ELSEVIER'S
DICTIONARY OF THE PRINTING
AND ALLIED INDUSTRIES

ELSEVIER'S

DICTIONARY OF THE PRINTING AND ALLIED INDUSTRIES

IN SIX LANGUAGES

English, French, German, Dutch, Spanish and Italian

COMPILED AND ARRANGED ON
AN ENGLISH ALPHABETICAL BASE BY

F. J. M. WIJNEKUS

OF THE
FORMER RESEARCH INSTITUTE FOR THE GRAPHIC AND ALLIED INDUSTRIES TNO,
AMSTERDAM
FORMER DIRECTOR OF A PRINTING AND PUBLISHING COMPANY

AND

E. F. P. H. WIJNEKUS

PRODUCTION MANAGER OF TOTAL DESIGN, AMSTERDAM

WITH A FOREWORD BY

† W. HOPE COLLINS C.B.E.

PRESIDENT OF THE INTERNATIONAL FEDERATION
OF MASTER PRINTERS

SECOND, REVISED AND ENLARGED EDITION

ELSEVIER
AMSTERDAM — LONDON — NEW YORK — TOKYO
1983

ELSEVIER SCIENCE PUBLISHERS B.V.
1 Molenwerf
P.O. Box 211, 1000 AE Amsterdam, The Netherlands

Distributors for the United States and Canada:

ELSEVIER SCIENCE PUBLISHING COMPANY INC.
52, Vanderbilt Avenue
New York, N.Y. 10017

First edition, October 1967
Second impression, May 1969
Second, revised and enlarged edition, October 1983

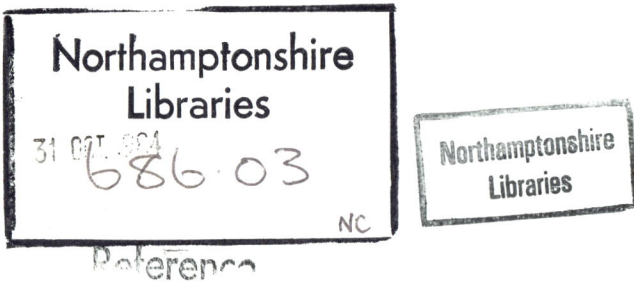
Library of Congress Cataloging in Publication Data
Main entry under title:

Elsevier's dictionary of the printing and allied
    industries in six languages.

    Bibliography: p.
    Includes index.
    1. Printing--Dictionaries--Polyglot. 2. Book
industries and trade--Dictionaries--Polyglot.
I. Wijnekus, F. J. M. II. Wijnekus, E. F. P. H.
Z118.E5 1983   686'.03   83-16586
ISBN 0-444-42249-8

ISBN 0-444-42249-8

Printed in The Netherlands

v

CONTENTS

FOREWORD TO THE FIRST EDITION

It gives me great pleasure to write a short foreword to this international dictionary of trade terms. It has been my experience that even in one's own language there are misunderstandings because of the difference of expression as well as difference in usage of certain technical terms. If there can be this misunderstanding within one country, then it is obvious on an international basis even more misunderstandings are likely to arise, if the exact translation of an expression or term is not clearly understood.

Looking through this very comprehensive dictionary the first thing which amazed me was the fantastic number of terms and expressions that we use in the printing and allied industries! It would take a real expert to know whether the author has left anything out and it seems to me a wonderfully comprehensive work. The second point I note is the very clear description of the terms. So far as the English version is concerned, this in itself is an invaluable guide to young people who have just entered the industry. There should be no possibility now of misunderstanding if one takes the trouble to check carefully the term or expression that is to be converted into another language.

I have no doubt this dictionary will be much used by anyone who is dealing with any matter in the very varied and wide field covered by the printing and allied industries throughout the world, and I wish the author and the publisher every success.

W. Hope Collins
President of the International Federation of Master Printers

ACKNOWLEDGEMENT

The author likes to express his thanks to all those who have given him advice, assistance and encouragement in his work. Of these he is particularly indebted to:

American Paper and Pulp Association, New York.
American Photoplatemakers Association, Chicago.
Anpa Research Institute, Easton (Penn.).
Association Française de Colorimétrie, Paris.
Alain Bargilliat, Saint-Mandé (Seine).
Georges Baudry, Paris.
British Standards Institution, London.
Champlain Company, Inc., Roseland (N.J.).
Eastman Kodak Company, Rochester (N.Y.).
E. Edelmann, Paris.
Eidgenössische Materialprüfungs- und Versuchsanstalt für Industrie, Bauwesen und Gewerbe,
    St. Gallen.
Encyclopædia Britannica, Chicago (Ill.).
Joh. Enschedé en Zonen Grafische Inrichting N.V., Haarlem.
Graphic Arts Technical Foundation, New York.
Her Britannic Majesty's Stationery Office, London EC1P 1BN.
Miss Ilse Hesse, Hannover.
The Institute of Printing Ltd., London.
International Typographic Composition Association, Washington DC.
Hans Kotte, Alfeld/Leine.
R. Laraignou, Copenhagen.
Ugo Leonardi, Zürich.
Lettergieterij 'Amsterdam' v/h N. Tetterode N.V., Amsterdam.
Miss Diana Liberatori, Rome.
Linotype & Machinery Ltd., London.
Marinoni S. A., Paris.
Gérard Martin, Paris.
† Ferrand Mees, Amsterdam.
The Monotype Corporation Ltd.
Ernest Müller, Brugg/Ag. (Switzerland).
National Association of Printing Ink Manufacturers, Inc., New York.
Nederlands Normalisatie Instituut, The Hague.
Organisation for Economic Co-operation and Development, Paris.
G. W. Ovink, Amsterdam.
Research and Engineering Council of the Graphic Arts Industry, Inc., Washington DC.
Royal 'Shell' laboratories, Amsterdam.
Schnellpressenfabrik Koenig & Bauer A.G., Würzburg.
Seybold Publications, Inc., Media (Penn.).
Howard Smith Paper Mills Ltd., Montreal.
Enrique Tormo Freixes, Barcelona.

Vereniging Tijdnormendienst voor de Grafische Industrie, Amsterdam.
Günter Wadewitz, Leipzig.
Waltwin Publishing Company, Inc., New York.
John Wiley & Sons, Inc., New York.

PREFACE TO THE FIRST EDITION

During the years in which I have worked in the graphic industry and especially in those years in which I had to provide the business world with information, I realized how extremely difficult it was for many of my colleagues to understand, in foreign languages, the literature pertaining to their trade. The vastly different technical words and expressions resulting from the tremendous development in the already so complicated printers trade and everything related to it, have come into the foreground in recent years, forming an unsurmountable barrier. Moreover, during the past forty years no extensive multilingual dictionary has been published on this subject. Existing dictionaries are incomplete and out of date. They have never dealt with printing, paper, or ink technology. Where two meanings exist, the lack of a clear definition sometimes leads to disastrous misunderstandings.

In the course of the years I have recorded alphabetically, for my own information, thousands of terms and expressions in the modern languages. When this fact became known – I was called upon for help on so many occasions – it was suggested that I compile my notes and have them published.

To aid in the completion of my work, I have been privileged in having recourse to the library of the Research Institute for the Printing and Allied Industries TNO. This library comprises most modern literature and prominent trade papers. My book, for which I feel there is a great need, is supported by all this valuable information. One may wonder at the fact that the book contains many physico-chemical terms and comments on what many people may consider an old or nearly outdated procedure. But the present stems from the past and this dictionary contains innumerable entries concerning both the present and the future. While the printer transfers ink to paper, either in text form or as a picture, the materials he needs are manufactured by others. It is just these materials which are affected by a number of properties, not immediately perceptible, but presenting problems at the most inopportune moments. It therefore speaks for itself that one should be familiar with the methods of producing these materials as well as their properties and defects. This work is absolutely necessary to eliminate such problems.

In recent years the printing industry has changed considerably. Doubtlessly it will continue to do so. The publisher and I are fully aware, therefore, that the day will come when this book will become outdated. In the meantime it will be a valuable aid in the exchange of data and technical information.

I shall be glad to receive any comments on, or additions to, my work.

Haarlem, (Holland), September 1967   F. J. M. Wijnekus

PREFACE TO THE SECOND EDITION

The first edition of this dictionary, published in October 1967, has received a very kind reception, some have even gone so far as to designate it as the 'Bible' or the 'Widge' of the printing industry. The available supply being soon exhausted, the book was reprinted at the end of 1968. In that edition some minor corrections have been made.

I am glad that my son Ernest has joined me in preparing the new edition. He has a broad experience in printing and publishing. In this revised and enlarged edition some changes have been made in the text and many new supplementary articles included.

We would like to express our thanks to Mr Georges Baudry, Ingénieur A. et M., of Paris, Mrs Dr Diana Liberatori of the Ente Nazionale per la Cellulosa e per la Carta, of Rome, who accepted the responsibility for checking and completing part of the Italian, to Mr Enrique Tormo Freixes, Conservador del Museo del Libro y Artes Gráficas, of Barcelona, who checked and completed part of the Spanish, and to Mr Günter Wadewitz, Dr phil, Dr habil, of the Institut für Graphische Technik, of Leipzig, who checked and completed part of the German.

Reader's suggestions for additions and amendments to the dictionary will be welcomed.

Santpoort
Haarlem   February 1982

F. J. M. Wijnekus
E. F. P. H. Wijnekus

ABBREVIATIONS

abbrev.	abbreviation	m	masculine
adj.	adjective	Me	Mexico
app.	appendix	misn.	misnomer
approx.	approximately	n	neuter
Ar	Argentina	NSA	Northern South America
arch.	archaic	obs.	obsolete term
Be	Belgium	Pa	Panama
CA	Central America	Pe	Peru
Ch	Chile	pl	plural
Co	Colombia	pop.	popular
coll.	colloquially	qv	quod vide (which see)
Cu	Cuba	RP	Rio de la Plata
depr.	deprecated term	SA	South America
Ec	Ecuador	sl.	slang
El Sal	El Salvador	SSA	Southern South America
Es	Spain	Sw	Switzerland
f	feminine	UK	United Kingdom
fam.	familiar	Ur	Uruguay
fl.	floruit	US	United States of America
GB	Great Britain	v	verb
gen.	general	≈	approximately equal to
Gu	Guatemala	*	see
LA	Latin America		

Language indications

f	French	(Français)
d	German	(Deutsch)
n	Dutch	(Nederlands)
e	Spanish	(Español)
i	Italian	(Italiano)

A SHORT HISTORY OF PRINTING

The East. – Printing as western civilization knows it began in the middle of the 15th century in Mainz, Germany. The Chinese, Japanese and Koreans knew and used printing long before that time. The earliest extant samples of printing are Buddhist charms produced on the order of the empress Shotoku before 770 in Japan, supposedly in the quantity of 1,000,000. It is not known what material was used for the blocks from which these were printed; copper, wood, but also stone (steatite) are among the materials suggested by various scholars. But the charms themselves are well authenticated and can be seen in several American museums and private collections. The British museum treasures another famous printing relic, known as the Diamond Sutra roll. This roll was found by Sir Aurel Stein in 1900 in a wall cave at Tunhuang in China and is the oldest dated book in existence. It is a block-printed roll, 4.87 m long and 30.5 cm wide, and bears a printing date corresponding to May 16, 868, at its end.

Block printing is the classical form of Chinese printing. In 969, playing cards are first mentioned; between 971 and 983, the Tripitaka, the Buddhist canon, was printed. This enormous undertaking required the cutting of 130,000 wooden blocks. During the Sung dynasty (960–1279), Chinese block printing reached its peak. The great dynastic histories and the Chinese classics printed in this period have never been surpassed in quality. Less propitious was the application of block printing to the production of paper money; at about 1100, China experienced its first great inflation. Printing therewith definitely established itself as a force in religion, philosophy, history and economics.

Block printing was not the only Chinese contribution to printing. Between 1041 and 1049, a 'common man' (often erroneously designated a smith), Pi Sheng, made movable types of earthenware. For printing, he covered an iron plate with a mixture of resin, wax and paper ashes. On this thermoplastic base, he assembled his crockery type in an iron frame, warmed the base until the adhesive mixture melted and leveled his types by pressing them in the base. Pi Sheng's invention was improved by others, but neither his earthen types nor another kind made from tin and kept in place by stringing on wire, became generally used, possibly because of the lack of a suitable ink.

Chinese efforts toward typography continued. Wooden type was used about 1300; movable metal type was first used in Korea in 1241, and a Korean book of 1377 states that it was printed with movable type. The art of type casting was advanced in Korea with great energy on the initiative of the crown and in 1403, a governmental type foundry was established. Ten founts of type were cast in Korea between 1403 and 1516. From Korea type casting spread to China and Japan. But in spite of all enthusiasm and governmental support, movable type did not dislodge block printing, which remained the Chinese printing technique par excellence until modern times.

Block printing and movable type. – Why did the Chinese prefer block printing to movable type? Several reasons are given. One is the Chinese form of writing; our alphabet consists of 26 letters, the Chinese use thousands of ideograms. Another reason is the Chinese love for calligraphy, the art of fine writing, which can best be expressed in block printing. Some scholars emphasize the different function of printing in Chinese civilization. There, printing served to preserve the classics in the approved and standardized version; western printing was more concerned with dissemination of knowledge than with its pristine pureness. Finally, it is often stated that the lack of a suitable ink was the main reason for the failure of movable metal types. But this argument is not convincing if one considers that many books were printed in Korea, China and Japan with movable metal types.

This is not to say that technological reasons should be excluded. On the contrary, block printing was excellently suited for a handicraft technology. The original writing is done by a calligrapher on thin, transparent paper. This original is pasted face down to the perfectly planed surface of a pear or jujube board large enough for two pages. The paper is rubbed off but the inked image remains on the board and guides the workman in cutting away the non-printing areas. The finished block is a relief printing plate. Chinese printing ink is made of lamp black and a water-soluble binder; it has brushing consistency. The worker inks the block with a brush, lays the paper on the inked surface and takes the impression by brushing gently with a dry brush. The paper is printed one side only, two pages to the sheet. For binding, it is folded in the centre, unprinted sides inward, printed sides out and bound along its edges. Chinese books bound in this manner can easily be mistaken for uncut books. Block printing goes fast; a printer can produce approximately 2,000 sheets in a full day's work.

If we compare Chinese block printing and western typographic printing, we find that both are relief printing, but there the similarity ends. Apart from the difference between movable type and the custom-made block, we notice two other striking differences; the absence of a printing press in block printing and the fact that the paper cannot be printed on both sides. Chinese and western printing both produce a printed sheet but they arrive at this result by an entirely different technique. The west had knowledge of Chinese block prints in the form of paper money and playing cards. But Thomas Francis Carter to whose studies we owe most of our first systematical knowledge of Chinese printing, sees no evidence for western knowledge of Chinese printing techniques and, in particular, not of Chinese movable types before the western invention of printing. China's main contribution to western printing was the invention of paper; printing itself developed in the west under fundamentally different conditions and consequently, along fundamentally different lines.

The invention of printing in the West. – The literature on the invention of printing is enormous. D. C. McMurtrie's bibliography on the invention, edited in Chicago, 1942, lists more than 3,000 items. The lack of clear and incontrovertible evidence, the high esteem in which this invention was held by men of learning from the very beginning, and last but not least, nationalism, provincialism and localism resulted in a steady controversy in which the honour of having invented printing was claimed for many people and many cities. Modern bibliographical research has eliminated most of these trumped-up claims and resulted in the more or less general acceptance of Johann Gutenberg (1398–1468) as the inventor of printing. But the fact remains that there is no direct and authentic report on the invention; bibliographical research has not changed this situation but devoted itself to the study of detail. The archives and libraries were searched for documents, books and other early products of the press. The material has been examined, evaluated and correlated with infinite care and ingenuity. The discussion of the origin of printing has thereby not ceased; it has only become more specialized.

Semantics plays an unfortunate role in this discussion. People, and scholars are people too, have different things in mind when they speak of the 'invention of printing'. Some see the crux of the invention in the printed book and point out that the Chinese had printed books when the west was still in the throes of the dark ages. Others put the emphasis on movable type and insist that the honour belongs to a Dutch printer whose work is much more primitive than that connected with Gutenberg and therefore assumedly preceded this. Because Laurens Janszoon Coster of Haarlem, Holland (1405–1484) is the only man whose name can be connected with these prints, he is considered by some scholars the first inventor of printing.

Theodore Low de Vinne (1828–1914), the great American printer, was the first to redefine the meaning of the so-called invention of printing. For him, the key of the invention lies neither in the printed page, nor in movable metal type but in the adjustable type mould. The accent is on adjustable and not on mould, because only by means of the adjustable type mould did it become possible to make type with the precision necessary for typography and printing. Those who share his point of view distinguish emphatically between the adjustable type mould and the mould used in sand casting. The Korean types were sand-cast, and so were the types used by Coster. Prof. Gustav Zedler has studied the early Dutch prints down to their very minutiae. He too concedes that Coster could never have produced the kind of type indispensable for printing. In the final analysis, Zedler accepts De Vinne's opinion and attributes the invention of the adjustable type mould and therewith of western printing to Gutenberg.

Once the crucial role of type became recognized, it was studied extensively, so extensively that it sometimes looks as if type were all that is needed for printing. This is a grave mistake. All of printing needs four elements without the existence and combination of which printing is simply impossible. In letterpress, these four elements are: type, press, printing ink and paper. Type converts the written word into printable form; the press is the machine in which the type image is transferred to the paper by means of printing ink. Of these four indispensable elements, Gutenberg found only one ready and available, paper. The other three did not exist and had to be invented.

Many students are so overawed by the importance of type that they are inclined to belittle the rest. Zedler, for example, believes that any intelligent craftsman must have been capable of inventing the press. Our knowledge of Gutenberg's press is indeed non existent, nor do we know to what extent later presses represent his original design or improvements. But it is known that some of the work printed in his lifetime and by men who worked with him are marvels of perfection. It is, in this context, of no consequence that presses for making cheese and wine and for bookbinding and papermaking were known at the time of Gutenberg. Nor can the invention of printing ink be reduced to almost nothing by the fact that artists' oil paints were already in existence. The invention of printing was a technical achievement without precedent. It was much more than the invention of a single object or a machine; it was the development of a complete manufacturing process. This development included artistic design (type face and page layout), engraving (punch cutting), metallurgy (type casting), mechanics (type mould and printing press), and chemistry (printing ink). Any one of the many inventions made by Gutenberg and the men who worked with him would fill the development department of a big contemporary industrial organization with pride.

In the light of our own mass-production society, we can recognize in the invention of printing the first example of our own manufacturing techniques: standardized mass production by means of assembling interchangeable parts designed and manufactured for precision fit. Detroit was anticipated in Mainz about 1450. Gutenberg was certainly the first modern industrial genius.

Early printers. – The timeliness of the invention of printing is evidenced by its rapid diffusion. From Mainz, it spread not only to other German cities, but also to Italy, Spain, France, Austria and even Turkey. By the end of the 15th century, printing presses were established in all major European cities. William Caxton (1422–1491) introduced printing in England in 1476. He was a business man who had spent most of his adult years abroad, mainly in the Low Countries and

France. All his life a book lover and collector, he became finally a translator and printer. Caxton learned the new art of printing between 1471 and 1472 in Cologne, Germany, and began printing at Bruges, Belgium, in 1474 or 1475 in association with Colard Mansion, a calligrapher. Then he returned to England and set up his press at Westminster. Caxton, in contrast to the German printers, printed in his vernacular language, English, and not in Latin. A gentleman of independent means, he ran his press neither for the church nor exclusively for profit, but rather for his own pleasure.

The centres of printing were, during the 15th century and well into the 16th century, in Germany at Mainz, Frankfurt, Nürnberg, Augsburg and Cologne; in France, at Paris and Lyons; in Italy, at Venice and Rome, and in Switzerland at Basle. The printer of the incunabula period, the time up to the end of the 15th century, was a man of many talents. He understood not only composition and printing but was also his own ink chemist and press builder. In addition, he was often type designer, cutter and caster. The new art of printing attracted men of different interests, ability and character. Some, like Aldus Manutius of Venice (1450–1515) and Henri Estienne (Stephanus) of Paris (1460–1520), were classical scholars; others, such as Peter Schoeffer in Mainz (1425–1502) and Nicolas Jenson in Venice (1420–1480), were great type and book designers. But there is also the financier, like Johann Fust (1400–1466), Gutenberg's backer, and the business man of great organizing ability like Anton Koberger of Nürnberg (1440–1513). Competition in the 15th century was already keen and ethics were sometimes rather questionable. The great Nürnberg Chronicle of 1493, for example, had more than 1,800 woodcuts but only about 600 wooden blocks. Cynically, kings and popes were here not only chips but even prints off the same block. Bartholomew Troll of Lyons earned the self-bestowed epithet of 'the honest book-seller' by being a less honest printer; he shamelessly pirated Aldus' editions of the classics.

The output of 15th-century printing consisted mainly of religious, theological, philological, legal and classical books. But the press worked also for the day's necessities. Letters of indulgence, papal bulls, treaties, calendars and commercial announcements are among the so-called ephemera of printing. The great social power of the new art was generally recognized at its very beginning, though not always with enthusiasm. In 1501 Pope Alexander VI issued a bull against the unlicensed printing of books; the Index Librorum Prohibitorum has been continuously published since 1559 and is still serving Roman Catholics as their reading guide.

The decline of printing. – The 16th century saw Europe in upheaval. In 1517 Martin Luther nailed his 95 Theses to the door of the castle church at Wittenberg, Germany. The Reformation began and with it the ascendance of printing to the role of a crucial political and cultural power. The high and mighty who were the patrons of printing during the 15th century became its scared enemies during the 16th century. The character of printing changed; its Golden Age was over. The decline of printing was nowhere worse than in France where it had reached its greatest height. There, the religious intolerance of the Sorbonne, which officially controlled the intellectual life of France, had a catastrophic effect on printing. Every printer was under suspicion as a potential heretic. The fate of Etienne Dolet, the Lyonnaise scholar-printer who was convicted of heresy, tortured and burned at the stake in 1546, was equally possible for every other printer. Many fled the country. The best-known of them is Christophe Plantin (1514–1589), who migrated to Antwerp, Belgium, and founded a printing business in 1549 that existed until 1876 and was finally converted by the city of Antwerp into a museum. Printing deteriorated so much in France that Cardinal Richelieu had to send to Holland for first-rate compositors, pressmen and inkmakers when he wanted to start the royal printing office at the Louvre in 1640. The lead in printing passed to the Dutch who kept it until the end of the 18th century, if not longer.

In England printing remained free for the first 80 years of its history. The chartering of the Stationers' company in 1556 was the first governmental device for controlling and licensing. The Star Chamber decree of 1586 further tightened the screws. All printing was now restricted to the two universities (Oxford and Cambridge) and the City of London; there it was frozen at its then existing level of 21 shops with a total of 53 presses. The worst Star Chamber decree was given in 1637. This time, type founding was recognized as a separate branch and specifically controlled. The result was that England became completely dependent for type on the Dutch. John Milton's famous 'Areopagitica' was directed against the partial renewal of this decree in 1643; it took more than 50 years longer, until 1695, before printing was again free in England.

The western hemisphere. – In the new world, printing began in Mexico City in 1530s. Lima, Peru, had its first press in 1584. More than 50 years later, in 1638, the first press was set up in North America at Cambridge, Massachussets, by Stephan Daye, assisted by his son Matthew. The first two products of this press, which came under the control of Harvard college, were the 'Freeman's Oath' and 'An Almanac for 1639, calculated for New England by Mr. William Pierce, Mariner'. There are no copies of these two; the 'Bay Psalm Book' (The whole Booke of Psalms, faithfully translated into English Metre) of 1640 is the Earliest extant North American book. Another interesting item is the Indian Bible, the first American printing of the scriptures, translated by the missionary Rev. John Eliot into Indian. Samuel Green and Marmaduke Johnson printed it at Cambridge in 1663.

Printing spread with colonization. William Bradford started in Philadelphia, Pennsylvania, in 1685 and moved to New York in 1693. William Nuthead began printing in Jamestown, Virginia, in 1682, but was not permitted to continue and moved to Maryland in 1685. The first printer of Connecticut was Thomas Short who set up a press at New London in 1709; William Bradford also is supposed to have started printing in Perth Amboy, New Jersey, and therewith became the first printer of three states. Rhode Island's first printer was Benjamin Franklin's brother, James (1727); printing began in 1731 in South Carolina; in 1749 in North Carolina and in 1756 in New Hampshire. Isaiah Thomas, himself a successful printer and the first historian of printing in America, tells us that in 1775, 50 printers were active in the 13 colonies. The stamp tax on newspapers and advertising was one of the main stimuli for the American independence movement. Printing proved its political power for the second time in the world's history; the book was the intellectual foundation of the Reformation; the newspaper and pamphlet prepared the American War of Independence.

The 18th century. – At the end of the 18th century, printing was in a state of transformation. The American and French revolutions freed the energies of millions; everything was in flux and printing too had to change if it wanted to satisfy the needs of the times. Of the four indispensable elements that are combined in printing, image carrier, press, ink and paper, the press was most in need of modernization. (Printing-image carrier is here used as a generic term for not only type but for plates, cuts, cylinders, etc.) The printing press on which all western printing was done from the middle of the 15th century to the beginning of the 19th century was a wooden structure equipped for repetitive performance of two related motions. All presses known before the printing press had one motion only: up and down. The printing press had in addition a second motion: in and out. Both motions are reciprocating and at a 90° angle to each other. The up-and-down motion supplied the pressure necessary for ink transfer from the type forme to the paper;

the in-and-out motion positioned the forme with the paper on top for the making of the impression. The impression itself came about by pulling a lever that turned a spindle and thereby pressed a platen against the inked forme-paper assembly underneath it. After pulling, the platen was raised and the forme-paper assembly cranked out from under the platen.

Feeding the paper was done by placing it on a frame covered with vellum and equipped with adjustable points. This frame was known as the tympan. The paper was laid on, and the points served as guides to ensure registration of the back, i.e., printing in the same area on both sides of the sheet. The frisket was hinged to the tympan and protected the paper from all non-printing parts of the forme. The forme was inked by beating it with two ink balls. Then the paper was put on the tympan, the frisket over it and both on top of the forme. The carriage was run in, the lever pulled, the carriage run out and the printed sheet removed. Hand printing was primitive and cumbersome, but not as inefficient as one might think. In the American colonial period, a team of two men produced 2,000 sheets 38 cm by 50.8 cm in a ten-hour day. Considering that each sheet required two pulls because the press was not strong enough to print the whole forme in one, we must look with respect at this production.

If we look at 18th-century printing in the light of contemporary printing techniques, we are inclined to believe that printing did not develop at all for the first 350 years of its existence. At the end of the 18th century, type was still cast with the same tools as had been used in very early times and the printing press was still the same old wooden-screw press. But there had nevertheless been constant though slow progress. It consisted in improving tools and methods rather than in revolutionary change. The fact that the output of a day's work at the press was 2,000 sheets at the end of the 18th century whereas it is assumed not to have been more than 300 in the 15th, speaks for itself.

Development of modern presses. – This wooden hand press was the point of departure in the development of modern high-speed printing machinery. In this development, we can distinguish two different avenues of approach. One group of inventors wanted to maintain the basic design of the hand press but to improve its power and therewith its speed and size, or expressed in modern terms, its general efficiency. Whether Willem Janszoon Blaeu (1571–1638) of Amsterdam, the famous Dutch map printer, greatly improved the wooden press as Joseph Moxon thought, and as many writers on printing history have since asserted, or whether the Dutch in general produced better presses, as Lawrence C. Wroth maintained, will not be decided here. The essential steps in the improvement of the hand press were to replace wood with metal and to design every part on the basis of theoretical mechanics. The first man to do so was Wilhelm Haas (1741–1800) of Basle in 1772. But the Basle guilds prevented the exploitation of this press, and it took a quarter of a century until different inventors developed workable iron hand presses. Etienne Anisson-Duperon in France, Lord Stanhope in England, George Clymer in America and many others did notable work in this field.

The other avenue of approach was much more radical. It tried to connect both motions of the press, the vertical of the platen and the horizontal of the carriage in such a manner that they would form a complete cycle. The first record of this idea exists in the notebooks of Leonardo da Vinci in the 16th century. Friedrich Koenig from Saxony more than 200 years later not only conceived it independently again but also developed it to the point of practical performance. (The English writer and inventor William Nicholson anticipated many later printing developments in his patent of 1790 but never executed his ideas). On November 29, 1814, 'The Times' (London) proudly announced that the issue in the reader's hand was the 'result of the greatest improvement connected with printing since the discovery of the art itself'; the issue was printed on Koenig's flat-bed steam press. The great revolution of printing had begun. The newspaper was the strongest force behind printing progress in the 19th century. It was obvious that rotary printing was an absolute necessity for the newspaper's insatiable hunger for speed and production. How the problem of rotary printing was solved is one of the great sagas of 19th-century industrial development. Paper, ink, but most of all, stereotyping had to be radically changed before a successful newspaper press could be achieved. In 1869, 55 years after the debut of Koenig's steam press, 'The Times' (London) could ring the gong again. In the Walter press, it had achieved the prototype of the modern newspaper press; it printed on a continuous roll of paper from curved stereotypes. In the next decade, Richard Hoe of New York improved and developed newspaper printing by making it possible to deliver the completed newspaper folded and counted, ready for sale.

The spectacular development of the newspaper press should not allow us to forget that of the job press. Job work was and still is the domain of the small print shop. It took longer until this field could support presses of its own. The first attempt in the direction of the specialized job press was made by Johann Gottfried Freytag of Gera, Saxony, in 1777 and 1778. He made a treadle press that functioned, but he could not overcome the opposition of the guilds. The first successful job press (completed 1839) was the work of a Boston printer, Stephen P. Ruggles. His 'engine press' took the vertical motion and the tympan from the hand press, but it was already equipped with inking rollers and treadle operated. The most important contribution made by Ruggles was the introduction of the vertical bed in 1851. George Phinias Gordon of Salem, New Hampshire, is another famous press inventor. His best-known job press was the 'Franklin' jobber. Gordon introduced the inkdisk (which improved ink distribution) on job presses and also made many other important improvements. His presses enjoyed great popularity in the 1860s and 1870s. The third great name in job presses is the Universal, patented by Merrit Galley in 1863. It was the first heavy-duty job press and had a truly parallel impression. All these job presses were hand-fed and foot-powered. In mid-20th century America, most job presses have become automatic machines that feed, print and deliver the sheet with great speed and efficiency.

Type casting. – The history of printing types is divided into two subjects, technology and design. Both are, of course, related to each other but both have also their own independent development. Our historical knowledge of typographic technology is meagre. For the incunabula period of printing (up to 1500), we rely on inference and comparative study of the printed image more than on tangible type. The only evidence of this latter kind consists of some type found in 1878 in the Saône river, near Lyons. Anatole Claudin, one of the foremost historians of French printing, attributes them to the infancy of printing and concludes that they were cast in a mould. Another kind of evidence on the shape of type, visible rather than tangible, is the so-called turned up type. This is type that came accidentally to lie on top of the forme and thereby transferred its side instead of its image to the printed page.

The first technical discussion of type casting known to us is by the Siennese architect and metallurgist, Vanucci Biringuccio, in his 'Pirotechnia', published in Venice in 1540. He gives a quantitative formula for type metal consisting of lead, tin and antimony and discusses punch, matrix and casting in the adjustable type mould. His descriptions are clear and informative but without any illustrations. Joseph Moxon's 'Mechanick Exercises', the first printer's manual (1683), treats type casting more extensively and is also illustrated.

Hand casting of type was slow and tedious; a day's production averaged 2,000 to 4,000 pieces. In 1838, David Bruce Jr. of New York achieved a practical typecasting machine that made hand-casting obsolete. His machine was still hand-operated but its production surpassed hand-casting many times. Not until the end of the 19th century was type automatically finished as well as cast.

Type design. – A strictly aesthetic approach to the history of type design misses its essentially utilitarian purpose: 'Letters are to be read, not to be used as practice models for designers, or to be moulded by caprice or ignorance into fantastic forms of uncertain meaning. They are not shapes made to display the skill of their designers, they are forms fashioned solely to help the reader'. The writer of these sentences was Frederic W. Goudy (1865–1947), the most prolific type designer of the United States. The history of type design can be understood only as part of the whole history of printing, its technological as well as its social aspects.

It was only natural that Gutenberg should pattern his Bible type on the style of writing used for sacred books in his part of the world. It was equally natural that the German printers who spread the new art to the rest of Europe should continue to use the kind of types they knew. It was not less natural that these types were modified and finally completely redesigned according to the conditions prevalent in other countries. The first phase of type design consists in the adaptation of calligraphy and handwriting to the exigencies of printing. The next is that of typography as an independent art. Both phases may be divided in the sense that one man is a clear example for one and another for the second, Gutenberg and Peter Schoeffer for example; or both phases may be part of a single man's development, as perhaps in William Caxton, between his type one (1475) and type five (1489).

A telling example of how type design changes with the country is the semiroman of Conrad Sweynheim and Arnold Pannartz at Subiaco in 1465 and 1467. The gothic type of the fatherland was blended with the humanistic hand of Italy. This humanistic book hand became our roman type through the design of Nicolas Jenson in Venice in 1470. The contemporary companion of all roman text type, our italic, was born in 1501. Aldus Manutius needed a compact small type and adapted the slanted hand-writing used by the humanists for his purpose. Type cutting became an independent business with Claude Garamond (1480–1561) in the middle of the 16th century in Paris. Religious and political persecution on the one hand, the development of government control and the guilds on the other wrought a change in the personnel of printing by the end of the 16th century. The artist- and scholar-printer died out; the tradesman dominated the field. Printing became dull, unimaginative, routine. Design was completely stagnant, and mechanical production varied with the Dutch at the top and the English at the bottom of quality.

With the 18th century, things improved. Printers regained their pride, and type of good quality became available in many European cities, e.g., that of Pierre Simon Fournier of Paris, of Bernhard Breitkopf of Leipzig, Germany, Wilhelm Haas, Basle, Switzerland. William Caslon (1692–1766) did for England what these men did for Europe. Caslon's types are certainly outstanding in comparison with contemporary English products; whether they surpass good quality type of other countries is open to doubt. The next great name in English type design is that of John Baskerville (1706–1776). His type is customarily designated as transitional, but Baskerville was certainly the first modern man of printing. He wanted paper, type, ink and presswork all to be of excellent quality and perfectly attuned to each other. This concept has become the generally accepted credo of quality printing. Giambattista Bodoni (1740–1813), the Italian master of type design, took over where Baskerville left off and created the kind of type that our texts call modern.

During the 19th century the art of printing and type design fell lower and lower, quite in contrast with the unprecedented progress of printing technology during the same time. The revival of printing is linked to William Morris (1834–1896), a wealthy gentleman, poet and ardent socialist, who set new standards of excellence in book production. His main merit was that of a gadfly; he stung printers and publishers and opened the door again for the artist in the printing industry. Type design broadened in the 20th century as never before; its greatest stimulus is now advertising. The 20th century is also the first in which US designers distinguished themselves in the field.

Composing machines. – Much more important than the improvements in foundry type was the development of composing machines. The strongest social pressure in this direction came from the newspaper industry. During the 19th century, the newspaper grew at a fantastic rate; in number, volume and paper size. It needed faster printing and faster composition. All elements of printing – press, paper and ink – were radically changed by the end of the 1860s; only composition retained its mediaeval character.

The history of typesetting machines begins in 1822 when William Church of Boston was granted the first English patent for a composing machine. He introduced the keyboard principle, but mechanized only assembly; the other phases of the cycle, justification in particular, remained to be done by hand. But Church was also the first to suggest melting down instead of distributing type, a principle later generally accepted. Several other machines, similar in kind, followed and were successful. The Kastenbein machine, developed about 1870, was used for years by 'The Times' (London); the Fraser, developed about 1872, supposedly for the ninth edition of the 'Encyclopædia Britannica'; the Wicks was in production at the 'London Morning Post' from 1905 to 1910. The output of these machines varied from 5,000 to 17,000 assembled letters (or approximately 2,000 to 8,000 ems) per hour. Some of them had companion machines for distribution, but most of these caused more trouble than anything else.

The composing machine inventing bug was epidemic. It spread to people in all countries and of many different callings, but it had its greatest victim in the Paige machine. The Paige compositor was the brain child of J. W. Paige of Rochester, N.Y. It set type, justified the set lines, distributed the dead matter with precision and was considered one of the most ingenious mechanisms ever devised by man. In September 1894, the machine was tested by the 'Chicago Herald' where 32 Linotype machines were working at the same time. The Paige compositor needed several radical changes, but nevertheless proved its value. The development of the machine was financed among others by Mark Twain, who even used his first royalties from 'Huckleberry Finn' for it. The inventor was as stubborn as he was ingenious, and stubbornness is a heavy liability, particularly with financiers. They decided to drop the whole project rather than battle the Linotype. The loss was at least $ 800,000 not to speak of the waste of time and effort.

With the invention of the Linotype in 1884 by Ottmar Mergenthaler and Tolbert Lanston's invention of the Monotype in 1885, two machines of lasting value came into existence. They mechanized body or text composition; the composition of display type was modernized by Washington J. Ludlow of Chicago, Illinois, who invented the machine bearing his name. When the first Ludlows were made in 1911, the modernization of composition was completed in principle. From then until mid-20th century, the history of mechanical type-setting is the history of Linotype, Monotype and Ludlow machines. The next chapter of typographic evolution began after World War II. Its theme is photocomposition.

Photographic composition. – Nomic composition can be divided into the following groups; photographic composing machines for body and display type, photographic composing machines for unjustified lines of display type and nomic composing machines of the typewriter variety. The whole field is so very active and so far from settled that, at this time, only a rough summary can be presented.

The Fotosetter is the first photographic composing machine for body and display type to be commercially available. It is basically a line-casting machine, the Intertype, whose casting mechanism was exchanged for a photographic one. The circulating matrix scheme of the Linotype is retained with the difference that the matrix is now a transparent image of the type. Justification, on the other hand, is different from the Linotype; the line is justified by increasing all spaces, those between letters as well as between words. The opinions on the desirability of this form of justification are divided. The Fotosetter can change the body size of the type photographically. This feature increases efficiency of the machine and reduces the capital expenditure for mats of various sizes. The uses of the Fotosetter are manifold; in newspaper printing it is employed for display advertising composition that is converted into relief plates by magnesium etching. Four or five other machines were in various stages of development in the late 1950s.

Several machines exist for photographic composition of display type. Most of them produce long bands of paper or film imprinted with unjustified type; all are manually operated. The operator selects, positions and exposes by hand and eye. The resulting print is developed and finally incorporated in the artwork. Machines of this nature are often used by art studios and they have established themselves as valuable aids in the printing industry.

The last group of nomic composing machines is that which takes the typewriter as its point of departure. These machines face several serious problems. The standard typewriter with only one set width for all letters is contrary to all rules of typographic design. Several machines were developed to improve this condition; some have a limited number of width groups, others go much closer to that required for good typography. Some machines have exchangeable type founts, others only one; some have a wider and some a less wide range of body sizes. In most cases, the problem of justification is solved by double typing. One system works on the Monotype principle and has two machines, one for punching the ribbon and another for producing justified type.

The product of nomic composition is either a typed sheet of paper or a direct image offset plate. The typed paper can be converted into an offset printing plate by photographing or xerographing. Direct image plates are of paper treated to be receptive to the impression of the typewriter key through a specially prepared ribbon. They can be used directly for offset printing. This short résumé barely indicates the liveliness and variety present in nomic composition. The field is far from settled but it, too, has definitely established itself in the economic and cultural life of the nation.

Photography. – Photography was discovered by no one man. It was the outcome of the early observations of the alchemists and chemists on the action of light, a subject that belongs strictly to the domain of photochemistry. Although the blackening of silver salts was known in 1565, it was not until 1727, when J. H. Schulze used a mixture of silver nitrate and chalk under stencilled letters, that it was definitely recognized that this darkening action was caused by light and not by heat. The experiments were also done by W. Lewis in 1763. These experiments were important in that, in conjunction with those of K. W. Scheele, 1777, they led to the experiments of T. Wedgewood with silver nitrate on paper and leather in 1802. It was also suggested that silver chloride was more sensitive. These were practically failures because of the very long exposures required, and no means were known of removing the unaffected material and thus stabilizing or 'fixing' the image against light.

Camera. – Leaving for the time being the question of the sensitive salts, we may turn to the evolution of the camera. This was the outcome of the old camera obscura, the invention of which is usually ascribed to Giovanni Battista della Porta, 1553, though its principle had been indefinitely described by Alhazen, 1100, Roger Bacon, 1267, and others. Leonardo da Vinci (1452–1519), described and pictured a camera obscura in an unpublished manuscript. In 1550 J. Cardan suggested the use of a speculum or concave mirror in front of the instrument, and D. Barbaro, 1568, proposed convex lenses and the use of a diaphragm to secure greater sharpness of the images. E. Danti, 1573, corrected the reversed image by means of a mirror behind the lens, a device still in use. F. Risner, who died in 1580, described in his works published in 1606 the methods of enlarging and reducing, and a portable box in lieu of the cumbrous fixed hut in use until then. Porta in his second edition, 1589, of which an English translation was published in 1658, was the first to introduce the use of the convex lens in the camera. Johann Kepler in his work on optical astronomy, 1604–1611, described the use of a concave lens behind the convex to obtain larger images, and thus anticipated the telephoto lenses of the present day. In J. Zahn's book, 1665, there was described a portable camera obscura with two or three lenses, to secure greater brilliancy of the image, and side wings to shield it from extraneous light. At the beginning of the 18th century the portable camera obscura had become a regular article of commerce and was used to obtain sketches from nature. It was encouraged by the growth of amateur artists.

The beginnings of photography. – Toward the end of the 18th century there grew a demand for pictures. Wood engraving was revived and lithography was invented. A simple way of producing pictures was the silhouette, made by tracing the outline of a projected image of the face and filling it in with black. In 1786 G. L. Chrétien invented the physionotrace in which the projected image of a head was traced by a stylus, and by a pantograph arrangement an engraving tool cut a copper plate which could be inked and printed. Alois Senefelder discovered lithography in 1796, but, though it was introduced in Paris in 1802, it was not until 1813 that it became a success and a fashionable hobby. From that date until 1817 Joseph Nicéphore Niepce was engaged in examination of natural stones to find one suitable for the process. Since he was unable to draw, his son, Isidore, undertook this work, and when the latter was called for military service, Niepce was impelled to produce the images automatically. In conjunction with his brother Claude, Niepce first tried photography with silver chloride on paper. The light-sensitiveness of iron, manganese, acids and other compounds was tested, stone, metal and paper being used as supports. Finally, guaiacum resin was tried and then asphaltum, or bitumen of Judea, which had been in common use as an etching ground since the days of Rembrandt van Rijn. This, becoming insoluble in its usual solvents by the action of light, gave not only a resist for the etching of metal plates but also transparent images on glass. A successful result was obtained in 1822, and thus it may be said that the first permanent photograph was made in that year by Niepce. Prior to this, in 1816, Niepce had obtained a positive on paper, probably in silver chloride, which he could not fix.

Three years were required to perfect the process of heliography, as it was called, and in 1826 an etched metal plate was sent to G. Lemaître, of Paris, to be printed. In order that the pictures should be more distinct, they were exposed to the vapours of iodine. The high lights of the plates consisted of hardened bitumen; the shadows, of the bare metal; thus, the contrasts were very poor, which led to the use of tin instead of copper. Lemaître suggested silvering copper plates.

Daguerreotype. – In 1826 L. J. M. Daguerre, a painter who had experimented with silver salts, heard of Niepce's work and approached him to form a partnership. This was consummated in 1829. The work which Daguerre had done with Niepce had drawn their attention to the light-sensitiveness of silver iodide, and Daguerre discovered accidentally that the effect produced by exposing an iodized silver plate in a camera would result in an image if the plate were fumed with mercury vapour. The inventions of Niepce and of Daguerre were published to the world simultaneously by the French home minister in a bill presented to the house of deputies proposing a national reward to the inventors. A full description of their methods was presented by François Arago on August 19, 1839, at a meeting of the Academy of Sciences.

The daguerreotype process was a complete success. In its operation, a silvered copper plate was buffed and burnished to a high polish. It was then iodized by fuming with iodine, and later (1840) resensitized with bromine and then exposed in a camera. At first, exposures of minutes in bright sunlight were necessary; but later exposures were secured in a few seconds under favourable conditions. The development was effected by placing the exposed plate over a cup of mercury heated to about 75 °C. The image was then fixed with a solution of thiosulphate of soda and toned by treatment with gold chloride. The results obtained were excellent when all the operations had been carried out correctly; and the process flourished, especially for portraiture, until it was superseded by the wet collodion process in 1851.

Calotype. – A preliminary notice of Daguerre's success was made by Arago on January 7, 1839. This caused William H. Fox Talbot, an Englishman, to claim priority in obtaining a picture in the camera, and rendering it permanent. Talbot began his experiments in 1834, using the camera obscura with silver chloride and common salt of potassium iodide as fixing agents. A communication was made to the Royal society on January, 31, 1839. But not until a month later did he disclose his working method, which then involved silver iodide with excess of the nitrate, and fixation with sodium thiosulphate. The solvent action of this for silver salts had been discovered by Sir J. W. F. Herschel in 1819. Talbot's process was called 'calotype', and the image was developed with gallic acid, the action of which seems to have been discovered independently by the Rev. J. B. Reade.

The developed image on calotype paper was the exact reverse, as far as light and shade were concerned, of the original image. Such a picture was termed by Herschel in 1841 a 'negative'. The transparency of paper can be increased by waxing or oiling, and so Talbot was able to obtain true copies or positives of any negative by simple contact printing upon another piece of sensitized paper. Talbot's process is the first stage in the real line of photographic development, the inventions of Daguerre and Niepce being bypaths whose chief importance was the stimulus they gave to photographic evolution. The disadvantages of calotype were the somewhat long time required for printing and the structure of the paper.

Herschel suggested the use of glass plates and the deposition of silver chloride thereon. In 1848 Niepce de Saint-Victor suggested albumen as the vehicle for silver iodide. This process, called 'Niepceotype', held its own until 1851, when Frederic Scott Archer, an English architect, published his wet collodion process.

Collodion process. – To a solution of pyroxylin in ether and alcohol, Archer added a soluble iodide, usually with the addition of a little bromide, and coated a clean glass plate with it. In the darkroom, the iodized collodion was sensitized by immersion in a bath of silver nitrate and formed silver iodide with excess of silver nitrate. The plate was exposed wet in the camera, developed by pouring on a solution of pyrogallol containing acetic acid, and fixed with a strong solution of thiosulphate of soda for which cyanide of potassium was later substituted. Archer's collodion process was not patented, and in three or four years it displaced both calotype and daguerreotype.

The necessity for preparing the plates immediately before exposure and developing them immediately after considerably limited the practice of photography. J. Spiller and W. Crookes suggested bathing the prepared plates in a solution of hygroscopic salt, so that they could be kept some time both before and after exposure. This was followed by the use of hygroscopic substances of all kinds.

Collodion emulsion. – In 1864 B. J. Sayce and W. B. Bolton described the preparation of an emulsion of silver bromide in collodion. In 1874 Bolton washed the emulsion in bulk to remove the soluble salts formed in precipitating the silver bromide, and his methods are still followed.

Collodion emulsion was a practical advance on wet collodion, but it was no more sensitive. A considerable increase of sensitiveness was obtained when C. Russell introduced alkaline development in 1862, but the full fruit of this was reaped only with the introduction of gelatino-silver bromide dry plates.

Gelatine emulsion. – In 1871 R. L. Maddox made an emulsion of silver bromide in essentially the same manner as that used for making collodion emulsions but he replaced collodion by gelatine. The matter was followed up by other experimenters, among whom may be mentioned J. Burgess and J. Kennett. Kennett placed on the market a dry, washed emulsion which photographers could dissolve in warm water and use as a coating on glass and thus produce their own plates.

A great amount of experimental work was at once commenced on gelatine emulsions, the records of which filled the photographic journals between 1873 and 1885. The by-product salts were removed by washing. Sir William Abney recommended the use of iodide in small quantity with the bromide and found that this made it possible to obtain faster emulsions with less fog. Digestion, or 'ripening', as it was called, came into use – long digestion at low temperatures being suggested by C. Bennett – in 1878, and digestion with ammonia was used in 1876 by J. Johnson and in 1879 by D. B. van Monckhoven, who employed precipitation in ammoniacal solution as the basis of a process of manufacturing dry plates.

In 1877 the commercial plates of the Liverpool Dry Plate Company, Wratten and Wainwright and B. J. Edwards were introduced; and by 1879 comparatively rapid dry plates were available on the market similar in type to the slower varieties of plates used today.

After this period, amateurs gradually ceased their researches, and mass manufacture became general. Considerable increases in sensitiveness were made between 1890 and 1900, and after 1930 improved emulsion-making techniques resulted in further advance in sensitivity without the undesirable qualities which previously accompanied attempts to increase speed.

It is interesting at this point to record the advance made in photography in terms of the relative sensitivity of the process.

Progress in speed represented by the exposure required to give a good negative at f/11 in sunlight	
Original daguerreotype, 1839	4,000 sec.
Bromized daguerreotype, 1840	80 sec.
Wet plates, 1864	8 sec.
Early dry plates, 1880	1/2 sec.
Dry plates of 1885	1/10 sec.
Dry plates of 1900	1/50 sec.
Dry plates of 1931	1/100 sec.
Modern fast film	1/500 sec.

Colour sensitivity. – The silver halides are sensitive chiefly to the blue, violet and ultra-violet rays; hence all other colours are reproduced as dark grays or blacks. In 1873, H. W. Vogel discovered that the addition of certain dyes to the emulsion, or immersion of the coated plates in a solution of dye, increased the sensitiveness to the less refrangible colours.

Following this, J. Waterhouse found that eosin sensitized collodion emulsion, and shortly afterwards J. Clayton and P. A. Tailfer found that eosin would sensitize gelatine emulsions. They obtained a patent for its use in England and France, and their plates were placed on the market under the name of 'isochromatic' plates. In 1884 eosin was replaced by erythrosine, which was found by Joseph M. Eder to be a better sensitizer, and for 50 years erythrosine was used almost exclusively for the use of the so-called 'isochromatic' or 'orthochromatic' materials. In 1902 A. Miethe and A. Traube found that ethyl red, an isocyanine dye, gave strong colour sensitiveness as far as the orange of the spectrum, and in 1905 B. Homolka discovered pinacyanol, a dye of structure somewhat similar to ethyl red but which sensitizes very powerfully throughout the red.

It was now possible to make 'panchromatic' plates, which would photograph the entire spectrum without difficulty, and such plates were placed on the market in 1906.

The early dyes were largely of the cyanine and isocyanine types. In 1920 W. H. Mills and F. M. Hamer found pinacyanol to be a carbocyanine dye. After that time there was great activity in the study of sensitizing dyes, resulting in the present great variety which permit high sensitivity in all parts of the visible spectrum and out into the infra-red. Most of the new dyes were found after 1929 in England, Germany and the United States, and the most important types are modifications of the carbocyanines and the merocyanines.

The first panchromatic film was made by the Eastman Kodak company in 1914, but it was not generally used (for motion-picture work) until 1925. In 1931 a new super-sensitive type of panchromatic film was introduced. The progress in sensitizing dyes made possible the great advance in colour photography since about 1930 and the introduction of plates and films of high speed and low graininess.

Developing agents. – In the calotype process gallic acid was used as the developer, but in 1851 H. V. Regnault of Paris and Justus Liebig simultaneously discovered the more energetic pyrogallol, which actually allowed shorter exposures to be given. This was used also at first for wet collodion, though it was later replaced by an iron salt, and for dry collodion plates. Until 1861 free silver nitrate was an important ingredient of developers, but in that year G. Wardley proved that pyrogallol alone could be used. C. Russell and J. Leahy independently announced alkaline pyrogallol development, the former suggesting also an alkaline bromide as restrainer. Even after the introduction of the gelatine plate, pyro-ammonia was the only developer used until 1884, when the alkaline carbonates became general. In 1880 Abney discovered the developing properties of hydroquinone. C. Egli and Arnold Spiller recommended the use of hydroxylamine in 1884. In 1888 M. Andresen of Berlin patented the use of paraphenylene diamine, paratoluidine diamine and xylidene diamine as developers. In 1891 he patented the use of paraminophenol and its derivatives, especially monomethyl-paraminophenol, known under the trade name of metol.

Photographic printing processes. – The first printing process was with silver chloride and excess of silver nitrate on plain paper, which gave prints with a rather sunken appearance, the image being more or less buried in the fibres of the paper, and daylight was essential for printing. Louis-Désiré Blanquart-Evrard, in 1847, suggested the use of albumen to keep the sensitive salt more on the surface. In the same year Romieu proposed salting with a mixture of alkaline bromides and chlorides containing gelatine and subsequent floating on silver nitrate solution. A few years later the first coat of albumen was coagulated and a second applied, which gave a more brilliant glossy surface. The colour of prints obtained on these papers was foxy, and to darken this an acid was added to the fixing bath, causing sulphur toning. G. le Gray, in 1850, proposed the use of a gold salt to improve the colour, and an alkaline gold bath was recommended by Waterhouse in 1858. E. de Caranza, 1856, recommended the use of platinum in lieu of gold. Albumenized paper made in this way and toned with gold remained the standard photographic printing medium until 1890.

A. Gaudin had suggested an emulsion of silver chloride in collodion for printing-out. This was revived by Wharton Simpson in 1865. Paper thus prepared was introduced commercially by J. B. Obernetter three years later. Humbert de Molard, 1855, proposed to precipitate silver chloride, wash and suspend it in solution of starch or gelatine and paint the mixture onto paper. The same process was revived by W. H. Smith and J. E. Palmer in 1865 and by Abney in 1882. From that date gelatino-chloride or printing-out paper came into general use and eventually completely displaced albumen. Printing-out papers are little used now except for proofs.

Silver bromide with excess silver nitrate was used by Fox Talbot, and J. W. Swan in 1879 patented the use of paper coated with silver bromide emulsion not containing excess silver for development. It did not, however, come into general use until marketed by Eastman and by E. Just in 1883. J. M. Eder and G. Pizzighelli published, 1881, an exhaustive paper on gelatino-chloride of silver with development, and the former, two years later, described a chlorobromide emulsion. Chloride developing paper was introduced by A. Marion in England under the name of Alpha in 1889, and by Leo Baekeland in America as Velox. Since then the use of this paper has increased, special types, including a great variety of surface, being used for amateur and professional work. Modern papers are of the bromide and chlorobromide types mainly for enlargements.

Non-silver printing processes. – Robert Hunt was the first to utilize the light-sensitiveness of ferric oxalate in combination with a platinum salt in the hope of obtaining images in that metal. But the process was a failure because he did not recognize that the ferrous salt, formed by the action of light, must form a complex salt in order to reduce the platinum salt to the metallic state. W. Willis in 1878 introduced the first workable platinum process. The advantage of this process is that the pictures are permanent. The high cost of platinum and the ease with which development papers are worked has resulted in disuse of the process.

Louis Vauquelin in 1798 had discovered the light-sensitiveness of silver chromate; G. I. Suckow in 1832 observed that the bichromates were reduced by light in the presence of organic matter. Mungo Ponton, 1839, used paper treated with potassium bichromate. Edmond Becquerel 1840, and R. Hunt, 1843, made improvements; but Fox Talbot first pointed out, 1852, that bichromated gelatine became insoluble in light, and this was patented as a resist for photogravure, for which it is still used.

Paul Pretsch, Alphonse Poitevin, Testud de Beauregard, H. Garnier and A. Salmon and John Pouncy, 1859, devised carbon processes, in which carbon itself in a finely divided state was used as the pigment in bichromated gelatine. But they failed to recognize the important point that, as the light acted first on the outer surface of the gelatine, those parts underlying the half tones and high lights would still be soluble; hence, on developing with warm water, they would have no anchorage to the support and would be washed away. J. W. Swan in 1864 patented the production of carbon or pigment prints by transfer of the exposed pigmented tissue to a temporary support, on which it was developed from the back. His process is in use today, and its long scale of gradation, the possibility of using pigments of any colour and the stability of the pictures are much in its favour, though the necessity of day- or arc-light for printing is against it.

T. Manly, 1905, patented a pigment process in which a developed silver print was squeegeed into contact with pigmented tissue saturated with potassium ferricyanide and bichromate. These salts migrated to the silver image, bleached it, and the reduction products wandered back to the tissue, rendering it insoluble in ratio to the amount of silver in the picture, as though the tissue had been exposed to light under the negative. A modification of this, known as carbro, was introduced later.

In 1905 G. E. Rawlins reverted to an old process suggested by E. Mariot in 1866. Gelatinized paper is sensitized with bichromate, dried, exposed under a negative, freed from excess salt and inked up with a greasy ink, which adheres only to the light-affected parts. Somewhat similar is the bromoil process first suggested in 1907 by E. J. Wall and worked out by Welborne Piper. This utilizes the principle stated by Howard Farmer in 1889 that finely divided silver imbedded in gelatine reacts with bichromates and renders the contiguous colloid insoluble. A bromide print is treated with a hardening and bleaching solution, fixed, washed and dried, then soaked in water and inked up with a greasy ink which takes only on those parts where there was silver.

A number of printing processes have depended upon the light-sensitivity of salts of iron. The most used is the blueprint process for copying drawings, introduced by Herschel in 1842. Paper is sensitized with a mixture of potassium ferricyanide and ferric ammonium citrate. On exposure to light the latter is changed to the ferrous compound which, on wetting the paper, combines with the ferricyanide to form Prussian blue. In the ferrogallic process the exposed iron salts are developed in gallic acid to give a positive image in dark ink. The ferrous salts formed by exposure to light of ferric compounds can reduce other metal salts to the metal. Platinotype depends essentially on this reaction, but in the argentotype (Herschel, 1842) and kallitype (1889) and the more recent Vandyke processes, a silver image is formed by reduction of silver nitrate by the ferrous salt. In the ferro-gelatine process (A. Tellkampf and A. Traube, 1905), an unwashed blueprint is squeegeed in contact with a moist gelatine surface, then removed and the surface rolled up with greasy ink which gives a positive image which can be transferred to paper.

Light-sensitive organic compounds have formed the basis of many printing processes, the most used being the diazotype process for copying drawings (G. Kögel, 1917, 1920). In it, paper is coated with diazo compounds which are destroyed on exposure to light, development being with a 'coupler' which combines with the diazo compounds not exposed to form a dye.

Colour photography. – In 1810 J. T. Seebeck, in 1840 Sir J. Herschel and later E. Becquerel and Claude Niepce de Saint-Victor, observed that silver chloride darkened by exposure can record the spectrum of the sun by its direct action, in colours similar to those of the spectrum itself. The colours can not be fixed and the method has no practical value.

The first fixed colour photographs from nature were made by G. Lippmann in 1891, using the interference principle, the possibility of which had been pointed out by W. Zenker (1868), Lord Rayleigh (1887) and O. Wiener (1890). In the method, a transparent grainless panchromatic emulsion is coated on glass and exposed with the emulsion in contact with a reflecting surface, usually of mercury. The reflected light gives an interference effect, setting up standing waves in the emulsion. The exposed and developed silver is thus arranged in laminae half a wave length apart, and reflects light of double the separation, giving bright colours. Beautiful colour reproductions can be obtained, but the plates are insensitive and the viewing is not convenient.

The basic principles of modern colour photography are derived from an experiment made in 1861 by James Clerk Maxwell in a study of the theory of colour vision. He photographed coloured ribbons through red, green and blue filters, from the negatives made three positive black-and-white transparencies and projected these in register onto a screen using three lanterns, each transparency being projected by light of the colour by which its negative was taken. The image on the screen was a colour reproduction of the ribbons.

In 1869 Ducos du Hauron published a book in which he laid down the principles which provide the basis for all modern processes of three-colour photography. At about the same time C. Cros published an article with much the same conclusions. Du Hauron's book gives a clear account of the two fundamental processes of colour photography, the additive and the subtractive, with suggestions for carrying them out. The practical development of the processes was delayed for many years until panchromatic plates became available through the introduction of colour sensitizing by Vogel in 1873 and the introduction of isocyanine and carbocyanine sensitizers early in the 20th century.

Modern colour processes are divided into the additive process, which is used only for transparencies for projection and direct viewing, and the subtractive processes which can be used for transparencies and also for paper prints to be viewed by reflected light. In both cases the first step is to make three separation negatives of the subject by the red, green and blue light from it. It is also possible to make colour photographs by the two-colour process, in which the separation negatives are made by half the spectrum, blue-green and orange, but the best colour rendering is obtained by the three-colour process.

Photoengraving. – Photoengraving is the generic term being applied chiefly to the making of etched metal printing plates or blocks (copper, zinc, brass, etc.), the design being in relief for typographical (letterpress) printing. Parts of the process are also used in the photo-lithographic and rotogravure processes for the production of plates and cylinders.

The first specimens were produced, as far as can be ascertained, during the years 1859–1862 by Col. James in Southampton, by Gillot in Paris, Angerer in Vienna and Husnik in Prague. The name gillotype was for many years applied in France and England to line etchings.

Photography was not originally used in their production, the image on metal being obtained either by manual design or by lithographic transfer. The introduction of the photographic negative permitted the reduction or enlargement of the original design for reproduction.

'Nature Autotypy', an early reproduction process, furnished copies of natural objects which, while imperfect reproductions in the sense of our present standards, were still good enough to convey the image in multiple editions. Leonardo da Vinci describes the method in his 'Codex Atlanticus' (1490–1519): 'This paper must be coated with lamp black mixed with sweet oil. The colour of the leaves is thinned by applying white lead dissolved in oil, as printers do with type faces and then it is printed as usual. The leaves (i.e. the impression) will appear dark in the depression and light in the raised parts . . . .'. Several botanical publications of the late 17th and of the 18th centuries were illustrated by this method and are still preserved.

Niepce's early attempts to etch photographic intaglio images on metal sensitized with asphaltum initiated but did not hasten progress in the photomechanical processes. As early as 1841 daguerreotypes were etched in galvanic baths for printing on gravure presses. Donné was the first to show proofs of etched daguerreotypes; Fizeau, Claudet, Berres in Vienna and Grove in England also etched them. Paul Pretsch in London etched them in 1856. He was the originator of the later spray or blast system employed today in etching and was also the first to discover the swelled gelatine process of which many splendid examples are still in existence. The negatives for his work were usually made by the photographer Roger Fenton, using the wet collodion process; the printing plates were electrotypes moulded from the hardened gelatine reliefs.

As early as 1862 Pretsch exhibited at the London Exhibition halftone plates for letterpress printing. Poitevin, independent of Pretsch's work, pursued the same idea and produced relief electrotype printing plates from chromated gelatine originals. George Scamoni in Russia experimented and succeeded independently in the same process. Swan and Woodbury in England produced the Woodbury-type (1863).

In the development of reproduction processes, it was soon realized that the prime necessity for faithful reproduction consisted in finding a means for breaking up the continuous tones of photographs or drawings, reproducing in effect the quarter, half, and three-quarter tones of the subject. Graining and dusting of the printing surfaces, long employed, did not furnish the answer. Talbot recommended as early as 1852 in his gravure process the use of a gauze netting for breaking up the continuous tones into separate printing elements.

As in most inventions, the credit for the invention or discovery of the halftone process of photoengraving cannot be awarded to any one man.

A one-line glass screen in which parallel lines were scratched on an opaque background is described in a French patent of December 14, 1857, by Berchtold. Burnett published a method of single- and crossline screens for photography. Screens are mentioned in a patent of 1865 by the brothers Bullock. Egloffstein's patent of the same year can hardly be cited as his lines were engraved on steel and his process met with no success. Swan in 1866 also used line screens and so did Leggo in the United States in 1871, Jaffe in Vienna in 1877, Stephan H. Horgan in 1880. Meisenbach (1882) used a single-line screen which he turned at 90° after half the exposure was made.

To Frederic E. Ives of Philadelphia, Pennsylvania, is due the idea of the optical V which led to the present crossline screen which he used in his halftone plates, the printing surface of which was a bichromated glue enamel coating.

To the brothers Louis F. and Max Levy, also of Philadelphia, must be accorded the perfected halftone screen (grating) which today is universally used and has not been excelled since 1888.

The Levy brothers coated selected plates of high quality optical glass with an etching ground, in which parallel lines were cut by means of a ruling machine. The ruled lines were etched with hydrofluoric acid and filled in with Canadian balsam. The lines appeared in sharp contrast. The two diagonally ruled glass plates were faced at an angle of 90°, sealed and later bound together with aluminium frames.

One of the important contributions to the halftone process was the invention of the highlight process. F. J. M. Gerland patented such a method in 1883 and the Bassani Company a mechanical improvement in 1925. The outstanding difficulty in the early days of the process was its inability to meet the demands for printing in quantity, rapidity, and the danger of the ink filling up the spaces between dots. This required the use of highly coated, hard-surfaced and expensive paper. In the last twenty-five years, however, the process has been adapted to almost any kind of specification. Newspaper and poster plates are made through screens as coarse as 16 lines to the centimetre and pulp paper periodicals carry illustrations, sharp and full of colour, printed from plates up to 44 lines to the centimetre.

Colour plates. – Photoengraved colour plates are made to represent the subjects in the colours of the original colour photographs of coloured designs or from objects in their natural colours.

The early attempts to reproduce colour effects were the manual painting in of the different tints on black and white prints. Later stencils were used, but seldom for superimposed and colour combinations. The simplest form of colour reproduction followed the manual process and used photoengraved key plates and different plates for each colour or shade. Later shadings and overlapping plates were used to achieve colour nuances and combinations by overprinting.

The three- and four-colour processes used in 1940 are based on a Latin Treatise of 1611 by Dominis who described the additive system of complementary colours. Isaac Newton described light in its colour spectrum (1666). The German Le Blon in 1722, living in Paris, was the first to use three-colour printing with red, yellow and blue printing plates. Clerk Maxwell, the English physicist, was the first to publish (1861) the theory of colour reproduction by means of three-colour light filters. Ducos du Hauron and Charles Cross, independently of each other, published in 1868 their three-colour processes. The discovery (1873) by H. W. Vogel of optical sensitization led to a great advance in the reproduction of colour copy in monochrome by the use of orthochromatic plates.

The brothers Lumière of Lyons, France, gave the craft the autochrome which with other colour records by similar processes often serve as a colour guide when the original painting is not obtainable. Modern panchromatic plates are the result of the successive work of Vogel, König, Homolka, Eder, Miethe, Mees, Traube, Lehman, and other modern workers in the field of optical sensitizers. Colour separation negatives are made today on colour-sensitized gelatino-bromide dry plates or on colour-sensitized collodion emulsion. The old-fashioned hand press in the engraver's shop is now replaced by power presses. Colour reigns and is in steady ascendancy.

Paper. – The art of making paper (excluding papyrus) from fibrous matter appears to have been practiced by the Chinese at a very distant period. Different writers have traced it back to the 2nd century B.C. It is said that it was reputedly discovered about A.D. 104 by the eunuch T'sai Lun, in the district Lu-Yang, province Hunan. It was made from a mixture of mulberry and bamboo bark, rags, fishing nets and probably other vegetable fibres. It is assumed that this paper was made by spreading the pulp on a woven or plaited cloth on which it was left to dry. Two centuries later a change had been made to trays consisting of closely laid bamboo laths joined into matting by hairs or silken threads woven into them at regular intervals. The matting was laid on a groundwork of stronger bamboo rods and was dipped in a vat of pulp. On removal, the water was allowed to run off and the layer of paper was laid out to dry, the flexible matting being rolled away from it.

Paper first became available for the rest of the world in the middle of the 8th century. In 751 the Arabs, who had occupied Samarkand early in the century, were attacked there by Chinese. The invasion was repelled by the Arab governor, who in the pursuit, it is related, captured certain prisoners who were skilled in papermaking and who imparted their knowledge to their new masters. Hence began the Arabian manufacture, which rapidly spread to all parts of the Arab dominions.

The extent to which it was adopted for literary purposes is proved by the comparatively large number of the 9th-century Arabic manuscripts on paper which have been preserved. The material of the Arab paper was apparently substantially linen.

It seems that the Arabs and the skilled Persian workmen whom they employed at once resorted to flax, which grows abundantly in Khurasan, as their principal material, later also making use of rags, supplemented, as the demand grew, with any vegetable fibre that would serve; and that cotton, if used at all, was used very sparingly. Paper of oriental manufacture in the middle ages was usually distinguished by its stoutness, glossy surface and absence of watermarks. Paper was probably first brought into Greece from Asia. There is a record of its use by the empress Irene at about the end of the 11th century, but with one doubtful exception there are no extant Greek manuscripts on paper before the middle of the 13th century.

Paper in Europe. – The manufacture of paper in Europe was first established by the Moors in Spain in the middle of the 12th century, the headquarters of the industry being Xativa, Valencia and Toledo. But on the fall of the Moorish power the manufacture, passing into the hands of the less skilled Christians, declined in quality.

In Italy also the art of papermaking was no doubt established through the Arab occupation of Sicily. But the paper which was made both there and in Spain was in the first instance of oriental quality. In the laws of Alphonso X of 1263 it is referred to as cloth parchment, a term which well describes its stout substance. The first mention of rag paper occurs in the tract of Peter, abbot of Cluny (A.D. 1122–1150), 'adversus Iudaeos', cap. 5.

A few words may be said respecting manuscripts written in European countries on oriental paper or paper made in the oriental fashion. The oldest recorded document on paper was a deed of King Roger of Sicily, of the year 1102; and there are others of Sicilian kings, of the 12th century. A notarial register on paper, at Geneva, dates from 1154. The oldest known imperial deed on the same material is a charter of Frederic II to the nuns of Goess in Styria, of the year 1228. In 1231, however, the same emperor forbade further use of paper for public documents, which were in future to be inscribed on vellum. In Venice the 'Liber plegiorum', the entries in which begin with the year 1223, is made of rough paper; and similarly the registers of the Council of Ten, beginning in 1325, and the register of the emperor Henry VII (1308–1313), preserved at Turin, are written on a like substance. In the British Museum there is an older example in a manuscript (Arundel 268) which contains some astronomical treatises written on an excellent paper in an Italian hand of the first half of the 13th century. In the Public Record Office there is a letter on paper from Raymond, son of Raymond duke of Narbonne and count of Toulouse, to Henry III of England, written within the years 1216–1222. The letters addressed from Castile to Edward I, in 1279 and following years, are instances of Spanish-made paper.

In Italy the first place which appears to have become a great centre of the papermaking industry was Fabriano in the marquisate of Ancona, where mills were first set up in 1276. They rose to importance on the decline of the manufacture in Spain. The earliest known watermarks in paper from this factory are of 1293 and 1294. In 1340 a factory was established at Padua, another rose later at Treviso; and others followed in the territories of Florence, Bologna, Parma, Milan, Venice and other districts. From the factories of northern Italy the wants of southern Germany were supplied as late as the 15th century. But in Germany also factories were rapidly founded. The earliest are said to have been set up between Cologne and Mainz about 1320. England is said to have obtained paper at first from France and Burgundy. France owed the establishment of its first paper mills to Spain.

In the second half of the 14th century the use of paper for all literary purposes had become well established in all western Europe; and in the course of the 15th century it gradually superseded vellum. In manuscripts of the latter period it is not unusual to find a mixture of vellum and paper, a vellum sheet forming the outer, or the outer and inner, leaves of a quire, while the rest are of paper.

Paper in England. – With regard to the early use of paper in England, there is evidence that at the beginning of the 14th century it was a not uncommon material, particularly for registers and accounts. Under the year 1310, the records of Merton college, Oxford, show that paper was purchased 'pro registro'. There is, however, in the British Museum a paper manuscript, written in England, of even earlier date than the one recorded in the Merton archives. This is a register of the hustings court of Lyme Regis, the entries in which begin in the year 1309. The paper, of a rough manufacture, is similar to the kind which was used in Spain.

The knowledge, however, which is available of the history of papermaking in England is extremely scanty. The first maker whose name is known is John Tate, who is said to have set up a mill in Hertford late in the 15th century; and Sir John Spilman, jeweller to Elizabeth I, erected a paper mill at Dartford, and in 1589 obtained a licence for ten years to make all sorts of white writing paper and to gather, for the purpose, all manner of linen rags, scrolls or scraps of parchment, old fishing nets, etc. But it is incredible that no paper was made in the country before the time of the Tudors. The comparatively cheap rates at which it was sold in the 15th century in inland towns suggest that there was at that time a native industry in this commodity.

Paper in America. – The early printers of colonial America imported their paper from Europe, chiefly from the continent. The first paper mill was built in 1690 at Germantown, Pennsylvania, resulting from the combination of the needs of the Philadelphia printer, William Bradford, and the arrival of an ambitious German papermaker, William Rittenhouse. Two other mills were established in Pennsylvania in 1710 and 1729, one in Elizabethtown, N.J., in 1728, and the first in Massachussets at Milton in the same year. Virginia's first paper mill was built at Williamsburg in 1744 by its first newspaper publisher, William Parks. The first in New York was built at Hempstead, Long Island, in 1768. With imports cut off during the Revolutionary War and increased needs for newspapers, broadsides, pamphlets, records, correspondence, etc., an acute paper famine developed. Under these conditions additional mills sprang into existence, and there were probably 80 or 90 when the war ended. Paper manufacturing was protected in the first tariff. In 1810 there were more than 200 mills in operation making about $ 2,000,000 worth of products.

Bookbinding. – Bookbinding began in the Christian era with the change from the continuous roll or volume to the book made up of separate sheets. Early books are composed of single sheets of vellum at first, of paper later – folded once and collected into gatherings or sections of convenient size. The leaves were held together in the section form by sewing through the centre fold. The sections were held together in proper order by sewing them on to flexible bands or thongs at right angles to the backs. Later books only differ in that the section is usually a large single sheet folded several times so that the outer folds require cutting.

To keep the leaves flat and uninjured, early books, which were large, were placed between thin wooden boards. Soon it was found as convenient as it was simple to join book and boards together, by fixing to the boards the ends of the bands holding together the sections. By the time a leather covering had been added to hide and protect the back of the sections, overlapping or completely covering the boards, all the elements of the modern book, half bound or fully bound, had been evolved. A greater variety of materials is now used, but the principle of construction remains the same.

Early book decoration. – The covers of the book bound lent themselves readily to ornamentation and decoration. Already in the letters of St. Jerome reference is made to jewelled bindings, so that books were being sumptuously ornamented by the 4th century of the Christian era. Costly bindings were often destroyed for their valuable materials; but examples survive, mostly in churches and museums, of books covered or decorated with precious metals, enamels, jewels or carved ivory panels. The earliest is the 7th century Gospels of Theolinda at Monza in northern Italy. Famous examples are the so-called Gospels of Charlemagne in the Victoria and Albert Museum, London, and the Lindau Gospels in the J. Pierpont Morgan collection, New York. Others exist in libraries such as the British Museum and the Bibliothèque Nationale.

These precious bindings are, and always were, unusual; they are mostly found on devotional books intended for royal personages or for the service of the church. The ordinary book, covered wholly or partly with leather over boards, was decorated with patterns of lines or stamps, or both. The earliest surviving decorated leather binding on a book, belonging to the same period as the earliest known precious binding, is on St. Cuthbert's Gospel-book, taken from his coffin when it was transferred to the new Durham cathedral in 1104, and now preserved at Stonyhurst college. This is ornamented with repoussé and painted line-work, and stands quite by itself. The other early decorated bindings are impressed with small stamps in blind (i.e., ungilt) in more or less elaborate patterns, and, apart from isolated examples, date from the century in which St. Cuthbert's Gospel book was found. Fine examples of these bindings were made during that century at Winchester and Durham, and later at Oxford, Cambridge and London. Such bindings with small stamps, supplemented at the very end of the period by roll stamps, were the prevailing fashion in all the European countries from the 12th to the 15th century; but in the Germanic countries, cut leather bindings also were produced by incising a pattern in the leather, the outline being sometimes emphasized by stippling the background.

With the introduction of printing into Europe about the middle of the 15th century the number of books produced suddenly increased enormously, involving a corresponding increase of the number of people employed in binding. The making and binding of books was transferred from the monasteries to the houses of printers and binders, and soon the names, initials or devices of printers or binders are found stamped on book covers. Advances were made in the art of binding, styles of decoration developed and as books circulated freely, were copied in other countries.

Famous bookbinding styles. – About a quarter of a century after the invention of printing, the greatest advance was made in artistic bookbinding in Europe by the introduction, probably through Venice, of gold-tooling from the East, where it had been practiced much earlier. The art quickly developed in Italy and spread to other countries. The celebrated Venetian printer, Aldus Manutius, was the first to give his name to a style in this new art; but in general, the early styles are called not after their producers, who are unknown, but after famous collectors, or reputed collectors. 'Canevari' bindings, which have in the centre a cameo stamp of Apollo driving a chariot, were so-called after their supposed collector Demetrio Canevari, physician to Pope Urban VIII. Their real collector has later been shown to have been Pier Luigi Farnese, son of the succeeding Pope, Paul III. Many of the finest Italian and French bindings were made for Jean Grolier, viscount d'Aguisy, treasurer of France in 1545, and bear upon them the legend 'Portio mea domine sit in terra viventium' and 'Io. Grolierii et Amicorum'. Although not uniform in origin or appearance, these are known as Grolier or Grolieresque bindings. 'Maioli' bindings are named after Thomas Maiolus, another famous collector of the period, who used similar inscriptions to Grolier's. Until quite recently he was considered to be an Italian, Tommasso Maioli, but he is now claimed to be a Frenchman, Thomas Mahieu, and identified with the secretary of Catherine de Medici.

Italian gold-tooled bindings were imitated in other countries. In England, Thomas Berthelet, printer and binder to Henry VIII, was amongst the first to produce gold-tooled bindings 'in the Venetian manner', while Thomas Wotton, as a collector, is an English counterpart to Grolier. In Germany, on the other hand, blind stamping, especially with panels on pigskin, continued in general use. In Italy itself fine bindings long continued to be made for great patrons like popes and cardinals, but recent investigations suggest that the supremacy in binding was passing to France earlier than is generally supposed. Grolier and Maioli bindings were produced in France under those patrons of the arts, Francis I and Henry II. The royal printer and binder for the former, Geoffrey Tory, who also worked for Grolier, designed a decoration for bindings made for his books, which includes his device of the pot cassé. Fine bindings were made for Henry II, for his queen, Catherine de Medici, and for his mistress, Diane de Poitiers, and from this time onwards French binders and families of binders have excelled in technical skill and initiative. Later royal binders, Nicholas and Clovis Eve, developed 'fanfare' binding usually associated with their name. Le Gascon in the early 17th century developed the pointillé style where the dotted line replaces the solid line; and the Padeloup and Derome families of the late 17th and 18th century developed the dentelle binding, so called from their lace-work borders.

In England, after Berthelet's time, fine bindings were made for royal and noble patrons, and usually decorated with their arms or badges. Meantime velvet and embroidered bindings, common to most countries, increased in vogue and became especially popular during the Stuart period. In gold-tooled leather bindings a characteristic native style was not evolved until the late 17th century, when Samuel Mearne, royal binder to Charles II, devised the 'cottage' design, so-called from its walls and its roof appearance. Along with the richly decorated binding of Mearne and his followers, the characteristic blind-tooled black leather binding, with dark, instead of gilt edges, became fashionable for religious books in England, for some half a century. English binding deteriorated during the 18th century, but it was redeemed towards the end of the century by a brilliant and original artist, Roger Payne, who with his fine small tools and original designs, with their proper appreciation of blank spaces, is the most inspiring of the English book binders. During his time John Edwards of Halifax worked on different lines; he was famous for his transparent vellum bindings covering delicate paintings, and, with John

Whitaker, for 'Etruscan' bindings, so called from their classical borders and other ornamentations that were carried out in the classical tradition.

Modern work. – The 19th century witnessed the development of decoration by machinery, whole covers being impressed in blind or gold by means of metal dies, a practice which was greatly extended with the introduction of machine-made cloth bindings. But these developments hardly affected high-class bindings. In France, from the beginning of the century onwards, binding and decoration have been more remarkable than ever for their technical perfection, at the hands of a long line of artists; Bozérian, Thouvenin, Bauzonnet, Trautz, Lortic, Niedrée, Duru, Capé, Chambolle, Cuzin, Michel and others.

In England the most original binder of last century was Charles Lewis (Thomas Grenville's binder), while others followed traditional styles, sometimes rather mechanically, notably Kalthoeber, Staggemeier, Walther, Hering, Bedford and the existing firms of Rivière and Zaehnsdorf. But towards the end of the century an artistic revival took place, inspired by William Morris, who was responsible for the modern revival in printing. The practical founder of the modern school was T. J. Cobden Sanderson, who established the Doves Press and bindery; a series of his fine bindings is in the British Museum. His most successful pupil is Douglas Cockerell, whose increasing output, revealing originality, combined with a sense of the craft's historic background, confirms his position as the head of the bookbinders of the 1960s. Mention may also be made of Charles Ricketts, especially for his work on the Vale Press books and of a recent convert to the decoration of bindings, Glyn W. Philpot R.A. A number of women bookbinders, miss Adams (Mrs Webb), miss E. M. MacColl, miss Sarah Prideaux, miss Sybil Pye, miss Mary Robinson and others, have helped to increase the prestige of the modern English school, which has influenced the course of artistic binding on the Continent and in the United States.

In France and Germany, and to a lesser extent elsewhere, besides work on sounder lines, there is an increasing output of bindings which exceeds the true limits of book decoration, partly under the influence of modernist tendencies in other branches of art.

BASIC TABLE

1 abbreviate *v* — To reduce, as a word or phrase, to a shorter form intended to stand for the whole. See also hyphenation, ligature.
f abréger — **d** abkürzen; abbreviieren — **n** abbreviëren; afkorten — **e** abreviar; condensar; dividir; cortar; acortar — **i** abbreviare
2 abbreviation — Letter or group of letters representing a word or words. Also a shortened or contracted form of a word. See also hyphenation, ligature.
f abréviation *f* — **d** Abkürzung *f*; Abbreviatur *f* — **n** afkorting *f*; abbreviatie *f* — **e** abreviatura *f* — **i** abbreviazione *f*; abbreviatura *f*
3 ABC picture book — See picture primer.
4 abend — Abnormal end of computer task due to non-recoverable error condition.
5 aberration — Any of several optical errors contained in photographic lenses and which prevent the instrument from giving perfect definition. See astigmatism, chromatic aberration, spherical aberration, optical aberration.
f aberration *f* — **d** Aberration *f*; Verzerrung *f*; Abweichung *f* — **n** aberratie *f*; afwijking *f* — **e** aberración *f* — **i** aberrazione *f*
6 abietic acid; abietinic acid — Major active ingredient of *rosin where it occurs with other acids of closely related structure and properties, i.e., the resin acids. The term is often applied to these mixtures, separation of which is difficult and not achieved in technical grade material. Abietates (resinates) of heavy metals are used as varnish driers; esters in lacquers and varnishes. It is a yellowish resinous powder, soluble in alcohol, ether, chloroform, and benzene, insoluble in water. Formula: $C_{19}H_{29}COOH$.
f acide *m* abiétique — **d** Abietinsäure *f* — **n** abiëtinezuur *n* — **e** ácido *m* abiético; ácido *m* abietínico — **i** acido *m* abietico
7 abietinic acid — See abietic acid.
8 able to work
f capable de travailler; apte au travail; utilisable — **d** arbeitsfähig — **n** arbeidsgeschikt; geschikt tot werken; geschikt om werk te doen — **e** capaz de obrar — **i** abile al lavoro
9 abort *v* — Abortion of a program or process. To terminate the procedure, presumably because of detection, by hardware or software, of an error condition which prevents further processing or which, if it permitted it, would produce invalid results.
10 abrasion — Damage caused by the *scuffing or friction of a part against its package, or of a package against an external object.
f abrasion *f*; effet *m* abrasif — **d** Abreibung *f*; Abschürfung *f*; Abscheuerung *f*; Abschabung *f*; Verschleiß *m*; Reibung *f*; Abschleifen *n* — **n** afschuring *f*; afslijting *f*; afslijten *n*; slijtage *f* — **e** abrasión *f*; desgaste *m*; raspadura *f*; arañazo *m* — **i** logorio *m*; usura *f*; abrasione *f*

11 abrasion resistance; scuff resistance; abrasive resistance; abrasiveness — Ability to withstand the effects of repeated rubbing, scuffing and scratching.
f résistance *f* à l'abrasion; résistance *f* au frottement; résistance *f* à l'usure; résistance *f* aux abrasifs — **d** Abriebfestigkeit *f*; Scheuerfestigkeit *f*; Reibfestigkeit *f*; Abreibwiderstand *m* — **n** schuurweerstand *m*; slijtvastheid *f* — **e** resistencia *f* al frote; desgaste *m* por roce; desgaste *m* por rozamiento; resistencia *f* al rozamiento; resistencia *f* a la abrasión — **i** resistenza *f* all'abrasione; resistenza *f* allo sfregamento; resistenza *f* allo strofinio
12 abrasion resistant
f résistant à l'abrasion — **d** abriebfest — **n** slijtvast — **e** resistente al desgaste por abrasión; resistente al desgaste por roce — **i** resistente all'abrasione
13 abrasion test — Test designed to determine the ability to withstand the effects of rubbing and scuffing of printing ink.
f essai *m* de résistance au frottement — **d** Scheuerfestigkeitsversuch *m*; Reibfestigkeitsversuch *m*; Scheuertest *m* — **n** slijtvastheidsproef *fm* — **e** ensayo *m* al rozamiento; prova *f* di abrasività; prova *f* di abrasione; saggio *m* di abrasione
14 abrasive; abrasive material — Substance, e.g., aluminium oxide, crushed quartz, sand, garnet, used for rubbing or grinding down surfaces.
f abrasif *m* — **d** Schleifmittel *n*; Scheuermittel *n*; Abschabemittel *n*; Schleifmaterial *n* — **n** schuurmiddel *n*; slijpmiddel *n* — **e** lijante *m*; raedor *m*; abrasivo *m*; sustancia *f* raspadora; sustancia *f* abrasiva; corroyente *m* — **i** abrasivo *m*
15 abrasive material — See abrasive.
16 abrasiveness
See abrasion resistance.
17 abrasive paper; emery paper; sandpaper — Good grade of manila or other strong paper, coated with glue and an abrasive material, such as sand, emery, carborundum.
f papier *m* abrasif; papier *m* émeri — **d** Schleifpapier *n*; Putzpapier *n*; Polierpapier *n* — **n** schuurpapier *n*; slijppapier *n*; polijstpapier *n* — **e** papel *m* de lija; papel *m* (de) esmeril; papel *m* raspador — **i** carta *f* abrasiva; carta *f* smerigliata
18 abrasive resistance — See abrasion resistance.
19 abridged edition
f édition *f* abrégée; édition *f* réduite — **d** gekürzte Ausgabe *f*; Kurzausgabe *f*; kleine Ausgabe *f* — **n** verkorte uitgave *f*; kleine uitgave *f* — **e** edición *f* abreviada — **i** edizione *f* ridotta
20 abridgement; summary — Short condensation of a book or a long text or other printed work, emphasizing the most important points.
f abrégé *m*; précis *m*; résumé *m*; accourci *m*; aperçu *m*; sommaire *m* — **d** Abriß *m*; Zusammen-

fassung *f*; Übersicht *f*; Auszug *m* — **n** samenvatting *f*; resumé *n*; overzicht *n* — **e** resumen *m*; sumario *m*; abreviación *f*; compendio *m* — **i** sunto *m*; riassunto *m*; sommario *m*; ricapitolazione *f*

21 absciss — See abscissa.

22 abscissa; abscisse; absciss — X-axis (horizontal axis) in a diagram (*pl* abscissae; abscissas).
f abscisse *f*; axe *m* des abscisses — **d** Abszisse *f*; Abszissenachse *f* — **n** abscis *fm* — **e** abscisa *f* — **i** ascissa *f*; asse *m* dell'ascissa

23 abscisse — See abscissa.

24 absence-of-bloom gloss — See gloss.

25 absolute address; machine address; specific address — Pattern of characters that identifies a unique store location without further modification.
f adresse *f* absolue — **d** absolute Adresse *f* — **n** absoluut adres *n* — **e** dirección *f* absoluta; dirección *f* real — **i** indirizzo *m* assoluto

26 absolute addressing — See absolute coding.

27 absolute alcohol — *Ethyl alcohol containing not more than 0.2% of water.

28 absolute coding; absolute addressing; specific coding; specific addressing; actual coding — Coding which uses machine instructions in which absolute addresses are employed.
f codage *m* en absolu; adressage *m* en absolu — **d** absolute Adressierung *f* — **n** absolute codering *f*; absolute adressering *f* — **e** codificación *f* absoluta; codificación *f* real; direccionamiento *m* absoluto — **i** codifica *f* in assoluto; indirizzamento *m* in assoluto

29 absolute humidity — The amount of water vapour present in the air. At a temperature of 20 °C the air can contain 17.3 g water vapour per m^3.
f humidité *f* absolue (de l'air) — **d** absolute Feuchtigkeit *f*; absolute Luftfeuchtigkeit *f* — **n** absolute vochtigheid *f* (van de lucht) — **e** humedad *f* absoluta — **i** umidità *f* assoluta

30 absolute temperature — The temperature which is measured in relation to the absolute zero. See Kelvin.
f température *f* absolue — **d** absolute Temperatur *f* — **n** absolute temperatuur *f* — **e** temperatura *f* absoluta — **i** temperatura *f* assoluta

31 absorb *v*
f absorber — **d** absorbieren; aufsaugen; aufnehmen; einsaugen; saugen; verschlingen; zurück halten — **n** absorberen; opzuigen; opslorpen — **e** absorber — **i** assorbire

32 absorbency — That property of a material, such as paper, which causes it to take up liquids or vapours (e.g., moisture) with which it is in contact.
f pouvoir *m* absorbant — **d** Saugfähigkeit *f*; Absorptionsfähigkeit *f* — **n** absorberend vermogen *n*; absorptie *f* — **e** poder *m* absorbente; propiedad *f* absorbente; propiedad *f* de absorción; succibilidad *f* — **i** potere *m* assorbente; assorbenza *f*

33 absorbent paper — General term representing a class of bulky papers, spongy and bibulous in character, such as blotting, filter and towelling paper.
f papier *m* hydrophile; papier *m* absorbant — **d** saugfähiges Papier *n* — **n** absorberend papier *n*; opzuigend papier *n* — **e** papel *m* absorbente; papel *m* secante — **i** carta *f* assorbente

34 absorption — Penetration of a liquid into pore space. Absorption should be distinguished from *adsorption, the latter being a surface phenomenon.
f absorption *f*; absorptivité *f* — **d** Absorption *f*; Wegschlagen *n* (der Farbe); Eindringen *n* (der Farbe) — **n** absorptie *f*; wegslaan *n* (van de inkt) — **e** absorción *f* — **i** assorbimento *m*

35 absorption — Spectral absorption is illustrated when certain wavelengths of sunlight are absorbed by ordinary glass, thereby warming it slightly (the light energy is changed to heat) and depriving the transmitted light of certain wavelengths, particularly those of the ultraviolet spectrum. Transparent liquids are often identified and analysed by passing a beam of light through them and noting the extent of absorption of original light.
f absorption *f* — **d** Absorption *f*; Aufsaugung *f* — **n** absorptie *f*; opslorping *f*; opneming *f*; opname *fm* — **e** absorción *f* (óptica) — **i** assorbimento *m* (ottico)

36 absorption band; absorption range — Region of the *absorption spectrum in which the absorption passes through a maximum or inflection.
f bande *f* d'absorption; plage *f* d'absorption — **d** Absorptionsbande *f*; Absorptionsbereich *m* — **n** absorptieband *m* — **e** banda *f* de absorción — **i** banda *f* d'assorbimento; intervallo *m* d'assorbimento; fascia *f* d'assorbimento

37 absorption coefficient; absorption rate; absorption value
f coefficient *m* d'absorption — **d** Absorptionskoeffizient *m* — **n** absorptiecoëfficiënt *m* — **e** coeficiente *m* de absorción — **i** coefficiente *m* d'assorbimento

38 absorption ink test — This paper-porosity test follows the principle of applying a relative thick layer of a special coloured absorption ink to the paper and removing the surplus after some time has elapsed. The extent to which the paper is coloured gives an indication of its porosity.
f essai *m* de porosité à l'encre — **d** Farbabsorptionstest. *m*; Absorptionstest *m* — **n** inktabsorptieproef *fm* — **e** ensayo *m* de la absorción de la tinta; prueba *f* de la absorción de la tinta — **i** misura *f* dell'assorbenza

39 absorption range — See absorption band.

40 absorption rate — See absorption coefficient.

41 absorption spectrum — Spectrum consisting of dark lines or bands on a coloured background, formed when a substance is placed between a white light source and the spectroscope. The lines or bands are caused by the absorption of the incident light by the substance at certain wavelengths and their position is characteristic of the substance.

f spectre *m* d'absorption — **d** Absorptionsspektrum *n* — **n** absorptiespectrum *n* — **e** espectro *m* de absorción — **i** spettro *m* d'assorbimento

42 absorption tester — Instrument for determining the mean pore radius of a paper. When a strip of paper is hanged in a liquid, the level of the absorbed liquid can be seen. The rate of absorption can be timed. In this way an average pore radius can be measured. Viscosity and surface tension of the liquid must be known, and the contact angle must be 0 or very low. According to Lucas/Washburn the absorption is: $h_t^2 = \gamma rt/2\eta$ in which: h_t = height of column of liquid in cm after t seconds; γ = surface tension of liquid in dynes/cm; η = viscosity of liquid in poises; r = radius of capillary in cm; t = time in seconds. When γ and η are known, the radius of the capillary is: $r = 2\eta/\gamma \cdot h_t^2/t$ cm.

f altimètre *m* d'aspiration; appareil *m* pour l'essai à l'aptitude à absorber — **d** Saughöhenmesser *m*; Gerät *n* zur Bestimmung der Saughöhe — **n** opzuighoogtemeter *m*; opzuighoogtetoestel *n* — **e** aparato *m* para determinar la absorción; ensayador *m* de la absorción — **i** apparecchio *m* per la determinazione dell'assorbimento

43 absorption value — See absorption coefficient.

44 absorptive capacity — Volume of the pores and interstices of a paper or board capable of being filled with a fluid. It is determined under the conditions defined in the standard method of test.

f porosité *f* — **d** Porenvolumen *n* — **n** absorptievermogen *n*; opzuigend vermogen *n*; absorberend vermogen *n*; capillair volume *n* — **e** porosidad *f*; capacidad *f* de absorción; poder *m* absorbente — **i** porosità *f*; capacità *f* assorbente

45 abutted rules — Two rules put together to form a joint or a right angle as opposed to mitred rules.

f filets *mpl* assemblés sans biseautage — **d** zusammengestoßene Linien *fpl* (im Gegensatz zu mit Gehrung gesetzten Linienverbindungen) — **n** lijnen *fmpl* zonder verstek — **e** filetes *mpl* contrapeados; filetes *mpl* sin chaflán — **i** filetti *mpl* uniti di testa; filetti *mpl* senza smusso

46 AC — See alternating current.

47 academic dissertation — See thesis.

48 accelerate *v*; **speed** *v* **up**

f accélérer — **d** beschleunigen — **n** versnellen — **e** acelerar — **i** accelerare

49 accelerated test

f essai *m* accéléré — **d** Kurzprüfung *f*; Kurzzeitprüfung *f* — **n** versnelde proef *fm* — **e** ensayo *m* acelerado — **i** prova *f* accelerata

50 accelerating incentive; steepening incentive — See also differential piecework, geared incentive, multiple time plan.

f prime *f* de production très stimulante — **d** progressiver Leistungslohn *m* — **n** progressieve loonprikkel *m* — **e** prima estimuladora *f* a la producción; estímulo *m* acelerante — **i** incentivo *m*; incentivazione *f*

51 accelerating speed — Constantly increasing speed.

f vitesse *f* accélérée — **d** zunehmende Geschwindigkeit *f* — **n** toenemende snelheid *f* — **e** velocidad *f* acelerante; aumento *m* de velocidad — **i** velocità *f* accelerante

52 acceleration — Change in velocity; increasing in speed.

f accélération *f* — **d** Beschleunigung *f* — **n** versnelling *f* — **e** aceleración *f* — **i** accelerazione *f*; acceleramento *m*

53 accelerator — Chemical constituent of photographic developers which activates the developing agent and swells the gelatine to hasten penetration of the solution. See sodium borate, sodium carbonate, sodium hydroxide.

f accélérateur *m* — **d** Beschleunigungsmittel *n* — **n** versnellingsmiddel *n* — **e** medio *m* acelerador; producto *m* acelerador; acelerador *m* — **i** accelerante *m*; acceleratore *m*

54 accent — Diacritical marks usually made above sometimes below alphabetic characters in some languages to indicate correct pronunciation. The standard accents, shown here on an appropriate character are: acute (é), grave (è), circumflex (ê), diaeresis or umlaut (ü), cedilla (ç), and tilde (ñ). In mathematics a symbol used to distinguish similar quantities which differ in value, as in b', b", b''' (called b prime, b second or b double prime, b third or b triple prime, respectively). Accents are cast on the same type body as the characters for text but "floated" for display lines. See also floating accent.

f accent *m* — **d** Akzent *m* — **n** accent *n* — **e** acento *m*; ápice *m* — **i** accento *m*

55 accented letters — Those letters with the various added marks used to indicate pronunciation.

f lettres *fpl* accentuées — **d** Akzentbuchstaben *mpl* — **n** letters *fmpl* met accenten — **e** letras *fpl* con acento; letras *fpl* acentuadas; caracteres *mpl* con acento — **i** lettere *fpl* accentate

56 accented letters case

f casseau *m* de caractères accentuées — **d** Akzentkasten *m*; Akzentbuchstabenkasten *m* — **n** accentenkast *fm* — **e** caja *f* para letras acentuadas — **i** cassa *f* per lettere accentate

57 accentuate *v*

f accentuer — **d** akzentuieren — **n** accenten aanbrengen — **e** acentuar — **i** accentare

58 accentuated with bold

f avec du mi-gras — **d** mit Halbfetter ausgezeichnet — **n** met halfvet — **e** (acentuado) en seminegra; (acentuado) en negrilla — **i** in grassetto; (posto) in evidenza in grassetto; in neretto; (posto) in evidenza in neretto

59 accentuated with italic
f avec de l'italique — d mit Kursiv ausgezeichnet — n met cursief — e (acentuado) en cursiva — i in corsivo; (posto) in evidenza in corsivo

60 accentuated with small caps; accentuated with smalls
f avec des petites capitales — d mit Kapitälchen ausgezeichnet — n met kleinkapitaal — e (acentuado) en versalitas — i in maiuscoletto; (posto) in evidenza in maiuscoletto

61 accentuated with smalls — See accentuated with small caps.

62 accentuation — Placing of the grammatical accents.
f accentuation f — d Akzentuieren n — n accentuering f; accentuatie f; accentueren n — e acentuación f — i accentazione f; accentatura f

63 acceptable quality level
f niveau m de qualité acceptable — d zulässiges Qualitätsniveau n; annehmbare Qualität f der Lieferung; annehmbare Qualitätslage f; Annahmegrenze f; Gutgrenze f — n aanvaardbaar fabricageniveau n; aanvaardbare grenskwaliteit f; aanvaardbare kwaliteitsgrens fm — e calidad f aceptable — i livello m di qualità accettabile

64 acceptance number
f nombre m limite d'acceptation — d Gutzahl f; Annahmezahl f — n goedkeur-criterium n — e cantidad f límite de aceptación — i numero m d'accettazione

65 accepted stock; accepts — *Stock after it has been subjected to stock cleaning.
f pâte f épurée — d büttenfertiger Stoff m; Gutstoff m — n gezuiverde stof fm — e pasta f depurada — i impasto m accettato

66 accepts — See accepted stock.

67 access — To locate and process an area of main or auxiliary storage. See also random access, serial access.
f accès m — d Zugang m; Zugriff m — n toegang m — e acceso m — i accesso m

68 accessibility — The degree of approachability or reach. Good accessibility is an important consideration in the design of machinery having great bearing on the economic running of machines.
f accessibilité f; commodité f d'accès — d Zugänglichkeit f — n toegankelijkheid f — e accesibilidad f — i accessibilità f

69 accessories — Tools or parts necessary to the operation of machinery, such as a printing press, a paper machine, etc.
f accessoires mpl — d Zubehörteile fpl; Zubehör n — n toebehoren n; accessoires npl — e accesorios mpl; aditamentos mpl; utensilios mpl — i accessori mpl; pertinenze fpl; parti fpl ausiliarie

70 access time; read time — Computer-controlled typesetting term for the time interval between the call for the transfer of data to or from a computer store and the instant the operation is completed. The call is made by the control unit. The access time is the sum of the transfer time and the waiting time. The access speeds have an important bearing on the efficiency of a computer installation. Not to be confused with cycle time.
f temps m d'accès — d Zugriffszeit m — n toegangstijd m; accestijd m — e tiempo m de acceso — i tempo m di avviamento; tempo m di accesso

71 accidental error — In repeated observations of a quantity which in principle constant, it is in general found that slightly different values are obtained.
f erreur f accidentelle — d zufälliger Fehler m — n toevallige fout f — e error m accidental — i errore m occasionale

72 accordion fold — See accordion pleat.

73 accordion pleat GB; **accordion fold** US; **concertina fold; fan fold; over and back fold** — Map or large insert spread, folded to fit into a book, with folds parallel and in opposite direction for each successive fold.
f pliure m en accordéon; planche f en paravent; pliage m en accordéon; pliage m en zigzag — d Leporellofalz m; Harmonikafalz m; Zickzackfalz m — n harmonika-vouw fm; Leporello-vouw fm — e plegado m de acordeón; plegado m en zigzag; pliegue m en abanico; doblado m en acordeón; doblado m en fuelle; doblez m en zigzag; doblez m de acordeón — i piegatura f a soffietto; piegatura f a Leporello; piegatura f ad organetto; piegatura f a organetto; piegatura f a fisarmonica

74 account — In an *advertising agency the client whose advertising is handled.
f client m — d Kunde m — n klant m; cliënt m — e cliente m; cuenta m — i cliente m

75 accountant — One who is skilled in the practice of accounting or in charge of public or private accounts.
f comptable m — d Buchrevisor m; Bücherrevisor m; Rechnungsrevisor m — n accountant m — e contador m (público); perito m de contabilidad; revisor m de libros — i perito m contabile

76 account book; ledger — Book in which personal or commercial accounts are recorded.
f livre m de comptes; registre m de comptabilité — d Kontobuch n — n kantoorboek n — e libro m de cuentas; libro m de comercio — i libro m d'ufficio

77 account-book binders
f fabrique f de registres — d Geschäftsbücherfabrik f; Kontobuchbinderei f — n kantoorboekenfabriek f — e fábrica f de libros rayados — i fabbrica f di libri in bianco

78 account-book paper — See ledger paper.

79 account director — See advertising agency.

80 account executive — In an *advertising agency, contact man who provides liaison between client and agency, interprets client's desires into agency practice and advises client.
f directeur m de la clientèle — d Kontakter m —

e jefe *m* de cuenta; jefe *m* de órdenes; coordinador *m* publicitario; ejecutivo *m* de cuenta; ejecutivo *m* de órdenes — i responsabile *m* dei rapporti con la clientela

81 account mark — Sign used in commercial work, as @ or a/c (a conto).
f symbole *m* comptable — **d** Akontozeichen *n* — **n** a conto-teken *n* — **e** signo *m* de cuenta — i simbolo *m* contabile

82 account rule — The vertical lines (usually coloured) in ledgers, invoices, etc., to accommodate for prices, discounts, totals, etc.
f filet *m* de comptabilité — **d** Kontobuchlinie *f*; Spaltenlinie *f* (in Rechnungsformularen); Potenzlinie *f* — **n** boekhoudlijn *fm* — **e** filete *m* vertical de estados — i rigatura *f* per partite contabili

83 accumulator *GB*; **storage cell** *US*; **storage battery** — A cell or connected group of cells that converts chemical energy into electrical energy by reversible chemical reactions and that may be recharged by passing a current through it in the direction opposite to that of its discharge.
f accumulateur *m* — **d** Akkumulator *m* — **n** accumulator *m*; accu *m* — **e** acumulador *m* — i accumulatore *m*

84 accumulator — A computer device including an arithmetic register which stores a number (the *augend) and which on receipt of a second number (the *addend) adds them and stores the sum in place of the augend.
f accumulateur *m* — **d** Akkumulator *m* — **n** accumulator *m* — **e** acumulador *m* — i accumulatore *m*

85 accuracy
f précision *f*; exactitude *f* — **d** Genauigkeit *f*; Richtigkeit *f*; Präzision *f*; Vollständigkeit *f* — **n** nauwkeurigheid *f*; precisie *f* — **e** precisión *f*; exactitud *f* — i precisione *f*; accuratezza *f*; esattezza *f*

86 acetaldehyde; acetic aldehyde; ethyl aldehyde; ethanal; acetyl hydride *arch.* — Oxidation product of ethyl alcohol. An intermediate for acetic acid. Important raw material for organic compounds. It is a colourless, flammable liquid, miscible with water, alcohol, ether, benzene, gasoline, solvent naphtha, toluene, xylene, turpentine and acetone. Used in photography. Formula: CH_3CHO. TL-value: 200 ppm.
f aldéhyde *m* acétique; acétaldéhyde *m*; aldéhyde *m* éthylique; éthanal *m* — **d** Azetaldehyd *m*; Äthylaldehyd *m*; Äthanal *m* — **n** acetaldehyd(e) *m*; ethanal *m* — **e** acetaldehído *m*; aldehído *m* acético — i acetaldeide *f*; aldeide *f* acetica

87 acetaldehyde resin
f résine *f* acétaldéhyde — **d** Azetaldehydharz *n* — **n** acetaldehyd(e)hars *mn* — **e** resina *f* de acetaldehído — i resina *f* di acetaldeide

88 acetate — A compound derived from *acetic acid by replacing the acid hydrogen by a metal or a radical, so that the resulting compound contains the acetate radical or group

(CH_3COO^-).
f acétate *m* — **d** Azetat *n* — **n** acetaat *n* — **e** acetato *m* — i acetato *m*

89 acetate base — Common transparent plastic sheeting used as a base for photographic films or as a clear base for stripping.
f support *m* d'acétate — **d** Azetatträgerschicht *f*; Azetatunterlage *f* — **n** drager *m* uit acetaat; drager *m* van acetaat — **e** soporte *m* de acetato; base *f* de acetato — i supporto *m* in acetato

90 acetate film; triacetate film; safety film — Cellulose acetate film base, which replaced highly inflammable cellulose nitrate.
f film *m* acétate; film *m* triacétate; film *m* de sécurité — **d** Azetatfilm *m*; Triazetatfilm *m*; Sicherheitsfilm *m* — **n** acetaatfilm *m*; triacetaatfilm *m* — **e** película *f* de acetato; película *f* de triacetato — i pellicola *f* in acetato; pellicola *f* in triacetato

91 acetate foil — Transparent sheet made from cellulose acetate by spreading esterified cellulose onto a highly polished cylinder, so forming a thin film which can be stripped off. It is waterproof, and does not burn but melts. Increasingly used for packing purposes.
f feuille *f* d'acétate; pellicule *f* d'acétate — **d** Azetatfolie *f*; Zellglasfolie *f* — **n** acetaatfo(e)lie *fm*; celluloseacetaatfo(e)lie *fm* — **e** película *f* de acetato; lámina *f* de acetato; hoja *f* de acetato — i pellicola *f* di acetato; foglio *m* di acetato

92 acetate of lead — See lead acetate.
93 acetate of soda — See sodium acetate.
94 acetate proof — An inked reproduction proof pulled from type on thin acetate sheeting. Can be used as a positive for making contact negatives, or directly for stripping up a positive flat, and also for checking register or making a transfer of an etched engraving to another sheet of metal.
f cello *m* — **d** Andruck *m* auf Azetatfolie — **n** afdruk *m* op film — **e** prueba *f* en acetato — i bozza *f* su acetato

95 acetic acid; ethanoic acid; methanecarboxylic acid; vinegar acid — Clear, colourless, acid liquid miscible with water, alcohol, glycerol, and ether, insoluble in carbon disulphide. Contact with chromic acid, nitric acid, ethylene glycol and permanganates is dangerous. Latin: acidum aceticum. TL-value: 10 ppm or 25 mg/m^3.
Formula: CH_3COOH. See also glacial acetic acid.
f acide *m* acétique; alcool *m* de vinaigre; esprit *m* de vinaigre; acide *m* pyroligneux — **d** Essigsäure *f* — **n** azijnzuur *n*; methaancarbonzuur *n* — **e** ácido *m* acético — i acido *m* acetico

96 acetic acid anhydride — See acetic anhydride.
97 acetic aldehyde — See acetaldehyde.
98 acetic anhydride; acetic oxide; acetyl oxide; acetic acid anhydride — Colourless, very mobile, strongly refractive liquid. Miscible with alcohol, ether and acetic acid, decomposes in water, forming acetic acid. Latin: acidum aceticum anhydricum. TL-value: 5 ppm.
Formula: $(CH_3CO)_2O$.

f anhydride *m* acétique; anhydride *m* d'acide acétique — **d** Essigsäureanhydrid *m* — **n** azijn-zuuranhydride *n* — **e** anhídrido *m* acético — **i** anidride *f* acetica

99 acetic ether — See ethyl acetate.

100 acetic oxide — See acetic anhydride.

101 aceto-acetanilide; acetyl-acetanilide; acetyl acetic anilide — White, crystalline solid, slightly soluble in water, soluble in dilute sodium hydroxide, alcohol, ether, acids, chloroform, and hot benzene. It is used in organic synthesis; dyestuffs (intermediate in the manufacture of the dry colours generally referred to as Hansa and benzidine yellows).
Formula: $CH_3COCH_2CONHC_6H_5$.

f acéto-acétanilide *m*; acétyl-acétanilide *m*; anilide *m* de l'acide acétyl-acétique — **d** Azetylazet-anilid *n*; Azetoazetanilid *n*; Azetessigsäureanilid *n* — **n** aceto-acetanilide *n*; acetyl-acetanilide *n* — **e** acetoacetanilida *f*; acetil acetanilido *m* — **i** aceto-acetoanilide *f*

102 acetone; dimethyl ketone; ketopropane; methylacetyl; propanone; pyroacetic ether — Very active solvent used mainly in gravure inks. The fastest drying solvent in the ketone family. It is a colourless, flammable, volatile liquid with a characteristic smell, miscible with water, alcohol, ether, chloroform and most oils. Contact with concentrated mixtures of nitric acid, sulphuric acid and chloroform is dangerous. TL-value: 1000 ppm or 2400 mg/m^3.
Formula: CH_3COCH_3.

f acétone *f*; diméthylcétone *f*; méthylacétyle *m*; cétopropane *f*; éther *m* pyroacétique; propanone *f* — **d** Azeton *n*; Dimethylketon *n*; Ketopropan *n*; Propanon *n*; Essiggeist *m* — **n** aceton *nm*; dimethylketon *n*; propanon *n* — **e** acetona *f*; dimetilcetona *f* — **i** acetone *m*; dimetilchetone *m*

103 acetyl-acetanilide — See aceto-acetanilide.

104 acetyl acetic anilide — See aceto-acetanilide.

105 acetyl cellulose — See cellulose acetate.

106 acetylene black — See carbon black.
f noir *m* d'acétylène — **d** Azetylenschwarz *n*; Azetylenruß *m* — **n** acetyleenzwart *n* — **e** hollín *m* de acetileno; negro *m* de humo de acetileno — **i** nero *m* d'acetilene

107 acetyl hydride — See acetaldehyde.

108 acetyl oxide — See acetic anhydride.

109 achromatic — Term used in colour measurement. For primary light sources, the colour of the equi-energy system ($x = y = z = 1/3$). For surface colours, the light source serving as illuminant.
f achromatique — **d** achromatisch — **n** achromatisch — **e** acromático — **i** acromatico; cromaticamente corretto

110 achromatic colour — Colour which gives visual sensation, without hue (white-grey black).
f couleur *f* achromatique — **d** neutrale Farbe *f*; unbunte Farbe *f* — **n** neutrale kleur *fm*; achromatische kleur *fm* — **e** color *m* acromático — **i** colore *m* acromatico; colore *m* neutro

111 achromatic lens — A photographic lens corrected for chromatic aberration or one bringing visual and actinic rays to the same focus.
f objectif *m* achromatique — **d** achromatische Linse *f*; Achromat *n*; farbenkorrigierte Linse *f* — **n** achromatische lens *fm*; achromaat *m* — **e** objetivo *m* acromático; lente *m* acromático; lente *f* acromática — **i** obiettivo *m* acromatico; lente *f* acromatica

112 acicular — Needle-shaped. Term used in describing crystals or the particles in powders.
f aciculaire; en forme d'aiguilles; aiguillé — **d** nadelförmig — **n** naaldvormig; aciculair — **e** acicular; aciculado; aguzado — **i** aciculare; aghiforme

113 acid; acidic — Having the properties of an acid. The opposite of alkaline.
f acide — **d** sauer — **n** zuur — **e** ácido — **i** acido

114 acid — Any of various typically water-soluble and sour compounds capable of reacting with a *base yielding a salt that are hydrogen-containing molecules or ions able to give up a proton to a base or substances able to accept an unshared pair of electrons from a base.
f acide *m* — **d** Säure *f* — **n** zuur *n* — **e** ácido *m* — **i** acido *m*

115 acid anhydride — An oxide of a non-metallic element or an organic radical which is capable of forming an acid when united with water, or which can be formed by the abstraction of water from the acid molecule, or which can unite with basic oxides to form salts.
f anhydride *m* d'acide — **d** Säureanhydrid *n* — **n** zuuranhydride *n* — **e** anhídrido *m* (de) ácido — **i** anidride *f* di acido

116 acid bath
f bain *m* acidifère — **d** Säurebad *n* — **n** zuur-bad *n* — **e** baño *m* ácido; baño *m* de ácido — **i** bagno *m* d'acido

117 acid blast — Sprays of acid forced against the metal plates in an etching machine.
f jet *m* d'acide — **d** Säurestrahl *m*; Säuresprüh-regen *m* — **n** zuurstraal *fm* — **e** rociada *f* de mordiente — **i** getto *m* d'acido

118 acid colours — See acid dyes.

119 acid concentration; acid density
f concentration *f* d'acide; densité *f* de l'acide — **d** Säuredichte *f* — **n** zuurconcentratie *f*; soortelijk gewicht *n* van het zuur — **e** concentración *f* ácida; densidad *f* del ácido — **i** concentrazione *f* in acido

120 acid density — See acid concentration.

121 acid developer
f développeur *m* acide — **d** Säureentwickler *m* — **n** zure ontwikkelaar *m* — **e** revelador *m* ácido — **i** sviluppatore *m* acido

122 acid dyes; acid dyestuffs; acid colours; adjective dyestuffs; indirect dyestuffs — Dyes that contain SO_3H or CO_2H groups and form salts with alkalis.
f colorants *mpl* acides; couleurs *fpl* acides; teintures *fpl* acides — **d** saure Farbstoffe *mpl* —

n zure kleurstoffen *fmpl* — **e** colorantes *mpl* ácidos — **i** tinte *fpl* acide; coloranti *mpl* acidi
123 acid dyestuffs — See acid dyes.
124 acid fixer; acid fixing bath — Solution for the removal of undeveloped silver salts from photographic plates, films, or papers which is acidified for preservation of the solution and for immediate stopping of the developing action on transfer from alkaline developer.
f bain *m* de fixage acide — **d** saures Fixierbad *n* — **n** zuur fixeerbad *n* — **e** baño *m* fijador ácido — **i** bagno *m* di fissaggio acido
125 acid fixing bath — See acid fixer.
126 acid free — Not containing acids or corrosive elements. Term used in connection with paper, paste, glues, lacquers, etc.
f exempt d'acide; non acide — **d** säurefrei — **n** zuurvrij — **e** libre de ácidos; sin ácido — **i** esente da acidi
127 acid-free paper — Paper containing no acid or corrosive matter such as anti-tarnish tissue or bristol.
f papier *m* sans acide; papier *m* sans acidité; papier *m* exempt d'acide — **d** säurefreies Papier *n* — **n** zuurvrij papier *n* — **e** papel *m* sin ácido; papel *m* libre de ácido; papel *m* exento de ácido — **i** carta *f* senza acido; carta *f* esente da acidi
128 acid hardening fixing bath — An acid fixing bath which contains a hardening compound for tanning the emulsion during fixing.
f bain *m* de fixage tannant acide — **d** saures Härtefixierbad *n* — **n** hardend zuurfixeerbad *n* — **e** baño *m* fijador-endurecedor ácido — **i** bagno *m* di fissaggio acido induritore
129 acidic — See acid.
130 acidification
f acidification *f* — **d** Ansäuern *n*; Säuerung *f*; Säurebildung *f* — **n** aanzuren *n* — **e** acidulación *f* — **i** acidificazione *f*
131 acidify *v* — To make acid.
f acidifier — **d** ansäuern; einsäuern; säuern; sauer machen — **n** aanzuren — **e** acidular — **i** acidificare
132 acidimetry — Determination of the amount of acid in a solution by titration with a solution of base of known strength (standard solution); an indicator is used to establish the end point. See also pH.
f acidimétrie *f* — **d** Azidimetrie *f* — **n** acidimetrie *f* — **e** acidometría *f* — **i** acidimetria *f*
133 acidity — In aqueous solutions, the condition wherein the concentration of hydrogen ions exceeds that of hydroxyl ions. In paper, the condition which results in an acid solution when the paper is treated or extracted with water. In testing paper for acidity, the specimen is extracted with water at room temperature or at 100 °C, and the extract is tested to determine its pH or is titrated to determine the total amount of acid extracted from the paper. See also hydrogenion concentration.
f acidité *f*; degré *m* d'acidité — **d** Azidität *f*; Säure-

grad *m*; Säure *f* — **n** zuurgraad *m* — **e** acidez *f*; grado *m* de acidez — **i** acidità *f*
134 acid number; acid value — Measure of the free acids present. The number of milligrams of potassium hydroxide required to neutralize the free acids present in one gram of oil, wax, or resin. The determination is done by titrating the sample in hot 95% ethyl alcohol and using phenolphthalein as an indicator.
f indice *m* d'acidité; indice *m* d'acide; valeur *f* de l'acidité — **d** Säurezahl *f* — **n** zuurgetal *n* — **e** índice *m* de acidez — **i** numero *m* d'acidità; indice *m* d'acidità
135 acid of sugar — See oxalic acid.
136 acid potassium oxalate — See potassium binoxalate.
137 acid potassium sulphate — See potassium bisulphate.
138 acid proof — See acid resisting.
139 acid-proof paper — Paper which is not discoloured by acids.
f papier *m* résistant aux acides; papier *m* à l'épreuve des acides; papier *m* résistant à l'acide — **d** säurefestes Papier *n*; säureechtes Papier *n*; säurebeständiges Papier *n* — **n** zuurvast papier *n*; zuurbestendig papier *n*; tegen zuur bestand papier *n* — **e** papel *m* inerte al ácido; papel *m* resistente al ácido — **i** carta *f* acidoresistente; carta *f* resistente agli acidi
140 acid-proof test
f essai *m* de résistance aux acides — **d** Säurebeständigkeitsprobe *f* — **n** zuurbestendigheidsproef *f*; onderzoek *n* naar de zuurbestendigheid — **e** ensayo *m* de resistencia al ácido; prueba *f* de resistencia al ácido — **i** prova *f* di resistenza agli acidi
141 acid pump — Pump for handling corrosive liquids, frequently made of glass, earthenware, rubber, or special metal alloys.
f pompe *f* d'acide — **d** Säurepumpe *f* — **n** zuurpomp *fm* — **e** bomba *f* de ácido — **i** pompa *f* resistente agli acidi
142 acid resist — An acid-proof protective coating applied to metal plates prior to etching designs thereon. Bichromated solutions employed in photoengraving as sensitizers provide acid resists through the action of light on the sensitized surface.
f réserve *f* anti-acide — **d** Säureschutzschicht *f* — **n** zuurbestendige laag *fm* — **e** capa *f* protectora; capa *f* resistente al ácido; reserva *f* — **i** strato *m* acidoresistente; strato *m* antiacido
143 acid resistance — The quality of a material to be unaffected by corrosives and acid liquids.
f résistance *f* aux acides — **d** Säureechtheit *f*; Säurebeständigkeit *f*; Säurefestheit *f*; Säurefestigkeit *f* — **n** zuurvastheid *f*; bestandheid *f* tegen zuur; zuurbestendigheid *f* — **e** resistencia *f* al ácido; estabilidad *f* a los ácidos; solidez *f* a los ácidos — **i** resistenza *f* agli acidi
144 acid resistant — See acid resisting.
145 acid resisting; acid proof; acid resistant

f résistant à l'acide; inattaquable par les acides; inattaquable aux acides; réfractaire aux acides — d säurebeständig; säurefest; säureecht — n zuurbestendig; bestand tegen zuur; zuurvast — e resistente a los ácidos; inatacable por los ácidos; inatacable por el ácido; antiácido; inalterable a los ácidos — i resistente agli acidi; acidoresistente; a prova d'acido

146 acid salt; hydrogen salt — Salt in which not all of the hydrogen ions of the acid have been replaced by metal ions.
f sel *m* acide — d saures Salz *n*; Sauersalz *n* — n zuur zout *n* — e sal *f* ácida — i sale *m* acido

147 acid size — A rosin size which contains a considerable portion of unsaponified but emulsified free rosin. If dilution of such size produces a milky emulsion, it is known as white size.
f colle *f* acide; collage *m* acide — d Freiharzleim *m* — n zure lijm *m* — e cola *f* ácida; cola *f* con resina libre — i colla *f* acida; colla *f* di resina acida

148 acid value — See acid number.

149 acierage — See steel facing.

150 acknowledgement GB; **acknowledgment** US
f remerciements *mpl* — d Dank *m* an die Mitarbeiter — n dankwoord *n* voor de medewerking; dankwoord *n* aan de medewerker(s) — e agradecimiento *m* a los colaboradores — i ringraziamento *m* ai collaboratori

151 acknowledgment — See acknowledgement.

152 AC-motor — See alternating-current motor.

153 acorn nut — See cap nut.

154 across the grain — See cross direction.

155 across the page
f en travers de la page — d über die Seitenbreite — n over de breedte van de pagina — e a todo despliegue; a todo plana; a todo extensión; a todo columna — i trasversalmente alla pagina

156 acrylate resin — See acrylic resin.

157 acrylic acid — Colourless liquid. Polymerizes readily. Miscible with water, alcohol and ether. Formula: $CH_2CHCOOH$.
f acide *m* acrylique — d Akrylsäure *f* — n acrylzuur *n* — e ácido *m* acrílico — i acido *m* acrilico

158 acrylic resin; acrylate resin — Thermoplastic or thermosetting resin formed by polymerizing esters of amides of acrylic or methacrylic acid. Used chiefly where transparency is desired (surface coatings, adhesives, paper sizes, and in the methacrylate resin perspex). The repeating structure is
$-CH_2CH(COOR)-$.
f résine *f* acrylique — d Akrylharz *n*; Akrylsäureharz *n*; Akrylatharz *n* — n acrylhars *mn*; acrylaathars *mn* — e resina *f* acrílica; resina *f* de acrilo — i resina *f* acrilica

159 actinic — The property of light to affect a photographic surface in a normal length of time. Ultraviolet and blue rays possess the greatest actinism.
f actinique — d aktinisch; lichtempfindliche

Körper beeinflussend — n actinisch — e actínico — i attinico

160 actinic effect
f effet *m* actinique — d Strahlenwirkung *f* — n actinische werking *f* — e efecto *m* actínico — i effetto *m* attinico

161 actinic light — Chemically active light from arc lamps, mercury vapour lamps, photo-flood bulbs (gas filled tungsten filament incandescent lamps), to harden light-sensitive plate coating solutions in photographic platemaking techniques.
f lumière *f* actinique — d aktinisches Licht *n*; chemisch wirksames Licht *n* — n actinisch licht *n* — e luz *f* actínica — i luce *f* attinica

162 actinic rays — Rays of light which cause chemical changes; images appear on sensitized films, plates and paper. They render bichromated gelatine and bitumen insoluble.
f rayons *mpl* chimiques; rayons *mpl* actiniques — d aktinische Strahlen *mpl* — n actinische stralen *mfpl* — e rayos *mpl* actínicos — i raggi *mpl* attinici; radiazioni *fpl* attiniche

163 actinism — See actinometer.

164 actinometer — Instrument to measure the actinism or chemical power of light.
f actinomètre *m* — d Aktinometer *m* — n actinometer *m* — e actinómetro *m* — i attinometro *m*

165 activated carbon — See active carbon.

166 activated charcoal — See active carbon.

167 activation — The act or process of activating.
f activation *f* — d Aktivierung *f*; Aktivieren *n* — n activering *f* — e activación *f* — i attivazione *f*

168 active acid — See strong acid.

169 active carbon; activated carbon; activated charcoal; adsorptive carbon — Form of carbon characterized by high adsorptive capacity for gases, vapours and colloidal solids. The internal surface area is about 330 m^2 (3552 sq ft) per gram. Latin: carbo adsorbens.
f charbon *m* actif; charbon *m* activé — d Aktivkohle *f* — n actieve kool *fm*; adsorberende kool *fm* — e carbón *m* activado; carbón *m* adsorbente — i carbone *m* attivo; carbone *m* assorbente

170 activol — See rodinal.

171 actual coding — See absolute coding.

172 actual humidity — See humidity of the air.

173 actual weight — The actual weight as contrasted with the *nominal weight of a ream or a number of reams of paper or a bundle or a number of bundles of board.
f poids *m* réel — d wirkliches Gewicht *n* — n werkelijk gewicht *n* — e peso *m* real; peso *m* verdadero — i peso *m* reale

174 acutance — Sharpness of an image measured in an objective manner. See also sharpness.
f acutance *f*; netteté *f* apparente — d gemessene Schärfe *f* — n gemeten scherpte *f*; subjectieve scherpte *f* — e nitidez *f* medida — i acutezza *f*; nitidezza *f* misurata

175 acute accent — E.g., é, used especially in French.
f accent *m* aigu — d Akut *m*; spitzer Akzent *m*;

Accent *m* aigu — **n** acutus *m*; accent *m* aigu; kuut *fm coll.* — **e** acento *m* agudo; agudo *m* — **i** accento *m* acuto

176 ad — Short for *advertisement, usually applied to newspaper and magazine advertising. *pl* ads.

177 ad alley — See advertisement composing room.

178 adapt *v* — To adjust, reshape or make suitable, in connection with stories (adapted for children), books, descriptions.
f adapter — **d** bearbeiten; anpassen; adaptieren — **n** bewerken; aanpassen — **e** adaptar; apropiar — **i** adattare

179 adaptation — The adaptation of a text.
f adaptation *f* — **d** Bearbeitung *f*; Umarbeitung *f* — **n** bewerking *f*; aanpassing *f* — **e** adaptación *f* — **i** adattamento *m*

180 adapter — Frame or attachment for small plates in plateholders for large process cameras.
f adaptateur *m* — **d** Adapter *m*; Einlagehalter *m*; Einlage *f* — **n** adapter *m* — **e** adaptador *m*; marco *m* intermediario — **i** adattatore *m*; adattamento *m*

181 ad compositor — See advertisement compositor.

182 add *v* — To make additions to a manuscript or to the author's proof.
f ajouter — **d** hinzufügen; einfügen; ergänzen — **n** toevoegen — **e** añadir; agregar — **i** aggiungere

183 addend — A number to be added to another. See also accumulator.
f addende *m* — **d** Summand *m* — **n** opteller *m*; digitale opteller *m* — **e** sumador *m* — **i** addizionatore *m*

184 addendum — Addition to a book made after the main text has been printed (Latin; *pl* addenda).
f addendum *m* — **d** Anhang *m*; Nachtrag *m*; Ergänzung *f* — **n** addendum *n* — **e** adición *f*; añadidura *f*; apéndice *m* — **i** addendo *m*

185 ad department — See advertisement department.

186 adding-machine rolls — Paper in small rolls of standard width for adding, calculating and other machines.
f rouleaux *mpl* de papier pour machines comptables — **d** Rechenmaschinenpapier *n*; Addiermaschinenrollen *fpl* — **n** rekenmachinepapier *n*; rollen *mpl* papier voor rekenmachines — **e** papel *m* para máquina calculadora; rollos *mpl* de papel para máquina de calcular — **i** rotoli *mpl* di carta per macchine calcolatrici

187 additional keyboard
f clavier *m* auxiliaire — **d** Zusatztastatur *f* — **n** hulptoetsenbord *n* — **e** teclado *m* auxiliar; teclado *m* adicional — **i** tastiera *f* complementare

188 additive — Examples of additives for a damping solution are phosphoric acid and phosphates, gum arabic, alcohol, glycerol and glycol, bichromate, cobalt and manganese salts.
f adjuvant *m*; additif *m*; agent *m* additif — **d**

Zusatzmittel *n* — **n** toevoegingsmiddel *n* — **e** aditivo *m*; adición *f*; aditamento *m*; aditamiento *m Ec* — **i** additivo *m*

189 additive colour synthesis — See additive synthesis.

190 additive process — See additive synthesis.

191 additive synthesis; additive colour synthesis; additive process — Three-colour process wherein coloured lights are blended (added) to form the sensation of white.
f synthèse *f* additive (des couleurs) — **d** additive Farbsynthese *f*; additive Farbmischung *f*; additives Farbverfahren *n*; additive Dreifarbenmethode *f* — **n** additieve kleurmenging *f*; additieve kleursynthese *f* — **e** procedimiento *m* aditivo; síntesis *f* aditiva; procedimiento *m* tricrómico — **i** sintesi *f* additiva; procedimento *m* additivo

192 address — Computer-controlled typesetting term. There are quite a number of variations of the term "address" but generally it is a particular location in a store or some other data source or destination.
f adresse *f* — **d** Adresse *f* — **n** adres *n* — **e** dirección *f* — **i** indirizzo *m*

193 addressable — Accessible, as a location in a main storage or a record on an auxiliary storage device.
f accessible — **d** adressierbar; aufrufbar — **n** oproepbaar; toegankelijk — **e** accesible — **i** indirizzabile

194 addressable location — Computer term for a specific place where an address may be stored and found.

195 address card — See business card.

196 addressing machine — Machine for printing addresses on letters, packaging, etc., for mailing. "Addressograph" is a trade name.
f machine *f* à adresser — **d** Adressiermaschine *f* — **n** adresseermachine *f* — **e** máquina *f* para imprimir direcciones; máquina *f* para escribir direcciones — **i** macchina *f* per indirizzi; indirizzatrice *f*

197 address modification — Operation by a stored program to alter an address in a presribed manner.
f modification *f* d'adresse — **d** Adreßmodifizierung *f* — **n** adresmodificatie *f* — **e** modificación *f* de dirección — **i** modifica *f* d'indirizzo

198 Addressograph — See addressing machine.

199 address register — Computer term for a register in which an address is stored.
f registre *m* adresse — **d** Adressenregister *n* — **n** adresregister *n* — **e** registro *m* (de) dirección — **i** registro *m* indirizzo

200 add *v* **thin space** — To achieve justification when expansion of linecaster spacebands will not fill the line to the correct measure. A thin space is added to some or all spaceband positions. In TTS code the add-thin code is the sum of the thin and space codes.

201 ad face
f caractère *m* d'annonce — d Anzeigenschrift *f* —
n advertentieletter *fm* — e tipo *m* publicitario;
tipo *m* para publicidad; tipo *m* de carácter publicitario; tipo *m* de dibujo publicitario — i carattere
m per annunci pubblicitari
202 ad galley
f galée *f* pour annonces — d Anzeigenschiff *n* —
n advertentiegalei *fm* — e galera *f* para anuncios;
volandera *f SA* — i vantaggio *m* per composizioni
pubblicitarie
203 adhere *v*
f adhérer; coller — d haften; anhaften — n
kleven; plakken; aankleven; aanhangen — e pegarse — i aderire; fare presa
204 adherence — The quality of adhering.
f adhérence *f* — d Anhaftung *f*; Anhaften *n*; Haft-
vermögen *n* — n aanhechten *n*; hechting *f*; aan-
hechting *f* — e adherencia *f*; adhesividad *f* — i
aderenza *f*
205 adhesion — Adherence of two different
materials such as a printing ink film to a solid
surface. Also the sticking of two surfaces together
due to molecular attraction for each other.
f adhésion *f* — d Adhäsion *f* — n adhesie *f* — e
adhesión *f* — i adesione *f*
206 adhesive — Term sometimes used in ink
making for ink *vehicle or binding agent.
207 adhesive — See adhesive paste.
**208 adhesive binding; thermoplastic binding;
flexiback binding** — Term preferable to unsewn
binding since it indicates clearly the method of
binding. A special type of plastic glue is used on
the back of the book and on the linings. The
binding remains flexible and allows the work to
open flat.
f reliure *f* sans couture — d Klebebindung *f*;
fadenlose Bindung *f* — n garenloze binding *f* — e
encuadernación *f* sin cosido; encuadernación *f* sin
coser; encuadernación *f* sin costura; encuadernación *f* pegada; encuadernación *f* encolada — i
legatura *f* senza cucitura; legatura *f* in colla plastica
209 adhesive dope — Alcohol or other chemical
solution to wash imitation leather and facilitate
the pasting of book ends.
f solvant *m* (de reliure); mordant *m* — d Grundier-
mittel *n* — n grondeermiddel *n* — e mordiente *m*
para encuadernación (en falsa piel) — i solvente
m per legatoria
210 adhesive envelope
f enveloppe *f* gommée — d gummierte Briefhülle *f*; gummierter Briefumschlag *m* — n
gegomde envelop(pe) *fm*; envelop(pe) *fm* met
gegomde klep — e sobre *m* engomado — i busta *f*
(da lettere) gommata
211 adhesive foil
f feuille *f* adhésive; ruban *m* adhésif — d Klebefolie *f* — n plakband *n* — e hoja *f* adhesiva; cinta
f adhesiva — i foglio *m* adesivo
212 adhesive paper — See gummed paper.
213 adhesive paste; adhesive — Adhesive
prepared from flour or starch. Used in bookbinding, makeready, laminating sheets of paper or
board, etc.
f colle *f* (de pâte); adhésif *m*; matière *f* gluante —
d Kleb(e)stoff *m*; Klebemittel *n* — n kleefstof *fm*;
kleefmiddel *n*; plakmiddel *n* — e adhesivo *m*;
pegamento *m*; conglutinante *m* — i colla *f*; adesivo *m*
214 adhesive strength
f pouvoir *m* adhésif; force *f* adhésive — d Kleb(e)-
kraft *f* — n kleefkracht *fm*; klevend vermogen *n*
— e fuerza *f* adhesiva; poder *m* de adhesión — i
forza *f* adesiva
215 adhesive tape
f ruban *m* (de papier) gommé; ruban *m* adhésif —
d Klebeband *n* — n kleefband *nm*; plakband *nm*
— e cinta *f* adhesiva; banda *f* adhesiva — i nastro
m adesivo
216 adjacency effect — When two contiguous
areas of photographic material receive exposures
of different magnitude, their respective densities
and the position of their common boundary may
not reflect the exact distribution of light flux
within the emulsion at the time of exposure. As a
result there are various adjacency effects, generally caused by the chemical changes produced in
the developer or in the halide grain by the
development process, the extent of these changes
increasing with the amount of silver developed.
The enhancement of density at the edge of the
dense area is the *border effect and the depression of density immediately outside is the
fringe effect.
f effet *m* de proximité — d Nachbar-Effekt *m* —
n nabuureffect *n* — e efecto *m* de proximidad — i
effetto *m* di adiacenza
217 adjective dyestuffs — See acid dyes.
218 adjust *v*
f régler; ajuster — d abgleichen; ausrichten; ju-
stieren — n justeren; regelen; reguleren; stellen;
afstellen — e regular; ajustar; arreglar; acomodar
— i regolare; aggiustare; mettere a punto
219 adjustable back gauge
f équerre *f* réglable — d verstellbarer Sattel *m* —
n verstelbaar zadel *n* — e escuadra *f* graduable;
escuadra *f* de corredera; escuadra *f* regulable — i
squadra *f* posteriore regolabile; arresto *m*
posteriore regolabile
220 adjustable flange
f bride *f* réglable — d verstellbarer Facettenhalter *m*; verstellbarer Plattenschuh *m* — n
verstelbare flens *m* — e sujeta *m* regulable;
sujeta *m* ajustable — i portamusso *m* regolabile
221 adjustable speed — Speed which can be
regulated at will by a special driving mechanism.
f vitesse *f* réglable — d einstellbare Geschwindigkeit *f*; variabele Geschwindigkeit *f* — n regelbare
snelheid *f* — e velocidad *f* regulable; velocidad *f*
variable — i velocità *f* regolabile
222 adjusting nut — Nut to regulate the pressure
on a machine or the distance between two parts
of a machine.

f écrou *m* de réglage — **d** Einstellmutter *f*; Stellmutter *f*; Nachstellmutter *f* — **n** stelmoer *fm* — **e** tuerca *f* para graduación de precisión; tuerca *f* de precisión — **i** madrevite *f* di regolazione; dado *m* di regolazione

223 adjusting screw — Screw to control or regulate a machine or part of a machine.
f vis *f* de réglage — **d** Stellschraube *f*; Einstellschraube *f*; Regulierschraube *f*; Nachstellschraube *f*; Verstellschraube *f*; Justierschraube *f* — **n** stelschroef *fm* — **e** tornillo *m* de ajuste; tornillo *m* de regulación; tornillo *m* de movimiento — **i** vite *f* di regolazione; vite *f* di regolamento

224 adjusting screw — Screw to regulate flow of ink from ink fountain.
f vis *f* de l'encrier — **d** Zonenschraube *f* — **n** stelschroef *fm* van de inktbak; inktbakschroef *fm* — **e** tornillo *m* alimentador; perno *m* de ajuste del tintero — **i** vite *f* di regolazione

225 adjusting wedge
f cale *f* d'ajustage — **d** Stellkeil *m* — **n** stelwig *fm* — **e** cuña *f* de ajustar — **i** cuneo *m* di regolazione

226 adjustment; setting — Fitting part of a machine to another one or making changes on a machine to obtain better operation, e.g., adjustment of the ink duct on a press to regulate the flow of ink.
f réglage *m*; ajustage *m*; ajustement *m* — **d** Anordnung *f*; Einstellung *f*; Regelung *f* — **n** stellen *n*; afstellen *n*; regelen *n* — **e** puesta *f* a punto; graduación *f*; regulación *f*; ajuste *m* — **i** regolazione *f*; aggiustamento *m*; messa *f* a punto; registrazione *f*

227 adjustment notch — Notch to control the speed of a machine, the flow of a liquid (ink), etc.
f encoche *f* de réglage — **d** Raste *f*; Einstellkerbe *f* — **n** tand *m*; steltand *m* — **e** muesca *f* de graduación — **i** tacca *f* di regolazione

228 ad maker — See advertisement compositor.

229 ad man; ad writer; copy writer — Text writer for newspaper advertisements.
f spécialiste *m* de la publicité — **d** Werbetexter *m*; Anzeigenfachmann *m* — **n** copywriter *m*; tekstschrijver *m* voor advertenties — **e** redactor *m* publicitario; redactor *m* de textos (para anuncios); publicitario *m*; redactor *m* de anuncios; técnico *m* publicitario

230 ad man — See advertisement compositor.

231 ad matter — See advertisement composition.

232 ads — See ad.

233 ad side — See advertisement composing room.

234 ad smith — See advertisement compositor.

235 adsorbate — An adsorbed substance.
f produit *m* d'adsorption; substance *f* adsorbée — **d** Adsorbat *n*; adsorbierte Substanz *f*; Adsorptiv *n* — **n** geadsorbeerde stof *fm*; adsorbaat *n* — **e** adsorbato *m*; sustancia *f* adsorbida — **i** adsorbato *m*; sostanza *f* adsorbita

236 adsorbent — Substance able to condense or hold other substances on its surface, e.g., activated carbon, silica gel.
f adsorbant *m*; agent *m* adsorbant — **d** Adsorbens *n*; Adsorptionsmittel *n* — **n** adsorberend middel *n*; adsorbens *n* — **e** adsorbente *m*; medio *m* de adsorción — **i** adsorbente *m*; medio *m* de adsorbimento

237 adsorption — Concentration of molecules of a particular kind at the interface between two phases such as the pigment and the vehicle in printing inks. Can effectively remove a component such as the drier from an ink vehicle. It is not a chemical reaction and the binding force is a loose one. To be distinguished from *absorption.
f adsorption *f* — **d** Adsorption *f* — **n** adsorptie *f* — **e** adsorción *f* — **i** adsorbimento *m*

238 adsorption value — The per cent diphenylguanidine removed from a solution of 0.1 g of diphenylguanidine in 50 cc of alcohol by 1 g of carbon black.
f valeur *f* d'adsorption — **d** Adsorptionswert *m* — **n** adsorptiewaarde *f* — **e** valor *m* de adsorción — **i** valore *m* di adsorbimento

239 adsorptive carbon — See active carbon.

240 advance copy — Copy of a book or any printed work issued before the regular edition, often intended for review.
f exemplaire *m* anticipé; exemplaire *m* de lancement; exemplaire *m* précédent la mise en vente — **d** Vorausexemplar *n* — **n** vooruit gereed gemaakt exemplaar *n*; reisexemplaar *n*; exemplaar *n* ter bespreking — **e** ejemplar *m* anticipado; ejemplar *m* anticipado a la puesta en venta; ejemplar *m* preliminar de muestra; ejemplar *m* enviado a críticos antes de la fecha de publicación; ejemplar *m* de prueba — **i** esemplare *m* preliminare; esemplare *m* anticipato; esemplare *m* per la recensione

241 advance copy — Copy sent to the composing room ahead of time.
f copie *f* en avance — **d** Vorausmanuskript *n* — **n** vooruit gezonden kopij *f* — **e** material *m* anticipado; original *m* anticipado; copia *f* anticipada — **i** copia *f* anticipata; manoscritto *m* anticipato

242 advanced feed hole — See advanced sprocket.

243 advanced sprocket; advanced feed hole — Descriptive term for the sprocket holes in 6-level punched paper tape, which are in line with the leading edge of the larger code holes to indicate the leading and trailing ends of the tape. Usually required for special purpose TTS equipment.

244 advance sheet
f feuille *f* témoin — **d** Aushängebogen *m*; Aushänger *m* — **n** persvel *n*; uithouder *m*; vooruit gezonden vel *n* — **e** capilla *f*; pliego *m* de máquina — **i** foglio *m* anticipato; stampone *m*

245 advertise *v*
f insérer une annonce; passer une annonce — **d** anzeigen; inserieren — **n** adverteren — **e** anunciar; avisar; poner un anuncio — **i** avvisare; annunziare

246 advertisement — Announcement, as of goods for sale, in newspapers or magazines. Runs from an *agate line to full page displays.

f annonce *f* — **d** Anzeige *f*; Annonce *f*; Inserat *n* — **n** advertentie *f* — **e** anuncio *m* (publicitario); anuncio *m* (de publicidad); inserción *f* publicitaria; aviso *m SA, Co* — **i** annunzio *m* (pubblicitario); inserzione *f*

247 advertisement *gen.* — Public announcement of an event, of a sale, or request for services or goods.

f avis *m* — **d** Ankündigung *f* — **n** kennisgeving *f*; aankondiging *f* — **e** publicidad *f*; reclamo *m* — **i** avviso *m*

248 advertisement composing room; ad alley; ad side

f atelier *m* des annonces — **d** Anzeigensetzerei *f*; Inseratensetzerei *f* — **n** advertentiezetterij *f* — **e** sección *f* de anuncios; sala *f* de composición de los anuncios — **i** reparto *m* composizione della pubblicità; sala *f* di composizione di annunzi pubblicitari

249 advertisement composition; advertisement matter; ad matter

f composition *f* des annonces — **d** Anzeigensatz *m*; Inseratensatz *m* — **n** advertentiezetsel *n* — **e** composición *f* de anuncios; composición *f* publicitaria; composición *f* de avisos *SA* — **i** composizione *f* d'annunzi; composizione *f* della pubblicità

250 advertisement compositor; advertisement setter; ad compositor; ad man; ad maker; ad smith *coll.* — Compositor who sets advertisements for newspaper pages.

f annoncier *m* — **d** Anzeigensetzer *m*; Inseratensetzer *m* — **n** advertentiezetter *m* — **e** compositor *m* de anuncios; avisero *m Ar*; remendista *m Me*; avisero-cajista *m Me*; cajista *m* de anuncios — **i** compositore *m* della pubblicità

251 advertisement contract to be wound up within one year

f contrat *m* de publicité à utiliser dans l'année — **d** Anzeigenabschluß *m* innerhalb eines Jahres abzuwickeln — **n** advertentiecontract *n* voor één jaar — **e** contrato *m* de publicidad de un año — **i** contratto *m* pubblicitario annuale

252 advertisement department; advertising department; ad department

f service *m* des annonces — **d** Anzeigenabteilung *f* — **n** advertentieafdeling *f* — **e** departamento *m* de anuncios; departamento *m* de avisos; departamento *m* de publicidad — **i** reparto *m* pubblicità

253 advertisement layout

f maquette *f* de composition des annonces — **d** Anzeigenspiegel *m*; Anzeigenlayout *n* — **n** advertentieopmaak *m*; advertentielayout *m*; advertentie-indeling *f* — **e** diseño *m* anuncio; esbozo *m* anuncio; bosquejo *m* anuncio; maqueta *f* anuncio — **i** bozzetto *m* annunzio; bozzetto *m* inserzione; bozzetto *m* pubblicitario

254 advertisement maker-up

f metteur *m* des annonces — **d** Anzeigenmetteur *m* — **n** advertentieopmaker *m* — **e**

compaginador *m* de anuncios; ajustador *m* de anuncios — **i** impaginatore *m* degli annunzi

255 advertisement matter — See advertisement composition.

256 advertisement page; advertising page

f page *f* d'annonces — **d** Anzeigenseite *f* — **n** advertentiepagina *f* — **e** página *f* publicitaria; página *f* de publicidad; página *f* anunciadora; página *f* de propaganda; página *f* de anuncios; plana *f* publicitaria; plana *f* de publicidad; plana *f* de anuncios — **i** pagina *f* pubblicitaria

257 advertisement section

f partie *f* annonces; partie *f* publicitaire — **d** Anzeigenteil *m*; Inseratenteil *m* — **n** advertentiegedeelte *n* — **i** sezione *f* pubblicitaria; parte *f* destinata alla pubblicità

258 advertisement setter — See advertisement compositor.

259 advertisement to the reader — See notice to the reader.

260 advertisement wrapper — See publicity wrapper.

261 advertiser

f annonceur *m* — **d** Inserent *m*; Werbetreibender *m* — **n** adverteerder *m* — **e** anunciador *m*; anunciadora *f*; anunciante *mf*; avisador *m SSA* — **i** inserzionista *m*

262 advertiser — See advertising journal.

263 advertising agency — Commercial concern that provides copy, artwork, layouts, etc., and does all the business with an advertiser. The man who has the contact with the client is the *account executive, the account director supervises all the work.

f agence *f* de publicité; bureau *m* de publicité — **d** Werbebureau *n*; Werbestelle *f* — **n** reclamebureau *n* — **e** agencia *f* de publicidad; agencia *f* anunciadora; agencia *f* de propaganda; agencia *f* de anuncios; agencia *f* de avisos; agencia *f* publicitaria — **i** agenzia *f* di pubblicità; agenzia *f* pubblicitaria; agenzia *f* d'annunzi

264 advertising artist — See advertising designer.

265 advertising band — See publicity wrapper.

266 advertising blotters — Small coated or uncoated blotting papers on which advertising messages are printed.

f buvards *mpl* de publicité — **d** Reklamelöschblätter *npl* — **n** reclamevloeibladen *npl* — **e** secantes *mpl* publicitarios — **i** carte *fpl* asciuganti per messaggi pubblicitari

267 advertising campaign

f campagne *f* de publicité; campagne *f* publicitaire — **d** Werbekampagne *f*; Werbefeldzug *m* — **n** reclamecampagne *fm* — **e** campaña *f* anunciadora; campaña *f* publicitaria; campaña *f* de publicidad; campaña *f* de propaganda — **i** campagna *f* pubblicitaria

268 advertising conditions

f conditions *fpl* d'insertion — **d** Insertionsbedingungen *fpl* — **n** advertentievoorwaarden *fpl* — **e** condiciones *fpl* para la inserción de anuncios — **i** condizioni *fpl* d'inserzioni

269 advertising contract
f contrat *m* de publicité; engagement *m* de publicité — **d** Anzeigenabschluß *m* — **n** advertentiecontract *n* — **e** contrato *m* de publicidad; contrato *m* de propaganda — **i** contratto *m* d'annunzi

270 advertising department — See advertisement department.

271 advertising designer; advertising artist
f dessinateur *m* des annonces — **d** Anzeigengestalter *m*; Werbegestalter *m*; Werbegrafiker *m* — **n** reclameontwerper *m*; advertentieontwerper *m*; reclametekenaar *m* — **e** diseñador *m* publicitario; creador *m* de publicidad; artista *m* publigráfico; dibujante *m* gráfico; dibujante *m* publicitario; dibujante *m* maquetista; dibujante *m* comercial — **i** progettista *m* di pubblicità; disegnatore *m* pubblicitario

272 advertising journal; advertiser
f feuille *f* d'annonces; feuille *f* d'avis; moniteur *m*; indicateur *m* — **d** Anzeigenblatt *n*; Anzeiger *m*; Inseratenblatt *n* — **n** advertentieblad *n* — **e** diario *m* de anuncios; anunciador *m* — **i** foglio *m* pubblicitario

273 advertising material
f matériel *m* publicitaire; articles *mpl* publicitaires — **d** Werbematerial *n*; Reklamematerial *n* — **n** reclamemateriaal *n* — **e** material *m* de publicidad; material *m* de propaganda — **i** materiale *m* pubblicitario

274 advertising office
f office *m* de publicité — **d** Anzeigenbüro *n* — **n** advertentiebureau *n* — **e** oficina *f* de publicidad — **i** ufficio *m* pubblicitario

275 advertising page — See advertisement page.

276 advertising rates
f tarif *m* de publicité — **d** Anzeigentarif *m* — **n** advertentietarief *n* — **e** tarifa *f* de anuncios — **i** tariffa *f* d'annunzi

277 advertising stickers — See poster stamps.

278 advertising supplement
f supplément *m* publicitaire — **d** Reklamebeilage *f*; Anzeigenbeilage *f*; Werbebeilage *f* — **n** advertentiebijvoegsel *n*; reclamebijlage *fm* — **e** suplemento *m* publicitario — **i** supplemento *m* pubblicitario

279 advertising wrapper — See publicity wrapper.

280 ad writer — See ad man.

281 aerial photograph; air photograph; aerial picture
f photo *f* aérienne — **d** Luftbildaufnahme *f*; Luftbild *n* — **n** luchtfoto *fm* — **e** aerofoto *f*; fotografía *f* aérea — **i** fotografia *f* aerea

282 aerial picture — See aerial photograph.

283 aerograph — See air brush.

284 aerosol — Originally a fine division of fluid or solid particles. By extension a packaging for under pressure packed products, a pressurized packaging.
f aérosol *m* — **d** Aerosol *n* — **n** aërosol *n*; spuitbus *fm* — **e** aerosol *m* — **i** aerosol *m*

285 aesculin — See esculin.

286 aesthetic — Having to do with beauty as distinguished from utility.
f esthétique — **d** ästhetisch — **n** esthetisch — **e** estético — **i** estetico

287 affinity — Natural attraction for, as salt for moisture.
f affinité *f* — **d** Affinität *f* — **n** affiniteit *f* — **e** afinidad *f* — **i** affinità *f*

288 à froid — See blind tooling.

289 after tack — Tack that develops after an ink has apparently dried or after a heat-drying operation, or due to pressure in the delivery pile.
d Nachkleben *n* — **n** nakleven *n* — **e** repintado *m* — **i** appicciosità *f* ritardata

290 after treatment — See post treatment.

291 afterword — See epilogue.

292 against the grain — The direction opposite to that of the fibres.
f en sens travers — **d** gegen die Laufrichtung — **n** tegen de pool in; tegen de vleug in — **e** a contrafibra; a contrahilo — **i** in direzione contro-fibra

293 agalite — Grey, natural filler similar to talc although less soapy, used in papermaking to give stiffness and gloss. Hydrated magnesium silicate. Talc and asbestos powder are sometimes called agalite.
Formula: $Mg_3Si_4O_{11}·H_2O$.
f agalithe *f*; poudre *f* d'amiante — **d** Agalit *m* — **n** agalit *n* — **e** agalita *f*; polvo *m* de amianto — **i** agalite *f*; polvere *f* d'amianto

294 agar; agar agar; Japanese gelatine; Japanese isinglass; Chinese gelatine; Chinese isinglass — Gelatine-like product of some seaweeds, soluble in boiling water, insoluble in cold water and organic solvents. Used as a substitute for gelatine, in adhesives, as an emulsifier and as a sizing for paper. Sometimes erroneously named vegetable glue.
f agar-agar *m*; gélose *f* — **d** Agar-Agar *m* — **n** agar-agar *fm* — **e** agar-agar *m*; cola *f* de Bengala; gelosa *f* — **i** agar-agar *m*

295 agar agar — See agar.

296 agate; bloodstone; agate polisher — Hand tool used by gilders for *burnishing.
f agate *f*; agate *f* à polir; polissoir *m* en agate — **d** Achatstift *m*; Achatglättrolle *f* — **n** agaat *m*; agaatsteen *m*; agaatrol *fm* — **e** ágata *f*; rodillo *m* de ágata; bruñidor *m* de ágata — **i** agata *f*; politore *m* alla pietra; lisciatore *m* alla pietra

297 agate — American name for a size of type, used to measure the advertising space in a newspaper or a magazine. One agate is about 2 mm (14 lines to the inch). See app. no. 1.

298 agate line *US* — Unit of measurement for newspaper advertising space, 1/14 of a column inch. See also app. no. 1.
f ligne *f* d'agate — **d** Achatzeile *f* — **n** agaatregel *m* — **e** línea *f* ágata — **i** linea *f* d'agata

299 agate polisher — See agate.

300 agave — American aloe from which long, strong papermaking fibres can be obtained.
f agave *m* — **d** Agave *f* — **n** agave *fm* — **e** pita *f*; agave *fm*; henequen *m*; maguey *n*; cabuya *f* — **i**

agave *f*

301 ageing *GB*; **aging** *US*
f vieillissement *m* — **d** Altern *n*; Alterung *f*; Altwerden *n* — **n** veroudering *f*; verouderen *n* — **e** envejecimiento *m* — **i** invecchiamento *m*

302 ageing test *GB*; **aging test** *US* — Test to estimate the life expectancy of a product, paper, ink, etc.
f essai *m* de vieillissement — **d** Alterungsprüfung *f*; Alterungsversuch *m* — **n** verouderingsproef *fm* — **e** ensayo *m* de envejecimiento; prueba *f* de envejecimiento; verificación *f* de envejecimiento — **i** prova *f* di invecchiamento; esame *m* di invecchiamento

303 agglomerate — Cluster of undispersed particles.
f agglomérat *m* — **d** Agglomerat *n* — **n** agglomeraat *n* — **e** aglomerado *m* — **i** agglomerato *m*

304 aggregation — Group of two or more particles. If the particles are "cemented" together a hard aggregate is formed. Soft flocculates are also classified as aggregates, but they do not function rheologically as single particles like hard aggregates.
f agrégation *f* — **d** Aggregation *f*; Zusammenballung *f*; Zusammenschluß *m* — **n** aggregatie *f*; conglomeratie *f* — **e** agregación *f* — **i** aggregazione *f*

305 aging — See ageing.

306 aging test — See ageing test.

307 agitate *v* — See stir *v*.

308 agitation — Stirring or motion of solutions such as developers and etching baths during mixing and use.
f agitation *f* — **d** Bewegung *f*; Badbewegung *f*; Rühren *n*; Schütteln *n* — **n** in-beweging-brenging *f* — **e** agitación *f* — **i** agitazione *f*

309 agitator; stirrer — Mechanism to stir a solution or a mixture, e.g., in an ink duct.
f agitateur *m* — **d** Rührwerk *n*; Rührapparat *m* — **n** roerwerk *n* — **e** batidor *m*; agitador *m* — **i** agitatore *m*

310 aher — Coating from an aqueous solution of egg albumin and alum with which in the 16th century paper was coated to obtain a better writing surface.

311 air bells; air bubbles — Small bubbles of air occurring in glass and film supports, also those forming on photographic surfaces during development. In paper these surface defects are sometimes called *blisters or *foam spots.
f bulles *fpl* d'air — **d** Luftblasen *fpl* — **n** luchtblaasjes *npl*; luchtbelletjes *npl* — **e** burbujas *fpl* de aire; burbujas *fpl* de espuma — **i** bolle *fpl* d'aria

312 air blower — See fan.

313 air brake; air pressure brake; compressed-air brake — Braking arrangement on machinery in which the brake block is pressed against the moving part by compressed air or by a vacuum.
f frein *m* à air comprimé; frein *m* pneumatique — **d** Druckluftbremse *f*; pneumatische Bremse *f* — **n** luchtdrukrem *fm* — **e** freno *m* de aire comprimido; freno *m* neumático — **i** freno *m* ad aria compressa; freno *m* pneumatico

314 air brush; aerograph — Pencil-like apparatus for spraying liquid colours with compressed air.
f aérographe *m*; pinceau *m* à air; pistolet *m*; pinceau *m* pneumatique; pinceau-vaporisateur *m* — **d** Spritzapparat *m*; Aerograph *m*; Retuschiergerät *n*; Luftpinsel *m*; Farbspritzgerät *n* — **n** verfspuit *fm*; retoucheerspuit *fm* — **e** aerógrafo *m*; pistola *f* neumática; pincel *m* neumático; brocha *f* de aire; pulverizador *m* de aire comprimido; pulverizador *m* neumático — **i** aerografo *m*; pistola *f* pneumatica; pennello *m* pneumatico

315 air-brush coating — See air-knife coating.

316 air brushing — Method of placing smooth tints on photographs or drawings with an *air brush.
f travail *m* à l'aérographe; retouche *f* à l'aérographe — **d** Spritzen *n*; Spritzretusche *f*; Spritztechnik *f* — **n** spuiten *n*; spuitretouche *fm* — **e** aerografía *f*; retocado *m* al aerógrafo; retoque *m* al aerógrafo — **i** lavorazione *f* con aerografo; ritocco *m* a spruzzo

317 air bubbles — See air bells.

318 air buffer — See also buffer.
f amortisseur *m* à air — **d** Luftpuffer *m* — **n** luchtbuffer *m* — **e** amortiguador *m* de aire — **i** ammortizzatore *m* ad aria

319 air circulation system — Drying or ventilating system in which the same air is used repeatedly.
f système *m* de circulation d'air — **d** Umluftsystem *n* — **n** luchtcirculatiesysteem *n* — **e** sistema *m* de circulación de aire — **i** sistema *m* di ricircolazione d'aria

320 air compressor — See compressor.

321 air conditioner; air conditioning system — Equipment to control temperature and humidity of air.
f appareil *m* de climatisation; installation *f* de conditionnement d'air — **d** Klimaanlage *f*; Bewetterungsanlage *f* — **n** klimatiseringsinrichting *f* — **e** instalación *f* de aire acondicionado; instalación *f* de acondicionamiento de aire; acondicionador *m* de aire — **i** condizionatore *m* d'aria; impianto *m* di condizionamento dell'aria; climatizzatore *m*

322 air conditioning — Controlling the temperature and humidity of the air. It helps to overcome certain troubles in the printing plant.
f climatisation *f*; conditionnement *m* de l'air; conditionnement *m* d'air — **d** Luftkonditionierung *f*; Bewetterung *f* — **n** klimatisering *f*; klimaatregeling *f* — **e** climatización *f*; acondicionamiento *m* del aire — **i** climatizzazione *f*; condizionamento *m* dell'aria

323 air conditioning system — See air conditioner.

324 air cooling
f refroidissement *m* à l'air — **d** Luftkühlung *f* — **n** luchtkoeling *f* — **e** enfriamiento *m* de aire — **i**

raffreddamento *m* ad aria

325 air cushion

f coussin *m* d'air; matelas *m* d'air — **d** Luft-polster *n*; Luftkissen *n* — **n** luchtkussen *n* — **e** colchón *m* de aire — **i** cuscino *m* d'aria; cuscino *m* pneumatico

326 air-cushion table; air table

f table *f* pneumatique; table *f* à coussin d'air; table *f* soufflante — **d** Lufttisch *m*; Luftpolster-tisch *m*; Luftkissentisch *m* — **n** luchttafel *fm* — **e** mesa *f* neumática — **i** tavolo *m* a cuscino d'aria; tavolo *m* aspirante

327 air doctor dampening system — System on offset presses consisting of a rubber roller which rotates very fast in a water fountain and throws a thick water film on the plate. An air doctor regulates the film of water remaining on the plate.

f système *m* de mouillage à lame d'air — **d** Sprüh-feuchtwerk *n* mit Luftrakel; Schleuderfeucht-werk *n* — **n** sproeivochtwerk *n* met luchtrakel — **e** sistema *m* de mojado con cuchilla de aire — **i** sistema *m* di bagnatura a lama d'aria

328 air-dried brown — Coarse wrapping paper usually made from rope or similar fibres with a rough finish. The normal range of basis weights is 13.5-40 kg (24 x 36 - 500), i.e., 30-90 pounds.

f papier *m* d'emballage séché à l'air — **d** luft-getrocknetes Verpackungspapier *n* — **n** lucht-gedroogd bruinpak; luchtgedroogd bruin pak-papier *n* — **e** papel *m* de embalaje secado al aire — **i** carta *f* da imballo essiccata all'aria

329 air-dried paper — Paper dried by contact with air, at normal or elevated temperature, as distinguished from machine-dried paper, where drying is accomplished by contact with heated rolls.

f papier *m* séché à l'air — **d** luftgetrocknetes Papier *n* — **n** luchtgedroogd papier *n*; natuurlijk gedroogd papier *n* — **e** papel *m* secado al aire — **i** carta *f* asciugata all'aria; carta *f* essiccata con aria

330 air dryer — Part of a drying system on paper machines. Air drying is used on tub-sized papers which are not put through the loft-dryer.

f sécheur *m* à air (chaud) — **d** Lufttrockner *m* — **n** hete-luchtdroger *m* — **e** secador *m* de aire — **i** essiccatore *m* ad aria

331 air-drying — Method for drying pulp, paper or board, generally carried out by suspending sheets in freely circulating air, or in the web by contact with heated air in a room or in a tunnel.

f séchage *m* à l'air — **d** Lufttrocknung *f* — **n** luchtdroging *f*; droging *f* aan de lucht — **e** secado *m* por aire; secado *m* al aire — **i** essiccamento *m* ad aria

332 air-dry paper — Paper with a moisture content nearly in equilibrium with that of the air.

f papier *m* sec à l'air — **d** lufttrockenes Papier *n* — **n** luchtdroog papier *n* — **e** papel *m* seco al aire — **i** carta *f* secca all'aria

333 air-dry pulp — Pulp with a moisture content approximately in equilibrium with that of the sur-rounding atmosphere.

f pâte *f* sèche à l'air — **d** lufttrockener Zell-stoff *m*; lufttrockener Halbstoff *m* — **n** lucht-gedroogde celstof *fm* — **e** pasta *f* secada al aire — **i** pasta *f* secca all'aria

334 air eraser — Miniature sand-blasting hand appliance to remove superfluous lithographic images from the plate. Also used to erase art with-out destroying the texture of the medium.

f pistolet *m* sableur — **d** Sandstrahlgebläse *n* — **n** zandstraalapparaat *n* — **e** pistola *f* de chorro de arena; soplador *m* de chorro de arena — **i** appa-recchio *m* di sabbiatura

335 air extractor — Ventilating device for the con-tinuous extraction of air from a room.

f aspirateur *m*; extracteur *m* d'air — **d** Ventilator *m*; Exhaustor *m*; Luftabsaugung *f* — **n** ventilator *m* — **e** extractor *m* de aire — **i** estrat-tore *m* d'aria

336 air hole — Small hole in the surface of any casting, as in an inking roller, type, etc., caused by the pressure of air bubbles during casting.

f soufflure *f* — **d** Luftblase *f* — **n** gietgal *fm* — **e** burbuja *f* de aire; sopladura *f*; escarabajo *m* — **i** soffiatura *f* d'aria

337 air-jet coating — See air-knife coating.

338 air-knife coating; air-brush coating; air-jet coating — Roll coating method in which the applied coating slip is levelled and controlled, and the excess removed by a stream of compressed air issuing from an orifice placed close to the coated surface while the web is supported by a backing roll. Operated as a machine coating or separate coating method.

f couchage *m* par lame d'air; couchage *m* à lame d'air — **d** Luftbürstenstreichverfahren *n*; Luft-messerstreichverfahren *n*; Luftpinselstreich-verfahren *n*; Luftdüsenstreichverfahren *n*; Streich-verfahren *n* mit der Luftrakel — **n** strijken *n* met een luchtmes; strijken *n* met een luchtrakel — **e** estucado *m* con chorro de aire dosificador; estucado *m* por cepillo de aire; estucado *m* por labio soplante; estucado *m* por cuchilla de aire — **i** patinatura *f* a lama d'aria

339 airmail bond — See airmail paper.

340 airmail envelope

f enveloppe *f* poste aérienne — **d** Luftpost-umschlag *m* — **n** luchtpostenvelop(pe) *fm* — **e** sobre *m* aéreo; sobre *m* de avión; sobre *m* para vía aérea — **i** busta *f* di posta aerea

341 airmail paper; airmail bond — Lightweight writing paper for letters and circulars to be trans-ported by airplanes. The basis weight is 2.25-4 kg (17 x 22 - 500), i.e., 5-9 pounds. Its light weight is its most significant property; opacity is also impor-tant.

f papier *m* avion; écriture *f* pour poste-aérienne — **d** Luftpostpapier *n*; Flugpostpapier *n*; Dünn-postpapier *n* — **n** luchtpostpapier *n* — **e** papel *m* aéreo; papel *m* avión; papel *m* para correo aéreo; papel *m* vía aérea; papel *m* para correspondencia aérea; papel m cebolla — i carta f per posta aerea

342 air moistener — See humidifier.

343 air passage
f passage m d'air — d Luftkanal m; Luftdurchlaß m — n luchtkanaal n; luchtbuis fm — e
vientos mpl — i canale m d'aria

344 air permeability — According to *Tappi the
average number of seconds required for the
displacement of 100 ml of air through an area of
6.45 cm² (1 sq in) of the paper. Incorrect to use the
term air porosity.
f perméabilité f à l'air — d Luftdurchlässigkeit f
— n luchtdoorlatendheid f — e permeabilidad f al
aire — i permeabilità f all'aria

345 air permeability tester
f mesureur m de la perméabilité à l'air — d Luftdurchlässigkeitsprüfer m — n luchtdoorlatendheidsmeter m — e comprador m de la permeabilidad al aire — i apparecchio m per prova di
permeabilità all'aria; misuratore m di permeabilità all'aria

346 air photograph — See aerial photograph.

347 air pin; air piston — Part of the Monotype
machine.
f piston m à air — d Luftstift m; Luftkolben m —
n luchtpen fm; luchtklep fm — e pistón m de
aire; émbolo m del aire — i pistone m ad aria

348 air pin block — Part of the Monotype caster.
f bloc m de pistons — d Luftstiftblock m — n
penblok n — e bloque m de pistones de aire — i
blocco m dei pistoni

349 air piping
f conduite f soufflerie — d Luftleitungen fpl — n
luchtleiding(en) f(pl) — e conducto m de aire;
tubería f de aire — i tubazioni fpl dell'aria

350 air piston — See air pin.

351 air pollution
f pollution f de l'air — d Luftverunreinigung f —
n luchtverontreiniging f — e ensuciamiento m del
aire; polución f del aire — i contaminazione f
dell'aria; inquinamento m dell'aria

352 air porosity — See air permeability.

353 air pressure brake — See air brake.

354 air pump — Apparatus for drawing in, compressing or exhausting air.
f pompe f à air; compresseur m — d Luftpumpe f;
Luftpresser m; Druckluftpumpe f — n luchtpomp
fm; luchtperspomp fm; compressor m — e bomba
f de aire; compresor m de aire — i pompa f
d'aria; pompa f pneumatica; compressore m

355 air table — See air-cushion table.

356 air tight
f étanche à l'air; hermétique — d luftdicht; hermetisch — n luchtdicht; hermetisch — e hermético; a
prueba de aire — i ermetico; a tenuta

357 air vent — Outlet for outgoing or inlet for in-
coming air.
f aspirail m; évent m; ventouse f — d Luftventil n
— n luchtklep fm — e válvula f de aire;
ventosa f; chupón m de aire — i valvola f d'aria;
apertura f di sfogo

358 -al — Suffix indicating that a compound con-

tains an aldehyde group.

359 Alaska seal — Ground sheep or cowskin
which imitates sealskin.
f simili veau-marin m — d imitiertes Seehunds-
fell n; imitiertes Seal n — n imitatie-zeehonden-
leer n — e imitación f de piel de foca — i
similfoca m; cuoio m imitazione foca

360 albardine pulp — See esparto pulp.

361 Albert; Albert note — A size of writing (note)
paper. See app. no. 3.

362 Albert effect — In 1899, E. Albert found that
if a collodion plate was heavily exposed, treated
with nitric acid, washed and re-exposed to diffuse
light, a positive image of the first exposure was
obtained. This is known as the Albert effect and
H. Lüppo Cramer showed that the effect could be
observed on a silver bromide emulsion made in
gelatine. He used chromic acid or ammonium persulphate in place of nitric acid. Lüppo Cramer
found that to produce the Albert effect the first exposure must be heavy but need not produce
solarization. The effect can be obtained in
emulsions which contain sodium nitrite (compare
solarization) when they are developed with non-
solvent chemical or prefixation physical
developers. Chemical developers which dissolve
silver bromide give a negative image, and no
reversal can be observed with postfixation
physical development.
f effet m Albert — d Albert-Effekt m — n Albert-
effect n — e efecto m de Albert — i effetto m di
Albert

363 Albert-Fischer electroplate — See Albert
galvano.

364 Albert-Fischer galvano — See Albert galvano.

**365 Albert galvano; Albert-Fischer galvano;
Albert-Fischer electroplate** — Kind of electro
made by *lead-moulding for close register work
and multi-colour printing. Process developed by E.
Albert (Munich), modified in cooperation with G.
Fischer.
f galvano m Albert; empreinte f au plomb;
galvano m au plomb — d Albert-Fischer-
Galvano n; Galvano n nach Bleiabprägung des
Originals — n loodgalvano m — e galvano m
Albert-Fischer; electrotipia f extradura moldeada
en plomo — i galvano m Albert-Fischer

366 Albert note — See Albert.

367 albertype — See collotype process.

368 Albion press — Small iron handpress in
which the type forme slides under the platen and
pressure is exerted by a toggle-jointed lever,
forcing the platen on the type. Devised by
Richard Whittacker Cope, London (1823). Used by
William Morris at the Kelmscott Press.
f presse f à platine Albion; presse f à genouillère
— d Albionpresse f; Handpresse f; Kniehebelpresse f — n Albionpers f; kniehevelpers f — e
prensa f Albion — i torchio m Albion; torchio m
a mano

369 album — Binding in which the loose leaves
are separated by "stubs" at the binding edge to

take up bulk of pictures or other material mounted on the leaves.
f album *m* — **d** Album *n*; Sammelbuch *n* — **n** album *n* — **e** álbum *m* — **i** album *m*
370 album board — See album paper.
371 albumen — Natural protein, soluble in water and most commonly found in the white of eggs, the colloid used for certain bichromated sensitizers employed in photomechanics. It is also used as size in gold stamping and application of gold leaves to edges. Latin: albumen ovi siccum. (albumen is a general scientific name; chemical name: albumin.)
f albumine *f* du blanc d'œuf; blanc *m* d'œufs — **d** Albumin *n*; Eiweiß *n*; Eialbumin *n*; Eiweißstoff *m* — **n** (droog) eiwit *n*; ei-albumine *fm*; albumine *fm* — **e** albúmina *f* de huevo; clara *f* de huevo — **i** albume *m* (d'uovo); albumina *f*
372 albumen copy
f copie *f* à l'albumine — **d** Eiweißkopie *f*; Albuminkopie *f* — **n** eiwitkopie *f* — **e** copia *f* a la albúmina; pasado *m* a la albúmina — **i** copia *f* all'albumina
373 albumenized paper — See albumen paper.
374 albumen paper; albumenized paper — Paper used in photography, coated with albumen (albumin) from the whites of eggs mixed with ammonium chloride and treated with silver salts. Sensitive to light.
f papier *m* albuminé; papier *m* albumineux; papier *m* glacé albumineux — **d** Albuminpapier *n*; Eiweißpapier *n* — **n** albuminepapier *n* — **e** papel *m* (de) albúmina; papel *m* albuminado — **i** carta *f* all'albumina
375 albumen plate — *Surface plate, the image area of which is hardened albumen.
f plaque *f* albumine — **d** Eiweißplatte *f* — **n** eiwitplaat *f* — **e** plancha *f* a la albúmina — **i** lastra *f* all'albumina
376 albumen process — Procedure of photomechanics utilizing a coating of bichromated albumin as a sensitized surface on which images are made by exposure under a line or halftone negative followed by development of the inked image with water.
f procédé *m* à l'albumine — **d** Albuminverfahren *n*; Eiweißkopierverfahren *n* — **n** eiwitprocédé *n* — **e** procedimiento *m* a la albúmina — **i** procedimento *m* all'albumina; processo *m* all'albumina
377 albumen solution
f solution *f* d'albumine — **d** Eiweißlösung *f* — **n** eiwitoplossing *f* — **e** solución *f* de albúmina — **i** soluzione *f* d'albumina
378 album paper; album board — Black, brown or grey cover paper for albums, suitable for pasting photographs or other material.
f papier *m* pour albums; carte *f* pour albums — **d** Albumpapier *n*; Albumkarton *n* — **n** papier *n* voor albums; albumpapier *n*; opzetpapier *n*; opzetkarton *n* — **e** cartulina *f* para álbumes — **i** carta *f* per album

379 Alcogravure — Proprietary name for a gravure process for production of depth-and-area-variable gravure printing. Use is made of a continuous tone master positive, a contact screen, a high speed photographic film, and an improved etching procedure.
f procédé *m* Alco — **d** Alco-Verfahren *n* — **n** Alco-procédé *n* — **e** procedimiento *m* Alco; método *m* Alco — **i** procedimento *m* Alco
380 alcohol — In common parlance, term for *ethyl alcohol. In chemical sense a generic term for a series of hydroxyl compounds, the simplest of which has the formula $C_nH_{2n+1}OH$.
f alcool *m* — **d** Alkohol *m* — **n** alcohol *m* — **e** alcohol *m* — **i** alcool *m*; alcole *m*
381 alcohol resistant; resistant to alcohol
f résistant à l'alcool; inattaquable par l'alcool — **d** spritecht — **n** bestand tegen alcohol; alcoholecht — **e** resistente al alcool; alcoholresistente — **i** resistente all'alcool
382 alcohols — Family of organic solvents containing the grouping C-OH. The most common members of this group are methyl (wood) alcohol, ethyl alcohol, propyl alcohol, isopropyl alcohol.
f alcools *mpl* — **d** Alkohole *mpl* — **n** alcoholen *mpl* — **e** alcoholes *mpl* — **i** alcoles *mpl*
383 alcohol wash; spirit wash — A liquid, e.g., anhydrous industrial methylated spirit or anhydrous isopropyl alcohol, used in the production of reversal plates to remove developing and deep-etch solutions before application of lacquer to the image areas.
f solution *f* alcoolique de lavage — **d** Waschalkohol *m*; Waschspiritus *m*; Sprit *m* — **n** afwasspiritus *m*; spiritus *m* — **i** soluzione *f* di lavaggio ad alcool
384 Alco process — Gravure process using a continuous tone master positive, a contact screen, a new high speed photographic film and an improved etching procedure.
f procédé *m* Alco — **d** Alco-Verfahren *n* — **n** Alco-procédé *n* — **e** procedimiento *m* Alco; método *m* Alco — **i** procedimento *m* Alco
385 aldehyde — Generic term for a class of organic compounds. If primary alcohols are oxidized carefully, an aldehyde can be produced. To lithographers only *formaldehyde is of interest.
f aldéhyde *m* — **d** Aldehyd *mn* — **n** aldehyd(e) *n* — **e** aldehído *m* — **i** aldeide *f*
386 aldehyde resin — High polymer obtained as condensation product of aldehydes by treating these with strong caustic soda. Important in the plastics industry.
f résine *f* aldéhydique — **d** Aldehydharz *n* — **n** aldehydhars *mn*; aldehyde hars *mn* — **e** resina *f* aldehídica; resina *f* de aldehído — **i** resina *f* aldeidica
387 alder — Hardwood from which soda pulp can be made. The bark was used as a source for a black dye with copperas or iron liquor.
f aulne *m*; aune *m* — **d** Erlenholz *n*; Erle *f* — **n**

elzenhout *n* — **e** madera *f* de chopo; madera *f* de aliso; aliso *m* — **i** legno *m* di ontano; legno *m* di alno

388 Aldine; Aldine edition; Aldus — Edition of Greek and Latin classics printed by Aldus Manutius (Teobaldo Mannucci or Manuzio) and his family (1490-1597), in compact editions.
f Alde *m* — **d** Aldine *f*; Aldine-Ausgabe *f* — **n** Aldine *f* — **e** Aldino *m*; edición *f* aldina — **i** Aldino *m*; Aldina *f*

389 Aldine edition — See Aldine.

390 Aldus — See Aldine.

391 alfa — See esparto.

392 alfa pulp — See esparto pulp.

393 algebraical signs; algebraic signs — See also mathematical signs.
f signes *mpl* algébriques — **d** algebraische Zeichen *npl* — **n** algebratekens *npl*; algebraïsche tekens *npl* — **e** signos *mpl* algebraicos — **i** segni *mpl* d'algebra

394 algebraic oriented language — See algorithmic language.

395 algebraic signs — See algebraical signs.

396 algebraic sum
f somme *f* algébrique — **d** algebraische Summe *f* — **n** algebraïsche som *fm* — **e** suma *f* algebraica — **i** somma *f* algebraica; somma *f* algebrica

397 alginate — Metal salt of algin, i.e., the dried, bleached, gelatinous form of various seaweeds.
f alginate *m* — **d** Alginat *n* — **n** alginaat *n* — **e** alginato *m* — **i** alginato *m*

398 algol — Acronym for algorithmic language or algebraic oriented language. A high-level programming language initiated in 1958 and developed in Europe, previously called international algebraic language. Procedure oriented and machine independent. Used in scientific applications to express numerical values and processes in a form suitable for automatic translation to machine code with the use of a compiler.
f algol *m* — **d** Algol *n* — **n** algol *n* — **e** algol *m* — **i** algol *m*

399 algorithm — Set of rules for the solution of a problem in a finite number of steps, e.g., a full statement of an arithmetical procedure for evaluating sin x to a stated precision.
f algorithme *m* — **d** Algorithmus *m* — **n** algoritme *n*; rekenschema *n* — **e** algoritmo *m* — **i** algoritmo *m*

400 algorithmic language — Computer language in which information is expressed in algebraic notation according to the rules of Boolean algebra. See also algol.
f langage *m* algorithmique — **d** algorithmische Sprache *f* — **n** algorithmische taal *fm* — **e** lenguaje *m* algorítmico — **i** linguaggio *m* algoritmico

401 algorithmic translation; algorithm translation — Specific, effective, essentially computational method to obtain a translation from one language into another.
f traduction *f* algorithmique — **d** algorithmische

Übersetzung *f* — **n** algorithmische vertaling *f* — **e** traducción *f* algorítmica — **i** traduzione *f* algoritmica

402 algorithm translation — See algorithmic translation.

403 algraphy — See aluminography.

404 align *v*; **line** *v* **up** — To place letters, words, designs, etc., on the same horizontal line.
f aligner — **d** in Linie bringen; ausrichten; geraderichten; alinieren; alineieren — **n** in de lijn brengen — **e** alinear; poner en línea; nivelar — **i** allineare

405 alignment; alinement — The position of printer's characters on the type body to ensure that the base of all characters without descenders are in alignment across the line. Characters of the same body size in different type faces do not necessarily align if set together in the same line; differences of more than 0.0381 mm (0.0015 in) are detectable. Most film setter characters are designed with the same alignment for different type sizes.
f alignement *m* — **d** Ausrichtung *f* — **n** lijning *f* — **e** alineación *f*; alineamiento *m*; alineo *m* — **i** allineamento *m*

406 alignment chart — See nomogram.

407 alinement — The spelling alinement is not usual, but is permissable. See alignment.

408 alinement — See alignment.

409 aliphatic solvents — Organic liquids with an open chain hydrocarbon structure and KB values below 40. Relatively poor solvents for printing ink resins. Examples are *VM & P naphtha, textile spirits and mineral oils.
f solvants *mpl* aliphatiques — **d** aliphatische Lösungsmittel *npl* — **n** alifatische oplosmiddelen *npl* — **e** disolventes *mpl* alifáticos — **i** solventi *mpl* alifatici

410 alive matter — See standing matter.

411 alizarin assistant — See turkey red oil.

412 alizarin dye — Dye formerly obtained from madder, now from anthraquinone and used for red and blue inks.
Formula: $C_6H_4(CO)_2 \cdot C_6H_2(OH)_2$.
f colorant *m* d'alizarine — **d** Alizarinfarbstoff *m* — **n** alizarinekleurstof *fm* — **e** colorante *m* de alizarina — **i** colorante *m* d'alizarina

413 alizarin oil — Neutralized product obtained by treating olive oil, castor oil, cottonseed oil or other glycerides of fatty acids with sulphuric acid.
f huile *f* d'alizarine — **d** Alizarinöl *n* — **n** alizarineolie *fm* — **e** aceite *m* de alizarina — **i** olio *m* di alizarina

414 alkali — Chemical compound able to form a salt with an acid. Lime, soda, potash are alkalis. Ammonia and ammonium hydroxide are sometimes referred to as alkalis.
f alcali *m* — **d** Alkali *n* — **n** alkali *n* — **e** álcali *m* — **i** alcale *m*; alcali *m*

415 alkali blue; reflex blue — Strong alkali blue toner with a bronzy or reflex cast made of the sodium salt of monosulphonic acid of aniline blue.

Alkali blue, on a weight basis, has the highest tinting strength of all blue pigments.

f bleu *m* alcalin; bleu *m* alcalin reflexe — **d** Alkaliblau *n*; Reflexblau *n* — **n** alkaliblauw *n*; reflexblauw *n* — **e** azul *m* de álcali — **i** blu *m* alcalino; blu *m* riflesso

416 alkali fast; alkali resistant; alkali proof

f inattaquable par les alcalis; résistant aux alcalis — **d** alkaliecht; alkalibeständig; alkalifest; laugenfest — **n** alkali-echt; alkalivast; alkalibestendig; bestand tegen alkaliën — **e** resistente a los álcalis; álcalirresistente — **i** resistente agli alcali

417 alkaline — Relating to, or having the properties of an alkali. Having a pH of more than 7.

f alcalin — **d** alkalisch; laugenartig — **n** alkalisch — **e** alcalino — **i** alcalino

418 alkalinity — The state or quality of being alkaline. For paper cooking liquors the term causticity is sometimes used.

f alcalinité *f* — **d** Alkalinität *f*; Alkalität *f* — **n** alkaliniteit *f*; alcaliniteit *f* — **e** alcalinidad *f* — **i** alcalinità *f*

419 alkali proof — See alkali fast.

420 alkali-proof paper — Paper with a high degree of resistance to alkali, used for wrapping and packaging of alkaline materials, such as soaps, adhesives, etc. An important characteristic is the stability of colour without appreciable discoloration when wetted with 1% sodium hydroxide (caustic soda) or 40 °Baumé sodium silicate. Made from a variety of pulps, primarily semibleached and fully bleached chemical wood pulps.

f papier *m* résistant aux alcalis; papier *m* à l'épreuve des alcalis — **d** alkalifestes Papier *n*; alkalibeständiges Papier *n*; laugenfestes Papier *n* — **n** alkalibestendig papier *n*; alkalivast papier *n* — **e** papel *m* a prueba de álcalis; papel *m* resistente a los álcalis — **i** carta *f* resistente agli alcali

421 alkali resistance — The ability to withstand the action of alkalis. To be distinguished from soap resistance.

f résistance *f* à l'alcali; résistance *f* aux alcalis — **d** Alkaliechtheit *f*; alkalibeständigkeit *f* — **n** alkalibestendigheid *f* — **e** resistencia *f* al álcali; estabilidad *f* a los álcalis; solidez *f* a los álcalis — **i** resistenza *f* agli alcali

422 alkali resistant — See alkali fast.

423 alkali violet — Violet variety of alkali blue.

f violet *m* d'alcali — **d** Alkaliviolet *n* — **n** alkaliviolet *n* — **e** violeta *f* de álcali — **i** violetto *m* di alcali

424 alkanes; paraffins; paraffin hydrocarbons — Homologous series of hydrocarbons. The first four members of the series (methane, ethane, propane, butane) are gases; the next eleven are liquids and form the principal constituents of various mineral oils; the higher members are solids, forming the chief constituents of paraffin wax.

Formula: C_nH_{2n+2}.

f alcanes *fpl*; hydrocarbures *mpl* saturés; hydrocarbures *mpl* paraffiniques; paraffines *fpl* — **d** Alkane *npl*; Paraffinkohlenwasserstoffe *mpl*; Grenzkohlenwasserstoffe *mpl* — **n** alkanen *mpl*; paraffine-koolwaterstoffen *fmpl*; paraffinen *fmpl* — **e** alcanos *mpl*; hidrocarburos *mpl* parafínicos; parafinas *fpl* — **i** idrocarburi *mpl* di paraffina

425 alkyd resin; glyptal resin *trade mark* — Thermosetting resin, obtained as reaction product from glycerol and phthalic acid. When modified with certain vegetable oils (e.g., linseed, tung, soya bean) alkyds form a very important group of vehicles for letterpress or litho inks. Also used for paper coatings and adhesives.

f résine *f* alkyde; composé *m* glycérophtalique; alkyde *f*; résine *f* glycérophtalique — **d** Alkydharz *n* — **n** alkydhars *mn* — **e** resina *f* alcídia; resina *f* alquídica; resina *f* alcohílica — **i** resina *f* alchílica

426 all along stitching — Method of binding on tapes or cords, in which the thread passes from end to end of each section, i.e., from kettle-stitch to kettle-stitch inside the fold of each section.

f cousure *f* à un cahier; couture *f* à un cahier — **d** Durchausheften *n* — **n** geheel doornaaien *n*; naaien *n* over de gehele lengte — **e** cosido *m* con puntada seguida; cosido *m* a punto seguido — **i** cucitura *f* a punto appiccicato

427 Allen press — One of the first automatic presses, of American origin, for rapid printing of envelopes.

428 Allen wrench — Wrench formed from a piece of hexagonal bar stock bent to a right angle, for turning allen screws (i.e. a screw turned by means of an axial hexagonal hole in its head. Formerly a trade-mark.

f clé *f* Allen — **d** Stiftschlüssel *m* — **n** stiftsleutel *m* — **e** llave *f* Allen — **i** chiave *f* per viti Allen

429 alley — The space between two type frames, stands, or cabinets, where the compositor works.

f lézarde *f* — **d** Gasse *f*; Setzergasse *f* — **n** gangetje *n* — **e** pasillo *m* — **i** canaletto *m*

430 all in; all in hand — Term used when the complete copy has been given to the composing room.

f copie toute entrée; copie complète — **d** das gesamte Manuskript ist zum Satz gegeben — **n** gehele manuscript in de zetterij — **e** pasado a la sala de composición — **i** copia completa in sala compositori

431 all in hand — See all in.

432 "all in" time — The time including all allowances.

f temps *m* compensé — **d** Vorgabezeit *f* — **n** berekende tijd *m*; gecalculeerde tijd *m* — **e** tiempo *m* presupuestado — **i** tempo *m* compensato

433 all metal camera — Process camera whose stand and camera body, except the bellows, are made entirely from metal.

f chambre *f* photographique tout métal — **d** Ganzmetallkamera *f* — **n** geheel metalen camera *f* — **e** cámara *f* toda de metal — **i** apparecchio *m* fotografico metallico

434 allotter — Computer controlled device to distribute paper tape to line casters, avoiding manual tape handling. Paper tape is generated at a punch station near the linecaster operating unit to create a buffer of tape between the linecaster and the punch. The computer determines the sequencing and availability of linecasters for the particular size and type faces required. See also torntape system.

435 all out — When the types of a fount have all been used up in a composition.
f casse f épuisée; police f épuisée (être envoyé à la balade. Aller à la balade, à proprement parler, aller se promener. C'est manquer de copie ou de distribution, pour le compositeur; attendre après les formes ou le papier, pour les imprimeurs. "Le prote nous envoie à la balade", c'est un chomage momentané, occasionnel.) — **d** die Schrift ist versetzt; es fehlt an Manuskript (ausgesetzter Kasten m; leerer Kasten m) — **n** wachten — **e** agotado (el tipo) — **i** ci manca il carattere; cassa f esaurita

436 allowed time — See incentive time.

437 alloy — Metallic compound obtained by melting two or more metals together. Type metal is an alloy of lead, tin and antimony. There is much divergency as to the casting range in the various countries. See also composing machine metal, stereotype metal, type metal.
f alliage m; alliage m d'imprimerie — **d** Legierung f; Metallegierung f; Schriftlegierung f — **n** legering f; alliage f — **e** aleación f (de metal de imprenta) — **i** lega f (di metalli)

438 all-rag paper — See rag paper.

439 all rights reserved
f reproduction interdite; tout droit de reproduction réservé; tous les droits de traduction (de reproduction, de représentation, d'exécution) réservés — **d** alle Rechte vorbehalten; Nachdruck verboten — **n** alle rechten voorbehouden; nadruk verboden — **e** reservados todos los derechos; derechos reservados por ...; es propiedad de ...; prohibida la reproducción — **i** tutti i diritti riservati; riproduzione vietata

440 all the way across the page
f tout en travers de la page — **d** quer über die ganze Seite; über die ganze Seitenbreite — **n** over de gehele breedte van de pagina — **e** de todo el ancho de la página — **i** attraverso tutta la pagina; traversalmente alla pagina

441 Alltone process — Method of newspaper plate-making involving etching of type matter and half-tone illustrations as a single unit on zinc plates, the ordinary non-printing areas of the plate bearing a formation of small halftone dots, which serve as bearers for the inking rollers when printing direct from the plates on rotary newspaper presses.
f procédé m Alltone — **d** Alltone-Verfahren n — **n** Alltone-procédé n — **e** procedimiento m Alltone; método m Alltone — **i** procedimento m Alltone

442 all up — When the composition work is completed.

f tout composé — **d** ausgesetzt — **n** klaar met zetten; afgezet — **e** terminado — **i** tutto composto

443 allyl alcohol; propenyl alcohol; propenol — Colourless liquid, miscible with water, alcohol, chloroform, ether, petroleum ether. Poisonous. TL-value: 5 ppm.
Formula: C_3H_6O.
f alcool m allylique; vinylcarbinol m — **d** Allyl-alkohol m — **n** allylalcohol m — **e** alcohol m alílico — **i** alcool m allilico

444 allyl chloride — Colourless liquid with unpleasant purgent odour, insoluble in water, miscible with alcohol, chloroform, ether and petroleum ether. TL-value: 5 ppm.
Formula: C_3H_5Cl.
f chlorure m d'allyle; chlorure m allylique — **d** Allylchlorid n; Chlorallyl n — **n** allylchloride n — **e** cloruro m de alilo — **i** cloruro m allilico; cloropropilene m

445 allyl isosulphocyanate — See allyl isothiocyanate.

446 allyl isothiocyanate; allyl isosulphocyanate — Colourless to light yellow, oily liquid, soluble in alcohol, ether, carbon disulphide, slightly soluble in water. Causes blistering of skin and is harmful to lungs.
Formula: C_3H_5NCS.
f isothiocyanate m d'allyle; isosulfocyanate m d'allyle — **d** Allylisothiozyanat n; Isosulfozyanallyl n — **n** allylisothiocyanaat n — **e** isotiocianato m de alilo — **i** isotiocianato m di allile

447 allyl resin — Thermosetting transparent abrasion-resistant synthetic resins derived from allyl alcohol.
f résine f allylique — **d** Allylharz n — **n** allylhars mn — **e** resina f alílica — **i** resina f allilica

448 allyl sulphocarbamide — See allyl thiourea.

449 allyl sulphourea — See allyl thiourea.

450 allyl thiourea; allyl sulphocarbamide; allyl sulphourea; thiosinamine — White crystalline solid. Toxic, soluble in water, ether, and solutions of borax, benzoates, urethane, insoluble in benzene, slightly soluble in 70% alcohol.
Formula: $C_3H_5NHCSNH_2$.
f thiourée f d'allyle — **d** Allylthioharnstoff m — **n** allylthioureum n — **e** aliltiourea f; alilsulfocarbamida — **i** tiourea f allilica

451 almanac — Book, booklet, or pamphlet with calendar and other information; usually published as an annual.
f almanach m; annuaire m — **d** Almanach m; Kalender m — **n** almanak m; kalender m — **e** almanaque m; calendario m — **i** almanacco m

452 aloe — Plant with long, strong fibres, similar to sisal hemp, used in papermaking.
f aloès m — **d** Aloe f — **n** aloë fm — **e** áloe m; aloe m — **i** aloe m

453 alpha- — Prefix denoting position of an atom or group in a compound.
f alpha- — **d** Alpha- — **n** alfa- — **e** alfa- — **i** alfa-

454 alphabet — See also German alphabet, Greek alphabet, French alphabet, Italian, Spanish, Dutch.

f alphabet *m* — **d** Alphabet *n* — **n** alfabet *n* — **e** alfabeto *m* — **i** alfabeto *m*

455 alphabetical index; alphabetical table
f table *f* alphabétique — **d** alphabetisches Verzeichnis *n* — **n** alfabetische inhoudsopgave *fm*; alfabetisch register *n* — **e** índice *m* alfabético — **i** indice *m* alfabetico

456 alphabetical order — In the order of the letters of the alphabet. See especially Spanish.
f classement *m* alphabétique; ordre *m* alphabétique — **d** alphabetische Ordnung *f* — **n** alfabetische volgorde *fm* — **e** orden *m* alfabético — **i** ordine *m* alfabetico; ordinamento *m* alfabetico

457 alphabetical table — See alphabetical index.

458 alphabetize *v* — To put or arrange in alphabetical order. Also to express by or furnish with an alphabet. See especially Spanish.
f classer par ordre alphabétique; ranger par ordre alphabétique — **d** in alphabetischer Ordnung einreihen; alphabetisieren — **n** in alfabetische volgorde plaatsen; alfabetiseren; alfabetizeren — **e** colocar por orden alfabético; ordenar alfabéticamente; alfabetizar — **i** mettere in ordine alfabetico; alfabetare

459 alphabet length — The horizontal measurement, in points, of the lower case alphabet set in type of a particular face and size.
f longueur *f* de l'alphabet — **d** Alphabetlänge *f* — **n** alfabetlengte *f* — **e** longitud *f* del alfabeto — **i** lunghezza *f* d'alfabeto

460 alphameric — See alphanumeric.

461 alphanumeric *GB*; **alphameric** *US* — Contraction of alphabetic-numeric, the characters which include letters of the alphabet, numerals and other symbols such as punctuation or mathematical symbols. An alphanumeric instruction can be used equally well with alphabetic or numeric kinds of fields of data.
f alphanumérique — **d** alphanumerisch — **n** alfanumeriek — **e** alfanumérico — **i** alfanumerico

462 alter *v*
f changer — **d** ändern — **n** veranderen — **e** alterar — **i** cambiare

463 alteration — Change from the manuscript copy introduced in proof, distinguished from a *correction made to eliminate a printer's error. Billed as a separate item above the charge for original composition. Alterations made by the author (author's alterations, colloquially called AA), or part of them, are customarily charged against his royalties.
f modification *f*; changement *m* — **d** Änderung *f*; Abänderung *f*; Veränderung *f*; Umänderung *f* — **n** verandering *f*; wijziging *f* — **e** alteración *f*; modificación *f*; cambio *m*; corrección *f* — **i** alterazione *f*; modificazione *f*

464 alternate standard — Standard for an alternative method due to change in tools, equipment or machinery.
f standard *m* de rechange — **n** aangepaste norm *fm* — **e** tiempo *m* standard de recambio — **i** standard *m* di ricambio

465 alternating current; AC — Supply or signal (in) which (the voltage) is continuously reversing in direction. One pair of reversals is called a cycle. The domestic mains voltage alternates at 50 or 60 c/sec.
f courant *m* alternatif; CA — **d** Wechselstrom *m* — **n** wisselstroom *m* — **e** corriente *f* alterna — **i** corrente *f* alternata

466 alternating-current motor; AC-motor — Motor that converts alternating current into mechanical energy.
f moteur *m* à courant alternatif; moteur *m* à CA; alternomoteur *m* — **d** Wechselstrommotor *m* — **n** wisselstroommotor *m* — **e** motor *m* de corriente alterna; alternomotor *m* — **i** motore *m* a corrente alternata

467 alternative title — Title given as an alternative to the main title of a book. Printed below the main title on the title page and linked to that title by the word "or".
f variante *f* de titre — **d** Wechseltitel *m* — **n** ondertitel *m* — **e** subtítulo *m* — **i** titolo *m* alternativo; secondo titolo *m*

468 alum — Term commonly but incorrectly applied by papermakers to various qualities of aluminium sulphate. Used for precipitating the rosin size onto the pulp to give water resistancy to paper. Chemically the name refers to compounds similar to the double salt $K_2SO_4 \cdot Al_2(SO_4)_3 \cdot 24H_2O$. In the paper industry, papermaker's alum is $Al_2(SO_4)_3 \cdot 14H_2O$ or $Al_2(SO_4)_3 \cdot 18H_2O$, or a mixture of these hydrates. It is known as alum because it is used for the purposes for which alum was formerly used. See also aluminium sulphate, aluminium potassium sulphate.
f alun *m*; sulfate *m* d'alumine; sulfate *m* d'aluminium — **d** Alaun *m*; Aluminiumsulfat *n* — **n** aluin *m*; aluminiumsulfaat *n* — **e** alumbre *m*; sulfato *m* alumínico — **i** allume *m*; solfato *m* d'alluminio

469 alum — See aluminium potassium sulphate.

470 alumina white — See aluminium hydrate.

471 aluminium *GB*; **aluminum** *US* — Silvery, ductile metal, soluble in strong acids and alkalis. Symbol: Al. Latin: aluminium.
f aluminium *m* — **d** Aluminium *n* — **n** aluminium *n* — **e** aluminio *m* — **i** alluminio *m*

472 aluminium acetate — Normally known only in solution.
Formula: $Al(C_2H_3O_2)_3$. Basic aluminium acetate is an amorphous white powder, soluble in water.
Formula: $Al(C_2H_3O_2)_2OH$.
f acétate *m* d'aluminium; acétate *m* d'alumine; mordant *m* au rouge — **d** Aluminiumazetat *n*; essigsaures Aluminium *n*; essigsaure Tonerde *f* — **n** aluminiumacetaat *n* — **e** acetato *m* de aluminio; acetato *m* alumínico — **i** acetato *m* d'alluminio

473 aluminium ammonium sulphate; ammonia alum; ammonium alum — Colourless crystals, soluble in water, glycerol, insoluble in alcohol. Used for sizing paper, tanning, etc.

Formula: $Al_2(SO_4)_3(NH_4)_2SO_4·24H_2O$.
f sulfate *m* double d'aluminium et d'ammonium; alun *m* ammoniacal; alun *m* d'ammonium — **d** Aluminium-Ammoniumsulfat *n*; Ammoniumalaun *m*; Ammoniakalaun *m*; Ammonalaun *m* — **n** aluminium-ammoniumsulfaat *n*; ammonium-aluminiumsulfaat *n*; ammoniakaluin *m* — **e** sulfato *m* alumínico-amónico; alumbre *m* de amonio — **i** solfato *m* ammonico d'alluminio

474 aluminium bronze — Alloy containing a high percentage of copper with 5-11% aluminium and varying amounts of iron, nickel, manganese and other elements.
f bronze *m* d'aluminium — **d** Aluminiumbronze *f* — **n** aluminiumbrons *n* — **e** bronce *m* al aluminio — **i** bronzo *m* d'alluminio

475 aluminium chloride — Yellowish-white crystalline powder, soluble in water and alcohol. Formula: $AlCl_3·6H_2O$.
f chlorure *m* d'aluminium — **d** Aluminiumchlorid *n* — **n** aluminiumchloride *n* — **e** cloruro *m* de aluminio; cloruro *m* alumínico — **i** cloruro *m* d'alluminio

476 aluminium foil — Aluminium sheet rolled to a thickness of less than 0.15 mm (0.006 in). There are three forms: 1. rolls of foil, i.e., foil in coiled form with trimmed edges; 2. sheets of foil, i.e., foil in rectangular form, sheared to size; 3. foil stock, i.e., a reroll stock for manufacture in foil. See also defects of aluminium foil.
f feuille *f* (mince) d'aluminium — **d** Aluminiumfolie *f*; Blattaluminium *n* — **n** aluminiumfo(e)lie *fm* — **e** lámina *f* de aluminio; hoja *f* de aluminio; película *f* de aluminio — **i** foglio *m* d'alluminio; alluminio *m* sottile

477 aluminium hydrate; aluminium trihydrate; aluminium hydroxide; aluminium trihydroxide; hydrated aluminium; alumina white; hydrate — White inorganic pigment used as an extender in inks and noted for its transparency. Also used for paper coatings. Formula: $Al(OH)_3$.
f hydrate *m* d'alumine, hydroxyde *m* d'aluminium, alumine *f* hydratée; blanc *m* d'alumine — **d** Aluminiumhydroxyd *n*; Tonerdehydrat *n*; Transparentweiß *n* — **n** toonaardehydraat *n*; aluinaardehydraat *n*; aluminiumhydroxyde *n*; transparantwit *n*; hydroxyde — **e** hidrato *m* de alúmina; hidrato *m* alumínico; hidróxido *m* de aluminio — **i** idrossido *m* d'alluminio

478 aluminium hydroxide — See aluminium hydrate.

479 aluminium ink — Ink with aluminium powder as pigment.
f encre *f* d'aluminium — **d** Aluminiumdruckfarbe *f* — **n** aluminiuminkt *m* — **e** tinta *f* de aluminio; tinta *f* plateada — **i** inchiostro *m* d'alluminio

480 aluminium oxide — Formula: Al_2O_3. The source of commercial aluminium oxide used in the production of metallic aluminium is bauxite which is a hydrated aluminium oxide with a somewhat variable proportion of water and the oxides of iron, silicon and titanium as principal impurities. The mineral *corundum is natural aluminium oxide and emery, ruby and sapphire are impure crystalline varieties.
f oxyde *m* d'aluminium — **d** Aluminiumoxyd *n* — **n** aluminiumoxyde *n* — **e** óxido *m* de aluminio; óxido *m* alumínico — **i** ossido *m* d'alluminio

481 aluminium paper — Base paper of ordinary wrapping weight coated with aluminium powder, sometimes made by incorporating the powder in the beater or a size press. Used for wrapping, particularly for wrapping food products and tobacco.
f papier *m* aluminium — **d** Aluminiumpapier *n* — **n** aluminiumpapier *n* — **e** papel *m* aluminio; papel *m* plateado — **i** carta *f* d'alluminio

482 aluminium plate — Thin plate of aluminium used in lithography for some press plates, image applied photographically used for both surface-type and deep-etch offset plates.
f plaque *f* d'aluminium; planche *f* d'aluminium — **d** Aluminiumplatte *f* — **n** aluminiumplaat *fm* — **e** plancha *f* de aluminio; lámina *f* de aluminio; placa *f* de aluminio *Me* — **i** lastra *f* d'alluminio

483 aluminium potassium sulphate; potash alum; potassium aluminium sulphate; alum; burnt alum — White crystals, soluble in water, insoluble in alcohol. Solutions in water are acid. Latin: sulfas kalico-aluminicus.
Formula: $Al_2(SO_4)_3·K_2SO_4·24H_2O$.
f sulfate *m* d'aluminium et de potassium; sulfate *m* double d'aluminium et de potassium; alun *m* ordinaire; alun *m* de potassium — **d** Kalium-Aluminiumsulfat *n*; Kalium-Aluminiumalaun *m*; Kaliumalaun *m*; Kalialaun *m* — **n** kalium-aluminiumsulfaat *n*; kalium-aluminiumaluin *m*; kalialuin *m*; aluin *m* — **e** sulfato *m* de aluminio potásico; sulfato *m* alumínico-potásico — **i** solfato *m* di potassio e alluminio; solfato *m* doppio d'alluminio e potassio; allume *m* di potassio

484 aluminium powder — Metallic aluminium ground into very fine powder, used for dusting over prints or as a pigment in silver inks.
f poudre *f* d'aluminium — **d** Aluminiumbronze *f*; Aluminiumpulver *n*; Silberbronze *f* — **n** aluminiumpoeder *n* — **e** aluminio *m* en polvo — **i** polvere *f* d'alluminio

485 aluminium silicate — Chemical name for *China clay. Synthetic, colourless crystals, insoluble in water.
Formula: $Al_2O_3·2SiO_2·H_2O$.
f silicate *m* d'aluminium — **d** Aluminiumsilikat *n* — **n** aluminiumsilicaat *n* — **e** silicato *m* de aluminio; silicato *m* alumínico — **i** silicato *m* d'alluminio.

486 aluminium sodium sulphate — See sodium alum.

487 aluminium sulphate; papermaker's alum — White crystals, soluble in water, insoluble in alcohol. Used for sizing paper. Latin: sulfas aluminicus.
Formula: $Al_2(SO_4)_3·18H_2O$. See also alum.

f sulfate *m* d'aluminium — **d** Aluminiumsulfat *n*; Tonerdesulfat *n*; schwefelsaure Tonerde *f*; schwefelsaures Aluminium *n* — **n** aluminium-sulfaat *n* — **e** sulfato *m* de aluminio; sulfato *m* alumínico — **i** solfato *m* d'alluminio; allume *m* dei cartai

488 aluminium trihydrate — See aluminium hydrate.

489 aluminium trihydroxide — See aluminium hydrate.

490 aluminography — Offset process employing an aluminium plate instead of a stone. Invented in 1894 by Jos Scholz, Mainz. The print of such a plate was originally called algraphy.

f aluminographie *f*; algraphie *f*; impression *f* avec plaque aluminium — **d** Aluminiumdruck *m*; Algraphie *f* — **n** aluminiumdruk *m*; algrafie *f* — **e** aluminografía *f*; algrafía *f*; impresión *f* algráfica; impresión *f* con planchas de aluminio — **i** allumin-grafia *f*; algrafia *f*; stampa *f* con lastre d'alluminio

491 aluminotype — Reproduction of engravings or type formes in aluminium, cast from a plaster of Paris mould.

f aluminotypie *f* — **d** Aluminiumätzung *f*; Aluminium-Duplikatdruckform *f* von Ätzungen oder Satzformen — **n** aluminiumtypie *f* — **e** aluminotipia *f* — **i** alluminotipia *f*

492 aluminum — See aluminium.

493 AM — See amplitude modulation.

494 amalgam — Alloy of mercury with another metal, solid or liquid at room temperature depending on the proportion of mercury.

f amalgame *m* — **d** Amalgam *n* — **n** amalgaam *n* — **e** amalgama *f* — **i** amalgama *m*

495 amateur binding; collector's binding — Usually a half-leather binding made for the owner of the books.

f reliure *f* d'amateur; reliure *f* amateur — **d** Lieb-habereinband *m* — **n** bibliofiele band *m*; bibliofiele uitgave *f* — **e** encuadernación *f* de aficionado — **i** legatura *f* da amatore

496 ambient temperature — Temperature of the surrounding medium.

f température *f* ambiante; température *f* d'ambiance; température *f* de l'air ambiant — **d** Umgebungstemperatur *f* — **n** temperatuur *f* der omgeving — **e** temperatura *f* ambiente; ambiente *m* — **i** temperatura *f* ambiente

497 American groove — Similar to *French groove but of a more rounded nature, of an approximate standard of 3 mm (1/8 in) for all books.

498 American height to paper — See also height to paper.

f hauteur *f* américaine — **d** amerikanische Schrift-höhe *f*; amerikanische Höhe *f* — **n** Amerikaanse letterhoogte *f*; Amerikaanse hoogte *f* — **e** altura *f* americana — **i** altezza *f* americana

499 American Russia; imitation Russia — Strong, split cowhide for book covers.

f imitation *f* cuir de Russie — **d** imitiertes Jucht(en)leder *n*; imitierter Juchten *m* — **n**

imitatie-juchtleer *n* — **e** imitación *f* de cuero de Rusia; imitación *f* de piel de Rusia — **i** imitazione *f* cuoio di Russia

500 amide — Organic compound formed by replacing a hydrogen atom of ammonia (NH_3) by organic acid radicals, e.g., acetamide (CH_3CONH_2) and urea (NH_2CONH_2).

f amide *m* — **d** Amide *n*; Amid *n* — **n** amide *n* — **e** amida *f* — **i** ammide *f*

501 amido — Adjective form of *amide; also a synonym for amino.

502 amidol — See diaminophenol hydrochloride.

503 amine — Organic compound derived from ammonia by replacement of hydrogen by one or more univalent alkyl or aryl radicals, e.g., phenyl-amine, $C_6H_5NH_2$.

f amine *f* — **d** Amine *n*; Amin *n* — **n** amine *fm* — **e** amina *f* — **i** ammina *f*

504 amino — See amido.

505 aminobenzene — See aniline.

506 amino plastic — See amino resin.

507 amino resin; amino plastic — Thermosetting resin made from a substance containing one or more amino (NH_2) groups, usually aldehyde urea, thiourea, melamine, dicyanodiamide or aniline products. Used in adhesives, protective coatings, paper manufacture, etc.

f résine *f* aminique; résine *f* amino — **d** Amino-harz *n*; Aminoplastharz *n* — **n** aminohars *mn* — **e** resina *f* amínica — **i** resina *f* amminica

508 ammeter — See ampere meter.

509 ammonia — Product (a gas) of the destructive distillation of soft coal. Also a solution of this material in water. Very soluble in water, alcohol and ether. Formula: NH_3. TL-value: 25 ppm or $18mg/m^3$. See also ammonium hydroxide.

f ammoniac *m*; gaz *m* ammoniac — **d** Ammoniak *n* — **n** ammoniak *m* — **e** amoníaco *m* — **i** ammoniaca *f*

510 ammonia alum — See aluminium ammonium sulphate.

511 ammonia process — Two-component diazo-type process in which both the diazo and the coupler are on the base, and development is achieved by neutralizing the acidic stabilizers with vapours derived from evaporating aqueous ammonia. See also diazoprint.

512 ammonia water — See ammonium hydroxide.

513 ammonium acetate — White, deliquescent, crystalline mass, soluble in water, alcohol. Formula: $NH_4C_2H_3O_2$.

f acétate *m* d'ammonium; acétate *m* d'ammoniaque — **d** Ammoniumazetat *n*; essig-saures Ammonium *n* — **n** ammoniumacetaat *n* — **e** acetato *m* de amonio; acetato *m* amónico — **i** acetato *m* d'ammonio

514 ammonium acid carbonate — See ammonium bicarbonate.

515 ammonium acid fluoride — See ammonium bifluoride.

516 ammonium alum — See aluminium

ammonium sulphate.

517 ammonium bicarbonate; ammonium acid carbonate; ammonium hydrogen carbonate — White crystals, soluble in water, insoluble in alcohol. Formula: NH_4HCO_3.

f bicarbonate m d'ammonium — **d** Ammoniumbikarbonat n; doppeltkohlensaures Ammonium n — **n** ammoniumbicarbonaat n — **e** bicarbonato m de amonio; bicarbonato m amónico — **i** bicarbonato m d'ammonio

518 ammonium bichromate — See ammonium dichromate.

519 ammonium bifluoride; ammonium acid fluoride; ammonium hydrogen fluoride — White crystals, deliquescent, decomposed by heat, poisonous, soluble in cold water and alcohol, decomposes in hot water. Formula: NH_4HF_2.

f bifluorure m d'ammonium — **d** Ammoniumbifluorid n — **n** ammoniumbifluoride n — **e** fluoruro m de amonio ácido; bifluoruro m amónico — **i** bifluoruro m d'ammonio

520 ammonium bromide; bromide of ammonia — Colourless crystals or a white powder, soluble in water and alcohol. Used for precipitating silver salts for photographic plates. Formula: NH_4Br. Latin: brometum ammonicum.

f bromure m d'ammonium; bromhydrate m d'ammoniaque — **d** Ammoniumbromid n; Bromammonium n — **n** ammoniumbromide n; broomammonium n — **e** bromuro m de amonio; bromuro m amónico — **i** bromuro m d'ammonio

521 ammonium carbonate; carbonate of ammonia — Colourless crystal plates, unstable in air, being converted into the bicarbonate, soluble in water, decomposes in hot water, yielding ammonia and carbon dioxide. Formula: $(NH_4)_2CO_3$.

f carbonate m d'ammonium (neutre); carbonate m d'ammoniaque; sel m volatile — **d** Ammoniumkarbonat n; Ammoniumbikarbonat n; kohlensaures Ammonium n; Hirschhornsalz n; Geistersalz n — **n** ammoniumcarbonaat n; koolzure ammoniak m; hertshoornzout n — **e** carbonato m de amonio; carbonato m amónico — **i** carbonato m d'ammonio

522 ammonium chloride; sal ammoniac; ammonium muriate — White crystals, soluble in water and glycerol, slightly soluble in alcohol. Formula: NH_4Cl. Latin: chloretum ammonicum.

f chlorure m d'ammonium; chlorhydrate m d'ammoniaque; sel m ammoniaque — **d** Ammoniumchlorid n; Chlorammonium n; salzsaures Ammoniak n — **n** ammoniumchloride n; chloorammonium n; salmiak m — **e** cloruro m de amonio; cloruro m amónico; sal f amoniaca; sal f amónica — **i** cloruro m d'ammonio

523 ammonium chromate — Yellow crystals, soluble in cold water, insoluble in alcohol. Used as a photographic sensitizer for gelatine coatings. Formula: $(NH_4)_2CrO_4$.

f chromate m d'ammonium — **d** Chrom-

ammonium n; Ammoniumchromat n — **n** ammoniumchromaat n — **e** cromato m de amonio; cromato m amónico — **i** cromato m d'ammonio

524 ammonium citrate; dibasic ammonium citrate — White granules, soluble in water, very slightly soluble in alcohol. Formula: $(NH_4)_2HC_6H_5O_7$.

f citrate m d'ammonium — **d** Ammoniumzitrat n — **n** ammoniumcitraat n — **e** citrato m de amonio; citrato m amónico — **i** citrato m d'ammonio

525 ammonium citrate of iron — See ferric ammonium citrate.

526 ammonium dichromate; ammonium bichromate — Salt formed by neutralizing chromic acid with ammonia. Soluble in water, insoluble in alcohol. Sensitive to actinic light rays; used with a colloid as a sensitizer for printing plate coatings. Latin: bichromas ammonicus; ammonium bichromicum. Formula: $(NH_4)_2Cr_2O_7$.

f bichromate m d'ammonium; bichromate m d'ammoniaque; ammonium m bichromate — **d** Ammoniumbichromat n; Ammoniumdichromat n; Ammoniumhydrogenkarbonat n; doppelchromsaures Ammonium n; saures Ammoniumkarbonat n — **n** ammoniumbichromaat n; ammoniumdichromaat n; dubbelchroomzure ammoniak m — **e** dicromato m de amonio; dicromato m amónico; bicromato m de amonio; bicromato m amónico — **i** bicromato m d'ammonio

527 ammonium ferric citrate — See ferric ammonium citrate.

528 ammonium ferric oxalate — See ferric ammonium oxalate.

529 ammonium ferric sulphate — See ferric ammonium sulphate.

530 ammonium fluoride — White crystals, soluble in cold water. Poisonous. Used for etching of glass. Formula: NH_4F.

f fluorure m d'ammonium — **d** Ammoniumfluorid n; Fluorammonium n — **n** ammoniumfluoride n — **e** fluoruro m de amonio; fluoruro m amónico — **i** fluoruro m d'ammonio

531 ammonium hydrogen carbonate — See ammonium bicarbonate.

532 ammonium hydrogen fluoride — See ammonium bifluoride.

533 ammonium hydroxide; ammonia water; aqua ammonia — Solution of the gas ammonia (NH_3) in water, used by lithographers for clearing surface plates after development. Formula: NH_4OH. Latin: liquor ammonii caustici; ammonium hydricum; ammonia liquida.

f ammoniaque m; eau f ammoniacale; alcali m volatil — **d** Ammoniakwasser n; Ammoniaklösung f; Ammoniumhydroxyd n; Ätzammoniak n; Salmiakgeist m — **n** ammonia m — **e** hidróxido m de amonio; agua f amoniacal; agua f de amoníaco; solución f amoniacal — **i** ammoniaca f

liquida; acqua *f* ammoniacale; idrossido *m* d'ammonio

534 ammonium hyposulphite — See ammonium thiosulphate.

535 ammonium iodide — White crystals or white granular powder, soluble in water or alcohol. Sublimes with decomposition. Formula: NH_4I. Latin: iodetum ammonicum.
f iodure *m* d'ammonium; iodhydrate *m* d'ammoniaque — **d** Ammoniumjodid *n*; Jodammonium *n* — **n** ammoniumjodide *n* — **e** yoduro *m* de amonio; yoduro *m* amónico — **i** ioduro *m* d'ammonio

536 ammonium iron citrate — See ferric ammonium citrate.

537 ammonium lactate — Colourless to yellowish syrupy liquid, soluble in water and alcohol. Formula: $NH_4C_3H_5O_3$.
f lactate *m* d'ammonium — **d** Ammoniumlaktat *n*; milchsaures Ammonium *n* — **n** ammoniumlactaat *n* — **e** lactato *m* de amonio; lactato *m* amónico — **i** lattato *m* d'ammonio

538 ammonium muriate — See ammonium chloride.

539 ammonium nickel sulphate — See nickel ammonium sulphate.

540 ammonium nitrate — Colourless crystals, soluble in water, alcohol and alkalis. Explosive but not readily detonating. Latin: nitras ammonicus. Formula: NH_4NO_3.
f nitrate *m* d'ammoniaque — **d** Ammoniumnitrat *n*; Ammonsalpeter *m* — **n** ammoniumnitraat *n* — **e** nitrato *m* de amonio; nitrato *m* amónico — **i** nitrato *m* d'ammonio

541 ammonium oxalate — Colourless crystals, soluble in water. Poisonous. Decomposed by heat. Latin: oxalas ammonicus. Formula: $(NH_4)_2C_2O_4·H_2O$.
f oxalate *m* d'ammonium — **d** Ammoniumoxalat *n* — **n** ammoniumoxalaat *n* — **e** oxalato *m* de amonio; oxalato *m* amónico — **i** ossalato *m* d'ammonio

542 ammonium oxalate of iron — See ferric ammonium oxalate.

543 ammonium persulphate — White crystals, soluble in water. Strong oxidizing agent. Latin: ammonium persulfuricum. Formula: $(NH_4)_2S_2O_8$.
f persulfate *m* d'ammoniaque; persulfate *m* d'ammonium — **d** Ammoniumpersulfat *n*; überschwefelsaures Ammonium *n* — **n** ammoniumpersulfaat *n* — **e** persulfato *m* de amonio; persulfato *m* amónico — **i** persolfato *m* d'ammonio; persolfato *m* ammonico

544 ammonium sulphate — Brownish-grey to white crystals according to degree of purity. Soluble in water, insoluble in alcohol and acetone. Latin: sulfas ammonicus. Formula: $(NH_4)_2SO_4$.
f sulfate *m* d'ammonium; sulfate *m* d'ammoniaque — **d** Ammoniumsulfat *n*; schwefelsaures Ammonium *n* — **n** ammoniumsulfaat *n* — **e**

sulfato *m* de amonio; sulfato *m* amónico; vitriolo *m* amoniacal — **i** solfato *m* d'ammonio

545 ammonium sulphate of iron — See ferrous ammonium sulphate.

546 ammonium sulphide — Yellow crystals, soluble in water, alcohol and alkalis. Latin: ammonium sulfuratum. Formula: $(NH_4)_2S$.
f sulfure *m* d'ammonium — **d** Ammoniumsulfid *n*; Ammoniumsulfür *n*; Schwefelammonium *n* — **n** ammoniumsulfide *n*; zwavelammonium *n* — **e** sulfuro *m* de amonio; sulfuro *m* amónico — **i** solfuro *m* d'ammonio

547 ammonium sulphite — Colourless crystals used in photography. Hygroscopic. Must be well stoppered. Formula: $(NH_4)_2SO_3·H_2O$.
f sulfite *m* d'ammonium — **d** Ammoniumsulfit *n* — **n** ammoniumsulfiet *n* — **e** sulfito *m* de amonio; sulfito *m* amónico — **i** solfito *m* d'ammonio

548 ammonium sulphocyanate — See ammonium thiocyanate.

549 ammonium sulphocyanide — See ammonium thiocyanate.

550 ammonium thiocyanate; ammonium sulphocyanate; ammonium sulphocyanide — Colourless, deliquescent crystals, soluble in water, alcohol, acetone and ammonia. Latin: ammonium sulfocyanatum; ammonium rhodanatum. Formula: NH_4SCN.
f thiocyanate *m* d'ammonium; sulfocyanate *m* d'ammonium; sulfocyanure *m* d'ammonium — **d** Ammoniumthiozyanat *n*; Ammoniumsulfozyanat *n*; Ammoniumsulfozyanid *n*; Rhodanammonium *n*; Ammoniumrhodanid *n*; Schwefelzyanammonium *n* — **n** ammoniumthiocyanaat *n*; ammoniumsulfocyanide *n*; sulfocyaanammonium *n*; ammoniumrhodanide *m*; rhodaanammonium *n* — **e** tiocianato *m* de amonio; tiocianato *m* amónico; sulfocianuro *m* amónico; sulfocianuro *m* de amonio; rodanuro *m* de amonio; rodanuro *m* amónico — **i** tiocianato *m* d'ammonio; solfocianuro *m* d'ammonio

551 ammonium thiosulphate; ammonium hyposulphite — White crystals, decomposed by heat. Very soluble in water, pH of 60% solution 6.5-7.0. Certain photographic emulsions, particularly those with large grains, will clear much faster when ammonium thiosulphate is used instead of sodium thiosulphate. Formula: $(NH_4)_2S_2O_3$.
f thiosulfate *m* d'ammonium; hyposulfite *m* d'ammonium — **d** Ammoniumthiosulfat *n*; Ammoniumhyposulfit *n* — **n** ammoniumthiosulfaat *n* — **e** tiosulfato *m* de amonio; tiosulfato *m* amónico; hiposulfito *m* de amonio; hiposulfito *m* amónico — **i** tiosolfato *m* d'ammonio; iposolfito *m* d'ammonio

552 amorphous — Having no crystalline structure, e.g., glass.
f amorphe — **d** amorph(isch); gestaltlos — **n**

amorf; niet kristallijn; vormloos — **e** amorfo — **i** amorfo

553 amperage — The strength of an electric current measured in amperes.

f ampérage *m*; intensité *f*; débit *m* — **d** Stromstärke *f* — **n** amperage *n*; stroomsterkte *f* — **e** amperaje *m*; intensidad *f* en amperios — **i** intensità *f* di corrente; amperaggio *m* depr.

554 ampere — Unit of electrical current intensity, being that produced by one volt acting through a resistance of one ohm. The power (actinicity) of arc lamps for photoengraving usually is indicated in amperes. Symbol: A (single and plural), or amp (no point). In the SI one international ampere is the constant current which in one second, precipitates electrolytically 1.118 mg of silver from a silver nitrate solution. Thus:
$$1 A_{int} = 1 V_{int}/\Omega_{int} \approx 0.999850 \text{ A}.$$
f ampère *m* — **d** Ampere *n* — **n** ampère *f* — **e** amperio *m* — **i** ampère *m*

555 ampere-hour — Quantity of electricity conveyed when a current of 1 A flows for 1 h. In gravure printing 1 ampere-hour used with an acid copper solution at 100% cathode efficiency will deposit 1.186 g of copper, 2800 ampere-hours will deposit 3320 g or 117 ounces of copper.

f ampère-heure *f* — **d** Amperestunde *f* — **n** ampère-uur *n* — **e** amperio-hora *m* — **i** amperora *f*

556 ampere meter; ammeter — Instrument to measure the intensity of an electric current.

f ampèremètre *m* — **d** Amperemesser *m*; Amperemeter *n*; Ammeter *n*; Stromstärkenmesser *m* — **n** ampèremeter *m* — **e** amperímetro *m*; amperómetro *m* — **i** amperometro *m*

557 ampersand; short and — The sign "&" which is a kind of logotype for Latin "et", or English "and". Used in names of companies, e.g., Smith & Co., but preferably not at the beginning or end of a line. Should not be used in text matter to replace "and", unless the author specially marks it so. The abbreviation etc. should be used in preference to &c. The word ampersand is a corruption of the expression "and per se and", supposed to have been used in reading out the letters of the alphabet and other characters.

f perluète *f*; "et" *m* commercial; esperluète *f* obs.; pirlouète *f* obs. — **d** Et-Zeichen *n*; Und-Zeichen *n* — **n** en-teken *n*; et-teken *n* — **e** signo *m* & — **i** segno *m* &; congiunzione *f* commerciale

558 amplified and revised edition; enlarged and revised edition

f édition *f* amplifiée et corrigée — **d** vermehrte und verbesserte Auflage *f*; vermehrte und verbesserte Ausgabe *f* — **n** vermeerderde en verbeterde druk *m* — **e** edición *f* ampliada y aumentada; edición *f* corregida y aumentada — **i** edizione *f* ampliata e migliorata; ristampa *f* corretta ed arricchita di nuove aggiunte

559 amplified edition — See enlarged edition.

560 amplifier — Device to amplify an electric signal. It receives the signal at a low level and sends it out at a high level in identical or nearly identical form.

f amplificateur *m* — **d** Verstärker *m* — **n** versterker *m* — **e** amplificador *m*; ampliador *m* — **i** amplificatore *m*

561 amplitude — If a quantity is varying in an oscillatory manner about an equilibrium value, the maximum departure from that value is called the amplitude; e.g., in the case of a pendulum the amplitude is half the legth of the swing. For a wave motion, e.g., electromagnetic waves or sound waves, the amplitude of the wave determines the amount of energy carried by the wave.

f amplitude *f* — **d** Amplitude *f* — **n** amplitudo *f*; amplitude *f*; slingerwijdte *f* — **e** amplitud *f* — **i** ampiezza *f*

562 amplitude modulation — The form of modulation in which the amplitude of the carrier is varied in accordance with the instantaneous value of the modulating signal. In amplitude modulation technics the amplitude of an alternating current voltage (or current) contains the information. Abbrev.: AM.

f modulation *f* d'amplitude — **d** Amplituden-Modulation *f* — **n** amplitudemodulatie *f* — **e** amplitud *f* de modulación — **i** modulazione *f* d'ampiezza

563 amyl acetate; amyl acetic ether; isoamyl acetate — Colourless liquid, slightly soluble in water, miscible with alcohol and ether, used in printing ink compounds, as a solvent for celluloid, nitrocellulose, ethyl cellulose. Latin: amylum aceticum. TL-value: 100 ppm or 525 mg/m^3.
Formula: $CH_3COOC_5H_{11}$.

f acétate *m* d'amyle; éther *m* amylacétique; acétate *m* d'isoamyle; essence *f* de poires — **d** Amylazetat *n*; essigsaurer Amylester *m*; Isoamylazetat *n*; Birnenäther *m* — **n** amylacetaat *n*; isoamylacetaat *n* — **e** acetato *m* de amilo; acetato *m* de isoamilo — **i** acetato *m* d'amile; acetato *m* di isoamile

564 amyl acetic ether — See amyl acetate.

565 amyl alcohol; fusel oil — Colourless, oily mixture obtained as a by-product in the alcoholic fermentation of sugars; contains chiefly isoamyl alcohol, $C_5H_{11}OH$. Latin: alcohol amylicus. TL-value: 100 ppm.

f alcool *m* amylique; huile *f* de fusel — **d** Amylalkohol *m*; Amyloxyhydrat *n*; Fuselöl *n* — **n** amylalcohol *m*; foezelolie *fm* — **e** alcohol *m* amílico; aceite *m* de fusel — **i** alcool *m* amilico; olio *m* di flemma

566 amylum — See starch.

567 anachromat — See anachromatic lens.

568 anachromatic lens; anachromat

f anachromat *m*; objectif *m* anachromatique — **d** Anachromat *n*; anachromatisches Objektiv *n* — **n** anachromaat *m*; anachromatische lens *fm* — **e** anacromático *m*; lente *f* anacromática; objetivo *m* anacromático — **i** lente *f* anacromatica; obiettivo *m* anacromatico

569 anaglyph — Illustration giving a stereoscopic or relief effect when viewed through proper colour filters or spectacles.
f anaglyphe *m* — **d** Anaglyph *f* — **n** anaglyf *m* — **e** anáglifo *m* — **i** anaglifo *m*

570 anagram — Word or phrase made by transposing the letters of another word or phrase.
f anagramme *f* — **d** Anagramm *n* — **n** anagram *n* — **e** anagrama *m* — **i** anagramma *m*

571 analogue channel — Channel on which the information transmitted can take any value between the limits defined by the channels.
f voie *f* analogique — **d** Analogkanal *m*; Analogübertragungskanal *m* — **n** analoog transmissiekanaal *n* — **e** canal *m* analógico — **i** canale *m* analogico

572 analogue computer — Computer in which numerical processes are undertaken by measuring physical quantities such as voltage, resistance or current; in contrast with digital computer which operates directly on numbers. Analogue computers are not used in typesetting.
f ordinateur *m* analogique; calculateur *m* analogique — **d** Analogcomputer *m*; Analogrechner *m* — **n** analoge computer *m* — **e** calculador *m* analógico; computadora *f* analógica — **i** calcolatore *m* analogico

573 analogue signal — In data communication systems, a continuous electrical signal that varies in direct correlation to a signal impressed on a transducer. The frequency or amplitude of the signal may vary, for instance, in response to changes in phenomena or characteristics such as sound, light, heat, position or pressure. Generally, voice transmission is in analogue form.
f signal *m* analogique — **d** analoges Signal *n* — **n** analoog signaal *n* — **e** señal *f* analógica — **i** segnale *m* analogica

574 analysing head — See scanning head.

575 analysis
f analyse *f* — **d** Analyse *f*; Analisierung *f* — **n** analyse *f* — **e** análisis *f* — **i** analisi *f*

576 analyst
f analyste *m* — **d** Analytiker *m* — **n** analist *m*; analiste *f* — **e** analizador *m* — **i** analista *m*

577 analytical method — Method to find the nature and/or the amounts of ingredients in a substance.
f procédé *m* analytique — **d** analytisches Verfahren *n* — **n** analytische methode *f* — **e** método *m* analítico; procedimiento *m* analítico — **i** metodo *m* analitico

578 analytical rapid balance
f balance *f* analytique à pesée rapide — **d** Analyse-Schnellwaage *f*; Analysenschnellwaage *f* — **n** analytische snelbalans *fm* — **e** balanza *f* rápida de análisis — **i** bilancia *f* analitica rapida

579 anastatic printing — Process of printing from slightly raised metallic surfaces, reproducing a series of new impressions from an old print.
f impression *f* anastatique; réimpression *f* anastatique — **d** anastatisches Druckverfahren *n*; ana-statischer Druck *m* — **n** anastatisch drukprocédé *n* — **e** impresión *f* anastática — **i** stampa *f* anastatica; ristampa *f* anastatica

580 anastigmat — See anastigmatic lens.

581 anastigmatic lens; anastigmat — Photographic lens corrected for astigmatism as well as other distortions. Process lenses are anastigmats.
f objectif *m* anastigmatique; anastigmat *m* — **d** anastigmatisches Objektiv *n*; Anastigmat *n* — **n** anastigmatische lens *fm*; anastigmaat *m* — **e** objetivo *m* anastigmático; lente *m* anastigmático — **i** obiettivo *m* anastigmatico; lente *f* anastigmatica

582 anchor *v* — To mount printing plates on wood or metal mounts when no flanges are available. A thin bolt and nut is let through the mount and secures the plate.
f monter à griffes — **d** befestigen mit Schraubenbolzen — **n** monteren met schroefbouten — **e** asegurar con tornillos; atornillar; anclar al zócalo; fijar al zócalo — **i** fissare mediante viti; avvitare

583 anchoring — Bonding inks to the material on which they are printed.
f accrochage *m* des encres — **d** Haften *n* der Farbe — **n** hechten *n* van de inkt — **e** ligamiento *m* de las tintas en superposición — **i** ancoraggio *m*

584 ancillary equipment
f matériel *m* auxiliaire — **d** Hilfsmaterial *n* — **n** hulpstukken *npl* — **e** material *m* auxiliar — **i** fornitura *f* ausiliaria

585 and — Logic operator where more than one thing, fact or condition must be present for the statement to be effective. A and B must both be true, for example; or, A, B, C and D must all be true. See also gate.

586 anepigraph; anepigraphon — Work without title or inscription.
f anépigraphe *m* — **d** Anepigraphon *n*; Werk *n* og; Werk *n* ohne Titel; unbetitelte Schrift *f*; unbetiteltes Werk *n* — **n** anepigraaf *m*; geschrift *n* zonder titel — **e** anepígrafo *m*; obra *f* sin título; obra *f* que carece de portada — **i** anepigrafo *m*; opera *f* senza titolo

587 anepigraphon — See anepigraph.

588 angle bar; turner bar; turning bar — Metal bar in a rotary press, laid horizontally at 45 ° angle form the direction of the press. Used to turn the web when feeding from the side, or to bypass the former in ribbon folding. Usually filled with air and perforated to reduce friction from web travel.
f barre *f* de retournement; barre *f* de renversement — **d** Wendestange *f* — **n** keerstang *fm* — **e** barra *f* de inversión; recodo *m* desviador; barra *f* de inflexión — **i** barra *f* d'inversione; barra *f* per volgere; barra *f* di rovesciamento; barra *f* diagonale

589 angle-bar fold — See ribbon fold.

590 angle clamps — Angular pieces of metal to hold the corners of rules to keep them properly joined. They are of the same height as quads and are made in 6 and 12 point.

f blocs *mpl* d'anglets — **d** Winkelstücke *npl* — **n** hoekstukken *npl* — **e** piezas *fpl* angulares; piezas *fpl* de esquina; esquinazos *mpl* — **i** squadre *fpl* per unione filetti ad angolo

591 angle cut paper — Paper cut to special angles usually for the manufacture of envelopes.

f papier *m* coupé oblique — **d** schräggeschnittenes Papier *n*; diagonal geschnittenes Papier *n*; Schrägschnittpapier *n*; Diagonalschnittpapier *n* — **n** diagonaal gesneden papier *n*; schuin gesneden papier *n* — **e** papel *m* cortado al sesgo; papel *m* cortado en diagonal — **i** carta *f* con taglio diagonale

592 angle cutter; diagonal cutter — Machine for cutting paper at an angle to the machine direction of the paper. Envelope blanks are usually cut in this manner.

f coupeuse *f* à coupe oblique — **d** Diagonalschneider *m*; Schrägschneider *m*; Schrägschneidemaschine *f* — **n** diagonaalsnijmachine *f* — **e** cortadora *f* diagonal — **i** tagliarina *f* diagonale

593 angle modulation — See phase modulation.

594 angle of contact test — Test to measure the surface wettability of paper by aqueous solutions, such as writing and ruling inks. Measured is the tangential angle of a drop of water placed on the surface of the paper.

f essai *m* de l'angle de contact — **d** Randwinkelmessung *f* — **n** randhoekmeting *f* — **e** ensayo *m* del ángulo de contacto — **i** prova *f* dell'angolo di contatto

595 angle of incidence — Angle that a straight line, ray of light, etc., makes with the normal to the surface at the point of meeting the surface.

f angle *m* d'incidence — **d** Einfallswinkel *m* — **n** hoek *m* van inval — **e** ángulo *m* de incidencia — **i** angolo *m* d'incidenza

596 angle of ninety degrees — See right angle.

597 angle of reflection — Angle that a ray of light, reflected from a surface, makes with the normal to the surface at the point of reflection.

f angle *m* de réflexion — **d** Prallwinkel *m*; Reflexionswinkel *m* — **n** hoek *m* van terugkaatsing — **e** ángulo *m* de reflexión — **i** angolo *m* di riflessione

598 angle of refraction — Angle between a ray of light after refraction at an interface between the two media and the normal to the interface.

f angle *m* de réfraction — **d** Brechungswinkel *m* — **n** brekingshoek *m* — **e** ángulo *m* de refracción — **i** angolo *m* di rifrazione

599 angle of view — Angle between the two extreme rays passing through the centre of a lens for which the lens is calculated to give a correct image.

f angle *m* optique — **d** Bildwinkel *m* — **n** beeldhoek *m* — **e** ángulo *m* de campo — **i** angolo *m* di campo

600 angle of wipe — In rotogravure printing the angle between the doctor blade and the engraved cylinder.

f inclination *f* de la racle — **d** Rakelwinkel *m* — **n** rakelstand *m*; rakelhoek *m* — **e** inclinación *f* de la cuchilla; inclinación *f* de la racleta; ángulo *m* de la raspadora *SA* — **i** inclinazione *f* della racla

601 ångström; ångström unit — The unit of measurement for wavelengths, equal to 1/6 x 438,4696 of the wavelength of the red cadmium line in dry air at standard atmospheric pressure, 15 °C and 0.03 per cent by volume of carbon dioxide. The ångström is the former unit used in optics, in spectroscopy and in electron microscopy. In the SI (1978) the symbol is replaced by Å. Thus: 1 Å = 10^{-10} m (approx. 254.000.000 ångström units in an inch). Name derived from Anders Jöns Ångström (1814-1874), Swedish astronomer and physicist. In the SI the ångström is replaced by the nanometre.

f ångström *m*; unité *f* ångström — **d** Ångström *fn*; Ångström-Einheit *f* — **n** ångström *m*; ångström-eenheid *f* — **e** ångström *m*; unidad *f* ångström — **i** ångström *m*; unità *f* ångström

602 ångström unit — See ångström.

603 angular acceleration — The time rate of change of angular velocity of a rotating body.

f accélération *f* angulaire — **d** Winkelbeschleunigung *f*; Radialbeschleunigung *f* — **n** hoekversnelling *f* — **e** aceleración *f* angular — **i** accelerazione *f* angolare

604 angular velocity — The time rate of change of angular position of a rotating body, usually expressed in radians per second or radians per minute.

f vitesse *f* angulaire — **d** Winkelgeschwindigkeit *f* — **n** hoeksnelheid *f* — **e** velocidad *f* angular — **i** velocità *f* angolare

605 anhydride — The anhydride of a substance, when combined with water, gives the substance; e.g., sulphur trioxide SO_3 is the anhydride of sulphuric acid H_2SO_4.

f anhydride *m* — **d** Anhydrid *n* — **n** anhydride *n* — **e** anhídrido *m* — **i** anidride *f*

606 anhydrite — See calcium sulphate.

607 anhydrous; dehydrated; water free — Applied to chemicals: free from water, especially the water of crystallization.

f anhydre — **d** wasserfrei; entwässert — **n** watervrij; anhydrisch — **e** anhidro; sin agua; exento de agua; desprovisto de agua — **i** anidro

608 anhydrous alcohol — Water-free ethyl alcohol used for several purposes, e.g., to remove the deep-etch solution in offset platemaking.

f alcool *m* pur; alcool *m* anhydre — **d** wasserfreier Alkohol *m* — **n** watervrije alcohol *m* — **e** alcohol *m* anhidro; alcohol *m* puro — **i** alcool *m* anidro

609 anhydrous plate wash — Anhydrous alcohol used in lithographic platemaking to wash the plate before applying the lacquer image base.

f solution *f* de lavage anhydre — **d** wasserfreies Druckplatten-Reinigungsmittel *n* — **n** watervrij wasmiddel *n* — **e** líquido *m* anhidro para limpieza de planchas — **i** soluzione *f* anidra per il lavaggio delle lastre

610 anhydrous sodium carbonate — See sodium carbonate.

611 anhydrous sodium iodide — See sodium iodide.

612 anhydrous sodium sulphate — See sodium sulphate.

613 aniline; aminobenzene; phenylamine — Basic oily liquid, originally a coal-tar product, now generally made from nitrobenzene. Used in making dyes for printing inks and printing ink pigments. Soluble in alcohol, ether and benzene, slightly soluble in water. Poisonous. TL-value: 5 ppm.
Formula: $C_6H_5NH_2$.
f aniline *f*; phénylamine *f* — **d** Anilin *n*; Aminobenzol *n*; Phenylamin *n* — **n** aniline *fm*; aminobenzeen *n*; aminobenzol *n depr.*; fenylamine *n* — **e** anilina *f*; fenilamina *f* — **i** anilina *f*; fenilamina *f*

614 aniline dyes — See coal-tar dyes.

615 aniline fudge box — Small printing unit on newspaper machines for inserting latest news by printing with aniline inks. See also fudge box.
f élément *m* d'impression flexographique; élément *m* d'impression à l'aniline — **d** Flexoeindruckwerk *n*; Anilineindruckwerk *n* — **n** flexo-indrukwerk *n*; aniline-indrukwerk *n* — **e** elemento *m* de impresión a la anilina — **i** elemento *m* per stampa flessografica; elemento *m* per stampa all'anilina

616 aniline ink — See flexographic ink.

617 aniline point — The lowest temperature at which a hydrocarbon solvent is completely soluble in an equal volume of freshly distilled aniline. Below this point, the mixture is cloudy and separates into two layers. Used as a measure of solvent power of hydrocarbon solvents.
f point *m* d'aniline — **d** Anilinpunkt *m* — **n** anilinepunt *n* — **e** punto *m* de anilina — **i** punto *m* d'anilina

618 aniline press — See flexo press.

619 aniline print — Insolubilized image of glue or cold enamel stained with an aniline dye during or after development, the dyed print promoting greater visibility.
f copie *f* à l'aniline — **d** Anilinkopie *f* — **n** anilinekopie *f* — **e** copia *f* a la anilina — **i** copia *f* all'anilina

620 aniline printing — See flexography.

621 aniline rubber-plate printing — See flexography.

622 aniline sulphate — Solution of 2 g in 100 ml of water containing one drop of sulphuric acid gives a deep yellow colour with mechanical wood fibres.
f sulfate *m* d'aniline — **d** Anilinsulfat *n*; Anilinsulfatlösung *f* — **n** anilinesulfaat *n*; zwavelzure anilineoplossing *f* — **e** sulfato *m* de anilina — **i** solfato *m* d'anilina

623 animal glue; animal size; gelatine size — *Glue obtained from hides, bones and hoofs of animals; used in bookbinding, in surface sizing of paper to increase its strength and durability, in sizing textiles, in photo type and photography and in the manufacture of composition rollers.
f colle *f* animale; colle *f* forte; gélatine *f* — **d** tierischer Leim *m*; Tierleim *m*; animalischer Leim *m*; Gelatineleim *m* — **n** dierlijke lijm *m*; gelatinelijm *m* — **e** cola *f* animal; cola *f* fuerte; cola *f* de gelatina; gelatina *f* — **i** colla *f* animale; gelatina *f*

624 animal size — See animal glue.

625 animal sized paper — Paper that has passed through a solution of animal size (i.e., a substance containing glue or gelatine made from hides or hoofs) to enhance the quality of the sheet after it leaves the driers of a papermaking machine.
f papier *m* collé en cuve à la gélatine — **d** mit Tierleim geleimtes Papier *n* — **n** papier *n* met dierlijke lijming — **e** papel *m* encolado con cola animal; papel *m* bañado en gelatina; papel *m* encolado animal; papel *m* de gelatina — **i** carta *f* con collatura animale

626 animal tub sized; surface sized — Adjective used for paper sized by passing by sheet through a bath (tub or vat) of animal size (gelatine).
f collé à la cuve à la gélatine; avec collage en surface — **d** mit Tierleim geleimt; mit Oberflächenleimung — **n** nagelijmd met dierlijke lijm; met oppervlakte-lijming — **e** encolado en tina; encolado en la superficie (con gelatina) — **i** collata in superficie con colla animale

627 anion — Ion with a negative charge migrating to the positive pole if a current is sent through the solution. A *cation possesses a positive charge and therefore migrates to the negative pole if a current is sent through the solution. See also non-ionics.
f anion *m* — **d** Anion *m* — **n** anion *m* — **e** anión *m* — **i** anione *m*

628 anionics
f agents *mpl* de surface anioniques — **d** anionaktive Verbindungen *fpl* — **n** anion-actieve stoffen *fmpl* — **e** agentes *mpl* de superficie aniónica — **i** agenti *mpl* anionici

629 anisotropy — Property of a material with unequal optical properties along different axes.
f anisotropie *f* — **d** Anisotropie *f* — **n** anisotropie *f* — **e** anisotropía *f* — **i** anisotropia *f*

630 annaline *obs.* — Natural fibrous form of gypsum, which after grinding is used as a filler for paper. See also calcium sulphate.
f annaline *f* — **d** Annaline *f*; Weißerde *f* — **n** annalinwit *n* — **e** analina *f*; tierra *f* blanca — **i** annalina *f*

631 anneal *v* — To heat and then cool hardened steel dies to soften and render them less brittle to make them ready for cutting alterations in the text or image.
f tremper — **d** tempern; anlassen; nachlassen — **n** temperen; ontlaten; aanlaten — **e** templar — **i** temperare

632 annotate *v* — To supply with critical or

explanatory notes; remark upon in notes.
f commenter — **d** kommentieren — **n** commentariëren; van commentaar voorzien — **e** comentar — **i** commentare

633 annotate *v* — To make annotations or notes.
f annoter — **d** annotieren; mit Anmerkungen versehen — **n** annoteren; van verklarende aantekening voorzien — **e** anotar — **i** annotare

634 annotated edition
f édition *f* annotée — **d** Ausgabe *f* mit Anmerkungen; mit Anmerkungen versehene Ausgabe *f* — **n** van opmerkingen voorziene uitgave *f* — **e** edición *f* anotada — **i** edizione *f* commentata

635 annotation — Notes or comments written in the margin of a printed work. See also centre notes, cut-in notes, foot notes, marginal notes.
f annotation *f*; commentaires *mpl* — **d** Anmerkung *f* — **n** kanttekening *f*; aantekening *f*; annotatie *f* — **e** anotación *f* (marginal) — **i** annotazione *f*

636 announcement — Official printed notice of an event to take place or having taken place.
f avis *m* — **d** Anzeigenkarte *f* — **n** aankondigingskaart *f*; kennisgeving *f* — **e** tarjeta *f* de anuncio — **i** annuncio *m*

637 announcement papers — See announcements.

638 announcements; announcement papers — Paper or cards, plain or paneled, cut to size or folded to fit envelopes made from the same paper and sold in sets. Greeting cards, business and social stationery, weddings, etc.
f papiers *mpl* pour faire-part; faire-parts *mpl* — **d** Anzeigenpapiere *npl* — **n** papier *n* voor gelegenheidsdrukwerk — **e** tarjetas *fpl* de participación — **i** carte *fpl* per annunci; annunci *mpl*

639 annual — See year-book.

640 annual turnover
f chiffre *m* d'affaires annuel — **d** Jahresumsatz *m*; jährlicher Umsatz *m* — **n** jaaromzet *m*; jaarlijkse omzet *m* — **e** giro *m* de ventas anual — **i** giro *m* d'affari annuale

641 anode — Electrode connected to the positive side of supply voltage.
f anode *f*; électrode *f* positive — **d** Anode *f*; Pluspol *m* — **n** anode *f*; positieve pool *m* — **e** ánodo *m*; polo *m* positivo — **i** anodo *m*; polo *m* positivo

642 anode bag — Fabric container for anodes in electroplating installations to prevent anode mud getting into the electrolyte.
f sac *m* à anodes — **d** Anodenbeutel *m* — **n** anodezak *m* — **e** bolsa *f* de ánodo — **i** guaina *f* anodica; sacco *m* filtro per anodi

643 anode current *GB*; **plate current** *US* — Electron current flowing from the cathode to the anode in an electronic tube.
f courant *m* anodique — **d** Anodenstrom *m* — **n** anodestroom *m* — **e** corriente *f* anódica — **i** corrente *f* anodica

644 anode current resistance — See anode resistance.

645 anode differential resistance — See anode resistance.

646 anode efficiency — The ratio of ampere-hours to the amount of metal dissolved in the solution.
f rendement *m* anodique — **d** anodische Stromausbeute *f* — **n** anodestroomrendement *n* — **e** rendimiento *m* anódico — **i** rendimento *m* anodico

647 anode resistance; anode current resistance; anode differential resistance; plate resistance — The quotient of a small change in anode voltage and the corresponding small change of the anode current, all the other electrode voltages being maintained constant. It is equal to the reciprocal of the anode conductance.
f résistance *f* interne — **d** innerer Widerstand *m*; Innenwiderstand *m* — **n** inwendige weerstand *m* — **e** resistencia *f* interna; resistencia *f* de placa — **i** resistenza *f* interna; resistenza *f* differenziale

648 anode voltage *GB*; **plate voltage** *US* — Voltage applied to the anode of an electronic tube.
f tension *f* anodique — **d** Anodenspannung *f* — **n** anodespanning *f* — **e** tensión *f* de ánodo; tensión *f* anódica — **i** tensione *f* anodica

649 anodic oxidation; anodizing — Production of a protective oxide film on aluminium or other light metals by passing a high voltage electric current through a bath in which the metal is suspended. The bath usually contains sulphuric, chromic or oxalic acid. The metal serves as the anode.
f oxydation *f* anodique; traitement *m* anodique — **d** anodische Oxydation *f*; Eloxierung *f* — **n** anodische oxydatie *f*; anodisering *f*; anodiseren *n* — **e** oxidación *f* anódica; anodización *f* — **i** ossidazione *f* anodica

650 anodized aluminium plate — Aluminium plate for offset lithography with a specially prepared surface. See also anodic oxidation.
f plaque *f* d'aluminium anodisée — **d** eloxierte Aluminiumplatte *f* — **n** geanodiseerde aluminiumplaat *fm* — **e** plancha *f* de aluminio anodizada — **i** lastra *f* d'alluminio anodizzata

651 anodizing — See anodic oxidation.

652 anonymous
f anonyme — **d** anonym — **n** anoniem; zonder naam; naamloos — **e** anónimo — **i** anonimo

653 anopistograph — Document, parchment, manuscript or book with pages written or printed on one side of the leaves.
f anopistographe *f* — **d** Anopistographie *f*; anopistographischer Druck *m* — **n** anopistografie *f*; anopistografische druk *m* — **e** anopistografía *f*; anopistógrafo *m*; impresión *f* anopistográfica — **i** anopistografo *m*

654 anthology — Collection of selected literary pieces or passages.
f anthologie *f*; florilège *m* — **d** Florilegium *n*; Blumenlese *f* — **n** anthologie *f*; bloemlezing *f* — **e** antología *f*; florilegio *m* — **i** antilogia *f*; florilegio *m*

655 anthracene — Crystalline cyclic hydrocarbon,

obtained from coal-tar distillation. Soluble in alcohol and ether, insoluble in water.
Formula: $C_6H_4(CH)_2C_6H_4$.
f anthracène *m*; anthracine *f* — **d** Anthrazen *n* — **n** anthraceen *n* — **e** antraceno *m* — **i** antracene *m*
656 anthracene oil — Coal-tar fraction boiling in the range 270-360 °C, used as a source of anthracene and similar aromatics.
f huile *f* d'anthracène — **d** Anthrazenöl *n* — **n** anthraceenolie *fm* — **e** aceite *m* de antraceno; aceite *m* antracénico — **i** olio *m* d'antracene
657 anti-blocking agent — An anti-stick agent. See also blocking.
f agent *m* anti-bloquage; agent *m* anti-collant — **d** verblockungshemmendes Mittel *n*; verklebungshemmendes Mittel *n* — **n** anti-blokmiddel *n* — **e** antibloqueador *m* — **i** agente *m* antiadesivo
658 antichlor — Chemical to remove traces of free chlorine or hypochlorite from materials bleached with these substances. Sodium bisulphite ($NaHSO_3$) and sodium thiosulphate ($Na_2S_2O_3$) are common antichlors. Many years ago the term was only used for sodium thiosulphate, but now also for sodium bisulphite, etc.
f antichlore *m* — **d** Antichlor *m* — **n** antichloor *m* — **e** anticloro *m* — **i** anticloro *m*
659 anti drier
f anti-siccative *f* — **d** Antitrockner *m*; trocknungsverzögerndes Mittel *n* — **n** anti-droger *m*; antidroogstof *fm*; droging vertragende stof *fm* — **e** antisecante *m*; producto *m* antisecativo — **i** antisiccativo *m*; antiessicante *m*
660 anti-foam agent; anti-foaming agent; defoaming agent; defoamer; anti-froth agent; antifrothing agent — Product to reduce foaming, which often interferes with processing operations.
f agent *m* antimousse; antimousse *m*; agent *m* anti-écume — **d** Antischaummittel *n*; Entschäumungsmittel *n*; Entschäumer *m*; Schaumverhütungsmittel *n*; Demulgator *m* — **n** antischuimmiddel *n*; schuimwerend middel *n* — **e** agente *m* antiespuma; eliminador *m* de espuma; antiespumante *m* — **i** agente *m* antischiuma; sostanza *f* antischiuma
661 anti-foaming agent — See anti-foam agent.
662 antifoggant
f antivoile *m* — **d** Antischleiermittel *n* — **n** antisluiermiddel *n* — **e** antivelo *m* — **i** antivelo *m*
663 anti-friction metal — See babbitt metal.
664 anti-froth agent — See anti-foam agent.
665 anti-frothing agent — See anti-foam agent.
666 anti-halation — Property of a film or plate, usually with an opaque backing which prevents halation.
f antihalo *m* — **d** Lichthoffreiheit *f* — **n** antihalo *m* — **e** antihalo *m* — **i** antialo *m*
667 anti-halation backing; anti-halo layer — Coating on the back of a film containing a dye or pigment to absorb light rays, thus preventing their reflection from the back surface of the film base.
f dorsale *f* anti-halo; sous-couche *f* anti-halo; couche *f* anti-halo — **d** Rückenschutzschicht *f*; Lichthofrückschicht *f*; Lichthofschutzschicht *f*; lichthoffreie Rückschicht *f* — **n** antihalo-laag *fm*; antireflex-laag *fm* — **e** capa *f* antihalo; respaldo *m* antihalo; dorso *m* antihalo — **i** strato *m* antialo
668 anti-halo layer — See anti-halation backing.
669 antimony — Silver white, lustrous, hard, brittle metal. Soluble in hot concentrated sulphuric acid, insoluble in dilute acids. Symbol: Sb. Latin: stibium. Used in the manufacture of type metal alloys to harden it and in the manufacture of anti-friction bearing alloys, etc. Melting point about 630.5 °C. TL-value: 0.5 mg/m^3.
f antimoine *m*; stibine *(nom chimique)* — **d** Antimon *n* — **n** antimoon *n*; antimonium *n* — **e** antimonio *m* — **i** antimonio *m*
670 antimony oxide — See antimony trioxide.
671 antimony trioxide; antimony oxide; antimony white — White, odourless, crystalline powder, insoluble in water, soluble in concentrated hydrochloric and sulphuric acids. Formula: Sb_2O_3.
f trioxyde *m* d'antimoine; oxyde *m* d'antimoine; blanc *m* d'antimoine — **d** Antimontrioxyd *n*; Antimonoxyd *n*; Antimonweiß *n* — **n** antimoniumtrioxyde *n*; antimoniumoxyde *n*; antimoonwit *n* — **e** trióxido *m* de antimonio; óxido *m* de antimonio; blanco *m* de antimonio — **i** triossido *m* d'antimonio; ossido *m* antimonioso; bianco *m* d'antimonio
672 antimony trisulphide — The mineral stibnite, consisting of antimony trisulphide is the chief source of *antimony. The sulphide is insoluble in water, soluble in concentrated hydrochloric acid and sulphide solutions. Used for pigments and for fire-proofing fabrics and paper. Formula: Sb_2S_3.
f trisulfure *m* d'antimoine — **d** Antimontrisulfid *n*; Antimonsulfür *n*; Schwefelantimon *n*; Stibnit *m* — **n** antimoontrisulfide *n*; antimoniumtrisulfide *n* — **e** trisulfuro *m* de antimonio — **i** trisolfuro *m* d'antimonio
673 antimony white — See antimony trioxide.
674 anti-oxidant — Agent which retards the oxidation of drying oils and other substances.
f antioxydant *m* — **d** Antioxydationsmittel *n*; Antioxydant *m*; Oxydationsverzögerer *m* — **n** antioxydant *m* — **e** antioxidante *m*; agente *m* antioxidante — **i** antiossidante *m*
675 antiquarian — Owner or manager of an antiquarian bookshop.
f antiquaire *m*; bouquiniste *m* — **d** Antiquar *m*; Antiquariatsbuchhändler *m* — **n** antiquaar *m* — **e** anticuario *m* — **i** antiquario *m*; libraio *m* antiquario
676 antiquarian — A size of writing paper, recognized but not standard in the UK. The term is rarely used. See app. no. 3.
677 antiquarian bookshop — Shop where valuable old or rare books are sold.
f librairie *f* antiquaire; librairie *f* de livres anciens — **d** Antiquariatsbuchhandlung *f*; Antiquariat *n* — **n** antiquariaatsboekhandel *m*; antiquariaat *n*

— **e** librería *f* anticuaria; librería *f* de lance; librería *f* de viejo — **i** libreria *f* antiquaria

678 antique — See Lineals, app. no. 2.

679 antique — See blind tooling.

680 antique binding — Style of bookbinding with a decoration of *blind tooling.

f reliure *f* à l'antique — **d** Antik-Einband *m* — **n** antieke band *m* — **e** encuadernación *f* a la antigua — **i** legatura *f* antica

681 antique book paper — See antique paper.

682 antique paper; antique book paper — Originally applied to machine-made paper to denote colour and finish. Now any good bulking paper with a rough surface.

f papier *m* d'édition antique; papier *m* à grain; papier *m* non apprêté — **d** Antikpapier *n*; Antik-druckpapier *n*; Werkpapier *n* — **n** romandruk *m*; opdikkend romandruk papier *n* — **e** papel *m* áspero — **i** carta *f* tipo antico

683 antiques — See Lineals.

684 anti-rust paper — See anti-tarnish paper.

685 anti set-off apparatus — See anti set-off spray.

686 anti set-off liquid

f liquide *m* anti-macule — **d** Druckbestäubungs-flüssigkeit *f* — **n** sproeivloeistof *fm* — **e** líquido *m* antimaculador; líquido *m* antimancha — **i** liquido *m* antiscartino; agente *m* antiscartino

687 anti set-off powder; anti-spray powder — Powder sprayed at the delivery of a printing press onto the freshly printed sheet to prevent damage to the print.

f poudre *f* anti-maculage — **d** Druckbestäubungs-puder *m* — **n** anti-smetpoeder *n* — **e** polvo *m* antimaculador — **i** polvere *f* antiscartino

688 anti set-off spray; anti set-off apparatus; dusting apparatus — Device on the delivery end of the printing machine to prevent set-off by projecting a fine spray of liquid or powder at the sheet.

f dispositif *m* antimaculateur — **d** Bestäubungs-apparat *m*; Druckbestäuber *m*; Besprüh-einrichtung *f* — **n** anti-smetapparaat *n* — **e** anti-maculador *m*; equipo *m* antimaculador — **i** apparecchio *m* antiscartino a spruzzo

689 anti-skinning agent — Chemical substance that retards the skin formation on the surface of an oxidizable oil or ink. Frequently an *anti-oxidant.

f retardeur *m*; agent *m* antipeaux; antipeaux *m*; antipelliculeur *m*; agent *m* protecteur — **d** Haut-verhütungsmittel *n*; Hautverhinderungsmittel *n*; Deckmasse *f* — **n** anti-velmiddel *n*; anti-vel-vormer *m*; velvorming-verhinderend middel *n* — **e** agente *m* antipellejo; agente *m* para evitar la formación de piel — **i** antipellicola *f*; agente *m* antipellicolare

690 anti-spray powder — See anti set-off powder.

691 anti-static agents; antistats — Materials which reduce static electrical charges on paper, textiles and resins. Such charges are often built up by friction and cause difficulty in handling and create a fire hazard. The antistat allows the charge to leak off, usually by retaining enough moisture to provide good electrical conduction by way of molecularly held water. High molecular weight fatty alcohols and medium-sized polymers are used as such.

f agents *mpl* antistatiques — **d** Antistatikmittel *npl* — **n** anti-static middelen *npl* — **e** agentes *mpl* antiestáticos; agentes *mpl* antiparasitarios — **i** agenti *mpl* antistatici

692 anti-static device — See static neutralizer.

693 antistats — See anti-static agents.

694 anti-tarnish paper; anti-rust paper — Acid, alkali and sulphur-free paper used for wrapping metals, especially silver articles, to prevent tarnishing. Term originally used for a lightweight tissue made from rags, later for wrapping paper made from sulphite and sulphate pulp.

f papier *m* anti-ternissure; papier *m* anti-rouille; papier *m* pour coutellerie — **d** Rostschutz-papier *n*; rostschützendes Papier *n*; Silberpack-papier *n* — **n** juwelierszijde *fm*; juweliersvloei *n* — **e** papel *m* anticorrosivo; papel *m* protector contra la oxidación — **i** carta *f* anticorrosiva; carta *f* antiruggine

695 ant oil — See furfural.

696 anvil — Part of a stapling machine on which the staples are bent.

f enclume *f* d'agrafeuse — **d** Amboß *m*; Heft-pfanne *f* — **n** aambeeld *n* — **e** yunque *m* — **i** incudine *f* (di macchina cucitrice)

697 anvil beater — See lock-up man.

698 aperture card; microcard — Punched card which contains an aperture covered by a photographic film.

f carte *f* à fenêtre — **d** Filmlochkarte *f* — **n** vensterkaart *fm* — **e** tarjeta *f* de ventana — **i** scheda *f* a finestra

699 aperture effect — See circle of confusion.

700 aperture, lens — See lens aperture.

701 aphorism — Brief sententious statement of a truth or principle. Also a proverb, maxim or precept.

f aphorisme *m* — **d** Aphorismus *m* — **n** aforisme *n* — **e** aforismo *m* — **i** aforismo *m*; aforisma *m*

702 aplanat — See aplanatic lens.

703 aplanatic lens; aplanat — Compound lens corrected for spherical aberration and coma.

f objectif *m* aplanétique; aplanat *m* — **d** aplanatisches Objektiv *n*; Aplanat *n* — **n** aplanatische lens *fm*; aplanatisch objectief *n*; aplanaat *m* — **e** objetivo *m* aplanático; lente *m* aplanático; lente *f* aplanática — **i** obiettivo *m* aplanatico; lente *f* aplanatica; lente *f* aplanetica

704 apochromat — See apochromatic lens.

705 apochromatic lens; apochromat — Lens corrected for spherical aberration at two wavelengths or colours and for chromatic aberration at three wavelengths. Especially designed for three-colour photography, it is the highest quality of lens.

f objectif *m* apochromatique; apochromat *m* — **d** apochromatisches Objektiv *n*; Apochromat *n* — **n** apochromatische lens *fm*; apochromaat *m* — **e** objetivo *m* apocromático; lente *m* apocromático; lente *f* apocromática — **i** obiettivo *m* apocromatico; lente *f* apocromatica

706 apostil — Marginal annotation. Also the form on which the annotation or comment is written.
f apostille *f* — **d** Randbemerkung *f*; Randglosse *f*; Apostill *m* — **n** apostille *fm*; apostil *fm* — **e** apostilla *f* — **i** postilla *f*

707 apostilb — Unit of luminance. Abbrev. asb. In the SI: 1 apostilb = $(1/\pi)$ cd/m^2 $\approx$ 0.318310 cd/m^2.
f apostilb *m* — **d** Apostilb *m* — **n** apostilb *m* — **e** apostilb *m* — **i** apostilb *m*

708 apostrophe — See punctuation marks.
f apostrophe *f* — **d** Apostrof *m*; Apostrophe *m*; Apostroph *m*; Auslassungszeichen *n* — **n** apostrof *fm*; weglatingsteken *n* — **e** apóstrofo *m*; virgulilla *f* — **i** apostrofo *m*

709 apothecaries' (weight) signs — See medical signs.

710 apparatus — See device.

711 apparent density — Weight of a unit volume of powder, usually expressed as g/cm^3, determined by a specific method.
f densité *f* apparente — **d** scheinbare Dichte *f*; Aufschüttdichte *f*; Rüttelgewicht *n*; Schüttgewicht *n* — **n** schijnbare dichtheid *f* — **e** densidad *f* aparente — **i** densità *f* apparente

712 apparent specific density — The number obtained by dividing the substance of paper or board expressed in g/m^2 by its thickness expressed in microns.
f densité *f* apparente; masse *f* volumique apparente — **d** Raumgewicht *n*; Rohwichte *f* — **n** schijnbaar soortelijk gewicht *n* — **e** peso *m* por unidad cúbica; peso *m* cúbico — **i** densità *f* apparente

713 apparent viscosity — Imaginary quantity applied to non-Newtonians. It is the viscosity the non-Newtonian would have (at a given rate of shear) if it were a Newtonian. It is obtained experimentally by dividing the shearing stress by the rate of shear and then multiplying by the instrumental constant.
f viscosité *f* apparente — **d** scheinbare Viskosität *f* — **n** schijnbare viscositeit *f* — **e** viscosidad *f* aparente — **i** viscosità *f* apparente

714 appear *v* — To come out (of a newspaper, magazine, etc.).
f paraître — **d** erscheinen (Zeitung); zum Vorschein kommen (beim Entwickeln) — **n** verschijnen; uitkomen — **e** aparecer; publicarse; salir a luz — **i** apparire; uscire alla luce

715 appear shortly, to — To be published soon; about to be published.
f pour paraître prochainement — **d** erscheint demnächst; erscheint in Kürze — **n** verschijnt binnenkort — **e** aparecerá próximamente; de próxima publicación — **i** in corso di pubblicazione; di prossima pubblicazione

716 appendix — Supplement at the end of a book (*pl* appendices).
f appendice *m*; supplément *m*; annexe *f* — **d** Anhang *m*; Appendix *m*; Zusatz *m* — **n** appendix *mn*; aanhangsel *n* — **e** apéndice *m*; suplemento *m*; anexo *m* — **i** appendice *f*

717 appliance — See device.

718 application program — Computer program for a specific use in contrast with a general or broad program for a variety of uses. It may apply to a program written by a user for his own purposes (rather than one supplied by a computer vendor), but also to a program written by a vendor specifically for a user or type of user.
f programme *m* d'application — **d** Anwendungsprogramm *n* — **n** toepassingsprogramma *n* — **e** programa *m* de aplicación — **i** programma *m* d'applicazione

719 applicator; ink applicator — Mechanism to apply ink uniformly to the surface of an engraved cylinder.
f applicateur *m* d'encre — **d** Farbverteiler *m* — **n** inktverdeler *m* — **e** distribuidor *m* de la tinta — **i** distributore *m* d'inchiostro

720 applied scientific research
f recherche *f* scientifique appliquée — **d** angewandte naturwissenschaftliche Forschung *f* — **n** toegepast (natuur)wetenschappelijk onderzoek *n* — **e** investigación *f* científica aplicada — **i** ricerca *f* scientifica applicata

721 apprentice — One who is learning by practical experience under skilled workers. See also printer's devil, cub.
f apprenti *m* — **d** Lehrling *m*; Lehrbursche *m*; Fachschüler *m* — **n** leerling *m* — **e** aprendiz *m*; alumno *m* — **i** apprendista *m*; allievo *m*

722 apprentice compositor
f apprenti *m* typographe; attrape-science *m sl.*; gosse *m sl.* — **d** Schriftsetzerlehrling *m*; Setzerlehrling *m* — **n** zettersleerling *m*; leerling-letterzetter *m* — **e** aprendiz *m* de tipógrafo — **i** apprendista *m* compositore

723 apprentice compositor, eldest
f prote *m* aux gosses *sl.* — **d** ältester Setzerlehrling *m* — **n** oudste zettersleerling *m* — **e** aprendiz *m* mayor de tipógrafo — **i** apprendista *m* compositore anziano

724 apprenticeship
f apprentissage *m*; temps *m* d'apprentissage — **d** Lehre *f*; Lehrzeit *f* — **n** leertijd *m* — **e** aprendizaje *m* — **i** apprendistato *m*; tirocinio *m*

725 appropriation — Amount of money apportioned for advertising or other particular purposes.
f budget *m* — **d** Budget *n*; Etat *m*; Voranschlag *m*; Plansumme *f*; Verwendung *f* — **n** toegestaan bedrag *n*; budget *n*; begroting *f* — **e** presupuesto *m* — **i** budget *m*

726 approval, on — Without obligation to buy unless satisfactory to the customer upon examination (and, otherwise, returnable).
f à l'examen — **d** zur Ansicht — **n** op zicht; ter

inzage — **e** para su inspección — **i** in esame

727 apron — Garment covering part of the front of the body and tied at the waist to protect the wearer's clothing.

f tablier *m* — **d** Schürze *f*; Lederschürze *f*; Schurz *m* — **n** voorschoot *mn*; schort *n* — **e** delantal *m* (de trabajo); mandil *m* — **i** grembiale *m*; grembiule *m*

728 apron — Rubber or rubberized sheet which extends from the headbox to the wire on the paper machine and carries the diluted stock from one to the other.

f tablier *m* — **d** Schürze *f*; Siebleder *n*; Spritzleder *n*; Siebschürze *f* — **n** schort *n* — **e** faldón *m* — **i** grembiale *m*; grembiule *m*

729 apron; pouring sheet; shield; pan; tailboard; tail piece — Pouring sheet of a casting box pasted to or placed over the end of a mat to prevent metal from running behind it when a stereotype cast is made.

730 aqua ammonia — See ammonium hydroxide.

731 aqua destillata — See distilled water.

732 aqua fortis — See nitric acid.

733 aquafortist — See etcher.

734 aqua regia; nitrohydrochloric acid; nitromuriatic acid; chlorazotic acid — Mixture of nitric and hydrochloric acids, usually one part of nitric acid and three or four parts of hydrochloric acid. Fuming yellow, corrosive, suffocating, volatile liquid.

f eau *f* régale; acide *m* chlorhydronitrique; acide *m* nitromuriatique; acide *m* chloroazotique — **d** Königswasser *n*; Salpetersalzsäure *f*; Goldscheidewasser *n* — **n** koningswater *n* — **e** agua *f* regia — **i** acqua *f* regia

735 aquatint; aquatinta — Tonal intaglio method of etching on copper or steel by means of nitric acid, by which prints imitating the broad flat tints of India ink or sepia drawings are produced. Also the print produced by the aquatint process. Important as a fine-art printing method but also as a forerunner of halftone printing in general and photogravure in particular. Aquatint produces the counterpart of mezzoprint cells, though with a difference. In mezzotint, the plate is mechanically roughened, in aquatint a corresponding effect is produced by etching. To this the metal plate is dusted with an acid proof resin powder and then heated until the resin melts and attaches itself as tiny globules to the plate. The etcher controls the image by stopping out, or by protecting an area with an acid-resist and by varying the etching time. See also sugar aquatint, sulphur aquatint.

f manière *f* de lavis — **d** Aquatinta *f* — **n** aquatint *fm* — **e** acuatinta *f*; aguatinta *f*; grabado *m* al aguatinta — **i** acquatinta *f*

736 aquatinta — See aquatint.

737 aquatone; aquatone printing — Printing method produced on offset presses, invented by R. John, New York, resembling collotype. Use is made of a very fine halftone screen of about 80 to 160 lines per cm (200 to 400 lines to the inch),

whilst in the collotype process use is made of a reticulated grain screen. The copy is transferred to a sensitized-gelatine coating on a zinc plate and developed in water and alcohol and bekaed.

f aquatone *f*; procédé *m* aquatone — **d** Aquatone *m*; Aquatone-Verfahren *n* — **n** aquatoneprocédé *n* — **e** acuatono *m*; procedimiento *m* de impresión acuatono — **i** aquatone *m*; procedimento *m* di aquatone

738 aquatone printing — See aquatone.

739 aqueous solution — A solution of, like, or containing water.

f solution *f* aqueuse; solution *f* délayée — **d** wässerige Lösung *f* — **n** waterige oplossing *f* — **e** solución *f* acuosa — **i** soluzione *f* acquosa

740 aqueous suspension

f suspension *f* aqueuse — **d** wässerige Aufschlämmung *f* — **n** waterige suspensie *f*; suspensie *f* in water — **e** suspensión *f* acuosa — **i** sospensione *f* acquosa

741 aqueous vapour — See water vapour.

742 aqueous vapour permeability — See water-vapour permeability.

743 arabesks — See arabesques.

744 arabesques; arabesks; mauresques — Ornaments or designs as in Arabian or Moorish architecture, employing patterns of intertwined scrollwork, conventionalized leaves or flowers, etc., used on bookbindings.

f arabesques *fpl* — **d** Arabesken *fpl*; Mauresken *fpl* — **n** arabesken *fmpl* — **e** arabescos *mpl* — **i** arabeschi *mpl*

745 arabic acid

f acide *m* arabique — **d** Arabinsäure *f* — **n** arabinezuur *n* — **e** ácido *m* arábigo; arabina *f* — **i** acido *m* arabico

746 arabic characters — See arabic type.

747 arabic figures; arabic numerals — The figures in common use, 1, 2, 3, 4, etc., introduced in Europe from Arabia in the 12th century, as distinct from Roman figures I, II, III, IV, etc.

f chiffres *mpl* arabes — **d** arabische Ziffern *fpl* — **n** Arabische cijfers *npl* — **e** cifras *fpl* arábigas; cifras *fpl* árabes; números *mpl* arábigos; números *mpl* árabes; guarismos *mpl*; números *mpl* de guarismo; letras *fpl* de guarismo — **i** numeri *mpl* arabi; cifre *fpl* arabe; cifre *fpl* arabiche

748 arabic numerals — See arabic figures.

749 arabic type; arabic characters — Some symbols have a root portion and an attached terminal portion to make up one character, and a set of type has symbols which print the terminal portions of more than one symbol.

f caractères *mpl* arabes — **d** arabische Schrift *f*; arabische Buchstaben *mpl* — **n** Arabisch schrift *n*; Arabische letters *fmpl* — **e** caracteres *mpl* arábigos — **i** caratteri *mpl* arabici

750 arbitrary sign; natural sign — Sign that is interpreted by the connections of cause and effect, such as # for number, © for copyright, ' for degree, £ for pound sterling, $ for dollar, etc.

f signe *m* naturel — **d** Einheitenzeichen *n* — **n**

natuurlijk teken *n* — **e** signo *m* natural; símbolo *m* natural — **i** segno *m* convenzionale; simbolo *m* convenzionale

751 archil — See orchil.

752 architect's soft eraser
f plastiline *f*; gomme *f* d'architecte — **d** weicher Radiergummi *m* — **n** zacht vlakgom *mn* — **e** goma *f* de borrar — **i** gomma *f* pane; gomma *f* (per cancellare)

753 architectural binding — See cathedral binding.

754 arc lamp — Powerful illuminant in which an electric current passes through a pair of slightly separated electrodes of carbon causing combustion (vaporization) of the carbon and emission of intensely bright light containing all the colours of the spectrum. Colour temperature 5000 K.
f lampe *f* à arc — **d** Bogenlampe *f*; Kohlenbogenlampe *f*; Lichtbogenlampe *f*; Aufnahmebogenlampe *f* — **n** booglamp *fm*; koolbooglamp *fm* — **e** lámpara *f* de arco (voltaico); arco *m* voltaico; lámpara *f* de carbón — **i** lampada *f* ad arco

755 arc lamp carbon
f charbon *m* pour lampe à arc — **d** Bogenlampenkohle *f*; Kohle *f* für Bogenlampen — **n** koolspits *fm* voor de booglamp — **e** carbón *m* para lámpara de arco; electrodo *m* de carbón para lámpara de arco — **i** carbone *m* per lampada ad arco

756 arc light — Light produced by an *arc lamp.
f arc *m* voltaïque — **d** Bogenlicht *n* — **n** booglicht *n* — **e** luz *f* de arco voltaico — **i** luce *f* di lampada ad arco

757 area evenness — Uniform distribution of densities over the area of a sensitive material.
f uniformité *f* en surface — **d** Gleichmäßigkeit *f* der Schwärzung über die Fläche eines Films; Tonruhe *f* — **n** oppervlakte-gelijkmatigheid *f*; gelijkmatigheid *f* van het oppervlak — **e** uniformidad *f* superficial — **i** uniformità *f* di superficie

758 area mask — See dodging mask.

759 area of the hysteresis loop
f aire *f* de la bouche d'hystérésis — **d** Fläche *f* der Hysteresis-Schleife; Inhalt *m* der Hysteresis-Schleife — **n** hysteresisgebied *n* — **e** área *f* de la curva histerética — **i** area *f* del ciclo d'isteresi

760 area-variable gravure — Gravure printed from a printing surface, the printing elements of which are cells with a varying area but even or near even depth.
f gravure *f* à surface variable; gravure *f* autotypique — **d** flächenvariabler Tiefdruck *m*; autotypischer Tiefdruck *m depr.* — **n** volautotypische diepdruk *m* — **e** huecograbado *m* autotípico — **i** rotocalco *m* a superficie variabile; rotocalco *m* autotipico

761 areometer — Floating instrument to measure the specific gravity of liquids.
f aréomètre *m*; pèse-liqueur *m*; pèse-bain *m* — **d** Aräometer *nm*; Areometer *m*; Dichtigkeitsmesser *m*; Dichtemesser *m*; Senkwaage *f*; Senkspindel *f*; Gradierwaage *f*; Säuremesser *m* — **n** areometer *m*; zuurweger *m* — **e** areómetro *m*; pesalicores *m*; pesa-ácidos *m* — **i** areometro *m*; densimetro *m*

762 Areopagite — Title of a pamphlet by John Milton (1644), advocating freedom of the press. Also a request for press liberties.
f aréopagite *m* — **e** areopagita *m*

763 argentometer — Hydrometer to test silver nitrate solutions, indicating the strength in grams per ounce of solution.
f argentomètre *m*; pèse-argent *m*; pèse-bain *m* — **d** Argentometer *m*; Silbermeter *m* — **n** argentometer *m* — **e** argentómetro *m* — **i** argentometro *m*

764 argument — Computer term for an independent variable which defines conditions to be obtained at a given time, or which, should they be obtained, would call for a certain action. For example, if a command is given for "change line length", the argument would indicate the specific line length desired.
f argument *m* — **d** Argument *n* — **n** argument *n* — **e** argumento *m* — **i** argomento *m*

765 arithmetical mean — The average obtained by dividing the sum of all measurements by the number of measurements.
f moyenne *f* arithmétique — **d** arithmetisches Mittel *n* — **n** rekenkundig gemiddelde *n* — **e** medio *m* aritmético — **i** media *f* aritmetica

766 arithmetical signs — Signs used in the most elementary branch of mathematics.
f signes *mpl* arithmétiques — **d** arithmetische Zeichen *npl* — **n** rekenkundige tekens *npl* — **e** signos *mpl* aritméticos — **i** segni *mpl* aritmetici

767 arithmetic and logical unit — See arithmetic unit.

768 arithmetic unit; arithmetic and logical unit — Unit of a computer system containing the circuits that perform arithmetic operations (adding, subtracting, dividing, multiplication) and makes logical decisions.
f unité *f* arithmétique — **d** Rechenwerk *n* — **n** rekenorgaan *n* — **e** unidad *f* aritmética — **i** unità *f* aritmetica

769 Arkansas stone — Finely grained natural stone used for grinding.
f pierre *f* à aiguiser Arkansas — **d** Arkansas-Schleifstein *m* — **n** Arkansas-slijpsteen *m*; Arkansas-steen *m* — **e** piedra *f* amoladora Arkansas — **i** pietra *f* da affilare Arkansas

770 arm — The projecting or unclosed horizontal or short upward stroke on the letters E, L, F. See also bar.

771 armature — The main current carrying winding of an electric motor.
f induit *m* — **d** Anker *m*; Läufer *m* — **n** anker *n* — **e** inducido *m* — **i** rotore *m*; indotto *m*

772 Armenian bole — Material used as a base for gilding book edges. Dense, dark-red powder or lumps, soluble in acids, not in water. Formula: Fe_2O_3. Sometimes called *gilder's delight.
f bol *m* d'Arménie — **d** armenischer Bolus *m* — **n**

Armeense bolus *m* — **e** bol *m* arménico; bol *m* de Armenia — **i** bolo *m* d'Armenia; bolo *m* armenico

773 arming press *GB* — Small hand-operated *blocking press, so called because of its original association with impressing armorial bearings. Used today for short runs.

774 armorial; book of heraldry — Book with pictures of coats of arms.
f armorial *m* — **d** Wappenbuch *n* — **n** armoriaal *n*; wapenboek *n* — **e** armorial *m* — **i** armerista *f*; stemmario *m*

775 armorial binding — Binding on which a coat of arms is printed.
f reliure *f* armoriée — **d** Wappeneinband *m* — **n** wapenband *m*; heraldische band *m* — **e** encuadernación *f* heráldica — **i** rilegatura *f* araldica

776 aromatic hydrocarbons — Hydrocarbons characterized by a molecular structure with one or more six-carbon atom rings, and with properties similar to those of benzene, the simplest member of this group. Toluene, xylene, naphthalene, anthracene and phenanthrene are other key members of this group.
f hydrocarbures *mpl* aromatiques — **d** aromatische Kohlenwasserstoffe *mpl*; Benzolkohlenwasserstoffe *mpl*; zyklische Kohlenwasserstoffe *mpl* — **n** aromatische koolwaterstoffen *fmpl*; benzeenkoolwaterstoffen *fmpl* — **e** hidrocarburos *mpl* aromáticos — **i** idrocarburi *mpl* aromatici

777 aromatic solvents — Organic liquids with a cyclic hydrocarbon structure and KB values over 40. Good solvents for printing ink resins. Examples are toluene and xylene.
f solvants *mpl* aromatiques — **d** aromatische Lösungsmittel *npl* — **n** aromatische oplosmiddelen *npl* — **e** disolventes *mpl* aromáticos — **i** solventi *mpl* aromatici

778 ARQ — See automatic request for repetition.

779 arrowheaded characters — See cuneiform writing.

780 arrowheaded writing — See cuneiform writing.

781 ars artis conservatrix — Latin for art preservative of all arts. A poetic expression.
f l'art qui préserve tous les autres — **d** die Kunst die alle Andere verwahrt; die Kunst die alle anderen Künste bewahrt — **n** kunst die alle andere kunsten bewaart — **e** arte conservatriz de todas las artes — **i** arte conservatrice dell'arte

782 arsenic yellow — See orpiment.

783 art — See art work.

784 art board — See coated board.

785 art canvas — Heavy woven material prepared like buckram but with a matt surface in imitation of canvas. Being penetrable by adhesives it is given a thin paper backing during manufacture.
f toile *f* artificielle — **d** Kunstleinen *n* — **n** kunstlinnen *n* — **e** tela *f* artificial — **i** tela *f* artificiale

786 art editor — Editor of a newspaper or magazine who selects the pictures to be published.

787 article; contribution — Literary composition forming an independent part of a publication.
f article *m*; contribution *f* — **d** Artikel *m*; Aufsatz *m*; Beitrag *m* — **n** artikel *n*; bijdrage *fm* — **e** artículo *m*; contribución *f* — **i** articolo *m*; contributo *m*

788 artificial ant oil — See furfural.

789 artificial gum — See dextrin.

790 artificial language — Language based on a set of prescribed rules that are established prior to its usage.
f langage *m* artificiel — **d** Kunstsprache *f* — **n** kunsttaal *fm* — **e** lenguaje *m* artificial — **i** linguaggio *m* artificiale

791 artificial leather; imitation leather — Cotton fabric, base dyed, surface coated with cellulose ester and embossed to give it the appearance of leather.
f similicuir *m*; cuir *m* artificiel — **d** Kunstleder *n* — **n** kunstleer *n*; imitatie-leer *n* — **e** cuero *m* artificial; imitación *f* cuero — **i** cuoio *m* artificiale; finto cuoio *m*; imitazione *f* pelle

792 artificial leather paper — Paper (or board) embossed with leather-like finishes; used for book covers.
f papier *m* similicuir; papier *m* imitation cuir — **d** Lederpapier *n*; Lederimitationspapier *n*; Kunstlederpapier *n* — **n** kunstleerpapier *n* — **e** papel *m* imitación cuero — **i** carta *f* similpelle; carta *f* uso pelle

793 artificial light — Illumination provided by incandescent, fluorescent, or flame sources as distinguished from natural light produced by solar radiation or self-luminous organisms.
f lumière *f* artificielle — **d** Kunstlicht *n*; künstliches Licht *n* — **n** kunstlicht *n* — **e** luz *f* artificial — **i** luce *f* artificiale

794 artificial malachite — See copper carbonate.

795 artificial parchment — See imitation parchment-paper.

796 artificial resin — See synthetic resin.

797 artificial rubber — See synthetic rubber.

798 artist's board — High grade drawing board used by artists for pen and ink drawings, wash drawings, water colours, etc.
f carte *f* d'artiste — **d** Malkarton *m*; Malpappe *f* — **n** tekenkarton *n* — **e** cartón *m* para pinturas — **i** cartone *m* da disegno

799 artist's paint brush — See paint brush.

800 Art Nouveau typography — Art Nouveau is a style of fine and applied art, current in the late 19th and early 20th centuries, characterized chiefly by curvilinear motifs derived from natural forms. The publishing of the magazine Pan (1895) by Drugulin at Leipsic was of great interest. Collaborators were Eckmann, Klinger, Heine, Weiß. Of the same interest was the publication of the magazine Jugend by Hirt at Munich in 1896, from which originates the German name Jugendstil (*Youth style). In typography it was used and influenced by William Morris.
f art *m* nouveau — **d** Jugendstil *m* — **n** Jugend-

stil *m*; slaolie-stijl *m sl* — **e** arte *m* (gráfico) modernista — **i** tipografia *f* Art Nouveau

801 art of bookbinding — The art or practise of bookbinding. In early Christian times bibliopegy.
f art *m* de reliure; art *m* de la reliure — **d** Buchbindekunst *f* — **n** boekbindkunst *f* — **e** arte *m* de la encuadernación; bibliópegia *f* — **i** arte *f* della legatoria

802 art of etching
f art *m* de graver à l'eau-forte — **d** Ätzkunst *m* — **n** etskunst *f* — **e** arte *m* de grabar al aguafuerte — **i** arte *f* d'incisione all'acquaforte; arte *f* ad acqua forte; acquaforte *f*

803 art of printing
f art *m* typographique — **d** Buchdruckerkunst *f*; Druckkunst *f* — **n** boekdrukkunst *f*; drukkunst *f* — **e** arte *m* gráfico; arte *m* tipográfico; arte *m* de la imprenta — **i** arte *f* tipografica; arte *f* impresoria; arte *f* della stampa

804 art of the book — The designing and manufacturing of books as objects of art.
f art *m* du livre — **d** Buchkunst *f* — **n** boekkunst *f* — **e** arte *m* del libro — **i** arte *f* del libro

805 art of wood engraving
f art *m* de la xylographie; xylographie *f* — **d** Holzschneidekunst *f* — **n** houtsneekunst *f* — **e** arte *m* de xilografía; xilografía *f* — **i** xilografia *f*; silografia *f*

806 artotype — See collotype process.

807 art paper — See coated paper.

808 art work; art — Hand drawn originals used for photomechanical reproduction and supplied by the customer, to be photographed, transferred to the printing forme or plate and printed.
f travail *m* d'artiste; maquette *f* — **d** Originalvorlage *f*; Originalentwurf *m*; Original *n*; Entwurf *m*; Aufsichtsvorlage *f* — **n** originele tekening *f*; tekenwerk *n*; opzichtmodel *n depr.* — **e** original *m*; maqueta *f* — **i** originale *m*; disegno *m* originale; disegno *m* eseguito a mano

809 art work — See copy.

810 asbestos board
f carton *m* amiante — **d** Asbestpappe *f* — **n** asbestkarton *n* — **e** cartón *m* de amianto; cartón *m* de asbesto — **i** cartone *m* d'amianto; cartone *m* d'asbesto

811 asbestos fibres
f fibres *fpl* d'amiante — **d** Asbestfasern *fpl* — **n** asbestvezels *fmpl* — **e** fibras *fpl* de amianto — **i** fibre *fpl* d'amianto

812 asbestos paper — Insulating paper made from asbestos fibres or a mixture of asbestos and cellulose fibres.
f papier *m* amiante; papier *m* d'asbeste — **d** Asbestpapier *n* — **n** asbestpapier *n* — **e** papel *m* de amianto; papel *m* de asbesto — **i** carta *f* d'amianto; carta *f* d'asbesto

813 ascender — Part of a character which extends above the common body height of the lower case character of a type face, as in the letters b, d, f, h, l and t. By extension the character itself (ascending letter).

f jambage *m* ascendant; montant *m* d'un caractère; longue *f* du haut; ascendante *f*; lettre *f* montante — **d** Oberlänge *f* des Buchstabens — **n** stok *m*; stokletter *fm* — **e** trazo *m*; alto *m*; palo *m* alto; cabeza *f*; rasgo *m* ascendente; letra *f* ascendente; letra *f* de palo ascendente; letra *f* estrecha — **i** asta *f* ascendente; ascendente *f*

814 ascender line — The implied line to which ascending letters normally reach, and which demarcates in most instances the uppermost limits of the characters of the fount. See also mean line.

815 ascending letter — See ascender.

816 ash content — The percentage residue of a sample of paper left after incineration to constant weight.
f teneur *f* en cendres; poids *m* de cendres — **d** Asche *f*; Asch(en)gehalt *m*; Aschegehalt *m*; Glührückstand *m* — **n** asgehalte *n* — **e** porcentaje *m* de cenizas; contenido *m* de cenizas — **i** contenuto *m* in ceneri

817 ashless paper — Paper that leaves a negligible residue after complete combustion.
f papier *m* sans cendre — **d** aschefreies Papier *n* — **n** asvrij papier *n* — **e** papel *m* sin ceniza — **i** carta *f* esente da ceneri; carta *f* senza ceneri; carta *f* priva di cariche

818 ash scale
f balance *f* à cendres — **d** Aschenwaage *f* — **n** asweegschaal *fm*; asweger *m* — **e** balanza *f* para cenizas; balanza *f* de ceniza — **i** bilancia *f* da cenere

819 asp — See aspen pulp.

820 aspen pulp — Pulp made of aspen, a type of broadleafed hardwood tree, used in the manufacture of soda pulp. Aspen is an adjective derived from asp, obsolete for the branching polar tree.
f pâte *f* de tremble — **d** Espenzellstoff *m*; Espenstoff *m* — **n** espehoutstof *fm* — **e** pasta *f* de álamo (temblón); celulosa *f* de álamo (temblón) — **i** pasta *f* di pioppo tremulo

821 asphalt; asphaltum — Naturally occurring bituminous mixture of hydrocarbons or obtained as residue in refining petroleum, used in various inks and varnishes and as an acid resist or protectant in photomechanics. Soluble in carbon disulphide of 69% (native), 99% (petroleum).
f asphalte *m* — **d** Asphalt *m* — **n** asfalt *n* — **e** asfalto *m* — **i** asfalto *m*

822 asphalt papers — Papers saturated, coated, or laminated with asphalt or bituminous material.
f papiers *mpl* asphaltés; papiers *mpl* bitumés — **d** Asphaltpapiere *npl* — **n** asfaltpapieren *npl*; gebitumeerde papieren *npl*; bitumenpapieren *npl* — **e** papeles *mpl* embreados; papeles *mpl* alquitranados; papeles *mpl* bituminados; papeles *mpl* asfaltados — **i** carte *fpl* asfaltate; carte *fpl* bitumate

823 asphalt powder
f asphalte *m* en poudre — **d** pulverisierter Asphalt *m* — **n** asfaltpoeder *n* — **e** asfalto *m* en polvo — **i** asfalto *m* polverizzato

824 asphalt process — The first method of producing a stable photograph (Nièpce, 1822).
f procédé *m* de copie au bitume — **d** Asphaltverfahren *n*; Asphaltkopierverfahren *n* — **n** asfaltprocédé *n* — **e** procedimiento *m* al asfalto; procedimiento *m* al betún — **i** procedimento *m* al bitume; processo *m* al bitume
825 asphaltum — See asphalt.
826 asphaltum, under — See under asphaltum.
827 asphaltum washout solution — Powdered asphaltum, or gilsonite in an organic solvent, used in lithography to make printing images on press plates permanently ink receptive.
f solution *f* d'asphalte; solution *f* d'enlevage — **d** Auswaschtinktur *f*; Asphalttinktur *f* — **n** asfalttinctuur *fm*; uitwastinctuur *fm depr.* — **e** solución *f* (a base) de asfalto — **i** soluzione *f* d'asfalto; soluzione *f* di preparazione; soluzione *f* di trattamento all'asfalto
828 asphalt varnish — Solution of bitumen in volatile solvents, usually turpentine and/or benzene, used to protect metal parts against the influence of corroding liquids in etching processes.
f vernis *m* de bitume; bitume *m* à couvrir — **d** Asphaltlack *m*; Asphalt-Benzollack *m*; Abdecklack *m* — **n** asfaltlak *mn* — **e** laca *f* de asfalto; barniz *m* de asfalto — **i** vernice *f* d'asfalto; vernice *f* all'asfalto
829 aspheric surface — Surface of a lens which is ground slightly different from spherical to reduce distortions.
f surface *f* asphérique — **d** asphärische Oberfläche *f*; kugelgestaltswidrige Oberfläche *f* — **n** asferisch oppervlak *n* — **e** superficie *f* asférica — **i** superficie *f* asferica
830 assemble *v* — Computer term. To translate a symbolic language source program into a machine language object program; usually produces one object program step from each source program step. Assembly involves translating the symbolic operation codes and addresses into machine language and may incorporate library subroutines.
f assembler — **d** assemblieren — **n** assembleren; samenvoegen — **e** ensamblar; compaginar — **i** insiemare
831 assemble *v* — See gather *v*.
832 assembler; assembling elevator — Part of a line composing machine to which matrices are conveyed from the magazine and assembled into a line of pre-determined measure.
f composteur *m* — **d** Sammler *m*; Sammelelevator *m* — **n** verzamelaar *m*; verzamelelevator *m* — **e** reunidor *m*; elevador-reunidor *m*; elevador *m* del aparato colector — **i** compositoio *m*; raccoglitore *m* (elevatore)
833 assembler; assembly program — Computer typesetting term. A translating routine which accepts or selects required subroutines, assembles parts of a program and makes the necessary adjustments to cross-references.
f programme *m* d'assemblage — **d**

Assemblierer *m*; Assemblierprogramm *n* — **n** assembleur *m*; assembleerprogramma *n* — **e** programa *m* de ensamblaje; programa *m* de compaginación — **i** programma *m* assiematore
834 assembler belt — Belt which transports the matrices from the magazine to the assembler in slug composing machines.
f courroie *f* d'assemblage; courroie *f* de l'assembleur — **d** Sammlerriemen *m*; Sammelriemen *m* — **n** verzamelriempje *n* — **e** correa *f* del reunidor; correa *f* del colector — **i** cinghia *f* del raccoglitore; cinghia *f* del compositoio
835 assembler channel — Part of the slug composing machine.
f canal *m* d'assemblage — **d** Sammlerkanal *m* — **n** verzamelkanaal *n* — **e** canal *m* colector — **i** canale *m* di raccolta
836 assembler chute finger — See chute finger.
837 assembler entrance — Part of the line composing machine.
f lamelles *fpl* de conduite des matrices — **d** Sammlereintritt *m* — **n** matrijzengeleiders *mpl* — **e** entrada *f* del reunidor; entrada *f* del colector — **i** entrata *f* del compositoio
838 assembler routines — Set of computer programs which translates or converts symbolic commands or instructions into the machine language for the computer in question. The symbolic commands are somewhat more abstract or general in nature than machine language, but not as abstract as a higher-level language such as fortran or cobol. Writing in assembler enables the programmer to take advantage of the inherent efficiences of his particular computer model, as would be the case if he were to write in the computer's own machine language, but the assembler routines will simplify the task, especially when programs are re-assembled, since the locations of the instructions are automatically re-sequenced to avoid patching or branching to a different section of core when new or improved subroutines are incorporated into the program. The subject program is thus written in assembler coding, while the object or converted program is assembled into machine language.
839 assembler slide — Part of the slug composing machine.
f traîneau *m* du composteur — **d** Sammlerschlitten *m* — **n** verzamelslede *fm* — **e** deslizadora *f* del reunidor; deslizadora *f* del colector; corredera *f* del reunidor; corredera *f* del colector; corredera *f* del componedor — **i** slitta *f* del compositoio
840 assembler star — See star wheel.
841 assembling elevator — See assembler.
842 assembly language — Machine oriented language for which a translating program (assembler) is supplied by the computer manufacturer. As an aid to programming operation codes often have symbolic names, and addresses may be given meaningful names by the programmer.

f langage *m* d'assemblage — **d** Assembliersprache *f* — **n** assembleertaal *fm* — **e** lenguaje *m* de compaginación; lenguaje *m* de ensamblaje — **i** linguaggio *m* assiematore

843 assembly line; in-line operation
f chaîne *f* d'assemblage — **d** Montagestrecke *f*; Montagelinie *f*; Fließband *n* — **n** montageband *m*; lopende band *m* — **e** correa *f* sin fin; cinta *f* continua; cinta *f* sin fin; operación *f* en línea — **i** linea *f* di montaggio

844 assembly line work
f travail *m* sur chaîne d'assemblage; assemblage *m* sur chaîne — **d** Fließarbeit *f*; Arbeit *f* am Fließband; Fließbandfertigung *f* — **n** lopende-bandwerk *n* — **e** trabajo *m* a la correa sin fin; trabajo *m* a la cinta sin fin; trabajo *m* a la cinta continua — **i** lavoro *m* alla linea di montaggio

845 assembly program — See assembler.

846 associate *v* — To bring together separate webs in a rotary press.
f assembler — **d** zusammenführen — **n** samenvoegen — **e** reunir — **i** riunire

847 association book — See association copy.

848 association copy; association book — Copy of a book once owned or annotated by the author, by someone otherwise associated with the book, or by some famous person.
f exemplaire *m* annoté — **d** Buch *n* aus persönlichem Besitz des Autors; Buch *n* mit Autor-Annotationen; Buch *n* aus Besitz einer berühmten Persönlichkeit; Erinnerungsexemplar *n* — **n** geannoteerd exemplaar *n* — **e** libro *m* con anotaciones (autográficos) — **i** libro *m* con annotazioni (autografe)

849 assorting — See sorting.

850 asterisk; star — Sign or reference mark (*) added to a word or sentence to refer the reader to a similar sign in another part, e.g., a foot note. When a number of reference signs are used, the asterisk comes first. In computer language one of the marks used to signify ellipses or matter not printed.
f astérisque *m*; étoile *f* — **d** Sternzeichen *n*; Stern *m*; Sternchen *n*; Notensternchen *n*; Anmerkzeichen *n*; Anmerkungszeichen *n* — **n** sterretje *n*; ster *f* — **e** asterisco *m*; signo *m* asteristico; estrella *f*; estrellita *f* — **i** asterisco *m*; stella *f*

851 asterisk signature; starred signature; star signature; inset signature — Numbers on the third page of each book sheet, indicated with an asterisk, i.e., on page 3 = 1*, page 19 = 2*, page 35 = 3*, etc. The asterisk denotes that the page is an inner page and this helps the binder for correct feeding of the folding machine.
f signature *f* seconde — **d** Sekunde *f* — **n** stersignatuur *f*; stercijfer *n* — **e** segunda signatura *f*; signatura *f* con asterisco — **i** segnatura *f* seconda

852 astigmatic aberration — Inability of an optical system to reproduce straight lines at various angles in focus at the same time. The image of a point in front of the lens with astigmatic aberration is not a point but two lines at right angles slightly separated from each other.
f aberration *f* astigmatique — **d** astigmatische Abweichung; astigmatischer Verzeichnungsfehler *m* — **n** astigmatische aberratie *f*; astigmatische afwijking *f* — **e** aberración *f* astigmática — **i** aberrazione *f* astigmatica

853 astigmatism — Inability of a photographic lens to sharply focus both vertical and horizontal lines, especially near the margin of the field image.
f astigmatisme *m* — **d** Astigmatismus *m* — **n** astigmatisme *n* — **e** astigmatismo *m* — **i** astigmatismo *m*

854 ASTM — See kerosine.

855 astrafoil — Sheets of polymerized vinyl chloride and metacrylic acid forming a transparent support, dimensionally stable under varying climatic conditions.
f astralon *m* — **d** Astralon *n* — **n** astralon *n* — **e** astralón *m* — **i** astralon *m*; supporto *m* di astralon

856 astronomical chart — See celestial chart.

857 astronomical signs; astronomical symbols — See app. no. 10.
f signes *mpl* astronomiques; symboles *mpl* d'astronomie — **d** astronomische Zeichen *npl* — **n** astronomische tekens *npl*; sterrenkundige tekens *npl* — **e** signos *mpl* de astronomía; símbolos *mpl* astronómicos — **i** segni *mpl* astronomici; simboli *mpl* astronomici

858 astronomical symbols — See astronomical signs.

859 asynchronous computer — Computer in which each circuit event initiates its successor(s). As opposed to synchronous processing where circuit events are clock controlled.
f calculateur *m* asynchrone — **d** asynchroner Rechner *m* — **n** asynchrone computer *m*; asynchrone rekenautomaat *m* — **e** computadora *f* asíncrona — **i** calcolatore *m* asincrono

860 asynchronous mode — System of working where a signal is generated to start the next operation; either the previous operation has been completed or peripheral equipment is available for the next operation and is awaiting data.
f mode *m* asynchrone — **d** asynchroner Modus *m*; asynchrone Betriebsart *f* — **n** asynchrone werkwijze *fm*; asynchrone modus *m* — **e** modo *m* asíncrono; modalidad *f* asíncrona — **i** modo *m* asincrono

861 atlas — Volume of maps usually bound together. Also any bound collection of plates or engravings showing systematically the development of a subject, or any work producing such effect by tabular arrangement.
f atlas *m* — **d** Atlas *m* — **n** atlas *m* — **e** atlas *m* — **i** atlante *m*

862 atlas — A size of writing and printing paper, recognized but not standard in the UK. See app. no. 3.

863 atmosphere — The air in any locality, including all its constituents, oxygen, nitrogen, carbon dioxide, the rare gases, dust and moisture. In the SI also unit of pressure. So 1 normal atmosphere = 101.325 x 10^3 Pa and 1 technical atmosphere = 1 kgf/cm^2 = 98.0665 x 10^3 Pa.

f atmosphère *f* — **d** Atmosphäre *f* — **n** atmosfeer *fm* — **e** atmósfera *f* — **i** atmosfera *f*

864 atmospheric conditions — Condition of the air, with particular regard to its temperature and relative humidity.

f conditions *fpl* atmosphériques; conditions *fpl* ambiantes; ambiance *f* — **d** atmosphärische Bedingungen *fpl*; Umgebung *f* — **n** atmosferische omstandigheden *fpl*; omgeving *f* — **e** condiciones *fpl* atmosféricas; condiciones *fpl* del ambiente; ambiente *m* — **i** condizioni *fpl* atmosferiche; condizioni *fpl* ambiente

865 atomic weight — Weight of an atom of an element, expressed on a scale in which the weight of the oxygen atom is exactly 16.

f poids *m* atomique — **d** Atomgewicht *n* — **n** atoomgewicht *n* — **e** peso *m* atómico — **i** peso *m* atomico

866 atomiser — Apparatus which by pressure or compressed air emits a stream of liquid in the form of minute separated droplets, forming a kind of mist.

f pulvérisateur *m*; pistolet *m* — **d** Zerstäuberdüse *f*; Sprühdüse *f* — **n** verstuiver *m* — **e** atomizador *m* — **i** atomizzatore *m*; polverizzatore *m*

867 attenuation — The decrease in amplitude that accompanies propagation or passage through equipment, lines or space, expressed in decibels.

f atténuation *f* — **d** Dämpfung *f*; Abschwächung *f* — **n** demping *f*; verzwakking *f* — **e** atenuación *f* — **i** attenuazione *f*; smorgamento *m*

868 attrition mill
f affleureuse *f* — **d** Stoffegalisator *n* — **e** molino *m* de afloras — **i** omogeneizzatore *m* dell'impasto

869 audio response — Recorded voice response to suitable input, or the acceptance of audible sound waves as input to a system. The ultimate is acceptance of the human voice; a number of manufacturers and research bodies have investigations at various stages of development.

f réponse *f* parlée — **d** Sprachausgabe *f* — **n** audio-antwoord *n* — **e** respuesta *f* hablada — **i** riposta *f* parlata

870 audit bureau of circulation
f office *m* de contrôle des tirages; office *m* de justification des tirages — **d** Auflagekontrollbureau *n* — **n** bureau *n* voor oplagecontrole; oplagecontrolebureau *n* — **e** oficina *f* de control de las tiradas — **i** ufficio *m* controllo delle tirature

871 audit report; circulation statement
f justificatif *m* de tirage — **d** Auflageangabe *f* — **n** opgave *f* van de oplage — **e** certificación *f* de circulación; certificado *m* de circulación; informe *m* sobre circulación — **i** certificato *m* della tiratura; attestato *m* della tiratura

872 augend — A number to which another is

added. See also accumulator.

d Augend *m* — **n** optelgetal *n* — **e** sumando *m* — **i** addendo *m*

873 augmented edition — See enlarged edition.

874 auramine — Yellow crystalline solid, soluble in water, alcohol and ether. Used as a yellow dye for paper, textiles, printing inks, leather and as a fungicide.
Formula: $(CH_3)_2NC_6H_4(C:NH)C_6H_4N(CH_3)_2{\cdot}HCl$.
f auramine *f* — **d** Auramin *m* — **n** auramine *f* — **e** auramina *f*

875 auric chloride — See gold trichloride.

876 auric oxide — See gold oxide.

877 aurines — Group of dyestuffs derived from para-rosolic acid.
f aurines *fpl* — **d** Aurine *npl* — **n** aurinen *fpl* — **e** aurinas *fpl* — **i** aurine *fpl*

878 auripigment — See orpiment.

879 aurora yellow — See cadmium sulphide.

880 aurous bromide — See gold bromide.

881 author; authoress (female) — The original composer of a document.
f auteur *m*; écrivain *m*; femme *f* auteur; femme *f* écrivain — **d** Verfasser *m*; Autor *m*; Schriftsteller *m*; Verfasserin *f*; Schriftstellerin *f* — **n** auteur *m*; schrijver *m*; schrijfster *f* — **e** autor *m*; escritor *m*; autora *f*; escritora *f* — **i** autore *m*; scrittore *m*; autrice *f*; scrittrice *f*

882 authoress — See author.

883 authorized translation
f traduction *f* autorisée — **d** autorisierte Übersetzung *f* — **n** geautoriseerde vertaling *f* — **e** traducción *f* autorizada — **i** traduzione *f* autorizzata

884 authorized version — English version of the bible prepared in England under James I and published in 1611. Also called King James' version or King James' Bible.
f version *f* autorisée — **d** autorisierte Version *f* — **n** geautoriseerde versie *f* — **e** versión *f* autorizada — **i** versione *f* autorizzata

885 author publisher — An author who publishes his own work.
f auteur-éditeur *m* — **d** Selbstverleger *m* — **n** schrijver-uitgever *m*; auteur-uitgever *m* — **e** autor-editor *m* — **i** autore-editore *m*; autoeditore *m*

886 author's alterations — See author's corrections.

887 author's corrections; author's alterations — Corrections which depart from original copy as distinguished from those made necessary by printer's mistakes. They are charged to the customer.
f corrections *fpl* d'auteur — **d** Autorkorrekturen *fpl*; Verfasserkorrekturen *fpl*; Änderungen *fpl* des Verfassers in der Korrektur — **n** auteurscorrectie *f* — **e** correcciones *fpl* del autor; alteraciones *fpl* del autor — **i** correzioni *fpl* d'autore; correzioni *fpl* dell'autore

888 author's error
f erreur *f* d'auteur — **d** Autorfehler *m* — **n** fout

fm van de schrijver; fout *fm* van de auteur — **e** yerro *m* del autor; error *m* del autor — **i** errore *m* sfuggito all'autore; errore *m* dell'autore; sbaglio *m* dell'autore

889 author's proof — Proof supplied to an author for approval or correction.
f épreuve *f* d'auteur; première *f* d'auteur — **d** Korrekturabzug *m* für den Autor; Autorkorrektur *f*; Verfasserkorrektur *f* — **n** proef *fm* voor de schrijver; proef *fm* voor de auteur — **e** prueba *f* del autor; prueba *f* de segunda capilla — **i** bozza *f* d'autore; prova *f* d'autore; prova *f* per correzione d'autore

890 author's royalty — Payment made to an author for copies of his work sold.
f honoraires *mpl* de l'auteur — **d** Autorhonorar *n*; Verfasserhonorar *n* — **n** auteurshonorarium *n*; honorarium *n* voor de auteur; royalties *pl* — **e** honorarios *pl* del autor; retribución *f* pecuniaria — **i** onorari *pl* dell'autore; retribuzione *f* dell'autore

891 autobiography — An account of a person's life written by himself.
f autobiographie *f* — **d** Autobiographie *f*; Selbstbiographie *f* — **n** autobiografie *f* — **e** autobiografía *f* — **i** autobiografia *f*

892 autochrome plate — Plate for direct colour photography on glass which must be viewed by transparency, the colour patterns being formed by starch grains dyed blue-violet, green and orange. The first successful tricolour screen plate for producing full-colour transparencies by a single exposure of objects.
f plaque *f* autochrome — **d** Autochromplatte *f*; Farbrasterplatte *f* — **n** autochroomplaat *fm* — **e** placa *f* (fotográfica) autocroma; placa *f* autocrómica — **i** lastra *f* autochrome; lastra *f* autocromatica

893 autochrome printing — See autochrome process.

894 autochrome process; autochrome printing — Obsolete colour printing process using a halftone block or lithographic plate as key, adding colours by full-tone plates.
f procédé *m* autochrome — **d** Autochromverfahren *n* — **n** autochroomprocédé *n* — **e** procedimiento *m* autocrómica; impresión *f* autocrómica — **i** procedimento *m* autochrome; impressione *f* autocromia

895 autocode — Code to write computer programs which require translation into machine code. The term has become ambiguous, as it is used to describe various levels of computer language.
f autocode *m* — **d** Autocode *m* — **n** autocode *m* — **e** autocódigo *m* — **i** autocodice *m*

896 autographed copy — Copy of a book signed by the author. See also presentation copy.
f exemplaire *m* autographié — **d** handsignierte Ausgabe *f*; signierte Ausgabe *f*; Widmungsexemplar *n* — **n** gesigneerd exemplaar *n*; getekend exemplaar *n* — **e** ejemplar *m* autógrafo; ejemplar *m* dedicado por el autor; ejemplar *m* autografiado — **i** esemplare *m* autografato; copia

f autografata

897 automatic bag — See automatic self-opening bag.

898 automatic blankets — Felt blankets with a top coating of rubber or plastics material to cover the impression cylinder.
f blanchets *mpl* automatiques — **d** automatische Drucktücher *npl* — **n** automatische drukdoeken *npl* — **e** mantillas *fpl* automáticas — **i** rivestimenti *mpl* automatici; abbigliamenti *mpl*

899 automatic brake
f frein *m* automatique — **d** automatische Bremsvorrichtung *f* — **n** automatische rem *fm* — **e** freno *m* automático — **i** freno *m* automatico

900 automatic centring and quadding device — See self quadder.

901 automatic check; built-in check — Hardware which performs accuracy checking automatically, to interrupt in case of an error or hardware failure.
f contrôle *m* automatique; contrôle *m* incorporé — **d** eingebaute Prüfung *f* — **n** automatische controle *fm* — **e** comprobación *f* automática; verificación *f* automática — **i** controllo *m* automatico

902 automatic control — Self-acting appliance to regulate the operation of a machine.
f commande *f* automatique — **d** automatische Steuerung *f* — **n** automatische besturing *f* — **e** mando *m* automático — **i** comando *m* automatico

903 automatic cutting and creasing machine
f machine *f* automatique de découpe et de rainage — **d** Stanzautomat *m* — **n** stansautomaat *m* — **e** troqueladora *f* automática — **i** macchina *f* automatica fustellatrice

904 automatic cylinder press
f machine *f* automatique à cylindre; presse *f* automatique à cylindre — **d** automatische Zylinderflachform-Druckmaschine *f*; Zylinderdruckautomat *m*; Zylinderautomat *m depr.* — **n** cilinderautomaat *m* — **e** máquina *f* cilíndrica automática — **i** macchina *f* da stampa pianocilindrica

905 automatic data processing; datamation — Series of operations which include calculating, filing, interpreting, recording, merging, sorting and logical processes, performed on data to obtain information with a minimum of human intervention. ADP equipment uses electro-mechanical, electric or electronic means of processing data. See also electronic data processing.
f transformation *f* automatique des données; traitement *m* automatique des données — **d** automatische Datenverarbeitung *f* — **n** automatische gegevensverwerking *f*; automatische verwerking *f* van gegevens; automatische dataverwerking *f* — **e** procesado *m* automático de datos — **i** elaborazione *f* automatica dei dati; trattamento *m* automatico dei dati

906 automatic delivery
f sortie *f* automatique; réception *f* automatique; receveur *m* automatique — **d** automatische Auslage *f*; selbsttätige Auslage *f*; selbsttätiger

Bogenausleger *m* — **n** automatische uitleg *m* — **e** entrega *f* automática; receptor *m* automático — **i** uscita *f* automatica

907 automatic engraving — Direct production of relief line and halftone printing plates from photographic images by electronically operated engraving machines.

f gravure *f* électronique — **d** elektronische Gravur *f* — **n** elektronisch graveren *n* — **e** grabado *m* electrónico — **i** incisione *f* elettronica

908 automatic exposure timer; exposure-time computer; exposure computer — Apparatus to determine the quantity of light for the main and the flash exposure from the maximum and minimum density of the copy and the rate of enlargement. With a trial exposure it is possible to adapt the required exposure to the speed and the contrast of the film-speed/developer combination.

f appareil *m* automatique d'insolation; intégrateur *m* de lumière — **d** Belichtungscomputer *m*; Belichtungszeitrechner *m* — **n** belichtingsautomaat *m* — **e** autómata *m* de exposición — **i** calcolatore *m* dell'esposizione; esposimetro *m* computerizzato

909 automatic feeder; automatic feeding apparatus — Mechanical appliance to lift a sheet of paper from the pile and carry it to the feeding table.

f margeur *m* automatique — **d** automatischer Anlegeapparat *m*; automatischer Anleger *m*; automatischer Bogenanlegeapparat *m*; Selbstanleger *m*; selbsttätiger Einleger *m*; Bogenanleger *m*; Anlegeapparat *m* — **n** automatisch (werkend) inlegapparaat *n* — **e** marcador *m* automático; alimentador *m* automático; ponepliegos *m* automático; ponedor *m* automático; tomador *m* automático *Ar*; introductor *m* automático — **i** mettifoglio *m* automatico; alimentatore *m* automatico

910 automatic feeding apparatus — See automatic feeder.

911 automatic focusing — Determination of the correct position of the focal plane on process cameras.

f mise *f* au point automatique — **d** automatische Bildeinstellung *f*; automatische Scharfeinstellung *f* — **n** automatische instelling *f*; automatische scherpstelling *f* — **e** enfoque *m* automático — **i** messa *f* a fuoco automatica

912 automatic fountain; spray fountain *depr.* — Automatic system of feeding ink to the distributing cylinders by pumps and a perforated rail arrangement.

f encrier *m* à projection — **d** Farbwerk *n* mit Farbpumpen; automatisches Farbwerk *n* — **n** inktwerk *n* met pompsysteem — **e** tintero *m* a inyección (por bomba) — **i** inchiostratore *m* a pompa

913 automatic line

f chaîne *f* automatique — **d** automatische Fertigungsstraße *f* — **n** automatische produktielijn *fm* — **e** cadena *f* automatizada de producción

continua; tren *m* automático de fabricación; tren *m* automático de producción — **i** linea *f* automatica

914 automatic platen — See automatic platen press.

915 automatic platen machine — See automatic platen press.

916 automatic platen press; automatic platen; automatic platen machine; autoplaten

f presse *f* à platine automatique; platine *f* automatique; autoplatine *f* — **d** Tiegelautomat *m*; Druckautomat *m* — **n** degelautomaat *m*; automatische degelpers *f* — **e** máquina *f* automática de platina; prensa *f* automática de platina; minerva *f* automática — **i** platina *f* automatica

917 automatic pot feeder — Arrangement on a slug casting machine for automatic charging of the melting pot with an adequate amount of fresh metal each time a slug is cast.

f alimenteur *m* automatique du creuset — **d** automatischer Metallzuführer *m*; automatischer Bleizuführer *m* — **n** automatische potvuller *m* — **e** dispositivo *m* alimentador del crisol; alimentador *m* automático del metal — **i** alimentatore *m* automatico del crogiuolo; caricamento *m* automatico del metallo

918 automatic processor — See processor.

919 automatic program interrupt; program priority interrupt *US* — Ability of computers to abandon one operation temporarily when a priority operation arises.

f arrêt *m* automatique du programme — **d** Programmstop *m* — **n** automatische programmaonderbreking *f* — **e** parada *f* automática del programa — **i** arresto *m* automatico del programma

920 automatic programmer; programmatic; programmed guillotine — Guillotine equipped with an auto-repeating mechanism.

f massicot *m* à (coupe) programme; programmeur *m* — **d** Programmschnellschneider *m*; Programmschneider *m*; Schneidemaschine *f* mit Programmsteuerung; Schneidautomat *m* — **n** programsnijmachine *f* — **e** guillotina *f* con programa de corte — **i** tagliacarte *f* a programma

921 automatic quadder — See self quadder.

922 automatic reel change — See flying paster.

923 automatic request for repetition — System employing an error detecting code and so arranged that a signal detected as being in error automatically initiates a request for retransmission. Abbrev.: ARQ.

f demande *f* automatique de répétition — **d** selbsttätiger Wiederholungsruf *m* — **n** automatisch verzoek *n* om herhaling — **e** petición *f* automática de repetición — **i** richiesta *f* automatica di ripetizione

924 automatic self-opening bag; automatic bag; self-opening satchel bag; SOS bag — Bag which can be opened with a quick flip of the wrist. The pressure of the air entering the top of the bag

automatically opens the bottom. The bag is constructed with tucks in the side and a preformed flat, square bottom which permits it to stand upright when empty.
f cornet *m* à ouverture automatique; sac *m* à fond formé SOS − **n** zelfopenende zak *m* − **e** saco *m* con válvula de apertura automática − **i** sacchetto *m* ad apertura automatica

925 automatic sensing − In magnetic ink printing according to the Bull system, automatic character recognition takes place by identification of internal lengths between adjacent strokes.
f reconnaissance *f* automatique − **d** automatische Abtastung *f* − **n** automatische aftasting *f* − **e** exploración *f* automática − **i** riconoscimento *m* automatico

926 automatic switch − Self-acting switch which breaks the current in case of an accident or fault in an electrically driven machine, such as a printing press.
f interrupteur *m* automatique − **d** automatischer Schalter *m*; Selbstschalter *m* − **n** automatische schakelaar *m* − **e** interruptor *m* automático − **i** interruttore *m* automatico

927 automatic tag stringer and knotter − See tag stringing and knotting machine.

928 automatic translation − See machine translation.

929 automation; automatization − Method in which manufacturing processes are automatically performed by self-operating devices. Term derived from the words "automatic operation".
f automatisation *f*; automation *f* − **d** Automatisierung *f*; Automation *f* − **n** automatisering *f*; automatie *f* − **e** automatización *f*; automación *f*; automatismo *m* − **i** automatizzazione *f*; automazione *f*

930 automatization − See automation.

931 autopaster − See flying paster.

932 autoplate − Machine for casting curved stereoplates for letterpress rotary machines. Models are made for producing two, three or four plates per minute. Cast plates require finishing in the autoshaver machine, which bores, trims and cools plates ready for the press at the rate of five per minute.
f autoplate *f*; clicheuse *f* journal − **d** Autoplate *f* − **n** autoplate *fm* − **e** fundidora *f* automática para planchas curvas; fundidora *f* de tejas de impresión; potrillo *m* Ch − **i** autoplate *f*; macchina *f* per forme stereotipiche curve

933 autoplaten − See automatic platen press.

934 autopolymerization − Polymerization reaction catalyzed by the polymers formed. See also polymerization.
f autopolymérisation *f* − **d** Selbstpolymerisierung *f* − **n** auto-polymerisatie *f* − **e** autopolimerización *f* − **i** autopolimerizzazione *f*

935 autopositive − See direct positive.

936 autoscreen film − Trade name for an orthochromatic film embodying a halftone screen. Exposed to a continuous-tone image, it produces automatically a dot pattern just as if a halftone screen had been used.
f film *m* prétramé; matériel *m* photosensible à trame incorporée − **d** Rasterfilm *m*; Autoscreenfilm *m*; vorgerasterter Film *m* − **n** voorgerasterde film *m* − **e** película *f* pretramada; película *f* con trama incorporada; película *f* para tramado directo − **i** pellicola *f* preretinata; pellicola *f* autotipica; pellicola *f* con preretinatura

937 autoshaver; hobelmachine *pop.* − Machine on which curved stereoplates cast in an *autoplate are trimmed, shaved and bevelled ready for use on the press.
f autoshaver *m* − **d** Rundstereo-Bearbeitungsmaschine *f* − **n** plaatfreesmachine *f* − **e** fresadora; máquina *f* de fresar − **i** fresatrice *f*

938 autotype − German and Dutch name for (halftone) photoengraving. See photoengraving.

939 autotype process − *Collotype process in which the plate is coated with a light-sensitive resin instead of gelatine.

940 autoxidation
f autoxydation *f* − **d** Autooxydation *f*; Autoxydation *f* − **n** auto-oxydatie *f* − **e** autooxidación *f* − **i** autossidazione *f*

941 auxiliary device − See auxiliary machine.

942 auxiliary machine; auxiliary device
f machine *f* auxiliaire; dispositif *m* auxiliaire − **d** Hilfsmaschine *f*; Hilfsgerät *n* − **n** hulpmachine *f*; hulpapparaat *n*; hulptoestel *n* − **e** máquina *f* auxiliar; aparato *m* auxiliar − **i** macchina *f* ausiliaria; apparecchio *m* ausiliario

943 auxiliary magazine − The extra holder of matrices of a composing machine.
f magasin *m* auxiliaire − **d** Seitenmagazin *n*; Zusatzmagazin *n* − **n** zijmagazijn *n* − **e** almacén *m* auxiliar; depósito *m* auxiliar − **i** magazzino *m* supplementare

944 auxiliary reel stand; auxiliary roll stand *depr.* − Second reel stand that can be mounted on top of another one. Reduces down time by permitting one stand to be loaded while the other is still inwinding. Cannot be used to feed two webs at the same time unless converted to a dual reel stand.
f porte-bobine *m* auxiliaire − **d** zusätzlicher Rollenständer *m* − **n** extra rollenstandaard *m*; extra rollenhouder *m* − **e** portabobina *m* superpuesto − **i** portabobine *m* supplementare

945 auxiliary roll stand − See auxiliary reel stand.

946 auxiliary storage − See auxiliary store.

947 auxiliary store; secondary store; backing store *GB*; **auxiliary storage; secondary storage** *US* − Store under control of the computer system, but ancillary to the main store. Greater store space is available in the auxiliary store, but access speed is usually slower than that of the main store (e.g., magnetic tape, drum, card or disk).
f mémoire *f* auxiliaire − **d** Ergänzungsspeicher *m*; Hintergrundspeicher *m* − **n** hulp-

geheugen *n* — **e** memoria *f* auxiliar — **i** memoria *f* ausiliaria

948 aux petits fers — Said of a book cover decorated by impressing small individual tools to build up complete patterns. See also petits fers.

f aux petits fers — **d** mit Stempelsatz — **n** aux petits fers; met kleine stempeltjes — **e** a pequeños hierros — **i** a piccoli ferri

949 available chlorine

f chlore *m* actif — **d** aktives Chlor *n* — **n** actief chloor *n* — **e** cloro *m* activo — **i** cloro *m* attivo

950 available time — The time during which a computer is available for use.

f temps *m* disponible — **d** verfügbare Zeit *f* — **n** beschikbare tijd *m* — **e** tiempo *m* disponible — **i** tempo *m* disponibile

951 average — See also arithmetical mean.

f moyenne *f* — **d** Durchschnitt *m*; Durchschnittswert *m*; Mittelwert *m*; Mittel *n* — **n** gemiddelde *n* — **e** promedio *m*; media *f* — **i** media *f*

952 average capacity — See average output.

953 average deviation — See mean deviation.

954 average error — See mean deviation.

955 average outgoing quality

f qualité *f* moyenne après inspection — **d** Durchschlupf *m* — **n** gemiddeld doorgelaten uitvalpercentage *n* — **e** calidad *f* media después de inspección — **i** qualità *f* media del prodotto finito

956 average output; average production; average capacity

f rendement *m* moyen — **d** Dauerdruckleistung *f*; Durchschnittsleistung *f* — **n** gemiddelde produktie *f* — **e** producción *f* media; rendimiento *m* medio; capacidad *f* media — **i** produzione *f* media; rendimento *m* medio

957 average particle size — Usually the average particle diameter, though it can refer to surface, volume or weight.

f grosseur *f* moyenne des particles — **d** mittlere Teilchengröße *f*; durchschnittliche Teilchengröße *f* — **n** gemiddelde deeltjesgrootte *f* — **e** tamaño *m* medio de las partículas — **i** dimensione *f* media delle particelle

958 average production — See average output.

959 average quality level

f qualité *f* moyenne — **d** mittlere Güte *f* — **n** gemiddeld kwaliteitsniveau *n* — **e** calidad *f* intermedia — **i** qualità *f* media

960 average sample number

f effectif *m* moyen de l'échantillon — **d** mittlerer Stichprobenumfang *m* — **n** gemiddelde steekproefgrootte *f* — **e** efectivo *m* medio de la muestra — **i** numero *m* medio di campioni

961 average speed

f vitesse *f* moyenne — **d** Durchschnittsgeschwindigkeit *f*; mittlere Geschwindigkeit *f* — **n** gemiddelde snelheid *f* — **e** velocidad *f* media — **i** velocità *f* media

962 awl — See bodkin.

963 axe — Imaginary line extending lengthwise through the centre of the press cylinder. See also axle.

f axe *m* — **d** Achse *f* — **n** as *fm* — **e** eje *m*; árbol *m* — **i** asse *f*; albero *m*

964 axis of abscissas — See axis of the abscisse.

965 axis of ordinates — See axis of ordination.

966 axis of ordination; axis of ordinates; y-axis — The vertical axis in a plane Cartesian coordinate system.

f axe *m* des ordonnées; axe *m* des y — **d** Ordinatenachse *f*; y-Achse *f* — **n** as *fm* van de ordinaat; y-as *fm* — **e** eje *m* de las ordenadas; eje *m* dell'y — **i** asse *m* dell'ordinata; asse *m* dell'y

967 axis of the abscisse; axis of abscissas; x-axis — The horizontal axis in a plane Cartesian coordinate system.

f axe *m* des abscisses; axe *m* des x — **d** Abszissenachse *f*; x-Achse *f* — **n** as *fm* van de abscis; x-as *fm* — **e** eje *m* de abscisas — **i** asse *m* dell'ascissa

968 axle — Spindle on or with which a wheel revolves.

f arbre *m* — **d** Welle *f*; Achse *f* — **n** as *fm* — **e** árbol *m*; eje *m*; flecha *f* *Me* — **i** asse *m*; albero *m*

969 axle bearing; journal bearing

f portée *f* d'axe; palier *m* d'axe — **d** Achsenlager *n*; Achsellager *n*; Achslager *n*; Achslagerung *f* — **n** aslager *n* — **e** cojinete *m* del eje — **i** cuscinetto *m* dell'albero; supporto *m* dell'albero

970 ayr stone — See snakeslip.

971 azerty layout — Typewriter keyboard layout with the a, z, e, r, t, y-keys at the upper left hand row of the keys. See also Dvorak keyboard, qwert layout.

972 aziminobenzene — See benzotriazole.

973 azo colours; azo dyes — Pigments containing the azo group N=N.

f colorants *mpl* azoïques — **d** Azofarben *fpl*; Azofarbstoffe *mpl* — **n** azokleurstoffen *fmpl* — **e** colorantes *mpl* azoicos — **i** azocoloranti *mpl*; coloranti *mpl* azoici

974 azo dyes — See azo colours.

975 azotate — See nitrate.

976 azotic acid — See nitric acid.

977 azure blue — See cobalt blue.

978 azured tool — Tool with parallel diagonal lines used in book cover decorating.

f fer *m* azuré — **d** schraffierter Stempel *m* — **n** gearceerd stempel *n* — **e** filete *m* de líneas; filete *m* rayado — **i** ferro *m* azzurrato

979 azure laid writing paper — Writing paper with laid marks and light blue colour. *Azure wove writing paper differs only in having a wove finish.

f papier *m* d'écriture vergé bleu — **d** azurblau geripptes Schreibpapier *n* — **n** blauw gevergeerd schrijfpapier *n* — **e** papel *m* de carta azul celeste verjurado — **i** carta *f* da scrivere vergata azzurrina

980 azure tooling — Decorating book covers by tooling on them a pattern of close parallel lines.

f décoration *f* à lignes azurées; décoration *f* à fers azurés — **d** Azureelinienverzierung *f* — **n** versiering *f* met assuré-lijnen — **e** decoración *f*

con líneas azuradas — **i** decorazione *f* a linee parallele; decorazione *f* delle rilegatura con ferri azzurrati
981 azure wove writing paper — See also azure laid writing paper.

f papier *m* d'écriture vélin bleu — **d** azurblau ungeripptes Schreibpapier *n* — **n** blauw velijn schrijfpapier *n* — **e** papel *m* de carta azul celeste velin — **i** carta *f* da scrivere velina azzurrina

982 babbitt metal; bearing metal; white metal; anti-friction metal — Soft alloy containing approx. 89% tin, 7% antimony and 4% copper, used for bearings in machinery. The main types are lead base babbitt, arsenical babbitt. Named after Isaac Babbitt, the American inventor (1799-1862).
f métal *m* antifriction; métal *m* blanc; régule *m* — **d** Weißmetall *n*; Lagermetall *n* — **n** babbitt-metaal *n*; witmetaal *n*; lagermetaal *n*; antifrictie-metaal *n* — **e** metal *m* babbitt; metal *m* blanco; metal *m* antifricción; régulo *m* — **i** metallo *m* bianco antifrizione
983 back — Part of a book or a cover which is sewn when the book is bound. There are three kinds: a *hollow back, a *tight back and a flexible back.
f dos *m* — **d** Buchrücken *m*; Rücken *m* — **n** rug *m*; boekrug *m* — **e** lomo *m* — **i** dorso *m*
984 back — Reverse of nick or belly of type.
f dos *m* — **d** Rücken *m*; Rückseite *f* — **n** achterkant *m* — **e** cara *f* posterior; dorso *m* — **i** dorso *m*
985 back — Space between the sides of mated or companion pages in the forme.
f blanc *m* de petit fond; petit fond *m*; blanc *m* de dos; blanc *m* de couture — **d** Bundsteg *m* — **n** rugwit *n* — **e** medianil *m* — **i** bianco *m* del dorso
986 back — See back margin.
987 back — See spine.
988 back *v* — To make the back of a book.
f endosser — **d** den Buchrücken machen — **n** de boekrug maken — **e** enlomar; formar el lomo — **i** formare il dorso (di un libro)
989 backbone — See spine.
990 back copy — See back number.
991 back cover — The board covering the end of a book.
f plat *m* de derrière; plat *m* inférieur; plat *m* verso; plat *m* arrière — **d** hinterer Buchdeckel *m* — **n** achterplat *n* — **e** contracubierta *f*; contratapa *f* — **i** piatto *m* inferiore; piatto *m* posteriore
992 back edge — See trailing edge.
993 backed plate — Photographic plate or film backed with an anti-halation layer.
f plaque *f* antihalo — **d** lichthoffreie Platte *f*; photographischer Film *m* (Platte) mit Lichthofschutzschicht — **n** antihalo-plaat *fm* — **e** placa *f* con respaldo antihalo; placa *f* antihalo — **i** lastra *f* antialo
994 backer — See backing machine.
995 backer — See back rounding machine.
996 back etching — Opening up or lessening the density in a negative. Used in lithography for colour correcting continuous-tone negatives.
f morsure *f* de descente d'intensité — **d** Auflichten *n*; Aufhellen *n*; Zurückkätzen *n* — **n** terugetsen *n* — **e** rebajamiento *m*; debilitamiento *m* — **i** indebolimento *m*

997 backfall — Raised part of the floor in the beater, behind the roll.
f saut *m* — **d** Kropf *m*; Sattel *m*; Berg *m* — **n** zadel *n* — **e** salto *m* — **i** salto *m*
998 back focal distance; back focal length — Distance from the vertex of the back surface of the lens to the back focal point.
f distance *f* focale image; longueur *f* focale postérieure — **d** Bildschnittweite *f*; hintere Brennweite *f* — **n** afstand *m* van beeld tot lens — **e** distancia *f* focal imagen — **i** lunghezza *f* focale posteriore; seconda distanza *f* focale
999 back focal length — See back focal distance.
1000 back gauge — Part of a guillotine
f équerre *f* — **d** Sattel *m*; Anlegesattel *m*; Sattelvorschub *m*; Vorschubsattel *m* — **n** zadel *n*; voorschuif *fm* — **e** escuadra *f*; guía *f* (posterior); indicador *m* de atrás; calibrador *m* de atrás — **i** squadra *f* posteriore
1001 background — Light, flat tint, with or without designs, used as a decorative base for type matter printed in a different colour.
f fond *m* — **d** Untergrund *m*; Fond *m* — **n** ondergrond *m*; fond *mn* — **e** fondo *m* — **i** fondo *m*; sottostampa *f* di fondo
1002 background — In photography or artwork, that which stands at the back of the main subject of a picture.
f fond *m*; arrière-plan *m* — **d** Hintergrund *m* — **n** achtergrond *m* — **e** fondo *m*; último plano *m*; último término *m* — **i** sfondo *m*
1003 background — The use of a computer to process a job of a low priority when the computer is not occupied in performing some higher priority task. See also foreground.
1004 background colour
See backing-up colour.
1005 background plate — See flat-tint plate.
1006 background plate — See tint block.
1007 background program — Program, usually of the batch processing type, that can be executed when the facilities of a multiprogramming computer system are not required by real-time programs or other programs of high priority.
1008 backing — Making of the shoulder or joint of a book into which the cover boards will fit. See also back rounding.
f endossure *f* — **d** Falzeinpressen *n*; Abpressen *n* — **n** knepen *n* — **e** formación *f* de cajos — **i** formazione *f* del morso; formazione *f* del falso
1009 backing — Sheet of resilient paper or other material, placed under the sheet to be printed to ensure a more even impression.
f feuille *f* de dessous — **d** Unterlagbogen *m* — **n** onderlegvel *n* — **e** hoja *f* debajo el papel de impresión — **i** maestra *f*; foglio *m* di maestra
1010 backing — The back of a book. See back.
1011 backing away from fountain; holding back — Said of ink when lacking in flow it will not

keep in contact with the fountain roller so that the latter can transfer ink to the ductor roller. The prints become uneven, streaky and weak.

f dormir (l'encre dort) — **d** Nichtmitgehen *n* der Farbe im Farbkasten; Stehenbleiben *n* der Farbe im Farbkasten — **n** stilstaan *n* van de inkt in de inktbak — **e** remanso *m* (la tinta se remansa en el tintero; tinta dormida) — **i** mancato trascinamento *m* dell'inchiostro nel calamaio

1012 backing doctor — See doctor blade support.

1013 backing felt; felt backing; mat backing — Material to be attached to the back of the blank spaces on stereotype mats.

f garnissage *m* — **d** Auslegefilz *m*; Auslegepappe *f* — **n** matrijzenvilt *n* — **e** cartón *m* de reforzar matrices; cartón *m* de reforzar blancos; cartón *m* fieltro; cartón *m* muelle; cartón *m* engomado; cartón *m* gris — **i** feltro *m* da taccheggio (per flani)

1014 backing hammer; beating hammer — Round-faced hammer to round and shape the backs of books.

f marteau *m* de relieur — **d** Buchbinderhammer *m* — **n** rondzethamer *m*; boekbindershamer *m*; rondklophamer *m* — **e** martillo *m* para redondear lomos; martillo *m* de encuadernador — **i** martello *m* da rigelatore

1015 backing machine; backing press; backer — Machine in which books are shaped, joints raised, and books backed automatically. See also hand backer.

f presse *f* à endosser; machine *f* à endosser — **d** Abpreßmaschine *f* — **n** rondzet- en kneepmachine *f* — **e** máquina *f* para sacar cajos — **i** macchina *f* per formare i dorsi

1016 backing metal — Alloy to back electro shells. Of no use as type metal.

f métal *m* d'endossage; métal *m* de doublage — **d** Hintergießmetall *n* für Galvanos; Metall *n* zum Hintergießen von Galvanos — **n** achtergietmetaal *n* — **e** metal *m* de relleno — **i** metallo *m* di riempimento

1017 backing press — See backing machine.

1018 backing roll mark — Excess of coating caused by a gouge mark in the backing roll. Defect in blade coated paper that occurs regularly in the machine direction at one constant deckle position.

f marque *f* de rouleau de soutien — **d** Markierung *f* der Stützwalze — **n** merk *n* van de steunwals — **e** marca *f* del rodillo de sostén — **i** marcatura *f* del rullo di sostegno; impronta *f* del cilindro d'appoggio

1019 backing roll spots — Spots, usually 0.25 in or larger in size, that have no coating in the centre. A butterfly usually has an excess of coating around the periphery or at the sides; occurs at regular intervals in the machine direction of the sheet.

f taches *fpl* de rouleau de soutien; papillons *mpl* — **d** Strichkrater *mpl* — **e** manchas *fpl* del rodillo de sostén — **i** macchie *fpl* del cilindro d'appoggio; farfalle *fpl*

1020 backing store — See auxiliary store.

1021 backing up — Printing on the back side of a sheet, the first side being already printed.

f retiration; impression *f* second côté — **d** Widerdruck *m* — **n** weerdruk *m*; tegendruk *m* *depr.* — **e** retiro *m* (en tipografía); retiración *f*; dorso *m* (en lito) — **i** stampa *f* in volta; stampa *f* del verso

1022 backing up — Application of electrotype metal to the copper shell. See also electrotype.

f doublage *m* d'un galvano; garnissage *m* de la coquille; garnissage *m* au dos — **d** Hintergießen *n* (des Galvanos); Unterguß *m* — **n** achtergieten *n* (van de galvano) — **e** relleno *m* del dorso (de la cascarilla del galvano) — **i** colata *f* di riempimento (della conchiglia)

1023 backing up — Deprecated term for the letterpress rotary printing term underlay.

f mise *f* de dessous — **d** Plattenzurichtung *f* — **n** onderlegsel *n* — **e** arreglo *m* de plancha; recorte *m* por debajo — **i** avviamento *m*; teccheggio *m*

1024 backing up — See underlay.

1025 backing-up colour; background colour; under-colour; ground tint — Colour printed underneath an ink film to increase the density.

f teinte *f* de fond — **d** Hintergrundfarbe *f*; Untergrundfarbe *f*; Grundfarbe *f*; untergedruckte Farbe *f* — **n** ondergrondkleur *fm*; grondkleur *fm*; steunkleur *fm* — **e** tinta *f* amortiguadora — **i** colore *m* di fondo; colore *m* dello sfondo; sottostampa *f* di fondo

1026 back in sections

f dos *m* à compartiments — **d** Rücken *m* in Feldern — **n** rug *m* in vakken — **e** lomo *m* en compartimientos — **i** dorso *m* a scomparti

1027 back issue — See back number.

1028 back lash — Slackness between two or more machine parts working together, permitting one of them to move a certain distance without being followed by the other(s).

f jeu *m* — **d** Spiel *n*; Zahnspiel *n*; toter Gang *m*; Schlupf *m* — **n** speling *f*; tandspeling *f*; tandradspeling *f* — **e** juego *m* entre diente; juego *m* muerto; juego *m* perdido; juego *m* inútil; huelgo *m*; huelga *f* — **i** gioco *m*; gioco *m* fra in fianchi dei denti

1029 back-lash gear; no-lash gear

f engrenage *m* sans jeu de dents — **d** spielfreier Antrieb *m* — **n** spelingvrije aandrijving *f* — **e** engranaje *m* rectificador (de holguras); engranaje *m* sin holguras; engranaje *m* de contragolpe — **i** ingranaggio *m* senza gioco

1030 back-lash ring — Adjustable narrow gear in a rotary press mounted on the main gear used to counteract back lash.

f bride *f* de suppression de jeu; dispositif *m* pour enlever le jeu; couronne de rattrapage de jeu — **d** Beiläufer *m* — **n** spelingstandwiel *n* — **e** brida *f* de supresión de juego; brida *f* eliminadora *f* de holguras — **i** corona *f* dentata registrabile per l'eliminazione del gioco

1031 back lighting — Illuminating a subject to be

photographed from behind so that the subject stands out against the background.

f contre-jour *m* — **d** Gegenlicht *n* — **n** tegenlicht *n* — **e** iluminación *f* a contraluz — **i** controluce *f*

1032 back lining — See back-lining paper.

1033 back lining — See hollows.

1034 back-lining paper; back lining; book-backing paper; book-back liner — Paper (or other material) to reinforce the back of sewed books.

f papier *m* garniture; garniture *f* — **d** Rückenbeklebepapier *n*; Hinterklebepapier *n* — **n** rugoverlijmpapier *n*; overlijmpapier *n* — **e** lomera *f*; papel *m* para contralomo; papel *m* de forro — **i** carta *f* di rinforzo sul dorso

1035 back margin; inner margin; back — Inside margin of a book or that nearest to the binding edge.

f marge *f* intérieure; marge *f* de fond; marge *f* de petit fond — **d** innerer Seitenrand *m*; innerer Papierrand *m*; Bund *m* — **n** rugmarge *fm* — **e** margen *fm* medianil; margen *fm* del lomo; margen *fm* de cosido; margen *fm* de fondo; margen *fm* del crucero; margen *fm* interior; margen *fm* lateral interior — **i** margine *m* di dorso; margine *m* di cucitura; margine *m* di piega

1036 back marks — See collating marks.

1037 back matter — See end matter.

1038 back number; back issue; back copy

f vieux numéro *m*; exemplaire *m* périmé — **d** alte Nummer *f*; alte Ausgabe *f* — **n** oud nummer *n*; vorig nummer *n* — **e** ejemplar *m* atrasado; número *m* atrasado; ejemplar *m* no reciente; número *m* no reciente — **i** copia *f* arretrata; numero *m* arretrato; fascicolo *m* arretrato

1039 back nut — See check nut.

1040 back page — Left-hand page of a book.

f page *f* verso — **d** Rückseite *f* — **n** pagina *fm* op de achterzijde — **e** dorso *m* de página — **i** pagina *f* al verso; verso *m*

1041 back page; back-page companion — Compositor's term for the companion working behind him; whence also *side page and front page.

f compagnon *m* de la même rangée — **d** Gassengespann *m* der einem den Rücken zuwendet; Arschgespann *m* *sl.* — **n** achterbuurman *m* — **e** compañero *m* de la misma calle — **i** compagno *m* della stessa fila di casse

1042 back-page companion — See back page.

1043 back planing — Finishing operation to the back of an electrotype after the copper shell has been filled with backing metal.

f dressage *m* — **d** Rückseitenbearbeitung *f*; Planschleifen *n* der Rückseite — **n** justeren *n* — **e** rectificado *m* de planitud del dorso — **i** rettifica *f* della forma elettrotipica

1044 back pressure — Squeeze pressure between the blanket (offset) cylinder and the impression cylinder. Sometimes called *impression pressure.

f pression *f* d'impression — **d** Gegendruck *m* — **n**

tegendruk *m* — **e** contrapresión *f* — **i** contropressione *f*

1045 back printing; reverse printing — Printing on a transparent film so that the printing is on the inside of the package and can be observed through the film. Permits a higher gloss package because no printing is on the outside, but usually places the printing in contact with the contents. Should not be used for food packages.

f impression *f* en transparence; impression *f* cello-émail — **d** Bedrucken *n* transparenten Materials von der Rückseite — **n** drukken *n* op de achterzijde van een transparante film; drukken *n* aan de binnenkant — **e** imprimir al revés — **i** stampa *f* in trasparenza; stampa *f* di materiali trasparenti sul lato rovescio; imprimar al rovescio

1046 back rounding — Forming the back of a book into a convex shape.

f arrondissure *m* du dos — **d** Rückenrunden *n*; Rückenrundklopfen *n* — **n** rondzetten *n* van de rug; rondkloppen *n* van de rug — **e** redondeamiento *m* del lomo — **i** arrotondamento *m* del dorso

1047 back rounding machine; backer

f machine *f* à arrondir le dos; arrondisseuse *f* — **d** Rückenrundemaschine *f*; Bücherrücken-Rundmaschine *f* — **n** rondzetmachine *f* — **e** máquina *f* de redondear — **i** macchina *f* arrotondadorsi

1048 back router

f planeuse *f*; rectifieuse *f* — **d** Rückenfräser *m* — **n** freesmachine *f* — **e** calibradora *f*; máquina *f* calibradora; rectificadora *f* — **i** rettificatrice *f*; fresatrice *f*

1049 back tender

f aide-conducteur *m* de machine à papier — **d** Papiermaschinengehilfe *m* — **n** hulp *m* aan de papiermachine — **e** ayudante *m* de conductor de la máquina de papel — **i** aiuto *m* conduttore di una macchina continua; aiutante *m* di macchina continua

1050 back title — Name of a book stamped on the back of the cover. It reads either across the short dimension of the back or from top to bottom, so that it reads normally when the book is laying face up on a table.

f titre *m* de dos — **d** Rückentitel *m* — **n** rugtitel *m* — **e** título *m* del lomo; tejuelo *m* — **i** titolo *m* del dorso; titolo *m* sul dorso

1051 back transfer — Increase in density of halftone dots printed on the first unit by additional ink picked up from the blanket on the second unit.

d Rückübertragung *f* von Druckfarbe — **e** reporte *m* de la mantilla — **i** trasporto *m* residuo

1052 back v up — To print the reverse side of a sheet when the first side has been completed.

f imprimer au verso — **d** widerdrucken — **n** weerdrukken; weerdruk aanbrengen — **e** retirar; imprimir el retiro — **i** stampare in volta

1053 back v up — To back up a stereotype mat with strips of backing felt.

f habiller; garnir — **d** auslegen — **n** opplakken;

achterplakken — **e** calzar la matriz; cartonear *Pe* — **i** taccheggiare un flano

1054 back-up computer — Computer used in cases of defects of the main computer.
f ordinateur *m* de réserve; ordinateur *m* de secours — **d** Reservesatzrechner *m*; Not-Rechner *m* — **n** reservecomputer *m* — **e** computadora *f* de reserva; computadora *f* de emergencia — **i** elaboratore *m* di riserva; elaboratore *m* d'emergenza

1055 back-up impression cylinder — Steel roller on a rotogravure machine that presses against the rubber impression roller, which in turn presses the paper against the printing cylinder.
f presseur *m*; rouleau *m* presseur — **d** Presseur *m*; Stahlpresseur *m*; Presseurwalze *f*; Stützwalze *f*; Stützpresseur *m* — **n** staalpresseur *m*; contrapresseur *m*; tegenpresseur *m*; steunpresseur *m* — **e** cilindro *m* de contrapresión — **i** cilindro *m* di contropressione

1056 back water — See white water.

1057 bad break — Incorrect division of a word.
f division *f* incorrecte — **d** falsche Worttrennung *f*; falsche Trennung *f*; falsche Abteilung *f* — **n** onjuiste woordafbreking *f*; onjuiste afbreking *f* — **e** división *f* incorrecta de palabra — **i** divisione *f* scorretta di parole

1058 bad break — See wrong overturn.

1059 bad copy — Copy that is badly written or where the sense is difficult to make out. Compositors were usually paid extra for the loss of time they incurred in deciphering it. See also fair copy.
f copie *f* défectueuse — **d** schlechtes Manuskript *n*; unleserliches Manuskript *n*; Geschmier *n fam.* — **n** slechte kopij *f*; slecht leesbare kopij *f*; geknoei *n fam.* — **e** manuscrito *m* de difícil lectura; original *m* deficiente; original *m* sucio; original *m* malo — **i** manoscritto *m* cattivo

1060 bad letter — See batter.

1061 badly legible copy — Copy trying the eyes.
f copie *f* difficile à déchiffrer — **d** schlecht lesbares Manuskript *n*; Augenpulver *n* — **n** slecht leesbare kopij *f*; slecht leesbaar manuscript *n* — **e** manuscrito *m* de difícil lectura; original *m* malo — **i** copia *f* malamente leggibile

1062 badly shaped reel; misshapen reel — Not truly cylindrical reel of paper.
f bobine *f* déformée — **d** deformierte Rolle *f*; eingedrückte Rolle *f* — **n** vervormde rol *fm*; ingedeukte rol *fm*; niet-rond zijnde rol *fm*; slechte rol *fm* — **e** bobina *f* deformada; bobina *f* aplastada — **i** bobina *f* deformata

1063 bad press — Bad review given to a book.
f mauvaise presse *f* — **d** schlechte Presse *f* — **n** slechte pers *f* — **e** mala prensa *f* — **i** cattiva recensione *f*; cattiva critica *f*

1064 baffle plates — Plates behind paper web in dryer section of a rotogravure press.
f déflecteurs *mpl*; persiennes *fm*; chicanes *fpl* — **d** Prallplatten *fpl*; Trockenhauben *fpl* — **n** prangplaten *fmpl* — **e** placas *fpl* de desviación; placas *fpl* deflectoras; deflectores *mpl*; persianas *fpl* (de cierre) — **i** lamiere *fpl* deflettrici

1065 Bafour system — See BBR system.

1066 bag — Preformed container with an opening that can be sealed after filling. Made of flexible material such as paper, metal foil, cellulose and plastic films, textiles, etc., any of which may be coated, laminated or treated in other ways to provide the property required for the packaging. The four basic standard styles of bags are: 1. *automatic self-opening bag; 2. satchel-bottom bag; 3. *flat bag; 4. square bag. The five basic standard types are: 1. grocery bag; 2. merchandise paper; 3. industrial sack; 4. textile sack; 5. paper shipping sack. See also sack.
f sac *m* en papier; cornet *m* de papier — **d** Papiersack *m*; Beutel *m*; Papierbeutel *m*; Papiertüte *f*; Tüte *f* — **n** papieren zak *m*; winkelzak *m*; zak *m*; puntzak *m* — **e** bolsa *f*; cucurucho *m*; cartucho *m*; saco *m* — **i** sacchetto *m*

1067 bagasse; sugar cane bagasse; sugar cane pulp — Crushed stalks of sugar cane from which the sugar has been extracted, used in papermaking.
f bagasse *f* — **d** Bagasse *f* — **n** bagasse *m*; ampas *m* — **e** bagazo *m* — **i** bagassa *f*

1068 bag cap — A UK size for wrapping paper. See app. no. 3.

1069 baggage tag — Strong tag with eyelet, for identification of personal baggage, made from manila tag board, tough check, etc.
f étiquette *f* pour bagage — **d** Anhängeetikett *n*; Anhängezettel *m*; Gepäckanhänger *m* — **n** label *m*; bagagelabel *m* — **e** etiqueta *f* para equipaje; etiqueta *f* colgante; etiqueta *f* de envío — **i** cartellino *m* per bagaglio; etichetta *f* per bagaglio

1070 baggy reel — Result of a non-uniform drawing web, where certain slack areas of the web are not under as much tension as the remaining areas.
f bobine *f* molle — **d** schlecht, ungleichmäßige oder zu lose gewickelte Papierrolle *f*; Rolle *f* mit Wassersäcken — **n** zakvormige rol *fm* — **i** rotolo *m* bombato

1071 bag liner; bag lining — Paper used within bags for protection of contents.
f doublure *f* pour sacs; papier *m* à doubler les sacs — **d** Beutelfutter *n* — **n** zakvoering *f*; papier *n* voor zakkenvoering — **e** papel *m* para forros de sacos — **i** carta *f* per foderare di sacchetti

1072 bag lining — See bag liner.

1073 bag machine; bag making machine — Machine on which web paper is cut and pasted into bags. On some machines the bags are also printed.
f machine *f* à fabriquer des sacs — **d** Tüten- und Beutelmaschine *f* — **n** zakkenmachine *f* — **e** máquina *f* para fabricar bolsas — **i** macchina *f* per fabbricare sacchetti

1074 bag making machine — See bag machine.

1075 bag paper — Paper used for making bags.

f papier *m* à sacs; papier *m* à sachets; papier *m* pour sachets; papier *m* pour cornets — **d** Tütenpapier *n*; Beutelpapier *n*; Sackpapier *n* — **n** zakkenpapier *n* — **e** papel *m* para bolsas; papel *m* para sacos — **i** carta *f* per sacchi; carta *f* per sacchetti

1076 bag sealer — See bag sealing machine.

1077 bag sealing machine; bag sealer
f machine *f* à fermer les sacs — **d** Beutelverschließmaschine *f* — **n** zakkensluitmachine *f* — **e** máquina *f* de cerrar bolsas — **i** macchina *f* per la chiusura di sacchi

1078 bails; tympan bails; tympan clamps; bales — The hinged metal bars that hold the tympan sheets against the platen of a job press.
f tringles *fpl* de tension de l'habillage — **d** Aufzugstäbe *mpl*; Aufzugstängchen *npl* — **n** leggerklemmen *fmpl* — **e** abrazaderas *fpl* del tímpano — **i** nastri *mpl* metallici per fissare il timpano

1079 bain marie — See water bath.

1080 baked; caked — Said of type when it has been rinsed and laid up for distribution and adheres so closely together that it can only be separated with difficulty. New type is particularly liable to this trouble. Saturation with soft soap and water before new type is laid in case will prevent this. It will also diminish the extreme brightness of new type which may strain the compositor's eyes.
f collé — **d** zusammengebacken — **n** aaneengekoekt — **e** pegado — **i** aderente

1081 baked types — See caked types.

1082 bakelite — Trade name of a group of thermosetting plastics formed by the reaction of phenol, formaldehyde and their derivatives, having a wide range of properties and uses. Named after the Belgian chemist Leo H. Baekeland (1863-1944).
f bakélite *f* — **d** Bakelit *n* — **n** bakeliet *n* — **e** baquelita *f* — **i** bachelite *f*

1083 baking soda — See sodium bicarbonate.

1084 baking varnish
f vernis *m* émail — **d** Heißemail-Kopierschicht *f*; Heißemail-Schicht *f* — **n** warmemail *n* — **e** cola *f* esmalte; esmalte *m* de pasado — **i** vernice *f* smalto

1085 balanced — Said of rotating parts which are in equilibrium when stationary and rotating (in the latter case "dynamic balance"). Such parts are free from vibrations.
f équilibré — **d** ausgewuchtet — **n** in balans; uitgebalanceerd — **e** equilibrado — **i** equilibrato; bilanciato

1086 balance of motions — Simultaneous and identical motions for left and right hands.
f équilibre *m* des mouvements — **n** gelijke en gelijktijdige beweging *f* — **e** movimientos *mpl* equilibrados — **i** equilibrio *m* dei movimenti

1087 balance time; unoccupied (cycle) time — Period during controlled machine time when an operator is neither working nor taking authorized rest.
e tiempo *m* muerto — **i** tempo *m* disoccupato

1088 balance weight; counter weight
f contre-poids *m*; masse *f* d'équilibrage — **d** Ausgleichsgewicht *n*; Balancegewicht *n*; Gegengewicht *n* — **n** contragewicht *n*; tegengewicht *n* — **e** contrapeso *m*; peso *m* balancín; peso *m* de quilibrado — **i** contrappeso *m*; peso *m* di bilanciamento

1089 balcony type press — Reel-fed rotary press built in such a way that the reel standards are near ground level and the printing units placed above, so that these are reached by a sort of balcony.
f rotative *f* à balcon — **d** Druckmaschine *f* in Etagenbauweise; Etagenmaschine *f* — **n** pers *f* volgens etagebouw — **i** rotativa *f* a balconata

1090 bale *obs.* — A packing of ten reams of paper.
f balle *f*; ballot *m* — **d** Ballen *m* — **n** baal *fm* — **e** bala *f*; fardo *m*; paca *f* — **i** balla *f*

1091 bales — See bails.

1092 baling board
f plateau *m* de balle; plateau *m* d'emballage (pour papier); planche *f* — **d** Ballenbrett *n* (zur Aufnahme von Papier) — **n** schot *n* (van een papierbaal) — **e** tabla *f* de embalaje (de los fardos) — **i** tavola *f* per balle (di carta)

1093 baling press
f presse *f* à ballots — **d** Ballenpackpresse *f*; Bündelpresse *f* — **n** balenpers *f* — **e** prensa *f* embaladora; prensa *f* de embalar; embaladora *f*; enfardadora *f*; máquina *f* embaladora; máquina *f* para embalar; prensa *f* de hacer balas — **i** pressa *f* per balle di cartaccia; pressa *f* imballatrice di refili

1094 ball-and-socket joint; ball joint *US*; **swivel joint** *GB* — Joint in which a ball moves within a socket to allow rotary motion in every direction within certain limits.
f joint *m* à rotule; genouillère *f*; articulation *f* à genouillère — **d** Kugelgelenk *n* — **n** kogelgewricht *n* — **e** junta *f* de rótula; rótula *f*; articulación *f* esférica; cabeza *f* giratoria — **i** giunto *m* sferico

1095 Ballard process — Process to recondition used intaglio printing cylinders for new etching. A thin copper shell of 0.1-0.25 mm (0.004-0.01 in) is electrodeposited over an insulating silver layer onto a copper cylinder. The skin is sufficiently adherent to the base to stay out during etching and printing but can afterwards be stripped off mechanically by peeling, leaving the base copper ready for repeat of the procedure. This method results in printing cylinders of constant diameters.
f procédé *m* Ballard; procédé *m* de cuivrage brillant — **d** Ballard-Verfahren *n*; Glanzkupferverfahren *n*; Ballard-Aufkupferungsverfahren *n*; Hautaufkupferungsverfahren *n* — **n** Ballardprocédé *n*; glans-koperprocédé *n* — **e** procedimiento *m* Ballard; procedimiento *m* al cobre brillante — **i** procedimento *m* Ballard; processo *m* Ballard; ramatura *f* Ballard

1096 ball bearing — Low friction bearing consisting of an inner ring on the shaft and an outer ring in the bearing housing between which steel balls are used to reduce friction.
f roulement *m* à billes; coussinet *m* à billes; palier *m* à billes — **d** Kugellager *m* — **n** kogellager *n* — **e** cojinete *m* de bolas; cojinete *m* a bobillas; corredera *f*; munición *f* SA; rodamiento *m* de bolas; rodamiento *m* a bolas — **i** cuscinetto *m* a sfere

1097 ball graining machine — See graining machine.

1098 ball, ink — See ink ball.

1099 ball joint — See ball-and-socket joint.

1100 ball mill — Dispersion equipment used in the production of certain printing inks. It consists of a rotating cylinder containing balls, which cascade and disperse the pigment by impact and attrition as the cylinder revolves.
f broyeuse *f* à billes; alsing *m* — **d** Kugelmühle *f* — **n** kogelinktmolen *m*; inktmolen *m*; kogelmolen *m* — **e** molino *m* de bolas — **i** mulino *m* a palle

1101 balloon former — Former in a rotary press mounted above other formers from which the folded webs are gathered to make up the sections in multi-sectioned newspapers or magazines. See also length fold collection.
f triangle *m* supérieur — **d** Obertrichter *m*; Doppeltrichter *m*; Ballonformer *m* — **n** ballontrechter *m* — **e** horma *f* superior; formador *m* superior; triángulo *m* superior — **i** cono *m* di piega superiore

1102 balloon linen
f toile-ballon *f* — **d** Ballonleinen *n* — **n** ballonlinnen *n* — **e** tela *f* de balón

1103 ball pen — See ball-point pen.

1104 ball-point pen; ball pen — Pen in which the point is a fine ball bearing.
f stylo *m* à bille — **d** Kugelschreiber *m* — **n** balpen *fm*; ballpoint(pen) *fm* — **e** bolígrafo *m* — **i** penna *f* a sfera

1105 ball rack — Arrangement to hang out paper for conditioning. It consists of strips of wood with slanting slots in which loose glass or metal balls are retained for automatic clamping of the inserted sheets.
f pendrier *m* à papier — **d** Trockenleiste *f*; Kugeltrockenleiste *f*; Papieraushängevorrichtung *f* mit Bogenhalterung durch Glaskugeldruck — **n** papier-uithanginrichting *f* — **e** tablero *m* de bolas — **i** dispositivo *m* a pinze per sospendere la carta; sbarre *f* a pinze per sospendere la carta

1106 ball stick — See ink ball.

1107 balsam fir — Resinous tree used in the production of wood pulp.
f sapin *m* baumier — **d** Balsamholz *n* — **n** balsahout *n* — **e** madera *f* de pino melis — **i** legno *m* di balsamo; balsamo *m*

1108 balsam of fir — See Canada balsam.

1109 bamboo — Arborescent grass, one of the early sources of papermaking fibres.
f bambou *m* — **d** Bambus *m* — **n** bamboe *mn* — **e** bambú *m*; bambuc *m* — **i** bambù *m*

1110 band — Group of recording tracks as on a magnetic storage drum.
f bande *f*; zone *f* — **d** Spurgruppe *f* — **n** sporengroep *fm* — **e** banda *f* — **i** banda *f*; gruppo *m* di piste

1111 band driver — See band nippers.

1112 band nippers; band driver — Tool used by bookbinders for squaring and shaping after covering.
f pince-nerfs *f*; pince *f* à nerfs; pince *f* de relieur — **d** Bundzange *f*; Bündezange *f* — **n** ribbentang *fm* — **e** alicates *mpl* de boca chata (para ceñir los nervios) — **i** pinze *fpl* da rilegatore

1113 bands — Strings on which the sheets of a volume are sewn. If these project from the back of the book they are called *raised bands.
f accolures *fpl* — **d** Heftschnüre *fpl*; Bünde *mpl*; Heftbünde *mpl* (am Kopf und Fuß: Fitzbund) — **n** touwen *npl* — **e** cordones *mpl*; cintas *fpl* — **i** nastri *mpl*; fasce *fpl*

1114 band saw — Machine with a continuous saw in the form of a narrow circulating blade to cut mounted printing plates accurately. See also circular saw, saw trimmer, jig saw.
f scie *f* à ruban; scie *f* sans fin — **d** Bandsäge *f* — **n** lintzaag *fm* — **e** sierra *f* de cinta; sierra *f* sin fin — **i** sega *f* a nastro

1115 B & W — See diazoprint.

1116 band width — The range of frequencies within the limits of the band.
f largeur *f* de bande — **d** Bandbreite *f* — **n** bandbreedte *f* — **e** ancho *m* de banda — **i** larghezza *f* di banda

1117 bang — See screamer.

1118 bang — See screamer.

1119 banjo — The end adjusting duct screw of a rotary press.
f banjo *m*; réglage *m* à distance des vis d'encrier; commande *f* de vis micrométriques — **d** Farbstellpult *m*; Farbregulierungspult *m*; Farbeinstellpult *m* — **n** inktregeltafel *fm* — **e** pupitre *m* de mando a distancia del entintado — **i** comando *m* a vite micrometrica

1120 bank — Table usually about 122 cm (4 ft) high on which to lay paper before or after printing.
f table *f* à papier — **d** Papiertisch *m*; Ablegetisch *m*; Tisch *m* — **n** papiertafel *fm*; tafel *fm* — **e** tablero *m* apilador del papel — **i** banco *m* da carta

1121 bank — Line of a newspaper headline. A headline consists of two, three or more banks.
f ligne *f* de titre — **d** Zeitungsüberschriftzeile *f* — **n** kopregel *m* — **i** riga *f* di un titolo di giornale

1122 bank — See dump.

1123 bank envelope — See banker envelope.

1124 banker — See banker envelope.

1125 banker envelope; bank envelope; banker — Envelope with opening and flap on the side with the greater dimension.

f enveloppe f américaine — d Umschlag m mit Klappe an der längeren Seite — n cabinet-envelop(pe) f — e sobre m banquero; sobre m apaisado; sobre m americano — i busta f all'americana

1126 bank-note — Promissory note issued by a bank payable to bearer on demand without interest and acceptable as money.
f billet m de banque — d Banknote f — n bank-biljet n — e billete m de banco; billete m bancario — i banconota f; biglietto m di banca

1127 bank-note paper — See currency paper.

1128 bank-note printing — See also security printing.
f impression f de billets de banque — d Banknotendruck m; Papiergelddruck m — n bank-biljettendruk m — e impresión f de billetes de banco; impresión f de papel moneda — i stampa f di banconote; stampa f di biglietti di banca

1129 bank paper — High-grade writing paper made of rag or chemical wood pulp with bond characteristics and durability.
f papier m (d'écriture) coquille; papier m machine à écrire; papier m poste; coquille f anglaise — d Bankpostpapier n — n bankpostpapier n; bank-post n — e papel m de escribir (apergaminado) — i carta f da scrivere di buona qualità

1130 banks — See thumb index.

1131 banner; banner line; banner head; banner headline — Large head extending across a newspaper page.
f titre m en caractères d'affiches; titre m à gros caractères — d Balkenüberschrift f — n grote kopregel m; grote kop m — e título m a toda plana; titular m a toda página; título m a todo lo ancho de la plana; cabecera f a todo plana; título m a toda página; título m en bandera CA; titular m en bandera CA; título m a todo despliegue CA; titular m a todo despliegue CA; cabeza f en bandera CA; cabeza f a todo despliegue CA; cintillo m Cu, Gu — i titolone m; titolo m su tutta la larghezza della pagina

1132 banner head — See banner.

1133 banner headline — See banner.

1134 banner line — See banner.

1135 bar; closed arm — The closed horizontal stroke in the letters A, H, E and e.
f barre f (de lettre); trait m latéral — d Querstrich m; Querbalken m; Balken m — n dwarsstreepje n — e línea f transversal (de la letra); travesaño m; trazo m horizontal (de la letra) — i tratto m traversale; asta f trasversale

1136 bar; devil's tail coll., US — Handle on handpresses which operates the press.
f barreau m — d Preßbengel m; Preßschwengel m; Bengel m der Handpresse — n perszwengel m; zwengel m — e barra f (del husillo de la prensa) — i leva f di torchio; leva f di pressione; braccio m di torchio; leva f a mano

1137 bar — See cross bar.

1138 bar — See vinculum.

1139 barge — Small wooden case with six or eight divisions to hold spaces when correcting type matter; also with ten divisions for time-table correcting.
f casseau m à espaces; casseau m pour espaces; casse f d'espaces — d Ausschlußkasten m; Materialkasten m — n spatiekast fm; witkast fm — e caja f blancos; caja f para blancos; cajuela f blancos; cajuela f para blancos — i cassettino m per spaziatura; cassa f per la spaziatura

1140 barged case — Type case in which some of the boxes are overflowing.
f casse f trop pleine — d übervoller Schriftkasten m; ungleich gefüllter Schriftkasten m — n overlopende kast fm; te volle kast fm; overvolle kast fm — e caja f demasiado llena — i cassa f troppo piena

1141 barite — See barytes.

1142 barium — Silver-white, slightly lustrous, somewhat malleable metal. Soluble in acids, decomposes in water. TL-value 0.5 mg/m^3. Symbol: Ba.
f baryum m — d Barium n — n barium n — e bario m — i bario m

1143 barium chromate; baryta yellow; ultramarine yellow — Heavy, yellow, crystalline powder, soluble in acids, insoluble in water. Poisonous. Formula: $BaCrO_4$.
f chromate m de baryum; jaune m d'outremer; jaune m citron; jaune m de baryte — d Bariumchromat n; Ultramaringelb n; Barytgelb n — n bariumchromaat n — e cromato m de bario; cromato m bárico; amarillo m de bario; amarillo m bárico — i cromato m di bario; cromato m barico; giallo m di barite

1144 barium dioxide — See barium peroxide.

1145 barium hydrate — See barium hydroxide.

1146 barium hydroxide; barium hydrate; barium octahydrate; caustic baryta — White powder or crystals. Poisonous. Must be kept well stoppered since it absorbs carbon dioxide from air. Soluble in water, alcohol and ether. Formula: $Ba(OH)_2$.
f baryte f (hydratée); hydrate m de baryum; baryte f caustique — d Bariumhydroxyd n; Bariumhydrat n; Bariumoxydhydrat n; Baryt-hydrat n; Ätzbaryt n — n bariumhydroxyde n — e hidróxido m de bario; hidróxido m bárico — i idrossido m di bario; idrossido m barico

1147 barium hyposulphite — See barium thiosulphate.

1148 barium monosulphide — See barium sulphide.

1149 barium monoxide — See barium oxide.

1150 barium nitrate — Lustrous, white crystals, soluble in water, insoluble in alcohol. Formula: $Ba(NO_3)_2$.
f nitrate m de baryum — d Bariumnitrat n — n bariumnitraat n — e nitrato m de bario; nitrato m bárico — i nitrato m di bario; nitrato m barico

1151 barium octahydrate — See barium hydroxide.

1152 barium oxide; barium monoxide; barium protoxide; calcined baryta — White to yellow-white powder. Absorbs carbon dioxide readily from air. Poisonous. Soluble in acids, reacts violently with water to form the hydroxide. Formula: BaO.

f oxyde *m* de baryum; baryte *f* caustique; baryte *f* anhydre — **d** Bariumoxyd *n* — **n** bariumoxyde *n* — **e** óxido *m* de bario; óxido *m* bárico — **i** ossido *m* di bario; ossido *m* barico

1153 barium peroxide; barium dioxide; barium superoxide — Greyish-white powder. Poisonous. Decomposes in water. Formula: BaO_2.

f peroxyde *m* de baryum; bioxyde *m* de baryum; oxylithe *m* — **d** Bariumperoxyd *n*; Bariumsuperoxyd *n* — **n** bariumperoxyde *n* — **e** peróxido *m* de bario; peróxido *m* bárico — **i** perossido *m* di bario; perossido *m* barico

1154 barium protoxide — See barium oxide.

1155 barium rhodanide — See barium thiocyanate.

1156 barium sulphate; barium white — Chemical compound obtained either from the natural mineral *barytes or by chemical reaction, used as a filler in the manufacture of paper or as an extender in printing ink in combination with other pigments. The artificial products are called blanc fixe, fast white, and permanent white. Formula: $BaSO_4$. Latin: sulfas baryticus.

f sulfate *m* de baryum; blanc *m* de baryte précipité sec; blanc fixe *f*; blanc *m* permanent — **d** Bariumsulfat *n*; Barytweiß *n*; Blanc fixe *m* — **n** bariumsulfaat *n*; barietwit *n*; blanc fixe *m*; permanent-wit *n* — **e** sulfato *m* de bario (artificial); blanco *m* fijo; blanco *m* permanente — **i** solfato *m* di bario; solfato *m* barico; bianco *m* fisso

1157 barium sulphide; barium monosulphide — Yellowish-green or grey powder or lumps. Poisonous. Soluble in water, decomposes to the hydrosulphide. Formula: BaS.

f sulfure *m* de baryum — **d** Bariumsulfid *n* — **n** bariumsulfide *n* — **e** sulfuro *m* de bario; sulfuro *m* bárico — **i** solfuro *m* di bario; solfuro *m* barico

1158 barium sulphite — White powder, decomposed by heat. Soluble in hydrochloric acid (dilute), insoluble in water. Formula: $BaSO_3$.

f sulfite *m* de baryum — **d** Bariumsulfit *n* — **n** bariumsulfiet *n* — **e** sulfito *m* de bario; sulfito *m* bárico — **i** solfito *m* di bario; solfito *m* barico

1159 barium sulphocyanide — See barium thiocyanate.

1160 barium superoxide — See barium peroxide.

1161 barium thiocyanate; barium sulphocyanide; barium rhodanide — White crystals, soluble in water and alcohol. Deliquescent. Poisonous. Used in photography. Formula: $Ba(SCN)_2$.

f sulfocyanure *m* de baryum — **d** Bariumsulfozyanid *n*; Bariumrhodanid *n* — **n** bariumrhodanide *n* — **e** sulfocianuro *m* de bario; sulfocianuro *m* bárico — **i** solfocianuro *m* di bario; solfocianuro *m* barico

1162 barium thiosulphate; barium hyposulphite — White, crystalline powder. Poisonous. Slightly soluble in water, insoluble in alcohol. Formula: BaS_2O_3.

f thiosulfate *m* de baryum; hyposulfite *m* de baryum — **d** Bariumthiosulfat *n*; Bariumhyposulfit *n* — **n** bariumthiosulfaat *n* — **e** tiosulfato *m* de bario; tiosulfato *m* bárico — **i** tiosolfato *m* di bario; tiosolfato *m* barico

1163 barium white — See barium sulphate.

1164 barker; barking machine — Machine to remove the bark from pulpwood logs.

f écorceuse *f* — **d** Entrindungsmaschine *f*; Rindenschälmaschine *f*; Schälmaschine *f*; Schäler *m* — **n** schilmachine *f*; schiller *m* — **e** descortezadora *f*; máquina *f* de descortezar — **i** scortecciatrice *f*

1165 barking — Removing bark from pulpwood before processing by means of a knife (disk), drum, abrasion, hydraulic barker, or by chemical means.

f écorçage *m* — **d** Entrindung *f*; Entrinden *n* — **n** schillen *n*; ontschorsen *n* — **e** descortezado *m* — **i** scortecciatura *f*

1166 barking drum — Large revolving cylinder in which pulpwood logs are thrown against each other, the bark being loosened and broken by repeated impacts.

f tambour *m* écorceur — **d** Entrindungstrommel *f* — **n** schiltrommel *fm* — **e** tambor *m* descortezador; bombo *m* descortezador — **i** tamburo *m* scortecciatore

1167 barking machine — See barker.

1168 baronet — A British size of cards. See app. no. 3.

1169 bar printer — Printer in which the printing element consists of one or more bars with a complete set of types.

f imprimante *f* à barre — **d** Stabdrucker *m* — **n** stang-afdrukeenheid *f*; afdrukmachine *f* met drukstang — **e** impresora *f* de barras — **i** stampatrice *f* a barra

1170 barrel; mount of a lens; focusing mount — Mechanical structure in which the lens of a camera is mounted.

f monture *f* de l'objectif — **d** Objektivfassung *f*; Linsenfassung *f* — **n** lensvatting *f*; vatting *f* van de lens; montuur *n* van de lens — **e** montura *f* de un objetivo; barrilete *m* del objetivo; tubo *m* del objetivo — **i** montatura *f* dell'obiettivo; tubo *m* dell'obiettivo

1171 barrel distortion — Optical distortion where the images of straight lines are curved lines concave to the centre of the plate or film.

f distorsion *f* en barillet; distorsion *f* négative — **d** tonnenförmige Verzeichnung *f* — **n** tonvormige vertekening *f* — **e** distorsión *f* esférica; distorsión *f* de convexidad; distorsión *f* de barrel — **i** distorsione *f* a botte; distorsione *f* con incurvamento convesso

1172 barrel printer — See drum printer.

1173 barrier layer photocell; photo-voltaic cell — Photoelectric cell in which the light causes the passage of electrons across the surface of contact between a conductor and semi-conductor.

f cellule *f* photoélectrique à couche d'arret; cellule *f* photovoltaique — **d** Sperrschichtfotozelle *f*; Photowiderstandszelle *f* — **n** sperlaagfotocel *f*; grenslaagfotocel *f*; keerlaagfotocel *f* — **e** célula *f* fotoeléctrica de bloqueo; célula *f* fotoeléctrica de detención — **i** fotocellula *f* a strato di sbarramento

1174 barring lever and hub — Manual device to turn the rotary press.
f levier *m* pour la commande à la main — **d** Handantrieb *m* — **n** handaandrijfinrichting *f* — **e** accionamiento *m* manual; accionamiento *m* a mano — **i** comando *m* a mano

1175 barring motor — See inching motor.

1176 bar roll; fillet roll; gilding roll; roll; roulette — Little roll or wheel used by gilders to press down the gold.
f roulette *f* — **d** Vergolderolle *f*; Rolle *f*; Rolleisen *n*; Räderstempel *n*; Roulett *n*; Linienrulette *f* — **n** rolfilet *mn* — **e** rueda *f* de dorar; rueda *f* fileteadora; ruleta *f* — **i** rullino *m* per dorare

1177 bar viscometer — Viscometer consisting of a cylindrical rod falling through a slightly wider cylindrical annulus in which the liquid is contained.
f viscosimètre *m* à tige — **d** Stabviskosimeter *n* — **n** staafviscosimeter *m* — **e** viscosímetro *m* de varilla; viscosímetro *m* a barra — **i** viscosimetro *m* a barra

1178 baryta board — Similar to *baryta paper, but of a heavier weight.
f carton *m* baryté — **d** Barytkarton *m* — **n** barietkarton *n* — **e** cartulina *f* baritada — **i** cartone *m* baritato

1179 baryta paper — Paper coated with barium sulphate to obtain a smooth high-reflectance finish for subsequent coatings, frequently used as base for photographic emulsions.
f papier *m* baryté; papier *m* à sulfate de baryte — **d** Barytpapier *n*; Kreidepapier *n* — **n** barietpapier *n* — **e** papel *m* de barita; papel *m* baritado — **i** carta *f* baritata; carta *f* alla barite

1180 baryta white — See barytes.

1181 baryta yellow — See barium chromate.

1182 barytes; barite; baryta white; heavy spar — Natural *barium sulphate used as an ink pigment and a white extender and as a filler in papermaking and in coating of paper. It is considerably more abrasive and gritty than synthetic barium sulphate.
f barytine *f*; sulfate *m* de baryte — **d** Barytweiß *n*; Baryt *n*; Schwerspat *m* — **n** barietwit *n*; zwaarspaat *n* — **e** barita *f*; baritina *f*; espato *m* pesado — **i** baritina *f*

1183 basan; basil — Material of large sheepskins, mostly used in account-book bindings. Bark tanned and dyed.
f basane *f* — **d** Schafleder *n* — **n** bezaan *n*; bazaan *n*; bezaansleer *n*; bazaansleer *n* — **e** badana *f*; piel *f* de carnero — **i** bazzana *f*; cuoio *m* di pecora

1184 base — Compound formed by the combination of any positive ion, except H$^+$, with the negative OH$^-$ ion (hydroxide ion). Neutralizes acids, forming a salt and water, and turns litmus blue.
f base *f* — **d** Base *f* — **n** base *f* — **e** base *f* — **i** base *f*

1185 base; base ink — In ink manufacture, a dispersion containing usually only one colouring matter, pigment or dye, properly dispersed in a vehicle, used for mixing to produce the desired end product.
f encre *f* de base — **d** Grundfarbe *f*; Stammfarbe *f*; Konzentratfarbe *f* — **n** grondinkt *m* — **e** tinta *f* de base — **i** inchiostro *m* base; base *f* monopigmentata

1186 base — Foundation for further operation, as a data base.

1187 base — See deep-etch lacquer.

1188 base — See mount.

1189 base board for the protection of frozen foods
f carton *m* support pour la protection des aliments congelés et surgelés — **d** Gefrierrohkarton *n*; Tiefkühlrohkarton *n* — **n** diepvrieskarton *n* — **i** cartone *m* supporto per la protezione di cibi congelati

1190 base colour — Colour used as a background on which other colours are printed.
f couleur *f* de dessous — **d** Grundfarbe *f* — **n** grondkleur *fm*; ondergrondkleur *fm* — **e** color *m* de fondo; color *m* básico — **i** colore *m* di fondo

1191 base copper — Copper shell which is electrolytically deposited onto the gravure cylinder, in the *Ballard process.
f cuivre *m* de base — **d** Grundkupfer *n* — **n** grondkoper *n* — **e** cobre *m* de base — **i** rame *m* di base

1192 base cylinder — Rotogravure printing cylinder prior to copperplating, polishing and etching.
f cylindre *m* non cuivré — **d** Stahlzylinder *m* — **n** stalen cylinder *m* — **e** cilindro *m* de acero por cobrear — **i** anima *f* d'acciaio dei cilindri rotocalco

1193 base eight — Method of expressing an octal counting system, as opposed to a decimal system. One counts from 0 through 7; 10 through 17, etc., so that 10 to base 8 is the decimal number 9.

1194 base ink — See base.

1195 base lacquer — See deep-etch lacquer.

1196 base line — The lowest of the three imaginary framework lines (base line, cap line, mean line) on which letters are constructed. Some characters which seem to rest on top of or at the optical base line actually drop somewhat below it, especially such characters as "o" or "c". Leading or primary leading is the distance from one base line to the next. See also x-height, ascender, descender.
f ligne *f* de base — **d** Grundlinie *f*; Schriftlinie *f*; untere Schriftlinie *f*; Normalschriftlinie *f* — **n** onderlijn *fm*; voetlijn *fm*; grondlijn *fm*; letterlijn

fm — **e** línea *f* básica; alineación *f* inferior — **i** allineamento *m* inferiore dell'occhio mediano

1197 base oils — See blown oils.

1198 base paper — See body paper.

1199 basic anhydride — Metallic oxides are often called basic anhydrides, meaning a *base minus the water.

f anhydride *m* basique — **d** basisches Anhydrid *n* — **n** basisch anhydride *n* — **e** anhídrido *m* básico — **i** anidride *f* basica

1200 basic bismuth nitrate — See Spanish white.

1201 basic colours — See basic dyes.

1202 basic dyes; basic colours; basic dyestuffs — Organic colouring agents in which the dye-base is the basic constituent of a salt. Used for tinting paper stock in the beaters. Usually not as fast to light as acid dyes, but of brighter hue. Generally marketed as colourless hydrochlorides or colourless double salts with zinc chloride.

f colorants *mpl* basiques; teintures *fpl* basiques; matières *fpl* colorantes basiques — **d** basische Farbstoffe *mpl* — **n** basische kleurstoffen *fmpl* — **e** colorantes *mpl* básicos — **i** coloranti *mpl* basici

1203 basic dyestuffs — See basic dyes.

1204 basicity — The state of being a base. Also the power of an acid to react with bases, dependent on the number of replaceable hydrogen atoms of the acid.

f basicité *f* — **d** Basizität *f* — **n** basiciteit *f*; basiditeit *f* — **e** basicidad *f* — **i** basicità *f*

1205 basic lead carbonate — See flake white.

1206 basic size — Sheet size recognized by buyers and sellers as the one from which its basis weight is determined. Initially, it was that size which printed, folded, and trimmed most advantageously. See app. no. 3.

f format *m* de base — **d** Grundformat *n* — **n** basisformaat *n*; grondformaat *n* — **e** tamaño *m* básico; tamaño *m* de base — **i** formato *m* base

1207 basic weight — See basis weight.

1208 basil — See basan.

1209 basis weight; basic weight — The weight in pounds of a ream (usually 500, sometimes 480 sheets of paper cut to its *basic size). 1. The US Government Printing Office uses a unit of 1000 sheets. This unit also appears in certain trade customs. For cultural papers the basis weight of 500 sheets 43.1 x 55.8 cm (17 x 22 in) is sometimes called substance or substance number. 2. In most foreign countries and in some domestic test procedures the standard size is a square metre and the weight is expressed in grams per square metre. N.B. For paperboards basis weight is frequently expressed in pounds per 100 square feet. The standard ream for boxboard is 500 sheets measuring 63.5 x 101.6 cm (25 x 40 in) and is also used in reporting basis weight.

f 1. poids *m* par rame; 2. poids *m* au mètre carré; force *f* du papier; grammage *m* — **d** 1. Riesgewicht *n*; 2. Flächengewicht *n*; Quadratmetergewicht *n*; Grammgewicht *n* — **n** 1. riemgewicht *n*; gewicht *n* per riem; 2. gramgewicht *n*;

vierkante-metergewicht *n* — **e** 1. peso *m* por resma; peso *m* elemental (del papel); peso *m* básico (del papel); peso *m* de base; 2. gramaje *m* del papel; peso *m* en gramos; peso *m* por metro cuadrado — **i** 1. peso *m* di base; peso *m* di risma; 2. grammatura *f*; peso *m* al metro quadrato

1210 Bassani process — Method of highlight halftone photography involving slight rotation of the halftone screen during part of the exposure.

f procédé *m* Bassani — **d** Bassani-Verfahren *n*; Bassani-Prozess *m* — **n** Bassani-procédé *n* — **e** procedimiento *m* Bassani; método *m* Bassani — **i** procedimento *m* Bassani

1211 basswood — Broadleaf, hardwood tree used for the manufacture of soda pulp.

f tilleul *m* — **d** Lindenholz *n* — **n** lindenhout *n* — **e** madera *f* de tilo americano; tilo *m* — **i** legno *m* di tiglio

1212 bastarda — Type face used until 1500 by French printers, showing sloping, pointed characters. Not to be confused with *bastard fount.

f lettre *f* bâtarde — **d** Bastardschrift *f* — **n** bâtarde *f* — **e** letra *f* inglesa — **i** lettera *f* inglesa

1213 bastard body — See bastard fount.

1214 bastard colour separation — See fake-colour process.

1215 bastard font — See bastard fount.

1216 bastard fount *GB*; **bastard font** *US*; **bastard body; bastard type** — Type cast on another body than a standard body, e.g., an 8-point on a 9-point body. Not to be confused with the French "bâtarde", a type used by the copperplate engravers in the middle of the 17th century, which lies between the "ronde" and the "anglaise" (Spanish: letra inglesa; Italian: lettera inglesa) and which was the precursor of the modern *italics.

f fonte *f* bâtarde — **d** Bleisatzschrift *f* bei der Schriftgrad und Kegelgröße nicht übereinstimmen (Nicht zu verwechseln mit Bastardschrift) — **n** bastaardletter *fm*; bastaardcorps *n* — **e** caracteres *mpl* bastardos; fundición *f* bastarda — **i** caratteri *mpl* bastardi

1217 bastard title; bas title; fly title; mock title; half title *depr.*; **focus** *GB* — Brief title on a separate page preceding the full title or the dividing sections of a book. It should be placed three eights of the page depth from the top of the page.

f faux-titre *m*; avant-titre *m* — **d** Vortitel *m*; Schmutztitel *m* — **n** voordehandse titel *m*; Franse titel *m* — **e** anteportada *f*; anteporta *f*; portadilla *f*; media portada *f*; falsa portada *f* — **i** occhietto *m*; falso frontespizio *m*

1218 bastard type — See bastard fount.

1219 bast fibres — Fibres from the phloem or inner bark of a woody plant but also fibres found in other outer portions of the plant, such as cortex and pericycle. Botanically, the term is superfluous, since the terms phloem, fibre, cortical fibre, and pericyclic fibre are accurate and specific. *Flax, *hemp, *jute, manila hemp, *ramie, mitsumata, and gampi are typical bast-fibre plants.

f fibres *fpl* libériennes — **d** Bastfasern *fpl* — **n** bastvezels *fmpl* — **e** fibras *fpl* de líber — **i** fibre *fpl* liberiane

1220 bas title — See bastard title.

1221 batch — The quantity of material prepared or required for one operation.

f lot *m* — **d** Partie *f* — **n** fabricagepartij *fm*; partij *fm* — **e** lote *m* — **i** partita *f*; lotto *m*

1222 batch processing — The use of a computer system to perform only one job at a time. The job cannot be interrupted, except to be aborted, once it has been undertaken, although there are some versions of batch processing in which the operating system itself can interrupt one batch program in order to start or continue the processing of another one. See also interactive program.

f traitement *m* par lot — **d** abschnittweise Datenverarbeitung *f*; schubweise Verarbeitung *f*; Stapelverarbeitung *f* — **n** stapelverwerking *f*; groepsgewijze verwerking *f* — **e** procedimiento *m* per lotes — **i** elaborazione *f* a blocchi

1223 bath — Chemical solution employed in photoengraving, but specifically applied to silver nitrate, developing, fixing, and etching solutions.

f bain *m* — **d** Bad *n* — **n** bad *n* — **e** baño *m* — **i** bagno *m*

1224 bath holder — Container for chemical solutions.

f cuve *f* — **d** Chemikalienbehälter *m*; Chemikalienflasche *f* — **n** tank *m*; reservoir *n* — **e** cubeta *f* para baño — **i** serbatoio *m*

1225 bath note — A size of note paper. See app. no. 3.

1226 batik paper; battik paper — Paper imitating the Indonesian hand-dyed fabrics, in which use is made of wax as a dye repellent to cover parts of a design and in which the uncovered fabric is dyed with colours, after which the wax is dissolved in boiling water.

f papier *m* batique — **d** Batikpapier *n*; Javakunstpapier *n* — **n** batikpapier *n* — **e** papel *m* artístico bático — **i** carta *f* batik

1227 bâtonné — Ruled with parallel lines.

f réglure *f* parallèle — **d** mit Parallellinien versehen — **n** bâtonné — **e** alineación *f* paralela — **i** rigato con linee parallele

1228 batter; battered type; bad letter — A damaged letter, character, or rule. See also worn-out type.

f caractère *m* endommagé; caractère *m* cogné — **d** defekter Buchstabe *m*; lädierter Buchstabe *m*; beschädigte Type *f* — **n** kapotte letter *fm*; beschadigde letter *fm*; defecte letter *fm* — **e** tipo *m* estropeado; estropeo *m*; letra *f* defectuosa; tipo *m* machacado; arañazo *m*; aplastamiento *m* — **i** carattere *m* deteriorato; carattere *m* danneggiato; lettera *f* guasta

1229 battered type — See batter.

1230 battik paper — See batik paper.

1231 baud — Originally the rate of transmission by telegraph, including a stop and start signal for serial transmission. Now used for bits per second in a train of binary signals. With a start and stop system the character transmission rate is less than the bauds divided by the number of bits per character. Hence 1200 bauds means 1200 bits, or perhaps 120 characters, can be transmitted per second.

1232 Baudot code — Standard five channel communications code, devised about 1867, from which 31 combinations, extended by letter and figure shifts, can be formed. Known as International Telegraph Alphabet no. 1 US.

f code *m* de Baudot — **d** Baudot-Kode *m*; Fernschreibkode *m* — **n** Baudot-code *m* — **e** código *m* de Baudot — **i** codice *m* di Baudot

1233 Baumé hydrometer; Baumé meter — Hydrometer for industrial use, invented by the French chemist Antoine Baumé (1728-1804), to determine the specific gravity of liquids. Of the two types available, the one measuring liquids heavier than water is most generally used in photomechanics and electrotyping. For lithographic plate coating solutions, the specific gravity is referred to as density of solution. This varies according to the temperature of the solution. Degrees B = 144.3 (SG - 1): SG. Specific gravity =

144.3 : (144.3 - degr. B).

f aréomètre *m* Baumé — **d** Baumé-Aräometer *mn* — **n** areometer *m* van Baumé; Baumé-meter *m* — **e** areómetro *m* Baumé — **i** areometro *m* di Baumé

1234 Baumé meter — See Baumé hydrometer.

1235 bayonet holder — See bayonet socket.

1236 bayonet mount — See bayonet socket.

1237 bayonet socket; bayonet holder; bayonet mount; swan socket — Socket with push-and-twist engagement of parts to prevent detachment by simple pull.

f douille *f* à baïonnette — **d** Bajonettfassung *f*; Swanfassung *f* — **n** bajonetsluiting *f*; bajonetfitting *f* — **e** portalámpara *m* de bayoneta; montura *f* a bayoneta — **i** portalampada *m* a baionetta; madrevirola *f* a baionetta; innesto *m* a baionetta

1238 bay windows — Openings in frames of a rotary press through which webs or folded webs from balloon formers pass on their way to the folding mechanism.

f dégagements *mpl* — **d** Seitenöffnungen *fpl* — **n** zij-openingen *fpl* — **e** aberturas *fpl* laterales — **i** svincoli *mpl*

1239 B box — See modifier register.

1240 B bristol — See bogus bristol.

1241 BBR system; Bafour system — Computer controlled typesetting system developed in 1954 by G. Bafour, Blanchard and Raymond of the Imprimerie Nationale at Paris.

f système *m* BBR — **d** BBR-System *n* — **n** BBR-systeem *n* — **e** sistema *m* BBR — **i** sistema *m* BBR

1242 bead — Small roll formed by the knots in the headband and tailband of bound books.
f talon *m* de la tranchefile — **e** vivo cordoncillo *m* de la cabezada — **i** cannoncino *m*
1243 bead — Raised score along the bottom of a cardboard box to ensure proper folding. See also bottoming.
f bourrelet *m* de refoulage — **d** Biegewulst *m* — **n** kraal *fm* — **e** hendido *m* para el doblado — **i** cordonatura *f* segnapiega
1244 beam of light
f faisceau *m* de lumière — **d** Lichtbündel *n*; Licht-büschel *m* — **n** lichtbundel *m* — **e** haz *m* de luz; haz *m* luminoso — **i** fascio *m* di luce
1245 beam press — Old type of handpress for letterpress printing in which pressure is applied by screw action.
f presse *f* à vis — **d** Spindelpresse *f* — **n** spindel-pers *f*; schroefpers *f* — **i** torchio *m* a vite
1246 beam splitter — Optical arrangement of prisms or mirrors that divides a light ray into several paths.
f séparateur *m* de rayons lumineux — **d** Strahl-teiler *m* — **n** lichtstraalverdeler *m* — **e** divisor *m* de luz
1247 beard — In GB the space between the foot of the type face and the bottom edge, i.e., the sum of bevel and shoulder of the type. In US the sloping part between the type face and the shoulder. See also bevel.
f blanc *m* — **d** Fleisch *n* — **n** baard *m*; vlees *n* — **e** blanco *m* del tipo; blanco *m* de la letra — **i** bianco *m* di spalla
1248 beard *sl.* — Metallic fuzz between or around line cast letters usually due to faulty or dirty matrices. See also hairing.
f bavure *f* — **d** Spieß *m* — **n** haartje *n* — **e** rebaba *f* (de fundición) — **i** bava *f* di fonderia
1249 beard — Outside shading in ornamental type faces.
f hachures *fpl* — **d** Außenschattierung *f* bei Ornamenttypen — **n** arcering *f* langs de buiten-zijde — **e** sombreado *m* en tipos ornamentales — **i** tratteggio *m*; ombreggiatura *f*
1250 beard — See hair-line.
1251 bearer — Clump or anything type-high to bear off the impression from the light parts of a forme when it is being printed.
f lingot *m* — **d** schrifthoher Steg *m*; Laufsteg *m*; Laufschiene *f*; Walzsteg *m* — **n** drager *m* — **e** imposición *f* de altura; camino *m* de carga — **i** lingotto *m* all'altezza del carattere
1252 bearer rings; bearers; cylinder bearers — Circumferential bands fixed to the ends of the plate and impression cylinder on a press or inte-gral with them. The diameter of these bands pro-vides a datum for cylinder packing and for the exact centre distance of the cylinders. On American offset presses the bearers make rolling contact for proper meshing of the driving gears. See also bed bearers.
f couronnes *fpl*; bandes *fpl* du cylindre; bagues

fpl du cylindre — **d** Zylinderlaufringe *mpl*; Schmitzringe *mpl*; Meßringe *mpl* — **n** smetringen *mpl* — **e** coronas *fpl* del cilindro; caminos *mpl* laterales del cilindro; cejas *fpl* del cilindro; rebordes *mpl* laterales del cilindro; chumaceras *fpl* — **i** anelli *mpl* di rotolamento; anelli *mpl* di controllo; corone *fpl* di controllo
1253 bearers — See bearer rings.
1254 bearing — In rotating machinery, support on which journal of axle or shaft rests. See also needle bearing, plain bearing, roller bearing, self-aligning roller bearing.
f palier *m*; coussinet *m* — **d** Lager *n* — **n** lager *n* — **e** cojinete *m* — **i** cuscinetto *m*
1255 bearing bush — See bushing.
1256 bearing bushing — See bushing.
1257 bearing cap
f chapeau *m* de palier — **d** Lagerdeckel *m*; Lager-schalendeckel *m* — **n** lagerdeksel *n* — **e** tapa *f* de cojinete; sombrerete *m* de cabeza — **i** cappello *m* del cuscinetto; coperchio *m* del cuscinetto
1258 bearing metal — See babbitt metal.
1259 beat — See scoop.
1260 beat *v* — To prepare a flong by laying a damp matrix paper over a type forme and beating the back of it with a brush.
f mouler à la brosse; prendre empreinte à la brosse — **d** mit der Bürste klopfen; mit der Bürste abformen — **n** een matrijs kloppen met een borstel — **i** prendere impronte con la spazzola
1261 beater; Dutch engine — Machine to confer by mechanical means on pulps in aqueous sus-pension the qualities necessary to make paper or board of the required characteristics. See also hollander, refiner, Jordan machine.
1262 beater chest — Storage tank in which the paper stock from two or more beaters is mixed to secure greater uniformity of the papermaking material.
1263 beater plate — See bed plate.
1264 beater-sized paper; engine-sized paper — Paper for which the pulp has been made more or less waterproof by the addition of rosin size and alum to the stock in the beater.
f papier *m* collé à la pile; papier *m* collé dans la pile; papier *m* collé à la pâte — **d** stoffgeleimtes Papier *n*; in der Masse geleimtes Papier *n*; Papier *n* mit Stoffleimung; Papier *n* mit Holländer-leimung — **n** papier *n* met stoflijming; in de stof gelijmd papier *n*; papier *n* met inwendige lijming — **e** papel *m* encolado en la pila — **i** carta *f* collata nell'olandese
1265 beater sizing; engine sizing — Process of sizing paper by application of sizing materials in the beater, or to the furnish prior to sheet formation, as distinguished from surface sizing or tub sizing. Rosin size and alum are usually used, but also other sizing agents.
f collage *m* en pile — **d** Holländerleimung *f*; Massenleimung *f*; Stoffleimung *f* — **n** stof-lijming *f*; lijming *f* in de stof; inwendige lijming *f* — **e** encolado *m* en la pila — **i** collatura *f*

nell'olandese

1266 beating — Mechanical treatment of fibrous materials in a beater or refiner to give them the properties necessary for the manufacture of a definite quality of paper or board.
f raffinage *m* (en pile) — **d** Mahlung *f*; Holländermahlung *f*; Mahlen *n*; Mahlarbeit *f*; Zerfaserung *f*; Zerfasern *n* — **n** maling *f*; hollandermaling *f* — **e** molturación *f*; trabajo *m* de desfibrado; refino *m* — **i** raffinazione *f*

1267 beating brush; letter brush; proof brush; printer's brush — Short brush with stiff bristles, in former times used to take proofs of composed type.
f brosse *f* à épreuves; brosse *f* pour faire des épreuves — **d** Abklopfbürste *f*; Abziehbürste *f*; Korrekturabziehbürste *f* — **n** klopborstel *m*; letterborstel *m*; proevenborstel *m* — **e** bruza *f* para tirar pruebas; bruza *f* para sacar pruebas; bruza *f* para pruebas — **i** spazzola *f* per tirare prove

1268 beating brush; matrix beating brush — Brush to beat the back of a damp matrix paper laid upon the forme to prepare a flong.
f brosse *f* à mouler — **d** Abschlagbürste *f*; Abklopfbürste *f* — **n** klopborstel *m*; stypborstel *m* — **e** bruza *f* para amoldar; escobilla *f* para golpear; cepillo *m* para batir matrices; bruza *f* de batir — **i** spazzola *f* per stereo

1269 beating hammer — See backing hammer.

1270 Becquerel effect — Photographic effect, discovered in 1841 by Edmond Becquerel, occurring with silver chloride paper. When this paper is exposed to bluish-violet rays until the layer is slightly coloured, the density can further be increased by exposing the layer to yellow or green light, although originally for this range of the spectrum no sensitivity was present. With silver bromide layers this effect arises only when they have got a flash exposure.
f effect *m* Becquerel — **d** Becquerel-Effekt *m* — **n** Becquerel-effect *n* — **e** efecto *m* Becquerel — **i** effetto *m* Becquerel

1271 bed; type bed; forme bed; type plate — Flat part or table of a press supporting the forme or plate during printing.
f marbre *m* — **d** Druckfundament *n*; Formbrett *n* — **n** drukfundament *n* — **e** platina *f* de una prensa; platina *f* del carro; padrón *m* Ch — **i** piano *m* portaforma

1272 bed — Foundations of a printing press.
f massif *m*; socle *m* — **d** Fundamentplatte *f* — **n** fundament *n* — **e** placa *f* de fundamento; zócalo *m* — **i** basamento *m*

1273 bed — Water solution of gum tragacanth used in bookbinding to marble book edges.
f solution *f* de gomme adragante — **d** Marmoriergrund *m*; Grund *m* — **n** marmerbad *n* — **e** baño *m* de goma tragacanto; cama *f* de goma tragacanto — **i** soluzione *f* acquosa di adragante

1274 bed-bearer height
f hauteur *f* des tasseaux — **d** Schmitzleistenhöhe *f* — **n** looplijsthoogte *f* — **e** altura *f* de los caminos — **i** altezza *f* della guida di rotolamento

1275 bed bearers — Metal bands on the forme bed of flat-bed printing machines, which in intimate contact with the cylinder bearers determine the gap between the bed and the cylinder. Sometimes called mackling rails.
f tasseaux *mpl*; bandes *fpl* du marbre — **d** Schmitzleisten *fpl*; Laufleisten *fpl*; Friktionsschienen *fpl* — **n** looplijsten *fmpl* — **e** bandas *fpl* laterales; reglas *fpl* laterales del carro; caminos *mpl* de la máquina; caminos *mpl* de la platina — **i** guide *fpl* di rotolamento del piano portaforma

1276 bed drive; carriage drive
f commande *f* du marbre; entraînement *m* du chariot — **d** Karrenantrieb *m* — **n** aandrijving *f* van de kar; karaandrijving *f* — **e** impulso *m* del carro; accionamiento *m* del carro; mando *m* del carro — **i** comando *m* del carro

1277 bed latch — Safety catch securing the platform between reel stands.
d Arretierung *f* der Schiebebühne — **n** veiligheidspal *m* — **e** trinquete *m* de seguridad — **i** dente *m* d'arresto

1278 bed plate — Bottom section of the frame on which the press is built.
f bâti *m* de fond; socle *m* — **d** Fußrahmen *m* — **n** grondplaat *fm*; machinefundering *f* — **e** zócalo *m*; placa *f* de base; placa *f* de asiento; placa *f* de basamento — **i** piastra *f* di base; piastra *f* di fondazione

1279 bed plate; beater plate — Metallic plate in which bronze or steel knives are fitted, set directly underneath the beater roll of the hollander. Pulp fibres are crushed between the bars of the roll and the knives of the bed plate.
f platine *f* de pile — **d** Grundwerk *n* — **n** grondwerk *n* — **e** platina *f* de la holandesa — **i** platina *f* dell'olandese

1280 bed plate — See platen.

1281 bed size
f format *m* de marbre — **d** Fundamentgröße *f* — **n** grootte *f* van het drukfundament — **e** tamaño *m* de la cama; tamaño *m* de la platina — **i** formato *m* del piano portaforma

1282 bed stone — The fixed horizontal member of a *kollergang.
f meule *f* de fond; pierre *f* gisante; meuleton *m*; gîte *m*; meule *f* de dessous — **d** Bodenstein *m* des Kollergangs — **n** grondsteen *m* van de kollergang — **e** muela *f* yacente; muela *f* de fondo — **i** pietra *f* di fondo nella molazza

1283 bed track
f glissière *f* — **d** Karrenbahn *f* — **n** karrebaan *fm* — **e** vía *f* del carro; vía *f* de la máquina de platina; carril *m*; riel *m*; caminos *mpl* — **i** corsa *f* del carro

1284 beech; beech wood — Broadleaf, hardwood tree used in the manufacture of soda pulp.
f hêtre *m* — **d** Buchenholz *m* — **n** beukehout *n* — **e** madera *f* de haya; haya *f* — **i** legno *m* di faggio; faggio *m*

1285 beech wood — See beech.

1286 beer mat — Mat made of coaster board, also called beer mat board, or beer plaques, a heavy paperboard made of wood pulp, absorbent and not warping when wet, with a thickness of 1.5-3.5 mm (0.060-0.135 in).

f dessous *m* de bocks; soucoupe *f* de verres; sous-bock *m* — **d** Bierfilz *m*; Bierglasuntersetzer *m*; Bieruntersatz *m* — **n** bierviltje *n* — **e** platillo *m* para cerveza; redondelito *m* para cerveza — **i** sottobicchiere *m* per birra

1287 beer mat board — See coaster board.

1288 beeswax — Solid yellowish substance secreted by bees, plastic when warm and melting at about 63 °C (145 °F).

f cire *f* d'abeilles — **d** Bienenwachs *n* — **n** bijenwas *mn* — **e** cera *f* de abejas — **i** cera *f* d'api

1289 begin *v* **even** — To begin a line without an indention.

f commencer sans enfoncement; commencer à l'alignement — **d** stumpf anfangen — **n** niet inspringen — **e** comiéncese sin sangría — **i** cominciare senza capoverso; cominciare il capoverso a filo; cominciare il capoverso allineato

1290 beginner — See cub.

1291 begin quote — See commence quote.

1292 bellows — The folding portions which unite the front and back sections of process cameras.

f soufflet *m* — **d** Balg *m*; Balgen *m*; Kamerabalgen *m* — **n** balg *m*; camerabalg *m* — **e** fuelle *m* — **i** soffietto *m*

1293 bellows; hand bellows — Hand compressor to blow out the dust from type cases, etc. Nowadays a small vacuum cleaner with a screened nozzle is used.

f soufflet *m* — **d** Blasebalg *m* — **n** blaasbalg *m* — **e** fuelle *m*; barquín *m*; barquinera *f* — **i** soffietto *m*

1294 bellows extension

f tirage *m* du soufflet — **d** Kameraauszug *m*; Balgenauszug *m* — **n** camera-uittrek *m*; balguittrek *m* — **e** tiro *m* del fuelle — **i** corsa *f* del soffietto

1295 bellows support

f support *m* de soufflet — **d** Balgenstütze *f* — **n** balgsteun *m*; balg-ondersteuning *f* — **e** soporte *m* del fuelle; apoyo *m* del fuelle — **i** supporto *m* del soffietto

1296 belly — The front or nick side of type. In France the nick is at the opposite side, i.e., at the side of the accents.

f front *m*; côté *m* du cran; devant *m* du caractère — **d** Vorderseite *f* — **n** voorkant *m*; kerfzijde *fm*; zijde *fm* van de kerf — **e** frente *m*; cara *f* anterior del tipo — **i** fronte *m* del carattere; faccia *f* anteriore del fusto del carattere

1297 belly — Raised portion of a printing plate not conform the cylinder surface on which it is mounted.

f bosse *f* — **d** Plattenwölbung *f*; Plattenaufbiegung *f* — **n** bolling *f* — **e** abollado *m* de plancha; abolladura *f* de plancha; combado *m* de plancha — **i** bombatura *f* di lastra

1298 belly — Concave distortion in a pile of sheeted paper.

f ventre *m* — **n** zak *m* — **e** panza *f*; bolsa *f* de papel; abombado *m* del papel — **i** pancia *f*

1299 belly — The lower edge of a graver.

f talon *m* — **n** buik *m* — **i** bordo *m* inferiore di un bulino

1300 belly — Bulge in a poorly locked-up forme.

1301 belt — Leather or rubberized fabric, cord or band passing around wheels, drums or pulleys to transmit motion from one to another.

f courroie *f* — **d** Riemen *m*; Treibriemen *m* — **n** drijfriem *m*; riem *m* — **e** correa *f* (de transmisión); banda *f* (de transmisión) — **i** cinghia *f* di trasmissione; correggia *f* di trasmissione

1302 belt drive

f entraînement *m* par courroie; transmission *f* par courroie — **d** Riemenantrieb *m* — **n** riemaandrijving *f* — **e** propulsión *f* por correa; impulso *m* por correa — **i** propulsione *f* a cinghia; comando *m* a cinghia

1303 belt fastener

f attache *m* de courroie; agrafe *f* de courroie — **d** Riemenklammer *f*; Riemenverbinder *m* — **n** riemverbinder *m* — **e** grapa *f* de unión de correas — **i** graffa *f* di giunzione della cinghia

1304 belt fork

f fourchette *f* de courroie — **d** Riemengabel *f* — **n** riemvork *fm* — **e** horquilla *f* de la correa — **i** forcella *f* della coreggia

1305 belt guard

f protège-courroie *m* — **d** Riemenschutzvorrichtung *f*; Riemenschutz *m*; Riemenabdeckung *f* — **n** riembeschermer *m* — **e** guarda *f* de correa — **i** protezione *f* di sicurezza della cinghia

1306 belt lacing

f attache *m* de courroie — **d** Riemenverbindung *f* — **n** riemveter *m* — **e** cordón *m* de la correa — **i** giunzione *f* della cinghia; legaccio *m* della cinghia

1307 belts; brake bands; moving belt braking; running belt tension — Metal, leather or composition strips at the outside of the reel of paper acting as a brake on the reel to increase the amount of pull or tension on the web. The belts may be stationary or running.

f courroies *fpl* de freinage — **d** Gürte *mpl*; Riemen *mpl*; Spannung *f* im laufenden Gurt — **n** remriemen *mpl* — **e** cinchos *mpl* de freno — **i** correggie *fpl* di frenatura

1308 belt sheave; pulley

f rouelle *f* de courroie; poulie *f* — **d** Riemenscheibe *f* — **n** riemschijf *fm*; poelie *f* — **e** polea *f* (de transmisión) — **i** puleggia *f*

1309 belt shifter — Machine part used in belt drives to shift the driving belt from a loose pulley to a fixed pulley to start or stop a machine.

f guide-courroie *m*; pose-courroie *m* — **d** Riemenabsteller *m*; Riemenausrückung *f* — **n** riemafsteller *m* — **i** spostacinghia *m*

1310 belts on — See brake down.

1311 bench; work bench; work table

f établi *m*; banc *m*; table *f* de travail — **d** Werkbank *f*; Arbeitstisch *m* — **n** werkbank *fm*; werktafel *fm* — **e** banco *m* de trabajo; banco *m* de taller; mesa *f* de trabajo — **i** banco *m*; bancone *m*

1312 bench ruling machine
f machine *f* à régler de table — **d** Tischliniiermaschine *f* — **n** tafellinieermachine *f* — **e** máquina *f* rayadora de mesa; rayadora *f* de mesa — **i** macchina *f* rigatrice a tavola; rigatrice *f* a tavola

1313 bend *v*
f cintrer; courber — **d** biegen; runden — **n** buigen; rondzetten — **e** curvar; encorvar — **i** curvare; incurvare

1314 benday — A *shading medium. See also benday process.

1315 benday artist; colour artist; film layer — Man skilled in laying benday tints on photoengravings.
f chromiste *m* benday — **n** filmlegger *m*; tintlegger *m* — **e** grisador *m*; aplicador *m* de fondos; operario *m* cromista — **i** cromista *m*

1316 bendayed plates — Relief etchings on zinc or copper, on which various tints or patterns are introduced through the medium of benday films.
f clichés *mpl* exécutés par benday — **d** Klischees *npl* mit Tangierraster — **n** clichés *npl* met filmraster — **e** grabados *mpl* bendéi; clisés *mpl* con grisado bendéi — **i** incisioni *fpl* in rilievo eseguite col procedimento benday

1317 benday machine — See shading machine.

1318 benday process — Method of mechanically transferring line, dot or texture patterns to paper, metal or glass by general or local pressure on the back of an inked benday screen (a relief on a gelatine film). Invented in 1879 by the American printer Benjamin Day, New Jersey (1838-1916), for introduction of shading effects in line drawings and reproduction therefrom. The same effects can be produced by working directly on the original artwork with trans parent adhesive printed screens such as Zip-a-tone, or with *Craftint.
f procédé *m* au benday — **d** Bendayverfahren *n*; Tangierverfahren *n* — **n** filmraster-procédé *n*; benday-procédé *n*; tangeer-procédé *n* — **e** procedimiento *m* bendéi; procedimiento *m* de grisado bendéi — **i** procedimento *m* benday

1319 bending machine
f machine *f* à cintrer — **d** Biegemaschine *f*; Biegepresse *f* — **n** buigmachine *f*; plaatbuigmachine *f* — **e** encorvadora *f* de planchas; máquina *f* de encorvar; máquina *f* de curvar; curvadora *f* — **i** macchina *f* curvatrice

1320 bending machine — In boxmaking a machine to bend sides and ends of base, lid, or tray.
f machine *f* à courber du carton — **d** Pappenbiegemaschine *f*; Biegemaschine *f*; Biegepresse *f* — **n** buigmachine *f*; kartonbuigmachine *f* — **i** macchina *f* curvatrice; macchina *f* per curvare il cartone

1321 bending of light

f diffraction *f* de lumière — **d** Diffraktion *f* des Lichtes; Beugung *f* des Lichtes — **n** lichtdiffractie *f*; lichtbuiging *f* — **e** difracción *f* de la luz — **i** diffrazione *f* della luce

1322 bending rollers; forming rollers — Rollers in a rotary press at the point or nose of the former.
f rouleaux *mpl* d'entrée — **d** Trichterfalzwalzen *fpl* — **n** trechtervouwrollen *fmpl*; trechterinvoerrollen *fmpl* — **e** rodillos *mpl* de entrada — **i** rulli *mpl* di piega al cono

1323 bending strength
f résistance *f* à la flexion — **d** Biegefestigkeit *f*; Biegewiderstand *m* — **n** buigweerstand *m*; buigsterkte *f* — **e** resistencia *f* a la flexión — **i** resistenza *f* alla flessione

1324 bending-strength tester — Instrument to measure the resistance to bending of boards. A test strip, 5 cm wide, is fixed in two superimposed clamps which are initially aligned, both revolving in such a way that the clamped strip is bent or folded over and eventually breaks or creases. The lower clamp only is moved and pulls over the upper clamp with the trial strip. This upper clamp is connected with a pendulum, the deviation of which being proportional to the force exerted, indicates on a scale the resistance of the trial strip for the various angles reached before breaking.
f appareil *m* pour déterminer la flexibilité des cartons — **d** Pappenbiegeprüfer *m* — **n** knik- en buigtoestel *n* voor karton — **e** ensayadora *f* (del esfuerzo) de flexión; aparato *m* para medir la flexión — **i** apparecchio *m* per determinare la flessibilità; apparecchio *m* per determinare la resistenza alla flessione

1325 bending test
f essai *m* de flexion — **d** Biegeprüfung *f* — **n** buigproef *fm* — **e** ensayo *m* de flexión — **i** prova *f* di flessione

1326 benzene — Clear, colourless, flammable, volatile spirit, obtained when soft coal is heated in the absence of air, used as a medium for carrying gravure inks. It is narcotic and toxic, miscible with alcohol, ether, acetone, carbon tetrachloride, carbon disulphide, acetic acid and slightly soluble in water. Can be absorbed by the skin. Forms an explosive mixture with air within limits of 1.5 to 8% by volume. Benzene has a ring structure. Formula: C_6H_6. TL-value: 25 ppm or 80 mg/m^3. See also benzol.
f benzène *m*; benzol *m*; benzine *f* de houille; benzine *f* cristallisable — **d** Reinbenzol *n*; Benzol *n* — **n** benzeen *n*; benzol *nm depr.* — **e** benceno *m*; benzol *m* — **i** benzene *m*; benzolo *m*

1327 benzene azimide — See benzotriazole.

1328 benzene nucleus — See benzene ring.

1329 benzene ring; benzene nucleus — Six-carbon ring in the molecular structure of benzene and in all the organic compounds derived therefrom. The graphic representation is a hexagon.
f noyau *m* benzénique; anneau *m* de benzène — **d**

Benzolkern *m*; Benzolring *m* − **n** benzeenring *m* − **e** anillo *m* de benceno − **i** anello *m* benzenico; formula *f* del benzene

1330 benzidine yellow − Organic azo pigment, approx. twice as strong as Hansa yellow but somewhat poorer in light permanency. Used as a strong yellow toner for many types of printing inks.
f jaune *m* de benzidine − **d** Benzidingelb *n* − **n** benzidinegeel *n* − **e** amarillo *m* de bencidina; amarillo *m* bencénico − **i** giallo *m* di benzidina

1331 benzine − Archaic and misleading name for a colourless, volatile, flammable, liquid mixture of various hydrocarbons, obtained in the distillation of petroleum. Term should not be used. See SBP spirit.

1332 benzol − Term still used commercially, but not in favour in modern nomenclature. See benzenes, solvent naphtha.

1333 benzoline − See SBP spirit.

1334 benzophenol − See phenol.

1335 benzoquinone − See quinone.

1336 benzotriazole; aziminobenzene; benzene azimide − White to light tan, crystalline compound, soluble in alcohol and benzene, slightly soluble in water. Used as a photographic restrainer.
Formula: $C_6H_4NHN_2$.
f benzotriazole *m* − **d** Benzotriazol *n*; Aziminobenzol *n* − **n** benzotriazol *n* − **e** benzotriazol *m*; aziminobenceno *m* − **i** benzotriazolo *m*

1337 Berlin blue − Name applied loosely to any of a number of the varieties of iron blue pigments. See also Prussian blue.
f bleu *m* de Prusse − **d** Berlinerblau *n* − **n** Berlijns blauw *n* − **e** azul *m* Berlín − **i** azzuro *m* di Berlino

1338 Berlin blue − See Prussian blue.

1339 Berlin red − Red pigment consisting essentially of red iron oxide. See also iron oxide reds.
f rouge *m* (de) Berlin − **d** Berlinerrot *n* − **n** Berlijns rood *n* − **e** rojo *m* Berlín − **i** rosso *m* di Berlino

1340 Berthold film setter − See Diatype.

1341 bestseller
f livre *m* à forte vente; livre *m* à succès; clou *m*; bestseller *m*; livre *m* le plus recherché; livre *m* qu'on s'arrache − **d** Verkaufsschlager *m*; Buchverkaufsschlager *m*; erfolgreiches Buch *n*; Bucherfolg *m*; Reißer *m coll.*; Bestseller *m* − **n** bestseller *m*; veel verkocht boek *n*; kasstuk *m* − **e** libro *m* de mucha venta; libro *m* de (gran) éxito; libro *m* de mayor venta − **i** libro *m* di grande successo; libro *m* maggiomente venduto

1342 beta- − Prefix to a chemical name denoting the second position from a particular group or atom in the structure of an organic molecule.

1343 beta rays − Stream of beta particles (charged particles) emitted from radioactive decay or fission.
f rayons *mpl* bêta − **d** Betastrahlen *mpl* − **n** bêtastralen *mfpl* − **e** rayos *mpl* beta − **i** raggi

mpl beta

1344 bevel − Sloping edge on a printing plate (electrotype, or stereotype) used to fasten the plate onto the block.
f biseau *m* − **d** Facette *f*; Abschrägung *f* − **n** facet *n*; spijkerrand *m* − **e** bisel *m*; pestaña *f*; faceta *f*; chaflán *m* − **i** smusso *m*; bisello *m*; faccetta *f*

1345 bevel − The angled edge on the rotary printing plate on which the dogs act.
f chanfrein *m* d'accrochage de cliché − **d** Facettenschräge *f*; Facettenwinkel *m*; Schrägkante *f* − **n** facet *n*; schuine kant *m* − **e** ángulo *m* del bisel; ángulo *m* de la pestaña; rebajo *m* − **i** smusso *m* sui bordi; sghembo *m*

1346 bevel; bevel of shoulder; neck − The sloping surface of a type running up from the shoulder of the face. See also beard.
f biseau *m*; talus *m* (de pied); facette *f* − **d** Konus *m*; Hals *m*; Schrägkante *f*; Facette *f*; Achselschrägung *f*; Gehrung *f* − **n** talud *n*; facet *n*; schuine kant *m* − **e** talud *m* del hombro − **i** fianco *m* inclinato del rilievo dell'occhio

1347 beveled rule − See bevelled rule.

1348 bevel gear; bevel gearing − Gears which connect two shafts at an angle with the centre lines of the shafts in the same plane.
f engrenage *m* conique − **d** Kegelgetriebe *n*; Kegelradantrieb *m* − **n** conische tandwielaandrijving *f* − **e** engranaje *m* cónico − **i** ingranaggio *m* conico

1349 bevel gear − Gear with teeth cut into a conical surface, usually meshing with a similar gear set at right angles.
f pignon *m* conique; couronne *f* conique − **d** Kegelrad *n*; konisches Zahnrad *n*; Winkelrad *n* − **n** conisch tandwiel *n* − **e** piñón *m* cónico; rueda *f* cónica − **i** pignone *m* conico

1350 bevel gearing − See bevel gear.

1351 beveling − See bevelling.

1352 bevelled rule *GB*; **beveled rule** *US*; **sidefaced rule** − Rule with face flush with one side and bevelled on the other side.
f filet *m* à épaulement; filet *m* œil de côté − **d** Linie *f* auf Kante − **n** lijn *fm* met het beeld aan één zijde − **e** filete *m* fundido al canto; raya *f* fundida al canto − **i** filetto *m* smussato

1353 beveller − See bevelling machine.

1354 bevelling *GB*; **beveling** *US* − Placing a bevel angle. In photoengraving, rectangular plates are put on a *bevelling machine.
f biseautage *m* − **d** Facettieren *n*; Abschrägen *n* − **n** facetteren *n* − **e** abiselar; biselar; achaflanar; chaflanar; facetar − **i** bisellatura *f*; smussatura *f*

1355 bevelling machine; beveller − Machine to cut a narrow rabbit or bevel around the edges of square or rectangular printing plates to provide an approx. 3.5 mm wide channel or flange for nailing plates on wooden or metal blocks. The flange is the bevel of the plate. See also bevelling.
f biseauteuse *f* − **d** Facettiermaschine *f*; Facetten-

fräsmaschine *f*; Facettenfräser *m*; Facettenhobel *m*; Kantenabschrägmaschine *f* — **n** facetteermachine *f*; facetschaaf *fm* — **e** biseladora *f*; abiseladora *f*; cepillo *m* biselador; achaflanadora *f* — **i** macchina *f* bisellatrice

1356 bevel of shoulder — See bevel.

1357 bias chase; biased chase — Chase with a tendency to bend.

f châssis *m* déformé — **d** schiefer Rahmen *m*; schiefer Formrahmen *m*; schiefer Schließrahmen *m*; verzogener Schließrahmen *m*; zum Durchbiegen neigender Schließrahmen *m* — **n** verbogen sluitraam *n*; scheef sluitraam *n* — **e** rama *f* sesgada — **i** telaio *m* deformato; telaio *m* inclinato

1358 biased chase — See bias chase.

1359 bible paper; Oxford bible paper; Oxford India paper; India paper; India bible paper; India Oxford paper — Light-weight, strong, opaque paper for printing of bibles, prayer books, dictionaries, etc. Basis weights normally 6.5-13.6 kg or 14-30 pounds (25 x 38 - 500).

f papier *m* bible; papier *m* pour bibles; papier *m* indien; papier *m* d'Oxford; papier *m* de Chine — **d** Bibeldruckpapier *n*; Bibelpapier *n*; Dünndruckpapier *n*; chinesisches Papier *n*; Indiapapier *n* — **n** bijbeldrukpapier *n*; bijbeldruk *n*; dundrukpapier *n* — **e** papel *m* biblia; papel *m* de China; papel *m* India — **i** carta *f* bibbia; carta *f* India; carta *f* sottile

1360 bibliographer — One who writes about books, especially in regard to their authorship, date, typography, etc.

f bibliographe *m* — **d** Bibliograph *m* — **n** bibliograaf *m* — **e** bibliógrafo *m* — **i** bibliografo *m*

1361 bibliography; references — List of sources or references which an author has consulted in preparation of a book or index of published works referred to by an author. Also list of publications on a particular subject. Usually at end of article, chapter, or book.

f bibliographie *f*, liste *f* bibliographique, références *fpl* — **d** Bibliographie *f*; Literaturverzeichnis *n*; Literaturnachweis *m*; Literaturhinweis *m*; Quellenangabe *f* — **n** bibliografie *f*; literatuuropgave *f*; geraadpleegde literatuur *f*; bronvermelding *f* — **e** bibliografía *f*; mención *f* de las fuentes — **i** bibliografia *f*; riferimento *m* alla fonti

1362 bibliomaniac — Person with a mania of collecting books, especially old and rare books.

f bibliomane *m* — **d** Büchernarr *m* — **n** bibliomaan *m*; boekengek *m* — **e** bibliómano *m* — **i** bibliomane *m*

1363 bibliopegy — See art of bookbinding.

1364 bibliophile; bibliophilist; book lover — One who seriously loves and collects books and knows how to discriminate between good and bad editions.

f bibliophile *m* — **d** Bibliophile *m*; Bücherfreund *m*; Bücherliebhaber *m* — **n** bibliofiel *m*; boekenliefhebber *m*; liefhebber *m* van boeken —

e bibliófilo *m*; amador *m* de libros; aficionado *m* a libros — **i** bibliofilo *m*; amatore *m* di libri

1365 bibliophilist — See bibliophile.

1366 biborate of soda — See sodium borate.

1367 bicarbonate of soda — See sodium bicarbonate.

1368 bichloroacetic acid — See dichloroacetic acid.

1369 bichromate — See dichromate.

1370 bichromated albumen — See dichromated albumen.

1371 bichromated coating — See dichromated coating.

1372 bichromated colloids — See dichromated colloids.

1373 bichromated fish glue — See enamel glue.

1374 bichromated gelatine — See dichromated gelatine.

1375 bichromated glue — See dichromated glue.

1376 bichromated gum — See dichromated gum.

1377 bichromate of potash — See potassium dichromate.

1378 bichromate of soda — See sodium dichromate.

1379 biconcave lens — See concave lens.

1380 biconvex lens — See convex lens.

1381 bifoliate *v* — To number two opposite pages of an account book with the same numbers (1 - 1; 2 - 2; 3 - 3; etc.), i.e., on the left- and right-hand pages.

f bifolioter — **d** foliieren — **n** foliëren — **e** foliar; bifoliar — **i** numerare

1382 big-ticket book — See edition de luxe.

1383 bilingual — Able to speak two languages with approximately equal facility. Also spoken or written in two different languages.

f bilingue — **d** zweisprachig — **n** tweetalig; in twee talen — **e** bilingüe — **i** bilingue

1384 billet note — A size of note paper. See app. no. 3.

1385 billhead — Ruled and printed form to render an account to a debtor.

f en-tête *m* de facture; facture *f* — **d** Rechnungskopf *m* — **n** rekening *f* — **e** cabecera *f* de factura; factura *f* — **i** fattura *f* intestata

1386 billion — In US, France, Belgium, the Netherlands and Italy a thousand million or 10^9. In GB, Germany and Spain a million millions or 10^{12}.

f milliard *m* — **n** miljard *n* — **i** milliardo *m*

1387 bill machine — Small rotary or flat-bed press to print news content bills.

1388 bill of fount — See fount scheme.

1389 bill of sorts

f formulaire *m* de commande des réassortiments — **d** Defektzettel *m*; Zusatzbestellung *f* — **n** bestelformulier *n* voor defecten; defectenpolis *fm* — **e** boletín *m* de letras incompletas — **i** modulo *m* per riordinazioni

1390 bill of type — See fount scheme.

1391 bimanual — Involving or requiring the use of both hands.

f bimanuel — **d** beidhändig — **n** tweehandig — **e** bimano; bímano; de ambas manos — **i** bimano

1392 bimetallic plate — See bimetal plate.

1393 bimetal plate; bimetallic plate — Lithographic plate in which the printing image base is formed of one metal and the non-printing area of a second metal. Generally, the printing area is copper, while the non-printing areas may be nickel, chromium, or stainless steel. Some plates employ a third metal as a base or backing and could be regarded as trimetallic, multimetallic or polymetallic. Both surface and deep-etch plate-making techniques are used to make such plates. With a plating of chromium on copper sheet, the resist stencil of deep-etch method permits etching chromium from image to reach copper for base of lithographic image; with plating of copper on stainless steel base, surface platemaking methods temporarily protect the image while copper is etched away to give stainless steel for non-printing areas of plate.

f plaque f bimétallique — **d** Bimetallplatte f — **n** bimetaalplaat f — **e** plancha f bimetálica — **i** lastra f bimetallica

1394 bi-monthly — Appearing every two months or appearing twice a month. Since it is not always clear which meaning of "bi-monthly" is intended, the use of *semi-monthly for twice a month is preferable, because it is unambiguous. Since there is no single, unambiguous term for "every two months", this is the least confusing.

1395 binary code — Code that makes use of exactly two distinct characters. Usually 0 and 1.

f code m binaire — **d** Binärkode m — **n** binaire code m — **e** código m binario — **i** codice m binario

1396 binary coded decimal representation — Method of number representation in which each decimal numeral is represented by some designated binary number. Each decimal digit is represented by a pattern of four binary digits. Thus the decimal number 93 is represented in a binary code with channel values of 8 - 4 - 2 - 1 by 1001, 0011. In pure binary different values for the code channels, for example 5 - 2 - 1 - 1 or 7 - 4 - 2 - 1. A 6-level binary coded decimal (BCD) code was devised as a tabulator card and paper and magnetic tape compatible code; in tape form it has 64 unique code values. Extended binary coded decimal (EBCDIC) was devised to extend the available codes to 256, by the use of eight levels, plus a parity bit.

f notation f décimale codifiée en binaire; numération f décimale codée en binaire; numération f décimale binaire — **d** binär verschlüsselte Dezimalzifferndarstellung f; binär kodierte Dezimalzifferndarstellung f — **n** binair gecodeerde decimale voorstelling f — **e** notación f decimal codificada en binario — **i** notazione f decimale codificata in binario

1397 binary digit — A numeral in the *binary system.

f chiffre m binaire — **d** Binärzahl f; Binärziffer f — **n** binair getal n — **e** cifra f binaria — **i** numero m binario

1398 binary information transfer — See bit.

1399 binary search — System to search a file or table by creating a series of midpoint indices; e.g. to search a table of 20 entries, in order to find the eighth entry one would first compare the argument to the first midpoint index 10. Eight is lower than 10, so the next comparison would be to the next lowest midpoint index, 5. Eight is higher than 5 so it is evident that the entry lies between 5 and 10. That area is then searched until the correct entry is found.

f recherche f binaire — **d** Halbierungsverfahren n — **n** halveringsmethode f; halveringsselectie f — **e** busqueda f binaria — **i** ricerca f binaria

1400 binary system — In this system use is made of merely two numerals, zero (0) or one (1), sometimes written as L to prevent confusion. The numeral may be equivalent to an on or off condition, a yes or a no. The two *binary digits must be organized into a code to represent numbers larger than 2, hence 4 is represented by 100, where the 1 in the first position represents 4, the middle zero represents two and is zero to indicate that this column has no value, and the last column is zero to indicate that no. 1 is present. Thus 0 = 0; 1 = 1; 2 = 10; 3 = 11; 4 = 100; 5 = 101; 6 = 110; 7 = 111; 8 = 1000; 9 = 1001; 10 = 1010; 11 = 1011; 12 = 1100; 13 = 1101; 14 = 1110; 15 = 1111; 16 = 10000; etc.

f système m binaire; numération f binaire — **d** Binärsystem n; binäres Zahlensystem n — **n** binair stelsel n; tweetallig stelsel n — **e** sistema m binario; numeración f binaria — **i** sistema m binario; sistema m a due fori

1401 bind v — To join pages of a book, etc., with thread, wire, adhesive or other means. Also to enclose them in a cover when so specified.

f relier; brocher — **d** einbinden; binden — **n** binden; inbinden — **e** encuadernar — **i** legare

1402 binder — Material in an ink film which holds the pigment to the printed surface. See also vehicle.

1403 binder — Temporary cover for ledger or loose leaf forms.

f classeur m pour feuillets mobiles — **d** Schnellhefter m; Aktendeckel m; Ordner m — **n** snelhechter m; ordner m — **e** carpeta f; cubierta f para hojas sueltas — **i** copertina f provvisoria a foglio sciolto

1404 binder; binder line; binder head — Single headline over a long article or series of articles in a newspaper, usually on an inside page.

f bandeau m général; titre m général — **d** Sammelüberschrift f — **n** verzameltitel m — **e** línea f cumbrera

1405 binder — See bookbinder.

1406 binder head — See binder.

1407 binder line — See binder.

1408 binder's board — Single-ply solid board

made on a wet machine from a base stock of mixed papers, ranging in thickness from 30 points (0.76 mm or 0.030 in) to 300 points.

f carton *m* (pour) reliure; carton *m* pour registres — **d** Buchbinderpappe *f*; Deckelpappe *f* — **n** bindersbord *n*; boekbindersbord *n*; grijsbord *n*; Zaans bord *n* — **e** cartón *m* para tapas; cartón *m* para encuadernaciones — **i** cartone *m* per rilegatura

1409 binder's cloth — See bookbinder's cloth.

1410 binder specks — Specks which give a grainy or textured appearance to the surface of blade coated paper; indicated by variations in ink receptivity or printed gloss and caused by non-uniform coating binder distribution. Specks appear as small dots (usually about 0.7 mm (1/32 in), uniformly distributed over all or much of the sheet area. Can be visualized as one railroad track, the entire width of the machine. This is a drying effect. Ink wipe or print test will verify the non-uniform ink receptivity and/or printed gloss. A light iodine burnout will show a similar pattern.

f taches *fpl* de liant — **d** Bindemittelflecken *mpl*; Binderflecken *mpl* — **n** bindmiddelvlekken *fmpl* — **i** macchie *fpl* dovute al legante di patina

1411 binder's stamp; book stamp — Design or lettering cut in brass, used for stamping or embossing back and flaps of a book.

f fer *m* à dorer — **d** Buchbinderstempel *m* — **n** bandstempel *n* — **e** plancha *f* para estampar libros; hierro *m* de dorar; tronquillo *m* para estampar libros — **i** ferro *m* per dorare

1412 bindery — Commercial establishment specializing in bookbinding. See also binding department.

f atelier *m* de reliure; service *m* de reliure — **d** Buchbinderei *f* — **n** boekbinderij *f*; binderij *f* — **e** taller *m* de encuadernación — **i** legatoria *f*

1413 bindery — See binding department.

1414 binding; bookbinding — The act of binding a book.

f reliure *f* — **d** Buchbinden *n*; Binden *n* — **n** boekbinden *n*; binden *n* — **e** encuadernación *f*; empastadura *f LA*; empastar *LA* — **i** legare

1415 binding — The cover of a book. See also case.

f reliure *f* — **d** Einband *m* — **n** boekband *m*; band *m* — **e** cubierta *f*; tapa *f* — **i** legatura *f*; rilegatura *f*

1416 binding — In locking up a forme, if the furniture is longer or wider than the type, it is said to bind, and the forme cannot be tightened up properly.

f d'une ligne ou d'une interligne trop longue on dit qu'il "commande" ou qu'il "force" — **d** die Form ist nicht systematisch und kann nicht einwandfrei geschlossen werden — **n** de vorm kan niet goed worden gesloten omdat het insluitwit niet past — **e** encaballar(se) el tipo al no estar justificada la línea — **i** non e possibile serrare la forma

1417 binding agent — See vehicle.

1418 binding à la fanfare — See fanfare binding.

1419 binding à l'éventail — See fan style binding.

1420 binding cloth — See bookbinder's cloth.

1421 binding department; bindery

f service *m* de reliure; atelier *m* de reliure — **d** Buchbindereiabteilung *f*; Buchbinderei *f* — **n** boekbinderij *f*; binderij *f*; afdeling *f* (boek)-binderij — **e** taller *m* de encuadernación; sección *f* de encuadernación — **i** reparto *m* legatura

1422 binding edge — Back of the book where the sections or signatures are sewed together.

f dos *m* — **d** Einheftkante *f*; Rücken *m* — **n** rug *m* — **e** lomo *m* — **i** dorso *m* del libro

1423 binding "en gist" — See Bradel binding.

1424 binding machinery

f machines *fpl* de reliure — **d** Buchbindereimaschinen *fpl* — **n** boekbinderijmachines *fpl*; binderijmachines *fpl* — **e** maquinaria *f* para encuadernación — **i** macchine *fpl* per legatoria

1425 binding medium — See vehicle.

1426 binding posts — Metal posts in loose-leaf covers to hold the perforated filler leaves in place.

f tiges *fpl*; poteaux *mpl*; bornes *fpl* à vis — **d** Buchbolzen *mpl* — **n** busbouten *mpl* — **e** pasadores *mpl* para hojas sueltas — **i** morsetti *mpl*

1427 binding press

f presse *f* de relieur; presse *f* à relier — **d** Buchbinderpresse *f* — **n** boekbinderspers *f* — **e** prensa *f* de encuadernación — **i** pressa *f* da legatore; pressa *f* per legatoria; strettoio *m* di legatore

1428 binding saw — Saw used by bookbinders to cut grooves across backbone of book to receive cords or bands used in binding.

f scie *f* à grecquer — **d** Fuchsschwanz *m* — **n** binderszaag *fm*; boekbinderszaag *fm* — **e** serrucho *m* de encuadernador — **i** sega *f* per grecare

1429 binding screws — Screws to lock the binding edge of a loose-leaf book.

f vis *fpl* à relier — **d** Buchschrauben *fpl* — **n** busschroeven *fmpl* — **e** tornillos *mpl* de libros de hojas cambiables — **i** viti *fpl* per legare

1430 binding thread; sewing thread

f fil *m* à brocher — **d** Heftfaden *m*; Faden *m*; Heftzwirn *m*; Zwirn *m* — **n** boekbindersgaren *n* — **e** hilo *m* bramante; hilo *m*; bramante *m*; hilo *m* de encuadernación — **i** filo *m* refe

1431 binding varnish — See vehicle.

1432 binding wire — See stitching wire.

1433 Bingham body — Ideal material that has a definite yield value, like putty, printing inks, and paints, i.e. pigment-vehicle suspensions where the pigment-vehicle ratio is sufficiently high. Term employed by Reiner.

f corps *m* de Bingham — **d** Bingham'scher Körper *m* — **n** Binghamse stof *fm*; Binghams materiaal *n* — **e** cuerpo *m* de Bingham — **i** corpo *m* di Bingham

1434 Bingham yield value — The shearing stress in dynes/cm^2 below which value flow will not take place.

f rigidité f selon Bingham — d Fließgrenze f eines Bingham'schen Körpers — n Binghamse vloeigrens fm — e límite m de fluencia según Bingham; límite m de estiraje según Bingham — i rigidità f secondo Bingham

1435 binocular viewer
f binoculaire m — d Binokel n; zweiaugiges Betrachtungsgerät n — n binoculair n — e binocular m; mirador m binocular — i binoccolo m

1436 biodegradable etching bath — Zinc, which is often used in photoengraving, is a pollutant and in some countries the maximum level in effluent discharge is 5-10 ppm. A biodegradable solution (capable of being broken down by microorganisms) eliminates the need for expensive anti-pollution equipment. Such equipment is necessary for the disposal of exhausted conventional powderless etching baths containing metal salts (zinc nitrates), unused nitric acid, petroleum solvents and non-biodegradable surfactants.
f bain m de morsure biodégradable — d biologisch abbaubares Ätzbad n — n biologisch afbreekbaar etsbad n — i bagno m d'incisione biodegradabile

1437 biography
f biographie f — d Biographie f; Lebensbeschreibung f — n biografie f; levensbeschrijving f — e biografía f — i biografia f

1438 birch; birchwood — Broadleaf, hardwood tree used in the production of special pulp by the sulphate process.
f bouleau m — d Birkenholz n; Birke f — n berkehout n; berk m — e madera f de abedul; abedul m — i legno m di betulla; betulla f

1439 birchwood — See birch.

1440 births, marriages, and deaths — In a jocular sense column of "hatches, matches, and dispatches". See also hotch.
f naissances, mariages, décès; faire-parts mpl — d Geburten, Hochzeiten und Todesfälle; kleine Familiennachrichten fpl — n geboorten, huwelijken, overlijden; familieberichten npl — e nacimientos, matrimonios, fallecimientos — i nascite, matrimoni, morti

1441 biscuit bag
f sac m à bisquit — d Zwiebackbeutel m — n beschuitzak m — e bolsa f para bizocho — i sacchetto m per biscotti

1442 bismuth — Greyish-white, hard, brittle metal with a reddish tinge. Soluble in hydrochloric acid (in presence of oxygen), hot concentrated sulphuric acid and nitric acid, insoluble in water. Symbol: Bi. Latin: bismuthum.
f bismuth m; étain m de glace — d Wismut n — n bismut n — e bismuto m — i bismuto m

1443 bismuth oxychloride — See pearl white.

1444 bismuth subnitrate — See Spanish white.

1445 bi-stable circuit — See flip-flop.

1446 bisulphite lye — Lye of sodium bisulphite in water.
f lessive f de bisulfite — d Bisulfitlauge f — n bisulfietloog fm — e lejía f de bisulfito — i lisciva f di bisolfito

1447 bisulphite of soda — See sodium bisulphite.

1448 bit — Acronym for binary information transfer; the information transferred is either a 0 or a 1. A data bit represents one pulse, one positive or negative charge, one hole or no hole in a paper tape, capable to describe one of two conditions. In US also slang term for an amount equivalent to 12.5 cents, used only in even multiples. Also any small coin.
f bit m; chiffre m binaire; digit m binaire — d Bit m; Binärzeichen n; Binärzahl f; Binärziffer f; Dualziffer f — n bit n; binair cijfer n; binair getal n; tweetallig cijfer n — e bit m; cifra f binaria; número m binario; dígito m binario — i bit m; cifra f binaria

1449 bit density — Number of bits recorded within a specified length.
f densité f de bits — d Bitdichte f — n bitdichtheid f — e densidad f de bits — i densità f di bit

1450 bite — The occasion and period of time during which a metal plate is subjected to etching by an acid or mordant.
f mordançage (en photogravure); morsure f (en gravure à l'eau-forte) — d Ätzung f; Ätzen n; Anätzung f — n etsing f; etsen n — e mordido m; mordedura f; acidulación f — i morsura f; mordenzatora f; incisione f con mordente; incisione f

1451 bite — Surface characteristic of paper enabling it to accept ink, pencil, or other impressions.
f nature f de la surface — d Oberflächenbeschaffenheit f; Farb-, Tinten- oder Bleistift-Annahmefähigkeit f — n aard m van het papieroppervlak; gesteldheid f van het papieroppervlak (bedrukbaarheid; beschrijfbaarheid) — e propiedad f superficial; calidad f de la superficie; naturaleza f de la superficie; propiedad f de la superficie — i caratteristica f superficiale (della carta)

1452 bite — White spot in an impression due to small pieces of paper or other foreign matter on the sheet. See also hickies.
f moine m; larron m — d Putzen m — n witte plek fm; spanjool m — e lardón m; ladrón m; fraile m — i macchiolina f bianca

1453 bite, gripper — See gripper margin.

1454 biting time — See etching time.

1455 bitone ink — See double-tone ink.

1456 bit rate — Speed at which a stream of bits is transmitted, usually in bits per second. See also baud.
f débit m binaire — d Bit/S; Übertragungsgeschwindigkeit f — n bitsnelheid f — e velocidad f de transferencia de bits — i velocità f di trasferimento di bit

1457 bit, router — See router bit.

1458 bitumen; mineral pitch; Judean pitch — Hydrocarbon material of natural or pyrogenous origin, or combinations of both, frequently accompanied by their non-metallic derivatives, which may be gaseous, liquid, semi-liquid, or solid, and which is completely soluble in carbon disulphide. Found in asphalt and mineral waxes. Also the components of coal that are soluble in organic solvents. Used as a binder in printing inks and in stop-off lacquers.

f bitume *m*; poix *f* de Judée — **d** Bitumen *n*; judäischer Asphalt *m* — **n** bitumen *n*; Syrisch asfalt *n* — **e** betún *m*; bitumen *m*; betún *m* de Judea — **i** bitume *m*; bitume *m* di Giudea

1459 bituminous varnish

f laque *f* à l'asphalte; vernis *m* asphaltique — **d** Asphaltlack *m* — **n** asfaltlak *m* — **e** laca *f* de asfalto; barniz *m* de asfalto — **i** lacca *f* all'asfalto

1460 bivalent; divalent — See also valence.

f bivalent; divalent — **d** zweiwertig — **n** tweewaardig — **e** bivalente — **i** bivalente

1461 bi-weekly — Appearing every two weeks or appearing twice a week. Since it is not always clear which meaning of bi-weekly is intended, the use of *semi-weekly for twice a week is preferable, because it is unambiguous. Since there is no single, unambiguous term for "every two weeks", this is the least confusing.

1462 black — See turned letter.

1463 black and white — Said of originals and reproductions displayed in monochrome (single colour), as distinguished from polychrome or multicolour.

f noir et blanc — **d** schwarz-weiß — **n** zwart/wit — **e** blanco y negro — **i** bianco e nero

1464 black and white line finish — Narrow black finish line, separated from the edge of a halftone plate by a white line of suitable width; can be mechanically introduced with a lining beveller.

f cadrage *m* par filet décollé; cadrage *m* par filet détaché — **d** schwarz-weiße Randlinie *f* — **n** zwart/wit kader — **e** recuadro *m* con raya negra — **i** inquadratura *f* in bianco e nero; inquadratura *f* a filetti staccati in bianco e nero

1465 black and white print — Print displaying only black and white tones, without colour.

f impression *f* en noir et blanc — **d** Schwarz-Weiß-Abdruck *m* — **n** zwart/wit-afdruk *m* — **e** impresión *f* en negro — **i** stampa *f* in bianco e nero

1466 black ash — Mixture of sodium carbonate, carbon and mineral matter, obtained in caustic recovery in paper manufacture.

1467 black body; perfect radiator; complete radiator; full radiator — Ideal body which would, if it existed, absorb all and reflect none of the radiation falling upon it. Its reflectivity would be zero and its absorptivity would be 100%.

f corps *m* noir; corps *m* noir parfait — **d** schwarzer Körper *m*; schwarzer Strahler *m* — **n** zwart lichaam *n*; zwartstraler *m* — **e** cuerpo *m* negro; radiador *m* completo — **i** corpo *m* nero; corpo *m* completo; corpo *m* perfecto

1468 black border — See mourning border.

1469 black-border(ed) paper — See mourning paper.

1470 black box — Any unit that forms part of an electronic circuit and that has its functions, but not its components, specified. See control unit.

1471 black box — See control unit.

1472 black copper oxide — See copper oxide.

1473 black edge — See mourning border.

1474 black-edged paper — See mourning paper.

1475 blackening — Darkening of paper by crushing at the calenders or supercalenders, which is associated with a decrease in the opacity of the sheet; caused by excessive pressure, excessive moisture in the paper, or a combination of these factors.

f noircissement *m*; plombage *m* (à la calandre) — **d** Schwärzung *f*; Vergrauung *f*; Transparenzverminderung *f*; Farbvertiefung *f* — **n** doodsatineren *n* — **e** enegricimiento *m* — **i** annerimento *m* da calandra

1476 black face type — See bold face type.

1477 black ferric oxide — See black iron oxide.

1478 blacking up — Printing of the non-image areas on a letterpress rotary.

f tableau *m* noir — **d** Grundeinfärbung *f*; Einfärbung *f* des Plattengrundes — **n** smetten *n* — **e** entintado *m* del fondo de la plancha — **i** stampa *f* di fondi

1479 black ink

f encre *f* noire — **d** schwarze Farbe *f*; Druckerschwärze *f*; Schwärze *f* — **n** zwarte inkt *m* — **e** tinta *f* negra — **i** inchiostro *m* nero

1480 black iron oxide; black ferric oxide; ferroso ferric oxide; ferro ferric oxide — Reddish-black, amorphous powder, soluble in acids, insoluble in water, alcohol and ether.

Formula: $FeO \cdot Fe_2O_3$.

f hydroxyde *m* ferrosoferrique; oxyde *m* magnétique artificiel; noir *m* de fer — **d** Ferroferrioxyd *n*; Eisenoxyduloxyd *n*; Eisenoxydschwarz *n*; Eisenschwarz *n* — **n** ijzeroxydezwart *n* — **e** óxido *m* de hierro negro — **i** ossido *m* doppio ferrosoferrico; nero *m* di ferro

1481 black lead — See graphite.

1482 black-leading — Applying graphite on the mould for making electrotypes to make it conductive.

f graphitage *m* — **d** Graphitieren *n* — **n** grafiteren *n* — **e** grafitar; plombaginar — **i** grafitatura *f*

1483 black-leading brush; graphite brush

f brosse *f* à graphiter — **d** Graphitierbürste *f* — **n** grafiteerborstel *m* — **e** cepillo *m* de grafitar; cepillo *m* para grafitar — **i** spazzola *f* per grafitatura

1484 blackleg — Worker who continues to work while his fellows are on strike.

f renard *m*; jaune *m*; non-gréviste *m*; faux frère *m* fam.; traître *m* fam. — **d** Streikbrecher *m*; nicht-zünftiger Arbeiter *m*; Arbeitswillige(r) *m* — **n** werkwillige *m*; stakingbreker *m*; onderkruiper *m* fam. — **e** esquirol *m*

1485 Black letter — Heavy-faced type in a style like that of early European handlettering and earliest printed books. See app. no. 2.

1486 black light — See ultra-violet light.

1487 black liquor — Liquor from the soda and sulphite processes from which the chemicals are removed.

f lessive *f* noire; liqueur *f* noire — **d** Schwarzlauge *f* — **n** donkere loog *fm* — **e** lejía *f* negra — **i** liscivio *m* nero

1488 black paper; black photopaper — Paper, usually of the duplex type, used to protect or wrap sensitized photographic materials. It must be free from pinholes, chemicals or other materials harmful to a photographic emulsion. The basis weight is about 40 pounds (24 x 36 - 500).

f papier *m* noir (pour plaques photographiques) — **d** Lichtschutzpapier *n*; Schwarzpapier *n* — **n** zwart papier *n* (voor fotografisch materiaal) — **e** papel *m* negro (protector de material fotográfico) — **i** carta *f* nera

1489 black photopaper — See black paper.

1490 black printer — Black plate for colour reproductions to give proper emphasis to the neutral tones and detail. Made frequently by exposure of copy on panchromatic emulsion through a yellow filter.

f négatif *m* du noir — **d** Schwarzplatte *f* — **n** zwartplaat *fm*; fotografische zwartplaat *fm* — **e** negativo *m* del negro — **i** lastra *f* del nero

1491 black printer — Image in a subtractive colour process which is to be printed in black ink.

f cliché *m* du noir — **d** Schwarzdruckplatte *f*; Schwarzplatte *f* — **n** krachtplaat *fm*; zwarte drukplaat *fm*; zwartplaat *fm* — **e** clisé *m* del negro; plancha *f* del negro — **i** lastra *f* del nero; cliscè *m* del nero; cliché *m* del nero

1492 black pull

f épreuve *f* en noir — **d** Schwarzabzug *m*; Schwarzandruck *m* — **n** proef *fm* in zwart — **e** prueba *f* en negro — **i** bozza *f* in nero; prova *f* in nero

1493 blacks — Unintended marks made on a printed sheet by raised leads, spaces or pieces or furniture. See also work up.

f marques *fpl* de blanc; blancs *mpl* levés — **d** Spiesse *mpl* — **n** gerezen wit *n* — **e** cascabelas *fpl*; blanco *m* levantado; zopilotes *mpl* — **i** spazio *m* all'aria; neri *mpl*; macchie *fpl* nere

1494 black shop — Formerly part of electrotyping where the wax mould was black-leaded as a conductor for the copper deposit forming the shell of the electrotype.

1495 black vignette; Egyptian vignette; Russian vignette — Vignetted halftone image, the borders of which are not gradually fading away to pure white, but to black.

f vignette *f* noire; vignette *f* égyptienne — **d** Schwarzvignette *f*; in Schwarz verlaufende Rasterätzung *f* — **n** autotypie *f* met verloop naar zwart — **e** esfumado *m* en negro — **i** vignetta *f* in nero

1496 blade coating; trailing blade coating; flex-ible blade coating — Method of coating a continuous paper web with a flexible blade close to a roll supported web and which allows a controlled thickness of coating slip from a pond contained by the blade to be applied to the web. Used as a *machine coating or a separate coating operation.

f couchage *m* à lame (trainante); procédé *m* à racle-traineuse; dispositif *m* enducteur à lame porteux — **d** Schlepprakelstreichverfahren *n*; Rakelstreichverfahren *n*; Glättschaberstreichverfahren *n* — **n** rakelstrijkprocédé *n* — **e** estucado *m* por cuchilla flexible de ángulo regulable — **i** patinatura *f* a racla mobile

1497 blade cut — Straight cut (or near cut) in paper running in a straight line, parallel to the direction of web travel on coated papers.

f fente *f* — **d** Schabermarkierung *f*; Rakeleinschnitt *m*; Kratzer *m* — **n** rakelsnee *fm*; rakelsnede *fm* — **i** taglio *n* nella carta

1498 blade folder — See knife folder.

1499 blanc fixe — See barium sulphate.

1500 blank; blank space — Blank space in a printed text, such as areas occupied by spaces, quadrats, leads, slugs, etc.

f blanc *m* — **d** freier Raum *m* — **n** wit *n*; open ruimte *f* — **e** espacio *m* en blanco; blanco *m*; aparte *m*; intervalo *m* — **i** bianco *m*; spazio *m* in bianco; finestra *f*

1501 blank — See blank page.

1502 blank — See carton blank.

1503 blank — See cut out.

1504 blank — See form.

1505 blank *v* — See leave *v* blank.

1506 blank book; ledger — Strong, flat and free opening binding usually canvas covered, with leather back and corners.

f livre *m* de comptabilité — **d** Register *n*; Geschäftsbuch *n*; Kontobuch *n* — **n** register *n*; kantoorboek *n* — **e** libro *m* de contabilidad; libro *m* rayado de contabilidad — **i** libro *m* contabile

1507 blanked — See blind blocked.

1508 blanket; cylinder blanket; impression blanket — Textile, rubber, plastics, cork or composition cover for the impression cylinder of a printing press.

f blanchet *m* — **d** Drucktuch *n* — **n** drukdoek *n* — **e** camisa *f*; mantilla *f* de la prensa; bayeta *f*; muletón *m*; pañete *m*; paño *m* Me; mantilla *f* del cilindro; hule *m* — **i** telo *m*; rivestimento *m* del cilindro

1509 blanket; offset blanket; rubber blanket — On offset presses a fabric-reinforced sheet of rubber to transfer the impression from the plate on to the paper.

f blanchet *m* (en caoutchouc); blanchet *m* de caoutchouc — **d** Gummituch *n* — **n** rubberdoek *n*; offsetdoek *n* — **e** mantilla *f* de caucho; tela *f* de caucho; franela *f* de caucho; mantilla *f* litográfica — **i** tessuto *m* gommato; cauccù *m*; telo *m* gommato; blanket *m*

1510 blanket — Upper cloth of a paper ruling machine.

f blanchet *m* de la régleuse — **d** Liniiertuch *n*; Liniertuch *n*; oberes Tuch *n* einer Liniiermaschine — **n** doek *n* van de linieermachine — **e** paño *m* superior de la rayadora — **i** tessuto *m* superiore della macchina lineatrice; telo *m* superiore della rigatrice

1511 blanket; drying blanket; drying felt — Moleskin or felt impregnated with hygroscopic salts to dry pigment paper for intaglio printing. Reconditioned in a drying oven for subsequent use.

f feutre *m* sécheur; molleton *m* sécheur — **d** Trockendecke *f*; Trockenfilz *m* — **n** droogvilt *n* — **e** fieltro *m* secador; fieltro *m* enjugador — **i** feltro *m* per asciugante; mollettone *m* asciugante

1512 blanket bar — See blanket wind.

1513 blanket clamp — Clamp to fix the end of a cylinder packing to the impression cylinder of a letterpress rotary press.

f dispositif *m* d'accrochage du blanchet — **d** Tuchklemmvorrichtung *f* — **n** klemlijst *fm*; doekklemlijst *fm* — **e** varilla *f* de la mantilla; barra *f* de la mantilla; vástago *m* de la mantilla — **i** morsetto *m* per il fissaggio del blanket

1514 blanket clamping shaft

f arbre *m* de serrage de blanchet — **d** Gummituchspannwelle *f* — **n** doekspanas *fm* — **e** eje *m* de la barra tensora de la mantilla — **i** albero *m* di serraggio del blanket

1515 blanket creep — Slight forward movement of that part of the blanket surface that is in contact with the plate or paper.

f glissement *m* du blanchet — **d** leichtes Vorwärts-Kriechen *n* des Gummituchs; Wandern *n* des Gummituchs — **n** kruipen *n* van het rubberdoek — **e** deslizamiento *m* de la mantilla — **i** scorrimento *m* del blanket

1516 blanket cylinder — On offset presses, the cylinder where the rubber blanket is mounted. It receives the inked design from the plate cylinder and offsets or transfers it onto the surface to be printed.

f cylindre *m* de blanchet; cylindre *m* porteblanchet; cylindre *m* porte-caoutchouc — **d** Gummituchzylinder *m* — **n** rubberdoekcilinder *m* — **e** cilindro *m* de la mantilla; cilindro *m* portamantilla — **i** cilindro *m* portablanket; cilindro *m* portacaucciù; cilindro *m* portatessuto gommato

1517 blanket embossing — See embossing.

1518 blanket pins — Pins to attach the blanket or canvas to another sheet on the blanket wind of a letterpress rotary.

f picots *mpl* de blanchet — **d** Tuchspannstifte *mpl* — **n** doekpennen *fmpl* — **e** pinchos *mpl* fijadores de la mantilla — **i** puntine *fpl* per fissare il blanket

1519 blanket scum — A whitish deposit built up on the blanket and wearing the offset plate. Frequent stops for cleaning are required. It is caused by lack of moisture resistance of the coating. It softens and adheres to the blanket.

f souillure *f* du blanchet par le papier; encrassement *m* du blanchet par le papier — **d** Offsettuch-

verschmutzung *f*; Verschmutzung *f* des Offsettuchs; Absetzen *n* von Staub auf Gummituch; Haften *n* des Staubes am Gummituch; Verunreinigtwerden *n* des Gummituches — **n** offsetdoekvervuiling *f*; vervuiling *f* van het rubberdoek — **e** ensuciamiento *m* de la mantilla; empolvamiento *m* de la mantilla — **i** insudiciamento *m* del tessuto gommato; insudiciamento *m* del blanket

1520 blanket sheet — Newspaper, larger than average, common in mid-19th-century England.

f journal *m* anglais grand format — **d** Zeitung *f* überdurchschnittlichen Formats — **n** krant *fm* van buitengewoon formaat — **e** diario *m* de tamaño extraordinario; periódico *m* de tamaño extraordinario — **i** giornale *m* più largo della norma

1521 blanket stretch

f allongement *m* du blanchet — **d** Tuchausdehnung *f*; Ausdehnung *f* des Gummituchs — **n** rek *m* van het (rubber)doek — **e** estiramiento *m* de la mantilla — **i** allungamento *m* del blanket

1522 blanket tail — Piece of the blanket sewn to the end of the blanket to attach it to the blanket bar on a letterpress rotary press.

f fourreau *m* (de blanchet) — **d** Drucktuchschlaufe *f*; angenähtes Stoffende *n* am Drucktuch — **n** staart *m* van de drukdoek — **e** cola *f* de la mantilla; rabo *m* de la mantilla; extremo *m* de la mantilla — **i** coda *f* del blanket

1523 blanket thickness gauge — Micrometer to measure the offset blanket under uniform pressure.

f micromètre *m* pour mesure des épaisseurs de blanchet — **d** Gummituchdickenmesser *m* — **n** rubberdoekdiktemeter *m* — **e** calibrador *m* para la mantilla — **i** spessimetro *m* per blanket

1524 blanket-to-blanket press; unit perfecting press — Perfecting press in which the web runs between two blanket cylinders, each of which acts as the impression cylinder for the other.

f presse *f* blanchet sur blanchet — **d** Gummi-gegen-Gummi-Presse *f* — **n** doek-op-doek-pers *f* — **e** máquina *f* caucho contra caucho; prensa *f* caucho contra caucho — **i** macchina *f* gomma (contro) gomma

1525 blanket, top — See top blanket.

1526 blanket wash — Solvent to clean ink from the offset blanket.

f solution *f* de lavage pour blanchet — **d** Tuchwaschmittel *n*; Offsettuchwaschmittel *n* — **n** doekwasmiddel *n*; offsetdoekwasmiddel *n* — **e** limpiamantillas *m*; solución *f* para limpiar mantillas; solución *f* lavamantilla — **i** detersivo *m* per blanket

1527 blanket wind; blanket bar — Bar to tension the loose end of a cylinder packing blanket, usually a rotating bar round which the blanket is wound.

f tringle *f* à blanchet — **d** Tuchspannstange *f*; Tuchstange *f* — **n** doekspanstang *fm*; doekstang *fm*; spanstang *fm* — **e** barra *f* tensora de la mantilla — **i** barra *f* di tensione del blanket

1528 blank form — See form.

1529 blanking — See blind tooling.

1530 blank line — See white line.

1531 blank material — Material lower than high to fill all non-printing areas of a letterpress forme.
f blancs *mpl* — **d** Blindmaterial *n*; blindes Material *n* — **n** witmateriaal *n* — **e** material *m* de blancos; blancos *mpl* — **i** bianchi *mpl* tipografici

1532 blank page; white page; blank — Unprinted page.
f page *f* blanche — **d** Leerseite *f*; Vakantseite *f*; Blankseite *f*; unbedruckte Seite *f* — **n** blanco pagina *fm*; onbedrukte pagina *fm*; witpagina *fm* — **e** página *f* en blanco; página *f* sin impresión — **i** pagina *f* bianca; pagina *f* non stampata

1533 blank paper — Unprinted or unwritten paper.
f papier *m* en blanco — **d** blanko Papier *n*; unbedrucktes Papier *n* — **n** blanco papier *n*; onbedrukt papier *n* — **e** papel *m* en blanco; papel *m* sin impresión — **i** carta *f* bianca

1534 blanks — Class of paperboard with a thickness of 0.3-2 mm (0.012-0.078 in), with corresponding basis weights of 54-350 kg (120-775 pounds 22 x 28 - 500). Either single-ply Fourdrinier board, multi-ply cylinder board, or laminations of these. The liner may be made of de-inked stock, clean shavings, bleached or unbleached groundwood, or chemical pulps. The surface may be either coated or uncoated; if uncoated, called plain mill blanks; if coated, coated blanks, or if coated on one side car card. Generally made to produce maximum stiffness and surface smoothness. Used for various purposes where stiffness and good printing qualities are required as in window displays, etc.
f cartons-pancartes *mpl*; cartons *mpl* pour affiches — **e** cartoncillos *mpl*; cartulinas *fpl* gruesas — **i** cartone *mpl* rigido per pubblicità

1535 blank sheet
f feuille *f* non-imprimée — **d** unbedruckter Bogen *m* — **n** onbedrukt vel *n*; niet-bedrukt vel *n* — **e** hoja *f* no impresa; hoja *f* en blanco — **i** foglio *m* non stampato

1536 blank space — See blank.

1537 blazed oils — See blown oils.

1538 bleach *v* — In photography, to convert the silver image of a negative or print to a silver halide, either to remove the image or to change its tone.
f blanchir — **d** bleichen; ausbleichen — **n** bleken; uitbleken — **e** blanquear — **i** sbiancare

1539 bleached kraft paper — Paper made from bleached kraft pulp.
f papier *m* kraft blanchi — **d** gebleichtes Kraft-papier *n* — **n** gebleekt kraft(papier) *n* — **e** papel *m* kraft blanqueado — **i** carta *f* kraft bianca

1540 bleached pulp — Pulp with a high degree of whiteness as a result of a chemical post-treatment.

f pâte *f* à papier blanchie; pâte *f* blanchie — **d** gebleichter Halbstoff *m*; gebleichter Zellstoff *m* — **n** gebleekte celstof *fm* — **e** pasta *f* blanqueada; pulpa *f* blanqueada — **i** pasta *f* bianchita

1541 bleacher — See potcher.

1542 bleaching — Whitening of photographic images during intensification, or for purpose of removing the image entirely.
f blanchiment *m* — **d** Bleichen *n*; Ausbleichen *n* — **n** bleken *n*; uitbleken *n* — **e** blanqueo *m*; blanqueado *m* — **i** imbianca *f*

1543 bleaching — Destruction, or modification to a greater or lesser extent, of the colour of pulps to improve their whiteness.
f blanchiment *m* — **d** Bleichung *f*; Bleichen *n* — **n** bleking *f*; bleken *n* — **e** blanqueado *m*; descolorado *m* — **i** sbianca *f*

1544 bleaching agent — Agent to bleach paper or textiles, such as hydrogen peroxide (the most common), sodium hypochlorite, sodium peroxide, sodium chlorite, calcium hypochlorite, and many organic chlorine derivatives.
f agent *m* de blanchiment; produit *m* de blanchiment — **d** Bleichmittel *n*; bleichendes Mittel *n* — **n** bleekmiddel *n* — **e** agente *m* blanqueador — **i** agente *m* di sbianca; sbiancante *m*

1545 bleaching bath — Chemical solution used in processing to convert the metallic silver to a soluble silver salt.
f bain *m* de blanchiment — **d** Bleichbad *n* — **n** bleekbad *n* — **e** baño *m* de blanqueo — **i** bagno *m* sbiancante

1546 bleaching engine — See potcher.

1547 bleaching liquor — In papermaking, a solution of a bleaching agent, usually calcium hypochlorite or sodium hypochlorite.
f liqueur *f* de blanchiment; solution *f* de blanchiment; liqueur *f* chlorurée — **d** Bleichflüssigkeit *f*; Bleichlösung *f*; Bleichlauge *f*; Bleichwasser *n* — **n** bleekloog *fm*; bleekvloeistof *fm* — **e** solución *f* de blanquear; lejía *f* de blanqueo — **i** liscivio *m* di sbianca

1548 bleaching powder — Mixture of calcium hydroxide, chloride, and hypochlorite, used to bleach. See also calcium hypochlorite.
f poudre *f* à blanchir — **d** Bleichpulver *n* — **n** bleekpoeder *n* — **e** polvo *m* de blanqueo — **i** cloruro *m* di calcio

1549 bleaching powder — See calcium hypochlorite.

1550 bleachout process — Method to make line drawings on photographs and silverprints with waterproof inks, the image serving as a guide to the artist and afterwards removed by bleaching, leaving only the drawing on the surface of the paper.
f procédé *m* de blanchiment; procédé *m* de blanchissage — **d** Ausbleichverfahren *n* — **n** uitbleekprocédé *n* — **e** procedimiento *m* de blanquear — **i** processo *m* d'imbiancamento

1551 bleach test *US*; **tint test** *GB* — Method to

measure tinctorial strength of an ink or toner, usually by mixing a small portion of the ink or toner with a large amount of white base and evaluating the tinting strength of the ink versus a control standard.

d Weißausmischungsprobe *f*; Weißausmischung *f* — **n** wit-uitmengproef *f*; wit-uitmenging *f* — **e** ensayo *m* de blanqueo; prueba *f* de blanqueo — **i** prova *f* della forza di colore; saggio *m* della forza di colore

1552 bled-off plates — Plates designed to come to the extreme edge of the leaf so that their edges are deliberately cut-off.

f illustrations *fpl* à fond perdu; clichés *mpl* à fond perdu; impressions *fpl* à plein papier; illustrations *fpl* franc-bord; illustrations *fpl* rognées à vif — **d** am Rande abfallende Bilder *npl*; am Rande angeschnittene Bilder *npl* — **n** aflopende beelden *npl*; aflopende illustraties *fpl* — **e** ilustraciones *fpl* decayentes; ilustraciones *fpl* a sangre; láminas *fpl* sin margen; ilustraciones *fpl* voladas *Gu*; grabados *mpl* a sangre; grabados *mpl* sin margen; clisés *mpl* a sangre; clisés *mpl* sin margen — **i** illustrazioni *fpl* tagliati al vivo; illustrazioni *fpl* a bordi rifilati; illustrazioni *fpl* che si estendono oltre i limiti di rifilatura

1553 bleed — Extra amount of tone added to edges and contours of the sketch that is later to be die-cut into a conforming shape. Window displays and contour cards are usually cut to shape to make them look more realistic. This bleed permits the slight variations that occur when the reproduction is being die-cut.

f fond *m* perdu — **d** Beschnitt *m*; Übergriff *m*; Schnittrand *m* — **n** vlees *n*; aangezette rand *m* — **e** borde *m* perdido; borde *m* a sangre; borde *m* guillotinado a sangre

1554 bleed — Slight extension or thickening of printed detail, usually of lighter colour or tint, to produce colour overlap zones, so that a white gap will not show in printing when slight variations in register occur throughout the press run.

f engrainement *m* des couleurs pour repérage — **d** Übergriff *m*; Farbüberstand *m* — **n** overvul *m* — **i** estensione *f* delle zone stampanti

1555 bleed — Term used by gravure printers when the ink trails from the back edge of a solid area, owing to low ink viscosity, high pressure or a combination of both, particularly on non-absorbent materials.

1556 bleed *v* — To dissolve in the fountain etch and cause tinting (of lithographic ink pigment).

f se laver — **d** ins Wasser gehen — **n** emulgeren; tinten; in het water gaan — **e** lavarse; emulsionar — **i** andare in acqua; emulsificarsi

1557 bleeding — Spreading or running of a pigment colour by the action of a solvent, e.g. water, alcohol.

f déteintage *m* — **d** Bluten *n*; Ausbluten *n*; Durchbluten *n*; Auslaufen *n* — **n** bloeden *n*; uitbloeden *n* — **e** correrse; descolosimiento *m* — **i** spandimento *m*; diffusione *f*

1558 bleeding colour — See bleeding ink.

1559 bleeding ink; bleeding colour — Ink which tends to spread or run when wetted by a solvent.

f couleur *f* soluble — **d** ausblutende Farbe *f*; auslaufende Farbe *f* — **n** bloedende inkt *m* — **e** tinta *f* que se corre; tinta *f* que se decolora — **i** inchiostro *m* che sanguina

1560 blend — A mixture, such as a mixture of solvents or a mixture of inks.

f mélange *m* — **d** Mischung *f* — **n** mengsel *n* — **e** mezcla *f*; mixtura *f*; mixtión *f* — **i** mescolanza *f*; miscela *f*

1561 blind — Dot on a photographic negative or positive, which lacks density and is so transparent that any light going through it falsifies the values desired. To be photographically effective all dots must be opaque and all transparent areas must be absolutely clear.

f blond — **d** blind — **n** ongedekt — **e** cegado — **i** accecato

1562 blind blocked; blind tooled; blanked; blind embossing — Embossed lettering on book covers which are not inked or gilded.

f doré à froid; gaufré à froid; tracé — **d** mit Blinddruck; mit Blindprägung — **n** met blinddruk; met blindstempel — **e** estampado en seco; timbrado en seco — **i** impresso a secco; stampato in rilievo a secco; goffrato

1563 blind blocking — In bookbinding, impression by hot tools only, without goldleaf or ink.

f empreinte *f* à chaud — **d** Blindprägung *f*; Heißprägung *f* — **n** blindstempeling *f*; blindpreging *f* — **e** estampado *m* a seco — **i** stampa *f* a secco; impressione *f* a caldo alla pressa

1564 blind embossing — See blind blocked.

1565 blind folio — A page number counted but not actually expressed in the make-up of a book.

f pagination *f* non indiquée — **d** Blindpaginierung *f*; gezählte, jedoch nicht paginierte Seite *f* — **n** ongenummerde pagina *fm*; nietgenummerde pagina *fm* — **e** página *f* sin numerar; página *f* sin número — **i** pagina *f* non numerata

1566 blind image — In lithography, an image that has lost its ink-receptivity.

f image *f* filée — **d** Blindbild *n* — **n** blind gelopen beeld *n* — **e** imagen *f* falta de afinidad para la tinta — **i** immagine *f* accecata

1567 blinding — Gradual weakening of the print on an offset plate during the run owing to the image base not accepting enough ink. This can be corrected.

f filage *m* (du report) — **d** Blindlaufen *n*; Blindwerden *n*; Schwinden *n* der Zeichnung auf der Druckplatte — **n** blindlopen *n* — **e** falta *f* de afinidad para la tinta; repelado *m* de la tinta; picado *m* de la tinta; cegado *m* — **i** accecamento *m*

1568 blinding — See blind tooling.

1569 blind keyboard; blind perforator — Input keyboard which usually produces a punched paper tape, but without visual character representation, such as a hard copy or soft copy display. Usually a blind keyboard produces a TTS six-level code.

f clavier *m* non renseigné — **d** Taster *m* ohne Klarschrift; Tastapparat *m* ohne Klartextwiedergabe — **n** blind toetsenbord *n* — **e** teclado *m* ciego — **i** tastiera *f* cieca

1570 blind P — See reversed P.

1571 blind perforator — See blind keyboard.

1572 blind print — See blind sheet.

1573 blind sheet; blind print — Sheet of paper printed by mistake on one side only.

f feuille *f* en blanc — **d** Schimmel *m*; Schimmelbogen *m*; unbedruckt gebliebener Bogen *m* — **n** eenzijdig bedrukt vel *n*; misdruk *m* — **e** impresión *f* de blanco; plana *f* — **i** bianca *f*

1574 blind stamp — Impression from a die on a cover, letterhead, certificate, or other piece of printing without the use of a colour.

f gaufrage *m*; dorure *f* à froid — **d** Blindprägung *f* — **n** blindstempel *n* — **e** estampado *m* en seco — **i** stampa *f* a secco

1575 blind tooled — See blind blocked.

1576 blind tooling; blanking; blinding; dumb tooling; à froid *misn.* — In bookbinding, impressing heated tools on leather by hand, without the use of goldleaf or ink. Sometimes referred to as antique.

f dorure *f* à froid; tirage *m* à froid; estampage *m* à froid — **d** Blindprägung *f* — **n** blindstempelen *n* — **e** estampado *m* en seco — **i** impressione *f* a secco

1577 blister — Small raised area, caused by expansion of trapped gas or other fluid beneath the paper surface, when the paper is dried too suddenly on the drying cylinder or when the felts are not in good condition.

f boursoufflure *f*; cloque *f* — **d** Aufblähung *f*; Blase *f* — **n** blaas *fm*; blaar *fm* — **e** ampolla *f*; burbuja *f*; sopladura *f* — **i** bolla *f*; vescica *f*

1578 blistered mat — In stereotype, mat which has been scorched by pouring too hot metal, the stereo obtained therefrom having a blistered appearance. See also cokey face.

f matrice *f* grésillée — **d** versengte Mater *f* — **n** verschroeide matrijs *f* — **e** flan *m* con burbujas; molde *m* con burbujas; cartón *m* con burbujas — **i** flano *m* bruciacchiato

1579 blistering

f formation *f* de soufflures; cloquage *m*; formation *f* de bulles — **d** Blasenbildung *f*; Blasenbilden *n* — **n** blaasvorming *f* — **e** ampollar — **i** formazione *f* di bolle; vescicatura *f*

1580 blistering of blanket — Local separation of the surface of an offset blanket from the main body of the material.

f décollement *m* d'un blanchet — **d** Abplätzen *n*; sich loslösen — **n** loslaten *n* van het rubber — **e** desprenderse — **i** sfaldamento *m* del blanket

1581 blister test — Test method for strongly beaten paper (e.g., greaseproof paper). By heating the paper in a flame, blisters will result because the water in the paper changes to steam.

f essai *m* de cloquage — **d** Blasenprobe *f*; Blasenversuch *m* — **n** blaasproef *fm* — **e** ensayo *m* de burbujas — **i** prova *f* delle bolle

1582 block *GB* — Letterpress printing plate.

f cliché *m* (de photogravure); bois *m* *obs.* — **d** Druckstock *m*; Klischee *n* — **n** cliché *n* — **e** clisé *m*; cliché *m*; plancha *f*; grabado *m*; fotograbado *m* — **i** cliché *m*; cliscè *m*

1583 block — Computer term. Group of consecutive bytes, words or records of convenient size, transferred as one unit for input, processing or output.

f bloc *m* — **d** Block *m* — **n** blok *n* — **e** bloque *m* — **i** blocco *m*

1584 block — Solid metal stamp to impress a design on a cover.

f plaque *f* à dorer — **d** Stempel *m* — **n** stempel *n*; bandstempel *n* — **e** bronce *m* de estampación; plancha *f* para estampar tapas; molde *m* para estampar tapas — **i** stampo *m* per impressioni

1585 block — See mount.

1586 block — See tablet.

1587 block bending apparatus

f machine *f* à cintrer des clichés; cintreuse *f* — **d** Klischeebiegeapparat *m* — **n** clichébuigmachine *f* — **e** máquina *f* curvadora de clichés — **i** macchine *f* curvatrice (di lastre di stampa)

1588 block book; xylographic book — Book printed from wooden blocks in which the text was cut in relief.

f incunable *m* xylographique; impression *f* tabellaire — **d** Blochbuch *n*; Holztafeldruck *m* — **n** blokboek *n* — **e** libro *m* impreso con grabados xilográficos — **i** libro *m* stampato in xilografia; libro *m* stampato in silografia

1589 block bottom bag — Bag made with a lengthwise lap joint, creased each side to fold flat, with the bottom made by first folding into a flat diamond shape, then over the points to meet in the centre. This gives a bag opening out into a square cross-section.

f sac *m* à fond carré — **d** Blochbeutel *m*; Klotzbodenbeutel *m*; Klotzbeutel *m*; Blockbodenbeutel *m* — **n** blokbodemzak *m*; klotsbodemzak *m* — **e** bolsa *f* de bloque; bolsa *f* de fondo cuadrangular — **i** sacchetto *m* con fondo piatto

1590 block calendar; tear-off calendar

f calendrier *m* bloc; calendrier *m* à effeuiller; calendrier *m* en effeuilles; calendrier *m* en feuilles; calendrier *m* éphéméride — **d** Abreißkalender *m* — **n** scheurkalender *m* — **e** calendario *m* en bloc; calendario *m* de taco; almanaque *m* de pared; almanaque *m* en hojas — **i** calendario *m* a blocco; calendario *m* da sfogliare

1591 block diagram; schematic diagram — Diagram in which the essential units of any system are drawn in the form of blocks, usually rectangles; their relation to each other is indicated by appropriate connecting lines.

f schéma *m* des liaisons; schéma *m* fonctionnel — **d** Blockdiagramm *n*; Blockschaltbild *n*; Blockschaltplan *m*; Funktionsschema *n* − **n** blok-diagram *n*; blokschema *n* − **e** esquema *m* de bloques; esquema *m* funcional − **i** schema *m* a blocchi; schema *m* funzionale

1592 block drawing − See working drawing.

1593 blocked-up type − Type that is composed but cannot be sent to machine for various reasons, because it is out on proof, awaiting author's corrections, scarcity of sorts, paper stock not yet reveived, etc.

f matière *f* en attente; matière *f* bloquée; bloqué *m* − **d** blockierte Schrift *f* − **n** geblokkeerd zetsel *n*; vast staand zetsel *n* − **i** composizione *f* bloccata in attesa di stampa

1594 block gap − See inter-block gap.

1595 block height gauge
f comparateur *m* (en pression) − **d** Klischeehöhen-messer *m*; Klischeeprüfgerät *n* − **n** cliché-hoogte-meter *m* − **e** medida *f* de la altura del cliché − **i** calibro *m* per misurare l'altezza dei clichè

1596 blocking; stamping − Impressing book covers, etc., by means of a hot die, brass types or blocks.

f empreinte *f* à la plaque − **d** Prägung *f*; Prägen *n* − **n** stempelen *n*; stempeling *f* − **e** estampado *m* con molde; estampado *m* con bronce − **i** stampigliatura *f*

1597 blocking − Mounting or nailing printing plates on permanent wood supports.

f montage *m* − **d** Aufklotzen *n*; Montieren *n*; Montage *f*; Aufholzen *n* − **n** monteren *n*; montage *f*; opspijkeren *n* − **e** montaje *m* de grabados − **i** montaggio *m*

1598 blocking − The sticking together between layers of material such as might occur under moderate pressure and/or temperature in storage or use, to the extent that damage to at least one surface is visible upon their separation.

f blocking *m*; adhérence *f* mutuelle; adhérence *f* de contact entre feuilles − **d** Verblocken *n*; Zusammenkleben *n* bedruckter Bogen im Stapel; Backen *n*; Zusammenbacken *n*; Blocken *n*; unerwünschtes Kleben *n* − **n** blokken *n*; kleven *n*; aan elkaar blijven plakken *n* − **e** adherencia *f*; pegajosidad *f*; blocking *m* − **i** aderenza *f* superficiale

1599 blocking factor − The number of logical units (records) in a physical unit (block) of data.

f facteur *m* de groupage − **d** Blockungsfaktor *m* − **n** blokkingsfactor *m* − **e** factor *m* de agrupa-miento − **i** fattore *m* d'aggruppamento

1600 blocking flush − Trimming block or mounted printing plate on one or more edges, so that printing surface is flushed with block.

f montage *m* pour rognage à vif − **d** Aufklotzen *n* ohne Facettenrand − **n** monteren *n* zonder facet; "gelijk-af" monteren *n* − **e** montaje *m* al ras de la letra − **i** montaggio *m* a filo

1601 blocking foil − Paper foil with a special coating of gold, coloured, pigments or metal, used for blocking. See also gold leaf.

f encre *f* sur film; encre *f* film; encre *f* en rouleau − **d** Prägefolie *f*; Abziehfolie *f* − **n** stempel-fo(e)lie *fm* − **i** foglio *m* per stampa

1602 blocking lumber − See mounting lumber.

1603 blocking nails − Small, flat-headed steel wire nails to fasten plates to blocks.

f pointes *fpl* pour montage des clichés − **d** Klischeemontiernägel *mpl*; Klischeenägel *mpl* − **n** clichéspijkers *mpl* − **e** clavos *mpl* para montar clichés; puntas *fpl* para montar clisés − **i** chiodi *mpl* di montaggio

1604 blocking press; stamping press − Heavily built press, often hydraulic, fitted with heating devices in which covers of books are stamped. See also arming press.

f presse *f* à dorer − **d** Vergoldepresse *f*; Gold-prägepresse *f*; Buchbinderprägepresse *f* − **n** verguldpers *f* − **e** prensa *f* de dorador; prensa *f* para dorar − **i** pressa *f* a dorare

1605 blocking slab − Cast iron plate about 2.5 cm (1 in) thick, which provides a solid bed or surface on which blocking of plates can be performed.

f marbre *m* de montage − **d** Aufklotzblock *m*; Montierblock *m* − **n** tafel *fm* voor het monteren van clichés − **e** platina *f* para montaje de clichés − **i** piano *m* per il montaggio delle lastre sulle basi

1606 blocking test − Test where two prints are placed face to face and put between two pieces of plate rubber each 6.45 cm^2 (1 sq in). One and one-half pounds of weight is applied to the top piece of rubber and the whole set-up is kept at 49 °C for 16 hours. When removed, the ink should not stick or pick.

f essai *m* de la résistance à l'adhérence en surface − **d** Blockpunktprobe *f*; Blocktest *m*; Block-festigkeitsprobe *f*; Prüfung *f* auf Blocken − **n** blokproef *fm* − **e** ensayo *m* de bloqueo − **i** prova *f* di resistenza all'autoadesione

1607 block length − The number of bytes or words in a block, fixed or variable according to the system used.

f longueur *f* de bloc − **d** Blocklänge *f* − **n** blok-lengte *f* − **e** longitud *f* de bloque − **i** lunghezza *f* di blocco

1608 block letters − Originally types cut from wood.

f caractères *mpl* en bois − **d** Holzbuchstaben *mpl*; Holzschrift *f* − **n** houten letters *fmpl* − **e** tipos *mpl* de madera − **i** caratteri *mpl* di legno

1609 block letters − Characters suggesting wood types, square cut and without serifs. Also writing imitating printing characters.

f lettres *fpl* moulées; caractères *mpl* bâtons; capitales *fpl* bâtons; moulé *m* − **d** Blockschrift *f*; Druckschrift *f* − **n** blokletters *fmpl* − **e** tipos *mpl* abastonados; tipos *mpl* de palo seco − **i** caratteri *mpl* bastoncini; caratteri *mpl* bastoni

1610 block leveler − See block leveller.

1611 block leveller *GB*; **block leveler** *US*; **type-high planer** − Machine for milling (planing) wooden and metal mounts of relief printing plates to perfect planity or evenness.

f planeuse-rectifieuse *f* — **d** Planfräsmaschine *f*; Planfräser *m*; Hobelmaschine *f*; Plattenschabemaschine *f* — **n** vlakfreesmachine *f* — **e** acepilladora *f*; cepilladora *f*; niveladora *f*; calibradora *f* — **i** spianatrice *f*

1612 block maker — See process engraver.

1613 block *v* **out** — See opaque *v*.

1614 block *v* **out** — See stop *v* out.

1615 blockout halftone — See outlined halftone.

1616 block pull — See engraver's proof.

1617 block stitching
f piqûre *f* à plat — **d** Blockheftung *f* — **n** hechting *f* door het plat; nieten *n* door het plat — **e** costura *f* a través de la tapa; cosido *m* en bloc — **i** cucitura *f* attraverso la copertina; cucitura *f* a blocco; cucitura *f* di traverso a punti metallici

1618 bloodstone — See agate.

1619 bloom — Surface film on glass or packaging materials resulting from attack by the atmosphere or from the deposition of smoke or other vapour, e.g. the grey, milky veil appearing on the surface of some flexo prints. Usually caused by too rapid evaporation of the solvent, resulting in condensation on the surface, or excess of acid in the ink varnish or excessive use of driers. See also gloss.
f altération *f* — **d** Beschlag *m* — **n** aanslag *m* — **e** empañamiento *m* — **i** appannamento *m*

1620 blotch — Smear or smudge caused by defective rollers, ink, paper, overlays, underlays, etc., in a print.
f tache *f* d'encre; tache *f* de couleur; pâté *m* — **d** Tintenklecks *m*; Tintenfleck *m* — **n** inktvlek *fm*; vuile vlek *fm* — **e** borrón *m*; mancha *f* — **i** chiazza *f* d'inchiostro

1621 blotter — Sheet of blotting paper often with advertising matter.
f buvard *m* — **d** Löschblatt *n* — **n** vloeiblad *n* — **e** teleta *f*; secante *m* — **i** foglio *m* di carta assorbente

1622 blotting paper — Unsized and absorbent paper made of specially prepared rag. The cheapest kinds are merely unsized printing papers. They range from 60 to 140 pounds (19 x 24 - 500).
f papier *m* buvard; papier *m* brouillard — **d** Löschpapier *n*; Fließpapier *n* — **n** vloeipapier *n*; inktvloei *n* — **e** papel *m* secante; papel *m* chúpon — **i** carta *f* assorbente; carta *f* suga; cartasuga *f*

1623 blower foot; gooseneck; second-sheet stop — Metal air-blowing device, right angle air nozzle to blow and hold down back edge of sheets.
f pied-de-biche *m*; pied *m* souffleur — **d** Trennsauger *m*; Drückerfuß *m*; Tastenbläser *m*; Stapeltaster *m* mit Blasfuß — **n** blaasvoet *m* — **e** boquilla *f* del soplador; chupador *m* Ar — **i** piedino *m* soffiatore

1624 blower Linotype — First commercially used Linotype (1886), so called because matrices were carried into the assembler by a blast of air.

1625 blown linseed oil — Linseed oil which is boiled, i.e., its viscosity is increased by having air bubbled through it while heated to 125 °C. The reaction is mainly oxidation followed by polymerization of the oxidized molecules. The resulting product dries to a harder film than heat-bodied oils.
f huile *f* de lin soufflée — **d** gesprühtes Leinöl *n* — **n** geblazen lijnolie *fm*; bruine lijnolie *fm* — **e** aceite *m* de linaza soplado — **i** olio *m* di semi di lino soffiato

1626 blown oils; blazed oils; base oils; thickened oils; oxidized oils; polymerized oils — Products obtained by blowing air through heated drying or semi-drying oils (linseed, rape, fish, whale, castor).
f huiles *fpl* soufflées — **d** Blasöle *npl*; gesprühte Öle *npl* — **n** geblazen oliën *fmpl* — **e** aceites *mpl* inyectados — **i** oli *mpl* soffiati

1627 blown-up halftone; halftone blowup; blowup — Coarse screen halftone, containing exaggerated lights and shadows, resulting from enlargement of halftone negative or photo-enlargement of proof from a halftone.
f simili *f* agrandie; reproduction *f* à grosse trame — **d** Rasterprojektion *f* — **n** reproduktie *f* naar een vergroot gerasterd model — **e** ampliación *f* autotípica; medio tono *m* amplificado; medio tono *m* de trama ampliada — **i** ingrandimento *m* da negativo retinato

1628 blowout — Highlight halftone, especially one made from a highlight negative.
f simili *f* à blancs purs — **d** Hochlichtaufnahme *f* — **n** hooglicht-opname *f* — **e** medio tono *m* recortado; medio tono *m* con partes en blanco; medio tono *m* directo siluetado — **i** mezzatinta *f* senza punti nelle alte luci

1629 blow *v* **out a case**
f souffler une casse — **d** einen Schriftkasten ausblasen; einen Setzkasten ausblasen — **n** een letterkast schoonblazen — **e** desempolvar una caja — **i** soffiare una cassa

1630 blowpipe — Tube to direct a current of air or gas into a flame for local concentration of heat and used in soldering.
f chalumeau *m* — **d** Blasrohr *n* — **n** blaaspijp *fm* — **e** soplete *m*; boquilla *f* sopladora — **i** cannello *m* ferruminatorio

1631 blow pit — See diffuser.

1632 blow torch — See fan.

1633 blowup — See blown-up halftone.

1634 blow *v* **up** — See enlarge *v*.

1635 blue black — Black with bluish highlights.
f noir *m* bleu; noir *m* bleuâtre — **d** Blauschwarz *n* — **n** blauwzwart *n* — **e** negro *m* azulado; negro *m* azuloso — **i** nero *m* bluastro

1636 blue filter — Filter to make the yellow plate, photographing the red and green.
f filtre *m* bleu; écran *m* bleu — **d** Blaufilter *n* — **n** blauwfilter *n* — **e** filtro *m* azul — **i** filtro *m* blu

1637 blue-green — See cyan.

1638 blue key — Blueprint on glass or a vinyl plastic sheet of a basic design containing all elements with register marks, used as a guide for stripping a flat of photographic elements of other colours to register. For deep-etch plates, positives

are usually stripped to flats made of vinyl plastics.
f faux décalque *m* en bleu — **d** Blauschlüssel *m* — **n** blauwkopie *f* — **e** plantilla *f* azul de montaje; plantilla *f* azul de registro — **i** tracciato *m* blu (per montaggio)

1639 blue lacquer — See cold enamel.

1640 blue-line — Blueprint on an offset plate used as a guide in applying tusche or crayon handwork.
f faux décalque *m* — **d** Klatsch *m*; Klatschdruck *m*; Abklatsch *m* — **n** klats *m* — **e** decalco *m* — **i** decalco *m* di guida

1641 blue plate — See cyan printer.

1642 blueprint — Photographic print, usually by contact with negative on paper, glass or metal. Guide for artists in making keyed art for multicolour. Also used to secure a copy of unit negatives or of flat for checking layout and imposition. Ozalid prints used for prints from photographic positives.
f bleu *m*; ozalide *m*; cyanotypie *f* — **d** Blaupause *f*; Lichtpause *f* — **n** blauwdruk *m*; blauw gekleurde diazotypie *f* — **e** copia *f* al ferroprusiato; copia *f* cianográfica; cianotipia *f* — **i** copia *f* cianografica; blu *m*; cianotipia *f*

1643 blue printer — See cyan printer.

1644 blueprint paper; cyanotype paper; ferroprussiate paper — Paper of good wet strength, usually of rag content, sensitized with ferric ammonium citrate and potassium ferricyanide, used for making copies of engineering drawings, etc., in white lines on a blue ground. The blueprint paper is placed under the tracing paper and exposed to bright light. The ferric ions are reduced to ferrous ions by the effect of the light, and the ferrous ions react with the ferricyanide ions forming Turnbull's blue, $Fe[Fe(CN)_6]_2$. Under the black lines on the tracing paper no ferrous ions are reduced and no blue colour is produced.
f papier *m* à bleus; papier *m* au ferroprussiate; papier *m* ferrogallique; papier *m* héliographique; papier *m* photographique au trait bleu — **d** Blaupauspapier *n*; Lichtpauspapier *n* — **n** blauwdrukpapier *n* — **e** papel *m* (al) ferroprusiato; papel *m* (de) ferroprusiato para copias; papel *m* cianográfico — **i** carta *f* cianografica

1645 blunt *v*
f émousser — **d** abstumpfen — **n** afstompen — **e** arromar; embotar — **i** spuntare; smussare

1646 blur — See blurred image.

1647 blurb — Brief, concise summary or laudatory estimate of a book, sometimes issued as advance information; often printed on a book jacket.
f annonce *f* sur le couvre-livre; annonce *f* avantageuse d'un livre sur le point de paraître; jus *m* fam. — **d** Klappentext *m*; Waschzettel *m* — **n** flaptekst *m* — **e** bombo *m*; elogio *m*; solapa *f*; texto *m* de aleta — **i** blurb *m*; presentazione *f* editoriale sulla copertina dei libri

1648 blurred image; blur — Slurred impression in printing or in stamping.

f image *f* papillotée — **d** verwischtes Bild *n*; verschwommener Druck *m* — **n** geveegd beeld *n*; uitgesmeerd beeld *n* — **e** imagen *f* borrosa; imagen *f* remosqueada; imagen *f* sombreada; imagen *f* empastelada; imagen *f* cegada; imagen *f* maculada — **i** immagine *f* sfocata

1649 blurring — Spreading of ink in printing beyond the face of the types.
f papillotage *m* typo; frisotage *m*; regiflage *m*; allongement *m* du point (en offset) — **d** Verwischen *n* — **n** vegen *n*; smeren *n* — **e** remosqueo *m* — **i** chiazzatura *f*; allungamento *m* dei punti

1650 blushing — Milky or foggy appearance in a transparent ink or coating due to precipitation or incompatability of one of the ingredients, commonly caused by excessive moisture condensation; most frequent in periods of high humidity. See also bloom.
f opalescence *f*; turbidité *f* — **d** Trübung *f*; Mattierung *f* — **n** troebeling *f*; wolkvorming *f* — **e** turbieza *f* — **i** torbidità *f*

1651 board — See cardboard.

1652 board — See letter board.

1653 board, baling — See baling board.

1654 board, beer mat — See coaster board.

1655 board, coaster — See coaster board.

1656 board, feed — See feedboard.

1657 board, feeding — See feedboard.

1658 board for dead matter
f ais *m* à distribuer — **d** Ablegebrett *n*; Brett *n* für Ablegesatz — **n** letterbord *n* voor distributiezetsel — **e** tablero *m* para distribuir; mesa *f* de distribución — **i** tavolo *m* da scomposizione

1659 board for pressing — Board specially prepared to form, by pressing between dies, a substantially three-dimensional article, e.g., the bottom or lid of a box.
f carton *m* pour emboutissage — **d** Ziehpappe *f* — **e** cartón *m* para embutición; cartón *m* estirado — **i** cartone *m* per imbutitura; cartone *m* per formatura

1660 board machine — See cylinder machine.

1661 board-making machine — See cylinder machine.

1662 board mill — Mill equipped with cylinder machines on which various types of boards are made.
f fabrique *f* de carton — **d** Pappenfabrik *f*; Kartonfabrik *f* — **n** kartonfabriek *f*; strokartonfabriek *f* — **e** fábrica *f* de cartón; cartonería *f*; fábrica *f* de cartón de paja — **i** fabbrica *f* di cartone; cartonificio *m*

1663 board, paper — See paperboard.

1664 board rack — Stand or rack in a composing room to store formes of type or standing matter.
f rayon *m*; rayon-layette *f* — **d** Brettregal *n* — **n** letterbordenregaal *n* — **e** estanteria *f* para moldes — **i** scaffale *m*

1665 boards — See cover boards.

1666 board shears — Hand-operated, shear-type knife used in cutting boards in small quantities.

f cisaille *f* (à main) — **d** Pappenschere *f*; Kartonschneider *m*; Kartonschere *f* — **n** bordschaar *fm* — **e** cizalla *f* — **i** cesoia *f* a mano per cartone; tagliarina *f* per cartone
1667 bobbin — See wire bobbin.
1668 bocasin — See buckram.
1669 bodied
f ayant de consistance — **d** dickflüssig — **n** viskeus; consistent; dik — **e** viscoso; consistente; espeso — **i** consistente; denso
1670 bodied linseed oil — See boiled linseed oil.
1671 bodkin — Pointed steel instrument in a round handle to lever up type characters when correcting type formes. Often called spike. Correcting type matter is mostly done with the *tweezers.
f pointe *f* à corriger; pointe *f* à correction; pointe *f* — **d** Ahle *f*; Korrekturahle *f* — **n** els *fm* — **e** punzón *m* de correcciones; punta *f* de correcciones; lezna *f*; lesna *f* — **i** punta *f* per correzioni
1672 bodkin; awl — Pointed steel instrument used by bookbinders to pierce holes and to trace lines on the cover.
f alêne *f*; poinçon *m*; perçoir *m* — **d** Einbindenadel *f*; Punkturnadel *f* — **n** priem *m* — **e** punzón *m*; lezna *f* de encuadernación; lesna *f* de encuadernación — **i** lesina *f*; punteruolo *m* per legatori
1673 bodkin handle
f porte-pointe *m*; manche *m* — **d** Ahlenheft *n*; Ahlengriff *m* — **n** elsklos *fm* — **e** mango *m* del punzón — **i** manico *m* della lesina
1674 bodkin point
f pointe *f* de la pointe à corriger — **d** Ahlenspitze *f* — **n** punt *m* van de els; elspunt *m* — **e** punta *f* de la lesna — **i** punta *f* della lesina
1675 Bodleian binding — Binding named after the books collected by Sir Thomas Bodley (1545-1613), for the Oxford University Library.
f reliure *f* Bodléeinal — **d** Bodley-Einband *m* — **n** Bodley-band *m* — **e** encuadernación *f* estilo Bodley — **i** legatura *f* di Bodley
1676 Bodoni dash — See French rule.
1677 body — Technical term synonymous with the rheological term concistency. For Newtonian liquids it is the same as *viscosity.
f corps *m*; consistance *f* — **d** Körper *m* — **n** consistentie *f* — **e** cuerpo *m*; consistencia *f* — **i** consistenza *f*
1678 body — See body size.
1679 body colour — Colour of unmoistened pigment before mixing with the vehicle.
f couleur *f* à sec — **d** Trockenfarbe *f* — **n** pigmentkleur *fm*; kleur *fm* van de droge kleurstof — **e** color *m* seco — **i** colore *m* a secco
1680 body gum; # 8 varnish — Linseed oil, heatpolymerized to a heavy, gummy state, commonly used as *bodying agent.
f mordant *m*; vernis *m* fort — **d** starker Leinölfirnis *m* — **n** mordantvernis *mn* — **i** vernice *f* mordente

1681 bodying agent; thickening agent; thickener — Material added to ink to increase its viscosity. See also body gum.
f épaississant *m*; agent *m* d'épaississement — **d** Verdickungsmittel *n*; Viskositätserhöher *m* — **n** verdikkingsmiddel *n* — **e** producto *m* espesante; espesador *m* — **i** prodotto *m* per ispessire gli inchiostri; agente *m* d'ispessimento
1682 body line — See type line.
1683 body matter — Text matter as distinct from display.
f labeur *m* — **d** glatter Satz *m*; Textsatz *m*; Paketsatz *m* — **n** plat zetsel *n*; platte tekst *m* — **e** composición *f* de texto; composición *f* corrida; composición *f* corriente; composición *f* seguida — **i** composizione *f* corrente
1684 body of the work — Text part of a book.
f corps *m* d'ouvrage — **d** Werksatz *m* — **n** platte tekst *m*; plat gedeelte *n* — **e** cuerpo *m* del texto — **i** testo *m* del libro
1685 body paper; base paper; body stock; raw stock; coating paper — Paper or board for conversion by coating or surface application or impregnation. In some countries paper to which a layer of other material (aluminium, plastics, etc.) is added.
f papier *m* support; papier *m* support pour couchage; support *m* (de couche); papier *m* de base — **d** Rohpapier *n*; Streichrohpapier *n*; Streichpapier *n*; Trägerpapier *n*; Grundpapier *n* — **n** grondpapier *n*; drager *m*; binnenpapier *n*; basispapier *n*; strijkpapier *n* — **e** papel *m* soporte; papel *m* soporte para estucar; cartulina *f* soporte — **i** carta *f* (di) supporto; carta *f* supporto alla patinatura
1686 body size; type size; point size; body — The measurement (or thickness) from top to bottom of a type, slug, rule, lead, etc. See also point system.
f force *f* de corps; corps *m* — **d** Kegelstärke *f*; Schriftkegel *m*; Kegel *m*; Schriftgrad *m*; Korpus *f* — **n** corps *n*; lettercorps *n*; lettergrootte *f* — **e** medida *f* del cuerpo; cuerpo *m* (de la letra); cuerpo *m* del tipo; fuerza *f* del cuerpo; tamaño *m* del cuerpo; grado *m* del cuerpo; calibre *m* del cuerpo — **i** corpo *m* (di un carattere)
1687 body stock — See body paper.
1688 body type — Type for text or reading matter, as distinguished from headings and display type for advertisements.
f caractères *mpl* de labeur; petits clous *mpl sl.* — **d** Brotschrift *f*; Textschrift *f* — **n** broodletter *fm* — **e** tipo *m* de texto; tipo *m* común; tipo *m* corriente; tipo *m* usual; caracteres *mpl* comunes; caracteres *mpl* de texto; caracteres *mpl* de obra — **i** caratteri *mpl* del testo; caratteri *mpl* correnti; caratteri *mpl* comuni
1689 body wise — See pointwise.
1690 boea gum — Fossil resin of the Manila copal class, used for durable oil varnishes.
1691 bogus — Paper and paperboard made from old paper or inferior or low-grade stock to imitate grades of higher quality.

f imitation *f* − **d** Ersatz *m* − **n** namaak *m* − **e** falso; espúreo; postizo *US fam.*; fictico; de imitación − **i** imitazione *f*

1692 bogus bristol; imitation bristol; B bristol − Bristol board usually made on a cylinder machine. It may be solid, or with different stocks for filler and liners. The furnish consists of overissue news, blank news, bleached sulphite, soft white shavings, and hard white shavings in varying amounts, according to the quality.

f simili-bristol *m* − **d** Imitationsbristolkarton *m*; Bristolersatzkarton *m* − **n** imitatie-bristolkarton *n* − **e** cartulina *f* bristol imitación − **i** cartoncino *m* uso bristol

1693 bogus manila − Paper used instead of sulphite or kraft manila paper when strength and quality are not essential, commonly made from reclaimed paper and coloured to represent manila.

f simili-kraft *m* − **d** Manila-Imitation *f* − **n** imitatie-manilla *n* − **e** papel *m* imitación manila − **i** carta *f* uso manilla

1694 bogus setting *US* − Matter set, by Union requirement, by a compositor and later discarded, duplicating the text of an advertisement for which a plate has been supplied or type set by another publisher. See also featherbedding.

n overzetten *n* van advertenties − **e** material *m* compuesto sin intención de publicarlo − **i** composizione *f* fatta senza intenzione di farne uso

1695 boiled linseed oil; bodied linseed oil; boiled oil − Linseed oil heated (not boiled) to a high temperature for a short time to increase the viscosity and drying rate. It usually contains a small amount of drier.

f huile *f* de lin raffinée (décapée disaient les vieux imprimeurs); huile *f* de lin cuite; huile *f* cuite; huile *f* de lin épaissie − **d** gekochtes Leinöl *n* − **n** gekookte lijnolie *fm*; gestookte lijnolie *fm* − **e** aceite *m* de linaza cocido − **i** olio *m* di lino cotto

1696 boiled oil − See boiled linseed oil.

1697 boiler − Large vessel for heating water or generating steam.

f chaudière *f* − **d** Dampfkessel *m* − **n** stoomketel *m*; ketel *m* − **e** caldera *f* de vapor; generador *m* de vapor − **i** caldaia *f* (a vapore)

1698 boiler − See digester.

1699 boiler plate; plate matter − A mat delivered from mat services.

f flans *mpl* fournis par des agences d'information − **d** Matern *fpl* geliefert von Korrespondenten − **n** matrijzen *fpl* geleverd door derden − **e** material *m* clisado; material *m* estereotipado; planchas *fpl* para impresión; grabados *mpl* para impresión

1700 boiler plate − In computer typesetting, a set of repetitive blocks of copy that may be picked up and included routinely without the need to recreate them, as the "fine print" in an insurance policy.

1701 boiling agent; digesting agent − Chemical used for cooking wood chips to produce chemical woodpulp.

f lessive *f* de cuisson − **d** Kochmittel *n* − **n** kookmiddel *n* − **i** agente *m* (additivo) di cottura

1702 boiling-point − The temperature at which the vapour pressure of a liquid is equal to the atmospheric pressure, e.g., 100 °C or 212 °F for water at sea level.

f température *f* d'ébullition; point *m* d'ébullition − **d** Siedepunkt *m* − **n** kookpunt *n* − **e** punto *m* de ebullición; temperatura *f* de ebullición − **i** punto *m* di ebollizione

1703 boiling-point thermometer

f thermomètre *m* à point d'ébullition − **d** Siedepunkt-Thermometer *n* − **n** kookpuntthermometer *m*; kookpuntmeter *m* − **e** termómetro *m* para punto de ebullición; ebuliómetro *m*; ebulioscopio *m* − **i** termometro *m* a punto d'ebollizione

1704 boiling range − Many solvents have no specific boiling point but distill over a range of temperatures, e.g. VM & P naphtha 250-300 °F.

f intervalle *m* d'ébullition − **d** Kochgebiet *n* − **n** kookpunttraject *n*; kooktraject *n*; kookgebied *n* − **e** intervalo *m* de ebullición − **i** intervallo *m* d'ebollizione

1705 bold − In type-face design characters that have been thickened to produce a blacker or more conspicuous selection of composition. See also semi-bold face, extra bold, weight of type.

f gras − **d** fett − **n** vet − **e** negrilla; negrita − **i** neretto; grassetto

1706 bold, accentuated with − See accentuated with bold.

1707 bold face type; black face type; fat face type − Heavy or thickened form of type face giving a blacker effect in printing. Indicated in manuscripts by wavy underlining. Used for emphasis, captions, subheadings, etc. See app. no. 5.

f caractère *m* gras − **d** fette Schrift *f*; vollfette Schrift *f*; fetter Buchstabe *m* − **n** vette letter *fm* − **e** letra *f* negra; letra *f* negrilla; letra *f* negrita; tipo *m* negro; tipo *m* negrillo; tipo *m* negrito − **i** carattere *m* neretto; carattere *m* grassetto

1708 bole, Armenian − See Armenian bole.

1709 Bologna chalk − See calcium carbonate.

1710 bolster − The high portion or edges of the blanket not compressed by the printing plate.

f hausse *f*; parties *fpl* non imprimées; bordure *f* du blanchet − **d** druckfreie Stelle *f* − **n** nietdrukkend deel *n* van het rubberdoek − **e** zonas *fpl* descargadas de presión − **i** cuscini *mpl*

1711 bolster − Recesses round the printing forme or mould.

f échancrure *m* − **d** Punzen *m* − **i** depressioni *fpl*

1712 bolster − In die cutting, the structural members that support the upper and lower die chases.

1713 bolster marks − Ink or grease in margins caused by contact of bolsters with the web. Also uneven impression produced by printing on previously formed bolsters.

f maculage *m* des marges − **d** Abschmieren *n*

durch druckfreie Stellen — **n** smetten *n* op niet te bedrukken plaatsen; smetten *n* van het drukdoek — **e** maculado *m* de los márgenes — **i** maculatura *f* dovuta a cuscini

1714 bolt — Strong fastening rod, pin or screw, threaded to receive a nut.
f boulon *m*; vis *f* munie d'un écrou — **d** Bolzenschraube *f*; Bolzen *m* — **n** bout *m*; schroefbout *m*; moerbout *m* — **e** perno *m*; tornillo *m*; bulón *m SA* — **i** bullone *m*

1715 bolt *v*; **tighten** *v*; **screw** *v* **down**
f boulonner; serrer — **d** festziehen; festschrauben; verbolzen; spannen — **n** aandraaien; vastschroeven; aanschroeven — **e** empernar; juntar con pernos; atornillar; apretar con tornillos; fijar — **i** avvitare; imbullonare; serrare

1716 bolts; closed bolts; closed folds; edge bolts — Uncut edges or foredges of a bound book; closed ends or signatures of untrimmed books (head bolt; foredge bolt; tail bolt).
f témoins *mpl* — **d** unaufgeschnittener Druckbogen *m*; Beschnittrand *m*; Faltkante *f* — **n** nietopengesneden kanten *mpl* — **e** hojas *fpl* intonsas; cantos *mpl* doblados — **i** bordi *mpl* non tagliati

1717 bolts out — To bind a book with bolts out means to leave the edges of the book uncut.
f laissé avec les témoins — **d** unbeschnitten — **n** onafgesneden — **e** no cortado — **i** non tagliato

1718 bond — Certificate of ownership of a portion of a debt due to be paid by a government or corporation to an individual holder and usually bearing a fixed rate of interest.
f obligation *f* — **d** Obligation *f*; Schuldverschreibung *f*; Anleihschein *m* — **n** obligatie *f* — **e** obligación *f* — **i** obbligazione *f*; titolo *m*

1719 bond — First page of a share, debenture bond, share certificate.
f manteau *m* — **d** Mantel *m* — **n** mantel *m* — **e** capa *f* — **i** primera pagina *f*

1720 bonding strength — The force with which fibres adhere to each other within a sheet, with which a coating and/or film adheres to the surface of a sheet, or with which plies in a board or laminated sheet adhere to each other.
f cohésion *f* des fibres — **d** Haftfestigkeit *f*; Haftigkeit *f* — **n** vezelbinding *f*; hechting *f* van de vezels — **e** cohesión *f* de fibras — **i** forza *f* di coesione; resistenza *f* allo scollamento

1721 bonding-strength device
f appareil *m* pour mesurer la liaison — **d** Gerät *n* zur Bestimmung der Adhäsion — **n** hechtingsmeter *m*; sealkrachtmeter *m* — **e** aparato *m* para medir la adhesión — **i** apparecchio *m* per misurare l'adesione

1722 bond ink — Ink used on hard sized bond and ledger papers; should be as concentrated and as heavy and tacky as the grade of paper and type of press permit. Should dry entirely by oxidation.
f encre *f* pour "bond" — **d** Hartpostfarbe *f* — **n** hardpostinkt *m* — **e** tinta *f* para papel cartas; tinta *f* para papel bond — **i** inchiostro *m* per

carta dura per corrispondenza

1723 bond paper — Originally glue-sized rag paper for printing of bonds and certificates. Now both rag and sulphite papers, sized for writing purposes, for letterheads and other commercial forms. Usually made in basis weights from 13-24 pounds (17 x 22 - 500).
f papier *m* bond — **d** Bankpostpapier *n*; Bondpapier *n*; Hartpostpapier *n*; Feinpostpapier *n* — **n** hard bankpostpapier *n*; hard bankpost *n*; bondpapier *n*; bond *n* — **e** papel *m* bond — **i** carta *f* per banche

1724 bone — See folding stick.

1725 bone black; hard black; Paris black — Black ink pigment made by calcining bones without air; also used as filtering medium and deodorizing agent.
f noir *m* d'os; noir *m* animal — **d** Knochenschwarz *n* — **n** beenzwart *n* — **e** carbón *m* de huesos — **i** nero *m* d'ossa

1726 bone dry — See oven dry.

1727 bone folder — See folding stick.

1728 bone glue — Adhesive product from animal bones, which is applied as a hot solution; formerly used for binding books. See also glue.
f colle *f* d'os — **d** Knochenleim *m*; Kuchenleim *m* — **n** beenderlijm *m* — **e** cola *f* de huesos — **i** colla *f* d'ossa

1729 book — Completely assembled and bound printed sheets which constitute the finished product. By extension sheets of blank writing paper bound together and used for making entries. Also a treatise, written or printed on any material and put together in any convenient form.
f livre *m* — **d** Buch *n* — **n** boek *n* — **e** libro *m* — **i** libro *m*

1730 book-backing paper — See back-lining paper.

1731 book-back liner — See back-lining paper.

1732 bookbinder; binder — Workman who assembles and binds together printed sheets to constitute a book.
f relieur *m* — **d** Buchbinder *m*; Kleistergraf *m fam.* — **n** boekbinder *m*; binder *m* — **e** encuadernador *m*; empastador *m LA* — **i** legatore *m*; rilegatore *m*

1733 bookbinder's bone — See folding stick.

1734 bookbinder's calico — See calico.

1735 bookbinder's cloth; binder's cloth; binding cloth; book cloth; book linen; linen — Cotton cloth, sized, glazed or impregnated with synthetic resins for book covers, available in a large variety of weights, finishes, colours and patterns.
f toile *f* à reliure; toile *f* — **d** Buchleinen *n*; Buchbindeleinen *n*; Leinen *n* — **n** boekbinderslinnen *n*; boeklinnen *n*; linnen *n* — **e** tela *f* de encuadernación; tela *f* para encuadernación — **i** tela *f* per legatoria; tela *f* per legatura

1736 bookbinder's glue
f colle *f* de relieur — **d** Buchbinderleim *m* — **n** boekbinderslijm *m* — **e** cola *f* para libros; pasta *f* para libros — **i** colla *f* per legatori

1737 bookbinder's needle — See stitching needle.
1738 bookbinding — See binding.
1739 bookbinding machinery — All machines for binding books, such as guillotines, sewing machines, stitchers, gold presses, perforators, presses.
f machines *fpl* pour la reliure — d Buchbinderei-maschinen *fpl* — n boekbinderijmachines *fpl*; binderijmachines *fpl* — e máquinas *fpl* de encuadernación — i macchine *fpl* per legatoria
1740 bookbinding press — See press.
1741 book block
f corps *m* d'ouvrage — d Buchblock *m* — n boek-blok *n* — e cuerpo *m* del libro; tripa *f* — i blocco *m* del libro
1742 book canvasser; canvasser — One who seeks orders for selling books.
f courtier *m* en librairie; démarcheur *m* en livres — d Buchhandlungsreisender *m* — n reiziger *m* voor de boekhandel — i piazzista *m* di libri
1743 book case — See case.
1744 book case — See slip case.
1745 book clamp — Vise used by bookbinders to hold the book during the backing operation.
f presse *f* à endosser; étau *m* à endosser — d Preßlade *f*; Klotzpresse *f*; Klemmbacke *f* — n boekbinderspers *f* — e prensa *f* para encua-dernar; burro *m*; mordaza *f* para encuadernar *LA* — i morsa *f* per la formazione del dorso
1746 book cloth — See bookbinder's cloth.
1747 book collector
f collectionneur *m* de livres — d Bücher-sammler *m* — n boekenverzamelaar *m* — e coleccionador *m* de libros — i collezionista *m* di libri; raccoglitore *m* di libri
1748 book designer
f graphiste *m* — d Buchkünstler *m*; Buch-gestalter *m* — n boekontwerper *m*; boek-kunstenaar *m*; typografisch verzorger *m* — e arquitecto *m* de libros; proyectista *m* de libros — i disegnatore *m* grafico
1749 book destroyer
f bibliocaste *m* — d Buchvernichter *m* — n boekenvernieler *m* — e bibliocasto *m*; destructor *m* de libros — i bibliocasta *m*
1750 book end paper — Paper used for inside of book covers and to attach the body of the book to the cover.
f papier *m* pour gardes — d Vorsatzpapier *n* — n schutbladenpapier *n* — e papel *m* para guardas — i carta *f* per risguardi
1751 book face; book fount — Fount of body type used mostly in printing of books, different from fancy or job types.
f caractères *mpl* d'édition — d Buchschrift *f*; Werkschrift *f*; Werksatzschrift *f* — n boekletter *fm* — e tipo *m* para libros — i caratteri *mpl* per libri; caratteri *mpl* di testo
1752 book fair; bookseller's fair — Meeting of book sellers and buyers.
f foire *f* de librairie; foire *f* du livre; foire *f* du commerce du livre — d Buchmesse *f*; Buch-

händlerbörse *f*; Bücherbörse *f* — n boekenbeurs *fm*; boekenjaarbeurs *fm* — e feria *f* del libro — i fiera *f* del libro
1753 book fount — See book face.
1754 bookholder — Copyholder accommodating open books during photography of pages.
f porte-livre *m* — d Buchgestell *n* — n model-houder *m* voor boeken — e portalibros *m* — i portalibri *m*
1755 book industry — All activities in the produc-tion of books.
f industrie *f* du livre — d Buchgewerbe *n*; buch-herstellende Industrie *f* — n boekbedrijf *n*; boek-wezen *n* — e industria *f* del libro — i industria *f* del libro
1756 book inks — Free flowing inks which dry partly by oxidation, partly by penetration; more drier is used in inks for supercalendered book paper.
f encres *fpl* labeurs — d Werkdruckfarben *fpl*; Buchdruckfarben *fpl* — n gewone zwarte inkten *mpl*; boekinkten *mpl* — e tintas *fpl* para obras — i inchiostri *mpl* per opere
1757 book jacket — See jacket.
1758 booklet — See also brochure.
f livret *m* — d Heftchen *n* — n boekje *n* — e librito *m*; librillo *m*; opúsculo *m* — i libretto *m*
1759 book linen — See bookbinder's cloth.
1760 book lover — See bibliophile.
1761 book mark — See book marker.
1762 book marker; book mark; flag; tassel; register ribbon — Ribbon fastened at one end to the spine of a book to mark a place when reading.
f signet *m* — d Lesebändchen *n*; Lesezeichen-band *n*; Zeichenband *n*; Lesezeichen *n*; Buch-zeichen *n*; Blattzeichen *n* — n leeslint *n*; blad-wijzer *m* — e señal *f* de libro; señal *f* de lectura; cinta *f* indicadora; registro *m* — i nastrolibro *m*; segnalibro *m*
1763 book number — See Standard Book Number.
1764 Book Number, Standard — See Standard Book Number.
1765 book of fables
f fablier *m* — d Fabelbuch *n* — n fabelboek *n* — e libro *m* de fábulas — i libro *m* di favole
1766 book of heraldry — See armorial.
1767 book of hours — Book with the prescribed order of prayers, readings from Scripture, and rites for the canonical hours.
f livre *m* d'heures — d Stundenbuch *n* — n getijdenboek *n* — e libro *m* de horas — i libro *m* d'ore
1768 book ornamentation
f décoration *f* du livre — d Buchschmuck *m*; Buchausstattung *f* — n boekversiering *f* — e adorno *m* del libro — i ornamento *m* del libro; decorazione *f* del libro
1769 book page puller — Device to measure the page strength of unsewn bindings in books, pam-phlets and other publications. Also to measure the tensile strength of some thin flexible materials

and of joins made between such materials.

1770 book paper — Originally soft sized or unsized paper used for printing of books; now also used for magazines, circulars, labels, etc. See also mechanical book paper.

f papier *m* d'édition; papier *m* pour livres; papier *m* pour impression de livres — **d** Bücherpapier *n* — **n** tekstpapier *n*; romandruk(papier) *n* — **e** papel *m* para libros; papel *m* para obras; papel *m* edición; papeles *mpl* de obra — **i** carta *f* per lavori editoriali; carta *f* da stampa per edizioni

1771 book-plate; ex-libris — A book owner's identification label usually pasted to the inside front cover of a book.

f ex-libris *m* — **d** Exlibris *n*; Bucheignerzeichen *n* — **n** ex-libris *n* — **e** ex-libris *m*; marca *f* de propiedad — **i** ex-libris *m*

1772 book pocket — Flat pouch or envelope, usually made from manila paper, pasted on the inside back or front cover of a library book, to insert the identifying card.

f pochette *f* de livre — **d** Buchtasche *f*; Kartentasche *f*; Tasche *f* — **n** kaarttasje *n*; tasje *n* — **e** bolsa *f* (para) libros — **i** taschina *f* entro il libro

1773 book review — Critical description and evaluation in a newspaper or magazine of a newly published book by a critic, journalist, etc.

f critique *f* d'un livre — **d** Buchbesprechung *f*; Buchrezension *f* — **n** boekbespreking *f*; recensie *f* van een boek — **e** reseña *f* de un libro; crítica *f* de un libro — **i** recensione *f* d'un libro; critica *f* d'un libro

1774 book reviewer — Critical estimator of a book. See also critic.

f critique *m* littéraire — **d** Buchbesprecher *m*; Buchrezensent *m*; Buchkritiker *m* — **n** boekrecensent *m*; recensent *m*; boekbespreker *m*; boekbeoordelaar *m*; boekcriticus *m* — **e** reseñante *m*; crítico *m* de libros — **i** recensore *m*; critico *m*

1775 bookseller — Person who sells books. See also bookshop.

f libraire *m* — **d** Buchhändler *m* — **n** boekhandelaar *m*; boekverkoper *m* — **e** librero *m*; bibliópola *m* — **i** libraio *m*

1776 bookseller's fair — See book fair.

1777 book sewing machine — See sewing machine.

1778 bookshop; book store — Store where books are sold.

f librairie *f*; boutique *f* de librairie — **d** Buchhandlung *f* — **n** boekwinkel *m*; boekhandel *m* — **e** librería *f*; comercio *m* de libros — **i** libreria *f*

1779 book size — Some untrimmed book sizes are: crown octavo 12.7 x 19 cm (5 x 7.5 in); crown quarto 19 x 25.4 cm (7.5 x 10 in); demy octavo 14.25 x 28.5 cm (5.625 x 8.75 in); demy quarto 22.2 x 29.8 cm (8.75 x 11.25 in); foolscap octavo 10.6 x 17 cm (4.25 x 6.75 in); medium octavo 14.6 x 22.8 cm (5.75 x 9 in); royal octavo 15.8 x 25.4 cm (6.25 x 10 in); royal quarto 25.4 x 31.7 cm (10 x 12.5 in).

f format *m* de livre — **d** Buchformat *n* — **n** boekformaat *n* — **e** tamaño *m* del libro; formato *m* del libro — **i** formato *m* del libro

1780 book stamp — See binder's stamp.

1781 book stock — For book stock sometimes *de-inked-paper stock is used.

1782 book store — See bookshop.

1783 bookwork — The printing of books as distinguished from magazine, newspaper or job printing.

f labeur *m*; travaux *mpl* d'édition — **d** Werkdruck *m* — **n** boekwerk *n*; drukken *n* van boekwerken — **e** impresión *f* de libros — **i** stampa *f* dei libri

1784 bookworm — Insect that feed on books.

f ver *m* — **d** Bücherwurm *m* — **n** boekenwurm *m* — **e** verme *m* de libros — **i** verme *m* dei libri

1785 bookworm — Person devoted to books and study.

f dévoreur *m* de livres; bouquineur *m*; liseur *m* acharné; rat *m* de bibliothèque — **d** Bücherwurm *m* — **n** boekenwurm *m* — **e** ratón *m* de biblioteca; rata *f* de biblioteca — **i** lettore *m* appassionato; topo *m* di biblioteca

1786 book wrapper — See publicity wrapper.

1787 Boolean logic — Named after George Boole (1815-1864), English/Irish professor of mathematics and logic who developed a series of logic propositions which may be expressed in symbolic terms, such as "and", "or", "not", "exclusive or", etc.

1788 bootstrap — Method to provide information at the beginning of a program to cause the balance of the program to be loaded and activated (as from paper tape).

f amorce *f* — **d** Selbststart *m* — **n** zelfstart *m* — **e** autocarga *f* — **i** autocarica *f*

1789 bootstrap loader — Loader whose first few instructions are sufficient to bring the rest of it into the computer's storage from an input device. Usually refers to routines used to initialize the computer for use.

1790 boracic acid — See boric acid.

1791 borax — See sodium borate.

1792 border; box — Printed line or design surrounding an illustration or other printed matter.

f cadre *m*; bordure *f*; encadrement *m* — **d** Rand *m*; Einfassung *f*; Umrandung *f* — **n** kader *n*; omlijsting *f*; rand *m* — **e** cierre *m* de rayas; cerco *m* de rayas; cierre *m* de filetes; cerco *m* de filetes; recuadro *m*; encuadrado *m*; contorno *m*; orla *f* de contorno; marco *m* — **i** cornice *f*; inquadratura *f*; contorno *m*; bordo *m*; quadro *m*

1793 border — Finishing line or design on printing edges of a plate.

f cadre *m* — **d** Umrandung *f*; Randlinie *f* — **n** kader *n* — **e** borde *m* — **i** riquadro *m*

1794 border effect — Faint dark line just within the high-density side of a boundary between a lightly exposed and heavily exposed region on a developed emulsion. The faint light line just within the low-density side of the margin is the fringe effect. These lines have been called *mackie lines. See also adjacency effect.

f effet *m* de bordure — **d** Kanteneffekt *m* — **n** rand-effect *n* — **e** efecto *m* borde — **i** effetto *m* bordo

1795 border-punched card — See edge-punched card.

1796 border rule

f filet *m* d'encadrement — **d** Einfassungslinie *f*; Randlinie *f*; Akzidenzlinie *f* — **n** kaderlijn *fm* — **e** raya *f* de contorno; orla *f* de contorno; filete *m* de contorno — **i** filetto *m* di inquadratura; filetto *m* di contorno

1797 border slide — See matrix slide.

1798 boric acid; boracic acid; orthoboric acid — Colourless, odourless scales or white powder. Stable in air, soluble in cold water, more soluble in boiling water, soluble in alcohol and glycerine. Acts as a buffer to limit the change of pH of the fixing bath solution and helps to prevent the precipitation of aluminium compounds which occurs when the pH changes too much. Formula: H_3BO_3. Latin: acidum boricum.

f acide *m* borique — **d** Borsäure *f* — **n** boorzuur *n* — **e** ácido *m* bórico — **i** acido *m* borico

1799 boss — Person who employs or superintends workmen.

f grand vingt-deux *m* — **d** Alte *m* — **n** baas *m* — **e** patrón *m*; maestro *m* — **i** anziano *m*

1800 bosses — The pieces of metal inset into the cover of a book for protection of the corners or edges or for decoration.

f ferrure *f*; cabochons *mpl* — **d** Buchbeschläge *mpl*; Bucheinbandbeschläge *mpl* — **n** boekbeslag *n*; beslag *n* — **e** bullones *mpl*; cabujones *mpl*; clavos *mpl* — **i** borchie *fpl*

1801 bosses — See propellers.

1802 botanical signs; botanical symbols — See app. no. 10.

f signes *mpl* botaniques; symboles *mpl* de botanique — **d** botanische Zeichen *npl* — **n** botanische tekens *npl* — **e** signos *mpl* botánicos; símbolos *mpl* botánicos — **i** segni *mpl* botanici

1803 botanical symbols — See botanical signs.

1804 botch; botched job; bungle — In the composing room, clumsy or poor piece of work, e.g., when a compositor spikes the composition with a bodkin, inserts cards between type, etc.

f travail *m* mal fait; travail *m* bousillé — **d** Pfusch *m* — **n** ondeugdelijk werk *n*; slecht werk *n* — **e** trabajo *m* inservible; trabajo *m* inferior — **i** lavora *m* rappezzato

1805 botched job — See botch.

1806 botcher *sl.*; **bungler** *sl.* — Clumsy and inefficient workman.

f gâcheur *m*; gâte-métier *m*; machurat *m* — **d** Pfuscher *m*; Hudler *m*; schlechter Arbeiter *m* — **n** knoeier *m*; beunhaas *m* — **e** chambón *m*; chapucero *m* — **i** guastamestieri *m*; ciabattino *m*

1807 bottle-arsed type — See bottle-bottom type.

1808 bottle-bottom type; bottle-arsed type — Type or slug the bottom of which is wider than the top.

f caractère *m* évasé — **d** Type *f* (oder Zeile) deren Dicke am Fuß größer ist als am Kopf — **n** letter *fm* (of regel) die aan de voet breder is dan aan de kop — **e** tipo *m* (o línea) cuyo pie es mas grueso que la cabeza — **i** carattere *m* (o riga) più grosso al piede che in testa

1809 bottle-necked type — Type or slug the top of which is wider than the bottom.

f caractère *m* aplati; caractère *m* écrasé — **d** Type *f* (oder Zeile) deren Dicke am Kopf größer ist als am Fuß — **n** letter *fm* (of regel) die aan de kop breder is dan aan de voet — **e** tipo *m* (o línea fundida) cuyo cabeza es mas grueso que el pie — **i** carattere *m* (o riga) più grosso in testa che al piede

1810 bottoming — When, in box making, score rule bottoms too hard against board and cutting plate, results may be a breaking or partial cutting of the board on the scores. Too high score rule or too much makeready under scores may also result in bottoming. Scores should bottom to some extent. If score height is maintained and difference between cutting rule and score rule is no more than thickness of board, a good bead will be formed, which will not crack. If score rule is not high enough and does not bottom properly, a crack will appear in the centre of the bead.

1811 bottom line — See foot line.

1812 bottom margin — See tail margin.

1813 bottom notes — See foot notes.

1814 boudoir — A size of note paper. See app. no. 3.

1815 bound book — Book on which boards of cover have been attached before the covering of leather, cloth or other material is secured to boards. See also cased book.

f livre *m* relié — **d** gebundenes Buch *n* — **n** gebonden boek *n* — **e** libro *m* encuadernado — **i** libro *m* rilegato

1816 bound in boards — See in boards.

1817 bourgeois — Old name for a size of type. The name is said to be derived from the famous French typographer Geofroy Tory (1480-1533), born at Bourges, or from the writing used in France by the bourgeois. See app. no. 1.

1818 boustrophedon; boustrophedon writing — The early Greek method of writing, alternately from right to left and from left to right. Literally "in the manner in which an ox turns" when ploughing.

f boustrophédon *m*; inscriptions *fpl* en boustrophédon — **d** Bustrophedon *n*; Furchenschrift *f* — **n** boustrofedon *n* — **e** bustrófedon *m* — **i** bustrofedone *m*

1819 boustrophedon writing — See boustrophedon.

1820 bow *v* — To bend a letter so that it cannot be used again.

f arquer — **d** verbiegen — **n** ombuigen; verbuigen; kapot buigen — **e** torcer — **i** storcere

1821 bowden cable
f câble *m* bowden — **d** Bowdenzugkabel *n*; Bowdenzug *m* — **n** bowdenkabel *m* — **e** cable *m* bowden — **i** cavo *m* bowden
1822 bowdlerize *v* — To expurgate in editing by deleting offensive words or passages, as did Dr. Thomas Bowdler (1754-1825) in his Shakespeare edition.
f expurger; corriger maladroitement; gâter en voulant corriger — **d** verballhornen; verballhornisieren; ballhornisieren (nach Johann Ballhorn, Buchdrucker zu Lübeck im 16. Jahrhundert, mit einer Fibel getan); verschlimmbessern; scheinbessern — **n** overdreven zuiveren (en bederven) — **e** recortar; mutilar; hacer una corrección mal hecha — **i** espurgare; correggere esageratamente
1823 bowl; cup — Full rounded oval or circular form in a letter, complete in an O or modified in D, B, d, b, a.
f panse *f* — **d** Rundung *f*; Kreisbogen *m*; Oval *n* — **n** boog *m* — **e** ojal *m* — **i** ovale *m*
1824 bowl; paper bowl — Non-metal roll, usually made of (or covered with) cotton, linen, paper or asbestos, used in supercalender machine.
f rouleau *m* (en) papier — **d** Papierwalze *f*; Kalanderwalze *f* — **n** papierwals *fm* — **i** rullo *m* di carta; cilindro *m* di carta
1825 bowl track — The runners on top of the frame of a letterpress machine on which the bowl carriage is moving.
f glissière *f*; bande *f* à galets; bandes *fpl* à glissières — **d** Rollenbahn *f*; Karrenbahn *f* — **n** rollenbaan *fm* — **e** camino *m* de rolletes — **i** binari *mpl* del carro portaforme
1826 box — Paperboard container with base and lid.
f boîte *f*; carton *m* — **d** Schachtel *f* — **n** doos *fm* — **e** caja *f* (de cartón) — **i** scatola *f*; scatolina *f*; scatoletta *f*
1827 box; case — Wooden packing case.
f caisse *f*; boîte *f* — **d** Kiste *f* — **n** kist *fm* — **e** caja *f* — **i** cassa *f*
1828 box — Receiving trough or magazine on a folding or gathering machine.
f poche *f* — **d** Tasche *f* — **n** tas *fm* — **e** bolsa *f*; bolsilla *f* — **i** tasca *f*
1829 box — Subdivision in a type case in which characters are kept separate.
f cassetin *m* — **d** Kastenfach *n*; Gefach *n* — **n** vakje *n* van de letterkast; kastvakje *n* — **e** cajetín *m*; gacetín *m Me*; compartimiento *m* — **i** cassettino *m*; scomparto *m*
1830 box — See border.
1831 boxboard — Paperboard used to fabricate boxes, made of wood pulp or paper stocks or any combinations of these; it may be plain, lined or clay coated. Classification of boxboard grades is based on the composition of the top liner, filler and back liner, e.g. patent coated news, kraft back, has a patent coated top, a news filler and a kraft back liner. Double manila-lined news has a manila furnish on the top and back sides with a news filler.
f carton *m* pour boîtes (pliantes); carton *m* pour cartonnages — **d** Schachtelkarton *m*; Schachtelpappe *f*; Faltschachtelkarton *m*; Kartonagenpappe *f*; Kartonagenkarton *m*; Karton *m* für Kartonagen — **n** vouwdozenkarton *n*; kartonnagekarton *n*; kartonnagebord *n* — **e** cartón *m* para cartonaje; cartón *m* para cajas — **i** cartone *m* per scatole; cartone *m* per cartonagi
1832 box cover paper — Paper coated, glazed or decorated on one side only, used for covering boxes. Plain paper is also used.
f papier *m* extérieur pour boîtes; extérieur *m* pour carton — **d** Schachtelüberzugpapier *n* — **n** dozenbeplakpapier *n* — **e** papel *m* para forrar cajas — **i** carta *f* per foderare le scatole
1833 boxed head — See box heading.
1834 box head — See box heading.
1835 box heading; box head; boxed head — Heading surrounded by rules and borders. Similar to a cut-in heading, but it has a rule around it, or it is a head for a column in a ruled table.
f titre *m* encadré — **d** umrandete Überschrift *f* — **n** omkaderde kop *m* — **e** título *m* encasillado; encabezamiento *m* encasillado; titular *m* encuadrado — **i** titolo *m* inquadrato
1836 box lining machine
f machine *f* à recouvrir les boîtes — **d** Schachtelbeklebemaschine *f*; Überziehmaschine *f* für Schachtel — **n** dozenbeplakmachine *f* — **e** forradora *f* de cajas — **i** foderatrice *f* per scatole
1837 box stayer — See corner stayer.
1838 box stitcher
f agrafeuse *f* pour boîtes — **d** Schachtelheftmaschine *f* — **n** dozenhechtmachine *f* — **e** cosedora *f* de cajas — **i** cucitrice *f* per scatole; cucitrice *f* a punti metallici per scatole
1839 boxwood — Dense, close-grained wood used by engravers for fine woodcuts.
f buis *m* — **d** Buchsbaumholz *n*; Palmholz *n* — **n** palmhout *n*; bukshout *n* — **e** boj *m*; madera *f* de boj — **i** bosso *m*; legno *m* di bosso
1840 braces — Typographical signs, cast on own body usually to a definite number of ems and used to enclose a note, reference or explanation and separate it from the context. Sectional braces are made of several parts which can be assembled to the required length. Vinculum is an old name for a brace.
f accolades *fpl* — **d** Akkoladen *fpl*; geschweifte Klammern *fpl*; Nasenklammern *fpl* — **n** accoladen *fpl* — **e** abrazaderas *fpl*; llaves *fpl*; corchetes *mpl* — **i** grappe *fpl*; graffe *fpl*
1841 bracket — The filling in at the angle or angles between the serif and the stem of letters.
f empattement *m* (triangulaire) — **d** Winkelfüllung *f* zwischen Serif und Abstrich — **n** opvulling *f* tussen schreef en stam — **e** base *f* triangular; base *f* filiforme — **i** raccordo *m* fra terminale
1842 bracket — Support for fixing two parts at an angle.
f console *f*; support *m* — **d** Konsole *f*; Bock *m*;

Bügel *m*; Träger *m* — **n** hoeksteun *m* — **e** apoyo *m* — **i** mensola *f*

1843 bracket *v* — To set a word between parentheses.

f mettre entre parenthèses; mettre entre crochets — **d** einklammern; in Klammern einschließen — **n** tussen haakjes plaatsen — **e** poner entre paréntesis; abrazar con corchetes — **i** mettere fra parentesi

1844 brackets; square brackets; crotchets *obs.* — Typographical signs used to enclose a note, reference or explanation and separate it from the context. See also parentheses, punctuation marks.

f crochets *mpl* — **d** eckige Klammern *fpl* — **n** vierkante haakjes *npl*; teksthaken *mpl* — **e** paréntesis *mpl* rectangulares; paréntesis *mpl* cuadrados; claudátur *m* — **i** parentesi *fpl* quadre; parentesi *fpl* quadrate

1845 Bradel binding; binding "en gist" — Bookbinding with a spine of leather or linen, uncut edges, sometimes top gilded. Said to be of German origin; named, however, after Alexis Pierre Bradel, France.

f reliure *f* à la Bradel; cartonnage *m* à la Bradel; reliure *f* en gist — **d** Bradel-Einband *m*; Einband *m* à la Bradel — **n** Bradel-band *m* — **e** encuadernación *f* à la Bradel; encuadernación *f* Bradel — **i** legatura *f* alla Bradel

1846 Braille — System of printing for the blind in which the characters consist of raised dots to be read by the fingers. Named after Louis Braille (1809-1852), French teacher of the blind.

f impression *f* du Braille (par points en relief); impression *f* pour aveugles — **d** Blindenschrift *f* — **n** Brailleschrift *n*; blindenschrift *n* — **e** escritura *f* Braille; escritura *f* para ciegos; escritura *f* en relieve — **i** caratteri *mpl* Braille; caratteri *mpl* in rilievo per i ciechi; scrittura *f* Braille

1847 Braille paper

f papier *m* Braille — **d** Braillepapier *n*; Blindenschriftpapier *n* — **n** Braillepapier *n*; blindenpapier *n*; papier *n* voor blindenschrift — **e** papel *m* Braille; papel *m* para ciegos; papel *m* de ciegos — **i** carta *f* Braille; carta *f* da stampa per ciechi

1848 brake; reel brake — The manual or automatic mechanism applied to the reel spindle or reel to control the tension of the web of paper when the printing machine is running.

f frein *m* — **d** Bremse *f* — **n** rem *fm* — **e** freno *m* — **i** freno *m*

1849 brake bands — Endless leather, metal or fabric strips applied to a drum or pulley on the reel shaft. See also belts.

1850 brake bands — See belts.

1851 brake down; belts on — The action of applying extra pressure to the brake to increase the tension of the web.

f freinage *m* — **d** Abbremsen *n* — **n** afremmen *n* — **e** frenado *m*; enfrenamiento *m*; refrenamiento *m* — **i** frenatura *f*

1852 brake drum — Wheel attached to one end of

a reel spindle upon which pressure is exerted to control the tension of the web.

f tambour *m* de frein — **d** Bremstrommel *f* — **n** remtrommel *fm* — **e** tambor *m* de freno — **i** tamburo *m* del freno

1853 brake hand — The operator on a rotary machine responsible for adjusting the reel brake and reel side margin control. He may also be the *controller hand.

f bobinier *m* — **d** Verantwortlicher *m* für Papierrollenbremse — **n** remmer *m* — **e** operario *m* guardafrenos; guardafrenos *m* — **i** responsabile *m* del freno bobina

1854 branch — Computer term for a set of instructions that are executed between two successive decision instructions.

f branchement *m* — **d** Verzweigung *f* — **n** vertakking *f*; sprong *m* — **e** bifurcación *f* — **i** salto *m*

1855 branching — Feature of a computer program which, when it determines that a certain condition exists, alters the processing cycle. E.g., a series of numbers may be deducted from a certain sum, and the remainder is then tested to ascertain whether or not the result is negative. If it is not, the program remains in its loop. If negative, the program branches to another routine.

1856 branch *v* **out** *obs.* — To space outlines with leads or reglets to open or extend the matter. See also lead *v*.

f jeter du blanc; interligner — **d** die Zwischenräume vergrößern — **n** interliniëren; uitdrijven — **e** interlinear; espaciar — **i** spaziare; interlineare

1857 branch point — See node.

1858 Brasil wax — See carnauba wax.

1859 brass — Zinc-copper alloy (30-40% zinc) harder than copper, which may be cast, rolled, cut, and drilled more readily than copper and zinc.

f laiton *m*; cuivre *m* jaune — **d** Messing *n* — **n** messing *n*; koper *n* *depr.*; geel koper *n* *depr.* — **e** bronce *m*; latón *m* — **i** ottone *m*

1860 brass blocking plate

f plaque *f* à empreindre de laiton — **d** Messingprägeplatte *f* — **n** messing preegstempel *n*; koperen preegstempel *n* *depr.* — **e** plancha *f* de bronce para estampar — **i** lastra *f* per rilievo in ottone

1861 brass border

f filet *m* de cuivre pour cadre — **d** Messingeinfassung *f* — **n** messing kaderlijn *fm*; koperen kaderlijn *fm* *depr.* — **e** filete *m* de bronce para encuadrar; marco *m* de bronce; orla *f* de bronce — **i** contorno *m* d'ottone

1862 brass corner

f coin *m* de laiton; coin *m* de cuivre jaune — **d** Messingecke *f* — **n** messing hoek *m*; koperen hoek *m* *depr.*; koperen hoekstuk *n* *depr.* — **e** esquinazo *m* de latón — **i** angolo *m* d'ottone

1863 brass matrix

f matrice *f* en laiton — **d** Messingmatrize *f* — **n** messing matrijs *fm*; koperen matrijs *fm* *depr.* — **e** matriz *f* de latón; matriz *f* de bronce — **i**

matrice *f* d'ottone
1864 brass plate

f plaque *f* en laiton — **d** Messingplatte *f* — **n** messing plaat *fm*; koperen plaat *fm depr.* — **e** plancha *f* de latón; lámina *f* de latón; placa *f* de latón; plancha *f* de bronce; lámina *f* de bronce; placa *f* de bronce — **i** lastra *f* d'ottone
1865 brass roller — Roller in the water fountain of an offset press.

f rouleau *m* de laiton; rouleau *m* de bassine en laiton — **d** Messingwalze *f* — **n** messing vochtrol *fm*; koperen vochtrol *fm depr.* — **e** rodillo *m* mojador de latón; rodillo *m* mojador de bronce — **i** rullo *m* umidificatore d'ottone
1866 brass rule — Thin, type-high strip of brass of various thicknesses and different faces.

f filet *m* en cuivre; filet *m* d'airain — **d** Messinglinie *f*; Abschnittlinie *f* — **n** messing lijn *fm*; koperen lijn *fm depr.* — **e** filete *m* de latón; filete *m* de bronce; pleca *f* de bronce; pleca *f* de latón; raya *f* de latón; raya *f* de bronce — **i** filetto *m* d'ottone
1867 brass space — See also space.

f espace *f* en laiton — **d** Messingspatie *f*; Messingausschuß *m*; Messingdurchschuß *m* — **n** messing spatie *f*; koperen spatie *f depr.* — **e** espacio *m* de latón; espacio *m* de bronce — **i** spazio *m* d'ottone
1868 brass type

f caractère *m* en laiton; caractère *m* en cuivre — **d** Messingschrift *f* — **n** messing letter *fm*; koperen letter *fm depr.* — **e** tipo *m* de latón; tipo *m* de bronce — **i** tipo *m* d'ottone
1869 brass width — Width of the individual matrices used in line casters. Character widths are usually expressed in thousandths of an inch but to facilitate calculation of line lengths some founts for use in Teletype setting have been designed in unit widths where each character in different type faces is given the same unit value proportional to 18ths of an em. Later, perforator counting magazines were developed with brass widths converted to the appropriate unit of measurement to allow the use of non-unit founts.

f largeur *f* de matrice — **d** Matrizenbreite *f*; Breite *f* des Matrizenkörpers — **n** matrijzenbreedte *f* — **e** anchura *f* de la matriz — **i** larghezza *f* delle matrici singole
1870 bray *v* — To distribute ink with a hand roller.

f encrer au rouleau — **d** auswalzen der Druckfarbe; einfärben — **n** oprollen; inrollen — **e** batir a mano; entintar con rodillo de mano — **i** distribuire l'inchiostro con rullo a mano
1871 brayer; hand roller — Hand ink-roller. Formerly a round wooden pestle, flat at one end and with a handle on the other, used to spread out ink for the inking balls.

f rouleau *m* à main; brayon *m* — **d** Handwalze *f*; Handauftragwalze *f*; Auftragwalze *f* — **n** handrol *fm* — **e** rodillo *m* de mano; rulo *m* de mano — **i** rullo *m* a mano
1872 Brazil wax — See carnauba wax.

1873 bread wrappers — Paper for wrapping bread, often made opaque with materials such as titanium dioxide, giving good opacity despite waxing. Beater filling and surface coating are used to supply opacity to the sheet.

f papier *m* pour boulangerie — **d** Broteinschlagpapier *n*; Broteinwickelpapier *n*; Broteinwickler *m* — **n** papier *n* voor broodwikkels — **e** papel *m* para envolver pan — **i** carta *f* per avvolgere il pane
1874 break — Tear in the web of a paper reel, spliced with gummed or rubber tape and marked by a protruding piece of coloured paper, called a flag. See also downer.

f cassure *f*; rupture *f* — **d** Bruch *m*; Reißer *m* — **n** breuk *fm*; scheur *m* — **e** rotura *f*; ruptura *f*; romperse — **i** rottura *f*; strappo *m*
1875 break — The end or the beginning of a *paragraph.

f fin *f* d'alinéa; alinéa *m*; paragraphe *m* — **d** Ausgang *m*; Ausgangszeile *f*; Endzeile *f*; Absatz *m*; Alinea *n* — **n** uitgang *m*; uitgangsregel *m*; alinea *f*; inspringing *f* — **e** aparte *m*; párrafo *m* aparte; nueva línea *f* — **i** linea *f* a capo; capoverso *m*; daccapo *m*
1876 break — In newspaper production, the amount of break in a print or the area not covered by ink showing the white of the paper surface, due to its roughness.

f manque *m* d'encre — **n** sneeuw *fm* (in krantenpapier) — **e** falta *f* de tinta — **i** stampa *f* mancata
1877 break — See paragraph.
1878 break *v* — To interrupt a communication circuit.

f déconnecter — **d** abschalten — **n** verbreken — **e** desconectar — **i** disconnettere
1879 break *v* **a colour** — To add a third ink to change the colour toward grey or black. Breaking inks are brown, olive or grey.

f rabattre une encre — **d** eine Farbe brechen; eine Druckfarbe brechen — **n** een kleur breken; een inkt versnijden — **i** tagliare un inchiostro; tagliare con tinte scure
1880 breakdown — The enforced stoppage of production due to a machine defect.

f panne *f*; dérangement *m* — **d** Panne *f*; Störung *f*; Maschinendefekt *m*; Stopfer *m*; Stopper *m* — **n** storing *f*; oponthoud *n*; stagnatie *f* — **e** pana *f* — **i** interruzione *f*; fermata *f*; incidente *m* meccanico; fermomacchina *f*; guasto *m*; avaria *f*
1881 breaker; breaker beater — Beater used to disintegrate rags and old papers, before refining.

f pile *f* défileuse; pile *f* désagrégante — **d** Auflöseholländer *m*; Halbstoffholländer *m*; Halbzeugholländer *m*; Stetigholländer *m* — **n** lompenhollander *m* — **e** pila *f* desfibradora; filocho *m* — **i** olandese *f* sfilacciatrice
1882 breaker beater — See breaker.
1883 breaker rollers — Metal rollers used in a rotary press to convey, support and direct the paper, cellophane, foil, glassine or cardboard web

through the press. Term stems from the action of breaking the web at an angle over the roller. Freedom of rotation, diameter and angle of break over roller are usually of great importance due to conditions of friction, checking, etc. See also carrier rollers.

f rouleaux *mpl* de renvoi — **d** Umleitwalzen *fpl*; Umleitrollen *fpl* — **n** draagrollen *fmpl*; geleiderollen *fmpl* — **e** rodillos *mpl* conductores — **i** rulli *mpl* di rinvio

1884 break-even performance; threshold performance — Work study term for the performance level at which bonus commences.

f seuil *m* de prime; seuil *m* d'allure — **n** drempelprestatie *f* — **i** soglia *f*

1885 break *v* **in** — To insert cuts in their proper position in the text, as marked on the proofs.

f insérer; intercaler — **d** einbauen — **n** inbouwen — **e** colocar; intercalar los grabados en el texto; embutir los grabados en el texto — **i** inserire; intercalare

1886 breaking — Sudden and distinct separation of a gelatinous mass from raw linseed or other oils.

f floculation *f* — **d** Brechen *n*; Schleimen *n* — **i** separazione *f*

1887 breaking away of liner — When a cutting rule is placed close to a score rule on a platen press, the score rule will act to take or pull board into the counter. When there is no extra board available, it takes some board from under the cutting rule before the cutting operation is completed. This causes the filler to break away from the liner and fracture the board.

1888 breaking length — Calculated limiting length of a strip of paper or board of uniform width, beyond which, if such a strip were suspended by one end, it would break by its own weight.

f longueur *f* de rupture — **d** Reißlänge *f* — **n** breeklengte *f* — **e** latitud *f* de ruptura; latitud *f* de rotura; latitud *f* de desgarro — **i** lunghezza *f* di rottura

1889 break-line; broken line — The last line of a paragraph. It should not begin a column or page, unless it fills out the measure.

f ligne *f* boiteuse; ligne *f* courte; bout *m* de ligne — **d** Ausgangszeile *f*; Ausgang *m*; Diebszeile *f* — **n** halve regel *m*; uitgangsregel *m* — **e** línea *f* quebrada; línea *f* corta; línea *f* de fin de párrafo — **i** righino *m*; linea *f* mozza

1890 break mark — See paragraph mark.

1891 break *v* **off** — When movable type breaks off, usually in isolated words and folio numbers. Swash characters and kerned are easily broken.

f casser — **d** abbrechen — **n** breken; afbreken — **e** romperse — **i** rompere

1892 break *v* **off the jet** — To remove the surplus metal from the foot of a type when casting.

f rompre le jet — **d** den Gußzapfen abbrechen — **n** de gietprop afbreken — **e** quebrar el cabo; romper el jito — **i** rompere il getto

1893 break *v* **the line** — To make a new paragraph.

f couper des lignes; couper des alinéas — **d** Zeile trennen; Absätze bilden; Zeile brechen — **n** een nieuwe alinea maken — **e** separar líneas; separar renglones; fraccionar — **i** iniziare un nuovo paragrafo

1894 break *v* **up** — To unlock a forme and distribute the furniture and type.

f désosser (la forme) — **d** Form aufschließen und aufräumen; ausschlachten; Satz auseinandernehmen — **n** de vorm opbreken; de vorm uiteen nemen — **e** desarmar; desmontar; deshacer la forma — **i** spaginare; scomporre una forma

1895 break *v* **up for colours** — See dissect *v* for colours.

1896 break, web — See web break.

1897 breast bar — See ink rail bar.

1898 breast box — See head box.

1899 breast-roll — Roll supporting the wire-cloth of a paper machine at the forward end where the pulp is delivered to the wire from the head or breast box.

f rouleau *m* de tête — **d** Brustwalze *f* — **n** borstwals *fm* — **e** rodillo *m* de cabeza; cilindro *m* mandil — **i** cilindro *m* capotela

1900 breathings — Signs used in Greek to indicate that initial vowels are or are not aspirated. The rough breathing or *spiritus asper (') is placed over the P (rho) when it is the beginning consonant of a word. The smooth breathing or *spiritus lenis (') is placed over the beginning vowel, which does not have the spiritus asper.

f esprits *mpl*; signes *mpl* d'aspiration — **d** Hauchlautzeichen *npl*; Zeichen *npl* für Hauchlaute — **n** aspiratietekens *npl*; tekens *npl* voor al of niet geaspireerde stemzet — **e** signos *mpl* espíritu; espíritus *mpl* — **i** segni *mpl* di aspirazione

1901 breve — See short accent.

1902 brevier — Old name for a size of type. See app. no. 1.

1903 brick-wall pattern — Screen with a brick-wall pattern.

f trame *f* mur de briques; trame *f* brique — **d** Backsteinraster *m* — **n** baksteenraster *n* — **e** trama *f* en forma de ladrillo; retícula *f* en forma de ladrillo — **i** retino *m* a mattoni; retino *m* con trama a muratura di mattoni

1904 bridge — The detachable support for the dividers or blade adjusting screws of some ink ducts.

f séparateur *m* à plomb d'encrier — **d** Brücke *f*; Farbkastenbrücke *f* — **n** inktbakschot *n* — **e** tabique *m* del tintero — **i** divisore *m* del calamaio

1905 bridge — A land or portion of the die that is not cut through.

f point *m* d'attache — **d** Verbindungssteg *m* — **n** bruggetje *n* — **e** punto *m* de sujeción; punto *m* de agarre — **i** parte *f* della fustella non tagliata

1906 bridge — See hand rest.

1907 brief — The size of a standard foolscap sheet when folded in half.

1908 brightening agent — See optical bleaching agent.

1909 brightness; subjective brightness — Attribute of sensation by which an observer is aware of differences of *luminance. The brightness of a paper for instance is measured relative to that of a baryt standard which is taken equal to 100%. Deprecated term for *luminosity. See also degree of whiteness.
f brillance f — d Helligkeit f — n helderheid f — e resplandor m; claridad f — i brillanza f

1910 brightness range — Range between the brightest highlights of a scene and the darkest shadows.
f gamme f de brillances — d Helligkeitsumfang m — n helderheidsomvang m — e alcances mpl de reflectividad — i intervallo m di riflettanza

1911 Brightype — Registered trade name for a conversion process (1954) for the production of type transparencies by direct photography of a prepared letterpress printing forme. Use is made of a special camera with a rotating lamp system. It is marketed by a Ludlow Co., but it has never enjoyed wide acceptance.

1912 brilliant — Old name for a size of type. See app. no. 1.

1913 brilliant green
f vert m brillant — d Brillantgrün n — n briljantgroen n — e verde m brillante — i verde m brillante

1914 brilliant lake — Concentrated red organic colour with light overtone and blue undertone.
f laque f brillante — d Brillantfarblack m — n briljantlak mn — e laca f brillante — i lacca f brillante

1915 brilliant schwarz
f noir m brillant — d Glanzschwarz n — n glanszwart n; prachtzwart n — e negro m de lustre — i nero m brillante

1916 brimstone — *Sulphur used in the preparation of the cooking acid in the sulphite process of papermaking.

1917 Brinell hardness — See hardness.

1918 Brinell hardness number — Number indicating the hardness of a material determined by a ball of chromic steel of 10 mm diameter, under a load of 500 kg, applied to a sample of the material (type alloy) for one minute; then the diameter of the depression is measured. See also Shore hardness.
f dureté f selon Brinell — d Härte f nach Brinell; Brinell-Härte f; Härtezahl f nach Brinell — n Brinell-hardheidsgraad m; Brinell-hardheid f; hardheid f volgens Brinell — e coeficiente m de dureza Brinell; grado m de dureza según Brinell; dureza f Brinell — i durezza f secondo Brinell; durezza f Brinell

1919 bring v into register — To effect exact correspondence in the position of pages on both sides of the sheet of paper.
f faire le registre — d Register machen — n in register brengen; pas maken — e hacer el registro; buscar el registro — i portare a registro; mettere a registro

1920 bring v up; build v up — To underlay or interlay a block to give it the correct printing height.
f mettre de hauteur — d unterlegen — n ophogen; op hoogte brengen — e calzar; nivelar la forma — i alzare; mettere alla giusta altezza

1921 bristol; bristol board — Cardboard made on a Fourdrinier or a cylinder machine in thicknesses of 0.15 mm (0.006 in) and more. The three principal types are index, mill, and wedding bristols. Cylinder-made bristols are designated as mill bristols to distinguish them from Fourdrinier bristols or index bristols.
f carte f bristol; bristol m; feuilleton m — d Bristolkarton m — n bristolkarton n; triplexkarton n; geplakt karton n — e cartulina f bristol — i cartoncino m bristol

1922 bristol board — See bristol.

1923 British gum — See dextrin.

1924 British thermal unit — Unit of energy. Abbrev. Btu. In the SI (1978) defined by the equation: $1 \text{ Btu/1b} = 2.326 \times 10^3 \text{ J/kg}$. This definition ensures that specific heats are numerically identical in the British system with the *calorie. Thus: $1 \text{ Btu} \approx 1.05506 \times 10^3 \text{ J}$.
f unité f de chaleur anglaise — d Britische Hitzeeinheit f — n Engelse warmte-eenheid fm — e unidad f de calor inglés; unidad f térmica británica — i unità f di calore inglese

1925 brittle
f cassant; fragile — d spröde; brüchig; hartbrüchig — n bros — e quebradizo; frágil — i fragile; friabile

1926 brittleness (of a film)
f tendance f à l'écaillage (d'un film) — d Sprödigkeit f (eines Films) — n brosheid f (van een film) — e fragilidad f (de una película) — i fragilità f (di una pellicola)

1927 broad obs. — Wooden furniture 4 ems (48 points) wide.
f lingot m en bois de 48 pts — d hölzerner Formatsteg m von 48 Punkt Breite — n houten wit n van 48 punten breed — e lingote m de madera de 48 puntos — i lingotto m di legno da 48 punti

1928 broad double demy — A size of paper. See app. no. 3.

1929 broad double medium — A size of paper. See app. no. 3.

1930 broadloaf tree — Generally, a deciduous tree, such as poplar or maple; distinguished from coniferous trees by relatively broad, flat leaves. The wood is usually termed *hardwood.
f arbre m feuillu — d Laubbaum m — n loofboom m — e árbol m de fronda; árbol m frondoso — i albero m a fronda latifoglia

1931 broadsheet — Traditionally the sheet before it is folded; sometimes the size of newspapers when run columns around the cylinder.
d Planobogen m — n plano formaat n; vol formaat n; ongevouwen formaat n — e hoja f

plana; hoja *f* in-plano — **i** foglio *m* non ripiegato

1932 broadside page — Page designed to read normally when a book is turned 90°. Wide tables and illustrations are often run broadside. Mostly the left side of a broadside table or illustration is at the bottom of the page.

f page *f* en travers; page *f* à l'italienne — **d** Querfolio *m*; gestürzte Seite *f* — **n** dwarse pagina *fm*; dwarspagina *fm* — **e** página *f* en folio; página *f* apaisada — **i** pagina *f* di traverso; pagina *f* trasversale; pagina *f* traversa

1933 brochure; pamphlet — Sewed or stitched book with or without paper cover and about 8-32 pages. See also booklet.

f brochure *f*; pamphlet *m*; opuscule *m* — **d** Broschüre *f* (Weichbroschüre; Kartonbroschüre; Steifbroschüre) — **n** brochure *fm* — **e** folleto *m*; opúsculo *m* — **i** brossura *f*; fascicolo *m*; opuscolo *m*

1934 broke — Paper or board discarded at any stage during manufacture. Usually repulped. There are two kinds: wet broke and dry broke. Not to be confused with the English term broken paper which is equivalent to the American *job lot paper.

f cassés *mpl* de fabrication; rejets *mpl* — **d** Ausschuß *m*; Maschinenausschuß *m*; Fabrikationsabfälle *mpl* — **n** uitval *m*; uitschot *n*; kas *n* — **e** recortes *mpl*; costeros *mpl*; merma *f* desperdicio — **i** fogliacci *mpl*; scarto *m*

1935 broken bastard — Old name for 32-point type. See app. no. 1.

1936 broken line — See break-line.

1937 broken matter — See pie.

1938 broken paper — See job lot paper.

1939 broken types — See classification of type design and especially note to app. no. 2.

f caractères *mpl* brisés; caractères *mpl* à fractures — **d** gebrochene Schriften *fpl* — **n** gebroken letters *fpl* — **e** tipos *mpl* quebrados — **i** caratteri *mpl* fratti

1940 bromic ether — See ethyl bromide.

1941 bromide — Salt of hydrobromic acid. Silver bromide is a light-sensitive salt used in photographic emulsions.

f bromure *m* — **d** Bromide *n* — **n** bromide *n* — **e** bromuro *m* — **i** bromuro *m*

1942 bromide *US sl.* — One who utters platitudes; also a platitude, a dull remark.

f homme *m* ennuyeux; raseur *m* — **d** Langweiler *m* — **n** vervelende man *m* — **e** hombre *m* trivial — **i** persona *f* noiosa

1943 bromide — See bromide print.

1944 bromide of ammonia — See ammonium bromide.

1945 bromide of potash — See potassium bromide.

1946 bromide paper — Photographic paper base coated with an emulsion coating AgBr, i.e., *silver bromide. It is the most popular printing paper for enlarging because its high sensitivity to light allows short exposure times.

f papier *m* au bromure; papier *m* au gélatinobromure; papier *m* bromure — **d** Bromsilberpapier *n* — **n** bromidepapier *n* — **e** papel *m* bromuro; papel *m* al bromuro de plata; papel *m* de bromuro — **i** carta *f* al bromuro

1947 bromide print; bromide — Photographic contact print or enlargement made on paper sensitized with a bromide emulsion.

f bromure *m* — **d** Bromsilberdruck *m*; Bromsilberkopie *f* — **n** bromideafdruk *m*; afdruk *m* op bromidepapier — **e** copia *f* en papel bromuro — **i** copia *f* al bromuro

1948 bromide proof — Monochrome photographic print to be submitted to a customer for approval.

f épreuve *f* bromure — **d** einfarbige Papierkopie *f*; Bromsilberpapierprobe *f* — **n** proef *fm* op bromidepapier — **e** prueba *f* en papel bromuro — **i** prova *f* al bromuro

1949 bromine — Non-metallic halogen element; deep red corrosive toxic liquid giving off an irritating reddish brown vapour of disagreeable odour. Symbol: Br. Latin: bromium.

f brome *m* — **d** Brom *n* — **n** broom *n* — **e** bromo *m* — **i** bromo *m*

1950 bromocresol green — Tetrabromo-metacresolsulphonphthalein, an acid base indicator, changes from yellow to blue between pH 3.8-5.4. Yellow crystals, slightly soluble in water, soluble in alcohol.

f vert *m* de bromocrésol — **d** Bromkresolgrün *n* — **n** broomcresolgroen *n* — **e** verde *m* de bromocresol — **i** verde *m* di bromocresolo

1951 bromocresol purple — Dibromo-ortho-cresolsulphonphthalein, an acid base indicator, changes from yellow to purple between pH 5.2-6.8. Yellow crystals, soluble in alcohol, insoluble in water.

f pourpre *m* de bromocrésol; rouge *m* de bromocrésol — **d** Bromkresolpurpur *n* — **n** broomcresolpurper *n* — **e** púrpura *f* de bromocresol — **i** porpora *f* di bromocresolo

1952 bromoethane — See ethyl bromide.

1953 bromoil process — Process to make an offset reproduction by making a photographic print on paper with a silver bromide emulsion, wetting it, and then using it as a lithographic plate; the lighter parts of the emulsion tend to repel the oil base of the ink while the darker parts tend to hold it.

f oléobromie *f*; bromoil *m*; bromoléotypie *f*; bromocollographie *f* — **d** Bromölverfahren *n* — **n** broomoliedruk *m* — **e** procedimiento *m* al bromóleo; procedimiento *m* al oleobromo; bromóleo *m*; oleobromía *f*; bromoleotipia *f* — **i** procedimento *m* al bromolio

1954 bromomethane — See methyl bromide.

1955 bromophenol blue — Tetrabromo-phenolsulphonphthalein, an acid base indicator, changes from yellow to purple between pH 3.0-4.6.

f bleu *m* de bromophénol — **d** Bromphenolblau *n* — **n** broomfenolblauw *n* — **e** azul *m* de bromofenol — **i** azzurro *m* di bromofenolo

1956 bronze — Alloy of tin and copper used for

machine parts, bearings, etc.

f bronze *m* — **d** Bronze *f* — **n** brons *n* — **e** bronce *m* — **i** bronzo *m*

1957 bronze — Metallic powder used in decorative work in printing and bookbinding.

f bronze *m*; poudre *f* à bronzer; poudre *f* de bronze — **d** Bronzepulver *n*; Bronzestaub *m* — **n** brons *n*; bronspoeder *n* — **e** bronce *m*; polvo *m* de bronce; purpurina *f* — **i** polvere *f* da bronzatura; porporina *f*

1958 bronze; iridescence — An appearance characteristic of printed films in which the colour of the print depends on the angles of viewing and illumination.

f mordoré *m*; impression *f* mordorée — **d** Bronzestich *m*; Goldkäferglanz *m*; Bronzelüster *m* — **n** bronskleur *m*; iridiscentie *f* — **e** impresión *f* irisada — **i** iridescenza *f*; stampa *f* iridata

1959 bronze blue — See Prussian blue.

1960 bronze blue ink — Ink made with a ferrocyanide pigment which gives it a bronzy tone.

f encre *f* bleue fluorée — **d** bronzeblaue Farbe *f* — **n** bronsblauwe inkt *m* — **e** tinta *f* azul bronce — **i** inchiostro *m* bronzo azzurro

1961 bronze bushing

f bague *f* de bronze — **d** Bronzebuchse *f* — **n** bronzen lagerbus *fm* — **e** casquillo *m* de bronce — **i** boccola *f* in bronzo

1962 bronze dust — See bronze powder.

1963 bronze dusting — See bronzing.

1964 bronze ink — Printing ink containing bronze powder or pigments with an exceptionally marked bronze cast.

f encre *f* bronze — **d** Bronzefarbe *f*; Bronzedruckfarbe *f* — **n** bronsinkt *m* — **e** tinta *f* de bronce; tinta *f* metálica de bronce; tinta *f* metálica de oro; purpurina *f* oro — **i** inchiostro *m* bronzo

1965 bronze powder; bronze dust — Metallic pigment for printing ink consisting of copper alloys in fine flakes.

f poudre *f* de bronze; bronze *m* en poudre — **d** Bronzepulver *n*; Bronzestaub *m*; Bronzepigment *n* — **n** bronspoeder *n* — **e** polvo *m* de bronce — **i** polvere *f* di bronzo; bronzo *m* pesto

1966 bronze printing

f impression *f* bronzée — **d** Bronzedruck *m*; Bronzierdruck *m* — **n** bronsdruk *m* — **e** estampación *f* en bronce — **i** stampa *f* bronzo

1967 bronzer

f bronzeur *m* — **d** Bronzierer *m* — **n** bronzer *m* — **e** bronceador *m* — **i** bronzista *m*

1968 bronzing; bronze dusting — Application of bronze powder to the press sheet, usually by a bronzing machine. A sticky base of clear or yellow varnish, called gold size, is applied by the press just before the sheets are fed to the bronzer. There the powder is sifted onto the sheet where it adheres to the sticky image.

f bronzage *m* — **d** Bronzieren *n* — **n** bronzen *n* — **e** bronceado *m* — **i** bronzatura *f*

1969 bronzing machine — Machine in which the bronze or other metallic powder is brushed over a sheet freshly printed with varnish or size.

f machine *f* à bronzer; bronzeuse *f*; bronzeuse-essuyeuse *f* — **d** Bronziermaschine *f*; Bronzierapparat *m* — **n** bronsmachine *f* — **e** máquina *f* de broncear; bronceadora *f* — **i** macchina *f* a bronzare; bronzatrice *f*

1970 bronzing medium; bronzing size — Stiff, yellow ink used for printing before bronzing with bronze powder.

f mordant *m* à bronzage; mordant *m* — **d** Bronzeunterdruckfarbe *f* — **n** bronsvoordrukinkt *m* — **e** mordiente *m* para broncear; sisa *f* para broncear — **i** mordente *m* per bronzatura; inchiostro *m* sottostampa per bronzatura

1971 bronzing plush — Fabric of silk, cotton, or wool, whose pile is more than 3 mm (1/8 in) high. Used in a *bronzing machine.

f peluche *f* à bronzer — **d** Bronzierplüsch *m* — **n** bronspluche *mn* — **e** felpa *f* para broncear; peluche *m* de broncear — **i** felpa *f* per bronzatura

1972 bronzing size — See bronzing medium.

1973 brownline — See brownprint.

1974 brown lined chipboard

f carton *m* gris de vieux papiers recouvert brun — **d** braungedeckte Graupappe *f* — **n** bruingedekt grijsbord *n*; bruingedekt grijs karton *n* — **e** cartón *m* gris — **i** cartone *m* grigio rivestito in bruno

1975 brown mechanical pulp board — Board made from brown mechanical woodpulp.

f carton *m* de pâte mécanique brun; carton *m* bois brun — **d** Braunschliffpappe *f*; braune Holzpappe *f* — **n** leerbord *n* uit bruinslijp — **e** cartón *m* compacto de pasta mecánica parda; cartón *m* de madera parda — **i** cartone *m* di pasta meccanica bruna

1976 brown mechanical woodpulp — Pulp obtained by grinding steamed or boiled logs.

f pâte *f* mécanique brune; pâte *f* de bois brune mécanique — **d** brauner Holzschliff *m*; Braunschliff *m* — **n** bruine houtslijp *m*; bruinslijp *m* — **e** pasta *f* (de) madera parda; pulpa *f* (de) madera parda — **i** pasta *f* meccanica bruna; pastalegno *m* bruno

1977 brown mixed pulp board — Board from waste papers sometimes with addition of pulps, brown on both sides.

f carton *m* de pâte brune mixte — **d** braune Mischpappe *f* — **e** cartón *m* gris mixto — **i** cartone *m* di pasta bruna mista

1978 brownprint; brownline; vandyke print; silverprint — Brown photographic contact print made on thin paper sensitized with a silver and iron compound. Requires intense light for exposure. Frequently used by offset lithographers as a proof for the customer to check layout or other corrections. Not to be confused with a sepia print, or black photograph, which has been chemically converted (toned) to a brown colour. Vandyke print is named after Sir Anthony van Dyck, Flemish painter, Antwerp 1599 - London 1641.

f diapose *m* — **d** Braunpause *f*

1979 bruise mark — Excessive pressure in relief stamping, causing distortion of the stamped material, which bruises the edges of the stamped area in the shape of the counterpart.
d Quetschmarkierung *f* — **n** gekneusde plek *fm* in het papier — **i** ammaccatura *f*; impressione *f* ammaccata
1980 brunak — Treatment for aluminium litho plates, using a solution of ammonium bichromate (27 av.oz.) and hydrofluoric acid (3 fl.oz.- 48%) in three gallons of water, to make them non-oxidizing. The solution is sometimes called patrol.
f procédé *m* brunak — **d** Brunak-Verfahren *n* — **n** brunak-procédé *n* — **e** procedimiento *m* brunak; método *m* brunak — **i** procedimento *m* brunak
1981 brush
f brosse *f* — **d** Bürste *f* — **n** borstel *m* — **e** bruza *f*; cepillo *m* — **i** setola *f*; spazzola *f*
1982 brush — See carbon brush.
1983 brush — See paint brush.
1984 brush — See paper brush.
1985 brush — See paste brush.
1986 brush *v* — To remove by brushing or by lightly passing over.
f brosser; donner un coup de brosse — **d** ab-bürsten; mittels Bürste reinigen — **n** afborstelen — **e** bruzar; brozar — **i** spazzolare
1987 brush, air — See air brush.
1988 brush, beating — See beating brush.
1989 brush, black-leading — See black-leading brush.
1990 brush, carbon — See carbon brush.
1991 brush, cleaning — See type brush.
1992 brush coated paper — Paper coated by brush-coating process.
f papier *m* couché brosse; papier *m* couché à la brosse — **d** bürstengestrichenes Papier *n*; Papier *n* mit Bürstenauftrag — **n** met de borstel gestreken papier *n* — **e** papel *m* estucado a cepillo — **i** carta *f* patinata a spazzola
1993 brush coating — Application of a semifluid mixture of pigment and binder to a paper web by means of a revolving cylindrical brush and smoothing this coating by means of oscillating flat brushes which contact the coated sheet as it is being drawn tightly over a moving rubber apron or a revolving drum.
f couchage *m* à la brosse; couchage *m* par brosses — **d** Bürstenstreichverfahren *n*; Bürstenstrich *m* — **n** borstelstrijkprocédé *n*; borstelcoating *m* *depr.*; brush coating *m* *depr.* — **e** estucado *m* con cepillos; procedimiento *m* de estucado por cepillos — **i** patinatura *f* a spazzola
1994 brush, cylinder — See cylinder brush.
1995 brush, developing — See developing brush.
1996 brush, dusting — See dusting brush.
1997 brush, dusting — See dusting brush.
1998 brush enamel paper — Paper coated on one or both sides and brush polished previous to calendering to produce a smooth, even, and brilliant surface, used largely for cigar labels, illus-trations and box coverings.
f papier *m* couché émail; papier *m* couché brillant — **d** gestrichenes Glacé-Papier *n*, satiniertes gebürstet — **e** papel *m* esmaltado a cepillo — **i** carta *f* patinata lucidata a spazzola
1999 brush etching
f morsure *f* au pinceau — **d** Pinselätzung *f* — **n** kwastetsing *f*; penseeletsing *f* — **e** grabado *m* a pincel — **i** morsura *f* al pennello
2000 brush, etching — See etching brush.
2001 brush finished paper — Coated paper, the surface of which has been given the required degree of finish by means of rotating brushes.
f papier *m* couché fini à la brosse — **d** gebürstetes Kunstdruckpapier *n* — **n** met de borstel gestreken kunstdrukpapier *n* — **e** papel *m* estucado abrillantado a cepillo — **i** carta *f* lucidata a spazzola
2002 brush finishing — Method to impart gloss to coated papers and boards by subjecting the coated surface (usually mineral based) to the action of rapidly rotating brushes.
f finissage *m* à la brosse — **d** Bürstenglättung *f* — **n** satinage *f* met borstels — **e** abrillantado *m* a cepillo — **i** lucidatura *f* a spazzole; finitura *f* a spazzole
2003 brush, flat — See flat brush.
2004 brush, graphite — See black-leading brush.
2005 brushing — Setting of the beater roll so that the fibres are crushed and refined, rather that cut.
f affleurage *m* — **d** Ausgleichung *f* — **n** stellen *n* — **e** afloramiento *m* — **i** stemperamento *m*
2006 brushing out — See fibrillation.
2007 brush lettering — See brush writing.
2008 brush, lye — See lye brush.
2009 brush, marbling — See marbling brush.
2010 brush, matrix beating — See beating brush.
2011 brush, pick — See type brush.
2012 brush proof
f morasse *f*; épreuve *f* à la brosse — **d** Bürsten-abzug *m*; Korrekturabzug *m* — **n** borstelproef *fm*; met de borstel geklopte proef *fm*; drukproef *fm* — **e** prueba *f* hecha con un cepillo; prueba *f* de plana — **i** bozza *f* alla spazzola
2013 brush, reel — See reel brush.
2014 brush, retouching — See retouching brush.
2015 brush, type — See type brush.
2016 brush, wire — See wire brush.
2017 brush work — Originals made by painting with brushes.
f peinture *f* au pinceau — **d** Pinseltechnik *f* — **n** penseelwerk *n* — **e** pintado *m* a pincel — **i** dipinto *m* a pennello
2018 brush writing; brush lettering
f écriture *f* au pinceau — **d** Pinselschrift *f* — **n** op met het penseel getekend lijkende letter *fm* — **e** escritura *f* de pincel — **i** scrittura *f* a pennello
2019 bubble coating — Coating on paper obtained by bringing air bells in the coating slip.
f couchage *m* mousse — **d** Luftbläschenstreich-verfahren *n*; Luftbläschenstrich *m* — **n** schuim-strijklaag *fm*; bubble coating *m* *depr.* — **i** patina-

tura *f* bubble coating

2020 bubble etcher; bubble powderless etcher — Powderless etching machine. Air is pumped in the air grid located at the bottom of the tank, where it escapes through small holes to create millions of bubbles which supply the bath activating force.
f machine *f* à graver par bullage — **d** Blasenätzmaschine *f* — **i** macchina *f* per morsura a bolle

2021 bubble powderless etcher — See bubble etcher.

2022 bubble-tube viscometer; bubble viscometer — Viscometer composed of a short glass tube closed at both ends that contains the material to be tested together with a substantial size air bubble. The tube is turned upside down and the time required by the bubble to travel the length of the tube is recorded. This time is compared with a set of standard tubes containing materials of known viscosity.
f viscosimètre *m* à bulle d'air — **d** Luftblasenviskosimeter *n* — **n** luchtbel-viscosimeter *m* — **e** viscosímetro *m* de burbuja — **i** viscosimetro *m* a bolla d'aria

2023 bubble viscometer — See bubble-tube viscometer.

2024 bubbling — Swelling of rubber blankets or rollers caused by oil or solvent on a rotary press.
f se gonfler — **d** quellen — **n** zwellen; opzwellen — **e** hincharse; abotargarse *fam.* — **i** gonfiarsi

2025 bubbling — The effect produced on a printing plate, caused by a defective mould in letterpress rotary printing.

2026 buckle — Defect of aluminium foil. A rough, short gathering or fullness, cross-wise the direction of rolling, usually at the centre or about one-quarter of the way from the edge of the rolled width.

2027 buckle folder machine; plate folding machine; pocket-folding machine — Machine for folding sheets of paper. The paper is run up against a stop by means of driving rollers. After reaching the stop, the driving rollers force the paper to buckle in the position where the fold is to be formed and another pair of rollers grip the buckle, complete the fold and eject the folded sheet.
f plieuse *f* à poches — **d** Taschenfalzmaschine *f*; Stauchfalzmaschine *f* — **n** tassenvouwmachine *f* — **e** plegadora *f* de bolsa — **i** piegatrice *f* a tasche; piegatrice *f* a castelli

2028 buckling — Tendency of paper to bend or become wavy with changes in atmospheric conditions.
f gondolage *m* — **d** Welligwerden *n*; Wölbung *f* — **n** golven *n* — **e** abarquillarse — **i** bombatura *f*

2029 buckling — Distortion of a letterpress rotary printing plate caused by excessive pressure from a compression type lock-up.
f ventre *m* — **d** Plattenverspannung *f*; Plattenaufwölbung *f* — **n** opkomen *n*; buigen *n* — **e** panza *f* — **i** pancia *f*

2030 buckram — Heavy woven material, used for bookbinding; the best quality has a basis of linen. It is finished with either a glazed or matt calendered surface. Bocasin is a fine quality buckram.
f bougran *m* — **d** Buckram *n*; Steifleinen *n* — **n** buckram *n* — **e** bucarán *m*; bocacé *m*; zángala *f* — **i** tela *f* buckram

2031 buckskin — Strong leather for bookbinding originally from deerskins, now usually from sheepskins.
f peau *f* de daim — **d** Buckskin *m*; Wildleder *n* — **n** bukskin *n*; wildleer *n* — **e** piel *f* de ante — **i** pelle *f* di daimo

2032 buffer; shock absorber — Machine part to absorb shocks. See also air buffer.
f amortisseur *m* de chocs; amortisseur *m* à piston — **d** Stoßdämpfer *m*; Dämpfer *m*; Puffer *m* — **n** schokbreker *m* — **e** parachoques *m*; paragolpes *m* *SA*; tope *m*; amortiguador *m* — **i** ammortizzatore *m*

2033 buffer; buffer store *GB*; **buffer storage** *US* — Store to compensate for the difference in the speeds of the central processor and peripheral equipment. It can receive data at different time intervals. Buffered input and output channels allow terminal operations to continue while the CPU continues processing.
f mémoire *f* tampon; mémoire *f* intermédiaire — **d** Pufferspeicher *m*; eingebauter Speicher *m*; Zusatzspeicher *m*; Puffer *m* (hardware); Zwischenspeicher *m* (software) — **n** buffer *m*; buffergeheugen *n* — **e** memoria *f* intermedia — **i** memoria *f* intermedia; memoria *f* di transito; memoria *f* ponte

2034 buffer solution — Substance, or mixture of compounds, that, added to a solution, can neutralize both acids and bases without appreciably changing the original acidity or alkalinity of the solution. A chemical agent used to control the activity of a developer.
f solution *f* tampon — **d** Pufferlösung *f* — **n** bufferoplossing *f* — **e** solución *f* tampón; solución *f* amortiguadora; solución *f* reguladora; solución *f* compensadora; amortiguador *m* — **i** soluzione *f* tampone

2035 buffer storage — See buffer.

2036 buffer store — See buffer.

2037 buffing; burnishing — Roughing out with emery wheel or polishing with soft fabric wheel.
f polissage *m* — **d** Polieren *n*; Schwabbeln *n*; Abglätten *n* — **n** polijsten *n*; glad maken *n* — **e** pulir — **i** politura *f*; lucidatura *f*

2038 buffing — Split cowhide. See also American Russia.
f peau *f* sciée de vache — **d** Spaltleder *n*; Rindspaltleder *n* — **n** spouwleer *n*; gespouwen rundleer *n* — **i** cuoio *m* spaccato

2039 buffing machine — Machine for buffing or polishing with rotating discs covered with canvas.
f polisseuse *f* — **d** Schwabbelpoliermaschine *f* — **n** polijstmachine *f* — **e** máquina *f* pulidora; bruñidora *f* mecánica — **i** politrice *f*

2040 bug — In computer controlled typesetting, a mistake or malfunction.

f erreur *f*; incident *m*; défaut *m*; imperfection *f* d'une programme — **d** Störung *f*; Defekt *m* (am Satzrechner, im Programm); Panne *f*; Programmierfehler *m* — **n** storing *f*; programmeringsfout *fm*; panne *fm* — **e** defecto *m*; falla *f*; incidencia *f* (en computadora, en programa) — **i** difetto *m*; guasto *m*

2041 bug *US sl.* — Printed matter produced by a union shop.

2042 build *v* **up** — See bring *v* up.

2043 built-in

f encastré; incorporé — **d** eingebaut; angebaut — **n** ingebouwd — **e** integrado; integrante; integral; incorporado; emportado; acoplado — **i** incorporato; incassato

2044 built-in check — See automatic check.

2045 bulk; bulk index; bulking index *GB*; **specific volume** *US* — The number obtained by dividing the thickness of paper or board expressed in microns (0.001 mm) by its substance expressed in g/m².

f indice *m* de bouffant; bouffant *m*; indice *m* de main — **d** Dickgriffigkeit *f*; Bauschigkeit *f*; spezifisches Volumen *n* — **n** opdikking *f*; opdikkendheid *f* — **e** cuerpo *m* (del papel) — **i** voluminosità *f*

2046 bulk *v* — Paper or board is said to bulk well (high) when the thickness seems high for its substance and to bulk low when the reverse is the case.

f avoir de la main — **d** aufbauschen; auftragen; Griffigkeit haben — **n** opdikken; opdikkend zijn — **i** mano *m* della carta

2047 bulk index — See bulk.

2048 bulking dummy — See dummy.

2049 bulking index — See bulk.

2050 bulking paper — See bulky paper.

2051 bulky paper; bulking paper — Thick, uncalendered or lightly calendered paper of relatively light weight. See also featherweight paper.

f papier *m* bouffant; papier *m* volumineux — **d** bauschiges Papier *n*; auftragendes Papier *n*; Dickdruckpapier *n*; Daunendruckpapier *n*; aufbauschendes Papier *n* — **n** opdikkend papier *n*; volumineus papier *n* — **e** papel *m* pluma; papel *m* voluminoso — **i** carta *f* grossa

2052 bulldog; bulldog edition — Early edition of a morning newspaper, often appearing the evening before the date of imprint and intended for distribution out of town. The Sunday edition is called the bull pup.

f édition *f* du matin — **d** Morgenausgabe *f* — **n** ochtend-editie *f* — **e** edición *f* matutina; edición *f* de la mañana; edición *f* de mula — **i** primera edizione *f* delle giornale del mattino

2053 bulldog edition — See bulldog.

2054 bullet — Instant dismissal (to get the bullet, to be fired).

f renvoi *m* immédiat — **d** fristlose Entlassung *f*; gefeuert werden — **n** ontslag *n* op staande voet — **e** despedida *f* en el acto; despedida *f* de inmediato; despedida *f* incontinenti; despedido *m* en jergo — **i** licenziamento *m* sull'istante; licenziamento *m* in tronco

2055 bullet — See centred dot.

2056 bull horn wrench — Wrench used by the letterpress rotary minder in connection with a particular form of instantaneous plate lock-up.

f clef *f* de serrage rapide; clef *f* de serrage instantané — **d** Plattenspannschlüssel *m* — **n** plaatsleutel *m* — **e** mordaza *f* de cierre; mordaza *f* rápida — **i** chiave *f* di serraggio rapido

2057 bull pup — See bulldog.

2058 bull screen — Coarse screen to remove slivers and knots in groundwood pulp as it comes from the grinders.

f trieur *m* de bûchettes — **d** Splitterfänger *m*; Schüttelsortierer *m*; Sauerkrautfang *m* — **e** recoge-astillas *m*; separa-astillas *m*; separador *m* de astillas — **i** separaschegge *m*

2059 bull's-eye — Magnifying lens mounted on a stand. See also linen tester.

f loupe *f* sur pied — **d** Vergrößerungsglas *n* mit Ständer — **n** vergrootglas *n* met voet — **e** lupa *f* de pie; lupa *f* montada — **i** lente *f* d'ingrandimento a piede

2060 bull's-eye — Formerly a glass balloon filled with water to which some droplets of copper sulphate were added to give the light from a paraffin lamp falling through the water a bluish tint. Formerly used by engravers and punch cutters to reduce shadows.

d Glasskugel *f* — **n** glazen bol *m*; glasbol *m*

2061 bumping — Treatment of underlaying halftone blocks to be printed from curved stereotype and electrotype plates. Practically confined to illustrations in newspapers and magazines, giving a varying height-to-paper surface to the plates, the solids being on the highest plane and the highlights on the lowest.

f bosselage *m* des clichés — **d** Unterlegen *n*; Klischeezurichtung *f*; Durchprägen *n* — **n** onderleggen *n* van clichés — **i** taccheggio *m*

2062 bump-up process — Method of etching and treating halftone plates to introduce *premakeready in the printing surface.

f pré-mise *f* des clichés — **d** Ätzausschnitt *m*; Herstellung *f* einer Kraftzurichtung im Ätzverfahren — **n** pikeersel *n* in het cliché — **e** arreglo *m* previo — **i** taccheggio *m* mediante morsura di una lastra retinata

2063 buna — Synthetic rubber made by polymerization of butadiene with other substances, as styrene. German trade-mark (contraction of butadiene and natrium).

f buna *f* — **d** Buna *n* — **n** bunarubber *mn*; kunstrubber *mn* — **e** buna *f*; caucho *m* buna; caucho *m* artificial; caucho *m* sintético — **i** buna *f*; gomma *f* sintetica

2064 bundle press; bundling press; bundle tying machine; bundle tyer; bundler; tying machine — Machine for laying a string around bundles of daily newspapers or other printed matter.

f fardeleuse *f*; presse *f* à paqueter — **d** Verschnür-automat *m*; Bündelmaschine *f*; Packpresse *f* — **n** bundelmachine *f*; bundelpers *f*; bundelapparaat *n*; pakpers *f* — **e** atadora *f*; máquina *f* de atar; amarradora *f* automática; enfajilladora *f*; máquina *f* de amarrar; prensa *f* para enfardar; máquina *f* de prensar; compresora *f* — **i** legatrice *f* impaccatrice

2065 bundler — See bundle press.

2066 bundle tyer — See bundle press.

2067 bundle tying machine — See bundle press.

2068 bundling press — See bundle press.

2069 bung — See reel cone.

2070 bungle — See botch.

2071 bungler — See botcher.

2072 burin — Tempered steel *graver or cutting tool with a lozenge-shaped point, used for line engraving on metal.

f burin *m* — **d** Stichel *m* — **n** burijn *m*; steker *m* — **e** buril *m* — **i** bulino *m*

2073 burn — Exposure made with an arc lamp.

f exposition *f* à l'arc — **d** Bogenlichtaufnahme *f* — **n** booglicht-opname *fm* — **e** exposición *f* con lámpara de arco — **i** esposizione *f* con lampada ad arco

2074 burning-in — Heating a developed glue print on copper or zinc to bake the enamel image and impart resistance to acids or mordants; in etching, application of heat to a plate dusted with etching powder, to fuse or melt the powder.

f cuisson *f* — **d** Einbrennen *n* der Emailkopie — **n** inbranden *n* — **e** calentado *m*; cocido *m*; fundido *m*; quemado *m*; tostado *m*; esmaltado *m* — **i** smaltare; cottura *f*

2075 burning-in oven; burning-in stove

f chaufferette *f*; four *m* à cuire — **d** Einbrennofen *m* — **n** inbrandoven *m* — **e** hornillo *m* para esmaltar planchas; aparato *m* de calefacción; horno *m* para calentar; estufa *f* para calentar — **i** forno *m* per la cottura dello smalto

2076 burning-in stove — See burning-in oven.

2077 burning-in tongs

f pince *f* à chauffer — **d** Einbrennzange *f* — **n** inbrandtang *m* — **e** tenazas *fpl* de fundido — **i** tenaglie *fpl* da cottura

2078 burnisher — In photoengraving, steel tool used by finishers to darken tints and tones on halftone plates by rubbing with pressure; also to remove rough spots from plates and edges.

f brunissoir *m* — **d** Polierstahl *m*; Glättstahl *m* — **n** polijststaal *n* — **e** pulidora *f* — **i** brunitore *m*

2079 burnisher — Binder's tool, sometimes steel, more often bloodstone or agate, used to burnish book edges, or leather.

f dent *f* de loup; dent *f* à brunir — **d** Glättzahn *m* — **n** bruineertand *m*; gladtand *m* — **e** bruñidor *m* (de diente de lobo) — **i** brunitoio *m*; brunitore *m*

2080 burnishing — Corrective treatment of a printing plate with a burnisher to darken areas by spreading the printing surface of lines and dots.

f fonçage *m* — **d** Polieren *n* — **n** polijsten *n* — **e** pulido *m* — **i** pulire; polire

2081 burnishing — Polishing the edges of a book, previously stained, marbled or gilded with a bloodstone or agate.

f brunissage *m* — **d** Glätten *n* — **n** bruineren *n*; bruneren *n*; glad maken *n* — **e** bruñir — **i** brunire; lisciare

2082 burnishing — See buffing.

2083 burnout; colour burnout — Objectionable change in the colour of printing ink which may occur either in bulk or on the printed sheet. In bulk it is associated primarily with tints and is caused by a chemical reaction between certain components in the ink formulation. On the printed sheet it is generally caused by heat generated in a pile of printed material during drying of an oxidizing type of ink.

f virage *m* — **d** Verfärbung *f* — **n** verkleuring *f* — **e** virado *m*; viraje *m* — **i** scolorazione *f*

2084 burn *v* **out** — To overexpose in such a way that no halftones come up. This is done on deep-etch plates where positives are used. Edges can be sharpened up, while lettering can be obtained in toned areas by burning out or by double shooting.

f surexposer — **d** überbelichten — **n** overbelichten — **e** sobreexponer excesivamente — **i** sovraesporre

2085 burn-out mask — See print-out mask.

2086 burnt alum — See aluminium potassium sulphate.

2087 burnt lime — See calcium oxide.

2088 burnt Sienna — Orange-brown pigment made by calcining raw sienna.

f terre *f* de Sienne brûlée — **d** gebrannte Siena *f*; gebrannte Sienna *f* — **n** gebrande siënna *f* — **e** tierra *f* de Siena calcinada — **i** terra *f* di Siena bruciata

2089 burnt umber — Deep brown pigment obtained by calcining umber, consisting chiefly of oxides of iron and manganese; used in lithographic inks.

f terre *f* d'ombre brûlée — **d** gebrannte Umbra *f* — **n** gebrande omber *f* — **e** tierra *f* de Umbría quemada — **i** terra *f* di ombra bruciata

2090 burr — Rough edge or shoulder on metal. In blockmaking caused by the routing machine, graver, saw or cutter; may print if not removed.

f barbe *f*; morfil *m* — **d** Grat *m*; rauhe Kante *f* — **n** braam *fm* — **e** rebaba *f*; pelo *m*; pelito *m Ar* — **i** bava *f* (metallica); filo *m* morto

2091 burst — Transmission of a group of signals as a continuous block at relatively high speeds.

f chaîne *f* de données fermée — **d** geschlossene Datenkette *f* — **n** gesloten gegevensketen *fm* — **e** cadena *f* de datos cerrada — **i** catena *f* di dati chiusa

2092 burst — Irregular separation or rupture in reels of paper. When in the web in the machine direction, evident as the web continues to unwind from the reel, it is called burst machine direction. When the rupture is shorter and runs across the web it is called burst cross machine direction.

f crevasse *f*; éclatement *m* — **d** Platzstelle *f*; Riß *m* — **n** barst *fm* — **e** reventamiento *m* — **i** scoppio *m*

2093 burst cross machine direction — See burst.

2094 burst factor — Factor expressing the *bursting strength of paper in terms of unit substance in oven-dry condition; expressed as bursting strength in g/cm^2 divided by substance (oven-dry) in g/m^2.

f indice *m* d'éclatement; facteur *m* de résistance à l'éclatement *obs.*; résistance *f* relative à l'éclatement *obs.* — **d** Berstfaktor *m*; relativer Berstdruck *m*; relativer Berstwiderstand *m*; relative Berstfestigkeit *f* — **n** relatieve berstdruk *m*; relatieve berstweerstand *m*; relatieve berststerkte *f*; berstfactor *m* — **e** factor *m* relativo de reventamiento; factor *m* de estallido — **i** fattore *m* relativo di scoppio

2095 bursting; continuous forms bursting — Separation of continuous business forms into separate items.

f séparation *f* des formules en continu — **d** Zerlegung *f* von Endlosformularen — **n** scheiding *f* van kettingformulieren — **e** separación *f* de formularios continuos — **i** separazione *f* di moduli continui

2096 bursting strength — Limiting resistance of a piece of paper to a uniformly distributed pressure applied at right angles to its surface, up to the point at which it breaks. It is expressed in kg/m^2 or lbs/in^2, for a surface of 7.245 cm^2 or of 1.2 in in diameter. The results of bursting strength tests in UK are commonly expressed as *burst ratio.

f résistance *f* à l'éclatement; résistance *f* à la crevaison; résistance *f* à la perforation absolue *obs.* — **d** effektiver Berstdruck *m*; effektiver Berstwiderstand *m*; effektive Berstfestigkeit *f* — **n** effectieve berstdruk *m*; effectieve berstweerstand *m*; effectieve berststerkte *f* — **e** resistencia *f* efectiva al reventamiento; resistencia *f* efectiva al estallidomiento — **i** resistenza *f* effettiva allo scoppio

2097 bursting strength tester — See also Mullen tester, pop tester.

f essayeur *m* de crevaison; éclatomètre *m* — **d** Berstdruckprüfer *m*; Berstdruckprüfgerät *n* — **n** berstweerstandsmeter *m* — **e** ensayador *m* de resistencia al reventamiento; ensayador *m* de resistencia al estallido — **i** apparecchio *m* per la prova di scoppio; scoppiometro *m*

2098 burst machine direction — See burst.

2099 burst mode — Mode of operation when a peripheral device is made ready, receives a block of data and is then quiescent until the next block.

f mode *m* continu — **d** Einpunktbetrieb *m* — **n** burst mode *f*; continue modus *m*; stootsgewijze verzendingsmethode *f* — **e** modalidad *f* continua — **i** modo *m* continuo

2100 burst ratio — Ratio obtained by dividing the bursting strength of paper in lbs/in^2 by the basis weight using the basic ream size (double crown, 20 x 30 inches - 480). When expressed in metric units, it can be confused with the burst factor.

f rapport *m* de résistance à l'éclatement — **d** Berstindex *m* — **n** berstindex *m* — **e** índice *m* de reventamiento; índice *m* de estallido — **i** fattore *m* di scoppio

2101 bus bars — Bars along an electrotyping tank, conducting the electric current.

f barres *fpl* conductrices du courant — **d** Stromzuführungsstangen *fpl* — **n** stroomrails *mpl* — **e** barras *fpl* ómnibus — **i** barre *fpl* collettrici

2102 bush — See bushing.

2103 bushing; bearing bushing; bush; bearing bush — Perforated tube, made from brass or soft metal, used on rotating parts of machinery to receive wear of journals or pivots.

f coussinet *m* — **d** Lagerbüchse *f*; Lagerschale *f*; Zapfenlager *m* — **n** lagerbus *fm*; lagervoering *f*; bus *fm* — **e** casquillo *m* de metal antifricción; casquillo *m* de bronce — **i** boccola *f*; bronzina *f*

2104 business — A size of cards. See app. no. 3.

2105 business card; address card; calling card — Small size card used for visiting purposes, with inscribed name and (for business calls) address. See also visiting card.

f carte *f* d'adresse; carte *f* de commerce; carte *f* de recommandation; carte *f* d'affaires — **d** Geschäftskarte *f*; Adreßkarte *f*; Besuchskarte *f* — **n** adreskaartje *n* — **e** tarjeta *f* comercial; tarjeta *f* de negocios — **i** carta *f* da visita commerciale; biglietto *m* di visita commerciale

2106 business forms; commercial forms

f formulaires *mpl* commerciaux; formulaires *mpl* de commerce; liasses *fpl* — **d** Geschäftsformulare *npl*; Formularsätze *mpl* — **n** formulieren *npl* — **e** formularios *mpl* comerciales; esqueletos *mpl* comerciales — **i** moduli *mpl* commerciali; moduli *mpl* d'affari

2107 business stationary — Letterheads, invoices, statements, envelopes, package labels, printed with the name of the firm.

f papeterie *f* commerciale — **d** Geschäftsdrucksache *f* — **n** handelsdrukwerk *n* — **e** impreso *m* comercial — **i** stampato *m* commerciale; cancellaria *f* per uso commerciale

2108 butadiene — Colourless gas, easily liquified, flammable, soluble in alcohol and ether, insoluble in water. The material polymerizes readily, particularly if oxygen is present. TL-value: 1000 ppm. Formula: $CH_2CHCHCH_2$.

f butadiène *m* — **d** Butadien *n* — **n** butadieen *n* — **e** butadieno *m* — **i** butadiene *m*

2109 butanol — See butyl alcohol.

2110 butchers paper — See dry-finish butchers wrap.

2111 butchers wrap — See dry-finish butchers wrap.

2112 butchers wrapping paper — See dry-finish butchers wrap.

2113 butted joint — See butted splice.

2114 butted slugs; butt slugs — Slugs placed end to end to form one line.

f lignes *fpl* raboutées; lignes-blocs *fpl* mises bout

à bout; lignes *fpl* en deux parties — **d** zusammen-gefügte Zeilen *fpl* — **n** regels *mpl* met de einden tegen elkaar — **e** lingotes *mpl* yuxtapuestos; lingotes *mpl* empalmados; lingotes *mpl* para unir — **i** linee *fpl* disposte testa a testa

2115 butted splice; butted joint; butt joint — Joint formed by trimming the ends of two webs of paper, placing them end to end and pasting a strip over and under to make a continuous web without overlapping.
f joint *m* bout à bout; joint *m* en bout — **d** Stumpflasche *f*; Stumpfstoß *m*; stumpfer Stoß *m* — **n** stuitlas *fm*; stompe las *fm*; las *fm* waarin de einden tegen elkaar sluiten — **e** empalme *m*; unión *f* a tope — **i** giunzione *f* testa a testa

2116 butt end; reel end; stub — The unused paper left on the reel centre after the reel has been run to its practicable limit.
f fin *f* de la bobine — **d** Rollenrest *m* — **n** klis *fm* — **e** rezago *m* — **i** rimanenza *f* di bobina; fine *f* bobina; residuo *m* bobina

2117 butterfly — See backing roll spots.
2118 butterfly nut — See wing nut.
2119 butter of zinc — See zinc chloride.
2120 butter paper — Greaseproof paper for wrapping butter, lard and other greasy substances. Usually vegetable parchment paper or un-calendered grease-resistant paper. The usual basis weights are 27 pounds (24 x 36 - 500) for vegetable parchment and 30 pounds for dry-waxed greaseproof paper.
f papier *m* à beurre; papier *m* de beurrerie — **d** Butterverpackungspapier *n*; Butterpergament-papier *n* — **n** boterwikkelpapier *n*; papier *n* voor boterwikkels — **e** papel *m* para manteca; papel *m* para mantequilla — **i** carta *f* burro; carta *f* da burro

2121 buttery ink — Short but not tacky printing ink. It may be stirred, but a knife or other implement is inclined to cut its way through, leaving channels which will not readily close up.
f encre *f* beurrée; encre *f* onctueuse — **d** buttrige (salbenartige) Farbe *f* — **n** boterachtige inkt *m* — **e** inta *f* untuosa — **i** inchiostro *m* burroso

2122 butt joint — See butted splice.
2123 button control — See push-button control.
2124 butt register — See kiss register.
2125 butt slugs — See butted slugs.

2126 butt splicer — Mechanism to make a splice with top and bottom surfaces of web on inline press equipment.
f dispositif *m* de collage bout à bout; dispositif *m* de collage butt splicer — **d** Anklebevorrichtung *f* für Stumpfläsche — **n** stuitlas-aanplakinrichting *f* — **e** empalmadora *f* de unión a tope — **i** dispositivo *m* per effettuare una giunzione testa a testa

2127 butyl acetate — Limpid, colourless liquid, soluble in alcohol, ether, and hydrocarbons, slightly soluble in water. Used as a solvent in production of lacquers and synthetic resins. TL-value: 150 ppm or 710 mg/m^3.
Formula: $CH_3COO(CH_2)_3CH_3$.
f acétate *m* de butyle; éther *m* butylacétique — **d** Butylazetat *n*; essigsaures Butyl *n* — **n** butyl-acetaat *n* — **e** acetato *m* de butilo — **i** acetato *m* di butile; butil acetato *m*

2128 butyl alcohol; butyric alcohol; butanol — Limpid, flammable, colourless liquid. Miscible with alcohol and ether. TL-value: 100 ppm.
Formula: $CH_3(CH_2)_2CH_2OH$.
f alcool *m* butylique; butanol *m* — **d** Butyl-alkohol *m*; Butanol *m* — **n** butylalcohol *m*; butanol *m* — **e** alcohol *m* butílico; butanol *m* — **i** alcool *m* butilico; butanolo *m*

2129 butyric alcohol — See butyl alcohol.
2130 byline — See by-line.
2131 by-line; byline — Line under the title line, indicating the name of the writer of a newspaper story.
f ligne *f* sous le titre; ligne *f* avec le nom de l'auteur — **d** Zeile *f* mit der Autorenangabe — **n** regel *m* met de naam van de auteur — **i** riga *f* sotto il titolo riportante il nome dell'autore

2132 by-product — Material or product incidental to a manufacturing process.
f sous-produit *m* — **d** Abfallprodukt *m*; Neben-produkt *m* — **n** bijprodukt *n*; nevenprodukt *n* — **e** producto *m* accesorio; subproducto *m* — **i** sotto-prodotto *m*; prodotto *m* secondario

2133 byte — String of bits, usually eight, sufficient to describe a number of different characters or symbols. An 8-bit byte can describe 256 such different characters.
f multiplet *m* — **d** Byte *f* — **n** byte *m* — **e** byte *m* — **i** byte *m*

2134 cab — Line negative or photographic image used in colour composing for accurate positioning of continuous tone separation positives.
f négatif *m* de montage — **d** Kontertype *f*; Montagescheibe *f* — **n** contournegatief *n* — **e** negativo *m* para montaje; diagrama *m* fotográfico — **i** negativo *m* di montaggio

2135 cab — Carriage or negative carrier used in the three-point bar system of colour composing.
d Plattenträger *m*; Plattenhalter *m* — **n** plaathouder *m* — **i** portanegativo *m*; portalastra *m*

2136 Cabannes-Hofmann effect — When one area of a light-sensitive layer is exposed to low intensity light and another one to high intensity light, and the times of exposure are adjusted so that the same density is obtained upon completion of development, the developed image will appear first on the area exposed with the lesser intensity.
f effet *m* Cabannes-Hofmann — **d** Cabannes-Hofmann-Effekt *m* — **n** Cabannes-Hofmann-effect *n* — **e** efecto *m* Cabannes-Hofmann — **i** effetto *m* Cabannes-Hofmann

2137 cabinet; type cabinet — Enclosed rack to hold type cases which slide in and out like drawers.
f rang *m* — **d** (geschlossenes) Regal *n*; Satzregal *n*; Schriftregal *n*; Akzidenzregal *n*; Schrank *m* — **n** regaal *n* — **e** comodín *m*; chibalete *m* — **i** armadio *m*; portacasse *m*

2138 cabinet — A size of cut cards. See app. no. 3.

2139 cadmium — Soft, blue-white, ductile, malleable metal or greyish-white powder. Tarnishes in moist air, soluble in acids and in ammonium nitrate solutions, insoluble in water; becomes brittle at 80 °C and burns when heated. Used as a white pigment. Symbol: Cd. TL-value of cadmium fumes: 0.1 mg/m³.
f cadmium *n* — **d** Kadmium *n* — **n** cadmium *n* — **e** cadmio *m* — **i** cadmio *m*

2140 cadmium bromide — White to yellowish, efflorescent crystalline powder, soluble in water, acetone, and alcohol, slightly soluble in ether. Used in photography, process engraving and lithography. Latin: brometum cadmicum.
Formula: $CdBr_2$ or $CdBr_2 \cdot 4H_2O$
f bromure *m* de cadmium — **d** Kadmiumbromid *n* — **n** cadmiumbromide *n* — **e** bromuro *m* de cadmio — **i** bromuro *m* di cadmio

2141 cadmium chloride — Small white crystals, odourless, soluble in water and alcohol. Used in photography and printing. Latin: chloretum cadmicum.
Formula: $CdCl_2$.
f chlorure *m* de cadmium — **d** Kadmiumchlorid *n*; Chlorkadmium *n* — **n** cadmiumchloride *n* — **e** cloruro *m* de cadmio — **i** cloruro *m* di cadmio

2142 cadmium green — Pigment consisting of a mixture of hydrated chromium oxide with cadmium sulphide, characterized by its strong green colour and slow drying rate.
f vert *m* de cadmium — **d** Kadmiumgrün *n* — **n** cadmiumgroen *n* — **e** verde *m* de cadmio — **i** verde *m* di cadmio

2143 cadmium iodide — White, flaky, odourless crystals, becoming yellow on exposure to air and light. Soluble in water, alcohol, ether, acetone and ammonia. Occurs in two allotropic forms. Used in photography, process engraving and lithography. Latin: jodetum cadmicum.
Formula: CdI_2.
f iodure *m* de cadmium — **d** Jodkadmium *n*; Kadmiumjodid *n* — **n** cadmiumjodide *n* — **e** yoduro *m* de cadmio — **i** ioduro *m* di cadmio

2144 cadmium red — Permanent and heat resistant inorganic red ink pigment; consists essentially of cadmium sulphide and selenide.
f rouge *m* de cadmium — **d** Kadmiumrot *n* — **n** cadmiumrood *n* — **e** rojo *m* de cadmio — **i** rosso *m* di cadmio

2145 cadmium sulphide; orange cadmium; Orient yellow; aurora yellow — Light yellow or orange powder, soluble in acids and ammonia, insoluble in cold water, forming a colloid in hot water. Used in printing inks.
Formula: CdS. See also cadmium yellow.
f sulfure *m* de cadmium — **d** Kadmiumsulfid *n*; Schwefelkadmium *n* — **n** cadmiumsulfide *n* — **e** sulfuro *m* de cadmio — **i** solfuro *m* di cadmio

2146 cadmium yellow — Permanent and heat resistant inorganic yellow ink pigment; consists essentially of *cadmium sulphide.
f jaune *m* de cadmium — **d** Kadmiumgelb *n* — **n** cadmiumgeel *n* — **e** amarillo *m* de cadmio — **i** giallo *m* di cadmio

2147 caesium
See cesium.

2148 cage; magazine frame — The matrix frame of the Typograph typesetting machine.
f corbeille *f* à matrices; aigle *m* — **d** Matrizenkorb *m* — **n** matrijzenkorf *m* — **e** cesta *f* de (las) matrices — **i** cesto *m* delle matrici; cesto *m* portamatrici

2149 cake box
f boîte *f* pâtissière; boîte *f* à gateaux — **d** Tortenschachtel *f*; Kuchenschachtel *f* — **n** gebaksdoos *fm* — **e** caja *f* de pastel — **i** scatola *f* per dolci

2150 caked — See baked.

2151 caked types; baked types — Types, especially new ones, which adhere to one another and can hardly be separated.
f caractères *mpl* collés — **d** zusammenbackende Typen *fm* — **n** aan elkaar gekoekte letters *fmpl* — **e** tipos *mpl* pegados — **i** caratteri *mpl* incollati

2152 caking — See piling.

2153 calcination — Process of heating a material to a high temperature, but below its fusing point, to cause it to lose moisture or other volatile material or be oxidized or reduced.

f calcination f — d Kalzinierung f; Kalzinieren n; Kalzination f; Glühung f; Glühen n; Brennen n; Verkalkung f — n calcinering f; calcineren n — e calcinación f — i calcinazione f

2154 calcined baryta — See barium oxide.

2155 calcined calcium sulphate — See crown filler.

2156 calcined calcium sulphate — See plaster of Paris.

2157 calcined magnesium oxide — See magnesium oxide.

2158 calcined soda — See sodium carbonate.

2159 calcite — Natural crystalline *calcium carbonate.

2160 calcium — Moderately soft, white metal with a brilliant crystalline surface when fresh. Soluble in acid, decomposes, liberating hydrogen gas. Must be kept dry, in well-stoppered bottles. Symbol: Ca.
f calcium m — d Kalzium n — n calcium n — e calcio m — i calcio m

2161 calcium acid sulphite — See calcium bisulphite.

2162 calcium bisulphite; calcium hydrogen sulphite; calcium acid sulphite — Main ingredient of the cooking acid used in the sulphite process of papermaking. Exists only in solutions; is a solution of calcium sulphite in an aqueous sulphur dioxide solution. Soluble in water and acids.
Formula: $Ca(SO_3H)_2$. See also sulphite acid liquor.
f bisulfite m de calcium; sulfite m acide de calcium — d Kalziumbisulfit n; saurer schwefligsaurer Kalk m; doppeltschwefligsaures Kalzium n — n calciumbisulfiet n — e bisulfito m de calcio; bisulfito m cálcico — i bisolfito m di calcio; bisolfito m calcico

2163 calcium carbonate; precipitated chalk; precipitated calcium carbonate; limestone; Bologna chalk — Pure calcium carbonate is a white powder or colourless crystals, insoluble in water and alcohol, soluble in acids with evolution of carbon dioxide. Chalk is natural calcium carbonate composed of the calcareous remainings of minute marine organisms. It may contain up to 99% calcium carbonate in the form of calcite with silica, etc. Calcium carbonate is used as a filler and coating pigment, by binders for removing grease spots and, when dusted onto wet letterpress printed sheets, facilitates the drying of ink. Black ink dusted in this way takes a grey, lifeless tone. Latin: carbonas calcicus.
Formula: $CaCO_3$. See also whiting.
f carbonate m de calcium; carbonate m de chaux; craie f de Bologna — d Kalziumkarbonat n; kohlensaurer Kalk m; gewöhnlicher Kalk m; Wiener Kalk m; Bologneser Kreide f; Kalkstein m — n calciumcarbonaat n; koolzure kalk m; Bolognees krijt n; kalkspaat n — e carbonato m de calcio; carbonato m de cal — i carbonato m di calcio; carbonato m calcico; creta f di Bologna

2164 calcium chloride — White, deliquescent crystals, granules, lumps or flakes, soluble in water and alcohol. Water solution is neutral or slightly alkaline. Latin: chloretum calcicum.
Formula: $CaCl_2·6H_2O$.
f chlorure m de calcium — d Kalziumchlorid n; Chlorkalzium n — n calciumchloride n; chloorcalcium n — e cloruro m de calcio; cloruro m cálcico — i cloruro m di calcio; cloruro m calcico

2165 calcium hydrate — See calcium hydroxide.

2166 calcium hydrogen sulphite — See calcium bisulphite.

2167 calcium hydroxide; calcium hydrate; lime hydrate; hydrated lime — Soft, white, crystalline powder with alkaline, slightly bitter taste. Soluble in water, glycerin, syrup and acids; insoluble in alcohol. Absorbs carbon dioxide from air. It is commonly called slaked lime. Latin: hydras calcicus.
Formula: $Ca(OH)_2$.
f hydroxyde m de calcium; chaux f hydratée; chaux f éteinte; chaux f délitée — d Kalziumhydroxyd n; Kalkhydrat n; gelöschter Kalk m — n calciumhydroxyde n; kalkhydraat n; gebluste kalk m — e hidrato m de calcio; hidróxido m cálcico; hidrato m de cal; agua f de cal — i idrossido m di calcio

2168 calcium hypochlorite; calcium oxychloride; chloride of lime; chlorinated lime; hypochlorite of lime; bleaching powder — White, crystalline solid, soluble in water. Latin: hypochloris calcicus.
Formula: $Ca(OCl)_2$.
f hypochlorite m de chaux; chlorure m de chaux; hypochlorite m de calcium; chaux f chlorée; poudre f à blanchir — d Kalziumhypochlorit n; Chlorkalk m; Bleichkalk m; unterchlorsaurer Kalk m; Bleichpulver n — n calciumhypochloriet m; onderchlorigzure kalk m; chloorkalk m; bleekpoeder n — e hipoclorito m de calcio; hipoclorito m cálcico; cloruro m de calcio; cloruro m de cal; cal f clorada; polvo m de blanqueo — i ipoclorito m di calcio; cloruro m di calcio; calce f clorurata

2169 calcium nitrate; lime nitrate; nitro calcite; lime saltpetre — White deliquescent mass, soluble in water, alcohol and acetone.
Formula: $Ca(NO_3)_2·4H_2O$ or $Ca(NO_3)_2$.
f nitrate m de calcium; nitrate m de chaux — d Kalziumnitrat n; salpetersaures Kalzium n; Kalksalpeter m — n calciumnitraat n — e nitrato m de calcio; nitrato m de cal; nitrato m cálcico — i nitrato m di calcio; nitrato m calcico

2170 calcium orthophosphate — See calcium phosphate.

2171 calcium oxide; burnt lime; quick lime; unslaked lime; caustic lime — White or greyishwhite hard lumps, sometimes with a yellowish or brownish tint, due to iron. Soluble in acid, very slightly soluble in water, forming calcium hydroxide. TL-value: 5 mg/m³.
Formula: CaO.
f oxyde m de calcium; chaux f anhydre; chaux f caustique; chaux f vive — d Kalziumoxyd m;

gebrannter Kalk *m*; Ätzkalk *m* — **n** calcium-oxyde *n*; ongebluste kalk *m* — **e** óxido *m* de calcio; óxido *m* de cal; óxido *m* cálcico; cal *f* viva — **i** ossido *m* di calcio; ossido *m* calcico; calce *f* viva

2172 calcium oxychloride — See calcium hypochlorite.

2173 calcium phosphate; tribasic calcium phosphate; tricalcium phosphate; calcium orthophosphate; precipitated calcium phosphate — The precipitated product is a white, amorphous powder, soluble in acids, insoluble in water, alcohol and acetic acid. Latin: phosphas calcicus.
Formula: $Ca_3(PO_4)_2$.
f phosphate *m* tricalcique; diphosphate *m* tricalcique — **d** Kalziumphosphat *n*; Trikalziumphosphat *n* — **n** calciumfosfaat *n* — **e** fosfato *m* cálcico — **i** fosfato *m* di calcio; fosfato *m* calcico

2174 calcium sulphate — Chemical compound; very slightly soluble in water; used primarily as a filler pigment. In nature, it may be in the form of anhydrite or as *gypsum. Precipitated calcium sulphate is known as *crown filler.
General formula: $CaSO_4·H_2O$; anhydrite: $CaSO_4$; gypsum: $CaSO_4·2H_2O$.
f sulfate *m* de calcium — **d** Kalziumsulfat *n*; schwefelsaures Kalzium *n* — **n** calciumsulfaat *n* — **e** sulfato *m* de calcio; sulfato *m* de cal — **i** solfato *m* di calcio; solfato *m* calcico

2175 calcium sulphite — Chemical prepared by the interaction of sulphurous acid and calcium hydroxide; used as a filler or coating pigment in the manufacture of paper. Soluble in sulphurous acid, slightly soluble in water.
Formula: $CaSO_3·2H_2O$.
f sulfite *m* de calcium — **d** Kalziumsulfit *n*; schwefelsaurer Kalk *m* — **n** calciumsulfiet *n* — **e** sulfito *m* de calcio; sulfito *m* cálcico — **i** solfito *m* di calcio; solfito *m* calcico

2176 calculating rule
f règle *f* à calcul — **d** Rechenschieber *m* — **n** rekenliniaal *f* — **e** regla *f* de cálculo — **i** regolo *m* calcolatore

2177 calculator — Device that performs arithmetic operations and can perform modifications upon data but not upon its own program.
f calculatrice *f* — **d** Rechengerät *n*; Rechenmaschine *f* — **n** rekenmachine *f*; telmachine *f* — **e** calculadora *f* — **i** calcolatrice *f*; macchina *f* calcolatrice

2178 calendar — Systematic arrangement of subdivisions of time, as years, months, weeks, days, etc. Illustrated calendars are one of the most popular forms of advertising.
f calendrier *m* — **d** Kalender *m*; Abreißkalender *m*; Dauerkalender *m* — **n** kalender *m*; scheurkalender *m*; altijddurende kalender *m* — **e** calendario *m* — **i** calendario *m*

2179 calendar back; calendar mount — Stiff cardboard printed with an attractive design to carry a wall calendar.
f tableau *m* de calendrier; dos *m* de calendrier —

d Kalenderrückwand *f*; Rückwand *f* für Kalender — **n** kalenderschild *n* — **e** dorso *m* del calendario — **i** dorso *m* di calendario

2180 calendar edge — See calendar slide.

2181 calendar mount — See calendar back.

2182 calendar rim — See calendar slide.

2183 calendar slide; calendar edge; calendar rim
f baguette *f* — **d** Leiste *f*; Metalleiste *f* für Kalender; Kalenderschiene *f* — **n** baget *fm* — **e** varilla *f* para ribete; ribete *m* de metal — **i** montatura *f* del calendario

2184 calender — Machine consisting of a number of superposed rolls (bowls) of which only one is power driven. See also calendering.
f calandre *f* — **d** Kalander *m*; Maschinenkalander *m*; Walzwerk *n*; Glättwerk *n*; Satinierwerk *n* — **n** kalander *fm*; machinekalander *fm* — **e** calandria *f*; calandra *f*; satinadora *f*; máquina *f* de calandria; planchadora *f* — **i** calandra *f*

2185 calender coloured paper — See calender dyed paper.

2186 calender creases; dry wrinkles — Creases in the web of paper brought about during the smoothing or calendering process at the paper mill. In extreme cases this may lead to an actual cut in the paper (*calender cuts).
f plis *mpl* de calandre — **d** Kalanderfalten *fpl*; Kalanderzugfalten *fpl*; Quetschfalten *fpl*; Satinierfalten *fpl* — **n** kalanderplooien *fmpl*; satinageplooien *fmpl* — **e** arrugas *fpl* de calandria; plegaduras *fpl* de calandria — **i** pieghe *fpl* dovute alla calandra

2187 calender cuts — Defects in paper caused by wrinkles in the paper as it passes through the calender rolls, appearing as straight, sharp cuts, running for a relative short distance at an angle to the direction of web travel. See also fibre cut.
f coupures *fpl* de calandre — **d** Satinierrisse *mpl*; Kalanderrisse *mpl*; Kalanderzugschlitten *mpl* — **n** kalanderscheuren *fmpl*; kalandersplitten *mpl* — **e** ranuras *fpl* de calandra — **i** tagli *mpl* dalla calandra

2188 calender dyed paper; calender coloured paper — Paper or paperboard dyed or stained at the calender rolls. The dye solution is supplied from calender boxes to the calender rolls, which transfer it to either one or both sides of the paper or the board.
f papier *m* colorié à la calandre; papier *m* colorié sur calandre — **d** auf dem Kalander gefärbtes Papier *n* — **n** op de kalander gekleurd papier *n* — **e** papel *m* colorado en la calandra — **i** carta *f* colorata in calandra

2189 calendered paper — See calender finished paper.

2190 calender finished paper; calendered paper — Paper with a surface glazed by means of calenders. It does not include plate finish but refers to machine finish, English finish, supercalendered, and calender-friction glazed. See also friction glazed paper.
f papier *m* satiné; papier *m* calandré; papier *m*

apprêté — **d** kalandriertes Papier *n*; satiniertes Papier *n*; geglättetes Papier *n* — **n** gesatineerd papier *n* — **e** papel *m* calandrado; papel *m* satinado — **i** carta *f* calandrata; carta *f* satinata

2191 calendering — Process carried out by means of a *calender on at least partially dried paper or board to improve its surface finish or smoothness. It may permit some control of calliper.
f calandrage *m* — **d** Kalandrieren *n*; Satinieren *n*; Satinage *f*; Maschinenglättung *f* — **n** kalanderen *n* — **e** calandrar; cilindrar; satinar; glasear; pasar por la calandra — **i** calandratura *f*; lisciatura *f* in macchina

2192 calender marks; calender spots — Defects or imperfections in paper appearing in the form of glazed or indented spots, often transparent; caused by small flakes or pieces of paper which adhere to the calender rolls or are carried through the rolls on the paper sheet. See also calender streaks, pocks.
f plages *fpl* de calandre; marques *fpl* de calandrage — **d** Kalanderflecken *mpl*; Kalandermarkierung *f*; Kalanderschmarren *fpl*; Satinierflecken *mpl* — **n** kalandervlekken *fmpl* — **e** marcas *fpl* por la calandra — **i** segni *mpl* di calandra

2193 calender scale; dryer scale — Material not strongly attached and randomly distributed on the surface of the web, appearing as shiny spots; can cause non-uniform ink absorbency or lead to picking.
f dépôt *m* de calandre — **d** Kalanderspäne *fpl*; Kalanderstaub *m* — **n** kalanderstof *n* — **i** sporco *m* di liscia; sporco *m* di essiccatore

2194 calender-sized paper — Paper surface-sized at the calenders by application of the sizing solution to one or more of the calender rolls which transfer it to the paper surface.
f papier *m* surfacé à la calandre; papier *m* surfacé sur calandre — **d** auf dem Kalander geleimtes Papier *n* — **n** kalandergelijmd papier *n*; papier *n* met kalanderlijming — **e** papel *m* encolado en la calandra — **i** carta *f* collata in liscia di macchina; carta *f* collata in pressa collante

2195 calender spots — See calender marks.
2196 calender stack — See stack.
2197 calender streaks — Continuous streaks of darkened paper occurring parallel to the grain, caused by uneven pressing and drying preliminary to calendering.
f plages *fpl* de calandrage — **d** Kalandrierstreifen *mpl* — **n** kalanderstrepen *fmpl* — **e** fallos *mpl* de calandra — **i** strisce *fpl* di calandra

2198 calender varnishing
f vernissage *m* à calandre — **d** Kalanderlackierung *f* — **n** kalandervernissen *n* — **e** barnizado *m* a calandra — **i** laccatura *f* su calandra; verniciatura *f* calandra

2199 calf; calf skin; calf leather — High-grade leather, in natural colour or dyed, from calf skin, in rough or smooth finish, used for bookbinding.
f cuir *m* de veau; veau *m* — **d** Kalbleder *n* — **n**

kalfsleer *n* — **e** becerillo *m*; piel *f* de becerro; becerro; vitela *f* — **i** cuoio *m* di vitello; pelle *f* di vitello; vitello *m*

2200 calf binding
f reliure *f* en veau (pleine) — **d** Kalbslederband *m* — **n** kalfsleren band *m* — **e** encuadernación *f* en becerro; encuadernación *f* en pasta francesa — **i** legatura *f* in pelle di vitello

2201 calf leather — See calf.
2202 calf skin — See calf.
2203 calf vellum — See vellum.
2204 calibration — Determination of the true values of division on a graduated scale. Establishment of the relative value of an arbitrary scale.
f étalonnage *m*; calibrage *m* (d'un tube); tarage *m* (d'un ressort) — **d** Eichung *f*; Kalibrierung *f*; Überprüfung *f* — **n** ijking *f* — **e** calibrado *m*; comparación *f* — **i** taratura *f*

2205 calibration error
f erreur *f* d'étalonnage — **d** Eichfehler *m* — **n** ijkfout *fm* — **e** error *m* de aforo — **i** errore *m* di taratura

2206 calico; bookbinder's calico — Originally a cotton cloth imported from India, a plain cotton fabric heavier than muslin.
f calicot *m* — **d** Kaliko *n*; Steifleinen *n* — **n** calico *n*; calicot *n* — **e** calicó *m*; zaraza *f* Me; augaripola — **i** calico *m*

2207 calico paper — Decorative paper with calico-imitating designs.
f papier *m* pour impression genre cotonnades — **d** Kalikopapier *n*; Kattundruckpapier *n* — **n** calico-papier *n* — **e** papel *m* calicó — **i** carta *f* tipo cotone

2208 calico printing machine; cloth printing machine — Machine for printing onto textiles, either surface or intaglio. Many units (one for each colour) are arranged around a common impression cylinder of very large diameter to facilitate good registering.
f machine *f* à imprimer sur tissus — **d** Zeugdruckmaschine *f*; Textildruckmaschine *f*; Stoffdruckmaschine *f* — **n** textieldrukmachine *f* — **e** máquina *f* de imprimir tejidos — **i** macchina *f* per stampa su tessuti

2209 California job case — Standard American type case, about 42.25 x 82 cm, containing upper and lower case letters, the upper case letters in compartments at the right-hand side.
f casse *f* californienne — **d** Californischer Schriftkasten *m* — **n** Californische kast *fm* — **e** caja *f* californiana — **i** cassa *f* californiana

2210 caliper — See calliper.
2211 caliper — See callipers.
2212 calipers — See callipers.
2213 calipers — See callipers.
2214 Callier coefficient — See Callier Q factor.
2215 Callier effect — When a negative is enlarged with specular light (focused condenser or point light lamp) the contrast is greater that when the negative is enlarged with diffuse light (diffuser in the light beam) or is printed by contact. Depends

upon light scatter by the developed silver particles, and the particle size. See also Callier Q factor.

f effet *m* Callier — **d** Callier-Effekt *m* — **n** Callier-effect *n* — **e** efecto *m* de Callier — **i** effetto *m* di Callier

2216 Callier Q factor; Callier coefficient — In 1909, A. Callier determined the relation between the density D″ measured by parallel light and the density D* measured by diffuse light. The ratio Q = D″/D* is commonly referred to as the Callier Q factor. According to Callier's measurements Q is a constant at all values of density for a given material.

f facteur *m* de Callier; coefficient *m* de Callier — **d** Callier-Faktor *m*; Callier-Koeffizient *m* — **n** Callier-factor *m*; Callier-coëfficiënt *m* — **e** factor *m* de Callier; coeficiente *m* de Callier — **i** fattore *m* di Callier; coefficiente *m* di Callier

2217 calligrapher — Fine writer and letterer.

f calligraphe *m* — **d** Kalligraph *m*; Schönschreiber *m* — **n** kalligraaf *m* — **e** calígrafo *m*; maestro *m* de escritura — **i** calligrafo *m*

2218 calligraphy — Art of penmanship.

f calligraphie *f* — **d** Kalligraphie *f* — **n** kalligrafie *f* — **e** caligrafía *f* — **i** calligrafia *f*

2219 calling card — See business card.

2220 calliper *GB*; **caliper** *US*; **thickness** — Thickness of a sheet of paper measured under specified conditions, usually expressed in thousandths of an inch (mils or points).

f épaisseur *f* du papier — **d** Dicke *f*; Papierdicke *f*; Stärke *f*; Papierstärke *f* — **n** dikte *f* van het papier; papierdikte *f* — **e** espesor *m* del papel — **i** spessore *m* di carta

2221 callipers *GB*; **calipers** *US*; **caliper** — Instrument to measure the bulk (i.e., thickness) of paper. See also micrometer, packing gauge.

f calibre *m*; jauge *f*; micromètre *m*; palmer *m* — **d** Stärkemesser *m*; Lehre *f*; Dickenmesser *m*; Mikrometer *n* — **n** diktemeter *m*; micrometer *m* — **e** compás *m* de espesor — **i** calibro *m* di spessore

2222 callipers *GB*; **calipers** *US* — Instrument to measure internal or external diameters inaccessible to a scale, consisting usually of a pair of adjustable pivoted legs. See also dividers.

f compas *m* d'épaisseur; maître *m* de danse — **d** Dickenmesser *m*; Tasterzirkel *m* — **n** krompasser *m* — **e** compás *m* de calibre; compás *m* de puntas curvas — **i** compasso *m*

2223 calomel — See mercury chloride.

2224 calomel electrode — Standard electrode of mercury, mercurous chloride, and potassium chloride.

f électrode *f* au calomel — **d** Kalomel-Elektrode *f* — **n** calomel-elektrode *f* — **e** electrodo *m* de calomelanos — **i** elettrodo *m* di calomelano

2225 calorie — Unit for the measurement of the quantity of heat, defined in 1978 by the SI as 1 calorie = 4.1868 J (joules), i.e. the amount of heat which will raise the temperature of 1 gram of water 1 centigrade. Abbrev. cal.

f calorie *f* — **d** Kalorie *f* — **n** calorie *f* — **e** caloría *f* — **i** caloria *f*

2226 calotype — See talbotype.

2227 cam — Rotating or sliding piece that imparts motion to a roller moving against its edge or to a pin free to move in a groove on its face or that receives motion from such a roller or pin.

f came *f*; rampe *f* — **d** Steuernocken *m*; Nocken *m*; Exzenter *m*; Steuerkurve *f*; Kurve *f*; Kurvenscheibe *f*; Exzenterscheibe *f* — **n** kam *m*; nok *fm*; excentriek *n*; excentriekschijf *fm*; excenter *m* — **e** leva *f*; camba *f*; excéntrica *f*; balancín *m Ch* — **i** eccentrico *m*; camma *f*

2228 camber, lateral — See lateral camber.

2229 camber, transverse — See transverse camber.

2230 camboge — See gamboge.

2231 cambogia — See gamboge.

2232 cambric — Thin, plain cotton or linen fabric of fine close weave, used for embossing linen-finished stationery paper and in bookbinding.

f batiste *f* (de lin); percale *f* (de coton) — **d** Batist *m*; Kambrikleinen *n* — **n** cambric *n* — **e** percal *m* — **i** percalle *m*

2233 cambric paper — Paper with a finish with a design resembling cambric cloth, produced by plating or embossing.

f papier *m* grainé toile — **d** Leinenpapier *n*; Papier *n* mit Leinenprägung; Kambrikpapier *n* — **n** postpapier *n* met linnen persing; papier *n* met linnenpersing — **e** papel *m* tela para escribir — **i** carta *f* batista; carta *f* telata

2234 Cambridge binding — Leather book cover decorated with double panels with a flower tool at each of the four corners, a style originating from a group of bookbinders at Cambridge (1610-1630).

f reliure *f* de Cambridge; reliure *f* à la Cambridge — **d** Cambridge-Einband *m*; Cambridgeband *m* — **n** Cambridge-band *m* — **e** encuadernación *f* de Cambridge — **i** legatura *f* di Cambridge

2235 cameo — Stamping on a book cover where the background is coloured with the design standing out in plain relief. See also cameo binding.

2236 cameo binding; plaquette binding — In the early 16th century book covers decorated with interlaced bands and small tooled ornaments, in the centre of which a cameo or medallion was inset.

f reliure *f* à camée; reliure *f* à plaquette — **d** Kamee-Einband *m*; Plaketteneinband *m*; Bucheinband *m* mit Kameeprägung — **n** cameoband *m*; cameeband *m* — **e** encuadernación *f* camafeo; encuadernación *f* en camafeo — **i** legatura *f* con cammeo; legatura *f* con medaglione

2237 cameo coated paper — See mat art paper.

2238 cameo paper — See mat art paper.

2239 camera — See process camera.

2240 camera back — Back of the camera which holds the photographic material. In a darkroom process camera the camera back holds the plateholder, the ground glass, and the halftone screen. Special purpose camera backs have special equipment such as micrometer adjustment of plateholder for step-and-repeat negatives.

f corps *m* arrière de l'appareil — **d** Kamerahinterkasten *m* — **n** achterkant *m* van de camera — **e** respaldo *m* de la cámara; caja *f* posterior de la cámara — **i** cassetta *f* posteriore della macchina fotografica

2241 camera-back mask — Mask in contact with, or close to, the sensitive film, through which the image to be modified is projected, also on the enlarger easel. Camera-back masks for originals are usually negative masks, the Kodak magenta masking method being one example.

f masque *m* dans la chambre noir; masque *m* à l'appareil — **d** Saugwand-Maske *f* — **n** cameramasker *m* — **e** máscara *f* en la cámara; máscara *f* para respaldo de cámara — **i** maschera *f* nella macchina fotografica

2242 camera copyboard; copy board; copy frame — Part of a process camera on which copy to be photographed is placed, frequently with a hinged glass cover to hold copy flat; can be tilted to horizontal position for placing copy, and may have a removable section in which a transparency holder can be positioned for back-lighting illumination. See also vacuum copyboard.

f porte-modèle *m* — **d** Originalhalter *m*; Reißbrett *n* — **n** modelbord *n* — **e** portaoriginal(es) *m*; tablero *m* portaoriginales — **i** portaoriginale *m*

2243 camera extension — In photomechanics, the distance between the lens diaphragm and photographic surface in process cameras at any definite scale of reproduction.

f tirage *m* de la chambre (photographique) — **d** Kameraauszug *m* — **n** camera-uittrek *m* — **e** tiro *m* de la cámara; extensión *f* de la cámara — **i** tiro *m* della camera; tiro *m* della macchina fotografica

2244 camera flare — See flare.

2245 camera lens — See lens.

2246 camera lucida; Lucy — Prism projection system producing an enlarged or reduced image on convenient easel for sketching or layout purposes.

f chambre *f* claire — **d** Camera Lucida *f* — **e** cámara *f* lúcida; cámara *f* clara

2247 camera objective — See lens.

2248 camera-ready copy — Artwork, type proofs, typewritten material, etc., ready to be photographed for reproduction.

f original *m* prêt à la reproduction — **d** aufnahmefertiges Original *n*; kamerafertige Vorlage *f* — **n** voor opname gereed model *n* — **e** copia *f* lista para cámara — **i** originale *m* pronto per la riproduzione

2249 camera stand — Heavy frame or support for process cameras to absorb or prevent vibration of the camera.

f pied *m* de chambre photographique — **d** Kamerastütze *f*; Kamerastativ *n* — **n** camerachassis *n* — **e** soporte *m* de cámara; pie *m* de la cámara — **i** sostegno *m* della macchina fotografica

2250 campeachy wood — See logwood.

2251 camphor — White, volatile, translucent, crystalline compound, distilled from the wood and bark of the camphor tree; also obtained by organic synthesis. Soluble in alcohol, ether, chloroform, carbon disulphide and solvent naphtha, insoluble in water.
Formula: $C_{10}H_{16}O$.

f camphre *m* — **d** Kampfer *m* — **n** kamfer *m* — **e** alcanfor *m* — **i** canfora *f*

2252 camphor oil; liquid camphor — Colourless, natural oil with characteristic odour, soluble in ether, chloroform, insoluble in alcohol.

f huile *f* de camphre; essence *f* de camphre — **d** Kampferöl *n*; Campheröl *n* — **n** kamferolie *fm* — **e** aceite *m* de alcanfor — **i** olio *m* di canfora

2253 cam roller — Part of the slug composing machine.

f galet *m* de came — **d** Kurvenrolle *f* — **n** excentriekrol *fm* — **e** rodillo *m* de excéntricas — **i** rullo *m* di eccentrico; rullo *m* a camma

2254 cam shaft — Shaft to which a cam is fastened or of which a cam forms an integral part.

f arbre *m* à cames — **d** Nockenwelle *f*; Exzenterwelle *f* — **n** nokkenas *fm* — **e** árbol *m* de levas; eje *m* de levas — **i** albero *m* di distribuzione; albero *m* a camma

2255 Canada balsam; Canada turpentine; Canadian turpentine; balsam of fir — Pale yellow or greenish-yellow, transparent, viscous liquid. Slowly dries to a transparent varnish when exposed to the air. Soluble in benzene, chloroform, and ether, insoluble in water. Used as a cement for lenses and halftone screens.

f baume *m* du Canada — **d** Kanadabalsam *m*; kanadischer Balsam *m* — **n** Canada-balsem *m* — **e** bálsamo *m* del Canadá; trementina *f* del Canadá — **i** balsamo *m* del Canada

2256 Canada turpentine — See Canada balsam.

2257 Canadian Freeness — See degree of beating.

2258 Canadian turpentine — See Canada balsam.

2259 canard — See hoax.

2260 cancel *v* — To cut out blank or printed pages of a book to correct a misprint or to make an alterion by replacing another leaf.

f découronner et remplacer la page — **d** fehlerhafte Blätter auswechseln — **n** foutieve pagina's verwijderen en vervangen — **e** suprimir una página errónea para substituirla con otra — **i** eliminare una pagina incorrecta ad una buona pagina; sostituire una pagina incorrecta ad una buona pagina

2261 cancel *v* — See strike *v* out.

2262 cancel *v* **a subscription**

f se désabonner — **d** ein Abonnement abbestellen; vom Bezug zurücktreten — **n** een abonnement opzeggen; bedanken — **e** cancelar una subscripción; desabonarse a una revista — **i** sospendere un abbonamento

2263 cancelled figure; scratch figure — Figure

with a horizontal line through its face.
f chiffre *m* barré — **d** durchgestrichene Ziffer *f* — **n** doorgestreept cijfer *n*; cijfer *n* met doorstreping — **e** número *m* tachado; número *m* barrado — **i** cifra *f* cancellata; numero *m* cancellato; cifra *f* con cancellatura; cifra *f* con barra diagonale
2264 cancels — New leaf or signature replacing a defective one or one containing errors; any material substituting for deleted material.
f feuillets *mpl* refaits — **d** Umdruckblätter *npl* — **n** vervangende vellen *npl* — **e** pliegos *mpl* de reposición — **i** fogli *mpl* rifatti
2265 candela — Unit of luminous intensity where the luminance of a black body emitter at the temperature of freezing platinum (2042 K) is 600.000 cd/m² (stilbs) or 60 cd/cm². Replaces the international candle.
f candela *f* — **d** candela *f* — **n** candela *fm* — **e** candela *f* — **i** candela *f*
2266 candle — Obsolete unit of luminous intensity; superseded by the *candela.
f bougie *f* — **d** Kerze *f* — **n** kaars *fm* — **e** bujía *f* — **i** candela *f*
2267 Canevari binding — Bookbinding with in the centre a cameo stamp of Apollo driving a chariot, named after print collector Demetrio Canevari, physician of Pope Urban VIII.
f reliure *f* Canevari — **d** Canevari-Einband *m* — **n** Canevari-band *m* — **e** encuadernación *f* Canevari — **i** legatura *f* da Canevari
2268 canon — Obsolete name for a size of type. See app. no. 1.
2269 canvas — Strong, heavy cotton cloth used for covering blank books and postbinders, sometimes called duck.
f canevas *m* — **d** Kanvas *m* — **n** canvas *n* — **e** lona *f* — **i** canovaccio *m*; canavaccio *m*
2270 canvasser — See book canvasser.
2271 caoutchouc — See rubber.
2272 cap — In bookbinding, the top (headcap or head of the back), and the bottom part (tail cap) of the book cover that lays over the headband.
f coiffe *f* — **d** Häubchen *n* — **n** kapje *n* — **e** cabezada *f* — **i** cuffia *f*
2273 cap — A size of wrapping (bag) paper. The term is also applied to writing paper. See app. no. 3.
2274 cap — Abbreviation sometimes used for *foolscap.
2275 capacitor — Electrical device consisting primarily of two conductors, separated by a dielectric.
f condensateur *m* — **d** Kondensator *m* — **n** condensator *m* — **e** condensador *m* — **i** condensatore *m*
2276 cap case — See upper case.
2277 capillarity — The rise of a liquid in capillary tubes or fine spaces between fibres due to surface tension, caused by the attraction between molecules.
f capillarité *f* — **d** Kapillarität *f* — **n** capillariteit *f* — **e** capilaridad *f* — **i** capillarità *f*

2278 capillary — Fine pore such as exist in most paper surfaces.
f capillaire *m* — **d** Kapillare *f*; Haarröhrchen *n* — **n** capillair *n* — **e** capilar *m* — **i** capillare *m*
2279 capillary action
f action *f* capillaire — **d** Kapillarwirkung *f* — **n** capillaire werking *f* — **e** acción *f* capilar — **i** azione *f* capillare
2280 capillary flow — Ascent or descent of liquids within tubes of very small calibre due to the relative attraction between the molecules of the liquid and the molecules of the glass.
f écoulement *m* capillaire; courant *m* capillaire — **d** Kapillarströmung *f* — **n** capillaire stroming *f* — **e** flujo *m* capilar — **i** flusso *m* capillare
2281 capillary force — Force causing a liquid to penetrate into the pores of the paper surface. It depends on the wetting power of the liquid and fineness of the pores.
f force *f* capillaire — **d** Kapillarkraft *f* — **n** capillaire druk *m*; capillaire kracht *fm* — **e** fuerza *f* capilar — **i** forza *f* capillare
2282 capillary penetration — See capillary rise.
2283 capillary rise; capillary penetration — Height to which a liquid rises in a test piece of paper or board under standard conditions of test.
f ascension *f* capillaire — **d** Kapillarsteigung *f*; Saughöhe — **n** capillaire opstijging *f*; opzuighoogte *f* — **e** ascensión *f* capilar — **i** ascensione *f* capillare
2284 capillary tube — Tube with a hairlike bore.
f tube *m* capillaire; vaisseau *m* capillaire — **d** Kapillarrohr *n* — **n** capillaire buis *fm* — **e** tubo *m* capilar — **i** tubo *m* capillare
2285 capillary viscometer — Viscometer where the flow takes place in a capillary tube. Either gravity or air pressure is used to produce the flow. Volume of flow per second and the activating pressure are the factors measured.
f viscosimètre *m* par capillarité — **d** Kapillarviskosimeter *n* — **n** capillair-viscosimeter *m* — **e** viscosímetro *m* capilar — **i** viscosimetro *m* capillare
2286 capital letters — See capitals.
2287 capitals; capital letters; caps — Capitals are usually indicated in manuscripts by three lines underneath the letters.
f capitales *fpl*; majuscules *fpl* — **d** Großbuchstaben *mpl*; Versalbuchstaben *mpl*; Majuskeln *mpl*; Versalien *mpl* — **n** kapitalen *fmpl*; hoofdletters *fmpl* — **e** capitulares *mpl*; mayúsculas *fpl*; versales *fpl*; letras *fpl* mayúsculas; letras *fpl* versales; letras *fpl* capitales; letras *fpl* de caja alta — **i** caratteri *mpl* maiuscoli; lettere *fpl* maiuscole; maiuscole *fpl*
2288 cap line — The topmost line of the three (imaginary) framework lines (base line, cap line, and mean line) on which letters are constructed.
f ligne *f* de tête — **d** Oberlinie *f*; obere Linie *f* — **n** bovenlijn *fm* — **e** línea *f* superior — **i** linea *f* superiore
2289 cap nut; acorn nut

f écrou *m* borgne — **d** Hutmutter *f* — **n** dopmoer *fm* — **e** tuerca *f* ciega — **i** dado *m* cieco

2290 capped box — Paperboard container with a lid.
f boîte *f* à calotte; boîte *f* à rabattues — **d** Kappenschachtel *f* — **n** doos *fm* met deksel — **e** caja *f* de tapa — **i** scatola *f* a cappuccio

2291 caps — English term for thin wrapping paper used in a variety of trades.
f papier *m* mince d'emballage; papier *m* soie d'emballage — **d** (holzhaltiges) Packseidenpapier *n*; dünnes Einwickelpapier *n* — **n** pakzijdepapier *n*; pakzijde *n* — **i** carta *f* da involgere sottile

2292 caps — See capitals.

2293 caps and lower case — Indication in composing instructions or on proofs to use capital and lower-case letters.
f capitales et bas de casse — **d** Kapital mit Unterkasten — **n** kapitaal met onderkast — **e** altas y bajas; alta y baja — **i** alta e bassa

2294 caps and small caps — See caps and smalls.

2295 caps and smalls; caps and small caps — Small capitals with initial letters in capitals. Abbreviated: c and sc.
f capitales et petites capitales *fpl*; grandes et petites capitales *fpl*; majuscules et médiuscules *fpl* — **d** Kapitälchen mit grossen Anfangsbuchstaben — **n** klein kapitaal met kapitaal; kapitaal met klein kapitaal — **e** mayúsculas y versalitas; mayúsculas y versalillas; mayúsculas y mayúsculas pequeñas — **i** maiuscole e maiuscolette *fpl*

2296 cap screw — Screw bolt with a long thread and generally with a hexagonal or square head.
f vis *f* à six pans — **d** Kopfschraube *f* — **n** kopschroef *fm* — **e** tornillo *m* de cabeza hexagonal — **i** vite *f* mordente

2297 caption — Title set above an illustration, chart, table, etc. To be distinguished from a *legend which usually appears below.
f en-tête *m* — **d** Überschrift *f* — **n** tekst *m* boven een illustratie; bovenschrift *n* — **e** cabeza *f*; encabezado *m*; encabezamiento *m* — **i** titolo *m*

2298 caption — Heading of a chapter, article or document.
f titre *m* de chapitre; titre *m* de colonne; titre *m* de page — **d** Kapitelüberschrift *f* — **n** hoofdstuktitel *m*; titel *m* van een hoofdstuk — **e** título *m* del capítulo — **i** titolo *m* del capitolo

2299 captive tapes — Tapes in a letterpress rotary which operate only at slow speeds to thread the web.
f cordons *mpl* d'alimentation; cordons *mpl* d'alimentation; courroies *fpl* d'alimentation — **d** Einziehbänder *npl*; Zuführbänder *npl* — **n** invoerbanden *mpl* — **e** cintas *fpl* de entrada — **i** nastri *mpl* da introduzione; nastri *mpl* di guida

2300 caput mortuum — Red variety of impure ferric oxide similar to Venetian red. See also iron oxide reds.
f caput mortuum *m* — **d** Totenkopf *m* — **n** dodekop *mn* — **e** colcótar *m*; caput mortuum *m* — **i** caput mortuum *m*

2301 carbohydrates — Organic compounds, usually of vegetable origin, such as cellulose, sugar, starch etc.
f hydrates *mpl* de carbone — **d** Kohlehydrate *npl* — **n** koolhydraten *npl* — **e** hidratos *mpl* de carbono; hidratos *mpl* carbónicos — **i** carboidrati *mpl*

2302 carbolic acid — See phenol.

2303 carboloy — Trade name for an extremely hard carbonized steel for tips and edges of cutting tools used in relief printing platemaking; especially in finishing operations on plastic plates.

2304 carbon — Non-metallic, chiefly tetravalent element found native (as in diamond and graphite) or as a constituent of coal, petroleum, and asphalt, of limestone and other carbonates, and of organic compounds; obtained artificially in varying degrees of purity, especially as carbon black, lamp black, active carbon, charcoal and coke. Soluble in some molten metals from which it crystallizes out as graphite, insoluble in common solvents. Symbol: C. Latin: carboneum.
f carbone *m* — **d** Kohlenstoff *m* — **n** koolstof *fm* — **e** carbono *m* — **i** carbonio *m*

2305 carbonate — Salt of carbonic acid. Formula: H_2CO_3.
f carbonate *m* — **d** Karbonat *n* — **n** carbonaat *n* — **e** carbonato *m* — **i** carbonato *m*

2306 carbonate of ammonia — See ammonium carbonate.

2307 carbonate of potash — See potassium carbonate.

2308 carbonate of soda — See sodium carbonate.

2309 carbon-backed forms — See carbonized forms.

2310 carbon bisulphide — See carbon disulphide.

2311 carbon black — All carbon products in the colloidal range of particle size produced by the thermal decomposition of hydrocarbon products. Each of the five principal processes develops its own characteristic carbon: channel or *impingement black, furnace black, lamp black, thermal black, and acetylene black.
f noir *m* de carbone — **d** Gasruß *m*; Ruß *m* — **n** carbon black *n*; roet *n* — **e** negro *m* de carbón — **i** nero *m* di carbone

2312 carbon black; carbon ink; carbonizing ink — Ink for *carbonized forms.
f encre *f* pour carbone — **d** Karbonfarbe *f*; Karbondruckfarbe *f*; Karbonisierfarbe *f*; Durchschreibefarbe *f* — **n** carboninkt *m* — **e** tinta *f* de carbonizar; tinta *f* para calcar — **i** inchiostro *m* per carbonatura

2313 carbon brush; brush — Electrical conductor made of a block of carbon (sometimes copper strips) that makes sliding contact between a stationary and a moving part of a generator or a motor.
f balai *m* de charbon; balai *m* de moteur — **d** Kohlebürste *f*; Bürste *f*; Stromabnehmer *m* — **n** koolborstel *m*; borstel *m* — **e** escobilla *f* de

carbón — **i** spazzola *f* di carbone
2314 carbon copy; copy — Copy of a letter, etc., made by means of carbon paper.
f copie *f* (à carbone) — **d** Durchschlag *m*; Kopie *f* — **n** doorslag *m*; kopie *f* — **e** copia *f* al carbón; copia *f* en papel carbónico — **i** copia *f* carbone; velina *f*
2315 carbon dioxide; carbonic acid; carbonic anhydride — Colourless, odourless gas, or heavy, volatile, colourless liquid, or white, snow-like solid. Soluble in water, alcohol and acetone. Latin: acidum carbonicum. TL-value: 5000 ppm. Formula: CO_2.
f bioxyde *m* de carbone; acide *m* carbonique; anhydride *m* carbonique — **d** Kohlendioxyd *n*; Kohlensäure *f*; Kohlenstoffsäureanhydrid *n*; Kohlensäureanhydrid *n* — **n** kooldioxyde *n*; koolzuur *n* — **e** dióxido *m* de carbono; dióxido *m* carbónico; ácido *m* carbónico; anhídrido *m* carbónico — **i** biossido *m* di carbonio; anidride *f* carbonica
2316 carbon disulphide; carbon bisulphide — Clear, colourless, flammable liquid, soluble in alcohol, benzene and ether, slightly soluble in water. Formula: CS_2.
f sulfure *m* de carbone; bisulfure *m* de carbone; anhydride *m* sulfocarbonique — **d** Schwefelkohlenstoff *m* — **n** zwavelkoolstof *fm* — **e** bisulfuro *m* de carbono; disulfuro *m* de carbono — **i** bisolfuro *m* di carbonio; solfuro *m* di carbonio; anidride *f* solfocarbonica
2317 carbon holder
f porte-charbon *m* — **d** Kohlenhalter *m* — **n** koolspitshouder *m* — **e** portacarbón *m* — **i** portacarbone *m*
2318 carbonic acid — See carbon dioxide.
2319 carbonic anhydride — See carbon dioxide.
2320 carbon ink — See carbon black.
2321 carbonized forms; carbon-backed forms — Paper in sets (sheets or continuous forms or unit books), the backs of which are coated with a pressure-transferable pigmented layer so that copies can be obtained without inserting separate sheets of carbon paper.
f formules *fpl* carbonées — **d** Durchschreib(e)formulare *npl*; Formulare *npl* mit Karbonaufdruck — **n** gecarboniseerde formulieren *npl* — **e** formularios *mpl* al carbón; formularios *mpl* con dorso de carbón copiativos — **i** moduli *mpl* a ricalco; moduli *mpl* con carbonatura sul verso
2322 carbonizing ink — See carbon black.
2323 carbonless copying paper — See carbonless paper.
2324 carbonless forms — *Carbonless copying paper assembled in multiple parts which are cut into unit sets, cut unit sets bound into books, or held in continuous form either in folded packs or rolls.
f liasses *fpl* non carbonées; liasses *fpl* sans carbone — **d** karbonfreier Formularsatz *m*; Selbstdurchschreibeformularsatz *m* — **n** carbonloze

formulieren *npl*; formulieren *npl* zonder carbon — **e** formularios *mpl* sin carbón — **i** moduli *mpl* senza carbone; moduli *mpl* senza carbonatura
2325 carbonless paper; carbonless copying paper; no carbon paper — These papers are of two types: chemical and physical. The chemical papers produce an image by means of a chemical transfer from sheet to sheet. The physical papers produce an image by means of a coating. The papers are pressure-responsive. See also no carbon required.
f papier *m* non carboné pour copie; papier *m* autocopiant — **d** karbonfreies Durchschreibepapier *n*; Selbstdurchschreibepapier *n* — **n** papier *n* zonder carbon — **e** papel *m* sin carbón; papel *m* para copias sin carbón — **i** carta *f* copiativa senza carbonatura; carta *f* autocopiante
2326 carbon paper — Paper coated with a pressure-transferable pigmented layer, to make copies at the same time as the original.
f papier *m* carbone; papier *m* communicatif — **d** Kohlepapier *n*; Karbonpapier *n* — **n** carbonpapier *n* — **e** papel *m* carbón; papel *m* de copias; papel *m* para copias — **i** carta *f* carbone
2327 carbon print — Photograph made on carbon tissue.
f copie *f* sur papier charbon — **d** Pigmentkopie *f*; Pigmentdruck *m* — **n** pigmentkopie *f* — **e** copia *f* en papel pigmento — **i** copia *f* su carta pigmento
2328 carbon rods; carbons — Long cylindrical rods of baked carbon, used as electrodes or light sources for arc lamps. See also cored carbons.
f charbons *mpl* — **d** Kohlestifte *mpl*; Bogenlampenkohlen *fpl* — **n** koolspitsen *fmpl*; koolstaven *fmpl* — **e** carbones *mpl* para lámparas de arco; electrodos *mpl* de carbón para lámpara de arco — **i** carboni *mpl* per lampade
2329 carbons — See carbon rods.
2330 carbon tetrachloride; tetrachloromethane; perchloromethane — Colourless liquid with peculiar odour, yielding heavy vapours; nonflammable, poisonous, miscible with alcohol, ether, chloroform, benzene, solvent naphtha, and most of the fixed and volatile oils, very slightly soluble in water. Latin: tetrachloretum carbonicum. TL-value: 10 ppm or 65 mg/m^3. Formula: CCl_4.
f tétrachlorure *m* de carbone; tétrachlorométhane *m* — **d** Tetrachlorkohlenstoff *m*; Tetrachlormethan *n*; Kohlenstofftetrachlorid *n* — **n** tetrachloorkoolstof *fm*; tetrachloormethaan *n*; tetra *m* — **e** tetracloruro *m* de carbono; tetracloruro *m* carbónico — **i** tetracloruro *m* di carbonio
2331 carbon tissue; pigment paper; gravure tissue — Paper coated with hardened dichromated gelatine, containing carbon or other pigments in suspension, for transfer work in the production of printing plates for rotogravure printing. Originally the pigment was carbon black, and the material was a thin tissue. The pigment is required to provide a visible image and to limit the diffusion

of light within the gelatine layer during exposure. Now a reddish-brown iron oxide is used.

f papier *m* charbon; papier *m* au charbon; papier *m* transfert − **d** Pigmentpapier *n*; Ätzpigmentpapier *n* − **n** pigmentpapier *n* − **e** papel *m* pigmentado; papel *m* pigmento; papel *m* de reportar − **i** carta *f* pigmento

2332 carbon tissue transfer − See also lay down.

f application *f* de la feuille de papier charbon insolée sur la forme de cuivre − **d** Übertragung *f* des Pigmentpapiers; Übertragung *f* der Pigmentkopie − **n** overdraging *f* van het pigmentpapier; overdraging *f* van de pigmentkopie − **e** reporte *m* del papel pigmento; aplicación *f* del papel pigmento − **i** trasporto *m* su carta pigmento

2333 carbon tissue transfer machine; laying machine − Machine for adhering pigment paper to the copper surface of gravure cylinder or forme by wetting the gelatine surface only with water. Used to avoid register difficulties caused by stretching of the pigment paper.

f machine *f* à appliquer le papier charbon; machine *f* à report du papier charbon − **d** Pigmentpapier-Übertragungsmaschine *f* − **n** overdraagmachine *f* voor pigmentpapier − **e** máquina *f* de aplicar el papel pigmento − **i** macchina *f* per il trasporto della carta al pigmento; macchina *f* trasportatrice della carta pigmento

2334 carborundum − Trade mark for *silicon carbide. The crystalline form ranges from small to massive crystals varying from transparent to opaque with colours from pale green to deep blue or black. Not affected by acids. Used for various abrasives.

f carborundum *m* − **d** Carborundum *n* − **n** carborundum *n* − **e** carborundo *m* − **i** carborundum *m*

2335 carboxyl group − Univalent COOH group, the radical characteristic of all organic acids.

f groupe *m* carboxyl; groupe *m* carboxylique; groupement *m* carboxyl − **d** Karboxylgruppe *f* − **n** carboxylgroep *fm* − **e** grupo *m* carboxílico − **i** gruppo *m* carbossilico

2336 carboxymethylcellulose − White granular, odourless, tasteless powder, readily dispersible in cold or hot water. Used as a surface active agent. Abbrev. CMC.

Formula: $(C_6H_7O_2(OH)_2OCH_2COOH)_n$.

f carboxyméthylcellulose *f* − **d** Karboxymethylzellulose *f* − **n** carboxymethylcellulose *fm* − **e** carboximetilcelulosa *f* − **i** carbossimetilcellulosa *f*

2337 carboy; demijohn − Large glass bottle encased in a wooden box or basketwork; container for acids and corrosive liquids; capacity about 4.5 to 45 liters (1-10 gallons).

f tourie *f*; dame-jeanne *f* − **d** Korbflasche *f*; Glasballon *m* − **n** mandfles *fm* − **e** damajuana *f*; garrafón *m*; bombona *f*; castaña *f* − **i** damigiana *f*

2338 carbro − See carbro print.

2339 carbro print; carbro − Reproduction colour print made by superimposition of colour pigments

transferred from carbon tissues whose differentially hardened image areas were made insoluble in hot water by contact with enlarged photographic bromide prints made from the colour-separation negatives. The name is a hybrid of carbon and bromide, distinguishing the process from regular carbon prints which were limited to contact print size of the separations because of the low sensitivity of the bichromate sensitizer used.

f ozobrome *m* − **d** Ozobrom-Druck *m*; Carbro-Druck *m* − **n** carbro *m*; ozobroomdruk *m* − **e** carbrotipia *f*; ozobromía *f*; ozobromo *m* − **i** carbrotipia *f*

2340 card *v* − To space out lines of text by relatively minute amounts so that a page may be filled, with fewer lines, to avoid an awkward page or column break. Cardboard strips are sometimes inserted between lines by a (hot-metal) make-up man. With photocomposition devices the ability of the machine to produce satisfactory pages in response to a carding command will be a function of the minimum leading increment which the machine is capable of providing, which may be 0.5 point, 0.25 point or perhaps even 0.1 point. Sometimes called faking.

f placer interlignes en carton; placer interlignes de Limoges − **d** austreiben; aussperren − **n** interliniёren met reepjes karton − **e** interlinear con cartulinas de justificación − **i** inserire strisce di carta

2341 cardan joint − See universal coupling.

2342 cardboard − Board of 0.15 mm (0.006 in) or more in thickness, where stiffness is the paramount characteristic. The word cardboard as used by the public is too vague to be technical. In the paper industry the term board is generally used in combination with words indicating its character or use (blanks; bristols; postcard bristol; railroad board; thick China; tough check; translucents; etc.).

f carton *m* − **d** Karton *m* − **n** karton *n* − **e** cartulina *f*; cartoncillo *m* − **i** cartoncino *m*

2343 cardboard bending machine

f machine *f* à cintrer le carton − **d** Pappenbiegemaschine *f* − **n** kartonbuigmachine *f* − **e** máquina *f* para doblar cartón − **i** piegatrice *f* per cartoncino

2344 cardboard box − Box made of thin, stiff pasteboard.

f caisse *f* carton; boîte *f* carton − **d** Pappschachtel *f*; Pappkarton *m*; Karton *m*; Pappkiste *f* − **n** kartonnen doos *fm* − **e** caja *f* de cartón − **i** scatola *f* di cartone

2345 cardboard cutter

f cisaille *f* à carton − **d** Kartonschere *f*; Pappschere *f*; Kartonschneider *m* − **n** bordschaar *fm* − **e** cizalla *f* para cortar cartón − **i** cesoia *f* per cartoni; tagliacartone *m*

2346 cardboard for playing cards − See playing card stock.

2347 cardboard machine − See intermittent

board machine.

2348 cardboard tube
f tube *m* de carton — **d** Papphülse *f*; Papprohr *n* — **n** kartonnen koker *m* — **e** tubo *m* de cartón — **i** tubo *m* di cartone

2349 card fount — The smallest complete fount of type sold by a type founder. See also minimum weight fount.
f police *f* minimale — **d** Minimum *n* — **n** minimum *n* — **e** mínimo *m* — **i** più piccolo assortimento *m*

2350 card image — One-to-one representation of the contents of a punched card, e.g., a matrix in which a 1 represents a punch and a 0 represents the absence of a punch.
f image *f* de carte — **d** Kartenbild *n* — **n** kaartbeeld *n* — **e** imagen *f* de tarjeta — **i** immagine *f* di scheda

2351 cardinal points — In a thick lens or system of lenses, the two nodal points and the two focal points.
f points *mpl* cardinaux; repères *mpl* — **d** Kardinalpunkte *mpl* — **n** kardinale punten *npl* — **e** puntos *mpl* cardinales — **i** punti *mpl* cardinali

2352 card index — Index on cards.
f cartothèque *f*; fichier *m*; classeur *m* — **d** Kartei *f* — **n** cartotheek *f*; kaartsysteem *n* — **e** fichero *m*; clasificador *m* — **i** cartoteca *f*; schedario *m*

2353 carding — See shimming.

2354 card pips — The fifty-three characters cast on bodies of about 36 point, used to print plays in card games, each character representing a card.
f figurines *fpl* de cartes à jouer — **d** Spielkartenzeichen *npl* — **n** kaartspeltekens *npl* — **e** viñetas *fpl* de naipes — **i** caratteri *mpl* delle carte di gioco

2355 card reader — Input device to a data processing system which senses data on punched cards, usually with 80 columns, but sometimes with 40 columns on each card.
f lecteur *m* de cartes (perforées) — **d** Lochkartenleser *m*; Kartenleser *m*; Kartenabfühler *m*; Kartenabfühleinheit *f* — **n** ponskaartlezer *m*; kaartlezer *m*; ponskaartenleesapparaat *n* — **e** lector *m* de tarjetas — **i** lettore *m* di schede; lettore *m* della scheda

2356 cards — Sizes cut from various kinds of boards, usually bristols. Their use is indicated by prefixing another word, such as business, postal, visiting, wedding, etc. The word card as used by the public is too vague to be technical. In the paper industry the term board is generally used in combination with words indicating its character or use. See also blanks, bristol.
f cartes *fpl* — **d** Kartenkarton *m*; Karten *fpl* — **n** kaartkarton *n* — **e** cartulina *f* para tarjetas — **i** biglietti *mpl*

2357 caret — Symbol like a reversed V, indicating that something is to be inserted where the caret is placed. See proof correction marks, app. no. 5.
f renvoi *m* (de marge); signe *m* d'intercalage;

signe *m* d'omission — **d** Einschaltungszeichen *n* — **n** verwijzingsteken *n*; correctieteken *n* — **e** carete *m*; signo *m* de intercalación; careta *f* CU — **i** segno *m* di inserimento; segno *m* d'omissione

2358 carload lot — Minimum amount of paper required for individual freight-car shipment at the carload rate of freight. It ranges from 36,000 to 100,000 pounds, depending upon the freight classification zone. The minimum for bulkier paper is less. Also the amount customarily shipped my mills in one freight car. Most shipments exceed the minimum carload lot. See also case lot.
f minimum *m* de livraison par voiture — **d** kleinste Papiermenge *f* für Lastwagentransport; kleinste Wagenladung *f*; Mindestfrachtmenge *f* an Papier — **n** kleinste hoeveelheid *f* per wagenlading — **e** cantidad *f* mínima de un camión de carga — **i** quantità *f* per vagone; carico *m* di carta di un vagone ferroviario

2359 Carlovingian minuscules — See Carolingian minuscules.

2360 carnauba wax; Brazil wax; Brasil wax — One of the hardest and most expensive commercial waxes, obtained from the South American pine. Soluble in ether, boiling alcohol and alkalis, insoluble in water. Used in carbon paper coating.
f cire *f* de carnauba — **d** Karnaubawachs *n* — **n** carnaubawas *mn* — **e** cera *f* de carnauba; cera *f* carnauba; cera *f* de carandai — **i** cera *f* di carnauba

2361 Caroline minuscules — See Carolingian minuscules.

2362 Carolingian minuscules; Carlovingian minuscules; Caroline minuscules — 9th century type face originating from the St. Martin's Abbey at Tours and revived in Italy in the 15th century.
f minuscules *fpl* carolines — **d** Karolingische Minuskeln *fpl* — **n** Karolingische minuskels *fmpl* — **e** minúsculas *fpl* carolingias — **i** minuscole *fpl* carolinge

2363 carrageen — See Irish moss.

2364 carriage — Part of a printing press on which the forme is laid and which runs under the cylinder.
f chariot *m* — **d** Karren *m* — **n** kar *fm* — **e** carro *m* — **i** carro *m*

2365 carriage drive — See bed drive.

2366 carriage movement
f mouvement *m* de marbre; mouvement *m* du chariot — **d** Karrenbewegung *f*; Schlittenbewegung *f* — **n** beweging *f* van de kar; karbeweging *f* — **e** movimiento *m* del carro — **i** movimento *m* del carro

2367 carriage rails
f chemins *mpl* de marbre — **d** Karrenleisten *fpl* — **n** karrebanen *fmpl* — **e** caminos *mpl* del carro — **i** guide *fpl* del carro

2368 carriage return — Computer term. The operation or symbol that causes the next character to begin a new line.

f retour *m* du chariot — **d** Wagenrücklauf *m* — **n** wagenterugloop *m* — **e** retorno *m* del carro — **i** ritorno *m* del carro

2369 carriage return movement
f mouvement *m* de retour du chariot — **d** Karrenrückgang *m* — **n** terugloop *m* van de kar — **e** movimiento *m* de retorno del carro — **i** movimento *m* di ritorno del carro

2370 carrier — Adjustable bars in a plateholder for process cameras.
f barres *fpl* de châssis — **d** Plattenhalterleisten *mpl* — **n** plaathouderrails *fmpl* — **e** dispositivo *m* conductor; dispositivo *m* portador — **i** barre *fpl* regolabili del portalastra

2371 carrier — In data communication a high-frequency current that can be modulated by voice or signalling impulses.
f courant *m* porteur — **d** Trägerwelle *f* — **n** draaggolf *fm* — **e** corriente *f* portadora — **i** corrente *f* portatore

2372 carrier bag — Strong bag, reinforced at the opening, fitted with or incorporating a carrying handle, usually of twine.
f sac *m* portable — **d** Tragbeutel *m*; Tragtasche *f*; Tragetasche *f* — **n** draagzak *m* — **e** bolsa *f* para compras; bolsa *f* portátil — **i** sacchetto *m* da asporto

2373 carrier rollers — Similar to *breaker rollers except that function is usually interpreted as applying to support a web in a rotary press; used where long, straight web leads exist.
f rouleaux *mpl* de support; rouleaux *mpl* porteurs — **d** Tragrollen *fpl*; Leitwalzen *fpl*; Papierleitwalzen *fpl* — **n** draagrollen *fmpl*; transportrollen *fmpl* — **e** rodillos *mpl* guía (de la hoja); rodillos *mpl* guía (del papel); rodillos *mpl* conductores (de la hoja) — **i** rulli *mpl* portanti; rulli *mpl* di guida

2374 carrier system — In data communication, a means of conveying a number of channels over a single path by modulating each channel on a different carrier frequency and demodulating at the receiving point to restore the signals to their original form.
f système *m* à courants porteurs — **d** Trägerwellensystem *n* — **n** draaggolfsysteem *n* — **e** sistema *m* de corrientes portadoras — **i** sistema *m* portante

2375 carrying medium — See vehicle.
2376 carte de visite — A size of cards. Not to confuse with *visiting card. See app. no. 3.
2377 cartographer
f cartographe *m* — **d** Kartograph *m*; Kartenzeichner *m* — **n** cartograaf *m*; kaarttekenaar *m* — **e** cartógrafo *m* — **i** cartografo *m*

2378 carton — 1. a folding paper box; 2. a rigid set-up box; 3. a fibreboard shipping container; 4. a folding paperboard box as distinguished from a setup or rigid box or a shipping container. 5. a shipping unit of paper which usually weighs 56-68 kg (125-150 pounds) equivalent to one-fourth of a case.

f boîte *f*; boîte *f* en carton — **d** Schachtel *f*; Pappschachtel *f* — **n** doos *fm*; kartonnen doos *fm* — **e** caja *f*; caja *f* de cartón — **i** scatola *f*; scatola *f* di cartone; astuccio *m* pieghevole

2379 carton — See folding carton.
2380 carton blank; blank — Piece of material from which a container or a part will be made by further operation. 1. In closures, the basic form of a cap as it is stamped in the press in the first fabricating operation and before forming the thread, wire or knurl. 2. In metal cans and other metal containers, the plain, flat piece of plate cut to size but not formed to shape. 3. Any die-cut, scored and corner-cut or otherwise partially prepared section of boxboard, in the flat, to be formed into a setup paper box or part thereof, e.g., base blank, lid blank, tray blank, etc. 4. The folding carton after cutting and scoring but before folding and gluing.
f découpe *f* à plat de boîte pliante — **d** Faltschachtelzuschnitt *m*; flachgefalteter Karton *m* — **n** vouwdoos *fm* in plano-vorm; niet-opgezette doos *fm* — **e** troquelado *m* en plano de una caja plegable — **i** pezzo *m* per astucci pieghevole; pezzo *m* tranciato in cartone

2381 carton for mechanical erecting cartoning machine — See also carton blank.
f boîte *f* pliante à verrouillage sur parois latérales — **d** Faltschachtel *f* für maschinelles Aufrichten — **n** machinaal opzetbare vouwdoos *fm* — **e** caja *f* plegable con enganches en los costados — **i** pezzo *m* di cartone per macchine per montare astucci

2382 carton gluer — See folding carton machine.
2383 carton gluing machine — See folding carton machine.
2384 carton with bandshape pull-through carton — See also carton blank.
f boîte *f* pliante à ceinture rapportée — **d** Durchzieh-Gürtelschachtel *f* — **n** tweedelige gordeldoos *fm* — **e** caja *f* solapas abrochamiento dos piezas con pestana inferior a toda altura y reborde — **i** astuccio *m* pieghevole con striscia di cartone a strappo

2385 carton with crash lock bottom — See also carton blank.
f boîte *f* pliante à fond automatique — **d** Faltschachtel *f* mit Schnapfbodenverschluß — **n** vouwdoos *fm* met zelfsluitende bodem — **e** caja *f* plegable con fondo automático — **i** astuccio *m* pieghevole con chiusura automatica a incastro del fondo

2386 carton with full plain ends and short side flaps — See also carton blank.
f boîte *f* pliante à pattes droites — **d** Faltschachtel *f* mit Schlitzverschluß — **n** vouwdoos *fm* met insteeksluiting — **e** caja *f* solapas ranuradas — **i** astuccio *m* pieghevole con falde laterali ridotte

2387 carton with hinged full fold over lid; carton with integrated full fold over lid — See also carton blank.
f boîte *f* pliante type coffret en seul pièce — **d**

Faltschachtel *f* mit Klappdeckel — **n** vouwdoos *fm* met vast deksel — **e** caja *f* plegable de una sola pieza — **i** astuccio *m* pieghevole tipo cofanetto in un sol pezzo

2388 carton with integrated full fold over lid — See carton with hinged full fold over lid.

2389 carton with reverse tuck-in ends — See also carton blank.

f boîte *f* pliante à pattes rentrantes alternés — **d** Faltschachtel *f* mit Einsteckverschluß — **n** vouwdoos *fm* met wisselende insteekkleppen — **e** caja *f* plegable con solapas alternas — **i** astuccio *m* pieghevole con falde rientranti alternate

2390 carton with straight tucks and film window — See also carton blank.

f boîte *f* pliante à pattes rentrantes opposées à fenêtre pelliculée — **d** Fensterfaltschachtel *f* mit gegenüberliegenden Klappen — **n** vouwdoos *fm* met insteekkleppen aan dezelfde zijde en met venster — **e** caja *f* solapas abrochamiento con ventana de celofán — **i** astuccio *m* pieghevole con falde rientranti opposte e finestra trasparente

2391 carton with tuck top and full plain end bottom — See also carton blank.

f boîte *f* pliante mixte — **d** Faltschachtel *f* mit kombiniertem Verschluß — **n** vouwdoos *fm* met verschillende sluiting — **e** caja *f* plegable mixta — **i** astuccio *m* pieghevole con falde superiori ridotte e inferiori complete

2392 cartoon — Satiric image or drawing.

f caricature *f*; portrait *m* caricaturé — **d** Karikatur *f*; Zerrbild *n*; Spottbild *n*; Fratze *f*; Fratzenbild *n* — **n** karikatuur *f*; spotprent *fm* — **e** caricatura *f* — **i** caricatura *f*

2393 cartoon — Design drawn on strong paper of the same size as that to be reproduced on a mural fresco, mosaic or tapestry.

f carton *m* — **d** Karton *m*; Entwurf *m* eines Gemäldes — **n** modelblad *n* — **e** cartón *m* — **i** cartone *m*

2394 cartouche — Small ornamental mostly bowed design at the top of commercial letterheads, bills, invoices, etc., representing the name of the firm.

f cartouche *m* — **d** Kartusche *f* — **n** cartouche *fm* — **i** cartoccio *m*

2395 cartridge paper — Hard, tough paper made with a rough surface in a number of grades. See also offset cartridge.

f papier *m* cartouche; papier *m* à cartouche; papier *m* pour cartouches — **d** Kartuschenpapier *n*; Patronenpapier *n* — **n** kardoespapier *n*; kardoes *n* — **e** papel *m* para cartuchos — **i** carta *f* per cartucce

2396 case; type case — Receptacle in which movable type is laid to compose from. When in pairs, defined as upper (containing caps and small caps mostly) and lower (containing lower-case mostly) case respectively.

f casse *f*; boîte *f* — **d** Schriftkasten *m*; Setzkasten *m*; Kasten *m* — **n** letterkast *fm*; kast *fm* — **e** caja *f* (tipográfica); caja *f* de tipo(s); caja *f* de imprenta — **i** cassa *f* per composizione

2397 case; book case — Ready-made cover for a book, i.e., before it is attached to the book block.

f couverture *f* — **d** Buchdeckel *m*; Buchdecke *f* — **n** band *m*; boekband *m* — **e** tapa *f* (confeccionada aparte); cubierta *f* — **i** coperta *f*; copertina *f*

2398 case — Box-like enclosure of machine parts.

f cage *m*; carter *m* — **d** Gehäuse *n*; Verkleidung *f* — **n** kast *fm*; bak *m* — **e** caja *f* — **i** scatola *f* di protezione

2399 case — See box.

2400 case bound book — See cased book.

2401 cased book; case bound book — Cover or case made independently of the book, which is finally fixed into the case by gluing or pasting down the end papers. See also bound book, combination style.

f livre *m* cartonné — **d** gebundenes Buch *n* — **n** gebonden boek *n* — **e** libro *m* (con tapa confeccionada) empastado; libro *m* de tapa montada — **i** libro *m* legato; libro *m* incassato nella copertina

2402 case department — See composing room.

2403 casein — White, amorphous solid, hygroscopic, stable when kept dry, rapidly deteriorating when damp. It is a protein usually obtained from skimmed milk (3% casein). Soluble in dilute alkalis and concentrated acids, almost insoluble in water; precipitates from weak acid solutions. Used to make sizings for coated papers, adhesive solutions, also as a binder in aqueous dispersions of pigments.

f caséine *f* — **d** Kasein *n* — **n** caseïne *fm* — **e** caseína *f* — **i** caseina *f*

2404 case *v* **in** — To apply paste or glue to end papers of a book, insert the book in the cover and set in press to dry.

f emboîter — **d** einhängen — **n** inhangen; in de band zetten — **e** meter en tapas; entrar tapas; colocar los pliegos cosidos en la cubierta — **i** incassatura *f*

2405 case lay; case layout

f modèle *m* de casse — **d** Setzkasteneinteilung *f*; Schriftkasteneinteilung *f* — **n** kastindeling *f*; letterkastindeling *f*; indeling *f* van de letterkast — **e** disposición *f* de los cajetines; esquema *m* de los cajetines; distribución *f* de los cajetines — **i** divisione *f* della cassa; schema *m* della cassa

2406 case layout — See case lay.

2407 case *v* **letters** — See lay *v* type.

2408 case-lining paper; casing paper — Strong, heavy-weight kraft wrapping paper to line the inside of packing cases. Also duplex asphalt and dry-waxed papers are used.

f papier *m* pour doublage de saisses; papier *m* d'emballage — **d** Kistenauslegepackpapier *n*; Kistenausschlagpapier *n*; Kistenfutterpapier *n*; Kistenauskleidepapier *n* — **n** papier *n* voor bekleding van de binnenzijde van kisten — **e** papel *m* para forrar cajas; papel *m* para recubrir cajas — **i** carta *f* per rivestimento interno di casse

2409 case lot — Quantity of flat paper or paperboard usually wrapped in packages and enclosed in a fibreboard or wooden box. Usually a quantity of 500-600 pounds or four cartons. See also carload lot.

f minimum *m* de livraison par caisse — **d** Packmenge *f* an Bogenpapier die eine Kiste füllt — **n** hoeveelheid *f* per kist — **e** cantidad *f* de caja — **i** quantità *f* per imballaggio in casse

2410 case maker — See case making machine.

2411 case making — Making the cover of a book.
d Decken machen — **n** banden maken — **e** confección *f* de tapas; hacer tapas — **i** confezionatura *f* delle copertine

2412 case making machine; case maker — Machine for the manufacture of book cases, essentially an automatic gluing for pasting the pre-cut cover boards to the cover linen in the correct position, with an attachment for folding over and sticking down the surplus linen.
f machine *f* à fabriquer les couvertures; machine *f* à faire les couvertures; machine *f* à faire les emboîtages — **d** Buchdeckenmaschine *f*; Buchdeckenherstellmaschine *f*; Deckenmachmaschine *f* — **n** bandenmaakmachine *f*; bandenmachine *f* — **e** máquina *f* de hacer tapas; máquina *f* de tapas sueltas; confeccionadora *f* de tapas — **i** macchina *f* confezionatrice di copertine

2413 case overseer — See composing-room overseer.

2414 case rack; case stand — Receptacle to hold cases when out of use.
f layette *f*; rayon-layette *f*; meuble *m* à casses — **d** Kastenregal *n* (ohne Setzpult) — **n** loket *n* — **e** chibalete *m*; burro *m* LA; comodín-chibalete *m*; estante *m* para cajas — **i** scaffale *m*; castello *m*

2415 caser-in — The workman who cases in books.
f emboîteur *m* — **d** Buchbindereiarbeiter *m* der Bücher einhängt — **n** inhanger *m* — **i** incassatore *m*

2416 case room — See composing room.

2417 case rules — See labour-saving rules.

2418 case stand — See case rack.

2419 case with strips of wood — Type case for storing expensive or precious types placed upright between thin wooden boards.
f casseau *m* à barrettes mobiles — **d** Steckschriftkasten *m* — **n** steekkast *fm* — **e** caja *f* con listones móviles — **i** cassa *f* con strice in legno

2420 cash-block making machine — See cash-pad printing machine.

2421 cash pad
f bloc-notes *m* de caisse — **d** Kassenblock *m* — **n** kassablok *n* — **e** talonario *m* de caja — **i** blocco *m* note di cassa

2422 cash-pad printing machine; cash-block making machine
f machine *f* à imprimer les blocs-caisses — **d** Kassenblockmaschine *f*; Kassenblockdruckmaschine *f* — **n** machine *f* voor het maken van kassabloks; kassablokmachine *f* — **e** máquina *f* de imprimir talonarios de caja — **i** macchina *f* per la stampa dei blocchi note di cassa

2423 cash purchase; purchase for cash
f achat *m* au comptant — **d** Bareinkauf *m*; Einkauf *m* gegen Barzahlung — **n** aankoop *m* tegen contante betaling — **e** adquisición *f* contante — **i** acquisto *m* a contanti

2424 casing — A UK size of wrapping paper. See app. no. 3.

2425 casing-in machine — Automatic machine which applies glue to the end papers of a book block and presses the prefabricated case around it, to finish the binding.
f machine *f* à emboîter — **d** Bucheinhängemaschine *f*; Einhängemaschine *f* — **n** inhangmachine *f* — **e** máquina *f* de poner tapas; máquina *f* de meter (libros) en tapas — **i** macchina *f* incassatrice; incassatrice *f*

2426 casing paper — See case-lining paper.

2427 Cassel brown — See vandyke brown.

2428 casse paper — See cassie paper.

2429 cassette — Holder for reels of magnetic tape, which can be easily clamped and detached from the tape deck of a recorder.
f cassette *f* — **d** Kassette *f* — **n** cassette *fm* — **e** caseta *f* — **i** cassetta *f*

2430 cassette — See plate holder.

2431 cassie paper; cassie quires; casse paper; cording quires — Old terms for the damaged sheets or quires at the *outsides of reams of paper. From the French "casser", to break.
f cassées *fpl* — **d** Ausschuß *m*; Abfall *m* — **n** beschadigde vellen *npl*; afval *n* — **i** carta *f* danneggiata

2432 cassie quires — See cassie paper.

2433 cast — See casting.

2434 cast *v* — In stereotype, running of the molten metal in the matrix.
f jeter — **d** abgießen — **n** afgieten — **e** fundir tejas — **i** eseguire una fusione; eseguire un getto; gettare

2435 cast *v* — In typography, running of the molten metal in the character moulds.
f fondre — **d** gießen — **n** gieten — **e** colar; fundir — **i** fondere

2436 cast-coated paper — Paper or board, the coating of which is allowed to harden or set while in contact with a finished casting surface. Cast-coated papers have, in general, a high gloss. For printing, they may be made (1) with greater ink receptivity than supercalendered coated papers, (2) suitable for gloss inks, or (3) with highly impervious surface coatings.
f papier *m* couché moulé — **d** gußgestrichenes Papier *n*; Papier *n* mit Höchstglanzstrich — **n** papier *n* met gegoten glanslaag — **e** papel *m* estucado moldeado — **i** carta *f* con patinatura per colata; carta *f* patinata cast-coated

2437 cast coating — Paper coating obtained by drying the coating slip in contact with a highly polished metal surface, either in the form of a continuous belt or on a cylinder. Cast coating may be carried out as a machine coating or separate coating operation.

f couche *f* à haut brillant; couchage *m* au glacis — **d** Höchstglanzschicht *f*; Gußschicht *f* — **n** hoogglanslaag *fm*; gietlaag *fm* — **e** capa *f* de alto brillo — **i** patina *f* per colata

2438 castellated nut — See castle nut.

2439 castelled nut — See castle nut.

2440 caster; casting machine — Machine for casting single type.

f fondeuse *f* — **d** Gießmaschine *f*; Gießapparat *m* — **n** gietmachine *f* — **e** fundidora *f* (de tipos) — **i** macchina *f* fonditrice; fonditrice *f*

2441 caster attendant

f fondeur *m* Monotype — **d** Monogießer *m*; Monotypegießer *m*; Gießer *m* — **n** Monotype-gieter *m*; gieter *m* — **e** fundidor *m* monotipista; operario *m* fundidor (de monotipia) — **i** fonditore *m* monotipista

2442 casting; cast — The cast object in relation to the mould in which it was cast.

f moulage *m*; pièce *f* moulée — **d** Abguß *m* — **n** gietsel *n* — **e** fundición *f*; vaciado *m* en molde; pieza *f* fundida; pieza *f* colada; clisé *m* estereotípico — **i** getto *m*; fusione *f*

2443 casting box — Cast-iron box, in which the matrix is set for casting stereotypes.

f moule *m* à stéréo; moule *m* à cliché — **d** Gießapparat *m* — **n** gietapparaat *n* — **e** máquina *f* fundidora; aparato *m* fundidor; fundidora *f*; molde *m*; caja-molde *f*; caja *f* de fundir *Me*; máquina *f* de fundir; estereotipadora *f* — **i** forma *f* a fondere

2444 casting box for curved stereos — See casting machine for curved stereos.

2445 casting box for flat stereos — See casting machine for flat stereos.

2446 casting frame

f équerre *f* de coulée — **d** Gießrahmen *m* — **n** gietraam *n* — **e** escuadra *f* para fundir — **i** squadra *f* di fusione

2447 casting ladle

f pochon *m* (de fonderie); pochon *m* à fondre; cuillière *f* à fondre; louche *f* à fondre; louche *f* à main — **d** Gießlöffel *m*; Gießkelle *f* — **n** gietlepel *m* — **e** cazo *m* de colada; cuchara *f* para fundir; cucharón *m* (para fundir) — **i** cucchiaio *m* (da fonditore)

2448 casting machine — See caster.

2449 casting machine for curved stereos; casting box for curved stereos — See also casting machine for flat stereos.

f moule *m* pour couler des stéréos ronds; moule *m* pour couler des stéréos cintrés; moule *m* pour couler des stéréos semi-cylindriques — **d** Rundstereo-Gießwerk *n*; Rundgießwerk *n* — **n** rondstyp-apparaat *n* — **e** fundidora *f* para estereotipias semicilíndricas; molde *m* para estereotipia curva; fundidora *f* curva; molde *m* de fundición curva; fundidora *f* de tejas — **i** fonditrice *f* per stereotipia curva

2450 casting machine for flat stereos; casting box for flat stereos — See also casting machine for curved stereos.

f moule *m* pour couler des stéréos à plat — **d** Gießgerät *n* für Flachstereotypie — **n** vlakstypapparaat *n* — **e** fundidora *f* para estereotipia plana; caja *f* de fundir *Me*; molde *m* para estereotipia plana — **i** fonditrice *f* per stereotipia piana

2451 casting mould — See mould.

2452 casting of rollers

f fonte *f* des rouleaux — **d** Walzenguß *m*; Walzengießen *n* — **n** gieten *n* van rollen; rollengieten *n* — **e** fundición *f* de rodillos — **i** colata *f* del rivestimento di gelatina sui rulli

2453 casting square — The side of a flat casting box limiting the size of the stereo plate.

f équerre *f* du moule; équerre *f* à fondre — **d** Gießwinkel *m*; Gießbacke *f* — **n** gietraam *n* — **e** escuadra *f* para fundir — **i** squadra *f* per fusione; forma *f* di fusione

2454 casting temperature

f température *f* de coulée — **d** Gießtemperatur *f* — **n** giettemperatuur *f* — **e** temperatura *f* de fundición; temperatura *f* de fundir — **i** temperatura *f* di fusione

2455 cast iron — Moulded iron, often used in making the framework of machinery. Iron-carbon alloy containing more than 2.0% carbon, usually 2.25-4.5%; usually also 0.05-0.15% sulphur, 0.5-3% silicon, 0.5-1% manganese and up to 1% phosphorus. Cannot be shaped by hammering, rolling, or pressing, except on an experimental or development scale.

f fer *m* coulé; fer *m* de fonte — **d** Gußeisen *n* — **n** gietijzer *n* — **e** hierro *m* fundido; hierro *m* colado; hierro *m* de fundición — **i** ghisa *f*

2456 castle nut; castellated nut; castelled nut — Tall lock nut with radial slits on its outer face for insertion of a cotter pin or wire through the nut and a hole in its bolt, to prevent the nut from coming loose.

f écrou *m* à créneaux — **d** Kronenmutter *f* — **n** kroonmoer *fm* — **e** tuerca *f* de corona; tuerca *f* entallada — **i** dado *m* a corona; dado *m* ad intagli

2457 cast *v* off — To estimate the amount of printed matter that a manuscript will make.

f estimer (une) copie; évaluer la copie; compter; calculer; piger — **d** Satzumfang schätzen; Text berechnen; Manuskript abschätzen; Manuskript berechnen; Umfang des Manuskriptes berechnen — **n** omvang berekenen — **e** calcular un manuscrito; avalorar un manuscrito; calcular un original; avalorar un original; hacer el cálculo de espacio; hacer el cálculo de material; hacer el cálculo tipográfico; hacer el cálculo de la composición; calcular el espacio de original(es); calcular el material — **i** stimare un originale per il lavoro di composizione

2458 castor oil; ricinus oil — A non drying oil. Pale-yellowish or almost colourless, transparent, viscous liquid. Soluble in alcohol, ether, benzene, chloroform and carbon disulphide. Latin: oleum ricini.

f huile *f* de ricin — **d** Rizinusöl *n*; Castoröl *n*; Wunderbaumöl *n* — **n** ricinusolie *fm*; castorolie *fm*; wonderolie *fm* — **e** aceite *m* de castor; aceite *m* de ricino; aceite *m* de palmacristi; carapato *m* — **i** olio *m* di ricino

2459 cast *v* **up** — To calculate the cost of composition.
f évaluer (le prix de composition) — **d** Satzkosten berechnen; Satzkosten kalkulieren; überschlagen — **n** zetkosten berekenen — **e** estimar el precio de la composición; calcular el precio de la composición — **i** calcolare il prezzo di composizione

2460 casual work — Small printing jobs, such as calling and announcement cards, ordered by casual customers.
f bilboquets *mpl*; bibelots *mpl* — **d** Gelegenheitsaufträge *mpl* — **n** familiedrukwerk *n*; gelegenheidsdrukwerk *n* — **e** remendería *f* — **i** lavoro *m* avventizi

2461 catalog — See catalogue.

2462 catalogue *GB*; **catalog** *US* — Essentially a booklet, but larger, and with many more pages. Can be case-bound (hard cover) or loose-leaf, and usually with an illustrated listing of merchandise, with or without prices.
f catalogue *m* — **d** Katalog *m* — **n** catalogus *m* — **e** catálogo *m* — **i** catalogo *m*

2463 catalyst — Material that speeds up a chemical reaction without itself being changed or used up.
f catalyseur *m* — **d** Katalysator *m* — **n** katalysator *m* — **e** catalizador *m* — **i** catalizzatore *m*

2464 catching up — In lithography, the printing of non-image areas of the plate, and overcome generally by increasing the amount of fountain solution applied to it. It may occur as a run is started before damping adjustment is correctly set.
f graissage *m* — **d** Tonen *n* — **n** tonen *n* — **e** taparse; tupirse; llenarse; empastarse; borrarse; cegarse — **i** velatura *f*

2465 catch line — Short line in small type set between large display lines.
f ligne *f* perdue — **d** Zwischenzeile *f* — **n** tussenregel *m* uit kleiner corps gezet — **e** línea *f* perdida — **i** linea *f* corta intermedia

2466 catch line — The line which contains the catchword.
f ligne *f* de réclame — **d** Stichzeile *f*; Leitzeile *f*; Schlagzeile *f* — **n** slagregel *m* — **e** línea *f* de reclamo

2467 catch stitch — See kettle stitch.

2468 catch title — Sub-title of each book in a set of volumes. See also title, back title, section title, sub-title.
f sous-titre *m* — **d** Untertitel *m*; Nebentitel *m*; Unterrubrik *f* — **n** ondertitel *m* — **e** título *m* de sección — **i** sottotitolo *m*

2469 catchword — Word at the top of each page in a dictionary or other reference book to indicate the first or last article on that page.
f mot *m* d'amorce — **d** Stichwort *n*; Suchwort *n* — **n** trefwoord *n* — **e** palabra *f* clave — **i**

esponente *m* di testa

2470 catchword; entry — Word, usually printed in bold face, at the beginning of a line, as in a dictionary.
f entrée *f* — **d** Stichwort *n* — **n** trefwoord *n* — **e** entradilla *f* — **i** esponente *m* di testa

2471 catchword — Isolated word at the bottom of a page in old books, inserted to connect the text with the beginning of the next page.
f réclame *f* — **d** Kustode *m*; Kustos *m*; Folgezeiger *m* — **n** wachter *m*; bladwachter *m*; custos *m* — **e** reclamo *m* — **i** richiamo *m*

2472 catchword index
f table *f* des mots-souches — **d** Stichwörterverzeichnis *n* — **n** trefwoordenregister *n* — **e** tabla *f* de entradillos; tabla *f* de palabras reclamas — **i** elenco *m* dei richiami

2473 catechol — See pyrocatechol.

2474 catechu — Modification of Malay: kachu. Astringent substance obtained from various tropical plants, especially from the wood of two East Indian acacias, acacia catechu and acacia Sumatra, used in dyeing and tanning.
f cachou *m* — **d** Katechu *n* — **n** cachou *m*; catechu *m* — **e** cachou *m*; catechu *m*; cachú *m*; catacú *m*

2475 catenati — See chain books.

2476 catenation — See concatenation.

2477 cater cornered — Said of paper that is accidentally cut diagonally.
d (fälschlich) diagonal geschnitten — **n** scheef gesneden; niet haaks gesneden — **i** tagliata non a squadra

2478 cathedral binding; cathédrale binding; architectural binding — Book-cover decoration style originating from the 16th century, suggesting columns supporting an arch and cathedral windows and destined for books on architecture. In the 19th century this style was revived by the French binder J. Thouvenin (1834).
f reliure *f* à la cathédrale; style *m* à la cathédrale — **d** Kathedral-Einband *m*; Einband *m* im Kathedralstil — **n** kathedraalband *m*; band *m* in kathedraalstijl — **e** encuadernación *f* catedral — **i** legatura *f* alla cattedrale

2479 cathédrale binding — See cathedral binding.

2480 cathode — In electronics, an electrode in valves connected to the negative side of supply voltage; emits electrons when heated to a high temperature. In electroplating and electrotyping, the wax or lead mould which acts as the negative pole on which the copper or nickel is deposited.
f cathode *f*; électrode *f* négative — **d** Kathode *m*; Katode *m*; negative Elektrode *f* — **n** kathode *f* — **e** cátodo *m*; polo *m* negativo — **i** catodo *m*; polo *m* negativo

2481 cathode ray tube — Generates characters or images by the bombardment of electrons against a phosphor coating on the inside of the tube. The stream of electrons is emitted by a cathode ray gun, focused into a narrow beam and directed to the proper location on the face of the tube by means of deflection yokes.

f tube *m* à rayons cathodiques — **d** Kathoden-strahlröhre *f* — **n** elektronenstraalbuis *fm*; kathodestraalbuis *fm*; beeldbuis *fm* — **e** tubo *m* de rayos catódicos — **i** tubo *m* da raggi catodici; tubo *m* catodico

2482 cathode ray tube display; CRT display; visual display unit; display unit — Cathode ray tube, similar to a television screen, used to display information as a means of communication with a computer system. Data in the computer store are displayed during processing and can be amended by the computer using a CRT display for operator monitored hyphenation. Other display applications include page make-up and display composition.

f unité *f* d'affichage — **d** Sichtgerät *n* — **n** beeld-station *n*; beeldbuisstation *n* — **e** pantalla *f*; unidad *f* de representación visual — **i** unità *f* di visualizzazione

2483 cation — Ion with a positive charge. Cations in a liquid subjected to electrical potential collect at the negative pole or cathode. See also anion.

f cation *m* — **d** Kation *m* — **n** kation *n* — **e** catión *m* — **i** catione *m*

2484 cationic reagents; cationics — Surface-active substances which have the active constituent in the positive ion.

f agents *mpl* de surface cationiques — **d** kationische Reagentien *npl* — **n** kationactieve stoffen *fmpl* — **e** reactivos *mpl* catiónicos superficiales — **i** agenti *mpl* cationici

2485 cationics — See cationic reagents.

2486 catoptric phenomenon — Formation of images by mirrors.

f phénomène *m* de catoptrie — **d** Reflexions-erscheinung *f*; Spiegelungserscheinung *f* — **n** reflectieverschijnsel *n* — **e** fenómeno *m* catóptrico — **i** fenomeno *m* di luce catodica riflessa

2487 caustic baryta — See barium hydroxide.

2488 causticity — See alkalinity.

2489 caustic lime — See calcium oxide.

2490 caustic lye — See sodium hydroxide.

2491 caustic potash — See potassium hydroxide.

2492 caustic soda — See sodium hydroxide.

2493 cave

f vingt-deux, voilà le prote — **d** Achtung, der Fax — **n** kijk uit, de chef — **e** ¡cuidado, el jefe! — **i** ecco, il proto

2494 cavitation — The continuous rapid formation and sudden collapse of microscopic bubbles in a liquid as a result of the reduction of total pressure.

f cavitation *f* — **d** Kavitation *f*; Hohlraum-bildung *f* — **n** cavitatie *f* — **e** cavitación *f* — **i** cavitazione *f*

2495 cedilla — Accent added to c, thus: ç, used in French to indicate that the c must be pronounced as s.

f cédille *f* — **d** Cedille *f* — **n** cedille *fm* — **e** zedilla *f*; cedilla *f*; virgulilla *f* — **i** cediglia *f*

2496 celestial blue — See Prussian blue.

2497 celestial chart; astronomical chart

f carte *f* céleste; carte *f* du ciel; carte *f* astronomique — **d** Himmelskarte *f*; Sternkarte *f* — **n** sterrenkaart *fm*; hemelkaart *fm*; astronomische kaart *fm* — **e** mapa *m* celeste; mapa *m* astronómico — **i** mappa *f* celeste; mappa *f* astronomica

2498 cell; store cell *GB*; **storage cell** *US* — Location of storage for a unit of information, as a binary cell it is one bit. Also a subdivision of storage space on a data cell drive, containing several strips of tape upon which data is recorded.

f cellule *f* de mémoire; mot *m* — **d** Speicher-element *n*; Zelle *f*; Streifenspeicher *m* — **n** geheugenelement *n*; cel *fm* — **e** célula *f* de memoria — **i** cella *f* di memoria

2499 cell — See ink cell.

2500 cell — See photocell.

2501 cellophane — Regenerated cellulose used in thin transparent sheets for wrapping and other purposes. It can be printed easily in flexography and gravure but the degree of adhesion is dependent on temperature and humidity.

f cellophane *f*; pellicule *f* cellulosique — **d** Zellophan *f*; Zellglas *n* — **n** cellofaan *n* — **e** celofán *m*; celofana *f* — **i** cellofan *m*

2502 cellophane bag

f sac *m* de cellophane — **d** Zellglasbeutel *m* — **n** cellofaanzakje *n* — **e** bolsa *f* de celofán — **i** sacchetto *m* di cellofan

2503 cellophane foil

f feuille *f* de cellophane — **d** Zellglasfolie *f* — **n** cellofaanfo(e)lie *fm* — **e** lámina *f* de celofán; lámina *f* celulósica — **i** foglio *m* di cellofan

2504 cellophane tape test — See Scotch tape test.

2505 cellophane window

f fenêtre *f* en cellophane — **d** Zellglasfenster *n* — **n** cellofaanvenster *n* — **e** ventanilla *f* de celofán — **i** finestra *f* di cellofan

2506 cellosolve — Trade name for *ethylene glycol monoethylether.

2507 celluloid — Trade-mark for *cellulose nitrate.

f celluloïd(e) *m* — **d** Zelluloid *n*; Zellhorn *n* — **n** celluloid *n*; celluloïde *n* — **e** celuloide *m* — **i** celluloide *f*

2508 cellulose — Chief component of the cell walls or woody structure of plants; the fibrous material remaining after the non-fibrous components of wood have been removed by pulping and bleaching operations, used in making paper.

Formula: $(C_6H_{10}O_5)_n$.

f cellulose *f* — **d** Zellstoff *m*; Zellulose *f* — **n** cel-stof *fm*; cellulose *fm* — **e** celulosa *f* — **i** cellulosa *f*

2509 cellulose acetate; acetyl cellulose — Reaction of cellulose with acetic anhydride forms cellulose acetate. Subsequently extruding and plasticizer addition yields the resultant, a thermoplastic material which is used in the graphic industries in thin transparent sheets for proofing, non-flammable photographic films, and other purposes. It is not extremely difficult to print in flexography, depending, however, on humidity and temperature. Soluble in acetone, ethyl acetate, cyclohexanol, nitropropene, ethylene dichloride, but subject to dimensional change due to cold flow, heat, or moisture absorption.

Formulas: $C_6H_5(CO_2H_3)_5$, $C_6H_6O(CO_2CH_3)_4$ or $C_6H_7O_2(CO_2CH_3)_3$.

f acétate *m* de cellulose; acétyl-cellulose *f*; acéto-cellulose *f* — **d** Zelluloseazetat *n*; Azetyl-zellulose *f*; Azetatzellulose *f* — **n** cellulose-acetaat *n* — **e** acetato *m* de celulosa; acetato *m* celulósico; acetil celulosa *f* — **i** acetato *m* di cellulosa; acetilcellulosa *f*

2510 cellulose acetate film — Flexible material from cellulose acetate.

f film *m* acétate — **d** Azetatfilm *m* — **n** cellulose-acetaatfilm *m* — **e** película *f* de acetato de celulosa — **i** film *m* di acetato di cellulosa

2511 cellulose acetate lacquer

f vernis *m* acétocellulosique — **d** Azetylzellulose-lack *m* — **n** acetylcelluloselak *mn* — **e** laca *f* celulósica; barniz *m* celulósico — **i** lacca *f* all'acetato di cellulosa

2512 cellulose ester

f ester *m* de cellulose; ester *m* cellulosique — **d** Zelluloseester *m* — **n** cellulose-ester *m* — **e** éster *m* de celulosa — **i** estere *m* di cellulosa

2513 cellulose fibres — Main constituents of trees and plants; remaining after the removal of the non-fibrous components of wood, used for making paper.

f fibres *fpl* de cellulose — **d** Zellulosefasern *fpl* — **n** cellulosevezels *fmpl*; celstofvezels *fmpl* — **e** fibras *fpl* de celulosa — **i** fibre *fpl* di cellulosa

2514 cellulose gum — Water-soluble gum from wood fibre cellulose, chemically designated as *carboxymethylcellulose. In lithography a substitute for gum arabic and synthetic gums.

f gomme *f* cellulosique — **d** Zellulosegummi *n* — **n** cellulosegom *m* — **e** goma *f* de celulosa — **i** gomma *f* di cellulosa

2515 cellulose lacquer; cellulose varnish — Solution of a cellulose ester or ether in an organic solvent, with modifying agents such as plasti-cizers, used for coating paper and board for protection and decoration.

f laque *f* cellulosique; vernis *m* cellulosique — **d** Zelluloselack *m* — **n** celluloselak *m*; cellulose-vernis *mn* — **e** laca *f* de celulosa — **i** vernice *f* alla cellulosa

2516 cellulose methyl ether — See methyl cellulose.

2517 cellulose nitrate — Highly inflammable sheet material plasticized with camphor. Formerly known under the trade name *celluloid. The name cellulose nitrate is technically correct although nitrocellulose is more commonly used.

f nitrate *m* de cellulose; cellulose *f* nitrique; nitro-cellulose *f* — **d** Zellulosenitrat *n*; Cellulose-nitrat *n*; Nitrozellulose *f*; Nitratzellulose *f* — **n**

cellulosenitraat *n*; nitrocellulose *fm* — **e** nitrato *m* de celulosa; nitrocelulosa *f* — **i** nitrato *m* di cellulosa; nitrocellulosa *f*

2518 cellulose nitrate — See nitrocellulose.

2519 cellulose sponge — See viscose sponge.

2520 cellulose varnish — See cellulose lacquer.

2521 cellulose wadding — Porous, absorbent paper used for sanitary purposes; it may also be made moisture-proof and used as a protective packaging material.

f ouate *f* de cellulose — **d** Zellstoffwatte *f*; Zell-watte *f* — **n** celstofwatten *pl* — **e** guata *f* de celulosa; guata *f* celulósica — **i** ovatta *f* di cellulosa

2522 cellusuède — Trade name for a *flock paper.

2523 cell wall; screen wall

f cloison *f* (de trame); muretin *m* — **d** Steg *m*; Tiefdrucksteg *m* — **n** rasterkam *m* — **e** barra *f* de la trama; puente *m* de la trama — **i** parete *f* fra gli alveoli

2524 Celsius — See centigrades.

2525 cement *v*

f cimenter — **d** verkitten — **n** kitten — **e** cemen-tar — **i** cementare

2526 censorship of the press

f censure *f* de la presse — **d** Pressezensur *f*; Zensur *f* der Presse — **n** perscensuur *f* — **e** censura *f* de (la) prensa — **i** censura *f* di stampa

2527 centered dot — See centred dot.

2528 centered sprocket hole — See in-line feed hole.

2529 center hole — See in-line feed hole.

2530 centigrades; degrees centigrade; degrees Celsius — Measuring unit where the interval between the two standard points, the freezing point and the boiling point (at 760 mm in baro-metric pressure) of water, is divided into 100 parts or degrees. Indicated by °C. To convert degrees Celsius to degrees Fahrenheit, multiply by 9, divide by 5, then add 32 to the result.

f degrés *mpl* centigrades; centigrades *mpl*; degrés *mpl* Celsius — **d** Grade *mpl* Celsius — **n** graden *mpl* Celsius; centigraden *mpl* — **e** grados *mpl* Celsius; centígrados *mpl* — **i** gradi *mpl* Celsius; centigradi *mpl*

2531 centimeter — See centimetre.

2532 centimetre *GB*; **centimeter** *US* — Metric unit of length. Abbrev. cm. See app. no. 7.

f centimètre *m* — **d** Zentimeter *m* — **n** centi-meter *m* — **e** centímetro *m* — **i** centimetro *m*

2533 centipoise — One one-hundredth of a *poise, the unit of viscosity. Abbrev. cP. Liquids of low vis-cosity are usually given in centipoise units.

$1 cP = 0.01 P = 10^{-3} Pa·s$.

f centipoise *f* — **d** Centipoise *fn*; Zentipoise *fn* — **n** centipoise *mf* — **e** centipoise *m* — **i** centi-poise *m*

2534 centistokes — One one-hundredth of a *stokes. Abbrev. cSt. In the SI: $1 cSt = 0.01 St = 10^{-6} m^2/s$.

2535 cent mark

f signe *m* ¢ — **d** Centzeichen *n* — **n** centteken *n*

— **e** signo *m* de centavo; signo *m* ¢ — **i** segno *m* centesimo; segno *m* ¢

2536 central heating

f chauffage *m* central; chauffage *m* au calorifère — **d** Zentralheizung *f* — **n** centrale verwarming *f* — **e** calefacción *f* central — **i** riscaldamento *m* centrale

2537 centralized lubrication; central lubrication; one-shot lubrication

f graissage *m* centralisé — **d** Zentralschmierung *f*; Eindruckschmierung *f* — **n** centrale smering *f* — **e** lubricación *f* central; engrase *m* central — **i** lubrificazione *f* centralizzata

2538 central lubrication — See centralized lubrication.

2539 central processing unit; central processor; main frame; main processor; CPU — Main component of a computer which includes arithmetic and logic to execute its instruction set.

f unité *f* centrale de traitement — **d** Zentraleinheit *f* — **n** centrale verwerkingseenheid *f*; centraal verwerkingsorgaan *n* — **e** equipo *m* central; unidad *f* central — **i** unità *f* elaborativa

2540 central processor — See central processing unit.

2541 centred dot *GB*; **centered dot** *US*; **bullet** *fam* — Heavy dot used as an ornament before a paragraph, or for marking it to call attention to a particular section. A lighter centred dot is used in mathematical composition as a multiplication sign.

f point *m* vignette — **d** fetter mittestehender Punkt *m* — **n** blikvanger *m* — **i** punto *m* centrato

2542 centred head — See centre head.

2543 centred sprocket hole — See in-line feed hole.

2544 centre head; centred head; cross head — Headline at equal distances from both margins of the page or column.

2545 centre hole — See in-line feed hole.

2546 centre margin ring; centre ring — Bevelled circumferential strip on a plate cylinder of a letterpress rotary, against which the printing plates are locked.

f cercle *m* d'accrochage — **d** Mittelfacette *f* — **n** middelste zijaanleg *m*; midden-zijaanleg *m* — **e** cincho *m* — **i** bordo *m* centrale smussato

2547 centre notes — Notes between columns.

f notes *fpl* centrales — **d** Anmerkungen *fpl* zwischen Textspalten; Anmerkungen *fpl* im Text — **n** noten *fmpl* tussen de kolommen — **e** notas *fpl* intercaladas — **i** note *fpl* al centro fra le colonne

2548 centre ring — See centre margin ring.

2549 centre spread — Middle opening of a journal, booklet, or folder where the design occupies the double-page area. See also double spread.

f centre *m* éployé; à livre ouvert — **d** Seitenpaar *n* in der Mitte eines Bogens, aus einem Blatt bestehend — **n** middelste pagina's *fmpl* in een katern — **e** plana *f* doble; página *f* doble; doble página *f* — **i** doppia pagina *f* nel centro di un quaderno

2550 centre stitching — Stitching of pamphlets with thread by working it in three places in the fold, like wire saddle stitching.

f cousure *f* à trois points — **d** Rückstichheftung *f* mit drei Stichen — **n** hechting *f* met drie steken door de rug — **e** costura *f* (por el lomo) con tres puntadas — **i** cucitura *f* a tre punti

2551 centre *v* the line — To compose a line in such a way that its centre coincides exactly with that of the type area.

f centrer la ligne — **d** die Zeile zur Mitte stellen; die Zeile auf die Mitte einstellen; die Zeile zentrieren — **n** de lijn in het midden plaatsen; de lijn centreren — **e** centrar la línea; justificar la línea al centro — **i** centrare

2552 centrifugal

f centrifugal — **d** zentrifugal — **n** centrifugaal; middelpuntvliedend — **e** centrífugo — **i** centrifugo

2553 centrifugal force — Force with which a rotating or whirling object tends to pull away from the central point of rotation.

f force *f* centrifuge — **d** Zentrifugalkraft *f*; Schwungkraft *f* — **n** centrifugale kracht *fm*; centrifugaalkracht *fm*; middelpuntvliedende kracht *fm* — **e** fuerza *f* centrífuga — **i** forza *f* centrifuga

2554 centring and quadding device — See quadder.

2555 centring tack; matrix centring pin — Part of the Monotype casting machine.

f plongeur-centreur *m*; pointeau *m* de centrage — **d** Zentrierstift *m* — **n** centreerpen *fm* — **e** aguja *f* centradora — **i** punta *f* di centratura

2556 ceramic ink

f encre *f* pour décalcomanie céramique — **d** keramische Farbe *f* — **n** keramische inkt *m*; ceramische inkt *m* — **e** tinta *f* cerámica — **i** inchiostro *m* per decalcomanie su ceramica

2557 ceramic transfer paper — See also decalcomania paper.

f papier *m* pour décalcomanies céramiques — **d** Papier *n* für keramische Abziehbilder — **n** papier *n* voor keramische transfers — **e** papel *m* para calcomanías cerámicas — **i** carta *f* per decalcomanie su ceramica

2558 ceramic transfer picture

f décalcomanie *f* céramique — **d** keramisches Abziehbild *n* — **n** keramisch(e) transfer *mn*; ceramisch(e) transfer *mn* — **e** calcomanía *f* cerámica — **i** decalcomania *f* su ceramica

2559 cerecloth — Cloth treated with melted wax or gummy matter sometimes used for book covers.

f toile *f* cirée — **d** Wachstuch *n* — **n** wasdoek *n* — **e** hule *m*; encerado *m* — **i** tela *f* cerata; tela *f* incerata

2560 ceresin — See ozokerite.

2561 ceresin wax — See ozokerite.

2562 ceric sulphate; cerium sulphate — White or reddish-yellow crystals, soluble in water (decomposes), soluble in dilute sulphuric acid.

Formula: $Ce(SO_4)_2 \cdot 4H_2O$.
f sulfate *m* cérique — **d** Cerisulfat *n* — **n** cerisulfaat *n* — **e** sulfato *m* cérico — **i** solfato *m* cerico
2563 cerin — See ozokerite.
2564 ceriphs — See serifs.
2565 cerium sulphate — See ceric sulphate.
2566 cerium sulphate — See cerous sulphate.
2567 cerotype; wax engraving — Process of engraving (cerography) in which the design is cut on a wax-coated metal plate, from which a printing surface is produced by electrotyping; used in map making.
f cérotype *m* — **d** Wachsradierung *f* — **n** cerotypie *f* — **e** cerotipia *f* — **i** impronta *f* cerografica; forma *f* cerografica
2568 cerous sulphate; cerium sulphate — White crystals or powder, soluble in water and in acids. Formula: $Ce_2(SO_4)_3 \cdot 8H_2O$.
f sulfate *m* céreux — **d** Cerosulfat *n* — **n** cerosulfaat *n* — **e** sulfato *m* ceroso — **i** solfato *m* ceroso
2569 certificate — In limited editions, the statement usually printed on the page facing the title page giving the number of books printed.
f justificatif *m* — **d** Auflagebeglaubigung *f* — **n** opgave *f* van de oplage — **e** certificado *m* — **i** certificato *m*
2570 certinal — See rodinal.
2571 ceruse — See flake white.
2572 cesium; caesium — Silver-white, soft, ductile metal, soluble in acids and alcohol. Used for photo-electric cells, etc. Symbol: Cs.
f césium *m* — **d** Zäsium *n* — **n** caesium *n* — **e** cesio *m* — **i** cesio *m*
2573 cetylacetic acid — See stearic acid.
2574 cetylic acid — See palmitic acid.
2575 CGO — See punk.
2576 chadless — Without *chads.
2577 chads — Small paper disks formed when holes are punched in paper tape. Chads remain attached by about one quarter of its circumference to the hole.
f confetti *m*; débris *mpl* de perforation — **d** Schnipsel *npl* — **n** confetti *m* — **e** pedacites *mpl* de papel; papel *m* picado — **i** coriandoli *mpl*
2578 chaff cutting engine
f hache-paille *f* — **d** Häckselmaschine *f*; Häckselbank *f*; Häcksellade *f* — **n** hakselmachine *f* — **e** corta-paja *m*; máquina *f* de cortar paja; cortadora *f* de paja — **i** trinciapaglia *m*
2579 chain books; chained books; catenati — Precious books in former times secured to the desk by a chain.
f livres *mpl* enchaînés — **d** Kettenbücher *npl* — **n** kettingboeken *npl* — **e** libros *mpl* encadenados — **i** libri *mpl* incatenati
2580 chain delivery device — Device on a printing machine that takes the sheets over from the grippers of the impression cylinder by a gripper bar, which in turn is fixed to a pair of chains to transport them to the delivery board.

f sortie *f* à chaîne; réception *f* à chaîne — **d** Kettenausleger *m*; Kettenauslage *f*; Kettengreiferauslage *f* — **n** kettinguitleg *m*; kettinggrijperuitleg *m* — **e** sacador *m* de cadena; entrega *f* por cadena; salida *f* por cadenas; salida *f* a cadena; sacapliegos *m* de cadena; sacapliegos *m* a cadena; receptor *m* de cadena — **i** uscita *f* con catene
2581 chain design
f chaînette *f* — **d** kettenartig verschlungene Verzierung *f* — **n** kettingvormige versiering *f* — **i** ornamento *m* a catenelle
2582 chain dot screen; elliptical dot screen — Halftone screen producing an elongated dot formation.
f trame *f* à points elliptiques; trame *f* à points en chaîne — **d** Kettenpunktraster *m*; Kettpunktraster *m*; Perlformraster *m* — **n** kettingpuntraster *n* — **e** trama *f* de puntos elípticos; retícula *f* de puntos elípticos — **i** retino *m* a punto ellittico
2583 chain drive — Arrangement in which the driving force is transferred by a chain running over sprocket wheels.
f commande *f* par chaînes; entraînement *m* par chaîne — **d** Kettenantrieb *m*; Kettentrieb *m* — **n** kettingaandrijving *f* — **e** impulsión *f* de cadena; propulsión *f* de cadena; transmisión *f* por cadena — **i** trasmissione *f* a catena
2584 chained binding — Book secured by a chain to the desk, formerly in monastic libraries.
f reliure *f* enchaînée — **d** Ketteneinband *m* — **n** kettingband *m* — **e** encuadernación *f* encadenada — **i** legatura *f* incatenata
2585 chained books — See chain books.
2586 chained list — List in which items which may not be sequential are linked together by means of a reference at the end of one to the next.
f liste *f* enchaînée — **d** verknüpfte Liste *f* — **n** keten *fm* — **e** lista *f* encadenada — **i** lista *f* concatenata
2587 chain gripper device; chain grippers
f dispositif *m* de pinces à chaînes — **d** Kettengreifersystem *n*; Kettengreiferauslage *f*; Kettengreifer *m* — **n** kettinggrijper *m* — **e** uña *f* de cadena; pinza *f* de cadena — **i** dispositivo *m* uscitafoglio con catene portapinze
2588 chain grippers — See chain gripper device.
2589 chain lines — In laid papers, the vertical, prominent widely spaced (approx. 2.5 cm) lines parallel to the machine direction or grain of the paper.
f chaînettes *fpl*; pontuseaux *mpl* — **d** Wasserlinien *fpl* — **n** kettinglijnen *fmpl*; staande lijnen *fmpl*; waterlijnen *fmpl* — **e** corondeles *mpl*; cadenetas *fpl*; líneas *fpl* de agua — **i** treccioli *mpl*; filoni *mpl*; ponticelli *mpl*; righe *fpl* d'acqua
2590 chain printer — High speed line printer which prints "on the fly". The character set is engraved on slugs which are fixed to a chain which travels across the face of the paper along the length of the printed line.

f imprimante *f* à chaîne — **d** Kettendrucker *m* —
n kettingdrukker *m* — **e** impresora *f* de cadena
— **i** stampatrice *f* a catena

2591 chain sprocket — See sprocket.

2592 chain stitch — See kettle stitch.

2593 chain wheel — See sprocket.

2594 chalcography — Art of engraving on copper.
f chalcographie *f*; gravure *f* sur cuivre; gravure *f*
en taille douce — **d** Chalkographie *f*; Kupferstech-
kunst *f* — **n** chalcografie *f*; kunst *f* der kopergravure; kopergraveerkunst *f*; kunst *f* van het
graveren in koper; plaatdrukkunst *f* — **e** calcografía *f* — **i** calcografia *f*

2595 chalk — See calcium carbonate.

2596 chalk — See lithographic chalk.

2597 chalk *v* — To apply magnesium carbonate
(chalk) to etched halftone plates, the material
filling the etched areas and imparting good visibility to the halftone image for study of tone
values and progress of etching.
f passer au blanc de magnésie — **d** Magnesium
einreiben — **n** inkrijten — **e** espolvorear con
blanco de magnesio — **i** applicare polvere di
magnesia

2598 chalk drawing — Drawing with crayon on a
lithographic surface.
f dessin *m* au crayon litho — **d** Kreidelitho-
graphie *f*; Kreidezeichnung *f* — **n** krijtlitho *fm*;
krijttekening *f* — **e** dibujo *m* litográfico a lápiz —
i disegno *m* con gesso litografico

2599 chalking — Condition of printing ink in
which the pigment is not properly bound to the
paper and can easily be rubbed off as a powder.
Caused by too rapid absorption of the vehicle into
the paper.
f poudrage *m* — **d** Abkreiden *n*; Abmehlen *n*;
Mehlen *n*; Kreiden *n*; Kreidigwerden *n* — **n**
poederen *n*; afpoederen *n*; afkrijten *n* — **e**
polvorizado *m* — **i** spolverio *m* dell'inchiostro;
sfarinatura *f*

2600 chalking — Condition of paper where fine
particles of pigment leave the sheet during the
funishing, converting, printing operation, or subsequent use.

2601 chalk offsets — Impressions on super paper,
dusted with red chalk, and pulled over on stone or
metal as a key or guide for the artist. This technique is limited to poster hand plates to be run on
a direct rotary lithographic press.
f faux décalques *mpl* lithographiques (à la Sienne)
— **d** Klatschdrucke *mpl* — **n** klatsen *mpl* — **e**
impresiones *fpl* reportadas — **i** falsi decalchi *mpl*
litografici

2602 chalk overlay — Interlay or overlay mechanically made for making ready line and halftone
blocks.
f découpage *m* à la craie; découpage *m* chimique;
mise *f* au relief de craie — **d** mechanische Kreidezurichtung *f* — **n** krijtpikeersel *n* — **e** alza *f*
mecánica de greda — **i** taccheggio *m* con rilievo
in gesso

2603 chalk overlay paper — Thick, dense paper,
coated with chalk, to make overlays.
f papier *m* porcelaine; papier *m* pour découpage à
la craie — **d** Kreidezurichtepapier *n* — **n** papier *n*
voor krijtpikeersels — **e** papel *m* para alzas
mecánicas; papel *m* para arreglo (mecánico) con
greda — **i** carta *f* per avviamento meccanico;
carta *f* per avviamento in creta

2604 chalk relief makeready — Use of a sheet of
paper coated with chalk printed on both sides and
developed in Javelle water which dissolves the
chalk on the non-printed areas. These areas
correspond with the lighter parts of the image.
f mise *f* en train au relief de craie; mise *f* en train
à la craie — **d** Kreidereliefzurichtung *f* — **n** toestellen *n* met krijtpikeersels — **e** arreglo *m*
mecánico con greda — **i** avviamento *m* di rilievo
in creta

2605 chalky — In photography: lacking in detail
due to extreme contrast.

2606 chalky
f crayeux — **d** kreidig — **n** krijtachtig — **e** cretáceo; gredoso — **i** cretoso; gessoso

2607 chalky ink
f encre *f* terreuse — **d** kreidige Farbe *f* — **n** krijtachtige inkt *m* — **e** tinta *f* gredosa — **i** inchiostro
m gessoso

2608 Challenge press — Small job press of the
Gordon style.

2609 chamfer *v*
f chanfreiner — **d** abschrägen; abfasen; abschärfen — **n** afschuinen; een schuine kant
maken — **e** achaflanar; biselar — **i** bisellare;
smussare

2610 chamfered rule — See mitred rule.

2611 chammy-leather — See chamois-leather.

2612 chamois-leather; **chammy-leather;**
shammy-leather — Soft, pliable leather dressed
with oil, especially fish oil, originally prepared
from the skin of the chamois.
f peau *f* de chamois; chamois *m*; peau *f*
chamoisée; peau *f* de daim — **d** Sämischleder *n*
— **n** zeemleer *n* — **e** piel *f* de gamuza; gamuza *f*
— **i** pelle *f* scamosciata

2613 champlevé enamel — Enamel embedded in
the cavities of a metal plate. See also enamel
binding.
f émail *m* champlevé — **d** Grubenschmelz — **n**
champlevé — **i** smallatura *f* ad incavo

2614 change *v*
f changer; remplacer — **d** ändern; wechseln; wenden — **n** veranderen; vervangen — **e** cambiar — **i**
cambiare; sostituire

2615 change of body size
f changement *m* du corps — **d** Kegelwechsel *m* —
n corpsverandering *f*; verandering *f* van corps —
e cambio *m* de cuerpo — **i** cambiamento *m* di
corpo

2616 change of measure
f changement *m* de la justification; changement
m du format — **d** Formatwechsel *m*; Format-
verstellung *f* — **n** formaatverandering *f*;
verandering *f* van het formaat — **e** cambio *m* de

formato; cambio *m* de medida — **i** cambiamento *m* di giustezza

2617 change over of a forme — Taking off a printing forme from the press and replacing it by another.

f changement *m* de forme — **d** Formwechsel *m* — **n** vormverwisseling *f* — **e** cambio *m* de forma — **i** cambio *m* della forma

2618 change-over time — Set-up time plus teardown time.

f temps *m* par série; temps *m* de préparation; temps *m* de réglage — **d** Umstellungszeit *f*; Rüstzeit *f* — **n** insteltijd *m*; omsteltijd *m* — **e** tiempo *m* de cambio — **i** tempo *m* di cambiamento; tempo *m* di preparazione

2619 changes in the forme

f changements *mpl* dans la forme — **d** Änderungen *fpl* in der Form — **n** veranderingen *fpl* in de vorm — **e** enmiendas *fpl* en la forma; enmiendas *fpl* en el molde — **i** cambiamenti *mpl* nella forma

2620 channel — One of the ninety channels in which the magazine of a line-casting machine usually is divided, each channel holding up to twenty identical matrices ready for assembling.

f canal *m* — **d** Kanal *m* — **n** kanaal *n* — **e** canal *m* — **i** canale *m*

2621 channel; track — One of a number of parallel paths across and along which signals representing data are encoded, normally by punched holes or magnetic pulses. Sometimes a method of connecting the CPU with I/O units.

f canal *m* — **d** Lochspur *f*; Spur *f* — **n** gatenspoor *n*; spoor *n* — **e** pista *f* de perforación — **i** pista *f* di perforazione; canale *m*

2622 channel black — Carbon black produced by impinging a natural-gas flame against an iron plate. It is characterized by lower pH, higher volatile content, small particle size and less structure between the particles. See also impingement black.

f noir *m* channel; noir *m* au tunnel — **d** Channelruß *m*; Channelschwarz *n*; Kanalruß *m* — **n** gasroet *n*; channel black *n* — **e** negro *m* de canal; negro *m* de túnel — **i** nerofumo *m* di gas

2623 channel entrance — Part of the slug composing machine through which the matrices pass after leaving the distributor bar and then into the magazine.

f entrée *f* de magasin — **d** Magazineintritt *m* — **n** kanaalingang *m* — **e** entrada *f* del depósito; guía *f* de entrada — **i** entrata *f* del canale

2624 channel escapement — See escapement.

2625 chapel — Association of journeymen compositors in a printing office. Also a meeting of that association. To "chapel" a person, means to report him to the chapel. See also father of the chapel.

f chapelle *f* — **d** Betriebsrat *m* — **n** personeelsraad *m* — **e** consejo *m* obrero — **i** consiglio *m* personal

2626 chapter — Main division of a book.

f chapitre *m* — **d** Kapitel *n*; Abschnitt *m* — **n** hoofdstuk *n* — **e** capítulo *m* — **i** capitolo *m*

2627 chapter heading

f tête *f* de chapitre; en-tête *m* — **d** Kapitelüberschrift *f*; Absatztitel *m* — **n** hoofdstuktitel *m* — **e** cabeza *f* de capítulo; cabecero *m* — **i** intestazione *f* di capitolo; testata *f*

2628 character — 1. Letter of the alphabet, a figure (number), or a punctuation symbol, sometimes called a sign as distinguished from a *graphic. 2. Symbol which may be assigned for information processing purposes, to an eight-bit byte. 3. Image created by a typewriter, typesetter, or computer line printer, of an alphanumeric symbol and hence a part of its repertoire of symbols. For example, a typesetter's fount disk may contain 16 founts of 100 characters each.

f caractère *m*; lettre *f* — **d** Buchstabe *m*; Schriftzeichen *n*; Zeichen *n*; Type *f* — **n** letter *fm*; letterteken *n*; teken *n* — **e** carácter *m*; letra *f* — **i** carattere *m*; lettera *f*

2629 character count — In copyfitting, count made by computing the number of characters and spaces in an average line of the manuscript and multiplying this by the number of lines.

f calibrage *m* de la copie — **d** Manuskriptberechnung *f*; Schriftmengenberechnung *f*; Umfangsberechnung *f* — **n** lettertelling *f* — **e** contado *m* de caracteres — **i** conteggio *m* dei caratteri

2630 character dimension — See also magnetic encoded cheques.

f dimensions *fpl* des caractères — **d** Zeichenabmessung *f* — **n** lettergrootte *f* — **e** tamaño *m* de los caracteres — **i** dimensioni *fpl* del carattere

2631 character direction — See character skew.

2632 character generation — Projection or construction of typographic images on the face of a cathode ray tube, usually in association with a high-speed computer-controlled photocomposition system.

f projection *f* du caractère — **d** Buchstabenerzeugung *f*; Schriftbildprojektion *f* — **n** beeldprojectie *f*; letterprojectie *f* — **e** generación *f* de caracteres; creación *f* del carácter — **i** proiezione *f* del carattere

2633 character generator — Cathode ray (TV) tube or similar device used to display characters in high speed photocomposition systems.

f générateur *m* de caractères — **d** Schriftzeichengenerator *m* — **n** beeldgenerator *m*; tekengenerator *m* — **e** generador *m* de caracteres — **i** generatore *m* di caratteri

2634 character grid — Plastic or glass master image used in most filmsetting equipment. A grid is assembled from individual character images, projected on film by an optical or cathode ray tube co-ordinate selector system. Some machines have a system of lenses to obtain different sizes, while others require a fount of separate master images for each point size.

f grille *f* — **d** Matrizenträger *n*; Matrizenrahmen *m* − **n** letterraster *n*; letterdrager *m* − **e** rejilla *f* de caracteres; cuadriculado *m* − **i** griglia *f* di caratteri

2635 characteristic; job factor; requirement − A work study term used in job evaluation.

f critère *m*; exigence *f* − **d** Bewertungsmerkmal *n* − **n** kenmerk *n* − **e** característica *f* − **i** fattore *m* lavoro

2636 characteristic − See feature.

2637 characteristic curve; sensitometric curve; density gradation curve; D log E curve − Characteristic curve for photosensitive material, showing the relation between the exposure of the material and the resulting density of the silver deposit.

f courbe *f* caractéristique; courbe *f* D log E − **d** Gradationskurve *f*; Schwärzungskurve *f*; D-log-E-Krumme *f*; D-log-E-Kurve *f* − **n** gevoeligheidskromme *f*; gradatiekromme *f*; gradatiecurve *f*; D log E-kromme *f*; D log E-curve *f* − **e** curva *f* característica; curva *f* D log E − **i** curva *f* caratteristica; curva *f* D log E

2638 characteristics

f caractéristiques *fpl* − **d** Eigenschaften *fpl*; Merkmale *npl*; Kennzeichen *npl* − **n** eigenschappen *fpl* − **e** características *fpl* − **i** caratteristiche *fpl*

2639 character recognition − Identification by machine of graphic characters and thus the conversion of visually observable characters into machine-readable data.

f reconnaissance *f* de caractères − **d** Zeichenerkennung *f*; maschinelles Lesen *n* − **n** schriftlezen *n*; schrifttekenlezen *n*; tekenherkenning *f*; mechanisch lezen *n* − **e** reconocimiento *m* de caracteres; lectura *f* mecánica − **i** riconoscimento *m* di caratteri; lettura *f* mecanica

2640 character set − Collection of characters (character repertoire) in an array for a *fount of type style, as in the case of a typesetter, or the array of characters displayable on a video terminal, available on a keyboard, or on a line printer.

f jeu *m* de caractères − **d** Zeichenmenge *f*; Zeichenvorrat *m* − **n** stel *n* tekens; tekensverzameling *f* − **e** conjunto *m* de caracteres; juego *m* de caracteres − **i** assortimento *m* di caratteri; insieme *m* di caratteri

2641 character skew − In magnetic ink printing the angle between a line called character direction and a line perpendicular to the bottom reference line. See also magnetic encoded cheques.

f inclinaison *f* de caractère − **d** Zeichenschräge *f* − **n** tekenschuinte *f* − **e** inclinación *f* de carácter − **i** inclinazione *f* di carattere

2642 characters per hour

f caractères *mpl* par heure − **d** Buchstaben *mpl* je Stunde; Buchstaben/h *mpl*; Zeichen/h *npl* − **n** letters *fmpl* per uur − **e** caracteres *mpl* por hora − **i** caratteri *mpl* all'ora

2643 characters per inch

f caractères *mpl* par pouce − **d** Buchstaben *mpl* je Zoll − **n** letters *fmpl* per inch − **e** caracteres *mpl* por pulgada − **i** caratteri *mpl* per pollice

2644 charcoal − Black, porous, odourless carbonaceous substance, burning with little or no flame, obtained by imperfect combustion of organic matter, as of wood. Used as a fuel, an absorbent, a filter, for fine grinding and polishing, and for making drawings. See also charcoal crayon, grinding charcoal.

f charbon *m* (de bois) − **d** Holzkohle *f* − **n** houtskool *fm* − **e** carbón *m* de leña; carbón *m* vegetal; carbón *m* dulce para emolar − **i** carbone *m* di legna

2645 charcoal crayon

f crayon *m* de charbon − **d** Zeichenkohle *f*; Reißkohle *f* − **n** houtskool *fm*; tekenhoutskool *fm* − **e** carboncillo *m* − **i** carbone *m* da disegnare; lapis *m* nero

2646 charcoal drawing − Drawing or sketch made with charcoal stick or pencil.

f dessin *m* au fusain; fusain *m* − **d** Kohlezeichnung *f* − **n** houtskooltekening *f* − **e** dibujo *m* al carbón − **i** disegno *m* al carbone; disegno *m* a carboncino

2647 charcoaling − Polishing a metal plate by rubbing with wet charcoal.

f polissage *m* au charbon de bois − **d** Schleifen *n* mit Kohle − **n** slijpen *n* met houtskool − **e** pulido *m* con carbón vegetal; pulido *m* con carbón de tilo − **i** politura *f* con carbone di legna

2648 chart; nautical chart − Hydrographic or marine *map.

f carte *f* marine − **d** Seekarte *f* − **n** zeekaart *fm*; kaart *fm* − **e** carta *f* marítima; carta *f* marina; carta *f* de navegar; carta *f* de marear − **i** carta *f* marina; carta *f* nautica

2649 chart paper − Extra strong, tub-sized paper of high folding endurance, for nautical charts or maps and automatic recording charts.

f papier *m* pour cartes maritimes − **d** Seekartenpapier *n* − **n** zeekaartenpapier *n*; papier *n* voor zeekaarten − **e** papel *m* para cartas marítimas; papel *m* para mapas − **i** carta *f* per mappe marittima

2650 chase − Frame of steel or cast or wrought iron in which type etc. is locked up for printing.

f châssis *m* − **d** Formrahmen *m*; Schließrahmen *m*; Rahmen *m* − **n** vormraam *n*; insluitraam *n*; raam *n* − **e** rama *f*; marco *m* − **i** telaio *m*

2651 chase − The negative (or positive) frame with glass face with register lines on which photographic film or plates are positioned to register in a photocomposing machine. It is attached to the register device, a jig-type fixture which establishes uniform register relationship between all images being photocomposed on offset pressplates by the step-and-repeat method.

f châssis *m* de montage; châssis *m* à négatif − **d** Montagerahmen *m*; Negativhalterrahmen *m* − **n** montageraam *n*; negatiefhouder *m* − **e** chasis *m*

portapelícula — **i** telaio *m* di montaggio
2652 chase bar — See cross bar.
2653 chased edges — See goffered edges.
2654 chase galley — Frame with easily detachable shank. After making up the forme is locked by simply turning two levers.
f galée *f* châssis — **d** Schließsetzschiff *n* — **n** snelsluitgalei *fm* — **e** galera *f* rama — **i** vantaggio *m* serraforma
2655 chaser — New slugs or correction material to be placed in the forme on the press.
f repiquage *m* — **d** Neusatz *m* — **n** loodcorrecties *fpl* — **i** nuova riga *f* per composizione in piedi
2656 chaser — Person who engraves metals.
f ciseleur *m* — **d** Ziseleur *m*; Graveur *m* — **n** ciseleur *m*; ciseleerder *m*; metaaldrijver *m* — **e** cincelador *m* — **i** cesellatore *m*
2657 chaser — See progress chaser.
2658 chase rack — See forme rack.
2659 chattering — Noise made by loosely fitting gears on a letterpress rotary. See also plate slap.
f claquement *m* — **d** Schlagen *n* — **n** ratelen *n*; rammelen *n*; kletteren *n*; slaan *n* — **e** tabletear; matraquar; rechinar; chirriar — **i** schiocco *m*
2660 cheap edition
f édition *f* populaire — **d** Volksausgabe *f*; billige Ausgabe *f* — **n** volksuitgave *f*; goedkope uitgave *f* — **e** edición *f* popular; edición *f* económica — **i** edizione *f* popolare; edizione *f* economica
2661 check — A control, test, or inspection that ascertains performance or prevents error. See also parity check.
f contrôle *m* — **d** Prüfung *f* — **n** controle *m* — **e** comprobación *f* — **i** controllo *m*
2662 check — See cheque.
2663 check *v* — See also control *v*.
f vérifier; contrôler — **d** überprüfen; nachsehen; kontrollieren — **n** nazien; controleren — **e** verificar; chequear — **i** verificare; controllare; riscontrare; rivedere
2664 check bit — Bit added to the data stream during machine verification to enable the machine to detect errors either in a single character (in case the bit has been dropped in recording, reading or transmission) or in a data block of a prescribed length.
f bit *m* de contrôle — **d** Prüfbit *n* — **n** controlebit *n* — **e** bit *m* de comprobación — **i** bit *m* di controllo
2665 check digit; check number — Digit to check the accuracy of a numeric field (such as record number, account number, etc.), usually created by some algorithm such as prime number division.
f chiffre *m* de contrôle — **d** Prüfziffer *f* — **n** controlecijfer *n* — **e** dígito *m* de verificación — **i** cifra *f* di controllo
2666 checker — See counter.
2667 checkerboard screen
f trame *f* en damier — **d** Schachbrettraster *m* — **n** schaakbordraster *n*; dambordraster *n* — **e** trama *f* cuadriculada — **i** retino *m* a scacchiera
2668 checkers — See draughtsmen.

2669 check folio — A size of writing paper. See app. no. 3.
2670 checking copy — See voucher copy.
2671 check number — See check digit.
2672 check nut; lock nut; jam nut; back nut — Nut, usually thinner than an ordinary nut, to keep the regular nut from loosening.
f contre-écrou *m* — **d** Gegenmutter *f*; Kontermutter *f* — **n** contramoer *fm*; tegenmoer *fm* — **e** contratuerca *f* — **i** controdado *m*
2673 check point; dump point — Place in a routine where a check, or a recording of data for restart purposes, is performed.
f point *m* de contrôle — **d** Prüfpunkt *m* — **n** controlepunt *n* — **e** punto *m* de comprobación — **i** punto *m* di controllo
2674 check royal — A size of writing paper. See app. no. 3.
2675 cheeks, knife — See knife cheeks.
2676 cheeks, side — See knife cheeks.
2677 cheese box
f boîte *f* à fromage — **d** Käseschachtel *f* — **n** kaasdoos *fm* — **e** caja *f* para queso — **i** scatola *f* per formaggio
2678 cheese cake *sl.*; **leg art** *sl.* — Photograph featuring an attractive woman's legs and body.
f image *f* affriolante *fam.* — **d** Photo *f* attraktiver Frauenkörper oder Frauenkörperteile — **n** blotebenen illustratie *f* — **i** illustrazione *f* piccante
2679 cheese resistance
f résistance *f* aux fromages — **d** Käseechtheit *f*; Echtheit *f* gegenüber Käse — **n** bestendigheid *f* tegen kaas; kaasbestendigheid *f* — **e** resistencia *f* al queso; solidez *f* al queso — **i** resistenza *f* al formaggio
2680 chemical fog
f voile *m* chimique — **d** chemischer Schleier *m* — **n** chemische sluier *m* — **e** velo *m* químico — **i** velo *m* chimico
2681 chemical formula — Expression of the constituents of a compound by symbols and figures.
f formule *f* de chimie — **d** chemische Formel *f* — **n** chemische formule *fm* — **e** fórmula *f* química — **i** formola *f* chimica
2682 chemical grain — Fine grain produced by the chemical etching of a metal plate.
f grain *m* chimique — **d** chemisches Korn *n* — **n** chemisch grein *n* — **e** grano *m* químico — **i** grana *f* chimica
2683 chemical graining — Producing a grained or rough surface on metal plates by means of chemicals, generally used on direct-image metal offset plates.
f grainage *m* chimique — **d** chemisches Körnen *n* — **n** chemisch greinen *n* — **e** granear químicamente — **i** granitura *f* chimica
2684 chemically pure
f chimiquement pur — **d** chemisch rein — **n** chemisch zuiver — **e** químicamento puro — **i** chimicamente puro
2685 chemical paper — Paper without mechanical wood pulp.

f papier *m* sans bois — **d** holzfreies Papier *n* — **n** houtvrij papier *n*; houtslijpvrij papier *n* — **e** papel *m* sin fibra de madera; papel *m* sin pasta mecánica — **i** carta *f* senza pastalegno

2686 chemical pulp; chemical woodpulp — Pulp obtained from wood or other material of vegetable origin by chemical treatment eliminating the greater part of the non-fibrous components.
f pâte *f* chimique; pâte *f* de bois chimique; pulpe *f* chimique — **d** Zellstoff *m*; Holzzellstoff *m* — **n** celstof *fm*; houtcelstof *fm* — **e** pasta *f* química; pulpa *f* química — **i** pasta *f* chimica; cellulosa *f* tecnica

2687 chemical pulp board — Board merely made of chemical pulp.
f carton *m* de pâte chimique — **d** Zellstoff-karton *m* — **n** celstofkarton *n*; cellulosekarton *n* — **e** cartón *m* de pasta química; cartón *m* de pulpa química — **i** cartone *m* di pasta chimica; cartone *m* di polpa chimica

2688 chemical resistance — Resistance to chemical reagents.
f résistance *f* chimique; résistance *f* aux agents chimiques — **d** chemische Beständigkeit *f*; Beständigkeit *f* gegen Chemikalien — **n** chemische bestendigheid *f*; bestendigheid *f* tegen chemische aantasting; bestendigheid *f* tegen chemicaliën — **e** resistencia *f* química; resistencia *f* frente a los productos químicos; estabilidad *f* frente a los productos químicos — **i** resistenza *f* agli agente chimici

2689 chemical reversal — Converting a negative to a positive (or vice versa) by chemically treating the photographic image.
f inversion *f* chimique — **d** chemisches Umkehr-verfahren *n* — **n** chemische omkering *f* — **e** inversión *f* química — **i** inversione *f* chimica

2690 chemicals
f produits *mpl* chimiques — **d** Chemikalien *npl*; chemische Stoffe *mpl*; chemische Produkte *npl* — **n** chemicaliën *pl*; chemische stoffen *fmpl* — **e** químicos *mpl*; productos *mpl* químicos — **i** prodotti *mpl* chimici

2691 chemical signs — See app. no. 10.
f signes *mpl* de chimie — **d** chemische Zeichen *npl* — **n** chemische tekens *npl* — **e** signos *mpl* químicos; símbolos *mpl* químicos — **i** segni *mpl* di chimica; abbreviazioni *fpl* di chimica

2692 chemical wood paper — Paper from pulp obtained by digestion of wood with solutions of various chemicals. The principal chemical processes are the sulphite, sulphate (kraft), and soda processes.
f papier *m* de pulp chimique; papier *m* de pâte chimique — **d** Zellstoffpapier *n*; Holzzellstoff-papier *n* — **n** papier *n* uit houtcelstof — **e** papel *m* de pasta química — **i** carta *f* di pasta chimica da legno

2693 chemical woodpulp — See chemical pulp.

2694 chemigraphy — Process of producing a printing surface on metals, plastics, etc., by the use of etching solutions or chemicals.
f chimigraphie *f* — **d** Chemigraphie *f* — **n** chemigrafie *f* — **e** quimigrafía *f* — **i** chimigrafia *f*

2695 cheque *GB*; **check** *US* — Written order directing a bank to pay money.
f chèque *m* — **d** Scheck *m* — **n** cheque *m* — **e** cheque *m* — **i** cheque *m*; assegno *m* bancario

2696 cheque-book — Book of bank cheques, usually with marginal stubs for date, amount, and name of payee.
f carnet *m* de chèques — **d** Scheckbuch *n* — **n** chequeboekje *n* — **e** libro *m* de cheques; libreta *f* de cheques; talonario *m* de cheques — **i** libretto *m* di cheques

2697 cheque line; combination rule
f filet *m* azuré — **d** Assureelinie *f*; Azureelinie *f*; Sicherheitslinie *f* — **n** assurélijn *fm*; wissellijn *fm* — **e** filete *m* azurado; filete *m* de seguridad; azurado *m* — **i** filetto *m* azzurrato

2698 cheque paper; safety paper — Writing paper chemically treated to betray any tampering with cheques or other documents.
f papier-chèque *m*; papier *m* à chèque; papier *m* pour chèques; papier *m* de sûreté — **d** Scheck-papier *n*; Sicherheitspapier *n* — **n** cheque-papier *n*; onvervalsbaar papier *n* — **e** papel *m* para cheques; papel *m* (de) seguridad; papel *m* infalsificable — **i** carta *f* per assegni; carta *f* di sicurezza

2699 chessmen — Types for the pieces used in the chess game.
f signes *mpl* pour jeu d'échecs; figurines *fpl* d'échecs — **d** Schachspielzeichen *npl*; Setz-material *n* zum Setzen von Schachproblemen — **n** schaakspeltekens *npl* — **e** tipos *mpl* para juego del ajedrez — **i** caratteri *mpl* dei pezzi degli scacchi; segni *mpl* per giuoco degli scacchi

2700 chest, machine — See stock chest.

2701 chestnut tree — Main source of tannin which is used for tanning leather. After extraction of the tannin, the residual fibres are sometimes used in the manufacture of chip and testboard.
f châtaignier *m* — **d** Kastanienholz *n* — **n** kastanjehout *n* — **e** castaño *m* — **i** castagno *m*

2702 chest, stock — See stock chest.

2703 chest, stuff — See stock chest.

2704 cheveril — See chevrotain.

2705 chevrotain; cheveril — Leather made from hides of the small guinea deer, used in the 15th century protecting covers (chemises) for fine decorated leather bound books.

2706 chewed plate — Imperfect etching in which lines or dots have been attacked and rendered ragged or broken by untoward action of the mordant or failure of the acid resist.
f cliché *m* panné — **d** verätztes Klischee *n* — **n** kapot geëtst cliché *n* — **e** clisé *m* pisado; clisé *m* machacado; clisé *m* magullado — **i** cliché *m* difettoso

2707 chiaroscuro — Proper representation of lights and shadows in a photograph, painting, drawing or other original, without regard to colour.

f clair-obscur *m* — **d** Clair-obscur *n*; Hell-
dunkel *n*; Verteilung *f* von Licht und Schatten —
n clair-obscur *n* — **e** claroscuro *m*; grabado *m* al
claroscuro — **i** chiaroscuro *m*
2708 chief editor — See editor in chief.
2709 children's book
f livre *m* d'enfants — **d** Kinderbuch *n* — **n** kinderboek *n* — **e** libro *m* infantil; libro *m* para niños
— **i** libro *m* per bambini; libro *m* per ragazzi
2710 chill roll — See cooling roller.
2711 China clay; porcelain clay; kaolin — Fine,
white variety of clay, used by papermakers to
obtain finish and consistency of surface, for
coating art and chromo papers and as an ink
extender. Chemical name: aluminium silicate.
Formula: $Al_2O_3 \cdot 2SiO_2 \cdot 2H_2O$.
f kaolin *m*; argile *f*; terre *f* à porcelaine — **d**
Kaolin *n*; Porzellanerde *f*; Porzellanton *m*; Tonerde *f* — **n** kaolien *n*; China klei *fm*; porselein-
aarde *fm* — **e** caolín *m*; arcilla *f*; tierra *f* arcillosa
blanca; arcilla *f* de aluminio — **i** caolino *m*;
argilla *f*
2712 China grass — See ramie.
**2713 China paper; Chinese paper; India proof
paper** — Soft, waterleaf (unsized) paper made in
China from bamboo fibre, with pale yellow colour
and very fine texture. The usual size is 68.5 x
144.75 cm (27 x 57 in). Used by (engraved) plate
printers to pull proof.
f papier *m* de Chine; papier *m* indien — **d** China-
papier *n*; chinesisches Papier *n* — **n** Chinees
papier *n* — **e** papel *m* China — **i** carta *f* Cina;
carta *f* China
2714 China wood oil — See tung oil.
2715 Chinese bean oil — See soya bean oil.
2716 Chinese binary — See column binary.
2717 Chinese blue — See Prussian blue.
2718 Chinese brush — Oriental writing brush for
application of ferric chloride solution in re-
etching copper halftone.
f pinceau *m* chinois — **d** chinesischer Pinsel *m*;
Chinapinsel *m* — **n** Chinees penseel *n*; Chinese
kwast *m* — **e** brocha *f* china; pincel *m* china — **i**
pennello *m* cinese
2719 Chinese gelatine — See agar.
2720 Chinese isinglass — See agar.
2721 Chinese paper — See China paper.
2722 Chinese white — See zinc oxide.
2723 Chinese wood oil — See tung oil.
2724 chinone — See quinone.
2725 chintz paper — Paper decorated or printed
to imitate chintz cloth. Originally chintz is a
printed calico from India.
f papier *m* Perse — **d** Zitskattunpapier *n* — **n**
chintzpapier *n* — **e** papel *m* zaraza; zaraza *f*;
papel *m* persa — **i** carta *f* chintz
2726 chipboard — Board from low-grade mixed
waste paper on a continuous machine, used with
or without a lining on one or both sides of a
furnish layer of different composition.
f carton *m* gris — **d** Graupappe *f* — **n** grijsbord *n*;
gedekt grijsbord *n* — **e** cartón *m* gris; cartón *m*

ordinario — **i** cartone *m* grigio
2727 chip crusher — Machine to break oversize
chips which have been rejected in the chip screen.
f broyeur *m* à copeaux — **d** Hackschnitzel-
Desintegrator *m* — **i** macchina *f* sminuzzatrice
2728 chipper; chopper; chopping machine —
Machine with rotating disk which carries knives
to cut the pulpwood logs into chips.
f déchiqueteuse *f*; coupeuse *f* à copeaux — **d**
Hackmaschine *f* — **n** hakselmachine *f* — **e**
troceadora *f* de madera; máquina *f* astilladora;
astilladora *f* — **i** sminuzzatrice *f*
2729 chips — Small pieces of pulpwood, approx.
6.5 cm² x 1.25 cm thick (1 sq in x 0.5 in),
produced by the chipper for the manufacture of
chemical woodpulp and fed into the digesters.
f copeaux *mpl* (de bois) — **d** Späne *mpl*; Hack-
späne *mpl* — **n** houtspanen *fmpl*; spanen *fmpl*;
spaanders *mpl* — **e** virutas *fpl* — **i** trucioli *mpl*;
chips *mpl*
2730 chip screen — Rotating or vibrating flat
screen to eliminate oversize and undersize chips.
f classeur *m* à copeaux; trieur *m* à copeaux; tamis
m à copeaux — **d** Sortiertrommel *f* für Hack-
späne; Schleudermühle *f* — **n** sorteerzeef *fm* — **e**
tambor *m* clasificador; tambor *m* cernedor — **i**
assortitore *m* di minuzzoli
2731 chisel — In photoengraving, tool used by
finishers to remove burrs and excrescenses from
surfaces of printing plates.
f échoppe *f* — **d** Stichel *m* — **n** steker *m* — **e**
cincel *m*; formón *m* — **i** cesello *m*
2732 chisel *v*
f enlever au ciseau — **d** grappeln; wegsticheln —
n wegbeitelen — **e** quitar con (el) escoplo; escoplear — **i** scalpellare
2733 chlorate of potash — See potassium
chlorate.
2734 chlorazotic acid — See aqua regia.
2735 chlorhydrate — See hydrochloride.
2736 chloric zinc iodide — See Herzberg's stain.
2737 chloride — Salt of hydrochloric acid.
f chlorure *m* — **d** Chlorid *n* — **n** chloride *n* — **e**
cloruro *m* — **i** cloruro *m*
2738 chloride of lime — See calcium hypochlorite.
2739 chloride paper — Slow photographic paper
with a coating of silver chloride emulsion, for
contact prints.
f papier *m* au chlorure (d'argent) — **d** Chlorsilberpapier *n* — **n** chloorzilverpapier *n* — **e** papel *m*
de cloruro (de plata) — **i** carta *f* al cloruro
(d'argento)
2740 chlorinated lime — See calcium
hypochlorite.
2741 chlorinated rubber — Chemical compound of
chlorine and latex; binder for gravure inks.
f caoutchouc *m* chloré — **d** Chlorkautschuk *m*;
chlorierter Kautschuk *m* — **n** chloorrubber *mn* —
e caucho *m* clorado — **i** gomma *f* clorurata;
cauccù *m* clorurato
2742 chlorine — Halogen element isolated as a
heavy greenish-yellow irritating gas of pungent

odour, used especially as a bleach and oxidizing agent. Symbol: Cl.

f chlore *m* — **d** Chlor *n* — **n** chloor *mn* — **e** cloro *m* — **i** cloro *m*

2743 chlorine-free paper

f papier *m* sans chlore — **d** chlorfreies Papier *n* — **n** chloorvrij papier *n* — **e** papel *m* sin cloro — **i** carta *f* esente da cloro

2744 chloro bromide paper — Photographic paper with an emulsion coating of silver chloride and silver bromide.

f papier *m* au chloro-bromuro — **d** Chlorbrom-silberpapier *n* — **n** broomzilverpapier *n* — **e** papel *m* al chlorobromuro — **i** carta *f* al cloro-bromuro

2745 chloroethane — See ethyl chloride.

2746 chloroform; trichloromethane — Clear, colourless, highly refractive, volatile liquid. Miscible with alcohol, ether, benzene, solvent naphtha, slightly soluble in water. Usually a mixture of 99% $CHCl_3$ with 1% absolute alcohol. Contact with acetone is dangerous. Latin: chloroformum. TL-value: 25 ppm or 120 mg/m^3. Formula: $CHCl_3$.

f chloroforme *m*; trichlorométhane *m* — **d** Chloroform *n*; Trichlormethan *n* — **n** chloroform *m*; trichloormethaan *n* — **e** cloroformo *m*; triclorometano *m* — **i** cloroformio *m*; triclorometano *m*

2747 chlorohydroquinone — White to light-tan fine crystals, soluble in water and alcohol, slightly soluble in ether. Used as a photographic developer and in dyestuffs. Formula: $ClC_6H_3(OH)_2$.

f chlorhydroquinone *f* — **d** Chlorhydrochinon *n* — **n** chloorhydrochinon *n* — **e** clorohidroquinona *f* — **i** cloroidrochinone *m*

2748 chloroplatinic acid — See platinum chloride.

2749 chocolate box

f carton *m* à chocolat; boîte *f* à chocolat — **d** Schokoladenschachtel *f* — **n** chocoladedoosje *n* — **e** caja *f* para chocolate; cajita *f* para chocolate — **i** scatola *f* per cioccolato; scatolina *f* per cioccolato; scatoletta *f* per cioccolato

2750 choice of type

f choix *m* du caractère; choix *m* des (styles de) caractères — **d** Schriftwahl *f* — **n** letterkeuze *fm* — **e** selección *f* del tipo — **i** scelta *f* del carattere

2751 choice of type area

f choix *m* de la mise en pages — **d** Wahl *f* des Satzspiegels — **n** pagina-opmaak *m*; zetspiegel-keuze *fm*; keuze *fm* van de bladspiegel — **e** selección *f* de la página caja — **i** scelta *f* della impaginazione

2752 choir book

f livre *m* de cantiques — **d** Gesangbuch *n*; Choralbuch *n* — **n** gezangenboek *n* — **e** libro *m* de coro; libro *m* de cánticos; libro *m* antifonal; antifonario *m* — **i** libro *m* corale

2753 choked type — Ink-filled or dirty type.

f caractère *m* empaté — **d** verschmierte Schrift *f* — **n** vuile letter *fm*; vervuilde letter *fm* — **e** tipo *m* sucio; carácter *m* tapado — **i** carattere *m* sporco; carattere *m* intasato

2754 chopper — Mechanism which accomplishes the chopper fold. Signature is conveyed from the first parallel fold in a horizontal plane, spine forward, until it passes under a reciprocating blade which forces it down between folding rollers to complete the fold.

f lame *f* de pli — **d** Falzmesser *n* — **n** vouwmes *n* — **e** cuchilla *f* plegadera — **i** coltello *m* piegatore

2755 chopper — See chipper.

2756 chopper fold; cross fold; quarter fold — In rotary printing, fold made by following the first parallel fold and at right angles to it. Produces signatures that are 16-page multiples of the number of webs in the press, 1/4 web width x 1/2 cut-off length.

f pli *m* croisé — **d** Querfalz *m*; Zylinderfalz *m* — **n** kruisvouw *fm*; dwarsvouw *fm* — **e** pliego *m* cruzado — **i** piega *f* a croce; piega *f* incrociata; piega *f* trasversale; piega *f* traversa

2757 chopping machine — See chipper.

2758 chord keyboard — Keyboard which either requires more than one key to be struck to obtain a character or enables several characters to be struck at the same time.

2759 Christmas card

f carte *f* postale de Noël; carte *f* (du jour) du Noël — **d** Weihnachtskarte *f* — **n** kerstkaart *fm* — **e** tarjeta *f* de Navidad; tarjeta *f* de felicitación para Navidad — **i** carta *f* di Natale

2760 chroma — Short for *Munsell chroma. Term not admitted by CIE. Attribute of a visual sensation which permits a judgment to be made of the amount of pure chromatic colour, irrespective of the amount of achromatic colour. The subjective variables of this lightness, hue and saturation or chroma are used in certain colour atlases, but then become objectively defined by the samples of the atlas.

f saturation *f* de la couleur — **d** Farbsättigung *f* — **n** verzadiging *f* van de kleur — **e** saturación *f* del color — **i** saturazione *f* di colore

2761 chromatic aberration — Inability of a photographic lens to bring yellow and red rays to the same focus as blue and violet; lack of colour correction.

f aberration *f* chromatique — **d** chromatische Aberration *f*; chromatische Abweichung *f*; chromatischer Farbfehler *m*; chromatische Farbenbrechung *f*; chromatische Farbabweichung *f*; chromatische Farbenzerstreuung *f* — **n** chromatische aberratie *f*; chromatische afwijking *f* — **e** aberración *f* cromática — **i** aberrazione *f* cromatica

2762 chromatic circle; colour circle; colour disk

f cercle *m* chromatique — **d** Farbenkreisel *m* — **n** kleurencirkel *m*; chromatische cirkel *m* — **e** círculo *m* cromático — **i** cerchio *m* cromatico

2763 chromaticity — In colour measurement, colour quality of a light definable by its chromaticity co-ordinates, or by its dominant (or complementary) wavelength and its purity taken together (CIE).

f chromaticité *f* — **d** Farbart *f*; Chromatizität *f* — **n** chromaticiteit *f* — **e** cromaticidad *f* — **i** cromaticità *f*

2764 chromaticity co-ordinates — In colour measurement, ratios of each of the tristimulus values of a light to their sum (CIE). In the standard 1931 CIE colorimetric system the symbols x, y, z are recommended for the chromaticity co-ordinates.

f coordonnées *fpl* trichromatiques; coefficients *mpl* trichromatiques — **d** Normfarbwertanteile *mpl* — **n** kleurcoördinaten *fmpl* — **e** coordenadas *fpl* tricromáticas; coordenadas *fpl* de cromaticidad — **i** coordinate *fpl* tricromatiche

2765 chromaticity diagram — See colour triangle.

2766 chromaticness — In colour measurement, attribute of visual sensation combining the hue and the saturation of a colour. Psychosensorial correlate of chromaticity.

f chromie *f* — **d** empfindungsmäßige Farbart *f* — **n** kleurindruk *m* — **e** cromía *f* — **i** cromia *f*

2767 chromatic sensitizer — See optical sensitizer.

2768 chromatography, paper — See paper chromatography.

2769 chrome — Abbreviated name for *chromium, used to designate alloys, for example ferrochrome.

2770 chrome-albumen process

f procédé *m* à l'albumine bichromatée — **d** Chromalbuminverfahren *n*; Chromeiweißverfahren *n* — **n** chroomeiwitprocédé *n* — **e** procedimiento *m* a la albúmina (cromatada) — **i** procedimento *m* all'albumina bicromatata; processo *m* all'albumina bicromatata

2771 chrome-albumen solution

f solution *f* d'albumine bichromatée — **d** Chromeiweißkopierlösung *f* — **n** chroomeiwitoplossing *f* — **e** solución *f* de albúmina bicromatada — **i** soluzione *f* di albumina bicromatata

2772 chrome alum — See chromium potassium sulphate.

2773 chrome bath — Chromium depositing bath.

f bain *m* de chromage; bain *m* de chrome — **d** Verchromungsbad *n*; Chrombad *n* — **n** verchroombad *n*; chroombad *n* — **e** baño *m* de cromatado — **i** bagno *m* di cromatura

2774 chrome bath — Solution of potassium or ammonium dichromate used for sensitizing pigment paper, process glue and other resist-forming materials.

f bain *m* bichromaté; bain *m* chromaté — **d** Bichromatbad *n*; Chrombad *n* — **n** bichromaatbad *n*; chroombad *n* — **e** baño *m* al bicromato; baño *m* cromatado — **i** bagno *m* di cromatura

2775 chrome coating

f couche *f* de chrome — **d** Chromschicht *f* — **n** chroomlaag *fm* — **e** capa *f* de cromo — **i** rivestimento *m* di cromo

2776 chrome facing — See chromium depositing.

2777 chrome green — Green pigment made by mixing freshly precipitated iron blue and chrome

yellow. A fairly permanent opaque pigment.

f vert *m* de chrome — **d** Chromgrün *n* — **n** chromaatgroen *n* — **e** verde *m* de cromo — **i** verde *m* di cromo

2778 chrome plating — See chromium depositing.

2779 chrome plating installation

f installation *f* de chromage — **d** Verchromungsanlage *f* — **n** verchroominstallatie *f* — **e** instalación *f* de cromado — **i** impianto *m* di cromatura

2780 chrome poisoning; chromic poisoning; dichromate poisoning

f empoisonnement *m* par le chrome — **d** Chromvergiftung *f* — **n** chroomvergiftiging *f* — **e** envenenamiento *m* de dicromato — **i** avvelenamento *m* da cromo

2781 chrome potash alum — See chromium potassium sulphate.

2782 chrome salt; chromic salt; chromium salt — Salt of chromium, usually stable, blue or violet in colour.

f sel *m* de chrome — **d** Chromsalz *n* — **n** chroomzout *n* — **e** sal *f* crómica — **i** sale *m* di cromo

2783 chrome yellow — Permanent yellow ink pigment composed essentially of *lead chromate.

f jaune *m* de chrome — **d** Chromgelb *n* — **n** chromaatgeel *n* — **e** amarillo *m* de cromo — **i** giallo *m* di cromo

2784 chrome yellow — See lead chromate.

2785 chromic acid; chromic anhydride; chromium trioxide — Dark, purplish-red crystals, soluble in water, alcohol, and mineral acids. Can explode in contact with organic substances. Contact with acetic acid, glycerol, turpentine, alcohol and inflammable liquids is dangerous. Formula: CrO_3. The name chromic acid is in common use, although the true chromic acid, H_2CrO_4, exists only in solution. TL-value of chromates: 0.1 mg/m^3.

f acide *m* chromique; anhydride *m* chromique; trioxyde *m* chromique — **d** Chromsäure *f*; Chromsäureanhydrid *n*; Chromtrioxyd *n* — **n** chroomzuur *n* — **e** ácido *m* crómico — **i** acido *m* cromico

2786 chromic anhydride — See chromic acid.

2787 chromic chloride; chromium chloride; chromium sesquichloride — Bluish-violet, crystalline powder. Anhydrous: insoluble in water, but a trace of chromous chloride will cause it to go into solution. Hydrated: soluble in alcohol. Formula: $CrCl_3$ or $CrCl_3 \cdot 6H_2O$.

f chlorure *m* chromique; perchlorure *m* de chrome; sesquichlorure *m* de chrome — **d** Chromchlorid *n*; Chromichlorid *n* — **n** chromichloride *n* — **e** cloruro *m* crómico — **i** cloruro *m* cromico

2788 chromic dermatitis — See chromium dermatitis.

2789 chromic fluoride; chromium fluoride; chromium trifluoride — Green crystalline powder: the hydrates are soluble in water and acids, insoluble in alcohol. Used for dyeing and printing woolens. Formula: $CrF_3 \cdot 4H_2O$.

f fluorure *m* de chrome; fluorure *m* chromique — **d** Chromfluorid *n* — **n** chroomfluoride *n*; chromifluoride n — **e** fluoruro m de cromo; fluoruro m crómico — **i** fluoruro m di cromo; fluoruro m cromico

2790 chromic hydrate — See chromic hydroxide.

2791 chromic hydroxide; chromium hydroxide; chromic hydrate; chromium hydrate — Green, gelatinous precipitate, decomposed to chromic oxide by heat. Insoluble in water, soluble in acids and strong alkalis.
Formula: $Cr(OH)_3$.
f hydrate m de chrome; oxyde m de chrome hydraté — **d** Chromhydroxyd n — **n** chroomhydroxyde n — **e** hidróxido m de cromo; hidróxido m crómico — **i** idrossido m di cromo; idrossido m cromico; idrato m di cromo

2792 chromic oxide; chromium oxide; chromium sesquioxide — Bright green, crystalline powder, insoluble in water, acids and alkalis.
Formula: Cr_2O_3.
f sesquioxyde m de chrome — **d** Chromsesquioxyd n; Chromoxyd n — **n** chromioxyde n — **e** óxido m de cromo; óxido m crómico — **i** ossido m di cromo; ossido m cromico

2793 chromic poisoning — See chrome poisoning.

2794 chromic salt — See chrome salt.

2795 chroming plant
f installation f de chromage — **d** Verchromungsanlage f — **n** verchromingsinstallatie f — **e** instalación f de cromado; instalación f para cromado — **i** impianto m di cromatura

2796 chromium — Hard, brittle, steel-grey metal. Does not tarnish in air. Resists very strong oxidizing agents due to passivity. Soluble in strong alkalis, in acids except nitric, insoluble in water. Symbol: Cr. Latin: chromium. See also chrome.
f chrome m — **d** Chrom n — **n** chroom n — **e** cromo m — **i** cromo m

2797 chromium chloride — See chromic chloride.

2798 chromium depositing; chromium plating; chrome plating; chrome facing — Electrodepositing of a very thin layer of hard chromium on the surface of printing plates to increase wear resistance.
f chromage m — **d** Verchromung f; Verchromen n — **n** verchroming f; verchromen n — **e** cromado m — **i** cromatura f

2799 chromium dermatitis; chromic dermatitis — Occupation disease in photoengraving, causing sores, ulcers and various afflictions by absorption of bichromates or chromium compounds through the skin, or by wounds on hands or body.
f dermatose f du chrome — **d** Chromekzem n; Chromhautentzündung f; Hautvergiftung f durch Chrom — **n** chroomeczeem n — **e** dermatosis f del cromo; eczema f de cromo — **i** dermatite f da cromo; eczema f da cromo

2800 chromium-faced plate — Printing plate coated with a deposit of chromium to improve its wearing qualities, used for long runs.
f cliché m chromé — **d** Chromplatte f; Chromklischee n — **n** verchroomd cliché n; verchroomde

plaat fm — **e** grabado m cromado — **i** lastra f cromata

2801 chromium fluoride — See chromic fluoride.

2802 chromium hydrate — See chromic hydroxide.

2803 chromium hydroxide — See chromic hydroxide.

2804 chromium iodide
f iodure m de chrome; iodure m chromique — **d** Chromjodid n; Chromjodür n — **n** chroomjodide n — **e** yoduro m de cromo; yoduro m crómico — **i** ioduro m di cromo; ioduro m cromico

2805 chromium oxalate
f oxalate m de chrome; oxalate m chromique — **d** Chromoxalat m — **n** chromioxalaat n — **e** oxalato m de cromo; oxalato m crómico — **i** ossalato m di cromo; ossalato m cromico

2806 chromium oxide — See chromic oxide.

2807 chromium plating — See chromium depositing.

2808 chromium potassium sulphate; potassium chromium sulphate — Chemical names for chrome alum or chrome potash alum. Dark, violet-red crystals, efflorescent, soluble in water.
Formula: $CrK(SO_4)_2 \cdot 12H_2O$.
f sulfate m de chrome et de potassium; alun m de chrome — **d** Kaliumchromsulfat n; Chromkaliumsulfat n; Kalichromalaun m; Chromalaun m — **n** kalium-chroomaluin m; kalium-chroomsulfaat n; chroomaluin m — **e** sulfato m crómico-potásico; alumbre m de cromo — **i** solfato m di cromo e potassio; solfato m doppio di cromo e potassio; allume m di cromo

2809 chromium salt — See chrome salt.

2810 chromium sesquichloride — See chromic chloride.

2811 chromium sesquioxide — See chromic oxide.

2812 chromium trifluoride — See chromic fluoride.

2813 chromium trioxide — See chromic acid.

2814 chromo blotting board
f carton m buvard chromo — **d** Chromolöschkarton m — **n** chromovloeikarton n — **e** cartulina f secante al cromo — **i** cartone m cromo assorbente

2815 chromogen developing
f développement m chromogène — **d** Chromogenentwicklung f — **n** chromogeenontwikkeling f — **e** revelado m cromógeno — **i** sviluppo m cromogeno

2816 chromolitho — See chromolithography.

2817 chromolithographer
f chromolithographe m — **d** Chromolithograph m; Farblithograph m — **n** chromolithograaf m — **e** cromolitógrafo m — **i** cromolitografo m

2818 chromolithography; chromolitho; colour lithography — Process of reproducing a coloured original in lithography by means of a series of hand drawn images. By extension picture printed in colours from stone.
f chromolithographie f — **d** Chromolithographie f;

Farblithographie *f*; lithographischer Farbendruck *m*; farbiger Steindruck *m* — **n** chromolithografie *f* — **e** cromolitografía *f* — **i** cromolitografia *f*; litografia *f* a colori

2819 chromometer — See colorimeter.

2820 chromo paper — Coated paper or board particularly suited for coloured printing with surface characteristics to take many colours like smoothness, uniformity of ink receptivity, high total reflectance, and neutrality of shade (true white rather than tinted). Term is not used in US.
f papier *m* chromo; papier *m* couché une face — **d** Chromopapier *n* — **n** chromopapier *n* — **e** papel *m* cromo — **i** carta *f* cromo

2821 chromo xylography; colour wood engraving
f chromoxylographie *f*; procédé *m* bois en couleurs — **d** Farbenholzschnitt *m*; Chromoxylographie *f* — **n** kleurenhoutsnede *fm*; chromoxylografie *f* — **e** cromoxilografía *f*; xilografía *f* a color — **i** cromosilografia *f*; cromoxilografia *f*

2822 chronological order — Order in which data are arranged in order of time.
f ordre *m* de dates; ordre *m* chronologique; classement *m* chronologique — **d** chronologische Ordnung *f*; zeitliche Reihenfolge *f* — **n** chronologische volgorde *fm* — **e** orden *m* cronológico — **i** ordine *m* cronologico

2823 chronological study; production study; overall study — See also time study.
f étude *f* diagnostique — **d** Fertigungsablaufstudie *f*; pauschale Arbeitsablaufstudie *f* — **n** produktietijdstudie *f* — **e** estudio *m* diagnóstico — **i** studio *m* dei tempi di produzione

2824 chuck — See reel cone.

2825 chute — Guiding channel, usually steel, to slide newspapers or magazines to the loading platform.
f gouttière *f*; canal *m* — **d** Rutsche *f* — **n** goot *fm*; glijbaan *fm* — **i** scivolo *m*

2826 chute finger — The assembler chute finger in a line composing machine.
f lamelle *f* guidante; lamelle *f* de guidage — **d** Matrizenführung *f*; Führungsblech *n* — **n** lange vinger *m* — **e** dedo *m* guiador; lengüeta *f* guiadora — **i** dito *m* di guida

2827 cicero — Typographic measure for 12 points Didot (4.51368 mm). Name derived from the types used in 1467 at Rome by Sweynheim and Pannartz for printing Cicero's letters. See also pica, point, point system, app. no. 1.
f douze *m*; corps *m* de douze; cicéro *m* (selon Fournier); Saint-Augustin *m* (selon Didot) — **d** Cicero *f* — **n** cicero *fm*; augustijn *m* — **e** cícero *m* — **i** cicero *m*

2828 cigar-band paper; cigar-label paper — One-side coated paper of good strength and finish to take gold and coloured printing, embossing, and die-cutting. The basis weight is 45-60 pounds (25 x 38 - 500).
f papier *m* pour bagues de cigare — **d** Papier *n* für Zigarrenetiketten — **n** papier *n* voor sigarebandjes — **e** papel *m* para sortijas de puro; papel

m para vitolas — **i** carta *f* per le fascette dei sigari

2829 cigaret box — See cigarette box.

2830 cigarette box; cigaret box
f boîte *f* à cigarettes — **d** Zigarettenschachtel *f*; Zigarettenumhüllung *f*; Zigarettenkästchen *n* — **n** sigarettendoosje *n*; sigarettendoos *fm* — **e** cajetilla *f* para cigarillos — **i** pacchetto *m* di sigarette; scatola *f* di sigarette

2831 cigarette paper — Strong tissue paper, unsized or very lightly sized, usually made from rags or flax straw pulp with calcium carbonate filler to make it burn evenly with little or no ash.
f papier *m* à cigarettes — **d** Zigarettenpapier *n* — **n** sigarettenpapier *n* — **e** papel *m* de cigarillos; papel *m* cigarillo; papel *m* de fumar — **i** carta *f* da sigarette

2832 cigar-label paper — See cigar-band paper.

2833 cinematic viscosity — See kinematic viscosity.

2834 cinnabar; natural vermilion — Natural mercuric sulphide with a red, scarlet, reddish-brown to black colour. Soluble in aqua regia. See also vermilion.
f cinabre *m* — **d** Zinnober *m* — **n** cinnaber *n* — **e** cinabrio *m* — **i** cinabro *m*

2835 circle — The circumference of a circle is $2\pi r$, whilst its surface is πr^2, in which $\pi = 3.14159$ or 22/7.
f cercle *m* — **d** Kreis *m* — **n** cirkel *m* — **e** círculo *m* — **i** cerchio *m*

2836 circle of confusion; aperture effect — Small circle which is the image of a point object due to imperfect imagery and to the fact that images of points at different object distances are commonly observed on a single image-plane.
f cercle *m* de confusion — **d** Zerstreuungskreis *m*; Unschärfering *m* — **n** verstrooiingscirkel *m*; diffusiecirkel *m*; uitvloeiingscirkel *m* — **e** círculo *m* de difusión — **i** circolo *m* di confusione

2837 circuit — Combination of electrical components.
f circuit *m* — **d** Stromkreis *m*; Umkreis *m*; Kreis *m*; Schaltung *f*; Leitungsnetz *n*; Wirkungskreis *m* — **n** stroomkring *m*; circuit *n*; stroomketen *fm* — **e** circuito *m* — **i** circuito *m*

2838 circuit — Computer term for a two way communications link, in contrast to a channel.

2839 circuit edges — See divinity edges.

2840 circuit switching; line switching — Method of handling traffic through a switching centre, either from local users, or from other switching centres, whereby a direct electrical connection is established between the calling and called stations.
f commutation *f* de circuit — **d** Leitungswählen *n* — **n** kringkeuze *fm* — **e** conmutación *f* de circuito — **i** commutazione *f* di circuito

2841 circular — See circular letter.

2842 circular grinder — See levigator.

2843 circular knives — See slitters.

2844 circular letter; circular — Letter,

memorandum, etc., printed in large quantities and addressed to a large number of persons or intended for general circulation.

f circulaire *f*; lettre *f* circulaire — **d** Zirkular *n*; Rundschreiben *n* — **n** circulaire *fm*; rondschrijven *n* — **e** circular *f*; carta *f* circular — **i** circolare *f*; lettera *f* circolare

2845 circular motion of arc lamps
f mouvement *m* circulaire des lampes à arc — **d** Kreisbewegung *f* der Bogenlampen — **n** rondgaande beweging *f* van de booglampen — **e** movimiento *m* circular de las lámparas de arco; movimiento *m* giratorio de las lámparas de arco — **i** movimento *m* circolare delle lampade ad arco

2846 circular saw — Disk-shaped saw, which rotates round its axle. See also band saw, jig saw, saw trimmer.
f scie *f* circulaire — **d** Kreissäge *f* — **n** cirkelzaag *fm* — **e** sierra *f* circular — **i** sega *f* circolare

2847 circular screen — Circular-shaped halftone screen that rotates in the vertical plane and which enables the camera operator to obtain the proper screen angles for colour halftones without disturbing the copy.
f trame *f* ronde; trame *f* circulaire — **d** Drehraster *m*; Rundraster *m*; Kreisraster *m* — **n** rond raster *n*; draairaster *n* — **e** trama *f* circular — **i** retino *m* circolare

2848 circular shears
f cisaille *f* circulaire — **d** Kreisschere *f* — **n** rondschaar *fm* — **e** cizalla *f* circular — **i** cesoia *f* circolare

2849 circulating ink system — Pumps, lines, applicator, return and tank through which ink flows continuously.
f système *m* de circulation d'encre — **d** Farbumwälzsystem *n* — **n** inktcirculatiesysteem *n* — **e** sistema *m* de circulación para el tintaje — **i** sistema *m* di circolazione dell'inchiostro

2850 circulating library — See lending library.

2851 circulating pump
f pompe *f* de circulation — **d** Umlaufpumpe *f*; Umwälzpumpe *f* — **n** circulatiepomp *fm* — **e** bomba *f* de circulación — **i** pompa *f* di circolazione

2852 circulation — The number of printed copies of a newspaper or a magazine.
f tirage *m* — **d** Auflage *f*; Auflagehöhe *f* — **n** oplage *f* — **e** tirada *f*; tiraje *m*; circulación *f*; cifra *f* de tirada — **i** tiratura *f*

2853 circulation statement — See audit report.

2854 circumference
f circonférence *f* — **d** Kreislinie *f*; Kreisumfang *m*; Umfang *m* — **n** omtrek *m*; omvang *m* — **e** circunferencia *f*; periferia *f* — **i** circonferenza *f*

2855 circumference of impression cylinder
f circonférence *f* du cylindre d'impression — **d** Druckzylinderumfang *m*; Zylinderumfang *m*; Umfang *m* des (Gegen)druckzylinders — **n** omtrek *m* van de (tegen)drukcilinder — **e** circunferencia *f* del cilindro impresor — **i** circonferenza *f* del

cilindro di pressione

2856 circumferential register — Register in the direction of the web on reel fed printing machines. Register along the circumference of the printing forme.
f repérage *m* circonférentiel; repérage *m* longitudinal — **d** Umfangsregister *n*; Passer *m* in Druckrichtung — **n** omvangsregister *n* — **e** registro *m* circunferencial — **i** registro *m* circonferenziale

2857 circumferential register adjustment — Adjustment of one operation in the rotary press in relation to another in the direction of the run of paper.
f réglage *m* du repérage circonférentiel; mise *f* en repérage circonférentiel — **d** Umfangsregister-Einstellung *f*; Verstellung *f* der Zylinder im Umfang — **n** regeling *f* van het omvangsregister — **e** puesta *f* a punto del registro circunferencial — **i** regolazione *f* del registro circonferenziale

2858 circumferential register screws — Adjustable screws in place of margin strips allowing slight movement of the printing plate around the cylinder of a rotary press.
f vis *fpl* de réglage du repérage circonférentiel — **d** Schrauben *fpl* für Umfangsregister-Einstellung — **n** stelschroeven *fmpl* voor het omvangsregister — **e** tornillos *mpl* del registro circunferencial — **i** viti *fpl* di regolazione del registro circonferenziale

2859 circumferential speed
f vitesse *f* périphérique — **d** Umfangsgeschwindigkeit *f*; Abwicklungsgeschwindigkeit *f* — **n** omtreksnelheid *f* — **e** velocidad *f* circunferencial — **i** velocità *f* periferica

2860 circumflex accent — The mark over a vowel, as ê, used specially in French.
f accent *m* circonflexe; flexe *m* — **d** Zirkumflexakzent *m*; Zirkumflex *mn* — **n** accent *m* circumflex; circumflexe *m*; kapje *n*; kap *fm coll.* — **e** acento *m* circunflejo; acento *m* circunflexo; circunflejo *m*; capucha *f* — **i** accento *m* circonflesso; circonflesso *m*

2861 cite *v*; **quote** *v*
f citer; alléguer — **d** zitieren — **n** citeren; aanhalen — **e** citar — **i** citare

2862 citol — See rodinal.

2863 citrate of iron — See ferric citrate.

2864 citric acid; oxytricarballylic acid — Colourless, translucent crystals or white powder with a strongly acid taste. Hydrated form is efflorescent in dry air. Soluble in water, alcohol and ether. Latin: acidum citricum.
Formula: $C_3H_4(OH)(CO_2H)_3 \cdot H_2O$.
f acide *m* citrique — **d** Zitronensäure *f*; Oxytricarballylsäure *f* — **n** citroenzuur *n* — **e** ácido *m* cítrico — **i** acido *m* citrico

2865 citron yellow — See zinc yellow.

2866 city editor *GB* — Newspaper or magazine editor in charge of financial and commercial news.
f rédacteur *m* de la rubrique financière — **d** Finanzschriftleiter *m*; Finanzredakteur *m*;

Wirtschaftsschriftleiter   *m*;   Wirtschaftsredakteur *m*; Handelsschriftleiter *m* — **n**
financieel redacteur *m*; redacteur *m* financiën —
e jefe *m* de información financiera; redactor *m* de
información financiera — **i** redattore *m*
finanziario
2867 city editor *US* — Newspaper editor in
charge of local news and assignments to
reporters.
f titulaire *m* de la rubrique locale — **d** Lokalredakteur *m* — **n** redacteur *m* stadsnieuws — **e**
jefe *m* de información local; redactor *m* de
información local — **i** redattore *m* della cronaca
locale
2868 city room — Room in which local news for a
newspaper is handled.
f salle *f* de rédaction — **d** Lokalredaktion *f* — **n**
stadsredactiezaal *fm*; stadsredactie *f* — **e** sala *f*
de redacción de información local — **i** sala *f* di
redazione
2869 clamp; plate clamp — Clamp to attach an engraving to a metal base.
f griffe *f* — **d** Facettenhalter *m*; Plattenklemme *f*
— **n** plaatklem *fm*; clichéklem *fm* — **e** uña *f*; gafa
f (de uña); grapa *f* — **i** morsetto *m*
2870 clamp — See clamping bar.
2871 clamp bar — Bar to hold the blanket of the
letterpress rotary in position on the pins of an impression cylinder.
f tringle *f* à blanchet — **d** Tuchstange *f* — **n** doekstang *fm* — **e** barra *f* de la tela; grapa *f* de la tela
— **i** barra *f* di bloccaggio (del rivestimento)
2872 clamp bar — Bar to lock or unlock the
mechanism that holds the plates on the rotary
press.
f barre *f* de calage — **d** Klemmstange *f* — **n** plaat-
klemmlijst *fm*; klemlijst *fm*; plaatopsluitlijst *fm* —
e barra *f* de uña; barra *f* de grapa — **i** barra *f* di
bloccaggio (del mecanismo di fissaggio)
2873 clamping bar; clamp — Heavy bar pressing
down the paper in a guillotine cutter.
f pressoir *m* — **d** Preßbalken *m*; Anpreßbalken *m*;
Preßleiste *f* — **n** persbalk *m*; dracht *fm* — **e**
pisón *m*; prensador *m* — **i** barra *f* premicarta;
pressino *m*
2874 clamping screw
f vis *f* de bloquage; vis *f* de serrage — **d** Klemmschraube *f* — **n** klemschroef *fm* — **e** tornillo *m*
de aprieto; tornillo *m* de bloqueo — **i** vite *f* di
pressione
2875 Clarendon — Condensed form of printing
type (designed about 1845) like roman in outline
but with thicker serifs. Name derived from the
Clarendon Press (founded 1713), Oxford, England.
See app. no. 2.
2876 clasp — Metal fastener or lock on books to
hold covers closed.
f fermoir *m*; fermail *m* — **d** Buchschloß *n*; Buch-
schließe *f*; Schließe *f* — **n** boekslot *n*; slot *n* — **e**
broche *m*; corchete *m*; cierre *m*; manecilla *f* — **i**
fermaglio *m*; serratura *f*
2877 clasp — Fastener to hold loose sheets to-

gether.
f agrafe *f*; attache-lettres *m* — **d** Klammer *f* — **n**
knijper *m* — **i** morsetto *m*
2878 clasp envelope — Envelope of which the flap
is closed with a metal clasp.
f enveloppe *f* à agrafe; pochette *f* — **d** Umschlag
m mit Klammerverschluß; Umschlag *m* mit
Patentverschluß — **n** envelop(pe) *fm* met pen-
sluiting — **e** bolsa *f* con grapa — **i** busta *f* con
lembo chiudibile con fermaglio
2879 classification of type design — Since 1954 attempts have been made to reconstruct the classification of type designs, of which the system of
Maximilian Vox of Lurs (France) has been
adopted by the A.Typ.I in 1962. This system is
divided in nine groups: 1. Humanistics; 2.
Garaldics; 3. Transitionals; 4. Didonics; 5.
Mechanistics; 6. Lineals; 7. Inciseds; 8. Scripts; 9.
Manuals. For Western Germany A.Typ.I adopted
in the same year the classification of Hermann
Zapf of Frankfurt am Main. This system is
divided in eleven groups. See especially note to
app. no. 2. In the American printing industry the
system set forth in the PIA composition manual is
widely accepted. This manual establishes eight
classes of type faces: 1. old style types; 2. modern-
face types; 3. transitional faces; 4. square-serif
types; 5. sans-serif types and "Gothics"; 6.
cursives and scripts; 7. text letters; 8. decorative
types.
f classification *f* des caractères — **d** Klassification
f der Druckschriften — **n** classificatie *f* van letter-
soorten — **e** clasificación *f* de tipos — **i**
classificazione *f* dei caratteri
2880 classified   ads   —   See   classified
advertisements.
2881 classified advertisements; classified ads —
Advertisements in a particular section of a newspaper or magazine, typically set in agate type in
one column, without illustrations, generally
dealing with jobs, houses, help wanted, used cars,
etc.
f annonces *fpl* groupées; annonces *fpl* anglaises
— **d** Kleinanzeigen *fpl*; kleine Anzeigen *fpl*;
rubrizierte Anzeigen *fpl* — **n** kleine advertenties
fpl — **e** anuncios *mpl* por palabras; anuncios *mpl*
económicos; pequeños avisos *mpl Ar* — **i** annunzi
mpl pubblicitari classificati
2882 claw lock carton — See also carton blank.
f boîte *f* pliante à pattes droites opposées — **d**
Faltschachtel *f* mit Greiferverschluß — **n** vouwdoos *fm* met lipsluiting — **e** caja *f* plegable con
solapas opuestas — **i** astuccio *m* pieghevole con
chiusura a lembi aggraffabili
2883 Claybourn process — Method for producing
letterpress plates, originating from US, in which
copper or nickel electros of the original type
forme are used. The formes are pressed hydraulically in such a way that unevennesses appear on
the back which later can easily be smoothed out.
This system reduces make-ready time.
f procédé *m* Claybourn — **d** Claybourn-

Verfahren *n* — **n** Claybourn-procédé *n* — **e** procedimiento *m* Claybourn — **i** procedimento *m* Claybourn

2884 clay, China — See China clay.

2885 Clayden effect — If a photographic emulsion is given a very brief exposure to light of high intensity, it is desensitized towards a subsequent longer exposure to light of moderate intensity. That is, the second exposure produces less effect that if the pre-exposure had not been given. Observed originally by A.W. Clayden in 1899 when photographing lightning flashes. It can be produced equally well by any type of light source, provided the intensity is sufficiently high and the duration short enough, e.g. 1/1000 sec or less, or by x-rays. The pre-exposure may be of longer duration if it is given at the temperature of liquid air, because all exposures made at this temperature behave as though they had been made at very high intensity. The density produced by the pre-exposure is not reduced by the subsequent low-intensity exposure. Thus, the normal Clayden effect is not a reversal but a desensitization caused by the first exposure.

f effet *m* Clayden — **d** Clayden-Effekt *m* — **n** Clayden-effect *n* — **e** efecto *m* de Clayden — **i** effetto *m* (di) Clayden

2886 clean *v* — To remove grease from dampers.

f dégraisser les mouilleurs — **d** die Feuchtwalzen reinigen — **n** de vochtrollen wassen; de vochtrollen schoonmaken — **e** desengrasar los mojadores — **i** sgrassare rulli bagnatori

2887 clean *v* — To remove the ink from the offset blanket.

f laver le blanchet — **d** das Gummituch waschen — **n** het rubberdoek wassen — **e** lavar la mantilla; limpiar la mantilla — **i** lavare il blanket

2888 cleaner's naphtha — See cleaning spirit.

2889 cleaning — Operation to eliminate foreign matter in pulp, paper or board by physical means from the stock or raw material in the form of a suspension in water, e.g., 1. cleaning by gravity; 2. centrifugal cleaning; 3. cleaning by passing through orifices of definite size.

f épuration *f* — **d** Reinigung *f*; Säuberung *f* — **n** reiniging *f* — **e** depuración *f* — **i** epurazione *f*

2890 cleaning brush — See type brush.

2891 cleaning cloth; cleaning rag

f chiffon *m* de nettoyage — **d** Putzlappen *m*; Putztuch *n* — **n** poetslap *m*; poetsdoek *mn* — **e** trapo *m* de limpieza — **i** strofinaccio *m*

2892 cleaning rag — See cleaning cloth.

2893 cleaning solvent — See cleaning spirit.

2894 cleaning spirit — An *SBP spirit specifically used for cleaning, including dry cleaning. Boiling ranges may vary locally, but generally lie between 80 and 160 °C. For technical purposes heavier fractions may be used. Sometimes called (dry) cleaning solvent; cleaner's naphtha.

f essence *f* de lavage — **d** Waschbenzin *n* — **n** wasbenzine *fm* — **i** benzina *f* per pulizie

2895 clean proof — Proof with few or no errors,

pulled after matter has been corrected.

f épreuve *f* peu chargé; épreuve *f* d'auteur; bonne épreuve *f* — **d** fehlerfreier Abzug *m*; sauberer Abzug *m*; korrigierter Abzug *m*; Revisionsbogen *m* — **n** schone proef *fm*; revisie *f* — **e** prueba *f* limpia; prueba *f* corregida; prueba *f* con pocos errores; prueba *f* en limpio; prueba *f* nítida — **i** bozza *f* corretta; prova *f* pulita; bozza *f* senza errori

2896 clean sheet; specimen sheet — Sheets set aside during printing to show progress of work.

f bonne feuille *f* — **d** Aushängebogen *m*; Ansichtsbogen *m*; Reindruckbogen *m* — **n** vel *n* van de pers — **e** capilla *f* de prensa — **i** prova *f* di stampa

2897 clean tape; corrected tape; pure tape — Punched tape on which all corrections and alterations are taken up and ready to be sent to the typesetting machine.

f bande *f* corrigée; bande *f* code finale — **d** korrigierter Lochstreifen *m*; korrigierter Streifen *m* — **n** schone band *m*; schone ponsband *m*; gecorrigeerde band *m* — **e** banda *f* corregida; cinta *f* correcta — **i** nastro *m* (perforato) corretto

2898 clean *v* **up** — To make an offset plate ink-repellent again when the non-printing areas have begun to become ink-receptive.

f dégraisser — **d** ätzen — **e** desengrasar — **i** sgrassare

2899 clear *v* — To delete all data in a computer store or location by bringing all its cells to a prescribed state.

f effacer — **d** löschen — **n** schoonmaken — **e** borrar — **i** cancellare

2900 clear *v* — Composing room term for distributing type, etc.

f distribuer — **d** ablegen; aufräumen — **n** opruimen; distribueren; wegmaken — **e** distribuir; desempastelar; arreglar — **i** scomporre; distribuire

2901 clear *v* — To apply very light flat bleaching etch to negatives or positives to remove a fog or scum (slight silver deposit).

f dévoiler — **d** aufhellen; auflichten; entschleiern; klären — **n** lichter maken; ontsluieren — **e** aclarar; desvelar — **i** schiarire; rischiarire

2902 clear *v* — To remove the light-hardened stencil from the non-image areas of a deep-etch press plate.

f dépouiller — **d** entschichten — **n** kopieerlaag verwijderen; ontschichten *depr.* — **e** decapar — **i** rimuovere del strato di riserva

2903 clear *v* — See reset *v*.

2904 clearance; play; free motion — Distance between two parts of machinery, one of them moving, the other stationary.

f jeu *m*; espace *m*; dégagement *m* — **d** Spiel *n*; Spielraum *m* — **n** speling *f*; ruimte *f*; speelruimte *f* — **e** juego *m*; espacio *m* muerto — **i** gioco *m*

2905 clear area — See clear band.

2906 clear band; clear area — In magnetic ink

printing, a 16 mm high unprinted area over the whole length and on both sides of the document.
f aire *f* d'exploration — **d** nicht-beschriftete Oberfläche *f*; nicht-beschriftete Fläche *f* — **n** onbedrukte ruimte *f*; onbedrukte strook *fm*; onbeschreven oppervlakte *f* — **e** área *f* de exploración; zona *f* no impresa — **i** area *f* d'esplorazione; area *f* non stampata

2907 clearer — See distributor.

2908 clearing forme — See distribution forme.

2909 cleat; finger — Strip of wood or metal to hold or support something. Sometimes used for a *gripper.
f taquet *m* (d'arrêt); taquet *m* de prise — **d** Anlage *f*; Marke *f*; Leiste *f* — **n** aanleg *m* — **e** guía *m*; marginera *f* — **i** squadra *f*

2910 clenching die — Part of a stitching mechanism which forms the ends of a wire stitch which can either be inward or outward as the wire is forced through the paper. The anvil of a stitcher.
d Klammerschließer *m* — **n** ombuiger *m* — **i** stampo *m* di chiusura

2911 clerical error; slip of the pen — Error made by clerk, copyist or typist which was not in original manuscript. Latin: lapsus calami.
f erreur *f* de plume; faute *f* de copiste; paragramme *m* — **d** Abschreibfehler *m*; Schreibfehler *m* — **n** schrijffout *fm*; typefout *fm* — **e** error *m* de pluma; errata *f* de copia; errata *f* de redacción — **i** errore *m* di scrittura

2912 clerical work; office work
f travail *m* de bureau — **d** Büroarbeit *f*; Bürotätigkeit *f* — **n** administratieve arbeid *m*; bureauwerk *n*; kantoorwerk *n* — **e** trabajo *m* de oficina — **i** lavoro *m* amministrativo

2913 Cleveland open cup — Standard size brass cup to determine the flash and fire points of inflammable liquids.
d Cleveland-Flammpunktbestimmungsbecher *m*; Cleveland-Flammpunktprüfer *m* — **n** Cleveland open cup *m*; Cleveland-vlampuntsbepaler *m* — **e** copa *f* Cleveland; comparador *m* Cleveland del punto de inflamación — **i** tazza *f* Cleveland

2914 clicker — Foreman compositor, or foreman maker-up, who receives copy and instructions direct from the overseer or principal, and is responsible to his companions for charging the work. English custom.
f prote *m*; chef-typo *m* — **d** Obersetzer *m*; Metteur *m* — **n** voormanzetter *m* — **e** primer cajista *m*; compaginador *m* — **i** primo compositore *m*; proto *m*

2915 client; customer
f client *m*; commettant *m* — **d** Kunde *m*; Auftraggeber *m* — **n** klant *m*; cliënt *m*; opdrachtgever *m* — **e** cliente *m* — **i** cliente *m*

2916 clipping — See news cutting.

2917 clipping agency; clipping bureau; clipping service
f bureau *m* de coupures de journeaux; agence *f* de coupures de presse — **d** Zeitungsausschnitt-

büro *n*; Ausschnittbüro *n* — **n** knipseldienst *m*; knipselbureau *n*; knipselagentschap *n* — **e** agencia *f* de recortes (de diarios); servicio *m* de recortes (de diarios) — **i** ufficio *m* di ritagli (di giornali)

2918 clipping bureau — See clipping agency.

2919 clipping service — See clipping agency.

2920 clips — Self-adjusting clamp-like devices to hold one edge of the sheet when hanging.
f crochets *mpl* de suspension (des feuilles) — **d** Papieraufhängeklammern *fpl*; Klammern *fpl*; Aufhänger *mpl* — **n** knijpers *mpl*; papierknijpers *mpl* — **e** colgadores *mpl*; tendedores *mpl* — **i** morsetti *mpl*

2921 clock card; time card — Card on which the working time of an employee is entered, used to compute his wages.
f carte *f* de pointage; carton *m* de pointage — **d** Stechkarte *f* — **n** klokkaart *fm*; prikkaart *fm*; tijdkaart *fm* — **e** tarjeta *f* horaria — **i** cartellino *m* di presenza

2922 clockwise — In the direction of the hands of a clock.
f en sens d'horloge; dans le sens des aiguilles d'une montre — **d** im Uhrzeigersinn; rechtsgängig — **n** volgens de wijzers van de klok; rechtsom — **e** según las agujas del reloj; hacia la derecha — **i** in direzione oraria; in senso orario; a destra

2923 clogging — See filling up.

2924 cloisonné — Multicoloured decoration of enamels, poured into the divided areas in an outlined book cover design. See also enamel binding.
f cloisonné — **d** Cloisonné *n*; Zellenschmelz *m* — **n** cloisonné *n* — **e** cloisonné *m*; alveolado — **i** cloisonné *m*

2925 close — The second member of any pair of signs, as ″,),].

2926 close-cut halftone — See outlined halftone.

2927 closed arc — See enclosed arc lamp.

2928 closed arm — See bar.

2929 closed bolts — See bolts.

2930 closed folds — See bolts.

2931 closed loop — Group of instructions in a program that are repeated indefinitely.
f boucle *f* fermée; boucle *f* de régulation — **d** geschlossener Kreis *m*; Regelkreis *m* — **n** gesloten lus *fm*; gesloten kring *m* — **e** anillo *m* cerrado; circuito *m* cerrado — **i** anello *m* chiuso; anello *m* di regolazione

2932 closed sections — See closed signatures.

2933 closed signatures; closed sections — Sections showing uncut folds.
f cahiers *mpl* à plis non coupés — **d** unaufgeschnittener Bogen *m* — **n** niet-opengesneden katerns *fmnpl* — **i** segnature *fpl* non refilate

2934 closed up — When all compositors have completed their work.
f composition complétée — **d** ausgesetzt — **n** met zetten gereed; klaar met zetten — **e** composición completa — **i** composizione completatta

2935 close formation — See well-closed formation.

2936 close leaders — Periods used in setting lines to lead the eye, spaced one dot to an en. See also open leaders.

f points *mpl* de conduite — **d** Führungspunkte *mpl* auf Halbgeviert; Halbgeviertführungspunkte *mpl* — **n** blokpunten *mpl* (op half vierkant) — **e** puntos *mpl* conductores; puntos *mpl* de conducción — **i** punti *mpl* di conduzione

2937 close leaders — See leaders.

2938 closely spaced line

f ligne *f* serrée — **d** dicht ausgesperrte Zeile *f* — **n** nauw gespatieerde regel *m* — **e** línea *f* poco espaciada — **i** linea *f* serrata

2939 close matter — See solid matter.

2940 close quote; end quotation mark — Mark used to end a quotation, usually an apostrophe or a pair of apostrophes. See also commence quote.

f guillemet *m* fermant — **d** Abführung *f*; Abführungszeichen *n*; Ausführungszeichen *n* — **n** aanhalingsteken *n* "sluiten"; sluitteken *n* — **e** comilla *f* final; fin *m* de comillas — **i** segno *m* di chiusura; virgoletta *f* di chiusura

2941 close register work — See tight register work.

2942 close to paper, print *v* — See print *v* close to paper.

2943 close *v* **up** — To push together, to remove spacing-out leads.

f rapprocher; gagner — **d** zusammenrücken; enger setzen; Zwischenraum vermindern; Zwischenraum herausnehmen; einlaufen — **n** wit uithalen; uitwinnen; inlopen — **e** aproximar; estrechar; quitar espacio; unir; juntar; suprimir espacio — **i** avvicinare; unire; diminuire lo spazio

2944 close-up mark — See proof correction marks, app. no. 5.

2945 closing hour for news — See deadline.

2946 cloth — See bookbinder's cloth.

2947 cloth back — Cover of a book with sides of paper and backstrips of cloth.

f dos *m* en toile — **d** Leinenrücken *m* — **n** linnen rug *m* — **e** lomo *m* en tela — **i** dorso *m* di tela

2948 cloth binding — Book bound with cloth.

f reliure *f* plein-toile; reliure *f* en toile — **d** Ganzleinenband *m*; Leinenband *m*; Ganzgewebeband *m* — **n** (geheel) linnen band *m* — **e** encuadernación *f* en tela — **i** legatura *f* in tela

2949 cloth boards — Book covers made from bookbinder's cloth pasted over stiff boards.

f couvertures *fpl* en toile; couvertures *fpl* entoilées — **d** Gewebedecke *fpl*; Leinendecke *fpl* — **n** linnen platten *npl* — **e** tapas *fpl* en tela — **i** piatti *mpl* in tela

2950 cloth bound — Protected by a rigid cover, usually cloth wrapped around boards. See also paper bound.

f relié toile — **d** in Leinen gebunden; Leineneinband *m*; Gewebeeinband *m* — **n** (in) linnen gebonden; in linnen band — **e** encuadernado en tela — **i** rilegato in tela

2951 cloth case — Bookbinding from bookbinder's cloth pasted over stiff boards.

f couverture *f* toile — **d** Leinenband *m*; Gewebeeinband *m* — **n** linnen band *m* — **e** cubierta *f* en tela; tapa *f* de tela — **i** copertina *f* in tela; copertina *f* telata

2952 cloth-faced paper — See cloth-lined paper.

2953 cloth hinge — Cloth joint to reinforce the back of the book, extending from the book to the cover.

f mors *m* en toile — **d** Leinenfalz *m*; Gewebefalz *m* — **n** linnen kneep *fm* — **e** charnela *f* de tela — **i** morso *m* in tela

2954 cloth hinge — Cloth joint to bind in heavy inserts.

f onglet *m* en toile — **d** Falz *m*; Fälzel *n*; Leinenfalz *m* — **n** linnen oortje *n*; linnen strookje *n* — **e** uña *f* de tela; bisagra *f* de tela; unglete *m* de percalina — **i** brachetta *f* in tela

2955 cloth lined — Backed with cloth.

f entoilé; toilé — **d** leinenkaschiert — **n** met linnen beplakt — **e** forrado de tela — **i** telato

2956 cloth-lined board — Paperboard pasted with cloth.

f carton *m* entoilé; carton *m* toilé — **d** Leinenkarton *m* — **n** karton *n* met linnen achterzijde; met linnen beplakt karton *n* — **e** cartón *m* forrado de tela — **i** cartone *m* telato

2957 cloth-lined paper; cloth-faced paper; reinforced paper — Paper lined with linen, canvas or other textile to reinforce it, used by map makers and envelope manufacturers.

f papier *m* entoilé (une face); papier *m* renforcé (de toile) — **d** Gewebepapier *n*; Papier *n* mit Gewebe; Leinwandpapier *n* — **n** met textielweefsel beplakt papier *n* — **e** papel *m* forrado de tela — **i** carta *f* telata

2958 cloth printing ink

f encre *f* pour imprimer sur textiles — **d** Textildruckfarbe *f*; Stoffdruckfarbe *f* — **n** textieldrukinkt *m* — **e** tinta *f* para la imprenta de tejidos — **i** inchiostro *m* per la stampa di tessuti

2959 cloth printing machine — See calico printing machine.

2960 cloth *v* **rollers** — To cast the composition on the roller stocks.

f couler les rouleaux — **d** Walzen gießen — **n** rollen gieten — **e** fundir rodillos — **i** rivestire i rulli

2961 cloudy effect — See cloudy formation.

2962 cloudy formation; cloudy effect; wild formation — Light and dark areas in a sheet of paper, more or less regularly distributed, seen when viewed by transmitted light.

f épair *m* nuageux; épair *m* irrégulier — **d** wolkige Durchsicht *f*; unregelmäßige Durchsicht *f*; unruhige Durchsicht *f* — **n** wolkig doorzicht *n*; onregelmatig doorzicht *n* — **e** formación *f* nubosa; transparencia *f* nubosa; transparencia *f* sin uniformidad — **i** spera *f* nuvolosa; speratura *f* nuvolosa; nuvolosità *f*

2963 club — A size of cut card. See app. no. 3.

2964 clump — Piece of metal line-spacing material thicker than 6-point. See also lead.

f lingot *m* — **d** Steg *m*; Reglette *f* — **n** reglet *fmn*

— e lingote *m*; regleta *f* (gruesa); imposición *f* plomada *Me* **— i** lingotto *m*; margine *m*

2965 clump — English name for a slug of the composing machine.

2966 clump — See fountain divider.

2967 cluster — Small group of liquid molecules with about the same spatial arrangement as in a solid crystal. Clusters are found in liquids at temperatures not far removed from the melting point.
f agglomérat *m* de molécules **— d** Molekülkomplex *m* **— n** moleculenopeenhoping *f* **— e** aglomerado *m* de moléculas **— i** agglomerato *m* di molecole

2968 clutch — See coupling device.

2969 CMC 7 — One of the two magnetic ink character recognition founts in current use on a large scale. The character set consists of numerals 0-9, alphabetic characters A-Z and four symbols. Characters are composed of seven vertical strokes each 0.15 mm wide and each character is unique in the spacing between the strokes. The fount which can be read at high speeds by a single reading head, was developed in France by Compagnie Française Bull. See also E 13 B.

2970 coagulate *v*
f (se) coaguler **— d** gerinnen; koagulieren; stocken **— n** coaguleren; stremmen **— e** coagular; cuajar **— i** coagulare; coagularsi

2971 coagulation — Process of converting a finely divided or colloidally dispersed suspension of a solid into particles of such size that reasonably rapid setting occurs. This is usually accomplished by adding the salt of a di- or trivalent metal.
f coagulation *f* **— d** Koagulation *f*; Koagulierung *f*; Koagulieren *n*; Gerinnung *f*; Gerinnen *n* **— n** coagulatie *f* **— e** coagulación *f*; coágulo *m*; espesamiento *m* **— i** coagulazione *f*

2972 coal tar — Black, viscous liquid or semisolid, denser that water, with a characteristic naphthalene-like odour and a sharp burning taste. Obtained in the destructive distillation of coal, soluble in ether, benzene, carbon disulphide, chloroform, partially soluble in alcohol, acetone, methanol, slightly soluble in water.
f goudron *m* de houille **— d** Kohlenteer *m*; Steinkohlenteer *m* **— n** koolteer *mn* **— e** alquitrán *m* de hulla **— i** catrame *m* di carbone

2973 coal-tar dyes; aniline dyes *obs.* — Organic colouring dyes, derived from coal-tar hydrocarbons, a residue of the distillation of coal, or their derivatives such as benzene, toluene, xylene, naphthalene, anthracene, aniline, etc.
f colorants *mpl* dérivés du goudron de houille; couleurs *fpl* dérivées du goudron de houille; colorants *mpl* d'aniline **— d** Teerfarbstoffe *mpl*; Teerfarben *fpl*; Anilinfarbstoffe *mpl depr.* **— n** teerkleurstoffen *fmpl* **— e** colorantes *mpl* de alquitrán de hulla; colorantes *mpl* de anilina *depr.* **— i** coloranti *mpl* da catrame di carbone; coloranti *mpl* all'anilina *depr.*

2974 coal-tar naphtha — See solvent naphtha.

2975 coarse grain — Grain sufficiently rough to produce a tonal effect worked on with lithographic crayons.
f grain *m* grossier; gros grain *m* **— d** grobes Korn *n* **— n** grof grein *n* **— e** grano *m* grueso **— i** grana *f* grossolana

2976 coarse screen — Halftone screen up to 30-40 lines per cm (85-100 lines to the inch), used on rough finished papers.
f grosse trame *f*; trame *f* grossière **— d** grober Raster *m*; Zeitungsraster *m* **— n** grof raster *n*; krantenraster *n* **— e** trama *f* gruesa; trama *f* ancha; trama *f* abierta; trama *f* con lineatura gruesa **— i** retino *m* a trama grossa

2977 coarse screen etching — See coarse screen halftone.

2978 coarse screen halftone; coarse screen etching — Printing plate produced by means of a halftone screen of 36 lines per cm (90 lines to the inch) or less, used on rough finished papers.
f simili *f* (à) grosse trame **— d** Grobrasterautotypie *f*; Grobrasterklischee *n*; Grobrasterätzung *f* **— n** grofrastercliché *n*; grofrasterautotypie *f* **— e** medio tono *m* de trama gruesa; grabado *m* de trama gruesa; fotograbado *m* de trama gruesa; medio tono *m* de trama ancha; autotipia *f* de trama gruesa **— i** cliscè *m* con retinatura grossolana; cliché *m* con retinatura grossolana; autotipia *f* con retinatura grossolana; autotipia *f* a trama grossa; autotipia *f* a retino grosso

2979 coaster board; beer mat board — Board for *beer mats.
f carton *m* feutre pour soucoupes pour verres **— d** Bierfilzpappe *f* **— n** karton *n* voor bierviltjes; bierviltjeskarton *n* **— e** cartón *m* para redondelitos de cerveza **— i** cartone *m* per sottobicchieri di birra

2980 coat *v*; **sensitize** *v* — To apply light sensitive solutions to a plate surface usually by means of a plate whirler.
f étendre la couche sensible; couler la couche sensible **— d** beschichten; eine Schicht auftragen; mit einer Schicht überziehen **— n** een kopieerlaag aanbrengen (prepareren, in chlichébedrijf, diepdruk) **— e** imprimir sensibilización; sensibilizar **— i** applicare il strato ricoprente; spalmare il strato ricoprente

2981 coated board *US*; **art board** *GB* — Clay coated paperboard, used for box making and other purposes.
f carton *m* couché **— d** Kunstdruckkarton *m* **— n** kunstdrukkarton *n* **— e** cartón *m* estucado; cartulina *f* estucada **— i** cartone *m* patinato; cartoncino *m* patinato

2982 coated lens — Lens covered with a thin layer of transparent material (e.g. magnesium fluoride) which reduces the amount of light reflected by the outer surface of the lens. A coated lens transmits more light than an uncoated lens.
f objectif *m* traité **— d** vergütetes Objektiv *n* **— n** gecoate lens *fm* **— e** lente *m* tratado **— i** obiettivo *m* trattato

2983 coated machine-glazed poster paper

f papier *m* couché litho frictionné — **d** gestrichenes Lithopapier *n* — **n** gestreken biljet-litho(-papier) *n* — **e** papel *m* estucado glaseado — **i** carta *f* monolucida patinata a macchina

2984 coated offset paper — Coated paper with a high resistance to picking, coated on two sides and sized like coated lithograph paper, suitable for use in offset printing. The basis weights usually range from 50-100 pounds (25 x 38 - 500).

f papier *m* offset couché (classique) — **d** gestrichenes Offsetpapier *n* — **n** gestreken offsetpapier *n*; gecoucheerd offsetpapier *n* — **e** papel *m* offset estucado — **i** carta *f* offset patinata

2985 coated on both sides — See two-sided coated paper.

2986 coated paper *US*; **art paper** *GB* — Paper coated with clay, other white pigments and a suitable binder. See also enamel paper.

f papier *m* couché (classique); papier *m* couché hors machine; papier *m* crayé *Sw* — **d** gestrichenes Papier *n*; Kunstdruckpapier *n*; Kreidepapier *n* — **n** gestreken papier *n*; gecoucheerd papier *n*; kunstdrukpapier *n* — **e** papel *m* estucado; papel *m* esmaltado — **i** carta *f* patinata; carta *f* spalmata

2987 coating; layer — Layer spread over a surface.

f couche *f*; pellicule *f* — **d** Schicht *f*; Film *m* — **n** laag *fm*; laagje *n*; film *m* — **e** capa *f*; película *f* — **i** strato *m*; pellicola *f*

2988 coating — Process of covering the surface of a paper or board with one or more layers of coating slip.

f couchage *m*; procédé *m* de couchage — **d** Streichverfahren *n*; Streichen *n* — **n** strijkprocédé *n*; strijken *n*; coucheren *n*; aanbrengen *n* van een strijklaag — **e** estucado *m*; procedimiento *m* de estucado — **i** patinatura *f*; processo *m* di patinatura

2989 coating — Mineral substances, such as China clay, blanc fixe, satin white, etc., to cover the surface of paper, thus making the coated surface of enamelled papers.

f couche *f* — **d** Strich *m*; Streichschicht *f* — **n** strijklaag *fm*; couche *fm*; coating *depr.* — **e** estucado *m* — **i** patina *f*

2990 coating — See layer.

2991 coating colour — See coating slip.

2992 coating-engraving technique — Engraving in an easily workable layer to obtain a copying layer, especially in cartography.

f gravure *f* de la couche — **d** Schichtgravur *f* — **n** graveren *n* in een laag — **e** grabado *m* en el strato — **i** incisione *f* dello strato

2993 coating, lens — See lens coating.

2994 coating lump — Discoloured, shiny, hard and brittle spot in the web.

f mâton *m*; pâton *m* — **d** Patzen *m*; Batzen *m*; Stoffbatzen *m* — **n** patzen *pl* — **e** grumo *m* (de pasta); grumo *m* (de estucado) — **i** macchia *f* di patina

2995 coating machine — Machine on which the coating slurry is applied to paper by means of rollers, the wet coating being evened on the paper surface by means of brushes. The wet coated paper is dried in festoon driers. In another type of machine, coating slurry is applied by means of a brush and evened out by a jet of air.

f coucheuse *f*; fonceuse *f*; machine *f* à coucher — **d** Streichmaschine *f* — **n** strijkmachine *f* — **e** máquina *f* de estucar; estucadora *f* — **i** macchina *f* patinatrice

2996 coating paper — See body paper.

2997 coating pick; micro pick

f micro-arrachage *m* — **d** Mikrorupfen *n*; Strichrupfen *n* — **n** micropluk *m* — **e** microrepelado *m*; repelado *m* del estucado — **i** microstrappo *m*; microdistacco *m*

2998 coating slip; coating colour; slurry — Mixture, generally of mineral base with an adhesive or a binder, for application to the surface of paper or board. Other materials, such as colouring matters, may also be incorporated in the mixture.

f lait *m* de couchage; pâte *f* liquide; sauce *f* de couchage — **d** Streichmasse *f* — **n** strijkmassa *fm* — **e** masa *f* para estucar; estuco *m* — **i** patina *f* fluida; miscuglio *m* di patinatura

2999 coating splash — See spits.

3000 coating thickness

f épaisseur *f* de la couche — **d** Schichtstärke *f*; Schichtdicke *f* — **n** dikte *f* van de strijklaag; strijklaagdikte *f* — **e** espesor *m* de la capa — **i** spessore *m* della patina

3001 coat of arms — Heraldic achievement of arms.

f armoiries *fpl* — **d** Wappen *n* — **n** wapenschild *n*; wapen *n* — **e** escudo *m* de armas; armas *fpl* — **i** stemma *m*

3002 coauthor; joint author — One or two or more authors, as of a book.

f coauteur *m* — **d** Mitverfasser *m*; Koautor *m*; Mitautor *m* — **n** medeauteur *m*; medeschrijver *m* — **e** coautor *m* — **i** coautore *m*

3003 coaxial cable — Cable consisting of one conductor, usually a small copper tube or wire, within and insulated from another conductor of larger diameter which is usually braided copper surrounded by a lead sheath. It is primarily used for wide band transmission.

f câble *m* coaxial — **d** koaxiales Kabel *n*; konzentrisches Kabel *n* — **n** coaxiale kabel *m*; concentrische kabel *m* — **e** cable *m* coaxial — **i** cavo *m* coassiale

3004 coaxial drive — Rotary press drive where a number of smaller powered motors drive the press units by means of couplings on a common shaft.

f commande *f* coaxiale; arbre *m* de transmission unique — **d** Parallelantrieb *m*; koachsialer Antrieb *m*; gemeinsamer Antrieb *m* — **n** gemeenschappelijke aandrijving *f* — **e** accionamiento *m* coaxial; accionamiento *m* en paralelo — **i** comando *m* coassiale

3005 cobalt — Steel-grey, slightly pinkish shining,

hard, ductile, somewhat malleable metal. Slowly soluble in dilute acids (hydrochloric and sulphuric acid), readily in nitric acid. Latin: cobaltum. TL-value: 0.5 mg/m^3. Symbol: Co.
f cobalt *m* — **d** Kobalt *m* — **n** kobalt *n* — **e** cobalto *m* — **i** cobalto *m*

3006 cobalt blue; azure blue — Blue pigment of variable composition, consisting essentially of mixtures of cobalt oxide and alumina, approximating cobaltous aluminate $Co(AlO_2)_2$. The most durable of all blue pigments, being completely unaltered by the atmosphere and only slightly subject to chemical reagents.
f bleu *m* de cobalt — **d** Kobaltblau *n* — **n** kobalt-blauw *n* — **e** azul *m* de cobalto; azul *m* cobalto — **i** blu *m* di cobalto; blu *m* cobalto

3007 cobalt drier — Material containing chemically combined cobalt used to accelerate oxidation and polymerization of an ink film. Very strong surface drier.
f siccatif *m* au cobalt; siccatif *m* cobaltique — **d** Kobalttrockner *m*; Kobaltsikkativ *m* — **n** kobalt-droger *m* — **e** secante *m* de cobalto; desecante *m* de cobalto — **i** essiccante *m* al cobalto

3008 cobalt green; cobalt zincate — Pigment consisting mainly of oxides of cobalt and zinc, characterized by its green colour, fast drying rate, permanence and lack of tinting strength. Formula: $CoZnO_2$.
f vert *m* de cobalt — **d** Kobaltgrün *n* — **n** kobalt-groen *n* — **e** verde *m* de cobalto; verde *m* cobalto — **i** verde *m* di cobalto; verde *m* cobalto

3009 cobalt linoleate; cobaltous linoleate — Brown, amorphous powder, soluble in alcohol, ether and acids, insoluble in water. Formula: $Co(C_{18}H_{31}O_2)_2$.
f linoléate *m* de cobalt; linoléate *m* cobalteux — **d** Kobaltlinoleat *n*; leinölsaures Kobalt *n* — **n** kobaltlinoleaat *n* — **e** linoleato *m* de cobalto; linoleato *m* cobaltoso — **i** linoleato *m* di cobalto; linoleato *m* cobaltoso

3010 cobalt naphthenate; cobaltous naphthenate — Brown, amorphous powder or bluish-red solid. Soluble in alcohol, ether, oils, insoluble in water. Composition indefinite. Used in varnish driers.
f naphténate *m* de cobalt; naphténate *m* cobalteux — **d** Kobaltnaphtenat *n* — **n** kobaltnaftenaat *n* — **e** naftenato *m* de cobalto; naftenato *m* cobaltoso — **i** naftenato *m* di cobalto; naftenato *m* cobaltoso

3011 cobaltous — Containing bivalent cobalt.
3012 cobaltous linoleate — See cobalt linoleate.
3013 cobaltous naphthenate — See cobalt naphthenate.
3014 cobaltous resinate — See cobalt resinate.
3015 cobalt resinate; cobaltous resinate — Brown-red powder, soluble in oils, insoluble in water. Principally cobalt abietate. Used as a varnish drier. Formula: $Co(C_{44}H_{62}O_4)_2$.
f résinate *m* de cobalt; résinate *m* cobalteux — **d** Kobaltresinat *n* — **n** kobaltresinaat *n* — **e** resinato *m* de cobalto; resinato *m* cobaltoso — **i** resinato *m* di cobalto; resinato *m* cobaltoso

3016 cobalt zincate — See cobalt green.
3017 Cobb's paper — Thin paper for the covering of book boards and for end papers. Named after Thomas Cobb (1796).
f papier *m* Cobb; papier *m* doublure — **d** Cobb's Vorsatzpapier *n* — **n** Cobb-papier *n* — **e** papel *m* Cobb — **i** carta *f* di Cobb

3018 Cobb test — Method for the determination of the water absorbency capacity of paper.
f essai *m* Cobb — **d** Cobb-Test *m*; Test *m* der Wasserabsorption des Papiers — **n** Cobb-test *m* — **e** ensayo *m* Cobb; prueba *f* de Cobb — **i** prova *f* di Cobb

3019 cobol — Acronym for common business oriented language. A powerful international procedure-oriented language devised for data processing for commercial users and written in meaningful English words and arithmetic symbols. Compilers for cobol have been written by most computer manufacturers. In this sense it is a machine independent language although not all implementations are the same. The language was initiated and its development directed by the Codasyl group (Conference on data systems languages).
f cobol *m* — **d** Cobol *m* — **n** cobol *n* — **e** cobol *m* — **i** cobol *m*

3020 cobwebbing — Application defect in gravure printing, exhibiting a filmy, web-like build-up of dried ink or clear material on the doctor blade, ends of impression roll or engraving.

3021 coccine — Red tar dyestuff.
f coccine *f* — **d** Coccin *n* — **n** coccine *fm* — **e** coccín *f*

3022 cochineal — Red dyestuff consisting of the dried bodies of female cochineal insects, which live on the cactuses in Central America, used as a stain and as an indicator.
f cochenille *f* — **d** Koschenille *f* — **n** cochenille *fm* — **e** cochinilla *f* — **i** coccinella *f*

3023 cock — The middle portion of a *brace. See also composite brace.
3024 cock and hens — See composite brace.
3025 cocking roller — See slewing roller.
3026 cockling; waviness — Warping effect occurring along the edges of paper, particularly across the grain, due to packing too tightly while it is immature, or to too rapid drying. May be overcome by exposing to air, by hanging in a uniform temperature, or by stacking under pressure.
f recoquillement *m*; gondolage *m*; ondulation *f* — **d** Welligwerden *n*; Werfen *n* — **n** golven *n* — **e** abarquillado *m* — **i** arricciatura *f*; conchigliatura *f*

3027 cockroach *coll.* — Display matter set entirely in lower-case type.
3028 cock robin shop — Small printing office which produces cheap printing from second-rate materials.
f boîte *f*; boutique *f*; rabaisien *m* — **d** Quetsche *f*; Waschküchendruckerei *f*; Feuerzeug *n* — **n**

tweederangs drukkerijtje *n* — **e** casa *f* de segunda fila — **i** stamperia *f* di seconda mano; stamperia *f* di scarsa qualità

3029 cock-up initial; stick-up initial; raised initial — Initial letter that aligns at the bottom with the first line of text but sticks up into the white space above.
f lettrine *f* — **d** Initiale *f* über eine Zeile; Initial *n* — **n** initiaal *n* over meer regels bovenuitstekend; bovenuitstekend initiaal *n* — **e** inicial *f* sobre una línea; inicial *f* saliente — **i** iniziale *f* di più righe

3030 code; conversion code — System of unique combinations of binary digits each defining a function or character. Also the implementation of a procedure in a form which the computer can accept and interpret.
f code *m* — **d** Kode *m* — **n** code *m* — **e** código *m* — **i** codice *m*

3031 code conversion — Process by which a code of a predetermined bit structure (for example 6, 7, 14 bits per character interval) is converted to a second code with more or less bits per character interval. No alphabetical significance is assumed in this process. In certain cases, such as the conversion from start/stop telegraph equipment, a code conversion process may only consist of discarding the stop and start bits and adding a sixth bit to indicate the stop and start condition. In other cases, it may consist of addition or deletion of control and/or parity bits.
f conversion *f* de code — **d** Kodekonversion *f*; Kodeumsetzung *f* — **n** codeconversie *f*; code-omzetting *f* — **e** conversión *f* de código — **i** conversione *f* di codice

3032 code converter; transcoder — Device to convert one set of code assignments into another.
f transcodeur *m*; convertisseur *m* de code — **d** Kodeumsetzer *m*; Kodeumwandler *m* — **n** code-omzetapparaat *n*; codeomzetter *m*; code-omvormer *m* — **e** convertidor *m* de código — **i** convertitore *m* di codice

3033 coder — Person who converts a program into instruction codes. See also encoder.
f codeur *m* — **d** Kodierer *m* — **n** codeur *m*; codist *m* — **e** codificador *m* — **i** codificatore *m*

3034 code set — Finite and complete set of representations defined by a code.
f liste *f* code — **d** Kodeliste *f* — **n** codelijst *fm* — **e** lista *f* código — **i** lista *f* codice

3035 code standard — Set of conventions whereby bytes are given a specific meaning for the purpose of information interchange.

3036 coding — System of symbols representing data, functions and instructions in a form in which they can be dealt with by computers.
f codage *m* — **d** Kodierung *f*; Verschlüsselung *f* — **n** codering *f* — **e** codificación *f* — **i** codificazione *f*

3037 cods; graining balls; graining marbles; marbles — Glass, porcelain, or steel marbles used in graining planographic printing surfaces.
f billes *fpl* (de grainoir); marbres *mpl*; billes *fpl* pour le grainage — **d** Schleifkugeln *fpl*; Kugeln *fpl* zum Körnen; Märbel *mpl*; Glasmärbel *mpl*; Glaskugeln *fpl* — **n** slijpkogels *mpl*; greinkogels *mpl*; greinknikkers *mpl*; knikkers *mpl* — **e** bolas *fpl* graneadoras; bolas *fpl* para granear; bolas *fpl* para amolar — **i** biglie *fpl*; sfere *fpl* per granitoio

3038 coefficient of expansion — The fractional change in length, area, or volume per unit change in temperature of a solid, liquid, or gas at constant pressure.
f coefficient *m* de dilatation — **d** Dehnungskoeffizient *m*; Ausdehnungskoeffizient *m*; Ausdehnungszahl *f* — **n** uitzettingscoëfficiënt *m* — **e** coeficiente *m* de expansión — **i** coefficiente *m* di dilatazione

3039 coefficient of friction — The ratio between the force required to move one surface along another and the force, normal to the surfaces, pressing them together.
f coefficient *m* de friction; coefficient *m* de frottement — **d** Reibungskoeffizient *m*; Reibungszahl *f*; Reibwert *f* — **n** wrijvingscoëfficiënt *m* — **e** coeficiente *m* de fricción; coeficiente *m* de frotamiento; coeficiente *m* de rozamiento — **i** coefficiente *m* di attrito

3040 coefficient of variation — In statistics, the ratio of the standard deviation to the mean, i.e., a measure of relative variability. Sometimes multiplied by 100 to give the percentage variation.
f coefficient *m* de variation — **d** Variationskoeffizient *m* — **n** variatiecoëfficiënt *m* — **e** coeficiente *m* de variación — **i** coefficiente *m* di variazione

3041 coefficient of velocity — The ratio of the mean flow velocity in a liquid jet to the theoretical velocity of discharge of a frictionless liquid, the latter being given by $(2gh)^{1/2}$.
f coefficient *m* de vitesse d'écoulement — **d** Ausströmungsgeschwindigkeits-Koeffizient *m* — **n** uitstromingssnelheidscoëfficiënt *m* — **e** coeficiente *m* de velocidad de gasto — **i** coefficiente *m* di velocità di efflusso

3042 coercitive force — See coercive force.

3043 coercive force; coercitive force — In magnetic printing ink manufacture the reverse magnetizing force required to reduce the residual induction (remanence) to zero.
f force *f* coercitive — **d** Koerzitivkraft *f*; Retentionskraft *f* — **n** coërcitiefkracht *fm*; coërcitiekracht *fm* — **e** fuerza *f* coercitiva — **i** forza *f* coercitiva

3044 coherence — See cohesion.

3045 coherency — See cohesion.

3046 cohesion; cohesive attraction; coherence; coherency — Attraction of molecules in a substance.
f cohésion *f*; cohérence *f*; adhérence *f* — **d** Kohäsion *f*; Zusammenhang *m* — **n** cohesie *f* — **e** cohesión *f* — **i** coesione *f*

3047 cohesive attraction — See cohesion.

3048 coincident current store — Store in which each storage element has the property of changing its state on receipt of a pulse from two coincident sources, i.e., x and y wires.

f mémoire *f* à entrée par coïncidence — **d** Koinzidenzstromspeicher *m* — **e** memoria *f* seleccionadora de corriente de coincidencia — **i** memoria *f* selezione di corrente di coincidenza

3049 cokey face — Bad surface on a stereo plate casting from metal due to release of gases from the flong through excessive heat. See also blistered mat.

f surface *f* grésillée — **d** blasige Oberfläche *f* — **n** oppervlak *n* met blaasjes — **e** superficie *f* picada; superficie *f* que tiene burbujas; superficie *f* remosqueada *Ch*; superficie *f* mosqueada *Ch* — **i** superficie *f* vaiolata; superficie *f* con soffiature

3050 colcothar — See Venetian red.

3051 cold blocking — Bookbinder's impression with cold tools, requiring much higher pressure than hot blocking.

f empreinte *f* à froid — **d** Kaltprägung *f* — **n** koud stempelen *n* — **e** estampado *m* en frío — **i** impressione *f* a freddo

3052 cold colour — Colour which is on the bluish side, that is green or blue, or colours which exhibit a predominance of these.

f couleur *f* froide — **d** kalte Farbe *f* — **n** koude kleur *fm* — **e** color *m* frío — **i** colore *m* freddo; tinta *f* fredda

3053 cold composition — See cold type.

3054 cold enamel; blue lacquer — Photoengraving sensitizer for metal plates consisting of a solution of bichromated shellac, the image being ready for etching after development and drying of the print.

f émail *m* à froid — **d** Kaltemail; Blaulack *m* — **n** koudemail *n*; blauwlak *mn* — **e** esmalte *m* en frío — **i** smalto *m* a freddo

3055 cold enamel process; cold top process — Printing down process where the layer is made for instance of shellac which resists the acid after developing and needs not to be baked.

f procédé *m* à l'émail à froid — **d** Kaltemailverfahren *n*; Emailkopierverfahren *n*; Emailverfahren *n*; Blaulackkopierverfahren *n* — **n** koudemailprocédé *n*; koudlakprocédé *n*; blauwlakprocédé *n* — **e** procedimiento *m* al esmalte en frío — **i** procedimento *m* allo smalto freddo

3056 cold glue

f colle *f* froide — **d** Kaltleim *m* — **n** koudlijm *m* — **e** cola *f* en frío — **i** colla *f* fredda

3057 cold pressed paper — Loft-dried paper pressed in a hydraulic press to remove roughness and give it a slight gloss.

f papier *m* apprêté à froid — **d** kaltgepreßtes Papier *n* — **n** koudgeperst papier *n* — **e** papel *m* prensado en frío — **i** carta *f* pressata a freddo

3058 cold-set inks; cold-setting inks — Solid inks which must be melted and applied on a hot press. They solidify again on contact with the relatively cold paper.

f encres *fpl* cold-set; encres *fpl* séchant par refroidissement; encres *fpl* à séchage par refroidissement; encres *fpl* fixées à froid — **d** Cold-set-

Farben *fpl*; Kalttrockenfarben *fpl*; kalttrocknende Farben *fpl* — **n** cold-set inkten *mpl* — **e** tintas *fpl* de fraguado al frío; tintas *fpl* de secado al frío — **i** inchiostri *mpl* cold-set; inchiostri *mpl* stabilizzati a freddo

3059 cold-setting inks — See cold-set inks.

3060 cold top process — See cold enamel process.

3061 cold type; cold composition — Composition method not involving hot metal type like typewriter composition, paper letters, photocomposition on photo paper, with mechanical lettering instruments and photographic composition on film.

f composition *f* à froid; composition *f* sans plomb; composition *f* photomécanique — **d** Kaltsatz *m*; bleiloser Satz *m*; nicht-metallischer Satz *m*; Lichtsatz *m* — **n** koud zetsel *n*; fotografisch zetsel *n*; fotozetsel *n* — **e** composición *f* fría; composición *f* en frío; composición *f* sin plomo; fotocomposición *f* — **i** composizione *f* a freddo; fotocomposizione *f*

3062 collapsible folding box — See folding carton.

3063 collate *v* — In bookbinding, to examine the folded signatures of a book, after gathering but before sewing, for proper sequence.

f collationner — **d** kollationieren; vergleichen — **n** collationeren; nakijken; controleren — **e** repasar; cotejar; colacionar — **i** collazionare

3064 collate *v*; **interpolate** *v* — Computer term. To combine two sequences of items in such a way that the same sequence is observed in the combined sequence.

f interclasser — **d** abgleichen; mischen; zuordnen — **n** op volgorde sorteren; tussensorteren — **e** intercalar; mezclar — **i** inserire; intercalare

3065 collate *v* — Misnomer for gathering, i.e., putting the sections of a book in right sequence, ready for binding. See also gather.

3066 collating marks; back marks; niggerheads — Black step marks printed on the back of folded sheets, to facilitate collating and checking of the sequence of book signatures.

f indices *mpl* de collationnement — **d** Flattermarken *fpl*; Falzzeichen *npl depr.* — **n** collationeertekens *npl* — **e** señales *fpl* escalonadas; marcas *fpl* guías; marcas *fpl* del pliego — **i** contrassegni *mpl* di raccolta; svolazzi *mpl*

3067 collating table; gathering table

f table *f* d'assemblage; table *f* à assembler — **d** Zusammentragtisch *m* — **n** verzameltafel *fm* — **e** mesa *f* alzadora; mesa *f* para alzar; mesa *f* de cotejar — **i** tavola *f* di raccolta

3068 collation — Comparison of the copy with the original manuscript.

f collation *f*; confrontation *f* de textes — **d** Vergleichen *n* — **n** vergelijken *n*; collationering *f* — **e** cotejo *m* — **i** collazione *f*

3069 collator; interpolator — Machine to collate (compare) punched cards, merge, select and check the sequence of them.

f interclasseuse *f* — **d** Kartenmischer *m*;

Mischer *m* — **n** collator *m* — **e** intercaladora *f* — **i** inseritrice *f*

3070 collect *v* — To gather sections or pages on a web-fed machine.
f assembler — **d** zusammentragen — **n** vergaren; verzamelen — **e** acoplar — **i** raccogliere

3071 collect *v* — See gather *v*.

3072 collecting cylinder; collecting drum; collector — Cylinder in a rotary press for collecting sheets or sections, prior to folding or delivery.
f cylindre *m* d'accumulation; cylindre *m* collecteur; cylindre *m* d'assemblage — **d** Sammelzylinder *m*; Sammeltrommel *f* — **n** verzamelcilinder *m* — **e** cilindro *m* colector — **i** cilindro *m* di accumulo; tamburo *m* raccoglitore

3073 collecting cylinder cam — Cam in a rotary press folder operating the collecting cylinder needles or grippers.
f came *f* de commande du cylindre assembleur — **d** Sammelzylinder-Steuerkurve *f* — **n** verzamelcilinderexcenter *m* — **e** camba *f* del cilindro colector; excéntrica *f* del cilindro colector — **i** camma *f* di comando del tamburo raccoglitore

3074 collecting drum — See collecting cylinder.

3075 collector — Device to accumulate current from contact conductors.
f collecteur *m* — **d** Kollektor *m* — **n** collector *m* — **i** collettore *m*

3076 collector — See collecting cylinder.

3077 collector's binding — See amateur binding.

3078 collect run — Method to increase the number of pages in a copy by collection before the final fold and delivery. The folder is so adjusted that one section of the paper is held back until the next section is ready to go with it around the folding cylinder, hence "collect". The total product is always an even number of sections.
f accumulation *f* — **d** einfache Produktion *f*; gesammelte Produktion *f*; Sammelproduktion *f* — **n** verzamelgang *m* — **e** tirada *f* combinada; tirada *f* con los productos coleccionados — **i** stampa *f* con accumulo

3079 collimate *v* — To direct the line of sight into a straight path perpendicular to the plane of the negative or flat. In this way printing details on two separate films or flats, that are over each other but not in contact, can be checked for exact alignment and register.
f collimater — **d** kollimationieren; in die optische Achse bringen — **n** collimeren — **e** colimar; poner en línea — **i** collimare

3080 collimation error
f erreur *f* de collimation — **d** Kollimationsfehler *m* — **n** collimatiefout *fm* — **e** yerro *m* de colimación — **i** errore *m* di collimazione

3081 collimator — Optical instrument that directs the line of sight down its axis, to check alignment for register between two or more photographic images that are separated from each other, as on superimposed flats. See also register tube.
f collimateur *m* — **d** Kollimator *m* — **n**

collimator *m* — **e** colimador *m* — **i** collimatore *m*

3082 collodion — Pale yellow, syrupy, viscous liquid, consisting of a solution of nitrocellulose (pyroxylin) in a mixture of alcohol and ether (nitrocellulose 40 g, ether 750 ml, alcohol 250 ml), formerly used in the *collodion process.
f collodion *m* — **d** Kollodium *n* — **n** collodium *n* — **e** colodión *m* — **i** collodio *m*

3083 collodion bromide — Solution of collodion sensitized with silver bromide.
f collodion *m* au bromure d'argent — **d** Bromsilberkollodium *n* — **n** broomzilver-collodion *n* — **e** colodio-bromuro *m* de plata — **i** collodio *m* al bromuro d'argento

3084 collodion chloride — Solution of collodion sensitized with silver chloride.
f collodion *m* au chlorure d'argent — **d** Chlorsilberkollodium *n* — **n** chloorzilver-collodium *n* — **e** colodio-cloruro *m* de plata — **i** collodio *m* al cloruro d'argento

3085 collodion cotton — See pyroxylin.

3086 collodion emulsion — Negative collodion emulsified with silver bromide, with or without colour sensitizing dyes.
f émulsion *f* au collodion — **d** Kollodiumemulsion *f* — **n** collodiumemulsie *f* — **e** emulsión *f* de colodión; emulsión *f* al colodión — **i** emulsione *f* al collodio

3087 collodion process; wet collodion process; wet-plate process — Photographic negative process invented about 1850 independently by Le Gray and F.S. Archer. A glass plate coated with a layer of *iodized collodion is sensitized in a silver nitrate solution and exposed wet in the camera. Development in an iron salt solution (pyrogallic acid or ferrous sulphate) must begin before the plate starts to dry. Universally used for photography until the appearance of gelatine dry plates in the 1870's. Sometimes still in use today for mechanical line work and halftone work to obtain sharp high contrast.
f procédé *m* au collodion (humide) — **d** Kollodiumverfahren *n*; Naßverfahren *n*; nasses Kollodiumverfahren *n* — **n** collodiumprocédé *n*; natte-collodiumprocédé *n*; natte-plaatprocédé *n* — **e** procedimiento *m* al colodión (húmedo); procedimiento *m* de placa húmeda — **i** procedimento *m* al collodio umido; processo *m* al collodio umido

3088 collodion wool — See pyroxylin.

3089 colloid — Water-soluble, non-crystalline substance of gelatinous nature (albumen, gelatine glue, gum arabic, dextrin), used as vehicle in photomechanical sensitizers; rendered light sensitive by addition of a bichromate. The particle size is within a range of 10^{-7} and 10^{-9} mm in diameter.
f colloïde *m* — **d** Kolloid *n*; Kolloide *f* — **n** colloïde *fmn* — **e** coloide *m* — **i** colloide *m*

3090 colloidal — Composed of particles that are very small but larger than most molecules.
f colloïdal — **d** kolloidal — **n** colloïdaal — **e** coloidal — **i** colloidale

3091 colloidal suspension — Suspension in which the dispersed phase consists of particles of colloidal size.

f suspension *f* colloïdale — **d** kolloidale Suspension *f* — **n** colloïdale suspensie *f* — **e** suspensión *f* coloidal — **i** sospensione *f* colloidale

3092 colloid mill — Machine for dispersion of pigments in some types of printing ink. The ink passes through the very narrow space between a stator and a rotor which is revolving at very high speed. This produces intense shearing stresses in the liquid and the solid particles flowing between these surfaces, due to viscous forces in the liquid rather than to a grinding action between the moving surfaces.

f broyeuse *f* colloïdale — **d** Kolloidmühle *f* — **n** colloïdmolen *m* — **e** molino *m* coloidal — **i** mulino *m* per dispersioni colloidali

3093 colloid resist — Colloid stencil used in making etched plates to protect certain areas of the metal from the corroding liquid used for the etching.

f réserve *f* colloïdale de morsure — **d** Kolloidätzwiderstand *m*; Ätzschutzschicht *f*; Ätzgrund *m* — **n** zuurbestendige laag *fm* — **e** reserva *f* colloidal — **i** riserva *f* colloidale

3094 collotype grain

f grain *m* à phototypie — **d** Runzelkorn *n*; Lichtdruckkorn *n* — **n** lichtdrukkorn *m* — **e** grano *m* de fototipia; grano *m* de heliotipia — **i** grana *f* fototipica; grana *f* d'eliotipia

3095 collotype ink; photogelatine ink — Lithographic ink with heavy body, buttery consistency and strong colour.

f encre *f* pour phototypie — **d** Lichtdruckfarbe *f* — **n** lichtdrukinkt *m* — **e** tinta *f* para fototipia; tinta *f* para heliotipia — **i** inchiostro *m* per fototipia; inchiostro *m* per eliotipia

3096 collotype paper — Paper of sufficient consistency and sizing to withstand the damp surface of the gelatine plate in printing collotype.

f papier *m* pour phototypie; papier *m* pour collotypie — **d** Lichtdruckpapier *n* — **n** lichtdrukpapier *n* — **e** papel *m* para heliografía; papel *m* para fototipia — **i** carta *f* per eliotipia; carta *f* per fototipia

3097 collotype plate

f dalle *f* pour phototypie — **d** Lichtdruckplatte *f*; Lichtdruckglasplatte *f* — **n** lichtdrukplaat *fm*; lichtdrukglasplaat *fm* — **e** plancha *f* de fototipia; plancha *f* de heliotipia; plancha *f* para fotogelatinografía; plancha *f* per fotocolografía — **i** lastra *f* fototipica; lastra *f* eliotipica

3098 collotype printing — See collotype process.

3099 collotype process; collotype printing; photogelatine printing *US*; **phototype process** — Semi-intaglio photomechanical non screen halftone process based on the action of light on bichromatized gelatine. Where the light has affected the glass plate the gelatine hardens; the hardened parts retain the ink, and the soft or watery portions repel it. Reticulations in the surface, finer or coarser as the action of the light has been lesser or greater, provide the "bite" and ensure reproduction of halftones without the need of a screen. Used for art reproductions in comparatively short runs. Collotype is known by a number of names: albertype, artotype, heliotype. Invented by Joseph Albert (1825-1886), an Austrian photographer.

f collotypie *f*; phototypie *f*; héliotypie *f*; photocollographie *f* — **d** Lichtdruck *m*; Phototypie *f*; Albertypie *f*; Albertochromie *f* — **n** lichtdruk *m*; albertypie *f* — **e** heliotipia *f*; fotogelatinografía *f*; colotipia *f*; fotocolotipia *f*; artotipia *f*; albertipia *f*; fototipia *f* de Albert; fotocolografía *f* — **i** fototipia *f*; eliotipia *f*; albertipia *f*; albertotipia *f*

3100 colombier — A British size of drawing paper. See app. no. 3.

3101 colon — Punctuation mark (:) used when the preceding matter leads on to that which follows. It is not necessary to add a metal rule to the colon. Used in mathematics to indicate ratio.

f deux-points *mpl* — **d** Doppelpunkt *m*; Kolon *n* — **n** dubbele punt *m* — **e** dos puntos *mpl*; colón *m*; colón *m* perfecto — **i** due punti *mpl*

3102 colophon — In early books, a final paragraph giving particulars of authorship and printer, where printed, date of completion, and possibly other details. Often arranged as a tail piece and ornament. Now sometimes replaced by the *title-page and *imprint, although some publishers often append a colophon to a book which has an imprint and title page.

f colophon *m*; achevé *m* d'imprimer; achevé *m* d'impression — **d** Kolophon *m*; Schlußschrift *f* — **n** colofon *mn* — **e** colofón *m* — **i** colophon *m*

3103 colophony — See rosin.

3104 color — See colour.

3105 coloration on the back side — Yellow spots at the back side of a sheet of paper, originating from gaseous products formed by oxidation, polymerization of the ink vehicle during drying. Not to be confused with setting-off.

f coloration *f* au verso; migration *f* du colorant — **d** Verfärbungserscheinungen *fpl* auf der Rückseite — **n** verkleuring *f* aan de achterzijde — **e** amarilleado *m* del reverso; descoloración *f* parcial — **i** colorazione *f* sul lato di volta

3106 Colorgraph — Trade name for an electronic colour scanner for the production of tone- and colour-corrected negatives or diapositives from reflection or transparent colour copy or separations. Manufactured in Kiel, Germany.

3107 colorimeter; chromometer — Device for comparing colours and analyzing them quantitatively with Nessler tubes and photoelectric devices.

f colorimètre *m* — **d** Kolorimeter *n*; Farbenmesser *m*; Farbenmeßapparat *m* — **n** colorimeter *m* — **e** colorímetro *m* — **i** colorimetro *m*

3108 colorimetric — Measuring by means of colour.

f colorimétrique — **d** kolorimetrisch — **n** colorimetrisch — **e** colorimétrico — **i** colorimetrico

3109 colorimetric analysis — See colorimetry.

3110 colorimetric purity — The ratio of the luminance of a spectrally homogeneous component to the luminance of the white light with which it must be mixed, to match the chromaticity of a sample of light. This ratio is L_d/L, where L is the luminance of the sample stimulus and L_d is the luminance of a spectral stimulus (or of a suitable combination of extreme special red and extreme special violet) which, by additive mixture with the adopted achromatic stimulus, can form a match with the sample stimulus in both luminance and chromaticity. Symbol: p_c. It is recommended (CIE 1948) that colorimetric purity be calculated by the formula: $p_c = p_e \times y_d/y$, where y_d and y are chromaticity co-ordinates defined to *excitation purity.

f pureté f colorimétrique — **d** farbmetrische Echtheit f; spektraler Leuchtdichteanteil m — **n** colorimetrische zuiverheid f — **e** pureza f colorimétrica — **i** purezza f colorimetrica

3111 colorimetry; colorimetric analysis — Determination of the amounts of substances by comparing the intensity of colour produced by them with specific reagents with the intensity of colour produced by a standard amount of the substance.

f colorimétrie f — **d** Kolorimetrie f; Farbenmessung f; Farbmessung f — **n** colorimetrie f; kleurmeting f — **e** colorimetría f; medida f de los colores — **i** colorimetria f; misura f del colore

3112 colors — See colours.

3113 colour *GB*; **color** *US* — 1. Perceived or subjective colour. Aspect of visual perception by which an observer may distinguish differences between two fields of view of the same size, shape and structure, such as may be caused by differences in the spectral composition of the radiation concerned in the observation. 2. Psychophysical or objective colour. Characteristic of a visible radiation by which an observer may distinguish differences between two fields of view of the same size, shape and structure, such as may be caused by differences in the spectral composition of the radiation concerned in the observation, e.g., a red light, a blue lamp, a white face, etc.

f couleur f — **d** Farbe f — **n** kleur fm — **e** color m — **i** colore m

3114 colour adjusting — Regulating the flow of ink.

f réglage m du débit d'encre — **d** Farbzufuhr-Einstellung f — **n** regeling f van de inkttoevoer; inktregeling f; inktgeving f — **e** arreglo m de tintaje; ajuste m del tintero; entintaje m — **i** regolazione f dell'inchiostrazione

3115 colour analyst; colour analyzer — Apparatus operating on the principle of additive synthesis and designed to show a full-colour picture from monochrome colour separation prints of halftone proofs as an aid in judging the results of colour photography and plate making.

f analyseur m de couleurs — **d** Farbanalyst m — **n** kleurmeter m — **e** analizador m de colores — **i** analizzatore m dei colori

3116 colour analyzer — See colour analyst.

3117 colour appraisal — Inspection of surfaces or objects under a suitable illuminant to make subjective judgements on their colour characteristics. These judgements may vary between the extremes of colour matching and aesthetic appreciation of an isolated colour.

f examen m des couleurs — **d** Farbabmusterung f; Farbbeurteilung f — **n** kleurbeoordeling f — **e** valuación f de colores — **i** valutazione f dei colori

3118 colour artist — See benday artist.

3119 colour balance — Relationship between the overall intensities of the different colour images in a colour reproduction.

f équilibre m chromatique; équilibre m des couleurs — **d** Farbgleichgewicht n — **n** kleurbalans fm; kleurevenwicht n — **e** balance m del color — **i** equilibrio m cromatico; equilibrio m di colori

3120 colour bank — The four printing units printing in succession yellow, magenta, cyan and black.

f séquence f quadrichrome — **d** Druckwerksfolge f — **n** kleurvolgorde fm — **i** ordine m dei colori successivamente

3121 colour bath; colouring bath — Vat or tank in which surface coloured papers are dipped.

f bain m de teinture — **d** Farbbad n — **n** kleurbad n — **e** baño m de color — **i** bagno m di coloritura; tino m di coloritura

3122 colour blind

f daltonien — **d** farbenblind — **n** kleurenblind — **e** daltoniano — **i** daltonico

3123 colour blindness; daltonism — Incapacity to perceive colours, independent of capacity to distinguish light and shade. In scientific terminology: defective colour vision, i.e., the condition in which colour discrimination is significantly reduced in comparison with the normal trichromat.

f daltonisme m; insensibilité f à la couleur; achromatopsie f — **d** Farbenblindheit f — **n** kleurenblindheid f — **e** daltonismo m; ceguedad f de los colores; acromatopsia f; discromatopsia f — **i** daltonismo m; acromatopsia f

3124 colour burnout — See burnout.

3125 colour chart — See colour-control chart.

3126 colour circle — See chromatic circle.

3127 colour collotype — Collotype printing in three or more colours. The best facsimile reproductions in this process are using as many as twelve and more colours. See also collotype process.

f phototypie f en couleurs — **d** Farbenlichtdruck m — **n** kleurenlichtdruk m — **e** fototipia f en colores — **i** fototipia f a colori

3128 colour control — Measures in all steps in the production of printed material to ensure matching and uniformity of colour.

f contrôle m des couleurs — **d** Farbkontrolle f — **n** kleurhouden n; kleurcontrole fm — **e** comprobación f de color — **i** controllo m del colore

3129 colour-control chart; colour chart; scale of colours — Chromatic colour standards against which coloured materials or printing may be compared, to aid in developing uniformity or consistency of colour tones.
f carte *f* de couleurs; carte *f* collection de teintes; comparateur *m* de couleurs — **d** Farb(en)skala *f*; Farb(en)karte *f*; Farbentafel *f*; Farb(en)musterkarte *f* — **n** kleurenkaart *f* — **e** carta *f* muestra del color — **i** carta *f* dei colori
3130 colour correcting mask — See colour-correction mask.
3131 colour correction — Method such as masking, dot etching, and re-etching, to improve colour rendition. Can be done on screened or continuous-tone negatives or by corrective work on the halftone printing plates.
f correction *f* de couleur — **d** Farbkorrektur *f* — **n** kleurcorrectie *f* — **e** rectificación *f* cromática; corrección *f* del color — **i** correzione *f* cromatica
3132 colour-correction mask; colour correcting mask
f masque *m* de correction de couleur — **d** Farbkorrekturmaske *f* — **n** kleurcorrectiemasker *n* — **e** máscara *f* de corrección del color; máscara *f* correctiva del color — **i** maschera *f* per la correzione dei colori
3133 colour deck — See single-colour unit.
3134 colour disk — See chromatic circle.
3135 coloured edges — The cut edges of a book stained with a solid colour.
f tranches *fpl* coloriées — **d** gefärbte Buchschnitte *mpl*; glatter Farbschnitt *m* — **n** geverfde sneden *fmpl* — **e** cortes *mpl* coloreados; cortes *mpl* pintados — **i** tagli *mpl* colorati
3136 coloured fibres
f fibres *fpl* colorées — **d** gefärbte Fasern *fpl*; Melierfasern *fpl* — **n** gekleurde vezels *fmpl*; mêleervezels *fmpl*; minuteervezels *fmpl*; minuutvezels *fmpl* — **e** fibras *fpl* coloreadas — **i** fibre *fpl* colorati
3137 coloured paper — There are two kinds of coloured paper: 1. stock coloured paper; 2. surface coloured paper.
f papier *m* de couleur (1. papier coloré dans la masse; 2. papier coloré en surface) — **d** farbiges Papier *n* (1. massegefärbtes Papier; 2. oberflächengefärbtes Papier) — **n** gekleurd papier *n* (1. in de stof geverfd papier; 2. oppervlakte-geverfd papier) — **e** papel *m* de color; papel *m* coloreado — **i** carta *f* colorata
3138 coloured sheepskin
f alude *f*; alute *f*; peau *f* de mouton teintée — **d** gefärbtes Schafleder *n* — **n** gekleurd schapeleer *n* — **e** badana *f* colorada; baldés *m* — **i** bazzana *f* colorata
3139 colour etcher — See colour process etcher.
3140 colour etching; gravure à la poupée — Method of printing many colours simultaneously from a single plate in which pads of rolled felt are used to ink and wipe each colour separately.
f impression *f* taille-douce dite à la poupée; gravure *f* à la poupée; gravure *f* à l'eau-forte en couleurs; eau-forte *f* en couleurs — **d** Farbradierung *f* — **n** kleurets *fm* — **e** impresión *f* con muñequilla al aguafuerte; grabado *m* colorido con muñeca — **i** incisione *f* all'acqua forte a colori; stampa *f* colorata al tampone
3141 colour fading — Gradual change in colour of paper or ink, usually by exposure to light and acid present in the atmosphere or in the paper.
f se décolorer — **d** verblassen; sich verfärben — **n** verkleuren; verschieten — **e** desvanecer del color; decolorar gradualmente; descolorarse — **i** scolorimento *m*
3142 colour fastness — Ability of a pigment to retain its original hue, chroma and value under conditions of storage and use. Generally refers to action by light, but it may also refer to other agents, e.g., colour-fastness to alkali, to acid, to ultra-violet light, etc.
f solidité *f* de la couleur — **d** Farbechtheit *f* — **n** kleurechtheid *f* — **e** permanencia *f* del color; solidez *f* del color — **i** solidità *f* di colore
3143 colour-fastness device — Instrument to test the fastness to light by measuring the degree of fading.
f appareil *m* pour examiner la solidité des couleurs à la lumière — **d** Lichtechtheitsprüfgerät *n*; Lichtechtheitsgerät *n* — **n** lichtechtheidsbepalingstoestel *n*; lichtechtheidstoestel *n* — **e** aparato *m* para examinar la sensibilidad a la luz — **i** apparecchio *m* per misurare la solidità alla luce
3144 colour film — Multi-layer photographic film to produce coloured pictures.
f film *m* sur la photo en couleurs — **d** Farbfilm *m* — **n** kleurenfilm *n* — **e** película *f* de color; película *f* en colores; película *f* fotocroma — **i** pellicola *f* a colori; film *m* a colori
3145 colour filter; colour screen; light filter — Coloured gelatine film cemented between glass, or coloured liquid in a transparent container, interposed between the camera lens and the object photographed during exposure, to filter out colours, i.e., to absorb some colours and allow others to pass. See also trichromatic process printing.
f filtre *m* coloré; écran *m* coloré — **d** Farbfilter *m* — **n** kleurfilter *n* — **e** filtro *m* de color; filtro *m* coloreado; filtro *m* cromofotográfico; filtrarayos *m*; filtrarrayos *m* — **i** filtro *m* colore; filtro *m* colorato
3146 colour foil — Foil for stamping in colour of book covers.
f feuille *f* oeser; oeser *f* — **d** Oeser-Folie *f*; Farbfolie *f*; Prägefolie *f* — **n** kleurfo(e)lie *fm* — **e** hoja *f* en color — **i** colore *m* in foglia; foglia *f* colorata
3147 colour forme — Forme to print each colour separately.
f forme *f* (pour impression) en couleur — **d** Farbenform *f* — **n** kleurvorm *m* — **e** forma *f* para imprimir en color — **i** forma *f* per la stampa a colori

3148 colouring bath — See colour bath.
3149 colouring book — Book of outline drawings for colouring in crayons or water colours by children.
f livre *m* à colorier; album *m* à colorier — **d** Kolorierbuch *n*; Ausmalebuch *n*; Ausmaleheft *n* — **n** kleurboek *n* — **e** libro *m* para pintar — **i** libro *m* da colorare
3150 colouring matter — See pigment.
3151 colourless
f non coloré; incolore — **d** farblos; ungefärbt — **n** kleurloos — **e** sin color; incoloro — **i** senza colore; incolore
3152 colour lithography — See chromolithography.
3153 colour lumps — *Hard lumps of coating left on the surface of a sheet of coated paper. See also lumps, soft lumps.
f mâtons *mpl* — **d** Farbklumpen *mpl*; Farbbatzen *mpl*; Farbpatzen *mpl* — **n** strijklaagknopen *mpl* — **i** macchie *fpl* di patina
3154 colour masking — See masking.
3155 colour matching — Visual comparison under a suitable illuminant of two materials or fields, usually contiguous, to judge their similarity in colour.
f contretypage *m* des couleurs — **d** Farbabstimmung *f*; Nachstellung *f* von Farben; Abmusterung *f* von Farben; Farbabmusterung *f* — **n** kleurmaken *n*; afstemmen *n* van de kleur; kleurafstemming *f* — **e** casado *m* de colores; contrahacer el color; matizar — **i** messa *f* a punto dei colori; accostamento *m* dei colori; accordo *m* dei colori; confronto *m* del colore
3156 colour of light source — Colour that cannot be classified due to its lightness, but to the intensity of its source. Grey and brown are lacking.
f couleur *f* de la lumière — **d** Lichtfarbe *f* — **n** kleur *fm* van het licht; lichtkleur *fm* — **e** color *m* de la luz — **i** colore *m* della luce
3157 colour patch — Small pre-printed card showing the inks being used for process colour work and attached to the original art when photographed for colour separations. An aid to the correction artist when analyzing his tones.
f charte *f* de couleurs — **d** Farbskala *f*; Farbkontrollfelder *npl* — **n** kleurkaart *fm* — **e** escala *f* de colores; gama *f* de colores — **i** carta *f* dei colori; scala *f* dei colori
3158 colour photo — See colour photograph.
3159 colour photograph; colour photo
f photographie *f* en couleurs; photo *f* en couleurs — **d** Farbphoto *n*; Farbphotographie *f*; Farbaufnahme *f* — **n** kleurenfoto *f* — **e** fotografía *f* en colores; fotografía *f* en color; foto *f* en colores; foto *f* en color; fotocromía *f*; fotocromografía *f* — **i** fotografia *f* a colori; foto *f* a colori
3160 colour photography — Photographic reproduction of a subject or original in its natural colours; photography of any coloured original to record one or more colours.
f photographie *f* en couleurs — **d** Farbenphoto-

graphie *f* — **n** kleurenfotografie *f* — **e** fotografía *f* en colores; fotografía *f* en color; cromofotografía *f*; fotocromía *f*; fotocolor *f* — **i** fotografia *f* a colori
3161 colour pigments — Pulverized, insoluble, coloured, inorganic substances used in ink and paper making.
f pigments *mpl* colorés — **d** Farbpigmente *npl* — **n** gekleurde pigmenten *npl* — **e** pigmentos *mpl* de color — **i** pigmenti *mpl* colorati
3162 colour print — Print in two or more colours.
f imprimé *m* polychrome; planche *f* en couleurs; impression *f* en couleurs — **d** Farbdruck *m* — **n** kleurendruk *m*; plaat *fm* in kleuren; kleurenplaat *fm* — **e** fotocromograbado *m*; fototipocromía *f*; fotocromotipograbado *m*; cromotipografía *f*; cromotipia *f*; impresión *f* en colores; reproducción *f* en colores; policromía *f*; autocromía *f* — **i** stampato *m* a colori; policromia *f*
3163 colour printing — Printing in two or more colours to simulate a design or picture in its original colours.
f impression *f* en couleurs — **d** Mehrfarbendruck *m*; Buntfarbendruck *m*; asynchromer Druck *m* — **n** kleurendruk *m* — **e** impresión *f* a colores; impresión *f* en colores; cromotipia *f*; cromotipografía *f*; tipocromía *f*; policromotipografía *f*; policromía *f* — **i** policromia *f*; stampa *f* a colori
3164 colour process etcher; colour etcher
f chromiste *m* — **d** Farbätzer *m*; Farbenätzer *m* — **n** kleuretser *m* — **e** cromista *m* — **i** cromista *m*
3165 colour process plates — Halftone colour plates, particularly those made from separation negatives.
f clichés *mpl* simili en couleurs — **d** Farbplatten *fpl* — **n** kleuren-autotypieclichés *npl* — **e** medios tonos *mpl* para policromía; grabados *mpl* para policromía; clisés *mpl* de punto para similigrabados *Ch*; láminas *fpl* para policromía — **i** lastre *fpl* dei colori
3166 colour process printing — See colour process work.
3167 colour process work; process colour work; colour process printing — Reproduction of colour made by photographic separations. Also the method or the copy requiring such operation.
f photogravure *f* en couleurs — **d** Mehrfarbendruck *m* mit Farbauszügen; Farbreproduktion *f* — **n** kleurenwerk *n* — **e** impresión *f* policroma; fotocromotipografía *f*; policromía *f*; fotograbado *m* en colores — **i** riproduzione *f* a colori; procedimento *m* a colori
3168 colour proof — Proof of combined and registered plates printed in proper colours.
f épreuve *f* en couleurs — **d** Probeabzug *m* in Farben — **n** proef *fm* in kleuren — **e** prueba *f* de color(es); prueba *f* en color(es) — **i** bozza *f* a colori
3169 colour register — The accurate aligning or superimposing of one colour over another in succession on the web.

f repérage *m* des couleurs — **d** Farbpasser *m*; Farbregister *n* — **n** kleurregister *n* — **e** registro *m* de los colores — **i** registro *m* dei colori

3170 colour rendering
f rendu *m* des couleurs — **d** Farbwiedergabe *f*; Farbtonwiedergabe *f* — **n** kleurweergave *fm* — **e** rendición *f* en color — **i** resa *f* dei colori

3171 colour reproduction — Print in colours made from photomechanically produced printing formes.
f reproduction *f* en couleurs — **d** Farbdruck *m*; Farbendruck *m*; Farbreproduktion *f* — **n** reproduktie *f* in kleuren — **e** reproducción *f* en colores — **i** riproduzione *f* a colori

3172 colour reproduction — The quality of rendering the colours of an original.
f reproduction *f* des couleurs — **d** Farbwiedergabequalität *f*; Farbreproduktionsqualität *f*; Farbwiedergabe *f* — **n** kleurweergave *f* — **e** rendición *f* de colores — **i** riproduzione *f* dei colori

3173 colour retouching — The manual application of masks, usually by brushing neutral grey dye solutions to certain parts of colour separations for reducing the transparency in these parts, resulting in a reduction of printing strength of the colour in the affected parts.
f retouche *f* de couleurs — **d** Farbretusche *f* — **n** kleurretouche *fm* — **e** retoque *m* de los colores — **i** ritocco *m* colore

3174 colour reversion — See yellowing.

3175 colour room — Room where the pigments or dyes are stored, mixed and prepared, usually located near the beater room of a paper mill.
f salle *f* des colorants; atelier *m* de coloration — **d** Farbküche *f* — **e** cocina *f* — **i** locale *m* di pigmenti e coloranti

3176 colours *GB*; **colors** *US* — Pigments, lakes or dyes used in the manufacture of coloured paper or ink. The shades of a given colour may vary widely in papers and inks by different manufacturers.
f couleurs *fpl* — **d** Farbstoffe *mpl* — **n** kleurstoffen *fmpl* — **e** colorantes *mpl*; materias *fpl* colorantes — **i** colori *mpl*; coloranti *mpl*; materie *fpl* coloranti

3177 colour scanner — Photoelectrical device for production and automatic colour correction of continuous tone separation negatives made from multicolour originals; intended for more accurate balancing of a set of four-colour images and the elimination of undercolours therefrom.
f sélecteur *m* électronique; correcteur *m* électronique — **d** Scanner *m* — **n** kleurenscanner *m* — **e** seleccionadora *f* para la separación tricroma y confección de los clisés — **i** apparecchio *m* per la selezione elettronica dei colori; analizzatore *m* elettronico dei colori; scanner *m*

3178 colour screen — See colour filter.

3179 colour selection negative — See colour separation negative.

3180 colour sensitiveness — See colour sensitivity.

3181 colour sensitivity; colour sensitiveness — Response of photosensitive material to coloured light.
f sensibilité *f* chromatique — **d** Farbenempfindlichkeit *f* — **n** kleurgevoeligheid *f* — **e** sensibilidad *f* cromática — **i** sensibilità *f* cromatica

3182 colour sensitizing — Increasing the colour sensitivity of emulsions on photographic plates and films by addition of dyes or colour sensitizers.
f sensibilisation *f* chromatique — **d** Farbsensibilisierung *f* — **n** kleurensensibilisatie *f* — **e** sensibilizado *m* cromático — **i** sensibilizzazione *f* cromatica

3183 colour separation — In colour photography, the isolation or division of the colours of an original into their primary hues, each record or negative used for the production of a colour plate. Manually separating or introducing colours in printing plates. Direct separations are made with the use of the halftone screen; indirect separations involve continuous tone separation negatives and screened positives made from these.
f sélection *f* des couleurs — **d** Farbauszug *m*; Farbausscheidung *f* — **n** kleurscheiding *f* — **e** selección *f* de color(es); selección *f* en colores; selección *f* fotomecánica de colores; selección *f* cromática; selección *f* tricroma; descarte *m* — **i** selezione *f* dei colori; separazione *f* dei colori

3184 colour separation negative; colour selection negative — Photographic negative exposed through a colour filter and recording only one of the primary colours in full colour work or one of the two colours in two-colour work. In platemaking, manual separation of colours by handwork directly on the printing surface. The negative is a grey tonal record of the intensity and the colour it is reproducing, being light where the colour is strong in the copy, dark where the colour is weak. Sometimes called a cut out.
f négatif *m* de sélection — **d** Farbauszugsnegativ *n*; Farbtrennungsnegativ *n* — **n** kleurdeelnegatief *n*; deelnegatief *n* — **e** negativo *m* de selección; negativo *m* seleccionado; negativo *m* de citocromía; negativo *m* de tricromía — **i** negativo *m* di selezione

3185 colour sequence — The order in which colours are printed.
f ordre *m* de passage des couleurs — **d** Reihenfolge *f* der Farben; Farbreihenfolge *f*; Farbenfolge *f*; Farbfolge *f* — **n** volgorde *fm* der kleuren; kleurvolgorde *fm* — **e** orden *m* de impresión — **i** sequenza *f* dei colori

3186 colours of the spectrum — If a beam of white light passes through a glass prism, it shows to exist of seven (visible) colours, red, orange, yellow, green, blue, indigo and violet. If the refracted rays are projected onto a screen the entire range is called the *spectrum.
f couleurs *fpl* spectrales — **d** Spektralfarben *fpl* — **n** spectrale kleuren *fmpl*; kleuren *fmpl* van het spectrum — **e** colores *mpl* del espectro — **i**

colori *mpl* dello spettro

3187 colour solid — In colour measurement, part of the colour space which is occupied by surface colours.
f solide *f* des couleurs — **d** Farbkörper *m* — **n** kleurvlak *n* — **e** solido *m* cromático — **i** solido *m* cromatico

3188 colour space — In colour measurement, manifold of three dimensions for the geometrical representation of colours.
f espace *m* chromatique — **d** Farbenraum *m* — **e** espacio *m* cromático — **i** spazio *m* cromatico

3189 colour strength; tinctorial strength; tinctorial power; tinting strength — In printing ink, the effective concentration of colouring material per unit of volume. The higher the colour strength, the thinner the film that can be printed and the greater the number of sheets resulting per volume of ink.
f pouvoir *m* colorant; puissance *f* colorante; intensité *f* de la couleur — **d** Farbenstrenge *f*; Farbstärke *f*; Farbintensität *f* — **n** kleurkracht *fm*; intensiteit *f* van de kleur — **e** fuerza *f* colorante — **i** forza *f* di colore

3190 colour swatch — Small, usually square solid print to identify the sketch, negative, positive or printing plate and furnish a sample of the actual ink colours used. A guide in colour separation and correction operations.
f plage *f* d'identification de couleur; témoin *m* de couleur; touche *f* de couleur — **d** Farbkontrollfeld *n*; Farbmarke *f* — **n** kleurvlakje *n* — **e** patrón *m* de color; probeta *f* de color — **i** campione *m* di colore

3191 colour temperature — Means of measuring and indicating the spectral quality of visible light by determining the relative amounts of blue and red rays in sources of illumination. Measurement is done with specially designed meters and is expressed in degrees Kelvin (K), which are in equivalent to the actual temperature of the light source plus 273 degrees Centigrade.
f température *f* de couleur — **d** Farbtemperatur *f* — **n** kleurtemperatuur *f* — **e** temperatura *f* de(l) color — **i** temperatura *f* di colore

3192 colour transparency — Full-colour photographic positive image on a transparent support, in natural colours.
f diapositif *m* en couleurs — **d** Farbdiapositiv *n*; Farbtransparenz *f* — **n** kleurendia *m*; kleurdiapositief *n* — **e** transparencia *f* en color; diapositiva *f* en color(es); diapositiva *f* de color; diapositiva *f* cromática — **i** diapositivo *m* a colori; trasparente *m* a colori

3193 colour triangle; chromaticity diagram; xyz system — System to define each colour corresponding to the sensation of the colour experienced by a conventional mean standard observer, defined by the CIE, based on the three coloured primary lights: primary red, wavelength 700.0 mμ; primary green, wavelength 546.1 mμ; and primary blue, wavelength 435.8 mμ. These primaries are in increasing order of frequency (in thousand millions of kilocycles).
f triangle *m* chromatique; diagramme *m* chromatique; diagramme *m* de chromaticité; triangle *m* des couleurs — **d** Farbendreieck *m*; Farbendiagramm *n* — **n** kleurendriehoek *m*; kleurendiagram *n* — **e** triángulo *m* de colores; diagrama *m* de cromaticidad — **i** triangolo *m* dei colori; triangolo *m* cromatico; diagramma *m* di cromaticità

3194 colour unit — Printing unit on a multicolour press.
f groupe *m* couleur — **d** Farbdruckwerk *n*; Farbdruckeinheit *f* — **n** kleureenheid *f*; kleurdrukwerk *n* — **e** unidad *f* de color — **i** elemento *m* colore

3195 colour woodcut
f gravure *f* (en taille d'épargne) en couleurs — **d** Farbenholzschnitt *m* — **n** kleurenhoutsnede *fm* — **e** xilografía *f* in colores — **i** incisione *f* di colori in legno

3196 colour wood engraving — See chromo xylography.

3197 colour work — Printing in two or more colours, such as line and tint work, two-colour halftone, three-colour halftone, planographic and intaglio work in colour.
f travail *m* en couleurs — **d** Farbdruckarbeit *f*; Mehrfarbendruck *m* (mehrere Farben übereinander); Buntdruck *m* (mehrere Farben nebeneinander) — **n** kleurenwerk *n*; kleuren-autotypiewerk *n*; kleurenlijnwerk *n* — **e** impresión *f* en colores; trabajo *m* en colores — **i** lavoro *m* a colori

3198 Columbian — Obsolete name for an American size of type. See app. no. 1.

3199 column — Vertical division of a printed page separated by a rule from an other vertical division, as in newspapers or by white spaces, as in magazines.
f colonne *f* — **d** Spalte *f*; Druckspalte *f*; Kolumne *f*; Kolonne *f* — **n** kolom *fm* — **e** columna *f* — **i** colonna *f*

3200 column — Regular department of a newspaper, periodical etc., usually with a readily identifiable heading and the name of the author.
f rubrique *f* — **d** Rubrik *f* — **n** vaste kolom *fm*; kolom *fm*; rubriek *f* — **e** sección *f* fija; columna *f*; rúbrica *f* — **i** rubrica *f*

3201 column — Character or digit position in a physical device such as an eighty column punched card. Also positions in a format where the characters appear in rows as in a table or in a listing of data.

3202 column binary; Chinese binary — Binary representation of data on punched cards in which adjacent positions in a column correspond to adjacent bits of data. As opposed to *row binary.
f carte *f* binaire par colonne — **d** Karte *f* binär pro Spalte — **n** kaart *fm* binair per kolom — **e** tarjeta *f* binaria en columna — **i** scheda *f* binaria in colonna

3203 column composition — See column matter.

3204 column galley — See slip galley.

3205 column gutter — See gutter.

3206 column inch — Measurement unit for newspaper advertising space, one inch deep and one column wide.

3207 columnist — Author who writes articles for a newspaper that are published regularly in a specially reserved column.

f rubriquart *m* — **d** Kolumnist *m* — **n** kolomschrijver *m*; vaste medewerker *m* — **e** columnista *m* titular de sección; colaborador *m* permanente — **i** columnist *m*

3208 column matter; column composition — In US the column width of periodicals is usually 12, 12.5 and 13 picas.

f tableaux *mpl*; composition *f* de(s) tableaux — **d** Spaltensatz *m*; zusammengesetzte Spalten *fpl* — **n** kolomzetsel *n*; zetsel *n* in kolommen — **e** composición *f* de ancho de columna — **i** composizione *f* su colonne

3209 column rule — Brass rule with a fine face, to separate columns of type matter.

f colombelle *f* — **d** Spaltenlinie *f* — **n** kolomlijn *fm* — **e** corondel *m*; columnaria *f Me*; raya *f* de columna *Pa*; raya *f* para columna; raya *f* entre columnas *Pe*; rayado *m Ur* — **i** filetto *m* fra due colonne; colombella *f sl*.

3210 columns across — Letterpress rotary term for printing with the column length imposed across the cylinder.

f colonnes *fpl* en sens traverse — **d** Spaltenquerlage *f* — **n** dwarskolommen *fmpl*; kolommen *fmpl* overdwars — **e** columnas *fpl* transversales — **i** colonne *fpl* in senso trasversale

3211 columns around — Letterpress rotary term for printing with the column length imposed around the cylinder.

f colonnes *fpl* dans le sens de la rotation — **d** Spaltenlängslage *f* — **n** kolommen *fmpl* rondom de cilinder — **e** columnas *fpl* en el sentido de rotación — **i** colonne *fpl* in senso periferico

3212 column screw — Duct adjusting screw on a letterpress rotary to regulate the flow of ink to one column of printed matter.

f vis *f* de réglage d'encrier; vis *f* d'encrier — **d** Zonenschraube *f* — **n** kolomschroef *fm* — **e** tornillo *m* de ajuste del tintero — **i** vite *f* di regolazione su una colonna

3213 column width; width of column — The width or measure of a printed column.

f justification *f* de colonne — **d** Spaltenbreite *f*; Zeilenbreite *f* — **n** kolombreedte *f*; regelbreedte *f* — **e** ancho *m* de la columna; medida *f* de la columna; ancho *m* de la composición; regleta *f Ur* — **i** larghezza *f* di colonna; giustezza *f* di colonna

3214 coma — Fault in lens performance that results in an asymmetrical light patch, flairing away like the tail of a comet.

f coma *m* — **d** Koma *n* — **n** coma *fm* — **e** coma *f* — **i** coma *m*

3215 comb — See lining tool.

3216 comb — See marbling comb.

3217 combed edges — Marbled edges with a pattern obtained by combing the colour.

f tranches *fpl* peignées — **d** Kammarmorschnitt *m* — **n** snede *fm* met kammarmer — **e** cantos *mpl* de barras de peine — **i** tagli *mpl* a pettine; tagli *mpl* marmorizzati

3218 combers; combing wheels — Beaded wheels which separate and buckle top sheets in the feeder apparatus.

f frotteurs *mpl*; dragues *fpl* — **d** Streichräder *npl* — **n** strijkraderen *npl* — **e** ruedas *fpl* alimentadoras; peiñadores *mpl*; esquinadores *mpl*; combadores *mpl* — **i** ruote *fpl* steccatore; ruote *fpl* a strisciamento

3219 combination; tooth combination — Number and arrangement of the teeth on a matrix of a slug composing machine, to distribute the matrix into its correct channel in the magazine.

f combinaison *f* — **d** Kombination *f* (der Matrizenzähne); Zahnung *f* — **n** combinatie *f*; tandcombinatie *f* — **e** combinación *f* de los dientes — **i** combinazione *f* della dentatura

3220 combination fraction — See piece fraction.

3221 combination newsboard — See newsboard.

3222 combination plate — See line-halftone combination (plate).

3223 combination rule — See cheque line.

3224 combination style — Binding process in which the sections are sewn together, the back reinforced by cloth hinges and the book made ready for insertion in the case, which has been built-up separately.

3225 combined head — See read/write head.

3226 combined read/write head — See read/write head.

3227 combing wheels — See combers.

3228 comb, perforating — See perforating comb.

3229 come *v* in — To occupy a designated space when set.

f rentrer — **d** hineinpassen; ausfüllen — **n** (precies) passen — **e** alcance *m*; entrada *f* — **i** rientrare

3230 come out — See appear *v*.

3231 comets — Extraneous ink film deposited in the shape of a comet. Caused by the printing cylinder having been gouged by a hard foreign particle freeing itself from under the doctor blade.

f comètes *fpl* — **d** Kometen *mpl* — **n** komeetjes *npl* — **e** cométas *fpl* — **i** comete *fpl*

3232 comic ink — Ink for colour printing, that is strong, free flowing, and dry by penetration.

f encre *f* en couleur à journal — **d** bunte Zeitungsrotationsfarbe *f* — **n** kleurinkt *m* voor rotatieboekdruk; gekleurde rotatieboekdrukinkt *m* — **e** tinta *f* de color para periódicos — **i** inchiostro *m* colorato da giornali

3233 comic strip — Sequence of drawings, either in colour or black and white, relating a comic incident, an adventure, or mystery story, etc., with dialogues printed in balloons. Often serialized, usually a horizontal strip in daily newspapers and in an uninterrupted block or longer sequence of such strips in Sunday newspapers and comic books.

f bande *f* dessinée — **d** Bildergeschichte *f*; Bilderstreifen *m*; Comic strip *m* — **n** beeldverhaal *n*; stripverhaal *n* — **e** tira *f* cómica; tira *f* de muñecos; tira *f* de muñequitos; historieta *f* gráfica; historieta *f* cómica; tirilla *f* cómica; tirilla *f* de muñecos; tirilla *f* de muñequitos — **i** romanzo *m* a fumetti

3234 comma — Punctuation mark (,) to indicate the briefest of the pauses that break sentences into work groupes, placed before a continuative relative clause. In the decimal system it is used as a decimal point instead of a dot.

f virgule *f* — **d** Komma *n*; Beistrich *m* — **n** komma *fm* — **e** coma *f*; vírgula *f* — **i** virgola *f*

3235 comma friend; ink slinger — By-name for a *proofreader.

f père *m* virgule — **d** Kommajäger *m*; Kommareiter *m* — **n** zifter *m*; haarklover *m* — **e** emborronador *m* de cuartilla; corrector *m* pedantesco — **i** correttore *m* pedantesco

3236 command — Control signal, or set of signals indicating one step in the performance of a circuit or device.

f signal *m* de commande — **d** Führungssignal *n* — **n** geleidesignaal *n* — **e** señal *f* de mando; mandato *m* — **i** segnale *m* di comando

3237 command — See instruction.

3238 command language — Sources language consisting primarily of procedural operators, each capable of invoking a function to be executed.

f langage *m* à instructions — **d** Befehlssprache *f* — **n** opdrachttaal *fm* — **e** lenguaje *m* de instrucciones — **i** linguaggio *m* ad istruzioni

3239 commence quote; begin quote — Mark used to denote the beginning of a quotation. See also close quote.

f guillemet *m* ouvrant — **d** Anführung *f*; Anführungszeichen *n* — **n** aanhalingsteken *n* "openen" — **e** comilla *f* de abertura; comilla *f* de escomienzo — **i** virgoletta *f* di apertura

3240 commercial artist; designer — Designer of graphic art for commercial uses, esp. for advertising, illustrations in magazines or books, or the like.

f dessinateur *m* en publicité; dessinateur *m* publicitaire — **d** Gebrauchsgraphiker *m* — **n** grafisch tekenaar *m*; grafisch ontwerper *m*; reclametekenaar *m* — **e** dibujante *m* de trabajos de publicidad; dibujante *m* comercial; dibujante *m* publicitario; dibujante *m* gráfico; dibujante *m* maquetista; artista *m* dibujante; publigráfico *m*; diseñador *m*; bocetista *m* — **i** disegnatore *m* grafico; bozzettista *m*; progettista *m* grafico

3241 commercial forms — See business forms.

3242 commercial letter; commercial post — A size of writing paper. See app. no. 3.

3243 commercial lithographer — Lithographer for mercantile work.

f lithographe *m* pour travaux de commerce — **d** Merkantillithograph *m*; Industrielithograph *m* —

n mercantiellithograaf *m* — **e** litógrafo *m* comercial — **i** litografo *m* per lavoro commerciale

3244 commercial note — A size of writing paper. See app. no. 3.

3245 commercial post — See commercial letter.

3246 common — Customary cash column rule used on account books and paper.

f réglure *f* comptable — **d** Betragsspaltenlinie *f* — **n** geldkolom *fm*

3247 common alcohol — See ethyl alcohol.

3248 common business oriented language — See cobol.

3249 common impression cylinder press; drum type press — Offset press with four or five printing units around one impression cylinder. The web passes through a dryer and proceeds to the second printing unit, where the opposite side of the web is printed and dried.

f machine *f* planétaire; machine *f* à tambour; machine *f* à groupe satellite — **d** Trommelzylindermaschine *f*; Maschine *f* in Trommelbauweise; Maschine *f* mit einem gemeinsamen Gegendruckzylinder — **n** satellietmachine *f* — **e** máquina *f* satélite — **i** macchina *f* satellite; macchina *f* a tamburo; macchina *f* tipo planetario

3250 common machine language

f langage *m* commun machine — **d** Standardmaschinenkodeschrift *f* — **n** gemeenschappelijke machinetaal *fm* — **e** lenguaje *m* de máquina comun — **i** linguaggio *m* macchina comune

3251 common salt — See sodium chloride.

3252 communicating time — See computer time.

3253 comp — See complimentary copy.

3254 comp — See compositor.

3255 companionship; 'ship — Company of compositors working together.

f équipe *f* — **d** Arbeitsgruppe *f* — **n** ploeg *fm* — **e** cuadrilla *f* (de obreros); equipo *m* — **i** gruppo *m* di lavoro; squadra *f*

3256 compare *v* — To examine two words to discover either their relative magnitude or their relative order in a sequence.

f comparer — **d** vergleichen — **n** vergelijken — **e** comparar — **i** comparare

3257 compartment carton

f emballage *m* à alvéoles — **d** Faltzellenverpackung *f* — **n** draagkarton *n* met aangestanst interieur — **i** scatola *f* d'imballaggio a compartimenti

3258 compatibility — In general the capability of existing together in the same subject. In computer language: 1. Hardware: the characteristics required of all the units in a computer configuration or range of systems to ensure that they are compatible mechanically, electronically and logically with each other. 2. Software: the characteristics required of a set of computer program commands to ensure that it can be translated or interpreted on more than one computer model. 3. Hardware and software: the characteristics required of computers, peripheral equipment and software to ensure that data and programs prepared for one computer system may be used in the same form on another.

f compatibilité — **d** Verträglichkeit *f*; Austauschbarkeit *f* — **n** uitwisselbaarheid *f* — **e** compatibilidad *f* — **i** compatibilità *f*

3259 compatible — Mutually soluble; get along well together. One ink cannot be mixed with another ink unless the two inks are compatible and of similar type.

f compatible — **d** verträglich — **n** elkaar verdragend — **e** compatible — **i** compatibile

3260 compendium — Abridgement of a work or treatise, giving the sense and substance, within smaller compass.

f compendium *m*; précis *m*; abrégé *m* — **d** Kompendium *n*; Handbuch *n*; kurzgefaßtes Lehrbuch *n* — **n** compendium *n*; handboek *n*; kort begrip *n* — **e** compendio *m* — **i** compendio *m*

3261 compensating filter; correction filter

f écran *m* compensateur — **d** Kompensativfilter *m* — **n** correctiefilter *n* — **e** filtro *m* compensador; filtro *m* de compensación; filtro *m* de corrección — **i** filtro *m* compensatore; filtro *m* di correzione

3262 compensating plate lock-up — Compression plate lock-up on a letterpress rotary with resilient means to compensate imperfections in the plate bevel.

f système *m* de calage à griffes régables — **d** Plattenspannvorrichtung *f* mit Ausgleich — **n** compenserende compressie-bevestiging *f* — **e** cierre *m* de planchas compensador — **i** dispositivo *m* di tensione con compensazione

3263 compensating relaxation allowance; CR allowance; personal allowance; personal needs allowance; rest allowance; relaxation allowance — Allowance for recovery and personal needs.

f allocation *f* pour besoins personnels — **d** (persönliche) Verteilzeit *f* (Refa) — **n** toeslag *m* voor persoonlijke verzorging — **e** sobresueldo *m* por cuidado personal — **i** permesso *m* per motivi personali

3264 compensating roller — See jockey roller.

3265 compensation guards; filling-in guards; guards; stubs — In bookbinding, narrow strips of paper sewed between the signatures to compensate for the thickness of inserted plates, maps, samples, photographs.

f onglets *mpl* (de remplissage) — **d** Füllfälze *mpl*; Ausgleichungsstreifen *mpl* — **n** opvulstrookjes *npl* — **e** cartivanas *fpl* de relleno; escartivanas *fpl* de relleno — **i** braghette *fpl* di compensazione; braghette *fpl* di riempimento

3266 compensator — See jockey roller.

3267 compilation

f compilation *f* — **d** Kompilation *f* — **n** compilatie *f* — **e** compilación *f* — **i** compilazione *f*

3268 compile *v* — To prepare a machine language program (or a program expressed in symbolic coding) from a program written in another programming language.

f compiler — **d** kompilieren — **n** compileren — **e** compilar — **i** compilare

3269 compiler — One who makes compilations, i.e., a hash-up of the work of others.

f compilateur *m* — **d** Kompilator *m* — **n** compilator *m*; kompilator *m* — **e** compilador *m*; recopilador *m* — **i** compilatore *m*

3270 compiler; compiler program — Translating program for a higher level language than that for which an assembler is written. The source program, usually written in a machine independent language, is translated into the machine coded object program. The compiler expands the source program by selecting subroutines and generates more than one machine coded instruction for each source program step.

f compilateur *m*; convertisseur *m* de programme — **d** Kompilierer *m*; Kompilierprogramm *n* — **n** compilator *m*; compileerprogramma *n* — **e** compilador *m* — **i** compilatore *m*

3271 compiler program — See compiler.

3272 complementary colours — Two opposite (or contrasting) light colours which, when combined, produce white or grey. A mixture of any two primary colours is the complement of the remaining primary. In printing, complementary colours have the power either to neutralize or to accentuate each other; thus to diminish or enhance the attention value of the print. They are also called subtractive colours because they subtract or remove portions of the white light reflected from or travelling to the paper.

f couleurs *fpl* supplémentaires; couleurs *fpl* complémentaires — **d** Komplementärfarben *fpl*; besondere Kompensativfarben *fpl*; Ergänzungsfarben *fpl* — **n** complementaire kleuren *fmpl* — **e** colores *mpl* complementarios — **i** colori *mpl* complementari

3273 complementary wavelength — Wavelength of the spectrum light that, when combined in suitable proportions with the light considered, yields a match with the specified achromatic light (CIE). Symbol: λ_c.

f longueur *f* d'onde complémentaire — **d** kompensative Wellenlänge *f* — **n** complementaire golflengte *f* — **e** longitud *f* de onda complementaria — **i** lunghezza *f* d'onda complementare

3274 complete radiator — See black body.

3275 complex compound — Compound formed by combination of two or more other compounds, usually involving co-ordinate valences, and usually capable of being easily converted back into the original compounds.

f composé *m* complex — **d** komplexe Verbindung *f* — **n** complexe verbinding *f* — **e** compuesto *m* complejo — **i** composto *m* complesso

3276 complicated composition; complicated setting

f composition *f* compliquée — **d** komplizierter Satz *m*; schwieriger Satz *m* — **n** moeilijk zetwerk *n*; lastig zetwerk *n*; ingewikkeld zetwerk *n*;

gecompliceerd zetwerk *n* — **e** composición *f* complicada; composición *f* intricada; composición *f* enrevesada — **i** composizione *f* complicata; composizione *f* difficile; composizione *f* complessa

3277 complicated setting — See complicated composition.

3278 complimentary copy; comp — Copy of a book given by the publisher or author as a mark of esteem. See also free copy.

f exemplaire *m* en hommage de l'éditeur; exemplaire *m* en hommage de l'auteur — **d** Freiexemplar *n* — **n** presentexemplaar *n*; auteursexemplaar *n* — **e** ejemplar *m* de obsequio — **i** esemplare *m* omaggio; copia *f* omaggio

3279 compo — Glue-glycerine composition of letterpress ink rollers. See also roller composition.

3280 compo — See compositor.

3281 component; constituent part — One of the minimum number of chemical substances required to state the composition of all phases of a system. In the absence of chemical reaction, any one of the substances in a mixture.

f composant *m*; constituant *m*; élément *m* constitutif — **d** Bestandteil *m*; Komponente *f*; Grundstoff *m* — **n** bestanddeel *n*; component *m* — **e** componente *m* — **i** componente *m*

3282 compose *v*; **set** *v* — To produce type matter ready for printing, either by hand or by machine. Typesetting.

f composer — **d** setzen; Schrift setzen — **n** zetten; letterzetten — **e** componer; levantar letra; formar; parar tipo *Me* — **i** comporre

3283 composing computer

f calculateur *m* de composition — **d** Satzrechner *m*; Satzcomputer *m*; Setzereielektronenrechner *m*; elektronische Verarbeitungsanlage für Satzherstellung — **n** zetcomputer *m* — **e** computador *m* para composición tipográfica — **i** calcolatore *m* della composizione

3284 composing frame; frame; composing rack; composing stand — Sloped stand to hold type cases, lower case on the lower part of the frame at an angle of about 30°; the upper case above this at an angle of about 45°. Usually with a rack underneath to hold cases not in use. Height about 106.6 cm (3 ft 6 in) in front.

f rang *m* — **d** Setzpult *n*; Setzregal *n* — **n** zetbok *m* — **e** chibalete *m*; comodín-chibalete *m*; burro *m LA* — **i** telaio *m* da composizione; banco *m* da composizione

3285 composing instructions — Instructions for the compositor.

f indications *fpl* pour la composition — **d** Satzanweisungen *fpl* — **n** zetinstructie *f*; zetaanwijzingen *fpl* — **e** indicaciones *fpl* para componer; instrucciones *fpl* para el compositor — **i** istruzione *f* da composizione

3286 composing machine; typesetting machine — Machine for hot-metal typesetting, not to confuse with the American for *step-and-repeat machine. See also Linotype, Intertype, Monotype.

f composeuse *f*; machine *f* a composer — **d** Setzmaschine *f*; eiserner Kollege *m coll.* — **n** zetmachine *f*; letterzetmachine *f* — **e** componedora *f*; máquina *f* compositora; máquina *f* de componer; máquina *f* de composición; colega *m* de hierro *coll.* — **i** compositrice *f*; macchina *f* compositrice; compagno *m* di ferro *coll.*

3287 composing machine alloy — See composing machine metal.

3288 composing-machine heater

f chauffage *m* de la machine à composer — **d** Setzmaschinenheizung *f* — **n** zetmachineverwarming *f* — **e** calentador *m* de la componedora — **i** riscaldamento *m* della compositrice

3289 composing machine metal; composing machine alloy

f métal *m* pour machines à composer — **d** Setzmaschinenmetall *n* — **n** zetmachinemetaal *n* — **e** metal *m* para máquinas de componer; aleación *f* para máquinas de componer — **i** metallo *m* per macchine compositrici

3290 composing machine operator — See machine compositor.

3291 composing rack — See composing frame.

3292 composing room; case department; case room; floor *US* — The portion of a printing office occupied by the compositors and their equipment.

f atelier *m* de composition (manuelle); section *f* composition; salle *f* de composition — **d** Setzerei *f*; Setzersaal *m* — **n** zetterij *f*; letterzetterij *f*; handzetterij *f* — **e** sala *f* de composición; sala *f* de componer; sala *f* da cajas; sección *f* de cajas; taller *m* de composición; departamento *m* de composición; sección *f* de componer *Cu*; sección *f* de emplane — **i** sala *f* composizione; reparto *m* composizione; sala *f* compositori

3293 composing-room foreman — See composing-room overseer.

3294 composing-room overseer; overseer of the composing room; case overseer; composing-room foreman — Man in charge of composing room who distributes the work amongst compositors and records the time taken for each job. See also working director.

f prote *m* (de la salle des compositeurs); cheftypo *m* — **d** Setzerfaktor *m*; Faktor *m*; Fax *m coll*; Setzereileiter *m*; Setzermeister *m* — **n** chef *m* van de zetterij; chef *m* zetterij — **e** encargado *m* de sección de cajas; capataz *m*; jefe *m* de armado; regente *m* de cajas — **i** proto *m*

3295 composing rule; setting rule — Piece of brass or steel in a composing stick against which the types are set, usually with a nib at one or both ends, to remove it from behind the line when the justification is completed.

f lève-ligne *m* — **d** Setzlinie *f*; Setzlatte *f* — **n** zetlijn *fm* — **e** pleca *f* (sacalíneas); sacalíneas *f*; filete *m* sacalíneas; regleta *f* sacalíneas — **i** cavarighe *m*

3296 composing stand — See composing frame.

3297 composing stick; setting stick — Adjustable

metal (formerly wooden) hand tray for receiving movable types as they are set.

f composteur *m*; truelle *f coll.* — **d** Winkelhaken *m*; Löffel *m coll.* — **n** zethaak *m*; letterhaak *m*; haak *m* — **e** cazuela *f*; componedor *m* de cazuela — **i** compositoio *m*

3298 composing time

f temps *m* de composition — **d** Setzzeit *f*; Satzzeit *f* — **n** zettijd *m* — **e** tiempo *m* de composición — **i** tempo *m* di composizione

3299 composite block — Plate made up of two or more originals; also a block combining halftone and line.

f cliché *m* composite; planche *f* composite — **d** zusammengesetztes Klischee *n* — **n** samengesteld cliché *n* — **e** fotograbado *m* combinado; fotograbado *m* compuesto; fotograbado *m* mixto; grabado *m* de combinación — **i** cliscè *m* composto; cliché *m* combinato

3300 composite brace — Brace composed of loose type pieces. Sometimes colloquially named cock and hens, the centre piece being called the *cock.

f accolade *f* combinée — **d** Nasenklammer *f* aus Mittel- und Endstücken — **n** samengestelde accolade *f* — **e** corchete *m* de tres piezas — **i** grappa *f* composta; graffa *f* composta; graffa *f* con naso centrale; graffa *f* con becco centrale

3301 composition; type matter — The assembly of characters into words, lines and paragraphs of text or body matter for reproduction by printing.

f composition *f* — **d** Satz *m*; Schriftsatz *m* — **n** zetsel *n*; letterzetsel *n* — **e** composición *f* — **i** composizione *f*

3302 composition board — See letter board.

3303 composition costs

f prix *m* de la composition; frais *mpl* de la composition; dépenses *fpl* de la composition — **d** Satzpreis *m*; Satzkosten *pl* — **n** zetkosten *pl* — **e** gastos *mpl* de composición; costes *mpl* de composición — **i** spesa *f* per la composizione; prezzo *m* della composizione

3304 composition formatting — The insertion into a text stream, or the concatenation with the text stream, of command codes or symbols that will cause the text to be arranged in lines, blocks or pages in a prescribed set of type faces, to give a measure, and perhaps also to give a given depth. Formatting of tabular material is included in this definition.

3305 composition of abbreviations

f composition *f* avec beaucoup d'abréviations — **d** Abbreviaturensatz *m* — **n** zetsel *n* met vele afkortingen — **e** composición *f* con muchas abreviaturas — **i** composizione *f* con molte abbreviazioni

3306 composition of a title page

f composition *f* du grand titre — **d** Titelsatz *m* — **n** zetten *n* van een titelpagina — **e** composición *f* de portada; composición *f* de carátula *LA* — **i** composizione *f* della pagina di titolo

3307 composition of scientific formulae

f composition *f* des formules scientifiques — **d** wissenschaftlicher Formel(n)satz *m*; Formelsatz *m* — **n** zetten *n* van wetenschappelijke formules — **e** composición *f* de fórmulas científicas — **i** composizione *f* di formule scientifiche

3308 composition rider rollers

f rouleaux *mpl* chargeurs en pâte; chargeurs *mpl* en pâte — **d** Masse-Reiterwalzen *fpl* — **n** ruiterrollen *fmpl* van specie; specie-ruiterrollen *fmpl* — **e** rodillos *mpl* cargadores de pasta (gelatinosa); rodillos *mpl* caballeros de pasta (gelatinosa) — **i** rulli *mpl* macinatori in pasta di gelatina

3309 composition roller; glue-glycerine roller; gelatine-composition roller — Ink(ing) roller consisting of a metal core coated with a resilient or somewhat elastic composition made from a mixture of gelatine, glycerine, molasses and other materials.

f rouleau *m* (en) pâte; rouleau *m* de pâte; rouleau *m* de gélatine; rouleau *m* de composition — **d** Massewalze *f* — **n** specierol *fm*; compositierol *fm* — **e** rodillo *m* de cola; rodillo *m* de pasta; rodillo *m* de gelatina — **i** rullo *m* di gelatina; rullo *m* (con rivestimento) in pasta

3310 composition, roller — See roller composition.

3311 composition shop

f salle *f* des composeuses; atelier *m* des machines à composer; section *f* des machines à composer — **d** Setzmaschinensaal *m*; Maschinensetzerei *f* — **n** machinezetterij *f* — **e** taller *m* de composición mecánica; sala *f* de las componedoras; sala *f* de máquinas de componer — **i** sala *f* di macchine compositrices; reparto *m* di macchine compositrices; compartimento *m* di macchine compositrices

3312 composition software — Program or set of programs to formate text into justified lines and produce output to drive a typesetting machine. Generally includes hyphenation, justification and pagination routines. May include correction routines, or editing programs, as well as an operating system and file management capabilities. It may be a subset of a larger set of integrated programs. See also software.

3313 compositor; type-setter; compo *coll.***; typo; comp; galley slave** *fam.* — Man who sets type or performs any operations incidental to preparing forme for the press. Sometimes called floor man; in the 19th century sometimes called a puddler.

f compositeur *m*; typograph *m*; typo *m*; bourreur *m* de lignes *coll.* — **d** Schriftsetzer *m*; Setzer *m* — **n** letterzetter *m*; zetter *m*; typograaf *m* — **e** cajista *m* (tipógrafo); cajista *m* de líneas; cajista *m* liniero; tipógrafo *m* cajista; tipógrafo *m* compositor; compositor *m* (tipográfico); esclavo *m* de galera *coll.* — **i** compositore *m*

3314 compositor of solid matter — Compositor who merely makes packets of solid matter.

f paquetier *m*; compositeur *m* en paquets; compositeur *m* de labeurs; labeurier *m* — **d** Paketsetzer *m*; Spaltensetzer *m*; Stücksetzer *m*; Werksetzer *m*; Bachulke *m* (aus dem Polnischen pascholek, d.h. Knecht) — **n** platzetter *m* — **e** paquetero *m*; cajista *m* especializado en la composición de páginas (de libros) — **i** compositore *m* di opere; compositore *m* di libri

3315 compositor's error; printer's error — Error in printed matter made by the compositor.

f erreur *f* de composition — **d** Satzfehler *m*; Druckereiversehen *n* — **n** zetfout *fm*; drukfout *fm* — **e** error *m* del cajista; error *m* del compositor; error *m* de imprenta; yerro *m* de imprenta; error *m* tipográfico — **i** errore *m* di stampa; errore *m* tipografico

3316 compound — Substance composed of atoms or ions of two or more elements, which may be represented by a chemical formula. Each compound has its own characteristic properties.

f combinaison *f* chimique; composé *m* — **d** Verbindung *f*; Bindung *f*; Zusammensetzung *f* — **n** chemische verbinding *f*; verbinding *f* — **e** compuesto *m* — **i** composto *m*

3317 compound; ink compound — Wax, grease or other material to be added to ink to improve laying, prevent offset, sticking and picking, and to shorten or reduce the ink. See also dope.

f correctif *m*; adjuvant *m* — **d** Farbzusatz *m*; Farbenzusatz *m*; Druckfarbezusatzmittel *n* — **n** toevoegingsmiddel *n*; inkttoevoegingsmiddel *n*; inktpasta *mn*; drukpasta *mn* — **e** pasta *f* (suavizante) de tinta — **i** additivo *m* per inchiostro

3318 compound fraction — See piece fraction.

3319 compound lens — Lens consisting of two or more pieces of glass, sometimes cemented together with Canada balsam.

f lentille *f* composée; objectif *m* composite — **d** zusammengesetzte Linse *f*; Verbundlinse *f*; mehrlinsiges Objektiv *n* — **n** samengestelde lens *fm*; samengesteld objectief *n* — **e** lente *f* compuesta; objetivo *m* compuesto — **i** lente *f* composta; obiettivo *m* composto

3320 compound word; link-word — A word compound of two or more words united either with a hyphen (hypheme) or without (solideme), and usually distinguished from a phrase by a reduction of stress on one of the elements and a shortening of the pause between the words, as greenhouse and a green house. See app. no. 6.

f mot *m* composé — **d** Kuppelwort *n*; Kompositum *n* — **n** koppelwoord *n* — **e** palabra *f* compuesta; partícula *f* copulativa; conjunción *f* — **i** parola *f* composta

3321 comprehensive — Detailed layout of an advertisement and the like, showing placement of illustrations, copy, etc., for presentation to a client.

f maquette *f* élaborée; maquette *f* complète — **d** Feinlayout *n*; Feinentwurf *m* — **n** uitgewerkte tekening *f* — **e** maqueta *f* detallada — **i** completo *m*; dettagliato *m*

3322 compressed air — Air compressed to a pressure higher than the surrounding atmospheric pressure.

f air *m* comprimé — **d** Preßluft *f*; Druckluft *f*;

komprimierte Luft *f* — **n** perslucht *fm*; samengeperste lucht *fm*; gecomprimeerde lucht *fm* — **e** aire *m* comprimido — **i** aria *f* compressa

3323 compressed-air brake — See air brake.

3324 compressibility — The percentage decrease in caliper of the sheet of paper or paperboard produced by an arbitrarily specified increase in load; of considerable importance in several uses of paper, notably in printing and bookbinding. The conditions under which the determinations are made must be completely specified. In engineering practice, it is the ratio of the fractional change in volume.

f compressibilité *f* — **d** Kompressibilität *f*; Zusammendrückbarkeit *f*; Verdichtbarkeit *f* — **n** samendrukbaarheid *f* — **e** compresibilidad *f* — **i** compressibilità *f*

3325 compression plate lock-up — Mechanism on a letterpress rotary for locking up printing plates on to the plate cylinder with movable dogs which operate on the angle edge of the plate from the side towards the centre of the cylinder. See also tension plate lock-up.

f calage *m* par compression des stéréos; fixation *f* du cliché — **d** Facettenverschluß *m*; Druckverschluß *m* — **n** facettenbevestiging *f*; compressiebevestiging *f* — **e** cierre *m* de planchas por compresión — **i** dispositivo *m* di bloccaggio per compressione; dispositivo *m* di staffatura

3326 compression spring

f ressort *m* de compression; ressort *m* travaillant à la compression — **d** Druckfeder *f* — **n** drukveer *fm* — **e** muelle *m* de compresión; resorte *m* de compresión — **i** molla *f* di compressione

3327 compressor; air compressor — Pump to compress air or gas into a container.

f compresseur *m* (d'air) — **d** Kompressor *m*; Luftkompressor *m*; Luftpresser *m*; Drucklufterzeuger *m*; Preßluftgerät *m*; Gebläse *n* — **n** compressor *m* — **e** compresor *m* — **i** compressore *m* (d'aria)

3328 compressor unit

f installation *f* du compresseur — **d** Kompressoranlage *f* — **n** compressorinstallatie *f* — **e** unidad *f* compresora — **i** impianto *m* di compressione

3329 compute bound — Data processing routines or programs in which the throughput is limited by the speed at which the central unit (CPU) can handle, manage or process the data, executing the various programming instructions. The limitation is not on the speed of the input or output devices (in which case the program would be I/O-bound) but by the program execution time.

3330 computer — Device capable of accepting information, applying prescribed processes to the information, and supplying the results of these processes, usually consisting of input and output devices, storage, arithmetic and logical units, and a control unit.

f ordinateur *m*; calculateur *m* électronique — **d** Computer *m*; Datenrechner *m*; Rechner *m*; Datenverarbeitungsanlage *f*; elektronische Rechenanlage *f*; Elektronenrechner *m*; Daten-verarbeitungsmaschine *f* — **n** computer *m*; elektronische rekenmachine *f*; rekenautomaat *m* — **e** computador *m* (electrónico); calculador *m* (de datos); computadora *f* — **i** calcolatore *m* elettronico; ordinatore *m*; elaboratore *m* di dati

3331 computer code; computer instruction code; machine code — Code to represent the elementary operations of a programming system.
f code *m* machine — **d** Maschinenbefehlskode *m* — **n** machinecode *m*; machineopdrachtcode *m* — **e** código *m* de máquina; código *m* de ordenador — **i** codice *m* d'istruzioni di macchina

3332 computer composition program — Set of instructions which can be loaded or brought into the memory of a computer or typesetting controller (intelligent typesetter) which enables the device to accept input of text in an unformatted sequence, break the text into lines which are justified to a specified measure, and otherwise arrange in desired order, with proper required spacing, type founts, type sizes and other features.

3333 computer configuration — See configuration.
3334 computer-controlled typesetting; computerized typesetting; computer typesetting — Range of operations by computers to assist the process of typesetting. In its simple application it is concerned with automatic justification, hyphenation and the advantages associated with stored data for typesetting different measures, type faces and sizes on a wide range of type-setting equipments. In a more sophisticated approach it embraces display, page make-up, sorting, updating, merging and the information processing of data generated for typesetting, thus avoiding re-keyboarding.
f composition *f* par ordinateur; composition *f* automatique (des textes); composition *f* programmée; emploi *m* du calculateur électronique pour l'automatisation de la composition — **d** Satzherstellung *f* mit elektronischen Datenverarbeitungsanlagen; Satzherstellung *f* mit EDV-Anlagen; Gebrauch *n* eines Elektronenrechners für das Schriftsetzen — **n** letterzetten *n* met behulp van een computer; zetten *n* met een computer; computerzetten *n* — **e** composición *f* programada; composición *f* con computadora — **i** composizione *f* programmata; composizione *f* con calcolatori elettronica; composizione *f* automatica comandata da calcolatore

3335 computer-independent language — See machine-independent language.
3336 computer input card — Sometimes called IBM or Hollerith card, usually with 80 columns, which may be punched by a keypunch to record data for computer input by means of a card reader.
3337 computer instruction; machine instruction — Instruction which can be obeyed by the computer directly without the use of an interpretive or translating routine.

f instruction *f* machine — **d** Maschinenbefehl *m* — **n** machineopdracht *fm* — **e** instrucción *f* de máquina; instrucción *f* de ordenador — **i** istruzione *f* di macchina
3338 computer instruction code — See computer code.
3339 computerized typesetting — See computer-controlled typesetting.
3340 computer language — See machine language.
3341 computer merging — See tape merging.
3342 computer peripheral — Device connected (interfaced) to a computer which passes information on to the computer, receives information from it, or both, such as a paper tape reader or punch, a magnetic tape station, a disk, a video terminal, or a line printer. See also peripheral equipment.
3343 computer program — Ordered set of instructions to be loaded into computer memory to cause the computer to perform pre-defined tasks.
f programme *m* du calculateur — **d** Computerprogramm *n*; Rechnerprogramm *n* — **n** computerprogramma *n* — **e** programa *m* de la computadora — **i** programma *m* del calcolatore
3344 computer run — Processing of a job or task through a computer, utilizing a given program for the performance of that task.
3345 computer system — See configuration.
3346 computer time — Reckoning of the utilization of the computer for the processing of a particular job or task. Time may be expressed as "connect time", which represents the time that a computer is required to pay attention to the possible demands of a user who has access to its facilities at the moment, or processing time, which is a measure of the actual use of the computer facilities in the performance of a given job or task. Processing time, in turn, may be divided into *cycle time, during which the main frame is actually at work executing instructions, and communicating time, during which the main frame is waiting for data to be brought in or disposed of, as, for example, from or to a magnetic tape or disk.
3347 computer typesetting — See computer-controlled typesetting.
3348 computer word; machine word — Unit of length of data, i.e., a set of digits commonly treated by the equipment as a unit.
f mot *m* machine — **d** Maschinenwort *n* — **n** machinewoord *n* — **e** palabra *f* de máquina; palabra *f* de ordenador — **i** parola *f* di macchine
3349 concatenation; catenation — The process of linking one independent data set to another for the performance of a task.
f enchaînement *m* — **d** Verkettung *f*; Zusammenfassung *f* — **n** vereniging *f* — **e** encadenamiento *m* — **i** concatenazione *f*
3350 concave lens; biconcave lens; concavo-concave lens; double-concave lens — A *lens, thinner at the centre than at the edges. See also

reducing glass.

f lentille *f* concave; lentille *f* biconcave — **d** Konkavlinse *f*; bikonkave Linse *f*; Bikonkavlinse *f* — **n** concave lens *fm*; biconcave lens *fm*; dubbel-concave lens *fm*; dubbel-holle lens *fm* — **e** lente *m* cóncavo; lente *m* bicóncavo; lente *f* cóncava; lente *f* bicóncava — **i** lente *f* concava; lente *f* biconcava

3351 concavo-concave lens — See concave lens.

3352 concentrate

f concentré *m*; substance *f* concentrée — **d** Konzentrat *n* — **n** concentraat *n* — **e** concentrado *m* — **i** concentrato *m*

3353 concentrate *v*

f concentrer — **d** eindampfen; eindicken; verdicken — **n** indampen; indikken — **e** concentrar — **i** concentrare

3354 concentrated acetic acid — See glacial acetic acid.

3355 concentrated acid

f acide *m* concentré — **d** konzentrierte Säure *f* — **n** geconcentreerd zuur *n* — **e** ácido *m* concentrado — **i** acido *m* concentrato

3356 concentration — Amount of a given substance in a stated unit of a mixture or solution. Common methods of stating concentrations are per cent by weight or by volume, normality, weight per unit volume, as grams per cubic centimetre or pounds per gallon.

f concentration *f*; enrichissement *m* — **d** Konzentration *f*; Lösungsstärke *f*; Anreicherung *f* — **n** concentratie *f* — **e** concentración *f* — **i** concentrazione *f*

3357 concentric — Having the same centre.

f concentrique — **d** konzentrisch — **n** concentrisch — **e** concéntrico; homocéntrico — **i** concentrico

3358 concertina fold — See accordion pleat.

3359 concurrent processing — See multiprogramming.

3360 concurrent working — Automatic data processing system in which more that one operation or sequence of operations is executed at the same time.

f exécution *f* simultanée d'instructions — **d** simultane Befehlsausführung *f* — **n** gelijktijdige verwerking *f* — **e** ejecución *f* simultánea de instrucciones — **i** esecuzione *f* simultanea d'istruzioni

3361 condensation

f condensation *f* — **d** Kondensation *f*; Verdichtung *f* — **n** condensatie *f* — **e** condensación *f* — **i** condensazione *f*

3362 condensed face — See condensed type.

3363 condensed letter — See condensed type.

3364 condensed type; condensed letter; condensed face — Modified type design in which the *width of each character is contracted (setwise) so that more characters will fit within specified width of the line measure. See also set.

f caractère *m* serré; caractère *m* étroit — **d** schmaler Buchstabe *m*; schmale Schrift *f*; enge Schrift *f* — **n** smalle letter *fm*; smal lopende letter *fm* — **e** tipo *m* condensato; tipo *m* alargado; tipo *m* esqueleto; tipo *m* estrecho; tipo *m* chupado; tipo *m* metido; tipo *m* compacto; letra *f* compacta; letra *f* chupada; letra *f* delgada; carácter *m* estrecho; carácter *m* chupado; carácter *m* apretado — **i** carattere *m* stretto

3365 condenser — See condensing lens.

3366 condenser lens — See condensing lens.

3367 condensing lens; condenser lens; condenser — Lens or lens system concerned with the even distribution of light and not with the formation of an image.

f lentille *f* condensatrice; condensateur *m* — **d** Sammellinse *f*; Kondensor *m* — **n** verzamellens *fm*; condensor *m* — **e** lente *m* condensador; lente *f* condensadora; condensador *m*; condensadora *f* — **i** lente *f* convergente

3368 conditioned paper; seasoned paper — Paper in which the moisture content is in equilibrium with the humidity of the atmosphere. See green paper.

f papier *m* conditionné; papier *m* saisonné — **d** klimatisiertes Papier *n* — **n** geconditioneerd papier *n* — **e** papel *m* acondicionado; papel *m* aclimatado; papel *m* ambientado — **i** carta *f* condizionata

3369 conditioner, paper — See paper conditioner.

3370 conditioning; paper conditioning — Essentially the same as seasoning, except that in general usage it refers to the exposure of paper to accurately controlled and specified atmospheric conditions, so that its moisture content is in equilibrium with the surrounding atmosphere. The properties of paper are measurably affected by its moisture content. The standard atmospheric conditions in US are 50% relative humidity and 23 °C (73 °F). In many other countries, the present standard for relative humidity is 65%.

f conditionnement *m* du papier — **d** Papierklimatisierung *f*; Konditionieren *n* des Papiers; Klimatisieren *n* des Papiers; Klimatisierung *f* des Papiers; Konditionierung *f* des Papiers — **n** conditioneren van papier; klimatiseren *n* van papier — **e** acondicionamiento *m* del papel; climatización *f* del papel; aclimación *f* del papel; ambientación *f* del papel — **i** condizionamento *m* della carta

3371 conditions of employment — See conditions of work.

3372 conditions of sale — See also delivery terms.

f conditions *fpl* de vente — **d** Verkaufsbedingungen *fpl* — **n** verkoopsvoorwaarden *fpl* — **e** condiciones *fpl* de venta — **i** condizioni *fpl* di vendita

3373 conditions of work; conditions of employment

f conditions *fpl* de travail — **d** Arbeitsbedingungen *fpl* — **n** arbeidsvoorwaarden *fpl* — **e** condiciones *fpl* de trabajo — **i** condizioni *fpl* di lavoro

3374 conductance — Readiness with which a

conductor transmits an electric current; the reciprocal of electrical resistance.

f conductance *f* — **d** Konduktanz *f*; Wirkleitwert *m*; Leitwert *m* — **n** geleiding *f*; geleidbaarheid *f*; conductantie *f* — **e** conductancia *f* — **i** conduttanza *f*

3375 conducting rollers

f rouleaux *mpl* conducteurs — **d** Leitwalzen *fpl* — **n** geleiderollen *fmpl*; leirollen *fmpl* — **e** rodillos *mpl* guía; guías *mpl*; rodillos *mpl* conductores; rodillos *mpl* de conducción — **i** rulli *mpl* di conduzione; rulli *mpl* di guida

3376 cone shaped — See conical.

3377 cone-shaped bag — See cornet bag.

3378 confectionary bag — Style of envelope in sizes from 8.9 x 11.5 cm or 17 x 24 cm (3.5 x 4.5 and 6.75 x 9.5 in).

3379 configuration; computer configuration; computer system — Interrelated equipment organized to form a complete system; it may include one central processor, or a number of compatible computers, with on-line and off-line peripheral equipment.

f configuration *f* — **d** Konfiguration *f*; Systemübersicht *f* — **n** configuratie *f*; groepopstelling *f* — **e** configuración *f* — **i** configurazione *f*

3380 conic — See conical.

3381 conical; conic; cone shaped

f conique; côné; en forme de cône — **d** konisch; kegelförmig — **n** conisch; kegelvormig — **e** cónico; coniforme — **i** conico

3382 conifer; coniferous tree — Gymnosperm tree or shrub belonging to the order coniferales, so called because the fruit of the tree is a cone, as in pines and firs. The wood is usually termed *softwood.

f conifère *m* — **d** Nadelholzbaum *m* — **n** naaldboom *m* — **e** conífera *f* — **i** conifera *f*

3383 coniferous tree — See conifer.

3384 coniferous wood — See softwood.

3385 coniferous wood pulp — See softwood pulp.

3386 conjugate foci; conjugate points — On a process camera, the respective distances from the lens to the copyboard and from the lens to the image or focal plane at any scale of reproduction.

f foyers *mpl* conjugués — **d** konjugierte Brennpunkte *mpl* — **n** geconjugeerde brandpunten *npl* — **e** focos *mpl* conjugados — **i** fuochi *mpl* coniugati; distanzi *fpl* focali

3387 conjugate points — See conjugate foci.

3388 connected dots — In negatives and plates, halftone dots joined together by a bridge.

f points *mpl* liés — **d** Kreuzlage *f*; zusammenhängende Rasterpunkte *mpl*; hängende Rasterpunkte *mpl* — **n** punten *mpl* met bruggetjes — **e** puntos *mpl* cegados; puntos *mpl* (reticulares) unidos; pantalla *f* cerrada *Me* — **i** punti *mpl* di retino uniti; punti *mpl* di retino chiusi

3389 connected structure — Structure in which the suspended particles are present in the form of a continuous network, giving a form of rigidity to the material, e.g., a flocculated-pigment sus-

pension where the pigment-vehicle ratio is sufficiently high.

f structure *f* réticulée; structure *f* agrégée — **d** zusammenhängende Struktur *f* — **n** samenhangende structuur *f*; netwerk *n* — **e** estructura *f* reticulada — **i** struttura *f* reticolata

3390 connect time — See computer time.

3391 consignment — In statistics, the bulk of material or collection of units delivered at one time to the user.

f livraison *f* — **d** Lieferung *f* — **n** levering *f* — **e** entrega *f* — **i** consegna *f*

3392 consignment on approval — See delivery on approval.

3393 consistency — The flow nature of a material by virtue of its internal structure. Consistency is represented by a curve (shearing stress versus rate of shear) and not by a single value except in the case of Newtonians. The consistency of a Newtonian is defined by its coefficient of viscosity. For Newtonian liquids, consistency and viscosity are synonymous. For non-Newtonian liquids, it qualitatively represents plastic flow. The newer term for consistency is body and is measured by inkometers.

f consistance *f* — **d** Konsistenz *f*; Dickflüssigkeit *f* — **n** consistentie *f* — **e** consistencia *f* — **i** consistenza *f*

3394 consistent grease

f graisse *f* consistante — **d** Maschinenfett *n*; Heißlagerfett *n*; Wasserpumpenfett *n*; konsistentes Fett *n* — **n** consistentvet *n*; smeervet *n* — **e** grasa *f* consistente — **i** grasso *m* denso

3395 console; control desk — Separate container in which electrical controls for a rotary press are grouped.

f pupitre *m* de commande; console *f* — **d** Schaltpult *n*; Bedienungspult *n*; Konsole *f* — **n** schakeltafel *fm*; schakelkast *fm*; bedieningslessenaar *m* — **e** pupitre *m* de mando; cuadro *m* de control; consola *f* — **i** pannello *m* di comando; banco *m* di comando

3396 console — Control panel or nerve centre of a computer system to control the machine manually, correct errors, determinate the status of the machine circuits, etc., determine the contents of storage, and revise the contents of store.

f console *f* — **d** Konsole *f* — **n** console *f* — **e** consola *f*

3397 constancy

f permanence *f*; constance *f* — **d** Beständigkeit *f* — **n** bestendigheid *f* — **e** constancia *f*; perseverencia *f*; firmeza *f*; fuerza *f*; fidelidad *f*; permanencia *f*; estabilidad *f* — **i** costanza *f*

3398 constant — Quantity which does not vary, e.g., π (pi).

f constante *f* — **d** Konstante *f* — **n** constante *fm* — **e** constante *f* — **i** costante *f*

3399 constant tension — Tension of the paper web under ideal conditions of non-varying pull in the direction of travel.

f tension *f* constante de la bande — **d** konstante

Bahnspannung *f* — **n** constante spanning *f* van de papierbaan — **e** tensión *f* constante de la bande — **i** tensione *f* costante del nastro

3400 constant tension rewind — Reel rewind with core to maintain automatically at the surface of the rewound reel, a speed always equal to the speed of the approaching web. Therefore, when drive works in conjunction with a dancer roller, constant tension is maintained in the web being rewound.

f rebobineuse *f* à tension constante — **d** Wiederaufwickler *m* mit konstanter Bahnspannung — **n** opwikkelinrichting *f* met handhaving van een constante baanspanning — **e** rebobinadora *f* a tensión constante — **i** ribobinatura *f* a tensione costante

3401 constant-voltage stabilizer — See stabilizer.

3402 constituent part — See component.

3403 constructional fault; construction defect; fault of construction

f défectuosité *f* (de construction); défaut *m* de fabrication — **d** Konstruktionsfehler *m* — **n** constructiefout *fm*; fabricagefout *fm* — **e** defecto *m* de construcción — **i** difetto *m* di costruzione

3404 construction defect — See constructional fault.

3405 contact angle — Internal angle made by a drop of liquid on a solid surface; it adjusts itself so that the cohesional force of the liquid and the cohesional force between liquid and solid balance each other.

f angle *m* de contact — **d** Kontaktwinkel *m*; Randwinkel *m* — **n** randhoek *m* — **e** ángulo *m* de contacto — **i** angolo *m* di contatto

3406 contact box; contact cabinet — Contact printer where the light source is in a box, the top of which is fitted to take the light-sensitive material together with the original.

f tireuse *f* — **d** Kontaktkopierapparat *m*; Kontaktkopierkasten *m* — **n** contactkast *fm* — **e** impresora *f* por contacto — **i** macchina *f* a contatto; bromografo *m*

3407 contact cabinet — See contact box.

3408 contact copy — See contact print.

3409 contact diapositive — Photographic image produced from a continuous tone negative by the contact method.

f diapositif *m* par contact; diapositive *f* de contact — **d** Kontaktdiapositiv *n* — **n** contactdiapositief *n* — **e** diapositiva *f* por contacto; diapositiva *f* al contacto — **i** diapositiva *f* a contatto

3410 contact mask — Mask used in contact with the image which is to be modified, usually made by contact printing. Contact masks for originals are usually negative masks combined with a positive colour transparency. Often a highlight premask is made first. Masks made with the aid of a highlight premask are often called principal masks. Contact masks for negatives are of several varieties. *Positive masks are used for colour correction or undercolour removal. Negative highlight masks are used for tone correction, but should not be confused with the highlight premasks used for the same purpose in making principal masks for colour transparencies. Dropout masks are a form of highlight mask. In the two-stage masking method, positive premasks are first made, and the principal masks made from combinations of these with the negatives are neither negative nor positive, but are purely colour-correcting masks. The use of a highlight premask is similar in principle to the two-stage method, but might cause confusion if classified as such.

f masque *m* par contact; masque *m* de contact — **d** Kontaktmaske *f* — **n** contactmasker *n* — **e** máscara *f* de contacto; máscara *f* por contacto; mascarilla *f* por contacto — **i** maschera *f* a contatto

3411 contact negative — Photographic image produced from a positive by the contact method.

f négatif *m* par contact; négatif *m* de contact — **d** Kontaktnegativ *n* — **n** contactnegatief *n* — **e** negativo *m* por contacto — **i** negativo *m* a contatto

3412 contact paper — Slow-speed paper specifically for printing down an image from a negative held in contact with the paper.

f papier *m* de tirage par contact — **d** Kontaktpapier *n*; Kopierpapier *n* — **n** contactpapier *n* — **e** papel *m* de contacto — **i** carta *f* per copia a contatto

3413 contact positive — Misnomer for a contact-diapositive.

f positif *m* par contact; positif *m* de contact — **d** Kontaktpositiv *n* — **n** contactpositief *n* — **e** positivo *m* por contacto — **i** positivo *m* a contatto

3414 contact print; contact copy — Photographic same size copy made by exposure of a sensitized emulsion in contact with the transparency, a negative, or a positive, the exposing light passing through the master image.

f copie *f* par contact; copie *f* de contact; contretype *m*; photo *f* par contact; épreuve *f* par contact — **d** Kontaktabzug *m*; Kontaktdruck *m*; Negativnutzen *m*; Positivnutzen *m*; Negativnutzenfilm *m*; Positivnutzenfilm *m*; Kontaktbild *n* — **n** contactafdruk *m*; contactkopie *f*; contactdruk *m depr.* — **e** contacto *m*; copia *f* por contacto — **i** copia *f* a contatto

3415 contact printing — Photographic operation in which a sensitized surface is exposed in contact with a negative or a positive while locked in a printing frame.

f copier par contact — **d** eine Kontaktkopie herstellen; kontaktkopieren — **n** een contactkopie maken; contacten — **e** copiar por contacto — **i** copiare per contatto

3416 contact screen; variable opacity screen; vignetted contact screen — Photographically made halftone screen with a dot structure of graded density and usually used in vacuum contact with the film or plate.

f trame *f* par contact; trame *f* de contact — **d**

Kontaktraster *m*; Skalenraster *m* — **n** contact-raster *n*; verloopraster *n* — **e** retícula *f* de contacto; trama *f* de contacto; pantalla *f* de contacto — **i** retino *m* a contatto

3417 contaminate *v* — To make impure.

f contaminer; souiller — **d** verunreinigen — **n** verontreinigen — **e** contaminar; manchar; corromper; contagiar; inficionar — **i** sporcare; insudiciare; imbrattare

3418 contamination of offset blanket

f encrassage *m* du blanchet offset — **d** Offsettuchverschmutzung *f*; Verschmutzung *f* des Offsettuchs — **n** offsetdoekvervuiling *f* — **e** ensuciamiento *m* de la mantilla offset — **i** insudiciamento *m* del blanket

3419 conté crayon; conté pencil — Pencil in which non-greasy agents (graphite and clay) are used to make temporary drawings or designs on lithographic plates. Named after N.J. Conté, 18th-century French chemist.

f crayon *m* Conté — **d** Conté-Stift *m* — **n** Conté-potlood *n* — **e** lápiz *m* Conté — **i** matita *f* Conté

3420 contention — Condition in which two or more demands are placed upon a given device or channel at the same time in an unregulated manner.

f engorgement *m* — **d** Wetteifer *m* — **n** rivaliteit *f* — **e** rivalidad *f* — **i** contesa *f*

3421 contents; table of contents — Table at the beginning of a book listing the chapter numbers, titles and subtitles of chapters and page numbers.

f table *f* des matières — **d** Inhaltsverzeichnis *n*; Inhalt *m* — **n** inhoudsopgave *fm* — **e** contenido *m*; tabla *f* de materias — **i** tavola *f* delle materie

3422 conté pencil — See conté crayon.

3423 continental horsepower — See metric horsepower.

3424 continuation sheets — Sheets of paper for additional pages of multiple-page letters, usually of the same size and quality as the letterhead, but not printed with the letterhead design.

f feuilles *fpl* de continuation — **d** Anschlußblätter *npl* — **n** vervolgbladen *npl*; vervolgvellen *npl* — **e** hojas *fpl* adicionales — **i** fogli *mpl* di continuazione

3425 continuing action of light; continuing reaction of light — Chemical reaction of bichromated colloids to become increasingly insoluble (tanned) between the exposure to light and development, continuing even in darkness.

f action *f* continuatrice; durcissement *m* dans l'obscurité — **d** Nachkopiereffekt *m*; Nachdunkeln *n* — **n** nakopiereffect *n*; nadonkeren *n* — **e** efecto *m* continuo — **i** indurimento *m* susseguente all'esposizione

3426 continuing reaction of light — See continuing action of light.

3427 continuous current — See direct current.

3428 continuous forms — Printed and perforated business forms, such as invoices, bill-heads, etc., supplied in rolls or fanfold form with carbon paper and duplicate sheets inserted.

f formulaires *mpl* en continu; formules *fpl* continues; liasses *fpl* sans fin; liasses *fpl* en continu — **d** Endlosformulare *npl* — **n** kettingformulieren *npl* — **e** formularios *mpl* continuos; formas *fpl* continuas — **i** moduli *mpl* continui

3429 continuous forms bursting — See bursting.

3430 continuous grinder — Machine in paper mills for grinding wood logs.

f défibreur *m* en continu — **d** Stetigschleifer *m* — **n** stetigslijper *m*; continuslijper *m* — **e** desfibrador *m* continuo — **i** sfibratore *m* continuo

3431 continuously undulating inking mechanism

f dispositif *m* d'encrage semi-continu — **d** halbkontinuierliches Farbwerk *n* — **n** halfcontinu inktwerk *n* — **e** batería *f* de tintaje semicontinua — **i** dispositivo *m* inchiostratore semicontinuo

3432 continuous spectrum — Unbroken continuity of wavelength over a wide range presented by light or any other radiation when analyzed with a spectroscope.

f spectre *m* continu — **d** kontinuierliches Spektrum *n* — **n** continu spectrum *n* — **e** espectro *m* continuo — **i** spettro *m* continuo

3433 continuous tone — Said of those images (wash drawings, oil paintings, photographic negatives and positives) in which the detail and tone values of the subject are reproduced by a varying deposit (density) of developed silver in the picture. Photogelatine or collotype reproductions duplicate continuous-tone copy without use of a screen and are said to be continuous-tone reproductions. In lithographic colour correction, continuous-tone colour separation negatives are stained locally by grey dyes to add density, or treated by chemicals to lessen density.

f à ton continu; à demi-ton; modelé — **d** Halbton- — **n** ongerasterd — **e** de tono continuo; de tono no reticulado — **i** a tono continuo

3434 continuous tone negative — Photographic negative produced without the use of a halftone screen.

f négatif *m* à tons continus — **d** Halbtonnegativ *n* — **n** ongerasterd negatief *n* — **e** negativo *m* de tono continuo; negativo *m* de tono no reticulado — **i** negativo *m* a tono continuo

3435 continuous tone subject

f modèle *m* à modelé continu — **d** Halbtonvorlage *f* — **n** ongerasterd model *n* — **e** original *m* a tono continuo; original *m* a tono no reticulado — **i** modello *m* a tono continuo

3436 continuous tone wedge

f gamme *f* de gris à modelé continu — **d** kontinuierlicher Graukeil *m* — **n** continue toonwig *fm* — **e** cuña *f* de tono continuo; cuña *f* de tono no reticulado — **i** cuneo *m* a tono continuo

3437 contour — Outline of a plate, illustration etc.

f contour *m* — **d** Kontur *f*; Umriß *m* — **n** contour *m*; omtreklijn *fm* — **e** contorno *m* — **i** contorno *m*

3438 contraries — Visible matter unwanted in the paper pulp.

f impuretés *fpl* (visibles) — **d** Verunreinigungen *fpl*; Fremdstoffe *mpl* — **n** ongerechtigheden *fpl*; ongewenste stoffen *fpl*; onreinheden *fpl* — **e** indeseados *mpl* — **i** impurezze *fpl*

3439 contrast — Measure of the degree of increased or decreased density differences between corresponding tones of grey in the original and in the reproduction. The contrast change ratio is measured as the gamma for the reproduction. Identical differences in value between the tones in the original and in the reproduction indicate a gamma of 1.00. Photomechanical films that are used for line and halftone photography have high contrast with a gamma of 9.00 and greater.

f contraste *m* — **d** Kontrast *m* — **e** contraste *m* — **i** contrasto *m*

3440 contrast gloss — See gloss.

3441 contrast index — Average gradient measured over the part of the curve commonly used in practice for continuous-tone, black and white, negative and positive films and plates and recommended as an aid in determining development times for these materials. Gamma, the slope of the straight-line portion of the D-log E curve, is not always an appropriate basis for selecting proper development times. It often fails when applied to films for which D-log E curves have unusually long or unusually short toes because it does not take into account the fact that a portion of the toe of the curve is normally involved in the exposure of a typical negative. The optimum development time is actually the time required to produce a certain average gradient measured over the used part of the curve.

f indice *m* de contraste — **d** Kontrastindex *m* — **n** contrastindex *m* — **e** índice *m* de contraste — **i** indice *m* di contrasto

3442 contrasting colour — Colour that appears strikingly different from another colour.

f couleur *f* contrasté; ton *m* opposé — **d** kontrastierende Farbe *f* — **n** contrasterende kleur *fm* — **e** color *m* contrastante — **i** colore *m* contrastante

3443 contrast range — Difference between the lowest and the highest density of an image.

f intervalle *m* de densité — **d** Schwärzungsumfang *m*; Kontrastumfang *m* — **n** contrastomvang *m* — **e** escala *f* de contraste — **i** gamma *f* di contrasto

3444 contrast ratio *depr.* — Many ink technologists determine *hiding power as a contrast ratio of black-backed reflectance to white-backed reflectance. See also opacity (white backing).

3445 contrast-reducing mask — Mask for an excessively contrasty image to reduce it to an easily reproducible density range.

f masque *m* de réduction de contraste — **d** kontrastvermindernde Maske *f* — **n** contrastverminderend masker *n* — **e** máscara *f* de reducción de contraste; máscara *f* reductora de contraste — **i** maschera *f* di riduzione del contrasto; maschera *f* per smorzare il contrasto

3446 contrast rule — See Oxford rule.

3447 contrasty — For photographic materials, when small changes in exposure cause large differences in density of the developed image.

f contrasté — **d** kontrastreich; hart — **n** contrastrijk; hard — **e** contrastado; vigoroso; duro; con mucho contraste; definido; de exagerado contraste — **i** contrastato

3448 contribution — See article.

3449 control *v* — The English "to control" and the French "contrôler" have different meanings. This not infrequently gives rise to trouble. In English it emplies a continuous regulation or direction, of persons or machines by some means. In French, German and Dutch it implies a brief inspection or check to ensure that everything is correct. Thus, your ticket is controlled at the barrier of the métro in Paris, but on the London underground it is inspected.

f commander; régler — **d** steuern — **n** regelen; sturen; besturen — **e** mandar; gobernar — **i** regolare; comandare

3450 control board — See control panel.

3451 control box — Case on the rotary press containing one or more push buttons or other control units.

f boîte *f* de boutons-poussoirs; boîtier *m* — **d** Druckknopfstation; Kontrollkasten *m*; Steuerkasten *m* — **n** drukknopstation *m* — **e** estación *f* de mando — **i** pulsantiera *f* di comando

3452 control character; functional character — Character or signal in a data stream to modify the meaning of that which follows.

f caractère *m* de commande; caractère *m* fonctionnel — **d** Steuerzeichen *n* — **n** stuurteken *n*; besturingsteken *n* — **e** carácter *m* de mando; carácter *m* funcional — **i** carattere *m* di comando; carattere *m* funzionale

3453 control current; controlling current

f courant *m* de commande — **d** Steuerstrom *m* — **n** stuurstroom *m* — **e** corriente *f* de mando — **i** corrente *f* di comando

3454 control desk — See console.

3455 controller — See starter controller.

3456 controller hand — Operator responsible for operating the warning system and stopping and starting a rotary machine from the main control position.

f conducteur *m* — **d** Maschinenmeister *m* — **n** voormandrukker *m* — **e** contramaestre *m* — **i** conduttore *m*

3457 controlling current — See control current.

3458 control motor

f moteur *m* régulateur *m* — **d** Steuermotor *m* — **n** regelmotor *m*; stuurmotor *m* — **e** motor *m* de mando — **i** motore *m* di comando

3459 control panel; control board — Panel, usually isolated, with switches, dial, and other equipment to regulate electrical devices, lights, etc.

f panneau *m* de commande; pupitre *m* de commande; tableau *m* de commande — **d** Schalttafel *f*; Bedienungstafel *f*; Steuerpult *n* — **n** bedieningspaneel *n*; bedieningstafel *m* — **e** botonera *f*; tablero *m* de mando; panel *m* de control; cuadro *m* de control — **i** pannello *m* di comando; quadro *m* di comando

3460 control transfer — See jump.

3461 control unit; black box — Part of an automatic data processing equipment which directs the sequence and timing of operations, interprets the coded instructions, and stimulates the proper circuits to execute the instruction.
f unité *f* de contrôle — **d** Leitwerk *n*; Steuereinheit *f* — **n** stuurorgaan *n*; stuureenheid *f*; besturingsorgaan *n*; besturingseenheid *f* — **e** unidad *f* de control — **i** unità *f* di controllo

3462 conventional gravure; depth variable gravure — Gravure printed from a printing surface in which the cells are of equal (square) size but different depth.
f héliogravure *f* conventionnelle; gravure *f* à profondeur variable — **d** konventioneller Tiefdruck *m*; tiefenvariabler Tiefdruck *m* — **n** conventionele diepdruk *m*; variabele diepdruk *m* — **e** huecograbado *m* convencional — **i** rotocalco *m* convenzionale; rotocalco *m* a profondità variabile

3463 conventional ink — See normal ink.

3464 converging lens; positive lens — Lens to converge light. Sometimes called converging meniscus.
f lentille *f* convergente; lentille *f* positive — **d** konvergierende Linse *f*; positive Linse *f*; Sammellinse *f*; Halbmondlinse *f* — **n** convergerende lens *fm*; positieve lens *fm*; verzamellens *fm* — **e** lente *mf* convergente; lente *m* positivo — **i** lente *f* convergente; lente *f* positiva

3465 converging meniscus — See converging lens.

3466 conversational mode — Mode of operation that implies a "dialogue" between a computer and its user, in which the computer program examines the input supplied by the user and formulates questions or comments which are directed back to the user.
f mode *m* de communication par conversation — **d** Sprachmodus *m* — **n** conversationele modus *m* — **e** modo *m* de comunicación por conversación — **i** modo *m* di comunicazione per conversazione

3467 conversion — Process of changing information from one form of representation to another, e.g., from the language of one machine to that of another, from magnetic tape to the printed page, from one data processing method to another, or from one type of equipment to another, e.g., from paper tape to magnetic tape equipment.
f conversion *f* — **d** Umsetzung *f*; Umwandlung *f* — **n** omzetting *f*; conversie *f* — **e** conversión *f* — **i** conversione *f*

3468 conversion — See extension.

3469 conversion coating; off-machine coating; separate coating — Application of one or more coatings to one or both sides of paper or paperboard, separate from papermaking.
f couchage *m* hors-machine; couchage *m* séparé — **d** Separatstreichverfahren *n*; Beschichtung *f* außerhalb der Maschine; Streichen *n* außerhalb der Papiermaschine — **n** klassieke strijkmethode *f*; strijken *n* buiten de papiermachine — **e** estucado *m* fuera máquina — **i** patinatura *f* fuori macchina

3470 conversion code — See code.

3471 conversion method — Transformation of a letterpress printing forme into negatives or positives for use in offset or gravure printing.
f procédé *m* de conversion — **d** Umwandlungsverfahren *n*; Konversionsverfahren *n* — **n** conversieprocédé *n* — **e** método *m* de conversión; procedimiento *m* de conversión — **i** metodo *m* di conversione; procedimento *m* di conversione

3472 conversion scale; conversion table — Tabular arrangement of the equivalent values of units of different systems.
f échelle *f* de conversion; table *f* de conversion — **d** Umrechnungstabelle *f* — **n** omrekeningstabel *fm*; omrekeningsschaal *f*; omrekeningstafel *fm* — **e** tabla *f* de conversión — **i** tabella *f* di conversione

3473 conversion table — See conversion scale.

3474 convert *v* — To change data from one format to another, from one code to another, or from paper tape to magnetic tape.
f convertir — **d** umsetzen — **n** omzetten; converteren — **e** convertir — **i** convertire

3475 converter — Device to convert data from one form to another (e.g., punched cards to magnetic tape, 8-level tape to 31-level paper tape).
f convertisseur *m* — **d** Umwandler *m*; Umsetzer *m* — **n** omzetter *m* — **e** convertidor *m*; unidad *f* de conversión — **i** convertitore *m*; unità *f* di conversione

3476 converting — Secondary operation in which paper or board is processed into a finished or more finished product.
f transformation *f*; façonnage *m* — **d** Verarbeitung *f*; Weiterverarbeitung *f* — **n** verwerking *f*; verdere verwerking *f* — **e** transformación *f* — **i** trasformazione *f*

3477 converting process; paper converting process — Process which transforms paper and board into paper products, such as envelopes, bags, etc.
f procédé *m* de transformation — **d** Papierverarbeitung *f* — **n** papierverwerking *f* — **e** transformación *f* del papel — **i** processo *m* di trasformazione

3478 convex lens; biconvex lens; convexo-convex lens; double-convex lens — *Lens, thicker at the centre than at the edges.
f lentille *f* convexe; lentille *f* biconvexe — **d** Konvexlinse *f*; bikonvexe Linse *f*; Bikonvexlinse *f* — **n** convexe lens *fm*; biconvexe lens *fm*; dubbelconvexe lens *fm*; dubbel-bolle lens *fm* — **e** lente *m* convexo; lente *m* biconvexo — **i** lente *f* convessa; lente *f* biconvessa

3479 convexo-convex lens — See convex lens.

3480 conveyer belt — See conveyor belt.

3481 conveyor — Automatic mechanical device to carry paper or other products from one machine or room to another.
f transporteur *m*; convoyeur *m* — **d** Förderbahn *f*; Transportanlage *f*; Transportweg *m* — **n** transportbaan *fm* — **e** transportadora *f*; mecanismo *m* de conducción; conveyor *m* Pe — **i** trasportatore *m*

3482 conveyor belt; conveyer belt — Endless belt to carry objects or materials short distances.
f bande *f* transporteuse; chaîne *f* porteuse; tapis *m* roulant — **d** Transportband *n*; Förderband *n*; Fördergurt *m*; laufendes Band *n*; Transportband *n* — **n** transportband *m*; lopende band *m* — **e** correa *f* transportadora; correa *f* portadora; cinta *f* continua — **i** nastro *m* trasportatore

3483 convoluted tube — Tube made of thin board coiled up longitudinally. See also cardboard tube.
f tube *m* en carton enroulé — **d** gewickelte Tube *f*; Wickeltube *f* — **n** gespiraliseerde koker *m* — **e** tubo *m* pegado longitudinal — **i** tubo *m* attorcigliato

3484 convolute tube winder — See convolute winding machine.

3485 convolute winding machine; convolute tube winder — Machine for making tube-shaped containers.
f machine *f* à fabriquer des tubes de carton; machine *f* à fabriquer des tubes en carton rondes — **d** Hülsenwickelmaschine *f* — **n** rechtwikkelmachine *f* — **e** máquina *f* para hacer tubos espirales para envases; máquina *f* de enrollar y hacer tubos en espiral — **i** macchina *f* fabbricare tubi

3486 cooking; digesting — Treatment of cellulosic raw material (rags, pulpwood, straw, etc.) with chemicals usually at a high pressure and temperature to remove impurities and produce a pulp suitable for making paper. The French "cuisson" can also indicate a treatment without chemicals.
f cuisson *f*; lessivage *m* — **d** Kochung *f*; Kochen *n*; Aufschluß *m*; Aufschliessung *f* — **n** ontsluiting *f*; koken *n* — **e** cocción *f*; lejiación *f* — **i** cottura *f*; bollitura *f*

3487 cooling
f refroidissement *m* — **d** Kühlung *f*; Abkühlung *f* — **n** koeling *f* — **e** refrigeración *f* — **i** raffreddamento *m*

3488 cooling cylinder — See sweat drier.

3489 cooling device
f dispositif *m* refroidisseur — **d** Kühlvorrichtung *f* — **n** koelinrichting *f*; koelinstallatie *f* — **e** enfriador *m*; dispositivo *m* de refrigeración — **i** dispositivo *m* refrigerante

3490 cooling roller; chill roll — Roller immediately after the drying oven of a rotary offset press, used to reduce the temperature of the web from the approx. 177 °C of the oven to the setting temperature of the heat-set inks, about 27-32 °C.
f rouleau *m* refroidisseur; refroidisseur *m* de nappe — **d** Kühlwalze *f*; Kühlzylinder *m* — **n** koelrol *fm*; koelcilinder *m*; koelwals *fm* — **e** rodillo *m* refrigerador; rodillo *m* refrigerante — **i** rullo *m* raffreddatore; rullo *m* di raffreddamento

3491 coordinates — Numbers to locate a point on a graph in relation to the x axis and y axis. Rectangular coordinates are named abscissa and ordinate.
f coordonnées *fpl* — **d** Koordinaten *fpl* — **n** coördinaten *mpl* — **e** coordenadas *fpl* — **i** coordinate *fpl*

3492 copal — See copal resin.

3493 copal resin; copal — Hard, transparent, amber-like resin, collected from various tropical trees. Used in varnish and ink vehicles and in photogravure for graining; soluble in oil of turpentine and linseed oil, after fusion.
f résine *f* (de) copal; gomme *f* copal; copal *m* — **d** Kopalharz *n*; Kopal *n* — **n** kopalhars *mn*; kopal *mn* — **e** resina *f* copal; goma *f* copal — **i** resina *f* coppale; coppale *m*

3494 copal varnish
f vernis *m* (au) copal; vernis *m* anglais — **d** Kopallack *m* — **n** kopalvernis *mn*; kopallak *mn* — **e** laca *f* de copal; barniz *m* de copal — **i** lacca *f* di coppale

3495 copolymer — Product formed by simultaneous polymerization of two or more substances. See also polymer.
f copolymère *m* — **d** Kopolymer *n*; Mischpolymerisat *n* — **n** copolymeer *m* — **e** copolímero *m* — **i** copolimero *m*

3496 copolymerization — See polymerization.

3497 copper — Reddish chiefly univalent and bivalent metallic element, ductile and malleable, dissolving in nitric acid and hot concentrated sulphuric acid. Latin: cuprum. TL-value: 1 mg/m^3. Symbol: Cu. See also brass.
f cuivre *m* — **d** Kupfer *n* — **n** koper *n* — **e** cobre *m* — **i** rame *m*

3498 copper acetate; cupric acetate — Greenish-blue, fine powder. Poisonous. Soluble in water, alcohol and ether. Formula: $Cu(C_2H_3O_2)_2$.
f acétate *m* de cuivre — **d** Kupferazetat *n* — **n** koperacetaat *n* — **e** acetato *m* de cobre — **i** acetato *m* di rame

3499 copperas; green copperas — Common name for *ferrous sulphate.
f couperose *f* verte; sulfate *m* ferreux — **d** Eisenvitriol *n*; Ferrosulfat *n* — **n** ijzervitriool *nm*; ferrosulfaat *n*; ijzersulfaat *n* — **e** caparrosa *f* verde; sulfato *m* ferroso; sulfato *m* de hierro — **i** copperosa *f* verde; solfato *m* ferroso

3500 copper bath — Vat with an electrolyte containing copper salts, anodes and an arrangement for placing objects to be copperplated by electro-deposition.
f bain *m* de cuivrage — **d** Kupferbad *n* — **n** koperbad *n* — **e** baño *m* de cobrizar — **i** bagno *m* di ramatura

3501 copper bromide; cupric bromide — Black, crystalline powder or crystals, soluble in acetone, alcohol, water. Used as a photographic intensifier. Formula: $CuBr_2$.

f bromure *m* cuivrique — **d** Kupferbromid *n*; Bromkupfer *n*; Cupribromid *n* — **n** koperbromide *n* — **e** bromuro *m* cúprico; bromuro *m* de cobre — **i** bromuro *m* rameico; bromuro *m* di rame

3502 copper bromide intensifier
f renforçateur *m* au bromure cuivrique — **d** Bromkupferverstärker *m* — **n** broomkoperversterker *m* — **e** reforzador *m* de bromuro cúprico — **i** rinforzatore *m* al bromuro rameico

3503 copper carbonate; cupric carbonate; mineral green; artificial malachite — Green poisonous powder, soluble in acids, insoluble in water. Used in pigments.
Formula: $Cu_2(OH)_2CO_3$. See also malachite green.
f carbonate *m* de cuivre; carbonate *m* cuivrique; malachite *f* artificielle; vert *m* minéral — **d** Kupferkarbonat *n*; künstliches Malachit *n*; kohlensaures Kupferkarbonat *n* — **n** kopercarbonaat *n* — **e** carbonato *m* de cobre — **i** carbonato *m* di rame

3504 copper chloride; cupric chloride — Brownish-yellow powder (formula $CuCl_2$) or green, deliquescent crystals (formula $CuCl_2 \cdot 2H_2O$). Used in sympathetic inks, preservation of pulpwood and ground pulp, and in photography. Latin: cuprum chloratum.
f chlorure *m* cuivrique; chlorure *m* de cuivre — **d** Kupferchlorid *n*; Kupferbichlorid *n*; Cuprichlorid *n*; Chlorkupfer *n* — **n** koperchloride *n*; cuprichloride *n* — **e** cloruro *m* de cobre; cloruro *m* cúprico — **i** cloruro *m* di rame; cloruro *m* rameico

3505 copper colorimeter — Device to determine the concentration or content of dissolved copper (cupric chloride) in used ferric chloride etching baths.
f colorimètre *m* à cuivre — **d** Kupferkolorimeter *m* — **n** koper-colorimeter *m* — **e** colorímetro *m* de cobre — **i** colorimetro *m* per la determinazione del rame

3506 copper content
f teneur *f* en cuivre — **d** Kupfergehalt *m* — **n** kopergehalte *n* — **e** contenido *m* de cobre — **i** tenore *m* di rame

3507 copper cyanide; cupric cyanide — Green powder, exceedingly poisonous, soluble in acids and alkalis, insoluble in water. Must be kept well stoppered.
Formula: $Cu(CN)_2$.
f cyanure *m* cuivrique; cyanure *m* de cuivre — **d** Kupferzyanid *n*; Cuprizyanid *n* — **n** kopercyanide *n* — **e** cianuro *m* de cobre — **i** cianuro *m* di rame

3508 copper deposit — Coating of copper deposited usually by an electric current.
f dépôt *m* de cuivre — **d** Kupferniederschlag *m* — **n** koperneerslag *m* — **e** depósito *m* de cobre; posos *mpl* de cobre — **i** deposito *m* di rame

3509 copper depositing; copper facing; copper plating — The depositing of pure copper onto another metal. Use is made of two types of bath. 1.

The alkaline or cyanide bath in which copper is present as the cyanide, used when copper is to be plated on die castings and on iron or steel; preferred for plating over brass or bronze. 2. The acid bath in which copper is present in the electrolyte as copper sulphate, used for plating copper over copper or nickel previously deposited on steel from the cyanide bath.
f cuivrage *m* — **d** Verkupferung *f*; Aufkupfern *n*; Verkupfern *n* — **n** verkopering *f*; verkoperen *n* — **e** cobreado *m*; cobreamiento *m*; encobreamiento *m*; revestimiento *m* electrolítico de cobre — **i** ramatura *f*

3510 copper depositing installation — Installation for the electro-deposition of copper, consisting of a direct current supply (motor-generator or rectifier), the regulators for the current and the voltage, vats and connection to the vats.
f installation *f* de cuivrage — **d** Aufkupferungsanlage *f*; Verkupferungsanlage *f* — **n** verkoperingsinstallatie *f* — **e** instalación *f* para cobrear — **i** impianto *m* di ramatura

3511 copper engraving — See copperplate engraving.

3512 copper etching — Copper plate with an etched image.
f cliché *m* sur cuivre — **d** Kupferklischee *n* — **n** kopercliché *n* — **e** fotograbado *m* en cobre — **i** cliscè *m* di rame; cliché *m* di rame

3513 copper etching — Act of etching line and halftone images in relief in copper.
f gravure *f* sur cuivre — **d** Kupferätzung *f* — **n** etsen *n* in koper — **e** grabado *m* al aguafuerte en cobre — **i** incisione *f* del rame

3514 copper-faced stereo
f stéréo *m* cuivré — **d** verkupfertes Stereo *n* — **n** verkoperde styp *m* — **e** estereotipia *f* cobreada; estéreo *m* cobreado — **i** stereo *m* ramato

3515 copper facing — See copper depositing.

3516 copper ferrocyanide; cupric ferrocyanide — Reddish-brown powder, insoluble in water and acids, soluble in ammonia and potassium cyanide solutions. As a pigment, copper ferrocyanide retains colour, light fastness, and chalking resistance and is compatible with high quality organic red and maroon pigments.
Formula: $Cu_2Fe(CN)_6 \cdot 7H_2O$.
f ferrocyanure *m* de cuivre; ferrocyanure *m* cuivrique — **d** Kupfereisenzyanid *n* — **n** ferrocyaankoper *n* — **e** ferrocianuro *m* de cobre — **i** ferrocianuro *m* di rame

3517 copper monoxide — See copper oxide.

3518 copper oxide; black copper oxide; cupric oxide; copper monoxide — Brownish-black, amorphous or crystalline powder, soluble in acids, insoluble in water.
Formula: CuO.
f oxyde *m* cuivrique; oxyde *m* de cuivre; bioxyde *m* de cuivre — **d** Kupferoxyd *n* — **n** koperoxyde *n* — **e** óxido *m* de cobre; óxido *m* cúprico — **i** ossido *m* di rame; ossido *m* rameico

3519 copper phthalocyanine blue — See

monastral blue.

3520 copper phthalocyanine green — See monastral blue.

3521 copperplate
f plaque *f* de cuivre — **d** Kupferplatte *f*; Kupfer-druckplatte *f* — **n** koperplaat *fm* — **e** lámina *f* de cobre; plancha *f* de cobre — **i** lastra *f* di rame

3522 copperplate engraver
f graveur *m* en taille-douce; taille-doucier *m* — **d** Kupferstecher *m* — **n** kopergraveur *m* — **e** tallador *m*; artista *m* grabador en cobre — **i** incisore *m* in rame; calcografo *m*

3523 copperplate engraving; copper engraving — Hand-engraved copper plate used in intaglio printing.
f gravure *f* en taille-douce; gravure *f* sur cuivre; taille-douce *f*; gravure *f* au burin — **d** Kupferstich *m*; Kupferstichplatte *f*; Kupfergravüre *f* — **n** kopergravure *fm* — **e** grabado *m* en cobre; grabado *m* en talla dulce — **i** incisione *f* su rame; taglio *m* dolce; incisione *f* in rame al bulino

3524 copperplate engraving ink
f encre *f* pour gravure sur cuivre; encre *f* taille-douce — **d** Kupferdruckfarbe *f* — **n** plaatdruk-inkt *m*; koperdrukinkt *m* — **e** tinta *f* para talla dulce — **i** inchiostro *m* per calcografia; inchiostro *m* per taglio dolce

3525 copperplate press; copperplate printing press; etching press — Press for printing etchings and hand engraved copper (and steel) plates, some-times called rolling press.
f presse *f* pour taille-douce; presse *f* en taille-douce; presse *f* taille-douce; machine *f* pour l'impression en taille-douce — **d** Kupferdruck-presse *f* — **n** plaatdrukpers *f*; etspers *f* — **e** prensa *f* de grabado en cobre; prensa *f* para impresión de grabados en cobre; tórculo *m* — **i** macchina *f* da stampa calcografia

3526 copperplate printer
f taille-doucier *m*; imprimeur *m* — **d** Kupfer-drucker *m* — **n** koperdrukker *m*; plaatdrukker *m* — **e** estampador *m* — **i** stampatore *m* in rame; calcografo *m*

3527 copperplate printing
f impression *f* en taille-douce; impression *f* en creux — **d** Kupferdruck *m* — **n** koperdruk *m*; plaatdruk *m* — **e** estampación *f* en talla dulce; impresión *f* en cobre — **i** stampa *f* calcografica; calcografia *f*

3528 copperplate printing press — See copperplate press.

3529 copper plating — See copper depositing.

3530 copper potassium ferrocyanide; potassium copper ferrocyanide; potassium cupric ferro-cyanide — Brownish-red powder, insoluble in water, used in pigments.
Formula: $K_2CuFe(CN)_6 \cdot H_2O$.
f cyanure *m* de potassium cuivrique — **d** Zyan-kupferkalium *n*; Kupferkaliumzyanid *n* — **n** kopercyaankalium *n* — **e** cianuro *m* de potasio cúprico — **i** cianuro *m* di potassio rameico

3531 copper shell — See shell.

3532 copper space
f espace *f* de cuivre — **d** Kupferspatium *n* — **n** koperen spatie *f*; koperspatie *f* — **e** cascarilla *f* de cobre; espacio *m* de cobre — **i** spazio *m* di rame

3533 copper sulphate; cupric sulphate; sulphate of copper — Blue crystals or blue crystalline granules or powder, soluble in water, methanol, slowly soluble in alcohol and glycerol. Latin: cuprum sulfuricum.
Formula: $CuSO_4 \cdot 5H_2O$.
f sulfate *m* de cuivre — **d** Kupfersulfat *n*; Cupri-sulfat *n*; schwefelsaures Kupfer *n* — **n** koper-sulfaat *n*; cuprisulfaat *n* — **e** sulfato *m* de cobre — **i** solfato *m* di rame

3534 copy; manuscript; matter — Matter to be reproduced in type. Abbrev. Ms, *pl* Mss.
f copie *f*; manuscrit *m* — **d** Manuskript *n*; Satz-vorlage *f* — **n** kopij *f*; manuscript *n* — **e** copia *f*; manuscrito *m* — **i** copia *f*; manoscritto *m*

3535 copy — Single number of a publication.
f exemplaire *m* — **d** Exemplar *n*; Heft *n*; Nummer *f* — **n** exemplaar *n*; nummer *n*; af-levering *f* — **e** ejemplar *m* — **i** esemplare *m*

3536 copy; original; art work — Although "copy" is widely applied to photographs and different types of art work for reproduction, "original" is better, because from this originates the reproduction. In photographic platemaking there are two kinds of copy, line and tone, photo-graphed without and with the halftone screen, respectively. In line copy the design or image of the original is composed of lines or dots of solid colour; in tone copy tones or shades of solid colour appear.
f original *m*; modèle *m* — **d** Vorlage *f*; Original *n* — **n** origineel *n*; model *n* — **e** original *m* — **i** originale *m*

3537 copy — See carbon copy.

3538 copy *v* — To duplicate by photographic contact printing.
f copier — **d** kopieren; eine Kopie machen — **n** af-drukken; contacten; een kopie maken — **e** copiar (por contacto) — **i** copiare

3539 copy board — See camera copyboard.

3540 copy deadline — See deadline.

3541 copy fitting — Calculations to determine the size of type and width of line to fit the copy into a given area of space.
f évaluation *f* de copie; estimation *f* de copie — **d** Umfangsberechnung *f* — **n** berekening *f* van de omvang — **e** cálculo *m* tipográfico; cálculo *m* previo; cálculos *mpl* previos; cálculo *m* de espacio; cálculo *m* de material; cálculo *m* de original(es); cálculo *m* de la composición — **i** tipoconteggio *m*

3542 copy-fitting table — Method of casting-off copy, based on character count.
f tableau *m* calibrage — **d** Manuskript-berechnungstabelle *f* — **n** tabel *fm* voor de berekening van het aantal letters — **e** tabla *f* de cálculo tipográfico — **i** tabella *f* per tipoconteggio

3543 copy frame — See camera copyboard.

3544 copyholder — Device for holding copy sometimes used by composers.

f porte-copie *m*; visorium *m obs.* — **d** Manuskripthalter *m*; Tenakel *n* — **n** kopijhouder *m* — **e** atril *m*; sujetacuartillas *m*; divisorio *m*; mordante *m* — **i** porta *m* manoscritto

3545 copyholder — Glass-covered copyboard.

f porte-modèle *m* — **d** Vorlagenhalter *m* — **n** modelbord *n* — **e** portaoriginal(es) *m*; portamodelo *m*; tablero *m* — **i** portaoriginale *m*

3546 copyholder — Person who reads the manuscript aloud to the proofreader, who checks the proof.

f lecteur *m* — **d** Vorleser *m* (für den Korrektor) — **n** tegenlezer *m*; voorlezer *m* — **e** atendedor *m*; lector *m* de pruebas; ayudante *m* del corrector — **i** assistente *m* del correttore; lettore *m*

3547 copy hook — Hook at the side of a typesetting machine or a copy desk, to file copy or proofs. See also spike.

f porte-copie *m* — **d** Manuskripthaken *m*; Haken *m* zum Aufspießen von Manuskripten — **n** kopijhaak *m* — **e** gancho *m* portaoriginales — **i** gancio *m*

3548 copying ink — Water soluble ink which can be transferred after drying by the application of moisture and pressure; may be applied either with a pen or printing press, or from carbon paper by typewriter.

f encre *f* communicative — **d** Kopierfarbe *f* — **n** kopieerinkt *m* — **e** tinta *f* de copiar; tinta *f* comunicativa — **i** inchiostro *m* per copiare; inchiostro *m* da copia

3549 copying press — Obsolete screw-type press for taking copies from letters written with aniline copying ink.

f presse *f* à copier — **d** Kopierpresse *f* — **n** kopieerpers *f* — **e** prensa *f* de copiar — **i** pressa *f* per copiatura

3550 copy in pristine — Copy having its original purity; uncorrupted or unsullied.

f exemplaire *m* impeccable — **d** tadelloses Exemplar — **n** onberispelijk exemplaar *n* — **e** ejemplar *m* impecable; ejemplar *m* excelente — **i** esemplare *m* perfetto

3551 copy preparation; copy styling — In typesetting, careful revision of copy to ensure a minimum of changes or corrections after type is set.

f préparation *f* de la copie; préparation *f* du manuscrit — **d** Manuskriptbearbeitung *f*; Manuskriptvorbereitung *f*; Manuskriptverbesserung *f*; Textbereinigung *f* — **n** kopijvoorbereiding *f* — **e** corrección *f* del original — **i** preparazione *f* del manoscritto; preparazione *f* della copia

3552 copy preparation — In photomechanical processes, the directions as to desired size and other details for illustration and the arrangement into proper position of the various parts of the page to be photographed for reproduction. Also the work of preparing copy in paste-up form of text and art as a unit, termed a *mechanical

paste-up.

f préparation *f* des originaux — **d** Vorlagenvorbereitung *f* — **n** voorbereiding *f* van de originelen — **e** preparación *f* de originales — **i** preparazione *f* di originali

3553 copyright — Exclusive right, secured by law, for authors and artists to publish and control their works during specified periods. Unpublished material is protected under common law copyright, published material under statutory copyright.

f droit *m* d'auteur; copyright *m*; droits *mpl* réservés — **d** Urheberrecht *n*; Autorenrecht *n*; Vervielfältigungsrecht *n*; Verlagsrecht *n*; Autorrecht *n* — **n** auteursrecht *n*; eigendomsrecht *n*; recht *n* op de geestelijke eigendom — **e** derechos *mpl* de autor; copyright *m*; propiedad *f* intelectual; propiedad *f* literaria; propiedad *f* artística — **i** copyright *m*; diritti *mpl* d'autore; proprietà *f* letteraria (riservata)

3554 copyright deposit

f dépôt *m* légal — **d** Ablieferungspflicht *f*; gesetzliche Pflichtlieferung *f*; Pflichtexemplar *n* — **n** wettelijk depot *n* — **e** depósito *m* legal — **i** deposito *m* legale

3555 copyright notice; copyright page — Verso of the title page of a book bearing publisher's imprint and often other information (the standard book number, the Library of Congress catalog card number). See © in app. no. 8.

f avis *m* de copyright; mention *f* de copyright; mention *f* d'interdiction; mention *f* de réserve — **d** Copyright-Vermerk *n* — **n** auteursrechtaanduiding *f* — **e** mención *f* de propiedad (intelectual) — **i** indicazione *f* del diritto d'autore; menzione *f* del diritto d'autore

3556 copyright page — See copyright notice.

3557 copy scaling — Determining and marking originals with the dimensions in which they are to be reproduced.

f détermination *f* des dimensions de reproduction — **d** Größenangabe *f*; Formatangabe *f* — **n** aangeven *n* van de grootte — **e** indicación *f* de medidas de reproducción — **i** determinazione *f* della scala di riproduzione

3558 copy styling — See copy preparation.

3559 copy writer — See ad man.

3560 cording quires — See cassie paper.

3561 cordovan; cordovan leather; cordwain *arch.*; **Spanish leather** — Soft, smooth leather originally made at Córdoba (Spain) of goatskin but later also of split horse hide, pigskin, etc., used for bookbinding.

f cordouan *m*; cuir *m* de Cordoba; cuir *m* d'Espagne — **d** Korduan *m*; Korduanleder *n* — **n** corduaan *n* — **e** cordobán *m*; cuero *m* cordobán — **i** cuoio *m* di Spagna

3562 cordovan leather — See cordovan.

3563 cord, page — See page cord.

3564 cordwain — See cordovan.

3565 core — In lithography and other photomechanical operations, the centre portion of the halftone dot.

f centre *m* — **d** Kern *m* — **n** kern *fm* — **e** núcleo *m*; alma *f* — **i** nucleo *m*

3566 core — See roller stock.

3567 cored arc lamp carbons — See cored carbons.

3568 cored carbons; cored arc lamp carbons

f charbons *mpl* métallisés; charbons *mpl* à âme métallique; charbons *mpl* minéralisés — **d** Docht-kohlestifte *mpl*; Dochtkohlen *fpl* — **n** koolspitsen *fmpl* met kern van metaalzouten; koolstaven *fmpl* met kern van metaalzouten — **e** carbones *mpl* mechados; carbones *mpl* de mecha — **i** carboni *mpl* a miccia

3569 cored types; hollow types — Large size types and spaces with a hollow feet to reduce the weight of metal.

f caractères *mpl* évidés — **d** Spargußschrift *f* — **n** holle letter *fm* — **e** tipos *mpl* ahuecados; tipos *mpl* mechados — **i** caratteri *mpl* cavi

3570 core, magnetic — See magnetic core.

3571 core memory and core size — The *memory of the computer, usually called core, is described by its size in terms of so many K of words or bytes. For example, an 8K, 16-bit word core memory actually provides 8,192 words or 16,384 bytes, or 131,072 information bits, for data storage. A "K" is not precisely 1,000, as is generally supposed, but the nearest binary number, equivalent, or 1,024. See also store, magnetic core store, magnetic disk store, magnetic drum store, magnetic tape store.

3572 core plug — See reel cone.

3573 core, reel — See reel core.

3574 core store; magnetic core store *GB*; **magnetic core storage** *US* — Immediate access memory device, composed of individual doughnut-shaped cores of ferromagnetic material and inter-laced with wire so that each represents one bit of binary information by clockwise or anti-clockwise magnetization. One core is placed at each intersection of a wire grid.

f mémoire *f* à ferrites; mémoire *f* à tores magnétiques; mémoire *f* à noyaux magnétiques — **d** Kernspeicher *m*; Magnetkernspeicher *m* — **n** kernengeheugen *n*; magneetkerngeheugen *n*; magnetisch kerngeheugen *n* — **e** memoria *f* de núcleo magnético; memoria *f* de ferritas — **i** memoria *f* a nuclei di ferrite; memoria *f* a nuclei magnetici

3575 core waste — See newsprint waste.

3576 cork blanket — Printing blanket for letterpress rotary machines.

f blanchet *m* de liège — **d** Korkdrucktuch *n* — **n** kurkdoek *n* — **e** mantilla *f* de corcho; revestimiento *m* de corcho; cama *f* de corcho; patrón *m* de corcho — **i** rivestimento *m* in sughero

3577 corking — In die cutting, all chase dies are corked to release or strip away the cut cartons from the cutting knives. Cork is cut in strips about 1/4 in wide and glued to the plywood on each side of all cutting knives. The height of this cork depends on the height at which cutting knives are used and the thickness of the plywood furniture. Several materials are used for corking, cork, composition of cork and rubber, or rubber are the more common ones.

3578 corner — Corner of a book cover, square or rounded.

f coin *m* — **d** Ecke *f* — **n** hoek *m* — **e** punta *f*; puntera *f*; cantonera *f* — **i** angolo *m*

3579 corner — See corner piece.

3580 cornering machine — See corner rounding machine.

3581 corner piece; corner rule; corner — In type borders, piece which forms the corner or right angle of the panel.

f coin *m* de cadre; angle *m* (d'ornement); cornier *m* — **d** Gehrungsstück *n*; Eckstück *n* — **n** hoekstuk *n*; hoekversiering *f* — **e** esquinazo *m*; esquina *f* — **i** gherone *m*

3582 corner rounder — See corner rounding machine.

3583 corner rounding machine; corner rounder; cornering machine; round-cornering machine — Machine for cutting round corners of pages or book covers.

f machine *f* à arrondir les coins; machine *f* à coins ronds — **d** Eckenrundstoßmaschine *f* — **n** rondhoekmachine *f*; rondehoekenmachine *f*; hoekenrondmachine *f* — **e** máquina *f* de redondear esquinas; cortadora *f* de esquinas — **i** macchina *f* per arrotondare angoli

3584 corner rule — See corner piece.

3585 corner stapling — See corner stitching.

3586 corner stayer; box stayer; staying machine; stayer — Machine to reinforce corners of set-up boxes.

f machine *f* à relier les coins — **d** Eckenverbindemaschine *f*; Eckenverstärkungsmaschine *f* — **n** dozenhechtmachine *f*; hoekenhechtmachine *f* — **e** máquina *f* para juntar esquinas de cajas — **i** macchina *f* per fermare spigoli

3587 corner stitching; corner stapling

f piquage *m* en coin; piquage *m* des coins; agrafage *m* des coins — **d** Eckenheftung *f* — **n** hoekhechting *f* — **e** engrapado *m* a la esquina; cosido *m* a la esquina — **i** agraffatura *f* degli angoli; cucitura *f* degli angoli

3588 cornet bag; cone-shaped bag — Cone-shaped paper wrapper.

f cornet *m* (de papier); sac *m* à pointe — **d** Spitztüte *f* — **n** puntzakje *n* — **e** cucurucho *m* — **i** sacchetto *m* conico; sacchetto *m* a punta

3589 corona — Visible glow produced by ionized air in the path of a high voltage electrical discharge. In electrostatics, the specific type of charging unit employed.

f corona *f* — **d** Korona *f* — **n** corona *f* — **e** corona *f* — **i** corona *f*

3590 corona discharge — Discharge caused by ionization of a gas surrounding a conductor, which occurs when the potential gradient exceeds a certain value but is not sufficient to cause sparking. Before printing in the screen process, polythene needs to be pre-treated to obtain good ink adhesion. This is usually done by exposing the surface to a corona discharge.

f décharge *f* corona — **d** Korona-Entladung *f* — **n** coronaontlading *f* — **e** descarga *f* corona — **i** scarica *f* corona

3591 correct *v* — In GB the expression correcting in the metal is used to distinguish the mechanical operation from the reading and marking of proofs. In US the term is used both for the marking of errors in proofs by the proofreader and the correction of these errors on the forme by the compositor. See also read *v*.

f corriger sur plomb — **d** korrigieren im Blei; auf dem Blei korrigieren — **n** loodcorrectie aanbrengen — **e** corregir (en la forma); corregir en el molde; corregir sobre el plomo; corregir en la platina — **i** correggere sul piombo; correggere in piombo

3592 correct *v* — See read *v*.

3593 corrected edition
f édition *f* corrigée; édition *f* améliorée — **d** verbesserte Ausgabe *f* — **n** verbeterde uitgave *f* — **e** edición *f* corregida; edición *f* enmendada — **i** edizione *f* corretta; edizione *f* emendata

3594 corrected line — See also correction line.
f ligne *f* corrigée — **d** korrigierte Zeile *f* — **n** gecorrigeerde regel *m* — **e** línea *f* corregida — **i** riga *f* corretta

3595 corrected tape — See clean tape.

3596 correcting compositor
f corrigeur *m* — **d** korrektur ausführender Setzer *m* — **n** loodcorrector *m* — **e** corrector *m* de metal — **i** correttore *m* in piombo; compositore-correttore *m*

3597 correction — Elimination of a printer's error.
f correction *f* — **d** Korrektur *f* — **n** correctie *f*; verbetering *f* — **e** corrección *f*; enmienda *f* — **i** correzione *f*

3598 correction filter — See compensating filter.
3599 correction line — See also corrected line.
f ligne *f* de correction — **d** Korrekturzeile *f* — **n** correctieregel *m*; gecorrigeerde regel *m* — **e** línea *f* con enmienda; línea *f* enmendada — **i** riga *f* di correzione

3600 correction marks; proof correction marks; proofreaders' marks; readers' marks — Symbols used by proofreader to indicate corrections to be made in composition. See app. no. 5.
f signes *mpl* de correction — **d** Korrekturzeichen *npl* — **n** correctietekens *npl* — **e** señales *fpl* de corrección; signos *mpl* de corrección; signos *mpl* para correcciones; señales *fpl* de atención; llamadas *fpl* de corrección; marcas *fpl* de corrección — **i** segni *mpl* di correzione; segni *mpl* per la correzione; simboli *mpl* di correzione

3601 correction metal — See reviving alloy.
3602 correction on film
f correction *f* sur film — **d** Filmkorrektur *f*; Korrektur *f* im Film; Filmsatzkorrektur *f* — **n** filmcorrectie *f*; correctie *f* op film — **e** corrección

f en película — **i** correzione *f* in pellicola
3603 correction on lead; type correcting
f correction *f* sur plomb; correction *f* sur mobile — **d** Bleikorrektur *f*; Bleisatzkorrektur *f*; Korrektur *f* im Bleisatz — **n** loodcorrectie *f* — **e** corrección *f* en plomo — **i** correzione *f* in piombo

3604 correction routine — Computer program to change or update a body of text by deleting, adding or substituting new text or command codes.

3605 corrections — Changes made in type after proof has been taken.
f corrections *fpl* — **d** Korrektur *f*; Korrekturen *fpl* — **n** correctie *f*; correcties *fpl* — **e** correcciones *fpl*; enmiendas *fpl* — **i** correzioni *fpl*

3606 correction tape
f bande *f* de correction — **d** Korrekturlochstreifen *m*; Korrekturstreifen *m* — **n** correctieband *m* — **e** banda *f* de corrección; cinta *f* de corrección — **i** nastro *m* (perforato) delle correzioni

3607 corrector of the press — See proofreader.
3608 correlation — The extent to which two variables are related.
f corrélation *f* — **d** Korrelation *f*; Beziehung *f*; Wechselbeziehung *f* — **n** correlatie *f*; wederzijdse betrekking *f* — **e** correlación *f* — **i** correlazione *f*

3609 correspondence — A size of cut card. See app. no. 3.
3610 correspondence card — Cards suitable for pen and ink writing, usually made of wedding or index bristol cut to envelope sizes, generally used for personal correspondence.
f carte *f* de correspondance — **d** Korrespondenzkarte *f* — **n** correspondentiekaart *fm* — **e** carta *f* de correspondencia — **i** cartolina *f* di corrispondenza

3611 correspondence envelope; correspondence pocket — Flat case, generally rectangular, made from one sheet of paper, folded to provide a plain front and a back consisting of four overlapping flaps. Generally three flaps (but occasionally only two) are stuck together, the fourth (gummed or ungummed) serving as a closure. This fourth flap may be either on the short side (pocket shape) of the rectangle. The front may have a transparent window.
f enveloppe *f* postale; pochette *f* postale — **d** Briefhülle *f*; Briefumschlag *m*; Kuvert *n* — **n** envelop(pe) *fm*; couvert *n*; briefomslag *n* — **e** sobre *m* de correspondencia — **i** busta *f* di corrispondenza

3612 correspondence pocket — See correspondence envelope.
3613 corrode *v* — Destruction of metals by atmospheric or similar influences, especially by chemical action.
f corroder; attaquer (le métal) — **d** korrodieren; anfressen — **n** corroderen; aantasten; aanvreten — **e** corroer — **i** corrodere

3614 corroded layer
f couche *f* corrodée — **d** zerfressene Schicht *f* —

n gecorrodeerde laag *fm* — **e** capa *f* corroída — **i** strato *m* corroso

3615 corrosion — The eating away of material or metal.

f corrosion *f*; attaque *f* — **d** Korrosion *f*; Ätzung *f*; Anfressen *n*; Zerfressen *n* — **n** corrosie *f*; aantasting *f*; aanvreting *f* — **e** corrosión *f*; picado *m* — **i** corrosione *f*

3616 corrosion test

f essai *m* de corrosion — **d** Korrosionsversuch *m*; Korrosionsprobe *f* — **n** corrosieproef *fm*; corrosie-onderzoek *n* — **e** ensayo *m* de corrosión — **i** prova *f* di corrosione

3617 corrosive mercury chloride — See mercury chloride.

3618 corrosive sublimate — See mercury chloride.

3619 corrugated board; corrugated fibreboard — Board consisting of one or more sheets of fluted paper stuck to a flat sheet of paper or board between several sheets. There are three main classifications: 1. single face corrugated board; 2. double face corrugated board; and 3. double-double face corrugated board.

f carton *m* ondulé; fibre *f* ondulée — **d** Well-pappe *f*; Wellpapier *n*; Riffelpapier *n* — **n** wel-bord *n*; golfkarton *n*; gegolfd karton *n* — **e** cartón *m* ondulado; cartón *m* acanalado; cartón *m* corrugado; papel *m* (embalaje) ondulado — **i** cartone *m* ondulato; carta *f* ondulata

3620 corrugated box; corrugated container; corrugated case

f boîte *f* en carton ondulé; caisse *f* en carton ondulé — **d** Wellpappfaltkiste *f*; Wellpapp-karton *m* — **n** doos *fm* van golfkarton; doos *fm* van welbord — **e** caja *f* de cartón ondulado — **i** scatola *f* in cartone ondulato; contenitore *m* in cartone ondulato; cassa *f* in cartone ondulato

3621 corrugated case — See corrugated box.

3622 corrugated container — See corrugated box.

3623 corrugated fibreboard — See corrugated board.

3624 corundum — Very hard mineral, used as abrasive, consisting of *aluminium oxide.

f corindon *m* — **d** Korund *m* — **n** korund *n* — **e** corindón *m* — **i** corindone *m*

3625 cosecant — The secant of the complement of an angle or an arc. See also trigonometric functions.

f cosécante *f* — **d** Kosekante *f* — **n** cosecans *f* — **e** cosecante *f* — **i** cosecante *f*

3626 cosine — The sine of the complement of a given angle or arc. See also trigonometric functions.

f cosinus *m* — **d** Kosinus *m* — **n** cosinus *m* — **e** coseno *m* — **i** coseno *m*

3627 co-solvent ink — Ink of a category of flexo inks, based on polyamide resins. Most polyamide inks, however, are modified with other resins, producing the most widely used flexo inks for plastic packaging films. Excellent gloss, adhesion, block resistance, water and wet pack resistance, and printability are characteristic, heat resistance is relatively poor.

f encre *f* co-solvant — **d** Co-solvent-Farbe *f* — **n** co-solvent inkt *m* — **e** tinta *f* co-solvente — **i** inchiostro *m* a co-solvente

3628 Costeriana — Books (chiefly grammars, as Donats, Abecedaria, etc.) printed in the 15th century by Dutch printers, probably at Utrecht. Sometimes attributed to Laurens Janszoon Coster, of Haarlem, Holland; however, the 7 or 8 different type designs in these books make it very improbable that they originate from one workshop.

3629 costing

f calcul *m* du prix de revient; postcalculation *f* — **d** Nachkalkulation *f*; Kostenberechnung *f* — **n** na-calculatie *f*; naberekening *f* — **e** cálculo *m* de los gastos — **i** calcolo *m* dei costi

3630 costing system

f système *m* de calcul des prix de revient — **d** Kostenberechnungssystem *n*; Kalkulations-system *n* — **n** kostprijssysteem *n* — **e** sistema *m* de computar el costo — **i** sistema *m* per la determinazione dei costi

3631 cost price; production costs — Price at which goods or merchandise are sold without profits, i.e. the price paid for them or at manufacturing cost.

f prix *m* de revient; prix *m* coûtant — **d** Selbst-kostenpreis *m*; Gestehungspreis *m*; Herstellungs-kosten *pl* — **n** kostprijs *m*; kosten *pl* van vervaardiging — **e** precio *m* de coste; precio *m* de costo; costo *m* de fabricación — **i** prezzo *m* di costo

3632 costs of printing — See printing costs.

3633 cotangent — The tangent of the complement of an angle. See also trigonometric functions.

f cotangente *f* — **d** Kotangente *f* — **n** cotangens *f* — **e** cotangente *f* — **i** cotangente *f*

3634 cottage style binding — Book cover decorated in a style like a cottage gable, used in the late 17th century by Samuel Mearne (1660-1683), binder to King Charles II of England.

f reliure *f* cottage — **d** Cottage-Einband *m* — **n** cottage-band *m* — **e** encuadernación *f* de alero de cabaña — **i** legatura *f* architettonica

3635 cotter pin — See split pin.

3636 cotton — Staple fibres, usually 19-63.5 mm (0.75-2.5 in) long, surrounding the seeds of various species of gossypium. Cotton is the major textile fibre and also an important source of cellulose, which constitutes 88-96% of the fibre. Subject to mildew.

f coton *m* — **d** Kattun *m*; Baumwolle *f* — **n** katoen *n* — **e** algodón *m* — **i** cotone *m*

3637 cotton content — Amount of cellulose fibres derived from raw cotton in paper, not to be confused with rag-content paper.

f teneur *f* en fibres de coton — **d** Baumwoll-gehalt *m* — **n** katoenstofgehalte *n* — **e** contenido *m* de algodón — **i** contenuto *m* in fibre di cotone

3638 cotton linters; linters — Short fibres adhering to cottonseed after ginning, including the stumps or bases of the longer textile fibres, or lint and an undercoat or coarse, short fuzz fibres in most upland varieties. Cut from the cottonseed by another saw gin operation. Linters may be removed in one operation to give a blend of fibre lengths called mill run linters, or the longer fibres may be removed first as first cut linters while the remaining shorter fibres will then be removed by a second operation as second cut linters. Linters are used in cotton content paper manufacture but mainly as raw material for the manufacture of cellulose derivatives.

f duvets *mpl* de coton; linters *mpl* — **d** Baumwollinters *mpl*; Linters *mpl* — **n** linters *mpl*; katoenlinters *mpl*; katoenhaartjes *npl* — **e** línters *mpl*; línteres *mpl*; borras *fpl* de algodón — **i** linters *mpl* (di cotone)

3639 cotton powder — See nitrocellulose.

3640 cotton-wool wad — See pledget of cotton.

3641 couch *v* — To press down a hand-made sheet of paper onto a couch felt.

f coucher — **d** gautschen; abgautschen — **n** koetsen; afkoetsen — **e** sacar el papel de la tela; desprender el papel del manchón — **i** porre; goffare

3642 coucher — In papermaking, the craftsman who transfers wet sheets of wet pulp to the couch (board or felt blanket) on which he builds up a post.

f coucheur *m* — **d** Gautscher *m*; Filzer *m* — **n** koetser *m* — **e** prensador *m* — **i** ponitore *m*

3643 couch felt — Woolen, tubular felt to cover the top couch roll, or the heavy felt to remove a sheet of paper from the hand mould on which the sheet is dried.

f feutre *m* coucheur; flôtre *m obs.* — **d** Gautschfilz *m* — **n** koetsvilt *n* — **e** fieltro *m* de manchón — **i** feltro *m* ponitore

3644 couch jacket — Tubular woven thick wool felt shrunk onto the top couch roll. Replaced by the suction couch roll.

f manchon *m* en feutre — **d** Filzmanchon *m*; Manchon *m*; Gautschwalzenbezug *m*; Filzschlauch *m* — **n** manchon *m*; viltmantel *m* — **e** manchón *m* de fieltro — **i** feltro *m* manicotto

3645 couch roll — Roll to separate wet paper web from mould or paper machine wire. On modern paper machines, the felt-jacketed couch roll has been replaced by a suction roll.

f rouleau *m* coucheur; cylindre *m* coucheur — **d** Gautschwalze *f* — **n** koetswals *fm*; koetswals *fm* — **e** prensa *f* de la tela; manchón *m*; prensa *f* húmeda — **i** cilindro *m* manicotto

3646 coulomb — Practical unit of the quantity of electricity. Abbrev. C. In the SI (1978):
$$1\ C_{int} \cdot s/\Omega_{int} \approx 0.999850\ C.$$
f coulomb *m* — **d** Coulomb *n* — **n** coulomb *m* — **e** coulomb *m* — **i** coulomb *m*

3647 coulombmeter — See coulometer.

3648 coulometer; coulombmeter — Electrolytic cell to measure the quantity of electricity by the chemical action produced in accordance with Faraday's law.

f coulombmètre *m*; voltamètre *m* — **d** Coulometer *n*; Voltameter *n* — **n** coulometer *m* — **e** culombímetro *m*; voltámetro *m* — **i** voltametro *m*

3649 coumaric resin — See coumarone-indene resin.

3650 coumarone-indene resin; coumaric resin; indene resin — Resin mixture of polymerized coumarone, C_8H_6O, and polymerized indene, C_9H_8, from solvent naphtha fractions of coal tar at 160-190 °C, used in printing inks, adhesives, etc.

f résine *f* (de) coumarone; résine *f* coumaronique — **d** Cumaronharz *n*; Kumaronharz *n* — **n** cumaronhars *mn* — **e** resina *f* de cumarón; resina *f* cumarona — **i** resina *f* cumaron-indenica

3651 counter — Space inside a letter, which does not print.

f intérieur *m* de l'œil; blanc *m* — **d** Punzen *m*; Bunzen *m*; Bunze *m* — **n** pons *m* — **e** cuenca *f* (del tipo); blanco *m* interior; blanco *m* interno; punzón *m*; profundidad *f* central; ojo *m* — **i** bianco *m* interno dell'occhio

3652 counter; checker — One who counts sheets of paper or copies.

f compteur *m*; compteuse *f* — **d** Zähler *m*; Zählerin *f* — **n** teller *m*; telster *f* — **e** contador *m*; contadora *f* — **i** contatore *m*; contatrice *f*

3653 counter — Counting device on the machine that records the number of copies produced for a straight run. See also tachometer.

f compteur *m* (de feuilles); compte-feuilles *m*; appareil *m* compteur — **d** Zähler *m*; Zählapparat *m*; Zählvorrichtung *f*; Zählgerät *n*; Bogenzähler *m*; Auflagenzähler *m* — **n** teller *m*; telapparaat *n*; vellentelapparaat *n*; telklok *fm*; klok *fm* — **e** contadora *f*; contador *m* automático; aparato *m* contador — **i** contatrice *f*; contatore *m*; apparecchio *m* contatore

3654 counter — See counter punch.

3655 counter-clockwise — In the direction opposite to that of the hands of a clock.

f en sens inverse d'horloge; dans le sens inverse des aiguilles d'une horloge — **d** linksum; linksgängig; linksdrehend; gegen den Uhrzeigersinn — **n** linksom; tegen de wijzers van de klok in — **e** contrario al de las agújas del reloj; de rotación inversa; de rotación a la izquierda — **i** in senso antiorario; in direzione antiorario; a sinistra

3656 counteretch *v* — To clean, before coating, a grained offset metal plate of dirt and oxides, without damaging the grain, with a weak acid solution such as 1 oz. hydrochloric acid in a gallon (4.54596 liters) of water. Sometimes this is referred to as sensitizing the plate, i.e., making the metal sensitive to grease.

f décaper — **d** entsäuern — **n** ontzuren; zuren — **e** decapar — **i** decappare; preparare

3657 counterfeit *v*

f contrefaire — **d** nachahmen — **n** namaken — **e** contrahacer — **i** contraffare

3658 counterfeiting

f contrefaçon *f*; contrefaction *f* — **d** Nachahmung *f* — **n** namaak *m* — **e** falsificación *f* — **i**

contraffazione *f*

3659 counterfoil — See stub.

3660 counter mask — Mask of the opposite sign (negative with positive or vice versa) useed with another mask to lessen its contrast-reducing effect.
f contremasque *m* — **d** Gegenmaske *f* — **n** tegenmasker *n* — **e** contramáscara *f* — **i** contromaschera *f*

3661 counter punch — Engraving in high relief of the counter or hollow part of a type design, i.e. the part which appears white when the letter or character is printed.
f contre-poinçon *m* — **d** Konterstempel *m*; Punzen *m*; Bunzen *m*; Bunze *m* — **n** pons *m*; ponsoen *m* — **e** contrapunzón *m* — **i** contropunzone *m*; controstampo *m*

3662 counter sample
f contre-type *m* — **d** Gegenmuster *n* — **n** tegenmonster *n* — **e** contraprueba *f*; contratipo *m*; contramuestra *f* — **i** controtipo *m*

3663 countersunk screw — Screw with a cone-shaped head sunk into a prepared depression to be flush with or below the surface.
f vis *f* à tête noyée; vis *f* à tête perdue — **d** Senkkopfschraube *f*; Senkschraube *f* — **n** schroef *fm* met verzonken kop — **e** tornillo *m* avellanado; tornillo *m* de cabeza avellanada — **i** vite *f* a testa svasata

3664 countersunk watermark — See shaded watermark.

3665 counter weight — See balance weight.

3666 counting — Operation in which the sheets of paper, signatures, books, etc., are counted to compare with ordered quantities.
f comptage *m* — **d** Zählung *f*; Abzählung *f*; Abzählen *n* — **n** telling *f*; aftellen *n* — **e** contado *m* — **i** contare

3667 counting perforator — Paper-tape perforating keyboard with a character width counting mechanism to determine the end of each line. The simplest boards indicate when justification zone is reached; the more sophisticated boards are linked to electronic calculating devices with indicator feedback, and give automatic shift and reduced key stroke facilities.

3668 couple — Two equal and opposite forces that act along parallel lines. See also torque.
f couple *m* — **d** Moment *n* — **n** koppel *n* — **e** par *m* — **i** coppia *f*

3669 coupling — The combination of an amine or phenol with a diazonium compound to give an azo compound, the reaction of which azo dyes are prepared, thus meta-phenylenediamine $C_6H_4(NH_2)_2$ couples with benzene diazonium chloride $C_6H_5N_2Cl$ to produce the dye $C_6H_5N_2C_6H_3(NH_2)_2$, which gives orange colours.
f copulation *f* — **d** Verbindung *f* — **n** koppeling *f* — **e** copulación *f* — **i** accoppiamento *m*

3670 coupling device; clutch — Part of a machine for connecting a driving.

f embrayage *m*; accouplement *m* — **d** Kupplung *f*; Kupplungsvorrichtung *f* — **n** koppeling *f* — **e** embrague *m*; acoplamiento *m*; cuplón *m*; unión *f* — **i** accoppiamento *m*; giunto *m*; innesto *m*

3671 coupon — Dated certificate attached to, e.g., a bond, representing interest accrued and payable at stated periods.
f coupon *m* de dividende; coupon *m* de titre — **d** Kupon *m*; Zinsschein *m* — **n** coupon *m*; dividendbewijs *n* — **e** cupón *m* — **i** cedola *f*; tagliando *m*; cupone *m*

3672 court — A UK size of cut card. See app. no. 3.

3673 cover — Outer covering of a book or pamphlet; may be made of paper, board, cloth-lined board, leather, etc.
f couverture *f* — **d** Umschlag *m*; Einband *m*; Überzug *m*; Einbanddecke *f*; Deckel *m*; Buchdeckel *m* — **n** omslag *n*; boekband *m*; band *m* — **e** cubierta *f*; tapa *f*; forro *m* — **i** coperta *f*; copertina *f*

3674 cover — See envelope.

3675 coverage; spreading rate; spreading capacity — See also ink coverage.
f pouvoir *m* couvrant en surface; rendement *m* en surface; couvrage *m* de l'encre — **d** Ausgiebigkeit *f* — **n** uitstrijkend vermogen *n*; rendement *n* van het uitstrijken — **e** capacidad *f* cubriente; capacidad *f* cubridora; poder *m* cubriente; poder *m* cubridor — **i** capacità *f* coprente

3676 cover boards; boards — Composition pasteboard covers on the sides of books. In former times made of wood.
f plats *mpl* de couverture; cartons *mpl* de couvertures — **d** Deckelpappen *fpl* — **n** platten *npl* — **e** tapas *fpl* — **i** piani *mpl* della coperta

3677 covering — In bookbinding done by hand, the pasting-on and drawing of the leather or cloth over the boards and around the edges and corners of the boards.
f couvrure *f* — **d** Überziehen *n*; Beziehen *n* — **n** omtrekken *n* — **e** cubrimiento *m* — **i** rivestimento *m* esterno della coperta

3678 covering power — See hiding power.

3679 covering varnish — See masking lacquer.

3680 cover ink — Heavy bodied ink, generally opaque, with a high pigment concentration.
f encre *f* couvrante — **d** Deckfarbe *f* — **n** dekinkt *m* — **e** tinta *f* cubriente — **i** inchiostro *m* coprente

3681 cover paper — Strong antique finished coloured paper suitable for brochure covers, book covers, magazines, catalogues, etc.
f papier *m* (à) couverture; papier *m* pour couvertures — **d** Umschlagpapier *n*; Beklebepapier *n*; Bezugspapier *n*; Deckpapier *n* — **n** omslagpapier *n* — **e** papel *m* para cubiertas; papel *m* para carátulas; papel *m* para tapas — **i** carta *f* per copertine

3682 cover royal — A British standard size of writing and printing papers. See app. no. 3.

3683 cover unit — Separate printing unit for a letterpress rotary coupled to the main press to produce the cover of a magazine or paper.
f groupe *m* (pour impression de la) couverture — **d** Umschlagdruckwerk *n*; Druckwerk *n* für Umschlagseite — **n** omslagdrukeenheid *f* — **e** unidad *f* de cubiertas — **i** gruppo *m* per la stampa della copertina
3684 cover with rounded back
f volume *m* avec dos arrondi — **d** Einband *m* mit rundem Rücken — **n** band *m* met ronde rug — **e** cubierta *f* con lomo redondo — **i** legatura *f* con dorso mezzo tondo
3685 cover with square back
f volume *m* avec dos plat; cartonnage *m* anglais — **d** Einband *m* mit geradem Rücken — **n** band *m* met rechte rug; band *m* met platte rug; Engelse band *m* — **e** cubierta *f* con lomo cuadrado — **i** legatura *f* con dorso piatto
3686 cow catcher — Etched area or bars across the leading edge of a gravure printing plate that clears the doctor blade of foreign particles before it passes over the printing area of the plate. Sometimes called scavenger lines.
f preneur *m* de pétouilles
3687 cowhide; neat's leather — Grained cowhide, used in best quality loose-leaf work.
f cuir *m* de vache; vache *f* — **d** Rindleder *n* — **n** rundleer *n* — **e** cuero *m* de vaca; vaqueta *f* — **i** cuoio *m* di vacca
3688 cps — See cycle/second.
3689 CPU — See central processing unit.
3690 crabs — See returns.
3691 cracking — Defect in coated paper, caused by separation of the coating layer or formation of fissures in the surface of the coating in printing or other converting processes.
f décollement *m*; se fendiller — **d** brechen — **n** barsten; splijten; breken — **e** arrancamiento *m* — **i** distacco *m* (della patina)
3692 cracking — Breaking of stocks when a box is folded 90 ° or 180 °.
f craquelure *f* — **d** Bruch *m* — **n** breuk *fm* — **e** cuarteado *m* — **i** screpolatura *f*
3693 cracking at the fold — See poor fold strength.
3694 crackle — Network of fine cracks on a smooth surface, as in the glaze of some kinds of porcelain.
f craquelure *f* — **d** Krakelüre *f*; Krakelée *n* — **n** craquelé *n* — **e** craquelado *m* — **i** crepolatura *f*
3695 cradle — Curved metal mould on which mats are placed for casting curved stereos.
f berceau *m* — **d** Gießschale *f* — **n** gietschaal *fm* — **i** culla *f*
3696 cradle book — See incunabulum.
3697 Craftint — Chemically treated drawing sheet; a modern form of shading sheet. See also benday process.
3698 CR allowance — See compensating relaxation allowance.
3699 crank handle — See rounce.

3700 crank shaft — Part of an engine to which the connecting rods are fitted.
f vilebrequin *m* — **d** Kurbelwelle *f* — **n** krukas *fm* — **e** cigüeñal *m* — **i** asse *m* a manovella; albero *m* a gomito
3701 crash — See mull.
3702 crash finish — Finish similar to linen finish but coarser in texture.
f fini *m* toile gros grain; apprêt *m* grosse toile; fini *m* grosse toile — **d** grobe Leinenprägung *f*; Grobleinenprägung *f*; Tuchprägung *f* — **n** grove linnenpersing *f*; jutepersing *f*; canvaspersing *f* — **e** acabado *m* imitación lona — **i** telatura *f* grossa
3703 crate — Box made of wooden slats used to ship paper.
f caisse *f* à claire-voie; plateau *m* à claire-voie; cadre *m* à claire-voie — **d** Lattenkiste *f*; Gitterschachtel *f* — **n** krat *n*; schot *n*; lattenkist *fm* — **e** jaula *f*; esqueleto *m*; guacal *m* — **i** cassa *f* a trafori; cassetta *f* a trafori; gabbia *f* (di legno)
3704 crate *v* — To pack in crates or in boxes.
f emballer dans une caisse à claire-voie; mettre en caisse — **d** in Lattenkisten verpacken; einballen; einballieren — **n** verpakken in kratten (of kisten) — **e** encajonar — **i** imballare in gabbie
3705 crawl *v* — The continuous turning of the rotary press at slow speed.
f marcher au ralenti — **d** auf Einziehgeschwindigkeit — **n** langzaam draaien — **e** girar lentamente — **i** girare al minimo
3706 crawling — The contraction of an ink film into drops, after printing on a surface which the ink does not wet properly, as glass, lacquer, cellophane.
f rétraction *f* — **d** Perlen *n* — **n** parelen *n*; samentrekken *n* — **e** agrumarse — **i** perlatura *f*
3707 crawling — In rotogravure printing, a "squeezing out" of ink from cells thereby producing a ragged print.
f galeux *m* — **d** Ausquetschen *n* — **n** wegpersen *n*
3708 crayon engraving — Copperplate engraving to reproduce the texture of chalk and pencil. As in *stipple engraving the etching ground is removed and further elaboration and completion are accomplished by working directly on the plate.
f manière *f* de crayon — **d** Crayonmanier *f*; Kreidemanierzeichnung *f* — **n** crayonmanier *fm*; op de wijze van een krijttekening — **e** dibujo *m* hecho al carbon — **i** incisione *f* in forma di disegno a matita; stampa *f* in forma di disegno a matita
3709 crayoning — Application of greasy litho crayon to halftone etchings or retouching on direct hand-made or photographic plates as a delicate and blendable acid resist.
f retouche *f* au crayon litho — **d** Bleistiftretusche *f*; Kreideretusche *f*; Crayonretusche *f*; Abdecken *n* mit Lithokreide — **n** potloodretouche *fm*; bijwerken *n* met potlood — **e** retoque *m* con lápiz litográfico — **i** ritocco *m* con gesso grasso litografico

3710 crazed drying; gas crazing; gas checking; wrinkling; throwing out; frosting — The wrinkling of a tung oil film under certain drying conditions, caused by rapid absorption of oxygen on the surface.

f gélification *f* — **d** Runzeln *n* — **n** rimpelen *n*; oprimpelen *n* — **e** encrespadura *f* — **i** increspamento *m*; increspatura *f*

3711 cream laid paper — Cream-coloured paper showing the wire marks when held up.

f papier *m* vergé crème — **d** cremegeripptes Papier *n*; gelblichweißes geripptes Papier *n* — **n** crème gevergeerd papier *n* — **e** papel *m* verjurado cremoso — **i** carta *f* vergata crema

3712 cream white paper — White paper with a cream cast.

f papier *m* blanc crème — **d** cremefarbiges Papier *n* — **n** crème getint papier *n* — **e** papel *m* crema; papel *m* ahuesado; papel *m* cremoso — **i** carta *f* crema

3713 cream wove paper — Cream-coloured paper without laid lines.

f papier *m* vélin blanc — **d** cremes Velinpapier *n*; cremes ungeripptes Papier *n* — **n** crème velijn papier *n* — **e** papel *m* velin cremoso — **i** carta *f* velina crema

3714 crease — Indention in board to give the line of fold.

f refoulage *m*; rainure *f*; tracé *m* refoulé — **d** Rille *f* — **n** rillijn *fm*; ril *fm* — **e** ranura *f*; raya *f*; estría *f*; acanaladura *f*; hendido *m* — **i** cordonatura *f*; linea *f* di cordonatura

3715 crease — Folding or rip in a sheet.

f pli *m* — **d** Falte *f* — **n** plooi *fm*; vouw *fm* — **e** pliegue *m* — **i** piega *f*; cordonatura *f* segnapiega

3716 crease *v* — To score a card or heavy paper so that it will not crack when folded.

f refouler (profond); rainer (peu profond) — **d** rillen — **n** rillen — **e** ranurar; acanalar; rayar; hender — **i** cordonare

3717 creasing

f plissage *m* (ne pas confondre avec pliage) — **d** Faltenbildung *f* — **n** plooien *n* (omslaan tijdens het drukken) — **e** formación *f* de arrugas; formación *f* de pliegues — **i** formazione *f* di pieghe

3718 creasing machine — Machine for scoring a sheet of board to be bent along a defined line.

f machine *f* à rainer — **d** Rillmaschine *f* — **n** rilmachine *f* — **e** ranuradora *f*; máquina *f* para ranurar; acanaladora *f* — **i** cordonatrice *f*; macchina *f* per cordonare

3719 creasing-resistance tester — Apparatus to test the resistance to creasing for folding. A rubber covered wheel with a sharp metal edge in the centre, is pressed in a groove between two metal wheels, which can be fixed any distance apart, permitting grooving or creasing to any depth to suit the thickness of the board being tested.

f appareil *m* pour déterminer la résistance au rainage; mesureur *m* de l'aptitude au rainage — **d** Rillfähigkeitsprüfgerät *n*; Rillbarkeitsprüfer *m* — **n** riltoestel *n* — **e** aparato *m* para determinar la resistencia al ranurado — **i** apparecchio *m* per determinare la resistenza alla cordonatura

3720 creasing rule — Brass rule to make indentions in board to give the line of fold.

f filet *m* à refouler; filet *m* raineur; filet *m* de rainure; filet *m* de refoulage — **d** Rillinie *f* — **n** rillijn *fm* — **e** filete *m* hendedor; filete *m* doblador; filete *m* para marcar dobleces; filete *m* para ranurar; pleca *f* para ranurar — **i** filetto *m* per cordonatura

3721 credit line — Line acknowledging the source of origin of published or exhibited material.

f référence *f* de source; indication *f* de la source — **d** Quellenangabe *f*; Herkunftsvermerk *m*; Urhebervermerk *m* — **n** bronvermelding *f*; bron *fm* — **e** mención *f* de la fuente — **i** menzione *f* della fonte

3722 creep *v* — Forward movement of the blanket surface during operation of offset press due to improper pressure or stretch of offset blanket.

f glisser — **d** wandern; kriechen — **n** kruipen; slippen — **e** desplazarse — **i** scorrimento *m*

3723 creep *v* — Flowing of ink beyond limits of the design after printing when ink is not viscous enough.

f étaler — **d** auslaufen — **n** uitvloeien; uitlopen — **e** derramarse; correrse — **i** spandersi

3724 creeper — Rubber blanket for rolling mats.

f blanchet *m* de prise d'empreinte — **d** Gummituch *n* — **n** preegdoek *m* — **e** mantilla *f* de caucho; mantilla *f* de respaldar; mantilla *f* de calzar — **i** telo *m* di gomma

3725 creeper belt delivery — See creeper delivery.

3726 creeper delivery; creeper belt delivery

f transporteur *m* de sortie à courroie; transporteur *m* de sortie à cordons — **d** Bänderauslage *f* — **n** banduitleg *m*; kruipbanduitleg *m* — **e** entrega *f* de cintas; transportador *m* de salida de cintas — **i** uscita *f* con nastri trasportatori

3727 Cremnitz white — See flake white.

3728 Crems white — See flake white.

3729 crêpe paper — Paper with a texture simulating crêpe such as paper towels or napkins, coloured decorative tissue. The crêpe effect is obtained by wrinkling the wet paper on a roll with a knife or doctor blade.

f papier *m* crêpé; papier *m* crêpe; papier *m* crepon — **d** Krepp-Papier *n*; Kreppapier *n* — **n** crêpepapier *n*; gecrêpt papier *n* — **e** papel *m* crêpe; papel *m* crêpé; papel *m* crespado — **i** carta *f* crespata

3730 crêping — Crinkling a sheet of paper to increase its stretch and softness.

f crêpage *m* — **d** Kreppen *n*; Kreppung *f* — **n** crêpen *n* — **e** crespado *m* — **i** crespatura *f*

3731 cresol; methyl phenol; hydroxymethylbenzene — Mixture of the ortho-, meta-, and/or para-cresols; used as a photographic developer. Rapidly absorbed through skin, causing severe

burns. TL-value: 5 ppm.
Formula: C_7H_8O.
f crésol *m* — **d** Kresol *n* — **n** cresol *n* — **e** cresol *m* — **i** cresolo *m*
3732 cretaceous — Having the characteristics of or abounding in chalk.
f crétacé; crayeux — **d** kreidig; kreideartig — **n** krijtachtig — **e** gredoso; cretáceo — **i** cretoso
3733 criblée — See manière criblée.
3734 critic — One who engages in the analysis, evaluation, or appreciation of literary works.
f critique *m* — **d** Rezensent *m*; Kritiker *m* — **n** recensent *m* — **e** crítico *m* — **i** critico *m*
3735 crocus; crocus martis — Powder consisting of ferric oxide used for polishing gravure cylinders.
3736 crocus martis — See crocus.
3737 cronak process — Process of treating zinc plates with a mixture of sodium bichromate and sulphuric acid to make them resistant to oxidation.
f procédé *m* cronak — **d** Cronak-Verfahren *n* — **n** cronak-procédé *n* — **e** procedimiento *m* cronak; método *m* cronak — **i** procedimento *m* cronak; processo *m* cronak; metodo *m* cronak
3738 Cronar — Trade-mark for a polyester photographic film.
3739 crooked lines
f lignes *fpl* bombées — **d** krumme Zeilen *fpl*; krummstehende Zeilen *fpl* — **n** kromme regels *mpl*; krom staande regels *mpl* — **e** renglones *mpl* torcidos; renglones *mpl* encaballados; líneas *fpl* torcidas; renglones *mpl* chuecos *Me* — **i** linee *fpl* (di composizione) incurvate
3740 crop *v* — To trim a book to a size smaller than intended or specified.
f rogner trop fort; réduire — **d** zu stark beschneiden — **n** te kort afsnijden; teveel afsnijden — **e** desmochar; recortar con exceso — **i** rifilare eccessivamente
3741 crop *v* — To opaque, mask, cut, or trim a negative or plate, an illustration, or its reproduction to the required size. See also crop marks.
f élaguer — **d** beschneiden — **n** afsnijden — **e** cortar; recortar; dar corte — **i** ridurre
3742 crop marks — Markings at edges of original or on guide sheet to indicate the area desired in reproduction, with negative or plate trimmed (cropped) at the markings.
f repères *mpl* de rogne; repères *mpl* de coupe — **d** Schnittzeichen *npl*; Schneidemarken *fpl*; Abstriche *mpl* — **n** snijtekens *npl*; afsnijtekens *npl* — **e** marcas *fpl* recorte; señales *fpl* de encuadre — **i** segni *mpl* (di riferimento) per la rifilatura
3743 cropper — In 1862 and 1867 the British firm H.S. Cropper & Co. commenced the manufacture of an improved Franklin platen press and called it Minerva. The company's name became so attached to this type of press that any platen machine operator was known as a "cropper

hand". This term occurs in the minutes of the Platen Machine Minders Society (formed in 1890 and not particularly concerned with any make of machine). See also treadle-press.
f minerve *f*; pédale *f* — **d** Drucktiegel *m* mit Fußbetrieb; Tiegeldruckpresse *f* mit Fußbetrieb; Tiegeltretpresse *f* — **n** trappers *f* — **e** minerva *f* a pedal — **i** pedalina *f*
3744 cropper hand — See cropper.
3745 cropper hand — See platen printer.
3746 cross — Proof correction mark indicating a faulty letter or any part of the printing surfaces that requires attention. See proof correction marks in app. no. 5.
f croix *f* — **d** Kreuz *n*; Kreuzchen *n* — **n** kruisje *n* — **e** cruz *f* — **i** croce *f*
3747 cross — See set mark.
3748 cross association; transverse collection — When the webs from one side of a double folder are taken across to the other side after the former but before cutting. This has the effect of doubling the number of pages in the copy and halving production.
f assemblage *m* transversal des bandes — **d** Zusammenlauf *m* der Papierstränge — **n** samenlopen *n* der papierbanen — **e** reunirse; encajarse — **i** riunione *m* trasversale dei nastri; concorso *m* trasversale dei nastri
3749 cross bar; chase bar; bar — Bar which divides some chases into sections; fixed in cast-iron chases, removable in wrought-iron and steel chases. Normally the short bar represents in the chase the first fold of the paper.
f barre *f* transversale — **d** eiserner Mittelsteg *m* — **n** kruis *n* — **e** crucero *m* de la rama; medianil *m* — **i** barra *f* trasversale
3750 cross bars; girders — Parts extending over the whole of a machine.
f traverses *fpl*; barres *fpl* de traverse — **d** Traversen *fpl* — **n** traversen *fpl*; langsverbindingen *fpl* — **e** travesaños *mpl*; vigas *fpl* — **i** traverse *fpl*
3751 cross-cut of cylinder; cross-section of cylinder
f coupe *f* en travers du cylindre — **d** Zylinderquerschnitt *m* — **n** doorsnede *fm* van de cilinder; dwarsdoorsnede *fm* van de cilinder — **e** sección *f* transversal del cilindro — **i** sezione *f* trasversale del cilindro
3752 cross direction; across the grain — Dimension of paper or board at right angles to the machine direction.
f sens *m* travers (du papier); sens *m* transversal — **d** Querrichtung *f*; Querlaufrichtung *f* — **n** dwarsrichting *f*; breedterichting *f* — **e** sentido *m* transversal; sentido *m* contrario; sentido *m* atravesado; grano *m* atravesado; dirección *f* transversal; hilo *m* cruzado; hilo *m* corto; papel *m* a contrafibra — **i** direzione *f* trasversale
3753 cross feints — See feint lines.
3754 cross fold — See chopper fold.
3755 cross grain wood — Wood like *pearwood

with the grain or fibres running transversely.
f bois *m* de bout — **d** Hirnholz *n* — **n** kopshout *n* — **e** madera *f* a contrafibra; madera *f* a contrahilo — **i** legno *m* di testa

3756 cross hatching — Drawing or tooling parallel lines to cross each other at certain angles for effect of tone or texture.
f recoupe *f* de hachures — **d** Kreuzschraffierung *f* — **n** kruisharcering *f*; kruisarcering *f* — **e** cuadrícula *f*; sombreado *m* de líneas cruzadas — **i** tratteggio *m* a spinapesce; tratteggio *m* incrociato

3757 crosshead — Metal block to which one end of a piston rod is secured which slides on parallel guides, and with a pin for attachment of the connecting rod.
f crosse *f* de tête — **d** Kreuzkopf *m* — **n** kruiskop *m* — **e** traviesa *f* móvil; barra *f* de entrada — **i** cappello *m*; testa *f* a croce

3758 cross head — See centre head.

3759 crossline screen; square-ruled screen — Standard halftone (glass) screen with the opaque lines crossing each other at right angles, thus forming transparent squares or screen apertures.
f trame *f* cristal; trame *f* quadrillée — **d** Kreuzlinienraster *m* — **n** kruislijnraster *n* — **e** trama *f* cuadriculada; retícula *f* de medio tono — **i** retino *m* a linee incrociate

3760 cross-lugging — When the lugs or ears of the matrices of typesetting machines are arranged so that the matrices project over both sides of the magazine channel instead of one side, as is generally the case. To accommodate larger faces in the magazine than it could otherwise contain.

3761 cross-marks — See register marks.

3762 cross *v* out — See strike *v* out.

3763 cross perforation — Perforations made at right angles to the direction of web travel to prevent bursting of signature during folding.
f perforation *f* transversale — **d** Querperforation *f*; Perforierschnitt *m* — **n** vouwperforatie *f*; dwarsperforatie *f* — **e** perforación *f* transversal — **i** perforazione *f* trasversale

3764 cross reference — An instruction to the user of an index or similar material to look under a synonym or an additional entry for further information. Also a reference from the text of a book to some other part of the book (figure, map, table, etc.) or to another page.
f double renvoi *m*; renvoi *m* réciproque; renvoi *m* — **d** Kreuzverweisung *f*; Verweisung *f*; Hinweisung *f*; Hinweis *m* — **n** kruisverwijzing *f*; verwijzing *f* — **e** referencia *f*; doble renvío *m*; renvío *m* recíproco — **i** rimando *m* reciproco; rimando *m* (in un libro)

3765 cross rules — See also with cross rules, with separate cross rules.
f réglure *f* intérieure — **d** Querlinien *fpl* — **n** contralijnen *fmpl* — **i** righe *fpl* trasversale

3766 cross rules, with — See with cross rules.

3767 cross-section of cylinder — See cross-cut of cylinder.

3768 cross section paper — See squared paper.

3769 cross stop — Inset in a photographic lens aperture with an opening in the form of a cross.
f diaphragme *m* en croix — **d** Kreuzblende *f* — **n** kruisdiafragma *n*; kruisstop *m* — **e** diafragma *m* en cruz; stop *m* en cruz — **i** diaframma *m* a croce

3770 crotchets — See brackets.

3771 crowd *v*; flood *v*; overink *v* — To ink the offset plate too heavily in an attempt to print a darker tone. Applying too heavy an ink film to plate.
f encrage *m* trop chargé; encrage *m* excessif; excès *m* d'encrage — **d** stark einfärben; laden — **n** teveel inkt geven; met inkt overladen — **e** sobrecargado *m* de tinta — **i** sovraccaricare con inchiostro

3772 crowdout; over matter; hold over — Excess of matter set over the space allotted to it.
f texte *m* en trop — **d** Übersatz *m* — **n** teveel *n* aan zetsel — **e** material *m* sobrante; sobrante *m* — **i** composizione *f* in eccesso

3773 crown — A size of printing paper. See app. no. 3.

3774 crown filler — Hydrated calcium sulphate ($CaSO_4 \cdot 2H_2O$) prepared by the interaction of calcium chloride and sodium sulphate, also known as pearl hardening. Used particularly in high-grade papeteries where a high white colour or delicate tint is desired. Now replaced largely by calcined calcium sulphate. *Plaster of Paris is the anhydrous commercial form.
f sulfate *m* de calcium; gypse *m*; plâtre *m* — **d** Kalziumsulfat *n*; schwefelsaures Kalzium *n*; Gips *m* — **n** calciumsulfaat *n*; gips *n* — **e** sulfato *m* de calcio; sulfato *m* cálcico; yeso *m* — **i** solfato *m* di calcio; solfato *m* calcico; gesso *m*

3775 crown octavo — A UK size of writing paper. See app. no. 3.

3776 CRT display — See cathode ray tube display.

3777 crucible — Porcelain or platinum container in which substances are burned to ash or melted.
f creuset *m*; pot *m* — **d** Tiegel *m* — **n** kroes *m*; smeltkroes *m* — **e** crisol *m* — **i** crogiolo *m*; crogiuolo *m*

3778 crucible — See metal pot.

3779 crude oil — See petroleum.

3780 crumpling apparatus; crumpling device — Hand operated apparatus which crumples the paper under pressure to investigate the wear by crumpling, e.g. flex tester, more especially the Gelbo flex tester.
f froisseuse; appareil *m* pour l'essai de froissement; appareil *m* à froisser — **d** Knitterprüfgerät *n* — **n** kreukeltoestel *n* — **e** aparato *m* para el ensayo del arrugado — **i** apparecchio *m* per prove di sgualcimento

3781 crumpling device — See crumpling apparatus.

3782 crumpling proof
f essai *m* de froissement — **d** Knitterprobe *f* — **n** kreukelproef *fm* — **e** ensayo *m* al frotamiento; ensayo *m* de resistencia al frote — **i** prova *f* di sgualcimento

3783 crushed core — Paper core squeezed out-of-round or completely collapsed as a result of excessive impact.
f mandrin *m* déformé; mandrin *m* écrasé — **d** zerquetschte Rollenhülse *f*; verunstaltete Rollenhülse *f*; verdrückte Rollenhülse *f* — **n** vervormde koker *m* — **e** alma *f* deformada; mandril *m* deformado — **i** anima *f* ovalizzata; anima *f* deformata

3784 crushed levant morocco — Levant pressed or stamped or hand polished with a hot polishing tool to smooth down its natural grain.
f maroquin *m* à grains écrasés; maroquin *m* écrasé — **d** Maroquin écrasé *n*; geglättetes Saffianleder *n* — **n** écrasé maroquin *n* — **e** tafilete *m* — **i** marocchino *m* di levante a grani schiacciati; marocchino *m* di levante cilindrato

3785 crushing — Defect in machine-made paper by excessive pressure or moisture at the dandy, couch, or press rolls, characterized by a mottled or cloudy appearance and giving local areas of greater translucency or holes. Crushing by excessive pressure in the calenders is *blackening.
f écrasage *m*; écrasé *m* — **d** Verdrücken *n*; Abquetschen *n* — **e** marca *f* de bayeta — **i** franatura *f*

3786 crushing strip — Strip sometimes fitted on fold rollers to increase the nip or pressure on the fold.
f garnissage *m* des rouleaux de plieuse — **d** Preßstreifen *m* — **n** knijpstrookje *n* — **e** guarnecido *m* de los rodillos de la plegadora — **i** striscia *f* di schiacciamento

3787 cryptics — Signs above or below all-capital characters in computer print-out to represent typographic changes, such as lower case, italic or change of type face.
f drapeaux *mpl* — **d** zusätzliche Zeichen *npl* — **n** additionele tekens *npl*; bijzondere tekens *npl*; geheimtekens *npl* — **e** signos *mpl* adicionales — **i** segni *mpl* accessori

3788 crystal diode; crystal rectifier — Electronic semi-conducting device with two electrodes. A diode permits current to flow in one direction with little resistance.
f diode *f* à cristal; redresseur *m* à cristal — **d** Kristalldiode *f*; Kristallgleichrichter *m* — **n** kristaldiode *f*; kristaldetector *m*; kristalgelijkrichter *m* — **e** diodo *m* de cristal; rectificador *m* de cristal — **i** diodo *m* a cristallo; raddrizzatore *m* a cristallo

3789 crystalline
f cristallin *f* — **d** kristallinisch — **n** kristallijn — **e** cristalino — **i** cristallino

3790 crystalline structure — Formation composed of crystals, for example, in type metals.
f structure *f* cristalline — **d** Kristallstruktur *f* — **n** kristallijne structuur *f* — **e** estructura *f* cristalina — **i** struttura *f* cristallina

3791 crystallization — The formation of crystals.
f cristallisation *f* — **d** Kristallisieren *n*; Kristallisation *f* — **n** kristallisatie *f* — **e** cristalización *f* — **i** cristallizzazione *f*

3792 crystallization *depr.*; **dry trap** — Condition in which a dried ink film repels a second ink which must be printed on top of it.
f refus *m* d'impression en superposition; cristallisation *f* — **d** Abstoßen *n* (nachfolgender Farben) — **n** ketsen *n*; afstoten *n* — **e** rechazo *m* de las sobreimpresiones — **i** rifiuto *m* di sovrastampa; rifiuto *m* dell'inchiostro di sovrastampa; rifiuto *m* dell'inchiostro

3793 crystallized — Solidified into crystals. Incorrect term for dried ink films that will not trap succeeding colours. See also crystallization, trapping.
f cristallisé — **d** kristallisiert — **n** gekristalliseerd — **e** cristalizado — **i** cristallizzato

3794 crystal rectifier — See crystal diode.

3795 c/s — See cycle/second.

3796 cub; cub reporter; beginner — Young person apprenticing as a newspaper reporter.
f apprenti *m*; blanc-bec *m*; jeune rédacteur *m*; débutant *m* — **d** junger Reporter *m*; Redaktionsvolontär *m*; Zeitungsvolontär *m* — **n** aankomend reporter *m*; aankomend verslaggever *m*; aankomend redacteur *m*; aankomend journalist *m* — **e** reportero *m* novicio; reportero *m* nuevo; reporter *m* novel; principiante *m*; aspirante *m* (a reportero); aprendiz *m*; aprendiza *f*; auxiliar *m* de redacción; currinche *m* — **i** cronista *f* principiante

3797 cube root — See root.

3798 cubic centimeter — See cubic centimetre.

3799 cubic centimetre *GB*; **cubic centimeter** *US* — Metric unit of volume. Abbrev. cm³ or cc. See app. no. 7.
f centimètre *m* cube — **d** Kubikzentimeter *nm* — **n** kubieke centimeter *m* — **e** centímetro *m* cúbico — **i** centimetro *m* cubo

3800 cubic decimetre *GB*; **dubic decimeter** *US* — Metric unit of volume. Abbrev. dm³. See app. no. 7.
f décimètre *m* cube — **d** Kubikdezimeter *m* — **n** kubieke decimeter *m* — **e** decímetro *m* cúbico — **i** decimetro *m* cubo

3801 cubic foot — Unit of volume. *pl* feet. Abbrev. ft³ or cu ft. See app. no. 7 and 7a.
f pied *m* cube — **d** Kubikfuß *m* — **n** kubieke (Engelse) voet *m* — **e** pie *m* cúbico — **i** piede *m* cubo

3802 cubic inch — Unit of volume. Abbrev. in³ or cu in. See app. no. 7 and 7a.
f pouce *m* cube — **d** Kubikzoll *m* — **n** kubieke inch *mn* — **e** pulgada *f* cúbica — **i** pollice *m* cubo

3803 cubic meter — See cubic metre.

3804 cubic metre *GB*; **cubic meter** *US*; **stere** — Metric unit of volume. Abbrev. m³. See app. no. 7.
f mètre *m* cube; stère *m* — **d** Kubikmeter *n*; Ster *m* — **n** kubieke meter *m*; stère *fm* — **e** metro *m* cúbico; estéreo *m* — **i** metro *m* cubo

3805 cubic yard — Unit of volume. Abbrev. yd³ or cu ya. See app. no. 7 and 7a.
f yard *m* cube — **d** Kubikyard *m* — **n** kubieke yard *m* — **e** yarda *f* cúbica — **i** iarda *f* cuba

3806 cub reporter — See cub.

3807 cuir bouilli binding — Style of book decoration used between the 9th and 14th century, in which the leather cover was soaked in hot water (not boiled), modeled and hammered to raise the design in relief.

f reliure f en cuir bouilli; reliure f en cuir estampé — d Cuir-bouilli-Einband m; Einband m mit Ledertreibarbeit — n cuir-bouilli-band m — e encuadernación f en cuir bouilli — i legatura f in cuoio sbalzato; legatura f in cuir bouilli

3808 cuir ciselé binding — Binding, widely practised in the 15th century in Germany, with a design in relief cut into the leather, instead of being tooled or stamped on it.

f reliure f cuir ciselé — d Lederschnittband m; Lederschnitteinband m — n cuir-ciselé-band m — e encuadernación f en cuir ciselé; encuadernación f cincelada — i legatura f in cuoio cesellato; legatura f in cuir ciselé

3809 cultural papers — Papers such as writing and printing used for cultural purposes.

3810 cumulative distribution function; cumulative probability function

f fonction f cumulative; distribution f cumulative — d kumulative Verteilungsfunktion f — n cumulatieve frequentieverdeling f — e función f acumulativa — i funzione f cumulativa; distribuzione f cumulativa

3811 cumulative probability function — See cumulative distribution function.

3812 cuneiform characters — See cuneiform writing.

3813 cuneiform writing; cuneiform characters; arrowheaded writing; arrowheaded characters — Wedge-shaped characters used in writing by the ancient Assyrians, Babylonians, Persians and others.

f écriture f cunéiforme; caractères mpl cunéiformes — d Keilschrift f — n spijkerschrift n — e escritura f cuneiforme — i scrittura f cuneiforme

3814 cup — See bowl.

3815 cup — See ink cell.

3816 cup board

f carton m à coupes — d Becherkarton n — n bekerkarton n — e cartulina f para cálices — i cartone m per bicchierini; cartone m per coppette

3817 cupboard lining paper — See shelf paper.

3818 cup-cast quads — See quotation quads.

3819 cupel — Shallow, absorbent vessel, generally of bone ash (a white, friable substance, composed mainly of calcium phosphate), used in assaying gold and silver from lead.

f coupelle f — d Kapelle f; Kupelle f; Schiffchen n; Näpfchen n — n cupel fm; kapel fm — e cupel m; copela f — i coppella f

3820 cupping — In flexography, the distortion caused by curving the rubber plate around the printing cylinder, the shrink inherent in the compound and the degree of control excercised by certain plate constructions combines to cause one of the most troublesome plate conditions. Cupping causes halos, hard edges on screens, and leads to excess impression, distortion and other printing problems.

3821 cuprammonium hydroxide; Schweizer's reagent — Solution of cupric hydroxide in aqueous ammonium hydroxide, used in paper-making for dissolving cellulose.

f hydroxyde m de cuivre ammoniacal — d Cuprammoniumhydroxyd n — n cuprammonium-hydroxyde n — e hidróxido m cuproamoniacal — i idrossido m di cuprammonio; idrossido m d'ammoniuro di rame

3822 cupric — Designation for *copper salts in which the copper is divalent, e.g., cupric chloride $CuCl_2$. See also cuprous.

f cuivrique — d Cupri- — n cupri- — e cúprico — i ramico

3823 cupric acetate — See copper acetate.

3824 cupric bromide — See copper bromide.

3825 cupric carbonate — See copper carbonate.

3826 cupric chloride — See copper chloride.

3827 cupric cyanide — See copper cyanide.

3828 cupric ferrocyanide — See copper ferrocyanide.

3829 cupric oxide — See copper oxide.

3830 cupric sulphate — See copper sulphate.

3831 cuprous — Pertaining to copper; chemically, designating a compound of monovalent copper. See also cupric.

f cuivreux — d Cupro- — n cupro- — e cuproso — i rameoso

3832 cuprous chloride — White cubical crystals, soluble in acids, ammonia, ether, insoluble in water. Absorbent for carbon monoxide. Formula: $CuCl$ or Cu_2Cl_2.

f chlorure m cuivreux; protochlorure m de cuivre — d Kupferchlorür n; Cuprochlorid n — n koperchloruur n — e cloruro m de cobre; cloruro m cuproso — i cloruro m di rame; cloruro m rameoso

3833 curl — Malformation or deformation of paper due to faulty manufacture or to changes in atmospheric conditions.

f gondolage m; recoquillement m — d Kräuselung f — n krulling f; krul m — e curvatura f del papel; ondulación f del papel; rizado m del papel; abarquillado m — i accartocciamento m; imbarcamento m; arricciamento m

3834 curl v — Not lying flat and tending form into cylindrical or wavy shapes.

f rouler; se rouler — d sich rollen; sich werfen — n krullen; omkrullen — e encurrujarse; abarquillarse — i imbarcarsi; accartocciarsi

3835 curlicue; curlycue — American name for an ornamented type face or other ornament. See ornament.

3836 curling test; skim test — Test where a strip of paper floated on water shows a serpentine curl, owing to the swelling of the fibres. The time from the contact of the paper with the water until the instant of uncurling is a measure of the degree of sizing.

f essai *m* de gondolage — **d** Kräuselprobe *f*; Kräuselversuch *m* — **n** krulproef *fm* — **e** ensayo *m* de abarquillado — **i** prova *f* d'accartocciamento; prova *f* d'imbarcamento; prova *f* d'arricciamento

3837 curlycue — See curlicue.

3838 curly ñ — See ñ.

3839 currency paper *GB*; **bank-note paper** *US* — Paper made from flax pulp, new linen and cotton rags for bank-notes. That used by the Bank of England is made from new unbleached linen.

f papier *m* monnaie; papier *m* pour billets de banque; papier *m* fiduciaire — **d** Banknotenpapier *n*; Wertpapier *n*; Wertzeichenpapier *n*; Werttitelpapier *n* — **n** bankbiljettenpapier *n*; waardepapier *n* — **e** papel *m* moneda; papel *m* de estado — **i** carta *f* per biglietti di banca; carta *f* per cartamoneta

3840 current; electric current; flow — Flow of electrons round a circuit.

f courant *m*; passage *m* (du courant); flux *m* — **d** Strom *m*; Stromart *f* — **n** stroom *m*; elektrische stroom *m* — **e** corriente *f* (eléctrica) — **i** corrente *f* (elettrica)

3841 current consumption

f consommation *f* de courant — **d** Stromverbrauch *m* — **n** stroomverbruik *n* — **e** consumo *m* de corriente; consumo *m* de energía eléctrica — **i** consumo *m* di elettricità; consumo *m* d'energia elettrica

3842 current density — In electro-deposition, the current flowing per unit area. Usually given as amperes per square decimetre or per square foot.

f densité *f* de courant — **d** Stromdichte *f* — **n** stroomdichtheid *f* — **e** densidad *f* de corriente — **i** densità *f* di corrente

3843 current intensity

f intensité *f* de courant — **d** Stromstärke *f* — **n** stroomsterkte *f* — **e** intensidad *f* de corriente — **i** intensità *f* di corrente

3844 cursor — Movable spot of light on the cathode ray tube display of a console or terminal indicating where the next action will occur.

3845 curve — Curve is divided in the shoe (in photographical materials for instance, the underexposed region), the straight-line portion, and the shoulder (in photographical materials the overexposed region).

f courbe *f* — **d** Kurve *f* — **n** kromme *fm*; curve *fm* — **e** curva *f* — **i** curva *f*

3846 curve *v* — To bend the printing plate to a desired arc.

f cintrer — **d** biegen — **n** buigen — **e** curvar — **i** curvare

3847 curved

f cintré — **d** gebogen; gerundet — **n** gebogen — **e** encorvado; curvo — **i** curvato

3848 curved plate — Semi-cylindrical printing plate of proper thickness, curved to fit cylinder or rotary press.

f cliché *m* cintré — **d** Rundplatte *f* — **n** gebogen plaat *fm* — **e** plancha *f* curva; plancha *f* curvada — **i** lastra *f* curva; lastra *f* cilindrica

3849 curved stereo; cylindrical casting — Stereotype cast in a curved flong to produce a printing plate for a rotary press.

f stéréo *m* cintré; stéréo *m* cylindrique — **d** Rundstereo *n* — **n** rondstyp *m* — **e** estereotipo *m* semicilíndrico; estereotipia *f* cilíndrica; estereotipia *f* curva(da); teja *f* de estereotipia — **i** stereo *m* curvo

3850 cushion-shaped distortion — See pincushion distortion.

3851 customer — See client.

3852 customs declaration

f déclaration *f* de douane; déclaration *f* en douane — **d** Zolldeklaration *f*; Zollangabe *f* — **n** douaneverklaring *f* — **e** declaración *f* de aduana — **i** dichiarazione *f* doganale

3853 cut — Old and inappropriate term for photoengraving; properly applied only to wood engravings and other surfaces manually engraved.

3854 cut — See cut out.

3855 cut *v* — To trim sheets of paper and edges of books.

f couper; massicoter; rogner — **d** schneiden; abschneiden; beschneiden — **n** snijden; afsnijden — **e** cortar; guillotinar — **i** tagliare; refilare; rifilare

3856 cut *v*; **thin** *v* — To dilute an ink, lacquer or varnish with solvents or with clear base.

f allonger; descendre — **d** verschneiden — **n** versnijden — **e** diluir — **i** diluire

3857 cut *v* **away** — To engrave in wood.

f champlever — **d** wegstechen; ausstechen — **n** wegsteken — **e** escoplear — **i** togliere; incidere

3858 cut *v* **bolts** — To open top and foredge of book signatures with a blunt blade or a folding stick.

f découronner — **d** beraufen — **n** snoeien — **e** cortar los cantos; desbarbar

3859 cutch — Tough paper sheets used by goldbeaters.

3860 cut edges — Smooth, trimmed edges of a book.

f tranches *fpl* rognées; tranches *fpl* coupées; tranches *fpl* ébarbées — **d** beschnittene Seiten *fpl*; Schnitt *m*; Buchschnitt *m* — **n** afgesneden randen *mpl*; afgesneden kanten *mpl*; sneden *fmpl* — **e** decantos *mpl* cortados — **i** tagli *mpl* rifilati; margini *mpl* rifilati

3861 cut *v* **flush** — To cut or trim edges of cover and sheets of a book together in one operation so that the cover is the same size as the body of the book.

f rogner à fleur; couper à vif — **d** Broschürenschnitt *m* — **n** gelijk-afsnijden — **e** recortar a ras; cortar a sangre — **i** rifilare a filo della copertina; refilare a filo della copertina

3862 cut in boards — See in boards.

3863 cut-in head — Head placed in a box of white space cut into the side of the type page, usually set in type different from that of the text and placed under the first two lines of the paragraph. Also, a head cutting across the body of a table.

f titre *m* rentrant — **d** eingebauter Kopf *m* — **n** ingebouwde kop *m* — **e** título *m* entrado — **i** titolo *m* inserito

3864 cut-in initial — See cut-in letter.

3865 cut-in letter; cut-in initial — Two or three line (or larger) initial letter at the beginning of a chapter or paragraph.

f initiale *f* rentrante; initiale *f* tombée — **d** eingebautes Initial *n* — **n** ingebouwd initiaal *n*; ingebouwd kapitaal *fm* — **e** letra *f* capitular entrada; inicial *f* en racada — **i** lettera *f* inserita; iniziale *f* inserita

3866 cut-in notes; let-in notes; incut notes — Notes set into the text at the outer edge of a paragraph, generally about halfway down, with white space forming a rectangle around them.

f notes *fpl* marginales insérées dans le texte — **d** eingezogene Marginalien *fpl*; eingezogene Randglossen *fpl*; Anmerkungen *fpl* die in den Satzspiegel hineinreichen — **n** ingebouwde noten *fmpl* — **e** notas *fpl* intercaladas (en el texto); anotaciones *fpl* intercaladas; ladillos *mpl* — **i** note *fpl* marginali rientranti nel testo; annotazioni *fpl* marginali rientranti nel testo

3867 cutlery paper — Thin white or brown paper, produced to avoid acidity to attack cutlery, etc.

f papier *m* pour coutellerie — **d** Papier *n* für Stahlwaren; säurefreies Papier *n* — **n** zuurvrij papier *n* voor staalwaren — **e** papel *m* para cuchillería — **i** carta *f* da coltelleria

3868 cutline — *Caption placed inside an illustration. See also legend.

3869 cut-off knife — See knife.

3870 cut-off rubber — Rubber strip set in a cylinder of a folder of a rotary press against which the serrated cut-off knife strikes when cutting the web into lengths.

f listeau *m* en caoutchouc; contre-partie *f* — **d** Gummileiste *f* — **n** rubber snijlat *fm* — **e** goma *f* contracorte — **i** controlama *f* di gomma

3871 cut-off rule — Printing rule to separate advertisements or different news items.

f filet *m* déséparation — **d** Trennlinie *f* — **n** scheidingslijn *fm* — **e** pleca *f* divisoria; raya *f* de separación; filete *m* de separación; filete *m* de corondel; columnaria *f* — **i** filetto *m* di separazione

3872 cut out; cut; blank — Fancy shaped printed novelty, cut into a circle, oval or irregular shape with steel knives, such as window displays, cartons, etc.

f découpe *f* (à plat) — **d** Zuschnitt *m*; gestanzter Ausschnitt *m*; Nutzen *m* — **n** uitgestanst model *n* — **e** troquelado *m* — **i** fustellato *m*; stampato *m* fustellato

3873 cut out — See colour separation negative.

3874 cut *v* out — See silhouet *v*.

3875 cut out background — Halftone image in which the screen dots and other details are eliminated.

f fond *m* détouré — **d** abgedeckter Hintergrund *m*

— **n** uitgedekte achtergrond *m* — **e** fondo *m* silueteado — **i** fondo *m* contornado

3876 cut-out halftone — See outlined halftone.

3877 cut-out in lead — See lead engraving.

3878 cut *v* reglets

f couper une reglette (un filet) — **d** Regletten (Linien) abhacken — **n** lijnen (regletten) afhakken — **e** cortar lingotes — **i** tagliare interlinee

3879 cutter — See guillotine.

3880 cutter — See router bit.

3881 cutter and creaser — See cutting and creasing press.

3882 cutter cam — Cam which operates a cutting device in the folder of a rotary press for severing the wire of the stitcher.

f came *f* de coupe du fil métallique — **d** Messerexzenter *m* — **n** mesexcentriek *n* — **e** cama *f* de la cuchilla; balancín *m* de la cuchilla *Ch* — **i** camma *f* cesoiatrice

3883 cutter creaser — See cutting and creasing press.

3884 cutter operator

f rogneur *m* de papier — **d** Papierschneider *m*; Schneider *m* — **n** papiersnijder *m*; snijder *m* — **e** guillotinista *m*

3885 cut *v* the line — Formerly when one or more of the companionship was out of copy or type, according to previous agreement, the whole companionship must stop work until there was a fresh supply. Now generally to cease work.

f cesser le travail; finir — **d** die Arbeit niederlegen — **n** uitscheiden; ophouden; eindigen; staken; kappen *sl.* — **e** acabar el trabajo — **i** cessare il lavorro

3886 cutting — Reduction of density or dot size.

f morsure *f* du point — **d** Ätzen *n* des Rasterpunktes — **n** puntetsen *n*; kleiner etsen *n* van de rasterpunt — **e** reducir (el tamaño de los puntos tramadas); reducir la densidad; rebajar puntos; afinar puntos — **i** morsura *f* del punto

3887 cutting — See news cutting.

3888 cutting and creasing platen press — Platen-type die cutter.

f presse *f* à platine à découper; découpeuse *f* à platine — **d** Stanztiegel *m*; Stanzpresse *f*; Stanzmaschine *f*; Tiegelstanzmaschine *f* — **n** stansdegel *m*; stanspers *f*; stansmachine *f* — **e** minerva *f* troqueladora — **i** macchina *f* fustellatrice a platina; pressa *f* a platina tagliatrice-cordonatrice

3889 cutting and creasing press; cutter creaser; cutter and creaser — Machine for cutting and creasing or scoring boxboard blanks.

f découpeuse-refouleuse *f*; machine *f* découpeuse-refouleuse — **d** Stanz- und Rill- und Ritzmaschine *f* — **n** stans- en rilmachine *f* — **e** troqueladora-hendidora *f*; prensa *f* de cortar y marcar; cortadora y marcadora *f*; cortadora y ranudora *f*; prensa *f* para cortar y hender — **i** tagliatrice-cordonatrice *f*; pressa *f* tagliatrice-cordonatrice

3890 cutting angle

f angle *m* de coupe — **d** Abschneidwinkel *m* — **n**

snijhoek *m* — **e** ángulo *m* de corte — **i** angolo *m* di taglio

3891 cutting buffer — See cutting strip.

3892 cutting colour — In letterpress rotary reducing or stopping the flow of ink.

f arrêt *m* de prise d'encre; réduire le débit d'encre; réduire la consommation d'encre — **d** Farbabstellung *f* — **n** de inktgeving verminderen — **e** disminuir el entintado; diminuir el consumo de tinta — **i** riduzione *f* del flusso d'inchiostro; arresto *m* del flusso d'inchiostro

3893 cutting cylinder — Small cylinder in the folder of a rotary press carrying the knives which, by pressure on the cutting stick in the folding roller, cut each separate copy from the web.

f cylindre *m* de coupe — **d** Schneidzylinder *m*; Messerzylinder *m* — **n** snijcilinder *m* — **e** cilindro *m* de corte — **i** cilindro *m* di taglio; cilindro *m* portalama

3894 cutting die — See die.

3895 cutting disks — See slitters.

3896 cutting dust

f poussière *f* de coupe — **d** Schneidstaub *m*; Schnittstaub *m* — **n** snijstof *n* — **e** polvo *m* de corte; polvo *m* cortante — **i** polvere *f* di tagliare

3897 cutting forme; forme cutter

f forme *f* de découpage — **d** Stanzform *f* — **n** stansvorm *m*; stansplank *fm* — **e** forma *f* troqueladora — **i** forma *f* fustellatrice

3898 cutting machine — See guillotine.

3899 cutting marks — Marks usually added at copy stage and carried through to printed work, to indicate positions for cutting, slitting, punching, etc.

f repères *mpl* de coupe — **d** Beschnittmarken *fpl*; Schneidlinien *fpl* — **n** snijtekens *npl* — **e** marcas *fpl* de corte — **i** marche *fpl* di taglio

3900 cutting of negatives — Making the density of a negative or positive less by etching, in halftone by reducing in dot size. See also reduction of density.

f descente *f* du point — **d** abschwächen — **n** afzwakken *n* (van de rasterpunt); etsen *depr.* — **e** debilitar; rebajar (el punto) — **i** indebolire

3901 cutting of negatives — Cleaning up of bad negatives by the opaquer with a needle or scraper.

f grattage *m* du négatif — **d** Aufhellen *n* eines Negativs — **n** schrapen *n*; afschrapen *n* van een negatief — **e** raspado *m* del negativo — **i** schiarimento *m* di un negativo

3902 cutting-out knife; overlay knife; makeready knife — A knife with which overlays and other materials are cut.

f couteau *m* de mise en train — **d** Zurichtemesser *n* — **n** toestelmesje *n*; pikeermesje *n* — **e** cuchilla *f* para arreglo; cuchilla *f* para emparejar; cuchilla *f* para recortes; cuchilla *f* para (cortar) alzas; lanceta *f*; cortante *m Ar* — **i** sgarzino *m*; lama *f* per avviamento; coltello *m* da frastagliamento; coltello *m* per taccheggio; lama *f* per taccheggio

3903 cutting rule — Steel strip, one edge of which is ground to centre or side face, used for cutting through boxboard stock.

f filet *m* coupeur; filet *m* de découpe — **d** Schneidelinie *f*; Schnittlinie *f* — **n** snijlijn *fm* — **e** filete *m* de cortar; filete *m* cortante; pleca *f* para cortar; pleca *f* de cortar; pleca *f* cortadora; pleca *f* cortante; troquel *m*; suaje *m Me* — **i** filetto *m* di taglio

3904 cutting stick — Wood or plastic strip on which the knife of guillotine descends when cutting paper. Also a strip of wood, vulcanite, or hard rubber, on which the knife in the cutting cylinder of a rotary web press makes contact to cut each copy from the web.

f règle *f* de coupe — **d** Schneidleiste *f* — **n** snijlat *fm* — **e** guardafilo *m*; palo *m* guardafilo; listón *m* de la guillotina — **i** listello *m* salvafilo

3905 cutting strip; cutting buffer — Strip of rubber, fibre or other material which operates in conjunction with the serrated cut-off blade to sever the section or sheet from the web in a rotary press.

f listeau *m*; contre-partie *f* de coupe; réglette *f* de coupe — **d** Schneidleiste *f* — **n** snijlat *fm*; snijstrip *m*; snijblok *n*; snijbuffer *m* — **i** striscia *f* divisoria; striscia *f* di rifilatura

3906 cutting width — Cutting length of a knife in a guillotine.

f longueur *f* de coupe — **d** Schnittlänge *f*; Schneidlänge *f* — **n** snijbreedte *f* — **e** luz *f* de corte; longitud *f* de corte — **i** lunghezza *f* di taglio

3907 cut to register — So cut that the watermark in each sheet of paper falls in the same position. See also localized watermark.

f coupé au repère — **d** wasserzeichengerecht geschnitten — **n** naar het watermerk gesneden — **e** recortado al registro — **i** taglio a registro

3908 cyan; blue-green — Minus-red subtractive primary colour in three-colour processes with an absorption band in the region 600-700 millimicrons and transmits or reflects light in the range 400-600 millimicrons.

f cyan *m*; bleu-vert — **d** Zyan *n*; blaugrün — **n** cyaan *n*; blauwgroen — **e** cian; cianógeno; cianea *Es*; azul verdoso — **i** ciano; verdazzurro

3909 cyanic acid — Strong, unstable, poisonous acid obtained as a mobile volatile liquid by heating cyanuric acid. Formula: HCNO.

f acide *m* cyanique — **d** Zyansäure *f* — **n** cyaanzuur *n* — **e** ácido *m* ciánico — **i** acido *m* cianico

3910 cyanide — Short name for *sodium cyanide or *potassium cyanide. Used to remove the product of the silver bleaching action caused either by ferricyanide or by iodine. It is a much better solvent than hypo but it has a softening effect on gelatine emulsions and is very poisonous. Extreme care must be taken when using it.

3911 cyanide of potash — See potassium cyanide.

3912 cyanide of potassium — See potassium cyanide.

3913 cyanide of silver — See silver cyanide.

3914 cyanide of sodium — See sodium cyanide.

3915 cyanogen iodide; iodine cyanide — Colourless needles, soluble in water, alcohol, and ether. Violent poison. Formula: CNI.

3916 cyanogen iodide reducer
f affaiblisseur *m* cyanure; eau *f* iodée — **d** Jod-Zyan-Abschwächer *m* — **n** jood-cyaan-afzwakker *m*

3917 cyanotype — Blueprint on paper sensitized with iron salts.
f bleu *m* — **d** Blaupause *f* — **n** blauwdruk *m* — **e** marión *m* azul — **i** cianografia *f*

3918 cyanotype paper — See blueprint paper.

3919 cyan printer; blue printer *depr.*; **blue plate** *depr.* — The image in a subtractive colour process to be printed in blue. The copy for this printing plate is photographed through a red (or orange) filter.
f cliché *m* du cyan; cliché *m* du bleu — **d** Zyanplatte *f*; Blauplatte *f*; Blaudruckplatte *f* — **n** cyaanplaat *fm*; blauwplaat *fm*; blauwe drukplaat *fm* — **e** plancha *f* del cian; plancha *f* del azul — **i** lastra *f* del cian; lastra *f* del blu; cliché *m* del cian; cliscè *m* del blue

3920 cybernetics — Study of human control functions and of mechanical and electric systems designed to replace them, involving the application of statistical mechanics to communication engineering.
f cybernétique *f* — **d** Kybernetik *f* — **n** cybernetica *f*; stuurkunde *f* — **e** cibernética *f* — **i** cibernetica *f*

3921 cycle — One complete repetition of an action or process capable of being repeated over and over in the same order.
f cycle *m* — **d** Periode *f*; Kreislauf *m* — **n** cyclus *m*; kringloop *m* — **e** ciclo *m* — **i** ciclo *m*

3922 cycle/second; cps; c/s — Unit of frequency, superseded by the hertz.
f cycle *m* par seconde — **d** Periode *f* pro Sekunde — **n** periode *f* per seconde — **e** ciclo *m* por segundo — **i** periodo *m* per seconda

3923 cycle time — The length of time that a computer requires to move a "word" from a given location to another location.
f temps *m* de cycle — **d** Zykluszeit *f* — **n** cyclustijd *m* — **e** tiempo *m* de ciclo — **i** tempo *m* di ciclo

3924 cyclic element; repetitive element — Element which occurs in every work cycle.
f élément *m* cyclique — **n** repeterend element *n* — **e** elemento *m* cíclico — **i** elemento *m* ciclico

3925 cyclohexanone; ketohexamethylene; pimelic ketone — Water-white to pale yellow liquid, miscible with most solvents. TL-value: 50 ppm or 200 mg/m³.
Formula: $CH_2(CH_2)_4CO$.
f cyclohexanone *m* — **d** Zyklohexanon *n* — **n** cyclohexanon *n* — **e** ciclohexanona *f* — **i** cicloesanone *m*

3926 cycloidal engine — Term sometimes used for a *guilloching machine.

3927 cyclostyle paper — See duplicating stencil.

3928 cylinder bearer gauge; packing gauge *misn.* — U-shaped steel block with a dial gauge mounted on one end. When clamped on plate cylinder of press, and the gauge button permitted to rest upon the bearer, a pointer indicates the difference in height between the two. To measure the amount of packing and blanket underpacking needed.
f jauge *f* de la hauteur des couronnes — **d** Schmitzringhöhemesser *m* — **n** smetringhoogtemeter *m* — **e** medidor *m* de la altura de coronas — **i** misuratore *m* dell'altezza degli anelli

3929 cylinder bearer height; undercut — In printing presses, the difference between the radius of the bearers and the radius of the cylinder body; the allowance for plate or blanket plus a margin for packing adjustment.
f hauteur *f* des couronnes; hauteur *f* des bandes du cylindre; profondeur *f* du logement de l'habillage; gorge *f* de cylindre — **d** Schmitzringhöhe *f* — **n** smetringhoogte *f*; loopringhoogte *f* — **e** altura *f* de la corona del cilindro — **i** altezza *f* degli anelli

3930 cylinder bearers — See bearer rings.

3931 cylinder bearing
f palier *m* de cylindre — **d** Zylinderlager *n* — **n** cilinderlager *n* — **e** cojinete *m* del cilindro — **i** cuscinetto *m* del cilindro

3932 cylinder blanket — See blanket.

3933 cylinder brake — The mechanism fitted to a fast running press to stop it quickly.
f frein *m* au cylindre; frein *m* de cylindre — **d** Zylinderbremse *f* — **n** cilinderrem *fm* — **e** freno *m* del cilindro — **i** freno *m* al cilindro

3934 cylinder bristol — See mill bristol.

3935 cylinder brush — Device for pressing the running web against the new reel when the join is made on an automatic reel change.
f brosse *f* du cylindre — **d** Zylinderbürste *f* — **n** cilinderborstel *m* — **e** cepillo *m* del cilindro; escobilla *f* del cilindro — **i** spazzola *f* del cilindro

3936 cylinder catch
f fourchette *f* du cylindre — **d** Fanggabel *f* des Zylinders; Auffanggabel *f* des Zylinders — **n** gaffel *fm* voor de cilinder; cilindergaffel *fm*; cilindervang *f* — **e** horquilla *f* del cilindro — **i** forcella *f* del cilindro; dispositivo *m* di presa (del cilindro); forchetta *f* di presa

3937 cylinder covering — See packing.

3938 cylinder diameter
f diamètre *m* de cylindre — **d** Zylinderdurchmesser *m* — **n** cilinderdiameter *m*; diameter *m* van de cilinder — **e** diámetro *m* del cilindro — **i** diametro *m* del cilindro

3939 cylinder dressing — See packing.

3940 cylinder dressing clamp — See dressing clamp.

3941 cylinder gap — Gap or space in cylinders of printing machines where the mechanism for plate clamps, blanket bar and tightening shaft and grippers is housed.

f évidement *m* du cylindre; vide *m* du cylindre — **d** Zylinderkanal *m*; Zylindergrube *f* — **n** cilinder-kanaal *n* — **e** vacío *m* del cilindro — **i** incavo *m* del cilindro

3942 cylinder gauge — Instrument to measure the differences of the distance between the bed and the cylinder of a flat-bed printing press, or between the cylinder bearer and the bed.

f jauge *f* de cylindre; jauge *f* d'écartement du cylindre — **d** Zylinderlehre *f* — **n** cilinderhoogte-meter *m* — **e** calibrador *m* del cilindro — **i** calibro *m* per il controllo dell'altezza del cilindro

3943 cylinder guide marks — Marks on the offset press-plate to match corresponding marks on the cylinder of the press, so that each plate will be positioned the same on the press.

f repères *mpl* de calage de la plaque — **d** Platt-führungsmarken *fpl*; Paßmarken *fpl* auf Offset-platte — **n** plaatmerken *npl*; aanlegtekens *npl* op de plaat — **e** marcas *fpl* de registro de la plancha — **i** marche *fpl* di guida della lastra

3944 cylinder machine; mould machine; board machine; board-making machine; vat machine *GB* — Paper machine (invented by John Dickinson in 1808) with rotating cylinder covered with wire cloth, partially immersed in a vat containing the pulp suspension, the paper web being formed around the cylinder. There may be a series of such cylinders, with vats filled with the same pulp or with different pulps. Thus solid or filled boards may be built to various thicknesses by combining or laminating the webs formed on each cylinder.

f machine *f* à forme ronde; passoire *f* ronde — **d** Rundsiebmaschine *f* — **n** rondzeefmachine *f* — **e** máquina *f* (a forma) redonda; máquina *f* de tamiz redondo — **i** macchina *f* in tondo

3945 cylinder machine — See cylinder press.

3946 cylinder packing — See packing.

3947 cylinder plate marks — Datum lines on the plate cylinder to assist the positioning of the print-ing plate.

f repères *mpl* de calage sur cylindre — **d** Zylinder-führungsmarken *npl* — **n** cilindermerken *npl* — **e** marcas *fpl* de registro del cilindro — **i** marche *fpl* di guida del cilindro

3948 cylinder press; cylinder machine — Printing machine which gives the impression by a cylinder.

f presse *f* à cylindre; presse *f* en blanc — **d** Zylinderpresse *f*; Schnellpresse *f* — **n** cilinder-pers *f*; snelpers *f* — **e** máquina *f* de cilindro; prensa *f* de cilindro; máquina *f* cilíndrica; prensa *f* cilíndrica; máquina *f* planocilíndrica; prensa *f* planocilíndrica; máquina *f* de cilindro y platina; prensa *f* de cilindro y platina — **i** macchina *f* a cilindro; pressa *f* a cilindro

3949 cylinder range — Minimum to maximum circumference of engraved cylinders.

f circonférence *f* minimum et maximum du cylindre gravé — **d** Umfang *m* des Druck-zylinders; Zylinderumfang *m* — **n** cilinder-omvang *m*; omvang *m* van de cilinder — **e** circun-ferencia *f* máxima y mínima del cilindro — **i** circonferenza *f* periferica del cilindro

3950 cylinder revolution

f tour *m* du cylindre — **d** Zylinderumdrehung *f* — **n** omwenteling *f* van de cilinder — **e** revolución *f* del cilindro — **i** rivoluzione *f* del cilindro; giro *m* del cilindro

3951 cylindrical casting — See curved stereo.

3952 Cyrillic — Noting or pertaining to a script derived from Greek uncials and traditionally sup-posed to have been invented by St. Cyril, first used for the writing of old Church Slavonic and adopted with minor modifications for the writing of Russian, Bulgarian, Serbian, and also Mongolian, Tajik, and many other USSR lan-guages.

f cyrillien; cyrillique — **d** kyrillisch — **n** cyrillisch — **e** cirílico — **i** cirillico

3953 czarina — A size of writing paper. See app. no. 3.

3954 dabber — Pad of cotton used by engravers to apply an acid resisting varnish to the plate before this is hand or machine engraved.
f tapette *f* — **d** Tampon *m* — **n** tampon *m* — **e** bala *f* de entintar — **i** tampone *m*

3955 dagger; obelisk — Second reference mark (†), after the asterisk. In English before, in Germany, the Netherlands and other countries after a person's name signifies dead or died.
f croix *f* (mortuaire); obèle *m* — **d** Kreuz *n*; Kreuzzeichen *n*; Kreuzchen *n* — **n** kruisje *n*; overlijdenskruisje *n* — **e** obelisco *m*; obelo *m*; cruz *f*; daga *f* — **i** croce *f*; obelisco *m*

3956 Daguerreotype — First practical photographic process, comprising an image on a silvered copperplate.
f Daguerréotype *m* — **d** Daguerreotypie *f* — **n** Daguerreotypie *f* — **e** daguerreotipia *f*; daguerrotipia *f*; daguerrotipo *m* — **i** dagherrotipo *m*

3957 daily newspaper — See daily paper.

3958 daily paper; daily newspaper; newspaper
f journal quotidien *m*; quotidien *m* — **d** Tageszeitung *f*; Tageblatt *n*; Zeitung *f* — **n** dagblad *n*; krant *fm* — **e** diario *m*; periódico *m* diario; periódico *m* cotidiano; cotidiano *m* — **i** giornale *m*; quotidiano *m*

3959 daily press
f presse *f* quotidienne — **d** Tagespresse *f* — **n** dagbladpers *fm* — **e** prensa *f* diaria; prensa *f* cotidiana — **i** stampa *f* quotidiana

3960 daltonism — See colour blindness.

3961 daluwang paper — See deluwang paper.

3962 damar — See dammar.

3963 damask paper — Fine writing paper.
f papier *m* écriture fin — **d** Damastpapier *n* — **n** fijn schrijfpapier *n*; damastpapier *n* — **e** papel *m* damasco — **i** carta *f* damascata

3964 dammar; dammar gum; dammar resin; damar — White to yellow, semi-transparent lumps obtained as an exudation from trees in the East Indies and Philippines. Dammar resins are soluble in alcohol, benzene, turpentine, and oils. In general they have lower acid numbers than the copals. Other physical properties are highly variable. Dammar gum is used as an ingredient of printing ink varnishes.
f gomme *f* dammar; résine *f* dammar; dammar *f* — **d** Dammarharz *n*; Dammar *n* — **n** damarhars *mn*; damar *mn* — **e** damar *m*; damara *f*; resina *f* damar — **i** dammara *f*; resina *f* dammara

3965 dammar gum — See dammar.

3966 dammar resin — See dammar.

3967 damning critique
f critique *f* violente; éreintement *m* — **d** Verriß *m* — **n** vernietigende kritiek *f* — **e** crítica *f* demoledora — **i** critica *f* fulminante

3968 damp *v GB*; **dampen** *v US* — To make wet. See also damping.
f mouiller — **d** feuchten — **n** vochten — **e** humedecer — **i** umettare

3969 dampen *v* — See damp *v*.

3970 dampeners — See dampers.

3971 dampening — The English for to make wet is to damp, but this is frequently enlarged to dampen. The meaning is exactly the same and the extra syllable is unnecessary, but is favoured particularly in the US, who seem to like making long words where adequate short words already exist. Dampening is not wrong and is in fact current in the US. Damping is sufficient and preferable.

3972 dampening — See damping.

3973 dampening etch
See damping solution.

3974 dampening rollers — See dampers.

3975 dampening solution — See damping solution.

3976 dampers; damping rollers *GB*; **dampeners; dampening rollers** *US* — Cloth-covered rollers that distribute the damping solution received from the ductor roller of the damping unit to the lithographic press plate.
f rouleaux *mpl* mouilleurs; rouleaux *mpl* de mouillage; mouilleurs *mpl* — **d** Feuchtwalzen *fpl*; Wischwalzen *fpl*; Feuchtauftragwalzen *fpl*; Wasserwalzen *fpl* — **n** vochtrollen *fmpl*; vochtopdraagrollen *fmpl*; waterrollen *fmpl* — **e** rodillos *mpl* mojadores; rodillos *mpl* humectadores; mojadores *mpl*; rodillos *mpl* de agua — **i** rulli *mpl* bagnatori; rulli *mpl* umidificatori

3977 damping *GB*; **dampening** *US* — Moistening non-design areas of lithographic plates by means of water-moistened rollers. See also dampening.
f mouillage *m* — **d** Feuchtung *f* — **n** vochtgeving *f* — **e** mojado *m*; mojadura *f* — **i** umettazione *f*; umidificazione *f*; bagnatura *f*

3978 damping rollers — See dampers.

3979 damping roller sleeve
f manchon *m* mouilleur — **d** Feuchtwalzenschlauch *m*; Wischwalzenschlauch *m* — **n** vochtrollenslang *fm* — **e** forro *m* de mojador; manchón *m* — **i** manicotto *m* dei rulli bagnatori; manicotto *m* dei rulli umidificatori

3980 damping solution *GB*; **fountain solution; dampening solution** *US*; **dampening etch** — Solution of gum arabic, and various etches in water, used for wetting the lithographic pressplate and keep the non-printing areas from accepting ink (the grease/water principle of lithography). Commonly called the water.
f eau *f* de mouillage; solution *f* de mouillage — **d** Feuchtwasser *n*; Wischwasser *n* — **n** vochtwater *n* — **e** solución *f* mojadora; solución *f* de mojador; solución *f* humectadora — **i** acqua *f* di bagnatura; soluzione *f* di bagnatura

3981 damping system; damping unit — Mechanism, usually comprising a succession of rollers, to convey moisture from the fountain to the printing plate.

f dispositif *m* de mouillage — **d** Feuchtwerk *n* — **n** vochtwerk *n* — **e** dispositivo *m* mojador; tren *m* de humectación; sistema *m* de humectación; mecanismo *m* de mojado — **i** sistema *m* bagnatore

3982 damping unit — See damping system.

3983 dancer — See jockey roller.

3984 dancing roller — See jockey roller.

3985 dandy; dandy roll; watermark dandy; watermark roll — Cylinder of wire gauze on the papermaking machine which comes into contact with the paper while it is in a wet and elementary stage. The dandy roll impresses a watermark. The ordinary dandy has the laid wires, which impress the ribs in laid papers, running transversely from end to end. On a spiral dandy they are arranged circularly. Spiral laid papers owe their distinctive appearance to the greater facility with which the liquid pulp flows under a spiral dandy. In a wove dandy the fabric is more or less ordinary wire gauze.

f rouleau *m* égoutteur; rouleau *m* filigraneur; cylindre *m* filigraneur — **d** Wasserzeichenwalze *f*; Egoutteurwalze *f*; Siebwalze *f*; Vordruckwalze *f*; Egoutteur *m* — **n** egoutteur *m*; dandywals *fm*; voordrukwals *fm*; voordrukrol *fm* — **e** rodillo *m* filigranador; filigranador *m*; rodillo *m* de afiligranar; rodillo *m* marcador; rodillo *m* para marcas de agua; rodillo *m* desgotador; egoutteur *m*; bailarín *m Ar* — **i** rullo *m* per filigrana; ballerino *m* filigranatore; rullo *m* ballerino; cilindro *m* ballerino

3986 dandy roll — See dandy.

3987 dark adaptation *GB*; **dark adaption** *US* — Conditioning of photoelectrostatic copying papers in the absence of light to permit the development or recovery of photoconductive properties.

f adaptation *f* à l'obscurité — **d** Dunklererholung *f*; Dunkeladaptation *f* — **n** donkeradaptatie *f* — **e** adaptación *f* a la obscuridad — **i** adattamento *m* all'oscurità

3988 dark adaption — See dark adaptation.

3989 darkening — Growing dark of printing inks.

f assombrissement *m* — **d** Nachdunkeln *n*; Verdunklung *f* — **n** nadonkeren *n*; donker worden *n* — **e** ensombrecimiento *m*; ennegrecimiento *m* — **i** scurimento *m*; imbrunimento *m*

3990 dark print — Overexposed print from a negative in which shadow details of the picture merge.

f épreuve *f* foncé — **d** überbelichteter Abzug *m* — **n** overbelichte afdruk *m* — **e** prueba *f* sobrexpuesta — **i** copia *f* sovraesposta

3991 dark reaction — With light-sensitive plate coatings, the hardening action which takes place without light. This action is greater with high humidity and temperature. Staling is the practical term.

f réaction *f* à l'obscurité; voilage *m* à l'obscurité; réaction *f* au noir — **d** Dunkelgerbung *f*; Dunkelreaktion *f* — **n** donkerreactie *f* — **e** reacción *f* de obscuridad — **i** indurimento *m* nell'oscurità

3992 darkroom — Chamber free from actinic light, in which photographic operations are carried out with light-sensitive materials.

f laboratoire *m* noir; laboratoire *m* obscur; chambre *f* noire; chambre *f* obscure — **d** Dunkelkammer *f* — **n** donkere kamer *fm* — **e** cuarto *m* o(b)scuro; cámara *f* obscura; cuartoscuro *m*; laboratorio *m* fotográfico — **i** camera *f* oscura

3993 darkroom camera; darkroom process camera — Process camera with the rear section permanently built into the wall of a darkroom.

f chambre *f* laboratoire — **d** Dunkelkammerkamera *f*; Zweiraumkamera *f*; Dunkelkammerreproduktionskamera *f* — **n** donkere-kamercamera *f* — **e** cámara *f* de cuarto o(b)scuro — **i** macchina *f* fotografica a camera oscura

3994 darkroom clock

f pendule *f* de laboratoire — **d** Kontrolluhr *f*; Laboruhr *f* — **n** donkere-kamerklok *fm* — **e** reloj *m* de cuarto obscuro; reloj *m* de cámara obscura — **i** orologio *m* controllo (per camera oscura)

3995 darkroom lamp; darkroom safelight; safelamp — Coloured source of illumination used in a photographic darkroom to which the photographic materials are relatively insensitive. Red light is usually used for orthochromatic, and faint green for panchromatic materials.

f lampe *f* à chambre obscur; lanterne *f* de laboratoire; lampe *f* inactinique — **d** Dunkelkammerlampe *f* — **n** donkere-kamerlamp *fm* — **e** lámpara *f* de laboratorio (fotográfico); farol *m* para cuarto oscuro; linterna *f* para cuarto oscuro; bombilla *f* para cámara oscura; lámpara *f* de securidad — **i** lampada *f* di scurezza (per camera oscura)

3996 darkroom process camera — See darkroom camera.

3997 darkroom safelight — See darkroom lamp.

3998 dark slide — Removable panel covering photographic surfaces in a plateholder.

f volet *m* (de châssis porte-plaque) — **d** Schiebejalousie *f* — **n** jaloezie *f* — **e** bastidor *m* — **i** schermo *m* scorrevole

3999 dark slide — See plate holder.

4000 darts — Term used in gravure printing for *comets.

4001 dash; em dash; em rule; dash rule — In type matter a horizontal line to indicate a rest in reading. The length of dashes is designated as en dashes, em dashes, 2-em dashes, 3-em dashes, jim dashes, Bodoni dashes, fancy dashes. See also punctuation marks.

f tiret *m* — **d** Geviertstrich *m* — **n** kastlijntje *n* — **e** pleca *f* de eme; pleca *f* cuadratín; raya *f* cuadratín; cuadratín *m* de raya; rayita *f*; bigote *m* — **i** quadrato *m* lineato; lineato *m*; quadrato *m* rigato

4002 dash; en dash; en rule — A dash with the width of an en space. Used almost exclusively in tabular work, as it is the same width as the figures and will range them. See also punctuation marks.

f tiret *m* sur demi-cadratin — **d** Halbgeviert-strich *m*; Halbgeviertgedankenstrich *m*; Strecken-strich *m* — **n** half kastlijntje *n* — **e** pleca *f* de ene; raya *f* de medio cuadratín; guión *m* de ene — **i** quadratino *m* rigato; quadratino *m* lineato; lineetta *f*; lineato *m* breve

4003 dash — In printed text a short horizontal line (a hyphen). This is not the word used in typographical sense.

f trait *m* d'union — **d** Bindestrich *m*; Trennungsstrich *m*; Abkürzungsstrich *m*; Abteilungszeichen *n* — **n** koppelteken *n*; afbrekingsteken *n*; verbindingsteken *n* — **e** guión *m*; división *f*; rayita *f* — **i** tratto *m* d'unione; lineetta *f* d'unione

4004 dash rule — See dash.

4005 data — Basic elements, involved in a process, which are capable of communicating information. Data is strictly the plural of datum but is often treated as a single collective noun.

f données *fpl* — **d** Daten *npl*; Unterlagen *fpl*; Informationen *fpl* — **n** data *npl*; gegevens *npl*; informatie *f* — **e** datos *mpl* — **i** dati *mpl*

4006 data bank — The mass storage of data which may be selected for processing. A number of sources might be involved in a particular application, such as magnetic tape files, magnetic drum, disk or card storage.

f banque *f* de données — **d** Datenbank *f* — **n** gegevensbank *fm*; informatiebank *fm* — **e** banco *m* de datos — **i** banca *f* di dati

4007 data collecting — See data collection.

4008 data collection; data collecting; data gathering — Bringing data from one or more points to a central point.

f collecte *f* de données; rassemblement *m* de données — **d** Datenerfassung *f*; Datensammeln *n* — **n** gegevensverzameling *f*; verzamelen *n* van gegevens — **e** recogida *f* de datos; recopilación *f* de datos — **i** raccolta *f* dati

4009 data dissemination; dissemination of data; dissemination of information — The distribution of information at a centre to one or more other places.

f distribution *f* d'informations; diffusion *f* d'informations — **d** Verbreitung *f* von Informationen; Verteilung *f* von Informationen — **n** informatieverspreiding *f*; informatieverdeling *f* — **e** distribución *f* de informaciones; diseminación *f* de informaciones — **i** distribuzione *f* d'informazioni; disseminazione *f* d'informazioni

4010 data gathering — See data collection.

4011 data link — Equipment and circuits to transmit data in digital format between communication terminals, to which computers or computer peripherals may be connected.

f connexions *fpl* des données — **d** Datenverbindungen *fpl* — **n** gegevensverbindingen *fpl* — **e** conexiones *fpl* de los datos — **i** connessioni *fpl* dei dati

4012 datamation — See automatic data processing.

4013 data processing — Method of manipulating information, and of outputting all or selected pieces of information in a systematic fashion by means of a computer.

f traitement *m* des données — **d** Datenverarbeitung *f* — **n** gegevensverwerking *f* — **e** tratamiento *m* de los datos — **i** trattamento *m* dei dati

4014 data reduction — Condensation of a large quantity of data into a smaller selected quantity.

f réduction *f* des données — **d** Datenreduktion *f*; Datenverdichtung *f* — **n** comprimeren *n* van gegevens — **e** reducción *f* de los datos; simplificación *f* de los datos — **i** riduzione *f* dei dati

4015 data set — Collection of data in one of several prescribed arrangements and described by control information to which the system has access.

f ensemble *m* de données; jeu *m* de données — **d** geordnete Datenmenge *f*; Datei *f* — **n** gegevensverzameling *f*; dataset *m* — **e** conjunto *m* de datos — **i** complesso *m* di dati; insieme *m* di dati

4016 data transmission — Transmission between remote points of data in coded form by means of signals.

f transmission *f* de données — **d** Datenübertragung *f* — **n** gegevenstransmissie *f*; datatransmissie *f* — **e** transmisión *f* de datos — **i** trasmissione *f* di dati

4017 date — See dateline.

4018 date block

f bloc *m* pour calendrier — **d** Kalenderblock *m*; Abreißkalenderblock *m* — **n** kalenderblok *n*; scheurkalenderblok *n* — **e** bloc *m* de calendario — **i** blocco *m* per calendario

4019 dateline; date — Line at the head of newspaper pages indicating the date of publication.

f ligne *f* de date — **d** Spitzmarke *f* — **n** datumregel *m* — **e** línea *f* de fecha; fecha *f*; línea *f* de procedencia; procedencia *f*; título *m* de fecha; título *m* de corrido *Cu* — **i** riga *f* per la data

4020 date marker — See date stamp.

4021 date stamp; date marker — Device for stamping dates and frequently the place of origin or receipt, as on postal matter.

f dateur *m*; timbre *m* à date — **d** Datumstempel *m* — **n** datumstempel *n* — **e** fechador *m* — **i** timbro *m* della data

4022 day-glo ink — See fluorescent ink.

4023 daylight — Light of the sun.

f lumière *f* du jour — **d** Tageslicht *n* — **n** daglicht *n* — **e** luz *f* diurna; luz *f* del día — **i** luce *f* diurna; luce *f* del giorno

4024 day shift — See day side.

4025 day side; day shift — Workmen who work during the daytime.

f équipe *f* du jour — **d** Tagschicht *f* — **n** dagploeg *fm* — **e** turno *m* diurno; turno *m* de día — **i** squadra *f* del giorno

4026 DC motor — See direct current motor.

4027 DDT; dichloro-diphenyl-trichloroethane —

Colourless crystals or white to slightly off-white powder, soluble in acetone, ether, benzene, carbon tetrachloride, kerosene, dioxane and pyridine, insoluble in water. Powerful insecticide effective on contact. Generic name accepted by the Entomological Society of America. TL-value: 1 mg/m^3. Formula: $(ClC_6H_4)_2CHCCl_3$.

4028 dead — Without electric current.

f sans courant — **d** stromlos — **n** stroomloos — **e** sin corriente — **i** senza corrente

4029 deadhead *v* — To have an advertisement inserted without paying.

f placer gratuitement — **d** kostenlos einrücken — **n** gratis plaatsen — **e** insertar gratis — **i** inserzione *f* gratuita

4030 dead horse — See horse flesh.

4031 dead letter — Type left after a case has been set out.

f lettre *f* au rebut — **d** übriggebliebene Schrift *f* — **n** overgebleven letter *fm*

4032 deadline; copy deadline; closing hour for news — The time after which copy is not accepted for a particular issue of a publication.

f heure *f* limite de la rédaction — **d** Redaktionsschluß *m* — **n** sluittijd *m* van de redactie; sluitingstijd *m* voor nieuws; tijd *m* van afsluiting — **e** hora *f* de(l) cierre; hora *f* de cerrar; hora *f* de clausura; momento *m* del cierre; tiempo *m* del cierre; tiempo *m* de entrega; cierre *m*; última hora *f* — **i** tempo *m* di consegna

4033 deadline — The time at which an edition of a newspaper or magazine must go to the press.

f heure *f* limite de calage — **d** Imprimatur *n* für letzte Zeitungsseite — **n** sluittijd *m* voor de zetterij — **e** al minuto — **i** ora *f* di andare in macchina

4034 dead line — Line or mark scratched on the bed of flat-bed cylinder press as a guide in placing the forme. If placed beyond this line the matter will be injured by the grippers.

f repère *m* du marbre — **d** Markierungslinie *f* — **n** aanslaglijn *fm*; aanslag *m* — **e** línea *f* de entrada de presión; línea *f* de imposición; límite *m* de imposición — **i** linea *f* di riferimento; marca *f* di riferimento

4035 dead matter — Type matter or plates which are not to be used for reprint and which may distributed or melted.

f matière *f* morte; matière *f* à distribuer; composition *f* à distribuer — **d** Ablegesatz *m* — **n** distributiezetsel *n*; afgedrukt zetsel *n* — **e** composición *f* a distribuir; material *m* muerto; material *m* usado; material *m* a deshacer — **i** scomposizione *f*; materiale *m* in scomposizione

4036 dead stone — Stone or table in the composing room to hold *dead matter.

f table *f* de distribution — **d** Ablegetisch *m* — **n** distributietafel *fm* — **e** tablero *m* para la distribución; depósito *m* para material usado — **i** piano *m* per la scomposizione

4037 dead stuck — See unsaleable book.

4038 debenture bond — Bond usually secured by an indenture with protective provisions but without a specific lieu or any asset.

f titre *m* d'obligation — **d** Obligation *f* — **n** obligatie *f* — **e** obligación *f* — **i** obbligazione *f*

4039 Debot effect — Photographic effect in which a red-sensitive emulsion is first exposed in a normal way and then treated with chromic acid to reduce or eliminate the latent image. When the layer receives only an after-exposure with red light or infra-red light and then is developed again an image arises. This is related to the *Herschel effect. It is caused by the transfer of the silver germs of the inner latent image to the surface of the silver bromide crystals.

f effet *m* Debot — **d** Debot-Effekt *m* — **n** Debot-effect *n* — **e** efecto *m* Debot — **i** effetto *m* Debot

4040 debugging — Searching for, isolating from and correcting errors in a computer program which is rarely error free during the first run. Sometimes also detection and correction of hardware faults.

f mise *f* au point d'une programme; mise *f* au point des programmes — **d** Suchen *n* von Fehlern; Kontrollieren *n* eines Programms; Kontrolle *f* der Computerprogramme — **n** opsporen *n* van fouten; ontstoren *n* van het programma; programmacontrole *fm* — **e** puesta *f* a punto de un programa; depuración *f* de un programa; eliminación *f* de fallas — **i** messa *f* a punto d'un programma; spulciatura *f* d'un programma

4041 decalcomania; decalcomania pictures; decals; transfers — Printed sheets from which the design or picture is transferred to glass, pottery, wood or metal.

f décalcomanies *fpl* (glissantes); images *fpl* à décalquer — **d** Abziehbilder *npl* — **n** transfers *mnpl*; schuiftransfers *mnpl* — **e** calcomanías *fpl*; decalcomanías *fpl* — **i** calcomanie *fpl*; decalcomanie *fpl*

4042 decalcomania paper; transfer paper — Temporary base paper on which designs are printed for transfer onto a permanent base. See also ceramic transfer paper.

f papier *m* à décalcomanie; papier *m* pour décalcomanie — **d** Abziehbilderpapier *n* — **n** decalcomaniepapier *n*; transferpapier *n* — **e** papel *m* para calcomanías; papel *m* decalco — **i** carta *f* per decalcomanie

4043 decalcomania pictures — See decalcomania.

4044 decalcomania printing — See transfer printing.

4045 decals — See decalcomania.

4046 decameter — Apparatus for the determination of the water content of paper or ink by means of the dielectric constant.

f décamètre *m* — **d** Dekameter *n*; DK-Meter *n* — **n** decameter *m* — **e** decámetro *m* — **i** decametro *m*

4047 decant *v* — To draw off without disturbing the sediment or the lower liquid layers. Also to pour from one vessel into another.

f décanter; transvaser — **d** dekantieren; abgießen — **n** decanteren; afgieten; afschenken — **e** decantar; trasegar; verter — **i** decantare

4048 decantation — Pouring off clear supernatant liquid from its sediment.

f décantation *f* — **d** Dekantieren *n*; Abgießen *n*; Abklären *n*; Abscheiden *n* — **n** decanteren *n*; afgieten *n*; afschenken *n* — **e** decantación *f*; trasiego *m* — **i** decantazione *f*

4049 decay time — Time taken for the light on a CRT screen to fall to about 10% after impact of the electron beam; applied to the phosphor coating on the face of the cathode ray tube which fluoresces under electron impact and continues to phosphoresce after the beam excitation has moved away.

f temps *m* de chute — **d** Abfallzeit *f*; Signalabfallzeit *f* — **n** afvaltijd *m* — **e** tiempo *m* de caída — **i** tempo *m* di discesa; tempo *m* di caduta

4050 decibel — Dimensionless unit for expressing the ratio of two values of power, the number of decibels being ten times the logarithm of the power ratio. Unit for measuring the relative loudness of sound.

f décibel *m* — **d** Dezibel *m* — **n** decibel *m* — **e** decibelio *m*; decibel *m* — **i** decibel *m*

4051 deciduous wood — See hardwood.

4052 deciduous wood fibres — Fibres from trees which lose their leaves annually.

f fibres *fpl* de bois feuillu — **d** Laubholzfasern *fpl* — **n** loofhoutvezels *fmpl* — **e** fibras *fpl* de madera de árboles foliculares — **i** fibre *fpl* di latifoglia

4053 deciduous wood pulp — See hardwood pulp.

4054 decimal classification — System of classifying publications into ten main classes of knowledge with subdivision in these classes by use of a decimal system.

f classification *f* décimale — **d** Dezimalklassifizierung *f* — **n** decimale classificatie *f* — **e** clasificación *f* decimal — **i** classificazione *f* decimale

4055 decimal comma; decimal point *depr.*; **separatrix** — In UK and US, decimal fractions are separated from the whole number to which they belong by a decimal point (i.e. 5.62345). A comma is used to divide each thousandth part of a whole number (i.e. 1,000,000 = one million). In January 1978, the Système International (SI) was accepted in Western Europe. This reverses the roles as stated above of the point and the comma (i.e. 5,62345 and 1.000.000). Whether Britain and the US will adopt the SI remains a matter of speculation. In GB the decimal comma is also used for amounts less than a pound. Express as (new) pence: 54 p (roman, no point); for amounts of one pound or more express as pounds and decimal fractions of a pound: £ 54; £ 54,65; £ 54,07 (omitting p). The (new) halfpenny should be expressed as a fraction: £ 54,07 1/2. See also punctuation marks.

f virgule *f* — **d** Dezimalzeichen *n* — **n** decimaalteken *n*; decimale komma *fm* — **e** coma *f* de decimales — **i** punto *m* decimale

4056 decimal point — See decimal comma.

4057 decimo octavo — See eighteenmo.

4058 decimo sexto — See sixteenmo.

4059 decision — Result of a logical operation.

f décision *f* — **d** Entscheidung *f* — **n** beslissing *f*; besluit *n*; decisie *f* — **e** decisión *f* — **i** decisione *f*

4060 deck — Complete unit for printing on a multi-unit press, when the units are arranged on different levels in the press.

f élément *m* d'impression — **d** Druckeinheit *f*; Druckwerk *n* (bei Etagenmaschinen mit mehreren Druckwerken) — **n** drukeenheid *f* — **e** unidad *f* de impresión — **i** elemento *m* di stampa; unità *f* stampante (a elementi sovrapposti)

4061 deck — Platform of a rotary press.

f passerelle *f* — **d** Plattform *f*; Laufsteg *m*; Rundgang *m* — **n** platvorm *n* — **e** piso *m*; plataforma *f* — **i** piattaforma *f*

4062 deck — See rostrum.

4063 deckle — Removable rectangular frame which fits on the wire mould used in hand-made paper manufacture to prevent the stock running of the mould.

f cadre *m* volant; couverte *f* — **d** Schöpfrahmen *m*; Auflaufrahmen *m*; Deckelrahmen *m* — **n** schepraam *n* — **e** bastidor *m* — **i** cornice *f* (della forma) per carta a mano; cascio *m*; casso *m*

4064 deckle boards — Stationary equipment to retain the stock laterally on the wire during the early part of drainage. This equipment can be adjusted laterally to obtain the required width on the Fourdrinier wire part.

f réglettes *fpl* — **d** Stauplatte *fpl*; Staubrette *npl*; seitliche Begrenzungslineale *npl* — **n** formaatstrippen *mpl*; formaatbanden *mpl* — **e** regletas *fpl* — **i** regoli *mpl* laterali

4065 deckle edge; featheredge — Feathery edge surrounding a sheet of hand-made paper formed by the rubber deckle straps at the sides of the paper machine, or by artificial means such as a water jet.

f barbe *f*; témoin *m* — **d** Schöpfrand *m*; Büttenrand *m* — **n** scheprand *m* — **e** orilla *f* barbada; barba *f* (del papel); canto *m* barbado; borde *m* greco *Ar*; borde *m* antique *Ar* — **i** barba *f*; sfrangiatura *f*

4066 deckle-edged paper — Paper with feathery edges produced by leaking of the pulp underneath the deckle, or by impinging small water jets at specified intervals on the wet web. See also hand-made paper.

f papier *m* similiforme; papier *m* barbé — **d** Papier *n* mit Büttenrand; Büttenpapier *n* — **n** papier *n* met scheprand — **e** papel *m* de tina; papel *m* de barba — **i** carta *f* barbata

4067 deckle of suction box — Stationary equipment inside suction boxes to restrict the suction area to the width of the web; can be adjusted laterally to the width of the web.

f tiroir *m* de caisse aspirante — **d** Stillwand *f* — **e**

cierre *m* lateral móvil de caja aspirante — **i** regolo *m* della cassa aspirante

4068 deckle straps — Endless belts, generally rectangular in cross section, that move with the wire and serve to retain the stock laterally on the wire during the early part of drainage.

f courroies-guides *mpl*; couvertes *fpl* — **d** Deckelriemen *mpl* — **n** dekkelriemen *mpl* — **e** tapas *fpl* de correa — **i** centiguide *fpl*

4069 deck of cards — See pack of cards.

4070 decks — Pairs of horizontal printing couples arranged one above the other in a letterpress rotary.

f dispositifs *mpl* straightline — **d** Doppeldruckwerke *npl* — **n** dubbele drukeenheden *fpl* — **e** doble unidades *fpl* impresoras; doble grupos *mpl* de impresión — **i** doppi gruppi *mpl* stampanti

4071 declutching device

f dispositif *m* de débrayage — **d** Kupplungs-Ausrückvorrichtung *f* — **n** ontkoppelings-inrichting *f* — **e** dispositivo *m* de desembrague; juego *m* de desembrague — **i** dispositivo *m* di disimbracamento; dispositivo *m* di disinnesto

4072 decode *v* — To reverse previous encoding.

f décoder — **d** entschlüsseln — **n** decoderen — **e** descodificar — **i** decodificare

4073 decorated with fleurs-de-lis

f fleurdelisé; orné de fleurs de Lys — **d** mit Lilienmuster verziert — **n** met Franse lelies versierd — **e** decorado con flores de lis — **i** gigliato

4074 dedication — Complimentary inscription addressed to one or more persons by the author. See also dedicatory title.

f dédicace *f*; épitre *m* dédicatoire; épitre *m* liminaire — **d** Widmung *f*; Dedikation *f*; Zueignung *f*; Zueignungsschrift *f*; Widmungsschrift *f* — **n** opdracht *fm*; dedicatie *f* — **e** dedicación *f*; dedicatoria *f* — **i** dedica *f*

4075 dedication — See dedicatory title.

4076 dedicatory title; dedication — Title in the *prelims of a book, dedicated to a person or cause, on the tail part of a right-hand page.

f titre *m* de dédicace; dédicace *f* — **d** Widmungstitel *m*; Dedikationstitel *m* — **n** opdrachtstitel *m*; opdracht *fm* — **e** título *m* dedicatorio; dedicatoria *f* — **i** titolo *m* di dedica

4077 deepen *v*

f foncer; donner une nuance (couleur) plus foncée — **d** abdunkeln (Farbton) — **n** donkerder maken — **e** obscurecer; oscurecer — **i** offuscare; scurire; incupire; rendere più scuro

4078 deep-etch *v* — To etch or sink down bare metal areas of relief plates to obtain necessary printing depth.

f graver en creux — **d** tiefätzen — **n** diepetsen — **e** regrabar — **i** incidere in profondità

4079 deep-etched halftone — See deeply-etched halftone plate.

4080 deep-etched offset plate; deep-etch plate — Offset printing plate with an intaglio image filled with a substance that attracks ink to make it planographic.

f plaque *f* d'offset en creux — **d** tiefgeätzte Offsetplatte *f* — **n** diepgelegde offsetplaat *f* — **e** plancha *f* al hueco-offset — **i** lastra *f* offset deep-etch; lastra *f* deep-etch

4081 deep etching — Extra etching or bites, after the rough etching and before the first round etching, required for proper printing depth of halftone blocks; specifically applied to combination plates on copper and zinc, and to etching highlight effect in halftones.

f morsure *f* de grands creux; acidulation *f* profonde — **d** tiefgeätztes Rasterklischee *n* — **n** diepgeëtst autotypiecliché *n* — **e** mordido *m* profundo; mordido *m* extrahondo — **i** incisione *f* profonda; incisione *f* all'incavo; incisione *f* d'approfondimento

4082 deep-etching — Making lithographic plates by the application of various etches, depending on metal of plate, to etch the metal (about 0.007 mm or 0.0003 in deep) in order to provide space for the lacquer base of the ink-receptive printing image.

f morsure *f* en creux — **d** Tieflegen *n*; Tiefätzung *f* — **n** diepleggen *n* — **e** mordido *m* en hueco; mordido *m* en profundo — **i** incidere in profondità

4083 deep-etch ink

f encre *f* pour morsure en creux — **d** Tiefätzfarbe *f* — **n** diepetsinkt *m* — **e** tinta *f* para mordido profundo — **i** inchiostro *m* deep-etch; inchiostro *m* per incisione in profondità

4084 deep-etch lacquer; base lacquer; base *depr.* — Solution applied to image areas of a plate to form the printing image.

f laque *f* pour plaque offset; laque *f* offset en creux — **d** Tiefätzlack *m* — **n** diepetslak *mn* — **e** laca *f* deep-etch; laca *f* al hueco offset; laca *f* para mordido extrahondo — **i** lacca *f* deep-etch

4085 deep-etch offset; intaglio offset — Offset printing with plates on which the printing areas are etched below the surface to compensate for wear of long runs and to allow transfer of a thicker film of ink.

f offset *m* en creux; offset *m* creux — **d** Offsettiefdruck *m* — **n** offsetdiepprocédé *n* — **e** hueco-offset *m* — **i** stampa *f* offset deep-etch; stampa *f* offset incisa

4086 deep-etch pad — Plush covered, wooden block to apply the deep-etch solution in platemaking. Separate pads for each operation.

f tampon *m* de peluche — **d** Plüschtampon *m*; Ätztampon *m* — **n** pluche tampon *m*; pluche borstel *m* — **e** tampón *m* de peluche; tampón *m* de felpa — **i** tampone *m* di peluche; tampone *m* di felpa; tampone *m* d'incisione

4087 deep-etch plate — See deep-etched offset plate.

4088 deep-etch process — Reversal process using a deep-etch solution. A reversal plate is not necessarily a deep-etch plate.

f procédé *m* offset creux — **d** Tiefätzverfahren *n*; Offsettiefverfahren *n* — **n** diepetsprocédé *n*; dieplegprocédé *n*; offsetdiepprocédé *n* — **e** procedimiento *m* al hueco-offset — **i** procedimento *m* deep-etch

4089 deep-etch solution — Solution to clean the image areas of a reversal printing plate.

f solution *f* de morsure en creux — **d** Tiefätzlösung *f* — **n** diepetsoplossing *f* — **e** solución *f* mordiente para hueco-offset — **i** soluzione *f* deep-etch

4090 deep-etch stencil — Light-hardened bichromated gum resist which protects the non-printing parts of the plate from the developing and deep-etching solutions.

f réserve *f* (en offset creux) — **d** Kopierschicht *f*; lichtgehärtete Kopierschicht *f* — **n** diepetslaag *fm* — **e** reserva *f* de goma — **i** strato *m* deep-etch; strato *m* per incisione in profondità

4091 deeply-etched halftone plate; deep-etched halftone — Plate in which the spaces between the dots have been etched deeper to allow printing on matt surface or poor quality paper.

f simili *m* grand creux — **d** tiefgeätztes Rasterklischee *n* — **n** diepgeëtst autotypiecliché *n* — **e** autotipia *f* de mordido extrahondo; medio tono *m* de mordido extrahondo — **i** cliché *m* autotipico inciso in profondità

4092 deep page — Usual page of a book, the vertical size being longer than the horizontal size.

f page *f* en hauteur; page *f* à la française — **d** Seite *f* im Hochformat — **n** staande pagina *fm* — **e** página *f* vertical — **i** pagina *f* verticale; pagina *f* di formato alto

4093 defective; incomplete — Having a defect; faulty.

f défectueux; incomplet — **d** defekt; fehlerhaft; unvollständig — **n** defect; incompleet; onvolledig — **e** defectuoso; incompleto — **i** difettoso; incompleto

4094 defective — Imperfect copy.

f pièce *f* défectueuse — **d** fehlerhaftes Exemplar *n* — **n** foutief exemplaar *n* — **e** ejemplar *m* defectuoso; ejemplar *m* adolecido — **i** esemplare *m* difettoso

4095 defective colour vision — See colour blindness.

4096 defect of image

f défaut *m* en imperfection des images — **d** Abbildungsfehler *m*; Bildfehler *m* — **n** afbeeldingsfout *fm* — **e** defecto *m* de imagen — **i** mancanza *f* d'immagine

4097 defects of aluminium foil — These are: belly, blister, broken mat, buckle, burred edge, chatter, coating streaks, dropped edge, ears, herringbone streaks, inclusions, kink, lateral camber, pinch marks, pinholes, scratches, slivers, stain, sticking, telescoping, transverse camber, trim inclusions, wavy edge.

4098 definition — Defining power of a lens, or its property to project sharp images.

f définition *f* — **d** Genauigkeit *f* — **n** scheidend vermogen *n* — **e** definición *f*; fineza *f* de detalles — **i** definizione *f*

4099 definition — Impression of clarity of detail perceived when viewing a photograph.

f netteté *f* — **d** Zeichnung *f*; Detailzeichnung *f*; Schärfe *f* — **n** scherpte *f* — **e** nitidez *f* — **i** nitidezza *f*

4100 deflection — Yield of a machine part under pressure. Deflexion is correct, but deflection is increasingly common, and usual in US.

f flexion *m*; déformation *f* par flexion; fléchissement *m* — **d** Durchbiegung *f* — **n** doorbuiging *f* — **e** flexión *f* — **i** incurvatura *f*; deformazione *f*

4101 deflection yoke *GB*; **sweeping coil** *US* — Electronic component surrounding or built in a cathode ray tube to direct the beam of electrons emitted from the tube's electron gun horizontally and vertically across the face of the tube. Accomplished by varying the directional intensity of electrostatic or electro-magnetic fields created by north, south, east and west changes emitted by the yoke.

f collier *m* de déviation — **d** Ablenkspule *f* — **n** afbuigspoel *fm*; deflectiespoel *fm* — **e** bobina *f* de deflexión; bobina *f* de desviación — **i** bobina *f* deflettrice

4102 deflocculation — Separating flocculated particles, e.g. by stirring. Reflocculation takes place when stirring stops. Deflocculation is the reverse of flocculation and can be accomplished by lowering the interfacial tension with a suitable wetting agent.

f défloculation *f* — **d** Deflockulation *f*; Entflockung *f*; Peptisierung *f* — **n** peptisatie *f* — **e** desfloculación *f*; peptización *f* — **i** deflocculazione *f*

4103 defoamer — See anti-foam agent.

4104 defoaming agent — See anti-foam agent.

4105 deformation

f déformation *f* — **d** Verformung *f*; Formänderung *f* — **n** vervorming *f*; deformatie *f*; vormverandering *f* — **e** deformación *f* — **i** deformazione *f*

4106 degrease *v* — To clean with a solvent or hot vapour, to remove grease.

f dégraisser — **d** entfetten — **n** ontvetten — **e** desengrasar — **i** sgrassare

4107 degree Kelvin — See Kelvin.

4108 degree mark — Typographic mark (°), used with figures, a contraction for degrees. Mark must be placed directly before the letter which follows, and separated from the figures.

f symbole *m* de degré; signe *m* de degré; zéro *m* supérieur — **d** Gradzeichen *n*; hochstehende Null *f* — **n** graadteken *n* — **e** signo *m* de grado; cerito *m* — **i** segno *m* di grado

4109 degree of beating — The degree of beating of the paper pulp is defined in degrees Schopper-Riegler (°SR) or in Canadian Freeness.

f indice *m* de raffinage; degré *m* de raffinage — **d** Mahlgrad *m* — **n** malingsgraad *m*; maalgraad *m* — **e** grado *m* de refino; grado *m* de engrasamiento — **i** grado *m* di raffinazione

4110 degree of esterification

f degré *m* d'estérification — **d** Veresterungsgrad m — **n** mate fm van verestering — **e** grado m del esterificación — **i** grado m di esterificazione

4111 degree of grinding — See fineness of grind.

4112 degree of hardness
f dureté f — **d** Härtegrad m — **n** hardheidsgraad m; hardheid f — **e** grado m de dureza; dureza f — **i** grado m di durezza; durezza f

4113 degree of humidity
f degré m d'humidité — **d** Feuchtigkeitsgrad m — **n** vochtigheidsgraad m; mate fm van vochtigheid — **e** grado m de humedad — **i** grado m d'umidità

4114 degree of proof — Method to designate the strength of ethyl alcohol/water mixtures, most frequently for taxation according to the number of gallons of "proof spirit" or "100 proof" alcohol that can be made from 100 gallons of the alcohol/water mixture. Proof spirit is defined by the regulation (in US) that it contains "one half its volume of alcohol of specific gravity 0.7939 (60 °F)". This latter is absolute alcohol, and pure alcohol is therefore 200 proof. The degree of proof is twice the per cent by volume of alcohol. Due to volume concentration upon mixing alcohol and water, it is necessary to add 53.73 volumes of water to 50 volumes of alcohol in order to produce 100 volumes of 100 proof alcohol.

4115 degree of whiteness
f degré m de blanc; degré m de blancheur — **d** Weißgrad m — **n** witheidsgraad m — **e** grado m de blancor; grado m de blancura; grado m del albura — **i** grado m di bianco; indice m riflettometrico nel blu

4116 degrees Celsius — See centigrades.

4117 degrees centigrade — See centigrades.

4118 degrees Fahrenheit — See Fahrenheit.

4119 dehumidification — Process in air-conditioning in which air is cooled below the *dew point so that part of the water vapour is condensed.
f déshydratation f — **d** Luftentfeuchtung f; Regelung f der Luftfeuchtigkeit — **n** luchtvochtigheidsregeling f — **e** deshumectación f — **i** deumidificazione f

4120 dehydrated — See anhydrous.

4121 dehydration — Loss of water from a substance or mixture either by drying, or by a decomposition process that produces water, by centrifugal force or hydraulic pressure. Usually not loss of water from a water solution by evaporation or boiling.
f déshydratation f — **d** Entwässerung f; Wasserentzug m; Wasserentziehung f; Dehydratisierung f — **n** dehydratatie f; dehydratie f; wateronttrekking f — **e** deshidratación f — **i** disidratazione f

4122 deinked-paper stock — Paper stock made of the pulp resulting from the deinking of waste paper (old books and magazines) and re-used in papermaking.
f pâte f de vieux papiers désencrés — **d** entschwärzter Altpapierstoff m; entfärbter Altpapierstoff m; druckfarbengereinigter Altpapierstoff m — **n** ontinkte papierstof m — **e** pasta f de recortes detintados — **i** pasta f di carta disinchiostrata

4123 deinking — Operation of reclaiming fibres from waste paper by removing ink, colouring materials, and fillers.
f désencrage m — **d** Entfärben n; Druckfarbenentfernung f; Deinking n — **n** ontinkten n — **e** detintado m — **i** disinchiostrazione f

4124 delamination — Separation of paper into layers caused by shearing when the paper leaves the printing nip. See also peeling.
f délamination f; délaminage m — **d** Delaminierung f; Schichttrennung f; Schichtspaltung f — **n** delaminatie f — **e** delaminación f — **i** delaminazione f

4125 delay line — Component or circuit to delay the transmission of a signal.
f ligne f à retard — **d** Verzögerungsleitung f; Verzögerungsstrecke f — **n** vertragingslijn fm — **e** línea f de retardo — **i** linea f di ritardo

4126 dele — See delete.

4127 deleatur — See delete.

4128 delete; dele; deleatur — Proofreader's mark signifying to take out the marked letter or word(s). See proofcorrection marks, app. no. 5.
f à supprimer; deleatur — **d** streichen; ausstreichen; herausnehmen; entfernen — **n** laten vervallen; weghalen — **e** borrar; dele; deleátur; ¡suprimir! — **i** da sopprimere; deleatur

4129 delete v — To remove type or to direct the compositor to do so. Also, in a manuscript, to strike out.
f supprimer — **d** löschen; streichen; beseitigen — **n** schrappen; verwijderen — **e** eliminar; suprimir; delear — **i** sopprimere

4130 delete button; kill button
f touche f d'annulation; touche f d'effaçage — **d** Löschtaste f — **n** uitwistoets m — **e** tecla f de anulación — **i** tasto m annullatore

4131 deletion mark
f signe m de suppression; deleatur m — **d** Deleaturzeichen n — **n** weghaalteken n — **e** señal f de quitar (letra); señal f de suprimir (letra); signo m de supresión — **i** segno m di soppressione (una lettera); segno m per sopprimere (una lettera)

4132 delimiter — Code or flag to indicate that the codes which follow are an instruction or argument. Differs from a function code in that delimiters appear in pairs, with an opening delimiter at the beginning of the command string and a closing delimiter at the end of the string.
f délimiteur m; symbole m de délimitation — **d** Begrenzungszeichen n; Begrenzungssymbol n — **n** begrenzingsteken n — **e** delimitador m; símbolo m de delimitación — **i** delimitatore m; simbolo m di delimitazione

4133 deliquescence — Tendency of chemicals to absorb atmospheric moisture.
f déliquescence f — **d** Hygroskopizität f;

Weichwerden *n* — **n** hygroscopiciteit *f*; zacht worden *n* — **e** delicuescencia *f* — **i** igroscopicità *f*

4134 delivery — System that carries the sheet from the impression cylinder to the front pile.
f sortie *f* (de feuilles); réception *f* — **d** Bogenausgang *m*; Ausgang *m*; Ablage *f*; Bogenauslage *f*; Ausführung *f* — **n** uitleg *m*; vellenuitleg *m* — **e** entrega *f*; salida *f*; sacapliegos *m*; dispositivo *m* de salida; mecanismo *m* de salida — **i** uscita *f*; uscita *f* fogli

4135 delivery belts — Driven tapes or belts which carry the product away from the press.
f courroies *fpl* de sortie; tapis *m* de sortie — **d** Auslagebänder *npl*; Auslegebänder *npl*; Ausführbänder *npl* — **n** uitlegbanden *mpl* — **e** correas *fpl* de entrega; correas *fpl* de salida — **i** nastri *mpl* d'uscita

4136 delivery board — See delivery table.

4137 delivery carriage — Oscillating carrier which transports the printed sheets to the pile and then drops them onto the pile.
f chariot *m* de sortie — **d** Auslegewagen *m* — **n** uitlegwagen *m* — **e** caro *m* de salida — **i** carrello *m* d'uscita

4138 delivery chain — Chain which carries gripper bars for the transfer of sheets from the impression cylinder to the delivery board.
f chaîne *f* de réception — **d** Auslegerkette *f* — **n** uitlegketting *f*; ketting *f* van de uitleg — **e** cadena *f* del receptor — **i** catena *f* d'uscita

4139 delivery channel — Part of the slug composing machine wich carries the assembled matrices from the assembler to the first elevator.
d Überführungskanal *m* — **n** overschuifkanaal *n* — **e** canal *m* de entrega — **i** canale *m* d'uscita

4140 delivery cylinder — See skeleton cylinder.

4141 delivery date — Date on which the printer delivers bound books to the publisher's warehouse.
f date *f* de livraison — **d** Lieferdatum *n* — **n** afleveringsdatum *m* — **e** término *m* de entrega; plazo *m* de entrega — **i** termine *m* di consegna

4142 delivery fly — See fly.

4143 delivery gripper
f pince *f* de sortie — **d** Auswerfgreifer *m*; Ausleiter *m* — **n** uitleggrijper *m* — **e** pinza *f* de entrega; pinza *f* de salida; uña *f* de entrega; uña *f* de salida; cogebocados *m* — **i** pinza *f* del mecanismo d'uscita

4144 delivery lever — Part of the slug composing machine.
f levier *m* du chariot d'envoi des matrices — **d** waagerechte Überführung *f* — **n** transporteurarm *m*; wegzendarm *m* — **e** leva *f* del separador — **i** leva *f* d'uscita

4145 delivery mechanism
f sortie *f* — **d** Ausführapparat *m* — **n** uitlegmechanisme *n* — **e** mecanismo *m* de entrega; mecanismo *m* de salida — **i** meccanismo *m* d'uscita

4146 delivery on approval; consignment on approval
f envoi *m* en communication; envoi *m* à condition; envoi *m* à l'examen — **d** Ansichtssendung *f*; Konditionssendung *f* — **n** zichtzending *f*; zending *f* op zicht — **e** envío *m* a condición; envío *m* para inspección; envío *m* a examen; envío *m* condicional — **i** invio *m* in esame; spedizione *f* in visione

4147 delivery pile
f pile *f* de sortie; pile *f* de recette — **d** Ablegestapel *m* — **n** uitlegstapel *m* — **e** pila *f* de entrega; pila *f* de salida; pila *f* del receptor; pila *f* (de las hojas impresas) — **i** pila *f* d'uscita

4148 delivery slide — Part of the slug composing machine for the transport of matrices.
f chariot *m* d'envoi des matrices — **d** Überführungsschlitten *m* — **n** transporteur *m* — **e** carro *m* de entrega; corredera *f* de salida; carro *m* transportador; carro *m* conductor — **i** carrello *m* per il trasferimento delle matrici

4149 delivery table; delivery board; fly board — Table on a press or cutting or folding machine on which the printed, cut, or folded sheets are piled automatically.
f table *f* de sortie; table *f* de réception; table *f* de recette — **d** Auslegetisch *m*; Ablegetisch *m*; Ablagetisch *m* — **n** uitlegtafel *fm* — **e** mesa *f* receptora; tablero *m* receptor; receptor *m*; sacador *m* — **i** tavola *f* d'uscita

4150 delivery tapes
f cordons *mpl* de sortie; cordons *mpl* à la sortie — **d** Ausführbänder *npl* — **n** uitlegbanden *mpl* — **e** cintas *fpl* sacapliegos; cintas *fpl* de arrastre; cintas *fpl* conductoras — **i** nastri *mpl* d'uscita

4151 delivery terms — See also conditions of sale.
f conditions *fpl* de livraison — **d** Lieferbedingungen *fpl*; Lieferungsbedingungen *fpl*; Lieferfrist *f* — **n** leveringsvoorwaarden *fpl* — **e** condiciones *fpl* de entrega — **i** condizioni *fpl* della fornitura

4152 delivery tray
f plateau *m* de sortie — **d** Maschinenschiff *n*; Auslagetrog *m*; Auslagekasten *m* — **n** machinegalei *fm* — **e** galera *f* receptora — **i** vantaggio *m* della macchina compositrice; piano *m* d'uscita

4153 delta modulation — See differential modulation.

4154 deluwang paper; daluwang paper — Writing and wrapping paper originating from Java, made by hand from the inner bark of the paper *mulberry tree.

4155 de luxe binding
f reliure *f* de luxe — **d** Prachtband *m* — **n** prachtband *m*; luxe band *m* — **e** encuadernación *f* espléndida — **i** legatura *f* splendida; coperta *f* splendida; copertina *f* splendida

4156 de luxe edition — See edition de luxe.

4157 demijohn — See carboy.

4158 demodulation — Process wherein a wave resulting from previous modulation is used to derive a wave with the characteristics of the original modulating wave.
f démodulation *f* — **d** Demodulation *f* — **n**

demodulatie *f* — **e** demodulación *f* — **i** demodulazione *f*

4159 demy — A British standard size for printing papers. The term is also used for boards. See app. no. 3.

4160 denaturate *v* — See denature *v*.

4161 denature *v*; **denaturate** *v* — To make unfit to be eaten by adding poisonous matter.

f dénaturer — **d** denaturieren; vergällen — **n** denatureren — **e** desnaturalizar — **i** denaturare

4162 denatured alcohol — Ethyl alcohol to which ingredients have been added to make it unfit for human consumption. See also methylated alcohol.

f alcool *m* dénaturé — **d** denaturierter Alkohol *m*; denaturierter Spiritus *m*; vergällter Alkohol *m*; vergällter Spiritus *m* — **n** gedenatureerde alcohol *m*; brandspiritus *m* — **e** alcohol *m* desnaturalizado; alcohol *m* de quemar — **i** alcool *m* denaturato

4163 dendrites — Crystals with branching, tree-like forms.

f dendrites *fpl* — **d** Dendrite *mpl*; Tannenbaum-kristalle *mpl* — **n** dendrieten *mpl* — **e** dendritas *fpl* — **i** dendrite *fpl*

4164 dendritic growths — Effect caused by age (oxidation) in the particles of copper or brass in papers that have been beaten with bronze roll bars. After a few years irregular lines spread from these particles through the fibres.

d Bronzeflecke *mpl* — **n** bronsvlekken *fmpl* — **e** manchas *fpl* de cobre; manchas *fpl* de bronze — **i** macchie *fpl* bronzo

4165 Dennison wax test — Method to determine the surface bonding strength of paper by melting a series of waxes with graded adhesive strength and sticking them to the surface of the sheet. The waxes are cooled 15 minutes, then pulled vertically with a quick jerk. The paper is said to pass the highest number wax which does not break or injure the surface of the paper, etc.

f essai *m* aux cires Dennison — **d** Dennison-Wachstest *m* — **n** wasproef *fm* volgens Dennison — **e** ensayo *m* de ceras Dennison — **i** prova *f* alla cera Dennison

4166 dense reel — See tight reel.

4167 densitometer — Electric instrument to measure optical density or tone values, or to check the uniformity of print colours. Two types: visual and photo-electric. *Transmission densitometers measure the full density range of negatives, and reflection densitometers the reflection range of opaque copy. With a photocell search unit the instrument can be used as an illumination meter on the ground glass of camera. See also reflectometer.

f densitomètre *m* — **d** Densitometer *n*; Schwärzungsmesser *m* — **n** densitometer *m*; zwartingsmeter *m* — **e** densitómetro *m* — **i** densitometro *m*

4168 densitometry — Technique to measure the amount of light absorbed or reflected by the blackening (density) on photographic materials

after exposure and processing.

f densitométrie *f* — **d** Densitometrie *f* — **n** densitometrie *f* — **e** densitometría *f* — **i** densitometria *f*

4169 density — Mass per unit volume, usually expressed in grams per cubic centimetre or in pounds per cubic foot or gallon. As a specific term it has been applied in electricity, photography and other areas with reference to length-force-time systems of units.

f densité *f* — **d** Dichte *f* — **n** dichtheid *f* — **e** densidad *f* — **i** densità *f*

4170 density — Amount of information which can be stored on a medium per unit of length, e.g., on magnetic tape; a common density is 800 bytes or characters per inch.

4171 density — See optical density.

4172 density gradation curve — See characteristic curve.

4173 density, ink — See ink density.

4174 density, print — See print density.

4175 density range — Difference in optical density of halftone films between the highlight and shadow areas of a photographic or other image.

f rangée *f* de la densité — **d** Dichteumfang *m*; Schwärzungsumfang *m*; Schwärzungsbereich *m* — **n** zwartingsomvang *m*; zwartingsbereik *n* — **e** grado *m* de ennegrecimiento — **i** intervallo *m* di densità

4176 densometer — Equipment to determine the porosity or resistance of paper to the passage of air.

f densomètre *m*; porosimètre *m* à air — **d** Densometer *n*; Luftdurchlässigkeitsprüfgerät *n* — **n** densometer *m*; luchtdoorlatendheidsmeter *m* — **e** porosímetro *m* (ad aria) — **i** porosimetro *m* (ad aria)

4177 dent; depression of blanket — Depression in an offset blanket.

f faible *m* d'un blanchet — **d** Einbeulung *f* eines Offsetgummituchs — **n** dunne plek *fm* in een offset-rubberdoek; kuil *m* in een offset-rubber-doek — **e** abolladura *f* — **i** ammaccatura *f*; depressione *f* in un blanket

4178 dent *v*

f enfoncer; abîmer — **d** einbeulen — **n** indeuken — **e** abollar — **i** ammaccare

4179 dentelle à la grecque binding — Book cover decorated in a lace-pattern style of so-called Greek design.

f reliure *f* à la dentelle à la grecque — **d** Mäander-Einband *m* — **n** meander-band *m* — **e** encuadernación *f* dentelle à la grecque — **i** legatura *f* a meandri

4180 dentelle à l'oiseau binding — Book cover decorated in the style used by Nicolas-Denis Derôme (Derôme le Jeune), 1731-1788. In the small dentelle ornaments often small birds were tooled.

f reliure *f* à la dentelle à l'oiseau — **d** Dentelle à l'oiseau-Einband *m* — **n** dentelle à l'oiseau-band *m* — **e** encuadernación *f* dentelle à l'oiseau

— **i** legatura *f* a merlotti

4181 dentelle binding — Book cover decoration style developed by A.M. Padeloup († 1758) and Antoine Derôme († 1788), famous bookbinder families of the 17th and 18th century. They used broad borders having an effect of lace edging tooled on the covers.

f reliure *f* dentelle; reliure *f* à la dentelle — **d** Dentelle-Einband *m*; Spitzenmuster-Einband *m* — **n** dentelle-band *m* — **e** encuadernación *f* dentelle — **i** legatura *f* dentelle

4182 deodorization

f désodorisation *f* — **d** Geruchlosmachung *f*; Geruchlosmachen *n* — **n** reukloos maken *n* — **e** hacer inodoro — **i** deodorizzazione *f*

4183 deoxidize *v* — To remove oxygen from a compound or a molten metal.

f désoxyder; décaper — **d** desoxydieren — **n** desoxyderen — **e** desoxidar; desoxigenar — **i** deossidare

4184 deoxidizer — Substance which removes oxygen from a compound or a molten metal.

f désoxydant *m* — **d** Desoxydationsmittel *n* — **n** desoxydatiemiddel *n* — **e** desoxidante *m* — **i** disossidante *m*; riducente *m*

4185 department — Separate part of a plant or an industry in which a specific process or function is carried out; bookbinding department; sales department, etc.

f atelier *m*; service *m*; section *f* — **d** Abteilung *f*; Werkabteilung *f* — **n** afdeling *f* — **e** servicio *m*; departamento *m*; sección *f* — **i** reparto *m*

4186 deposit; sediment — Substance which settles down from a solution or a suspension.

f dépôt *m*; sédiment *m* — **d** Niederschlag *m*; Ablagerung *f*; Sediment *n*; Bodensatz *m*; Abschlag *m* — **n** depot *n*; neerslag *m*; sediment *n*; bezinksel *n*; uitzaksel *n*; afzetting *f* — **e** depósito *m*; sedimento *m* — **i** deposito *m*; sedimento *m*

4187 deposit copy — See statutory copy.

4188 depression of blanket — See dent.

4189 depth — See etching depth.

4190 depth dial gauge; depth gauge — Instrument to measure depths.

f indicateur *m* de profondeur — **d** Tiefenmeßuhr *f* — **n** dieptemeetklok *fm*; dieptemeter *m* — **e** indicador *m* de profundidad; profundímetro *m*; medidor *m* a mostrador de profundidad — **i** comparatore *m* a quadrante per le misurazioni di profondità

4191 depth gauge — See depth dial gauge.

4192 depth gauge — See etching-depth meter.

4193 depth microscope — Optical instrument to measure the depth of relief printing plates by ocular focusing.

f microscope *m* pour mesure des creux — **d** Tiefenmeßmikroskop *n* — **n** microscoop *mn* met meetoculair — **e** microscopio *m* medidor (de profundidad) — **i** microscopio *m* per misure di profondità

4194 depth of bite — See etching depth.

4195 depth of focus — Region on either side of the exact focus where quality and sharpness of the image are up to a required standard.

f profondeur *f* de foyer — **d** Tiefenschärfe *f*; Schärfentiefe *f* — **n** scherptediepte *f*; dieptescherpte *f* — **e** profundidad *f* de foco; volumen *m* focal; nitidez *f* de profundidad — **i** profondità *f* focale; profondità *f* di fuoco

4196 depth of page; page depth

f hauteur *f* de page — **d** Seitenhöhe *f*; Satzhöhe *f* — **n** paginahoogte *f* — **e** altura *f* de la página; alto *m* de la página — **i** altezza *f* della pagina

4197 depthometer — See etching-depth meter.

4198 depth variable gravure — See conventional gravure.

4199 derivative — Chemical substance structurally so related to another one as to be theoretically derivable from it, or a substance that can be made from another one in one or more steps.

f dérivé *m*; dérivée *f* — **d** Derivat *n*; Abkömmling *m* — **n** derivaat *n* — **e** derivado *m*; producto *m* derivado — **i** prodotto *m* derivato

4200 dermatitis — Skin disease, characterized by an itching rash, swelling or roughening of the skin, or watery pustules; in the printing industry caused by photographic developers, chromium compounds, and solvents.

f dermatose *f*; maladie *f* de peau — **d** Hautkrankheit *f*; Hautentzündung *f*; Hauterkrankung *f* — **n** eczeem *f*; huiduitslag *m*; huidontsteking *f* — **e** dermitis *f*; dermatitis *f*; inflamación *f* cutánea; inflamación *f* de la piel — **i** dermatite *f*; malatitta *f* della pelle; malatitta *f* cutanea

4201 descender — Part of a type character that extends below the common body size (base line) of a fount of type, such as g, j, p, q, and y. By extension the letter itself.

f longue *f* du bas; jambage *m* descendant; lettre *f* descendante; descendante *f* — **d** Unterlänge *f* des Buchstabens; geschwänzter Buchstabe *m* — **n** staart *m* van de letter; staartletter *fm* — **e** cola *f*; trazo *m* bajo; trazo *m* descendiente; rasgo *m* descendiente; letra *f* con trazo bajo; letra *f* descendente; letra *f* de palo descendente; letra *f* de cola — **i** discendente *m*; asta *f* discendente; carattere *m* discendente

4202 descending figures — See hanging figures.

4203 desensitation — Destruction or diminishing of the sensibility to light of a photographic emulsion without affecting the latent image. Used for development of panchromatic materials in red light.

f désensibilisation *f* — **d** Desensibilisierung *f* — **n** desensibilisatie *f* — **e** desensibilización *f* — **i** desensibilizzazione *f*

4204 desensitize *v*; **etch** *v depr.* — To apply to the lithographic plate a solution of various chemicals (called an etch) to produce a surface on the non-printing areas capable of being wet by water and not by greasy inks and to increase its capacity to retain moisture. The solutions are for zinc gum arabic, chromic acid and phosphoric acid, for aluminium gum arabic and phosphoric acid.

f aciduler; préparer — **d** ätzen — **n** etsen — **e** desensibilizar

4205 desensitizer — In photography, an agent to decrease the colour sensitivity of a photographic emulsion to facilitate development under comparatively bright light, after exposure.

f désensibilisateur *m* — **d** Desensibilisator *m* — **n** desensibilisator *m* — **e** desensibilizador *m* — **i** desensibilizzatore *m*

4206 desiccate *v* — To dry up or cause to dry up.

f dessécher — **d** austrocknen — **n** uitdrogen — **e** desecar — **i** essiccare

4207 desiccator — Tight container with a strong drying agent such as concentrated sulphuric acid or anhydrous calcium chloride.

f dessicateur *m*; exsiccateur *m* — **d** Exsiccator *m*; Trockner *m* — **n** exsiccator *m* — **e** desecador *m* — **i** essiccatore *m*

4208 design — Preliminary sketch or plan for work to be done, used in preparation of dummy. See also layout.

f ébauche *f*; étude *f*; maquette *f* — **d** Entwurf *m*; Skizze *f*; Plan *m*; Musterzeichnung *f*; typographische Aufmachung *f*; Gestaltung *f* (einer Drucksache); Projekt *n* — **n** ontwerp *n*; schets *fm* — **e** dibujo *m*; diseño *m* — **i** abbozzo *m*; bozzetto *m*; progetto *m*; disegno *m*; traccia *f*

4209 designer — See commercial artist.

4210 desk calendar pad — Pad with the date and space for scribbling notes and appointments.

f bloc-notes *m* éphéméride; calendrier *m* éphéméride — **d** Schreibtischkalenderblock *m*; Umlegekalender *m* — **n** omlegkalender *m* — **e** calendario *m* de notas; calendario *m* de bloque de notas — **i** calendario *m* blocco da tavolo

4211 desorption — Removal of adsorbed material from the substance on which it is absorbed. Opposite of absorption.

f désorption *f* — **d** Desorption *f* — **n** desorptie *f* — **e** desorción *f* — **i** deassorbimento *m*

4212 destructive reading; destructive read-out — Type of read-out in which the data are no longer in store after reading.

f lecture *f* destructive — **d** löschendes Lesen *n*; zerstörendes Lesen *n* — **n** uitwissend lezen *n* — **e** lectura *f* destructiva — **i** lettura *f* distruttiva

4213 destructive read-out — See destructive reading.

4214 detachable — Removable.

f démontable; amovible — **d** abnehmbar — **n** afneembaar; demonteerbaar — **e** desmontable; desarmable — **i** smontabile

4215 detail — Defining power of a lens. In originals and reproductions, the minute subdivisions of an image. Good detail indicates accurate portrayal of all subdivisions in their proper tonal value or strength.

f détail *m* — **d** Detail *n*; Detailzeichnung *f* — **n** detailweergave *fm* — **e** detalle *m* — **i** dettaglio *m*

4216 detail exposure — See main exposure.

4217 detail rendering

f rendu *m* du détail — **d** Detailwiedergabe *f* — **n** detailweergave *fm* — **e** reproducción *f* de detalles — **i** resa *f* dei dettagli

4218 detail stop — Aperture in halftone photography to render middletones, as differentiated from shadows and highlights.

f diaphragme *m* des tons moyens — **d** mittlere Blendenöffnung *f* — **n** hoogbelichtings-diafragma *n*; diafragma *n* voor de middentonen — **e** diafragma *m* de los detalles; abertura *f* de media de diafragma — **i** apertura *f* di diaframma media

4219 detergent — Cleaning agent or plate cleaner.

f détergent *m* — **d** Reinigungsmittel *n* — **n** wasmiddel *n*; reinigingsmiddel *n* — **e** detergente *m*; detersivo *m*; detersorio *m* — **i** detergente *m*; detersivo *m*; solvente *m*

4220 determination — Finding of a quantity or value.

f détermination *f*; dosage *m* — **d** Bestimmung *f*; Ermittlung *f* — **n** bepaling *f* — **e** determinación *f* — **i** determinazione *f*

4221 develop *v* — To subject exposed photograph material to a chemical treatment to produce a visible image.

f développer; révéler — **d** entwickeln — **n** ontwikkelen — **e** revelar — **i** sviluppare

4222 developer — Chemical agent or solution to render photographic images visible by reduction of light-affected (exposed) silver particles to a form of metallic silver; any agent or solution for treating a photographic image after exposure.

f révélateur *m* (en sens chimique, une solution réductrice); développateur *m* (une substance réductrice, donc le constituant essentiel) — **d** Entwickler *m* — **n** ontwikkelaar *m* — **e** revelador *m* — **i** rivelatore *m*; sviluppatore *m*; sviluppo *m*

4223 developer streaks — Uneven areas of development of a negative or positive image due to lack of agitation, or improper handling of film during processing.

f trainées *fpl* de développement — **d** Entwicklerstreifen *mpl* — **n** ontwikkelstrepen *fmpl* — **i** striature *fpl* dello sviluppatore

4224 developing bath

f bain *m* de développement; bain *m* révélateur — **d** Entwickelbad *n* — **n** ontwikkelbad *n* — **e** baño *m* revelador; baño *m* de revelado — **i** bagno *m* rivelatore; bagno *m* di sviluppo

4225 developing brush — Brush for the development of an offset printing plate.

f brosse *f* à développer; tampon *m* à développer — **d** Entwicklungsbürste *f*; Bürste *f* — **n** ontwikkelborstel *m*; plaatborstel *m*; borstel *m* — **e** bruza *f* de revelado; cepillo *m* de revelado; cepillo *m* para revelar — **i** pennello *m* per lo sviluppo; tampone *m* per lo sviluppo

4226 developing ink — Greasy liquid applied to plate images in photolithography to protect the image and keep it ink-receptive while the plate is being developed, etched and gummed. On some surface plates used to make the image ink-receptive.

f encre *f* à report; encre *f* à développer — **d** Entwicklungsfarbe *f*; Entwicklerfarbe *f* — **n** ontwikkelinkt *m*; ontwikkelverf *fm depr.* — **e** tinta *f* de revelar; tinta *f* reveladora; tinta *f* para revelar; tinta *f* de reporte — **i** inchiostro *m* per sviluppo; inchiostro *m* prottetivo

4227 developing pad — Usually a pluche-covered wooden block to work the developing solution over the surface of a deep-etch lithographic plate to remove the unhardened image areas.

f patte *f* de développement; tampon *m* à développer — **d** Entwicklungstampon *m* — **n** ontwikkeltampon *m*; ontwikkelborstel *m* — **e** tampón *m* de revelado; muñequilla *f* de revelado — **i** tampone *m* di sviluppo

4228 developing sink

f évier *m* de développement — **d** Entwicklungs-rinnstein *m*; Rinnstein *m* — **n** ontwikkelgoot-steen *m*; gootsteen *m* — **e** pila *f* de revelado; cubeta *f* de revelar — **i** asquaio *m* di sviluppo

4229 developing tank — Plastic or metal light-tight container to process roll or 35 mm film in daylight. The processing solutions can be poured into the tank through a spout in the cover.

f cuve *f* de développement — **d** Entwicklungs-tank *m*; Entwicklungsdose *f* — **n** ontwikkel-tank *m* — **e** tanque *m* de revelado — **i** vasca *f* di sviluppo

4230 developing tray; dish

f cuvette *f* de développement — **d** Entwicklungs-schale *f*; Entwicklerschale *f* — **n** ontwikkelbak *m*; ontwikkelschaal *fm* — **e** cubeta *f* de revelar; cubeta *f* de revelado; cuba *f* de revelar — **i** bacinella *f* di sviluppo

4231 development — Rendering a latent (in-visible) image visible with a reducing agent or developer; in platemaking the removal of soluble (unexposed) particles of bichromated colloid from the image.

f développement *m* — **d** Entwicklung *f* — **n** ontwikkeling *f*; ontwikkelen *n* — **e** revelado *m* — **i** sviluppo *m*

4232 development latitude — Degree of tolerance of a developer system in relation to the variation in contrast of the effective speed of the photographic material.

f latitude *f* de développement — **d** Entwicklungs-spielraum *m* — **n** armslag *m* in de ontwikkeling — **e** latitud *f* de revelado — **i** portata *f* di sviluppo

4233 development time

f temps *m* de développement — **d** Entwicklungs-zeit *f* — **n** ontwikkeltijd *m* — **e** tiempo *m* de revelado — **i** tempo *m* di sviluppo

4234 deviation — Difference between an observation and the mean of all observations.

f déviation *f* — **d** Abweichung *f* — **n** afwijking *f* — **e** desviación *f* — **i** deviazione *f*

4235 deviation of light rays — Bending of rays of light away from a straight line.

f déviation *f* des rayons lumineux — **d** Licht-strahlenlenkung *f* — **n** afwijking *f* van de licht-stralen — **e** desviación *f* de rayos de luz — **i** deviazione *f* di raggi di luce

4236 device; apparatus; appliance — Piece of equipment.

f dispositif *m*; appareil *m*; outil *m* — **d** Vorrichtung *f*; Apparat *m*; Gerät *n* — **n** toestel *n*; apparaat *n*; instrument *n*; inrichting *f* — **e** dispositivo *m*; aparato *m*; mecanismo *m* — **i** dispositivo *m*; apparecchio *m*; meccanismo *m*

4237 device — Emblematic design, used especially as a heraldic bearing.

f devise *f* — **d** Devise *f*; Wappenspruch *m* — **n** devies *n*; wapenspreuk *fm* — **e** divisa *f*; mote *m*; lema *m* — **i** divisa *f*; motto *m*

4238 devil — See willow.

4239 devil, printer's — See printer's devil.

4240 devil's tail — See bar.

4241 devil stick — Beater to mix the glue used in bookbinding.

f agitateur *m* — **d** Rührstock *m*; Rührholz *n* — **n** roerstok *m*; stok *m* om te roeren; roerspaan *fm*; roerstaaf *fm* — **e** agitador *m* — **i** agitatore *m*

4242 dew point — Temperature at which air is saturated with moisture, or at which a gas is saturated with respect to a condensable component.

f point *m* de rosée; point *m* de condensation — **d** Taupunkt *m* — **n** dauwpunt *n* — **e** punto *m* de rocío; punto *m* de condensación — **i** punto *m* di rugiada; temperatura *f* di condensazione

4243 dextrin; artificial gum; British gum; starch gum; vegetable gum — Carbohydrate produced from starch by hydrolysis with dilute acids, enzymes, or dry heat. Intermediate product between starch and the sugars resulting from starch on hydrolysis. Dextrin is an amorphous, white or yellowish powder, soluble in water or dilute alcohol but precipitated by strong alcohol; used as a thickening agent in printing inks, for sizing paper and as an adhesive in the preparation of gummed paper.

f dextrine *f*; fécule *f* soluble; amidon *m* grillé — **d** Dextrin *n*; Klebestärke *f* — **n** dextrine *f* — **e** dextrina *f* — **i** destrina *f*; dextrina *f*

4244 dextroglucose — See dextrose.

4245 dextro-glucose — See glucose.

4246 dextrose; dextroglucose; glucose; grape sugar — Colourless crystals, or white, crystalline or granular powder, soluble in water, slightly soluble in alcohol. Latin: glucosum.
Formula: $C_6H_{12}O_6 \cdot H_2O$.

f dextrose *f*; glucose *m*; sucre *m* de raisin; sucre *m* d'amidon — **d** Dextrose *f*; Glukose *f*; Glykose *f*; Traubenzucker *m*; Stärkezucker *m* — **n** dextrose *fm*; glucose *fm*; glykose *fm*; druivesuiker *m* — **e** dextrosa *f*; glucosa *f*; azúcar *m* de uva — **i** destrosio *m*; glucosio *m*; zucchero *m* d'uva

4247 d-glucose — See glucose.

4248 dia — See diapositive.

4249 diacritic — See diacritical mark.

4250 diacritical mark; diacritic — Mark, point or

sign added or attached to a letter or character, as a cedilla, tilde, circumflex, macron, umlaut, etc., to give it a phonetic value, to indicate stress, etc.

f signe *m* diacritique — **d** diakritisches Zeichen *n*; Aussprache-Zeichen *n* — **n** diacritisch teken *n* — **e** signo *m* diacrítico; acento *m* diacrítico; señal *f* diacrítica; marca *f* diacrítica; símbolo *m* diacrítico; punto *m* diacrítico — **i** segno *m* diacritico

4251 diaeresis; dieresis — Accent (¨) placed over the second of two vowels to show that they are to be pronounced separately, e.g., naïve, omitted when the vowels are separated by a hyphen at the end of a line. See also Umlaut.

f diérèse *f*; tréma *m* — **d** Diärese *f*; Trema *n* — **n** diaeresis *f*; trema *n*; deelteken *n*; jeris *fm coll.* — **e** diéresis *f*; acento *m* diéresis; crema *f* — **i** dieresi *f*

4252 diagnostic program; diagnostic routing — Program to locate a fault in the equipment or an error in programming.

f programme *m* diagnostic — **d** Diagnostik-programm *n*; Diagnoseprogramm *n* — **n** diagnostisch programma *n* — **e** programa *m* diagnóstico — **i** programma *m* diagnostico

4253 diagnostic routing — See diagnostic program.

4254 diagonal — See solidus.

4255 diagonal cutter — See angle cutter.

4256 diagonal indentation — Stepwise arrangement of type lines, the first line beginning at the left, the following lines equally indented.

f renfoncement *m* en escalier — **d** schräger Einzug *m* — **n** trapsgewijze inspringing *f* — **e** sangría *f* diagonal; sangría *f* escalonada — **i** rientranza *f* diagonale

4257 diagram; graph — Drawing showing the relation between two varying values.

f diagramme *m*; graphique *m*; schéma *m* — **d** Diagramm *n*; Darstellung *f*; Schaubild *n*; Plan *m*; graphische Darstellung *f* — **n** diagram *n*; grafiek *f*; grafische voorstelling *f* — **e** diagrama *m*; gráfico *m* — **i** diagramma *m*; grafico *m*

4258 diagram of connection; wiring diagram

f schéma *m* de branchement; schéma *m* électrique; schéma *m* de câblage; plan *m* de câblage — **d** Leitungsschema *n*; Schaltschema *n*; Schaltplan *m*; Schaltbild *n* — **n** schakelschema *n*; bedradingsschema *n* — **e** esquema *m* de conexiones; dibujo *m* de conexiones; esquema *m* de cableado — **i** schema *m* di connessione; schema *m* di cablaggio

4259 dial — Face of a clock or a metering device marking speed, pressure, etc., of the machinery to which it is attached.

f cadran *m* — **d** Skala *f* — **n** schaal *fm*; wijzerplaat *fm* — **e** esfera *f*; cuadrante *m*; disco *m* — **i** quadrante *m*; scala *f*

4260 diameter of the impression cylinder

f diamètre *m* du cylindre d'impression — **d** Druckzylinderdurchmesser *m* — **n** diameter *m* van de drukcilinder — **e** diámetro *m* del cilindro

impresor — **i** diametro *m* del cilindro d'impressione

4261 diaminophenol chlorhydrate — See diaminophenol hydrochloride.

4262 diaminophenol hydrochloride; diaminophenol chlorhydrate; amidol *trade name* — Greyish-white crystals, soluble in water, slightly soluble in alcohol. Used as a photographic developer.

Formula: $C_6H_3(NH_2)_2OH \cdot 2HCl$.

f diaminophénol chlorhydrate *m*; chlorhydrate *m* de diaminophénol; amidol *m* — **d** Diaminophenol-hydrochlorid *n*; Diaminophenolchlorhydrat *n*; salzsaures Diaminophenol *n*; Amidol *m* — **n** diaminofenol-hydrochloride *n*; amidol *n*; dolmi *n* — **e** hidrocloruro *m* de diaminofenol; amidol *m*; clorhidrato *m* de diaminofenol — **i** idrocloruro *m* di diamminofenolo; amidol *m*

4263 diamond — Old name for a size of type. See app. no. 1.

4264 diamond — A British size of card. See app. no. 3.

4265 diamond — Term denoting imperfect but usable paper. See retrees.

4266 diamond — See retrees.

4267 diamond dash — See French rule.

4268 diamond shaped — See lozenge shaped.

4269 diamond tool — Turning tool, the cutting tip of which is a diamond, for fine surfacing of gravure cylinders.

f outil *m* de diamant — **d** Drehdiamant *m*; Diamantwerkzeug *n* — **n** diamant *n* — **e** diamante *m* de corto — **i** punta *f* di diamante

4270 diaper design — Design of a book cover decoration, tooled with a small, repeating all-over pattern.

f dessin *m* à répétition; dessin *m* courant — **d** repetierendes Muster *n*; repetierendes Rautenmuster *n* — **n** repeterend motief *n*; repeterend ruitmotief *n* — **e** estampa *f* (losanjeada) en repetición — **i** ornamento *m* a ripetizione

4271 diaphanic paper — See vitrauphanie paper.

4272 diaphanous ink

f encre *f* diaphane — **d** Diaphaniefarbe *f* — **n** vitrauphanie-inkt *m* — **e** tinta *f* para diáfanos — **i** inchiostro *m* per vetrofanie

4273 diaphragm; stop — Lens aperture.

f diaphragme *m* — **d** Diaphragma *n*; Blende *f*; Kamerablende *f* — **n** diafragma *n*; blende *f depr.*; stop *m depr.* — **e** diafragma *m*; agujero *m*; stop *m* — **i** diaframma *m*

4274 diaphragm control — Device for process cameras to co-ordinate all optical factors of line and halftone photography, and to indicate correct apertures for any camera extension and screen ruling.

f commande *f* de diaphragme — **d** Blendenkontrollsystem *m*; Blendeeinrichtung *f* — **n** diafragmaregeling *f* — **e** mando *m* del diafragma; regulador *m* del diafragma — **i** comando *m* del diaframma

4275 diaphragm indicator — Pointer or device

attached to the iris diaphragm of process lenses to ascertain the exact diameter of the aperture.

f indicateur *m* de diaphragme — **d** Blendeneinstellscheibe *f* — **n** diafragmainstelring *f* — **e** indicador *m* de la abertura del diafragma — **i** indicatore *m* dell'apertura di diaframma

4276 diaphragm plane
f plan *m* du diaphragme — **d** Blendenebene *f* — **n** diafragmavlak *n* — **e** plano *m* de diafragma — **i** piano *m* del diaframma

4277 diaphragm scale — See lens scale.

4278 diapositive; dia — Photographic transparency or image to be viewed and reproduced by transmitted light. The terms "transparent positive" or "negative-positive" are erroneous.
f diapositive *f*; diapositif *m* — **d** Diapositiv *n*; Bilddiapositiv *n*; Durchsichtsbild *n* — **n** diapositief *n*; dia *m* — **e** diapositiva *f* (se refiere a color); diapositivo *m* (se refiere a negro y en tono continuo); transparencia *f* positiva — **i** diapositiva *f*; diapositivo *m*

4279 diary — Book with printed dates on blank leaves for a daily record of events and appointments.
f agenda *m* — **d** Notizkalender *m*; Kalender *m* — **n** agenda *fm* — **e** diario *m*; carnet *m*; libreta *f* de apuntes; libro *m* de apuntes; libro *m* de memoria; agenda *f* — **i** diario *m*

4280 diatomaceous earth; diatomite; infusorial earth; fossil meal; fossil flour; kieselguhr; siliceous earth; terra silicea; tripolite — Soft early rock composed of the siliceous skeletons of small aquatic plants called diatoms. Resists all acids but hydrofluoric and is slowly dissolved by hot caustic alkali. Used as a filler in paper and as a powder for the cleaning of gravure cylinders.
f diatomite *f*; terre *f* de diatomées; terre *f* pourrie (d'Angleterre); terre *f* d'infusoires; terre *f* à infusoires; farine *f* fossile; kieselguhr *m*; tripoli *m* (silicieux) — **d** Diatomit *m*; Diatomeenerde *f*; Infusorienerde *f*; Kieselguhr *f*; Kieselerde *f*; Bergmehl *n*; Polierschleifer *m*; Tripel *m*; Tripelerde *f* — **n** diatomiet *n*; diatomeeënaarde *fm*; infusoriënaarde *fm*; kiezelgoer *n*; tripel *n* — **e** diatomita *f*; tierra *f* de diatomeas; tierra *f* de infusorios; tierra *f* infusoria; trípoli *m*; tripol *m*; quiselgur *m* — **i** diatomite *f*; farina *f* fossile; trippoli *m*

4281 diatomite — See diatomaceous earth.

4282 Diatype; Berthold film setter — Film setting device of the Berthold type foundry, Berlin, which, via a typedisk with 190 characters, provides a range of type faces in sizes from 4.5-38 point; the speed is limited to about 50 characters per minute.

4283 diazo compound — Light-sensitive compound applied to paper, plastic, or metal sheets. Requires exposure to intense light. Normally develops as an autopositive print with ammonia fumes or by damp development. Used for proofs of positive flats, for colourpack natural colour proofs, and as a coating for presensitized press plates.

f composé *m* diazoïque; diazoïque *m* — **d** Diazoverbindung *f*; Diazokörper *m* — **n** diazoverbinding *f* — **e** compuesto *m* de diazo; compuesto *m* diazoico — **i** composto *m* diazoico

4284 diazo paper; diazotype paper; dyeline paper; direct positive paper — Wet-strength paper with a light-sensitive layer to multiply text or drawings.
f papier *m* diazo — **d** Diazopapier *n*; Lichtpauspapier *n* — **n** diazopapier *n* — **e** papel *m* de compuesto diazoico; papel *m* para diazotipia; papel *m* diazo — **i** carta *f* diazo

4285 diazoprint; white print; B & W — Slow-speed direct positive photographic sensitizer applied to paper or film material, to obtain positive proofs in black or in single colour from positives. The diazo dye used is decomposed by exposure to an intense (arc) light. The unexposed design areas develop when treated with ammonia fumes or an alkaline solution. See also ammonia process, direct positive.
f copie *f* diazotypique — **d** Lichtpause *f* — **n** diazokopie *f* — **e** copia *f* a diazo — **i** diazocopia *f*

4286 diazotype — Print produced on paper treated with a diazo compound that disintegrates upon exposure to light and developing the unexposed areas by the use of diazo dyes.
f diazotypie *f* — **d** Diazotypie *f* — **n** diazotypie *f* — **e** diazotipia *f*; diazotipo *m* — **i** diazotipia *f*

4287 diazotype paper — See diazo paper.

4288 dibasic — Containing two replaceable or ionizable hydrogen atoms, (dibasic acid), or having two univalent, basic atoms, as dibasic sodium phosphate, Na_2HPO_4.
f dibasique; bibasique — **d** zweibasisch — **n** tweebasisch — **e** dibásico; bibásico — **i** bibasico

4289 dibasic ammonium citrate — See ammonium citrate.

4290 dibutyl phthalate — Colourless, non-volatile, non-toxic, stable, oily liquid. Miscible with the common organic solvents, slightly soluble in water. Used in paper coatings, printing inks, synthetic resins, adhesives, plastics, etc.
Formula: $C_6H_4(CO_2C_4H_9)_2$.
f phtalate *m* de butyle — **d** Dibutylphthalat *n* — **n** dibutylftalaat *n* — **e** ftalato *m* de butilo — **i** dibutilftalato *m*

4291 diced leather — Leather tooled with cube-like figures, for book covers.
f cuir *m* losangé; cuir *m* en losange — **d** Rautenleder *n* — **n** geruit leer *n* — **e** piel *f* losanjeada — **i** cuoio *m* con motivo a losanga

4292 dichloroacetic acid; bichloroacetic acid; dichloroethanoic acid — Colourless liquid, soluble in water, alcohol and ether.
Formula: $Cl_2CH \cdot COOH$.
f acide *m* dichloracétique — **d** Dichloressigsäure *f* — **n** dichloorazijnzuur *n* — **e** ácido *m* bicloroacético — **i** acido *m* bicloroacetico

4293 dichloro-diphenyl-trichloroethane — See DDT.

4294 dichloroethanoic acid — See dichloroacetic

acid.

4295 dichroic fog; dichroitic fog; green fog
f voile *m* dichroïque — **d** dichroitischer Schleier *m*; zweifarbiger Schleier *m* — **n** dichroïtische sluier *m*; tweekleurige sluier *m* — **e** velo *m* dicroico; velo *m* de doble coloración — **i** velo *m* dicroico

4296 dichroism — Phenomenon in which a secondary source shows a marked change in hue with change in the observing conditions. Instances are: 1. change in colour temperature of the illuminant; 2. change in concentration of an absorbing material; 3. change in thickness of an absorbing layer; 4. change in direction of illumination or viewing; and 5. change in conditions of polarization.
f dichroïsme *m* — **d** Dichroismus *m* — **n** dichroïsme *n* — **e** dicroísmo *m* — **i** dicroismo *m*

4297 dichroitic fog — See dichroic fog.

4298 dichromate; bichromate — Compound with two chromium atoms with a valency of six. In printing processes an orange-coloured aqueous solution of potassium dichromate (formula: $K_2C_2O_7$), or (rarely) ammonium dichromate which, when combined with gelatine or albumin produces a light-sensitive coating used as a resist in photoengraving and for offset plates, as well as for dye transfer, collotype, etc.
f bichromate *m* — **d** Bichromat *n* — **n** bichromaat *n* — **e** bicromato *m*; dicromato *m* — **i** bicromato *m*

4299 dichromated albumen; bichromated albumen
f albumine *f* bichromatée — **d** Eiweißbichromat *n* — **n** eiwitbichromaat *n* — **e** albúmina *f* bicromatada — **i** albumina *f* bicromatata

4300 dichromated coating; bichromated coating — Coating of a dichromated sensitizer used in making printing plates.
f couche *f* bichromatée — **d** Bichromatschicht *f* — **n** bichromaatlaag *fm* — **e** capa *f* bicromatada — **i** strato *m* bicromatato

4301 dichromated colloids; bichromated colloids — Various substances (albumen, glue, gum arabic, shellac) used as vehicles in photoengraving, offset platemaking, etc., and rendered sensitive in solution by the addition of ammonium dichromate.
f colloïdes *mpl* bichromatés — **d** bichromatete Kolloide *npl*; Bichromatkolloide *npl* — **n** gebichromateerde colloïden *fmnpl* — **e** coloides *mpl* bicromatados — **i** colloidi *mpl* bicromatati

4302 dichromated gelatine; bichromated gelatine — Gelatine sensitized to light by immersion in a solution of ammonium, potassium, or sodium dichromate. Light has a tanning effect on dichromated gelatine and makes the affected parts insoluble in hot water.
f gélatine *f* bichromatée; gélatine *f* chromatée — **d** Chromgelatine *f*; mit Bichromat sensibilisierte Gelatine *f* — **n** chroomgelatine *fm*; gebichromateerde gelatine *fm* — **e** gelatina *f* bicromatada; gelatina *f* cromatada — **i** gelatina *f* bicromatata; gelatina *f* al bicromato

4303 dichromated glue; bichromated glue — Glue for the glue deep-etch platemaking process in lithography. Process still used in Europe and Far East.
f colle *f* bichromatée — **d** Bichromatleim *m* — **n** gebichromateerde lijm *m* — **e** cola *f* bicromatada — **i** colla *f* bicromatata; colla *f* al bicromato

4304 dichromated gum; bichromated gum — Gum arabic made light sensitive with dichromate.
f gomme *f* bichromatée — **d** Chromgummi *n* — **n** gebichromateerde gom *m* — **e** goma *f* bicromatada — **i** gomma *f* bicromatata; gomma *f* al bicromato

4305 dichromate poisoning — See chrome poisoning.

4306 dictionary — Book with an alphabetically arranged list of words with their meanings.
f dictionnaire *m* — **d** Wörterbuch *n*; Lexikon *n* — **n** woordenboek *n* — **e** diccionario *m*; léxico *m* — **i** dizionario *m*

4307 dictionary — Computer term. Method of *hyphenation, often in conjunction with rules of logic. When a word needs breaking, a dictionary in store is searched and if the word is included the most appropriate hyphenation point is used to end the line, and the remainder of the word is turned over to the beginning of the next line. A dictionary can also be used to look up and correct spelling errors by automatic data processing means.
f dictionnaire *m* — **d** Verzeichnis *n* — **n** woordenlijst *fm* — **e** diccionario *m* — **i** dizionario *m*

4308 dictionary approach — In computer controlled typesetting, system where the *hyphenation point of a word is obtained from a stored list of prehyphenated words.
f division *f* syllabique au dictionnaire — **d** Wörterbuch-Trennprogramm *n* — **n** afbreking *f* met behulp van een woordenlijst — **e** división *f* silábica en el diccionario — **i** divisione *f* sillabica nel dizionario

4309 Didones — See Didonics.

4310 Didonics — Group of type faces with an abrupt contrast between thin and thick. The axis of the curves is vertical; the serifs of the ascenders of the lower-case are horizontal and there are often no brackets to the serifs. They include the Didot types and the Bodoni types. They are *Transitionals with rectilinear serifs, to which belong Egmont, Promotor, Orator, Arsis, Juno, Visite, Gras Vibert, Walbaum, Corvinus, Normande. Formerly called Modern. Before 1971 this group was named Didones. See also classification of type design.
f Didones *fpl* — **d** klassizistische Antiqua *f* — **n** Didonen *fmpl* — **e** Didodianos *mpl* — **i** Bodoniani *mpl*

4311 Didot system — System of type measurement (about 1775) used generally throughout the continent of Europe, originating from François

Ambroise Didot (1730-1804), and his son Firmin. Based on the "pied du roi" (King's foot or French foot). In 1879 the German Hermann Berthold was charged by the German type founders to improve this system and to bring it into accordance with the metre. He pointed out that one point Didot was 1/2660 of a metre. The attempts of Firmin Didot (1811) and others to make a decimal system out of the Didot system were a complete failure. See app. no. 1, point system.

f système *m* Didot — **d** Didotsches System *n*; Didot-System *n*; französisches System *n* — **n** Didot-systeem *n* — **e** sistema *m* Didot — **i** sistema *m* Didot

4312 die — Design, letters or pattern cut in metal, mostly brass, for stamping book covers or embossing.

f matrice *f* de timbrage — **d** Prägestempel *m*; Prägeplatte *f* — **n** stempel *n* — **e** plancha *f* para estampar las tapas; plancha *f* para timbrar las tapas; molde *m* para estampar las tapas — **i** stampo *m* a goffrare; timbro *m* a goffrare

4313 die; cutting die — Sharp edged device, usually made of steel rule, to cut paper, cardboard, etc., on a printing press.

f forme *f* de découpe — **d** Stanzmesser *n*; Stanzform *f*; Bandstahlform *f*; Bandstahlstanzform *f* — **n** stansmes *n*; stansvorm *m*; uitkapvorm *m* Be — **e** troquel *m* parar cortar; troquel *m* de pleca de acero; troquel *m* de fleje de acero; suaje *m* Me — **i** fustella *f*

4314 die case — See matrix case.

4315 die-cut cards — Small cards cut out with steel dies.

f cartes *fpl* découpées à la forme — **d** ausgestanzte Karten *fpl* — **n** uitgestanste kaarten *fmpl* — **e** tarjetas *fpl* troqueladas; tarjetas *fpl* cortadas a troquel — **i** cartoncini *mpl* fustellati

4316 die cutter — See die cutting machine.

4317 die cutting — To cut paperboard, metal, plastic or other material by dies on a percussion press.

f découper à l'emporte-pièce; couper à la forme; stamper — **d** stanzen; ausstanzen — **n** stansen; uitstansen; stampen; uitstampen — **e** sacaboquear; troquelar; cortar a troquel; cortar con troquel; estampar a troquel; estampar con troquel — **i** fustellare

4318 die cutting machine; die cutter

f machine *f* à découper; presse *f* de découpe — **d** Stanzmaschine *f*; Stanzpresse *f*; Stanze *f* — **n** stansmachine *f*; stanspers *f* — **e** máquina *f* para troquelar; prensa *f* para cortar a troquel; troqueladora *f* — **i** macchina *f* fustellatrice

4319 die-cutting rule

f filet *m* à découper; filet *m* de découpe — **d** Stanzlinie *f*; Schneidlinie *f* — **n** stanslijn *fm* — **e** filete *m* para troquelar — **i** filetto *m* per fustellare

4320 dielectric — See dielectricum.

4321 dielectric coefficient — See dielectric constant.

4322 dielectric constant; dielectric coefficient; spe-

cific inductive capacity; relative permittivity — Ratio of the capacitance of a two-plate electrical condenser when the space between the plates is filled with the test sample to the capacitance of the same condenser when the space between the plates is filled with air (evacuated). Of importance in condenser paper and in electrical insulating material.

f constante *f* diélectrique; pouvoir *m* inducteur spécifique — **d** dielektrische Konstante *f*; Dielektrizitätskonstante *f* — **n** diëlektrische constante *fm* — **e** constante *f* dieléctrica; poder *m* específico inductor — **i** costante *f* dielettrica; potere *m* induttore specifico

4323 dielectric paper — Paper free from metallic or other impurities that conduct electricity. Used as a dielectric material.

f papier *m* diélectrique — **d** dielektrisches Papier *n* — **n** diëlektrisch papier *n* — **e** papel *m* dieléctrico — **i** carta *f* dielettrica

4324 dielectric properties

f propriétés *fpl* diélectriques — **d** dielektrische Eigenschaften *fpl* — **n** diëlektrische eigenschappen *fpl* — **e** propiedades *fpl* dieléctricas; características *fpl* dieléctricas — **i** caratteristiche *fpl* dielettriche

4325 dielectric strength — Strength to resist the passage of an electric spark discharge. Specifically, the potential difference at which a spark passes through a medium of specified thickness under specified conditions, usually expressed as a voltage gradient in volts per mil thickness. Not to be confused with dielectric constant (specific inductive capacity).

f rigidité *f* diélectrique — **d** Durchschlagfestigkeit *f* — **n** doorslagvastheid *f*; doorslagsterkte *f* — **e** resistencia *f* dieléctrica — **i** rigidità *f* dielettrica

4326 dielectricum; dielectric — Material characterized by its poor electrical conductivity, an insulator.

f diélectrique *m* — **d** Dielektrikum *n* — **n** diëlektricum *n* — **e** dieléctrico *m* — **i** dielettrico *m*

4327 die making — Assembling and mounting cutting dies in plywood.

f montage *m* des formes de découpe — **d** Stanzformenbau *m* — **n** vervaardiging *f* van de stansvorm — **e** confección *f* de troqueles; confección *f* de suajes *Me* — **i** montaggio *m* delle fustelle

4328 diene number; diene value — Figure denoting the extent to which open-chain compounds are unsaturated.

f valeur *f* diène — **d** Dienzahl *f* — **n** dieengetal *n* — **e** valor *m* en dienos — **i** valore *m* di non-saturazione

4329 diene value — See diene number.

4330 dieresis — See diaeresis.

4331 die sheet; strike sheet — Accurate imprint made from the package manufacturer's dies to enable the package designer to secure accuracy of colour-register and design-register in his art work. Shows the exact location of cutting and

scoring rules and is made by inking the dies and proofing them on a sheet of parchment or other semi-non shrinking surface.
f feuille *f* modèle — **d** Musterbogen *m* — **n** stansvel *n* — **e** hoja *f* de muestra; hoja *f* modelo — **i** foglio *m* modello

4332 diesis — See double dagger.

4333 die stamp
f timbre *m* sec — **d** Trockenstempel *m* — **n** droogstempel *n* — **e** timbre *m* a seco; plancha *f* para estampar en seco — **i** timbro *m* a secco

4334 die stamping; relief stamping — Intaglio process of printing raised letters either coloured or blind (i.e., without colour) by means of engraved steel plates. Ink is smeared over the surface of the die, the surface is wiped off and the ink remaining in the design is printed under heavy pressure, wich also partially embosses the paper.
f impression *f* sur plaque d'acier; impression *f* en timbrage-relief — **d** Stahlstichdruck *m*; Stahldruck *m*; Stahlstich *m*; Prägedruck *m* — **n** staalstempeldruk *m*; reliëfdruk *m*; stempeldruk *m* — **e** estampado *m* en relieve; timbrado *m* en relieve; estampación *f* en acero; estampación *f* en relieve — **i** stampa *f* da lastre d'acciaio incise; stampa *f* incavorilievografica

4335 die stamping ink — See stamping ink.

4336 die stamping press
f presse *f* à timbrer — **d** Stahlstichprägepresse *f*; Stahlstichpresse *f*; Reliefdruckpresse *f* — **n** stempelpers *f*; staalstempelpers *f* — **e** prensa *f* para estampar (con grabados de acero); prensa *f* para timbrar — **i** macchina *f* per stampa con incisione in acciaio; macchina *f* per stampa rilievografica

4337 die-stamping press; stamping machine — In the manufacture of tin cans, a machine to punch circular disks (or other shapes) of tin plate which later are curled on the edge to form the can ends.
f estampeuse *f*; presse *f* à estamper — **d** Stampfwerk *n*; Stampfmaschine *f* — **n** stamppers *fm* — **e** embutidora *f*

4338 diethyl ether — See ether.
4339 diethyl oxide — See ether.
4340 die wipe — See plate-wiping paper.
4341 die-wiping paper — See plate-wiping paper.
4342 differential modulation — Modulation in which the choice of the significant condition for any signal element is dependent on the choice of the previous signal element, e.g., delta modulation.

4343 differential piecework — See also accelerating incentive, geared incentive, multiple time plan.
f salaire *m* aux pièces différentiel — **d** Differentialstücklohn *m* — **n** gedifferentieerd stukloon *n* — **e** jornal *m* a destajo — **i** salario *m* a cottimo differenziato

4344 diffraction — Deviation of light rays from a straight course when partially cut off by an obstacle, or in passage near the edges of a small opening or through a small hole.
f diffraction *f* — **d** Diffraktion *f*; Beugung *f* — **n** diffractie *f*; straalbuiging *f* — **e** difracción *f* — **i** diffrazione *f*

4345 diffraction theory — Theory of halftone photography that dot formations are due to the action of diffracted light.
f théorie *f* de la diffraction — **d** Diffraktionstheorie *f* — **n** diffractietheorie *f* — **e** teoría *f* de difracción — **i** teoria *f* della diffrazione

4346 diffused light
f lumière *f* diffuse; lumière *f* diffusée; lumière *f* parasite — **d** Streulicht *n*; zerstreutes Licht *n*; diffuses Licht *n* — **n** strooilicht *n*; diffuus licht *n* — **e** luz *f* difusa; luz *f* atenuada — **i** luce *f* diffusa; luce *f* parassita

4347 diffused lighting — Soft, indirect lighting by means of light reflected off or transmitted through a white matt material, used to soften shadows which result when illuminating with direct lighting.
f lumière *f* diffuse — **d** indirektes Licht *n* — **n** indirecte verlichting *f*; diffuse verlichting *f* — **e** iluminación *f* difusa — **i** illuminazione *f* diffusa

4348 diffused mask — See unsharp mask.

4349 diffuser; blow pit — Large tank with perforated bottom in which the cooked pulp, blown from the digester, is drained and washed of its spent liquor.
f diffuseur *m*; cuve *f* de lavage; fosse *f* de décharge (par soufflage) — **d** Diffuseur *m* — **n** diffuseur *m* — **e** difusor *m* — **i** diffusore *m*

4350 diffusion of light; light diffusion
f diffusion *f* de la lumière — **d** Diffusion *f* des Lichtes; Lichtzerstreuung *f*; Überstrahlung *f*; Verbreitung *f*; Zerstreuung *f* — **n** lichtdiffusie *f* — **e** difusión *f* de luz — **i** diffusione *f* della luce

4351 digallic acid — See tannic acid.

4352 digester; boiler; fibre boiler — Large vessel in which pulpwood is cooked under pressure with chemicals to eliminate lignin and other constituents and to obtain cellulose fibres. Also for cooking straw and other papermaking material.
f lessiveur *m* — **d** Zellstoffkocher *m*; Kocher *m* — **n** celstofkoker *m* — **e** lejiador *m* (de celulosa); lejiadora *f* — **i** bollitore *m* per fibre

4353 digesting — See cooking.
4354 digesting agent — See boiling agent.
4355 digging out — Removing small surface areas of metal in relief plates manually with tools.
f retouche *f* (des clichés) à l'outil — **d** Ausstechen *n* — **n** uitsteken *n*; wegsteken *n*; metaalretouche *f* — **i** incidere col bulino

4356 digit — Character to represent one of the non-negative integers smaller than the radix, e.g., in decimal notation one of the characters 0 to 9. See also check digit.
f chiffre *m*; digit *m* — **d** Ziffer *f* — **n** cijfer *n*; cijferteken *n* — **e** cifra *f*; dígito *m* — **i** cifra *f*

4357 digit — See fist.

4358 digital — As contrasted with analogue, a method of recording and transmitting information by sensing and conveying either the presence or absence of a condition, usually an electric current, i.e., on or off, positive or negative, rather than to record and transmit more or less current (amplitude and/or frequency). Digital observations are usually made intelligible by grouping them into an array or pattern of "on/off" signals which are expressed in binary terms, and the grouping may consist of a byte or a word.

f digital; numérique — **d** digital; ziffernmäßig — **n** digitaal — **e** digital; numérico — **i** digitale; numerico

4359 digital computer — Computer which operates by using numbers to express all the quantities and variables of a problem, the numbers in turn are expressed by electrical impulses.

f calculateur *m* digital; calculateur *m* numérique; compteur *m* digital — **d** Digitalrechner *m*; Ziffernrechner *m*; digitaler Rechner *m*; digitale Datenverarbeitungsanlage *f*; digitale Rechenanlage *f* — **n** digitale computer *m*; digitale rekenautomaat *m* — **e** computadora *f* digital; calculador *m* numérico — **i** calcolatore *m* digitale; calcolatore *m* numerico; ordinatore *m* digitale

4360 digital plotter — Device to record incremental steps, under program control, by the movement of a pen across and along graph paper, to produce straight and curved lines and symbols.

4361 digital signal — Nominally discontinuous electrical signal that changes from one state to another in discrete steps. It could, e.g., change its amplitude or polarity in response to outputs from computers, teletypewriters, etc.

f signal *m* digital; signal *m* numérique — **d** digitales Signal *n* — **n** digitaal signaal *n* — **e** señal *f* digital; señal *f* numérica — **i** segnale *m* numerico

4362 digital-to-analog — See digital-to-analogue.

4363 digital-to-analogue *GB*; **digital-to-analog** *US* — Electronic circuitry which converts digital information, represented as bytes, bits, or words, into pulses or fluctuations of current which can convey variable or qualitative information. For example, a light bulb may be on or off, and this would be a digital fact. But it could be brighter or dimmer, and this would be an analogue fact. Information can be transferred digitally, even on a voice-grade telephone line, as a serial stream of on/off signals, or in analogue fashion, as varying frequencies representing, for example, the effect of the human voice on an electromagnetic diaphragm. D-to-a or a-to-d circuitry can convert one kind of signal (e.g., a bit pattern) into another (e.g., varying intensities or frequencies), or vice versa.

f digital-analogique; numérique-analogique — **d** Digital-Analog — **n** digitaal-analoog — **e** digital-analógico; numérico-analógico — **i** digitale-analogico; numerico-analogico

4364 digit *v* **a word** — To place a word between inverted commas.

f guillemeter — **d** unterführen — **n** tussen aanhalingstekens plaatsen — **e** poner entre comillas — **i** mettere tra virgolette

4365 digitized information — Text or graphic material can be input, processed and output, as well as recorded and transmitted, as codes or signals. In some instances this information is converted into "characters" while in other cases it simply consists of observations as to the presence or absence of a black dot at the point of observation. Such digital data may sometimes be compacted for economy in storage or transmission.

4366 digitizer — Device to obtain from an analogue representation of a physical quantity of digital representation of the value of the quantity.

f codeur *m* digital; codeur *m* numérique — **d** Analog-Digital-Umsetzer *m*; Zifferndarsteller *m* — **n** analoog-digitaal omzetter *m*; nummeraar *m* — **e** codificador *m* digital; codificador *m* numérico — **i** digitalizzatore *m*; numerizzatore *m*

4367 dihydroxysuccinic acid — See tartaric acid.

4368 dilatancy — In rheology, the flow property of suspensions in which the resistance to flow increases at a greater rate than the increase in the rate of flow. In general the property of colloidal solutions to become solid, or set under pressure.

f augmentation *f* de la consistance — **d** Verfestigung *f* — **n** consistentievergroting *f* — **e** aumento *m* de consistencia; dilatancia *f* — **i** dilatanza *f*

4369 dilatant; shear thickening — Having the property of increasing in apparent viscosity with increasing shear. Dilatant materials are particle suspensions which are deflocculated and are in a settled state of minimum voids. An attempt to make such a system flow dilates the voids and thereby increases the resistance to flow. Also the material takes on a dry appeerence when pressure is applied to it.

f dilatant — **d** dilatant — **n** dilatant — **e** dilatante; dilatador — **i** dilatante

4370 dilatation — Increase in volume per unit volume of a continuous material.

f dilatation *f* — **d** Ausdehnung *f* — **n** dilatatie *f*; uitzetting *f* — **e** dilatación *f* — **i** dilatazione *f*

4371 diluent — See reducer.

4372 dilute *v* — To weaken or thin by the addition of a liquid.

f diluer; allonger; étendre — **d** verdünnen (Flüssigkeit); verschneiden (Farbe im Tiefdruck); strecken — **n** verdunnen; versnijden — **e** diluir (en agua); rebajar (un color) — **i** diluire

4373 diluting agent — See reducer.

4374 dilution ratio — A measure of amount of non-solvent that can be added to a solution with nitrocellulose, synthetic resin or other solute before precipitation occurs.

f tolérance *f* de dilution — **d** Verdünnungsverhältnis *n*; Verschnittfähigkeit *f* — **n** verdunning *f*; verdunningsverhouding *f* — **e** grado *m* de dilución — **i** tolleranza *f* alla diluizione

4375 dimension — Width and height of the final

reproduction.

f dimension *f*; format *m* — **d** Abmessung *f*; Größe *f*; Ausmaß *n* — **n** afmeting *f*; formaat *n*; maat *fm* — **e** dimensión *f*; medida *f*; tamaño *m*; formato *m* — **i** dimensione *f*; formato *m*

4376 dimensional stability; hygro-instability — Ability to maintain size; resistance of paper or photographic film to dimensional change with change in moisture content.

f stabilité *f* dimensionnelle — **d** Maßhaltigkeit *f*; Dimensionsstabilität *f*; Maßbeständigkeit *f* — **n** maatvastheid *f*; dimensionele stabiliteit *f* — **e** estabilidad *f* dimensional — **i** stabilità *f* dimensionale

4377 dimension marks — Points on an original outside image area to be reproduced, between which size of reproduction is marked and focussing performed.

f repères *mpl* de dimension — **d** Bildmaßzeichen *npl* — **n** beeldmaten *fmpl*; tekens *npl* voor de beeldmaat — **e** marcas *fpl* de dimensión; marcas *fpl* de tamaño — **i** segni *mpl* di dimensioni

4378 dimethylbenzene — See xylene.

4379 dimethyl carbinol — See isopropyl alcohol.

4380 dimethyl ketone — See acetone.

4381 diminishing glass — See reducing glass.

4382 dimmed lighting — In photography, reduced lighting of part of the subject to obtain contrast.

f éclairage *m* réduit — **d** abgeblendete Beleuchtung *f* — **n** gedeeltelijk afgeschermd licht — **e** luz *f* atenuada — **i** illuminazione *f* ridotta

4383 dingbat — American term for a small picture or vignette, used as an ornament. See ornament.

4384 dinky reel — See dinky sheet.

4385 dinky sheet; dinky reel — Web considerably narrower than the full width of the press. It produces usually two standard or four tabloid pages.

f bobine *f* petit format; bobine *f* étroite — **d** teilbreite Papierbahn *f* — **n** smalle papierbaan *fm*; bijloper *m* — **e** bobina *f* de papel de una (sola) página; bobina *f* de ancho reducido — **i** bobina *f* di piccolo formato; bobina *f* stretta

4386 DIN sizes — See international sizes.

4387 diols — See glycols.

4388 diopter — See dioptre.

4389 dioptre; diopter — Unit of power of a lens, the reciprocal of its focal length in metres, the power of a divergent lens being given a negative sign. Abbrev. dpt. In the SI: 1 dpt = 1 m^{-1}.

f dioptrie *f* — **d** Dioptrie *f* — **n** dioptrie *f* — **e** dioptría *f* — **i** diottria *f*

4390 dip coated paper — Paper that has been given a single or double sided coating by the dip coating process.

f papier *m* couché au trempé; papier *m* couché par trempage; papier *m* couché par immersion — **d** im Tauchstreichverfahren gestrichenes Papier *n*; im Tauchbad gestrichenes Papier *n* — **n** in dompelbad gestreken papier *n* — **e** papel *m* estucado por inmersión — **i** carta *f* patinata per immersione

4391 dip coating; immersion coating — Coating a continuous paper web by passing the web round a roll immersed in a pan of coating slip; may be a *machine coating or a separate coating operation.

f couchage *m* par immersion; couchage *m* par trempage — **d** Tauchstreichverfahren *n*; Tauchbeschichtung *f* — **n** dompelstrijkmethode *f*; dompelstrijkprocédé *n*; dompelcoating *m depr.* — **e** procedimiento *m* de estucado por inmersión; aprestado *m* por inmersión — **i** patinatura *f* per immersione

4392 diphthong — Two jointed characters cast on the same body (or in computer controlled typesetting a combination of two signs in one graphic) as in the case with ligatures, except that diphthongs occur in certain words, usually of Greek origin. Generally no ligatures are used in US. See also French alphabet.

f diphtongue *f* — **d** Diphthong *m*; Zweilaut *m*; Doppellaut *m*; Doppelvokal *m* — **n** diftong *fm*; tweeklank *m* — **e** diptongo *m* — **i** dittongo *m*

4393 diple — Critical mark, resembling a V. It is placed horizontally as (<), the opening to the right. It is used in old manuscripts to indicate a quotation from the Bible or Scriptures, etc. Predecessor of quotation marks.

f diple *m*; antilambda *m* — **d** Anführungszeichen *n* bei Bibelzitaten — **n** horizontaal V-teken *n* — **e** signo *m* de una V horizontal — **i** segno *m* simile ad una V orrizontale

4394 dipper — See vatman.

4395 dipping; immersion — Immersing photographic plates or images in chemical solutions. In papermaking, surface dyeing or colouring paper by passing it through a colour bath.

f immersion *f*; trempage *m* — **d** Eintauchung *f*; Eintauchen *n* — **n** onderdompeling *f*; indoping *f* — **e** inmersión *f* — **i** immersione *f*

4396 dipping roller

f rouleau *m* barboteur; barboteur *m* — **d** Tauchwalze *f* — **n** dompelrol *fm* — **e** rodillo *m* inmersor — **i** rullo *m* d'immersione; rullo *m* pescatore

4397 direct access — See random access.

4398 direct access storage — See direct access store.

4399 direct access store *GB*; **direct access storage** *US*; **immediate access storage** *US* — Store with insignificant switching time.

f mémoire *f* à accès direct — **d** Speicher *m* mit unmittelbarem Zugriff — **n** direct toegankelijk geheugen *n* — **e** memoria *f* de acceso directo — **i** memoria *f* d'accesso diretto

4400 direct current *DC*; **continuous current** — Supply or voltage in which one side of the supply is always positive and the other side always negative.

f courant *m* continu — **d** Gleichstrom *m* — **n** gelijkstroom *m* — **e** corriente *f* continua — **i** corrente *f* continua

4401 direct current motor; DC motor

f moteur *m* à courant continu; moteur *m* à CC — **d** Gleichstrommotor *m* — **n** gelijkstroommotor *m* — **e** motor *m* de corriente continua — **i** motore *m* a corrente continua

4402 direct dyes; substantive dyes — Class of aniline dyes, so named because of their high affinity for cellulose. While direct and acid dyes are both sodium salts of dye acids, the former do not require a mordant such as size and alum for efficient retention in stock dying, and hence useful in colouring unsized papers such as blotting, facial, and sanitary tissues. As a class they have less tinctorial value than basic dyes and are duller in shade than either basic or acid types. In general, they are much faster to light than the basic dyes and, in some cases, than the acid dyes.
f teintures *fpl* directes; colorants *mpl* directs; couleurs *fpl* substantives — **d** Direktfarbstoffe *mpl* — **n** directe verfstoffen *mpl*; directe kleurstoffen *mpl* — **e** colorantes *mpl* directos — **i** coloranti *mpl* diretti

4403 direct halftone negative — Halftone negative made by direct exposure of an object through a halftone screen.
f négatif *m* tramé direct — **d** direktes Rasternegativ *n* — **n** direct rasternegatief *n* — **e** negativo *m* de medio tono directo; negativo *m* reticulado directo; negativo *m* tramado directo — **i** negativo *m* retinato direttamente

4404 direct halftone process — Production of colour separation negatives from originals directly through colour filters and halftone screen in one exposure for each colour.
f sélection *f* directe — **d** direkter Rasterauszug-Prozess *m* — **n** directe rastermethode *f* — **e** procedimiento *m* de medio tono directo; procedimiento *m* tricromático directo; procedimiento *m* de selección y tramado simultáneos — **i** procedimento *m* di retinatura diretta; selezione *f* e retinatura simultanee

4405 direct-image plate — Paper or thin metal plate which may be typed on or drawn on directly.

4406 direct lighting — Lighting that is not reflected, diffused or refracted. In photography it gives sharpness of details.
f éclairage *m* direct — **d** direkte Beleuchtung *f* — **n** directe verlichting *f* — **e** alumbrado *m* directo — **i** illuminazione *f* diretta

4407 direct lithography — See lithography.

4408 directory — Book containing an alphabetical index, the names and addresses of persons, organizations, etc.
f répertoire *m* d'adresses; bottin *m*; annuaire *m* (des téléphones) — **d** Adreßbuch *m*; Adressenverzeichnis *n*; Telephonadreßbuch *n*; Telephonverzeichnis *n* — **n** adresboek *n*; adressengids *m*; telefoongids *m*; telefoonboek *n*; naamlijst *fm* — **e** guía *f* telefónica; guía *f* de direcciones — **i** libro *m* degli indirizzi; guida *f* commerciale; elenco *m* telefonico; indirizzario *m*

4409 direct photography — Making of halftone images by photographing the object to be reproduced rather than by re-photographing a continuous-tone picture of the object; frequently used in the reproduction of jewelry, cutlery, shoes, textiles, etc.
f photographie *f* directe — **d** Direktaufnahme *f* — **n** directe rasteropname *f*; directe rasterfotografie *f* — **e** fotografía *f* directa — **i** fotografia *f* diretta

4410 direct positive; autopositive *trade name* — Photographic film or paper material that provides a positive from a positive by single development after contact exposure to an intense yellow filtered light. By altering exposure techniques, the material can produce a negative from a positive, outline characters from block type, etc. See also diazoprint.
n direct positief *n*; autopositief *n* — **e** autopositiva *f*; autopositivo *m* directo; positivo *m* de inversión — **i** pellicola *f* autopositiva

4411 direct positive paper — See diazo paper.

4412 direct separation negative; screen separation negative — Separation negative made with halftone screen from copy.
f négatif *m* de sélection *f* directe — **d** Direktauszug *m*; Rasterauszug *m* — **n** direct deelnegatief *n*; direct rasternegatief *n* — **e** negativo *m* de selección directo; negativo *m* de selección reticulado; negativo *m* de citocromía directo — **i** negativo *m* di selezione diretta; negativo *m* di separazione diretta

4413 dirty-bottom etching — See ragged etching.

4414 dirty colour — Colour containing primary colours with black.
f couleur *f* rabattue — **d** gebrochene Farbe *f*; unreine Farbe *f* — **n** gebroken kleur *fm*; vuile kleur *fm* — **i** colore *m* non puro

4415 dirty proof — Proof marked with many corrections and alterations, or the first impression sent to the proofreader. See also reader's proof, first proof.
f épreuve *f* chargée — **d** Saukorrektur *f*; erste Korrektur *f* — **n** vuile proef *f* — **e** prueba *f* sucia; prueba *f* llena de erratas; prueba *f* llena de errores; prueba *f* llena de moscas; prueba *f* plegada de erratas; prueba *f* mentirosa; prueba *f* moteada *Ch*

4416 dis *v* — See distribute *v*.

4417 disc — See disk.

4418 discards — Soiled or defective printed sheets which are set aside.
f déchets *mpl* — **d** Ausschuß *m*; Makulatur *f*; Abfall *m* — **n** uitschot *n*; misdrukken *mpl*; afval *n* — **e** papeles *mpl* de deshecho; recortes *mpl* — **i** scarti *mpl*; maculature *fpl*

4419 discharge lamp — Lamp in which light is produced by an electric discharge through a gas.
f lampe *f* à décharge gazeuse — **d** Entladungslampe *f* — **n** ontladingslamp *fm*; gasontladingslamp *fm* — **e** lámpara *f* de descarga; lámpara *f* de destello — **i** lampada *f* a elettroluminescenza

4420 discoloration; discolouring — See also

burnout.

f décoloration *f*; changement *m* de couleur — **d** Verfärbung *f*; Entfärbung *f* — **n** verkleuring *f* — **e** descoloramiento *m*; decoloramiento *m* — **i** scolorimento *m*

4421 discoloured; faded

f décoloré; pâli; déteint — **d** verschossen; verblaßt; verfärbt — **n** verschoten; verbleekt; verkleurd — **e** descolorado; palidecido — **i** smorto; sbiadito

4422 discolouring — See discoloration.

4423 discontinued; not continued

f cessé de paraître — **d** eingegangen; nicht fortgesetzt; erscheint nicht mehr — **n** uitgave gestaakt; verschijnt niet meer; opgehouden te verschijnen — **e** discontinuado — **i** non più pubblicato

4424 discontinue v publication

f cesser de paraître; arrêter la publication — **d** das Erscheinen einstellen — **n** ophouden te verschijnen; de verschijning staken — **e** discontinuar la publicación — **i** interrompere la pubblicazione

4425 discount — Fixed rate off the retail, or list, price of a book which the publisher allows the bookseller or wholesaler.

f rabais *m* — **d** Abzug *m*; Rabatt *m* — **n** korting *f*; boekhandelskorting *f*; uitgeverskorting *f*; reductie *f* — **e** descuento *m*; rebaja *f* — **i** sconto *m*; ribasso *m*; riduzione *f*

4426 discount granted to colleagues

f remise *f* à titre de confrère — **d** Kollegenrabatt *m* — **n** collegiale korting *f* — **e** descuento *m* para colegas — **i** sconto *m* per colleghi

4427 discount price — See net price.

4428 discretionary hyphen — System hyphen inserted in long words by a keyboard operator producing unjustified data, or inserted by program, to indicate possible hyphenation points. During the subsequent computer run appropriate discretionary hyphens are used to terminate justified lines and the others are discarded. The discretionary hyphen and the hyphen used to connect two words or parts of a word are allocated different codes on input and a common code at the output stage. This method of hyphenation under operator control is used in some comparatively simple typesetting systems and does not require the continuous use of a display system or a hyphenation program.

4429 dish — See developing tray.

4430 dis hand — See distributor.

4431 dished — With a symmetrical, concave depression in the surface. Opposite of *domed.

f recoquillé; en cuvette — **d** tellerförmig; schüsselförmig — **n** schotelvormig — **e** en forma de plato; en forma de platillo — **i** accartocciato; deformato a coppa; imbarcato

4432 dished paper — Paper in sheets with a concave form due to atmospheric conditions.

f papier *m* recoquillé — **d** tellerförmiges Papier *n*; Papier *n* mit Randschrumpf — **n** schotelvormig papier *n*; papier *n* met een "zak" — **e** papel *m* en forma de plato; papel *m* en forma de platillo — **i** carta *f* imbarcata

4433 disk — Thin, flat, circular plate or object. Disc is now accepted for disk, which is the correct spelling. "Disc" is especially used in compounds.

f disque *m* — **d** Scheibe *f*; Teller *m* — **n** schijf *fm* — **e** disco *m* — **i** disco *m*

4434 disk — Magnetic computer component for the storage of and random access to data, managed by a disk controller which in turn interfaces with the computer main frame itself. It may be a moving head disk, or a fixed-head disk, with removable or non-removable platters, or a diskette or floppy disk.

f disque *m* — **d** Platte *f*; Scheibe *f* — **n** schijf *fm* — **e** disco *m* — **i** disco *m*

4435 disk — Glass or plastic carrier containing film fount masters for some second-generation typesetters. See also segmented disc.

4436 disk — See ink disk.

4437 disk barker — See barker.

4438 disk controller — Controller which acts as an interface between the main frame computer and one or more disk drives, regulates the way in which the disk drives access or record data and transfer the same to and from the computer.

4439 disk drive — System or device which physically reads information from or writes information on a magnetic disk.

f commande *f* de disque — **d** Plattenantrieb *m*; Scheibenantrieb *m* — **n** schijfaandrijving *f* — **e** transmisión *f* de disco — **i** comando *m* di disco

4440 disk driver — Electro-mechanical device in which magnetic storage disks (or disk packs) are mounted and which causes the disks to rotate and the disk heads to pick up or write information.

4441 diskette — See floppy disc.

4442 disk inking arrangement — Arrangement on small platen presses in which an inking disk applies the ink to the inking rollers.

f dispositif *m* d'encrage à plateau — **d** Tellerfarbwerk *n* — **n** schijf-inktwerk *n* — **e** tintaje *m* de disco; tintaje *m* cilíndrico — **i** inchiostrazione *f* a disco; inchiostrazione *f* a piatto

4443 disk knives — See slitters.

4444 disk pack; disk stack — Pile of magnetic disks contained in a movable loading device.

f chargeur *m* de disques; pile *f* de disques — **d** Plattenstapel *m*; Scheibenstapel *m* — **n** schijvenpakket *n* — **e** juego *m* de discos; pila *f* de discos — **i** pila *f* di dischi

4445 disk ruling machine — Machine for ruling lines with disks instead of steel pens.

f machine *f* de réglure à disques — **d** Rollenlin(i)iermaschine *f* — **n** schijvenlinieermachine *f*; rollenlinieermachine *f* — **e** máquina *f* rayadora de discos; rayadora *f* de discos — **i** rigatrice *f* a dischi; lineatrice *f* a dischi

4446 disk stack — See disk pack.

4447 disk store, magnetic — See magnetic disk store.

4448 disk track — Addressable section of a disk

pack with information which may be read from that location, or, conversely, to which information may be written. Only a finite amount of data may be stored on that track or a segment thereof, depending upon the density of the disk.

4449 dismantle *v*
f démonter — **d** abmontieren; demontieren; abbauen — **n** demonteren; uiteen nemen — **e** desmontar; desarmar — **i** smontare

4450 disperse *v* — To break down a substance into fine particles and suspend them uniformly in a liquid.
f disperser — **d** dispergieren — **n** dispergeren — **e** dispersar — **i** disperdere

4451 disperse phase — Phase in a colloidal system, composed of the distributed particles.
f phase *f* de dispersion — **d** Dispersionsphase *f* — **n** disperse fase *f* — **e** fase *f* dispersa — **i** fase *f* dispersa

4452 dispersing agent — Agent added in small amounts to facilitate dispersion of a solid substance into a liquid medium. See also wetting agent.
f agent *m* de dispersion; agent *m* dispersant; substance *f* dispersive — **d** Dispergiermittel *n*; Dispergierungsmittel *n*; Dispersionsmittel *n* — **n** dispergeermiddel *n* — **e** agente *m* dispersante; dispersivo *m* — **i** agente *m* disperdente; agente *m* di dispersione

4453 dispersing lens — See diverging lens.

4454 dispersion — Mixture of particles in a vehicle. The particles may contain hard aggregates and/or may be flocculated. A dispersion is good when the hard aggregates are ground down to fine particles. A good dispersion may be either flocculated or deflocculated. For this reason the term is a poor one and should never be used unless defined. It is usually applied to colloidal particles with a diameter of 1-100 millimicrons, suspended in a suitable medium.
f dispersion *f* — **d** Dispersion *f* — **n** dispersie *f* — **e** dispersión *f* — **i** dispersione *f*

4455 dispersion — Splitting of a light ray passing through a lens or prism into its component colours when the path of light is diverted or bent because of refraction.
f dispersion *f* — **d** Streuung *f* — **n** dispersie *f*; kleurschifting *f* — **e** dispersión *f* de la luz — **i** dispersione *f* della luce

4456 dispersional analysis — See particle size analysis.

4457 displaced line — See misplaced line.

4458 display — Visual representation of computer output, as on a cathode ray tube.
f affichage *m* — **d** optische Anzeige *f* — **n** visuele weergave *fm*; afbeelding *f* — **e** representación *f* visual — **i** visualizzazione *f*

4459 display *v* — To make prominent by use of heavy-face or large type, by spacing or by printing in colour.
f mettre en vedette; accentuer — **d** auszeichnen; hervorheben — **n** laten uitkomen; zetten op op-vallende wijze — **e** destacar; resaltar; poner en resalte; desplegar *Me* — **i** porre in evidenza

4460 display attachment
f appareil *m* pour gros corps — **d** Großkegelvorrichtung *f* — **n** grootkorps-apparaat *n*; smoutlettergietapparaat *n* — **e** aparato *m* para tipos grandes; componedor *m* para titulares — **i** accessorio *m* per grandi caratteri

4461 display box
f boîte *f* pliante presentoir — **d** Gürtelschachtel *f* — **n** driedelige gordeldoos *fm* — **e** caja *f* exposición — **i** scatola *f* da esposizione

4462 display caster — See display machine.

4463 display composition — See display matter.

4464 displayed in bold type
f en vedette — **d** gesetzt aus fetter oder Auszeichnungsschrift — **n** vet gezet; in vette letters — **e** destacado — **i** composto in caratteri grassetto

4465 display faces — See display type.

4466 display line
f ligne *f* aux caractères travaux de ville — **d** Auszeichnungszeile *f* — **n** regel *m* gezet uit grootkorps-letter — **e** línea *f* epigráfica; línea *f* de titulares; título *m* desplegado; título *m* grande; epígrafe *m*; rótulo *m* — **i** linea *f* con grandi caratteri

4467 display machine; display caster
f machine *f* à titre; machine *f* pour gros corps; fondeuse *f* display; fondeuse *f* gros corps — **d** Großkegelmaschine *f*; Großkegelgießmaschine *f* — **n** smoutlettermachine *f*; grootkorps-giet-machine *f* — **e** máquina *f* tituladora; máquina *f* de titulares; máquina *f* de componer para titulares (y anuncios); titulera *f* *Ch Co*; tituladora *f*; fundidora *f* de titulares — **i** macchina *f* compositrice per grandi caratteri; macchina *f* compositrice di titoli; fonditrice *f* display

4468 display matrix
f matrice *f* pour gros corps — **d** Großkegel-matrize *f* — **n** grootkorpsmatrijs *f* — **e** matriz *f* titular; matriz *f* de titulares — **i** matrice *f* per grande carattere

4469 display matter; display composition — Matter distinguished by being set in larger or different type and on lines by itself, e.g. the title page, chapter headings, and sub-headings.
f composition *f* d'intercalations; composition *f* gros corps — **d** Akzidenzsatz *m* — **n** smoutzetsel *n* — **e** composición *f* con titulares; composición *f* de tipos grandes; composición *f* de disposición variada; composición *f* de epigráfica — **i** composizione *f* pubblicitaria

4470 display tube — Cathode ray tube in front of a monitor, on which words which overset a line and require division, are displayed by pressing a button indicating one or several hyphenation or non-hyphenation alternatives.
f cadre *m* auquel paraît le dernier mot — **d** Bildschirm *m* — **n** beeldscherm *n* — **e** pantalla *f* del imagen — **i** schermo *m* (per la proiezione) di imaggini

4471 display type; display faces — Large, conspicuous type characters used in printing advertisements, etc., usually 18 point or larger.

f caractère *m* à vedette; vedette *f*; caractères *mpl* travaux de ville — **d** Akzidenzschriften *fpl*; Titelschrift *f*; Auszeichnungsschrift *f* — **n** smoutletter *fm*; grootkorpsletter *fm* — **e** tipos *mpl* para remiendos; tipos *mpl* grandes; tipos *mpl* (para) titulares; tipos *mpl* publicitarios; tipos *mpl* para títulos; letras *fpl* titular; letras *fpl* de epígrafe; capitulares *mpl* — **i** caratteri *mpl* di grande corpo; caratteri *mpl* pubblicitari; caratteri *mpl* per titoli

4472 display unit — See cathode ray tube display.

4473 display work — Type displayed, such as title pages, headings, jobbing work, as distinct from *solid matter.

f composition *f* en vedette — **d** Auszeichnungssatz *m* — **n** smoutwerk *n* — **e** composición *f* de disposición variada; composición *f* versiforme; composición *f* titular; composición *f* de tipos grandes; composición *f* epigráfica; composición *f* de epigrafía — **i** composizione *f* con caratteri pubblicitari

4474 dissect *v* **for colours; break** *v* **up for colours; skeletonize** *v* — A job to be printed in two or more colours is first set up as if for one colour. A proof is taken and the matter that is to print the other colour (or colours) is then removed and imposed in a separate chase, using the proof as a guide for position and spacing.

f séparer la forme en plusieurs parties pour impression en polychromie; démonter pour tirage en couleurs — **d** den Satz im Farbe stellen — **n** de drukvorm voor kleurwerk splitsen — **e** sacar contramoldes; decomponer un molde para impresión a colores; desglosar — **i** separare la forma per stampa a colori

4475 dissemination of data — See data dissemination.

4476 dissemination of information — See data dissemination.

4477 dissertation — See thesis.

4478 dissolve *v* — Complete mixing with a liquid of a gas or solid composed of different molecules.

f (se) dissoudre — **d** auflösen; lösen — **n** oplossen — **e** disolver — **i** sciogliere

4479 distilled water; aqua destillata — Water from which impurities are removed by boiling tap water in a water still.

f eau *f* distillée — **d** destilliertes Wasser *n* — **n** gedistilleerd water *n* — **e** agua *f* destilada — **i** acqua *f* distillata

4480 distinctness-of-image gloss — See gloss.

4481 distortion — In lenses, departure from the proper perspective of an image. In photography and platemaking, departure in size or change of shape of negative or reproduction as compared to the original, due to lack of larallelism in camera equipment, or errors in stripping and handling of negatives and printing plates. See also barrel distortion, pincushion distortion.

f distorsion *f* — **d** Verzerrung *f*; Verformung *f*; Verkrümmung *f*; Deformierung *f*; Verbiegung *f*; Verzeichnung *f*; Distorsion *f*; Abbildungsfehler *m* — **n** vertekening *f*; vervorming *f*; distorsie *f* — **e** distorsión *f* — **i** distorsione *f*

4482 distortion — Change of the shape of the picture or image toward the edges of a cathode ray tube since the electrons which bombard the interior phosphorized surface of the tube have farther to travel and thus strike the tube at an angle. To maintain a quality image it is desirable to correct for this distortion by altering the focus and deflection of the electron beam.

f distorsion *f* — **d** Verzerrung *f* — **n** vertekening *f*; vervorming *f* — **e** distorsión *f* — **i** distorsione *f*

4483 distortion camera — See rescaling camera.

4484 distribute *v*; **dis** *v coll.* — To put back each letter and space into its proper compartment in the case or each matrix into its magazine of the slug composing machine.

f distribuer; mettre en casse; disposer — **d** ablegen; abstecken — **n** distribueren; wegmaken *coll.* — **e** distribuir (la composición); distribuir los tipos; poner los tipos; deshacer — **i** scomporre; distribuire

4485 distributer — See distributor.

4486 distributer — See distributor.

4487 distribute *v* **the ink**

f distribuer l'encre — **d** die Farbe auf die Walze ausstreichen — **n** de inkt verdelen — **e** distribuir la tinta — **i** spalmare el inchiostro sul rullo

4488 distributing cylinder, ink — See ink drum.

4489 distributing rider — See rider roller.

4490 distributing roller — See distributor roller.

4491 distributing roller mechanism — Part of the slug composing machine.

f mécanisme *m* de distribution — **d** Förder- und Verteilmechanismus *m* für die Ablegung; Ablegemechanismus *m*; Ablegevorrichtung *f* — **n** distributiemechanisme *n* — **e** mecanismo *m* de distribución — **i** meccanismo *m* per la scomposizione; meccanismo *m* di distribuzione

4492 distribution — Returning type, etc. to their proper case or store after use.

f distribution *f* — **d** Ablegen *n* — **n** distributie *f*; afleggen *n* — **e** distribución *f* — **i** scomposizione *f*; distribuzione *f*

4493 distribution curve — In statistics, a curve representing ideally the form to which the frequency distribution tends as more and more observations are obtained.

f courbe *f* de distribution; courbe *f* de répartition — **d** Verteilungskurve *f* — **n** verdelingskromme *fm*; verdelingscurve *fm* — **e** curva *f* de distribución — **i** curva *f* di distribuzione

4494 distribution fault — See wrong fount.

4495 distribution forme; clearing forme

f forme *f* à distribuer — **d** Ablegeform *f* — **n** distributievorm *m* — **e** forma *f* para distribuir — **i** forma *f* da scomporre; forma *f* di scomposizione

4496 distributor; distributer; dis hand; clearer — Compositor who returns types to their proper location after use.

f distributeur *m* — **d** Ableger *m*; Aufräumer *m* — **n** distribueerder *m*; opruimer *m*; wegmaker *m* — **e** distribuidor *m*; cajista *m* distribuidor — **i** scompositore *m*

4497 distributor; distributer — Part of a slug composing machine which carries the matrices back into their proper channels in the magazine. The distributor screws the matrices along the distributor bar until the correct channel is reached, then the matrices are released and drop by gravity into the magazine.

f distributeur *m* — **d** Ableger *m* — **n** distributiemechanisme *n* — **e** distribuidor *m*; aparato *m* distribuidor — **i** distributore *m*

4498 distributor — See distributor roller.

4499 distributor bar — Part of slug composing machine. See also distributor.

f barre *f* de distribution — **d** Ablegezahnstange *f*; Ablegerstange *f*; Ablegestange *f* — **n** distributorbar *m* — **e** barra *f* distribuidora; barra *f* de distribución — **i** barra *f* di distribuzione; barra *f* di guida della scomposizione

4500 distributor box — Part of a typesetting machine.

f boîte *f* de distribution — **d** Ablegeschloß *n*; Ablegekasten *m* — **n** distributiebox *m* — **e** caja *f* de distribución; caja *f* del distribuidor; caja *f* distribuidora — **i** scatola *f* di distribuzione

4501 distributor box fount distinguisher — Part of the typesetting machine.

f sélecteur *m* des matrices — **d** Ablegeschloß-Kontrolleinrichtung *f* — **n** matrijzencontrolestift *fm* — **e** selector *m* de matrices — **i** selettore *m* delle matrici

4502 distributor bracket — Part of the slug composing machine.

f châssis *m* de la distribution — **d** Ablegeschloß-Träger *m* — **n** distributorframe *n* — **e** chasis *m* de distribución — **i** telaio *m* di sostegno del mecanismo di distribuzione

4503 distributor roller; distributing roller; distributor — In a set of inking or damping rollers, the metal or hard rollers which oscillate from side to side to distribute ink or water more evenly across inking table or forme or damping rollers.

f rouleau *m* distributeur; rouleau *m* bal(l)adeur; baladeur *m*; rouleau *m* broyeur; broyeur *m* — **d** Verreibwalze *f*; Reiber *m*; Reibwalze *f*; Ulmer *m*; Stahlwalze *f* — **n** distributierol *fm*; verdeelrol *fm*; verwrijfrol *fm depr.* — **e** rodillo *m* de distribución; rodillo *m* batidor Ar; rodillo *m* dador — **i** rullo *m* distributore; rullo *m* macinatore

4504 distributor shifter — Part of typesetting machine.

f glissière *f* du poussoir des matrices à la distribution — **d** Einschieber *m* zum Ableger — **n** geleider *m* van de matrijzenschuif van de distributor — **e** impulsora *f*; impeledora *f* Ar — **i** frusta *f* spingimatrici alla scatola di distribuzione

4505 ditto marks — Two inverted commas under the word that is to be repeated.

f signes *mpl* de répétition — **d** Unterführungszeichen *npl*; Wiederholungszeichen *npl* — **n** aanhalingstekens *npl* — **e** comillas *fpl*; virgulillas *fpl* — **i** segni *mpl* di ripetizione

4506 divalent — See bivalent.

4507 diverging lens; dispersing lens; negative lens — Lens which increases the divergence of already diverging light rays. Sometimes called diverging meniscus.

f lentille *f* divergente; lentille *f* negative — **d** divergierende Linse *f*; Zerstreuungslinse *f*; negative Linse *f* — **n** divergerende lens *fm*; verstrooiingslens *fm*; negatieve lens *fm* — **e** lente *mf* divergente; lente *mf* negativo — **i** lente *f* divergente; lente *f* negativa

4508 diverging meniscus — See diverging lens.

4509 divide *v* **a word**

f diviser un mot — **d** das Wort am Ende der Zeile brechen; ein Wort trennen — **n** een woord afbreken — **e** dividir una palabra; fraccionar — **i** suddividere una parola (in sillabe)

4510 divide *v* **into takes** — To separate copy to be set into several takes so that a number of compositors may work on it at the same time.

f répartir la copie — **d** in Schiebungen zerlegen — **n** in stukken verdelen — **e** original *m* separado para varios compositores — **i** dividere il manoscritto in più parti

4511 dividers — Pair of compasses for dividing lines, measuring, etc. When the two prongs are connected by a threaded rod on which a threaded nut operates to open or close the prongs, it is called spring dividers. When the two prongs are held together and operate by a wing-like curved bar it is called wing dividers.

f compas *m* à pointes sèches; compas *m* droit (à pointes); compas *m* de mesure — **d** Stechzirkel *m*; Handzirkel *m* — **n** steekpasser *m*; verdeelpasser *m* — **e** compás *m* de división — **i** compasso *m* a punte fisse

4512 dividing coating — See dividing layer.

4513 dividing layer; dividing coating — Insulation layer of inert material between two materials to separate these to prevent diffusion or to provide a means for mechanical separation of the two layers, as for example with gravure printing cylinders.

f couche *f* de séparation — **d** Trennschicht *f* — **n** scheidingslaag *fm* — **e** capa *f* divisora — **i** strato *m* di separazione

4514 dividing rule — Small rule between the heads and text in newspapers, magazines, etc., or between articles.

f couillard *m* — **d** Trennlinie *f*; Trennungslinie *f*; Abteilungslinie *f*; Abschnittlinie *f* — **n** scheidingslijn *fm* — **e** raya *f* de división; línea *f* de división; línea *f* de sección — **i** linea *f* di separazione; filetto *m* di separazione; lineetta *f* di divisione

4515 divinity binding — See yapp binding.

4516 divinity circuit — See divinity edges.

4517 divinity edges; circuit edges; divinity circuit — Wide overhanging edgings of limp leather bindings that come together when the book is closed, used on bibles and prayer books, usually decorated with gold. See also yapp binding.

f reliure *f* yapp; reliure *f* souple débordante — **d** übergreifende Kanten *fpl* — **n** slap overhangende kanten *mpl*; slap overhangend omslag *n* — **e** bordes *mpl* cerrados — **i** coperte *mpl* sporgenti

4518 division into degrees

f division *f* en degrés — **d** Gradeinteilung *f* — **n** graadverdeling *f*; verdeling *f* in graden — **e** división *f* en grados — **i** divisione *f* in gradi

4519 division of words; word division — Separation of words into syllables so that part of a word can appear at end of one line and part at beginning of next line. Never separate a group of letters representing a single sound, but divide a word so that each part retains its present sound.

f division *f* des mots — **d** Worttrennung *f*; Wortteilung *f*; Trennung *f* von Wörtern; Abbrechen *n* von Wörtern — **n** afbreken *n* van woorden; woordafbreking *f* — **e** separación *f* de las palabras; división *f* de las palabras; corte *m* de las palabras; partición *f* de las palabras — **i** divisione *f* della parola; spezzatura *f* della parola

4520 divisorium

f visorium *m* — **d** Divisorium *n*; Manuskripthalter *m* — **n** manuscripthouder *m* — **e** atril *m*; divisorio *m* — **i** divisorio *m*

4521 D log E curve — See characteristic curve.

4522 D-notice — Official request (denial) to a newspaper not to publish an item.

f interdiction *f* de publier — **d** offizielle Aufforderung *f* eine Meldung nicht zu bringen — **n** verzoek *n* om een bericht niet te publiceren — **i** richiesta *f* ufficiale di silenzio stampa

4523 doctor *v* — See dope *v*.

4524 doctor bearing grid — Crossline pattern forming the recessed cells in a gravure printing surface to keep the doctor blade in the plane of the cylinder surface over the etched areas.

f cloisons *fpl* de trame; ponts *mpl* de trame — **d** rakelführendes Rasterstegnetz *n*; Tiefdruckrakelstegnetz *n* — **n** rasternet *n* — **i** griglia *f* del portaracla

4525 doctor blade; wiping blade; knife — Steel edge to wipe excess ink from the surface of engraved cylinders. In inverted type relief stamping presses, a device to remove surplus ink from the surface of the die prior to being wiped.

f racle *f* (d'essuyage); essuyeur *m*; raclette *f* — **d** Rakel *f*; Rakelblech *n* — **n** rakel *m* — **e** raspadora *f*; hoja *f* raspadora; cuchilla *f* tangente; racleta *f*; raspador *m*; rasqueta *f* — **i** racla *f*; racletta *f*

4526 doctor blade angle

f inclination *f* de la racle — **d** Rakelanstellwinkel *m* — **n** rakelhoek *m* — **e** ángulo *m* de la raspadora — **i** inclinazione *f* della racla

4527 doctor blade carrier — See doctor blade holder.

4528 doctor blade grinder — See doctor blade grinding machine.

4529 doctor blade grinding machine; doctor blade grinder

f machine *f* à affûter les racles; affûteuse *f* — **d** Rakelschleifmaschine *f*; Rakelmesser-Schleifmaschine *f* — **n** rakelslijpmachine *f* — **e** amoladora *f* para raspadores; rectificadora *f* para raspadores — **i** macchina *f* affilaracle; affilatrice *f* per racle

4530 doctor blade holder; doctor blade carrier

f porte-racle *m* — **d** Rakelbalken *m*; Rakelträger *m*; Rakelhalter *m* — **n** rakelhouder *m* — **e** portarraspadora *m* — **i** portaracla *m*

4531 doctor blade streaks — Long ink streaks which appear on the printed web or sheet surface when the wiping action of the doctor blade is faulty.

f stries *fpl* de la racle — **d** Rakelstreifen *mpl* — **n** rakelstrepen *fmpl* — **e** rayas *fpl* de raspadora — **i** striature *fpl* di racla

4532 doctor blade stroke — Side movement of the doctor blade, usually between 6-25 mm (0.25-1 in).

f oscillation *f* de la racle; déplacement *m* latéral de la racle — **d** Rakelhub *m*; Rakelbewegung *f*; Hinundherbewegung *f* der Rakel — **n** rakelslag *m* — **e** oscilación *f* de la racleta; vaivén *m* de la cuchilla — **i** movimento *m* alternato laterale della racla

4533 doctor blade support; backing doctor — Piece of spring steel on intaglio presses to support the very thin doctor blade.

f support-racle *m*; contre-racle *f* — **d** Stützrakel *f* — **n** rakelsteun *m*; steunrakel *m* — **e** soporte *m* de la raspadora — **i** controracla *f*

4534 doctor fountain — Press fountain with a flexible doctor blade to control the flow of ink by means of adjustable thumb screws across the length.

f encrier *m* à lame — **d** Farbwerk *n* mit Duktorlineal — **n** inktbak *m* met inktmes — **e** tintero *m* con cuchilla — **i** calamaio *m* a lama

4535 document board — See tag board.

4536 document manila — See tag board.

4537 dodge *v* — To restrict partially the light reaching a section of a flat or negative during exposure, with a fixed or movable mask.

d abhalten; abdecken; abwedeln; abfächeln — **n** tegenhouden — **e** manipular localmente — **i** regolare il contrasto

4538 dodgers — See handbills.

4539 dodging — Holding back the projected image-forming light during part of the basic exposure to make that area of the print lighter.

f dodging *m* — **d** Kontraststeuerung *f* — **n** regeling *f* van het contrast — **e** regulación *f* local; modificación *f* local; control *m* local — **i** regolazione *f* del contrasto

4540 dodging mask — Very unsharp mask formerly called an area mask.

f masque *m* de détourage — **d** Kontraststeuerungsmaske *f* — **n** contrastscherpte verbeterend masker *n* — **e** máscara *f* de rebaje — **i** maschera *f* di regolazione del contrasto

4541 dog-eared book — See dog's eared book.

4542 dogs; plate dogs — Bevelled clips which slide in grooves on the plate cylinder of a letterpress rotary, holding the plates firmly in position.
f griffes *fpl* de serrage — **d** Facettenhaken *mpl*; Innenhaken *mpl*; Plattengreifhaken *mpl*; Plattenhaken *mpl* — **n** plaathaken *mpl* — **e** pinzas *fpl* de cierre — **i** staffe *fpl* intermedie di bloccaggio

4543 dog's ear — In bookbinding, folded corner of a sheet which has not been cut and which extends beyond the trimmed edges.
f larron *m*; corne *f* — **d** Eselsohr *n* — **n** ezelsoor *n* — **e** oreja *f*; punta *f* doblada; esquina *f* doblada; señal *f* — **i** orecchia *f*; becco *m*

4544 dog's eared book; dog-eared book — Book in which the corners of the leaves have been turned down or folded.
f livre *m* écorné — **d** Buch *n* mit Eselsohren — **n** boek *n* met ezelsoren — **e** libro *m* con orejas; libro *m* con puntas dobladas — **i** libro *m* con angoli di pagina ripiegati

4545 dog watch — See shift.

4546 dog ways — Slides or slots in the plate cylinder of a letterpress rotary in which dogs operate.
f rainures *fpl* de logement des griffes — **d** Facettennuten *fpl* — **n** plaathaakgroeven *fmpl* — **e** ranuradas *fpl* de pinzas — **i** scanalature *fpl* per le staffe intermedie

4547 dollar mark — The mark $, first used in 1779.
f symbole *m* du dollar — **d** Dollarzeichen *n* — **n** dollarteken *n* — **e** signo *m* de dólar — **i** simbolo *m* del dollaro

4548 dolly truck — See reel bogie.

4549 dolomite — Type of limestone containing calcium and magnesium carbonates, used in the preparation of the sulphite cooking liquor for papermaking.
Formula: $CaMg(CO_3)_2$.
f dolomite *f* — **d** Dolomit *m* — **n** dolomiet *n* — **e** dolomía *f*; dolomita *f*; caliza *f* lenta — **i** dolomia *f*

4550 domed; dome shaped — Symmetrical, convex protrusion in the surface. Opposite of *dished.
f bombé; arrondi; en forme de dôme; cloqué — **d** gewölbt; ausgewölbt — **n** bolvormig — **e** bombeado — **i** bombato

4551 dome shaped — See domed.

4552 dominant wavelength — Wavelength of a colour that indicates its *hue. The more this wavelength is accentuated the greater is the purity factor of the colour. Wavelength of the monochromatic light stimulus that, when combined in suitable proportions with the specified achromatic light stimulus, yields a match with the colour stimulus considered. Photometric correlate with hue. Symbol: λ_d.
f longueur *f* d'onde dominante — **d** farbtongleiche Wellenlänge *f* — **n** dominerende golflengte *f* — **e** longitud *f* de onda dominante — **i** lunghezza *f* d'onda dominante

4553 donat; donet; donatus — Schoolbook or grammar, named after Aelius Donatus, a famous grammarian and teacher in Rome (mid 4th century AD), who wrote a large and a small grammar, "Ars maior" and "Ars minor". The latter remained in use throughout the middle ages and the author's name became a common metonymy in the forms of "donat" and "donet" for grammar.
f donat *m* — **d** Donat *m* — **n** donaat *m* — **e** donato *m*

4554 donatus — See donat.

4555 donet — See donat.

4556 do not run keyboard — See sorts channel.

4557 don't have
f n'y en a pas — **d** ist nicht da; nicht da — **n** hebben we niet — **e** ¡no tiene!; ¡no hay! — **i** non ce n'è

4558 door — Dutch doors are half pages (more or less) bound horizontally into the magazine and occupying either the top half or the bottom half of the page. French doors are usually a half page cut vertically. The single gate fold is a full page that is folded at the outer edge of the book into the gutter, which can be opened out to form a spread out of a right-hand or left-hand page. There are also double gate folds, in which pages can be folded both to the right and to the left.

4559 dope — Ink conditioning compound, reducer or varnish.
f améliorant *m*; correctif *m*; adjuvant *m* — **d** Zusatzmittel *n* — **n** toevoeging *f* — **e** aditivo *m*; producto *m* auxiliar — **i** correttivo *m* per inchiostri; additivo *m* per inchiostri

4560 dope *v*; **doctor** *v* — To modify the characteristics of a printing ink in the pressroom with additives.
f améliorer; ajouter d'adjuvants; modifier; corriger — **d** die Druckfarbe anpassen; Zusätze machen — **n** toevoegingen aanbrengen; de drukinkt aanpassen — **e** modificar con algún ingrediente las tintas; alterar con algún modificante la tinta — **i** aggiuntare additivi all'inchiostro

4561 dope story; think piece — Article analyzing a news event and giving the background, often with the author's opinions and forecast for the future.
d Hintergrundbeitrag *m* — **e** información *f* especulativa; trabajo *m* especulativo; trabajo *m* teórico; material *m* para futura información — **i** pezzo *m* denso d'informazioni

4562 Doppler effect — Phenomenon evidenced by the change in the observed frequency of a wave in a transmission system caused by a time rate of change in the effective length of the path of travel between the source and the point of observation (Christian Doppler, 1803-1853, Austrian physicist).
f effet *m* Doppler — **d** Doppler-Effekt *m* — **n** Doppler effect *n* — **e** efecto *m* de Doppler — **i** effetto *m* di Doppler

4563 dos-à-dos binding — Bookbinding comprising two books bound in such a way that the foredge of one book meets the back of the other, the books being separated by a common back board.

f reliure *f* jumelée — **d** dos-à-dos-Einband *m* — **n** dos-à-dos-band *m* — **e** encuadernación *f* dos-à-dos — **i** legatura *f* dos-à-dos

4564 dot — See halftone dot.

4565 dot distortion — Fault in an eccentric shape of the dot in a halftone image.

f distorsion *f* des points de trame — **d** Punktverzerrung *f*; Punktverformung *f* — **n** vervorming *f* van de rasterpunt — **e** distorsión *f* del punto reticular — **i** deformazione *f* del punto di retino

4566 dot enlargement

f élargissement *m* de point — **d** Punktvergrößerung *f* — **n** vergroting *f* van de punt; puntvergroting *f* — **e** engrandecimiento *m* del punto — **i** allargamento *m* del punto

4567 dot etching — Tonal correction of halftone positives or negatives by judicious and controlled reduction of dot size with chemical reducers. Tray etching for all-over reduction or local etching done by an artist with a small soft brush.

f morsure *f* du point; retouche *f* par morsure (du point) — **d** Punktätzung *f* — **n** puntetsing *f* — **e** retoque *m* de puntos; retoque *m* de los medios tonos — **i** incisione *f* del punto

4568 dot formation — Arrangement and proper size of dots in halftone negatives and printing plates, necessary for accurate translation of detail and tone values.

f formation *f* de points — **d** Punktbildung *f*; Punktformation *f* — **n** puntvorming *f* — **e** puntillado *m* reticular; puntillado *m* de trama; formación *f* de puntos; punteado *m* reticular; punteado *m* de trama — **i** formazione *f* dei punti

4569 dot fringe; halo effect — Deposit of ink at the edges of halftone dots or printed letters, with lack of ink at the centres.

f bourrelet *m* d'encre; halo *m*; auréole *f*; pénombre *f* — **d** Quetschrand *m* — **n** kraalrandje *n* — **e** aureola *f*; halo *m* — **i** effetto *m* alone

4570 dot leaders — See leaders.

4571 dotless i — Lower-case letter "i" used in computer typesetting when an accent is to be positioned over the character, such as a circumflex, when the dot would interfere with the accent.

4572 dot matrix — Generation of characters by means of a pattern of dots which may be printed (as by an electrostatic or ink jet printer) or generated on the face of a TV tube. A 5 x 7 matrix is the minimum required to describe a basic upper and lower case alphabet, although characters with descenders can be accommodated only by displacing the dot pattern below the base line. A 7 x 9 dot matrix provides a better representation, and further addition of dots to the pattern will enhance the intelligibility of the image.

4573 dot printer — See matrix printer.

4574 dot reduction

f réduction *f* de point — **d** Punktverkleinerung *f* — **n** kleiner worden *n* van de punt; afzwakken *n* van de punt — **e** reducción *f* de los puntos reticulares — **i** riduzione *f* dei punti

4575 dotted line

f trait *m* ponctué; ligne *f* en pointillé; ligne *f* stipulée — **d** punktierte Linie *f* — **n** stippellijn *fm*; puntlijn *fm* — **e** linea *f* de puntos — **i** linea *f* punteggiata

4576 dotted manner — See manière criblée.

4577 dotted rule — Strips of brass (or lead) with a dotted face, ranging from fine dots to short dashes close together.

f filet *m* pointillé — **d** Punktlinie *f*; punktierte Linie *f* — **n** puntlijn *fm* — **e** filete *m* puntillado; filete *m* punteado; raya *f* puntillada — **i** filetto *m* punteggiato

4578 dotting wheel

f roue *f* à pointiller — **d** Punktierrädchen *n* — **n** roulette *fm* — **e** rueda *f* para puntillado — **i** rotella *f* per punteggiare

4579 double; doublet — Words repeated by mistake.

f doublon *m* — **d** Doppelsatz *m*; Dublette *f*; Hochzeit *f* *fam*; Doppelzeile *f* (im Maschinensatz) — **n** dubbelgezet woord *n*; dubbele regel *m* — **e** palabra *f* repetida; repetido *m*; duplicado *m*; letra *f* doble; doble letra *f*; ditografía *f* — **i** doppione *m*; gambero *m*

4580 double *v*

f répéter un ou plusieurs mots — **d** eine Hochzeit machen — **n** abusievelijk twee keer zetten; een woord dubbel zetten — **e** repetir palabras — **i** doppiare una parola

4581 double bag cap — A British size of wrapping paper. The term is becoming obsolete. See app. no. 3.

4582 double-body matrix — See double-letter matrix.

4583 double bound — See double linking.

4584 double cap — See double foolscap.

4585 double case — Case with a complete lower case and half a normal upper case; erroneously called *half case.

f casse *f* double — **d** Doppelsetzkasten *m* — **n** dubbele kast *fm* — **e** doble caja *f* — **i** cassa *f* doppia

4586 double coated paper — Paper heavily coated on one or two sides, for fine illustrations.

f papier *m* double émail; papier *m* couché double — **d** doppelgestrichenes Papier *n* — **n** dubbelgestreken papier *n*; tweemaal gestreken papier *n* — **e** papel *m* esmaltado doble; papel *m* estucado y esmaltado; papel *m* estucado doble; papel *m* estucado dos veces — **i** carta *f* patinata due volte

4587 double columbian — Old name for 32-point type. See app. no. 1.

4588 double column page — Page with two columns of type matter, extending across the full width. Sometimes called half measure.

f page *f* sur deux colonnes — **d** zweispaltige Seite *f* — **n** pagina *fm* over twee kolommen — **e**

página *f* a dos columnas; página *f* de doble columna; página *f* con corondeles — *i* pagina *f* su due colonne

4589 double-concave lens — See concave lens.

4590 double-convex lens — See convex lens.

4591 double copy — A size of writing and printing papers recognized but not standard in the UK. See app. no. 3.

4592 double crown — A British standard size for printing and wrapping papers. See app. no. 3.

4593 double cylinder machine — See perfecting press.

4594 double dagger; diesis — Third reference mark (‡), after the dagger.

f double-croix *f*; diésis *f* — **d** Doppelkreuz *n* — **n** dubbelkruisje *n* — **e** diesis *m*; cruz *f* doble; doble cruz *f* — **i** doppia crocetta *f*

4595 double delivery — Delivery on two piles which receive alternately a sheet. This permits the delivery to run at half the speed of the press.

f double sortie *f*; double recette *f* — **d** Doppelauslage *f*; Wechselauslage *f* — **n** dubbele uitleg *m*; wisseluitleg *m* — **e** salida *f* doble; receptora *f* doble; salida *f* alternativa — **i** uscita *f* in doppio

4596 double demy — A British standard size for printing and wrapping papers. See app. no. 3.

4597 double double cap — A size of writing paper. See app. no. 3.

4598 double-double face corrugated fibreboard; double-wall corrugated fibreboard; twin-flute corrugated fibreboard — Board made up of two sheets of fluted paper interposed and stuck between three sheets of paper or board.

f carton *m* ondulé double-double (face); carton *m* ondulé double cannelure — **d** Doppel-Doppel-Wellpappe *f*; Fünfbahnen-Wellpappe *f*; zweiwellige Wellpappe *f* — **n** dubbel-dubbelzijdig beplakt golfkarton *n*; dubbel-dubbelzijdig beplakt welbord *n* — **e** cartón *m* corrugado doble-doble; cartón *m* acanalado doble-doble — **i** cartone *m* ondulato doppio-doppio; cartone *m* ondulato a due onde; cartone *m* ondulato triplo

4599 double double imperial — A UK size of wrapping paper and of board. See app. no. 3.

4600 double double small hand — A size of wrapping paper. See app. no. 3.

4601 double elephant — A British standard size for writing and printing papers. See app. no. 3.

4602 double English — Old name for 28-point type. See app. no. 1.

4603 double exposure — In photography and photomechanics, a supplementary exposure to obtain special effects.

f double exposition *f* — **d** Doppelbelichtung *f* — **n** dubbele belichting *f* — **e** exposición *f* doble; pose *f* doble; insolación *f* doble; tiempo *m* doble; doble exposición *f* — **i** doppia esposizione *f*

4604 double exposure — Printing down from two or more positives or negatives to obtain a single image.

4605 double face corrugated fibreboard; single-wall corrugated fibreboard; single-flute corrugated fibreboard — Board made of a sheet of fluted paper stuck between two sheets of paper or board.

f carton *m* ondulé double face; carton *m* ondulé simple cannelure — **d** doppelseitige Wellpappe *f*; Dreibahnen-Wellpappe *f*; einwellige Wellpappe *f*; zweiseitige Wellpappe *f* — **n** dubbelzijdig beplakt golfkarton *n*; dubbelzijdig beplakt welbord *n* — **e** cartón *m* ondulado doble cara; cartón *m* corrugado doble; cartón *m* acanalado doble; cartón *m* ondulado forrado a dos caras *Es* — **i** cartone *m* ondulato doppio; cartone *m* ondulato ad una onda

4606 double-face self adhesive tape

f ruban *m* double collant; ruban *m* adhésif double face — **d** doppelseitiges Klebeband *n*; Doppelklebeband *n* — **n** tweezijdig klevend (plak)band *n*; tweezijdig zelfklevend band *n* — **e** cinta *f* con doble cara adhesiva; cinta *f* autoadhesiva de doble cara — **i** nastro *m* adesivo a doppia faccia; nastro *m* autoadesivo a doppia faccia

4607 double fine rule

f filet *m* double-maigre; filet *m* double-fin — **d** doppelfeine Linie *f* — **n** dubbelfijne lijn *fm* — **e** filete *m* de dos hilos; filete *m* de caña — **i** filetto *m* doppio chiaro; filetto *m* doppio filo

4608 double flat foolscap — A size of writing paper. See app. no. 3.

4609 double fold number — See folding endurance.

4610 double folio — A size of writing or printing paper. See app. no. 3.

4611 double foolscap; double cap — A British standard size of printing and writing papers. In some parts of the UK the term is also applied to writing paper. See app. no. 3.

4612 double frame — Stand to hold two pairs of type cases mounted in position for use, with space in between for trays, and rack for ten pairs of cases.

f double-layette *f* — **d** Doppelregal *n* — **n** dubbel regaal *n* — **e** chibalete *m* doble — **i** bancone *m* doppio; banco *m* doppio

4613 double globe — A size of printing paper. See app. no. 3.

4614 double great primer — Old name for 32-point type. See app. no. 1.

4615 double hambro — A size of wrapping paper. See app. no. 3.

4616 double-helical gear — See herring-bone gear.

4617 double imperial; double imperial cap — A British standard size for printing and writing papers. Also a UK size for wrapping paper. See app. no. 3.

4618 double imperial cap — See double imperial.

4619 double justification

f double justification *f* — **d** Doppelausschluß *m* — **n** dubbele uitvulling *f* — **e** doble justificación *f* — **i** doppia giustificazione *f*

4620 double keyboarding — Method to ensure the accuracy of input by keyboarding it twice, assuming that a second keyboarder would not make the same mistakes as the first. A character-by-character comparison is then done by a computer program and discrepancies are flagged so that a monitor can select the correct version. Sometimes the second keyboarding takes place on a verifying keyboard which locks if the operator strikes a different key than that used the first time.

4621 double knife — 1. Used as a verb, double knife means to surround a box with trim or scrap. All feed-up errors appear in this scrap area. Only used when size of box has to be perfect. More board is used in this method. On 90% of all commercial boxes double knifing is not required. 2. Used as a noun, double knife means a two-bladed tool to cut both sides of score at once.

4622 double large — A size of cut card. See app. no. 3.

4623 double large post — A British size of writing and printing or drawing paper. See app. no. 3.

4624 double leaded — Type matter set with double space between the lines by using two leads of two points each.
f à interligne double — **d** doppelter Durchschuß *m* — **n** dubbelgeïnterlinieerd; met twee tweepunts interline — **e** con dos interlíneas (entre cada renglón); con doble interlínea — **i** ad interlineatura doppia

4625 double letter — See ligature.

4626 double-letter matrix; double-body matrix; two-letter matrix; duplex matrix — Matrices from 4.75-24 pt, each of which has two type characters, as roman and italic or roman and bold.
f matrice *f* à deux œils; matrice *f* duplex — **d** Zweibuchstabenmatrize *f*; Zweikörpermatrize *f*; paarige Matrize *f* — **n** duplex-matrijs *fm*; tweelettermatrijs *fm*; tweeletterbeeldmatrijs *fm* — **e** matriz *f* de dos letras; matriz *f* doblada — **i** matrice *f* a doppia incisione

4627 double linking; double bound
f double liaison *f*; liaison *f* double — **d** Doppelbindung *f* — **n** dubbele binding *f* — **e** enlace *m* doble; ligadura *f* doble — **i** doppio legame *m*

4628 double long demy — A size of ledger paper, sometimes used in the US. See app. no. 3.

4629 double long medium — A size of ledger paper, sometimes used in the US. See app. no. 3.

4630 double lump — A size of wrapping paper. See app. no. 3.

4631 double medium — A British standard size for writing and printing papers. See app. no. 3.

4632 double medium-face rule
f filet *m* double-quart-gras; filet *m* gouttière quart-gras — **d** doppelstumpffeine Linie *f* — **n** dubbele stompfijne lijn *fm* — **e** filete *m* de doble descanterado; filete *m* descanterado doble — **i** filetto *m* doppio semiscuro

4633 double music — A size of printing paper. See app. no. 3.

4634 double nicanee — A size of wrapping paper. See app. no. 3.

4635 double octuple — Newspaper press of double width, four pages wide, eight perfecting units or decks.
f rotative *f* à 8 groupes 4 pages — **d** doppelbreite 8-Einheitenmaschine *f*; 64-seitige Rotationsmaschine *f* — **n** 64-zijdige rotatiepers *fm* — **e** rotativa *f* de 8 grupos de 4 páginas — **i** rotativa *f* a doppia larghezza; rotativa *f* a 4 pagine ed 8 gruppi stampanti

4636 double official — A British standard size of cards. See app. no. 3.

4637 double overlay mask — See two-stage mask.

4638 double paragon — Old name for 40-point type. See app. no. 1.

4639 double pica — Old name for 24-point type. See app. no. 1.

4640 double pinched post — A size of writing and printing papers, recognized but not standard in Great Britain. See app. no. 3.

4641 double plate — Two sets of plates on a letterpress rotary so that, for example, tabloid newspapers are produced on a full sheet press. See also two set.
f double plaque *f* — **d** doppelter Plattensatz *m* — **n** dubbele plaat *fm* — **e** plancha *f* doble; clisé *m* doble; lámina *f* doble — **i** lastra *f* doppia

4642 double plating — See two set.

4643 double post — A British standard size of writing, printing, and drawing papers. See app. no. 3.

4644 double pott — A size of writing or printing paper and of boards. See app. no. 3.

4645 double precision — Increased precision gained by the use of two computer words representing a number.
f double précision *f* — **d** doppelte Präzision *f* — **n** dubbele nauwkeurigheid *f* — **e** doble precisión *f*; precisión *f* larga — **i** doppia precisione *f*

4646 double print — Repeated impression on the same sheet.
f doublage *m* — **d** Doublieren *n*; Dublieren *n* — **n** doubleren *n* — **e** doble impresión *f*; remosqueo *m*; remosqueamento *m* — **i** stampa *f* doppia; doppia impressione *f*

4647 double printing — Combining printing details from two negatives (or complementary flats) into a single positive or press-plate. Such combinations usually fill in adjacent or surrounding matter as distinguished from overprinting or surprinting.
f copie *f* combinée — **d** Übereinanderkopieren *n* — **n** inkopiëren *n* — **e** sobreimpresión *f* fotomecánica — **i** copia *f* combinata

4648 double production — See two set.

4649 double quad pott — A size of writing or printing papers. See app. no. 3.

4650 double quotes — See punctuation marks.

4651 double royal — A British standard size of writing and printing paper. See app. no. 3.

4652 double rule — See parallel rule.

4653 double salt — Salt that crystallizes as a single substance but ionizes as two distinct salts when dissolved.
f sel *m* double — **d** Doppelsalz *n* — **n** dubbelzout *n* — **e** sal *f* doble — **i** sale *m* doppio

4654 double sextuple — Newspaper printing press of double width four pages wide, six perfecting units or decks.
f rotative *f* à 6 groupes 4 pages — **d** doppelbreite 6-Einheitenmaschine *f*; 48-seitige Rotationsmaschine *f* — **n** 48-zijdige rotatiepers *fm* — **e** rotativa *f* de 6 grupos de 4 páginas — **i** rotativa *f* a doppia larghezza; rotativa *f* a 4 pagine ed 6 gruppi stampanti

4655 double small — A standard size of cut card. See app. no. 3.

4656 double small cap — A size of wrapping paper. See app. no. 3.

4657 double small demy — A British standard size of writing and printing papers. See app. no. 3.

4658 double small foolscap — A British standard size of writing and printing papers. See app. no. 3.

4659 double small hand — A size of wrapping paper. See app. no. 3.

4660 double small pica — Old name for 22-point type. See app. no. 1.

4661 double small post — A size of writing and printing papers, recognized but not standard in Great Britain. See app. no. 3.

4662 double small royal — A size of writing paper. See app. no. 3.

4663 double spread; spread — Printing that extends across and fills a two-page opening of a brochure, book or folder; in the centre of a book or folder it is called a *centre spread.
f illustration *f* en double page; illustration *f* à livre ouvert — **d** Abbildung *f* auf zwei gegenüberstehenden Seiten — **n** afbeelding *f* over twee pagina's; illustratie *f* over twee pagina's — **e** ilustración *f* a doble plana; plana *f* doble; página *f* doble; doble página *f* — **i** illustrazione *f* su doppia pagina

4664 double super royal — A size of printing paper. See app. no. 3.

4665 doublet — See double.

4666 double-tone ink; duotone ink; bitone ink; duplex ink; varishade ink — Printing ink with soluble toner which bleeds out to produce the effect of two-colour printing caused by the colouring of the areas around the halftone dots.
f encre *f* double-ton; encre *f* deux-tons; encre *f* duotone — **d** Doppeltonfarbe *f*; Doppeltondruckfarbe *f*; Auslauffarbe *f* — **n** dubbeltooninkt *m* — **e** tinta *f* de doble tono; tinta *f* de dos tonos — **i** inchiostro *m* doppia tinta; inchiostro *m* a doppia tonalità

4667 double truck advertisement — Advertisement of two pages. Name derived from the double wheel truck with which the printing forme is transported. For the transport of a one-page forme a single truck is used, hence the term single truck advertisement.

4668 double-wall bag — See duplex bag.
4669 double-wall corrugated fibreboard — See double-double face corrugated fibreboard.

4670 double width press — Newspaper printing press four standard pages wide.
f rotative *f* (à) double largeur — **d** doppelbreite Maschine *f* — **n** dubbele pers *fm* — **e** rotativa *f* a doble ancho — **i** rotativa *f* a doppia larghezza

4671 doubling up — The emergency measure of having two craftsmen simultaneously engaged on the same work.
f emploi *m* simultane de deux ouvriers — **d** Doppelbesetzung *f*; Einsatz *m* von zwei Arbeitskräften für dieselbe Arbeit — **n** twee man *m* aan één werk — **e** empleo *m* simultáneo de dos operarios — **i** impiego *m* contemporaneo di due operai nello stesso lavoro

4672 doughnutting — Appearance of a screen dot in gravure printing that has printed the circumference of the dot but not the dot surface itself.

4673 dowel — Short register pin of plastic or metal, attached to a film support or plate, used to position a film or flat for double printing or for step-and-repeat exposures.
f pointure *f* — **d** Paßstift *m*; Registerstift *m* — **n** registerpen *fm*; paspen *fm* — **e** clavija *f* de ajuste; pasador *m* de ajuste (repetidora) — **i** spina *f* di riferimento; spina *f* di registro

4674 downdraught sink — Developing sink with facilities to extract noxious fumes downwards and away from the operator.
f bac-évier *m* à aspiration par en bas — **d** Entwicklungsrinnstein *m* mit Absaugung — **n** ontwikkelgootsteen *m* met afzuiginrichting — **e** artesa *f* de revelar con aspiración — **i** vasca *f* con aspirazione

4675 downer — Sudden breaking of the web while the press is in operation. See also break.
f casse *f* (de la bande) — **d** Papierreißer *m*; Papierbruch *m*; Bahnbruch *m* — **n** baanbreuk *fm*; breuk *fm* van de papierbaan — **e** rotura *f* de la banda; desgarro *m* de la banda — **i** rottura *f* del nastro in macchina

4676 downshift — Mode of accessing lower-case characters (and equivalent numerals) as on a typewriter. A downshift code may require that the succeeding bit patterns will represent lower-case characters, until further notice.

4677 down stroke of the knife
f mouvement *m* de descente du couteau — **d** Messerniedergang *m* — **n** neergang *m* van het mes — **e** movimiento *m* de descenso de la cuchilla; descenso *m* de la cuchilla — **i** corsa *f* discendente della lama

4678 down time; idle machine time — Time when a machine is stopped.
f temps *m* d'arrêt de machine — **d** Maschinenstillstandszeit *f*; Zeit *f* der Außerbetriebnahme — **n** machinestilstandtijd *m* — **e** período *m* de paro de la máquina; tiempo *m* inactivo — **i** tempo *m* d'arresto di macchina; tempo *m* inattivo di una macchina

4679 down time; fault time — Time when a computer system is not functioning properly. Mean time between failures (MTBF) is usually one of the figures recoreded to access the impact of down time.

f temps *m* de panne — **d** Störungszeit *f*; Ausfallzeit *f* — **n** storingstijd *m*; uitvaltijd *m* — **e** tiempo *m* de avería — **i** tempo *m* di guasto

4680 downward movement

f mouvement *m* descendant — **d** Niedergang *m*; Abwärtsbewegung *f* — **n** neergaande beweging *f*; neerwaartse beweging *f* — **e** movimiento *m* de descenso — **i** movimento *m* discendente

4681 drachm; fluid drachm — Apothecaries' fluid measure. See app. no. 7 and 7a.

4682 drachm — See dram.

4683 drafting board — See drawing board.

4684 drafting paper — See drawing paper.

4685 drag; draw — Register trouble in lithographic printing, when the dot is enlarged toward the back (non-gripper end) of the sheet.

f allongement *m* du point (vers l'arrière) — **d** Verlängerung *f* der Rasterpunkte — **n** uitrekking *f* van de rasterpunten — **e** punto *m* estirado — **i** allungatura *f* dei punti

4686 drag — See slur.

4687 drag *v* — A sheet of paper is said to drag when the end of it shows an unsharp image, due to the fact that it is not held close to the cylinder.

f entraîner — **d** schleppen — **n** slepen — **i** trascinare

4688 dragon's blood; gum dragon — Resinous substance, mostly red in colour, used in powder form in photoengraving to protect parts of the metal from the acid in the etching process. Soluble in alcohol, ether and volatile and fixed oils. See also etching powder.

f sang *m* dragon; sang *m* de dragon — **d** Drachenblut *n* — **n** drakenbloed *n* — **e** sangre *f* de dragón; sangre *f* de drago; sangre *f* drago; drago *m*; goma *f* de India — **i** sangue *m* di drago

4689 dragon's blood process — Method of relief etching in which the sides of lines and dots are protected against the mordant by dusting the plate on all four sides with dragon's blood or etching powder, then heating the dusted plate to melt the powder to form an acid resisting coating on the top and sides of the relief formations. Superseded by *powderless etching.

f procédé *m* au sang (de) dragon — **d** Drachenblutverfahren *n*; Anstaubverfahren *n* — **n** drakenbloedprocédé *n* — **e** procedimiento *m* de sangre de dragón; procedimiento *m* de drago — **i** procedimento *m* al salgue di drago

4690 drag-out — In gravure printing, a bead of excessive ink at the trailing edge of the print.

f flammèche *f* — **d** Farbwischer *m* — **i** sbavatura *f*

4691 drag roller — See roller top of former.

4692 drag tool — See draw tool.

4693 drainage rate

f vitesse *f* d'égouttage — **d** Entwässerungs-geschwindigkeit *f* — **n** ontwateringssnelheid *f* — **e** velocidad *f* de desgotaje

4694 drainer — Holder for glassware, photographic plates etc. to permit water to run off after washing.

f égouttoir *m* — **d** Abtropfständer *m* — **n** afdruiprek *n* — **e** caballete *m* para secar negativos; rejilla *f* para secar negativos; escurridor *m* — **i** scolapiatti *m*

4695 drainer; draining chest — Tiled or cement tank, with perforated bottom, in which the bleached and washed rag halftuff is left to drain and where it is stored for further use.

f caisse *f* d'égouttage — **d** Entwässerungsanlage *f*; Absetzbütte *f*; Eindickgrube *f* — **n** ontwateringsinrichting *f* — **e** instalación *f* de deshidratación; instalación *f* de desagüe — **i** cassa *f* di sgocciolamento

4696 draining chest — See drainer.

4697 dram; drachm — Unit of weight and apothecaries' fluid measure. Abbrev. dr (no point). Symbol: ℨ. See app. no. 7 and 7a.

f drachme *f* — **d** Drachme *f*; Dram *n* — **n** drachme *fn* — **e** dracma *f* — **i** dramma *f*

4698 draughtsman

f dessinateur *m*; dessinatrice *f* — **d** Zeichner *m*; Zeichnikus *m* fam. — **n** tekenaar *m*; tekenares *f* — **e** dibujante *mf* — **i** disegnatore *m*; disegnatrice *f*

4699 draughtsmen *GB*; **checkers** *US* — Types for the pieces in the game of draughts (game of checkers), cast on em bodies, usually 12-point or 24-point.

f signes *mpl* pour jeu de dames — **d** Damenspielzeichen *npl*; Damespielzeichen *npl* — **n** damspeltekens *npl* — **e** tipos *mpl* para juego de damas — **i** caratteri *mpl* dei pezzi degli dama; segni *mpl* per giuoco della dama

4700 draw — In trimming paper, the displacement of the sheet by the thickness of the knife which may result in irregular cutting.

f décalage *m* — **d** Verrutschen *n* des Papiers — **n** wegdrukken *n* van het papier — **e** resalto *m* en el corte del papel — **i** spostamento *m* del foglio nel taglio

4701 draw — Tension exerted on the paper web as it travels from section to section on web presses.

f tension *f* — **d** Zug *m*; Spannung *f* — **n** spanning *f* — **e** tracción *f* — **i** tensione *f*

4702 draw — See drag.

4703 drawdown; pull down; draw out — Ink chemist's method of roughly determining colour shade, colour strength, mass tone and undertone of an ink, by placing a small glob of ink on paper and spreading it with a spatula to get a thin film of ink.

f touche *f* étalée; touche *f* à spatule; essai *m* d'encre à la spatule; étalage *m* — **d** Abstrich *m*; Aufstrich *m*; Tupfprobe *f* — **n** uitstrijksel *n*; uitstrijkproef *fm* — **e** raya *f* de paleta — **i** prova *f* dell'inchiostro alla spatola

4704 drawer handle — Handle of a type case.
f poignée *f* (de la casse) — **d** Griff *m* (am Schrift-kasten) — **n** greep *fm*; kastgreep *fm*; greep *fm* van de letterkast — **e** tiradera *f* (de la caja) — **i** manico *m* (di cassa)

4705 drawing board — Paperboard for crayon or water colour drawings, made of wood pulp and waste paper stock of sufficient thickness to with-stand bending; sized with good texture, and finished without gloss.
f carton *m* à dessin — **d** Zeichenkarton *m* — **n** tekenkarton *n* — **e** cartón *m* de dibujo; cartón *m* para dibujo — **i** cartoncino *m* da disegno

4706 drawing board; drafting board — Smooth, flat board to which paper or canvas is attached for drawing.
f planche *f* à dessin; planche *f* à dessiner — **d** Zeichenbrett *n*; Reißbrett *n* — **n** tekenbord *n*; tekenplank *fm* — **e** tablero *m* de dibujo; mesa *f* de dibujar — **i** tavolo *m* da disegno

4707 drawing ink — See India ink.

4708 drawing ink for fountain pens
f encre *f* de Chine stylographique — **d** Füllhalter-tusche *f*; Füllhalterzeichentusche *f* — **n** tekeninkt *m* voor vulpenhouder — **e** tinta *f* China estilo-gráfica — **i** inchiostro *m* per penne stilografiche

4709 drawing paper; drafting paper — Heavily sized paper with rough or smooth finish, for pencil, pen-and-ink and crayon drawings.
f papier *m* à dessin — **d** Zeichenpapier *n* — **n** tekenpapier *n* — **e** papel *m* de dibujo; papel *m* para dibujo — **i** carta *f* da disegno

4710 drawing pen
f plume *f* à dessin — **d** Zeichenfeder *f* — **n** teken-pen *fm* — **e** pluma *f* de dibujo — **i** penna *f* da disegno; penna *f* da disegnare

4711 drawing pen; straight-line pen
f tire-ligne *f* — **d** Ziehfeder *f*; Reißfeder *f* — **n** trekpen *fm* — **e** tiralíneas *m* — **i** tiralinee *m*

4712 drawing pin *GB*; **thumb tack** *US* — Tack with a large, flat head, to be thrust into a board or other fairly soft object by the pressure of the thumb.
f punaise *f* — **d** Reißnagel *m*; Reißzwecke *f* — **n** punaise *f* — **e** chinche *f*; chincheta *f* — **i** cimice *f*; puntina *f* (da disegno)

4713 drawing rollers; draw rollers — Pair of rollers, usually both driven, to control the tension of the web in a rotary press.
f rouleaux *mpl* tendeurs; rouleaux *mpl* de traction; galets *mpl* tendeurs — **d** Zugwalzen *fpl* — **n** trekrollen *fmpl* — **e** rodillos *mpl* tiradores; rodillos *mpl* llamadores; rodillos *mpl* de tiro — **i** rulli *mpl* di trazione

4714 drawing rollers; pull rollers — Pair of ser-rated rollers on a rotary web press, placed in the folder a sufficient distance apart to hold several sheets of paper and pull them into position for folding and cutting.
f rouleaux *mpl* d'entraînement — **d** Einführungs-walzen *fpl*; Zugwalzen *fpl* — **n** invoerrollen *fmpl*; trekrollen *fmpl* — **e** rodillos *mpl* introductores —

i rulli *mpl* d'introduzione; rulli *mpl* di trazione

4715 drawing triangle
f équerre *f* en triangle — **d** Zeichendreieck *n*; Dreieck *n* — **n** tekendriehoek *m*; driehoek *m* — **e** escuadra *f* de dibujo; escuadra *f* de delineante; cartabón *m* — **i** triangolo *m* da disegno

4716 draw out; pull out — Piece of movable type pulled out from the forme by the suction of the forme rollers, due to poor locking up and bad justi-fication. See also fall out.
f blancs *mpl* qui lèvent — **d** Spiesse *mpl* — **n** gerezen wit *n*; uitgetrokken letter *fm* — **e** levanta-miento *m* de tipo; levantamiento *m* de letras; desprendimiento *m* de letra(s); salir(se); tipo *m* levantado; cascabel *m* — **i** carattere *m* rialzato; lettera *f* rialzata; carattere *m* all'aria; lettera *f* all'aria; spazio *m* all'aria

4717 draw out — See drawdown.

4718 draw rollers — See drawing rollers.

4719 draw sheet; top draw sheet; top sheet; tym-pan sheet — Top sheet of paper, usually kraft or manila paper, drawn over the makeready on the platen or on the printing cylinder.
f feuille *f* de mise; feuille *f* d'assise; feuille *f* de dessus; feuille *f* de couverture — **d** Deckbogen *m*; Spannbogen *m*; Straffer *m*; Straffen *n* — **n** span-vel *n* — **e** pliego *m* de cubrir; cubretímpano *m*; cubierta *f* del arreglo; capa *f* del arreglo; hoja *f* exterior de la cama — **i** foglio *m* di maestra

4720 draw tool; drag tool — Finisher's tool in the form of a hooked steel blade, for cutting white lines and borders into relief plates.
f griffe *f* — **d** Reißnadel *f*; Anreißnadel *f*; Abreiß-nadel *f* — **n** krasnaald *fm* — **e** cortarraya *f* — **i** truschino *m*

4721 dress *v* **a cylinder** — To cloth the cut away portion of a cylinder to provide the necessary resilience between the printing and impression surfaces. Usually consists of two parts: permanent packing, made of manila sheets held in place by a sheet of calico; temporary packing, remaining thickness made up of normal cylinder sheets and a top sheet of manila tightened down by a draw bar.
f habiller un cylindre — **d** den Aufzug machen — **n** een legger maken; een cilinderbekleding maken — **e** preparar la cama del cilindro; revestir; guarnecer — **i** rivestire un cilindro

4722 dress *v* **a forme** — To fit or place furniture to pages or chases, quoins, etc., before locking up.
f garnir la forme — **d** die Stege anlegen; Format-stege anlegen — **n** sluitwit aanbrengen — **e** guarnecer la forma — **i** guarnire la forma

4723 dressing clamp; cylinder dressing clamp
f mâchoire *f* de serrage d'habillage — **d** Aufzug-klappe *f* — **n** leggerklem *fm* — **e** agarradera *f* — **i** morsetto *m* per il rivestimento

4724 dressing stick — Straight edge used in type foundries for a metal plate with a sharp edge at the long side, which, when placed over a row of new-cast types and checked against a light source, shows whether the faces of the types are at the same level.

f jeton *m*; mire *f* à justifier — **d** Besehblech *n* — **n** ston *m* — **e** comprobante *m*; esquadra *f* comprobante — **i** squadretta *f* verificatrice

4725 drier — See dryer.

4726 drier — See drying agent.

4727 drier absorption — See drier dissipation.

4728 drier dissipation; loss of drier; drier absorption *depr.* — Loss in catalytic power of a drier due to a physical absorption or a chemical reaction with pigments.

f inhibition *f* du siccatif — **d** Trockenstoff-absorption *f* — **n** migreren *n* van de droogstof — **e** inhibición *f* del secante — **i** inibizione *f* dell'essiccante; perdita *f* del potere essiccante

4729 drier felt; felt — On a paper machine, the felt that keeps the paper web in contact with the drying cylinders.

f feutre *m* sécheur; feutre *m* à sécher — **d** Trockenfilz *m* — **n** droogvilt *n*; bovenvilt *n*; markeervilt *n* — **e** fieltro *m* secador; fieltro *m* superior; paño *m* superior de fieltro; fieltro *m* marcador — **i** feltro *m* essiccatore

4730 driers; drying cylinders — On a paper machine, steam-heated cylinders on which the paper web is dried.

f cylindres *mpl* sécheurs — **d** Trockenzylinder *mpl* — **n** droogcilinders *mpl* — **e** cilindros *mpl* secadores; tambores *mpl* para el secado — **i** cilindri *mpl* essiccatori; essiccatori *mpl*

4731 drill — Coarse cotton fabric in twill weave used for bookbinding.

f coutil *m*; treillis *m* — **d** Drell *m*; Drill *m*; Drilch *m*; Drillich *m* — **n** dril *n* — **e** dril *m*; lona *f* — **i** tela *f* di cotone grossolana

4732 drilling

f perçage *m*; forage *m* — **d** Bohren *n*; Ausbohren *n* — **n** boren *n* — **e** agujereado *m*; barrenado *m*; taladrado *m*; taladro *m* — **i** foratura *f*; trapanatura *f*

4733 drilling machine; paper drill — Machine to bore holes in paper or board.

f foreuse *f*; machine *f* à forer; perceuse *f* — **d** Bohrmaschine *f*; Papierbohrmaschine *f* — **n** boormachine *f*; papierboormachine *f* — **e** taladradora *f*; máquina *f* taladradora; máquina *f* de taladrar — **i** trapano *m* per carta; macchina *f* foratrice; foratrice *f*

4734 driography; dry planographic printing — Trade name for a lithographic process in which no use is made of water, invented by J.L. Curtin. A flat plate is coated with an ultra-violet light sensitive layer. Available in Europe since 1973. The production is stopped in 1977 due to high costs of platemaking.

f driographie *f* — **d** Driographie *f*; feuchtungsloser Flachdruck *m* — **n** driografie *f* — **e** driografía *f*; litografía *f* sin agua — **i** driografia *f*; stampa *f* driografica; stampa *f* senza bagnatura

4735 drip fountain — Fountain usually one page wide for running colour on a news press.

f encrier *m* sectionnel — **d** Farbkasten *m* für Eindruckwerk — **n** inktbak *m* voor indrukwerk — **e** tintero *m* para impresión posterior — **i** calamaio *m* a sezioni

4736 drive — Driving mechanism; shaft, pulleys, motors, etc.

f commande *f* — **d** Antrieb *m* — **n** aandrijving *f*; drijfwerk *n* — **e** accionamiento *m*; impulso *m*; mando *m*; impulsión *f*; transmisión *f*; propulsión *f* — **i** comando *m*; impulso *m*

4737 drive; shoulder — Space at the bottom or top of type which serves as a support for descenders or ascenders, and from which the bevel of the face begins.

f épaulement *m* — **d** Achsel *f*; Schulter *f* — **n** schouder *m* — **e** hombro *m*; rebaba *f*; quijada *f* Me — **i** spalla *f*

4738 drive — See strike.

4739 driven roller — Roller in a letterpress rotary geared to the press drive system to carry the web from one section of the machine to another and to maintain the tension of the web.

f rouleau *m* commandé — **d** angetriebene Walze *f* — **n** aangedreven rol *fm* — **e** rodillo *m* accionado — **i** rullo *m* comandato

4740 drive *v* **out** — To set matter widely word-spaced to increase the number of lines. See also take *v* in a line.

f chasser une ligne; espacer — **d** austreiben; (eine Zeile) ausbringen; erweitern — **n** (een regel) uitdrijven — **e** recorrer (une línea) — **i** rubare (una riga); allargare; allontanare

4741 drive *v* **out a double**

f chasser un doublure — **d** eine Hochzeit ausbringen — **n** een dubbelgezet woord weghalen — **e** suprimir una palabra repetida — **i** togliere un doppione

4742 driver program; driver tape — Software routine that produces output which drives or directs a specific typesetter and hence provides, with the text stream, the necessary coded instructions to cause the typesetter to perform functions, such as to advance film or paper, bring the appropriate type founts into play, size the type according to specifications, and arrange the derived characters into lines and text blocks on the page.

4743 driver tape — See driver program.

4744 drive side; far side; gear side; offside; off-lay — Part of the press on which the drive motor, shafts and gears are located, to differentiate from the *operating side.

f côté *m* commande — **d** Antriebseite *f* — **n** aandrijfzijde *fm* — **e** lado *m* de volante; lado *m* de la transmisión; costado *m* posterior — **i** lato *m* di comando; lato *m* comando; lato *m* comandi

4745 driving belt

f courroie *f* de transmission — **d** Antriebsriemen *m*; Treibriemen *m*; Riemen *m* für den Antrieb — **n** drijfriem *m* — **e** correa *f* de transmisión — **i** cinghia *f* di trasmissione

4746 driving by hand

f entraînement *m* manuel — **d** Handantrieb *m* —

n handaandrijving *f* — **e** impulso *m* a mano — **i** comando *m* a mano

4747 driving motor — Prime mover for driving a machine.
f moteur *m* de commande — **d** Antriebsmotor *m* — **n** aandrijfmotor *m*; drijfmotor *m* — **e** motor *m* de mando — **i** motore *m* di comando

4748 driving shaft
f arbre *m* de couche; arbre *m* de commande; arbre *m* de transmission — **d** Antriebswelle *f*; Treibwelle *f* — **n** drijfas *fm*; aandrijfas *fm* — **e** árbol *m* de transmisión; comunicador *m* — **i** albero *m* di comando

4749 driving wheel — Wheel which communicates motion to any part of a machine.
f roue *f* de transmission — **d** Antriebsrad *m* — **n** aandrijfrad *n*; drijfrad *n*; aandrijfwiel *n*; drijf-wiel *n* — **e** rueda *f* motriz — **i** ruota *f* di comando

4750 drop *v* — To unlock a forme and take the chase off.
f desserrer une forme — **d** aufschließen — **n** een vorm uitslaan — **e** abrir la forma; desacuñar la composición — **i** aprire la forma

4751 drop folio; dropped folio — Page number printed at the foot of the page.
f numéro *m* en bas de page — **d** Seitenzahl *f* am Fuß (der Seite); Kolumnenziffer *f*; Paginierung *f* — **n** paginacijfer *n* aan de voet (van de pagina); paginacijfer *n* onderaan (de pagina) — **e** folio *m* al pie (de la página) — **i** numero *m* al piede (di pagina)

4752 drop initial; drop letter — Large capital letter at the beginning of a chapter, which extends below the type line. See also two-line initial, three-line initial.
f lettrine *f* — **d** hängende Initiale *f* — **n** initiaal *n* over meer regels — **e** inicial *f* embutida — **i** lettera *f* iniziale estendentesi al disotto della linea

4753 drop in voltage — See voltage drop.

4754 drop letter — See drop initial.

4755 dropline indention
f renfoncement *m* en sommaire simple — **d** hängender Einzug *m*; Einzug *m* ab 2. Zeile — **n** inspringing *f* over twee regels — **e** sangría *f* escalonada; sangría *f* de escalón; sangría *f* en gradas — **i** rientranza *f* al disotto della linea

4756 dropout — See highlight halftone.

4757 drop-out halftone — See highlight halftone.

4758 drop-out halftone — See outlined halftone.

4759 drop-out negative — See highlight negative.

4760 dropped folio — See drop folio.

4761 dropped head — First page of a chapter, etc., beginning lower than other pages.
f page *f* de départ — **d** Seite *f* mit Vorschlag beginnen — **n** pagina *fm* van halve hoogte — **e** página *f* de partida — **i** pagina *f* con zona bianca in testa

4762 dropship *v* — To send a book to one address and the bill to another.
f envoyer un livre à une adresse et la facture à une autre — **d** Buch und Rechnung an getrennte Anschriften versenden — **n** boek en rekening naar verschillende adressen sturen — **e** despachar un libro a una dirección y la factura al otro — **i** spedire un libro ad uno indirizzo e il conto ad altro

4763 dross; skimmings — Scum which forms when melting metals.
f crasses *fpl*; scorie *f* — **d** Krätze *f*; Krätzmetall *n*; Bleiasche *f*; Metallasche *f* — **n** loodas *fm*; metaalas *fm* — **e** escoria *f* — **i** scoria *f*

4764 drum, drying — See drying drum.

4765 drum, ink — See ink drum.

4766 drum printer; barrel printer — Line printer which prints "on the fly" from a drum engraved with identical characters in each print position across the drum and the full character set engraved in each print position around the drum.

4767 drum type press — See common impression cylinder press.

4768 dry back — Change in colour of a print while drying, due to physical change in the surface; more noticeable in reds which become yellower because of the development of *bronze.
f virage *m* — **d** Farbveränderung *f* beim Trocknen — **n** kleurverandering *f* door het drogen — **e** viraje *m*; decoloración *f* al secarse — **i** viraggio *m*

4769 dry broke — Broke accumulated at any stage on the dry end of the paper or board machine, trimmings from the reeling, slitting and cutting operations, as well as paper rejected during sorting.
f cassés *mpl* (de fabrication) secs — **d** Trockenausschuß *m*; trockene Fabrikationsabfälle *mpl* — **n** droge uitval *m*; uitval *m* bij de droogpartij — **e** roturas *fpl* de fabricación; desconchados *mpl* — **i** fogliacci *mpl* secchi

4770 dry-brush drawing — Technique to create a distinctive stroke and contrast to a drawing with a relatively dry pigment.
f crachis *m* à la brosse — **d** Trockenpinselzeichnung *f* — **n** droge-penseel-tekening *f* — **e** dibujo *m* a pincel seco — **i** disegno *m* a pennello secco

4771 dry colour — Pigment in dry or powder form.
f couleur *f* sèche — **d** Trockenpigment *n*; Trockenfarbe *f* — **n** droog pigment *n*; droge kleurstof *fm* — **e** color *m* seco; color *m* en polvo — **i** pigmento *m* secco

4772 dry content; dryness — Relation between the mass remaining after drying by a standard method and the mass at the time of sampling, expressed as a percentage.
f siccité *f*; teneur *f* en matière sèche — **d** Trockengehalt *m*; Trockenheit *f* — **n** drooggehalte *n*; droogheid *f*; droogte *f* — **e** sequedad *f* — **i** contenuto *m* in secco; siccità *f*

4773 dry enamel process
f procédé *m* à émail à sec; émail *m* à sec — **d** Trockenemailverfahren *n* — **n** droog email-procédé *n* — **e** procedimiento *m* al esmalte seco — **i** procedimento *m* allo smalto secco

4774 dry end; dryer section — Drying section of the paper machine, consisting mainly of the driers, calenders, reels and slitters.
f section f des sécheurs; sécherie f; séchoir m — **d** Trockenpartie f — **n** droogpartij f — **e** sequería f; batería f de secadores — **i** seccheria f
4775 dryer; drying oven; dryer unit GB; **drier** US — Oven in the rotary offset press, through which the web passes after it leaves the last printing unit, used with heat-set inks. Heats the web about 177 °C using either gas, electricity, or steam to dry the vehicles. Air blasts are used to drive off volatile gases. Results in higher setting temperature for ink.
f unité f de séchage; séchoir m — **d** Trockenvorrichtung f; Trockenofen m — **n** drooginstallatie f — **e** unidad f de secado; dispositivo m de secado — **i** dispositivo m essiccatore
4776 dryer — See matrix dryer.
4777 dryer glazer — Equipment which automatically dries wet prints and give them a smooth, glossy surface (a glaze).
f sécheuse-glaceuse f — **d** Hochglanzpresse f — **n** droog-glansmachine f — **e** secadora abrillantadora f — **i** asciugatrice-smaltatrice f
4778 dryer scale — See calender scale.
4779 dryer section — See dry end.
4780 dryer unit — See dryer.
4781 dry-finish butchers wrap; butchers paper; butchers wrap; butchers wrapping paper — Well-sized, dry finished wrapping paper, made from mechanical and/or chemical wood pulp, adapted (waxed or paraffined) to the wrapping of meats; has a basis weight of 35-50 pounds (24 x 36 - 500) and is sold in standard size counter rolls and sheets.
f papier m (de) boucherie — **d** Fleischeinwickelpapier n; Fleischerpapier n — **n** vleesinwikkelpapier n — **e** papel m para carnicerías — **i** carta f per macelleria
4782 dry flong — See dry mat.
4783 drying — Conversion of an ink film to a solid state, accomplished by oxidation, evaporation, polymerization, penetration, gelation, precipitation, either singly or in combination.
f séchage m — **d** Trocknung f; Trocknen n — **n** droging f; drogen n — **e** secado m — **i** asciugatura f; asciugamento m; essiccamento m
4784 drying agent; siccative; drier depr. — Substance added to inks to hasten drying. They consist mainly of metallic salts which exert a catalytic effect on the oxidation and polymerization of the oil vehicles used. See also metallic soap.
f siccatif m; pâte f siccative — **d** Trockenstoff m; Trockner m; Trockenpasta f; Trockenmittel n; Sikkativ n — **n** droogstof fm; droogmiddel n; droger m — **e** secante m; secativo m — **i** pasta f essiccante
4785 drying blanket — See blanket.
4786 drying cabinet — Heated metal cabinet for drying negatives and other surfaces.
f étuve f; armoire m à sécher; cabine f sécheuse

— **d** Trockenschrank m; Wärmeschrank m — **n** droogkast f; droogstoof fm — **e** armario m para secar; armario m de secado; armario m secador; estufa f secadora; secadora f — **i** armadio m essiccatore; cabina f di asciugamento
4787 drying cylinders — See driers.
4788 drying drum
f tambour m sécheur; tambour m de séchange — **d** Trockentrommel f; Trockenzylinder m; Heiztrommel f — **n** droogcilinder m — **e** tambor m secador; cilindro m secador — **i** tamburo m essiccatore; cilindro m essiccatore
4789 drying felt — See blanket.
4790 drying in — Filling of cells in engraved cylinders with dry ink.
f séchage m de l'encre dans les alvéoles du cylindre gravé; bouchage m — **d** Eintrocknen n — **n** indrogen n — **e** secado m de la tinta en las alvéolos del cilindro — **i** essiccamento m dell'inchiostro negli alveoli dei cilindri
4791 drying loft — Loft or ventilated room for air drying paper.
f chambre f séchoir; séchoir m — **d** Trockenkammer f; Trockenboden m; Trockenraum n; Darrboden m — **n** droogkamer fm — **e** secadero m; tendedero m; tendalero m — **i** camera f di asciugamento
4792 drying marks — Imperfections on the surface of a photographic film caused by uneven drying or by chemicals dissolved in the washing water.
f taches fpl de séchage — **d** Trockenflecke mpl — **n** droogstrepen fmpl; droogvlekken fmpl — **e** trazos mpl de secado; rayas fpl de secado; manchas fpl de secado — **i** macchie fpl di asciugamento
4793 drying oil — Oily, organic liquid, occurring naturally, as linseed, soyabean, or dehydrated castor oil, or synthesized, which, on exposure to the air, absorbs oxygen and changes to a hard, tough, elastic substance.
f huile f siccative — **d** trocknendes Öl n — **n** drogende olie fm — **e** aceite m secante — **i** olio m essiccativo; olio m essiccante
4794 drying oven; drying stove
f étuve f — **d** Trockenofen m; Trockenschrank m — **n** droogstoof fm; droogkast fm — **e** estufa f secadora; horno m de secado — **i** stufa f per essiccare; stufa f essiccatore; armadio m di asciugamento; cabina f di asciugamento; forno m essiccatore
4795 drying oven — See dryer.
4796 drying process
f processus m de séchage; séchage m — **d** Trockenvorgang m; Trocknungsvorgang m — **n** droogproces n; droging f — **e** procedimiento m de secado — **i** essiccamento m; procedimento m di essicamento
4797 drying rack — Skeleton rack or frame on which printed sheets are spread to allow the ink to dry.
f râtelier m de séchage; séchoir m; claie f; rayon-

layette *f* — **d** Trockenregal *m*; Trockengestell *n*; Trockenständer *m* — **n** droogrek *n* — **e** secadero *m*; estante *m* para secar; anaquel *m* de secar — **i** essiccatoio *m*; rastrelliera *f* d'essiccazione; rastrelliera *f* per seccare; rastrelliera *f* per asciugare

4798 drying stove — See drying oven.

4799 drying through; through drying
f séchange *m* à cœur — **d** Durchtrocknung *f*; Durchtrocknen *n* — **n** doordrogen *n* — **e** secado *m* a fondo — **i** essiccazione *f* a fondo

4800 drying time — Time required for an ink film to become hard enough to handle. The degree of hardness required varies greatly from one case to another. The drying effect can be produced by oxidation, evaporation, penetration, precipitation, or solidification, or combinations of these. Many inks will set and be dry enough to handle or back up before they reach their final stage of hardness.
f temps *m* de séchage — **d** Trockenzeit *f*; Trockendauer *f* — **n** droogtijd *m* — **e** tiempo *m* de secado — **i** tempo *m* di essiccamento

4801 drying-time recorder — Instrument to determine the drying quality of an ink or paper, recording the time of offsetting of the ink printed on a strip of paper onto a sheet of offsetting paper. Replaced by the *rub-dryness tester.

4802 drying time test
f essai *m* de séchage — **d** Trocknungsversuch *m* — **n** droogproef *fm* — **e** medida *f* de la velocidad de secado — **i** prova *f* d'essiccamento

4803 drying tunnel; tunnel drier — Well-insulated sheet metal tunnel or large box, through which paper or board is passed for drying.
f tunnel *m* de séchage; gaine *f* de séchage — **d** Trockentunnel *m*; Trockenvorrichtung *f* — **n** droogtunnel *m* — **e** túnel *m* de secado — **i** tunnel *m* di essiccamento; tunnel *m* di essiccazione

4804 drying varnish
f vernis *m* siccatif — **d** Trockenfirnis *m* — **n** droogvernis *mn* — **e** barniz *m* secante — **i** vernice *f* essiccante; vernice *f* siccativa

4805 dry mat; dry flong — Ready formed matrix for the casting of stereotypes.
f flan *m* sec — **d** Trockenmater *f* — **n** droge matrijs *f* — **e** matriz *f* seca; matriz *f* de cartón; flan *m* *Es*; flan *m* estereotípico; flan *m* de estereotipia — **i** flano *m* secco

4806 dry mat — See flong.

4807 dry mat scorcher — See matrix dryer.

4808 dry mounting — Preparation of gravure cylinders to make the exposed pigment paper sink to the surface of the cylinder or plate before development, to prevent stretching of the pigment paper which would occur if the paper would be thoroughly wetted by only applying sufficient water to the gelatine surface to assure adherence and permitting no water to reach the paper.
f application *f* à sec — **d** Trockenübertragung *f*; Maschinenübertragung *f* — **n** droge overdraging *f* — **e** aplicación *f* en seco; reporte *m* en seco — **i** trasporto *m* a secco

4809 dry needle — See etching needle.

4810 dry-needle etching — See dry-point.

4811 dryness — See dry content.

4812 dry offset; dry relief offset; offset letterpress; letterset — Process in which a metal plate is etched to a depth of approx. 0.15 mm (0.006 in), making a right-reading relief plate, printed on the offset blanket and then to the paper without the use of water.
f offset *m* à sec; offset *m* sec; procédé *m* offset à sec; typographie *f* indirecte — **d** indirekter Buchdruck *m*; Lettersetdruck *m*; Trockenoffsetdruck *m*; Trockenoffsetverfahren *n*; Hochoffsetdruck *m* — **n** droge offsetdruk *m*; droge offset *m*; droog offsetprocédé *n*; indirecte boekdruk *m* — **e** tipoffset *m*; tipo-offset *m*; tipografía *f* indirecta; impresión *f* tipográfica indirecta; offset *m* tipográfico — **i** offset *m* a secco; tipografia *f* indiretta; stampa *f* tipografica indiretta; letterset *m*

4813 dry pick — Occurrence of picking in the absence of water introduced by the printing process.
f arrachage *m* à sec — **d** Trockenrupfen *n*; Rupfen *n* ohne Einwirkung von Wasser — **n** droge pluk *m* — **e** repelado *m* seco; picado *m* seco; arrancado *m* de fibras en seco — **i** strappo *m* superficiale a secco

4814 dry pint — Unit of capacity (dry measure). Abbrev. dry pt (no point). See app. no. 7 and 7a.
f pinte *f* sèche — **n** dry pint *fm* — **e** pinta *f* seca

4815 dry planographic printing — See driography.

4816 dry plate — Glass plate sensitized with a film of gelatine/silver emulsion, the plate exposed in a dry condition. See also wet plate.
f plaque *f* sèche — **d** Trockenplatte *f* — **n** droge plaat *fm* — **e** placa *f* seca — **i** lastra *f* secca

4817 dry-point; dry-point engraving; dry-needle etching — Process where the metal plate is scratched with a pointed hand tool. The intaglio image consists of the turned-up metal, the so-called burr, behind which the ink is deposited. The metal is usually copper, steel, brass, aluminium, or zinc. The cutting or scratching point may be steel, diamond, sapphire, or ruby jewel set in metal.
f gravure *f* à la pointe sèche; pointe-sèche *f*; gravure *f* directe — **d** Kaltnadelradierung *f* — **n** droge-naaldets *fm* — **e** grabado *m* a puntaseca — **i** incisione *f* a puntasecca; puntasecca *f*

4818 dry-point engraving — See dry-point.

4819 dry proving — Printing of succeeding colours after the previous colour has set or dried.
f épreuve *f* sur couleurs sèches — **d** Trocken-in-Trocken-Andruck *m* — **n** proeftrekken *n* nadat de laatste aangebrachte kleur droog is — **e** ensayo *m* sobre colores secos — **i** prova *f* di stampa su colori secchi

4820 dry quart — Unit of capacity. Abbrev. dr qt (no point). See app. no. 7 and 7a.

4821 dry relief offset — See dry offset.

4822 dry solids content — See solids content.

4823 dry spray — Anti set-off spray with a

powder.

f pulvérisateur *m* à poudre; appareil *m* à poudre — **d** Druckbestäuber *m* — **n** droogsproei-apparaat *n*; droogsproeier *m*; drukbestuivings-apparaat *n* — **e** pulverizador *m* (antirrepinte); aparato *m* pulverizador; antimaculador *m*; pulverizador-antimaculador *m*; rociador *m* anti-maculador — **i** antiscartino *m* secco

4824 dry spray powder

f poudre *f* anti-maculage — **d** Druckbestäubungs-pulver *n*; Druckbestäubungsmittel *n* — **n** anti-smetpoeder *n* — **e** polvo *m* antimaculador — **i** polvere *f* antiscartino

4825 dry stamping

f timbrage *m* à sec — **d** Trockendruck *m* — **n** droogstempelen *n* — **e** timbrado *m* a seco — **i** stampa *f* a secco

4826 dry-transfer letters — See self-adhering letters.

4827 dry trap — See crystallization.

4828 dry weight

f poids *m* à sec — **d** Trockengewicht *n* — **n** droog-gewicht *n* — **e** peso *m* en seco — **i** peso *m* a secco

4829 dry wrinkles — See calender creases.

4830 dual reel stand; dual roll stand *depr.* — Reel stand supporting two reels, one above the other, to feed two webs at the same time, or to reduce re-loading down time if only a single web is being used.

f dérouleur *m* double; porte-bobine *m* double — **d** doppelter Rollenständer *m*; doppelte Rollen-lagerung *f* — **n** twee-rollenstandaard *m* — **e** porta-bobina *m* doble; portarrollo *m* doble — **i** porta-bobine *m* doppio

4831 dual roll stand — See dual reel stand.

4832 dubic decimeter — See cubic decimetre.

4833 Ducali binding — Bookbinding and decoration style used in the 16th century for the decrees of the Doges of Venice. Combination of Western and Oriental styles. The bookboard was applied with a paper composition, then a thin leather pasted on it, the corners and the centre being recessed. After this, the background was coloured and decorated with arabesques of gold.

f reliure *f* Ducali — **d** Ducali-Einband *m* — **n** Ducali-band *m* — **e** encuadernación *f* Ducali; encuadernación *f* veneciana — **i** legatura *f* Ducali

4834 duchess — A British standard size of printer's card. See app. no. 3.

4835 duchess quarto — A British size of note paper. See app. no. 3.

4836 duck — See canvas.

4837 duckfoot quotes — See quotation mark.

4838 duct adjusting screws; fountain keys; fountain blade adjusting screws; feed screws; ductor keys; fountain screws — The screws which position the duct blad to the duct roller, to regulate the thickness of the ink film.

f vis *fpl* de l'encrier; vis *fpl* de réglage de l'encrier — **d** Zonenschrauben *fpl*; Farbregulierschrauben *fpl*; Stellschrauben *fpl* — **n** inktbakschroeven

fmpl; stelschroeven *fmpl* van de inktbak — **e** tornillos *mpl* graduadores del tintero; tornillos *mpl* reguladores del tintero — **i** viti *fpl* del calamaio

4839 duct blade; ductor blade; fountain blade — Bevelled blade attached to the ink duct with ad-justing screws, to regulate the amount of ink delivered by the duct or ductor roller by setting of the blade.

f lame *f* de l'encrier; couteau *m* de l'encrier — **d** Duktorlineal *n* — **n** mes *n* van de inktbak; bak-mes *n* — **e** cuchilla *f* del tintero — **i** lama *f* del calamaio

4840 duct, ink — See ink duct.

4841 ductor; ductor roller; transfer roller; ink-feed roller; feed roller; vibrator; vibrating roller — Roller in inking and damping mechanism which alternatively contacts fountain roller and drum roller. Length of contact or dwell of ductor can be adjusted. There is some confusion between authors. At one time it may be the ductor, at an-other time an oscillating distributor roller. In the US it is a metal or hard roller. See also lifter roller.

f rouleau *m* preneur; preneur *m* — **d** Leckwalze *f*; Farbheber *m*; Heber *m* — **n** likrol *fm* — **e** rodillo *m* tomador; tomador *m* — **i** rullo *m* prenditore

4842 ductor blade — See duct blade.

4843 ductor keys — See duct adjusting screws.

4844 ductor knife — See ink knife.

4845 ductor roller — See ductor.

4846 duct roller; ink-duct roller; fountain roller; ink-fountain roller — Roller which supplies the ink from the duct to the inking system by rotating through the ink and against the duct blade.

f rouleau *m* de l'encrier; rouleau *m* d'encrier; cylindre *m* d'encrier — **d** Farbkastenwalze *f*; Farb-zylinder *m*; Duktorwalze *f*; Duktor *m* — **n** inkt-bakrol *fm* — **e** rodillo *m* del tintero — **i** rullo *m* del calamaio; rullo *m* duttore

4847 duct roller — On flexographic printing machines a rubber covered or screened chromium-faced roller, rotating in the ink duct and transferring the ink onto the forme roller and plate cylinder.

f rouleau *m* barboteur; barboteur *m* — **d** Tauch-walze *f* — **n** dompelrol *fm* — **e** cilindro *m* del tintero; rodillo *m* del tintero; cilindro *m* de inmersión — **i** rullo *m* pescante nel calamaio

4848 dud — See unsaleable book.

4849 Dufay color — Colour film of the screen-plate type to produce direct colour transparencies from coloured objects.

4850 duff *v* out — To opaque certain regions of a negative for separating colours.

f boucher — **d** abdecken — **n** uitdekken — **e** opacar — **i** opacizzare

4851 duke — A British standard size of note paper; also a British card size. The latter is also known as *official, when used for postcards. See app. no. 3.

4852 dull-coated printing paper — Paper with a coated surface low in gloss, usually a free sheet base stock coated on two sides with calcium carbonate or blanc fixe and finished to flatten or smooth the surface and give minimum gloss or glare. Made in all standard book sizes and weights.
f papier *m* couché mat — **d** mattgestrichenes Papier *n* — **n** mat gestreken papier *n*; mat gecoucheerd papier *n* — **e** papel *m* estucado mate — **i** carta *f* patinata mat

4853 dull finish; mat finish; matt finish — Finish low in gloss, applied to papers coated with low gloss pigments, such as carbonates and blanc fixe, specifically to any coated box paper with a glare test less than 55%, as distinguished from a semi-dull finish which covers the field between dull and glossy.
f apprêt *m* mat; fini *m* mat — **d** mattsatiniert; mattgeglättet; Mattglanz *m*; matte Oberfläche *f* — **n** mat gesatineerd — **e** aprestado *m* mate — **i** finitura *f* mat

4854 dull ink — Ink which dries with little or no lustre.
f encre *f* mate — **d** Mattfarbe *f* — **n** matte inkt *m* — **e** tinta *f* mate; tinta *f* apagada — **i** inchiostro *m* mat

4855 Dultgen method — Method to make screen positives with unjoined shadow dots (without altering the camera setting), and low-contrast continuous-tone positives. These can be dot etched or retouched, respectively.
f procédé *m* Dultgen — **d** Dultgenverfahren *n* — **n** Dultgen-procédé *n* — **e** procedimiento *m* de Dultgen; método *m* de Dultgen — **i** procedimento *m* di Dultgen; metodo *m* di Dultgen

4856 dumb tooling — See blind tooling.

4857 dummy; dummy copy *GB*; **mock-up** *US*; **bulking dummy** — Model resembling the finished book in every respect except that the pages and cover are blank, used by the designer as a final check on the appearance and "feel" of the book as a guide for the size and position of elements on the jacket, and as a positive indication of the width of the spine (for the stamping).
f maquette *f* papier; livre *m* modèle; modèle *m* — **d** Blindband *m*; Blindmuster *n*; Probeband *m*; Musterband *m* — **n** dummy *m*; model *n* — **e** modelo *m*; maqueta *f* — **i** menabò *m*; modello *m*

4858 dummy copy — See dummy.

4859 dummy make-up — Dummy with galley proof cut and pasted in.
f maquette *f* truffée — **d** Blindband *m* mit eingeklebten Korrekturabzügen — **n** ingeplakte proef *fm* — **e** maqueta *f* acabada — **i** libro *m* in bianco con bozze incollate

4860 dummy plates — Blank plates below printing diameter used to fill the cylinder for lock-up a letterpress rotary.
f faux-clichés *mpl* — **d** Blindplatten *fpl* — **n** vulplaten *fmpl* — **e** planchas *fpl* falsas; falsas planchas *fpl* — **i** false lastre *fpl*

4861 dump; bank; dumping bank; proof bank — Usually metal covered table, bench or shelf in a composing room to hold newly set type matter in galleys or without galleys for correction, make-up, etc.
f table *f* pour la composition — **d** Tisch *m*; Bank *f* — **n** uitzettafel *fm* voor zetsel — **e** mesa *f* para composición (en galeras); galerón *m* (de pruebas); banco *m* de pruebas; tablero *m* de pruebas; portamoldes *m*; portagaleras *m* *Pa* — **i** banco *m* da composizione; tavola *f* per deposito della composizione

4862 dump — Computer term. Location of the data resulting from the dumping process.
f emplacement *m* de l'analyse-mémoire; emplacement *m* de l'image-mémoire — **d** Speicherauszugstelle *f* — **n** bergplaats *fm* — **e** situación *f* de memoria retirada; situación *f* de memoria transferida — **i** locazione *f* di memoria trasferita

4863 dump *v*; **preserve** *v* — To retain by storage elsewhere the contents of a set of locations either because the locations are temporarily required for another purpose or as a safeguard, e.g., against power failure, or for a check.
f faire une analyse-mémoire; prendre une analyse-mémoire — **d** aufbewahren; sicherstellen — **n** bergen; redden — **e** transferir — **i** emettere

4864 dumping bank — See dump.

4865 dump point — See check point.

4866 dump *v* **the stick** — See empty *v* the stick.

4867 duodecimo — See twelvemo.

4868 duograph — See duotone.

4869 duotone; duograph; duotype — Colour reproduction from monochrome original produced by making two halftone negatives for opposite ends of grey scale at proper screen angles, and plates etched to represent proper tone and colour values. Keyplate usually printed in dark colour for detail, while second plate is printed in light flat tints. See also duplex halftone.
f duotone *m*; simili *m* deux tons; camaïeu *m* — **d** Duoton *m*; Duplexdruck *m* — **n** duplex-reproductie *f* — **e** doble tono *m*; bitono *m* — **i** riproduzione *f* duotone

4870 duotone ink — See double-tone ink.

4871 duotype — See duotone.

4872 dupe — See duplicate plate.

4873 duplex bag; double-wall bag — Bag of two plies of packaging materials, i.e., with double walls which may or may not be spot glued.
f cornet *m* à double paroi; sac *m* (en papier) doublé — **d** doppelwandiger Papiersack *m*; zweilagiger Sack *m*; Zweilagenbeutel *m* — **n** dubbelwandige zak *m*; dubbele zak *m* — **e** saco *m* de papel duplicado; bolsa *f* de papel de doble hoja — **i** sacchetto *m* a parete doppia

4874 duplex board — See duplex cardboard.

4875 duplex cardboard; duplex board; two-layer board — Board of two layers of mechanical board couched one upon another and lined with paper.
f carton *m* duplex — **d** Duplexkarton *m* — **n** duplex-karton *n* — **e** cartulina *f* duplex — **i**

cartoncino *m* duplex

4876 duplexed fount — When two founts mounted on a typesetting machine are designed so that each character in one fount has the same width value as its equivalent character in the alternate fount, they are called duplexed. When three or more faces correspond character-for-character in set width values, they are called multiplexed founts. Unit count founts could be called omniplexed founts. Linecaster founts are almost always duplexed.

4877 duplex halftone — Two colour halftone blocks, made from a monochrome original, the second plate being used as a tint. See also duotone.

f simili-duplex *m* — **d** Duplexautotypie *f*; Duplexklischee *n*; Duplexrasterätzung *f* — **n** duplexcliché *n*; duplex-autotypie *f* — **e** autotipia *f* duplex — **i** cliscè *m* doppiatinta; cliché *m* doppiatinta; autotipia *f* duplex

4878 duplex ink — See double-tone ink.

4879 Duplex machine — Trade name for a reel-fed letterpress machine for production of small newspapers, using flat formes, namely the original type composition locked up in a chase.

f machine *f* Duplex; rotative *f* à formes plattes — **d** Duplex-Maschine *f*; Flachformrotationsmaschine *f* — **n** Duplex-machine *f* — **e** máquina *f* Duplex; máquina *f* rotativa Duplex; prensa *f* Duplex — **i** macchina *f* Duplex; macchina *f* rotativa Duplex; rotativa *f* Duplex

4880 duplex matrix — See double-letter matrix.

4881 duplex paper; two layer paper — Paper with different colours, textures, or finish on the two sides, made by pasting two different kinds of paper together on a cylinder machine or a cylinder/Fourdrinier machine.

f papier *m* bicolore; papier *m* duplex — **d** zweifarbiges Papier *n*; Duplexpapier *n*; doppelfarbiges Papier *n*; zweiseitiges Papier *n* — **n** tweezijdig gekleurd papier *n*; tweezijdig papier *n* — **e** papel *m* (de) doble faz; papel *m* duplo; papel *m* duplex; papel *m* bicolor — **i** carta *f* duplex; carta *f* bicolore

4882 duplex screen — Halftone screen for production of highlight effects and combination line-halftone negatives.

f trame *f* duplex — **d** Duplexraster *m* — **n** duplex-raster *n* — **e** trama *f* dúplice — **i** retino *m* duplex

4883 duplex set — Combination of barring and crawl motor; main driving motor to provide power for the rotary press at all speeds.

d Doppelantrieb *m* — **n** duplex-aandrijving *f* — **e** accionamiento *m* combinado — **i** comando *m* motori combinato

4884 duplicate plate; dupe — Extra printing plate made from the same negative, etched and finished in the same manner, used in lieu of electrotype or stereotype.

f réplique *f* — **d** Duplikat-Klischee *n* — **n** duplicaat-cliché *n*; extra cliché *n* — **e** clisé *m* duplicado — **i** cliscè *m* duplicato; cliché *m* duplicato; duplicato *m*

4885 duplicating stencil *GB*; **stencil paper; cyclostyle paper** *US*; **wax stencil** — Paper of a special structure impregnated and/or coated with a suitable preparation for reproduction of textual matter or patterns impressed on it to permit the passage of an appropriate ink. The impression is generally obtained with a typewriter, by hand with a special stylus or by photomechanical process.

f stencil *m* — **d** Schablone *f*; Dauerschablone *f*; Wachsmatrize *f*; Schreibmaschinematrize *f* — **n** stencil *n* — **e** estencil *m* — **i** stencil *m*

4886 durability — Property of papers used for documents which are to be preserved. Rag papers are best suited.

f durabilité *f* — **d** Haltbarkeit *f*; Dauerhaftigkeit *f* — **n** houdbaarheid *f*; duurzaamheid *f*; levensduur *m* — **e** durabilidad *f* — **i** durabilità *f*; conservabilità *f*

4887 durometer — See hardness meter.

4888 durometer hardness — See Shore hardness.

4889 dust box — See dusting box.

4890 dust cover — See jacket.

4891 duster; rag duster; dusting machine; thrasher; rag thrasher — Perforated rotating cylinder or drum in which old rags and papers are shaken to remove dust and foreign particles.

f époussiéreuse *f*; blutoir *m* (à chiffons) — **d** Siebtrommel *f*; Lumpenschüttelmaschine *f* — **n** lompenstuiver *m* — **e** batidor *m* de trapos; batán *m* de trapos; diablo *m*; rasgador *m*; cernedor *m*; cedazo *m* para trapos — **i** macchina *f* spolveratrice degli stracci

4892 dusting; powdering

f poudrer; talquer — **d** talkumieren — **n** poederen; inpoederen — **e** empolvar; empolvorar; empolvorizar — **i** spolverare

4893 dusting

f essuyage *m*; dépoussiérage *m* — **d** Abstäuben *n*; Entstäubung *f* — **n** afstoffen *n* — **e** despolvoreado *m*; despolvado *m* — **i** spolverare

4894 dusting — Removal of loose pigment material from the paper surface during printing. See also linting, fluffing.

f poussiérage *m* — **d** Stauben *n* — **n** stuiven *n* — **e** polvoreado *m* — **i** deposito *m* di fibre di carta

4895 dusting apparatus — See anti set-off spray.

4896 dusting box; dust box; grain box — Revolving cabinet containing resinous dust, used for depositing a dust grain on metal plates as a medium of tone translation; the cabinet or box containing the etching powder used in relief etching.

f boîte *f* à résine; boîte *f* à poudrer — **d** Staubkasten *m*; Einstaubkasten *m* — **n** stofkast *fm* — **e** caja *f* de espolvorear; caja *f* espolveadora; caja *f* de resina — **i** cassetta *f* per polvere; cassa *f* spolverizzatrice

4897 dusting brush — Brush to strew or sprinkle powder, dust or other fine particles, e.g., on photoengravings.

f brosse *f* à poudrer — **d** Einstaubpinsel *m*; Puderquaste *f* — **n** poederkwast *m* — **e** borla *f* para polvo — **i** spazzolino *m* da spolvero

4898 dusting brush — Brush to wipe dust or powder from a metal plate, a print, etc.

f pinceau *m* à épousseter; blaireau *m* à épousseter — **d** Abstaubpinsel *m* — **n** stofkwast *m*; afstofkwast *m* — **e** brocha *f* para despolvorear; brocha *f* para desempolvar — **i** spazzola *f* da spolverare

4899 dusting machine

f machine *f* à épousseter le bronze — **d** Abstaubmaschine *f* (an der Bronziermaschine) — **n** afstofmachine *f* — **e** máquina *f* para despolvorear — **i** spolveratrice *f*

4900 dusting machine — See duster.

4901 dusting of metal — See metal dust.

4902 dusting tape — Velvet tape in the bronzing machine.

f bande *f* d'essuyage (de bronzeuse) — **d** Abstaubband *n* — **n** stofband *n* — **e** cinta *f* para despolvorear — **i** nastro *m* spolveratore

4903 dust jacket — See jacket.

4904 dust, metal — See metal dust.

4905 dust, paper — See paper dust.

4906 dust proof — Capable of excluding dust.

f étanche à la poussière — **d** staubdicht; staubgeschützt — **n** stofdicht — **e** al abrigo del polvo; protegido contra el polvo — **i** a tenuta di polvere

4907 dust wrapper — See jacket.

4908 Dutch — Same as English but q and x only in borrowed foreign words. Accented letters are often used in stressed syllables. Marked letter ë and ö diaeresis. Ch must never be separated; for y use ij; for Y use IJ (not Ij, for the Dutch ij or IJ is one character).

f néerlandais *m*; hollandais *m*; néerlandais *adj.*; hollandais *adj.* — **d** Niederländische *n*; Holländische *n*; niederländisch *adj.*; holländisch *adj.* — **n** Nederlands *n*; Nederlands *adj.*; Hollands *n*; Hollands *adj.* — **e** neerlandés *m*; neerlandés *adj.*; holandés *m*; holandés *adj.* — **i** olandese *m*; olandese *adj.*

4909 Dutch bath — Synonym for *ferric chloride. It is the best mordant for softground etching.

4910 Dutch cut — Forme in which the work units are of different sizes and the final press sheet layout is not symmetrical.

4911 Dutch door — See door.

4912 Dutch engine — See beater.

4913 Dutch gold; Dutch leaf — Alloy of copper and zinc, used (only seldom), when beaten into leaf form, in tooling as an imitation of gold leaf; discolours quickly.

4914 Dutch leaf — See Dutch gold.

4915 Dutchman — Wooden pin driven into a type forme to tighten it.

f petite cale *f* en bois à l'intérieur d'une forme de mobile — **d** Holzspan *m* zum Austreiben schlecht ausgeschlossener Satzformen — **n** houten keg *fm* in een drukvorm; houten wig *fm* in een drukvorm — **i** zeppa *f*

4916 Dutch mordant — Mixture of potassium chlorate (2%) and hydrochloric acid (10%), used for fine etch work.

4917 Dutch paper — Originally Van Gelder's hand-made in Holland. Now any rough-surfaced paper, made of rags, with deckle edges.

f papier *m* de Hollande — **d** Holländisch-Bütten *n*; holländisches Papier *n* — **n** Hollands (geschept) papier *n*; oudhollands papier *n* — **e** papel *m* holandés; papel *m* de Holanda; papel *m* (de) barba; papel *m* imitación antiguo — **i** carta *f* d'Olanda

4918 Dutch top — In boxmaking, lid in which padded top is slightly less in length and width than lid, which is set-up with shell.

f couvercle *m* hollandais

4919 Dvorak keyboard — Keyboard arrangement, patented in 1932 by August Dvorak, an American of Czechoslovakian origin, to give better distribution of the characters than the *qwert layout.

4920 dwell — Length of time the ductor roller is in contact with the fountain roller.

f prise *f* d'encre — **d** Verweilzeit *f*; Verweilen *n*; Ruhen *n* — **n** contacttijd *m* — **e** punto *m* muerto — **i** tempo *m* di contatto

4921 dycril — Trade-mark for a photosensitive plastic bonded to steel, aluminium or Cronar base supports, used in printing plates. Exposure to an ultra-violet light source renders exposed areas insoluble to a subsequent mild alkaline washout solution so that a relief image of the exposed pattern remains. Depth of image is dependent on the thickness of the photosensitive plastic layer. There are eight types available.

4922 dye — Colloid material that imparts its colour to a surface by absorption or by chemical reaction. Dyes usually form colloid suspensions of particles too small to be seen with white-light microscopy. Pigment particles are considerably larger than dye particles. A dye can be exhausted from its bath, a true solution cannot.

f colorant *m*; matière *f* colorante — **d** Farbstoff *m* — **n** kleurstof *fm* — **e** colorante *m*; materia *f* colorante; tinte *m*; tintura *f* — **i** colorante *m*

4923 dye coupling developer

f révélateur *m* chromogène — **d** chromogener Entwickler *m*; Farbentwickler *m* — **n** kleurkoppelaar *m*; kleurontwikkelaar *m*; chromogeenontwikkelaar *m* — **e** acoplador *m* cromógeno — **i** rivelatore *m* cromogeno; sviluppatore *m* cromogeno

4924 dyeing — Application of a coloured or neutral dye to a photographic negative or positive to modify the density.

f retouche *f* par imbibition de colorant — **d** Anfärben *n* — **n** verfretouche *f* — **i** tintura *f*

4925 dyeline paper — See diazo paper.

4926 dye pigment — Dye insoluble in water that can be used as pigment without any chemical transformation.

f pigment-coloré *m* — **d** natürlicher Farbstoff *m*; Farbpigment *n* — **n** gekleurd pigment *n* — **i**

pigmento *m* colorato

4927 dye retouching — Retouching of continuous tone, line, or halftone negatives with solutions of red or black dyes.
f retouche *f* à la gouache — **d** Farbretusche *f* — **n** verfretouche *f* — **e** retoque *m* con tintes — **i** ritocco *m* con coloranti

4928 dyestuffs — Intensely coloured soluble organic compounds used for colouring by various processes and for production of many pigments.
f colorants *mpl*; matières *fpl* colorantes — **d** Farbstoffe *mpl* — **n** kleurstoffen *fmpl* — **e** materias *fpl* colorantes — **i** materie *fpl* coloranti; coloranti *mpl*

4929 dynamic smoothness of paper — When a glass roller is pressed against the paper and rolled over the paper surface it is possible to get an impression of the degree of contact between paper and roller, and thus of the smoothness of the paper. Contact is effected in the same way as in a print made on the press, i.e., this smoothness is measured under dynamic conditions. A light spot of 25 x 50 μm is projected in the contact zone, and with this the paper is scanned. The reflected signal is stored in the memory of a computer. From the frequency distribution the degree of contact and the spreading can be calculated, and an impression of the distribution of the smoothness over the paper surface can be obtained.
f lissé *m* dynamique du papier — **d** dynamische Papierglätte *f* — **n** dynamische gladheid *f* van papier — **e** lisura *f* dinámica del papel — **i** liscio *m* dinamico di carta

4930 dynamometer — Instrument to measure force exerted or power expended.
f dynamomètre *m* — **d** Dynamometer *m*; Kraftmesser *m* — **n** dynamometer *m* — **e** dinamómetro *m* — **i** dinamometro *m*

4931 dyne — Unit of force. The amount of force that, acting on one gram of matter for one second, gives it a velocity of one centimetre per second. Abbrev. dyn. In the SI:
1 dyn $= 1$ g·cm/s$^2 = 10^{-5}$ N, and
1 dyn/cm $= 10^{-3}$ N/m, and
1 dyn/cm$^2 = 0.1$ Pa.
f dyne *f* — **d** dyn *n* — **n** dyne *m* — **e** dina *f* — **i** dina *f*

4932 ear — Lug of a composing machine matrix.

f oreille *f* de la matrice — **d** Matrizenohr *n* — **n** oortje *n* van de matrijs; matrijzenoortje *n* — **e** oreja *f* de la matriz — **i** orecchia *f* della matrice; orecchietta *f* della matrice

4933 ear; ear space — Small box for notices, or for advertisements beside the front page nameplate of a newspaper.

f oreille *f* — **d** Anzeige *f* neben dem Zeitungskopf; Kurzinformation *f* neben dem Zeitungskopf — **n** kop-advertentie *f* — **e** anuncio *m* de cabecera; oveja *f* de cabecero — **i** occhiello *m*

4934 ear — In type, the finishing strokes of the g and the r.

4935 ear — See nib.

4936 early printed book — See incunabulum.

4937 early roman cursive

f vieux romain *m* courant — **d** ältere römische Kursiv *f* — **n** vroeg-Romeinse cursief *fm* — **e** cursivo *m* romano antiguo — **i** corsivo *m* romano antico

4938 ear space — See ear.

4939 earth pigment — Siennas, ochres and umber fallen into disuse owing to their coarseness and variability in colours and drying properties.

f couleur *f* terreuse; pigment *m* terreux; pigment *m* minéral naturel — **d** Erdfarbe *f*; natürliche Mineralfarbe *f* — **n** aardverfstof *fm* — **e** pigmento *m* de tierra; tinta *f* terrona — **i** pigmento *m* terroso

4940 easel — Device in which photographic paper is held flat during printing, generally equipped with an adjustable metal mask for framing.

f margeur *m* — **d** Vergrößerungsrahmen *m*; Vergrößerungskassette *f* — **n** afdrukraam *n* — **e** marginador *m*; marco *m* de copia — **i** marginatore *m*

4941 easel; support — Support pasted on the back of a showcard to sustain it in upright position for display.

f soutien *m* — **d** Rückenstütze *f* — **n** ezel *m* — **e** caballete *m*; atril *m* — **i** sostegno *m* per cartello; piedino *m*

4942 easer — Substance (oil or varnish) added to a printing ink to improve the absorption into the paper, accelerate drying, prevent picking, reduce set-off, etc.

f produit *m* d'addition — **d** Druckhilfsmittel *n* — **n** toevoegingsmiddel *n* — **e** suavizante *m* — **i** additivo *m*

4943 easy bleached pulp — Pulp that is not fully bleached.

f pulpe *f* légèrement blanchie; pâte *f* légèrement blanchie — **d** leicht gebleichter Stoff *m* — **n** licht gebleekte stof *fm* — **i** pasta *f* leggermente sbianchita

4944 eau de javel(le) — See javelle water.

4945 eau de Labarraque — See sodium hypochlorite.

4946 Eberhard effect — Effect of distribution of density in an image upon the development of specific areas of that image, named after G. Eberhard (1912), who exposed photographic plates behind a screen with circular openings of varying diameter from 0.3-30 mm. The densities of the developed plate were not uniform but the smaller images were denser than the larger, as if the smallest had received from one and a half times to twice as much exposure as the largest.

f effet *m* Eberhard — **d** Eberhard-Effekt *m* — **n** Eberhard-effect *n* — **e** efecto *m* de Eberhard — **i** effetto *m* di Eberhard

4947 ebonite
See hard rubber.

4948 eccentric — Disk mounted out of centre on a driving shaft, bound to it by a key, and surrounded by a collar or strap connected with a rod, effecting a crank motion.

f excentrique *m* — **d** Exzenter *m*; Exzentrik *f* — **n** excentriek *n* — **e** excéntrica *f* — **i** eccentrica *f*

4949 eccentric sheave

f roue *f* excentrique — **d** Exzenterscheibe *f* — **n** excentriekschijf *fm* — **e** disco *m* excéntrico — **i** rotella *f* eccentrica

4950 echoppe — See graver.

4951 écrasé leather — Leather mechanically crushed to produce a grained effect. See also crushed levant morocco.

f cuir *m* écrasé — **d** Ecrasé-Leder *n* — **n** écrasé *n*; écrasé leer *n* — **e** cuero *m* aplastado — **i** cuoio *m* cilindrato

4952 eddy current — Current set up in a substance by variation of an applied magnetic field, used in the driving, coupling and braking systems of some web splicers.

f courant *m* parasite; courant *m* de Foucault — **d** Wirbelstrom *m* — **n** wervelstroom *m* — **e** corriente *f* parásita; corriente *f* de Foucault — **i** corrente *f* parassita; corrente *f* di Foucault

4953 edge bolts — See bolts.

4954 edge effect — See mackie lines.

4955 edge fog

f voile *m* marginal — **d** Randschleier *m* — **n** randsluier *m* — **e** velo *m* marginal — **i** velo *m* marginale

4956 edge gilder

f doreur *m* de tranches — **d** Schnittvergolder *m* — **n** sneevergulder *m* — **e** dorador *m* de cantos; dorador *m* de cortes — **i** doratore *m* dei tagli

4957 edge mill — See kollergang.

4958 edge-notched card — See edge-punched card.

4959 edge-punched card; margin-punched card; edge-notched card; border-punched card — Standard-sized card to record information by punching a 5- to 8-level code along one edge of the card. Also called tape cards as they are supplied in sets of 25 cards and can be punched on suitable tape puncher. The term can also be applied to a variety of mechanical accounting cards.

f carte *f* à perforation marginale — **d** Randlochkarte *f*; Lochstreifenkarte *f* — **n** randponskaart *fm* — **e** tarjeta *f* de perforación marginal — **i** scheda *f* a perforazione marginale; scheda *f* perforata sui margini

4960 edge runner — See kollergang.

4961 edges — Outer extremity of folded sections.

f tranches *fpl* — **d** Schnitt *m*; Kanten *fpl*; Ränder *mpl* — **n** snede *fm*; kanten *mpl*; randen *mpl* — **e** cantos *mpl*; cortes *mpl* — **i** tagli *mpl*

4962 edge tearing resistance — See tearing strength.

4963 edging knife — See paring knife.

4964 Edison socket

f douille *f* Edison; douille *f* à vis — **d** Schraubfassung *f*; Schraubsockel *m* — **n** Edisonsluiting *f*; schroeffitting *f* — **e** portalámpara(s) *m* Edison; portalámpara(s) *m* roscado; portalámpara(s) *m* de rosca — **i** attacco *m* Edison

4965 edit *v* — To supervise or direct the preparation of a newspaper, magazine, etc.

f rédiger; diriger — **d** redigieren — **n** redigeren — **e** redactar — **i** redigere

4966 edit *v* — In data processing, to modify the form or format of data. May involve the rearrangement of data, the addition of data (e.g., insertion of dollar signs and decimal points), the deletion of data (e.g., suppression of leading zeros), code translation, and the control of layouts for printing (e.g., provision of headings and page numbers).

f éditer; mettre en forme — **d** aufbereiten zum Drucken — **n** voorbereiden; opmaken — **e** editar; preparar para la imprenta — **i** preparare per la stampa

4967 editing — Term used in data processing.

f mise *f* en forme — **d** Druckaufbereitung *f*; Aufbereiten *n* zum Drucken — **n** opmaak *m*; opmaken *n* — **e** preparación *f* para la imprenta — **i** preparazione *f* per la stampa

4968 editing program — Software routine to facilitate the revising of text material, implemented not only by software, but also by a particular hardware configuration, since the text may be displaced on a screen (VDT) or only in a printout. It may be revised only in a batch mode, or interactively. Revisions may be made sequentially or they may be made randomly. They may be extensive in size or very limited. They may be made only to the intellectual content or also to the typographic format. Changes may be made only specifically or globally (universally) or both.

4969 editing terminal — Usually a video terminal (VDT), but also a typewriter-like device producing only hard copy, to incorporate changes into a stream of textual material. Although called a terminal it may in fact be a "stand-alone" unit with its own "intelligence", or may be designed as a "shared-logic" device, whereby several terminals use the same intelligence to effect the editorial

revisions desired. It may in many cases be converted to a host computer. It may operate best in an editorial environment, as a tool for writing and performing other editorial tasks, or it may serve primarily as a device for implementing changes indicated by others (i.e., in a production environment, as a correction device).

4970 edition — All the copies of a book, etc., printed and published at one time, having regard to the type in which it is set and its general form, or to the number of copies produced at one time. It may be either an exact reprint or may be more or less revised.

f édition *f* — **d** Ausgabe *f* — **n** uitgave *f*; editie *f* — **e** edición *f*; publicación *f* — **i** edizione *f*; pubblicazione *f*

4971 edition bindery — Bookbinding plant where work in large quantities is produced.

f atelier *m* de reliure industrielle — **d** industrielle Buchbinderei *f*; Großbuchbinderei *f*; Verlagsbuchbinderei *f* — **n** uitgaafbinderij *f* — **e** taller *m* de encuadernación industrial — **i** legatoria *f* industriale

4972 edition binding; publisher's binding; trade binding — Ordinary machine bound book supplied by a publisher.

f reliure *f* d'éditeur — **d** Verlagseinband *m*; Verlegereinband *m* — **n** uitgeversband *m* — **e** encuadernación *f* de edición; encuadernación *f* de editorial; encuadernación *f* en serie — **i** rilegatura *f* editoriale; rilegatura *f* commerciale; rilegatura *f* industriale in serie

4973 edition de luxe; de luxe edition; fine edition; big-ticket book *fam.* — Special or limited edition of fine books. See also large-paper.

f édition *f* de luxe — **d** Luxusausgabe *f*; Prachtausgabe *f*; Luxusbuch *n* — **n** luxe uitgave *f*; uitgave *f* in prachtband — **e** edición *f* de lujo; edición *f* fina — **i** edizione *f* di lusso

4974 edition in sections

f édition *f* en fascicules; édition *f* en cahiers — **d** Heftausgabe *f*; Lieferungsausgabe *f*; Faßikelausgabe *f* — **n** uitgave *f* in afleveringen — **e** edición *f* en fascículos; edición *f* en cuadernos — **i** edizione *f* in fascicoli; edizione *f* in quaderni

4975 editio princeps — See first edition.

4976 editor *m*; **editress** *f* — One who conducts a newspaper or a magazine, or who prepares and superintends the printing of a literary work.

f rédacteur *m*; rédactrice *f* — **d** Schriftleiter *m*; Redaktor *m*; Schriftleiterin *f*; Redakteurin *f* — **n** redacteur *m*; redactrice *f* — **e** redactor *m*; redactora *f* — **i** redattore *m*; redattrice *f*

4977 editorial — Article in a newspaper or other periodical presenting the opinion of the publisher or editor(s).

f éditorial *m* — **d** Leitartikel *m* — **n** redactioneel artikel *n* — **e** editorial *m*; artículo *m* editorial; nota *f* editorial — **i** editoriale *m*

4978 editorial section; text part

f partie *f* rédactionnelle — **d** redaktioneller Teil *m* — **n** redactioneel gedeelte *n* — **e** parte *f* editorial

— **i** parte *f* redazionale
4979 editor in chief; chief editor — See also responsible editor.
f rédacteur *m* en chef — **d** Hautschriftleiter *m*; Chefredakteur *m* — **n** hoofdredacteur *m* — **e** redactor *m* en jefe; jefe *m* de (la) redacción — **i** redattore *m* capo
4980 editress *f* — See editor *m*.
4981 effective aperture
f ouverture *f* effective — **d** effektive Öffnung *f*; wirksame Öffnung *f* — **n** effectieve (lens)-opening *f*; werkzame opening *f* — **e** abertura *f* útil; abertura *f* eficaz — **i** apertura *f* effettiva
4982 effective value — Measured value of a property of paper or board.
f valeur *f* effective — **d** Effektivwert *f* — **n** effectieve waarde *f* — **e** valor *m* efectivo — **i** valore *m* effettivo
4983 efflux cup — Viscosimeter such as Zahn or Shell cup that allows rapid readings in seconds from full cup to empty through a standardized orifice.
f coupe *f* consistométrique — **d** Auslaufbecher *m* — **n** uitloopbeker *m* — **e** copa *f* de efusión — **i** becher *m* ad efflusso
4984 efflux time
f temps *m* d'écoulement — **d** Auslaufzeit *f* — **n** uitlooptijd *m* — **e** tiempo *m* de efusión — **i** tempo *m* d'efflusso
4985 efflux viscosimeter — Viscometer that discharges the test material through an orifice or through a capillary tube.
f viscosimètre *m* à écoulement — **d** Auslaufviskosimeter *n*; Ausflußviskosimeter *n* — **n** uitloopviscosimeter *m* — **e** viscosímetro *m* de efusión; viscosímetro *m* a derrame — **i** viscosimetro *m* ad efflusso
4986 effort-controlled cycle; unrestricted cycle — Cycle time determined by the effort of the operator.
f cycle *m* en travail libre — **d** Freiarbeitszyklus *m* — **n** ongebonden cyclus *m*; vrij werk *n* — **e** ciclo *m* en trabajo libre — **i** ciclo *m* in lavoro libero
4987 e.f. paper — See English finish paper.
4988 egg albumin — The white of a bird's egg often supplied dried, used for making albumin-type surface offset press-plate. Commercially, the white of a hen or duck egg.
f albumine *f* d'œuf; blanc *m* d'œuf; albumine *f* des œufs — **d** Eialbumin *n*; natürliches Albumin *n* — **n** eiwit *n* — **e** clara *f* de huevo; albúmina *f* de huevo — **i** albumina *f* d'uovo
4989 eggshell book paper — Book paper with an eggshell finish. The basis weight ranges from 45-80 pounds (25 x 38 - 500) It is considered to bulk between 440 and 640 pages per inch (2.54 cm) for a 45-pound basis.
f papier *m* d'édition coquille — **d** halbglänzendes Papier *n*; Eierschalen-Werkdruckpapier *n* — **n** halfmat papier *n*; papier *n* met eierschaaloppervlak — **e** papel *m* semibrillante; papel *m* obra de acabado ligero — **i** carta *f* pelle d'uovo; carta *f* semimat

4990 eggshell finish — Dull, soft, rough finish on paper, especially book paper, produced by omitting the calendering.
f apprêt *m* coquille; fini *m* coquille — **d** halbglänzend — **n** met eierschaal-oppervlak — **e** acabado *m* semimate; acabado *m* novela; acabado *m* semiáspero — **i** finitura *f* semimatta
4991 Egyptian — See app. no. 2.
4992 Egyptian vignette — See black vignette.
4993 eighteenmo; octodecimo; decimo octavo; sheet of eighteens — Book with eighteen leaves or 36 pages to the sheet. Abbrev. 18mo (no point).
f in-dix-huit *m*; in-18 *m* — **d** Achtzehnerformat *n*; Oktodezformat *n*; Oktodez *n* — **n** in achttienen — **e** en dieziochavo; en dieciochavo; en décimoctavo — **i** in diciottesimo
4994 eight-level code — Method to describe a combination of characters by the use of eight bits, turned "on" or "off", as in a frame of paper tape. In this fashion 256 code combinations are available and 256 different characters can be represented.
f alphabet *m* à huit unités — **d** Achtenalphabet *n* — **n** acht-eenhedenalfabet *n*; acht-eenheden-code *m* — **e** alfabeto *m* de ocho unidades — **i** alfabeto *m* ad otto unità
4995 eight points; 8 points — See app. no. 1.
f corps *m* huit; huit points *mpl* — **d** Korpus *f* acht; acht Punkte *mpl* — **n** corps *n* acht; acht punten *mpl* — **e** cuerpo *m* ocho; ocho puntos *mpl* — **i** corpo *m* otto; otto punti *mpl*
4996 eight-reel machine — See eight-reel press.
4997 eight-reel press; eight-reel machine
f machine *f* à huit bobines; machine *f* octuple — **d** Achtrollenmaschine *f* — **n** acht-rollenmachine *f* — **e** máquina *f* de ocho bobinas; prensa *f* de ocho bobinas — **i** macchina *f* ad otto bobine
4998 eights — See octavo.
4999 eight-to-pica — See app. no. 1.
f point *m* et demi de force de corps — **d** Achtel-cicero *f* — **n** anderhalve punt *m*; anderhalf punt *m* — **e** punto *m* y medio — **i** punto *m* e mezzo
5000 eight-to-pica leads
f interlignes *fpl* d'un point et demi; réglettes *fpl* d'un point et demi — **d** Achtelcicero-Durchschuß *m*; Achtelcicero-Regletten *fpl* — **n** anderhalfpunts interlinies *fpl* — **e** interlíneas *fpl* de un punto y medio — **i** interlinee *fpl* di un punto e mezzo
5001 ejector
f éjecteur *m* — **d** Auswerfer *m*; Ausstoßer *m*; Ausstoßmechanismus *m* — **n** uitstoter *m*; uitstootmechanisme *n*; uitwerper *m* — **e** expulsador *m*; expulsor *m* — **i** eiettore *m*
5002 ejector blade — Part of a slug composing machine that pushes the slug from the mould.
f lame *f* éjectrice; lame *f* d'éjecteur — **d** Ausstoßplatte *f* — **n** uitstoterblad *n* — **e** lámina *f* expulsora; lámina *f* expulsadora; lámina *f* del expulsor; lámina *f* del expulsador; lámina *f* de ex-

pulsión; botador *m Me* — **i** lama *f* d'espulsione

5003 ejector blade magazine — Part of the slug composing machine that contains the ejector plates.

f boîte *f* des éjecteurs — **d** Ausstoßplatten-Magazin *n*; Ausstoßplatten-Kassette *f* — **n** uitstotercassette *f* — **e** marquesina *f* de la lámina expulsora; botador *m Me* — **i** magazzino *m* della lama d'espulsione

5004 elastic deformation — Deformation in which no true flow takes place. Each molecule, though displaced from its initial position, still remains within the range of attraction of its original neighbouring molecules. When the stress is removed, the displaced molecules return to their original position. As a consequence, in elastic deformation, strain is proportional to stress.

f déformation *f* élastique — **d** elastische Verformung *f* — **n** elastische vervorming *f*; elastische deformatie *f* — **e** deformación *f* elástica — **i** deformazione *f* elastica

5005 elasticity — Property of a material which enables it to undergo deformation and to recover its original dimensions after removal of the deforming stress; determined more by ability to recover initial shape than by capacity to be deformed or extended. Should not be confused with extensibility or stretch.

f élasticité *f* — **d** Elastizität *f*; Spannkraft *f* — **n** elasticiteit *f*; veerkracht *fm* — **e** elasticidad *f* — **i** elasticità *f*

5006 elasticity — See resiliency.

5007 electrical etching — Etching of relief printing plates by means of electrolysis, really an inversion of electroplating.

f gravure *f* électrolytique — **d** elektrische Ätzung *f* — **n** elektrisch etsen *n* — **e** grabado *m* electrolítico; mordido *m* electrolítico — **i** elettro-incisione *f*

5008 electrical insulating board — See electrical insulating paper.

5009 electrical insulating paper; electrical insulating board — Paper or board with properties like high electric strength, durability, absence of conductive metallic particles and uniformity in thickness and formation, used for insulation. See also presspahn.

f papier *m* diélectrique; carton *m* diélectrique — **d** Isolierpapier *n*; Isolierpappe *f*; Elektroisolierpapier *n*; Elektroisolierpappe *f* — **n** condensatorpapier *n*; condensatorkarton *n* — **e** papel *m* aislante eléctrico; papel *m* de condensador; cartón *m* aislante eléctrico; cartón *m* condensador — **i** carta *f* per condensatori elettrici; carta *f* isolante; carta *f* per isolamento elettrico; cartone *m* per isolamento elettrico

5010 electrical pressboard — See presspahn.

5011 electrical resistance — See resistance.

5012 electric bulb

f ampoule *f* électrique — **d** Glühlampe *f* — **n** gloeilamp *fm* — **e** ampolla *f* eléctrica; bombilla *f* eléctrica — **i** lampadina *f* elettrica

5013 electric charge

f charge *f* électrique — **d** elektrische Ladung *f* — **n** elektrische lading *f* — **e** carga *f* eléctrica — **i** carica *f* elettrica

5014 electric current — Stream of electrons flowing through a conducting body, usually measured an expressed in amperes, coulombs, and volts.

f courant *m* électrique — **d** elektrischer Strom *m* — **n** elektrische stroom *m* — **e** corriente *f* eléctrica — **i** corrente *f* elettrica

5015 electric current — See current.

5016 electric drive

f commande *f* électrique — **d** elektrischer Antrieb *m* — **n** elektrische aandrijving *f* — **e** impulsión *f* eléctrica; accionamiento *m* eléctrico — **i** comando *m* elettrico; propulsione *f* elettrica

5017 electric field — Region where forces capable of acting on electrons can be found.

f champ *m* électrique — **d** elektrisches Feld *n* — **n** elektrisch veld *n* — **e** campo *m* eléctrico — **i** campo *m* elettrico

5018 electric field intensity — See field strength.

5019 electric force — See field strength.

5020 electric heating

f chauffage *m* électrique — **d** elektrische Heizung *f*; Elektroheizung *f* — **n** elektrische verwarming *f* — **e** calefacción *f* eléctrica — **i** riscaldamento *m* elettrico

5021 electric hygrometer — Hygrometer that depends for its readings on changes in conductivity of films containing hygroscopic salts such as lithium chloride. Requires an electric current, and is usually used where the RH is small. Not practical in printing work rooms.

f hygromètre *m* électrique — **d** elektrisches Hygrometer *n*; elektrischer Feuchtigkeitsmesser *m* — **n** elektrische hygrometer *m* — **e** higrómetro *m* eléctrico — **i** igrometro *m* elettrico

5022 electric intensity — See field strength.

5023 electric motor — See electromotor.

5024 electric power

f énergie *f* électrique; puissance *f* électrique — **d** Betriebskraft *f*; elektrische Leistung *f*; Elektroenergie *f* — **n** elektrische kracht *fm*; elektrische energie *f*; elektrisch vermogen *n*; krachtstroom *m* — **e** energía *f* eléctrica; fuerza *f* eléctrica — **i** energia *f* elettrica; forza *f* elettrica

5025 electric typewriter

f machine *f* à écrire électrique — **d** elektrische Schreibmaschine *f* — **n** elektrische schrijfmachine *f* — **e** máquina *f* de escribir eléctrica — **i** macchina *f* per scrivere elettrica

5026 electro — See electrotype.

5027 electrode — Conducting element that performs one or more of the functions of emitting, collecting or controlling by an electric field the movement of electrons or ions.

f électrode *f* — **d** Elektrode *f* — **n** elektrode *f* — **e** electrodo *m* — **i** elettrodo *m*

5028 electrodeposit — Material precipitated at an electrode as the result of the passage of an

electric current through a solution or suspension of the material, e.g., copper from copper sulphate solution.
f dépôt *m* électrolytique — **d** elektrolytische Abscheidung *f* — **n** galvanische neerslag *m* — **e** depósito *m* electrolítico — **i** deposizione *f* elettrolitica

5029 electrode potential — See electrode voltage.

5030 electrode voltage; electrode potential — Voltage between an electrode and the cathode or a specified point of a filamentary cathode, usually the negative end.
f tension *f* d'électrode — **d** Elektrodenspannung *f* — **n** elektrodespanning *f* — **e** tensión *f* de electrodo — **i** tensione *f* d'elettrodo

5031 electrofax — Electrostatic printing process resembling xerography.

5032 electrolysis — Decomposition of a chemical compound in aqueous solution by the passage of an electric current into positive and negative ions which migrate to and collect at the negative and positive electrodes.
f électrolyse *f* — **d** Elektrolyse *f* — **n** elektrolyse *f* — **e** electrólisis *f* — **i** elettrolisi *f*

5033 electrolyte — 1. Solution or bath in electric etching machines. 2. Substance which dissociates into ions when in a solution or a fused state and which will then conduct an electric current. Common examples are sodium chloride and sulphuric acid.
f électrolyte *f*; liquide *m* excitateur — **d** Elektrolyt *m* — **n** elektrolyt *m* — **e** electrólito *m* — **i** elettrolito *m*

5034 electrolytic bath — Tank in which electrolytic solutions are decomposed by the passage of an electric current, used in electrotyping and electroplating.
f bain *m* électrolytique — **d** elektrolytisches Bad *n* — **n** elektrolytisch bad *n* — **e** baño *m* electrolítico — **i** bagno *m* elettrolitico

5035 electrolytic etching — Etching of relief printing plates by means of electrolysis, really an inversion of electroplating.
f gravure *f* électrolytique — **d** elektrolytische Ätzung *f* — **n** elektrolytisch etsen *n* — **e** grabado *m* electrolítico — **i** incisione *f* elettrolitica

5036 electrolytic grain — Grain on an offset plate by electrolysis.
f grain *m* électrolytique — **d** elektrolytisches Korn *n* — **n** elektrolytisch grein *n* — **e** grano *m* electrolítico — **i** grana *f* elettrolitica

5037 electromagnet
f électro-aimant *m* — **d** Elektromagnet *m* — **n** elektromagneet *m* — **e** electroimán *m* — **i** elettromagnete *m*

5038 electromagnetic clutch — Electromagnetically operated coupling.
f embrayage *m* électromagnétique — **d** elektromagnetische Kupplung *f* — **n** elektromagnetische koppeling *f* — **e** embrague *m* electromagnético — **i** giunto *m* elettromagnetico

5039 electromagnetic thickness gauge
f déterminateur *m* de l'épaisseur par mesure électromagnétique — **d** elektromagnetischer Dickemesser *m* — **n** elektromagnetische (laag)diktemeter *m* — **e** medidor *m* de espesor electromagnético; calibre *m* medidor de espesor electromagnético — **i** misuratore *m* elettromagnetico di spessore; spessimetro *m* elettromagnetico; calibro *m* di spessore elettromagnetico

5040 electromagnetic unit — Abbrev. emu.
f unité *f* électromagnétique — **d** elektromagnetische Einheit *f* — **n** elektromagnetische eenheid *f* — **e** unidad *f* electromagnética — **i** unità *f* elettromagnetica

5041 electromotive force; emf — Source of electrical energy required to produce an electrical current in a circuit, measured in volts.
f force *f* électromotrice; f.e.m. — **d** Elektromotivkraft *f*; elektromotorische Kraft *f*; EMK — **n** elektromotorische kracht *fm*; emk — **e** fuerza *f* electromotriz; f.e.m. — **i** forza *f* elettromotrice; f.e.m.

5042 electromotor; electric motor
f électro-moteur *m*; moteur *m* électrique — **d** Elektromotor *m* — **n** elektromotor *m* — **e** electromotor *m*; motor *m* eléctrico — **i** elettromotore *m*; motore *m* elettrico

5043 electron — Basic element of an atom. Some electrons are only loosely bound to the atoms in all conductors of electricity, and it is the movement of the electrons from one atom to another which constitutes a flow of current.
f électron *m* — **d** Elektron *n* — **n** elektron *n* — **e** electrón *m* — **i** elettrone *m*

5044 electronic
f électronique — **d** elektronisch — **n** elektronisch — **e** electrónico — **i** elettronico

5045 electronic cutting — See electronic engraving.

5046 electronic data processing — Data processing by devices which are mainly electronic (e.g., digital computers and associated peripheral equipment). Abbrev. EDP. In GB the term *automatic data processing is frequently used.
f transformation *f* électronique des données; traitement *m* électronique des données — **d** elektronische Datenverarbeitung *f* — **n** elektronische gegevens-verwerking *f*; elektronische verwerking *f* van gegevens; elektronische data-verwerking *f* — **e** procesado *m* electrónico de los datos — **i** elaborazione *f* elettronica dei dati; trattamento *m* elettronico dei dati

5047 electronic engraver — See photoelectric engraver.

5048 electronic engraving; electronic cutting — Engraving in screened form, directly onto plastics or metal, with the cutting stylus of an electronic scanner.
f gravure *f* électronique — **d** elektronische Gravur *f* — **n** elektronisch graveren *n* — **i** incisione *f* elettronica

5049 electronic flash lamp — Discharge lamp giving a high light output for a very brief period.

f lampe f éclair électronique − d Blitzröhre f − n elektronenflitslamp f − e lámpara f de destello electrónico − i lampada f di luce elettronica

5050 electronic photoengraving machine − See photoelectric engraver.

5051 electronic register control − Device to control the accurate positioning of one colour with another or one operation with another or in-setting a pre-printed web automatically.
f repérage m électronique; repérage m photo-électrique; repérage m automatique; contrôle m électronique du repérage − d elektronische Passerkontrolle f; photoelektrische Register-steuerung f; photoelektrischer Registerregler m; elektronische Registersteuerung f − n elek-tronische registerregeling f; elektronische regeling f van het register − e regulación f de registro electrónica − i regolazione f elettronica del registro

5052 electronics − Branch of physics which deals with the emission, behaviour, and effects of elec-trons, and with electronic devices.
f électronique f − d Elektronik f − n elek-tronica f − e electrónica f − i elettronica f

5053 electronic scanning − See scanning.

5054 electronic sidelay control − Device to control automatically sidelay errors of the web.
f réglage m électronique latéral − d elektronische Seitenregistersteuerung f; vollautomatische Kantensteuerung f − n elektronische regeling f van het dwarsregister − e mando m electrónico; gobierno m electrónico; guíalateral m electrónico − i controllo m elettronico del registro laterale

5055 electronic stylus − See light pen.

5056 electronic tube; electron tube US; **electronic valve** GB − Electronic device in which con-duction takes place through a vacuum or a gaseous medium within a gas-tight envelope.
f tube m électronique − d Elektronenröhre f − n elektronenbuis f − e válvula f electrónica − i valvola f elettronica; tubo m elettronico

5057 electronic valve − See electronic tube.

5058 electron microscope − Microscope that pro-duces an image by means of a beam of electrons instead of a beam of light. The "lenses" are mag-netic and electric fields.
f microscope m électronique − d Elektronen-mikroskop n − n elektronenmicroscoop m − e microscopio m electrónico − i microscopio m elettronico

5059 electron optics − The laws of optics define the formation of images subjected to refraction and reflection by light passing through lenses or reflected by mirrors. Electron optics apply to similar phenomena obtained by passing a beam of electrons through magnetic or electrostatic fields which acts in the same manner as lenses on light rays.

5060 electron tube − See electronic tube.

5061 electrophoresis − Movement of suspended particles through a fluid under the action of an electromotive force applied to electrodes in contact with the suspension.
f électrophorèse f; cataphorèse f − d Elektro-phorese f; Kataphorese f − n elektroforese f − e electrofóresis f − i elettroforesi f

5062 electrophotography − Reproduction process where use is made of presensitized planographic plates with photoconductive layers that are so light-sensitive that exposure can follow in the reproduction camera.
f électrophotographie f − d Elektrophotographie f − n elektrofotografie f − e electrofotografía f − i elettrofotografia f

5063 electroplating − Deposition of a usually thin layer of metal on a base metal or conducting sur-face by electrolysis.
f galvanoplastie f; électrodéposition f − d Galvanisierung f − n galvaniseren n − e electro-chapeado m; electrogalvanización f; galvano-plastia f − i galvanoplastica f

5064 electroplating bath; electroplating vat
f bain m galvanoplastique; bain m pour galvano-plastie − d galvanisches Bad n; galvano-plastisches Bad n − n galvanisch bad n − e baño m galvanoplástico − i bagno m galvanico

5065 electroplating vat − See electroplating bath.

5066 electrostatic field − Electric field that is constant in time, produced by stationary charges.
f champ m électrostatique − d elektrostatisches Feld n − n elektrostatisch veld n − e campo m electrostático − i campo m elettrostatico

5067 electrostatic printer − Device to provide a hard copy of textual material from machine-readable input by forming matrix patterns of char-acters through electrostatic deposits of carbon sub-stances. An array of "nibs" is positioned hori-zontally across a paper carrier. As the paper moves forward, black dots or spots are deposited or created, forming letters and words.
f dispositif m de tirage électrostatique − d elektrostatische Kopiermaschine f − n elektro-statische kopieermachine f − e dispositivo m impresor electrostático − i dispositivo m stampatore elettrostatico

5068 electrostatic printing method − Repro-duction process in which the image on the paper or other surface is formed by applying an electric charge to the toner or pigment particles, bringing them into contact with an oppositely charged image area on the surface and treating the electro-statically held image to bind it to the surface.
f méthode f d'impression électrostatique; impression f électrostatique − d elektrostatisches Druckverfahren n; elektrostatischer Druck m − n elektrostatisch drukprocédé n − e impresión f electrostática − i stampa f elettrostatica

5069 electrostatics − Branch of physics that deals with electric phenomena not associated with electricity in motion. Of this fact use is made in some methods (*xerography) in which advantage is taken that there are materials which can be made light sensitive.
f électrostatique f − d Elektrostatik f − n elektro-

statica f — **e** electrostática f — **i** elettrostatica f

5070 electrostatic unit — Abbrev. esu.

f unité f électrostatique — **d** elektrostatische Einheit f — **n** elektrostatische eenheid f — **e** unidad f electrostática — **i** unità f elettrostatica

5071 electrotype; electro — Curved or flat printing plate (the original of which is a halftone, line engraving, type, or combination) made by the electrotyping process, consisting of a copper or a nickel-and-copper shell, (sometimes faced with a coating of chromium) backed up with an alloy of lead, tin and antimony, known as electrotype metal; used when it is desirable to retain the original printing plate in good condition and when long press runs or a second printing are required.

f galvano m; galvanotype m; électro m — **d** Galvano n — **n** galvano m — **e** electrotipo m; electro m; electrotipia f; galvano m; molde m electrotipado; plancha f electrotípica; cascarilla f electrotipada; plancha f galvánica; clisé m electrotípico — **i** galvano m

5072 electrotype metal — Alloy containing approx. 92% lead, 3.6% antimony and 4.4% tin, used to back up the copper shell.

f matière f à galvano — **d** Hintergießmetall n für Galvanos — **n** achtergietmetaal n voor galvano's; metaal n voor het achtergieten van galvano's — **e** metal m de electrotipia; metal m de galvanotipia; metal m de galvanoplástica — **i** lega f per galvanotipia; metallo m per galvanotipia

5073 electrotyper

f galvanoplaste m; galvanotypeur m — **d** Galvanoplastiker m — **n** galvaniseur m — **e** electrotipador m; electrotipista m — **i** galvanotipista m; elettrotipista m

5074 element; elementary substance — Simple substance that cannot be separated into simpler substances; one of the parts of which anything is made up.

f élément m; corps m simple — **d** Element n — **n** element n — **e** elemento m — **i** elemento m

5075 elementary — Made up of a single chemical element.

f élémentaire — **d** elementar — **n** elementair — **e** elementar; elemental — **i** elementare

5076 elementary substance — See element.

5077 elephant — A British size of wrapping paper. The term is rarely used. See app. no. 3.

5078 elevating truck — See lift truck.

5079 elevator, first — See first elevator.

5080 elevator, second — See second elevator.

5081 elevator transfer slide — Part of the slug composing machine.

f poussoir m des matrices — **d** Schlitten m zur Matrizen- und Spatienkeilüberführung — **n** transportslede fm — **e** carro m de traslación del elevador — **i** carrello m per il trasferimento delle matrici

5082 eleven point metal — Metal for relief printing plates for use on patent base with a thickness of 3.86 mm (0.152 in).

f métal m épais pour clichés (onze points) — **d** Elfpunktmetall n — **n** clichémetaal n ter dikte van elf punten — **e** lámina f de once puntos — **i** lastra f metallica di undici punti

5083 elision — The striking out of a letter, as in "o'er" for "over".

f élision f — **d** Elision f; Ausfall m einer Silbe; Verkürzung f; Auslassung f; Verschleifung f — **n** elisie f; weglating f van een letter; uitlating f van een letter — **e** elisión f — **i** elisione f

5084 elite type — Typewriter face with 12 characters or 6 lines to the inch.

5085 ellipse — A plane curve such that the sums of the distances of each point in its periphery from two fixed points, called the foci, are equal. It is a conic section formed by the intersection of a right circular cone by a plane which cuts obliquely the axis and the opposite side of the cone. Equation:

$x^2/a^2 + y^2/b^2 = 1$. Not to confuse with *ellipsis.

f ellipse f — **d** Ellipse f — **n** ellips fm — **e** elipse f — **i** ellisse f

5086 ellipsis — Figure of syntax by which a word or words left out are implied. pl ellipses. To this three spaced-out full points or asterisks are used, thus … or ***. A fourth point should not be added at the end of an incomplete sentence. Not to confuse with *ellipse.

f ellipse f — **d** Ellipse f — **n** ellips fm; beletselteken n — **e** elipsis f — **i** ellissi f

5087 ellipsograph; elliptograph — Machine or trammel to draw or cut ellipses and circles on originals, masks, negatives and printing plates.

f ellipsographe m — **d** Ellipsenzirkel m — **n** ellipsograaf m; ellipspasser m — **e** elipsógrafo m — **i** ellissografo m

5088 elliptical dot screen — See chain dot screen.

5089 elliptograph — See ellipsograph.

5090 Elmendorf test — Method to determine the resistance of paper to tearing. See also tearing strength, tearing test.

5091 elon — Trade name for photograde methyl paraminophenol sulphate.

5092 elon — See methyl-paraminophenol sulphate.

5093 elongated format

f format m allongé — **d** verlängertes Format n; Schmalformat n — **n** langwerpig formaat n — **e** tamaño m alargado — **i** formato m allungato

5094 elongated type

f caractère m allongé; caractère m effilé — **d** verlängerte Schrift f — **n** verlengde letter fm — **e** tipo m estrecho; letra f chupada — **i** lettera f allungata; carattere m allungato

5095 elongation at break; elongation of rupture

f allongement m à la rupture; allongement m de rupture — **d** Bruchdehnung f — **n** rek m bij breuk — **e** alargamiento m de rotura — **i** allungamento m a rottura

5096 elongation of rupture — See elongation at break.

5097 Elrod — Machine for casting rules and

borders between 1 to 36 point, manufactured since 1920 by the Ludlow Typograph Co., Chicago.

5098 em — Printer's unit of area measurement equal in width and height to the height of any selected type body size. So named because the letter m was at one time designed to fit on a square body. Now used as an abbreviation of pica em, where the em is equivalent to 12 point (approx. one-sixth in) units. See also pica.

f em *m* (12 points anglo-américains) — **d** em *n* (12 englisch-amerikanische Punkte) — **n** em *m* (12 Engels-Amerikaanse punten) — **e** eme *f* (12 puntos angloamericanos) — **i** em *m* (12 punti anglo-americano)

5099 emboss *v* — To raise in relief a design or letters already printed on card or tough paper by an uninked block or die, in rubber and plastic platemaking usually by heat.

f gaufrer — **d** prägen — **n** pregen — **e** gofrar — **i** goffrare

5100 embossed glassine — *Glassine paper decorated with a continuous formal design by embossing rolls.

f papier *m* cristal gaufré — **d** geprägtes Pergamin-papier *n*; geprägtes Pergamin *n* — **n** geperst pergamijn *n* — **e** papel *m* apergaminado y gofrado — **i** glassina *f* goffrata; carta *f* pergamena goffrata

5101 embossed paper — Paper printed with a deeply engraved plate.

f papier *m* gaufré — **d** geprägtes Papier *n* — **n** geperst papier *n*; gepreegd papier *n* — **e** papel *m* gofrado — **i** carta *f* goffrata

5102 embossing — Raised surface obtained by stamping or tooling.

f gaufrure *f* — **d** Prägung *f* — **n** preging *f*; preeg *m* — **e** gofrado *m* — **i** goffratura *f*

5103 embossing; blanket embossing — Swelling of the image on an offset blanket due to its absorbing solvents from the ink.

f gonflement *m* (d'un blanchet) — **d** Quellen *n* (des Gummituchs) — **n** zwellen *n* (van het rubber-doek) — **e** hinchazón *f* (de la mantilla) — **i** rigonfiamento *m* (del blanket); rigonfiamento *m* sul tessuto gommato

5104 embossing plate — Plate etched or engraved below its surface (intaglio), into which paper is forced to produce a raised (embossed) design on surface of sheet.

f plaque *f* à gaufrer; plaque *f* à embosser — **d** Prägeplatte *f* — **n** preegstempel *n*; preegplaat *fm* — **e** plancha *f* para el estampado en relieve — **i** lastra *f* incisa in incavo per imprimere in rilievo; forma *f* incavografica

5105 embossing press — Machine for embossing paper or cardboard.

f presse *f* à gaufrer — **d** Prägepresse *f*; Gaufrier-presse *f* — **n** preegpers *fm*; gaufreerpers *fm* — **e** máquina *f* gofradora — **i** pressa *f* goffratrice; pressa *f* per la goffratura

5106 embossment — In magnetic ink printing, the distance between the average paper surface and the ink surface. It has an influence on the signal waveform. See also magnetic encoded cheques.

f foulage *m* du papier — **d** Prägetiefe *f* — **n** moet *fm*; inpersing *f* — **e** hondura *f* de la presión; diente *m* — **i** profondità *f* dell'impronta

5107 embroidery transfer ink — Ink for printing needlework patterns on paper to be transferred on to a fabric with a hot iron.

f encre *f* transfert pour broderies — **d** Abbügel-farbe *f* — **n** inkt *m* voor strijkpatronen — **i** inchiostro *m* trasferibile a caldo

5108 em dash — See dash.

5109 emerald — Obsolete name for a size of type. See app. no. 1.

5110 emerald green — A clear, deep green colour.

f vert *m* d'émeraude — **d** Smaragdgrün *n*; Smaragden *n* — **n** smaragdgroen *n* — **e** esmaragdino *m*; esmeraldino *m*; color *m* verde esmeraldo — **i** smeraldo *m* verde

5111 emery paper — Abrasive paper, the base stock of which is a kraft sheet in basis weight of 65 to 70 pounds (24 x 36 - 500), coated with emery powder with a glue or resin bond, used in sheets and in narrow rolls for hand-polishing operations on metal.

f papier *m* à l'émeri; papier *m* d'émeri; papier *m* émeri; papier *m* émerisé — **d** Schmirgelpapier *n* — **n** amarilpapier *n*; schuurpapier *n* — **e** papel *m* de esmeril; papel *m* esmeril — **i** carta *f* smerigliata; carta *f* smeriglio; carta *f* vetrata; carta *f* abrasiva

5112 emery paper — See abrasive paper.

5113 emery powder — Abrasive powder of alumina.

f poudre *f* d'émeri — **d** Schmirgelpulver *n* — **n** schuurpoeder *n* — **e** polvo *m* de esmeril — **i** polvere *f* di smeriglio

5114 emf — See electromotive force.

5115 emission spectrum — Spectrum produced by radiation from an emitting source, as distinguished from absorption spectra.

f spectre *m* d'émission — **d** Emissionsspektrum *n* — **n** emissiespectrum *n* — **e** espectro *m* de emisión — **i** spettro *m* di emissione

5116 emperor — A size of ledger paper, sometimes used in the US. See app. no. 3.

5117 empire — A size of note paper. See app. no. 3.

5118 empirical — Based on practical experience or observation without regard to science and theory.

f empirique — **d** empirisch — **n** empirisch; proef-ondervindelijk — **e** empírico — **i** empirico

5119 empty *v* **the stick; dump** *v* **the stick** — To take out the composed lines from the composing stick and place them on a galley.

f vider le composteur — **d** den Winkelhacken ausheben — **n** de haak leeg maken; de zethaak leeg maken — **e** bajar el componedor; vaciar el componedor — **i** estrarre; portare una riga sul vantaggio

5120 em quad; em quadrat; em space; mutton

quad *fam.*; **mutton** *fam.*; **mutt**; **mut** — A space the set width of the type face. An 8-point 8-set type face has an em with a square cross section of 8 points each side; an 8-point 8 1/2-set type face has an em quad 8 points deep and 8 1/2 points wide in cross section of the type body. In the Monotype system the em is 18 units wide. A measure of 25 ems of its own is 25 ems of set of the type face used. The em space is normally used before the first word of a new paragraph.
f cadratin *m* — **d** Geviert *n*; Ganzgeviertstück *n*; Quadrätchen *n* — **n** vierkant *n* — **e** cuadratín *m*; espacio *m* de eme — **i** quadratone *m*; quadrato *m*; quadrato *m* da una; quadrato *m* tondo

5121 em quadrat — See em quad.

5122 em rack
f crémaillère *f* des ems — **d** Einheitenzangstange *f* — **n** em-rek *n* — **e** cremallera *f* del tipó metro; anaquel *m* para cuadratines; depósito *m* para cuadratines — **i** cremagliera *f* del tipometro

5123 em rule — See dash.

5124 em scale — Part of the Monotype keyboard.
f échelle *f* des ems — **d** Formatskala *f*; Geviertskala *f* — **n** em-liniaal *f m* — **e** tipómetro *m* (del monotipo) — **i** tipometro *m*

5125 em space — See em quad.

5126 emulsification — Process of dispersing one liquid in another when they normally do not mix.
f émulsification *f*; émulsionnage *m* — **d** Emulgierung *f* — **n** emulgeren *n* — **e** emulsificación *f* — **i** emulsificazione *f*

5127 emulsification — See tinting.

5128 emulsifier; emulsifying agent — Material to facilitate the preparation of an emulsion and to improve its stability.
f agent *m* émulsifiant; émulsif *m*; émulsier *m*; agent *m* émulsif — **d** Emulgierungsmittel *n*; Emulgator *m*; Emulgiermittel *n* — **n** emulgeermiddel *n*; emulgator *m* — **e** agente *m* emulgente; emulgente *m*; emulsionante *m* — **i** agente *m* emulsionante; emulsificante *m*

5129 emulsify *v* — To form an emulsion as when lithographic inks emulsify with the fountain solution.
f émulsionner — **d** emulgieren — **n** emulgeren — **e** emulsionar; emulsificar — **i** emulsionare

5130 emulsifying — See tinting.

5131 emulsifying agent — See emulsifier.

5132 emulsion — System consisting of a liquid dispersed with or without an emulsifier in an immiscible liquid in droplets of larger than colloidal size. See also photographic emulsion.
f émulsion *f* — **d** Emulsion *f* — **n** emulsie *f* — **e** emulsión *f* — **i** emulsione *f*

5133 emulsion coated paper — Paper coated by an emulsion coating process.
f papier *m* couché à l'émulsion — **d** im Emulsionsstreichverfahren gestrichenes Papier *n* — **n** met een emulsielaag gestreken papier *n* — **e** papel *m* estucado por emulsión — **i** carta *f* patinata a emulsione

5134 emulsion coating — Method of coating a continuous paper web, in which the plastics or resin material is applied as an emulsion by the roll, brush or air jet coating operation.
f couchage *m* par émulsion; couchage *m* à l'émulsion — **d** Emulsionsstreichverfahren *n*; Emulsionsbeschichtung *f* — **n** emulsiestrijkprocédé *n* — **e** estucado *m* por emulsión — **i** patinatura *f* a emulsione

5135 emulsion ink — Ink consisting of colouring matters; modifying agents and a stabilized dispersion of one liquid in another in which it is insoluble. One of the liquids is usually water.
f encre *f* à l'eau — **d** Emulsionsfarbe *f* — **n** emulsieinkt *f* — **e** tinta *f* de emulsión — **i** inchiostro *m* d'emulsione; inchiostro *m* ad acqua; inchiostro *m* all'acqua

5136 emulsion side — The side of a photographic film to which the emulsion is applied and on which the image is developped.
f côté *m* de la couche — **d** Schichtseite *f* — **n** emulsiezijde *f*; zijde *f* van de gevoelige laag; kant *m* van de gevoelige laag — **e** cara *f* sensible; lado *m* de la capa sensible; cara *f* del emulsión — **i** lato *m* emulsionato

5137 en — Printer's unit of area measurement equal to the same height but half the width of the *em, sometimes used to specify the area of composition as its value closely approximates the number of characters in the text.
f largeur *f* moyenne des caractères — **d** durchschnittliche Buchstabenbreite *f* — **n** gemiddelde letterbreedte *f* — **e** ancho *m* medio de las letras — **i** spessore *m* medio delle caratteri

5138 enamel — Originally an abbreviation of glue enamel, but now applied to cold enamel, the sensitized coating on the surface of the metal plate (engravings) to receive the image by light transference through a line or halftone negative.
f émail *m* (à froid) — **d** Email *n*; Emaille *n*; Kaltemail *n* — **n** email *n*; koudemail *m* — **e** esmalte *m* (frío) — **i** smalto *m*; smalto *m* freddo

5139 enamel binding; enamelled binding — Bookbinding decorated with enamels. See also champlevé enamel, cloisonné.
f reliure *f* émaillée — **d** Emaileinband *m* — **n** geëmailleerde band *m* — **e** encuadernación *f* en esmalte — **i** legatura *f* in smalto

5140 enamel blotting paper — See enamelled blotting paper.

5141 enamel glue; bichromated fish glue
f colle-émail *f*; colle *f* de poisson — **d** Fischleim *m* — **n** vislijm *m* — **e** cola *f* esmalte; cola *f* de pescado; esmalte *m* a la cola; esmalte *m* ordinario — **i** colla *f* di pesce

5142 enamelled binding — See enamel binding.

5143 enamelled blotting paper; enamel blotting paper — Blotting paper to which a coated book paper or card has been pasted to give on one side a smooth, hard surface suitable for printing and lithographing.
f buvard *m* émaillé; buvard *m* chromo — **d** Chromolöschkarton *m* — **n** chromo-inktvloei *n*;

chromovloeikarton *n* — **e** papel *m* secante esmaltado; teleta *f* esmaltada — **i** carta *f* smalto sugante; carta *f* smalto asciugante

5144 enamelled paper — See enamel paper.

5145 enamel paper; enamelled paper — Originally one-side coated paper with a high polish, white or coloured, used for box covers; now any coated paper.
f papier *m* émaillé; papier *m* couché une face — **d** Hochglanzpapier *n*; Kartonagenglacé *n* — **n** cartonnage glacé *n*; chromopapier *n* — **e** papel *m* esmaltado (para cartonaje) — **i** carta *f* smaltata

5146 enamel print — Print on metal with glue or cold enamel.
f copie *f* à l'émail — **d** Emailkopie *f* — **n** email-kopie *f* — **e** copia *f* al esmalte; copia *f* a la cola — **i** copia *f* allo smalto

5147 enamel top — Acid resisting image of either glue or cold enamel.
f couche *f* d'émail — **d** Emailschicht *f* — **n** email-laag *fm*; emaillelaag *fm* — **e** capa *f* de esmalte — **i** strato *m* smalto

5148 encaustic ink — Ink which penetrates the body of the sheet of paper, rendering erasure much more difficult.
f encre *f* indélébile; encre *f* ineffaçable — **d** Sicherheitsfarbe *f* — **n** cheque-inkt *m* — **e** tinta *f* encáustica — **i** inchiostro *m* indelebile

5149 enclosed arc lamp; closed arc *depr.* — Electric arc lamp in which the carbons are enclosed in a glass cylinder and burn in a partial vacuum.
f lampe *f* à arc (électrique) fermé; lampe *f* à arc en vase clos — **d** Bogenlampe *f* mit ein-geschlossenem Lichtbogen — **n** gesloten booglamp *fm* — **e** lámpara *f* de arco (voltaico) cerrada; lámpara *f* voltaico cerrada — **i** lampada *f* ad arco chiusa

5150 enclosed fountain — In a rotogravure press, an ink fountain sealed from the outside atmosphere by a series of shields which allow only a small area (at the impression point) to be exposed; inhibits evaporation and drying of ink in the etched surface of the engraved cylinder; reduces explosion hazard present due to the use of highly volatile inks; prevents ink splash.
f système *m* de circulation d'encre entièrement fermé — **d** völlig eingeschlossenes Farbumwalz-system *n* — **n** geheel gesloten inktbak *m* — **e** tintero *m* cerrado — **i** calamaio *m* chiuso

5151 enclosed motor; enclosed-type motor; protected motor
f moteur *m* blindé — **d** geschlossener Motor *m*; gekapselter Motor *m* — **n** ingesloten motor *m* — **e** motor *m* acorazado; motor *m* blindado — **i** motore *m* chiuso

5152 enclosed-type motor — See enclosed motor.

5153 enclosure — Letter or paper enclosed with another one in an envelope.
f pièce *f* jointe — **d** Beilage *f*; Anlage *f* — **n** bij-lage *fm* — **e** incluída *f*; encerrada *f* — **i** allegato *m*

5154 encode *v* — To apply a code.

f coder; codifier — **d** kodieren; verschlüsseln — **n** coderen — **e** codificar; poner en código — **i** codificare

5155 encoder; magnetic tape encoder; magnetic data inscriber — Input keyboard device which sets up a series of signals in code form on magnetic tape.
f codeur *m* — **d** Verschlüssler *m* — **n** codeer-orgaan *n* — **e** codificadora *f* — **i** codificatore *m*

5156 encrust — Impurity in paper, originating from wood, as pectin, resin, but especially lignin; accelerates the ageing of paper.
f incrustation *f* — **d** Inkruste *f*; inkrustierender Bestandteil *m* — **n** incrust *f* — **e** incrustación *f*; sustancia *f* incrustante; substancia *f* incrustante — **i** incrostazione *f*

5157 end *v* **a break** — To end a composition with a line not being full. Opposite to *end *v* even.
n een korte regel uitvullen met vierkanten — **e** llenar la línea con cuadratines; ajustar la última línea; justificar la línea corta — **i** terminare la composizione giustificando con quadratoni l'ultima linea

5158 end adjusting screws — Mechanical arrangement on a letterpress rotary to adjust the column screws, fountain, or duct adjusting screws from the side of the press instead of between the units. See also banjo.
f vis *fpl* d'encrier (avec commande latéral) — **d** Zonenschrauben *fpl*; mechanische Farb-verstellung *f*; Zonenfarbeinstellung *f* — **n** inktstel-schroeven *fmpl*; inktbakschroeven *fmpl*; bak-schroeven *fmpl* — **e** tornillos *mpl* reguladores del tintero; tornilladores *mpl* de ajuste automático — **i** viti *fpl* di regolazione dell'inchiostrazione (con comando laterale)

5159 en dash — See dash.

5160 end *v* **even; make** *v* **even** — To end a composition with a full line; opposite to *end *v* a break.
f tomber en ligne — **d** stumpf halten; stumpf aus-gehen lassen; mit einer vollen Zeile ausgehen — **n** volle regels maken; met een volle regel laten ein-digen — **e** terminar con línea llena; alcanzar con línea llena; terminar un alcance con línea llena; terminar a final de párrafo; llenar la última línea — **i** terminare a linea piena; terminare con linea completa

5161 end-leaf paper; fly-leaf paper; end paper; end sheet — White or coloured chemical wood pulp paper (50-80 pounds 25 x 38 - 500) or cotton fibre, used at the beginning and the end of books, with sufficient strength to withstand tearing at the fold; pastes smoothly on the cover.
f papier *m* de garde — **d** Vorsatzpapier *n*; Schutz-blattpapier *n* — **n** schutbladenpapier *n* — **e** papel *m* para guardas — **i** carta *f* per bisguardie

5162 end leaves; end papers; end sheets — Two blank fly leaves at the beginning and end of a book. One leaf of each two is pasted down to the cover. See also paste down, fly leaf.
f gardes *fpl*; feuilles *fpl* de garde — **d** Vorsatz-

blätter *npl* — **n** schutbladen *npl* — **e** guardas *fpl* — **i** risguardi *mpl*; sguardie *fpl*; guardie *fpl*; fogli *mpl* di risguardo

5163 endless forms
f formulaires *mpl* en continu; liasses *fpl* sans fin — **d** Endlosformulare *npl* — **n** kettingformulieren *npl* — **e** formularios *mpl* continuos; formularios *mpl* sin fin — **i** moduli *mpl* continui; moduli *mpl* su nastro continuo

5164 endless screw — See worm.

5165 end matter; back matter; reference matter; subsidiaries — Material at the end of a book, after the proper text, including appendices, bibliography, glossary, indexes, colophon, etc.
f pages *fpl* de fin — **d** Anhang *m*; ergänzende Buchteile *mpl* am Buchende — **n** pagina's *fmpl* achterin een boek — **e** material *m* suplementario — **i** materiale *m* in fondo al libro

5166 end paper — See end-leaf paper.

5167 end papering machine
f machine *f* à coller les gardes — **d** Vorsatz-einklebemaschine *f*; Vorsatzblätter-Anleim-maschine *f* — **n** schutbladenplakmachine *f* — **e** máquina *f* de pegar guardas — **i** macchina *f* per incollare i risguardi

5168 end papers — See end leaves.

5169 end-piece — See tail-piece.

5170 end quotation mark — See close quote.

5171 end sheet — See end-leaf paper.

5172 end sheets — See end leaves.

5173 enfant de la balle — French term, denoting a son of a compositor, who is a compositor on turn.

5174 engineering — A size of wrapping paper. See app. no. 3.

5175 engine-sized paper — See beater-sized paper.

5176 engine sizing — See beater sizing.

5177 Engler degree — Unit in engineering instrumentation to measure viscosity.
f degré *m* Engler — **d** Engler-Grad *m* — **n** graad *m* Engler — **e** grado *m* Engler — **i** grado *m* Engler

5178 english — Obsolete name for a size of type. See app. no. 1.

5179 English finish; velvet finish — Smooth finish without gloss, intermediate between machine and supercalender finishes.
f apprêt *m* satiné; fini *m* satiné; fini *m* à l'anglaise — **d** mattsatiniert; halbglänzend — **n** mat gesatineerd — **e** acabado *m* semimate; acabado *m* inglés; acabado *m* aterciopelado; satinado *m* común — **i** satinatura *f* mat

5180 English finish paper; velvet finished paper; velvet paper; e.f. paper
f papier *m* satiné semi-mat — **d** halbglänzendes Papier *n*; mattsatiniertes Papier *n* — **n** halfmat papier *n*; mat gesatineerd papier *n* — **e** papel *m* (de acabado) semimate; papel *m* satinado mate — **i** carta *f* vellutata; carta *f* satinata

5181 English red — See Venetian red.

5182 engrave *v* — To cut or incise lines or designs in metal, wood and other surfaces manually with tools. Often but improperly applied to etching.
f graver (à la main) — **d** gravieren; stechen (in Kupfer usw.); radieren; schneiden (in Holz) — **n** graveren; steken; snijden (in hout) — **e** grabar; burilar — **i** incidere; bulinare

5183 engraved blanket — Blanket where the image area surface has sunk below the rest of the surface due to disintegration by ink ingredients.
f blanchet *m* gravé — **d** durch Farbe vertieftes Gummituch *n* — **n** rubberdoek *n* met een reliëf — **e** mantilla *f* hundida — **i** blanket *m* abbassato

5184 engraved cylinder; forme cylinder — Intaglio cylinder, etched and ready for printing on a rotogravure press.
f cylindre *m* gravé; cylindre *m* imprimeur — **d** Tiefdruckzylinder *m*; Formzylinder *m* — **n** diep-drukcilinder *m*; cilinder *m* — **e** cilindro *m* de huecograbado — **i** cilindro *m* rotocalco

5185 engraved roller — Screened roller in flexographic and gravure inking systems.
f rouleau *m* tramé — **d** Rasterwalze *f* — **n** raster-rol *fm* — **e** rodillo *m* reticulado; rodillo *m* grabador — **i** rullo *m* con retinatura incisa

5186 engraver — One who engraves. The term is loosely applied to *photoengravers, regardless of branch or activity. See also plate engraver, wood engraver.
f graveur *m* — **d** Graveur *m* — **n** graveur *m* — **e** grabador *m* — **i** incisore *m*; intagliatore *m*; bulino *m*

5187 engraver, photo — See photoengraver.

5188 engraver's acid — See nitric acid.

5189 engraver's pad — Round leather, sand-filled pad to support copper or steel plates during hand engraving.
f coussinet *m* du graveur — **d** Gravierkissen *n* — **n** graveerkussen *n* — **e** almohada *f* de grabado — **i** cuscinetto *m* d'appoggio per incisori

5190 engraver's proof; block pull — In photoengraving, an impression of a line or halftone etching on good quality paper.
f épreuve *f* de photogravure — **d** Klischee-andruck *m*; Andruck *m* der Klischeeanstalt — **n** proef *fm* van de clichémaker; afdruk *m* van de clichémaker — **e** prueba *f* de grabador; prueba *f* de ensayo — **i** bozza *f* del cliscè; bozza *f* del cliché

5191 engraving — Illustration printed by means of an engraved metal plate.
f estampe *f* — **d** Gravüre *f* — **n** gravure *fm* — **e** grabado *m* — **i** incisione *f*

5192 engraving — Plate produced by engraving and intended for use on a printing press.
f cliché *m* (de photogravure) — **d** Druckstock *m*; Klischee *n*; Ätzung *f* — **n** cliché *n*; drukplaat *fm* — **e** grabado *m*; lámina *f* grabada; plancha *f*; clisé *m* (de fotograbado) — **i** lastra *f* incisa; cliscè *m*; cliché *m*

5193 engraving head — Part of an electronic engraving machine.

f tête *f* à graver — **d** Gravierkopf *m* — **n** graveer-kop *m* — **e** cabeza *f* de grabar; cabeza *f* grabadora — **i** testa *f* per incidere

5194 engraving in lead — See lead engraving.

5195 engraving machine
f machine *f* à graver — **d** Graviermaschine *f* — **n** graveermachine *f* — **e** máquina *f* de grabar; grabadora *f* — **i** macchina *f* da incidere

5196 engraving negatives — See scriber.

5197 engraving, photo — See photoengraving.

5198 enlarge *v*; **blow** *v* **up** — To produce an image larger than the original.
f agrandir — **d** vergrößern — **n** vergroten — **e** agrandar; ampliar; amplificar — **i** ingrandire

5199 enlarged and revised edition — See amplified and revised edition.

5200 enlarged edition; amplified edition; augmented edition
f édition *f* augmentée — **d** erweiterte Ausgabe *f*; vermehrte Ausgabe *f* — **n** vermeerderde uitgave *f*; vermeerderde druk *m* — **e** edición *f* aumentada — **i** edizione *f* ampliata; edizione *f* accresciuta

5201 enlargement — Reproduction larger than the original.
f agrandissement *m* — **d** Vergrößerung *f* — **n** vergroting *f* — **e** ampliación *f*; amplificación *f* — **i** ingrandimento *m*

5202 enlarger; enlarging camera — Equipment to make photographic enlargements, consisting of a lamphouse, negative carrier, lens and adjustment to vary the distances between lens, negative and original.
f agrandisseur *m*; appareil *m* d'agrandissement — **d** Vergrößerungsapparat *m*; Vergrößerungskamera *f*; Vergrößerungsgerät *n* — **n** vergrotingsapparaat *n* — **e** ampliadora *f*; cámara *f* ampliadora; amplificadora *f* — **i** ingranditore *m*; apparecchio *m* ingranditore

5203 enlarging camera — See enlarger.

5204 enlarging paper — Relatively fast paper, as compared with contact paper, for printing and enlarged image projected from a negative.
f papier *m* d'agrandissement — **d** Vergrößerungspapier *n* — **n** vergrotingspapier *n* — **e** papel *m* de ampliación — **i** carta *f* per ingrandimento

5205 en quad; nut; nut quad; en space — Space usually half the em width of a type body, but alternatively equivalent to the width of the fount figures.
f demi-cadratin *m* — **d** Halbgeviert *n* — **n** pasje *n* — **e** medio cuadratín *m*; cuadratín *m* de ene — **i** quadratino *m*

5206 en rule — See dash.

5207 en space — See en quad.

5208 entry — See catchword.

5209 envelope; cover; wrapper — Paper covering for mailing or packaging with an open side or open end. Flaps may be gummed or ungummed, or closed with metal or string attachments.
f enveloppe *f*; pli *m* — **d** Briefumschlag *m*; Umschlag *m*; Briefhülle *f* (Klappe an der langen Seite); Brieftasche *f* (Klappe an der kurzen Seite); Kuvert *n* — **n** enveloppe *fm*; envelop *fm*; couvert *n*; briefomslag *mn* — **e** sobre *m*; cubierta *f* de carta — **i** busta *f* (da lettere)

5210 envelope filler — See stiffener.

5211 envelope making machine
f machine *f* à enveloppes — **d** Briefumschlagmaschine *f* — **n** enveloppenmachine *f* — **e** máquina *f* para la confección de sobres — **i** macchina *f* per la manifattura di buste

5212 envelope paper — Paper for envelopes, such as kraft, manila, stationery, etc.
f papier *m* pour enveloppes (postales) — **d** Briefumschlagpapier *n*; Briefhüllenpapier *n* — **n** enveloppenpapier *n* — **e** papel *m* para sobres — **i** carta *f* per buste; carta *f* da buste

5213 envelope stuffer — See stuffer.

5214 enveloping machine — Machine which inserts printed matter into envelopes.
f machine *f* à mettre sous enveloppe — **d** Kuvertiermaschine *f* — **n** envelopinsteekmachine *f* — **e** máquina *f* para meter en sobres — **i** macchina *f* per confezionare in buste

5215 eosin(e) — Red organic dyestuff mainly used for process inks.
Formula: $C_{20}H_8Br_4O_5$.
f éosine *f* — **d** Eosin *n* — **n** eosine *fmn* — **e** eosina *f* — **i** eosina *f*

5216 EPC-process — Essentially simple and potentially less expensive than other printing methods the EPC-process (Electrostatic Printing Corporation of America) uses a thin flexible printing element with finely screened openings defining the image to be printed. An electric field is established between this image element and the surface to be printed. Finely divided "electroscopic" ink particles are metered through the image openings and attracted to the surface to be printed where they are firmly held by electrostatic attraction until fixed by heat or by chemical means.

5217 epigram — Short poem, often satirical, dealing concisely with single subject and usually ending with a witty or ingenious turn of thought.
f épigramme *m* — **d** Epigramm *n*; kurzes Sinngedicht *n* — **n** epigram *n*; puntdicht *n* — **e** epigrama *m* — **i** epigramma *m*

5218 epilog — See epilogue.

5219 epilogue *GB*; **epilog** *US*; **afterword** — Concluding part added to a literary work.
f épilogue *m*; postface *f* — **d** Epilog *m*; Nachwort *n*; Nachruf *m* — **n** epiloog *m*; nawoord *n*; naschrift *n*; slotwoord *n*; narede *fm* — **e** epílogo *m*; postdata *f* — **i** epilogo *m*

5220 epoxy(d) resin; ether resin — Flexible, usually thermosetting resin made by polymerization of an epoxide and used chiefly in coatings and adhesives. An epoxide is a compound containing an oxygen atom bound to two already connected, usually carbon, atoms, thus forming a ring, as in ethylene oxide or epoxy-ethane, $H_2C(O)CH_2$.
f résine *f* époxyde; résine *f* époxydrique; résine *f*

épikote; résine *f* d'épichlorhydrique — **d** Eponharz *n*; Epoxydharz *n*; Epichlorhydrinharz *n*; Ätherharz *n*; Äthoxylinharz *n* — **n** epoxyhars *mn*; epikotehars *mn*; ethoxylinehars *mn* — **e** resina *f* epóxido; resina *f* epoxídica; resina *f* epiclorhidrina; resina *f* epoxi — **i** resina *f* epossidica

5221 Epsom salt — See magnesium sulphate.

5222 equalization — The process of reducing frequency and/or phase distortion of a circuit by the introduction of networks to compensate for the difference in attenuation and/or time delay at the various frequencies in the transmission band.
f compensation *f*; correction *f*; égalisation *f* — **d** Ausgleichung *f* — **n** gelijkmaking *f*; effening *f* — **e** compensación *f*; igualación *f* — **i** compensazione *f*; equilibrazione *f*

5223 equalizer rod; Meyer rod; Meyer bar — Metal rod wound with a fine wire around its axis so that an ink can be drawn down evenly and at a given thickness across a piece of paper, sometimes also called film applicator.
f étaleur *m* d'encre — **d** Streichrakel *f*; Ziehdreieck *n* für Farbauftrag — **n** strijkmes *n* — **e** barra *f* rascadora; rascador *m* — **i** racla *f* di Meyer

5224 equal mark; equal sign — Symbol (=) in mathematical expressions to indicate that the terms it separates are equal.
f signe *m* d'égalité — **d** Gleichheitszeichen *n*; Gleichheitsstrich *m* — **n** gelijkteken *n*; is-gelijkteken *n* — **e** signo *m* de igualdad; signo *m* de ecuación — **i** segno *m* d'uguaglianza

5225 equal sign — See equal mark.

5226 equation — Expression or proposition, often algebraic, asserting the equality of two quantities, often used in determining a value of an unknown included in one or both quantities.
f équation *f* — **d** Gleichung *f* (im Formelsatz) — **n** vergelijking *f* — **e** ecuación *f* — **i** equazione *f*

5227 equilibrium — State of balance. Paper is said to be in equilibrium with the atmosphere, when it neither takes up nor gives off moisture in that particular atmosphere.
f équilibre *m* — **d** Gleichgewichtslage *f*; Gleichgewichtausgleich *m* — **n** evenwicht *n*; evenwichtstoestand *m* — **e** equilibrio *m* — **i** equilibrio *m*

5228 equilibrium moisture content — The percentage moisture content of paper after bringing it into balance with the surrounding atmosphere.
f teneur *f* en eau d'équilibre — **d** Feuchtigkeitsgehalt *m* — **n** vochtgehalte *n* — **e** equilibrio *m* de la humedad relativa — **i** contenuto *m* d'umidità d'equilibrio

5229 equivalent exposure — Change in exposure time required when the camera lights are moved from their normal position.
f pose *f* équivalente — **d** gleichwertige Belichtung *f*; equivalente Belichtung *f* — **n** gelijkwaardige belichting *f* — **e** exposición *f* equivalente — **i** esposizione *f* equivalente

5230 equivalent weight — Weight of an odd size sheet of paper figured from the standard basis weight by multiplying the basis weight or substance by the odd size and dividing by the standard size.
f poids *m* correspondant — **d** Äquivalentgewicht *n* — **n** verhoudingsgewicht *n* — **e** peso *m* equivalente — **i** peso *m* equivalente

5231 erasability; erasing property — Property of a sheet concerned with the ease of removing typed and/or written characters and impressions from the sheet by mechanical erasure, the cleanliness or the amount of abrading on the erased portion, and the suitability of the erased portion for re-use.
f résistance *f* à l'effaçure; résistance *f* au grattage — **d** Radierbarkeit *f*; Radierfestigkeit *f* — **n** radeerbaarheid *f* — **e** resistencia *f* al rozamiento — **i** resistenza *f* allo sfregamento; resistenza *f* alla cancellatura; cancellabilità *f*

5232 erase *v* — To obliterate information from a magnetic surface or some other memory unit by returning the magnetic state of a cell to a uniform null condition.
f effacer; mettre à zero — **d** löschen; ausblenden — **n** uitwissen; wissen — **e** borrar — **i** annullare; cancellare

5233 eraser — Piece of rubber to erase marks. See also ink eraser.
f gomme-grattoir *f*; grattoir *m*; gomme *f* à effacer — **d** Radiergummi *m* — **n** vlakgum *mn*; vlakgum *mn* — **e** goma *f* de borrar; goma *f* elástica; hule *m* Me — **i** gomma *f* per cancellare

5234 erasing knife
f grattoir *m*; grattoir *m* pour gravures — **d** Radiermesser *n*; Schabernadel *f* — **n** radeermes *n* — **e** cuchillo *m* raspador; desbarbador *m* — **i** raschietto *m*

5235 erasing property — See erasability.

5236 erect *v* — To build up printing presses or other machinery.
f installer; monter — **d** aufstellen — **n** opstellen — **e** montar — **i** montare

5237 erector
f monteur *m* (de machines); ajusteur-monteur *m* (de machines) — **d** Fabriksmonteur *m*; Maschinenmonteur *m*; Maschinenaufsteller *m* — **n** machinemonteur *m*; fabrieksmonteur *m* — **e** montador *m* de máquinas — **i** montatore *m* di macchine

5238 erg — Unit of work or energy, being the work when a steady force of 1 dyne produces a displacement of 1 centimetre in the direction of the force.
In the SI: 1 erg = 1 dyne cm = 10^{-7} J.
f erg *m* — **d** Erg *f* — **n** erg *m* — **e** erg *m* — **i** erg *m*

5239 ergonomics — Study of man in relationship to his working environment. See also human engineering.
f ergologie *f* — **n** ergonomie *f* — **e** ergología *f* — **i** ergonomia *f*

5240 errand boy
f galopin *m*; garçon *m* de magasin — **d** Laufbursche *m* — **n** loopjongen *m*; boodschappen-

jongen *m* — **e** muchacho *m*; chico *m* — **i** fattorino *m*; galoppino *m*; piazzista *m*

5241 errata — List of author's or printer's errors usually inserted at the begining of a book between the title page and the preface. Plural of *erratatum.

f errata *pl*; liste *f* d'erratas; liste *f* des fautes d'impression — **d** Druckfehlerverzeichnis *n*; Errata *fpl* — **n** errata *npl* — **e** erratas *fpl*; fé *f* de erratas — **i** errata *mpl*; errata corrige *m*

5242 erratum — Error discovered and corrected after printing.

f erratum *m* — **d** Erratum *n*; Druckfehlerberichtigung *f* — **n** erratum *n* — **e** errata *f* — **i** erratum *m*

5243 error burst — In data transmission, a sequence of signals containing one or more errors but counted as only one unit in accordance with some specific criterion or measure.

f série *f* de signaux à erreurs — **d** Fehlerpaket *n*; Signalreihe *f* — **n** foutenpakket *n*; signaalserie *f* — **e** serie *f* de señales de errores — **i** serie *f* di segnali ad errori

5244 error condition — Indication that the data being processed or the instructions being received are internally inconsistent so that either the processing run must be aborted or, if completed, will contain defects not initially anticipated. The error may be attributable to hardware failure or to software deficiencies.

5245 error correcting code — Error-detecting code using additional code elements so that the mutilated representation resembles more closely the original than any other valid representation.

f code *m* converteur d'erreurs — **d** Fehlerkorrekturkode *m* — **n** fouten-verbeterende code *m*; fouten-corrigerende code *m*; fouten-herstellende code *m* — **e** código *m* corrector de errores — **i** codice *m* autocorrettivo

5246 error detecting code — Code in which each representation of a character conforms to specific rules of construction, so that for certain errors the mutilated representation corresponds to no valid character; the presence of these errors can be detected without reference to the original message.

f code *m* détecteur d'erreurs — **d** Fehlererkennungskode *m* — **n** fouten-aangevende code *m*; fouten-signalerende code *m* — **e** código *m* de detección de errores — **i** codice *m* rivelatore d'errori

5247 error rate — Measure of quality of circuit or system; the number of erroneous bits or characters in a sample, frequently taken per 100,000 characters.

f fréquence *f* d'erreurs — **d** Fehlerhäufigkeit *f* — **n** foutenfrequentie *f* — **e** frecuencia *f* de errores — **i** frequenza *f* d'errori

5248 erythrosine — Brown powder made by iodination of fluorescein, used for colouring; soluble in alcohol.
Formula: $C_{20}H_6I_4Na_2O_5$.

f érythrosine *f* — **d** Erythrosin *n* — **n** erytrosine *f* — **e** eritrosina *f* — **i** eritrosina *f*

5249 escapement; channel escapement; magazine escapement — Mechanism on a line-caster which releases a matrix from a channel when the operator presses a key, or, in case of a teletypesetter, when the tape read by the line-caster operating unit signals the appropriate character code.

f échappement *m* de magasin — **d** Sperrkegel *m* (am Magazin) — **n** magazijnuitlaat *m* — **e** escape *m* — **i** scappamento *m*

5250 escapement — Process of moving a mirror, lens, prism, or an electronic beam, to avoid superimposing one character on top of another. The escapement mechanism must move over by the width of the character which has just been set or which is just about to be set. Some typesetters escape and then flash while others flash and then escape. In either event, the typesetting machine needs to know or be told how far to escape for each character. In the case of a typewriter, escapement (usually of monospaced or non-proportional characters) is provided mechanically by the movement of the typing element or the typewriter platen, or roller.

5251 esculin; aesculin — White, crystalline, slightly water-soluble glucoside, obtained from the bark of the common horse chestnut, used chiefly in skin preparations as a protective.
Formula: $C_{15}H_{16}O_9$.

f esculine *f* — **d** Äskulin *n*; Schillerstoff *m* — **n** esculine *f* — **e** esculina *f* — **i** esculina *f*

5252 esparto; esparto grass; Spanish grass; alfa — Coarse grass from Southern Spain and Northern Africa, obtained from two species, lygeum spartum and stipa tenacissima, with short and relatively fine and wiry fibres. Generally cooked with caustic soda alone or with sodium sulphide. Used mainly by English and Scottish papermakers for the production of better grades of book paper. Not to be confused with alpha cellulose.

f sparte *m*; alfa *m* — **d** Esparto *n*; Espartogras *n*; Spartogras *n*; Alphagras *n* — **n** esparto *n*; espartogras *n*; alfagras *n* — **e** esparto *m*; alfa *m* — **i** sparto *m*; alfa *f*

5253 esparto grass — See esparto.

5254 esparto paper

f papier *m* d'alfa; papier *m* alfa; papier *m* de sparte — **d** Espartopapier *n*; Alfapapier *n* — **n** espartopapier *n* — **e** papel *m* alfa — **i** carta *f* di alfa; carta *f* di sparto

5255 esparto pulp; alfa pulp; albardine pulp — Pulp obtained from esparto grass.

f pâte *f* d'alfa — **d** Espartozellstoff *m*; Albardinzellstoff *m* — **n** espartocelstof *fm* — **e** pulpa *f* de alfa; pasta *f* de alfa — **i** pasta *f* di sparto; pasta *f* di alfa

5256 essence of lavender — See lavender oil.

5257 essential oil — Volatile oil, usually nearly colourless when fresh, but becoming darker and thick on exposure to air. Unsaponifiable, except when containing organic ethers.

f huile *f* essentielle; huile *f* volatile; essence *f* — **d** ätherisches Öl *n* — **n** etherische olie *fm*; essence *fm* — **e** aceite *m* esencial; aceite *m* volátil; esencia *f* — **i** olio *m* essenziale

5258 established wages — See stab wages.

5259 ester — Organic liquid, product of the reactions of organic acids and alcohols; used as solvent. Examples are ethyl acetate, isopropyl acetate, dioctyl phthalate.

f ester *m* — **d** Ester *m* — **n** ester *m* — **e** éster *m*; éter sal *m* — **i** estere *m*

5260 ester gum; rosin ester gum — Hard resin produced by the esterification of a natural resin, especially rosin, with a polyhydric alcohol, chiefly glycerol. Used as an ingredient of some printing ink varnishes.

f gomme *f* ester; résine *f* estérifiée — **d** Harzester *m*; Esterharz *n*; Estergummi *m* — **n** harsester *m*; esterhars *mn* — **e** resina *f* esterificada; resina *f* de éster — **i** resina *f* esterificata; gomma *f* esterificata

5261 esterification — Formation of an ester by the direct action of an acid on alcohol in the presence of hydrogen ions.

f estérification *f* — **d** Veresterung *f*; Esterifikation *f* — **n** verestering *f* — **e** esterificación *f* — **i** esterificazione *f*

5262 esterify *v* — To make or change into an ester.

f estérifier — **d** veresteren; esterifizieren — **n** veresteren — **e** esterificar — **i** esterificare

5263 ester number — The difference, expressed in milligrams of potassium hydroxide per gram of oil, between the acid number and saponification number.

f indice *m* d'ester — **d** Esterzahl *f* — **n** estergetal *n* — **e** índice *m* de éster — **i** indice *m* d'estere

5264 estimation — Computation or approximate evaluation of the cost of the work to be done, on which a *quotation may be based.

f devis *m* (provisoire); évaluation *f* — **d** Preisberechnung *f*; Voranschlag *m*; Kostenvoranschlag *m*; Kalkulation *f*; Vorauskalkulation *f* — **n** calculatie *f*; voorcalculatie *f*; prijsberekening *f* — **e** presupuesto *m*; cálculo *m* (de precios) — **i** preventivo *m*; calcolo *m* preventivo; stima *f*

5265 estimator

f préposé *m* aux devis; chiffreur *m*; calculateur *m* — **d** Kalkulator *m*; Vorkalkulator *m* — **n** calculator *m* — **e** estimador *m*; computista *m*; calculador *m* — **i** addetto *m* ai preventivi

5266 etch — Mixture for *desensitizing offset printing plates.

f mordant *m* — **d** Ätzmittel *n*; Ätze *f* — **n** etsvloeistof *fm* — **e** corrosivo *m* — **i** corrosivo *m*

5267 etch *v* — See desensitize *v*.

5268 etch *v* **away** — To remove by etching.

f enlever par morsure — **d** abätzen — **n** wegetsen — **e** grabar — **i** incidere

5269 etcher; aquafortist — Artist who makes fine prints by etching copper or steel plates. See also halftone etcher, engraver.

f aquafortiste *m*; graveur *m* à l'eau-forte — **d** Radierer *m*; Ätzer *m* — **n** etser *m* — **e** aguafortista *m*; grabador *m* al agua fuerte — **i** acquafortista *m*; incisore *m* ad acquaforte

5270 etching — Proof taken from an etched plate.

f gravure *f* à l'eau-forte; eau-forte *f* — **d** Radierung *f* — **n** ets *fm* — **e** grabado *m* al agua fuerte; aguafuerte *m* — **i** incisione *f* all'acqua forte; incisione *f* ad acquaforte

5271 etching — Biting a design or image into a metal plate by chemical or electrolytical action, to make some areas considerably lower than the surface of the plate.

f morsure *f*; gravure *f* — **d** Ätzen *n* — **n** etsen *n* — **e** morder; grabar (al agua fuerte) — **i** incidere

5272 etching — Relief or intaglio printing plate produced by etching.

f gravure *f* (à l'eau-forte) — **d** Ätzplatte *f* — **n** ets *fm*; etsplaat *fm* — **e** grabado *m* al agua fuerte — **i** incisione *f* (per acqua forte)

5273 etching basin

f cuve *f* de morsure — **d** Ätzschale *f*; Ätzwanne *f* — **n** etsbak *m* — **e** cubeta *f* para el mordido — **i** bacinella *f* per gli acidi

5274 etching bath — Solution or mordant used for etching.

f basin *m* de morsure — **d** Ätzbad *n* — **n** etsbad *n* — **e** baño *m* de mordido; baño *m* mordiente; baño *m* para grabar; baño *m* ácido; mordiente *m* para grabar — **i** bagno *m* d'incisione

5275 etching brush — Flat brush for etching.

f blaireau *m* à morsure; queue *f* de morue — **d** Ätzpinsel *m* — **n** etskwast *m* — **e** brocha *f* para agua fuerte; paletina *f* para mordida — **i** pennello *m* per mordanzare; pennello *m* da incisione

5276 etching depth; depth of bite; printing depth; depth — In relief plates, the vertical distance from the actual printing surface to the bottom of any low (etched) area. The etching depth, for instance, for newspaper halftone engravings should produce the best results: highlight dots 55 lines 0.008 in, highlight dots 65 lines 0.0065 in, highlight dots 75 lines 0.005 in, middle tone dots 55 lines 0.005 in, middle tone dots 65 lines 0.0045 in, middle tone dots 75 lines 0.0035 in, shadow dots 55 lines 0.003 in, shadow dots 65 lines 0.0035 in, shadow dots 75 lines 0.002 in.

f profondeur *f* de gravure; creux *m* (de gravure) — **d** Ätztiefe *f*; Tiefe *f* — **n** etsdiepte *f*; diepte *f* — **e** profundidad *f* de mordido; profundidad *f* de grabado; profundidad *f* de mordedura — **i** profondità *f* d'incisione

5277 etching-depth meter; halftonometer; depthometer; depth gauge — Calibrated gauge for micrometrically measuring the depth of printing plates.

f appareil *m* pour la mesure des creux — **d** Ätztiefemeßgerät *n*; Rastertiefenmesser *m*; Tiefenmaß *n* — **n** etsdieptemeter *m*; rasterdieptemeter *m*; clichédieptemeter *m* — **e** medidor *m* de profundidad de clisés — **i** apparecchio *m* per misurare la profondità d'incisione

5278 etching ground — Thin acid resistant coating of beeswax, resin, and asphaltum, etc.
f réserve *f* de morsure — **d** Ätzgrund *m* — **n** etsgrond *m* — **e** sisa *f* mordiente; sisa *f* de asfalto; recubridor *m* — **i** vernice *f* antiacido di protezione per morsura

5279 etching ink — Greasy ink of resinous constituency used as an acid resist in conjunction with etching powders applied to the inked print or plate.
f encre *f* à morsure — **d** Ätzschutzfarbe *f*; Ätzfarbe *f* — **n** bescherminkt *m* — **e** tinta *f* para mordido; tinta *f* de grabar; tinta *f* para fotograbado — **i** inchiostro *m* per incisione; inchiostro *m* per morsura; inchiostro *m* per mordenzatura

5280 etching machine — In chemical engraving, mechanically operated apparatus to accelerate the process of etching by agitation of the mordant and discharge of the solution against the surface of a metal plate.
f machine *f* à morsure; machine *f* à mordancer; machine *f* à graver (à l'eau-forte) — **d** Ätzmaschine *f*; Klischeeätzmaschine *f* — **n** etsmachine *f* — **e** máquina *f* para morder; máquina *f* de grabar al ácido; máquina *f* de grabar al agua fuerte; grabadora *f* — **i** macchina *f* da incidere; macchina *f* per incisione; macchina *f* a incidere

5281 etching needle; etching point; etching stylus; dry needle — Sharp pointed steel instrument to draw lines into an acid resisting surface.
f échoppe *f*; aiguille *f* (de graveur) — **d** Radiernadel *f*; Ätznadel *f*; Bolzer *m* (für Kupferstecher) — **n** etsnaald *f*; graveernaald *f* — **e** aguja *f* del fotograbador; punta *f* del fotograbador — **i** punta *f* per mordenzatura

5282 etching point — See etching needle.

5283 etching powder; topping powder — Finely ground mixture of resins used for dusting relief plates during etching to protect the top and sides of lines and dots. See also dragon's blood.
f résine *f* en poudre — **d** Ätzpulver *n*; Einstaubpuder *m* — **n** harspoeder *n* — **e** polvo *m* para grabar; polvo *m* (resinoso) para reserva; polvo *m* de resina — **i** polvere *f* resinosa

5284 etching press — See copperplate press.
5285 etching rate — Ratio between the vertical etching depth and the horizontal etching width of a halftone block; used to express the lateral effect of the etching. The greater this factor the less is the side etching.
f facteur *m* de morsure — **d** Ätzfaktor *m*; Ätzverhältnis *n* — **n** etsfactor *m* — **e** factor *m* de mordido; factor *m* de mordadura; factor *m* de grabado — **i** fattore *m* d'incisione

5286 etching solution; mordant — In photoengraving, an acid or other corrosive liquid able to eat into and dissolve a metal surface.
f solution *f* de morsure; acide *m* (de morsure); mordant *m* — **d** Ätzlösung *f*; Ätzwasser *n*; Ätz-

flüssigkeit *f*; Ätze *f*; Ätzmittel *n* — **n** etsvloeistof *fm* — **e** mordiente *m* (para grabar); agua *f* fuerte — **i** soluzione *f* d'incisione; soluzione *f* per mordenzatura

5287 etching stylus — See etching needle.
5288 etching time; biting time — Time during which a metal plate is exposed to the biting action of an acid.
f temps *m* de morsure; durée *f* de la morsure — **d** Ätzdauer *f* — **n** etstijd *m*; etsduur *m* — **e** tiempo *m* de mordido; tiempo *m* de grabado — **i** tempo *m* d'incisione; durata *f* d'incisione

5289 etching trough — See etching tub.
5290 etching tub; etching trough — Stoneware or acid proof tray or trough mechanically oscillated to cause an acid solution to flow back and forth over a metal plate during etching.
f cuve *f* de morsure; bac *m* de morsure — **d** Ätztrog *m*; Ätzkasten *m*; Ätzschale *f* — **n** etsbak *m* — **e** cubeta *f* para grabar; artesa *f* para grabar; cubeta *f* basculante; cubeta *f* de balanceo — **i** vasca *f* d'incisione

5291 etch *v* **in relief**
f mordre en relief — **d** hochätzen — **n** hoogetsen — **e** grabar en relieve — **i** incidere in rilievo

5292 ethanal — See acetaldehyde.
5293 ethanethiolic acid — See thioacetic acid.
5294 ethanoic acid — See acetic acid.
5295 ethanol — See ethyl alcohol.
5296 ethene — See ethylene.
5297 ether; ethyl ether; diethyl ether; ethyl oxide; diethyl oxide; sulphuric ether — Light, transparent, colourless volatile, exceedingly flammable, mobile liquid. Soluble in alcohol, chloroform, benzene, solvent naphtha and oils, slightly soluble in water. The vapour of ether mixed with air explodes when ignited. Strongly narcotic. Latin: aether.
Formula: $(C_2H_5)_2O$.
f éther *m* ordinaire; éther *m* éthylique; éther *m* sulfurique; oxyde *m* di-éthylique — **d** gewöhnlicher Äther *m*; Diäthyläther *m*; Äthyläther *m*; Schwefeläther *m* — **n** ether *m*; diethylether *m* — **e** éter *m*; éter *m* etílico — **i** etere *m*; etere *m* etilico; etere *m* solforico

5298 ether resin — See epoxy(d) resin.
5299 E 13 B — One of the two magnetic ink recognition founts, essentially numerical, developed in the US by the American Bankers Association. E stands for the fifth design proposed, and in this case accepted, 13 for the character width (0.33 mm or 0.013 in), and B for the second revision of the design. See also CMC 7.
5300 ethyl acetate; acetic ether — Colourless, flammable liquid, soluble in chloroform, alcohol and ether, slightly soluble in water. Used for lacquer type inks for flexography and rotogravure printing. TL-value: 400 ppm or 1400 mg/m³.
Formula: $CH_3CO_2 \cdot C_2H_5$.
f acétate *m* d'éthyle; éther *m* acétique; éther *m* éthylacétique — **d** Äthylazetat *n*; Essigäther *n*; Essigsäureäthylester *m* — **n** ethylacetaat *n* — **e**

acetato *m* de etilo; acetato *m* etílico; éter *m* acético — i acetato *m* d'etile; acetato *m* etilico; etil acetato *m*

5301 ethyl alcohol; ethylic alcohol; common alcohol; ethyl hydroxide; ethanol — Colourless limpid, volatile liquid, soluble in water, methyl alcohol, ether and chloroform. TL-value: 1000 ppm or 1900 mg/m³.

Formula: C_2H_5OH.

f alcool *m* éthylique; éthanol *m*; alcool *m* ordinaire — **d** Äthylalkohol *m*; Äthanol *m*; Spiritus *m*; Sprit *m* — **n** ethylalcohol *m*; ethanol *m*; spiritus *m* — **e** alcohol *m* etílico; etanol *m* — i alcool *m* etilico; etanolo *m*

5302 ethyl aldehyde — See acetaldehyde.

5303 ethyl bromide; bromoethane; bromic ether; hydrobromic ether — Colourless, flammable, volatile liquid, soluble in alcohol and ether, sparingly soluble in water.

Formula: C_2H_5Br.

f bromure *m* d'éthyle — **d** Äthylbromid *n*; Bromäthan *n*; Bromäthyl *n* — **n** ethylbromide *n* — **e** bromuro *m* de etilo — i bromuro *m* d'etile; bromuro *m* etilico

5304 ethyl carbinol — See propyl alcohol.

5305 ethyl cellulose — Ethyl ether of cellulose. White, granular, thermoplastic solid. Soluble in most organic liquids, and compatible with resins, waxes, oils, and plasticizers; inert to alkalis and dilute acids. Insoluble in water and glycerol. Used as a film former in gravure inks and lacquers, in adhesives, etc.

f éthylcellulose *f* — **d** Äthylzellulose *f* — **n** ethylcellulose *fm* — **e** etilcelulosa *f* — i etilcellulosa *f*

5306 ethyl chloride; chloroethane; hydrochloric ether; muriatic ether — Gas at ordinary temperature. Compressed, a colourless, highly flammable, volatile liquid. Miscible with most of the commonly used solvents, slightly soluble in water. TL-value: 1000 ppm.

Formula: C_2H_5Cl.

f chlorure *m* d'éthyle; chloréthane *m*; éther *m* éthylchlorhydrique; éther *m* chlorhydrique — **d** Chloräthyl *n*; Äthylchlorid *n*; Chloräthan *n* — **n** ethylchloride *n*; monochloorethaan *n* — **e** cloruro *m* de etilo; cloruro *m* etílico — i cloruro *m* d'etile; cloruro *m* etilico

5307 ethylene; ethene — Colourless, inflammable gas with a sweetish smell. Slightly soluble in water, alcohol and ethyl ether.

Formula: C_2H_4.

f éthylène *m*; éthène *m*; gas *m* oléifiant — **d** Äthylen *n*; Äthen *n*; ölbildendes Gas *n* — **n** ethyleen *m*; etheen *m* — **e** etileno *m* — i etilene *m*

5308 ethylene alcohol — See ethylene glycol.

5309 ethylene glycol; ethylene alcohol; glycol; glycol alcohol — Clear, colourless, syrupy liquid. Hygroscopic. Soluble in water, alcohol and ether; used as an anti-freeze solution.

Formula: $CH_2OH \cdot CH_2OH$.

f éthylène glycol *m*; glycol *m*; monoéthylène glycol *m*; éthanediol *m*; dialcol *m*; diol *m* — **d** Äthylenglykol *n*; Glykol *n*; Äthylenalkohol *m*; Äthandiol *n* — **n** ethyleenglycol *m*; etheenglycol *m*; glycol *m* — **e** etilenglicol *m*; glicol *m*; etilenoglicol *m*; etanodiol *m* — i etilenglicol *m*; glicol *m* etilenico; glicol *m*; etandiolo *m*

5310 ethylene glycol monoethyl ether *GB*; **methyl glycol** *US* — Relative slow drying flexographic ink solvent frequently used as a retarder. Also used as a solvent for natural and synthetic resins, cellulose derivatives, and dyestuffs. See also cellosolve.

f mono-méthyléther *m* d'éthylène-glycol; éther *m* éthylglycolique; méthylglycol *m* — **d** Äthylenglykol-monoäthyläther *m*; Glykolmonomethyläther *m*; Methylglykol *m* — **n** ethyleenglycolmonomethylether *m*; methylglycol *m* — **e** monometiléter *m* de etilenglicol; éter *m* de metílico del etilenglicol — i etere *m* monometilico del glicol etilenico

5311 ethyl ether — See ether.

5312 ethyl hydroxide — See ethyl alcohol.

5313 ethylic alcohol — See ethyl alcohol.

5314 ethyl iodide; iodoethane — Clear, colourless liquid. Turns brown on exposure to light. Soluble in alcohol and ether, slightly soluble in water.

Formula: C_2H_5I.

f iodure *m* d'éthyle; iodo-éthane *m* — **d** Jodäthyl *n*; Äthyljodid *n* — **n** ethyljodide *n*; monojoodethaan *n* — **e** yoduro *m* de etilo; yodoetano *m* — i ioduro *m* d'etile; ioduro *m* etilico

5315 ethyl oxide — See ether.

5316 ethyl red — Iso-chinocyanin. Photographic sensitizer.

f rouge *m* iso-chinocyanine — **d** Äthylrot *n* — **n** ethylrood *n* — **e** rojo *m* de etilo — i rosso *m* di etile

5317 Etruscan binding — Bookbinding with a decoration of classical ornaments, devised in England in the late 18th century by John Edwards of Halifax and John Whitaker.

f reliure *f* étrusque — **d** etruskischer Einband *m* — **n** Etruskische band *m* — **e** encuadernación *f* etrusco — i legatura *f* etrusco

5318 eutectic — Lowest melting point of all alloys of, e.g., antimony and lead. Eutectic alloys melt at a temperature below the melting point of their constituents. For example, the melting point of lead is 327.5 °C, of tin 231.9 °C and of antimony 630.5 °C. A type metal alloy of 4 tin and 12 antimony has an eutectic of 239 °C. However, the eutectic of an alloy of 13 antimony and 87 lead is 248 °C.

f eutectique *m* — **d** Eutektikum *n* — **n** eutecticum *n* — **e** eutéctico *m* — i eutettico *m*

5319 evaluation — Finding the value or measuring the quantity of.

f évaluation *f* — **d** Bewertung *f* — **n** waardebepaling *f*; waardering *f* — **e** evaluación *f*; valuación *f*; tasa *f* — i valutazione *f*

5320 evaporation — Changing from the liquid to the gaseous or vapour stage, as when the solvent leaves the printed ink film.

f évaporation *f* — **d** Verdünstung *f*; Abdünstung *f*; Vergasung *f*; Verdampfung *f*; Verdampfen *n* — **n** verdamping *f*; verdampen *n* — **e** evaporación *f* — **i** evaporazione *f*

5321 evaporator — Apparatus used in paper-making to concentrate the spent liquors in the soda and kraft pulp processes.

f évaporateur *m* — **d** Evaporator *m*; Verdampfer *m* — **n** vaporisator *m*; evaporator *m*; verdamper *m* — **e** evaporador *m* — **i** vaporizzatore *m*

5322 Eve binding — Bookbinding in the style used by Nicolas Eve († 1582) and his son or nephew Clovis († 1635), binders and booksellers to the king of France, showing an ornamental design with fleurs-de-lis, sometimes with a centre piece.

f reliure *f* selon Eve; reliure *f* d'Eve — **d** Eve-Einband *m* — **n** Eve-band *m*; band *m* van Eve — **e** encuadernación *f* de Eve — **i** legatura *f* di Eve

5323 evening gazette; evening paper; evening newspaper

f gazette *f* du soir; journal *m* du soir; feuille *f* du soir — **d** Abendblatt *n*; Abendzeitung *f* — **n** avondblad *n* — **e** papel *m* de la tarde; papel *m* de la noche; diario *m* de la tarde; diario *m* de la noche; hoja *f* de la noche; diario *m* vespertino — **i** giornale *m* della sera

5324 evening newspaper — See evening gazette.

5325 evening paper — See evening gazette.

5326 even leaves — See even pages.

5327 even pages; left-hand pages; even leaves — The left-hand or verso pages, these usually bearing the even number 2, 4, 6, etc.

f pages *fpl* paires; fausses pages *fpl*; pages *fpl* de gauche — **d** gerade Seiten *fpl*; linke Seiten *fpl* — **n** even pagina's *fmpl*; linker pagina's *fmpl*; linkse pagina's *fmpl* — **e** páginas *fpl* de izquierda; páginas *fpl* izquierdas; páginas *fpl* pares; páginas *fpl* verso; vueltos *mpl* — **i** pagine *fpl* pari; pagine *fpl* a sinistra

5328 even small caps — See even smalls.

5329 even smalls; even small caps; level small caps — Small caps without full-size initials.

f petites capitales *fpl* — **d** Kapitälchen *npl* ohne Initiale — **n** klein kapitaal *n* — **e** versalitas *fpl*; mayúsculas *fpl* — **i** lettere *fpl* in maiuscoletto; maiuscolette *fpl*

5330 everdamp — See transfer paper.

5331 everdamp transfer paper — *Transfer paper which remains limp by having a hygroscopic substance in its coating.

5332 every two months — See bi-monthly.

5333 every two months — See two-months publication.

5334 every two weeks — See bi-weekly.

5335 excelsior — Old American name of a size of type. See app. no. 1.

5336 exception dictionary — List of words, word roots or word segments stored in a computer to be accessed prior to hyphenating a word by a hyphenation algorithm, because the logic of the hyphenation routine would produce the wrong or undesirable results. They are "exceptions" to the particular hyphenation algorithm.

f dictionnaire *m* d'exceptions — **d** Ausnahmenverzeichnis *n* — **n** uitzonderingswoordenboek *n* — **e** diccionario *m* de excepciones — **i** dizionario *m* delle eccezioni

5337 excess feed; making paper — Letterpress rotary term for the increase in the amount of paper drawn through the press when the peripheral speed of a web driving component is greater than that of the next in line ahead. Its antinym is *insufficient feed.

f trop grand débit *m* de papier — **d** Mehrförderung *f*; Voreilung *f* — **n** te grote afwikkeling *f* — **e** avance *m*; adelanto *m* — **i** alimentazione *f* eccessiva

5338 excess material

f matière *f* en excès — **d** überschüssiges Material *n* — **n** overtollig materiaal *n* — **e** exceso *m* de material — **i** materiale *m* in eccesso

5339 excess pressure — Squeeze pressure that causes distortion or undue tension, e.g., between plate and offset blanket.

f excès *m* de pression — **d** übermäßiger Druck *m*; Überdruck *m* — **n** overmatige spanning *f*; overmatige drukkracht *fm* — **e** presión *f* excesiva — **i** pressione *f* eccessiva

5340 excess shrinkage — Cross-direction shrinkage of paper web between colour impressions, causing misregister.

f rétrécissement *m* excessif; retrait *m* excessif — **d** zu starke Querschrumpfung *f* — **n** overmatige krimp *m* in dwarsrichting — **i** eccesso *m* di restringimento; eccesso *m* di ritiro; eccesso *m* di contrazione; ritiro *m* del nastro in senso trasversale

5341 exchangeable disk store — See magnetic disk.

5342 exchange copy

f exemplaire *m* d'échange — **d** Tauschexemplar *n*; Austauschexemplar *n* — **n** ruilexemplaar *n* — **e** ejemplar *m* de canje; ejemplar *m* de cambio; periódico *m* de cambiar — **i** esemplare *m* di scambio

5343 exchange list; stock exchange list

f cote *f* (de la bourse); bulletin *m* financier; bulletin *m* de cours — **d** Kurszettel *m*; Kursblatt *n*; Börsenbericht *m* — **n** koerslijst *fm*; prijslijst *fm*; beurs *fm* — **e** hoja *f* de las cotizaciones; cotización *f* de la bolsa; cuadro *m* de las cotizaciones; listín *m*; boletín *m* de la bolsa — **i** bollettino *m* della bolsa valori; listino *m* dei corsi

5344 exchange list — List of duplicates for exchange.

f liste *f* (de doubles) pour échange — **d** Tauschliste *f* — **n** ruillijst *fm* — **e** lista *f* (de duplicados) para canje — **i** elenco *m* (di duplicati) per cambio

5345 exchange list figures

f chiffres *mpl* de bulletin des changes; chiffres *mpl* de bulletin des cours — **d** Kurszettelziffern *fpl*; Kursziffern *fpl* — **n** cijfers *npl* voor de beurs-notering — **e** cifras *fpl* para la nota del cambio — **i** numeri *mpl* del bollettino di cambio

5346 exchange of experience
f échanges *mpl* de vue — **d** Erfahrungsaustausch *m* — **n** uitwisseling *f* van ervaring(en) — **e** cambio *m* de experiencia — **i** scambio *m* d'esperienza

5347 excitation purity — The ratio of the distances, measured on the CIE chromaticity diagram, from the adopted achromatic stimulus to the sample stimulus and to the stimulus lying on the spectrum locus (or the straight line joining its extremes), which by additive mixture with the adopted achromatic stimulus, can form a match with the sample stimulus. **f** Pureté *f* d'excitation — **d** spektraler Farbanteil *m* — **n** verzadigingsgraad *m* — **e** pureza *f* de excitación — **i** purezza *f* d'eccitazione

5348 exclamation mark; exclamation point; note of exclamation; strike'm stiff — Punctuation mark (!) after an exclamatory passage, or (occasionally) after a striking or unexpected sentence. In Spanish, an inverted exclamation mark is placed at the beginning of the exclamatory sentence.
f point *m* d'exclamation; point *m* admiratif — **d** Ausruf(e)zeichen *n*; Ausrufungszeichen *n* — **n** uitroepteken *n* — **e** signo *m* de admiración; signo *m* admirativo; punto *m* de admiración; punto *m* admirativo; admiración *f* — **i** punto *m* esclamativo

5349 exclamation point — See exclamation mark.
5350 exclusive — See scoop.
5351 execute *v* — To perform the operations indicated by a machine instruction; also to run a program.
f exécuter — **d** ausführen — **n** uitvoeren — **e** ejecutar — **i** eseguire

5352 executive — A size of cut card. See app. no. 3.
5353 executive program — See supervisor.
5354 Exeltype — Early and original form of screenline and shadowgraph procedures.
5355 exhaust duct — Part of the dryer which carries solvent laden air to the outside atmosphere.
f conduite *f* d'évacuation — **d** Luftabzugrohr *n* — **n** afzuigbuis *fm*; afzuigleiding *f* — **e** escapes *mpl* de aire; aires *mpl* — **i** condotto *m* d'evacuazione

5356 ex-libris — See book-plate.
5357 exotic types — See foreign types.
5358 expanded type; extended type — Type of wide face. See also width, set.
f caractère *m* large — **d** breite Schrift *f*; breitlaufende Schrift *f*; breite Buchstaben *mpl* — **n** verbrede letter *fm*; brede letter *fm*; breedlopende letter *fm* — **e** tipo *m* ancho; tipo *m* de ojo ancho; tipo *m* extendido; tipo *m* extenso *CU*; tipo *m* abierto *Me* — **i** tipo *m* largo; lettera *f* larga; carattere *m* largo

5359 expander roll; swimming roller; floating roller — Roller rotating on a stationary shaft on self aligning bearings. A space is provided between the shaft and the inside roller, divided into two compartments by two sealing strips sliding radially across the shaft and rubbing on the inner face of the roller. Hydraulic pressure is applied to the roller evenly across the inside face of the roller, which prevents run out or deflection at the nip. Installed in 1962 in a paper machine these rollers are adapted to rotary presses. Sometimes called Mount-Hope roller.
f cylindre *m* flottant — **d** Breitstreckwalze *f*; schwimmende Walze *f*; Küsterswalze *f*; Mount-Hope-Walze *f* — **n** zwemmende rol *fm* — **e** rodillo *m* expansivo; rodillo *m* flotador; rodillo *m* de flotación — **i** rullo *m* rotante su albero fisso con allineamento automatico

5360 expanding roller — Roller to tension a belt.
f galet-tendeur *m* — **d** Spannrolle *f* — **n** spanrol *fm* — **e** rollo *m* tensor — **i** rotolo *m* di tensione

5361 expansion — Change in the dimensions of a sheet of paper or board caused by atmospheric changes. See also hygroexpansivity, stretch.
f allongement *m*; jeu *m* — **d** Feuchtdehnung *f*; Ausdehnung *f*; Dehnung *f* — **n** uitzetting *f*; rek *m* — **e** alargamiento *m* — **i** espansione *f*; allungamento *m*

5362 expansion — Process in which a constant mass of a substance undergoes an increase in size.
f dilatation *f*; expansion *f* — **d** Ausdehnung *f* — **n** dilatatie *f*; uitzetting *f* — **e** dilatación *f*; expansión *f* — **i** dilatazione *f*; espansione *f*

5363 expansion, coefficient of — See coefficient of expansion.
5364 expansion wrinkles — Wrinkles which form on the top layers of a reel of paper when the paper absorbs moisture from the air.
f rides *fpl* d'expansion; cordons *mpl* — **d** Dehnfalten *fpl* — **n** uitzettingsplooien *fmpl* — **e** arrugas *fpl* de expansión — **i** grinze *fpl* d'espansione

5365 expansivity, hygro — See hygroexpansivity.
5366 exploded view — Photo or drawing showing the separate components of an assembly as they would appear if removed radially.
5367 expose *v* — To subject a sensitive photographic film, plate or paper to the action of radiant energy.
f exposer; insoler — **d** belichten; exponieren; die Aufnahme machen — **n** belichten; exponeren; opnemen — **e** exponer; insolar; posar; pasar — **i** esporre

5368 exposure; insolation — Act and period of time during which a sensitized surface is subjected to the action of actinic light, either in a camera or printing frame.
f exposition *f*; pose *f*; insolation *f* — **d** Belichtung *f*; Exposition *f* — **n** belichting *f*; exponeren *n* — **e** exposición *f*; pose *f*; insolación *f* — **i** esposizione *f*; posa *f*

5369 exposure chart — In lithographic platemaking, chart of one or more variables to determine the exposure required under various conditions of temperature and humidity. See also exposure table.

f diagramme *m* des expositions; tableau *m* de temps de pose — **d** Belichtungstabelle *f* — **n** belichtingstabel *fm* — **e** carta *f* de exposición; carta *f* de pose; cuadro *m* de tiempo; calculador *m* de exposiciones — **i** tabella *f* d'esposizione

5370 exposure computer — See automatic exposure timer.

5371 exposure control unit — See light integrator.

5372 exposure latitude; latitude of exposure

f latitude *f* d'exposition; latitude *f* de pose — **d** Belichtungsspielraum *m* — **n** speelruimte *f* in de belichtingstijd; armslag *m* in de belichtingstijd; speling *f* in de belichtingstijd — **e** margen *fm* de exposición; margen *fm* de pose; latitud *f* de exposición — **i** latitudine *f* di posa

5373 exposure meter; light meter; photometer — Instrument to measure light intensity and to provide a calculator for determining proper exposure. Calibrated for different emulsion speeds.

f compte-pose *m*; photomètre *m*; posemètre *m* — **d** Belichtungsmesser *m*; Lichtmesser *m*; Lichtmeßgerät *n* — **n** belichtingsmeter *m* — **e** exposí metro *m*; fotómetro *m*; posómetro *m*; medidor *m* de luz; posofotómetro *m* — **i** esposimetro *m*; fotometro *m*

5374 exposure setting — The lens opening and shutter speed selected to expose film or paper.

f réglage *m* d'exposition — **d** Belichtungseinstellung *f* — **n** belichtingsinstelling *f*; instelling *f* van de belichting — **e** ajuste *m* de exposición — **i** regolazione *f* dell'esposizione

5375 exposure table — Table indicating the main exposure time and the flash exposure time for use in halftone photography. See also exposure chart.

f tableau *m* de temps de pose; barème *m* d'exposition — **d** Belichtungstabelle *f* — **n** belichtingstabel *fm* — **e** tabla *f* de pose; tabla *f* de exposición — **i** tabella *f* d'esposizione; tabella *f* dei tempi d'esposizione

5376 exposure time; time of exposure; length of exposure

f temps *m* d'exposition; temps *m* de pose; pose *f* — **d** Belichtungszeit *f*; Belichtungsdauer *f* — **n** belichtingstijd *m* — **e** tiempo *m* de exposición; tiempo *m* de insolación; tiempo *m* de pose — **i** tempo *m* d'esposizione; tempo *m* di posa

5377 exposure-time computer — See automatic exposure timer.

5378 expurgated edition — Edition of a book from which morally harmful, offensive, or erroneous parts have been eliminated.

f édition *f* expurgée — **d** gesäuberte Ausgabe *f* — **n** gezuiverde uitgave *f* — **e** edición *f* espurgada — **i** edizione *f* purgata

5379 exsiccated sodium sulphate — See sodium sulphate.

5380 extended colour — Extra area of colour necessary for trimming or die-cutting when the printed image goes to the very edge, to avoid a white area due to inaccurate cutting of paper.

f fonds *m* perdu — **d** überstehender Farbrand *m* — **n** vlees *n*; leven *n* — **e** fondo *m* a sangre — **i** fondo *m* perduto

5381 extended cover; overhang cover; extension cover — In pamphlet binding, cover which extends beyond the trimmed edges of a book.

f couverture *f* à chasses — **d** überstehender Buchdeckel *m*; Umschlag *m* mit überstehenden Kanten — **n** overstekend omslag *n*; overhangend omslag *n* — **e** cubierta *f* saliente; cubierta *f* de cejas; cubierta *f* sobresaliente — **i** copertina *f* con unghiatura

5382 extended type — See expanded type.

5383 extender — Any colourless or white inorganic material used in printing inks. They usually give relatively low opacity when ground in varnish and are used to build up pigment content for proper working properties and to reduce colours to the proper strength.

f charge *f*; matière *f* de charge — **d** Extender *m* — **n** extender *m*; versnijpigment *n* — **e** carga *f*; material *m* de carga — **i** carica *f*; bianco *m* di riduzione

5384 extender varnish; gravure varnish — Varnish usually composed of resins and/or film formers dissolved in a solvent or a blend of solvents.

f blanc *m* à atténuer; vernis *m* d'allongement — **d** Verschnitt *m* — **n** versnijmiddel *n*; versnit *m depr.* — **e** barniz *m* para rebajar — **i** diluente *m*; agente *m* diluente

5385 extension; conversion; normalizing — Calculation of basic time from observed time.

5386 extension cover — See extended cover.

5387 external label — Label at the outside of a file media holder identifying the file.

5388 external storage — See external store.

5389 external store *GB*; **external storage** *US* — Store not permanently linked to a computer but holding data in a form acceptable to it.

f mémoire *f* externe — **d** peripherer Speicher *m*; externer Speicher *m* — **n** extern geheugen *n* — **e** memoria *f* externa — **i** memoria *f* esterna

5390 extra; extra binding — Binding of a book with gilt ornaments on side and back, silk headbands, etc., and on which all work is performed by hand. Not to confuse with half binding which can be bound extra, whilst a full binding, i.e., same material all over, can be bound half extra.

f reliure *f* à la main — **d** Handeinband *m* — **n** met de hand vervaardigde boekband *m* — **e** encuadernación *f* manual y de esmeralda calidad — **i** legatura *f* a mano

5391 extra antiquarian — A size of boards. See app. no. 3.

5392 extra atlas — A size of boards. See app. no. 3.

5393 extra binding — See extra.

5394 extra bold — Heavy bold-face type. See also weight of type.

f gras — **d** fett — **n** vet — **e** supernegro; muy negro; grueso pesado; fuerte; negrísimo — **i** nero

5395 extra charge — Additional charge for difficult or badly written copy.
f surcharge *f* — **d** Aufschlag *m*; Zuschlag *m*; Sprach(en)aufschlag *m*; Manuskriptentschädigung *f*; Entschädigung *f* für schlechtes Manuskript; Erschwerungszuschlag *m* — **n** toeslag *m*; opslag *m*; extra kosten *pl* — **e** recargo *m*; sobreprecio *m*; aumento *m* — **i** sovrapprezzo *m*; aumento *m* di prezzo
5396 extract — Printer's term for block quotation, a long quotation marked off from the text by being set in smaller type or on narrower measure than the body copy.
f extrait *m* — **d** Extrakt *m* — **n** extract *n*; uittreksel *n* — **e** extracto *m* — **i** estratto *m*
5397 extraction — Process of separating the soluble and insoluble components of a substance.
f extraction *f* — **d** Extraktion *f*; Ausziehung *f*; Ausziehen *n*; Gewinnung *f*; Förderung *f* — **n** extractie *f* — **e** extracción *f* — **i** estrazione *f*
5398 extractor — Device to separate the pigment from the vehicle when analyzing ink samples. In most cases a Soxhlet extractor is used.
f extracteur *m*; Soxhlet appareil *m* à extraction — **d** Extraktor *m*; Extraktionsapparat *m* (nach Soxhlet); Soxhlet-Extraktionsapparat *m*; Soxhlet *m* — **n** extractie-apparaat *n*; Soxhlet *m* — **e** extractor *m*; aparato *m* extractor; extractor *m* de Soxhlet — **i** estrattore *m*; apparato *m* estrattore; estrattore *m* di Soxhlet
5399 extra double crown — A size of wrapping paper. See app. no. 3.
5400 extra large — A British size of cards. See app. no. 3.
5401 extra large atlas — A size of drawing paper and of boards. See app. no. 3.
5402 extra large casing — A British size for wrapping and casing papers. See app. no. 3.
5403 extra large lump — A size of wrapping paper. See app. no. 3.
5404 extra large post — A size of writing paper. See app. no. 3.
5405 extraneous ink back — Magnetic ink on the reverse side of the *clear band. See also magnetic encoded cheques.
f tache *f* d'encre au verso du document — **d** Farbspritzer *m* auf der Hinterseite — **n** inktspat *fm* op de achterzijde — **e** mancha *f* de tinta en el reverso — **i** macchia *f* d'inchiostro sul lato in volta
5406 extraneous ink front — Magnetic ink outside the printed edge zones and within the *clear band in the area that should be ink free. See also magnetic encoded cheques.
f tache *f* d'encre au recto du document — **d** Farbspritzer *m* auf der Vorderseite — **n** inktspat *fm* op de voorzijde — **e** mancha *f* de tinta en la cara — **i** macchia *f* d'inchiostro sul lato in bianca
5407 extras — Items not covered in an estimate, which are charged extra or over the estimated price.
f surcharges *fpl* — **d** Mehrkosten *pl* — **n** meerkosten *pl*; extra kosten *pl* — **e** recargo *m*; gastos

mpl adicionales — **i** extra costi *mpl*; maggiorazioni *fpl* di costo
5408 extra small — A British size of cards. See app. no. 3.
5409 extra thirds — A size of cut cards. See app. no. 3.
5410 extra work
f travail *m* supplémentaire — **d** Mehrarbeit *f* — **n** meerdere werk *n* — **e** trabajo *m* adicional — **i** lavoro *m* extra
5411 extruder — Device to press thermoplastics through a narrow orifice under a high temperature.
f extrudeuse *f*; boudineuse *f* — **d** Strangpresse *f*; Spritzmaschine *f* — **n** extruder *m*; spuitmachine *f* — **e** estrujadora *f*; budinadora *f* — **i** estrusore *m*; trafila *f*
5412 extrusion coated paper — Paper coated by the extrusion coating process.
f papier *m* couché par extrusion — **d** extrusionbeschichtetes Papier *n* — **n** papier *n* met extrusie-laag; papier *n* met extrusie-coating; papier *n* met extrusie-couche — **e** papel *m* estucado por extrusión — **i** carta *f* patinata per estrusione; carta *f* trattata per estrusione
5413 extrusion coating — Method of coating a continuous web with resins, plastics or similar compounds. The coating is applied through an extender die on to a suitably supported and cooled paper web. Extrusion coating is a separate *coating operation.
f couchage *m* par extrusion — **d** Extrudierstreichverfahren *n*; Extrusionsbeschichtung *f* — **n** extrusiecoating *m* — **e** estucado *m* por extrusión — **i** patinatura *f* per estrusione; trattamento *m* superficiale per estrusione
5414 eye catcher
f tire l'œil *m*; éclat *m* — **d** Blickfang *m* — **n** blikvanger *m* — **e** señuelo *m*; ilustración *f* señuelo — **i** attrazione *f*
5415 eyelets — Metal, celluloid, plastic or gummed paper rings to protect apertures from tearing out where cords or fasteners are inserted.
f œillets *mpl* — **d** Ösen *fpl* — **n** ringetjes *npl* — **e** ojillos *mpl*; ojetes *mpl* — **i** occhielli *mpl*
5416 eyeletting — To make an eyelet in or to insert metal eyelets.
f œilleter — **d** ösen; Ösen einsetzen — **n** ringetjes aanbrengen — **e** ojetear; colocar ojetes — **i** occhiellare
5417 eyeletting machine
f machine *f* à poser des œillets; œilleteuse *f* — **d** Ösenmaschine *f*; Ösmaschine *f*; Ösenhefter *m*; Öseneinsetzmaschine *f*; Ösenapparat *n* — **n** ringetjesmachine *f* — **e** máquina *f* para ojetear; máquina *f* de poner ojillos; ojeteadora *f* — **i** macchina *f* occhiellatrice; occhiellatrice *f*
5418 eyepiece; ocular — Lens or combination of lenses at the eye and of an optical instrument.
f oculaire *m*; œilleton *m* — **d** Okular *n* — **n** oculair *n* — **e** ocular *m* — **i** oculare *m*
5419 eyes, fish — See fish eyes.

5420 eye shade — Visor worn on the head or forehead to shield the eyes from overhead light. See also lamp shade.

f garde-vue *m* — **d** Augenschirm *m*; Lichtschirm *m* — **n** oogkap *fm*; lichtkap *fm*; oogscherm *n*; lichtscherm *n*; lichtklep *fm* — **e** visera *f*; parasol *m*; caperuza *f*; pantalla *f* — **i** para-occhi *m*; paralume *m*; guardavista *m*

5421 fac — See factotum initial.

5422 face — Sensitized side of a photographic surface; working surface of a photographic image or printing plate.

f face *f* — **d** Schichtseite *f* — **n** zijde *fm* van de gevoelige laag; emulsiezijde *fm* — **e** lado *m* de la capa sensible; cara *f* del emulsión — **i** lato *m* di emulsione (fotosensibile); lato *m* dello strato sensibile

5423 face; type face — Printing surface of type consisting of stem and (sometimes) serifs and kern.

f œil *m* du caractère; œil *m* du lettre — **d** Schriftbild *n*; Buchstabenbild *n*; Bild *n* des Buchstabens — **n** letterbeeld *n*; beeld *n* — **e** ojo *m* del tipo; ojo *m* de la letra; cara *f*; estilo *m* de letra; tipo *m* de letra — **i** occhio *m* del carattere

5424 face-plate frame — Part of a slug composing machine or other machines.

f devanture *f* — **d** Frontplatte *f* — **n** frontplaat *fm* — **e** plato *m* frontal — **i** piastra *f* frontale

5425 facet edge — Impression left on paper by the ininked edge of a hand engraved or etched copper or steel plate. The term is a pleonasm. Meant is a plate mark.

f facette *f* — **d** Facette *f*; Fazette *f* — **n** moet *fm* van het facet; indruk *m* van het facet; facetafdruk *m* — **e** faceta *f*; impresión *f* de la faceta — **i** faccetta *f*; smusso *m*

5426 facsimile — Reproduction the same size of the original, in which all detail and tone values are reproduced closely resembling those of the original. To apply the term to a reproduction of an oil painting is meaningless; a facsimile reproduction would entail furnishing an exactly duplicated printing. A highlight halftone plate in which the highlight dots have been eliminated by local etching.

f fac-similé *m* — **d** Faksimile *n* — **n** facsimile *n* — **e** facsímil *m* — **i** facsimile *m*

5427 facsimile transmission — Scanning of originals and transmitting them to a remote place, e.g. large newspaper pages.

f transmission *f* par fac-similé — **d** Faksimile-Übertragung *f* — **n** facsimile-overbrenging *f* — **e** transmisión *f* facsímile — **i** trasmissione *f* di facsimile

5428 factory manager — See works manager.

5429 factotum initial; fac — Initial letter with ornamentation in various patterns. See also former.

f initiale *f* passe-partout; initiale *f* encastrée; initiale *f* ornée — **d** verzierter Versal *m*; verzierter Versalbuchstabe *m*; verzierter Anfangsbuchstabe *m*; Kasseteninitial *n* — **n** versierde initiaal *fm* — **e** inicial *f* orlada; inicial *f* en orla; inicial *f* encastrada; inicial *f* encuadrada; inicial *f* encuadrada en ornamento; capitular *f* adornada; letra *f* de fantasía; letra *f* historiada; letra *f*

florida — **i** iniziale *f* incorniciata; iniziale *f* arabescata; iniziale *f* rabescata; iniziale *f* incastrata in un ornamento; lettera *f* ornata

5430 faded
See discoloured.

5431 faded colour — Colour which has lost its brightness due to exposure to light, heat or chemical agents.

f couleur *f* passée — **d** verschossene Farbe *f*; verblassene Farbe *f* — **n** verschoten kleur *fm*; verbleekte kleur *fm* — **e** color *m* quebrado; color *m* muerto; color *m* pasado — **i** colore *m* appassito

5432 fadeometer — Instrument in which prints are exposed to intense light to measure in a short time the light fastness of their pigments. The light from an electric arch passes through a glass globe to accelerate the fading.

f fadéomètre *m* — **d** Fadeometer *n*; Lichtechtigkeitsprüfgerät *n*; Lichtechtheitmesser *m* — **n** lichtechtheidsmeter *m* — **e** fadeómetro *m* — **i** fadeometro *m*; misuratore *m* di decolorazione

5433 fading — Loss of colour by destruction of colouring matter caused by exposure to light, heat or other agents (or insufficient elimination of hyposulphite).

f se décolorer; se déteindre; se faner — **d** verblassen; verschießen; sich verfärben; ausbleichen — **n** verkleuren; verschieten; verbleken — **e** desteñirse; devanecimiento *m*; decoloración *f*; pérdida *f* de color — **i** sbiadire; scolorimento *m* dell'immagine

5434 fading test — Test to determine the degree of resistance to discoloration of paper or ink, when exposed to sunlight or to artificial light.

f essai *m* de décoloration — **d** Lichtechtheitsprüfung *f* — **n** lichtechtheidsbepaling *f* — **e** ensayo *m* de pérdida del color — **i** prova *f* di solidità alla luce

5435 Fahrenheit; degrees Fahrenheit — Indicated by F. The Fahrenheit thermometer on which, under standard atmospheric pressure, the boiling point of water is at 212 ° and the freezing point at 32 ° above the zero of its scale. To convert degrees F to degrees C (see centigrades), subtract 32 from F value, multiply by 5 and divide by 9.

f degrés *mpl* Fahrenheit — **d** Grade *mpl* Fahrenheit — **n** graden *mpl* Fahrenheit — **e** grados *mpl* Fahrenheit — **i** gradi *mpl* Fahrenheit

5436 faint impression — Grey appearance of a print; may be due to poor makeready, soft packing, worn out or shallow cut, lack of ink, poor ink distribution, ink not dense enough, paper with irregular finish or varying thickness.

f impression *f* faible; impression *f* grise — **d** schwacher Druck *m*; weicher Druck *m* — **n** fletse afdruk *m*; slappe afdruk *m*; grijze afdruk *m*; krachteloze afdruk *m* — **e** impresión *f* débil — **i** stampa *f* morbida

5437 faint lines — See feint lines.

5438 faint ruled — See feint ruled.

5439 fair calf — See law calf.

5440 fair copy — Transcript free from corrections. See also bad copy.

f bonne copie f — d sauberes, einwandfreies Manuskript n; Reinschrift f — n goede kopij f; onberispelijke kopij f — e buen original m — i buona copia f; buono manoscritto m

5441 fake; fake story; fals story — Newspaper story with a false character.

f article m faux; article m truqué; article m maquillé; article m camouflé — d Falschmeldung f — n verzonnen verhaal n; verzinsel n — e infundio m; noticia f falsa — i falso giornalistico m

5442 fake-colour process; bastard colour separation — Production of halftone colour plates from monochrome originals, whereby the colour effects are introduced by retouching of photoprints, and by re-etching and finishing of the halftone plates. For lithographic reproduction, colours are pre-separated by mechanical overlays, with grey tones, to provide tone copy for the camera.

f procédé m d'interprétation de couleurs par bromure — d Fake-Colour-Verfahren n — n fake-colour-procédé n — e seudotricromía f; tricromía f falsa — i pseudotricromia f; procedimento m fake-colour

5443 faked photograph — Photographic print in which detail or tone values have been deliberately changed or altered by retouching.

f retouche f photographique — d retuschierte Photographie f — n geretoucheerde foto f — e fotografía f retocado — i fotografia f ritoccata

5444 fake news — See hoax.

5445 fake story — See fake.

5446 faking — See card v.

5447 falling-ball viscometer — Viscometer composed of a large tube containing a metal ball. The material to be measured is placed in the tube; no air bubbles are allowed to enter. Viscosity is measured by the rate of fall of the ball through the material. Stoke's law is used for the calculation. The radius and density of the ball, and the density of the liquid must be known.

f viscosimètre m à bille — d Kugelfallviskosimeter n — n kogelviscosimeter m — e viscosímetro m a bola — i viscosimetro m a sfera

5448 fall out — Type that drops from the forme when this is lifted.

f caractère m mal croché; type m mal croché; sonnette f; sentinelle f — d loser Buchstabe m; Glöckchen n (aus der Form fallender Buchstabe) — n losse letter fm; uitgevallen letter fm — e letra f caída; cascabel m — i campanella f

5449 false back — See hollow back.

5450 false bands — Leather bands or strips of board glued to the spine of books before attaching the cover. See also raised bands.

f faux nerfs mpl — d falsche Bünde mpl — n onechte ribben fmpl; valse ribben fmpl — e falsos nervios mpl — i nervi mpl finti; nervi mpl falsi;

rialzi mpl finti

5451 false body — Technical expression in rheology with various meanings. To some investigators, it means yield values, to others thixotropy.

f faux corps m — e falso cuerpo m — i corpo m falso

5452 false motions — The movement of the compositor, in which he over and over again bumps the types against the border of the composing stick.

f battre le briquet — d beim Satzen klappern — n de letters telkens onnodig tegen de zethaak tikken — e golpear innecesariamente el componedor con el tipo — i sbattere i caratteri di continuo contro il compositoio

5453 false motions — The unnecessary movements made by some compositors during typesetting.

f pomper — d beim Setzen den Oberkörper stark bewegen; pumpen wie ein Spritzenmann — n tijdens het zetten het bovenlichaam onnodig bewegen — e movimientos mpl innecesarios; movimientos mpl falsos — i movimento m non necessario

5454 fals story — See fake.

5455 family, type — See type family.

5456 fan; air blower; blow torch — Apparatus to create air circulation by blowing air.

f séchoir m électrique; foehn m — d Föhn m — n (elektrisch) droogapparaat n; föhn m — e soplador m de aire; secador m — i ugello m; soffiatore m d'aria

5457 fanal — Registered trade name (BASF) for complex compounds of basic colours (rhodamine, auramin, Victoria blue), complexed with phosphomolybdic acid or phosphotungstic acid or mixture of both. They are characterized by brilliancy and good fastness to light.

5458 fancy finish — Embossed, plated or decorative finish on paper for book covers or box liners.

f apprêt m fantaisie; fini m fantaisie — d Phantasiemuster n — n fantasieoppervlak n; fantasiepreging f — i finitura f fantasia

5459 fancy gift-wrapping paper — See fancy paper.

5460 fancy paper; fancy gift-wrapping paper; gift wrapping paper

f papier m fantaisie — d Phantasiepapier n; Ausstattungspapier n — n fantasiepapier n — e papel m (de) fantasía — i carta f fantasia

5461 fancy types

f caractères mpl fantaisie — d Phantasieschriften fpl; Künstlerschriften fpl; Zierschriften fpl; Schmuckschriften fpl — n fantasieletter fm — e tipos mpl de fantasía; caracteres mpl de fantasía; caracteres mpl de adorno; caracteres mpl de historiados; caracteres mpl ornamentados — i caratteri mpl fantasia

5462 fan delivery — Water-wheel type rotary unit to transfer folded signatures from various folding sections to conveyors that carry them to the press delivery.

f sortie *f* à moulinet; moulinet *m* — **d** Schaufelrad-ausleger *m*; Staffelausleger *m* — **n** schoepenrad-uitleg *m* — **e** salida *f* de palas; dispositivo *m* de salida — **i** uscita *f* a mulinello

5463 fan *v* **dry** — To dry wet offset plates or litho stones by moving piece of board or zinc over them or with an electric fan.

f éventer; sécher au ventilateur — **d** fächeln; fächern; wedeln; trocknen — **n** droog waaieren; drogen — **e** secar a ventilador — **i** asciugare ventilando

5464 fanfare binding; binding à la fanfare — Book decoration style developed by Nicholas and Clovis Eve, binders to the king of France. The cover bears in the centre the arms of the owner, and the surrounding small areas are fully tooled with small flowers. The major areas show spirals with leaves or leaf tendrils. Name derived from Charles Nodier (1788-1844), a book collector who charged bookbinder J. Thouvenin to bind and decorate a book "Les Fanfares et Courvées Abbadesques des Roule-Bontemps de la haute et basse Cocquaigne et Dépendances selon un reliure d'Eve" (1613).

f reliure *f* à la fanfare; style *m* à la fanfare — **d** Fanfare-Einband *m*; Fanfares-Einband *m*; Fanfare-Stil *m* — **n** fanfare-band *m* — **e** encuadernación *f* a la fanfarria; encuadernación *f* vistosa — **i** legatura *f* a fanfara

5465 fan fold — See accordion pleat.

5466 fanning out; shuffling — Separating sheets by gently rubbing them to form a fan-shaped pile.

f mettre en éventail; étager — **d** staffeln; auffächern; ausfächern — **n** uitstrijken; uitwaaieren; staffelen — **e** poner en abanico; formar abanico; formar escalerilla; despegar esparciendo; despegar escalonando; abrir — **i** mettere a ventaglio; disporre a squama

5467 fanning (out) — Spreading or separation of printing subjects, usually along the back edge of the press sheet, caused by uneven moisture absorption and expansion of the sheet.

f allongement *m* en éventail — **d** Strecken *n* eines Offsetbogens beim Durchgang durch die Presse — **n** uitwalsen *n* van het papier — **e** ensanchamiento *m* de la hoja en abanico — **i** deformazione *f* a ventaglio

5468 fan style binding; fers à l'éventail binding; binding à l'éventail — Book cover decoration style of the 17th century with a tooling on the covers with a fan-like ornament.

f reliure *f* à l'éventail — **d** Fächermuster-Einband *m*; Fächerstil-Einband *m* — **n** éventail-band *m*; band *m* met waaier-ornament — **e** encuadernación *f* de abanico — **i** legatura *f* a ventaglio; legatura *f* al stile a ventaglio

5469 farad — Unit of capacitance equal to the capacitance of a capacitor between whose plates a potential appears of one volt when it is charged by one coulomb of electricity. Abbrev. F. In the SI: $1 \text{ F} = 10^{-9}$ emu (electromagnetic unit) or $9 \cdot 10^{11}$ esu

(electrostatic unit). Named after Michael Faraday (1791-1867), English physicist and chemist.

f farad *m* — **d** Farad *m* — **n** farad *m* — **e** faradio *m* — **i** farad *m*

5470 faraday — The quantity of electricity transferred in electrolysis per equivalent weight of an element or ion. Abbrev. F. In the SI: $1 \text{ F} \approx 96.4846 \times 10^3$ C.

f faraday *m* — **d** Faraday *m* — **n** faraday *m* — **e** faraday *m* — **i** faraday *m*

5471 Farmer's reducer — Solution to reduce the density of developed negatives, containing principally potassium ferricyanide and sodium thiosulphate, invented by Howard Farmer. Tends to increase the contrast of the negative. Strong solution is used to dissolve and remove the black silver of negatives from unwanted areas.

f affaiblisseur *m* de Farmer; réducteur *m* de Farmer; liqueur *f* de Farmer — **d** Farmerscher Abschwächer *m* — **n** Farmerse afzwakker *m* — **e** rebajador *m* (de) Farmer; reductor *m* Farmer; debilitador *m* Farmer — **i** indebolitore *m* Farmer

5472 farm *v* **out** — To sublet or give a printing job to sub-contractor.

f donner en sous-contrat — **d** eine Druckarbeit aus dem Haus geben; unterbringen; in Submission begeben — **n** een werk uitbesteden — **e** subencargar; subcontratar; compartir — **i** dare in subappalto; subappaltare; accottimare

5473 far side — See drive side.

5474 fascicle; fascicule — Part of printed work published in sections. Also small number of printed or written pages bound together.

f fascicule *m* — **d** Heft *n*; Lieferung *f* — **n** aflevering *f*; deel *n* — **e** fascículo *m*; entrega *f* — **i** fascicolo *m*

5475 fascicule — See fascicle.

5476 fashion journal — See fashion paper.

5477 fashion magazine — See fashion paper.

5478 fashion paper; fashion journal; fashion magazine

f feuille *f* de modes; journal *m* de modes — **d** Modezeitschrift *f*; Modeblatt *n* — **n** modeblad *n* — **e** revista *f* de modas — **i** giornale *m* della moda; rivista *f* di moda

5479 fast back — See tight back.

5480 fast beating — See free beating.

5481 fast colour; resistant colour — Coloured material which retains its colour under the influence of aging, water, light, heat, acids and alkalis, and does not bleed.

f couleur *f* solide; couleur *f* inaltérable — **d** dauerhafte Farbe *f*; haltbare Farbe *f* — **n** sterke kleur *fm*; resistente kleur *fm* — **e** color *m* estable; color *m* firme; color *m* indeleble — **i** colore *m* durevole

5482 fast drying; quick drying

f à séchage rapide — **d** schnelltrocknend — **n** sneldrogend — **e** de secado rápido — **i** a rapido essiccamento

5483 fast-drying ink — See quick-drying ink.

5484 fastness to light; light fastness; resistance to

light — Resistance of paper, ink or other materials to the fading action of light, especially sunlight. Essential in poster work.

f solidité *f* à la lumière; résistance *f* à la lumière — **d** Lichtechtheit *f*; Lichtbeständigkeit *f* — **n** lichtechtheid *f* — **e** solidez *f* a la luz; resistencia *f* a la luz; estabilidad *f* a prueba de luz — **i** solidità *f* alla luce; resistenza *f* alla luce

5485 fastopake ink — Ink for printing on waxed surfaces.

f encre *f* pour surfaces cireuses — **d** Druckfarbe *f* zum Druck auf gewachste Oberflächen — **n** inkt *m* voor het drukken op was — **i** inchiostro *m* per la stampa su superfici paraffinate

5486 fast pulp — See free stock.

5487 fast solvent — Organic solvent of low boiling point which evaporates rapidly.

f solvant *m* rapide — **d** schnellverdünstendes Lösungsmittel *n* — **n** snel verdampend oplosmiddel *n* — **e** solvente *m* rápido — **i** solvente *m* a basso punto d'ebollizione

5488 fast white — See barium sulphate.

5489 fat face type — See bold face type.

5490 father of the chapel — Person selected by the organized employees of a composing, printing or other department to represent them in matters related to welfare. Abbrev. FOC. See also chapel.

f homme *m* de confiance; chef *m* de chapelle — **d** Vertrauensmann *m*; ≈ Betriebsratmitglied *n* — **n** ≈ vertrouwensman *m* — **e** hombre *m* de confianza — **i** uomo *m* di coscienza (degli operai)

5491 fatique allowance — See compensating relaxation allowance.

5492 fat matter; phat matter — Blank or short pages, light, open matter, charged at the same rate as solid matter. Also an easy job in a composing department.

f beurre *m*; feuilletage *m* — **d** Specksatz *m*; Speck *m*; Vorteilsatz *m*; vorteilhafter Satz *m*; Süß *m*; hoch zu berechnender Satz *m* — **n** voordelig zetwerk *n*; gemakkelijk zetwerk *n* — **e** momio *m*; forma *f* con mucho blanco; forma *f* ventajosa — **i** composizione *f* vantaggiosa

5493 fatty acid — Acid component of oils and fats, e.g. oleic, palmitic and stearic acids. Formula: $C_nH_{2n}O_2$.

f acide *m* gras — **d** Fettsäure *f* — **n** vetzuur *n* — **e** ácido *m* graso — **i** acido *m* grasso

5494 fault — In data processing, a malfunction that is reproducible, as contrasted to an error, which is not reproducible. A malfunction is reproducible if it occurs consistently under the same circumstances.

f défaillance *f* — **d** Störung *f* — **n** storing *f* — **e** fallo *m* — **i** guasto *m*

5495 fault of construction — See constructional fault.

5496 fault time — See down time.

5497 fax — Term sometimes used in computer controlled typesetting for *facsimile transmission.

5498 Faxi-Foto — See Morisawa.

5499 featherbedding — Requiring an employer to hire unnecessary employees or to limit production according to a union rule or safety statute.

n werk laten verrichten zonder dat dit direct noodzakelijk is — **e** trabajo *m* innecesario; trabajo *m* inútil — **i** lavoro *m* inutile

5500 featheredge — See deckle edge.

5501 featherweight paper — Light-handling antique, laid or woven book paper, made chiefly from esparto, so loosely woven that 75% of the bulk is air space. See also high bulk.

f papier *m* bouffant; papier *m* plume — **d** Federleichtpapier *n*; Dickdruckpapier *n*; Daunendruckpapier *n*; auftragendes Papier *n*; bauschiges Papier *n* — **n** opdikkend papier *n*; papier *n* met grote opdikking — **e** papel *m* liviano; papel *m* ligero; papel *m* fofo; papel *m* pluma *El Sal* — **i** carta *f* voluminosa

5502 feature; property; characteristic

f propriété *f*; caractéristique *f* — **d** Eigenschaft *f*; Merkmal *n* — **n** eigenschap *f*; kenmerkende eigenschap *f*; kenmerk *n* — **e** rasgo *m*; característico *m* — **i** caratteristica *f*

5503 feed *v* — To lay on.

f marger — **d** anlegen; einlegen — **n** inleggen — **e** marcar (a mano); alimentar; marginar — **i** alimentare

5504 feedboard; feeding board; laying-on board — Part of a printing press from which the paper is fed in.

f table *f* de marge; table *f* à papier — **d** Anlegebrett *n*; Anlegeplatte *f*; Einlegetisch *m* — **n** inlegtafel *fm* — **e** tablero *m* de alimentación; tablero *m* marcador; tablero *m* ponepliegos; mesa *f* de alimentación; mesa *f* de alimentar; mesa *f* de marcar; mesa *f* para marginar; mesa *f* para marcar; ponepliegos *m*; ponedor *m*; metedor *m* — **i** tavola *f* d'alimentazione; tavola *f* per marginare; tavola *f* a carta; piano *m* di carta

5505 feedboard pile

f pile *f* de marge — **d** Anlegestapel *m* — **n** inlegstapel *m* — **e** pila *f* de entrada; pila *f* del ponepliego; pila *f* del marcador — **i** pila *f* mettifoglio

5506 feed edge — See gripper edge.

5507 feeder; layer-on; layer-on girl — On handfed presses, the person who puts the sheets to be printed in position.

f margeur *m*; margeuse *f* — **d** Anleger *m*; Anlegerin *f*; Einleger *m*; Einlegerin *f* — **n** inlegger *m*; inlegster *f* — **e** alimentador *m*; marcador *m*; ponedor *m*; alimentadora *f*; marcadora *f*; marginadora *f*; ponedora *f* — **i** operaio *m* addetto al mettifoglio; operaia *f* addetta al mettifoglio

5508 feeder; feeder mechanism; feeding apparatus — Device to send the sheet into the press for printing.

f margeur *m* (automatique); appareil *m* de marge; dispositif *m* de marge — **d** Bogenanleger *m*; Bogenanlegeapparat *m*; Anlegeapparat *m*; Anleger *m* — **n** inlegapparaat *n*; inlegmechanisme *n* — **e** mecanismo *m* alimentador; alimentadora *f*; ponepliegos *m*; ponepliegos *m* automático; marcador *m* — **i** mettifoglio *m* (automatico)

5509 feeder, automatic — See automatic feeder.

5510 feeder, automatic pot — See automatic pot feeder.

5511 feeder mechanism — See feeder.

5512 feeder pump
f pompe *f* du margeur — **d** Anlegerpumpe *f* — **n** pomp *fm* van het inlegapparaat — **e** bomba *f* alimentadora — **i** pompa *f* di alimentazione

5513 feeder side — See operating side.

5514 feeder's platform; footboard — Platform on the side of a printing press, or the step on the rear of a typesetting machine.
f marche-pied *m* — **d** Fußbrett *n*; Trittbrett *n*; Laufbrett *n* — **n** bordes *n* — **e** estribo *m* — **i** piattaforma *f* d'entrata

5515 feed guide — See front lay.

5516 feed guide stop — Device against which fed sheets come to a stop and which helps correct feeding.
f taquet *m* (de marge) — **d** Anlegemarke *f* — **n** aanleg *m* — **e** guía *f* (de entrada); guía *f* capuchina — **i** squadra *f* d'arresto

5517 feed holes; sprocket holes — Holes punched in a paper tape to enable it to be driven or indexed longitudinally. See also in-line feed hole.
f trous *mpl* d'entraînement — **d** Transportlöcher *npl*; Führungslochungen *fpl* — **n** transportgaten *npl*; geleidegaten *npl* — **e** perforaciones *fpl* de arrastre — **i** fori *mpl* di trascinamento

5518 feeding apparatus — See feeder.

5519 feeding board — See feedboard.

5520 feed roller — See ductor.

5521 feed rollers — Usually steel or rubber covered rollers or a series of wheels or rollers which, as parts of a feed or metering unit, activate the movement of the web to a printing or fabricating unit.
f rouleaux *mpl* d'introduction — **d** Einführwalzen *fpl*; Zuführwalzen *fpl*; Vorschubrollen *fpl* — **n** invoerrollen *fmpl* — **e** rodillos *mpl* de alimentación; rodillos *mpl* alimentadores; rodillos *mpl* de avance; rodetes *mpl* — **i** rulli *mpl* d'alimentazione

5522 feed screws — See duct adjusting screws.

5523 feed unit — See metering unit.

5524 feel *v*; **fimble** *v* — To judge finish and general quality of paper by touching and handling.
f toucher — **d** den Griff prüfen — **n** aanvoelen; voelen — **e** tacto *m*; cuerpo *m*; característica *f* de la superficie del papel — **i** provare al tatto

5525 feet — See foot.

5526 feint lines; faint lines *depr.*; **feint rules; cross feints** — Writing lines ruled horizontally on paper in feint blue or grey ink, as in account books.
f lignes *fpl* de conduite; guide-âne(s) *mpl* — **d** (schwache) Führungslinien *fpl*; Unterdrucklinien *fpl* — **n** flauwlijnen *fmpl*; contralijnen *fmpl* — **e** rayas *fpl* tenues — **i** righe *fpl* di fondo

5527 feint ruled; faint ruled *depr.* — Paper ruled with very thin ink, leaving barely discernible lines.

5528 feint rules — See feint lines.

5529 Fell types — Old-Face types bough by the agents in Holland of Dr. John Fell, bishop of Oxford (1625-1686) and presented by him to the Oxford University Press. Probably from the Christoffel van Dijck Foundry, Amsterdam.

5530 felt — See drier felt.

5531 felt backing — See backing felt.

5532 felt drier
f sécheur *m* de feutre; cylindre *m* chauffeur pour feutres — **d** Filztrockner *m* — **n** viltdroogcilinder *m* — **e** secador *m* de fieltro; secador *m* de paño — **i** cilindro *m* asciugafeltro

5533 felt finish — Finish applied on the paper machine by using a special marking felt which differs from the usual machine felt.
f apprêté au feutre — **d** mit Filzprägung — **n** met viltpersing — **e** con acabado de fieltro — **i** con finitura mediante feltri marcatori

5534 felt mark — Mark left on the surface of paper by the coarse texture of the felt or by foreign material imbedded in the felt. See also wire mark.
f marque *f* de feutre — **d** Filzmarkierung *f*; Filzmarke *f* — **n** viltmarkering *f*; viltmerk *n* — **e** marca *f* del fieltro; marca *f* del manchón — **i** marcatura *f* del feltro; segno *m* del feltro

5535 felt side — See top side.

5536 female compositor
f compositrice *f* typographe; typote *f* — **d** Setzerin *f* — **n** letterzetster *f*; vrouwelijk zetter *f* — **e** compositora *f* — **i** compositrice *f*

5537 female thread — See internal thread.

5538 Fenchel expansion tester — Apparatus to determine the expansion of paper due to change in the amount of moisture.
f appareil *m* Fenchel — **d** Fenchale-Apparat *m* — **n** Fenchel-vochtrektoestel *n* — **e** aparato *m* Fenchel — **i** apparecchio *m* di Fenchel

5539 fender — See gauge.

5540 ferric — Designation for iron salts in which the iron is trivalent, e.g., ferric chloride, $FeCl_3$. See also ferrous.
f ferrique — **d** Ferri- — **n** ferri- — **e** férrico — **i** ferrico

5541 ferric ammonium alum — See ferric ammonium sulphate.

5542 ferric ammonium citrate; iron ammonium citrate; ammonium ferric citrate; ammonium iron citrate; ammonium citrate of iron — Green scales, insoluble in alcohol and ether, used for cyanotype paper.
Formula: $(NH_4)_3Fe_2(C_6H_5O_7)_2$.
f citrate *m* de fer ammoniacal; ammonio-citrate *m* de fer — **d** Ferri-Ammoniumzitrat *n*; Eisen-Ammoniumzitrat *n*; zitronensaures Eisenoxydammonium *n*; Ammoniumferrizitrat *n* — **n** ferriammoniumcitraat *n*; ijzer-ammoniumcitraat *n*; citroenzuur-ijzeroxyde-ammoniak *m* — **e** citrato

m férrico-amónico; citrato *m* de hierro amoniacal
— **i** citrato *m* ferrico d'ammonio; citrato *m* ferrico
ammoniacale; citrato *m* di ferro ammoniacale

5543 ferric ammonium oxalate; iron ammonium oxalate; ammonium oxalate of iron; ammonium ferric oxalate — Green crystals, soluble in water and alcohol. Used in blueprint photography.
Formula: $(NH_4)_3Fe(C_2O_4)_3 \cdot 3H_2O$.

f oxalate *m* ferrique d'ammonium; oxalate *m* de fer ammoniacal — **d** Ferri-Ammoniumoxalat *n*; Eisen-Ammoniumoxalat *n* — **n** ferri-ammonium-oxalaat *n*; ijzer-ammoniumoxalaat *n* — **e** oxalato *m* férrico-amónico; oxalato *m* de hierro amoniacal — **i** ossalato *m* ferrico d'ammonio; ossalato *m* ferrico ammoniacale; ossalato *m* di ferro ammoniacale

5544 ferric ammonium sulphate; iron ammonium sulphate; ammonium ferric sulphate; ferric ammonium alum — Lilac to violet, efflorescent crystals, soluble in water, insoluble in alcohol. Latin: sulfas ammonico ferricus.
Formula: $FeNH_4(SO_4)_2 \cdot 12H_2O$.

f alun *m* de fer ammoniacal; alun *m* ferrique ammoniacal — **d** Ammonium-Eisenalaun *m*; Ferri-Ammoniumsulfat *n*; Eisen-Ammoniak-alaun *m* — **n** ferri-ammoniumsulfaat *n*; ijzer-ammoniakaluin *n* — **e** sulfato *m* férrico-amónico; sulfato *m* de hierro amoniacal — **i** solfato *m* ferrico d'ammonio; solfato *m* ferrico ammoniacale; solfato *m* di ferro ammoniacale

5545 ferric bromide; iron bromide — Dark-red, deliquescent crystals, soluble in water, alcohol and ether.
Formula: $FeBr_3$.

f bromure *m* ferrique — **d** Ferribromid *n*; Eisenbromid *n* — **n** ferribromide *n*; ijzerbromide *n* — **e** bromuro *m* férrico; bromuro *m* de hierro — **i** bromuro *m* ferrico; bromuro *m* di ferro

5546 ferric chloride; ferric chloride hydrate; ferric perchloride; ferric trichloride; iron chloride; iron perchloride; iron trichloride; iron sesquichloride; perchloride of iron — Orange-yellow, very deliquescent crystals. Decomposes to yield hydrochloric acid if exposed to moist air or light. Used in the manufacturing of pigments and in etching processes. Latin: ferrum sesquichloratum.
Formula: $FeCl_3 \cdot 6H_2O$. See also Dutch bath.

f chlorure *m* ferrique; perchlorure *m* de fer; sesquichloride *m* de fer — **d** Eisenchlorid *n*; Eisensesquichlorid *n*; salzsaures Eisen *n*; Ferrichlorid *n*; Chloreisen *n* — **n** ijzerchloride *n*; ferrichloride *n* — **e** cloruro *m* ferrico; cloruro *m* de hierro; percloruro *m* de hierro — **i** cloruro *m* ferrico; cloruro *m* di ferro; percloruro *m* di ferro

5547 ferric chloride hydrate — See ferric chloride.

5548 ferric citrate; iron citrate; citrate of iron — Reddish brown scales, soluble in water, insoluble in alcohol. To be kept away from light. Used for blueprint paper.
Formula: $FeC_6H_5O_7 \cdot 5H_2O$.

f citrate *m* de fer; citrate *m* ferrique; citrate *m* de sesquioxyde de fer — **d** Ferrizitrat *n*; Eisen-

zitrat *n* — **n** ferricitraat *n*; ijzercitraat *n* — **e** citrato *m* férrico; citrato *m* de hierro — **i** citrato *m* ferrico; citrato *m* di ferro

5549 ferric hydrate — See ferric hydroxide.

5550 ferric hydroxide; iron hydroxide; ferric hydrate; iron hydrate; hydrated ferric oxide; hydrated iron oxide — Brown flocculent precipitate which dries to the oxide. Soluble in acids, insoluble in water, alcohol and ether.
Formula: $Fe(OH)_3$.

f hydrate *m* ferrique; hydrate *m* d'oxyde ferrique; hydrate *m* de sesquioxyde de fer — **d** Ferrihydroxyd *n*; Eisenhydroxyd *n*; Eisenoxydhydrat *n* — **n** ferrihydroxyde *n* — **e** hidróxido *m* férrico — **i** idrossido *m* ferrico; idrato *m* ferrico

5551 ferric oxalate; oxalate of iron — Pale yellow, amorphous, odourless scale or powder. Soluble in water and acids, insoluble in alkali. Used in photographic printing papers.
Formula: $Fe_2(C_2O_4)_3$.

f oxalate *m* ferrique; sesquioxalate *m* de fer; oxalate *m* de fer — **d** Ferrioxalat *n*; oxalsaures Eisenoxyd *n* — **n** ferri-oxalaat *n*; ijzeroxalaat *n* — **e** oxalato *m* férrico; oxalato *m* de hierro — **i** ossalato *m* di ferro; ossalato *m* ferrico

5552 ferric oxide; red ferric oxide; red iron oxide; red iron trioxide; ferric trioxide — Pigment formerly consisting of hematite, now made by calcining ferrous sulphate. Soluble in acids, insoluble in water. There are many synonyms, applying to relatively impure materials used as pigments, some naturally occurring others referring to synthetic materials.
Formula: Fe_2O_3.

f oxyde *m* ferrique; oxyde *m* rouge de fer; rouge *m* vénitien; hématite *f* rouge — **d** Ferrioxyd *n*; Eisenoxyd *n*; Eisenglanz *n* — **n** ferri-oxyde *n*; ijzeroxyde *n* — **e** óxido *m* férrico; óxido *m* de hierro — **i** ossido *m* ferrico; ossido *m* di ferro

5553 ferric perchloride — See ferric chloride.

5554 ferric potassium sulphate; iron potassium sulphate; iron alum — Pale violet crystals, soluble in water, insoluble in alcohol.
Formula: $FeK(SO_4)_2 \cdot 12H_2O$.

f sulfate *m* ferrique de potassium; alun *m* de fer potassique — **d** Kalium-Eisensulfat *n*; Kalium-Eisenalaun *m*; Eisenalaun *m* — **n** kalium-ijzer-sulfaat *n*; kalium-ijzeraluin *m* — **e** sulfato *m* férrico-potásico — **i** solfato *m* ferrico di potassio

5555 ferric salt

f sel *m* ferrique — **d** Ferrisalz *n* — **n** ferrizout *n* — **e** sal *f* férrica — **i** sale *m* ferrico

5556 ferric sulphate; ferric trisulphate; iron sulphate; iron persulphate; persulphate of iron — Yellow crystals or greyish-white powder. Formula: $Fe_2(SO_4)_3$, which is slightly soluble in water, or $Fe_2(SO_4)_3 \cdot 9H_2O$, which is very soluble in water.

f sulfate *m* ferrique; persulfate *m* de fer; sesquisulfate *m* de fer — **d** Ferrisulfat *n*; Eisenoxydulsulfat *n*; schwefelsaures Eisenoxyd *n*; Eisensesquisulfat *n* — **n** ferrisulfaat *n*; bruin ijzersulfaat *n* — **e** sulfato *m* férrico — **i** solfato *m* ferrico

5557 ferric trichloride — See ferric chloride.
5558 ferric trioxide — See ferric oxide.
5559 ferric trisulphate — See ferric sulphate.
5560 ferricyanide of potash — See potassium ferricyanide.
5561 ferrite core — Minute ring of magnetic material used as a unit of core storage. The ring is pulsed or polarized by wires wound through it, and remains in one of two states, until induced to change, acting as a switch or store. Ferrites are the inorganic salts of the formula MFe_2O_4, where M represents a bivalent metal.
f tore m en ferrite — **d** Ferritkern m; Magnetkern m — **n** ferrietkern fm — **e** núcleo m en ferrita — **i** nucleo m in ferrite
5562 ferrocyanide — Salt of ferrocyanic acid, containing the quadrivalent radical $Fe(CN)_6$, as potassium ferricyanide, $K_3Fe(CN)_6$ or potassium ferrocyanide, $K_4Fe(CN)_6$. Used in the making of iron-blue inks.
f ferrocyanure m — **d** Eisenzyanid n; Ferrozyanid n — **n** ferrocyaan n — **e** ferrocianuro m — **i** ferrocianuro m
5563 ferro ferric oxide — See black iron oxide.
5564 ferromagnetics — Science dealing with the magnetic polarization properties of materials.
f ferromagnétisme m — **d** Ferromagnetismus m — **n** ferromagnetisme n — **e** ferromagnetismo m — **i** ferromagnetismo m
5565 ferroprussiate — See potassium ferrocyanide.
5566 ferroprussiate paper — See blueprint paper.
5567 ferroso ferric oxide — See black iron oxide.
5568 ferrotype; tintype — Photographic image on a collodion coated thin black sheet of iron.
f ferrotypie f — **d** Ferrotypie f — **n** ferrotypie f — **e** ferrotipo m; ferrotipia f — **i** ferrotipo m; ferrotipia f
5569 ferrotype v — To put a glossy surface on a photographic print by pressing it while wet on a metal sheet.
f glacer — **d** mit Hochglanz trocknen — **n** glanzen; drogen en glanzen — **e** abrillantar — **i** lucidare; dare lucido
5570 ferrotype plate — Plate with a highly polished surface used to give a photographic print on paper a gloss by squeezing face down while wet and allow to dry.
f plaque f d'essorage — **d** Glanzplatte f; Hochglanztrockenplatte f — **n** glansplaat fm — **e** abrillantadora f; brillantadora f; satinadora f; plancha f satinadora; plancha f de abrillantar
5571 ferrotype tin — Material for the support upon which formerly after immersion in the sensitizing solution, the carbon tissue was squeezed; now plexiglass is recommended.
f dalle f d'essorage — **d** Trockenscheibe f; Trockenblech n — **n** droogplaat fm
5572 ferrous — Designation for iron salts in which the iron is divalent, e.g., ferrous chloride, $FeCl_2$. See also ferric.
f ferreux — **d** Ferro-; Eisen-II- — **n** ferro — **e**

ferroso — **i** ferroso
5573 ferrous ammonium sulphate; iron ammonium sulphate; ammonium sulphate of iron — Light-green crystals, soluble in water, insoluble in alcohol.
Formula: $Fe(SO_4)(NH_4)_2 \cdot 6H_2O$.
f sulfate m ferreux ammoniacal — **d** Ferroammonsulfat n; schwefelsaures Eisenoxydulammoniak n — **n** ferro-ammonium-sulfaat n — **e** sulfato m ferroso-amónico — **i** solfato m ferroso di ammonio
5574 ferrous bromide; iron bromide — Green crystalline powder, very deliquescent. Soluble in water and alcohol. To be kept tightly closed and protected from light.
Formula: $FeBr_2 \cdot 6H_2O$.
f bromure m ferreux — **d** Eisendibromid n — **n** ferrobromide n; ijzerbromide n — **e** bromuro m ferroso — **i** bromuro m ferroso
5575 ferrous carbonate; iron carbonate — Greenish-brown crystals, soluble in acids, insoluble in water.
Formula: $FeCO_3$ or $FeCO_3 \cdot H_2O$.
f carbonate m de fer — **d** Eisenkarbonat n — **n** ijzercarbonaat n; ferrocarbonaat n — **e** carbonato m ferroso; carbonato m de hierro — **i** carbonato m ferroso; carbonato m di ferro
5576 ferrous chloride; iron chloride; iron dichloride; iron protochloride; ferrous protochloride — Greenish-white crystals, soluble in alcohol and water.
Formula: $FeCl_2$ or $FeCl_2 \cdot 4H_2O$.
f chlorure m ferreux; protochlorure m de fer — **d** Eisendichlorid n; Eisenchlorid n; Ferrochlorid n; Eisenchlorür n — **n** ferrochloride n; ijzerchloruur f — **e** cloruro m ferroso; cloruro m de hierro — **i** cloruro m ferroso; cloruro m di ferro
5577 ferrous oxalate; oxalate of iron; iron oxalate — Pale yellow, odourless crystalline powder, soluble in acids, insoluble in water. Used in photographic developers.
Formula: $FeC_2O_4 \cdot 2H_2O$.
f oxalate m ferreux — **d** Ferrooxalat n — **n** ferrooxalaat n; ijzeroxalaat n; zuringzuur ijzeroxydule n — **e** oxalato m ferroso; oxalato m de hierro — **i** ossalato m ferroso; ossalato m di ferro
5578 ferrous protochloride — See ferrous chloride.
5579 ferrous salt
f sel m ferreux — **d** Ferrosalz n — **n** ferrozout n — **e** sal f ferrosa — **i** sale m ferroso
5580 ferrous sulphate; iron sulphate; protosulphate of iron — Greenish crystals, often brownish-yellow in colour from oxidation and efflorescence, soluble in water, insoluble in alcohol. Latin: sulfas ferrosus; ferrum sulfuricum oxydulatum.
Formula: $FeSO_4 \cdot 7H_2O$. See also copperas.
f sulfate m ferreux; sulfate m de fer; protosulfate m de fer — **d** Ferrosulfat n; Eisensulfat n; schwefelsaures Eisenoxydul n; Schweinfürter Grün n — **n** ferrosulfaat n; zwavelzuur ijzer-

oxydule *n* — **e** sulfato *m* ferroso; sulfato *m* de hierro — **i** solfato *m* ferroso; solfato *m* di ferro

5581 fers à l'éventail binding — See fan style binding.

5582 fers pointillé, à — Method of tooling of book covers, showing dotted lines and curves.
f à fers pointillés — **d** mit Stempelbild in punktierten Linien — **n** à fers pointillés — **e** con hierros puntillados — **i** a mille punti

5583 fers-pointillé binding — See pointillé binding.

5584 festoon dryer — Dryer which carries the paper web in loops, about 2.5 m (10 ft) high, over travelling poles through the drying chamber. Used for drying surface sized and coated papers.
f sécheur *m* à festons — **d** Trockenhänge-Vorrichtung *f*; Trockenhänge *f* — **n** festoendroger *m* — **e** secador *m* de tendido; secadero *m* de suspensión — **i** essiccatoio *m* a festoni

5585 feuilleton — Part of a newspaper, marked off by a rule, devoted to light literature, fiction, criticism, etc.
f feuilleton *m* — **d** Feuilleton *n* — **n** feuilleton *mn* — **e** folletín *m*; folletón *m* — **i** feuilleton *m*

5586 feuilleton section, in the
f au rez-de-chaussée — **d** unter dem Strich; unterm Strich; im Feuilleton — **n** in het feuilleton-gedeelte — **e** folletín — **i** parte *f* destinata a romanzi d'appendice

5587 fiber — See fibre.

5588 fibre *GB*; **fiber** *US* — Unit cell of vegetable growth which is many times longer than its diameter and which is the unit of paper pulps. Fibres are sometimes divided into two classes, bast and wood, but they are best designated by the tissue or region in which they occur, as cortical fibres, pericyclic fibres, phloem fibres, wood fibres, etc.
f fibre *f* — **d** Faser *f* — **n** vezel *fm* — **e** fibra *f* — **i** fibra *f*

5589 fibre boiler — See digester.

5590 fibre bonding
f liaison *f* des fibres — **d** Faserbindung *f* — **n** vezelbinding *f* — **e** ligado *m* de fibras; trabazón *f* de las fibras — **i** legame *m* fibroso

5591 fibre composition — See fibre content.

5592 fibre content; fibre composition — Fibrous constituents of paper or board and their proportions in it, usually expressed in percentage figures by weight, taking the total fibrous material of the paper or board as 100 parts. The fibres may be identified under the microscope by their particular structure and their reaction to various staining reagents. See app. no. 4 for the French classification.
f composition *f* fibreuse; suspension *f* fibreuse — **d** Stoffzusammenstellung *f*; Faserstoff-Suspension *f*; Faserzusammensetzung *f*; Faserstoffaufschwemmung *f* — **n** vezelstofgehalte *n* — **e** composición *f* fibrosa; composición *f* de fibras; composición *f* de la pasta; composición *f* de la pulpa — **i** composizione *f* fibrosa

5593 fibre cut — Short, straight, fairly smooth, randomly located cut, which occurs in a paper web where fibres larger than normal paper-making fibres are contained in the web and go through a calender or press nip. See also calender cuts.
f poil *m*; coupure *f* de fibre — **d** Faserriß *m* — **n** splinterbreuk *fm* — **e** corte *m* de fibra — **i** taglia *f* da fibra

5594 fibre furnish — The fibrous constituents of the paper and their various proportions in the paper. The fibre furnish is usually expressed in percentage figures, taking the total fibrous material of the paper as 100 parts.
f composition *f* de fabrication; composition *f* fibreuse — **d** Faserstoffzusammensetzung *f* — **n** stofsamenstelling *f* — **e** composición *f* fibrosa; composición *f* de la pasta de fibras — **i** composizione *f* fibrosa; composizione *f* di fabbricazione

5595 fibre knots — Not properly disintegrated bunches of fibres in paper which may be seen by transparence.
f nœuds *mpl*; agglomérats *mpl* de fibres; mâtons *mpl* — **d** Knoten *mpl* — **n** knopen *mpl* — **e** nudos *mpl* de fibras — **i** nodi *mpl*

5596 fibre optics — Plastics of suitable dimensions which exhibits the property of complete internal reflection. A "light pipe" is made from a bundle of these fibres and has the property of transmitting an image from one end to the other independent of any distortion of the path.

5597 fibre stock
f pâte *f* de fibres — **d** Faserstoff *m* — **n** vezelstof *fm*; vezelmassa *fm* — **e** pasta *f* de fibras; materia *f* fibrosa; sustancia *f* fibrosa; substancia *f* fibrosa — **i** pasta *f* fibrosa; impasto *m* fibroso

5598 fibrillae *GB*; **fibrils** *US* — Threadlike elements of the wall of the native cellulose fibre visible with a microscope.
f fibrilles *fpl* — **d** Fibrillen *fpl* — **n** fibrillen *fmpl* — **e** fibrilas *fpl* — **i** fibrille *fpl*

5599 fibrillate *v* — To beat, crush, smash, and split lengthwise paper fibres to form *fibrils.
f fibriller — **d** fibrillieren — **n** fibrilleren — **e** fibrilar — **i** fibrillare

5600 fibrillation; brushing out — Loosening of threadlike elements from the fibre wall to provide greater surface for forming fibre-to-fibre bonds, associated with refining of pulp.
f fibrillation *f* — **d** Fibrillierung *f* — **n** fibrillering *f* — **e** fibrilación *f*; descomposición *f* en fibrilas — **i** fibrillazione *f*

5601 fibrils — See fibrillae.

5602 field — Group of related characters treated as a unit in computer operations, generally used in connection with punched cards or recording on magnetic tape.
f zone *f* — **d** Feld *n* — **n** veld *n* — **e** campo *m* — **i** campo *m*; zona *f*

5603 field intensity — See field strength.

5604 field of force — Electric charges exert force upon other electric charges, magnets exert forces upon other magnets, matter exerts gravitational force upon other matter. All these action-at-a-distance phenomena are conveniently described in terms of the force field set up by a source.

f champ *m* d'intensité — **d** Kraftfeld *n* — **n** krachtveld *n* — **e** campo *m* de intensidad — **i** campo *m* d'intensità

5605 field strength; electric field intensity; electric force *GB*; **electric intensity; field intensity** *US* — The magnitude of a field, expressed as a vector quantity in volts per unit length at a given point.

f intensité *f* du champ électrique — **d** Feldstärke *f*; elektrische Feldstärke *f* — **n** veldsterkte *f*; elektrische veldsterkte *f* — **e** intensidad *f* de campo eléctrico — **i** intensità *f* di campo elettrico

5606 figure; text figure — Illustration or diagram inserted in the text, in distinction to a plate, which is printed separately.

f figure *f*; illustration *f*; diagramme *m* — **d** Bild *n*; Abbildung *f*; Diagramm *n* — **n** afbeelding *f*; diagram *n* — **e** grabado *m*; figura *f*; imagen *f*; diagrama *m* — **i** illustrazione *f*; diagramma *m*

5607 figure — A single number (0-9). The numbers may be conventional or *"old style" figures. See also arabic figures, Roman figures.

f chiffre *m*; nombre *m* — **d** Ziffer *f*; Zahl *f* — **n** cijfer *n*; getal *n* — **e** cifra *f*; número *m*; guarismo *m* — **i** cifra *f*; numero *m*

5608 figure case — Case with ten boxes for figures and some for spaces and quads.

f casseau *m* à chiffres — **d** Ziffernkasten *m* — **n** cijferkast *fm* — **e** caja *f* para números — **i** cassa *f* da numeri

5609 figure space — In computer controlled typesetting, "fixed" space (en space), which takes on the width value of the figure of the fount in question.

5610 file — Stiff, box-like cover for filing documents, letters, etc. See also paper file.

f classeur *m* — **d** Ordner *m*; Rechnungsordner *m* — **n** ordner *m* — **e** clasificador *m* — **i** raccoglitore *m*; classificatore *m*

5611 file — Collection of related records, usually, but not necessarily, arranged in sequence according to a key in each word. Computer term.

f fichier *m* — **d** Datei *f*; Informationsreihe *f* — **n** bestand *n* — **e** fichero *m* — **i** archivio *m*

5612 file — Tool of hardened steel with cutting ridges to smooth and form metal surfaces.

f lime *f* — **d** Feile *f*; Raspel *f* — **n** vijl *fm*; rasp *fm* — **e** lima *f* — **i** lima *f*

5613 file *v*

f classer — **d** ablegen — **n** afleggen; opbergen — **e** archivar

5614 file maintenance — Keeping a file up to date by adding, changing or deleting data.

f tenue *f* de fichier — **d** Dateifortschreibung *f*; Fortschreibung *f*; Bestandsführung *f* — **n** bestandsbijwerking *f* — **e** actualización *f* de ficheros — **i** aggiornamento *m* d'archivio

5615 file management — Program which keeps track of tasks processed simultaneously by a computer so that each task may be quickly retrieved and reprocessed or passed on for the performance of some other function.

f traitement *m* du fichier — **d** Dateibearbeitung *f* — **n** bestandsbehandeling *f* — **e** tratamiento *m* de fichero — **i** trattamento *m* d'archivio

5616 file management program — Program or series of programs which keep track of the files of various system users, enabling them to fetch their files and work on them when they desire, editing and correcting them, and linking their components together. Essential ingredient of any system which functions in a multi-user environment.

5617 file number — Number of a job entered in the order-book by the printer and reproduced on the printed job, to determine conditions under which the work was produced.

f numéro *m* de grebiche; grebiche *f* — **d** Bestellnummer *f* (eines Druckauftrages); Kommissionsnummer *f*; Registraturnummer *f* — **n** ordernummer *n*; commissienummer *n*; opdrachtnummer *n* — **e** número *m* de pedido — **i** numero *m* di commissione

5618 file *v* **off**

f enlever à la lime — **d** abfeilen — **n** afvijlen — **e** limar; quitar con la lima — **i** limare; raspare

5619 filing cabinet

f meuble-classeur *m*; cartonnier *m*; fichier *m* — **d** Karteikasten *m*; Kartei *f*; Registraturschrank *m* — **n** cartotheekkast *fm*; kast *fm* voor een kaartsysteem — **e** fichero *m*; archivador *m*; clasificador *m* — **i** cartoteca *f*; schedario *m*

5620 filled paper — Paper filled or loaded with mineral pigment.

f papier *m* chargé — **d** aschehaltiges Papier *n* — **n** papier *n* met vulstof; gevuld papier *n* — **e** papel *m* cargado — **i** carta *f* caricata

5621 filler; plugger; stopgap — Stop advertisement in a newspaper.

f bouche-trou *m*; bouchon *m*; bourrage *m* — **d** Füllinserat *n*; Füller *m*; Lückenbüßer *m* *coll.* — **n** stopper *m* — **e** anuncio *m* de relleno — **i** riempitivo *m*; riempimento *m*; stelloncino *m*

5622 filler — Short matter kept ready to fill space in a journal or magazine. See also punk.

f article *m* explétif; article *m* de remplissage — **d** Füllartikel *n* — **n** stopartikel *n*; stopper *m* — **e** relleno *m*; artículo *m* para llenar; fiambre *m* *Ur*; cuña *f* *Me, Cu* — **i** articolo *m* espletivo; articolo *m* di riempimento

5623 filler; loading material — Finely divided, relatively insoluble material, such as clay, calcium carbonate, etc., incorporated in papermaking composition, usually prior to sheet formation, to modify characteristics of the finished sheet like opacity, texture, printability, finish, weight, etc.

f charge *f*; matière *f* de charge — **d** Füllstoff *m* — **n** vulstof *fm* — **e** carga *f*; material *m* de carga; material *m* de relleno — **i** carica *f*; sostanza *f* di carica

5624 filler — Interior of the sheet of paper or board made on a multi-cylinder machine, as distinguished from the liners or outside plies.
f intérieur *m* (de carton) — **d** Einlage *f*; Mittelschicht *f* — **n** binnenlaag *fm* — **e** strato *m* interno; interno *m* (del cartone)

5625 filler — See punk.

5626 filler — See weft.

5627 filler block — Block used in the stick of the Ludlow.
f bloc *m* de remplissage — **d** Füllstück *n*; Ausschlußstück *n*; Zwischenstück *n* — **n** vulstuk *n*; uitvulstuk *n* — **e** bloque *m* de relleno; cuadrador *m* de justificación — **i** blocco-spazio *m*; blocco *m* di riempire

5628 fillet — Face-lined tool in the shape of a segment and a short handle, used for book cover decoration.
f palette *f* — **d** Filete *n* — **n** filet *mn* — **e** filete *m*; fileteador *m*; paleta *f*; tronquillo *m* — **i** filetto *m* a paletta

5629 fillet roll — See bar roll.

5630 filling — See weft.

5631 filling and dosing machine
f machine *f* de remplissage-dosage — **d** Abfüll- und Dosiermaschine *f* — **n** vul- en doseermachine *f* — **e** llenadora-dosificadora *f* — **i** macchina *f* dosatrice e riempitrice

5632 filling-in guards — See compensation guards.

5633 filling up; clogging; plate filling — Condition in the printing of halftones by letterpress or lithography, where the ink fills areas between the dots, and produces a solid rather than a sharp halftone dot.
f bouchage *m*; empâtement *m* — **d** Zusetzen *n* — **n** vollopen *n* — **e** taparse; llenarse; empastarse; borrarse; tupirse — **i** riempimento *m*

5634 fill-in powder — Thermosetting moulding material in granular form, used with sheet thermosetting matrix material to solidify the low spaces of original subject in a matrix making operation.
f poudre *f* à garnir; poudre *f* à flan — **d** Matrizenpulver *n* — **n** matrijzenpoeder *n* — **e** polvo *m* para patrices — **i** carica *f* termoindurente

5635 fill up — Unimportant news used for filling only.
f chien *m* écrasé — **d** unbedeutende Nachricht *f* — **n** onbelangrijk bericht *n* — **i** riempitivo *m*

5636 film; photographic film — Thin transparent plastic sheet coated with a photographic emulsion. After exposure it is developed and processed to produce a negative or a positive.
f film *m*; film *m* photographique; pellicule *f* photographique — **d** Film *m*; photographischer Film *m* — **n** film *m*; fotografische film *m* — **e** película *f*; film *m*; película *f* fotográfica — **i** pellicola *f*; pellicola *f* fotografica; film *m*

5637 film — Thin layer of light-sensitive silver emulsion applied to plates, films and papers.
f pellicule *f* — **d** Schicht *f*; Film *m* — **n** laag *fm*;

film *m* — **e** capa *f* — **i** strato *m*

5638 film — Thin skin or pellicle formed on drying ink.
f pellicule *f*; peau *f* — **d** Häutchen *n* — **n** velletje *n* (in de inktbus); huidje *n*; filmpje *n*; laagje *n* — **e** película *f*; piel *f* — **i** pellicola *f*

5639 film — See layer.

5640 film applicator — See equalizer rod.

5641 film clip — Clip to hang processed roll films for drying, either with a hook for hanging and attached to the top of the film, or weighted and attached to the bottom of the roll to keep the film from curling.
f pince *f* à film — **d** Filmklammer *f* — **n** filmklem *fm*; filmclip *m* — **e** pinza *f* para película — **i** pinzetta *f* per pellicola

5642 film former; film forming matter — Resin with qualities to form a tough, continuous film, usually plastics like nitrocellulose, chlorinated rubber, and vinyl.
f filmogène *m*; matière *f* filmogène; substance *f* filmogène — **d** Filmbildner *m*; Schichtbildner *m* — **n** filmvormer *m* — **e** formador *m* de películas; substancia *f* filmógena — **i** filmogeno *m*

5643 film forming matter — See film former.

5644 film holder — Support in the dark slide of a reproduction camera to which a piece of film is fixed to carry it flat in the focal plane.
f porte-film *m* — **d** Filmträger *m* — **n** filmdrager *m*; filmplaat *m* — **e** portapelículas *n* — **i** portafilm *m*; portapellicola *m*

5645 film layer — See benday artist.

5646 filmogen — Capable of being converted into a film. See also film former.
f filmogène — **d** filmbildend — **n** filmvormend — **e** filmógeno — **i** filmogeno

5647 film printer — Reel fed press with special tension, cutting, re-reeling, tension-control equipment, etc., to print on films or foils.
f rotative *f* pour impression sur pellicules — **d** Foliendruckmaschine *f* — **n** filmdrukmachine *f* — **e** máquina *f* para impresión de películas — **i** macchina *f* per la stampa di pellicole

5648 film processing unit — See processor.

5649 film processor — See processor.

5650 film setter — See photo-typesetting machine.

5651 filmsetting — See photo-typesetting.

5652 filmsetting machine — See photo-typesetting machine.

5653 film speed — See speed of an emulsion.

5654 filter — Porous wad of absorbent cotton or sheet of special paper through which photographic or other solutions are passed (filtered) to remove precipitates and foreign matter held in suspension.
f filtre *m* — **d** Filter *mn* (in der Technik wird der Gebrauch des Neutrums bevorzugt) — **n** filter *mn* — **e** filtro *m*; separador *m*; colador *m* — **i** filtro *m*

5655 filter — See knot screen.

5656 filter action — Action of a dye incorporated in a translucent layer to limit penetration of light.
f effet *m* de filtre — **d** Filterwirkung *f* — **n** filter-

werking *f* — **e** efecto *m* del filtro — **i** filtraggio *m*

5657 filter, colour — See colour filter.

5658 filter factor; multiplying factor — Number indicating, by multiplication, the increased exposure required when a particular colour filter is used during the camera exposure.

f coefficient *m* de filtre; facteur *m* de filtre — **d** Filterfaktor *m* — **n** filterfactor *m* — **e** factor *m* del filtro; factor *m* filtro; coeficiente *m* del filtro — **i** fattore *m* filtro

5659 filter holder

f porte-écran *m*; porte-filtre *m* — **d** Filterhalter *m* — **n** filterhouder *m* — **e** chasis *m* portafiltro; chasí *m* portafiltro; portafiltro *m* — **i** portafiltri *m*

5660 filter paper; Josef paper — Unsized, absorbent and porous paper for filtration.

f papier *m* filtre; papier *m* à filtrer; papier *m* Joseph — **d** Filtrierpapier *n* — **n** filtreerpapier *n* — **e** papel *m* filtro(s); papel *m* de filtro(s); papel *m* para filtrar; papel *m* de filtrar — **i** carta *f* da filtro; carta *f* da filtrare; carta *f* filtro

5661 filter ratio scale; ratiometer — Device to determine the exposure factors for colour filters used with colour-sensitive photographic negative materials.

f échelle *f* de coefficients de filtres colorés — **d** Filterverhältnisskala *f*; Filterfaktorbestimmungsgerät *n* — **n** filterfactormeter *m* — **e** escala *f* de filtros de color; cuña *f* de coeficientes de color — **i** scala *f* di rapporto dei fattori filtro; misuratore *m* dei fattori filtro

5662 filtrate — Clear liquid after filtration. Substance which has been filtered, containing no suspended matter.

f filtrat *m* — **d** Filtrat *n* — **n** filtraat *n* — **e** filtrado *m*; líquido *m* filtrado; líquido *m* colado — **i** filtrato *m*

5663 filtrate *v*

f filtrer — **d** filtrieren; abfiltern; filtern; seihen; durchseihen — **n** filtreren — **e** filtrar; posar por un filtro; colar — **i** filtrare; passare per il filtro

5664 fimble *v* — See feel *v*.

5665 final letter; terminal — Letter with a decorative flourish used at the end of words.

f lettre *f* finale — **d** Finalbuchstabe *m*; Schlußbuchstabe *m* — **n** sluitletter *fm* — **e** letra *f* final — **i** lettera *f* finale

5666 final s — See also final letter, ligature.

f s *m* rond — **d** Schluß-s *m* — **n** sluit-s *m* — **e** s *m* final — **i** esse *f* rotonda

5667 Finch device — Instrument to test the wet tensile breaking strength of paper.

f pince *f* selon Finch — **d** Finch-Gerät *n* — **n** Finch-klem *fm* — **e** pinza *f* Finch — **i** apparecchio *m* di Finch

5668 fine edition — See edition de luxe.

5669 fine etching; re-etching; re-biting — Supplementary or local etching of halftone plates to render the detail, tones and colour values of the original in the printed reproduction. Lightening a tone or local area by additional etching.

f remorsure *f*; réacidulation *f*; morsure *f* par couverture; morsure *f* d'effets — **d** Feinätzung *f*; Nachätzung *f*; Tiefätzung *f* — **n** fijnetsing *f*; reinetsing *f* — **e** refinado *m*; regrabado *m* — **i** morsura *f* finale

5670 fine-face rule — See hair-line rule.

5671 fine grain — Designation of film emulsions in which the grain size is small.

f grain *m* fin — **d** Feinkorn *n* — **n** fijne korrel *m* — **e** grano *m* fino — **i** grana *f* fine

5672 fine-grain developer — Photographic developer which produces negatives of low granularity.

f révélateur *m* à grain fin — **d** Feinkornentwickler *m* — **n** fijnkorrel-ontwikkelaar *m* — **e** revelador *m* de grano fino — **i** sviluppo *m* a grana fine

5673 fine-grain emulsion

f émulsion *f* à grain fin — **d** Feinkornemulsion *f* — **n** fijnkorrel-emulsie *f*; fijnkorrelige emulsie *f* — **e** emulsión *f* de grano fino — **i** emulsione *f* a grana fine

5674 fine line; hair line — Thin black finishing line enclosing the image on a printing plate.

f trait *m* fin — **d** Haarstrich *m*; Haarlinie *f* — **n** haarlijntje *n*; fijn lijntje *n* — **e** trazo *m* fino; perfil *m* fino; raya *f* delgada — **i** linea *f* nera sottile d'inquadratura

5675 fineness gauge — See fineness-of-grind gauge.

5676 fineness of grind — Degree of grinding or dispersion of a pigment in a printing ink. Poorly ground inks cause fill-up and piling.

f finesse *f* de broyage; finesse *f* de grain — **d** Feinheit *f*; Mahlfeinheit *f*; Kornfeinheit *f* — **n** fijnheid *f*; maalfijnheid *f* — **e** fineza *f* de moledura — **i** grado *m* di macinazione

5677 fineness-of-grind gauge; fineness gauge; grind gauge — Metal block with two troughs graduated numerically from the greatest depth to zero depth. Inks are placed in the troughs and drawn toward the zero mark with a finely machined scraper blade. The point at which scratches or interruptions of the smooth ink film appear is the designation of the fineness of grind.

f jauge *f* de broyage; appareil *m* pour mesurer la finesse de broyage — **d** Mahlfeinheitsmeßgerät *n* — **n** fijnheidsmeter *m*; inktfijnheidsmeter *m* — **e** medidora *f* de molturación — **i** misuratore *m* della finezza di macinazione

5678 fineness-of-grind measurement

f mesure *f* de la finesse de broyage — **d** Mahlfeinheitsmessung *f*; Farbfeinheitsmessung *f* — **n** fijnheidsmeting *f*; inktfijnheidsmeting *f*; meting *f* van de maalfijnheid — **e** medición *f* de la fineza — **i** misurazione *f* della finezza

5679 fine paper — Writing paper which good pen and ink writing characteristics, such as bond, ledger, mimeo, duplicator and manifold papers.

f papier *m* fin — **d** Feinpapier *n* — **n** fijn papier *n* — **e** papel *m* fino — **i** carta *f* fine

5680 fine screen — In photoengraving, a screen usually above 33 lines per cm (85 lines to the inch).

5680 f trame *f* fine; trame *f* serrée — **d** Feinraster *m* — **n** fijn raster *n* — **e** trama *f* fina; lineatura *f* fina — **i** retino *m* fine; trama *f* sottile

5681 finger — See cleat.

5682 finger — See gripper.

5683 fingernails — See parentheses.

5684 finger nail test — Test to assess the hardness of a varnish or film by scratching with a finger nail.
f essai *m* de rayure à l'ongle — **d** Fingernagelprobe *f* — **n** nagelvastheidsproef *fm* — **e** ensayo *m* con la uña de dedo; prueba *f* con la uña de dedo — **i** saggio *m* della resistenza all'unghia; prova *f* della resistenza all'unghia

5685 finger-prints — Impression of fingertips.
f empreints *fpl* digitales — **d** Fingerabdrücke *mpl* — **n** vingerafdrukken *mpl* — **e** impresiones *fpl* digitales — **i** impronte *fpl* digitali

5686 finish — Surface property of a sheet determined by its surface contour and gloss.
f apprêt *m*; fini *m* — **d** Oberfläche *f*; Glätte *f*; Satinage *f* — **n** oppervlakte *f*; gladheid *f*; satinage *f* — **e** acabado *m*; aprestado *m*; lisura *f* — **i** finitura *f*

5687 finish *v* — In newspaper printing, to stop sending new copy to the composing room.
f boucler; terminer — **d** aussetzen; den Satz beendigen — **n** eindigen met zetten; afsluiten — **e** entregar; cerrar a su hora — **i** terminare; finir la copia

5688 finisher — He who performs final re-etching, engraving, tooling and other operations on printing plates before proving.
f retoucheur *m* à l'outil — **d** Nachschneider *m* — **n** graveur *m* — **e** abridor *m* de láminas; grabador *m* — **i** ritoccatore *m* intagliatore

5689 finishing — Final work performed on a printing plate to remove defects and promote the best possible reproduction.
f finissage *m* — **d** letzte Bearbeitung *f*; Nachschneiden *n* — **n** afwerking *f* — **e** acabo *m* — **i** rifinitura *f*

5690 finishing line — Border line on the outer edges of halftone plates.
f cadre *m* — **d** Randlinie *f* — **n** kader *n*; kaderlijn *fm* — **e** recuadro *m* — **i** linea *f* di inquadratura; linea *f* di cornice

5691 finishing press — See lying press.

5692 finishing room — Department of a paper mill, where the paper is cut, trimmed, sorted, counted, packed and prepared for shipments.
f salle *f* de triage — **d** Sortiersaal *m* — **n** sorteerzaal *fm* — **e** sala *f* de escogido; sala *f* de clasificación — **i** reparto *m* assortimento

5693 fir — Coniferous tree, sometimes used mixed with spruce, in the manufacture of sulphite pulp.
f sapin *m*; bois *m* de sapin — **d** Tanne *f* — **n** sparrehout *n*; zilverspar *m* — **e** abeto *m*; madera *f* de abeto — **i** abete *m*; legno *m* di abete

5694 fired, to be — See also to get the *bullet.
f être renvoyé; être congédié; être mis à la porte

— **d** entlassen werden; den Sack bekommen — **n** ontslagen worden; de zak krijgen *fam.* — **e** despedir a una persona; despachar; desacomodar; dar (el) pasaporte a quedar cesante — **i** ricevere la licenza

5695 fire eater — See whip hand.

5696 fire point — See flash point.

5697 fireproof paper; flameproof paper — Paper treated with chemicals so that it will not support combustion. Not actually fireproof but it will not carry a flame.
f papier *m* incombustible; papier *m* ininflammable; papier *m* réfractaire; papier *m* ignifuge — **d** feuerfestes Papier *n*; unverbrennbares Papier *n*; flammsicheres Papier *n* — **n** onbrandbaar papier *n* — **e** papel *m* incombustible; papel *m* refractario; papel *m* ininflamable — **i** carta *f* incombustibile

5698 fire-red toner — Brilliant orange red ink toner pigment.
f toner *m* rouge feu — **d** Brillantrot *n* — **n** briljantrood *n* — **e** modificador *m* de color de fuego — **i** pigmento *m* rosso fuoco

5699 firm order
f commande *f* ferme; commande *f* à compte-ferme — **d** feste Bestellung *f* — **n** vaste bestelling *f*; vaste order *fm*; vaste opdracht *fm* — **e** pedido *m* en firme — **i** ordinazione *f* fissa

5700 first bite — See flat etch.

5701 first down colour — In multicolour printing, the first colour printed on the substrate to be overprinted by other colours.
f première couleur *f* — **d** erste Farbe *f*; zuerst gedruckte Farbe *f* — **n** eerste kleur *fm* — **e** primer color *m* — **i** primo colore *m*

5702 first edition; editio princeps; original edition — All the copies of a book or other work as first printed and published, including the first impression and subsequent impressions exactly similar to the first. A new edition implies that there are changes or additions to the text, that the type has been reset, or that the format has been changed.
f édition *f* princeps; édition *f* originale; première édition *f*; livre *m* princeps — **d** Erstausgabe *f*; Originalausgabe *f* — **n** eerste uitgave *f*; oorspronkelijke uitgave *f* — **e** edición *f* príncipe; edición *f* original; primera edición *f* — **i** prima edizione *f*; edizione *f* originale

5703 first elevator — Part of the slug composing machine which receives the matrices from the assembler, carries them to the casting position, and after the casting is made elevates them to the first transfer position. Has become out of use in some machines.
f premier élévateur *m* — **d** erster Elevator *m* — **n** eerste elevator *m* — **e** primer elevador *m* — **i** primo elevatore *m*

5704 first elevator jaw
f tête *f* de l'élévateur — **d** Elevatorkopf *m* — **n** elevatorkop *m* — **e** cabeza *f* del primer elevador; quijada *f* del primer elevador — **i** testa *f* dell'ele-

vatore; testa *f* del primo elevatore

5705 first etching — See flat etch.

5706 first form — See first forme.

5707 first forme *GB*; **first form** *US*
f première forme *f*; forme *f* de l'impression recto — **d** Schöndruckform *f* — **n** schoondrukvorm *m* — **e** primera forma *f*; forma *f* de blanco (tipografía); forma *f* de cara (litografía) — **i** prima forma *f*; forma *f* di bianca

5708 first generation computer — Computer utilizing vacuum tube components, operating in milliseconds.
f calculateur *m* de première génération — **d** Computer *m* der ersten Generation; Rechner *m* der ersten Generation — **n** computer *m* van de eerste generatie — **e** computadora *f* de primera generación — **i** calcolatore *m* di prima generazione

5709 first generation typesetter — Photocomposition device patterned after hot-metal machines, specifically the Intertype Fotosetter, which used the circulating matrix principle (with a photo image on the matrix rather than an intaglio mould), and the original Monophoto (which caused type to be photographed from a matrix resembling the original Monotype grid.
f composeuse *f* de première génération — **d** Setzmaschine *f* der ersten Generation — **n** zetmachine *f* van de eerste generatie — **e** componedora *f* de primera generación — **i** compositrice *f* di prima generazione

5710 first impression — All the copies of a book or other work printed at the first printing, before alterations or additions have been made to the text or before reprints have been made.
f premier tirage *m*; première impression *f* — **d** Erstdruck *m* — **n** eerste druk *m* — **e** tirada *f* (de la) primera; tiraje *m* primero; tiro *m* primero; primera tirada *f*; primera impresión *f* — **i** prima stampa *f*; prima tiratura *f*

5711 first impression — The first printing operation on one side of the sheet of paper or web.
f impression *f* en blanc; impression *f* sur premier côté — **d** Schöndruck *m* — **n** schoondruk *m* — **e** tirada *f* en blanco; impresión *f* de blanco; tirada *f* de blanco (tipografía); tirada *f* de primera cara (litografía) — **i** stampa *f* in bianca

5712 first impression offset — The offsetting of ink on the first impression side of the sheet from the second impression backing cylinder. Common in newspaper printing.
f maculage *m* au recto — **d** Verschmieren *n* der Schöndruckseite; Abliegen *n* der Schöndruckseite; Abziehen *n* der Schöndruckseite — **n** overzetten *n* op de schoondrukzijde — **e** repintado *m* a la cara de blanco — **i** maculatura *f* della bianca

5713 first parallel fold; tabloid fold — Fold made in the *jaw folder immediately following the former fold when the web has been slit in half longitudinally. Results in 8-page multiples of the number of webs in the press, signature size 1/2

cut-off lengths x 1/2 web width.
f premier pli *m* parallèle — **d** erster Parallelfalz *m* — **n** eerste parallelvouw *fm* — **e** primero doblado *m* en paralelo — **i** prima piega *f* parallela

5714 first proof — See reader's proof.

5715 fish eyes — Small, round, glazed, or transparent spots, caused by slime, undefibred portions of stock, or foreign materials which are crushed in calendering the sheet. Also round, transparent spots in the coated surface of coated board, caused by excess defoamer of an oil-base type.
f yeux *mpl* — **d** glasige Flecke *mpl*; Löcher *npl* — **n** glazige plekken *fmpl* — **e** motas *fpl*; manchas *fpl* de agua — **i** macchioline *fpl* vetrose

5716 fish glue — Glue made from the skins and refuse of cod fish, used as a vehicle in glue enamel and rendered light sensitive with ammonium bichromate. Also used as an adhesive for stamps and paper boxes.
f colle *f* de poisson; colle-émail *f* — **d** Fischleim *m* — **n** vislijm *m* — **e** colapez *f*; cola *f* de pescado; cola *f* de pez; cola *f* para fotograbado — **i** colla *f* di pesce

5717 fish oil — Oil derived from the bodies of fish such as menhaden, by cooking or pressing and further purification. Sometimes used in printing ink vehicles.
f huile *f* de poisson — **d** Fischöl *n* — **n** visolie *fm* — **e** aceite *m* de pescado — **i** olio *m* di pesce

5718 fist; hand; index; index mark; digit — Index or fist type character.
f main *f* — **d** Hand *f*; Hinweiszeichen *n*; Handzeichen *n*; Flosse *f coll.* — **n** handje *n*; hand *fm*; index *m* — **e** manecilla *f*; mano *m*; llamada *f* — **i** manina *f*; mano *m*

5719 fitting — The degree of closeness each letter has to its adjacent letters.
f approche *f* — **d** Zwischenraum *m* — **n** breedte *f* van het vlees — **i** avvicinamento *m*

5720 five-line nonpareil — Old name for 30-point type. See app. no. 1.

5721 five-line pica — Old name for 60-point type. See app. no. 1.

5722 fix *v* — To stabilize the photographic image on film or paper after development by dissolving or neutralizing the remaining unexposed silver salts in the emulsion.
f fixer — **d** fixieren — **n** fixeren — **e** fijar; virar — **i** fissare

5723 fix *v* — To spray a fixative over a pencil or charcoal drawing to prevent it from being rubbed off.
f fixer — **d** fixieren — **n** fixeren — **e** fijar — **i** fissare

5724 fixation; fixing — Chemical removal of undeveloped silver salts from photographic images by treatment in a fixing bath.
f fixage *m* — **d** Fixierung *f*; Fixieren *n* — **n** fixeren *n* — **e** fijación *f*; fijado *m* — **i** fissaggio *m*

5725 fixative — Plastic or varnish coating sprayed over charcoal, chalk, pastel, and other art work to protect and to prevent smudging.

f fixatif *m* — **d** Fixativ *n* — **n** fixatief *n*; fixeermiddel *n* — **e** fijativo *m* — **i** fissativo *m*

5726 fixed cut-off folder — Folder of a rotary press, in which the length of the product cannot be changed.

f plieuse *f* à format fixe — **d** festformatiger Falzapparat *m* — **n** niet-variabel vouwapparaat *n* — **e** plegadora *f* de formato fijo — **i** piegatrice *f* a formato fisso; piegatrice *f* invariabile

5727 fixed disk store — See magnetic disk.

5728 fixed-head disk — For explanation of the difference between a fixed-head disk and a moving head disk, see moving head disk.

5729 fixed length record — Extent of a record in terms of column numbers of a card, or number of locations on punched or magnetic tape. The length of a fixed record remains constant although the amount of data contained within it may not fill the locations.

f enregistrement *m* à longueur fixe — **d** Satz *m* mit fester Länge — **n** record *n* met vaste lengte — **e** registro *m* de longitud fija — **i** registrazione *f* a lunghezza fissa

5730 fixed point — In mathematics the system in which all numbers, whole or fractional, are processed as whole numbers. It is the users' responsibility to maintain decimal point location. Contrast with *floating point.

f virgule *f* fixe — **d** Festkomma *n* — **n** vaste komma *fmn*; vast decimaalteken *n* — **e** coma *f* fija; punto *m* fijo — **i** virgola *f* fissa

5731 fixed space — Space providing a gap between words or characters of a given width, such as an em, en or thin space. Opposed to spaceband or justifying space. See also figure space.

5732 fixed store *GB*; **read-only storage** *US*; **read-only memory** — Solid-state memory for storing programs or information which cannot be altered or written over by the user. They can take the place of either hardwired logic or of random access stores for the storage of computer programs and are cheaper than random access stores, but since the information can never be changed they are relatively inflexible.

f mémoire *f* morte — **d** Festspeicher *m*; Festwertspeicher *m* — **n** dood geheugen *n* — **e** memoria *f* muerta; memoria *f* de sólo lectura — **i** memoria *f* protetta

5733 fixed word length — Pertaining to a machine word or operand that always has the same number of bits or characters. Contrast with variable word length.

f longueur *f* de mot fixe — **d** feste Wortlänge *f* — **n** vaste woordlengte *f* — **e** longitud *f* fija de palabra — **i** lunghezza *f* fissa di parola

5734 fixer — See fixing bath.

5735 fixing — See fixation.

5736 fixing bath; fixer — Solution that makes the developed image on a film permanent by dissolving or neutralizing the remaining unexposed emulsion.

f bain *m* de fixage; bain *m* de fixation — **d** Fixierbad *n* — **n** fixeerbad *n* — **e** baño *m* fijador; baño *m* de fijado; fijador *m* — **i** bagno *m* di fissaggio; bagno *m* fissatore

5737 fixing salt — Salt for making up fixing baths, commonly referred to as *hypo. Sodium thiosulphate, $Na_2S_2O_3$, easily dissolves silver bromides, silver chlorides and silver iodides, forming soluble thiosulphates of silver which is removed by subsequent washing.

f sel *m* de fixage; sel *m* fixateur — **d** Fixiersalz *n* — **n** fixeerzout *n* — **e** sal *f* fijadora; sal *f* para fijar — **i** sale *m* di fissaggio

5738 fix *v* **the blanket** — To put on the blanket.

f mettre le blanchet en place; fixer le blanchet — **d** das Gummituch befestigen; das Gummituch festmachen — **n** het rubberdoek aanbrengen — **e** fijar la mantilla — **i** montare il blanket; fissare il blanket

5739 flabby — Said of paper with little or no internal stiffness.

f mou — **d** lappig — **n** slap; zacht — **e** flojo; blando; en girones — **i** floscia

5740 flabbyness

f mollesse *f* — **d** Weichheit *f* — **n** weekheid *f*; zachtheid *f*; soepelheid *f* — **e** blandura *f*; mullido *m*; blando *m* — **i** mollezza *f*

5741 flag; splice tag — Small piece of paper or similar material, placed in a reel of paper, extending beyond the end to denote a splice or a defect. See also ream marker.

f pavillon *m*; marqueur *m* — **d** Streifen *m*; Papierstreifen *m* — **n** strookje *n* (papier); papierstrookje *n*; lasstrookje *n* — **e** registro *m*; banderita *f*; señal *f* — **i** bandierina *f*; strisciolina *f* di carta

5742 flag; sentinel; pointer; marker — Programmed indicator usually consisting of one bit, used to test a condition set at a previous point in the program. Also an indicator of the beginning of a group of data.

f drapeau *m* — **d** Hinweissymbol *n*; Kennzeichen *n* — **n** vlag *fm*; sein *n* — **e** bandera *f* — **i** indicatore *m*

5743 flag — American term for the leading illustration on the first page of a daily paper.

5744 flag — See book marker.

5745 flag — See mast-head.

5746 flag — See ream marker.

5747 flag code — Code which changes the meaning of the following single character code in data input.

5748 flake albumen

f albumine *f* en paillettes — **d** Trockeneiweiß *n*; Flockeneiweiß *n* — **n** eiwitvlokken *fmpl* — **e** albúmina *f* en escamas — **i** albumina *f* in scaglie; albumina *f* secca

5749 flake white; white lead; lead subcarbonate; Cremnitz white; Kremnitz white; Crems white; ceruse — Basic lead carbonate, soluble in acids, insoluble in water, used as a pigment. Poisonous; characterized by a fugitive white colour, covering power and tough, flexible film forming properties.

Cremnitz white is an old name for white lead produced in Europe by the chamber process, named after Kremnitz (Czechoslovakia). Latin: carbonas plumbicus. Formula: $2PbCO_3 \cdot Pb(OH)_2$.

f carbonate *m* de plomb (basique); hydrocarbonate *m* de plomb; céruse *f* en lamelles; blanc *m* d'argent; blanc *m* de plomb; blanc *m* de Clichy; blanc *m* de Gênes; blanc *m* de Hambourg; blanc *m* de Hollande; blanc *m* de Crems; blanc *m* de Vénise — **d** basisches Bleikarbonat *n*; Kremserweiß *n*; Bleiweiß *n* — **n** basisch loodcarbonaat *n*; Kremserwit *n*; loodoxydwit *n* — **e** carbonato *m* básico de plomo; blanco *m* de plomo; blanco *m* de Krems; cerusa *f*; albayalde *m* — **i** carbonato *m* di piombo basico; bianco *m* di piombo; biacca *f* di Krems

5750 flameproof paper — See fireproof paper.

5751 flammable — See inflammable.

5752 flange — Rib or rim for strength, for guiding, or for attachment to another object.
f bride *f* — **d** Flansch *m*; Anschraubflansch *m* — **n** flens *m* — **e** brida *f* — **i** flangia *f*

5753 flange connection
f joint *m* à brides — **d** Flanschverbindung *f* — **n** flensverbinding *f* — **e** conexión *f* abordonado; conexión *f* de bridas — **i** collegamento *m* a flangia

5754 flannel — Loosely-woven fabric of cotton, or of cotton and wool, with soft, nap-like surface, used for damping rollers.
f flanelle *f* — **d** Flanell *n*; Walzenstoff *m* — **n** flanel *n* — **e** franela *f*; bayeta *f* — **i** flanella *f*

5755 flap — Side end or part of the envelope which is folded over to close and seal the envelope.
f patte *f* d'enveloppe — **d** Schlußklappe *f*; Klappe *f* — **n** klep *fm* van een envelop(pe) — **e** aleta *f*; solapa *f*; cartera *f* (de sobre) — **i** falda *f*

5756 flap — See front cover.

5757 flap lid — Lid on boxes without sides or ends, hinged at the base.
f couvercle *m* rabattant — **d** Klappdeckel *m* — **n** klapdeksel *n* — **e** tapa *f* con alas — **i** coperchio *m* cerniera

5758 flare — Appearance of a circular disc of light in the centre of photographic images when viewed on a focusing screen, due either to optical errors in the camera lens or to mechanical causes. Camera flare is due to scattered light within the apparatus and causes fogged images, particularly in the shadow areas of halftone negatives.
f tache *f* de lumière — **d** Reflektionsfleck *m*; Spiegelfleck *m*; Nebenlicht *n*; Glanzlicht *n* — **n** lichtreflectie *f*; lichtplek *fm*; overstraling *f*; weerschijn *m*; blink *m*; weerspiegeling *f* — **e** ráfaga *f*; mancha *f* de luz; relumbre *m* — **i** macchia *f* di luce

5759 flare *v* the balls — To heat the ink balls over a flame to make them tacky.
f chauffer les tampons (d'encrage) — **d** die Farbballen anwärmen — **n** de tampons aanwarmen —

e calentar las balas (de entintado); calentar los tampones — **i** scaldare i tampone (d'inchiostro)

5760 flash — In computer controlled typesetting, "exposing" a photographic image of a type character usually from a xenon light source. The flash is timed to expose the character when it is properly positioned before the light source and in relation to the desired position in a line of type.

5761 flashed opaque glass — See opal glass.

5762 flash exposure; fogging exposure; padding *coll.* — Supplementary exposure (about 45 sec) in halftone photography with a small lens stop to a sheet of white paper placed over the original or to the raya of a flash lamp, to strengthen the dots in the shadows of the negative.
f exposition *f* sur papier blanc; exposition *f* flash; pose *f* flash; pose *f* auxiliaire; exposition *f* voile — **d** Flash-Belichtung *f*; Nachbelichtung *f*; kontrastvermindernde Belichtung *f*; Schleierbelichtung *f* — **n** sluierbelichting *f*; witbelichting *f*; nabelichting *f*; voorbelichting *f*; contrastverminderende belichting *f*; schaduwbelichting *f*; wit geven *n*; wit nageven *n* — **e** exposición *f* en blanco; exposición *f* auxiliar; preexposición *f* — **i** esposizione *f* flash; esposizione *f* in bianco; preesposizione *f*

5763 flashing; papering — Giving the *flash exposure in halftone photography.

5764 flash lamp — Electric lamp at the front of a process camera to transmit light through the lens during the flash exposure.
f lampe *f* flash — **d** Vorbelichtungslampe *f* — **n** voorbelichtingslamp *fm*; nabelichtingslamp *fm* — **e** lámpara *f* de preexposición; lámpara *f* para exposición en blanco — **i** lampo *m* flash; lampada *f* per la preesposizione

5765 flash point; fire point — Minimum temperature at which an inflammable liquid will burn for five seconds upon exposure to flame in a standard cup. For the higher flash points use is made of the *Cleveland open cup and the Tag closed cup.
f point *m* d'inflammation — **d** Flammpunkt *m* — **n** vlampunt *n* — **e** punto *m* de ignición; punto *m* de inflamación; punto *m* de inflamabilidad; temperatura *f* de inflamación; centelleo *m* *Me*; punto *m* centelleo — **i** punto *m* d'infiammabilità; temperatura *f* d'infiammabilità

5766 flat — Glass plate on which a number of stripfilm negatives have been transferred and arranged to promote photoprinting and etching as a single unit.
f glace *f* de montage; verre *m* de montage — **d** Montageplatte *f*; Glasplatte *f* (für Sammelmontage) — **n** montageruit *fm*; montageplaat *fm* — **e** luna *f* de montaje; plantilla *f* de montaje; vidrio *m* de montaje; plancha *f* de montaje — **i** lastra *f* di montaggio

5767 flat — In lithography, the assembly of photographic negatives or positives on goldenrod paper, glass or vinyl acetate for exposure in vacuum frame in contact with sensitized metal press plate.

Equivalent to a typographic forme and containing text as well as art. Types of flat are: l. goldenrod flat, an assembly of film negatives on goldenrod paper. The paper serves as a layout, mask, and film support. 2. blueline flat, an assembly of films or stripfilm sections in register with blueline images printed on a glass or plastic sheet. 3. complementary flat, one of a pair or set of flats, each of which contributes its printing detail by successive exposures made to the same pressplate.

f montage *m* − **d** Montage *fm*; montierte Vorlageform *f* − **n** montage *f* − **e** lámina *f* de montaje − **i** montaggio *m*

5768 flat − Zinc or copper plate with a number of line or halftone images and photoprinted from a flat.

f planche *f* (de métal) copiée − **d** Sammelplatte *f*; Tableau *n*; Ganzstückdruckplatte *f* − **n** metaalkopie *f*; verzamelkopie *f* op metaal − **e** plancha *f* insolada; plancha *f* pasada − **i** lastra *f* copiata (di metallo)

5769 flat − Lengthwise indentation on a rubber or composition roller of a letterpress rotary.

f plat *m* − **d** Abplattung *f* − **n** afplatting *f* − **e** aplanamiento *m*; achatamiento *m* − **i** appiattimento *m*; schiacciamento *m*

5770 flat − Width of contact of a rubber or composition roller when set on an ink drum in a rotary press.

f largeur *f* de contact; bande *f* de contact − **d** Berührungsstreifen *m* − **n** breedte *f* van de contactlijn − **e** ancho *m* de contacto; banda *f* de contacto − **i** larghezza *f* di contatto

5771 flat; in the flat − Paper or board supplied in sheets, not folded, as distinct from reels.

f papier *m* à plat − **d** Planopapier *n* − **n** plano papier *n*; plano vellen *npl*; papier *n* aan vellen; ongevouwen papier *n* − **e** papel *m* en hojas; papel *m* plano − **i** carta *f* in fogli

5772 flat back − See square back.

5773 flat bag − Flat tube with the bottom closed by turning up this tube and gumming the turn-up to the outside of the bag so formed.

f sac *m* plat; sachet *m* plat − **d** Flachbeutel *m* − **n** platte zak *m* − **e** bolsa *f* plana − **i** sacchetto *m* piatto

5774 flat-bed cylinder press − See flat-bed press.

5775 flat-bed press; flat-bed cylinder press − Printing press in which a flat bed holding the printing forme moves against a revolving cylinder which carries the paper.

f presse *f* plate; presse *f* à plate − **d** Flachformdruckpresse *f*; Flachformmaschine *f* − **n** drukpers *f* voor vlakke vormen − **e** prensa *f* plana; máquina *f* plana; prensa *f* cilíndrica; máquina *f* cilíndrica; máquina *f* de (im)presión planocilíndrica; máquina *f* de cama plana; máquina *f* planocilíndrica; prensa *f* de (im)presión planocilíndrica; prensa *f* de cama plana − **i** macchina *f* da stampa per forme piane; macchina *f* piana

5776 flat-bed rotary press; flat-bed web press

f presse *f* rotative à plat; rotative *f* à plat − **d** Flachformrotationsmaschine *f* − **n** vlakke-vorm rotatiepers *f* − **e** rotaplana *f*; máquina *f* rotaplana *Es*; prensa *f* rotaplana; máquina *f* semirrotativa; prensa *f* semirrotativa; semirrotativa *f*; prensa *f* plana con papel continua − **i** macchina *f* da stampa rotativa per forme piane; macchina *f* (da stampa) pianocilindrica

5777 flat-bed web press − See flat-bed rotary press.

5778 flat brush − Wide brush used by stereotypers to spread the pulp, and by engravers to apply the varnish.

f queue *f* de morue − **d** Flachbürste *f* − **n** platte kwast *m* − **e** brocha *f* plana; palatina *f* − **i** spazzola *f* piatta

5779 flat cap − A size of ledger paper, sometimes used in the US. See app. no. 3.

5780 flat colour plate − See line colour plate.

5781 flat colour printing − See line colourwork.

5782 flat colourwork − See line colourwork.

5783 flat copy − See reflection copy.

5784 flat cut index − See thumb index.

5785 flat cutter − See guillotine.

5786 flat delivery − Delivery on a printingpress supplying the sheets unfolded.

f sortie *f* à plat − **d** Planoauslage *f* − **n** planouitleg *m* − **e** salida *f* plana − **i** uscita *f* in piano

5787 flat etch; flat etching; first etching; first bite; pre-etching − In photoengraving the first or initial etch given to a halftone plate to promote the required printing depth.

f première morsure *f*; morsure *f* préliminaire − **d** Anätzung *f*; Anätzen *n*; Vorätzung *f*; Vorätzen *n* − **n** aanetsing *f*; aanetsen *n* − **e** mordedura *f*; primer mordido *m*; mordido *m* preliminar − **i** morsura *f* preliminare; incisione *f* preliminare; premordenzatura *f*

5788 flat etching − See flat etch.

5789 flat foolscap − A size of writing paper. See app. no. 3.

5790 flat graver − Engraving tool with a flat face at right angle to the cutting edge.

f échoppe *f* plate − **d** Flachstichel *m* − **n** vlaksteker *m* − **e** buril *m* plano − **i** scalpello *m* piatto; bulino *m* piatto

5791 flat image − Image which lacks contrast.

f image *f* sans contraste − **d** kontrastloses Bild *n*; flaches Bild *n* − **n** contrastloos beeld *n*; krachteloos beeld *n* − **e** imagen *f* con poco contraste; imagen *f* con falta de contraste − **i** immagine *f* piatta; immagine *f* priva di contrasto

5792 flat impression; flat pull; rough pull − Proof without over- or underlaying, used only for checking.

f épreuve *f* brute; première épreuve *f* − **d** Abzug *m* ohne Zurichtung − **n** niet-toegestelde proef *fm*; afdruk *m* van een niet-toegestelde drukvorm; ruwe proef *fm* − **e** prueba *f* sin arreglo en la forma; prueba *f* de máquina − **i** bozza *f* senza avviamento della forma

5793 flat, in the − See flat.

5794 flat lighting − In photography, lighting that

produces a minimum of shadows and very little contrast or modeling on the subject.

f éclairage *m* flou; éclairage *m* plat − **d** flache Beleuchtung *f* − **n** contrastlose belichting *f*; vlakke belichting *f*; vlakke verlichting *f* − **e** iluminación *f* uniforme − **i** illuminazione *f* senza contrasti; illuminazione *f* uniforme

5795 flat negative − Negative with little contrast.

f négatif *m* plat − **d** flaches Negativ *n*; kontrastarmes Negativ *n* − **n** vlak negatief *n*; contrastloos negatief *n* − **e** negativo *m* sin contraste; negativo *m* sin relieve − **i** negativo *m* piatto

5796 flatness − Condition of paper or board when it has no curl, cockle or wave.

f plat *m* − **d** Flachliegen *n* − **n** vlakliggen *n* − **e** aplanado *m* − **i** planarità *f*

5797 flatness meter − Device to measure the unevenness of the bed of a printing press.

f mesureur *m* de la planéité (du marbre) − **d** Planheitsmesser *m* − **n** vlakheidsmeter *m* − **e** medidor *m* del plano (del mármol) − **i** misuratore *m* di planarità

5798 flat paper − Paper which comes from the mill in flat sheets, without fold or crease; especially writing paper sold by manufacturers and dealers in packages of flat sheets of standard sizes and finishes.

f papier *m* plano; papier *m* à plat; papier *m* inerte − **d** flachliegendes Papier *n* − **n** vlakliggend papier *n*; plano papier *n* − **e** papel *m* apilado; papel *m* en hojas − **i** carta *f* stesa; carta *f* in piano

5799 flat plate − Etched halftone on which no re-etching or finishing has been performed. Halftone plate, lacking in contrast.

f cliché *m* plat sans contraste − **d** flache Platte *f*; kontrastlose Ätzung *f* − **n** vlakke plaat *fm*; contrastloze plaat *fm* − **e** grabado *m* blando; grabado *m* con poco contraste; medio tono *m* blando; medio tono *m* con poco contraste; plancha *f* blanda; plancha *f* con poco contraste − **i** lastra *f* senza contrasti; lastra *f* a toni appiattiti

5800 flat printed news; overissue news − Unsold newspapers. See also returns.

5801 flat proof; rough proof − Impression made without makeready from an unfinished or finished printing plate. A proof lacking in quality.

f épreuve *f* nature; épreuve *f* d'état; épreuve *f* en première − **d** erster Abzug *m* − **n** eerste proef *fm* − **e** prueba *f* de primeras − **i** prima bozza *f*

5802 flat proofs − See progressive colour proofs.

5803 flat pull − See flat impression.

5804 flat ratings − Set of ratings in which the variations in the rate of working have been underestimated. See also steep ratings.

5805 flat stereo

f stéréo *m* plat; stéréotype *m* plan − **d** Flachstereo *n* − **n** vlakstyp *m* − **e** estereotipia *f* plana; plancha *f* plana estereotípica − **i** stereotipia *f* piana; stereo *m* piano

5806 flat-stitched booklet; side-stitched booklet − Booklet in which the stitches are going

through its total thickness.

f livret *m* piqué à plat − **d** Broschüre *f* mit Blockheftung; Heft *n* mit Blockheftung − **n** door het plat geniete brochure *fm*; door het plat geniet boekje *n* − **e** libro *m* cosido con alambre; folleto *m* cosido por el costado − **i** opuscolo *m* cucito di piatto (a punti metallici)

5807 flat stitching − See side stitching.

5808 flat tint

f teinte *f* en aplat − **d** ungerasteter Ton *m* − **n** vlakke tint *fm*; vlakke toon *m*; ongerasterde tint *fm*; ongerasterde toon *m* − **e** color *m* liso; color *m* sin trama; tinta *f* plana − **i** tinta *f* a tono continuo; tinta *f* a tono pieno

5809 flat-tint plate; tint block; background plate; solid block − Solid piece of metal, usually zinc, cut to a required shape to apply a flat tint.

f cliché *m* d'aplat; cliché *m* de teinte − **d** Tonplatte *f*; Untergrundplatte *f* − **n** toonplaat *fm*; ongerasterde tintplaat *fm* − **e** plancha *f* para fondo plano; plancha *f* de tinta plana − **i** lastra *f* per fondo a tono pieno

5810 flaw − A defect. See also air hole.

f soufflure *f* de fonte − **d** Gußblase *f*; Lunker *m*; schädlicher Hohlraum *m* in Gußstücken − **n** gietgal *fm* − **e** sopladura *f*; paja *f*; burbuja *f*; escarabajo *m*; poro *m* − **i** soffiatura *f*

5811 flax − Flax fibres from flax straw and linen rags are used in strong durable papers, like banknote papers.

f lin *m* − **d** Flachs *m* − **n** vlas *n* − **e** lino *m* − **i** lino *m*

5812 flaxseed oil − See linseed oil.

5813 flesh colour − See flesh tone.

5814 flesh tint − See flesh tone.

5815 flesh tone; flesh colour; flesh tint

f ton *m* chair; couleur *f* de chair − **d** Fleischfarbe *f*; Fleischton *m* − **n** vleeskleur *fm*; vleestint *fm* − **e** color *m* carne − **i** tono *m* carne; tinta *f* carne

5816 fleuron − Printer's mark or design on title page or cover of a book.

f fleuron *m* − **d** Druckerzeichen *n*; Druckermarke *f* − **n** drukkersmerk *n*; vignet *n* van de drukker; tjap *mn* van de drukker − **e** marca *f* de impresor − **i** rosone *m*; fiorone *m*; sigla *f* dello stampatore

5817 fleurons − See flowers.

5818 flexiback binding − See adhesive binding.

5819 flexibility − Flexible quality or state.

f aptitude *f* à la flexion; aptitude *f* au pliage − **d** Biegsamkeit *f*; Falzfähigkeit *f* − **n** buigzaamheid *f*; geschiktheid *f* tot vouwen − **e** flexibilidad *f* − **i** flessibilità *f*

5820 flexible

f flexible; souple; pliable − **d** biegsam; geschmeidig − **n** buigzaam; flexibel; soepel − **e** flexible − **i** flessibile; pieghevole

5821 flexible back − See back.

5822 flexible binding − Binding with back sewed on raised band with thread passing entirely around each band. Also applied to books with flexible covers.

f reliure *f* flexible — **d** biegsamer Einband *m*; biegsame Bindung *f* — **n** slappe band *m* — **e** encuadernación *f* flexible; encuadernación *f* en material plástico — **i** legatura *f* flessibile

5823 flexible binding — Term sometimes used for *unsewn binding.

5824 flexible blade coating — See blade coating.

5825 flexographic ink; spirit ink; aniline ink *obs.* — Quick drying printing ink used on kraft paper, cotton fabric, cellophane, polyethylene etc. Originally solutions of coal-tar dyes in organic solvents (alcohols, esters, ketones, ethers), now with pigments rather than dyes and of two types, spirit inks, containing organic solvent as vehicle, and *emulsion inks, in which water is the main vehicle.

f encre *f* flexographique; encre *f* à l'alcool; encre *f* d'aniline *obs.* — **d** Flexographiefarbe *f*; Flexofarbe *f*; flexographische Farbe *f*; Gummidruckfarbe *f*; Anilindruckfarbe *f* *obs.*; Anilinfarbe *f* *obs.*; Teerfarbe *f* *obs.* — **n** flexografische inkt *m*; flexo-inkt *m*; flexografie-inkt *m*; aniline-inkt *m* *obs.* — **e** tinta *f* flexográfica; tinta *f* de flexografía; tinta *f* al alcohol; tinta *f* a la anilina *obs.*; tinta *f* de anilina *obs.* — **i** inchiostro *m* flessografico; inchiostro *m* all'anilina *obs.*

5826 flexographic press — See flexo press.

5827 flexographic printing — See flexography.

5828 flexography; flexographic printing; rubberplate printing; aniline printing *obs.*; **aniline rubber-plate printing** *obs.* — Method of rotary letterpress printing using flexible rubber plates and rapid drying fluid inks. The name originates from a decision made in 1952 by the American Packaging Institute.

f flexographie *f*; impression *f* flexographique; impression *f* à l'aniline *obs.* — **d** Flexographie *f*; Flexodruck *m*; Anilindruck *m* *obs.*; Anilingummidruck *m* *obs.* — **n** flexografie *f*; flexodruk *m*; anilinedruk *m* *obs.* — **e** flexografía *f*; impresión *f* con clisés flexibles; impresión *f* con clisés plásticos; impresión *f* a la anilina *obs.* — **i** flessografia *f*; stampa *f* flessografica; stampa *f* all'anilina *obs.*

5829 flexo press; flexographic press; aniline press *obs.* — Rotary printing press usually employing rubber stereos and aniline inks. Simple in construction comprising a duct roller, an inking roller, a printing cylinder and an impression cylinder. All cylinders except the steel impression roller are made from rubber. See also flexography.

f machine *f* d'impression flexographique; machine *f* à imprimer à l'aniline *obs.* — **d** Flexodruckmaschine *f*; Anilindruckmaschine *f* *obs.* — **n** flexopers *f*; anilinepers *f* *obs.* — **e** prensa *f* flexográfica — **i** macchina *f* flessografica; macchina *f* per la stampa all'anilina *obs.*

5830 flex tester — A *crumpling apparatus, more especially the Gelbo flex tester.

5831 flex tester — See crumpling apparatus.

5832 flimsy paper — Very thin paper such as tissues, manifolds, onion skin, etc.

f papier *m* mince — **d** dünnes Papier *n* — **n** dun papier *n* — **e** papel *m* delgado; papel *m* débil; papel *m* sutil — **i** carta *f* sottile

5833 flint glass — Heavy brilliant glass of high dispersion and relatively high index of refraction, composed of alkalis, lead oxide and silica, with or without other bases. Used for optical purposes.

f flint *m* — **d** Flintglas *n*; Kieselglas *n* — **n** flintglas *n* — **e** cristal *m* de roca; flintglas *m*; cristal *m* óptico; vidrio *m* de potasio y plomo — **i** vetro *m* flint

5834 flint glazing — Method of imparting a hard, brilliant polish to paper, especially to coated paper, by rubbing or rolling with a smooth stone or stone burnisher on a flint glazing machine.

f lissage *m* à la pierre; lissage *m* à l'agate — **d** Steinglättung *f* — **n** gladmaking *f* met de (agaat)-steen — **e** satinado *m* a la piedra (de ágata) — **i** lisciatura *f* a pietra

5835 flint paper — White or coloured coated paper on a thin body paper, with a hard burnished surface produced by a stone or flint burnisher travelling backwards and forwards across the coated surface as it leaves the calenders.

f papier *m* lissé une face — **d** Flintpapier *n*; Glanzpapier *n*; Glacépapier *n* — **n** glacépapier *n*; glacé *n* — **e** papel *m* charol — **i** carta *f* rasata

5836 flip-flop — Bi-stable circuit or device that can be in only one of two possible conditions at a time, remaining in one condition until forced by a electronic impulse to change to the other.

f bascule *f* — **d** Flipflop *m* — **n** flip-flop *m* — **e** sistema *m* flip-flop; sistema *m* de cambio brusco — **i** circuito *m* flip-flop

5837 floatation test — Test in which a small piece of paper is floated on the surface of a liquid; the time taken by the liquid to penetrate through determines the resistance of the paper to the penetration of liquid. Used to determine the resistance to penetration of water, writing ink, or oil.

f essai *m* de flottage — **d** Schwimmprobe *f* — **n** opdrijfproef *fm* — **e** ensayo *m* por flotación — **i** prova *f* di flottazione

5838 floating accent; loose accent *GB*; **piece accent** *US* — Separate diacritical mark produced on a separate piece of metal and placed over the type character to which it belongs. In photocomposition, most devices can set floating accents, by using the non-accented character (except for the dotless "i") and positioning the desired accent above or below it, either by keyboarding convention and/or by software subroutine.

f accent *m* séparé; accent *m* superposé — **d** Akzent *m* zum Übersetzen; Akzent *m* zum Ansetzen — **n** los accent *n*; los teken *n* — **e** acento *m* postizo; acento *m* separado — **i** accento *m* separato; accento *m* sovrapposto

5839 floating point — In mathematics, the system whereby numbers are handled as characteristic (exponent) and mantissa (decimal portion). It requires special hardware or software on most computers. Allows handling of very large and very small numbers efficiently in scientific calculations. See also fixed point.

f virgule *f* flottante — **d** Gleitkomma *f* — **n** drijvende komma *fmn* — **e** coma *f* flotante — **i** virgola *f* mobile

5840 floating position — See full position.

5841 floating roller — See expander roll.

5842 flocculation — In rheology, the clustering of suspended particles if there is a decrease in free energy of the system when particles collide.

f floculation *f* — **d** Ausflockung *f*; Ausflocken *n*; Flockenbildung *f*; Flockung *f* — **n** uitvlokking *f* — **e** floculación *f* — **i** flocculazione *f*

5843 flocculation tendency

f aptitude *f* à la floculation — **d** Ausflockbarkeit *f* — **n** neiging *f* tot uitvlokken; uitvlokneiging *f* — **e** aptitud *f* a la floculación — **i** tendenza *f* alla flocculazione

5844 flocking — The deposition (usually electrostatically) of short fibres or pile or other particles onto a design printed with an adhesive ink to which they adhere.

f flockage *m* — **d** Beflockung *f*; Beflocken *n* — **n** bevlokken *n* — **e** aterciopelado *m* — **i** decorazione *f* per flocculazione; flocculazione *f*

5845 flock paper — Cover paper sized, either over the whole surface or over special parts constituting the pattern only, and then powdered with specially dyed flock (powdered wool, cotton, or rayon). Originally used to imitate tapestry and Italian velvet brocades.

f papier *m* floqué; papier *m* suédé; papier *m* velours — **d** Samtpapier *n*; Velourspapier *n*; Plüschpapier *n* — **n** velourspapier *n* — **e** papel *m* aterciopelado; papel *m* paño; papel *m* de raso; papel *m* afelpado — **i** carta *f* vellutata

5846 flong *GB*; **stereotype dry mat**; **mat** *US*; **dry mat** *US*; **matrix**

f flan *m*; matrice *f* — **d** Mater *f* (vor dem Prägen); Stereomater *f*; Matrize *f* — **n** matrijs *fm*; flan *m* — **e** estereomatriz *f*; matriz *f* de estereotipia; matriz *f* estereotípica; matriz *f* de papel; matriz *f* (de cartón); flan *m* (estereotípico); mat *f Pe* — **i** flano *m*

5847 flong drying box — See matrix dryer.

5848 flong paper; **matrix board**; **matrix paper** — Pulplike coated board for casting stereotypes.

f carton *m* à matrice; carton *m* pour matrices; carton *m* pour flans (de clicherie) — **d** Maternkarton *m*; Maternpappe *f*; Matrizenpappe *f*; Stereotypiepappe *f* — **n** matrijzenkarton *n* — **e** cartón *m* (de) matriz; cartón *m* de estereotipia; cartón *m* para matrizar — **i** cartone *m* per flani

5849 flood *v* — See crowd *v*.

5850 flood coater — In screen process printing, a blade behind the squeegee. At the end of the printing stroke the squeegee rises and the flood coater moves into position to return the ink. Flood coating gives heavier ink deposits and prevents quick-drying inks from drying out on the screen during printing.

f contre-raclette *f* de ramassage — **d** Farbrechen *m*; Vorflutrakel *f* — **n** voorrakel *m* — **e** raspadora *f* de retorno — **i** racla *f* di ritorno

5851 flooding — Excess of ink on type or forme caused by the ink fountain being open too wide. See also crowd *v*.

f excès *m* d'encre — **d** Überfärben *n* — **n** teveel inkt geven *n*; overladen *n* met inkt — **e** sobrecargado *m* de tinta; excesiva *f* de tinta — **i** eccesso *m* d'inchiostro

5852 flooding — Separation of one pigment from the others on the surface of a printing ink.

f flottage *m* — **d** Ausschwimmen *n* — **n** opdrijven *n* — **i** affioramento *m*

5853 floor — Composing room without the composing machines.

5854 floor — See composing room.

5855 floor hand — See make-up hand.

5856 floor man — See compositor.

5857 floor man — See hand compositor.

5858 floor man — See make-up hand.

5859 floor space; **ground space**

f encombrement *m* — **d** Raumbedarf *m*; Bodenfläche *f* — **n** vloeroppervlak *n*; plaatsruimte *f* — **e** encumbramiento *m*; lugar *m* solar — **i** spazio *m*; ingombro *m*

5860 flop *v* — To reverse a film or flat to bring the other side out on top.

f tourner — **d** umkippen — **n** over de kop keren — **e** dar vuelta; girar — **i** rovesciare

5861 floppy disc; **diskette** — A method of storing data and providing access thereto on inexpensive, magnetically-coated plastic patters about the size and shape of a 45 rpm record. Access time is considerably slower than conventional moving head discs (about 1 sec maximum access time as compared to 1/10 sec) and the unit of data storage is much less (about 240,000 characters as compared to several million or more). But the floppy discs are cheap, and may be stored in job envelopes. Unlike the read/write heads on conventional discs, the read/write head on a floppy disc is in actual physical contact with the disc. Therefore floppy discs (which are removable and inexpensive) will wear out with repeated use.

5862 florets — See flowers.

5863 flourishes — Ornaments of curved lines or sweeping strokes, cast in brass, used in lines of type.

f traits *mpl* de plume; parafes *mpl* — **d** Federzüge *mpl* — **n** pennekrullen *mpl* — **e** plumadas *fpl*; rasgos *mpl* (de pluma); viñetas *fpl* de rasgos caprichosos — **i** ornamenti *mpl* a svolazzi e ghirigori

5864 flow — Property of an ink to level out as a true liquid. Inks of poor flow are classed as short or buttery in body, while inks of good flow are said to be long.

f écoulement *m* — **d** Fließen *n*; Fließeigenschaft *f* — **n** vloei *fm*; vloeiend vermogen *n* — **e** derramamiento *m*; fluido *m*; flujo *m* — **i** fluidità *f*; scorri-

mento *m*
5865 flow — See current.
5866 flow box — See head box.
5867 flow chart; flow diagram; flow sheet; process chart *US* — A step-by-step graphical representation of a logical procedure. A program flow is prepared by the system analyst (outline flow chart) and often rewritten in more detailed form by the programmer before coding (detailed flow chart). Flow charts consist of standard symbols corrected by flow lines indicating the direction of the flow supported by commentary and documentation. The standard symbols represent equipment, processes, decisions, or data.
f graphe *m* de fluence; ordinogramme *m*; organigramme *m* — **d** Betriebsfolgediagramm *n*; Datenflußplan *n*; Flußbild *n*; Programmablaufplan *n*; Flußdiagramm *n* — **n** blokschema *n*; informatiestroomschema *n*; stroomschema *n*; organigram *n*; werkinstructie *f* — **e** ordinograma *m*; organigrama *m* — **i** diagramma *m* di flusso; schema *m* di flusso
5868 flow curve — Graph of shearing force versus rate of shear. In the flow curve of a printing ink it is the slope of a tangent.
f courbe *f* d'écoulement — **d** Fließkurve *f* — **n** vloeikromme *fm*; vloeicurve *fm* — **e** curva *f* de flujo — **i** curva *f* di fluidità; curva *f* di scorrimento
5869 flow diagram — See flow chart.
5870 flowers; fleurons; florets — Printer's type ornament copied from designs used by the early bookbinders for ornamenting bindings.
f fleurons *mpl*; vignettes *fpl* (typographiques) — **d** Aldusblätter *npl*; Vignetten *fpl*; Verzierungen *fpl*; Röslein *npl* — **n** vignetjes *npl*; bloem-vignetjes *npl*; blad-ornamenten *npl* — **e** florones *mpl*; viñetas *fpl*; adornos *mpl*; ornamentos *mpl*; ornamentos *mpl* florales — **i** ornamenti *mpl* decorazioni; tipi *mpl* ornamentali
5871 flowers of sulphur — See sulphur.
5872 flower tissue paper — Intensively vat coloured, unsized, machine finished, woodfree tissue paper, used for making artificial flowers and for decoration.
f papier *m* soie pour fleurs artificielles — **d** Blumenseidenpapier *n* — **n** bloemenzijdepapier *n* — **e** papel *m* de seda para hacer flores — **i** carta *f* seta per fiori
5873 flow line production
f production *f* en ligne — **d** Fließproduktion *f*; Fließfertigung *f* — **n** lopende-band-produktie *f*; produktie *f* aan de lopende band — **e** línea *f* de fabricación; producción *f* por línea de fabricación; tren *m* de fabricación — **i** produzione *f* continua
5874 flow meter
f appareil *m* pour mesurer l'étalement; fluidmètre *m*; débitmètre *m* — **d** Fließprüfer *m*; Strömungsmesser *m*; Durchflußmeßgerät *n* — **n** vloeimeter *m* — **e** medidor *m* del caudal — **i** misuratore *m* di portata
5875 flow properties — See rheological properties.

5876 flow sheet — See flow chart.
5877 flubdub — American slang term for an insignificant stock *ornament.
5878 fluff — Dust from paper which gathers on the rollers of a printing press or on the rolls or doctors of a paper machine. See also paper dust, lint.
f poussière *f* (de papier); duvet *m*; peluche *f* — **d** Papierstaub *m*; Staub *m* — **n** papierstof *n*; stof *n* — **e** pelusa *f* del papel — **i** polvere *f* di carta; peluria *f* di carta; lanugine *f* di carta
5879 fluff free paper
f papier *m* non pelucheux — **d** nicht-staubendes Papier *n* — **n** niet-stuivend papier *n* — **e** papel *m* sin pelusa — **i** carta *f* senza polvere; carta *f* esente di polvere
5880 fluffing — Adhering of paper fibres on the non-printing areas of the blanket. See also linting.
f peluchage *m* — **d** Stauben *n* — **n** stuiven *n* — **e** repelón *m* — **i** spolvero *m*; spolverio *m*
5881 fluffing tendency
f tendance *f* au peluchage — **d** Staubungsneigung *f* — **n** neiging *f* tot stuiven — **e** tendencia *f* a la pelusa — **i** tendenza *f* allo spolverio
5882 fluffy paper
f papier *m* pelucheux — **d** staubendes Papier *n*; haariges Papier *n* — **n** stuivend papier *n*; harig papier *n* — **e** papel *m* polvoriente — **i** carta *f* che spolvera
5883 fluid drachm — See drachm.
5884 fluidics — Methods of machine control for hydraulic and pneumatic equipment, using small volumes of liquid or gas flowing through intricate channels to control the action of larger volumes in the functioning of the equipment. These channels in a solid component or circuit perform logic, computation, and control functions analogous to the flip-flop, and, nand, or, nor, and not gates, triggers, etc., in electronic computers and control devices. Printing applications are still in an exploratory stage. The word is a synthesis of "fluids" and "logic".
5885 fluidity — Rate of shear induced by a unit shearing strength; reciprocal of the coefficient of viscosity.
f fluidité *f* — **d** Fluidität *f*; Flüssigkeit *f*; Flüssigkeitszahl *f* — **n** vloeibaarheid *f* — **e** fluidez *f* — **i** fluidità *f*
5886 fluid ounce — Unit of apothecaries' fluid measure. Abbrev. fl oz (no points) or f℥ fl. See app. no. 7 and 7a.
f once *f* fluide — **d** Unze *f* — **n** ons *n* — **e** onza *f* — **i** oncia *f*
5887 fluid resin — See tall oil.
5888 fluorescence — Emission of visible light by some substances under the influence of ultraviolet radiation or visible light of short wavelength. See also luminescence, phosphorescence.
f fluorescence *f* — **d** Fluoreszenz *f*; Fluoreszieren *n* — **n** fluorescentie *f* — **e** fluorescencia *f* — **i** fluorescenza *f*

5889 fluorescence process; fluorographic process
— Process in which fluorescent water-colour pigments are used to simplify separation of water-colour art by photographic methods using special lights in the camera colour-separation work.
f procédé *m* de dessin de maquette aux couleurs fluorescentes — **d** Fluoreszenzverfahren *n* — **n** fluorescentieprocédé *n* — **e** procedimiento *m* de fluorescencia — **i** procedimento *m* fluorografico
5890 fluorescent — Capable to convert invisible light rays such as ultra-violet into visible light.
f fluorescent — **d** fluoreszierend — **n** fluorescerend — **e** fluorescente — **i** fluorescente
5891 fluorescent ink; day-glo ink — Ink which exhibits fluorescence. This effect can be very brilliant.
f encre *f* fluorescente — **d** Fluoreszenzfarbe *f*; fluoreszierende Farbe *f*; Leuchtfarbe *f*; Tagesleuchtfarbe *f* — **n** fluorescerende inkt *m* — **e** tinta *f* fluorescente — **i** inchiostro *m* fluorescente
5892 fluorescent lamp; fluorescent tube — Mercury vapour illuminant emitting a high degree of cool and visible light, the radiation equalling daylight in spectral quality or colour temperature. Sometimes used for illumination or light source in offset platemaking. The colour temperature ranges from 3500-4500 °K.
f lampe *f* fluorescente; tune *m* fluorescent — **d** Fluoreszenzlampe *f*; Leuchtstofflampe *f*; Leuchtröhre *f*; Leuchtstoffröhre *f* — **n** fluorescentielamp *fm*; TL-buis *fm*; fluorescentiebuis *fm*; gasontladingslamp *fm* — **e** lámpara *f* fluorescente; tubo *m* fluorescente — **i** lampada *f* fluorescente; tubo *m* fluorescente
5893 fluorescent paper — One-side coated paper, the coating of which contains materials which convert ultra-violet radiation into visible radiation.
f papier *m* fluorescent — **d** Leuchtfarbenpapier *n* — **n** fluorescerend papier *n* — **e** papel *m* fluorescente — **i** carta *f* fluorescente
5894 fluorescent tube — See fluorescent lamp.
5895 fluorescent whitening agent — See optical bleaching agent.
5896 fluorhydric acid — See hydrofluoric acid.
5897 fluorographic process — See fluorescence process.
5898 flush — Absence of indention. Flush lines of type begin at the left margin. "Flush right" indicates that type aligns at the right.
5899 flush blocking — In photoengraving, mounting a cut on a wooden block with a double-sided adhesive flush with the printing surface. This has replaced the older method of tacking, which required a margin of wood all around.
f montage *m* à vif; montage *m* à fleur — **d** Montieren *n* ohne Facette — **n** monteren *n* zonder facet — **e** montaje *m* a ras; montaje *m* sin bisel; montaje *m* sin chaflán — **i** montaggio *m* a filo
5900 flush colours — See flushed colours.
5901 flushed colours; flush colours — Pigments dispersed in oil, varnish, etc. The transfer from

the water phase to the oil phase is effected without the usual drying and grinding of the dye pigment.
f pigments *mpl* flushés — **d** geflushte Pigmente *npl*; geflushte Farbstoffe *mpl* — **n** geflushte pigmenten *npl* — **e** pigmentos *mpl* flush — **i** inchiostri *mpl* preparati col procedimento flushing
5902 flush edge — In photoengraving, a plate with no excess metal on one or more edges of the printing surface.
f bord *m* rogné à vif — **d** randlose Montage *f*; Flush-Montage *f* — **n** cliché *n* zonder facet — **e** orilla *f* del clisé montado a ras; sin márgenes in biseles — **i** senza margine; bordo *m* a filo
5903 flush head — Heading of which the line(s) begin at the extreme left or end at the extreme right of any measure; usually extreme left.
f sans enfoncement — **d** stumpfer Zeilenanfang *m* — **n** niet-ingesprongen regel *m* — **e** título *m* sin sangría; título *m* alineado — **i** titolo *m* non rientrante; titolo *m* allineato con le linee di testo
5904 flushing — Method of transferring dispersions of pigments in water to dispersions in oil by displacement, with compounds known as flushing agents.
f flushing *m*; broyage *m* à l'huile — **d** Flushing *m*; Flushing-Verfahren *n*; Fluschen *n* — **n** flushing-procédé *n* — **e** procedimiento *m* flushing — **i** procedimento *m* flushing
5905 flushing — See flush trimming.
5906 flush left — See quad left.
5907 flush left and right — See quad middle.
5908 flush paragraph — Paragraph in which the first line is not indented.
f paragraphe *m* carré; alinéa *m* à fleur de marge — **d** stumpf anfangender Absatz *m*; Alinea *n* ohne Einzug; Absatz *m* ohne Einzug — **n** niet-ingesprongen alinea *f*; niet-inspringende alinea *f* — **e** párrafo *m* sin sangría; párrafo *m* sin entrado — **i** capoverso *m* non rientrante; capoverso *m* senza a capo; capoverso *m* allineato con le linee di testo
5909 flush right — See quad right.
5910 flush trimming; flushing — In photoengraving, trimming unmounted plate flush with printing surface.
f rognage *m* à vif; rognage *m* à fleur — **d** bündig beschneiden — **n** afsnijden zonder facet — **e** cortar los lados al ras del tipo — **i** rifilatura *f* della lastra a filo
5911 flute — The wave-shaped paper layer of *corrugated fibreboard. There are the following kinds of flutes:
A-flute, approx. 118 flutes per metre (36 per foot) 5 mm high;
B-flute, approx. 164 flutes per metre (51 per foot) 3 mm high;
C-flute, approx. 138 flutes per metre (42 per foot) 4 mm high;
E-flute, approx. 300 flutes per metre (90 per foot) 4.75 mm high. The jumbo flute is 70 flutes per

metre and 9 mm high.

f cannelure *f* (A-cannelure, grande cannelure, grosse cannelure; B-cannelure, petite cannelure; C-cannelure, moyenne cannelure; E-cannelure, micro cannelure, cannelure fine; jumbo cannelure, macro cannelure) — **d** Welle *f*; Flute *f* (A-Welle, Grobwelle, B-Welle, Feinwelle; C-Welle, Mittelwelle; E-Welle, ultrafeine Welle, Mikrowelle; D-Welle, Übergrobwelle) — **n** golf *fm* (A-golf, grove golf; B-golf, fijne golf; C-golf, middengolf; E-golf, microgolf, minigolf; D-golf, Goliath-golf) — **e** ondulado *m* (ondulado A; ondulado B; ondulado C; ondulado E; ondulado D, ondulado *m* jumbo) — **i** onda *f* (onda A, onda grossa; onda B, onda fina; onda C, onda media; onda E, onda ultrafina; onda D, onda jumbo)

5912 flux; fluxing agent — Additive in melting metals to separate the impurities as slag.

f pâte *f* à nettoyer l'alliage; flux *m* — **d** Fluß-mittel *n*; Schmelzpaste *f* — **n** vloeimiddel *n* — **e** pasta *f* detergente para metal de fundición — **i** fondente *m*

5913 fluxing — Melting an ink and a coated film such as moisture-proof cellophane, usually with short burst of flame from a gas burner.

f flammage *m* — **d** Anschmelzen *n*; Einbrennen *n* — **n** aansmelten *n*; aansmelting *f* — **i** fusione *f*

5914 fluxing agent — See flux.

5915 fly; delivery fly; flyer; sheet flyer — Device on a power printing press, which receives the printed sheet from the cylinder and places it on the delivery board or table.

f raquette *f* (de sortie); abat-feuille *m*; raquettes *fpl* — **d** Bogenfänger *m*; Bogenauswerfer *m*; Stab-ausleger *m*; Ausleger(r)echen *m*; Rechen *m* — **n** waaier *m*; waaieruitleg *m* — **e** abanico *m*; saca-pliegos *m* de abanico — **i** levafoglio *m*

5916 fly — Part of the rotary press which delivers the copy from the folder to the delivery belts.

f moulinet *m*; araignée *f* — **d** Fächer *m*; Schaufel-rad *n* — **n** uitlegster *fm* — **e** salida *f* de palas; rueda *f* de palas — **i** uscita *f* a mulinello; disposi-tivo *m* con ruota a palette

5917 fly — See fly hand.

5918 fly board — See delivery table.

5919 fly boy — See fly hand.

5920 flyer — See fly.

5921 flyer laths — See flyer sticks.

5922 flyers — See handbills.

5923 flyer sticks; flyer laths

f verges *fpl* de raquette — **d** Auslegerstäbe *mpl* — **n** waaierlatten *fmpl* — **e** barillas *fpl* saca-pliegos — **i** stecche *fpl* del levafoglio

5924 fly hand; fly boy; fly — He who removes the printed product from the press or conveyor system of a rotary press.

f receveur *m* — **d** Abnehmer *m* — **n** afnemer *m* — **e** tomador *m* de pliegos — **i** operaio *m* addetto a levare

5925 flying — See ink misting.

5926 flying cam — Cam that reduces the shock in the operation of the fold blade in the folder of a rotary press.

f came *f* de commande de lame plieuse — **d** Falz-messer-Steuerkurve *f* — **n** vouwmesexcentriek *n* — **e** excéntrica *f* de la cuchilla plegadora; regulador *m* de la cuchilla plegadora — **i** camma *f* di comando della lama piegatrice; regolo *m* di piega

5927 flying paster; autopaster; automatic reel change; web paster — Device on rotary presses to attach a new web of paper to the tail end of the expiring reel without stopping the press.

f autocolleur *m*; dérouleur *m* à collage en marche; dérouleur *m* à collage automatique; changeur *m* de bobine en marche; système *m* de collage en marche; flying paster *m* — **d** Autopaster *m*; Selbstankleber *m* für fliegenden Rollenwechsel; automatische Anklebevorrichtung *f*; selbsttätiger Rollenwechsel *m* — **n** autopaster *m*; flying paster *m* — **e** empalmador *m* de la banda (de papel) a toda velocidad; cambio *m* automático de las bobinas — **i** incollatore *m* automatico; cambiatore *m* di bobine automatico; dispositivo *m* per le giunte al volo delle bobine; autopaster *m*

5928 flying spot scanner — In optical character recognition, a device using a moving spot of light to scan a sample space, the intensity of the trans-mitted or reflected light being sensed by a photo-electric transducer.

f explorateur *m* à tache mobile — **d** Abtaster *m* mit bewegendem Lichtfleck — **n** scanner *m* met bewegende lichtvlek; aftaster *m* met bewegende lichtvlek — **e** explorador *m* de punto móvil — **i** analizzatore *m* a punto mobile

5929 flying tuck; wheel and pinion folder — A gear driven continuously rotating folding blade mounted in the cylinder of the folding mechanism of a rotary press.

f commande *f* de lames engageantes — **d** Räder-falzmesser *n*; Tuckerfalzmesser *n* — **n** radervouw-mes *n* — **e** cuchilla *f* plegadora de ruedas — **i** lama *f* piegatrice a movimento cicloidale

5930 fly leaf; inner end-paper — Blank leaf at the beginning or end of a book. See also paste down.

f feuille *f* de garde — **d** fliegendes Vorsatzblatt *n*; inneres Vorsatzblatt *n*; fliegendes Blatt *n* — **n** vrij-hangend deel *n* van het schutblad — **e** hoja *f* de guarda exterior — **i** guardia *f*; sguardia *f*; ris-guardo *m*

5931 fly-leaf paper — See end-leaf paper.

5932 fly nut — See wing nut.

5933 fly press — Hand screw press with a weighted lever which can be swung horizontally, to make relief marks in paper.

f presse *f* à balancier; balancier *m* — **d** Balancier-presse *f* — **n** balanceerpers *fm* — **i** bilanciere *m* a mano

5934 fly title — See bastard title.

5935 flywheel — Heavy disc or wheel rotating on a shaft so that its momentum gives almost uniform rotational speed to the shaft and to all connected machinery.

f volant *m* — **d** Schwungrad *n*; Schwungscheibe *f*

— **n** vliegwiel *n* — **e** volante *m*; rueda *f* volante — **i** volano *m*

5936 flywheel shaft
f arbre *m* du volant — **d** Schwungradwelle *f* — **n** as *fm* van het vliegwiel; vliegwielas *fm* — **e** árbol *m* del volante — **i** albero *m* di volano

5937 f/numbers; focal ratio — The numerals or divisions on an iris diaphragm, indicating the diameter of the aperture in relation to the focal length of the lens, written following the symbol f, a smaller number indicating a larger lens diameter for a specific focal length and hence a smaller time of exposure. f/1.4 signifies that the focal length of the lens is 1.4 times as great as the diameter.
f nombres *mpl* du diaphragme — **d** Blendezahlen *fpl* — **n** diafragmagetallen *npl* — **e** números *mpl* del diafragma; valores *mpl* del diafragma; coeficientes *mpl* del diafragma — **i** numeri *mpl* del diaframma

5938 foam; froth; lather; suds — Foam is a mass of fine bubbles formed on liquids by agitation. Froth are stable bubbles produced at the air/liquid interface, due to agitation, aeration or ebullition. Lather is a foam caused when a detergent is agitated in water or other liquid. Suds is a foam generated on or in a detergent solution, especially on a soapy water. Foam and froth may be used interchangeably.
f écume *f*; mousse *f* — **d** Schaum *m* — **n** schuim *n* — **e** espuma *f* — **i** spuma *f*

5939 foaming
f moussage *m* — **d** Schäumen *n* — **n** schuimen *n* — **e** espumadura *f* — **i** formazione *f* di schiuma

5940 foam killer — Compound to reduce the foam in pulp suspensions and coating mixtures to prevent depressed spots and pinholes in the sheet.
f antimousse *m* — **d** Schaumbekämpfungsmittel *n* — **n** anti-schuimmiddel *n* — **e** antiespumante *m*

5941 foam marks — See foam spots.

5942 foam rubber — See sponge rubber.

5943 foam spots; foam marks; froth pits — Depressed spots and pinholes caused by foam bubbles in the paper stock on the paper machine wire.
f vésicules *fpl* de mousse — **d** Schaumblasen *fpl*; Schaumlöcher *npl* — **n** schuimblaasjes *npl*; patsen *mpl* (in papierfabriek) — **e** barbujitas *fpl* — **i** schiumini *mpl*

5944 focal distance — See focal length.

5945 focal length; focal distance — The distance of the principal focus of a lens from its optical centre; determines the size of the image at a given distance from the subject or original.
f distance *f* focale; focale *f*; longueur *f* focale; foyer *m depr.* — **d** Brennweite *f* — **n** brandpuntsafstand *m* — **e** distancia *f* focal; longitud *f* focal; longitud *f* de foco — **i** distanza *f* focale; lunghezza *f* focale

5946 focal plane — Plane surface on which the image transmitted by a lens is brought to sharpest focus; on process cameras, the position interchangeably occupied by the focusing screen and the photographic plate or film.
f plan *m* focal — **d** Brennebene *f* — **n** brandvlak *n* — **e** plano *m* focal; plano *m* de enfoque — **i** piano *m* focale

5947 focal plane shutter — Shutter in a camera in the form of a blind running in close proximity to the light-sensitive film.
f obturateur *m* à rideau — **d** Schlitzverschluß *m* — **n** gordijnsluiter *m*; spleetsluiter *m* — **e** obturador *m* de cortina — **i** otturatore *m* a tendina

5948 focal point — See focus.

5949 focal ratio — See f/numbers.

5950 focus; focal point — Point at which the rays of light transmitted by a lens converge to form a sharp image of the subject or original. *pl*: foci.
f foyer *m* — **d** Brennpunkt *m*; Fokus *m* — **n** brandpunt *n*; focus *n* — **e** foco *m*; punto *m* focal; enfoque *m* — **i** fuoco *m*; centro *m* focale

5951 focus — See bastard title.

5952 focusing — Bringing a camera image to correct size and proper sharpness.
f mise *f* au point (optique) — **d** Bildeinstellung *f*; Scharfeinstellung *f*; Einstellen *n* des Bildes — **n** instellen *n* (van de camera) — **e** enfocar; focar; poner en foco; hallar el foco; acomodar el foco — **i** messa *f* a fuoco; mettere a fuoco; focallizzare

5953 focusing glass — Small achromatic magnifying glass for ocular examination of camera images on focusing screens to promote sharpest possible focus.
f loupe *f* de mise au point — **d** Einstellupe *f* — **n** instelloep *fm*; instelloupe *fm* — **e** lupa *f* de enfoque; lente *fm* para enfocar; enfocador *m*; pantalla *f* para enfocar — **i** lente *f* per messa a fuoco

5954 focusing mount — See barrel.

5955 focusing scale; scale — Graduated scale on a process camera for bringing images to correct size without image measurement and ocular focusing.
f échelle *f* de mise au point; échelle *f* graduée de mise au point — **d** Einstellskala *f*; Scharfeinstellungs-Skala *f* — **n** instelschaal *fm* — **e** escala *f* de enfoque; escala *f* enfocadora; escala *f* para enfocar; escala *f* de distancia — **i** scala *f* per messa a fuoco; scala *f* a fuoco

5956 focusing screen; ground glass — Surface on which camera images are viewed and brought to correct size and sharpness. Usually consists of a sheet of ground glass, preferably with a square or circular patch of transparency in the centre to facilitate closer examination of line images and delicate detail.
f verre *m* dépoli; dépoli *m*; glace *f* dépolie; vitre *f* dépolie — **d** Mattscheibe *f*; Visierscheibe *f*; Einstellscheibe *f*; Glasscheibe *f* — **n** matschijf *fm*; matruit *fm*; matglas *n* — **e** vidrio *m* despulido; vidrio *m* esmerilado; vidrio *m* deslustrado — **i** schermo *m* smerigliato; schermo *m* per messa a fuoco

5957 fog; veil — Photographic defect in which the image is completely or locally veiled by a deposit of silver of varying density, due either to the action of extraneous actinic light or through improper chemical action. Can sometimes be removed by *clearing.
f voile *m* — **d** Schleier *m*; Ton *m*; Grauschleier *m* — **n** sluier *m* — **e** velo *m* — **i** velo *m*
5958 fog *v* — To get a hazy effect on a developed negative or positive, caused by light or other than that forming the image, by improper handling during development, or by the use of excessively old film.
f voiler — **d** schleiern — **n** sluieren — **e** velar — **i** velarsi
5959 fogged negative
f négatif *m* voilé — **d** geschleiertes Negativ *n* — **n** gesluierd negatief *n* — **e** negativo *m* velado — **i** negativo *m* velato
5960 fogging exposure — See flash exposure.
5961 foil — Thin metal membrane less than 0.15 mm (0.006 in) thick. When thicker than 0.15 mm it is called sheet. In many European countries the French "feuille" is used to describe any thin material, cellophane or plastic films as well as thin metals. See also aluminium foil.
f feuille *f* (métallique) — **d** Folie *f* — **n** fo(e)lie *fm* — **e** lámina *f* metálica; hoja *f* metálica — **i** foglio *m*; lamina *f*
5962 fold — Line at which a sheet of paper has been folded.
f pli *m* — **d** Falz *m*; Falte *f*; Bruch *m* — **n** vouw *fm* — **e** doblez *m*; pliegue *m* — **i** piega *f*
5963 fold *v* — To bend and crease a sheet of paper to form a folder, brochure, or signature of a book.
f plier — **d** falzen — **n** vouwen — **e** plegar; doblar — **i** piegare
5964 folded in quires — A quire is a section of printed leaves in proper sequence after folding.
f pliage *m* en portefeuilles — **d** Lagenfalzung *f* — **n** in katerns gevouwen — **e** doblado en cuadernos — **i** piegatura *f* a quinterni; piegatura *f* a intercalare
5965 folder; folder unit — Section of the rotary press where the webs are associated, folded, cut, delivered and sometimes stitched.
f plieuse *f* — **d** Falzwerk *n*; Falzaggregat *m*; Falzapparat *m*; Falzer *m*; Falzvorrichtung *f* — **n** vouwwerk *n*; vouwapparaat *n* — **e** doblador *m*; dobladora *f*; plegadora *f*; aparato *m* plegador; unidad *f* plegadora — **i** piegatore *m*; gruppo *m* di piega
5966 folder; folding machine — Machine for folding printed sheets of paper and book sections.
f machine *f* à plier; plieuse *f* — **d** Falzmaschine *f*; Falzapparat *m* — **n** vouwmachine *f*; vouwapparaat *n* — **e** plegadora *f* (mecánica); máquina *f* de plegar; máquina *f* plegadora; dobladora *f* Me; formadora *f* Cu — **i** piegatrice *f*; macchina *f* piegatrice
5967 folder — Small printed piece of paper that is

folded once or several times without being stitched or bound.
f dépliant *m*; prospectus *m* plié — **d** Faltblatt *n*; Faltprospekt *m*; Faltbroschüre *f* — **n** vouwblad *n*; folder *m* — **e** porfolio *m*; prospecto *m* plegado; tríptico *m* — **i** pieghevole *m* pubblicitario
5968 folder — See folding stick.
5969 folder blade — See folding blade.
5970 folder creases — Wrinkles or creases caused by carelessness on the folding machine or in hand folding.
f becs *mpl*; faux plis *mpl* à la pliure — **d** Falten *fpl* — **n** plooien *fmpl* — **e** arrugas *fpl* — **i** grinze *fpl* di piegatura
5971 folder unit — See folder.
5972 fold flattening machine
f machine *f* à écraser les plis — **d** Falzniederdruckmaschine *f* — **n** vouwneerdrukpers *f* — **e** prensa *f* de pliegos; prensa *f* de picado; máquina *f* prensadora de pliegos — **i** macchina *f* pressatrice delle pieghe
5973 folding blade; folder blade; tucker blade; tucking blade; tucker — Metal strip producing the fold by thrusting the web or webs into jaws or rollers in the folder of a rotary press.
f lame *f* plieuse — **d** Falzmesser *n* — **n** vouwmes *n* — **e** cuchilla *f* plegadora; regla *f* plegadora — **i** lama *f* piegatrice; coltello *m* di piega; stecca *f* per piegare
5974 folding boxboard — Usually vat lined board used in the manufacture of folding boxes, with special folding qualities.
f carton *m* pour boîte pliante — **d** Faltschachtelkarton *m* — **n** vouwdozenkarton *n* — **e** cartoncillo *m* para cajas plegables; cartón *m* para estuches *SA* — **i** cartone *m* per astucci pieghevoli
5975 folding carton *US*; **carton** *GB*; **folding (paper) box**; **collapsible folding box** *depr.* — Closed container made of bending grades of paperboard (0.4-1.15 mm or 0.016-0.045 in thick), plain or printed, cut and creased, in a variety of sizes and shapes, folded and delivered in flat, or glued and collapsed form by the maker and to be set up, filled and closed by the user. To package consumer-size units weighing not more than 4.5 kg (10 lbs). There are many types, e.g., tuck-end carton, seal-end carton, reverse tuck, straight tuck, two-piece, display carton. Distinct from set-up boxes and corrugated and solid fibre boxes. See also carton blank.
f boîte *f* pliante; carton *m* pliant — **d** Faltschachtel *f*; Pappschachtel *f*; Karton *m*; Aufrichtschachtel *f* — **n** vouwdoos *fm*; samenvouwbare doos *fm* — **e** caja *f* plegadiza; caja *f* de cartón semirrigida; caja *f* plegable — **i** astuccio *m* pieghevole; scatola *f* da montare
5976 folding carton machine; carton gluing machine; carton gluer — Machine to form, glue or stitch, prepared carton blanks into containers or parts thereof.
f machine *f* à coller des boîtes pliantes; plieuse-encolleuse *f* — **d** Faltschachtelklebemaschine *f* —

n vouwdozenplakmachine *f*; dozenplakmachine *f* — **e** máquina *f* de cajas plegadizas; máquina *f* para plegar cajas plegables; máquina *f* para encolar cajas plegables — **i** macchina *f* incollatrice per astucci pieghevoli

5977 folding chase; twin chase — Pair of chases used instead of one large one to lock up large book formes.

f châssis *m* à feuillure — **d** Doppelschließrahmen *m* — **n** dubbel raam *n*; dubbel insluitraam *n*; dubbel vormraam *n* — **e** rama *f* dúplex

5978 folding cylinder — Cylinder which holds the folder blade or jaw for making a fold in rotary printing.

f cylindre *m* plieur — **d** Falzzylinder *m* — **n** vouwcilinder *m* — **e** cilindro *m* plegador; cilindro *m* doblador — **i** cilindro *m* di piega; cilindro *m* piegatore

5979 folding endurance; folding strength; double fold number — The number of folds under specified conditions in a specified instrument which a specimen will withstand before failure. Tested by subjecting a specimen under tension repeatedly to double folds through a wide angle.

f résistance *f* au pliage; résistance *f* au pli; nombre *m* doubles plis; double-pli *m* — **d** Falzfestigkeit *f*; Falzwiderstand *m*; Faltfestigkeit *f*; Doppelfalzung *f*; Falzzahl *f*; Doppelfalzzahl *f* — **n** vouwsterkte *f*; vouwweerstand *m*; weerstand *m* tegen vouwen; vouwgetal *n*; dubbelvouwgetal *n* — **e** resistencia *f* al plegado; resistencia *f* al doblado — **i** resistenza *f* alla piegatura; resistenza *f* alla doppie pieghe

5980 folding-endurance tester; fold tester — Laboratory folding machine (Schopper; Köhler-Molin, etc.), to determine the *folding endurance of paper.

f pliagraphe *m*; appareil *m* pour l'essai de résistance au pliage — **d** Falzfestigkeitsprüfer *m*; Doppelfalzer *m*; Falzgerät *n*; Falzer *m* — **n** dubbelvouwtoestel *n* — **e** ensayador *m* de plegado; ensayador *m* de resistencia al doblez; ensayador *m* de doblez — **i** apparecchio *m* per prova di resistenza alle doppie pieghe

5981 folding jaws — Gripping section on a nip and tuck folder.

f mâchoires *fpl* — **d** Falzklappen *fpl* — **n** vouwkleppen *fmpl* — **e** quijadas *fpl* plegaderas — **i** battenti *mpl* di piega; ganasce *fpl* di piega

5982 folding machine — See folder.

5983 folding (paper) box — See folding carton.

5984 folding pins — See needles.

5985 folding quality — See pliability.

5986 folding rollers — Pair of serrated rollers on a rotary web press immediately below the folding cylinder, which receives the paper as thrust down by the folding blade, giving the sheets their last fold before passing them on to the delivery straps or fly. See also nipping rollers.

f rouleaux *mpl* plieurs — **d** Falzwalzen *fpl* — **n** vouwrollen *fmpl* — **e** rodillos *mpl* plegadores — **i** rulli *mpl* piegatori

5987 folding stick; bookbinder's bone; bone folder; bone; folder — Piece of polished, white bone or wood, about 15 cm (6 in) long and rounded on both ends and on its sides, to fold sheets of paper by hand, providing firm edges without damaging the paper; also to apply pressure on waxed type to make this adhere to paste-up.

f plioir *m* — **d** Falzbein *n* — **n** vouwbeen *n* — **e** plegadora *f* — **i** piegacarta *m*; stecca *f* (piegacarta)

5988 folding strength — See folding endurance.

5989 fold marks — Marks added to a flat for printing along the margins of a press sheet, as in book signature, to guide the folding operations after printing.

f repères *mpl* de pliage — **d** Falzmarken *fpl* — **n** vouwtekens *npl*; vouwmerken *npl* — **e** signos *mpl* de plegado — **i** marche *fpl* di piegatura; linee *fpl* di piegatura

5990 fold-out — See throw-out.

5991 fold tester — See folding-endurance tester.

5992 fold wrapping

f emballage *m* pliant — **d** Falteinschlag *m* — **n** vouwverpakking *f* — **e** cubierta *f* plegadizada; envoltura *f* plegadizada — **i** avvolgimento *m* ripiegato

5993 foliation; pagination — Allotting of folio numbers to pages. See also bifoliate *v*.

f foliotage *m*; pagination *f* — **d** Foliieren *n*; Paginierung *f*; Foliierung *f* — **n** paginering *f*; pagineren *n* — **e** paginación *f*; numeración *f* de las páginas — **i** numerazione *f* delle pagine

5994 folio — Book, etc., composed of sheets folded once, thus having two leaves or four pages to the sheet.

f in-folio *m* — **d** Folioformat *n* (einmal gefalzt) — **n** folioformaat *n*; in tweeën (eenmaal gevouwen) — **e** en folio — **i** in folio

5995 folio; page number — Number of a page at top or bottom (for preparatory matter lower case Roman numerals), centred or flush left on left pages and flush right on right pages, often with the *running headline. See also drop folio.

f numéro *m*; folio *m*; nombre *m* de page — **d** Seitenzahl *f*; Seitennummer *f*; Nummer *f* — **n** paginacijfer *n*; paginanummer *n* — **e** folio *m*; número *m* de la página — **i** numero *m* di pagina

5996 folio; folio post — A size of writing paper. See app. no. 3.

5997 folio *v* — To number the pages of a book. See also bifoliate *v*.

f folioter; paginer; numéroter — **d** paginieren — **n** pagineren; nummeren — **e** paginar; foliar; numerar — **i** numerare

5998 folio note — A size of note paper. See app. no. 3.

5999 folio post — See folio.

6000 follow copy — Order to set-up the matter exactly as it appears on copy, with no changes whatever.

f ne rien changer; ne varietur; aller chou pour

chou — **d** Zeile auf Zeile (setzen); Männchen auf Männchen (setzen) — **n** zetten volgens de kopij — **e** iojo!; sígase; componer progresivo — **i** comporre riga per riga; comporre linea a linea; attenersi esattamente al manoscritto

6001 font — See fount.

6002 font case — See fount case.

6003 food package

f emballage *m* de produits alimentaires — **d** Lebensmittel(ver)packung *f*; Nahrungsmittel-(ver)packung *f* — **n** verpakking *f* voor levensmiddelen — **e** envase *m* de alimentos — **i** imballaggi *mpl* alimentari; confezione *m* di prodotto alimentare

6004 foolscap — A British size of writing and printing papers. Formerly the paper showed a watermark of a jester's head with cap and bells. Abbrev. cap. See app. no. 3.

6005 foolscap and half — See foolscap 1 1/2 sheet.

6006 foolscap and third — See foolscap 1 1/3 sheet.

6007 foolscap folio — A size of writing paper. See app. no. 3.

6008 foolscap long folio — A size of writing paper. See app. no. 3.

6009 foolscap 1 1/2 sheet; foolscap and half — A British standard size for writing and printing papers. See app. no. 3.

6010 foolscap 1 1/3 sheet; foolscap and third — A British standard size for writing and printing papers. See app. no. 3.

6011 foolscap 4to — A size of writing paper. See app. no. 3.

6012 foot; tail — Space at the bottom of a book page.

f blanc *m* de pied — **d** Fußsteg *m* — **n** staartwit *n* — **e** blanco *m* de pie — **i** bianco *m* al piede

6013 foot — Bottom part of a type opposite to the face. *pl*: feet. Due to the *groove the letter is said to stand on its feet.

f talon *m*; pied *m* — **d** Fuß *m* — **n** voet *m* — **e** pie *m*; base *f*; culata *f*; pata *f* — **i** piede *m*; pie *m*

6014 foot — Unit of length. *pl* feet. Abbrev. ft (no point). See app. no. 7 and 7a.

f pie *m* (anglais) — **d** (englischer) Fuß *m* — **n** (Engelse) voet *m* — **e** pie *m* (inglès) — **i** piede *m* (inglese)

6015 footboard — See feeder's platform.

6016 foot-candle — Unit of illuminance or luminous flux density when the foot is taken as the unit length. For colour appraisal and matching the minimum of illumination should be about 200 foot-candles.

f bougie-pied *f* — **d** Footcandle *f* — **n** footcandle *f* — **e** bujía-pie *f*; bujía-metro *f* — **i** footcandle *f*

6017 foot-lambert — Unit of luminance to $1/\pi$ candle per sq ft, or to the uniform luminance of a perfectly diffusing surface emitting or reflecting light at the rate of one lumen per sq ft. 1 ft la = 10.76391 asb.

6018 foot line; bottom line — Bottom line of a page, especially the line with the folio, or the line that is blank except for signature of sheet.

f ligne *f* de pied; ligne *f* de bas de page — **d** Fußzeile *f* — **n** onderste regel *m*; voetregel *m* — **e** línea *f* del pie — **i** linea *f* al piede

6019 foot margin — See tail margin.

6020 footnote callout reference — Symbol such as an asterisk, a superior letter or number which indicates that an additional comment or a citation (as of the source of the statement) will be found at the end of the column, page, chapter or article. The presence of such references poses a problem for page make-up, especially if the text of the footnote must also be located on the same page.

6021 foot notes; bottom notes — Notes explanatory to the text in smaller type at the foot of pages. See also marks of reference.

f notes *fpl* en bas de page; notes *fpl* de bas de page — **d** Fußnoten *fpl* — **n** voetnoten *fmpl*; noten *fmpl* onder aan de pagina; noten *fmpl* aan de voet van de pagina — **e** notas *fpl* al pie (de la página); notas *fpl* explicativas — **i** note *fpl* a piede di pagina; note *fpl* a pie di pagina

6022 foot of page

f bas *m* de page — **d** unterer Rand *m* — **n** voet *m* van de pagina — **e** pie *m* de página — **i** pie *m* di pagina; piede *m* della pagina

6023 foot rule

f filet *m* de pied — **d** Fußlinie *f* — **n** voetlijn *fm* — **e** raya *f* de pie; filete *m* de pie — **i** filetto *m* al piede

6024 foot stick — See also sticks.

f blanc *m* de pied — **d** Fußsteg *m* — **n** staartwit *n* — **e** regleta *f* de pie; imposición *f* de pie — **i** marginatura *f* al piede

6025 foot stick — See sticks.

6026 forced drying

f séchage *m* accéléré — **d** beschleunigte Trocknung *f* — **n** versnelde droging *f*; geforceerde droging *f* — **e** secado *m* acelerado; secado *m* forzado — **i** essiccamento *m* forzato

6027 forced lubrication

f graissage *m* sous pression; graissage *m* à pression (forcé) — **d** Druckschmierung *f*; Hochdruckschmierung *f* — **n** druksmering *f*; smering *f* onder druk — **e** lubricación *f* a presión; engrase *m* forzado — **i** lubrificazione *f* forzata

6028 forced motion

f mouvement *m* commandé — **d** Zwangsbewegung *f*; zwangsläufige Bewegung *f*; zwangsweise Bewegung *f* — **n** gedwongen beweging *f* — **e** movimiento *m* forzado — **i** movimento *m* forzato; movimento *m* coatto

6029 force per unit area — See shearing stress.

6030 Ford cup — Cup-type viscometer, consisting of a reservoir and orifice (about 10 diameters long at most), with or without a temperature-control jacket and a receiving flask.

f coupe *m* Ford; Ford coupe *f* consistométrique — **d** Fordbecher *m*; Ford-Auslaufbecher *m*; Ford-topf *m* — **n** Ford-cup *m*; Ford-beker *m* — **e** copa *f* de Ford; probeta *f* Ford — **i** coppetta *f* Ford; bicchierino *m* Ford

6031 foredge *GB*; **fore edge** *US* — Edge of a book opposite to the back. (Remark: a hyphen is used in non-technical uses.)

f gouttière *f*; tranche *f* extérieure — **d** Vorderschnitt *m* — **n** voorsnee *fm*; voorsnede *fm* — **e** corte *m* delantero; canto *m* delantero; delantera *f*

6032 foredge bolt — See bolts.

6033 foredge margin; outer margin — Blank space at the outside of a printed book page, opposite to the folds of the sections.

f marge *f* extérieure — **d** äußerer Papierrand *m*; äußerer Seitenrand *m* — **n** zijmarge *fm*; voormarge *fm* — **e** margen *fm* exterior; margen *fm* lateral (exterior); margen *fm* externo — **i** margine *m* di taglio

6034 fore edge — See foredge.

6035 foreground — Area between the camera and the subject on a photograph.

f premier plan *m* — **d** Vordergrund *m* — **n** voorgrond *m* — **e** primer plano *m* — **i** primo piano *m*

6036 foreground — In computer systems capable to process more than one task simultaneously, jobs performed in the foreground are those which receive more computer attention, either because the response time is more critical or because they deserve higher priority. Background programs are executed when there is not enough call upon the computer's processing capability to occupy it exclusively on foreground tasks, as is frequently the case since foreground programs are usually I/O bound rather than compute bound.

6037 foreign language composition

f composition *f* en langues étrangères — **d** Fremdsprachensatz *m*; fremdsprachlicher Satz *m* — **n** zetten *n* van vreemde talen — **e** composición *f* de lenguas extranjeras — **i** composizione *f* in lingue stranieri

6038 foreign types; exotic types — To this group of modern type design classification belong all the types not of Roman origin, Greek, Cyrillic, Hebrew, Arabic, etc. See app. no. 2.

f caractères *mpl* étrangers — **d** fremde Schriften *fpl* — **n** vreemde letters *fpl* — **e** tipos *mpl* extranjeros — **i** caratteri *mpl* stranieri

6039 forel; forrel *depr.*; **forril** *depr.* — Parchment of poor quality made from split sheepskins, used in its natural colour for making book covers.

f parchemin *m* de peau de mouton — **d** Pergament *n* aus Schaffell — **n** perkament *n* uit schapevel — **e** pergamino *m* de piel de carnero — **i** pergamena *f* di pelle di montone

6040 forel — Case made of *forel formerly used to hold books or manuscripts.

f étui *m* (de parchemin) — **d** Buchhülle *f*; Schuber *m* — **n** (perkamenten) foedraal *n* — **e** estuche *m* (de pergamino) — **i** coperta *f* (di pergamena)

6041 foreman — Man in charge of a department in a printing plant.

f prote *m*; contre-maître *m* — **d** Faktor *m*; Faks *m coll.*; Abteilungsleiter *m* — **n** chef *m* — **e** capataz *m* — **i** capo *m* reparto; proto *m*

6042 foreman printer — See press-room overseer.

6043 foreword — Introductory remarks on the purpose, aim or content of a book, usually written by a person other than the author. Sometimes called *preface but foreword was substituted for it when restoration of native terms was in vogue. See also introduction.

f avant-propos *m*; préface *f*; avertissement *m* (préface de peu d'étendue) — **d** Vorwort *n*; Geleitwort *n*; zum Geleit — **n** voorwoord *n*; woord *n* vooraf; ten geleide — **e** prefacio *m* — **i** prefazione *f*

6044 fork-lift truck — See fork truck.

6045 fork truck; fork-lift truck — Electrically driven car with two parallel horizontal arms to lift, displace and transport loaded pallets.

f chariot *m* élévateur (à fourche) — **d** Gabelhubwagen *m*; Gabelstapler *m* — **n** vorkhefwagen *m*; vorkheftruck *m*; vorktruck *m* — **e** apiladora *f* de horquilla — **i** carrello *m* elevatore a forca; carrello *m* elevatore a forcella

6046 form; blank form; blank — Printed form or document with blanks to be filled in. See also questionary.

f formule *f*; formulaire *m* — **d** Formular *n*; Vordruck *m* — **n** formulier *n* — **e** formulario *m* (en blanco); forma *f* (en blanco); fórmula *f*; modelo *m* (en blanco); esqueleto *m* *Me*; machote *m* *Me* — **i** formulario *m*; modulo *m*

6047 form — See forme.

6048 formaldehyde; formic aldehyde; formol; formalith; oxymethylene; methanal — Colourless gas, soluble in water, alcohol and ether. One of the chemicals in silver spray solutions used for metallic coating of electrotype moulds and in certain plastics for rubber and plastic platemaking. Usually handled as an aqueous solution, with or without methanol. TL-value: 2 ppm or 3 mg/m^3.
Formula: HCHO.

f formaldéhyde *f*; aldéhyde *m* formique; méthanal *m* — **d** Formaldehyd *m*; Formylhydrat *n*; Methylaldehyd *m*; Ameisensäurealdehyd *m* — **n** formaldehyd(e) *n*; methanal *n* — **e** formaldehído *m*; aldehído *m* fórmico — **i** formaldeide *f*; aldeide *f* formica

6049 formaldehyde resin

f résine *f* de formaldéhyde — **d** Formaldehydharz *n* — **n** formaldehydhars *mn* — **e** resina *f* de formaldehído — **i** resina *f* di formaldeide

6050 formalin; formol — Clear, colourless, 40% aqueous solution of *formaldehyde. Formerly a trade mark.

f formaline *f*; formol *m* — **d** Formalin *n*; Formol *n* — **n** formaline *fm*; formol *mn* — **e** formalina *f*; formol *m* — **i** formalina *f*; formol *m*

6051 formalith — See formaldehyde.

6052 format — Size and shape of a page. In a broader meaning, the size, shape, style and general appearance of a book or similar work, including the quality of paper, style of type, and kind of binding. Formerly, the size and proportion

of a book as determined by the number of times the sheets were folded, as folio, quarto, octavo, etc.
f format *m* — **d** Format *n*; Buchformat *n* — **n** formaat *n*; boekformaat *n* — **e** formato *m*; tamaño *m* — **i** formato *m*

6053 format — Use of relatively complex command codes or strings of such codes to achieve specific typographical formatting effects. To assure that these effects are created reliably, when used repetitively, they may be stored, usually in random access storage, at the keyboard, or, in a stand-alone typesetter, during one of the composing passes. One of the criteria for evaluating the capability of a stand-alone typesetter is the amount of core or other storage area available for formats, and the efficiency with which they may be stored.
f disposition *f*; modèle *m* — **d** Form *f*; Format *n* — **n** indeling *f*; vorm *m* — **e** formato *m* — **i** formato *m*

6054 format *v* — To store in a computer predefined styles and repetitive data pertaining to a particular job. The formats may be called by a computer program upon the recognition of a unique pattern or by an operator typing a character or command.
f disposer — **d** einteilen — **n** indelen — **e** formar

6055 formation — Disposition and distribution of the fibres in a sheet of paper, observed by looking through the sheet. See also well-closed formation, cloudy formation, look-through.
f formation *f* de la feuille — **d** Blattbildung *f* — **n** vervilting *f* — **e** formación *f* del papel; estructura *f* del papel; trama *f* del papel; afieltrado *m* — **i** formazione *f* del foglio; formazione *f* della carta

6056 forme; type forme *GB*; **form; type form** *US* — Type matter or type and block with its accompanying spacing material secured in the forme called a chase.
f forme *f* (d'impression) — **d** Form *f*; Druckform *f*; Satzform *f* — **n** vorm *m*; drukvorm *m*; lettervorm *m* — **e** forma *f* de imprenta; forma *f* tipográfica; molde *m* de imprenta; molde *m* de impresión; molde *m* tipográfico — **i** forma *f* di stampa; forma *f* stampante

6057 forme bed — See bed.

6058 forme board — See letter board.

6059 forme carriage — See forme trolley.

6060 forme cutter — See cutting forme.

6061 forme cylinder — See engraved cylinder.

6062 forme proof — Proof pulled after imposition of the forme.
f épreuve *f* en seconde; tierce *f* — **d** Presseabzug *m* — **n** persproef *fm*; proef *fm* van de pers — **e** prueba *f* de lanzado — **i** prova *f* di macchina

6063 former; kite *coll.* — Triangular device on a folder, slanting at approx. 55 ° from the horizontal, point downward, over which the web travels to be folded in half longitudinally before entering the jaw folder. A roller at the top keeps the web smooth, a rounded nose at the point cushions the web, and tiny air jets at the edges and nose

reduce web friction. May be equipped with pasting mechanism.
f triangle *m*; cône *m* plieur *depr.*; cône *m* de premier pli *depr.*; cône *m* triangulaire *depr.*; plieur *m* triangulaire *depr.*; entonnoir *m* de pliage *depr.*; entonnoir *m* de pliure *depr.*; cornet *m* depr. — **d** Falztrichter *m*; Trichter *m*; Falzkegel *m* — **n** vouwtrechter *m*; trechter *m* — **e** embudo *m*; embudo *m* plegador; horma *f* triangular; cono *m* triangular; triángulo *m*; triángulo *m* formador; formador *m*; recodo *m* — **i** imbuto *m* piegatore; cono *m* piegatore

6064 former — Ornamentation surrounding an initial. See also factotum initial.

6065 former — Enlarged model of a letter or character from which the punch is made.
f patron *m*; modèle *m* — **d** Modell *n* — **n** model *n* — **e** patrón *m* — **i** modello *m*

6066 former — See matrix dryer.

6067 forme rack; chase rack — Rack consisting of two sets of grooves to house formes, sometimes beneath imposing stones.
f râtelier *m* à formes; râtelier *m* à châssis; meuble *m* pour formes serrées; porte-châssis *m* — **d** Form(en)regal *n* (für geschlossene Formen) — **n** vormenrek *n* — **e** portarramas *m*; armario *m* portarramas; armario *m* para ramas — **i** rastrelliera *f* per forme; bancone *m* per forme

6068 former angle — Angle of the *former relevant to the bending rollers.
f angle *m* du triangle — **d** Trichterwinkel *m* — **n** trechterhoek *m* — **e** ángulo *m* de triángulo — **i** angolo *m* del cono piegatore

6069 former carriage — Adjustable frame on which the *former is mounted.
f bâti *m* du triangle — **d** Trichteraufhängung *f* — **n** trechterwagen *m*; trechterkar *fm* — **e** bastidor *m* del triángulo — **i** telaio *m* del cono piegatore

6070 former fold; newspaper fold — Long fold made by the former folder as the web passes over the former.
f pli *m* longitudinal; pli *m* (au) cornet — **d** Trichterfalz *m*; Längsfalz *m* — **n** trechtervouw *fm*; langsvouw *fm* — **e** pliego *m* longitudinal — **i** piega *f* al cono; prima piega *f* longitudinale

6071 former folder — Folder which uses a former to fold the web in half, in the direction of the travel, before it enters the jaw folder.
f plieuse *f* en entonnoir — **d** Trichterfalzapparat *n* — **n** trechtervouwapparaat *n* — **e** dobladora *f* triangular — **i** piegatrice *f* a cono

6072 former nose; nose of former — Lowest part, sometimes spring loaded, of the triangular *former over which the paper runs to give the fold.
f bec *m* de cornet; bec *m* d'entonnoir — **d** Trichternase *f*; Trichterspitze *f* — **n** trechterneus *m* — **e** nariz *f* del embudo; nariz *f* del triángulo; nariz *f* del formador; nariz *f* de la horma; cabo *m* del embudo — **i** naso *m* del piegatore; naso *m* del cono di piega

6073 forme rollers — Ink rollers which contact

and supply ink to the printing forme in a letter-press machine or a rotary press. Deprecated in offset printing.

f rouleaux *mpl* toucheurs; toucheurs *mpl*; rouleaux *mpl* encreurs — **d** Auftragwalzen *fpl*; Farbauftragwalzen *fpl*; Formwalzen *fpl* — **n** opdraagrollen *fpl*; letterrollen *fpl*; vormrollen *fpl* — **e** rodillos *mpl* dadores; rodillos *mpl* entintadores; rodillos *mpl* de entintación — **i** inchiostratori *mpl*; rulli *mpl* inchiostratori della forma

6074 former sections — Detachable sections of the *former.

6075 forme trolley; forme truck; forme carriage — Small wheeled device to move type formes in vertical position.

f chariot *m* porte-formes; chariot *m* de transport de formes — **d** Formentransportwagen *m*; Formentransporttisch *m* — **n** vormenwagen *m* — **e** carretilla *f* portaformas; carretilla *f* para formas; carro *m* portaformas; mesa *f* portátil para formas — **i** carrello *m* portaforma; carrello *m* trasportatore

6076 forme truck — See forme trolley.

6077 forme wear

f usure *f* de la forme — **d** Formenverschleiß *m*; Abnutzung *f* der Druckform — **n** vormslijtage *f*; slijtage *f* van de drukvorm — **e** desgaste *m* de la forma — **i** usura *f* della forma

6078 formic acid; hydrogen carboxylic acid; methanoic acid — Colourless, fuming liquid, soluble in water, alcohol and ether. Latin: acidum formicum.
Formula: HCOOH.

f acide *m* formique — **d** Ameisensäure *f*; Hydrokarbonsäure *f*; Methansäure *f* — **n** mierenzuur *n* — **e** ácido *m* fórmico — **i** acido *m* formico

6079 formic aldehyde — See formaldehyde.

6080 forming rollers — See bending rollers.

6081 formol — See formaldehyde.

6082 formol — See formalin.

6083 formula — In chemistry: an expression of the constituents of a compound by symbols and figures. In mathematics: a rule or principle frequently expressed in algebraic symbols.

f formule *f* — **d** Formel *f* — **n** formule *fm* — **e** fórmula *f* — **i** formula *f*

6084 formulate *v*

f formuler; donner une formule — **d** formulieren — **n** formuleren — **e** formular — **i** formulare

6085 forrel — See forel.

6086 forril — See forel.

6087 fortnightly — See semi-monthly.

6088 fortran — Acronym for formula translator, an international machine-independent procedure-oriented, high-level language, introduced by IBM. Most computer manufacturers have produced fortran compilers, although not all are implemented identically. Statements are written in explicit English, algebraic and arithmetic notation. Intended as a language for scientific application but has been adopted by some commercial users. Codification of fortran IV has been undertaken by the American Standards Institution.

f fortran *m* — **d** Fortran *m* — **n** fortran *n* — **e** fortrán *m* — **i** fortran *m*

6089 fortuitous distortion — See jitter.

6090 forty-eightmo — Book size with 96 pages to the sheet. Abbrev. 48mo (no point).

f in-quarante-huit — **d** Achtundvierzigerformat *n* — **n** in achtenveertigen — **e** en cuaranta-y-ochoavo — **i** in ventiquattresimo

6091 forwarding — In bookbinding, the processes between folding the sheets and casing in, such as rounding and backing, putting on headbands, reinforcing backs, etc.

f façonnage *m* — **d** buchbinderische Weiterverarbeitung *f*; Fertigmacherei *f* — **n** verdere werkzaamheden *fpl*; verdere verwerking *f* — **e** obras *fpl* de la encuadernación después la costura — **i** finitura *f*; allestimento *m*

6092 forwarding sucker — Rubber suction device on a printing press or a folding machine, which leads the sheets of paper to the feed board.

f aspirateur *m* — **d** Sauger *m* — **n** zuiger *m* — **e** aspirador *m* adelantador — **i** ventosa *f* aspirante; aspiratore *m* a ventosa

6093 forward reading

f lecture *f* avant — **d** lektorieren — **n** voorlezing *f* — **e** prelectura *f* — **i** lettura *f* anticipata

6094 fossil flour — See diatomaceous earth.

6095 fossil meal — See diatomaceous earth.

6096 fotomat — Brass matrix in a typesetting machine, bearing a photographic negative (type design) for making photo-type composition.

f fotomat *f*; filmatrice *f*; matrice *f* fotografique — **d** Photomatrize *f* — **n** fotomatrijs *f* — **e** matriz *f* fotográfica; matriz *f* de fotocomposición — **i** fotomatrice *f*; fotomat *f*

6097 Fougeadoire — See rubber reduction.

6098 foul case — Type case in which the types have been distributed into the wrong compartments, causing faulty hand composition.

f casse *f* mélangée — **d** verfischter Setzkasten *m*; Setzkasten *m* mit falsch abgelegter Schrift — **n** vuile kast *fm* — **e** caja *f* empastelada — **i** cassa *f* sporca

6099 founders' type — See foundry type.

6100 foundry — See type foundry.

6101 foundryman — See stereotyper.

6102 foundry type; founders' type — Type produced by manufacturers for use in typography set by hand, usually reserved for display material or other small jobs.

f caractère *m* de composition à la main; lettre *f* de casse — **d** Handsatztype *f* — **n** handletter *fm*; kastletter *fm* — **e** tipo *m* de fundición; tipo *m* suelto; tipo *m* de caja; tipo *m* a mano; tipo *m* de mano — **i** carattere *m* per composizione a mano

6103 fount; type fount *GB*; **font; type font** *US* — Complete assortment of characters, numbers, points, etc., for one size and type face design as used in the composition of text matter. For hot-metal typesetting a separate fount is required for each size of the same style, whereas in phototypesetting by means of photographic enlargement or reduction, one master fount may be used for different sizes. See also set of matrices.

f fonte *f*; assortiment *m* complet (d'un certain œil); police *f* (de caractère) — **d** ganzer Satz *m*; Schriftsatz *m*; Schnitt *m*; Schriftgrad *m* (alle Figuren eines Grades) — **n** font *n*; compleet stel *n* — **e** fundición *f* (de tipos); póliza *f* de tipos; fuente *f* (de tipos) *Cu, Me*; juego *m* de tipos *Me* — **i** assortimento *m* completo (di caratteri)

6104 fountain blade — See duct blade.

6105 fountain blade adjusting screws — See duct adjusting screws.

6106 fountain block — See fountain divider.

6107 fountain clump — See fountain divider.

6108 fountain divider; fountain clump; clump; fountain block — Piece of lead to separate the ink duct.

f plomb *m* d'encrier — **d** Farbklotz *m* (im Farbkasten); Farbbrocken *m* — **n** inktlood *n* — **e** plomo *m* del tintero; divisor *m* del tintero — **i** blocco *m* del calamaio; separatore *m* del calamaio

6109 fountain, ink — See ink duct.

6110 fountain keys — See duct adjusting screws.

6111 fountain roller — See duct roller.

6112 fountain screws — See duct adjusting screws.

6113 fountain solution — See damping solution.

6114 fountain splitting

f division *f* de l'encrier — **d** Farbkastenteilung *f* — **n** inktbakverdeling *f*; verdeling *f* van de inktbak — **e** división *f* del tintero — **i** divisione *f* del calamaio

6115 fountain stops — Devices (e.g., squeegees), applied to the surface of a fountain roller, to limit the amount of moisture conveyed.

f mangeurs *mpl* d'eau — **d** Wasserabstreifer *mpl*; Wasserblöcke *mpl* — **n** waterafscheiders *mpl* — **e** separadores *mpl* de agua — **i** arresti *mpl*

6116 fountain, water — See water fountain.

6117 fount case *GB*; **font case** *US*; **sorts case** — Case to hold superfluous sorts which cannot be contained in the ordinary cases; used as a reserve.

f casse *f* de réserve(s); casse *f* d'assortiments; bardeau *m* — **d** Reservekasten *m*; Ausraffkasten *m*; Ausraffekasten *m* — **n** reservekast *fm* — **e** caja *f* de suertes sobrantes — **i** cassa *f* di riserva

6118 fount combination — See fount scheme.

6119 fount list — See fount scheme.

6120 fount scheme; fount list; fount synopsis; fount combination; bill of fount; bill of type — Plan or ratio by which founts of type or matrices are made up to provide the correct proportion of each letter or character, as ascertained by past experience or calculation as to probable requirements.

f police *f* — **d** Gießzettel *m* — **n** letterpolis *fm*; gietpolis *fm* — **e** póliza *f* (de tipos); lista *f* de tipos; esquema *m* de la póliza de tipos — **i** polizza *f* (di caratteri)

6121 fount synopsis — See fount scheme.

6122 four-colour machine — See four-colour press.

6123 four-colour press; four-colour machine

f presse *f* à quatre couleurs; machine *f* à quatre couleurs — **d** Vierfarbenmaschine *f* — **n** vierkleurenpers *f* — **e** máquina *f* de cuatro colores; prensa *f* de cuatro colores — **i** macchina *f* a quattro colori

6124 four-colour process; quadricolour process; quadrichromie — Extension of three-colour reproduction in which a fourth or black plate is used with those for the primary colours.

f impression *f* en quatre couleurs; quadrichromie *f* — **d** Vierfarbendruck *m* — **n** vierkleurendruk *m*; vierkleurenprocédé *n* — **e** cuatricromía *f*; citocromía *f*; tetracromía *f*; policromía *f* — **i** stampa *f* a quattro colori; quadricromia *f*

6125 four-cylinder perfecting press — Press with four printing cylinders to print two colours on each side of the sheet or one colour on one side and three colours on the other.

f presse *f* recto-verso à quatre couleurs — **d** Vierfarben-Schön- und Widerdruckmaschine *f* — **n** vierkleuren schoon- en weerdrukpers *f* — **e** máquina *f* de tiro y retiro de cuatro cilindros; prensa *f* de tiro y retiro de cuatro cilindros — **i** macchina *f* per la stampa in bianca e volta a quattro colori

6126 Fourdrinier bristol — Bristol including index bristols and sometimes mill bristols. The stock may be rag or chemical wood pulps in varying proportions. Used principally for index, record, business and commercial cards, social announcements, invitations, specialties, etc.

f bristol *m* Fourdrinier; carton *m* bristol à la machine Fourdrinier — **d** Naturkarton *m* — **n** natuurkarton *n* — **e** cartulina *f* bristol fabricada en la máquina Fourdrinier — **i** cartoncino *m* bristol fabbricato su macchina a tavola piana

6127 Fourdrinier machine; Fourdrinier paper machine — Paper machine, named after the Fourdrinier brothers, who financed the development of the machine invented by Nicolas Louis Robert, a Frenchman, in 1798. It consists of a moving and endless web of wire cloth on which the sheet of paper is formed from the pulp suspension. The continuous web sheet is then pressed and dried. The wire part and the presses are known as the *wet end of the machine.

f machine *f* à papier Fourdrinier; machine *f* à papier à table plate; passoire *f* longue — **d** Langsiebmaschine *f*; Langsiebpapiermaschine *f* — **n** langzeefmachine *f*; Fourdrinier-machine *f* — **e** máquina *f* Fourdrinier; máquina *f* plana; máquina *f* de tamiz largo — **i** macchina *f* a tavola piana

6128 Fourdrinier paper machine — See Fourdrinier machine.

6129 Fourdrinier wire — See machine wire.

6130 four-em quad; 4-em quad

f quatre-douzes — **d** Quadrat *n* auf vier Gevierte — **n** kwadraat *n* op vier vierkanten; stuk *n* van vier — **e** cuadrado *m* de cuatro emes — **i** spazio *m* a quattro quadratoni

6131 four-line brevier — Old name for 32-point type. See app. no. 1.

6132 four-line pica — Old name for 48-point type. See app. no. 1.

6133 Fournier system — The Fournier type-body system (of 1764) is based on the Parish foot and one point equal 0.34718 mm (1/864 foot or 0.0137 in). This system is practically out of use. The French names and their sizes are as follows: perle 4 points (1.391 mm); parisienne 5 points (1.739 mm); nonpareille 6 points (2.087 mm); mignonne 7 points (2.435 mm); petit-texte 8 points (2.783 mm); gaillarde 9 points (3.131 mm); petit-romain 10 points (3.479 mm); philosophie 11 points (3.827 mm); cicéro 12 points (4.174 mm); Saint-Augustin 14 points (4.870 mm); gros-texte 16 points (5.566 mm); gros-romain 18 points (6.262 mm); petit-paragon 20 points (6.940 mm); gros-paragon 22 points (7.645 mm); palestine 24 points (8.349 mm); petit canon 28 points (9.740 mm); trismégiste 36 points (12.523 mm); double canon 56 points (19.479 mm); grosse-nonpareille 96 points (33.396 mm). See app. no. 1.

6134 Fourth Estate — Name given to the newspaper press, the newspaper publishers and writers collectively.

6135 fourth-generation typesetter — Photographic typesetting machine in which a laser beam is used.
f composeuse *f* de quatrième génération — **d** Setzmaschine *f* der vierten Generation — **n** zetmachine *f* van de vierde generatie — **e** fotocomponedora *f* de cuarta generación — **i** compositrice *f* di quarto generazione

6136 four-to-em space — See middle space.

6137 four-to-pica — See app. no. 1.
f corps *m* de trois; trois points *mpl* — **d** Viertelcicero *f*; drei Punkte *mpl* — **n** corps *n* drie; drie punten *mpl* — **e** cuerpo *m* tres; cuerpo *m* de tres puntos — **i** corpo *m* tre; tre punti *mpl*

6138 four-way powdering — In the dragon's blood process of relief etching, brushing of etching powder in four different directions on the plate, thereby protecting all sides of lines and dots. Process out of use.
f poudrage *m* exécuté sur les quatre côtés — **d** Vierweg-Pinselverfahren *n*; Anstaubverfahren *n* — **n** drakenbloedprocédé *n* — **e** espolvoreado *m* en cuatro direcciones — **i** impolveratura *f* in quattro direzioni

6139 foxed leaves — Spotted or stained leaves of paper, found in old books, as the result of mildew or other organic agents. The spots are called foxing. See also mildew spots.
f feuilles *fpl* rousselées; taches *fpl* de mouillure(s); taches *fpl* de moisissure; piqûres *mpl* — **d** Meltaufflecken *mpl*; Stockflecken *mpl*; Schimmelflecken *mpl* — **n** vellen *npl* papier met vochtvlekken — **e** hojas *fpl* rosáceas; hojas *fpl* con manchas de humedad — **i** fogli *mpl* con macchie d'umidità; fogli *mpl* con macchie di muffa; fogli *mpl* macchiati; fogli *mpl* danneggiati

6140 foxing — See foxed leaves.

6141 fraction — In the decimal system a comma is used, but in GB no naught is placed at the begin of a fraction, only a point, for example: .572.
f fraction *f*; nombre *m* fractionnaire — **d** Bruch *m*; Bruchziffer *f* — **n** breuk *fm*; breukcijfer *n* — **e** fracción *f*; quebrado *m*; número *m* quebrado; número *m* fraccionario — **i** frazione *f*

6142 fraction case
f casse *f* à fractions; casse *f* à nombres rompus — **d** Bruchzifferkasten *m* — **n** breukkast *fm*; breukcijferkast *fm* — **e** caja *f* de cifras quebradas — **i** cassa *f* per frazioni

6143 fraction mark
f signe *m* de fraction — **d** Bruchzeichen *n* — **n** breukteken *n* — **e** signo *m* de fracción — **i** segno *m* di frazione

6144 frame — Metal structure on a printing press or a machine to hold the cylinder bearings or other fitments.
f bâti *m* — **d** Maschinengestell *n*; Gestell *n* — **n** frame *n* — **e** armadura *f*; bancada *f*; armazón *f* — **i** intelaiatura *f*; incastellatura *f*

6145 frame — Area one recording position long extending across the width of a magnetic or paper tape perpendicular to its movement.
f cadre *m*; trame *f* — **d** Bandsprosse *f*; Rahmen *m* — **n** raster *n* — **e** cuadro *m*; trama *f* — **i** quadro *m*

6146 frame — Wooden frame to protect paper in transit.
f cadre *m* — **d** Lattenkiste *f* — **n** krat *n* — **e** jaula *f* — **i** quadro *m*; telaio *m*

6147 frame — See composing frame.

6148 franker — See franking machine.

6149 franking machine; franker; postage-meter machine; postage meter
f machine *f* à affranchir — **d** Frankiermaschine *f*; Freistempelmaschine *f*; Freistempler *m* — **n** frankeermachine *f* — **e** máquina *f* de franquear; franqueadora *f*; estampilladora *f* — **i** macchina *f* per affrancare

6150 Fraunhofer's lines; spectrum lines — Series of fixed transverse dark lines intersecting the solar spectrum, identifying visible spectral colours according to their wavelength.
f raies *fpl* de Fraunhofer; raies *fpl* noires du spectre — **d** Fraunhofer Linien *fpl* — **n** Fraunhoferse lijnen *fmpl*; lijnen *fmpl* van het spectrum — **e** rayas *fpl* de Fraunhofer; líneas *fpl* negras del espectro — **i** linee *fpl* di Fraunhofer

6151 fray *v* — To make raveled threads at the end of the cords of a book, to fasten them to the book cover.
f effilocher; ouvrir les ficelles — **d** aufschaben; auffächern — **n** uitrafelen — **e** deshilachar los clavos; deshilachar los cordeles — **i** sfilacciare

6152 frayed slip — Frayed ends of the cords projecting from the back of a sewed but uncovered book.
f bout *m* de ficelle effiloché — **d** aufgeschabenes Bundende *n* — **n** uitgerafeld touweinde *n* — **e**

mecha *f* deshilada — **i** estremità *f* sfiaccilata

6153 free — Said of paper not containing ground-wood or mechanical pulp.

f maigre — **d** rösch — **n** droog — **e** magro — **i** magro

6154 free acid

f acide *m* libre — **d** freie Säure *f* — **n** vrij zuur *n* — **e** ácido *m* libre — **i** acido *m* libero

6155 free beating; fast beating — Treatment of paper pulp in which the fibres are not fibrillated.

f raffinage *m* maigre — **d** rösche Mahlung *f*; riesche Mahlung *f* — **n** droge maling *f*; riese maling *f* — **e** refino *m* magro — **i** raffinazione *f* magra

6156 free copy — Copy of a book given by the publisher or author for publicity etc. See also complimentary copy.

f exemplaire *m* gratuit — **d** Freiexemplar *n* — **n** gratis exemplaar *n* — **e** ejemplar *m* gratuito; ejemplar *m* suministrado gratuitamente; ejemplar *m* no venal; ejemplar *m* de regalo; ejemplar *m* fuera de venta — **i** esemplare *m* gratuito; copia *f* gratuita

6157 freedom of opinion

f liberté *f* d'opinion — **d** Meinungsfreiheit *f* — **n** vrijheid *f* van meningsuiting; vrije meningsuiting *f* — **e** libertad *f* de expresión; libertad *f* de opinión; libertad *f* de comentario; libertad *f* de opinar — **i** libertà *f* di opinione

6158 freedom of the press; liberty of the press; press freedom — Right to publish newspapers, magazines, and other printed matter without governmental restriction.

f liberté *f* de la presse — **d** Pressefreiheit *f* — **n** persvrijheid *f*; vrijheid *f* van drukpers — **e** libertad *f* de (la) prensa; libertad *f* de (la) imprenta — **i** libertà *f* di stampa

6159 free motion — See clearance.

6160 freeness tester — Apparatus to measure the degrees of hydration or refining of a pulp stock, by determining how fast water drains from it.

f essayeur *m* de raffinage; essayeur *m* pour raffinage — **d** Mahlungsgradprüfgerät *n*; Mahl-gradprüfer *m* — **n** maalgraad-bepalingstoestel *n* — **e** ensayador *m* del grado de refino; aparato *m* para determinar el grado de refino; aparato *m* para ensayo de refino — **i** raffinometro *m*

6161 freeness value — Proportion of dilution water recovered from a sample of paper stock by drainage measured by a standard method.

f indice *m* d'égouttage — **d** Mahlgrad *m*; Mahlungsgrad *m* — **n** malingsgraad *m*; maalgraad *m* — **e** grado *m* de refino — **i** indice *m* di scolantezza

6162 free press

f presse *f* libre — **d** freie Presse *f* — **n** vrije pers *f*; onafhankelijke pers *f* — **e** prensa *f* libre; prensa *f* independiente — **i** stampa *f* libera

6163 free silver — Solution of silver nitrate, as e.g., on wet collodion plates.

f argent *m* libre — **d** freies Silber *n* — **n** vrij zilver *n* — **e** plata *f* soluble — **i** argento *m* libero

6164 free stock; fast pulp — Pulp suspension from which the water drains freely. Papermaker's term.

f pâte *f* maigre — **d** rösch gemahlener Stoff *m*; röscher Stoff *m* — **n** droog gemalen stof *fm* — **e** pasta *f* magra; pulpa *f* magra — **i** pasta *f* magra; pasta *f* poco refinada

6165 free water — See white water.

6166 freeze *v* — To change from a liquid substance to a solid form.

f (se) congeler; geler — **d** erstarren; gefrieren; frieren — **n** stollen — **e** congelar — **i** rassegare; rassegarsi; congelarsi

6167 French alphabet — Same as English. The acute accent (´) is used only over e: when two e's come together the first always has acute accent, as née; adjectives of nationality, the first personal pronoun, months and days of week, do not take initial cap, as anglais, je, mars, lundi; cedilla c (ç) only used before a, o, and u; circumflex accent (â) is used over any vowel; diaeresis as in English; grave accent (à) used over a and e; the diphthongs æ and œ not to be separated (except in writing).

f alphabet *m* français — **d** französisches Alphabet *n* — **n** Franse alfabet *n* — **e** alfabeto *m* francés — **i** alfabeto *m* francese

6168 French chalk — Soft, finely ground variety of steatite, basically hydrated magnesium silicate. See also talc.

6169 French curve — Instrument to draw continuous sweeps and bowls.

f courbe *f*; pistolet *m* (de dessinateur) — **d** Kurvenlineal *n* — **n** tekenmal *m*; mal *m* — **e** plantilla *f* de curvas (múltiples) — **i** curvalinee *m*

6170 French door — See door.

6171 French fold(er) — Folder with an advertising message or announcement, printed on one side of a sheet folded once horizontally and once vertically. Thus, while there are eight pages in the folder, only the four outside ones (1, 4, 5, 8) are printed.

f dépliant *m* français; pli *m* à la française; pli *m* français — **d** Respectform *f* — **n** Amerikaanse vouwbrief *m*; Amerikaanse verzendbrief *m* — **e** pliego *m* a la francesa — **i** pieghevole *m* alla francese

6172 French folio — Thin, smooth finished paper for typewriter copies, light circulars, proofing, overlays and underlays. Thicker than tissue paper.

f papier *m* pelure; folio *m* français; manifold *m* — **d** Seidenpapier *n* — **n** pelure *fm* — **e** papel *m* delgado; papel *m* de seda — **i** carta *f* velina francese

6173 French furniture — Metal furniture in the smaller sizes sometimes also called quotations. Large hollow quads with an exact point measurement. See also quotation quads.

f garnitures *fpl* en plomb (creuses) — **d** Hohl-würfel *mpl* — **n** holwit *n* — **e** cuadrados *mpl* huecos — **i** quadrati *mpl* vuoti

6174 French groove; French joint; grooved joint; sunk joint — Space between the backing shoulder and the inside edge of the board used in the cover of a book, the depth and the width of the groove varying with the book dimension. See also American groove, groove.

6175 French height to paper — See height to paper.

6176 French horsepower — See metric horsepower.

6177 French joint — See French groove.

6178 French lid — In boxmaking, a lid with extension edge and with sides and ends of lesser height than sides and ends of base, and fitting outside of sides and ends of base.

6179 french morocco — Low-grade goatskin, sheepskin or cowhide with small grain, used in bookbinding. (Do not use capital letters.)
f maroquin *m* chagrin; chagrin *m* — **n** Frans marokijn *n* — **e** marroquín *m* frances — **i** marocchino *m* francese

6180 French nick — See nick at the back.

6181 French quotes — See quotation mark.

6182 French rule; swelled rule; diamond dash; Bodoni dash; tapered dash; swell dash — Rule of brass or type metal, designed as a diamond in the middle of a straight line.
f filet *m* couillard; filet *m* anglais — **d** englische Linie *f*; Zierlinie *f*; verzierte Trennlinie *f* — **n** Engelse lijn *fm*; fileet *m* — **e** bigote *m*; pleca *f* — **i** filetto *m* inglese

6183 French sewing; plain sewing — Sewing either by hand or machine which does not include the attaching of a movable tape or cord.
f brochure *f* à points; cousure *f* en brochure — **d** Holländern *n* — **n** hollanderen *n* — **e** cosido *m* a la francesa — **i** cucitura *f* a punto corrente

6184 French sewn
f à points de brochure — **d** geholländert (mit Buchfadenheftmaschine) — **n** gehollanderd — **e** cosido a la francesa — **i** cucito a punti correnti

6185 French transfer paper — See transfer paper.

6186 frequency — In statistics, the number of observations with a value between two specified limits.
f fréquence *f* — **d** Häufigkeit *f*; Frequenz *f* — **n** frequentie *f* — **e** frecuencia *f* — **i** frequenza *f*

6187 frequency distribution — In statistics, the relation between the magnitude of an observed variable characteristic and the frequency of its occurrence, expressed in tabular form by grouping the observed data according to the magnitude of the variable characteristic, and represented graphically as a histogram or frequency-curve or as a mathematical function.
f distribution *f* de fréquence — **d** Häufigkeits-verteilung *f* — **n** frequentieverdeling *f* — **e** distribución *f* de frecuencia — **i** distribuzione *f* di frequenza

6188 frequency modulation — Impressing a signal on a radio carrier wave by varying its frequency (distinguished from amplitude modulation). In frequency-modulation technics the frequency of an electric current signal represents the information.
f modulation *f* de fréquence — **d** Frequenz-

modulation *f* — **n** frequentiemodulatie *f* — **e** modulación *f* de frecuencia — **i** modulazione *f* di frequenza

6189 freshly printed
f fraîchement imprimé — **d** frisch gedruckt; frisch bedruckt — **n** vers gedrukt — **e** recién impreso — **i** stampa fresca

6190 fret — Ornament or ornamental work (Greek or Japanese), consisting of small straight bars intersecting one another in right or oblique angles, used as decoration of book pages and book covers since it can be made up from type. See also Greek fret.

6191 fretted — Said of rollers when cut (or fretted) by formes of sharp new type or brass rule.
f éraillé; coupé; éraflé — **d** zerschnitten; abgenutzt — **n** ingesneden — **e** arrugado — **i** frastagliati

6192 friar — Unprinted area caused by a defect in the forme roller of a letterpress machine.
f feinte *f*; faiblesse *f*; faible *m*; manque *m* d'encrage — **d** Mönch *m*; Mönchschlag *m*; Mönchs-bogen *m*; blinder Bogen *m* — **n** vel *n* met kale partijen; kale plek *fm* — **e** fraile *m*; fallo *m* — **i** frate *m*

6193 friction calender; friction glazing calender — Calender consisting of three rolls: the bottom and top roll of chilled iron, bored to admit steam, the intermediate roll of cotton. These rolls are so geared that the smaller has the higher peripheral speed. It has a burnishing action on the paper.
f calandre *f* frictionneuse; calandre *f* à friction — **d** Glanzkalander *m*; Friktionskalander *m* — **n** frictiekalander *m*; glanskalander *m* — **e** calandria *f* a fricción; satinadora *f* — **i** calandra *f* a frizione

6194 friction clutch; friction gearing — Clutch in which connection is made through sliding friction, as in a slug composing machine.
f embrayage *m* à friction; friction *f* — **d** Friktion *f*; Friktionskupplung *f*; Friktions-getriebe *n* — **n** frictiekoppeling *f*; frictie *f* — **e** embrague *m* de fricción; acoplamiento *m* a fricción — **i** accoppiamento *m* di frizione; frizione *f*

6195 friction feeder
f margeur *m* à friction — **d** Reibanleger *m* — **n** strijkinlegapparaat *n* — **i** mettifoglio *m* per attrito

6196 friction gearing — See friction clutch.

6197 friction glazed paper — Paper with a very high finish, secured by passing the sheet through chilled iron rolls revolving at different peripheral speeds, e.g., coated box-lining paper, waterproof paper, bronzed and silver paper, etc. See also flint glazing.
f papier *m* glacé par friction; papier *m* glacé à la calandre frictionneuse; papier *m* glacé sur calandre à friction — **d** friktionsgeglättetes Papier *n*; Hochglanzpapier *n* — **n** gefrictioneerd papier *n*; hooggeglansd papier *n* — **e** papel *m* de alto brillo; papel *m* charol — **i** carta *f* lisciata alla calandra a frizione

6198 friction glazing calender — See friction

calender.

6199 frilling — Wrinkling and separation of photographic images on plates or films during development, fixing or washing.

d Runzelung *f*; Runzelbildung *f* — **n** reticulatie *f*; oprimpelen *n* — **e** despegamiento *m* — **i** raggrinzamento *m*

6200 fringe, dot — See dot fringe.

6201 fringe effect — See adjacency effect.

6202 fringe effect — See border effect.

6203 fringe effect — See mackie lines.

6204 frisket — Thin iron frame jointed to the tympan of a hand-press to prevent the sheet of paper being dirtied or blackened.

f frisquette *f*; braie *f* — **d** Einlagedeckel *m*; Rähmchen *n* (an der Handpresse) — **n** verschet *n*; frisket *n* — **e** frasqueta *f*; bastidor *m* — **i** fraschetta *f*

6205 frisket — Paper mask or stencil placed over certain areas of photographs during air-brushing.

f cache *m* — **d** Schablone *f*; Papiermaske *f* — **n** schabloon *n* — **e** máscara *f*; mascarilla *f* — **i** maschera *f* di carta; sagoma *f* protettiva di carta

6206 froid, à — See blind tooling.

6207 front cover; recto; flap — Outside front cover of a book.

f plat *m* de devant; plat *m* avant; plat *m* recto — **d** Einbandvorderseite *f*; Umschlagvorderseite *f*; Vorderdeckel *m* — **n** voorplat *n*; eerste pagina *fm* van het omslag — **e** delante *m*; carátula *f*; primera página *f* de cubierta — **i** piatto *m* anteriore

6208 front delivery — Delivery of sheets from a two-revolution printing press with the printed side up, to facilitate easy pile delivery, to increase the drying time, and to enable the pressman to inspect the sheets.

f sortie *f* frontale; réception *f* frontale; réception *f* à l'avant — **d** Frontbogenauslage *f*; Frontbogenausleger *m*; Frontbogenausgang *m* — **n** frontuitleg *m* — **e** salida *f* frontal (de las hojas) — **i** uscita *f* frontale (dei fogli)

6209 front edge — See leading edge.

6210 front guide — See front lay.

6211 frontispiece — Illustration or picture, often a tip-in plate, facing the title page in a book.

f frontispiece *m* — **d** Titelbild *n*; Frontispiz *n* — **n** titelplaat *fm*; frontispice *n* — **e** contraportada *f*; frontis *m*; frontispicio *m obs.* — **i** frontespizio *m*

6212 front lay; front guide; gripper guide; feed guide — Collapsible or otherwise movable stop (usually only a pair) against which the leading edge of the sheet is presented.

f taquet *m* de front; taquet *m* de pince — **d** Vordermarke *f*; Vorderanschlag *m* — **n** vooraanleg *m* — **e** guía *f* de frente; guía *f* delantera — **i** squadra *f* frontale

6213 front margin

f marge *f* de tête — **d** vorderster Papierrand *m* — **n** voormarge *f* — **e** margen *m* delantero; margen *m* exterior — **i** margine *m*

6214 front matter — See preliminary matter.

6215 front page — First page of a daily newspaper, etc.

f première page *f* — **d** Vorderseite *f* — **n** voorpagina *f* — **e** página-portada *f*; portada *f*; página *f* frontal; primera página *f*; plana *f* primera — **i** prima pagina *f*

6216 front page — See back page.

6217 frosted glass; ground glass — Frosted glass plates used as transparent table tops for the examination and retouching of negatives and positives.

f verre *m* dépoli — **d** Mattglas *n* — **n** matglas *n*; matruit *fm depr.* — **e** vidrio *m* despulido; vidrio *m* esmerilado; vidrio *m* mate — **i** vetro *m* smerigliato

6218 frosting — See crazed drying.

6219 froth — The constant agitation of paper making stock tends to generate foam, which may be the cause of serious troubles on the paper machine. See foam.

6220 froth — See foam.

6221 froth pits — See foam spots.

6222 frottoir — French name for a wooden stick with a concave end to clean glue off the backs of books. No English equivalent.

6223 fudge — Part of a newspaper reserved for extra late news, sometimes known as *stop press.

f dernières nouvelles *fpl* — **d** letzte Meldung *f*; letzte Nachrichten *fpl* — **n** laatste nieuws *n*; laatste berichten *npl* — **e** alcance *m*; últimas noticias *fpl* — **i** attualità *f*

6224 fudge box — Device on a rotary newspaper press to insert extra news or to print a small portion of matter in colour.

f dispositif *m* de dépêches de dernière heure; fudge *m*; unité *f* d'impression complémentaire — **d** Depeschen-Eindruckwerk *n*; Eindruckwerk *n* für letzte Nachrichten — **n** indrukwerk *n* voor het laatste nieuws — **e** dispositivo *m* de impresión de últimas noticias — **i** dispositivo *m* per la stampa delle ultime notizie

6225 fudge cylinder — Cylinder in the newspaper printing press holding the late news, fudge boxes or plates.

f cylindre *m* fudge — **d** Depeschen-Druckzylinder *m* — **n** indrukcilinder *m* — **e** cilindro *m* de cambio — **i** cilindro *m* per le ultime notizie

6226 fudge plate — Printing plate to fit on, or be cast in the fudge box or to fit direct on the fudge cylinder.

f cliché *m* pour fudge; stéréo *m* de dernière heure — **d** Depeschen-Klischee *n* — **n** plaat *fm* met het laatste nieuws; styp *m* met het laatste nieuws — **e** plancha *f* de cambio para las noticias de última hora — **i** lastra *f* per la stampa delle ultime notizie

6227 fudge shaft — Shaft on which the fudge cylinder or head is mounted.

f arbre *m* (de) fudge — **d** Formzylinderachse *f* — **n** as *fm* van de indrukcilinder — **e** árbol *m* del cilindro de cambio — **i** albero *m* del cilindro per le ultime notizie

6228 fudge unit — Unit on the newspaper press comprised of a framework for carrying the fudge ink duct and fudge shaft(s).
f groupe *m* fudge — **d** Depeschen-Druckwerk *n* — **n** indrukwerk *n* voor het laatste nieuws — **e** grupo *m* de cambio para las últimas noticias — **i** gruppo *m* stampante delle ultime notizie
6229 fugitive — Having poor stability, or, of a dye or pigment, very poor light-fastness.
f fugitif — **d** nicht beständig; flüchtig; unbeständig — **n** onbestendig; niet sterk; niet houdbaar — **e** versátil; fugitiva — **i** a bassa resistenza; labile
6230 fugitive colour — Colour changing or disappearing rapidly when exposed to light, heat, moisture or other weather conditions.
f couleur *f* instable; couleur *f* non-résistante; couleur *f* fugitive — **d** unbeständige Farbe *f*; unechte Farbe *f* — **n** onbestendige kleur *fm*; zwakke kleur *fm* — **e** color *m* inestable; color *m* no resistente — **i** colore *m* non resistente; colore *m* labile
6231 full-bound; whole-bound — Said of a book when the entire back and sides are covered with the same material whether leather or cloth.
f en plein reliure — **d** ganzleder gebunden; ganzleinen gebunden — **n** geheel leer gebonden; geheel linnen gebonden — **e** encuadernación a toda piel; encuadernación a toda tela — **i** rilegatura intera
6232 full-face rule; heavy-face rule
f filet *m* plein; filet *m* gras — **d** fette Linie *f* — **n** vette lijn *fm*; volvette lijn *fm* — **e** raya *f* (de) luto; filete *m* (de) luto — **i** filetto *m* nero; filetto *m* scuro
6233 full-face type — See titling.
6234 full measure — Full measure refers to copy set full width.
6235 full *v* out — To start matter flush without indention.
f sans renfoncement — **d** stumpfer Zeilenanfang *m* — **n** zonder inspringen; niet ingesprongen; over de volle breedte — **e** sin sangría; sin sangrar — **i** iniziare senza rientranza
6236 full page ad — See full page advertisement.
6237 full page advertisement; full page ad — Advertisement occupying the whole of a page.
f annonce *f* en pleine page — **d** ganzseitige Anzeige *f*; ganzseitige Annonce *f* — **n** advertentie *f* over de gehele pagina — **e** anuncio *m* a toda página; aviso *m* a toda página; anuncio *m* a plana entera — **i** annunzio *m* a piena pagina; aviso *m* a piena pagina; pubblicità *f* a piena pagina; inserzione *f* a piena pagina
6238 full point; full stop; point — Punctuation mark (.), to indicate the end of a sentence and abbreviations.
f point *m* — **d** Punkt *m*; Schlußpunkt *m* — **n** punt *m* — **e** punto *m* final; punto *m* redondo — **i** punto *m*
6239 full position; floating position; island position; solus — Placement of an advertising sur-rounded by reading matter.
f mise en vedette — **d** alleinstehend — **n** vrijstaand tussen de tekst — **e** emplazamiento *m* aislado (en el texto); emplazamiento *m* aislado (entre texto) — **i** pubblicità *f* attorniata da testo
6240 full radiator — See black body.
6241 full size — Unfolded sheet of paper.
f in-plano *m*; feuille *f* à plat; non plié — **d** Planoformat *n*; ungefalzt — **n** plano; ongevouwen — **e** en plano — **i** formato *m* pieno
6242 full stop — See full point.
6243 full title; main title
f grand-titre *m* — **d** Haupttitel *m* — **n** hoofdtitel *m* — **e** título *m* principal — **i** titolo *m* generale; titolo *m* principale
6244 full-title page — Page of a book with the complete title (sometimes amplified), the names of author and publisher, date of publishing, place of origin, and sometimes the printer's name and address.
f page *f* du grand-titre — **d** Haupttitelseite *f* — **n** titelpagina *fm*; pagina *fm* met de hoofdtitel — **e** portada *f* — **i** pagina *f* a grandi titoli
6245 fully automatic
f entièrement automatique; pleinement automatique — **d** vollautomatisch — **n** vol-automatisch; geheel automatisch — **e** completamente automático — **i** completamente automatico
6246 fumaric resin — Synthetic hard resin formed by the reaction of fumaric acid and rosin.
f résine *f* fumarique — **d** Fumarharz *n* — **n** fumaarhars *mn* — **e** resina *f* fumárica — **i** resina *f* fumarica
6247 fuming nitric acid — Nitric acid with a concentration higher than 86% HNO_3.
f acide *m* nitrique fumant — **d** rauchende Salpetersäure *f* — **n** rokend salpeterzuur *n* — **e** ácido *m* nítrico fumante — **i** acido *m* nitrico fumante
6248 functional character — See control character.
6249 function code *GB*; operational code *US* — A code, generally in the text stream, which acts as a flag or signal to indicate that a certain number of succeeding codes convey format information (e.g., measure, type size, face, indent and the like). In some cases the code itself conveys the necessary information, e.g., a quad left or a quad centre command. In some cases a code can stand for a text character or signal a differentiating fuction depending upon the context.
f code *m* fonctionnel — **d** Funktionskode *m* — **n** functiecode *m* — **e** código *m* de función; código *m* de operación — **i** codice *m* d'operazione
6250 fundamental colours — See primary colours.
6251 fungicide — Agent that destroys or is hostile to fungi.
f fongicide *m*; agent *m* fongicide; anti-pourriture *f* — **d** Pilzvertilgungsmittel *n*; fungizider Stoff *m*; Fungizid *n*; pilzetötender Stoff *m* — **n** fungicide *n*; schimmeldodende stof *fm*; verdelgingsmiddel *n* voor schimmels — **e**

fungicida *f* — **i** fungicida *f*; sostanza *f* anticrittogamica

6252 fungoid growth; fungoid moulding

f végétation *f* mycélienne — **d** Pilzbefall *m*; Schimmelwachstum *n* — **n** schimmelvorming *f*; schimmelen *n*; beschimmelen *n* — **e** crecimiento *m* fungoideo — **i** vegetazione *f* micelica; proliferazione *f* di funghi

6253 fungoid moulding — See fungoid growth.

6254 funnel

f entonnoir *m* — **d** Trichter *m*; Einfülltrichter *m* — **n** trechter *m* — **e** embudo *m*; tolva *f* — **i** imbuto *m*

6255 furbish *v*; **furbish** *v* **up** — To restore shopworn books to freshness of appearance or condition.

f mettre à neuf; remettre à neuf — **d** aufarbeiten; erneuern; ausbessern; auffrischen — **n** restaureren; opknappen — **e** acicalar — **i** ristorare; lucidare

6256 furbish *v* **up** — See furbish *v*.

6257 furfural; furfuraldehyde; furol; furfurol *misn.*; **ant oil; artificial ant oil** — Colourless, mobile liquid when very pure, changing to reddish brown upon exposure to light and air. Soluble in alcohol, ether, and benzene; 8.3% soluble in water at 20 °C. TL-value: 5 ppm. Formula: C_4H_3COOH.

f furfurol *m*; furfural *m*; furol *m* — **d** Furfurol *n*; Furol *n*; Furaldehyd *n* — **n** furfurol *n*; furfural *n*; furol *n* — **e** furfurol *m* — **i** furfurolo *m*

6258 furfuraldehyde — See furfural.

6259 furfurol — See furfural.

6260 furnace black — Carbon black obtained by decomposing natural gas under controlled conditions in a furnace and precipitating the pigment. It has a pleasing blue tone as opposed to the strong brown tone of *channel black. It has a long flow and high opacity and is low of costs.

f noir *m* de fourneau; noir *m* furnace — **d** Furnaceruß *m*; Ofenruß *m* — **n** ovenroet *n*; vlamroet *n*; carbon black *n* — **e** negro *m* de horno; negro *m* de fogón — **i** nerofumo *m* fornace

6261 furnish — Mixture of various materials that are blended in the stock suspension from which paper or board is made. The chief constituents are fibrous materials (pulp), sizing materials, additives, fillers and dyes. See also fibre furnish.

6262 furniture — Material to make *margins etc., for a printed sheet, also to fill up the space left in a chase after the type matter has been inserted.

f garnitures *fpl*; garnitures *fpl* en métal; lingots *mpl* — **d** Blindmaterial *n*; eiserne Stege *mpl*; Bleistege *mpl*; Formatstege *mpl*; Schließstege *mpl*; Hogestege *mpl* — **n** gerief *n*; blindmateriaal *n*; formaatwit *n*; sluitwit *n*; insluitwit *n*; holwit *n* — **e** guarnición *f*; guarniciones *fpl*; imposiciones *fpl* (de metal); lingotes *mpl* de plomo; fornitura *f depr.* — **i** marginatura *f*; bianchi *mpl* tipografici

6263 furniture cabinet; furniture rack

f lingotier *m* — **d** Stegregal *n* — **n** geriefkast *fm* — **e** lingotero *m*; armario *m* para imposiciones; estante *m* para imposiciones — **i** armadio *m* per marginatura

6264 furniture rack — See furniture cabinet.

6265 furol — See furfural.

6266 fusel oil — See amyl alcohol.

6267 fuzz; lint — Loose fibres on surface of paper. See also paper dust.

f poussière *f* (de papier); poils *mpl* — **d** Staub *m*; Papierstaub *m* — **n** stof *n*; papierstof *n*; haartjes *npl* — **e** pelusa *f*; polvillo *m* — **i** polvere *f*; polvere *f* di carta

6268 fuzzing

f poussiérage *m* — **d** Stauben *n* — **n** stuiven *n* — **e** empolvado *m* — **i** spolverio *m*; spolvero *m*

6269 fuzzy paper — Paper with hairiness on the surface, caused by individual fibres standing out. It may cause trouble on the printing press, when the detached fibres mix with the ink. See also paper dust.

f papier *m* pelucheux — **d** staubendes Papier *n*; haariges Papier *n* — **n** stuivend papier *n*; harend papier *n* — **e** papel *m* velloso — **i** carta *f* che spolvera

6270 gage — See gauge.

6271 gag law — Law restricting freedom of the press, free speech, or the right of petition.

f loi *f* sur la presse — **d** Preßgesetz *n*; Knebelgesetz *n* — **n** wet *fm* op de vrijheid van meningsuiting; wet *fm* op de persvrijheid; persbreidel *m* — **e** ley *f* de imprenta — **i** legge *f* sulla stampa

6272 galena; galenite; lead glance — Mineral consisting of lead sulphide, which is the principal source of lead, soluble in nitric acid, also in excess of hot hydrochloric acid. Formula: PbS. Chemical name: natural lead sulphide.

f galène *f* — **d** Galenit *n*; Bleiglanz *m*; Schwefelblei *n*; Bleisulfid *m* — **n** loodglans *n* — **e** galena *f*; sulfuro *m* de plomo; sulfuro *m* vírgen de plomo; alquifol *m* — **i** galena *f*

6273 galenite — See galena.

6274 galipot; galipot gum — White resin similar to Burgundy pitch used by engravers on surface of plate to be engraved.

f galipot *m*; poix *f* de Bourgogne — **d** Galipot *n*; Gallipot *n*; Scharrharz *n* — **n** pijnhars *mn*; dennehars *mn* — **e** galipodio *m*; trementina *f* de Burdeos — **i** galipot *m*

6275 galipot gum — See galipot.

6276 gallery — Photographic department of a photoengraving plant.

f département *m* photographique — **d** Kameraabteilung *f* — **n** afdeling *f* fotografie; fotoafdeling *f* — **e** taller *m* de fotografía; galería *f* fotográfica — **i** reparto *m* fotografico

6277 gallery camera — Reproduction camera in which the photographic material is transported in a dark slide from the darkroom to the camera and vice versa.

f chambre *f* laboratoire — **d** Atelierkamera *f* — **n** ateliercamera *fm*; reproduktiecamera *fm* — **e** cámara *f* de galería; cámara *f* de retratos — **i** macchina *f* fotografica da laboratorio

6278 galley — Flat oblong tray into which composed type matter is put and kept until made up into pages in the forme. Also a similar tray on a slug composing machine which receives the slugs as they are ejected.

f galée *f* — **d** Setzschiff *n*; Schiff *n*; Satzschiff *n*; Zeilenschiff *n* — **n** galei *fm*; zetmachinegalei *fm* — **e** galera *f*; galerín *m* (a single column galley); galerada *f* (a tray filled with type) — **i** vantaggio *m*

6279 galley boy — Apprentice compositor charged with proofing and the care of the galleys.

f grouillot *m* — **d** Abzieher *m* — **n** proeventrekker *m* — **e** lingotero *m* — **i** apprendista *m* incaricato di tirare le prove

6280 galley cabinet — See galley rack.

6281 galley flange

f bride *f* de galée — **d** Randleiste *f* (am Schiff); Schiffsrand *m* — **n** rand *m* van de galei; galeirand *m* — **e** borde *m* de la galera; barrote *m* de

la galera — **i** listella *f* del vantaggio

6282 galley press — Proof press consisting of a base and a heavy roller on which galleys of type are proved. Not to be confused with a *Gally press.

f presse *f* à galée; presse *f* à épreuves — **d** Abziehpresse *f*; Korrekturabziehpresse *f* — **n** proefpers *f* — **e** prensa *f* para pruebas de galeradas — **i** torchio *m* tipografico per tirare bozze in colonna

6283 galley proof — See slip proof.

6284 galley rack; galley cabinet — Rack, sometimes consisting of sloped partitions to avoid pieing, used for storing galleys.

f meuble *m* à galées; classeur *m* à galées — **d** Setzschiff-Regal *n* — **n** galeien-regaal *n* — **e** armario *m* portagaleras; comodín *m* para composición en galeras — **i** mensola *f* per vantaggi

6285 galley slave — See compositor.

6286 galley slice; slice — Removable false bottom in certain old-fashioned galleys. See also slice galley.

f planchette *f* de galée à coulisse; plateau *m* (de galée) à coulisse — **d** Schiffszunge *f* — **n** schuif *fm* van de galei; galeischuif *fm* — **e** pala *f* (de la galera); volandera *f* (de la galera) — **i** paletta *f* (del vantaggio)

6287 galley slug; take slug — Temporary slug to identify the end of the assembled set matter on a galley.

f ligne *f* d'identification — **d** Satzzeile *f* zur (vorübergehenden) Kennzeichnung von Satzspalten — **n** galeiregel *m* — **e** lingote *m* indicador; guía *f* (de ajuste); línea *f* de guía; línea *f* de identificación *Ch* — **i** lingotto *m* di vantaggio

6288 galley stick — Long sloping piece of wood at the side of a galley to give a temporary lock-up for pulling a proof.

6289 gallic acid — Colourless or slightly yellow, crystalline needles or prisms, soluble in alcohol and glycerol, sparingly soluble in water and ether. Used in process engraving and lithography. Latin: acidum gallicum.

Formula: $C_6H_2(OH)_3CO_2H \cdot H_2O$.

f acide *m* gallique; acide *m* pyrogallolcarbonique — **d** Gallussäure *f*; Trioxybenzoesäure *f* — **n** galluszuur *n* — **e** ácido *m* gálico; ácido *m* agálico — **i** acido *m* gallico

6290 gallon — Unit of capacity. Abbrev. gal. See app. no. 7 and 7a.

f gallon *m* — **d** Gallon *f* — **n** gallon *mn* — **e** galón *m* — **i** gallone *m*

6291 gallotannic acid

See tannic acid.

6292 Gally press — Platen press invented by the American Merritt Gally, in 1869. Not to be confused with a *galley press.

f presse *f* (de) Gally — **d** Gally-Tiegeldruckmaschine *f* — **n** Gally-pers *f* — **e** prensa *f* de

Gally; máquina *f* Gally — **i** platina *f* Gally

6293 galvanometer — Instrument to detect, compare, or measure small electric currents, sometimes calibrated in amperes; requires calibration when an actual current measurement is needed. See also ampere meter.
f galvanomètre *m* — **d** Galvanometer *m* — **n** galvanometer *m* — **e** galvanómetro *m* — **i** galvanometro *m*

6294 gamboge; cambogia; camboge; gum camboge — Gum resin, soluble in acetone and benzene, used as a yellow pigment and for stopping out areas on printing plates to lay tints.
f gomme *f* gutte — **d** Gummigutt *n* — **n** guttegom *mn* — **e** cambogia *f*; goma *f* guta; gomaguta *f*; gutagamba *f*; gutiambar *m* — **i** gommagutta *f*

6295 gamboging — Application of an aqueous emulsion of a gamboge to areas of prints on metal before laying tints with benday films.
f gommage *m* (benday) — **d** Schablonendecken *n* zum Tangieren — **n** afdekken *n* voor tangeren — **e** aplicación *f* (de la emulsión acuosa) de gutagamba — **i** ricoprire a gommagutta

6296 gamma — Photographic term for negative *contrast resulting not from development and the contrast of the subject itself; a numerical measure of contrast in the development of a negative.
f gamme *f* — **d** Gamma *n*; Gammawert *m* — **n** gamma *fmn* — **e** gama *f* — **i** gamma *m*

6297 gamma- — Prefix added to a chemical name, denoting the third in position in the structure of an organic molecule from a particular group or atom.

6298 gampi — See bast fibres.

6299 gang — Group of formes or imposition in the same forme of different jobs arranged and positioned to be printed together.
f mariage *m*; amalgame *m* — **d** Sammelform *f* — **n** verzamelvorm *m* — **e** grupo *m*; juego *m*; serie *f* — **i** forma *f* composta; forma *f* multipla

6300 gang negative; multiple negative; multinegative — In photomechanics, a negative with a number of properly positioned images. To handle many small images as one unit in the step-and-repeat machine used in lithographic platemaking for making multiple images.
f négatif *m* multiplié; négatif *m* à plusieurs poses — **d** Sammelnegativ *n*; Negativ *n* mit mehreren gleichen Bildern (Nutzen) — **n** verzamelnegatief *n* — **e** negativo *m* en serie; negativo *m* en múltiple — **i** negativo *m* multiplo

6301 gang plate — Printing plate with a number of duplicate images as an integral unit; plate made from a gang negative.
f cliché *m* multiplié — **d** Sammelplatte *f* — **n** verzamelplaat *fm* — **e** clisés *mpl* en serie — **i** lastra *f* di stampa multipla

6302 gang punch — Device to punch identical data into a succession of punched cards.
f perforateur *m* en bloc — **d** Folgestanzer *m*; Schnellstanzer *m* — **n** stansmachine *f* — **e** multiperforador *m* desde el emisor — **i** riproduttrice *f*

6303 gang-punch *v* — To punch all or part of the information from one punched card into succeeding cards.
f poinçonner — **d** folgestanzen — **n** stansen — **e** punzonar — **i** punzonare

6304 gang stitcher; gathering and wire stitching machine; gatherer and wire stitcher
f encarteuse-piqueuse *f* — **d** Sammeldrahtheftmaschine *f*; Sammelheftmaschine *f*; Sammelhefter *m*; Zusammentrag- und Heftmaschine *f* — **n** verzamelhechtmachine *f* — **e** alzadora *f* con alambre; alzadora *f* cosedora; cosedora *f* de alimentación múltiple; cosedora *f* para cosido en serie — **i** macchina *f* raccoglitrice-cucitrice a filo metallico; raccoglitrice-cucitrice *f* a punti metallici

6305 gap — See cylinder gap.

6306 gap — See gutter.

6307 Garaldes — See Garaldics.

6308 Garaldics — Group of type faces, representing the pure Renaissance style. The axis of the curves is inclined to the left. There is generally a greater contrast in the relative thickness of the strokes than in Humanistic designs. The serifs are bracketed; the bar of the lower case e is horizontal. The serifs of the ascenders in the lower case are oblique. To this group belong all the Garamonts, Bembo, Caslon, Arrighi, Poliphilus, De Roos, Romulus and the modern Astrée and Vendôme. Formerly known as Old Face or Old Style. Before 1971 this group was named Garaldes. See also classification of type design.
f Garaldes *fpl* — **d** französische Renaissance Antiqua *f* — **n** Garalden *fmpl* — **e** Garaldos *mpl* — **i** Elzeviri *mpl*

6309 garbage *sl.* — In computer controlled typesetting, unnecessary and unwanted information on tape or in store. For bad input producing bad output the term "gigo" (garbage in, garbage out) is used.

6310 gas — Sulphur dioxide formed by the burning of sulphur in the sulphite pulp process.
f gaz *m* — **d** Gas *n* — **n** gas *n* — **e** gas *m* — **i** gas *m*

6311 gas black — Black pigment produced by a incomplete combustion of natural gas, used for the manufacture of black printing ink.
f noir *m* de gaz — **d** Gasruß *m* — **n** gasroet *n* — **e** negro *m* de gas — **i** nero *m* da gas

6312 gas burner — Illuminating-gas jet used on cylinder presses as a heating medium to partly dry the ink and prevent offsetting.
f brûleur *m* à gaz — **d** Brennerrampe *f* — **n** gasbrander *m* — **e** mechero *m* a gas — **i** bruciatore *m* a gas

6313 gas burst agitation — See nitrogen burst.

6314 gas checking — See crazed drying.

6315 gas crazing — See crazed drying.

6316 gaseous
f gazeux — **d** gasartig; gasförmig; luftförmig — **n** gasachtig; gasvormig — **e** gaseoso — **i** gassoso

6317 gas heating

f chauffage *m* au gaz — **d** Gasheizung *f* — **n** gasverwarming *f* — **e** calefacción *f* a gas — **i** riscaldamento *m* a gas

6318 gasket; packing ring — Rubber or metal ring to place around a joint to make it watertight.

f anneau-joint *m*; joint *m* presse-étoupe — **d** Dichtring *m* — **n** pakkingring *m*; afsluitring *m* — **e** anillo *m* de junta — **i** anello *m* di guarnizione a tenuta

6319 gaslight paper — Light-sensitive paper for contact printing coated with a relatively slow emulsion of silver chloride.

f papier *m* pour copie par contact (au gélatino-chlorure d'argent) — **d** Kontaktkopierpapier *n*; Gaslichtpapier *n*; Chlorsilberpapier *n* — **n** gaslichtpapier *n* — **e** papel *m* para luz de gas — **i** carta *f* fotosensibile per stampa a contatto; carta *f* al cloruro d'argento

6320 gasoline *US*; **petrol** *GB*; **motor spirits** — Mixture of volatile hydrocarbons derived from petroleum with boiling range of 60-200 °C (140-390 °F). TL-value: 500 ppm.

f essence *f*; essence *f* de pétrole — **d** Benzin *n* — **n** benzine *fm* — **e** gasolina *f*; esencia *f* de petróleo — **i** benzina *f*

6321 gas permeability — See vapour permeability.

6322 gate; gate circuit — Circuit which gives an output logically dependent upon the input; used in digital computers for logical and arithmetic operations. An "and" gate gives an output only if there is an input on all input wires in the logic module, whereas an "or" gate gives an output if there is an input on any input wire; other elements are nor, not-and, and exclusive-or gates.

f circuit *m* porte; porte *f* — **d** Torschaltung *f*; Tor *n* — **n** poortschakeling *f*; poort *fm* — **e** circuito *m* puerta; puerta *f*; compuerta *f* — **i** circuito *m* porta; porta *f*

6323 gate circuit — See gate.

6324 gatefold — See throw-out.

6325 gather *v*; **assemble** *v*; **collect** *v* — To gather the sheets after drying and folding in single copies of complete books for binding. See also collate *v*.

f assembler — **d** zusammentragen; sammeln — **n** vergaren — **e** alzar; alzar y acoplar; juntar; reunir; coleccionar — **i** raccogliere; riunire

6326 gatherer — See gathering machine.

6327 gatherer and wire stitcher — See gang stitcher.

6328 gathering — Section of a book. See also signature.

6329 gathering and wire stitching machine — See gang stitcher.

6330 gathering machine; gatherer — Machine with a moving belt or a turning circular table, which associates and registers in correct sequence the sheets. See also collating machine.

f machine *f* à assembler; assembleuse *f* — **d** Zusammentragmaschine *f*; Zusammentraganlage *f* — **n** vergaarmachine *f* — **e** alzadora *f* (de hojas); reunidora *f* de hojas; alzadora *f* de pliegos; re-unidora *f* de pliegos; ensambladora *f* — **i** macchina *f* raccoglitrice; raccoglitrice *f*; macchina *f* per raccogliere fogli singoli

6331 gathering table — See collating table.

6332 gatling — See gun.

6333 gauffered edges — See goffered edges.

6334 gauge *GB*; **gage** *US* — Instrument to measure thickness of paper, of metal sheets, diameter of pipes and wires, height and volume of liquids in tanks, gas and liquid pressures, etc. See also type gauge.

f appareil *m* vérificateur; calibre *m*; jauge *f*; micromètre *m*; vérificateur *m* — **d** Meßgerät *n*; Prüfgerät *n*; Prüfer *m*; Lehre *f*; Maß *n* — **n** meet-instrument *n*; kaliber *n*; micrometer *m* — **e** calibre *m* — **i** calibro *m*; apparecchio *m* di misura

6335 gauge; fender — Small piece of board, metal or wood glued to tympan of a platen press, to hold printed sheet in position. See also lay guide.

f capucin *m*; queue *f* de rat — **d** Frosch *m*; Fröschchen *n*; Kapuziner *m* — **n** aanleg *m* — **e** capuchina *f*; guía *f* — **i** taccheggio *m*

6336 gauge — Depth of the face of a type letter from the top serif to the bottom serif.

f hauteur *f* de l'œil; hauteur *f* d'œil — **d** Höhe *f* des Buchstabenbildes; Schriftgröße *f*; Bildfläche *f*; Schriftfläche *f* — **n** beeldhoogte *f* — **e** altura *f* del ojo — **i** altezza *f* dell'occhio

6337 gauge — Unit of measurement for photoengraving metal. See also thin gauge metal.

6338 gauge pin — Spring tongue metal pin on a platen press to hold the sheet in printing position.

f capucin *m* — **d** Frosch *m*; Fröschchen *n*; Anlegefrosch *m*; Kapuziner *m* — **n** aanlegspeld *fm* — **e** guía *f* de lengüeta; guía *f* de resorte; guía *f* para minervas — **i** squadra *f* di marginatura

6339 gauss — Term used in magnetic ink printing processes, abbreviated G. It is the CGS unit of magnetic induction obtained from the law which relates magnetic induction with the electromotive force induced in a conductor which is moving through a magnetic field. The law is: $(v \times 1)B = E$, in which v being the velocity in centimetres per second when a conductor of length 1 in centimetres is moving through a magnetic field where the magnetic induction in gausses is B.

f gauss *m* — **d** Gauss *n* — **n** gauss *m* — **e** gauss *m* — **i** gauss *m*

6340 Gautschbrief — German term for a certificate of qualification, stating that an apprentice has been taken up into the corps of printers and that he has undergone the customary baptism.

f certificat *m* de baptême d'apprenti allemand — **d** Gautschbrief *m* — **e** certificado *m* de aprendizaje

6341 Gautschen — German term for a baptism on the nomination of an apprentice to pressman, to which he is set down on a wet sponge. See also couch *v*.

f baptême *m* — **d** Gautschen *n*; Gautschung *f* — **n** dopen *n*; doop *m* — **e** bautizar — **i** battesimo *m*

6342 gauze — Loosely woven mesh from textile

fibres or wire.

f gaze *f* — **d** Gaze *f* — **n** gaas *n* — **e** gasa *f* — **i** garza *f*

6343 gazette; official gazette — In GB, official government journal containing lists of government appointments and promotions, bankruptcies, etc.

f journal *m* officiel; gazette *f* officielle; moniteur *m* — **d** Staatsanzeiger *m*; Amtsblatt *n* — **n** staatscourant *fm*; staatsblad *n* — **e** boletín *m* oficial; diario *m* oficial; gaceta *f* oficial — **i** gazzetta *f* ufficiale

6344 gazette — In US, a newspaper.

f gazette *f* — **d** Zeitung *f* — **n** gazet *fm* — **e** gaceta *f* — **i** gazzetta *f*

6345 gear — Appliance connected with the moving part of a machine, to transmit motion; usually consists of a wheel with teeth on its circumference which engage similar teeth on another wheel.

f engrenage *m* — **d** Getriebe *n* — **n** aandrijf-mechanisme *n*; tandwieloverbrenging *f* — **e** engranaje *m*; propulsión *f* de engranajes — **i** ingranaggio *m*

6346 gear; tooth wheel

f pignon *m*; roue *f* dentée — **d** Zahnrad *n* — **n** tandrad *n*; tandwiel *n* — **e** rueda *f* dentada; piñón *m* — **i** ruota *f* dentata

6347 geared incentive — Plan where the rate of payment is not directly proportional to the rate of working. See also accelerating incentive, differential piecework, multiple time plan.

6348 geared rider — Gear driven roller which normally reciprocates. See also rider roller.

f chargeur *m* commandé — **d** im Getriebe laufende Reiterwalze *f* — **n** aangedreven ruiterrol *fm* — **e** rodillo *m* cargador accionado; cargador *m* accionado — **i** rullo *m* cavaliere comandato

6349 gear marks — See gear streaks.

6350 gear side — See drive side.

6351 gear streaks; gear marks — Parallel streaks appearing across printed sheet at same interval as gear teeth on cylinder. Caused by improper under-packing or defective press conditions resulting in difference or surface speed between cylinders and pitch diameter of gears.

f marques *fpl* de dents — **d** Zahnstreifen *mpl* — **n** tandstrepen *fmpl* — **e** franjas *fpl* (en la impresión) por franquicia de engranajes — **i** strisce *fpl* testimoni nella stampa del gioco tra i denti degli ingranaggi

6352 gel — Colloid in a more solid form than a *sol. State or condition in which an ink or vehicle has a jelly-like consistency.

f gel *m* — **d** Gel *n* — **n** gel *mn* — **e** gel *m* — **i** gel *m*

6353 gel *v* — To change from a liquid to a gel.

f se coaguler; se gélifier; se prendre en gelée — **d** gelieren — **n** geleren — **e** coagular — **i** gelare

6354 gelatin — See gelatine.

6355 gelatin coating — See gelatine coating.

6356 gelatine *GB*; **gelatin** *US* — Albuminous material in bones, ligaments and skins, which may be extracted by boiling water (the water may be slightly acidified with hydrochloric acid); differs from glue in its purity and in the care observed in its manufacture. Soluble in hot water, glycerol and acetic acid, insoluble in alcohol, chloroform and other organic solvents.

f gélatine *f* — **d** Gelatine *f* — **n** gelatine *fm* — **e** gelatina *f* — **i** gelatina *f*

6357 gelatine coating *GB*; **gelatin coating** *US*

f couche *f* de gélatine — **d** Gelatineschicht *f* — **n** gelatinelaag *fm* — **e** capa *f* de gelatina — **i** strato *m* di gelatina

6358 gelatine-composition roller — See composition roller.

6359 gelatine duplicating — See hectography.

6360 gelatine filter — Gelatine covered glass plate or film which allows certain kinds of light to pass through but absorbs light of a different colour, used to obtain separation plates.

f écran *m* de gélatine; filtre *m* de gélatine — **d** Gelatinefilter *m* — **n** gelatinefilter *n* — **e** filtro *m* de gelatina — **i** filtro *m* di gelatina

6361 gelatine printing — See hectography.

6362 gelatine relief — Film with a relief pattern for inking and subsequent transfer of the pattern by pressure. See also shading medium.

6363 gelatine relief plate — Printing plate or surface consisting of a relief image of light-hardened gelatine; a cheap substitute for photoengravings.

f cliché *m* à relief de gélatine — **d** Gelatinerelief-platte *f* — **n** gelatinereliëfplaat *fm* — **e** gelatino-grafía *f*; clisé *m* al fotorrelieve; clisé *m* con relieve de gelatina — **i** lastra *f* di stampa per gelatinografia

6364 gelatine size — See animal glue.

6365 gelatinochloride

f gélatine-chlorure *m* (d'argent) — **d** Chlorsilber *n* — **n** chloorzilver *n* — **e** gelatinocloruro *m* (de plata) — **i** gelatino-cloruro *m* (d'argento)

6366 gelatinous; jelly-like — Resembling gelatine or jelly, or relating to or containing gelatine.

f gélatineux — **d** gelatinös; gallertartig; gelatine-artig — **n** gelatineus; gelei-achtig — **e** gelatinoso — **i** gelatinoso

6367 gem — Old name for a size of type. See app. no. 1.

6368 general purpose computer — Device with a hardwired set of instructions which can be executed according to the program and coded into its memory to enable it to perform a wide variety of processing tasks involving the use of arithmetic and logic. There are arithmetic and logic circuits, a memory, or core for the storing of programs and data, and some peripheral devices to accept input and to provide output. Almost invariably binary logic is used. They are digital, rather than analogue devices.

f calculateur *m* universel; ordinateur *m* universel — **d** Allzweckrechner *m*; Universalrechner *m* — **n** computer *m* voor algemeen gebruik; universele computer *m* — **e** computadora *f* de uso general;

computador *m* universal — **i** calcolatore *m* universale; elaboratore *m* universale

6369 generating program — See generator.

6370 generation — See first generation computer.

6371 generation — See first generation typesetter.

6372 generation — See fourth-generation typesetter.

6373 generation — See second-generation computer.

6374 generation — See second-generation typesetter.

6375 generation — See third-generation computer.

6376 generation — See third-generation typesetter.

6377 generator; generating program — Problem-oriented routine used to write programs, for a range of problems, from a set of parameters specified by the user.
f générateur *m* — **d** Generierer *m*; Generierprogramm *n* — **n** generator *m*; genereerprogramma *n* — **e** generador *m* — **i** generatore *m*

6378 genol — Trade mark for para-methyl aminophenol sulphate.

6379 gentlemen's card — A size of cut card. See app. no. 3.

6380 genuine leatherboard — See leatherboard.

6381 geometrical lathe — Term sometimes used for a *guilloching machine.

6382 geometrical signs — See geometrical symbols.

6383 geometrical symbols; geometrical signs — See app. no. 10.
f symboles *mpl* géométriques; signes *mpl* de géométrie — **d** geometrische Zeichen *npl* — **n** meetkundige tekens *npl* — **e** signos *mpl* de geometría — **i** segni *mpl* di geometria

6384 German alphabet — Same as English, plus Ä, Ö, Ü, ä, ö, ü, and logotypes for ch, ck, ss. There are elaborate rules for capitalization, division of words, compounding, etc. The German characters were Fraktur or Schwabacher, the roman Antiqua, today Latin types are used. Fraktur has no small caps or italics; emphasis being given by letter-spacing. Ae, Oe, Ue, except in proper names are now rendered Ä, ö, ü. The ligatures æ, œ are not generally used; t is now used for th in all but proper names and foreign words. Division of words by sound, but prefixed nouns, prepositions, etc., remain intact, also suffixes beginning with a consonant, such as -chen, -keit, -lein, -ling, -nis. All nouns and words used as such have capital initials. So also have the personal pronouns of the second person, Ihnen, Ihr, Sie, and adjectives from names of places and persons, as Leipziger Messe, Kölnisches Wasser, Kantische Philosophie, but adjectives of countries lower-case, as russische Sprache, deutsche Industrie, etc. See also ligature.
f alphabet *m* allemand — **d** deutsches Alphabet *n* — **n** Duitse alfabet *n* — **e** alfabeto *m* alemán — **i** alfabeto *m* tedesco

6385 German height to paper — See also height

to paper.
f hauteur *f* allemande — **d** deutsche Höhe *f* — **n** Duitse hoogte *f* — **e** altura *f* alemana — **i** altezza *f* tedesca

6386 get *v* **in** — To set more matter in a line, page or forme, that is in the printed copy from which the compositor works, or to set manuscript copy so that it makes less space than estimated; to take in; to set close; to thin space a line to make it come within the required space.
f rentrer (le texte); serrer — **d** einbringen — **n** inwinnen; nauwer zetten — **e** ganar; comer; comer línea — **i** guadagnare spazio; rientrare; serrare

6387 get *v* **position**
f mettre les blancs — **d** Standmachen *n* — **n** standmaken — **e** posición *f* adelantada — **i** stabilere i bianchi

6388 get *v* **to bed** — See put *v* to bed.

6389 g.h. — London compositor's term signifying "Queen Anne is dead", after the initials of George Horne, an individual who specialized in the retailing of stale news.
f c'est connu — **d** alte Kamellen *pl*; überholte Nachrichten *fpl* — **n** oud nieuws *n* — **e** noticias *fpl* añejas — **i** notizie *fpl* superate

6390 ghost etching — In rotogravure a silver-grey appearance sometimes noticed just before a tone is brought-in, quite distinct from the dark brown associated with etching. There is no evidence that the copper is attacked. The cause is still obscure.

6391 ghosting — Image which appears as a lighter area on a subsequent print due to local blanket depressions from previous image areas on a letterpress rotary machine as well as on an offset press.
f impression *f* fantôme; image *f* fantôme — **d** Geistererscheinung *f*; Geistereffekt *m* — **e** impresión *f* fantasma; imagen *f* fantasma — **i** falsa immagine *f*; impressione *f* fantasma

6392 ghosting — Marring a print by an image on it of work printed on the reverse side which has interfered with its drying so that differences in the trapping of some colours or gloss variations are apparent. Letterpress rotary term.
d Geistererscheinung *f* — **n** doorslaan *n*

6393 ghosting — See repeat.

6394 ghosting — See second impression set-off.

6395 ghost writer — One who writes articles, speeches, books, etc., for another person who is named as or presumed to be the author.
f collaborateur *m* anonyme — **d** anonymer Mitarbeiter *m* — **n** anoniem medewerker *m* — **e** colaborador *m* anónimo; colaborador *m* oculto — **i** collaboratore *m* anonimo

6396 gift wrapping paper — See fancy paper.

6397 giggering; jiggering — Moving back and forth of a tool in blind tooling.
f berçage *m* — **d** Hinundherbewegung *f* — **n** heen- en weerbewegen *n* — **e** movimiento *m* de vaivén — **i** movimento *m* di andirivieni

6398 gilbert — CGS unit of magnetomotive force or magnetic potential on the electromagnetic

system. A point has a magnetic potential of 1 gilbert if the work done in bringing a unit positive pole up to that point is 1 erg. Abbrev. Gb.
f gilbert *m* — **d** Gilbert *n* — **n** gilbert *m* — **e** gilbert *m* — **i** gilbert *m*
6399 gilder — The workman who gilds books. See also press gilder, edge gilder.
f doreur *m* — **d** Vergolder *m* — **n** vergulder *m* — **e** dorador *m*; estampador *m* — **i** doratore *m*
6400 gilder's delight — Size for book edges, especially made for gilders. See also Armenian bole.
6401 gilder's press — See gilding press.
6402 gilder's size; size; glaire; glair — Egg albumen composition used on tooled book covers and edges of books to make gold adhere.
f glaire *f*; blanc *m* du doreur; blanc *m* d'œuf; colle *f* d'or — **d** Vergoldeeiweiß *n*; Eiweißmischung *f*; Eiweiß *n* — **n** eiwit *n* — **e** cola *f* de dorar; sisa *f* de dorar; mordiente *m*; clara *f* de huevo — **i** sottostrato *m* per doratura; albume *m* d'uovo
6403 gilder's skewings — See skewings.
6404 gilder's tip — Flat brush for lifting gold leaf and depositing it on the surface to be gilded.
f couchoir *m* — **d** Anschießer *m*; Aufträger *m* — **n** verguldkwast *m*; opbrengkwast *m*; goudopdraagkwastje *n* — **e** palillo *m* de levantar (el pan de oro al dorar); polonesa *f* — **i** pennellessa *f*
6405 gilding; hand gilding — Adhesion of gold leaf to the edges of a book with a liquid agent made permanent with various burnishing tools.
f dorure *f* à la main — **d** Handvergoldung *f* — **n** handvergulden *n* — **e** dorado *m* a mano — **i** indoratura *f* di mano; indoratura *f* a mano; doratura *f* di mano; doratura *f* a mano
6406 gilding; machine gilding — Simulation of the hand process by the use of gold foils applied with heat.
f dorure *f* à presse — **d** Preßvergoldung *f* — **n** persvergulden *n* — **e** dorado *m* a prensa — **i** doratura *f* alla pressa
6407 gilding press; gilder's press; gold blocking press — Press to hold a book between two wooden boards when gilding.
f presse *f* à dorer — **d** Vergoldepresse *f* — **n** verguldpers *f* — **e** prensa *f* para dorar; prensa *f* de dorar — **i** pressa *f* a dorare
6408 gilding roll — See bar roll.
6409 gill — Unit of liquid measure. Abbrev. gi. See app. no. 7 and 7a.
6410 gilling — Cylinder glazing of paper with a gill machine, named after the inventor.
6411 gilsonite — Black asphaltic resinous material used in black printing ink. Soluble in all proportions of carbon disulphide. It is one of the purest (99.9%) natural bitumens.
f gilsonite *f* — **d** Gilsonit *m*; Gilsonitasphalt *m* — **n** gilsoniet *m* — **e** gilsonita *f* — **i** gilsonite *f*
6412 gilt edge — Bound book with edges cut and gilded to prevent soiling from dust.
f tranche *f* dorée; doré sur tranche — **d** Goldschnitt *m*; mit Goldschnitt — **n** verguld op snee; goud op snee — **e** corte *m* dorado; canto *m*

dorado — **i** taglio *m* dorato
6413 gilt top — The gilt top edge of a book.
f tranche *f* supérieure dorée — **d** Kopfgoldschnitt *m* — **n** vergulde kopsnee *fm*; goud op snee — **e** corte *m* dorado de cabecera; de corte de cabecera dorado — **i** taglio *m* superiore dorato
6414 gingerbread — Heavily, gaudily and superfluously ornamented book decoration.
f décoration *f* de mauvais goût — **d** übertriebene Verzierung *f*; geschmacklose Verzierung *f* — **n** overdreven versiering *f*; smakeloze versiering *f* — **e** adorno *m* cursi; cursi *m*; churrigueresco *m* — **i** decorazione *f* esagerata; decorazione *f* di cattivo gusto
6415 girders — See cross bars.
6416 girth tape — Metal tape with scale to measure cylinder circumferences (deprecated method).
f mètre-ruban *m* — **d** Meßband *n* — **n** rolmaat *fm* — **e** medida *f* arrollable — **i** metro *m* flessibile; nastro *m* graduato
6417 give *v* **the OK** — See pass *v* for press.
6418 glacial acetic acid; concentrated acetic acid — Pure acetic acid (99.8%) in distinction to the common water solution, known as acetic acid. Latin: acidum aceticum glaciale.
Formula: CH_3COOH.
f acide *m* acétique cristallisable; acide *m* acétique glacial; acide *m* acétique concentré; vinaigre *m* glacial — **d** Eisessig *m*; kristallisierte Essigsäure *f* — **n** ijsazijn *m*; geconcentreerd azijnzuur *n*; sterk azijnzuur *n* — **e** ácido *m* acético glacial — **i** acido *m* acetico glaciale; acido *m* acetico cristallizzabile
6419 glair — See gilder's size.
6420 glair *v*; **glaire** *v* — To coat book covers and the like with glaire, before applying gold.
f glairer — **d** grundieren mit Eiweiß — **n** eiwit aanbrengen; gronderen met eiwit — **e** aplicar el mordiente (para dorar) — **i** applicare il mordente per la doratura
6421 glaire — See gilder's size.
6422 glaire *v* — See glair *v*.
6423 glaire pencil — See glair pencil.
6424 glair pencil; glaire pencil — Small bookbinder's brush made of camel's hair to apply glair.
f pinceau *m* à glairer — **d** Grundierpinsel *m*; Pinsel *m* — **n** grondeerkwastje *n*; kwastje *n* — **e** pincel *m* para aplicar el mordiente de dorar — **i** pennellino *m* per applicare il sottostrato di doratura
6425 glare — Specular light or reflection from a polished or glossy surface, such as supercalendered or glazed paper.
f éclat *m* — **d** Glanzlicht *n* — **n** schittering *f*; blink *m* — **e** deslumbre *m*; deslumbramiento *m* — **i** abbagliamento *m*; riflesso *m* di luce
6426 glarimeter — Instrument to measure the glare, sometimes to estimate the finish or polish of paper.
f glarimètre *m* — **d** Glanzmesser *m*; Glanzprüfer *m* — **n** glansmeter *m* — **e** glasímetro *m*; aparato *m* para determinar el brillo — **i**

apparecchio *m* per misurare l'abbagliamento

6427 glass — See magnifier.

6428 glass electrode — Thin-walled glass membrane, separating a solution of known pH from another solution whose pH is to be determined.
f électrode *f* de verre; demi-cellule *f* en verre — **d** Glaselektrode *f*; Glashalbzelle *f* — **n** glaselektrode *f* — **e** electrodo *m* de vidrio; semicelda *f* de vidrio — **i** elettrodo *m* a vetro

6429 glass fibres
f fibres *fpl* de verre — **d** Glasfaser *fpl* — **n** glasvezels *fmpl* — **e** fibras *fpl* de vidrio — **i** fibre *fpl* di vetro

6430 glass grinding marbles
f billes *fpl* de verre pour grainage — **d** Glaskugeln *fpl*; Glasmärbel *mpl*; Körnglaskugeln *fpl* — **n** glazen knikkers *mpl*; glaskogels *mpl* — **e** bolas *fpl* de vidrio para amolar; bolas *fpl* para granear planchas litográficas — **i** biglie *fpl* di vetro per granitura

6431 glassine paper — Transparent paper, produced from well beaten sulphite pulp and by heavy calendering of the sheet, used for packaging, envelopes, bags, covers, etc.
f papier *m* cristal — **d** Pergamin *n*; Pergaminpapier *n* — **n** pergamijn *n* — **e** papel *m* glasín; papel *m* cristal — **i** pergamine *m* sottile; carta *f* pergamena; glassina *f*

6432 glass master — Photosensitive glass plate used in flexography to mould rubber plates. Customer art work is prepared as for a conventional plate, then imposed photographically in the glass and finally etched. Steps not necessary to the process are applying photo-resist and making a matrix.
f matrice *f* de verre — **d** Glasmater *f* — **n** glasmatrijs *f* — **e** matriz *f* de vidrio — **i** matrice *f* di vetro

6433 glass negative
f négatif *m* sur verre — **d** Glasnegativ *n*; negatives Glasbild *n* — **n** glasnegatief *n*; negatief *n* op glas — **e** negativo *m* sobre vidrio — **i** negativo *m* su lastra di vetro

6434 glass paper — Abrasive paper. See also emery paper.
f papier *m* de verre; papier *m* verré — **d** Glaspapier *n*; Schleifpapier *n* — **n** schuurpapier *n* — **e** papel *m* (de) vidrio; papel *m* de lija — **i** carta *f* vetrata

6435 glass plate — Glass plate with a coating sensitized for photographic reproduction.
f plaque *f* de verre; verre *m* — **d** Glasplatte *f* — **n** glasplaat *fm* — **e** placa *f* de vidrio; luna *f* — **i** lastra *f* di vetro

6436 glass positive — A *diapositive on glass.
f positif *m* sur verre; diapositif *m* — **d** Glaspositiv *n*; positives Glasbild *n*; Diapositiv *n* — **n** glaspositief *n*; diapositief *n* op glas — **e** positivo *m* sobre vidrio — **i** positiva *f* su lastra di vetro

6437 glass rod; glass stirring rod — Small stick of glass to stir liquids.
f baguette *f* en verre; agitateur *m* en verre — **d** Glasstab *m* (zum Rühren) — **n** (glazen) roerstaaf *fm*; glasstaaf *fm* — **e** barra *f* de vidrio; varilla *f* de agitar — **i** agitatore *m* in vetro; paletta *f* in vetro

6438 glass screen
f trame *f* en verre — **d** Glasraster *m* — **n** glasraster *n*; glazen raster *n* — **e** trama *f* de vidrio — **i** retino *m* di vetro

6439 glass stirring rod — See glass rod.

6440 Glauber's salt — See sodium sulphate.

6441 glaze — In offset, hard, shiny surface on rollers or blanket due to oxidation.
f patine *f* — **d** glasige Oberfläche *f* — **n** glad oppervlak *n*; glazig oppervlak *n* — **e** glaseadura *f*; endurecimiento *m* lustroso — **i** superficie *f* indurita lucida

6442 glazed coloured paper; glazed coloured printing paper — Coloured chemical or mechanical paper for cheap printing matter.
f papier *m* d'impression couleur — **d** farbiges Druckpapier *n*; farbig Druck *m* — **n** couverture *n* — **e** papel *m* coloreado; papel *m* de colores — **i** carta *f* satinata colorata

6443 glazed coloured printing paper — See glazed coloured paper.

6444 glazed imitation parchment — Calendered, translucent, unbleached, woodfree paper, between 25 and 50 g/m².
f papier *m* parchemin imité; papier *m* opaline — **d** Florpostpapier *n*; Florpost *f* — **n** floorpostpapier *n*; floorpost *n* — **e** papel *m* cebolla para cartas — **i** carta *f* pergamena imitata

6445 glazed paper — Paper which has been highly calendered to obtain a smoother surface and a higher gloss than machine-finished paper.
f papier *m* satiné; papier *m* glacé; papier *m* supercalandré — **d** satiniertes Papier *n*; Glanzpapier *n* — **n** gesatineerd papier *n* — **e** papel *m* glaseado; papel *m* satinado — **i** carta *f* satinata

6446 glazed roller — Roller with a smooth surface caused during printing.
f rouleau *m* glacé — **d** glatt gewordene Walze *f* — **n** glad geworden rol *fm* — **e** rodillo *m* glaseado; rodillo *m* endurecido — **i** rullo *m* indurito; rullo *m* lucido

6447 glazed roller — Offset nap roller with a specially prepared surface for short colour runs. It saves much time in cleaning and scraping when changing colours, as opposed to an ordinary roller.
f rouleau *m* lisse — **d** glatte Walze *f* — **n** gladde rol *fm* — **e** rodillo *m* liso — **i** rullo *m* liscio

6448 glazing cylinder — Large drying cylinder, with highly polished surface, on which machine-glazed papers are dried and glazed at the same time.
f cylindre *m* frictionneur — **d** Glättwalze *f*; Glättzylinder *m*; Glättpresswalze *f* — **n** satineercilinder *m*; eenzijdigglad-cilinder *m* — **e** cilindro *m* satinador; rodillo *m* alisador — **i** cilindro *m* lisciatore

6449 glazing of a blanket

f glaçage *f* d'un blanchet — **d** Glattwerden *n* des Drucktuches — **n** gladworden *n* van het rubberdoek — **e** glaseado *m* de la mantilla — **i** lucidatura *f* del blanket

6450 global search — Search by a computer program through a text stream or data base to discover and report the occurrences of a certain string of characters.

6451 global substitution — As in the case of global search, location of certain unique strings of characters or codes, and the automatic substitution of these by another string of data.

6452 gloss — The perceptual (appearance) aspects are for (1) specular gloss: shininess, brilliance of highlights (such as medium gloss surfaces of book paper, paint, plastics, etc.); (2) sheen: shininess at grazing angles (such as low gloss surfaces of paper, paint, etc.); (3) contrast gloss: contrast between specularly reflecting areas and other areas (such as low gloss surfaces of newsprint, bond paper, paint, textile cloth, etc.); (4) absence-of-bloom gloss: absence of haze, or milky appearance adjacent to reflected highlights (such as high, and semi-gloss surfaces in which reflected highlights may be seen); (5) distinctness-of-image gloss: the distinctness and sharpness of mirror images (such as high gloss surfaces in which mirror images may be seen); (6) surface-uniformity gloss: surface uniformity, freedom from visible non-uniformities (such as medium to high-gloss surfaces of all types).
f lustre *m*; brillant *m*; éclat *m*; luisance *f*; glacé *m*; verni *m*; poli *m*; satiné *m* — **d** Glanz *m*; Feuer *m* (der Farbe) — **n** glans *m* — **e** brillo *m*; lustre *m* — **i** brillantezza *f*; lucido *m*; lustro *m*; lucidezza *f*

6453 gloss ink; glossy ink — Ink with a special high finish resin varnish. It dries quickly to a high lustre with a minimum of penetration into the stock.
f encre *f* brillante; encre *f* lustrée — **d** Glanzdruckfarbe *f*; Glanzfarbe *f*; Emailfarbe *f* — **n** glansinkt *m* — **e** tinta *f* brillante; tinta *f* de alto brillo; tinta *f* lustrosa — **i** inchiostro *m* brillante; inchiostro *m* lucido

6454 gloss meter — Instrument to measure light reflected by a surface as a percentage of the incident light, where the angle of reflectance is the same as the angle of illumination. This is called *specular reflection. Gloss is measured as a function of both specular and diffuse reflections of light. The relation between these two factors can be expressed by the formula:
$L = I_s/(I_s + I_d)$, where L is the gloss, I_s is the specular reflection, and I_d is the diffuse reflection.
f mesureur *m* de brillant; brillancemètre *m*; lampomètre *m* — **d** Glanzmesser *m*; Glanzmeßgerät *n* — **n** glansmeter *m* — **e** medidor *m* de brillo — **i** misuratore *m* di brillantezza; apparecchio *m* per prova di brillantezza

6455 gloss varnish; glossy lacquer
f vernis *m* luisant — **d** Glanzfirnis *m*; Glanz-lack *m* — **n** glansvernis *mn*; glanslak *mn* — **e** barniz *m* lustroso; barniz *m* brillante; barniz *m* de brillo — **i** vernice *f* brillante

6456 gloss white — White mineral pigment used as an ink extender, made by co-precipitation of alumina hydrate and blanc fixe.
f blanc *m* brillant — **d** Mischweiß *n* — **n** mengwit *n* — **e** blanco *m* lustroso — **i** bianco *m* brillante

6457 glossy — Having a superficial lustre or brightness.
f éclatant; brillant — **d** glänzend — **n** glanzend — **e** brillante — **i** brillante

6458 glossy — See glossy print.

6459 glossy coated paper
f papier *m* couché brillant — **d** hochglanzgestrichenes Papier *n* — **n** hoogglanzend kunstdrukpapier *n* — **e** papel *m* estucado brillante — **i** carta *f* patinata brillante

6460 glossy ink — See gloss ink.

6461 glossy lacquer — See gloss varnish.

6462 glossy print; glossy — Photograph with a hard, very shiny finish, preferred for reproduction work.
f bromure *m* glacé — **d** Hochglanzabzug *m*; Hochglanzphotographie *f* — **n** glansafdruk *m* — **e** fotografía *f* brillante; copia *f* brillante; positivo *m* brillante; positiva *f* brillante; prueba *f* positiva sobre papel brillante

6463 glucose; dextro-glucose; d-glucose — Identical with *dextrose. It is the term preferred by biochemists.

6464 glucose — See dextrose.

6465 glue — This term should actually apply only to the raw *animal glue and *fish glue. It is commonly used, however, as a synonym for the word adhesive. See also bone glue, enamel glue.
f colle *f*; colle *f* forte — **d** Leim *m* — **n** lijm *m* — **e** cola *f*; pasta *f* adhesiva — **i** colla *f*

6466 glued-on cover
f couverture *f* collée au dos — **d** am Blockrücken angeleimter Umschlag *m* — **n** aangeplakt omslag *n* — **e** cubierta *f* encolada — **i** coperta *f* incollata

6467 glue-enamel process; glue process — Use of a solution of bichromated fish glue as a sensitizer for metal plates, development of the image with water, and burning-in of the print to convert the image into an acid-resistant.
f procédé *m* à la colle — **d** Heißemail-Prozess *m*; Heißemail-Kopierverfahren *n*; Leimemailverfahren *n* — **n** warmemailprocédé *n*; vislijmprocédé *n* — **e** procedimiento *m* al esmalte (ordinario); procedimiento *m* a la cola esmalte; esmalte *m* a la cola; cola *f* esmalte — **i** procedimento *m* allo smalto caldo; procedimento *m* alla colla di pesce

6468 glue-glycerine roller — See composition roller.

6469 glue pot — Small, double, copper boiler, electrically heated, to melt glue.
f pot *m* à colle (forte); jatte *f* sl. — **d** Leimtopf *m*;

Leimkessel *m*; Leimbehälter *m*; Leimkocher *m*;
Kleistertopf *m* — **n** lijmpot *m*; plakselpot *m* — **e**
cazo *m* de cola; cazo *m* para cola; olla *f* para cola;
colero *m* — **i** recipiente *m* della colla
6470 glue process — See glue-enamel process.
6471 gluer — See gluing machine.
6472 glue seam; gluing edge; gluing seam
f bord *m* de collage — **d** Klebenaht *f*; Kleberand *m*; Klebelasche *f* — **n** plakrand *m*; gomrand *m*; kleefrand *m* — **e** borde *m* de encolado —
i bordo *m* collato; bordo *m* d'incollatura
6473 glue top — Bichromated glue solution, especially commercial mixtures (by extension a burnt-
in glue print).
f couche *f* de colle — **d** Leimschicht *f*; Chromleim-
schicht *f* — **n** lijmlaag *fm* — **e** capa *f* de cola bicromatada — **i** strato *m* di colla bicromatada
6474 gluing edge — See glue seam.
6475 gluing machine; gluer
f encolleuse *f*; machine *f* à encoller — **d** Anleim-
maschine *f*; Leimmaschine *f*; Klebemaschine *f* —
n aanlijmmachine *f* — **e** encoladora *f*; máquina *f*
encoladora — **i** macchina *f* incollatrice; macchina
f per incollare
6476 gluing seam — See glue seam.
6477 gluing up — Application of glue to the backs
of books, the first step in bookbinding after
sewing.
f passure *f* en colle — **d** Anschmieren *n*; Ver-
leimen *n*; Ableimen *n* — **n** aanlijmen *n* (van de
rug) — **i** applicazione
f della colla (al dorso)
6478 glyceric acid — Colourless, sirupy compound, formed during alcoholic fermentation and
by oxidizing glycerol with nitric acid.
Formula: $C_3H_6O_4$.
f acide *m* glycérique — **d** Glyzerinsäure *f* — **n**
glycerinezuur *n* — **e** ácido *m* glicérico
6479 glyceride — Ester obtained from glycerol by
the replacement of one, two, or three hydroxyl
groups with a fatty acid.
f glycéride *m* — **d** Glyzerid *n* — **n** glyceride *m* —
e glicérido *m* — **i** gliceride *m*
6480 glycerin — See glycerol.
6481 glycerine — See glycerol.
6482 glycerol; glycerine *GB*; **glycerin** *US*; **glycyl
alcohol** — Clear, colourless or pale yellow, hygroscopic liquid, soluble in water and alcohol, insoluble in ether, benzene, chloroform and fixed
and volatile oils. Aqueous solutions are neutral.
Used in graphic arts, as plasticizers or softening
agent in paper, in composition inking rollers, in
some printing inks, in photography, in bookbinding glues, etc. The name "glycerol" is preferred over "glycerine" since it indicates its
alcohol structure.
Formula: $C_3H_5(OH)_3$.
f glycérol *f*; glycérine *f*; propane-triol *f* — **d**
Glyzerin *n*; Propantriol *n*; Trioxytripan *n*; Ölsüß *n* — **n** glycerol *fm*; glycerine *fm* — **e**
glicerol *m*; glicerina *f* — **i** glicerolo *m*; glicerina *f*
6483 glycerol tristearate — See stearin.

6484 glyceryl trinitrate — See nitroglycerine.
**6485 glycine; para-hydroxyphenylglycine; photo-
glycine** — White to buff crystals or amorphous
powder, slightly soluble in water, soluble in
alkaline solutions. Used in photographic
developers, cellulose and nitrocellulose acetate
lacquers. Poisonous.
Formula: $HOC_6H_4NHCH_2COOH$.
Not to be confused with aminoacetic acid or glyco-
coll which has the formula NH_2CH_2COOH.
f glycine *f*; para-oxyphénylglycine — **d** Glyzin *n*;
para-Oxyphenylglyzin *n*; para-Oxyphenyl-Amido-
Essigsäure *f*; para-Amidophenolglyzin *n* — **n**
glycine *f*; para-oxyfenyl glycine *f* — **e** glicina *f*;
para-hidroxifenilglicina *f* — **i** glicina *f*; para-
idrossifenilglicina *f*
6486 glycol — See ethylene glycol.
6487 glycol alcohol — See ethylene glycol.
6488 glycolic acid; hydroxyacetic acid — Acid, in
the juice of cane sugar and unripe grapes, also
made synthetically. Used in adhesives, in electroplating and as a preservative added to developers
of amidol.
Formula: $C_2H_4O_3$.
f acide *m* glycolique; acide *m* hydroacétique — **d**
Glykolsäure *f*; Hydroxyessigsäure *f* — **n** glycolzuur *n*; hydroxy-azijnzuur *n* — **e** ácido *m*
glicólico; ácido *m* hidroacético — **i** acido *m*
glicolico; acido *m* idrossiacetico
6489 glycols; diols — Colourless, odourless, non-
toxic, not inflammable and non-volatile solvents,
used in the manufacture of printing inks. Compounds with two hydroxyl groups attached to separate carbon atoms in an aliphatic carbon chain.
The saturated glycols may be represented by the
formula $C_nH_{2n}(OH)_2$, and as such, they are a class
of alcohols. When a second hydroxyl group is
introduced into ethyl alcohol, C_2H_5OH, the result
is *ethylene glycol.
f glycols *mpl* — **d** Glykolen *npl* — **n** glycolen *mpl*
— **e** glicolas *fpl* — **i** glicoli *mpl*
6490 glycyl alcohol — See glycerol.
6491 Glyphics — See Inciseds.
6492 glyptal resin — See alkyd resin.
6493 goatskin — Leather used for bookbindings,
made from the skin of goats, e.g., Arabian goat,
domestic goat, East India goat, French goat,
Levant, Niger goat, Persian goat, Turkey goat.
f peau *f* de chèvre; chevreau *m* — **d** Ziegenleder *n* — **n** geiteleer *n* — **e** piel *f* de cabra;
marroquín *m* de cabra — **i** pelle *f* di capra
6494 goffered edges; gauffered edges *depr.*;
chased edges — Gilded edges of a book, decorated
with heated tools, by painted designs or illustrations.
f tranches *fpl* ciselées; tranches *fpl* orientales —
d ziselierter Buchschnitt *m*; ziselierter Schnitt *m*;
bemalter Buchschnitt *m* — **n** geciseleerde sne(d)e
fm — **e** cortes *mpl* cincelados; cortes *mpl*
gofrados — **i** tagli *mpl* cesellati; tagli *mpl* goffrati
6495 goldbeater's skin — Prepared outside membrane of the large intestine of the ox, used by

goldbeaters to lay between the leaves of the metal while they beat it into gold leaf. It looks like cellophane. Used also for repairing parchment bindings and documents since it is akin to parchment.

f baudruche *f* — **d** Goldschlägerhaut *f* — **n** goudslagershuid *fm*; goudslagersvlies *n* — **e** película *f* de batihoja; tripa *f* de buey — **i** membrana *f* intestinale del bue

6496 gold blocking; gold stamping — Design on book covers, etc., stamped with gold leaf using heated tools or dies.

f empreinte *f* dorée — **d** Golddruck *m*; Vergoldung *f*; Goldprägung *f* — **n** goudstempel *n*; gouden opdruk *m* — **e** estampación *f* en oro — **i** impressione *f* in oro

6497 gold blocking press — See gilding press.

6498 gold bromide; aurous bromide; gold monobromide — Yellowish-grey mass. Insoluble in water.
Formula: AuBr.

f bromure *m* d'or — **d** Goldbromid *n* — **n** goudbromide *n* — **e** bromuro *m* de oro — **i** bromuro *m* d'oro

6499 gold bronze

f bronze *m* d'or — **d** Goldbronze *f* — **n** goudbrons *n* — **e** bronce *m* al oro — **i** bronzo *m* d'oro

6500 gold chloride — See gold trichloride.

6501 gold cushion — Cushion of sheepskin or calfskin on which gold leaf is laid out for cutting.

f coussin *m* à l'or; coussinet *m* à l'or — **d** Goldkissen *n*; Vergoldepolster *n* — **n** goudkussen *n*; kussen *n* — **e** almohadilla *f* de dorar — **i** cuscinetto *m* per la doratura

6502 golden mean — See golden section.

6503 goldenrod flat *US* — The method of assembling and positioning lithographic negatives or positives for exposure in contact with light-sensitized press-plate. The goldenrod paper used is translucent enough to see pencilled layout on underside, or master layout on separate white paper beneath, so film negatives can be attached in proper position with red Scotch tape. The goldenrod paper beneath image areas is cut away before flat is reversed to place emulsion-side of negatives to emulsion on metal plate. Flat is also used for making blueprint of forme for checking imposition. Sometimes referred to as a forme. A lithographic flat corresponds to a typographic forme.

f tracé *m* sur papier inactinique — **d** Montage *f* auf nicht-aktinischem Papier — **n** montage *f* op niet-actinisch papier — **e** montaje *m* sobre papel inactínico — **i** montaggio *m* su carta inattinica

6504 goldenrod paper — Instead of the transfer impression, the line or halftone cameraman provided a photographic image on film. This was used by a stripper, who assembled the films onto a ruled out sheet, called the layout. The negatives were placed in position and taped down. This sheet usually an opaque yellow (from which it got its name goldenrod) was called a *flat. The golden-

rod paper serves as a base for drawing the layout and attaching the film negatives. When exposure openings are cut through it, the remainder serves as an exposure mask, since its yellowish-orange colour does not transmit actinic light.

f papier cache *m*; cache *m* en papier — **d** Montageunterlagepapier *n* — **n** montagepapier *n* — **e** papel *m* para cubrir; papel *m* enmascar — **i** carta *f* inattinica per montaggi

6505 golden section; golden mean; sectio aurea; sectio divina — A ration between two portions of a line, or the two dimensions of a plane figure, in which the lesser of the two is to the greater as the greater is to the sum of both: a ratio of approximately 0.616 to 1.000.

f section *f* d'or — **d** goldener Schnitt *m* — **n** gulden snede *fm* — **e** divina proporción *f*; proporción *f* áurea; formato *m* áureo; canon *m* áureo; rectángulo *m* de proporciones áureas — **i** sezione *f* aurea

6506 gold foil — See gold leaf.

6507 gold ink — Ink of golden colour consisting of special varnish and bronze powder used as a substitute for gold sizing and subsequent bronzing; usually mixed just before printing.

f encre *f* d'or directe; encre *f* dorée — **d** Golddruckfarbe *f* — **n** goudinkt *m* — **e** tinta *f* para imprimir en oro — **i** inchiostro *m* oro; inchiostro *m* per stampa oro

6508 gold knife — Knife with a long, flat blade sharpened on both sides, used by gilders for cutting gold leaf from the gold cushion.

f couteau *m* à or; couteau *m* de doreur — **d** Vergoldemesser *n*; Goldmesser *n* — **n** goudmes *n* — **e** espátula *f* de dorar — **i** coltellino *m* da doratore

6509 gold leaf; gold foil — Thin sheets of gold used for stamping book covers, etc., with a heated die. See also blocking foil.

f feuille *f* (mine) d'or — **d** Blattgold *n*; Goldfolie *f* — **n** bladgoud *n*; goudfo(e)lie *fm* — **e** oro *m* (batido) en hojas; pan *m* de oro; pan *m* para dorar — **i** foglia *f* d'oro; oro *m* in foglie

6510 gold monobromide — See gold bromide.

6511 gold oxide; auric oxide; gold trioxide — Brownish powder, insoluble in water, soluble in hydrochloric acid.
Formula: Au_2O_3.

f oxyde *m* d'or — **d** Goldoxyd *n* — **n** goudoxyde *n* — **e** óxido *m* de oro — **i** ossido *m* d'oro

6512 gold paper — Metallic bronze-coated paper with many qualities and weights ranging from light weights used in the paper-box industry to heavy bristols.

f papier *m* doré — **d** Goldpapier *n* — **n** goudpapier *n* — **e** papel *m* dorado — **i** carta *f* dorata

6513 gold potassium bromide — See potassium gold bromide.

6514 gold potassium chloride — See potassium gold chloride.

6515 gold powder

f poudre *f* de bronze — **d** Goldbronzepuder *m* —

n goudbronspoeder *n* — **e** polvo *m* dorado — **i** oro *m* in polvere
6516 gold stamping — See gold blocking.
6517 gold trichloride; auric chloride; gold chloride; trichloride of gold — Yellow to red crystals, decomposed by heat, soluble in water, alcohol and ether.
Formula: $AuCl_3$ or $AuCl_3 \cdot 2H_2O$.
f chlorure *m* d'or; aurichlorure *m* — **d** Gold-chlorid *n*; Aurichlorid *n* — **n** goudchloride *n* — **e** cloruro *m* de oro; cloruro *m* áurico — **i** cloruro *m* d'oro; cloruro *m* aurico
6518 gold trioxide — See gold oxide.
6519 golf-ball typewriter
f machine *f* à écrire à (avec) tête imprimante — **d** Kugelkopf-Schreibmaschine *f* — **n** kogelschrijf-machine *f* — **e** máquina *f* de escribir con cabeza impresora — **i** macchina *f* per scrivere a sfera
6520 golf-ball typing head
f tête *f* en balle de golf; tête *f* en sphère — **d** Schreibkugel *f*; Kugelkopf *m* — **n** schrijfkogel *m*; letterkogel *m* — **e** cabeza *f* impresora — **i** sfera *f* di battuta; testa *f* scrivente a sfera; pallina *f* porta-caratteri
6521 gone to bed — In printing offices, passed to press; there is no time left to make corrections or alterations.
f sous presse — **d** in die Maschine (gegangen) — **n** ter perse; reeds op de pers — **e** en prensa; en impresión — **i** andato in macchina
6522 good — Said of type matter when it is likely to be required again, and should not be distributed.
f à conserver — **d** stehen lassen — **n** laten staan; bewaren — **e** a guardar — **i** da conservare
6523 goose — See wayzgoose.
6524 gooseneck — See blower foot.
6525 gooseneck — See throat.
6526 Gordon press — Platen press invented by G.P. Gordon (1858).
f presse *f* Gordon — **d** Gordon-Presse *f*; Gordon-Maschine *f* — **n** Gordon-pers *fm* — **e** máquina *f* Gordon — **i** pressa *f* Gordon; macchina *f* Gordon
6527 Gothic — See Gothic type.
6528 Gothic type; Gothic — Formerly all bold sans-serif and grotesque type faces. Originally Black letter (or Old English), type faces based on the manuscripts of the scribes. Originates from the North of France; after 1500 superseded but continuing in Germany in the Schwabacher and the Fraktur, in France as bâtarde, and in the Netherlands as flamande or Old Dutch.
f caractère *m* gothique — **d** gotische Schrift *f*; Gotisch *f* — **n** Gothische letter *fm* — **e** carácter *m* gótico; letra *f* gótica — **i** carattere *m* gotico; lettera *f* gotica
6529 go *v* **to press** — To set the forme on a press and get ready for press operation.
f mettre sous presse — **d** einheben — **n** ter perse nemen — **e** entrar en prensa; meter en prensa; sentar la forma; echar forma; lanzar — **i** mettere in macchina

6530 gouache painting — Picture painted with opaque colours that have been ground in water and mingled with a preparation of gum.
f peinture *f* à la gouache; gouache *f* — **d** Guasche *f*; Gouache *f* — **n** gouache *fm* — **e** gouache *m*; guache *m* — **i** pittura *f* a guazzo; guazzo *m*
6531 gouge — Finishing tool, used in hand engraving, the belly of which is shaped like a circle segment. See also graver.
f échoppe *f* ronde — **d** Rundstichel *m* — **n** rond-steker *m* — **e** escoplo *m* redondo — **i** bulino *m* tondo
6532 gouge — In book decoration, brass tool to impress curved lines.
f gouge *f* — **d** Bogenstempel *m*; Bogensatz-stempel *m* — **n** boogstempel *n* — **e** gubia *f* — **i** filetto *m* a curva
6533 gouge index — See thumb index.
6534 gradation — Gradual passage of tones from one to another in originals, negatives and reproductions.
f gradation *f* — **d** Gradation *f*; Abstufung *f*; Steilheit *f* — **n** gradatie *f*; steilheid *f* — **e** gradación *f* — **i** gradazione *f*
6535 gradient — Rate of increase or decrease. Steepness of a slope.
f gradient *m*; inclinaison *f*; pente *f* — **d** Gradient *m*; Gefälle *n*; Neigung *f* — **n** gradient *m*; verloop *n* — **e** pendiente *f* (rising gradient); declive *m* (falling gradient) — **i** gradiente *m*
6536 grain; photographic grain — Distribution of silver particles in photographic emulsions and images.
f grain *m*; granulation *f* — **d** Korn *n* — **n** korrel *m* — **e** grano *m*; granillo *m*; granulado *m* — **i** grana *f*
6537 grain — Direction of the fibres in a sheet of paper, governing paper properties such as increased size change with relative humidity across the grain, and better folding qualities along the grain.
f grain *m*; direction *f* des fibres; sens *m* des fibres — **d** Faserlauf *m*; Faserrichtung *f* — **n** vezel-richting *f* — **e** grano *m* del papel; dirección *f* de las fibras; dentillado *m* granoso; flor *m* — **i** direzione *f* della fibra; direzione *f* di fibra
6538 grain — Roughened or irregular surface of an offset printing plate, which assists the retention of moisture and control of the image. See also chemical grain, coarse grain, fine grain, electrolytic grain.
f grain *m* — **d** Korn *n* — **n** grein *n*; korn *m* — **e** grano *m* — **i** grana *f*
6539 grain — Unit of weight avoirdupois. Abbrev. gr (no point). See app. no. 7 and 7a.
f grain *m* (anglais) — **d** (englischer) Grain *m*; (englisches) Gran *n* — **n** (Engelse) grein *n* — **e** grano *m* (inglés) — **i** grano *m* (inglese)
6540 grain, against the — See against the grain.
6541 grain box — See dusting box.

6542 grain direction — See machine direction.
6543 grained paper — Embossed or decorated paper with a surface imitating various grains such as wood, marble, alligator, Spanish leather, usually made in cover or box-cover weights.
f papier *m* grainé; papier *m* chagriné; papier *m* granulé — **d** gekörntes Papier *n*; genarbtes Papier *n*; Maserpapier *n* — **n** gekornd papier *n*; gekorreld papier *n* — **e** papel *m* veteado; papel *m* granulado — **i** carta *f* granulata
6544 grained roller — See nap roller.
6545 grain halftone — Halftone reproduction made with a dust grain or with a grainy screen.
f similigravure *f* à grain — **d** Kornrasterreproduktion *f* — **n** reproduktie *f* met een kornraster — **e** autotipia *f* de grano — **i** riproduzione *f* con retino granulato
6546 graining; pebbling — Preparing the surface of a lithographic stone or metal plate by the oscillation of a plate-graining machine containing glass or porcelain marbles (or cods) of about 2.5 cm (1 in) diameter, to impart greater water retention to the otherwise non-porous surface. The marbles rotate and cut up the surface of the plate or stone with the aid of sand, glass or pumice powder, giving it the appearance of ground glass.
f grainage *m*; grenage *m* — **d** Körnung *f*; Körnen *n* — **n** greinen *n*; kornen *n* — **e** graneado *m* — **i** granitura *f*
6547 graining — See missing dots.
6548 graining balls — See cods.
6549 graining bath — Dilute nitric acid-alum solution to impart in photoengraving a matt surface to zinc plates before sensitization of the metal.
f bain *m* de décapage — **d** Anrauhbad *n*; Ankörnbad *n* — **n** greinbad *n* — **e** baño *m* de granear; baño *m* de graneador — **i** bagno *m* d'irruvidimento
6550 graining machine; ball graining machine; plate graining machine — Machine to produce the grain on lithographic plates by oscillating the plate under wet abrasive and marbles or steel ball-bearings to give weight to abrasive.
f grainoir *m* (à billes); machine *f* à grainer des plaques offset — **d** Kornmaschine *f*; Plattenkornmaschine *f* — **n** greinmachine *f*; knikkerbak *m* — **e** graneadora *f* (de planchas); máquina *f* de granular; máquina *f* de granear; máquina *f* graneadora — **i** granitoio *m*; macchina *f* granitrice (a biglie)
6551 graining marbles — See cods.
6552 grain screen; mezzograph screen; metzograph screen — Halftone screen embodying a grain rather than a ruled line formation.
f trame *f* à grain; trame *f* mezzographe; trame *f* metzographe; écran *m* au grain — **d** Kornraster *n* — **n** kornraster *n*; korrelraster *n*; mezzograph raster *n* — **e** trama *f* graneada; grisado *m* de resina; trama *f* de resina; trama *f* mezzograph; trama *f* metzograph — **i** retino *m* a grana; retino

m granulato
6553 grain shading — Shading effect obtained by dusting and melting resinous or bituminous powder on a plate before etching. The molten grains of resin act as a resist and leave a design on the etched cut.
f teinte *f* au grain de résine — **d** Staubkorn *n* — **n** kornraster *n* — **e** grano *m* de resinada — **i** tono *m* granuloso
6554 grainy printing — Printing characterized by unevenness, particularly of halftones.
f impression *f* poivrée — **d** körniger Druck *m*; ungleichmäßiger (Halbton)Druck *m* — **n** ongelijkmatige afdruk *m* — **e** impresión *f* granulosa — **i** stampa *f* granulosa
6555 gram — See gramme.
6556 gramme *GB*; **gram** *US* — Metric unit of mass. Abbrev. g (no point). See app. no. 7.
f gramme *m* — **d** Gramm *n* — **n** gram *n* — **e** gramo *m* — **i** grammo *m*
6557 gramme-molecule — See mol.
6558 gram-molecule — See mol.
6559 grand total — See newsprint waste.
6560 grangerize *v* — To augment the illustrative content of a book by inserting additional prints, drawings, engravings, etc., not included in the original volume.
f truffer — **d** übermäßig mit Bildern aus dritten Büchern illustrieren; trüffeln; durch zusätzliche Bilder erweitern — **n** opvullen met illustraties uit andere boeken — **e** agregar con ilustraciones — **i** aumentare il contenuto di un libro inserendo altro materiale illustrato
6561 grangerize *v* — To mutilate books to get illustrative material for other books.
f décortiquer — **d** Illustrationen aus Büchern entnehmen — **n** illustraties uitnemen — **e** cortar los grabados de un libro para ilustrar otro — **i** ritagliare illustrazioni da un libro per inserirle in un altro
6562 grape sugar — See dextrose.
6563 graph — See diagram.
6564 graphic — In computer controlled typesetting, *sign displayed in a particular size and type fount.
f signe *m* graphique — **d** Schriftzeichen *n* — **n** schriftteken *n* — **e** representación *f* gráfico — **i** segno *m* grafico
6565 graphic arts — The entire graphic arts industry (printing, lithography, engraving, bookbinding, electrotyping, stereotyping, and allied processes and trades).
f industrie *f* graphique — **d** graphisches Gewerbe *n* — **n** grafische industrie *f* — **e** artes *fpl* gráficas — **i** arti *fpl* grafiche
6566 graphic arts quality; typographic quality — Quality equivalent to letterpress composition which must be achieved in type design, reproduction and format by equipment that produces good quality film or paper output at fast speeds. Distinct from the reproduction obtained from typewriters and line printers used with computer out-

put.

6567 graphic designer
f graphiste *m*; artiste *m* graphique — **d** Graphiker *m* — **n** grafisch ontwerper *m* — **e** grafista *m*; creativo *m* gráfico — **i** disegnatore *m* grafico

6568 graphic image — "Sign" with a particular and specific appearance, such as a Time roman 8 point lower case a. Also any illustrative material, such as a halftone or a line drawing.

6569 Graphics — See Manuals.

6570 graphic transcriber — Device to digitize diagrams or maps with an operator following the lines of the drawing with an electronic device which records the changing position of the stylus as digital codes.

6571 graphite; black lead — Finely powdered form of carbon, used (1) in electrotype moulding for the free release (without sticking) of the wax after the mould is made, and to help prevent sideslipping during moulding; (2) in coating the case to provide a conductive film to the surface of the mould; (3) in coating the surface of the matrix in plastic platemaking to effect a free release of the plate from the matrix; (4) as coating wax as a lubricant for spacebands of typesetting machines; (5) in the manufacture of pencils.
f graphite *m*; mine *f* de plomb — **d** Graphit *m*; Graphitschwärze *f*; Reißblei *n* — **n** grafiet *n* — **e** grafito *m*; negro *m* de plomo — **i** grafite *f*

6572 graphite brush — See black-leading brush.

6573 graph paper; plotting paper; metric rule paper — Paper for charts and graphs, ruled in both directions with parallel, equally spaced lines, used by layout men, scientists, etc.
f papier *m* pour graphiques; papier *m* millimétrique; papier *m* millimétré — **d** Millimeterpapier *n* — **n** grafiekpapier *n*; millimeterpapier *n* — **e** papel *m* cuadriculado; papel *m* milimetrado — **i** carta *f* millimetrata

6574 grasshand *sl.* — Compositor employed casually by the day.
f journalier *m* — **d** Springer *m*; Wanderer *m* — **n** tijdelijk zetter *m* — **e** compositor *m* pasajero — **i** giornaliero *m*

6575 grass pulp and similar pulp — Pulp obtained from the leaves and stems of grasses, reeds, etc.
f pâte *f* de monocotylédones — **d** Halbstoff *m* aus einkeimblättigen Pflanzen — **n** celstof *fm* uit gras, riet, enz. — **e** pasta *f* de monocotiledóneos; pulpa *f* de monocotiledóneos — **i** pasta *f* a base di cellulosa di foglie (canne, ecc.); pasta *f* di erba e similar pasta

6576 graticule *v* — To divide a picture or plan into small squares for reproduction or enlargement.
f graticuler — **d** in Quadrate *npl* einteilen; mit einem Netz versehen — **n** in vierkanten verdelen — **e** cuadricular — **i** quadrettare; tracciare un reticolo

6577 grave accent — E.g., à, used especially in French and Italian.
f accent *m* grave — **d** Gravis *m*; Accent grave *m* — **n** gravis *m*; accent *m* grave; graaf *fm coll.* — **e** acento *m* grave — **i** accento *m* grave

6578 graver; echoppe — Cutting or shaping tool like, e.g., scorper, scoop, scooper, *gouge, spitzsticher, used by engravers and finishers in cutting design in plates. See also burin, sculper.
f échoppe *f*; burin *m*; échoppe *f* ronde; onglette *f*; gouge *f*; gravoir *m*; ciselet *m*; style *f*; pied-de-biche *m*; outil *m* à ton vélo — **d** Stichel *m* (Grabstichel; Bollstichel; Boltstichel; Flachstichel; Spitzstichel; Messerstichel; Fadenstichel; Drahtstichel; Vierkantstichel) — **n** steker *m* (graafsteker; vlaksteker; diepsteker; fadensteker) — **e** buril *m* (de grabar) (buril-rastrillo; buril con cuatro cortes; buril aguisado; herramienta de múltiples líneas; chaple) — **i** bulino *m* (lingua di gatto; cesello)

6579 graveyard — See morgue.

6580 gravity headbox
f caisse *f* d'arrivée (de pâte) à gravité — **d** Schwerstoffauflauf(kasten) *m* — **e** caja *f* de entrada por gravedad; caja *f* de alimentación por gravedad — **i** cassa *f* d'afflusso a gravità

6581 gravure — See photogravure.

6582 gravure à la poupée — See colour etching.

6583 gravure cell — See ink cell.

6584 gravure coating — Roll coating process in which the applicator roll consists of (or is supplied with coating material by) a metal roll engraved with small, closely spaced cells or depressions over its entire surface. Generally a separate coating operation.
f couchage *m* au rouleau gravé — **d** Tiefdruckbeschichtung *f*; Streichen *n* im Tiefdruckverfahren — **n** strijkprocédé *n* met gerasterde rol — **e** estucado *m* semejante el rotograbado — **i** patinatura *f* a rullo inciso

6585 gravure cylinder
f cylindre *m* de gravure; cylindre *m* porte-image — **d** Tiefdruckzylinder *m* — **n** diepdrukcilinder *m* — **e** cilindro *m* de huecograbado — **i** cilindro *m* rotocalco

6586 gravure ink — Thin, quick drying, low viscosity ink based on volatile solvents, used in intaglio printing methods. See also copperplate engraving ink, rotogravure ink.

6587 gravure machine; gravure press
f presse *f* en creux — **d** Tiefdruckmaschine *f*; Tiefdruckpresse *f* — **n** diepdrukpers *f*; diepdrukmachine *f* — **e** máquina *f* de huecograbado; máquina *f* para huecograbado — **i** macchina *f* rotocalco

6588 gravure paper — See rotogravure paper.

6589 gravure press — See gravure machine.

6590 gravure printing — See intaglio printing.

6591 gravure screen
f trame *f* hélio — **d** Tiefdruckraster *m* — **n** diepdrukraster *n* — **e** trama *f* de huecograbado — **i** retino *m* per rotocalco

6592 gravure tissue — See carbon tissue.

6593 gravure varnish — See extender varnish.

6594 gray ... — See grey ...
6595 gray bookcloth — See grey bookcloth.
6596 gray cloth — See grey bookcloth.
6597 Gray code — See reflected binary code.
6598 gray scale — See grey scale.
6599 gray wedge — See grey-wedge.
6600 grazing incidence light — Light with angles of incidence close to 90°. It becomes convenient to speak of grazing incidence light, that is the complement of the usual angle of incidence, which is the angle subtended by the incident ray and the normal to the reflection surface.
f lumière *f* rasante; lumière *f* arasante — **d** Schräglicht *n*; schrägeinfallendes Licht *n* — **n** strijklicht *n*; schuin invallend licht *n* — **e** luz *f* de incidencia rasante — **i** luce *f* incidente radente
6601 grease — See also consistent grease.
f graisse *f* — **d** Schmierfett *n* — **n** smeervet *n* — **e** grasa *f* — **i** grasso *m*
6602 greaseproof ink — Ink resistant to the action of fats, oils and greases.
f encre *f* résistante aux graisses — **d** fettechte Druckfarbe *f*; fettbeständige Druckfarbe *f* — **n** vetbestendige inkt *m* — **e** tinta *f* resistente al grasa — **i** inchiostro *m* resistente ai grassi
6603 greaseproof paper; oilproof paper — Paper free from mechanical pulp, with a high resistance to penetration by grease and an appearance of vegetable parchment, both obtained by heavy beating.
f papier *m* simili-sulfurisé; simili-sulfurisé *m* — **d** fettdichtes Papier *n*; Butterbrotpapier *n* — **n** vetdicht papier *n* — **e** papel *m* a prueba de grasa; papel *m* para grasas — **i** carta *f* impermeabile ai grassi; carta *f* resistente ai grassi
6604 grease repellent; grease resisting
f refusant les corps gras — **d** fettabstoßend — **n** vetafstotend — **e** rechazando los cuerpos grasos — **i** repellente ai grassi
6605 grease-resistant paper — Paper which cannot be penetrated by grease to any appreciable extent.
f papier *m* ingraissable; papier *m* imperméable à la graisse — **d** fettechtes Papier *n*; fettbeständiges Papier *n* — **n** vetbestendig papier *n* — **e** papel *m* impermeable a la grasa — **i** carta *f* resistente ai grassi
6606 grease resisting — See grease repellent.
6607 grease-resisting area; ink-resisting area; non-printing area
f région *f* refusant le corps gras; région *f* non-imprimante — **d** fettabstoßende Fläche *f*; farbabweisende Fläche *f*; nicht-druckende Fläche *f* — **n** vetafstotend gedeelte *n*; inktafstotend gedeelte *n*; niet-drukkende partij *f* — **e** parte *f* rechazanda la tinta — **i** zona *f* inchiostrorepellente; zona *f* non stampante
6608 grease stain; spot of grease
f tache *f* de graisse — **d** Fettfleck *m* — **n** vetvlek *fm* — **e** mancha *f* de grasa; mancha *f* de pringue; churrete *m* — **i** macchia *f* di grasso; macchia *f* d'unto

6609 greasing — Sensitization of the non printing areas of a lithographic plate resulting in undesirable adhesion of the ink that cannot be removed by gumming up. Generally starts as a spreading of the image areas. See also tinting.
f graissage *m* (en offset) — **d** Tonen *n* — **n** tonen *n* — **e** engrasado *m* (en offset) — **i** velatura *f*; ingrassamento *m*
6610 greasing — See also oiling.
f graissage *m* — **d** Schmierung *f* — **n** smering *f* — **e** engrase *m*; engrasado *m* — **i** ingrassaggio *m*
6611 greasing — See scumming.
6612 greasy ink — In offset, ink which is likely to cause smudging.
f encre *f* grasse — **d** fette Farbe *f* — **n** vette inkt *m* — **e** tinta *f* grasa — **i** inchiostro *m* grasso
6613 great-capacity bag
f sac *m* de grande contenance — **d** großer Papiersack *m*; großvolumiger Papiersack *m* — **n** grote papierzak *m* — **e** bolsa *f* grande — **i** sacco *m* di grande capacità
6614 great indulgence — Entrance fee expected to be paid by a compositor when he newly entered an office, usually in the form of treating the companions. Now used to imply "good time", "bust up", or "razzle".
f payer sa bienvenue; droit *m* de bienvenue; droit *m* de quatre heures — **d** Introitum *n*; den Einstand geben; Einstandgeld geben — **n** entree betalen; tol betalen — **e** pagar la entrada; entrada *f* — **i** pedaggio *m*; diritto *m* d'entrata
6615 great primer — Old name for a size of type. See app. no. 1.
6616 grecs du roi — Greek cursive type with many ligatures, cut (1541-1549) in three sizes by Claude Garamond for Francis I of France, after the designs of Angelos Vergetios of Crete and under the supervision of Robert Estienne.
6617 Greek — See Greek characters.
6618 Greek alphabet — Classical and modern printed the same; alphabet 17 consonants, seven vowels, two breathings, three accents (acute, grave, and circumflex), one apostrophe, one diaeresis; note of interrogation same as English semicolon; the colon or semicolon same as turned point (·); comma, exclamation point, and period as in English. There are many rules for composing. Grave accent only on last syllable; diphthongs, accents, and breathings on second vowel, all vowels or diphthongs commencing a word have either asper (ʽ) or lenis (ʼ) breathings; the sigma (σ) when final is always ς; in dividing words ending in κτος the κ is turned over. The Greek letters, their names and, in parentheses their English equivalents are listed as follows: Α, α alpha (a); Β, β beta (b); Γ, γ gamma (g); Δ, δ delta (d); Ε, ε epsilon (ĕ); Ζ, ζ zeta (z); Η, η eta (ē); Θ, θ theta (th); Ι, ι iota (i); Κ, κ kappa (k); Λ, λ lambda (l); Μ, μ mu (m); Ν, ν nu (n); Ξ, ξ xi (x); Ο, ο omicron (ŏ); Π, π pi (p); Ρ, ρ rho (r); Σ, σ sigma (s); Τ, τ tau (t); Υ, υ ypsilon (u); Φ, φ phi (ph); Χ, χ chi (ch); Ψ, ψ psi (ps); Ω, ω omega (o).

f alphabet *m* grec — d griechisches Alphabet *n* —
n Griekse alfabet *n* — e alfabeto *m* griego — i
alfabeto *m* greco

6619 Greek characters; Greek letters; Greek —
See also Greek alphabet.
f caractères *mpl* grecques; lettres *fpl* grecques —
d griechische Buchstaben *mpl* — n Griekse
letters *fmpl*; Griekse lettertekens *npl* — e
caracteres *mpl* griegos; tipos *mpl* griegos — i
caratteri *mpl* greco

6620 Greek fret — Design made up of bands or
fillets or borders for decoration of pages, or to be
tooled on book covers; characterized by angular
alterations and interlocking lines in various
patterns. Common motif for typographic borders.
f greque *f* — n Griekse rand *m* — e orla *f* de
greca; adorno *m* de greca; greca *f* — i greca *f*;
contorno *m* a greca; cornice *f* grecata

6621 Greek letters — See Greek characters.

6622 green copperas — See copperas.

6623 green filter — Filter that photographs the
red and blue rays. This plate is printed with the
magenta ink.
f filtre *m* vert; écran *m* vert — d Grünfilter *m* —
n groenfilter *n* — e filtro *m* verde — i filtro *m*
verde

6624 green fog — See dichroic fog.

6625 green paper — Recently made paper, not sufficiently conditioned to equalize moisture content.
See also mature paper.
f papier *m* jeune; papier *m* non mûri; papier *m*
frais — d nicht-ausgereiftes Papier *n* — n on-
bestorven papier *n*; niet-bestorven papier *n* — e
papel *m* inmaduro — i carta *f* non stagionata

6626 greeting cards — Cards, often with an illustration or decorations, to send a message of
holiday greetings, congratulations, good wishes, or
other sentiments.
f cartes *fpl* de vœux — d Glückwunschkarten *fpl*;
Gratulationskarten *fpl*; Grußkarten *fpl* — n
felicitatiekaarten *fmpl*; gelukwensen *mpl* — e
tarjetas *fpl* de felicitación — i biglietti *mpl*
d'augurio; cartoncini *mpl* d'augurio; biglietti *mpl*
augurali

6627 grey bookcloth; grey cloth *GB*; **gray bookcloth; gray cloth** *US* — Grey or greyish cloth used
in commercial and office bindings.
f bisonne *f* — d graues Buchbinderleinen *n*; graue
Leinwand *f* — n grijs boekbinderslinnen *n*; grijs
linnen *n* — e tela *f* gris para encuadernación — i
tela *f* grigia per rilegatura

6628 grey cloth — See grey bookcloth.

6629 grey scale *GB*; **gray scale** *US*; **step wedge;
halftone step scale; sensitivity guide** — Step scale
of increasing densities prepared on a photographic film or paper, used as a control in photography to measure reproduction contrast changes
or to time development.
f gamme *f* de gris (échelonnée); échelle *f* de gris
— d Grauskala *f*; Stufengraukeil *m*; Grauleiter *m*;
Rasterstufenkeil *m*; Empfindlichkeitskontroll-
skala *f* — n grijstrap *m* — e escala *f* de grises;

gama *f* de grises; guía *f* de densidades; guía *f* de
sensibilidad — i scala *f* dei grigi; scala *f* controllo

6630 grey-wedge *GB*; **gray wedge** *US* —
Unstepped photographic film or paper with continuously increasing densities.
f gamme *f* de gris continue — d Graukeil *m* — n
grijswig *f* — e cuña *f* de grises — i cuneo *m* dei
grigi; scala *f* dei grigi

6631 grid — See character grid.

6632 grid current — Current which moves within
the vacuum tube from the grid to the cathode.
f courant *m* de grille; intensité *f* de grille — d
Gitterstrom *m* — n roosterstroom *m* — e
corriente *f* de rejilla — i corrente *f* di griglia

6633 grid potential — See grid voltage.

6634 grid voltage; grid potential
f tension *f* de grille; potentiel *m* de grille — d
Gitterspannung *f* — n roosterspanning *f* — e
tensión *f* de rejilla — i tensione *f* di griglia

6635 grind; swot *GB sl.* — Tedious and laborious
work.
f pièce *f* de bœuf; labeur *m* monotone et continu
— d tüchtiges Stück *n* Arbeit; quälende Arbeit *f*
— n eentonig en langdurig Werk *n*; karwei *n*;
kluif *fm fam.* — e trabajo *m* aburrido

6636 grind *v* — The grinding of a lithographic
stone.
f meuler; rectifier — d abschleifen; schleifen — n
slijpen; afslijpen; wegslijpen; uitslijpen — e
amolar — i levigare; rettificare; affilare

6637 grind *v* **a colour**
f broyer — d anreiben; mahlen; vermahlen; ver-
reiben — n malen; (de inkt) aanwrijven — e
moler (la tinta); preparar (la tinta) — i macinare
l'inchiostro

6638 grinder — Machine to produce mechanical
wood pulp or groundwood by defibring pulpwood
logs against a rotating stone.
f défribreur *m* — d Schleifer *m* — n slijper *m* —
e desfibrador *m* — i sfibratore *m*

6639 grinder — See grinding machine.

6640 grind gauge — See fineness-of-grind gauge.

6641 grinding — Producing mechanical wood pulp
or groundwood by defibring pulpwood logs
against a rotating stone.
f défibrage *m*; râpage *m* — d Schleifen *n* — n
slijpen *n* — e desfibrado *m* — i sfibratura *f*

6642 grinding charcoal
f charbon *m* à polir — d Schleifkohle *f* — n houtskool *fm* om te slijpen; slijpkool *fm* — e carbón *m*
dulce para emolar — i carbone *m* (di salice) per
levigare

6643 grinding machine; grinder — Machine to
remove metal by the action of fast rotating wheels
of abrasive material, resulting in a very smooth
polishable surface.
f machine *f* à meuler; affûteuse *f* — d Schleif-
maschine *f* — n slijpmachine *f* — e amoladora *f*
— i macchina *f* affilatrice

6644 grinding medium — See vehicle.

6645 grinding mill; groundwood mill
f râperie *f* — d Schleiferei *f*; Holzschleiferei *f* —

n slijperij *f* — **e** desfibraduría *f*; fábrica *f* de pasta mecánica; instalación *f* de pasta mecánica; instalación *f* de pasta desfibrada — **i** fabbrica *f* di pastalegno; fabbrica *f* di pasta meccanica; sfibratura *f*

6646 grinding sand — See sand.

6647 grinding stone; grindstone — Natural granite, basalt or lava stone or artificial carborundum stone with burred surface used to defibre pulpwood logs and produce groundwood.
f meule *f* — **d** Schleifstein *m* — **n** maalsteen *m* — **e** muela *f*; amoladera *f*; piedra *f* amoladera — **i** pietra *f* abrasiva; pietra *f* da affilare; affilatoio *m*; cote *f*

6648 grindstone — See grinding stone.

6649 grip — See hand grip.

6650 gripper — In a sheet-fed press, the mechanical fingers that take hold of the press sheet and carry it through the press for printing. See also cleat.
f pince *f* — **d** Greifer *m* — **n** grijper *m* — **e** pinza *f* (de enganche); uña *f* — **i** pinza *f*

6651 gripper allowance — See gripper margin.

6652 gripper bar — Metal bar carrying a row of grippers.
f tringle *f* de pinces; barre *f* de pinces — **d** Greiferstange *f* — **n** grijperstang *fm* — **e** portauñas *m*; barra *f* de las pinzas; varilla *f* de la pinzas; biela *f* conductora para las lengüetas; árbol *m* conductor para las lengüetas — **i** barra *f* delle pinze; barra *f* portapinze

6653 gripper bite — See gripper margin.

6654 gripper edge; feed edge — Edge of the paper which is fed to the press gripper and into which the image must not extend, about 9.5 mm (3/8 in).
f bord *m* pinces; côté *m* pinces; côté *m* des pinces — **d** Greiferkante *f*; Greiferrand *m*; Anlegekante *f* — **n** grijperkant *m* — **e** orilla *f* de entrada; margen *fm* de entrada de pinzas; blanco *m* de pinzas — **i** lato *m* pinze

6655 gripper guide — See front lay.

6656 gripper margin; gripper allowance; gripper bite; gripper pad — Unprinted area between the edge of the sheet and lead edge of the printing area, allotted for the press grippers to hold the sheet.
f blanc *m* de pinces; marge *f* de pinces; prise *f* de pince; blanc *m* de prise — **d** Greiferkante *f*; Greifersteg *m*; Anlagesteg *m*; Kapitalsteg *m* — **n** grijperwit *n* — **e** margen *fm* de uña(s); margen *fm* para las uñas; agarre *m* de la uña; blanco *m* de entrada — **i** margine *m* principale; margine *m* di pinza; bianco *m* per presa pinze

6657 gripper pad — See gripper margin.

6658 grit — Hard particles in loading materials of paper.
f gravier *m* — **d** Sandkörner *npl* — **n** zanderige deeltjes *npl* — **e** granos *mpl* de arena; partículas *fpl* ásperas y duras — **i** sabbiosità *f*; granelli *mpl* di sabbia

6659 grocers 4-lb — A size of wrapping paper. See app. no. 3.

6660 grocers 6-lb — A size of wrapping paper. See app. no. 3.

6661 grocery bag — See bag.

6662 grocery bag paper — See grocery paper.

6663 grocery paper; grocery bag paper — Paper used by the grocery and provision trade.
f papier *m* d'épicerie; papier *m* de denrées — **d** Tütenschrenz *m*; Materialwarenpack *m* — **n** zakkenschrens *fm* — **e** estraza *f* para bolsas — **i** carta *f* per generi alimentari

6664 Grolier binding; Grolieresque binding — Books bound for Jean Grolier, viscount d'Asguisy (1479-1565), treasurer of France in 1545. There are three main styles: 1. with simple interlaced strapwork, small flowers, and tools cut solid; 2. plain or coloured ground with an intricate interlaced fillet; 3. without a fillet, decorated with azured stamps. Grolier was one of the first collectors to have the title lettered on the spine of his books.
f reliure *f* Grolier — **d** Grolier-Einband *m* — **n** Grolier-band *m* — **e** encuadernación *f* Grolier — **i** legatura *f* Grolier

6665 Grolieresque binding — See Grolier binding.

6666 groove; heelnick — Hollow between the feet of a type.
f gouttière *f*; gouttière *f* au pied — **d** Fußrille *f*; Fußausschnitt *m* — **n** breuk *fm*; groef *fm* — **e** muesca *f*; hendidura *f*; canal *m*; ranura *f* — **i** canale *m*

6667 groove; grooved joint — Valley or channel at either side or parallel to the backbone joints of a cased-in book. See also French groove, American groove.
f mors *m*; charnière *f* — **d** Falz *m* — **n** kneep *fm* — **e** surco *m* — **i** morso *m*

6668 groove *v* — To make grooves in heavy weight board to allow it to bend. The material is cut in with two knives and the chip between the cuts is taken away so that a gutter arises.
d nuten — **n** groeven; een groef maken — **e** acanalar; hacer ranuras — **i** scanalare

6669 groove cylinder — Forme cylinder on some letterpress machines covered with grooves to anchor the plates in the appropriate positions.
f cylindre *m* à rainures — **d** Nutenzylinder *m*; Plattenzylinder *m* mit eingefrästen Nuten — **n** groefcilinder *m* — **e** cilindro *m* ranurado — **i** cilindro *m* a scanalature

6670 grooved joint — See French groove.

6671 grooved joint — See groove.

6672 gross hundredweight — See long hundredweight.

6673 gross ton — See long ton.

6674 gross weight
f poids *m* brut — **d** Bruttogewicht *n* — **n** brutogewicht *n* — **e** peso *m* bruto — **i** peso *m* lordo

6675 Grotesque — Early sans-serif types. On account of *modern type design classification, this term is deprecated. See app. no. 2.
f caractères *mpl* grotesques — **d** Grotesk *f* — **n** Grotesk *f* — **e** Grotesca *f*; caracteres *mpl* grotescos — **i** Grotesca *f*

6676 ground glass — See focusing screen.

6677 ground glass — See frosted glass.

6678 ground space — See floor space.

6679 ground tint — See backing-up colour.

6680 groundwood free — Not containing mechanical wood pulp, generally meaning less than 5%.

f sans pâte mécanique — **d** holzfrei — **n** houtvrij — **e** sin fibra de madera; sin pasta (mecánica); sin mecánica; sin pasta (de madera); sin madera — **i** esente da pastalegno

6681 groundwood-free paper; woodfree paper — Paper not containing mechanical pulp.

f papier *m* sans bois; papier *m* sans pâte mécanique — **d** holzfreies Papier *n* — **n** houtvrij papier *n* — **e** papel *m* sin (fibra de) madera; papel *m* sin pasta mecánica de madera — **i** carta *f* senza legno

6682 groundwood mill — See grinding mill.

6683 groundwood pulp — See mechanical woodpulp.

6684 group — Quantity of identical images on a negative or positive.

f groupement *m*; groupage *m* — **d** Nutzenfilm *m* — **n** verzamelnegatief *n*; verzamelpositief *n* — **e** negativo *m* de imágenes combinadas; positivo *m* de imágines combinadas — **i** raggruppamento *m* (di negativi; di positivi); pellicola *f* per più pose

6685 group drive — Drive to two or more printing couples ganged together of a rotary press.

f entraînement *m* de groupe d'impression; commande *f* par groupe — **d** Gruppenantrieb *m* — **n** groepsaandrijving *f* — **i** comando *m* a gruppi

6686 grouping machine — Miniature *photocomposing machine to make multiple negatives or positives.

f machine *f* à multiplier sur film — **d** Addiermaschine *f*; Nutzenkopiermaschine *f* — **n** kopieermachine *f* — **e** copiadora *f* de repetición — **i** macchina *f* fotoripetitrice; macchina *f* fotoaddizionatrice

6687 grow *v* hollow — To wear hollow.

f se creuser — **d** aushöhlen — **n** uitslijten; uithollen — **e** rabajarse — **i** incavarsi

6688 GSA system — Computer controlled typesetting method, abbreviation of Güttinger Satz Automation, of Swiss origin; a 6-level TTS tape is provided.

f système *m* GSA — **d** GSA-System *n* — **n** GSA-systeem *n* — **e** sistema *m* GSA — **i** sistema *m* GSA

6689 guard — Mechanical means preventing the reaching or touching of moving machine parts.

f garde-main *m*; étrier *m* de protège-main; protecteur *m* — **d** Handschutz *m*; Handschutzbügel *m*; Schutzblech *n*; Handabweiser *m* — **n** handbeschermer *m*; beschermplaat *fm* — **e** salvamanos *f*; protector *m* de las manos — **i** salvamano *m*; protezione *f*

6690 guard *v* — To add *guards.

f placer des onglets — **d** fälzeln; Fälzelstreifen ankleben — **n** oortjes aanbrengen — **e** aplicar cartivanas; aplicar escartivanas — **i** applicare brachette

6691 guard *v* — To mount on *guards.

f coller sur onglets — **d** an Fälze hängen — **n** op oortjes plakken — **e** encolar sobre cartivanas; pegar sobre cartivanas — **i** incollare alle brachette

6692 guards — Strips of paper or muslin, etc., inserted in the back of books for plates, or additional leaves, to be pasted on. The extra bulk in the back allows for insertion.

f onglets *mpl* — **d** Fälze *mpl*; Fälzel *npl*; Fälzelstreifen *mpl*; Heftränder *mpl* — **n** oortjes *npl* — **e** cartivanas *fpl*; escartivanas *fpl* — **i** striscioline *fpl*

6693 guards — See compensation guards.

6694 guide bars — See guides.

6695 guide edge — In some paper tape equipment, the edge of a tape used to determine its transverse position.

f bord *m* guide — **d** Führungsrand *m*; Bezugskante *f* — **n** referentierand *m* — **e** borde *m* de guía — **i** canto *m* di guida

6696 guide mark — Crossline marks on the offset press plate to indicate trim centring of sheet, centring of plate, etc., as well as press register in multicolour work. Not to be confused with register marks used for stripping elements to register.

f repère *m* de calage — **d** Ausrichtemarke *f*; Leitmarke *f*; Führungsmarke *f* — **n** uitrichtteken *n* — **e** signo *m* de calado — **i** segno *m* di riferimento per il registro

6697 guider — On proof presses, means to fix the exact position of paper to obtain register.

f taquet *m* — **d** Registermarke *f* — **n** aanleg *m* — **e** guía *m*; tope *m*; tacón *m* — **i** tacca *f* di registro; marca *f* di registro

6698 guider — Preliminary sketch, coloured proof or drawing to serve as guidance in making colour plates.

f tracé *m* — **d** Modell *n*; Vorlage *f* — **n** model *n*; gids *m* — **e** guía *m* trazado para comprobación de registro — **i** modello *m*

6699 guide roller — See slewing roller.

6700 guides; guide bars — Stationary members in a *folder which retain or direct webs or copies as required.

f guides *mpl* — **d** Führungen *fpl*; Leitspindeln *fpl* — **n** geleiders *mpl*; geleidestangen *fmpl*; leistangen *fmpl* — **e** guías *mpl*; barras *fpl* de guía — **i** guide *fpl*; sbarre *fpl* di guida

6701 guilloche — Intricated, machine-engraved design as for borders, ornaments, backgrounds, used in security printing to prevent counterfeiting and falsification.

f guilloche *m* — **d** Guilloche *f* — **n** guilloche *m* — **e** guilloque *m*

6702 guillochee *v* — To adorn with guilloches.

f guillocher — **d** guillochieren — **n** guillocheren — **e** guilloquear — **i** guillosciare

6703 guilloching machine — See also geometrical lathe, cycloidal engine, rose engine machine.

f machine *f* à guillocher; tour *m* à guillocher — **d** Guillochiermaschine *f* — **n** guillocheermachine *f*

— **e** máquina *f* de guilloches — **i** macchina *f* da guillosciare

6704 guillotine; paper cutter; cutter *US*; **cutting machine; power cutter; trimmer** *US* — Machine with a long heavy removable knife to trim paper with a downward slicing action. "Guillotine" commemorates the name of the French physician Dr. J.I. Guillotin. See also handwheel cutter, lever cutter.
f massicot *m*; coupe-papier *m*; massiquot *m depr.* — **d** Schneidemaschine *f*; Schnellschneider *m*; Papierschneidemaschine *f* — **n** snijmachine *f* — **e** guillotina *f* — **i** tagliacarte *m* a ghigliottina; taglierina *f* ghigliottina

6705 guillotine; flat cutter — Machine with a heavy treadle-operated steel blade, to cut sheets of copper and zinc and to trim finished etchings.
f cisaille *f* (à métal) — **d** Metallschere *f* — **n** metaalschaar *fm*; plaatschaar *fm* — **e** cizalla *f* — **i** ghigliottina *f* per il taglio di materiale metallico

6706 gum — In lithography, water-soluble colloid used for coating the metal plate to make the non-image areas ink repellent and to preserve the plate for future use. See also gum arabic, cellulose gum.
f gomme *f* — **d** Gummi *n* — **n** gom *mn* — **e** goma *f* — **i** gomma *f*

6707 gum arabic — Gum obtained from either of two species of acacia trees, used in all branches of the graphic arts. There are a number of varieties known in commerce, e.g., Turkey gum, Egyptian gum. It forms a large part of the fountain solutions used on lithographic presses; it is also used as an adhesive for stamps, labels, cigarette paper, etc. Gum arabic solutions are used to desensitize or remove any affinity for ink in the non-printing areas of lithographic plates.
f gomme *f* arabique — **d** Gummiarabicum *n*; arabisches Gummi *n* — **n** Arabische gom *mn* — **e** goma *f* arábiga — **i** gomma *f* arabica

6708 gum camboge — See gamboge.
6709 gum dragon — See dragon's blood.
6710 gum elastic — See rubber.
6711 gum etch
f préparation *f* à la gomme arabique; gomme *f* arabique — **d** Gummiätze *f* — **n** gom-ets *m* — **e** goma *f* desensibilizadora; goma *f* mordiente; goma *f* desoxidante; laca *f* desensibilizadora — **i** preparazione *f* alla gomma arabica
6712 gum layer
f couche *f* de gomme — **d** Gummischicht *f* (auf der Druckplatte); Gummierschicht *f* (auf Papier) — **n** gomlaag *fm* — **e** capa *f* de goma — **i** strato *m* di gomma
6713 gummed all over flap — Abbrev. g.a.o.f.
f gommée sur toute la patte — **d** mit vollgummierter Schlußklappe — **n** met gegomde klep — **e** engomados (los sobres) a toda la solapa — **i** gommata su tutta la falda
6714 gummed label
f étiquette *f* gommée — **d** gummiertes Etikett *n* — **n** gegomd etiket *n* — **e** etiqueta *f* engomada —

i etichetta *f* gommata
6715 gummed paper; adhesive paper — Paper coated on one side with adhesive gum, the adhesive being a dextrin, fish or animal glue, and resin or a blend of any of these, used for stickers, labels, seals, stamps, splices, tapes, etc. The usual basis weight is in the range of 38 - 45 pounds (24 x 36 - 500).
f papier *m* gommé; papier *m* adhésif — **d** gummiertes Papier *n*; Klebepapier *n*; Haftpapier *n* — **n** gegomd papier *n* — **e** papel *m* engomado — **i** carta *f* gommata; carta *f* adesiva
6716 gummed paper tape — See gummed tape.
6717 gummed sealing tape — See gummed tape.
6718 gummed tape; gummed paper tape; gummed sealing tape — Kraft paper, coated on one side with glue, cut in narrow rolls, for sealing or reinforcing.
f ruban *m* de papier gommé — **d** Klebeband *n*; Klebestreifen *m* — **n** plakband *n*; gegomd band *n* — **e** cinta *f* engomada — **i** nastro *m* adesivo; nastro *m* gommato; nastro *m* di gomma
6719 gumming — Applying a layer of adhesive to the whole or a part of one side of paper or board.
f gommage *m*; engommage *m* — **d** Gummieren *n*; Gummierung *f* — **n** gommen *n*; voorzien *n* van gom — **e** engomado *m*; engomadura *f* — **i** gommatura *f*
6720 gumming; gumming up — In lithography, the treating of surfaces with a thin coating of gum arabic as a protection against oxidation and ink-receptive coatings for image, and as an aid to desensitizing the plate.
f gommage *m* — **d** Gummierung *f* — **n** gommen *n*; in de gom zetten *n* — **e** recubrimiento *m* con goma — **i** gommatura *f*
6721 gumming machine — Machine for the application of gum or adhesive.
f gommeuse *f*; machine *f* à gommer — **d** Gummiermaschine *f* — **n** gommeermachine *f* — **e** engomadora *f* — **i** macchina *f* ingommatrice; macchina *f* gommatrice
6722 gumming up — See gumming.
6723 gum rosin — See rosin.
6724 gum sandarac; sandarac resin — Yellow, brittle, translucent, amorphous lumps or powder. Soluble in alcohol, ether, amyl alcohol and hot caustic alkali, partially soluble in volatile oils, carbon disulphide, chloroform and oil of turpentine, insoluble in light petroleum hydrocarbons, benzene and water. Used in the manufacture of lacquers and varnishes.
f gomme *f* sandaraque — **d** Sandarak *m* — **n** sandarak *n*; sandrak *n* — **e** goma *f* sandáraca; sandáraca *f* — **i** sandracca *f*
6725 gum stencil — The acid-resistant stencil formed in deep-etch platemaking in non-copy areas of the plate when gum arabic is a constituent of the coating solution; it protects non-image areas while image areas are developed and etched.
f pellicule *f* de gomme — **d** Gummischicht *f*;

säurewiderstandsfähige Schicht *f* — **n** gomlaag
fm; kopieerlaag *fm*; zuurbestendige laag *fm* — **e**
película *f* de goma — **i** strato *m* di gomma (resistente agli acidi)

6726 gum streaks — Streaks, particularly in halftones, produced by the uneven gumming up of
litho plates.
f marques *fpl* de gomme — **d** Streifenbildung *f*
durch unregelmäßiges Gummieren — **n** gomstrepen *fmpl* — **e** rayas *fpl* de goma — **i** striature
fpl di gomma

6727 gum tragacanth — See tragacanth gum.

6728 gum turpentine — See turpentine.

6729 gun; gatling — Device with moulds, used for
clothing printing rollers with composition.
f appareil *m* pour fondre les rouleaux — **d** Walzen-
gießapparat *m*; Walzengießform *f* — **n** rollengiet-
apparaat *n* — **e** molde *m* para fundir rodillos — **i**
forma *f* per la fusione dei rulli

6730 gusset — Fold in a *gusset bag.
f soufflet *m* — **d** Tütenfalte *f*; Falte *f*; Seitenfalte *f*; Zwickel *m* — **n** soufflet *n*; tussenvouw *fm*;
zijvouw *fm* — **e** fuelle *m*; pliegue *m* lateral — **i**
soffietto *m*; piega *f* laterale a soffietto

6731 gusset bag — Bag with three folds on each
side to make a pleat, to allow it to expand and
take up a rectangular cross section. The capacity
is measured with the gusset unfolded.
f sac *m* à soufflet — **d** Faltentüte *f*; Falten-
beutel *m*; Seitenfaltenbeutel *m* — **n** zak *m* met
soufflet — **e** bolsa *f* con pliegue lateral — **i**
sacchetto *m* con soffietto

6732 gusset envelope — Envelope in pocket or
banker shape with pleats for expansion.
f enveloppe *f* à soufflet — **d** Versandbeutel *m* (mit
seitlichen Einschlägen); Musterbeutel *m* — **n**
envelop(pe) *fm* met soufflet — **e** sobre *m* con
pliegue lateral — **i** busta *f* a soffietto

6733 gutta-percha — Substance similar to rubber
but containing more resin and changing less
during vulcanization, used to take impressions on
uneven or hard surfaces. Partly soluble in carbon
disulphide, chloroform, solvent naphtha and warm
benzene.
Formula: $(C_{10}H_{16})_n$.

f gutta-percha *m* — **d** Guttapercha *n*; Gutta-
Percha *n* — **n** guttapercha *mn*; getah pertsja *mn*
— **e** gutapercha *f* — **i** guttaperga *f*

**6734 gutter; gap; river; street; lizard; hound's
teeth** *US coll.* — White vertical line in a page
occurring when spacing of words in consecutive
lines fall in a straight line.
f colombier *m* (entre les mots); rue *f*; ruelle *f*;
lézarde *f*; cheminée *f* — **d** Gasse *f*; Strasse *f*; Gießbach *m*; Ritze *f* — **n** straatje *n*; naadje *n*;
gootje *n* — **e** calle *f*; callejón *m*; corral *m* — **i**
canaletto *m*

6735 gutter — Space between pages in the printing forme of a book, etc. for folding (double
foredge) and binding, i.e., the inside margin towards the back or binding edge. The space between columns on multi-column pages is sometimes called a gutter, but would be better defined
as column gutter. The name in other languages
depends on the number of pages imposed.
f blanc *m* de fond (blanc de grand fond, grand
fond, grand blanc; blanc de petits fonds, petit
fond; blanc transversal) — **d** Steg *m* (Randsteg,
Bundsteg, Kopfsteg, Seitensteg, Beschnittsteg,
Abschneidesteg, Schneidsteg, Außensteg, Mittelsteg, z.B. bei 8 Seiten, Kreuzsteg, z.B. bei 16 Seiten
im rechten Winkel zum Mittelsteg) — **n** wit *n*
(aanslagwit, middenwit, kopwit, kruiswit, staartwit, rugwit) — **e** medianil *m* — **i** bianco *m* tipografico (bianco di cucitura, bianco interno; bianco
di testa, bianco di superiore, bianco di pieda,
bianco di inferiore; bianco di taglio, bianco di
esterno)

6736 gutter paper — See yellow paper.

6737 gutter stick — See sticks.

6738 gypsum — Hydrous form of calcium sulphate occurring in nature. The pigment is used as
a filler in paper. Sometimes the material is
referred to as pearl filler, terra alba, etc. Insoluble
in water, soluble in ammonium salts, acids and
sodium chloride.
Formula: $CaSO_4 \cdot 2H_2O$.
f gypse *m*; plâtre *m* — **d** Gips *m* — **n** gips *n* — **e**
yeso *m*; escayola *f* — **i** gesso *m*

6739 H2K — See head to come.

6740 hachure; hatching — The drawing and also the pattern of fine, closely spaced lines chiefly to give an effect of shading.
f hachure *f*; système *m* de hachures — **d** Schraffur *f*; Schraffierung *f*; Schraffe *f*; Strichelung *f* — **n** arcering *f*; harcering *f* — **e** rayado *m*; azurado *m*; azurada *f*; sombra *f*; agrisado *m* — **i** tratteggio *m*

6741 Hadego — Photographic display machine, developed during World War II by H.J.A. de Goey (Haarlem, Netherlands). Use was made of plastic photo-matrices, assembled by hand, as in the Ludlow system, and then photographed, producing type or film from 4-82 point. Production of the machine ceased in 1973. Licences sold to America.

6742 hair hygrometer
f hygromètre *m* à cheveu — **d** Haarhygrometer *m* — **n** haarhygrometer *m* — **e** higrómetro *m* de caballo — **i** igrometro *m* a capello

6743 hairing — Fine black lines between the printed types due to damage of the matrix side walls. See also beard.
f bavures *fpl* — **d** Spießen *n* — **n** haren *n* van matrijzen — **e** rebabas *fpl* — **i** bave *fpl*

6744 hair-lead — Normally a one point thick lead for spacing lines of type.
f interligne *f* mince — **d** dünne Reglette *f* — **n** eenpunts interlinie *f* — **e** interlínea *f* extra fina — **i** interlinea *f* molto sottile

6745 hair line — See fine line.

6746 hair-line; hair-stroke — Thin stroke of a type face.
f délié *m*; terminaison *f* — **d** Haarstrich *m* — **n** haarlijntje *n* — **e** perfil *m* fino; trazo *m* fino; palo *m* fino — **i** asta *f* capillare

6747 hair-line; matrix hair-line; beard — Fine black line between printed characters caused by the breakdown of the side walls of the matrices of line casting machines, and metal adhesion to spacebands.
f bavure *f* — **d** Spieß *m* — **n** haartje *n* — **e** rebaba *f*; pelito *m*; pelo *m Ar* — **i** bava *f*

6748 hair-line — The narrowest or finest black or white line that can be etched or engraved on a relief printing plate. The finest normal printing line, approx. 0.07-0.1 mm (0.003-0.004 in) thick.
f trait *m* fin; trait *m* de l'épaisseur d'un cheveu — **d** Haarstrich *m*; Haarlinie *f* — **n** haarlijntje *n*; fijn lijntje *n* — **e** línea *f* fina — **i** tratto *m* sottilissimo

6749 hair-line register — In multicolour printing, perfect or nearly perfect register of successive colours.
f repérage *m* précis — **d** feiner Passer *m* — **n** nauwkeurig register *n* — **e** registro *m* exacto — **i** registro *m* perfetto

6750 hair-line rule; fine-face rule — Brass rule of hair-line thickness.

f filet *m* maigre; filet *m* fin — **d** Haarlinie *f*; feine Linie *f* — **n** haarlijn *fm*; fijne lijn *fm* — **e** filete *m* fino; filete *m* extrafino; filete *m* superfino; filete *m* de pelo — **i** filetto *m* finissimo; filetto *m* capillare

6751 hair-line space
See hair-space.

6752 hair-space; hair-line space — The thinnest space, about one point thick (wide).
f espace *f* fine; espace *f* mince; espace *f* de un point — **d** Haarspatium *n*; Haarspatie *n*; dünnes Spatium *n*; Punktspatium *n*; Achtelgeviertspatium *n*; Achtelpetitspatium *n*; Achtelgeviert *n* — **n** vliesje *n*; vliesspatie *f*; haarspatie *f*; 1-punts spatie *f* — **e** espacio *m* entrefino; espacio *m* de pelo; espacio *m* de un punto — **i** spazio *m* finissimo; spazio *m* di uno punto

6753 hair-stroke — See hair-line.

6754 halation — In photography, spreading of light action beyond proper boundaries in negatives, particularly in the highlight areas of the image. The dots on every negative or positive, when shot through the halftone screen, have a soft, fuzzy perimeter known as a halo, due to the fact that less light has reached the edge than the centre of the dot. The silver deposit is therefore weaker and shows a corresponding transparency. It permits chemical reduction, as in dot-etching.
f halo *m* — **d** Lichthofbildung *f* — **n** halo *m*; halovorming *f* — **e** halo *m*; aureola *f* — **i** alone *m*; aureola *f*

6755 half binding; half-leather binding — Bookbinding with back and corners and partly the sides in leather.
f reliure *f* demi-peau; demi-reliure *f* en cuir; demi-reliure *f* (à petits coins); reliure *f* demi-cuir; demi-reliure *f* en peau — **d** Halbband *m*; Halblederband *m*; Halbfranzband *m* — **n** halfband *m*; halfleren band *m*; halffranse band *m* — **e** media pasta *f*; encuadernación *f* media piel; encuadernación *f* a la francesa; encuadernación *f* a la holandesa — **i** mezza legatura *f*; legatura *f* in mezza pelle

6756 half bound — Said of a book with leather or cloth back and corners and cloth or paper sides.
f en demi-reliure à coins — **d** im Halbband — **n** halfgebonden — **e** encuadernado a medio — **i** mezzo legato

6757 half case; single case — Type case the same width (front to rear) as an ordinary case, but half the length (side to side).
f demi-casse *f* — **d** halber Kasten *m* — **n** halve kast *fm* — **e** caja *f* media; media caja *f* — **i** media cassa *f*

6758 half cloth binding — Bookbinding with back and corners in cloth and sides covered with paper.
f demi-reliure *f* en toile (à petits coins) — **d** Halbleinenband *m* — **n** halflinnen band *m* — **e** encuadernación *f* en media tela; media tela *f* — **i**

legatura *f* in mezza tela

6759 half-duplex channel — Channel that can transmit and receive signals, but in only one direction at a time.

f canal *m* duplex à sens unique — **d** Einrichtungs-duplexkanal *m*; Halbduplexkanal *m* — **n** een-richtingsduplexkanaal *n*; halfduplexkanaal *n* — **e** canal *m* duplex de sentido único — **i** canale *m* duplex a senso unico

6760 half-duplex working; two-wire system — Working at two stations which can send and receive messages to and from each other but not at the same time.

6761 half large — A size of cut card. See app. no. 3.

6762 half-leather binding — See half binding.

6763 half-line screen; single-line screen; one-way screen — Screen consisting of opaque parallel lines with clear interfaces to produce tonal values by forming a pattern in lines of varying thickness.

f trame *f* lignée — **d** Linienraster *m* — **n** lijnraster *n*; enkellijns raster *n* — **e** retícula *f* lineal — **i** retino *m* lineare

6764 half measure — See double column page.

6765 half-monthly — See semi-monthly.

6766 half nonpareil — Old name for a 3-point type. See app. no. 1.

6767 half sheetwork — See work and turn.

6768 half sized; medium sized — See also hardsized paper.

f mi-collé — **d** halbgeleimt — **n** halfgelijmd — **e** semiencolado; medio encolado — **i** mezza colla; mezzo collata

6769 half small — A size of cut cards. See app. no. 3.

6770 halfstuff — Pulp in condition to be charged into the beater. After beating it is called *whole stuff or stuff. Expression normally used in rag paper manufacture.

f demi-pâte *f* — **d** Papierhalbstoff *m*; Halbstoff *m*; Halbzeug *n* — **n** halfstof *fm* — **e** semipasta *f* — **i** mezza pasta *f*

6771 half super royal — A size of drawing paper and of boards. See app. no. 3.

6772 half title — See bastard title.

6773 halftone — Tone gradation by an image composed of dots of various sizes with equidistant centres.

f demi-teinte *f* — **d** Halbton *m* — **n** halftoon *m*; halftint *fm* — **e** medio tono *m*; media tinta *f*; reticulado *m* — **i** mezzatinta *f*

6774 halftone block; halftone plate; halftone engraving — Block in which the various tones are made by dots. Printing plate in relief produced by a photo-mechanical etching process, employing the principle of the halftone screen, or by an electronically controlled engraving needle, where all gradations of tone values in the original photograph or drawing are reproduced in the plate by variations in the size of minute dots of geometric arrangement, obtained either in the negative by the interposition of a cross-ruled screen or in the

action of the engraving needle controlled from the electronic scanning head.

f cliché *m* de similigravure; similigravure *f*; cliché *m* simili; simili *m*; cliché *m* tramé; planche *f* simili; autotypie *f* — **d** Rasterätzung *f*; Rasterklischee *n*; Autotypie *f*; Autotypieätzung *f*; Halbtonätzung *f*; Autotypieplatte *f*; Rasterplatte *f* — **n** autotypie *f*; autotypie-cliché *n*; rastercliché *n* — **e** clisé *m* de medio tono; clisé *m* de medios tonos; clisé *m* de medias tintas; clisé *m* de trama; clisé *m* tramado; autotipia *f*; similgrabado *m*; clisé *m* directo; fotograbado *m* directo; grabado *m* reticulado; clisé *m* de puntos *Pe* — **i** cliscè *m* a mezzatinta; cliché *m* a mezzatinta; autotipia *f*

6775 halftone blowup — See blown-up halftone.

6776 halftone comb — See lining tool.

6777 halftone copy — Continuous tone original, or that best reproduced with a halftone screen.

f document *m* de demi-ton — **d** Rasterkopie *f*; Rastervorlage *f* — **n** halftoon-origineel *n*; halftoon-model *n* — **e** original *m* de medio tono — **i** modello *m* a mezzatinta

6778 halftone diapositive — Positive transparency the tone values of which are produced by dots of varying size all being completely opaque.

f diapositif *m* tramé — **d** Rasterdiapositiv *n* — **n** rasterdiapositief *n* — **e** diapositivo *m* reticulado — **i** diapositivo *m* retinato

6779 halftone dot; dot — The individual formation or element in a halftone negative, printing plate and final impression.

f point *m* de trame — **d** Rasterpunkt *m* — **n** rasterpunt *f* — **e** punto *m* reticular; punto *m* de la trama; puntito *m* — **i** punto *m* del retino

6780 halftone engraving — See halftone block.

6781 halftone etcher

f similiste *m* — **d** Autotypieätzer *m*; Ätzer *m* — **n** auto-etser *m*; etser *m* — **e** grabador *m* de medio tonos — **i** fotoincisore *m*

6782 halftone film

f film *m* à modelé continu — **d** Halbtonfilm *m* — **n** lith-film *m* — **e** película *f* de medio tono; película *f* de tono continuo — **i** pellicola *f* a mezzatinta

6783 halftone gravure — Intaglio printing process in which the printing forme consists of etched recesses of various size and equal depth, as against ordinary gravure where the recesses are all of the same size but of varying depth. Advantageous for practical printing on account of the depth to which the highlight dots are etched, allowing long runs. The difficulty of the process is the production of doctor bearing lines in shadows.

f similigravure *f* en creux; héliogravure *f* autotypique; héliogravure *f* tramée — **d** autotypischer Tiefdruck *m* — **n** autotypische diepdruk *m* — **e** huecograbado *m* autotípico — **i** rotocalco *m* autotipico

6784 halftone image

f image *f* à demi-tons; image *f* de modelé continu — **d** Halbtonbild *n* — **n** halftoonbeeld *n* — **e** imagen *f* de medio tono — **i** immagine *f* a mezzatinta

6785 halftone ink — Printing ink made from carbon black with a blue toner, mixed in a heavy-bodied vehicle. Should be dense, clear working, quick setting and hard drying.
f encre *f* vignette — **d** Illustrationsfarbe *f*; Autotypiefarbe *f*; Halbtonfarbe *f* — **n** illustratie-inkt *m*; illustratiezwart *n* — **e** tinta *f* (negra) para ilustraciones — **i** inchiostro *m* da illustrazione

6786 halftone negative; screen negative — Negative obtained by insertion of a ruled screen between lens and photographic negative or between the film and the continuous tone positive transparency.
f négatif *m* tramé — **d** Rasternegativ *n* — **n** rasternegatief *n* — **e** negativo *m* tramado — **i** negativo *m* retinato; negativo *m* in mezzatinta

6787 halftone news — Newsprint grade paper, loaded and calendered, for printing of cheap magazines.
f papier *m* journal pour vignettes — **d** Illustrationszeitungsdruckpapier *n* — **n** krantenpapier *n* voor illustratiewerk; gesatineerd courantdruk *n* — **e** papel *m* de periódico para ilustraciones; papel *m* prensa para ilustraciones — **i** carta *f* da giornale per illustrazioni

6788 halftone no line — Halftone without the usual thin border line.
f simili *f* à claire-voie — **d** Autotypie *f* ohne Randlinie — **n** autotypie *f* zonder kader — **e** medio tono *m* sin recuadro — **i** illustrazione *f* senza (linea di) inquadratura

6789 halftone paper — Supercalendered or coated paper for printing halftones.
f papier *m* pour similigravure; papier *m* pour illustrations; papier *m* pour vignettes — **d** Illustrationsdruckpapier *n* — **n** illustratiedrukpapier *n*; illustratiedruk *n* — **e** papel *m* para fotograbados; papel *m* para autotipias; papel *m* para medio tono; papel *m* arte; papel *m* ilustración; papel *m* cícero *Es* — **i** carta *f* da stampa per illustrazioni; carta *f* per autotipie

6790 halftone photography — Reproduction of originals by photography with a halftone screen.
f photographie *f* tramée — **d** Rasterphotographie *f* — **n** rasterfotografie *f* — **e** fotografía *f* de medio tono — **i** fotografia *f* a mezzatinta

6791 halftone plate — See halftone block.

6792 halftone positive; screen positive — Positive made in the camera from a continuous tone negative with a halftone screen interposed between plate and lens or a positive made in contact from a halftone negative.
f positif *m* tramé — **d** Rasterpositiv *n* — **n** rasterpositief *n* — **e** positivo *m* punteado; positivo *m* de trama; positivo *m* tramado; positivo *m* de medio tono — **i** positivo *m* retinato; positivo *m* in mezzatinta

6793 halftone process — Method of photomechanical reproduction. Originals are photographed with a halftone screen, which translates details and tones in the negative in the form of a geometric and varying dot formation; the negative is then printed on sensitized copper or zinc and the resulting image subjected to etching to bring the individual dots to the required relief for successful letterpress printing.
f similigravure *f* — **d** Autotypieverfahren *n*; Rasterverfahren *n* — **n** autotypie-procédé *n* — **e** procedimiento *m* de medio tono — **i** procedimento *m* a mezzatinta

6794 halftone screen; screen — Series of ruled, etched, pigmented, evenly spaced lines ranging from 20-150 lines per cm or 50-400 lines to the inch, the lines etched on two glass plates which are then cemented together so that the opaque lines cross each other at right angles and form transparent square apertures. Halftone screens are also made with a grain formation and in the form of photographic films (variable opacity and contact screens) with a formation of halftone dots.
f trame *f* cristal; trame *f* de similigravure; trame *f* pour similigravure; trame *f* simili — **d** Autotypieraster *m*; Raster *m*; Bildraster *m* — **n** autotypieraster *n*; raster *n* — **e** trama *f* (de medio tono); retícula *f* (de medio tono); retícula *f* para autotipia — **i** retino *m* autotipico

6795 halftone step scale — See grey scale.

6796 halftone stop — Lens aperture used in halftone photography.
f diaphragme *m* pour similigravure — **d** Rasterblende *f*; Blende *f* — **n** diafragma *n*; insteekdiafragma *n*; stop *m* — **e** diafragma *m* de medio tono — **i** diaframma *m* per fotografia con retino; diaframma *m* da inserire

6797 halftonometer — See etching-depth meter.

6798 half-uncial — Combination of *uncials with cursives and ligatures, used in the 5th and 6th century in Ireland and England.
f semi-onciale *f* — **d** Halbunzial *f*; Halbuncial *f* — **n** halfunciaal *fm* — **e** semiuncial *f* — **i** semiunciale *f*

6799 half unit — See single-colour unit.

6800 half-weekly — See semi-weekly.

6801 halo — The circle or aurea of lesser density around the core of the halftone dot in negatives or positives.
f halo *m* — **d** Lichthof *m*; Hof *m*; Ring *m*; Kreis *m*; Schleier *m*; Überstrahlung *f* — **n** halo *m*; krans *m*; kring *m* — **e** halo *m*; aureola *f* — **i** alone *m*

6802 halo effect — In work study, tendency on the part of a rater to be overinfluenced by particularly dominating characteristics of the person being rated.
f effet *m* de halo — **d** Lichthofeffekt *m* — **n** haloeffect *n* — **e** efecto *m* de halo; efecto *m* de la aureola — **i** effetto *m* alone

6803 halo effect — See dot fringe.

6804 halogen — The chemically related elements fluorine, chlorine, bromine and iodine.
f halogène *m* — **d** Halogen *n* — **n** halogeen *n* — **e** halógeno *m* — **i** alogeno *m*

6805 halogen lamp

f lampe f halogène — d Halogenlampe f — n halogeenlamp fm — e lámpara f halógena — i lampada f alogena

6806 halogen silver
f argent m halogénique; halogénure m d'argent — d Halogensilber n — n halogeenzilver n — e plata f halógena — i alogenuro m d'argento

6807 hambro — See double hambro, single hambro.

6808 Hamburg blue — See Prussian blue.

6809 hammer finish paper — Paper with a surface that is smooth and rough in patches, with indentations where the smooth patches are, as if beaten with a hammer.
f papier m à surface martelée — d gehämmertes Papier n — n gehamerd papier n — e papel m martillado — i carta f martellata

6810 hammering — See repoussage.

6811 hand — See fist.

6812 hand, all in — See all in.

6813 hand backer — Hand-operated *backing machine.
f tas m à endosser; presse f à endosser — d Handabpreßmaschine f — n kneepapparaat n — e máquina f para sacar cajos de mano — i macchina f per formare i dorsi a mano

6814 hand bellows — See bellows.

6815 handbills; dodgers US**; flyers; throwaways** — Printed announcements for distribution by hand on the streets or to be taken away by customers in a store.
f prospectus mpl; circulaires fpl; papillons mpl — d Flugzettel mpl; Flugblätter npl; Handzettel mpl — n strooibiljetten npl — e carteles mpl; hojas fpl sueltas — i volantini mpl

6816 handbook; manual — Reference book of convenient size.
f manuel m; aide-mémoire m — d Handbuch n — n handboek n — e manual m — i manuale m; prontuario m

6817 hand coloured
f colorié à la main — d handkoloriert — n handgekleurd; met de hand gekleurd — e coloreado a mano; pintado a mano — i colorato a mano

6818 hand composition — Setting of type by hand.
f composition f à la main; composition f manuelle — d Handsatz m — n handzetsel n — e composición f a mano; composición f manual — i composizione f a mano

6819 hand compositor; floor man US
f compositeur m à la main — d Handsetzer m — n handzetter m — e compositor m a mano; cajista m de líneas; cajista m liniero — i compositore m a mano

6820 hand control
f commande f à main — d Handbedienung f; manuelle Steuerung f — n handbesturing f; handbediening f — e mando m a mano; control m manual; manejo m — i comando m a mano

6821 hand-cut overlay — Small pieces of paper pasted to a sheet under the tympan to com-pensate for depressions in the forme or to increase the pressure at certain points.
f béquet m — d Handausschnitt m (bei der Kraftzurichtung) — n handpikeersel n — e alza f — i alco m a mano

6822 H & D curve; Hurter & Driffield curve — Curve representing the response of a sensitive material, as shown by the relationship between density and exposure, in which density is plotted against the logarithm of the exposure. See also characteristic curve.
f courbe f H & D — d H & D-Kurve f — n H & D-kromme f; H & D-curve f; zwartingskromme f volgens H & D — e curva f H & D; curva f de sensibilidad — i curva f H & D; curva f della sensibilità; curva f di annerimento

6823 H & D speed number — Sensitivity curve for photographic material, named after Hurter and Driffield. See also H & D curve.
f degré m H & D — d HD-Einheiten fpl — n H & D-snelheid f — e grado m H & D — i grado m H & D

6824 H & D stiffness test — Test to predict the combined board column and box compression strengths. Abbrev. for Hinde and Dauch, Ltd., Toronto (Ontario), Canada.
f essai m de rigidité H & D — d H & D Steifigkeitsprüfung f; H & D Steifigkeitsmessung f — n H & D-stijfheidsproef fm; H & C-stijfheidsmeting f — e ensayo m de rigidez H & D; medición f de la rigidez H & D — i prova f di rigidità H & D

6825 hand fed — Said of sheets of paper when placed by hand up to the guides.
f alimenté à la main — d mit Handanlage — n met handinleg; met de hand ingelegd — e alimentado a mano; a marcado manual — i alimentato a mano

6826 handful of matter; handful of type
f poignée f — d Griff m — n greep m; handvol fm — e tomada f — i manata f

6827 handful of type — See handful of matter.

6828 hand gilding — See gilding.

6829 hand grip; grip
f poignée f — d Handgriff m; Griff m — n handgreep fm; greep fm — e puño m; empuñadura f — i impugnatura f; manopola f

6830 hand impression
f épreuve f à la main; impression f manuelle — d Handabzug m; Handdruck m — n handafdruk m; met de hand vervaardigde afdruk m — e impresión f a mano — i stampa f a mano

6831 handle of a bodkin
f manche m de la pointe à correction — d Ahlenheft n; Ahlheft n — n elsklos fm — e mango m de la punta — i manico m della punta

6832 handling — Process of getting the paper to the press, including unpacking, counting and conveying to the pressroom.
f manipulations fpl du papier — d An- und Abfahren n des Papiers; Handhabung f des Papiers — n af- en aanvoer m van het papier; transport n

van het papier — **e** manejo *m*; manipulado *m* del papel — **i** trattamento *m* e trasporto di carta

6833 handling and transit waste — See newsprint waste.

6834 hand-made paper; vat paper — Paper made by hand moulds in single sheets with rough or deckle edges on four sides. The mould, of size required, is dipped into the vat containing the stock and is lifted with a peculiar motion, forming the sheet. It is sometimes called *deckle-edged paper. In GB also paper made on a cylinder machine.

f papier *m* à la forme; papier *m* à la cuve; papier *m* à la main; papier *m* de cuve; papier *m* de Hollande — **d** handgeschöpftes Papier *n*; Handpapier *n*; Büttenpapier *n*; Schöpfpapier *n* — **n** handgeschept papier *n* — **e** papel *m* hecho a mano; papel *m* de tina — **i** carta *f* a mano

6835 hand mold — See hand mould.

6836 hand mould *GB*; **hand mold** *US* — Device in two parts used in old times to cast movable type.

f moule *m* à arçon; moule *m* à main — **d** Handgießform *f* — **n** handgietvorm *m* — **e** molde *m* a mano; molde *m* de fundición manual — **i** forma *f* a mano

6837 hand-out *US* — Newspaper copy obtained from a press agency.

6838 hand pasting — Attaching manually a new reel of paper to the expiring web.

f collage *m* à la main — **d** Handklebung *f* — **n** aanplakken *n* met de hand; aaneenplakken *n* met de hand — **e** empalmar a mano; encolar a mano; pegar a mano — **i** incollatura *f* a mano

6839 hand press — Printing press that is worked by hand, used for small jobs. See also lithographic handpress.

f presse *f* à main; presse *f* à bras — **d** Handpresse *f* — **n** handpers *f* — **e** prensa *f* a mano; prensa *f* de brazo; prensa *f* a brazo — **i** torchio *m* a mano; pressa *f* a mano

6840 handpress — See lithographic handpress.

6841 hand rest; bridge — Thin board slightly elevated above a printing plate, used by etchers when spotting, staging and re-etching. Also used by lithographers.

f appui-main *m*; support *m* — **d** Handstützbrett *n*; Handstützschiene *f* — **n** dekplank *fm* — **e** apoyamanos *m* — **i** appoggiamano *m*

6842 hand roller — See brayer.

6843 hand safety guard

f appareil *m* pare-main; appareil *m* protège-main — **d** Händeschutzvorrichtung *f*; Handschutzbügel *m* — **n** handbeschermer *m* — **e** guardamanos *m*; protector *m* de manos — **i** salvamani *m*

6844 Handsel day — See Lost Monday.

6845 hand set

f composé à la main — **d** handgesetzt — **n** met de hand gezet — **e** compuesto a mano; compuesto de caja — **i** composto a mano

6846 hand sewing

f couture *f* à la main — **d** Handheftung *f* — **n** naaien *n* met de hand — **e** cosido *m* a mano — **i** cucitura *f* a mano

6847 handshaking — Interface of one component or peripheral with the main processor.

f affirmation *f* de connexion — **d** Anschlußbestätigung *f* — **n** aansluitingsbevestiging *f* — **e** acuso *m* de conexión — **i** confirmazione *f* di connessione

6848 hand tooling — In photoengraving, hand finishing of a printing plate.

f burinage *m* — **d** Nachschneiden *n*; Nachstechen *n* — **n** nasteken *n* — **e** burilado *m* a mano — **i** rifinitura *f* al bulino

6849 hand tooling — In bookbinding, impressing ornamentations by hand on book covers.

f gaufrage *m* à la main — **d** Handprägung *f*; Handverzierung *f* — **n** handstempelen *n* — **e** gofrado *m* a mano — **i** goffratura *f* a mano

6850 handwheel cutter

f massicot *m* à volant — **d** Radschneider *m* — **n** handwiel-snijmachine *f* — **e** guillotina *f* de volante — **i** taglierina *f* con volantino a mano

6851 handwork — Manual effort, e.g., carried out on originals, negatives and printing plates, such as drawing, opaquing, re-etching or finishing.

f travail *m* à la main — **d** Handarbeit *f* — **n** handwerk *n*; handarbeid *m* — **e** trabajo *m* a mano; trabajo *m* manual — **i** lavoro *m* a mano; lavoro *m* manuale

6852 hang *v*

f suspendre (au moyen de pinces) — **d** aufhängen (des Papiers) — **n** uithangen (van papier) — **e** colgar (el papel) — **i** sospendere (la carta)

6853 hanging figures; descending figures; old-style figures; non-lining figures — Type faces which extend above and below defined limits as distinct from *lining figures.

f chiffres *mpl* elzéviriens — **d** Mediävalziffern *fpl*; Normalziffern *fpl* — **n** onder de lijn uithangende cijfers *npl*; onderuithangende cijfers *npl*; mediaeval-cijfers *npl* — **e** cifras *fpl* bajas; números *mpl* elzevirianos — **i** numeri *mpl* sotto la linea; cifre *fpl* discendenti

6854 hanging indention; hanging paragraph; reverse indention — Paragraph with first line set full length and all subsequent lines being indented.

f alinéa *m* en sommaire; composition *f* en sommaire — **d** Absatz *m* bei dem alle Zeilen außer der ersten eingezogen sind — **n** eerste regel voluit, de volgende regels ingesprongen — **e** párrafo *m* francés; sangrado *m* a la francesa; sumario *m* Ar — **i** capoverso *m* a sommario

6855 hanging paragraph — See hanging indention.

6856 Hansa yellow — Permanent, semi-opaque, organic yellow pigment, derived from coal tar.

f jaune *m* de Hansa; jaune *m* Hansa — **d** Hansagelb *n* — **n** Hansageel *n* — **e** amarillo *m* Hansa — **i** giallo *m* Hansa

6857 Hansel Monday — See Lost Monday.

6858 hard black — See bone black.

6859 hard copy; monitor copy; typescript-proof —

Product of an electric typewriter or line printer at various stages in data processing, in filmsetting operations often used for proof reading prior to correction of data, when the system does not incorporate a verification stage.

f document *m* en clair — **d** gedruckte Kopie *f* — **n** gedrukte kopij *f* — **e** copia *f* legible — **i** copia *f* leggibile

6860 hard dot — Halftone dot, negative or positive, characterized by a sharp, clean cut edge. Photographically, "hard" denotes excessive contrast. See also lateral hard dot.

f point *m* dur — **d** harter Punkt *m* — **n** harde punt *m* — **e** punto *m* duro; punto *m* recortado — **i** punto *m* duro

6861 harden *v* — To make harder by heating and subsequent quick cooling or by adding a tanning agent.

f durcir; tanner — **d** härten; gerben — **n** harden; looien — **i** indurire; tannare; conciare

6862 hardener; hardening agent; hardening solution — In photography, a chemical for raising the melting point of an emulsion.

f durcisseur *m*; agent *m* durcissant; durcissant *m* — **d** Erhärter *m*; Härtemittel *n*; Härtungsmittel *n* — **n** harder *m*; hardingsmiddel *n* — **e** endurecedor *m* — **i** induritore *m*

6863 hardening agent — See hardener.

6864 hardening bath — Solution to toughen photographic images, e.g., alum baths for negatives and chromic acid or bichromate mixtures for glue prints on zinc.

f bain *m* de durcissement — **d** Härtebad *n* — **n** hardingsbad *n* — **e** baño *m* endurecedor; baño *m* de endurecimiento; baño *m* curtiente — **i** bagno *m* d'indurimento

6865 hardening fixer — Fixing solution containing a chemical to harden the emulsion against damage during washing, drying and handling.

f fixateur *m* tannant — **d** Härtefixierbad *n* — **n** hardend fixeermiddel *n* — **e** fijador-endurecedor *m* — **i** fissatore-induritore *m*

6866 hardening of blanket

f durcissement *m* d'un blanchet — **d** Härten *n* des Gummituchs — **n** hardworden *n* van het rubberdoek — **e** endurecimiento *m* de la mantilla — **i** indurimento *m* del blanket

6867 hardening solution — See hardener.

6868 hard impression; overimpression; heavy impression — Firm impression between the paper and forme.

f foulage *m* — **d** Drucken *n* mit zu starkem Anpreßdruck; Schattierung *f* — **n** moet *fm*; drukken *n* met te grote drukkracht — **e** impresión *f* excesiva; exceso *m* de presión; clavo *m*; diente *m*; huella *f* — **i** impressione *f* eccessiva; impressione *f* dura; impressione *f* pesante

6869 hard lumps — Not completely disintegrated small pieces of dried pulp or of paper broke, which show up as white dense spots in the sheet and are crushed into shiners on calendering. See

also lumps, soft lumps, coating lump, colour lumps.

f pâtons *mpl* durs — **d** harte Knoten *mpl* — **n** harde stofknopen *mpl*

6870 hard negative — Photographic negative with a steep gradation curve.

f négatif *m* contrasté; négatif *m* vigoureux — **d** hartes Negativ *n* — **n** hard negatief *n* — **e** negativo *m* duro — **i** negativo *m* duro; negativo *m* fortemente contrastato

6871 hardness — To determine hardness, several methods are used, each being particularly suited for certain types of materials. Brinell hardness is widely used for metals, Rockwell hardness is used for metals and plastics. *Shore hardness is suitable for metals, not usually for plastics.

f dureté *f* — **d** Härte *f*; Härtezahl *f* — **n** hardheid *f* — **e** dureza *f* — **i** durezza *f*

6872 hardness meter; durometer; sclerometer — Instrument to measure hardness.

f duromètre *m*; scléromètre *m* — **d** Härtemesser *m*; Härteprüfer *m*; Härteprüfgerät *n* — **n** hardheidsmeter *m* — **e** durómetro *m*; medidor *m* de la dureza — **i** durometro *m*; apparecchio *m* per misurare la durezza

6873 hard packing — Hard sheets of paper and cardboard used to cover the platen or cylinder of a printing press in making a forme ready, required when printing on hard, smooth, dry papers from engravings, new type, etc., to obtain a sharp, clear impression.

f habillage *m* sec; garnissage *m* sec — **d** harter Aufzug *m* — **n** harde legger *m* — **e** cama *f* dura; patrón *m* duro; padrón *m* duro — **i** rivestimento *m* duro; abbigliamento *m* duro

6874 hard rubber; ebonite; vulcanite; vulcanized rubber — Rubber with a sulphur content of 18-47%, usually 25%, used in acid-proof material for photographic receptacles. Not to confuse with European *vulcanite fibre.

f caoutchouc *m* durci; ébonite *f*; vulcanite *f* — **d** Hartgummi *m*; Ebonit *n* — **n** hard rubber *n*; eboniet *n* — **e** caucho *m* endurecido; ebonita *f*; caucho *m* vulcanizado — **i** gomma *f* dura; ebanite *f*; vulcanite *f*

6875 hard-sized paper — Paper with a maximum of sizing to resist the penetration of ink and other aqueous solutions. Lesser degrees are indicated by *half-sized and *quarter-sized, *slack-sized.

f papier *m* fortement collé — **i** très collé — **d** stark geleimtes Papier *n*; vollgeleimtes Papier *n* — **n** hard gelijmd papier *n*; sterk gelijmd papier *n*; volgelijmd papier *n* — **e** papel *m* de encolado duro — **i** carta *f* fortemente collata

6876 hard vignette — Vignette effect with a delicate but definite printing edge instead of graduating to pure white.

f dégradé *m* défectueux — **d** abgebrochener Verlauf *m* — **n** afgebrokkeld verloop *n*; afgebroken verloop *n*; slecht verloop *n* — **e** esfumado *m* áspero — **i** sfumatura *f* troncata

6877 hardware — Electronic components of a computer or of a photocomposition system, exclusive of any programs (software) necessary to perform particular tasks.

f matériel *m* de traitement — **d** Maschinenausrüstung *f*; Geräteausstattung *f* — **n** apparatuur *f*; hardware *m* — **e** componentes *mpl* físicos; máquinas *fpl* y equipo *m*; hardware *m* — **i** componenti *mpl* di macchina; hardware *m*

6878 hardware module — Unit of equipment. Modules can be assembled into different configurations to form a variety of systems to suit different requirements.

6879 hard-wired — Said of a system with an unchangeable program, i.e., a circuit, program or system built in by the manufacturer that cannot be changed by programming.

f à programme fixe; à cablage fixe — **d** festprogrammiert; festverdrahter — **n** met ingebouwd programma

6880 hard-wired logic — The use of electronic components and circuit diagrams to make logical decisions which could otherwise be performed by software programs. Hard-wired devices are not programmable, but are constructed to perform only certain dedicated tasks.

6881 hardwood; deciduous wood — Wood from broad-leaved deciduous trees.

f bois *m* feuillu; bois *m* non-résineux — **d** Laubholz *n* — **n** loofhout *n*; hardhout *n* — **e** madera *f* de árboles foliculares — **i** legno *m* di latifoglie

6882 hardwood pulp; deciduous wood pulp — Pulp obtained from deciduous trees.

f pâte *f* de bois feuillus; pâte *f* de feuillus — **d** Laubholzzellstoff *m*; Laubholzhalbstoff *m*; Halbstoff *m* aus Laubholz — **n** loofhoutcelstof *fm*; loofhoutpulp *fm*; halfstof *fm* uit loofhout — **e** pasta *f* de árboles foliculares; pulpa *f* de árboles foliculares — **i** pasta *f* di latifoglie

6883 Harleian binding — Book cover decoration style used for the private collection of Robert Harley, first Earl of Oxford (1661-1724), showing an ornate centre piece composed of small tool forms, usually arranged in a lozenge-shaped design, and a broad border or a narrow tooled roll border.

f reliure *f* Harléienne — **d** Harley-Einband *m* — **n** Harley-band *m* — **e** encuadernación *f* Harley — **i** legatura *f* Harley

6884 harsh lines — Heavy lines in an engraving or etching.

f aigreurs *fpl* — **d** grobe Linien *fpl* — **n** zware lijnen *fmpl*; dikke lijnen *fmpl*; grove lijnen *fmpl* — **e** líneas *fpl* ásperas — **i** asprezze *fpl*; linee *fpl* aspre

6885 hatch *v* — To mark with fine, closely spaced lines for shading in drawing or engraving.

f hacher; hachurer — **d** schraffieren; schlummern (in der Kartographie) — **n** arceren; harceren — **e** sombrear; poner sombras (en la pintura); cruzar líneas (en el grabado) — **i** trattaggiare

6886 hatches, matches, and dispatches — See births, marriages, and deaths.

6887 hatching — See hachure.

6888 haven cap — A size of wrapping paper. See app. no. 3.

6889 head — Space between the pages of a printing forme, which are placed head to head.

f blanc *m* de tête; têtière *f* — **d** Kopfsteg *m* — **n** kopwit *n* — **e** blanco *m* de cabecera; blanco *m* de cabeza; testera *f* — **i** bianco *m* di testa; testa *f*

6890 head — The top of a page.

f tête *f* (de la page) — **d** Paginakopf *m*; Seitenkopf *m*; Kopf *m* (der Pagina) — **n** paginakop *m*; kop *m* (van de pagina) — **e** cabeza *f* (de la página) — **i** testa *f* (della pagina); capo *m* (della pagina)

6891 head — See heading.

6892 head, analysing — See scanning head.

6893 head and tail strength tester — Instrument for measuring the binding strength of adhesive bindings. A book page is pulled at its head or tail out a bound book at an angle of 45 °. The pulling force is registered and gives an indication of the binding quality.

f mesureur *m* de l'adhésion des feuilles de reliures collées — **d** Kopf- und Fußstärkemeßgerät *m*

6894 headband — In bookbinding, an ornamental silk, linen or cotton covered band stretching across the head edge and tail edge (*tailband) of a book and resting along the contour of the back of the book. It can be made by hand, but is obtainable in lengths ready for cutting and gluing to the spine. See also tranchefile chapiteau.

f tranchefile *f* (supérieure) — **d** (oberes) Kapitalband *n* — **n** (bovenste) kapitaalbandje *n*; (bovenste) besteekbandje *n* — **e** cabezada *f*; capitel *m* Ar — **i** capitello *m* superiore; capitello *m* di testa

6895 head bar — See margin bar.

6896 head bar — See stagger bar.

6897 head bolt — See bolts.

6898 head box; breast box; flow box — Tank in which the head or level of the papermaking stuff is regulated to provide uniform flow on the paper machine wire.

f bac *m* de tête; caisse *f* d'arrivée; tête *f* de machine — **d** Stoffauflaufkasten *m*; Stoffeinlauf *m*; Maschinenbütte *f* — **n** oploopkast *fm*; stofoploopkast *fm*; stofoploop *m* — **e** caja *f* de entrada; caja *f* de alimentación — **i** cassa *f* d'afflusso; serbatoio *m* a livello di carico regolato

6899 headcap — See cap.

6900 header; message header — First part of a message containing all necessary information for directing the message to the destination(s).

f titre *m* de message — **d** Kopfanschrift *f* — **n** koptitel *m* — **e** rótulo *m* de encabezamiento — **i** titolo *m* di messaggio

6901 header label — Machine-readable record at the beginning of a file containing data identifying the file and data used in file control.

f label *m* de tête — **d** Vorsatz *m* — **n** koplabel *m* — **e** rótulo *m* inicial — **i** etichetta *f* iniziale

6902 heading; head — Type set apart (usually above) a section of the text, serving as a title or description of what follows. Not to confuse with *headline. See also box heading, centre head, chapter heading, shoulder heading, side heading, sub heading.
f titre *m*; en-tête *m* — **d** Überschrift *f* — **n** titel *m* — **e** título *m* — **i** titolo *m*

6903 heading — Brief summary of a chapter in a book or introductory paragraph of a newspaper article, at the beginning of the chapter or article.
f chapeau *m* — **d** Vorspann *m* — **n** korte inleiding *f* onder de kop — **e** cabecera *f* — **i** cappello *m*

6904 heading chase — Long narrow chase to lock up formes for account book headings.
f châssis *m* oblong — **d** Schließrahmen *m* für Tabellenkopf — **n** tabellensluitraam *n*; tabellenraam *n* — **e** rama *f* para encabezados — **i** telaio *m* per la composizione di tabelle

6905 headline — Line of type at the top of a page or mass of text matter. See also running headline.
f ligne *f* de tête — **d** Kopfzeile *f*; Hauptzeile *f*; Überschriftlinie *f* — **n** kopregel *m* — **e** título *m*; cabeza *f*; encabezamiento *m*; titular *m*; encabezado *m*; cabecera *f*; rótulo *m*; epígrafe *m* — **i** titolo *m*; riga *f* principale di un titolo

6906 headliner; headline setter; photohead setter
f composeuse *f* de titres; titreuse *f*; photo-titreuse *f* — **d** Titelsetzgerät *n* — **n** koppenzetter *m*; fotografische koppenzetter *m* — **e** máquina *f* fototituladora — **i** fototitolatrice *f*; macchina *f* fotocompositrice di titoli

6907 headline setter — The compositor who sets headlines into type.
f compositeur *m* de titres — **d** Titelsetzer *m* — **n** koppenzetter *m* — **e** cabecero *m* — **i** compositore *m* di titoli

6908 headline setter — See headliner.

6909 head margin — Upper margin of a book page.
f marge *f* supérieure; marge *f* de tête — **d** oberer Seitenrand *m* — **n** bovenmarge *fm*; kopmarge *fm*; marge *fm* aán de kop — **e** margen *fm* de cabeza — **i** margine *m* di testa

6910 head of the back — See cap.

6911 head of the sheet — Edge of the sheet which is fed to the guides.
f bord *m* pince — **d** Vorderkante *f* des Bogens — **n** grijperkant *m*; vooraanlegkant *m*; vooraanleg *m* — **e** orilla *f* de entrada — **i** bordo *m* del lato pinza

6912 head piece — Decorative block, usually a rectangular line block, at the top of the first page of a book, chapter, etc.
f bandeau *m*; tête *f* de page; tête *f* de chapitre — **d** Kopfleiste *f*; Zierleiste *f*; Kopfstück *n* — **e** orla *f* de cabecera — **i** fregio *m* di testa

6913 head rule
f filet *m* de tête — **d** Kopflinie *f* (einer Tabelle) — **n** koplijn *fm* — **e** filete *m* de cabeza — **i** filetto *m* in testa

6914 head, scanning — See scanning head.
6915 head, sensing — See scanning head.
6916 head stick — See sticks.
6917 head to come; HTC; HTK; H2K — An American newspaper term informing the composing room that the headline for the article will follow.
n kop volgt — **e** título *m* por venir — **i** titolo *m* a venire

6918 heat capacity — Heat required to raise the temperature of a substance 1 °C.
f capacité *f* calorifique; capacité *f* thermique — **d** Wärmeeinhalt *m*; Wärmekapazität *f* — **n** warmtecapaciteit *f* — **e** capacidad *f* de calor; capacidad *f* térmica; capacidad *f* calorífica — **i** capacità *f* termica

6919 heat conductivity
f conductibilité *f* calorifique; conductibilité *f* thermique — **d** Wärmeleitfähigkeit *f*; Wärmeleitung *f*; Wärmeleitvermögen *n* — **n** warmtegeleidend vermogen *n*; warmtegeleiding *f*; warmtegeleidendheid *f* — **e** conductividad *f* de calor — **i** conducibilità *f* termica

6920 heater; heating unit; heating apparatus
f appareil *m* de chauffage; calorifère *m*; réchauffeur *m* — **d** Heizkörper *m*; Heizapparat *m*; Heizgerät *n* — **n** verwarmingsinstallatie *f*; verwarmingstoestel *n* — **e** calentador *m* — **i** apparecchio *m* di riscaldamento; impianto *m* di riscaldamento

6921 heat flow
f flux *m* de chaleur — **d** Wärmestrom *m* — **n** warmtestroom *m* — **e** corriente *f* de calor — **i** flusso *m* di calore

6922 heating apparatus — See heater.

6923 heating element
f élément *m* de chauffage; élément *m* chauffant — **d** Heizelement *n*; Heizkörper *m*; Heizapparat *m* — **n** verwarmingselement *n* — **e** elemento *m* de calefacción; elemento *m* de caldeo — **i** elemento *m* di riscaldamento; elemento *m* riscaldatore

6924 heating rollers — Rollers containing heating elements for aiding in setting of inks on difficult materials such as foils.
f rouleaux *mpl* chauffants — **d** Heizrollen *fpl*; Heizwalzen *fpl* — **n** verwarmingsrollen *fmpl* — **e** rodillos *mpl* de calefacción — **i** rulli *mpl* di riscaldamento

6925 heating unit — See heater.

6926 heat of fusion — Heat absorbed by a unit mass of a given solid at its melting point which completely converts the solid to a liquid at the same temperature, equal to the heat of solidification.
f chaleur *f* de fusion — **d** Schmelzwärme *f* — **n** smeltwarmte *f* — **e** calor *m* de fusión — **i** calore *m* di fusione

6927 heat of solidification — See heat of fusion.

6928 heat of vaporisation — Heat absorbed per unit mass of a given material at its boiling point which completely converts the material to a gas at the same temperature. See also latent heat.

f chaleur *f* de vaporisation — **d** Verdampfungs-wärme *f* — **n** verdampingswarmte *f* — **e** calor *m* de vaporización; calor *m* de evaporación — **i** calore *m* di evaporazione

6929 heat resistance

f résistance *f* à la chaleur — **d** Wärmebeständig-keit *f*; Hitzebeständigkeit *f*; Hitzeechtheit *f* — **n** hittebestendigheid *f*; bestendigheid *f* tegen hitte — **e** resistencia *f* al calor — **i** resistenza *f* al calore

6930 heat resistant

f résistant à la chaleur — **d** hitzebeständig — **n** hittebestendig; bestand tegen hitte — **e** resistente al calor — **i** resistente al caldo; resistente al calore

6931 heat sealing — Uniting two or more surfaces by fusion, either of the coatings or of the base materials, under controlled conditions of temperature, pressure and time (dwell).

f thermosoudage *m*; thermoscellage *m* — **d** Heiß-schweißen *n*; Anschmelzen *n*; Heißsiegeln *n* — **n** (warm) aaneensmelten *n*; warm lassen *n*; sealen *n* — **e** soldadura *f* por calefacción; sellado *m* al calor; sellado *m* en caliente — **i** termosaldatura *f*; termosigillatura *f*; sigillatura *f* a caldo

6932 heat-sealing paper; heat-seal paper — Sur-face-coated paper, usually waxed paper with a suf-ficient coating of paraffin to permit sealing upon the application of heat.

f papier *m* collant à la chaleur — **d** Heißklebe-papier *n*; Heißsiegelpapier *n*; Plyadhäsivpapier *n* — **n** heat-seal-papier *n* — **e** papel *m* para sellado térmico; papel *m* para fraguado térmico — **i** carta *f* termosaldante

6933 heat-seal paper — See heat-sealing paper.

6934 heat-set inks — Inks which dry under the action of heat by evaporation of a solvent. The re-maining resin and pigment on the stock, being dry, cannot migrate or strike through and can be backed up immediately. See also thermosetting ink.

f encres *fpl* heat-set; encres *fpl* séchant par la chaleur; encres *fpl* prenant à la chaleur — **d** hitze-abbindende Farben *fpl*; Heat-Set-Farben *fpl*; Heißtrockenfarben *fpl*; heißtrocknende Farben *fpl* — **n** heat-set inkten *mpl*; warmdrogende inkten *mpl* — **e** tintas *fpl* de secado en caliente; tintas *fpl* de fraguado al calor — **i** inchiostri *mpl* termostabilizzanti; inchiostri *mpl* heat-set

6935 heavy-bodied ink — Ink with a high vis-cosity or stiff consistency.

f encre *f* forte — **d** steife Druckfarbe *f*; stark pigmentierte Druckfarbe *f*; hochviskose Druck-farbe *f* — **n** stijve inkt *m*; hoogviskeuze inkt *m* — **e** tinta *f* de alta viscosidad — **i** inchiostro *m* ad alta viscosità

6936 heavy-duty machine — Machine that can withstand unusual strain.

f machine *f* à grand rendement; machine *f* à fort débit; appareil *m* soumis à un travail très dur — **d** Hochleistungsmaschine *f* — **n** machine *f* voor zwaar werk; zware machine *f* — **e** máquina *f* de alto rendimiento — **i** macchina *f* per lavorazioni pesanti; macchina *f* di struttura robusta

6937 heavy-face rule — See full-face rule.

6938 heavy impression — See hard impression.

6939 heavy ink — Ink with a high proportion of pigment or with a high specific gravity.

f encre *f* lourde — **d** schwere Farbe *f* — **n** zware inkt *m* — **e** tinta *f* pesada — **i** inchiostro *m* pesante

6940 heavy metal — Photoengraving metal rolled to 11 point or 3.865 mm (0.152 in) thick.

f métal *m* épais (de photogravure de hauteur de 11 points) — **d** schweres (Klischee-)Metall *n* — **n** dik clichémetaal *n* — **e** plancha *f* de 3.865 mm de espesor — **i** lastra *f* di 3.865 mm di spessore

6941 heavy naphtha — Mixture of xylene and higher homologs, derived from coal tar by frac-tional distillation. See also SBP spirit.

f essence *f* lourde — **d** Schwerbenzin *n* — **n** zware benzine *fm* — **e** nafta *f* pesada — **i** benzina *f* pesante

6942 heavy spar — See barytes.

6943 heavy-weight forme — Letterpress printing forme with a printing zone of 50-70% of the dis-tance between the two bed bearers. See also pre-tension.

f forme *f* lourde — **d** schwere Druckform *f* — **n** zware drukvorm *m*; zware vorm *m* — **e** forma *f* pesada; forma *f* densa — **i** forma *f* pesante

6944 Hebrew — Hebrew has 22 letters in the basic alphabet, but many "pointed" characters are used in certain kinds of text. It is read from right to left. Set with the nicks downwards.

f hébreu *m* — **d** Hebräisch *n* — **n** Hebreeuws *n* — **e** hebreo *m* — **i** ebraico *m*

6945 hectography; gelatine printing; gelatine duplicating — Duplicating process utilizing a gel-atine pad which receives copy of a specially prepared image using a special type of aniline ink and then transfers it to suitable paper. Sometimes refers to the spirit process.

f impression *f* hectographiée; impression *f* à la gélatine — **d** hektographischer Druck *m* — **n** hectografische druk *m* — **e** impresión *f* hecto-gráfica — **i** impressione *f* ettografica

6946 heelnick — See groove.

6947 height of capital letter

f hauteur *f* des capitales — **d** Versalhöhe *f* — **n** kapitaalhoogte *f* — **e** alto *m* de las mayúsculas — **i** corpo *m* della maiuscola

6948 height of Frankfort — See also height to paper.

f hauteur *f* de Franckfurt — **d** Frankfurter Höhe *f* — **n** Frankforter hoogte *f* — **e** altura *f* de Francfort — **i** altezza *f* di Francoforte

6949 height of shank; height of shoulder

f hauteur *f* à l'épaule; hauteur *f* de moule; hauteur *f* socle — **d** Schulterhöhe *f*; Achselhöhe *f* — **n** schouderhoogte *f* — **e** altura *f* del árbol; altura *f* del hombro (de tipo) — **i** altezza *f* di spalla; altezza *f* del fusto

6950 height of shoulder — See height of shank.

6951 height of type — See height to paper.

6952 height of type face — Vertical dimension of the face of a character expressed in points.

f hauteur *f* de l'œil; hauteur *f* d'œil — **d** Höhe *f* des Buchstabenbildes; Schriftgröße *f*; Bildfläche *f*; Schriftfläche *f* — **n** beeldhoogte *f* (van de letter); letterbeeldhoogte *f* — **e** altura *f* de ojo — **i** altezza *f* d'occhio; altezza *f* dell'occhio

6953 height to paper; type height; height of type — Standard height of types, blocks or any letterpress printing material, in UK and most American countries 23.3167 mm (0.918 in), in France 23.56 mm, Germany 23.56 mm (Old German 24.90 mm *obs.*, Leipsic 24.82 mm *obs.*), Netherlands 24.85 mm, Belgium 23.68 mm (Flemish 24.85 mm, Old Flemish 24.78 mm), Austria 23.56 mm, Russia 25.10 mm and bookbinders type high 6.5 mm.

f hauteur *f* en papier; hauteur *f* du caractère — **d** Schrifthöhe *f* — **n** letterhoogte *f* — **e** altura *f* tipográfica; altura *f* del tipo; altura *f* de los tipos — **i** altezza *f* tipografica; altezza *f* del carattere

6954 helical gear — Gear wheel the teeth of which follow the pitch surface in a helical manner.

f pignon *m* à taille hélicoïdale; roue *f* à denture hélicoïdale — **d** Schraubenrad *n*; schrägverzahntes Rad *n* — **n** tandrad *n* met schuine vertanding — **e** rueda *f* helicoidal; engranaje *m* helicoidal — **i** ruota *f* con dentatura elicoidale; ingranaggio *m* elicoidale

6955 heliogen — Registered trade name (BASF) for phthalocyanine dyestuffs in manufacturing of printing inks, characterized by outstanding fastness to light as well as brilliancy of shade. See also monastral blue.

6956 heliogen pigments — See monastral blue.

6957 heliotype — See collotype process.

6958 hellbox; metal box; shoe; old shoe — Receptacle in which defective or worn out types are thrown for remelting.

f boîte *f* à défets; cassetin *m* au diable; caisse *f* pour la refonte; sabot *m obs.* — **d** Zeugkiste *f*; Defektenkasten *m*; Hölle *f sl.* — **n** helbak *m*; hel *m* — **e** cajetín *m* del pastel; cajetín *m* del diablo; cajetín *m* de (las) ánimas; cajón *m* (de metal); cajón *m* para echar el plomo; cajón *m* para botar el plomo; zapato *m* — **i** cassa *f* del piombo di rifondere; cassettino *m* del diavolo; cassettino *m* degli accordi; cassettino *m* dei refusi

6959 hematite — See ferric oxide.

6960 hemicellulose — Cellulose which can be converted by hydrolysis with dilute acids into their individual sugars.

f hémicellulose *f* — **d** Hemizellulose *f*; Halbzellulose *f* — **n** hemicellulose *fm* — **e** hemicelulosa *f*; semicelulosa *f*; seudocelulosa *f* — **i** emicellulosa *f*

6961 hemlock fir; hemlock spruce — Resinous wood used in the manufacture of sulphite pulp.

f pruche *f*; sapin *m* du Canada — **d** Schierling *m*; Schierlingstanne *f*; Hemlockstanne *f* — **n**

hemlockspar *m*; Canadese hemlock *n* — **e** pinabete *m*; abeto *m* americano; pino *m* de Virginia — **i** abete *m* del Canada

6962 hemlock spruce — See hemlock fir.

6963 hemp — Soft, white fibres, coarser than flax, but stronger, more glossy and more durable than cotton, used in strong papers and boards. The main source of hemp fibres are old ropes, and old fishing nets.

f chanvre *m* — **d** Hanf *m* — **n** hennep *m* — **e** cáñamo *m* — **i** canapa *f*

6964 hemp paper — See manilla paper.

6965 Henderson process — Process for newspaper and commercial gravure, either in single or in multi-colour work, based on the use of halftone positives (films) that are exposed by means of special equipment on cylinders sensitized with cold enamel (bichromated shellac).

f procédé *m* Henderson — **d** Henderson-Verfahren *n* — **n** Henderson-procédé *n* — **e** procedimiento *m* Henderson; método *m* Henderson — **i** procedimento *m* Henderson

6966 henry — Electromagnetic unit of inductance. Abbrev. H (no point). Equal to the inductance of a circuit in which an electromotive force of one volt is produced by a current in the circuit which varies at the ratio of one ampere per second.

f henry *m* — **d** Henry *n* — **n** henry *m* — **e** henry *m* — **i** henry *m*

6967 herring-bone gear; double-helical gear — Helical gear with teeth that lie on the pitch cylinder in a V-shaped form so that one half of each tooth is on a right-handed helix and the other half on a left-handed helix.

f engrenage *m* à chevrons — **d** Pfeilradgetriebe *n* — **n** V-vormige vertanding *f* — **e** engranaje *m* doble helicoidal; engranaje *m* doble angular — **i** ingranaggio *m* (con dentatura) a cuspide

6968 herring-bone streaks — Defect of aluminium foil. Surface imperfections in a series of V-formations.

6969 Herschel effect — In 1840 Sir John F.W. Herschel projected the image of a spectrum onto silver chloride paper which was exposed at the same time to diffuse daylight. This discoloured the paper by forming print-out silver but, on the area that was exposed to light at the red end of the spectrum, the discoloration was destroyed. Colours could be seen in the bleached region which corresponded to the colour of the light causing the bleaching. In particular, a "full and fiery" red was observed where the red light fell. Now "Herschel effect" is usually applied to the destruction of latent image rather than of print-out silver by subsequent exposure to light of long wavelength. It differs from solarization in that it is produced by light that is not effective (or only effective) for forming a latent image.

f effet *m* Herschel — **d** Herschel-Effekt *m* — **n** Herschel-effect *n* — **e** efecto *m* de Herschel — **i** effetto *m* di Herschel

6970 hertz — In 1965 the International Electrotechnical Commission has recommended that hertz be the international unit of frequency. One hertz is equal to 1 cycle per second. The alternative c/s (cycle/second) is allowed. Symbol: Hz (no point). After Heinrich Hertz, German physicist (1857-1894).

f hertz *m* — **d** Hertz *n* — **n** hertz *m* — **e** hertz *m* — **i** hertz *m*

6971 Herzberg's stain; chloric zinc iodide — Mixture of 25 ml, at 20 °C of a saturated solution of zinc chloride with a solution of 0.25 g of iodine and 5.25 g of potassium iodide in 12.5 ml of water. It gives red with rags, purple blue with chemical pulp containing a little lignin, and yellow with fibres containing lignin.

f chlorure *m* de zinc jodé; réactif *m* de Herzberg — **d** Chlorzinkjodlösung *f* — **n** chloorzinkjodium *n*; chloorzinkjood *n* — **e** solución *f* de cloruro de cinc yodado — **i** cloroioduro *m* di zinco; reattivo *m* di Herzberg

6972 hexadecanoic acid — See palmitic acid.

6973 hexadecimal system — Method of counting to base sixteen. Binary hexadecimal counting proceeds from 9 through A, B, C, D, E, and F, and then to the value 10. A congenial counting and coding system for an eight-bit byte oriented computer. A hexadecimal digit can be expressed in four bits.

f système *m* hexadécimal — **d** hexadezimales System *n* — **n** hexadecimaal stelsel *n* — **e** sistema *m* hexadecimal — **i** sistema *m* essadecimale

6974 hexagonal screen — See Schulze screen.

6975 hickeys — See hickies.

6976 hickies *GB*; **hickeys** *US* — Imperfection in presswork due to, e.g., dirt on press, hardened specks of ink or any dry hard particle working into the ink, forme, plate or offset blanket.

f mouches *fpl*; larrons *mpl*; puces *fpl sl.* — **d** Butzen *mpl*; Partisanen *mpl* — **n** spanjolen *mpl*; vlekjes *npl* — **e** ladrones *mpl*; lardones *mpl*; lunares *mpl* — **i** ladri *mpl*

6977 hiding power; covering power — Ability of ink to hide the material beneath (substrate) and to produce a uniform, opaque surface. See also opacity.

f pouvoir *m* couvrant; pouvoir *m* colorant; pouvoir *m* opacifiant; puissance *f* couvrante; couvrant *m*; pouvoir *m* masquant — **d** Deckfähigkeit *f*; Deckkraft *f*; Deckvermögen *n* — **n** dekkend vermogen *n*; dekvermogen *n*; dekkracht *fm* — **e** poder *m* cubriente — **i** potere *m* coprente

6978 hiding-power test — Test where a small portion of the ink sample is drawn down on the test paper with a black stripe and the film is scraped off with a spatula. The drawdown is inspected in transmitted light and compared with the control sample.

f essai *m* de pouvoir couvrant — **d** Deckfähigkeitsprüfung *f* — **n** dekkrachtonderzoek *n* — **e** ensayo *m* del poder cubriente; prueba *f* del poder cubriente — **i** prova *f* del potere coprente; saggio

m del potere coprente

6979 hieroglyphics — Ideographic symbols developed by the ancient Egyptians, Mayas, Incas, and others in the early history of writing. These symbols tended to become phonetic but never approached the simplicity of an alphabet.

f hiéroglyphes *mpl*; signes *mpl* hiéroglyphiques — **d** Hieroglyphen *mpl* — **n** hiërogliefen *fmpl*; hiëroglyfen *fmpl* — **e** jeroglíficos *mpl*; hieroglíficos *mpl* — **i** geroglifici *mpl*; scrittura *f* geroglifica

6980 high bulk — A book paper of relatively great thickness in relation to substance weight has a high bulk. Synonymous with the English term featherweight, which in US is also applied to extra light weight enamels and manifold papers.

6981 high contrast — Relationship of highlights to shadows on a negative, whether continuous tone or halftone, where the highlights are very black and the shadows very open according to the tonal scale.

f grand contraste *m*; contraste *m* élevé — **d** hoher Kontrast *m*; kontrastreich — **n** hoog contrast *n*; contrastrijk — **e** contraste *m* alto — **i** contrasto *m* alto

6982 higher level language — Programming language one or more steps removed from basic machine language, in which it is easier to write programs since the language makes more frequent use of familiar terms and assumes that certain tedious chores will be performed automatically when the program is re-compiled into machine language for the appropriate computer. Often "portable", i.e., can be translated by compiler programs to run on computers utilizing different machine languages.

6983 high-frequency welding

f soudage *m* par haute fréquence — **d** Hochfrequenzschweißen *n*; Hochfrequenzverschweißung *f*; Hochfrequenzschweißung *f* — **n** hoogfrequent lassen *n* — **e** soldadura *f* por alta frecuencia — **i** saldatura *f* ad alta frequenza

6984 high-gloss ink — See also gloss ink.

f encre *f* lustrée — **d** Hochglanzdruckfarbe *f*; Hochglanzfarbe *f* — **n** hoogglanzende inkt *m* — **e** tinta *f* de alto lustre — **i** inchiostro *m* ad alta brillantezza

6985 high key — Picture with only light and some middle tones; dark tones are absent.

f high key *m* — **d** Highkey *m* — **n** high key *m* — **e** imagen *f* pálida — **i** a soli toni

6986 highlight dots — Small dots representing the highlights in a halftone negative and printing plate.

f points *mpl* de hautes lumières — **d** Hochlichtpunkte *mpl* — **n** hogelicht-punten *mpl* — **e** puntillado *m* en las altas luces — **i** punti *mpl* nelle alte luci

6987 highlight halftone; drop-out halftone; dropout — Halftone plate in which the highlight areas are devoid of dots to accentuate the contrast of the reproduction. Made from a highlight negative or by special etching of the halftone print on metal.

f similigravure *f* à blancs purs; simili *f* à hautes lumières — **d** Hochlichtautotypie *f*; Hochlichtrasterätzung *f* — **n** hogelicht-autotypie *f*; hooglicht-autotypie *f* — **e** grabado *m* con blancos; autotipia *f* de blancos puros — **i** autotipia *f* di alte luci; mezzatinta *f* di alte luci

6988 highlighting exposure — Additional exposure when making a screen negative if the density range of the original must be extended.

f pose *f* des grandes lumières — **d** Hochlichtexposition *f* — **n** hogelicht-belichting *f*; lijnbelichting *f* — **e** exposición *f* de altas luces; posa *f* de altas luces — **i** esposizione *f* per alte luci

6989 highlight mask — Mask to retain or increase highlight contrast in the reproduction.

f masque *m* des lumières — **d** Hochlichtmaske *f* — **n** hogelicht-masker *n* — **e** máscara *f* de claros; máscara *f* acladora; máscara *f* aclaradora — **i** maschera *f* per le alte luci

6990 highlight negative; drop-out negative — Halftone negative in which the dot formation in the highlights has been plugged up or eliminated during exposure of the original by various methods, including the use of special stops and employment of originals specially prepared for highlight effects.

f négatif *m* à blancs purs; négatif *m* à hautes lumières — **d** Hochlichtnegativ *n* — **n** hooglicht-negatief *n* — **e** negativo *m* autotípico sin puntos en las altas luces — **i** negativo *m* senza punti nelle alte luci

6991 highlight reducer — Super-proportional reducer which attacks the heavier densities of silver first, thus lowering contrast as well as density. It consists of ammonium persulphate, sulphuric acid and water.

f affaiblisseur *m* surproportionnel — **d** überproportionaler Abschwächer *m* — **n** hogelicht-afzwakker *m* — **e** rebajador *m* de altas luces; reductor *m* de altas luces — **i** indebolitore *m* superproporzionale

6992 highlights — Lightest or whitest areas of an original or reproduction, represented by the densest portions of a continuous tone negative and by the smallest dot formations in a halftone negative and printing plate.

f hautes lumières *fpl*; grandes lumières *fpl*; grands blancs *mpl* — **d** Hochlichter *npl* — **n** hogelicht-partijen *fpl*; hoog-contrastrijke partijen *fpl* — **e** realces *mpl*; altas luces *fpl*; grandes luces *fpl*; claros *mpl*; blancos *mpl*; luces *fpl*; zona *f* de blancos — **i** alte luci *fpl*

6993 highlight stop — Lens aperture used in halftone photography to join up the highlight dots in the negative and to record highlight detail.

f diaphragme *m* des lumières — **d** Hochlichtblende *f* — **n** diafragma *n* voor de sluitbelichting; moordenaar *m* *sl.* — **e** diafragma *m* para acentuar los blancos — **i** diaframma *m* per le alte luci

6994 high-M.f. — Abbrev. for high machine finish, a finish between Mf and Sc.

6995 high-pressure mercury vapour lamp — Electric lamp emitting mainly blue, violet and ultraviolet rays from an arc burning between electrodes in an atmosphere of mercury vapour. Very efficient for all printing down processes but unsuited for colour work and direct negative making because of the line spectrum of the rays.

f lampe *f* à vapeur de mercure à haute pression — **d** Quecksilberhochdrucklampe *f* — **n** hoge-druk kwikdamplamp *fm* — **e** lámpara *f* de vapor de mercurio a alto presión — **i** lampada *f* a vapori di mercurio ad alto pressione

6996 high quad — See high spaces.

6997 high spaces — Quads and leads used in high-spaced type formes for duplicate relief plate-making or to support letters which overhang their body such as swash italic letters. They are higher than ordinary printing standard to provide reinforcement for accurate moulding (21.64 mm or 0.852 in); also of much value for the mounting of *16-gauge zinc and copper photoengraving plates, which are 1.65 mm (0.065 in) thick, thus avoiding the necessity of mounting the engravings on wood or metal for insertion in the page. High spacing for duplicate platemaking is also made 22.23 mm (0.875 in) high.

f espaces *fpl* hautes — **d** hoher Ausschluß *m*; achselhoher Ausschluß *m*; schulterhoher Ausschluß *m* — **n** hoog wit *n*; steunwit *n* — **e** espacios *mpl* de altura del hombro; espacios *mpl* altos — **i** spazi *mpl* di altezza uguale a quella della spalla

6998 high-speed cutting machine

f massicot *m* rapide — **d** Schnellschneidemaschine *f*; Schnellschneider *m* — **n** snelsnijmachine *f*; snelsnijder *m* — **e** guillotina *f* rápida; cortadora *f* rápida — **i** taglierina *f* rapida

6999 high-speed etching — See powderless etching.

7000 high-speed machine — Machine turning at hight speed.

f machine *f* à grand vitesse — **d** schnellaufende Maschine *f*; Schnellaufer *m* obs. — **n** snellopende machine *f*; snelloper *m* — **e** máquina *f* rápida; máquina *f* de gran velocidad; máquina *f* de alta velocidad — **i** macchina *f* rapida; macchina *f* ad alta velocità

7001 high-speed printer — Output printer which prints at great speed.

f imprimante *f* ultrarapide — **d** Schnelldrucker *m*; Hochleistungsbelegdrucker *m* — **n** snelle regeldrukker *m* — **e** impresora *f* de alta velocidad — **i** macchina *f* da stampa ad alta velocità; stampatrice *f* rapida

7002 high-speed reader

f lecteur *m* ultrarapide — **d** Hochleistungsleser *m* — **n** snellezer *m* — **e** lectora *f* de alta velocidad — **i** lettore *m* ad alta velocità

7003 high-task rating method — Procedure in which the time study man compares the actual rate of working of an operator with his concept of the rate necessary to earn the normal incentive earnings. See also low-task rating method.

7004 high tension
f haute tension *f* — **d** Hochspannung *f* — **n** hoogspanning *f* — **e** alta tensión *f* — **i** alta tensione *f*

7005 high tension switch
f interrupteur *m* à haute tension — **d** Hochspannungsschalter *m* — **n** hoogspanningsschakelaar *m* — **e** interruptor *m* de alta tensión — **i** interruttore *m* ad alta tensione

7006 high yield pulp — Chemical pulp produced from vegetable materials and giving a higher output.
f pâte *f* à haut rendement; pulpe *f* à haut rendement — **d** Hochausbeutezellstoff *m* — **n** celstof *fm* met hoge opbrengst — **e** pulpa *f* de alta rendición; pasta *f* de alta rendición — **i** pasta *f* ad alta resa

7007 Hilite lens — Process lens with integral apertures to produce highlight effects in halftone negatives.

7008 hinged lever — See toggle lever.

7009 hinged plate-holder
f châssis *m* à volet pivotant — **d** Schwenkkassette *f* — **n** scharnierende cassette *fm* — **e** portaplaca *m* fotográfico con bisagra; chasis *m* portaplacas con bisagra; portaplaca *m* giratorio — **i** telaio *m* portalastra con cerniera

7010 histogram — In statistics, a diagram representing a frequency distribution, so drawn that area in the diagram corresponds to frequency in the table.
f histogramme *f* — **d** Histogramm *n*; Treppenlinie *f* — **n** histogram *n* — **e** histograma *m* — **i** istogramma *m*

7011 historiated — Decorated with figures of people or animals, as the initials of paragraphs or the borders of the pages of a book, manuscript, etc.
f historié — **d** durch figürlichen Schmuck belebt — **n** met figuren versierd — **e** historiado — **i** istoriato; eccessivamente ornato

7012 hit-on-the-fly printer — Computer lineprinter in which the type carrier does not stop at the time of impression. An impression is obtained on the paper when the character is in the required position.
f imprimante *f* à volée — **d** Drucker *m* für fliegenden Abdruck — **n** vliegende-afdruk machine *f* — **e** impresora *f* al vuelo — **i** stampatrice *f* continua

7013 hoax; canard; fake news — False or unfounded report or story, especially a fabricated report.
f canard *m* — **d** Ente *f*; Zeitungsente *f*; falsche Nachricht *f*; Falschmeldung *f* — **n** canard *m* — **e** noticia *f* falsa; noticia *f* absurda; noticia *f* extravagante; infundio *m*; embuste *m*; ánade *m*; canard *m LA* — **i** notizia *f* falsa; notizia *f* inventata

7014 hobelmachine — See autoshaver.

7015 holding back — See backing away from fountain.

7016 hold over — See crowdout.

7017 hole — A hole in a paper web can come from various causes: the most common being slime, stock lumps, coating splashes and wire holes.
f trou *m* — **d** Loch *n* — **n** gat *n* — **e** agujero *m* — **i** foro *m*

7018 hole punching machine — See punching machine.

7019 Holland cloth — Smooth starch coated fabric 1.25-1.75 mm (0.005-0.007 in) thick. Used as a separator on moulding press platens in thermosetting matrix and rubber platemaking.

7020 hollander — The most commonly used type of *beater, developed by the Dutch papermakers in the 17th century. Oval shaped tub with a middle partition in which the papermaking fibres are circulated, disintegrated and cut, and in which the sizing, colouring and loading of the paper stock are carried out.
f pile *f* hollandaise — **d** Holländer *m*; Mahlholländer *m* — **n** hollander *m*; maalbak *m* — **e** pila *f* holandesa; máquina *f* holandesa; holandesa *f* — **i** olandese *f*

7021 hollerith card — See also hollerith code, punched card.
f carte *f* hollerith — **d** Hollerithkarte *f* — **n** hollerithkaart *fm* — **e** tarjeta *f* hollerith — **i** scheda *f* hollerith

7022 hollerith code — In computer technology, a system for coding data into punch cards, in which each horizontal row is assigned a different value, and letters, numbers, or special characters are encoded as combinations of these values in a vertical column. Named after Herman Hollerith (1890). Successfully used in the US census of 1890.
f code *m* Hollerith — **d** Hollerithkode *m* — **n** hollerithcode *m* — **e** código *m* hollerith — **i** codice *m* hollerith

7023 hollow back; open back; loose back; spring back; false back — When a book with a hollow back is opened the spine portion of the cover moves away from the back of the book. The normal cased book is a hollow back but this is assumed and therefore not specified.
f dos *m* brisé; dos *m* à ressort — **d** Sprungrücken *m*; Hohlrücken *m* — **n** springrug *m* — **e** lomo *m* destacado; lomo *m* despegado; lomo *m* libre; lomera *f* destacada; lomera *f* despegada; lomera *f* libre — **i** dorso *m* con salti

7024 hollow (cast) quads — See quotation quads.

7025 hollow patch — See indention.

7026 hollows; back lining — Strips of paper glued to the back and the boards of a book.
f papier *m* garniture — **d** Hinterklebepapier *n*; Rückenbeklebepapier *n* — **n** rugoverplakpapier *n*; rugoverlijmpapier *n* — **e** lomeras *fpl* — **i** carta *f* del lomo

7027 hollow types — See cored types.

7028 hologram — Product of a process of lensless

photography which utilizes lasers as a light source. The point light source gives the hologram its three-dimensional effect.

7029 holography — Method of reproducing illustrations with a photographic film, which takes the position of an imaginated "fence". The light-rays, coming from the different points of the object cross this fence under different angles and reach the observer. The film has the effect of a window but not as the illustration itself.

7030 home organ — See house organ.

7031 hone *v* — To grind a cutting edge fine.

f aiguiser — **d** honen; abziehen — **n** honen; afnemen met de oliesteen — **e** esmerilar — **i** smerigliare

7032 honeycomb base — See honeycomb mount.

7033 honeycomb mount; honeycomb base

f bloc *m* nid d'abeilles — **d** Wabenfundament *n*; wabenförmige Plattenunterlage *f* — **n** honingraatfundament *n* — **e** platina *f* de celdillas; platina *f* alveolada — **i** base *f* a nido d'ape

7034 honeycomb screen

f trame *f* en nid d'abeilles — **d** Wabenraster *m* — **n** honingraat-raster *n* — **e** trama *f* de nido de abeja; trama *f* hexagonal — **i** retino *m* a nido d'ape

7035 hoof oil — See neat's foot oil.

7036 Hooke's joint — See universal coupling.

7037 Hooke's law — The law of deformation for a perfectly elastic body. A term used for example in rheology stating the stress/strain relationship for a Hooke body. When a load is applied to any elastic body so that the body is deformed or strained, then the resulting stress (the tendency of the body to resume its normal condition) is proportional to the strain. Stress is measured in units of force per unit area, strain is the extent of the deformation. For example, when a bar of metal is subjected to a stretching load, the extent of the increase in length of the bar is directly proportional to the force per unit area, i.e., to the stretching load or stress. In general, Hocke's law applies only up to a certain stress called the yield strength.

f loi *f* de Hooke — **d** Hooke'sches Gesetz *n* — **n** wet *fm* van Hooke — **e** ley *f* de Hooke — **i** legge *f* di Hooke

7038 hook *v* **in; hook** *v* **up** — To place, when there are too many words for one line, those at the end to the end of the preceding or succeeding line, preceded by a square bracket, as in poetry, etc.

f rejeter — **d** überschließen; unterschließen — **n** erboven zetten en achterin houden; eronder zetten en achter inhouden — **e** amortiguar — **i** agganciare

7039 hook *v* **up** — See hook *v* in.

7040 horizontal camera — Camera in which the main movements and structure are in a horizontal plane.

f appareil *m* (de reproduction) horizontal; caméra *f* (de reproduction) horizontale — **d** Horizontal-(Reproduktions)Kamera *f* — **n** horizontale camera *f*; horizontale reproduktie-camera *f* — **e** cámara *f* (fotomecánica) horizontal; cámara *f* de reproducción horizontal — **i** macchina *f* fotografica orrizontale; apparecchio *m* orrizontale

7041 horizontal lines — Lines on a bill form, writing paper, etc., running in the same direction as the written text. See also feint lines.

f lignes *fpl* transversales — **d** Querlinien *fpl* — **n** contralijnen *fpl*; dwarslijnen *fpl*; horizontale lijnen *fpl* — **e** rayas *fpl* transversales — **i** linee *fpl* trasversali; linee *fpl* orrizontali

7042 horse *v* — To read proofs without the assistance of a copyholder.

f corriger sans lecteur — **d** korrekturlesen ohne Vorleser — **n** corrigeren zonder tegenlezen — **e** leer sin atendedor — **i** correggere senza lettore

7043 horse flesh — Surplus of work if a journeyman sets down in his bill more work than he has done, and he abates it in his next bill, the unprofitable remainder, being dead horse.

f travail *m* payé d'avance; salé *m sl* — **d** Sauer *n coll.*; Sauerkraut *n coll.*; vorgegessenes Brot *n coll.* — **n** voorschot *n*; vooruitbetaling *f* — **e** avance *m* (del sueldo); adelanto *m* — **i** anticipo *m*

7044 horsepower — Unit of power. Abbrev. hp. In the SI (1978):

1 hp = 550 ft·lbf/s ≈ 745.700 W. See also metric horsepower.

f cheval-vapeur *m*; puissance *f* en chevaux; force *f* en chevaux — **d** Pferdestärke *f* — **n** paardekracht *fm* — **e** caballo *m* de fuerza; caballo *m* de potencia; caballo *m* de energía; caballo *m* de vapor — **i** cavallo *m* di forza; cavallo-vapore *m*

7045 host computer — Main frame or central processing unit which provides the intelligence to support various peripheral devices and to execute the basic programs. If several computers are involved in one system, one tends to serve as the "host" or main computer.

7046 hot calendering

f calendrage *m* à chaud — **d** Heißsatinieren *n* — **n** warm kalanderen *n* — **e** calandrado *m* con calor; calandrado *m* en caliente — **i** calandratura *f* a caldo

7047 hotch — In news offices (probably originated in the Times office, London) the small announcements under the headings "*births, marriages and deaths".

7048 hot-melt adhesive; hot-melt glue — Material composed of combinations of resins and/or waxes which must be liquefied by heat for adhesion, and set immediately on cooling, forming a tough, strong mildew resistant bond; unaffected by weather conditions. Generally used in the range 65-205 °C (150-400 °F) for bookbinding, particularly paperbound books.

f adhésif *m* à chaud; colle *f* hot-melt; adhésif *m* hot-melt — **d** Heißschmelzleim *m*; Heißschmelzkleber *m*; Schmelzkleber *m*; Hot-Melt-Klebstoff *m*; Hot-Melt-Kleber *m* — **n** heet smeltende lijm *m*; smeltlijm *m* — **e** adhesivo *m* termosellable — **i** adesivo *m* hot-melt

7049 hot-melt coated paper — Paper coated by a hot melt coating method, which can be either a roll, gravure or extrusion coating process.
f papier *m* couché par fusion — **d** hotmelt-beschichtetes Papier *n*; Papier *n* mit Schmelz-strich — **n** papier *n* met smeltlaag — **e** papel *m* estucado por fusión — **i** carta *f* spalmata hot-melt; carta *f* spalmata con hot-melt

7050 hot-melt coating — Method of coating a con-tinuous paper web with a 100% solids compound of wax, resins, or polymers heated to a fluid state and applied to the substrate by a roll or gravure coating process with a subsequent chilling device.
f couchage *m* par fusion; couchage *m* par coulage à chaud — **d** Heißschmelzstreichverfahren *n*; Heißschmelzkleberbeschichtung *f* — **n** smeltlaag-procédé *n* — **e** estucado *m* por fusión — **i** spalmatura *f* hot-melt; spalmatura *f* con hot-melt

7051 hot-melt glue — See hot-melt adhesive.

7052 hot pressed — Glazed. Originally applied to a process of applying pressure and heat to paper during manufacture; now to papers finished by plate glazing.
f pressé à chaud — **d** heißgepreßt; heißsatiniert — **n** warm gesatineerd — **e** prensado en caliente — **i** pressato a caldo

7053 hot rolling — Glazing by means of steam heated cylinders.
f calandrage *m* à chaud — **d** Heißkalandrierung *f* — **n** warm kalanderen *n* — **e** calandrado *m* en caliente — **i** calandratura *f* a caldo

7054 hot-setting adhesive — Adhesive which re-quires a temperature at or above 100 °C for setting, generally after the addition of a curing agent or catalyst.
f adhésif *m* à chaud — **d** heißabbindender Kleb-stoff *m* — **n** warmdrogende kleefstof *fm*; warm-drogende lijm *m* — **e** adhesivo *m* en caliente — **i** adesivo *m* a caldo

7055 hound's teeth — See gutter.

7056 hours per hour; time on daywork — Hours paid at hourly rate.
f heures *fpl* à l'heure — **d** Zeitlohnstunden-anteil *m* — **n** betaalde werkuren *npl* — **e** sueldo *m* a horas — **i** ore *fpl* pagate all'ora

7057 house corrections — Corrections made neces-sary by printer's mistakes, such as typographical errors.
f première correction *f*; correction *f* en première — **d** Hauskorrektur *f*; erste Korrektur *f* — **n** huis-correctie *f* — **e** corrección *f* de primeras — **i** correzione *f* in prima; prima correzione *f*; correzione *f* preliminare

7058 house journal — See house organ.

7059 house magazine — See house organ.

7060 house organ; house journal; house paper; house magazine; home organ — Magazine or other publication published by a business or manufacturing firm for the information of its employees and customers.
f revue *f* d'entreprise; journal *m* d'entreprise — **d** Hauszeitschrift *f*; Hauszeitung *f*; Werkzeitschrift *f* — **n** huisorgaan *n* — **e** revista *f* de comunicación (interna); revista *f* órgano; revista *f* para uso del personal; boletín *m* para el personal — **i** periodico *m* aziendale

7061 house paper — See house organ.

7062 house style; style of the house — Typo-graphic style and special rules used in a printing or publishing establishment. See also follow copy.
f style *m* maison — **d** Hausstil *m*; typo-graphischer Hausstil *m* — **n** huisstijl *m*; stijl *m* van het huis — **e** estilo *m* tipográfico — **i** stile *m* tipografico

7063 HTC — See head to come.

7064 HTK — See head to come.

7065 hub — See nave.

7066 hue — Attribute of visual sensation which has given rise to colour names, such as blue, green, yellow, red, purple, etc. Psychosensorial correlate of dominant wavelength.
f teinte *f*; tonalité *f* chromatique — **d** Farb-tönung *f*; Farbton *m* depr. — **n** kleurtoon *m* — **e** matiz *m*; tonalidad *f* cromática — **i** tono *m* di colore

7067 human engineering — Management of human being and affairs especially in the industry; also a science that deals with the design of mechanical devices for efficient use by human beings. See also ergonomics.
f adaptation *f* de l'homme au travail — **d** Anpassung *f* der Arbeitsbedingungen an den Menschen — **n** aanpassing *f* van de werk-omstandigheden aan de mens — **e** adaptación *f* del trabajo al hombre — **i** adattamento *m* delle condizioni di lavoro all'uomo

7068 Humanistics — Group of type faces e.g., Jenson, Veronese, Goudy, Hollandse mediaeval, Erasmus, Palatino. The cross-stroke of the lower case e is oblique; the axis of the curves is inclined to the left; no great contrast between thick and thin strokes; serifs are bracketed; serifs of the ascenders in the lower case are oblique. Originate from the Roman type, renewed during the renais-sance. Formerly known as Venetian. Before 1971 this group was named Humanists. See also classifi-cation of type design.
f Humanes *fpl* — **d** venezianische Renaissance-Antiqua *f* — **n** Humanen *fmpl* — **e** Humanísticos *mpl* — **i** Veneziani *mpl*

7069 Humanists — See Humanistics.

7070 human relations
f relations *fpl* humaines — **d** menschliche Beziehungen *fpl* — **n** menselijke verhoudingen *fpl* — **e** relaciones *fpl* humanas — **i** relazioni *fpl* umane

7071 humidification — Adding moisture. Increasing the *relative humidity.
f humidification *f* — **d** Befeuchtung *f*; Befeuchten *n*; Feuchtegradeerhöhung *f* — **n** bevochtiging *f*; vochtig maken *n*; verhoging *f* van de vochtigheidsgraad — **e** humidificación *f*; humectación *f* — **i** umidificazione *f*

7072 humidifier; air moistener — Device to moisten the air by means of water sprays.
f appareil *m* humidificateur; humidificateur *m*; appareil *m* à humecter; humecteur *m* — **d** Luftbefeuchter *m* — **n** bevochtigingsinstallatie *f*; bevochtigingsapparaat *n* — **e** humectador *m*; humedecedor *m*; cajón *m* humectante; aparato *m* humedecedor; humectadora *f* de chorritos; mojadora *f* — **i** umidificatore *m* d'aria
7073 humidity — Moisture condition of the air. As applied to conditions of printing papers, most favourable percentage is 65. See also relative humidity.
f humidité *f* — **d** Feuchtigkeit *f* — **n** vochtigheid *f* — **e** humedad *f* — **i** umidità *f*
7074 humidity cabinet — Climate cabinet in which an ambient relative humidity up to 95% at a definite temperature (adjustable from 20-45 °C) can be established to imitate the climates of tropical and subtropical areas.
f armoire *f* de climatisation; armoire *f* tropicale — **d** Klimaschrank *m*; Klimakammer *f*; Klimaraum *m* — **n** klimaatkast *fm*; vochtkast *fm* — **e** cámara *f* de acondicionamiento de aire — **i** condizionatrice *f*
7075 humidity effect — Influence of humidity or atmospheric conditions on sensitized surfaces and materials used in photography.
f influence *f* de l'humidité — **d** Feuchtigkeitseffekt *m* — **n** vochtigheidseffect *n* — **e** efecto *m* de la humedad — **i** effetto *m* dell'umidità
7076 humidity expansion — See hygroexpansivity.
7077 humidity of the air — Moisture condition of the air. Actual humidity is the number of grains of moisture in the air at a given time. See also relative humidity.
f humidité *f* atmosphérique — **d** Luftfeuchtigkeit *f* — **n** luchtvochtigheid *f*; vochtigheid *f* van de lucht — **e** humedad *f* del aire — **i** umidità *f* dell'aria
7078 humidor — Cabinet to keep dry mats slightly moist.
f armoire *f* d'humidification — **d** Feuchtkasten *m* — **n** vochtkast *fm* — **e** cajón *m* humectante; cajón *m* humectativo — **i** armadio *m* condizionato per flani
7079 hundred twenty-eightmo; 128mo
f en cent-vingt huit — **d** Hundertachtundzwanziger Format *m* — **n** in honderdachtentwintigen — **e** en cientoveinte-y-ochoavo — **i** in centoventottesimo
7080 hundredweight, gross — See long hundredweight.
7081 hundredweight, long — See long hundredweight.
7082 hundredweight, short — See short hundredweight.
7083 hunger for reading
f soif *f* de lecture — **d** Lesehunger *m* — **n** leeshonger *m* — **e** afán *m* de leer; voracidad *f* lectora — **i** fame *f* di lettura
7084 hung on guards

f monté sur onglets — **d** auf Fälzel geklebt — **n** op oortjes geplakt — **e** encolado sobre cartivanas — **i** incollato su brachette
7085 hunt — In rotogravure printing, tendency of rollers to move laterally back and forth.
f déplacement *m* latéral — **d** seitliches Verlaufen *n* — **n** zijdelings afwijken *n* — **e** desplazamiento *m* lateral — **i** spostamento *m* laterale
7086 Hurter & Driffield curve — See H & D curve.
7087 Hycar — Trade mark for types of synthetic rubber.
7088 hydrargyrum — See mercury.
7089 hydrate — Compound formed by the combination of water with another substance, in which the water retains its molecular state as H_2O. The water combines in a definite weight ratio and the hydrate may be represented by a chemical formula. Most hydrates are decomposed by gentle heating.
f hydrate *m* — **d** Hydrat *n* — **n** hydraat *n* — **e** hidrato *m* — **i** idrato *m*
7090 hydrate — See aluminium hydrate.
7091 hydrated aluminium — See aluminium hydrate.
7092 hydrated ferric oxide — See ferric hydroxide.
7093 hydrated iron oxide — See ferric hydroxide.
7094 hydrated lime — See calcium hydroxide.
7095 hydration — Process of absorption or combination of water with another substance, involving both chemical reaction and mere absorption, usually not applied in cases where a liquid solution results. In paper industry, a prolonged treatment in the beater whereby a viscous pulp is produced that gives water-resistance and crackle to the finished paper.
f hydratation *f* — **d** Hydratation *f* — **n** hydratering *f*; hydrateren *n* — **e** hidratación *f* — **i** idratazione *f*
7096 hydraulic flow — See turbulent flow.
7097 hydraulic press — Machine in which a great force is communicated to a large plunger or ram by means of a liquid forced into the cylinder in which it moves, by a forcing pump of small diameter. Used in electrotyping, stereotyping, rubber and plastic platemaking.
f presse *f* hydraulique — **d** hydraulische Presse *f* — **n** hydraulische pers *f* — **e** prensa *f* hidráulica — **i** pressa *f* idraulica; torchio *m* idraulico
7098 hydraulic pressure — Pressure applied evenly by means of a liquid (oil) under pressure.
f pression *f* hydraulique — **d** hydraulischer Druck *m* — **n** hydraulische druk *m* — **e** presión *f* hidráulica — **i** pressione *f* idraulica
7099 hydrobromic ether — See ethyl bromide.
7100 hydrocarbons — Compounds which consist solely of carbon and hydrogen.
f carbures *mpl* d'hydrogène; hydrocarbures *mpl* — **d** Kohlenwasserstoffe *mpl* — **n** koolwaterstoffen *fmpl* — **e** hidrocarburos *mpl* — **i** idro-

carburi *mpl*

7101 hydrochinone — See hydroquinone.

7102 hydrochloric acid; hydrogen chloride; muriatic acid — Clear, colourless or slightly yellow, fuming liquid. Poisonous. Soluble in water, alcohol and ether. Formula: HCl (in aqueous solution). Latin: acidum hydrochloricum. TL-value: 5 ppm or 7 mg/m^3.
f acide *m* chlorhydrique; acide *m* hydrochlorhydrique; acide *m* muriatique — **d** Salzsäure *f*; Chlorwasserstoffsäure *f* — **n** zoutzuur *n*; chloorwaterstofzuur *n* — **e** ácido *m* clorhídrico; cloruro *m* de hidrógeno; ácido *m* muriático — **i** acido *m* cloridrico; acido *m* muriatico

7103 hydrochloric ether — See ethyl chloride.

7104 hydrochloride; chlorhydrate — Salt formed when a basic dye combines with hydrochloric acid.
f chlorhydrate *m* — **d** Hydrochlorid *n*; Chlorhydrat *n* — **n** chloorhydraat *n* — **e** clorhidrato *m*; hidroclorato *m* — **i** cloroidrato *m*

7105 hydrocyanic acid; hydrogen cyanide; prussic acid — Water-white liquid at temperature below 26.5 °C. Soluble in water and ether, miscible in all proportions with water and alcohol. Vapours very poisonous. TL-value: 5 ppm.
Formula: HCN.
f acide *m* cyanhydrique; acide *m* prussique — **d** Zyanwasserstoffsäure *f*; Blausäure *f* — **n** cyaanwaterstofzuur *n*; blauwzuur *n* — **e** ácido *m* hidrociánico; ácido *m* cianhídrico; ácido *m* prúsico — **i** acido *m* cianidrico; acido *m* prussico

7106 hydrofluoric acid; hydrogen fluoride; fluorhydric acid — Colourless, fuming, mobile, corrosive liquid. Poisonous. Gives terrible sores when in contact with the skin. Formula: HF (in aqueous solution).
f acide *m* fluorhydrique; acide *m* hydrofluorique — **d** Fluorwasserstoffsäure *f*; Flußsäure *f* — **n** fluorwaterstofzuur *n*; vloeispaatzuur *n* — **e** ácido *m* fluorhídrico; fluoruro *m* de hidrógeno — **i** acido *m* fluoridrico

7107 hydrogen — Non-metallic univalent element that is the simplest and lightest of the elements; a colourless, odourless, highly flammable diatomic gas. Symbol: H. Latin: hydrogenium.
f hydrogène *m* — **d** Wasserstoff *m* — **n** waterstof *fm* — **e** hidrógeno *m* — **i** idrogeno *m*

7108 hydrogen carboxylic acid — See formic acid.

7109 hydrogen chloride — See hydrochloric acid.

7110 hydrogen cyanide — See hydrocyanic acid.

7111 hydrogen dioxide — See hydrogen peroxide.

7112 hydrogen fluoride — See hydrofluoric acid.

7113 hydrogen-ion concentration — Determines the acidity or alkalinity of a liquid. The scale runs from pH 0-14. pH 7.0 is the neutral point. Values below 7.0 indicate an acid condition, above 7.0 alkalinity. See also pH value.
f concentration *f* en ions-hydrogène; concentration *f* ionique d'hydrogène — **d** Wasserstoffionen-Konzentration *f* — **n** waterstofionen-concentratie *f* — **e** concentración *f* hidrogen-iónica; concentración *f* iónica de hidrógeno; concentración *f* de los iones de hidrógeno — **i** concentrazione *f* in ioni idrogeno

7114 hydrogen nitrate — See nitric acid.

7115 hydrogen peroxide; hydrogen dioxide; peroxide; perhydrol — Colourless, heavy liquid, usually sold in aqueous solution of various strengths; TL-value: 1 ppm.
Formula: H$_2$O$_2$.
f peroxyde *m* d'hydrogène; eau *f* oxygénée; perhydrol *m* — **d** Wasserstoffperoxyd *n*; Wasserstoffsuperoxyd *n* — **n** waterstofperoxyde *n*; waterstofsuperoxyde *n* — **e** peróxido *m* hidrogenado; peróxido *m* de hidrógeno; agua *f* oxigenada — **i** perossido *m* d'idrogeno; acqua *f* ossigenata

7116 hydrogen salt — See acid salt.

7117 hydrogen sulphide; sulphuretted hydrogen — Colourless, flammable gas. Poisonous. Soluble in water and alcohol. Latin: acidum hydrosulfuricum. TL-value: 20 ppm.
Formula: H$_2$S.
f acide *m* sulfhydrique; hydrogène *m* sulfuré — **d** Schwefelwasserstoff *m*; Wasserstoffsulfid *n* — **n** zwavelwaterstof *fm* — **e** sulfuro *m* de hidrógeno; hidrógeno *m* sulfurado — **i** idrogeno *m* solforato; acido *m* solfidrico

7118 hydrographer
f ingénieur *m* hydrographe; hydrographe *m* — **d** Hydrograph *m* — **n** hydrograaf *m* — **e** hidrógrafo *m* — **i** idrografo *m*

7119 hydrolysis — Chemical reaction in which by the action of water one or more new substances are formed.
f hydrolyse *f* — **d** Hydrolyse *f* — **n** hydrolyse *f* — **e** hidrólisis *f* — **i** idrolisi *f*

7120 hydrometer — Instrument to measure the specific gravity or density of liquids, extensively used in graphic arts and papermaking.
f aréomètre *m*; hydromètre *m* — **d** Aräometer *mn*; Dichtigkeitsmesser *m*; Dichtemesser *m*; Hydrometer *m* — **n** areometer *m* — **e** areómetro *m*; densímetro *m*; hidrómetro *m* — **i** areometro *m*; densitometro *m*

7121 hydrophile — See hydrophilic.

7122 hydrophilic; hydrophile; water receptive — Having a strong affinity for water. Refers to colloids which swell in water and are not easily coagulated.
f hydrophile; absorbant l'eau; retenant l'eau — **d** hydrophil; wasserannehmend; wasseranziehend; wasserfreundlich — **n** hydrofiel; wateraannemend — **e** hidrófilo — **i** idrofilo

7123 hydrophobe — See hydrophobic.

7124 hydrophobic; hydrophobe; water repellent — Literally water fearing or water hating. Refers to materials that are not wetted by water. Hydrophobic pigments flocculate in water.
f hydrophobe; repoussant l'eau — **d** hydrophob; wasserabstoßend; wasserabweisend — **n** hydrofoob; waterafstotend — **e** hidrófobo; repelente al agua — **i** idrofobo; idrorepellente

7125 hydrophobic colloids; lyophobic colloids; ir-

reversible colloids − Colloids that will not redissolve after removal of solvent at atmospheric temperature.
f colloïdes *mpl* hydrophobes − **d** hydrophobe Kolloide *npl*; wasserabstoßende Kolloide *npl* − **n** hydrofobe colloïden *fmnpl*; waterafstotende colloïden *fmnpl* − **e** coloides *mpl* hidrófobos − **i** colloidi *mpl* idrofobi

7126 hydroquinol − See hydroquinone.

7127 hydroquinone; quinol; hydroquinol; hydrochinone; para-hydroxybenzene − White crystals, soluble in water, alcohol and ether. Used in photographic developers. TL-value: 2 mg/m³.
Formula: $C_6H_4(OH)_2$.
f hydroquinone *m*; para-dioxybenzène *m* − **d** Hydrochinon *n*; para-Dioxybenzol *n* − **n** hydrochinon *n*; chinol *n*; para-dioxybenzeen *n* − **e** hidroquinona *f*; quinol *m* − **i** idrochinone *m*

7128 hydroquinone developer
f révélateur *m* hydroquinone − **d** Hydrochinon-Entwickler *m* − **n** hydrochinon-ontwikkelaar *m* − **e** revelador *m* a la hidroquinona − **i** rivelatore *m* all'idrochinone

7129 hydrosodic carbonate − See sodium bicarbonate.

7130 hydrosulphite − See sodium hydrosulphite.

7131 hydroxyacetic acid − See glycolic acid.

7132 hydroxybenzene − See phenol.

7133 hydroxyl group; OH-group − Univalent radical consisting of one hydrogen and one oxygen atom, forming a part of a molecule.
f groupe *m* hydroxyle; groupe *m* OH − **d** Hydroxylgruppe *f*; OH-Gruppe *f* − **n** hydroxylgroep *fm*; OH-groep *fm* − **e** grupo *m* hidróxilo; grupo *m* oxhidrilo; grupo *m* oxidrilo; grupo *m* OH − **i** gruppo *m* idrossilico; gruppo *m* OH

7134 hydroxymethylbenzene − See cresol.

7135 hygroexpansivity; humidity expansion − Change in dimension of paper resulting from a change in the relative humidity of the air, expressed as a percentage and usually several times higher for the cross direction than for the machine direction. Of great importance when the dimensions of paper sheets and cards are critical.
f allongement *m* à l'humidité − **d** Feuchtdehnung *f*; Ausdehnung *f* bei Feuchtigkeitseinwirkung *f* − **n** uitzetting *f* door vocht; vochtrek *m* depr. − **e** alargamiento *m* por humedad − **i** dilatazione *f* d'umidità

7136 hygro-instability − See dimensional stability.

7137 hygrometer − Device to measure the moisture or relative humidity in the atmosphere. There are three types: wet-and-dry bulb, hygroscopic element, and electric. See also psychrometer, hair hygrometer.
f hygromètre *m* − **d** Hygrometer *n*; Feuchtigkeitsmesser *m* − **n** hygrometer *m*; vochtigheidsmeter *m* − **e** higrómetro *m* − **i** igrometro *m*; misuratore *m* dell'umidità

7138 hygroscope − Direct reading instrument to indicate relative humidity.

f hygroscope *m* − **d** Hygroskop *mn*; Feuchtigkeitsanzeiger *m* − **n** hygroscoop *m* − **e** higroscopio *m* − **i** igroscopio *m*

7139 hygroscopic − Of materials such as paper and acetate films, able to absorb or release moisture, and expanding or contracting their dimensions.
f hygroscopique − **d** hygroskopisch; wasseranziehend; feuchtigkeitsempfindlich − **n** hygroscopisch; vochtaantrekkend − **e** higroscópico − **i** igroscopico; sensibile all'umidità

7140 hygroscopicity − Property to gain or loose moisture according to surrounding atmospheric conditions.
f hygroscopicité *f* − **d** Hygroskopizität *f* − **n** hygroscopiciteit *f* − **e** higroscopicidad *f* − **i** igroscopicità *f*

7141 hygrostability
f inertie *f* à l'eau − **d** Feuchtstabilität *f* − **n** vochtbestendigheid *f* − **e** higroestabilidad *f* − **i** igrostabilità *f*; stabilità *f* ad umido

7142 hygro-thermograph − Instrument that automatically records the humidity and temperature of the air on a graph card attached to the instrument.
f hygro-thermographe *m* − **d** Hygrothermograph *m* − **n** hygrothermograaf *m* − **e** higrotermógrafo *m* − **i** igrotermografo *m*

7143 hyphen − Punctuation mark to connect two words or two parts of a word, and also to divide a word at the end of a line. See also punctuation marks, dictionary, discretionary hyphen.
f trait *m* d'union; division *f* − **d** Bindestrich *m*; Trennungsstrich *m*; Abkürzungsstrich *m*; Abkürzungszeichen *n*; Abteilungszeichen *n*; Divis *n* − **n** koppelteken *n*; afbrekingsteken *n*; verbindingsteken *n*; divisie *f* − **e** guión *m*; división *f*; raya *f*; rayita *f* − **i** tratto *m* d'unione; divisione *f*

7144 hyphenation − In *computer-controlled typesetting, word division at the end of a line. Most printers and publishers in the US use the division of Webster's New International Dictionary. See also dictionary, discretionary hyphen, app. no. 6, ligature.
f coupure *f* syllabique; division *f* syllabique; justification *f* et coupure *f* des mots − **d** Worttrennung *f*; Silbentrennung *f*; Abbrechen *n* von Wörtern; Ausschließen *n* und Abkürzen *n* von Wörtern − **n** woordafbreking *f*; afbreken *n* van woorden; afbreken *n* en spatiëren *n* van woorden − **e** división *f* de silabas; división *f* de palabras; separación *f* de silabas − **i** divisione *f* delle parole in sillabe

7145 hyphen perforating − See perforating rule.

7146 hypo; sodium thiosulphate *misn.* − White, translucent crystals or crystalline powder, soluble in water and oil of turpentine, insoluble in alcohol. Used to fix the image on a photographic plate after development. One of the ingredients in the hardening bath. In reducing or etching solutions such as Farmer's solution, it dissolves

and removes the silver ferricyanide formed when the plate is placed in a bleaching bath of potassium ferricyanide. Latin: natrium thiosulfuricum. Formerly wrongly named hyposulphite.
Formula: $Na_2S_2O_3 \cdot 5H_2O$.
f hypo *m*; thiosulfate *m* de sodium — **d** Natriumthiosulfat *n*; unterschwefligsaures Natrium *n*; Fixiernatron *n* — **n** hypo *n*; natriumthiosulfaat *n* — **e** tiosulfato *m* de sodio; tiosulfato *m* sódico — **i** tiosolfato *m* di sodio; tiosolfato *m* sodico

7147 hypochlorite of lime — See calcium hypochlorite.

7148 hypochlorous acid — Strong oxidizer and bleaching substance known only in salts and solution.
Formula: HClO.
f acide *m* hypochloreux — **d** Unterchlorsäure *f*; unterchlorige Säure *f* — **n** onderchlorig zuur *n* — **e** ácido *m* hipocloroso — **i** acido *m* ipocloroso

7149 hyposulphite — See hypo.

7150 hysteresis; lag — Difference between the ascending and descending curves in a diagram. With paper, for example, it is the lag effect characterized by never getting quite back to where it started as far as moisture content is concerned.

With magnetic ink printing, magnetizing force and magnetic induction have a cause and effect relationship. Here, hysteresis refers to the lag in build up and fall off of magnetic induction when a magnetic substance is subject to a changing magnetizing force. If this lag is plotted in relation to the value of the magnetizing force, a hysteresis curve (magnetization curve) will be formed. This property is very useful in measuring and defining magnetic materials. Other examples arise in the viscosity or resistance to flow of certain colloids when subjected to deforming forces.
f hystérésis *f*; hystérèse *f* — **d** Hysteresis *f*; Hysterese *f* — **n** hysteresis *f* — **e** histéresis *f* — **i** isteresi *f*

7151 hysteresis loop — In a diagram, the part between two adjacent nodes. For example, the consistency curve of a thixotropic material, obtained on a rotational viscometer and formed by plotting the up and down curves together. The extent of the area indicates the amount of thixotropic breakdown.
f boucle *f* d'hystérésis — **d** Hysteresis-Schleife *f* — **n** hysteresislus *fm* — **e** curva *f* histerética — **i** ciclo *m* d'isteresi

7152 idiot tape — See unjustified (paper) tape.

7153 idle machine time — See down time.

7154 idler — See idler roller.

7155 idler roller; idler; idling roller; pipe roller — Free turning grated roller to support and guide the web as it travels through the press. Often used interchangeably with web lead rollers.

f rouleau *m* libre; rouleau *m* non commandé; tube *m* de tension (du papier) non commandé — **d** Mitläuferwalze *f*; freilaufende Walze *f*; Papierleitwalze *f* — **n** vrijlopende rol *fm* — **e** rodillo *m* libre — **i** rullo *m* libero; rullo *m* folle

7156 idling roller — See idler roller.

7157 illegible — Unreadable. See also undecipherable (inscription).

f illisible — **d** unleserlich — **n** onleesbaar — **e** ilegible — **i** illeggibile

7158 illuminant — See light source.

7159 illuminated line-up table — See register table.

7160 illumination; lighting — Light falling on the copy on the copyboard of the process camera.

f éclairement *m*; lumination *f* — **d** Beleuchtung *f* — **n** verlichting *f* — **e** iluminación *f*; alumbrado *m* — **i** illuminazione *f*

7161 illumination — Printed or hand applied decoration in different colour of initial letter, chapter heads, tail pieces, vignettes, etc.

f enluminure *f* — **d** Illuminierung *f*; Ausmalung *f*; Buchmalerei *f* — **n** verluchting *f* — **e** iluminación *f* — **i** illustrare

7162 illustrate *v*

f illustrer — **d** illustrieren — **n** illustreren — **e** ilustrar — **i** illustrare

7163 illustrated article

f article *m* illustré — **d** illustrierter Beitrag *m*; bebilderter Aufsatz *m* — **n** geïllustreerd artikel *n* — **e** artículo *m* ilustrado — **i** articolo *m* illustrato

7164 illustrated edition — Book with text and illustrations.

f édition *f* illustrée — **d** illustrierte Ausgabe *f* — **n** geïllustreerde uitgave *f* — **e** edición *f* ilustrada — **i** edizione *f* illustrata

7165 illustration — Pictorial representation in a book or other printed work to elucidate the text.

f illustration *f*; image *f*; figure *f* — **d** Illustration *f*; Abbildung *f*; Bild *n* — **n** illustratie *f*; afbeelding *f* — **e** ilustración *f*; figura *f*; grabado *m*; dibujo *m* — **i** illustrazione *f*

7166 image — Optical counterpart of an original projected by a lens on the focusing screen of a camera. Picture reflected by a mirror or prism. Photographic or printed reproduction of an original.

f image *f* — **d** Bild *n*; Abbild *n* — **n** beeld *n* — **e** imagen *f* — **i** immagine *f*

7167 image area

f partie *f* imprimante — **d** Druckfläche *f* — **n** beeldpartij *f* — **e** parte *f* impresora; elemento *m* impresor; superficie *f* impresora — **i** zona *f* stampante

7168 image definition — Sharpness and resolution of an image.

f netteté *f* de l'image — **d** Scharfzeichnung *f*; Durchzeichnung *f* — **n** beeldscherpte *f* — **e** nitidez *f* de la imagen — **i** nitidezza *f* dell'immagine

7169 imbrication — Book cover decoration style, in which the pattern consists of overlapping scales or leaves.

f imbrication *f*; écailles *fpl* de poisson — **d** Schuppenmuster *n* — **n** schubornament *n* — **e** escamas *fpl* de pescado; imbricación *f* — **i** ornamento *m* a squama di pesce

7170 imitation art paper — Highly finished printing paper prepared by adding a high percentage of China clay to the pulp, as distinct from art paper where the clay is spread over the surface after the paper is made. In the other it is mixed in the pulp. Imitation art is water finished, i.e., sprinkled from water jets immediately before passing through the calender rolls.

f papier *m* surglacé; papier *m* similicouché — **d** Naturkunstdruckpapier *n*; Illustrationsdruckpapier *n* — **n** natuurkunstdrukpapier *n*; natuurkunstdruk *n*; imitatie-kunstdrukpapier *n* — **e** papel *m* satinado sin estucar; imitación *f* de papel estucado; papel *m* imitación estucado — **i** carta *f* uso patinata

7171 imitation bristol — See bogus bristol.

7172 imitation embossing
See thermography.

7173 imitation Japanese vellum — Strong printing paper of wild formation imitating Japanese vellum.

f papier *m* simili Japon — **d** imitiertes Japanpergamentpapier *n* — **n** imitatie-Japans pergamentpapier *n* — **e** papel *m* vitela imitación Japón — **i** carta *f* imitazione pergamena giapponese

7174 imitation leather — See artificial leather.

7175 imitation parchment-paper; artificial parchment — Well-hydrated sulphite paper imitating *vegetable parchment paper. There are four grades: 1. bleached, calendered imitation parchment, usually called *glassine paper, 2. unbleached, calendered artificial parchment; 3. bleached, uncalendered imitation parchment; 4. unbleached, uncalendered imitation parchment.

f papier *m* parcheminé — **d** Pergamentersatzpapier *n* — **n** imitatie-perkamentpapier *n*; perkament-ersatzpapier *n*; perkament-ersatz *n* — **e** papel *m* apergaminado; papel *m* simil pergamino — **i** carta *f* pergamenacea; carta *f* pergamena imitata; carta *f* uso pergamena

7176 imitation relief printing — See thermography.

7177 imitation relief stamping — See

thermography.

7178 imitation Russia — See American Russia.

7179 imitation steel-die embossing — See thermography.

7180 imitation watermark; simulated watermark — Watermark produced by printing dry paper with transparentizing compounds.

f filigrane *m* (à) sec — **d** künstliches Wasserzeichen *n*; imitiertes Wasserzeichen *n* — **n** imitatie-watermerk *n*; namaak-watermerk *n* — **e** filigrana *f* imitada; marca *f* de agua imitada; falsa filigrana *f* — **i** filigrana *f* a secco; filigrana *f* imitata

7181 immediate access storage — See direct access store.

7182 immersion — See dipping.

7183 immersion coating — See dip coating.

7184 immersion heater — Electric heating element immersed in the material to be heated.

f élément *m* chauffant à immersion; thermoplongeur *m*; plongeur *m* chauffant — **d** Tauchsieder *m*; elektrischer Tauchsieder *m* — **n** elektrische dompelaar *m*; elektrisch dompelelement *n* — **e** termoinmersor *m* eléctrico — **i** riscaldatore *m* ad immersione; elemento *m* riscaldante ad immersione

7185 immiscible — Incapable of mixing or attaining homogenity.

f inmiscible; non miscible — **d** nicht mischbar; unmischbar — **n** niet mengbaar; onvermengbaar; niet te mengen — **e** inmiscible — **i** immiscibile

7186 impact — In mechanics, the action of two bodies in collision, whereby the velocity of one or both bodies is changed.

f choc *m*; impact *m*; collision *f* — **d** Stoß *m*; Zusammenstoß *m*; Prall *m*; Auftreffen *n* — **n** stoot *m*; slag *m* — **e** choque *m* — **i** urto *m*

7187 impact strength

f résistance *f* à la traction dynamique — **d** dynamische Zugfestigkeit *f* — **n** dynamische breukenergie *f* — **e** fuerza *f* de impacto — **i** resistenza *f* all'impatto

7188 impalpable — Extremely fine so that no grit can be felt.

f impalpable — **d** unfühlbar; nicht spürbar — **n** niet voelbaar; uiterst fijn verdeeld — **e** impalpable — **i** impalpabile

7189 impedance — Apparent resistance of an alternating current circuit or path, equal to the vector sum of the resistance and reactance of the path.

f impédance *f* — **d** Impedanz *f*; Scheinwiderstand *m* — **n** impedantie *f* — **e** impedancia *f* — **i** impedenza *f*

7190 imperial — A British standard size of writing and printing papers. See app. no. 3.

7191 imperial cap — A British size of wrapping paper. See app. no. 3.

7192 imperial card — A British size of cards. See app. no. 3.

7193 imperial quarto — A British size for drawing paper; although used, it is not regarded as a standard. See app. no. 3.

7194 impermeability — The quality or state of being impermeable.

f imperméabilité *f* — **d** Undurchlässigkeit *f* — **n** ondoorlatendheid *f* — **e** impermeabilidad *f* — **i** impermeabilità *f*

7195 impingement black — Another name for *carbon black, so called to distinguish it from the newer surface blacks and thermal decomposition blacks, which are also produced from natural gas by impinging the gas flames against a metal surface.

7196 impose *v* — To arrange pages of type in a forme so that they will read consecutively when the printed sheet is folded.

f imposer — **d** ausschießen — **n** inslaan — **e** casar; imponer — **i** impostare

7197 imposing scheme; imposition scheme — Scheme for the arrangement of the pages of a book etc., so that they follow the correct sequence when the sheets are folded.

f maquette *f* de mise en pages — **d** Ausschießschema *n* — **n** inslagschema *n* — **e** esquema *m* del lanzado; modelo *m* de imposición; plantilla *f* de imposición; plan *m* de armada — **i** schema *m* d'impostazione

7198 imposing stone; imposing surface; imposing table; stone *coll.* — Table with a planed iron (formerly stone) surface, on which formes are imposed.

f marbre *m* de serrage; marbre *m* d'imposition — **d** Schließplatte *f*; Schließtisch *m*; Ausschießplatte *f*; Ausschießbrett *n* — **n** insluitsteen *m*; insluittafel *fm*; insluitplaat *fm*; steen *m* — **e** mármol *m*; piedra *f* de imposición; mesa *f* de imponer; mesa *f* de acuñar; mesa *f* de imposición; platina *f* de imposición — **i** tavolo *m* d'impostazione; piano *m* d'impostazione

7199 imposing surface — See imposing stone.

7200 imposing table — See imposing stone.

7201 imposing to quire — Imposition of pages in such a way that they inset into each other like the sheets of a quire of note paper.

f imposition *f* en cahiers — **d** Ausschießen *n* in Lagen — **n** inslaan *n* als boekvellen — **e** imposición *f* en cuadernos — **i** imposizione *f* intercalare

7202 imposition — Arrangement of type formes so that they print correctly on the press sheet, and the pages are in proper order when the sheets are folded.

f imposition *f* — **d** Ausschießen *n* — **n** inslag *m* — **e** imposición *f*; casado *m*; lanzado *m* — **i** imposizione *f*; impostazione *f*

7203 imposition — Arranging and fastening negatives or positives to a supporting flat for use in offset lithography platemaking and intaglio printing. Multiple imposition: the exposure of the same flat in two or more positions on the press plate or carbon tissue sheet.

f imposition *f* — **d** Formatausrichtung *f* — **n** indeling *f* van het vel — **e** casado *m* — **i** imposizione *f* a formato

7204 imposition scheme — See imposing scheme.

7205 impositor — See lock-up man.

7206 impregnated paper — Paper treated with a liquid, such as oil, or with a molten substance such as wax or asphalt, or with a solution of resinous or other material in a volatile or aqueous solvent.

f papier *m* imprégné — **d** imprägniertes Papier *n*; getränktes Papier *n* — **n** geïmpregneerd papier *n* — **e** papel *m* impregnado — **i** carta *f* impregnata

7207 impressed watermark; press-mark; rubber-stamp watermark — Watermark made with a rubber letterpress-type marking disc on the top side of the sheet at a point beyond the press section of the paper machine where the water content of the paper has been reduced below the level required for a true watermark.

f filigrane *m* à la molette; filigrane *m* par pression; marque *f* (à la) molette — **d** Molette-Wasserzeichen *n* — **n** perswatermerk *n*; geperst watermerk *n*; persmerk *n*; molette-watermerk *n* — **e** filigrana *f* en seco; filigrana *f* sobre prensa; filigrana *f* sobre secador; marca *f* de agua molette — **i** filigrana *f* alla moletta

7208 impression — All the copies of a book or other work printed at one time from the same type or plates, constituting second, etc., impressions but not a new edition.

f tirage *m*; impression *f* — **d** Auflage *f* — **n** oplage *f*; oplaag *f* — **e** tirada *f*; tiraje *m* — **i** tiratura *f*

7209 impression — Product resulting from one cycle of the printing machine.

f impression *f* — **d** Abdruck *m* — **n** afdruk *m* — **e** impresión *f*

7210 impression blanket — See blanket.

7211 impression cylinder — On a letterpress machine, the cylinder that carries the paper into contact with the type.

f cylindre *m* d'impression — **d** Druckzylinder *m* — **n** drukcilinder *m* — **e** cilindro *m* impresor; cilindro *m* empresor; cilindro *m* de impresión; bombo *m* de impresión; bombo *m* de presión; tambor *m* impresor; tambor *m* de presión; tambor *m* de impresión — **i** cilindro *m* d'impressione; cilindro *m* di pressione; cilindro *m* impressore

7212 impression cylinder — On an offset press, the cylinder that carries the paper sheet into contact with the blanket.

f cylindre *m* de contre-pression — **d** Gegendruckzylinder *m* — **n** tegendrukcilinder *m* — **e** cilindro *m* de impresión — **i** cilindro *m* di contropressione

7213 impression cylinder; impression roller; pressure roller; printing cylinder — Cylinder, usually hydraulically or mechanically operated, which is brought into contact with the reverse side of the web in rotogravure (or flexographic) presses to ensure proper pressure on the web surface as it passes between the impression cylinder and the surface of the engraved cylinder. Usually covered with rubber of hardness suited for mate-

rial being printed. Softer coverings are used for light materials such as cellophane, and harder coverings for materials such as antique stock or cardboard. See also back-up impression cylinder.

f rouleau *m* presseur; cylindre *m* presseur; presseur *m* — **d** Presseur *m*; Gummiwalze *f*; Gummizwischenwalze *f*; Gummizylinder *m* — **n** presseur *m* — **e** rodillo *m* de presión — **i** cilindro *m* pressore; cilindro *m* di pressione

7214 impression eccentric — A means of impression adjustment by rotating the bearing or bearing housing in which the cylinder shaft of a letterpress rotary is mounted out of centre.

f excentrique *m* de cylindre d'impression — **d** Exzenterbüchse *f* zur Druckverstellung — **n** drukcilinder-excentriek *n* — **e** excéntrica *f* del cilindro impresor — **i** eccentrico *m* del cilindro di pressione

7215 impression indicator — Device on a letterpress rotary to indicate the degree of pressure, or distance between the plate and the impression cylinder.

f indicateur *m* de pression; vernier *m* de pression — **d** Druckanzeiger *m* — **n** drukkrachtaangever *m*; drukaangever *m depr.* — **e** indicador *m* de presión — **i** indicatore *m* di pressione

7216 impression line — See impression zone.

7217 impression point — See impression zone.

7218 impression pressure — Amount of pressure (usually gauged in pounds per lineal inch) applied by the impression cylinder to the reverse surface of a web as the web is held against the rotating surface of the engraved cylinder.

f pression *f* du cylindre presseur — **d** Druck *m* des Presseurs — **n** druk *m* van de presseur — **e** presión *f* de rodillo de presión — **i** pressione *f* del cilindro pressore

7219 impression pressure — See also back pressure.

f pression *f* d'impression — **d** Gegendruck *m* (im Offsetdruck); Druckkraft *f* (im Flachdruck); Druckspannung *f depr.* — **n** druk *m* (in offsetdruk); drukkracht *fm* (in boekdruk); drukspanning *f depr.* — **e** fuerza *f* de impresión; presión *f* de impresión — **i** forza *f* di pressione; pressione *f* di stampa

7220 impression pressure — See printing pressure.

7221 impression roller — See impression cylinder.

7222 impression setting — Adjusting printing pressure.

f réglage *m* de pression — **d** Druckeinstellung *f* — **n** instellen *n* van de drukkracht — **e** ajuste *m* de la presión — **i** regolazione *f* della pressione di stampa

7223 impressions per hour — Number of sheets per hour multiplied by the numbers of colours the machine prints at one pass of the sheet.

f feuilles/heure — **d** Bogen pro Stunde — **n** druks per uur; vellen per uur — **e** pliegos por hora; impresiones por hora; ejemplares por hora — **i**

copie/ora

7224 impression throw-off — Arrangement for the quick release of pressure between the printing forme and the impression cylinder.

f mise *f* hors pression — **d** Druckabsteller *m* — **n** drukafsteller *m* — **e** mecanismo *m* salvapliegos; dispositivo *m* salvapliegos; salvapliegos *m*; aparato *m* de levantamiento del cilindro; quita-presión *m* — **i** dispositivo *m* per l'arresto; disinseritore *m* di pressione; dispositivo *m* di distacco; dispositivo *m* d'arresto

7225 impression wedge — Tapered element on a letterpress rotary, which is movable to locate the cylinder to give the desired pressure.

f cale *f* de réglage de pression; coin *m* de réglage de pression — **d** Druckkeil *m* — **n** stelwig *fm* — **e** cuña *f* de ajuste de la presión — **i** cuneo *m* di regolazione della pressione

7226 impression with the letter — See proof with the letter.

7227 impression zone; impression line; impression point; printing zone — Line of contact between forme and printing cylinder, or plate and the sheet of paper on an offset press. Also the point where the web is brought into contact with the impression cylinder and the engraved cylinder in a rotogravure press.

f ligne *f* d'impression; point *m* d'impression — **d** Drucklinie *f*; Druckpunkt *m* — **n** druklijn *fm* — **e** línea *f* de impresión; línea *f* de presión — **i** linea *f* di pressione; punto *m* d'impressione

7228 imprimatur — Latin for "let it be printed", usually printed at back of title page with signature and date on which it was granted.

f imprimatur *m*; peut être imprimé; approbation *f* — **d** Imprimatur *n*; Druckgenehmigung *f*; Druckerlaubnis *f* — **n** imprimatur *n*; het worde gedrukt; fiat voor afdrukken — **e** imprimátur *m*; imprímase — **i** visto *m* per la stampa; buono a stampare; buono a tirare

7229 imprint; printer's imprint; printer's note; printer's mark — Line of type or small section of printed matter added to a job to provide abbreviated reference information concerning the job. Sometimes incorrectly called a *colophon.

f marque *f* d'imprimerie; marque *f* d'imprimeur — **d** Impressum *n*; Druckvermerk *m*; Druckermarke *f*; Druckerzeichen *n* — **n** drukkersmerk *n*; merk *n* van de drukker; naam *m* van de drukker; tjap *mn* van de drukker — **e** lema *m* del impresor; pie *m* de imprenta; marca *f* del impresor; marca *f* tipográfica; marca *f* de imprenta — **i** nome *m* dello stampatore; soscrizione *f* dello stampatore

7230 imprint; publisher's imprint — Subject matter on the lower part of a title page, concerning the place and date of publication, the publisher's name and sometimes the name and address of the printer. Also a *colophon. In GB essential by Act of Parliament. If omitted, the printer cannot recover payment from the customer by law.

f marque *f* d'éditeur — **d** Impressum *n*; Verlegerzeichen *n* — **n** uitgeversmerk *n*; naam *m* van de uitgever(ij) — **e** lema *m* del editor; marca *f* del editor — **i** nome *m* dell'editore; soscrizione *f* dell'editore

7231 imprint *v* — To insert text by printing.

f imprimer en repérage; repiquer — **d** eindrucken — **n** indrukken — **e** hacer una impresión secundaria — **i** sovrastampare

7232 imprinting — Printing of a price mark, suppliers name, addresses, etc., on formerly printed matter. Also the overprinting of a dried ink film.

7233 imprint unit — Device to print imprints on one side of the web from rubber plates.

f dispositif *m* d'impression supplémentaire; appareil *m* d'impression supplémentaire — **d** Eindruckwerk *n* — **n** indrukwerk *n* — **e** dispositivo *m* de impresión suplementaria; grupo *m* de impresión suplementaria — **i** gruppo *m* stampante supplementare

7234 improver — Apprentice who has not served his full time to become a journeyman.

f apprenti *m* en cours de perfectionnement — **d** Lehrling *m*; Stift *m* — **n** leerling *m* — **e** aprendiz *m* — **i** apprendista *m*

7235 impurities

f impuretés *fpl* — **d** Verunreinigungen *fpl*; Unreinigkeiten *fpl* — **n** verontreinigingen *fpl* — **e** impurezas *fpl*; suciedades *fpl* — **i** insudiciamenti *mpl*; impuritas *fpl*

7236 inactinic filter — Glass or other coloured filter, which stops the actinic rays of light, used in dark rooms while handling sensitive photographic material.

f écran *m* inactinique — **d** inaktinischer Filter *m* — **n** niet-actinisch filter *n*; inactinisch filter *n* — **e** filtro *m* inactínico — **i** filtro *m* inattinico

7237 inactive acid — See weak acid.

7238 in and in — See interlapped.

7239 in-board forwarding — The term in-board is derived from the technique of cutting the edges of a book after it has been rounded and backed, laced-in and pressed, etc.

7240 in boards — Said of a book with plain board sides and cloth back. Also of a book trimmed after board sides have been affixed. When edges of the book are cut after boards are affixed or laced on, it is bound in boards or cut in boards.

f cartonné — **d** kartoniert; in Pappband — **n** gekartonneerd — **e** en cartoné; encartonado en tapas; cortado con tapas — **i** cartonato; incartonato

7241 incandescent lamp — Lamp the light of which is derived from a material becoming luminous with heat (tungsten). Colour temperature ranges from about 2800-3400 °K.

f lampe *f* à incandescence — **d** Wolframlampe *f*; Glühlampe *f* — **n** wolfram-draadgloeilamp *fm*; (gewone) gloeilamp *fm* — **e** lámpara *f* incandescente — **i** lampada *f* ad incandescenza

7242 incapable of work; unfit for work

f impropre au travail; incapable de travailler — **d** arbeitsunfähig — **n** arbeidsongeschikt — **e** incapaz de trabajar — **i** inabile a lavorare

7243 incentive time; allowed time — See also standard time, target time.

f temps *m* alloué — **d** Vorgabezeit *f*; Akkordzeit *f* — **n** tarieftijd *m*; toegestane tijd *m* — **e** tiempo *m* incentivo — **i** tempo *m* incentivo

7244 inch — Unit of length. Abbrev. in (no point). See app. no. 7 and 7a.

f pouce *m* — **d** Zoll *m*; Inch *m* — **n** inch *mn*; duim *m obs.* — **e** pulgada *f* — **i** pollice *m*

7245 inch *v* — To turn the cylinder of a printing machine by inches at a time.

f marcher par à-coups; marcher coup par coup — **d** zentimeterweise vorrücken; langsam gehen — **n** tornen — **e** arrancar poco a poco — **i** marciare ad impulsi

7246 inching motor; barring motor — Small motor to drive the rotary press at inch and crawl speeds.

f moteur *m* de ralenti — **d** Hilfsantriebmotor *m* — **n** tornmotor *m* — **e** motor *m* de arranque — **i** motore *m* ad impulsi

7247 incidental element; intermittent element; occasional element; non-cyclic element — Element which may occur at regular or irregular intervals, in work cycles.

f élément *m* intermittent; élément *m* acyclique; élément *m* occasionnel; irrégularité *f* propre au travail normal — **n** neven-element *n*; niet-cyclische gebeurtenis *f*; incidentele gebeurtenis *f* — **e** elemento *m* intermitente; elemento *m* a-cíclico — **i** elemento *m* intermittente; elemento *m* aciclico

7248 incident light

f lumière *f* incidente — **d** einfallendes Licht *n*; Schräglicht *n* — **n** invallend licht *n* — **e** luz *f* incidente; luz *f* de incidencia — **i** luce *f* incidente

7249 incinerator — Furnace or container to incinerate waste materials, for laboratory work.

f incinérateur *m* — **d** Verascher *m* — **n** asoventje *n* — **e** incinerador *m* — **i** inceneratore *m*

7250 Inciseds — Group of type faces, e.g., Pascal, Othello, Albertus, Latin, Spartan, Bavo, Optima, which are chiselled rather than calligraphic in form. New name for the Latin types, executed by the tool of wood cutter, wood engraver or stone engraver. Before 1971 they were named Glyphics. See also classification of type design.

f Incises *fpl* — **d** (sonstige) Antiquavarianten *fpl* — **n** Inciezen *mfpl* — **e** Incisos *mpl* — **i** Lapidari *mpl*

7251 inclined at an angle of 30 ° (thirty degrees)

f incliné à 30 ° (trente degrés) — **d** unter einem Winkel von 30 ° (dreißig Grad) — **n** onder een hoek van 30 ° (dertig graden) — **e** inclinado en ángulo de 30 ° (treinta grados) — **i** inclinato a 30 ° (trenti gradi)

7252 inclined-plane viscometer — Viscometer with an inclined plane and a rolling device. The test material is spread on the plane, and the time required for a cylinder or sphere to roll down the plate over the film is a measure of its viscosity.

f viscosimètre *m* à plan incliné — **d** schiefe Ebene-Viskosimeter *n* — **n** viscosimeter *m* met hellend vlak — **e** viscosímetro *m* a plano inclinado — **i** viscosimetro *m* a piano inclinato

7253 inclusions — Defect of aluminium foil. Foreign material in metal or rolled-in surface dirt.

7254 incomplete — See defective.

7255 incunable — See incunabulum.

7256 incunabulum; incunable; cradle book; early printed book — Book printed from the invention of printing up to and including the year 1500. Name derived from Benghem's "Incunabula Typographica", printed in Amsterdam in 1688. About 450,000 incunabula have been saved.

f incunable *m*; impression *f* ancienne; impression *f* primitive — **d** Inkunabel *f*; Wiegendruck *m*; Frühdruck *m*; Erstlingsdruck *m* — **n** incunabel *m*; wiegedruk *m* — **e** incunable *m*; libro *m* incunable — **i** incunabulo *m*; incunabolo *m*

7257 incut notes — See cut-in notes.

7258 indanthrene pigment — Trade-mark for a series of vat dyes of blue, maroon and pink shades used as printing ink pigment. Some are characterized by exceptional light fastness even in extremely reduced tints. Strong blue powder, soluble in dilute alkaline solution. Formula: $C_{28}H_{14}N_2O_4$.

f colorant *m* d'indanthrène — **d** Indanthrenfarbstoff *m*; Indanthrenpigment *n* — **n** indantreenkleurstof *fm*; indantreenpigment *n* — **e** colorante *m* indantreno; colorante *m* indantrénico — **i** colorante *m* d'indantrene

7259 indelible ink — Marking ink, water and alkali fast, used on cloth to withstand laundering.

f encre *f* indélébile — **d** unauslöschliche Farbe *f*; unzerstörbare Farbe *f* — **n** onuitwisbare inkt *m*; merkinkt *m* — **e** tinta *f* indeleble — **i** inchiostro *m* indelebile

7260 indemnity for sickness

f indemnité *f* de maladie — **d** Krankenunterstützung *f*; Krankengeld *n* — **n** ziekengeld *n*; ziekte-uitkering *f* — **e** socorro *m* a los enfermos — **i** sussidio *m* per malattia; indennità *f* malattia

7261 indene resin — See coumarone-indene resin.

7262 indent *v* — To insert one or more em quads at beginning of a line, usually the first line of a paragraph.

f rentrer; renfoncer — **d** (weiter) einziehen; einrücken — **n** inspringen — **e** sangrar; entrar; indentar *Gu* — **i** rientrare

7263 indentation — See indention.

7264 indention; indentation — Blank space at the beginning of a line, setting it back from the other lines in the left-hand margin.

f renfoncement *m*; rentrée *f* — **d** Einzug *m* — **n** inspringing *f*; inspringen *n* — **e** sangría *f*; sangrado *m*; entrada *f*; acápite *m* — **i** rientranza *f*

7265 indention; hollow patch — Low spot in a rub-

f initialiser — **d** initialisieren — **n** initialiseren — **e** inicializar — **i** inizializzare

7321 initial letter; initial — Large capital or decorated letter at the beginning of a chapter or work.
f lettre *f* initiale; initiale *f*; lettrine *f* — **d** Initial *n*; Initiale *f*; Versalbuchstabe *m*; großer Anfangsbuchstabe *m* — **n** initiaal *n* — **e** capital *f* inicial; inicial *f* mayúscula; letra *f* inicial; inicial *f*; capitular *m* — **i** iniziale *m*; lettera *f* iniziale

7322 initial line
f première ligne *f*; ligne *f* de tête — **d** Anfangszeile *f* — **n** beginregel *m*; eerste regel *m*; bovenste regel *m*; kopregel *m* — **e** línea *f* inicial; línea *f* de cabeza; cabecera *f* — **i** riga *f* di daccapo; accapo *m*

7323 initial tearing strength — See tearing strength.

7324 initial tearing tester
f appareil *m* pour déterminer la résistance au déchirement — **d** Einreißfestigkeitsprüfer *m* — **n** inscheurtoestel *n* — **e** aparato *m* para ensayar la resistencia al rasgado; aparato *m* para ensayar la resistencia al desgarramiento — **i** apparecchio *m* per misurare la resistenza alla lacerazione iniziale

7325 ink; printing ink; typographic ink — Mixture of finely divided pigments such as carbon black, suspended in a drying oil such as heat-bodied linseed oil, alkyds, phenol-formaldehyde or other synthetic resins. Cobalt, manganese and lead soaps are added to achieve rapid drying by oxidation and polymerization. Sometimes mineral oils are present and there are types of ink which dry by evaporation of a volatile solvent rather than by oxidation and polymerization of a drying oil or resin. For coloured inks, pigments such as chrome yellows, are used. See also letterpress ink, offset ink, flexographic ink, gravure ink, screen printing ink, writing ink, drawing ink.
f encre *f* d'impression; encre *f* — **d** Druckfarbe *f*; Farbe *f* — **n** drukinkt *m*; inkt *m*; verf *fm depr.* — **e** tinta *f* de imprenta; tinta *f* de imprimir; tinta *f* — **i** inchiostro *m* da stampa; inchiostro *m*

7326 ink; writing ink
f encre *f* à écrire; encre *f* à écriture; encre *f* — **d** Schreibtinte *f*; Tinte *f* — **n** schrijfinkt *m*; inkt *m* — **e** tinta *f* para escribir; tinta *f* — **i** inchiostro *m* per scrivere; inchiostro *m* da scrivere; inchiostro *m*

7327 ink absorption — Absorption of the ink vehicle by the paper.
f absorption *f* (du papier); pénétration *f* de l'encre dans le papier — **d** Wegschlagen *n* der Farbe — **n** wegslaan *n* van de inkt; penetratie *f* van de inkt in het papier; absorptie *f* van de inkt door het papier — **e** absorción *f* de tinta; absorción *f* del papel — **i** assorbimento *m* dell'inchiostro; penetrazione *f* dell'inchiostro; assorbimento *m* della carta

7328 ink applicator — See applicator.

7329 ink ball; tampon — Round leather, canvas or felt cushion, stuffed with wool, hair or similar material and fastened to a handle called a ball stick. Used in pairs before the invention of rollers

to distribute the ink on the type forme. Formerly also called pelt ball or *pelt.
f balle *f* (d'encrage); balle *f* (d'imprimeur) — **d** Druckerballen *m*; Farbballen *m*; Ballen *m*; Tampon *m* — **n** tampon *m* — **e** tampón *m* (de entintado); bala *f* — **i** tampone *m* (inchiostratore)

7330 ink can; ink tin
f boîte *f* à encre — **d** Farbbüchse *f*; Farbdose *f* — **n** inktbus *fm*; bus *fm* — **e** lata *f* de tinta; bote *m* de tinta — **i** latta *f* d'inchiostro; barattolo *m* d'inchiostro

7331 ink cell; cell; cup; gravure cell — Depression in surface of engraved cylinder to retain ink.
f alvéole *mf* (en héliogravure); godet *m* (de trame); cellule *f* — **d** Farbnäpfchen *n* (Tiefdruck) — **n** rasternapje *n*; inktnapje *n* (in koperdiepdruk) — **e** alveolo *m*; alvéolo *m*; cazuelita *f* (de la retícula) — **i** cella *f*; alveolo *m*

7332 ink compound — See compound.

7333 ink consumption — See ink coverage.

7334 ink coverage; ink consumption; mileage — Number of square centimetres or square inches covered by one kilogram or one pound of ink.
f consommation *f* d'encre; rendement *m* de l'encre; couvrage *m* de l'encre; poids *m* d'encre; taux *m* d'étendage — **d** Farbverbrauch *m*; Farbdeckung *f* — **n** inktverbruik *n* — **e** capacidad *f* de entintado; poder *m* de cubrimiento de la tinta; consumo *m* de tinta — **i** rendimento *m* dell'inchiostro; consumo *m* d'inchiostro

7335 ink cylinder — Metal cylinder contained within the ductor.
f cylindre *m* de l'encrier — **d** Farbzylinder *m* — **n** bakrol *fm*; inktbakrol *fm* — **e** rodillo *m* del tintero — **i** cilindro *m* del calamaio

7336 ink density — Colour strength of an ink for a given film thickness.
f intensité *f* de l'encre — **d** Farbdichte *f* — **n** inktdichtheid *f* — **i** densità *f* dell'inchiostro

7337 ink disk; disk — Flat circular plate on some platen machines upon which the ink is distributed.
f disque *m* d'encrage; plat *m* d'encrage — **d** Farbteller *m* — **n** inktplaat *fm* — **e** plato *m* entintador — **i** piastra *f* inchiostrante

7338 ink distributing cylinder — See ink drum.

7339 ink distribution
f distribution *f* d'encre — **d** Farbverteilung *f* — **n** inktdistributie *f*; inktverdeling *f* — **e** distribución *f* de tinta; entintado *m*; tintaje *m* — **i** distribuzione *f* dell'inchiostro

7340 ink dot scum — On aluminium plates, oxidation scum characterized by scattered pits that print sharp, dense dots.
f oxydation *f* par points — **d** Platterton *f*; Tonen *n* — **n** plaattoon *m* — **e** oxidación *f* por puntos — **i** velatura *f* su lastra per ossidazione

7341 ink drum; inking drum; ink distributing cylinder — Solid or cored metal drum, part of an inking mechanism. Used to break down the ink and transfer it to the forme rollers. On some proof presses used in lieu of a fountain. The ink is supplied to the drum manually. On an offset press ink drums are oscillating rollers (move back and forth sideways) and aid in ink distribution.

f table *f* d'encrage — **d** Farbtrommel *f*; Farbzylinder *m*; Stahlzylinder *m*; Reibezylinder *m* aus Stahl; Nacktzylinder *m* — **n** inktcilinder *m*; stalen distributierol *m*; stalen inktrol *m* — **e** cilindro *m* batidor de la tinta; mesa *f* cilíndrica; mesa *f* distribuidora; cilindro *m* distribuidor — **i** cilindro *m* d'inchiostrazione; cilindro *m* macinatore

7342 ink drying — Property of an ink which determines its rate of drying.

f siccité *f* de l'encre — **d** Trocknung *f* der Farbe — **n** droging *f* van de inkt — **e** secado *m* de la tinta — **i** siccatività *f* dell'inchiostro

7343 ink duct; ink fountain *depr.* — Trough on a press in which ink is placed for distribution to the forme inking system.

f encrier *m*; bac *m* à encre — **d** Farbkasten *m*; Farbbehälter *m* — **n** inktbak *m* — **e** tintero *m* (de prensa) — **i** calamaio *m*

7344 ink-duct roller — See duct roller.

7345 ink eraser

f gomme *f* à encre — **d** Tintengummi *m* — **n** inktgum *mn*; inktgom *mn* — **e** borrador *m* de tinta; quitatinta *f*; goma *f* de borrar; borratintas *f* — **i** gomma *f* per inchiostro

7346 inkers — See inking rollers.

7347 inker unit — See inking mechanism.

7348 ink feed; ink supply

f débit *m* d'encre; commande *f* de l'encrage — **d** Farbzufuhr *f*; Farbzuführung *f* — **n** inkttoevoer *m*; inktgeving *f*; inktregeling *f* — **e** regulación *f* de entintado; regulación *f* de la cantidad de tinta — **i** alimentazione *f* dell'inchiostro

7349 ink-feed roller — See ductor.

7350 ink film *GB*; **ink layer** *US*

f pellicule *f* d'encre; film *m* d'encre — **d** Farbschicht *f*; Farbfilm *m*; Farbhäutchen *n* — **n** inktlaag *fm*; inktlaagje *n*; inktfilm *m* — **e** película *f* de tinta — **i** pellicola *f* d'inchiostro; film *m* d'inchiostro

7351 ink-film thickness; ink-layer thickness — Thickness of the ink layer carried on the inking system or transferred to the plate or blanket or printed material.

f épaisseur *f* de la pellicule d'encre — **d** Farbeschichtdicke *f* — **n** inktlaagdikte *f* — **e** espesor *m* de la capa de tinta — **i** spessore *m* del strato d'inchiostro

7352 ink fly — See ink misting.

7353 ink flying — See ink misting.

7354 ink fountain — See ink duct.

7355 ink-fountain roller — See duct roller.

7356 inking — Distribution of ink on the forme by the inking rollers.

f touche *f*; encrage *m* — **d** Farbgebung *f*; Farbabgabe *f*; Auftragung *f* der Farbe; Auftragen *n* der Farbe; Einfärbung *f*; Einfärben *n*; Einwalzen *n* — **n** inktgeving *f*; oprolling *f*; ininkten *n* — **e** entintado *m*; cobertura *f*; tintaje *m* — **i** inchiostrazione *f*

7357 inking arrangement — See inking mechanism.

7358 inking drum — See ink drum.

7359 inking mechanism; inking unit; inker unit; inking system; inking arrangement — On a printing press, the ink fountain and all the parts used to meter, transfer, breakdown, distribute, cool or heat, and supply the ink to the printing member.

f dispositif *m* d'encrage; dispositif *m* encreur; système *m* d'encrage — **d** Farbwerk *n*; Farbübertragungsmechanismus *n* — **n** inktwerk *n* — **e** mecanismo *m* de entintado; grupo *m* de entintado — **i** sistema *m* inchiostratore; gruppo *m* inchiostratore

7360 inking rollers; ink rollers; inkers — Rollers which carry and distribute ink from fountain to printing forme. See also distributor roller, rider roller, forme rollers, vibrator.

f rouleaux *mpl* encreurs; encreurs *mpl*; rouleaux *mpl* d'encre; tables *fpl* d'encrage — **d** Farbwalzen *fpl* — **n** inktrollen *fmpl* — **e** rodillos *mpl* de tinta; rodillos *mpl* de entintar; rodillos *mpl* entintadores; rodillos *mpl* de tintaje; rodillos *mpl* de batido — **i** rulli *mpl* inchiostratori

7361 inking system — See inking mechanism.

7362 inking table — See ink slab.

7363 inking unit — See inking mechanism.

7364 ink-in-water emulsion — Emulsion of ink and water in which the ink is broken up into fine droplets surrounded by water.

f émulsion *f* encre dans eau — **d** Farbe-Wasser-Emulsion *f* — **n** inkt-in-water-emulsie *f* — **e** emulsión *f* tinta en agua — **i** emulsione *f* acqua-inchiostro

7365 ink jet printer — Device which deposits droplets of ink across a line of paper, as it moves past the ink jet heads - or as the single ink jet nozzle traverses the width of paper - to create a dot matrix pattern forming a line of type, as by the nibs or styli of an electrostatic printer. Droplets of ink may be charged so that they are deflected into a repository for recirculation, or non-charged, in which case they reach the paper directly opposite the jet. Illustrative material is also generated in this fashion; the pattern must then first be stored in a computer buffer area, with the location of each dot being defined.

7366 ink knife; ductor knife — Long blade in the ductor which regulates with screws the amount of ink to be given to each impression.

f lame *f* de l'encrier — **d** Farblineal *n*; Farbmesser *n*; Abstreichmesser *n* — **n** inktbakmes *n*; mes *n* van de inktbak; bakmes *n* — **e** cuchilla *f* del tintero — **i** lama *f* del calamaio

7367 ink knife; ink pallet; palette knife; pulldown knife; spatula; slice *sl.*; **ink slice** *sl.* — Small hand knife to facilitate the handling of ink during grinding, mixing and testing procedures and to fill the ink duct from a tin of ink.

f couteau *m* à encre; spatule *f* — **d** Farbmesser *m*; Farbspachtel *n*; Spachtel *n*; Spatel *m*

− n inktmes *n*; inktspatel *fm*; spatel *fm*; tempermes *n* **− e** espátula *f* de tinta; espátula *f* para tinta; espátula *f*; paleta *f* de tinta **− i** spatola *f* per inchiostro; spatola *f*

7368 ink layer − See ink film.

7369 ink-layer thickness − See ink-film thickness.

7370 ink maker
f fabricant *m* d'encre (d'impression) **− d** Farbenfabrik *f*; Druckfarbenfabrik *f* **− n** inktfabriek *f*; drukinktfabriek *f* **− e** fábrica *f* de tinta **− i** fabbrica *f* d'inchiostri (da stampa)

7371 ink mill − Machinery for pigment dispersion in the manufacture of printing inks. See also ball mill, colloid mill, roller mill.
f broyeuse *f* d'encre **− d** Farbenanriebmaschine *f*; Farbereibmaschine *f*; Farbmühle *f* **− n** inktmolen *m* **− e** moledora *f* de tintas; molino *m* de tintas **− i** mulino *m* per la macinazione degli inchiostri

7372 ink mist − See ink misting.

7373 ink misting; ink mist; ink flying; flying; misting; spitting; spraying; ink fly − Condition in which ink is reduced to a fine mist. It occurs only between two contacting surfaces rotating in opposite directions under friction, long inks being more susceptible than short inks. Electrostatic charges of the ink due to low relative humidity of the air are sometimes a contributory cause.
f voltige *m*; brouillard *m* d'encre; crachage *m* de l'encre **− d** Spritzen *n*; Fliegen *n*; Umherspritzen *n* **− n** stuiven *n*; inktmist *m* **− e** vaho *m*; emanación *f* de tinta **− i** nebbia *f* d'inchiostro

7374 ink mixer − See mixer.

7375 inkometer − Instrument to measure certain properties (tack and length) of printing inks, consisting of a series of ink rollers. The ink exerts a viscous drag on one of these, pulling it away from its normal position. The restoring force is measured at several speeds from which the calculations are made.
f inkomètre *m* **− d** Inkometer *n* **− n** inkometer *m* **− e** tintómetro *m* **− i** inkometro *m*

7376 ink pallet − See ink knife.

7377 ink pipette − Laboratory instrument for measuring small quantities of ink to determine the specific gravity or in tests for which an ink layer of prescribed thickness is indispensable.
f pipette *f* à encre **− d** Farbenpipette *f* **− n** inktpipet *fmn* **− e** pipeta *f* de tinta; aspirador *m* de tinta **− i** pipetta *f* per inchiostro

7378 ink plate − See ink table.

7379 ink pump − Pump-operated ink system.
f pompe *f* à encre **− d** Farbpumpe *f*; Pumpenfarbwerk *n*; Farbpumpenaggregat *n* **− n** inktpompwerk *n* **− e** bomba *f* para subir la tinta **− i** pompa *f* per inchiostro

7380 ink pump − Circulating pump to agitate liquid inks.
f pompe *f* à encre de circulation **− d** Umwälzpumpe *f*; Farbumlaufpumpe *f*; Farbpumpe *f*; Farbezirkulationspumpe *f* **− n** inktcirculatiepomp *fm*; inktpomp *fm*; circulatiepomp *fm* **− e** bomba *f* de tinta; bomba *f* de circulación de la tinta **− i** pompa *f* per la circolazione dell'inchiostro

7381 ink quantity
f quantité *f* d'encre **− d** Farbemenge *f* **− n** inkthoeveelheid *f*; hoeveelheid *f* inkt **− e** cantidad *f* de tinta **− i** quantità *f* d'inchiostro

7382 ink rail − Part of a pump-operated ink duct which extends across the width of the machine to apply ink to the ink drum.
f bâti *m* d'encrier; réglette *f* d'encrage **− d** Farbverteilungsbalken *m*; Farbeverteilbalken *m*; Farbbalken *m* **− n** inktbalk *m* **− e** barra *f* de distribución de tinta **− i** regolo *m* del calamaio

7383 ink rail bar; breast bar − Removable portion of the ink rail on pump fed inking systems.
d Farbverteilerleiste *f* **− i** asta *f* del regolo del calamaio

7384 ink receptive − Having the property of being wet by greasy ink in preference to water.
f amoureux de l'encre **− d** farbeannahmefähig; farbanziehend **− n** inktaannemend **− e** propio a tomar tinta **− i** ricettivo all'inchiostro

7385 ink receptivity − Facility with which a printed film of ink accepts an additional colour.
f réceptivité *f* à l'encre **− d** Farbaufnahmefähigkeit *f* **− n** inktaanneming *f*; inktaannemend vermogen *n* **− e** recibo *m* de tinta; aptitud *f* para tomar tinta **− i** ricettività *f* all'inchiostro

7386 ink remaining on the forme
f encre *f* restant sur la forme **− d** auf der Druckform zurückgebliebene Farbe *f* **− n** op de vorm achtergebleven inkt *m* **− e** tinta *f* rezagada en la forma **− i** inchiostro *m* rimasto sulla forma

7387 ink repellent − Property of the non-image area of the lithographic plate which, when damped, enables it to repel ink.
f refusant l'encre; repoussant l'encre **− d** farbabstoßend **− n** inktafstotend **− e** repelente la tinta; rechazando la tinta **− i** inchiostrorepellente; rifiuta l'inchiostro

7388 ink resistance − Property of paper which prevents the penetration of printing or writing ink.
f résistance *f* à l'encre **− d** Farbbeständigkeit *f* **− n** bestendigheid *f* tegen inkt **− e** resistencia *f* a la tinta **− i** resistenza *f* all'inchiostro

7389 ink resistant
f repoussant l'encre **− d** farbbeständig **− n** bestand tegen inkt **− e** resistente a la tinta **− i** resistente all'inchiostro

7390 ink-resisting area − See grease-resisting area.

7391 ink return blade − Appliance in a screen printing press.
f contre-raclette *f* de ramassage **− d** Vorrakel *f* **− n** voorrakel *m*

7392 ink rollers − See inking rollers.

7393 ink slab; ink stone; inking table − Slab of marble, glass or slate (sometimes a lithographic stone), upon which small quantities of ink are mixed or distributed by hand.

f pierre *f* à encre; table *f* à encre; table *f* au noir — **d** Farbstein *m*; Farbmischstein *m*; Stein *m*; Farbplatte *f*; Farbtisch *m* — **n** inktsteen *m*; steen *m* — **e** plato *m* de batir la tinta; mesa *f* para la tinta; piedra *f* para mesclar la tinta; plato *m* de la tinta — **i** tavoletta *f* di macinazione d'inchiostro

7394 ink slice — See ink knife.

7395 ink sling — Accumulation of ink, usually at the end of the rollers, built up during running of the press and then flung from the rollers on to the various parts of the web or press.

7396 ink slinger — See comma friend.

7397 ink splitting; splitting of ink film

f scission *f* du film d'encre — **d** Druckfarbenspaltung *f*; Spalten *n* der Farbe — **n** inktsplitsing *f*; splitsing *f* van de inktlaag — **e** escisión *f* de la tinta — **i** scissione *f* (in strati) della pellicola d'inchiostro; separazione *f* del film d'inchiostro

7398 ink stone — See ink slab.

7399 ink supply — See ink feed.

7400 ink table; ink plate — Flat, usually iron surface, on which the ink is rolled before being applied to the forme on a letterpress machine.

f table *f* d'encrage — **d** Farbtisch *m* — **n** inkttafel *fm* — **e** tabla *f* de la tinta; plato *m* de la tinta; mesa *f* distribuidora; mesa *f* tintera; mesa *f* de distribución — **i** tavola *f* di macinazione d'inchiostro

7401 ink taking — Amount of ink taken from the printing plate.

f prise *f* d'encre — **d** Farbabnahme *f* — **n** inktafname *fm* — **e** toma *f* de tinta — **i** presa *f* d'inchiostro

7402 ink tank — Main ink storage unit.

f réservoir *m* d'encre de stockage — **d** Farbtank *m* — **n** hoofdinkttank *m*; hoofdinktreservoir *n* — **e** estanque *m* principal de tinta; tanque *m* de tinta *SA* — **i** serbatoio *m* d'inchiostro

7403 ink tank — Small reserve tank used in conjunction with a circulating pump.

f réservoir *m* d'encre — **d** Farbbehälter *m* — **n** inkttank *m*; inktreservoir *n* — **e** tanque *m* de tinta; depósito *m* de tinta; recipiente *m* de tinta — **i** contenitore *m* dell'inchiostro

7404 ink tin — See ink can.

7405 ink transfer

f transfert *m* d'encre — **d** Farbübertragung *f*; Übergabe *f* der Druckfarbe; Farbtransport *m* — **n** inktoverdracht *fm*; inktoverdraging *f* — **e** traspaso *m* de tinta — **i** trasferimento *m* d'inchiostro

7406 ink transfer coefficient

f coefficient *m* de transfer d'encre — **d** Farbübertragungszahl *f*; Farbabgabeverhältnis *n* — **n** inktoverdrachts-coëfficiënt *n* — **e** coeficiente *m* de traspaso de tinta — **i** coefficiente *m* di trasferimento d'inchiostro

7407 ink *v* up — See roll *v* up.

7408 ink/water balance

f équilibre *m* encre-eau — **d** Farb/Wasser-Abgleich *m*; Farb/Wasser-Balance *f* — **n** evenwicht *n* tussen inkt en water — **e** equilibrio *m* tinta/agua — **i** equilibrio *m* inchiostro/acqua

7409 ink/water fountain — Damping unit separate from the inking system, in which the damping solution on the plate is controlled by a gap between an intermediate ink cylinder with a lipophilic surface and a damping roller, onto which ink is also supplied.

f mouillage *m* sur rouleau encreur — **d** farbfreudiges Feuchtwerk *n*

7410 inlaid binding — See mosaic binding.

7411 inlay; onlay — Panel of paper, cloth, leather or other material set into a book cover flush with surface to create a design.

f mosaïque *f* — **d** aufgelegte Arbeit *f* — **n** mozaïek *n* — **e** mosaico *f* — **i** mosaico *m*; intarsio *m*

7412 inlet and exit wire guides — Flexible or rigid steel guides which lead stitching wires from the wire reel to the stitcher head. Also guides leading the formed staples from the stitcher head.

f guides *mpl* du fil métallique — **d** Heftdrahtführungselemente *npl* — **n** hechtdraadgeleiders *mpl* — **e** guías *fpl* del hilo metálico — **i** guide *fpl* del filo metallico

7413 in-line — Connecting operation of a machine or appliance in conjunction with the operation of another one.

7414 in-line feed hole; centre hole *GB*; **center hole** *US*; **centered sprocket hole** *US*; **centred sprocket hole** *US* — Small hole longitudinally in the centre of punched tape for feeding by sprocket wheel.

f trou *m* central d'entraînement — **d** zentrales Transportloch *n* — **n** centraal transportgat *n*; centraal geleidegat *n* — **e** perforación *f* central de arrastro — **i** foro *m* centrale di trascinamento

7415 in-line operation — See assembly line.

7416 inner chase width; inside chase dimension

f dimension *f* intérieure du châssis de serrage; format *m* intérieur du châssis de serrage — **d** innere Rahmenweite *f*; innere Schließrahmenweite *f* — **n** grootte *f* van het vormraam binnenwerks; binnenraamwijdte *f* — **e** luz *f* interior de la rama; ancho *m* interior del chasis — **i** luce *f* interna del telaio; luce *f* entro il telaio

7417 inner end-paper — See fly leaf.

7418 inner forme; inside forme — Forme containing the pages of a type forme which fall on the inside of a printed sheet in sheet work. The reverse of *outer forme.

f forme *f* intérieure; forme *f* du second côté — **d** innere Form *f*; Widerdruckform *f* — **n** binnenvorm *m*; weerdrukvorm *m* *depr.* — **e** retiro *m* — **i** forma *f* interna

7419 inner margin — See back margin.

7420 inorganic pigments — Class of pigments used in printing ink manufacture consisting mainly of compounds of metals. Some occur naturally such as iron oxides but they are more frequently manufactured by chemical reaction.

f pigments *mpl* inorganiques — **d** anorganische Pigmente *npl* — **n** anorganische pigmenten *npl* — **e** pigmentos *mpl* inorgánicos — **i** pigmenti *mpl* inorganici

7421 in print — Still on sale.

f en librairie — **d** lieferbar; erhältlich — **n** nog leverbaar; nog verkrijgbaar — **e** ya en venta; abastecido — **i** in vendita; in libreria

7422 input — Current, voltage, power, or driving force applied to a circuit or device.

f entrée *f* — **d** Eingang *m* — **n** ingang *m* — **e** entrada *f* — **i** ingresso *m*

7423 input — Stream of data that is fed into a computer for processing or analysis. Data may be "input" in many ways: on-line from sensing devices, from optical scanning, from paper or magnetic tape, or from on-line keyboards, optical character readers, or video display terminals.

f entrée *f* — **d** Eingabe *f* — **n** ingang *m* — **e** entrada *f* — **i** ingresso *m*

7424 input bound — Said of a computer or other similar device (or a typesetter) when restricted in its processing speed by the physical or mechanical limitations of reading the input (especially from punched cards or paper tape). If it is limited by the output speeds, as of a paper tape punch, it is said to be *output bound. If limited at both ends it is input/output bound (I/O bound).

7425 input/output; I/O — Techniques, devices and media to communicate with data processing equipment and the data involved in these communications. Depending upon the context, the term may mean either "input and output" or "input or output".

f entrée/sortie *f* — **d** Eingabe/Ausgabe *f*; Ein/Ausgabe *f* — **n** invoer/uitvoer *m* — **e** entrada/salida *f* — **i** ingresso/uscita *f*

7426 input/output bound — See input bound.

7427 input/output control — Standard routine(s) to initiate and control the input and output processes of a computer system, making it unnecessary to prepare detailed coding for these processes.

f commande *f* d'entrée/sortie — **d** Ein/Ausgabesteuerung *f* — **n** in- en uitvoerbesturing *f* — **e** mando *m* de entrada/salida — **i** comando *m* d'ingresso/uscita

7428 input signal — Electrical signal received from the analysing head, which is fed into the computer of an electronic scanner or engraver, and which initiates action.

f signal *m* d'entrée — **d** Eingangssignal *n* — **n** ingangssignaal *n* — **e** señal *f* de entrada — **i** segnale *m* d'ingresso

7429 input terminal — Device to create input for a computer system or a typesetter; may be "free standing", i.e., not on-line to the computer or typesetter, and recording its input on paper tape, or on magnetic tape, floppy disc, magnetic cassette

etc.

7430 in quires — See in sheets.

7431 inscriber, magnetic data — See encoder.

7432 insecticide — Powder etc. to kill insects.

f insecticide *m*; poudre *f* insecticide — **d** Insektizid *n*; Schädlingsbekämpfungsmittel *n*; Insektenvertilgungsmittel *n* — **n** insekticide *n*; insektendodend middel *n*; insektenverdelgingsmiddel *n*; insektenpoeder *n* — **e** insecticida *m*; polvo *m* insecticida — **i** insetticida *m*; polvere *f* insetticida

7433 insert — In printing, a page, etc., that is printed separately and then placed into or bound with the main publication.

f encart *m* — **d** Beilage *f* — **n** bijlage *f* — **e** encartado *m*; encartaje *m* (galicismo) — **i** inserto *m*; allegato *m*

7434 insert — Additional sentence or paragraph added to a proof to be inserted in the revise or final proof.

f insertion *f* — **d** Einfügung *f* — **n** inlas *m*; inlassing *f* — **e** intercalación *f*; intercaladura *f*; intercalo *m*; intercalado *m*; inserción *f*; añadido *m*; adición *f*; agregado *m*; añadidura *f*; metido *m*; insertación *f Cu* — **i** inserzione *f*

7435 insert — Extra original and negative therefrom, introduced into some area of the main negative before printing and etching the plate.

7436 insert *v* — To place sections within one another after the press run.

f encarter — **d** einlegen; beilegen; ineinanderfalzen; einstecken — **n** insteken — **e** insertar; intercalar; meter; encajar; embuchar — **i** inserire; intercalare

7437 inserter — See inserting machine.

7438 inserting machine; inserter — Machine to insert a folded product into the opening of another separately printed one.

f encarteuse *f*; machine *f* à encarter — **d** Einsteckmaschine *f*; Einlegemaschine *f* — **n** insteekmachine *f* — **e** máquina *f* insertadora — **i** macchina *f* inseritrice

7439 insertion — Publication of a notice or advertisement in a newspaper or magazine.

f insertion *f* — **d** Textanzeige *f* — **n** advertentie *f* tussen de tekst; ingezonden mededeling *f* — **e** colocación *f* aislada — **i** inserzione *f*

7440 insertion mark

f signe *m* d'intercalation — **d** Einschaltungszeichen *n* — **n** tussenvoegingsteken *n* — **e** señal *f* de intercalación — **i** segno *m* per interpolazione; marca *f* per interpolazione

7441 insert *v* **letter; take** *v* **out turns** — To replace turned-for-sorts letter by proper letter in set forme. See also turn for sorts.

f débloquer — **d** deblockieren; Fliegenköpfe berichtigen; Fliegenköpfe herausnehmen; Fliegenköpfe richtigstellen — **n** deblokkeren — **e** desbloquear — **i** rettificare

7442 inset — Advertisement on a separate leaf inserted in a magazine or booklet.

f encart *m*; papillon *m* — **d** Einlage *f*; Beilage *f* —

n bijlage *fm* — **e** encarte *m*; encaje *m* — **i** inserto *m* sciolto

7443 inset — Section which when insetted becomes a permanent part of the book.

f encart *m*; signature *f* encartée; cahier *m* encarté; carton *m* (de 4 pages) — **d** Einsteckbogen *m*; Einsatz *m*; eingesteckter Bogen *m* — **n** insteekvel *n*; insteker *m*; ingestoken vel *n* — **e** pliego *m* de encaje; encaje *m*; encarte *m* — **i** inserto *m* fissato; segnatura *f* da inserire

7444 inset plate

f illustration *f* hors texte; hors-texte *m* — **d** eingeklebte Beilage *f*; Abbildung *f* außerhalb des Textes — **n** buitentekst-plaat *fm*; ingeplakte plaat *fm* — **e** lámina *f* fuera de texto — **i** illustrazione *f* fuori testo

7445 inset signature — See asterisk signature.

7446 insetter — Device on a reel fed press to run a pre-printed web into the press for additional printing or to run a complete printed additional web into the folder.

f insetter *m* — **d** Insetter *m* — **n** insetter *m* — **e** insetter *m*; máquina *f* encartadora — **i** inseritore *m*; insetter *m*

7447 insetting — In rotary printing, the association of a number of webs before the spine fold. The introduction of an additional web which may be pre-printed into the main product or folder.

7448 in sheets; in quires *obs.* — When a book is printed, or printed and folded, but not bound.

f en feuilles; en cahiers — **d** in Rohbogen; roh — **n** aan vellen; aan plano vellen; ongevouwen; in katerns — **e** en hojas — **i** in fogli; in quaderni

7449 inside back cover; third cover

f troisième page *f* de couverture; troisième couverture *f*; contreplat *m* verso — **d** hintere Umschlaginnenseite *f*; Innenseite *f* des hinteren Deckels — **n** pagina *fm* drie van de omslag; binnenzijde *fm* van het achterplat — **e** tercera cubierta *f*; interior *m* de contracubierta; interior *m* de contratapa; tercera página *f* de cubierta — **i** terza copertina *f*; copertina *f* del dorso interno

7450 inside chase dimension — See inner chase width.

7451 inside forme — See inner forme.

7452 inside front cover; second cover

f deuxième page *f* de couverture; deuxième couverture *f*; contreplat *m* recto — **d** vordere Umschlaginnenseite *f*; Innenseite *f* des vorderen Deckels — **n** pagina *fm* twee van de omslag; binnenzijde *fm* van het voorplat — **e** segunda cubierta *f*; interior *m* de cubierta — **i** seconda copertina *f*; copertina *f* frontale interna

7453 inside mortise — Opening cut entirely inside a mounted printing plate for insertion of type or other matter.

f ajour *m*; fenêtre *f*; passe-partout *m* — **d** Ausschnitt *m* — **n** binnenwerkse uitsparing *f* — **e** mortaja *f* interior; entalladura *f*; muesca *f* interior; escopladura *f* al centro; escopleadura *f* interior; escopleadura *f* al centro; escopleadura *f* adentro — **i** finestra *f*

7454 insides — English term applied to the 18 inside quires of a (twenty quire) ream of paper, the top and bottom quires being *outsides. A ream of insides has its outsides removed and insides added to make a complete ream. It is charged at 7.5% more than an ordinary ream.

f intérieur *m* — **d** innere Lagen *fpl* — **n** binnenvellen *npl* — **e** hojas *fpl* interiores — **i** prima scelta *f* di una risma

7455 in slips — Composed matter in galleys from which proofs are printed.

f en placards — **d** in Spalten; in Kolumnen — **n** aan stroken; in slips — **e** en columnas — **i** in colonna

7456 insolation — See exposure.

7457 insoluble

f insoluble — **d** unlöslich; unauflöslich — **n** onoplosbaar; niet oplosbaar — **e** insoluble; indisoluble — **i** insolubile

7458 inspection

f inspection *f* — **d** Prüfung *f* — **n** keuring *f* — **e** inspección *f*; ensayo *m*; cata *f* — **i** ispezione *f*

7459 instable — Opposite of *stable.

f instable — **d** unbeständig — **n** onstabiel; niet stabiel; instabiel; onbestendig — **e** inestable — **i** instabile

7460 instantaneous plate lock-up — Method of rapidly operating a compression or tension plate lock-up on a letterpress rotary.

f serrage *m* rapide (de clichés); calage *m* instantané (de stéréos); calage *m* rapide (des plaques) — **d** Plattenschnellverschluß *m*; Schnellverschluß *m* — **n** snelsluitinrichting *f* — **e** acuñado *m* instantáneo — **i** bloccaggio *m* rapido

7461 instant lettering — System of dry transfer of characters printed on an adhesive backed film. The lettering can either be photographed direct or transferred to the layout.

f lettrage *m* instantané (par report direct); lettrage *m* avec lettres à décalquer par pression; lettrage *m* avec lettres à décalques — **d** Beschriften *n* mit Schriftfolien; Beschriften *n* mit Haftfolien — **n** aanbrengen *n* van plakletters — **e** colocación *f* de letras transferibles; colocación *f* de letras adhesivas — **i** scrittura *f* per riporto a pressione da fogli adesivi

7462 instruction *GB*; **command** *US* — Phrase, sentence or set of related characters which when translated to machine-readable format, direct the computer to perform a given function.

f instruction *f* — **d** Befehl *m*; Instruktion *f* — **n** instructie *f*; opdracht *fm* — **e** instrucción *f* — **i** istruzione *f*

7463 instruction book

f mode *m* d'emploi; livret *m* d'instructions — **d** Bedienungsanleitung *f*; Bedienungshandbuch *n* — **n** gebruiksaanwijzing *f* — **e** libro *m* de indicaciones; instrucciones *fpl* de manejo — **i** libro *m* delle istruzioni

7464 insufficient feed — Restricted ink flow to the printing plate.

f débit *m* d'encre insuffisant — **d** ungenügende

Farbgebung *f* — **n** onvoldoende inktgeving *f* — **e** arribo *m* insuficiente de tinta — **i** alimentazione *f* insufficiente dell'inchiostro

7465 insufficient feed; losing paper — Decrease in the amount of paper in a web press drawn through the press, which results in an increase in web tension between two units or components. See also excess feed.

f débit *m* de papier insuffisant — **d** Minderförderung *f*; Nacheilung *f* — **n** onvoldoende papieraanvoer *m* — **e** retraso *m* de papel — **i** alimentazione *f* insufficiente della carta

7466 insulating board — See presspahn.

7467 intaglio etching — Image etched into a plate instead of appearing in relief, usually made from a photographic positive.

f gravure *f* en creux; héliogravure *f* — **d** Tiefätzung *f*; Tiefdruckätzung *f* — **n** diepetsing *f*; diepdruketsing *f* — **e** grabado *m* en hueco; grabado *m* calcográfico — **i** incisione *f* (chimica) in incavo

7468 intaglio ink — Ink for *intaglio printing. There are two types: 1. *copperplate engraving ink and *photogravure ink; 2. *gravure ink and *rotogravure ink.

7469 intaglio offset — See deep-etch offset.

7470 intaglio paper — Paper suitable for intaglio printing. See also rotogravure paper, plate paper, steel-plate paper.

7471 intaglio printing; gravure printing — Printing process employing minute engraved wells. Based on the principle that deeply etched wells carry more ink than a raised surface, hence print darker values. Shallow wells are used to print light values. A doctor blade wipes excess of ink from the cylindrical printing surface. Rotogravure employs etched cylinders and web-fed stock.

f impression *f* en creux; impression *f* en héliogravure; héliogravure *f*; procédé *m* chalcographique — **d** Tiefdruck *m*; Kupfertiefdruck *m* — **n** diepdruk *m*; koperdiepdruk *m* — **e** huecograbado *m*; calcografía *f*; estampación *f* fotocalcográfica — **i** rotocalco *m*; rotocalcografia *f*; stampa *f* rotocalco

7472 intaglio proof — Impression from an intaglio plate. Produced by depositing ink in the etched incisions or lines and wiping the surface of the plate clean before taking the impression on paper.

f épreuve *f* en creux — **d** Tiefdruckandruck *m* — **n** proef *fm* van de plaat; plaatproef *fm* — **e** prueba *f* de huecograbado — **i** bozza *f* da forma ad incavo

7473 intaglio watermark — See shaded watermark.

7474 integral cover; self-contained cover; self cover — Cover printed on same kind of paper as body of pamphlet or booklet. May be made also of other paper stuff.

f couverture *f* à même — **d** Integral-Einband *m*; Integral-Buchdecke *f* — **n** integraal-band *m* — **e** autocubierta *f*; tapa *f* integral; cubierta *f* integral — **i** copertina *f* integrale; autocopertina *f*

7475 integral mask — Mask on the same support

as the image which they are to modify. Could have been included under contact masks. Integral masks for originals might be considered to include the Kodak fluorescence process, in which fluorescent pigments are used in preparing the copy. Integral masks for negatives used in Kodacolor, Ektacolor, and Eastman color negative film.

f masque *m* intégral — **d** eingebaute Maske *f* — **n** automatisch masker *n* — **e** máscara *f* integral — **i** maschera *f* integrale; automaschera *f*

7476 integral sign — See also mathematical signs.

f intégrale *m*; signe *m* d'intégration — **d** Integral-Zeichen *n* — **n** integraal-teken *n* — **e** signo *m* integral; señal *f* de integración — **i** segno *m* di integrale

7477 integrated circuit; monolithic circuit — Array of elements integrated with or deposited on a semi-conductor substrate and able to undertake the functions of electronic circuits. A hybride integrated circuit consists of an integrated circuit together with one or more discrete components.

f circuit *m* intégré — **d** integrierter Schaltkreis *m*; integrierte Schaltung *f* — **n** geïntegreerde schakeling *f*; geïntegreerd circuit *n* — **e** circuito *m* integrado — **i** circuito *m* integrato

7478 integrated halftone density — Light stopping power of an area in a halftone photograph. It is the combined measure of the dense and clear areas within the field of the densitometer aperture.

f densité *f* tramée intégrée — **d** integrierte Dichte *f*; integrierte Punktdichte *f* — **n** geïntegreerde dichtheid *f*; geïntegreerde densiteit *f* — **e** densidad *f* integrada — **i** densità *f* integrata di un retino

7479 integrating exposure meter — See light integrator.

7480 integrating light meter — See light integrator.

7481 intelligent — Said of a typesetting device if it is able to perform certain logical or arithmetic functions of a decision-making character which formerly had to be performed independently either by a person or by a general-purpose computer. Also a typesetting device that can accept a stream of data, break it into lines, justify these lines and perform hyphenation if required. A video terminal that contains logic to perform certain complex editing or counting functions without recourse to a computer. In many systems intelligence is "distributed" among various devices, each of which may perform some but not all functions previously performed only in one central computer, so that many tasks can be performed without placing demands upon the main frame or host computer which is freed to perform more vital or demanding tasks.

7482 intelligent typesetter — See computer composition program.

7483 intensification — Increasing the opacity of developed and fixed photographic negatives by chemical treatment of the image.

f renforcement *m* — **d** Verstärkung *f*; Verstärken *n*; Schwärzung *f* — **n** versterking *f*; versterken *n* — **e** reforzado *m*; intensificación *f*; refuerzo *m* — **i** rinforzo *m*; rafforzamento *m*

7484 intensifier — Chemical agent capable of intensifying photographic negatives.

f renforçateur *m* — **d** Verstärker *m* — **n** versterker *m* — **e** intensificador *m*; reforzador *m* — **i** rinforzatore *m*

7485 intensify *v* — To increase the density or contrast of each silver particle in the film emulsion by adding silver compounds of heavier metals such as chromium or mercury.

f renforcer — **d** verstärken — **n** versterken — **e** reforzar — **i** rinforzare

7486 intensity — The intensity of coloured light coming directly from its source or by transmission is its *luminance; term not admitted by CIE; in common speech its *luminosity.

f intensité *f* — **d** Intensität *f*; Leuchtkraft *f* — **n** intensiteit *f* — **e** intensidad *f* — **i** intensità *f*

7487 intensity of light

f intensité *f* lumineuse — **d** Lichtstärke *f* — **n** lichtsterkte *f* — **e** intensidad *f* lumínica; potencia *f* luminosa; potencia *f* lumínica; poder *m* lumínico — **i** intensità *f* della luce

7488 interactive program — Program or system which permits the user to make independent decisions during the processing run which will affect the outcome of the process. The user may add to, delete from, or alter the characteristics of the task while it is in process.

7489 inter-block gap; block gap — Space required to enable magnetic tape to stopped and started between data blocks, necessary because magnetic tape cannot be stopped and started between codes except on incremental decks.

f intervalle *m* de blocs — **d** Blockzwischenraum *m*; Zwischenraum *m* — **n** interblokspatie *f*; bloktussenruimte *f*; blokspouw *fm* — **e** intervalo *m* de bloques — **i** intervallo *m* tra blocchi

7490 interchangeability

f interchangeabilité *f* — **d** Auswechselbarkeit *f*; Verwechselbarkeit *f* — **n** verwisselbaarheid *f* — **e** recambiabilidad *f*; intercambiabilidad *f* — **i** intercambiabilità *f*

7491 interchangeable lens — Removable lens which permits substitution of lenses of different focal lengths and/or apertures in some cameras.

f objectif *m* interchangeable — **d** Wechselobjektiv *n*; auswechselbares Objektiv *n* — **n** verwisselbaar objectief *n* — **e** objetivo *m* intercambiable — **i** obiettivo *m* intercambiabile

7492 interface — Link between computer systems or parts of a configuration. It may be a standard of plug used to link peripheral devices, through data and control channels, to a range of central processors; in a more detailed definition, it involves compatibility between a complex of systems and may be concerned with the design of the system and equipment, codes, transfer speed, tape format or other points of possible difference;

it also exists between man and machine, and much attention is paid to development of systems involving enquiry points, display consoles and data terminals.

f interface *f*; jonction *f* — **d** Verbindung *f* — **n** schakel *mf* — **e** acoplamiento *m* mutuo; zona *f* interfacial — **i** interfaccia *f*; interficie *f*

7493 interfacial activity

f activité *f* interfaciale — **d** Grenzflächenaktivität *f* — **n** grensvlakactiviteit *f* — **e** actividad *f* interfacial — **i** attività *f* all'interfacie

7494 interfacial tension — Contractile force existing at the interfaces of two immiscible liquids.

f tension *f* interfaciale — **d** Grenzflächenspannung *f* — **n** grensvlakspanning *f* — **e** tensión *f* interfacial; tensión *f* de la superficie de límite — **i** tensione *f* interfacciale

7495 interference phenomenon

f phénomène *m* d'interférence — **d** Interferenzerscheinung *f* — **n** interferentie-verschijnsel *n* — **e** fenómeno *m* de interferencia — **i** fenomeno *m* d'interferenza

7496 interlapped; in and in — Method of packing large sheets of paper for handling or shipment. The packages are divided in half and the two portions then *lapped one into the other.

f plié en trois; plié à l'anglaise — **d** zweimal umgeklappt; ineinandergeklappt; ineinandergefaltet — **n** tweemaal toegeslagen; tweemaal dichtgeslagen; tweemaal omgeslagen; tweemaal ingeslagen — **e** gualdrapeado; doblado en tres; doblado en zigzag; plegado en zigzag; en tres dobleces — **i** interposto

7497 interlay — Sheet of paper or other material between a printing plate and its mount to raise the surface of the plate to proper height of paper.

f mise *f* entre cuir et chair; mise *f* en train entre cuir et chair; découpage *m* entre cuir et chair — **d** Plattenzurichtung *f*; Zwischenzurichtung *f* — **n** pikeersel *n* tussen cliché en voet — **e** calzo *m*; calza *f*; alza *f*; recorte *m*; pliego *m* de cama intermedio — **i** taccheggio *m* tra cliché e piede

7498 interleave *v* — To insert extra leaves, usually blank or ruled, between regular leaves of a book.

f interfolier; entrefeuilleter — **d** einschießen — **n** doorschieten — **e** interfoliar; interpaginar — **i** interfogliare; intercalare

7499 interleave *v* — To insert leaves of paper between freshly printed sheets to prevent set-off.

f intercaler; mettre en macules — **d** Makulaturbogen einschießen; durchschießen mit Makulatur — **n** tussenschieten — **e** intercalar (de maculaturas entre los impresos); intercalar (hojas de papel entre los impresos) — **i** intercalare; inserire scartini; tramezzare (con carta)

7500 interleaves — Extra leaves, usually blank, inserted between the regular leaves of a book, e.g., to make notes.

f pages *fpl* interfoliées — **d** Zwischenblätter *npl*; eingeschossene Bogen *npl* — **n** doorschietvellen *npl* — **e** hojas *fpl* para interfoliar — **i** fogli *mpl* intercalati

7501 interleaves — See slip sheets.

7502 interleaving — See slip sheeting.

7503 interleaving paper; interleaving tissue
f feuille *f* de séparation — **d** Einschießpapier *n* — **n** tussenschietpapier *n* — **i** scartini *mpl*

7504 interleaving tissue — See interleaving paper.

7505 interlineation — Increasing the spacing between lines to cover specified area.
f interlignage *m* — **d** Durchschuß *m*; Kolumne Austreiben *n*; Zeilen Austreiben *n*; Durchschießen *n* — **n** interliniëring *f* — **e** interlineado *m*; entrerrenglón *m*; entrerrenglonadura *f* — **i** interlineatura *f*

7506 interlock
f verrouillage *m* mutuel — **d** gegenseitige Verriegelung *f* — **n** wederzijdse vergrendeling *f*; wederzijdse blokkering *f* — **e** enclavamiento *m* mutuo; intercierre *m* — **i** bloccaggio *m* reciproco; interblocco *m*

7507 intermediate — Contact print from on original negative or positive required for printing down duplicates by a photographic process.

7508 intermediate channel — See transfer channel.

7509 intermediate coat
f couche *f* intermédiaire — **d** Zwischenschicht *f*; Zwischenlage *f* — **n** tussenlaag *fm* — **e** capa *f* intermedia — **i** strato *m* intermedio

7510 intermediate colours — Colours obtained by mixing adjacent primary and secondary colours.
f couleurs *fpl* intermédiaires — **d** Zwischenfarben *fpl*; Mittelfarben *fpl* — **n** tussenliggende kleuren *fmpl*; tussenkleuren *fmpl* — **e** colores *mpl* intermedios — **i** colori *mpl* intermedi

7511 intermediate dogs — Clips between plates on a letterpress rotary machine. See also dogs.
f griffes *fpl* intermédiaires — **d** Zwischenhaken *mpl* — **n** tussenhaken *mpl* — **e** grapas *fpl* intermediarias — **i** staffe *fpl* intermedie

7512 intermediate rollers — Rollers transferring ink from one ink drum to another.
f rouleaux *mpl* intermédiaires — **d** Zwischenwalzen *fpl*; Verreibwalzen *fpl*; Heberwalzen *fpl* — **n** tussenrollen *fmpl* — **e** rodillos *mpl* intermediarios — **i** rulli *mpl* intermedi

7513 intermittency effect — Failure of photographic material to integrate correctly an interrupted exposure. In the early years of photography, sensitometric tests were frequently made by using rapidly rotating sector wheels to modulate the exposure on a photographic material. Beginning with the studies of Abney in 1893, extensive investigations were conducted to determine the conditions under which this procedure was reliable. In general, it was found that a continuous exposure of a given intensity and time did not produce the same photographic density as a second, equal-energy exposure, given in a number of discrete increments of the same intensity.
f effet *m* d'intermittente — **d** Intermittenz-

Effekt *m* — **n** intermitterings-effect *n* — **e** efecto *m* de intermitencia — **i** effetto *m* d'intermittenza

7514 intermittent — Coming and going at intervals.
f périodique; intermittent — **d** zeitweise; intermittierend; aussetzend — **n** intermitterend; met tussenpozen — **e** intermitente — **i** intermittente

7515 intermittent board machine; wet board machine; cardboard machine — Machine to produce thick sheets of board. It consists of a Fourdrinier wire or one or more cylinder moulds or vats. The web is wound round a making roll to form a sheet consisting of several layers, of which the thickness increases with the number of turns of the making roll.
f enrouleuse *f* (pour carton) — **d** Handpappenmaschine *f*; Wickelpappenmaschine *f*; Handpappenwickelmaschine *f* — **n** handbordmachine *f* — **e** máquina *f* plana de cartón; máquina *f* de cartón a la enrolladura — **i** macchina *f* discontinua in tondo per cartoni

7516 intermittent element — See incidental element.

7517 internal memory — See internal store.

7518 internal storage — See internal store.

7519 internal store *GB*; **internal storage** *US*; **internal memory** *US* — Store built into a computer and directly controlled by it.
f mémoire *f* interne — **d** Zentralspeicher *m* — **n** intern geheugen *n* — **e** memoria *f* interna — **i** memoria *f* interna

7520 internal thread; female thread
f filetage *m* intérieur; filetage *m* femelle — **d** Innengewinde *n*; Muttergewinde *n* — **n** inwendige draad *m*; inwendige schroefdraad *m* — **e** rosca *f* interna; rosca *f* interior; rosca *f* hembra — **i** filetto *m* interno; filettatura *f* femmina

7521 international sizes; standard sizes; DIN sizes — The international sizes of paper are divided into four groups of sizes, known as A, B, C and D. The printing sizes are categorized as A sizes, with the B, C and D sizes referring to intermediate sizes. The C and D sizes are practically out of use. See app. no. 3.
f formats *mpl* internationaux; formats *mpl* normalisés; formats *mpl* étalons; formats *mpl* (de papier) standardisés — **d** Normalformate *npl*; normalisierte Formate *npl*; DIN-Formate *npl* — **n** normaalformaten *npl*; genormaliseerde formaten *npl*; eenheidsformaten *npl*; DIN-formaten *npl* — **e** tamaños *mpl* internacionales; tamaños *mpl* (de papel) normalizados; formatos *mpl* (de papel) normalizados — **i** formati *mpl* unificati (della carta); formati *mpl* (di carte) normalizzati; formati *mpl* internazionali

7522 international sizes of envelopes — See app. no. 4.
f formats *mpl* internationaux d'enveloppes — **d** internationale Briefhüllenformate *npl*; Normal-Kuvertformate *npl* — **n** eenheidsformaten *npl* voor enveloppen — **e** tamaños *mpl* internacionales de sobres — **i** formati *mpl* inter-

nazionali delle buste

7523 International Telegraph Alphabet — See Baudot code.

7524 interpolate *v* — See collate *v*.

7525 interpolator — See collator.

7526 interrecord gap; record gap — Blank, unused space between unblocked records. See also inter-block gap.
f entre-enregistrement *m* — **d** Satzzwischenraum *m* — **n** recordtussenruimte *f*; interrecordspatie *f* — **e** separación *f* entre registros — **i** intervallo *m*

7527 interrogation mark; mark of interrogation; interrogation point; question mark; note of interrogation — Punctuation mark (?) placed at the end of a question. In Spanish an inverted interrogation mark is placed at the beginning of a question.
f point *m* d'interrogation — **d** Fragezeichen *n* — **n** vraagteken *n* — **e** punto *m* de interrogación; punto *m* interrogante; signo *m* de interrogación; signo *m* interrogativo; interrogación *f* — **i** punto *m* interrogativo

7528 interrogation point — See interrogation mark.

7529 interrupt — Temporarily disrupt of the normal operation of a routine by a signal from the computer. See also automatic program interrupt.
f interrupteur *m* — **d** Unterbrecher *m* — **n** ingreep *m* — **e** interruptor *m* — **i** interruttore *m*

7530 interrupt signal — Signal generated in a computer to cause an interruption.
f signal *m* d'interruption — **d** Eingriffssignal *n* — **n** ingreepsignaal *n*; onderbrekingssignaal *n* — **e** señal *f* de interrupción — **i** segnale *m* d'interruzione

7531 interspacing — See letter spacing.

7532 Intertype — Slug-composing machine originally based on expired Linotype patents but manufactured by another company. Matrices for these two machines are interchangeable.
f Intertype *f* — **d** Intertype *f* — **n** Intertype *mf* — **e** Intertipo *m*; Intertipia *f Ar* — **i** Intertipia *f*

7533 Intertypesetter — Solid-state unit to increase the speed of productivity of the Monarch linecasting machine (Febr. 1966). With this machine the normal operating speed was increased at about five times.

7534 interview — Meeting at which a reporter obtains information from a person, also a report or reproduction of information so obtained.
f interview *mf* — **d** Interview *n* — **n** interview *n*; vraaggesprek *n* — **e** interview *m*; interviú *fm*; entrevista *f* — **i** intervista *f*

7535 interword knob — Knob used to end a line ahead of a word which cannot be properly divided and which appears on the display, and to carry this word to the next line providing the line is justifiable at that point. If not, the word will remain on the display.
f touche *f* d'abréviation — **d** Abkürzungstaste *f* —
n afbreektoets *m*; scheidingstoets *m* — **e** tecla *f* de abreviaturas — **i** tasto *m* per parole che non possono essere divise su due righe

7536 in the flat — See flat.

7537 in the press — Of work which is being printed and which is advertised as a coming publication.
f sous presse; à l'impression — **d** im Druck — **n** ter perse; in druk — **e** en prensa; en curso de impresión — **i** in pressione; sotto stampa; in corso di stampa

7538 intimation — A size of cut card. See app. no. 3.

7539 introduction — Author's statement on the matter of the contents of a book, placed after the *foreword.
f introduction *f* — **d** Einführung *f*; Einleitung *f* — **n** inleiding *f* — **e** introducción *f*; preámbulo *m*; exordio *m*; prólogo *m*; introito *m*; iniciación *f*; isagoge *f* — **i** introduzione *f*

7540 inversely proportional; in inverse ratio to
f inversement *m* proportionnel; en raison inverse de — **d** umgekehrt proportional — **n** omgekeerd evenredig — **e** en razón inversa de — **i** inversamente proporzionale a

7541 inverted commas — See quotation mark.

7542 inverted image — Positive image changed into a negative one, e.g., white characters on a blank background.
f image *f* inversée — **d** Kehrbild *n* — **n** negatief beeld *n* — **e** imagen *f* invertida — **i** immagine *f* inversa

7543 inverted letter — See turned letter.

7544 inverter; power inverter — Device to transform a direct current into an alternating current.
f inverseur *m* continu-alternatif — **d** Umformeraggregat *n*; Wechselrichter *m* — **n** omvormer *m* — **e** convertidor *m*; grupo *m* convertidor — **i** invertitore *m*

7545 inverting prism — See reversing prism.

7546 invisible ink; sympathetic ink — Ink which cannot be seen under ordinary conditions but can be made visible by special techniques, e.g., wetting the sheets.
f encre *f* invisible; encre *f* sympathique — **d** unsichtbare Druckfarbe *f*; sympathetische Druckfarbe *f* — **n** onzichtbare inkt *m* — **e** tinta *f* invisible; tinta *f* simpática — **i** inchiostro *m* invisibile

7547 I/O — See input/output.

7548 I/O bound — See input bound.

7549 iodic acid — Colourless, rhombic crystals or white, crystalline powder. Moderately strong acid, soluble in water. Latin: acidum iodicum. Formula: HIO_3.
f acide *m* iodique — **d** Jodsäure *f* — **n** joodzuur *n* — **e** ácido *m* yódico — **i** acido *m* iodico

7550 iodide — Salt of hydriodic acid. Formula: HI.
f iodure *m* — **d** Jodid *n*; Jodür *n* — **n** jodide *n* — **e** yoduro *m* — **i** ioduro *m*

7551 iodimetry — See iodometry.

7552 iodine — Heavy, greyish-black plates or

granules. Poisonous. Readily sublimed. Corrosive. Soluble in alcohol, carbon disulphide, chloroform, ether, glycerol, alkaline iodide solutions and carbon tetrachloride, insoluble in water. Symbol: I; in German and Dutch: J; in Spanish: Y.

f iode *m* — **d** Jod *n* — **n** jodium *n*; jood *n* — **e** yodo *m* — **i** iodio *m*

7553 iodine cyanide — See cyanogen iodide.

7554 iodine intensifier

f renforçateur *m* à l'iode — **d** Jodverstärker *m* — **n** joodversterker *m* — **e** reforzador *m* de yodo — **i** rinforzatore *m* allo iodio

7555 iodine number; iodine value — The number of centigrams of iodine absorbed by the double bonds in one gram of a drying oil or resin. A high iodine number is generally indicative of high reactivity and an ability to dry rapidly.

f indice *m* d'iode — **d** Jodzahl *f* — **n** joodgetal *n* — **e** índice *m* de yodo — **i** indice *m* di iodio; numero *m* di iodio

7556 iodine-potassium intensifier

f renforçateur *m* à l'iodure de potassium — **d** Jodkalium-Verstärker *m* — **n** joodkali-versterker *m* — **e** reforzador *m* de yoduro de potasio — **i** rinforzatore *m* all'ioduro di potassio

7557 iodine trichloride — Orange-yellow, deliquescent, crystalline powder, with pungent, irritating odour. Poisonous. Soluble in water with decomposition, alcohol and benzene. Formula: ICl_3.

f trichlorure *m* d'iode — **d** Jodtrichlorid *n*; Jodchlorid *n* — **n** joodtrichloride *n* — **e** tricloruro *m* de yodo — **i** tricloruro *m* di iodio; tricloruro *m* iodico

7558 iodine value — See iodine number.

7559 iodize *v* — To treat, impregnate, or affect with iodine or iodide.

f ioder — **d** jodieren — **n** joderen — **e** yodurar — **i** iodurare

7560 iodized collodion

f collodion *m* iodé — **d** jodiertes Kollodium *n*; Jodkollodium *n* — **n** gejodeed collodium *n* — **e** colodión *m* yodado — **i** collodio *m* iodato

7561 iodobenzene; phenyl iodide — Formula: C_6H_5I.

f iodobenzène *m* — **d** Jodbenzol *n* — **n** joodbenzeen *n* — **e** yodobenceno *m* — **i** iodobenzene *m*

7562 iodoethane — See ethyl iodide.

7563 iodometry; iodimetry — Volumetric analytical procedure making use of iodine.

f iodométrie *f* — **d** Jodometrie *f* — **n** jodometrie *f* — **e** yodometría *f* — **i** iodometria *f*

7564 iodo-potassium iodide — Solution of iodine in aqueous potassium iodide.

f solution *f* d'iode dans l'iodure de potassium — **d** Jod-Jodkalium-Lösung *f* — **n** jodium-oplossing *f*; jood-joodkali *m* — **e** solución *f* de yodo yoduro de potasio — **i** soluzione *f* di iodio in ioduro di potassio

7565 ion — An electrically charged atom or group of atoms formed by the loss or gain of one or more electrons, as a cation (positive ion), formed by electron loss and attracted to the cathode in electrolysis, or as an anion (negative ion), formed by electron gain and attracted to the anode. The valence of an ion is equal to the number of electrons lost or gained and is indicated by a plus sign for cations and a minus sign for anions, e.g., Na^+, Ca^{++}, Cl^-.

f ion *m* — **d** Ion *n* — **n** ion *n* — **e** ión *m* — **i** ione *m*

7566 ion exchange — Reversible chemical reaction between a solid (*ion exchanger) and a fluid mixture (usually an aqueous solution) to interchange ions.

f échange *m* d'ions; échange *m* ionique — **d** Ionenaustausch *m* — **n** ionenuitwisseling *f* — **e** cambio *m* de iones; cambio *m* iónico — **i** scambio *m* d'ioni; scambio *m* ionico

7567 ion exchanger

f appareil *m* pour échanger des ions; échangeur *m* d'ions — **d** Ionenaustauscher *m* — **n** ionenuitwisselaar *m* — **e** intercambiador *m* de iones — **i** scambiatore *m* ionico

7568 Ionic — See app. no. 2.

7569 ionization — Formation of ions.

f ionisation *f* — **d** Ionisation *f*; Ionisierung *f* — **n** ionisatie *f*; ionisering *f* — **e** ionización *f* — **i** ionizzazione *f*

7570 iridescence — See bronze.

7571 iridescent paper — See mother-of-pearl paper.

7572 iris diaphragm — In photographic lenses, an adjustable diaphragm consisting of a number of thin flat metal tongues fastened to a ring on the lens barrel, the aperture made larger or smaller by turning the ring backward or forward, causing the tongues to enlarge or contract the opening.

f diaphragme *m* iris — **d** Irisblende *f* — **n** irisdiafragma *n* — **e** diafragma *m* iris — **i** diaframma *m* a iride

7573 Irish gum — See Irish moss.

7574 Irish moss; Irish gum; carrageen — Dried carrageen, seaweed of the Atlantic coasts of Europe and North America, used as a size in marbling of paper. Name derived from Carragheen, near Waterford (Ireland).

f mousse *f* perlée d'Irlande; carragheen *m* — **d** irländisches Moos *n*; Perlmoos *n*; Karragheenmoos *n*; Karragheenmoos *n*; Karrageenmoos *n* — **n** Iers mos *n* — **e** musgo *m* de Irlanda; carrageen *m* — **i** muschio *n* d'Irlanda

7575 iron — Silvery white, soft, ductile, magnetic metal, properties greatly affected by relative small percentage of carbon (steel). Symbol: Fe. Latin: ferrum.

f fer *m* — **d** Eisen *n* — **n** ijzer *n* — **e** hierro *m* — **i** ferro *m*

7576 iron alum — See ferric potassium sulphate.

7577 iron ammonium citrate — See ferric ammonium citrate.

7578 iron ammonium oxalate — See ferric ammonium oxalate.

7579 iron ammonium sulphate — See ferric ammonium sulphate.

7580 iron ammonium sulphate — See ferrous ammonium sulphate.

7581 iron black — Finely divided black antimony prepared by the reduction of an antimony salt solution by zinc or a similar metal. It contains no iron. Used to impact the appearance of polished steel to papier maché and plaster of Paris.

7582 iron bromide — See ferric bromide.

7583 iron bromide — See ferrous bromide.

7584 iron carbonate — See ferrous carbonate.

7585 iron chloride — See ferric chloride.

7586 iron chloride — See ferrous chloride.

7587 iron citrate — See ferric citrate.

7588 iron developer
f révélateur *m* au sulfate de fer; développateur *m* au sulfate de fer — **d** Eisenentwickler *m* — **n** ijzerontwikkelaar *m* — **e** revelador *m* al hierro — **i** sviluppatore *m* al ferro

7589 iron dichloride — See ferrous chloride.

7590 iron furniture
f lingots *mpl* en fonte — **d** eiserne Stege *mpl* — **n** ijzerwit *n* — **e** imposiciones *fpl* de hierro; guarniciones *fpl* de hierro — **i** lingotti *mpl* in ghisa

7591 iron hydrate — See ferric hydroxide.

7592 iron hydroxide — See ferric hydroxide.

7593 iron oxalate — See ferrous oxalate.

7594 iron oxide pigments — Pigments of which the basic colours are determined by chemical composition. The synthetic oxides are up to 96-99.5% pure. The natural oxides show a wide range of iron oxide content, and vary in colour through red, yellow and brown. See also iron oxide reds, iron oxide yellows.
f pigments *mpl* d'oxyde de fer — **d** Eisenoxydpigmente *npl* — **n** ijzeroxydepigmenten *npl* — **e** pigmentos *mpl* de óxido de hierro — **i** pigmenti *mpl* all'ossido di ferro

7595 iron oxide reds — Ferric oxide pigments known under many names according to their content of relative impure materials.
Formula: Fe_2O_3.
f oxydes *mpl* de fer rouges — **d** Eisenoxydrot *n* — **n** ijzeroxyderood *n* — **e** rojos *mpl* de óxido de hierro — **i** rossi *mpl* all'ossido di ferro

7596 iron oxides — Series of compounds of oxygen and iron occurring naturally or manufactured, used as printing ink pigments. They vary in hue from yellow to brown, to red, to black. Some have special properties that make them useful in magnetic ink printing. TL-value of iron oxide fumes: 15 mg/m³.
f oxydes *mpl* de fer — **d** Eisenoxyden *npl* — **n** ijzeroxyden *npl* — **e** óxidos *mpl* de hierro; óxidos *mpl* férricos; óxidos *mpl* ferrosos — **i** ossidi *mpl* di ferro

7597 iron oxide yellows — Precipitated pigments (hydrated ferric oxide, formula $Fe_2O_3 \cdot H_2O$) of much finer particle size and much greater tinctorial strength than the naturally occurring oxides such as ochre. Very useful for producing cream and buff-coloured tints and with excellent light-fastness and resistance to alkali.
f oxydes *mpl* de fer jaunes — **d** Eisenoxydgelb *n* — **n** ijzeroxydegeel *n* — **e** amarillos *mpl* de óxido de hierro — **i** gialli *mpl* all'ossido di ferro

7598 iron perchloride — See ferric chloride.

7599 iron persulphate — See ferric sulphate.

7600 iron potassium sulphate — See ferric potassium sulphate.

7601 iron protochloride — See ferrous chloride.

7602 iron pyrite — See pyrite.

7603 iron sesquichloride — See ferric chloride.

7604 iron sulphate — See ferric sulphate.

7605 iron sulphate — See ferrous sulphate.

7606 iron to iron — See metal to metal.

7607 iron trichloride — See ferric chloride.

7608 irregularity — See magnetic encoded cheques.

7609 irreversible colloids — See hydrophobic colloids.

7610 ISCC colour system — Standard designation for colours, using words commonly understood, developed by the Inter-Society Colour Council.

7611 island position — See full position.

7612 iso- — Prefix to the name of one compound to denote another isomeric with it, e.g. isoamyl alcohol.

7613 isoamyl acetate — See amyl acetate.

7614 isocyanate resin
f résine *f* d'isocyanate — **d** Isozyanatharz *n* — **n** isocyanaathars *mn* — **e** resina *f* isociánica — **i** resina *f* di isocianato

7615 isoelectric point — Hydrogen ion concentration at which a colloid is unchanged, e.g., will not migrate in an electric field.
f point *m* isoélectrique — **d** isoelektrischer Punkt *m* — **n** isoelektrisch punt *n* — **e** punto *m* isoeléctrico — **i** punto *m* isoelettrico

7616 isography — Anastatic printing process.

7617 isoprene — Colourless, volatile water-insoluble liquid of the terpene class, usually obtained from rubber or from oil of turpentine by pyrolysis, used chiefly in the manufacture of synthetic rubber by polymerization.
Formula: C_5H_8.
f isoprène *m* — **d** Isopren *n* — **n** isopreen *mn* — **e** isopreno *m* — **i** isoprene *m*

7618 isopropanol — See isopropyl alcohol.

7619 isopropyl alcohol; isopropanol; dimethyl carbinol — Colourless, clear, mobile, flammable liquid, soluble in water, alcohol and ether; used as a surface active agent in offset damping solutions. TL-value: 200 ppm or 200 mg/m³.
Formula: $(CH_3)_2CHOH$.
f alcool *m* isopropylique; alcool *m* d'isopropyle — **d** Isopropylalkohol *m*; Äthylkarbinol *m* — **n** isopropylalcohol *m* — **e** alcohol *m* isopropílico; isopropanol *m* — **i** alcool *m* isopropilico; isopropanolo *m*

7620 Italian — The Italian alphabet is the same as English, however omitting k, w, x, y. Pronunciation nearly as French. There are two accents, grave and acute; any vowel may have either. The apostrophe is frequently used for the vowel at the end of a word. When the next word begins with a vowel a space is put between the apostrophe and the next word although there is a growing tendency to close up to the following word as in French. Follow copy in this respect. Divisions of words as in French.

f italien *m/adj.* — **d** Italienische *n*; italienisch *adj.* — **n** Italiaans *n/adj.* — **e** italiano *m/adj.* — **i** italiano *m/adj.*

7621 italic — Of or relating to a type style with characters that slant upward to the right.

f italique — **d** kursiv — **n** cursief; kursief — **e** de bastardilla; en bastardilla; en cursiva — **i** corsivo; italico

7622 italic, accentuated with — See accentuated with italic.

7623 italicize *v*

f mettre en italiques; imprimer en italiques — **d** mit Kursiv drucken; in Kursiv setzen; in Kursiv stellen — **n** cursief zetten — **e** poner en cursiva; poner de bastardilla; poner en letras bastardillas — **i** mettere in corsivo

7624 italicized

f en italiques — **d** kursiv — **n** cursief; in cursief gezet — **e** acursivado — **i** in corsivo

7625 italics; italic types — Style of type in which the letters slope upwards to the right, distinct from roman types, used to emphasize a word or sentence. Invented by Aldus Manutius (1500) and copied from an everyday writing (chancery) of the period. Marked in manuscripts by one line underneath.

f italiques *mpl*; caractères *mpl* italiques — **d** Kursivschrift *f*; Kursiv *f*; Schrägschrift *f* — **n** cursieve letters *fmpl*; cursief *f* — **e** letras *fpl* cursivas; tipos *mpl* cursivos; letras *fpl* bastardillas; bastardillas *fpl*; letras *fpl* itálicas; acursivado *m* — **i** caratteri *mpl* corsivi; corsivo *m*; italico *m obs.*

7626 italic types — See italics.

7627 ivory board — In England ivory board applies to thin bristol used for visiting and business cards.

f carte *f* ivoirine; carton *m* ivoire; carte *f* porcelaine — **d** Elfenbeinkarton *m* — **n** ivoorkarton *n*; opalinekarton *n* — **e** cartulina *f* marfil; cartón *m* marfil; cartón *m* ahuesado — **i** cartoncino *m* avorio; cartoncino *m* opalino

7628 ivory board *US*; **menu board; translucent** — Highly finished, non-bending bristol coated on both sides, used for hotel and restaurant menus and similar work where halftone printing is required. Fourdrinier or cylinder machine product.

f carte *f* à menus — **e** cartón *m* marfil; cartulina *f* marfil — **i** cartone *m* avorio

7629 jacket; book jacket; dust jacket; dust cover; dust wrapper — Detachable paper cover or wrapper of a bound book to protect it from dust and soiling, often printed with attractive design to promote the sale of the book.

f fausse couverture *f*; couverture *f* de protection; chemise *f*; jaquette *f* — **d** Schutzumschlag *m*; Schmutzumschlag *m* — **n** stofomslag *m*; boekomslag *n* — **e** sobrecubierta *f*; camisa *f*; chaqueta *f*; guardapolvo *m*; cubretapas *m*; sobretapas *f* — **i** sovraccopertina *f*; copertina *f* di protezione

7630 jacket; job ticket; job envelope; work ticket — Large envelope form that carries the instructions for preparing and producing the printed job. Dummies, proofs and other information may be enclosed.

f dossier *m* de fabrication — **d** Auftragzettel *m*; Laufzettel *m*; Arbeitstasche *f* — **n** orderzak *m*; commissiezak *m* — **e** boleta *f* de trabajo; bolsa *f* de ruta — **i** busta *f* delle istruzioni; guida *f* lavoro

7631 jacket, couch — See couch jacket.

7632 jack screw; screw jack — Screw-operated jack to lift or to exert pressure.

f cric *m* à vis; vérin *m* à vis — **d** Schraubhebel *m*; Schraubspindel *f* — **n** vijzel *fm* — **e** gato *m* de tornillo — **i** binda *f* a vite

7633 jaconet — Medium cotton cloth to line up book spines.

f jaconas *m* — **d** Jakonett *m* — **n** jaconnet *m* — **e** chaconada *f* — **i** giaconetta *f*

7634 jam nut — See check nut.

7635 Jansenist binding — Book binding of the late 17th century, named after the Roman-Catholic Jansenists, with a sober decoration with untooled outsides. Much attention was paid to the leather and the binding, but the covers show often a rich tooling.

f reliure *f* Janséniste — **d** Jansenist-Einband *m*; Jansenisten-Einband *m* — **n** Jansenist-band *m*; Jansenisten-band *m* — **e** encuadernación *f* jansenista — **i** legatura *f* giansenista

7636 Japan drier — Liquid drier that acts as reducer and siccative, consists of organic lead and manganese salts dissolved in linseed oil with or without rosin or gum, and diluted to proper consistency with turpentine. Used in softer black inks, which dry partly by oxidation, partly by penetration and partly by evaporation.

f siccatif *m* Japon; Vernis *m* Japon — **d** Japan-Trockner *m* — **n** Japan-droger *m* — **e** secante *m* de barniz (para tintas); secante *m* Japón — **i** siccativo *m* giapponese

7637 Japanese colour print — For old Japanese colour prints (1650-1700) special paper size names are used, e.g., for oblong (landscape) sizes (yokoy e): koban 16 x 25 cm, chuban 19 x 25.5 cm, aiban 22 x 33 cm, oban 25.5 x 37.5 cm, obosho 37.5 x 51 cm; for upright sizes (tatey e) also nagay e 13 x 65 cm, hosoy e 15 x 30 cm, o tanzaku 17 x 37.5 cm, kakemonoy e 25.5 x 75 cm. Prints coloured with beni (carmine) are called beniy e, those coloured with tan (cinnabar) tany e; those merely in black sumiy e and prints of which some parts are covered urushi (lacquer) urushiy e. Multicolour prints are named nishikiy e, and those merely tinted in blue aizuriy e and loose prints ichimai. All these Japanese names are used for primitive prints.

f estampe *f* japonaise — **d** japanische Holzschnitt *m* — **n** Japanse prent *fm*; Japanse houtsnede *fm* — **e** estampa *f* japonesa — **i** stampa *f* giapponese

7638 Japanese copying paper — Thin, strong paper made in Japan from long fibres, such as mitsumata and paper mulberry, and used for copying. The fibres are pulped by hand and the sheets made on moulds of bamboo or hair.

f papier *m* de soie du Japon; pelure *f* japonaise — **d** Japan-Seidepapier *n*; Japan-Kopierpapier *n* — **n** Japans zijdepapier *n*; Japans papier *n* — **e** papel *m* japonés para copiar; Japón *m* para copias; papel *m* seda Japón; papel *m* seda japonés — **i** carta *f* da copia Giappone

7639 Japanese gelatine — See agar.

7640 Japanese isinglass — See agar.

7641 Japanese parchment paper
See Japanese vellum.

7642 Japanese tallow — See sumac.

7643 Japanese tissue paper
See Japan paper.

7644 Japanese vellum; Japanese parchment paper — Thick paper made in Japan from long fibres. The formation is very cloudy. Usually has a cream or a natural colour. Finished with a good surface and used for certificates or for other purposes where a tough durable paper is necessary. See also Japan.

f papier *m* vélin Japon; vélin *m* Japon; Japon-vellum *m* — **d** Japan-Papier *n* — **n** Japans papier *n* — **e** papel *m* japonés — **i** carta *f* Giappone vellutata

7645 Japan paper; Japanese tissue paper — Light-weight paper made by hand from kozo fibres and commonly used for mimeograph stencils and lens papers, etchings and engravings.

f papier *m* Japon — **d** Japan-Papier *n* — **n** Japans papier *n* — **e** papel *m* Japón — **i** carta *f* Giappone; carta *f* giapponese

7646 Japan wax — See sumac.

7647 Japon — French imitation of Japanese vellum.

7648 jargon — Debased language.

f jargon *m* — **d** Jargon *m*; Sondersprache *f*; Kauderwelsch *n*; Kaudersprache *f*; Rotwelsch *n* — **n** jargon *n*; koeterwaals *n fam.* — **e** jerga *f*; jerigonza *f*; algarabía *f* — **i** gergo *m*; linguaggio *m* corrotto

7649 javelle water; eau de javel(le) — Aqueous solution of potassium hypochlorite and potassium chloride, occasionally used as bleaching agent in photography. Name derived from Javel, former village in France.
Formula: KOCl.
f eau *f* de Javelle; eau *f* javellisée; solution *f* de hypochlorite de potasse — **d** Eau *m* de Javelle; Javellewasser *n*; Javellesche Lauge *f*; Kaliumhypochloritlösung *f*; Kaliumhypochloritlauge *f* — **n** kalium-hypochlorietoplossing *f* — **e** agua *f* de Javel — **i** acqua *f* di Javel

7650 jaw — Part of the slug composing machine.
f mâchoire *f* de la tête d'élévateur — **d** Backe *f* — **n** eerste elevatorhaak *m* — **e** quijada *f*; mordaza *f*; abrazadera *f* — **i** ganascia *f*

7651 jaw fold; parallel fold; tucker fold — Fold in the web made by a jaw folder, in the same direction as the fold immediately preceding, i.e., parallel to the leading edge of the sheet as it enters the folding machine.
f pli *m* à couteau — **d** Klappenfalz *m*; Tuckerfalz *m* — **n** kleppenvouw *fm* — **e** pliegue *m* a cuchilla — **i** piega *f* fra ganasce

7652 jaw folder *US*; **nip and tuck folder** *GB*; **tuck and grip folder** — Folder that consists of three cylinders between which the web passes to make one or two parallel folds at right angle to the direction of web travel. The lead edge of the web is caught on pins which carry it around the first cylinder. Half way around tucker blades on this cylinder force the centre of the signature-to-be into folding jaws on the second cylinder. At the same time a cut-off knife separates the tail of the signature from the web. The signature is carried around and released by the jaws and the cycle continues. The signature can be passed to the third cylinder in a similar manner to make the second parallel fold.
f plieuse *f* à mâchoires — **d** Klappenfalzapparat *m* — **n** kleppenvouwapparaat *n* — **e** plegadora *f* de quijada — **i** piegatrice *f* a ganasce; gruppo *m* piegatore a ganasce

7653 jef *v* — To gamble with nine one-em quadrats, i.e., to throw quads like dice, using the nick side, appearing uppermost, representing one and the other sides blanks. Very old custom, now almost out of practice. See also Mary.
f jouer aux (six) cadratins — **d** quadräteln — **n** dobbelen met kwadraten — **e** jugar de los cuadratines — **i** giocare coi quadrati; giocare con nove quadrati

7654 jelly — Substance almost always composed of submicroscopic particles, which form a firm network showing marked elasticity. Not to confuse with *gel.
f gelée *f* — **d** Gallerte *f* — **n** gelei *mf* — **e** jalea *f* — **i** gelatina *f*

7655 jelly-like — See gelatinous.

7656 jerks — Violent intermittent pulls of paper through a rotary press or folder usually due to incorrect or worn drives.

f à-coups *mpl* (de tension) — **d** Rucke *mpl*; Stöße *mpl* — **n** rukken *mpl* — **e** golpes *mpl* — **i** strappi *mpl*

7657 jet; tang — Projection at bottom of type formed by cooling metal in opening of mould. It is broken off and a groove is made which forms foot of type.
f jet *m*; rompure *f* — **d** Gußzapfen *m*; Anguß *m* — **n** gietprop *fm* — **e** cabo *m* de fundición; jito *m*; apéndice *m* — **i** attacco *m* di colata; picciolo *m* di colata

7658 jet-black — Very deep, neutral black.
f noir de jais; noir comme du jais — **d** tiefschwarz — **n** diep zwart — **e** negro como el azabache; negro azabache intenso; negro intenso — **i** nero giaietto

7659 jiffy bag — Trade name for a padded paper sack used for shipment of books, etc.
f sac *m* kraft capitonné — **d** Jiffy-Beutel *m*; Jiffy-Tasche *f*; gepolsterter Beutel *m* — **n** gewatteerde zak *m*; gewatteerde envelop(pe) *fm* — **e** sobre *m* acolchado; cubierta *f* de expedición acolchada — **i** busta *f* imbottita jiffy

7660 jig die — Die made from one piece of plywood. Job is laid out and a jig saw is used to cut cutting and score slots. A bridge in these slots then holds die together.
f forme *f* de découpe monté en bois contreplaqué — **d** in Sperrholz montiertes Stanzmesser *n* — **n** plankstansmes *n* — **e** troquel *m* en madera — **i** stampo *m* i legno compensato

7661 jigger — American term for an *ornament.

7662 jiggering — See giggering.

7663 jig saw — Machine carrying a straight slender saw blade, vertically mounted between the ends of two arms, to which a rapid up and down motion is imparted. Used for mortising mounted printing plates and for irregular sawing. See also circular saw, band saw, saw trimmer.
f scie *f* sauteuse; sauteuse *f*; scie *f* à découper; scie *f* anglaise — **d** Dekupiersäge *f*; Laubsäge *f* — **n** decoupeerzaag *fm* — **e** sierra *f* de marquetería; sierra *f* de vaivén; sierra *f* de calar; caladora *f*; sierra *f* perfiladora — **i** sega *f* per trafori

7664 jim dash — *Dash, often three ems long, used within a headline, between the headline and the main body of printed matter, between items in a single column, or between related but different material within a story.

7665 jitter; fortuitous distortion — Short time instability of a signal, especially on a cathode ray tube screen, due to either amplitude, or phase, or both.
f distorsion *f* irrégulière de ligne — **d** unregelmäßige Verzerrung *f* — **n** onregelmatige vervorming *f*; onregelmatige vertekening *f* — **e** distorsión *f* fortuita — **i** distorsione *f* anormale

7666 jobbing compositor — See job compositor.

7667 jobbing department; job department
f atelier *m* des travaux de ville — **d** Akzidenzsetzerei *f*; Akzidenzabteilung *f*; Werksatzabteilung *f*; Akzidenzzimmer *n* — **n** smout-

zetterij *f* — **e** departamento *m* de remendería; departamento *m* de obra suelta *Gu*; departamento *m* de obras *Pe* — **i** reparto *m* lavori avventizi; reparto *m* di composizione per lavori commerciali

7668 jobbing ink — See job ink.

7669 jobbing work — See job printing.

7670 job case — Case to hold types, both upper-case and lower-case letters. See also California job case.

f casse *f* de caractères pour travaux de ville — **d** Akzidenzkasten *m* — **n** smoutkast *fm* — **e** caja *f* de tipos para remiendos

7671 job chase — Small chase without cross bars.

f châssis *m* à bilboquets; ramette *f*; petit châssis *m* — **d** Akzidenzschließrahmen *m* — **n** vormraam *n* zonder kruis — **e** rama *f* para remiendos — **i** telaio *m* senza crociera; telaio *m* per lavori avventizi

7672 job-composing room foreman — Head of typographers working on jobs paid by the hour.

f chef-typo *m*; chef *m* de conscience; prote *m* — **d** Setzereifaktor *m* — **n** chef *m* van de smout-zetterij

7673 job compositor; jobbing compositor — Hand compositor who makes type formes for commercial work.

f compositeur *m* de travaux de ville; conscience *f*; bibelotier *m* — **d** Akzidenzsetzer *m* — **n** smout-zetter *m* — **e** cajista *m* remendista; cajista *m* de remiendos; moldista *m* de remiendos — **i** compositore *m* di lavori avventizi

7674 job department — See jobbing department.

7675 job envelope — See jacket.

7676 job factor — See characteristic.

7677 job ink; jobbing ink — Usually heavy bodied, short, buttery ink, suitable for various types of work on different grades of paper, for use on platen presses.

f encre *f* labeur; encre *f* pour travaux de ville; couleur *f* pour travaux de ville — **d** Akzidenz-druckfarbe *f*; Werkdruckfarbe *f*; Akzidenzfarbe *f* — **n** smoutinkt *m* — **e** tinta *f* para remiendos — **i** inchiostro *m* per lavori avventizi

7678 job lot paper *US*; broken paper *GB* — Paper rejected only because of a mechanical error at the mill, e.g., cutting to the wrong size or imperfect paper, sold at a discount.

f lot *m* désassorti; papier *m* non conforme aux spécifications — **d** Papier *n* zweiter Wahl — **n** tweede keus *f*; retiré *m* — **e** papel *m* quebrado; quebrado *m* de segunda clase — **i** carta *f* di seconda scelta; carta *f* scartata

7679 job printer — Printer doing mostly contract or casual work as opposed to publishing of material for resale.

f bibelotier *m*; imprimeur *m* de travaux de ville — **d** Akzidenzdrucker *m* — **n** handelsdrukkerij *f* — **e** imprenta *f* comercial; impresor *m* remendista; remendista *m* — **i** stamperia *f* di lavori avventizi; stampatore *m* di lavori avventizi

7680 job printing; jobbing work — Printing of commercial work, such as forms, folders, circulars, letterheads, etc.

f bilboquets *mpl*; travaux *mpl* de ville; ouvrages *mpl* de ville — **d** Akzidenzen *fpl*; Akzidenz-arbeiten *fpl*; Akzidenzdruck *m* — **n** handelsdruk-werk *n* — **e** trabajo *m* comercial; trabajo *m* de remiendo; trabajo *m* de remendería; trabajo *m* de obra pacotilla; remendería *f* — **i** stampa *f* di lavori avventizi; stampa *f* di lavori commerciali

7681 job ticket; time ticket — Slip on which a workman records the time spent on a specific job or jobs.

f fiche *f* de travail — **d** Arbeitszettel *m*; Arbeits-stundenzettel *m*; Stundenzettel *m* — **n** werk-briefje *n*; urenbriefje *n* — **e** hoja *f* de trabajo; boleta-horario *f* — **i** cartellino *m* di lavoro

7682 job ticket — See jacket.

7683 jockey roller; compensating roller; compensator; monkey roller; dancing roller; dancer; looping roller; spring loaded idler; tension roller — Roller, usually the first to be traversed by the web, to compensate for any uneven tension as the reel unwinds.

f rouleau *m* compensateur; rouleau *m* de tension — **d** Pendelwalze *f*; Schwingwalze *f*; Tänzer-walze *f*; Ausgleichswalze *f*; Spannwalze *f* — **n** danserrol *fm*; danser *m*; spanrol *fm* — **e** rodillo *m* tensor; rodillo *m* compensador — **i** rullo *m* tenditore; rullo *m* di tensione; rullo *m* compen-satore; rullo *m* ballerino

7684 jog *v* — To inch a press, making it go and stop.

f faire marcher par à-coups — **d** rückweiser Lauf *m* — **n** tornen — **e** mover (intermitente-mente); dar vuelta parcial (a una prensa); correr vuelta a marcha lente — **i** avviare a colpi (una macchina); avviare a piccoli impulsi

7685 jog *v*; knock *v* up — To shake or oscillate gently, i.e., to push together sheets of paper into a compact pile.

f égaliser; dresser — **d** aufstoßen (Papier); gerade-stoßen (Bogen); glattstoßen — **n** gelijkstoten; op-stoten — **e** enderezar (el papel); emparejar; igualar — **i** pareggiare (la carta)

7686 jogger — Arrangement attached to the deliv-ery board of a press to keep sheets in order as they are delivered.

f égalisateur *m* de feuilles — **d** Bogenglattstoß-vorrichtung *f*; Bogengeradleger *m* — **n** vellen-gelijkstoter *m*; gelijkstoter *m* — **e** enderezador *m* de pliegos; igualador *m* de pliegos; empare-jadora *f*; escuadra *f* — **i** pareggiatore *m*

7687 jogger; jogging machine — Vibrating, sloping platform or through where stock is placed on end to even up the edges before feeding, trimming, folding, padding, etc.

f taqueuse *f*; machine *f* à égaliser — **d** Bogenglatt-stoßmaschine *f*; Bogenschüttelmaschine *f*; Bogen-geradstoßmaschine *f*; Glattstoßmaschine *f*; Rüttler *m*; Rütteltisch *m* — **n** gelijkstoot-machine *f*; vellen-gelijkstootmachine *f* — **e** igualadora *f*; aparato *m* de igualar; empare-jador *m*; emparejadora *f*; máquina *f* para enderezar pliegos — **i** pareggiatrice *f*

7688 jogging machine — See jogger.

7689 joint; pasting joint — Adhesion between the ends of two webs to ensure the continuous run of paper through the printing machine.
f collage *m*; endroit *m* collé — **d** Klebelasche *f*; Klebestelle *f*; Lasche *f* — **n** las *fm*; papierlas *fm* — **e** cubrejunta *f* — **i** giunto *m*; giunta *f*; parte *f* giunta; giuntura *f*

7690 joint — Hinge or part of a book cover where it joins the back on the inside.
f mors *m* — **d** Falz *m* — **n** kneep *fm* — **e** cajo *m* — **i** morso *m*

7691 joint — Line or point of contact between joined negatives and plates.
f joint *m*; ligne *f* de contact — **d** Blendlinie *f*; Stoßlinie *f*; Berührungslinie *f* — **n** contactlijn *fm*; aansluitlijn *fm* — **e** línea *f* de contacto — **i** linea *f* di contatto; punto *m* di contatto

7692 joint author — See coauthor.

7693 Jordan machine; Jordan refiner — Refining machine consisting of a conical rotor and housing between which the fibre slurry is passed, to shorten the fibres and improve sheet formation.
f raffineuse *f* Jordan; Jordan *m*; raffineur *m* conique; moulin *m* conique; moulin *m* à cônes — **d** Kegelstoffmühle *f*; Stoffmühle *f*; Jordanmühle *f*; Kegelmühle *f* — **n** kegelstofmolen *m*; stofmolen *m*; kegelmolen *m*; Jordanmolen *m*; Jordan *m*; refiner *m* — **e** batidora *f* Jordán; pila *f* Jordán; refinadora *f* Jordán; molino *m* cónico — **i** raffinatore *m* Jordan; raffinatore *m* conico

7694 Jordan refiner — See Jordan machine.

7695 Josef paper — See filter paper.

7696 joule — Basic unit of energy. Abbrev. J (no point). In the SI: 1 J_{int} = 1 international Joule = 1 $V^2_{int} \cdot s / \Omega_{int} \approx 1.00019$ J.
f joule *m* — **d** Joule *m* — **n** joule *m* — **e** julio *m* — **i** joule *m*

7697 journal — Part of shaft or axle of rotating machine which rests on bearings.
f tourillon *m* — **d** Zapfen *m*; Lagerzapfen *m* — **n** tap *m* — **e** gorrón *m*; muñón *m*; pivote *m*; muñequilla *f* de eje — **i** perno *m* dell'albero; parte *f* sopportata dell'albero

7698 journal; magazine; review; periodical — Technical or trade magazine.
f journal *m*; revue *f*; recueil *m*; périodique *m*; publication *f* périodique — **d** Journal *n*; Zeitschrift *f*; Magazin *n*; Revue *f* — **n** tijdschrift *n* — **e** revista *f*; periódico *m* — **i** rivista *f*; periodico *m*

7699 journal bearing — See axle bearing.

7700 journalist
f journaliste *mf* — **d** Journalist *m*; Zeitungsschreiber *m* — **n** journalist *m*; journaliste *f* — **e** periodista *m*; diarista *m*; gacetero *m*; publicista *m* — **i** giornalista *m*; gazzettiere *m*; pubblicista *m*

7701 journeyman — Compositor or pressman who has completed his apprenticeship. Derived from French journée, a day, day's work or day's pay.
f ouvrier *m* qualifié — **d** Geselle *m* — **n** gezel *m* — **e** oficial *m* (cajista); oficial *m* (tipógrafo) — **i** operaio *m* qualificato

7702 Judean pitch — See bitumen.

7703 jumbled type — See pie.

7704 jumbo roll — See mill roll.

7705 jump; control transfer; transfer of control — Departure from the normal sequence of obeying instructions in a computer.
f saut *m* — **d** Sprung *m* — **n** sprong *m* — **e** salto *m*; bifurcación *f* — **i** salto *m*

7706 jump line — Line of type identifying the page on or from which a newspaper story is continued.
f ligne *f* de tourne — **d** Verweiszeile *f*; Hinweiszeile *f* — **n** verwijzingsregel *m* — **e** línea *f* de continuación; línea *f* de traslado — **i** linea *f* identificante

7707 junction point — See node.

7708 justification — Allocation of spaces between words to fit a line into a pre-determined measure or width, to align at the right- and left-hand margins.
f justification *f* — **d** Ausschließen *n* — **n** uitvullen *n*; uitvulling *f* — **e** justificación *f*; justificado *m* — **i** giustificazione *f*

7709 justification key
f touche *f* de justification — **d** Ausschlußtaste *f*; Ausschließtaste *f* — **n** uitvultoets *m* — **e** tecla *f* de espaciado — **i** tasto *m* di giustificazione

7710 justification lever — Part of the slug composing machine.
f levier *m* de justification — **d** Ausschließhebel *m* — **n** uitvularm *m* — **e** palanca *f* de justificación — **i** leva *f* di giustificazione

7711 justification pointer — See justifying scale pointer.

7712 justification wedge — Part of the Monotype machine.
f coin *m* de justification — **d** Ausschlußkeil *m* — **n** uitvulwig *fm* — **e** cuña *f* justificadora; cuña *f* de justificación — **i** cuneo *m* di giustificazione

7713 justification zone — Area at the end of a line in which the composing equipment can automatically justify the line. In line casters it corresponds to the amount the line can be expanded by the spacebands between words.

7714 justify *v* a line — To space out a line of type to a specified width or measure so that both the left- and right-hand margins of the printed matter are aligned, by adjusting the spacing between words, or between words and characters (letter spacing) so as to fill the measure.
f justifier une ligne — **d** eine Zeile ausschließen — **n** een regel uitvullen — **e** justificar una línea — **i** giustificare una linea; giustificare le righe; chiudere le righe

7715 justifying scale — Part of the Monotype keyboard.
f tambour *m* de justification; échelle *f* justificative — **d** Settrommel *f*; Ausschlußtrommel *f* — **n** uitvultrommel *fm* — **e** tambor *m* de justificación — **i** tamburo *m* di giustificazione

7716 justifying scale pinion

f pignon *m* du tambour de justification — d Set-trommelantrieb *m* — n aandrijfmechanisme *n* van de uitvultrommel — e piñón *m* del bombo de justificación — i pignone *m* del tamburo di giustificazione

7717 justifying scale pointer; justification pointer — Part of the Monotype keyboard.

f indicateur *m* du tambour justification — d Ausschlußzeiger *m* — n trommel-aanwijsnaald *fm* — e puntero *m* de espaciado (de la scala indicadora) — i lancetta *f* del tamburo

7718 justify *v* the stick — See set *v* a stick.

7719 just published

f vient de paraître — d soeben erschienen; neu herausgekommen — n zojuist verschenen — e acaba de aparecer; acabo de publicarse; novedad *f* — i appena pubblicato; appena uscito; viene de essere pubblicato; novità *f* editoriale

7720 jute — Indian bast fibre for the manufacture of coarse sacking and bags and gunny. Old gunny and sacks are used as raw material in paper-making. Now also a furnish consisting substantially of paper stock reclaimed from waste papers.

f jute *m* — d Jute *f* — n jute *fm* — e yute *m* — i iuta *f*

7721 jute liner — See test linerboard.

7722 jute paper — Paper made from jute fibres or burlap waste or with various proportions of kraft or sulphite pulp, used extensively for envelopes, folders, tag stock, wrappers, cover stock, bristols, pattern papers, hydrated lime and cement bags, flour sacks, etc. The basis weight is 20-300 pounds (24 x 36 - 500).

f papier *m* jute — d Jutepapier *n* — n jute-papier *n* — e papel *m* de yute — i carte *f* di iuta

7723 juxtaposition, in — Literally side by side, or being placed in nearness or contiguity.

f juxtaposé; en juxtaposition — d nebeneinanderstehend; gegenüberstehend — n ten opzichte van elkaar geplaatst — e yuxtapuesto — i giustapposto

7724 kakemono — Picture on paper or silk attached to a roller, used as a wall hanging. From Japanese kake = hang, mono = thing. About 30.5 cm wide and up to 89 cm long. See also Japanese colour print.
f kakémono *m* — **d** Kakemono *m* — **n** kakemono *m* — **e** kakemono *m* — **i** kakemono *m*
7725 kalium — See potassium.
7726 kalogen — See rodinal.
7727 K & N test — Test indicating the rate of printing-ink varnish absorption at the surface of paper or paperboard. Name originates from K & N laboratories, Melrose Park (Ill.).
f essai *m* K & N — **d** K & N-Versuch *m* — **n** K & N-proef *fm* — **e** prueba *f* de K & N — **i** prova *f* K & N; saggio *m* K & N
7728 kaolin — See China clay.
7729 kappa number of pulp — The permanganate number as a measure of bleachability.
f indice *m* kappa — **d** Kappazahl *f* — **n** kappagetal *n* — **e** índice *m* kappa; coeficiente *m* kappa — **i** numero *m* kappa
7730 Kassel brown — See vandyke brown.
7731 Kassel earth — See vandyke brown.
7732 kauri; kauri gum; kauri resin — Fossil copal resin derived from the kauri pine of New Zealand, used in oleoresinous varnishes and lacquers.
f kauri *m* — **d** Kauri *n*; Kauriharz *n* — **n** kauri *mn*; kaurigom *mn* — **e** kauri *m*; goma *f* de kauri — **i** kauri *f*; resina *f* di kauri
7733 kauri-butanol value; kb-value — Measure of the solvency action of a hydrocarbon solvent. It is the number of millilitres of the thinner required to cause cloudiness when added to 20 g of a solution of kauri gum in butyl alcohol. The solution is prepared in the proportion of 100 g of kauri in 500 g of butyl alcohol. The values range from 20, which is a poor solvent, to a high of 105, which is an excellent solvent. Less favoured test than the aniline point test.
f valeur *f* kauri-butanol; indice *m* de kb — **d** Kauri-Butanolwert *m*; KB-Wert *m* — **n** kauri-butanolwaarde *f*; kb-waarde *f*; kb-getal *n* — **e** valor *m* kauri butanol; índice *m* de kauri butílico; índice *m* de kb — **i** valore *m* di kauri-butanolo; indice *m* di kauri-butanolo
7734 kauri gum — See kauri.
7735 kauri resin — See kauri.
7736 kb-value
See kauri-butanol value.
7737 keep *v* down — To use capitals sparingly. See also keep *v* up.
f épargner les majuscules — **d** in Gemeinen setzen; in Kleinbuchstaben setzen; klein drucken — **n** uit onderkast zetten — **e** componer con minúsculas — **i** risparmiare le maiuscole
7738 keep *v* in — To set matter closely spaced so that it makes as few lines as possible.
f serrer la composition — **d** eng anschließen — **n**

nauw zetten; nauw spatiëren — **e** apretar (la composición); componer con espacios medianos — **i** stringere
7739 keep *v* in alignment
f aligner; s'aligner — **d** Ausrichtung einhalten; Zeile halten — **n** lijn houden — **e** alinear — **i** allineare
7740 keeping quality — Property of photographic plates, films, chemicals, solutions, etc., to remain in good condition for some time.
f qualité *f* de conservation — **d** Lagerfähigkeit *f*; Haltbarkeit *f* — **n** houdbaarheid *f* — **e** estabilidad *f*; conservabilidad *f*; solidez *f* — **i** qualità *f* di conservazione; conservabilità *f*
7741 keep *v* out — To set matter widely spaced so that it makes as many lines as possible, a vile practice.
f blanchir à l'excés — **d** Satz weit halten — **n** wijd zetten — **e** componer con espacios gruesos; extender la composición — **i** spaziare al massimo; comporre con spazi molto larghi
7742 keep standing — Order not to distribute type pending possible reprinting.
f gardez debout; à conserver — **d** stehen lassen — **n** laten staan; bewaren — **e** ¡conserve!; conservado; conservada; guardar la composición; conservar — **i** conservare in piedi la composizione
7743 keep *v* up — To use capitals liberally. See also keep *v* down.
f prodiguer les majuscules — **d** in Großbuchstaben setzen — **n** kapitaal zetten; uit bovenkast zetten — **e** emplear abundancia de mayúsculas — **i** comporre con molte maiuscole
7744 Kelvin; degree Kelvin — Unit of thermodynamic temperature defined by the Conférence Générale des Poids et Mesures. O °C = 273.16 K; 100 °C = 373.16 K.
f degré *m* Kelvin — **d** Grad *m* Kelvin — **n** graad *m* Kelvin — **e** grado *m* Kelvin — **i** grado *m* Kelvin
7745 Kemart process — Patented process for highlighting or dropouts and silhouetting. A fluorescent white pigment is applied to dropout areas, or a fluorescent coated illustrator's board is used for silhouetting. When copy is photographed, special flash lights cause the pigment to fluoresce and burn-out the coated areas of copy.
f procédé *m* Kemart — **d** Kemart-Verfahren *n* — **n** Kemart-procédé *n* — **e** procedimiento *m* Kemart; método *m* Kemart — **i** procedimento *m* Kemart
7746 Kent cap — A British size for wrapping paper. See app. no. 3.
7747 kern — Part of a movable type projecting beyond the body, as the tail of an italic f. See also kerned letter.
f crénage *m*; saillie *f* — **d** Überhang *m* (am unterschnittenen Buchstaben) — **n** overhang *m* — **e** parte *f* sobresaliente — **i** crenatura *f*

7748 kerned letter; kern letter — Type with part of the face projecting beyond the body. Kerns are easily broken, except in slug composing machine matter, where they are cast as part of the slug and cannot break off. See also kern.
f lettre *f* crénée; crénée *f*; lettre *f* débordante — **d** unterschnittener Buchstabe *m*; überhängender Buchstabe *m* — **n** overhangende letter *fm*; créne *fm* — **e** tipo *m* sobresaliente; tipo *m* saledizo; letra *f* desbordante; hombro *m* saliente; letra *f* sobresaliente; letra *f* de perfil saliente — **i** lettera *f* crenata

7749 kern letter — See kerned letter.

7750 kerosene — See kerosine.

7751 kerosine; kerosene — Mixture of hydrocarbons derived from petroleum, with a boiling range of approx. 150-300 °C, used for cleaning and washing formes and inking rollers on printing presses, in papermaking sometimes as a foam killer. The spelling kerosine is officially adopted by the ASTM and the British Institute of Petroleum, in preference to kerosene, to bring it into line with gasoline. Paraffin or paraffin oil are undesirable as applied to kerosine owing to the confusion with paraffin wax, paraffin hydrocarbons and medicinal paraffin.
f kérosine *f*; kérosène *m*; pétrole *m* (lampant) — **d** Kerosin *n*; Petroleum *n*; Leuchtpetroleum *n*; Leuchtöl *n* — **n** kerosine *f*; petroleum *m*; lampolie *fm* — **e** queroseno *m*; kerosén *m*; kerosena *f*; keroseno *m*; petróleo *m* de lámpara — **i** cherosene *m*; petrolio *m* raffinato

7752 ketohexamethylene — See cyclohexanone.

7753 ketone — Organic compound with a carbonyl group attached to two carbon atoms, e.g., acetone, methyl ethyl ketone, methyl isobutyl ketone. Specific for the vinyl type inks and widely used in gravure inks.
f cétone *f* — **d** Keton *n* — **n** keton *n* — **e** cetona *f* — **i** chetone *m*

7754 ketopropane — See acetone.

7755 kettle — Utensil for heating or boiling liquids, and for melting composition for covering ink rollers.
f chaudière *f* — **d** Kessel *m*; Schmelzkessel *m* — **n** smeltketel *m*; ketel *m*; smeltpot *m* — **e** caldera *f*; calderón *m* (large kettle); calderico *m* (small kettle); calderilla *f* (minor kettle) — **i** caldaia *f*

7756 kettle stitch; chain stitch; catch stitch — Stitch at the head and tail of sewed books.
f chaînette *f* — **d** Fitzbund *m* — **n** kettingsteek *m* — **e** cadeneta *f*; punto *m* de cadeneta — **i** nodo *m* di catenella; punto *m* di catenella

7757 key; key plate — Layout, film, block, forme, plate, or stone which acts as the guide for position and register for other colours or for standard signatures in the same colour.
f mise *f* en place; cliché-clé *m*; cliché *m* de base — **d** Konturplatte *f*; Paßplatte *f*; Paßform *f*; Klatsch *m* — **n** blauwkopie *f*; basisplaat *fm*; contourplaat *fm*; klats *m* — **e** grabado *m* de base; grabado *m* de registro — **i** lastra *f* chiave; lastra *f* di riferimento

7758 key — In photography the emphasis on lighter or darker tones in a negative or print. High key indicates prevalence of light tones, low key prevalence of darker tones.
f effet *m* (dans les blancs); effet *m* (dans les ombres) — **d** Helligkeitsumfang *m* — **n** helderheidsbereik *n*; helderheidsomvang *m* — **e** efecto *m* en los blancos y en las sombras — **i** effetto *m* nelle ombre

7759 key — Small instrument to close and open metal quoins.
f clef *f*; clé *f* — **d** Schlüssel *m*; Formschlüssel *m* — **n** sleutel *m*; vormsleutel *m*; kooisleutel *m* — **e** llave *f* — **i** chiave *f*

7760 key — Wedge-shaped piece ot steel.
f clavette *f* — **d** Keil *m* — **n** spie *fm* — **e** chaveta *f* — **i** chiavetta *f*

7761 key; key button — Component part of a keyboard.
f touche *f* — **d** Taste *f*; Tastenknopf *m* — **n** toets *m* — **e** tecla *f*; botón *m* (de la tecla) — **i** tasto *m*

7762 key — One or more digits used to identify an item or record.
f clef *f*; clé *f* — **d** Schlüssel *m*; Schlüsselzahl *f*; Kennzahl *f* — **n** sleutel *m* — **e** clave *f* — **i** chiave *f*

7763 key *v* — To identify positions of art or copy in a dummy by symbols, usually letters.
f indiquer les couleurs — **d** Farben anheben — **n** kleur aangeven (met letters) — **e** indicar los colores (con letras) — **i** indicare il colore (con numeri o simboli)

7764 key bar frame — Part of the Monotype keyboard.
f intermédiaire *m* — **d** Zwischenrahmen *m* — **n** toetsstangraam *n* — **e** intermediario *m* — **i** telaio *m* intermedio

7765 keyboard — In typesetting or typewriting, instrument or device operated by the fingers (often with a "touch" system so that the fingers can access the appropriate keys without visual assistance), to create words by determining the sequence in which the keys are stuck. It may present some visual feed back, in the form of hard or soft copy or it may be entirely blind, presenting no feed back at all.
f clavier *m* — **d** Taster *m*; Tastatur *f*; Tastapparat *m*; Klaviatur *f* — **n** toetsenbord *n* — **e** teclado *m*; aparato *m* teclado — **i** tastiera *f*

7766 keyboard cam — Cam connected with the key rod of the slug composing machine, which makes one complete revolution every time a key button is depressed and releases a matrix.
f came *f* de clavier — **d** Klaviaturexzenter *m* — **n** toetsenbord-excentriek *n* — **e** cama *f* del teclado — **i** eccentrico *m* di tastiera

7767 keyboard cam rubber roll — Part of the slug composing machine.
f rouleau *m* (en caoutchouc) du clavier — **d**

Gummiwalze *f* der Tastatur — **n** rubberrol *fm* van het toetsenbord — **e** rodillo *m* de goma del teclado — **i** rullo *m* caucciù della tastiera

7768 keyboarding convention — Specific method of arranging a keyboard or utilizing combinations of keystrokes to achieve desired objectives. Implies some "tailor-made" improvisations to meet user requirements.

7769 keyboard key rod — Part of the slug composing machine.
f tringle *f* d'échappement — **d** Klaviaturauslösestab *m* — **n** uitlaatstang *fm* — **e** uña *f* de escape — **i** bacchetta *f* di sgancio

7770 keyboard layout — See also Dvorak keyboard, azerty layout, qwert layout.
f plan *m* du clavier — **d** Tastaturplan *m* — **n** toetsenbord-indeling *f* — **e** plano *m* de teclado — **i** schema *m* della tastiera; disposizione *f* dei tasti

7771 keyboard operator — See also machine compositor.
f claviste *mf* — **d** Taster *m*; Tasterin *f* — **n** toetsenbordbewerker *m*; tikker *m*; toetsenbordbewerkster *f*; tikster *f* — **e** teclista *mf*; monotipista *mf*; mecanotipista *f* — **i** tastierista *mf*; compositore *m* monotipista; operatore *m* alla tastiera; dattilografo *m* monotipista

7772 keyboard operator — See operator.
7773 keyboard paper — See Monotype paper.
7774 keyboard punch — See key punch.
7775 keyboard shorthand — Method of reducing key strokes on input data by allocating a single code, requiring one key stroke, to frequently used words and their associated spaces.

7776 keyboard spaceband cam — Part of the slug composing machine.
f excentrique *m* des espaces-bandes — **d** Spatienkeilexzenter *m*; Keilexzenter *m* — **n** spatie-excentriekje *n* — **e** excéntrica *f* del escape de los espacios-cuna — **i** eccentrico *m* dello spazio mobile

7777 key button — See key.
7778 key-button bank — Part of the Monotype keyboard.
f dessus *m* de clavier — **d** Tastenbrett *n* — **n** toetsenbank *fm* — **e** superficie *f* del teclado — **i** mensola *f* della tastiera

7779 key drawing; key line drawing — Outline drawing containing only guide lines or those necessary for separation of colours on printing plates.
f tracé *m* (de contours) — **d** Umriß-Zeichnung *f*; Konturzeichnung *f* — **n** contourtekening *f*; tekening *f* van de omtrek — **e** perfil *m*; contornos *mpl* para separación de colores — **i** tracciato *m* di guida

7780 key groove — See key way.
7781 key line drawing — See key drawing.
7782 keylining — See mechanical colour separations.
7783 key plate — See key.
7784 key punch; keyboard punch — Keyboard actuated device that punches holes in a card to represent data.
f perforateur *m* — **d** Locher *m*; Tastaturlocher *m* — **n** ponsapparaat *n*; ponsmachine *f* — **e** perforador *m* — **i** perforatore *m*

7785 key stroke — Pushing down of a key of a keyboard.
f frappe *f* — **d** Tastenanschlag *m*; Anschlag *m* — **n** toetsaanslag *m*; aanslag *m* — **e** pulsación *f* — **i** battuta *f*

7786 key way; key groove — Recess in a shaft or boss to accommodate a flat piece of metal connecting a wheel to a shaft.
f rainure *f* de clavette; rainure *f* de clavetage; logement *m* de clef; logement *m* de clavette — **d** Keilnute *f* — **n** spiebaan *fm* — **e** muesca *f* de chaveta; ranura *f* para chaveta; chavetera *f*; cajera *f* — **i** scanalatura *f* per chiavetta; cava *f* per chiavetta

7787 kick copy — Copy displaced by the kicker on a rotary press, to indicate the completion of a count.
f exemplaire *m* décalé — **d** Abzählexemplar *n*; zur Seite verschobenes Exemplar *n* — **n** terzijde geschoven exemplaar *n*; aftelexemplaar *n* — **e** ejemplar *m* echado — **i** copia *f* estratta

7788 kicker — Mechanism on a rotary press folder which indicates a counted series of product by displacing at the delivery one copy askew (skew kick), sideways or with parallel projection. See also quire spacing.
f sortie *f* en paquets — **d** Abzähler *m*; Paketausleger *m*; Abteilvorrichtung *f* — **n** aftelinrichting *f* — **e** sacador *m* en postetas — **i** mecanismo *m* d'espulsione a intermittenza

7789 kicker cam — Cam which operates the *kicker.
f came *f* d'espacement — **d** Steuerkurve *f* für Abzähler; Steuernocken *m* für Abzähler; Verschieber-Exzenter *m*; Verschieber-Exzentrik *f* — **n** excenter *m* voor de telvinger; telvinger-excentriek *n* — **e** excéntrica *f* de desplazamiento — **i** camma *f* dell'espulsore

7790 kicker roller — Timed roller to remove cut cartons from the continuous web and to synchronize the web to the stripper.
f rouleau *m* arracheur — **d** Ausstoßwalze *f*; Ausstoßer *m* — **n** uitstootrol *fm* — **e** rodillo *m* de expulsión — **i** rullo *m* espulsore

7791 kieselguhr — See diatomaceous earth.
7792 kill *v* — To indicate in a proof the text or illustrations which are not to be reprinted.
f enlevez; supprimez — **d** streichen; löschen; tilgen; Deleatur — **n** wegnemen; laten vervallen — **e** suprimir -- **i** eliminare

7793 kill *v* — To distribute (or to remelt) type matter when no longer required.
f distribuer — **d** ablegen — **n** distribueren; wegmaken; weggooien — **e** distribuir; deshacer — **i** scomporre; distribuire

7794 kill button — See delete button.
7795 kilocycle — One thousand cycles per second. Abbrev. kc, however, use kHz for kc/s. See also

hertz.

f kilohertz *m* — **d** Kilohertz *n* — **n** kilohertz *m* — **e** kilohertz *m* — **i** kilohertz *m*

7796 kilogram — See kilogramme.

7797 kilogramme *GB*; **kilogram** *US* — Metric unit of mass. Abbrev. kg. See app. no. 7.

f kilogramme *m* — **d** Kilogramm *m* — **n** kilogram *nm*; kilo *nm* — **e** kilogramo *m*; quilogramo *m*; kilo *m*; quilo *m* — **i** chilogramma *f*; chilogrammo *m*; chilo *m*

7798 kilogramme-force — Metric unit of force. Abbrev. kgf. In the SI: 1 kgf = 9.80665 N. Thus: 1 kgf/cm² = 98.0665 × 10³ Pa.

f kilogramme-force *f* — **d** Kilopond *n* — **n** kilogramkracht *fm* — **e** kilolibra *f* — **i** kilogrammoforza *f*

7799 kilowatt — Unit of power. Abbrev. kW. One kilowatt = 1,000 Watts.

f kilowatt *m*; kilovolt-ampère *m* — **d** Kilowatt *n* — **n** kilowatt *m* — **e** kilovatio *m* — **i** chilowatt *m*

7800 kilowatt-hour — The energy consumption involved by the steady use of the electrical power of 1 kilowatt for 1 hour. Abbrev. kWh. In the SI: 1 kWh = 3.6 × 10⁶ J.

f kilowatt-heure *f* — **d** Kilowattstunde *f* — **n** kilowattuur *n* — **e** kilovatiohora *f* — **i** chilowattora *f*

7801 kinematic viscosity; cinematic viscosity — Viscosity of a material divided by its density.

f viscosité *f* cinématique — **d** kinematische Viskosität *f* — **n** kinematische viscositeit *f* — **e** viscosidad *f* cinemática — **i** viscosità *f* cinematica

7802 kinetic energy — Energy associated with a body when in motion with respect to another; equal to one-half the mass times the square of the relative velocity.

f énergie *f* cinétique — **d** kinetische Energie *f*; Arbeitsvermögen *n* — **n** kinetische energie *f* — **e** energía *f* cinética — **i** energia *f* cinetica

7803 King James' Bible — See authorized version.

7804 King James' version — See authorized version.

7805 kiss impression; kiss printing; kiss pressure — Printing when there is no continuation of pressure or dwell in the motion of a press, of the paper against the printing surface while the impression is being made, usually involving hard packing. In offset, a contact squeeze between blanket and plate, or blanket and paper of 0.7-0.1 mm (0.003-0.004 in) for transfer of impressions.

f pression *f* minimale; impression *f* par effleurage; impression *f* au minimum de pression — **d** Kußdruck *m*; ganz leichte Druckpressung *f*; minimaler Druck *m*; geringe Druckspannung *m* — **n** uiterst lichte druk *m*; minimale druk *m*; lichtste drukspanning *f* — **e** presión *f* mínima; presión *f* de contacto; presión *f* suave; presión *f* de leve contacto; impresión *f* con grueso mínimo — **i** stampa *f* con poca pressione; stampa *f* con pressione minime

7806 kissing the binder's daughter — Joke played on apprentices. A new apprentice would be told that it is a recognized custom to kiss the binder's daughter. When he entered the binding department he would receive a flop in the face with a well-charged brush of paste.

7807 kiss pressure — See kiss impression.

7808 kiss printing — See kiss impression.

7809 kiss register; butt register — Registering of two or more colours, meeting without overlapping.

7810 kit — Adapter. A nest of interfitting series of wooden or metal frames used as a transparency holder on process cameras.

f adaptateur *m* — **d** Satz *m* auswechselbarer ineinanderpassender Diahalterrahmen — **n** diahouder *m* met vaste maten — **e** adaptador *m* — **i** adattatore *m*

7811 kite — See former.

7812 Klischograph — Electronic engraving machine to produce halftone printing plates, devised by Rudolph Hell, Kiel (Germany).

7813 knee — See sliding bar.

7814 knife; cut-off knife — Blade(s) on an automatic reel change that sever the old web after the join has been made.

f couteau *m* — **d** Abschlagmesser *n*; Trennmesser *n* — **n** kapmes *n*; mes *n* — **e** cuchilla *f* — **i** coltello *m*

7815 knife — See doctor blade.

7816 knife barker — See barker.

7817 knife blade — Sharp tool-steel blade used on paper cutters and trimmers.

f lame *f* de couteau — **d** Messer *n* — **n** mes *n*; blad *n* van het mes; mesblad *n* — **e** hoja *f* de(l) cuchillo — **i** lama *f* di coltello

7818 knife block — Part of a typesetting machine.

f couteau-bloc *m* — **d** Messerblock *m* — **n** mesblok *n* — **e** bloque *m* de cuchillas — **i** blocco *m* dei coltelli

7819 knife box *coll.* — Knife sections assembled in the knife case on a cutting cylinder of a rotary press.

f porte-lame *m* — **d** Messerbalken *m* — **n** meshouder *m* — **e** barra *f* portacuchilla — **i** portacoltelli *m*; portalama *m*

7820 knife cheeks; side cheeks; woods — Spring-loaded fibre or synthetic substance on either side of the knife which grip the web while it is being severed.

f réglettes *fpl* de coupe; bois *mpl* de scie; listeaux *mpl* du cylindre coupeur — **d** Messerleisten *mpl*; Schneidmesserschienen *fpl*; Fiberleisten *mpl*; Kunststoffleisten *mpl*; Holzleisten *mpl* — **n** snijlatten *fmpl*; mesbuffers *mpl* — **e** listones *mpl* de cuchilla — **i** guance *fpl*

7821 knife coating — Coating process in which a doctor, knife, or a straight edge is used to spread and control the amount of coating on the paper.

f couchage *m* à la lame; couchage *m* à la racle — **d** Schaberstreichverfahren *n*; Messerstreichverfahren *n*; Rakelstreichverfahren *n* — **n** rakelstrijkprocédé *n* — **e** procedimiento *m* de estucado a cuchilla — **i** patinatura *f* a lama raschiante

7822 knife drum

f cylindre *m* à lames — **d** Holländerwalze *f* — **n** messenwals *fm* van de hollander — **e** cilindro *m* de la pila holandesa — **i** cilindro *m* a coltelli

7823 knife edge — Sharpened edge of a knife.
f arête *f* du couteau; fil *m* du couteau; tranchante *m* de la lame — **d** Schneide *f* — **n** scherp *n* — **e** corte *m*; tajo *m*; filo *m* — **i** filo *m* della lama

7824 knife folder; blade folder — Machine folder in which a knife edge is pressed down over the fold to be and forces it in the nip of two rollers which complete the fold and transport the sheet of paper at the same time.
f plieuse *f* à lame — **d** Schwertfalzmaschine *f*; Messerfalzmaschine *f* — **n** messenvouwmachine *f* — **e** máquina *f* dobladora con cuchillo — **i** macchina *f* piegatrice a coltello

7825 knife grinder — See knife grinding machine.

7826 knife grinding machine; knife grinder
f machine *f* à affûter les couteaux; affûteuse *f*; machine *f* à aiguiser — **d** Messerschleifmaschine *f* — **n** messenslijpmachine *f* — **e** amoladora *f* de cuchillas; máquina *f* afiladora de cuchillas; afiladora *f* — **i** macchina *f* affilalama; macchina *f* per affilare le lame; affilatrice *f* per lame

7827 knife section — One of the series of hard steel serrated perforating blades which comprise the knife.
f élément *m* de scie — **d** Messersektion *f* — **n** messtuk *n* — **e** elemento *m* sustentador de la hoja cortante — **i** sezione *f* del coltello

7828 knife wiper — Part of the typesetting machine which removes all shavings from the knives after a slug is trimmed.
f essuyeur *m* de couteaux — **d** Messerputzfahne *f*; Messerputzer *m* — **n** meswisser *m* — **e** limpiacuchillas *m*; limpiador *m* de (las) cuchillas; banderita *f* — **i** pulisci-lama *m*

7829 knock *v* **up** — See jog *v*.

7830 knot screen; strainer; screen — Filtering device in pulp and paper mills to remove knots and large dirty particles from the fibrous suspension.
f trieur *m* de nœuds; classeur *m*; capteur *m* de nœuds — **d** Knotenfänger *m*; Astfänger *m* — **n** knopenvanger *m*; kwastenvanger *m* — **e** recogenudos *m* — **i** separanodi *m*

7831 knurled screw — Screw with small ridges on the edge or surface; milled.
f vis *f* moletée — **d** Rändelschraube *f*; Kordelschraube *f* — **n** gekartelde schroef *m* — **e** tornillo *m* moleteado — **i** vite *f* a filettatura tonda

7832 Kodachrome — Trade name for a colour transparency, manufactured by the Eastman Kodak Comp.

7833 Kodacolor — Trade name for a coloured negative material with hue couplers of complementary colours, such as green for red areas of copy. Also colour prints with inherent dye couplers in emulsion, which are developed out after exposure to Kodacolor negative.

7834 kodzu — See mulberry tree.

7835 kollergang; edge runner; edge mill — German type of beater to grind, crush and macerate pulp, waste paper or broke. It consists of two heavy stone rolls (runners) which roll around on the bottom stone (bed stone) of an annular trough or pan. Its refining action is due to the crushing and abrasion of the fibres between the rolls and the bed stone. A small model made of metal is sometimes used in laboratories for pulp processing tests.
f meuleton *m* broyeur; raffineuse *f* à meuletons — **d** Kollergang *m*; Kollermühle *f*; Trottmühle *f* — **n** kollergang *m*; kollermolen *m* — **e** molino *m* de muelas; molino *m* de rulos; molino *m* de ruedas verticales — **i** molazza *f*

7836 Kostinsky effect — Form of the border and fringe effects in which an increase in the separation of two small nearby images is produced by a shift in the geometric centre of the images which is caused by the increased concentration of the development products between the two images leading to an asymmetrical growth of the image. Discovered by the Russian Kostinsky (1906).
f effet *m* Kostinsky — **d** Kostinsky-Effekt *m* — **n** Kostinsky-effect *n* — **e** efecto *m* de Kostinsky — **i** effetto *m* di Kostinsky

7837 kozo — See mulberry tree.

7838 kraft liner — See kraft linerboard.

7839 kraft linerboard; kraft liner — Linerboard made on a cylinder or Fourdrinier machine from a furnish containing 85% or more virgin kraft woodpulp. Nominal grade weights range from 26-90 lb/M sq ft and thicknesses from 9-30 points.
f couverture *f* kraft; kraftliner *m* — **d** Kraftliner *m* — **n** kraftliner *m* — **e** papel *m* kraft para cartón corrugado — **i** copertina *f* in carta kraft

7840 kraft paper — Paper of high mechanical strength made entirely from unbleached, sulphate, softwood chemical pulp. In some European countries the terms sulphate kraft paper and sodium kraft paper are used interchangeably.
f papier *m* kraft — **d** Kraftpapier *n*; Natronpapier *n* — **n** kraft *n*; kraftpapier *n*; natronkraft *n*; natronkraftpapier *n* — **e** papel *m* kraft; papel *m* a la sosa — **i** carta *f* kraft

7841 kraft pulp — Softwood sulphate pulp used especially for the manufacture of kraft paper.
f pâte *f* kraft — **d** Kraftzellstoff *m*; Natronzellstoff *m* — **n** kraftcelstof *fm*; natroncelstof *fm* — **e** pasta *f* kraft; pasta *f* al sulfato; pulpa *f* kraft; pulpa *f* al sulfato — **i** pasta *f* kraft

7842 Kremnitz white — See flake white.

7843 Kromekote — Trade mark for a cast coated paper which is given a very high gloss with a highly polished, chromium-faced cylinder, used for high quality printing jobs and for reproduction type proofs.

7844 kutch — Package of parchment interspersed with gold sheets for the initial steps of gold beating.

7845 Labarraque's solution — See sodium hypochlorite.

7846 label — Printed slip of paper for affixing to something for identification or description.

f étiquette *f* — **d** Etikett *n*; Aufklebezettel *m*; Anklebezettel *m*; Zettel *m* — **n** etiket *n*; wikkel *m* — **e** etiqueta *f*; marbete *m*; rótulo *m* — **i** etichetta *f*

7847 label — See panel.

7848 label *v* — To paste labels on finished goods or packages.

f étiqueter — **d** etikettieren; bezetteln — **n** etiketteren; etiketten opplakken; wikkels opplakken — **e** etiquetar — **i** etichettare

7849 labeller — See labelling machine.

7850 labelling machine; labeller — Machine for affixing labels to boxes, bottles, etc.

f machine *f* à étiqueter; étiqueteuse *f* — **d** Etikettiermaschine *f* — **n** etiketteermachine *f* — **e** máquina *f* etiquetadora; máquina *f* de etiquetar; etiquetadora *f* — **i** macchina *f* per etichettare; macchina *f* etichettatrice; etichettatrice *f*

7851 label paper; litho label paper; litho paper; lithograph label paper; lithograph paper — Mf or Ef paper supercalendered or coated on one side and especially sized for gumming, printing, and/or lithographing, used for all kinds of labels.

f papier *m* pour étiquettes; papier *m* d'étiquettes — **d** Etikettenpapier *n*; Konservenetikettenpapier *n* — **n** billetlitho *n*; lithopapier *n*; etikettenpapier *n*; papier *n* voor buswikkels — **e** papel *m* para etiquetas; papel *m* para marbetes; papel *m* para rótulos — **i** carta *f* per etichette

7852 label printing and punching machine

f machine *f* à imprimer et découper les étiquettes — **d** Etikettendruck- und Stanzmaschine *f* — **n** etiketten-automaat *m*; etikettendruk- en stansmachine *f* — **e** máquina *f* de imprimir y troquelar etiquetas — **i** macchina *f* per stampare e fustellare etichette

7853 label varnish

f vernis *m* pour étiquettes — **d** Etikettenlack *m* — **n** etikettenvernis *mn*; etikettenlak *mn* — **e** barniz *m* para etiquetas; barniz *m* para etiquetar — **i** vernice *f* per etichette

7854 laboratory apparatus — See laboratory device.

7855 laboratory balance

f balance *f* de laboratoire — **d** Laboratoriumwaage *f* — **n** laboratorium-balans *fm* — **e** balanza *f* de laboratorio — **i** bilancia *f* da laboratorio

7856 laboratory device; laboratory instrument; laboratory apparatus

f appareil *m* de laboratoire — **d** Laborgerät *n* — **n** laboratorium-instrument *n*; laboratorium-toestel *n*; laboratorium-apparaat *n* — **e** aparato *m* de laboratorio; dispositivo *m* de laboratorio — **i** apparecchio *m* da laboratorio

7857 laboratory instrument — See laboratory device.

7858 laboratory sample

f échantillon *m* pour laboratoire — **d** Laboratoriumprobe *f* — **n** laboratoriummonster *n* — **e** muestra *f* (de ensayo) de laboratorio — **i** campione *m* da laboratorio

7859 laboratory test

f essai *m* de laboratoire — **d** Laboratoriumsversuch *n*; Laboratoriumsprüfung *f*; Laborversuch *m*; Laborprüfung *f* — **n** laboratoriumproef *fm*; laboratorium-onderzoek *n* — **e** ensayo *m* en laboratorio; ensayo *m* de laboratorio; prueba *f* de laboratorio — **i** prova *f* di laboratorio; saggio *m* di laboratorio

7860 labourer — See workman.

7861 labour-saving rules; case rules — Rules cut into regular sizes, ready for use.

f filets *mpl* systématiques — **d** Stücklinien *fpl*; systematisch geschnittene Linien *fpl* — **n** op maat gesneden lijnen *fmpl* — **e** plecas *fpl* sistemáticas; filetes *mpl* economizadores — **i** filetti *mpl* sistematici

7862 lace *v* **in** — To draw the cords on which the book is sewed through the holes previously pierced in the book boards, to flat them down and glue them to the boards. See also frayed slip.

f passer les nerfs en carton — **d** durchziehen; einschnüren — **n** touwen doorhalen — **e** entrar en tapas; meter en tapas — **i** fissare delle corde

7863 lacquer — Solution in an organic solvent of a natural or synthetic resin, a cellulose ester, such as cellulose nitrate or cellulose acetate, or a cellulose ether, such as methyl or benzyl cellulose, together with modifying agents, such as plasticizers, resins, waxes and pigments. The solvent evaporates after application of the lacquer, leaving the dissolved material as a shiny, more or less continuous protective film on the surface of the material. Used for coating paper to give them functional qualities such as decreased water-vapour transmission rates, heat-sealing properties, grease resistance, gloss and decorative effects.

f vernis-laque *m*; laque *f* — **d** Lack *m*; Papierlack *m*; Lackfirnis *m* — **n** lak *mn*; vernis *mn* — **e** laca *f*; barniz *m* — **i** lacca *f*; vernice *f*

7864 lacquer, base
See deep-etch lacquer.

7865 lacquer, deep-etch — See deep-etch lacquer.

7866 lacquering; varnishing

f laquage *m*; vernissage *m* — **d** Lackierung *f*; Lackieren *n* — **n** vernissen *n* — **e** barnizado *m*; lacado *m* — **i** verniciatura *f*; laccatura *f*

7867 lacquering machine — See varnishing machine.

7868 lacquer ink — Pigmented lacquer used in the gravure process for high gloss publication work or in the packaging field for speciality wrapper such as foil, plastic, cellophane, etc.

f encre *f* transparente; laque *f* — **d** Lackfarbe *f* —

n lakinkt *m*; lak *m* — **i** inchiostro *m* lacca

7869 lactic acid; milk acid — Yellow or colourless, thick liquid, miscible with water, alcohol, glycerol, soluble in ether, insoluble in chloroform, petroleum ether, carbon disulphide. Latin: acidum lacticum.
Formula: $CH_3CHOH \cdot CO_2H$.
f acide *m* lactique — **d** Milchsäure *f* — **n** melkzuur *n* — **e** ácido *m* láctico — **i** acido *m* lattico

7870 ladies' cards — A size of cut card. See app. no. 3.

7871 ladle — Deep-bowled, long-handled spoon especially for dipping up and conveying liquids. See casting ladle, skimming ladle.
f pochon *m*; louche *f* à main — **d** Löffel *m* — **n** lepel *m* — **e** cuchara *f*; cucharón *m* — **i** cucchiaio *m*

7872 lag — See hysteresis.

7873 laid antique paper — See laid paper.

7874 laid dandy roll; laid roll — Dandy roll made with wires parallel to the axis of the roll and attached to a frame and kept in position by chain wires evenly spread and encircling the circumference of the roll.
f rouleau *m* vergeur — **d** Wasserlinienzylinder *m* — **n** vergeerrol *fm*; waterlijnenrol *fm* — **e** rodillo *m* verjurador; rodillo *m* de verjuras — **i** rullo *m* per vergatura

7875 laid lines — Closely spaced light lines in laid papers, produced by the laid wires of the mould or dandy roll, usually running across the grain of the paper. Spiral laid paper has lines parallel with the grain or in the machine direction. See also chain lines.
f vergeures *fpl*; verjures *fpl* — **d** Egoutteurrippung *f*; Rippung *f* — **n** vergure *f* — **e** puntizones *mpl* — **i** vergatura *f*

7876 laid paper; laid antique paper — Originally, hand-made paper bearing the impression of the laid mould wire; now produced with a laid dandy roll on the paper machine.
f papier *m* vergé — **d** geripptes Papier *n*; Papier *n* mit Wasserlinien — **n** gevergeerd papier *n* — **e** papel *m* acanillado; papel *m* de línea marca; papel *m* verjurado; papel *m* estriado; papel *m* vergueteado; papel *m* vergé (galicismo) — **i** carta *f* vergata

7877 laid roll — See laid dandy roll.

7878 laid writing paper — Writing paper with a laid mark.
f papier *m* d'écriture vergé — **d** geripptes Schreibpapier *n*; Schreibpapier *n* mit Wasserlinien — **n** gevergeerd schrijfpapier *n* — **e** papel *m* de escribir verjurado — **i** carta *f* da scrivere vergata

7879 Lainer effect — Photographic effect where the image appears sooner and the emulsion layer is lesser inclined to fogging by the treatment of the layer with a solution of potassium iodide (or an admixture of potassium bromide to the developer).
f effet *m* Lainer — **d** Lainer-Effekt *m* — **n** Lainer-effect *n* — **e** efecto *m* Lainer — **i** effetto *m* Lainer

7880 lake — Pigment consisting of an organic, soluble colouring matter combined with an inorganic base, characterized by a bright colour and a pronounced translucency, used in the manufacturing of printing inks. There are three types: 1. The original, a precipitation of a soluble colour on an inorganic base, such as alumina, barium sulphate, to render it insoluble in water. Obsolete. 2. The modern and more extensive type made by precipitating from various coal-tar colours by means of a metallic salt, or other suitable reagent, upon a base either previously prepared or coincidentally formed. 3. A combination of both types.
f laque *f* — **d** Farblack *m*; verlackter Farbstoff *m*; Farbstofflack *m*; Substratfarbe *f*; verlacktes Pigment *n* — **n** lak *m*; substraatpigment *n* — **e** laca *f*; colorante *m* lacado — **i** lacca *f*

7881 laketine — Transparent reducer (magnesia in linseed oil) for extending letterpress or lithographic inks.
f blanc *m* transparent — **d** Farbtonabschwächer *m* — **n** transparantwit *n*

7882 lambert — Unit of luminance equal to $1/\pi$ candle per cm^2, and, therefore equal to the uniform luminance of a perfectly diffusing surface emitting or reflecting light at a rate of 1 lumen per cm^2. Abbrev. L (no point). In the SI:
one lambert = $(1/\pi)$ $cd/cm^2 \approx 3.18310 \times 10^3 cd/m^2$.
f lambert *m* — **d** Lambert *m* — **n** lambert *m* — **e** lambert *m* — **i** lambert *m*

7883 lambs vellum
f parchemin *m* d'agneau — **d** Lammpergament *n* — **n** perkament *n* van lamshuid — **e** pergamino *m* de cordero — **i** pergamena *f* d'agnello

7884 laminar flow — Orderly flow; flow in parallel layers, sometimes called telescopic flow. Opposite of *turbulent flow.
f écoulement *m* laminaire — **d** Laminarströmung *f*; laminare Strömung *f*; laminares Fließen *n* — **n** laminaire stroming *f* — **e** flujo *m* laminar — **i** corrente *f* laminare

7885 laminate *v* — To cement a transparent plastic film to printed sheets for protection and to enhance gloss. Also to unite layers of materials with adhesives.
f laminer — **d** kaschieren — **n** lamineren — **e** tallar en láminas; cortar en láminas; tallar en hojas; cortar en hojas; recubrir con película plástica; contrapegar — **i** accoppiare; laminare

7886 laminated — Composed of two or more thin layers cemented together.
f laminé — **d** kaschiert; beschichtet — **n** gelamineerd — **e** laminado; de láminas — **i** accoppiato

7887 laminated backing — Blocking lumber consisting of several layers of wood glued together with alternate crossing of wood grain, used for mounting blocks.
f contreplaqué *m* pour montage de clichés; bloc *m* de montage en contreplaqué — **d** Sperrholz *n*; Klischeeholz *n* — **n** multiplex *n* voor clichévoeten — **e** madera *f* contrachapeada; madera *f* laminada

m Lainer

– i legno *m* laminato; legno *m* compensato

7888 laminated paper — Paper built up to a desired thickness or a given desired surface by joining two or more webs or sheets.

f papier *m* doublé; papier *m* contrecollé; papier *m* laminé — **d** Schichtpapier *n*; kaschiertes Papier *n* — **n** gelamineerd papier *n* — **e** papel *m* laminado; papel *m* revestido — **i** carta *f* rivestita; carta *f* accoppiata

7889 laminated wood — See plywood.

7890 laminated-wood block — Wood block made from thin layers of wood glued together, grain of the layers being crossed alternatively.

f base *f* en bois contreplaqué — **d** Sperrholz-Klischeefuß *m* — **n** multiplex voet *m* — **e** bloque *m* de madera laminada; bloque *m* de madera contrachapeada; bloque *m* de madera terciada — **i** zoccolo *m* di legno laminato; zoccolo *m* di legno compensato

7891 laminating machine; laminator

f machine *f* à laminer — **d** Kaschiermaschine *f*; Folienkaschiermaschine *f* — **n** lamineermachine *f* — **e** máquina *f* de cubrir; máquina *f* de forrar — **i** macchina *f* accoppiatrice

7892 laminator — See laminating machine.

7893 laminé paper — Paper or board that has been passed through a machine suitable for giving it uniform thickness. No such term is used in English.

i carta *f* calibrata

7894 lamp-black — Carbon printing ink pigment prepared by the incomplete combustion of tar, vegetable oils, petroleum, or asphaltic materials. Used to achieve a dull, smooth-finish, black ink. See also carbon black.

f noir *m* de lampe; noir *m* de fumée — **d** Lampenruß *m*; Lampenschwarz *n*; Kienruß *m* — **n** lamproet *n*; lampzwart *n* — **e** negro *m* de humo; negro *m* de lámpara; hollín *m* de lámpara — **i** nerofumo *m* di lampada

7895 lampoon — Harsh satire, usually directed against an individual.

f pasquinade; libelle *m*; satire *f*; brocard *m* — **d** Satire *f* — **n** satire *fm*; hekelend geschrift *n* — **e** sátira *f*; libelo *m* — **i** satira *f*

7896 lamp shade — Shade, usually translucent or opaque, to shield the glare of a light-source in a lamp or to direct the light over a particular area. See also eye shade.

f abat-jour *m* — **d** Lampenschirm *m* — **n** lampekap *fm* — **e** pantalla *f* de lámpara — **i** paralume *m*

7897 land — See bridge.

7898 landscape; oblong size — Page, booklet, or book which is wider than it is deep.

f format *m* à l'italienne; format *m* en largeur; format *m* oblong; format *m* en travers; format *m* atlas — **d** Querformat *n*; Langformat *n*; breites Format *n* — **n** oblong formaat *n*; liggend formaat *n* — **e** formato *m* apaisado; formato *m* oblongo; tamaño *m* oblongo — **i** formato *m* oblungo

7899 language — Set of representations, conventions and rules to convey information. See also machine language, natural language, object language, programming language, source language.

f langage *m* — **d** Sprache *f* — **n** taal *fm* — **e** lenguaje *m* — **i** linguaggio *m*

7900 language processor — See processor.

7901 lanolin; lanum — Purified grease of sheep's wool. Yellowish-white, or nearly white mass, soluble in ether, chloroform, insoluble in water.

f lanoléine *f*; lanoline *f*; graisse *f* de laine purifiée — **d** Lanolin *n*; gesäubertes Wollfett *n* — **n** lanoline *fm*; gereinigd wolvet *n* — **e** lanolina *f*; grasa *f* de lana purificada — **i** lanolina *f*; grasso *m* di lana purificato

7902 lanum — See lanolin.

7903 lap — Slight extension of areas of printing surfaces to assure register of colour plates.

f engrainement *m* des couleurs pour repérage — **d** Übergriff *m*; Farbüberstand *m* — **n** overvul *m* — **e** recubrimiento *m*; sobreposición *f* de colores — **i** estensione *f* delle zone stampanti

7904 lap — In poster printing, the margin allowed for one sheet of a poster to cover the edge of an adjacent sheet.

7905 lap *v* — To grind optics before polishing.

f roder — **d** läppen — **n** leppen — **e** lapidar — **i** lappare

7906 lapis lazuli — Deep blue mineral compound mainly of lazurite (sodium aluminium silicate and sulphide, $Na_5Al_3Si_3O_{12}S_3$) with small quantities of other minerals, formerly used as a pigment. Now superseded by the artificial product.

f lapis-lazuli *m* — **d** Lapis lazuli *m* — **n** lapis lazuli *m* — **e** lapislázuli *m* — **i** lapislazzuli *m*; lapislazzolo *m*

7907 lapped; soft fold — Said of large sheets of paper: when folded by hand, no crease results when opened up after handling or shipment. See also interlapped.

f plié en portefeuille — **d** umgeklappt; zugeklappt; umgeschlagen; zugeschlagen; nicht-scharfer Falz *m*; nicht-scharf gebrochener Falz *m* — **n** (eenmaal) toegeslagen; (eenmaal) dichtgeslagen; (eenmaal) omgeslagen — **e** gualdrapeado — **i** sovrapposto; piegato; piegatura *f* non marcata

7908 larch gum — See Venetian turpentine.

7909 larch turpentine — See Venetian turpentine.

7910 large — A size of cut card and of board. See app. no. 3.

7911 large atlas — A size of drawing paper and of board. See app. no. 3.

7912 large court 8vo — A size of cut card. See app. no. 3.

7913 large double loaf — A size of wrapping paper. See app. no. 3.

7914 large foolscap — A size of printing paper. See app. no. 3.

7915 large half royal — A size of boards. See app. no. 3.

7916 large imperial — A British size for boards

and wrapping paper. See app. no. 3.

7917 large middle — A size of boards. See app. no. 3.

7918 large official postcard — A size of card. See app. no. 3.

7919 large-paper — Special copy of a book with wide margins. Also termed *edition de luxe.

f édition *f* grand-format; édition *f* sur grands papiers; édition *f* de luxe — **d** Luxusausgabe *f*; Prachtausgabe *f*; Luxuseinband *m* — **n** luxe uitgave *f*; uitgave *f* in prachtband; prachtuitgave *f* — **e** edición *f* de lujo — **i** edizione *f* di lusso; rilegatura *f* di lusso

7920 large post — A British standard size of writing and printing papers. See app. no. 3.

7921 large post 8vo — A size of writing paper. See app. no. 3.

7922 large post 4to — A size of writing paper. See app. no. 3.

7923 large royal — A British size of paper. See app. no. 3.

7924 large-scale integrated circuitry — Computer or logic device constructed of minute elements to perform computer-line functions by means of miniaturized hardware. Makes use of metal oxide silicon wafers treated with impurities which govern the direction and characteristics of the electronic circuits. The scope of the logic is extensive rather than specific or particularized.

7925 large single — A size of wrapping paper. See app. no. 3.

7926 large size for driers — A size of board. See app. no. 3.

7927 large whole royal — A size of board. See app. no. 3.

7928 laser — Acronym for "light amplification by stimulated emission of radiation". Light of extraordinary high energy density, in a very narrow beam, and capable of providing a continuous output that can be modulated. The potential of laser extends into all facets of graphic arts, including composition.

f laser *m* — **d** Laser *m* — **n** laser *m* — **e** láser *m* — **i** laser *m*

7929 latency — See waiting time.

7930 latent heat — Heat required to change a solid to a liquid or a liquid to gas. Also called *heat of fusion for solids and *heat of vaporisation for liquids. When a liquid solidifies or a gas changes to a liquid, latent heat is released.

f chaleur *f* latente — **d** latente Wärme *f*; bleibende Wärme *f*; gebundene Wärme *f* — **n** latente warmte *f* — **e** calor *m* latente; calor *m* desarrollado — **i** calore *m* latente

7931 latent image — Invisible image produced by exposure of a photographic surface and rendered perceptible by development.

f image *f* latente — **d** latentes Bild *n* — **n** latent beeld *n* — **e** imagen *f* latente — **i** immagine *f* latente

7932 lateral camber — Defect of aluminium foil. Deviation of edge from parallelism with a straight line between the two ends of the sheet of the edge. See also transverse camber.

7933 lateral displacement

f déplacement *m* latéral — **d** seitliche Verschiebung *f* — **n** zijdelingse verplaatsing *f* — **e** desplazamiento *m* lateral — **i** spostamento *m* laterale

7934 lateral hard dot — Gravure process in which the shadow dots are square and sides of the dots are lined up parallel to each other. The dots get proportionally smaller as they represent middle tones and highlights. Different from letterpress, offset, or the original Dultgen process where shadow dots meet at the corners instead of the sides.

7935 lateral movement

f mouvement *m* latéral — **d** Seitwärtsbewegung *f* — **n** zijwaartse beweging *f*; zijdelingse beweging *f* — **e** movimiento *m* lateral; movimiento *m* de lado — **i** movimento *m* laterale

7936 lateral reversal — Turning of a photographic image, so that words appear reversed in reading direction (mirror image), achieved either with optical reversing devices, by flopping the negative for stripping, or by placement of the image in a transparency holder during photography. Frequently used to get emulsion-to-emulsion in contact printing of lithographic press plates. Not to be confused with transpose.

f retournement *m* (de l'image) — **d** Kontern *n* — **n** omkeren *n*; rechts/links maken *n* — **e** inversión *f* lateral — **i** rovesciamento *m* dell'immagine

7937 late watch — See shift.

7938 lathe — Machine for working metals, wood and other substances by causing the material to turn against a fixed tool, used for evening and smoothing rotogravure cylinders, press rolls, calender rolls, etc.

f tour *m* — **d** Drehbank *m* — **n** draaibank *fm* — **e** torneadora *f*; torno *m* — **i** tornio *m*

7939 lather — See foam.

7940 Latin alphabet — Same as English without w. Accents and ligatures falling into disuse; most scholars do not differentiate the letter j from i, and many use u for v; V usually stands for both U and V.

f alphabet *m* latin — **d** lateinisches Alphabet *n* — **n** Latijnse alfabet *n* — **e** alfabeto *m* latino — **i** alfabeto *m* latino

7941 latitude of exposure — See exposure latitude.

7942 lavender flower oil — See lavender oil.

7943 lavender oil; lavender flower oil; essence of lavender — Essential oil. Colourless, yellowish or greenish-yellow. Characteristic odour. One volume oil can be dissolved in 4 volumes 70% alcohol. Latin: oleum lavandulae.

f essence *f* de lavande — **d** Lavendelöl *n* — **n** lavendelolie *fm* — **e** aceite *m* de lavanda; aceite *m* de espliego; esencia *f* de lavanda — **i** olio *m* essenziale di lavanda

7944 law calf; fair calf — Bark tanned calf leather

not artificially coloured, used for the binding of law books. See also leather.
f veau *m* foncé — **d** feines Kalbleder *n* — **n** fijn kalfsleer *n* — **e** piel *f* (superior) de becerro — **i** vitello *m* fino

7945 law cap — A size of writing paper. See app. no. 3.

7946 lay — Arrangement of printing subjects on a sheet of paper, a flat, or a press-plate, for printing, cutting, trimming, folding, backing, etc.
f position *f* — **d** Satzanordnung *f*; Stand *m* — **n** stand *m*; indeling *f* — **e** colocación *f* — **i** disposizione *f*

7947 lay — See lay guide.

7948 lay *v* **down** — The act of transferring the carbon tissue (image) to the cylinder. Dry laydown is the method of transferring the carbon tissue to the cylinder in which the tissue is dry. Water is applied during the laydown process between the gelatine side of the tissue and the copper face of the cylinder or plate. Pressure of a rubber roller and softening of the gelatine by water cause the carbon tissue to adhere to the copper surface. See also carbon tissue transfer.

7949 lay edge — Edge of a sheet of paper being fed into a printing press.
f côté *m* de la marge (de la feuille) — **d** Anlagekante *f* (des Bogens) — **n** aanlegkant *m* van het vel papier — **e** margen *mf* de la guía de pecho — **i** lato *m* di marginazione; bordo *m* lato squadra

7950 layer; film; coating
f couche *f*; pellicule *f* — **d** Schicht *f*; Film *m* — **n** laag *fm*; film *m* — **e** capa *f*; película *f*; film *m* — **i** strato *m*; pellicola *f*

7951 layer — See coating.

7952 layered; stratified
f stratifié — **d** schichtenförmig — **n** gelaagd; in lagen — **e** estratificado — **i** stratificato

7953 layer-on — See feeder.

7954 layer-on girl — See feeder.

7955 lay guide; lay — Side guide on a cylinder press, to hold the sheet of paper in printing position.
f taquet *m* (de marge) — **d** Anlegemarke *f* — **n** aanleg *m*; zij-aanleg *m* — **e** guía *f* de lengüeta; guía *f* de resorte — **i** squadra *f* di marginatura; squadra *f* laterale

7956 laying machine — See carbon tissue transfer machine.

7957 laying-on board — See feedboard.

7958 laying-on side — See operating side.

7959 laying press — See lying press.

7960 lay of the case — Order in which type is laid in a case.
f modèle *m* de casse; mise *f* en casse — **d** Kasteneinteilung *f*; Kastenschema *n* — **n** kastindeling *f* — **e** disposición *f* de la distribución de las cajas; diagrama *m* de la subdivisión de las cajas — **i** schema *m* di cassa per composizione; modello *m* di cassa per composizione

7961 layout — Preliminary sketch or arrangement showing the size, position and colours of the illustrations and text matter in advertisements and other printed matter and also including special instructions to compositors, platemakers, printers and binders.
f croquis *m*; tracé *m*; maquette *f* (selon le degré de précision, le croquis est le plus simple, puis vient le tracé et enfin la maquette) — **d** Layout *n*; Entwurf *m*; Gestaltung *f*; Aufmachung *f*; Anordnungsskizze *f*; Skizze *f*; Aufmachungsmuster *n* — **n** layout *m*; indeling *f*; opmaak *m* — **e** boceto *m*; diseño *m*; esbozo *m*; bosquejo *m*; croquis *m* esquemático; croquis *m* detallado; dibujo *m* acabado; dibujo *m* definitivo; confección *f*; maqueta *f*; diagrama *m* — **i** progetto *m* (grafico); bozzetto *m*; layout *m*

7962 layout man; typotect *US* — Man who designs the arrangements of type and cuts for works to be printed.
f maquettiste *m* — **d** Drucksachengestalter *m*; Entwerfer *m*; Layouter *m* — **n** layout-man *m* — **e** abocetador *m*; bocetista *m*; proyectista *m* gráfico; confeccionador *m*; diagramador *m*; maquetista *m* — **i** progettista *m* grafico

7963 layout of folding positions
f maquette *f* de pliage; schéma *m* de pliage; polichinelle *f* *fam.* — **d** Falzschema *n*; Falzmuster *n* — **n** vouwschema *n* — **e** maqueta *f* de plegada — **i** schema *m* delle piegature

7964 lay *v* **type; case** *v* **letters** — To distribute new type in the case.
f mettre en casse; dresser une casse — **d** Schrift einlegen — **n** een kast inleggen — **e** echar fundición; distribuir tipo nuevo en las cajas — **i** porre i caratteri nuovi in cassa

7965 lazy fingers — Web holders used on box making machines to keep the web flat against the cutting plate while the cutter is rising. Formed from doctor blade stock and bent to shape, then usually screwed into the large panels of a die. If not used and the web lifts because of a sudden rush of air from the side, carton can break under the die.
f lames-ressorts *fpl* — **d** Blattfedern *fpl* — **n** springveren *fmpl*; bladveren *fmpl* — **e** resortes *mpl* de lámina; resortes *mpl* de hoja; resortes *mpl* de planos — **i** molle *fpl*

7966 Lazy Susan — Revolving tray placed on a table for serving sheets.

7967 LC number — See Library of Congress Catalog Card Number.

7968 lead — Heavy, malleable, ductile, grey, soft metal, soluble in dilute nitric acid, insoluble in water but slowly soluble in water containing weak acid. Symbol: Pb. Latin: plumbum. TL-value: 0.15 mg/m^3.
f plomb *m* — **d** Blei *n* — **n** lood *n* — **e** plomo *m* — **i** piombo *m*

7969 lead — Thin strip of lead less than type high, used to separate type lines and as a general spacing material. Thicknesses: 1-point, 1.5-point ("thin"), 2-point and 3-point ("thick"). Supplied in 60 cm lengths (24 in), or cut to size as wanted. A

unit of spacing beyond these limits is called a
*clump (lead cast in 6-points or 12-points) and a
*reglet (strips of wood 6-points to 18-points).
f interligne *f* — **d** Durchschußlinie *f*; Durchschuß *m*; Zeilendurchschuß *m* — **n** interlinie *f* —
e interlínea *f*; lujo *m Pe* — **i** interlinea *f*
7970 lead — Boxmaker's term of measurement
used when rule or panel is moved in die. A shim
is 0.2 mm (0.008 in) thick.
7971 lead *v*; **lead** *v* **out** — To separate the lines of
type by interposing leads, or to set on a larger
body so as to increase the space between lines.
See also branch *v* out.
f interligner; jeter du blanc — **d** die Zwischenräume vergrößern; durchschießen; aussperren —
n interliniëren; doorschieten met interlinies;
uitdrijven — **e** interlinear; colocar interlíneas; entrerrenglonar; entremeter; regletear; espaciar — **i**
interlineare; spaziare
7972 lead acetate; acetate of lead; sugar of lead
obs. — White crystals, soluble in water, soluble in
alcohol, freely soluble in glycerol. Poisonous. Used
in the manufacturing of varnishes, lead driers,
and in the manufacture of chrome pigments.
Latin: acetas plumbicus; plumbum aceticum.
Formula: $Pb(H_3C_2O_2)_2 \cdot 3H_2O$.
f acétate *m* (neutre) de plomb; acétate *m* de
plomb cristallisé; diacétate *m* de plomb; sucre *m*
de plomb; sel *m* de saturne — **d** Bleiazetat *n*;
essigsaures Blei *n*; Bleizucker *m* — **n** lood-
acetaat *n*; azijnzuur lood *n*; loodsuiker *m* — **e**
acetato *m* de plomo — **i** acetato *m* di piombo
7973 lead alloy — In printing, the lead, tin and
antimony alloys used in type casting and platemaking.
f alliage *m* de plomb — **d** Bleilegierung *f* — **n**
loodalliage *fn*; loodlegering *f*; loodsamenstelling *f*
— **e** aleación *f* de plomo — **i** lega *f* di piombo
**7974 lead and leave edges; leading and trailing
edges** — The first and last printed edges respectively of a copy or print.
f blanc *m* de tête et de queue d'impression — **d**
Druckanfang *m* und Druckende *n* — **n** op- en
afloopkanten *mpl* — **e** orillas *fpl* de entrada y
salida — **i** bianchi *mpl* di testa e di coda
7975 lead base; lead mount
f pied *m* de plomb — **d** Bleifuß *m*; Bleiunterlage *f*
— **n** loden voet *m* — **e** pie *m* de plomo; base *f* de
plomo — **i** base *f* di piombo
7976 lead carbonate — White powdery crystals,
soluble in acids, insoluble in water and alcohol.
Poisonous.
Formula: $PbCO_3$.
f carbonate *m* de plomb — **d** Bleikarbonat *n* — **n**
loodcarbonaat *n* — **e** carbonato *m* de plomo — **i**
carbonato *m* di piombo
**7977 lead chromate; neutral lead chromate;
chrome yellow; Paris yellow; Leipzig yellow; permanent yellow** — Yellow crystals, soluble in acids,
insoluble in water, used as a pigment.
Formula: $PbCrO_4$.
f chromate *m* (neutre) de plomb; jaune *m* de

chrome; jaune *m* de Paris; jaune *m* de Leipzig —
d (neutrales) Bleichromat *n*; Chromgelb *n*; Pariser
Gelb *n* — **n** chromaatgeel *n*; loodchromaat *n* — **e**
cromato *m* de plomo; amarillo *m* de cromo — **i**
cromato *m* di piombo; giallo *m* di cromo
7978 lead cutter; slug clipper — Small machine
with short strong knife and a movable gauge, for
cutting leads or rules and slugs.
f coupoir *m* à lignes; coupoir *m* à interlignes — **d**
Zeilenschneideapparat *m* — **n** lijnen-afhak-
machine *f*; regelhakmachine *f* — **e** cortadora *f* de
líneas; cortadora *f* de interlíneas; cortadora *f* de
regletas; cortalingotes *m*; cortante *m Ch* — **i**
taglialinee *m*; cesoia *f* per interlinee
7979 lead drier — Chemical combinations of lead
with various organic acids, used as driers for printing ink.
f siccatif *m* au plomb — **d** Bleitrockner *m*; Bleisikkativ *n* — **n** looddroger *m*; loodsiccatief *n* — **e**
secante *m* de plomo — **i** essiccante *m* al piombo
7980 leaded matter — Type matter with lines separated by leads, or cast on a larger body to increase the space between them.
f composition *f* interlignée — **d** durchschoßener
Satz *m* — **n** geïnterlinieerd zetsel *n* — **e**
composición *f* interlineada; composición *f*
regleteada — **i** composizione *f* interlineata
**7981 lead engraving; engraving in lead; cut-out
in lead**
f gravure *f* sur plomb — **d** Bleischnitt *m* — **n** lood-
gravure *fm* — **e** grabado *m* en plomo — **i** incisione *f* in piombo
7982 leader — See leading article.
7983 leaderette — Short editorial paragraph,
printed in the same type as the leaders in a newspaper.
7984 leaders; dot leaders — Periods or dots cast
to different multiples of the body size, used to
trace lines (close leaders) or in making tables or
indexes (*open leaders).
f points *mpl* de conduite; points *mpl* cadratinés;
pointillés *mpl* de conduite; signes *mpl*
conducteurs; gros points *mpl* — **d** Ausführpunkte
mpl; Ausführungspunkte *mpl*; Führungspunkte
mpl; Leitzeichen *npl* — **n** blokpunten *fmpl* — **e**
puntos *mpl* conductores; guiones *mpl*
conductores; puntos *mpl* de conducción; guiones
mpl de conducción; puntos *mpl* suspensivos;
puntos *mpl* corridos — **i** punti *mpl* di conduzione;
puntini *mpl* di guida
7985 leader writer
f auteur *m* des articles de fond; éditorialiste *m*;
ténor *m fam.* — **d** Verfasser *m* von Leitartikeln;
Leitartikler *m* — **n** schrijver *m* van het hoofd-
artikel; hoofdartikelschrijver *m* — **e** escritor *m*
del artículo de fondo; editorialista *m* — **i** scrittore
m dell'articolo di fondo
7986 lead fumes
f vapeurs *fpl* de plomb — **d** Bleidämpfe *mpl* — **n**
looddampen *mpl* — **e** vapores *mpl* del plomo — **i**
vapori *mpl* di piombo
7987 lead glance — See galena.

7988 leading and trailing edges — See lead and leave edges.

7989 leading article; leader — The most important news story in a newspaper or magazine.
f article *m* de fond; article *m* de tête; article *m* principal — **d** Hauptartikel *m*; Spitzenmeldung *f* — **n** hoofdartikel *n* — **e** artículo *m* de fondo; fondo *m* — **i** articolo *m* di fondo

7990 leading edge; front edge — Edge of the printing plate nearest to the first point of the impression.
f bord *m* avant du cliché — **d** Druckanfang *m* der Platte — **n** oploopkant *m* van de plaat — **e** entrada *f* de la plancha — **i** lato *m* anteriore della lastra

7991 lead intensifier
f renforçateur *m* au plomb — **d** Bleiverstärker *m* — **n** loodversterker *m* — **e** reforzador *m* de plomo — **i** rinforzatore *m* al piombo

7992 lead monoxide — See litharge.

7993 lead mould — Mould for electrotypes.
f empreinte *f* sur plomb — **d** Bleimater *f*; Bleimatrize *f* — **n** loodmatrijs *fm* — **e** matriz *f* de plomo — **i** matrice *f* da piombo; flano *m* in piombo; impronta *f* su piombo

7994 lead moulding — Process used in making electros as distinct from the wax-mould method. See also Albert galvano.
f moulage *m* en plomb — **d** Bleiprägung *f* — **n** loodafpersing *f* — **e** moldeado *m* en plomo; estampación *f* en plomo — **i** procedimento *m* con impronta in piombo

7995 lead mount — See lead base.

7996 lead nitrate; nitrate of lead; plumbic nitrate — White crystals, soluble in water and alcohol. Poisonous. Used as a sensitizer in photography, process engraving and lithography. Latin: nitras plumbicus; plumbum nitricum. Formula: $Pb(NO_3)_2$.
f nitrate *m* de plomb; azotate *m* de plomb — **d** Bleinitrat *n*; salpetersaures Blei *n*; Bleisalpeter *m* — **n** loodnitraat *n*; salpeterzuur lood *n* — **e** nitrato *m* de plomo — **i** nitrato *m* di piombo

7997 lead of the web — Start of a reel of paper after it has been torn diagonally for threading through the press. A wrongly threaded path may be referred to as a lost lead or wrong lead.
f passage *m* de papier; engagement *m* de la bande de papier — **d** Papierführung *f* — **n** doorvoer *m* van de papierbaan — **e** conducción *f* de la hoja de papel; guiado *m* de la hoja de papel — **i** passaggio *m* del nastro di carta

7998 lead *v* out — See lead *v*.

7999 lead oxide — See litharge.

8000 lead poisoning; saturnism — Poisoning caused by the slow, continuous absorption of lead by the tissues of the body. When chronic: plumbism.
f saturnisme *m*; empoisonnement *m* par le plomb; affection *f* saturnine — **d** Bleivergiftung *f* — **n** loodvergiftiging *f* — **e** saturnismo *m*; intoxicación *f* saturnina; cólicos *mpl* saturninos — **i** satur-nismo *m*; avvelenamento *m* da piombo

8001 lead protoxide — See litharge.

8002 lead rack — Receptacle to hold various sizes of leads.
f lingotière *f* — **d** Durchschußregal *n*; Durchschuß-kasten *m* — **n** geriefkast *fm* — **e** lingotero *m*; armario *m* para imposiciones — **i** lingottiera *f*; scaffale *m* per interlinee

8003 lead rule
f filet *m* de plomb — **d** Bleilinie *f* — **n** loden lijn *fm* — **e** filete *m* de plomo; filete *m* de metal fuerte — **i** filetto *m* di piombo

8004 lead subcarbonate — See flake white.

8005 lead sulphate — White, rhombic crystals, slightly soluble in hot water, insoluble in alcohol. Poisonous. Used as a pigment. Formula: $PbSO_4$.
f sulfate *m* de plomb — **d** Bleisulfat *n* — **n** loodsulfaat *n* — **e** sulfato *m* de plomo — **i** solfato *m* di piombo

8006 lead tapes — See tapes.

8007 leaf — One of a number of folds (each containing two pages) which compose a book or manuscript; a folio. Hence the matter printed or written thereon. Two pages back to back.
f feuillet *m* (recto-verso) — **d** Blatt *n* — **n** blad *n* — **e** hoja *f* (impresa a dos caras); foja *f Ch* — **i** foglio *m*

8008 leafing — Phenomenon of metallic flakes (metallic pigments), when mixed with oil or varnish, rising to the surface and forming a continuous film resembling that of fallen leaves.
d Blättern *n*; Schwimmen *n* von Metallpigmenten auf dem Bindemittel — **i** potere *m* fogliante

8009 lean matter — Matter that is difficult to set because of complexity or intermixed founts.
f composition *f* compliquée — **d** schwerer Satz *m*; schwierig abzusetzender Satz *m* — **n** ingewikkeld zetsel *n*; moeilijk zetsel *n*; lastig zetsel *n* — **e** composición *f* complicada — **i** composizione *f* difficile

8010 learner — See trainee.

8011 leather — Material for bookbinding from four animals: the goat giving morocco leather; the calf providing natural calf or law calf, dyed calf and vellum; the sheep providing basil and roan for account books and second quality books, and parchment and forel for less robust work; and pig which provides a skin which theoretically is the strongest.
f cuir *m*; peau *f* — **d** Leder *n* — **n** leder *n*; leer *n* — **e** cuero *m* (de encuadernar); piel *f* — **i** cuoio *m*; pelle *f*

8012 leather binding
f reliure *f* peau (pleine) — **d** Lederband *m*; Leder-einband *m*; Ganzlederband *m* — **n** geheel leren band *m*; leren band *m* — **e** encuadernación *f* cuero — **i** legatura *f* in cuoio; legatura *f* in pelle

8013 leatherboard; genuine leatherboard — Board made from pulped scrap leather, formed into sheets of specified thickness on a wet machine.

f carton *m* cuir — **d** Lederkarton *m*; Lederpappe *f* — **n** leerbord *n* — **e** cartón *m* cuero — **i** cartone *m* cuoio

8014 leathercloth — Heavy substance woven material imitating leather. Rendered waterproof and hard wearing by the use of cellulose and oil coatings.
f similicuir *m*; cuir *m* artificiel — **d** Kunstleder *n* — **n** kunstleer *n*; imitatie-leer *n* — **e** cuero *m* artificial; cuero *m* sintético — **i** similcuoio *m*; similcuoio

8015 leather cover
f reliure *f* en cuir; reliure *f* en peau — **d** Ledereinband *m*; Lederband *m* — **n** leren band *m* — **e** tapa *f* de piel; encuadernación *f* en piel — **i** legatura *f* in cuoio; legatura *f* in pelle

8016 leatherette — Trade-mark for a strong paper, coloured, finished and embossed in imitation of grained leather, used in bookbinding.
f papier *m* couché similicuir — **d** Kunstlederpapier *n* — **n** imitatie-leerpapier *n* — **e** papel *m* similcuero; cuerina *f*; cueroide *m*; cuero *m* artificial — **i** carta *f* imitazione cuoio

8017 leatheroid — Trade-mark for an artificial leather consisting of chemically treated paper combined with rubber and sandarac, used for bookbindings.
f leatheroïd *m*; armite *m* — **d** Kunstleder *n*; Lederimitationspapier *n* — **n** imitatie-leer *n* — **e** cueroide *m*; cuero *m* artificial — **i** similpelle *f*

8018 leatherpulp board — Board generally manufactured on an intermittent board machine and containing more than 50% of leather.
f carton *m* de pâte de cuir — **d** Lederfaserpappe *f*; Lederfaserkarton *m* — **n** leerbord *n* — **e** cartón *m* de pasta de cuero; cartón *m* de fibras de cuero — **i** cartone *m* da cascami di cuoio

8019 leather roller
f rouleau *m* en cuir; rouleau *m* de cuir — **d** Lederwalze *f* — **n** leren rol *fm*; lederrol *fm* — **e** rodillo *m* de cuero — **i** rullo *m* di cuoio

8020 leave *v* **blank; blank** *v* — To let out some type matter in a composition.
f laisser en blanc — **d** blankschlagen; blindschlagen; Raum freilassen — **n** open laten; onbedrukt laten — **e** dejar (en) blanco; dejar espacio — **i** lasciare uno spazio bianco; lasciare in bianco; lasciare una finestra nella composizione

8021 leave *v* **out a word**
f omettre un mot; élider un mot — **d** ein Wort auslassen; ein Wort weglassen — **n** een woord weglaten — **e** omitir una palabra; saltar — **i** omettere una parola

8022 ledger — See account book.

8023 ledger — See blank book.

8024 ledger paper; register paper; account-book paper; record paper — High quality, strong, well-finished hand-made or machine-made writing paper for records or documents which must be preserved for long periods.
f papier *m* pour registre(s); papier *m* pour documents; papier *m* pour livres comptables — **d** Buchungspapier *n*; Geschäftsbücherpapier *n*; Geschäftsbuchpapier *n*; Kontobücherpapier *n*; Registerpapier *n*; Kanzleipapier *n* — **n** registerpapier *n*; boekhoudpapier *n* — **e** papel *m* registro; papel *m* para contabilidad — **i** carta *f* registro; carta *f* per registri

8025 ledger royal — A British standard size of writing and printing papers. See app. no. 3.

8026 ledger super royal — A British standard size of ledger paper. See app. no. 3.

8027 left-hand pages — See even pages.

8028 left-hand thread — Helical or spiral member, as a gear tooth or screw thread, that twists counter-clockwise.
f filet *m* (de vis) avec pas à gauche; filet *m* à gauche; filetage *m* à gauche — **d** linksgängiges Gewinde *n*; Linksgewinde *n* — **n** linkse (schroef)draad *m* — **e** filete *m* (de mano) izquierda; filete *m* a la izquierda; rosca *f* a izquierdas — **i** filetto *m* sinistrorso; filetto *m* sinistro

8029 leg art — See cheese cake.

8030 legato touch — Smooth touch of keyboard keys.

8031 legend; underline — Line(s) under an illustration. Often erroneously called a *caption.
f légende *f* — **d** Legende *f*; Bildunterschrift *f*; Unterschrift *f*; Bildertext *m* — **n** onderschrift *n*; bijschrift *n* — **e** leyenda *f*; epígrafe *m*; pie *m* de grabado — **i** leggenda *f*; didascalia *f*; dicitura *f*

8032 legibility
f lisibilité *f* — **d** Lesbarkeit *f* — **n** leesbaarheid *f* — **e** legibilidad *f* — **i** leggibilità *f*

8033 Leipsic height to paper — See Leipzig height to paper.

8034 Leipzig height to paper; Leipsic height to paper — See also height to paper.
f hauteur *f* de Leipzig — **d** Leipziger Höhe *f* — **n** Leipziger hoogte *f* — **e** altura *f* de Leipzig — **i** altezza *f* di Lipsia

8035 Leipzig yellow — See lead chromate.

8036 lending library; circulating library; rental library — Small library maintained by a commercial establishment, comprising mainly current books which are lent to customers for a minimal daily fee.
f bibliothèque *f* de prêt; bibliothèque *f* de location de livres — **d** Leihbibliothek *f*; Ausleihbibliothek *f*; Leihbücherei *f* — **n** uitleenbibliotheek *f*; leesbibliotheek *f* — **e** biblioteca *f* circulante; librería *f* circulante — **i** biblioteca *f* circolante

8037 length — Property of ink to stretch out into a long thread without breaking. Long inks have good flow in the fountain.
f longueur *f* — **d** Länge *f* — **n** kortheid *f* — **e** lungitud *f* — **i** lunghezza *f*

8038 length fold collection — Bringing together folded products from a series of formers sometimes mounted one above another. See also balloon former.
f assemblage *m* de cahiers — **d** längsgefalzte Lagensammlung *f* — **n** langsverzameling *f* — **e** encajado *m* de cuadernos; encaballado *m* de

pliegos — **i** raccolta *f* di quinterni

8039 length of a page — See page length.

8040 length of exposure — See exposure time.

8041 lens

f lentille *f* — **d** Linse *f*; Glaslinse *f* — **n** lens *fm* — **e** lente *fm* — **i** lente *f*

8042 lens; camera lens; camera objective; objective — Photographic lens made up of several elements, both negative and positive, separated in the lens barrel at definite distances by an air space. A negative lens element is double concave (resembling two prisms placed apex to apex); a positive lens element is double convex (resembling two prisms placed base to base) which bring the light rays to a focus. See also process lens.

f objectif *m* — **d** Objektiv *n*; Aufnahmeobjektiv *n*; Kameralinse *f* — **n** objectief *n*; cameralens *fm*; lens *fm* — **e** objetivo *m* (fotográfico); lente *fm* — **i** obiettivo *m*; lente *f*

8043 lens aperture; lens opening — Effective opening to a lens. See also diaphragm.

f ouverture *f* de l'objectif — **d** Objektivöffnung *f*; Öffnung *f* des Objektivs; Linsenöffnung *f* — **n** lensopening *f*; opening *f* van de lens — **e** abertura *f* del objetivo — **i** apertura *f* del'obiettivo

8044 lens cap — Light-tight cover to be placed over a lens to exclude light from the camera.

f couvre-objectif *m* — **d** Objektivdeckel *m* — **n** lenskap *fm* — **e** tapa *f* de objetivo — **i** copriobiettivo *m*

8045 lens coating — Deposition of a very thin film of evaporated metallic fluorides on the uncemented glass elements of a lens, to increase the speed of the objective and reduce reflections from glass to air surfaces. With process lenses, it results in a slight increase of contrast and somewhat sharper line negatives.

f traitement *m* de surface de l'objectif — **d** Objektivvergütung *f*; Objektivbeschichtung *f*; Linsenvergütung *f* — **n** lenscoating *f* — **i** ricoprimento *m* della lente

8046 lens opening — See lens aperture.

8047 lens scale; diaphragm scale — Chart or scale mounted above process lenses on the camera to aid quick and accurate setting of iris diaphragm apertures, particularly in halftone photography.

f indicateur *m* de diaphragme — **d** Objektivskala *f*; Blendenskala *f*; Blendenreihe *f* — **n** diafragmaschaal *fm* — **e** escala *f* del objetivo — **i** scala *f* del'obiettivo

8048 lens turret — Part of a photo-typesetting machine with a number of lenses to obtain phototype composition in different point sizes.

f tourelle *f* porte-lentille — **d** Linsensystem *n*; optisches System *n* — **n** lenzenschijf *fm* — **e** portaobjetivos *m* giratorio — **i** portaobiettivo *m* multiplo

8049 let-in notes — See cut-in notes.

8050 let it stand — See stet.

8051 letter — A *character of the alphabet. See also type.

f lettre *f* — **d** Buchstabe *m*; Letter *f* — **n** letter *fm* — **e** letra *f* — **i** lettera *f*

8052 letter — Style of type.

f caractère *m* — **d** Schriftstil *m*; Stil *m* — **n** letterstijl *m*; stijl *m*; karakter *n* — **e** carácter *m*; estilo *m* — **i** carattere *m*; stile *m*

8053 letter — See type.

8054 letter balance

f pèse-lettre(s) *m* — **d** Briefwaage *f* — **n** briefweger *m*; brieveweger *m* — **e** pesacartas *m* — **i** pesalettere *m*

8055 letter board; composition board; forme board; board — Wooden tray on which standing type matter is kept in racks.

f ais *m* pour forme — **d** Satzbrett *n*; Stehsatzbrett *n*; Form(en)brett *n* — **n** letterbord *n* — **e** tablero *m* para guardar la composición; tablero *m* corredizo para formas — **i** balestra *f*; piattaforma *f* portaforme

8056 letter brush — See beating brush.

8057 letter file — Folder with a fastening device for securing papers.

f classeur *m* de lettres — **d** Briefordner *m*; Ordner *m* — **n** briefordner *m*; ordner *m*; brievenmap *fm* — **e** clasificador *m*; archivador *m* de cartas; bibliorrato *m* — **i** raccoglitore *m* di corrispondenza

8058 letterhead — Printed heading, usually on bond paper, giving the name, address, etc., of the writer, a business concern, etc.

f en-tête *m* de lettre; tête *f* de lettre — **d** Briefkopf *m* — **n** briefhoofd *n*; briefpapier *n* — **e** cabecera *f* de carta; membrete *m* — **i** intestazione *f*

8059 lettering piece — See panel.

8060 letter paper — Writing paper cut to proper size for correspondence.

f papier *m* à lettres — **d** Briefpapier *n*; Postpapier *n* — **n** briefpapier *n*; postpapier *n* — **e** papel *m* de carta — **i** carta *f* da lettere

8061 letterpress — Relief printing from type, blocks, or other raised surface, as distinct from planographic (lithographic, collotype, etc.), and intaglio (gravure, engravings, etc.) printing.

f typographie *f* — **d** Buchdruck *m* — **n** boekdruk *m* — **e** tipografía *f* — **i** tipografia *f*

8062 letterpress ink; typographic ink — *Ink used in letterpress printing. The viscosity, fluidity, length and tack vary according to the type of work to be done and the grade of paper to be used: bond ink, job ink, etc.

f encre *f* typographique; encre *f* typo — **d** Buchdruckfarbe *f* — **n** boekdrukinkt *m* — **e** tinta *f* tipográfica — **i** inchiostro *m* tipografico

8063 letterpress machine; letterpress printing press — Printing press in which the impression is taken from the raised surface of type.

f presse *f* typographique; presse *f* typo — **d** Buchdruckpresse *f*; Buchdruckmaschine *f* — **n** boekdrukpers *f* — **e** máquina *f* tipográfica; máquina *f* para la impresión tipográfica; prensa *f* para tipografía — **i** macchina *f* tipografica

8064 letterpress printer
f imprimeur *m* typographe — **d** Buchdrucker *m* — **n** boekdrukker *m* — **e** maquinista *m* tipógrafo; impresor *m* tipógrafo — **i** impresore *m* tipografo

8065 letterpress printing — Relief printing process by means of types, electrotypes, stereotypes, rubber or plastic plates.
f impression *f* typo; impression *f* typographique — **d** Buchdruck *m*; Hochdruck *m*; Hochdruckverfahren *n* — **n** boekdruk *m*; hoogdruk *m* — **e** impresión *f* tipográfica; tipografía *f* — **i** stampa *f* tipografica; tipografia *f*

8066 letterpress printing press — See letterpress machine.

8067 letterpress rotary — See newspaper press.

8068 letterset — Printing from a rubber blanket on which the print image has been deposited from a relief printing plate. Modern term for *dry offset.

8069 letterset — See dry offset.

8070 letter spacing; interspacing — The insertion of spaces between the letters of a word or words for lengthening the measure, improving the appearance of the spacing or for emphasis.
f espacement *m* — **d** Spationieren *n*; Spatiieren *n*; Sperren *n*; Sperrung *f* — **n** spatiëring *f* — **e** espaciado *m* de (las) letras; espaciado *m* entre letras; espaciatura *f* Me — **i** spazieggiatura *f*

8071 lettre bâtarde — See bastarda.

8072 lettre cadeau — French name for old capital letters, placed at the head of books.

8073 lettre de forme — French name for the Gothic pointed letter of the primitive xylographs, used by the printers of the heroic times for the composition of their first books, in the 42-lines Bible, the Mainz Psalter, etc.
f lettre *f* de forme; lettre *f* Gothique anguleuse — **d** spitzgotische Schrift *f* — **n** lettre *f* de forme — **e** letra *f* gótica congulosa — **i** lettera *f* di testo; scrittura *f* formata; scrittura *f* gotica

8074 lettre de Somme — *Lettre de forme did not satisfy the men of the Renaissance. They amended it bit by bit, disencumbered it of unnecessary appendages and so it became the lettre de Somme. It is said that the name of this type originates from the fact that it was used for the first time to print the "Summa Theologica" of Thomas Aquinas, and which prepared the coming of the roman type. Thus it was a character of the decline.
f lettre *f* de Somme; lettre *f* Gothique arrondie — **d** rundgotische Schrift *f* — **n** lettre *f* de Somme — **e** letra *f* de Tortis (named after the Venetian printer Battista de Tortis, who used it for his law books, in 1490); letra *f* de Suma; letra *f* gótica redondeada — **i** scrittura *f* gotica rotonda

8075 lettre tourneure — Exclusive French name for a rounded (tournante) capital type face, which occupied the chapter heads. This character seems to have been the beginning of the *lettre de Somme of which it recalled the rounded form.

8076 levant morocco leather — High grade *morocco leather with large grain. Do not use capital letters.

8077 level small caps — See even smalls.

8078 lever — Rigid bar turning round a fixed pivot to transmit pressure or motion to machine parts.
f levier *m* — **d** Hebel *m*; Bedienungshebel *m* — **n** hendel *nm*; handel *nm*; hefboom *m* — **e** palanca *f* — **i** leva *f*

8079 lever — See sliding bar.

8080 lever cutter — Lever-operated cutting machine.
f massicot *m* à levier — **d** Hebelschneidemaschine *f* — **n** hefboom-snijmachine *f* — **e** guillotina *f* de palanca; cortadora *f* de palanca — **i** taglierina *f* a leva

8081 levigator; stone planing machine; circular grinder — Circular steel abrasive plate with a weight of about 15 kg (30 pounds) and 7.5-10 cm thick (3-4 in) with an excentrically mounted handle, used for grinding lithographic stones by hand.
f bourriquet *m* — **d** Handschleifplatte *f*; Handschleifer *m*; Steinschleifer *m* — **n** handslijper *m* — **e** lijadora *f* — **i** levigatrice *f*

8082 liberty of the press — See freedom of the press.

8083 library
f bibliothèque *f* — **d** Bibliothek *f*; Bücherei *f*; Büchersammlung *f* — **n** bibliotheek *f*; boekerij *f* — **e** biblioteca *f*; librería *f* — **i** biblioteca *f*; libreria *f*

8084 library — See morgue.

8085 Library of Congress — The national library of the US in Washington, D.C., established by Congress in 1800 for service to its members. It now also serves government agencies, other libraries, and the public. This library is yearly adding about 500,000 volumes to its collections, many through deposits of the copyright law and exchanges of official publications with foreign governments and learned societies. In US the copyright is secured by publication of the work with the required notice, after which copies of the work are deposited in the US copyright office together with an application and fee for registration of the copyright claim. The copies deposited are made available to the Library of Congress.
f ≈ dépôt *m* légal — **e** ≈ depósito *m* legal — **i** deposito *m* legale

8086 Library of Congress Catalog Card — Card printed by the US Library of Congress, giving bibliographical information on a publication and assigning call numbers, made available to libraries for their own catalogues. See also Library of Congress Catalog Card Number.

8087 Library of Congress Catalog Card Number; LC number *fam.* — Number assigned by the US Library of Congress upon application of the publisher, usually printed on the copyright page of a book. Not to be confused with the call number used by libraries to show where a book is shelved. See also Standard Book Number.

8088 library program; library routine — Program or subroutine stored in executable format on an auxiliary storage device and available on call.

f programme *m* de bibliothèque; routine *f* de bibliothèque — **d** Bibliotheksprogramm *n*; Bibliotheksroutine *f* — **n** bibliotheekprogramma *n*; bibliotheekroutine *f* — **e** programa *m* de biblioteca; rutina *f* de biblioteca; programa *m* standard; rutina *f* standard — **i** programma *m* di biblioteca

8089 library routine — See library program.

8090 lid — Upper or covering portion of a paper box.

f couvercle *m* — **d** Deckel *m* — **n** deksel *nm* — **e** tapa *f* — **i** coperchio *m*

8091 lift — Number of sheets that can be readily handled, usually 30-50, hung together in a paper conditioning machine. Also a pack of sheets being cut or trimmed together.

f poignée *f* — **d** Stoß *m* — **n** pak *n* — **e** posteta *f* — **i** manata *f*; pacchetto *m*

8092 lift — A compositor who produces type matter in a handy way is said to have a good lift.

f être un habile leveur — **d** einen guten Griff haben — **e** tener buen tacto — **i** avere la mano svelta

8093 lift — Device to elevate (chiefly British term).

f ascenseur *m* (pour personnes); monte-charge(s) *m* (pour des charges); élévateur *m* (pour les bobines de papier); monte-bobines *m* — **d** Aufzug *m*; Lift *m*; Fahrstuhl *m* — **n** lift *m*; personenlift *m*; goederenlift *m*; rollenlift *m* — **e** ascensor *m* — **i** ascensore *m*

8094 lift *v* — To remove a forme from the press or platen, usually before the job is finished.

f décaler; relever la forme de la presse — **d** die Form ausheben — **n** de vorm opbreken; de vorm van de pers nemen; wegnemen — **e** levantar (la forma); levantar la forma de la máquina — **i** rimuovere la forma della macchina; togliere la forma della macchina

8095 lifted matter; pickup — Type matter taken from one forme for uses in another.

f matière *f* levée — **d** herausgezogenes Setzmaterial *n* — **n** uitgetrokken materiaal *n*; uitgenomen materiaal *n*; materiaal *n* uit een andere vorm — **e** material *m* levantado — **i** materiale *m* sollevato

8096 lifter roller — Ink carrying roller in a rotary press which moves between the duct roller and the distribution system. See also ductor.

8097 lifting — Proper adhering of one colour (ink) to a previously printed colour, or to the sheet.

f prise *f* — **d** Haften *n* — **n** hechten *n* — **e** adherencia *f* (entre las tintas superpuestas) — **i** presa *f*

8098 lifting — The lifting of surface fibres from paper and paperboard during printing is one of the printer's most serious problems. It is variously

known as picking, lifting, fibre lift, whiskers and fluffing.

8099 lifting — See picking.

8100 lifting property — Susceptibility of ink to removal from the plate surface or blanket at the time of the impression.

f capacité *f* de transfert — **d** Farbübertragung *f* — **n** inktoverdracht *fm*; inktafname *fm* — **e** capacidad *f* de transferencia — **i** trasferibilità *f* (dell'inchiostro); capacità *f* di trasporto

8101 lift-off — In the screen printing process, space between the inked stencil and the stock. Critical factor in high quality printing.

f dégagement *m* — **d** Absprung *m* — **n** afsprong *m*

8102 lift *v* **off the rollers** — See throw *v* off the rollers.

8103 lift *v* **out** — To lift a freshly imposed printing forme from the imposing stone to check whether a type falls down which indicates that the type is not locked up well. If not the forme is said to lift out. See also fall out.

8104 lift truck; elevating truck; paper truck; truck carriage — See also fork truck.

f chariot *m* élévateur; transporteur *m* à papier — **d** Stapler *m*; Hubwagen *m* — **n** heftruck *m* — **e** carretilla *f* montacargas; carretilla *f* de elevación; carretilla *f* elevadora — **i** carrello *m* elevatore; carrello *m* accatastore

8105 ligature; double letter — Two or more letters joined together and forming one type character as fi, ffi, œ, æ, etc. For metal type the word should be *logotype. In German type composition ligatures must always be used as both characters belong to the stem of the word. The flexional yet does not influence the ligature, as e.g., in: Stoff, Stoffe, Hoffnung, hoffen, streifig, flämisch, Haft, haften, künftig, Schrift, Schriften, Sitz, sitzen. On the other hand a ligature may not be used when it stands between two stems of words belonging together, as, e.g., in: Schaf/fell, Kauf/interesse, höf/lich, Auf/lage, Hof/tor, Auf/trag, Ent/ziehung, ent/zwei. A particularity is found in the conjuncture of three characters each of which may form a ligature. In these cases the limit of the syllable is decisive, as in: Raf/finade, auf/fliegen, muf/fig. In the word Schiffahrt the ligature must always be used, because one of the three consonants f is left out. In compounds of words as Sauerstoff-flasche, two ligatures ff and fl are set. In words like Geschafft the use of the ligatures ff and ft is permitted. In those cases the ligature is placed in the end of the word, because this gives a better look. Thus this word is composed as Geschaf/ft. In interspaced words, set in Lineals (Antique type) no ligatures are used. In broken types (Fraktur, see app. no. 2) on the contrary, ch, ck, tz and st are mostly not interspaced. However, in view of a better look of interspaced slugs of broken types (Fraktur), the ligatures are spaced, especially for the major body sizes, that is to say, the ligature st with lesser space, in order to avoid a bad composition style. œ for single sounds, is in GB generally employed in English, French, Latin and old English words. But the modern tendency (as in the O.E. dictionary) is to use separate letters. See also quaint character, diphthong.

f ligature *f*; lettres *fpl* accouplées — **d** Ligatur *f*; Doppelbuchstabe *m* (ß ist keine Ligatur. Ist diese Type nicht vorhanden, wird dafür ss gesetzt) — **n** ligatuur *f* — **e** ligadura *f*; ligado *m*; letras *fpl* ligadas; letra *f* doble — **i** ligatura *f*

8106 light — Form of radiant power (energy flux), which is capable of acting upon the eye in such manner as to render visible the object from which it proceeds. All colours depend on light.

f lumière *f* — **d** Licht *n* — **n** licht *n* — **e** luz *f* — **i** luce *f*

8107 light absorption — Property of absorbing light or any of its constituent rays, the latter causing the formation of colour.

f absorption *f* de la lumière — **d** Lichtabsorption *f* — **n** lichtabsorptie *f* — **e** absorción *f* de luz — **i** assorbimento *m* della luce

8108 light-alloy — See light metal.

8109 light diffusion — See diffusion of light.

8110 light face — See light-face type.

8111 light-face type; light face — Type face of a fine appearance. See also weight of face.

f caractère *m* maigre; maigre *m* — **d** magere Schrift *f* — **n** magere letter *fm* — **e** letra *f* fina; letra *f* blanca; carácter *m* blanco; carácter *m* fino; carácter *m* de trazo fino — **i** carattere *m* magro

8112 light fastness — See fastness to light.

8113 light filter — See colour filter.

8114 light flux — See luminous flux.

8115 light forme — See light-weight forme.

8116 light hardening

f tannage *m* par la lumière — **d** Lichthärtung *f*; Lichtgerbung *f* — **n** verharding *f* door licht; lichtharding *f* — **e** endurecimiento *m* por la luz — **i** indurimento *m* per effetto della luce

8117 lighting — See illumination.

8118 light integrator; integrating light meter; integrating exposure meter; exposure control unit — Instrument to measure a predetermined amount of light by taking into consideration both intensity and time of exposure.

f intégrateur *m* de lumière; posemètre *m* à intégration — **d** Lichtdosiergerät *n*; Lichtintegrator *m*; integrierender Lichtmesser *m* — **n** lichtdoseringsmeter *m*; integrerende belichtingsmeter *m* — **e** dosificador *m* de luz; exposímetro *m* integrador — **i** dosatore *m* di luce; esposimetro *m* integratore; apparecchio *m* per dosare la luce

8119 light ligroine — See ligroine.

8120 light ligroine — See SBP spirit.

8121 light-lock — Baffle type of entrance to photographic darkrooms, preventing entry of stray light during traffic from the chamber.

f chicane *f* d'entrée; entrée *f* en chicanes; chicane *f* d'accès — **d** Lichtschleuse *f* — **n** lichtsluis *fm* — **e** paraluz *m*; compuerta *f* de luz — **i** entrata *f* a labirinto

8122 light metal; light-alloy

f alliage *m* léger — **d** Leichtmetall *n*; Leichtmetall-Legierung *f* — **n** lichtmetaal *n*; legering *f* van lichtmetaal; lichtmetaal-alliage *fn* — **e** metal *m* ligero; aleación *f* ligera — **i** metallo *m* leggero; lega *f* leggera

8123 light meter — See exposure meter.

8124 lightness — Attribute of visual sensation in accordance with which a body seems to transmit or reflect diffusely a greater or smaller fraction of the incident light. Psychosensorial correlate of luminance factor.

f clarté *f*; leucie *f* — **d** Helligkeit *f* — **n** helderheid *f* — **e** claridad *f*; grado *m* de claridad — **i** chiarezza *f*

8125 light pen; light pencil; light stylus; electronic stylus — Pen-like tube with a photocell which, when directed at a cathode ray tube display, reacts to light from the display. The response is transmitted to the computer and text in the data store may be deleted or inserted.

f crayon *m* à cellule photoélectrique; crayon *m* cathodique; marqueur *m* lumineux — **d** Leuchtstift *m*; Lichtfeder *m*; Lichtgriffel *m* — **n** lichtpen *fm* — **e** marcador *m* luminoso; pincel *m* luminoso — **i** marcatore *m* luminoso; penna *f* luminosa

8126 light pencil — See light pen.

8127 light print — In photography, an underexposed or underdeveloped print which lacks contrast and will not give a good reproduction.

f épreuve *f* faible; épreuve *f* floue — **d** weicher Abzug *m*; weicher Photoabzug *m* — **n** zachte afdruk *m*; zachte foto-afdruk *m* — **e** prueba *f* débil — **i** fotografia *f* debole

8128 light-ray — Beam of light of very small diameter.

f rayon *m* lumineux — **d** Lichtstrahl *m* — **n** lichtstraal *fm* — **e** rayo *m* de luz — **i** raggio *m* di luce; raggio *m* luminoso

8129 light scattering — See also scattered light.

f diffusion *f* de la lumière — **d** Lichtstreuung *f* — **n** lichtverstrooiing *f*; lichtstrooiing *f* — **e** dispersión *f* de la luz — **i** dispersión *f* della luce; diffusione *f* della luce

8130 light sensitive — See sensitive to light.

8131 light-sensitive coating

f couche *f* sensible à la lumière — **d** lichtempfindliche Schicht *f* — **n** lichtgevoelige laag *m* — **e** capa *f* sensible a la luz — **i** strato *m* sensibile alla luce

8132 light setting — See photo-typesetting.

8133 light source; illuminant

f source *f* lumière; source *f* de lumière — **d** Lichtquelle *f*; Belichtungsquelle *f* — **n** lichtbron *fm* — **e** fuente *f* de luz; manantial *m* de luz — **i** sorgente *f* luminosa

8134 light spots — See sun spots.

8135 light stimulus — Radiation entering the eye and producing a sensation of light.

f stimulus *m* lumineux; stimulus *m* de lumière — **d** Lichtreiz *m* — **n** lichtprikkel *m* — **e** estímulo *m* de luz; estímulo *m* luminoso — **i** stimolo *m* luminoso

8136 light struck — Said of photographic material accidentally fogged by action of actinic light, or by extraneous light creeping into faulty camera equipment.

f voilé — **d** geschleiert — **n** gesluierd — **e** velado por la luz — **i** velato

8137 light stylus — See light pen.

8138 light table — See register table.

8139 light-weight forme; light forme — Letterpress printing forme with a printing zone of 10-30% of the distance between the two bed bearers. See also pre-tension.

f forme *f* légère — **d** leichte Druckform *f* — **n** lichte drukvorm *m*; lichte vorm *m* — **e** forma *f* ligera — **i** forma *f* leggera

8140 light-weight papers — Papers having a substance of less than 40 g/m².

f papiers *mpl* minces — **d** Dünndruckpapiere *npl*; Dünnpapiere *npl* — **n** dun papier *n*; dundrukpapier *n*; licht papier *n* — **e** papeles *mpl* ligeros — **i** carte *fpl* leggere; carta *f* a bassa grammatura

8141 lignin — Amorphous polymeric substance related to cellulose that with cellulose forms the woody cell walls of plants and cementing material between them. In papermaking considered as an impurity which is eliminated in the chemical cooking processes. Used in the production of synthetic resins. See also encrust.

f lignine *f* — **d** Lignin *n* — **n** lignine *fm* — **e** lignina *f* — **i** lignina *f*

8142 lignite wax — See montan wax.

8143 ligroin — See ligroine.

8144 ligroine; ligroin — In US a flammable mixture of hydrocarbons that boils at from 20 to 135 °C (58-275 °F), obtained from petroleum by distillation and used as a solvent for general laboratory use. In Europe an obsolete name for a gasoline fraction boiling in the range of approximately 100 to 150 °C. See also SBP spirit.

f ligroïne *f* — **d** Ligroin *n* — **n** ligroïne *fm*; ligroïen *fm* — **e** ligroína *f* — **i** ligroina *f*

8145 lime — In a specific sense, *calcium oxide; generally chemical and physical form of quicklime, hydrated lime, and hydraulic lime.

f chaux *f* — **d** Kalk *m* — **n** kalk *m* — **e** cal *f* — **i** calce *f*

8146 lime hydrate — See calcium hydroxide.

8147 lime nitrate — See calcium nitrate.

8148 lime saltpetre — See calcium nitrate.

8149 limestone — See calcium carbonate.

8150 limited edition

f édition *f* à tirage limité; tirage *m* limité — **d** beschränkte Auflage *f*; begrenzte Auflage *f* — **n** beperkte oplage *f* — **e** edición *f* limitada; edición *f* de pocos ejemplares — **i** edizione *f* (a tiratura) limitata

8151 limit switch — Switch to reverse or cut off the current by a movement at the end of the stroke.

f interrupteur *m* de fin de course; contacteur *m* de fin de course — **d** Endschalter *m* — **n** eind-

schakelaar *m* — **e** interruptor *m* de fin de carrera; ruptor *m* de fin de carrera; contactor *m* de fin de carrera — **i** interruttore *m* di fine corsa

8152 limit value

f valeur *f* limite — **d** Schwellenwert *m*; Grenzwert *m* — **n** grenswaarde *f* — **e** valor *m* límite — **i** valore *m* limite

8153 limp binding — Book case made of leather or cloth with thin card or paper in place of stiff board. On occasions even this is omitted.

f reliure *f* souple; reliure *f* flexible — **d** weicher Buchumschlag *m*; flexibler Buchumschlag *m*; flexibler Einband *m* — **n** slappe band *m*; soepele band *m*; flexibele band *m* — **e** encuadernación *f* flexible — **i** rilegatura *f* flessibile

8154 line — Row of written or printed letters, words, figures, etc., which extends across a column or page.

f ligne *f* — **d** Zeile *f*; Linie *f* — **n** regel *m* — **e** línea *f*; renglón *m*; riga *f* — **i** linea *f*; riga *f*

8155 line

f trait *m* — **d** Strich *m* — **n** lijn *m*; streep *fm* — **e** raya *f*; línea *f* — **i** linea *f*

8156 line *v* — If the serifs or ends of characters at the feet form a perfectly straight horizontal line when they are placed side by side, they are said to line.

f aligner — **d** Linie halten — **n** lijnen; in de lijn staan — **e** alinear — **i** allineare

8157 Lineals — Group of type faces without serifs, e.g., Noble, Futura, Erbar, Paris, Mercator, Annonce, Grotesque, Folio, Univers, Cairoli, Placard, Headline, Iris, Gill, (the former antiques). Before 1971 this group was called Sans Serifs. See also classification of type design.

f Linéales *fpl* — **d** serifenlose Linear-Antiqua *f* — **n** Linearen *mfpl* — **e** Lineales *mpl* — **i** Lineari *mpl*

8158 line-at-a-time printer — See line printer.

8159 line block; line engraving; line etching; line cut; line plate — Letterpress printing block consisting of solid areas and lines (with occasionally mechanical tint or stipple), reproduced direct from a black-and-white drawing without any intermediate tones other than tints.

f cliché *m* de trait; cliché *m* au trait — **d** Strichätzung *f*; Strichklischee *n*; Strichplatte *f* — **n** lijncliché *n* — **e** fotograbado *m* de línea; fotograbado *m* pluma; fotograbado *m* de trazo; grabado *m* pluma; grabado *m* de trazo; clisé *m* pluma; clisé *m* de trazo; zincografía *f*; zincotipia *f*; fotozincotipia *f* — **i** cliscè *m* al tratto; cliché *m* al tratto; zincotipia *f*

8160 line by itself

f ligne *f* pour elle-même — **d** Zeile *f* für sich — **n** afzonderlijke regel *m*; aparte regel *m* — **i** riga *f* per se stessa

8161 line caster; line-casting machine — Casting machine for the production of leads and rules.

f fondeuse *f* d'interlignes — **d** Liniengießmaschine *f* — **n** lijnengietmachine *f* — **e** máquina *f* para fundir por líneas — **i** fonditrice *f* di linee

8162 line caster — See slug-composing machine.

8163 line-casting machine — See line caster.

8164 line-casting machine — See slug-composing machine.

8165 line colour plate; flat colour plate
f cliché *m* couleur au trait — **d** (ungerasterte) Farbplatte *f*; (ungerasterte) Strichplatte *f*; (ungerasterte) Tonplatte *f* — **n** kleurenlijncliché *n*; kleurenlijnplaat *f* — **e** grabado *m* de línea para colores — **i** cliscè *m* colore al tratto

8166 line colourwork; flat colourwork; flat colour printing — Colour effects produced by means of line images, bendaying and the use of shading sheets.
f impression *f* en couleurs au trait — **d** Strich-farbenreproduktion *f* — **n** kleurenlijnwerk *n* — **e** impresión *f* con tintas planas; policromía *f* de línea — **i** policromia *f* al tratto

8167 line-composing machine — See slug-composing machine.

8168 line copy — Copy composed of lines or dots, suitable for reproduction without using a screen.
f document *m* de trait; original *m* de trait; original *m* au trait — **d** Strichvorlage *f*; Strich-zeichnung *f*; Schwarz-Weiß-Original *n* — **n** lijn-model *n*; lijntekening *f*; lijnwerk *n* — **e** original *m* de pluma; modelo *m* de pluma; original *m* de dibujo a la pluma — **i** originale *m* al tratto

8169 line cut — See line block.

8170 lined board — Board lined with a paper different from the body of the board, e.g., mill-lined board, sheet-lined board, vat-lined board.
f carton *m* doublé; carton *m* revêtu de papier — **d** kaschierte Pappe *f*; beklebte Pappe *f* — **n** beplakt karton *n* — **e** cartón *m* forrado — **i** cartone *m* rivestito

8171 line drawing — Drawing or sketch in solid black ink, produced with pen or brush, or occasionally pencil, from which a line block can be made.
f dessin *m* au trait; dessin *m* de trait — **d** Strich-zeichnung *f*; Linienzeichnung *f* — **n** lijn-tekening *f* — **e** dibujo *m* de línea; dibujo *m* de líneas; dibujo *m* de trazo; dibujo *m* de pluma; dibujo *m* a la pluma — **i** disegno *m* al tratto

8172 line engraving — Engraving in which the effect is obtained by lines or combinations of lines by direct incision of the graver on the steel or copper plate. See also steel-plate engraving, copperplate engraving.
f gravure *f* à l'outil — **d** Gravüre *f*; Stich *m* — **n** gravure *fm* — **e** grabado *m* al buril — **i** incisione *f* al bulino

8173 line engraving — See line block.

8174 line etching — Etching a line reproduction.
f morsure *f* de trait — **d** Strichätzung *f* — **n** lijn-etsen *n* — **e** grabado *m* de línea; grabado *m* de trazo; grabado *m* de pluma — **i** incisione *f* al tratto

8175 line etching — See line block.

8176 line for line; page on page — To compose a facsimile of an originally printed text.
f aller chou pour chou; aller ligne pour ligne — **d** Männchen auf Männchen setzen; Zeile-auf-Zeile gehen — **n** letter voor letter overzetten; op iden-tieke wijze overzetten — **e** componer a plana (y) renglón — **i** comporre linea a linea; comporre riga per riga

8177 line gage — See type gauge.

8178 line gauge — See type gauge.

8179 line-halftone combination (plate); combination plate — Printing plate comprising both line and halftone images.
f cliché *m* simili et trait; trait simili *m* combiné — **d** Strich-Auto-Kombination *f*; (kombinierte) Strich-Rasterätzung *f*; Strichklischee *n* mit ein-kopiertem Raster — **n** auto-lijncliché *n*; auto-lijn-combinatie *f* — **e** grabado *m* de combinación; medio tono *m* con zonas al trazo; clisé *m* de línea y trama; clisé *m* combinado; clisé *m* mixto; clisé *m* compuesto — **i** cliscè *m* misto al tratto e retinata

8180 line holder — Oblong iron clamp in which small blocks or mounted plates are tightly held during trimming.
f mâchoire *f* à découper — **d** Anreißhalter *m*; Anreißlineal *n* — **n** snijliniaal *fmn* — **e** mordaza *f* — **i** reggi-originale *m*

8181 line length — See measure.

8182 linen — See bookbinder's cloth.

8183 line negative — Photographic image made directly from a line original without interposition of a halftone screen, the image represented more or less in the form of a stencil, with opaque back-ground and completely transparent lines.
f négatif *m* de trait — **d** Strichnegativ *n* — **n** lijn-negatief *n* — **e** negativo *m* de línea; negativo *m* de trazo — **i** negativo *m* al tratto

8184 linen finish — Paper finish produced 1. by compressing the sheets between alternate sheets of linen cloth so that the pattern of the cloth is im-pressed upon the surface of the paper; 2. by pressing the continuous web of paper between two endless belts of linen cloth by means of press rolls; or 3. by embossing a continuous web of paper with a steel roll which has been knurled or engraved to simulate the surface of linen cloth.
f apprêt *m* toile; apprêt *m* similitoile; fini *m* toile; gaufré *m* toile — **d** Leinenprägung *f*; Leinen-pressung *f* — **n** linnenpersing *f* — **e** acabado *m* imitación de tela — **i** telatura *f*

8185 linen paper — Paper which has a linen finish.
f papier *m* fini toile; papier *m* apprêt similitoile; papier *m* toilé — **d** Leinenpapier *n*; leinen-geprägtes Papier *n* — **n** papier *n* met linnen-persing — **e** papel *m* tela; papel *m* imitación tela — **i** carta *f* lino; carta *f* telata

8186 linen paper — Paper made wholly or partly from linen rags.
f papier *m* de chiffons — **d** Papier *n* aus Leinen-lumpen — **n** papier *n* uit linnen lompen — **e** papel *m* de hilo — **i** carta *f* da stracci di lino

8187 linen tester — Small magnifying glass

mounted above its base at a distance equal to the focal length of the lens. Originally for counting threads in linen. Modern achromatic glasses of this type are widely used for examination of negatives, plates and proofs.

f compte-fils *m* — **d** Fadenzähler *m*; Lupe *f*; Vergrößerungsglas *n* — **n** loep *fm*; loupe *fm*; vergrootglas *n*; dradenteller *m* — **e** cuentahilos *m*; lupa *f*; lente *f* de aumento — **i** contafili *m*; lupa *f*

8188 line number — In computer typesetting, computer generated consecutive number before each line in a print-out or filmset proof, used as reference number in the correction system.

8189 line photography — Direct photography of line originals on negative material of high contrast, without use of halftone screen.

f photographie *f* de trait — **d** Strichaufnahme *f* — **n** lijnopname *f* — **e** fotografía *f* para grabados de línea — **i** fotografia *f* al tratto; foto *f* al tratto

8190 line plate — See line block.

8191 line positive — Transparency or diapositive of a line original.

f positif *m* de trait — **d** Strichpositiv *n* — **n** lijnpositief *n* — **e** positivo *m* de línea; positivo *m* de trazo — **i** positivo *m* al tratto

8192 line printer; line-at-a-time printer — Printout device to provide a hard copy of information which is in machine-readable form. Frequently operates at high speeds, from a computer interface. The character repertoire may provide 64, 96, 128 or 256 different images. The printer may be of the impact variety, with a revolving belt containing one or more alphabets (or a revolving drum with the entire repertoire available for each horizontal drum position), or it may be an electrostatic printer. Some printers may operate remotely or off-line (from paper or magnetic tape input).

f imprimeur *m* en lignes — **d** Zeilendrucker *m* — **n** regeldrukker *m* — **e** impresor *m* por líneas — **i** stampatrice *f* in parallelo

8193 liner — Piece of steel inserted at each end of the opening in the mould of a slug composing machine to determine the length and body of the slug.

f cale *f* de justification (et d'épaisseur) — **d** Einsatzstück *n*; Zeilenschwinder *m*; Gießformeinsatzstück *n* — **n** kaliberplaatje *n*; inzetstukje *n* — **e** alineador *m*; cala *f* Es; regleta *f* Ar; cala *f* móvil — **i** piastrina *f* di giustezza

8194 linerboard — Paperboard made on a Fourdrinier or cylinder machine, used as facing material in corrugated and solid fibre shipping containers. Usually classified according to furnish and method of web formation as, e.g., Fourdrinier kraft linerboard, cylinder kraft linerboard, jute linerboard.

8195 line screen

f trame *f* lignée — **d** Linienraster *m*; Strichraster *m* — **n** lijnraster *n* — **e** trama *f* rayada; trama *f* lineal — **i** retino *m* a linee; retino *m* lineare

8196 line spectrum — Spectrum produced by radiation in which the energy values of the property being measured, such as energy, mass, etc., cluster about one or more discrete values, in contact with a continuous spectrum.

f spectre *m* de lignes; spectre *m* de raies — **d** Linienspektrum *n* — **n** lijnenspectrum *n* — **e** espectro *m* de líneas — **i** spettro *m* di righe

8197 line speed — See transmission speed.

8198 lines to the inch — Classification method of crossline halftone screens according to the number of opaque lines ruled at right angles to each other of which they are made up. The 120- and 133-line screens are the ones most commonly used. In large size work the screen ruling may be 40 lines to the inch, in very fine detail work 250 lines or more. The number of dots to the square inch is the square of the lines to the inch of the screen. A 133-line screen, for example, produces 17,689 dots per square inch.

f linéature *f* de trame — **d** Anzahl *f* der Rasterlinien je Zoll — **n** aantal *n* lijnen per inch — **e** número *m* de líneas por pulgada — **i** numero *m* di linee per pollice

8199 line switching — See circuit switching.

8200 line *v* **up** — To insert leads at top and/or bottom, of one size of type to make it align with a larger size in the same line.

f parangonner — **d** ausgleichen; parangonieren — **n** gelijkmaken; opvullen — **e** parangonar — **i** parangonare

8201 line *v* **up** — See align *v*.

8202 line-up sheet

f feuille *f* de registre; feuille *f* tracée — **d** Standbogen *m* — **n** standvel *n* — **e** hoja *f* modelo — **i** foglio *m* di registro; foglio *m* di norma

8203 line width — See measure.

8204 line-width variator

f convertisseur *m* de l'épaisseur des traits — **d** Strichstärkenwandler *m*; Strichstärkeumwandler *m* — **n** lijnbreedte-variator *m*; lijndikte-variator *m* — **e** variador *m* del grueso de líneas — **i** variatore *m* dello spessore dei tratti

8205 line work — Copy or reproduction consisting of solid elements only, as distinct from halftone.

f travail *m* de trait; travail *m* au trait — **d** Strichoriginal *n*; Strichreproduktion *f* — **n** lijnwerk *n* — **e** trabajo *m* de línea — **i** lavoro *m* al tratto

8206 lining figures; ranging figures; modern figures; modernized figures — Figures that align horizontally at top and bottom, instead of some extending above or below others in the line (old style figures).

f chiffres *mpl* alignants; chiffre *mpl* Didot; chiffres *mpl* Didot — **d** liniehaltende Normalziffern *fpl*; liniehaltende Ziffern *fpl*; Antiquaziffern *fpl* — **n** in de lijn staande cijfers *npl*; lijnende cijfers *npl* — **e** cifras *fpl* alineadas — **i** cifre *fpl* allineate al piede e in testa

8207 lining paper — Paper pasted on boxes or for other lining purposes. Not to confuse with end papers in books.

f papier *m* doublure; papier *m* à doubler; extérieur *m* pour carton — **d** Klebepapier *n*; Überzugpapier *n*; Kaschierungspapier *n*; Kaschierpapier *n* — **n** beplakpapier *n* — **e** papel *m* para forrar; papel *m* para forros — **i** carta *f* per rivestimento

8208 lining tool; multiple graver; shooter; halftone comb; comb — Graver with a number of cutting points matching the rulings of standard halftone screens, used by finishers for shooting or cutting between the dots of a halftone plate.

f griffe *f* à filets — **d** Fadenstichel *m* — **n** fadensteker *m* — **e** buril *m* rayador; buril *m* agrisado — **i** bulino *m* rigatore

8209 lining up — Pasting the back of a book with a strip of paper.

f doublure *f*; contrecollage *m*; pose *f* d'un papier au dos — **i** incollatura *f*

8210 lining-up table — See register table.

8211 link; return — Instruction or address to leave a closed subroutine on its completion to return to some desired point in the return from which the subroutine was entered.

f instruction *f* d'enchaînement; liaison *f*; lieu *m* — **d** Anschluß *m*; Rückkehrinformation *f* — **n** terugkeeropdracht *fm*; terugkeeradres *n*; terugkeermodificator *m* — **e** conexión *f* — **i** vincolo *m*

8212 linkage — Interconnection between a main routine and a closed routine, i.e., entry and exit for a closed routine from the main routine.

f liaison *f*; lien *m* — **d** Verbindung *f* — **n** koppeling *f*; verbinding *f* — **e** vinculación *f* — **i** agganci *pl*; agganciamenti *pl*

8213 link-word — See compound word.

8214 linocut — Method of producing a line block without process engraving. Material similar to ordinary cork linoleum cut by hand after being mounted. Broad flat masses and bold lettering are applicable to the process.

f gravure *f* (sur) linoléum — **d** Linolschnitt *m*; Linoleumschnitt *m* — **n** linoleumsnede *fm*; lino *fm* — **e** grabado *m* en linóleo — **i** linoleumgrafia *f*; incisione *f* su linoleum

8215 linoleate; linoleate drier — Generally the salt of soaps of linseed fatty acids, more properly the salt of soaps of linoleic acid. Cobalt, lead, and manganese linoleates are widely used as driers in printing inks.

f linoléate *m* — **d** Linoleat *n*; Ölsikkativ *n* — **n** linoleaat *n*; oliesiccatief *n* — **e** linoleato *m*; aceite *m* secante — **i** linoleato *m*

8216 linoleate drier — See linoleate.

8217 linoleic acid; linolic acid — Unsaturated fatty acid containing two double bonds, which occurs as a glyceride in all drying oils, such as linseed and cottonseed. Insoluble in water, soluble in alcohol and most organic solvents.
Formula: $C_{17}H_{31}COOH$.

f acide *m* linoléique — **d** Linolsäure *f* — **n** linolzuur *n* — **e** ácido *m* linoleico — **i** acido *m* linoleico

8218 linolenic acid — Unsaturated fatty acid which occurs as the glyceride in many seed fats. Must by present in drying oils in appreciable proportion to produce effective film formation and therefore the natural oils, such as linseed, perilla and hempseed with a linolenic acid content of 25-65 % are in demand. Used in the manufacture of synthetic resins, varnishes, printing inks, etc.
Formula: $C_{18}H_{30}O_2$.

f acide *m* linolénique — **d** Linolensäure *f* — **n** linoleenzuur *n* — **e** ácido *m* linolénico — **i** acido *m* linolenico

8219 linoleum — Mixture of oxidized linseed oil, rosin and powdered cork which is heated and pressed hot onto canvas.

f linoléum *m* — **d** Linoleum *n* — **n** linoleum *n* — **e** linóleo *m*; linóleum *m* — **i** linoleum *m*

8220 linolic acid — See linoleic acid.

8221 Linotype — Machine operated by means of a keyboard for setting and casting lines or slugs or words, etc., instead of setting movable types by hand. Named after the US inventor Ottmar Mergenthaler, born in Germany (1854-1899). First machine installed July 1886, New York Tribune; first machine in GB, 1889, Newcastle Evening Chronicle.

f Linotype *f* — **d** Linotype *f* — **n** Linotype *fm* — **e** Linotipia *f*; Lineotipia *f*; Linotipo *m*; máquina *f* Linotipia — **i** Linotype *f*; macchina *f* Linotype; macchina *f* linocompositrice

8222 Linotype composition

f composition *f* linotypique — **d** Linotypesatz *m* — **n** Linotype-zetsel *n* — **e** composición *f* linotípica — **i** composizione *f* linotipica

8223 Linotype operator

f compositeur *m* linotypiste; linotypiste *m*; opérateur *m* de la Linotype — **d** Linotypesetzer *m* — **n** Linotype-zetter *m* — **e** compositor *m* linotipista; linotipista *m*; oficial *m* linotipista; operador *m* de linotipo; linógrafo *m* *Ch* — **i** linotipista *f*

8224 linseed oil; flaxseed oil; stand oil *GB* — Drying oil obtained by pressing flax seeds. Used as the starting point in the production of varnishes and as printing ink vehicle, which thickens and hardens on the exposure to the air. Soluble in ether, chloroform, carbon disulphide, benzene and turpentine, slightly soluble in alcohol.

f huile *f* de lin — **d** Leinöl *n* — **n** lijnolie *fm* — **e** aceite *m* de linaza; aceite *m* de lino — **i** olio *m* di semi di lino; olio *m* di lino; standolio *m* di lino

8225 linseed oil varnish — Varnish obtained from boiled linseed oil, used especially in litho or offset inks.

f vernis *m* de lin; vernis *m* d'huile de lin — **d** Leinölfirnis *m* — **n** lijnolievernis *mn* — **e** barniz *m* de aceite de linaza — **i** vernice *f* ad olio di seme di lino

8226 lint; whiskers — Particles of fibres that separate or dust off from paper during manufacturing or converting operations. See also fuzz, paper dust.

f poussière *f* (de papier); peluche *f* — **d** Papierstaub *m*; Faserstaub *m* — **n** papierstof *n*; vezelstof *n*; stof *n* — **e** granillo *m* del papel; pelusa *f* del papel; polvo *m* de papel; polvillo *m* de papel — **i** polvere *f* di carta

8227 lint — See fuzz.

8228 linters — See cotton linters.

8229 linting — Removal of completely loose fibrous materials from the paper surface during printing. In Europe, this phenomenon is referred to as *fluffing. See also dusting.

8230 lipids — See lipophilic.

8231 lipophilic — Having a strong affinity for lipids, a group of organic compounds, greasy to the touch, insoluble in water, soluble in alcohol, ether and other fat solvents.

8232 liquefaction — The process of making or becoming fluid.

f liquéfaction *f* — **d** Verflüssigung *f* — **n** vloeibaarmaking *f*; vloeibaarwording *f* — **e** licuefacción *f* — **i** liquefazione *f*

8233 liquefying stress — Ease with which concentrated printing ink can be liquefied by remilling after storage.

f contre-thixotropie *f* — **d** Thixotropiegrad *m* — **e** fuerza *f* licuante — **i** facilità *f* di manipolazione

8234 liquid camphor — See camphor oil.

8235 liquid drier

f siccatif *m* liquide — **d** flüssiges Sikkativ *n*; Sikkativlösung *f* — **n** vloeibaar droogstof *fm* — **e** secador *m* líquido; secante *m* líquido — **i** essiccante *m* liquido; siccativo *m* liquido

8236 liquid glass — See sodium silicate.

8237 liquid paraffin — See paraffin oil.

8238 liquid phase

f phase *f* liquide — **d** flüssige Phase *f* — **n** vloeibare fase *f* — **e** fase *f* líquida — **i** fase *f* liquida

8239 liquid pint — Unit of capacity (liquid measure). Abbrev. liq pt (no point). See app. no. 7 and 7a.

f pinte *f* liquide — **d** englische Pinte *f* — **n** Engelse pint *fm* — **e** pinta *f* líquida — **i** pinta *f* liquida

8240 liquid quart — Unit of capacity (liquid measure). Abbrev. liq qt (no point). See app. no. 7 and 7a.

f quart *m* de gallon; quarte *f* — **n** kwart gallon *mn* — **e** cuarta *f* líquida — **i** quarta *f* liquida

8241 liquid rosin — See tall oil.

8242 list of abbreviations — See app. no. 8.

f table *f* des abréviations — **d** Verzeichnis *n* der Abkürzungen — **n** lijst *fm* van afkortingen — **e** lista *f* de abreviaturas — **i** elenco *m* delle abbreviazioni

8243 list of illustrations

f table *f* des illustrations — **d** Verzeichnis *n* der Abbildungen; Verzeichnis *n* der Illustrationen — **n** lijst *fm* van afbeeldingen — **e** índice *m* de ilustraciones — **i** elenco *m* delle illustrazioni

8244 liter — See litre.

8245 literal; literal error — Error made by the compositor in substituting one character for another, including turns, wrong founts and defective types.

f coquille *f* — **d** Satzfehler *m*; Buchstabenfehler *m*; versetzter Buchstabe *m*; fehlerhafter Buchstabe *m* — **n** zetfout *fm* — **e** error *m* del cajista; yerro *m* del cajista; falta *f* tipográfica — **i** errore *m* di compositore; refuso *m*

8246 literal error — See literal.

8247 literal translation — Translation following the words of the original exactly.

f traduction *f* littérale — **d** wortgetreue Übersetzung *f* — **n** letterlijke vertaling *f* — **e** traducción *f* literal — **i** traduzione *f* letterale

8248 litharge; yellow lead oxide; plumbous oxide; lead oxide; lead monoxide; lead protoxide — Lead oxide made by heating metallic lead to 550 °C or formed at very high temperature under the throat of the casting equipment in stereotyping, causing failures in the forme. Used as a raw material for making pigments. Soluble in acids and alkalis. See also massicot.

f litharge *f*; protoxyde *m* de plomb — **d** Lithargyrum *n*; Bleioxyd *n*; Bleimonoxyd *n*; Bleiglätte *f*; Glätte *f*; Silberglätte *f* — **n** loodoxyde *n*; loodglit *n* — **e** litargirio *m*; monóxido *m* de plomo; almártaga *f*; almártega *f* — **i** litargirio *m*

8249 lith film — Photographic film of extreme contrast.

f film *m* dit à effet lith; film *m* lith — **d** Lith-Film *m* — **n** lith-film *m* — **e** película *f* lith — **i** pellicola *f* lith; film *m* lith

8250 lithium hydrate — See lithium hydroxide.

8251 lithium hydroxide; lithium hydrate — Colourless crystals, soluble in water, slightly soluble in alcohol. Used in photographic developers. Formula: LiOH.

f hydroxyde *m* de lithium; lithine *f* caustique — **d** Lithiumhydroxyd *n*; Lithiumhydrat *n*; Lithionhydrat *n* — **n** lithium-hydroxyde *n* — **e** hidróxido *m* de litio — **i** idrossido *m* di litio

8252 litho chalk — See lithographic chalk.

8253 litho crayon — See lithographic chalk.

8254 litho etching — Litho stone or metal plate with a design drawn with a crayon. Very seldom used now.

f crayon *m* litho(graphique) — **d** Lithographie *f* — **n** lithografie *f* — **e** litografía *f* — **i** litografia *f*

8255 lithograph — Print produced from a lithographic stone, on which the design is drawn with fatty ink or crayon.

f lithographie *f*; dessin *m* d'artiste lithographique — **d** Lithographie *f*; Kunststeinzeichnung *f* — **n** lithografie *f*; litho *f* — **e** litografía *f*; dibujo *m* litográfico de artista — **i** litografia *f*; disegno *m* litografico dell'artista

8256 lithographer; lithographic artist — Crayon artist who draws on litho stones or metal plates. Negative retoucher who works on glass or film negatives and positives. Artist or stainer who works on continuous-tone negatives with dyes to adjust density of selected areas as a method of colour correction.

f lithographe *m*; artiste *m* lithographe — **d** Lithograph *m*; Steinzeichner *m* — **n** lithograaf *m* — **e**
litógrafo *m*; artista *m* litógrafo — **i** litografo *m*;
artista *f* litografico
8257 lithographic artist — See lithographer.
**8258 lithographic chalk; litho chalk; lithographic
crayon; litho crayon; chalk** — Pencil-shaped stick
consisting of soap, tallow, shellac, wax and lamp-
black, used for crayon sketches on grained paper,
for direct drawing on litho stone and metal plates,
and as a delicate staging medium in halftone
relief etching.
f crayon *m* litho(graphique); craie *f* grasse — **d**
Lithographiekreide *f*; Fettkreide *f*; litho-
graphische Kreide *f*; Lithostift *m* — **n** litho-
grafisch krijt *n*; vet krijt *n* — **e** lápiz *m* litográfico
— **i** gesso *m* litografico
8259 lithographic crayon — See lithographic
chalk.
8260 lithographic handpress; handpress — Printing press operated by hand for pulling proofs
from lithographic stones.
f presse *f* à bras lithographique; presse *f* lithographique à bras — **d** Steindruckhandpresse *f* —
n steendrukhandpers *f* — **e** prensa *f* litográfica a
brazo — **i** pressa *f* litografica a mano
8261 lithographic image — Ink-receptive image
on the lithographic press plate, either photographic or direct hand or transfer image. The
design or drawing on stone or metal plate.
f image *f* lithographique; report *m* lithographique
— **d** Bild *n* (auf einer Flachdruckplatte) — **n**
beeld *n* op de plaat; lithografie *f* — **e** imagen *f*
litográfica; litografía *f* — **i** immagine *f* litografica
8262 lithographic ink — See litho ink.
8263 lithographic press
f presse *f* litho(graphique) — **d** Steindruckpresse *f*; Steindruckmaschine *f* — **n** steendruk-
pers *f* — **e** prensa *f* litográfica — **i** macchina *f* da
stampa litografica
8264 lithographic roller — Properly, leather-
covered (nap) roller for hand rolling (inking) of
litho printing surfaces, but also rubber and metal
inking roller suitable for lithographic inks.
f rouleau *m* à main — **d** Steindruckwalze *f* — **n**
steendrukhandrol *fm*; handrol *fm* — **e** rodillo *m*
de mano — **i** rullo *m* manuale
8265 lithographic stone; litho stone — Slab of
limestone, free from veins and spots, 5-10 cm
thick (2-4 in). It must be compact enough to take
a good polish and yet sufficiently porous to
absorb the grease of the crayon. Liver-coloured
stones are the best.
f pierre *f* lithographique — **d** Lithographie-
stein *m*; lithographischer Stein *m*; Druckstein *m*;
Stein *m* — **n** lithografische steen *m*; steen *m* — **e**
piedra *f* litográfica — **i** pietra *f* litografica
8266 lithographic transfer — Proof or design or
type matter pulled with special transfer ink on
transfer paper from a lithographic image or a
relief image, used to transfer an image to the
press plate or stone.

f épreuve *f* à report — **d** Umdruck *m* — **n** overdruk *m* — **e** reporte *m* litográfico; transporte *m*
litográfico — **i** trasporto *m* litografico
8267 lithographic varnish — See litho varnish.
8268 lithograph label paper — See label paper.
8269 lithograph paper — See label paper.
8270 lithography — Planographic method of printing using the chemical repulsion between water
and grease to separate the printing from the non-
printing areas. The printing design is greasy and
ink-receptive, while the balance of the stone or
plate is dampened with water to make it ink
repellent. Name derived from litho and graphy, or
stone-writing, as limestone slabs were originally
used as lithographic printing formes. Invented in
1796 by Alois Senefelder (1771-1834). In direct
lithography the paper is printed from a reversed
design on a stone or plate; in *offset lithography
the design is first printed on a rubber-covered cylinder and offset on to the paper.
f lithographie *f*; procédé *m* lithographique — **d**
Steindruck *m*; Steindruckverfahren *n* — **n** steendruk *m*; steendrukprocédé *n* — **e** litografía *f*;
procedimiento *m* litográfico — **i** litografia *f*;
stampa *f* litografica
8271 litho ink; lithographic ink — Highly concentrated, tacky, water repellent and acid resistant ink, used for printing by lithographic stones
or offset presses. The principal characteristic is its
ability to resist excessive emulsification by the
fountain solution.
f encre *f* litho(graphique); encre *f* offset — **d**
Offsetfarbe *f* — **n** offsetinkt *m* — **e** tinta *f* lito-
(gráfica); tinta *f* offset — **i** inchiostro *m* litografico; inchiostro *m* offset
8272 lithol — Registered trade name for azo
lacquers and difficult solvable lacquable azo dyestuffs.
8273 litho label paper — See label paper.
8274 lithol red — Relatively brilliant, moderately
light-fast organic red pigment derived from a dyestuff by treatment with metallic salts. It varies
from orange to deep maroon and is the most widely used red pigment.
f rouge *m* lithol — **d** Litholrot *n* — **n** litholrood *n*
— **e** rojo *m* litol — **i** rosso *m* litolo
8275 litho paper — See label paper.
**8276 lithopone; zinc baryta white; zinc sulphide
white** — Formerly widely used white pigment
consisting of varying percentages of barium sulphate and zinc sulphide. Now rarely used in printing, but as a filler in paper.
Formula: $BaSO_4 \cdot ZnS$.
f lithopone *m*; blanc *m* anglais; blanc *m* de lithopone — **d** Lithopone *f*; Zinksulfidweiß *n*; Schwefelzinkweiß *n*; Patenzinkweiß *n* — **n** lithopoon *n* —
e litopón *m*; blanco *m* inglés — **i** litopone *m*
8277 litho stone — See lithographic stone.
8278 Lithotine — Lithographic solvent developed
by LTF to replace turpentine; is less toxic and
irritating to the skin; its use prevents dermatitis.
8279 litho varnish; lithographic varnish —

Varnish obtained by heating linseed oil until it reaches a certain degree of viscosity.

f vernis *m* litho(graphique) — **d** Lithofirnis *m* — **n** lithografische vernis *mn* — **e** barniz *m* lito-(gráfico) — **i** vernice *f* litografica

8280 litmus paper — Chemical test paper made by soaking filter paper in tincture of litmus (a blue, amorphous powder obtained from certain lichens), used in laboratories and industrial operations to detect acidity or alkalinity of solutions. Acids turn the paper red at pH 4.5, alkalis turn it blue at pH 8.3.

f papier *m* réactif; papier *m* tournesol; papier *m* de tournesol — **d** Lackmuspapier *n*; Reagenz-papier *n* — **n** lakmoespapier *n* — **e** papel *m* tornasol; papel *m* de tornasol — **i** cartina *f* di tornasole

8281 litre *GB*; **liter** *US* — Unit of capacity in the metric system. Abbrev. l (no point). Until 1964 the litre was defined as the volume of 1 kg of pure water at its maximum density and a pressure of 1 normal atmosphere. In the SI: 1 litre = 10^{-3} m^3.

f litre *m* — **d** Liter *nm* — **n** liter *m* — **e** litro *m* — **i** litro *m*

8282 little Neddy — See Neddy.

8283 live matter — See standing matter.

8284 livering — Irreversible increase in the body of inks as a result of gelation or chemical change during storage or hot milling.

f caoutchoutage *m*; épaississement *m*; épais-sissage *m* — **d** Eindicken *n* — **n** indikken *n*; taai worden *n* — **e** espesamiento *m* — **i** ispessi-mento *m*

8285 liver of antimony — See sodium-thioantimonate.

8286 liver of sulphur — See potassium trisulphide.

8287 live web — In a reel-fed printing press, amount of web between the impression point of one colour unit and the first subsequent offset roller.

f longueur *f* de band d'un groupe imprimeur à l'autre — **d** Bahnlänge *f* zwischen einem Druck-werk und dem Nächsten — **n** lengte *f* van de baan papier tussen twee drukeenheden — **e** longitud *f* de la banda de papel entre dos grupos impresores — **i** lunghezza *f* nastro fra gruppi stampanti contigui

8288 lizard — See gutter.

8289 load *v* — To place data into internal store.

f charger — **d** laden — **n** laden — **e** cargar — **i** caricare

8290 load-and-go — Computer operating technique without stops between the loading and execution phases of a program and which may in-clude assembling or compiling.

f charger et exécuter — **d** umwandeln und aus-führen — **n** laden en uitvoeren — **e** cargar y comenzar — **i** caricare e va

8291 load at break

f charge *f* de rupture — **d** Bruchlast *f* — **n** belasting *f* bij breuk; breukbelasting *f* — **e** carga *f* de rotura — **i** carico *m* di rottura

8292 loader — Service routine to read programs into internal storage in preparation for their execution.

f chargeur *m* — **d** Ladevorrichtung *f* — **n** lader *m*; laadprogramma *n* — **e** cargador *m* — **i** caricatore *m*

8293 loading material — See filler.

8294 loading platform

f quai *m* d'embarquement — **d** Ladebühne *f*; Lade-stelle *f* — **n** laadbordes *n*; laadperron *n* — **e** plata-forma *f* de carga; sitio *m* de carga — **i** piatta-forma *f* di carico e scarico

8295 loaf — See small double loaf, large double loaf.

8296 loan paper; security paper — Tough, tub sized, wove paper, made of the strongest materials and finished with a good writing surface, not highly glazed. The materials and treatment (hydration) are similar to those employed for bank and bond papers, but it is tougher and the basis weights are heavier. Used chiefly for bonds and shares, etc., which possibly gave rise to the name.

f papier *m* (pour) valeurs; papier *m* pour titres — **d** Werttitelpapier *n*; Wertzeichenpapier *n*; Wert-schriftenpapier *n*; Wertpapier *n*; Dokumenten-papier *n*; Aktienpapier *n* — **n** waardepapier *n*; aandelenpapier *n*; papier *n* voor aandelen — **e** papel *m* para acciones; papel *m* para valores; papel *m* para títulos; papel *m* para documentos — **i** carta *f* di sicurezza; carta *f* per carta valori

8297 lobe — Small projection of a negative be-yond its normal trimmed rectangular form to retain positioning marks or other data as an integral part of the negative.

f marge *f* — **n** marge *fm* voor de paskruizen — **e** margen *fm* para marcas de registro — **i** margine *m* per crocette per il registro

8298 lobster shift — See shift.

8299 lobster trick — See shift.

8300 local corrections — Corrections applied by hand to isolated parts of the picture, as distinct from overall corrections.

f corrections *fpl* locales — **d** örtliche Korrekturen *fpl* — **n** plaatselijke correcties *fpl* — **i** correzioni *fpl* locali

8301 localized watermark — Watermark appear-ing regular intervals in a sheet of paper. See also cut to register.

f filigrane *m* localisé — **d** placiertes Wasser-zeichen *n*; zentriertes Wasserzeichen *n*; Format-wasserzeichen *n* — **n** regelmatig geplaatst water-merk *n* — **e** filigrana *f* coordinada a los formatos — **i** filigrana *f* localizzata; filigrana *f* collocata regolare

8302 location — Position in a store which holds a word or part of a word.

f emplacement *m* — **d** Speicherzelle *f* — **n** geheugenplaats *fm*; locatie *f*; lokatie *f* — **e** situación *f* de memoria; ubicación *f*; posición *f*; emplazamiento *m* — **i** locazione *f*

8303 lock — Safety device on a rotary press in most control boxes, to prevent the press from being moved under power or, when the press is running, to prevent any further increase in speed. The press can only be made free running by cancelling the lock.

f arrêt *m* de sécurité — **d** Sperrvorrichtung *f* — **n** blokkeerinrichting *f* — **e** paro *m* de seguridad — **i** dispositivo *m* di sicurezza; blocco *m*

8304 locking condition — See precedence code.

8305 locking device

f dispositif *m* de verrouillage — **d** Verriegelung *f*; Sperre *f* — **n** grendel *m* — **e** dispositivo *m* de bloqueo; mecanismo *m* de inmovilización — **i** dispositivo *m* di arresto; dispositivo *m* di chiusura; dispositivo *m* di sicurezza

8306 lock nut — See check nut.

8307 lock-up; plate lock-up — Mechanical arrangement to hold the printing plates or formes on a rotary press.

f dispositif *m* de calage (des plaques); dispositif *m* de verrouillage (des plaques) — **d** Plattenverschluß *m* — **n** opsluitmechanisme *n*; plaatopsluitmechanisme *n* — **e** cierre *m*; acuñado *m*; ajuste *m* — **i** dispositivo *m* per il fissaggio di lastre

8308 lock *v* **up** — In letterpress printing, to wedge the type pages firmly within the chase or frame, by means of quoins opened and closed with a quoin key (formerly with shooter and mallet).

f serrer — **d** aufschließen; schließen — **n** insluiten; opkooien — **e** acuñar; cerrar; ajustar la forma; apretar — **i** serrare; chiudere

8309 lock *v* **up** — To lock the printing plates or formes on a rotary press. See also compression plate lock-up, tension plate lock-up.

f caler; verrouiller — **d** schließen — **n** opsluiten — **e** fijar la plancha al cilindro; sujetar (la plancha); montear la plancha; bloquear — **i** bloccare; staffare

8310 lock-up finger — Small lever engaging the notches or grooves in the underside of the plate in a tension plate lock-up of a rotary press.

f griffe *f* de verrouillage; doigt *m* d'accrochage (de cliché) — **d** Spannklaue *f* — **n** opsluitklauw *mf* — **e** uña *f* de cierre; uña *f* de montaje — **i** griffa *f* di bloccaggio; staffa *f* di bloccaggio

8311 lock-up man; stone man; stone hand; impositor; anvil beater *coll.* — Person who imposes the formes and fastens them securely in a chase.

f imposeur *m*; monteur *m* de formes; cogneur *m* — **d** Formenschließer *m* — **n** inslaander *m*; insluiter *m* — **e** platinero *m*; imponedor *m*; impositor *m*; acuñador *m* (de las formas) — **i** serra-forma *m*

8312 loft-dried paper — Tub-sized paper dried by hanging it over poles in a drying shed or in the loft dryer. Usually of better quality than machine-dried paper.

f papier *m* séché en feuilles; papier *m* séché à l'air — **d** an der Luft getrocknetes Papier *n*; luft-getrocknetes Papier *n* — **n** aan de lucht gedroogd papier *n* — **e** papel *m* secado al aire — **i** carta *f* asciugata stesa all'aria

8313 loft dryer — Heated room in which tub-sized sheets are hung over poles and dried. Used only for more expensive rag papers, nearly obsolete in commercial practice.

f séchoir *m* à feuilles; étendoir *m* — **d** Trockenboden *m*

8314 log *v* — To record the occurrence of an event, such as beginning and end of a program, on some output device, usually the console typewriter.

f enregistrer — **d** aufzeichnen — **n** optekenen — **e** registrar — **i** registrare

8315 logarithm — Exponent of the power to which a base number must be raised to equal a given number, e.g., 2 is the logarithm of 100 to the base 10. Symbol: log (no point).

f logarithme *f* — **d** Logarithmus *m* — **n** logaritme *fm* — **e** logaritmo *m* — **i** logaritmo *m*

8316 logic — The logic and arithmetic functions of a computer consist of electronic circuitry designed to execute, upon command, each of a repertoire of instructions, such as compare, add, subtract, multiply, store etc. These instructions have many of the properties of Boolean logic, such as "and", "or", "not", and "exclusive or".

f logique *f* — **d** Logik *f* — **n** logica *f* — **e** lógica *f* — **i** logica *f*

8317 logical display — Computer typesetting in which the type size can be controlled by logical rules applied to the data.

8318 logic decision — Choice or ability to choose between alternatives.

f décision *f* logique — **d** logischer Entschluß *m* — **n** logische beslissing *f* — **e** decisión *f* lógica — **i** decisione *f* logica

8319 logic hyphenation program — Routine for the analysis of words which must be divided into syllables.

f programme *m* de coupure logique — **d** logische Silbentrennung *f* — **n** taalkundige woordafbreking *f* — **e** programa *m* de corte lógico — **i** programma *m* di divisione delle parole in sillabe

8320 logo — Identification or trade-mark of the packaging customer, which is part of the over-all design.

8321 logo — See logotype.

8322 logotype; logo *fam.* — Several letters or entire word cast on one type or as a single machine matrix, such as rd, st, av, or a uniquely drawn letter within a circle, such as ©, ®, etc., used in printing of directories or other works in which they are repeated frequently. Not to confuse with *ligature.

f logotype *m*; polytype *m* — **d** Logotype *f*; Silbentype *f*; Mehrbuchstabenletter *f* — **n** logotype *f* — **e** logotipo *m* — **i** logotipo *m*

8323 logs — Pulpwood cut in 60-120 cm (2-4 ft) lengths, before shipment to the pulp mill.

f rondins *mpl* — **d** Klötze *mpl*; geschältes Rundholz *n*; Prügel *m* — **n** papierhout *n* — **e** leños *mpl* — **i** tondelli *mpl*

8324 logwood; campeachy wood — Heavy, brownish-red heartwood of a West Indian and Central America tree, formerly used as a dye in printing inks, mainly in the manufacture of bleachable black inks.
f extrait *m* de campêche — **d** Blauholz *n* — **n** campêchehout *n*; blauwhout *n* — **e** campeche *m*; madera *f* de campeche; palo *m* de campeche — **i** legno *m* di campeggio

8325 long — A size of board. See app. no. 3.

8326 long accent; long mark; straight accent; macron — Quality mark (ˉ), placed over a vowel as in ā, ē, etc., indicating a long sound.
f accent *m* long; marque *f* de longueur (d'une voyelle) — **d** Längezeichen *n* — **n** lang accent *n* — **e** acento *m* largo — **i** segno *m* di allungamento (su una vocale)

8327 long colombier — A size of board. See app. no. 3.

8328 long descenders; long letters — Alternative characters e.g., g, j, p, q, y, in Caledonia, Electra, Garamond, Garamond bold, Janson, and Times roman with their respective italics.
f lettres *fpl* descendantes; lettres *fpl* longues — **d** Buchstaben *mpl* mit Unterlängen; lange Buchstaben *mpl* — **n** staartletters *fmpl*; lange letters *fmpl* — **e** letras *fpl* con trazo bajo; letras *fpl* largas (con rasgo descendiente) — **i** lettere *fpl* discendenti; lettere *fpl* lunghe

8329 long direction — See machine direction.

8330 long double demy — A size of paper. See app. no. 3.

8331 long double elephant — A size of board. See app. no. 3.

8332 long double medium — A size of paper. See app. no. 3.

8333 long elephant — Paper (usually wallpaper base), in reels of various lengths. See app. no. 3.

8334 long fold — Folded lengthwise, with the grain. In bristol boards the grain runs lengthwise of the sheet.
f plié en sens machine — **d** Faltung *f* in der Längsrichtung; in der Längsrichtung gefältet — **n** in langsrichting gevouwen — **e** doblado a lo largo — **i** piega *f* longitudinale

8335 long-grain paper
f papier *m* sens machine — **d** Schmalbahn-papier *n*; Papier *n* in Schmalbahn — **n** langlopend papier *n* — **i** carta *f* a fibre longitudinali

8336 long-grain wood; long-tailed wood
f bois *m* de fil — **d** Langholz *n* — **n** langshout *n* — **e** madera *f* de hilo — **i** legno *m* a fibre longitudinali

8337 long hundredweight; gross hundredweight — Unit of weight. Abbrev. cwt. See also short hundredweight, app. no. 7 and 7a.

8338 long ink — Ink which stretches considerably before breaking, and has good flow in the fountain. High viscosity, no yield value. See also length.

f encre *f* longue; encre *f* fluante — **d** lange Farbe *f* — **n** lange inkt *m* — **e** tinta *f* blanda dúctil — **i** inchiostro *m* filante

8339 longitudinal parity — Summation check that the number of bits within a block or within each channel in a block are of the nominated odd or even parity.
f parité *f* horizontale — **d** horizontale Parität *f* — **n** horizontale pariteit *f* — **e** paridad *f* horizontal — **i** parità *f* orizzontale

8340 longitudinal parity check; track parity check — Parity check performed on the bits in each track of magnetic tape or punched tape by recording them simultaneously at the end of each block in the form of a "longitudinal check character", which is regenerated and checked when the block is read.
f contrôle *m* de parité horizontale — **d** horizontale Paritätskontrolle *f* — **n** horizontale pariteitscontrole *fm* — **e** comprobación *f* de paridad horizontal — **i** controllo *m* di parità orizzontale

8341 long letters — See long descenders.

8342 long mark — See long accent.

8343 long page — Page with more lines than the standard number, that must be reduced accordingly.
f page *f* longue — **d** lange Seite *f*; Seite *f* auf Handtuchformat — **n** lange pagina *fm* — **e** página *f* larga — **i** pagina *f* lunga

8344 long primer — Old name for a size of type. See app. no. 1.

8345 long royal — A size of board. See app. no. 3.

8346 long s — Old form of the "s", only distinguishable from the f by the fact that on the long s the horizontal cross-stroke projects to the left side only of the vertical stroke. Discarded by John Bell about 1775.
f s *m* long — **d** langes S *n* — **n** lange s *m* — **e** s *f* larga — **i** s *f* lunga

8347 long-storage paper — Paper for long storage documents.
f papier *m* pour documents de longue conservation — **d** Dokumentenpapier *n*; Papier *n* für langdauernde Belege — **n** documentpapier *n* — **e** papel *m* de larga duración; papel *m* de carpetas — **i** carta *f* per documenti di lunga conservazione

8348 long-tailed wood — See long-grain wood.

8349 long thin — A size of board. See app. no. 3.

8350 long ton; gross ton — Unit of weight. Abbrev. UK ton. See also short ton, app. no. 7 and 7a.
f tonne *f* forte — **d** Tonne *f* — **n** long ton *fm* — **e** tonelada *f* larga — **i** tonnellata *f* lunga

8351 long varnish — Varnish to thin down the ink.
f vernis *m* fort; vernis *m* long; vernis *m* à réduire — **d** Drucköl *n* — **n** slappe vernis *mn* — **e** barniz *m* blando dúctil — **i** vernice *f* di allungamento

8352 long vowel
f voyelle *f* longue — **d** langer Laut *m*; langer Vokal *m*; lang auszusprechender Vokal *m* — **n** lange klinker *m* — **e** vocal *f* larga — **i** vocale *f*

lunga

8353 look-through — Structural appearance of a sheet of paper observed when viewed by transmitted light. See also formation.

f épair *m* — **d** Durchsicht *f* — **n** doorzicht *n* — **e** trasluz *m* — **i** speratura *f*; spera *f*; sfera *f*

8354 loop — Rounded part of a letter, not of a circular or formal oval character, as in the lower part of g.

f boucle *f* — **d** Schleife *f* — **n** lus *fm* — **e** ojal *m* — **i** gancio *m*

8355 loop — Series of computer program steps which are repeated until a predetermined condition is reached. E.g., in computer typesetting each input code is read and identified, character widths are extracted from a look-up table and added to the cumulative line width. This is repeated until the justification zone is reached.

f boucle *f* — **d** Schleife *f*; Meßschleife *f*; Regelschleife *f* — **n** lus *fm* — **e** bucle *m*; lazo *m* — **i** giro *m*; giro *m* controllato

8356 looper — Special needle in a book sewing machine.

f loupier *m* — **d** Hakennadel *f* — **n** haaknaald *fm* — **e** aguja *f* de ganchillo — **i** uncinetto *m*; ago *m* a gancio

8357 looping roller — See jockey roller.

8358 loose accent — See floating accent.

8359 loose back — See hollow back.

8360 loose fibers — See loose fibres.

8361 loose fibres *GB*; **loose fibers** *US*

f fibres *fpl* lâches — **d** lockere Fasern *fpl* — **n** losse vezels *fmpl* — **e** fibras *fpl* sueltas — **i** fibre *fpl* sciolte

8362 loose leaf — Sheet trimmed on four sides, plain ruled or printed with punched holes in the back margin for insertion in a loose-leaf binder.

f feuillet *m* mobile — **d** loses Blatt *n* — **n** los blad *n* — **e** hoja *f* suelta — **i** foglio *m* mobile

8363 loose-leaf binder — See loose-leaf binding.

8364 loose-leaf binding; loose-leaf binder — Binding case with screws, posts or rings to hold punched sheets in place between the covers.

f reliure *f* à feuillets mobiles — **d** Lose-Blätter-Buch *n*; Buch *n* in loser Blattform — **n** losbladig boek *n*; losbladige band *m* — **e** encuadernación *f* de hojas sueltas; encuadernación *f* de hojas movibles — **i** legatura *f* a fogli mobili; legatura *f* a fogli sciolti

8365 loose-leaf ledger

f registre *m* à feuillets mobiles — **d** Lose-Blätter-Geschäftsbuch *n* — **n** losbladig register *n*; register *n* met losse bladen — **e** libro *m* de hojas cambiables; libro *m* de hojas movibles; libro *m* de hojas sueltas — **i** registro *m* a fogli mobili

8366 loose line

f ligne *f* faible — **d** zu schwache Zeile *f*; zu schwache Matrizenzeile *f* — **n** slappe regel *m* — **e** línea *f* floja — **i** linea *f* troppo debole

8367 loosely wound reel; poorly wound reel

f bobine *f* peu serrée; bobine *f* molle — **d** lose gewickelte Rolle *f*; lose aufgewickelte Rolle *f*; weiche Rolle *f*; weichgewickelte Rolle *f*; schlecht gewickelte Rolle *f* — **n** slappe rol *fm*; slecht gewikkelde rol *fm* — **e** bobina *f* floja — **i** bobina *f* molle

8368 loosen a forme — See unlock *v.*

8369 loosened binding — See slippery book.

8370 loose water — See white water.

8371 losing paper — See insufficient feed.

8372 loss in voltage — See voltage drop.

8373 loss of density

f perte *f* de densité — **d** Schwärzungsverlust *m* — **n** zwartingsverlies *n* — **e** pérdida *f* de densidad — **i** perdita *f* di densità

8374 loss of drier — See drier dissipation.

8375 loss of paper — To determine the loss of paper in reels due to damage, the following formula can be used: percentage of paper loss = $400 \times D \ (R - D)/(R^2 - C^2)$, in which: D = depth of damage in cm; R = diameter of paper reel in cm; C = diameter of reel cone in cm.

f perte *f* de papier — **d** Papierverlust *m* — **n** papierverlies *n* — **e** pérdida *f* de papel — **i** perdita *f* di carta

8376 lost lead — See lead of the web.

8377 Lost Monday; Handsel day; Hansel Monday — A handsel is a gift or token for good luck or as an expression of good wishes, as at the beginning of the new year, or when entering upon a new state, situation or enterprise.

f lundi *m* perdu — **d** Blauer Montag *m* — **n** Koppermaandag *m* (eerste maandag na Driekoningen); Verloren Maandag *m Be*; Egyptische Maandag *m Be* — **i** prima giornata *f* di lavoro; giorno *m* d'inaugurazione

8378 lot tolerance per cent defective

f proportion *f* de pièces défectueuses tolérées dans le lot — **d** Schlechtgrenze *f*; Rückweisegrenze *f* — **n** grenskwaliteit *f* voor de afnemer — **e** porcentaje *m* de piezas defectuosas — **i** percentuale *f* di scarti tollerata nel lotto

8379 lower case — Case which holds the lower case letters.

f bas *m* de casse — **d** Unterkasten *m* — **n** onderkast *fm* — **e** caja *f* baja — **i** cassa *f* bassa; bassa cassa *f*

8380 lower-case letters — Smaller characters of a type face as distinguished from the upper case or capital letters, so named because they are contained in the lower of two type cases as used by the hand compositor when setting type.

f lettres *fpl* bas de casse; lettres *fpl* minuscules — **d** Kleinbuchstaben *mpl*; kleine Buchstaben *mpl*; Gemeine *mpl* — **n** onderkastletters *fmpl*; kleine letters *fmpl* — **e** letras *fpl* de caja baja; letras *fpl* minúsculas; minúsculas *fpl* — **i** lettere *fpl* bassa cassa; lettere *fpl* minuscole

8381 lower deck — Lower printing unit on a multi-unit press in which the units are placed at different levels.

f groupe *m* inférieur — **d** unteres Druckwerk *n* — **n** onderste drukwerk *n* — **e** grupo *m* de impresión inferior — **i** gruppo *m* inferiore

8382 lower magazine — Magazine on a line casting machine. An associated unique code is required for TTS operation when changing from upper magazine to lower magazine.
f magasin *m* inférieur — **d** unteres Magazin *n* — **n** onderste magazijn *n*; ondermagazijn *n* — **e** depósito *m* bajo; almacén *m* bajo — **i** magazzino *m* inferiore

8383 lower rail — See rail.

8384 low-intensity desensitization — This term should be restricted to the case where the low-intensity exposure is given first, but it should be broadened to include post-exposure by light at low temperature.

8385 low key — Picture with predominately dark tones.
f low key *m* — **d** Lowkey *n*; Bild *n* mit vorwiegend dunklen Tönen — **n** low key *m* — **e** imagen *f* oscura — **i** a toni scuri

8386 low negative — Too thin halftone negative with no opaque dots in the shadows.
f négatif *m* mou — **d** dünnes Negativ *n*; unterbelichtetes Negativ *n* — **n** zacht negatief *n* — **e** negativo *m* débil — **i** negativo *m* sottoesposto

8387 low spot — Depression in the cylinder of a printing press.
f faible *m* — **d** Vertiefung *f* — **n** kuil *m* — **e** depresión *f* — **i** depressione *f*

8388 low-task rating method — Procedure in which the time study man compares the actual rate of working of an operator with his concept of the rate necessary to earn the basic wage. See also high-task rating method.
n tempo-waardering *f* (op basis van de redelijke minimum prestatie) — **i** metodo *m* di calcolo delle prestazioni minime

8389 low voltage
f basse tension *f* — **d** Niederspannung *f* — **n** lage spanning *f*; laagspanning *f* — **e** baja tensión *f* — **i** bassa tensione *f*

8390 lozenged — See lozenge shaped.

8391 lozenge shaped; lozenged; lozengy; diamond shaped
f losangé; en losange — **d** rautenförmig — **n** ruitvormig; in ruitvorm — **e** romboidal; de figura de romboide — **i** a forma di losanga; rombico; romboidale

8392 lozengy — See lozenge shaped.

8393 lube oil — See lubricating oil.

8394 lubricant — Grease or oil used on the bearings or moving parts of a machine to reduce friction. Ensures smoother and more efficient operation and longer life of the parts.
f lubrifiant *m*; agent *m* de lubrification — **d** Schmiermittel *n*; Schmierstoff *m* — **n** smeermiddel *n* — **e** lubricante *m* — **i** lubrificante *m*

8395 lubricate *v*; **oil** *v*
f lubrifier; graisser — **d** schmieren; ölen — **n** smeren; oliën — **e** lubricar; engrasar; aceitar — **i** lubrificare; ingrassare

8396 lubricate *v* — See oil *v*.

8397 lubricating oil; lube oil; machine oil — Selected fractions of refined mineral oils used for lubrication of moving surfaces; also transformer oils used for electrical insulation and cooling. Usually with small amounts of additives to impart properties such as viscosity index and detergency. They range in consistency from thin liquids to grease-like substances. In contrast to lubricating greases, lube oils do not contain solid or fibrous materials.
f huile *f* de graissage; huile *f* de machine; huile *f* de lubrication — **d** Schmieröl *n*; Maschinenöl *n* — **n** smeerolie *fm*; machine-olie *fm* — **e** aceite *m* de engrase; aceite *m* para maquinaria; aceite *m* de máquina; aceite *m* lubricante; engrase *m* líquido; óleo *m* lubrificante; lubrificante *m* — **i** olio *m* lubrificante; olio *m* da macchina

8398 lubricating point; oil hole
f trou *m* de graissage; orifice *m* de graissage; trou *m* graisseur — **d** Schmierloch *n*; Ölloch *n* — **n** smeergat *n* — **e** punto *m* de engrase — **i** buco *m* da oliare

8399 lucca oil — Oil cold pressed from ripe olives.
f huile *f* d'olives de Lucques — **d** Olivenöl *n* — **n** olijfolie *f* uit lucca — **e** aceite *m* de oliva — **i** olio *m* d'oliva de Lucca

8400 Lucy — See camera lucida.

8401 Ludlow — Slug-casting machine for composing lines of type, principally in display sizes of 18 points or larger. Matrices are assembled by hand and locked in a frame, then placed over a slot in the machine; molten metal is brought in contact with the matrices, thereby casting a solid line (slug).
f Ludlow *f* — **d** Ludlow-Setzmaschine *f* — **n** Ludlow *fm* — **e** Ludlow *f* — **i** Ludlow *f*

8402 Ludlow operator
f opérateur *m* Ludlow — **d** Ludlow-Setzer *m* — **n** Ludlow-zetter *m* — **e** ludlista *m*; titulero *m* Ar — **i** operatore *m* alla Ludlow

8403 lug — See ear.

8404 lumbard — Term used during the seventeenth century for a size of paper. Now replaced by *demy.

8405 lumbeck *v* — Trade name for an adhesive binding process.
f relier par collage Lumbeck — **d** lumbecken; klebebinden — **n** lumbecken — **e** encuadernar sin costuras Lumbeck — **i** legare con il procedimento di Lumbeck

8406 lumber hand — A size of wrapping paper. Name derived from lombard, a paper size name and the hand in the watermark. See app. no. 3.

8407 lumen — The luminous flux emitted within unit solid angle (steradian) by a point source having a uniform intensity of 1 candle.
f lumen *m* — **d** Lumen *n* — **n** lumen *n* — **e** lumen *m* — **i** lumen *m*

8408 lumerg — Unit of luminous energy; a contraction of lumen-erg. If the luminous efficiency of a source emitting 1 erg of radiant energy is L lumens per watt, it has a luminous of L lumergs.
f lumen-erg *m* — **d** Lumen-erg *m* — **n** lumen-

erg *m* − **e** lumen-erg *m* − **i** lumen-erg *m*
8409 luminance *US*; **photometric brightness** *GB*
− Unit of luminous *intensity. A light source with a luminous intensity of 1 candela and a luminous surface of 1 m^2 has a luminance of 1 nit (cd/m²). Symbol: L. The luminance of a colour (the mean of the ordinates of its transmission or reflexion curve) can be modified by the addition of white of black.
f luminance *f* − **d** Leuchtdichte *f* − **n** luminantie *f* − **e** luminancia *f* − **i** luminanza *f*
8410 luminance factor − Ratio of the luminance of the body to that of a perfect reflecting diffuser identically luminated. Photometric correlate of lightness.
f facteur *m* de lumineux − **d** Leuchtdichte-faktor *m* − **n** luminantie-factor *m* − **e** factor *m* de luminancia − **i** fattore *m* di luminanza
8411 luminescence − Emission of light (visible or invisible) from a body (both organic and inorganic substances) from any cause other than high temperature. A distinction is made between *fluorescence and *phosphorescence. Fluorescent substances cease to shine when the exciting source is removed, while phosphorescent substances continue to radiate light for a characteristic period of time.
f luminescence *f* − **d** Lumineszenz *f* − **n** luminescentie *f* − **e** luminescencia *f*; luminiscencia *f* − **i** luminescenza *f*
8412 luminescent ink
f encre *f* luminescente − **d** lumineszierende Farbe *f*; Lumineszenzdruckfarbe *f* − **n** luminescerende inkt *m* − **e** tinta *f* de luminiscencia; tinta *f* de luminescencia − **i** inchiostro *m* luminescente
8413 luminosity − In common speech the luminosity of a colour is determined by the presence in the colour of more or less black. See also brightness.
f luminosité *f* − **d** Helligkeit *f*; Helligkeitswert *m* − **n** helderheid *f* − **e** luminosidad *f* − **i** luminosità *f*
8414 luminous flux; light flux − Flow of radiant energy commonly measured in lumens.
f flux *m* lumineux − **d** Lichtstrom *m* − **n** lichtstroom *m* − **e** flujo *m* luminoso − **i** flusso *m* luminoso
8415 luminous intensity − In any direction, the ratio of the luminous flux emitted by any element of a source, in an infinitesimal solid angle containing this direction, to the solid angle.
f intensité *f* lumineuse − **d** Lichtstärke *f* − **n** lichtsterkte *f* − **e** intensidad *f* luminosa − **i** intensità *f* luminosa
8416 Lumitype − Name used in Europe for the *Photon composing machine. With the demise of Photon, and the Photon products prior to the pacesetter model, this name is of purely historical interest.
8417 lumps − Bunches of fibres or other material which produce spots in a sheet higher than the surface of the paper. See also colour lumps, hard lumps, soft lumps.
f pâtons *mpl*; mâtons *mpl*; boutons *mpl*; andouilles *fpl* − **d** Knoten *mpl*; Batzen *mpl*; Patzen *mpl* − **n** stofknopen *mpl*; knopen *mpl* − **e** grumos *mpl* − **i** grumi *mpl*
8418 luster − See lustre.
8419 lustre *GB*; **luster** *US* − Reflection of light from the surface of an object or material, and the appearance of the surface by that reflected light.
f lustre *m* − **d** Glanz *m* − **n** glans *m* − **e** lustre *m* − **i** lucido *m*
8420 lux − Unit of illuminance in the metric system, defined as 1 lumen per m²; equivalent to the metre candle.
f lux *m* − **d** Lux *m* − **n** lux *fm* − **e** lux *m* − **i** lux *m*
8421 lye − Preparation of sodium hydroxide or potassium hydroxide, used by printers to clean type formes before introduction of gasoline.
f lessive *f* − **d** Lauge *f*; Waschlauge *f* − **n** loog *fmn* − **e** lejía *f* − **i** liscivio *m*
8422 lye brush
f brosse *f* à lessive − **d** Laugenbürste *f* − **n** loogborstel *m* − **e** broza *f* para (limpiar con) lejía; cepillo *m* para lejía − **i** spazzola *f* per liscivi
8423 lying press; finishing press; laying press *depr.* − Horizontally lying screw press, to be placed on a binder's bench to hold a book for backing, edge gilding and other bookbinding operations.
f presse *f* allemande; presse *f* à billot − **d** Klotzpresse *f* − **n** blokpers *fm* − **e** prensa *f* para el acabado de libros a mano − **i** torchietto *m*
8424 lyonnaise − Style of book binding with painted covers and a dotted background, and large corner ornaments. Though Lyons in the 16th century was a centre of bookbinding it is said that these styles are not peculiar to Lyons bindings.
f reliure *f* à la lyonnaise − **d** Lyoneser Einband *m* − **n** boekband *m* à la lyonnaise − **e** encuadernación *f* a la Lionesa − **i** legatura *f* alla lionese
8425 lyophilic − Characterizing a material which readily goes into colloidal suspension; if into water, it may be termed *hydrophilic. The colloid is stabilized by the formation of a protective layer of molecules of the dispersing medium about the suspended particles. Examples: glue, gelatine.
f lyophile; aisément soluble − **d** lyophil − **n** lyofiel − **e** liofílico − **i** liofilo
8426 lyophobe − See lyophobic.
8427 lyophobic; lyophobe − Characterizing a material which exists in the colloidal state without having any significant affinity for the medium, generally stabilized by the adsorption of ions and coagulating when the charge is neutralized.
f lyophobe − **d** lyophob − **n** lyofoob − **e** liófobo − **i** liofobo
8428 lyophobic colloids − See hydrophobic colloids.

8429 macerate *v* — To mash or crush into a soft mass, usually in water.

f macérer — **d** erweichen; einweichen — **n** ontsluiten; weken — **e** macerar — **i** macerare

8430 macerated straw pulp — Pulp obtained by fermentation of straw soaked in an alkaline solution without external heating.

f pâte *f* de paille macérée — **d** erweichter Strohzellstoff *m* — **n** gemacereerde strocelstof *fm* — **e** pasta *f* química de paja; pulpa *f* química de paja — **i** pasta *f* di paglia macerata

8431 machine address — See absolute address.

8432 machine chest — Large tank with agitator in which the prepared papermaking fibres are stored before being pumped to the paper machine.

f cuvier *m* de machine; cuvier *m* de tête — **d** Maschinenbütte *f* — **e** tina *f* de (la) máquina; tina *f* de alimentación; tina *f* de cabeza — **i** tino *m* della macchina; tina *f* della macchina

8433 machine chest — See stock chest.

8434 machine coated board — High-grade bending board, the top of which has been coated on the paper machine with a mixture of a mineral pigment and a suitable adhesive to improve the printability and appearance.

f carton *m* couchémachine; carton *m* couché sur machine — **d** maschinengestrichener Karton *m* — **n** machinegestreken karton *m* — **e** cartón *m* estucado a máquina — **i** cartone *m* patinato in macchina

8435 machine coated paper; mc paper — Paper coated while running on the paper machine. See also coated paper.

f papier *m* couché-machine; papier *m* couché sur machine — **d** maschinengestrichenes Papier *n* — **n** machinegestreken papier *n* — **e** papel *m* estucado a máquina — **i** carta *f* patinata in macchina

8436 machine coating — Applying coating to paper or paperboard with equipment which is an integral part of the paper machine. Sometimes called *on-machine coating; distinguished from *off-machine coating or *conversion coating.

f couchage *m* sur machine — **d** Maschinenstreichverfahren *n*; Maschinenstrich *m* — **n** machinestrijkprocédé *n*; strijken *n* op de machine — **e** estucado *m* en la máquina; estucado *m* a máquina; aplicación *f* en la máquina — **i** patinatura *f* a macchina

8437 machine code — See computer code.

8438 machine composition — Composition done with a typesetting machine.

f composition *f* mécanique — **d** Maschinensatz *m* — **n** machinezetsel *n* — **e** composición *f* mecánica; composición *f* de máquina — **i** composizione *f* meccanica

8439 machine compositor; composing machine operator — See also keyboard operator.

f claviste *m* — **d** Maschinensetzer *m* — **n** machinezetter *m* — **e** mecanotipista *m*; operador *m* mecanotipista — **i** compositore *m* a macchina; operatore *m* di macchina; tastierista *m*

8440 machine compositor
See operator.

8441 machine correction

f changements *mpl* sous presse; correction *f* sous presse — **d** Änderungen *fpl* in der Druckmaschine; Maschinenrevision *f* — **n** machinecorrectie *f*; perscorrectie *f* — **e** cambio *m* en la prensa — **i** correzione *f* di macchina

8442 machine dependent — In computerized typesetting, function governed by the speed or characteristics of a specific device.

8443 machine direction; making direction; long direction; grain direction — Dimension of paper or board corresponding to the direction of the flow on the paper machine.

f sens *m* machine; sens *m* du papier; sens *m* de fabrication; sens *m* du grain; sens *m* marche; sens *m* de coulée — **d** Maschinenrichtung *f*; Faserlaufrichtung *f*; Laufrichtung *f*; Papierlaufrichtung *f*; Längsrichtung *f*; Arbeitsrichtung *f* — **n** looprichting *f*; machinerichting *f*; langsrichting *f* — **e** dirección *f* de la máquina; dirección *f* de fibra; sentido *m* de fabricación; sentido *m* de fibra — **i** direzione *f* di macchina; direzione *f* di fabbricazione

8444 machine dried paper — Paper dried by passing over steam heated cylinder(s) which constitute the drying part of the paper machine.

f papier *m* séché à la machine — **d** maschinengetrocknetes Papier *n* — **n** op de machine gedroogd papier *n* — **e** papel *m* secado a máquina — **i** carta *f* essiccata in macchina

8445 machine etching — Etching printing plates in a machine by mechanical application of the mordant to the metal surface, or by electrical action.

f morsure *f* à la machine — **d** Maschinenätzung *f* — **n** machinaal etsen *n*; machine-etsing *f* — **e** grabado *m* a máquina; mordido *m* a máquina — **i** morsura *f* a macchina

8446 machine finish; mill finish — Finish obtained on a paper machine, of the sheet as it leaves the last drier or as it leaves the calender stack. It may be a dry or a water finish.

f fini *m* sur machine; apprêté sur machine — **d** maschinenglatt — **n** machineglad — **e** acabado de máquina; alisado de máquina — **i** lisciomacchina *f*; lisciato in macchina

8447 machine-finished paper; mill-finished paper; mf paper — Paper treated mechanically on a paper machine to obtain a smooth and uniform appearance on both sides.

f papier *m* apprêté (sur machine); papier *m* apprêté à la machine — **d** maschinenglattes Papier *n* — **n** machineglad papier *n* — **e** papel *m* alisado; papel *m* semisatinado; papel *m* supercalandrado — **i** carta *f* lisciata in macchina; carta *f* liscia di macchina

8448 machine gilding — See gilding.

8449 machine-glazed paper or board; mg-paper; mg-board — Paper or board made smooth and glossy on one side by drying on a heated, polished metal cylinder which forms part of the drying section of the machine.
f papier *m* frictionné (sur machine); papier *m* frictionné d'un côté; carton *m* frictionné; carton *m* frictionné d'une côté — **d** einseitig glattes Papier *n*; einseitig glatter Karton *m* — **n** eenzijdig glad papier *n*; eenzijdig glad karton *n* — **e** papel *m* monolúcido; papel *m* glaseado a máquina; papel *m* satinado por una cara; cartón *m* monolúcido; cartón *m* glaseado a máquina; cartón *m* satinado por una cara — **i** carta *f* monolucida; cartone *m* monolucido

8450 machine-glazed sulphite paper
f papier *m* au bisulfite frictionné — **d** einseitig glattes Sulfitpapier *n* — **n** perkament sigaretpapier *n* — **e** papel *m* sulfito satinado — **i** carta *f* al solfito lisciata a macchina

8451 machine-independent language; computer-independent language — Programming language developed without consideration of the characteristics of a particular data processing system. In practice such languages are more powerful than most machine-oriented languages and require compilers to translate from source program to machine coded object program (e.g., fortran, algol, cobol).
f langage *m* indépéndant de la machine — **d** maschinenunabhängige Programmiersprache *f* — **n** machine-onafhankelijke taal *fm* — **e** lenguaje *m* independiente de la máquina — **i** linguaggio *m* indipendente dalla macchina

8452 machine instruction — See computer instruction.

8453 machine language; computer language — The manner of addressing or commanding a computer to execute its instructions, at the simplest and most efficient level the computer can "understand", but in a manner requiring more complex and less flexible programming than is true of higher-level languages.
f langage *m* machine — **d** Maschinensprache *f* — **n** machinetaal *fm* — **e** lenguaje *m* de máquina — **i** linguaggio *m* di macchina

8454 machine-made — Made by means of a machine, instead of by hand.
f fait à la machine — **d** maschinell hergestellt; maschinengemacht — **n** machinaal vervaardigd — **e** fabricado a máquina — **i** fatto a macchina

8455 machine minder — See pressman.

8456 machine-minder's assistant
f aide-conducteur *m*; aide-conductrice *f*; manœuvre *m* — **d** Hilfsarbeiter *m* (an der Druckmaschine); Helfer *m* des Maschinenmeisters — **n** hulparbeider *m*; hulp *m* van de drukker — **e** ayudante *m* mecánico; auxiliar *m* del prensista — **i** assistente *m* di macchina

8457 machine oil — See lubricating oil.

8458 machine plate — See press-plate.

8459 machine-readable data; machine-readable information; machine-sensible information — Data in a form to be read into a computer, e.g., punched tape, magnetic tape, disc and drum, and input for optical character recognition and magnetic ink character recognition equipment. When optical character recognition techniques can be applied to printers' type faces all printed data will be potentially machine readable.
f information *f* lisible par la machine — **d** maschinenlesbare Information *f* — **n** door de machine leesbare informatie *f* — **e** información *f* legible por la máquina — **i** informazione *f* leggibile per la macchina

8460 machine-readable information — See machine-readable data.

8461 machine room — See press-room.

8462 machine-room manager — See press-room overseer.

8463 machine-sensible information — See machine-readable data.

8464 machine sewing — Continuous sewing of the signatures of books to tapes in a sewing machine, the book blocks being separated afterwards.
f couture *f* mécanique — **d** Maschinenheftung *f* — **n** naaien *n* op de machine — **e** cosido *m* a máquina — **i** cucitura *f* a macchina

8465 machine shop — Department equipped with lathes, planers, drilling machines, etc., in which parts of machines can be made or repaired.
f atelier *m* de machinerie — **d** Werkstätte *f*; Schlosserei *f* — **n** machinewerkplaats *fm*; bankwerkerij *f* — **e** sección *f* de talleres — **i** officina *f* meccanica

8466 machine speed; speed of machine — The number of printing cycles made by the cylinders of a printing machine per hour.
f vitesse *f* de la machine — **d** Maschinengeschwindigkeit *f* — **n** machinesnelheid *f*; snelheid *f* van de machine — **e** velocidad *f* de la máquina — **i** velocità *f* della macchina

8467 machine translation; automatic translation; mechanical translation — Translation from one form to another.
f traduction *f* à machine; traduction *f* mécanique — **d** maschinelle Übersetzung *f*; automatische Übersetzung *f* — **n** machinale vertaling *f*; mechanische vertaling *f* — **e** traducción *f* mecánica — **i** traduzione *f* mecanica

8468 machine width — Width of material that can be made or processed on a machine.
f largeur *f* de machine — **d** Maschinenbreite *f* — **n** machinebreedte *f* — **e** ancho *m* de la máquina; anchura *f* de la máquina — **i** larghezza *f* di macchina

8469 machine wire; Fourdrinier wire; wire gauze — Endless woven wire cloth on which the diluted paper stock flows and the paper web is formed. The water drains through the wire, depositing the fibres as a felted mass.

f toile *f* de la machine; table *f* de fabrication; toile *f* métallique; table *f* de Fourdrinier; table *f* plate; tamis *m* — **d** Maschinensieb *n*; Sieb *n* — **n** machinezeef *fm*; zeef *fm* van de papiermachine; zeef *fm*; zeefdoek *n* — **e** tela *f* de máquina; tela *f* Fourdrinier; tela *f* sin fin de la máquina de papel — **i** tela *f* di macchina; tavola *f* di fabbricazione; setaccio *m* della macchina per carte

8470 machine word — See computer word.

8471 machine word length — See word length.

8472 mackie lines — *Adjacency effect which occurs at the edge of a dark area in a developed negative image as dark or light lines, caused by changes of developer activity in the edge areas. The edge of the image may be darker than the rest (edge effect) and the adjacent lighter area may be lighter at the edge than the rest of the area (fringe effect).

8473 mackle — Spot or blemish in printing. Blurred impression.

f impression *f* maculée — **d** Schmitz *m*; Schmitzen *n*; Schmieren *n*; Dublieren *n* — **n** misdruk *m* — **e** remosqueo *m*; repintado *m*; repinte *m*; maculado *m* — **i** chiazza *f*; macchia *f* di stampa; maculatura *f*

8474 mackling rails — See bed bearers.

8475 macro — Prefix denoting large, in contrast to micro, i.e., small.

8476 macro — Computer instruction or a program to execute such instruction. Macro coding provides names for subroutines which can be invoked by reference to the appropriate "macro".

8477 macro instruction — Single instruction which is expanded to a predetermined sequence of instructions during assembly of the program. The number of instructions in the expansion may vary depending on information which may or may not be passed along by the original instruction.

f macro-instruction *f* — **d** Makrobefehl *m*; Makroinstruktion *f* — **n** macro-opdracht *fm*; macroinstructie *f* — **e** macroinstrucción *f* — **i** macroistruzione *f*

8478 macron — See long accent.

8479 magazine — Receptacle to store slug composing matrices ready to be assembled in lines. A magazine has 91 channels, containing up to 20 identical matrices for founts from 6-18 points, however, there are machines having 72-channel magazines for larger type sizes.

f magasin *m* — **d** Magazin *n*; Ablegekasten *m* — **n** magazijn *n* — **e** almacén *m*; depósito *m* Pe — **i** magazzino *m*; cassetta *f* di scomposizione

8480 magazine — See journal.

8481 magazine escapement — See escapement.

8482 magazine frame — Part of the slug composing machine.

f châssis *m* de magasin; châssis *m* des magasins — **d** Magazinrahmen *m* — **n** magazijnframe *n*; magazijnfreem *n* — **e** portaalmacenes *m*; soporte *m* para almacenes — **i** telaio *m* dei magazzini

8483 magazine frame — See cage.

8484 magazine frame operating equipment —

Part of a typesetting machine.

f mécanisme *m* de changement de magasin — **d** Magazinrahmen-Umschaltmechanismus *m*; Ausrüstung *f* zur Magazinrahmenbewegung — **n** magazijnframe-verstelling *f* — **e** mecanismo *m* de cambio de almacén — **i** meccanismo *m* di commutazione del magazzino

8485 magazine paper — Paper for periodicals. A wide range of grades and finishes are used, coated and uncoated. There are no definite quality requirements.

f papier *m* pour revues; papier *m* pour périodiques — **d** Zeitschriftenpapier *n*; Werkdruckpapier *n*; Druckpapier *n* — **n** tijdschriftenpapier *n*; papier *n* voor tijdschriften — **e** papel *m* para revistas — **i** carta *f* per riviste

8486 magenta — Colour between red and violet, formerly called red, absorbing green light strongly, used in multi-colour printing.

f magenta *m* — **d** Magenta *n* — **n** magenta *n* — **e** magenta *f* — **i** magenta *f*

8487 magenta contact screen — Contact film screen composed of magenta dyed dots of variable density used for making halftone negatives in the camera, mostly for lithographic negatives.

f trame *f* magenta — **d** Magentakontaktraster *m* — **n** magenta-contactraster *n*; magenta-raster *n* — **e** trama *f* magenta de contacto; retícula *f* magenta de contacto — **i** retino *m* a contatto magenta

8488 magenta filter; red filter *depr.* — Filter to photograph the blue and green colours. This makes the cyan plate.

f filtre *m* magenta; écran *m* magenta; filtre *m* rouge; écran *m* rouge — **d** Magentafilter *n*; Rotfilter *n* — **n** magenta-filter *n*; rood filter *n depr.* — **e** filtro *m* magenta; filtro *m* rojo — **i** filtro *m* magenta; filtro *m* rosso

8489 magenta ink — Standard red printing ink, used in multi-colour printing.

f encre *f* magenta — **d** Magentadruckfarbe *f*; Magentafarbe *f* — **n** magenta-inkt *m* — **e** tinta *f* magenta — **i** inchiostro *m* magenta

8490 magenta masking — Method in which the mask image consists of magenta dye.

f masquage *m* magenta — **d** Magentamaskierverfahren *n* — **n** magenta-maskerprocédé *n*; magenta-maskermethode *f* — **e** enmascarado *m* magenta — **i** mascheratura *f* magenta

8491 magenta printer; red printer *depr.*; **red plate** *depr.* — Image in a subtractive colour process to be printed in magenta ink. The copy for this printing plate is photographed through a green filter.

f cliché *m* du magenta; cliché *m* du rouge *depr.* — **d** Magentaplatte *f*; Magentadruckplatte *f*; Rotdruckplatte *f depr.*; Rotplatte *f depr.* — **n** magenta-plaat *fm*; rode drukplaat *fm*; roodplaat *fm* — **e** plancha *f* del magenta; plancha *f* del rojo *depr.* — **i** lastra *f* del magenta; lastra *f* del rosso *depr.*

8492 magic ink — Ink which develops a secondary colour such as red, blue, green or yellow.

Prints made with these inks are developed by application of heat or water.

f encre *f* magique — **d** Zauberfarbe *f*; magische Druckfarbe *f* — **n** toverinkt *m* — **e** tinta *f* mágica — **i** inchiostro *m* magico

8493 magnesia — This term is incorrectly used in the trade for *magnesium oxide.

8494 magnesia — See magnesium oxide.

8495 magnesite — See magnesium carbonate.

8496 magnesium — Silvery, malleable, moderately hard metal. Oxidizes and tarnishes in moist air, stable in dry air. Soluble in acids, insoluble in water. Contact with halogenated hydrocarbons as, e.g., carbon tetrachloride and trichloro-ethylene, hydrochloric acid and sulphuric acid, is dangerous. Symbol: Mg. Latin: magnesium metallicum.

f magnésium *m* — **d** Magnesium *n* — **n** magnesium *n* — **e** magnesio *m* — **i** magnesio *m*

8497 magnesium bisulphite — Chemical used in wood-cooking process, similar to calcium bisulphite, except that the material used is pure magnesium oxide instead of lime stone or milk of lime. The process lends itself to the recovery of the magnesium and sulphur.

f bisulfite *m* de magnésium — **d** Magnesiumbisulfit *n* — **n** magnesiumbisulfiet *n* — **e** bisulfito *m* de magnesio — **i** bisolfito *m* di magnesio

8498 magnesium block

f cliché *m* magnésium — **d** Magnesiumklischee *n* — **n** magnesiumcliché *n* — **e** clisé *m* de magnesio — **i** cliscè *m* di magnesio; cliché *m* di magnesio

8499 magnesium carbonate — Very light, white powder, soluble in acids, insoluble in water and alcohol. Often called *magnesia, common name magnesite. It is used as a transparent white extender in printing inks. Latin: carbonas magnesicus.

Formula: Mg_3CO.

f carbonate *m* de magnésium; carbonate *m* de magnésie — **d** Magnesiumkarbonat *n*; kohlensaures Magnesium *n* — **n** magnesiumcarbonaat *n* — **e** carbonato *m* de magnesio; carbonato *m* magnésico — **i** carbonato *m* di magnesio

8500 magnesium oxide; calcined magnesium oxide; magnesia *depr.* — White powder, soluble in acids and ammonium salt solutions, insoluble in water and alcohol. TL-value of magnesium oxide fumes: 15 mg/m³.

Formula: MgO.

f oxyde *m* de magnésium; magnésie *f* calcinée; magnésie *f* — **d** Magnesiumoxyd *n*; gebrannte Magnesia *f*; Magnesia *f* — **n** magnesiumoxyde *n*; magnesia *fm* — **e** óxido *m* de magnesio; óxido *m* magnésico; magnesia *f* — **i** ossido *m* di magnesio *f*; magnesia *f*

8501 magnesium plate — Halftone or line plate made of magnesium. Light in weight, strong, and adaptable to high-speed etching, made 11 points thick. See also magnesium block.

f plaque *f* de magnésium — **d** Magnesiumplatte *f* — **n** magnesiumplaat *fm* — **e** plancha *f* de magnesio; lámina *f* de magnesio; placa *f* de magnesio — **i** lastra *f* di magnesio

8502 magnesium sulphate; Epsom salt — Colourless crystals, soluble in glycerol, very soluble in water, sparingly soluble in alcohol. Latin: sulfas magnesicus.

Formula: $MgSO_4·7H_2O$.

f sulfate *m* de magnésium; sel *m* anglais — **d** Magnesiumsulfat *n*; schwefelsaures Magnesium *n*; Bittersalz *n* — **n** magnesiumsulfaat *n*; bitterzout *n*; Engels zout *n* — **e** sulfato *m* de magnesia; sulfato *m* magnésico; sal *f* de Epsom; epsomita *f*; sal *f* de la Higuera — **i** solfato *m* di magnesio

8503 magnetic card — Plastic card coated with material which can retain magnetic charges, to receive and store data that are written into, or read from, the card by word processing typewriters.

f carte *f* magnétique — **d** Magnetkarte *f* — **n** magneetkaart *f*; magnetische kaart *f* — **e** tarjeta *f* magnética — **i** scheda *f* magnetica

8504 magnetic core — Main frame memory whereby bits are designated, usually as "on" or "off", "O" or "I", by changing the polarity of a magnetizable material.

f tore *m* magnétique — **d** Magnetkern *m*; Kern *m* — **n** magneetkern *fm*; kern *fm* — **e** núcleo *m* magnético; toro *m* magnético; toroide *m* magnético — **i** nucleo *m* magnetico

8505 magnetic core storage — See core store.

8506 magnetic core store — See core store.

8507 magnetic data inscriber — See encoder.

8508 magnetic disk — Type of random access store. Continuously rotating circular plate, magnetically coated on both sides and mounted horizontally or vertically with other disks to form a pack. Data are stored as magnetically polarized spots representing binary digits in concentric tracks. Read-write heads for each side of disks are mounted on arms to give quick access to the correct data track, which is read sequentially when accessed. Disks may be mounted permanently (fixed disk store), or temporarily with the facility to remove and substitute another disk pack (exchangeable disk store).

f disque *m* magnétique — **d** Magnetscheibe *f*; Magnetplatte *f*; Plattenspeicher *m* — **n** magneetschijf *fm*; magnetische plaat *fm* — **e** disco *m* magnético — **i** disco *m* magnetico

8509 magnetic disk storage — See magnetic disk store.

8510 magnetic disk store *GB*; **magnetic disk storage** *US* — Magnetic store in which the magnetic medium is on the surface of a rotating disk.

f mémoire *f* à disque magnétique — **d** Magnetscheibenspeicher *m*; Magnetplattenspeicher *m* — **n** magnetisch schijfgeheugen *n* — **e** memoria *f* de disco magnético — **i** memoria *f* a disco magnetico

8511 magnetic drive station — See magnetic tape drive.

8512 magnetic drum — Type of random access store. Consists of a magnetic coating on the surface of a spinning magnetic drum. Small magnetized spots represent binary data in tracks around the drum. A read/write head is mounted above each track which is read sequentially when accessed. It provides faster access time but less storage capacity than disk.

f tambour *m* magnétique — **d** Magnettrommel *f* — **n** magneettrommel *fm* — **e** tambor *m* magnético — **i** tamburo *m* magnetico

8513 magnetic drum storage — See magnetic drum store.

8514 magnetic drum store *GB*; **magnetic drum storage** *US* — Metal cylinder with a sensitized surface which spins inside a jacket with read/write heads. The address of all data on the drum is known and the data can be placed or removed by the read/write head. Data can be placed into or called from store without sorting input data or searching the entire memory as is necessary with tape reels.

f mémoire *f* à tambour magnétique — **d** Magnettrommelspeicher *m*; Trommelspeicher *m* — **n** magnetisch trommelgeheugen *n*; trommelgeheugen *n* — **e** memoria *f* de tambor magnético — **i** memoria *f* a tamburo magnetico

8515 magnetic encoded checks — See magnetic encoded cheques.

8516 magnetic encoded cheques *GB*; **magnetic encoded checks** *US* — The following terms are standard with all electronic machine companies in the US for use with magnetic encoded cheques: Format, refers to placement of characters on a cheque. Space, refers to alignment (right to left) of characters. Character dimension, size and shape of characters. Irregularity, saw tooth or fuzziness in excess or permissable range. Void, a term denoting absence of ink in either large or small spots. Embossment, term used to indicate impression. If too heavy it can lead to trouble. Extraneous ink front, specking or splatter. Extraneous ink back, splatter or speck on back of document which can be read by machine. Uniformity, amount of electronic impulse. Skew, relationship of the character to bottom edge.

f chèques *mpl* à marquage magnétique — **d** magnetisch kodierte Schecks *mpl* — **n** magnetisch gecodeerde cheques *mpl* — **e** cheques *mpl* con codificación magnética — **i** assegni *mpl* a marcazione magnetica; assegni *mpl* numerati con inchiostri magnetici

8517 magnetic field — Region around a magnet with forces capable of attracting ferrous metals.

f champ *m* magnétique — **d** magnetisches Feld *n* — **n** magnetisch veld *n* — **e** campo *m* magnético — **i** campo *m* magnetico

8518 magnetic field strength

f intensité *f* de champ magnétique — **d** magnetische Feldstärke *f* — **n** magnetische veldsterkte *f* — **e** intensidad *f* del campo magnético — **i** intensità *f* di campo magnetico

8519 magnetic ink — Ink to enable a computer to "read" certain stylized founts by magnetic

scanning. These founts are also legible to people. Used on bank cheques. Use is made of Fe_2O_3 or Fe_3O_4 in Υ-crystal structure.

f encre *f* magnétisable — **d** magnetisierbare Druckfarbe *f* — **n** magnetiseerbare inkt *m*; magnetische inkt *m depr.* — **e** tinta *f* magnética — **i** inchiostro *m* magnetico

8520 magnetic ink character recognition — Recognition of characters printed or typed in magnetic ink by interpreting the distinct wave form induced in a reading head by a material in the ink which can be magnetized. There are a number of figure series specially designed to conform to the very stringent requirements of electronic cheques-sorting machines. The printing of these characters is known as MICR printing but for simplicity the term magnetic ink printing is used. See also E 13 B, CMC 7.

f reconnaissance *f* magnétique de caractères — **d** magnetische Zeichenerkennung *f* — **n** magnetisch schriftlezen *n* — **e** reconocimiento *m* de caracteres de tinta magnética; reconocimiento *m* magnético de caracteres; lectura *f* magnética — **i** riconoscimento *m* magnetico di caratteri

8521 magnetic ink printing; MICR printing

f impression *f* à l'encre magnétique — **d** Drucken *n* mit magnetische Druckfarbe — **n** drukken *n* met magnetische inkt — **e** impresión *f* con tinta magnética — **i** impressione *f* con inchiostro magnetico; stampa *f* con inchiostri magnetizzabili

8522 magnetic storage — See magnetic store.

8523 magnetic store *GB*; **magnetic storage** *US* — Device using magnetic materials as a medium for storage of data: magnetic disk, film, tape, drum, core etc.

f mémoire *f* magnétique — **d** magnetischer Speicher *m*; Magnetspeicher *m* — **n** magnetisch geheugen *n* — **e** memoria *f* magnética; almacenamiento *m* magnético — **i** memoria *f* magnetica

8524 magnetic tape; mag tape — Although cartridges and cassettes may be described as magnetic tape, the term generally refers to computer compatible tape onto which data are written as magnetic charges (in the form of bits) across the tape width. In this way each character can be read or written at one time rather than serially. Industry compatible magnetic tape is usually written at a density of 800 bpi (bits per inch), which really means 800 eight-level (plus parity) characters per inch, although 1600 bpi tape is increasingly common. Older computers often made use of 7 tracks (6 bits plus parity) tape at densities of 250, 300 or perhaps 500 bpi. Magnetic tape transports read and/or write on this tape at a rate expressed in inches per second (ips). Dec tape (frequently used on Digital Equipment systems) is not industry compatible and offers certain "indexable" features not otherwise available.

f bande *f* magnétique; ruban *m* magnétique — **d** Magnetband *m* — **n** magneetband *m*; magnetische band *m* — **e** cinta *f* magnética;

banda *f* magnética — **i** nastro *m* magnetico

8525 magnetic tape drive — Tape transport or station which winds or unwinds computer tape and passes it by a read/write head at a constant speed so that data may be transferred from computer memory to tape or vice versa. Sometimes called magnetic drive station.

8526 magnetic tape encoder — See encoder.

8527 magnetic tape format — Method of writing data onto magnetic tape, format covering all aspects of the recording, including tape width, number of tape channels, packing format and density, code, number of characters per record and number of records per block.

8528 magnetic tape reader — Tape transport mechanism with a reading head and associated electrical circuits to read magnetic tape.
f lecteur *m* de bande magnétique — **d** Magnetbandlesegerät *n* — **n** magneetbandlezer *m* — **e** lector *m* de cinta magnética — **i** lettore *m* di nastro magnetico

8529 magnetic tape storage — See magnetic tape store.

8530 magnetic tape store *GB*; **magnetic tape storage** *US* — Computer term for reels of metallic or plastic tape with sensitized surface which is magnetized to hold data. Data can be read, erased, entered or replaced by recording heads.
f mémoire *f* à bande magnétique — **d** Magnetbandspeicher *m* — **n** magneetbandgeheugen *n* — **e** memoria *f* de cinta magnética — **i** memoria *f* a nastro magnetico

8531 magnetic track; track — Part of a data storage medium that is influenced by (or influences) one head, e.g., the ring-shaped portion of the surface of a drum associated with non-movable head, or one of several divisions (most commonly nine) running parallel to the edges of the magnetic tape.
f piste *f* magnétique — **d** magnetische Spur *f* — **n** magnetisch spoor *n* — **e** pista *f* magnética — **i** pista *f* magnetica; traccia *f* magnetica

8532 magnification — Ratio of the linear size of the image with respect to that of the original copy.
f grossissement *m*; agrandissement *m* — **d** Vergrößerung *f* — **n** vergroting *f* — **e** ampliación *f* — **i** ingrandimento *m*

8533 magnification sign
f signe *m* de multiplication — **d** Vervielfältigungszeichen *n*; Multiplikationszeichen *n* — **n** teken *n* van vermenigvuldiging; vermenigvuldigingsteken *n* — **e** signo *m* de multiplicación — **i** segno *m* di moltiplicazione

8534 magnifier; magnifying glass; glass — Small convex lens or combination of lenses capable of magnification, to examine dot and line structures in negatives and printing plates.
f loupe *f* — **d** Lupe *f*; Vergrößerungsglas *n* — **n** loep *fm*; loupe *fm*; vergrootglas *n* — **e** lupa *f*; amplificador *m*; vidrio *m* de aumento; cristal *m* de aumento; lente *f* de aumento — **i** lente *f* di ingrandimento

8535 magnifying glass — See magnifier.

8536 mag tape — See magnetic tape.

8537 mail circulation
f édition *f* étrangère; édition *f* postale — **d** Postauflage *f* — **n** posteditie *f* — **e** edición *f* para el correo — **i** edizione *f* dell'estero

8538 mail fold — See second chopper fold.

8539 mailing card — Postcard with a printed announcement or advertisement.
f carte *f* publicitaire — **d** Werbepostkarte *f* — **n** reklamekaart *fm* — **i** cartolina *f* postale pubblicitaria

8540 mailing room — See mail room.

8541 mail room; mailing room; shipping department
f salle *f* d'expédition — **d** Versandraum *m*; Versandabteilung *f* — **n** expeditie-afdeling *f*; afdeling *f* expeditie — **e** departamento *m* de empaque; departamento *m* de embarques; departamento *m* de expedición; paquete *m* *Ch* — **i** reparto *m* spedizione

8542 main drive — Main shaft or gearing, driving the rotary press, to which all sections are connected. Sometimes the main motor or motors driving the press.
f arbre *m* de commande principale — **d** Hauptantriebswelle *f* — **n** hoofdas *fm*; hoofdaandrijfas *fm* — **e** impulso *m* mayor — **i** albero *m* principale; albero *m* del comando principale

8543 main exposure; detail exposure — Exposure when making a screen negative without additional exposures.
f pose *f* principale; pose *f* des détails — **d** Hauptbelichtung *f* — **n** hoofdbelichting *f* — **e** pose *f* principal; exposición *f* principal; insolación *f* principal — **i** esposizione *f* principale

8544 main frame — See central processing unit.

8545 main line — See stem.

8546 main motor
f moteur *m* principal — **d** Hauptmotor *m* — **n** hoofdmotor *m* — **e** motor *m* principal; motor *m* mayor — **i** motore *m* principale

8547 main processor — See central processing unit.

8548 mains — See town mains.

8549 main shaft bearing
f coussinet *m* de l'arbre principal — **d** Hauptwellenlager *n* — **n** hoofdaslager *n* — **e** cojinete *m* del árbol principal; soporte *m* del árbol principal — **i** cuscinetto *m* dell'albero principale; supporto *m* dell'albero principale

8550 main storage — See main store.

8551 main store *GB*; **main storage** *US* — Usually the fastest storage device in a computer system in which program steps are undertaken. Usually core store or thin-film store.
f mémoire *f* principale — **d** Hauptspeicher *m* — **n** hoofdgeheugen *n* — **e** memoria *f* principal — **i** memoria *f* principale

8552 main stroke — See stem.

8553 maintenance — Up-keep of machinery in

good running order.

f entretien *m* — **d** Pflege *f* (von Maschinen) — **n** onderhoud *n* — **e** mantenimiento *m* — **i** manutenzione *f*

8554 main title — See full title.

8555 Maioli binding — Binding named after the French book collector Thomas Mahieu. Considered to be Italian (Tommaso Maioli), he is identified with the secretary of Catherine de Medici.

f reliure *f* Maioli — **d** Maioli-Einband *m* — **n** Maioli-band *m* — **e** encuadernación *f* Maioli — **i** legatura *f* da Maioli

8556 major defect

f défaut *m* majeur; défaut *m* principale — **d** Hauptfehler *m* — **n** ernstige fout *fm* — **e** yerro *m* de bulto — **i** difetto *m* principale (a more serious defect); difetto *m* grave (a less serious defect)

8557 majuscule — Capital letter.

f majuscule *f*; capitale *f* — **d** Versal *n*; Großbuchstabe *m* — **n** hoofdletter *fm*; kapitaal *fm* — **e** mayúscula *f* — **i** maiuscola *f*; lettera *f* maiuscola

8558 make — See making order.

8559 make and hold order — Order to make a lot of paper and hold it awaiting shipping instructions of the customer.

f ordre *m* d'appel — **d** Abrufauftrag *m* — **n** afroeporder *fm* — **e** pedido *m* de entrega diferida; pedido *m* escalonado — **i** commessa *f* su appello; ordinazione *f* su appello

8560 make *v* **even** — See end *v* even.

8561 make *v* **margins** — To arrange the spaces between the pages of a forme so that the margins will be correctly proportioned.

f mettre la garniture; garnir la forme — **d** Format machen — **n** formaat maken — **e** guarnecer la forma — **i** guarnire; marginare

8562 make over — Repetition of manufacture caused by spoilage.

f réfection *f* — **d** Wiederholung *f* (einer Arbeit) nach fehlerhafter Ausführung — **n** overmaking *f* — **e** rehacimiento *m*; reposicíon *f* de faltas — **i** rifacimento *m*

8563 makeready — See makeready sheet.

8564 make *v* **ready** — To prepare a forme for printing by patching up with paper or cutting away on the impression cylinder or the platen bed, also by underlaying or interlaying the blocks. See also pre-makeready.

f faire la mise en train — **d** zurichten; druckfertig machen — **n** toestellen — **e** arreglar; hacer el arreglo; calzar; nivelar — **i** avviare

8565 makeready knife — See cutting-out knife.

8566 makeready sheet; makeready

f feuille *f* de mise (en train) — **d** Zurichtebogen *m* — **n** toestel *n* — **e** arreglo *m*; padrón *m* *Ar*; patrón *m*; pliego *m* de arreglo — **i** foglio *m* d'avviamento

8567 makeready tissue — Tissue paper of standardized caliper, usually of 0.025, 0.038 and 0.05 mm (0.001, 0.0015 and 0.002 in) thickness, used to adjust diameter or plate of offset cylinder by placing under plate or offset blanket in making ready offset press for operation. Used only when necessary to compensate for variables.

f papier *m* de soie pour la mise en train — **d** Zurichteseidenpapier *n* — **n** toestelvloei *n*; onderlegvloei *n* — **e** papel *m* de seda para arreglo; papel *m* de seda para recorte — **i** carta *f* per avviamento

8568 maker up — See make-up hand.

8569 make-up — Arrangement of printing elements to compose a page or other printing formes.

f mise *f* en page(s); empagement *m* — **d** Umbruch *m* — **n** opmaak *m* — **e** compaginación *f*; ajuste *m* tipográfico; armado *m*; armazón *f*; formación *f* *Me*; armadura *f* *Pe*; armada *f* *Co; Pe; Ur; Ec*; emplane *m*; emplanado *m*; emplanaje *m*; emplanamiento *m* *Cu* — **i** impaginazione *f*

8570 make *v* **up** — To arrange set type, cuts, etc., into columns or pages.

f mettre en pages — **d** umbrechen — **n** opmaken — **e** armar; compaginar; formar; ajustar; confeccionar — **i** impaginare

8571 make-up assistant

f homme *m* de bois; fonctionnaire *m* de bois — **d** Hilfsmetteur *m* — **n** hulpopmaker *m*; assistentopmaker *m* — **e** ayudante *m* del compaginador — **i** impaginatore *m* d'aiuto; bardotto *m* *fam.*

8572 make-up editor — Person in charge of arranging the news and editorial matter of a daily newspaper in the formes.

f rédacteur *m* au marbre; rédacteur *m* qui fait la marbre; rédacteur *m* chargé de la mise en pages — **d** Umbruchredakteur *m*; Umbruchschriftleiter *m* — **n** redacteur-opmaker *m* — **e** redactor *m* de platina; armador *m*; diagramador *m*; confeccionador *m*; compaginador *m* de platina; jefe *m* de armada *Co*; emplanador *m* *Cu*; encargado *m* de armadura *Pe* — **i** redattore *m* di impaginazione

8573 make-up hand; make-up man; maker up — Sometimes called floor man or floor hand.

f metteur *m* en pages; pageux *m* — **d** Metteur *m* — **n** opmaker *m* — **e** armador *m*; compaginador *m*; ajustador *m*; cajista montador *m*; montador *m*; platinero *m*; emplanador *m* *Cu*; formador *m* *Me* — **i** impaginatore *m*

8574 make-up man — See make-up hand.

8575 make-up pay — Work study term for an amount by which the earnings on incentive fall below guaranteed minimum pay.

8576 make-up section

f section *f* de mise en page(s) — **d** Umbruchabteilung *f* — **n** opmaakafdeling *f*; opmakerij *f* — **e** departamento *m* de formación — **i** reparto *m* dell'impaginazione

8577 make-up table — See making-up table.

8578 making direction — See machine direction.

8579 making order; make — Order that cannot be supplied from stock but is made to the purchaser's specifications.

f commande *f* de fabrication; fabrication *f* sur

commande — **d** Anfertigungsmenge *f*; Anfertigung *f* — **n** aanmaakpartij *f*; aanmaakorder *fm* — **e** partida *f* de fabricación — **i** commessa *f* di fabbricazione

8580 making paper — See excess feed.

8581 making-up table; make-up table
f marbre *m* de mise en pages — **d** Umbruchtisch *m* — **n** opmaaktafel *fm* — **e** mesa *f* de armado; mesa *f* de imposición; mesa *f* de imponer; mesa *f* de montaje; mesa *f* de compaginación *Ur*; mesa *f* de armada *Co;Pe*; mesa *f* de armadura *Pe*; mesón *m* Ch; plancha *f* Gu; mármol *m*; platina *f* — **i** tavolo *m* di imposizione; tavolo *m* di impaginazione

8582 malachite green; Victoria green — Green basic *copper carbonate.
Formula: $Cu_2CO_3(OH)_2$.
f vert *m* malachite — **d** Malachitgrün *n* — **n** malachietgroen *n* — **e** verde *m* malaquita — **i** verde *m* malachite

8583 male-female die — Precision punch and die set in which either the product or the scrap is sheared exactly according to the cross section of the punch and separated from the sheet, allowing the remaining material to pass after the punch has been withdrawn.
f moule *m* à mâle et femelle — **d** Patrize-Matrize Form *f* — **n** stempel *m* met contrastempel — **i** stampo *m* e controstampo

8584 maleic resin — Synthetic hard resin formed by the reaction of maleic acid (formula HOOCCH:CHCOOH) and modified rosin.
f résine *f* maléique — **d** Maleinharz *n*; Maleinsäureharz *n* — **n** maleïnehars *mn* — **e** resina *f* maleica — **i** resina *f* maleica

8585 malleable — Capable of being extended by rolling or hammering.
f malléable; ductile — **d** walzbar; hämmerbar — **n** walsbaar; smeedbaar — **e** maleable; dúctil — **i** malleabile

8586 mallet — Wooden hammer to strike the shooting stick when locking up a forme and when planing the forme.
f maillet *m* de serrage — **d** Holzhammer *m*; Holzschlegel *m*; Schlegel *m* — **n** houten hamer *m*; dresseerhamer *m* — **e** maceta *f*; mazo *m*; mallete *m* — **i** mazzuola *f*

8587 mandrel; mandril *depr.* — Shaft on which machine parts are fixed to be rotated in bearings.
f mandrin *m* — **d** Spindel *f*; Drehstift *m*; Spanndorn *m* — **n** opspan-as *fm* — **e** mandril *m* — **i** mandrino *m*

8588 mandril — See mandrel.

8589 manganese — Reddish-grey or silvery, brittle metallic element. Decomposes water. Slowly dissolves in dilute acids. Symbol: Mn. TL-value: 6 mg/m³.
f manganèse *m* — **d** Mangan *n* — **n** mangaan *n* — **e** manganeso *m* — **i** manganese *m*

8590 manganese drier — Compound of manganese and various fatty acids to accelerate the oxidation and polymerization of printing ink films.
f siccatif *m* au manganèse — **d** Mangantrockner *m*; Mangansikkativ *mn* — **n** mangaandroger *m* — **e** secante *m* de manganeso — **i** essiccante *m* al manganese

8591 manganese sulphate; manganous sulphate — Translucent, pale rose-red, efflorescent prisms, soluble in water, insoluble in alcohol.
Formula: $MnSO_4 \cdot 4H_2O$
f sulfate *m* de manganèse; sulfate *m* manganeux — **d** Mangansulfat *n*; Manganosulfat *n*; schwefelsaure Mangan *n* — **n** mangaansulfaat *n* — **e** sulfato *m* de manganeso; sulfato *m* manganoso — **i** solfato *m* di manganese; solfato *m* manganoso

8592 manganous — Containing divalent manganese, e.g., manganous chloride, $MnCl_2$.

8593 manganous sulphate — See manganese sulphate.

8594 mangle — See mat roller.

8595 mangle *v* — To print with very heavy impression.
f imprimer avec excès de foulage — **d** drucken mit hoher Druckspannung — **n** drukken met te grote drukkracht; drukken met te zware drukspanning; mangelen *coll.* — **e** imprimir en exceso de huella — **i** stampare con eccesso di pressione

8596 mangle rack motion — Drive for the bed of a two revolution press consisting of a continuously turning cog wheel, engaging alternatively in two cog bars causing a forth and aft movement of the bed.
f mouvement *m* à pignon et crémaillères — **d** Doppelrechenantrieb *m* — **i** movimento *m* di va e veni a doppia cremagliera

8597 manière criblée; dotted manner — Copperplate printing process of the 15th century. Lines and shadows were produced by short flicks of the engraver's burin, without biting the plate with acid. Ancestor of *stipple engraving and *crayon engraving.
f manière *f* criblée; gravure *f* au criblé; gravure *f* à fond criblé — **d** Schrottschnitt *m*; Schrottmanier *f*; Metallschnitt *m* — **n** manière *f* criblée; schrootmanier *fm*; schrootblad *n* — **e** manera *f* punteada; plancha *f* punteada — **i** maniera *f* criblée; incisione *f* a fondo punteggiato; stampa *f* a fondo punteggiato

8598 manière noire — See mezzotint.

8599 manifold paper; manifold tissue — Paper used for additional copies by interleaving with carbon paper on the typewriter; manufactured in substance weight of 3-4 kg or 7-9 lbs (17 x 22 - 500), with an unglazed or machine glazed finish, sometimes a cockle finish.
f papier *m* pelure; papier *m* pour copies multiples — **d** Durchschreibpapier *n*; Durchschlagpapier *n*; Vervielfältigungspapier *n*; Abzugpapier *n* — **n** doorslagpapier *n* — **e** papel *m* para copias — **i** carta *f* per copie multiple

8600 manifold tissue — See manifold paper.

8601 manila — See manilla.

8602 manila tag — See tag board.

8603 manila writing paper — See railroad manila.

8604 manilla *GB*; **manila** *US* — Colour and finish comparable to that formerly obtained from manilla hemp (rope) stock. Unless definitely stated, no manilla fibre is present. Under the present usage, it has no definite significance as to furnish.

8605 manilla paper; hemp paper — Strong, coloured paper made from rope (hemp), manilla hemp, sisal, jute, coir, flax (cotton) and chemical wood pulp. Rope manilla paper is used for envelopes and is made of manilla hemp (commonly called rope) and has a basis weight range from 60-140 pounds (24 x 36 - 600).

f papier *m* corde; papier *m* bulle — **d** Tauenpapier *n*; (braun) Tauenpackpapier *n*; Patentpackpapier *n*; Hanfpapier *n* — **n** casing *n*; manillapapier *n* — **e** papel *m* manila — **i** carta *f* manila

8606 man/machine chart — See also multiple activity chart.

f graphique *m* homme-machine — **d** Mann-Maschinenkarte *f*; Unterweisungskarte *f* bei Maschinenarbeit — **n** man/machine kaart *fm* — **e** gráfica *f* hombre/máquina — **i** carta *f* uomo/macchina

8607 mantissa — Decimal, always positive, portion of a common logarithm.

f mantisse *f* — **d** Mantisse *f* — **n** mantisse *f* — **e** mantisa *f* — **i** mantissa *f*

8608 manual — See handbook.

8609 Manuals — Group of type faces, e.g., Gothics, Libra, Jacno, Banco, Studio, Flambard, Old Dutch, Polka, Lasso, Raffia, Cartoon, Klang, whose characters suggest that they have been designed individually by drawing in contradiction to the freer style of writing. Before 1971 this group was named Graphics. See also classification of type design.

f Manuaires *fpl* — **d** handschriftliche Antiqua *f* — **n** Manuaren *mfpl* — **e** Manuales *mpl* — **i** Fantasie *mpl* medioevali

8610 Manul process — Reflex printing process in which the sensitized material consists of dichromated gelatine on glass. After developing in water and colouring, and due to an intermediate layer, the film can be removed from the glass and used as a negative in offset printing. Name is the reverse of that of the inventor, F. Ullmann (Germany).

f procédé *m* Manul — **d** Manulverfahren *n*; Manuldruck *m* — **n** manul-procédé *n* — **e** procedimiento *m* Manul; método *m* Manul — **i** procedimento *m* Manul

8611 manuscript — See copy.

8612 map — See also nautical chart, road map.

f carte *f* géographique — **d** Landkarte *f*; Karte *f* — **n** landkaart *fm*; kaart *fm* — **e** mapa *m* geográfico — **i** mappa *f*; carta *f* geografica

8613 map maker — See also cartographer, hydrographer.

f cartographe *m* — **d** Kartograph *m* — **n** cartograaf *m* — **e** cartógrafo *m* — **i** cartografo *m*

8614 map paper — Paper of good folding strength, clearliness and writing surface, made of part rag or all rag stock, suitable for lithography. Groundwood or unbleached wood pulp is excluded. See also chart paper.

f papier *m* pour cartes géographiques — **d** Landkartenpapier *n* — **n** landkaartenpapier *n* — **e** papel *m* para mapas (geográficos); papel *m* para atlas — **i** carta *f* per carte geografiche; carta *f* per cartografia

8615 marbled edges — Decorated edges of a book, usually wavy lines or spots in one or more colours.

f tranches *fpl* marbrées; tranches *fpl* jaspées; tranches *fpl* peignées; marbré sur tranche — **d** marmorierte Buchschnitte *mpl*; Marmorschnitt *m*; mit marmoriertem Schnitt *m* — **n** gemarmerde sneden *fmpl*; marmersnee *fm*; op snee gemarmerd — **e** cortes *mpl* jaspeados; cortes *mpl* veteados — **i** tagli *mpl* marmorizzati

8616 marbled paper — See marble paper.

8617 marble paper; marbled paper — Tinted or printed paper resembling marble, for lining book covers and fancy boxes.

f papier *m* marbré — **d** Marmorpapier *n*; marmoriertes Papier *n* — **n** gemarmerd papier *n*; marmerpapier *n* — **e** papel *m* jaspeado — **i** carta *f* marmorizzata

8618 marbler — Man who stains paper or edges of books to give them the appearance of marble.

f marbreur *m* — **d** Marmorierer *m* — **n** marmeraar *m* — **e** jaspeador *m* — **i** marmorizzatore *m*

8619 marbler's comb — See marbling comb.

8620 marbles — See cods.

8621 marbling — Staining the edges of a book or the surface of paper to give it the appearance of marble. The colours are floated on a solution composed of gum tragacanth or carrageen, boiled in water, mixed with ox gall and poured into a flat tray. The edges of the book are treated with a solution of concentrated alum, before they are tipped against the surface of the marbling solution.

f marbrure *f*; marbrage *m*; jaspage *m* (des tranches) — **d** Marmorierung *f* — **n** marmeren *n* — **e** marmoración *f*; jaspeado *m* — **i** marmorizzazione *f*

8622 marbling brush — Brush to spatter colours in a vat. See also marbling.

f brosse *f* à jasper — **d** Spritzbürste *f*; Sprengbürste *f* — **n** sprenkelborstel *m* — **e** cepillo *m* para jaspear — **i** spazzolino *m* per spruzzare

8623 marbling colour

f encre *f* pour marbrure — **d** Marmorierfarbe *f* — **n** marmerverf *fm* — **e** color *m* para jaspear — **i** colore *m* da marmorizzatura

8624 marbling comb; marbler's comb; comb; peg rake — Coarse comb used by marblers to make patterns.

f peigne *m* à marbrer — **d** Marmorkamm *m* — **n** marmerkam *m* — **e** peine *m* para el jaspeado — **i** pettine *m* per marmorizzatura

8625 marbling grill; wire screen — Wire mesh device used in sprinkling and marbling edges of books.
f grille *f* à jasper; grille *f* à marbrer — **d** Sprenggitter *n*; Spritzgitter *n* — **n** sprenkelraam *n* — **e** rejilla *f* para jaspear — **i** griglia *f* per marmorizzare; griglia *f* per spruzzare

8626 marbling vat — Vat in which coloured patterns are floated on a sized surface of water to transfer them to paper or some other surface.
f cuve *f*; cuvette *f*; baquet *m* — **d** Marmorbecken *n* — **n** marmerbak *m* — **i** bagno *m* di marmorizzazione

8627 margin — Unprinted spaces around a book page. The four margins are: *head, *foredge, *back margin, and *tail margin or *foot margin. An acceptable ratio for the size of margins, in the order as above, is 1 : 1 1/2 : 2 : 2 1/2.
f marge *f* — **d** Papierrand *m*; Rand *m* — **n** marge *fm*; rand *m*; wit *n* — **e** margen *mf*; borde *m* — **i** margine *m*

8628 marginal figure — See runner.

8629 marginal note; side note — Explanatory note in the margin of a page. See also shoulder note.
f addition *f* (marginale); manchette *f*; note *f* en marge; note *f* marginale; marginale *f*; glose *f* marginale — **d** Marginalnote *f*; Marginale *f* (*pl* Marginalien); Randbemerkung *f*; Randglosse *f* — **n** kantnoot *fm*; kanttekening *f*; zijnoot *fm*; marginale *fm* — **e** nota *f* marginal; anotación *f* en el margen; ladrillo *m*; arracada *f*; acotación *f*; apostilla *f* — **i** nota *f* marginale; nota *f* di margine; postilla *f*

8630 margin bar; head bar; margin strip — Metal strip across the cylinder against which the printing plates are positioned in a letterpress rotary.
f barrette *f* d'appui; barrette *f* de tête — **d** Blindsteg *m* — **e** barra *f* de cabeza — **i** listello *m* d'appoggio

8631 margin-punched card — See edge-punched card.

8632 margin strip — See margin bar.

8633 mark *v* — To correct mistakes.
f marquer les fautes — **d** Fehler anzeichnen; anstreichen — **n** fouten aanstrepen — **e** marcar; indicar erratas; corregir — **i** indicare; marcare; segnare errori

8634 marker — See flag.

8635 marker, book — See book marker.

8636 marking felt — Felt to impart a distinctive pattern or mark to a sheet of paper.
f feutre *m* marqueur — **d** Markierfilz *m* — **n** markeervilt *n* — **e** filtro *m* marcador — **i** feltro *m* marcatore

8637 marking ink — Ink to mark cloth to withstand laundering. See also indelible ink.
f encre *f* à marquer — **d** Wäschezeichenfarbe *f* — **n** merkinkt *m* — **e** tinta *f* para marcar — **i** inchiostro *m* per stampigliatura

8638 mark of interrogation — See interrogation mark.

8639 mark of multiplication; multiplication sign; point of multiplication; multiple mark — According to the *SI the point of multiplication must be placed in printed matter halfway above the line. In manuscripts typed out on a typewriter it may be placed on the line. See also mathematical signs.
f signe *m* de multiplication; marque *f* de multiplication; symbole *m* de multiplication — **d** Malzeichen *n*; Multiplikationszeichen *n* — **n** maalteken *n*; teken *n* van vermenigvuldiging; vermenigvuldigingsteken *n* — **e** signo *m* de multiplicación — **i** segno *m* di moltiplicazione

8640 mark sensing — Mark detection system employing printed cards on which pencil marks are made by hand. The marks are detected electrically by brushes passing a small current through the graphite from the pencil. Up to 27 marks can be accommodated on each side of the card. Recent mark sensing readers are photosensitive.
f lecture *f* de marques sensibles — **d** Markierungslesung *f*; Zeichenabfühlung *f* — **n** aanstreepmethode *f* — **e** lectura *f* de marcas sensibles — **i** lettura *f* di marche sensibili

8641 marks of reference — In order: *, †, ‡, §, ¶. Today mostly arabic figures are used. See also foot notes.
f pieds *mpl* de mouche — **d** Hinweiszeichen *npl*; Verweisungszeichen *npl* — **n** verwijzingstekens *npl*; noottekens *npl* — **e** llamadas *fpl* (indicativas); llamadas *fpl* marginales; señales *fpl* indicativas; signos *mpl* de referencia; signos *mpl* de llamada — **i** segni *mpl* di richiamo

8642 mark *v* **up** — To write up instructions as on a dummy.
f noter les instructions; préparer l'original — **d** versehen mit Anweisungen; versehen mit Anmerkungen — **n** aanwijzingen aanbrengen; gegevens aanbrengen — **e** introducir indicaciones — **i** scrivere l'istruzioni

8643 marriage advertisement
f annonce *f* matrimoniale — **d** Heiratsanzeige *f* — **n** huwelijksadvertentie *f* — **e** anuncio *m* matrimonial — **i** annunzio *m* di matrimonio

8644 Mary; Mollie — The throw in *jeffing if none of the nicks appear uppermost in throwing.

8645 mask — Controlled low density continuous-tone positive or negative to correct the tone range or colour errors in a negative of an illustration. The negative and its mask(s) are registered together and then photographed as a unit. See also contact mask.
f masque *m* — **d** Maske *f* — **n** masker *n* — **e** máscara *f* — **i** maschera *f*

8646 mask — Cut out sheet of paper to protect parts of sheet to be printed against smearing of the ink.
f cache *m* — **d** Schablone *f*; Abdeckschablone *f* — **n** schabloon *n*; masker *n* — **e** plantilla *f* de enmascarar; máscara *f*; mascarilla *f* — **i** maschera *f* di copertura

8647 masking; masking process; colour masking
— Use of a mask on certain areas of originals, negatives or sensitized metal for better reproduction. Correction of colour separation negatives by attachment of corrective photographic masks to the images. See also contact mask.
f masquage *m*; correction *f* par masquage — **d** Maskieren *n*; Maskerverfahren *n*; Maskierverfahren *n*; Maskenverfahren *n*; Maskierung *f* — **n** maskeren *n*; masker-procédé *n*; maskermethode *f* — **e** enmascarado *m*; procedimiento *m* de enmascarado; corrección *f* con mascarillas; corrección *f* por mascarillas — **i** mascheratura *f*; procedimento *m* di mascheratura

8648 masking lacquer; covering varnish
f vernis *m* à couvrir; vernis *m* à masquer — **d** Abdecklack *m* — **n** afdeklak *mn* — **e** barniz *m* para tapar; barniz *m* para cubrir; laca *f* para tapar — **i** vernice *f* protettiva

8649 masking process — See masking.

8650 masking tape — Opaque black or red cellulose tape to mask photographic negatives.
f ruban *m* opaque — **d** Abdeckklebeband *n*; rotes Klebeband *n*; roter Klebstreifen *m* — **n** afdekband *n* — **e** cinta *f* de enmascarar — **i** nastro *m* per mascheratura

8651 Massey coating — Method of *roll coating.
f couchage *m* Massey — **d** Massey-Streichverfahren *n* — **n** Massey-strijkprocédé *n* — **e** estucado *m* Massey — **i** patinatura *f* Massey

8652 massicot — Oxide of lead, PbO, formed by oxidation of a bath of metallic lead at about 345 °C. If the oxide is melted it is converted into *litharge.
f massicot *m* — **d** Massicot *n* — **n** koningsgeel *n* — **e** masicote *m* — **i** massicot *m*

8653 mass production
f fabrication *f* en (grande) série; travail *m* en (grande) série; production *f* de masse — **d** Massenfabrikation *f*; Massenherstellung *f*; Massenfertigung *f* — **n** massa-produktie *f* — **e** producción *f* en masa; producción *f* en serie; producción *f* en cadena — **i** produzione *f* di massa; produzione *f* in grande quantità

8654 mass tone; self tone — Colour of an ink in bulk. Opaque inks print nearly the bulk colour. Transparent inks are much darker in bulk than the printing colour.
f couleur *f* dans la masse — **d** Farbton *m* der Farbe im Farbehälter — **n** kleur *fm* van de inkt in de bus — **e** color *m* en la masa — **i** colore *m* dell'inchiostro nel recipiente; colore *m* dell'inchiostro prima della stampa

8655 master file — File that is relatively permanent, or that is treated as an authority in a particular job. Computer term.
f fichier *m* principal — **d** Hauptdatei *n* — **n** stambestand *n*; referentieverzameling *f* — **e** fichero *m* maestro — **i** archivio *m* principale

8656 master image — 1. Type character image on a film strip or disk from which an exposure is created during photographic typesetting. 2. Art design of a character from which the *photographic master is derived. 3. Digital representation from which a third generation cathode ray tube typesetter is able to create a character image. 4. Art design of a character which may be digitized to provide the representation referred to in 3.
f image *f* maîtresse; image *f* matrice — **d** Mutterbild *n*; Originalbild *n* — **n** moederbeeld *n* — **e** imagen *f* original — **i** immagine *f* originale

8657 master instrument — Instrument used to calibrate others.
f instrument *m* d'étalonnage — **d** Eichinstrument *n* — **n** ijkmaat *fm*; ijkinstrument *n* — **e** aforo *m* — **i** strumento *m* di taratura

8658 master plate; pattern plate — Well-hardened and well-routed electrotype or stereotype made direct from the original type and kept not for printing but for making duplicate plates.
f cliché *m* matrice; planche *f* maîtresse — **d** Originalgalvano *n*; Originalplatte *f* — **n** moederplaat *fm* — **e** placa *f* original — **i** lastra *f* originale; lastra *f* madre; galvano *m* originale; galvano *m* madre

8659 master printer — Director, manager or owner of a printing plant.
f maître *m* imprimeur — **d** Buchdruckerprinzipal *m*; Buchdruckereibesitzer *m* — **n** drukkerspatroon *m*; drukker *m* — **e** maestro *m* impresor; imprentero *m* Ar; imprentario *m* Ch — **i** industriale *m* tipografo

8660 master roll — See mill roll.

8661 mast-head; name plate; flag — Name, address, etc., of a newspaper, usually published on the editorial page.
f en-tête *m* — **d** Zeitungsimpressum *n* — **n** vaste kop *m* — **e** cabecera *f* (del periódico); título *m* del periódico; membrete *m*; marca *f* de cabecera; cabecera-marca *f*; machote *m* Cu; machón *m* Gu — **i** testata *f*

8662 mat — Abbrev. for matrix, the papier mâché mould made from relief plates and formes used for casting stereotypes. See also flong.

8663 mat — Lustreless and dull in surface. Matt is displacing it in technical use.
f mat — **d** matt — **n** mat — **e** mate — **i** mat

8664 mat — See flong.

8665 mat art paper; mat finished art paper; mat coated paper; cameo paper US; cameo coated paper US — Art paper with a mat finish.
f papier *m* couché mat — **d** mattgestrichenes Papier *n*; Mattkunstdruckpapier *n* — **n** mat kunstdrukpapier *n*; mat gestreken papier *n* — **e** papel *m* estucado mate — **i** carta *f* patinata non lucida

8666 mat backing — See backing felt.

8667 matching, colour — See colour matching.

8668 mat coated paper — See mat art paper.

8669 mat dryer — See matrix dryer.

8670 mat finish — See dull finish.

8671 mat finished art paper — See mat art paper.

8672 mathematical formula
f formule *f* de mathématiques — **d** mathematische

Formel *f* — **n** wiskundige formule *fm* — **e** fórmula *f* matemática — **i** formola *f* matematica

8673 mathematical setting — Setting of type for mathematical formulas, etc.

f composition *f* de mathématiques — **d** mathematischer Satz *m* — **n** zetten *n* van wiskundige formules — **e** composición *f* de matemáticas — **i** composizione *f* di matematica

8674 mathematical signs — See also algebraical signs, arithmetical signs, app. no. 10.

f signes *mpl* de mathématiques — **d** mathematische Zeichen *npl* — **n** wiskunde-tekens *npl* — **e** signos *mpl* de matemáticos; signos *mpl* de matemático — **i** segni *mpl* matematici

8675 mat inks — Inks which dry with a dull surface, mostly water-colour inks.

f encres *fpl* mates — **d** Mattfarben *fpl* — **n** matte inkten *mpl* — **e** tintas *fpl* mates; tintas *fpl* apagadas — **i** inchiostri *mpl* non lucidi

8676 mat paper — Coated or uncoated paper with a low, dull finish.

f papier *m* mat — **d** mattes Papier *n* — **n** mat papier *n* — **e** papel *m* mate — **i** carta *f* non lucida

8677 matrix — On typesetting machines, a flat brass plate having on one edge a character (or characters) in intaglio from which type is cast, and in its upper end a series of teeth to select and distribute it to its proper channel in the magazine of the machine. The Monotype system requires only one matrix for each letter whereas line casters require about 20 matrices of each letter to assemble a complete line, allowing for circulating matrices.

f matrice *f* — **d** Matrize *f* — **n** matrijs *fm* — **e** matriz *f* — **i** matrice *f*

8678 matrix — Copper mould struck by a punch from which type is cast.

f matrice *f* — **d** Matrize *f* — **n** matrijs *fm* — **e** matriz *f* — **i** matrice *f*

8679 matrix — In mathematics, two-dimensional rectangular array of quantities which is manipulated according to defined rules.

f matrice *f* — **d** Matrix *f* — **n** matrix *f* — **e** matriz *f* — **i** matrice *f*

8680 matrix — See flong.

8681 matrix — See mat.

8682 matrix beating brush — See beating brush.

8683 matrix block; matrix slide block — Base to carry matrix slides from which rules, borders, etc., are cast on the slug composing machine.

f bloc *m* pour filets-glissières — **d** Liniengießblock *m* — **n** lijnengietblok *n* — **e** bloque *m* para fundir orlas y filetes — **i** blocco *m* per fondere filetti

8684 matrix board — See flong paper.

8685 matrix case; die case — Part of the Monotype casting machine containing the matrices.

f châssis *m* porte-matrices — **d** Matrizenrahmen *m* — **n** matrijzenraam *n* — **e** chasis *m* portamatrices; cuadro *m* de matrices — **i** telaio *m* portamatrici

8686 matrix centring pin — See centring tack.

8687 matrix cleaning device

f machine *f* à nettoyer les matrices — **d** Matrizenreinigungsgerät *n* — **n** matrijzenreiniger *m* — **e** máquina *f* para limpiar matrices — **i** macchina *f* per pulire matrici

8688 matrix conducting wires — Wires which conduct the matrices in the Typograph typesetting machine.

f conduite *f* des matrices — **d** Matrizenführung *f* — **n** matrijzengeleiders *mpl* — **e** conducta *f* de las matrices — **i** conduzione *f* delle matrici

8689 matrix delivery belt

f courroie *f* de l'assembleur — **d** Sammlerriemen *m*; Sammelriemen *m* — **n** transportriempje *n*; matrijzentransportriempje *n* — **e** correa *f* transportadora de matrices; bajada *f* de las matrices — **i** cinghia *f* trasportamatrici

8690 matrix dryer; matrix drying box; mat dryer; dry mat scorcher; scorcher; flong drying box; former; dryer

f séchoir *m* de flans; séchoir *m* aux flans; séchoir *m* d'empreintes; appareil *m* à sécher les flans — **d** Materntrockenkasten *m*; Materntrockner *m*; Materntrockenapparat *m* — **n** matrijzendroogkast *fm*; matrijzendroger *m*; matrijzendrooginrichting *f*; droogkast *fm* voor matrijzen — **e** secadora *f* para estereomatrices; secadora *f* de matrices estereotípicas; secador *m* de matrices (estereotípicas); secadora *f* (de matriz); aparato *m* secador; estufa *f* secadora (con tambor rotativo) — **i** essiccatoio *m* per flani; essiccaflani *m*

8691 matrix drying box — See matrix dryer.

8692 matrix hair-line — See hair-line.

8693 matrix paper — See flong paper.

8694 matrix printer; matrix wire printer; wire printer; mosaic printer; stylus printer; dot printer — Serial printer which prints the representation of characters in dot formation by selecting the correct pattern of wire ends from a matrix.

f imprimante *f* à bloc; imprimante *f* à fils — **d** Matrixdrucker *m*; Stiftdrucker *m* — **n** mozaïekdrukker *m* — **e** impresora *f* por puntos — **i** stampatrice *f* a fili

8695 matrix rake — Part of the Typograph typesetting machine.

f râteau *m* à matrices; peigne *m* (pour recueillir des matrices) — **d** Matrizenrechen *m*; Rechen *m* — **n** matrijzenrek *n* — **e** peine *m* para las matrices — **i** rastello *m* delle matrici

8696 matrix slide; border slide; rule slide — Strip of brass on which is engraved a design of a rule, border, etc. It slides into a base or block ready for the casting of slugs on a typesetting machine.

f moule *m* à filets — **d** Liniengießform *f* — **n** lijnengietvorm *m* — **e** molde *m* para fundir orlas y filetes; matriz *f* corrediza — **i** forma *f* per fondere filetti

8697 matrix slide block — See matrix block.

8698 matrix wire printer — See matrix printer.

8699 mat roller; mangle *coll.* — Cylinder press for taking deep impression on a flong from a raised printing surface to produce a matrix for stereo-typing.

f presse *f* à empreindre — **d** Matrizenpräge-presse *f* — **n** preegpers *f*; mangel *m coll.* — **e** matrizador *m*; matrizadora *f*; máquina *f* de matrizar; arrollador *m* de matrices; prensa *f* para estampar matrices; calandria *f Es*; calandra *f*; cartonera *f Ch*; matricera *f Ch*; máquina *f* de prensar *Gu* — **i** calandra *f* per flani

8700 mat surface
f surface *f* mate — **d** matte Oberfläche *f*; Matt-schicht *f* — **n** mat oppervlak *n* — **e** superficie *f* mate — **i** superficie *f* non lucida

8701 matt — See mat.

8702 matter — Type that is composed.
f composition *f*; matière *f* — **d** Satz *m* — **n** zetsel *n* — **e** composición *f* — **i** composizione *f*

8703 matter — See copy.

8704 matt finish — See dull finish.

8705 maturation — Developing of the quality of freshly made paper by proper storing.
f maturation *f*; mûrissage *m* — **d** Reifung *f* — **n** laten besterven *n* — **e** maduración *f* — **i** matura-zione *f*

8706 mature paper — Paper kept in stock for a considerable period before use for quality improve-ment. Newly made papers tend to fluff, cockle, and stretch; also electricity generated by the friction of manufacture cause sheets to adhere and waste time on the printing machine. See also green paper.
f papier *m* maturé — **d** reifes Papier *n*; abgelager-tes Papier *n* — **n** bestorven papier *n* — **e** papel *m* maduro — **i** carta *f* stagionata

8707 mat varnish — Type of varnish in which the natural glossiness of the coating has been di-minished. Used as a pre-treatment for the retouching of photographic negatives with a crayon pencil. Solution of *dammar in turpentine.
f vernis *m* mattolin; mattolin *m* — **d** Mattlack *m*; Mattfirnis *m* — **n** matlak *mn*; matte vernis *nm*; matvernis *nm*; mattoleïen *m*; mattoleïne *m* — **e** barniz *m* mate; laca *f* mate — **i** vernice *f* non lucida

8708 mauresques — See arabesques.

8709 maximal allowable concentration — See threshold limit value.

8710 maximum sheet size
f format *m* maximum du papier — **d** größtes Papierformat *n*; Maximalformat *n* des Papiers — **n** grootste papierformaat *n* — **e** formato *m* máximo de papel; tamaño *m* máximo de papel — **i** formato *m* massimo del foglio

8711 maxwell — SI unit of magnetic flux. Abbrev. Mx (no point). 1 Mx = 10^{-8} Wb.
f maxwell *m* — **d** Maxwell *m* — **n** maxwell *m* — **e** maxwell *m* — **i** maxwell *m*

8712 Mazarin bible — The 42-line first book of 1282 printed pages in two columns, printed from movable type by Johann Gutenberg. The French cardinal Mazarin (1602-1661) had 25 copies.

8713 mc paper — See machine coated paper.

8714 mealiness — See missing dots.

8715 mean deviation; average error; average de-viation — In statistics, the mean of the deviations when all are given a positive sign.
f écart *m* moyen; déviation *f* moyenne — **d** durch-schnittliche Abweichung *f*; mittlere Abweichung *f* — **n** gemiddelde afwijking *f* — **e** desviación *f* media — **i** deviazione *f* media

8716 mean edge — In magnetic ink printing, the centre line between the two straight lines parallel to the character direction. It divides the irregular-ities of the printed edge in the printed edge zone in such a way that the sum of the white areas on the stroke side is equal to the sum of the black areas on the space side.
i linea *f* principale

8717 mean line — The middle of the three main imaginary framework lines (base line, cap line, mean line) on which letters are constructed. See also ascender line.
f ligne *f* médiane — **d** Mittellinie *f*; Zenterlinie *f* — **n** middenlijn *fm* — **e** línea *f* central — **i** linea *f* mediana

8718 mean time between failures — See down time.

8719 mean value — In mathematics, the ratio of the integral of a given function over a closed inter-val to the length of the interval.
f valeur *f* moyenne — **d** Mittelwert *m* — **n** ge-middelde waarde *f* — **e** valor *m* medio — **i** valore *m* medio

8720 measure; line width; line length — Width to which type is set, i.e., the maximum length of the lines, usually expressed in picas or ciceros.
f justification *f*; format *m* — **d** Formatbreite *f*; Satzbreite *f*; Satzformat *n*; Spaltenbreite *f*; Zeilen-länge *f*; Zeilenformat *n* — **n** zetbreedte *f*; regel-breedte *f*; kolombreedte *f*; lengte *f* van de regel — **e** justificación *f*; ancho *m* de composición; formato *m*; medida *f* — **i** giustezza *f* (di riga)

8721 measured daywork — Work study term for a wage plan based on fixed rates of pay but employee productivity and output are closely controlled through sound time standards.
n getarifieerde arbeid *m* — **i** lavoro *m* giornaliero misurato

8722 measurement; measuring
f mesure *f*; jaugeage *m* — **d** Messung *f*; Messen *n* — **n** meting *f*; meten *n* — **e** medida *f*; medición *f* — **i** misura *f*; misurazione *f*

8723 measuring — See measurement.

8724 measuring eyepiece — Finely-divided scale in the focal plane of the eyepiece of a microscope to measure the dimensions of viewed objects.
f oculaire *m* — **d** Meßokular *n* — **n** meet-oculair *n* — **e** ocular *m* de medida — **i** oculare *m* per misurare

8725 mechanical — Having to do with machinery.
f machinal; à la machine — **d** mechanisch; maschi-nell — **n** mechanisch; machinaal — **e** mecánico; a máquina — **i** meccanico; macchinale; a macchina

8726 mechanical — See mechanical paste-up.

8727 mechanical binding — See spiral binding.

8728 mechanical book paper — Low grade, wood containing paper, thicker than newsgrade, with a rough surface, or sometimes machine finished or machine glazed. See also book paper.

f papier *m* labeur; papier *m* ordinaire — **d** Werkdruckpapier *n* — **n** werkdrukpapier *n* — **e** papel *m* para obras — **i** carta *f* ordinaria per edizioni; carta *f* di qualità inferiore per edizioni

8729 mechanical chalk overlay — Overlay made by taking an impression with resist ink on a sheet covered with chalk, which is then etched with an acid solution to remove the chalk in so far as it is not protected by the ink.

f découpage *m* chimique — **d** mechanische Kreidereliefzurichtung *f* — **n** krijtpikeersel *n* — **e** arreglo *m* mecánico con greda — **i** avviamento *m* con rilievo in creta; avviamento *m* chimico

8730 mechanical characteristics

f caractéristiques *fpl* mécaniques — **d** mechanische Eigenschaften *fpl* — **n** mechanische eigenschappen *fpl* — **e** características *fpl* mecánicas — **i** caratteristiche *fpl* meccaniche

8731 mechanical colour separations — Pre-separation of colours for line or fake-colour art by the artist, executing the art for each colour in black or grey tones, usually separate overlays of key copy, ready for the camera. Either line or tone separations. Keylining is a method of separating each colour area by a line for copy executed in one piece; and opaquing areas to 0.3 mm (1/8 in) of division line. Colour tissue overlay is the colour guide for the opaquer who, with duplicate negatives or positives, makes a negative or positive for each colour. The width of the dividing line (common to adjoining colour areas) is the overlap of the inks.

f couleurs *fpl* par masque manuel — **d** Handfarbauszüge *mpl* — **n** met de hand vervaardigde kleurdeelplaten *fmpl* — **e** selecciones *fpl* de colores a mano — **i** selezioni *fpl* meccaniche

8732 mechanical composition — Means of producing a type surface, or its equivalent in proof form, without the aid of hand-set type.

f composition *f* mécanique — **d** Maschinensatz *m* — **n** machinezetsel *n* — **e** composición *f* mecánica; composición *f* a máquina; composición *f* mecanotípica; composición *f* de máquina — **i** composizione *f* meccanica

8733 mechanical focusing — Focusing of process cameras by means of scales.

f mise *f* au point automatique — **d** mechanische Einstellung *f*; Einstellen *n* nach Skala — **n** automatische scherpstelling *f*; automatische instelling *f* — **e** enfoque *m* mecánico — **i** messa *f* a fuoco automatica

8734 mechanically set

f composé à la machine — **d** gesetzt mit der Maschine; maschinell gesetzt — **n** gezet op de machine; machinaal gezet — **e** compuesto a máquina — **i** composto a macchina

8735 mechanical paper; wood paper — Paper made of mechanical wood pulp.

f papier *m* avec bois — **d** holzhaltiges Papier *n* — **n** houthoudend papier *n* — **e** papel *m* mecánico; papel *m* leñoso; papel *m* hecho a base de pulpa mecánica; papel *m* de pasta de madera; papel *m* de pasta mecánica — **i** carta *f* con pasta legno; carta *f* con legno; carta *f* da pasta meccanica

8736 mechanical paste-up; mechanical — Method of assembling all copy elements into a unit for photographic platemaking (copy ready for the camera). May include all copy except text, or be complete with text as well as line and tone copy, proportioned and positioned. The illustrations may be veloxes or blocked photoprints made to scale to aid in the subsequent stripping in of halftone inserts.

f groupement *m* — **d** Vorlagemontage *f* — **n** montage *f*; verzamelmodel *n* — **e** combinados *m* — **i** montaggio *m* per fotografia grafica

8737 mechanical pulp — See mechanical woodpulp.

8738 mechanical pulp board — Board made principally from mechanical woodpulp.

f carton *m* de pâte mécanique; carton *m* bois blanc — **d** weiße Holzpappe *f*; Weißschliffpappe *f* — **n** houtbord *n* — **e** cartón *m* de pasta mecánica; cartón *m* de madera blanca — **i** cartone *m* da pasta meccanica

8739 mechanical resistance — Opposition from a material to forces which tend to produce motion.

f résistance *f* mécanique — **d** mechanischer Widerstand *m* — **n** mechanische weerstand *m* — **e** resistencia *f* mecánica — **i** resistenza *f* meccanica

8740 mechanical stability — Property of a body to develop forces in opposition to any position or motion disturbing influence.

f stabilité *f* mécanique — **d** mechanische Stabilität *f* — **n** mechanische stabiliteit *f* — **e** estabilidad *f* mecánica — **i** stabilità *f* meccanica

8741 mechanical tint — Tint (dot, line or other pattern) added by the engraver to a line block as distinct from one which is on the drawing. See also benday process.

f grisé *m* photomécanique — **d** einkopierter Raster *m* — **n** filmraster *n* — **e** tinta *f* mecánica; grisado *m*

8742 mechanical translation — See machine translation.

8743 mechanical woodpulp; groundwood pulp; mechanical pulp — Pulp obtained by grinding cleaned logs with revolving pulp stones. There are several varieties, used principally in newsprint, the lower grade printing papers and boards.

f pâte *f* mécanique; pulpe *f* mécanique — **d** Holzschliff *m*; Holzstoff *m* — **n** houtslijp *m*; houtstof *fm* — **e** pasta *f* mecánica (de madera); pasta *f* de madera preparada mecánicamente — **i** pasta *f* meccanica (di legno); pastalegno *m*

8744 mechanics — Branch of physics dealing with the behaviour of matter under the action of force.

f mécanique *f* — **d** Mechanik *f* — **n** mechanica *f*

— **e** mecánica *f* — **i** meccanica *f*

8745 Mechanistics — Group of type faces, e.g., Atlas, Beton, Luxor, Egyptienne, Clarendon, Egizio, Ideal, Rockwell (the former Egyptians), showing heavy, square-ended serifs with or without brackets. Before 1971 they were named Slab Serifs. See also classification of type design.
f Mécanes *fpl* — **d** serifenbetonte Linear-Antiqua *f* — **n** Mechanen *mfpl* — **e** Mecanos *mpl* — **i** Egiziani *mpl*

8746 median — In statistics, the value of a variable characteristic which is greater than one half the observations and less than the other half.
f médiane *f* — **d** Zentralwert *m*; Medianwert *m* — **n** mediaan *fm* — **e** mediana *f* — **i** mediana *f*

8747 medical signs; medical symbols; pharmaceutical symbols; apothecaries' (weight) signs — See app. no. 10.
f signes *mpl* médicaux; symboles *mpl* de médecine; symboles *mpl* pharmaceutiques; symboles *mpl* de pharmacie; signes *mpl* de pharmacie; abréviations *fpl* médicales — **d** Apothekerzeichen *npl* — **n** farmaceutische tekens *npl* — **e** signos *mpl* de farmacia — **i** segni *mpl* di medicina

8748 medical symbols — See medical signs.

8749 medieval binding
f reliure *f* médiévale — **d** mittelalterlicher Einband *m* — **n** middeleeuwse band *m* — **e** encuadernación *f* medioeval — **i** legatura *f* medioevale

8750 medium — A British standard size for writing, printing and wrapping papers. See app. no. 3.

8751 medium — See weight of face.

8752 medium — See width.

8753 medium, binding — See vehicle.

8754 medium-bold face — See semi-bold face.

8755 medium, carrying — See vehicle.

8756 medium copying — A size of writing paper. See app. no. 3.

8757 medium-face rule
f filet *m* demi-gras; filet *m* mi-gras; filet *m* demi-maigre — **d** halbfette Linie *f*; stumpffeine Linie *f* — **n** halfvette lijn *fm*; stompfijne lijn *fm* — **e** filete *m* seminegro — **i** filetto *m* scuretto; filetto *m* medio nero; filetto *m* mezzo scuro

8758 medium, grinding — See vehicle.

8759 medium 8vo — A size of writing paper. See app. no. 3.

8760 medium post — A size of writing paper. See app. no. 3.

8761 medium 4to — A size of writing paper and paper for drawing. See app. no. 3.

8762 medium sized — See half sized.

8763 mega — Prefix in the metric system of measurement denoting one million. See also prefixes for metric units.
f még(a)- — **d** Mega- — **n** mega- — **e** megá- — **i** mega-

8764 megacycle; megahertz — A million cycles per second. See also prefixes for metric units.

f megahertz *m* — **d** Megahertz *m* — **n** megahertz *m* — **e** megahertz *m* — **i** megahertz *m*

8765 megahertz — See megacycle.

8766 melamine formaldehyde resin — Thermosetting resin with good temperature and moisture resistance and dielectric strength.
f résine *f* de mélamine-formol — **d** Melamin-Formaldehydharz *n* — **n** melamine-formaldehydehars *mn* — **e** resina *f* de melamínica formaldehída; resina *f* melamina formaldehída — **i** resina *f* melammino-formaldeidica

8767 melamine resin — Thermosetting resin formed by the interaction of melamine (formula: $C_3N_3(NH_2)_3$) and formaldehyde (formula: HCHO). Used as coating for paper, adhesive for laminated materials, plastics and textiles.
f résine *f* mélamine — **d** Melaminharz *n* — **n** melaminehars *mn* — **e** resina *f* de melamina; resina *f* melamínica — **i** resina *f* alla melammina

8768 mellow paper — Term is confusing. Mellow may mean well-matured (*mature paper), but it is also said to refer to a sheet of paper which is relatively pliable and compressible, in comparison with the normal paper of its particular grade.

8769 melting-point — Temperature at which a solid substance begins to change from solid to fluid form, specifically when under standard pressure.
f point *m* de fusion — **d** Schmelzpunkt *m* — **n** smeltpunt *n* — **e** punto *m* de fusión — **i** punto *m* di fusione

8770 melting pot; metal furnace — Receptacle or crucible in which type or stereotype casting metal is melted.
f chaudière *f* à fusion; pot *m* à fusion — **d** Schmelzkessel *m* — **n** smeltketel *m*; smeltpot *m* — **e** caldera *f* (de refundición); caldera *f* de refundir; horno *m* para refundir; horno *m* de refundición; marmita *f* — **i** caldaia *f* per fondere; crogiuolo *m* di fusione

8771 memo pad — Small size tablet used for scribbling notes, sometimes with an advertisement.
f bloc-notes *m*; aide-mémoire *m* — **d** Schreibblock *m*; Notizblock *m* — **n** schrijfblok *n*; notitieblok *n* — **e** taco *m* de papel (blanco); libretilla *f* para anotaciones; papel *m* en bloc — **i** blocco *m* per appunti

8772 memory — Term used in the earlier days of the computer and superseded by *store (GB), or storage (US).

8773 meniscus — In general, a crescent or crescent-shaped body, as the curved surface of a liquid in a vessel. If the contact angle between the liquid and the wall of the vessel is less than 90 °, the meniscus is concave; if greater, the meniscus is convex. See also converging lens, diverging lens.
f ménisque *m*; sommet *m* — **d** Meniskus *m*; Kuppe *f* — **n** meniscus *m* — **e** menisco *m* — **i** menisco *m*

8774 menu board — See ivory board.

8775 mercerize *v* — To treat cotton yarns or fabric with caustic alkali under tension to in-

crease its strength, luster and affinity for dye. Named after John Mercer (1791-1866), English calico printer.
f merceriser — **d** mercerisieren; merzerisieren — **n** merceriseren — **e** mercerizar
8776 merchandise paper — See bag.
8777 mercuric — Designation for mercury salts where the mercury is divalent, e.g., mercuric chloride, $HgCl_2$. See also mercurous.
f mercurique — **d** Merkuri- — **n** mercuri- — **e** mercúrico — **i** mercurico
8778 mercuric bromide — See mercury bromide.
8779 mercuric chloride — See mercury chloride.
8780 mercuric sulphocyanate — See mercury thiocyanate.
8781 mercuric sulphocyanide — See mercury thiocyanate.
8782 mercuric thiocyanate — See mercury thiocyanate.
8783 mercurography — Treatment of partially etched copper or zinc plates with a mercuric chloride solution to form an amalgam on the metal, the amalgam repelling greasy ink and permitting the plate to be rolled up with ink without depositing it on the sides and bottom of the etching.
8784 mercurous — Designation for mercury salts in which the mercury is monovalent, e.g., mercurous chloride, HgCl. See also mercuric.
f mercureux — **d** Merkuro- — **n** mercuro- — **e** mercurioso — **i** mercuroso
8785 mercurous bromide — See mercury bromide.
8786 mercurous chloride — See mercury chloride.
8787 mercurous iodide — See mercury iodide.
8788 mercury; quicksilver — Silvery, liquid, metallic element. Poisonous. Insoluble in hydrochloric acid, soluble in sulphuric acid upon boiling, readily and completely soluble in nitric acid, insoluble in water, alcohol and ether. Symbol: Hg. Latin: hydrargyrum. TL-value: 0.1 mg/m^3.
f mercure m; vif argent m — **d** Quecksilber n — **n** kwikzilver n; kwik n — **e** mercurio m — **i** mercurio m
8789 mercury bichloride — See mercury chloride.
8790 mercury bromide; mercuric bromide — White, rhombic crystals, sensitive to light, soluble in alcohol and ether, sparingly soluble in water. Poisonous.
Formula: $HgBr_2$.
f bromure m de mercure — **d** Merkurobromid n; Quecksilberbromid n; Bromquecksilber n — **n** kwikbromide n; mercuro-bromide n — **e** bromuro m de mercurio; bromuro m de mercúrico — **i** bromuro m di mercurio; bromuro m mercurico
8791 mercury bromide; mercurous bromide — White powder or colourless crystals, soluble in fuming nitric acid, hot concentrated sulphuric acid, hot ammonium carbonate or ammonium succinate solutions, sparingly soluble in water, insoluble in alcohol and ether.
Formula: HgBr.

f bromure m mercureux; bromure m de mercure — **d** Merkurobromid n; Quecksilberbromid n; Bromquecksilber n — **n** kwikbromide n; mercuro-bromide n — **e** bromuro m de mercurio; bromuro m mercurioso — **i** bromuro m di mercurio; bromuro m mercuroso
8792 mercury chloride; mercury bichloride; mercury dichloride; mercury perchloride; mercuric chloride; corrosive sublimate; corrosive mercury chloride; perchloride of mercury — White crystals or powder, soluble in water, alcohol, ether, pyridine, glycerol and acetic acid ester. Poisonous. Latin: chloretum hydrargyricum.
Formula: $HgCl_2$.
f chlorure m mercurique; bichlorure m de mercure; sublimé m corrosif — **d** Quecksilberchlorid n; Quecksilbersublimat n; (doppelt) Chlorquecksilber n; Merkurichlorid n — **n** kwikchloride n; mercuri-chloride n; sublimaat n — **e** cloruro m mercúrico; sublimado m (corrosivo); bicloruro m de mercure — **i** cloruro m mercurico; sublimato m (corrosivo); bicloruro m di mercurio
8793 mercury chloride; mercury monochloride; mercury protochloride; mercury subchloride; mild mercury chloride; calomel; mercurous chloride — White, rhombic crystals or crystalline powder, soluble in water, alcohol and ether. Not poisonous.
Formula: HgCl.
f chlorure m mercureux; protochlorure m de mercure; mercure m doux; calomel m — **d** Quecksilberchlorür n; Quecksilberchlorid n; Merkurochlorid n; Kalomel n — **n** kwikchloruur f; mercuro-chloride n; kalomel n — **e** cloruro m mercurioso; protocloruro m de mercurio; calomel m; calomelanos mpl — **i** cloruro m mercuroso; calomelano m
8794 mercury dichloride — See mercury chloride.
8795 mercury intensifier
f renforçateur m au mercure; renforçateur m au chlorure mercurique; renforçateur m au bichlorure de mercure — **d** Quecksilberverstärker m — **n** kwikversterker m — **e** reforzador m de sales de mercurio — **i** rinforzatore m al mercurio
8796 mercury iodide; mercurous iodide; mercury protoiodide — Bright yellow, amorphous powder, becoming greenish on exposure to light, soluble in castor oil, liquid ammonia, aqua ammonia, insoluble in water, alcohol, ether.
Formula: Hg_2I_2.
f iodure m de mercure — **d** Jodquecksilber n; Quecksilberjodid n — **n** kwikjodide n — **e** yoduro m de mercurio; yoduro m mercurioso — **i** ioduro m mercuroso
8797 mercury iodide intensifier
f renforçateur m à l'iodure de mercure — **d** Jodquecksilber-Verstärker m — **n** joodkwikversterker m — **e** reforzador m al yoduro mercúrico — **i** rinforzatore m all'ioduro di mercurio
8798 mercury monochloride — See mercury chloride.

8799 mercury perchloride — See mercury chloride.

8800 mercury protochloride — See mercury chloride.

8801 mercury protoiodide — See mercury iodide.

8802 mercury rhodanide — See mercury thiocyanate.

8803 mercury subchloride — See mercury chloride.

8804 mercury sulphocyanate — See mercury thiocyanate.

8805 mercury sulphocyanide — See mercury thiocyanate.

8806 mercury thiocyanate; mercury sulphocyanate; mercury sulphocyanide; mercury rhodanide; mercuric thiocyanate; mercuric sulphocyanate; mercuric sulphocyanide — White powder. Poisonous. Slightly soluble in alcohol, insoluble in water. Used in photography. Formula: $Hg(CNS)_2$.
f thiocyanate *m* de mercure; sulfocyanure *m* de mercure — **d** Quecksilbersulfozyanat *n*; Merkurirhodanid *n*; Quecksilberrhodanid *n* — **n** mercurithiocyanaat *n*; mercuri-rhodanide *n* — **e** tiocianato *m* mercúrico — **i** tiocianato *m* mercurico; solfocianuro *m* di mercurio

8807 mercury vapour lamp — Lamp which gives blue-white light containing no red rays. The light is produced by passing an electric current through vapour of mercury. The colour temperature ranges from 4500-5000 K. Used in photographic work in place of arc lamp.
f lampe *f* à vapeur de mercure — **d** Quecksilberdampflampe *f* — **n** kwikdamplamp *fm* — **e** lámpara *f* (de vapor(es) de) mercurio — **i** lampada *f* a vapore di mercurio

8808 merge *v* — To combine two or more sets of items into one, usually in a specified sequence.
f fusionner; intercaler — **d** mischen — **n** samenvoegen; ineenvoegen; invoegen — **e** fusionar; refundir — **i** inserire

8809 meridian — Old type-body name, listed in 1822 by Bruce's New York Type Foundry, for seven-line pica (72 points). See app. no. 1.

8810 merit rating
f évaluation *f* du mérite personnel; notation *f* du personnel — **d** Persönlichkeitsbewertung *f*; Persönlichkeitsbeurteilung *f* — **n** prestatiebeoordeling *f*; prestatie-beloning *f*; merit rating *m* — **e** evaluación *f* del mérito personal — **i** valutazione *f* di merito

8811 mesh *v* — To interact (of the teeth on spur wheels, etc.).
f engrener — **d** eingreifen; ineinandergreifen; im Eingriff stehen — **n** ineengrijpen — **e** engranar — **i** ingranare; ingranarsi

8812 mesh count — Classification of screen meshes according to the thickness of the threads, the mesh opening and the number of threads per centimetre or inch.
f grosseur *f* de trame — **d** Fadenzahl *f*; Gewebezahl *f*; Gewebenummer *f* — **n** draadgetal *n*; draad-

nummer *n* — **i** titolo *m* di un tessuto

8813 mesh opening — Because there is a relationship between the mesh opening of a fabric used in screen printing process and the size of the ink pigment, the mesh opening of screen printing fabrics should be at least 2 1/2 - 3 times greater than the size of the ink.
f ouverture *f* de la maille — **d** Maschenweite *f*; offene Maschenfläche *f*; offene Siebfläche *f* — **n** maaswijdte *f* — **e** abertura *f* de la malla — **i** apertura *f* di maglia

8814 message header — See header.

8815 meta — Prefix referring to inorganic acids that are derived from the ortho- or ordinary form by loss of one molecule of water from each molecule of acid.

8816 meta-dihydroxybenzene — See resorcinol.

8817 metal base — Block of solid type metal on which relief plates are mounted type high in substitution of a wooden base.
f bloc *m* matière — **d** Metallfuß *m*; Metallklotz *m* — **n** metalen voet *m* — **e** base *f* de metal; base *f* metálica — **i** zoccolo *m* di metallo

8818 metal box — See hellbox.

8819 metal decorating; tin printing *depr.* — Printing by offset lithography on sheet metal such as tinplate or aluminium. The work is specialty and is done on special built presses.
f impression *f* sur fer-blanc; impression *f* sur tôle; impression *f* sur métal — **d** Blechdruck *m* — **n** blikdruk *m* — **e** impresión *f* en hojalata; impresión *f* sobre hojalata; impresión *f* sobre metal; impresión *f* metalográfica; metalografía *f* — **i** stampa *f* su latta; lattografia *f*; metallografia *f*

8820 metal decorating ink; tin printing ink *depr.*
f encre *f* à impression sur fer-blanc; encre *f* à impression sur tôle; encre *f* à impression sur métal — **d** Blechdruckfarbe *f* — **n** blikdrukinkt *m* — **e** tinta *f* para impresión sobre hojalata; tinta *f* para impresión sobre metal — **i** inchiostro *m* per stampa su latta

8821 metal decorating machine; tin-plate printing machine *depr.* — Machine for printing on sheet metal.
f machine *f* à imprimer le fer-blanc; machine *f* à impression sur fer-blanc; machine *f* à impression sur tôle; machine *f* à impression sur métal — **d** Blechdruckmaschine *f* — **n** blikdrukpers *f*; blikdrukmachine *f* — **e** máquina *f* para la impresión en hojalata; máquina *f* para imprimir hojalata — **i** macchina *f* per la stampa su latta

8822 metal dust; dusting of metal — Fine invisible metal dust floating in the air near typesetting machines.
f poussière *f* métallique — **d** Müllern *n* — **n** stuiven *n* van metaal — **e** polvo *m* metálico — **i** polvere *f* metallica

8823 metal edger — See metal edging machine.

8824 metal edging machine; metal edger
f machine *f* à baguetter; baguetteuse *f* — **d** Randleistenandrückmaschine *f*; Randleistenmaschine *f*; Metallkantmaschine *f*; Metallschienenandrück-

maschine *f* — **n** bagetteermachine *f* — **e** ribeteadora *f*; envarilladora *f* — **i** blindatrice *f*

8825 metal feeder; metal pot feeder; ingot feeder — Device which, by the movement of the metal pot of the type-casting machine, automatically feeds a long ingot into the pot when the level of the metal demands it.
f alimentateur *m* de métal — **d** Metallzuführer *m*; Metallzuführung *f* — **n** potvuller *m* — **e** alimentador *m* de metal — **i** alimentatore *m* del metallo

8826 metal furnace; metal melting furnace — See also melting pot.
f four *m* à refondre — **d** Schmelzofen *m*; Metallschmelzofen *m* — **n** smeltoven *m* — **e** horno *m* para metal; caldera *f* para metal; horno *m* de refundición; caldera *f* de refundir — **i** forno *m* per la fusione metallica

8827 metal furnace — See melting pot.

8828 metallic ink — Finely divided metal powders mixed with appropriate ink vehicles. Gold and silver are the usual descriptions, the former being an alloy of copper and zinc and the latter aluminium.
f encre *f* métallisée — **d** Metalldruckfarbe *f* — **n** metaalinkt *m* — **e** tinta *f* metálica — **i** inchiostro *m* metallizzato

8829 metallic oxide — Binary compound of a metal and oxygen.
f oxyde *m* métallique — **d** Metalloxyd *n* — **n** metallisch oxyde *n* — **e** óxido *m* metálico — **i** ossido *m* metallico

8830 metallic paper — Paper coated with aluminium or bronze powder.
f papier *m* métallisé — **d** Metallpapier *n* — **n** metaalpapier *n* — **e** papel *m* metálico; papel *m* metalizado — **i** carta *f* metallizzata

8831 metallic salt
f sel *m* métallique — **d** Metallsalz *n* — **n** metaalzout *n* — **e** sal *f* metálica — **i** sale *m* metallico

8832 metallic slush — Formations of crystals around the opening of the metal pot containing much tin and antimony. To mix these crystals into the alloy, a heat of more than 275 °C is needed. See also dross.
f crasses *fpl* — **d** Metallkrätze *f*; Garschaum *m*; Krätzeansatz *m* — **n** metaalkristallen *npl* — **i** scorie *fpl* metalliche

8833 metallic soap — Paste drier for printing inks, commonly metal salts as acetates, borates, or oxalates dispersed in a drying oil.
f savon *m* métallique — **d** Metallseife *f* — **n** metaalzeep *fm* — **e** jabón *m* metálico — **i** sapone *m* metallico

8834 metal melting furnace — See metal furnace.

8835 metal mounted — See mounted on metal.

8836 metal pot; crucible; pot crucible; pot — In a typesetting machine a crucible containing molten metal, which is pumped through the throat of the crucible into the mould, thereby forming a slug or letter.
f creuset *m* — **d** Gießtopf *m* — **n** loodpot *m*; gietpot *m* — **e** crisol *m*; caldera *f* de fundición *Es*;

fundición *f Ar* — **i** crogiolo *m*; crogiuolo *m*

8837 metal pot feeder — See metal feeder.

8838 metal print — Line or halftone image on a copper or zinc plate for relief etching.
f copie *f* sur métal — **d** Metallkopie *f* — **n** metaalkopie *f* — **e** copia *f* sobre metal; pasado *m* de plancha — **i** copia *f* su lastra metallica

8839 metal-printing room
f salle *f* de copie sur métal — **d** Kopierabteilung *f*; Kopierraum *m* — **n** kopieerafdeling *f* — **e** sala *f* de insolación — **i** reparto *m* copia

8840 metal rimming
f baguettage *m* — **d** Randleisten anbringen; Blechleisten anbringen; Blechbeleistung *f* — **n** bagetteren — **e** montar varillas; poner varillas — **i** modanatura *f* metallica

8841 metal to metal; iron to iron; steel to steel — Distance between the surfaces of an undressed impression cylinder and an undressed plate cylinder of an offset press.
f intervalle *m* entre cylindres — **d** Zwischenmaß *n* Eisen auf Eisen — **n** afstand *m* tussen de onbeklede cilinders — **e** intervalo *m* entre cilindros — **i** intervallo *m* tra i cilindri; distanza *f* tra i cilindri

8842 metameric — Said of colours, when appearing to be different in hue in different lights.
f métamère — **d** metamer — **n** metameer — **e** metamérico — **i** metamerico

8843 metameric colours — Colours which appear as a satisfactory match under one kind of illumination and a relatively poor match (or mismatch) under other. They would have different spectrophotometric curves.
f couleurs *fpl* métamères — **d** metamere Farben *fpl* — **n** metamere kleuren *fmpl* — **e** colores *mpl* metaméricos — **i** colori *mpl* metamerici

8844 metamerism — Phenomenon, whereby spectrally different radiations produce the same colour under the same viewing conditions.
f métamérisme *m* — **d** Metamerie *f*; bedingtgleiche Farbvalenzen *fpl* — **n** metamerie *f* — **e** metamería *f* — **i** metamerismo *m*

8845 meteorological signs — See meteorological symbols.

8846 meteorological symbols; meteorological signs; signs of meteorology — See app. no. 10.
f signes *mpl* de météorologie — **d** meteorologische Zeichen *npl*; Witterungszeichen *npl* — **n** meteorologische tekens *npl* — **e** signos *mpl* de meteorología — **i** segni *mpl* di meteorologia

8847 meter — Instrument to measure flow of solids, liquids, or gases or to count articles, such as number of sheets or impressions.
f mesureur *m*; compteur *m*; jaugeur *m*; appareil *m* de mesure; dispositif *m* de mesure — **d** Meßgerät *n*; Lehre *f*; Maß *n*; Meßvorrichtung *f* — **n** meetinstrument *n*; meetinrichting *f* — **e** instrumento *m* de medición — **i** apparecchio *m* di misura

8848 meter — See metre.

8849 meter *v* — On reel-fed rotary presses and in-line machines, to feed a web at a regulated speed to a printing or a fabricating unit.

f alimenter — **d** speisen — **n** doorvoeren — **e** alimentar — **i** alimentare

8850 metering unit; meter unit; feed unit; in-feed rollers — Series of three rollers (two driven, one free) mounted on a reel stand. Used to smooth the web and control its tension and speed as it feeds from reel into first printing unit.

f groupe *m* d'introduction; rouleaux *mpl* d'introduction — **d** Einführungswerk *n*; Einlaufwalzen *fpl* — **n** invoerinrichting *f*; invoerrollen *fmpl* — **e** rodillos *mpl* de introducción; rodillos *mpl* de entrada — **i** gruppo *m* d'alimentazione

8851 meter unit — See metering unit.

8852 methanal — See formaldehyde.

8853 methanecarboxylic acid — See acetic acid.

8854 methanoic acid — See formic acid.

8855 methanol; methyl alcohol; wood alcohol — Clear, colourless, mobile, volatile flammable liquid. Poisonous. Soluble in water, alcohol and ether. TL-value: 200 ppm or 260 mg/m³.
Formula: CH_3OH.

f méthanol *m*; alcool *m* méthylique; alcool *m* de bois; esprit *m* de bois; carbinol *m* — **d** Methanol *m*; Methylalkohol *m*; Holzgeist *m*; Holzspiritus *n*; Holzalkohol *m*; Karbinol *m* — **n** methanol *n*; methylalcohol *m*; houtgeest *m* — **e** metanol *m*; alcohol *m* metílico; alcohol *m* de madera; espíritu *m* de madera — **i** metanolo *m*; alcool *m* metilico; spirito *m* di legno

8856 methylacetyl — See acetone.

8857 methyl alcohol — See methanol.

8858 methylated alcohol; methylated spirit — Ethyl alcohol rendered undrinkable by means of methyl alcohol. See also denatured alcohol.

8859 methylated spirit — See methylated alcohol.

8860 methyl benzene — See toluene.

8861 methyl benzol — See toluene.

8862 methyl bromide; bromomethane — Colourless gas or volatile liquid. Poisonous. Miscible with most organic solvents; from a voluminous crystalline hydrate with cold water.
Formula: CH_3Br.

f bromure *m* de méthyle — **d** Methylbromid *n*; Bromethan *n*; Monobrommethan *n*; Brommethyl *n* — **n** methylbromide *n* — **e** bromuro *m* de metilo — **i** bromuro *m* di metile

8863 methyl cellulose; cellulose methyl ether — Greyish, fibrous powder resulting when methyl chloride reacts with cellulose. Capable of forming a film that may be printed very simply; however, due to the nature of the film, moderate heat must be used in drying. Swells in water and produces a clear to opalescent, viscous, colloidal solution. Aqueous suspensions are neutral to litmus. Insoluble in alcohol, ether and chloroform, soluble in glacial acetic acid, unaffected by oils and greases, flammable when ignited, stable to light.
Formula: $[C_6H_7O_2(OH)_2OCH_5]_n$.

f méthylcellulose *m* — **d** Methylzellulose *f* — **n** methylcellulose *fm* — **e** metilcelulosa *f* — **i** metilcellulosa *f*

8864 methyl ethylene glycol — See propylene glycol.

8865 methyl glycol — See ethylene glycol monoethyl ether.

8866 methyl methacrylate — Important monomer for the production of a number of plastics and resins; commonly known under the trade-name *plexiglas.

f méthacrylate *m* de méthyle — **d** Methylmethakrylat *n* — **n** methyl-methacrylaat *n* — **e** metacrilato *m* de metilo — **i** metacrilato *m* di metile; metilmetacrilato *m*

8867 methyl-paraminophenol sulphate; para-methyl-aminophenol sulphate — White, soluble powder, derivative of cresol, soluble in water and alcohol, insoluble in ether. Used as a photographic developer and known by a number of trade names, e.g., metol, the original German name, and elon, the name used by the Eastman Kodak Company.
Formula: $HOC_6H_4NHCH_3 \cdot 1/2H_2SO_4$.

f sulfate *m* de monométhyl-paraminophénol; métol *m* — **d** Methylaminophenolsulfat *n*; schwefelsaures Monomethyl-Paraminophenol *n*; Metol *n* — **n** monomethyl-paraminofenolsulfaat *n*; metol *mn*; genol *mn*; rhodal *m* — **e** sulfato *m* de monometil paraminofenol; metol *m* — **i** solfato *m* di metil-paramminofenolo; metolo *m*

8868 methyl phenol — See cresol.

8869 methyl violet — Basic organic violet dyestuff, derived from rosaniline, which is used as an oil soluble toner for black ink. When properly precipitated, it produces bright violet pigments which are used as purple colours or as toners for blacks. Also used to colour printing down on metal in making letterpress printing plates and as an indicator in chemistry. There is a confusing series of names for closely related and overlapping derivatives of para rosaniline, a triphenyl methane type of dye.

f méthylviolet *m*; violet *m* de méthyle — **d** Methylviolett *n* — **n** methylviolet *n* — **e** violeta *f* de metilo — **i** violetto *m* di metile

8870 metol — See methyl-paraminophenol sulphate.

8871 metre *GB*; **meter** *US* — Fundamental unit of length in the metric system, originally intended to be, and being very nearly, equal to one ten-millionth of the distance from the equator to the pole, measured on a meridian. Defined from 1889 to 1960 as the distances between two lines on a platinum/iridium bar at the International Bureau of Weights and Measures at Sèvres (near Paris), and now defined as 1.65076373 wavelengths of the orange-red radiation of the isotope krypton 86 under specified conditions. Abbrev. m (no point). See app. no. 7.

f mètre *m* — **d** Meter *nm* — **n** meter *m* — **e** metro *m* — **i** metro *m*

8872 metre candle — See lux.

8873 metric horsepower; continental horsepower;

French horsepower — The power required to raise 75 kg against the force of gravity through a distance of 1 metre in 1 second. In the SI See also horsepower.

8874 metric rule paper — See graph paper.

8875 metric system — Decimal system of weights and measures established in France in 1799 and legalized in the US in 1866. Based on the metre as the unit of length, the litre being the standard of capacity or volume and the gram serving as the standard of weight. For the UK, see app. no. 7.
f système *m* métrique — **d** Dezimalsystem *n* — **n** decimaal stelsel *n*; tiendelig stelsel *n*; tientallig stelsel *n*; metrieke stelsel *n* — **e** sistema *m* métrico; sistema *m* decimal — **i** sistema *m* metrico (di misura); sistema *m* metrico decimale

8876 metric thread
f filet *m* métrique; filetage *m* métrique — **d** metrisches Gewinde *n* — **n** metrische draad *m*; metrische schroefdraad *m* — **e** rosca *f* métrica; filete *m* (de rosca) métrica — **i** filettatura *f* metrica

8877 metzograph — See mezzograph.

8878 metzograph screen — See grain screen.

8879 Meyer bar — See equalizer rod.

8880 Meyer rod — See equalizer rod.

8881 mezzograph; metzograph — Printing plate with a grained screen, invented by the Englishman James Wheeler (1897), as a substitution of the crossline screen. See also grain screen.

8882 mezzograph screen — See grain screen.

8883 mezzotint; manière noire — Intaglio print made from ground (e.g., crossed in several directions with knife-edged cuts) copper plate; the burred surface is scraped away to create the design. The ground is applied with a *rocker.
f mezzo-tinto *m*; gravure *f* en manière noire; manière *f* noire; manière *f* anglaise; estampe *f* en manière noire — **d** Mezzotinto *m*; Schab(e)kunst *f*; Schwarzkunst *f*; Mezzotintostich *m*; Mezzotintodruck *m* — **n** mezzotint *fm*; mezzo tinto *fm*; manière *f* noire; zwartekunstplaat *fm*; zwartekunstprent *fm* — **e** grabado *m* a la maneranegra; grabado *m* en negro; estampa *f* mediatinta — **i** mezzotinto *m*; manière *f* noire; maniera *f* nera

8884 mf paper — See machine-finished paper.

8885 mg-board — See machine-glazed paper or board.

8886 mg-paper — See machine-glazed paper or board.

8887 micelle — Particle of colloidal dimensions, e.g., a colloidal ion.
f micelle *f* — **d** Mizelle *f* — **n** micel *fm* — **e** micela *f* — **i** micella *f*

8888 micelle formation
f formation *f* de micelles — **d** Mizellbildung *f* — **n** micelvorming *f* — **e** formación *f* de micelas — **i** formazione *f* di micelle

8889 micro — See prefixes for metric units.

8890 microcard — See aperture card.

8891 microcomputer — Tiny, solid-state chip of LSI circuitry, which contains its own instruction set and, when combined with a memory to handle data and some method of receiving and disposing of data, can perform arithmetic and logical functions pursuant to a program. Hence, depending upon the instructions it may be a general purpose computer, although it requires additional elements to make it so function.

8892 micro-contour ink wipe test — See micro-contour test.

8893 micro-contour test; micro-contour ink wipe test — Paper smoothness test in which an ink of particular formation is applied to the surface of the test piece. With a soft cloth most of the ink is removed. This is continued until no further lightening of the colour of the surface takes place. Pits in coatings will be seen as discrete coloured dots and surface fibres are white lines against the background. Regularly repeating areas of light and dark would be likely to give rise to mottling.
f test *m* du micro-contour; micro-contour test *m* — **d** Mikrokonturtest *m* — **n** micro-contourtest *m* — **e** ensayo *m* micro-contor — **i** saggio *m* micro-contour; prova *f* micro-contour; micro-contour-test *m*

8894 microcrystalline wax — Petroleum wax of high molecular weight, characterized by minute crystals and distinguished by its solid, wax-like appearance at room temperature. Used for laminating and surface coating in preference to paraffin wax which is less ductile and more brittle. The term amorphous wax is incorrect.
f cire *f* micro-cristalline — **d** mikrokristallinisches Wachs *n* — **n** microkristallijne was *mn* — **e** cera *f* microcristalina — **i** cera *f* microcristallina

8895 microfiche — Sheet of microfilm, in a form suitable for filing, containing several images.
f microfiche *f*; microfilm *m* en feuille — **d** Mikroplanfilm *m*; durchsichtige Mikrokarte *f* — **n** microfiche *m* — **e** microficha *f* transparente — **i** microscheda *f* trasparente

8896 microfilm — Reproduction of text or illustrations on 16 mm of 35 mm film to provide an information record that may later be retrieved.
f microfilm *m* — **d** Mikrofilm *m* — **n** microfilm *m* — **e** microfilm *m*; micropelícula *f* — **i** microfilm *m*

8897 micro-integrated circuit — See micro-miniaturized circuit.

8898 micrometer; micrometer gauge; thickness gauge — Instrument to measure the thickness of paper, electrotype shells, printing plates, cylinder packings etc. See also callipers, packing gauge.
f micromètre *m*; palmer *m* — **d** Mikrometer *n*; Mikrometerlehre *f*; Dickenmesser *m*; Meßbügel *m*; Feinmesser *m* — **n** micrometer *m*; diktemeter *m* — **e** micrómetro *m*; medidor *m* de gruesos; calibrador *m*; lengüeta *f* calibradora — **i** micrometro *m*

8899 micrometer — See micron.

8900 micrometer gauge — See micrometer.

8901 micrometer scale
f échelle *f* micrométrique — **d** Mikrometerskala *f*

— **n** micrometerschaal *fm* — **e** escala *f* micrométrica — **i** scala *f* micrometrica

8902 micrometer screw — Screw with fine and very accurately cut threads and a circular, graduated head, which shows the amount of advancement or retraction of the screw. Used in making fine measurements often to 0.025 mm (0.001 in).
f vis *f* micrométrique — **d** Mikrometerschraube *f* — **n** micrometerschroef *fm* — **e** tornillo *m* micrométrico; microtornillo *m* — **i** vite *f* micrometrica

8903 micrometre — See micron.

8904 micro-miniaturized circuit; micro-integrated circuit — Electronic circuit using techniques of integrated, hybrid, thin-film monolithic circuitry.
f circuit *m* microminiaturisé — **d** mikrominiaturisierte Schaltung *f* — **n** micro-miniatuurschakeling *f* — **e** circuito *m* microminiaturizado — **i** circuito *m* micro-miniaturizzato

8905 micron; micrometre *GB*; **micrometer** *US* — In the SI: 1 micron = 1 μm = 1 micrometre = 10^{-6} m. See also prefixes for metric units.
f micron *m* — **d** Mikron *n* — **n** micron *nm* — **e** micrón *m* — **i** micron *m*

8906 microphotograph — Photograph requiring optical enlargement to render it visible in details.
f microphotographie *f* — **d** Mikrobild *n* — **n** microfoto *f* — **e** microfotografía *f* — **i** microfotografia *f*

8907 micro pick — See coating pick.

8908 microprogramming — Operation of the control unit of a computer in which each instruction, instead of initiating control signals directly, starts the execution of a sequence of micro instructions at a more elementary level. The micro instructions are usually stored in a special read-only storage unit.
f microprogammation *f* — **d** Mikroprogrammierung *f* — **n** microprogrammering *f* — **e** microprogramación *f* — **i** microprogrammazione *f*

8909 micro reciprocal degree — See mired.

8910 microscope — See also electron microscope, pocket microscope.
f microscope *m* — **d** Mikroskop *m* — **n** microscoop *mn* — **e** microscopio *m* — **i** microscopio *m*

8911 microsecond — One millionth part of a second (10^{-6} sec). Abbrev. μs or μsec. See also prefixes for metric units.
f microseconde *f* — **d** Mikrosekunde *f* — **n** microseconde *fm* — **e** microsegundo *m* — **i** microsecondo *m*

8912 microwaves — Frequencies whose wavelengths are sufficiently short to exhibit some properties of light, in that they can be easily focused, reflected or refracted.
f micro-ondes *fpl* — **d** Mikrowellen *fpl* — **n** microgolven *fmpl* — **e** microondas *fpl* — **i** micro-onde *fpl*

8913 MICR printing — See magnetic ink printing.

8914 middle — Size of board, recognized but not standard in the UK. Also known as small demy. See app. no. 3.

8915 middle demy — A size of board. See app. no. 3.

8916 middle hand — A size of writing paper. See app. no. 3.

8917 middle leads — Leads two points thick that come between the thin (1 point) and thick (6 point) leads.
f interlignes *fpl* moyennes; interlignes *fpl* de deux points — **d** 2-Punkt-Regletten *fpl* — **n** 2-punts interlinies *fpl* — **e** interlíneas *fpl* de dos puntos — **i** interlinee *fpl* di due punti; interlinee *fpl* medie

8918 middles — A size of paper. See app. no. 3.

8919 middle space; four-to-em space — Space cast four to the em of its own body.
f espace *f* moyenne — **d** Viertelspatium *n*; Viertelgeviertspatium *n*; Viertelgeviert *n* — **n** spatie *f* vier op vierkant — **e** espacio *m* medio; espacio *m* mediano — **i** spazio *m* mezzano

8920 middletone — Neutral tone intermediate between the lightest and darkest tones of an original, negative or reproduction therefrom.
f ton *m* moyen — **d** Mittelton *m* — **n** middentoon *m* — **e** tono *m* medio; media tinta *f* — **i** tono *m* intermedio

8921 middletone — Refers to a *drawdown of a wet ink sample. This is the portion at the top of the drawdown where the knife is held at 45 degree angle and medium pressure is applied. This is considered comparable to the line printed effect on the substrate.

8922 middletone stop — Lens aperture used in halftone photography to register the middletones, the diameter of which is midway between that of the highlight and detail stops.
f diaphragme *m* des tons moyens; diaphragme *m* des demi-tons — **d** Mitteltonblende *f* — **n** diafragma *n* voor de middentonen — **e** diafragma *m* para los medios tonos — **i** diaframma *m* per toni intermedi

8923 middle-weight forme — Letterpress printing forme with a printing zone of 30-50% of the distance between the two bed bearers. See also pre-tension.
f forme *f* moyenne — **d** mittlere Druckform *f* — **n** middelzware drukvorm *m* — **e** forma *f* de peso medio — **i** forma *f* di peso medio

8924 midget — A size of cut card. See app. no. 3.

8925 migration of the plasticizer — Movement or diffusion of chemical ingredients in their supporting medium.
f migration *f* du plastifiant — **d** Weichmacherwanderung *f* — **n** migratie *f* van de weekmaker — **e** migración *f* del plastificante — **i** migrazione *f* del plastificante

8926 mil — Confusing abbreviation for a thousandth.
f millième *m* — **d** Tausendstel *n* — **n** duizendste *n* — **e** milésimo *m* — **i** millesimo *m*

8927 mil — In the SI: milli-inch or 25.4×10^{-6} m. Also abbreviation for *millilitre.

8928 mildew spots — Yellow-brown spots on old

paper, a superficial, usually whitish growth produced on organic matter or living plants by fungi. See also foxed leaves.

f taches *fpl* de moisissure; taches *fpl* de mouillure(s); piqûres *fpl* — **d** Schimmelflecken *mpl*; Meltauflecken *mpl*; Stockflecken *mpl* — **n** schimmelvlekken *fmpl* — **e** manchas *fpl* de moho — **i** macchie *fpl* di muffa

8929 mild mercury chloride — See mercury chloride.

8930 mileage — See ink coverage.

8931 milk acid — See lactic acid.

8932 millboard — Homogeneous board made usually of mixed *waste papers on an intermittent board machine, thicker than 1 mm (0.040 in). Most countries accept that millboard is not made thinner than 1 mm, but thinner material is produced in the UK.

f carton *m* à l'enrouleuse — **d** Handpappe *f*; Wickelpappe *f*; Maschinenpappe *f* — **n** handbord *n*; wikkelbord *n* — **e** cartón *m* grueso — **i** cartone *m* a mano

8933 mill bristol; cylinder bristol; printing bristol — Bristol usually made on a cylinder machine. The basis weights range from 90 to 200 pounds (22 1/2 x 28 1/2 - 500). Normally for printing purposes.

f bristol *m* fabriqué sur forme ronde — **d** auf der Rundsiebmaschine hergestellter Bristolkarton *m* — **n** op de rondzeefmachine vervaardigd bristolkarton *n* — **e** cartulina *f* bristol fabricada en la máquina a forma redonda — **i** cartoncino *m* bristol fabbricato su macchina in tondo

8934 mill finish — See machine finish.

8935 mill-finished paper — See machine-finished paper.

8936 milliliter — See millilitre.

8937 millilitre *GB*; **milliliter** *US* — Metric unit of capacity. Abbrev. ml or mil. In the SI: 10^{-6} m^3. See app. no. 7.

f millilitre *m* — **d** Milliliter *nm* — **n** milliliter *m* — **e** mililitro *m* — **i** millilitro *m*

8938 millimicron — A unit for measuring wavelengths of light equivalent to the millionth part of a millimetre (written mµ). Preference is given to *nanometre. See also prefixes for metric units.

f millimicron *m* — **d** Millimikron *n* — **n** millimicron *nm* — **e** milimicrón *m* — **i** millimicron *m*

8939 milling — Bevelling a plate, also routing away dead metal from the printing area of the plate.

f toupillage *m* — **d** Fräsen *n*; Abfräsen *n*; Rauten *n* — **n** frezen *n* — **e** fresado *m* — **i** fresatura *f*; smussatura *f*

8940 millisecond — One thousandth part of a second (10^{-3} sec). Abbrev. ms or msec. See also prefixes for metric units.

f milliseconde *f* — **d** Millisekunde *f* — **n** milliseconde *fm* — **e** milisegundo *m* — **i** millisecondo *m*

8941 mill join — Adhesion between the ends of two similar webs during manufacture to ensure continuous run of web as the reel unwinds. May be marked by the mill for the guidance of the printer.

f joint *m* de fabrication; collage *m* de fabrication — **d** Klebestelle *f* von der Papierfabrik — **n** fabriekslas *fm* — **e** junta *f* de fabricación — **i** giunta *f* di fabbricazione

8942 mill roll; master roll; jumbo roll — Reel stock as received from the paper mill.

f bobine *f* brute; bobine *f* mère — **d** Fabriksrolle *f*; Maschinenrolle *f* — **n** fabrieksrol *fm*; machinerol *fm* — **i** bobina *f* di fabbrica

8943 mill wrappers — See ream wrappers.

8944 mill wrapping — Packing paper for wrapping reels or bundels of paper to protect them in transit.

f emballage *m* de la bobine — **d** Verpackung *f* der Papierrolle — **n** verpakking *f* van de papierrol; rolverpakking *f* — **e** embalaje *m* de la bobina (de papel) — **i** imballo *m* della bobina (de carta)

8945 milori blue — Brilliant green shade iron blue pigment, greener than Prussian blue, produced for the first time by Milori (Paris) in the second half of the 19th century.

f bleu *m* milori — **d** Miloriblau *n* — **n** moriblauw *n* — **e** azul *m* milori — **i** blu *m* milori; azzurro *m* milori

8946 milori green — Yellowish green pigment, chiefly lead chromate. See also milori blue.

f vert *m* milori — **d** Milorigrün *n* — **n** milorigroen *n* — **e** verde *m* milori — **i** verde *m* milori

8947 mimeograph — Trade mark for a device used in offices, in which a waxed paper stencil with text that has been cut on a typewriting machine, or text or a drawing done by hand with a stylus, is fastened to a drum which is inked on the inside, so that the ink penetrates the cut areas and is deposited on a new sheet of paper with each revolution of the drum.

8948 mimeograph ink — Ink for stencil duplicating processes, made from carbon black or lamp black mixed in mineral or castor oil and a toner. They dry chiefly by absorption into the paper.

f encre *f* duplicateur; encre *f* miméographe; encre *f* multigraphe — **d** Druckfarbe *f* für Schablonendruck — **n** stencilinkt *m* — **e** tinta *f* (para) mimeó grafo; tinta *f* duplicativa — **i** inchiostro *m* ciclostile; inchiostro *m* mimeografico

8949 mimeograph paper — Writing paper used for making copies on the Mimeograph machine. A wide variety of fibres ranging from cotton fibre and bleached chemical wood pulps to those containing mechanical pulp are used in varying proportions. Opacity, finish, absorbency, lack of fuzz and mimeographing qualities are significant proporties. The basis weight is usually 20 pounds (17 x 22 - 500) but may range from 16 to 24 pounds. Originally a slack-sized paper unsuitable for pen and ink signature, now practically all mimeograph papers have good writing qualities. Name derived from the invention (1884) of Albert Blake Dick.

f papier *m* (pour) duplicateur-cyclostyle; papier *m* miméographe; papier *m* multigraphe — **d** Stencilpapier *n*; Saugpostpapier *n* — **n** cyclostylepapier *n*; stencilpapier *n* — **e** papel *m* (para) mimeógrafo — **i** carta *f* per ciclostilare

8950 minder — Operator responsible for the general supervision of the rotary press or part of it.
f conducteur *m* de rotative — **d** Rotationsmaschinenmeister *m*; Rotationer *m* — **n** rotatiepersdrukker *m* — **e** maquinista *m* de la rotativa — **i** rotativista *m*

8951 mineral blue — See Prussian blue.

8952 mineral green — See copper carbonate.

8953 mineral oil — Oil of petroleum origin consisting of high molecular weight hydrocarbons. Common ingredient of news ink vehicles. Official name in US.
f huile *f* minérale — **d** Mineralöl *n* — **n** minerale olie *fm* — **e** aceite *m* mineral — **i** olio *m* minerale

8954 mineral pigment — Pulverized inorganic substance, natural or artificial, such as ochres, Prussian blue, etc., used in the manufacture of coloured papers and inks.
f pigment *m* minéral — **d** anorganisches Pigment *n*; Erdpigment *n*; Mineralpigment *n*; Mineralfarbstoff *m* — **n** anorganisch pigment *n* — **e** pigmento *m* mineral; pigmento *m* inorgánico — **i** pigmento *m* minerale; pigmento *m* inorganico

8955 mineral pitch — See bitumen.

8956 mineral turpentine — See white spirit.

8957 mineral wax — See ozokerite.

8958 Minerva — See cropper.

8959 minikin — Old name for a size of type. See app. no. 1.

8960 minim — Smallest unit of liquid measure. Symbol: ℳ. Abbrev. min (no point). See app. no. 7 and 7a.

8961 minimum — Size of photoengraving or printing plate below which the price remains fixed because of manufacturing costs. The lowest price charged for a plate.
f format *m* minimum — **d** Minimum *n*; Minimalformat *n* — **n** minimum *n*; minimumformaat *f* — **e** tamaño *m* mínimo; formato *m* mínimo — **i** formato *m* minimo

8962 minimum coded tape — See unjustified (paper) tape.

8963 minimum weight fount; weight fount — Smallest assortment of type of one body and design in which the number of each character is proportioned according to the frequency of its use.
f poids *m* des minima — **d** Schriftminimum *n* — **n** minimum *n* gewicht aan letter — **e** mínimo *m*; peso *m* mínimo — **i** minimo *m* (polizza del carattere); quantità *f* minima

8964 minion — Name of an old type size. See app. no. 1.

8965 minionette — Old name for a size of type. See app. no. 1.

8966 minium — See red lead.

8967 minuscule — A small or lower-case letter.

f minuscule *f*; lettre *f* bas de casse — **d** Minuskel *f*; Gemeine *m*; Kleinbuchstabe *m* — **n** kleine letter *fm*; onderkastletter *fm* — **e** minúscula *f* — **i** minuscola *f*; carattere *m* basso

8968 minus-ing — See minussing.

8969 minus sign — See also mathematical signs.
f moins *m* — **d** Minuszeichen *n*; Subtraktionszeichen *n* — **n** minteken *n* — **e** menos *m*; signo *m* de menos; signo *m* negativo — **i** meno *m*; segno *m* di sottrazione

8970 minussing; minus-ing — In gravure printing, reducing exposure in carbon printing by blocking out or protecting too light areas during over-all exposure of positive or forme of positives.
f sous-copie *f*; souscopie *f* — **d** Zurückhalten *n* — **n** terughouden *n* — **e** corto *m* de exposición; subexpuesto *m* — **i** riduzione *f* dell'esposizione

8971 minute — In time, 1/60 part of an hour; in geometry, 1/60 part of a degree.
f minute *f* — **d** Minute *f* — **n** minuut *fm* — **e** minuto *m* — **i** minuto *m*

8972 minute mark — Symbol (') for minutes or angle (or feet).
f symbole *m* des minutes; signe *m* minute — **d** Minutenzeichen *n* — **n** minuutteken *n* — **e** signo *m* de minuto — **i** segno *m* di minuto

8973 mired — Unit of measurement (short for micro reciprocal degree) for colour temperature, defined by the formula:
$$1 \text{ mired} = 10^6/\text{K}.$$
Thus 3200 K correspond to 313 mireds. The notation in mireds has the advantage of a nominal correction constant.

8974 mirror — Optical device in the form of a special metallic mirror attached to process lenses to laterally reverse the position of a negative image on a film or plate.
f miroir *m* (de retournement) — **d** Spiegel *m*; Umkehrspiegel *m* — **n** spiegel *m*; omkeerspiegel *m* — **e** espejo *m* — **i** specchio *m*; specchio *m* d'inversione

8975 mirror writing — Backward writing that resembles a mirror image of ordinary script.
f écriture *f* en miroir; écriture *f* spéculaire — **d** Spiegelschrift *f* — **n** spiegelschrift *n* — **e** escritura *f* de espejo — **i** scrittura *f* a specchio

8976 miscibility — Property of being miscible.
f miscibilité *f* — **d** Mischbarkeit *f* — **n** mengbaarheid *f* — **e** miscibilidad *f* — **i** miscibilità *f*

8977 miscible — Capable of being mixed without any tendency to separate.
f miscible — **d** mischbar — **n** mengbaar — **e** miscible — **i** miscibile

8978 misplaced line; displaced line
f ligne *f* déplacée — **d** verhobene Zeile *f* — **n** verkeerd aangebrachte regel *m* — **i** riga *f* spostata

8979 misprint — Typographical error.
f faute *f* typographique; erreur *f* typographique — **d** Druckfehler *m* — **n** drukfout *fm* — **e** errata *f*; error *m* de imprenta; falta *f*; falla *f* Me — **i** errore *m* di stampa; errore *m* tipografico

8980 misprint devilries
f diableries *fpl* d'erreur de composition — **d** Druckfehlerteufeleien *fpl* — **n** drukfouten-duiveltjes *npl* — **e** diablerías *fpl* de las erratas — **i** errori *mpl* di composizione
8981 misprints; spoils; spoilage — See also spoils.
f maculatures *fpl*; rebuts *mpl*; gaspillage *m* — **d** Makulaturen *fpl*; Fehldruck *m* — **n** misdrukken *mpl*; misdruk *m* — **e** desperdicio *m* — **i** scarto *m*
8982 misrolling — See slip.
8983 miss — Sheet not printed by mistake on the printing press.
f tour *m* sans feuille; manque *m* de feuille — **d** Fehlbogen *m* — **n** misser *m* — **i** foglio *m* mancato
8984 miss — A size of cut card. See app. no. 3.
8985 misshapen reel — See badly shaped reel.
8986 missing dots; skipped dots; skips; graining; mealiness — White speckles in the halftone areas of a rotogravure print, appearing when the cells have failed to transfer ink to the paper, especially pronounced in light or middle tones. See also speckle.
f points *mpl* manquants; galeux *m* — **d** fehlende Rasterpunkte *mpl*; nicht-ausdruckende Punkte *mpl* — **n** witte punten *mpl*; niet-uitdrukkende punten *mpl* — **e** puntos *mpl* fallados — **i** punti *mpl* mancanti
8987 misspelling — See spelling mistake.
8988 mistake in spelling — See spelling mistake.
8989 misting — See ink misting.
8990 miter — See mitre.
8991 miter *v* — See mitre *v*.
8992 mitered rule — See mitred rule.
8993 miterer — See mitring machine.
8994 mitering machine — See mitring machine.
8995 mitre *GB*; **miter** *US* — Bevel at ends of rules or borders to insure a close joint in geometric design of frames.
f anglet *m* — **d** Gehrung *f* — **n** verstek *n* — **e** inglete *m*; chaflán *m*; bisel *m*; colilla *f* *Co*; refilón *m* *Gu* — **i** angolo *m*
8996 mitre *v* *GB*; **miter** *v* *US* — To cut the ends of rules or borders at an angle of 45 ° so that when a complete border is made the corners will join flush.
f couper des anglets — **d** Gehrung schneiden; ausschneiden nach der Gehrung — **n** verstekken snijden — **e** ingletear; achaflanar; chaflanar; colillar *Co*; esquinar *Pa* — **i** tagliare angoli
8997 mitred rule *GB*; **mitered rule** *US*; **chamfered rule** *obs.* — Brass rule with ends cut at an angle of 45 ° so that when a complete border is made the corners will join flush.
f filet *m* à anglets — **d** Gehrungslinie *f* — **n** lijn *fm* met verstek — **e** raya *f* ingleteada; raya *f* achaflanada; chaflán *m*; esquina *f* *Ar* — **i** filetto *m* tagliato ad angolo
8998 mitring machine *GB*; **mitering machine** *US*; **miterer**
f machine *f* à couper les anglets — **d** Gehrungschneidmaschine *f* — **n** verstekmachine *f* — **e**

máquina *f* para hacer ingletes; achaflanadora *f*; caja *f* de ingletes; ingleteadora *f*; cuadrante *m*; cizalla *f* biseladora; colilladora *f* *Co*; esquinero *m* *Pa* — **i** macchina *f* per tagliare ad angolo
8999 Mitscherlich pulp — Slow-cooked sulphite pulp of superior strength, easily bleached. Named after Dr. Eilhardt Mitscherlich, the German inventor (1876).
f pâte *f* Mitscherlich — **d** Mitscherlich-Zellstoff *m* — **n** Mitscherlich-celstof *fm* — **e** pasta *f* de Mitscherlich; pasta *f* al bisulfito de calcio — **i** pasta *f* Mitscherlich
9000 mitsumata — See bast fibres.
9001 MIT-test — Folding endurance test (double fold) of Massachussets Institute of Technology.
f essai *m* pliagraphe MIT — **d** MIT-Falz *m* — **n** dubbelvouwproef *fm* volgens MIT; MIT-proef *fm* — **e** ensayo *m* MIT — **i** saggio *m* MIT; prova *f* di doppie pieghe MIT
9002 mixed composition — See mixed matter.
9003 mixed matter; mixed composition — Composition containing a variety of different types.
f composition *f* lardée — **d** gemischter Satz *m*; Mischsatz *m* — **n** gemengd zetsel *n* — **e** composición *f* mezclada; composición *f* mixta — **i** composizione *f* mista
9004 mixed papers — Waste papers or paper stock not sorted or from which the better grades have been taken out. They are collected from department stores, offices, schools, etc.
f papiers *mpl* mélangés — **d** unsortierte Papierabfälle *mpl* — **n** bont papierafval *n*; bont *n* — **e** papel *m* de desecho sin surtido — **i** carte *fpl* non assortite
9005 mixed rags; unsorted rags
f drilles *fpl* — **d** unsortierte Lumpen *fpl*; Hadern *mpl* — **n** bonte lompen *fmpl*; ongesorteerde lompen *fmpl* — **e** trapos *mpl* sin clasificar; trapos *mpl* sin escoger — **i** stracci *mpl*
9006 mixed straw paper or board — Paper or board containing a major proportion of unbleached straw pulp. This term is rarely used in certain countries.
f papier *m* paille mixte; carton *m* paille mixte — **d** Strohmischpapier *n*; Strohmischpappe *f* — **n** stropapier *n*; strokarton *n* — **e** papel *m* paja; cartón *m* paja; cartón *m* amarillo — **i** carta *f* paglia; cartone *m* paglia
9007 mixer; ink mixer — Small tank provided with agitator, in which the vehicle and pigment of printing inks are stirred and mixed, before being processed in the ink mill.
f malaxeur *m* (d'encre); mélangeur *m* (d'encre) — **d** Mischmaschine *f*; Mischer *m* — **n** inktmengmachine *f*; mengmachine *f* — **e** mezcladora *f* de tinta — **i** mescolatore *m* d'inchiostro; miscelatore *m* d'inchiostro
9008 mixer — Line-casting machine with which matrices from adjacent magazines may be mixed in one line to introduce a wider range of type founts. It has a double distributor mechanism to ensure distribution to the correct magazine. On

certain models mixing facilities are extended to adjacent side magazines but these are not driven by tape.

f mixer *m*; machine *f* lignes-blocs à double distribution — **d** Mixermodell *n* — **n** mixer *m* — **e** modelo *m* mixer; máquina *f* mezcladora — **i** modello *m* mixer; compositrice *f* monolineare a distribuzione multipla

9009 mixer — See mixing chest.

9010 mixing

f mélange *m* (des couleurs) — **d** Mischung *f*; Mischen *n*; Farbmischung *f* — **n** menging *f*; mengen *n*; vermenging *f*; vermengen *n* — **e** mezcla *f*; mixtura *f*; mixtión *f* — **i** miscelazione *f*

9011 mixing chest; mixer

f cuvier *m* mélangeur; caisse *f* de mélange; cuvier *m* de mélange; mélangeur *m*; pile *f* à mêler — **d** Rührbütte *f*; Mischbütte *f*; Mischkasten *m* — **n** mengkuip *fm*; menger *m* — **e** tina *f* de mezcla; tina *f* mezcladora; caja *f* de mezcla — **i** mescolatore *m*

9012 mixing of type faces

f mélange *m* des caractères — **d** Schriftmischung *f* — **n** lettermenging *f*; menging *f* van letters — **e** mezcla *f* de caracteres — **i** mescolanza *f* di caratteri

9013 mixing ratio

f proportion *f* de mélange; proportion *f* du mélange — **d** Mischungsverhältnis *n* — **n** mengverhouding *f* — **e** proporción *f* de mezcla — **i** rapporto *m* di miscelazione

9014 mixing white — Transparent or opaque white ink to make tints.

f blanc *m* de mélange — **d** Mischweiß *n* — **n** mengwit *n* — **e** blanco *m* para mezclar; blanco *m* transparente; laca *f* transparente — **i** bianco *m* d'allungamento; bianco *m* diluente

9015 mixture

f mélange *m* — **d** Mischung *f* — **n** mengsel *n* — **e** mezcla *f* — **i** miscela *f*

9016 mnemonic — Literally assisting human memory. Name for a condition, instruction, routine or format, usually consisting of a few unique characters (such as MPY for multiply and SUB for subtract), which suggest the meaning or purpose of the function.

f mnémonique — **d** mnemonisch — **n** mnemonisch — **e** mnemónico — **i** mnemonico

9017 mobility — Rate of shear induced by a shearing stress of one dyne in excess of the yield value. Reciprocal of plastic viscosity. Analogous to "fluidity" for liquids.

f mobilité *f* — **d** Beweglichkeit *f* — **n** beweeglijkheid *f* — **e** movilidad *f* — **i** mobilità *f*

9018 mobilometer — Viscometer composed of a metal tube and a plunger. The plunger is a perforated plate and the tested liquid is squeezed through the perforations as the plunger descends in the tube. Different weights can be applied to the plunger and a consistency curve is obtained by plotting these weights against the reciprocal of the time it takes the plunger to descend.

f mobilomètre *m* — **d** Mobilometer *n* — **n** mobilometer *m* — **e** moviliómetro *m* — **i** mobilometro *m*

9019 mock title — See bastard title.

9020 mock-up — See dummy.

9021 mode — Way of operation, for example, the binary mode, the interpretive mode, the alphameric mode, etc.; also the most frequent value in the statistical sense.

f mode *m* — **d** Betriebsart *f*; Modus *m* — **n** modus *fm*; werkwijze *fm* — **e** modalidad *f*; modo *m* — **i** modo *m*

9022 modem — Acronym for modulator-demodulator, terminal to convert the representation of data to achieve compatibility between processing and transmission systems.

f convertisseur *m* de signal — **d** Signalumsetzer *m* — **n** signaalomzetter *m* — **e** modem *m*; modulador-demodulador *m* — **i** modulatore-demodulatore *m*

9023 Modern — See Didonics.

9024 Modern Face *depr.* — A Roman letter design, e.g., Bodoni, characterized by considerable difference between thick and thin strokes, serifs without brackets, and vertical shading. See classification of type design, Didonics, app. no. 2.

9025 modern figures — See lining figures.

9026 modernized figures — See lining figures.

9027 modifier register *GB*; **index register** *US*; **B box** — Register or location which holds a modifier in a computer with facilities for automatic modification.

f registre *m* index — **d** Indexregister *n*; Indexzelle *f* — **n** indexregister *n*; modificatorregister *n* — **e** registro *m* indicativo — **i** registro *m* indice

9028 modular system — Building-up system in which the component parts can be combined in a large number of possibilities.

f système *m* modulaire — **d** Modularsystem *n*; auslauffähiges System *n* — **n** modulair systeem *n*; voor uitbreiding vatbaar systeem *n* — **e** sistema *m* modular — **i** sistema *m* modulare

9029 modulation — Process of varying some characteristics of the carrier wave in accordance with the instantaneous value, or samples of the intelligence to be transmitted.

f modulation *f* — **d** Modulation *f* — **n** modulatie *f* — **e** modulación *f* — **i** modulazione *f*

9030 modulator — Device to produce modulation.

f modulateur *m* — **d** Modulator *m*; Sender-Signalumsetzer *m* — **n** modulator *m* — **e** modulador *m* — **i** modulatore *m*

9031 module — Packaged functional hardware unit to be used with other components. In case of technical trouble it simplifies changing groups of parts or enables the machine to do other functions or to increase storing capacity.

f module *m* — **d** Modul *m*; Baustein *m* — **n** moduul *m* — **e** módulo *m* — **i** modulo *m*

9032 moiré; screen pattern; pattern — Undesirable symmetrical checkered or watered silk formations produced when making a halftone reproduction from a halftone proof or from steel engravings, caused by diffraction or conflict between the dot formation or lines of the original and the ruling of the halftone screen. In colour reproduction when two or more halftone colour plates are made and printed at improper screen angles or when slight errors of register occur in printing the plates.

f moiré *m*; moirage *m* — **d** Moiré *mn* — **n** moiré-effect *n*; moiré *n* — **e** moaré *m*; muaré *m* — **i** moiré *m*

9033 moisture content — Percentage of water in pulp, paper, or paperboard, determined by completely drying the sample at 100-105 °C (212-221 °F). The result is expressed as a percentage of the original weight of the sample unless otherwise specified. If the basis for determining moisture content is not specified, "when sampled" is implied.

f humidité *f* (sur brut); teneur *f* en eau (sur brut); teneur *f* en humidité; teneur *f* en eau; degré *m* hygrométrique — **d** Feuchtigkeit *f*; Feuchtigkeitsgehalt *m*; Feuchtgehalt *m*; Feuchtigkeitsgrad *m* — **n** vochtgehalte *n*; vochtigheidstoestand *m*; vochtigheidsgraad *m*; watergehalte *n* — **e** contenido *m* de humedad; grado *m* de humedad; porcentaje *m* de humedad; tanto por ciento *m* de humedad; grado *m* higrométrico — **i** contenuto *m* d'umidità; grado *m* d'umidità; percentuale *f* d'umidità

9034 moisture content on a bone-dry basis — See moisture content on an oven-dry basis.

9035 moisture content on an oven-dry basis; moisture content on a bone-dry basis — Loss in weight in a sample of paper or board, after drying to constant weight under standard conditions, expressed as a percentage of the constant weight so obtained.

f humidité *f* sur sec; teneur *f* en eau (sur sec) — **d** Feuchtigkeitsgehalt *m* berechnet auf absolutem Trockengewicht — **n** vochtgehalte *n* berekend op absoluut droog gewicht — **e** contenido *m* de humedad a peso seco absoluto — **i** contenuto *m* d'umidità calcolato sul secco

9036 moisture-free — See oven dry.

9037 moisture pick-up — Absorption of moisture by paper from the atmosphere; also from the pressplate or the blanket in offset.

f absorption *f* d'humidité — **d** Feuchtaufnahme *f* — **n** vochtopname *f* — **e** absorción *f* de humedad — **i** assorbimento *m* d'umidità

9038 moisture pipes — See moisture welts.

9039 moisture-set ink — Ink that dries or sets principally by precipitation. The vehicle consists of a water insoluble resin dissolved in a hygroscopic solvent. Drying occurs when the hygroscopic solvent has absorbed sufficient moisture either from the atmosphere, substrate or external application to precipitate the binder. An important characteristic is low odour.

f encre *f* moisture-set; encre *f* fixée à la vapeur; encre *f* fixée par l'humidité; encre *f* séchant part apport d'humidité — **d** feuchtigkeitsabbindende Farbe *f*; Dampftrockenfarbe *f* — **n** moisture-set inkt *m* — **e** tinta *f* moisture-set; tinta *f* fijandose por humedad; tinta *f* fijandose por vapor — **i** inchiostro *m* moisture-set; inchiostro *m* stabilizzante ad umido

9040 moisture-vapour transmission rate — See water-vapour transmission rate.

9041 moisture welts; moisture pipes — Narrow bands that form as raised wells in the making direction of paper caused by a too high moisture content of the air.

f lignes *fpl* d'eau — **d** Feuchtschwielen *fpl* — **n** vochtrillen *fmpl*

9042 mol; mole; gramme-molecule *GB*; **gram-molecule** *US* — Gram-molecular weight of a substance; the weight of it in grams numerically equal to its molecular weight.

f mole *m*; molécule-gramme *f* — **d** Mol *n*; Grammol *n*; Grammolekül *n* — **n** mol *fm*; grammolecule *fmn* — **e** mol *m*; molécula-gramo *m* — **i** mole *m*; grammo *m* molecola

9043 mold — See mould.

9044 mold — See mould.

9045 mold — See mould.

9046 mold *v* — See mould *v*.

9047 mold disk slide — See mould disk slide.

9048 mold, hand — See hand mould.

9049 mold-made paper — See mould-made paper.

9050 mold slide — See mould disk slide.

9051 mole — See mol.

9052 molecular alignment — Alignment in the direction of flow taken by long molecules, when subjected to a shearing force. It accounts for the curvilinear nature of the pseudoplastic consistency curve.

f alignement *m* moléculaire — **d** molekulare Ausrichtung *f* — **n** moleculaire oriëntering *f* — **e** alineación *f* molecular — **i** allineamento *m* molecolare

9053 molecular solution — Term used in distinction to colloidal suspension. True solution in which the dissolved particles are molecules and not micelles. Unlike a dye, a molecularly dissolved material cannot be exhausted from solution.

f solution *f* moléculaire — **d** molekulare Lösung *f*; echte Lösung *f*; Molekularlösung *f*; molekulardisperse Lösung *f* — **n** moleculair-disperse oplossing *f*; echte oplossing *f* — **e** solución *f* molecular — **i** soluzione *f* molecolare

9054 molecular weight — Weight of one molecule of a compound in relation to hydrogen, where H (i.e., hydrogen) = 1.

f poids *m* moléculaire; masse *f* molaire — **d** Molekulargewicht *n* — **n** moleculair gewicht *n*; molair gewicht *n* — **e** peso *m* molecular — **i** peso *m* molecolare

9055 molecule — The smallest particle into which a substance can be divided without chemical change.

f molécule *f* — **d** Molekül *n* — **n** molecule *mfn*; molekuul *mfn* — **e** molécula *f* — **i** molecola *f*

9056 moleskin — Stout, twilled cotton with a

short thick velvet nap on one side, used in book-binding.

f moleskine *f*; molesquine *f* — **d** Moleskin *m* — **n** moleskin *n*; bevertien *n* — **e** molesquina *f* — **i** moleskin *m*; fustagno *m*

9057 molleton — Thick cotton fabric similar to flannel, with a long nap and used on damping forme rollers.

f molleton *m* — **d** Molton *m*; Wischwalzenstoff *m*; Feuchtwalzenstoff *m*; Feuchtfilz *m* — **n** molton *n* — **e** moletón *m*; muletón *m*; manga *f*; franela *f* de algodón — **i** mollettone *m*

9058 Mollie — See Mary.

9059 molybdate chrome orange — See molybdate orange.

9060 molybdated pigment — Pigment precipitated with molybdic acid to give it permanence and insolubility.

f pigment *m* de molybdène — **d** mit Molybdän-säure gefälltes Pigment *n* — **n** met molybdeen-zuur neergeslagen pigment *n* — **e** pigmento *m* molibdeno — **i** pigmento *m* al molibdeno

9061 molybdate orange; molybdenum orange; molybdate chrome orange — Manufactured mineral pigment to produce opaque orange inks or to brighten red inks by increasing opacity. It consists of a mixture of lead chromate and molyb-date.

f orange *m* de molybdène — **d** Molybdatorange *n* — **n** molybdeen-oranje *n*; molybdaat-oranje *n* — **e** anaranjado *m* molibdeno; anaranjado *m* de molibdato — **i** arancio *m* molibdato

9062 molybdenum orange — See molybdate orange.

9063 moment of inertia — The moment of inertia I of a body about any axis is the sum of the products of the mass dm of each element of the body and the square of r, its distance from that axis. Thus: $I = \Sigma r^2 dm$.

f moment *m* d'inertie — **d** Trägheitsmoment *m* — **n** traagheidsmoment *n* — **e** momento *m* de inercia — **i** momento *m* d'inerzia

9064 monastic binding — Binding with a decoration with repeated figures in heavy patterns.

f reliure *f* monastique — **d** Kloster-Einband *m* — **n** kloosterband *m* — **e** encuadernación *f* monástica — **i** legatura *f* monastica

9065 monastral blue — Trade-mark (1935) for in-soluble phthalocyanine pigments producing extremely bright shades of blue and green with excellent fastness properties and high tinctorial strength. Made from phthalic anhydride orig-inating from the oxidation of naphthalene in the middle heavy oils. First made in 1935 in GB and US. Copper phthalocyanines, blue and green, are sometimes named heliogen pigments.

f bleu *m* monastral; bleu *m* de phtalocyanure — **d** Monastralblau *n* — **n** monastraalblauw *n* — **e** azul *m* ftalociánico — **i** blu *m* Monastral

9066 monastral green — Chlorinated copper phthalocyanine. See also monastral blue.

f vert *m* monastral — **d** Monastralgrün *n* — **n** monastraalgroen *n* — **e** verde *m* ftalociánico — **i** verde *m* Monastral

9067 monitor — General computer program which keeps track of what data the computer is pro-cessing and what specific computer application programs are in use, and which serves to optimize the computer's resources by bringing in new input or programs or data when they are required or available, often by "polling" peripherals or responding to interrupt messages.

9068 monitor copy — See hard copy.

9069 monk — Ink blob or patch with too much ink appearing on the printed sheet.

f moine *m* — **d** verschmierte Stelle *f* — **n** zwarte, vette plek *fm* in een bedrukt vel papier — **e** mancha *f*; borrón *m* (de tinta) — **i** monaco *m*; macchia *f* nera di stampa per eccesso d'inchiostro

9070 monkey — See monkey wrench.

9071 monkey roller — See jockey roller.

9072 monkey wrench; monkey — See also wrench.

f clé *f* réglable; clé *f* à molette — **d** verstellbarer Schlüssel *m*; Franzose *m*; Engländer *m* — **n** Engelse sleutel *m*; sleutel *m* — **e** llave *f* inglesa — **i** chiave *f* regolabile; chiave *f* inglese

9073 mono- — Prefix denoting single element or radical.

9074 monobasic acid — Acids with one, two, and three displaceable hydrogen atoms per molecule are termed mono-, di-, and tribasic acids respectively. Monobasic, dibasic and tribasic salts are salts which are formed with displacement of one, two and three hydrogen atoms respectively from the acid. These terms are commonly applied to salts of tribasic acids, e.g. the orthophosphates.

f acide *m* monobasique; monoacide *m* — **d** ein-basische Säure *f* — **n** eenbasisch zuur *n* — **e** ácido *m* monobásico — **i** acido *m* monobasico

9075 mono-caster attendant

f fondeur *m* Monotype — **d** Monogießer *m* — **n** Monotype-gieter *m* — **e** fundidor *m* monotipista — **i** fonditore *m* monotipista

9076 monochromatic — In optics of, producing, or pertaining to one colour or to a very limited range of wavelengths.

f monochromatique — **d** monochromatisch — **n** monochromatisch — **e** monocromático — **i** mono-cromatico

9077 monochromatic light — Light consisting of vibrations of the same or nearly the same fre-quency. Light of one colour.

f lumière *f* monochromatique — **d** mono-chromatisches Licht *n*; einfarbiges Licht *n* — **n** monochromatisch licht *n* — **e** luz *f* mono-cromática — **i** luce *f* monocromatica

9078 monochrome — Having different shades of one single colour.

f monochrome — **d** einfarbig — **n** monochroom; in één kleur; eenkleurig — **e** monocromo; de un solo color — **i** monocromo; monocolore

9079 monochrome print; single-colour print

f impression *f* monochrome — **d** Einfarbendruck *m* — **n** afdruk *m* in één kleur; plaat *fm* in één kleur; afbeelding *f* in één kleur — **e** impresión *f* monocromo; impreso *f* monocromo; monocromía *f*; impresión *f* a un solo color — **i** riproduzione *f* monocroma; riproduzione *f* ad un colore; monocromia *f*

9080 monoclinic — Crystal form in which there are three unequal axes with one oblique intersection.
f monoclinique — **d** monoklin; monoklinisch — **n** monoclien — **e** monoclínico — **i** monoclino

9081 monolayer
f couche *f* monomoléculaire — **d** monomolekulare Schicht *f* — **n** monomoleculaire laag *fm* — **e** capa *f* monomolecular — **i** strato *m* monomolecolare

9082 monolithic circuit — See integrated circuit.

9083 monomer — Chemical compound consisting of single molecules. As opposed to a *polymer, the molecules of which are built up by the repeated union of monomer molecules. See also polymerization.
f monomère *m* — **d** Monomer *n* — **n** monomeer *n* — **e** monómero *m* — **i** monomero *m*

9084 Monophoto — Photo-typesetting machine of British origin (1957) adapted from the Monotype composing machine; the matrices contain photographic negatives.

9085 monoscope — Electronic tube to generate characters for CRT display. Characters are printed on a target inside the tube. Scanning is modulated by secondary emission from the target so that selected characters can be displayed on a CRT.

9086 monospace — Computer term, assigning an equal width to all type characters and spaces (as with most typewriters and line printers), as opposed to proportionally spaced characters as in conventional printing. Monospaced characters are usually as 10 or 12 characters (or spaces) to the inch.

9087 Monotype — Hot-metal typesetting machine comprising a separate keyboard and a casting machine, invented (first patent 1885) by Tolbert Lanston (1844-1913), an American lawyer. The Monotype unit system is based on a width of 1/18th of a point (0.0007685 in).
f Monotype *f* — **d** Monotype *f* — **n** Monotype *fm* — **e** Monotipia *f*; Monotipo *m* — **i** Monotipia *f*

9088 Monotype caster; Monotype casting machine
f fondeuse *f* Monotype — **d** Monotypegießmaschine *f* — **n** Monotype-gietmachine *f* — **e** máquina *f* fundidora de la Monotipia; máquina *f* fundidora del Monotipo; fundidora *f* Monotype — **i** fonditrice *f* Monotype

9089 Monotype casting machine — See Monotype caster.

9090 Monotype composition
f composition *f* Monotype — **d** Monotypesatz *m* — **n** Monotype-zetsel *n* — **e** composición *f* monotípica — **i** composizione *f* monotipica

9091 Monotype keyboard
f clavier *m* Monotype — **d** Monotypetastenbrett *n*; Monotypetaster *m* — **n** Monotype-toetsenbord *n*

— **e** teclado *m* Monotype — **i** tastiera *f* Monotype

9092 Monotype paper; keyboard paper — White or coloured sulphite book-paper in rolls to be perforated with small holes and used on a Monotype keyboard and casting machine.
f papier *m* pour Monotype; papier *m* pour clavier — **d** Monotypepapier *n* — **n** Monotype-papier *n* — **e** papel *m* para monotipo; papel *m* para monotipia — **i** carta *f* per monotipia

9093 monovalent — Capable of combination with one atom of hydrogen or the equivalent. See also valence.
f monovalent — **d** einwertig — **n** eenwaardig — **e** monovalente — **i** monovalente

9094 montage — See photomontage.

9095 montan wax; lignite wax — Dark-brown, bituminous wax, soluble in carbon tetrachloride, benzene and chloroform, insoluble in water, used for paper-sizing compositions and carbon papers.
f cire *f* de montan; cire *f* de lignite — **d** Montanwachs *n* — **n** montaanwas *mn* — **e** cera *f* montana — **i** cera *f* montana; cera *f* di lignite

9096 monthly — Periodical published once a month.
f revue *f* mensuelle; publication *f* mensuelle — **d** Monatsschrift *f*; Monatsheft *n*; monatliche Zeitschrift *f* — **n** maandblad *n*; maandelijks verschijnend blad *n* — **e** revista *f* mensual; publicación *f* mensual — **i** rivista *f* mensile; pubblicazione *f* mensile

9097 mood picture — Picture in colours which has no relation to the natural colours of the scene but is descriptive of human mood.
f image *f* en couleurs d'interprétation — **d** Stimmungsbild *n* — **n** stemmingsbeeld *n* — **i** illustrazione *f* di manera

9098 mordant — See etching solution.

9099 more hug — Increase in the wrap of the paper (web) round the cylinder.
f excès *m* d'enveloppement — **d** Vergrößerung *f* der Umschlingung — **n** grotere omvatting *f* — **e** exceso *m* en enrollado — **i** aumento *m* dell'avvolgimento

9100 more space
f aérer — **d** mehr Raum; mehr Zwischenraum — **n** meer wit; meer ruimte; meer spatie — **e** agregar espacio — **i** più spazio

9101 morgue *coll.* US; **graveyard** *coll.* US — Library of a daily paper, a place for storage of photographs, editorial material, clippings, mats, etc., that may be used at a later time.
f archives *fpl* pour photos et matériel de rédaction — **d** Archivraum *m* für Redaktionsmaterial — **n** redactie-archief *n* — **e** archivo *m* de la redacción — **i** archivio *m* della redazione

9102 Morisawa — Photographic lettering machine of Faxi-Foto, Japan, which can set type up to 60 point.

9103 morning daily — See morning newspaper.

9104 morning newspaper; morning paper; morning daily
f journal *m* du matin; gazette *f* du matin; feuille *f*

du matin — **d** Morgenblatt *n*; Morgenzeitung *f* — **n** ochtendblad *n* — **e** diario *m* de la mañana; diario *m* matinal; diario *m* matutino; papel *m* matutino; papel *m* de la mañana; hoja *f* de la mañana — **i** giornale *m* del mattino

9105 morning paper — See morning newspaper.

9106 morocco leather — Tanned skin of the goat. There are three distinct headings. The finest and most expensive is that of the Cape goat, when finished known as *levant morocco. The second variety is a hard grained morocco or goatskin. The third is niger. They are shipped partly tanned, re-tanned with sumach, dyed and finished, and sold as oasis morocco. Do not use capital letters. See also leather.

f maroquin *m* — **d** Maroquin *m*; Marokkoleder *n*; Saffian *m*; Saffianleder *n* — **n** marokijn *n*; maroquin *n*; saffiaan *n*; saffiaanleer *n*; marokaan *n*; marokijnleer *n* — **e** marroquín *m*; marroqué *mf*; tafilete *m* — **i** marocchino *m*

9107 mortise — Cavity or enclosed aperture in a mounted printing plate into which type or printing matter may be inserted. See also inside mortise, outside mortise.

f encoche *f*; ajour *m* — **d** Ausschnitt *m*; Aussparung *f* — **n** uitsparing *f* — **e** mortaja *f*; cotana *f*; escopleadura *f*; escopladura *f*; muesca *f*; entalladura *f* — **i** apertura *f*; feritoia *f*; finestra *f*

9108 mortise *v* — To cut out sections of mounted printing plates, usually with a drill and jig saw.

f encocher — **d** einen Ausschnitt machen; aussparen — **n** een uitsparing aanbrengen; uitsparen — **e** mortajar; escoplear — **i** scanalare; praticare una finestra

9109 mortised letter

f caractère *m* à entaille; caractère *m* à rentrure — **d** ausgeklinkter Buchstabe *m* — **n** letter *fm* met een uitsparing — **e** letra *f* escotada; carácter *m* hendido; tipo *m* en hueco — **i** lettera *f* scanalata; carattere *m* rientrante

9110 mosaic binding; inlaid binding — Book decoration style of the 18th century with inlaid leather designs, initiated by the French bookbinder J.C.H. le Monnier (1757).

f reliure *f* mosaïquée — **d** Mosaik-Einband *m*; Einband *m* mit Mosaikmuster; mit Mosaik eingelegter Einband *m* — **n** mozaïekband *m* — **e** encuadernación *f* mosaico; encuadernación *f* a mosaico; encuadernación *f* con mosaico; encuadernación *f* a mosaico embutido — **i** legatura *f* con lavoro mosaicato

9111 mosaic printer — See matrix printer.

9112 mother-of-pearl paper; iridescent paper — Glazed paper immersed in a bath of silver or lead salts and fixed with collodion after drying and exposing to the vapour of hydrogen sulphide, so giving iridescent effects.

f papier *m* nacré; papier *m* irisé — **d** Perlmutterpapier *n*; irisierendes Papier *n* — **n** papier *n* met parelmoerglans; parelmoerpapier *n* — **e** papel *m* nacarado; papel *m* irisado; papel *m* iridescente — **i** carta *f* madreperla; carta *f* iridescente

9113 motor spirits — See gasoline.

9114 mottle; mottling; pebbling — Uneven appearance in solid areas (mostly), small dark and light areas appearing in surface which should be perfectly smooth, caused by ink, paper or presswork. Also a photographic defect in rotogravure printing characterized by non-uniform density differences, usually in the pattern of tiny circular areas, due to unsufficient contact between diapositive and carbon tissue. See also speckle.

f moutonnage *m*; perlage *m*; jaspage *m*; madrure *f*; moutonnement *m* — **d** Perlen *n* (der Farbe) — **n** parelen *n* (van de inkt) — **e** agrumarse — **i** chiazzatura *f*; perlatura *f*

9115 mottled paper — Paper with variegated surface produced by irregular distribution of the coloured fibres in the web.

f papier *m* marbré — **d** meliertes Papier *n*; Melierpapier *n* — **n** gemarbreerd papier *n*

9116 mottled print — Spotty print due to poor distribution of the ink. Some coloured inks have a tendency to mottle, such as browns and greens, especially on solids.

f impression *f* moutonnée; impression *f* jaspée; impression *f* madrée; impression *f* perlée — **d** geperlter Druck *m* — **n** geparelde afdruk *m* — **e** impresión *f* grumosa — **i** stampa *f* chiazzata; stampa *f* perlate

9117 mottling — See mottle.

9118 mould *GB*; **mold** *US*; **casting mould** — Part of a typesetting machine in which the slug is cast.

f moule *m* — **d** Gießform *f* — **n** gietvorm *m* — **e** molde *m* de fundir; fundidor *m*; caja *f* de fundir *Me* — **i** forma *f* a fondere; forma *f* di fusione

9119 mould *GB*; **mold** *US* — Framed screen on which hand-made paper is formed.

f forme *f* au papier — **d** Schöpfform *f*; Schöpfsieb *m*; Papierform *f* — **n** schepvorm *m* — **e** molde *m* (para papel hecho a mano); forma *f* — **i** forma *f* a mano

9120 mould *GB*; **mold** *US* — Superficial often wooly growth produced on damp or decaying organic matter or on living organisms.

f moisissure *f* — **d** Schimmel *m* — **n** schimmel *m* — **e** moho *m* — **i** muffa *f*; marciume *m*

9121 mould *v GB*; **mold** *v US*

f prendre empreinte; mouler — **d** abformen — **n** pregen; afpersen — **e** moldear; cartonear — **i** formare

9122 mould-blade cam

f excentrique *m* pour lame du moule — **d** Kernexzenter *m* — **n** gietvormblad-excentriek *n* — **e** excéntrica *f* de las cuchillas — **i** eccentrico *m* della lama della forma a fondere

9123 mould cleaner — See mould wiper.

9124 mould cooling attachment

f refroidisseur *m* des moules — **d** Gießformkühlung *f* — **n** gietvormkoeling *f* — **e** refrigerador *m* del molde — **i** dispositivo *m* raffreddatore della forma

9125 mould disk; mould wheel — Part of a typesetting machine which contains four or more pockets and carries the moulds.

f roue-moule *f* — **d** Gießrad *n* — **n** gietwiel *n*; gietrad *n* — **e** disco *m* de moldes; rueda *f* (de) molde; rueda *f* fundidora; molde *m* de rueda; rueda *f* de fundición — **i** ruota *f* di fusione

9126 mould disk slide; mould slide *GB*; **mold disk slide; mold slide** *US*

f glissière *f* de la roue-moule — **d** Gießrad-schlitten *n* — **n** gietwielslede *fm* — **e** deslizador *m* del disco; deslizadora *f* del disco; corredera *f* del disco — **i** slitta *f* della ruota portaforme

9127 moulded-pulp container — See pulp moulding.

9128 mould, hand — See hand mould.

9129 moulding press — Hydraulic press to make a *mat.

f presse *f* à empreindre; presse *f* à empreinte — **d** Matrizenprägepresse *f*; Maternprägepresse *f*; Kalander *m* — **n** preepers *f*; kalander *fm*; matrijzenkalander *fm* — **e** matrizador *m*; máquina *f* denatrizar; prensa *f* hidráulica de matrizar; prensa *f* de matrices estereotipias; prensa *f* de estereotipar; prensa *f* para matrizar; prensa *f* para estampar matrices; matricera *f*; calandra *f*; calandria *f*; cartonera *f* *Ch*; estampadora *f* *Pe*; compresora *f* *Gu*; moldeadora *f* — **i** pressa *f* per matrici; pressa *f* a formare

9130 mould machine — See cylinder machine.

9131 mould-made paper *GB*; **mold-made paper** *US* — Deckle-edged paper resembling that made by hand but produced on a cylinder or cylindrical mould revolving in a vat of pulp, the various sizes being arrived at by dividing the surface with rubber bands to imitate the thinning of the deckle edge by hand-made paper or by cutting the web by means of a jet of water.

f papier *m* à la cuve à la machine; papier *m* genre cuve — **d** Maschinenbüttenpapier *n*; Bütten-papier *n*; Schöpfpapier *n* — **n** imitatie-hand-geschept papier *n*; machinegeschept papier *n*; machinaal geschept papier *n* — **e** imitación *f* de papel a mano; papel *m* imitación barba — **i** carta *f* a mano-macchina; carta *f* uso mano

9132 mould-pulp fitting — See pulp moulding.

9133 mould slide — See mould disk slide.

9134 mould wheel — See mould disk.

9135 mould wiper; mould cleaner — Felt wiper to clean the mould on the Linotype and the Inter-type machines.

f essuyeur *m* de moule — **d** Gießformwischer *m* — **n** gietvormwisser *m* — **e** limpiador *m* del molde — **i** scovolino *m* per pulire la forma a fondere

9136 mount; block; base — Wood or metal base on which a printing plate is permanently fastened for use on a press.

f bloc *m* de montage; semelle *f* — **d** Fuß *m*; Plattenfuß *m*; Klotz *m*; Klischeeunterlage *f*; Klischeefuß *m*; Unterlage *f*; Untersatz *m*; Platten-schuh *m* — **n** clichévoet *m*; voet *m* — **e** base *f*; piso *m*; zócalo *m*; pie *m* (de grabado); montura *f*;

taco *m*; montaje *m* *Ch* — **i** zoccolo *m*; zoccola-tura *f*; base *f* per il montaggio; blocco *m* di montaggio; base *f* di montaggio

9137 mount *v* — To attach the printing plate to a wood or metal base to bring it to the required height for printing or platemaking.

f monter — **d** montieren; aufklotzen — **n** mon-teren — **e** montar; bloquear — **i** montare

9138 mounted flush — Stereo or engraving plate, without bevel or margin between the edges of the plate and the base.

f monté à fleur — **d** auf Fuß, ohne Facette — **n** ge-monteerd zonder facet — **e** montado a flor; montado a ras — **i** montato al vivo

9139 mounted on canvas; mounted on linen

f collé sur toile; monté sur toile — **d** auf Lein-wand aufgezogen — **n** op linnen geplakt — **e** en-colado sobre tela de lino; encolado sobre lienzo — **i** montato su tela

9140 mounted on linen — See mounted on canvas.

9141 mounted on metal; metal mounted

f monté sur matière; monté sur plomb — **d** auf Metall montiert; auf Blei montiert; auf Bleifuß montiert — **n** gemonteerd op metaal; gemonteerd op lood — **e** montado en metal — **i** montato su piombo

9142 Mount-Hope roller — See expander roll.

9143 mounting board — Rigid paper or board lined with plain or decorated cover paper, used for mounting specimens, pictures, etc.

f carton *m* pour montage — **d** Aufziehkarton *m* — **n** opzetkarton *n* — **e** cartón *m* para monturas — **i** cartone *m* da montaggio

9144 mounting lumber; blocking lumber; mounting wood — Specially prepared lumber or wooden panels of good quality, used as a base or support for printing plates to bring them type high.

f bois *m* de montage (de clichés) — **d** Klischee-montierholz *n*; Aufmontierholz *n*; Aufklotzholz *n*; Klischeeholz *n* — **n** clichéhout *n*; clichémontage-hout *n*; hout *n* voor clichévoeten — **e** madera *f* para montaje (de clisés); madera *f* para zócalos; tablero *m* de montaje — **i** legno *m* da zoccolatura

9145 mounting machine — Apparatus to apply ad-hesives to wooden or metal blocks and mount printing plates thereon with heat and pressure.

f machine *f* à monter les clichés — **d** Montier-maschine *f*; Klischeemontiermaschine *f* — **n** monteermachine *f*; clichémonteermachine *f* — **e** máquina *f* de montaje — **i** macchine *f* per montare i clichè

9146 mounting tape — Double-coated tape to hold engravings, etc., on their bases.

f adhésif *m* double face pour montage — **d** Klebe-band *n* für Druckstockmontage — **n** clichéplak-band *n* — **e** cinta *f* para montar grabados — **i** nastro *m* di montaggio

9147 mounting wood — See mounting lumber.

9148 mount of a lens — See barrel.

9149 mourning border; black border; black edge

f bord *m* de deuil — **d** Trauerrand *m* — **n** rouwrand *m* — **e** borde *m* de luto; orla *f* negra — **i** orlo *m* di lutto

9150 mourning paper; black-border(ed) paper; black-edged paper
f papier *m* deuil; papier *m* de deuil — **d** Trauerpapier *n*; Trauerrandpapier *n*; Papier *n* mit Trauerrand — **n** rouwcirculaires *fmpl*; papier *n* voor rouwdrukwerk — **e** papel *m* de luto; esquelas *fpl* de defunción; esquelas *fpl* funerarias; esquelas *fpl* mortuorias; esquelas *fpl* de muerte — **i** carta *f* da lutto

9151 mouthpiece — Part of an instrument or machine from which a gas or liquid is squirted.
f embouchure *f* — **d** Düse *f* — **n** mondstuk *n* — **e** boquilla *f*; embocadura *f* — **i** imboccatura *f*; ugello *m*

9152 mouthpiece — See pot mouth.

9153 movable inking carriage
f chariot *m* d'encrage mobile — **d** fahrbares Farbwerkgestell *n* — **n** verwisselbaar inktwerk *n*; los inktwerk *n* — **e** tintero *m* móvil — **i** carro *m* d'inchiostrazione mobile

9154 movable types — Single types as distinguished from slugs or blocks.
f caractères *mpl* mobiles — **d** Einzelbuchstaben *mpl*; bewegliche Typen *fpl*; bewegliche Buchstaben *mpl* — **n** losse letters *fmpl* — **e** tipos *mpl* sueltos; caracteres *mpl* movibles; caracteres *mpl* móviles — **i** caratteri *mpl* mobili; tipi *mpl* mobili

9155 moving belt braking — See belts.

9156 moving head disk — Computer peripheral which provides on-line data storage which can be accessed in a random (e.g., non-sequential) fashion. The read/write head of the disk moves across the disk platter to the track upon which the desired data are stored, according to a "map" or "index" which indicates the location of the data. The map or index is either stored in core or on a section of the disk itself. The read/write head on the disk floats on a cushion of air above the surface of the disk. It is never in contact with the disk itself unless the air cushion is broken and the head crashes on to the surface of the disk. Moving head disks provide slower access than *fixed head disks, since in the former there is only one read/write head per surface and in the latter instance there is one read/write head per track, and consequently no movement of the read/write head from track to track is necessary. Moving head disks usually provide more storage per dollar of investment than fixed head disks. Information is filed on a disk by track and by sector within a track. A sector is always a fixed physical size, as 400 bytes in length. Information which cannot be stored on one sector does not necessarily flow over to the following sector. The operating system or file management program may write an address at the end of one sector indicating where to go to find the continuation of the data stream. This technique of typing information together is called index sequential.

9157 mr. — A size of cut card. See app. no. 3.
9158 mr. and mrs. — A size of cut card. See app. no. 3.
9159 mrs. — A size of cut card. See app. no. 3.

9160 mucilage — Gelatinous substance formed by disintegration and hydration of cellulose fibres in the beater. It is the principal material which cements the fibres together in paper.
f mucilage *m* — **d** Schleimstoff *m* — **n** slijm *n*; losgeraakte gefibriliseerde vezels *fmpl* — **e** mucílago *m*; pasta *f* mucilaginosa; pegamento *m*; engrudo *m* — **i** mucillagine *f*

9161 muddy print — Halftone reproduction in which details are lost due to poor quality of ink, excessive use or poor distributing of ink, causing fill-in of the small non-printing areas.
f impression *f* galeuse; impression *f* empâtée; impression *f* beurrée; épreuve *f* galeuse; épreuve *f* empâtée; épreuve *f* beurrée — **d** unsauberer Druck *m*; unsauberer Rasterdruck *m*; Rasterdruck *m* mit zugesetzten Rastertönen — **n** slordige afdruk *m*; slechte afdruk *m* — **e** impresión *f* empastada — **i** stampa *f* impestata; stampa *f* retinata impestata

9162 mudéjar binding — Spanish strapwork book cover decoration style made by Moorish binders. (Mudejares were moslims living under the protection of a Christian king, especially during the 8th to 11th centuries, but retaining their religion, laws and customs.) According to Josefina Diez Lassaleta this style originates from mixing of Arabian and Gothic styles.
f reliure *f* mudéjar (hispano-mauresque) — **d** Mudéjar-Einband *m* — **n** mudéjar-band *m* — **e** encuadernación *f* mudéjar — **i** legatura *f* mudéjar

9163 mulberry tree; paper mulberry tree; kozo; kodzu — Tree (broussonetia papyrifera) of Japan which grows to a height of about 2.75 m (9 ft) in 4-5 years; the inner bark is used for papermaking. The bark is boiled with alkali and beaten by hand.
f mûrier *m* (à papier); arbre *m* à papier; broussonétie *f* à papier; mûrier *m* de Chine — **d** Maulbeerbaum *m*; Maulbeere *f* — **n** kodzu *m*; kozo *m*; moerbeiboom *m*; moerbesboom *m* — **e** morera *f*; moral *m* — **i** albero *m* di gelso

9164 mull; scrim *GB***; crash; supercrash; super** *US* — Coarse muslin material glued to backs of books and pads for strengthening.
f mousseline *f* — **d** Gaze *f*; Heftgaze *f* — **n** gaas *n*; hechtgaas *n* — **e** gasa *f* para encuadernar; glasilla *f*; tarlatana *f* — **i** garza *f*

9165 Mullen paper tester — See Mullen tester.
9166 Mullen test — See pop test.
9167 Mullen tester; Mullen paper tester; pop tester — Instrument to test the resistance of paper to perforation. See also bursting strength.
f appareil *m* Mullen; appareil *m* pour l'essai de résistance à la perforation selon Mullen; essayeur *m* de crevaison selon Mullen — **d** Mullens Berstdruckprüfer *m*; Mullenprüfer *m*; Berstfestigkeitsprüfgerät *n* — **n** berstweerstandsmeter *m* volgens

Mullen — **e** ensayador *m* Mullen; aparato *m* Mullen; aparato *m* de ensayo Mullen — **i** strumento *m* di misura (di resistenza allo scoppio) Mullen; apparecchio *m* di Mullen; apparecchio *m* per prova di resistenza allo scoppio

9168 muller — Pestle-like instrument usually of glass used to grind pigments in varnish for testing. Also made to operate automatically. The power driven automatic muller consists of two circular glass grinding surfaces which can be made to rotate a given number of turns. It is displacing hand mulling.
f broyeur *m* muller — **d** Tellerreiber *m*; Engelmann-Apparat *m*; Farbreiber *m* — **n** muller *m* — **e** moleta *f* — **i** mescolatore *m* a molazza; molazza *f*

9169 multicolour — Having two or more colours.
f multicolore; polychrome — **d** mehrfarbig; vielfarbig; polychromatisch — **n** meerkleurig; in meer kleuren; polychroom — **e** multicolor; policromo — **i** multicolore; policromo

9170 multicolour — See polychrome.

9171 multicolour printing — See process printing.

9172 multicolour unit — Section of a letterpress rotary embodying two or more printing couples or two or more plate cylinders with a common impression cylinder to print one side of the web in two or more colours.
f unité *f* de couleurs; groupe *m* couleur — **d** Buntdruckgruppe *f* — **n** kleurenwerk *n* — **e** cuerpo *m* impresor en colores — **i** gruppo *m* di stampa a colori; gruppo *m* di stampa colore

9173 multicylinder machine — *Cylinder machine consisting of two or more vats arranged in tandem formation to facilitate the manufacture of thick papers, duplex papers, bristols and filled boards.
f machine *f* à formes rondes — **d** Mehrrundsiebmaschine *f* — **n** rondzeefmachine *f* — **e** máquina *f* redonda — **i** macchina *f* in tondo a più tamburi

9174 multi-layer board
f carton *m* multijet — **d** mehrlagiger Karton *m*; Multiplex-Karton *m*; Multiplex-Pappe *f* — **n** multiplex karton *n* — **e** cartulina *f* multicapa; cartón *m* multicapa — **i** cartone *m* multistrato; cartoncino *m* a più strati

9175 multi-layer film — Emulsion with two or more separate photosensitive layers. Colour film is a multi-layer emulsion.
f film *m* à émulsions superposés; film *m* multicouches — **d** Mehrschichtenfilm *f* — **n** gelaagde film *m*; meerlagige film *m* — **e** película *f* de varias capas — **i** pellicola *f* a più strati

9176 multi-level address — See indirect address.

9177 Multilith — Printing machine, often used as an office duplicator which operates on the offset principle.

9178 multi-negative — See gang negative.

9179 multiple access — Simultaneous search of a computer store by two or more independent searchers or search devices (on-line equipment).
f accès *m* multiple — **d** Mehrfachzugriff *m* — **n**

meervoudige toegang *m* — **e** acceso *m* múltiple — **i** accesso *m* multiplo

9180 multiple activity chart — Chart showing the interrelationship of the activities of men and for materials and/or equipment on a common time scale, e.g., gauge chart, *man/machine chart.
f graphique *m* d'activité multiple — **i** grafico *m* d'attività multipla

9181 multiple bath etching — Gravure etching method in which a sequence of acids of diminishing specific gravity is used.
f morsure *f* en plusieurs bains — **d** Mehrbadätzung *f* — **n** etsing *f* met verschillende baden — **e** grabado *m* con baños graduales; mordido *m* con baños graduales — **i** incisione *f* a più bagni

9182 multiple graver — See lining tool.

9183 multiple imposition — See imposition.

9184 multiple mark — See mark of multiplication.

9185 multiple negative — See gang negative.

9186 multiple time plan — Work study term for a stepped bonus arrangement. See also accelerating incentive, differential piecework, geared incentive.

9187 multiplex *v* — To interleave or simultaneously transmit two or more messages on a single channel.

9188 multiplexed fount — See duplexed fount.

9189 multiplexor — Device to transfer data from a number of peripherals with comparatively slow transfer speeds to a storage medium capable of operation at a high transfer speed, by splitting the channel into narrower bands or by allotting time to each device in turn.
f multiplexeur *m* — **d** Mehrfachschalter *m*; Multiplexer *m*; Datenübertragungs-Steuereinheit *f*; Steuereinheit *f* für Datenvermittler — **n** multiplexer *m* — **e** multiplexor *m* — **i** dispositivo *m* d'esecuzione d'operazioni concorrenti

9190 multiplication sign — See mark of multiplication.

9191 multi-ply board — Board consisting of several furnish layers of different composition felted together during manufacture by pressure while still moist without the use of adhesive.
f carton *m* multiplex; carton *m* multijets — **d** Multiplex-Pappe *f*; mehrschichtige Pappe *f* — **n** meerlagig karton *n*; meerlagig gekoetst karton *n* — **e** cartón *m* múltiple; cartón *m* de varias capas — **i** cartone *m* a più strati

9192 multiplying factor — See filter factor.

9193 multi-point circuit — *Circuit interconnecting several locations which makes information transmitted over the circuit available at all locations simultaneously.
f circuit *m* multipoint — **d** Mehrstationenschaltung *f*; Übertragungsleitung *f* mit mehreren Stationen — **n** meerstationsschakeling *f* — **e** circuito *m* multipunto — **i** circuito *m* multinodale

9194 multi-positive — Multiple image in positive form, usually the product of a step and repeat camera or machine.
f positif *m* multiplié — **d** Sammelpositiv *n* — **n** verzamelpositief *n*

9195 multi processor — Computer that can execute one or more computer programs employing two or more processing units under integrated control of programs or devices.
f multi-ordinateur *m* — **d** Mehrrechneranlage *f* — **n** computer *m* met multi-verwerking — **e** multicomputadora *f* — **i** molticalcolatore *m*

9196 multi-programming; concurrent processing — Method of operating a computer system so that several different tasks, even tasks requiring different programs, may be processed concurrently under the control of an operating system which keeps track of the various assignments according to demands upon the system and the priorities assigned. Usually some tasks are performed in the foreground and others in the background, depending upon the nature of the processing requirements, and whether or not real-time responses are desired.
f multiprogrammation *f* — **d** Multiprogramming *f*; Programmverzahnung *f*; verzahnt ablaufende Verarbeitung *f*; Simultanverarbeitung *f* — **n** multiprogrammering *f* — **e** multiprogramación *f* — **i** multiprogrammazione *f*

9197 multivat machine — See multicylinder machine.

9198 multiwall paper sack; multiwall paper shipping sack — Paper sack of from three to six-ply construction made in tubular form with each tube properly nested within each other. The material for the various liners or inner plies depends on the requirements of the product packed and the method of transportation. The paper is usually of heavy-duty kraft from 40-70 pound basis weight per ply (24 x 36 - 500). Linings may be waxed paper, glassine, grease-proof or vegetable parchment, or a bitumen-treated sheet, laminated cellophane-kraft sheet, bentonite-coated sheet or formed film, etc. The innermost wall is often moisture-proof or waterproof. The paper may be treated with resins to provide high wet-strength. In describing the construction of a multiwall shipping sack it is customary to begin with the innermost ply. There are 10 types of paper shipping sacks: 1. sewn, gusseted, valve-type sack; 2. sewn, gusseted, valve-type sack with outer sleeve (tuck-in sleeve); 3. sewn bottom, gusseted, open-mouth type sack; 4. sewn bottom, flat tube, open-mouth type sack; 5. sewn bottom, sewn top, open corner sack; 6. pasted valve-type sack; 7. pasted valve-type sack with outer sleeve (tuck-in sleeve); 8. pasted satchel bottom, open-mouth sack; 9. automatic (SOS) pasted bottom, open-mouth sack, to balers.
f sac *m* à parois multiple — **d** mehrwandiger Papiersack *m*; mehrlagiger Papiersack *m*; Mehrlagensack *m* — **n** meerwandige papierzak *m* — **e** bolsa *f* con foro múltiple; saco *m* con foro múltiple; bolsa *f* de varias capas; saco *m* de varias capas — **i** sacco *m* a parete multipla

9199 multiwall paper shipping sack — See multiwall paper sack.

9200 Munsell chroma — Estimated pure chromatic colour content of a surface colour on a scale of equal sensation intervals extending from grey (*chroma = O), as specified objectively by the samples of the Munsell atlas (or the smoothed data specified in CIE units).

9201 Munsell hue — *Hue of a surface colour on a scale of 100 equal sensation intervals round a colour circle of constant chroma, as specified objectively by the samples of the Munsell atlas (or the smoothed data specified in CIE units).

9202 Munsell system — System of colour classification in which colours are defined in terms of hue, chroma and value. The Munsell system presents the closest attempt at representing the colour solid of surface colours by samples spaced at equal sensation intervals, and therefore the closest correlation with the subjective variables, which are chroma, lightness (called value) and hue. See also Munsell chroma, Munsell value, Munsell hue.
f système *m* Munsell — **d** Munsell-System *n* — **n** Munsell-systeem *n* — **e** sistema *m* de color Munsell; sistema *m* de Munsell — **i** sistema *m* Munsell

9203 Munsell value — Estimated lightness of a surface colour on a scale of 10 equal sensation intervals extending from ideal black (*value = O) to ideal white (value = 10), as specified objectively for values from 1 to 9 in the Munseel atlas (or the smoothed data specified in CIE units).

9204 muriatic acid — See hydrochloric acid.

9205 muriatic ether — See ethyl chloride.

9206 mushroom dot — See undercut dot.

9207 music composition
f composition *f* de musique — **d** Notensatz *m* — **n** muziekzetsel *n* — **e** composición *f* de música — **i** composizione *f* di musica

9208 music demy — A size of printing and cartridge papers. See app. no. 3.

9209 music engraver
f graveur *m* en musique — **d** Notenstecher *m* — **n** muziekgraveur *m* — **e** grabador *m* de música — **i** incisore *m* di musica; intagliatore *m* di note musicali

9210 music engraving — Metal plate on which music is engraved. Formerly pewter was used, later an alloy of tin and lead, now most music is printed in offset lithography.
f gravure *f* de musique — **d** Notenstich *m* — **n** muziekgravure *f* — **e** grabado *m* de música — **i** incisione *f* di musica; incisione *f* di note musicali

9211 music paper — Paper to print or lithograph sheet music, made largely of bleached chemical wood pulp. It is well formed and sized and processed to give a good pick strength and minimum curl. A flat English finish is desired and stiffness to stand in a music rack. The basis weights in most common use range from 60 to 90 pounds (25 x 33 - 500).
f papier *m* à musique; papier *m* de musique — **d** Notendruckpapier *n*; Notenpapier *n* — **n** muziek-

papier *n*; notenpapier *n* — **e** papel *m* para
música; papel *m* de música — **i** carta *f* da musica
9212 music pen
f patte *f* — **d** Rastral *n* — **n** rastraal *n*; notenbalkentrekker *m*; lijnentrekker *m* — **e** pauta *f* —
i rastello *m*
9213 music printing
f impression *f* de musique — **d** Musikdruck *m*;
Notendruck *m* — **n** muziekdruk *m* — **e** impresión
f de música — **i** stampa *f* di musica; stampa *f* di
note musicali
9214 music publisher
f éditeur *m* de musique — **d** Musikverleger *m*;
Musikverlag *m* — **n** muziekuitgever *m* — **e** editor
m de música — **i** editore *m* di musica
9215 music type composition; note composition
f composition *f* de musique — **d** Musiknotensatz *m*; Notensatz *m* — **n** muziekzetsel *n* — **e**
composición *f* de tipos para música — **i** composizione *f* di segni di musica; composizione *f* di
segni musicali
9216 muslin — Plain-woven sheer to coarse cotton fabric.

f mousseline *f* — **d** Musselin *m*; Mull *mn* — **n**
mousseline *fmn* — **e** muselina *f* — **i** mussola *f*;
mussolina *f*
9217 must — Instruction on copy of set matter to
indicate that it must be inserted in a certain
edition without fail.
f forcé; article *m* de rigueur — **d** zwingendes
Erscheinen in bestimmter Zeitungsausgabe — **n**
moet worden opgenomen — **e** no puede quedarse
— **i** istruzione *f* ad indicare cosa da fare assoluta-
mente
9218 mut — See em quad.
9219 mutt — See em quad.
9220 mutton — See em quad.
9221 mutton quad — See em quad.
9222 mutton tallow — Mixture of the animal fat
of mutton. Used to grease the scraper leather and
the tympan in lithographic printing.
f suif *m* de mouton; graisse *f* de mouton — **d**
Hammeltalg *m* — **n** schapevet *n*; talg *m*; talk *m*
— **e** grasa *f* de carnero; gordura *f*; sebo *m* — **i**
sego *m*

9223 ñ; curly ñ; Spanish ñ; ñ with the tilde — Used in Spanish, called ñ con tilde, giving the value of ni in onion. The sign begins at the left bottom corner. ñ follows all other n's in Spanish.
f ñ *m* espagnol; ñ *m* avec tilde; ñ *m* tilde — **d** ñ *n* mit Tilde; spanisches ñ *n* — **n** Spaanse ñ *m*; ñ *m* met tilde; ñ *m* met tildeteken — **e** eñe *fm* con tilde — **i** ñ *m* spagnuola

9224 nacreous pigments — Pigments available in the form of suspensions in a liquid vehicle, which may be water, a plasticizer, resin, or lacquer, depending on the application. Pearl essence, also known as natural essence, is the original nacreous pigment, in which the particles are genuine crystals obtained from fish scales and fish skin. The crystals are extremely thin plates in the form of elongated hexagons. Synthetic nacreous pigments, also known as synthetical pearl essence, are generally inorganic substances crystallized in the form of thin plates (generally lead or bismuth compounds). The nacreous effect is obtained from the simultaneous reflection of light from the many parallel microscopic layers.
f pigments *mpl* nacrés — **d** Perlmutterpigmente *npl*; perlmutterartige Pigmente *npl* — **n** paarlemoerachtige pigmenten *npl* — **e** pigmentos *mpl* nacarados; pigmentos *mpl* nacarinos — **i** pigmenti *mpl* madreperlacei

9225 nail
f clou *m* — **d** Nagel *m* — **n** spijker *m*; spijkertje *n* — **e** clavo *m* — **i** chiodo *m*

9226 nailing — The nailing of printing relief plates on wooden or metal bases.
f clouage *m* — **d** Aufnagelung *f*; Aufnageln *n*; Aufklotzen *n*; Aufholzen *n* — **n** opspijkeren; spijkeren — **e** clavar — **i** chiodatura *f*; inchiodatura *f*

9227 nailing apparatus — See nailing machine.

9228 nailing machine; nailing apparatus — Mechanical apparatus to nail printing plates on wood blocks. See also tacker.
f machine *f* à clouer (les clichés) — **d** Nagelmaschine *f*; Nagelapparat *m* — **n** spijkerapparaat *n* — **e** clavadora *f*; máquina *f* de clavar; aparato *m* para clavar — **i** apparecchio *m* per inchiodare

9229 name index
f index *m* de noms propres — **d** Namenregister *n* — **n** naamregister *n* — **e** índice *m* de nombres propios — **i** indice *m* di nomi propri

9230 name plate — See mast-head.

9231 nano — Prefix in the metric system of measurement, denoting 0,000,001. See also prefixes for metric units.
f nano — **d** Nano- — **n** nano- — **e** nanó — **i** nano

9232 nanometre — Modern unit of measurement for wavelength, etc. (0.000001 mm or 10⁻⁹ m). In the SI it replaces the ångström. Abbrev. nm (no point). See also billion.

f nanomètre *m* — **d** Nanometer *n* — **n** nanometer *m* — **e** nanómetro *m* — **i** nanometro *m*

9233 nanosecond — Abbrev. ns. In US one billionth part, 10^{-9}, of a second. See also prefixes for metric units, billion.
f nanoseconde *f* — **d** Nanosekunde *f* — **n** nanoseconde *fm* — **e** nanosegundo *m* — **i** nanosecondo *m*

9234 nap — Surface and direction of the grain of a leather, velvet or plush roller.
f grain *m* — **d** Strich *m*; Narbung *f* — **n** vleug *fm* — **e** flor *m* (del cuero); grano *m* (de la piel) — **i** grana *f*

9235 naphtha; petroleum naphtha — Liquid fractions boiling below about 200 °C obtained from petroleum shale oil, coal tar and similar sources. Term is replaced by more precise terms. TL-value of naphtha or destillates: about 500 ppm or about 2000 mg/m³.
f naphta *m* — **d** Naphtha *n* — **n** nafta *m* — **e** nafta *f* — **i** nafta *f*

9236 naphthenate
f naphténate *m* — **d** Naphthenat *n* — **n** naftenaat *n* — **e** naftenato *m* — **i** naftenato *m*

9237 naphthenate drier — Compounds of naphthenic acid with metals, usually lead, cobalt, or manganese, used to accelerate the oxidation of printing ink films.
f siccatif *m* au naphténate — **d** Naphthenattrockner *m* — **n** naftenaat-droger *m* — **e** secante *m* de naftenato; secante *m* naftalino — **i** essiccante *m* al naftenato

9238 napkin paper
See napkin tissue.

9239 napkin tissue; napkin paper; serviette paper — Bleached sulphite or sulphate, white or coloured paper for making paper napkins. It may be plain, crêped, semi crêped or embossed.
f papier *m* à serviettes; papier *m* mousseline pour serviettes — **d** Serviettenpapier *n* — **n** servettenpapier *n*; papier *n* voor servetten — **e** papel *m* para servilletas; papel *m* servilleta — **i** carta *f* per salviette; carta *f* per tovaglioli

9240 nap roller; grained roller — Grained leather roller.
f rouleau *m* à grain; rouleau *m* (en cuir) grainé — **d** gerauhte Druckwalze *f*; gerauhte Lederwalze *f* — **n** ruwe leerrol *fm* — **e** rodillo *m* graneado — **i** rullo *m* felpato

9241 narrow *obs.* — Furniture 3 ems (36-point) wide.

9242 narrow-cut filter — Colour filter in which the spectral band transmission is restricted.
f filtre *m* à bande (de transmission) étroite — **d** Schmalbandfilter *m* — **n** smalband-filter *n* — **i** filtro *m* a banda (spettrale) ristretta

9243 narrow measure — Type composed in narrow widths compared with the body of the type, e.g., a line with less than 25 characters (including spaces).

f justification *f* courte — **d** schmaler Satz *m*; schmalspaltiger Satz *m* — **n** smalle breedte *f* — **e** medida *f* corta — **i** colonna *f* stretta

9244 narrow set type

f composition *f* serrée — **d** enger Satz *m* — **n** nauw zetsel *n*; nauw gespatieerd zetsel *n* — **e** composición *f* amontonada; composición *f* apiñada; composición *f* estrecha — **i** composizione *f* a spaziatura serrata; composizione *f* serrata; composizione *f* a spaziatura stretta

9245 natrium — See sodium.

9246 natural language — Language whose rules reflect and describe current usage rather than prescribed usage.

f langage *m* naturel — **d** natürliche Sprache *f* — **n** natuurlijke taal *fm* — **e** lenguaje *m* natural — **i** linguaggio *m* naturale

9247 natural lead sulphide — See galena.

9248 natural pigment — Pigment occurring naturally, e.g., a few of the iron oxide colours, such as ochres, umbers, siennas, etc. All others whether organic or inorganic are manufactured.

f pigment *m* naturel — **d** natürliches Pigment *n* — **n** natuurlijk pigment *n* — **e** pigmento *m* natural — **i** pigmento *m* naturale

9249 natural resin — See resin.

9250 natural resins — Resinous materials obtained or produced from plants or animals.

f résines *fpl* naturelles — **d** Naturharze *npl* — **n** natuurlijke harsen *mnpl* — **e** resinas *fpl* naturales — **i** resine *fpl* naturali

9251 natural rubber — See rubber.

9252 natural sign — See arbitrary sign.

9253 natural size — Picture or reproduction in the same size as the original or copy.

f grandeur *f* nature — **d** natürliche Größe *f* — **n** ware grootte *f*; natuurlijke grootte *f* — **e** tamaño *m* natural — **i** grandezza *f* naturale

9254 natural vermilion — See cinnabar.

9255 nautical chart — See chart.

9256 nave; hub — Central part of a wheel.

f moyeu *m*; bossage *m* — **d** Nabe *f* — **n** naaf *fm* — **e** cubo *m* de rueda — **i** mozzo *m* della ruota

9257 NCR paper — See no carbon required.

9258 nearside — See operating side.

9259 neat — Said of bookbindings when they are in good order and clean.

f en bon état — **d** in gutem Stande — **n** in goede staat — **e** en buen uso; en perfecto estado — **i** pulito; a posto

9260 neat's foot oil *GB*; **neet's foot oil** *US*; **hoof oil** — Pale yellow oil obtained from the feet of cattle and used as a wetting agent for leather treatment. Soluble in alcohol, ether, chloroform, and kerosine.

f huile *f* de pied de bœuf; huile *f* de pattes de bœuf — **d** Ochsenklaueöl *n*; Klauenöl *n*; Rinderklauenöl *n* — **n** ossepootolie *fm* — **e** aceite *m* de pezuña; aceite *m* de manitas — **i** olio *m* di piede di bue

9261 neat's leather — See cowhide.

9262 neb — See nib.

9263 neck — Part of the type character between the shoulder and the face. Usually called the *bevel.

9264 neck — See bevel.

9265 Neddy — Nickname for National Economic Development Council (NEDC); sub councils as for the newspaper, printing and publishing councils are called little Neddy.

9266 needle — Sharp pointed, steel needle-like tool, used by lithographers to make corrections on plates, glass or film negatives or positives.

f grattoir *m* litho — **d** Korrekturnadel *f* — **n** krabnaald *fm* — **e** rascador *m* para dibujo litográfico; buril *m* rascador — **i** raschietto *m* litografico

9267 needle bearing

f roulement *m* à aiguilles — **d** Nadellager *n* — **n** naaldlager *n* — **e** cojinete *m* de agujas — **i** cuscinetto *m* ad aghi

9268 needle cam — Cam that controls the movement of the *needles in a rotary press.

f came *f* des picots — **d** Punkturnadelnexzenter *m* — **n** naalden-excentriek *n* — **e** excéntrica *f* de las agujas; cama *f* de las agujas — **i** camma *f* degli aghi di trascinamento

9269 needles; pins; folding pins — Removable points, usually with screwed mounts, either fixed or retractable under cam action and used to control and convey the cut leading edge of the web or copy until it is severed, transferred, folded or stitched, as required. They can be mounted in the folding, transfer or cutting, and collection cylinder.

f picots *mpl*; ardillons *mpl* — **d** Punkturen *fpl*; Punkturnadeln *fpl* — **n** punctuurnaalden *fmpl*; naalden *fmpl* — **e** agujas *fpl*; punturas *fpl* — **i** aghi *mpl* di trascinamento

9270 needles — Needles to obtain register on hand presses when laying.

f pointures *fpl* — **d** Nadeln *fpl* — **n** naalden *fmpl* — **e** agujas *fpl* — **i** puntine *fpl* di registro

9271 needle setting — Adjustment of the projection of the *needles in their holders.

f réglage *m* de picots — **d** Punktureinstellung *f* — **n** stellen *n* van de puncturen — **e** ajuste *m* de las punturas — **i** regolazione *f* degli aghi di trascinamento

9272 neet's foot oil — See neat's foot oil.

9273 neg — See negative.

9274 negative; neg *abbrev.* — Photographically prepared image of a subject in reverse shades of brightness, usually prepared on a film or glass base that has been sensitized with an emulsion coating containing finely dispersed silver salts. The clear or bright areas of the original transmit more light to the corresponding areas on the sensitized emulsion, activating it so that on development the greatest density of silver or blackening occurs when the exposure to light is also greatest.

f négatif *m* — **d** Negativ *n* — **n** negatief *n* — **e** negativo *m* — **i** negativo *m*; negativa *f*

9275 negative carrier; negative holder — Part of the mechanism of a photocomposing machine which holds and positions the photographic image being exposed, and includes the carriage saddle, the register device and the negative frame.
f porte-négatif *m* — **d** Negativträger *m* — **n** negatiefhouder *m* — **e** chasis *m* portanegativos; portanegativos *m* — **i** portanegativo *m*
9276 negative holder — See negative carrier.
9277 negative lens — See diverging lens.
9278 negative paper — Photographic paper coated with a contrasty or process emulsion for *paper negatives.
f papier *m* à négatif — **d** Negativpapier *n* — **n** negatiefpapier *n* — **e** papel *m* para negativos; papel *m* negativo — **i** carta *f* per negativi
9279 negative varnish — Solutions applied to the face of photographic negatives as a protecting film.
f vernis *m* de protection d'un négatif — **d** Negativschutzschicht *f* — **n** negatieflak *mn* — **e** barniz *m* para negativos — **i** lacca *f* per negativi; lacca *f* di protezione per negativi
9280 neighbourhood effect — See Eberhard effect.
9281 neon lamp — Electric light bulb in which a gas discharge is caused through a filling of neon gas at very low pressure. Used as indicator lamp on account of the low luminosity and low current consumption.
f lampe *f* de néon — **d** Neonlampe *f* — **n** neonlamp *fm* — **e** lámpara *f* de neón — **i** lampada *f* al neon
9282 neoprene — Trade name for chloroprene rubber, resistant to oils, heat and oxidation. See also synthetic rubber.
9283 Nessler tube — Glass tube to compare and analyze colours, with standardized diameter and colour. Named after Julius Nessler (1827-1905), German chemist.
f tube *m* de Nessler — **d** Nessler-Rohr *n* — **n** Nessler-buis *fm* — **e** tubo *m* de Nessler — **i** tubo *m* di Nessler
9284 nested indents — The setting of one or more indention condition(s) inside another (or others), as to set a centred indent within a left indent measure, and a left indent within the centred indent. Nested indents are helpful when setting text in outline form. Computerized typesetting term.
9285 nesting — Positioning irregular shapes on an offset press plate by means of a photocomposing machine, to avoid waste of stock for die cut forms such as envelope blanks.
f placement *m* en quinconces — **d** unregelmäßige Einteilung *f* — **n** onregelmatige plaatsing *f*; onregelmatige indeling *f* — **e** postetas *fpl* alternadas — **i** disposizione *f* quinconce
9286 net price; discount price
f prix *m* net; prix *m* faible — **d** Nettopreis *m* — **n** nettoprijs *m* — **e** precio *m* neto; precio *m* con descuento — **i** prezzo *m* netto
9287 net weight
f poids *m* net — **d** Reingewicht *n* — **n** netto-

gewicht *n* — **e** peso *m* neto — **i** peso *m* netto
9288 network — Series of points interconnected by communication channels.
f réseau *m* — **d** Netz *n*; Netzwerk *n* — **n** net *n*; netwerk *n* — **e** red *f* — **i** rete *f*
9289 neutralize *v* — To counteract or overcome acidity or alkalinity.
f neutraliser — **d** neutralisieren — **n** neutraliseren — **e** neutralizar — **i** neutralizzare
9290 neutralizer — See static neutralizer.
9291 neutral lead chromate — See lead chromate.
9292 neutral potassium chromate — See potassium chromate.
9293 neutral sulphite pulp — Pulp obtained by cooking fibrous raw materials with a liquor containing a monosulphite.
f pâte *f* au sulfite neutre — **d** Neutralsulfitzellstoff *m* — **n** neutrale sulfietcelstof *fm* — **e** pasta *f* de sulfito neutro — **i** pasta *f* al solfito neutro
9294 neutral tints — See neutral tones.
9295 neutral tones; neutral tints — Tones that have no apparent colour or hue, such as white, grey and black.
f tons *mpl* neutres; teintes *fpl* neutres — **d** neutrale Töne *mpl*; unbunte Farben *fpl*; Neutraltinten *fpl* — **n** neutrale tonen *mpl*; neutrale tinten *fmpl* — **e** tonos *mpl* neutros — **i** toni *mpl* neutri; tinte *fpl* neutre; colori *mpl* neutri
9296 neutron — Uncharged elementary particle with a mass nearly equal to that of the proton, present in all atomic nuclei, except the hydrogen nucleus.
f neutron *m* — **d** Neutron *m* — **n** neutron *m* — **e** neutrón *m* — **i** neutrone *m*
9297 new edition — Re-edition of a book or printed work, usually with changes.
f nouvelle édition *f* — **d** Neuausgabe *f*; Neudruck *m* — **n** nieuwe uitgave *f* — **e** nueva edición *f* — **i** edizione *f* nuova
9298 news — See newsprint.
9299 news-agency; press-agency — Office collecting and distributing news.
f agence *f* d'information(s) — **d** Nachrichtenbüro *n*; Nachrichtenagentur *f* — **n** persbureau *n*; nieuws-agentschap *n* — **e** agencia *f* de publicidad; agencia *f* noticiosa; agencia *f* de noticias; oficina *f* de prensa — **i** agenzia *f* di stampa
9300 news agent; news dealer — Retailer who buys newspapers and periodicals and resells them at wholesale, usually to news-stands.
f dépositaire *m* de journaux — **d** Zeitungshändler *m*; Zeitungsgrossist *m* — **n** tijdschriftengroothandel *m* — **e** revendedor *m* de periódicos; agente *m* de periódicos — **i** commercio *m* di periodicos; rivenditore *m* di periodicos
9301 newsboard — Stiff paperboard mainly for setup boxes; made on a cylinder machine from old newspapers, generally 0.4 to 1.4 mm thick. Combination newsboard has a news base or centre and is lined on one or both sides with a higher grade of stock. Solid newsboard is made entirely from printing news.

f carton *m* gris — **d** Graupappe *f*; Feingrau-pappe *f* — **n** grijs bord *n* — **e** cartón *m* gris; cartoncillo *m* gris — **i** cartone *m* grigio

9302 news colour plates — Colour plates for news-paper printing.

f clichés *mpl* couleurs pour journaux — **d** Farb-klischees *npl* für Zeitungen — **n** kleurenclichés *npl* voor kranten — **e** planchas *fpl* para impresión multicolor de periódicos — **i** cliscè *mpl* a colori per giornali

9303 news compositor; news hand

f canardier *m* — **d** Zeitungssetzer *m* — **n** kranten-zetter *m* — **e** compositor *m* de periódicos — **i** compositore *m* per quotidiani; compositore *m* di giornale

9304 news cutting; cutting *GB*; **newspaper cut-ting; clipping** *US*; **newspaper clipping; press clipping** — Cutting from a newspaper or perio-dical.

f coupure *f* de journal; coupure *f* de presse — **d** Zeitungsausschnitt *m*; Ausschnitt *m* — **n** krante-knipsel *n*; knipsel *n* — **e** recorte *m* de periódico; recorte *m* de prensa — **i** ritaglio *m* di giornale; taglio *m* da un giornale

9305 news dealer — See news agent.

9306 news fractions — Fractions without diagonal or horizontal strokes, as these: ^{132}, ^{332}, ^{532}, to ob-tain a larger size of figure on small bodies.

f fractions *fpl* sans trait; nombres *mpl* fraction-naires sans trait — **d** Bruchziffern *fpl* ohne Schrägstrich; Bruchziffern *fpl* ohne Querstrich — **n** breukcijfers *npl* zonder deelstreep — **e** quebrados *mpl* sin barra; números *mpl* quebrados sin barra — **i** frazioni *fpl* senza linea; frazioni *fpl* senza barra

9307 news hand — See news compositor.

9308 news ink — Printing ink to run on news-print, consisting basically of carbon black or coloured pigments dispersed in mineral oil vehicles. Dries entirely by penetration of the vehicle into the paper stock.

f encre *f* (à) journal — **d** Zeitungsdruckfarbe *f*; Zeitungsfarbe *f* — **n** kranteninkt *m*; krantinkt *m* — **e** tinta *f* para periódico(s) — **i** inchiostro *m* per giornali; inchiostro *m* da giornale

9309 newspaper; sheet *sl.* — Publication printed on newsprint, appearing at regular intervals, and containing news, editorials, feature columns, and advertising. See also daily paper.

f journal *m*; feuille *f* périodique; feuille *f* publique — **d** Blatt *n*; Zeitschrift *f*; Zeitung *f* — **n** tijd-schrift *n*; nieuwsblad *n*; krant *fm* — **e** papel *m*; hoja *f* periódica; periódico *m* (de información) — **i** foglio *m*; rassegna *f*; periodico *m*

9310 newspaper — See daily paper.

9311 newspaper advertising — Publicity in news-papers for advertising.

f publicité *f* par la presse — **d** Zeitungswerbung *f* — **n** krantenreclame *fm* — **e** publicidad *f* perio-dística; publicidad *f* de prensa; propaganda *f* perio-dística — **i** pubblicità *f* di giornali

9312 newspaper clipping — See news cutting.

9313 newspaper column

f colonne *f* de journal — **d** Zeitungsspalte *f*; Zeitungskolumne *f* — **n** krantenkolom *f* — **e** columna *f* periodística — **i** colonna *f* di giornale

9314 newspaper cutting — See news cutting.

9315 newspaper fold — See former fold.

9316 newspaper fount

f caractères *mpl* pour journaux — **d** Zeitungs-schrift *f* — **n** krantletter *fm*; krantenletter *fm* — **e** tipos *mpl* para diarios; tipos *m* para periódicos — **i** caratteri *mpl* da giornale; tipi *mpl* da giornale

9317 newspaper kiosk; newsstand — Small light structure with one or more open sides used as a newsstand.

f kiosque *m* à journaux — **d** Zeitungskiosk *m*; Kiosk *m* — **n** krantenkiosk *fm*; kiosk *fm* — **e** quiosco *m* de periódicos; kiosco *m* de periódicos; kiosko *m* de periódicos; puesto *m* de periódicos; parada *f* (de venta) *Ar* — **i** edicola *f*

9318 newspaper press; letterpress rotary; rotary letterpress machine

f rotative *f* à journal — **d** Zeitungsrotations-maschine *f*; Zeitungspresse *f*; Zeitungsdruck-maschine *f* — **n** krantenrotatiepers *f*; kranten-pers *f*; boekdrukrotatiepers *f* — **e** rotativa *f* de periódicos — **i** rotativa *f* per giornali

9319 newspaper size — In GB and US about 45 x 66 cm. In Germany Hamburger Format, some-times called nordisches Format 40 x 57 cm, Ber-liner Format 34 x 48 cm, Rheinländer Format or rheinisches Format 37.5 x 53 cm, Pariser Format 45 x 60 cm, Schweizer Format 33 x 49 cm. See also tabloid format.

f format *m* de journaux — **d** Zeitungsformat *n* — **n** krantformaat *n*; krantenformaat *n* — **e** tamaño *m* de diario; formato *m* de diario — **i** formato *m* di giornali

9320 newsplate — Line, halftone or combination etching for newspaper printing and newspaper advertisements.

f cliché *m* de presse zur zinc — **d** Zeitungs-klischee *n* — **n** krantencliché *n* — **e** plancha *f* para impresión de periódicos — **i** lastra *f* di stampa per giornali

9321 newsprint; news *abbrev.* — Paper generally used for newspapers; machine-finished and slack sized, with little or no mineral loading. The furnish is largely mechanical wood pulp, with some chemical wood pulp. Made in basis weights varying from 30 to 35 pounds (24 x 36 - 500), mainly 32 pounds.

f papier *m* journal; papier *m* à journal — **d** Zeitungsdruckpapier *n*; Zeitungspapier *n* — **n** krantenpapier *n*; courantdruk *n* — **e** papel *m* diario(s); papel *m* para diario; papel *m* periódico; papel *m* para periódicos; papel *m* (de) prensa; papel *m* para prensa — **i** carta *f* giornale; carta *f* da giornali; carta *f* per quotidiani

9322 newsprint waste — In US classified as follows: 1. Wrappers. The weight of all wrapper protection stripped from the rolls, including the heads and body wrapper but not including white paper which may be removed during stripping. 2. Handling and transit waste. The weight of all white paper taken from the roll up to the minute it is started through the press. 3. White press waste. Includes all white (unprinted) paper taken from the roll, from the time the paper is started through the press to the time the core is removed from the reel. 4. Core waste. The weight of the white paper remaining on the core at the time it is removed from the reel, but not including the weight of the core itself. 5. Total white. The sum of handling and transit plus white press plus core waste. 6. Printed waste. The weight of all printed paper which does not leave the pressroom, including the weight of printed and folded papers. 7. Production waste. The sum of white press plus core plus printed waste. 8. Grand total. The sum of wrapper plus handling and transit plus white press plus core plus printed waste.

f bouillons *mpl* — **d** Zeitungspapiermakulatur *f* — **n** oud krantenpapier *n*; oude kranten *fmpl* — **e** diarios *mpl* viejos — **i** scarti *mpl* di carta giornale; carte *fpl* di recupero da carta da giornale

9323 newsstand — See newspaper kiosk.

9324 newstone — Halftone up to 100-line, etched on zinc, for newspaper illustrations or for printing on cheap paper.

f simili *f* de presse à grosse trame — **d** Grobraster-klischee *n*; Zeitungsklischee *n*; Grobraster-ätzung *f*; Grobrasterautotypie *f* — **n** grofraster-cliché *n*; krantencliché *n* — **e** clisé *m* de prensa — **i** fotoincisione *f* (su zinco) con retinatura grossolama

9325 newton — Unit of force, defined as the force which will accelerate a mass of 1 kg 1 m per s². Abbrev. N. In the SI: 1 N = kg·m/s².

f newton *m* — **d** Newton *m* — **n** newton *m* — **e** newton *m* — **i** newton *m*

9326 Newton concept — For liquids, the rate of flow is proportional to the shearing stress.

f concept *m* de Newton — **d** Newtonsches Gesetz *n*; Newtonsches Prinzip *n* — **n** stromings-wet *fm* van Newton — **e** teoría *f* de Newton — **i** principio *m* di flusso newtoniano

9327 Newtonian flow — In rheology, flow occurring in a liquid system where the rate of shear is directly proportional to the shearing force.

f écoulement *m* newtonien; flux *m* newtonien — **d** Newtonsche Strömung *f*; Newtonsches Fließen *n* — **n** newtonse stroming *f* — **e** flujo *m* Newtoniano — **i** flusso *m* newtoniano

9328 Newtonian liquid — Liquid conforming to the law that the homogeneous shearing stress is the product of the coefficient of viscosity and rate of shear.

f liquide *m* newtonien — **d** Newtonsche Flüssig-keit *f* — **n** newtonse vloeistof *fm* — **e** líquido *m* Newtoniano — **i** liquido *m* newtoniano

9329 Newtonian rings; Newton rings — Rainbow-like rings or formations of colour between closely spaced polished surfaces (e.g., when negatives are locked into contact with sensitized metal plates in printing formes) caused by interference of light waves. Popularly assumed to indicate perfect contact, but really indicative of a film of air between the two surfaces. They sometimes cause an unwanted mottling of halftones or tint areas.

f anneaux *mpl* de Newton — **d** Newtonsche Ringe *mpl* — **n** newtonse ringen *mpl* — **e** anillos *mpl* de Newton — **i** anelli *mpl* di Newton

9330 Newton rings — See Newtonian rings.

9331 nib; neb; ear — Projecting part of a composing rule by which it is pulled out from between lines.

f oreille *f* — **d** Ohr *n* — **n** oortje *n* — **e** oreja *f* del sacalíneas — **i** naso *m*

9332 nicanee — A size of wrapping paper. See app. no. 3.

9333 nick — Notch or notches on the shank of a movable type to set it correctly, and to distinguish different founts. In France and Spain always at the side of the accents. See also nick at the back.

f cran *m* — **d** Signatur *f* — **n** kerf *fm* — **e** cran *m* — **i** tacca *f*

9334 nick at the back; French nick — Type carrying the groove of the shank at the side of the accents. In France and Spain the nick is always at the side of the accents.

f cran *m* français — **d** französische Signatur *f* — **n** kerf *fm* aan de zijde van de accenten — **e** cran *m* francés — **i** tacca *f* francesa; tacca *f* dorsale

9335 nickel — Bluish white, hard, dense metal, to reinforce copper, zinc or stereotype plates by electrotyping. Symbol: N. Latin: niccolum.

f nickel *m* — **d** Nickel *n* — **n** nikkel *n* — **e** níquel *m* — **i** nichel *m*

9336 nickel ammonium sulphate; ammonium nickel sulphate — Green crystals, soluble in water, less soluble in ammonium sulphate solution, insoluble in alcohol. Decomposed by heat. Used in electroplating. Formula: $NiSO_4 \cdot (NH_4)_2SO_4 \cdot 6H_2O$.

f sulfate *m* de nickel ammoniacal — **d** Nickel-Ammoniumsulfat *n*; Ammonium-Nickelsulfat *n* — **n** ammonium-nikkelsulfaat *n*; nikkel-ammonium-sulfaat *n* — **e** sulfato *m* amónico de níquel; sulfato *m* de níquel y amonio — **i** solfato *m* di nichel e ammonio

9337 nickel deposition — See nickel-facing.

9338 nickel-faced

f nickelé — **d** vernickelt — **n** vernikkeld — **e** niquelado — **i** nichelato

9339 nickel-faced stereo plate — Stereo plate faced with a nickel alloy to increase its durability.

f stéréo *m* nickelé — **d** vernickeltes Stereo *n* — **n** nikkel-styp *m* — **e** estereotipo *m* niquelado; estereotipia *f* niquelada; plancha *f* niquelada; estereoníquel *m* — **i** stereotipia *f* nichelata

9340 nickel-facing; nickel-plating; nickel deposition — Facing (stereo)plates with a nickel alloy to increase durability.

f nickelage *m* — **d** Vernicklung *f*; Vernickelung *f*;

Nickelüberzug *m* — **n** vernikkeling *f* — **e** niquelado *m* — **i** nichelatura *f*

9341 nickel-plating — See nickel-facing.

9342 nickel stype — Electrotype faced with nickel, used for long runs.

f galvano *m* nickelé — **d** Nickelgalvano *n* — **n** nikkelgalvano *m* — **e** galvano *m* niquelado — **i** galvano *m* nichelato

9343 Nigerian morocco — See niger morocco.

9344 niger morocco; Nigerian morocco — Strong leather from Nigeria, used for bookbinding.

f maroquin *m* de Nigeria — **d** Marokkoleder *n* aus Nigeria — **n** Nigeriaans marokijn *n* — **e** marroquín *m* nigeriano — **i** marocchino *m* di Nigeria

9345 niggerheads — See collating marks.

9346 night-shift; night-side — Work force scheduled to work during night-time.

f équipe *f* de nuit — **d** Nachtschicht *f* — **n** nachtploeg *fm* — **e** turno *m* de noche — **i** squadra *f* di notte

9347 night-side — See night-shift.

9348 nigrosine — Deep blue or black coal tar dyestuff, obtained by oxidation of aniline.

f nigrosine *f* — **d** Nigrosin *n* — **n** nigrosine *f* — **e** nigrosina *f* — **i** nigrosina *f*

9349 nigrosine test — Test to determine that the body paper is sufficiently hard-sized to resist penetration of glue or water which may be applied to the label when it is being stuck on the can, or to make sure that the paper can resist the penetration of water if the label is applied on to a wet bottle in a bottling plant.

f essai *m* de résistance à la colle — **d** Nigrosintest *m*; Anilinschwarztest *m* — **i** prova *f* di resistenza all'acqua della collatura di un'etichetta

9350 nihil obstat — Certification by an official censor of the Roman Catholic Church that a book has been examined and found to contain nothing contrary to faith and morals.

9351 ninety-sixmo; 96mo

f in-quatre-vingt-seize — **d** Sechsundneunzigerformat *n* — **n** in zesennegentigen — **e** en noventa-y-seisavo — **i** in novantasei

9352 nip — Point of contact between superimposed rolls or cylinders or printing cylinder and forme.

f ligne *f* de contact — **d** Drucklinie *f*; Berührungslinie *f*; Berührungsstelle *f*; Druckspalt *m* — **n** druklijn *fm*; contactlijn *fm* — **e** línea *f* de contacto — **i** linea *f* di contatto

9353 nip *v* — See smash *v*.

9354 nip and tuck folder — See jaw folder.

9355 nipping rollers; pinching rollers *US*; **pinchers** *US*; **nips** *US* — Pair of adjustable rollers to exert pressure to set a fold. See also folding rollers.

n knijprollen *fmpl* — **e** rodillos *mpl* de pinza — **i** rulli *mpl* pressatori

9356 nipping sections — Areas of increased diameter in drawing, nipping or folding rollers of a rotary press to grip the fold and margin of the sheet.

d Knippvorrichtung *f* — **n** knijpnokken *fpl* — **e** dispositivos *mpl* de pinza — **i** sezioni *fpl* ingrossate

9357 nips — See nipping rollers.

9358 nip *v* **up** — To shape the bands or ribs of a book cover with *band nippers.

f nerver — **d** Bünde anbringen — **n** ribben zetten — **e** ceñir — **i** nervare i dorsi

9359 nit — 1. Unit of luminance equal to one candela per m^2; 2. unit of information equal to 1.44 bits; 3. concentration of nepit (naperian digit). Abbrev. nt (no point). In the SI: 1 nt = 1 cd/m^2.

f nit *m* — **d** Nit — **n** nit *m* — **e** nit *m* — **i** nit *m*

9360 niter — See potassium nitrate.

9361 nitrate; azotate — Salt or ester of nitric acid, or a compound containing the univalent group ONO_2 or ONO_3, e.g., sodium nitrate $NaNO_3$.

f nitrate *m*; azotate *m* — **d** Nitrat *n* — **n** nitraat *n* — **e** nitrato *m* — **i** nitrato *m*; azotato *m*

9362 nitrate of lead — See lead nitrate.

9363 nitrate of silver — See silver nitrate.

9364 nitration pulp — Usually sulphite pulp obtained by long cooking to eliminate hemicellulose, which is sometimes dissolved in caustic soda after cooking.

f plate *f* obtenue par nitration; pulp *f* obtenue par nitration — **d** Nitrierzellstoff *m* — **i** pasta *f* per nitrazione

9365 nitre — See potassium nitrate.

9366 nitric acid; azotic acid; engraver's acid; hydrogen nitrate; aqua fortis — Transparent, colourless or yellowish, fuming, suffocating, caustic and corrosive liquid. Used as an etchant for zinc and also added to gum arabic when etching a lithographic stone. Latin: acidum nitricum. TL-value: 2 ppm or 5 mg/m^3. Contact with acetic acid, inflammable gases and liquids is dangerous. Formula: HNO_3.

f acide *m* nitrique; acide *m* azotique; eau *f* forte — **d** Salpetersäure *f*; Ätzwasser *n*; Ätze *f*; Aqua fortis *n*; Scheidewasser *n*; Ätzflüssigkeit *f* — **n** salpeterzuur *n*; etsvloeistof *fm*; sterk water *n* — **e** ácido *m* nítrico; ácido *m* azótico; aguafuerte *m*; hidrógeno *m* nítrico; mordiente *m*; ácido *m* nitromuriático — **i** acido *m* nitrico; acquaforte *f*

9367 nitric acid/soda pulp — Chemical pulp obtained by treatment of vegetable material with nitric acid, followed by an alkaline treatment.

f pâte *f* nitrosodique; pulpe *f* nitrosodique — **d** natronsalpetersaurer Zellstoff *m*; salpetersaurer Natronzellstoff *m* — **e** pasta *f* nitrosódica — **i** pasta *f* alla nitro-soda

9368 nitrobenzene; nitrobenzol — Bright yellow crystals or yellow, oily liquid. Poisonous. Soluble in alcohol, benzene and ether, very slightly soluble in water. TL-value: 1 ppm. Formula: $C_6H_5NO_2$.

f nitrobenzène *m* — **d** Nitrobenzol *n* — **n** nitrobenzeen *n* — **e** nitrobenceno *m*; nitrobenzol *m* — **i** nitrobenzene *m*

9369 nitrobenzol — See nitrobenzene.
9370 nitro calcite — See calcium nitrate.
9371 nitrocellulose; nitrocotton; cotton powder; cellulose nitrate — Ester formed by the action of nitric acid on cellulose, as paper, linen, or cotton. Used as film forming material in flexographic and gravure inks. The name cellulose nitrate is technically correct although nitrocellulose is more commonly used.
f nitrocellulose *f*; coton-poudre *m* — **d** Nitrozellulose *f*; Zellulosenitrat *n*; Schießbaumwolle *f*; Schießwolle *f* — **n** nitrocellulose *fm*; cellulosenitraat *n*; schietkatoen *n* — **e** nitrocelulosa *f*; nitrato *m* de celulosa; nitroalgodón *m*; algodón *m* pólvora — **i** nitrocellulosa *f*; nitrato *m* di cellulosa; nitrocotone *m*; fulmicotone *m*
9372 nitrocellulose lacquer — Film-forming lacquer containing nitrocellulose dissolved in organic solvents, used to produce protective, grease-proof and water-resistant coating, with high gloss.
f laque *f* de nitrocellulose; vernis *m* nitrocellulosique; laque *f* nitrocellulosique — **d** Nitrozelluloselack *m*; Nitrolack *m* — **n** nitrocelluloselak *mn* — **e** laca *f* de nitrocelulosa; laca *f* nitrocelulósica; nitrobarniz *m*; nitrolaca *f* — **i** lacca *f* nitrocellulosica
9373 nitrocotton — See nitrocellulose.
9374 nitrogen — Colourless, odourless, tasteless, diatomic gas, soluble in water, slightly soluble in alcohol. Symbol: N. Latin: nitrogenium.
f azote *m* — **d** Stickstoff *m* — **n** stikstof *fm* — **e** nitrógeno *m*; ázote *m* — **i** azoto *m*
9375 nitrogen burst; gas burst agitation — Method of agitation in which nitrogen gas is released under the surface of photographic chemical solutions to keep the solutions agitated.
f agitation *f* par bullage d'azote; bullage *m* d'azote — **d** Stickstoffstoß *m*; Badbewegung *f* durch Gasblasen — **n** stikstof-beweging *f*; stikstof-agitatie *f* — **e** agitación *f* de ázote — **i** agitazione *f* in corrente d'azoto
9376 nitroglycerine; trinitroglycerine; glyceryl trinitrate — Pale yellow, thick, flammable, explosive liquid. Soluble in alcohol and ether, slightly soluble in water. TL-value: 0.5 ppm.
Formula: $CH_2NO_3CHNO_3CH_2NO_3$.
f nitroglycérine *f*; trinitroglycérine *f*; nitroglycérol *f*; trinitrine *f* — **d** Nitroglyzerin *n* — **n** nitroglycerol *n*; nitroglycerine *fm*; glyceroltrinitraat *n* — **e** nitroglicerina *f* — **i** nitroglicerina *f*
9377 nitrohydrochloric acid — See aqua regia.
9378 nitromuriatic acid — See aqua regia.
9379 nitronaphthalene — Yellow crystals, soluble in alcohol and ether, insoluble in water. Intermediate for dyestuffs.
Formula: $C_{10}H_7NO_2$.
f nitronaphtaline *f* — **d** Nitronaphthalin *n* — **n** nitronaftaleen *n*; nitronaftaline *fm depr.* — **e** nitronaftaleno *m*; nitronaftalina *f* — **i** nitronaftalene *m*; nitronaftalina *f*
9380 nitrous acid

f acide *m* nitreux — **d** salpetrige Säure *f*; Salpetrigsäure *f* — **n** salpeterig zuur *n* — **e** ácido *m* nitroso — **i** acido *m* nitroso
9381 no carbon paper — See carbonless paper.
9382 no carbon required; NCR paper — Paper for statements, invoices, bank deposit slips and other business forms where duplicating without carbon paper is desired. Typewritten or pen pressure releases microscopic globules of a chemical which forms colour when it strikes a clay-like substance on the duplicate sheet, reproducing the impression of the original. See also carbonless paper.
f nul carbone requis; papier *m* sans carbone; papier *m* autocopiant — **d** ohne Kohle-Papier *n*; kohlefreies Durchschlagpapier *n*; Selbstdurchschreibepapier *n* — **n** papier *n* zonder carbon — **e** ningún carbón requerido; papel *m* autocopiador — **i** carta *f* senza carbonatura; carta *f* da copia senza carbone
9383 nodal point — Two point on the lens axis such that a ray of light going into the lens aiming at one comes out of the lens aiming away from the other and parallel to its ingoing direction.
f point *m* nodal — **d** Knotenpunkt *m* — **n** knooppunt *n* — **e** punto *m* nodal — **i** punto *m* nodale
9384 node; junction point; branch point; vertex — Data communication term. Terminal of any branch of a network or a terminal common to two or more branches of a network.
f branchement *m* — **d** Verzweigungspunkt *m*; Verzweigung *f*; Scheitelpunkt *m* — **n** vertakkingspunt *n* — **e** bifurcación *f* — **i** punto *m* di salto
9385 noise abatement; noise suppression
f suppression *f* du bruit; réduction *f* du bruit — **d** Lärmbekämpfung *f* — **n** lawaaibestrijding *f* — **e** lucha *f* contra el ruido; insonorización *f* — **i** lotta *f* contro il rumore; riduzione *f* del rumore
9386 noiseless program paper — Paper used as the name indicates. This refers to a paper suitable for a program, so treated that it is free from rattle.
f papier *m* pour programmes — **d** Programmpapier *n*; geräuschloses Papier *n* — **n** geruisloos programmapapier *n*; geruisloos papier *n*; programmapapier *n* — **e** papel *m* para programas — **i** carta *f* per programmi; carta *f* silenziosa
9387 noise suppression — See noise abatement.
9388 no-lash gear — See back-lash gear.
9389 nom de plume — See pen-name.
9390 nominal weight — Weight stated or prescribed for a quantity of paper in sheets, usually a *ream of 1000 sheets, at which weight the paper is billed, whether the ream or 1000 sheets actually weights, within customary tolerances, more or less than the weight stated. See also actual weight, ream weight.
f poids *m* par rame; poids *m* de mille feuilles — **d** Riesgewicht *n*; Tausendbogengewicht *n* — **n** gewicht *n* per riem; riemgewicht *n*; gewicht *n* per duizend vel — **e** peso *m* por resma; peso *m* por mil hojas — **i** peso *m* di risma; peso *m* di mille fogli

9391 nomogram; nomograph; alignment chart — Graph enabling by means of a straight-edge to read off the value of a dependent variable when the value of the independent variable is given. Also a graphic representation of numerical relations.
f nomogramme *m* (à points alignés); abaque *m* (de conversion) — **d** Nomogramm *n*; Leitertafel *f* — **n** nomogram *n* — **e** nomograma *m* — **i** nomogramma *m*

9392 nomograph — See nomogram.

9393 non actinic — Said of light rays which do not affect photographic surfaces.
f inactinique — **d** nicht aktinisch — **n** niet actinisch — **e** no actíneo; inactínico — **i** inattinico

9394 non actinic light — Light of the wavelength or spectral band which does not affect a light-sensitive layer.
f lumière *f* inactinique — **d** nicht-aktinisches Licht *n* — **n** niet-actinisch licht *n* — **e** luz *f* inactínica — **i** luce *f* inattinica

9395 non-counting perforator; non-justifying perforator — Perforating keyboard which produces paper tape without line-end codes (primary, raw or unjustified, or idiot tape). The tape provides input to a computer or autojustifier for justification and hyphenation, although the latter may be introduced by operator or may be avoided by increasing word or letter spacing.
f perforateur *m* non justifiant; perforateur *m* au kilomètre — **d** nicht-ausschließender Perforator *m*; nicht-zählender Perforator *m* — **n** niet-tellende ponsbandperforator *m* — **e** perforador *m* sin contador

9396 non-curling gummed paper — Paper the gum of which has been broken into small particles after drying to present a discontinuous surface, and to prevent curling.
f papier *m* gommé inerte — **d** nicht-kräuselndes gummiertes Papier *n*; nicht-rollendes gummiertes Papier *n* — **n** vlakliggend gegomd papier *n*; niet-krullend gegomd papier *n* — **e** papel *m* engomado inerte — **i** carta *f* gommata inerte; carta *f* gommata inarricciabile

9397 non-cyclic element — See incidental element.

9398 non destructive reading — Type of readout in which data are retained in store after the reading operation.
f lecture *f* non destructive — **d** nicht-zerstörendes Lesen *n* — **n** niet-uitwissend lezen *n* — **e** lectura *f* no destructiva — **i** lettura *f* non distruttiva

9399 non drying oil — Oil which does not form dry surface films on standing when exposed to the atmosphere in thin layers.
f huile *f* non siccative — **d** nicht-trocknendes Öl *n* — **n** niet-drogende olie *fm* — **e** aceite *m* no secante — **i** olio *m* non siccativo

9400 non flammable; non inflammable
f ininflammable — **d** unentzündbar; nicht entzündbar — **n** onontvlambaar; niet brandbaar — **e** no inflamable; ininflamable — **i** ininfiammabile; non infiammabile

9401 non image areas; non printing areas — Blanc areas on a printing forme which do not accept ink.
f régions *fpl* non imprimantes; blancs *mpl* — **d** nicht-druckende Flächen *fpl* — **n** niet-drukkende partijen *fpl* — **e** zonas *fpl* no impresas; blancos *mpl* — **i** zone *fpl* non stampanti

9402 non inflammable — See non flammable.

9403 non ionics — Compounds that exhibit practically no ionization when dissolved in water and are usually unaffected by the presence of salts and other substances which frequently precipitate ionic wetting agents.
f agents *mpl* de surface non-ioniques — **d** nicht-ionogene Verbindungen *fpl* — **n** niet-ionogene stoffen *fmpl*; non-ionogene stoffen *fmpl* — **e** elementos *mpl* superficiales no iónicos — **i** agenti *mpl* non ionici

9404 non-justifying perforator — See non-counting perforator.

9405 non-lining figures — See hanging figures.

9406 non-locking condition — See precedence code.

9407 non-Newtonian liquid
f liquide *m* non newtonien — **d** nicht-Newtonsche Flüssigkeit *f* — **n** niet-newtonse vloeistof *fm* — **e** líquido *m* non Newtoniano — **i** liquido *m* non newtoniano

9408 non-offset spray — See offset spray.

9409 nonpareil — Obsolete name for a size of type. See app. no. 1.

9410 non-printing area — See grease-resisting area.

9411 non printing areas — See non image areas.

9412 non return-to-reference recording — See non return-to-zero recording.

9413 non return-to-zero recording; non return-to-reference recording — Method of recording data on a magnetic surface in which the signal does not return to the same reference point after recording each bit; only changes in binary sequences are detected.
f enregistrement *m* sans retour à l'état de référence — **d** Aufzeichnung *f* ohne Rückkehr zum Bezugszustand — **n** optekenen *n* zonder terugkeer naar de referentietoestand — **e** registro *m* sin retorno al estado de referencia — **i** registrazione *f* senza ritorno allo stato di riferimento

9414 non scratch — Said of inks which dry to films or relatively high abrasion or rub resistance, and which acquire mar resistance only after they have become thoroughly dry.
f résistant à l'abrasion; résistant aux égratignures — **d** kratzfest; nagelfest — **n** nagelvast — **e** a prueba de rascado; resistente a la abrasión — **i** resistente alla scalfittura; resistente alla raschiatura

9415 non-scratch ink — Ink which has high abrasion and mar resistance when dry.
f encre *f* résistante à l'abrasion; encre *f* résistante au frottement — **d** nagelfeste Farbe *f*; kratzfeste

Farbe f — n nagelvaste inkt m — e tinta f prueba de rascado — i inchiostro m resistente alla raschiatura

9416 non serif — See sans serif.

9417 non volatile storage — See non volatile store.

9418 non volatile store GB; **non volatile storage** US — Store which retains data when the power source is absent, e.g., magnetic tape or drum.
f mémoire f permanente — d Strukturspeicher m — n niet-vluchtig geheugen n — e memoria f estable; memoria f permanente — i memoria f permanente

9419 noon edition — See noon paper.

9420 noon paper; noon edition
f feuille f de midi; journal m de midi — d Mittagsblatt n; Mittagsausgabe f — n middag-editie f — e edición f de la tarde — i edizione f di mezzogiorno

9421 no-pack mat — See packless mat.

9422 no place no date; place nor date — Said of books without imprints. Latin: sin loco et anno (s.l.e.a.).
f sans lieu ni date — d ohne Ort und Jahr — n zonder plaats en jaar — e sin lugar ni año — i senza luogo nè data

9423 normal- — Prefix to designate those hydrocarbons or hydrocarbon radicals whose molecules contain a single unbranched chain of carbon atoms. Thus normal butane or n-butane is the compound whose molecular structure is indicated by the formula $CH_3CH_2CH_2CH_3$.
f normal — d normal- — n normaal — e normal — i normale

9424 normal ink; conventional ink
f encre f ordinaire; encre f classique — d klassische Druckfarbe f — n gewone drukinkt m — e tinta f (de impresión) normal — i inchiostro m classico

9425 normalization — See standardization.

9426 normalizing — See extension.

9427 normal performance — The work study term for this level of performance has not yet been defined.
f allure f normal — d normale Leistung f — n normale prestatie f — e prestación f normal; marcha f normal — i prestazione f normale; comportamento m normale

9428 normal stop — Detail stop in halftone photography.
f diaphragme m normal — d Normalblende f — n normaal diafragma n; normaal insteek-diafragma n — e diafragma m para detalles — i diaframma m normale

9429 nose of former — See former nose.

9430 no-sheet detector — Device to trip the press when a sheet fails to come up to the front guides.
f détecteur m de manque de feuille; dispositif m de déclanchement en cas de non arrivée des feuillets aux taquets — d Fehlbogenmelder m; Fehlbogenschalter m — n drukafsteller n voor het ontbreken van een aan te leggen vel — e salvapliegos m de fallo de hoja — i dispositivo m rivelatore foglio mancante

9431 notch — Mortise cut into edge of mounted letterpress printing plate.
f encoche f — d Einschnitt m; Ausschnitt m — n uitsparing f — e ahorro m — i scanalatura f

9432 notch — Half "V" cutouts along the margins of a goldenrod flat, to locate the gripper line and centre line of the flat in platemaking.
f encoche f — d Einschnitt m — n inkeping f; keep fm — e muesca f; rebajo m; ranura f; escopleadura f; escopladura f; mortaja f — i intaccatura f; tacca f

9433 notch — Nick in a doctor blade.
f brèche f — d Scharte f — n kerf fm; schaarde fm; schaar fm — e mella f — i scheggiatura f

9434 not continued — See discontinued.

9435 note — Reference at the end of a book, which is too long to place as a *footnote.

9436 note — See folio note, pocket note.

9437 notebook
f carnet m (de poche); cahier m (grand format); calepin m; agenda m — d Notizbuch n; Taschenbuch n — n notitieboekje n; zakboekje n; agenda fm; zakagenda fm — e libro m de apuntes; agenda f; libreta f de notas; librito m de notas; cuaderno m de notas — i libro m di appunti; libro m per appunti; taccuino m

9438 note composition — See music type composition.

9439 note of exclamation — See exclamation mark.

9440 note of interrogation — See interrogation mark.

9441 note pad — See tablet.

9442 note paper — Flat writing paper of various qualities and finishes, usually folded, for personal correspondence.
f papier m à lettres — d Briefpapier n; Postpapier n; Billetpapier n — n briefpapier n; postpapier n — e papel m de carta(s); papel m para cartas — i carta f da lettere

9443 not for sale — See privately printed.

9444 notice to the reader; advertisement to the reader obs.
f avertissement m au lecteur; avis m au lecteur — d Vorbericht m; zur Beachtung — n voorwoord n; inleiding f; bericht n aan de lezer — e prólogo m; prefacio m; introducción f; preámbulo m; advertencia f al lector — i avviso m al lettore; avviso m per il lettore

9445 nozzle — Part of the dryer in a rotogravure press which directs air against the web; generally a small opening in a tube through which liquid or gaseous material is released to the atmosphere.
f buse f (de soufflage) — d Düse f; Blasdüse f; Luftdüse f; Ausspritzöffnung f — n blaaspijp fm; straalpijp fm — e soplador m; boquilla f — i bocca f di soffiaggio; ugello m

9446 nuller; nullo — Compositor's term for a figure 0 or zero, as opposed to a cap 0.

9447 nullo — See nuller.

9448 numbered copy — Copy of a limited or de luxe edition which contains a printed certificate: "250 copies of this book have been printed, of which this is no. ..."; the number is written in ink.

f exemplaire *m* numéroté — **d** numeriertes Exemplar *n* — **n** genummerd exemplaar *n* — **e** ejemplar *m* numerado — **i** esemplare *m* numerato; copia *f* numerata

9449 numbering

f chiffraison *m*; numérotage *m* — **d** Numerierung *f*; Numerieren *n* — **n** nummering *f* — **e** numeración *f* — **i** numerazione *f*

9450 numbering box

f numéroteur *m* — **d** Numerierwerk *n*; Ziffernwerk *n*; Numeroteur *m* — **n** nummerklokje *n*; numeroteur *m* — **e** caja *f* numeradora — **i** numeratore *m*

9451 numbering frame

f châssis *m* numérateur — **d** Numerierrahmen *m* — **n** nummerraam *n* — **e** rama *f* numeradora — **i** telaio *m* per numeratori

9452 numbering machine — Hand or foot operated or automatic mechanism to print consecutive numbers, principally in loose-leaf sheets, blankbooks, cheques, etc.

f machine *f* à numéroter; numéroteuse *f* — **d** Numeriermaschine *f*; Nummerndruckwerk *n*; Ziffernwerk *n* — **n** nummermachine *f*; numeroteur *m* — **e** máquina *f* de numerar; máquina *f* numeradora; numeradora *f* — **i** macchina *f* numeratrice; numeratore *m* (automatico)

9453 number of copies — Copies cut from one sheet.

f nombre *m* d'exemplaires (à la feuille); jeu *m*; découpe *f* — **d** Nutzen *mpl* — **n** aantal *n* exemplaren (uit één vel); aantal *n* exemplaren per vel — **e** número *m* de ejemplares — **i** numero *m* di copie (che si ottengono da un foglio)

9454 number of phases

f nombre *m* de phases — **d** Phasenzahl *f* — **n** aantal *n* fasen — **e** número *m* de fases — **i** numero *m* di fasi

9455 number of revolutions

f nombre *m* de tours — **d** Umdrehungszahl *f*; Tourenzahl *f*; Drehzahl *f* — **n** toerental *n*; aantal *n* toeren; aantal *n* omwentelingen — **e** número *m* de revoluciones — **i** numero *m* di giri

9456 number of slots

f nombre *m* d'encoches — **d** Nutenzahl *f* — **n** aantal *n* gleuven — **e** número *m* de ranuras — **i** numero *m* di scanalature

9457 number one — Placed in front of the grade name of paper or board, indicating that this is of the highest grade, e.g., number 1 bond stands for bond paper made from 100% new white rag stock.

f numéro *m* un; catégorie *f* un — **d** erstklassig; erste Qualität *f* — **n** eerste kwaliteit *f* — **e** primera calidad *f* — **i** prima qualità *f*

9458 numeral — Conventional symbol representing a number. Correct typographers' name for a figure.

f chiffre *m* — **d** Ziffer *f* — **n** cijfer *n* — **e** cifra *f*; guarismo *m*; número *m* — **i** numero *m*

9459 nut — Block, usually of metal, perforated with a threaded hole to be screwed down on a bolt to hold together objects through which the bolt passes.

f écrou *m* — **d** Mutter *f*; Schraubenmutter *f* — **n** moer *fm* — **e** tuerca *f* — **i** dado *m*

9460 nut — See en quad.

9461 nut quad — See en quad.

9462 ñ with the tilde — See ñ.

9463 Nyblin effect — Photographic effect in which the image is reversed when a heavily superexposed layer, which is not sensitive for red light, is developed in red light for a long time. It is evident that this effect has relations to the *Herschel effect and it is even supposed to be identical with it.

f effet *m* Nyblin — **d** Nyblin-Effekt *m* — **n** Nyblin-effect *n* — **e** efecto *m* Nyblin — **i** effetto *m* Nyblin

9464 nylon — Long-chain synthetic polymeric amide which has recurring amide groups -COHN- as an integral part of the main polymer chain and which is capable of being formed into a filament whose structural elements are oriented in the direction of the axis.

f nylon *m* — **d** Nylon *n* — **n** nylon *n* — **e** nylon *m*; nilón *m* — **i** nailon *m*; nylon *m*

9465 nylon printing plate

f cliché *m* nylon — **d** Nylonklischee *n*; Auswaschreliefdruckstock *m* — **n** nylon cliché *n* — **e** clisé *m* de nilón — **i** cliscè *m* di nailon; cliché *m* di nailon

9466 oasis goat — Leather made from the goat-skin from the Cape of Good Hope, used for the binding of books.
f chèvre *f* du Cap — **d** Oasenziegenleder *n*
9467 obelisk — See dagger.
9468 objective — See lens.
9469 object language; target language — Language that is an output from a translation process; contrast with source language.
f langage *m* d'exécution — **d** Zielsprache *f* — **n** doeltaal *fm* — **e** lenguaje *m* absoluto — **i** linguaggio *m* oggettivo
9470 object program; target program — Program created by a computer program from a source program, where the source is a higher-level language and the object program is either in machine language or in a language or coding structure close thereto. Normally changes to the program require going back to the source documentation and coding.
f programme *m* résultant — **d** Maschinenprogramm *n*; Zielprogramm *n*; Objektprogramm *n* — **n** doelprogramma *n*; werkprogramma *n* — **e** programa *m* objeto — **i** programma *m* oggettivo
9471 oblique lighting
f éclairage *m* oblique — **d** schräge Beleuchtung *f* — **n** scheve belichting *f*; zijdelingse belichting *f* — **e** iluminación *f* oblicua — **i** illuminazione *f* obliqua
9472 oblique serifs
f empattements *mpl* obliques — **d** schräge Serife *mpl* — **n** schuine schreven *fmpl* — **e** remates *mpl* oblicuos; bases *fpl* oblicuas — **i** grazie *fpl* inclinate
9473 oblique stroke — See solidus.
9474 oblong double foolscap — A British size for writing and printing papers. See app. no. 3.
9475 oblong double small foolscap — A British standard size for writing and printing papers. See app. no. 3.
9476 oblong page
f page *f* oblongue; page *f* à l'italienne; page *f* en largeur — **d** Querformatseite *f*; Seite *f* im Querformat — **n** liggende pagina *f* — **e** página *f* apaisada — **i** pagina *f* oblunga
9477 oblong size — See landscape.
9478 occasional element — See incidental element.
9479 ocerite — See ozokerite.
9480 ocher — See ochre.
9481 ochre *GB*; **ocher** *US* — Naturally occurring dirty iron oxide pigments, varying in shade from yellow to brown to red, used in tinting buff papers and in the manufacture of fast-to-light inks. See iron oxide reds.
f ocre *m* — **d** Ocker *m* — **n** oker *m* — **e** ocre *m* — **i** ocra *f*
9482 OCR — See optical character recognition.

9483 octadecanoic acid — See stearic acid.
9484 octal — Number system with a base of eight. Data are represented by the numbers 0 to 7. Successive positions are given values of the consecutive powers of base eight. Thus octal 125 represents the decimal (denary) number 85 (i.e., $1 \times 8^2 + 2 \times 8^1 + 5 \times 8^0$). The same number in binary coded octal is represented by 001, 010, 101, and in pure binary by 1010101 which has the same sequence of digits, except that the two leading zeros have been suppressed.
f octal — **d** Oktal; zur Basis 8 — **n** octal — **e** octal — **i** ottale
9485 octavo; eights *fam.* — Book made from sheets which have been folded three times, each sheet forming eight leaves or sixteen pages. Abbrev. 8vo (no point). Sometimes applied to any book measuring about 15 x 22.5 cm (6 x 9 in).
f in-octavo; quaternion *m obs.* — **d** Oktavformat *n*; Oktav *n* — **n** octavo *n*; octavo-formaat *n*; in achten — **e** en octavo; folio *m* francés — **i** in ottavo; in ottavo comune; in formato dei classici
9486 octoate — Metal salt of 2-ethyl-hexoic acid.
f octoate *m* — **d** Oktoat *n* — **n** octoaat *n* — **e** octoato *m* — **i** ottoato *m*
9487 octodecimo — See eighteenmo.
9488 ocular — See eyepiece.
9489 odd and even parity — In an odd parity system the parity bit is "1" if the recorded number of "1"s in the information channels is even and "0" if the number of "1"s is odd. In the case of an even parity system the parity bit is "1" if the number of "1"s in the information channels is odd and "0" if the number of "1"s is even.
9490 odd/even bit — See parity bit.
9491 odd/even check — See parity check.
9492 odd folios — See odd pages.
9493 oddments — Parts of a book separate from the main text, such as contents, index. See also preliminary matter.
9494 oddments
See preliminary matter.
9495 odd pages; uneven pages; odd folios — Right-hand pages of a book, numbered 1, 3, 5, etc.
f belles pages *fpl*; pages *fpl* impaires — **d** ungerade Seiten *fpl*; rechte Seiten *fpl*; Schauseiten *fpl* — **n** rechtse pagina's *fmpl*; rechter pagina's *fmpl*; oneven pagina's *fmpl* — **e** páginas *fpl* impares; páginas *fpl* de la derecha; páginas *fpl* nones — **i** pagine *fpl* dispari; pagine *fpl* a destra
9496 odd volume
f volume *m* dépareillé — **d** Einzelband *m* (von mehreren) — **n** los deel *n*; enkel deel *n* — **e** volumen *m* suelto — **i** volume *m* scompagnato
9497 odor — See odour.
9498 odour *GB*; **odor** *US* — Solvent remaining in the dried printed ink which affects the sense of smell.

f odeur *f* — **d** Geruch *m* — **n** reuk *m*; lucht *fm* — **e** olor *m* — **i** odore *m*

9499 odour free; odourless
f inodore; sans odeur — **d** geruchlos; geruchfrei — **n** reukloos; reukeloos — **e** inodoro; sin olor — **i** inodore; senza odore

9500 odourless — See odour free.

9501 odourless inks
f encres *fpl* inodores; encres *fpl* sans odeur — **d** geruchlose Farben *fpl*; geruchfreie Farben *fpl* — **n** reukloze inkten *fpl* — **e** tintas *fpl* inodoras; tintas *fpl* sin olor — **i** inchiostri *mpl* inodori; inchiostri *mpl* senza odore

9502 oersted — Unit of magnetic field strength. Abbrev. Oe. In the SI: 1 Oe $\approx$ 79.5775 A/m. Name derived from Hans C. Oersted (1777-1851), Danish physicist.
f oersted *m* — **d** Oersted *m* — **n** oersted *m* — **e** oersted *m* — **i** oersted *m*

9503 offcut — Piece of the sheet of paper removed during processing, smaller than the size ordered but large enough for use other than for re-pulping.
f à-côté — **d** verwendbarer Abschnitt *m* — **n** nog bruikbaar afsnijsel *n* — **e** recortes *mpl* — **i** sotto-formato *m*

9504 offcuts — Pieces cut off from sheets to reduce them to proper size.
f chutes *fpl*; rognures *fpl* — **d** Papierabschnitte *mpl*; Beschnitt *m* — **n** rijfels *pl*; papierafsnijsels *npl*; papierafsnijdsels *npl* — **e** recortaduras *fpl* — **i** ritagli *mpl*

9505 office — Term sometimes used for a printing workshop.
f officine *f* — **d** Offizin *f* — **n** drukkerij *f*; werkplaats *fm* van de drukker; drukkerswerkplaats *fm* — **e** oficina *f*; despacho *m* — **i** officina *f*

9506 office manager — See supervisor.

9507 office work — See clerical work.

9508 official — A British standard size for postcards. See app. no. 3.

9509 official gazette — See gazette.

9510 official paper — See petition paper.

9511 off its feet — Type not standing squarely on its feet and therefore printing imperfectly.
f couché; caractères *mpl* couchés; caractères *mpl* taqués — **d** stürzender Satz *m*; liegender Satz *m*; abgefallene Buchstaben *mpl* — **n** omliggend zetsel *n* — **e** echado; caída; composición *f* echada; composición *f* inclinada; composición *f* ladeada; composición *f* tumbada — **i** composizione *f* coricata

9512 off-lay — See drive side.

9513 off-line — Performance of a function of an electronic device in such a way that its input and/or its output must be manually carried to or from the device. Example: OCR scanner producing paper tape output which in turn will be mounted onto a paper tape reader of a computer typesetting system. Line printer which reads a paper tape mounted on its paper tape reader and produces a printout of the same. Typesetter which runs from a paper tape or cassette produced from a keyboard (also an off-line device). (Opposite of an on-line device or system.)
f autonome; indirect; non connecté (à l'ordinateur) — **d** getrennt (vom übrigen System); unabhängig; off-line — **n** niet gekoppeld; gescheiden; afzonderlijk — **e** indirecto; fuero de línea — **i** fuori linea

9514 off-machine coated paper — See separately coated paper.

9515 off-machine coating — See conversion coating.

9516 offprint; separate — Separate printed copy or a small edition which was originally part of a larger publication.
f tiré *m* à part; tirage *m* à part — **d** Separatabdruck *m* — **n** overdruk *m* — **e** reimpresión *f* aparte; separata *f*; tirada *f* aparte — **i** stampa *f* a parte

9517 offset — See offset lithography.

9518 offset — See offsetting.

9519 offset blanket — See blanket.

9520 offset cartridge — Paper with a rough surface that can be printed on an offset press. Name derived from *cartridge paper which was used during the introduction of offset printing to demonstrate that the rough surface was not a disadvantage.

9521 offset deep printing
f procédé *m* offset creux; offset *m* en creux — **d** Offsettiefdruck *m*; Offsettiefverfahren *n* — **n** offsetdiepdruk *m*; offsetdiepprocédé *n* — **e** impresión *f* hueco-offset — **i** offset *m* inciso

9522 offset flat-bed press
f presse *f* offset à plat — **d** Flachoffsetpresse *f* — **n** vlakpers *f*; vlakoffsetpers *f* — **e** prensa *f* offset plana; máquina *f* offset plana — **i** macchina *f* offset con forma in piano

9523 offset ink — Short, tacky, water repellent ink made with strong, acid resistant pigments.
f encre *f* offset; encre *f* pour offset — **d** Offsetfarbe *f*; Offsetdruckfarbe *f* — **n** offsetinkt *m* — **e** tinta *f* offset — **i** inchiostro *m* offset; inchiostro *m* da stampa offset

9524 offset letterpress — See dry offset.

9525 offset lithography; offset — Indirect printing method in which the inked image on the pressplate is first printed onto a rubber blanket, then in turn offsets the inked impression on to the sheet of paper. Although it almost certainly began with Voirin in France, several claims have been made. But it is quite certain that the first branch of the trade to benefit was metal decoration. As early as 1875, R. Barday was responsible in England for converting a direct flat-bed machine into an offset machine by adding an extra cylinder. This machine took the impression from the stone on to the surface of the old impression cylinder covering, and then transferred it to the sheet of metal which was pressed against it by the new cylinder. This mixture of old and new, flat-bed offset and direct rotary, continued till 1904 when George Mann & Company produced a rotary offset tin printing machine which was simply a direct rotary machine with an additional cylinder. Ira Rubel was at this time designing a rotary offset machine for paper printing in the US, so that the turn of the century was a period of intensive activity and interest.

f offset *m*; impression *f* offset; rotocalcographie *f*; rotocalco *m* — **d** Offsetdruck *m*; Flachdruck *m* — **n** offsetdruk *m*; offsetdruktechniek *f*; vlakdruk *m* — **e** procedimiento *m* offset; lito-offset *m*; lit-offset *m*; rotocalco *m* — **i** stampa *f* offset

9526 offset machine — See offset press.

9527 offset-machine minder — See pressman.

9528 offset paper — Paper suitable for offset lithography.

f papier *m* offset; papier *m* pour offset; papier *m* pour impression (en) offset — **d** Offsetpapier *n* — **n** offsetpapier *n* — **e** papel *m* offset; papel *m* para offset — **i** carta *f* offset; carta *f* da stampa offset; carta *f* per offset

9529 offset perfecting press — See offset perfector.

9530 offset perfector; offset perfecting press

f machine *f* offset à retiration — **d** Schön- und Widerdruck-Offsetmaschine *f* — **n** schoon- en weerdruk-offsetpers *f* — **e** máquina *f* de retiración offset; máquina *f* de reacción offset; máquina *f* de doble impresión offset — **i** macchina *f* offset in bianca e volta

9531 offset plate; printing plate; press plate; plate — Plate mostly made of metal but also of paper or plastic, used for offset lithography printing.

f plaque *f* offset — **d** Offsetplatte *f*; Offsetdruckplatte *f*; Druckplatte *f*; Platte *f*; Offsetblech *n* — **n** offsetplaat *fm*; offsetdrukplaat *fm*; drukplaat *fm*; persplaat *fm* — **e** plancha *f* offset; plancha *f* litográfica — **i** lastra *f* offset

9532 offset press; offset machine — Press which consists of a plate cylinder, a transfer or blanket cylinder and a printing cylinder. The inked image is transferred from the plate on to the rubber blanket and therefrom on the paper held on the printing cylinder.

f presse *f* offset; presse *f* rotocalco; machine *f* offset; machine *f* rotocalcographique; rotocalco *f* — **d** Offsetmaschine *f*; Offsetpresse *f*; Offsetdruckmaschine *f* — **n** offsetpers *f*; offsetmachine *f* — **e** máquina *f* offset *Es*; prensa *f* offset — **i** macchina *f* da stampa offset; macchina *f* per la stampa offset

9533 offset press — See smoothing press.

9534 offset pressman — See pressman.

9535 offset printer — See pressman.

9536 offset printing

f impression *f* offset; impression *f* rotocalcographique — **d** Offsetdruck *m*; Offsetdruckverfahren *n* — **n** offsetdruk *m*; offsetprocédé *n* — **e** impresión *f* offset; procedimiento *m* offset; impresión *f* litográfica — **i** stampa *f* offset

9537 offset proof press — See proof press.

9538 offset rotary press — See rotary offset press.

9539 offset spray; non-offset spray — Substance such as paraffin sprayed on freshly printed sheets to prevent offsetting of the ink.

f poudre *f* anti-maculage; liquide *m* anti-maculage — **d** Druckbestäubungspuder *m*; Druckbestäubungsflüssigkeit *f*; Druckbestäubungsmittel *n* — **n** anti-smetpoeder *n*; anti-smetvloeistof *fm*; anti-smetmiddel *n* — **e** rocío *m* anti-maculador — **i** polvere *f* antiscartino; liquido *m* antiscartino; prodotto *m* antiscartino

9540 offsetting; offset *US* — Transference or smearing of ink from freshly printed matter to another surface with which the undried prints come into contact. See also set *v* off.

f maculage *m* — **d** Abliegen *n*; Abschmieren *n*; Abschmutzen *n* — **n** overzetten *n* — **e** repintado *m*; repinte *m*; maculado *m* — **i** controstampa *f*; maculatura *f*

9541 offset zinc plate

f plaque *f* de zinc offset; zinc *m* de machine offset — **d** Offsetzinkplatte *f* — **n** offsetzinkplaat *fm*; zinkplaat *fm*; zinken offsetplaat *fm* — **e** plancha *f* de zinc offset; plancha *f* de cinc offset — **i** lastra *f* offset di zinco

9542 offshade — Adjective applying to a paper or board the colour of which does not conform to an agreed specimen.

f écart *m* de teinte — **d** Tonabweichung *f*; Farbtonabweichung *f* — **n** kleurafwijking *f* — **e** diferencia *f* de tinte; diferencia *f* de matíz — **i** fuoritinta

9543 offsheets — See slip sheets.

9544 offside — See drive side.

9545 off the strip — Condition on a letterpress rotary when a printing plate has moved round the cylinder away from the margin bar.

f glissement *m* du cliché; cliché *m* qui a glissé — **d** Plattenwandern *n*; Plattenverschiebung *f* — **n** plaatverschuiving *f*; verschuiving *f* van de plaat — **e** desplazamiento *m* del clisé; clisé *m* desplazado — **i** slittamento *m* del cliscè; scivolamento *m* del cliché

9546 OH-group — See hydroxyl group.

9547 ohm — Unit of electrical resistance. Symbol: Ω. In the SI: 1 international ohm is the resistance of a column of quicksilver with a mass of 14.4521 g, a length of 106.300 cm and a constant cross section at the temperature of melting ice. So: 1 $\Omega_{int} \approx 1.00049\ \Omega$. Name derived from G.S. Ohm (1787-1854), German physicist.

f ohm *m* — **d** Ohm *n* — **n** ohm *nm* — **e** ohmio *m*; ohm *m* — **i** ohm *m*

9548 oil *v*; **lubricate** *v*

f huiler; lubrifier — **d** ölen — **n** oliën; smeren met olie — **e** aceitar; lubricar; lubrificar — **i** lubrificare

9549 oil *v* — See lubricate *v*.

9550 oil absorption — Quantity of oil required to wet completely a definite weight of a given pigment.

f absorption *f* d'huile — **d** Ölabsorption *f*; Ölaufnahme *f* — **n** olieabsorptie *f* — **e** absorción *f* del aceite — **i** assorbimento *m* d'olio; presa *f* d'olio

(dei pigmenti)

9551 oil absorption test; oil penetration test —
Test in which the rate of absorption of a small
drop of oil into paper is determinded, as an indication of the printing ink absorbency.
f essai *m* d'absorption de l'huile — **d** Ölabsorptionsprobe *f* — **n** olieabsorptieproef *fm*; olieabsorptiebepaling *f* — **e** ensayo *m* de absorción
de aceite — **i** prova *f* dell'assorbimento d'olio

9552 oil can — Can with a long spout through
which oil is poured or squirted to lubricate
machinery or the like.
f bidon *m* d'huile; burette *f* — **d** Ölkanne *f*;
Öler *m*; Ölkännchen *n* -- **n** oliespuit *fm*; oliekan
fm; oliekannetje *n* — **e** aceitera *f*; bidón *m* de
aceite — **i** oliatore *m*

9553 oiled board — Board treated with oil for use
as stencils, dividers, etc.
f carton *m* huilé — **d** Ölkarton *m* — **n** geolied
karton *n* — **e** cartulina *f* aceitada — **i** cartone *m*
oleato

9554 oiled paper — Wrapping paper treated with
mineral oil to increase the protective features.
f papier *m* huilé — **d** Ölpapier *n* — **n** geolied
papier *n*; oliepapier *n* — **e** papel *m* aceitado;
papel *m* encerado — **i** carta *f* oleata

9555 oil groove
f gorge *f* de graissage; patte *f* d'araignée — **d**
Schmiernut *f*; Ölnut *f* — **n** smeergroef *fm* — **e**
ranura *f* de engrase; pata *f* de araña — **i**
scanalatura *f* per lubrificazione

9556 oil hole — See lubricating point.

9557 oiling — See also greasing.
f lubrification *f* — **d** Ölen *n* — **n** oliën *n*;
smeren *n*; smering *f* — **e** lubricación *f*;
lubricación *f* — **i** lubrificazione *f*

9558 oil, lube — See lubricating oil.

9559 oil, lubricating — See lubricating oil.

9560 oil, machine — See lubricating oil.

9561 oil of turpentine — See turpentine oil.

9562 oil of vitriol — See sulphuric acid.

9563 oil penetration test — See oil absorption
test.

9564 oil permeability
f perméabilité *f* à l'huile — **d** Öldurchlässigkeit *f*
— **n** oliedoorlatendheid *f* — **e** permeabilidad *f* al
aceite — **i** permeabilità *f* all'olio

9565 oilproof paper — See greaseproof paper.

9566 oil repellent
f refusant l'huile — **d** ölabweisend; ölabstoßend —
n olieafstotend — **e** repelente al aceite — **i**
repellente all'olio; oleorepellente

9567 oil resistance — Essential characteristic of
paper used to wrap greasy or fatty food products;
determined by measuring the rate of penetration
of a mineral or vegetable oil through the paper.
f résistance *f* à l'huile — **d** Ölechtheit *f*; Ölbeständigkeit *f* — **n** oliebestendigheid *f* — **e**
resistencia *f* al aceite — **i** resistenza *f* all'olio

9568 oilstone — Small rectangular slab of fine
grained stone to sharpen finisher's tools.
f pierre *f* à huile — **d** Ölstein *m*; Wetzstein *m* —

n oliesteen *m*; wetsteen *m* — **e** piedra *f* aceitada
para afilar — **i** pietra *f* a olio

9569 oiticica oil — Light yellow drying oil from
the seeds of the oiticica tree of Brasil, used in
printing inks. Similar to China wood oil.
f huile *f* d'oiticica — **d** Oitikikaöl *n* — **n** oiticica-
olie *fm* — **e** aceite *m* de oiticica — **i** olio *m*
d'oiticica

9570 -ol — Suffix used in the names of chemical
derivatives, representing alcohol (glycerol, phenol,
naphthol) or sometimes phenol or less definitely
assignable phenol derivatives (resorcinol).

9571 old binding — In France to be understood as
anterior to 1800.
f reliure *f* ancienne — **d** alter Einband *m* — **n**
oude band *m*; oude boekband *m*; antieke band *m*
— **e** encuadernación *f* antigua; vieja encuadernación *f* — **i** legatura *f* antica

9572 old face *GB*; **old style** *US* — Type face
based on originals of the period 1500-1700, characterized by bracketed serifs and oblique stress in
the curved strokes, e.g., Caslon. On account of the
modern *classification of type design use of these
terms is deprecated. See Garaldics, app. no. 2.

9573 old papers
f vieux papiers *mpl* — **d** Altpapier *n* — **n** oud
papier *n* — **e** recortes *mpl* — **i** carta *f* vecchia;
cartaccia *f*; carta *f* di recupero

9574 old rag — A size of card. See app. no. 3.

9575 old rags — See rags.

9576 old shoe — See hellbox.

9577 old style — See old face.

9578 old-style figures — See hanging figures.

9579 oleic acid; oleinic acid — Yellowish, oily
liquid, darkening on exposure to air, soluble in
alcohol, ether and most organic solvents, fixed
and volatile oils, fatty acids and oil-soluble materials, insoluble in water.
Formula: $CH_3(CH_2)_7CH:CH(CH_2)_7COOH$.
f acide *m* oléique — **d** Ölsäure *f*; Oleinsäure *f*;
Oktadezensäure *f* — **n** oliezuur *n*; oleïne *fm*
(technisch); oleïen *fm* (technisch) — **e** ácido *m*
oleico — **i** acido *m* oleico

9580 oleinic acid — See oleic acid.

9581 oleography — Lithography which simulates
oil painting. The finished print is mounted on
canvas, sized and varnished, and the irregularities
of oil painting and canvas are reproduced by embossing.
f oléographie *f* — **d** Oleographie *f* — **n** oleografie *f* — **e** oleografía *f* — **i** oleografia *f*

9582 oleophilic — Wettable by greasy substances.
f oléophile; qui accepte l'huile; recevant les corps
gras — **d** fettempfänglich — **n** olieaannemend; vet-
aannemend; oleofiel — **i** oliofilia; affine al grasso

9583 oleophobic
f oléophobe — **d** fettabstoßend; oleophob — **n** olieafstotend; oleofoob — **e** oleofobo — **i** oleofobo

9584 omission; out — Typographical error where
a word, sentence or paragraph has been dropped
from the copy.
f bourdon *m* — **d** Auslassung *f*; Weglassung *f*;

Schusterjunge *m*; Schuster *m*; Leiche *f sl.* — **n** weglating *f*; wegvalling *f* — **e** bordón *m*; omisión *f*; olvidado *m*; manchego *m fam.*; mochuelo *m fam.*; letra *f* omitida; salto *m* — **i** omissione *f*; parola *f* omessa

9585 omit *v* — To leave out words or sentences from manuscripts.
f omettre; passer — **d** weglassen; auslassen — **n** weglaten — **e** omitir — **i** omettere

9586 omniplexed fount — See duplexed fount.

9587 one-bath etching — See single-bath etching.

9588 one-bite etching — See powderless etching.

9589 one-shot lubrication — See centralized lubrication.

9590 one-side coated paper
f papier *m* couché une face; papier *m* couché un côté — **d** einseitig gestrichenes Papier *n* — **n** eenzijdig gestreken papier *n*; chromopapier *n* — **e** papel *m* estucado de una cara; papel *m* estucado por una cara; papel *m* esmaltado de una cara — **i** carta *f* patinata da un lato; carta *f* patinata su una superficie; carta *f* patinata su una faccia; carta *f* monopatinata

9591 one-side coloured paper
f papier *m* coloré une face — **d** einseitig gefärbtes Papier *n* — **n** eenzijdig gekleurd papier *n* — **e** papel *m* coloreado por una cara — **i** carta *f* monocolorata

9592 one-side printed paper — See blind sheet.

9593 one-stage etching — See powderless etching.

9594 one-time carbon paper — Carbon paper that will lose its copying layer after having been used one time.
f papier *m* carbone une fois — **d** Einmalkohlepapier *n* — **n** eenmalig carbonpapier *n* — **e** papel *m* carbón para (usar) una vez; papel *m* carbón para uso una vez — **i** carta *f* carbone per una copia

9595 one-way screen — See half-line screen.

9596 one-way tint; straight-line tint — Halftone tint consisting of parallel lines of uniform width, usually produced by photography through a standard halftone screen with a slit stop in the camera lens. A one-way halftone in a plate in which the tones are translated by means of a single series of parallel lines.
f cliché *m* tramé en lignes — **d** Klischee *n* mit Linienraster — **n** cliché *n* met lijnraster — **e** grisado *m* rayado; rayado *m* — **i** cliscè *m* a retino lineare; cliché *m* a retino lineare

9597 onion skin paper — Thin glazed or unglazed, lightweight bond-type paper for copies, made from rags or sulphite pulp.
f papier *m* pelure; coquille *f* pelure; pelure *f* d'oignon; pelure *f* d'ognon — **d** Florpost *f*; Pelürepapier *n*; Zwiebelschalenpapier *n* — **n** pelure *fm*; floorpost *n* — **e** papel *m* cebolla — **i** carta *f* pelle d'aglio

9598 onlay — See inlay.

9599 on-line — Pertaining to equipment used for text processing or typesetting which accept or pass on information as electronic signals without recourse to an intervening medium, such as paper tape. Computer peripherals (line printers for example) are on-line to the computer. A typesetting machine can also operate on-line to a computer so that the text which is generated or formatted by the computer need not be recorded on a paper or magnetic tape which in turn must be inserted into a tape reader associated with the typesetter.
f connecté à l'ordinateur; direct — **d** angeschlossen an das übrige System; on-line — **n** gekoppeld — **e** directo — **i** in linea

9600 on-machine coating
f couchage *m* sur machine — **d** Streichen *n* in der Maschine — **n** strijken *n* op de machine — **e** estucado *m* en máquina — **i** patinatura *f* in macchina

9601 on-the-fly printer, hit- — See hit-on-the-fly printer.

9602 on the slow — Running a rotary press at a very slow speed.
f marche *f* lente; au ralenti — **d** Andruckgeschwindigkeit *f* — **n** langzaam draaien — **e** marcha *f* lenta; andar lento — **i** marcia *f* lenta

9603 opacimeter; opacity tester; opacity indicator — Instrument to measure the degree of opacity of paper.
f opacimètre *m* — **d** Opazitätsmeßgerät — **n** opaciteitsmeter *m* — **e** opacómetro *m*; metro *m* de opacidad; medidor *m* de opacidad — **i** opacimetro *m*; misuratore *m* dell'opacità

9604 opacity — Property of paper which prevents dark objects on or in contact with the back side of sheet from being seen through it. It is decreased by the penetration of oil from the printing ink, with the amount of calendering or with increased finish. It is increased with the weight and with the bulk of the sheet, and by using opaque pigments, like titanium dioxide and calcium carbonate as fillers. See also opacity (paper backing), opacity (white backing), Tappi opacity, show-through.
f opacité *f* — **d** Opazität *f*; Lichtundurchlässigkeit *f* — **n** opaciteit *f*; ondoorschijnendheid *f* — **e** opacidad *f* del papel — **i** opacità *f*

9605 opacity indicator — See opacimeter.

9606 opacity (paper backing) — Ratio expressed as a percentage of the amount of light reflected from a single sheet of paper with a black backing to the amount of light reflected from the same sheet of paper backed by an effectively opaque pile (of sufficient thickness that the addition of more sheets does not affect the reading obtained) of the same paper.
f opacité *f* sur fond papier — **d** Opazität *f* gegen Papierunterlage; Lichtundurchlässigkeit *f* gegen Papierunterlage — **n** opaciteit *f* tegen papier — **e** opacidad *f* al fondo papel — **i** opacità *f* su fondo carta

9607 opacity tester — See opacimeter.

9608 opacity (white backing) — Ratio expressed as a percentage of the amount of light reflected

from a single sheet of paper with a black backing to the amount of light reflected by the same sheet of paper backed by a standard white.
f opacité *f* sur un fond blanc — **d** Opazität *f* gegen weiße Unterlage; Lichtundurchlässigkeit *f* gegen weiße Unterlage — **n** opaciteit *f* tegen witte ondergrond — **e** opacidad *f* al fondo blanco — **i** opacità *f* su fondo bianco
9609 opal glass — Translucent or opaque glass usually of a milky white hue for light diffusion. A sheet of transparent glass carrying a thin coating of white opaque glass is called flashed opaque glass (In German: Überfangglas). Glass which is white coloured right through is called pot opal.
f verre *m* opalin — **d** Opalglas *n*; Milchglas *n*; Überfangglas *n* — **n** opaalglas *n*; melkglas *n* — **e** vidrio *m* opalino — **i** vetro *m* opalino
9610 opaline — Very dense white sheet of paper about 0.38 mm thick and with an exceptionally clear formation, used in the manufacture of lamp shades.
f opaline *f* — **d** Opalinekarton *m* — **n** opalinekarton *n*; opaline *fmn* — **e** cartulina *f* opalina — **i** carta *f* opalina
9611 opaque — Not transmitting light.
f opaque — **d** undurchsichtig; deckend; opak — **n** ondoorschijnend; opaak — **e** opaco — **i** opaco
9612 opaque; opaquing medium — Water-soluble reddish paint that dries quickly and has good coverage, to paint out defects or areas in a film where no light transmission is desired.
f encre *f* opacifiante; opacifiant *m* — **d** Deckfarbe *f*; Abdeckfarbe *f* — **n** uitdekverf *fm*; dekverf *fm* — **e** tinta *f* de cubrir; tinta *f* de opacar; laca *f* para opacar; opaco *m*; tinta *f* opaca — **i** inchiostro *m* coprente
9613 opaque *v*; **block** *v* **out** — To make an area opaque, i.e., unable to permit light to pass through, as of negatives and transparencies.
f boucher; pocher — **d** abdecken; ausdecken — **n** afdekken; uitdekken — **e** opacar — **i** coprire; opacizzare
9614 opaque ink — Ink that prevents transmission of light and has good hiding power. It does not permit the paper or previous printing to show through.
f encre *f* opaque; encre *f* couvrante — **d** Deckfarbe *f*; undurchsichtige Farbe *f* — **n** dekinkt *m*; dekkende inkt *m* — **e** tinta *f* opaca; tinta *f* cubriente — **i** inchiostro *m* opaco
9615 opaque paper — Paper which, because of its ingredients or the method used in its manufacture, has more opacity for a given basis weight, colour, etc.
f papier *m* opaque — **d** undurchsichtiges Papier *n*; opakes Papier *n* — **n** ondoorschijnend papier *n*; opaak papier *n* — **e** papel *m* opaco; papel *m* no transparente; papel *m* adiáfeno — **i** carta *f* opaca
9616 opaquing medium — See opaque.
9617 open arc lamp — Arc lamp in which the electrodes burn in free air.

f lampe *f* à arc libre — **d** Bogenlampe *f* mit offenem Lichtbogen — **n** open booglamp *fm* — **e** lámpara *f* de arco abierto — **i** lampada *f* ad arco libero
9618 open back — See hollow back.
9619 open ended — Computer term. Pertaining to a process or system that can be augmented.
f extensible — **d** dehnbar — **n** elastisch — **e** ampliable — **i** espansibile; estendibile
9620 open-end wrench — See also wrench.
f clé *f* plate — **d** Stechschlüssel *m*; Maulschlüssel *m*; Schlüssel *m* — **n** steeksleutel *m*; sleutel *m* — **e** llave *f* fija — **i** chiave *f* a boca
9621 open-faced types — See outline letters.
9622 open fly — Delivery on a rotary press for hand take off only.
f sortie *f* à la main — **d** offener Exemplarausgang *m* — **n** open uitleg *m* — **e** salida *f* manual — **i** uscita *f* a mano
9623 open leaders — Periods used in setting tables or lines to lead the eye, spaced one dot to an em. See also leaders, close leaders.
f points *mpl* de conduite; gros points *mpl* — **d** Führungspunkte *mpl* auf Geviert; Geviertführungspunkte *mpl* — **n** blokpunten *mpl* één op vierkant — **e** puntos *mpl* conductores; puntos *mpl* de conducción; puntos *mpl* suspensivos; guiones *mpl* conductores; guiones *mpl* de conducción; guiones *mpl* suspensivos — **i** punti *mpl* di conduzione
9624 open letters — See outline letters.
9625 open matter — Matter widely leaded or containing many short lines.
f matière *f* blanchie; matière *f* espacée — **d** splendider Satz *m* — **n** open zetsel *n*; wijd zetsel *n*; gemakkelijk zetsel *n* — **e** composición *f* de ancho interlineado; composición *f* quebrada; composición *f* abierta; composición *f* espaciado ancho; forma *f* amplia — **i** composizione *f* abbondantemente interlineata
9626 open press — Blanket-to-steel press.
f presse *f* à trois cylindres (plaque, blanchet, contre-pression) — **d** Dreizylindermaschine *f* (Platte, Gummituch, Gegendruck) — **n** driecilindermachine *f* (plaat, rubberdoek, tegendruk) — **e** máquina *f* de tres cilindros (plancha, mantilla, impresión) — **i** macchina *f* a tre cilindri (lastra, blanket, contropressione)
9627 open *v* **up** — To render less opaque so that more light may pass through areas intended to appear more open.
f éclaircir — **d** aufhellen — **n** lichter maken; oplichten; afzwakken — **e** aclarar — **i** schiarire
9628 operand — Quantity upon which a mathematical operation is performed.
f opérande *f* — **d** Operand *m* — **n** operand *m*; opdrachtelement *n* — **e** operando *m* — **i** operando *m*
9629 operating characteristic
f courbe *f* d'efficacité — **d** Operations-Charakteristik *f*; Annahmekennlinie *f*; Prüfkennlinie *f* — **n** keuringskarakteristiek *f* — **e** curva *f*

de eficacia — **i** caratteristica *f* operativa

9630 operating side; nearside; laying-on side *depr.*; **feeder side** — Side of the press on which most of the controls are located. Opposite of *drive side.

f côté *m* de conduite; côté *m* opérateur; côté *m* de manœuvre — **d** Bedienungsseite *f* — **n** bedieningszijde *f* — **e** lado *m* de guía; lado *m* de atender; costado *m* anterior — **i** lato *m* operatore; lato *m* di servizio; lato *m* d'avviamento

9631 operating system — Program or series of programs which control the manner in which work is processed through a computer. A basic operating system brings into a core the program desired to perform the next task when it ascertains that that program is required. This avoids the necessity for bootstrap loading of each separate program. A more sophisticated operating system will also handle all computer input and output, and may even permit different jobs to be run in the computer apparently simultaneously (multi-programming) or may even permit multiple users to access the same computer concurrently, running different programs (time sharing). See also supervisor.

f système *m* d'exploitation — **d** Betriebssystem *n* — **n** hoofdbesturingssysteem *n* — **e** sistema *m* operativo — **i** sistema *m* operativo

9632 operating time — Interval between the time a circuit is seized manually or automatically and the time that actual transmission of the message begins. It is variable dependent upon type of service, type of selection and number of stations on a circuit.

f temps *m* de fonctionnement — **d** Betriebszeit *f* — **n** bedrijfstijd *m* — **e** tiempo *m* de funcionamiento — **i** tempo *m* di funzionamento

9633 operation — Condition of a machine when carrying out that for which it was built.

f fonctionnement *m* — **d** Betrieb *m* — **n** bedrijf *n*; werking *f* — **e** funcionamiento *m* — **i** funzionamento *m*

9634 operation — In general, a defined action. As a computer term, the deviation of an item or items, called the result, from one or more given items called operands, according to defined rules which specify the result for any permissible combination of values for the operands.

f opération *f* — **d** Operation *f* — **n** bewerking *f* — **e** operación *f* — **i** operazione *f*

9635 operational code — See function code.

9636 operational research; operations research — Application of mathematical methods to the solution of management problems.

f préparation *f* scientifique des décisions; recherche *f* opérationnelle — **e** preparación *f* científica de las decisiones — **i** ricerca *f* operativa

9637 operations research — See operational research.

9638 operator — Man in charge of the operation of a machine.

f conducteur *m*; opérateur *m*; ouvrier *m* — **d** Maschinenmeister *m* — **n** bedienaar *m* — **e** operador *m*; manipulador *m* — **i** operatore *m*

9639 operator; keyboard operator; machine compositor — Person who sets copy at the keyboard in one of the mechanical typesetting processes.

f claviste *m*; opérateur *m*; linotypiste *m*; monotypiste *m* — **d** Maschinensetzer *m* — **n** machinezetter *m* — **e** teclista *f* tipográfico; linotipista *m*; monotipista *m* — **i** tastierista *m*; compositore *m* a macchina

9640 operator — Computer term, used in the description of a process, that which indicates the action to be performed on operands.

f opérateur *m* — **d** Operationssymbol *n* — **n** bewerkingsteken *n*; operator *m* — **e** operador *m*; signo *m* de operación — **i** operatore *m*

9641 opposite page — In double page spreads, the odd folio page.

f page *f* en regard — **d** gegenüberliegende Seite *f*; Gegenseite *f* — **n** tegenoverliggende pagina *fm* — **e** página *f* opuesta — **i** pagina *f* opposta

9642 opposition paper

f feuille *f* d'opposition — **d** Oppositionsblatt *n* — **n** oppositieblad *n* — **e** periódico *m* rival; periódico *m* competencia — **i** giornale *m* d'opposizione

9643 optical aberration — Failure of an optical system to form an image of a point as a point, of a straight line as a straight line, and of an angle as an equal angle.

f aberration *f* optique — **d** optische Aberration *f*; optischer Abbildungsfehler *m*; optische Verzerrung *f*; optische Abweichung *f* — **n** optische aberratie *f*; optische afwijking *f* — **e** aberración *f* óptica — **i** aberrazione *f* ottica

9644 optical axis — Line through the foci and the vertices of the optical surfaces.

f axe *m* optique — **d** optische Achse *f* — **n** optische as *fm* — **e** eje *m* óptico — **i** asse *m* ottico

9645 optical bleaching — Incorporation in the paper of an almost colourless substance that can convert ultra-violet into visible light.

f blanchiment *m* optique — **d** optische Aufhellung *f* — **n** optische witmaking *f*; optisch witmaken *n* — **e** blanqueo *m* óptico — **i** sbianca *f* ottica

9646 optical bleaching agent; optical whitening agent; optical brightener; optical dye; brightening agent; fluorescent whitening agent — Bleaching agent that is able to absorb invisible violet rays and remit them as visible blue rays. Generally derivations of coumarin, diaminostilbenzene, benzimidazol, diaminobenzol, diaminophenyl-urea, diphenylimidazolan, phenylbenzothiazol, triaminophenyl ether, etc. Added to the pulp in the beater. They differ from chemical bleaching agents.

f agent *m* de blanchiment optique; agent *m* optique de blanchiment; agent *m* d'azurage optique; blanchissant *m* optique; azureur *m*; azurant *m* optique; éclaircisseur *m* optique; blancophore *m* — **d** optisches Bleichmittel *n*; optischer Aufheller *m*; Weißtöner *m*; Weißmacher m; Fluoreszenzfarbstoff m — **n** optisch bleekmiddel n; optische witmaker m — **e** agente m de azulamiento óptico; agente m de blanqueo óptico; blanqueante m óptico; blanqueador m óptico — **i** sbiancante m ottico; azzurrante m ottico

9647 optical brightener — See optical bleaching agent.

9648 optical centre; optic centre — Point about 10% above the exact centre of a printed page, but appearing central to normal vision.
f centre m optique — **d** optische Mitte f; optischer Mittelpunkt m — **n** optisch midden n; optisch middelpunt n — **e** centro m óptico — **i** centro m ottico

9649 optical character reader — Device for automatically reading written characters by means of an optical process.
f lecteur m optique de caractères — **d** optisches Lesegerät n; optisches Schriftlesegerät n — **n** optische schriftlezer m; optische lezer m; optisch leesapparaat n — **e** lectora f óptica de caracteres — **i** lettore m ottico dei caratteri

9650 optical character recognition; OCR — Recognition of printed or typed character by means of an optical reading head which senses the light reflected from the printed image and provides impulses to recognition circuits which identify the character.
f lecture f optique de caractères; reconnaissance f optique du caractère — **d** optisches Lesen n; optische Zeichenerkennung f — **n** optisch lezen n; optisch schriftlezen n; optische letterherkenning f — **e** lectura f óptica de caracteres; reconocimiento m óptico de caracteres — **i** lettura f ottica dei caratteri; ricognizione f ottica dei caratteri; riconoscimento m dei caratteri ottici

9651 optical character scanner — Device to recognize a type-written or printed character and translate it into a form suitable for electronic computer input.
f lecteur m de caractères en clair — **d** Klarschriftleser m — **n** optische letteraftastinrichting f — **e** lector m óptico de caracteres — **i** analizzatore m ottico dei caratteri; lettore m ottico dei caratteri

9652 optical contact — Condition of two surfaces in such intimate contact that all intervening air is excluded.
f contact m optique — **d** optischer Kontakt m — **n** optisch contact n — **e** contacto m absoluto; contacto m hermético — **i** contatto m ottico

9653 optical density; density — Degree of darkening of the developed photographic image, or of the ink coverage in printing. Density values are logarithmic in scale so that a density of 0.00 is completely transparent, while a density of 0.30 transmits one-half the incident light. Similarly a density of 1.00 transmits only one-tenth, and a density of 2.00 transmits only one-hundredth of the incident light.
f densité f optique; densité f du noir; noircissement m — **d** optische Dichte f;

Schwärzung f — **n** optische dichtheid f; zwarting f; densiteit f — **e** densidad f óptica — **i** densità f ottica

9654 optical depth gauge
f appareil m de mesure optique de profondeur — **d** optisches Tiefenmeßgerät n — **n** optisch dieptemeetinstrument n — **e** medidor m óptico de hondura — **i** apparecchio m per la misura ottica della profondità

9655 optical distortion
f distorsion f optique — **d** optische Verzeichnung f; optische Verzerrung f — **n** optische vertekening f — **e** distorsión f óptica — **i** distorsione f ottica

9656 optical dye — See optical bleaching agent.

9657 optical flat — Colour filter consisting of a sheet of dyed gelatine cemented between two sheets of optically surfaced glass.
f filtre m coulé sur glace — **d** optisch flacher Farbfilter m — **n** optisch kleurfilter n — **e** superficie f planoparalela de precisión — **i** filtro m colorato ottico

9658 optically spaced
f approches fpl rectifiées — **d** optisch ausgeglichen — **n** optisch gespatieerd — **e** espaciado m óptico — **i** spaziato otticamente; avvicinato m otticamente

9659 optical reversal — Lateral inversion of camera images with optical reversing devices.
f renversement m optique; retournement m optique — **d** optische Umkehrung f — **n** optische omkering f — **e** inversión f lateral óptica — **i** inversione f ottica

9660 optical scanner; visual scanner — Device that scans optically and usually generates an analogue or digital signal. Also a device that optically scans printed or written data and generates their digital representation.
f explorateur m optique — **d** optischer Abtaster m — **n** optische aftaster m — **e** explorador m óptico — **i** dispositivo m ottico d'esplorazione

9661 optical sensitizer; chromatic sensitizer — Dye for colour sensitization.
f sensibilisateur m chromatique — **d** chromatischer Sensibilisator m — **n** kleursensibilisator m — **e** sensibilizador m óptico — **i** sensibilizzatore m ottico; sensibilizzatore m cromatico

9662 optical sensitizing — Rendering photographic emulsions colour sensitive by addition of sensitizing dyes.
f sensibilisation f chromatique — **d** optisch empfindlich machen — **n** sensibiliseren n; kleurgevoelig maken n — **e** sensibilizado m óptico; sensibilizado m para determinar colores — **i** sensibilizzazione f ottica; sensibilizzazione f cromatica

9663 optical whitening agent — See optical bleaching agent.

9664 optic centre — See optical centre.

9665 optics — Branch of physical science relating to the nature and properties of light, its modification by opaque and transparent substances, and the laws of vision.

f optique *f* — **d** Optik *f* — **n** optica *f* — **e** óptica *f* — **i** ottica *f*

9666 OR — Logic operator in which the proposition is advanced that either (or perhaps all) of two or more propositions must be present for the statement to be effective. There are two forms: "inclusive or" requires that either A or B, or both A and B must obtain, whereas "exclusive or" requires that either A or B, but not A and B, must obtain. See also gate.

f organe *m* OU — **d** Oder-Glied *n*; logischer Operator *m* OR — **n** OR-element *n*; OR-poort *fm* — **e** órgano *m* OR — **i** elemento *m* di somma logica; elemento *m* OR; elemento *m* O

9667 orange cadmium — See cadmium sulphide.

9668 orange filter

f filtre *m* orange; écran *m* orange — **d** Orangefilter *n* — **n** oranje-filter *n* — **e** filtro *m* anaranjado — **i** filtro *m* arancio

9669 orange mineral — See red lead.

9670 orange peel — Mottle that occurs in flexographic, gravure and tin printing.

f peau *f* d'orange — **d** Orangenschaleneffekt *m*; Apfelsinenschalenhaut *f* — **i** marezzatura *f*; buccia *f* d'arancia

9671 orchil; archil; orseille — Natural occurring matter, obtained from lichens, containing orcin $C_6H_3(CH_3)(OH)_2$ which is white, crystalline prisms, soluble in water, alcohol, and ether.

f orseille *f* — **d** Orseillenfarbe *f*; Persio *n* — **n** orseille *n* — **e** orchilla *f*; orcina *f* — **i** orcina *f*

9672 order

f commande *f*; ordre *m*; commission *f* — **d** Auftrag *m*; Bestellung *f* — **n** bestelling *f*; opdracht *fm*; order *m* — **e** pedido *m*; encargo *m*; orden *f* — **i** ordinazione *f*; commissione *f*

9673 order *v* — To place an order.

f passer une commande; commander — **d** eine Bestellung aufgeben; bestellen — **n** een opdracht geven; een order opgeven; een bestelling opgeven; bestellen — **e** pedir; encargar — **i** ordinare; commissionare

9674 order department

f service *m* des commandes — **d** Auftragsabteilung *f* — **n** orderafdeling *f* — **e** departamento *m* de pedidos; sección *f* de encargos — **i** reparto *m* ordinazioni; sezione *f* ordinazioni

9675 ordinary terminals — Terminals of a letter as in J, j, c, a, t. The curved lines at the beginning and ending of italics are called pothooks. The thick ends at the stems of the lower case letters f, j, y are called tail dots.

f extrémités *fpl* normales — **n** gewone uiteinden *npl* — **i** terminali *mpl* comuni

9676 ordinate — Number indicating the distance of a point from the axis of abscissas. Also the distance between points and this.

f ordonnée *f* — **d** Ordinate *f* — **n** ordinaat *fm* — **e** ordenada *f* — **i** ordinata *f*

9677 organic dye; organic dyestuff

f colorant *m* organique — **d** organischer Farb-

stoff *m* — **n** organische kleurstof *fm* — **e** colorante *m* orgánico — **i** colorante *m* organico

9678 organic dyestuff — See organic dye.

9679 organic matter — Chemical compound containing carbon; material made up of carbon compounds.

f composé *m* organique — **d** organische Substanz *f*; organische Verbindung *f* — **n** organische stof *fm*; organische verbinding *f* — **e** materia *f* orgánica — **i** sostanza *f* organica

9680 organic pigments *GB*; **toners** *US* — Pigments manufactured from coal tar and its derivatives. Compared with inorganic pigments they are in general brighter, less light fast, and more prone to bleeding in oils and solvents.

f pigments *mpl* organiques — **d** organische Pigmente *npl*; Pigmentfarbstoffe *mpl* — **n** organische pigmenten *npl*; organische kleurstoffen *fmpl* — **e** pigmentos *mpl* orgánicos — **i** pigmenti *mpl* organici

9681 organic solvent — Solvent of an organic chemical structure.

f solvant *m* organique — **d** organisches Lösungsmittel *n* — **n** organisch oplosmiddel *n* — **e** solvente *m* orgánico — **i** solvente *m* organico

9682 organosol — Suspension of particles in an organic solvent most usually made with vinyl resins, solvents and plasticizers.

f organosol *m* — **d** Organosol *m* — **n** organosol *m* — **e** organosol *m* — **i** organosolo *m*

9683 Orient yellow — See cadmium sulphide.

9684 orifice — Small opening through which a substance may pass.

f orifice *m*; embouchure *f*; orifice *m* de coulée — **d** Mündung *f*; Ausguß *m*; Gießloch *n*; Gußloch *n*; Öffnung *f*; Düse *f* — **n** mondstuk *n* — **e** boquilla *f*; embocadura *f* — **i** bocchino *m*; orifizio *m*

9685 original — Photograph, drawing, painting, design, print or other matter submitted for photomechanical reproduction, usually referred to as *copy.

f original *m*; modèle *m* — **d** Original *n*; Vorlage *f*; Aufsichtsbild *n* — **n** origineel *n*; model *n* — **e** original *m*; modelo *m* — **i** originale *m*; modello *m*

9686 original — See copy.

9687 original edition — See first edition.

9688 ornament — Rule, border or conventional decoration cast in type and used for the embellishment of printed matter.

f ornement *m* — **d** Verzierung *f*; Zierstück *n*; Ornament *n*; Schmuck *m*; Vignette *f* — **n** ornament *n*; versiering *f*; versiersel *n*; vignet *n* — **e** adorno *m*; ornamento *m*; friso *m* — **i** ornamento *m*

9689 ornamental border

f bandeau *m*; bordure *f* — **d** Zierrand *m*; Schmuckeinfassung *f* — **n** versierde rand *m*; sierrand *m* — **e** orla *f* de decoración — **i** riquadro *m* ornamentale

9690 ornamental rule

f filet *m* orné — **d** Zierlinie *f*; Schmucklinie *f* — **n**

sierlijn *fm* — **e** orla *f*; raya *f* ornamental; raya *f* de adorno — **i** fusello *m*; fusetto *m*; filetto *m* ornamentale

9691 orpiment; auripigment; arsenic yellow; yellow arsenic sulphide — Orange to lemon yellow arsenic trisulphide, As_2S_3, found as a mineral or produced artificially and used as a pigment and as a dyestuff.
f orpiment *m*; orpin *m* doré; orpin *m* jaune; trisulfure *m* d'arsenic; sulfure *m* d'arsenic; arsenic *m* jaune — **d** Auripigment *n*; Operment *n*; Arsentrisulfid *n*; Königsgelb *n*; gelbes Arsensulfid *n* — **n** orpiment *n*; auripigment *n*; koningsgeel *n* — **e** oropimente *m*; génuli *m*; rejalgar *m* — **i** orpimento *m*

9692 orseille — See orchil.

9693 orthamine — See ortho-phenylenediamine.

9694 ortho- — Prefix designating a compound where the maximum number of hydroxyl groups is attached to a particular atom, e.g. $Si(OH)_4$ is called ortho-silicic acid. Also to indicate benzene disubstitution products when substitution is at adjacent carbons, e.g., ortho-dibromo benzene (o-dibromo benzene).
f ortho- — **d** Ortho- — **n** ortho- — **e** orto- — **i** orto-

9695 orthoboric acid — See boric acid.

9696 orthochromatic — Said of photographic plates and films sensitive to yellow and green as well as blue and violet, but insensitive to red rays. A negative or reproduction of a varicoloured original showing correct monochrome rendition of the colour values and natural tones of the subject.
f orthochromatique — **d** orthochromatisch — **n** orthochromatisch — **e** ortocromático — **i** ortocromatico

9697 orthochromatic film — Photographic film that is sensitive to all colours except red.
f film *m* orthochromatique — **d** orthochromatischer Film *m*; Orthofilm *m* — **n** orthochromatische film *m*; orthofilm *m* — **e** película *f* ortocromática — **i** pellicola *f* ortocromatica

9698 ortho-diaminobenzene — See ortho-phenylenediamine.

9699 ortho-dihydroxybenzene — See pyrocatechol.

9700 orthography — See spelling.

9701 ortho-hydroxybenzoic acid — See salicylic acid.

9702 ortho-phenylenediamine; orthamine; ortho-diaminobenzene — Colourless crystals, soluble in alcohol, ether, water, and chloroform. Used for the manufacture of dyes and as a photographic developer.
Formula: $C_6H_4(NH_2)_2$.
f ortho-phénylène-diamine *f*; diaminobenzène *f* — **d** Ortho-Phenylendiamin *n* — **n** ortho-fenyleendiamine *fmn* — **e** orto-fenilendiamina *f* — **i** orto-fenilendiammina *f*

9703 ortho-phthalic acid — See phthalic acid.

9704 oscillating ink drum — See reciprocating ink drum.

9705 oscillating ink roller — See reciprocating ink drum.

9706 oscillating motion — Non-uniform motion changing its direction periodically.
f oscillation *f*; mouvement *m* alternatif; mouvement *m* de va-et-vient — **d** Oszillation *f*; Wechselbewegung *f*; Hinundherbewegung *f* — **n** oscillerende beweging *f*; oscillatie *f*; heen- en weergaande beweging *f* — **e** oscilación *f*; movimiento *m* oscilante — **i** oscillazione *f*; movimento *m* oscillante

9707 oscillating roller — Roller in an inking ir damping system with a side to side as well as a rotary motion.
f rouleau *m* bal(l)adeur; bal(l)adeur *m* — **d** Verreibwalze *f*; Reiber *m* — **n** (heen- en weergaande) distributierol *fm*; (heen- en weergaande) verdeelrol *fm* — **e** rodillo *m* oscilador; rodillo *m* oscilante; rodillo *m* oscilatorio; rodillo *m* batidor; rodillo *m* de vaivén; cilindro *m* oscilante — **i** rullo *m* oscillante

9708 oscillation — One complete period of vibratory or periodic motion, e.g., the whole succession of states that take place before the motion begins to repeat itself.
f oscillation *f* — **d** Schwingung *f*; Oszillation *f* — **n** slingering *f*; oscillatie *f* — **e** oscilación *f* — **i** oscillazione *f*

9709 oscilloscope — Instrument in which the variations in a fluctuating electrical quantity appear temporarily as a visible wave from on the fluorescent screen of a cathode ray tube.
f oscilloscope *m* — **d** Oszilloskop *n* — **n** oscilloscoop *m* — **e** osciloscopio *m* — **i** oscilloscopio *m*

9710 osmotic pressure — Force that a dissolved substance exerts on a semi-permeable membrane through which it cannot penetrate, when separated by it from pure solvent.
f pression *f* osmotique — **d** osmotischer Druck *m* — **n** osmotische druk *m* — **e** presión *f* osmótica — **i** pressione *f* osmotica

9711 ounce — Unit of weight avoirdupois. Abbrev. oz (no point), or sometimes ℥ av. In the SI: 1 oz = 1/16 lb ≈ 28.3495 x 10^{-3} kg. See app. no. 7 and 7a, fluid ounce, troy ounce.
f once *f* avoirdupois; once *f* anglaise; once *f* du commerce — **d** englische Unze *f*; Avoirdupois-Unze *f* — **n** Engels ons *n*; ons *n* avoirdupois — **e** onza *f* avoirdupois — **i** oncia *f* avoirdupois

9712 out — See omission.

9713 outdated — See out of date.

9714 outer forme — Commonly the last printing couple. Traditionally the forme carrying the first and last pages will be on the outside of the folded section.
f forme *f* du premier côté — **d** äußere Druckform *f*; äußere Form *f*; Schöndruckform *f* — **n** buitenvorm *m*; schoondrukvorm *m* depr. — **e** forma *f* en blanco; tiro *m* en blanco; forma *f* de cara — **i** prima forma *f*; forma *f* in bianca

9715 outer margin — See foredge margin.

9716 outline — Line defining or bounding a

figure.

f contour *m* — **d** Kontur *f* — **n** contour *m*; contourlijn *fm*; omtreklijn *fm* — **e** contorno *m* — **i** contorno *m*; linea *f* di contorno

9717 outline — Rough sketch of a drawing of a project or of a display. See also layout.

f croquis *m*; ébauche *f*; esquisse *f* — **d** Skizze *f* — **n** schets *fm* — **e** esquema *m*; bosquejo *m*; corquis *m*; trazo *m* — **i** schizzo *m*; bozetto *m*; traccia *f*

9718 outline *v* — To box in an illustration with a line border usually engraved through the negative emulsion.

f entourer (d'un cadre); encadrer — **d** einfassen; mit Rand versehen; rändern; rändeln — **n** omkaderen — **e** encuadrar — **i** inquadrare; incorniciare

9719 outline *v* — To opaque out the background around the margin of a principal object in a halftone negative.

f détourer — **d** freistellen; abdecken — **n** uitdekken — **e** recuadrar — **i** coprire

9720 outlined and vignetted halftone — Halftone plate on which part of the background is cut away and part vignetted.

f simili *f* détourée et dégradée — **d** Rasterätzung *f* mit Verlauf — **n** gedeeltelijk uitgedekte autotypie *f* met verloop — **e** grabado *m* directo degradado — **i** autotipia *f* schermata con vignette; autotipia *f* con fondo sfumato

9721 outlined halftone; blockout halftone; dropout halftone; silhouette halftone; cut-out halftone; close-cut halftone — Halftone illustration plate from which the screen surroudning any part of the image has been cut or etched away.

f simili *f* détourée; simili *f* découpée — **d** freistehende Autotypie *f*; freistehende Rasterätzung *f*; freistehendes Halbtonbild *n* — **n** uitgedekte autotypie *f*; vrijstaande autotypie *f* — **e** medio tono *m* silueteado; grabado *m* silueteado; fotograbado *m* silueteado; grabado *m* recortado; fotograbado *m* recortado; autotipia *f* silueteada — **i** autotipia *f* con fondo eliminato; autotipia *f* decontornata; autotipia *f* schermata

9722 outline letters; open-faced types; open letters — Types which comprise only the outline of the letter, in opposition to solid types which print in full face.

f lettres *fpl* blanches; caractères *mpl* à jour — **d** Konturschrift *f* — **n** opengesneden letters *fmpl* — **e** caracteres *mpl* fileteados; letras *fpl* fileteadas; letras *fpl* caladas — **i** caratteri *mpl* contornati

9723 out of alignment

f désaligné; sortante — **d** nicht Linie haltend; schlecht ausgerichtet — **n** uit de lijn; niet in de lijn — **i** fuori allineamento

9724 out of date; outdated

f périmé — **d** veraltet; altmodisch — **n** verouderd; uit de tijd; ouderwets — **e** anticuado; en desuso — **i** sorpassato; passato di moda; fuori di moda

9725 out of print — Said of a book when the publisher has no more copies for sale and there is no plan to reprint.

f épuisé — **d** vergriffen — **n** uitverkocht — **e** agotado — **i** esaurito

9726 out of register — Failure of a colour or bracketed sheet to coincide with the other colours or the other sheet.

f repérage *m* défectueux; registre *m* défectueux; non repéré; bamboché; de guingois — **d** hält nich Register; schlechter Passer *m*; ungenauer Passer *m*; nicht registerhaltig — **n** misregister *n*; niet in register; niet passend — **e** fuera de registro; fuera de puntura — **i** fuori registro; non tiene il registro

9727 out-of-round reel — Reel identified by a flat area at edge of the paper reel and a starred pattern on the end of the reel caused by the disruption in the circular pattern of the winding. It will hump as it rotates and cause excessive web vibration.

f bobine *f* excentrée — **d** unrunde Rolle *f* — **n** niet-ronde rol *fm*; rol *fm* met platte kant; platgedrukte rol *fm* — **e** bobina *f* no redonda; bobina *f* decentrada — **i** rotolo *m* ovalizzato; bobina *f* ovalizzata

9728 out of stock

f pas en magasin; stock *m* temporairement épuisé — **d** nicht auf Lager; nicht vorrätig — **n** niet in voorraad; niet op magazijn — **e** sin existencias en depósito; agotado — **i** scorta *f* esaurita

9729 output — Production per hour.

f production *f* horaire; tirage *m* à l'heure — **d** Stundenleistung *f* — **n** produktie *f* per uur; uurproduktie *f* — **e** producción/hora *f*; tirada *f* efectiva — **i** produzione *f* oraria

9730 output — Current, voltage, power, or driving force delivered by a circuit or device.

f sortie *f* — **d** Ausgang *m* — **n** uitgang *m* — **e** salida *f* — **i** uscita *f*

9731 output — Process of transferring data from an internal store to an external store or to a peripheric equipment. Also the transferred data.

f sortie *f* — **d** Ausgabe *f* — **n** uitvoer *m* — **e** salida *f* — **i** uscita *f*

9732 output bound — Computer program which is limited in its throughout speed by the paper tape punch or other method of recording or passing on the product, such as the speed of recording onto paper tape, or writing on a disc, or transferring information to an on-line typesetter. See also input bound, compute bound.

9733 output device — Unit that translates computer results into usable or final form.

f équipement *m* de sortie — **d** Ausgabegerät *n* — **n** uitvoereenheid *f* — **e** dispositivo *m* de salida — **i** dispositivo *m* d'uscita

9734 output tube — Tube to display results of computer processing (cathode ray tube or similar device) or to compare type (e.g., a CRT typesetter).

9735 outsert; wraparound; wrap round

f encarté *m*; encartage *m* à l'extérieur — **d** umgelegter Bogen *m* — **n** omsteker *m*; omsteekvel *n* — **i** inserto *m* imbarato esternamente

9736 outside mortise — Notch cut into one or more sides of a mounted printing plate from the outer edges, but in such manner that the aperture is not completely enclosed.
f encoche *f* — **d** Einschnitt *m*; Ausschnitt *m* — **n** uitsparing *f* aan de buitenkant — **e** muesca *f* exterior; escopladura *f* exterior; escopladura *f* al borde; escopleadura *f* exterior; escopleadura *f* al borde; mortaja *f* al borde — **i** tacca *f* esterna
9737 outsides — Top and bottom quires or imperfect sheets of a ream. Formerly callded *cassie paper.
f feuilles *fpl* de protection; feuilles *fpl* de rebut — **d** Decklage *f* — **n** buitenvellen *npl*; dekvellen *npl* — **e** hojas *fpl* costeras; costeras *fpl*; papel *m* costero; papel *m* quebrado — **i** fogli *mpl* esterni; fogli *mpl* di protezione; fogli *mpl* di scarto
9738 outturn — See outturn sheet.
9739 outturn sample — See outturn sheet.
9740 outturn sheet; outturn sample; outturn — Sample taken from a run on the paper machine and sent to the customer as representative of the paper made.
f feuille *f* échantillon type; échantillon *m* référence — **d** Ausfallmuster *n*; Vorlagemuster *n*; Muster *n* — **n** uitvalmonster *n*; monster *n* uit de partij — **e** muestra *f* de fabricación; muestra *f* de referencia — **i** foglio *m* campione tipo
9741 oval — Figure having the lengthwise outline of an egg with one end larger than the other. Popularly but erroneously used in photoengraving to designate an *ellipse.
f ovale *m* — **d** Oval *n* — **n** ovaal *n* — **e** óvalo *m* — **i** ovale *m*
9742 oval *v* — To cut a printing plate to the shape of an ellipse. See also ellipsograph, oval.
f cadrer en ovale — **d** oval anreißen; oval fräsen — **n** ovaal maken; een ovale vorm geven — **e** ovalar — **i** contornare ovale
9743 oval head screw
f vis *f* à tête ronde — **d** Linsenschraube *f* — **n** lenskopschroef *fm* — **e** tornillo *m* de cabeza ovalada — **i** vite *f* con testa a calotta
9744 oven dry; bone dry; moisture-free — Containing no moisture. A paper which has been dried to a constant weight at a temperature of 100 to 105 °C (212 to 221 °F) in a completely dry atmosphere. In practice, the atmosphere in the oven usually is not completely dry, and the paper retains a few tenths of one per cent of moisture even when constant weight is attained.
f absolument sec; sec à l'absolu — **d** absolut trocken — **n** absoluut droog — **e** seco absoluto — **i** secco assoluto
9745 overall corrections — See local corrections.
9746 over-all study — See chronological study.
9747 over and back fold — See accordion pleat.
9748 overcast binding — See whip stitching.
9749 overcasting — See whip stitching.
9750 over development
f excès *m* de développement — **d** Überentwicklung *f* — **n** overontwikkeling *f* — **e**

sobrerrevelado *m*; exceso *m* de revelado — **i** sviluppo *m* eccessivo
9751 over-etched — In relief etching said of lines of dots that have been damaged or partially etched away by excessive etching or action beyond normal period of time.
f trop mordu — **d** verätzt — **n** kapot geëtst; veretst — **e** sobremordido; grabado con exceso — **i** sovrainciso; mordenzato eccessivamente
9752 overexpose *v* — See overexposure *v*.
9753 overexposed
f surexposé — **d** überbelichtet — **n** overbelicht — **e** sobreexpuesto; con exceso de exposición — **i** sovraesposto
9754 overexposure — Excessive exposure of a sensitized photographic surface to the action of actinic light.
f surexposition *f* — **d** Überbelichtung *f* — **n** overbelichting *f* — **e** sobreexposición *f*; exceso *m* de exposición; exceso *m* de mucha exposición — **i** sovraesposizione *f*
9755 overexposure *v*; **overexpose** *v* — To overexpose a film in photography.
f surexposer — **d** überbelichten — **n** overbelichten — **e** sobreexponer — **i** sovraesporre
9756 overfold — Lip or overhang formed by the leading edge of the section or copy when the fold is out of centre. Adjustment of the folding mechanism can have the reverse effect which is termed *underfold.
f faux pli *m* — **d** Überfalz *m* — **n** overvouw *fm* — **e** falso doblez *m* — **i** lembo *m* sporgente
9757 overhang cover — See extended cover.
9758 overhead — In computing cost of a job, fixed charges and general operating expenses.
f frais *mpl* généraux — **d** Gemeinkosten *fpl*; Regiekosten *fpl* — **n** opslagkosten *pl*; algemene kosten *pl* — **e** gastos *mpl* generales — **i** costi *mpl* generali
9759 overhead camera; suspended camera — Camera suspended from an overhead track permitting easy access to copyboard, etc.
f chambre *f* suspendue; appareil *m* suspendu — **d** hängende Kamera *f*; Überkopfkamera *f*; Brückenkamera *f* — **n** hangende camera *f* — **e** cámara *f* colgante; cámara *f* de suspensión — **i** macchina *f* fotografica sospesa; apparecchio *m* sospeso; apparecchio *m* a ponte
9760 overhead conveyor
f transporteur *m* aérien — **d** Schwebeförderer *m*; Überkopfförderer *m* — **n** hangende transportband *m*; hangende transportbaan *fm* — **e** transportadora *f* de suspensión — **i** trasportatore *m* sospeso; trasportatore *m* aereo
9761 overheating — Raising the temperature of a liquid above its boiling point.
f surchauffement *m*; surchauffage *m*; surchauffe *f* — **d** Überhitzung *f* — **n** oververhitting *f* — **e** achicharramiento *m*; sobrecalentamiento *m*; calentamiento *m* excesivo — **i** surriscaldamento *m*
9762 overimpression — See hard impression.

9763 overink *v* — See crowd *v*.

9764 overissue news — See flat printed news.

9765 overissues — See returns.

9766 overlap — Overlapping of one colour into another in colour printing.

f recouvrement *m* — **d** Überlappung *f* — **n** overlapping *f* — **e** rebaba *f* — **i** sovrapposizione *f*

9767 overlapping edges — See squares.

9768 overlay — Piece of thin paper cut out and pasted onto sheet of paper underneath tympan draw sheet over those areas which require heavier impression.

f béquet *m*; découpage *m*; épaisseur *f* — **d** Zurichtungsausschnitt *m*; Zurichtepapier *n* — **n** pikeersel *n* — **e** alza *f*; calzo *m*; recorte *m* — **i** taccheggio *m*; spessore *m*; alzo *m*

9769 overlay — Original art work in which one or more colour components are drawn on a transparent sheet attached to the art work.

f superposition *f* de calques

9770 overlaying — Transferring segments of a program from auxiliary store into internal store for execution, so that two or more segments occupy the same store locations at different times.

f recouvrement *m* — **d** Überlagerung *f* — **n** superponeren *n* — **e** recubrimiento *m* — **i** sovrapposizione *f*

9771 overlaying — Increasing pressure on the solids or dark tones of a printing plate and decreasing it on the lighter tones and highlights, by means of pasting bits of thin paper onto a sheet of the cylinder packing underneath tympan draw sheet.

f mise *f* de dessus; mise *f* en train sur le cylindre — **d** oberes Zurichten *n*; Zylinderzurichtung *f* — **n** pikeren *n* — **e** calzar; poner alzas; calzado *m*; colocación *f* de alzas; alza *f* — **i** avviamento *m*

9772 overlay knife — See cutting-out knife.

9773 overlay sheet

f feuille *f* de mise en train — **d** Zurichtebogen *m* — **n** toestel *n* — **e** hoja *f* de arreglo; primera hoja *f* de la cama — **i** foglio *m* di avviamento

9774 over matter — See crowdout.

9775 overpacking — Packing the cylinder covering, the plate or blanket to a level that is excessively above the level of the cylinder bearers.

f excès *m* d'habillage — **d** übermäßiger Aufzug *m* — **n** te dikke legger *m*; te dik onderlegsel *n* (in offset) — **e** sobrearreglo *m*; arreglo *m* excesivo — **i** abbigliamento *m* eccessivo

9776 overplus — See overs.

9777 overprint — See surprint.

9778 overprinted — Said of images which have been overexposed during contact printing.

f surexposé — **d** überbelichtet — **n** overbelicht — **e** sobreexpuesto — **i** sovraesposto

9779 overprinting — Impressing a printed surface with additional printings, as when printing one colour over another.

f surimpression *f* — **d** Aufdruck *m*; Überdruck *m*; Eindruck *m* — **n** opdruk *m*; overdruk *m*; indruk *m* — **e** sobreimpresión *f* — **i** sovrastampa *f*

9780 overprint varnish — Broadly: transparent coating, including spirit varnishes and lacquers, applied by sprayer, roller coating, or printing methods to beautify or protect printed surfaces. Specifically: varnish applied to the printed surface by ordinary printing methods. Must be pale-coloured, hard drying, scuff resistant and glossy, both on the unprinted paper and the printed areas. It usually becomes yellowish upon ageing of the package to which it is applied.

f vernis *m* à surimpression; vernis *m* de surlaquage — **d** Überdruckfirnis *m*; Überdrucklack *m*; Überzugslack *m* — **n** overdrukvernis *mn* — **e** barniz *m* para sobreimpresión; laca *f* para sobreimpresiónes — **i** vernice *f* a finire; vernice *f* per soprastampa; lacca *f* sovrastampa

9781 over-punching — Use of the upper curtate to represent a digit independently of the use of the lower curtate.

f perforation *f* de zone — **d** Überlochung *f* — **n** bovenponsing *f* — **e** perforación *f* de zona — **i** sopraelevazione *f*

9782 overrun — Surplus of copies printed.

f surplus *m*; exemplaires *m* excédentaires; excédentaires *mpl* — **d** Überschuß *m*; Überlieferung *f*; Übermenge *f*; Überdruck *m* — **n** teveel gedrukte exemplaren *npl*; meerlevering *f* — **e** exceso *m* de ejemplares — **i** copie *fpl* stampate in più

9783 overrun *v* — To carry over words from the end of one line to the beginning of the next one to allow insertion of corrections or to get a better disposition of the next.

f remanier (quelques lignes); aller en Germanie — **d** neu umbrechen (Zeilen); in die nächste Zeile hinübernehmen (Wörter) — **n** verlopen (regels) — **e** recorrer (la composición) — **i** rimaneggiare (riga)

9784 overs; overplus; over sheets; plus sheets; spoils; waste sheets — Extra sheets of paper allowed on all quantities as spoils to be expected in a process.

f passe *f* (du papier); feuilles *fpl* de passe; chaperon *m* obs.; chaperons *mpl* obs.; passes *fpl* — **d** Zuschuß *m*; Zuschußbogen *mpl* — **n** inschiet *m*; inschietvellen *npl*; vellen *npl* voor de inschiet; overvellen *npl* — **e** mano *f* perdida; perdido *m*; hojas *fpl* de falta; hojas *fpl* de más; aumento *m* — **i** aggiunta *f* (di fogli); fogli *mpl* aggiunti; soprappiù *m*; aggiunte *fpl*

9785 overscore *v* — The opposite of *underscore *v*.

9786 overseer of the composing room — See composing-room overseer.

9787 overseer of the machine room — See pressroom overseer.

9788 oversewing — See whip stitching.

9789 over sheets — See overs.

9790 overshot duct; overshot (ink) fountain — Ink duct using a flexible steel blade mounted above a roller which rotates in a trough of ink.

f encrage *m* pelliculaire supérieur — **d** ober-

schlächtiger Farbkasten *m*; Farbkasten *m* mit obenliegendem Farbmesser; Filmfarbwerk *n* — **n** inktbak *m* met bovenliggend mes; filminktwerk *n* — **e** tintero *m* de cuchilla entintado pelicular — **i** calamaio *m* con trasferimento pellicolare continuo
9791 overshot (ink) fountain — See overshot duct.
9792 overstrike *v* — To substitute one character for another, on a video display tube, by positioning the cursor at a particular location and then depressing the key representing the desired character while the video terminal is in an overstrike mode.
9793 overtime — Time in excess of a standard day or week.
f veillée *f*; heures *fpl* supplémentaires — **d** Überstunden *fpl*; Überstich *m*; Überzeitarbeit *f* — **n** overuren *mpl*; overwerk *n* — **e** horas *fpl* especiales; horas *fpl* extraordinarias; tiempo *m* suplementario — **i** lavoro *m* straordinario; ore *fpl* straordinarie
9794 overweight
f excédent *m* de poids; surpoids *m* — **d** Übergewicht *n* — **n** overgewicht *n*; overwicht *n* — **e** sobrepeso *m*; exceso *m* de peso — **i** sovrapeso *m*
9795 oxalate of iron — See ferric oxalate.
9796 oxalate of iron — See ferrous oxalate.
9797 oxalic acid — Transparent, colourless crystals, soluble in water, alcohol and ether. Poisonous. Chemical name for acid of sugar. Latin: acidum oxalicum. TL-value: 1 mg/m^3.
Formula: $CO_2HCO_2H·2H_2O$.
f acide *m* oxalique — **d** Oxalsäure *f*; Kleesäure *f*; Äthandisäure *f* — **n** oxaalzuur *n*; zuringzuur *n* — **e** ácido *m* oxálico; ácido *m* sacarino — **i** acido *m* ossalico
9798 Oxford bible paper — See bible paper.
9799 Oxford hollow — Hollow back of a book, which consists of a flat paper tube attached to the spine.
9800 Oxford India paper — See bible paper.
9801 Oxford rule; Scotch rule; shaded rule; contrast rule — Brass or metal rule with one thick and one thin line running parallel with each other.
f filet *m* (de) cadre; filet *m* gras-maigre — **d** fettfeine Linie *f* — **n** kaderlijn *fm*; vet-fijne lijn *fm*; carréijn *fm* — **e** mediacaña *f*; filete *m* mediacaña; raya *f* de media caña; filete *m* negro-fino — **i** filetto *m* chiaroscuro; filetto *m* Oxford
9802 ox gall — Bitter fluid secreted by the liver of the ox, which, after boiling and filtration, is mixed with ruling inks and water colours to make them flow better. Also used in *marbling to set the colours.
f fiel *m* de bœuf — **d** Ochsengalle *f* — **n** ossegal *fm* — **e** hiel *f* de toro; hiel *f* de buey — **i** fiele *m* di bue; fiele *m* di bovo *depr.*
9803 oxidation — Action of oxygen and air on metals, causing discoloration and marring of surfaces, together with the formation of rust or other coatings. Treatment of a zinc plate with a graining bath or dilute acid solution, the chemical ef-

fect taking place during etching. Also, the combination of oxygen with the vehicle of the ink, yielding a dry film.
f oxydation *f* — **d** Oxydation *f*; Oxydierung *f* — **n** oxydatie *f* — **e** oxidación *f* — **i** ossidazione *f*
9804 oxidation/reduction potential — See redox potential.
9805 oxide — Combination of oxygen with an element, such as lead or tin or zinc, copper, aluminium, etc.
f oxyde *m* — **d** Oxyd *n* — **n** oxyde *n* — **e** óxido *m* — **i** ossido *m*
9806 oxide film
f pellicule *f* d'oxyde — **d** Oxydhäutchen *n* — **n** oxydlaagje *n*; oxydhuidje *n* — **e** capa *f* de óxido — **i** pellicola *f* di ossido
9807 oxidize *v* — 1. To convert an element into its oxide; combine with oxygen. 2. To cover with a coating of oxide, or rust. 3. To take away hydrogen, as by the action of oxygen; add oxygen or any non-metal. 4. To increase the valence of an element. 5. To remove electrons from. To become oxidized.
f oxyder — **d** oxydieren — **n** oxyderen — **e** oxidar — **i** ossidare
9808 oxidized oils — See blown oils.
9809 oxidizing agent
f oxydant *m* — **d** Oxydationsmittel *n* — **n** oxydatiemiddel *n* — **e** medio *m* de oxidación; oxidante *m* — **i** ossidante *m*
9810 oxygen — Colourless, odourless, tasteless diatomic gas, slightly soluble in water, soluble in molten silver. Symbol: O. Latin: oxygenium.
f oxygène *m* — **d** Sauerstoff *fm* — **n** zuurstof *fm* — **e** oxígeno *m* — **i** ossigeno *m*
9811 oxymethylene — See formaldehyde.
9812 oxyphenic acid — See pyrocatechol.
9813 oxytricarballylic acid — See citric acid.
9814 ozachrome — Process of making photographic colour proofs from separation negatives with diazotype (ozalid) materials. These materials are developed in ammonia fumes and give visual transparency approximation of colour copy.
f ozachrome *m* — **d** Ozachrom *n* — **n** ozachroom — **e** ozacromo *m*
9815 ozalid paper — Light-sensitive paper for the multiplication of drawings, coated with a diazo compound, which forms dyes under the influence of ammonia vapour, but loses this quality when exposed to light.
f papier *m* ozalid — **d** Ozalidpapier *n*; Lichtpauspapier *n* — **n** ozalid-papier *n* — **e** papel *m* ozalid — **i** carta *f* ozalid
9816 ozalid process — Photographic process employing materials sensitized with diazo compounds, the exposed images usually developed with gaseous ammonia. Ozalid is a trade-mark.
f procédé *m* ozalid — **d** Ozalidverfahren *n* — **n** ozalid-procédé *n* — **e** procedimiento *m* ozalid — **i** procedimento *m* ozalid
9817 ozobrome — Positive carbon process in which the tanning of the pigmented gelatine is

accomplished by the chemical action of a bleach bath on a print made on bromide paper.

9818 ozokerite; ocerite; ceresin; ceresin wax; cerin; mineral wax — Mineral wax to mould electrotypes. Soluble in alcohol, benzene, chloroform, naphta, insoluble in water. Used for sizing and glossing papers, acid-proof coating for electrotypers' plates matrix compositions; in admixture with rosin and sulphur for making printing forms.
f ozokérite *f*; ozocérite *f*; cérésine *f*; cire *f* cérésine; cire *f* minérale — **d** Ozokerit *m*; Zeresin *n*; Zeresinwachs *m*; Bergwachs *n*; Erdwachs *n*; Mineralwachs *n* — **n** ozokeriet *n*; aardwas *mn*; gezuiverde aardwas *mn*; ceresine *fm*; ceresinewas *mn*; minerale was *mn* — **e** ozoquerita *f*; cera *f* de ceresina; cera *f* mineral — **i** ozocherite *f*; ozocerite *f*; ceresina *f*; cera *f* minerale

9819 ozone — Allotropic form of oxygen. Powerful bleaching agent. TL-value: 0.1 ppm or 0.2 mg/m^3. Formula: O$_3$.
f ozone *m* — **d** Ozon *n* — **n** ozon *nm* — **e** ozono *m* — **i** ozono *m*

9820 pack *v* — To combine more than one item of data into one unit of store to reduce the total amount of store required. Some computer systems require decimal numbers to be packed before arithmetic functions are performed upon them.
f comprimer; condenser — **d** packen — **n** inpakken; samenpakken; comprimeren — **e** empaquetar; condensar — **i** condensare; comprimere

9821 package engineering; packaging engineering — Application of scientific and engineering principles to solving the problem of functional design, formation, filling, closing and/or preparation for shipment of containers regardless of type or kind, or the product enclosed therein.
f technique *f* de l'emballage — **d** Verpackungstechnik *f* — **n** verpakkingstechniek *f* — **e** técnica *f* de embalaje — **i** tecnica *f* dell'imballaggio

9822 packaging engineering — See package engineering.

9823 packaging industry
f industrie *f* des emballages — **d** Verpackungsindustrie *f* — **n** verpakkingsindustrie *f* — **e** industria *f* de materiales de embalaje — **i** industria *f* dell'imballaggio

9824 packed cylinder
f cylindre *m* habillé; cylindre *m* garni — **d** Zylinder *m* mit Aufzug — **n** beklede cilinder *m* — **e** cilindro *m* con revestimiento; cilindro *m* vestido — **i** cilindro *m* rivestito; cilindro *m* abbigliato

9825 packing; cylinder covering; cylinder dressing; cylinder packing — In letterpress printing, rubber blanket, tympan paper, or other material covering the printing cylinder to bring it up to printing height. See also hard packing, soft packing.
f habillage *m* (du cylindre) — **d** Aufzug *m*; Zylinderaufzug *m* — **n** legger *m* — **e** cama *f*; cama *f* fija *Es*; mantilla *f* del cilindro; recubrimiento *m* del cilindro; revestimiento *m* del cilindro; colchón *m*; empaque *m* *Ar* — **i** rivestimento *m* del cilindro

9826 packing; under-packing — In lithography, the paper used to underlay a blanket, plate or proof to bring the surface to the desired height, the method of adjusting squeeze pressure.
f papier *m* d'habillage — **d** Zurichtepapier *n* — **n** onderlegpapier *n*; onderlegsel *n* — **e** papel *m* para arreglo; papel *m* para alzar — **i** carta *f* d'avviamento

9827 packing — Material to pack, to cushion or protect goods packed in a container.
f emballage *m* — **d** Verpackung *f*; Packmaterial *n* — **n** emballage *f*; verpakking *f*; verpakkingsmateriaal *n* — **e** embalaje *m* — **i** imballo *m*

9828 packing — Replaceable material forming a seal about a rotating or reciprocating element of a hydraulic or lubricated device. Replaceable portion of a gland or stuffing box assembly. See also gasket.
f presse-étoupe *m*; joint *m*; garniture *f* — **d** Packung *f*; Dichtung *f* — **n** pakking *f*; afdichting *f* — **e** empaquetadura *f*; junta *f*; guarnición *f*; rondana *f* *Me* — **i** guarnizione *f*

9829 packing density — Amount of data stored in a given unit of dimension. In magnetic tape, the number of bits, code rows, code frames, or characters recorded per linear inch.
f densité *f* d'enregistrement — **d** Packungsdichte *f* — **n** opslagdichtheid *f* — **e** densidad *f* de registro — **i** densità *f* di registrazione

9830 packing gauge
See also micrometer.
f jauge *f* d'habillage — **d** Aufzugmaß *n*; Aufzugsprüfer *m* — **n** leggerdiktemeter *m* — **e** calibre *m* del espesor de revestimiento; calibre *m* del espesor de la cama *Es* — **i** calibro *m* per il controllo dello spessore del rivestimento

9831 packing gauge — See cylinder bearer gauge.

9832 packing machine
f machine *f* à emballer; machine *f* à empaqueter — **d** Verpackungsmaschine *f* — **n** verpakkingsmachine *f* — **e** máquina *f* para embalar; máquina *f* empaquetadora — **i** macchina *f* impaccatrice

9833 packing ring — See gasket.

9834 packing thickness
f épaisseur *f* d'habillage — **d** Aufzugsdicke *f*; Aufzugstärke *f* — **n** leggerdikte *f* — **e** espesor *m* de la cama — **i** spessore *m* del rivestimento

9835 packless mat; packless stereo mat; no-pack mat; patchless mat; patchless stereo mat
f flan *m* sans garnissage; matrice *f* sans revêtement — **d** Nopak-Mater *f*; Mater *f* ohne Zurichtung — **n** no-pack matrijs *f* — **e** matriz *f* que no necesita relleno; matriz *f* sin respaldo — **i** flano *m* senza taccheggio

9836 packless stereo mat — See packless mat.

9837 pack of cards *GB*; **deck of cards** *US* — Computer term. In punched card usage, a collection of punched cards bearing data for a particular run.
f paquet *m*; jeu *m* — **d** Kartensatz *m*; Kartenstapel *m*; Stapel *m* — **n** kaartenpakket *n*; stapel *m* — **e** paquete *m* — **i** pacco *m*

9838 pad — Round, flat, sand-filled leather cushion on which engravers and finishers support printing plates while working on them.
f coussinet *m*; coussin *m* — **d** Stechkissen *n*; Kissen *n* — **n** graveerkussen *n*; kussen *n* — **e** cojín *m* de grabador — **i** cuscinetto *m* per incisori

9839 pad — Piece of cardboard, about 8 x 8 cm, with a wooden grip and covered with flanel, used by gilders to pick up gold leaf from the gold cushion.
f couchoir *m* — **d** Goldauftrager *m* — **n** goudoplegger *m*

9840 pad — Wadding which is sometimes placed between the board and the leather of a book cover.

f rembourrage *m* — **d** Füllung *f*; Füllsel *n* — **n** opvulsel *n* — **i** imbottitura *f*

9841 pad — See tablet.

9842 pad *v* — To increase the size of a book, etc., by inserting blank pages or pages of unnecessary matter. Also to add words so that type fill the prescribed space.
f remplir — **d** ausstopfen; auftragen — **n** opvullen — **e** rellenar — **i** riempire

9843 pad *v* — To glue sheets of paper together on one edge.
f mettre en tablette — **d** Blocks leimen — **n** aanlijmen (aan de kop); bloks aanlijmen; bloks maken — **e** fabricar blocs — **i** incollare blocchi

9844 padded sides — Covers of a book that are filled up with cotton wool.
f plats *mpl* capitonnés; plats *mpl* rembourrés — **d** wattierte Decken *fpl* — **n** gecapitonneerde platten *npl*; gewatteerde platten *npl*; opgevulde platten *npl* — **e** cantos *mpl* acolchados; cantos *mpl* colchados; cantos *mpl* acojinados — **i** coperte *fpl* imbottite

9845 padding; pad stapling
f mise *f* en tablette; piquage *m* de tablettes — **d** Blockheftung *f* — **n** hechten *n* van bloks; maken *n* van bloks; nieten *n* van bloks — **e** costura *f* de blocs; costura *f* en bloque — **i** cucitura *f* dei blocchi

9846 padding — Computer term. Filling out a short field or record with fill (or pad) characters, usually spaces for alphabetic data and zeros for numeric.
f remplissage *m* — **d** Auffüllen *n* — **n** opvullen *n* — **e** relleno *m* — **i** riempimento *m*

9847 padding — See flash exposure.

9848 paddles — Rotating curved or shaped fingers of the fly which transfer the copies from the folder of a rotary press to the delivery belt.
f branches *fpl* du moulinet — **d** Schaufeln *fpl* — **n** schoepen *fmpl* — **e** paletas *fpl*; palas *fpl* — **i** pale *fpl*

9849 pad stapling — See padding.

9850 page — One side of a leaf. Not to confuse with leaf or leaves: each leaf being printed on both sides consists of two pages. A single page should not contain less than five lines.
f page *f* — **d** Seite *f* (eines Blattes); Pagina *f* — **n** pagina *fm*; bladzijde *fm* — **e** página *f*; plana *f*; carilla *f* — **i** pagina *f*

9851 page — Type matter or type and blocks, arranged for printing on one side of a book leaf.
f paquet *m* — **d** Seite *f* — **n** pagina *fm*; pagina-zetsel *n* — **e** plana *f* (de composición) — **i** pagina *f* (di composizione)

9852 page, across the — See across the page.

9853 page, all the way across the — See all the way across the page.

9854 page-at-a-time printer; page printer — Teletype term for a printer with which the character pattern for the entire page is determined before printing.
f imprimante *f* par page — **d** Seitendrucker *m*; Blattschreiber *m* — **n** bladdrukker *m* — **e** impresora *f* por páginas — **i** stampatrice *f* pagina per pagina

9855 page cord; string — Fine cord to tie up pages, etc., of type before imposition or distribution.
f ficelle *f* (à lier les pages); ficelle *f* à colonnes — **d** Ausbindeschnur *f*; Kolumnenschnur *f*; Schnur *f* — **n** paginakoord *n*; opbindtouw *n* — **e** bramante *m*; hilo *m* (bramante); cordel *m*; piolín *m* Ar — **i** corda *f* per la legatura delle pagine

9856 page cut-off — Control on the ink duct of a letterpress rotary which allows the ink feed to be cut off from one page.
f mise *f* au repos de l'encrage d'une page; arrêt *m* de l'encrage d'une page — **d** plattenbreite Farbabstellung *f* — **n** inktgeving *f* per pagina; pagina-inktgeving *f* — **e** entintado *m* a la página; alimentación *f* de tinta a la página — **i** esclusione *f* di una pagina dall'inchiostrazione

9857 page depth — See depth of page.

9858 page length; length of a page
f longueur *f* de page — **d** Kolumnenlänge *f*; Kolumnenmaß *n* — **n** paginalengte *f* — **e** largo *m* de la columna — **i** lunghezza *f* della pagina

9859 page number — See folio.

9860 page on page — See line for line.

9861 page paper; page shoe — Sheets of paper or board used by compositors to support tied-up pages before imposition or for storage.
f porte-page *m* — **d** Portepage *f*; Papierunterlage *f* — **n** paginapapier *n* — **e** portapáginas *m* — **i** portapagina *m*; carta *f* per appoggio pagine

9862 page printer — See page-at-a-time printer.

9863 page proofs — Proofs from made-up pages, in opposition to galley or slip proofs.
f épreuves *fpl* en page; épreuves *fpl* en première — **d** Seitenabzüge *mpl* — **n** proeven *fmpl* aan pagina's; proeven *fmpl* aan opgemaakte pagina's; opgemaakte proeven *fmpl* — **e** pruebas *fpl* de página; pruebas *fpl* de la plana entera; páginas-muestras *fpl* — **i** bozze *fpl* impaginate

9864 page shoe — See page paper.

9865 paginate *v* — To indicate the page sequence by numbers or other characters on each leaf of a book. See also bifoliate *v*.
f paginer — **d** paginieren — **n** pagineren — **e** paginar — **i** numerare le pagine

9866 pagination — Numbering of the pages of a book, usually at head of outside margin or centred in bottom margin. Generally omitted on opening pages of chapters, main sections, etc.
f pagination *f* — **d** Paginierung *f* — **n** paginering *f* — **e** paginación *f*; numeración *f* de páginas — **i** numerazione *f* delle pagine

9867 pagination — See foliation.

9868 paging — Disposing type matter in pages of uniform length.
f mise *f* en page; empagement *m* — **d** Umbruch *m*; Umbrechen *n* — **n** opmaak *m*; opmaken *n* van pagina's — **e** compaginación *f* — **i** impaginazione *f*

9869 paging machine
f pagineuse f; machine f à paginer — d Paginiermaschine f — n pagineermachine f — e máquina f para paginar; máquina f paginatrice — i paginatrice f; macchina f paginatrice; macchina f a paginare

9870 paint brush; brush — An artist's paint brush.
f pinceau m; brosse f d'artiste — d Pinsel m; Malerpinsel m — n penseel n; kwast m — e pincel m; brocha f — i pennello m

9871 paint v in — To apply acid resisting varnish with a brush to certain areas of line and halftone plates before etching.
f reboucher — d abdecken — n afdekken — e aplicar tinta de reserva; aplicar el barniz — i ricoprire

9872 pair of scales — See scale.

9873 palaeography — See paleography.

9874 paleography; palaeography — Science of deciphering and reading ancient scriptures and hieroglyphic writings.
f paléographie f — d Paläographie f; Altschriftkunde f — n paleografie f — e paleografía f — i paleografia f

9875 palette knife — See ink knife.

9876 palimpsest — Parchment from which writing has been partially or completely erased to make room for another text. By extension a book made of these sheets.
f palimpseste m — d Palimpsest mn — n palimpsest m — e palimpsesto m — i palinsesto n

9877 palladium — Silver-white, ductile, malleable, metallic element of the platinum group, used especially as a catalyst in alloys, and as an amalgam leaf resembling silver, used in tooling bindings.
f palladium m — d Palladium n — n palladium n — e paladio m — i palladio m

9878 pallet — Load board with two decks separated by bearers or a single deck supported by bearers, for transport and stacking, with the overall height reduced to a minimum compatible with handling by fork lift trucks and pallet trucks. See also stillage.
f plateau m; palette f — d Palette f — n laadbord n; stapelbord n — e bandeja f estibadora — i paletta f

9879 pallet, type — See type holder.

9880 palmitic acid; palmitinic acid; cetylic acid; hexadecanoic acid — White crystals, soluble in alcohol and ether, insoluble in water. Fatty acid occurring in natural fats and oils and in tall oil, and in large amounts in commercial grade stearic acid. Formula: $CH_3(CH_2)_{14}COOH$.
f acide m palmitique; blanc m de blanc — d Palmitinsäure f; Zetylsäure f; n-Hexadezylsäure f — n palmitinezuur n — e ácido m palmítico — i acido m palmitico

9881 palmitinic acid — See palmitic acid.

9882 palm kernel oil; palm-nut oil; palm oil — Yellowish fatty oil, free of fatty acids when fresh and rapidly becoming rancid in air. Soluble in alcohol, ether, chloroform and carbon disulphide.
f huile f de palmiste — d Palmkernöl n — n palmpittenolie fm — e aceite m de nuez de palma — i olio m di noce di palma

9883 palm-nut oil — See palm kernel oil.

9884 palm oil — See palm kernel oil.

9885 pamphlet — See brochure.

9886 pan — See apron.

9887 panchromatic — Said of photographic plates and films sensitive to all visible colours of the spectrum.
f panchromatique — d panchromatisch; farbenempfindlich — n panchromatisch — e pancromático — i pancromatico

9888 panchromatic film — Photographic film sensitive to all visible colours of the spectrum.
f pellicule f panchromatique — d panchromatischer Film m; Panfilm m — n panchromatische film m; panfilm m; panchromatisch materiaal n — e película f pancromática — i pellicola f pancromatica

9889 panel; title piece; lettering piece; label — Panel on the spine of a book with the name of the author and the title of the book, usually made of coloured leather.
f entrenerfs m; pièce f de titre — d Rückenschild n; Rückenfeld n; Titelfeld n — n rugschildje n — e etiqueta f del lomo; recuadro m; panel m; entrenervios m; tejuelo m; tejo m Ar — i casella f del titolo

9890 panorama drawing
f croquis m panoramique; croquis m perspectif — d Panoramazeichnung f — n perspectieftekening f — e dibujo m panorámico — i disegno m panoramico

9891 panoramic printing — Continuous printing over two pages, only important for the inner pages, since in this way an impression is obtained without a white gutter.
f impression f en double page — d Panoramadruck m — n druk m over twee pagina's — i stampa f panoramica

9892 pantograph — Instrument for mechanically copying drawings, engravings, designs, plans, etc., either on the same scale or on reduced or enlarged scale.
f pantographe m — d Pantograph m; Storchschnabel m — n pantograaf m — e pantógrafo m — i pantografo m

9893 Pantone process — Photoengraving process, where a printing plate bearing an image that is not in relief is made repellent to ink in the bare parts by treating these with mercury, while the printing parts will take up ink. Such plates can be printed on any machine that will give printing pressure and will yield thousands of good prints on any kind of paper.
f procédé m Pantone — d Pantoneverfahren n; Pantonedruck m — n pantone-procédé n — e procedimiento m Pantone — i procedimento m Pantone

9894 paper — The two divisions of paper are

paper as a general term, and paperboard. In Europe paper has a maximum weight of approx. 165 g/m², in US paper is less than 12 points (0.30 mm or 0.012 in) thick. See also cardboard.

f papier *m* — **d** Papier *n* — **n** papier *n* — **e** papel *m* — **i** carta *f*

9895 paperback *GB*; **pocket book** *US* — Small, especially paperback book that can be carried in the pocket. See also hard-back book.

f livre *m* de poche — **d** Taschenbuch *n* — **n** pocketboek *n* — **e** libro *m* de bolsillo — **i** libro *m* tascabile; tascabile *m*

9896 paperback — British name for *pocket book, that is a book with a paper back (same paper as the cover), as distinctive from books which have a hard back and which are called hard-back books.

9897 paper backed — See paper bound.

9898 paper bag — See also bag.

f sac *m* en papier — **d** Papiertüte *f*; Papierbeutel *m*; Papiersack *m* — **n** papieren zak *m*; winkelzak *m* — **e** bolsa *f* de papel; saco *m* de papel — **i** sacchetto *m* di carta

9899 paperboard — One of the two broad subdivisions of *paper, more than 0.3 mm (0.012 in) thick and made from various fibre furnishes. Examples: container board; boxboard; setup boxboard; tube board; special food board.

f carton *m* mince — **d** Karton *m* — **n** karton *n* — **e** cartulina *f* — **i** cartoncino *m*

9900 paper bound; paper backed — Bound with a paper cover, sometimes with a distinctive pattern. A paper bound book is called a *paperback. See also cloth bound.

f broché — **d** broschiert — **n** gebrocheerd; met papieren omslag — **e** encuadernado en rústica — **i** legato in carte

9901 paper bowl — See bowl.

9902 paper box maker — Manufacturer of paper boxes and cartons.

f cartonnier *m* — **d** Kartonagenfabrik *f* — **n** kartonnagefabriek *f* — **e** cartonería *f*; fábrica *f* de cajas de cartón; cartonaje *m* — **i** fabbrica *f* di cartonaggi; scatolificio

9903 paper brush; brush — Device to remove surface dust from paper, used in a printing press.

f dispositif *m* de dépoussiérage — **d** Entstaubungseinrichtung *f* — **n** stofverwijderingsapparaat *n* — **i** dispositivo *m* per rimuovere la polvere

9904 paper cement — Adhesive made from rubber latex and a volatile solvent or vehicle.

f colle *f* caoutchouc à papier — **d** Papierkleber *m* aus Gummilösung — **n** rubberlijn *m* — **e** cola *f* de caucho — **i** adesivo *m* per carta in soluzione gommosa

9905 paper chromatography — Micro type of *chromatography. A drop of the liquid to be investigated is placed near one end of a strip of paper. This end is immersed in solvent, which travels down the paper and distributes the materials present in the original drop selectively. Comparison with known substances makes identification possible.

f chromatographie *f* sur papier — **d** Papierchromatographie *f* — **n** papierchromatografie *f* — **e** cromatografía *f* sobre papel — **i** cromatografia *f* su carta

9906 paper clip — Flat wire clip that can clasp sheets of paper between two of its projecting parts.

f agrafe *f*; happeur *m* (pour papiers); attache-papiers *m* — **d** Papierklammer *f*; Klammer *f*; Büroklammer *f* — **n** paperclip *m* — **e** sujetador *m*; sujetapapeles *m* — **i** agrafe *f*; fermaglio *m*

9907 paper conditioner; paper conditioning machine; paper seasoning machine — Equipment to bring paper into equilibrium with the printing room temperature and humidity.

f conditionneuse *f*; installation *f* de conditionnement de papier; appareil *m* de conditionnement de papier — **d** Klimaanlage *f*; Klimatisierungsanlage *f*; Bewetterungsanlage *f* — **n** conditioneerinrichting *f*; conditioneerinstallatie *f*; klimatiseerinrichting *f* — **e** instalación *f* de acondicionamiento para papel — **i** impianto *m* per il condizionamento della carta; condizionatore *m* per carta

9908 paper conditioning — See conditioning.

9909 paper conditioning machine — See paper conditioner.

9910 paper consumption

f consommation *f* de papier — **d** Papierverbrauch *m* — **n** papierverbruik *n* — **e** consumo *m* de papel — **i** consumo *m* di carta; fabbisogno *m* di carta

9911 paper converting process — See converting process.

9912 paper cutter — See guillotine.

9913 paper drill — See drilling machine.

9914 paper dust — Fibre (fluff) or loading (dust) which leaves the web as it passes through the printing press or slitters.

f poussière *f* (de papier) — **d** Staub *m*; Papierstaub *m* — **n** stof *n*; papierstof *n* — **e** polvo *m*; polvo *m* de papel — **i** polvere *f*; polvere *f* di carta

9915 paper embossment — Term used in magnetic ink printing. See also magnetic encoded cheques.

f embouti *m* du papier — **d** Prägen *n* der Schrift — **n** moet *fm* in het papier — **e** huella *f* en el papel; diente *m* en el papel — **i** rilievo *m* della carta

9916 paper factory — See paper mill.

9917 paper fibres

f fibres *fpl* de papier — **d** Papierfasern *fpl* — **n** papiervezels *fmpl* — **e** fibras *fpl* de papel — **i** fibre *fpl* di carta

9918 paper file — Cover formed by folding a sheet of paper or thin cardboard to keep loose papers together. See also file.

f chemise *f* — **d** Umschlag *m*; Ablegemappe *f*; Hülle *f*; Heftmappe *f*; Schnellhefter *m* — **n** map *fm*; omslag *mn* — **e** papelera *f*; guardapapeles *m* — **i** cartellina *f*

9919 paper hygroscope — See sword hygroscope.

9920 papering — See flashing.

9921 paper in reels — See reel paper.

9922 paper in sheets — Paper supplied in reams of sheets cut to size.

f papier *m* en feuilles — **d** Bogenpapier *n*; Formatpapier *n* — **n** papier *n* aan vellen — **e** papel *m* en hojas — **i** carta *f* in fogli

9923 paper machine — Machine on which cellulose fibres and other constituents of paper are formed into a sheet, which is pressed, dried, calendered and wound into reels or cut into sheets. See also Fourdrinier machine, cylinder machine.

f machine *f* à papier — **d** Papiermaschine *f* — **n** papiermachine *f* — **e** máquina *f* de papel — **i** macchina *f* per carta; macchina *f* continua (per carta)

9924 papermaker

f papetier *m*; fabricant *m* de papier — **d** Papiermacher *m*; Papiermüller *m*; Papierer *m* — **n** papiermaker *m*; papierfabrikant *m* — **e** papelero *m*; papelista *m*; fabricante *m* de papel — **i** cartaio *m*

9925 papermaker's alum — See aluminium sulphate.

9926 paper making

f fabrication *f* du papier — **d** Papierfabrikation *f*; Papierherstellung *f* — **n** papierfabricage *f* — **e** fabricación *f* de papel — **i** fabbricazione *f* della carta

9927 paper mill; paper factory — Factory where paper is manufactured. Termed mill, because the crushing of various materials to pulp usually takes place here.

f fabrique *f* de papier; usine *f* à papier; moulin *m* à papier; papeterie *f* — **d** Papierfabrik *f*; Papiermühle *f* — **n** papierfabriek *f*; papiermolen *m obs* — **e** fábrica *f* de papel; molino *m* de papel; fábrica *f* papelera; papelera *f* — **i** fabbrica *f* di carta; cartiera *f*

9928 paper mulberry tree — See mulberry tree.

9929 paper napkin — Soft, light weight, absorbent paper, plain, crêped, embossed or printed with a design, cut to napkin sizes.

f serviette *f* en papier — **d** Papierserviette *f* — **n** papieren servet *n* — **e** servilleta *f* de papel — **i** salvietta *f* di carta; tovagliolo *m* di carta; tovagliuolo *m* di carta

9930 paper negative — Negative on negative paper instead of plates or films, not suitable for fine work because of the grain of the paper.

f négatif *m* sur papier — **d** Papiernegativ *n* — **n** papieren negatief *n* — **e** negativo *m* de papel — **i** negativo *m* di carta

9931 paper permeability — See permeability.

9932 paper plate — Impregnated or presensitized paper offset plate on which typing and art work can be done directly. Also available presensitized. Not suitable for fine work, long runs, or close register.

f plaque *f* offset en papier — **d** Papieroffsetplatte *f* — **n** papieren offsetplaat *fm* — **e** plancha *f* offset de papel — **i** lastra *f* offset di carta

9933 paper pores

f pores *mpl* du papier — **d** Papierporen *fpl* — **n** papierporiën *fpl* — **e** poros *mpl* de papel — **i** pori *mpl* della carta

9934 paper pulp — See woodpulp.

9935 paper reel bogie — See reel bogie.

9936 paper reel conveying system — Mechanical arrangement to transport paper reels from store to the reel stands of a press.

f transporteur *m* de bobines — **d** Papierrollenförderanlage *f*; Papierrollentransportanlage *f* — **n** papierrollen-transportinrichting *f* — **e** transportadora *f* de bobinas de papel; transportadora *f* de rollos de papel — **i** sistema *m* trasportatore per bobine di carta

9937 paper reel dolly — See reel bogie.

9938 paper reel truck — See reel bogie.

9939 paper scale — See quadrant paper scale.

9940 paper scissors

f ciseaux *mpl* à papier — **d** Papierschere *f* — **n** papierschaar *fm* — **e** tijeras *fpl* para cortar papel — **i** forbici *fpl* da carta

9941 paper seasoning machine — See paper conditioner.

9942 paper shipping sack — See bag.

9943 paper shredder — See shredder.

9944 paper stack — See pile.

9945 paper storage — See storage of paper.

9946 paper store

f magasin *m* à papiers — **d** Papierlager *n* — **n** papiermagazijn *n* — **e** almacén *m* de papel; bodega *f* para papel *CA* — **i** magazzino *m* carta

9947 paper stretch — See stretch.

9948 paper surface

f surface *f* du papier — **d** Papieroberfläche *f* — **n** papieroppervlak *n*; oppervlakte *f* van het papier — **e** superficie *f* del papel — **i** superficie *f* della carta

9949 paper tape — Usually a 7/8 in or 1 in wide reel of paper several hundred feet long, used in paper tape perforators or mounted in computer driven paper tape punches. By extension the coded output recorded on this paper tape (5, 6, 7 or 8-level). Other forms of paper tape include a wider version for 15-level Linofilm keyboards and typesetters or still wider tapes for 31-level Monotype or Monophoto typesetters.

f bande *f* de papier; ruban *m* de papier — **d** Papierstreifen *m*; Papierband *n*; Lochstreifen *m* — **n** papierband *m*; ponsband *m* — **e** cinta *f* de papel; banda *f* de papel — **i** nastro *m* di carta

9950 paper tape punch — Mechanical device associated with a keyboard or a computer which punches or perforates holes in paper tape.

f perforateur *m* de bande de papier — **d** Papierstreifenlocher *m*; Streifenlocher *m* — **n** ponsbandponser *m*; papierbandponser *m* — **e** perforador *m* de cinta de papel — **i** perforatore *m* di nastro di carta

9951 paper-tape reader — See tape reader.

9952 paper test
f analyse *f* de papier; essai *m* du papier — **d** Papierprüfung *f* — **n** papieronderzoek *n* — **e** análisis *mf* del papel — **i** saggio *m* della carta; analisi *f* della carta
9953 paper tower — Part of the Monotype machine, keyboard as well as caster.
f tour *m* à papier — **d** Papierturm *m* — **n** papiertoren *m* — **e** torre *f* del papel — **i** torre *f* della carta
9954 paper tower cam — Part of the Monotype machine.
f excentrique *m* pour tour du papier — **d** Papierturmexzenter *m* — **n** papiertoren-excentriek *n* — **e** excéntrica *f* de la torre del papel — **i** eccentrico *m* della torre della carta
9955 paper truck — See lift truck.
9956 paper web — See web.
9957 papier mâché — Waste papers repulped, mixed with mineral matter, and moulded to serve numerous purposes.
f papier *m* mâché; carton *m* pierre *obs*; carton *m* pâte *obs* — **d** Papiermaché *n*; Pappmaché *n* — **n** papiermâché *n* — **e** cartón *m* piedra; papel *m* mascado; papel *m* majado — **i** cartapesta *f*
9958 papyroline paper — Paper with a linen fabric pasted on one or both sides, or with a fabric between two layers of paper. See also reinforced paper.
f papyrolin *m*; papier *m* entoilé — **d** Papyrolin *n* — **n** papyrolin *n* — **e** papirolina *f*; papel *m* entelado — **i** carta *f* tela; carta *f* foderata di tela
9959 papyrus — Tall sedge (cyperus papyrus) native to the Nile region, the pitch of which was sliced and pressed into matted sheets and used for writing material by the ancient Egytians, Greeks and Romans. Forerunner of paper and origin of the word.
f papyrus *m* — **d** Papyrus *m* — **n** papyrus *m* — **e** papiro *m* — **i** papiro *m*
9960 para- — Prefix added to a chemical name meaning the involving substitution at or characterized by two opposite positions in the benzene ring that are separated by two carbon atoms, e.g., paradichlorobenzene.
9961 para-aminophenol; para-hydroxyaniline — White or reddish-yellow crystals; turns violet on exposure to light, soluble in water and alcohol. Used as a photographic developer. Formula: $C_6H_4NH_2OH$.
f paraminophénol *m* — **d** Paraminophenol *n*; para-Aminophenol *n* — **n** para-aminofenol *n* — **e** para-aminofenol *m*; paraminofenol *m* — **i** para-amminofenolo *m*
9962 para-aminophenol hydrochloride — Light-brown to white crystals, soluble in water and alcohol. Used as dye and photographic chemical. Formula: $C_6H_7ON \cdot HCl$.
f chlorhydrate *m* de paraminophénol; paramino-phénol-chlorhydrate *m* — **d** para-Aminophenol-hydrochlorid *n*; salzsaures Paraminophenol *n* — **n** paraminofenol-hydrochloride *n* — **e** clorhidrato

m de paraminofenol — **i** idrocloruro *m* di para-amminofenolo
9963 para-diaminobenzene — See para-phenylenediamine.
9964 para-dichlorobenzene — White, volatile, crystalline compound, soluble in alcohol, benzene and ether, insoluble in water, widely used as an insecticide and in the manufacturing of dyestuffs. Formula: C_6H_4Cl.
f para-dichlorobenzène *m* — **d** para-Dichlorbenzol *n* — **n** para-dichloorbenzeen *n* — **e** para-diclorobenceno *m* — **i** para-diclorobenzene *m*
9965 paraffin hydrocarbons — See alkanes.
9966 paraffin oil; liquid paraffin — Oil dry-distillated from paraffin distillate. Liquid petrolatum is also known as paraffin oil. See also kerosene.
f huile *f* de paraffine — **d** Paraffinöl *n*; Vaselin-öl *n* — **n** paraffine-olie *fm* — **e** aceite *m* de parafina; parafina *f* liquida — **i** olio *m* di paraffina
9967 paraffin paper; waxed paper — Paper treated with paraffin wax to render it waterproof.
f papier *m* paraffiné; papier *m* ciré — **d** paraffiniertes Papier *n*; Wachspapier *n*; Paraffin-papier *n* — **n** geparaffineerd papier *n* — **e** papel *m* parafinado; papel *m* encerado; papel *m* de cera; papel *m* parafina — **i** carta *f* paraffinata; carta *f* cerata
9968 paraffins — See alkanes.
9969 paraffin wax — Mixture of high-boiling hydrocarbons derived from petroleum, solid at room temperature, classified by their melting points which range from approx. 40 to 90 °C, soluble in benzene, ligroin, warm alcohol, chloroform, turpentine, carbon disulphide, insoluble in water and acids. Used in printing inks, for sizing of paper, in coatings for carbon paper, and as an impregnating material for waxed papers. See also kerosine.
f paraffine *f*; cire *f* de paraffine — **d** Paraffin *n*; Paraffinwachs *n* — **n** paraffine *fm* — **e** parafina *f*; cera *f* de parafina — **i** paraffina *f*; cera *f* di paraffina
9970 paraformaldehyde — White solid with slight odour of formaldehyde, insoluble in alcohol and ether, soluble in strong alkali solution. Formula: $(HCOH)_n$.
f paraformaldéhyde *f* — **d** Paraformaldehyd *n* — **n** paraformaldehyde *n* — **e** paraformaldehído *m* — **i** paraformaldeide *f*
9971 paragon — Old name for a size of type. See app. no. 1.
9972 paragraph; break; section — Matter between one break line and the next, usually identified by an indention of the first line and a break in the last line. Last line should have more than five letters, except in narrow measures. Never to begin a page.
f alinéa *m*; paragraphe *m* — **d** Abschnitt *m*; Absatz *m*; Alinea *n* — **n** alinea *f* — **e** párrafo *m*; acápite *m*; apartado *m*; aparte *m*; inciso *m* — **i** capoverso *m*; alinea *m*; paragrafo *m*

9973 paragraph mark; break mark — Typographical sign ¶ showing where a new paragraph is to begin. Used when for some reason, such as for typographic effect, the whole of the matter is set solid, with no break-lines. Sometimes indicated NP (new paragraph).
f pied *m* de mouche; signe *m* de paragraphe; signe *m* d'alinéa — **d** Absatzzeichen *n*; Alineazeichen *n* — **n** alineateken *n* — **e** señal *f* de párrafo; señal *f* de apartado; párrafo *m* aparte; punto y aparte; punto y acápite; antígrafo *m*; calderón *m* — **i** segno *m* di capoverso; segno *m* di paragrafo

9974 para-hydroxyaniline — See para-aminophenol.

9975 para-hydroxybenzene — See hydroquinone.

9976 para-hydroxyphenylglycine — See glycine.

9977 parallax — Shift in location of a photographic image or its register mark with the angle of sight when they are separated from the key to which they are positioned as in sighting through two negatives that are separated by the thickness of the flat.
f parallaxe *f* — **d** Parallaxe *f* — **n** parallax *m* — **e** paralaje *f* — **i** parallasse *f*

9978 parallel connection — Arrangement of the components, as resistances, of a circuit in such a way that all positive terminals are connected to one point and all negative terminals are connected to a second point.
f accouplement *m* parallèle; couplage *m* en parallèle — **d** Parallelschaltung *f*; Nebeneinanderschaltung *f* — **n** parallelschakeling *f*; parallelverbinding *f* — **e** acoplamiento *m* en paralelo; conexión *f* en paralelo — **i** collegamento *m* in parallelo

9979 parallel fold — See also jaw fold.
f pli *m* parallèle — **d** Parallelfalz *m* — **n** parallelvouw *fm* — **e** plegado *m* en paralelo; doblado *m* en paralelo — **i** piega *f* parallela

9980 parallel fold — See jaw fold.

9981 parallel folding
f pliage *m* parallèle — **d** Parallelfalzung *f* — **n** parallelvouwen *n* — **e** doblar en paralelo; plegar en paralelo — **i** piegatura *f* parallela

9982 parallelism — Condition in process cameras when the copyboard, lensboard, halftone screen holder and focal plane are all exactly parallel and square with each other.
f parallélisme *m* — **d** Parallelismus *m* — **n** parallellisme *f*; evenwijdigheid *f* — **e** paralelelismo *m* — **i** parallellismo *m*

9983 parallel mark — Printer's sign // , used as the fifth reference mark for *footnotes.
f signe *m* de barres parallèles — **d** Parallelzeichen *n* — **n** parallelteken *n* — **e** signo *m* de paralelo — **i** segno *m* di parallela

9984 parallel motion — Impression mechanism to maintain a tight web during raising and lowering of impression cylinder.
f mouvement *m* parallèle — **d** gleichlaufende Bewegung *f* — **n** evenwijdige beweging *f*; parallelle beweging *f* — **e** movimiento *m* paralelo — **i** movimento *m* parallelo

9985 parallel operation — Performing a number of actions at the same time; transmitting or storing a computer word or part of a word simultaneously. Generally faster than serial operation in which all bits are handled sequentially.
f opération *f* parallèle — **d** Paralleloperation *f* — **n** parallelle operatie *f* — **e** operación *f* paralela — **i** operazione *f* parallela

9986 parallel-plate viscometer — Plastometer in which the flow is radial and takes place between two parallel plates that can be either pulled apart or squeezed together.
f viscosimètre *m* à disques parallèles — **d** Parallelplattenviskosimeter *n* — **n** viscosimeter *m* met evenwijdige platen — **e** viscosímetro *m* con placas paralelas — **i** viscosimetro *m* a dischi paralleli

9987 parallel rule; double rule — Rule with two parallel lines of equal weight.
f filet *m* double; filet *m* gouttière — **d** doppelfeine Linie *f* — **n** dubbelfijne lijn *fm* — **e** caña *f* (sencilla); filete *m* caña; raya *f* de caña sencilla; filete *m* de doble pelo; doble sencillo *m Ch* — **i** filetto *m* doppio

9988 parameter — In general, a quantity constant under specific conditions. In computer typesetting, for example, the specifications which may be stipulated by the use of a computer program (or the hardwired logic of a similar, non programmable device, where values may be "dialed in" for reference purposes). These specifications provide conditions which will govern calculations or logical operations, and may indicate, for example, the minimum interword space (prior to an attempt at hyphenation), whether or not letterspacing will be permitted, and if so, to what extent. Parameters for point size, leading, and for measure may also be stipulated. Parameters for page make-up (pagination), such as page depth, the expansion or contraction of secondary leading, or the use of carding, may also be specified in some systems of programs.
f paramètre *m* — **d** Parameter *n* — **n** parameter *m* — **e** parámetro *m* — **i** parametro *m*

9989 para-methyl-aminophenol sulphate — See methyl-paraminophenol sulphate.

9990 para-nitraniline red — See para red.

9991 paranol — See rodinal.

9992 para-phenylenediamine; para-diaminobenzene — White to light purple crystals, soluble in alcohol, ether, slightly soluble in cold water and chloroform. It is a non staining, clear working developer for tank use, gives fine-grained negatives.
Formula: $C_6H_4(NH_2)_2$.
f para-phénylène-diamine *f*; para-diaminobenzène *m* — **d** para-Phenylendiamin *n* — **n** para-fenyleendiamine *fmn* — **e** para-fenilendiamina *f* — **i** para-fenilendiammina *f*

9993 para red; para-nitraniline red — Organic red ink pigment. Also a group of lakes based on this dye.

Formula: $NO_2C_6H_4 \cdot NC_{10}H_6OH$.

f rouge *m* para — **d** Pararot *n* — **n** pararood *n* — **e** rojo *m* para — **i** rosso *m* para

9994 parchment — Skin of goat, sheep or other animal prepared with lime, like vellum, for writing and printing, and also for bookbindings.

f parchemin *m* — **d** Pergament *n* — **n** perkament *n* — **e** pergamino *m* — **i** pergamena *f*

9995 parchment binding; vellum binding

f reliure *f* plein parchemin; reliure *f* en parchemin — **d** Pergamentband *m* — **n** perkamenten band *m* — **e** encuadernación *f* de pergamino — **i** legatura *f* in pergamena

9996 parchment paper — See vegetable parchment paper.

9997 parchment-paper, imitation — See imitation parchment-paper.

9998 parentheses; round brackets; fingernails *coll* — Brackets to enclose words, clauses, or sentences inserted in a passage to which they are not grammatically essential, or which are explanatory. In GB () are known as brackets, [] as square brackets. See also punctuation marks.

f parenthèses *fpl* — **d** Parenthesen *fpl*; (runde) Klammern *fpl* — **n** parenthesen *fpl*; parentheses *fpl*; (ronde) haakjes *npl* — **e** paréntesis *mpl* (redondos) — **i** parentesi *fpl* tonde

9999 paring — Thinning of the edges of leather for book covers.

f parage *m* — **d** Schärfen *n*; Ausschärfen *n*; Abschärfen *n* — **n** dunnen *n* — **e** devastar; chiflar — **i** scarnitura *f*

10000 paring knife; edging knife — Knife for paring leather.

f couteau *m* à parer; paroir *m*; couteau *m* à dobler — **d** Schärfmesser *n*; Abschärfmesser *n* — **n** dunmes *n*; boekbindersmes *n* — **e** chifla *f*; cuchilla-serpeta *f* — **i** coltello *m* per scarnire

10001 paring stone — Usually a piece of lithographer's stone used as a base for hand-paring.

f pierre *f* à parer — **d** Schärfstein *m* — **n** dunsteen *m* — **e** piedra *f* para rebajar (pieles) — **i** pietra *f* per scarnire

10002 Paris black — See bone black.

10003 Paris blue — See Prussian blue.

10004 parisienne — The French name for the type case mostly used in France. See app. no. 1.

10005 Paris white — See whiting.

10006 Paris yellow — See lead chromate.

10007 parity — Term used to indicate that two integers are both odd or both even. In an odd parity system the parity bit is 1 if the recorded number of 1's in the information channels is even and 0 if the number of 1's is odd. In the case of an even parity system the parity bit is 1 if the number of 1's in the information channels is odd and 0 if the number of 1's is even.

f parité *f* — **d** Parität *f* — **n** pariteit *f* — **e** paridad *f* — **i** parità *f*

10008 parity bit; odd/even bit — Bit position added to a group of binary digits to make the total number of 1's odd or even. Thus on paper or magnetic tape the code row for a 6-bit character would be in 6 channels plus an additional parity channel to provide an odd or even number of 1's in each code row.

f bit *m* de parité — **d** Paritätsbit *n* — **n** pariteitsbit *n*; even/oneven bit *n* — **e** bit *m* de paridad — **i** bit *m* di parità

10009 parity check; odd/even check — Testing of parity to detect recording or transmission errors.

f contrôle *m* de parité — **d** Paritätsprüfung *f*; Bitzahlprüfung *f* — **n** pariteitscontrole *fm*; even/oneven controle *fm*; pariteitstoets *m*; even/oneven toets *fm* — **e** control *m* de paridad — **i** controllo *m* di parità

10010 pars — Compositors' term for *parentheses.

10011 part

f tome *m* — **d** Teil *m* — **n** deel *n* — **e** tomo *m* — **i** tomo *m*

10012 part

f fascicule *m* — **d** Heft *n* — **n** aflevering *f* — **e** fascículo *m* — **i** fascicolo *m*; dispensa *f*

10013 particle shape

f forme *f* de particule; forme *f* granulaire — **d** Teilchenform *f*; Korngestalt *f* — **n** deeltjesvorm *m* — **e** forma *f* de la partícula — **i** forma *f* della particella

10014 particle size; size of particle — Arbitrary measurement such as particle volume, particle surface, and horizontal diameter. In the case of regular materials such as cubes, it can be designated by the edge, the diagonal of the face, or the diagonal through the centre of the particle. Not to be confused with average particle size.

f grosseur *f* des particules — **d** Teilchengröße *f* — **n** deeltjesgrootte *f* — **e** tamaño *m* de la partícula — **i** grandezza *f* delle particelle; dimensione *f* delle particelle

10015 particle size analysis; dispersional analysis

f analyse *f* granulométrique; analyse *f* dispersoïdale — **d** Feinheitsanalyse *f*; Dispersoidanalyse *f* — **n** fijnheidsanalyse *f*; dispersieanalyse *f*; deeltjesgrootte-bepaling *f* — **e** análisis *mf* dispersoidal — **i** analisi *f* granulometrica; analisi *f* dispersionale

10016 partition *v* — See segment *v*.

10017 partly bleached pulp — Pulp having, as a result of a moderate chemical post-treatment, a degree of whiteness lower than when bleached.

f pâte *f* à papier semi-blanchie; pâte *f* chimique mi-blanchie — **d** halbgebleichter Halbstoff *m*; halbgebleichter Zellstoff *m* — **n** halfgebleekte celstof *fm* — **e** pasta *f* semiblanqueada; pulpa *f* semiblanqueada — **i** pasta *f* chimica semibianchita

10018 pascal — Unit of pressure. Symbol: Pa. In physics, the law that an external pressure applied to a fluid in a closed vessel, is uniformly transmitted throughout the fluid. In the SI:
1 Pa = 1 N/m^2.

10019 pass — Processing run through a computer to accomplish a specific result, as an editing or correction pass, or an "h and j" pass, more usually associated with a batch-processing computer programming procedure.

f passe *f* — **d** Arbeitsgang *m*; Durchgang *m*; Durchlauf *m* — **n** doorgang *m*; gang *m* — **e** pasada *f* — **i** passata *f*

10020 pass *v* **for press; give** *v* **the OK**

f donner le bon à tirer — **d** Gut zum Druck geben — **n** toestemming tot afdrukken geven; fiatteren; fiat geven — **e** dar el visto bueno; dar el V°.B°.; aprobar; dar el imprímase — **i** dare alla stampa

10021 paste — See starch glue.

10022 pasteboard — Both paperboard and cardboard that are formed by pasting a liner on stock of a different grade; popularly any stiff board or cardboard of medium thickness.

f carton *m* collé — **d** geklebter Karton *m*; Klebekarton *m*; Schichtenpappe *f*; Pappdeckel *m* — **n** geplakt karton *n* — **e** cartulina *f* pegada; cartón *m* pegado; cartón *m* de varias capas — **i** cartone *m* a più strati incollati

10023 paste brush; brush

f brosse *f* à colle; pinceau *m* à colle — **d** Kleisterpinsel *m*; Leimpinsel *m* — **n** stijfselkwast *m*; lijmkwast *m* — **e** brocha *f* de la cola; pincel *m* de la cola — **i** pennello *m* da colla

10024 pasted — Said of covers, bristols or paperboards, in which two or more thicknesses (which may or may not be of the same stock) have been pasted together as an operation separate from their manufacture on the paper machine, such as pasted blanks, pasted blotting, pasted bristol, etc.

f collé; doublé — **d** geklebt — **n** geplakt — **e** pegado — **i** collato

10025 pasted bristol — Bristol in which two or more thicknesses of paper, which may or may not be of the same stock, are pasted together as a secondary operation, generally by a machine but in the case of some wedding bristols hand pasting is used.

f carte *f* bristol collée — **d** geklebter Bristolkarton *m* — **n** geplakt bristolkarton *n* — **e** cartulina *f* bristol pegada

10026 pasted cover paper — Cover paper prepared by pasting two or more sheets together, often called double thick, triple thick, or double double thick, depending on the number of thicknesses pasted together. In many cases the individual plies have a basis weight of 65 pounds (20 x 26 - 500).

f papier *m* pour couvertures collé — **d** geklebtes Umschlagpapier *n* — **n** geplakt omslagpapier *n* — **e** papel *m* de cubiertas de varias capas — **i** carta *f* da copertina accoppiata con adesivo

10027 paste down — Part of the *end leaves which is pasted onto the board of a book. See also fly leaf.

f feuille *f* de garde attachée — **n** aangeplakt deel *n* van het schutblad

10028 paste drier — Ink drier, usually a combination of lead and manganese, which tends to dry throughout with tough surface film but less gloss than cobalt driers. Manganese driers are used alone only for non scratch inks. Some paste driers are known as trimetal driers, and contain cobalt, manganese and lead.

f siccatif *m* en pâte — **d** Trockenpasta *f* — **n** droogpasta *mn* — **e** secante *m* de pasta; pasta *f* secante — **i** siccativo *m* in pasta

10029 paste duct — Container for adhesive on a rotary press.

f bac *m* à colle — **d** Klebstoffkasten *m*; Klebekasten *m* — **n** kleefstofbak *m* — **e** artesa *f* de la cola; pastera *f* — **i** vaschetta *f* dell'adesivo

10030 paste gun — Hand-operated appliance to facilitate the even distribution of adhesive to the free end of a reel of paper.

f distributeur *m* de colle à main; appareil *m* de collage à main — **d** Handklebeapparat *m* — **n** handaanlijmapparaat *n*; handlijmapparaat *n*; lijmapparaat *n* — **e** distribuidor *m* manual de cola; empastador *m* — **i** apparecchio *m* per incollatura a mano

10031 paste lines — Shadows and marking created by illustrations and type proofs pasted into an original as additions.

f marques *fpl* de collage — **d** Kleisterstreifen *fpl* — **n** plakstrepen *fmpl*; lijmstrepen *fmpl* — **e** trazos *mpl* de pegado — **i** segni *mpl* di incollatura; ombre *fpl* di incollatura

10032 pastel inks — Light tints, like those obtained with coloured crayons.

f couleurs *fpl* pastel — **d** Pastellfarben *fpl* — **n** pastelinkten *mpl* — **e** tintas *fpl* al pastel — **i** inchiostri *mpl* (ad tonalità) pastello

10033 pastel shades — Pale, delicate tints in ink.

f pastel *m*; couleurs *fpl* tendres — **d** Pastellfarben *fpl* — **n** pasteltinten *fmpl*; pastelkleuren *fmpl* — **e** colores *mpl* pastel; colores *mpl* delicados; colores *mpl* suaves; colores *mpl* tiernos — **i** tinte *fpl* pastello; colori *mpl* pastello

10034 paste *v* **on** — To fasten or stick with past or the like.

f coller sur — **d** aufkleben — **n** opplakken — **e** encolar — **i** incollare

10035 paster — Device to apply fine line of paste on either or both sides of the web, in the direction of web travel, to produce finished booklets in which paste replaces stitching. See also flying paster.

f colleuse *f* — **d** Beleimvorrichtung *f*; Beleimer *m*; Klebeapparat *m*; Klebevorrichtung *f*; Anleimvorrichtung *f* — **n** rug-lijmapparaat *n* — **e** empalmador *m* — **i** dispositivo *m* incollatore

10036 paster disk; paster head — Device to apply a strip or film of adhesive onto the paper web in a rotary press.

f molette *f* encolleuse; tête *f* encolleuse — **d** Klebescheibe *f*; Klebekopf *m* — **n** kop *m* van het aanlijmapparaat; kleefstofschijf *fm* — **e** cabeza *f* encoladora; boquilla *f* encoladora — **i** testa *f* incollatrice

10037 paster head — See paster disk.

10038 paste royal — A British standard size for board. See app. no. 3.

10039 paster pump — Device to feed adhesive to a paster disc or head in a rotary press.
f pompe f à colle — **d** Klebstoffpumpe f — **n** kleefstofpomp fm — **e** bomba f de cola — **i** pompa f della colla

10040 paster tab — Strip of gummed paper with a weakened section or other device used to secure the prepared end of the new reel before joining it to the old reel on an automatic reel change.
f bande f de collage; étiquette f de collage — **d** Klebestreifen m; Klebemarke f — **n** plakband n — **e** cinta f engomada — **i** nastro m gommato

10041 pasteup — Preparation of copy, putting each component (veloxes, type glazed proofs) in proper position before photographing. The usual method of assembling copy elements including text for reproduction.
f maquette f collée en pages; maquette f collée de mise en pages — **d** geklebter Umbruch m; geklebte Montage f; Klebeumbruch m; Klebemontage f — **n** ingeplakte kopij f — **i** montaggio m incollato

10042 pasting joint — See joint.

10043 pasting paper
f papier m à doubler; papier m pour collage; papier m de collage pour cartonnages — **d** Beklebepapier n; Bezugspapier n; Kaschierpapier n — **n** beplakpapier n; papier n voor het beplakken — **e** papel m para forrar; papel m para pegar; papel m de forro — **i** carta f di rivestimento; carta f per accoppiamento

10044 pasting royal — A size of printing paper. See app. no. 3.

10045 patch — Hand cut overlay.
f béquet m — **d** Musche f; Handausschnitt m — **n** handpikeersel n; pikeersel n — **e** parche m; recorte m; alza f — **i** taccheggio m a mano; ritaglio m a mano

10046 patch — Temporary electrical connection.
f connexion f provisoire — **d** Notverbindung f — **n** noodverbinding f; tijdelijke verbinding f; provisorische verbinding f — **e** conexión f provisional — **i** connessione f provisoria

10047 patch v — To modify a computer program by adding or substituting subroutines, usually without re-assembling the source program.
f raccorder — **d** korrigieren — **n** corrigeren — **e** remender — **i** raccordare

10048 patch board — See plug board.

10049 patching sheet — See sticking-up sheet.

10050 patchless mat — See packless mat.

10051 patchless stereo mat — See packless mat.

10052 patch panel — See plug board.

10053 patch v **the blanket** — To lay paper under the offset rubber blanket.
f coller des béquets au blanchet — **d** das Gummituch unterlegen mit Papier oder Karton — **n** onderlegsels aanbrengen onder het rubberdoek — **e** pegar una alza — **i** disporre fogli di carta sotto il rivestimento di gomma

10054 patée — See pattée.

10055 patent base — See patent bloc.

10056 patent bloc; patent base — Sectional metal blocks used as support for printing plates and provided with means to hold the plates in position on the press.
f bloc m systématique — **d** Patentunterlage f — **n** patentvoet m — **e** base f de celdillas; platina f de celdillas — **i** blocco m sistematico; zoccolo m sistematico

10057 pat-out — Colour swatch made by tapping the ink out on paper with fingers. See also tap v out.
f essai m d'encre par tache — **d** Austüpfel m — **n** uitstrijk m; uitstrijksel n — **e** ensayo m de tinta por manchas — **i** macchia f di prova (spalmata a mano)

10058 pat v **out** — See tap v out.

10059 patrol — See brunak.

10060 patrol solution — For an example of a patrol solution, see brunak.

10061 pattée; patée; paty — Spreaded toward the extremity as a cross with arms of equal length, each expanding outward from the centre, used to describe book cover decoration styles.
f patté; patée — **n** naar buiten gespreid — **i** aperto alle estremità

10062 pattern — See moiré.

10063 pattern paper — Strong Fourdrinier paper used in the cutting out of cloth.
f papier m pour patrons (découpe); papier m de coupe; papier m de couture — **d** Schnittmusterpapier n; Modellpapier n — **n** patronenpapier n; patroonpapier n; papier n voor knippatronen — **e** papel m para patrones de labores — **i** carta f da modelli (per abiti); carta f per modelli

10064 pattern plate — See master plate.

10065 pattern recognition — Automatic identification of shapes, forms or configurations.
f reconnaissance f de configuration — **d** Strukturerkennung f — **n** vaststelling f van de configuratie — **e** reconocimiento m de configuración — **i** riconoscimento m di configurazione

10066 paty — See pattée.

10067 pawl — Pivoted bar adapted to engage with the teeth of a ratchet wheel to prevent movement or to impart motion.
f rochet m d'arrêt; cliquet m d'arrêt; chien m d'arrêt — **d** Sperrklinke f; Sperrhaken m; Klemmklinke f — **n** pal m; klink fm — **e** gatillo m de trinquete; linguete m; retén m; retenedor m — **i** nottolino m (d'arresto); dente m d'arresto; scatto m

10068 pay day — Day on which wages are given.
f jour m de paie; jour m de paiement; (jour m de) banque f sl. — **d** Löhnungstag m; Lohntag m; Zahltag m; Zahlungstag m — **n** betaaldag m — **e** día m de paga — **i** giorno m di paga

10069 peacock blue — Blue organic pigment used especially in inks for multi-colour printing; a lake of acid glaucine blue dye on alumina hydrate. Sometimes applied to other blue pigments, such as Prussian blue which has been treated with phosphotungstic acid. The hue is often referred to as cyan.

f bleu *m* paon — **d** Pfaublau *n* — **n** pauwblauw *n* — **e** azul *m* pavo real — **i** blu *m* pavone

10070 peak value
f valeur *f* de crête — **d** Scheitelwert *m*; Scheitelpunkt *m* — **n** topwaarde *f* — **e** valor *m* de vértice — **i** valore *m* di punta

10071 pearl — Obsolete name for a body size of type. See app. no. 1.

10072 pearl ash — See potassium carbonate.

10073 pearl hardening — See crown filler.

10074 pearl white; bismuth oxychloride — White pigment.
Formula: BiOCl.
f blanc *m* de perle; oxychlorure *m* de bismuth — **d** Perlweiß *n*; Wismutoxychlorid *n* — **n** bismutoxychloride *n* — **e** oxicloruro *m* de bismuto — **i** ossicloruro *m* di bismuto

10075 pearwood — Wood used by engravers as substitute for boxwood. Too soft to be used for fine woodcuts.
f poirier *m* — **d** Birnbaumholz *n* — **n** perehout *n* — **e** peral *m*; madera *f* de peral — **i** pero *m*; legno *m* di pero

10076 pebbling — See graining.

10077 pebbling — See mottle.

10078 peck — Unit of dry measure. Abbrev. pk. See app. no. 7 and 7a.

10079 pectin — Water-soluble substance in plant tissues that yields a gel. See also encrust.
f pectine *f* — **d** Pektin *n* — **n** pectine *fmn* — **e** pectina *f*

10080 peculiars — Out-of-the-way type characters, i.e., accents, etc., and any sort not included in a standard fount of type.
f caractères *mpl* spéciaux — **d** Sonderzeichen *npl*; Punkturen *fpl* — **n** speciale tekens *npl* — **e** signos *mpl* especiales — **i** segni *mpl* speciali

10081 peel adhesion test — Quantitative test of adhesion of ink to a treated surface. It is a modified Chapman test which is rather complicated.
f essai *m* de délaminage — **d** Abschältest *m* — **n** pelproef *fm*; afpelproef *fm*

10082 peeling — Coming off of the coating from the base paper. See also delamination.
f décollage *m* de la couche — **d** Abschälen *n* (des Strichs) — **n** loslaten *n* van de strijklaag; loslaten *n* van de couche — **e** soltura *f* del estuco — **i** spelamento *m*; spellatura *f*

10083 peeper — See register tube.

10084 peg rake — See marbling comb.

10085 pelt — Literally doll, a name for the small colour dabber to ink up handengraved plates.
f poupée *f* — **d** (kleiner) Tampon *m* (für Kupferdruck); Ballen *m*; Farbballen *m* — **n** tampon *m* — **e** tampón *m* — **i** tampone *m*

10086 pelt *v* — To beat or descend with violence, sometimes in connection with the dabbers used by copperplate printers.
f être en colère; avoir son bœuf; gober son bœuf; prendre son bœuf — **d** böse sein — **n** boos zijn; kwaad zijn; nijdig zijn — **e** estar disgutado — **i** essere adiarto

10087 pelt ball — See ink ball.

10088 penalty copy — Copy difficult to compose (heavily corrected, faint, foreign language, etc.), for which the typesetter charges extra.
f pensum *m* — **d** schwer abzusetzendes Manuskript *n*; schlechtes Manuskript *n* — **n** moeilijk te zetten kopij *f* — **i** originale *m* difficile

10089 pen and ink drawing; pen drawing — Black-and-white drawing made by pen and India ink, in lines, dots, or stipples.
f dessin *m* à la plume — **d** Federzeichnung *f* — **n** pentekening *f* — **e** dibujo *m* de pluma (y tinta); dibujo *m* a pluma — **i** disegno *m* a penna; schizzo *m* a penna

10090 pen and ink keys — Key tracing of a subject made on a transparent plastic sheet with pen and tusche. Sometimes used for transferring key outline of artwork to lith plates for making hand plates for posters.
f tracé *m* de contours — **d** Konturzeichnung *f*; Vorzeichnung *f* — **n** contourtekening *f* — **e** diseño *m* de contorno — **i** disegno *m* di contorno

10091 pen and ink test — Test to ascertain ink resistance of writing papers, made by crosshatching small areas of paper with pen and ink. A poorly sized paper absorbs the ink and shows feathering around the drawn or written lines.
f essai *m* à la plume — **d** Tintenfestigkeitsversuch *m*; Tintenprobe *f* — **n** beschrijfbaarheidsproef *f* — **e** ensayo *m* de tinta a la pluma — **i** prova *f* del tratto di penna

10092 pencil drawing
f dessin *m* au crayon — **d** Bleistiftzeichnung *f* — **n** potloodtekening *f* — **e** dibujo *m* de lápiz; dibujo *m* a lápiz — **i** disegno *m* a lapis; disegno *m* a matita

10093 pencil, glair — See glair pencil.

10094 pencil reproduction — Line or highlight halftone reproduction of a drawing made with a lead pencil.
f reproduction *f* d'un dessin au crayon — **d** Reproduktion *f* einer Bleistiftzeichnung — **n** reproduktie *f* naar een potloodtekening — **e** reproducción *f* de un dibujo a lápiz — **i** riproduzione *f* di un disegno a matita

10095 pen drawing — See pen and ink drawing.

10096 Penelope — Pitt Natural Language Processor, a computer language designed to simplify research on collecting large amounts of text in computer compatible form.

10097 penetrated ink
f encre *f* absorbée — **d** weggeschlagene Farbe *f* — **n** weggeslagen inkt *m*; gepenetreerde inkt *m*; geabsorbeerde inkt *m* — **e** tinta *f* penetrada; tinta *f* absorbida; tinta *f* bebida — **i** inchiostro *m* penetrato; inchiostro *m* assorbito

10098 penetration — Ability of a liquid (ink, varnish, or solvent) to be absorbed into the paper or other printing media.
f pénétration *f* — **d** Penetration *f*; Durchdringung *f*; Eindringfähigkeit *f* — **n** penetratie *f*;

indringing f — e penetración f — i penetrazione f

10099 penetration ink — Special ink for paper absorption tests.
f encre f porométrique — d Einschlagfarbe f — n wegslaginkt m — e tinta f porométrica; tinta f de penetración — i inchiostro m di penetrazione

10100 penetration volumeter — Instrument to determine the quantity of oil taken up by the paper at any moment in the course of the penetration process. A disc of paper of 25 cm^2 is clamped in a holder and a viscous oil applied to one side. As the oil penetrates, the displaced air volume is measured in a capillary tube. A plot of this volume against time characterizes the paper's ink absorbency.
f volumètre m de pénétration — d Penetrationsvolumeter nm — n penetratievolumeter m — e volúmetro m de penetración — i volumenometro m di penetrazione

10101 pen-name; pseudonym — Assumed name of the author. The term nom de plume is to be avoided.
f nom m de lettres; nom m de guerre (d'un journaliste); pseudonyme m; nom m de plume (post-War French) — d Pseudonym n; Deckname m — n pseudoniem n; schuilnaam m — e seudónimo m; nombre m de pluma — i pseudonimo m

10102 pennyweight — Unit of weight in the troy system of weights and measures. Abbrev. dwt. See app. no. 7 and 7a.

10103 pen ruling machine
f régleuse f à plumes — d Federliniiermaschine f; Federliniermaschine f — n pennenlinieermachine f — i macchina f rigatrice a penna

10104 penta- — Prefix, denoting five, five fold, in chemistry containing five atoms, groups, or equivalents.

10105 pentachlorophenol — White, crystalline powder, insoluble in water, soluble in dilute alkali, alcohol, acetone, ether, pine oil, benzene, slightly soluble in hydrocarbons. Used in fungicides and in preservatives.
Formula: C_6Cl_5OH.
f pentachlorophénol m — d Pentachlorphenol n — n pentachlorofenol n — e pentaclorofenol m — i pentaclorofenolo m

10106 pentane — Volatile, liquid hydrocarbons of the alkane series, C_5H_{12}, two of which are contained in petroleum and similar compounds. They differ from one another in behaviour with reagents.
f pentane m — d Pentan n — n pentaan n — e pentano m — i pentano m

10107 penumbra; penumbral shadow — In halftone photography theory, the halftone dot formation is presumed to be formed by shadows cast on the plate or film by the opaque lines of the standard halftone screen.
f pénombre f — d Halbschatten m — n halfschaduw fm — e penumbra f — i penombra f

10108 penumbral shadow — See penumbra.

10109 peptization — Spontaneous dispersion of a solid substance in a liquid on adding a small amount of a third substance, the peptizing agent.
f peptisation f — d Peptisation f — n peptisatie f — e peptización f — i peptizzazione f

10110 per- — Prefix denoting, in chemical nomenclature, an excess or the largest possible content or a relatively large proportion of a (specific) chemical element, e.g., peroxide.

10111 per cent defectives
f pourcentage m de défectueux — d Ausfallprozentsatz m — n uitvalpercentage n — e porcentaje m de desechos; porcentaje m de desperdicios; tanto por ciento m de residuos — i percentuale f di scarto

10112 per cent dot area — The black exposed and processed area in a halftone area expressed as per cent of the total area seen.
f pourcentage m de point — d Prozent n bedruckte Fläche; Flächendeckungsgrad m; Rastertonwert m — n percentage n rasteroppervlak — i superficie f percentuale di punto

10113 per cent mark — Typographical character (%) used as a contraction for per cent.
f symbole m de pourcentage; signe m de pourcentage — d Prozentzeichen n — n percentteken n — e signo m de porcentaje; signo m de tanto por ciento — i segno m di percentuale; segno m di per cento

10114 perchloride of iron — See ferric chloride.

10115 perchloride of mercury — See mercury chloride.

10116 perchloromethane — See carbon tetrachloride.

10117 percussion press — Press used in diecutting of envelopes, labels, etc. On pressing a lever or foot treadle, a pawl engages a dog on a rotating shaft which by a toggle drives one platen up to another with a quick sharp motion.
f presse f à percussion — d Schlagstanzmaschine f; Schlagstanzpresse f — n stampers fm — e máquina f de percusión — i pressa f a percussione

10118 perfect — Said of paper free from defects.
f parfait; papier m parfait — d erster Wahl; Papier n erster Wahl — n eerste keus; eerste kwaliteit f papier — e primeras fpl; de primera calidad; papel m de primera calidad; papel m de primer escogido — i prima scelta; carta f di prima scelta

10119 perfect binder — Machine to bind books, where the backs of the sections are cut off, roughened and glued together, and hung in a cover. See also perfect binding.
f brocheuse f automatique sans couture — d Klebebinder m; Klebebindemaschine f — n garenloze bindmachine f — e máquina f de encuadernación rústica; máquina f de encuadernación sin costura; máquina f para encuardernar sin cosido; máquina f de encuadernación a la americana — i macchina f legatrice senza cucitura; macchina f legatrice in colla (plastica)

10120 perfect binding — This term should not be

used since it was applied to, and is the property of, the first machine, invented about 1930.

10121 perfecting — Printing both sides of the paper (or other material) at the same pass through the printing machine.

10122 perfecting press; perfector; perfector press; two-cylinder press; double cylinder machine *obs* — Press with two printing cylinders on which both sides of the sheet or web are printed in one operation. See also blanket-to-blanket press.
f presse *f* à retiration; presse *f* recto-verso — **d** Schön- und Widerdruckpresse *f* — **n** schoon- en weerdrukpers *f* — **e** máquina *f* de retiración; prensa *f* de tiro y retiro; prensa *f* de doble impresión — **i** macchina *f* in bianca e volta

10123 perfectly set page — See perfect page.

10124 perfector — See perfecting press.

10125 perfector press — See perfecting press.

10126 perfect page; perfectly set page — Page or solid type composition without printing errors.
f page *f* sans faute — **d** Jungfer *f* (fehlerfreier Satz; see also app. no. 1) — **n** foutloze pagina *f*; foutloos zetwerk *n*; onberispelijk zetwerk *n* — **e** página *f* perfecta; columna *f* (compuesta) sin errores — **i** pagina *f* perfetta; colonna *f* (composta) senza errori

10127 perfect radiator — See black body.

10128 perforated plate
f plaque *f* perforée — **d** Lochplatte *f*; perforierte Platte *f* — **n** geperforeerde plaat *fm* — **e** placa *f* perforada — **i** lastra *f* perforata

10129 perforated tape — See punched tape.

10130 perforating — Punching small holes or slits in a sheet of paper or cardboard to facilitate tearing along a desired line. See also perforating rule, round hole perforating.
f perforage *m* — **d** Perforierung *f*; Perforieren *n* — **n** perforatie *f*; perforeren *n* — **e** perforación *f* — **i** perforazione *f*

10131 perforating comb — Bar-like perforating tool consisting of two beams. The top beam, which is movable in relation to the bottom beam, carries a set of perforating needles which engage in appropriate holes in the bottom beam and punches the holes in the sheets inserted between the two beams.
f peigne *m* de perforation — **d** Perforierkamm *m* — **n** perforeerkam *m* — **e** peine *m* de perforar; peine *m* perforador — **i** pettine *m* perforatore

10132 perforating machine; perforator — Machine equipped with needles mounted on blade or rotating disc used to perforate sheets of paper.
f perforeuse *f*; machine *f* à perforer — **d** Perforiermaschine *f* — **n** perforeermachine *f*; perforator *m* — **e** perforadora *f*; máquina *f* perforadora — **i** perforatrice *f*; macchina *f* perforatrice

10133 perforating rule — Dotted rule which forces itself into the paper and makes an impression for tearing purposes. Hyphen perforating is also done on machines provided with revolving notched discs.
f filet *m* perforateur — **d** Perforierlinie *f* — **n**

perforeerlijn *fm* — **e** pleca *f* perforadora; pleca *f* para perforar; filete *m* perforador — **i** filetto *m* perforatore; filetto *m* per perforare

10134 perforator; perforator keyboard — Keyboard unit to produce punched paper tape. Each character and function is given a unique code which is punched across the tape. Control of typesetting and film setting equipment requires 6, 7, 8, 15 or 31-level paper tape.
f clavier *m* à perforateur; clavier *m* perforateur — **d** Lochbandlocher *m*; Streifenlocher *m*; Satzlocher *m* — **n** ponsbandponser *m*; ponsbandperforator *m* — **e** teclado *m* perforador; perforador *m*; perforadora *f* — **i** tastiera *f* perforatrice del nastro; perforatore *m*

10135 perforator — See perforating machine.

10136 perforator keyboard — See perforator.

10137 perforator tape — See punched tape.

10138 perfumed inks — Regular printing inks to which a small percentage of concentrated perfume or scent has been added to impart a desired aroma or fragance to the printed sheet.
f encres *fpl* parfumées — **d** parfümierte Druckfarben *fpl* — **n** geparfumeerde drukinkten *mpl* — **e** tintas *fpl* perfumadas — **i** inchiostri *mpl* profumati

10139 perhydrol — See hydrogen peroxide.

10140 perilla oil — Light yellow liquid obtained from the seeds of mints of the genus Perilla. Soluble in alcohol, ether, chloroform, benzine and carbon disulphide. Substitute for linseed oil; used in the manufacture of varnish, printing ink, and artificial leather.
f huile *f* de périlla — **d** Perillaöl *n* — **n** perillaolie *fm* — **e** aceite *m* de perilla — **i** olio *m* di perilla

10141 period — In typography called the *full point, or point.

10142 periodical — See journal.

10143 peripheral equipment — The input/output units and secondary storage units of a computer system.
f équipement *m* périphérique — **d** periphere Einheiten *fpl*; Peripheriegeräte *npl* — **n** randapparatuur *f*; periferie-apparatuur *f* — **e** equipo *m* de periférico; equipo *m* auxiliar — **i** equipaggiamento *m* ausiliario

10144 permanent ink — Ink that does not readily fade or change colour when exposed to light and weather. Also used incorrectly to describe resistance to acids, alkalis, soaps, waxing, alcohols, spirit varnishes, etc.
f encre *f* permanente; encre *f* solide à la lumière — **d** lichtechte Farbe *f* — **n** lichtechte inkt *m* — **e** tinta *f* fija; tinta *f* resistente a la luz; tinta *f* inalterable — **i** inchiostro *m* resistente alla luce

10145 permanent violet — Light-resistant pigment used in printing inks. A tungstated or molybdated methyl violet.
f violet *m* permanent — **d** Fanalviolet *n* — **n** fanaalviolet *n* — **e** violeta *f* fanal — **i** violetto *m* fanal

10146 permanent white — See barium sulphate.

10147 permanent yellow — See lead chromate.

10148 permanganate — Salt of permanganic acid. The term is commonly applied to *potassium permanganate, $KMnO_4$.

f permanganate *m* — **d** Permanganat *n* — **n** permanganaat *n* — **e** permanganato *m* — **i** permanganato *m*

10149 permanganate of potash — See potassium permanganate.

10150 per mark; per sign — The mark ℣ .

10151 permeability; paper permeability — Readiness with which paper or board allow penetration from one surface to the other by a liquid. In strictly scientific sense this definition is not entirely accurate. Permeability is determined under the conditions defined in the standard method of test. It is to use the expression air porosity to designate air permeability.

f perméabilité *f* — **d** Permeabilität *f*; Durchlässigkeit *f* — **n** permeabiliteit *f*; doorlatendheid *f*; doorlaatbaarheid *f* — **e** permeabilidad *f* — **i** permeabilità *f*

10152 peroxide — See hydrogen peroxide.

10153 perpendicular — At right angles, making an angle of 90 °.

f perpendiculaire — **d** senkrecht — **n** loodrecht; haaks — **e** perpendicular — **i** perpendicolare

10154 Perry Higgings process — Method of assembling negatives for newspaper reproduction into page form, then etching plates therefrom as integral units on sheets of magnesium metal, followed by direct newsprinting from the etched plates. Modified form of *Alltone process.

f procédé *m* Perry-Higgins — **d** Perry-Higgins-Verfahren *n* — **n** Perry-Higgins-procédé *n* — **e** procedimiento *m* Parry-Higgins; método *m* Perry-Higgins — **i** procedimento *m* Perry-Higgins

10155 persian morocco leather — Leather usually finished on the grain side. (Do not use capital letters.)

f maroquin *m* grainé artificiellement — **e** marroquín *m* persiano — **i** marocchino *m* persiano

10156 per sign — See per mark.

10157 personal — A size of cut card. See app. no. 3.

10158 personal allowance — See compensating relaxation allowance.

10159 personal binding — Binding on which the initials or the name of the owner are stamped or printed.

f reliure *f* avec chiffre du possesseur — **d** Einband *m* mit Initialen des Besitzers — **n** boekband *m* met de naam van de eigenaar — **e** encuadernación *f* con el nombre del propietario — **i** legatura *f* di nome del proprietario

10160 personal needs allowance — See compensating relaxation allowance.

10161 perspex — Trade-mark for a clear plastic prepared by polymerization of methyl methacrylate. See also acrylic resin.

10162 persulphate of iron — See ferric sulphate.

10163 petition paper; official paper

f papier *m* écu; papier *m* ministre — **d** Kanzleipapier *n* — **n** gezegeld papier *n* — **e** papel *m* ministro — **i** carta *f* da protocollo; carta *f* da cancelleria

10164 petits fers — Small bookbinding tools to impress designs on bindings and by analogy used to describe printers' flowers. See also aux petits fers.

f petits fers *mpl* — **d** kleine Buchbinderstempel *mpl* — **n** kleine boekbinderstempels *npl* — **e** pequeños hierros *mpl* — **i** petits fers *mpl*; piccoli ferri *mpl*

10165 petits fers, aux — See aux petits fers.

10166 petrol — See gasoline.

10167 petrolatum; petroleum jelly; vaseline *trade mark* — Purified mixture of semi-solid hydrocarbons, e.g., one-third microcrystalline wax, one-third resins and one-third oil. Soluble in chloroform, benzene, carbon disulphide and oil of turpentine, slightly soluble in hot alcohol.

f pétrolatum *m*; vaseline *f* — **d** Petrolatum *m*; Vaseline *f*; Vaselin *n* — **n** petrolatum *n*; vaseline *fm* — **e** petrolato *m*; vaselina *f* — **i** petrolato *m*; vaselina *f*

10168 petroleum; crude oil — Liquid mixture of paraffin, naphthene and aromatic hydrocarbons of fossile origin with small amounts of nitrogen and oxygen compounds.

f pétrole *m*; huile *f* brute — **d** Erdöl *n*; rohes Erdöl *n*; Rohöl *n* — **n** aardolie *fm*; ruwe olie *fm* — **e** petróleo *m*; aceite *m* bruto — **i** petrolio *m*; petrolio *m* greggio

10169 petroleum ether — Misleading but widely used term, with specifications supplied by several American scientific societies and government organizations. It is also the legal label name used in the ICC and IATA shipping regulations (CG uses benzine). Latin: aether petrolei.

f éther *m* de pétrole — **d** Petroläther *m*; Petroleumäther *m* — **n** petroleumether *m* — **e** éter *m* de petróleo — **i** etere *m* di petrolio

10170 petroleum fraction — Product of the distillation of crude oils which separate then into many individual fractions, such as gasolines, mineral oils, lubricating oils, etc.

f fraction *f* de distillation du pétrole — **d** Erdölfraktion *f* — **n** petroleumfractie *f* — **i** frazione *f* della distillazione del petrolio

10171 petroleum jelly — See petrolatum.

10172 petroleum naphtha — See naphtha.

10173 petroleum spirit — In GB very volatile hydrocarbon mixture with a flash point below 32 °F. See also SBP spirit.

10174 petroleum spirit — See SBP spirit.

10175 Petzval lens — Lens consisting of two widely separated groups of lenses, each group approximately achromatic, very popular for projection work. Namend after J.M. Petzval (1840), of Vienna.

f objectif *m* Petzval — **d** Petzval-Objektiv *n* — **n** Petzval-objectief *n* — **e** objetivo *m* Petzal — **i** obiettivo *m* Petzval

10176 pewter plate — Tin alloy with 5-15% antimony, 0-3% copper, and 0-15% lead, formerly used in music engraving.
f potin *m*; plaque *f* d'étain — **d** Hartzinn *n*; Zinnplatte *f* für Notenstich — **n** tinplaat *f*; peauter *n*; piauter *n* — **e** peltre *m*; plancha *f* de peltre — **i** peltro *m*; piastra *f* di stagno; piastra *f* per incisione di musica

10177 pH — Means of expressing the degree of acidity or alkalinity of a solution, originally defined as the logarithm of the reciprocal of the hydrogen ion concentration in gram equivalents per litre of solution pH = -log(H^+) and this is in many cases approximately correct. In some cases it is seriously in error. Actually pH values are obtained by measuring the potential E of galvanic cells of the type (Pt)H_2; solution x, saturated KCl, reference electrode; and using the value of E in the equation
pH = (E - E_o)/(2.306 RT/F) in which E_o is a constant depending upon the nature of the reference electrode, and R, T and F are respectively the fundamental gas constant, the absolute temperature and the faraday. Commercial pH meters, however, are calibrated to read pH directly and no calculation is necessary. pH is the abbreviation for the Latin pondus hydrogenii, i.e., weight or mass of hydrogen. The term is said to originate from Sören Peter L. Sörensen (1868). Originally the H was an inferior character.
f pH *m* — **d** pH *m* — **n** pH *m* — **e** pH *m* — **i** pH *m*

10178 pH-control — Measuring the acidity or alkalinity of a solution or substance by determining its hydrogen ion content. The pH of paper, for example, is measured by cutting the paper into small pieces which are soaked in water and then testing the extract (1 g of paper is extracted with 70 ml of water). The extraction time specified is one hour at room temperature. For measuring the pH a calomel and a glass electrode are put into the extract.
f mesure *f* d'un pH — **d** pH-Messung *f* — **n** pH-meting *f*; pH-bepaling *f*; pH-onderzoek *n* — **e** medición *f* de pH; determinación *f* de pH — **i** misurazione *f* del pH

10179 pH-meter
f pH-mètre *m* — **d** pH-Meter *n*; pH-Meßgerät *n* — **n** pH-meter *m* — **e** medidor *m* de pH — **i** misuratore *m* del pH; misuratore *m* di pH

10180 pH-paper strip — Strip of paper containing some test colours in different ranges and a central indicator, to estimate the pH-value of solutions.
f fiche *f* du pH — **d** pH-Papierstreifen *m* — **n** pH-strookje *n* — **e** tira *f* de papel para pH; banda *f* de papel para pH — **i** cartina *f* per il pH; cartina *f* di tornasole per il pH

10181 pH-value — Value on the pH scale of acidity and alkalinity of a liquid or solution; 7.0 is neutral; 0-6.9 is acid; 7.1-14.0 is alkaline. The pH-value of a solution is pH = -log(H^+). See also hydrogen ion concentration.

f valeur *f* du pH — **d** pH-Wert *m* — **n** pH-waarde *f* — **e** valor-pH *m* — **i** valore *m* pH

10182 phantom — Original or reproduction in which certain details are depicted in ghostlike manner to direct attention to other parts.
f image *f* fantôme; image *f* truquée — **d** Abbildung *f* mit durch Retusche zurückgedrängtem Hintergrund — **n** afbeelding *f* met verdoezelde achtergrond — **e** imagen *f* fantasma; impresión *f* fantasma — **i** illustrazione *f* truccata

10183 pharmaceutical symbols — See medical signs.

10184 phase — 1. In chemistry, a mechanically separate, homogeneous part of a heterogeneous system (the solid, liquid, and gaseous phases of a system). 2. In physics, a particular stage or point of advancement in a cycle; the fractional part of the period through which the time has advanced, measured from some arbitrary origin. 3. In electrotechnics, each of the subsequent states of changes in an alternating current. An alternating-current motor may have one, two or three separate windings, each of which is called a phase. A single-phase motor has one such winding and is supplied by two wires. A two-phase motor has two such windings and is supplied by four wires (the two windings may or may not be interconnected). A three-phase motor has three separate windings which are interconnected and is supplied by three wires.
f phase *f* — **d** Phase *f* — **n** fase *f*; faze *f* — **e** fase *f* — **i** fase *f*

10185 phase angle
f angle *m* de phase — **d** Phasenwinkel *m* — **n** fasehoek *m*; fazehoek *m* — **e** ángulo *m* de fase — **i** angolo *m* di fase

10186 phase modulation; angle modulation — Modulation in which the angle of a sine wave carrier is the characteristic varied. In computer typesetting a method of transmitting or recording digital data in which a phase shift corresponds to a ″1″ or ″0″ signal. Abbrev. PM.
f modulation *f* de phase — **d** Phasemodulation *f* — **n** fasemodulatie *f* — **e** modulación *f* de fase — **i** modulazione *f* di fase

10187 phat matter — See fat matter.

10188 phenic acid — See phenol.

10189 phenol; carbolic acid; phenic acid; benzophenol; hydroxybenzene; phenylic acid — White, crystalline mass which turns pink or red if not perfectly pure or if under influence of light; absorbs water from air. Soluble in alcohol, water, ether, chloroform, glycerol, carbon disulphide, petrolatum, fixed or volatile oils and alkalis. Latin: phenolum; acidum carbolicum; acidum phenylicum. TL-value: 5 ppm or 19 mg/m^3. Phenol can be absorbed by the skin.
Formula: C_6H_5OH.
f phénol *m*; acide *m* carbolique; acide *m* phénique — **d** Phenol *m*; Karbolsäure *f* — **n** fenol *m*; carbolzuur *n* — **e** fenol *m*; ácido *m* carbólico; ácido *m* fénico — **i** fenolo *m*; acido *m* fenico

10190 phenol-formaldehyde resin — Plastic material made from phenol and formaldehyde or compounds in which these ingredients are predominant.
f résine *f* (à base de) phénol-formaldéhyde — **d** Phenol-Formaldehydharz *n* — **n** fenolformaldehyde hars *mn* — **e** resina *f* de fenol-formaldehído — **i** resina *f* fenolformaldeide

10191 phenolic resin — Synthetic thermosetting resin obtained by the condensation of phenol or substituted phenols with aldehydes such as formaldehyde, acetaldehyde, and furfural. Used to contribute gloss, hardness and drying in printing ink vehicles. *Phenol-formaldehyde resins are typical and constitute the chief class of phenolics.
f résine *f* phénolique — **d** Phenolharz *n* — **n** fenolhars *mn* — **e** resina *f* fenólica; resina *f* fénica — **i** resina *f* fenolica

10192 phenolphthalein — Pale yellow, crystalline powder, soluble in alcohol, ether and alkalis, insoluble in water.
Formula: $(C_6H_4OH)_2COC_6H_4CO$.
f phénol-phtaléine *f*; phtaléine *f* du phénol — **d** Phenolphtalein *n* — **n** fenolftaleïne *fm* — **e** fenolftaleína *f* — **i** fenolftaleina *f*

10193 phenomenon
f phénomène *m* — **d** Phänomen *n*; Vorgang *m*; Erscheinung *f* — **n** verschijnsel *n* — **e** fenómeno *m* — **i** fenomeno *m*

10194 phenylamine — See aniline.

10195 phenylethylene — See styrene.

10196 phenylic acid — See phenol.

10197 phenyl iodide — See iodobenzene.

10198 phenylmethane — See toluene.

10199 phloroglucin — See phloroglucinol.

10200 phloroglucinol; phloroglucin — White to yellow crystals, soluble in ether, alcohol and pyridine, slightly soluble in water. Solution of 4 g of phloroglucinol in a mixture of 100 ml of alcohol and 50 ml of hydrochloric acid gives a high red colour with wood or other lignified fibres. Latin: phloroglucinum.
Formula: $C_6H_3(OH)_3 2H_2O$.
f phloroglucine *f*; trioxybenzène *m* — **d** Phlorogluzin *n*; Phlorogluzinlösung *f* — **n** floroglucinol *fm*; floroglucine *fm*; floroglucine-oplossing *f* — **e** floroglucina *f*; floroglucinol *m* — **i** floroglucinolo *m*

10201 phloxine — Bluish-red pigment for process red inks. The hue is often referred to as *magenta. Applied to eosine lake. See also eosine.
f phloxine *f* — **d** Phloxine *f* — **n** floxine *f* — **e** floxina *f* — **i** floxina *f*

10202 phosphate — Salt of *phosphoric acid.

10203 phosphor — In electronics, a chemical powder which becomes phosphorescent when bombarded with electrons used for the inside face of a cathode ray tube. The nature will determine its colour, e.g., white or green, and its latency, the duration of the image.
f corps *m* phosphorescent — **d** Phosphor *m*; Leuchtschirmsubstanz *f* — **n** fosfor *n* — **e**

fósforo *m* — **i** fosforo *m*

10204 phosphorescence — *Fluorescence that continues for more than a very short time (10^{-6} sec) after the exciting radiation is stopped. See also luminescence.
f phosphorescence *f* — **d** Phosphoreszenz *f*; Nachleuchten *n* — **n** fosforescentie *f* — **e** fosforescencia *f* — **i** fosforescenza *f*

10205 phosphoric acid — Clear, colourless, sparkling liquid or transparent crystalline solid depending on the concentration and the temperature. Soluble in water and alcohol, very corrosive to ferrous metals and alloys. Latin: acidum phosphoricum. TL-value: 1 mg/m³.
Formula: H_3PO_4.
f acide *m* phosphorique — **d** Phosphorsäure *f* — **n** fosforzuur *n* — **e** ácido *m* fosfórico — **i** acido *m* fosforico

10206 phosphorous chloride — See phosphorus trichloride.

10207 phosphorus — Non metallic element. Vapour condenses at 280 °C to liquid white phosphorus, which solidifies at 44 °C to solid white phosphorus, which is soft, waxy, colourless, soluble in carbon disulphide and benzene, insoluble in water and alcohol. Very poisonous; causes severe burns. At room temperature it exhibits phosphorescence (slow oxidation) in air. Symbol: P.
f phosphore *m* — **d** Phosphor *m* — **n** fosfor *n* — **e** fósforo *m* — **i** fosforo *m*

10208 phosphorus oxychloride; phosphoryl chloride — Colourless, fuming liquid; decomposed by water and alcohol. Dangerous. Causes severe burns. TL-value: 0.5 ppm.
Formula: $POCl_3$.
f oxychlorure *m* de phosphore; chlorure *m* de phosphoryle — **d** Phosphoroxychlorid *n*; Phosphorylchlorid *n* — **n** fosforoxychloride *n*; fosforoxylchloride *n* — **e** oxicloruro *m* de fósforo — **i** ossicloruro *m* di fosforo

10209 phosphorus trichloride; phosphorous chloride — Clear, colourless, fuming liquid, soluble in ether, benzene, carbon disulphide and carbon tetrachloride.
Formula: PCl_3.
f trichlorure *m* de phosphore — **d** Phosphortrichlorid *n* — **n** fosfortrichloride *n* — **e** tricloruro *m* de fósforo — **i** tricloruro *m* di fosforo

10210 phosphoryl chloride — See phosphorus oxychloride.

10211 phot — Unit of light flux; the centimetre-candle. Illumination produced on a surface 1 cm from a standard candle.
f phot *m* — **d** Phot *m* — **n** foot *m* — **e** fot *m*; phot *m*; foto *m* — **i** fot *m*

10212 photistor — See phototransistor.

10213 photo — See photograph.

10214 photocathode; photoelectric cathode — Cathode which emits electrons when subjected to light.
f photocathode *f*; cathode *f* photoélectrique — **d**

Photokathode *f* — **n** fotokathode *f*; fotokatode *f*; foto-elektrische kathode *f* — **e** fotocátodo *m*; cátodo *m* fotoelectrónico — **i** fotocatodo *m*; catodo *m* fotoelettrico

10215 photocell; photoelectric cell; cell; photo-multiplier tube — Device in which measurable electrical effects occur when it is exposed to light.
f cellule *f* photoélectrique; œil *m* électrique; cellule *f* photorésistante — **d** Photozelle *f*; Fotozelle *f*; photoelektrische Zelle *f*; lichtelektrische Zelle *f* — **n** fotocel *f*; foto-elektrische cel *f* — **e** célula *f* fotoeléctrica — **i** cella *f* fotoelettrica; foto-cella *f*; cellula *f* fotoelettrica; fotocellula *f*

10216 photochemistry — Branch of chemistry that deals with the effect of light. The most important examples are in natural photosynthesis, in the production of a photographic image, and in the reaction of chlorine on hydrocarbons and other organic compounds.
f photochimie *f* — **d** Photochemie *f* — **n** foto-chemie *f* — **e** fotoquímica *f* — **i** fotochimica *f*

10217 photocomposer — See photo-typesetting machine.

10218 photocomposer — See step-and-repeat machine.

10219 photocomposing — See photo-typesetting.

10220 photocomposing machine — US term for *step-and-repeat machine. To avoid confusion with a step-and-repeat machine, which is also known as a photocomposing machine, it is recommended to use the term photo-typesetting machine.

10221 photocomposing machine — See photo-typesetting machine.

10222 photocomposing machine — See step-and-repeat machine.

10223 photo-composition — See photo-typesetting machine.

10224 photodiode — Two types of germanium or of silicon fused together to produce a junction layer which is sensitive to light. Used as a sensing device to detect the presence, absence or the amount of light at a given point and to transmit this information electronically as variable data, which is subsequently converted into digital information (i.e., light or no light). It can be quite minute so that an array of photodiodes may be necessary. Also used to clock pulses as timing slots (as on a rotating film disk) move past a light source opposite a photodiode.
f photodiode *f*; diode *f* photoélectronique — **d** Photodiode *f* — **n** fotodiode *f*; foto-elektrische diode *f* — **e** fotodiode *m*; diodo *m* fotoelectrónico — **i** fotodiodo *m*; diodo *m* fotoelettronico

10225 photoelectric — Relating to or utilizing electrical effects due to the interaction of light or other radiation with matter.
f photoélectrique — **d** photoelektrisch — **n** foto-elektrisch — **e** fotoeléctrico — **i** fotoelettrico

10226 photoelectric cathode — See photocathode.

10227 photoelectric cell — See photocell.

10228 photoelectric effect — Action or effect of light in decreasing the electrical resistance of metals and other substances when exposed to illumination of certain wavelengths.
f effet *m* photo-électrique — **d** photoelektrischer Effekt *m*; lichtelektrischer Effekt *m* — **n** foto-elektrisch effect *n* — **e** efecto *m* fotoeléctrico — **i** effetto *m* fotoelettrico

10229 photoelectric engraver; electronic photo-engraving machine; electronic engraver — Automatic engraving machine.
f machine *f* à graver électronique — **d** elektronische Klischiermaschine *f*; elektronische Klischeegraviermaschine *f*; elektronische graveermachine *f*; elektronische clicheer-machine *f* — **e** fotograbadora *f* electrónica; grabadora *f* electrónica; máquina *f* electrónica de grabar; máquina *f* para grabado electrónico — **i** macchina *f* per l'incisione elettronica

10230 photoelectricity — Electricity generated under the influence of light.
f photo-électricité *f* — **d** Photoelektrizität *f* — **n** foto-elektriciteit *f* — **e** fotoelectricidad *f* — **i** foto-elettricità *f*

10231 photoengraver — Man or firm supplying line and halftone blocks.
f photograveur *m* — **d** Klischeemacher *m* — **n** clichémaker *m* — **e** fotograbador *m*; foto-mecánico *m*; operador *m* fotograbador — **i** foto-incisore *m*

10232 photoengraving — Making printing plates by exposure of line and halftone negatives on sensitized metal, converting the image into an acid resist, and etching the print to the relief required for letterpress printing. See also process engraving.
f photogravure *f* (typographique) — **d** Klischeeherstellung *f* — **n** clichéfabricage *f* — **e** foto-grabado *m* — **i** fotoincisione *f*

10233 photoengraving; photo-etching — Etched relief printing plate produced by photography.
f cliché *m* de similigravure; cliché *m* simili; simili *f*; cliché *m* tramé — **d** Autotypie *f*; Rasterklischee *n*; Rasterätzung *f*; Autotypieätzung *f*; Halbtonätzung *f* — **n** autotypie *f*; autotypie-cliché *n* depr; rastercliché *n* — **e** fotograbado *m*; clisé *m* autotípico; clisé *m* fotograbado — **i** autotipia *f*; cliscè *m* a retino; cliché *m* a retino; foto-incisione *f*; fotoformatura *f* rilievografica

10234 photo-etching — See photoengraving.

10235 photogelatine ink — See collotype ink.

10236 photogelatine printing — See collotype process.

10237 photogenic property — Tendency of a pigment to darken or change colour on exposure to sunlight.
f propriété *f* photogénique — **d** Nachdunkeln *n* der Pigmente im Licht — **n** verdonkeren *n*; donker worden *n* — **e** propiedad *f* fotogénica — **i** proprietà *f* fotogenica

10238 photo-glycine — See glycine.

10239 photoglyphy — See Woodbury type.

10240 photograph; photo — Image or picture

produced by photography.

f photographie *f*; photo *f* — **d** Photographie *f*; Photo *n*; Lichtbild *n*; Aufnahme *f* — **n** fotografie *f*; foto *f* — **e** fotografía *f*; foto *f* — **i** fotografia *f*; foto *f*

10241 photographer
f photographe *m* — **d** Photograph *m* — **n** fotograaf *m* — **e** fotógrafo *m* — **i** fotografo *m*

10242 photographic department
f section *f* photographique; atelier *m* de photographie — **d** Photoabteilung *f*; Abteilung *f* Photographie; photographische Abteilung *f* — **n** fotoafdeling *f*; afdeling *f* fotografie — **e** departamento *m* de fotografía — **i** reparto *m* fotografico

10243 photographic emulsion — Viscous, light-sensitive solution of gelatine holding in suspension microscopic grains of silver salts, especially silver chloride or silver bromide.
f émulsion *f* pour plaques photographiques — **d** photographische Emulsion *f* — **n** fotografische emulsie *f* — **e** emulsión *f* fotográfica — **i** emulsione *f* fotografica

10244 photographic film — See film.

10245 photographic grain — See grain.

10246 photographic master — Image of a graphic design or character, either directly exposed to create type images on a photographic film (which often may be enlarged or reduced), or scanned to create digital information which in turn will cause a similar image to be represented on a CRT output tube. See also master image.
f image *f* maîtresse photographique; image *f* matrice photographique — **d** photographisches Mutterbild *n*; photographisches Originalbild *n* — **n** fotografisch moederbeeld *n* — **e** imagen *f* original fotográfica — **i** immagine *f* originale fotografica

10247 photographic paper — See photopaper.

10248 photographic plate — Photographic product in which the light-sensitive layer is supported on glass.
f plaque *f* photographique — **d** photographische Platte *f*; Photoplatte *f* — **n** fotografische plaat *f* — **e** placa *f* fotográfica — **i** lastra *f* fotografica

10249 photographic print — See photoprint.

10250 photographic proofs — Photographic prints made from negatives or positives. Blueprints and silverprints are used to check layout and imposition; silverprints approximate quality of halftones; various forms of diazo prints in colour are used to approximate quality of process colour plates sometimes termed print proofs.
f épreuves *fpl* photographiques — **d** Photodrucke *mpl* — **n** fotoproeven *fmpl*; fotografische proeven *fmpl* — **e** pruebas *fpl* fotográficas — **i** bozze *fpl* fotografiche

10251 photographic retouching — Corrective treatment of photographic negatives and prints by means of pencils, reducers, transparent shading sheets, air brush or application of aniline dyes for removal of blemishes and improvement of tone value, detail and photographic quality.

f retouche *f* photographique — **d** photographische Retusche *f* — **n** fotografische retouche *f* — **e** retoque *m* fotográfico; retocado *m* fotográfico — **i** ritocco *m* fotografico

10252 photography — Science and art of obtaining images of objects by the action of light on sensitized surfaces.
f photographie *f* — **d** Photographie *f* — **n** fotografie *f* — **e** fotografía *f* — **i** fotografia *f*

10253 photogravure; gravure — Producing by photographic action an incised metal surface for printing from. Also the printing process and the picture so produced.
f photogravure *f*; héliogravure *f* — **d** Heliogravüre *f* — **n** fotogravure *fm*; heliogravure *fm* — **e** fotohuecograbado *m*; fotograbado *m* en hueco; heliograbado *m* — **i** eliotipia *f*; fotoformatura *f* incavografica

10254 photogravure ink — Strong intaglio ink, used with photogravure plates in printing portraits, art calendars, greeting cards, and illustrations for fine books.
f encre *f* hélio au grain de résine — **d** Kupferdruckfarbe *f* — **n** plaatdrukinkt *m* — **e** tinta *f* para huecograbado (a grano de resina) — **i** inchiostro *m* calcografico

10255 photohead setter — See headliner.

10256 photolettering — Variety of devices and hand-operated methods for the production of display matter.
f photolettrage *m*; photocomposition *f* manuelle — **d** Fotohandsatz *m* — **n** handfotozetsel *n*; fotografisch smoutzetsel *n* — **e** fotocomposición *f* a mano — **i** fotocomposizione *f* manuale

10257 photolithography — Branch of lithographic printing in which photography is employed for production of the image on the final printing surface. The original printing surface, lithographic stone, has been almost completely displaced by thin and flexible sheets of metal (aluminium, zinc, stainless steel, bimetallic and polymetallic plates). The oldest photolithographic procedure was the albumen process in which the light-hardened and inked images of bichromated albumen served as the actual printing surface. Deep-etch plates are a modern form of photolithographic surface, as are also bimetallic plates and those of the plastic type.
f photolithographie *f*; photolitho *f* — **d** Photolithographie *f* — **n** fotolithografie *f* — **e** fotolitografía *f* — **i** fotolitografia *f*; fotolito *f*

10258 photomechanical — Generic and broadly applied term for any reproductory process in which photography is employed in the production of a printing surface. The term embraces collotype, photoengraving, rotogravure, screen photostencil printing, etc.
f photomécanique — **d** photomechanisch — **n** fotomechanisch — **e** fotomecánico — **i** fotomeccanico

10259 photometer — See exposure meter.

10260 photometric brightness — See luminance.

10261 photomontage — Combination or blending of several photographic images into a single print to present a variety of views or subjects.

f photo-montage *m*; assemblage *m* (photo-graphique) — **d** Photomontage *f* — **n** foto-montage *f* — **e** fotomontaje *m*; montaje *m* foto-gráfico; fotografía *f* de composición — **i** foto-montaggio *m*

10262 photomultiplier tube — See photocell.

10263 Photon — Photo-composing machine, patented in 1944 by the French engineers Louis Moyroud and René Higonnet, and known in Europe as *Lumitype. Not based on any of the hot-metal composing machines, but, using the binary counting system and a disk carrying 1400 characters which revolves at high speed, it is able to position, correctly expose a microfilm, and space characters which have originally been com-posed on a keyboard with a standard typewriter layout.

10264 photo-optical disc store — Transparent disc with microscopic opaque areas, forming binary coded digits.

10265 photo page — See picture page.

10266 photopaper; photographic paper — Paper sensitized with a photographic emulsion for con-tact or projecting printing, made of a high grade rag base stock. See also sensitized paper.

f papier *m* photographique — **d** Photopapier *n*; photographisches Papier *n* — **n** fotopapier *n*; foto-grafisch papier *n* — **e** papel *m* fotográfico; papel *m* para fotografía — **i** carta *f* fotografica

10267 photopic vision — Vision experienced by the normal eye when adapted to the higher levels of luminance. The cone receptors in the retina are considered to be the active elements under these conditions and the spectrum appears coloured.

f vision *f* photopique — **d** Tagessehen *n* — **n** fotopisch zien *n* — **e** visión *f* fotópica — **i** visione *f* fotopica

10268 photopolymer — Polymer or plastic made sensitive to light, used for printing and litho-graphic plates, for photographic prints and micro-film copying. One application involves the use of esters of polyvinyl alcohol which crosslink and so become insoluble, whereas unexposed portions of the material remain soluble.

f photopolymère *m* — **d** Photopolymer *n* — **n** foto-polymeer *m* — **e** fotopolímero *m* — **i** foto-polimero *m*

10269 photopolymer plate — Plate made of plastic mounted on steel (flexible type) and aluminium (rigid type), used as letterpress printing plates.

f planche *f* photopolymère — **d** Photopolymer-platte *f* — **n** fotopolymeerplaat *fm* — **e** plancha *f* fotopolímera — **i** lastra *f* fotopolimerica; lastra *f* di fotopolimeri

10270 photoprint; photographic print — Print made on a photographically sensitized paper by contact or projection exposure through a negative; used as illustration sections in paste-ups, as proofs or as an intermediate for reproduction.

f copie *f* photographique — **d** Photokopie *f* — **n** fotokopie *f*; fotoafdruk *m* — **e** copia *f* fotográfica

— **i** fotocopia *f*; stampa *f* fotografica; copia *f* foto-grafica; copia *f* a contatto

10271 photoprinting — See print *v*.

10272 photoresist — Acid-resisting image pro-duced on metal plates by photography or the hardening action of light on sensitized surfaces. Enamel print.

f réserve *f* anti-acide — **d** Schutzschicht *f*; Email-schutzschicht *f* — **n** beschermlaag *fm* — **e** capa *f* protectora — **i** strato *m* resistente agli acidi

10273 photostat; stat — Cheap and quickly made photographic reproduction in the form of a nega-tive or positive print, made with special camera apparatus.

f photostat *m*; photocopie *f* — **d** Photokopie *f* — **n** fotokopie *f* — **e** fotocopia *f*; copia *f* fotostática; fotostato *m* — **i** fotocopia *f*

10274 photo-textyping — Composing or setting of type or text matter with photographic apparatus. See also photo-typesetting.

10275 photothermographic — Said of processes in-volving the use of heat to produce an image. The heat changes the light sensitive material that has received exposure to a visible image.

10276 phototransistor; photistor — Transistor in which one of the junctions is sensitive to light.

f transistor *m* à effet photo-électrique — **d** licht-empfindlicher Transistor *m*; Kristall-Licht-verstärker *m* — **n** lichtgevoelige transistor *m* — **e** fototransistor *m* — **i** fototransistor *m*

10277 phototype process — See collotype process.

10278 photo-typesetting; filmsetting; light setting — Composing text matter directly onto a photo-graphic or other light-sensitive material. Note: the term photocomposing has been deprecated because in some countries it is used to describe what is known in the United Kingdom as "step and repeat".

f composition *f* photo (du texte) — **d** Photosatz *m*; Lichtsatz *m*; photomechanischer Satz *m*; Kalt-satz *m* — **n** fotografisch zetten *n*; fotozetten *n* — **e** composición *f* fotográfica; composición *f* foto-mecánica; composición *f* fototipográfica; composición *f* sobre película; fotocomposición *f* — **i** fotocomposizione *f*; composizione *f* fotografica; composizione *f* a freddo

10279 photo-typesetting machine; photo-composing machine *depr.*; **photocomposer; film-setting machine; film setter** — Machine for making photo-composition.

f photocomposeuse *f*; photocompositrice *f*; composeuse *f* photographique — **d** Photosetz-maschine *f*; Lichtsetzmaschine *f*; Filmsetz-maschine *f* — **n** fotozetmachine *f*; fotografische zetmachine *f*; lichtzetmachine *f* — **e** máquina *f* para fotocomposición; máquina *f* de foto-composición; máquina *f* de fototipocomposición; compositora *f* fotográfica; fotocompositora *f*; foto-componedor *m*; fotocomponedora *f* — **i** foto-compositrice *f*; compositrice *f* fotografica

10280 photo-typesetting machine — See photocomposing machine.

10281 phototypography — Art or technique of making printing surfaces by light or photography.
f phototypographie *f* — **n** fototypografie *f* — **e** fototipografía *f* — **i** fototipografia *f*

10282 photo-voltaic cell — See barrier layer photocell.

10283 phozy paper — Featherweight paper that is fluffy or too loosely manufactured.
f papier *m* duveteux; papier *m* spongieux — **d** lockeres Papier *n*; lappiges Papier *n* — **n** los papier *n*; slap papier *n* — **e** papel *m* fofo — **i** carta *f* spugnosa

10284 phthalic acid; ortho-phthalic acid — Colourless crystals, soluble in alcohol, sparingly soluble in water and ether, used for dyes. Not used so widely, since phthalic anhydride is the primary material.
Formula: $C_6H_4(CO_2H)_2$.
f acide *m* phtalique — **d** Phthalsäure *f* — **n** ftaalzuur *n* — **e** ácido *m* ftálico — **i** acido *m* ftalico

10285 phthalocyanine pigment — Blue and green ink pigment characterized by extreme light fastness, brilliance of shade, and resistance to solvents, acid and alkali.
Formula: $C_{32}H_{18}N_8$.
f pigment *m* de phtalocyanine — **d** Phthalozyaninpigment *n* — **n** ftalocyanine-pigment *n* — **e** pigmento *m* ftalociánico — **i** pigmento *m* di ftalocianina

10286 physical characteristics
f caractéristiques *fpl* physiques — **d** physikalische Eigenschaften *fpl* — **n** fysische eigenschappen *fpl* — **e** características *fpl* físicas — **i** caratteristiche *fpl* fisiche

10287 pi — See pie.

10288 pica — Standard of measurement, approx. 1/6 in. In the *point system 0.166 in. 4 to pica, or 6, 8, 10 or 12 to pica means that 4, 6, 8, 10 or 12 leads or brass rules equal one pica (11,215 point Didot).
In the SI:
1 pica = 12 points ≈ 4.21752 x 10^{-3} m. See also cicero, em, point, point system, app. no. 1.
f pica *m*; un-douze *m*; corps *m* douze; douze points *mpl* anglais-américains — **d** Pica *m*; Schriftgrad *m* 12; zwölf englisch-amerikanische Punkte *mpl* — **n** pica *fm*; corps *n* 12; twaalf Engels-Amerikaanse punten *mpl* — **e** pica *f*; cuerpo *m* 12; doce puntos *mpl* inglés-americanos — **i** pica *f*; corpo *m* 12; dodici punti *mpl* inglese-americano

10289 picamaster — Band saw or plate trimmer to accurately trim cemented printing plates to pica or point measurement without leaving a burr on the trimmed plates.
f scie *f* de précision — **d** Präzisionssäge *f* — **n** precisie-zaagmachine *f*; precisie-zaag *fm* — **e** sierra *f* de precisión — **i** sega *f* di precisione

10290 pica rule — See type gauge.

10291 pica type — The old name for a type size. Also typewriter type that runs ten characters to the inch. See app. no. 1, elite type.

10292 pick — Letter of which the face is choked with dirt. The compositor picks it out with the point of a needle.
f lettre *f* sale; lettre *f* à nettoyer — **d** verschmierter Buchstabe *m*; Putzen *fpl* — **n** vuile letter *fm* — **e** letra *f* sucia — **i** sudiciume *m*

10293 pick brush — See type brush.

10294 pick *v* **for sorts** — To take out short sorts from a standing forme to avoid distribution.
d Buchstaben ziehen — **n** letters trekken — **e** robar; picar un molde; picar una forma; picar composición — **i** rubare caratteri

10295 picking; plucking; lifting — Removal of parts of the surface of the paper during printing, when the pulling force (tack) of the ink is greater than the surface strength of the coated or uncoated paper. In stereotyping also the defect in the surface of the matrix whereby in the removal of the moulded plate from the matrix, small pieces of the matrix face adhere to the face of the duplicate plate. See also dry pick, wet pick.
f arrachage *m*; piquetage *m* — **d** Rupfen *n* — **n** plukken *n* — **e** repelado *m* (del papel); picado *m* (del papel); arrancamiento *m* (de tinta) — **i** strappo *m* (superficiale) della carta

10296 picking resistance — See pick resistance.

10297 picking-resistance tester — See pick tester.

10298 pickle — Weak acid or other chemical solution, in which metal objects, e.g. electros and engravings, are dipped to remove oxide scale of other adhering substances.
f solution *f* de décapage — **d** Pickelbad *n* — **e** solución *f* de casca para curtir

10299 picklet — In computer-controlled typesetting, a unit of measurement representing 1/64th of a pica.

10300 pick resistance; picking resistance
f résistance *f* à l'arrachage — **d** Rupffestigkeit *f* — **n** plukweerstand *m* — **e** resistencia *f* al repelado; resistencia *f* al arrancamiento — **i** resistenza *f* allo strappo

10301 pick-start viewer — Instrument to determine the moment at which a strip of paper starts picking. The test strip is observed in a special direction as light is striking the paper at a fixed angle. See also esculin.
f mireur *m* du commencement de l'arrachage — **d** Betrachtungsgerät *n* zur Beurteilung des Rupf-Begins — **n** plukbeginbeoordelingskast *fm*

10302 pick tester; picking-resistance tester — Instrument to measure the pick resistance of paper through the use of inks or oils with varying standardized tack.
f appareil *m* d'essai d'arrachage — **d** Rupffestigkeitsmeßgerät *n*; Rupffestigkeitsprüfer *m*; Rupffestigkeitsprüfgerät *n* — **n** plukweerstandsmeter *m* — **e** ensayador *m* de la resistencia al repelado; ensayador *m* de la resistencia al arrancamiento — **i** apparecchio *m* per prove di resistenza allo strappo

10303 pick *v* **the pie**
f trier les pâtes — **d** Zwiebelfische auslesen — **n** pastei uitzoeken; uitzoeken uit de pastei — **e** des-

empastelar — **i** separare il pasticcio; separare il refuso

10304 pickup — See lifted matter.

10305 pick *v* **up** — To re-use previously printed matter as part of a new work, either by printing from the original type or by photo-offset, sometimes abbreviated PU.

f reprendre — **d** Stehsatz wiederverwenden — **n** gebruik maken van staand zetsel — **i** servirsi di composizione già uso

10306 pickup system — Method of halftone photography employing three successively smaller lens apertures for highlight, middletone, and detail exposures. Three-stop exposure.

f procédé *m* aux trois diaphragmes — **d** Dreiblenden-Exposition *f* — **n** belichtingssysteem *n* met drie diafragma's — **e** sistema *m* a tres diafragmas — **i** fotografia *f* a tre esposizioni; fotografia *f* a tre aperture di diaframma

10307 picosecond — One million-millionth of a sec, or 10^{-12} sec. Abbrev. ps; psec.

10308 pictorial postcard — See picture postcard.

10309 picture ABC — See picture primer.

10310 picture book — Book consisting mainly of pictures, especially for children who cannot yet read. See also Struwwelpeter.

f livre *m* d'images — **d** Bilderbuch *n* — **n** prentenboek *n* — **e** libro *m* ilustrado; libro *m* con estampas; libro *m* de grabados; libro *m* de imágenes — **i** libro *m* illustrato

10311 picture page; photo page — Page with illustrations.

f page *f* d'illustrations — **d** Abbildungsseite *f*; Bildseite *f*; Illustrationsseite *f* — **n** fotopagina *f* — **e** página *f* gráfica; página *f* ilustrada — **i** pagina *f* illustrata

10312 picture postcard; pictorial postcard; postcard *US*

f carte *f* postale illustrée — **d** Ansichtskarte *f* — **n** prentbriefkaart *fm* — **e** tarjeta *f* postal ilustrada; tarjeta *f* postal con vistas — **i** cartolina *f* illustrata

10313 picture primer; picture ABC; ABC picture book — Elementary book with illustrations to teach children to read.

f abécédaire *m* illustré; abécédaire *m* avec images et figures — **d** Bilderfibel *m*; ABC-Buch *n* mit Bildern; Struwwelpeter *m* — **n** geïllustreerd abcboek *n* — **e** abecedario *m* pictórico; abecedario *m* con figuras; abecedario *m* ilustrado; abecedario *m* con imágenes y grabados — **i** abbecedario *m* illustrato; sillabario *m* illustrato

10314 picture stamps — See poster stamps.

10315 pie; jumbled type; broken matter; pi *depr.* — Type matter in disorder, e.g., before distribution. See also wrong fount.

f paté *m*; pâte *f*; coquille *f*; soleil *m*; composition *f* tombée — **d** Zwiebelfisch *m*; Fisch *m*; Eierkuchen *m*; gequirlter Satz *m*; zusammengefallener Satz *m*; Fischhaufen *m*; Zwiebelfischhaufen *m* — **n** pastei *fm*; ineengevallen zetsel *n* — **e** pastel *m*; empastelado *m*; empastelamiento *m*; encaballado *m*; encaballamiento *m*;

pato *m* *Ch* — **i** caratteri *mpl* in fascio; baracca *f*

10316 pie; pye — In the pre-Reformation English church a book of ecclesiastical rules to find the particulars of the service for the day.

10317 pie *v* — To mix up type accidentally. Also to make a bad mistake, or blunder, caused by the compositor or maker-up (e.g., misplaced captions or headlines of articles).

f mettre en pâte; faire du mastic; mêler les lettres dans une casse — **d** Zwiebelfische aufsetzen; den Kasten verfischen; Satz einwerfen; quirlen — **n** pastei maken — **e** empastelar — **i** mettere a fascio; mandare in baracca

10318 pie and sorts mechanism — See sorts stacker.

10319 piece accent — See floating accent.

10320 piece fraction; compound fraction; split fraction; combination fraction — Fraction made up of more than one piece of type. See also sloping fraction.

f fraction *f* composée; fraction *f* avec barre horizontale — **d** gebrochene Ziffer *f*; Bruchziffer *f* mit waagerechtem Strich; Bruchziffer *f* mit Horizontalstrich — **n** samengesteld breukcijfer *n*; breukcijfer *n* met horizontale deelstreep — **e** fracción *f* compuesta; quebrado *m* postizo; quebrado *m* seccional — **i** frazione *f* composta; frazione *f* con barra orizzontale

10321 piecework — Work paid for on the basis of output.

f ouvrage *m* à la tâche; ouvrage *m* à la pièce — **d** Akkordarbeit *f* — **n** stukwerk *n* — **e** trabajo *m* a piezas — **i** lavoro *m* a cottimo; cottimo *m*

10322 piecework composition — Composing of slips or packets.

f composition *f* aux pièces; composition *f* en paquets — **d** Paketsatz *m*; im Akkord gesetzter Satz *m*; berechneter Satz *m* — **n** stuk-zetten *n*; zetten *n* volgens stukwerk — **e** composición *f* en paquete — **i** composizione *f* a cottimo

10323 pie characters — Assortment of odd or non-standard characters for the setting of a particular job. In a line-casting machine they are inserted by hand by the operator, since they are not in the magazine which provides matrix storage. A phototypesetter fount disk or film strip may offer a limited selection of such special characters, and/or provide the facility for manual insertion. In some cases the system provides for on-line access to a virtually unlimited repertoire of such characters. See also sorts.

10324 pierced block — Blocks with an aperture cut through to let in type, etc.

f cliché *m* à fenêtre — **d** ausgeklinktes Klischee *n* — **n** cliché *n* met een uitsparing — **e** clisé *m* con ventana; clisé *m* calado — **i** cliscè *m* con feritoia; cliché *m* con feritoia; lastra *f* con feritoia

10325 piercer — Needle in a book sewing machine to make a hole in the back of each section.

f grecque *f* — **d** Vorstecher *m* — **n** voorsteeknaald *fm*

10326 pie stacker — See sorts stacker.

10327 pig — Moulded bar or cake of typographic or stereotype metal, melted and used for casting types or stereos.

f saumon *m*; lingot *m* — **d** Bleibarren *m*; Bleistange *f*; Metallzufuhrstange *f* — **n** loodstaaf *f m* — **e** barra *f* de metal; lingote *m*; bloque *m* de metal; pan *m* de metal; pastilla *f* — **i** barra *f*; lingotto *m* greggio; pane *m*

10328 pig dolly — Cart for the transport of bars of composing metal.

f chariot *m* à lingots — **d** Barrentransportwagen *m* — **n** loodtransportwagen *m* — **e** carro *m* portalingotes; carro *m* para transportar lingotes; carro *m* para transportar barras de metal — **i** carrello *m* per il trasporto di barre di metallo

10329 piggery — See press-room.

10330 pigment; colouring matter — Manufactured chemical colour, inorganic or organic, to impart colour (hue), usually to a surface. The former generally opaque and produced from basic materials, the latter include coal tar dye lakes formed by precipitation on alumina hydrate and are widely used for printing inks. Not soluble in its vehicle and in water. Its particle size is usually large enough to be seen with a microscope. The particles are larger than dye particles.

f pigment *m* — **d** Pigment *n*; Farbkörper *m* — **n** pigment *n* — **e** pigmento *m* — **i** pigmento *m*

10331 pigmented aniline ink — Spirit soluble printing ink, the dyeing matter of which is a pigment and not a soluble dye, used for rubber stereo printing.

f encre *f* d'aniline couvrante; encre *f* flexo couvrante — **d** deckende Flexofarbe *f*; deckende Anilinfarbe *f*; pigmentierte Flexodruckfarbe *f* — **n** gepigmenteerde flexo-inkt *m*; gepigmenteerde aniline-inkt *m* — **e** tinta *f* flexográfica pigmentada; tinta *f* de anilina pigmentada — **i** inchiostro *m* flessografico pigmentato

10332 pigment paper — See carbon tissue.

10333 pigment paper sensitizing

f préparation *f* du papier charbon — **d** Preparieren *n* des Pigmentpapiers — **n** prepareren *n* van het pigmentpapier — **e** preparado *m* del papel pigmento — **i** sensibilizzazione *f* della carta pigmento

10334 pig mould; ingot mould — Mould to cast bars of composing metal.

f moule *m* à lingots — **d** Barrengießform *f* — **n** gietvorm *m* voor loodstaven — **e** molde *m* para panes de metal; molde *m* de fusión — **i** forma *f* a fondere barre di metallo

10335 pigs — Nickname for the pressman of a letterpress machine.

10336 pigs — See pressman.

10337 pigskin — Strong leather with distinctive grain used for heavy bookbindings.

f truie *f*; peau *f* de truie — **d** Schweinsleder *n* — **n** varkensleer *n*; zwijnsleer *n* — **e** cerdo *m*; piel *f* de cerdo — **i** cuoio *m* di porco

10338 pigsty — See press-room.

10339 pilcher — In hand-made papermaking, the pad of three or four felts sewn together and put on top of a *post before pressing.

f matelas *m* de feutres — **d** Deckfilz *n* — **n** dekvilt *n* — **i** copriposta *f*

10340 pile; stack; paper stack — Sheets of paper placed one upon another in the delivery section of a printing press.

f pile *f* (de papier) — **d** Stapel *m*; Papierstapel *m* — **n** stapel *m*; papierstapel *m* — **e** pila *f*; montón *m* — **i** pila *f* (di carta); pila *f* di fogli

10341 pile height — Height of paper pile desirable for operation.

f hauteur *f* de la pile — **d** Stapelhöhe *f*; Einsatzhöhe *f* — **n** stapelhoogte *f* — **e** altura *f* de carga — **i** altezza *f* di pila; altezza *f* d'impilamento

10342 pile hoist; pile lifter; pile shift — Device to raise or lower the pile stock.

f élévateur *m* de pile — **d** Stapelhubeinrichtung *f*; Stapelheber *m*; Stapelhubgerät *n* — **n** stapellift *m*; stapelhefapparaat *n* — **e** elevador *m* de pila; ascensor *m* de pila — **i** elevatore *m* della pila

10343 pile lifter — See pile hoist.

10344 pile shift — See pile hoist.

10345 piling; caking — Collecting of ink particles upon plates, rollers, blankets, or ink table, caused primarily by the inability of the vehicle to hold the pigment in suspension.

f plaquage *m* — **d** Aufbauen *n*; Anhäufen *n*; Pelzen *n*; Antrocknen *n* — **n** koeken *n*; oplopen *n* — **e** amontonado *m* — **i** accumulo *m* d'inchiostro

10346 pillow distortion — See pincushion distortion.

10347 pilot lamp — Electric lamp, used in association with a control, which by means of position of colour indicates the functioning of the control; an indicator light or control light.

f lampe-témoin *f* — **d** Kontrollampe *f* — **n** controlelampje *n* — **e** lámpara *f* piloto; lámpara *f* testigo — **i** lampada *f* spia

10348 pilot plant — Experimental factory in which processes or techniques planned for use in full-scale manufacturing are tested.

f usine *f* pilote — **d** Versuchsanlage *f*; Probeanlage *f*; Pilotbetrieb *m*; Pilotanlage *f* — **n** proefbedrijf *n* — **e** ramo *m* experimental — **i** impianto *m* pilota

10349 pimelic ketone — See cyclohexanone.

10350 pinacyanol — Green, glossy crystals, easily dissolving in water and alcohol to a bluish colour, used as a photographic sensitizer for red ranging from 580-650 nm. Discovered in 1905 by Homolka.

f pinacyanol *m* — **d** Pinazyanol *n* — **n** pinacyanol *m* — **e** pinacianol *m*

10351 pinaflavol — Basic sensitizer for green, directly linked to the natural sensitivity of silver bromide. Formerly used in panchromatic emulsions in combination with pinacyanol, also for orthochromatic sensitizing with erythrosine. Flavol has the formula:

$$HOC_6H_3(CH)_2C_6H_3OH.$$

f pinaflavol *m* — **d** Pinaflavol *n* — **n** pinaflavol *m* — **e** pinaflavol *m*

10352 pinakryptol green — Dark, dull green powder or tablets. Strong photographic desensitizer which can be added to developers that do not contain hydroquinone, pyrocatechol or pyrogallol.
f pinacryptol *m* vert — **d** Pinakryptolgrün *n* — **n** pinakryptolgroen *n* — **e** pinacriptol *m* verde

10353 pinakryptol yellow — Light yellow powder or tablets. Strong photographic desensitizer, especially used for infra-red plates and panchromatic plates of high sensitivity. Made inactive by the sulphite of the developer.
f pinacryptol *m* jaune — **d** Pinakryptolgelb *n* — **n** pinakryptolgeel *n* — **e** pinacriptol *m* amarillo

10354 pinatypie — Relief coloured by impregnation of a non-pigmented, bichromated gelatine.
f pinatypie *f* — **d** Pinatypie *f* — **n** pinatypie *f* — **e** pinatipia *f*; pinotipia *f* — **i** pinatipia *f*

10355 pinch *v* — To steal type matter, sorts, etc., from a colleague. See also slum *v*.
f barboter — **d** mardern; hamstern; fuchsen *sl.* — **n** hamsteren; stelen; gappen — **e** acaparar — **i** accaparrare

10356 pinched post; small post — A British standard size of writing paper. See app. no. 3.

10357 pincher — Thief of type matter, sorts, etc., from a fellow compositor. Term not often used in English.
f barboteur *m* — **d** Marder *m*; Materialdieb *m*; Materialhamsterer *m*; unangenehmer Setzerkollege *m* — **n** hamsteraar *m*; materiaaldief *m* — **e** acaparador *m* — **i** accaparratore *m*

10358 pinchers — See nipping rollers.

10359 pinching rollers — See nipping rollers.

10360 pinch marks — Defect of aluminium foil. Pressed-in folds parallel to the direction of folding.

10361 pincushion distortion; pillow distortion — Cushion-shaped distortion of screen dots. Also *distortion of a lens when the images of straight lines are curved lines convex towards the centre of the plate.
f distorsion *f* en croissant; distorsion *f* en coussinet; distorsion *f* positive — **d** kissenförmige Verzerrung *f* — **n** kussenvormige vertekening *f* — **e** distorsión *f* de concavidad — **i** distorsione *f* a forma di cuscino

10362 pine — Resinous wood used in the manufacture of sulphate and sulphite pulps.
f pin *m* — **d** Kiefer *f*; Kiefernholz *n* — **n** den *m*; grove den *m*; dennehout *n* — **e** pino *m*; madera *f* de pino — **i** pino *m*; legno *m* di pino

10363 pinholes — Small, unwanted, transparent areas in the developed emulsion of a negative or positive, usually due to dust or other defects on the copy, copy-board, glass, or on the film, or to chemical action.
f pointes *fpl* d'aiguilles — **d** Nadelstiche *mpl* — **n** speldepuntjes *npl*; speldeprikjes *npl* — **e** puntos *mpl* — **i** vaiolature *fpl*; punte *fpl* di spillo; piccole soffiature *fpl* superficiali puntiformi

10364 pinholes — Minute and almost imperceptible pits in the surface of papers.
f piqûres *fpl* d'épingle; trous *mpl* d'épingle; trous *mpl* d'aiguille; cratères *mpl* — **d** Sandlöcher *npl*; kleine Löcher *npl* in Papieroberfläche — **n** speldegaatjes *npl*; speldeprikjes *npl*; speldepuntjes *npl* — **e** agujetas *fpl*; puntitos *mpl* — **i** forellini *mpl*

10365 pinhole theory — Theory of halftone photography in which each tiny aperture of the ruled halftone screen is supposed to act as a pinhole lens and form a vignetted image (dot) of the lens aperture.
f théorie *f* des ouvertures de la trame — **d** Rasteröffnung-Theorie *f* — **n** rastertheorie *f*; theorie *f* van de rasteropeningen — **e** teoría *f* de la abertura de la retícula — **i** teoria *f* della formazione dei punti del retino

10366 pinholing — Failure of printed ink to form a completely continuous film. Visible in the form of small holes in the printed area.
f formation *f* de piqûres d'aiguilles; formation *f* de petits cratères (circulaires) — **d** Kraterbildung *f*; Porenbildung *f*; Nadelstiche *mpl* — **n** optreden *n* van gaatjes — **e** formación *f* de puntitos; formación *f* de agujeritos — **i** pinholing *m*; rifiuto *m* di stampa a punta di spillo

10367 pinholing — Stopping out of minute holes in negatives or positives.
f repique *f* — **d** Ausflecken *n*; Ausdecken *n* kleiner Löcher — **n** uitdekken *n* van gaatjes — **i** stuccatura *f* di minutissimi fori

10368 pinion; tooth wheel; tooth gear — Small gear.
f pignon *m*; roue *f* dentée — **d** Ritzel *n*; Zahnrad *n* — **n** tandrad *n*; tandwiel *n*; rondsel *n* — **e** piñón *m* — **i** pignone *m*

10369 pin mark — Slight mark on the near side of a type near the top of the shank, made in casting.
f marque *f* du fondeur; poinçon *m* du fondeur — **d** Gießmarke *f* — **n** gietmerk *n* — **e** marca *f* del fundador — **i** marchio *m* della fonderia

10370 pinolin — See tall oil.

10371 pins — Small sharply pointed spikes sometimes used to locate and attach packing sheets to impression cylinders of a rotary press.
f picots *mpl*; ardillons *mpl* — **d** Aufzugnadeln *fpl* — **n** leggerpennen *fmpl* — **e** punturas *fpl* — **i** puntine *fpl* di fissaggio

10372 pins — See needles.

10373 pint — Unit of capacity. See also dry pint, liquid pint, app. no. 7 and 7a.
f pinte *f* anglaise — **d** Pinte *f*; englische Pinte *f* — **n** pint *fm*; Engelse pint *fm* — **e** pinta *f* inglés — **i** pinta *f* inglese

10374 pipe roller — See idler roller.

10375 pipette — Calibrated glass tube to measure small quantities of fluids. See also ink pipette.
f pipette *f* — **d** Pipette *f* — **n** pipet *fmn* — **e** pipeta *f*; bombilla *f*; gotero *m*; cuentagotas *m* — **i** pipetta *f*

10376 piston block — Part of the Monotype casting machine.
f bloc *m* de pistons — **d** Luftkolbenblock *m* — **n** zuigerblok *n* — **e** bloque *m* de pistón — **i** blocco *m* degli stantuffi; blocco *m* dei pistoni
10377 piston block base — Part of the Monotype casting machine.
f base *f* du bloc de pistons — **d** Luftventilblock *m* — **n** zuigerblokbasis *m* — **e** base *f* del bloque de pistón — **i** base *f* del blocco degli stantuffi; base *f* del blocco dei pistoni
10378 piston lever group
f cadre *m* support des leviers de pistons — **d** Luftkolbenhebelblock *m* — **i** gruppo *m* delle leve degli stantuffi
10379 pit — Cavity in the floor under a machine for accessibility to the underparts.
f fosse *f* — **d** Grube *f*; Wanne *f* — **n** put *m*; kuil *m* — **e** foso *m*; foso *m* de instalación — **i** fossa *f*
10380 pitch — Solid or semi-solid material of asphaltic nature used in ink to add flow or length.
f poix *f* — **d** Pech *n*; Teer *m* — **n** pek *nm*; teer *mn* — **e** pez *f*; brea *f*; alquitrán *m* — **i** pece *f*
10381 pitch — Resinous material in mechanical and unbleached chemical pulps, which at times separates out of the pulp suspension on the paper machine.
f brai *m* — **d** Zellstoffharz *n*; Harzausscheidung *f*; Pech *n* — **n** hars *mn*; papierhars *mn* — **e** pez *f*; brea *f* — **i** pece *f*
10382 pitch *v* — To take away the felts from the felt board and lay them upon the sheets of paper on a *post.
f monter la porse — **d** auflegen des Filzes — **n** vilten opleggen — **i** levare
10383 pitch circle; pitch line; primitive circle — Circle representing diameter of the gear. The circle at which gear device is most nearly perfect.
f cercle *m* primitif — **d** Teilkreis *m* — **n** steekcirkel *m* — **e** circunferencia *f* primitiva — **i** circonferenza *f* primitiva
10384 pitch diameter — Rolling diameter of a gear. On some printing presses the same basic diameter as the cylinder bearers.
f primitif *m* — **d** Teilkreisdurchmesser *m* — **n** diameter *m* van de steekcirkel; middellijn *m* van de steekcirkel — **e** diámetro *m* del círculo primitvo — **i** diametro *m* primitivo
10385 pitch line — Scribed line on the bed of a printing machine to position the forme correctly in relation to the gripper edge of the cylinder.
f repères *mpl* du marbre; gorge *f* — **d** Anfangslinie *f*; Druckanfang *m* — **n** aanslaglijn *m*; aanslag *m* — **e** principio *m* de impresión; línea *f* del principio de impresión — **i** linea *f* di principio della composizione; linea *f* di riferimento
10386 pitch line — See pitch circle.
10387 PIV control — Positive-infinitely-variable speed transmission.
f réglage *m* PIV — **d** PIV-Getriebe *n* — **n** PIV-aandrijving *f* — **e** variador *m* de velocidad PIV —

i regolazione *f* PIV
10388 place nor date — See no place no date.
10389 plagiarism — Copying of another person's writings and publishing the material copied as original matter.
f plagiat *m* — **d** Plagiat *n* — **n** plagiaat *n* — **e** plagio *m* — **i** plagio *m*
10390 plain bearing — Bearing not containing rolling elements, that presents to the shaft or axle it supports broad areas of corresponding form, usually segments of a cylinder.
f palier *m* lisse — **d** Gleitlager *n* — **n** glijlager *n* — **e** cojinete *m* liso — **i** cuscinetto *m* liscio
10391 plain sewing — See French sewing.
10392 plane *v* — To level off (and make type high) the rules, plates or blocks with a plane(r).
f raboter; raboter; aplanir au rabot — **d** abhobeln; hobeln; planschleifen — **n** afschaven; vlak schaven — **e** acepillar; cepillar; quitar con el cepillo — **i** levigare; piallare; lisciare; spianare; sbassare
10393 plane *v* **down** — To even out composed type by gently tapping the surface with a *planer and a mallet.
f taquer la forme; dresser la forme — **d** die Forme klopfen; abklopfen; die Forme gleichmäßig klopfen — **n** de vorm dresseren; opkloppen — **e** tamborilear; golpear; palmear; palmetear; palmotear; nivelar; igualar; asentar; aplanar *SA* — **i** appianare la forma; livellare la forma; appianare la composizione
10394 planer; type planer — Flat, smooth piece of hardwood to tap down and level type before locking up the printing forme.
f taquoir *m* — **d** Klopfholz *n* — **n** dresseerplank *fm* — **e** tamborilete *m*; palmeador *m*; asentador *m*; emparejador *m Me*; aplanador *m RP, SA* — **i** battitoio *m*
10395 planer — Machine used by printers and engravers to smooth and level back of stereos, electros, etc.
f planeuse *f*; rabotoir *m* — **d** Hobelmaschine *f* — **n** vlakschaafmachine *f* — **e** cepilladora *f*; acepilladora *f* — **i** macchina *f* piallatrice; macchina *f* spianatrice
10396 planet motion — Combination of geared mechanical parts to transform a circular motion into an oscillating motion.
f mouvement *m* planétaire — **d** Planetenbewegung *f*; Kreisbewegung *f* — **n** planeetwielbeweging *f* — **e** movimiento *m* de planeta — **i** movimento *m* planetario; movimento *m* ipocicloidale
10397 planning — See also production control.
f préparation *f* du travail — **d** Arbeitsvorbereitung *f*; Planung *f* — **n** werkvoorbereiding *f*; planning *f* — **e** programación *f* del trabajo; planing *f* — **i** preparazione *f* del lavoro; pianificazione *f*
10398 planning board
f tableau *m* de planning; tableau *m* d'ensemble des commandes — **d** Auftragübersichtstafel *f* — **n**

planbord *n*; planningbord *n* — **e** pizarra *f* de programación; tablero *m* de programación — **i** tavola *f* di pianificazione

10399 plano-concave lens — Lens which is flat on one side and concave (or hollow) on the other.
f lentille *f* plan-concave — **d** plankonkave Linse *f* — **n** planconcave lens *fm* — **e** lente *m* plano-cóncavo — **i** lente *f* piano-concava

10400 plano-convex lens — Lens which is flat on one side and convex on the other.
f lentille *f* plan-convexe — **d** plankonvexe Linse *f* — **n** planconvexe lens *fm* — **e** lente *m* plano-convexo — **i** lente *f* piano-convessa

10401 planographic printing; planography — Printing process in which a flat surface is used, i.e., where the image sections are in substantially the same plane as the non-image sections (such as direct and offset lithography, or collotype).
f impression *f* planographique — **d** Flachdruck *m* — **n** vlakdruk *m* — **e** impresión *f* planográfica; sistema *m* planográfico de impresión; impresión *f* plana; planografía *f* — **i** stampa *f* planografica; stampa *f* piana; stampa *f* in piano

10402 planography — See planographic printing.

10403 plant — Building in which a process or manufacturing is carried out. See also printing plant.
f usine *f*; atelier *m*; installation *f*; fabrique *f*; établissement *m* — **d** Anlage *f*; Betrieb *m*; Werk *n*; Fabrik *f* — **n** bedrijf *n*; bedrijfsgebouw *n*; fabriek *f*; fabrieksgebouw *n* — **e** establecimiento *m*; planta *f* — **i** stabilimento *m*; impianto *m*; fabbrica *f*

10404 plant manager — See works manager.

10405 plaquette binding — See cameo binding.

10406 plaster of Paris — Anhydrous commercial form of calcined calcium sulphate. See also crown filler.

10407 plastic — Not fluid but capable of being moulded or shaped.
f plastique — **d** plastisch — **n** plastisch — **e** plástico — **i** plastico

10408 plastic binding — Plastic material, shaped in various forms or styles, ring, spiral, tubular and others, inserted in the back binding edge of a book, so that the book lays flat for reading when opened. See also adhesive binding.
f reliure *f* plastique — **d** Plastikheftung *f* — **n** plastic band *m* — **e** encuadernación *f* con lomo plástico; encuadernación *f* a la americana — **i** legatura *f* con dorso plastico; legatura *f* con dorso di materiale plastico

10409 plastic block — See plastic plate.

10410 plasticizer; softening agent; softener — Agent or compound added to plastic materials, lacquers, paper stock, etc., to impart softness or flexibility. In adhesives a material or chemical incorporated or added to soften the wet or dry film.
f plastifiant *m* — **d** Weichmacher *m*; Weichmachungsmittel *n*; Weichhaltungsmittel *n* — **n** weekmaker *m* — **e** plastificante *m* — **i** plastificante *m*

10411 plastic plate; plastic stereo; plastic block — Duplicate printing plate made by moulding a thermoplastic moulding material into a thermosetting resin matrix under pressure and heat, then chilled in a cooling press. Also duplicate printing plate made from thermosetting plastic moulding materials in a thermosetting matrix under heat and pressure.
f cliché *m* en (matière) plastique — **d** Kunststoffklischee *n*; Kunstharzklischee *n*; Plaststereo *n*; Kunststoffstereo *n* — **n** plastic cliché *n*; plastic styp *m*; kunststof-cliché *n* — **e** clisé *m* de plástico; clisé *m* en plástico; plancha *f* de plástico; molde *m* de plástico — **i** cliscè *m* di plastica; cliché *m* di plastica; stereo *m* di plastica; forma *f* duplicata di plastica

10412 plastics — Substances capable of being moulded, or shaped, and includes any of a large group of organic synthetic materials.
f matières *fpl* plastiques; plastiques *mpl* — **d** Kunststoffe *mpl*; Plaste *mpl* — **n** kunststoffen *fmpl*; plastics *mpl* — **e** plásticos *mpl* — **i** materie *fpl* plastiche

10413 plastic stereo — See plastic plate.

10414 plastic viscosity — In rheology the excess of the shearing force of thixotropic materials over the yield value divided by the rate of shear.
f viscosité *f* plastique — **d** plastische Viskosität *f* — **n** plastische viscositeit *f* — **e** viscosidad *f* plástica — **i** viscosità *f* plastica

10415 plastisol — Suspension of particles in an organic liquid, similar to an organosol, but containing no solvents.
f plastisol *m* — **d** Plastisol *n* — **n** plastisol *m* — **e** plastisol *m* — **i** plastisol *m*

10416 plastometer — Viscometer in which Bingham bodies can be run to obtain their consistency curves.
f plastomètre *m* — **d** Plastometer *n* — **n** plastometer *m* — **e** plastómetro *m* — **i** plastometro *m*

10417 plastometer — Instrument to measure resilience of softness of moulded vulcanized rubber as in printing rollers, blankets, and printing plates.
f duromètre *m* — **d** Härtemesser *m*; Härteprüfer *m* — **n** hardheidsmeter *m* — **e** durómetro *m*; comprobador *m* de dureza — **i** durometro *m*

10418 plate; printing plate — Reproduction of type or cuts in metal, plastics, rubber, or other material, to form a plate bearing a relief, planographic or intaglio printing surface. See also offset plate.
f cliché *m* — **d** Druckstock *m*; Klischee *n*; Druckplatte *f*; Platte *f* — **n** cliché *n*; drukplaat *fm* — **e** clisé *m*; cliché *m*; plancha *f* — **i** cliscè *m*; cliché *m*; lastra *f* di stampa; forma *f* stampante

10419 plate — Printed illustration.
f illustration *f*; planche *f* — **d** Abbildung *f*; Illustration *f*; Bild *n* — **n** plaat *fm*; afbeelding *f*; illustratie *f*; prent *fm* — **e** ilustración *f*; grabado *m*; imagen *f*; figura *f* — **i** illustrazione *f*; immagine *f*; tavola *f*

plate 452

10420 plate — Photographically, a senzitized sheet of glass on which negatives or positives are made.
f plaque *f* de verre; plaque *f* — **d** Glasplatte *f*; Platte *f* — **n** glasplaat *fm*; plaat *fm* — **e** placa *f* de cristal — **i** lastra *f* di cristallo

10421 plate — See offset plate.

10422 plate cementing — Attachment or mounting of relief printing plates on wooden or metal supports by means of adhesives, usually with heat and pressure.
f fixation *f* des clichés par collage — **d** Aufkleben *n* von Druckstöcke — **n** plakken *n* van clichés op voeten — **e** fijación *f* de clisés con adhesivos; montado *m* adhesivo — **i** fissaggio *m* dei cliscè allo zoccolo mediante adesivo

10423 plate chattering — See plate slap.

10424 plate clamp — See clamp.

10425 plate cleaner — Solution or detergent to remove ink, etching powder and acid resisting images from the surface of relief etchings.
f solution *f* de lavage des clichés — **d** Klischeereinigungsmittel *n*; Plattenreinigungsmittel *n* — **n** clichéwasmiddel *n* — **e** solución *f* detergente para limpiar — **i** detersivo *m* per cliscè

10426 plate cling — See plate nip.

10427 plate coating — See also coat *v*.
f couchage *m* d'une plaque; couchage *m* des plaques — **d** Plattenbeschichtung *f* — **n** aanbrengen *n* van de kopieerlaag op de plaat — **e** cobertura *f* de la plancha — **i** stratificazione *f* della lastra

10428 plate coating machine — See whirler.

10429 plate cooler — Apparatus to cool heated plates during relief etching.
f marbre *m* à refroidir — **d** Plattenkühler *m*; Plattenkühlvorrichtung *f* — **n** plaatkoeler *m* — **i** raffreddatore *m* per lastre

10430 plate current — See anode current.

10431 plate curving — Process by which electrotypes and stereotypes are curved to fit on the plate cylinder.
f cintrage *m* du cliché — **d** Biegung *f* der Platte; Biegen *n* der Platte — **n** buigen *n* van de plaat; rondzetten *n* van de plaat — **e** curvado *m* de la plancha — **i** curvatura *f* della lastra

10432 plate cylinder — Cylinder on a rotary press to which the printing plates are attached; on an offset press the cylinder covered with the printing plate.
f cylindre *m* de plaque; cylindre *m* porte-plaque — **d** Plattenzylinder *m* — **n** plaatcilinder *m* — **e** cilindro *m* de plancha; cilindro *m* portaplancha(s); cilindro *m* de las planchas; cilindro *m* portaforma — **i** cilindro *m* portalastra

10433 plated — Covered by an electrolytic deposit of a stronger or better metal, like nickel-type.
f plaqué; étamé — **d** galvanisiert — **n** gegalvaniseerd — **e** galvanizado — **i** ricoperto elettroliticamente

10434 plate dogs — See dogs.

10435 plate engraver
f graveur *m* à l'outil; graveur *m* au burin — **d** Kupferstecher *m*; Stahlstecher *m* — **n** metaalgraveur *m*; kopergraveur *m*; staalgraveur *m* — **e** grabador *m* a buril; tallador *m* — **i** incisore *m* al bulino; incisore *m* in rame; incisore *m* in acciaio; calcografo *m*

10436 plate filling — See filling up.

10437 plate finish — Smooth, polished surface on paper obtained by placing sheets of paper between polished zinc or copper plates and passing a pile of these under high pressure and slight friction between rollers of a plating machine or platen. The finish may be a smooth or fancy finish, such as linen, ripple or coarse finish.
f apprêt *m* à la plaque; fini *m* à la plaque; glacé *m* à la plaque; laminé *m* à la plaque — **d** mittels Platten satiniertes Papier *n*; Plattensatiniertes Papier *n*; Hochglanzpapier *n* — **n** papier *n* met plaatpersing — **e** satinado *m* entre planchas de cinc — **i** lucidatura *f* a smalto

10438 plate folding machine — See buckle folder machine.

10439 plate graining machine — See graining machine.

10440 plate gravure machine — Sheet-fed gravure machine using etched plates as a printing forme which are fixed round a cylinder.
f machine *f* hélio à plaque; presse *f* hélio à plaque — **d** Plattentiefdruckmaschine *f* — **n** plaatdiepdrukmachine *f*; vellendiepdrukmachine *f* — **e** máquina *f* de huecograbado a planchas — **i** macchina *f* rotocalco a lastre

10441 plate hand — Operator responsible for laying and fixing the plates on the printing cylinders on a letterpress rotary.
f conducteur *m* chargé du calage; clicheur *m* — **d** Bedienungsmann *m* für Platten — **n** plateman *m* — **e** operario *m* encargado del calado de las planchas — **i** operaio *m* incaricato del montaggio delle lastre in macchina

10442 plate holder *US*; **dark slide** *GB*; **plate holding frame** *GB*; **cassette** — Light-tight frame or case to transport a photographic plate to and from the camera and to hold the sensitized material in position during exposure.
f châssis *m* porte-film; châssis *m* porte-plaque — **d** Plattenkassette *f*; Kassette *f*; Plattenhalter *m* — **n** cassette *fm*; plaathouder *m* — **e** chasis *m* portaplaca; chasis *m* portapelícula; chasis *m* portaplancha(s); chasis *m* cargapelícula; portaplaca *m* fotográfico; portaplanchas *m*; portaplacas *m*; chasis *m*; bastidor *m* — **i** chassis *m* telaio; telaio *m* portalastra

10443 plate holding frame — See plate holder.

10444 plate key — Key to operate the lock-up dogs or clips.
f clef *f* du calage; clef *f* de serrage — **d** Plattenschlüssel *m* — **n** plaatsleutel *m*; sleutel *m* — **e** llave *f* de plancha — **i** chiave *f* per lastra

10445 plateless engraving — See thermography.

10446 plate lock-up — See lock-up.

10447 platemaker
f copiste *m* offset — **d** Kopierer *m*; Kopist *m*;

Plattenhersteller *m* — **n** kopiïst *m* — **e** copista *m* — **i** copista *m*; fabbricante *m* lastre; addetto *m* alla preparazione delle lastre

10448 platemaking — Making a printing plate from a film or flat. Includes preparation of the plate surface, sensitizing, exposure through the flat, development or processing and finishing. **f** confection *f* de plaques offset; exécution *f* de la plaque; clichage *m* — **d** Plattenherstellung *f* — **n** plaatvervaardiging *f*; vervaardiging *f* van de drukplaat — **e** confección *f* de planchas; preparación *f* de planchas; copiado *m* de planchas — **i** preparazione *f* delle lastre

10449 platemaking department — Department responsible for the manufacture of the lithographic printing surfaces by photographic means. The following steps are involved: graining of the plate, making of photographic negatives or positives, coating the light-sensitive layer to the plate, and finishing after exposure. **f** service *m* de préparation des plaques — **d** Kopierabteilung *f*; Plattenherstellungsabteilung *f* — **n** kopieerafdeling *f* — **e** sección *f* de preparación de la copia; departamento *m* de preparación de la copia; ramo *m* de preparación de la copia — **i** reparto *m* preparazione lastre

10450 plate matter — See boiler plate.

10451 platen; bed plate — Part of a job printing press on which the makeready, tympan and guides are placed and on which the sheet takes the impression from the forme. **f** platine *f* — **d** Tiegel *m* — **n** degel *m* — **e** platina *f* — **i** platina *f*; piano *m* portaforma

10452 platen; platen machine; platen press — Press in which both forme and paper are on flat surfaces, which are brought against one another to give the impression. **f** presse *f* à platine; platine *f*; minerve *f* — **d** Tiegeldruckpresse *f*; Tiegelpresse *f* — **n** degelpers *fm*; degel *m* — **e** prensa *f* de platina; máquina *f* de platina; máquina *f* de presión plana; minerva *f* — **i** macchina *f* a platina; platina *f*; pressa *f* a platina

10453 plate nip; plate cling — Tight fit of the metal plate on the cylinder of a rotary press. **d** Plattensitz *m* — **n** passen *n* van de plaat; aanliggen *n* van de plaat — **e** asentado *m* de las planchas; asiento *m* de la plancha — **i** sede *f* della lastra sul cilindro

10454 platen machine — See platen.

10455 platen press — See platen.

10456 platen printer; cropper hand — See also cropper. **f** pédalier *m*; minerviste *m* — **d** Tiegeldrucker *m*; Tretpressendrucker *m*; Tiegelmüller *m coll.* — **n** degelpersdrukker *m*; degeldrukker *m*; trappersdrukker *m* — **e** minervista *m*; pedalinista *m* — **i** pedalista *m*

10457 plate paper — Thick, soft, high-quality printing paper, lightly sized, having a smooth flat surface without a high gloss. The thicker kinds are made by pressing two or more webs together

in the wet state. Used to take impressions from engraved copper and steel plates and also for woodcuts and lithographic printing. **f** papier *m* pour taille-douce; papier *m* pour gravure — **d** Kupferdruckpapier *n* — **n** plaatdrukpapier *n*; koperdrukpapier *n* — **e** papel *m* para talla dulce; papel *m* para grabados — **i** carta *f* per la stampa calcografica; carta *f* da stampa calcografica

10458 plate printing — Printing from engraved copper plates as distinguished from relief printing. The lines of the plate are filled with ink, the surface wiped clean, and the impression is taken by pressing the plate under the roller of a plate press. **f** taille-douce *f* — **d** Kupferdruck *m* — **n** plaatdruk *m* — **e** impresión *f* en talladulce; impresión *f* con plancha de cobre; impresión *f* con plancha de acero — **i** stampa *f* in rame; calcografia *f*

10459 plate resistance — See anode resistance.

10460 plate ribs — Ridges on the back of the printing plate for a letterpress rotary press. **f** nervures *fpl* (au dos du stéréo) — **d** Plattenrippen *fpl* — **n** plaatribben *fmpl* — **e** nervios *mpl*; nervuras *fpl* — **i** nervature *fpl* sul dorso (della lastra)

10461 plate slap; plate chattering — Noise resulting from the impact of loose or badly fitting plates against the cylinder on a rotary press. **f** claquement *m* de la planche — **d** Plattenschlag *m*; Knattern *n* der Platte — **n** kletteren *n* van de plaat — **e** golpeteo *m* de la plancha — **i** sbattimento *m* della lastra

10462 plate stagger — Arrangement of the plates on the printing cylinder of a rotary press where the heads of the plates across the cylinder are not in line. **f** mauvais calage *m* des cylindres — **d** Plattenversetzung *f* — **n** verspringen *n* van de platen — **e** calado *m* defectuoso — **i** impreciza disposizione *f* delle lastre sul cilindro

10463 plate thickness — Thickness of a printing plate. For example, the thickness of an electrotype is 3.86 mm (0.152 in) or 11 points. There are 150 standards for the thickness of litho plates. See also calliper, thin gauge metal. **f** épaisseur *f* du cliché — **d** Plattenstärke *f* — **n** plaatdikte *f* — **e** espesor *m* de la plancha — **i** spessore *m* di lastra

10464 plate v up — To secure plates to the printing cylinder of a rotary press. **f** poser les clichés; fixer les clichés; caler les plaques — **d** Platten umlegen — **n** platen opleggen — **e** colocar las planchas — **i** montare lastre in macchina

10465 plate voltage — See anode voltage.

10466 plate whirler — See whirler.

10467 plate-wiping paper; die-wiping paper; die wipe; wiping-off paper; wipe-off paper; wiping paper — Machine-crêped, soft, unsized paper (usually Mg kraft) with low stretch and good strength, used primarily by engravers and

printers for wiping plates. The basis weights are generally 40 to 60 pounds (24 x 36 - 500).
f papier *m* essuie-planche; macule *f* pour estampage — **d** Wischpapier *n*; Abwischpapier *n* — **n** afwispapier *n* — **e** papel *m* para limpiar — **i** carta *f* per asportare

10468 platform, feeder's — See feeder's platform.

10469 plating — Coating of a metal plate or object with another metal by immersion in an electrolytic bath.
f étamage *m*; placage *m*; plaquage *m* — **d** Galvanisierung *f* — **n** galvanisatie *f*; galvanisering *f* — **e** galvanización *f*; galvanizado *m* — **i** galvanizzazione *f*; rivestimento *m* mediante deposito elettrolitico

10470 plating sequence — Order in which printing plates are supplied and secured to the press.
f ordre *m* des plaques — **d** Plattenfolge *f* — **n** plaatvolgorde *fm* — **e** orden *m* de las planchas — **i** sequenza *f* lastre

10471 platinic chloride — See platinum chloride.

10472 platinum — Heavy, precious, greyish-white, non corroding, ductile, malleable metallic element that fuses with difficulty; used especially in chemical ware and apparatus, as a catalyst and in jewelry alloys. Symbol: Pt.
f platine *m* — **d** Platin *n* — **n** platina *n* — **e** platino *m* — **i** platino *m*

10473 platinum chloride; platinum tetrachloride; platinic chloride — Red crystals, soluble in water and alcohol. The commercial platinum chloride is usually chloroplatinic acid. Latin: platinum chloratum
Formula: $PtCl_4 \cdot 5H_2O$.
f chlorure *m* de platine — **d** Platinchlorid *n* — **n** platinachloride *n*; platinochloruur *f* — **e** cloruro *m* de platino — **i** cloruro *m* di platino

10474 platinum potassium chloride — See potassium chloroplatinite.

10475 platinum tetrachloride — See platinum chloride.

10476 play — See clearance.

10477 playing cards
f cartes *fpl* à jouer — **d** Spielkarten *fpl* — **n** speelkaarten *fmpl* — **e** naipes *mpl*; cartas *fpl*; barajas *fpl* — **i** carte *fpl* da gioco

10478 playing card stock; cardboard for playing cards — Stiff pasted cardboard, coated and glazed for printing of playing cards.
f carte *f* pour jeux de cartes; carton *m* pour jeux de cartes — **d** Spielkartenkarton *m* — **n** speelkaartenkarton *n* — **e** cartulina *f* para naipes — **i** cartoncino *m* per carte da gioco

10479 please turn over; p.t.o. — Indication generally at the foot of a single sheet to direct the reader's attention to the other side.
f tournez s'il vous plaît; t.s.v.p. — **d** bitte, wenden; b.w. — **n** omslaan a.u.b.; z.o.z. — **i** vedi retro; vedi a tergo; voltare la pagina

10480 pledget of cotton; cotton-wool wad
f tampon *m* d'ouate — **d** Wattebausch *m* — **n** prop *fm* watten — **e** muñequilla *f* de algodón;

almohadilla *f* de algodón; tampón *m* de guata; restregador *m* — **i** tampone *m* di cotone

10481 plexiglas — Trade-mark for a thermoplastic polymer of methyl methacrylate with a light weight, resistant to weathering; can be bent when hot but returns to its original shape when reheated.

10482 pliability; folding quality — Property of papers and artificial leathers to fold smoothly and without breaking at the fold.
f qualité *f* de pliure; aptitude *f* au pliage; aptitude *f* à la flexion — **d** Falzfähigkeit *f*; Biegsamkeit *f* — **n** vouwbaarheid *f*; geschiktheid *f* tot vouwen; buigzaamheid *f* — **e** plegabilidad *f* — **i** flessibilità *f*; attitudine *f* alla piegatura

10483 plotting paper — See graph paper.

10484 plough *GB*; **plow** *US*; **plow and press** *US* — Wooden implement with a knife used by hand bookbinders for cutting and trimming the edges of books.
f presse *f* à rogner; presse *f* de relieur et pointe à rogner — **d** Buchbindehobel *m*; Buchhobel *m*; Beschneidehobel *m*; Zungenhobel *m*; Hobel *m* — **n** ploeg *fm* — **e** ingenio *m* (para encuadernadores) — **i** pressa *f* per tranciare a mano; tagliacarte *m* a mano

10485 plow — See plough.

10486 plow and press — See plough.

10487 plucking — See picking.

10488 plug
f fiche *f*; fiche *f* de connexion; prise *f* de courant — **d** Stecker *m*; Netzstecker *m* — **n** stekker *m* — **e** ficha *f* tomacorriente — **i** spina *f* (della rete)

10489 plug *v* — To advertise or publicize insistently.
f faire de la publicité — **d** Reklame machen — **n** reclame maken — **e** dar bombo; elogiar — **i** propagandare insistentemente

10490 plug board; patch board; patch panel; problem board — Component of some data processing machines similar to a manual telephone exchange switchboard.
f panneau *m* de raccordement — **d** Programmierfeld *n*; Steckfeld *n*; Schalttafel *f*; Stecktafel *f* — **n** schakelbord *n*; schakelpaneel *n* — **e** panel *m* de acoplamiento — **i** pannello *m* di connessione

10491 plug-board coding — Method to program punched card accounting machines and some computers. Electrical leads are connected to terminals in a predetermined order and represent the computer program.

10492 plug flow — Part of the flow where the shearing stress is less than the yield value.
f écoulement *m* en bouchon

10493 plugger — See filler.

10494 plugging — Sign of underdevelopment of offset printing plate. Water on the plate or skimping on developer may be the cause. Developing on uneven table or sink may leave underdeveloped areas. For correct condition developer must be added and redeveloped.
n brugvorming *f*; samenklonteren *n* van de raster-

punten — **e** agrumarse de los puntos — **i** sotto-sviluppo *m* della lastra offset

10495 plumbic nitrate — See lead nitrate.

10496 plumbism — See lead poisoning.

10497 plumbous oxide — See litharge.

10498 plus-ing — See plussing.

10499 plus mark; plus sign — The sign . denoting addition or a positive quantity.
f plus *m* — **d** Pluszeichen *n* — **n** plusteken *n* — **e** signo *m* de adición; signo *m* positivo — **i** segno *m* d'addizione; segno *m* del più

10500 plus sheets — See overs.

10501 plus sign — See plus mark.

10502 plussing; plus-ing — In gravure printing, adding time applied to exposure of carbon tissue of positive or part of positive is too heavy or dense in relation to others on same forme.
f sur-copie *f*; surcopie *f* — **d** Anschleiern *n* — **n** langer belichten *n* — **e** largo de exposición; sobre-expuesto *m* — **i** aumento *m* dell'esposizione

10503 plutarch — A size of wrapping paper. See app. no. 3.

10504 ply — Fibrous web of homogeneous composition formed on the wire of the papermaking machine.
f jet *m* — **d** Lage *f*; Faserstofflage *f* — **n** laag *fm* — **e** capa *f* — **i** getto *m*

10505 plywood; laminated wood — Material formerly used for block mounting.
f contreplaqué *m*; bois *m* contreplaqué — **d** Sperrholz *n*; Klischeesperrholz *n* — **n** multiplex *n* — **e** madera *f* contrachapeada; madera *f* compensada; madera *f* terciada — **i** legno *m* compensato

10506 pneumatic post
f poste *f* par tubes pneumatiques — **d** Rohrpost *f* — **n** buizenpost *fm* — **i** posta *f* pneumatica

10507 pocket atlas
f atlas *m* de poche — **d** Taschenatlas *m* — **n** zakatlas *m* — **e** atlas *m* de bolsillo; atlas *m* portátil — **i** atlante *m* tascabile

10508 pocket book — Small, especially paperback book, that can be carried in the pocket. See paperback.
f livre *m* de poche — **d** Taschenbuch *n* — **n** pocketboek *n* — **e** libro *m* de bolsillo — **i** libra *m* tascabile

10509 pocket book — See paperback.

10510 pocket edition — Book with a small format, usually foolscap octavo.
f édition *f* de poche — **d** Taschenausgabe *f* — **n** zakuitgave *fm*; uitgave *fm* op zakformaat — **e** edición *f* de bolsillo — **i** edizione *f* tascabile

10511 pocket envelope — Envelope in which the opening and flap are on the smaller side, with seam or seams parallel to the longer side.
f pochette *f* — **d** Umschlag *m* mit Klappe an der kürzeren Seite; Versandtasche *f*; Versandumschlag *m* — **n** zak-envelop(pe) *fm* — **e** bolsa *f* de envío; bolsa *f* de expedición — **i** busta *f* a sacchetto

10512 pocket-folding machine — See buckle folder machine.

10513 pocket grinder — Mechanical woodpulp grinder in which the logs are held under pressure against the grinding stones in individual pockets (usually three pockets).
f défibreuse *f* à presses; défibreuse *f* à chambres — **d** Pressenschleifer *m* — **n** persslijper *m* — **e** desfibrador *m* de prensas — **i** sfibratore *m* a pressione

10514 pocket microscope
f microscope *m* de poche — **d** Taschenmikroskop *n* — **n** zakmicroscoop *mn* — **e** microscopio *m* de bolsillo — **i** microscopio *m* da tasca

10515 pocket note — A size of writing paper. See app. no. 3.

10516 pocket size
f format *m* de poche — **d** Taschenformat *n* — **n** zakformaat *n* — **e** tamaño *m* de bolsillo; formato *m* de bolsillo — **i** formato *m* tascabile

10517 pocks — Darkened or "burned" areas in paper which occur in calendering due to excessive moisture content in the web. See also calender marks.

10518 point — In reference to score rule, a thickness of 0.355 mm (0.014 in). To express the thickness of paper or board stock or any other material, a thickness of 0.0254 mm (0.001 in).
f point *m* — **d** Punkt *m* — **n** punt *m* — **e** punto *m* — **i** punto *m*

10519 point — A thousandth of an inch (or 0.0254 mm).

10520 point; typographic point — In the pica system a point is 0.351457 mm (1/72 or 0.013837 of an inch). In the Didot system a point is 0.376065 mm (at a temperature of 20 °C). However, since January 1st, 1978, the point Didot is fixed at 1000.333/2660 mm $\approx$ 0.376065 x 10^{-3} m. See also point system, app. no. 3.
f point *m*; point *m* typographique — **d** Punkt *m*; typographischer Punkt *m*; Achtelpetit *f obs.* — **n** punt *m*; typografische punt *m* — **e** punto *m*; punto *m* tipográfico — **i** punto *m*; punto tipografico

10521 point — See full point.

10522 point common line — See standard line.

10523 pointer — Part of the unit indicator of the Monotype keyboard.
f indicateur *m* des unités — **d** Einheitenzeiger *m* — **n** eenhedennaald *fm* — **e** aguja *f* de las unidades; indicador *m* de las unidades; puntero *m* de las unidades — **i** indicatore *m* di unità; indicatore *m* delle unità

10524 pointer — Needle of a speedometer or watch.
f aiguille *f* (indicatrice); index *m* — **d** Zeiger *m*; Anzeiger *m*; Weiser *m* — **n** wijzer *m*; naald *fm* — **e** indicador *m* — **i** lancetta *f*; indicatore *m*; indice *m*

10525 pointer — See flag.

10526 pointillé binding; fers-pointillé binding — Book cover decoration style originating from Le Gascon, Paris († 1730), characterized by the use of tool forms in dotted outline, while the cover is divided in panels, sometimes with yellow, green, or brown leather, and bordered with a narrow line ornament.

f reliure f pointillé; reliure f en style pointillé — d Spitzmuster-Einband m; Einband m in Spitzmusterstil; Einband m mit Spitzenstil — n pointillé-band m — e encuadernación f pointillé — i legatura f sile a mille punti; legatura f sile filigrana

10527 pointing — See punctuation.

10528 point light — For point light can be taken that it decreases with the square of the distance of the light source from the illuminated subject.

f lumière f punctiforme; lumière f ponctuelle — d Punktlicht n — n puntlicht n — e luz f puntiforme; alumbrado m concentrado — i luce f puntiforme

10529 point of multiplication — See mark of multiplication.

10530 point set — See unit set.

10531 point size — See body size.

10532 points of suspension — See also ellipsis.

f points mpl de suspension; points mpl suspensifs — d Auslassungspunkte mpl — n suspensietekens npl — e puntos mpl suspensivos — i puntini mpl sospensivi; punti mpl di reticenza

10533 points per pound — Term sometimes applied to the bursting strength of a paper.

10534 point system — The three principal point systems in use today differ basically and in decimal detail. The British-American system, universally adopted by English speaking countries, has for its standard of measurement the 0.166 pica and the 0.01383 point or one twelfth of the pica (thus 1,000 lines of pica or 12-point matter measure 166 inches). The Didot system used in most countries of continental Europe has the cicero as its basic unit. The cicero equals 12 "corps" or 0.178 in and the Didot "corps" or point measures exactly 0.01483. The mediaan system sometimes encountered has a "corps" or point measurement of 0.01374, the mediaan em or cicero measures 0.165. For metal type the 12-point system is the most ideal, because 12 can be divided by 2, 3, 4, 6 and 8. In the metric system dividing is only possible by 2 and 5. For photographic typesetting reducing or enlarging has no limits, so here a *typographic point system is not important. However, Eurograf (Organisation of printing federations in the EEC) accepted a unit based on the *SI. Since January 1978 the Didot point is fixed at 0.375 mm. About the increments of 0.1 mm (ISO) and 0.25 mm (Eurograf) there was up to now no consensus of opinion. In the pica system 12 pt $\approx$ 4.21752 x 10^{-3} m. See also cicero, point, app. no. 1.

f système m de point (typographique) — d (typographisches) Punktsystem n — n (typografisch) puntenstelsel n — e sistema m de medidas (tipográfico) en puntos — i sistema m di misura in punti

10535 point-to-point transmission — Trans-

mission of data directly between two points without the use of an intermediate terminal.

f transmission f entre deux terminaux — d Übertragung f zwischen zwei Stationen — n transmissie f tussen twee stations — e transmisión f de punto a punto — i trasmissione f tra due terminali

10536 pointwise; body wise — Vertical dimension or height of type as opposed to its horizontal dimension of *width, which is termed *setwise.

10537 poise — Unit of absolute viscosity of a fluid, signifying that a force of one gram will maintain unit rate of shear of a film of unit thickness between surfaces of unit area. Abbrev. P. In the SI:
1 P = 1 dyn·s/cm^2 = 0.1 Pa·s. See also centipoise.

f poise f — d Poise f — n poise m — e poise m — i poise f

10538 Poiseuille flow — Laminar flow that takes place in a capillary tube when streamline motion prevails; applied to Newtonians and Bingham bodies.

f écoulement m de Poiseuille — d Poiseuillesches Fließen n; Poiseuillesche Strömung f — n stroming f volgens Poiseuille — e corriente f según Poiseuille; flujo m según Poiseuille — i flusso m di Poiseuille

10539 polarisation — Action or process of affecting light or other radiation so that the vibrations of the wave assume a definite form.

f polarisation f — d Polarisation f — n polarisatie f — e polarización f — i polarizzazione f

10540 polarisation — Condition in a primary cell, in which the voltage falls off after a period of working, because of an accumulation of reaction products at the electrodes.

f polarisation f électrolytique — d elektrolytische Polarisation f — n polarisatie f — e polarización f electrolítica — i polarizzazione f elettrolitica

10541 polarized light — Light transmitted by a polarizing filter. If two polarizing filters are parallel the light transmitted by the first is transmitted unchanged by the second; if the second is turned through a right angle the light transmitted by the first is completely extinguished by the second.

f lumière f polarisée — d polarisiertes Licht n — n gepolariseerd licht n — e luz f polarizada — i luce f polarizzata

10542 polarizing filter — Filter placed on a camera lens or lights to eliminate undesirable reflections from the subject. It darkens a blue sky by transmitting light travelling in one plane while absorbing light travelling in other planes. Also used when photographing water, glass, or other objects with shiny surfaces.

f filtre m polarisant — d Polarisationsfilter n — n polarisatiefilter n — e filtro m polarizador — i filtro m polarizzatore

10543 polar materials — Materials whose molecules have ends that are opposite in character like the poles of a magnet.

f matériaux *mpl* polaires — **d** polare Stoffe *mpl*; polar entgegengesetzt wirkende Materialien *npl* — **n** polaire stoffen *fmpl* — **e** materiales *mpl* polares — **i** materiali *mpl* polari

10544 polar solvents — Solvents comprised of electrically non-symmetrical molecules, e.g., in gravure inks those made up of organic molecules which contain oxygen (alcohols, ketones, esters, ethers).
f solvants *mpl* polaires — **d** polare Lösungsmittel *npl* — **n** polaire oplosmiddelen *npl* — **e** solventes *mpl* polares — **i** solventi *mpl* polari

10545 Pola screen — Neutral grey light filter, transmitting plane polarized light of all visible colours, but absorbing ultra-violet. Used on camera lenses and lights to eliminate or subdue undesirable reflections from originals.
f trame *f* (simili) Pola — **d** Pola-Raster *m*; Pola-Autotypieraster *m* — **n** Pola-raster *n*; Pola-autotypieraster *n* — **e** trama *f* (de medio tono) Pola — **i** retino *m* (autotipico) Pola

10546 policy allowance — Work study term for an addition made to work values to increase earnings artificially.
f majoration *f* d'opportunité — **i** maggiorazione *f* per opportunità

10547 polish *v* **out** — To erase a design image or unwanted marks on a lithographic plate with a snake-stone or similar abrasive.
f effacer — **d** abglätten — **n** uitslijpen; wegslijpen — **e** alisar; rebajar; pulimentar — **i** lisciare

10548 polling — Flexible, systematic technique by which each of the terminals sharing a communication line is periodically interrogated to determine whether it requires servicing.
f appel *m* sélectif — **d** Abrufen *n* — **n** selectieve roep *m* — **e** llamada *f* selectiva — **i** consultazione *f* selettiva

10549 polyacrylate — Polymer formed from acrylic acid esters.
f polyacrylate *m* — **d** Polyakrylat *n* — **n** polyacrylaat *n* — **e** poliacrilato *m* — **i** poliacrilato *m*

10550 polyamide — Polymer having the structural units linked by amide groups.
f polyamide *m* — **d** Polyamid *n* — **n** polyamide *n* — **e** poliamida *f* — **i** poliamide *f*

10551 polyamide resin — Polymer in which the monomer units are linked together by the amide group CONH, e.g., nylon. Made by reaction of a dibasic acid with a diamine and used for plastic printing plates. It may be printed without any prior treatment; fairly low heat must be used for drying.
f résine *f* polyamide — **d** Polyamidharz *n* — **n** polyamide-hars *mn* — **e** resina *f* poliamida — **i** resina *f* poliammidica

10552 polychromatic — See polychrome.

10553 polychrome; polychromic; polychromatic; multicolour — Having many colours.
f polychrome; multicolore — **d** polychromatisch; vielfarbig; mehrfarbig — **n** meerkleurig; in meer kleuren; polychroom — **e** policromo; poli-

cromático; multicolor — **i** policromo; multicolore

10554 polychromic — See polychrome.

10555 polyester resin — Plastic material used in place of acetate for preparing photographic film tape, and stripping bases, characterized by dimensional stability and strength. The reaction of acids and alcohol from esters and the resultant polymer finds great use. The most commonly used polyester film is a polymer formed between dimethyl terephthalate and ethylene glycol and is called polyethylene terephthalate. It may be printed easily.
f résine *f* polyester; polyester *m* — **d** Polyesterharz *n*; Polyester *m* — **n** polyester-hars *m*; polyester *m* — **e** resina *f* poliéster; resina *f* de poliéster; poliéster *m* — **i** resina *f* poliesterica; poliestere *m*

10556 polyethylene — Synthetic polymer of high molecular weight resulting from the polymerization of ethylene gas under pressure. Milky white and wax-like; has good flexibility under extreme cold conditions, low moisture permeability, excellent resistance to acids, alkalis and inorganic chemicals and has no known solvents at room temperatures. Greatest application is in packaging and electrical insulation. Contains no plasticizer. Used as a coating on paper and applied in thicknesses of 0.01-0.1 mm (0.0005-0.005 in). Paper so coated is used as a protective packing material for wrapping foodstuffs, chemicals, tobacco, etc. Also available in film form. Extremely difficult to print due to its lack of adhesion.
Formula: $(C_2H_4)_n$. See also polythene.
f polyéthylène *m*; polythène *m* — **d** Polyäthylen *n* — **n** polyethyleen *n*; polytheen *n* — **e** polietileno *m* — **i** polietilene *m*

10557 polyhydric alcohol — Alcohol that contains more than one hydroxyl group, e.g., $CH_2OH \cdot CH_2OH$.
f alcool *m* polyatomique — **d** mehrwertiger Alkohol *m* — **n** meerwaardige alcohol *m* — **e** alcohol *m* polivalente — **i** alcool *m* polivalente

10558 polymer — Compound formed by the linking of simple and identical molecules having functional groups that permit their combination to proceed to higher molecular weights under suitable conditions. May be formed by polymerization (addition polymer) or polycondensation (condensation polymer). When two or more different monomers are involved, the product is called a *copolymer.
f polymère *m* — **d** Polymer *m*; Polymerisationsprodukt *n* — **n** polymeer *n* — **e** polímero *m*; cuerpo *m* polímero — **i** polimero *m*

10559 polymerization — Chemical reaction in which the molecules of a monomer are linked together to form large molecules whose weight is a multiple of that of the original substance. When two or more different monomers are involved, the process is called copolymerization. In the oxidation of drying oils, it is accompanied by the change from the liquid to the solid state.

f polymérisation f − **d** Polymerisation f − **n** polymerisatie f − **e** polimerización f − **i** polimerizzazione f

10560 polymerized oils − See blown oils.

10561 polystyrene; polystyrenic resin − Transparent thermoplastic synthetic resin, polymer of styrene, usually prepared from ethylene and benzene, soluble in aromatic hydrocarbons, chlorinated hydrocarbons and esters. It may be printed without any prior treatment. Formula: $(C_6H_5CHCH_2)_n$.

f polystyrolène m; résine f polystyrénique; résine f polystyrolique; résine f polystyrolénique − **d** Polystyrolharz n − **n** polystyreen n; polystyreenhars mn − **e** poliestireno m; polistireno m; resina f de estireno − **i** polistirene m; polistirolo m

10562 polystyrenic resin − See polystyrene.

10563 polythene − Generic name for *polyethylene. The word is no longer current in the US, but is still used in European countries.

10564 polyurethan resin − Thermoplastic or thermosetting synthetic polymer containing $NHCO_2$ groups. Polyurethan coatings are characterized by thoughness, hardness, mar-resistance, flexibility, and resistance to chemical action. There is a tendency to yellow with age.

f résine f de polyuréthane − **d** Polyurethanharz n − **n** polyurethaan n; polyurethaanhars nm − **e** resina f de poliuretano − **i** resina f di poliuretani

10565 polyvinyl acetal resin − See polyvinyl acetate.

10566 polyvinyl acetate; polyvinyl acetal resin; PVAc − Colourless synthetic resin. Transparent, thermoplastic solid, soluble in low molecular weight alcohols, esters, benzene, chlorinated hydrocarbons, ketones, insoluble in water, gasoline, oils and fats. Formula: $(H_2CCHOOCCH_3)_x$.

f acétate m de polyvinyle; acétate m polyvinylique − **d** Polyvinylazetat n − **n** polyvinylacetaat n − **e** acetato m de polivinilo − **i** acetato m di polivinile; acetato m polivinilico; polivinilacetato m

10567 polyvinyl alcohol; PVA; PVOH − Compound made when polyvinyl acetate is hydrolized with acid or alkali catalysts. May be printed easily. Formula: $(CH_2CHOH)_n$.

f alcool m polyvinylique − **d** Polyvinylalkohol m − **n** polyvinylalcohol m − **e** alcohol m de polivinilo; alcohol m polivinílico − **i** alcool m di polivinile; alcool m poliviniolico

10568 polyvinyl chloride; PVC − White powder which can be converted into colourless sheets or films. Insoluble in most organic solvents, somewhat soluble in methyl ethyl ketone and phorone. Formula: $(H_2CCHCl)_n$.

f chlorure m de polyvinyle; polychlorure m de vinyle; chlorure m polyvinylique − **d** Polyvinylchlorid n − **n** polyvinylchloride n − **e** cloruro m de polivinilo; cloruro m polivinílico − **i** cloruro m di polivinile; cloruro m polivinilico; polivinilcloruro m

10569 polyvinylidene chloride − See saran.

10570 polyvinyl process − Method based on the use of polyvinyl alcohol as a substitute for albumen and glue in photomechanical sensitizers.

f procédé m polyvinilique; procédé m à la colle synthétique − **d** Polyvinylverfahren n; Polyvinylalkoholverfahren n − **n** polyvinyl-procédé n − **e** proceso m polivinílico − **i** procedimento m al polivinile

10571 pony − Diminutive for small mechanical devices (pony bar, pony roll, pony cylinder, etc.).

10572 pony bar − One of a pair of turning bars in a rotary press to lead the paper web twice through a double wide printing unit.

f barre f de retournement pony − **d** Pony-Wendestange f; Kehrwendestange f; Parallelwendestange f − **n** ponystang fm − **e** barra f de inversión − **i** barra f diagonale pony

10573 poor fold strength − Inability of a web of paper to withstand folding, evident when the web cracks or ruptures when folded, providing the web has not been overdried. Could be termed cracking at the fold or split backs.

f mauvaise aptitude f au pliage − **d** ungenügende Falzfestigkeit f − **n** onvoldoende sterkte f tot vouwen; onvoldoende vouwweerstand m − **e** resistencia f insuficiente al plegado − **i** bassa resistenza f alla piegatura

10574 poorly wound reel − See loosely wound reel.

10575 pop envelope − Post Office Preferred range of envelopes which will qualify for the lowest rates of postage for inland letters and papers. Should be at least 90 x 140 mm (3.5 x 5.5 in) and not larger than 120 x 235 mm (4.75 x 9.25 in); oblong in shape, with the longer side at least 1.414 times the shorter side; made from paper weighing at least 63 g/m^2. The two most popular international size envelopes DL 109 x 220 mm (4.3 x 8.7 in) and C6 114 x 162.5 mm (4.5 x 6.4 in) are within the pop range.

10576 pope roll − Roll at the end of the papermaking machine to wind the finished product.

f enrouleuse f pope − **d** Poperoller m − **n** poperoller m − **e** enrolladora f de bobinas; enrolladora f pope − **i** arrotolatore m pope

10577 poplar − Broadleaf hardwood for the manufacture of soda pulp.

f peuplier m − **d** Pappelholz n − **n** populierehout n − **e** álamo m; madera f de álamo − **i** pioppe m; legno m di pioppo

10578 pop test; Mullen test − See also Mullen tester, bursting strength.

f essai m de résistance à la perforation; essai m de résistance à la crevaison; essai m de résistance à l'éclatement − **d** Berstfestigkeitsversuch m; Berstdruckprobe f − **n** berstweerstandsproef fm; berstdrukproef fm; berstproef fm − **e** ensayo m Mullen; ensayo m al reventamiento; ensayo m de rotura; ensayo m de resistencia (del papel) bajo presión; ensayo m de resistencia al estallido − **i**

prova *f* di resistenza allo scoppio di Mullen
10579 pop tester — See Mullen tester.
10580 porcelain clay — See China clay.
10581 porosity — Property of a substance containing connected air voids, dependent upon the number of the voids and their distribution in size, shape, and orientation; commonly evaluated by determining the *air permeability of the specimen.
f porosité *f* — **d** Porosität *f*; Porigkeit *f* — **n** porositeit *f*; poreusheid *f* — **e** porosidad *f* — **i** porosità *f*
10582 porosity test
f essai *m* porométrique; essai *m* de porosité — **d** Porositätsprüfung *f* — **n** porositeitsproef *f*; porositeitsonderzoek *n* — **e** ensayo *m* de porosidad — **i** saggio *m* porometrico; saggio *m* della porosità; prova *f* della porosità
10583 porous
f poreux — **d** porös; porig — **n** poreus — **e** poroso; burbujoso — **i** poroso
10584 porous slug
f ligne-bloc *f* poreuse; fonte *f* poreuse — **d** poröse Zeile *f* — **n** poreuze regel *m* — **e** lingote *m* poroso — **i** riga *f* porosa
10585 portable colour duct; portable colour fountain *depr.* — Interchangeable ink duct that can be attached to the inking system of a printing unit to provide an additional colour.
f encrier *m* portatif — **d** auswechselbare Farbwanne *f* — **n** verwisselbare inktbak *m* — **e** tintero *m* portátil para colores — **i** calamaio *m* intercambiabile
10586 portable colour fountain — See portable colour duct.
10587 portable fountain — Fountains for temporary attachment, usually one page in width.
f encrier *m* mobile — **d** auswechselbares Farbwerk *n*; tragbares Farbwerk *n* — **n** verwisselbaar inktwerk *n* — **e** tintero *m* portátil — **i** calamaio *m* portatile
10588 portable program — Computer program which can be made to work (without modification) on a computer of a different manufacturer or a different model.
10589 portfolio — A size of wrapping paper. See app. no. 3.
10590 portrait size; upright size
f format *m* en hauteur — **d** Hochformat *n* — **n** staand formaat *n* — **e** formato *m* vertical — **i** formato *m* verticale
10591 positive — Photographic image giving a natural representation of the light and shades of the subject or original. The reverse of a negative. A positive on a opaque support (paper, metal) is a print whereas one on a transparent support (glass or film) is a transparency. See also diapositive.
f positif *m* — **d** Positiv *n* — **n** positief *n* — **e** positivo *m* — **i** positivo *m*
10592 positive carbon process — See ozobrome.
10593 positive infinitely variable
f positif infiniment variable — **d** positif ideal-

veränderlich — **n** positief-ideaal veranderlijk — **e** positivo infinitamente variable — **i** positivo infinitamente variabile
10594 positive lens — See converging lens.
10595 positive mask; single-overlay mask *obs.* — See also contact mask.
f masque *m* à simple recouvrement — **d** einstufige Maske *f* — **n** positief masker *n* — **e** máscara *f* positiva — **i** maschera *f* positiva
10596 positive spacing — In computerized typesetting, spaces of a definite thickness, to justify tabular and multiple justification work without variable spaces, making up to even ems with the positive spaces.
10597 post — Pile of felts and sheets of paper, about 45 cm high, built up in hand-made papermaking. See also pilcher.
f porse *f* — **d** Pauscht *m* — **i** posta *f*
10598 post — A British standard size for writing paper. See app. no. 3.
10599 postage meter — See franking machine.
10600 postage-meter machine — See franking machine.
10601 postage stamp printing
f impression *f* timbres-poste; impression *f* des timbres-poste — **d** Briefmarkendruck *m* — **n** postzegeldruk *m*; drukken *n* van postzegels — **e** impresión *f* de sellos de correo — **i** stampa *f* di francobolli
10602 postal — A British standard size for pulp and pasteboards. See app. no. 3.
10603 postal card — A size of cut card. See app. no. 3.
10604 postal card — See postcard.
10605 postcard *GB*; **postal card** *US* — Card on which a message may be written for mailing without an envelope, in most countries printed with a postage stamp (in US in sheets of 48 cards). First issued in Austria (1869), in US in 1873.
f carte *f* postale — **d** Postkarte *f* — **n** briefkaart *fm* — **e** tarjeta *f* postal — **i** cartolina *f* postale
10606 postcard — A size of card. See app. no. 3.
10607 postcard — See picture postcard.
10608 postcard bristol — Coated or uncoated bristol for printing of postcards.
f carton *m* pour carte postale — **d** Postkartenkarton *m* — **n** briefkaartkarton *n* — **e** cartulina *f* para tarjetas postales; cartulina *f* para postales — **i** cartoncino *m* da cartoline postali
10609 poster; sign — Printed advertising matter or sign displayed on walls, boards, etc.
f affiche *f*; placard *m* (d'affiche); panneau-réclame *m* — **d** Plakat *n*; Aushang *m* — **n** affiche *nfm*; aanplakbiljet *n*; plakkaat *n* — **e** cartel *m* — **i** affisso *m*; manifesto *m*; cartellone *m*
10610 poster paper; sign paper — Machine-finished, white or coloured news or book paper with good non-fading qualities.
f papier *m* pour affiche(s); papier *m* pour panneaux-réclames; papier *m* d'affiche; papier *m* affiche — **d** Plakatpapier *n*; Anschlagpapier *n*; Zettelpapier *n* — **n** affichepapier *n*; aanplak-

biljettenpapier *n*; plakkaatpapier *n* — **e** papel *m* para carteles; papel *m* de carteles; papel *m* para reclamos — **i** carta *f* per affisi; carta *f* per manifesti; carta *f* per cartelloni

10611 poster stamps; picture stamps; advertising stickers — Small advertising illustrations in the form of perforated and detachable stamps.
f timbres-réclames *mpl*; timbres-vignettes *mpl*; timbres *mpl* de publicité — **d** Siegelmarken *fpl*; Verschlußmarken *fpl*; Reklamemarken *fpl* — **n** sluitzegels *mpl*; reclamezegels *mpl*; reclame-etiketten *npl* — **e** estampillas *fpl* de publicidad; estampillas *fpl* publicitarias; sellos *mpl* de reclamo — **i** francobolli *mpl* pubblicitari

10612 poster type — Large sized type for printing posters, made of wood, when over 72-points body size.
f caractères *mpl* pour affiches — **d** Plakatschrift *f*; Plakatbuchstaben *mpl* — **n** biljetletter *fm* — **e** letra *f* de cartel; letras *fpl* para carteles; tipo *m* de cartel — **i** caratteri *mpl* per affisi; caratteri *mpl* per manifesti; caratteri *mpl* per cartelloni

10613 post 8vo — A size of writing paper. Also a British size of cards. See app. no. 3.

10614 post 4to — A size of writing paper. See app. no. 3.

10615 posts, binding — See binding posts.

10616 post treatment; after treatment — Treatment of the non-image areas of the metal plate after image development.
f traitement *m* ultérieur; traitement *m* d'après — **d** Nachbehandlung *f* — **n** nabehandeling *f* — **e** tratamiento *m* ulterior — **i** trattamento *m* suppletivo; dopo trattamento *m*

10617 pot — See metal pot.

10618 potash — Originally applied to *potassium carbonate recovered from wood ashes, now used in trade in connections with any material containing potassium, irrespective of its composition; the potash value is expressed as the equivalent amount of the oxide K_2O.
f potasse *f* — **d** Pottasche *f* — **n** potas *fm* — **e** potasa *f* — **i** potassa *f*

10619 potash alum — See aluminium potassium sulphate.

10620 potash bromide — See potassium bromide.

10621 potash muriate — See potassium chloride.

10622 potassium — Soft, wax-like, silvery metal, soluble in liquid ammonia, aniline, mercury and sodium. Rapidly oxidizing in moist air. Symbol: K. Latin: kalium.
f potassium *m*; potasse *f* — **d** Kalium *n* — **n** kalium *n* — **e** potasio *m* — **i** potassio *m*

10623 potassium acetate — White, crystalline, deliquescent powder, soluble in water and alcohol, insoluble in ether. Solutions are alkaline to litmus but not to phenolphthalein. Must be kept well stoppered.
Formula: $KC_2H_3O_2$.
f acétate *m* de potassium; acétate *m* de potasse — **d** Kaliumazetat *n*; essigsaures Kalium *n* — **n**

kaliumacetaat *n* — **e** acetato *m* de potasio; acetato *m* potásico; acetato *m* de potasa — **i** acetato *m* di potassio; acetato *m* potassico

10624 potassium acid oxalate — See potassium binoxalate.

10625 potassium acid sulphate — See potassium bisulphate.

10626 potassium acid tartrate — See potassium bitartrate.

10627 potassium aluminium sulphate — See aluminium potassium sulphate.

10628 potassium auribromide — See potassium gold bromide.

10629 potassium aurichloride — See potassium gold chloride.

10630 potassium bichromate — See potassium dichromate.

10631 potassium binoxalate; potassium acid oxalate; acid potassium oxalate — White crystals, soluble in water, insoluble in alcohol. Bitter taste, hygroscopic, poisonous, decomposes when heated; used in photography.
Formula: $KHC_2O_4 \cdot 1/2H_2O$.
f bioxalate *m* de potassium; bioxalate *m* de potasse; sel *m* d'oseille — **d** Kaliumbioxalat *n*; saures Kaliumoxalat *n*; oxalsaures Kali *n*; Kleesalz *n* — **n** kaliumbioxalaat *n*; zuur kaliumoxalaat *n* — **e** bioxalato *m* de potasio; bioxalato *m* potásico; bioxalato *m* de potasa — **i** biossalato *m* di potassio; biossalato *m* potassico

10632 potassium bisulphate; acid potassium sulphate; potassium hydrogen sulphate; potassium acid sulphate — Colourless crystals, soluble in water, yielding a solution with acid reaction. Decomposes in alcohol.
Formula: $KHSO_4$.
f bisulfate *m* de potassium; bisulfate *m* de potasse; sulfate *m* acide de potassium — **d** Kaliumbisulfat *n*; saures schwefelsaures Kalium *n*; Kaliumhydrogensulfat *n*; Kaliumhydrosulfat *n* — **n** kaliumbisulfaat *n* — **e** bisulfato *m* de potasio; bisulfato *m* potásico; bisulfato *m* de potasa — **i** bisolfato *m* di potassio; bisolfato *m* potassico

10633 potassium bitartrate; potassium acid tartrate — White crystals, soluble in water, insoluble in alcohol. Latin: kalium bitartaricum.
Formula: $KHC_4H_4O_6$.
f bitartrate *m* de potassium; bitartrate *m* de potasse; tartrate *m* acide de potassium — **d** Kaliumbitartrat *m*; weinsaures Kalium *n*; Weinsteinrahm *m* — **n** kaliumbitartraat *n*; wijnsteen *m* — **e** bitartrato *m* de potasio; bitartrato *m* potásico; bitartrato *m* de potasa — **i** bitartrato *m* di potassio; bitartrato *m* potassico

10634 potassium bromide; potash bromide; bromide of potash — White, crystalline granules or powder, soluble in water and glycerol, slightly soluble in alcohol and ether. Used in process engraving and lithography, in photography for gelatine bromide papers and plates. Latin: brometum kalicum; kalium bromatum.

Formula: KBr.

f bromure *m* de potassium − **d** Bromkalium *n*; Kaliumbromid *n* − **n** broomkali *m*; broomkalium *n*; kaliumbromide *n* − **e** bromuro *m* de potasio; bromuro *m* potásico; bromuro *m* de potasa − **i** bromuro *m* di potassio; bromuro *m* potassico

10635 potassium bromoaurate − See potassium gold bromide.

10636 potassium carbonate; carbonate of potash; pearl ash − White, deliquescent, granular powder, soluble in water, insoluble in alcohol. Latin: carbonas kalicus.

Formula: K_2CO_3. See also potash.

f carbonate *m* (neutre) de potassium − **d** Kaliumkarbonat *n*; kohlensaures Kali *n* − **n** kaliumcarbonaat *n* − **e** carbonato *m* de potasio; carbonato *m* potásico; carbonato *m* de potasa − **i** carbonato *m* di potassio; carbonato *m* di potassa

10637 potassium chlorate; chlorate of potash − Transparent, colourless crystals or white powder, soluble in water and alkalis. Poisonous. Latin: chloras kalicus.

Formula: $KClO_3$.

f chlorate *m* de potassium; chlorate *m* de potasse; sel *m* de Bertholet − **d** Kaliumchlorat *n*; chlorsaures Kali(um) *n* − **n** kaliumchloraat *n*; chloorkali *n* − **e** clorato *m* de potasio; clorato *m* potásico; clorato *m* de potasa − **i** clorato *m* di potassio; clorato *m* potassico

10638 potassium chloride; potassium muriate; potash muriate − Colourless or white crystals or powder, soluble in water, slightly soluble in alcohol, insoluble in absolute alcohol. Latin: kalium chloratum.

Formula: KCl.

f chlorure *m* de potassium − **d** Kaliumchlorid *n* − **n** kaliumchloride *n* − **e** cloruro *m* de potasio; cloruro *m* potásico; cloruro *m* de potasa − **i** cloruro *m* di potassio; cloruro *m* potassico

10639 potassium chloroaurate − See potassium gold chloride.

10640 potassium chloroplatinite; potassium platinichloride; platinum potassium chloride − Small, orange-yellow crystals or powder, insoluble in alcohol, very slightly soluble in water.

Formula: K_2PtCl_6.

f chloroplatinite *m* de potassium; chlorure *m* double de platine et de potassium − **d** Kaliumplatinchlorür *n*; Platinchlorürkalium *n* − **n** kalium-platinachloride *n* − **e** cloroplatinito *m* de potasio; cloroplatinito *m* potásico; cloroplatinito *m* de potasa − **i** cloruro *m* doppio di platino e potassio

10641 potassium chromate; yellow potassium chromate; neutral potassium chromate − Yellow crystals, soluble in water, insoluble in alcohol. Latin: chromas kalicus.

Formula: K_2CrO_4.

f chromate *m* neutre de potassium − **d** (gelbes) Chromkali *n*; Kaliumchromat *n*; (neutrales) chromsaures Kali *n*; gelbes chromsaures Kalium *n* − **n** kaliumchromaat *n* − **e** cromato *m* de potasio; cromato *m* potásico; cromato *m* de potasa − **i** cromato *m* di potassio; cromato *m* potassico

10642 potassium chrome alum − See chromium potassium sulphate.

10643 potassium chromium sulphate − See chromium potassium sulphate.

10644 potassium citrate − Colourless or white crystals or powder, soluble in water and glycerol. Almost insoluble in alcohol. Latin: citras kalicus.

Formula: $K_3C_6H_5O_7 \cdot H_2O$.

f citrate *m* de potassium − **d** Kaliumzitrat *n*; zitronensaures Kali(um) *n* − **n** kaliumcitraat *n* − **e** citrato *m* de potasio; citrato *m* potásico; citrato *m* de potasa − **i** citrato *m* di potassio; citrato *m* potassico

10645 potassium copper ferrocyanide − See copper potassium ferrocyanide.

10646 potassium cupric ferrocyanide − See copper potassium ferrocyanide.

10647 potassium cyanide; cyanide of potassium; cyanide of potash − White, amorphous, deliquescent lumps or crystalline mass. Extremely poisonous. Soluble in water, alcohol and glycerol. Formula: KCN. Latin: kalium cyanatum.

f cyanure *m* de potassium; cyanure *m* de potasse; prussiate *m* de potasse − **d** Kaliumcyanid *n*; Cyankali(um) *n*; blausaures Kalium *n* − **n** kaliumcyanide *n*; cyaankali *m*; blauwzure kali *m* − **e** cianuro *m* de potasio; cianuro *m* potásico; cianuro *m* de potasa − **i** cianuro *m* di potassio; cianuro *m* potassico

10648 potassium dichromate; potassium bichromate; bichromate of potash; red potassium chromate − Bright, yellowish red transparent crystals, soluble in water, insoluble in alcohol. Used with eggs and water to sensitize zinc and copper plates ready for negative printing. Latin: bichromas kalicus.

Formula: $K_2Cr_2O_7$.

f bichromate *m* de potassium; bichromate *m* de potasse; chromate *m* rouge de potassium − **d** Kaliumbichromat *n*; Kaliumdichromat *n*; Chromkali *n*; doppeltchromsaures Kali(um) *n*; Kaliumpyrochromat *n* − **n** kaliumbichromaat *n*; dubbelchroomzure kali *m* − **e** dicromato *m* de potasio; dicromato *m* potásico; dicromato *m* de potasa; bicromato *m* de potasio; bicromato *m* potásico; bicromato *m* de potasa − **i** bicromato *m* di potassio; bicromato *m* potassico

10649 potassium ferricyanide; red prussiate of potash; red potassium prussiate; ferricyanide of potash − Bright red, lustrous crystals or powder. Poisonous. Soluble in water, slightly soluble in alcohol. Used to produce sensitive coatings on blueprint paper. Latin: kalium ferricyanatum; ferricyanetum kalicum.

Formula: $K_3Fe(CN)_6$.

f ferricyanure *m* de potassium; prussiate *m* rouge (de potassium); cyanure *m* rouge (de potassium); ferricyanhydrate *m* (de potassium) − **d** Kaliumferrizyanid *n*; rotes Blutlaugensalz *n*; Ferrizyan-

kalium *n*; Kaliumeisenzyanid *n*; Kaliumzyanoferrat-III *n* — **n** kaliumferricyanide *n*; ferricyaankali *m*; rood bloedloogzout *n* — **e** ferricianuro *m*
de potasio; ferricianuro *m* potásico; ferricianuro
m de potasa; ferricianuro *m* rojo; prusiato *m* rojo
— **i** ferricianuro *m* di potassio; ferricianuro *m*
potassico; prussiato *m* rosso

**10650 potassium ferrocyanide; yellow prussiate of
potash; yellow potassium prussiate; ferroprussiate** — Lemon yellow crystals or powder,
soluble in water, insoluble in alcohol. Used for dry
colours, dyeing, process engraving and lithography. Solution of ferroprussiate salt is used as a
coating medium to obtain a blue key by photomechanical means. Latin: kalium ferrocyanatum;
ferrocyanetum kalicum.
Formula: $K_4Fe(CN)_6 \cdot 3H_2O$.

f ferrocyanure *m* de potassium; prussiate *m* jaune
(de potassium) — **d** Kaliumferrozyanid *n*; gelbes
Blutlaugensalz *n*; gelbblausaures Kali *n*; Ferrozyankalium *n*; Kaliumeisenzyanür *n*; Kaliumzyanoferrat-II *n* — **n** kaliumferrocyanide *n*; ferrocyaankali *m*; geel bloedloogzout *n* — **e** ferrocianuro *m* de potasio; ferrocianuro *m* potásico;
ferrocianuro *m* amarillo; prusiato *m* de potasa;
prusiato *m* amarillo; ferrocianuro *m* de potasa —
i ferrocianuro *m* di potassio; ferrocianuro *m*
potassico; prussiato *m* giallo

**10651 potassium gold bromide; potassium auri-
bromide; potassium bromoaurate; gold potassium
bromide** — Violet crystals, soluble in alcohol and
water.
Formula: $AuBr_3 \cdot KBr \cdot 2H_2O$.

f bromure *m* d'or et de potassium — **d** Goldkaliumbromid *n* — **n** kalium-auribromide *n* — **e**
bromuro *m* de oro y potasio — **i** bromuro *m* d'oro
e potassio

**10652 potassium gold chloride; potassium auri-
chloride; potassium chloroaurate; gold potassium
chloride** — Yellow crystals, soluble in water,
alcohol and ether. Component of photographic
chemicals.
Formula: $AuCl_3 \cdot KCl \cdot 2H_2O$.

f chlorure *m* d'or et de potassium — **d** Goldkaliumchlorid *n* — **n** kalium-aurichloride *n* — **e** cloruro
m de oro y potasio — **i** cloruro *m* d'oro e potassio

10653 potassium hydrate — See potassium
hydroxide.

10654 potassium hydrogen sulphate — See
potassium bisulphate.

**10655 potassium hydroxide; caustic potash; potas-
sium hydrate** — White, deliquescent pieces, lumps,
sticks, soluble in water, alcohol, glycerol, slightly
soluble in ether. Must be kept well stoppered, absorbs water and carbon dioxide from the air.
Latin: kalium hydricum; hydras kalicus; kalium
causticum.
Formula: KOH.

f hydroxyde *m* de potassium; potasse *f* caustique;
hydrate *m* de potasse; lessive *f* de potasse — **d**
Kaliumhydroxyd *n*; Ätzkali *n*; Kaliumhydrat *n*;
Kalilauge *f*; kaustisches Kali *n* — **n** kalium-

hydroxyde *n*; kaliloog *fm*; etskali *m*; bijtende
kali *m*; kaliumhydraat *n* — **e** hidróxido *m* de
potasio; hidróxido *m* potásico; hidróxido *m* de
potasa; potasa *f* cáustica — **i** idrossido *m* di
potassio; idrossido *m* potassico; idrato *m* di
potassio; potassa *f* caustica

10656 potassium iodide — White crystals,
granules or powder, soluble in water, alcohol and
glycerol. Used for precipitating silver in photography. Latin: iodetum kalicum; kalium iodatum.
Formula: KI.

f iodure *m* de potassium — **d** Kaliumjodid *n*; Jod-
kalium *n* — **n** kaliumjodide *n*; joodkali *m*; jood-
kalium *n* — **e** yoduro *m* de potasio; yoduro *m*
potásico; yoduro *m* de potasa — **i** ioduro *m* di
potassio; ioduro *m* potassico

10657 potassium lactate

f lactate *m* de potassium — **d** Kaliumlaktat *n* — **n**
kaliumlactaat *n* — **e** lactato *m* de potasio; lactato
m potásico; lactato *m* de potasa — **i** lattato *m* di
potassio; lattato *m* potassico

**10658 potassium metabisulphite; potassium pyro-
sulphite** — White granules or powder, slightly
soluble in water and alcohol. Used as a developing
agent in photography. Latin: kalium metabisulfurosum; kalium pyrosulfurosum.
Formula: $K_2S_2O_5$.

f métabisulfite *m* de potassium — **d** Kaliummetabisulfit *n*; metaschwefligsaures Kali *n*; pyroschwefligsaures Kali *n* — **n** kalium-metabisulfiet *n* — **e** metabisulfito *m* de potasio; metabisulfito *m* potásico; metabisulfito *m* de potasa — **i**
metabisolfito *m* di potassio; metabisolfito *m*
potassico

10659 potassium muriate — See potassium
chloride.

10660 potassium nitrate; nitre *GB*; **niter** *US*; **saltpetre** — Transparent, colourless or white crystalline powder or crystals, soluble in water, slightly
soluble in alcohol and glycerol. Latin: nitras
kalicus.
Formula: KNO_3.

f nitrate *m* de potassium; azotate *m* de potassium;
salpêtre *m*; nitre *m* — **d** Kaliumnitrat *n*; salpetersaures Kali(um) *n* — **n** kaliumnitraat *n*; salpeterzure potas *fm*; kalisalpeter *m* — **e** nitrato *m* de
potasa; nitrato *m* potásico; nitrato *m* de potasa —
i nitrato *m* di potassio; nitrato *m* potassico

10661 potassium oxalate — Colourless, transparent crystals, soluble in water. Used as a
bleaching agent and stain remover, and in photography. Latin: kalium oxalicum.
Formula: $K_2C_2O_4 \cdot H_2O$.

f oxalate *m* de potassium — **d** Kaliumoxalat *n* —
n kaliumoxalaat *n* — **e** oxalato *m* de potasio;
oxalato *m* potásico; oxalato *m* de potasa — **i**
ossalato *m* di potassio; ossalato *m* potassico

10662 potassium percarbonate — Granular, white
mass, soluble in water. Used as an oxidizing agent
in photography and textile printing.
Formula: $K_2C_2O_6 \cdot H_2O$.

f percarbonate *m* de potassium — **d** Kaliumperkarbonat n − **n** kaliumpercarbonaat n − **e** percarbonato m de potasio; percarbonato m potásico; percarbonato m de potasa − **i** percarbonato m di potassio; percarbonato m potassico

10663 potassium perchlorate − Colourless crystals or white, crystalline powder, soluble in water, insoluble in alcohol. Used as an oxidizing agent and in photography. Latin: perchloras kalicus.
Formula: $KClO_4$.
f perchlorate m de potassium − **d** Kaliumperchlorat n; überchlorsaures Kali(um) n − **n** kaliumperchloraat n; overchloorzure kali m; perchloorzure kali m − **e** perclorato m de potasio; perclorato m potásico; perclorato m de potasa − **i** perclorato m di potassio; perclorato m potassico

10664 potassium permanganate; permanganate of potash − Dark purple crystals with a blue sheen, soluble in water and methanol, decomposed by alcohol. Powerful oxidizing agent. Contact with glycerol, ethylene glycol, sulphuric acid is dangerous. Latin: kalium permanganicum; permanganas kalicus.
Formula: $KMnO_4$.
f permanganate m de potassium − **d** Kaliumpermanganat n; übermangansaures Kali n; Kaliumhypermanganat n − **n** kaliumpermanganaat n; overmangaanzure kali m − **e** permanganato m de potasio; permanganato m potásico; permanganato m de potasa − **i** permanganato m di potassio; permanganato m potassico

10665 potassium peroxydisulphate − See potassium persulphate.

10666 potassium persulphate; potassium peroxydisulphate − White crystals, soluble in water, insoluble in alcohol. Used for defiberizing wet-strength paper, bleaching; oxidizing agent, reducent agent in photography. Latin: persulfas kalicus.
Formula: $K_2S_2O_8$.
f persulfate m de potassium − **d** Kaliumpersulfat n; überschwefelsaures Kali n − **n** kaliumpersulfaat n − **e** persulfato m de potasio; persulfato m potásico; persulfato m de potasa − **i** persolfato m di potassio; persolfato m potassico

10667 potassium platinichloride − See potassium chloroplatinite.

10668 potassium pyrosulphite − See potassium metabisulphite.

10669 potassium rhodanide − See potassium thiocyanate.

10670 potassium silicate; water-glass − Colourless to turbid liquid in various ratios of SiO_2 and K_2O, used as a catalyst and an ingredient in adhesives, as a binder and structural water-proofing agent and in textile chemistry.
f silicate m de potassium; silicate m de potasse − **d** Kaliumsilikat n; Kaliwasserglas n − **n** kaliumsilicaat n; kaliwaterglas n − **e** silicato m de potasio; silicato m potásico; silicato m de potasa − **i** silicato m di potassio; silicato m potassico

10671 potassium sulphate − Colourless or white, hard crystals or powder, soluble in water, insoluble in alcohol. Latin: sulfas kalicus; kalium sulfuricum.
Formula: K_2SO_4.
f sulfate m de potassium − **d** Kaliumsulfat n − **n** kaliumsulfaat n − **e** sulfato m de potasio; sulfato m potásico; sulfato m de potasa − **i** solfato m di potassio; solfato m potassico

10672 potassium sulphide − Red, crystalline mass, deliquescent in air, soluble in water, alcohol and glycerol, insoluble in ether.
Formula: K_2S.
f sulfure m de potassium − **d** Kaliumsulfid n; Schwefelkalium n − **n** kaliumsulfide n; zwavelkalium n − **e** sulfuro m de potasio; sulfuro m potásico; sulfuro m de potasa − **i** solfuro m di potassio; solfuro m potassico

10673 potassium sulphite − White crystals or powder, soluble in water, sparingly soluble in alcohol. Used in photography.
Formula: $K_2SO_3 \cdot 2H_2O$.
f sulfite m de potassium; sulfite m de potasse − **d** Kaliumsulfit n − **n** kaliumsulfiet n − **e** sulfito m de potasio; sulfito m potásico; sulfito m de potasa; pirosulfito m de potasio; pirosulfito m potásico; pirosulfito m de potasa − **i** solfito m di potassio; solfito m potassico

10674 potassium sulphocyanate − See potassium thiocyanate.

10675 potassium sulphocyanide − See potassium thiocyanate.

10676 potassium thiocyanate; potassium sulphocyanate; potassium sulphocyanide; potassium rhodanide − Colourless crystals, soluble in water, alcohol and acetone. Used as a photographic restrainer and intensifier. Latin: sulfocyanetum kalicum.
Formula: KCNS.
f sulfocyanure m de potassium; thiocyanate m de potassium; rhodanate m de potassium − **d** Kaliumthiozyanat n; Kaliumthiozyanid n; Schwefelzyankalium n; Kaliumrhodanid n; Rhodankalium n − **n** kaliumthiocyanaat n; kaliumsulfocyanide n; kaliumrhodanide n; rhodaankalium n obs. − **e** tiocianato m de potasio; tiocianato m potásico; tiocianato m de potasa; sulfocianato m de potasio; sulfocianato m potásico; sulfocianato m de potasa − **i** solfocianuro m di potassio; solfocianuro m potassico

10677 potassium trisulphide; sulphurated potash; liver of sulphur − Hard, brittle, liver-brown substance with nauseous alkaline taste. Soluble in water and odour of hydrogen sulphide. Latin: trisulfuretum kalicum.
Formula: K_2S_3.
f polysulfure m de potassium; foie m de soufre − **d** Kaliumsulfid n; Kaliumpolysulfid n; Schwefelleber m; Schwefelkalium n − **n** kaliumpolysulfide n; zwavellever fm − **e** trisulfuro m de potasio; trisulfuro m potásico; trisulfuro m de potasa − **i** polisolfuro m di potassio; polisolfuro m potassico

10678 pot cassé — Book cover decoration style characterized by a broken urn with arabesques and the words "Non Plus" for books printed (not bound) by Geoffroy Tory (1480-1533), printer to the king of France.

10679 potcher; bleaching engine; bleacher — Breaker for washing and bleaching of rag and pulp wood. For washing, a perforated cylinder is partially immersed in the pulp to the extract liquid continuously.

f pile *f* blanchisseuse; pile *f* laveuse; cuve *f* de blanchiment; chambre *f* de blanchiment — **d** Bleichholländer *m*; Bleicher *m*; Waschholländer *m* — **n** bleekhollander *m*; bleek- en washollander *m* — **e** pila *f* blanqueadora; blanqueador *m* — **i** olandese *f* di sbianca

10680 pot crucible — See metal pot.

10681 potentiometer — Device to measure electromotive force or potential difference by comparison with a known voltage.

f potentiomètre *m* — **d** Potentiometer *mn* — **n** potentiometer *m* — **e** potenciómetro *m* — **i** potenziometro *m*

10682 pot feeder, automatic — See automatic pot feeder.

10683 pothook — See ordinary terminals.

10684 pot mouth; mouthpiece — On typesetting machines, part of the metal pot through which the metal flows from the throat of the pot into the mould.

f bouche *f* (de coulée); bouche *f* du creuset — **d** Gießmund *m* — **n** gietmond *m* — **e** boca *f* del crisol; boca *f* de molde; boca *f* de fundir; boca *f* de fundición; boquilla *f* del crisol — **i** bocchetta *f* di fusione; bocchetta *f* del crogiolo; bocchello *m* del crogiolo

10685 pot opal — See opal glass.

10686 pot pump

f pompe *f* du creuset — **d** Bleitopfpumpe *f* — **n** potpomp *fm*; pomp *fm* van de loodpot — **e** bomba *f* del crisol — **i** pompa *f* di crogiolo

10687 pot pump plunger

f piston *m* du creuset — **d** Pumpenkolben *m* — **n** pompkolf *fm* — **e** émbolo *m* (de la bomba) del crisol; pistón *m* (de la bomba) del crisol — **i** stantuffo *m* del crogiolo

10688 pott — A size of writing or printing paper. See app. no. 3.

10689 pouch

f petit sac *m* — **d** Beutel *m*; Beutelchen *n* — **n** zakje *n* — **e** saco *m* pequeño; bolsillo *m*; faltriquera *f* — **i** sacchetto *m*

10690 pound — Unit of mass. Abbrev. lb av. In the SI: 1 pound avoirdupois ≈ 0.453592 kg. See also troy pound, app. no. 7 and 7a.

f livre *f* anglaise — **d** Pfund *n*; englisches Pfund *n* — **n** pond *n*; Engelse pond *n* — **e** libra *f* inglés — **i** libbra *f* inglesa

10691 pound mark; pound sterling mark — The symbol £.

f signe *m* de la livre sterling; symbole *m* de la livre sterling — **d** Pfundzeichen *n*; Pfund Sterling-zeichen *n* — **n** pond-sterlingteken *n* — **e** signo *m* del libra sterling — **i** segno *m* della lira sterlina

10692 pound sterling mark — See pound mark.

10693 poupée, gravure à la — See colour etching.

10694 pour *v*

f verser — **d** gießen; schütten — **n** gieten; storten — **e** vaciar — **i** versare; colare

10695 pouring sheet — See apron.

10696 pour *v* **off** — See decant *v*.

10697 powdering — Separation of particles from the dry sheet of paper by agitation or rubbing. See also chalking.

10698 powdering — Application or brushing of dragon's blood or etching powder on plates during relief etching, especially in the dragon's blood process.

f poudrage *m* — **d** Bestäubung *f*; Einstäubung *f* — **n** poedering *f*; poederen *n* — **e** empolvado *m*; empolvoramiento *m*; espolvoreado *m*; aplicación *f* de drago — **i** spolverizzatura *f*

10699 powdering — To prevent offsetting by powdering freshly printed proofs by hand with talcum.

f talquer; saupoudrage *m* avec de la poudre calcaire — **d** Pudern *n* mit Talkum — **n** poederen *n* met talkpoeder — **e** empolvoreado *m* con talco — **i** antiscartinatura *f*

10700 powdering — See dusting.

10701 powdering machine — Power-driven apparatus for mechanically applying powder to plates.

f machine *f* à poudrer — **d** Einstaubmaschine *f* — **n** poedermachine *f*; stuifkast *fm* — **e** cajón *m* de polvos — **i** macchina *f* spolverizzatrice

10702 powder leaf — A size of wrapping paper. See app. no. 3.

10703 powderless etching; one-bite etching; one-stage etching; high-speed etching; rapid etching; quick etch — Etching plates with a solution which does not undercut the printing surface, thus dispensing with the process of building up with dragon's blood.

f morsure *f* rapide; gravure *f* sans poudrage; gravure *f* en une étape; morsure *f* en continue — **d** Schnellätzverfahren *n*; Einstufenätzverfahren *n* — **n** snel-etsprocédé *n*; eenfase-etsprocédé *n*; snel-etsen *n* — **e** procedimiento *m* de mordido rápido; procedimiento *m* de mordido sin polvo; procedimiento *m* rápido de mordido; grabado *m* rápido (sin escalar); grabado *m* rápido (sin espolvorear); grabado *m* rápido sin escalas; grabado *m* continuo; mordido *m* continuo — **i** incisione *f* senza polvere; incisione *f* senza coperture; incisione *f* ad una fase

10704 powderless etching machine

f machine *f* à morsure continue; machine *f* à morsure rapide; machine *f* à morsure sans poudrage — **d** Einstufenätzmaschine *f*; Schnellätzmaschine *f* — **n** eenfase-etsmachine *f*; snel-etsmachine *f* — **e** máquina *f* de mordido rápido; máquina *f* de mordido sin polvo; máquina *f* de mordido continuo; grabadora *f* de mordido continuo; grabadora *f* rápida; grabadora *f* continua — **i** macchina *f* a incidere senza coperture

10705 power consumption — Power required to move a machine.

f consommation *f* de force motrice; consommation *f* de courant — **d** Energieverbrauch *m* — **n** krachtverbruik *n*; energieverbruik *n* — **e** consumo *m* de fuerza; consumo *m* de energía — **i** consumo *m* di forza motrice

10706 power cutter — See guillotine.

10707 power inverter — See inverter.

10708 prayer-book; service-book — Book with prayers and often other forms and directions for worship.

f livre *m* d'office; livre *m* de prières; livre *m* d'heures; paroissien *m* — **d** Gebetbuch *n*; Kirchenbibel *f* — **n** kerkboek *n*; gebedenboek *n* — **e** libro *m* de oraciones; libro *m* de devociones; libro *m* de rezo — **i** libro *m* delle preghiere

10709 precedence code — Signal in the input stream (usually a unique character or code, but also a standard one out of context) which denotes a change of mode, so that what follows represents a different condition. For example, a "shift" precedence code signals that until further notice all characters will be identified as in upper case. A supershift precedence code signifies that one or more following codes are to be identified as non-standard or unique characters grouped together more or less arbitrarily to extend the fount character repertoire. A precedence code may signify a locking or a non-locking condition. In the latter case the precedence code identifies a unique character only for the next code in the text stream or until a predefined condition obtains, whereas a "locking" precedence code signals that the special case will govern until further notice.

10710 precipitate — Relatively insoluble material produced by precipitation.

f précipité *m* — **d** Niederschlag *m*; Fällung *f*; Präzipität *n* — **n** neerslag *nm*; precipitaat *n* — **e** precipitado *m* — **i** precipitato *m*

10711 precipitate *v* — To separate from solution as a solid substance. Also, the settling out of small particles in either a liquid or gaseous medium.

f précipiter — **d** niederschlagen; ausfällen; fällen; Niederschlag bilden — **n** neerslaan; precipiteren; neerslag vormen; bezinken — **e** precipitar — **i** precipitare

10712 precipitated barium sulphate — See barium sulphate.

10713 precipitated calcium carbonate — See calcium carbonate.

10714 precipitated calcium phosphate — See calcium phosphate.

10715 precipitated calcium sulphate — See calcium sulphate.

10716 precipitated chalk — See calcium carbonate.

10717 precipitation — Reaction where two or more clear solutions or suspensions when mixed cause a solid product to drop out of the mixture.

f précipitation *f* — **d** Niederschlag *m*; Fällen *n* — **n** precipitatie *f*; neerslaan *n* — **e** precipitación *f* — **i** precipitazione *f*

10718 pre-curved printing plate

f cliché *m* pré-cintré — **d** vorgebogene Druckplatte *f* — **n** voorgebogen drukplaat *fm*; voorgebogen cliché *n* — **e** plancha *f* precurvada — **i** lastra *f* precurvata; cliscè *m* precurvato; cliché *m* precurvato; forma *f* stampante precurvata

10719 pre-etching — See flat etch.

10720 preface — Preface written by the author, introducing the book and stating its purpose. See also foreword, introduction, preliminary matter.

10721 prefixes for metric units — E = exa = 10^{18}; P = peta = 10^{15}; T = tera = 10^{12}; G = giga = 10^9; M = mega = 10^6; k = kilo = 10^3; h = hecto = 10^2; da = deka = 10; d = deci = 10^{-1}; c = centi = 10^{-2}; m = milli = 10^{-3}; μ = micro = 10^{-6}; n = nano = 10^{-9}; p = pico = 10^{-12}; f = femto = 10^{-15}; a = atto = 10^{-18}. Examples of usage: GHz = gigahertz; Hz = hertz (cycles/second); kHz = kilohertz; mHz = megahertz; pF = picofarad. In general, symbols for units are written in lower-case letters, except when the name of the unit is derived from a proper name, thus, hertz, Hz when abbreviated.

f préfixes *mpl* pour unités — **d** Präfixe *npl* für metrische Einheiten — **n** voorvoegsels *npl* voor metrische eenheden — **e** prefijos *mpl* para unidades métricas — **i** prefissi *mpl* per unità metriche

10722 preheater

f réchauffeur *m* — **d** Vorwärmer *m* — **n** voorverwarmer *m* — **e** precalentador *m*

10723 preliminaries — See preliminary matter.

10724 preliminary matter; prelims; preliminary pages; preliminaries; front matter; oddments — In bookwork, printed matter preceding the text. The order is: half title, frontispiece, title (on the back of this, number of editions, impressions, imprint, etc.), dedication, foreword, preface, contents, list of illustrations, introduction. All except frontispiece (which faces title) to come or begin on right-hand pages. See also Roman figures.

f texte *m* préliminaire; feuilles *fpl* de titre; folios *mpl* de la préface; préliminaires *fpl*; pièces *fpl* liminaires; feuillets *mpl* liminaires; pages *fpl* liminaires — **d** Titelei *f*; Titelbogen *m* — **n** voorwerk *n* — **e** páginas *fpl* preliminares; pliego *m* de principios — **i** preliminari *mpl*

10725 preliminary pages — See preliminary matter.

10726 prelims — See preliminary matter.

10727 pre-loader — Platform and sheet stacking guides to build up a pile of stock before positioning in the automatic sheet feeder.

f préchargeur *m* pour margeur — **d** Vorstapeleinrichtung *f* — **n** voorstapelaar *m*; voorstapelinrichting *f* — **i** precaricatore *m*

10728 pre-makeready — Operation on a relief printing plate to form in the plate different printing levels for the varying tone densities. Usually, the solids are made slightly above standard height with lighter tone densities progressively lower.

f pré-mise *f* en train − **d** Vorzurichtung *f* − **n** vormvoorbereiding *f* − **e** prearreglo *m* − **i** pre-avviamento *m*

10729 preprint − Advanced issue of a paper, or an article to be published in a journal or in book form.

f tirage *m* préliminaire − **d** Vorabdruck *m* − **n** afdruk *m* vooraf; voor-afdruk *m* − **e** impresión *f* anticipada − **i** stampa *f* preliminare

10730 preprinted reel − Printed re-reeled paper web.

f bobine *f* préimprimée − **d** vorgedruckte Rolle *f* − **n** voorbedrukte rol *fm* − **e** bobina *f* preimprimada − **i** bobina *f* prestampata

10731 preprinted web − Web printed and rewound as a separate operation.

f bande *f* préimprimée − **d** vorgedruckte Papierbahn *f* − **n** voorbedrukte papierbaan *fm* − **e** hoja *f* de papel preimpresa − **i** nastro *m* prestampato

10732 presensitized plate − Precoated plastic or aluminium plate which can be printed photographically from line or halftone negatives. Used mainly on small offset presses.

f plaque *f* précouchée; plaque *f* précoulée; plaque *f* présensibilisée; plaque *f* sensibilisée à l'avance − **d** vorbeschichtete Offsetplatte *f*; vorbeschichtete Druckplatte *f* − **n** gepresensibiliseerde drukplaat *fm*; voorgesensibiliseerde drukplaat *fm* − **e** plancha *f* presensibilizada − **i** lastra *f* presensibilizzata

10733 presentation copy

f exemplaire *m* gratuit − **d** Freiexemplar *n*; Gratisexemplar *n* − **n** presentexemplaar *n* − **e** ejemplar *m* gratuito; ejemplar *m* de regalo − **i** esemplare *m* gratuito

10734 presentation copy − Copy of a book with the compliments of the author.

f envoi *m* d'auteur − **d** Widmung *f* des Verfassers − **n** eigenhandig geschreven opdracht *fm* van de schrijver − **e** autógrafo *m*; dedicatoria *f* autógrafa − **i** invio *m* autografo dell'autore

10735 preserve *v* − See dump *v.*

10736 press − Brief for printing press or bookbinding press. In lithography commonly the offset printing press, though some direct lithographic presses were still in use in 1960.

f presse *f* (à imprimer) − **d** Presse *f*; Druckerpresse *f*; Druckmaschine *f* − **n** pers *fm*; drukpers *fm* − **e** prensa *f* (de imprenta) "Prensa" is originally used for a hand-operated press; "máquina" for a mechanically operated press. − **i** macchina *f* da stampa; stampatrice *f*

10737 press − In a paper machine a pair of rolls between which the paper web is passed for: 1. water removal at the wet press; 2. smoothing and levelling of the sheet surface at the *smoothing press; 3. application of surface treatments to the sheet at the *size press. The rolls are located between the paper machine wire and the dryers.

The bottom roll is usually rubber-covered, while the top one is of polished metal or granite. There are usually three sets of press rolls, constituting the first, second and third press.

f presses *fpl* − **d** Pressenpartie *f* − **n** perspartij *f* − **e** sección *f* de prensas − **i** parte *f* di presse

10738 Press, the

f presse *f*; journaux *mpl* − **d** Presse *f*; Zeitungswesen *n*; Zeitungswald *m* *fam.*; Blätterwald *m* *fam.* − **n** pers *f*; dagbladpers *f*; dagbladen *npl* − **e** prensa *f*; prensa *f* diaria − **i** stampa *f*; giornali *mpl*

10739 press-agency − See news-agency.

10740 pressboard; pressing board − Wooden board placed between books when in the standing press. Also used by printers to smooth printed sheets.

f ais *m* − **d** Preßbrett *n*; Preßplatte *f* − **n** bord *n* (in de pletpers); plank *fm* − **e** tabla *f* de prensar − **i** asse *f*; tavola *f*

10741 pressboard − See presspahn.

10742 press-book − Work published by private presses.

10743 press button − See push button.

10744 press-button envelope

f enveloppe *f* à bouton à pression; enveloppe *f* à bouton fermoir − **d** Umschlag *m* mit Druckknopfverschluß − **n** envelop(pe) *fm* met drukknoopsluiting − **e** sobre *m* de botón de presión − **i** busta *f* con bottone a pressione

10745 press clipping − See news cutting.

10746 press-copy − See review copy.

10747 press freedom − See freedom of the press.

10748 press gilder

f doreur *m* à presse − **d** Preßvergolder *m* − **n** persvergulder *m* − **e** dorador *m* a prensa − **i** doratore *m* alla pressa; indoratore *m* al torchio

10749 press-gilding − Applying gold to book covers. Also applying gold to book edges with a machine.

f dorage *m* à presse − **d** Preßvergoldung *f* − **n** persvergulden *n* − **e** dorado *m* de prensa; dorado *m* a volante − **i** doratura *f* alla pressa

10750 pressing board − See pressboard.

10751 press, in the − See in the press.

10752 press-libel

f délit *m* de presse − **d** Pressevergehen *n* − **n** persdelict *n* − **e** delito *m* de prensa

10753 pressman; machine minder; pigs *sl.* − Man in charge of one or more presses in a printing plant who makes ready the forme and supervises the running.

f conducteur *m* (typo); conducteur *m* (de machine); pressier *m*; imprimeur *m*; ouvrier *m*; ours *m* *sl.* − **d** Drucker *m*; Maschinenmeister *m* − **n** drukker *m*; man *m* aan de pers − **e** maquinista *m* (impresor); prensista *m* (tipográfico); tirador *m*; conductor *m* − **i** macchinista *m*; stampatore *m*; impressore *m*

10754 pressman; offset pressman; offset-machine minder; offset printer

f conducteur *m* (de la machine) offset; off-

settiste *m* — **d** Offsetdrucker *m*; Offsetmaschinenmeister *m* — **n** offsetdrukker *m* — **e** offsetista *m*;
maquinista *m* offset — **i** macchinista *m* per offset; stampatore *m* offset

10755 press-mark — See impressed watermark.

**10756 presspahn; pressboard; insulating board;
transformer board; electrical pressboard** *US* —
Very glossy pressboard with a density greater
than 1.15, chiefly characterized by its electrical
insulating properties and high mechanical
strength.
f presspahn *m*; carte *f* de Lyon; carton *m* isolant;
carton *m* simili-vulcanisé; carton *m* d'apprêt — **d**
Preßspahn *m*; Preßpappe *f*; Glanzpappe *f*; Glanzdecke *f* — **n** pletbord *n*; presspaan *n*; glansbord *n*
— **e** prespan *m*; cartón *m* prensado; cartón *m*
brillo; cartón *m* hidráulico — **i** presspan *m*;
cartone *m* presspan; cartone *m* pressato

10757 press plate — See offset plate.

10758 press-plate; machine plate
f plaque *f* machine — **d** Maschinenplatte *f*;
Maschinendruckplatte *f* — **n** machineplaat *fm*;
persplaat *fm* — **e** plancha *f* de máquina — **i**
lastra *f* per macchina

10759 press-proof — Actual press sheet to show
image, tone values and colours as well as imposition of forme or press-plate.
f tierce *f*; épreuve *f* en tierce; épreuve *f* sur
machine — **d** Revisionsabzug *m*; Preßrevision *f*;
letzte Korrektur *f* — **n** persproef *fm*; proef *fm*
van de pers — **e** pliego *m* de prensa; pliego *m* de
prueba; pliego *m* de máquina — **i** bozza *f* di
macchina; bozza *f* definitiva; terza bozza *f*

10760 press roll — See press.

**10761 press-room; machine room; printing depart-
ment; printing shop; piggery** *sl.*; **pigsty** *sl.* —
Room in which printing work is done or in which
the printing presses are.
f salle *f* des machines; secteur *m* impression; salle
f des impressions; service *m* des impressions — **d**
Druckersaal *m*; Maschinensaal *m* — **n**
drukkerij *f*; afdeling *f* drukkerij — **e** sala *f* de
prensas; sala *f* de máquinas; taller *m* de
impresión; departamento *m* de prensas; departamento *m* de máquinas — **i** sala *f* macchine; sala *f*
stampa

**10762 press-room overseer; overseer of the
machine room; press-room superintendent;
machine-room manager; foreman printer** — Man
in charge of the press-room.
f prote *m* aux machines; prote *m* de la salle des
presses; chef *m* des presses — **d** Druckereifaktor *m*; Druckerfaktor *m*; Faktor *m*; Fax *m*
coll.; Druckereileiter *m*; Saalmeister *m*; Obermaschinenmeister *m* — **n** chef *m* van de
drukkerij; chef-drukker *m* — **e** jefe *m* de
máquinas; jefe *m* de prensas; jefe *m* de la sección
de prensas; jefe *m* del taller de imprenta; regente
m de imprenta — **i** capo *m* macchinista

10763 press-room superintendent — See press-
room overseer.

10764 press-sheet — Full size sheet of paper se-

lected for the job to be printed on a sheet-fed
press, usually slightly larger than required for the
job to allow for grippers and trim margins. The
flat is prepared so that its subjects will print within the limits of this sheet.
f feuille *f* de tirage — **d** Auflagebogen *m* — **n** te
bedrukken vel *n* papier; persvel *n* — **e** hoja *f* de
máquina — **i** foglio *m* di pressa; foglio *m* di
macchina; foglio *m* di stampa

10765 pressure, impression — See printing
pressure.

10766 pressure, printing — See printing pressure.

10767 pressure roller — See impression cylinder.

10768 pressure-sensitive adhesive — Adhesive
which requires only little pressure at room temperature for adherence to a surface, e.g. wax adhesives. Generally rubber or silicone complexes
with small amounts of a non volatile solvent to remain tacky.
f colle *f* auto-adhesive — **d** selbstklebender Kleb-
stoff *m* — **n** zelfklevend plakmiddel *n* — **e** cola *f*
autoadhesiva — **i** adesivo *m* pressosensibile

10769 pressure-sensitive paper — See self-
adhesive paper.

10770 pressure spring
f ressort *m* de compression — **d** Druckfeder *f* —
n drukveer *f* — **e** resorte *m* de presión; muelle *m*
de presión — **i** molla *f* di compressione

10771 pressurized packaging — See aerosol.

10772 press-viscosity — Viscosity at which an ink
prints satisfactorily on a press, cylinder, paper,
etc.
f viscosité *f* sur presse — **d** Fortdruckviskosität *f*
— **n** drukviscositeit *f* — **e** viscosidad *f* de prensa
— **i** viscosità *f* di stampa

10773 press-work — In bookmaking, the actual
printing of the book, as distinct from composition,
make-up, and binding.
f travail *m* sur presse — **d** Druckarbeit *f* — **n**
werk *n* aan de pers — **e** trabajo *m* de impresión
— **i** lavoro *m* alla macchina

10774 pre-tension — The cylinder of a printing
press is set somewhat lower than the height of
the bed bearers. When fitting the bearers, the
frame of the press is stretched and the cylinder is
pressed to the bearers with a certain pre-tension.
If this is insufficient the cylinder may lift from
the bed bearers, producing slurs, guttering,
register troubles, and wear of printing formes.
The pre-tension is measured in kilograms,
according to the following formulae: for light-
weight formes 3 $LD^{1/2}$ for middle-weight formes 6
$LD^{1/2}$, and for heavy-weight formes 9 $LD^{1/2}$, where
L is the distance between the bed bearers in cm
and D the cylinder diameter in cm.
f bridage *m*; précontrainte *f*; pression *f* préalable;
tension *f* préalable — **d** Vorspannung *f* — **n** voor-
spanning *f*; voorspankracht *fm* — **e** fuerza *f*
mecánica tensora — **i** pre-tensione *f*

10775 pre-tension meter — Hydraulic jack with a
pressure-measuring device and micrometer
gauges, to be inserted between the bed and the

cylinder of a printing press to measure the change in the distance between theses two while the pressure is increased.

f indicateur *m* du bridage — **d** Vorspannungs-messer *m* — **n** voorspanningsmeter *m* — **e** manó metro *m* tensor — **i** apparecchio *m* di misura della pre-tensione

10776 pre-treatment — Treatment of the offset lithographic plate before graining to remove old ink and image substance, e.g., counter-etching, pre-etching, cronak, and brunak in offset zinc platemaking.

f pré-traitement *m*; traitement *m* préparatoire — **d** Vorbehandlung *f* — **n** voorbehandeling *f* — **e** pretratamiento *m*; tratamiento *m* preliminar — **i** pretrattamento *m*; trattamento *m* preliminare

10777 preventive maintenance — Tests, measurements, replacements and repairs carried out prevent faults to occur during operating.

f maintenance *f* préventive — **d** vorbeugende Wartung *f* — **n** preventief onderhoud *n* — **e** entretemiento *m* preventivo — **i** manutenzione *f* preventiva

10778 price cutter

f gâcheur *m* de prix — **d** Schmutzkonkurrent *m* — **n** prijsbederver *m*; deloyale concurrent *m* — **e** malbaratador *m* — **i** concorrente *m*; competitore *m*; quastatore *m* dei prezzi

10779 price list — Index of articles for sale with prices, in leaflet, brochure or book form.

f prix-courant *m* — **d** Preisliste *f*; Preis-verzeichnis *n* — **n** prijslijst *fm*; prijscourant *m* — **e** lista *f* de precios; tabla *f* de precios — **i** listino *m* dei prezzi

10780 price per sheet

f prix *m* à la feuille — **d** Bogenpreis *m* — **n** vel-prijs *m*; prijs *m* per vel — **e** precio *m* por hoja — **i** prezzo *m* di foglio; spesa *f* per foglio

10781 primary acute accent; prime — Accent (') placed immediately after a letter usually to show which syllable must be emphasized.

f accent *m* tonique; prime *f* — **d** Betonungs-zeichen *n* — **n** klemtoonteken *n* — **i** accento *m* acuto primario

10782 primary colours; fundamental colours — Fundamental colours to explain the process of colour vision, and to reproduce coloured originals by printing. Applied to light or light phenomena, the true primary colours are violet, red and green, but the primary colours of printing inks or those involved in subtractive three-colour reproduction are termed yellow, magenta (process red), and cyan (process blue), green formed by a mixture of cyan and yellow.

f couleurs *fpl* primaires — **d** Primärfarben *fpl* — **n** primaire kleuren *fmpl* — **e** colores *mpl* primarios; colores *mpl* primitivos; colores *mpl* fundamentales; colores *mpl* simples; colores *mpl* normales — **i** colori *m* primari

10783 primary input — Initial preparation of data in machine readable form. Primary input to the computer can be magnetic character recognition,

mark sensing, optical character recognition, tabulator cards, punched tape, magnetic tape and signals from on-line keyboards.

10784 primary keyboarding — The first occasion of which data is generated in machine readable form by keyboard operation.

10785 primary leading — Amount of vertical depth associated with the setting of a line of type. If ten point type is to be spread twelve points apart, then the primary leading is twelve points, whereas secondary leading could be additional white space between certain text blocks but not directly associated with the type in question. Whether the leading is above or below the base line of the type depends upon the manner in which the typesetting device functions.

10786 prime — See primary acute accent.

10787 primed letter — Letter followed by a primary acute accent.

10788 primitive circle — See pitch circle.

10789 princeps — A size of note paper. See app. no. 3.

10790 princess — A size of note paper. See app. no. 3.

10791 principal masks — See contact mask.

10792 print — Impression from an engraving, a plate, etc.

f impression *f* — **d** Abdruck *m* — **n** afdruk *m*; druk *m* *depr.* — **e** impreso *m*; impresión *f* — **i** stampa *f*

10793 print — Photographic image from a negative (or positive) made on metal or other opaque supports (paper). The line or halftone images on metal for relief etching or lithographic plate-making. See also photoprint.

f copie *f* (photographique) — **d** Kopie *f* — **n** kopie *f* — **e** copia *f* (fotográfica) — **i** copia *f* (foto-grafica)

10794 print *v* — To reproduce a design with ink on a printing press.

f imprimer; tirer — **d** drucken; abdrucken — **n** drukken; afdrukken — **e** imprimir; tirar; im-prentar *Ch* — **i** stampare; imprimere

10795 print *v* — To make photographic prints by the aid of light, usually through a negative, more popular called photoprinting.

f tirer — **d** abdrucken — **n** afdrukken; een afdruk maken — **e** copiar — **i** copiare

10796 printability of paper — Property required of all components in a printing process, judged by uniformity of colour of the printed areas, legibility of the printed matter and show-through, related to ink receptivity and its uniformity, compressibility, smoothness and opacity.

f imprimabilité *f* du papier; aptitude *f* du papier à l'impression — **d** Bedruckbarkeit *f* des Papiers; Druckeignung *f* — **n** bedrukbaarheid *f* van papier; geschiktheid *f* van het papier tot bedrukken — **e** imprimabilidad *f* del papel — **i** stampabilità *f*; attitudine *f* alla stampa della carta

10797 printability tester — Laboratory instrument, imitating a printing press, to investigate the characteristics of paper or ink and their combinations.

f appareil *m* d'essai d'imprimabilité — **d** Bedruckbarkeitsprüfgerät *n* — **n** bedrukbaarheidstoestel *n*; proefdruktoestel *n* — **e** aparato *m* para ensayos (del papel o tinta) — **i** apparecchio *m* per prove di stampabilità

10798 print *v* **close to paper; print** *v* **to the paper** — To print with a relatively small quantity of ink.

f imprimer sec — **d** mager drucken; mit sparsamer Farbgebung drucken — **n** schraal drukken — **i** stampare con relativemente piccole quantità d'inchiostro

10799 print density — Reflection density of a print.

f densité *f* d'impression — **d** optische Dichte *f* des Druckbildes — **n** drukdichtheid *f* — **i** densità *f* di stampa

10800 printed circuit — Electrical circuit produced by a printing process, whereby the image of the circuit is deposited or etched. Used in electronic equipment to achieve miniaturization, increased reliability, and savings in costs, weight and space.

f circuit *m* imprimé; schéma *m* imprimé — **d** gedruckte Schaltung *f*; gedruckte Leiterplatte *f* — **n** gedrukte schakeling *f*; gedrukte bedrading *f*; voorgedrukte bedrading *f* — **e** circuito *m* impreso — **i** circuito *m* stampato

10801 printed fabric — Fabric decorated by a printing process.

f tissu *m* imprimé — **d** bedruckter Stoff *m*; bedrucktes Gewebe *n* — **n** bedrukt textiel *n*; bedrukte stof *fm* — **e** textiles *mpl* estampados; tejidos *mpl* estampados — **i** tessuto *m* stampato

10802 printed matter — Matter to be mailed at special concession rates.

f imprimé *m* — **d** Drucksache *f* — **n** drukwerk *n* — **e** impresos *mpl* — **i** stampato *m*; stampa *f*

10803 printed on the ...

f imprimé sur les presses de ...; tirage *m* du ... — **d** Druck *m* von ... — **n** gedrukt op de persen van ...; gedrukt bij ...; druk *m* van ... — **e** impreso en el taller de ... — **i** stampato per i tipi di ...

10804 printed waste — See newsprint waste.

10805 printer — Computer term for a machine which produces a printed record of the data with which it is fed, usually in the form of discrete graphic characters, as opposed to a plotting table.

f imprimante *f* — **d** Drucker *m* — **n** afdrukmachine *f*; afdrukeenheid *f*; drukker *m* — **e** impresora *f* — **i** stampatrice *f*

10806 printer-computer — In printing industry, computer for typesetting.

10807 printer, master — See master printer.

10808 printer's brush — See beating brush.

10809 printer's devil — *Errand boy in a printing shop. The first stage of apprenticeship. See also apprentice.

f apprenti *m* (imprimeur); arpète *m*; mulet *m* — **d** Buchdruckerlehrling *m*; Lehrling *m* — **n** leerlingdrukker *m*; drukkersleerling *m*; leerling *m* — **e**

aprendiz *m* de impresor; aprendiz *m* de tipógrafo; aprendiz *m* gráfico; mandadero *m* de imprenta — **i** apprendista *m* tipografo; fattorino *m*; apprendista *m* tipografico

10810 printer's error — See compositor's error.

10811 printer's imprint — See imprint.

10812 printer-slotter — Machine for the simultaneous printing, slotting (the cuts between adjacent flaps), creasing and trimming corrugated and solid fibre box blanks.

f slotteur-imprimeur *m*; slotteur-imprimeuse *f*; imprimeuse *f* encocheuse; imprimeuse *f* refendeuse — **d** Druckerschlitzer *m*; Druckslotter *m* — **n** printer-slotter *m* — **e** impresora-troqueladora *f*; ranudora-impresora *f* — **i** printer-slotter *m*

10813 printer's mark — See imprint.

10814 printer's metal — See type metal.

10815 printer's note — See imprint.

10816 print, in — See in print.

10817 printing — Making a print.

f impression *f*; tirage *m* — **d** Drucken *n* — **n** drukken *n*; druk *m* — **e** impresión *f* — **i** tiratura *f*

10818 printing arc lamp

f lampe *f* à arc à copier — **d** Kopierbogenlampe *f* — **n** kopieerlamp *fm*; booglamp *fm* — **e** arco *m* voltaico para copiar; lámpara *f* de arco para copiar — **i** lampada *f* ad arco per copiare

10819 printing area — Extent of surface to be printed.

f surface *f* d'impression; partie *f* imprimante; région *f* imprimante — **d** Druckfläche *f* — **n** drukkende partij *f*; beeldpartij *f* — **e** superficie *f* de impresión; área *f* de impresión; zona *f* de mancha — **i** superficie *f* di stampa; zona *f* stampante

10820 printing bristol — *Mill bristol manufactured primarily for printing.

10821 printing bristol — See mill bristol.

10822 printing costs; costs of printing

f frais *mpl* d'impression; frais *mpl* du tirage — **d** Druckkosten *pl*; Kosten *pl* der Drucklegung; Druckpreis *m* — **n** drukkosten *pl*; kosten *pl* van het drukken; drukprijs *m* — **e** gastos *pl* de impresión — **i** costi *pl* di stampa; spese *pl* della stampa

10823 printing cylinder — See impression cylinder.

10824 printing department — See press-room.

10825 printing depth — See etching depth.

10826 printing down frame — See printing frame.

10827 printing edge zone — In magnetic ink printing a zone contained between the two straight lines delimiting one printing edge, and the mean edge as the central axis of this zone.

f zone *f* du bord d'impression — **d** Druckrandbereich *m* — **n** beeldgrens *fm* — **e** borde *m* de impresión — **i** campo *m* del bordo di stampa

10828 printing frame; printing down frame — Heavy glass-covered frame in which images are locked in contact with sensitized surfaces for exposure to light.

f châssis *m* à copier; châssis *m* de copie; châssis *m* à copie; tireuse *f* — **d** Kopierrahmen *m* — **n** kopieerraam *n* — **e** prensa *f* neumática; prensa *f* de vacio; prensa *f* de copiar; prensa *f* de positivar; marco *m* de imprimir; chasis *m* de imprimir; prensa *f* fotográfica; positivadora *f*; copiadora *f* — **i** telaio *m* da copia

10829 printing house; printing works; printing shop; printing office; printing plant — Generally the entire building or locals in which printing work is carried out.
f imprimerie *f*; maison *f* de l'impression; atelier *m* d'imprimerie — **d** Druckerei *f*; Druckhaus *n*; Druckereibetrieb *m*; graphische Anstalt *f*; graphischer Betrieb *m* — **n** drukkerij *f*; drukkerij-bedrijf *n* — **e** imprenta *f*; casa *f* de imprenta; casa *f* impresora; establecimiento *m* gráfico; establecimiento *m* tipográfico; oficina *f* tipográfica — **i** officina *f* tipografica; stamperia *f*

10830 printing impression — See printing pressure.

10831 printing industry; printing trade
f industrie *f* graphique — **d** graphisches Gewerbe *n*; Druckgewerbe *n* — **n** grafische industrie *f* — **e** industria *f* gráfica; artes *fpl* gráficas — **i** industria *f* grafica

10832 printing ink — See ink.

10833 printing ink factory
f fabrique *f* d'encres d'imprimerie — **d** Druck-farbenfabrik *f* — **n** drukinktfabriek *f* — **e** fábrica *f* de tinta de imprenta — **i** fabbrica *f* d'inchiostri da stampa

10834 printing lamp — Arc lamp or other high-powered illuminant (*xenon lamp), furnishing the source of light when photoprinting line and half-tone negatives on sensitized metal or other surfaces. See also printing arc lamp.
f lampe *f* de copie — **d** Kopierlampe *f* — **n** kopieerlamp *fm* — **e** lámpara *f* de impresión — **i** lampada *f* per copiare

10835 printing office — See printing house.

10836 printing opacity *depr.* — See also opacity (paper backing).
f opacité *f* d'impression *depr.* — **n** drukopaciteit *f depr.* — **e** opacidad *f* de impresión *depr.* — **i** opacità *f* di stampa *depr.*

10837 printing-out paper; print-out paper — Photographic paper, the image of which is appearing without the necessity of developing.
f papier *m* pour copie sans développement — **d** Auskopierpapier *n* — **n** uitkopieerpapier *n* — **e** papel *m* para positivado directo — **i** carta *f* (foto-grafica) che non necessita di sviluppa

10838 printing paper — Paper suitable for printing, such as book paper, bristols, newsprint, writing paper, etc.
f papier *m* pour impressions; papier *m* d'impression — **d** Druckpapier *n* — **n** druk-papier *n* — **e** papel *m* de imprenta; papel *m* de impresión; papel *m* de la prensa — **i** carta *f* da stampa

10839 printing plant — See printing house.

10840 printing plate — See offset plate.
10841 printing plate — See plate.
10842 printing press — See press.
10843 printing pressure; impression pressure; squash *GB*; **printing impression** *GB* — Force at the interface of the printing member and the paper, required to transfer the ink from the printing member to the paper. Taking the mean of all results, printing pressures with normal and average conditions in letterpress printing are found to be as follows: type 33.5 kg/cm^2 (300 lb/sq in), halftones 67 kg/cm^2 (600 lb/sq in), and solids 100 kg/cm^2 (1000 lb/sq in). These pressures must be considered as a standard for the stability of presses.
f pression *f* d'impression — **d** Anpreßdruck *m*; Druckkraft *f*; Druckspannung *f depr.* — **n** druk-kracht *f depr.* — **e** fuerza *f* de presión; presión *f* — **i** pressione *f* di stampa

10844 printing process — The main printing processes are: letterpress printing or relief printing, planography (lithographic printing, litho offset printing), intaglio printing, flexoggraphy, and screen printing. See also die stamping, collotype process.
f procédé *m* d'impression — **d** Druckverfahren *n* — **n** drukprocédé *n*; druktechniek *f* — **e** procedimiento *m* de impresión; método *m* de imprimir — **i** procedimento *m* di stampa; metodo *m* di stampa; sistema *m* di stampa

10845 printing result
f résultat *m* d'impression — **d** Druckergebnis *n* — **n** drukresultaat *n* — **e** resultado *m* de impresión — **i** risultato *m* di stampa

10846 printing shop — See press-room.
10847 printing shop — See printing house.
10848 printing speed — See printing velocity.
10849 printing telegraph — See Teletype.
10850 printing trade — See printing industry.
10851 printing types
f caractères *mpl* d'imprimerie — **d** Drucktypen *fpl* — **n** drukletters *fmpl* — **e** tipos *mpl* de imprenta — **i** caratteri *mpl* da stampa

10852 printing unit — One of several printing machines working together on the same product.
f élément *m* d'impression; groupe *m* imprimant — **d** Druckeinheit *f*; Druckwerk *n* — **n** druk-eenheid *f* — **e** elemento *m* impresor; unidad *f* impresora; grupo *m* impresor; grupo *m* de impresión; cuerpo *m* impresor; mecanismo *m* de impresión — **i** elemento *m* di stampa; gruppo *m* di stampa; gruppo *m* stampante; unità *f* di stampa

10853 printing velocity; printing speed
f vitesse *f* d'impression — **d** Druckgeschwindig-keit *f* — **n** druksnelheid *f* — **e** velocidad *f* de impresión — **i** velocità *f* di stampa; velocità *f* di tiratura

10854 printing works — See printing house.
10855 printing zone — See impression zone.
10856 Printon — Trade name for a multilayer material to make full-colour prints direct from colour film transparencies by a reversal procedure

which permits processing of the pictures by the photographer himself.

10857 printone — See undertone.

10858 printout — See print-out.

10859 print-out; printout — Hard copy representation which has been stored in machine-readable form, generated (usually) on fan-fold paper by a computer line printer.
f liste *f* de sortie — **d** Ausgabeliste *f* — **n** uitvoerlijst *fm* — **e** lista *f* de salida — **i** lista *f* d'uscita

10860 print-out effect — Appearance of metallic silver in a silver halide emulsion after prolonged exposure to light.
f procédé *m* de copie — **d** Auskopierprozeß *m* — **n** fotografie *f* met vol-uitdrukkend materiaal — **e** fotografía *f* de imagen directa — **i** fotografia *f* a stampa diretta

10861 print-out mask; burn-out mask — Mask to protect portions of an exposed plate when exposing it a second time, to eliminate unwanted work.
f masque *m* de détourage — **d** vollflächige Abdeckmaske *f*; vollgedeckte Abdeckmaske *f*; Ausdeckmaske *f* — **i** maschera *f* per sovraposizione

10862 print-out paper — See printing-out paper.

10863 print proofs — See photographic proofs.

10864 print *v* **the first side** — To make the first press run at one side of the sheet of paper.
f imprimer le premier coté — **d** Schöndruck ausführen — **n** de schoondruk aanbrengen; schoondrukken — **e** imprimir el blanco; tirar el blanco — **i** stampare in bianca

10865 print-through — Degree to which a print on one side of the sheet can be seen from the reverse, determined by measuring the percentage of light from the area in question compared with that from the unprinted page. May be due to show-through or strike-through, or more usually a combination of both. See also opacity.
f transvision *f*; perception *f* au verso — **n** inslagdoorschijnendheid *f* — **e** transparencia *f* de un impreso — **i** controstampa *f* per penetrazione; visibilità *f* della stampa sul retro; controstampa *f* per riflessione

10866 print *v* **to the paper** — See print *v* close to paper.

10867 priority indicator — Character in the *header of a message to define the order of transmitting messages over a communication channel.
f indicateur *m* de priorité — **d** Prioritätsanzeiger *m*; Vorranganzeiger *m* — **n** prioriteitsaanwijzer *m*; voorrangsaanwijzer *m* — **e** indicador *m* de prioridad — **i** indicatore *m* di priorità

10868 priority interrupt — The level of priority given to a program, routine or equipment. Each program in a multi-programming system, or subroutine within a program may be allocated a level of priority. Requests for interruption of the central processor by high priority interrupt are given consideration before all lower priority interrupts. In a time sharing system the priority given

to interrupts may be linked to the time taken to complete the task. In an on-line system stations may be given different levels of priority for central processor time.
d Vorrangunterbrecher *m* — **n** voorrangs-onderbreker *m*

10869 prism — Triangular prism made of material transparent to the light being used, e.g., glass for visible light, quartz for ultra-violet rays and near infra-red rays. See also reversing prism.
f prisme *m* — **d** Prisma *n* — **n** prisma *n* — **e** prisma *m* — **i** prisma *m*

10870 prism — See reversing prism.

10871 private edition
f édition *f* privée; édition *f* particulière; édition *f* hors commerce — **d** Privatausgabe *f*; Privatdruck *m* — **n** privé-uitgave *f*; particuliere uitgave *f*; eigen uitgave *f* — **e** edición *f* no oficial; edición *f* fuera de comercio; edición *f* no venal — **i** edizione *f* privata; edizione *f* particolare

10872 private library
f bibliothèque *f* privée — **d** Privatbibliothek *f* — **n** particuliere bibliotheek *f*; privé-bibliotheek *f* — **e** biblioteca *f* particular — **i** biblioteca *f* privata; biblioteca *f* particolare

10873 privately printed; not for sale
f hors commerce; non mis dans la commerce — **d** nicht im Handel; nicht zum Verkauf — **n** niet in de handel — **e** fuera de comercio; no puesta a la venta; no puesta en venta; no venal — **i** fuori commercio; fuori vendita; non in vendita

10874 private press — Small printing or publishing house which issues special editions for limited distribution that are usually not for sale.
f presse *f* privée — **d** Privatpresse *f* — **n** eigen pers *f*; eigen drukkerij *f*; particuliere drukkerij/uitgeverij *f* — **e** imprenta *f* privada; imprenta *f* particular — **i** stampa *f* particolare

10875 probability of rejection
f probabilité *f* de rejet — **d** Rückweisewahrscheinlichkeit *f* — **n** afkeurkans *fm* — **e** probabilidad *f* de lanzamiento — **i** probabilità *f* di essere respinto

10876 probable error — In statistics, a measure of variation, not necessarily associated with errors in estimation, equal to $0.6745 \times$ standard deviation. Connected with the properties of the normal distribution curve.
f erreur *f* probable — **d** wahrscheinlicher Fehler *m* — **n** waarschijnlijkheidsfout *fm* — **e** error *m* de probabilidades — **i** errore *m* probabile

10877 problem board — See plug board.

10878 problem-oriented language — Programming language devised for easier specification of a range of problems, as distinct from autocodes which are devised for ease of translation into machine code.
f langage *m* orienté vers le problème — **d** problemorientierte Sprache *f* — **n** probleemgerichte taal *fm*; op probleem gerichte taal *fm* — **e** lenguaje *m* orientado hacia el problema — **i** linguaggio *m* applicativo al problema

10879 procedure — Steps leading to the solution of a programming problem.

f procédure *f* — **d** Prozedur *f* — **n** procedure *fm* — **e** procedimiento *m* — **i** procedura *f*

10880 procedure-oriented language — Programming language (often machine independent) for the easier expression of the process of solving a wide range of problems, e.g., algol and fortran.

f langage *m* de procédure — **d** Prozedursprache *f*; verfahrensorientierte Programmiersprache *f* — **n** procedure-taal *fm*; procedure-gerichte taal *fm* — **e** lenguaje *m* de procedimiento — **i** linguaggio *m* di procedura

10881 process — Term noting, pertaining to, or involving photomechanical or photoengraving methods.

10882 process camera; reproduction camera; camera — Heavy copying camera designed for the particular requirements of line and halftone photography and especially for colour reproduction.

f appareil *m* de reproduction; chambre *f* photographique; banc *m* de reproduction — **d** Reproduktionskamera *f*; Kamera *f* — **n** reproduktiecamera *fm*; camera *fm* — **e** cámara *f* fotográfica de reproducción; cámara *f* de fotomecánica; cámara *f* de reproducción; aparato *m* fotográfico; máquina *f* para la reproducción — **i** macchina *f* fotografica; apparecchio *m* per riproduzione; reprocamera *f*

10883 process chart — See flow chart.

10884 process colour plates — See process plates.

10885 process colours — See process inks.

10886 process colour work — See colour process work.

10887 process developer — *Developer to give high contrast to line and halftone negatives on dry plates and films.

10888 process engraver; block maker — Craftsman who makes printing blocks.

f clicheur *m*; photograveur *m* — **d** Klischeur *m*; Klischeehersteller *m*; Chemigraph *m* — **n** cliché maker *m*; chemigraaf *m*; clichéfabrikant *m* — **e** grabador *m*; fotograbador *m*; fabricante *m* de clisés — **i** fotoincisore *m*; zincografo *m*

10889 process engravers — Business manufacturing line and halftone blocks.

f atelier *m* de photogravure; clicherie *f* — **d** Klischeeanstalt *f*; Reproduktionsanstalt *f* — **n** clichéfabriek *f*; clichébedrijf *n* — **e** fábrica *f* de clisés; taller *m* de fotograbado; fotomecánica *f* — **i** fotoincisore *m*

10890 process engraving — Making process halftone blocks for multi-colour printing.

f photogravure *f* (typographique) — **d** Klischeeherstellung *f* — **n** clichéfabricage *f*; cliché vervaardiging *f* — **e** fotograbado *m*; grabado *m* fototipográfico; fototipografía *f* — **i** fotoincisione *f*; zincografia *f*

10891 process film; process plate — Photographic surface sensitized with very contrasty emulsions to promote sharp line and dot formations.

f film *m*; plaque *f* process — **d** Reproduktionsfilm *m*; Reproduktionsplatte *f* — **n** reproduktiefilm *m*; fotomateriaal *n* voor de reproduktietechniek — **e** película *f* de fotomecánica; película *f* de extremo contraste — **i** pellicola *f* fotosensibile; lastra *f* fotosensibile

10892 process glue — Clarified *fish glue for photoengraving.

f colle *f* de copie; colle *f* de poisson — **d** Kopierleim *m*; Chromfischleim *m* — **n** kopieerlijm *m*; vislijm *m* — **e** cola *f* bicromatada; cola *f* de pescado — **i** colla *f* bicromatata; colla *f* di pesce

10893 processing — Photographic procedure during which exposed film or paper is developed, fixed, and washed to produce a negative or a positive image (transparency, or print).

f traitement *m* — **d** Entwicklung *f* — **n** afwerking *f*; behandeling *f* — **e** procesado *m* — **i** trattamento *m*

10894 processing

f traitement *m* — **d** Bearbeitung *f*; Verarbeitung *f* — **n** verwerking *f* — **e** tratamiento *m*; proceso *m* — **i** lavorazione *f*

10895 processing section — Computer unit that does the actual changing of input into output, including the arithmetic section and intermediate storage. Usually referred to as the central processor, which is based on the arithmetic unit.

f équipement *m* de traitement — **d** Verarbeitungsteil *m* — **n** verwerkingseenheid *f* — **e** equipo *m* de tratamiento — **i** organo *m* elaboratore

10896 processing time — See computer time.

10897 process inks; process colours; tricolour inks; standard inks — Printing inks, usually in sets of four colours. The most frequent combination is yellow, magenta, cyan and black, which are printed over one another in that order to obtain a coloured print with the desired hues, whites, blacks and greys.

f encres *fpl* pour polychromie; encres *fpl* trichromes; encres *fpl* normalisées — **d** Druckfarben *fpl* für Mehrfarbendruck; Mehrfarbendruckfarben *fpl*; Dreifarbendruckfarben *fpl*; Normalfarben *fpl*; Normaldruckfarben *fpl*; Normdruckfarben *fpl* — **n** inkten *mpl* voor drie- en vierkleurendruk; driekleureninkten *mpl*; normaalinkten *mpl* depr. — **e** tintas *fpl* tricromáticas; tintas *fpl* de tricromía; tintas *fpl* para policromía; tintas *fpl* para tricromía; tintas *fpl* primarias sustractivas; tintas *fpl* normalizadas — **i** inchiostri *mpl* per policromia; inchiostri *mpl* per tricolori; inchiostri *mpl* per più colori

10898 process lens — Camera lens to photograph flat copy. Used for line and halftone photography, continuous tone and colour separation for line; usually not sufficiently corrected for process-colour separation.

f objectif *m* de reproduction — **d** Reproduktionsobjektiv *n* — **n** reproduktielens *fm* — **e** objetivo *m* para fotomecánica; objetivo *m* fotomecánico; objetivo *m* de fotomecánico; objetivo *m* de fotograbado; objetivo *m* de reproducción — **i** obiettivo *m* da riproduzione

10899 process monger — Person who attempts to peddle or sell secret procedures of supposedly high values and efficiency, but which usually are either worthless or nothing but modifications of known methods.

10900 processor; automatic processor; film processor; film processing unit — Device to develop and fix and sometimes wash and dry a photographic emulsion.
f machine *f* (automatique) à développer — **d** Entwicklungsmaschine *f*; Entwicklungsautomat *m* — **n** (automatische) ontwikkelmachine *f*; ontwikkelautomaat *m* — **e** procesadora *f* (automática); máquina *f* reveladora de películas — **i** processor *m*; sviluppatrice *f* automatica; unità *f* di sviluppo (automatico); sviluppatore *m* (automatico)

10901 processor — Device or system capable of performing operations upon data. Note: the terms may refer to either hardware (*central processor) or software. An assembler or compiler is sometimes referred to as a "language processor".

10902 process plate — See process film.

10903 process plates; process colour plates — Line or halftone etchings in two or more colours, particularly those made from separation negatives.
f clichés *mpl* en couleurs — **d** Dreifarbenklischees *npl*; Dreifarbenplatten *fpl* — **n** kleurenclichés *npl* — **e** clisés *mpl* (tramado) para tricromía; clisés *mpl* para impresión multicolor — **i** lastre *fpl* ottenute per fotoformatura; lastre *fpl* in due colori; lastre *fpl* in più colori

10904 process printing; multicolour printing — Method of printing in the three primary colours from halftone plates. Sometimes a black plate or other colour plates are added.
f polychromie *f*; impression *f* en plusieurs couleurs; chromotypographie *f* — **d** Mehrfarbendruck *m* — **n** meerkleurendruk *m* — **e** policromía *f*; impresión *f* policroma; impresión *f* multicolor; fotocromotipografía *f*; citocromía *f* — **i** stampa *f* a (più) colori; policromia *f*

10905 process white — Water-colour preparation used to cover India ink or drawing or to retouch photographs from which blocks are to be made.
f gouache *f*; blanc *m* opaque; blanc *m* couvrant — **d** Retuschefarbe *f*; Deckweiß *n*; Wasserfarbe *f* — **n** retoucheerverf *fm*; dekwit *n*; dekverf *fm* — **e** tinta *f* de retoque; blanco *m* para retoque — **i** bianco *m* per ritocchi

10906 process work — Process of producing a printing surface on metals with etching solutions.
f photogravure *f* — **d** Chemigraphie *f* — **n** chemigrafie *f*; clichévervaardiging *f* — **e** quimicografía *f* — **i** fotoformatura *f*

10907 process work — Finished product of photoengraving.
f cliché *m* — **d** Druckstock *m*; Klischee *n*; Druckplatte *f* — **n** cliché *n* — **e** clisé *m*; cliché *m* — **i** cliscè *m*; cliché *m*; forma *f* stampante

10908 production control
f planning *m*; établir le programme d'ordonnancement — **d** Fertigungsplanung *f* — **n** produktieregeling *f* — **e** programación *f* de la producción — **i** controllo *m* di produzione; pianificazione *f*

10909 production costs — See cost price.

10910 production line
f chaîne *f* de fabrication — **d** Fertigungsstraße *f*; Fertigungslinie *f* — **n** produktie-straat *fm*; produktie-lijn *fm* — **e** tren *m* de producción — **i** linea *f* di produzione

10911 production study — See chronological study.

10912 production waste — See newsprint waste.

10913 professional — A size of cut card. See app. no. 3.

10914 program — Set of instructions or steps that tells the computer how to handle a complete problem such as type justification, production scheduling, payroll or other application. Most programs include alternate steps or routines to take care of variations. General program steps form a complete cycle. Each incoming unit of information sets off the complete cycle from beginning to end.
f programme *m* — **d** Programm *n* — **n** programma *n* — **e** programa *m* — **i** programma *m*

10915 program *v* — To produce a series of step-by-step instructions to control a particular computer run. The whole sequence covers system analysis, outline flow charts, detailed flow charts, coding the instructions, writing and calling subroutines, calling library routines, specifying format of input, debugging and integrating the computer run into the whole system.
f programmer — **d** programmieren — **n** programmeren — **e** programar — **i** programmare

10916 program execution — The carrying out or inplementation by a computer run of a specific program.

10917 program library — Organized collection of tested programs of general utility, with sufficient descriptive text to allow them to be used by others than their authors.
f bibliothèque *f* de programmes — **d** Programmbibliothek *f* — **n** programmabibliotheek *f* — **e** biblioteca *f* de programas — **i** biblioteca *f* di programmi

10918 programmable read-only memory; PROM — While read-only memories (roms) have information to be stored "built into them" at the time they are manufactured, proms are programmed at a later date by a special-program loader. This makes them particularly useful for customizing a product to the user's specifications. Generally speaking, proms like roms, cannot be "unprogrammed" once they have been loaded; however, a new type of prom which is programmed by exposure to ultraviolet light can, in fact, be "reprogrammed" by someone with the right equipment.

10919 programmatic — See automatic programmer.

10920 programmed cutting
f coupe f programmée — d Programmschnitt m — n geprogrammeerd snijden n — e corte m programado — i taglio m programmato

10921 programmed guillotine — See automatic programmer.

10922 programmer — Person who prepares a sequence of instructions in a computer language to perform operations on input to get a desired output. Programmers use languages, such as cobol, fortran, assembler, etc., as opposed to typesetting mark-up personnel who use commands supplied first by a programmer to achieve their desired results. This is a common mistake in nomenclature in the typesetting field. A programmer is sometimes called a system analyst although the latter may deal more specifically with problems of definition and the former with problems of implementation.
f programmateur m — d Programmierer m — n programmeur m — e programador m — i programmatore m

10923 programming instructions — Repertoire of commands (i.e., the instruction set) available for writing programs for a specific computer.

10924 programming language — Set of syntactical rules for the writing of computer instructions in other than machine language. The result is input to a compiler or assembler which translates the input and outputs an object program.
f langage m de programmation — d Programmiersprache f — n programmeertaal fm — e lenguaje m de programación — i linguaggio m di programmazione

10925 program priority interrupt — See automatic program interrupt.

10926 progress chaser; chaser
f suiveur m (des pièces) — d Terminjäger m — n planner m; man m van de planning — e controlador m de la producción; controlador m de plazos — i controllore m della produzione; controllore m dei termini

10927 progress control
f contrôle m des délais — d Terminkontrolle f — n voortgangscontrole fm — e control m de la producción; control m de plazos — i controllo m dei termini (di consegna)

10928 progressive colour proofs; progressive proofs; progressives; progs — Set of colour proofs consisting of proofs of each plate, single and in combination with other proofs as the job will be printed. For example: yellow by itself; magenta; yellow and magenta; cyan; yellow, magenta and cyan; black by itself; and all four colours combined. Sometimes called flat proofs.
f épreuves fpl gammes; gamme f d'épreuves trichromes — d Teilfarbenandrücke mpl; Farbskalen fpl; Skalendrücke mpl — n scala-drukken mpl; scala's fmpl; schaaldrukken mpl; deeldrukken mpl — e pruebas fpl progresivas; pruebas fpl de gama; pruebas fpl de estado; pruebas fpl de tricromía — i prove fpl progressive

10929 progressive mean — In statistics, the mean of all observed values in a series up to a given one.
f moyenne f progressive — d fortschreitendes Mittel n — n progressief gemiddelde n — e media f progresiva — i media f progressiva

10930 progressive proofs — See progressive colour proofs.

10931 progressives — See progressive colour proofs.

10932 progs — See progressive colour proofs.

10933 projected mask — Projected masks for originals include the Gamble process in which reflection copy is photographed and the resulting mask image is projected back onto the copy. Projected masks for negatives include automasking, in which the copy itself is used as a mask, its image being projected onto the negatives when the positives are made.
f masque m par projection — d Projektionsmaske f — n projectie-masker n — e máscara f proyectada; máscara f por proyección — i maschera f proiettata

10934 projection — Projecting a photographic image onto a sensitized surface or viewing screen.
f projection f — d Projektion f — n projectie f — e proyección f — i proiezione f

10935 PROM — See programmable read-only memory.

10936 proof — Impression from composed type or blocks, taken for checking and correction, or from a lithographic plate to check accuracy of layout, type matter, tone and colour reproduction.
f épreuve f — d Probedruck m; Probeabzug m; Abzug m; Korrekturabzug m; Andruck m; Probe f — n proef fm; drukproef fm — e prueba f; plana f — i bozza f di stampa; prova f di stampa

10937 proof v; **proove** v; **prove** v; **prooving; proving** — To pull proofs of plates for proofreading, revising, trial, approval of illustrations etc. before production printing.
f tirer d'épreuves; prendre d'épreuves — d andrucken; Probedrucke anfertigen — n proeven trekken — e sacar pruebas — i tirare una prova

10938 proof v — See pull v a proof.

10939 proof bank — See dump.

10940 proof before letters; proof before the letter — Proof from a hand-engraved plate, which does not show the legend or other indications, to check the progress of engraving.
f épreuve f avant la lettre; épreuve f premier état — d Abdruck m vor der Schrift; Druck m vor der Schrift; Andruck m vor der Schrift; Abzug m vor der Schrift; Abdruck m avant la lettre — n proef fm vóór (het aanbrengen van) de tekst; afdruk m vóór (het aanbrengen van) de tekst — e prueba f antes de la letra — i prova f avanti-lettere; prova f anti-lettere

10941 proof before the letter — See proof before letters.

10942 proof brush — See beating brush.

10943 proof correction marks — See correction marks.

10944 proofer — See proofing hand.

10945 proofing hand; proofer; proof puller
f tireur *m* d'épreuves; pressier *m* — **d** Abzieher *m* — **n** proeftrekker *m*; proeventrekker *m* — **e** sacapruebas *m*; sacador *m* de pruebas; tirador *m*; pruebero *m*; chonguero *m ch* — **i** tiratore *m* di bozze

10946 proofing paper; proof paper — Cheap book or newsprint paper to make galley proofs.
f papier *m* à épreuves; papier *m* pour épreuves — **d** Abziehpapier *n*; Abzugpapier *n*; Andruckpapier *n* — **n** proefdrukpapier *n*; drukproefpapier *n*; papier *n* voor drukproeven — **e** papel *m* para pruebas (de imprenta) — **i** carta *f* per prove; carta *f* per bozze

10947 proofing press — See proof press.

10948 proof on art paper
f épreuve *f* sur papier couché — **d** Kunstdruckabzug *m* — **n** proef *fm* op kunstdrukpapier; afdruk *m* op kunstdrukpapier — **e** prueba *f* sobre papel estucado — **i** prova *f* sopra carta patinata

10949 proof paper — See proofing paper.

10950 proof press; proofing press — Printing machine to make proofs, usually having most of the elements of a production machine but not for automatic sustained production.
f presse *f* à épreuves — **d** Andruckpresse *f*; Abziehpresse *f*; Abzugpresse *f* — **n** proefpers *f* — **e** prensa *f* de pruebas; prensa *f* sacapruebas; prensa *f* para (sacar) pruebas; máquina *f* sacapruebas; sacatipia *f Ch*; rodón *m Ch* — **i** macchina *f* tiraprove; tiraprove *m*; tirabozze *m*

10951 proof press; offset proof press — Offset press to transfer printed image to offset blanket and from that to the paper. Hand and power models, single and multicolour, as well as special presses for transparent proofs are available.
f presse *f* à contre-épreuves — **d** Offsetandruckpresse *f* — **n** offsetproefpers *f* — **e** prensa *f* de pruebas offset; prensa *f* para pruebas offset; prensa *f* sacapruebas offset — **i** tiraprove *m* offset

10952 proof puller — See proofing hand.

10953 proofread *v* — See read *v*.

10954 proofreader; reader; corrector of the press *obs.* — Person who reads printers' proof against the original copy and indicates corrections to be made. See also horse *v*, comma friend.
f correcteur *m* — **d** Korrektor *m* — **n** corrector *m* — **e** corrector *m* (de pruebas tipográficas); corrector *m* de imprenta; corrector *m* tipográfico; corrector *m* tipografo — **i** correttore *m* (di bozze)

10955 proofreaders' marks — See correction marks.

10956 proofreader's proof — See reader's proof.

10957 proofroom — See reading-room.

10958 proof with the letter; impression with the letter — First impression of a copperplate engraving or etching which shows the legend merely by the contours.

f épreuve *f* avec la lettre grise — **d** Abdruck *m* mit der Schrift; Andruck *m* mit der Schrift; Abdruck *m* mit ausgeführter Schrift — **n** proef *fm* met de letter; afdruk *m* met de letter — **e** prueba *f* con la letra — **i** prova *f* dopo lettera

10959 proove *v* — See proof *v*.

10960 prooving — See proof *v*.

10961 propanol — See propyl alcohol.

10962 propanone — See acetone.

10963 propellers; bosses — Individually mounted rollers, draw rollers or bosses, with independent adjustment and peripheries of rubber or synthetic substance or of knurled metal. They usually operate in conjunction with a driven roller to assist it to control the webs in a rotary press.
f rouleaux *mpl* cannelés — **d** Zugrollen *fpl*; Zugringe *mpl* — **n** trekrollen *fmpl* — **e** rodillos *mpl* de tracción; rodillos *mpl* de tiro — **i** rulli *mpl* di trazione

10964 propenol — See allyl alcohol.

10965 propenyl alcohol — See allyl alcohol.

10966 property — See feature.

10967 proportion — Relation of the dimensions of a subject or original, and that existing between the dimension of the original and the scale or size of the reproduction.
f proportion *f* — **d** Verhältnis *n*; Größenverhältnis *n* — **n** verhouding *f* — **e** proporción *f* — **i** proporzione *f*

10968 proportionally spaced characters — Characters of a typewriter which vary in width, as opposed to those in which capitals and lower-case are all the same width. Printer's term is *set width or unit width of the characters.

10969 proportional reducer — Reducer to reduce the silver uniformly throughout all tone values. The constituents are potassium permanganate, sulphuric acid, ammonium persulphate and water. A negative thus treated will retain its original contrast range but will lose over-all density.
f réducteur *m* proportionnel — **d** proportionaler Abschwächer *m* — **n** proportionele afzwakker *m* — **e** debilitador *m* proporcional — **i** indebolitore *m* proporzionale

10970 proportion rule; proportion scale — Device to determine the exact proportions or dimensions of an original at a given scale (size) or enlargement or reduction.
f échelle *f*; règle *f* proportionnelle — **d** Rechenscheibe *f* — **n** rekenschijf *fm*; rekenliniaal *fm*; proportieliniaal *fm* — **e** regla *f* de cálculo — **i** regolo *m* proporzionale

10971 proportion scale — See proportion rule.

10972 proprietary solvent alcohol — A completely denatured ethyl alcohol which is mixed to a set of US Government specifications and sold by the vendor under his own trade name. May be either anhydrous or 95% alcohol.

10973 propyl alcohol; ethyl carbinol; propanol — Clear, colourless liquid, soluble in water, alcohol and ether.
Formula: C_3H_7OH.

f alcool *m* propylique; propanol *m* — **d** Propyl-
alkohol *m*; Äthylkarbinol *m*; Propanol *m* — **n**
propylalcohol *m*; propanol *m* — **e** alcohol *m*
propílico; propanol *m* — **i** alcool *m* propilico;
propanolo *m*

10974 propylene glycol; methyl ethylene glycol —
Colourless, viscous, stable liquid, miscible with
water, alcohols and many organic solvents.
Formula: $CH_3CHOHCH_2OH$.
f propylène-glycol *m* — **d** Propylenglykol *n* — **n**
propyleen-glycol *m* — **e** propilenglicol *m* — **i**
glicolo *m* di propilene

10975 pro rata — In proportion.
f au prorata — **d** im Verhältnis — **n** in verhouding
tot; naar verhouding van — **e** en proporción a;
proporcional a; en comparación con; en
comparación de; en relación a; en relación en; a
razón de — **i** in proporzione di

10976 prospectus — Descriptive leaflet giving details of forthcoming publications, etc.
f prospectus *m* — **d** Prospekt *m* — **n** prospectus *n*
— **e** prospecto *m* — **i** prospetto *m*

10977 protected motor — See enclosed motor.

10978 protected store — Store reserved for special
purposes in which data cannot be stored or
changed without undergoing a screening procedure.

10979 protective colloid — Stable colloid which
protects another colloid, usually unstable, from
precipitation by coagulating influences.
f colloïde *m* protecteur — **d** Schutzkolloid *n* — **n**
beschermend colloïd *nfm* — **e** coloide *m*
protector; coloide *m* de protección — **i** colloide *m*
protettore

10980 protective cover
f couvercle *m* de protection — **d** Schutzkappe *f*;
Schutzdeckel *m*; Schutzabdeckung *f* — **n**
beschermkap *fm* — **e** tapa *f* protectora; funda *f*
protectora — **i** copertura *f* di protezione

10981 protein — Complex organic compound of
large molecular weight containing nitrogen, in the
cells of plant and animal matter.
f protéine *f* — **d** Protein *n* — **n** proteïne *fmn* — **e**
proteína *f* — **i** proteina *f*

10982 protocol — Observance of a required procedure for data transmission and reception so
that the input to a particular system or compound
will be what is anticipated, with respect to code
structure but especially identification of the text
stream so that the system will know what to do
with it.

10983 protometry — All the work done for the
checking of the separate parts of a printing forme
before printing.
f protométrie *f*; science et mesure du service du
prote — **d** Kenntnisse, Regeln und Meßverfahren
die dem Faktor zur Verfügung stehen — **n** kennis
en middelen ten dienste van de chef voor een
efficiente bedrijfsvoering — **i** scienza e misura al
servicio del proto

10984 proton — Elementary particle that is a fundamental constituent of all atomic nuclei, having

a positive charge equal in magnitude to that of
the electron. The atomic number, or the number
of protons, in a nucleus is different for each
element.
f proton *m* — **d** Proton *n* — **n** proton *n* — **e**
protón *m* — **i** protone *m*

10985 protosulphate of iron — See ferrous
sulphate.

10986 protractor — Half-circle divided into 180°
used in setting angle on bevelling machines, etc.
f rapporteur *m* — **d** Gradbogen *m* — **n** graadboog *m*; gradenboog *m depr.* — **e**
transportador *m*; trasportador *m* — **i** rapportatore *m*; goniometro *m*

10987 prove *v* — See proof *v*.

10988 provincial paper — Paper generally
restricted in its area of circulation.
f régional *m*; feuille *f* régionale — **d** Provinz-
blatt *n* — **n** provincieblad *n* — **e** periódico *m*
provinciano; periódico *m* de provincia — **i**
giornale *m* provinciale

10989 proving — See proof *v*.

**10990 proving and reversing press; reversing
press** — Flat-bed lithographic printing machine to
pull proofs by direct (countering) or indirect (offset) printing.
f presse *f* à contre-épreuves — **d** Konterpresse *f*
— **n** offsetproefpers *fm*; proefpers *fm* — **e**
máquina *f* para hacer contra-pruebas — **i**
macchina *f* tiraprove offset; tiraprove *m* offset;
tiraprove *m* litografico

10991 Prussian — A size of wrapping paper. See
app. no. 3, double lump.

10992 Prussian blue — Most common and best
known name for blue iron ferrocyanide (iron blue)
pigments made by a variety of procedures. Known
under a variety of names, as Berlin blue, Paris
blue, Chinese blue, mineral blue, Hamburg blue,
bronze blue, celestial blue.
Formula: $[Fe(CN)_6]_3Fe_4$.
f bleu *m* de Prusse; bleu *m* de Berlin; bleu *m* de
Paris; bleu *m* de Chine; bleu *m* acier — **d**
Preußischblau *n*; Berlinerblau *n*; Pariserblau *n*;
chinesisch Blau *n* — **n** Pruisisch blauw *n*;
Berlijns blauw *n* — **e** azul *m* de Prusia; azul *m*
de Paris; azul *m* chino; azul *m* acero — **i** blu *m* di
Prussia; blu *m* di Cina

10993 prussic acid — See hydrocyanic acid.

10994 pseudonym — See pen-name.

10995 pseudoplastic liquid — Non linear, non
Newtonian liquid that gives a consistency curve
which starts at the origin and curves away from
the force axis. Pseudoplastics have no Bingham
yield value.
f liquide *m* pseudo-plastique — **d** pseudo-
plastische Flüssigkeit *f* — **n** pseudo-plastische
vloeistof *fm* — **e** líquido *m* seudoplástico — **i**
liquido *m* pseudo-plastico

10996 psychrometer — Wet-and-dry bulb type of
hygrometer. The two thermometers with swivel
handle are whirled vertically in area for reading.
The most accurate instrument for industrial plant

use to determine *relative humidity.

f psychromètre *m* − **d** Psychrometer *m* − **n** psychrometer *m* − **e** psicrómetro *m* − **i** psicrometro *m*

10997 psychrometry − Measurement of the relative humidity of the atmosphere.

f psychrométrie *f* − **d** Luftfeuchtigkeitsmessung *f* − **n** psychrometrie *f* − **e** psicrometría *f* − **i** psicrometria *f*

10998 p.t.o. − See please turn over.

10999 publication date − Date upon which booksellers are authorized by a publisher to offer a book to the public. See also delivery date.

f date *f* de publication − **d** Erscheinungsdatum *n* − **n** verschijningsdatum *m* − **e** fecha *f* de la publicación − **i** data *f* di pubblicazione

11000 publicity magazine

f périodique *m* de propagande − **d** Werbezeitschrift *f* − **n** reclametijdschrift *n* − **e** revista *f* de publicidad − **i** rivista *f* pubblicitaria

11001 publicity matter

f imprimés *mpl* publicitaires − **d** Werbedrucksachen *fpl* − **n** reclamedrukwerk *n* − **e** impresos *mpl* publicitarios − **i** stampati *mpl* pubblicitari

11002 publicity wrapper; book wrapper; advertising band; advertising wrapper; advertisement wrapper − Strip of paper wrapped around a book like a dust jacket, measuring about one third of this, containing publicity matter.

f bande *f* de nouveauté; bande *f* de lancement; bande *f* de publicité; bande *f* publicitaire; bandeau *m* − **d** Bauchbinde *f coll.*; Buchbinde *f*; Streifband *n* mit Werbetext − **n** manchet *fm*; buikband *m* − **e** faja *f* de publicitaria − **i** striscione *m* pubblicitario

11003 publish *v* − To have books, newspapers or periodicals produced and placing them for sale to the public.

f publier; éditer − **d** verlegen; herausgeben − **n** uitgeven − **e** editar − **i** pubblicare

11004 published by the author

f chez l'auteur; à compte d'auteur − **d** im Selbstverlag − **n** uitgegeven door de schrijver; bij de schrijver − **e** en casa del autor − **i** pubblicato presso l'autore

11005 publisher; publishing house − Person or company who issues and sells books, magazines or newspapers. A publisher may not necessarily be a printer or bookbinder.

f éditeur *m*; maison *f* éditrice − **d** Verleger *m*; Herausgeber *m*; Verlag *m*; Verlagshaus *n* − **n** uitgever *m*; uitgeefster *f*; uitgeverij *f*; uitgeversmaatschappij *f* − **e** editor *m*; casa *f* editora; casa *f* editorial − **i** editore *m*; casa *f* editrice

11006 publisher's binding − See edition binding.

11007 publisher's catalogue − See stock catalogue.

11008 publisher's imprint − See imprint.

11009 publisher's list − See stock catalogue.

11010 publisher's reader; reader − One who reports on manuscripts to a publisher.

f lecteur *m* (dans une maison d'édition) − **d** Verlagslektor *m*; Verlagsberater *m*; Lektorat *m* − **n** lector *m* (van een uitgeversmaatschappij); uitgevers-adviseur *m*; adviseur *m* van de uitgever − **e** lector *m* de originales − **i** lettore *m* (in una casa editrice)

11011 publishing house − See publisher.

11012 puddler − See compositor.

11013 puff − Exaggerated commendation of a book.

f puff *m*; réclame *f* tapageuse − **d** lobende Kritik *f*; übertrieben lobende Kritik *f* − **n** overdreven lof *m*; overdreven goede kritiek *f* − **e** bombo *m* − **i** montatura *f*

11014 pugillares; pugillaria − Small Roman writing tablets made of wood, metal or ivory and covered with wax.

11015 pugillaria − See pugillares.

11016 pull − Impression taken on a hand press from a relief plate. See also proof.

11017 pull *v* − To pull apart the sections of a book before rebinding.

f débrocher − **d** ausreißen − **n** losmaken − **e** desencuadernar − **i** strappare; separare i blocchi incollati

11018 pull *v* **a proof; proof** *v* − Taking an impression for checking or verification.

f lever une épreuve; tirer une épreuve; tirer une impression; tirer les épreuves d'essai − **d** einen Abzug machen; andrucken; abziehen − **n** een proef trekken; een afdruk maken − **e** sacar una prueba; tirar impresión − **i** tirare; fare le bozze

11019 pull down − See drawdown.

11020 pulldown knife − See ink knife.

11021 pulley − See belt sheave.

11022 pull for position − Guide sheet for the positioning of type, blocks, etc.

f tierce *f* des blancs − **d** Standbogen *m* − **n** standvel *n* − **e** hoja *f* de registro; trazado *m* − **i** foglio *m* di registro; foglio *m* di norma

11023 pull out − See draw out.

11024 pull-out − See throw-out.

11025 pull rollers − See drawing rollers.

11026 pulp − Fibrous material of natural vegetable origin, prepared for the manufacture of paper and board.

f pâte *f* (matière première); pulpe *f* − **d** Halbstoff *m*; Papierfaserstoff *m* − **n** halfstof *fm*; stof *fm* (tijdens het fabricageproces) − **e** pasta *f*; pulpa *f* − **i** pasta *f*

11027 pulp board − Coarse board made from mixed papers, groundwood or mixtures of both.

f carton *m* bois − **d** Zellstoffpappe *f*; Zellstoffkarton *m* − **n** cellulosekarton *n* − **e** cartón *m* de celulosa; cartulina *f* de celulosa − **i** cartone *m* di cellulosa

11028 pulp engine − See refiner.

11029 pulper − Machine to disintegrate broke or waste paper for use in the beaters.

f broyeur *m*; triturateur *m* − **d** Zerfaserer *m*; Stoffauflöser *m*; Öffner *m* − **n** pulper *m* − **e** machacadora *f* de pulpa − **i** spappolatore *m*; idroapritore *m*

11030 pulp moulding; moulded-pulp container; mould-pulp fitting — Moulded object made from paper pulp.
f cartonnage *m* moulé — **d** Fasergußbehälter *m*; Pappengußbehälter *m* — **n** gietverpakking *f* — **e** cartón *m* moldeado — **i** contenitore *m* in cellulosa modellata

11031 pulp royal — A British standard size for board. See app. no. 3.

11032 pulp, slush — See slush pulp.

11033 pulp water — See white water.

11034 pulpwood — Wood suitable for the manufacture of chemical wood pulp, in the form of logs as they come from the forest or cut into shorter lengths suitable for the chipper or the grinder.
f bois *m* à pâte; bois *m* de pulpe — **d** Papierholz *n* — **n** papierhout *n* — **e** madera *f* (para la fabricación) de papel — **i** legno *m* per carta

11035 pulse — Electrical signal of short duration.
f impulsion *f* — **d** Impuls *m* — **n** impuls *m* — **e** impulso *m*; impulsión *f* — **i** impulso *m*

11036 pumice *v* — To polish plates with pumice stone.
f poncer; passer à la pierre ponce — **d** abbimsen; bimsen; abschleifen — **n** puimen; afpuimen; met puimsteen schuren; puimstenen — **e** apomazar — **i** passare alla pietra pomice; polire

11037 pumice stone — Highly vesicular lava, containing 65-75% silicon dioxide (SiO_2) and 10-20% aluminium oxide (Al_2O_3). Insoluble in water, not attacked by acids, used as an abrasive.
f pierre *f* ponce; ponce *f* — **d** Bimsstein *m*; Bims *m* — **n** puimsteen *nm* — **e** piedra *f* pómez; piedra *f* para apomazar — **i** pietra *f* pomice; pomice *f*

11038 pump cam
f excentrique *m* pour pompe — **d** Pumpenexzenter *m* — **n** pomp-excentriek *n* — **e** leva *f* de (la) bomba — **i** eccentrico *m* della pompa

11039 pump stop — See short-line preventer.

11040 punch — Piece of steel engraved at the end with a type character and hardened, for making matrices.
f poinçon *m* — **d** Stempel *m*; Schriftstempel *m* — **n** stempel *n*; letterstempel *n* — **e** punzón *m* — **i** punzone *m*

11041 punch card board; punch card paper stock
f papier *m* pour tickets de caisses enregistreuses; papier *m* à cartes statistiques — **d** Lochkartenkarton *m* — **n** ponskaartenkarton *n* — **e** cartón *m* para tarjetas perforadas — **i** cartoncino *m* per schede perforate

11042 punch card paper stock — See punch card board.

11043 punch cutter; type cutter
f graveur *m* de poinçons — **d** Stempelschneider *m*; Schriftschneider *m* — **n** stempelsnijder *m* — **e** grabador *m* de punzones — **i** punzonista *m*; incisore *m* di punzoni

11044 punch cutting — Process of engraving punches.
f gravure *f* des poinçons — **d** Stempelschneiden *n*; Schriftschneiden *n* — **n** stempelsnijden *n*; snijden *n* van stempels — **e** grabado *m* de punzones — **i** incisione *f* dei punzoni

11045 punched card — Card in which holes are punched in rows and columns to represent data. See also edge-punched card, hollerith card.
f carte *f* perforée — **d** Lochkarte *f*; gelochte Karte *f* — **n** ponskaart *fm* — **e** tarjeta *f* perforada; ficha *f* perforada — **i** scheda *f* perforata

11046 punched tape; perforated tape; perforator tape — Tape on which a pattern of holes or cuts is punched to represent data.
f bande *f* perforée; ruban *m* perforé — **d** Lochstreifen *m*; Lochband *n*; gelochter Streifen *m* — **n** ponsband *m* — **e** cinta *f* perforada; banda *f* perforada — **i** nastro *m* perforato

11047 punched-tape reader — Input device to a data processing system which senses the data on paper tape by electro-mechanical, photo-electric, dielectric or air-flow means.

11048 punching machine; hole punching machine — Hand-operated or motor-driven machine to punch or cut out holes in paper.
f poinçonneuse *f*; perceuse *f*; encocheuse *f* — **d** Lochmaschine *f*; Lochstanze *f* — **n** ponsmachine *f* — **e** punzonadora *f* — **i** macchina *f* punzonatrice

11049 punctuation; pointing — Act, practice or system of inserting standardized marks or signs in written matter to clarify the meaning and separate structural units. See also punctuation marks.
f ponctuation *f* — **d** Interpunktion *f*; Zeichensetzung *f* — **n** interpunctie *f*; plaatsen *n* van leestekens — **e** puntuación *f* — **i** punteggiatura *f*

11050 punctuation marks — Standardized marks or signs used in punctuation.
apostrophe '
brackets; square brackets [ ]
colon :
comma ,
double quotes " "
dash (en rule meaning "to") -
dash (em rule) —
ellipsis …
exclamation point !
hyphen -
inverted comma '
parentheses ( )
period, full stop, full point .
question mark, query, interrogation mark ?
semi-colon ;
single quotes ' '.
f signes *mpl* de ponctuation — **d** Satzzeichen *npl*; Interpunktionszeichen *npl* — **n** leestekens *npl* — **e** signos *mpl* de puntuación — **i** segni *mpl* di interpunzione; segni *mpl* di punteggiatura

11051 puncture tester — Device to test the bursting strength more particularly of fibre boards.
f appareil *m* pour l'essai à la résistance à la perforation — **d** Durchstoßprüfer *m*; Durchstoßprüfgerät *n*; Puncture-Tester *m* — **n** doorslag-

toestel *n*; puncture-tester *m* — **e** aparato *m* de ensayo de perforación — **i** apparecchio *m* per prova di perforazione

11052 punk; when room; red letter; filler; CGO — American term for newspaper copy which may be used at a later time when there is any need. CGO means can go over.

11053 purchase for cash — See cash purchase.

11054 pure tape — See clean tape.

11055 purified stand oils — See tekaols.

11056 purity — Generic for 1. excitation and 2. colorimetric purity, also short for excitation purity. Same as chroma (Munsell) and as intensity.
f pureté *f* (colorimétrique) — **d** Echtheit *f* — **n** zuiverheid *f* — **e** pureza *f* (colorimétrica) — **i** purezza *f* (colorimetrica)

11057 Purkinje effect — Decrease in luminosity of reds and oranges relative to blues and greens as the illumination is reduced; associated with the transition from the photopic to the scotopic state.
f effet *m* Purkinje — **d** Purkinje-Effekt *m* — **n** Purkinje-effect *n* — **e** efecto *m* Purkinje — **i** effetto *m* Purkinje

11058 purple no. 3 — A size of wrapping paper. See app. no. 3.

11059 purple no. 4 — A size of wrapping paper. See app. no. 3.

11060 push button; press button
f bouton *m* de contact; bouton *m* poussoir; poussoir *m*; poussette *f* — **d** Druckknopf *m*; Schaltknopf *m* — **n** drukknop *m*; knop *m* — **e** botonera *f*; botón *m* de contacto; botón *m* pulsador; pulsador *m*; interruptor *m* de botón — **i** pulsante *m*

11061 push-button control; button control
f commande *f* automatique par bouton; commande *f* à bouton-poussoir — **d** Druckknopfsteuerung *f*; Knopfsteuerung *f*; Druckknopfschaltung *f* — **n** drukknopbesturing *f* — **e** mando *m* a botón (pulsador); mando *m* por botón (pulsador) — **i** comando *m* a pulsante; comando *m* pulsantiera

11062 push-button switch
f interrupteur *m* à bouton-poussoir — **d** Druckknopfschalter *m*; Druckschalter *m* — **n** drukknopschakelaar *m* — **e** interruptor *m* de botón; llave *f* a botón; pulsador *m* — **i** interruttore *m* a pulsante

11063 push *v* down a risen space
f baisser une espace — **d** einen Spieß niederdrucken — **n** een spatie naar beneden drukken — **e** bajar un blanco subido — **i** abbassare uno spazio

11064 put *v* a strike through — See strike *v* out.

11065 put *v* down — To alter from caps to lower case.
f bas de casse; b.d.c. — **d** in Gemeinen setzen — **n** uit onderkast zetten; in onderkast zetten — **e** poner en minúsculas — **i** comporre in minuscole

11066 put *v* in figures — To print in a number in figures instead of spelling out.
f composer en chiffres — **d** Zahl in Ziffern aus-

drücken — **n** in cijfers zetten — **e** poner con cifras; poner con números — **i** comporre in cifre; mettere in cifre

11067 put *v* to bed; get *v* to bed; send *v* to bed; send *v* to press — To secure a printing forme on to the bed of a printing press.
f caler (la forme) sur le marbre — **d** Druckform auf Fundament befestigen — **n** op de pers nemen; ter perse nemen — **e** entrar en prensa; meter en prensa; poner en prensa — **i** assicurare in macchina

11068 put *v* up — To alter from lower case to caps.
f en capitales; cap — **d** in Versalien setzen — **n** kapitaal zetten — **e** poner de mayúsculas; poner en versales; poner de versales; poner en (letras) mayúsculas; poner de (letras) mayúsculas — **i** comporre in maiuscole

11069 putz pomade — Metal polish used by finishers and etchers on copper after etching or tooling.
f pâte *f* à polir — **d** Putzpomade *f* — **n** poetspommade *fm* — **e** pasta *f* para pulir — **i** pasta *f* da pulire

11070 PVA — See polyvinyl alcohol.

11071 PVAc — See polyvinyl acetate.

11072 PVC — See polyvinyl chloride.

11073 PVDC — See saran.

11074 PVOH — See polyvinyl alcohol.

11075 pycnometer — Device to determine the density of a liquid or solid, having a specific volume and often provided with a thermometer to indicate the temperature of the contained substance.
f pycnomètre *m* — **d** Pyknometer *n* — **n** pyknometer *m* — **e** picnómetro *m* — **i** picnometro *m*

11076 pye — See pie.

11077 pyramid roller — See steel rider.

11078 pyrite; iron pyrite — Mineral containing iron disulphide, used as a source of sulphur by some sulphite pulp manufacturers.
Formula: FeS_2.
f pyrite *f* — **d** Pyrit *m*; Schwefelkies *m*; Eisenkies *m* — **n** pyriet *n* — **e** pirita *f* — **i** pirite *f*

11079 pyroacetic ether — See acetone.

11080 pyrocatechin — See pyrocatechol.

11081 pyrocatechinic acid — See pyrocatechol.

11082 pyrocatechol; pyrocatechinic acid; pyrocatechin; catechol; oxyphenic acid; ortho-dihydroxybenzene — Colourless crystals, soluble in water, alcohol, ether, benzene and chloroform, used in photography, dyestuffs, electroplating, specialty inks, anti-oxidants.
Formula: $C_6H_4(OH)_2$. See also tannic acid.
f pyrocatéchine 1-2 *f*; ortho-dioxybenzène *m* — **d** Pyrokatechin *n*; ortho-Dioxybenzol *n*; Brenzkatechin *n* — **n** pyrocatechine *n*; pyrocatechol *n*; catechol *n*; ortho-dioxybenzeen *n*; dinol *n* — **e** pirocatequina *f*; pirocatecol *m*; catecol *m* — **i** pirocatecolo *m*; pirocatechina *f*; orto-diossi-benzolo *m*

11083 pyrocellulose — See pyroxylin.

11084 pyrogallic acid — See pyrogallol.

11085 pyrogallol; pyrogallic acid; trihydroxy-

benzene — White, lustrous crystals, soluble in water, alcohol, and ether. Used in photography and process engraving. Latin: acidum pyrogallicum.

Formula: $C_6H_3(OH)_3$. See also tannic acid.

f pyrogallol *m*; acide *m* pyrogallique; trioxybenzène *m* — **d** Pyrogallol *n*; Pyrogallussäure *f*; Trioxybenzol *n*; Brenzgallussäure *f* — **n** pyrogallol *n*; pyrogalluszuur *n* — **e** pirogalol *m*; ácido *m* pirogálico — **i** acido *m* pirogallico

11086 pyrometer — Highly sensitive instrument to measure small variations in temperature. Usually used on metal or other surfaces (gravure cylinders).

f pyromètre *m* — **d** Pyrometer *m*; Wärmemeßgerät *n* — **n** pyrometer *m* — **e** pirómetro *m* — **i** pirometro *m*

11087 pyroxylin; pyrocellulose; collodion cotton; collodion wool — Light yellow filaments, soluble in a mixture of ether and alcohol, or other organic solvents precipitated from solution by water. Used in making collodion. It is *nitrocellulose consisting chiefly of cellulose tetranitrate, formula $C_{12}H_{16}(NO_3)_4O_6$. The term is sometimes used for nitrocellulose of relatively low nitrogen content, also for solutions of nitrocellulose in ether/alcohol mixtures.

f pyroxyline *f* — **d** Pyroxylin *n* — **n** pyroxyline *f* — **e** piroxilinia *f*; coloxilina *f* — **i** pirossilina *f*

11088 quad — Placed before a paper size implies multiplying by four, e.g., crown 38.1 x 50.8 cm (15 x 20 in); quad crown 76.2 x 101.6 cm (30 x 40 in). **f** quatre fois — **d** viermal — **n** vier maal — **e** cuádruple; cuatro veces — **i** quadruplo

11089 quad — See quadrat.

11090 quad bag cap — A British size of wrapping paper. The term is becoming obsolete. See app. no. 3.

11091 quad crown — A UK size of printing paper. See app. no. 3.

11092 quad demy — A British standard size for printing paper. See app. no. 3.

11093 quadder; quadding and centring device; centring and quadding device — Device on a slug casting machine to close the right or left vice jaw independently or simultaneously so that a short line will be automatically quadded out either to the left, to the right or centred. **f** dispositif *m* de décadratinage et centrage; appareil *m* à centrer et à cadratiner — **d** Ausschluß- und Zentriervorrichtung *f*; Zeilenfüll- und Zentriervorrichtung *f* — **n** uitvul- en centreerapparaat *n* — **e** centrador *m*; mecanismo *m* cuadrador y centrador; justificador *m* automático; sangrador *m* (automático) — **i** centratore *m* e aggiustatore automatico; dispositivo *m* per la centratura e spaziatura

11094 quadding — Quadding left, right or centre means the quadding, or spacing, functions available on some line casting and film setting equipment. Synonymous with range left, range right, whilst quad centre give a line cast on the centre of the measure of the line. **f** cadratinage *m* — **d** Zeilenfüllen *n* — **n** regels uitvullen *n* — **e** cuadratinado *m* — **i** completamento *m* con quadrati

11095 quadding and centring device — See quadder.

11096 quad foolscap — A British standard size for writing and printing papers. See app. no. 3.

11097 quad foolscap 1 1/2 sheet — A size of writing and printing papers recognized but not standard in Great Britain. See app. no. 3.

11098 quad globe — A size of printing paper. See app. no. 3.

11099 quad imperial — A size of wrapping paper. See app. no. 3.

11100 quad large — A size of cut card. See app. no. 3.

11101 quad large post — A size of writing and drawing papers. See app. no. 3.

11102 quad left; flush left; range left — Typographical command code which signals a typesetting machine to place all material at the left of the line and to fill out the balance of the line with white space. In hot-metal composition, all non printing space within a line is filled out with em quads. In photocomposition, it is merely necessary to expose the graphic images and then to proceed to compose the next line. But the "quad left" concept still remains.

11103 quad medium — A size of printing paper. See app. no. 3.

11104 quad middle; flush left and right — Typographical command code, inserted at a specific location within a line of type, which causes all of the text material to the left of the signal or code to be flushed left and all of the text material to the right of the code to be flushed right, with all extra space in the centre of the line. See also quad left, quad right.

11105 quad post — A size of printing paper. See app. no. 3.

11106 quad pott — A size of printing paper, recognized but not standard in Great Britain. See app. no. 3.

11107 quadrant paper scale; paper scale — Scale to indicate the weight of a sheet of paper in grams per square metre, or of a ream (500 sheets) or of 1000 sheets of the same size as the sample being weighed. **f** balance *f* à papier; balance *f* à secteur — **d** Papierwaage *f*; Quadrant-Papierwaage *f*; Quadrantenwaage *f* — **n** papierweegschaal *fm*; papierweger *m* — **e** balanza *f* de cuadrante para papel — **i** bilancia *f* a settore per carta

11108 quadrat; quad — In hand composition, a piece of metal lower than type, used for spacing or filling in larger areas of white space. Usual sizes: en, em, 2, 3 and 4 ems. In mechanical composition, metal blanks or non-imaged brass matrices. **f** cadrat *m* — **d** Quadrat *n* — **n** kwadraat *n* — **e** cuadrado *m* — **i** quadrato *m*

11109 quadrichromatic; quadricolour — Consisting of four colours. **f** en quadrichromie — **d** vierfarbig — **n** in vier kleuren; vierkleurig — **e** en cuatro colores — **i** in quattro colori

11110 quadrichromie See four-colour process.

11111 quadricolour — See quadrichromatic.

11112 quadricolour process — See four-colour process.

11113 quad right; flush right; range right — Opposite of *quad left, placing the text to the right and filling out the beginning of the line with white space.

11114 quadrillé paper — See squared paper.

11115 quadrillé ruled paper — See squared paper.

11116 quadrivalent — See tetravalent.

11117 quad royal — A size of printing and wrapping paper. See app. no. 3.

11118 quad small — A size of cut card. See app. no. 3.

11119 quad small demy — A British standard size of writing and printing papers. See app. no. 3.

11120 quaint character — Ligature or type with a decorative form.

f ligature *f* décorative — **d** verzierte Ligatur *f* — **n** decoratieve ligatuur *f* — **i** ligatura *f* decorativa

11121 quality control — Use of statistical methods to evaluate the quality levels and variations during manufacturing of a product.

f contrôle *m* de la qualité — **d** Qualitätskontrolle *f*; Gütekontrolle *f*; Beschaffenheitsprüfung *f* — **n** kwaliteitszorg *fm* — **e** control *m* de calidad — **i** controllo *m* di qualità

11122 quality control testing; routine control testing — Examination and testing of in-coming materials (paper, ink, blocks, etc.) before printing.

f contrôle *m* de réception — **d** Eingangskontrolle *f*; Materialprüfung *f* — **n** ingangscontrole *f* — **e** control *m* de recepción de los materiales; ensayo *m* de materiales; comprobación *f* de materiales — **i** controllo *m* dei materiali in entrata; controllo *m* di qualità dei materiali in entrata

11123 quantity of heat

f quantité *f* de chaleur — **d** Wärmemenge *f* — **n** hoeveelheid *f* warmte — **e** cantidad *f* de calor — **i** quantità *f* di calore

11124 quantity of light

f quantité *f* de lumière — **d** Lichtmenge *f* — **n** hoeveelheid *f* licht; lichthoeveelheid *f* — **e** cantidad *f* de luz — **i** quantità *f* di luce

11125 quantize *v* — To subdivide the range of values of a variable into a finite number of non-overlapping and not necessarily equal, subranges or intervals, each of which is represented by an assigned value within the subrange.

f quantifier — **d** quantisieren — **n** quantiseren — **e** cuantizar — **i** quantizzare

11126 quart, dry — See dry quart.

11127 quarter — Unit of weight avoirdupois. See app. no. 7 and 7a.

11128 quarter binding with raised bands — See combination style.

11129 quarter bound — Books bound with back only in leather with cloth or board sides.

f relié en demi-cuir — **d** im Halblederband — **n** halfleer gebonden; in halfleren band — **e** encuadernado a media piel; encuadernación *f* holandesa — **i** legato in mezza pelle

11130 quarter cloth binding

f reliure *f* demi-toile — **d** Halbleinenband *m* — **n** halflinnen band *m* — **e** encuadernación *f* semi-tela; encuadernación *f* con lomo de tela — **i** legatura *f* mezzatela; legatura *f* in mezza tela

11131 quarter flush — In bookbinding, covering 1/5-1/8 width of board. Materials usual cloth for backs and paper for sides.

11132 quarter fold — See chopper fold.

11133 quarter-joint case — Case of which the dimensions of the board and backliner are such that the width of each of the unattached cloth strips left between these components is equal to one quarter of the total thickness of the book to obtain a flat-opening book. The strips are not pasted down to the book.

11134 quarter leather — In bookbinding, leather covering 1/5-1/8 in width of board, the remainder being cloth or paper.

f demi-cuir; mi-cuir — **d** halbleder — **n** halfleer — **e** media piel; de media pasta — **i** in mezza pelle

11135 quarter leather binding

f reliure *f* demi-peau — **d** Halblederband *m*; Halbledereinband *m* — **n** halfleren band *m* — **e** encuadernación *f* de media pasta — **i** legatura *f* in mezza pelle

11136 quarterly — Periodical issued once every three months.

f revue *f* trimestrielle; revue *f* par trimestre — **d** Quartalschrift *f* — **n** kwartaaluitgave *f*; kwartaalblad *n*; eens per drie maanden verschijnend blad *n* — **e** publicación *f* trimestral; revista *f* trimestral — **i** rivista *f* trimestrale

11137 quarterly

f trimestriel — **d** vierteljährlich — **n** per kwartaal; driemaandelijks — **e** trimestral — **i** trimestrale

11138 quarter page folder — Supplementary device to the *folder to give a third fold in line with the run of the web.

f plieuse *f* trois plis — **d** Einrichtung *f* für den dritten Falz — **n** inrichting *f* voor de derde vouw — **e** plegadora *f* de tres pliegues — **i** dispositivo *m* di terza piega

11139 quarter sized — See also hard-sized paper.

f peu collé — **d** schwachgeleimt; viertelgeleimt — **n** zwak gelijmd — **e** encolado ligero — **i** un quarto colla

11140 quarter tone — In advertising, to imitate scraperboard work for reproduction in newspapers or in commercial magazine printing on poor stock. The original copy, usually an artist's sketch or a monochrome bromide print, is photographed with a coarse screen. A negative is made without using the right angle prism and to a higher contrast (50%) than for normal halftone negatives. An enlarged bromide print is then made from the screen negative which is retouched. This is then reproduced as a normal line block.

f simili *f* à grosse trame imprimé comme cliché au trait — **d** Grobrasterautotypieklischee *n* gedruckt als Strichätzung — **n** grofrasterautotypie *f* als lijncliché gedrukt — **e** grabado *m* directo de trama gruesa; autotipia *f* de periódicos de trama gruesa — **i** lastra *f* di stampa retinata grossolanamento; cliscè *m* retinato grossolanamento

11141 quarter vellum binding

f reliure *f* demi-parchemin — **d** Halbpergamentband *m* — **n** halfperkamenten band *m* — **e** encuadernación *f* de medio pergamino — **i** legatura *f* in mezza pergamena

11142 quart, liquid — See liquid quart.

11143 quarto — Old term for a book made from sheets which have been folded twice, each sheet forming four leaves or eight pages. Sometimes applied to any book measuring about 22 x 30 cm (9 x 12 in).

f in-quarto; in-4 — **d** Quart *n*; Quartformat *n* — **n** kwarto *n*; kwarto-formaat *n*; in vieren — **e** en cuarto — **i** in quarto; quartino

11144 quartz lamp — Lamp consisting of an ultra-violet source, as mercury vapour, contained in a fused silica bulb that transmits ultra-violet light with little absorption. (Quartz, formula SiO_2, is crystallized silicon dioxide.)
f lampe *f* (à tube) de quartz; tube *m* de quartz — **d** Quarzlampe *f* — **n** kwartslamp *fm* — **e** lámpara *f* de cuarzo — **i** lampada *f* al quarzo

11145 quaternion — Four gathered sheets folded in two for binding together. See also ternion.
f quaternion *m* — **d** Quaternio *n*; Lage *f* von vier Bogen — **n** katern *fm* van vier — **e** cuaderno *m* de cuatro hojas — **i** quaternio *m*

11146 queen's — A size of note paper. See app. no. 3.

11147 query — On manuscript or proof, a question addressed to the author or editor. Abbrev. qy.
f interrogation *f* — **d** Frage *f* — **n** vraag *fm* — **i** domanda *f*

11148 questionary; questionnaire — Form with questions to be answered. In fact questionary is an adjective; questionnaire is a list of questions.
f questionnaire *m* — **d** Fragebogen *m*; Umfrageblatt *n* — **n** vragenlijst *fm* — **e** cuestionario *m* — **i** questionario *m*

11149 question mark — See interrogation mark.

11150 questionnaire — See questionary.

11151 quick drying — See fast drying.

11152 quick-drying ink; quick-setting ink; quickset ink; fast-drying ink — Inks for letterpress and offset that dry by filtration, coagulation, selective absorption or often a combination of these with other drying methods. The vehicles are generally special resin/oil combinations which, after the ink has been printed, separate into a solid material which remains on the surface as a dry film and an oily material which penetrates rapidly into the stock.
f encre *f* à séchage rapide; encre *f* à prise rapide; encre *f* très siccative; encre *f* quick-set — **d** Schnelltrockenfarbe *f*; schnelltrocknende Farbe *f*; schnellabbindende Farbe *f* — **n** sneldrogende inkt *m*; quick-set ink *m* — **e** tinta *f* de secado rápido; tinta *f* de fraguado rápido — **i** inchiostro *m* a rapida essiccazione; inchiostro *m* a rapida presa; inchiostro *m* quick-set

11153 quick-drying varnish
f vernis *m* à séchage rapide — **d** Schnelltrockenlack *m*; schnelltrocknender Lack *m* — **n** sneldrogende lak *mn* — **e** barniz *m* de secado rápido — **i** vernice *f* a rapido essiccamento

11154 quick etch — See powderless etching.

11155 quick lime — See calcium oxide.

11156 quick-set ink — See quick-drying ink.

11157 quick-setting ink — See quick-drying ink.

11158 quicksilver — See mercury.

11159 quinhydrone electrode — Half cell with a platinum electrode, immersed in a saturated quinhydrone, formula $C_6H_4O_2C_6H_4(OH)_2$, solution; used for measuring pH in neutral or acid solutions.
f demi-cellule *f* à la quinhydrone; électrode *f* à la quinhydrone — **d** Chinhydronhalbzelle *f* — **n** chinhydron-elektrode *f* — **e** electrodo *m* de quinhidrona; semicelda *f* de quinhidrona — **i** elettrodo *m* al chinidrone

11160 quinol — See hydroquinone.

11161 quinone; benzoquinone; chinone — Yellow crystals, soluble in alcohol, ether and alkalis, slightly soluble in hot water. Formed by oxidizing aniline or hydroquinone. Used in the manufacturing of dyes and in photography. TL-value: 0.1 ppm.
Formula: $C_6H_4O_2$.
f quinone *f* — **d** Chinon *n* — **n** chinon *m* — **e** quinona *f* — **i** chinone *m*

11162 quinternion — Five gathered sheets folded in two for binding together. See also ternion.
f quinternion *m* — **d** Quinternio *n*; Lage *f* von fünf Bogen — **n** katern *fm* van vijf — **e** quinterno *m* — **i** quinterno *m*

11163 quire — In UK the twentieth part of a ream, 24 sheets and one "outside" sheet making 25. In US 25 sheets in the case of a 500-sheet ream of fine papers, and 24 sheets in the case of a 480-sheet ream of coarse papers.
f main *f* de papier — **d** Lage *f* — **n** boek *n* — **e** mano *f* de papel — **i** quinterno *m* di carta; mazzetta *f* di fogli

11164 quire — A pre-determined number of copies, usually 26.

11165 quired paper — Reams folded in quires, not flat (in sections of twelve).
f papier *m* plié par mains; papier *m* façonné — **d** in Lagen gefalztes Papier *n*; fassoniertes Papier *n* — **n** opgemaakt papier *n*; in katerns gevouwen papier *n* — **e** papel *m* empaquetado en postetas dobladas — **i** carta *f* piegata in quinterni; carta *f* piegata in mazzette

11166 quires, in — See in sheets.

11167 quire spacing — Indication of a pre-determined number of copies by separation on a rotary press.
f espacement *m* — **d** Zählabstand *m* — **n** aftelafstand *m* — **e** separación *f* en postetas contadas — **i** conteggio *m* copie con distanziamento gruppi

11168 quod vide — Latin for "which see". Abbrev. qv.
f voir ce mot — **d** welches nachzusehen — **n** zie aldaar — **e** véase (el original) — **i** ecco qua; quod vide

11169 quoin — Metal wedges, formerly of wood, or expanding metal boxes, to tighten or lock up printing formes.
f coin *m* de serrage; cale *f* — **d** Schließkeil *m*; Schließzeug *n* — **n** sluitstuk *n* — **e** cuña *f* — **i** cuneo *m* di serraggio; cuneo *m* di serrare; cuneo *m* serraforma

11170 quoin key — T-shaped piece of iron to operate metal quoins in locking up type formes.
f clef *f* à béquille — **d** Formenschlüssel *m* — **n**

vormsleutel *m*; kooisleutel *m* — **e** llave *f* de acuñar; llave *f* de tuerca — **i** chiave *f* per serratura di chiusura

11171 quoin *v* **up** — To lock up a printing forme.
f serrer — **d** schließen; antreiben — **n** opsluiten; opkooien — **e** acuñar; cerrar — **i** serrare

11172 quotation — Statement or passage cited or repeated as the utterance of some other speaker or writer.
f citation *f* — **d** Zitat *n*; Anführung *f* — **n** aanhaling *f*; citering *f*; citaat *n* — **e** cita *f* — **i** citazione *f*

11173 quotation — Current or estimated selling price of commodities.
f devis *m* — **d** Kostenanschlag *m* — **n** prijsnotering *f*; notering *f* — **e** cotización *f* — **i** quotazione *f*

11174 quotation mark; quotes; inverted commas; turned commas; French quotes; duckfoot quotes — In English composition, two up-turned commas at the beginning and two apostrophes at the end of a quotation. In French composition, two right-handed hooks ≪ at the beginning and two left-handed hooks ≫ at the end. The apostrophes at the end of the quotation should come before all punctuation marks, unless these form part of the quotation itself. Quotes are to be used when citing titles of articles in magazines, chapters of books, essays, poems and songs. See also punctuation marks.
f guillemets *mpl* — **d** Anführungszeichen *npl*; Gänsefüßchen *npl*; englische Anführungszeichen *npl* — **n** aanhalingstekens *npl* — **e** vírgulas *fpl*; virgulillas *fpl*; entrecomillas *fpl*; comillas *fpl* francesas; comillas *fpl* de estilo francés; comillas *fpl* invertidas; comillas *fpl* de trinchante; guillemets *mpl*; comas *fpl* vueltas; comas *fpl* invertidas — **i** virgolette *fpl*

11175 quotation quads; quotations; cup-cast quads; hollow (cast) quads — Hollowed out pieces of metal to fill open spaces in a page of type. Made in various sizes, 24 x 36, 24 x 48, 36 x 48 points, etc. See also French furniture.
f cadrats *mpl* creux; cadrats *mpl* évidés; lingots *mpl* à noyaux; lingots *mpl* à cavités — **d** Hohlstege *mpl*; Hohlgußquadrate *npl*; Nüsse *fpl* — **n** holwit *n*; holle kwadraten *npl* — **e** cuadrados *mpl* huecos; cuadrados *mpl* de imposición — **i** quadrati *mpl* vuoti; lingotti *mpl* cavi

11176 quotations — See quotation quads.

11177 quote *v* — To enclose within quotation marks. When such a mark has to be set at the beginning of a sentence, the copyholder, when reading to the proofreader, must say "quote" or "commence quote", and at the end of the matter "close quote".
f mettre entre guillemets; guillemeter — **d** anführen; in Gänsefüßchen setzen; mit Anführungszeichen versehen — **n** aanhalen; tussen aanhalingstekens plaatsen — **e** poner entre vírgulas; poner entre virgulillas; poner entre comillas; poner entre comillado; entrecomillar — **i** mettere tra virgolette

11178 quote *v* — See cite *v*.

11179 quotes — See quotation mark.

11180 qwert layout; qwerty layout — Typewriter keyboard layout, with the q, w, e, r, t-keys at the upper left hand row of the keys. See also azerty layout, Dvorak keyboard.
f disposition *f* standard du clavier (de machine à écrire) — **d** qwert-Einteilung *f* — **n** qwertindeling *f* — **e** colocación *f* qwert; indicación *f* tipológica de los textos — **i** collocazione *f* qwert; collocazione *f* standard dei tasti

11181 qwerty layout — See qwert layout.

11182 rack — Framework with sloping top to hold type cases.
f rang *m* — **d** Setzpult *n*; Setzregal *n* — **n** zetbok *m*; bok *m* — **e** chibalete *m* — **i** bancone *m* da composizione; banco *m* da composizione; telaio *m* da composizione

11183 rack; rack bar — Bar with teeth on one face for gearing with a pinion or worm gear.
f crémaillère *f* — **d** Zahnstange *f* — **n** tandreep *m*; tandheugel *m* — **e** cremallera *f*; sierra *f* — **i** cremagliera *f*

11184 rack — See tool rack.

11185 rack *v* — To place thin lifts of paper on racks for easy access of air.
f étendre — **d** wegsetzen — **n** uitleggen op rekken — **e** poner en el tendedor; tender en el ambientador — **i** stendere ad aciugare

11186 rack and pinion — Bar with teeth on one of its sides, adapted to engage with the teeth of a pinion or the like, as for converting circular into rectilinear motion or vice versa.
f crémaillère *f* et pignon *m* — **d** Zahnstange *f* und Ritzel *n* — **n** heugel *m* met tandwiel — **e** cremallera *f* y piñón *m* — **i** cremagliera *f* e pignone *m*

11187 rack bar — See rack.

11188 rack, drying — See drying rack.

11189 rack hanger — Frame on a printing press which carries the driving mechanism bolted to the underside of the bed.
f porte *m* crémaillère — **d** Zahnstangerahmen *m* — **n** tandheugelframe *n*; tandheugelfreem *n* — **e** bastidor *m* de la cremallera — **i** telaio *m* della cremagliera

11190 radial — Arranged as lines from centre to circumference of a circle.
f radial — **d** radial — **n** radiaal; straalsgewijze — **e** radial — **i** radiale

11191 radiant energy — Energy which is transferred by electromagnetic waves without a corresponding transfer of matter.
f énergie *f* radiante; énergie *f* de rayonnement — **d** Strahlungsenergie *f* — **n** stralingsenergie *f* — **e** energía *f* radiante; energía *f* de radiación — **i** energia *f* di radiazione

11192 radiant flux; radiant power — Time rate of transfer of radiant energy.
f flux *m* énergétique; flux *m* de rayonnement — **d** Strahlungsfluß *m* — **n** stralingsstroom *m*; stralingsflux *m* — **e** flujo *m* radiante; flujo *m* de radiación — **i** flusso *m* di radiazione

11193 radiant power
See radiant flux.

11194 radiation — Emission of energy as light, heat, etc. The transfer of energy through space by electromagnetic waves.
f radiation *f*; rayonnement *m* — **d** Strahlung *f*; Ausstrahlung *f* — **n** straling *f*; uitstraling *f* — **e** radiación *f* — **i** radiazione *f*

11195 radical — Group of atoms which, though not normally able to exist in a separate state, can pass unchanged through a reaction or a series of reactions. Ethyl, C_2H_5 (valence 1), and sulphate, SO_4 (valence 2), are typical organic and inorganic radicals, respectively.
f radical *m* — **d** Radikal *n* — **n** radicaal *m* — **e** radical *m* — **i** radicale *m*

11196 radical sign — See root sign.

11197 radioactive — Having the ability to emit alpha, beta and gamma rays.
f radioactif — **d** radioaktiv — **n** radio-actief — **e** radioactivo — **i** radioattivo

11198 radius — Line segment extending from the centre of a circle to the curve.
f rayon *m* — **d** Radius *m*; Halbmesser *m* — **n** straal *fm* — **e** radio *m* — **i** raggio *m*

11199 rag; rag paper; trumpery paper — Insignificant journal. Newspaper of low repute.
f feuille *f* de chou; canard *m* — **d** Käseblatt *n*; Käseblättchen *n*; Wurstblatt *n*; Wurstblättchen *n* — **n** prulkrant *fm*; prulblad *n*; plaatselijk blaadje *n* — **e** diarucho *m*; periodicucho *m*; periodiquito *m* — **i** gazzetta *f* poco importante; giornalucolo *m*

11200 rag-content paper — See rag paper.

11201 rag duster — See duster.

11202 ragged etching; dirty-bottom etching — Defect resulting when the acid resistant powder adheres to the sides of the halftone dot during etching; in powderless etching caused by improper paddle speeds.
f gravure *f* galeuse — **d** unsaubere Ätzung *f*; Ätzung *f* mit rauhem Ätzgrund; Ätzung *f* mit unsauberem Rasterpunktprofil — **i** incisione *f* a fondo sporco

11203 ragged old book
f bouquin *m* — **d** Schmöker *m*; altes Buch *n* — **n** oud boek *n* — **e** libraco *m*; libracho *m* — **i** vecchio libro *m*

11204 ragged-right setting — See unjustified setting.

11205 rag paper; all-rag paper — Paper made entirely from rags, especially cotton rags. See app. no. 4. See also rag-content paper.
f papier *m* pur chiffon; papier *m* pur fil; papier *m* de chiffons — **d** Hadernpapier *n*; Lumpenpapier *n*; reines Hadernpapier *n* — **n** papier *n* uit zuiver lompen; lompenpapier *n* — **e** papel *m* puro de trapos — **i** carta *f* di puri stracci

11206 rag paper; rag-content paper — Paper containing a minimum of 25% rag or cotton fibre, generally made in the following grades: 25, 50, 75, 100% and extra no. 1 (100%). Used for bonds, currency, writing, ledgers, manifold and onion skin, papeteries and wedding, index, carbonizing, blueprint, and other reproduction papers, maps and charts and other industrial specialties. See app. no. 4.

f papier *m* mi-chiffons; papier *m* de chiffons — **d** hadernhaltiges Papier *n*; Hadernpapier *n*; Lumpenpapier *n* — **n** lompenhoudend papier *n*; lompenpapier *n*; papier *n* gedeeltelijk uit lompen — **e** papel *m* de trapos — **i** carta *f* con stracci; carta *f* di straccio

11207 rag paper — See rag.

11208 rag pulp — Pulp made by disintegrating new or old cotton or linen rags, and cleaning and bleaching the fibres; used principally in making high grade bond, ledger and writing papers, and papers required for permanent record purposes.
f pâte *f* de chiffons — **d** Halbstoff *m* aus Lumpen; Lumpenstoff *m* — **n** lompencelstof *fm*; lompenstof *fm* — **e** pasta *f* de trapos; pulpa *f* de trapos — **i** pasta *f* di stracci

11209 rag-right setting — See unjustified setting.

11210 rags — Cotton or linen fabrics used for paper making. New rags from textile manufacturers are known as rags cuttings, while rags gathered by rag dealers are called old rags, which have to be sorted into grades and quality. See also mixed rags.
f chiffons *mpl*; peille *f* — **d** Lumpen *mpl*; Hadern *mpl* — **n** lompen *fmpl* — **e** trapos *mpl* — **i** stracci *mpl*

11211 rags cuttings — See rags.

11212 rag sorting
f triage *m* des chiffons — **d** Hadernsortierung *f* — **n** sorteren *n* van lompen — **e** escogido *m* de trapos; clasificación *f* de trapos — **i** cernita *f* degli stracci

11213 rag thrasher — See duster.

11214 rag thrasher — See willow.

11215 rag willow — See willow.

11216 rail — Matrices for linecasting usually contain two (duplexed) characters. Duplex characters are referred to a upper rail and lower rail, either of which may be selected by the operator. On tape controlled line casters the first upper rail character must be preceded by a unique code and the last one followed by a lower rail code except at the end of a line where the lower rail condition is automatic. The lower rail character is usually roman face and the upper rail character is the italic or bold face counterpart of the roman face. This necessarily restricts the width of the italic or bold face characters. Small capitals are often duplexed with figures, reference marks and ligatures.

11217 railroad *v* — To send unedited copy to the composing room with great or undue speed, without having read it. Newspaper term.
f expédier la copie au marbre sans la relire ou la préparer — **d** ungelesenes, unredigiertes Manuskript eiligst in Setzerei schicken — **n** nietgelezen kopij doorgeven — **e** usar sin corregir — **i** inoltrare urgentemente alla sala composizione

11218 railroading — Improper printing of a continuous mark or line in the plain area of a design, often caused by a definite marking or scratching of the engraved cylinder, sometimes by particles

lodged behind the doctor blade.
f raies *fpl* d'impression; stries *fpl* de la racle — **d** Rakelstreifen *mpl* — **n** rakelstrepen *fmpl* — **e** mellas *fpl* de la racleta limpiadora de huecograbado — **i** rigature *fpl* di stampa rotocalco

11219 railroad manila; railroad writing; manila writing paper — Paper containing a substantial quantity of mechanical pulp (book grade) mixed with long-fibred chemical pulp, sized for pen and ink writing and usually supplied in a canary colour. Commonly used in school tablets, for printing business forms, for advertising throw-aways and flyers, for second sheets in typing, and in manufacturing sales books, order blanks, etc. The usual weights are 32 to 45 pounds (24 x 36 - 500).
f papier *m* bulle administration — **d** Manilafarbiges Schreibpapier *n* — **n** manilla-gekleurd schrijfpapier *n* — **e** papel *m* manila para escribir — **i** carta *f* da scrivere manila

11220 railroad tracking — Paper drying defect. Surface defect of blade coated paper, indicated by long striations in the machine direction when viewed in low angle light. It frequently shows up as variations in ink receptivity when printed. Striations may vary in width, wander back and forth or interconnect. Iodine or NH_4Cl burnout, ink wipe or print tests show the striations as a variation in ink receptivity, or as a variation in binder concentration at the surface of the sheet.
f trainées *fpl* de couchage — **d** Linien *fpl* geringen Druckfarbeaufnahmevermögens infolge Übertrocknung — **i** striature *fpl* della carta patinata a lama

11221 railroad writing — See railroad manila.

11222 rainbow printing — See split-colour printing.

11223 raised bands; ribs — Raised horizontal ridges across backbone of book.
f nerfs *mpl*; nervures *fpl* — **d** Rippen *fpl*; erhebene Bünde *mpl* — **n** ribben *fmpl* — **e** nervuras *fpl*; nervios *mpl*; nervaduras *fpl* — **i** nervi *mpl*; nervature *fpl*; rialzi *mpl*

11224 raised initial — See cock-up initial.

11225 raised printing — See thermography.

11226 raised space
f espace *f* levée — **d** gestiegenes Ausschlußstück *n* — **n** gerezen spatie *f* — **e** espacio *m* levantado — **i** spazio *m* alzato; spazio *m* sollevato

11227 RAM — See random access store.

11228 ramie — Plant of the nettle family native to tropical Asia, cultivated in other warm regions. The botanical name is Boehmeria nivea, especially important is variety tenacissima. The best fibres from the decorticated material is commercially known as China grass which is used as a textile fibre. Potential source of papermaking fibres, the tensile strength being four times that of flax.
f ramie *f* — **d** Ramie *f*; Ramiefaser *f* — **n** ramie *m*; ramee *m* — **e** ramio *m* — **i** ramié *m*

11229 random — Composing frame (desk) used by compositors when making up, and also for keeping standing lines of matter for distribution.

f rang *m* plat — **d** Metteur-Regal *n*; Umbruchtisch *m* — **n** loket *n*; opmaaktafel *fm* — **e** tablero *m* portamoldes — **i** bancone *m* da composizione

11230 random — Having no definite aim or purpose; made, done, occuring at haphazard.

f aléatoire — **d** zufällig — **n** toevallig — **e** aleatorio — **i** aleatorio

11231 random access; direct access — A method of finding information in a body of data without having to search the data base sequentially (from front to back or back to front) but to go more directly to the location of the desired information (the file or segment thereof) according to an index or map that enables the computer to bring the desired data into memory. Any amount of data can be accessed in this manner from a relatively small element (e.g., track and sector of a disc) to the entire body of stored information.

f accès *m* sélectif — **d** direkter Zugriff *m* — **n** direkte toegang *m*; rechtstreekse toegang *m* — **e** acceso *m* directo; acceso *m* al azar — **i** accesso *m* diretto

11232 random access storage — See random access store.

11233 random access store *GB*; **random access storage** *US*; **RAM** — Solid-state substitute for core memory, unlike core memory, generally with provision for standby power to keep the memory alive even when the machine is turned off, or with ROM (read only memory) bootstraps which can reload programs at the touch of a button when power is restored.

f mémoire *f* à libre accès — **d** Randomspeicher *m* — **n** random-geheugen *n*; uniform toegankelijk geheugen *n* — **e** memoria *f* de acceso al azar — **i** memoria *f* d'accesso casuale

11234 random display — Typography in which the type size and face varies and is not logically related to the data.

11235 random sample — In statistics, a sample selected without bias, in a haphazard manner, so that every item has an equal chance of inclusion.

f échantillon *m* aléatoire; échantillon *m* au hasard — **d** Zufallstichprobe *f* — **n** steekproefmonster *n* — **e** muestra *f* de escogido al alzar — **i** prova *f* a caso; campione *m* prelevato a caso

11236 range — In statistics, the difference between the maximum and the minimum observations, i.e. the distance between extreme observations.

f étendue *f* — **d** Variationsweite *f*; Spannweite *f*; Variationsspreite *f* — **n** bereik *n* — **e** alcance *m* — **i** intervallo *m*; portata *f*

11237 range *v* — To align types of different sizes.

f parangonner — **d** in Linie bringen; unterlegen — **n** in de lijn brengen; in de lijn plaatsen — **e** parangonar — **i** paragonare

11238 range left — See quad left.

11239 range of press — Minimum to maximum width of web and length of repeat.

f format *m* de la presse; format *m* de la rotative

— **d** Maschinenformat *n* — **n** machineformaat *n* — **e** tamaño *m* de la máquina — **i** formato *m* della macchina

11240 range right — See quad right.

11241 ranging figures — See lining figures.

11242 rape oil — See rapeseed oil.

11243 rapeseed oil; rape oil — Pale yellow, very viscous liquid, soluble in ether, chloroform and carbon disulphide. A slow drying oil used as a printing ink vehicle.

f huile *f* de colza; (huile *f* de) navette *f* — **d** Rüböl *n*; Rapsöl *n* — **n** raapolie *fm* — **e** aceite *m* de colza — **i** olio *m* di colza; olio *m* di ravizzone

11244 rapid etching — See powderless etching.

11245 raster — Pattern of scanning lines covering the area upon which the image is projected in the cathode ray tube of a television screen.

d Raster *m*; Rasternetz *n* — **n** raster *n*

11246 raster scanner — Hardware required to perform the scanning function.

11247 ratched wheel — See ratchet wheel.

11248 ratchet — See ratchet wheel.

11249 ratchet wheel; ratched wheel; ratchet — Wheel with teeth on the edge, into which a pawl drops or catches to prevent reversal of motion or convert reciprocating motion into rotatory motion.

f roue *f* à rochet; roue *f* à cliquet; roue *f* à chien; roue *f* encliquetée; roue *f* d'encliquetage — **d** Sperrad *n*; Ratsche *f* — **n** palwiel *n*; palrad *n* — **e** rueda *f* de trinquete — **i** ruota *f* a cricchetto; ruota *f* a nottolino

11250 rate of shear — Rheological term. The velocity of one plane with respect to another plane divided by the perpendicular distance between them. If the rate of shear between the planes is not constant then it is expressed by the ratio dv/dr. The Newton model is an example of the first case. Couette flow (rotational viscometer) is an example of the second case.

f vitesse *f* de cisaillement — **d** Schergeschwindigkeit *f*; Schubgeschwindigkeit *f* — **n** afschuifsnelheid *n* — **e** velocidad *f* de cizallamiento — **i** velocità *f* di cesoiamento

11251 rating scale — See scale of rating.

11252 ratio — Relative size or value. The ratio of 6 to 10 is 0.6 or 6/10. The ratio of 10 to 6 is 1.66 or 10/6.

f rapport *m* — **d** Verhältnis *n* — **n** verhouding *f* — **e** razón *f* — **i** rapporto *m*

11253 ratiometer — See filter ratio scale.

11254 rattle; ring — Crackling sound produced by crumpling or snapping paper to ascertain its stiffness.

f sonnant *m*; carteux *m* — **d** Klang *m*; Rascheln *n* — **n** klank *m* — **e** sonido *m* del papel; carteo *m* del papel; crujido *m* del papel — **i** sonorità *f* della carta; incarto *m* della carta; fruscio *m*

11255 rattling — Shaking of the rollers in their bearers.

f ballottement *m* — **d** Rattern *n* — **n** slaan *n*; trillen *n* — **e** golpeteo *m* — **i** scuotimento *m*;

sbattimento *m*; vibrazione *f*

11256 raw data — Data which have not been processed in a computer.

f données *fpl* brutes — **d** Ursprungsdaten *pl* — **n** oorspronkelijke gegevens *npl* — **e** datos *mpl* sin procesar — **i** dati *fpl* bruti

11257 raw linseed oil — Untreated raw oil from the flaxseed presses which is filtered through duck and flanel filter cloths in a plate and frame press. Yellow-brown or amber.

f huile *f* de lin crue; huile *f* de lin brute — **d** Rohleinöl *n*; rohes Leinöl *n* — **n** rauwe lijnolie *fm*; ongekookte lijnolie *fm* — **e** aceite *m* de linaza crudo; aceite *m* de lino crudo — **i** olio *m* di lino grezzo

11258 raw materials — Products in the manufacture of finished goods, e.g., in paper: cellulose fibres, sizing compounds, colouring substances, fillers; in printing: paper and ink.

f matières *fpl* premières; matériaux *mpl* bruts — **d** Rohstoffe *mpl* — **n** grondstoffen *fmpl* — **e** materia *f* prima — **i** materia *f* prima

11259 raw stock — See body paper.

11260 raw tape — See unjustified (paper) tape.

11261 reaction — Chemical action between two substances forming one or more different substances.

f réaction *f* — **d** Reaktion *f* — **n** reactie *f* — **e** reacción *f* — **i** reazione *f*

11262 read *v*; **proofread** *v GB*; **correct** *v US* — To mark errors on a proof. See also correct *v*.

f corriger — **d** korrekturlesen; berichtigen; beheben — **n** corrigeren; proeven lezen — **e** corregir; enmendar; rectificar; revisar; leer; cotejar; apuntar — **i** correggere (bozze); leggere le bozze

11263 read *v* — To extract or copy data from a record or signal.

f lire — **d** lesen — **n** lezen — **e** leer — **i** leggere

11264 readability

f lisibilité *f* — **d** Lesbarkeit *f* — **n** leesbaarheid *f* — **e** legibilidad *f* — **i** leggibilità *f*

11265 reader — Device to convert information in one form of storage to information in another form. See also card reader, magnetic tape reader, tape reader.

f lecteur *m*; appareil *m* de lecture — **d** Leser *m*; Lesegerät *n*; Leseapparat *m* — **n** lezer *m*; leesapparaat *n* — **e** lectora *f*; unidad *f* lectora; aparato *m* lector — **i** lettore *m*; apparecchio *m* lettore

11266 reader — See proofreader.

11267 reader — See publisher's reader.

11268 reader's box — See reading-room.

11269 readers' marks — See correction marks.

11270 reader's proof; proofreader's proof; first proof — First proof sent to the proofreader.

f épreuve *f* en première — **d** Hauskorrektur *f*; erste Korrektur *f*; erster Abzug *m* — **n** vuile proef *fm*; eerste proef *fm*; huiscorrectie *f* — **e** prueba *f* de imprenta; prueba *f* de primera; prueba *f* tipográfica — **i** prima bozza *f* preliminare

11271 read *v* **in** — To input data into a computer system, as from paper tape, magnetic tape, cassette, or disk, or to take observations which can be processed in machine-readable form.

n inlezen

11272 reading for press — Final reading before printing.

f lecture *f* en tierce; correction *f* en tierce — **d** letzte Korrektur *f* — **n** laatste correctie *f* — **e** última corrección *f*; corrección *f* de las pruebas finales — **i** correzione *f* ultima

11273 reading rate; reading velocity — Velocity with which a reading station of a computer senses data.

f vitesse *f* de lecture — **d** Lesegeschwindigkeit *f* — **n** leessnelheid *f* — **e** velocidad *f* de lectura — **i** velocità *f* di lettura

11274 reading-room; proofroom; reader's box — Room in newspapers and printing offices where the proofreaders work.

f salle *f* de correction (des épreuves) — **d** Korrektorzimmer *n*; Lesesaal *m* — **n** correctiekamer *fm* — **e** oficina *f* de correctores; sala *f* de corrección — **i** sala *f* di correttori; camera *f* di correttori

11275 reading-room — Room for reading as in a library or club.

f salle *f* de lecture — **d** Lesesaal *m* — **n** leeszaal *fm* — **e** sala *f* de lectura; salón *m* de lectura — **i** sala *f* di lettura

11276 reading station; sensing station — Part of a card track where the data on a punched card is read.

f station *f* de lecture — **d** Abfüllstation *f*; Lesestation *f* — **n** leesorgaan *n*; leesstation *n* — **e** estación *f* de lectura; estación *f* palpadora — **i** stazione *f* di lettura; stazione *f* di esplorazione

11277 reading velocity — See reading rate.

11278 read only memory — See programmable read-only memory.

11279 read-only memory — See fixed store.

11280 read-only storage — See fixed store.

11281 read *v* **out; roll** *v* **out** — Computer term for a counter which counts modulo n, to read its content by causing it to count a sequence of n pulses, determining at what stage in the sequence the content passes through zero.

f extraire les retenues — **d** auswalzen — **n** rondtellen — **e** extraer informes — **i** estrarre il riporto

11282 read right image

f image *f* droite; image *f* à l'endroit — **d** seitenrichtiges Bild *n* — **n** recht-leesbaar beeld *n* — **e** imagen *f* al directo — **i** immagine *f* diritta

11283 read time — See access time.

11284 read/write head; combined head; combined read/write head — Electro-magnetic device which reads information from a magnetic tape or disk, or writes information to a storage device or medium, such as disk store or magnetic tape. Some such devices merely read and do not write data.

f tête *f* de lecture/écriture — **d** Schreib/Lese-kopf *m* — **n** lees/schrijfkop *m* — **e** cabeza *f* de lectura/grabación; cabeza *f* lectora/grabadora — **i** testina *f* di lettura/iscrizione

11285 read wrong image
f image *f* à l'envers — **d** seitenverkehrtes Bild *n* — **n** omgekeerd-leesbaar beeld *n* — **e** imagen *f* al revés; imagen *f* de revés — **i** immagine *f* rovesciata

11286 ready for press
f bon à tirer; prêt pour l'impression — **d** druck-fertig — **n** drukklaar — **e** dispuesto para la impresión — **i** buono a tirare; buono a stampare; visto per la stampa; pronto per la stampa; sta bene

11287 ready reckoner
f barème *m* — **d** Rechentabelle *f*; Rechentafel *f* — **n** rekentabel *fm*; rekentafel *fm*; barema *n* — **e** baremo *m*; tabla *f* de calcular — **i** libro *m* dei conti fatti; prontuario *m* dei conti fatti

11288 reagent — Chemical to detect the presence of a substance or a particular quality, in the material being examined.
f réactif *m* — **d** Reagenz *n* (Reagenzien *npl*) — **n** reagens *n* (reagentia *npl*) — **e** reactivo *m* — **i** reagente *m*

11289 real image — Image formed in such a way that all of the light which passes through an optical system from a point of the object actually passes through a point of the image.
f image *f* réelle — **d** reelles Bild *n* — **n** reëel beeld *n* — **e** imagen *f* real — **i** immagine *f* reale

11290 real time — Operation of a computer system in an interactive mode whereby answers, solutions or information are provided, on a video display terminal or communicating typewriter, quickly so that the user can await the result. A right time response provides a less instantaneous answer.
f temps *m* réel — **d** Echtzeit *f* — **n** ware tijd *m* — **e** tiempo *m* real — **i** tempo *m* reale

11291 ream — Pack of 500 identical sheets of paper. In US either 480 or 500 sheets according to grade. See also short ream.
f rame *f* — **d** Ries *n*; Neuries *n* (1000 Bogen) — **n** riem *m* — **e** resma *f* — **i** risma *f*

11292 ream mark — See ream marker.

11293 ream marker; ream mark; slip; flag — Slip of paper placed in a pile of paper to mark its division into reams.
f séparateur *m* de rame; pavillon *m*; marqueur *m* — **d** Papierstreifen *m*; Rieszahlstreifen *m* — **n** riemstrookje *n*; telstrookje *n*; strookje *n*; vlaggetje *n* — **e** marca *f* de separación de resma — **i** strisciolina *f* di conteggio; strisciolina *f* di carta

11294 ream weight — Weight of one ream of paper, either actual or nominal. See also actual weight, nominal weight.
f poids *m* à la rame; poids *m* par rame — **d** Ries-gewicht *n* — **n** riemgewicht *n*; gewicht *n* per riem — **e** peso *m* por resma; peso *m* de la resma — **i** peso *m* per risma

11295 ream wrappers; mill wrappers — Papers to wrap reams of paper.
f enveloppes *fpl* de rames — **d** Riesumschläge *mpl*; Rieseinschlagpapier *n* — **n** kappen *fmpl*; riemkappen *fmpl*; papierkappen *fmpl* — **e** envol-torios *mpl* de las resmas — **i** carta *f* impacco risme

11296 rebind *v* — To sew an old book and make a new cover for it.
f relier en nouveau; relier à neuf — **d** aufs neue einbinden; umbinden; aufbinden — **n** overbinden; opnieuw inbinden; overnieuw binden — **e** en-cuadernar de nuevo — **i** rilegare di nuovo; legare di nuovo

11297 re-biting — See fine etching.

11298 receiving perforator — See reperforator.

11299 recess mould — Mould of the slug composing machine, the cap of which is so shaped that the body of a slug cast therefrom is recessed to conserve type metal and lighten formes.
f moule *m* squelette — **d** Spargießform *f* — **n** spaargietvorm *m* — **e** molde *m* hueco; molde *m* esqueleto — **i** forma *f* a fondere stretta

11300 recipe mark; response — Symbol, ℞ , in music printing, or in medical and liturgical books.
f signe *m* de répons; répons *m* — **d** Responsum-Zeichen *n*; Responsorium-Zeichen *n*; Antwort-Zeichen *n* — **n** reponsie-teken *n* — **e** responsorio *m*; signo *m* responsorio — **i** responsorio *m*; segno *m* di responsorio

11301 reciprocal — The reciprocal of a quantity is 1 divided by the quantity, e.g., the reciprocal of 5 is 1/5.
f réciproque *m* — **d** Kehrwert *m* — **n** reciproque waarde *f* — **e** recíproco *m* — **i** reciproco *m*

11302 reciprocating ink drum; oscillating ink roller; reciprocating ink roller; oscillating ink drum; vibrating ink drum *depr.***; vibrating ink roller** *depr.* — Ink drum or roller which moves from side to side while rotating.
f rouleau *m* bal(l)adeur; bal(l)adeur *m*; rouleau *m* distributeur — **d** Verreibwalze *f*; Reiber *m* — **n** heen- en weergaande distributierol *fm*; heen- en weergaande verdeelrol *fm* — **e** rodillo *m* de vaivén; rodillo *m* distribuidor — **i** rullo *m* macinatore di va e vieni; rullo *m* a movimento assiale alternativo

11303 reciprocating ink roller — See reciprocating ink drum.

11304 reciprocating motion
f mouvement *m* de va-et-vient; mouvement *m* alternatif — **d** Hinundherbewegung *f* — **n** heen-en weergaande beweging *f* — **e** movimiento *m* de vaivén; movimiento *m* de atras para adelante; movimiento *m* recíproco; vaivén *m* — **i** movi-mento *m* su e giú; movimento *m* alternativo

11305 reciprocating rider roller
f chargeur *m* bal(l)adeur — **d** Wechselreiber *m*; seitlich bewegbare Reiterwalze *f* — **n** heen- en weergaande ruiterrol *fm* — **e** cargador *m* de vaivén; cargador *m* oscilante; cargador *m*

oscilador — **i** rullo *m* intermittente macinatore; rullo *m* cavaliere a movimento assiale alternativo

11306 recognition, optical character — See optical character recognition.

11307 recognition, pattern — See pattern recognition.

11308 recompose *v* — See reset *v*.

11309 record — Group of related facts or fields of information, regarded as a unit. Records may be unbatched (single record per block), or batched (more than one record per block).
f enregistrement *m* — **d** Datensatz *m*; Satz *m* — **n** record *n*; registratie *f* — **e** registro *m* — **i** registrazione *f*

11310 recorder — Instrument usually attached to a measuring apparatus, such as a pressure or flow gauge, which registers the measurements continuously and permanently.
f enregistreur *m* — **d** Schreiber *m*; Aufzeichner *m*; Schreibwerk *n*; Registrierapparat *m*; Registriergerät *n*; Registriervorrichtung *f* — **n** registreerinrichting *f*; registreerinstrument *n* — **e** registrador *m*; dispositivo *m* registrador — **i** registratore *m*; strumento *m* registratore

11311 record gap — See interrecord gap.

11312 recording density — Number of useful storage cells per unit of length or area, e.g., the number of rows (or characters) per inch on a magnetic tape or punched tape, or the number of bits per inch on a single track of a tape or drum.
f densité *f* d'enregistrement — **d** Aufzeichnungsdichte *f*; Schreibdichte *f* — **n** optekendichtheid *f*; schrijfdichtheid *f* — **e** densidad *f* de registro — **i** densità *f* di registrazione

11313 record paper — See ledger paper.

11314 recovery
f récupération *f* — **d** Rückgewinnung *f*; Wiedergewinnung *f* — **n** terugwinning *f* — **e** recuperación *f* — **i** recupero *m*; ricupero *m*

11315 rectangular
f rectangulaire; rectangle; à angle droit — **d** rechtwinklig; rechtwinkelig — **n** rechthoekig; onder een rechte hoek — **e** rectángulo; rectangular — **i** rettangolare

11316 rectangular gauge thread
f filet *m* (de vis) rectangulaire; filet *m* (de vis) rectangle; filet *m* (de vis) orthogonal — **d** rechtwinkliges Gewinde *n* — **n** rechthoekige draad *m*; rechthoekige schroefdraad *m* — **e** filete *m* (de tornillo) rectangular; rosca *f* (de tornillo) rectangular — **i** filettatura *f* di vite rettangolare

11317 rectifier — Device to make a DC current (voltage) out of an AC current (voltage).
f redresseur *m* (de courant); rectificateur *m* (de courant) — **d** Gleichrichter *m* — **n** gelijkrichter *m* — **e** rectificador *m* — **i** rettificatore *m*; raddrizzatore *m*

11318 recto — First side of a processed sheet as distinct from the verso which is the reverse side.
f recto *m* — **d** Vorderseite *f* — **n** voorzijde *fm*; voorkant *m* — **e** anverso *m*; cara *f*; recto *m* — **i** lato *m* di bianca

11319 recto — Right-hand page of an open book having an odd page number. To start recto is to begin a recto page, as a preface or an index does.
f belle page *f*; page *f* impaire; recto *m* — **d** rechte Seite *f*; ungerade Seite *f*; Schauseite *f* — **n** rechter pagina *f*; rechtse pagina *f*; oneven pagina *f* — **e** página *f* de derecha; página *f* impar — **i** pagina *f* di destra; pagina *f* dispari; recto *m*

11320 recto — See front cover.

11321 recuperating plant
f installation *f* de récupération — **d** Rückgewinnungsanlage *f* — **n** terugwinningsinstallatie *f* — **e** planta *f* de recuperación — **i** impianto *m* di recupero; impianto *m* di ricupero

11322 red — See magenta.

11323 red button stop — Emergency stop button, usually red, in the control box.
f bouton *m* d'arrêt (en cas d'urgence) — **d** roter Druckknopf *m* (bei Notfällen) — **n** rode knop *m* (in noodgevallen); stopknop *m* — **e** botón *m* rojo (en caso de emergencia); botón *m* de presión; pulsador *m* rojo — **i** pulsante *m* rosso (di emergenza)

11324 red edges
f tranches *fpl* rouges — **d** Rotschnitt *m* — **n** rood op snee — **e** cantos *mpl* rojos — **i** tagli *mpl* rossi

11325 red ferric oxide — See ferric oxide.

11326 red filter — See magenta filter.

11327 red iron oxide — See ferric oxide.

11328 red iron trioxide — See ferric oxide.

11329 red lake C — Family of organic acid azo pigments.
f rouge *m* pour laque C — **d** Lack *m* für Rot C — **n** lakrood C *n* — **e** rojo *m* para laca C — **i** rosso *m* per lacca C

11330 red lead; red lead oxide; orange mineral; minium — Orange-red oxide of lead, used as an opaque printing ink pigment.
Formula: Pb_3O_4.
f minium *m* (de plomb); oxyde *m* (salin) de plomb — **d** Mennige *f*; Bleimennige *f*; Bleioxyd *n*; rotes Bleioxyd *n* — **n** menie *fm*; loodmenie *fm*; loodoxyde *n*; rood loodoxyde *n* — **e** minio *m* (de plomo); óxido *m* rojo de plomo; almártaga *f* — **i** minio *m* (di piombo); ossido *m* di piombo; rosso *m* di piombo

11331 red lead oxide — See red lead.

11332 red letter — See punk.

11333 red oil — Name for the commercial grade of *oleic acid.

11334 redox potential; oxidation/reduction potential — The redox potential measurement (in millivolts) can be carried out using a voltmeter. It consists of measuring the potential difference across two electrodes.
f potentiel *m* rédox; potentiel *m* d'oxydoréduction; potentiel *m* d'oxyréduction — **d** Redoxpotential *n* — **n** redox-potentiaal *m*; reductie/oxydatiepotentiaal *m* — **e** potencial *m* de óxidorreducción — **i** redopotenziale *m*; potenziale *m* redox

11335 ⸱ ⸱ plate — See magenta printer.

11336 red potassium chromate — See potassium

dichromate.
11337 red potassium prussiate — See potassium ferricyanide.
11338 red printer — See magenta printer.
11339 red prussiate of potash — See potassium ferricyanide.
11340 reduce *v* — To decrease the density of a photographic negative.
f affaiblir — **d** abschwächen — **n** afzwakken — **e** debilitar — **i** indebolire
11341 reduce *v* — To produce an image smaller than the original.
f réduire — **d** verkleinern — **n** verkleinen — **e** reducir — **i** ridurre
11342 reduced small — A size of cut card. See app. no. 3.
11343 reduce *v* **ink**
f allonger l'encre — **d** Farbe verdünnen — **n** inkt verdunnen; inkt dunner maken — **e** diluir la tinta — **i** diluire l'inchiostro; allungare l'inchiostro
11344 reducer; reducing agent — Chemical solution to dissolve part or all of the black silver image on a negative or positive. When used for clearing, it removes the background fog in the transparent areas of a negative or positive. When used for dot etching in colour correction, it progressively reduces the halftone dot size.
f affaiblisseur *m*; réducteur *m*; faiblisseur *m* — **d** Abschwächer *m* — **n** afzwakker *m*; afzwakmiddel *n*; afzwakkingsmiddel *n* — **e** debilitador *m*; reductor *m*; rebajador *m* — **i** indebolitore *m*
11345 reducer; reducing agent; diluent; diluting agent; thinner — Varnish, solvent, or oily or greasy compound to bring ink or varnish to a softer tack and consistency for use on the press.
f diluant *m*; allongement *m*; vernis *m* à réduire; fluidogène *m* — **d** Verdünnungsmittel *n*; Verdünner *m*; kurzmachender Zusatz *m* — **n** verdunningsmiddel *n*; verdunner *m* — **e** aditivo *m*; suavizante *m*; diluyente *m*; rebajador *m*; solvente *m*; adelgazante *m* — **i** additivo *m* diluente; diluente *m*
11346 reducing agent — See reducer.
11347 reducing agent — See reducer.
11348 reducing glass; diminishing glass — Double concave lens for viewing originals in reduced sizes.
f verre *m* diminuant — **d** Verkleinerungsglas *n*; Verkleinerungslinse *f* — **n** verkleinglas *n* — **e** vidrio *m* de disminución — **i** lente *f* di riduzione; lente *f* d'impicciolimento
11349 reduction — Reduction of size.
f réduction *f* — **d** Verkleinerung *f* — **n** verkleining *f* — **e** reducción *f* — **i** riduzione *f*
11350 reduction — Chemical reaction in which hydrogen combines with another substance or in which oxygen is lost from a substance. More generally a chemical change in which the valence state of an atom of an element is decreased, due to gain of one or more electrons by the atom.
f réduction *f*; désoxydation *f* — **d** Reduktion *f*;

Desoxydation *f*; Desoxydierung *f* — **n** reductie *f* — **e** rebajado *m*; rebajamiento *m* — **i** riduzione *f*
11351 reduction in surface tension — See surface tension lowering.
11352 reduction of density — Lessening the opacity of a photographic negative by chemical removal of some of the developed silver forming the image. The term reduction is unfortunate because the reactions employed are oxidations, and it would be better to refer to the process as weakening, the term used in most languages other than English.
f baissage *m*; affaiblissement *m* — **d** Abschwächen *n*; Abschwächung *f* — **n** afzwakken *n*; afzwakking *f* — **e** rebajado *m* de la densidad; debilitamiento *m* de la densidad; debilitación *f* de la densidad — **i** riduzione *f* di densità; riduzione *f* di opacità
11353 redundancy check — Check based on the transfer of more bits or characters than the minimum number required to express the message itself, the added bits or characters having been inserted systematically for checking.
f contrôle *m* de redondance — **d** Redundanzprüfung *f* — **n** redundantie-controle *fm* — **e** comprobación *f* por redundancia — **i** controllo *m* di ridondanza
11354 reel — Roll of paper wound on a shaft.
f bobine *f* (grand format); bobineau *m* (petit format); bobinette *f* (petit format) — **d** Papierrolle *f*; Rolle *f* — **n** papierrol *fm*; rol *fm* — **e** bobina *f* de papel; bobina *f* de papel continuo; bobina *f* de papel de periódico; rollo *m* de papel — **i** bobina *f* di carta; bobina *f*
11355 reel arms — Parts of the reel stand that support the centre or spindle of the reel of paper.
f bras *mpl* (de porte-bobine) — **d** Rollensternarme *mpl* — **n** armen *mpl* (van de rollenster) — **e** brazos *mpl* (del estante portabobinas) — **i** bracci *mpl* (del portabobine)
11356 reel bogie; paper reel bogie; paper reel truck; paper reel dolly; dolly truck; reel truck — Truck to transport a reel of paper.
f chariot *m* à bobine — **d** Rollenhunt *m*; Rollenwagen *m*; Papierrollenwagen *m*; Papierrollentransportwagen *m* — **n** rollenwagen *m* — **e** carretilla *f* para transportar bobinas (de papel) — **i** carrello *m* per il trasporto delle bobine
11357 reel brake — See brake.
11358 reel braking — Applying tension to the web when the machine is running.
f freinage *m* de la bobine (de papier) — **d** Abbremsen *n* der Papierrolle — **n** afremming *f* van de (papier)rol; afremmen *n* van de (papier)rol — **e** enfrenamiento *m* de la bobina (de papel); refrenamiento *m* de la bobina (de papel) — **i** frenatura *f* della bobina (di carta)
11359 reel brush — Device to press the running web against the new reel when the join is made on an automatic reel change.
f brosse *f* à bobines — **d** Rollenbürste *f* — **n** rollenborstel *m* — **e** bruza *f* de bobinas — **i**

spazzola *f* di bobine

11360 reel centre — See reel core.

11361 reel change — Changing of paper reels.
f changement *m* de bobine; remplacement *m* de la bobine — **d** Rollenwechsel *m*; Auswechseln *n* der Papierrolle — **n** rolwisseling *f* — **e** cambio *m* de bobina — **i** cambio *m* di bobina

11362 reel cone *GB*; **core plug; chuck; bung** *US* — Metal core inserted into the ends of the reel core to make it run concentrically on the reel spindle or between the reel arms.
f cône *m* de bobine — **d** Rollenkonus *m*; Rollkone *m depr.* — **n** conus *m*; rolconus *m* — **e** cono *m* de bobina; cono *m* portabobinas; taco *m* del alma; tarugo *m*; guarda *f* del alma — **i** cono *m* della bobina

11363 reel core; reel centre — Tube on which paper is wound.
f mandrin *m* — **d** Rollenhülse *f*; Papphülse *f*; Wickelhülse *f*; Hülse *f* — **n** koker *m*; rolkoker *m*; rolhuls *fm* — **e** alma *f* (de la bobina); mandril *m* — **i** anima *f* (della bobina)

11364 reel end — See butt end.

11365 reel end damage — Cuts, punctures, deep indentations on the paper reel ends which will cause a nick or tear on the web edge.
f lésion *f* du bord de la bobine; bord *m* abîmé — **d** Beschädigung *f* der Papierrollen-Stirnseite — **n** beschadiging *f* van de zijkant — **e** deterioración *f* del canto del bobina — **i** danneggiamento *m* del canto bobina

11366 reel fed — See web-fed.

11367 reel-fed bag making machine
f machine *f* à sac à marge en bobine — **d** von der Rolle arbeitende Beutelmaschine *f* — **n** van de rol werkende zakkenmachine *f* — **e** máquina *f* de bolsas en bobina — **i** macchina *f* per fabbricare sacchetti alimentata da bobina

11368 reel-fed press — See web machine.

11369 reel hand; reel tender; roll tender *depr.* — Operator responsible for preparing and loading the reel of paper to run on the printing press.
f bobinier *m* — **d** Rollenmixer *m*; Bedienungsmann *m* an der Rollenlagerung — **n** rollenman *m* — **e** bobinero *m* — **i** addetto *m* al carico delle bobine

11370 reel hoist — Mechanism in a rotary press to lift reels of paper.
f palan *m* à bobines — **d** Rollenaufzug *m*; Rollenanhebevorrichtung *f* — **n** rollenlift *m* — **e** cabria *f* de bobinas; pescante *m* de bobinas — **i** paranco *m* di sollevamento delle bobine

11371 reel of board — See reel of paper.

11372 reel of paper; reel of board; roll of paper — Continuous strip of paper (or board) rolled around itself; in some countries a reel may be wound round a core.
f rouleau *m* de papier; rouleau *m* de carton — **d** Rolle *f* Papier; Papierrolle *f*; Rolle *f* Karton; Rolle *f* Pappe; Kartonrolle *f*; Papperolle *f* — **n** rol *fm* papier; papierrol *fm*; rol *fm* karton — **e** rollo *m* de papel; rollo *m* de cartón — **i** rotolo *m* di carta;

rotolo *m* di cartone

11373 reel paper; paper in reels — Long webs of paper tightly rolled round a centre.
f papier *m* en bobines — **d** Rollenpapier *n* — **n** papier *n* aan rollen — **e** papel *m* en bobinas — **i** carta *f* in rotoli

11374 reel room — Floor on which the reel stands are located directly under a unit type press.
f étage *m* des bobines — **d** Rollensternraum *m* — **n** rollenkelder *m*; papierrollenkelder *m* — **e** sótano *m* de las bobinas — **i** piano *m* delle bobine

11375 reel stand; roll stand *depr.* — Part of a reel-fed inline printing press or fabricating machine which supports reel material that is subsequently fed to the machine or press.
f porte-bobine *m* — **d** Rollenlagerung *f*; Rollenstuhl *m*; Rollenständer *m* — **n** rollenstandaard *m* — **e** estante *m* portarrollos; portarrollos *m*; portabobinas *m* — **i** portabobine *m*

11376 reel tender — See reel hand.

11377 reel truck — See reel bogie.

11378 reel width — See width.

11379 re-etching — See fine etching.

11380 reference — Note at the bottom of page, explaining text or giving source of information.
f renvoi *m*; référence *f* — **d** Anmerkung *f*; Hinweis *m* — **n** verwijzing *f* — **e** indicación *f*; advertencia *f*; nota *f* — **i** riferimento *m*; richiamo *m*

11381 reference marks — Signs to direct the reader from the text to a footnote or to the bibliography at the end of chapter or book. Used in the following order: * (asterisk), † (dagger; obelisk), ‡ (double dagger), § (section mark), // (parallel mark), ¶ (paragraph mark), and then repeated in duplicate. Also figures in numerical sequence are used.
f signes *mpl* de référence; appels *mpl* de note; renvois *mpl*; pieds *mpl* de mouche — **d** Anmerkzeichen *npl*; Hinweiszeichen *npl* — **n** noottekens *npl*; noten *fmpl*; verwijzingstekens *npl* — **e** llamadas *fpl*; citas *fpl* — **i** segni *mpl* di riferimento; riferimenti *mpl*; segni *mpl* di richiamo; richiami *mpl*

11382 reference matter — See end matter.

11383 references — See bibliography.

11384 reference sample
f feuille-échantillon *f* — **d** Vergleichsmuster *n*; Vergleichsprobe *f* — **n** bestelmonster *n*; verwijzingsmonster *n* — **e** muestra *f* de referencia — **i** campione *m* di riferimento

11385 refiner; pulp engine — Machine to treat and cut the pulp, usually after it has left the beater; also used in pulp mills to refine coarse pulp fibres retained to the screens. See also Jordan machine.
f raffineur *m*; raffineuse *f*; pile *f* raffineuse; pulp-engine *f* — **d** Mahlgerät *n* — **n** maalwerk *n* — **e** machacadora *f* (de pulpa); pila *f* refinadora; pila *f* batidora; refinadora *f*; refinador *m*; batidora *f* — **i** raffinatrice *f*; raffinatore *m*

11386 refining — Mechanical treatment of pulp

fibres in a water suspension to develop the paper-making properties of the fibres. See also Jordan machine, beating.

f raffinage *m* (en raffineur) — **d** Refiner-Mahlung *f*; Mahlvorgang *m* — **n** maling *f* — **e** molturación *f*; refinaje *m*; refinado *m* — **i** raffinazione *f*; lavorazione *f* dell'impasto

11387 reflectance — Gloss or light reflection of ink or paper.

f reflet *m*; réflexion *f* — **d** Reflex *m*; Reflexion *f* — **n** reflectie *f*; reflex *m* — **e** reflejo *m*; reflexión *f* — **i** riflessione *f*; potere *m* riflettente

11388 reflected binary code *GB*; **Gray code** *US* — Binary code, devised by Elisha Gray in 1876. A type of cyclic unit-distance binary code built up from the four-word, two-bit unit distance code (00, 01, 11, 10) according to the following rule: To construct an (n.1) bit reflected binary code from an n-bit reflected binary code, write the n-bit code twice in sequence, first in forward, then in reverse sequence of code words. Prefix extra bits each word, taking the value 0 for the forward version of the n-bit code, and the value 1 for the backward version.

f code *m* binaire-cyclique; code *m* de Gray — **d** reflektierter Binärkode *m*; Gray-Verschlüsselung *f* — **n** gespiegelde binaire code *m*; Gray-code *m* — **e** código *m* cíclico binario; código *m* de Gray — **i** codice *m* ciclico binario; codice *m* di Gray

11389 reflection — Change of direction experienced by a ray of light falling upon a surface and thrown back into the medium from which it comes. Reflexion is etymologically correct, but use reflection.

f réflexion *f* — **d** Reflexion *f* — **n** reflectie *f*; terugkaatsing *f* — **e** reflexión *f* — **i** riflessione *f*

11390 reflection copy; flat copy — Copy viewed or photographed by reflected light.

f document *m* opaque — **d** Aufsichtsvorlage *f* — **n** model *n*; opzichtmodel *n* depr. — **i** originale *m* per riflessione; bozzetto *m*

11391 reflection densitometer — See reflectometer.

11392 reflection meter — See reflectometer.

11393 reflectometer; reflection meter; reflection densitometer — Photoelectric device to measure the amount of light reflected from an opaque surface. Used to control ink colour density on printing surfaces.

f réflectomètre *m*; densitomètre *m* par réflection — **d** Reflexionsdensitometer *n*; Aufsichtsdensitometer *n*; Aufsichtsdichtemesser *m* — **n** reflectiedensitometer *m* — **e** reflectómetro *m*; densitómetro *m* de reflexión — **i** riflettometro *m*; densitometro *m* per riflessione

11394 reflex blue — See alkali blue.

11395 refraction — Deviation of a light ray from a straight path when passing obliquely from one medium into another of different density or in transversing a medium whose density is not uniform.

f réfraction *f* — **d** optische Brechung *f*; Refraktion *f*; Lichtbrechung *f* — **n** refractie *f*; breking *f*; lichtbreking *f* — **e** refracción *f*; refracción *f* de la luz — **i** rifrazione *f*

11396 refractive index — Measure of refracting power; degree to which a light-ray is bent when it passes from one medium into another.

f indice *m* de réfraction — **d** Brechungsindex *m*; Brechungszahl *f*; Brechzahl *f*; Brechungsquotient *m*; Refraktionsindex *m* — **n** brekingsindex *m*; refractie-index *m* — **e** índice *m* de refracción — **i** indice *m* di rifrazione

11397 refractometer — Device to measure the angle at which a light-ray entering a substance is fully reflected. The angle is known as the critical or grazing angle. Used in ink-making to analyse oils, resins, or pigments.

f réfractomètre *m* — **d** Refraktometer *n* — **n** refractometer *m* — **e** refractómetro *m* — **i** rifrattometro *m*

11398 refresh *v* — To regenerate an image on a cathode ray tube (especially one used for editing or reviewing) with sufficient frequency that the image appears relatively constant. Necessary because the phosphor coating on the inside of the tube (unless it is a storage tube) has very little latency or persistency.

11399 refresh rate — Rate at which an image will be refreshed on a cathode ray tube; usually 30, 40, or 60 times per second. However, some tubes refresh the image on an interlaced basis so that alternate locations on the tube will be refreshed each cycle. An interlaced 60 second refresh rate will refresh a given location only thirty times per second.

11400 regenerated cellulose — See cellophane.

11401 regina — A size of note paper. See app. no. 3.

11402 register — Exact correspondence in the position of pages or other printed matter on two sides of a sheet or its relation to other matter already ruled or printed on the sheet.

f repérage *m*; registre *m*; enlignement *m* — **d** Passer *m*; Register *n*; Stand *m* — **n** passen *n*; register *n*; stand *m* — **e** registro *m* — **i** registro *m*

11403 register — Area in the central processor, used by the computer for accumulating, addressing, indexing etc. Access to registers is usually much faster than access to core.

f registre *m* — **d** Register *n* — **n** register *n* — **e** registro *m* — **i** registro *m*

11404 register control

f contrôle *m* du repérage — **d** Registersteuerung *f*; Passersteuerung *f*; Registerregelung *f* — **n** registerregeling *f*; registercontrole *f* — **i** controllo *m* del registro

11405 register differences

f irrégularités *fpl* de repérage; défauts *mpl* de repérage — **d** Passerdifferenzen *fpl* — **n** pasverschillen *npl* — **e** diferencias *fpl* del registro; diferencias *fpl* de puntura — **i** differenzas *fpl* di registro; difetti *mpl* di registro

11406 registered trade mark — Mark of ownership officially listed at the registration office to protect it from imitation by competitors.
f marque *f* déposée — **d** eingetragene Marke *f*; Schutzmarke *f* — **n** gedeponeerd handelsmerk *n*; ingeschreven handelsmerk *n* — **e** marca *f* registrada; marca *f* comercial registrada — **i** marca *f* di fabbrica registrata; marchio *m* depositato

11407 register error
f erreur *f* de repérage — **d** Registerfehler *m*; Paßfehler *m* — **n** pasfout *fm*; pasverschil *n* — **e** falta *f* de registro; fuera *f* de puntura — **i** errore *m* di registro

11408 register glass — See register tube.

11409 register holes and studs
f trous *mpl* et goujons *mpl* de registre — **d** Paßlöcher *npl* und Paßstifte *mpl* — **n** pasgaten *npl* en paspennen *fmpl* — **e** trepanados *mpl* y conos *mpl* (de puntura) — **i** fori *mpl* e chiodini di registro

11410 register lock-up — Mechanism to position the plates on the plate cylinder of a rotary press correctly.
f butées *fpl* réglables de mise en registre — **d** Registerstellvorrichtung *f* für die Platten — **n** register-stelinrichting *f* — **e** punturas *fpl* regulables de registro — **i** sistema *m* di messa a registro delle lastre

11411 register marks; register ticks; cross-marks — Guides in the form of small crosses or marks placed or drawn on an original before photography to facilitate registration in platemaking and proving; in process work, printed in the margin of the sheet to help in registering successive colours.
f croix *fpl* de repère; traits *mpl* de repère; repères *mpl* en croix; repères *mpl* croix — **d** Paßkreuze *npl*; Passerkreuze *npl*; Paßmarken *fpl*; Paßzeichen *npl*; Registermarken *fpl* — **n** paskruizen *npl*; pasmerken *npl*; registermerken *npl* — **e** cruces *fpl* de registro; marcas *fpl* de registro — **i** crocette *fpl* di registro; segni *mpl* di registro; marche *fpl* di registro

11412 register motor — Optional attachment to enable circumferential and lateral adjustments from a remote control station.
f moteur *m* pour le réglage du repérage; servomoteur *m* de correction — **d** Registermotor *m*; Steuermotor *m*; Stellmotor *m*; Korrekturservomotor *m* — **n** registermotor *m*; servomotor *m* — **e** motor *m* de ajuste del registro; servomotor *m* — **i** motore *m* per la regolazione del registro

11413 register paper — See ledger paper.

11414 register ribbon — See book marker.

11415 register roller — Adjustable roller to vary the web length between one unit on the rotary press and another.
f rouleau *m* de registre — **d** Registerwalze *f*; Registerspindel *f* — **n** registerwals *fm* — **e** cilindro *m* de registro — **i** rullo *m* di registro

11416 register rule — Metal measuring rule with

vernier to measure in thousandths of millimetres or inches, designed especially for checking size and register on a printing forme, a flat, pressplate or press sheet.
f règle *f* (pour le contrôle du repérage); règle *f* de repérage; équerre *f* de montage — **d** Registerlineal *n* — **n** registerliniaal *fm* — **e** regla *f* de registro — **i** regolo *m* di registro

11417 register table; lining-up table; light table; shining table; shining-up table; illuminated line-up table — Frosted glass-topped table illuminated from beneath and mounted with two precision straight edges, either movable at right angles to the other, to assure alignment of paste-ups, mechanicals and stripped-up negatives.
f table *f* de montage (lumineuse); table *f* lumineuse — **d** Leuchttisch *m*; Montagetisch *m*; Leuchtpult *n*; Lichttisch *m* — **n** montagetafel *fm*; lichtbak *m*; lichttafel *fm* — **e** mesa *f* luminosa (de montaje); mesa *f* de montaje; mesa *f* de alineación — **i** tavolo *m* luminoso; tavolo *m* di montaggio; tavolo *m* per la messa a registro

11418 register ticks — See register marks.

11419 register tube; register glass; peeper *US coll* — 20 cm long metal tube with a cross or opening at the base, for mounting negatives or positives in register.
f lunette *f* de visée; loupe *f* de mise au point — **d** Einpaßrohr *n*; Richtrohr *n*; Montagelupe *f* — **n** paskijker *m*; pasloep *fm*; pasloupe *fm* — **e** lente *fm* de registro — **i** cannocchiale *m* per messa a registro

11420 registration — Correct relative position and act of successively bringing negatives, plates and impressions into such position to form a perfect image or reproduction.
f repérage *m* — **d** Passer *m*; Paßgenauigkeit *f* — **n** passen *n*; register *n* — **e** alineación *f*; coincidencia *f* — **i** messa *f* a registro

11421 reglet — Strip of wood usually 6 points or 12 points thick, to make blanks between lines of type, and to lock up a forme. See also lead.
f réglette *f* — **d** Reglette *f* — **n** reglet *fmn* — **e** regleta *f* — **i** regoletto *m*

11422 regulating box — Box with a density measuring apparatus in which the consistency of the paper stock going to the paper machine is adjusted and maintained at a constant level.
f caisse *f* régulatrice — **d** Stoffdichteregler *m*; Stoffregulierkasten *m* — **n** regelkast *fm* — **e** caja *f* reguladora — **i** cassa *f* di regolazione; regolatore *m* della densità dell'impasto

11423 reimpose *v* — To impose again a composition in which corrections had to be made or which had been set aside.
f réimposer — **d** erneut ausschießen — **n** opnieuw inslaan — **e** reimponer — **i** impostare nuovamente

11424 reinforced binding — Binding reinforced by, e.g., pasting cambric around the signatures, at back of book, inside cover, etc.
f reliure *f* renforcée — **d** verstärkter Einband *m*

— **n** versterkte boekband *m*; versterkte band *m* — **e** encuadernación *f* reforzada — **i** legatura *f* rinforzata

11425 reinforced concrete — Concrete in which metal is embedded so that the two materials together resist forces.
f béton *m* armé; ciment *m* armé — **d** Eisenbeton *m*; Stahlbeton *m* — **n** gewapend beton *n* — **e** hormigón *m* armado; cemento *m* armado — **i** cemento *m* armato

11426 reinforced paper — Paper reinforced with threads or cloth to improve its mechanical strength. See also papyroline paper.
f papier *m* renforcé (de toile); papier *m* renforcé de tissus; papier *m* avec armature textile; papier *m* entoilé — **d** textilverstärktes Papier *n*; Gewebepapier *n*; Leinwandpapier *n*; Papier *n* mit Gewebe — **n** met textielgaas versterkt papier *n* — **e** papel *m* reforzado; papel *m* entelado; papel *m* con tejido — **i** carta *f* rinforzata (con tessuto); carta *f* telata

11427 reinforced paper — See cloth-lined paper.

11428 reinforced union paper; tarred thread paper — Paper composed of two sheets, stuck together with bitumen or tar, between which there is a lining (reinforcement with threads or cloth) which increases the mechanical strength.
f papier *m* brun goudronné renforcé — **d** textilverstärktes Bitumenpapier *n*; Teerpapier *n* mit Gewebe; Gewebepackpapier *n*; Gewebepack *n* — **n** met weefsel versterkt, geteerd papier *n*; teerpapier *n* met weefsel — **e** papel *m* embetunado entelado — **i** carta *f* doppia bitumata rinforzata

11429 reissue — Republication at a different price, or in a different form, of an impression which is already on the market.
f réédition *f*; nouvelle édition *f*; nouveau tirage *m* — **d** Neuausgabe *f* — **n** herdruk *m*; heruitgave *f*; nieuwe druk *m* — **e** edición *f* nueva; reimpresión *f* — **i** edizione *f* nuova; ristampa *f*

11430 relative aperture — Ratio of the focal length of an optical system to the diameter of the entrance pupil.
f ouverture *f* relative — **d** relative Öffnung *f* — **n** relatieve opening *f* — **e** abertura *f* útil relativa — **i** apertura *f* relativa

11431 relative density — See specific gravity.

11432 relative frequency — In statistics, ratio of the number of times an event occurs to the number of occasions on which it might occur in the same period.
f fréquence *f* relative — **d** relative Häufigkeit *f* — **n** relatieve frequentie *f* — **e** frecuencia *f* relativa — **i** frequenza *f* relativa

11433 relative humidity — Actual amount of moisture in the atmosphere expressed as a percentage of the total amount needed to saturate it at the same temperature. See also absolute humidity.
f humidité *f* relative (de l'air); humidité *f* relative de l'atmosphère; degré *m* hygrométrique — **d** relative Feuchtigkeit *f*; relative Luftfeuchtigkeit *f*

— **n** relatieve vochtigheid *f* (van de lucht); relatieve vochtgehalte *n* (van de lucht); relatieve luchtvochtigheid *f* — **e** humedad *f* relativa (del aire) — **i** umidità *f* relativa (dell'aria)

11434 relative permittivity — See dielectric constant.

11435 relative tack — Tack of the sample divided by the tack of a standard sample. At high velocities, it approaches the value of plastic viscosity.
f poisseux *m* relatif — **d** relativer Tack *m*; relative Zugigkeit *f*; relativer Zug *m* — **n** relatieve tack *m*; relatieve kleefkracht *f* — **e** pegajosidad *f* relativa — **i** peciosità *f* relativa

11436 relaxation allowance — See compensating relaxation allowance.

11437 relay — Device, usually consisting of an electromagnet and an armature, by which a change of current in one circuit can be made to produce a change in the electric condition of another circuit.
f relais *m* — **d** Relais *n* — **n** relais *n* — **e** relevador *m*; relais *m*; relé *m* — **i** relè *m*

11438 release coating — Coating applied to one side of sack kraft paper to prevent the paper surface sticking to the packed product.
f couche *m* anti-adhésif — **d** Trennbeschichtung *f* — **n** scheidingslaag *fm* — **e** capa *f* de separación; capa *f* antiadhesivo — **i** patina *f* antiadesiva

11439 releasing liquid — Base for printed matter, which later can be removed by treatment with a solvent, e.g., water or alcohol.

11440 reliability — Ability of a component, device, unit of equipment or functional section of a system to perform to a specified standard when required, without remedial action.
f fiabilité *f* — **d** Zuverlässigkeit *f* — **n** betrouwbaarheid *f* — **e** fiabilidad *f*; seguridad *f* — **i** attendibilità *f*; margine *m* di sicurezza

11441 relief, etch *v* in — See etch *v* in relief.

11442 relief etching
f gravure *f* en relief — **d** Hochätzung *f* — **n** hoogetsing *f*; reliëfetsing *f* — **e** grabado *m* en relieve — **i** incisione *f* in rilievo

11443 relief etching — See relief plate.

11444 relief map — Map showing a raised surface of the differences in elevation.
f carte *f* en relief — **d** Reliefkarte *f* — **n** reliëfkaart *fm*; kaart *fm* in reliëf — **e** mapa *m* en relieve — **i** carta *f* (geografica) in rilievo; plastico *m* geografico

11445 relief plate; relief etching — Printing plate in which the non-printing areas have been etched or cut below the surface of the material, leaving the relief presentation of the design in positive form.
f cliché *m* en relief — **d** Reliefdruckplatte *f*; Reliefplatte *f*; Reliefätzung *f*; Klischee *n* — **n** reliëfplaat *fm*; cliché *n* — **e** clisé *m* tipográfico en relieve — **i** lastra *f* di rilievo; cliscè *m*; cliché *m*; forma *f* stampante in rilievo

11446 relief-printing — Printing done from raised surfaces, like type, woodcuts, line and halftone plates.

f impression *f* en relief; impression *f* typo-(graphique) — **d** Hochdruck *m*; Buchdruck *m* — **n** hoogdruk *m*; boekdruk *m* — **e** impresión *f* de relieve — **i** stampa *f* in rilievo; impressione *f* in rilievo

11447 relief stamping — See die stamping.

11448 relocatable — Said of coding that can be loaded and executed in any available region of a computer's internal storage.

f transférable — **d** verschiebbar — **n** verplaatsbaar — **e** reubicable — **i** rilocabile

11449 relocate *v* — In programming, to move a routine from one portion of storage to another and to adjust the necessary address references so that the routine, in its new location, can be executed.

f transférer — **d** verschieben — **n** verplaatsen — **e** reubicar — **i** rilocare

11450 remainder — Part of an edition which is unsaleable at its original price.

f solde *m* (d'édition) — **d** Ramsch *m* — **n** ramsj *m* — **e** rezagos *mpl*; saldos *mpl*; sobrantes *mpl*; residuo *m* — **i** resto *m*; rimanenze *fpl*

11451 remainders — Imperfect books, sorted out in the binding department, before delivery to the publisher.

f déchets *mpl* — **d** Defekten *mpl* — **n** defecten *npl*; defecte exemplaren *npl* — **e** defectos *mpl* — **i** difetti *mpl*; libri *mpl* difetti

11452 remanence — Residual induction (or flux density) which remains when a magnetizing force is reduced to zero from a value sufficient to saturate a material.

f remanence *f* — **d** Remanenz *f* — **n** remanentie *f*; remanente magnetisatie *f*; remanent magnetisme *n* — **e** remanencia *f*; magnetismo *m* remanente; magnetismo *m* residual — **i** rimanenza *f*; magnetismo *m* residuo

11453 remedy

f remède *m* — **d** Abhilfe *f* — **n** remedie *fn*; tegenmiddel *n* — **e** remedio *m* — **i** rimedio *m*

11454 remelt *v*

f refondre — **d** umschmelzen — **n** omsmelten — **e** refundir — **i** rifondere

11455 remelter — See remelting furnace.

11456 remelting furnace; remelting pot; remelter — Furnace for melting composing machine metal.

f four *m* à refondre le métal de machine à composer; fondeuse *f* — **d** Umschmelzofen *m* für Setzmaschinenmetall — **n** omsmeltoven *m* voor zetmachine-metaal — **e** horno *m* de refundición; caldera *f* de refundir; crisol *m* de refundición — **i** crogiuolo *m* di rifusione

11457 remelting pot — See remelting furnace.

11458 remote control — System of operating or controlling a device or apparatus at some distance from this device or apparatus.

f commande *f* à distance; réglage *m* à distance — **d** Fernsteuerung *f*; Fernbedienung *f*; Fernschaltung *f*; Fernbetätigung *f* — **n** afstandsbediening *f* — **e** mando *m* a distancia; control *m* a distancia; control *m* remoto — **i** comando *m* a distanza; telecomando *m*

11459 remote station — Data terminal equipment for communicating with a data processing system from a distant location.

f téléterminal *m* — **d** Außenstelle *f*; entfernt aufgestellte Datenstation *f* — **n** buitenstation *n*; op afstand opgestelde terminal *m* — **e** teleterminal *m* — **i** teleterminale *m*

11460 Renck process — Dry offset process based on the phenomenon that certain amalgams are strongly grease repellent. Not practical on account of the poisonous character of mercury.

f procédé *m* Renck; procédé *m* mercurographique — **d** Renck-Verfahren *n*; Amalgam-Druckverfahren *n* — **n** Renck-procédé *n*; amalgaamdrukprocédé *n* — **e** procedimiento *m* Renck; técnica *f* de imprimir por amalgamación — **i** procedimento *m* Renck; metodo *m* di stampa d'amalgazione

11461 rental library — See lending library.

11462 repair *v*

f dépanner — **d** reparieren; ausbessern; beseitigen; wiederherstellen — **n** repareren; herstellen; maken — **e** reparar — **i** riparare

11463 repeat; ghosting *depr.* — Ink distribution fault resulting in an area or areas of varying ink film thickness on the finished print. These areas are related to the size and shape of the image being printed.

d Dublieren *n* — **n** doubleren *n* — **e** doblamiento *m* — **i** raddoppio *m*

11464 repel *v* — When a plate or the paper will not take the printing ink, they are said to repel the ink.

f refuser; repousser — **d** abstoßen — **n** afstoten — **e** repeler; rechazar — **i** rifiutare

11465 reperforator — Contraction of receiving perforator. Tape punch which automatically converts coded electrical signals into perforations in tape, as opposed to a perforator which punches tape by a keyboard operation. Output device on computer typesetting systems.

f réperforateur *m* — **d** Reperforator *m*; Lochstreifenempfänger *m*; Empfangslocher *m* — **n** reperforator *m*; ponsbandreproduceermachine *f* — **e** reperforador *m*; reperforadora *f* — **i** riperforatore *m*

11466 repertoire — Logical selection of characters or character images, sufficient to cope with the intellectual and aesthetic requirements of a given communication task. It may correspond with the characters on a typewriter keyboard, or a type fount.

f répertoire *m* — **d** Zeichenrepertoire *m*; Zeichenvorrat *m* — **n** repertoire *n* — **i** repertorio *m*

11467 repetitive element — See cyclic element.

11468 rep finish — Ribbed or corded finish or surface, like coarse linen or the weave of felt, produced by passing paper through grooved steel rollers or by felts with heavy warp threads, or in platers in which the books are made up with rep cloth similar to corduroy.

f apprêt *m* repsé; fini *m* repsé — **d** gereppt; Rips-finish *m*; Ripsoberfläche *f* — **n** geribbeld papier *n* — **e** acabado *m* pana; acabado *m* acordonado — **i** (carta *f*) telata; (carta *f*) uso telata

11469 reply card
f carte *f* de réponse — **d** Antwortkarte *f* — **n** antwoordkaart *f* — **e** tarjeta *f* de respuesta — **i** carta *f* di risposta

11470 report program generator — High-level programming language for easy generation of reports from a variety of file types.
f générateur *m* de programmes pour lister des informations — **d** Listprogrammgenerator *m* — **n** generator *m* van programma's voor het lijsten van gegevens — **e** generador *m* de programas para listar informes — **i** generatore *m* di programmi per listare informazioni

11471 repoussage — Flattening of the hollow areas of an etching or engraving plate by hammering it gently on the reverse side.
f repoussage *m* — **d** Zurücktreiben *n*; Auftreiben *n*; Treiben *n* — **n** terugdrijven *n*; terugkloppen *n* — **e** repujado *m* — **i** ricalcatura *f*; martellatura *f*

11472 repoussé — Formed in relief, as a design in metal, or adorned with such designs.
f repoussé — **d** mit erhabener Verzierung — **n** gedreven; geciseleerd — **e** repujado

11473 repped paper — Paper with a ribbed surface, obtained by embossing between engraved steel rolls.
f papier *m* repsé — **d** Ripspapier *n* — **n** papier *n* met persing — **e** papel *m* acordonado — **i** carta *f* a pressione

11474 representative — Salesman for a manufacturer or merchant.
f représentant *m*; vendeur *m* représentant — **d** Vertreter *m* — **n** vertegenwoordiger *m*; verkoper *m*; reiziger *m* *obs.* — **e** vendedor *m*; viajante *m* de comercio — **i** rappresentante *m*; agente *m*

11475 representative sample — In statistics, a sample in the selection of which planned action is necessary to ensure that a specified proportion of its contents is drawn from different subportions of the whole aggregation. The sampling from within each subportion may be random.
f échantillon *m* représentatif — **d** repräsentative Stichprobe *f* — **n** representatief monster *n* — **e** muestra *f* representativa — **i** campione *m* rappresentativo

11476 reprimand
f réprimande *f* — **d** Hering *m*; Tadel *m*; Rüge *f*; Rüffel *m*; Zurechtweisung *f* — **n** berisping *f*; terechtwijzing *f*; standje *n*; reprimande *f*; slechte beurt *fm coll.* — **e** represión *f*; reprimenda *f* — **i** ramanzina *f*; riprensione *f*; rabuffo *m*

11477 reprint — Second or new impression or edition of a printed work.
f réimpression *f* — **d** Nachdruck *m*; Neudruck *m* — **n** herdruk *m*; nieuwe druk *m* — **e** re-impresión *f* — **i** ristampa *f*

11478 reprint — Printed copy that must be reset.
f manuscrit *m* imprimée — **d** gedrucktes Manuskript *n* — **n** gedrukte kopij *f* — **e** manuscrito *m* impreso; copia *f* impresa — **i** copia *f* stampata

11479 reproducibility
f reproductibilité *f* — **d** Reproduzierbarkeit *f* — **n** reproduceerbaarheid *f* — **e** reproductibilidad *f* — **i** riproducibilità *f*

11480 reproduction — Process of duplicating an original by photographic or photomechanical means. The copy or final impression obtained from the original by photoengraving or other processes.
f reproduction *f* — **d** Reproduktion *f* — **n** reproduktie *f* — **e** reproducción *f* — **i** riproduzione *f*

11481 reproduction camera — See process camera.

11482 reproduction process
f procédé *m* de reproduction — **d** Reproduktionsverfahren *n* — **n** reproduktieprocédé *n* — **e** procedimiento *m* de reproducción — **i** procedimento *m* di riproduzione

11483 reproduction proofs; repro proofs; repros — Carefully printed proofs from type formes that are used directly, or in paste-ups, as camera copy for photographic reproduction.
f contre-épreuves *fpl*; faux décalques *mpl* — **d** Reproandrucke *mpl* — **n** proeven *fmpl* voor reproduktie; reproduktieproeven *fmpl* — **e** pruebas *fpl* de reproducción; pruebas *fpl* para reproducción — **i** stampe *fpl* per riproduzione

11484 reproduction ratio — Amount of enlargement or reduction in scaling copy when photographed; defined as any linear distance on the image divided by the corresponding distance on the copy. Same-size copy would be 1.0 expressed decimally, a reduction would be less than one.
f échelle *f* de reproduction — **d** Reproduktionsverhältnis *n* — **n** reproduktieverhouding *f* — **e** escala *f* de reproducción — **i** scala *f* della riproduzione; rapporto *m* di riproduzione

11485 reprography — Facsimile reproduction, as by photocopying of graphic matter.
f reprographie *f* — **d** Reprographie *f* — **n** reprografie *f* — **e** reprografía *f* — **i** riprografia *f*

11486 repro proofs — See reproduction proofs.

11487 repros — See reproduction proofs.

11488 request *v* on approval
f demander en communication — **d** zur Ansicht bestellen — **n** op zicht vragen; ter inzage vragen — **e** pedir para inspección — **i** richiedere in esame

11489 requirement — See characteristic.

11490 re-reeler — See rewind.

11491 rescaling camera; distortion camera — Phstographic camera with special lenses to alter the proportions of art work to compensate for the mechanical needs of printing reproduction processes. It can alter the proportions of either or both dimensions as well as enlarge or reduce the finished photocopy and also alter the relative "weight" of parts of the original art work.

f chambre *f* photographique à distorsion — **d** Distorsionskamera *f* — **n** distorsie-camera *fm* — **e** cámara *f* fotográfica de distorsión — **i** macchina *f* fotografica a distorsione

11492 research institute

f institut *m* de recherche — **d** Forschungsinstitut *n* — **n** researchinstituut *n*; onderzoekingsinstituut *n*; speurwerkinstituut *n* — **e** instituto *m* de investigaciones — **i** istututo *m* di ricerca

11493 reset *v*; **recompose** *v*; **set** *v* **up again** — To make a new composition.

f recomposer; composer à nouveau; aller à Saint-Jacques — **d** neu setzen; zum wiederholten Male setzen — **n** opnieuw zetten; overzetten; overnieuw zetten — **e** componer de nuevo; recomponer; hacer una nueva composición — **i** comporre di nuovo; ricomporre

11494 reset *v*; **clear** *v* — To restore a storage device to a prescribed initial state. Also to place a binary cell into the state denoting zero.

f remettre — **d** rücksetzen — **n** terugstellen; op nul stellen — **e** borrar — **i** azzerare; ripristinare

11495 residual; residual stencil — Very thin film of plate coating left on the metal of a lithographic plate after development. Post-treatment of the plate with various solutions removes this film and prepares the non printing areas of an albumin plate after development.

f réserve *f* — **d** Rückstand *m* — **n** achtergebleven kopieerlaag *fm* — **e** capa *f* de cola residual — **i** residuo *m*

11496 residual reclamation — Recovery of silver salts and other materials from exhausted fixing baths, photographic solutions and wastes by chemical or electrical treatment of the residue-containing substances.

f récupération *f* — **d** Rückstandgewinnung *f* — **n** terugwinning *f* — **e** recuperación *f* — **i** ricupero *m*

11497 residual stencil — See residual.

11498 residue; residuum — Remainder of material after the process of chemical or physical separation.

f résidu *m* — **d** Rückstand *m*; Rest *m* — **n** residu *n*; overblijfsel *n* — **e** residuo *m*; resto *m*; sobrante *m* — **i** residuo *m*; resto *m*

11499 residuum — See residue.

11500 resiliency; elasticity; springiness — Ability to regain original form after being bent compressed, or stretched.

f résilience *f*; élasticité *f* — **d** Elastizität *f* — **n** elasticiteit *f*; veerkracht *fm* — **e** elasticidad *f*; reacción *f* elástica; resiliencia *f* — **i** elasticità *f*

11501 resin; natural resin; true resin — Solid or semi-solid organic substance or synthetic or natural origin; amorphous, colourless to dark, and transparent or translucent. Insoluble in water, soluble in ether, alcohol, benzene and carbon disulphide, and softens on heating. Used in suitable solvents to produce many types of printing ink vehicles. See also rosin.

f résine *f* — **d** Harz *n* — **n** hars *mn* — **e** resina *f* — **i** resina *f*

11502 resin bonding — Cementing fibres together in papermaking with a synthetic resin, which takes the place of mucilage formed by hydration of the fibres.

f collage *m* à la résine — **d** Harzleimung *f* — **n** harslijming *f* — **e** encolado *m* a la resina; colaje *m* a la resina — **i** collatura *f* alla resina

11503 resin compound

f composé *m* résineux — **d** Harzverbindung *f* — **n** harsverbinding *f* — **e** compuesto *m* resinífero — **i** composto *m* a base di resina

11504 resinification

f résinification *f* — **d** Verharzung *f* — **n** verharsing *f* — **e** resinificación *f* — **i** resinificazione *f*

11505 resinous wood

f bois *m* résineux — **d** harziges Holz *n*; harzhaltiges Holz *n* — **n** harsachtig hout *n*; harshoudend hout *n* — **e** madera *f* resinosa — **i** legno *m* resinoso

11506 resin size — See rosin size.

11507 resist — In deep-etch offset platemaking, the non-image plate coating hardened by the action of light on bichromated gum or other coating solution; keeps developing solution from contacting these areas. Such solutions etch metal slightly. See also acid resist.

f réserve *f* — **d** Schutzschicht *f*; Schutzlack *m* — **n** kopieerlaag *fm*; beeldlaag *fm* — **e** reserva *f* — **i** riserva *f*

11508 resistance; electrical resistance — Ratio of the potential difference between the ends of a conductor to the current flowing in the conductor. The practical unit of resistance is the *ohm.

f résistance *f* (électrique) — **d** (elektrischer) Widerstand *m* — **n** (elektrische) weerstand *m* — **e** resistencia *f* (eléctrica) — **i** resistenza *f* (elettrica)

11509 resistance strain gauge; strain gauge

f jauge *f* de contrainte (à variation de résistance); extensomètre *m* — **d** Dehnungsmeßstreifen *m*; Widerstands-Dehnungsmeßstreifen *m*; Dehnmeßstreifen *m* — **n** rekstrookje *n*; weerstands-rekstrookje *n* — **e** banda *f* extensimétrica; extensímetro *m* — **i** nastro *m* estensimetro; estensimetro *m*

11510 resistance to light — See fastness to light.

11511 resistant; resisting

f résistant à — **d** widerstandsfähig gegen; widerstehend — **n** bestand tegen — **e** a prueba de — **i** resistente a

11512 resistant colour — See fast colour.

11513 resistant to alcohol — See alcohol resistant.

11514 resisting — See resistant.

11515 resist mask — Mask consisting of a mask image printed on a resist coated over, or incorporated in, the image which is to be modified. This permits local treatment of the areas unprotected by the resist. In the Hausleiter process, image-wise hardening of the gelatine of a positive, after dichromate treatment, permits local dot-etching; in the Hensel process, an ink-albumen mask image on a photoengraving permits local etching; the Hausleiter-Sportelli method employs double exposures on carbon tissue. These methods can hardly be classified as masking, although their object is similar.

f masque *m* réserve — **d** Retuschemaske *f* — **n** retouche-masker *n* — **e** máscara *f* de reserva — **i** maschera *f* da ritocco

11516 resolution — Ability of a microscope to separate two objects lying close together. It is given approximately by the equation d = λ/2 NA, where d is the smallest distance between two particles or edges that can just be resolved by light of wavelength λ and an objective of numerical aperture NA.

f pouvoir *m* séparateur — **d** Auflösevermögen *n*; Auflösung *f* — **n** scheidend vermogen *n* — **e** poder *m* separador — **i** potere *m* di separazione; risolvenza *f*

11517 resolving power — Ability of a lens and the sensitized surface of a photographic material to sharply transmit and record very fine detail or lines.

f pouvoir *m* résolvant; résolution *f* — **d** Auflösevermögen *n*; Auflösungsvermögen *n* — **n** oplossend vermogen *n*; scheidend vermogen *n* — **e** poder *m* resolvente; poder *m* resolutivo; poder *m* de resolución; poder *m* separador — **i** potere *m* risolvente; potere *m* risolutivo

11518 resorcin — See resorcinol.

11519 resorcinol; resorcin; meta-dihydroxybenzene — Very white crystals, becoming pink on exposure to light when not perfectly pure. Soluble in water, alcohol, ether, glycerol, benzene and amyl alcohol, slightly soluble in chloroform. Latin: resorcinum.
Formula: $C_6H_4(OH)_2$.

f résorcine *f*; métadioxybenzène *m*; métaoxyphénol *m* — **d** Resorzin *n* — **n** resorcinol *n*; meta-dioxybenzeen *n* — **e** resorcina *f* — **i** resorcina *f*

11520 Respi Rototype screen — See Respi screen.

11521 Respi screen; Respi Rototype screen — Gravure screen consisting of two distinct screen rulings and ranges, used to increase the tonal range in direct gravure printing.

11522 response — See recipe mark.

11523 response time — Elapsed time between the generation of a message at a terminal and the receipt of a reply in case of an inquiry or receipt of message by addressee.

f temps *m* de réponse — **d** Antwortzeit *f* — **n** antwoordtijd *m* — **e** tiempo *m* de respuesta — **i** tempo *m* di risposta

11524 response time — Time required for the absolute value of the difference between the output and its final value to become and remain less than a specified amount, following the application of a step input or disturbance.

f temps *m* d'établissement; temps *m* de réponse — **d** Ansprechzeit *f*; Einstellzeit *f* — **n** aanlooptijd *m*; responsietijd *m* — **e** tiempo *m* de

respuesta — **i** tempo *m* d'assestamento; tempo *m* di risposta

11525 response time; transient time — Time required for the output to first reach a definite value after the application of a step input or disturbance.

f réponse *f* transitoire — **d** Übergangsantwort *f* — **n** overgangsantwoord *n*; overgangsresponsie *f* — **e** respuesta *f* transitoria — **i** risposta *f* transitoria

11526 responsible editor

f rédacteur *m* responsable; rédacteur *m* gérant — **d** verantwortlicher Redakteur *m* — **n** verantwoordelijke redacteur *m* — **e** redactor *m* responsable; redactor *m* gerente — **i** redattore *m* responsabile

11527 rest allowance — See compensating relaxation allowance.

11528 restrainer — Ingredient of a photographic developer which prevents too rapid development and chemical fog, usually potassium bromide.

f retardateur *m* — **d** Verzögerer *m* — **n** antisluiermiddel *n* — **e** retardador *m*; moderador *m* — **i** ritardatore *m*

11529 retention — Power of retaining.

f retenue *f* — **d** Aufbewahren *n* — **n** retentie *f*; bewaren *n*; opslaan *n* — **e** retención *f* — **i** ritenuto *m*

11530 retention — Amount of filler or other material which remains in the finished paper expressed as a percentage of that added to the furnish before sheet formation. This depends on the nature of the stock, the degree of beating and the speed of the paper machine.

f rétention *f* — **d** Retention *f*; Zurückhaltung *f* — **n** retentie *f* — **e** retención *f* — **i** ritenzione *f*

11531 reticulated screen — Usually associated with collotype, where the sensitized and exposed gelatine in drying on the glassplate contracts and forms an irregular pattern which acts as a screen.

f trame *f* réticulée — **d** Runzelkornraster *m* — **n** korrelraster *n*; kornraster *n* — **e** trama *f* reticulada — **i** retino *m* granulare

11532 reticulation — Blemish in the form of an irregular pattern or formation of small wrinkles occurring in gelatine images, when subjected to solutions of high temperature, or to successive solutions of different temperatures. Caused by rapid expansion and shrinkage of the gelatine film. Reticulation is done intentionally in the *collotype process.

f réticulation *f* — **d** Kräuseln *n*; Kräuselung *f*; Runzelkornbildung *f*; Kräuselkorn *n*; Schrumpfkorn *n* — **n** reticulatie *f* — **e** reticulado *m* — **i** reticulazione *f*

11533 retoucher — The workman who retouches.

f retoucheur *m* (à la plume); plumiste *m*; retoucheur *m* (à l'outil) — **d** Retuscheur *m* — **n** retoucheur *m* — **e** retocador *m* — **i** ritoccatore *m*

11534 retouching; touching up — Corrective treatment performed on photographic negatives and prints with pencils, crayons, reducers, transparent shading sheets, air brush and dyes for elimination of flaws and imperfections, and for general improvement of tone value, detail and photographic quality.

f retouche *f* — **d** Retusche *f*; Retuschearbeit *f*; Retuschieren *n* — **n** retouche *fm* — **e** retoque *m* — **i** ritocco *m*

11535 retouching brush

f pinceau *m* à retouche; pinceau *m* pour retouche — **d** Retuschierpinsel *m* — **n** retoucheerkwast *m* — **e** pincel *m* de retoque — **i** pennello *m* di ritocco

11536 retouching desk

f pupitre *m* à retouches; pupitre *m* de retouche — **d** Retuschierpult *n*; Retuschiertisch *m* — **n** retoucheertafel *fm* — **e** mesa *f* de retoque; pupitre *m* de retoque; pupitre *m* para retocar — **i** tavolo *m* da ritocco

11537 retouching dye — Black or red dye used for retouching and spotting of negatives and prints.

f gouache *f* de retouche — **d** Retuschierfarbe *f*; Keilitzfarbe *f* — **n** retoucheerverf *fm* — **e** tinta *f* de retoque; laca *f* para retoque — **i** colore *m* da ritocco; tinta *f* da ritocco

11538 retouching knife — See scraper.

11539 retouching pencil

f crayon *m* à retoucher — **d** Retuschierbleistift *m* — **n** retoucheerpotlood *n* — **e** lápiz *m* de retoque; lápiz *m* de retocar — **i** lapis *m* da ritocco

11540 re-transfer — Impression taken from an existing lithographic stone, by using a special ink and specially coated paper, for duplicating onto another lithographic surface.

f double décalque *m* — **d** Umdruck *m* — **n** overdruk *m* — **e** transporte *m* (litográfico); reporte *m* — **i** trasporto *m*; riporto *m*

11541 retrees; seconds; diamond — Slightly defective sheets of paper (mostly hand-made sheets) as distinguished from perfect sheets, but merchantable at lesser value. Removed by *sorting at the mill.

f second choix *m* — **d** zweite Wahl *f*; (Papier *n*) zweiter Qualität — **n** tweede keus *fm*; retiré *mn* — **e** hojas *fpl* de papel defectuosas; papel *m* defectuoso; papel *m* imperfecto; papel *m* de segunda (clase); quebrado *m* de segunda (clase) — **i** seconda scelta *f*

11542 retrieval — Process of recovering information or data by means of a computer system, usually from magnetic tape or disc storage, either by sequential search or by indexing the data so that it can be accessed randomly (i.e., non sequentially).

f recouvrement *m* — **d** Wiederauffinden *n* — **n** terugvinden *n*; ontsluiting *f*; informatie-ontsluiting *f* — **e** recuperación *f*; localización *f* — **i** reperimento *m*

11543 retroussage — Technique or action in etching or engraving, of drawing ink up from within the incised lines of an inked plate by deftly passing a soft cloth across its surface to spread ink to the adjacent areas.

11544 return — See link.

11545 return key

f clef *f* de renversement — **d** Rückstelltaste *f* — **n** terugstelltoets *m* — **e** tecla *f* de restitución — **i** tasto *m* di restituzione

11546 returns; overissues; crabs — Copies of books, newspapers or magazines which have not been sold. See also flat printed news.

f invendus *mpl*; exemplaires *mpl* invendus; retour *m* des invendus; retour *m* en librairie; bouillon *m* (pour les journaux et périodiques) — **d** Remittenden *fpl*; Restauflage *f*; Krebse *mpl* — **n** onverkochte exemplaren *npl*; overgehouden exemplaren *npl*; restanten *npl*; kreeften *mpl* — **e** devoluciones *fpl*; remanente *m* de ejemplares; ejemplares *mpl* sobrantes; sobrantes *mpl*; residuo *m* — **i** copie *fpl* invendute; resa *f* di copie non vendute; rimanenze *fpl*

11547 return to zero — Method of recording data on a magnetic surface in which binary 1 and 0 are represented by opposite directions of magnetism and the signal returns to zero after recording each bit. Superseded by the non-return to zero mode of recording.

11548 reversal — Phenomenon that, if areas of a photographic emulsion layer are exposed to increasing amounts of light, the densities produced after development will not increase indefinitely but will reach a limit and then, with continued exposure, will diminish somewhat. This is known as reversal or *solarization. (The term solarization was originally given to highly over-exposed print-out images in which a change of reflectance of the image occurred, producing bronzed shadows.) Thus, after a certain degree of exposure has been reached, further exposure of the photographic grain tends to destroy the developable state induced by the earlier part of the exposure. The term reversal should include all positive-working photographic effects.

f inversion *f* — **d** Umkehrung *f* — **n** omkering *f* — **e** inversión *f*; reversión *f* — **i** inversione *f*; reversione *f*

11549 reversal of image; reversion of image

f inversion *f* de l'image — **d** Bildumkehrung *f*; Umkehrung *f* — **n** beeldomkering *f*; omkering *f* — **e** inversión *f* de la imagen — **i** inversione *f* dell'immagine

11550 reversal process — Processing of an exposed film so that it will become a positive instead of a negative. Reversing a screened negative allows the retoucher on four-colour work to match his plates while working with a positive.

f procédé *m* d'inversion; procédé *m* de retournement; traitement *m* reversible — **d** Umkehrverfahren *n* — **n** omkeerprocédé *n* — **e** procedimiento *m* de inversión; método *m* de inversión; proceso *m* reversible — **i** procedimento *m* di inversione; trattamento *m* di inversione

11551 reversed negative — Negative being right-left reversed as made through a prism or lens.

f négatif *m* inversé — **d** umgekehrtes Negativ *n*; seitenverkehrtes Negativ *n* — **n** omgekeerd negatief *n* — **e** negativo *m* invertido — **i** negativo *m* invertito; negativo *m* con immagine rovesciata

11552 reversed P; blind P — Paragraph mark ¶ , to imply the beginning of a new paragraph, usually when this is run on without a new line. Also, sixth reference mark for footnotes.
f pied *m* de mouche — **d** Alineazeichen *n*; neuer Absatz *m*; Hinweiszeichen *n* — **n** alineateken *n*; verwijzingsteken *n*; nootteken *n* — **e** signo *m* de párrafo; llamada *f* — **i** segno *m* di alinea; segno di riferimento

11553 reversed plate — See reverse plate.

11554 reverse indention — See hanging indention.

11555 reverse leading — Ability of the typesetting device to move the film or paper in the opposite direction to achieve certain typographic effects, such as to raise an image above the base line of the type, or to lay down a second column of text alongside a first column which has already been composed.

11556 reverse lettering; white-out lettering — Open lettering on a printed background.
f caractères *mpl* noir-au-blanc — **d** negative Schrift *f* — **n** negatief schrift *n* — **e** caracteres *mpl* negativos; letras *fpl* negativas; letras *fpl* blancas sobre fondo negro — **i** caratteri *mpl* in negativo; caratteri *mpl* bianci su fondo stampato

11557 reverse plate; reversed plate — Printing plate photomechanically reversed from type, decoration, or illustration so that black design on white paper becomes white design against background.
f cliché *m* inversé; cliché *m* noir au blanc — **d** Negativklischee *n* — **n** negatief cliché *n*; cliché *n* met witte letters op zwarte ondergrond — **e** clisé *m* negativo — **i** cliscè *m* negativo; cliché *m* negativo; lastra *f* di stampa invertita; lastra *f* di stampa negativa

11558 reverse printing — See back printing.

11559 reverse roll coating — Roll coating process in which the surface of the applicator roll is travelling in a direction opposite to that of the paper; generally a separate coating operation.
f couchage *m* au rouleau inversé — **d** Gegenlaufwalzen-Streichverfahren *n*; Umkehrwalzen-Streichverfahren *n* — **e** procedimiento *m* de estucado con rodillo invertido — **i** patinatura *f* a rullo invertito

11560 reversible unit — Printing unit with reversible drives which can print in either direction of rotation.
f groupe *m* réversible — **d** umschaltbares Druckwerk *n*; umsteuerbares Druckwerk *n* — **n** omkeerbare drukeenheid *f* — **e** grupo *m* impresor reversible; unidad *f* impresora reversible; elemento *m* impresor reversible; grupo *m* de impresión reversible — **i** gruppo *m* stampante reversibile

11561 reversing mirror; reversion mirror
f miroir *m* d'inversion — **d** Umkehrspiegel *m* — **n** omkeerspiegel *m* — **e** espejo *m* inversor — **i** specchio *m* a rovesciare

11562 reversing press — See proving and reversing press.

11563 reversing prism; inverting prism; prism — Optical instrument in the form of a triangular piece of glass, silvered on one side. Used in combination with the camera lens to reverse the image from right to left, or to make it read right when it would read wrong with the lens alone. The copyboard must be placed at right angles to the plate or film in the camera.
f prisme *m* de retournement — **d** Umkehrprisma *n*; Prisma *n* — **n** omkeerprisma *n*; prisma *n* — **e** prisma *m* inversor; prisma *m* — **i** prisma *m* di rovesciamento; prisma *m* d'inversione

11564 reversion mirror — See reversing mirror.

11565 reversion of image — See reversal of image.

11566 review — See journal.

11567 review *v* a book
f faire la critique — **d** ein Buch rezensieren; ein Buch beurteilen; ein Buch besprechen — **n** een boek recenseren; een boek beoordelen; een boek bespreken — **e** reseñar un libro — **i** criticare un libro

11568 review copy; reviewer's copy; press-copy — Copy of a publication sent to a reviewer for sales promoting.
f exemplaire *m* de critique; exemplaire *m* de presse; exemplaire *m* de service de presse — **d** Rezensionsexemplar *n*; Besprechungsexemplar *n*; Besprechungsstück *n* — **n** recensie-exemplaar *n*; exemplaar *n* ter bespreking — **e** ejemplar *m* para la crítica de la prensa; ejemplar *m* de publicidad — **i** esemplare *m* per recensione

11569 reviewer's copy — See review copy.

11570 revise; revise proof; revised proof; second proof — Second or subsequent proof with corrections made from the previous proof.
f révision *f*; revision *f* — **d** Revision *f*; zweiter Abzug *m* — **n** revisie *f*; tweede proef *fm* — **e** segunda prueba *f*; contraprueba *f*; prueba *f* posterior — **i** seconda bozza *f*

11571 revise *v* — To compare a marked proof with corrections made from the previous proof.
f réviser; reviser — **d** Revision lesen — **n** nakijken — **e** revisar — **i** rivedere

11572 revised and enlarged edition
f édition *f* corrigée et amplifiée; édition *f* corrigée et augmentée; édition *f* revue et augmentée — **d** verbesserte und vermehrte Auflage *f*; verbesserte und erweiterte Auflage *f*; durchgesehene und vermehrte Ausgabe *f* — **n** verbeterde en vermeerderde druk *m*; herziene en vermeerderde druk *m* — **e** edición *f* corregida y aumentada; edición *f* emendada y ampliada; edición *f* revisada y aumentada — **i** edizione *f* migliorata e ampliata; edizione *f* riveduta e accresciuta; ristampa *f* corretta ed arricchita di nuove aggiunte

11573 revised edition — Second or subsequent edition with amendments and/or corrections.
f édition *f* revisée — **d** revidierte Ausgabe *f*; verbesserte Ausgabe *f*; verbesserte Auflage *f* — **n**

herziene uitgave f; verbeterde uitgave f; herziene druk m; verbeterde druk m — **e** edición f corregida — **i** edizione f riveduta

11574 revised proof — See revise.

11575 revise proof — See revise.

11576 reviver — Chemical added to metal to restore the qualities lost by oxidation.
f régénérateur m du métal — **d** Metallreaktivierungsmittel n — **n** metaalreinigingsmiddel n — **e** depurador m de metal — **i** rigeneratore m del metallo

11577 reviving alloy; correction metal — Reviver of type to restore tin and antimony after frequent meltings.
f alliage m de remise au titre; alliage m de retitrage — **d** Zusatzmetall n — **n** toevoegingsmetaal n — **e** metal m de adición; metal m de regeneración — **i** aggiunta f di rititolazione

11578 revolution; turn
f révolution f; rotation f; tour m — **d** Umdrehung f; Tour f — **n** omwenteling f; toer m — **e** vuelta f; revolución f — **i** rivoluzione f; giro m

11579 revolutions per minute — Number of times per minute the main shaft of an engine revolves. Abbrev. RPM, in scientific work rev/min (no points).
f tours mpl à la minute — **d** Umdrehungen fpl je Minute — **n** omwentelingen fpl per minuut; toeren mpl per minuut — **e** revoluciones fpl por minuto — **i** rivoluzioni fpl per minuto

11580 rewind; rewinder; re-reeler — Delivery mechanism, frequently used with reel-fed printing presses, to rewind the printed web after it has passed through the press.
f rebobineuse f; enrouleuse f; système m de rebobinage; rembobineur m; groupe m de rebobinage; dispositif m de réembobinage — **d** Aufwickelvorrichtung f; Wiederaufwickler m; Wiederaufrollvorrichtung f; Aufwickler m; Aufroller m; Umroller m — **n** opwikkelinstallatie f; opwikkelinrichting f; overwikkelmachine f — **e** rebobinador m; rebobinadora f; máquina f rebobinadora — **i** ribobinatrice f

11581 rewind v — To take several butt ends of reels of paper and wind them into a reel for use in the press.
f rembobiner — **d** wiederaufwickeln — **n** weer oprollen; omrollen — **e** rebobinar; reenrollar — **i** ribobinare

11582 rewinder — See rewind.

11583 rex — A British standard size of note paper. Also a British standard card size. See app. no. 3.

11584 rheological properties; flow properties
f propriétés mpl rhéologiques — **d** rheologische Eigenschaften fpl — **n** reologische eigenschappen fpl; vloei-eigenschappen fpl — **e** propiedades fpl reológicas — **i** proprietà fpl reologiche

11585 rheology — Science of the flow properties of materials, such as printing ink.
f rhéologie f — **d** Rheologie f; Fließkunde f; Fließ-

lehre f — **n** reologie f; stromingsleer fm; vloeiingsleer fm — **e** reología f — **i** reologia f

11586 rhodamine; rhodamine B — Green crystals or reddish violet powder. Very soluble in water and alcohol, forming a bluish red, fluorescent solution, slightly soluble in acids or alkalis. Class of clean organic reds possessing good light-fastness, often called magenta in process printing. Also used as a dye in paper. Formula: $C_{28}H_{31}ClN_2O_3$.
f rhodamine B f; rose m brillant B — **d** Rhodamin B n — **n** rhodamine B f — **e** rodamina B f — **i** rodamine B f

11587 rhodamine B — See rhodamine.

11588 rhombohedral screen — See Schulze screen.

11589 ribbed back
f dos m à nerfs — **d** Rücken m mit (erhabenen) Bünden — **n** rug m met ribben — **e** lomo m con nervios — **i** dorso m a nervi

11590 ribbon face — See typescript.

11591 ribbon fold; angle-bar fold — Fold that does not require folding by a former folder for the first fold. The web is slit into multiple ribbons, or narrow webs, which use angle bars to by-pass the former. The ribbons are brought together at the jaw folder for holding and cut-off into desired signatures.
f coupe f à demie-circonférence — **d** Buchfalz m — **n** boekvouw m — **e** plegado m de bandas — **i** piega f alla barra diagonale

11592 ribs — See raised bands.

11593 ribs, plate — See plate ribs.

11594 rice paper misn. — Sheet material cut from the pith of a small tree (Aralia papyrifera) which grows in the swampy forests of Formosa. It is white and smooth and much used by the Chinese for painting.
f papier m de riz — **d** Reispapier n; Reisstrohpapier n — **n** rijstpapier n — **e** papel m de arroz — **i** carta f riso

11595 rice straw — Straw for the manufacture of chip and newsboard grades and of butcher's paper in Europe.
f paille f de riz — **d** Reisstroh n — **n** rijststro n — **e** paja f de arroz — **i** paglia f di riso

11596 ricinus oil — See castor oil.

11597 rider — Additional manuscript added to a proof.
f béquet m — **d** Einschaltung f — **n** inlas m; inlassing f; aanvulling f — **e** banderilla f — **i** aggiunta f; manoscritto m supplementare inserito nelle bozze

11598 rider — See rider roller.

11599 rider gauge — Device to check accuracy of cylinder diameter.
f calibre m à patins — **d** Reiterlehre f — **n** opzetkaliber n; cilinder-meetbrug fm — **e** calibre m de puente; calibre m para cilindro — **i** calibro m a cavaliere

11600 rider roller; riding roller; rider; distributing rider; riding changer US — Freely mounted, metal or rigid plastic roller on top of the other rollers in the inking mechanism which contact one or more soft (glue-glycerine, rubber, etc.) rollers and serve to break down, transfer and distribute the ink. Soft rollers are sometimes used as riders on large metal ink drums to break down the ink.

f rouleau *m* chargeur; chargeur *m* — **d** Reiterwalze *f*; Reiter *m*; Beschwerwalze *f* — **n** ruiterrol *f* — **e** rodillo *m* cargador; rodillo *m* de carga; cargador *m* — **i** rullo *m* cavaliere; rullo *m* caricatore

11601 ridging — Formation of peripheral ridges in ink films on rollers due to insufficient sidewise distribution.

f formation *f* de bourrelets — **d** wulstartiges Aufbauen *n* von Druckfarbe auf den Farbwalzen — **n** optreden *n* van inktringen — **e** formación *f* de anillos de tinta — **i** accumulo *m* di depositi d'inchiostro sui rulli

11602 riding changer — See rider roller.

11603 riding roller — See rider roller.

11604 riffler; sand trap; sand table; sand shifter — Trough or channel at bottom of paper machine through which a very dilute suspension of stock flows, to eliminate the heavy impurities from the suspension by gravity; sometimes fitted with submerged baffles.

f sablier *m* — **d** Sandfang *m* — **n** zandvang *m*; zandvanger *m* — **e** separador *m* de arena; arenero *m* — **i** sabbiere *m*

11605 right angle; angle of ninety degrees — An angle whose sides are perpendicular to each other. An angle of 90 ° (or $\pi/2$ radius).

f angle *m* droit; angle *m* de quatre-vingt-dix degrés — **d** rechter Winkel *m*; Winkel *m* von neunzig Grad — **n** rechte hoek *m*; hoek *m* van negentig graden — **e** ángulo *m* recto; ángulo *m* de noventa grados — **i** angolo *m* retto; angolo *m* di novanta gradi

11606 right-angled triangle

f triangle *m* rectangle — **d** rechteckiges Dreieck *n* — **n** rechthoekige driehoek *m* — **e** triángulo *m* rectángulo; triángulo *m* ortogonio — **i** triangolo *m* rettangolare

11607 right-angle folding

f pliage *m* en cahiers; plis *mpl* croisés; croisés *mpl* — **d** Kreuzbruchfalzung *f* — **n** kruisvouw *fm* — **e** plegado *m* de ángulo recto — **i** piega *f* incrociata

11608 right-hand pages — See odd pages.

11609 right-hand thread

f filet *m* (de vis) avec pas à droite; filet *m* à droite — **d** Rechtsgewinde *n*; rechtsgängiges Gewinde *n* — **n** rechtse schroefdraad *m*; rechtse draad *m* — **e** filete *m* (de mano) derecho; filete *m* al derecho; rosca *f* a derechas — **i** filettura *f* destrorsa

11610 right of translation

f droit *m* de traduction — **d** Übersetzungsrecht *n* — **n** vertalingsrecht *n*; vertaalrecht *n* — **e** derecho *m* de traducción — **i** diritto *m* di traduzione

11611 right time — See real time.

11612 rigid box — See set-up box.

11613 rigidity; stiffness; inflexibility — Resistance to bending. The resistance against deformation of a flat-bed printing press. This to be measured with the *pre-tension meter.

f rigidité *f* — **d** Steifigkeit *f*; Biegesteifigkeit *f*; Starrheit *f*; Strammheit *f*; Stärke *f*; Steife *f*; Widerstand *m* gegen Verbiegungen — **n** stijfheid *f*; onbuigzaamheid *f*; sterkte *f* — **e** rigidez *f* — **i** rigidezza *f*; rigidità *f*

11614 rigid paper container — See set-up box.

11615 rimmed letter

f caractère *m* encadré; caractère *m* cerné — **d** umstochene Schrift *f*; musierte Schrift *f* — **n** opengestoken letter *fm* — **e** letra *f* adornada — **i** carattere *m* incorniciato

11616 Rinco process — Direct method for making positives for magazine work.

11617 ring — See rattle.

11618 ring binder — Loose-leaf binder in which the sheets are held by two or more rings which can be opened.

f reliure *f* à anneaux — **d** Ringbinder *m*; Ringhefter *m* — **n** ringband *m* — **e** carpeta *f* de argollas; carpeta *f* de anillas — **i** legatrice *f* ad anello

11619 rinse *v* — To give a short wash in water.

f rincer — **d** abspülen; spülen; auswaschen — **n** afspoelen; spoelen; uitwassen — **e** enjuagar; lavar en chorro — **i** risciacquare; sciacquare

11620 rising; working up — Loosening of the printing forme, due to faulty lock up, allowing the spacing material to rise, which results in blurs on the printed sheet.

f levage *m* des blancs — **d** Spießen *n*; Hochgehen *n* der nicht druckenden Teile einer Druckform — **n** witrijzen *n* — **e** levantamiento *m* de blancos; levantamiento *m* de espacios; alzamiento *m* de blancos — **i** sollevamento *m* degli spazi

11621 river — See gutter.

11622 road map

f carte *f* routière — **d** Wegenkarte *f* — **n** wegenkaart *fm* — **e** mapa *m* caminero — **i** carta *f* stradale

11623 road-map paper — Sulphite bond or offset paper for printing road maps.

f papier *m* pour cartes routières — **d** Wegekartenpapier *n*; Landkartenpapier *n* — **n** papier *n* voor wegenkaarten; landkaartenpapier *n*; kaartpapier *n* — **e** papel *m* para mapas camineros; papel *m* para mapas geográficos; papel *m* para atlas — **i** carta *f* per carte stradali

11624 roan — Fine book leather made from sheepskin, given a sumac tannage and finished with a straight grain or hard (pinhead) grain surface. See also rutland.

f mouton *m* — **d** Schafleder *n*; Schaf *n* — **n** schapeleer *n*; schaapsleer *n* — **e** cuero *m* de carnero — **i** cuoio *m* di pecora

11625 Rochelle salt — See sodium-potassium tartrate.

11626 rocker — Serrated edged rocking tool used in mezzotint engraving, forming small recesses over the whole plate surface. It throws up a burr of displaced metal. Gradation of tone is obtained by using scrapers and burnishers to remove the burr and reduce the depth of the recesses.
f berceau *m* — **d** Mattoir *n*; Schaukeleisen *n*; Wiegemesser *n*; Wiegestahl *m*; Graniierstahl *m* — **n** wieg *fm*; wiegijzer *n* — **e** graneador *m* — **i** granitore *m*

11627 Rockwell hardness — See hardness.

11628 rodinal — Trade name for a photographic developer consisting of an alkaline solution of *para-aminophenol with *sodium bisulphite. Other trade names are: citol, activol, certinal, paranol, kalogen.
Formula: $NH_2C_6H_4OH$.
f rodinal *m*; développeur *m* au paraminophénol — **d** Rodinal *m*; Paraminophenol-Entwickler *m* — **n** rodinal *m*; paraminofenol-ontwikkelaar *m* — **e** rodinal *m*; revelador *m* de paraminofenol — **i** rodinal *m*; rivelatore *m* di para-amminofenolo; sviluppatore *m* di para-amminofenolo

11629 roll — See bar roll.

11630 roll coated paper — Paper coated by a roll coating process.
f papier *m* couché au rouleau — **d** walzengestrichenes Papier *n*; walzenbeschichtetes Papier *n* — **n** papier *n* met rolstrijklaag; papier *n* met rolcouche; papier *n* met rolcoating — **e** papel *m* estucado a rodillo — **i** carta *f* patinata a rullo

11631 roll coating — Process in which coating colour is applied to one or both sides of a paper web by transfer from a rubber applicator roll onto which the coating colour has been metered. It may be carried out on or off the paper machine.
f couchage *m* par rouleaux — **d** Walzenstreichverfahren *n*; Walzenstrich *m* — **n** rolstrijkprocédé *n*; rolcoating *f depr.* — **e** procedimiento *m* de estucado por rodillos; estucado *m* por rodillos — **i** patinatura *f* a rulli

11632 rolled Scotch — See transfer paper.

11633 roller — Composition coated cylinder for inking type formes or printing plates in printing operations.
f rouleau *m* — **d** Walze *f* — **n** rol *fm* — **e** rodillo *m* — **i** rullo *m*

11634 roller bearing — Bearing consisting of an inner and outer steel race between which a number of rollers are fitted which are held in place with a cage.
f roulement *m* à rouleaux — **d** Rollenlager *n* — **n** rollager *n* — **e** cojinete *m* de rodillos; rodamiento *m* de rodillos — **i** cuscinetto *m* a rulli

11635 roller black — Black pigment used in the manufacture of printing inks.
f noir *m* roller — **d** Roller-black *n* — **n** rollerzwart *n* — **e** negro *m* roller — **i** nero *m* roller

11636 roller bracket — See roller fork.

11637 roller carriage — Adjustable bearing bracket carrying an inking roller.
f peigne *m* — **d** Walzenschloß *n* — **n** rollenslot *n*

— **e** peine *m*; cierre *m* de los rodillos — **i** forca *f* portarulli

11638 roller composition — Stiff, jelly-like mixture of glue, sugar, glycerol, etc., that is moulded on letterpress ink rollers.
f pâte *f* à rouleaux — **d** Walzenmasse *f* — **n** rollerspecie *f* — **e** composición *f* de pasta; composición *f* de cola; pasta *f* de rodillos; pasta *f* para rodillos; pasta *f* gelatinosa — **i** pasta *f* per rulli; pasta *f* da rulli

11639 roller fork; roller bracket — Attachment on a printing press in which the ink rollers rest.
f fourchette *f* de rouleau — **d** Walzenlager *n* — **n** rolhouder *fm* — **e** horqueta *f* de rodillo; horquilla *f* de rodillo; portarrodillos *m* — **i** forcella *f* portarulli

11640 roller journals — Ends of a roller stock which revolve in bearings.
f tourillons *mpl* des rouleaux — **d** Walzenzapfen *mpl* — **n** roltappen *mpl*; tappen *mpl* van de rol — **e** gorrones *mpl* del rodillo; muñones *mpl* del rodillo; espigas *fpl* del rodillo — **i** perni *mpl* del rullo

11641 roller marks; roller streaks
f marques *fpl* de rouleaux; stries *fpl* de rouleaux — **d** Walzenstreifen *mpl* — **n** rollenstrepen *fmpl* — **e** ráfagas *fpl* en la impresión; barras *fpl* de rodillos — **i** strisci *fpl* dei rulli

11642 roller mill — Mill consisting of 2-5 horizontal steel rollers, mounted in a frame and geared to rotate consecutively in opposite directions, with the first pair of rollers rotating toward each other. Progressing from the first roller to the last, each roller rotates at a higher rate of speed than the preceding roller, this producing a grinding or shearing action.
f broyeuse *f* à rouleaux — **d** Walzenstuhl *m* — **n** inktwals *fm* — **e** molino *m* a rodillos; molino *m* de cilindros — **i** mulino *m* a rulli

11643 roller mould — See also gun.
f moule *m* à rouleaux — **d** Walzengießhülse *f*; Walzengießform *f*; Walzengießrohr *n*; Gießhülse *f* — **n** gietkoker *m* voor rollen; rollengietkoker *m* — **e** molde *m* para fundir rodillos — **i** forma *f* per la fusione dei rulli

11644 roller rack
f raquette *f* porte-rouleau — **d** Walzenständer *m* — **n** rollenstandaard *m*; rollenstander *m* — **e** portarrodillos *m* — **i** sostegno *m* per rulli

11645 roller reciprocation — Lateral movement of roller(s) in the inking system to improve distribution.
f mouvement *m* de va-et-vient des rouleaux — **d** Hinundherbewegung *f* der Walzen — **n** heen- en weergaande beweging *f* van de rollen — **e** movimiento *m* de vaivén de los rodillos; movimiento *m* alternativo de los rodillos — **i** movimiento *m* di va e vieni dei rulli

11646 roller setting — Adjusting the amount of contact area between inking rollers or between roller and forme, plate, or drum.
f réglage *m* des rouleaux — **d** Einstellen *n* der

Walzen; Walzeneinstellung f — **n** stellen n van de rollen — **e** colocación f de los rodillos — **i** regolazione f dei rulli

11647 roller spindle — See roller stock.

11648 roller star; triple reel stand; triple roll stand — Reel stand on a rotary press, with three arms.
f porte-bobine m en trèfle; porte-bobine m triple; porte-bobine m à trois positions; dispositif m en trèfle; trèfle m; disposition f en étoile — **d** Rollenstern m — **n** rollenster fm; rollenstandaard m; drierollenster m — **e** portarrollos m de estrella; estrella f; trébol m; portarrollo-tensorempalmador m automático — **i** portabobine m a stella; portabobine m triplo

11649 roller stock; roller spindle; core — Cylindrical metal bar upon and around which the composition is moulded in the making of an ink roller.
f mandrin m de rouleau — **d** Walzenspindel f; Spindel f — **n** rolspil fm; rollenspil fm; spil fm — **e** alma f del rodillo; alma f para rodillo; ánima f del rodillo; ánima f para rodillo; armazón f del rodillo; husillo m del rodillo; eje m del rodillo; mandril m del rodillo — **i** anima f del rullo

11650 roller streaks — See roller marks.

11651 roller stripping; stripping — Lithographic term denoting that the ink does not stick to the metal ink rollers on the press.
f blanchissement m des tables; refus m d'encre — **d** Blanklaufen n der Walzen; Blankwerden n der Walzen; Abstoßen n der Druckfarbe auf Metallwalzen — **n** kaallopen n van de rollen — **e** descarga f de los rodillos metálicos — **i** rifiuto m dell'inchiostro da parte dei rulli

11652 roller top of former; drag roller *US* — Roller immediately above the former in a rotary press. Abbrev. RTF.
f cannelé m (supérieur) de triangle — **d** Trichtereinführwalze f — **n** trechterinvoerrol fm; trechterrol fm — **e** rodillo m superior del triángulo — **i** rullo m d'introduzione nel cono

11653 roller trolley — Small trolley on a rail which facilitates the entry and withdrawal of a large ink roller.
d Walzeneinhebevorrichtung f — **n** inktrollenwagen m — **i** dispositivo m per l'introduzione di rulli

11654 roller trolley — Movable carriage which allows the roller or inking system to be moved clear of the printing unit.
f trolley m — **d** Walzentransportwagen m — **n** inktrollen-transportwagen m; inktrollenwagen m — **e** carro m de los rodillos — **i** carrello m portarulli

11655 roller washing machine
f dispositif m laveur de rouleaux — **d** Walzenwaschmaschine f — **n** rollenwasmachine f — **e** dispositivo m para lavar rodillos; lavador m de rodillos; aparato m para lavar rodillos; aparato m lavarrodillos; dispositivo m lavador; aparato m limpiarrodillos; limpiarrodillos m — **i** lavarulli m;

macchina f per lavare i rulli; macchina f per il lavaggio dei rulli

11656 roll film — Photographic film supplied in a light-tight roll.
f film m en bobine — **d** Rollfilm m — **n** rolfilm f — **e** película f en rollo — **i** pellicola f in bobina; rotolo m di pellicola

11657 roll-film camera — Process camera for use of films in roll form, accommodating rolls of different width and provided with a shearing device for cutting exposures from the rolls.
f chambre f pour film en rouleaux; chambre f pour pellicule en bobines — **d** Rollfilm-reproduktionskamera f — **n** rolfilm-camera fm (voor reproduktietechniek) — **e** cámara f (fotográfica) de película en rollo; máquina f de película enrollada — **i** apparecchio m (fotografico) a rullo di pellicole; macchina f fotografica per pellicole in bobine

11658 roll-film holder
f châssis m pour pellicule en bobine — **d** Rollfilmkassette f; Rollkassette f — **n** rolfilm-cassette f — **e** portador m de películas en rollo; portador m de películas enrolladas; portapelículas m — **i** telaio m per pellicola in bobina

11659 rolling — True rolling of a printing press requires that the speed of the sheet fitted on the packing of the cylinder equals the speed of the printing forme.
f déroulement m — **d** Abwicklung f; Abwickelung f — **n** afwikkeling f — **e** desarrollo m — **i** rotolamento m

11660 rolling defects
f défauts mpl de déroulement — **d** Abwicklungsfehler mpl — **n** afwikkelingsfouten $fmpl$ — **e** defectos mpl del desarrollo — **i** difetti mpl di rotolamento

11661 rolling press — See copperplate press.

11662 roll of paper — See reel of paper.

11663 roll v **out** — See read v out.

11664 roll stand — See reel stand.

11665 roll tender — See reel hand.

11666 roll v **up; ink** v **up** — To apply ink to metal plates with a roller either to create an acid resist, or to deposit ink on the surface of a printing plate to take an impression therefrom.
f encrer au rouleau — **d** auftragen; aufwalzen; einwalzen; Farbe auftragen; anfärben — **n** oprollen; inrollen — **e** entintar; dar tinta; batir la tinta — **i** inchiostrare con rullo a mano

11667 roll-up process — Method of etching halftone plates by rolling up the heated plate after the flat etch with a roller charged with etching ink, the heat of the plate causing the applied ink to melt and run down the sides of the dots.
f gillotage m; procédé m de gillotage — **d** Einwalzverfahren n — **n** inrol-methode f — **e** procedimiento m de entintaje — **i** processo m rolup

11668 ROM — See programmable read-only memory.

11669 romain du roi — Type face designed by Philippe Grandjean for the exclusive use in the

Imprimerie Royale. The first book printed with this type appeared in 1702. The work was completed in 1745 by Jean Alexandre and Louis Luce, prominent punch cutter (1695-1774), the son-in-law of Alexandre.

11670 roman; roman type — Normal upright letter as distinct from italic, bold or fancy type, etc.

f romain *m*; caractère *m* romain — **d** geradestehende Schrift *f*; aufrecht; gewöhnlich; nicht kursiv — **n** romein — **e** romano; redondo; letra *f* romana; letra *f* redonda — **i** tondo *m*; tipo *m* tondo

11671 Roman figures; Roman numerals — The caps I, V, L, D, M are not to be followed by a point. Lower case i, ii, etc., should be used for the pagination of preliminary matter. I = 1; II =2; V = 5; X = 10; L = 50; C = 100; D = 500; M = 1000. If the lesser number is placed after the greater, the lesser is added to the greater, XI = 11. If the lesser is placed before the greater, the lesser is deducted from the greater, IX = 9; XC = 90. Sometimes figures are represented by the contraction of other characters as in CIƆ, (1000), or in IƆCXXVi, (1626); in which CIƆ is a typographic form for the Roman M and IƆ for the Roman D. Thus: DC or IƆC = 600; CM = DCCC = IƆCCCC = 900; MM = CIƆCIƆ = 2000.

f chiffres *mpl* romains — **d** römische Ziffern *fpl*; römische Zahlen *fpl*; römische Zahlenzeichen *npl* — **n** Romeinse cijfers *npl* — **e** cifras *fpl* romanas — **i** cifre *fpl* romane; numeri *mpl* romani

11672 Roman numerals — See Roman figures.

11673 romantic binding — Book binding which shows a decoration in a non classical style.

f reliure *f* romantique — **d** Romantiker-Einband *m* — **n** romantieke band *m* — **e** encuadernación *f* romántica; encuadernación *f* al estilo romántico — **i** legatura *f* romantica; legatura *f* al stile romantico

11674 roman type — See roman.

11675 roof mirror — Combination of two mirrors meeting at exactly 90 ° which show a correct and not reversed image.

f miroir *m* en toit — **d** Dachspiegel *m* — **n** dakspiegel *m* — **i** prisma *m* a tetto

11676 room temperature — Many of the paper troubles occurring on the printing press are occasioned by insufficient conditioning of the paper to the room temperature. Paper should be left to stand at least 24 hours in the press-room before being printed.

f température *f* ambiante — **d** Raumtemperatur *f* — **n** temperatuur *f* der omgeving; kamertemperatuur *f* — **e** temperatura *f* ambiente — **i** temperatura *f* ambiente; temperatura *f* ambientale

11677 root — In mathematics, a quantity which, when multiplied by itself a certain number of times produces a given quantity. The number 2 is the square root of 4, the cube root of 8 and the fourth root of 16.

f racine *f* (racine carrée; racine cubique; racine quatrième) — **d** Wurzel *f* (Quadratwurzel, zweite Wurzel; Kubikwurzel, dritte Wurzel; vierte Wurzel) — **n** wortel *m* (vierkantswortel, tweedemachtswortel; derdemachtswortel; vierdemachtswortel) — **e** raíz *f* (raíz cuadrada; raíz cúbica; raíz bicuadrada) — **i** radice *f* (radice quadrata; radice cubica)

11678 root mean square — Square root of the arithmetic mean of the squares of the numbers in a given set of numbers, used as a measure of surface roughness.

11679 root sign; square root sign; radical sign — Radical sign $\sqrt{}$ or $\sqrt[2]{}$ used in mathematical work. See also mathematical signs.

f racine *f*; racine *f* carrée; radical *m*; signe *m* radical — **d** Wurzelzeichen *n*; Quadratwurzelzeichen *n* — **n** wortelteken *n*; vierkantworteltaken *n* — **e** radical *m*; signo *m* radical; signo *m* de raíz (cuadrada) — **i** radicale *m*; segno *m* di radice (quadrata)

11680 ROP colour — Abbrev. for "run-of-paper colour", sometimes "run of print" or "run of press". Any colour, other than black, applied to newsprint during the normal press run. There are two types, i.e., spot colour and full colour or process colour. Spot colour normally consists of a black key drawing with a second colour being printed as a solid or tint, to emphasize the advertisement. This type of colour work is normally produced through the use of overlays on a base drawing. Full colour or process colour reproduces an original in its full normal colours. This is normally produced from the original coloured art or photograph by a technique of colour separation.

f rop-colour *m*; impression *f* de deux ou trois couleurs d'un journal quotidien — **d** Rop-Farbdruck *m* — **n** rop-colour *m*; twee of meer kleuren-krantenrotatiedruk *m* — **e** rop-color *m*; impresión *f* dos o tres colores en la prensa diaria — **i** rop-color *m*; stampa *f* occasionale a colori sui giornali

11681 rope manilla paper — See manilla paper.

11682 rose bottom bag — Block-bottom bag with a small tag uniting the abutting points of the folded bottom. The open bag has a rectangular cross-section.

f sac *m* à fond croisé — **d** Kreuzbodenbeutel *m* — **n** kruisbodemzak *m* — **e** bolsa *f* con fondo cruzado — **i** sacchetto *m* a fondo incrociato

11683 rose engine machine — Term sometimes used for a *guilloching machine.

11684 rosin; gum rosin; colophony — Resin in a solid state obtained as a residue after the distillation of oil of turpentine from crude turpentine, which comes from the gum of the southern pine (chiefly the long-leaf species). It consists mainly of abietic acid. Freely soluble in alcohol, benzene, ether, glacial acetic acid, oils, carbon disulphide, dilute solutions of fixed alkali hydroxides, insoluble in water.

f colophane *f*; résine *f* naturelle (de pin) — **d** Kolophonium *n* — **n** colofonium *n* — **e** colofonia *f*; pez *f* griega — **i** colofonia *f*; pece *f* greca

11685 rosin content
f pourcentage *m* de résine — **d** Harzgehalt *m* — **n** harsgehalte *n* — **e** contenido *m* de resina; contenido *m* en resina — **i** contenuto *m* in resina naturale; contenuto *m* in resina; tenore *m* in resina

11686 rosin ester gum — See ester gum.

11687 rosin oil; rosinol — Viscous oil obtained from the destructive distillation of rosin, used in cheaper printing inks as an adulterant for boiled linseed oils and for impregnating paper. Soluble in ether, chloroform, fatty oils and carbon disulphide, slightly soluble in alcohol, insoluble in water.
f huile *f* de résine — **d** Harzöl *n* — **n** harsolie *fm* — **e** aceite *m* de resina — **i** olio *m* di resina

11688 rosinol — See rosin oil.

11689 rosin size — Soap solution obtained by cooking rosin with caustic soda or soda ash; added to the pulp furnish in the beater to render the paper or board water and ink resistant.
f colle *f* de résine — **d** Harzleim *m* — **n** harslijm *m* — **e** cola *f* de resina; cola *f* resinosa — **i** colla *f* di resina; sapone *m* di resina

11690 rosin sized paper — Paper made from cellulose fibres, which have been made more or less water resistant by the addition of rosin size and alum in the beater.
f papier *m* collé à la colophane — **d** harzgeleimtes Papier *n* — **n** papier *n* met harslijming; harsgelijmd papier *n* — **e** papel *m* encolado a la resina — **i** carta *f* collata alla colofonia

11691 Ross effect — Photographic effect which can be measured when determining densities with an interference device. Use is made of the capability of two light rays to damp down or to intensify (light interference) for the measurement of small film thicknesses or changes in film thicknesses, under predetermined conditions. The determination exists of making out whether heavily exposed image areas after drying have shrunk more than other less heavily exposed image areas. This shrinkage is explained by the tanning of the gelatine due to the concentrated oxidation products on the exposed areas. The tanning causes the gelatine to heavy concentration during drying.
f effet *m* Ross — **d** Ross-Effekt *m* — **n** Ross-effect *n* — **e** efecto *m* Ross — **i** effetto *m* Ross

11692 rostrum; deck *US*; **upper deck** *US* — Raised platform from which the composing room overseer or clicker can obtain a clear view of his staff.
f bureau *m* du prote — **d** Aufsichtspult *n*; Aufsichtsbühne *f* — **n** kantoortje *n* van de chef; beun *fm*; schavotje *n* — **e** regencia *f* — **i** ufficio *m* del proto

11693 rotary disks — Disks usually constructed following the steel rule die principle and utilize the anvil. The cutting edges, however, are machined and ground on special pantograph

machines to produce the compound curvature required. The male-female die principle is used on rotary machines also but is restricted to work requiring the punching of round holes of small diameter.
f poinçons *mpl* rotatifs — **d** Rotationsstanzformen *fpl* — **n** roterende stansvormen *mpl* — **i** fustelle *fpl* rotative

11694 rotary letterpress machine — See newspaper press.

11695 rotary letterpress printing
f impression *f* typo rotative — **d** Rotationsbuchdruck *m*; rotativer Buchdruck *m* — **n** rotatieboekdruk *m* — **e** tipografía *f* rotativa — **i** tipografia *f* rotativa

11696 rotary minder — See also rotary printer.
f rotativiste *m* — **d** Rotationsdrucker *m*; Rotationer *m* — **n** rotatiepersdrukker *m*; rotatiedrukker *m* — **e** rotativista *m*; obrero *m* rotativista — **i** operatore *m* alla rotativa; addetto *m* alla rotativa

11697 rotary offset machine — See rotary offset press.

11698 rotary offset press; offset rotary press; rotary offset machine; web offset press; web-fed offset rotary (press) — High speed offset press on which paper in continuous web is printed.
f rotative *f* offset à bobines; presse *f* offset à bobines — **d** Offsetrollenrotationsmaschine *f*; Rotationsoffsetmaschine *f*; Rollenoffsetmaschine *f* — **n** offset-rotatiepers *f*; rotatie-offsetpers *f*; rollen-offsetpers *f* — **e** rotativa *f* offset; máquina *f* rotativa offset (de bobina); máquina *f* para offset rotativo a bobinas; máquina *f* roto offset; prensa *f* roto offset — **i** macchina *f* offset rotativa; macchina *f* offset da bobina; rotativa *f* offset da bobina

11699 rotary press — Printing press on which the paper is printed by passing it between two rotating cylinders, one of which carries the curved plates. Used for all newspaper printing (except the *Duplex machine), periodicals of large circulation, and occasionally for books of large run.
f rotative *f*; machine *f* rotative; presse *f* rotative — **d** Rotationsmaschine *f* — **n** rotatiemachine *f*; rotatiepers *f* — **e** máquina *f* rotativa; rotativa *f* — **i** macchina *f* rotativa; rotativa *f*

11700 rotary printer — See also rotary minder.
f maître *m* machiniste (de la rotative) — **d** Rotationsmaschinenmeister *m*; Rotationsmaschinenführer *m* — **n** rotatiepersdrukker *m*; rotatiedrukker *m*; voorman *m* rotatie(pers)-drukker — **e** maquinista *m* de la rotativa; conductor *m* rotativista — **i** caposquadra *m* alla rotativa

11701 rotary printing — See also rotary press.
f impression *f* (sur) rotative; impression *f* en continu — **d** Rotationsdruck *m* — **n** rotatiedruk *m* — **e** impresión *f* rotativa — **i** stampa *f* rotativa; stampa *f* in rotativa

11702 rotational viscometer — Viscometer composed of a cylindrical cup and bob. The bob is suspended in the cup and the test material is put between the bob and the cup. Either the bob or cup can be made to rotate. The speed of rotation and the torque imposed by the viscous drag resulting therefrom are the measurements from which the coefficient of viscosity and the yield value can be calculated.

f viscosimètre *m* à rotation — **d** Rotationsviskosimeter *n* — **n** rotatieviscosimeter *m* — **e** viscosímetro *m* a rotación — **i** viscosimetro *m* a rotazione

11703 roto burner — Gas burner in a rotary offset press. It rotates 90 ° and provides 45 cm (18 in) of clearance for web protection when the press stops. Flames shut off immediately.

f sécheur *m* à flammes de gaz — **d** Gastrockner *m* — **n** gasbrander *m* — **i** bruciatore *m* a gas

11704 Rotofilm — Trade name for a product replacing carbon tissue. It looks like a negative, has a film backing instead of paper, and can be read on a densitometer after exposure and development.

11705 rotogravure — Process of printing with a thin, quick-drying ink from a cylindrical surface having an etched (recessed) design. Opposite of letterpress printing, since the design areas are recessed into the plate instead of being in relief. Industrial reel-fed gravure printing.

f héliogravure *f* rotative; rotogravure *f* — **d** Rotationstiefdruck *m*; Kupfertiefdruck *m*; Tiefdruck *m*; Rakeltiefdruck *m* — **n** rotatiediepdruk *m*; koperdiepdruk *m*; diepdruk *m* — **e** huecograbado *m* — **i** rotocalco *m*; rotocalcografia *f*

11706 rotogravure ink — Thin ink used for rotogravure printing. It consists of solvents, binders and colour, but contains no dryer as drying occurs entirely by evaporation.

f encre *f* hélio; encre *f* pour rotogravure; encre *f* hélio rotative — **d** Tiefdruckfarbe *f* — **n** diepdrukinkt *m*; rotatiediepdrukinkt *m* — **e** tinta *f* para huecograbado — **i** inchiostro *m* per rotocalco

11707 rotogravure paper; gravure paper — Unsized or lightly sized paper usually supercalendered, used mostly for printing of catalogues, magazines, and newspaper supplements.

f papier *m* hélio; papier *m* pour héliogravure; papier *m* pour rotogravure — **d** Tiefdruckpapier *n* — **n** diepdrukpapier *n* — **e** papel *m* para huecograbado — **i** carta *f* per rotocalco; carta *f* da stampa rotocalco

11708 roto iron — Slang term for *ferric chloride, used in etching rotogravure cylinders.

11709 rough breathing — See spiritus asper.

11710 roughing — Smoothing out and dressing a stereo or an electrotype before mounting on a block.

f dressage *m* — **d** Planschaben *n* und Geraderichten *n* — **n** justeren *n* — **e** debastadora *f* de planchas — **i** spianatura *f* e raddrizzatura *f*

11711 roughness; rugosity — For the surface of paper it is better to use the term *smoothness.

f rugosité *f* — **d** Rauheit *f*; Rauhigkeit *f*; Unebenheit *f* — **n** ruwheid *f*; oneffenheid *f* — **e** rugosidad *f* — **i** ruvidità *f*; ruvidezza *f*; scabrosita *f*

11712 roughness tester — See smoothness tester.

11713 rough proof — See flat proof.

11714 rough pull — See flat impression.

11715 rough sketch — First elementary drawing of a *picture or *layout.

f ébauche *f* — **d** Skizze *f* — **n** schets *fm* — **e** apunte *m*; proyecto *m* borrador — **i** schizzo *m*; traccia *f*; progetto *m* di massima

11716 roulet *v* — To indent the surface of a printing plate with a roulette to lighten or modify tone values.

f travailler à la roulette; pointiller à la roulette — **d** roulettieren — **n** bewerken met de roulette; bijwerken met de roulette; werken met de roulette — **e** grabar a la ruleta — **i** ritoccare con roulette

11717 roulette — Finisher's tool comprising a handle bearing on one end a rotating wheel or knurl, this possessing a serrated cutting pattern corresponding to the ruling of halftone screens.

f roulette *f* — **d** Roulette *f* — **n** roulette *fm* — **e** ruleta *f* — **i** roulette *f*

11718 roulette — See bar roll.

11719 rounce; crank handle — In early printing presses the handle to turn the roller that draws the bed from under the platen by means of a pair of leather straps.

f manivelle *f* — **d** Kurbel *f*; Handkurbel *f* — **n** slinger *m*; kruk *fm* — **e** manivela *f* — **i** manovella *f*

11720 round back

f dos *m* rond — **d** runder Einbandrücken *m* — **n** ronde rug *m* — **e** lomo *m* redondeado — **i** dorso *m* tondo

11721 round brackets — See parentheses.

11722 round chisel — See sculper.

11723 round cornering — Making round corners on the cover boards of books, cards, etc.

f arrondissure *m* des coins — **d** Ecken abrunden; Ecken rundstoßen — **n** rondhoeken; ronde hoeken aanbrengen — **e** redondear esquinas — **i** arrotondamento *m* degli angoli

11724 round-cornering machine — See corner rounding machine.

11725 rounded corners; rounded edges

f à coins arrondis — **d** mit runden Ecken — **n** met ronde hoeken; met (af)geronde hoeken — **e** puntas redondeadas — **i** angoli arrotondati

11726 rounded edges — See rounded corners.

11727 round hole perforating

f perforage *m* à trous ronds — **d** Rundlochperforation *f*; Rundlochung *f* — **n** perforatie *f* met ronde gaatjes — **e** perforación *f* redonda — **i** perforatura *f* a buchi circolari

11728 rounding, back — See back rounding.

11729 round-robin test — Test, the report of which bearing a number of signatures, as to a petition, written in a circle to avoid giving prominence to any name.

f rapport *m* revêtu de signatures en rond, ou en cercle pour ne pas révéler le chef de bande — **d** Ringversuch *m* — **n** anonieme test *m* — **e** ensayo *m* anónimo — **i** prova *f* anonima

11730 round sculper — See sculper.

11731 round thread
f filet *m* rond; filet *m* arrondi; filetage *m* rond; filetage *m* arrondi — **d** Rundgewinde *n* — **n** ronde draad *m*; ronde schroefdraad *m* — **e** rosca *f* redonda; filete *m* (de rosca) redonda — **i** filettatura *f* tonda

11732 router; routing machine — Machine with a rapidly revolving cutter held suspended in a vertical and mobile spindle head, to deepen the etched areas of plates, to remove superfluous metal therefrom, and to cut away part of the wood or metal block on which the plate is mounted.
f toupilleuse *f*; fraiseuse *f* — **d** Fräsmaschine *f* — **n** freesmachine *f* — **e** fresadora *f*; máquina *f* fresadora; rebajadora *f*; rauteadora *f*; taladradora *f* — **i** fresatrice *f*

11733 router bit; cutter — Ground steel blade or cutter used on routers.
f mèche *f* — **d** Fräse *f* — **n** frees *fm* — **e** broca *f* (para taladradora); fresa *f*; taladro *m*; fresadora *f* *Pe* — **i** fresa *f*

11734 routine — The whole or part of a program that has some general or frequent use. A routine or sub-routine consists of a series of logically-related processes, to accomplish a specific purpose (such as to add two numbers) or related purposes, such as to add a series of numbers and compare this result with the total of similar additions.
f programme *m*; routine *f* — **d** Programmteil *m*; Routine *f* — **n** programmadeel *n*; routine *f* — **e** programa *m*; rutina *f* — **i** programma *m*

11735 routine control testing — See quality control testing.

11736 routing — Removal of the unwanted parts of a printing plate, with a graver or gouge.
f échoppage *m* — **d** Ausstechen *n*; Wegstechen *n* — **n** wegsteken *n* — **e** grabado *m* a gubia — **i** asportare col bulino

11737 routing — The operation of a router or guiding the whirling cutter over an area of the printing plate for the removal of unessential metal, and of lowering the area of non printing surfaces and mounts to the required degree.
f toupillage *m*; fraisage *m* — **d** Fräsen *n*; Abfräsen *n*; Ausfräsen *n* — **n** frezen *n*; wegfrezen *n*; uitfrezen *n*; affrezen *n* — **e** fresado *m* — **i** fresatura *f*

11738 routing — To send along a certain way (goods, etc.).
f acheminement *m*; routage *m* — **n** routing *m* — **e** encaminado *m* — **i** convogliamento *m*

11739 routing indicator — Address or group of characters in the header of a message defining the final circuit or terminal to which the message has to be delivered.
f indicateur *m* d'acheminement — **d** Leitweganzeiger *m* — **n** route-aanwijzer *m* — **e** indicador *m* de enrutado — **i** indicatore *m* di via

11740 routing machine — See router.

11741 row binary — Binary representation of data on punched cards in which adjacent positions in a row correspond to adjacent bits of data, e.g., each row in an 80 column card may be used to represent 80 consecutive bits or two 40 bit words.
f carte *f* binaire par ligne — **d** Karte *f* binär pro Zeile — **n** kaart *fm* binair per regel — **e** tarjeta *f* binaria por línea — **i** scheda *f* binaria per riga

11742 royal — A British standard size of writing and printing papers. The term also refers to a size for wrapping papers. See app. no. 3.

11743 royal copying — A size of paper. See app. no. 3.

11744 royal hand — A size of wrapping paper. See app. no. 3.

11745 royal library
f bibliothèque *f* royale — **d** Königliche Bibliothek *f* — **n** koninklijke bibliotheek *f* — **e** biblioteca *f* real; biblioteca *f* del rey — **i** biblioteca *f* reale

11746 royal 4to — A size of drawing paper. See app. no. 3.

11747 rubber; natural rubber; India rubber; India gum; gum elastic; caoutchouc — Stereospecific polymer obtained from rubber trees and plants. Elastic solid obtained from the sap (latex) of the rubber tree (Hevea brasiliensis) by coagulation and drying, or similar sources. Soluble in carbon disulphide, petroleum and coaltar hydrocarbons and essential oils, used in rubber platemaking materials.
Formula: $(C_5H_8)_x$. See also buna, hard rubber, synthetic rubber.
f caoutchouc *m* (naturel de plantation) — **d** Kautschuk *m*; Naturkautschuk *m*; Naturgummi *m*; Rohgummi *m* — **n** rubber *mn*; natuurrubber *mn*; caoutchouc *m* — **e** hule *m*; caucho *m* (natural); goma *f* elástica — **i** caucciù *m*; gomma *f* naturale

11748 rubber blanket — See blanket.

11749 rubber block — See rubber plate.

11750 rubber impression roller — When the rubber impression cylinder of a rotogravure press is backed up by a steel roller, the impression cylinder gets in some languages another name. See also back-up impression cylinder.
f rouleau *m* de pression — **d** Zwischenwalze *f* — **n** presseur *m* — **e** cilindro *m* de presión — **i** cilindro *m* pressore

11751 rubber plate; rubber block; rubber stereo — Printing plate consisting of an engraved or moulded surface of rubber, made either by manual engraving or by moulding from relief etchings.
f cliché *m* de caoutchouc; cliché *m* en caoutchouc — **d** Gummiklischee *n*; Gummidruckstock *m*; Gummiplatte *f*; Gummistereo *n* — **n** rubber cliché *n*; rubber styp *m* — **e** clisé *m* de caucho; clisé *m* elástico; plancha *f* de caucho — **i** cliscè *m* di gomma; cliché *m* di gomma; stereo *m* di

gomma
11752 rubber-plate printing — See flexography.
11753 rubber reduction *obs.* — Method to reduce drawings on lithographic stones by making a print on a sheet of rubber, which is stretched with screws in a surrounding metal frame, called a Fougeadoire (patented 1877). By means of the screws the rubber contracts after which the print is transferred to another stone.
f réduction *f* à Fougeadoire — **d** Verkleinern *n* mit dem Gummipantographen — **n** verkleinen *n* met de rubber-pantograaf — **i** riduzione *f* mediante stampa su foglio teso di gomma
11754 rubber rider; rubber rider roller
f chargeur *m* en caoutchouc; chargeur *m* de caoutchouc — **d** Gummireiterwalze *f* — **n** rubber ruiterrol *fm* — **e** rodillo *m* cargador de caucho; rodillo *m* de carga de caucho — **i** rullo *m* cavaliere di gomma
11755 rubber rider roller — See rubber rider.
11756 rubber roller — Inking roller consisting of metal core covered with synthetic rubber.
f rouleau *m* en caoutchouc; rouleau *m* en caoutchouc; rouleau *m* de caoutchouc — **d** Gummiwalze *f* — **n** rubberrol *fm*; rubber rol *fm* — **e** rodillo *m* de caucho; rodillo *m* de goma — **i** rullo *m* di caucciù
11757 rubber stamp — Design or character moulded in vulcanized rubber, mounted on a wood or metal base, used for stamping by hand.
f étampe *f* en caoutchouc; cachet *m* en caoutchouc; timbre *m* en caoutchouc — **d** Gummistempel *m*; Kautschukstempel *m* — **n** rubberstempel *n* — **e** sello *m* de caucho; sello *m* de hule; sello *m* de goma — **i** timbro *m* in gomma; timbro *m* di cauccìu
11758 rubber-stamp moulding press — See vulcanizing press.
11759 rubber-stamp watermark — See impressed watermark.
11760 rubber stereo — See rubber plate.
11761 rubbing — See scuffing.
11762 rub-dryness tester — Instrument that replaced (1965) the *drying-time recorder. Since the composition of inks has changed, an instrument has been developed (1977) to indicate the point at which an ink film will be rub-dry. It consists mainly of a soft plastic foil over which a sheet of offsetting paper is placed. The test strips (12 strips, 35 mm wide), taken from a press sheet, are clamped to a metal plate and placed with their face against the set-off paper, without contacting it. The plastic foil is pressed with a bar against the offsetting paper when the instrument is switched on, the bar travelling at intervals. During bar transport there is no contact.
f mesureur *m* de temps de frottage d'une encre; mesureur *m* de temps de séchage; appareil *m* pour la mesure du temps de séchage — **d** Meßgerät *n* zur Bestimmung der Abreibungstrockenzeit; Trockendauermeßgerät *n*; Trockenzeitprüfer *m* — **n** afwrijf-droogtijdbepalingstoestel *n*

— **e** medidor *m* del tiempo de frotamiento de la tinta; medidor *m* del secado — **i** apparecchio *m* per la misura del sfregamento del'inchiostro; apparecchio *m* per la misura del tempo d'essiccamento
11763 rubometer — See rub-proof tester.
11764 rub proof — Said of ink when it reaches maximum dryness and the printed surface resists all normal abrasion.
f résistant au frottement — **d** reibfest — **n** wrijfvast — **e** resistente al frotamiento — **i** resistente allo strofinio
11765 rub-proof ink — Ink made of a compounded vehicle of synthetic resins, carnauba wax and linseed or China wood oil, used for printing magazine covers, labels and cartons.
f encre *f* inusable; encre *f* résistante au frottement — **d** scheuerfeste Farbe *f* — **n** slijtvaste inkt *m*; nagelvaste inkt *m* — **e** tinta *f* resistente al frote; tinta *f* antifricción — **i** inchiostro *m* resistente allo strofinio
11766 rub-proof tester; rubometer *US* — Instrument to measure rub or scuff resistance of a printed design.
f abrasimètre *m*; abrasiomètre *m*; mesureur *m* de la résistance à l'abrasion — **d** Scheuerfestigkeitsprüfgerät *n* — **n** slijtvastheidsmeter *m*; slijtweerstandsmeter *m*; wrijfvastheidsmeter *m*; slijttoestel *n* — **e** abrasímetro *m*; ensayador de la resistencia a la abrasión — **i** abrasimetro *m*; apparecchio *m* per determinare la resistenza all'abrasione
11767 rub resistance — Degree to which a print resists smearing by friction.
f résistance *f* au frottement — **d** Scheuerfestigkeit *f* — **n** slijtvastheid *f*; wrijfvastheid *f* — **e** resistencia *f* al rozamiento — **i** resistenza *f* allo sfregamento; resistenza *f* allo strofinio
11768 rubric; rubricated matter — Words, lines or symbols in a book or text printed in red to draw the attention of the reader.
f rubrique *f* — **d** Rubrik *f*; rot Gedruckte *n*; roter Titelkopf *m* — **n** (in) rood gedrukte *n*; rubriek *f* *obs* — **e** rúbrica *f*; rótulo *m*; rubro *m* — **i** rubrica *f*
11769 rubricated matter — See rubric.
11770 ruby — Old name for a size of type. See app. no. 1.
11771 rugosity — See roughness.
11772 rule — Strip of brass or steel, (sometimes of lead), type-high, used in printing lines, borders, etc., and named according to face or use, hair-line, double dotted, make-up, composing, column, etc., specified furthermore by its thickness, 2-point, 6-point, etc.
f filet *m* — **d** Linie *f* — **n** lijn *fm*; messing lijn *fm*; koperen lijn *fm* *depr.* — **e** pleca *f*; filete *m* — **i** filetto *m*
11773 rule *v* — To draw or to print parallel lines or account rules on paper.
f régler — **d** linieren; liniieren — **n** liniëren — **e** rayar — **i** rigare; tracciare linee; lineare

11774 rule case — Case to hold assortments of brass rules.

f casse *f* à filets — **d** Linienkasten *m* — **n** lijnenkast *fm* — **e** caja *f* para rayas; caja *f* para filetes; caja *f* para plecas — **i** cassa *f* da filetti

11775 ruled paper

f papier *m* réglé — **d** liniertes Papier *n*; liniiertes Papier *n* — **n** gelinieerd papier *n*; (batonné) gelijnd papier *n* — **e** papel *m* rayado — **i** carta *f* lineata; carta *f* rigata

11776 rule forme — Letterpress printing forme containing rules.

f forme *f* de filets — **d** Linienform *f* — **n** lijnenvorm *m* — **e** molde *m* de filetes; molde *m* de estado; composición *f* de estados — **i** forma *f* con molti filetti

11777 ruler

f régleur *m*; rayeur *m* — **d** Liniierer *m* — **n** linieerder *m* — **e** rayador *m* — **i** lineatore *m*

11778 ruler — See straight-edge.

11779 rule slide — See matrix slide.

11780 rule-work — Tabular or similar work in which brass rules are used to print borders, columns, etc. See also tabular work.

f tableautage *m* — **d** Liniensatz *m* — **n** lijnenwerk *n* — **e** filetaje *m*; filetería *f*

11781 ruling

f réglure *f* — **d** Liniierung *f*; Linierung *f* — **n** liniëring *f* — **e** rayadura *f* — **i** rigatura *f*

11782 ruling distance — See screen ruling.

11783 ruling ink — Water solution of dyes for ruling ledgers, etc.

f encre *f* pour réglure — **d** Linierfarbe *f*; Liniierfarbe *f* — **n** linieerinkt *m* — **e** tinta *f* para rayar — **i** inchiostro *m* per rigature; inchiostro *m* per lineare

11784 ruling machine — There are two types of ruling machines, the pen and the disk types. The pen ruling machines are used on the more intricated formes, such as ledger sheets, while the disk type is used for ruling cheaper and simpler work, such as index cards, notebooks, etc.

f régleuse *f*; machine *f* à régler — **d** Liniermaschine *f*; Liniiermaschine *f* — **n** linieermachine *f* — **e** máquina *f* de rayar; máquina *f* para rayar; rayadora *f* — **i** rigatrice *f*; macchina *f* tiralinee

11785 ruling pen — Pen in a ruling machine.

f plume *f* de réglure — **d** Linierfeder *f*; Liniierfeder *f* — **n** linieerpen *fm* — **e** pluma *f* de rayar — **i** penna *f* rigatrice

11786 ruling pen — Pen used by an artist to make designs of line work.

f tire-ligne *f* — **d** Ziehfeder *f*; Reißfeder *f*; Linienziehfeder *f* — **n** trekpen *fm* — **e** tiralíneas *m*; tirador *m* — **i** tiralinee *m*

11787 run — Number of copies of a job to be printed.

f tirage *m* — **d** Auflage *f* — **n** oplage *fm*; oplaag *fm* — **e** tiraje *m*; tirada *f* — **i** tiratura *f*

11788 run — Production of paper or board considered as one quantity.

f fabrication *f* — **d** Anfertigung *f* — **n** aanmaak *m* — **e** fabricación *f* — **i** fabbricazione *f*

11789 run — See pass.

11790 runability

f aptitude *f* au roulage; (bon) roulage *m*; (bon) passage *m* — **d** Verdruckbarkeit *f*; Lauffähigkeit *f*; störungsfreies Laufen *n* — **n** (gemakkelijke) verwerkbaarheid *f* op de pers — **e** marcha *f* sin dificultades — **i** macchinabilità *f*

11791 runaround — Arrangement of running text around other material, such as a half-column illustration, so that the normal measure of the text block is reduced temporarily.

11792 run *v* **around** — To set cut or illustration with text around it.

f intercaler dans le texte; dépaginer — **d** herumsetzen — **n** cliché tussen de tekst zetten; inbouwen — **e** intercalar — **i** comporre attorno ad un'illustrazione

11793 run-around block — Cut or illustration surrounded by type matter.

f cliché *m* intercalé dans le texte — **d** eingebautes Klischee *n* — **n** ingebouwd cliché *n*; cliché *n* tussen de tekst — **e** arracada *f*; ladillo *m* — **i** cliscè *m* inserito nella composizione; cliché *m* inserito nella composizione

11794 run *v* **back** — In reading proofs, to move material from the beginning of one line to the end of one above it. Abbrev. rb.

f retourner à la ligne précédente — **d** Text aus einer Zeile herumnehmen in die vorhergehende Zeile — **n** naar vorige regel overbrengen; terugverlopen — **i** riportare alla riga precedente

11795 run *v* **down** — In reading proofs, to move material from the end of one line to the beginning of the text. Abbrev. rd.

f reporter à la ligne suivante — **d** Text aus einer Zeile herumnehmen in die folgende Zeile — **n** naar volgende regel overbrengen; verlopen — **i** portare alla riga successiva

11796 run *v* **in** — To insert new copy (whether an omission of the operator or an author's addition) into the text.

f rentrer — **d** einschalten (von zusätzlichem Text in Manuskript); einfügen (in Satzteile) — **i** introdurre nuovo materiale nel testo

11797 run-in sheets — Sheets of paper run through the machine at the beginning of the run, other than for reproduction. Sometimes called wasters.

f feuilles *fpl* de mise en route — **d** Andruckmakulatur *f* — **n** voorlopers *mpl* — **i** scarti *mpl* d'avviamento

11798 run *v* **keyboard** — To fill the magazines of a composing machine with matrices.

f garnir en matrices — **d** Matrizen in Setzmaschinenmagazin einlaufen lassen — **n** matrijzen laten inlopen in het magazijn — **e** correr a teclado — **i** rifornire di matrici i magazzini

11799 run manual — Manual documenting the processing system, program logic controls, program changes, and operating instructions associated with a computer run.

11800 runner; marginal figure — Figure or letter placed down the length of a page to indicate the particular number of position of any given line.
f chiffre *m* en marge — **d** Randziffer *f*; Marginalziffer *f*; Zeilenzähler *m* — **n** verscijfer *n*; versnummer *n* — **e** cifra *f* marginal — **i** cifra *f* marginale

11801 runner — Rotating member(s) of a *kollergang.
f meule *f* courante — **d** Lauferstein *m* — **n** lopersteen *m*; loper *m*

11802 running belt tension — See belts.

11803 running foot — Information at the bottom of the page. Opposite of *running headline.

11804 running head — See running headline.

11805 running headline; running head; running title — Line of type or a title printed on the top of each page of a book or publication, with which folios are incorporated; odd folios to the right, even folios to the left.
f titre *m* courant — **d** lebender Kolumnentitel *m*; lebender Titelkopf *m* — **n** sprekende hoofdregel *m* — **e** folio *m* explicativo; título *m* de cabacera; título *m* corrido; cabeza *f* de página — **i** titolo *m* corrente; testatina *f*

11806 running schedule
f cédule *f* de marche; planning *m* de marche; tableau *m* de marche — **d** Fabrikationsprogramm *n*; Fabrikationsplan *m* — **n** fabricageprogramma *n*; fabrikatieprogram *n* — **e** programa *m* de fabricación; plan *m* de fabricación — **i** programma *m* di fabbricazione; programma *m* di marcia

11807 running tapes — Tapes run at the press speed, to lead or convey the web.
f cordons *mpl* de guidage du papier — **d** Führungsbänder *npl* — **n** geleidebanden *mpl*; leibanden *mpl* — **e** cintas *fpl* conductoras; cintas *fpl* de conducción — **i** nastri *mpl* conduttori

11808 running through; shooting through — Cutting through connected dot formations in etched halftone plates with a graver to lighten tone values or to bring the dots to proper size, also to introduce dots in solid spots or blemishes in the etching.
f filage *m* — **d** Nachschneiden *n*; Durchreißen *n*; Durchstoßen *n* — **n** nasteken *n*; open steken *n* — **e** retoque *m* (de la autotipia) a buril — **i** ritocco *m* al bulino (di lastre retinate)

11809 running title — See running headline.

11810 run *v* **on** — To commence printing after a job has been finally checked and made ready.
f rouler — **d** kann laufen; anlaufen lassen — **n** afdrukken; kan afgedrukt worden — **e** arrancar — **i** cominciare a stampare; andare in macchina

11811 run *v* **on** — No new paragraph, although the matter may be separated in the copy. Indicated by a line running from the end of one group to the beginning of the next, with the words "run on" in the margin. See app. no. 5.

f suivre sans alinéa; alinéa *m* à supprimer — **d** anhängen — **n** achter elkaar doorgaan; géén nieuwe alinea — **e** poner de seguido; poner sin párrafo; punto y seguido — **i** andare di seguito; sopprimere il capoverso

11812 run *v* **on solid** — To print in continuous line or paragraph without breaking or leading.
f composer plein — **d** paketsetzen — **n** plat zetten — **e** composición *f* sólida; composición *f* seguida — **i** comporre a dilungo

11813 run *v* **out and indent** — The first line to be full out and the subsequent lines indented.
f rentrer en sommaire — **d** Absatz *m*, dessen erste Zeile voll ausgeht, alle folgenden Zeilen aber eingezogen sind — **n** eerste regel over de volle breedte gezet, de volgende regels ingesprongen — **e** sangrado *m* a la francesa — **i** comporre a sommario

11814 runover — Large amount of reset material.
f recomposé *m* — **d** Neusatz *m* — **n** overgehouden zetsel *n* — **i** grande quantità *f* di materiale ricomposto

11815 runover — Continuation of a heading on a second line.
f continuation *f* d'un titre sur une deuxième ligne — **d** umlaufende Überschrift *f*; mehrzeilige Überschrift *f* — **n** kop *m* over twee regels — **i** testata *f* su due righe

11816 run *v* **sheets through the machine** — To condition sheets of paper to atmospheric conditions in the press-room.
f passer les feuilles dans la machine; passer en blanc — **d** Bogen im Blinddruck durch die Maschine laufen lassen; Streckgang *m* — **n** doorhalen; een strekgang maken — **e** pasar en blanco — **i** passare la carta in macchina

11817 run *v* **up the ink**
f introduire l'encre dans l'encrier — **d** die Farbe im Farbkasten laufen lassen — **n** de inkt in de bak doen — **e** batir la tinta en el tintero — **i** introdurre l'inchiostro nel calamaio

11818 rush job; rush order — Job which has to be finished with speed. The copy for such a job is marked with "rush".
f ouvrage *m* pressant; travail *m* urgent; coup *m* de feu — **d** Schnellschuß *m*; Eilarbeit *f* — **n** spoedorder *fm*; haastkarweitje *n* — **e** trabajo *m* urgente; trabajo *m* de emergencia; pedido *m* urgente — **i** lavoro *m* urgente; lavoro *m* d'urgenza

11819 rush order — See rush job.

11820 Russell effect — Photographic effect, discovered in 1899 by W.J. Russell, bringing about fogging of a light-sensitive layer. Produced by chemicals, e.g., hydrogen peroxide, penetrating through the packaging material.
f effet *m* Russell — **d** Russell-Effekt *m* — **n** Russell-effect *n* — **e** efecto *m* Russell — **i** effetto *m* Russell

11821 Russia leather — Willow bark tanned calfskin for bookbindings, dyed with sandal wood and soaked in birch oil.
f cuir *m* de Russie; youfte *m* — **d** Juchtleder *n*;

Juchtenleder *n*; Juchten *m* — **n** juchtleer *n* — **e** cuero *m* de Rusia; piel *f* de Rusia — **i** cuoio *m* di Russia

11822 Russian vignette — See black vignette.

11823 rust spots — Reddish-brown spots in paper caused by pressure of iron particles.

f taches *fpl* de rousseur; mouillures *fpl*; taches *fpl* roussâtres — **d** Rostflecken *mpl* — **n** roestvlekken *fmpl* — **e** manchas *fpl* de herrumbre — **i** macchie *fpl* ruggine

11824 rutland — Trade name for sheepskin. See also roan.

11825 Sabattier effect — If an exposed, incompletely developed, and washed, but not fixed film is given a second uniform exposure and developed again, a reversal of the original image may be obtained. The reversal may be partial or complete, depending on the relative magnitude of the first and second exposures. Discovered by C.P.D.A. Sabattier in 1860.

f effet *m* Sabattier — **d** Sabattier-Effekt *m* — **n** Sabattier-effect *n* — **e** efecto *m* Sabattier — **i** effetto *m* Sabattier

11826 sack — Although often used as a synonym for *bag, the term sack generally refers to the heavier duty or shipping bags.

f sac *m* (grand) — **d** Beutel *m*; Sack *m* — **n** (grote) zak *m* — **e** bolsa *f*; saco *m* (de papel) — **i** sacco *m*

11827 saddle *v* — To round-off the back of a book and to crease the hinges.

f arrondissure *m* et endorsure *m* — **d** Rückenrunden *n* und Falzeinpressen *n* — **n** rondzetten *n* en knepen *n* — **e** rondeado *m* y formación *f* del lomo — **i** arrotondare il dorso e imprimere la piega

11828 saddleback — A British size of wrapping paper. See app. no. 3.

11829 saddle gang stitcher

f encarteuse-piqueuse *f* — **d** Sammeldrahtheftmaschine *f* — **n** verzamelhechtmachine *f* — **e** alzadora-cosedora *f* — **i** raccoglitrice-cucitrice *f* a punti metallici

11830 saddle stitcher — Stitcher to insert a staple through the centre of the folded sheet from the back and clinch it in the fold.

f piqueuse *f* à cheval — **d** Rückstichheftmaschine *f* — **n** zadelhechtmachine *f* — **e** cosedora *f* de hilo metálico a caballo — **i** cucitrice *f* a sella; macchina *f* cucitrice a sella

11831 saddle stitching — Stitching where the wire staples pass through the spine from the outside and are clinched in the centre. Only used with folded sections, either single sections or two or more sections inset to form a single section.

f piqûre *f* à cheval; agrafage *m* à cheval — **d** Rückstichheftung *f* — **n** nieten *n* door de rug — **e** engrapado *m* de cuaderno; cosido *m* a galápago; cosido *m* a diente de perro — **i** cucitura *f* a sella

11832 Safawid binding; Savafid binding — Persian style of book cover decoration of the 16th century. The leather bindings showed medallions or were decorated over the whole surface with arabesques, birds, flowers etc., which were impressed from engraved steel or copper plates. The impression of the design was usually combined with gilding; gilding was sometimes done with a heated die before the binding was impressed. The bindings had various coloured areas of leather or paper and the back cover had a protecting flap, overlapping the front cover.

f reliure *f* séfévide; reliure *f* safavide — **d** Safawid-Einband *m* — **n** Safawid-band *m* — **e** encuadernación *f* persa — **i** legatura *f* persiana

11833 safe — Said of the rotary press when locked. See also lock.

f en position de sécurité — **d** sicher — **n** geblokkeerd; veilig — **e** seguro — **i** sicuro

11834 safe edge — Strips of paper or cut out sheets of paper to mask against light in photographic reproduction or to protect parts of the sheet to be printed against spoilage.

f cache *m* — **d** Abdeckstreifen *m* — **n** afdekband *n* — **e** marco-frasqueta *m* — **i** striscia *f* di copertura

11835 safelamp — See darkroom lamp.

11836 safety film — See acetate film.

11837 safety ink — Water colour or oil type ink which will change colour or bleed when ink eradicator or water is applied to prints. Also ink which will erase easily. Used for printing backgrounds on cheques, etc.

f encre *f* de sécurité; encre *f* pour chèques — **d** Sicherheitsfarbe *f*; Scheckfarbe *f* — **n** chequeinkt *m* — **e** tinta *f* de seguridad (para cheques) — **i** inchiostro *m* per la stampa di carte di sicurezza

11838 safety lights — Coloured indicators to show the setting of the electrical supply controls to the motors.

f lampes *fpl* témoins — **d** Kontroll-Leuchten *fpl*; Kontroll-Lampen *fpl* — **n** controlelichtjes *npl*; controelampjes *npl* — **e** lámparas *fpl* de seguridad — **i** luci *fpl* di sicurezza; lampade *fpl* spia

11839 safety paper
See cheque paper.

11840 safety ticket (paper) — Paper especially watermarked, or otherwise distinguished with dyes, etc., to make counterfeiting difficult.

f papier *m* de sécurité — **d** Sicherheitspapier *n* — **n** chequepapier *n* — **e** papel *m* de seguridad — **i** carta *f* di sicurezza

11841 Saint Bride — Irish abbes (453-523), a patron saint of Ireland. Name of a church in Fleetstreet, London, which is known as the printers' church, because here the bones rest of Wynkyn de Worde (1535), the assistant, foreman and successor of William Caxton, the first English printer (1491). Wynkyn de Worde brought printing to Fleetstreet, London's printing centre. He arrived in the parish of St. Bride in 1500 and since the church's close association with printing began. The St. Bride printing library in the St. Bride Institute has the most important British collection of books on the history of printing. It was opened in 1895 and was estimated, in 1977, to have 40,000 volumes.

11842 Saint-Jean — French term for the tools and smock which the tourist printer (trimardeur) took with him in old times. See also working clothes.

f Saint-Jean *m* — **d** Reisebündel *n*; Felleisen *n*; Berliner *m* — **n** reistas *fm* met werkkleding en gereedschap — **e** maleta *f* con traje de trabajo y trebejos — **i** borsa *f* con vesti di lavoro e arnesi; borsa *f* con abiti di lavoro e arnesi

11843 Saint John's day; Saint John's eve — Feast day, 6th May, of Saint John, the patron of the French compositors.

f Saint-Jean *m*; fête *f* de Saint-Jean-Porte-Latine — **d** Johannisfest *n* — **n** feestdag *m* van Sint Jan — **e** fiesta *f* de San Juan — **i** festa *f* di San Giovanni

11844 Saint John's eve — See Saint John's day.

11845 Saint Martin's day — In France, the feast of the pressmen (11th November).

f Saint-Martin *m*; fête *f* de Saint-Martin — **d** Martinsfest *n*; Martinustag *m* — **n** feestdag *m* van Sint Maarten — **e** fiesta *f* de San Martín — **i** festa *f* di San Martino

11846 sal ammoniac — See ammonium chloride.

11847 sale catalogue

f catalogue *m* de vente aux enchères — **d** Auktionskatalog *m*; Versteigerungskatalog *m* — **n** veilingcatalogus *m*; auctiecatalogus *m* — **e** catálogo *m* de subasta — **i** catalogo *m* di vendita pubblica

11848 sales department

f service *m* des ventes — **d** Verkaufsabteilung *f* — **n** verkoopafdeling *f*; afdeling *f* verkoop — **e** departamento *m* de ventas — **i** reparto *m* vendite

11849 salesman — A size of cut card. See app. no. 3.

11850 salesman — See representative.

11851 salicylic acid; ortho-hydroxybenzoic acid — White needle crystals or powder, soluble in acetone, oil of turpentine, alcohol, ether, benzene, very slightly soluble in water. Latin: acidum salicylicum. Formula: $C_6H_4(OH)(COOH)$.

f acide *m* salicylique — **d** Salizylsäure *f*; Spirsäure *f*; Spiroylsäure *f* — **n** salicylzuur *n* — **e** ácido *m* salicílico — **i** acido *m* salicilico

11852 salt — Compound formed by the reaction of an acid with a base, named after the acid and the metal from which it is derived, e.g., copper sulphate is a salt derived from sulphuric acid and copper.

f sel *m* — **d** Salz *n* — **n** zout *n* — **e** sal *f* — **i** sale *m*

11853 saltpetre — See potassium nitrate.

11854 sample — Portion of material or group of units taken from a large mass or aggregation (or population), to give informations as to the quality of the larger quantity.

f échantillon *m*; prélèvement *m* — **d** Stichprobe *f* — **n** monster *n* — **e** muestra *f* — **i** campione *m*

11855 sample book — Book, often loose-leaf type, in which samples of paper, textiles, leathers, etc. are displayed. See also swatch pad.

f album *m* d'échantillons; collection *f* d'échantillons — **d** Musterbuch *n*; Mustersammlung *f* — **n** monsterboek *n*; monstercollectie *f* — **e**

muestrario *m*; colección *f* de muestras — **i** campionario *m*

11856 sampling — Selection of representative samples of raw materials, paper, etc., for analysis and testing.

f échantillonnage *m*; prélèvement *m* d'échantillons — **d** Probeentnahme *f*; Musterentnahme *f* — **n** monsterneming *f*; monstername *fm* — **e** selección *f* de probetas; muestrario *m* — **i** campionatura *f*; campionamento *m*; prelievo *m* di campioni

11857 sampling instruction

f prescription *f* concernant l'échantillonnage — **d** Stichproben-Entnahmevorschrift *f* — **n** steekproefvoorschrift *n* — **e** instrucción *f* para escogidar pruebas al azar; instrucciones *fpl* para muestreo; instrucciones *fpl* para mostreo — **i** istruzione *f* per la campionatura; istruzione *f* per il prelevamento dei campioni

11858 sand; grinding sand — Finely sieved natural sand (or pulverized glass) used as an abrasive in plate graining.

f sable *m* à grainer — **d** Kornsand *m*; Schleifsand *m* — **n** zand *n*; slijpzand *n* — **e** sablón *m* — **i** sabbia *f*; polvere *f* per granitura

11859 sandarac resin — See gum sandarac.

11860 sand blasting — Producing a grain by impelling abrasive particles against the metal plate.

f projection *f* de sable; grainage *m* au jet de sable — **d** Sandstrahlen *n*; Körnung *f* mit Sandstrahlgebläse — **n** zandstralen *n*; greinen *n* met een zandstraal — **e** graneado *m* por chorro de arena — **i** granulazione *f* con sabbia

11861 sandpaper — Emery paper. See also abrasive paper.

f papier *m* sablé; papier *m* de verre; papier *m* d'émeri; papier *m* abrasif — **d** Sandpapier *n*; Schmirgelpapier *n*; Glaspapier *n* — **n** schuurpapier *n* — **e** papel *m* arenado; papel *m* de arena; papel *m* de lija — **i** carta *f* sabbia

11862 sandpaper — See abrasive paper.

11863 sand shifter — See riffler.

11864 sand table — See riffler.

11865 sand trap — See riffler.

11866 sandwiching — Use of ink of two or more types in multicolour printing. Rotogravure term.

f sandwiching *m* — **d** Verwendung *f* von zwei oder mehr Druckfarbentypen beim Mehrfarbendruck — **i** uso *m* di due o più tipi di inchiostro nella stampa in policromia

11867 sanguine — Red crayon or chalk, made from iron sulphate and clay, used by artists and by lithographers. Also a drawing executed with red chalks.

f sanguine *f* — **d** Rötel *m*; roter Kreidestift *m*; Rötelzeichnung *f*; Röteldruck *m* — **n** rood krijt *n* — **e** sanguina *f*; yeso *m* rojo; almazarrón *m*; almagre *m* — **i** sanguigna *f*; matita *f* rossa; gesso *m* rosso

11868 sans serif; non serif; unseriffed — Without serifs. Earlier forms were adaptional of the Roman types, but eliminating the serifs. Later sans serifs achieved wide acceptance with the introduction of Futura types. The word is probably derived from the Dutch schreef, i.e., stroke or line.

f sans empattement — **d** serifenlos — **n** schreefloos — **e** abastonado; sin palo (de bastón) — **i** senza grazie; senza terminali; bastoncino; bastone

11869 Sans Serifs — See Lineals.

11870 saponifiable — Capable of being saponified. See also saponification.

f saponifiable — **d** verseifbar — **n** verzeepbaar — **e** saponificable — **i** saponificabile

11871 saponification; saponifying — Hydrolysis of an ester. Often confined to the hydrolysis of an ester using an alkali, thus forming a salt (a soap in the case of the higher fatty acids) and the free alcohol.

f saponification *f* — **d** Verseifung *f* — **n** verzeping *f* — **e** saponificación *f* — **i** saponificazione *f*

11872 saponification number — Number of milligrams of potassium hydroxide required to saponify one gram of a sample of an ester (glyceride, fat), or mixture.

f indice *m* de saponification; nombre *m* de saponification — **d** Verseifungszahl *f* — **n** verzepingsgetal *n* — **e** índice *m* de saponificación; número *m* de saponificación — **i** numero *m* di saponificazione

11873 saponifying — See saponification.

11874 Saracenic binding — Binding with a decorative design which originates from the Mohammedans of North-Western Africa and the area of the Mediterranean.

f reliure *f* sarracénique — **d** sarazenischer Einband *m* — **n** saraceense band *m* — **e** encuadernación *f* sarracena — **i** legatura *f* saracena

11875 saran; polyvinylidene chloride; PVDC — Thermoplastic resin obtained by polymerization of vinylidene chloride or copolymerization of vinylidene chloride with not more than an equal weight of unsaturated compounds. Generally crystalline, orientable resins. May be printed easily; since the film is soft, low drying heats are recommended.

f saran *m*; chlorure *m* de polyvinylidène — **d** Saran *n*; Polyvinylidenchlorid *n* — **n** saran *n*; polyvinylideenchloride *n* — **e** cloruro *m* de vinilideno — **i** saran *m*; cloruro *m* di polivinilidene

11876 satchel bag — Gusseted tube with the bottom closed by turning up this tube and gumming the turn-up to the outside of the bag so formed.

f sac *m* à fond plat — **d** Bodenbeutel *m* — **n** bodemzak *m* — **e** bolsa *f* con fondo plano — **i** sacchetto *m* con fondo piatto

11877 satchel-bottom bag — See bag.

11878 satin white — White pigment made by the interaction of aluminium sulphate and slaked lime, actually a mixture of alumina hydrate and calcium sulphate; used in coating mixtures for the better grades of enamels.

f blanc *m* satin; pâte *f* satin — **d** Satinweiß *n*; Atlasweiß *n* — **n** satijnwit *n* — **e** blanco *m* satín — **i** bianco *m* satin

11879 saturated colour — See also saturation.

f couleur *f* saturée — **d** gesättigte Farbe *f*; Optimalfarbe *f* — **n** verzadigde kleur *fm* — **e** color *m* saturado; color *m* saciado — **i** colore *m* saturo

11880 saturated fatty acid — Saturated fatty acids have the general formula $C_nH_{2n}O_2$.

f acide *m* gras saturé — **d** gesättigte Fettsäure *f* — **n** verzadigd vetzuur *n* — **e** ácido *m* graso saturado — **i** acido *m* grasso saturo

11881 saturated solution — Solution which, at a fixed temperature, will not dissolve any more of a given substance.

f solution *f* saturée — **d** gesättigte Lösung *f* — **n** verzadigde oplossing *f* — **e** solución *f* saturada — **i** soluzione *f* satura

11882 saturation — Attribute of a visual sensation to judge the promotion of pure chromatic colour in the total sensation. Psychosensorial correlate of excitation purity. See also chroma.

f saturation *f* de la couleur — **d** Farbsättigung *f* — **n** verzadiging *f* van de kleur — **e** saturación *f* del color — **i** saturazione *f* del colore

11883 saturation — State of a solution when it holds the maximum equilibrium quantity of dissolved matter at a given temperature. Also the state in which all available valence bonds of an atom (esp. carbon) are attached to other atoms. The straight-chain paraffins are typical saturated compounds.

f saturation *f* — **d** Sättigung *f* — **n** verzadiging *f* — **e** saturación *f* — **i** saturazione *f*

11884 saturation — Process of impregnating a waterleaf sheet such as felt paper, with a waterproofing or preserving compound.

f imprégnation *f* — **d** Imprägnierung *f*; Durchtränkung *f* — **n** impregneren *n* — **e** impregnación *f* — **i** impregnazione *f*

11885 saturation bath

f bain *m* de saturation — **d** Sättigungsbad *n* — **n** verzadigingsbad *n* — **e** baño *m* de saturación — **i** bagno *m* di saturazione

11886 saturation force — Magnetizing force that will yield a magnetic flux density of approximately maximum value for a given sample of material, i.e., the saturation magnetization.

f niveau *m* de signal nominal — **d** Sättigungsgrad *m* — **n** verzadigingsgraad *m* — **e** fuerza *f* de saturación — **i** forza *f* di saturazione

11887 saturation point — Point at which a substance will receive no more of another substance in solution, chemical combination, etc.

f point *m* de saturation — **d** Sättigungspunkt *m* — **n** punt *n* van verzadiging; verzadigingspunt *n* — **e** punto *m* de saturación — **i** punto *m* di saturazione

11888 saturnism — See lead poisoning.

11889 sauerkraut — British rotary press term for

paper trimmings produced by slitters when the sidelay is incorrect. Also the trimmings at the cutting cylinder when using cylinder collection.

f rognures *fpl* — **d** Schnipsel *m*; Papierschnipsel *m*; Schnitzelbildung *f* — **n** snippers *mpl*; papiersnippers *mpl* — **e** trizas *fpl* — **i** ritagli *mpl*

11890 Savafid binding — See Safawid binding.

11891 save-all tray — Apparatus to reclaim fibres and fillers from *white water, usually on a filtration, sedimentation, flocculation or floatation principle.

f bacholle *f*; ramasse-pâte *m* — **d** Siebabwassertisch *m*; Siebwasserschiff *n*; Siebwasserauffangrinne *f*; Stoffänger *m* — **n** stofvanger *m* — **e** recoge *m* pasta — **i** gronda *f*

11892 save-all water — See white water.

11893 sawbind *v* — To cut grooves for the cords or bands across the backbone of a book.

f grecquer — **d** einsägen — **n** inzagen — **e** serrar; aserrar — **i** grecare

11894 saw blade

f lame *f* de scie — **d** Sägeblatt *n* — **n** zaagblad *n* — **e** hoja *f* de sierra — **i** lama *f* a sega

11895 sawn-in back — Back of a book block in which grooves are sawn, to receive the cords.

f dos *m* grecqué — **d** eingesägter Rücken *m* — **n** ingezaagde rug *m* — **e** lomo *m* aserrado — **i** dorso *m* grecato

11896 saw trimmer — Circular saw or a *band saw used by printers and engravers to saw and trim slugs, plates, blocks, etc. See also jig saw.

f scie *f* rectifieuse de précision; machine *f* à équerrer; dispositif *m* de coupe — **d** Präzisionssäge *f*; Sägeaggregat *n* — **n** precisie-zaagmachine *f*; zaaginrichting *f*; cirkelzaag *fm* — **e** sierra-cepillo *f*; sierra *f* de precisión — **i** sega *f* rifilatrice

11897 SBP spirit; petroleum spirit — The term SBP spirit, abbreviation for special boiling spirit, refers to a group of hydrocarbon mixtures derived from petroleum and usually having narrow boiling ranges. These mixtures are generally well-refined and may contain a particular type of hydrocarbons as the main constituent. The various commercial products are usually classified according to boiling range, e.g., SBP spirits 62/82, 100/140, 160/180 °C. The Dutch lakbenzine (170/200 °C), and the German Testbenzin (130/200 °C), Lackbenzin (140/190 °C), Klaffenbacher Benzin (100/140 °C) and Putzöl, also belong to this group. In ordering an SBP solvent the boiling range should be mentioned.

f essence *f* spéciale — **d** Siedegrenzenbenzin *n*; Lösungsbenzin *n* — **n** (speciale) kookpuntbenzine *fm* — **e** gasolina *f* a punto de ebullición especial — **i** benzina *f* a punto di ebollizione speciale; benzina *f* speciale

11898 scale; pair of scales — Apparatus to weigh chemicals, etc.

f balance *f* — **d** Waage *f* — **n** balans *fm*; weegschaal *fm* — **e** balanza *f* — **i** bilancia *f*

11899 scale — Rule of graduated dimensions.

f règle *f* divisée — **d** Meßlatte *f* — **n** meetlat *fm*; maatlat *fm* — **e** regla *f* graduada — **i** scala *f* graduata

11900 scale — Chart or means determining the cost of line and halftone etchings, with a special scale employed for estimating the cost of colour process plates.

f tarif *m* — **d** Tarif *m*; Preistarif *m*; Preisliste *f* — **n** tarief *n* — **e** tarifa *f* (de precios) — **i** tariffario *m*

11901 scale — See focusing scale.

11902 scale focusing — Bringing an image to correct size on process cameras by use of focusing scales.

f mise *f* au point par échelles graduées — **d** Scharfeinstellung *f* nach Skalen — **n** scherpstelling *f* met behulp van een schaal — **e** enfoque *m* según escala — **i** messa *f* a fuoco mediante scala graduata

11903 scale of colours — See colour-control chart.

11904 scale of rating; rating scale — Series of numerical indices given to various rates of working.

f échelle *f* d'allure — **d** Bewertungsskala *f* — **n** waarderingsschaal *fm* — **e** escala *f* de valuación — **i** scala *f* di valutazione

11905 scale of reproduction — Ratio of enlargement or reduction of an original to the final reproduction.

f échelle *f* de reproduction — **d** Abbildungsmaßstab *m*; Abbildungsverhältnis *n*; Verhältnis *n* — **n** verhouding *f*; reproduktieverhouding *f*; schaal *fm*; vergrotingsschaal *fm*; verkleiningsschaal *fm* — **e** escala *f* de reproducción — **i** scala *f* di riproduzione

11906 scaling — Measurement of pictures or illustrations to determine their proportional enlargement or reduction for reproduction.

f report *m* à l'échelle — **d** Bildgrößenberechnung *f*; Bestimmung *f* der verhältnisgleichen Veränderung *f* — **n** formaatberekening *f*; formaatbepaling *f* — **e** cálculo *m* de las dimensiones — **i** traduzione *f* in scala

11907 scan *v* — To move a light beam or flying spot to transmit data for electronic analysis, as for character recognition, or to define the elements of a character or picture so that these elements may be stored and subsequently reproduced, as in the case of facsimile transmission.

f balayer; explorer — **d** abtasten — **n** aftasten — **e** explorar — **i** esplorare; scandire

11908 Scan-a-graver — Automatic engraving machine for photoelectrically producing relief halftone printing plates on sheets of plastic by an electrically actuated and heated stylus, the action of the stylus controlled according to the tone values and gradation of the photographic image or original undergoing reproduction. The machine was made in 1948 by the Fairchild Camera & Instrument Corp. (US).

11909 scandal sheet — See yellow paper.

11910 scanner — Equipment with a device which electrically scans photographic copy or transparency, activating impulses which react upon a plastic plate or sensitized film, making a halftone plate on one type of machine, and a set of process colour negatives on another. See also flying spot scanner, optical scanner, visual scanner, Klischograph, Scan-a-graver.

f scanner *m*; appareil *m* de sélection électronique; sélecteur *m* électronique — **d** Scanner *m*; Abtaster *m*; Abtastvorrichtung *f*; Selektions- und Maskiermaschine *f*; Abfragevorrichtung *f* — **n** scanner *m*; aftaster *m*; aftastinrichting *f* — **e** escudriñador *m* electrónico; lector *m* electrónico — **i** scanner *m*; analizzatore *m* elettronico; apparecchio *m* per la selezione elettronica

11911 scanning; electronic scanning — Analysis of a picture through filters and simultaneous exposure of a photographic film in a separate position.

f exploration *f* de l'original — **d** Abtastung *f*; Abtasten *n* — **n** aftasten *n*; scannen *n depr.* — **e** escudriñamiento *m* electrónico; lectura *f* electrónica — **i** esplorazione *f* elettronica

11912 scanning head; sensing head; analysing head — Part of an electronic scanning or engraving machine which supplies information derived from the original copy or negative.

f tête *f* chercheuse; tête *f* d'analyse; tête *f* de lecture — **d** Abtastkopf *m*; Tastkopf *m*; Abtaster *m* — **n** tastkop *m* — **e** cabeza *f* escudriñadora; cabeza *f* lectora — **i** testina *f* esploratrice

11913 s.caps — See small capitals.

11914 scarlet lake — Fugitive red lake pigment, made by precipitation of an organic dye on a base of barium sulphate and aluminium hydrate or orange mineral or red lead. Used for transparent and glossy inks.

f laque *f* écarlate — **d** Scharlachlack *m*; Scharlachrot-Lack *m* — **n** scharlakrood *n* — **e** laca *f* escarlata — **i** lacca *f* scarlatta

11915 scattered light
f lumière *f* parasite; lumière *f* diffuse — **d** Streulicht *n* — **n** strooilicht *n* — **e** luz *f* dispersa; luz *f* parasitaria — **i** luce *f* parassita

11916 scavenger lines — See cow catcher.

11917 scented paper
f papier *m* parfumé — **d** parfümiertes Papier *n* — **n** geparfumeerd papier *n* — **e** papel *m* perfumado — **i** carta *f* profumata

11918 schematic diagram — See block diagram.

11919 Schlippe's salt — See sodium-thioantimonate.

11920 school book
f livre *m* scolaire — **d** Schulbuch *n*; Lehrbuch *n* — **n** schoolboek *n*; leerboek *n*; studieboek *n* — **e** libro *m* escolar; libro *m* de texto — **i** libro *m* scolastico; libro *m* di testo

11921 Schopper folding machine — Device to test the folding resistance of strips of paper of a standard width, and to count the number of double folds the inserted strip can withstand before breaking. The test is made both in the cross- and the machine direction.

f pliagraphe *m* Schopper — **d** Schopper-Falzer *m*; Schopper-Falzgerät *n* — **n** dubbelvouwtoestel *n* volgens Schopper — **e** ensayadora *f* plegadora según Schopper; ensayadora *f* a la flexión según Schopper — **i** apparecchio *m* Schopper per prove di piegatura

11922 Schopper-Riegler — See degree of beating.

11923 Schulze screen; rhombohedral screen; hexagonal screen
f trame *f* Schulze; trame *f* rhomboédrique — **d** Schulzeraster *m*; rhomboederraster *m*; hexagonaler Raster *m* — **n** Schulze-raster *n*; zestallig raster *n* — **e** trama *f* de Schulze; trama *f* romboedro; trama *f* hexagonal — **i** retino *m* di Schulze; retino *m* romboedrico; retino *m* a trama esagonale

11924 Schwabacher type — The German for a black-letter type used in printing books, newspapers, etc., in Germany. See also German alphabet, Gothic, app. no. 2.

f caractère *m* dit de Schwabach — **d** Schwabacher *f*; Schwabacherschrift *f* — **n** Schwabacher *m* — **e** Schwabacher *m* — **i** carattere *m* detto di Schwabach

11925 Schwarzschild effect — In photography, reciprocity failure reported by the astronomer J. Scheiner in 1889, who found an inefficiency in the photographic effect at low light intensity, named after Schwarzschild who made a detailed study of it.

f effet *m* Schwarzschild — **d** Schwarzschild-Effekt *m* — **n** Schwarzschild-effect *n* — **e** efecto *m* Schwarzschild — **i** effetto *m* Schwarzschild

11926 Schweizer's reagent — See cuprammonium hydroxide.

11927 science of journalism
f science *f* du journalisme — **d** Zeitungswissenschaft *f*; Journalistikwissenschaft *f* — **n** dagbladwetenschap *f* — **e** ciencia *f* diaria — **i** giornalismo *m*

11928 scientific research — See also applied scientific research.

f recherche *f* scientifique — **d** wissenschaftliche Forschung *f* — **n** wetenschappelijk onderzoek *n* — **e** investigación *f* científica — **i** investigazione *f* scientifica; ricerca *f* scientifica

11929 scissors — Cutting instrument for paper and thin metal sheets consisting of two knives hinged to each other. In trade, all such instruments less than 15 cm (6 in) long are called scissors, while all exceeding that length are termed *shears.

f ciseaux *mpl* — **d** Schere *f* — **n** schaar *fm* — **e** tijeras *fpl* — **i** forbici *fpl*

11930 sclerometer — See hardness meter.

11931 scoop — Degree of cut or removal of material affected by a router bit.

f creux *m*; excavation *f* — **d** Austiefung *f* — **n** diepte *f*; uitdieping *f* — **e** hueco *m* — **i** profondità *f* di fresatura; profondità *f* di passata della fresa

11932 scoop; exclusive; beat — In journalistic parlance, obtaining news items or data to the exclusion of competitors.

f primeur *f* d'une grosse nouvelle; exclusivité *f* — **d** Primeur *f* — **n** primeur *fm* — **e** antipación *f*; anticipo *m*; pisotón *m*; primicia *f* informativa *Es*; primicia *f* periodística; golpe *m* (periodístico) *Ch*; palo *m Cu* — **i** primizia *f*

11933 scoop — See graver.

11934 scorcher — See matrix dryer.

11935 score — Impression or cut in flat material to facilitate bending etc. In folding carton, two types of score are used: 1. folding score in which the fibres are compressed but not cut, to ensure that a fold or bend takes place on the score line; 2. tearing score in which the fibres are cut approximately half way through the board to permit tearing along the score line. In set-up paper boxes, scores are cut half through the board but corners formed at the score are usually reinforced by gummed paper stays.

f tracé *m* (coupant) — **d** Ritze *f*; Ritzlinie *f* — **n** ritslijn *fm*; rits *fm* — **e** trazo *m* cortante; raja *f* cortante — **i** tracciato *m*

11936 score *v* — To make an impression or a partial cut in flat material to facilitate bending, creasing, folding or tearing.

f tracer (au couteau) — **d** ritzen; anritzen — **n** ritsen — **e** rajar — **i** tracciare

11937 scoring machine

f mitrailleuse *f*; tracette *f* — **d** Ritzapparat *m*; Ritzmaschine *f* — **n** ritsapparaat *n*; ritsmachine *f* — **e** máquina *f* para rajar — **i** macchina *f* tracciatrice; tracciatrice *f*

11938 scoring rule — Steel or brass rule to crease covers, folders, etc., on a printing press.

f filet *m* à tracer — **d** Ritzlinie *f* — **n** ritslijn *fm* — **e** filete *m* para rayar; filete *m* para rasguñar; pleca *f* de estriar; pleca *f* estriadora — **i** filetto *m* per tracciare

11939 scorper — See graver.

11940 Scotch rule — See Oxford rule.

11941 Scotch stone — See snakeslip.

11942 Scotch tape test; cellophane tape test; sensitive tape test — For printing on plastics, test where a piece of pressure-sensitive tape, such as Scotch cellophane tape, is firmly pressed on the printed and dried film surface, then drawn back slowly over about half its length and rapidly pulled of the remaining area. The amount of ink removed by the tape is a measure of how well the ink is adhering to the printed film as it comes off the press.

f test *m* au Scotch — **d** Scotch-tape-Versuch *m* — **n** proef *fm* met Scotch tape — **e** ensayo *m* de cinta adhesiva de celulosa; ensayo *m* de cinta adhesiva al tacto de marca Scotch — **i** test *m* al nastro adesivo; prova *f* al nastro adesivo; saggio *m* al nastro adesivo

11943 s.c. paper — See super-calendered paper.

11944 scrap book — Blank book in which items

as newspaper clippings or pictures may be pasted or inserted.

f album *m* de coupures — **d** Einklebealbum *n*; Einklebeheft *n*; Einklebebuch *n* — **n** plakboek *n*; knipselboek *n* — **e** libro *m* de recortes; álbum *m* de recortes; libro *m* de extractos — **i** libro *m* di ritagli; album *m* di ritagli; raccolta *f* di ritagli

11945 scrap chopper — Usually a rotary mechanism consisting of a series of small cutting blades which, revolving together, chop waste trip or stripped material into fine pieces to aid in scrap disposal. Usually a part of a printing or fabricating press.

11946 scrape *v* — To scrape a leather roller against (or the smooth way) of the grain.

f gratter un rouleau dans le sens inverse (ou dans le sens) du grain — **d** eine Walze gegen den Strich (oder nach dem Strich) abreiben — **n** een rol schrapen tegen de vleug in (of met de vleug mee) — **e** raspar un rodillo en el sentido (de la flor) — **i** raschiare il cuoio dal rullo in direzione contraria alla fibra

11947 scrape *v* — To remove burrs from copper and steel plate engravings; by extension also around stereos.

f ébarber — **d** schaben — **n** wegschrapen; wegschaven — **e** raspar — **i** sbavare; raschiare; sbarbare (uno stereo)

11948 scrap ejector — See scrapings ejector.

11949 scraper; scraping tool — Three-faced steel tool, hollow ground, used by finishers to remove burrs and small defects from etched or engraved plates.

f grattoir *m* (pour graveur) — **d** Schaber *m*; Schabemesser *n* — **n** schraapstaal *n* — **e** buril *m* raspador; raspador *m*; desbarbador *m* — **i** raschietto *m*

11950 scraper; retouching knife — Pencil-like tool, or a sharp knife blade, to remove small surface defects in the emulsion of a negative or positive by abrasion.

f grattoir *m* à retouche; canif *m* — **d** Radiermesser *n*; Schabklinge *f*; Schabemesser *n* — **n** schraper *m*; retoucheermesje *n*; pennemesje *n* — **e** raspador *m* de retoque; cuchillo *m* de retoque — **i** coltello *m* da ritocco

11951 scraper

f râteau *m* (d'une presse lithographique à bras); râteau *m* de pression — **d** Reiber *m* (der Steindruckhandpresse) — **n** rijver *m* (van de steendrukhandpers) — **e** rastrillo *m*; raspador *m* — **i** rastrello *m* (per torchio litografico)

11952 scraper board — Board coated with China clay, in which a design can be scraped with a pen knife.

f carte *f* à gratter; papier *m* procédé — **d** Schabpapier *n*; Schabkarton *m* — **n** schaafkarton *n* — **e** cartoncino *m* para reproducciones por grabado — **i** cartone *m* per raschiare

11953 scrapings ejector; scrap ejector — Part of the Typograph typesetting machine.

f expulseur *m* de bouts — **d** Abfallausstoßer *m* —

n afval-uitstoter *m* — **e** expulsador *m* de residuos; expulsor *m* de residuos — **i** espulsore *m* degli scarti

11954 scraping tool — See scraper.

11955 scratch — See solidus.

11956 scratch comma — See solidus.

11957 scratches — Defect of aluminium foil. Visible fine lines marring the surface.

11958 scratch figure — See cancelled figure.

11959 scratching — See scuffing.

11960 scratching tendency tester
f appareil *m* à déterminer le pouvoir abrasif — **d** Kratzprüfer *m* — **n** krasproeftoestel *n* — **e** aparato *m* de ensayo del arañazo — **i** apparecchio *m* per misurare la tendenza alla raschiatura

11961 scratching test; scratch test — Test where the printed material is placed on a smooth, hard surface and scratched with a sharp object such as a knife or a fingernail, employing as much pressure as possible, without stretching or distorting the stock. Proper adhesion of the ink is indicated by no flaking or scratching of the ink surface.
f essai *m* de rayures; essai *m* de rayage — **d** Kratzprobe *f*; Ritzprobe *f* — **n** krasproef *fm*; nagelvastheidsproef *fm* — **e** ensayo *m* de arañazo — **i** prova *f* di raschiatura

11962 scratch test — See scratching test.

11963 scrawler; scribbler — Author who turns out badly written copy.
f griffonneur *m* — **d** Schlechtschreiber *m*; Schmierfink *m*; Schmierax *m* — **n** knoeipot *m*; knoeier *m*; krabbelaar *m* — **e** chafallón *m*; chapucero *m*; garabateador *m*; mal obreo *m* — **i** scarabocchiatore *m*

11964 scrawling
f gribouillage *m*; griffonnage *m*; pattes *fpl* de mouche — **d** unleserliches Geschmier *n*; unleserliche Handschrift *f*; Hieroglyphen *mpl* — **n** geknoei *n*; gekrabbel *n* — **e** garrapatos *mpl*; garabatos *mpl* — **i** scarabocchio *m*

11965 screamer; bang *US* — Big headline for sensational news in a daily paper. See also banner.

11966 screamer; bang — Compositor's expression for an *exclamation mark.

11967 screen — Stencil in (silk) screen printing.
f patron *m*; écran *m*; pochoir *m* — **d** Siebdruckschablone *f*; Schablone *f* — **n** zeef *fm*; zeefdrukschabloon *n*; schabloon *n* — **e** pantalla *f*; malla *f* de seda *obs.* — **i** setaccio *m* serigrafico; schermo *m* serigrafico

11968 screen — Chemically coated inside surface of the large end of a cathode ray tube which becomes luminous when struck by an electron beam.
f écran *m* — **d** Bildschirm *m*; Schirm *m* — **n** beeldscherm *n*; scherm *n* — **e** pantalla *f* — **i** schermo *m*

11969 screen — Brief for gravure screen, knot screen, halftone screen, wire screen.

11970 screen — See halftone screen.

11971 screen — See knot screen.

11972 screen *v* — To expose carbon tissue to a gravure screen.
f tramer; faire copie de trame — **d** rastern; Rasterbelichtung vornehmen; Rasterkopie vornehmen — **n** rasteren — **e** tramar — **i** retinare

11973 screen angle — In colour reproduction angle at which the halftone screen (or original) is placed for each of the colour plates of the set, to avoid pattern or moiré in the completed impression.
f orientation *f* de trame; angle *m* de la trame — **d** Rasterwinkel *m*; Rasterwink(e)lung *f* — **n** rasterhoek *m* — **e** angulación *f* de trama — **i** angolazione *f* del retino

11974 screen aperture — Clear space in a halftone screen.
f ouverture *f* de trame — **d** Rasteröffnung *f*; Rasterfenster *n* — **n** rasteropening *f* — **e** abertura *f* de (los cuadritos de) la retícula — **i** apertura *f* del retino

11975 screen carrier — See screen holder.

11976 screen compensator — Piece of glass positioned in front of the process camera lens when making combination line and halftone negative with two exposures. When halftone screen is removed, it is used to allow for difference in light refraction.
f glace *f* de compensation de trame — **d** Rasterdicken-Ausgleichsscheibe *f* — **n** rastercompensator *m* — **e** superretícula *f* — **i** vetro *m* di compensazione del retino

11977 screen distance; screen separation — In halftone photography, separation or space between the surface of a glass halftone screen of specific ruling and the photographic surface during exposure. It permits the rays of light passing through the screen to diffuse before striking the film or plate. Extent of diffusion of rays is in proportion to their intensity which in turn is determined by the reflection of light from the original copy. The variation in diffusion is what produces dots of different sizes on the film or plate.
f écart *m* de trame — **d** Rasterabstand *m* — **n** rasterafstand *m* — **e** distancia *f* de (la) trama; separación *f* de (la) trama — **i** distanza *f* del retino

11978 screened photoprint — Photoprint made through the halftone screen giving a screened print instead of a continuous-tone print. Can be photographed as line copy.
f copie *f* tramée — **d** gerasterter Photo-Abzug *m* — **n** gerasterde foto-afdruk *m* — **e** copia *f* fotográfica tramada; prueba *f* fotográfica tramada — **i** copia *f* fotografica retinata

11979 screen holder; screen carrier — In process cameras, mechanism or carrier for rectangular and circular halftone screens and for adjusting the screen distance.
f porte-trame *m*; châssis *m* porte-trame — **d** Rasterkassette *f*; Rasterhalter *m*; Rasterträger *m*

— n rasterhouder *m* — e portarretícula *m*; porta-trama *m*; portapantalla *m* — i portaretino *m*

11980 screen indicator — Ruled or lined instrument or device to determine the ruling of the screen employed for any halftone reproduction.
f linéomètre *m* — d Rasterlinienzähler *m* — n rastermeter *m* — e indicador *m* de (la lineatura de) la retícula — i lineometro *m*

11981 screening — Improper print condition in gravure printing where the ink flowout between cells is such that an uneven solid is formed, usually in a screen pattern, usually caused by high viscosity ink, or ink drying too rapidly.
f trame *f* apparente — d ungenügende Stegüberflutung *f*; Struktur *f* in den Tiefen des Drucks; Sichtbarwerden *n* der Rasterstege — n indrogen *n* — e secado *m* demasiado rápido de la tinta — i essiccamento *m* troppo rapido dell'inchiostro

11982 screenings — Paper in which pulp screenings are largely used, characterized by a profusion of specks or dark spots and used for wrappings and occasionally for stunt printing such as odd greeting cards.
f papier *m* gris pour emballages — d Schrenz *n* — n schrens *n* — e papel *m* de straza

11983 screen negative — See halftone negative.
11984 screen pattern — See moiré.
11985 screen positive — See halftone positive.
11986 screen printer; silk screen printer *depr.*
f imprimeur *m* sérigraphe — d Siebdrucker *m* — n zeefdrukker *m* — e serigrafo *m* — i serigrafo *m*

11987 screen printing ink; silk screen ink *depr.* — Ink for the screen printing process, made by grinding strong coloured pigments in a compounded varnish or an aqueous vehicle. Should be fairly short, non tacky, easy flowing, and non oily.
f encre *f* pour sérigraphie; encre *f* à pochoir sur soie *depr.*; encre *f* pour impression à trame de soie *depr.*; encre *f* pour impression à écran de soie *depr.* — d Siebdruckfarbe *f* — n zeefdrukinkt *m* — e tinta *f* de serigrafía; tinta *f* para impresión con trama de seda *depr.* — i inchiostro *m* per serigrafia; inchiostro *m* serigrafico

11988 screen printing machine — See screen process press.

11989 screen printing process; screen process; silk screen process *depr.*; **serigraphy** — Process of printing through the unblocked areas of a metal or fibre screen, with a free flowing ink which is spread and forced throught the screen by means of a squeegee. Used in the production of coloured posters, showcards, decalcomania, printed circuits, etc. Originally the screen was made almost exclusively of silk. Natural silk meshes, however, are woven from stranded threads and have irregularities and a rough surface structure. Most widely used materials today are nylon, terylene and metal. Serigraphy is literally drawing on silk (sericus and graphia).
f sérigraphie *f*; impression *f* sérigraphique; impression *f* à la trame de soie *depr.*; procédé *m* à l'écran de soie *depr.* — d Siebdruck *m*; Serigraphie *f*; Seidenrasterdruckverfahren *n* depr. — n zeefdruk *m*; serigrafie *f* — e impresión *f* serigráfica; impresión *f* con trama de seda *depr.*; impresión *f* con pantalla de seda *depr.*; impresión *f* con malla de seda *depr.*; impresión *f* por estarcido — i serigrafia *f*; procedimento *m* di stampa con retina di seda *depr.*

11990 screen process — See screen printing process.

11991 screen process press; screen printing machine; silk screen press *depr.*; **silk screen printing press** *depr.* — See also screen printing process.
f machine *f* pour impression sérigraphique; machine *f* à l'impression à trame de soie *depr.*; machine *f* à l'impression à l'écran de soie *depr.*; machine *f* à l'impression au pochoir sur soie *depr.* — d Siebdruckmaschine *f*; Siebdruckpresse *f* — n zeefdrukpers *f* — e prensa *f* serigráfica; prensa *f* de serigrafía; prensa *f* para impresión con trama de seda *depr.* — i macchina *f* serigrafica; pressa *f* serigrafica

11992 screen ruling; screen width; ruling distance — Distance between the screen lines of a *halftone screen.
f lignage *m* de trame; grosseur *f* de trame; finesse *f* de la trame; distance *f* des lignes de la trame — d Rasterweite *f*; Rasterfeinheit *f* — n rasterwijdte *f*; rasterfijnheid *f* — e lineatura *f* de la trama — i lineatura *f* del retino; finezza *f* del retino; frequenza *f* di retino

11993 screen separation — See screen distance.
11994 screen separation negative — See direct separation negative.
11995 screen tint — Screen tints or screen sheet usually used directly in stripping. See also shading medium.
11996 screen wall — See cell wall.
11997 screen width — See screen ruling.
11998 screw
f vis *f* — d Schraube *f* — n schroef *fm* — e tornillo *m* — i vite *f*
11999 screw closure — Closure for a container which screws into a thread provided on the neck.
f fermeture *f* à vis — d Schraubdeckel *m* — n schroefdop *m*; schroefsluiting *f* — e tapón *m* de rosca — i chiusura *f* a vite
12000 screw *v* **down** — See bolt *v*.
12001 screwdriver
f tournevis *m* — d Schraubenzieher *m* — n schroevedraaier *m* — e destornillador *m*; sacatornillos *m*; desarmador *m* *Me* — i cacciavite *m*
12002 screw jack — See jack screw.
12003 screws, binding — See binding screws.
12004 scribbler — See scrawler.
12005 scribbling paper
f papier *m* à brouillon — d Kladdepapier *n*; Konzeptpapier *n* — n kladpapier *n* — e papel *m* para borradores; borrador *m* — i carta *f* da minuta; carta *f* comune

12006 scribe machine — Device to mark copper plates or gravure cylinders accurately for correct register.
f machine *f* à tracer — **d** Anreißmaschine *f* — **n** aftekenmachine *f*; afschrijfmachine *f* — **i** macchina *f* tracciatrice
12007 scriber — Small pencil-like hand tool with a shaped point (scratch awl) to draw lines on the emulsion of an exposed photographic negative, the tool scraping off the black emulsion. The operation is termed engraving negatives and is used for fine ruled-form work.
f pointe *f* à tracer — **d** Reißzahle *f* — **n** krasnaald *fm* — **e** raedera *f* — **i** stilo *m*; punta *f* per tracciare
12008 scrim — See mull.
12009 scrinium — Cylindrical container used by the early Romans to hold scrolls.
12010 scriptorium — Writing room of a monastery, where records, annals, and manuscripts were written, copied or illustrated.
12011 Scripts — Ground of type faces that imitate writing, e.g., all scripts from English script, Choc, Rondo, Legende, Reiner script, Amazone, Excelsior, Pepita, Slogan, Gracia, Aigrette, Mistral. See also classification of type design.
f Scriptes *fpl* — **d** Schreibschriften *fpl* — **n** Scripten *fmpl* — **e** Escrituras *fpl* — **i** Scritti *mpl*
12012 script type — Type resembling handwriting. See also app. no. 2.
f caractère *m* d'écriture; caractère *m* anglais — **d** Schreibschrift *f* — **n** schrijfletter *fm*; Engels schrift *n* — **e** tipo *m* escritura; escritura *f* inglesa; letra *f* inglesa; inglesa *f*; plumilla *f* (inglesa); caracteres *fpl* de escritura — **i** carattere *m* di scrittura (a mano)
12013 scrubber — In the sulphite process, a large steel chamber in which the combustion of the sulphur is completed, the gas intimately mixed, and dust particles eliminated.
f purificateur *m*; laveur *m* (de gaz) — **d** Gaswäscher *m*; Gaswaschturm *m* — **e** lavador *m* de gas — **i** torre *f* di lavaggio; camera *f* di combustione
12014 scruple — Unit of weight in the apothecaries' units system. Abbrev. ℈ or ℥ . See app. no. 7 and 7a.
f scrupule *m* — **d** Skrupel *n* — **n** scrupel *n*; schrupel *n* — **e** escrúpulo *m*
12015 scuffing — Raising of the fibres on the surface of paper or paperboard when one piece is rubbed against another or comes in contact with a rough surface, especially when wet. Scuffing of inks is almost identical with rubbing and synonymous with scratching. Inks scuff off in subsequent fabricating operations. If used on containers, the inks scuff off during shipment, filling or handling of the filled containers. If the ink is used for printing halftones, unsightly scratch marks are formed if sheets are rubbed against the other or in folding operations.
f usure *f* — **d** Verschleiß *m* — **n** slijtage *f* — **e** arañamiento *m* — **i** abrasione *f*; graffiamento *m*
12016 scuff resistance — See abrasion resistance.
12017 sculper; round sculper; round chisel
f échoppe *f* ronde — **d** Boltstichel *m*; Bollstichel *m* — **n** rondsteker *m* — **e** buril *m* redondo — **i** bulino *m* tondo
12018 scum — Traces of ink which temporarily adhere to the non-image area of the offset plate due to its inability to repel ink.
f voile *m* — **d** Ton *m*; Plattenton *m* — **n** toon *m*; plaattoon *m* — **e** velo *m* — **i** velo *m* d'inchiostro; tracce *fpl* d'inchiostro
12019 scumming — In photoengraving, application of scum removers to glue prints on metal to restore billiancy and good etching properties to the image.
f enlevage *m* de voile — **d** Entschleiern *n* — **n** ontsluieren *n* — **e** aplicación *f* de quitavelo — **i** rimozione *f* della velatura
12020 scumming; greasing — Unwanted inking on the surface on the offset plate in non image areas not due to insufficient water or washing, basically due to spots or areas not remaining desensitized, often causing filling in of halftone dots, spreading of image, streaks. The scum will tend to adhere to the plate and become established if not quickly removed, e.g. by etching.
f graissage *m* (en offset) — **d** Tonen *n* — **n** tonen *n* — **e** formación *f* de velo; engrasado *m* — **i** ingrassamento *m*; scumming *m*
12021 scumming — In rotogravure printing, 1. a deposit of ink on the non-printing surface of a cylinder, very light in character, almost a haze; 2. an ink condition in an open fountain press where agitation is minimal. Solvent evaporates from the surface of the ink and leaves a film similar to a skin on letterpress ink. This scum will redissolve upon agitation.
f 1. voile *m*; 2. écume *f* — **d** 1. Tonen *n*; 2. Hautbildung *f* — **n** 1. tonen *n* — **i** 1. scumming *m*; 2. pellicolazione *f*
12022 scumming test; tone test — Test to check tendency of offset papers to scumming. A strip of wetted offset paper is pressed against a strip of aluminium offset plate. Then the damp strip of aluminium is inked up on the *printability tester by means of an inked rubber printing disk. Papers with a tendency to produce scumming will cause the metal to be inked, the non-scumming sample will not.
f essai *m* de graissage — **d** Prüfung *f* auf Tonen — **n** toonproef *fm* — **e** ensayo *m* de engrasado — **i** prova *f* d'ingrassamento; prova *f* di scumming
12023 scum remover — Acidified solution of common salt or chromic acid used in photoengraving to clean scummy glue prints on metal to restore brilliancy and good etching properties to the image.
f agent *m* d'enlevage de voile — **d** Schleierlösungsmittel *n* — **n** ontsluiermiddel *n* — **e** quitavelo *m*; solución *f* para quitar la capa de cola residual — **i** solvente *m* del velo

12024 seal — Small printing plate to indicate the edition of a newspaper.

d Siegel *n* — **n** editie-ster *f* — **i** sigillo *m*

12025 seal — Grained leather made from seal skin, used for limp book covers. It has a coarse grain and is soft in touch.

f peau *f* de phoque; phoque *m* — **d** Seehund-leder *n* — **n** zeehondeleer *n* — **e** piel *f* de foca — **i** cuoio *m* di foca

12026 sealing tape — See also gummed tape.

f bande *f* gommée; ruban *m* gommé — **d** Verschlußklebestreifen *m* — **n** plakband *n*; sluit-band *n* — **e** cinta *f* adhesiva; cinta *f* engomada — **i** nastro *m* di sigillatura; nastro *m* adesivo

12027 sealing-wax — Resinous preparation, soft when heated, to seal letters, documents, etc.

f cire *f* à cacheter; cire *f* d'Espagne — **d** Siegel-lack *mn* — **n** zegellak *mn* — **e** lacre *m* (para sellar) — **i** ceralacca *f*; cera *f* di Spagna

12028 seal unit — Small auxiliary printing unit to print the *seal, usually in a second colour on the front page.

f élément *m* de repiquage; fudge *m* — **d** (kleines) Eindruckwerk *n* — **n** drukeenheid *f* voor de editie-ster — **i** unità *f* supplementare per la stampa del sigillo

12029 search *v* — Examine a set of items for those that have a desired property. Computer term. See also binary search.

f chercher; compulser — **d** suchen — **n** zoeken — **e** buscar; investigar — **i** ricercare

12030 seasoned paper — See conditioned paper.

12031 seasoning — See conditioning.

12032 secant — In trigonometry, the ratio of the hypotenuse to the adjacent side of an angle of a right triangle. See also trigonometric functions.

f sécante *f* — **d** Sekante *f* — **n** secans *fm* — **e** secante *f* — **i** secante *f*

12033 secondary colours — Green, orange and violet, obtained by mixing two of the primary colours.

f couleurs *fpl* secondaires — **d** Sekundärfarben *fpl*; Mischfarben *fpl* — **n** secondaire kleuren *fmpl*; mengkleuren *fmpl* — **e** colores *mpl* secundarios; colores *mpl* binarios — **i** colori *mpl* secondari

12034 secondary leading — Computer term. Additional space between blocks of text, of a specified and sometimes variable amount. See also primary leading.

12035 secondary storage — See auxiliary store.

12036 secondary store — See auxiliary store.

12037 second bite

f deuxième morsure *f* — **d** Zwischenätzung *f*; zweite Ätzung *f* — **n** tweede etsing *f* — **e** mordedura *f* segunda — **i** secondo attacco *m*; seconda incisione *f*

12038 second chopper fold; mail fold — Fold accomplished in same manner as chopper fold, immediately following and parallel to it. Produces long, narrow signatures that are 32-page multiples of number of web used, 1/8 web width x 1/2 cut-off length.

f deuxième pli *m* parallèle — **d** zweimal Parallel-falz *m*; auf halbe Bahnbreite geschnitten; Zeitungsfalz *m* — **n** tweemaal parallelvouw *fm*; op halve papierbaan gesneden; krantenvouw *fm* — **e** segundo pliegue *m* paralelo — **i** seconda piega *f* parallela

12039 second cover — See inside front cover.

12040 second elevator — Arm on the slug composing machine, which swings down from the end of the distributor bar, and receives the matrices from the first elevator and carries them up to the distributor bar.

f second élévateur *m*; bras *m* preneur — **d** zweiter Elevator *m* — **n** tweede elevator *m* — **e** segundo elevador *m*; brazo *m* elevador; tomador *m Es* — **i** secondo elevatore *m*

12041 second-generation computer; solid-state computer — Computer utilizing solid state components. They operate in microseconds or a millionth of a second.

f calculateur *m* de la deuxième génération; ordinateur *m* de la deuxième génération — **d** Computer *m* der zweiten Generation; Rechner *m* der zweiten Generation — **n** computer *m* van de tweede generatie — **e** computadora *f* de la segunda generación — **i** calcolatore *m* della seconda generazione

12042 second-generation typesetter — Photo-composition device which composes type based upon photo-mechanical principles, but which is not an imitation of hot-metal machines as in the case with first-generation photo-typesetters. Type is created by flashing a light through a negative film representation of a character on a character-by-character basis. Such characters are laid down, according to their widths (*escapement) by means of a moving mirror, lens or prism, or by moving the film or character image (e.g., the disc of type images) a given distance after or before each character is exposed.

f composeuse *f* de la deuxième génération — **d** Setzmaschine *f* der zweiten Generation — **n** zet-machine *f* van de tweede generatie — **e** componedora *f* de la segunda generación — **i** compositrice *f* della seconda generazione

12043 second-hand — Said of goods offered for sale which have been in use.

f d'occasion — **d** aus zweiter Hand; gebraucht — **n** tweedehands; gebruikt — **e** de segunda mano; de lance; de ocasión; de viejo — **i** di seconda mano; usato

12044 second-hand book

f livre *m* d'occasion — **d** Buch *n* aus zweiter Hand — **n** tweedehands boek *n* — **e** libro *m* de segunda mano; libro *m* de lance; libro *m* de ocasión — **i** libro *m* di seconda mano; libro *m* d'occasione

12045 second-hand-bookseller — Dealer in old books or books of lesser value.

f bouquiniste *m* — **d** Büchertrödler *m* — **n** handelaar *m* in tweedehands boeken; handelaar *m* in oude boeken — **e** librero *m* de viejo — **i** venditore *m* di libri usati

12046 second-impression cylinder
f cylindre *m* de retiration; cylindre *m* de seconde; cylindre *m* de deux — **d** Widerdruckzylinder *m* — **n** weerdrukcilinder *m* — **e** cilindro *m* de retiración — **i** cilindro *m* di pressione (della stampa) in volta

12047 second impression set-off; ghosting — Condition in letterpress printing on a perfecting machine when the wet ink on one print transfers to the next impression cylinder and back on the paper on succeeding revolutions. This results in a blurred or an overemphasized appearance to the print.
f maculage *m* en retiration — **d** Abschmieren *n* der Farbe auf dem Widerdruck-Druckzylinder *m* — **n** overzetten *n* op de drukcilinder van de tweede drukeenheid — **e** mácula *f* de retiración — **i** controstampa *f* sul cilindro della stampa in volta

12048 second parallel fold — Fold made in *jaw folder immediately after the first parallel fold to it. Results in 16-page multiples of the number of webs in the press, signature size 1/2 web x 1/4 cutt-off length.

12049 second proof — See revise.

12050 seconds — See retrees.

12051 second sheets — Typewriter paper without letterhead, used for carbon copies.
f feuille *f* à copie — **d** Durchschlagpapier *n* — **n** doorslagpapier *n* — **e** papel *m* de copia — **i** carta *f* per dattilografia; carta *f* per copie dattiloscritte

12052 second-sheet stop — See blower foot.

12053 secret printing
f impression *f* clandestine — **d** heimlich hergestellter Druck *m* — **n** clandestiene uitgave *f* — **e** impresión *f* clandestina — **i** stampa *f* clandestina

12054 sectio aurea — See golden section.

12055 sectio divina — See golden section.

12056 section — See paragraph.

12057 section — See signature.

12058 sectional engraving — Line or halftone etching made in sections from a large original, dividing the reproduction into sectional negatives and plates because the camera or platemaking equipment will not accommodate the reproduction as a whole unit.
f cliché *m* fractionné — **d** Ätzung *f* in Teilen (wegen Überformat) — **n** cliché *n* uit delen samengesteld — **e** clisé *m* fraccionado — **i** incisione *f* frazionata

12059 section mark — Fourth reference mark, §, for footnotes.
f signe *m* de paragraphe — **d** Paragraphenzeichen *n* — **n** paragraafteken *n* — **e** signo *m* de párrafo; calderón *m* — **i** segno *m* di paragrafo; paragrafo *m*

12060 section title
f titre *m* divisionnaire — **d** Abteilungstitel *m* — **n** deeltitel *m* — **e** título *m* parcial — **i** titolo *m* parziale

12061 sector *v* — See segment *v*.

12062 security paper — See loan paper.

12063 security printing — Printing of securities making fraudulent reprinting difficult by employing guilloche backgrounds, multicolour printing with hues which are photographically very close not to be separated by filtering and hidden marks and signs.
f impression *f* de papiers-valeur; impression *f* fiduciaire — **d** Wertpapierdruck *m*; Wertzeichendruck *m* — **n** drukken *n* van waardepapieren — **e** impresión *f* de papeles de valor — **i** stampa *f* di carte-valori

12064 sediment — See deposit.

12065 sedimentation — Separation of suspended solid particles from a liquid by the action of gravity or other force.
f sédimentation *f* — **d** Sedimentation *f*; Absetzen *n* — **n** sedimentatie *f*; neerslaan *n* — **e** sedimentación *f* — **i** sedimentazione *f*

12066 see copy — Term used in correcting proofs. See app. no. 5.
f voyez copie; voy cop. — **d** siehe Manuskript — **n** zie (de) kopij — **e** vea original; véase el original — **i** vedi originale; vedi manoscritto

12067 segment *v*; **sector** *v*; **partition** *v* — To divide a program into parts. Some parts are retained in internal store during processing and contain the instructions necessary to call in other segments when required; used to economize on the amount of internal store for a program.
f segmenter — **d** segmentieren — **n** segmenteren — **e** segmentar; subdividir — **i** segmentare

12068 segmented disc — Combination of several founts on one disc, so that the user may combine certain desired faces. Segmentation may be by placing one smaller disc within a larger one, or by dividing the circular disc into segments or pie-shaped pieces, so that one or more founts will be contained within that particular segment.

12069 segregation — Separation of constituents during the freezing of an alloy which causes local variations in the composition of the solid alloy.
f ségrégation *f* — **d** Seigerung *f*; Entmischung *f* — **n** segregatie *f*; ontmenging *f* — **e** segregación *f* — **i** segregazione *f*

12070 Seignette salt — See sodium-potassium tartrate.

12071 selector channel — Hardware device connecting the central processing unit with a high speed input/output device. See also channel.
f canal *m* simple — **d** Selektorkanal *n* — **n** kieskanaal *n* — **e** canal *m* selector — **i** canale *m* selettore

12072 selenium — Non-metallic element, used in photographic cells, etc. Symbol: Se. TL-value: 0.1 mg/m^3.
f sélénium *m* — **d** Selen *n* — **n** selenium *n*; seleen *n* — **e** selenio *m* — **i** selenio *m*

12073 selenium cell — Photoelectric cell, depending for its action on the photo-conductive effect, or the photo-voltaic effect.

f cellule *f* au sélénium — **d** Selenzelle *f* — **n** selenium-cel *fm*; seleen-cel *fm* — **e** célula *f* de selenio — **i** cella *f* al selenio

12074 self-adhering letters; self-adhesive letters; dry-transfer letters — Sharp, opaque letters on transfer sheets, self-adherent to any surface. They are transferred by rubbing them from the sheet onto a surface.

f caractères *mpl* adhésifs; lettres *fpl* à transférer; lettres *fpl* transfert — **d** selbstklebende Lettern *fpl*; Abreibebuchstaben *fpl*; Sofortschrift *f* — **n** plakletters *fmpl* — **i** caratteri *mpl* trasferibili a secco; lettere *fpl* per riporto a pressione

12075 self-adhesive letters — See self-adhering letters.

12076 self-adhesive paper; pressure-sensitive paper; self-sticking paper
f papier *m* auto-collant; papier *m* adhésif à la pression — **d** Haftklebepapier *n*; Selbstklebepapier *n*; selbstklebendes Papier *n*; Trockenklebepapier *n* — **n** zelfklevend papier *n* — **e** papel *m* autoadhesivo; papel *m* autoadherente; papel *m* autocolante; papel *m* autopegante; pegatina *f* — **i** carta *f* autoadesiva

12077 self-adhesive tape; sticking tape
f ruban *m* auto-collant; ruban *m* auto-adhésif — **d** Selbstklebeband *n*; Haftklebeband *n* — **n** zelfklevend band *n* — **e** banda *f* autoadhesiva — **i** nastro *m* autoadesivo

12078 self-aligning roller bearing
f roulement *m* à rouleaux auto-centreurs — **d** Pendelrollenlager *n* — **n** zelfinstellend lager *n* — **e** cojinete *m* de rodillos de alineación automática — **i** cuscinetto *m* a rulli oscillanti

12079 self centring
f centrage *m* automatique — **d** automatische Zentrierung *f* — **n** automatisch centreren *n* — **e** autocentrado *m*; centrado *m* automático — **i** autocentratura *f*

12080 self-contained cover — See integral cover.

12081 self-copy paper
f papier *m* autocopiant — **d** Selbstkopierpapier *n* — **n** zelfkopiërend papier *n* — **i** carta *f* autoricalcante

12082 self cover — See integral cover.

12083 self-opening satchel bag; sos bag — Gusseted bag with rectangular bottom folded flat against one plain side. Filling opens the bag into a rectangular shape without any further creasing.

f sac *m* sos; sac *m* à fond rectangulaire et à soufflets latéraux; sachet *m* automatique à fond carré — **d** Selbstöffnerbeutel *m* — **n** zelfopengaande zak *m*; sos-zak *m* — **e** bolsa *f* con fondo rectangular y fuelles laterales — **i** sacchetto *m* con fondo quadrato ad apertura automatica

12084 self-opening satchel bag — See automatic self-opening bag.

12085 self quadder; automatic quadder; automatic centring and quadding device
f cadratinage *m* automatique — **d** automatische Zentriereinrichtung *f* — **n** automatisch centreer-

en uitvulapparaat *n* — **e** autocentrador y cuadrador *m* — **i** dispositivo *m* di rientranza e centratura

12086 self-sealing paper — Waxed, gummed or coated paper which adheres to another surface by cold or hot pressure.

f papier *m* auto-adhésif; papier *m* auto-collant — **d** siegelfähiges Papier *n* — **n** zelfplakkend papier *n*; zelfklevend papier *n* — **e** papel *m* autocierre; papel *m* autoadhesivo — **i** carta *f* autoadesiva

12087 self-sticking paper — See self-adhesive paper.

12088 self tone — See mass tone.

12089 selling price — Price at which a product is sold to customer.

f prix *m* de vente — **d** Verkaufspreis *m* — **n** verkoopsprijs *m*; verkoopprijs *m* — **e** precio *m* de venta — **i** prezzo *m* di vendita

12090 selvage; selvedge *depr.* — Edging of bookbinder's cloth different from the body of the fabric, to prevent ravelling.

f lisière *f* — **d** Salkante *f*; Salband *n*; Salleiste *f* — **n** zelfkant *m* — **e** orillo *m*; vendo *m*; hirma *f* — **i** vivagno *m*; cimosa *f*

12091 selvedge — See selvage.

12092 semée binding; semis binding — Sprinkled book cover with a narrow border of tooled lines around, and intervalling small monograms, fleurs-de-lis, etc. The centre of the cover shows usually the coat of arms of the owner. Originates from Nicolas and Clovis Eve, bookbinders to various kings of France of the 16th century.

f reliure *f* semée; reliure *f* semis — **d** Einband *m* mit Semisdekoration; Streumuster-Einband *m* — **n** semée-band *m* — **e** encuadernación *f* sembrado — **i** legatura *f* seminata

12093 semi-annual — Appearing every six months or twice a year.

f semestriel *d* halbjährlich — **n** halfjaarlijks; twee keer per jaar — **e** semestral — **i** semestrale

12094 semi-bleached pulp
f pâte *f* semi-blanchie — **d** halbgebleichter Halbzellstoff *m* — **n** halfgebleekte celstof *fm* — **e** pasta *f* semiblanqueada; pulpa *f* semiblanqueada — **i** pasta *f* semibianchita

12095 semi-bold face; medium-bold face — See also weight of face.

f caractère *m* demi-gras; caractère *m* mi-gras; mi-gras *m* — **d** halbfette Schrift *f* — **n** halfvette letter *fm* — **e** tipo *m* seminegro; letra *f* seminegra — **i** carattere *m* neretto; neretto *m*

12096 semi-chemical process — Process similar to standard wood cooking processes but with less active chemicals such as soda or lime and for shorter cooking periods. Suitable for making some paper boards and papers, depending on the mechanical refining treatment after cooking.

f procédé *m* semi-chimique — **d** halbchemisches Verfahren *n* — **n** halfchemisch procédé *n* — **e** procedimiento *m* semiquímico — **i** procedimento *m* semichimico

12097 semi-chemical pulp — Pulp obtained from wood or other material of vegetable origin by a light chemical treatment which eliminates only part of the non fibrous components. To separate the fibres, a further mechanical treatment is necessary.

f pâte *f* mi-chimique; pâte *f* semi-chimique — **d** Halbzellstoff *m*; halbchemischer Zellstoff *m* — **n** gedeeltelijk ontsloten celstof *fm*; halfchemische stof *fm* — **e** pasta *f* semiquímica — **i** pasta *f* semichimica

12098 semicolon — See also punctuation marks.

f point-virgule *m*; point et virgule *m* — **d** Strichpunkt *m*; Semikolon *n* — **n** punt-komma *fm*; komma-punt *fm* — **e** punto y coma *m*; semicolon *m* — **i** punto e virgola *m*

12099 semi-conductor — Material such as germanium, silicon, or selenium, whose electrical conductivity depends on the presence of a suitable impurity.

f semiconducteur *m* — **d** Halbleiter *m* — **n** halfgeleider *m* — **e** semiconductor *m* — **i** semiconduttore *m*

12100 semi-drying oil — Oil with the characteristics of a drying oil but to a lesser degree. There is no definite line of demarcation between drying and semi-drying oils.

f huile *f* semi-siccative — **d** halbtrocknendes Öl *n* — **n** halfdrogende olie *fm* — **e** aceite *m* semisecante — **i** olio *m* semisiccativo

12101 semi-monthly; half-monthly; fortnightly — Publication appearing twice a month. See also bi-monthly.

f publication *f* bimensuelle; publication *f* semi-mensuelle — **d** vierzehntägiges Blatt *n*; vierzehntägige Ausgabe *f*; halbmonatlich erscheinendes Blatt *n*; Halbmonatsschrift *f* — **n** veertiendaags blad *n*; halfmaandelijks (verschijnend) blad *n*; twee keren per maand verschijnend blad *n*; om de veertien dagen verschijnend blad *n* — **e** publicación *f* bimensual; publicación *f* quincenal; revista *f* quincenal; papel *m* quincenal; quincenario *m*; cuaderno *m* quincenal — **i** pubblicazione *f* quindicinale; rivista *f* quindicinale; quindicina *f*

12102 semis binding — See semée binding.

12103 semi-transparent pellicle mirror — See transmitting pellicle mirror.

12104 semi-weekly; half-weekly — Publication appearing twice a week. See also bi-weekly.

f publication *f* semi-hebdomadaire; publication *f* bihebdomadaire — **d** zweimal wöchentlich erscheinendes Blatt *n*; zweimal wöchentlich erscheinende Ausgabe *f* — **n** halfwekelijks blad *n*; twee keren per week verschijnend blad *n* — **e** publicación *f* bisemanal; bisemanario *m*

12105 send *v* **to bed** — See put *v* to bed.

12106 send *v* **to press** — See put *v* to bed.

12107 sensing head — See scanning head.

12108 sensing station — See reading station.

12109 sensitiveness — See sensitivity.

12110 sensitive tape test — See Scotch tape test.

12111 sensitive to light; light sensitive

f sensible à la lumière — **d** lichtempfindlich — **n** lichtgevoelig — **e** sensible a la luz — **i** sensibile alla luce

12112 sensitivity; sensitiveness — Response of a photographic material to incident light.

f sensibilité *f*; sensitivité *f* — **d** Empfindlichkeit *f*; Sensibilität *f*; Sensitivität *f* — **n** gevoeligheid *f* — **e** sensibilidad *f* — **i** sensibilità *f*

12113 sensitivity guide — See grey scale.

12114 sensitize *v* — To cause light sensitivity in a colloidal layer by bathing it in a solution of dichromate.

f sensibiliser; rendre sensible — **d** sensibilisieren; lichtempfindlich machen — **n** sensibiliseren; lichtgevoelig maken — **e** sensibilizar — **i** sensibilizzare

12115 sensitize *v* — See coat *v*.

12116 sensitized layer — Light-sensitive material spread out in a thin layer on a flat support for taking photographs or making photographic copies.

f couche *f* sensible — **d** lichtempfindliche Schicht *f* — **n** lichtgevoelige laag *fm* — **e** capa *f* sensibilizada — **i** strato *m* sensibilizzato

12117 sensitized paper — Paper for photographic printing, e.g., blueprint paper, bromide photographic paper, chloride photographic paper, photostat paper. The emulsions are applied to the surface of the paper or the paper passes through a solution of sensitive salts.

f papier *m* sensibilisé; papier *m* sensible — **d** lichtempfindliches Papier *n* — **n** lichtgevoelig papier *n*; gevoelig papier *n* — **e** papel *m* sensible; papel *m* sensibilizado — **i** carta *f* sensibile; carta *f* fotosensibile

12118 sensitizer — Chemical compound (e.g., salts of iron, silver and chromium, diazo compounds and dyes) to render photographic surfaces sensitive to light and colour. With lithography, also used to make plate coatings light sensitive.

f sensibilisateur *m* — **d** Sensibilisator *m* — **n** sensibilisator *m* — **e** sensibilizador *m* — **i** sensibilizzatore *m*

12119 sensitizing — See counteretch *v*.

12120 sensitometric curve — See characteristic curve.

12121 sensitometry — Study of the reproduction of a tonal scale by sensitive films, plates and paper.

f sensitométrie *f* — **d** Sensitometrie *f* — **n** sensitometrie *f* — **e** sensitometría *f* — **i** sensitometria *f*

12122 sentinel — See flag.

12123 separate — See offprint.

12124 separate coating — See conversion coating.

12125 separate cross rules

f à réglure separé — **d** mit Quersatz *m* für sich — **n** met contravorm apart; met lijnenvorm apart — **e** con reglaje separado — **i** filetti *mpl* trasversali separati

12126 separate cross rules, with — See with separate cross rules.

12127 separately coated paper; off-machine coated paper — Paper coated by a coating machine separate from the paper machine.

f papier *m* couché hors machine — **d** in der Streichmaschine gestrichenes Papier *n* — **n** niet op de papiermachine gestreken papier *n*; op de strijkmachine gestreken papier *n* — **e** papel *m* estucado fuera máquina — **i** carta *f* patinata fuori macchina

12128 separation, colour — See colour separation.

12129 separation filter

f filtre *m* de sélection; écran *m* de sélection; écran *m* sélectif — **d** Teilauszugsfilter *m* — **n** deelfilter *n* — **e** filtro *m* de selección — **i** filtro *m* di selezione

12130 separation negative — Image of a colour set, especially one taken through a colour filter.

f négatif *m* sélectionné; négatif *m* de sélection — **d** Auszugsnegativ *n*; Negativfarbauszug *m*; Teilauszugnegativ *n* — **n** deelnegatief *n* — **e** negativo *m* de (la) selección; negativo *m* seleccionado; negativo *m* de separación — **i** negativo *m* di selezione

12131 separatrix — Correction mark on a proof. The term is also used for a *decimal comma. See app. no. 5.

12132 separatrix — See decimal comma.

12133 sepia print — Print or photograph of a brown colour resembling sepia. Term also but incorrectly used for a *brownprint.

12134 sequence *v* — Sequencing means arranging in complete order two or more strings of data (or piles of cards) that are already within their own proper sequence, whereas sorting means to arrange them in order from a randomly-arranged collection.

f mettre en séquence — **d** in natürliche Reihenfolge bringen — **n** in numerieke volgorde brengen; op rangnummer schikken — **e** poner en secuencia — **i** riordinare in sequenza

12135 sequential access — See serial access.

12136 sequential processing — Processing of a file in sequential order (front to back, or back to front) where, in order to access a given record, all records previous to the required one must be passed. Opposed to random or direct access.

f traitement *m* séquentiel — **d** starr fortlaufende Verarbeitung *f* — **n** sequentiële verwerking *f* — **e** procesamiento *m* secuencial — **i** elaborazione *f* sequenziale

12137 serendipity — Capability of finding valuable or agreeable things without actually looking for them. Coined by Horace Walpole in 1754 after the three princes of Serendip who in their travels always gained by chance things they did not seek.

f combinaison *f* aléatoire de sujets — **d** Kopplung *f* aufs Geratewohl von Satzgegenständen — **n** toevallig ontdekken *n*; bij toeval ontdekken *n*; lukrake koppeling *f* van onderwerpen — **e** combinación *f* al azar de asuntos — **i** combinazione *f* aleatoria di soggetti

12138 serial — Handling the elements of a word or message (e.g., the bits or characters) one after another. Opposite to parallel.

f en série; sériel — **d** seriell; Serien- — **n** serie- — **e** en serie — **i** in serie; seriale

12139 serial access; sequential access — Access of an area or file in serial order. See also sequential processing.

12140 serial number

f numéro *m* de série; numéro *m* d'ordre; numéro *m* de référence — **d** Seriennummer *f* — **n** serienummer *n* — **e** número *m* de serie; número *m* de orden; número *m* de fábrica — **i** numero *m* di serie

12141 serial operation — Performing actions one after another; transmitting or storing the bits of a word, or part of a word, in a time sequence. Contrast with parallel operation. Magnetic tape provides serial access.

f opération *f* en série — **d** Serienoperation *f* — **n** serie-operatie *f* — **e** operación *f* en serie — **i** operazione *f* seriale

12142 serial printer — Printer in which the selected characters are printed one by one. Computer term.

f imprimante *f* série — **d** Seriendrucker *m*; Reihendrucker *m* — **n** serie-afdrukmachine *f* — **e** impresora *f* serie — **i** stampatrice *f* serie

12143 series connection — End-to-end arrangement of the components of a circuit e.g., resistances, so that the same current flows through each component.

f branchement *m* en série; couplage *m* en série — **d** Serienschaltung *f*; Reihenschaltung *f* — **n** serieschakeling *f* — **e** conexión *f* en serie; acoplamiento *m* en serie — **i** collegamento *m* in serie

12144 serifs; ceriphs — Short lines drawn at right angles to or obliquely across the ends of stems and arms of letters. See also slab serif, wedge serif.

f empattements *mpl* (at foot of letter); obits *mpl* (at top of letter); apices *mpl* (at top of letter) — **d** Serife *mpl*; Schraffen *fpl* — **n** schreven *fmpl* — **e** remates *mpl*; bigotillos *mpl*; trazos *mpl* de pie; rasgos *mpl* — **i** terminazioni *fpl*; grazie *fpl*

12145 serigraphy — See screen printing process.

12146 serrated roller — Roller with knurling or milling to give better pull to the paper which passes round it. Also used to minimize ink pick-up.

f rouleau *m* cannelé; cannelé *m*; tube *m* rainé — **d** Riffelwalze *f* — **n** geribbelde rol *fm*; geribbelde trekrol *fm* — **e** rodillo *m* dentellado; rodillo *m* acanalado — **i** rullo *m* scanalato; rullo *m* seghettato

12147 serviceable time — See up-time.

12148 service-book — See prayer-book.

12149 serviette paper — See napkin tissue.

12150 servomotor — Power-driven mechanism that supplements a primary control operated by a comparatively feeble force.

f servomoteur *m*; moteur *m* asservi — **d** Servomotor *n* — **n** servomotor *m* — **e** servomotor *m* —

i servomotore *m*

12151 sesqui — Prefix signifying one-half more. Often used for salts in which the proportions of metal oxide to acid anhydride are 2:3.

12152 sesqui — See sodium sesquicarbonate.

12153 set — Width of a character measured across the shank. On the Monotype unit system the fount set is the point width of a character which is 18 units wide. The narrowest characters are usually 5 units wide.

f set *m* — **d** Set *m* — **n** set *m* — **e** set *m*; prosa *f* — **i** set *m*

12154 set *v* — Inks are said to have set when the printed sheets, though not fully dry, can be handled without smudging.

f prendre — **d** nicht durch und durch trocken — **n** handdroog worden — **e** fraguar; quedar semiseco — **i** fare presa

12155 set *v* — See compose *v*.

12156 set *v* **a roller**

f régler un rouleau — **d** eine Walze abrichten — **n** een rol stellen — **e** graduar un rodillo; colocar un rodillo — **i** regolare un rullo

12157 set *v* **a stick; justify** *v* **the stick** — To adjust the *composing stick to the required measure.

f ajuster le composteur — **d** den Winkelhaken stellen — **n** de zethaak op maat stellen — **e** justificar el componedor — **i** justificare il compositoio; prendere la giustezza

12158 set-back — In offset platemaking, the distance from the front edge of the press-plate to the image area to allow for clamping to cylinder and also for gripper margin. Varies with different makes and press sizes.

f prise *f* de la plaque en pinces — **d** bildfreie Kante *f* der Druckplatte — **n** vrije kant *m* van de (druk)plaat — **e** margen *f* sujetaplanchas — **i** striscia *f* anteriore non stampante delle lastre

12159 set flush — Matter set without indention.

f composition *f* alignée — **d** ohne Einzug — **n** niet ingesprongen — **e** componer sin sangría; componer sin sangrar — **i** composto senza rientranze

12160 set *v* **in motion** — To throw into gear.

f mettre en marche; faire aller; faire marcher — **d** anstellen; in Gang setzen — **n** op gang brengen; in beweging brengen — **e** poner en marcha; disparar — **i** mettere in moto; mettere in funzione

12161 set mark; cross — Mark for wrong slugs: ⌖.

f étoile *f* croisée — **d** Kreuzstern *m*; August *m coll.*; Rauswerfer *m* — **n** ster *f* (voor een sterregel) — **e** estrella *f* crucera — **i** stella *f* crosciata; croce *f*; crocetta *f*

12162 set *v* **off** — To transfer or smear ink from freshly printed matter to another surface with which the undried print comes in contact; particularly from the face of one sheet onto the back of the sheet on top of it in the delivery pile or rewind roll. See also setting-off.

f maculer — **d** abschmieren; abliegen; abschmutzen; abziehen; absetzen; schmitzen — **n**

overzetten — **e** repintar — **i** sporcare; controstampare; macchiare

12163 set-off paper — Paper to remove the excess ink from an inked forme or from ink rollers. See also sheet *v* the rollers.

f papier *m* de décharge — **d** Abschmutzbogen *n*; Abziehpapier *n* — **n** afrolpapier *n* — **e** maculatura *f* de descarga — **i** carta *f* di scarica

12164 set-off paper — See slip sheets.

12165 set of matrices — All the matrices needed for the characters of a given body and style. See also fount.

f frappe *f* — **d** Matrizensatz *m* — **n** stel *n* matrijzen — **e** juego *m* de matrices; fuente *f* de matrices — **i** assortimento *m* di matrici

12166 set of plates — All the plates required to complete a job.

f jeu *m* de clichés — **d** Plattensatz *m* — **n** stel *n* platen — **e** juego *m* de planchas — **i** serie *f* di lastre

12167 set of rollers — All the inking rollers needed on a press.

f jeu *m* de rouleaux — **d** Walzensatz *m*; Walzenpaar *n*; Rollenpaar *n* — **n** stel *n* rollen — **e** juego *m* de rodillos; batería *f* de rodillos; tren *m* de rodillos — **i** serie *f* dei rulli; muta *f* dei rulli

12168 set-plate — The plate completing a set of curved stereos on a rotary press.

f plaque *f* complétant (un jeu) — **d** letzte Platte *f* eines Plattensatzes — **n** laatste plaat *fm* — **e** última plancha *f* — **i** lastra *f* completante

12169 set size — See width.

12170 set *v* **swiftly**

f mettre du zèle — **d** draufstechen; schnell drauflossetzen — **n** snel zetten — **e** componer rápidamente — **i** comporre con sveltezza

12171 setting — See adjustment.

12172 setting a headband — Gluing the *headband to the edges of the book block.

f placement *m* de tranchefile — **d** Ankleben *n* des Kapitalbands — **n** aanplakken *n* van de kapitaalband — **i** incollatura *f* del capitello

12173 setting blind — Composing solid matter without due allowance for the result.

f composer à l'aveugle — **d** pinnen — **n** in den blinde weg zetten; maar raak zetten — **e** componer a ciegas — **i** comporre alla cieca

12174 setting of abbreviations

f composition *f* des abréviations — **d** Setzen *n* von Abkürzungen; Abbreviatursatz *m* — **n** zetten *n* van afkortingen — **e** composición *f* de abreviaturas — **i** composizione *f* delle abbreviazioni

12175 setting-off — This is correct American, generally understood in Europe. The English term is *set-off; off-set being reserved for that process of lithographic printing which makes use of a rubber blanket.

12176 setting of ink — See also set.

f prise *f* de l'encre — **d** Abbinden *n* der Druckfarbe — **n** handdroog worden *n* — **e** asentamiento *m* de la tinta — **i** stabilizzazione *f*; fissaggio *m* dell'inchiostro

12177 setting rule — See composing rule.

12178 setting stick — See composing stick.

12179 setting time
f temps *m* de prise — **d** Härtezeit *f*; Abbindezeit *f* — **n** tijd *m* om handdroog te worden; tijd *m* om te zetten — **e** tiempo *m* de semisecado; tiempo *m* de fraguado — **i** tempo *m* di presa; tempo *m* di stabilizzazione

12180 setting to the trade — See trade setting.

12181 set *v* up again — See reset *v*.

12182 set-up box; set-up paper box; set-up carton; rigid box; rigid paper container — Box manufactured in the form and shape in which it is to be used, as distinguished from folding cartons, which are manufactured in a collapsed form and not set up until used.
f boîte *f* montée; boîte *f* à monter; boîte *f* rigide — **d** Aufrichtschachtel *f*; steife Schachtel *f*; formfeste Schachtel *f* — **n** verzenddoos *fm*; opgezette doos *fm*; doos *fm* in inelkaar gezette toestand — **e** caja *f* rigida; caja *f* montada — **i** scatola *f* rigida; scatola *f* a montare

12183 set-up boxboard — Non-bending board for rigid boxes.

12184 set-up carton — See set-up box.

12185 set-up paper box — See set-up box.

12186 set-up time — Time required to prepare a job to run: to load the program, load and ready input/output devices, etc.
f temps *m* de mise en place — **d** Rechenschaltzeit *f* — **n** insteltijd *m* — **e** tiempo *m* de preparación — **i** tempo *m* di preparazione

12187 set width — See width.

12188 setwise — See width.

12189 seven-level code — Method to describe characters or symbols (up to 128 different ones) by the use of a frame consisting of combinations of up to seven information bits.

12190 seven-line nonpareil — Old name for 42-point type. See app. no. 1.

12191 seventy-twomo — Signature of a book which contains 72 leaves or 144 pages. Abbrev. 72mo (no point).
f in-soixante-douze — **d** Zweiundsiebzigerformat *n* — **n** in tweeënzeventigen — **e** en setenta-y-dosavo — **i** en settanta e due

12192 sewing — Hand or machine operation in which the sections of a book are fixed securely to one another by means of threads drawn through the back fold of each signature.
f cousage *m* — **d** Nähen *n* — **n** naaien *n* — **e** cosido *m*; costura *f* — **i** cucitura *f*

12193 sewing bench — See sewing frame.

12194 sewing cords — Cords in a *sewing frame around which the sections are sewn.

12195 sewing frame; sewing bench — Large, flat, wooden board (the bed) with two upright posts with threaded tops and screws, holding a crossbar, to sew a book by hand. The cords, bands, or tapes are stretched between the bed and the cross-bar, while the book is placed between the upright posts.
f cousoir *m* — **d** Heftlade *f* — **n** naaibank *f* — **e** telar *m*; bastidor *m* para coser libros; bastidor *m* para cosido — **i** cucitoio *m*; telaio *m* di cucitura

12196 sewing heart — Piece of cardboard, in the form of a heart with in the middle of it a cut-in flap, used in sewing by hand, to find easily the middle spread of a signature.
f cœur *m* — **d** Heftherz *n* — **n** hartje *n* — **e** corazón *m* para cosido — **i** cuore *m* della cucitura

12197 sewing machine; book sewing machine — Machine for sewing the signatures of a book together with thread.
f couseuse *f* (au fil textile); machine *f* à coudre — **d** Fadenheftmaschine *f* — **n** garennaaimachine *f*; naaimachine *f* — **e** máquina *f* de coser; cosedora *f* con hilo; cosedora *f* de libros — **i** cucitrice *f* a filo refe; macchina *f* cucitrice a filo refe

12198 sewing on zip fastener
f cousage *m* à fermeture glissière; cousage *m* à fermeture à curseur; cousage *m* à fermeture éclair — **d** Annähen *n* des Reißverschlußes — **n** aannaaien *n* van ritssluitingen; aannaaien *n* van treksluitingen — **e** costura *f* sobre cierre de cremallera — **i** cucitura *f* a chiusura lampo

12199 sewing thread — See binding thread.

12200 sewn on hemp cords
f cousu sur ficelles; cousu sans grecques — **d** auf aufgedrehte Schnüre geheftet; Schnurheftung *f* — **n** genaaid op touw — **e** cosido con cordeles — **i** cucito su corde

12201 sewn on tapes
f cousu sur rubans; cousu sur chevillières — **d** auf Bänder geheftet; Heftung *f* auf Bänder — **n** genaaid op band — **e** cosido *m* con cintas — **i** cucito su nastri; cucito su fettucce

12202 sewn through linen
f cousu (au fil) sur gaze; cousu (au fil) sur canevas — **d** (mit Faden) auf Gaze geheftet — **n** genaaid op gaas — **e** cosido *m* con gasa — **i** cucito su filo con garza

12203 sextern — See sexternion.

12204 sexternion; sextern — Six gathered sheets folded in two for binding together. See also ternion.
f sexternion *m* — **d** Sexternio *n* — **n** katern *fm* van zes; sextern *fmn*

12205 sexto; sixmo — Book in which the sheets are folded into six (a half sheet of twelve). Abbrev. 6mo (no point).
f in-six *m*; in-6 *m* — **d** Sechserformat *n*; halber Duodezbogen *m* — **n** in zessen — **e** en sexto; sexterno — **i** in sei; in sesto

12206 sextodecimo — See sixteenmo.

12207 sextuple machine; sextuple press — Double-width press with three units.
f rotative *f* à 3 groupes 8 pages — **d** 24-seitige Rotationsmaschine *f*; doppelbreite 3-Einheitenmaschine *f* — **n** 24-zijdige rotatiepers *f*; 24-zijdige pers *f* — **e** rotativa *f* con 3 grupos de 8 páginas — **i** rotativa *f* a doppia larghezza a tre gruppi stampanti

12208 sextuple press — See sextuple machine.

12209 shade — Gradation of a colour resulting from the addition of a colour adjacent to it in the spectral scale, e.g., red may be shaded to a bluish tone by addition of violet; a colour toned with black or with a complementary colour having a dulling effect.

f ton *m* rompu; ton *m* rabattu; dégradation *f* de couleur — **d** Farbton *m*; Nuance *f*; getrübte Farbe *f*; Schwarzmischung *f* — **n** tint *fm*; gebroken kleur *fm*; zwart-uitmenging *f* — **e** matiz *m*; obscurecimiento *m* del color — **i** tonalità *f*; degradazione *f* di colore

12210 shade — See value.

12211 shadecraft watermark — See shaded watermark.

12212 shaded colours — Wash drawing in which the colours fade out gradually.

f teintes *fpl* dégradées — **d** abgetönte Farben *fpl*; verlaufende Farben *fpl* — **n** verlopende tinten *fmpl* — **e** colores *mpl* esfumados; colores *mpl* difumados — **i** tinte *fpl* sfumate; tinte *fpl* digradanti

12213 shaded rule — See Oxford rule.

12214 shaded watermark; shadecraft watermark; shadow mark — Watermark produced with a dandy roll in which a design is cut. Sometimes called intaglio watermark or countersunk watermark.

f filigrane *m* ombré; filigrane *m* enfoncé — **d** Schattenwasserzeichen *n* — **n** schaduwwatermerk *n* — **e** filigrana *f* obscura — **i** filigrana *f* in scuro

12215 shading machine; benday machine — Benday machine or apparatus to hold and adjust shade mediums while transferring tints or patterns to various surfaces.

f appareil *m* pour benday — **d** Tangiermaschine *f* — **n** tangeerapparaat *n*; filmapparaat *n* — **e** aparato *m* para grisar — **i** macchina *f* per applicare fondi benday

12216 shading medium; shading sheet — Photomechanical drawing material with a visible or latent (developable) pattern of lines or dots, used in the preparation of line originals for introduction of tint or tone effects. Distinguished from screen tints, or tint sheets that are usually used directly in stripping. See also benday process.

f relief *m* de gélatine; benday *m* — **d** Tangierfell *n*; Tangierfilm *m*; Einkopierraster *m* — **n** tangeerfilm *m*; rasterkopieerfilm *m*; filmraster *n*; benday-film *m* — **e** lámina *f* de grisar; película *f* de grisar; película *f* para grisado; grisado *m* bendéi; película *f* (de grisar) bendéi; tinta *f* mecánica — **i** pellicola *f* retinata in rilievo; pellicola *f* benday per fondi

12217 shading sheet — See shading medium.

12218 shadow areas — See shadows.

12219 shadow dots — Heavy dots in the dark areas of a plate which appear as a solid with evenly spaced white dots.

f grandes ombres *fpl* — **d** Tiefepunkte *mpl*; Tiefenpunkte *mpl* — **n** schaduwpunten *fmpl* — **e** puntos *mpl* de sombra; puntos *mpl* de oscuro — **i** punti *mpl* di retino nelle zone d'ombra

12220 shadow mark — Mark printed on one of the two layers of paper pasted together. See also shaded watermark.

d Schattenzeichen *n* — **n** schaduwmerk *n* — **e** filigrana *f* sombreada; filigrana *f* de relieve — **i** filigrana *f* in chiaroscuro

12221 shadow mark — See shaded watermark.

12222 shadow mask — Contact mask used with a positive to increase shadow contrast.

f masque *m* des ombres — **d** Schattenmaske *f* — **n** schaduwmasker *n* — **e** máscara *f* de sombra; máscara *f* para sombras — **i** maschera *f* per zone di ombra

12223 shadows; shadow areas — Darkest areas on an illustration or positive and the lightest areas on a negative. Good printing reproduction usually requires holding all tone values and the shadows open, by maintaining a pin-point clear dot in an otherwise solid printing ink area.

f ombres *fpl* — **d** Tiefen *fpl*; Tiefenpartien *fpl*; Schatten *mpl*; Schattenpartien *fpl*; Schattenstellen *fpl*; Schattenbereich *m* — **n** schaduwpartijen *fpl* — **e** sombras *fpl*; oscuros *mpl*; negros *mpl*; zona *f* de negros; zona *f* de sombras; claroscuros *mpl* — **i** zone *fpl* d'ombra; zone *fpl* di ombra; aree *fpl* d'ombra

12224 shadow stop — Flash stop in halftone photography to strengthen or introduce small opaque dots in the shadows of halftone negatives.

f ouverture *f* de piquage — **d** Tiefenblende *f*; kleinste Blende *f* — **n** klein insteekdiafragma *n* — **e** diafragma *m* para reforzar los negros — **i** apertura *f* di diaframma per accentuare le ombre

12225 shaft — Long revolving bar in an engine for the transmission of a movement.

f arbre *m* — **d** Welle *f*; Achse *f* — **n** as *fm* — **e** árbol *m*; eje *m*; flecha *f Me* — **i** asse *f*; albero *m*

12226 shagreen — Strong, granulated leather, originally made in Armenia from horse, mule or donkey hides, now generally imitated.

f chagrin *m*; peau *f* de chagrin — **d** Chagrin *m*; Chagrinleder *n* — **n** segrijn *n*; segrijnleer *n*; chagrijnleer *n* — **e** chagrén *m*; chagrín *m*; cuero *m* granilloso — **i** zigrino *m*

12227 shake — Device that causes oscillation of the Fourdrinier wire in the plane of the wire, at right angles to the machine direction, to secure the desired formation of the sheet. The oscillations may vary in frequency and length of stroke to obtain the desired result.

f branlement *m*; secouage *m* — **d** Schüttelbock *m*; Erschütterung *f*; Schüttelgerät *n*; Schüttelwerk *n* — **n** schudwerk *n* — **e** mecanismo *m* sacudidor; aparato *m* sacudidor — **i** scuotitore *m*; sistema *m* di scuotimento della tavola piana continua

12228 shale oil — Oil obtained from oil-bearing shale by destructive distillation, used in ink making.

f huile *f* de schiste — **d** Schieferöl *n* — **n** leisteenolie *fm* — **e** aceite *m* de esquistosa — **i** olio *m* di schisto

12229 shallow plate — Printing plate in which the non printing areas are not sufficiently far below the printing surface to give maximum print quality or to prevent blacking up.
f cliché *m* pas assez creux — **d** Platte *f* mit zu seichten Punzen — **n** onvoldoend diepe plaat *fm*; ondiepe plaat *fm* — **e** clisé *m* poco profundo — **i** lastra *f* troppo poco profunda; lastra *f* a rilievo insufficiente

12230 shammy-leather — See chamois-leather.

12231 shank — Body of a type on which are the nick, pin mark, belly and back.
f tige *f*; tronc *m* — **d** Schaft *m*; Säule *f* — **n** schacht *f* — **e** hombro *m*; vástago *m* del tipo; tronco *m* del tipo — **i** fusto *m*

12232 share — One of the equal fractional parts into which the capital stock of a joint-stock company or a corporation is divided.
f action *f*; titre *m* — **d** Aktie *f* — **n** aandeel *n* — **e** acción *f* — **i** azione *f*

12233 shared logic — The use of the "logic" or "intelligence" of a computer, or the electronic circuitry of a special-purpose "logic box" by more than one device, more or less simultaneously.

12234 sharpen *v* — To decrease in strength. When the image works too sharp, fine lines and dots become gradually smaller and tend to disappear. Lithography becomes weak and streaked looking. Opposite of thicken.
f baisser — **d** spitzer ätzen; kleiner ätzen — **n** spits worden — **e** debilitarse; archicarse los puntos — **i** assottigliare

12235 sharp image
f image *f* nette — **d** scharfes Bild *n* — **n** scherp beeld *n* — **e** imagen *f* nítida — **i** immagine *f* nitida

12236 sharp mask — Mask that accurately delineates the detail of the master from which it was taken.
f masque *m* net — **d** scharfe Maske *f* — **n** scherp masker *n* — **e** máscara *f* nítida — **i** maschera *f* nitidissima

12237 sharpness — In scientific sense, the sharpness of an image, subjectively estimated, as distinct from *acutance, i.e., the sharpness of an image in an objective manner.
f netteté *f* (d'une image photographique) — **d** Schärfe *f*; Bildschärfe *f* — **n** (visuele) scherpte *f*; subjectieve scherpte *f*; beeldscherpte *f* — **e** nitidez *f* — **i** nitidezza *f*

12238 shavings; trimmings — Strips cut from the sides of the web on a paper machine or strips from the trimming of sheets, which are returned to the beater. See also offcut.
f rognures *fpl* (à pertes); chutes *fpl* — **d** Abschnitt *m*; Papierabschnitt *m*; Beschnitt *m* — **n** afsnijdsel *n*; afsnijsel *n*; papierafsnijdsel *n*; papierafsnijsel *n*; rijfels *pl* — **e** recorte *m*; recortes *mpl* — **i** refili *mpl*; ritagli *mpl*; sfridi *mpl*

12239 shaving the counter — In paperbox making the planing procedure, done so that cutting knives will not cause board to bend sharply around counter. Corners of counter are sometimes shaved to reduce blisters. If not, outline of counter will emboss itself into the carton. Shaving may also refer to opening up or "skimming" one side of a score with a knife to take out a shift.

12240 shearing force — Force applied by a viscometer to produce flow.
f force *f* de cisaillement — **d** Schubkraft *f*; Scherkraft *f* — **n** afschuifkracht *fm* — **e** fuerza *f* de cizallamiento; fuerza *f* cortante — **i** forza *f* di taglio

12241 shearing stress — Rheological term for shearing force per unit area.
f effort *m* de cisaillement; contrainte *f* de cisaillement — **d** Schubspannung *f*; Scherspannung *f* — **n** schuifspanning *f*; afschuifspanning *f* — **e** tensión *f* de cizallamiento; tensión *f* de cortadura; fuerza *f* de empuje — **i** sollecitazione *f* al taglio

12242 shears — Device to cut boards, metal sheets, etc. by hand. See also scissors.
f cisaille *f* — **d** Pappschere *f*; Plattenschere *f* — **n** bordschaar *fm*; plaatschaar *fm* — **e** tijeras *fpl* para cartón — **i** cesoia *f*; tagliacartone *f*

12243 shear strain — Alteration in shape due to stress.
f déformation *f* au cisaillement — **d** Schubdeformation *f*; Schubdehnung *f* — **n** afschuifdeformatie *f*; afschuifvervorming *f* — **e** deformación *f* por cizallamiento — **i** deformazione *f* al taglio; deformazione *f* per cesoiamento

12244 shear thickening — See dilatant.

12245 shear thinning — See thixotropy.

12246 sheaves — See shives.

12247 sheen — See gloss.

12248 sheep's foot — Iron claw hammer formerly used by pressmen to lock up printing formes.
f marteau *m* d'imposer — **d** Klauenhammer *m* — **n** klauwhamer *m* — **i** martelletto *m* per chiudere le forme

12249 sheepskin — Hide of a sheep, specially tanned, used in de luxe bookbinding.
f mouton *m* — **d** Schafleder *n*; Schaf *n* — **n** schapeleer *n* — **e** badana *f* (blanca); blanquillo *m* — **i** pelle *f* di montone; pelle *f* di pecora

12250 sheet — Piece of paper or board, generally of rectangular shape.
f feuille *f* (de papier) — **d** Bogen *m*; Papierbogen *m* — **n** vel *n* (papier) — **e** hoja *f* — **i** foglio *m*

12251 sheet — See newspaper.

12252 sheet — See web.

12253 sheet and half demy usual — A size of printing paper. See app. no. 3.

12254 sheet and half foolscap — A British standard size for writing and printing papers. See app. no. 3.

12255 sheet and half imperial — A British size for writing and printing paper. See app. no. 3.

12256 sheet and half post — A size of printing

paper. See app. no. 3.

12257 sheet and half quad foolscap — A size of writing and printing paper, recognized but not standard in Great Britain. See app. no. 3.

12258 sheet and half small foolscap — A British standard size for writing and printing papers. See app. no. 3.

12259 sheet and third foolscap — A British standard size for writing and printing papers. See app. no. 3.

12260 sheet and third small foolscap — A British standard size for writing and printing papers. See app. no. 3.

12261 sheet calender
f calandre *f* pour feuilles; calandre *f* à feuilles — **d** Bogenkalander *m* — **n** vellenkalander *fm* — **e** calandria *f* para hojas; calandra *f* para hojas; calandria *f* para satinar el papel en hojas — **i** calandra *f* per fogli

12262 sheet calendered — A process of applying a finish or glaze to sheets of paper by passing them through a calender stack (but not in a continuous web) with the aid of a sheet fed gear. The stack consists of three to five rolls; chilled iron and cotton rolls are alternated in the stack.
f calandré en feuilles — **d** Bogenkalandriert — **n** aan vellen gesatineerd — **e** satinado en hojas; glaseado en hojas; calandrado en hojas; pasado en hojas — **i** calandrato in fogli

12263 sheet cutter — Large single knife, guillotine type of cutting machine, to cut and trim reams of paper to size.
f massicot *m* — **d** Bogenschneider *m*; Bogen-schneidemaschine *f* — **n** vellensnijmachine *f* — **e** cortadora *f* en hojas — **i** taglierina *f* per fogli; tagliafogli *f*

12264 sheet cutter — Cutting device which slits the paper reels into smaller widths and cuts the web transversally into sheets.
f découpeuse *f* en feuilles; découpeuse *f* en long et en travers — **d** Längs- und Quer-Schneide-vorrichtung *f* — **n** langs- en dwarssnij-inrichting *f*; snij-inrichting *f* — **e** cortadora *f* duplex — **i** taglierina *f* longitudinale e trasversale

12265 sheet drier — Drier in which the paper or board is dried in sheets that are laid flat on racks or hung up.
f séchoir *m* à plat; séchoir *m* accrocheur — **d** Bogentrockner *m* — **n** vellendrooginrichting *f* — **e** secador *m* de hojas — **i** essiccatore *m* per carta in fogli

12266 sheeter — Rotary unit over which the web passes to be cut into individual sheets for stacking if desired.
f coupeuse *f* en feuilles; coupe-feuilles *f* — **d** Schneidwerk *n*; Bogenschneidvorrichtung *f*; Kapp-vorrichtung *f* — **n** vellensnijwerk *n* — **e** cortadora *f* (de papel continuo) en hojas; cortadora *f* de hojas — **i** dispositivo *m* per il taglio in fogli

12267 sheet fed — The feeding of individual sheets into a printing machine.
f alimenté par feuilles; alimenté à feuilles — **d** mit

Bogenanlage — **n** met vellen-inleg — **e** alimen-tado con hojas — **i** alimentato da foglio

12268 sheet-fed gravure rotary (press)
f rotative *f* en creux en feuilles — **d** Tiefdruck-bogenrotationsmaschine *f* — **n** diepdruk-vellen-rotatiepers *f* — **e** rotativa *f* de huecograbado en hojas — **i** macchina *f* rotocalco da foglio

12269 sheet-fed machine — Press which receives its paper in separate sheets, as distinct from web or reel feed.
f machine *f* à feuilles; machine *f* à marge en feuilles — **d** Bogendruckmaschine *f*; Maschine *f* mit Bogenanlage — **n** vellenpers *f* — **e** máquina *f* de alimentación en hojas; prensa *f* para imprimir en hojas — **i** macchina *f* a foglio; macchina *f* da foglio

12270 sheet-fed offset rotary (press)
f rotative *f* offset à feuilles — **d** Offsetbogen-rotationsmaschine *f* — **n** offset-vellenrotatiepers *f* — **e** rotativa *f* offset en hojas — **i** rotativa *f* offset da foglio

12271 sheet flyer — See fly.

12272 sheet former — Laboratorium device to make proof sheets.
f formette *f* — **d** Blattbildner *m* — **n** blad-vormingstoestel *n* — **e** formador *m* de hojas

12273 sheeting paper — Paper to line paperboard such as chipboard after it has been cut into sheets.
f (papier *m*) extérieur pour carton — **d** Klebe-papier *n*; Kaschierpapier *n* — **n** beplakpapier *n* — **e** papel *m* para forrar (cartón) — **i** copertina *f* per cartoni

12274 sheet lined — Said of board to which a liner has been pasted after being sheeted.
f doublé — **d** beklebt — **n** beplakt — **e** pegado; forrado — **i** accoppiato (con materiale in fogli)

12275 sheet of eighteens — See eighteenmo.

12276 sheet of its own — Piece of paper of the size, colour, quality and substance on which the job is to be printed.
f feuille *f* (de papier) du tirage — **d** Auflage-bogen *m*; Auflagepapier *n* — **n** vel *n* uit de partij; te bedrukken vel *n* — **e** papel *m* para la tirada — **i** foglio *m* della carta di tiratura

12277 sheet severer — Automatic device to cut the web to prevent wrapping around.
f couteau *m* automatique; coupe-papier *m* — **d** Abschlagvorrichtung *f*; Kappvorrichtung *f*; Abschneidvorrichtung *f* — **n** kapinrichting *f* — **e** cuchilla *f* de toma rápida — **i** dispositivo *m* auto-matico per taglio in fogli

12278 sheets, in — See in sheets.

12279 sheet size — See size.

12280 sheets per hour
f feuilles *fpl* à l'heure — **d** Bogen *mpl* je Stunde — **n** vellen *npl* per uur; vel *n* per uur — **e** hojas *fpl* por hora — **i** fogli *mpl* per ora

12281 sheet v the rollers — To roll a sheet of paper between the inking rollers to take off sur-plus of ink. See also set-off paper.
f décharger (les rouleaux sur des maculatures) —

d die Farbwalzen abziehen; die überschüssige Farbe von den Walzen abziehen — **n** de inktrollen afrollen — **e** enjugar los rodillos; limpiar los rodillos — **i** staccare l'eccedente del colore dai rulli; asportare dai rulli l'eccesso di inchiostro

12282 sheet wander — Undesirable side movement of the running web in a rotary press.

f flottement *m* (latéral) de la bande — **d** seitliches Verlaufen *n* der Papierbahn — **n** slingeren *n* van de papierbaan — **e** desviación *f* de la banda de papel — **i** sbandamento *m* laterale del nastro

12283 sheetwise forme — See work and back.

12284 sheetwork — See work and back.

12285 shelfback — See spine.

12286 shelf life; storage life — Length of time that a container, or a material in a container, will remain in a saleable or acceptable condition under specified conditions of storage.

f durée *f* de conservation — **d** Haltbarkeit *f*; Lagerhaltbarkeit *f* — **n** houdbaarheid *f* — **e** solidez *f* gen.; conservabilidad *f*; consistencia *f*; estabilidad *f* (chemistry) — **i** conservabilità *f*; durata *f*

12287 shelf paper; cupboard lining paper

f papier *m* d'armoire; papier *m* pour étagères — **d** Schrankpapier *n*; Auslegepapier *n* — **n** kastpapier *n* — **e** papel *m* para anaqueles — **i** carta *f* per (foderare) armadi

12288 shell; copper shell — Deposit of copper or nickel and copper on the mould in the plating bath in the electrotyping process, 0.15-2 mm (0.006-0.008 in) thick.

f coquille *f* (galvanoplastique) — **d** Kupferhäutchen *n*; galvanische Kupferhaut *f*; Kupferniederschlag *m* — **n** koperhuidje *n* — **e** cascarilla *f* (galvánica) — **i** conchiglia *f* di rame; pellicola *f* di rame; pellicola *f* galvanica

12289 shellac — Resin secreted by the insect Laccifer lacca (coccus lacca). Insoluble in water; used in flexographic inks because of its solubility in alcohol.

f gomme-laque *f* — **d** Schellack *m* — **n** schellak *m* — **e** goma *f* laca — **i** gomma *f* lacca

12290 shellac varnish — Varnish made by dissolving *shellac in alcohol or a similar solvent.

f vernis *m* à la gomme laque — **d** Schellackfirnis *m* — **n** schellakvernis *mn* — **e** barniz *m* de goma laca — **i** vernice *f* di gomma lacca

12291 shell and slide carton — See also carton blank.

f boîte *f* pliante à coulisse et tiroir — **d** Schiebe-Faltschachtel *f*; Schiebeschachtel *f* — **n** schuifdoosje *n* — **e** caja *f* corrediza (de cartón) — **i** astuccio *m* con interno scorrevole; astuccio *m* accassetto e manicotto

12292 shell gold — Gold waste mixed with gum of honey, used by gilders and sold in shells.

f or *m* coquille; or *m* en coquille — **d** Muschelgold *n* — **n** schelpgoud *n* — **i** oro *m* in conchiglia

12293 shell removal — Operation by which the electro-deposited copper or nickel plate is removed from the mould.

f démoulage *m* — **d** Abnehmen *n* des Kupferhäutchens — **n** verwijdering *f* van het koperhuidje — **e** desmoldeo *m* de la cascarilla (galvánica) — **i** rimozione *f* della conchiglia

12294 shield — See apron.

12295 shift — Displacement, change, alteration in kind, character, position, direction or the like.

f changement *m* — **d** Abänderung *f*; Änderung *f*; Verstellung *f*; Umstellung *f* — **n** verandering *f* — **e** cambio *m* — **i** spostamento *m*

12296 shift; trick — In many works a working day is divided into three shifts of eight hours each. At the end of each shift a new crew takes over. The lobster shift, lobster trick or dog watch is an early morning shift between press time of the last edition of a newspaper and the next day's work. The late watch or the hours between 1 to 4 p.m. is the staff which handles the late editions of a newspaper.

f équipe *f* — **d** Schicht *f* — **n** ploeg *fm* — **e** equipo *m*; tanda *f* — **i** turno *m*; squadra *f*

12297 shift — Work in shifts.

f travail *m* par équipes — **d** Schichtarbeit *f* — **n** ploegendienst *m* — **e** trabajo *m* a turnos; trabajo *m* por equipos — **i** lavoro *m* in équipe

12298 shift *v* — In computer processing, to move data right or left. If the data are numerical digits the shift multiplies or divides by the power of the radix for each place moved. A circular shift returns dropped-off figures to the other end of the register, 78143621 shifted three places to the left becomes 43621781.

f décaler — **d** schieben — **n** schuiven; verschuiven — **e** desplazar — **i** scorrere; spostare

12299 shift code — The method used to extend the number of code combinations obtainable from 5, 6, 7 and 8-level tape. In the case of TTS coding only 64 are available from the 6-level tape used, but 90 channels plus function codes must be recorded on the tape. Two codes are allocated as shift and unshift functions; when used to proceed the remaining to codes a total of 124 combinations is obtained. Lower-case and ligatures are usually unshift while capitals and figures are on shift. For ease of operation certain points and function codes are on both shift and unshift. In practice the character set is further extended by upper rail and lower rail codes to access italic or bold characters duplexed with the roman face.

12300 shilling mark — See solidus.

12301 shilling stroke — See solidus.

12302 shim — See lead.

12303 shimming; carding — In rotary printing, the practice of putting thin sheets of paper between the plate and the head bar, or the plate and the centre ring, to register the plate.

d Einpassen *n* der Kartonstreifen — **n** opvullen *n* met strookjes karton — **i** registrazione *f* della lastra con striscioline di carta

12304 shiners — Lumps in paper which are crushed into transparent spots on calendering.

f paillettes *fpl* — **d** Glimmerverunreinigungen *fpl*;

glänzende Fehlerstellen *fpl*; Glimmerblättchen *npl* — **n** glimmerdeeltjes *npl* — **e** lunares *mpl* transparentes — **i** luccichini *mpl*

12305 shining table — See register table.

12306 shining-up table — See register table.

12307 'ship — See companionship.

12308 shipping department — See mail room.

12309 ship's newspaper
f gazette *f* du bord; journal *m* du bord — **d** Bordzeitung *f*; Schiffszeitung *f* — **n** scheepskrant *fm* — **i** giornale *m* di bordo

12310 shives — Undigested or undisintegrated splinters of wood resulting from incomplete resolution during pulping, which show up as imperfections in paper, particularly in groundwood grades. Sometimes called sheaves. See also sliver.
f faisceaux *mpl*; bouchons *mpl*; bûchettes *fpl* — **d** Schäben *fpl*; Schewen *fpl*; Schäwen *fpl*; Splitter *mpl*; Spreu *f* — **n** scheven *fmpl* (bij vlas); niet geheel ontsloten celstofdeeltjes *npl* — **e** grumos *mpl* de fibras; astillas *fpl* — **i** schegge *mpl*

12311 shock absorber — See buffer.

12312 shoe — See curve.

12313 shoe — See hellbox.

12314 shoot board; shooting board — Metal table fitted with a squaring plane, to square and trim plates.
f planche *f* à cadrer — **d** Facettenhobel *m*; Bestoßhobel *m* — **n** facetschaaf *fm* — **e** máquina *f* cepillo; cepillo *m* (para biselar); biseladora *f* — **i** pialla *f* da smusso; pialla *f* da sfaccettatura

12315 shooter — See lining tool.

12316 shooter — See shooting stick.

12317 shooting board — See shoot board.

12318 shooting stick; shooter — Implement, usually metal, formerly used with a mallet to lock up formes.
f décognoir *m* — **d** Keiltreiber *m* — **e** botador *m*; atacador *m*; taco *m*; descuñador *m* — **i** cacciacunei *m*

12319 shooting through — See running through.

12320 Shore durometer — Instrument to measure resilience or softness of moulded vulcanized rubber as in printing rollers, blankets, printing plates, etc. It has a flat tipped needle 1.5 mm diameter tapering to 0.75 mm. The hardness scale (soft to hard) is from 0 to 100 units. See also hardness meter.
f duromètre *m* Shore — **d** Shore-Härtemesser *m* — **n** Shore-hardheidsmeter *m* — **e** durómetro *m* de Shore — **i** durometro *m* Shore

12321 Shore hardness; durometer hardness — Degree of resilience or hardness of a material measured by a *Shore durometer. See also hardness.
f dureté *f* Shore — **d** Shore-Härte *f* — **n** Shore-hardheid *f*; hardheid *f* in Shore; hardheid *f* volgens Shore — **e** dureza *f* según Shore — **i** durezza *f* secondo Shore

12322 short — A size of board. See app. no. 3.

12323 short accent; short mark; breve — Curved mark to indicate a short vowel or syllable, as in ă, ĕ, etc.
f signe *m* brève; signe *m* de la prononciation brève — **d** Kürzezeichen *n* — **n** kort accent *n* — **e** signo *m* de brevedad; breve *m* — **i** segno *m* breve; segno *m* di pronunzia breve

12324 short and — See ampersand.

12325 short circuit — Connection between a wire and earth or between two wires, generally unintentional and caused by faulty insulation. (When used as a verb, with a hyphen.)
f court-circuit *m* — **d** Kurzschluß *m* — **n** kortsluiting *f* — **e** cortocircuito *m* — **i** cortocircuito *m*

12326 short-grain paper
f papier *m* sens travers — **d** Breitbahnpapier *n*; Papier *n* in Breitbahn — **n** breedlopend papier *n* — **i** carta *f* a fibre trasversali

12327 short hundredweight — Unit of weight. Abbrev. sh cwt (no point). See app. no. 7 and 7a.

12328 short ink — Ink with a relatively high yield value which does not flow freely and cannot be drawn into long threads.
f encre *f* courte; encre *f* non filante — **d** kurze Farbe *f*; strenge Farbe *f*; nicht-zügige Farbe *f* — **n** korte inkt *m*; strenge inkt *m* — **e** tinta *f* corta; tinta *f* espesa; tinta *f* poco dúctil — **i** inchiostro *m* a filo corto

12329 short letters — Letters which lie between the base and the mean line. Height of short lower-case letters is usually referred to as *x-height.
f lettres *fpl* courtes — **d** kurze Buchstaben *mpl* — **n** korte letters *fmpl*; kleine letters *fmpl* — **e** minúsculas *fpl* cortas; letras *fpl* cortas — **i** caratteri *mpl* di occhio mediano; lettere *fpl* di occhio mediano

12330 short-line preventer; short-line safety; pump stop — Device on the slug composing machine to automatically stop the operation of the plunger if a short line is sent away for casting.
f arrêt *m* de pompe — **d** Pumpenabstellvorrichtung *f* — **n** pompstop *m* — **e** corte *m* automático de la bomba; seguro *m* de la fundición; seguro *m* de la bomba de fundición; corte *m* de seguridad contra líneas cortas — **i** arresto *m* automatico della pompa

12331 short-line safety — See short-line preventer.

12332 short make — See underrun.

12333 short mark — See short accent.

12334 short page
f composition *f* en cul-de-lampe; page *f* de fin; queue *f* de page — **d** Spitzkolumne *f*; Ausgangsseite *f*; Ausgangskolumne *f*; Schlußkolumne *f* — **n** staartpagina *f* — **e** birlí *m*; página *f* corta; página *f* incompleta — **i** pagina *f* mozza; mozzino *m*

12335 short ream — Pack of identical sheets of paper denoting usually 480 sheets. See also ream.
f rame *f* de 480 feuilles — **d** Ries *n*; kleines Ries *n* — **n** riem *m* van 480 vel — **e** resma *f* de 480 hojas — **i** risma *f* ridotta

12336 short run; small run
f petit tirage *m* — d Kleinauflage *f*; niedrige Auflage *f* — n kleine oplage *fm*; kleine oplaag *fm* — e tiraje *m* bajo; tirada *f* baja — i bassa tiratura *f*

12337 short stack — Slang term for *disk pack.

12338 short-stop — See stop bath.

12339 short varnish
f vernis *m* faible — d schwacher Firnis *m* — n slappe vernis *mn* — e barniz *m* flojo — i vernice *f* debole

12340 short vowel — Vowel like ă, ĕ, etc.
f voyelle *f* brève — d kurzer Vokal *m*; kurzer Laut *m* — n korte klinker *m* — e vocal *f* breve — i vocale *f* breve

12341 short weight; underweight — Weight less than the ordered, normal or requisite amount.
f manque *m* de poids — d Untergewicht *n* — n ondergewicht *n*; onderwicht *n* — e falta *f* de peso; manco *m* de peso; merma *f* — i sottopeso *m*

12342 shoulder — In relief etching, the projecting ledge left at the sides of lines and dots after four-way powdering and etching. Excessive shoulder is considered a defect.
f talus *m* (de morsure) — d Ätzstufe *f* — n stoepje *n* — e hombro *m* — i gradino *m* sul fianco

12343 shoulder — Non-printing area below and above outline of face of type on upper end of body of type, from which the bevel of the face begins.
f épaulement *m* — d Achsel *f*; Achselfläche *f*; Schulter *f* — n schouder *m* — e hombro *m*; rebaba *f*; quijada *f* Me — i spalla *f*

12344 shoulder — See curve.

12345 shoulder — See drive.

12346 shoulder case — See Solander case.

12347 shoulder head — See shoulder heading.

12348 shoulder heading; shoulder head — Heading which precedes a paragraph; a text heading ranged left on a separate line to the text and set in bold face with the text run-on in the same line. This definition is not agreed by all printers, some of whom call them *side heading.
f manchette *f* — d nach links ausgeschossene Absatztitel *m* — n vóórin gehouden kopje *n* — e título *m* a ras con el texto; título *m* no sangrado; título *m* no centrado — i titolo *m* di paragrafo

12349 shoulder note — *Marginal note at the top outer corners of the paragraph.

12350 showcard — Advertising message printed on rigid board for display in stores or other public places.
f pancarte *f* — d Aussteller *m*; Stellplakat *n* — n etalagekaart *f*; showcard *m* — e cartel *m* de escaparate; tarjetón *m* de anuncio; tarjetón *m* publicitario; tarjeta *f* para avisos; letrero *m* — i cartello *m* reclamistico

12351 show-through — Condition where the printing on one side of the sheet can be seen from the other side when the latter is viewed by reflected light, due to low opacity of the paper.
f transparence *f* — d Durchscheinen *n* — n door-schijnen *n* — e traspintado *m*; transparencia *f* — i trapasso *m* ottico; trasparire

12352 shredder; shredding machine; paper shredder — Machine to destroy documents and files.
f machine *f* à détruire — d Aktenvernichtungsmaschine *f*; Papierwollschneidemaschine *f*; Aktenzerreißwolf *m*; Aktenvernichter *m* — n machine *f* voor het vernietigen van documenten — e trituradora *f* de papel — i macchina *f* trituratrice; trituratrice *f*; macchina *f* per distruggere documenti

12353 shredding machine — See shredder.

12354 shrink v
f (se) rétrécir — d schrumpfen — n krimpen; samentrekken — e contraerse — i ritirarsi; contrarsi

12355 shrinkage — Decrease in the dimensions of paper due to loss of moisture.
f rétrécissement *m* (du papier); retrait *m* (du papier) — d Papierschrumpfung *f*; Schrumpfen *n* des Papiers; Einschrumpfen *n*; Schrumpf *m*; Schrumpfung *f* — n krimp *m* — e contracción *f* (del papel); diminución *f* del papel; encogimiento *m* del papel — i restringimento *m* (della carta); restrazione *f* della carta; ritiro *m* della carta; contrazione *f* della carta

12356 shrinkage — Dimensional change from the original subject in the making of the matrix or the plate in duplicate relief platemaking. In stereotype moulding, the control of this shrinkage is an important feature of newspaper platemaking, and also in the making of stereotype plates for colour or other register printing.
f retrait *m* (du métal) — d Schwinden *n* (des Metalls beim Erkalten); Schwund *m* — n krimpen *n* (van het metaal) — e merma *f* (del metal) — i ritiro *m* (del metallo)

12357 shrinkage — Loss in weight between amount of papermaking materials furnished in the beater and the paper produced.
f perte *f* — d Verlust *m* — n verlies *n* — e pérdida *f*; merma *f* — i perdita *f*

12358 shuffling — See fanning out.

12359 shut v down — To stop a machine or press for washing or cleaning, for alterations and repairs, or for lack of orders.
f arrêter — d anhalten — n stoppen — e parar — i arrestare

12360 shut-down time
f temps *m* de rangement (à fin de poste) — d Abrüstzeit *m* (beim Arbeitsschluß) — n opruimtijd *m*; afflooptijd *m* — e tiempo *m* de ordenación — i tempo *m* di sistemazione (a fine lavoro)

12361 shutter — Camera part, placed in the way of the light rays to admit or exclude light at will for certain preselected lengths of time.
f obturateur *m* — d Verschluß *m*; Objektivverschluß *m* — n sluiter *m* — e obturador *m* — i otturatore *m*

12362 shutter — Blind made with horizontal slats fixed at an angle, in the cassette of a process camera.

d Schließschuber *m*; Verschluß *m* — **n** jaloezie *f*

12363 siccative — See drying agent.

12364 side bearing — Distance between one letter and another. It is the sum of the shoulders at the side of two characters.

f approche *f* — **d** Achselflächenbreite *f*; Punzenweite *f* — **n** afstand *m*; ruimte *f* — **e** espacio *m*; anchura *f* del punzón — **i** spalla *f* laterale

12365 side cheeks — See knife cheeks.

12366 side etching — See undercut.

12367 side-faced rule — See bevelled rule.

12368 side frame — Side members of a press or folder which carry the cylinder and other mechanism.

f bâti *m* de côté — **d** Seitengestell *n* — **n** zijframe *n*; zij-freem *n*; zij-geraamte *n* van de pers — **e** marco *m* lateral; costado *m* — **i** telaio *m* laterale

12369 side heading — In computerized typesetting, text heading ranged left on a separate line to the text. This definition is not agreed by all printers, some of whom call them *shoulder heading.

12370 side index — See thumb index.

12371 side lay; side mark — Mechanism to move the sheet to the datum point to position the side of the stock correctly in relation to the printed image.

f taquet *m* latéral; taquet *m* de côté; rectificateur *m* — **d** Seitenmarke *f*; Seitenanlegemarke *f*; Seitenanschlag *m*; Anstoßmarke *f* — **n** zijaanleg *m* — **e** tacón *m*; guía *f* lateral; guía *f* de lado; guía *f* de tope; guía *f* pecho; marco *m* lateral; escuadra *f* lateral — **i** taccheggio *m* laterale

12372 side lay — Side margin of the work controlled by the lateral position of the web or reel in the press.

d Seitenregister *n* — **n** zijregister *n* — **e** registro *m* lateral — **i** registro *m* laterale

12373 side-lay control; side (margin) control — Means of moving the web or reel laterally.

f réglage *m* latéral (de la bobine); guidage *m* latéral — **d** Seitenregister-Verstellung *f*; Seitenregister-Steuerung *f*; Kantensteuerung *f*; seitliche Verstellung *f* — **n** zijdelingse verstelling *f*; besturing *f* van het zij-register — **e** ajuste *m* del registro lateral (de la bobina) — **i** regolazione *f* del registro laterale; controllo *m* del registro laterale

12374 side lighting

f diffusion *f* latérale — **d** Unterstrahlung *f* — **n** onderkopiëren *n* — **e** iluminación *f* lateral — **i** irraggiamento *m* laterale; irradiazione *f* laterale

12375 side magazine — 34-channel magazine on some line casters. Requires an additional keyboard. Used for display setting, usually up to 36 or 48 point. Not under tape control.

f magasin *m* auxiliaire; magasin *m* latéral — **d** Seitenmagazin *n* — **n** zijmagazijn *n* — **e** depósito *m* auxiliar; magacén *m* auxiliar — **i** magazzino *m* ausiliario

12376 side margin — Amount of margin or space on either side of the web in a rotary press.

f marge *f* latérale — **d** Seitenrand *m* — **n** zijmarge *fm*; zijwit *n* — **e** margen *fm* lateral (exterior); margen *fm* lateral (externo) — **i** margine *m* laterale

12377 side (margin) control — See side-lay control.

12378 side mark — See side lay.

12379 side note — See marginal note.

12380 side page; side page companion — One or more compositors working in the same group of frames. See also back page.

f compagnon *m* de la même rangée; confrère *m* dans la même rangée — **d** Gespan *m*; Genosse *m*; Kollege *m*; Arbeitskamerad *m*; Gassengespan *m* — **n** buurman *m*; buurman-zetter *m* — **e** compañero *m* de sección — **i** collega *m* compositore vicino

12381 side page companion — See side page.

12382 side protection resist — Side-wall protecting film.

f solution *f* de protection des talus — **d** Flankenschutzmittel *n* — **i** protettore *m* dei fianchi; soluzione *f* di protezione dei fianchi

12383 side reel stand; side roll stand *depr.* — Reel stand located at the side of the press for most efficient space utilization. Web is guided into line by angle bars.

f support *m* de bobine latéral — **d** seitlich angestellter Rollenstern *m* — **n** naast de pers opgestelde rollenster *f* — **e** soporte *m* lateral de bobina — **i** portabobine *m* laterale

12384 side register — Horizontal location of a web in relation to its position across the length of printing cylinder with reference to location of a printed design or fabricated pattern.

f repérage *m* latéral — **d** Seitenregister *n*; seitlicher Passer *m* — **n** dwarsregister *n* — **i** registro *m* laterale

12385 siderography — Art of engraving on steel plates.

f sidérographie *f*; gravure *f* sur acier; aciérographie *f* — **d** Siderographie *f*; Stahlstichdruck *m*; Stahlstichkunst *f* — **n** siderografie *f*; staalgraveerkunst *f* — **e** siderografía *f* — **i** siderografia *f*

12386 side roll stand — See side reel stand.

12387 side stick — Wood or metal furniture of a tapered shape to lock up galleys and chases and inserted on their sides. See also sticks.

f blanc *m* de marge — **d** Randsteg *m*; Beschnittsteg *m* — **n** sneewit *n* — **e** regleta *f* marginal — **i** margine *m* del taglio

12388 side-stitched booklet — See flat-stitched booklet.

12389 side stitching; side wire stitching; side wire binding; flat stitching; stab stitching *misn.* — Stitching where the wire staples pass through the pile of sections or leaves gathered upon each other and are clinched on the underside. See also stabbing.

f piquer à plate (métallique); piquer à travers; agrafage *m* à plat — **d** Querheftung *f*; Klammerheftung *f* — **n** nieten *n* door het plat — **e** engrapado *m* talonario; cosido *m* en bloque; cosido *m* a diente de perro — **i** cucitura *f* di piatto a punti metallici; cucitura *f* di traverso a punti metallici; agraffatura *f* di piatto

12390 side wall — Nearly vertical part of the raised dots or lines on a letterpress photoengraving plate.

f talus *m* — **d** Flank *f*; Ätzflanke *f* — **n** talud *n* — **e** tabique *m*; costado *m*; talud *m* — **i** fianco *m*

12391 side wire binding — See side stitching.

12392 side wire stitching — See side stitching.

12393 siena — See sienna.

12394 sienna; siena — Ferruginous earth used as a yellowish-brown pigment, as raw sienna, or after roasting in a furnace, as a reddish-brown pigment, *burnt-sienna. Used for plate and stamping ink.

Formula: FeO_3.

f Sienne *f*; terre *f* de Sienne — **d** Siena *f*; Sienaerde *f* — **n** siënna *fm*; terrasiënna *fm*; terra *fm* di Siena — **e** siena *f*; tierra *f* de Siena — **i** Siena *f*; terra *f* di Siena

12395 sign — In computer controlled typesetting, an alphanumeric character for the intelligence it conveys, regardless of its size or graphic characteristics. A lower-case letter "a" is a sing, whereas an eight-point Times Roman lower-case letter "a" is a graphic.

12396 sign — See poster.

12397 signal converter — Converter that changes a signal coded in one form to the same signal coded another form.

f convertisseur *m* de signal — **d** Signalumsetzer *m*; Signalwandler *m* — **n** signaalomzetter *m* — **e** convertidor *m* de señal — **i** convertitore *m* di segnale

12398 signal level — Amplitude of the wave form delivered by a reading head scanning in a uniformly magnetized character.

f niveau *m* de signal — **d** Leseamplitude *f* — **n** leesamplitude *f* — **e** amplitud *f* de leer — **i** ampiezza *f* del segnale

12399 signature; section — Printed sheet (or its flat) that consists of a number of pages of a book, so layed out that they will fold and bind together as a section of a book. The printed sheet after folding.

f cahier *m* — **d** Lage *f* — **n** katern *fmn* — **e** cuadernillo *m*; cuaderno *m*; signatura *f*; pliego *m* — **i** quaderno *m*; quinterno *m*; segnatura *f*; sezione *f*

12400 signature mark — See signature number.

12401 signature number; signature mark — Number indicating the order of the sheets as a guide in gathering, placed in the tail margin near the back of the first page of each section, i.e., on the pages 1, 17, 33, 49, etc. For a multiplicity of sections of a book, various combinations may be used, e.g., A, AA, AAA, 1A, 2A, etc. See also asterisk signature.

f chiffre *m* de la signature — **d** Prime *f*; Primzahl *f*; Bogenzahl *f*; Bogenziffer *f* — **n** velcijfer *n*; velnummer *n* — **e** número *m* de la signatura — **i** numero *m* della segnatura; segnatura *f*; tacca *f*

12402 signature title — Short title in small type at the foot of the first page of each sheet, i.e., on the pages 1, 17, 33, 49, etc. See also asterisk signature.

f signature *f*; titre *m* de la signature — **d** Bogensignature *f*; Norm *f* — **n** signatuur *f* — **e** signatura *f* — **i** segnatura *f* ragionata

12403 sign of infinity; infinity sign — The sign ∞, used in mathematics.

f signe *m* d'infini — **d** Unendlichkeitszeichen *n* — **n** teken *n* van oneindigheid — **e** signo *m* de infinidad; signo *m* de infinito — **i** segno *m* infinito; segno *m* di infinito

12404 sign paper — See poster paper.

12405 signs of meteorology — See meteorological symbols.

12406 signs of the zodiac — See zodiacal signs.

12407 silhouet *v*; **cut** *v* **out** — Opaquing out the background around a subject on a halftone or a continuous-tone negative. Can be achieved on a positive by staging the subject and by flat-etching the background until it is entirely transparent.

f détourer — **d** freistellen; silhouettieren; ausdecken; abdecken; undurchsichtig machen — **n** uitdekken; vrijstaand maken — **e** opacar (ciertas partes del negativo); enmascarar (una parte de un negativo) — **i** eliminare del fondo attorno

12408 silhouette halftone — See outlined halftone.

12409 silica — Silicon dioxide occurring especially as quartz, sand flint and agate. Used in paper, as a plasticizer in resins and sometimes in ink to give a soft dull finish. Also used in the form of white powder in the manufacture of glass, water glass, ceramics and abrasives.

Formula: SiO_2.

f silice *f*; terre *f* siliceuse — **d** Kiesel *m*; Kieselerde *f* — **n** silica *n* — **e** sílice *f* — **i** terra *f* silicea

12410 silica gel — Porous material consisting of pure SiO_2, used as a dehumidifying and dehydrating agent.

f gel *m* de silice — **d** Silikagel *m* — **n** silicagel *mn* — **e** gel *m* de sílice — **i** gelo *m* di silice; gel *m* di silice

12411 silicated board — Paperboard used largely for making packages for foods. Any grade of board can be silicated (i.e., given a surface coating of sodium silicate) on one or both sides, either through the use of calender boxes or by a separate operation. It is slightly grease resistant.

f carton *m* silicaté — **d** silikonisierter Karton *m*; mit Wasserglas gestrichener Karton *m* — **n** gesiliconiseerd karton *n* — **e** cartón *m* siliconizado — **i** cartone *m* siliconato

12412 silicate of soda — See sodium silicate.

12413 siliceous earth — See diatomaceous earth.

12414 silicic acid — Amorphous, gelatinous mass,

formed when alkaline silicates are treated with acids, and easily decomposed into silica and water. Sometimes added to gravure and flexographic inks to promote drying properties.
f acide *m* silicique — **d** Kieselsäure *f* — **n** kieselzuur *n* — **e** ácido *m* silícico — **i** acido *m* silicico
12415 silicon — Tetravalent, non-metallic element that occurs combined as the most abundant element next to oxygen in the earth's crust and is used especially in alloys. Symbol: Si.
f silicium *n* — **d** Silizium *n* — **n** silicium *n* — **e** silicio *m* — **i** silicio *m*
12416 silicon carbide — Bluish black iridescent crystals, insoluble in water and alcohol, soluble in fused alkalis.
Formula: SiC. See also carborundum.
f carbure *m* de silicium; siliciure *m* de carbone — **d** Siliziumkarbid *n* — **n** siliciumcarbide *n* — **e** carburo *m* de silicio — **i** carburo *m* di silicio
12417 silicon dioxide — See silica.
12418 silicon treated paper
f papier *m* silicaté; papier *m* silicatisé — **d** silikonisiertes Papier *n* — **n** gesiliconiseerd papier *n* — **e** papel *m* siliconizado — **i** carta *f* siliconata
12419 silk screen ink — See screen printing ink.
12420 silk screen press — See screen process press.
12421 silk screen printer — See screen printer.
12422 silk screen printing press — See screen process press.
12423 silk screen process — See screen printing process.
12424 silver — White metallic, ductile, very malleable element chiefly univalent in compounds, with the highest thermal and electric conductivity of any substance. Symbol: Ag. Latin: argentum.
f argent *m* — **d** Silber *n* — **n** zilver *n* — **e** plata *f* — **i** argento *m*
12425 silver bath — Acidified silver nitrate solution to sensitize plates in wet-collodion photography.
f bain *m* d'argent — **d** Silberbad *n* — **n** zilverbad *n* — **e** baño *m* de plata — **i** bagno *m* d'argento
12426 silver bromide — Pale yellow crystals or powder, darkening on exposure to light, finally turning black. Soluble in potassium bromide, potassium cyanide and sodium thiosulphate solutions, very slightly soluble in ammonia water, insoluble in water. Used in the preparation of sensitive emulsion coatings for photographic materials.
Formula: AgBr.
f bromure *m* d'argent — **d** Silberbromid *n*; Bromsilber *n* — **n** zilverbromide *n*; broomzilver *n* — **e** bromuro *m* de plata — **i** bromuro *m* d'argento
12427 silver chloride — White granular powder, which darkens on exposure to light, finally turning black. Soluble in ammonium hydroxide, concentrated sulphuric acid and sodium thiosulphate and potassium bromide solutions; very

slightly soluble in water. Used in photography, photometry and optics. Latin: argentum chloratum
Formula: AgCl.
f chlorure *m* d'argent — **d** Silberchlorid *n*; Chlorsilber *n* — **n** zilverchloride *n*; chloorzilver *n* — **e** cloruro *m* de plata — **i** cloruro *m* d'argento
12428 silver cyanide; cyanide of silver — White, odourless, tasteless powder which darkens on exposure to light. Poisonous. Soluble in ammonium hydroxide, dilute boiling nitric acid and potassium cyanide and sodium thiosulphate solutions, insoluble in water. Used in silver plating.
Formula: AgCN.
f cyanure *m* d'argent — **d** Silberzyanid *n*; Zyansilber *n* — **n** zilvercyanide *n* — **e** cianuro *m* de plata — **i** cianuro *m* d'argento
12429 silver image
f image *f* argentique — **d** Silberbild *n* — **n** zilverbeeld *n* — **e** imagen *f* de plata — **i** immagine *f* d'argento
12430 silver ink — Common name for aluminium ink.
f encre *f* argentée — **d** Silberdruckfarbe *f* — **n** zilverinkt *m* — **e** tinta *f* de plata; tinta *f* argentina — **i** inchiostro *m* argento
12431 silver iodide — Pale yellow powder, darkening on exposure to light. Soluble in hydriodic acid, potassium iodide, potassium cyanide, ammonium hydroxide, sodium chloride and sodium thiosulphate solutions. Insoluble in water. Used in photography.
Formula: AgI; in German and Dutch AgJ.
f iodure *m* d'argent — **d** Silberjodid *n*; Jodsilber *n* — **n** zilverjodide *n*; joodzilver *n* — **e** yoduro *m* de plata — **i** ioduro *m* d'argento
12432 silver nitrate; nitrate of silver — Colourless, transparent, rhombic crystals, becoming grey or greyish-black on exposure to light in the presence of organic matter, soluble in cold water, more soluble in hot water, glycerol and hot alcohol, slightly soluble in ether. Used in photography. Latin: argentum nitricum; nitras argenticus.
Formula: $AgNO_3$.
f nitrate *m* d'argent; azotate *m* d'argent — **d** Silbernitrat *n*; salpetersaures Silber *n*; Silbersalpeter *m* — **n** zilvernitraat *n*; salpeterzuur zilver *n* — **e** nitrato *m* de plata — **i** nitrato *m* d'argento
12433 silver powder
f poudre *f* d'aluminium — **d** Aluminiumbronze *f*; Silberbronze *f* — **n** zilverpoeder *n* — **e** polvo *m* de plata; purpurina *f* plata — **i** argento *m* in polvere
12434 silverprint — See brownprint.
12435 silver sulphide — Greyish black, heavy powder, soluble in concentrated sulphuric and nitric acids, insoluble in water.
Formula: Ag_2S.
f sulfure *m* d'argent — **d** Silbersulfid *n* — **n** zilversulfide *n*; zwavelzilver *n* — **e** sulfuro *m* de plata — **i** solfuro *m* d'argento

12436 silver tissue — Rag, sulphite, or sulphate paper with a basis weight of eight pounds (20 x 30 - 480) or (24 x 36 - 480). Free from chemical impurities which would cause tarnishing. Generally made on a Fourdrinier machine.
f mousseline f pour orfèvrerie; papier m pour argenterie — d Silberwarenseidenpapier n; Goldwarenseidenpapier n — n juwelierszijde fm; juwelierszijdepapier n; juweliersvloei n; zijdevloei n — e papel m para joyería; papel m talco — i carta f per argenteria

12437 simo chart — Abbreviation for simultaneous motion cycle chart.
f simogramme m; graphique m simultané — d Simogramm n — n simokaart fm — e simograma m; gráfico m simultaneado — i diagramma m simultaneo

12438 simplex — Pertaining to a communications link that is capable of transmitting data in only one direction. Contrast with full duplex and half duplex.

12439 simplex working — Mode of operation of a data transmission station that can either send or receive messages but do not both.

12440 simulate v — To use a computer to imitate another model or type of computer. Also to represent, through computer operation, a problem or situation, e.g., traffic simulation.
f simuler — d nachahmen; simulieren — n nabootsen; simuleren — e simular — i simulare

12441 simulated watermark — See imitation watermark.

12442 simultaneous mode of working — Method of working closely similar to that of a simultaneous computer.
f mode m simultané de travail — d Simultanbetrieb m — n werkwijze fm met gelijktijdige bewerkingen — e modo m simultáneo de trabajo — i modo m simultaneo di lavoro

12443 sine — In trigonometry, the relationship between the side opposite an acute angle of a right-angle triangle and its hypotenuse. Sine equals side opposite over hypotenuse. See also trigonometric functions.
f sinus m — d Sinus m — n sinus m — e seno m — i seno m

12444 sine shaped — In the form of a sine.
f sinusoïdale — d sinusförmig — n sinusvormig; sinusoïdaal — e sinusoidal — i sinusoidale

12445 single-bath etching; one-bath etching — Etching gravure cylinders with only one solution of iron perchloride of a certain gravity.
f morsure f en un seul bain — d Einbadätzung f; Einbadätzverfahren n — n eenbad-etsmethode f; eenbad-etsing f — e mordido m a un solo baño; grabado m a un solo baño — i incisione f con bagno unico; incisione f in una sola fase

12446 single case — See half case.

12447 single-character machine — Composing machine casting single letters and arranging them automatically to justified lines which are then assembled to columns. The only machine in general use is Monotype.
f machine f à caractères séparés — d Einzelbuchstabenmaschine f — n losse-letterzetmachine f — i compositrice f a caratteri mobili; fonditrice f di caratteri mobili

12448 single coated paper — Paper or board coated once, either on one or both sides. Sometimes (incorrectly) paper or board coated on one side only, which should be called coated one side. See also one-side coated printing paper.
f papier m couché une fois — d einmal gestrichenes Papier n — n eenmaal gestreken papier n — e papel m estucado simple — i carta f patinata su una sola faccia; carta f patinata su uno solo lato; carta f monopatinata

12449 single colour — See monochrome.

12450 single-colour offset machine
f machine f offset une couleur; presse f offset une couleur — d Einfarben-Offsetmaschine — n eenkleur-offsetpers f — e prensa f offset de un color — i macchina f offset monocolore

12451 single-colour print — See monochrome print.

12452 single-colour unit; half unit; colour deck — Section of the rotary press embodying one printing couple to print one side of the web in one colour.
f groupe m couleur — d Einfarbendruckwerk n — n eenkleur-drukeenheid f — e unidad f de un color — i gruppo m colore

12453 single column
f sur une colonne — d einspaltig — n over één kolom — e a una columna; columna f sencilla — i su una colonna

12454 single-cylinder machine — See Yankee machine.

12455 single-face corrugated fibreboard — Board made up of one sheet of fluted paper stuck to one sheet of paper or board.
f carton m ondulé simple face — d einseitige Wellpappe f; Zweibahnen-Wellpappe f — n eenzijdig beplakt golfkarton n; eenzijdig beplakt welbord n — e cartón m ondulado por una cara; cartón m acanalado por una cara; cartón m corrugado por una cara — i cartone m ondulato semplice

12456 single-face self adhesive tape
f ruban m auto-adhésif simple face — d einseitiges Haftklebeband n; einseitiges Selbstklebeband n — n eenzijdig zelfklevend band n; eenzijdig kleefband n — e cinta f autoadhesiva por una cara; cinta f autocolante por una cara — i nastro m autoadesivo su una sola faccia

12457 single-flute corrugated fibreboard — See double face corrugated fibreboard.

12458 single hambro — A size of wrapping paper. See app. no. 3.

12459 single-line effect — One-way tint or halftone illustration showing detail represented by a single series of parallel lines.
f tramé m en lignes — d Linienraster-Effekt m — n enkellijnsraster-effect n — e efecto m de tramado lineal — i effetto m di retino lineare

12460 single-line screen — See half-line screen.

12461 single loaf — A size of wrapping paper. See app. no. 3.

12462 single lump — A size of wrapping paper. See app. no. 3.

12463 single-overlay mask — See positive mask.

12464 single plated — Plating the rotary press to produce one complete copy per revolution.

f cylindre *m* une plaque par tour — **d** eine Platte in Umfang — **n** inrichten *n* van de pers voor enkele produktie — **e** cilindro *m* de una sola plancha — **i** cilindro *m* con una lastra per giro

12465 single quotes — See punctuation marks.

12466 single-revolution press — Press with cylinder revolving once for each impression.

f presse *f* en blanc un tour; presse *f* mono-tour — **d** Eintourenpresse *f* — **n** eentoerpers *f* — **e** prensa *f* de una revolución; prensa *f* de gran cilindro; prensa *f* de tambor — **i** macchina *f* da stampa ad un giro; macchina *f* da stampa piano-cilindrica a un giro

12467 single small hand — A size of wrapping paper. See app. no. 3.

12468 single truck — See double truck advertisement.

12469 single truck advertisement — See double truck advertisement.

12470 single-wall corrugated fibreboard — See double face corrugated fibreboard.

12471 single width press — Newspaper printing machine two pages wide.

f rotative *f* (à) simple largeur — **d** einfachbreite Maschine *f* — **n** enkele pers *f*; enkelbrede pers *f* — **e** rotativa *f* a un solo ancho — **i** rotativa *f* a larghezza monomodulare

12472 sink — Depression in the printing surface of a plate.

f faiblesse *f* — **d** schlechte Stelle *f* — **n** kuil *m* — **e** abolladura *f* — **i** depressione *f*

12473 sinkage — Amount by which a chapter beginning may be dropped from the top of the page. Computer term.

12474 sink, developing — See developing sink.

12475 sinks — Results of air pockets immediately below the surface on type.

f soufflures *fpl* — **d** Gußblasen *fpl* — **n** gietgallen *fmpl* — **e** sopladuras *fpl*; pajas *fpl*; burbujas *fpl*; escarabajos *mpl* — **i** soffiature *fpl*

12476 sisal hemp — Plant (Agave siselana) and the fibre obtained from its leaves, used for hard fibre cordage. Native to Central America, it is grown extensively in the West Indies and Africa. Some obtained from cordage waste, is used in rope papers. (Sisal former seaport in Yucatan.)

f chanvre *m* de sisal — **d** Sisalhanf *m* — **n** sisalhennep *m* — **e** cáñamo *m* de sisal; cáñamo *m* de henequén; cáñamo *m* de pita — **i** canapa *f* sisal; canapa *f* sisalana

12477 six-level code — Method or describing characters or symbols (up to 64 different ones) by the use of a frame consisting of combinations of up to six information bits.

f alphabet *m* à six unités — **d** Sechsenalphabet *n* — **n** zes-eenhedenalfabet *n*; zes-eenhedencode *m* — **e** alfabeto *m* de seis unidades — **i** alfabeto *m* ad sei unità

12478 six-line pica — Old name for 72-point type. See app. no. 1.

12479 sixmo — See sexto.

12480 six-mould disc mechanism — Part of the slug composing machine.

f roue-moule *f* à six moules — **d** Sechs-Gieß-formen-Gießradmechanismus *m*; Sechsformengießrad *n* — **n** gietwiel *n* met zes gietmonden — **e** mecanismo *m* de seis moldes (de rueda); disco *m* de seis moldes — **i** meccanismo *m* della ruota di fusione a sei forme

12481 sixteenmo; sextodecimo; decimo sexto — Book each sheet of which forms 16 leaves or 32 pages. Abbrev. 16mo.

f in-seize *m*; in-16 *m* — **d** Sedezformat *n* — **n** in zestienen — **e** en diez-y-seisavo; dieciseisavo; décimosexto — **i** in sedicesimo; in sedici

12482 six-to-pica — See app. no. 1.

f deux points *mpl* — **d** Viertelpetit *f* — **n** twee punten *fmpl* — **e** dos puntos *mpl* — **i** due punti *mpl*

12483 six-to-pica leads — See two-points leads.

12484 sixty-fourmo — Book each sheet of which formes 64 leaves or 128 pages. Abbrev. 64mo.

f in-soixante-quatre; in-64 *m* — **d** Vierundsechzigerformat *n* — **n** in vierenzestigen — **e** en sesenta-y-cuatroavo — **i** in sessantaquattresimo

12485 size — Sticky ink printed as a base to hold bronze powder or flock.

f mordant *m* pour bronzage — **d** Bronzieruntergrundfarbe *f*; Goldunterdruckfarbe *f*; Grundierfarbe *f* — **n** voordrukinkt *m*; bronzvoordrukinkt *m*; goudvoordrukinkt *m* — **e** sisa *f*; tinta *f* mordiente — **i** mordente *m* per bronzare

12486 size — Material added or applied to paper to affect its ink or water absorbency (rosin size; animal size). Starch, alginates and glue are used in surface sizing. Transparent white ink can be printed as a size to minimize linting, to increase ink hold-out, to dry ink previously printed, or to overcome chalking.

f colle *f* — **d** Leim *m* — **n** lijm *m* — **e** cola *f* — **i** colla *f*

12487 size; sheet size — Dimensions of an unprocessed sheet of paper or board expressed in the following order: width, length, the width being the smaller dimension. (Also exact dimensions indicated on originals to designate the size to which the reproduction is to be made.) See app. no. 3 and no. 7.

f format *m* d'une feuille; dimensions *fpl* d'une feuille — **d** Bogenformat *n* — **n** velformaat *n*; formaat *n* van een vel (papier); afmetingen *fpl* van een vel (papier) — **e** tamaño *m* de la hoja — **i** formato *m* foglio

12488 size — See gilder's size.

12489 sized and supercalendered — Term less used than formerly denoting supercalendered book paper with the ordinary strength of sizing.

12490 sized paper
f papier *m* collé — **d** geleimtes Papier *n* — **n** gelijmd papier *n* — **e** papel *m* encolado — **i** carta *f* collata

12491 size of particle — See particle size.

12492 size press — In papermaking, press consisting of two rolls, usually rubber covered, to apply size on the surface of the paper. Located usually at the dry end of the paper machine between two sections of driers.
f presse *f* encolleuse; presse *f* à encoller; size presse *f*; mouilloir *m* — **d** Leimpresse *f* — **n** lijmpers *f*; lijmwals *fm* — **e** prensa *f* encoladora; size-press *m*; máquina *f* para encolar — **i** pressa *f* collante; size press *f*

12493 size press coated paper — Paper coated by the size press coating process.
f papier *m* couché à la presse encolleuse — **d** auf der Leimpresse gestrichenes Papier *n* — **n** op de lijmpers gestreken papier *n* — **e** papel *m* estucado en la prensa encoladora; papel *m* estucado en el size press — **i** carta *f* patinata in pressa collante; carta *f* patinata in size press

12494 size press coating — Roll coating method for light applications (about 5 g/m²), in which the coating material is transferred to the paper at the nip of a size press. Generally a machine coating operation.
f couchage *m* par presse encolleuse; couchage *m* à la size press — **d** Leimpressenstreichverfahren *n*; Streichen *n* in der Leimpresse — **n** strijken *n* op de lijmpers — **e** estucado *m* en prensa encoladora; estucado *m* en size press; estucado *m* en la máquina para encolar — **i** patinatura *f* in pressa collante; patinatura *f* in size press; patinatura *f* alla pressa collante

12495 sizes of board — See app. no. 3.
f formats *mpl* de carton — **d** Kartonformaten *npl* — **n** kartonformaten *npl*; formaten *npl* van karton — **e** tamaños *mpl* de cartones; formatos *mpl* de cartones — **i** formati *mpl* del cartone

12496 sizes of books
f formats *mpl* de livres — **d** Buchformaten *npl* — **n** boekformaten *npl* — **e** tamaños *mpl* de libro; formatos *mpl* de libro — **i** formati *mpl* dei libri

12497 sizes of cards — See app. no. 3.
f formats *mpl* de cartes — **d** Kartenformaten *npl* — **n** kaartformaten *npl*; formaten *npl* van kaarten — **e** tamaños *mpl* de carta; formatos *mpl* de carta — **i** formati *mpl* delle cartoline

12498 sizes of envelopes — See app. no. 3.
f formats *mpl* d'enveloppes — **d** Briefhüllenformaten *npl*; Hüllformaten *npl*; Umschlagformaten *npl* — **n** enveloppenformaten *npl*; formaten *npl* van enveloppen — **e** tamaños *mpl* de sobrecartas; formatos *mpl* de sobrecartas — **i** formati *mpl* delle buste

12499 sizes of ledger papers — See app. no. 3.
f formats *mpl* de papier registre — **d** Formaten *npl* van Registerpapier — **n** formaten *npl* van registerpapier — **e** tamaños *mpl* de papel registro; formatos *mpl* de papel registro — **i** formati *mpl* della carta per registri

12500 sizes of paper — In early times the two mostly used paper sizes were forma regalis, 50 x 70 cm, and forma mediana or forma communis, 30 x 50 cm. See app. no. 3.
f formats *mpl* de papier — **d** Papierformaten *npl* — **n** papierformaten *npl* — **e** tamaños *mpl* del papel; formatos *mpl* del papel — **i** formati *mpl* della carta

12501 size tub — Container for the sizing material used in the tub-sizing process.
f bac *m* à colle; cuve *f* à colle — **d** Leimbütte *f*; Leimtrog *m* — **n** lijmtrog *m*; lijmbak *m* — **e** tina *f* de encolado; artesa *f* de encolado — **i** gelatinatrice *f*

12502 size water — See white water.

12503 sizing — Addition of materials either to the stock (engine sizing) or to the surface of the paper or board (surface sizing), to increase its resistance to the spontaneous penetration of aqueous liquids, particularly writing ink, and the resistance to the surface spreading of such liquids.
f collage *m* — **d** Leimung *n*; Leimen *n* — **n** lijming *f*; lijmen *n* — **e** encolado *m* (del papel); encoladura *f*; encolamiento *m* — **i** collatura *f*; incollatura *f*; incollamento *m*

12504 skeleton cylinder; delivery cylinder — Cylindrical framework used in the transfer and delivery mechanism of a printing press, usually having rings or star wheels or similar shape, which contact the printed sheet in the margins.
f cylindre *m* squelette; tambour *m* de transfert; cylindre *m* de dégagement — **d** Skelettzylinder *m*; Übergabezylinder *m* — **n** uitlegcilinder *m* — **e** cilindro *m* esqueleto — **i** cilindro *m* uscitafogli; tamburo *m* uscitafogli

12505 skeletonize *v* — See dissect *v* for colours.

12506 sketching paper — Paper for rough drawings and plans. Almost any type or writing or low-grade drawing paper is suitable, but it should have a surface suitable for pencil marks.
f papier *m* à croquis; papier *m* à esquisse — **d** Skizzierpapier *n* — **n** schetspapier *n*; detailtekenpapier *n* — **e** papel *m* para croquis — **i** carta *f* per schizzi

12507 skew angle — In magnetic ink printing the angle between the vertical lines of the character and lines perpendicular to the reference line.
f inclinaison *f* — **d** Schrägstellung *f* — **n** inclinatie *f* — **e** ángulo *m* de través — **i** inclinazione *f*

12508 skewings; gilder's skewings — Excess of gold leaf on a stamped surface.
f déchets *mpl* d'or; rognures *fpl* d'or — **d** Abfälle *mpl* des Vergoldeprozeßes — **n** goud-afval *n* — **i** scarti *mpl* di doratura; residui *mpl* della doratura

12509 skew kick — See kicker.

12510 skid — See stillage.

12511 skim *v* — To remove the scum from the surface of molten metal or of glue.
f écumer; dégraisser — **d** abschäumen; abkrätzen

(Schriftmetall) — **n** afschuimen — **e** retirar le escoria (del crisol); espumar; despumar — **i** schiumare la scoria; scoriare

12512 skimmer — See skimming ladle.

12513 skimming — See shaving the counter.

12514 skimming ladle; skimmer
f écumoire *f*; cuiller *f* à écumer — **d** Schaumlöffel *m*; Abschaumlöffel *m*; Krätzlöffel *m* — **n** schuimlepel *m*; schuimspaan *fm* — **e** espumadera *f*; cuchara *f* para espumar — **i** cucchiaio *m* per scoriare; cucchiaio *m* schiumatore

12515 skimmings — See dross.

12516 skim test — See curling test.

12517 skin
f pellicule *f* — **d** Häutchen *n* — **n** huidje *n*; filmpje *n* — **e** piel *f*; costra *f*; película *f* — **i** pelle *f*; pellicola *f*

12518 skin depositing plant — Plant for the electro-deposition of thin copper skins onto the base copper of gravure cylinders. It consists of a low voltage DC supply and a plating vat, containing acidified copper sulphate solution together with the copper anodes and the cylinder to be plated. The cylinder is horizontally placed, submerged in the electrolyte for about 1/3 of its circumference, and rotated during plating to allow the use of high current densities. The skins are plated in one to two hours. See also Ballard process.
f installation *f* de cuivrage — **d** Hautaufkupferungsanlage *f*; Aufkupferungsanlage *f* — **n** verkoperingsinrichting *f* — **e** planta *f* de encobreamiento — **i** impianto *m* di ramatura

12519 skinning — Formation of a dried layer or film on the surface of ink after a period of standing.
f formation *f* d'une peau — **d** Hautbildung *f* — **n** velvorming *f*; huidvorming *f* — **e** encostradura *f* de la tinta; formación *f* de películas — **i** formazione *f* di pellicola

12520 skip — See speckle.

12521 skip *v* — To move with light springing steps.
f sauter — **d** springen; hüpfen — **n** verspringen; overspringen; overslaan — **e** pasar saltando; saltar — **i** saltare

12522 skipped dots — See missing dots.

12523 skipping — Intermittent failure to print of individual cells in rotogravure printing caused by drying in or by low ink supply. Sometimes known as snowflaking. See also speckle.

12524 skips — Randomly patterned surface areas in blade coated paper devoid of coating, usually elongated in the machine direction; may vary from a fraction of a centimetre to several centimetres in length and/or width. Skips may be found by print tests, black light, NH_4Cl burnout, ink wipe or any standard coating weight test which will verify that area is devoid of coating.
f endroits *mpl* non couchés — **d** (beim Streichen) ausgelassene Stellen *fpl* — **i** salti *mpl* di patinatura

12525 skips — See missing dots.

12526 skip slitter — *Cam-operated slitter giving intermediate cuts, whereby both broad-sheets and tabloid sections can be incorporated in a single copy by means of cylinder collection.
f molette *f* de coupe — **d** gesteuertes Rundschneidmesser *n* — **n** intermitterend mes *n*; bestuurd mes *n* — **e** cortadora *f* intermitente — **i** lama *f* per tagli intermedi (operata a camma)

12527 skiver — Outer or grain side of sheets in which sheepskin has been split, used for title pieces, book back labels, and cheap book bindings.
f mouton *m* scié côté fleur — **d** gespaltenes Schafleder *n* — **n** gespouwen schapeleer; skiver *n* — **e** piel *f* de carnero chiflado — **i** pelle *f* spaccata di montone; pelle *f* intagliata di montone; spaccato *m* di montone

12528 slab — In phototype printing, the heavy glass plate on which the gelatine is poured.
f dalle *f* — **d** Glasplatte *f*; Druckplatte *f* — **n** glasplaat *fm*; drukplaat *fm* — **e** plancha *f* de cristal — **i** lastra *f* di cristallo

12529 slabbing — Stage in the electrotyping process, when the metal used to fill in the copper shell is hammered with special punches to make the face of the electrotype perfectly even before planing the back to type height.
f dressage *m* — **d** Richten *n* des Galvanos — **n** justeren *n* van de galvano — **e** enderazamiento *m* del gálvano — **i** raddrizzatura *f* dei galvano

12530 slab, ink — See ink slab.

12531 slab serif — Serif which consists of a horizontal stroke without being bracketed. See also serifs.
f empattement *m* en forme de plaque — **d** plattenförmiger Serif *m* — **n** rechthoekige schreef *fm* — **e** terminal *m* cuadrado — **i** terminale *m* rettangolare

12532 Slab Serifs — See Mechanistics.

12533 slack edges — In reels of paper, edges of the web that are not wound as tinghtly as the centre, due to the web having one edge longer than the other.
f bords *mpl* détendus — **d** Flatterränder *mpl* — **n** slappe kanten *mpl* — **e** bordes *mpl* flojos — **i** bordi *mpl* molli

12534 slack edges — Warped or wavy edges of sheets.
f bords *mpl* ondulés — **d** wellige Kanten *fpl* — **n** gegolfde randen *mpl*; golvende randen *mpl* — **e** lados *mpl* ondulados — **i** bordi *mpl* ondulati

12535 slack reel; slack roll — Loosely wound reel of paper.
f bobine *f* desserrée; bobine *f* peu serrée — **d** lose gewickelte Rolle *f* — **n** zachte rol *fm*; slappe rol *fm*; niet-strak gewikkelde rol *fm* — **e** bobina *f* floja — **i** bobina *f* lenta

12536 slack roll — See slack reel.

12537 slack sheet — Loose web in a rotary press caused by incorrect tension control.
f feuille *f* non tendue; bande *f* mal tendue — **d** Durchhängen *n* der Papierbahn; Sackbildung *f* —

n slappe papierbaan *fm*; doorzakkende papierbaan *fm*; baan *fm* (papier) met een zak — **e** banda *f* (de papel) con comba; cinta *f* (de papel) con comba; tira *f* (de papel) con comba — **i** nastro *m* mal teso; foglio *m* non teso

12538 slack sized; soft sized — Applied to a lightly sized somewhat water-absorbent sheet of paper. Also used when the degree of water resistance is below standard.
f peu collé; légèrement collé — **d** schwachgeleimt — **n** zwak gelijmd — **e** poco encolado — **i** mezzacollata

12539 slag — Dross or scoria of metal.
f scorie *f* — **d** Schlacke *f* — **n** slak *fm*; metaalslak *fm* — **e** escorilla *f* — **i** scoria *f*

12540 slaked lime — See calcium hydroxide.

12541 slamming block — Old method of fixed cam operating the folding blades of the rotary press folder.
d Falzmesser-Steuerkurve *f* — **n** vouwmessen-excentriek *n* — **e** sujetacuchillas *m* — **i** camma *f* per comando delle lame piegatrice; comando *m* delle lame piegatrice

12542 slant — See solidus.

12543 slash — See solidus.

12544 slasher — Circular saw used in papermaking to cut logs to required length.
f tronçonneur *m* — **d** Kreissäge *f* (zum Ablängen) — **n** afkortzaag *fm* — **i** sega *f* per tondelli

12545 slave equipment — Device which has a master/slave relationship with another device. A slave computer is subsidiary to, and can be interrupted, by the linked master computer, for transmission of data.
f installation *f* secondaire — **d** Nebenanlage *f* — **n** neven-apparatuur *f*

12546 slewing roller; cocking roller; guide roller — Roller adjustable angularly to alter the run of the web in a rotary press, located on reel stand between reel of paper and dancer roll. Can be cocked to compensate for slight paper variation. It is another method to achieve sidelay.
f rouleau *m* dégauchisseur; rouleau *m* dégauchissable; rouleau *m* à inclinaison variable — **d** Papierwerfwalze *f*; exzentrisch gelagerte Leitwalze *f*; Führungswalze *f*; Leitwalze *f* — **n** stelrol *fm* voor het dwarsregister; geleiderol *fm* — **e** rodillo *m* guía; cilindro *m* guía — **i** rullo *m* di guida; rullo *m* conduttore

12547 slice — See galley slice.

12548 slice — See ink knife.

12549 slice galley — Galley with sliding false bottom to move matter to and from composing stone or to page shoe.
f galée *f* à coulisse — **d** Zungenschiff *n*; Satzschiff *n* mit Zunge — **n** schuifgalei *fm* — **e** galera *f* (de) volandera; galera *f* con lengüeta; galera *f* con pala — **i** balestra *f*

12550 slide — Photographic transparency, mounted for projection.
f diapositive *f* de projection; positif *m* pour projection — **d** Diapositiv *n*; Durchsichtspositiv *n*;

Projektionsbild *n* — **n** dia *m*; lantaarnplaatje *n*; lichtbeeld *n* — **e** diapositiva *f* para proyección; transparenza *f* — **i** diapositivo *m* per proiezione

12551 slide — See sliding bar.

12552 slide box — Box with a lid in the form of a shell into which the tray is slide.
f boîte *f* à glissière; boîte *f* coulissante — **d** Schiebeschachtel *f* — **n** schuifdoos(je) *n* — **e** caja *f* corredera — **i** scatola *f* con interno scorrevole

12553 slide fastener — See zip fastener.

12554 slide rule; slip stick *sl.* — Mathematical scale, made of two rules sliding against each other, to carry out mathematical operations rapidly.
f règle *f* à calcul — **d** Rechenschieber *m* — **n** rekenliniaal *fm* — **e** regla *f* de cálculo; regla *f* deslizante — **i** regolo *m* calcolatore

12555 sliding bar; slide; knee *US*; **lever** — Movable part of a composing stick.
f talon *m* justificateur; butée *f* mobile; clavette *f* — **d** Frosch *m* — **n** klavier *n* — **e** corredera *f*; corrediza *f*; zapatilla *f*; palanca *f* — **i** limite *m* mobile del compositoio

12556 slime holes; slime spots — Defect in paper caused by fungi or bacterial growths in the pulp stock. Although they are sterilized by the dryers, they are unsightly and may leave holes or patches in the paper.
f taches *fpl* de vase — **d** Pilzflecke *mpl* — **n** schimmelvlekken *fmpl* — **e** manchas *fpl* de moho — **i** macchie *fpl* di muffa

12557 slime spots — See slime holes.

12558 sling psychrometer
f psychromètre *m* tourbillon — **d** Schwingpsychrometer *n* — **n** slinger-psychrometer *m* — **e** psicrómetro *m* de torbellino — **i** psicrometro *m* ad imbragamento

12559 slip; misrolling — Symptom of misrolling between the sheet fitted on the packing of the cylinder of a printing press and the printing forme.
f dérapage *m*; glissement *m* — **d** Schlupf *m* — **n** slip *fm* — **e** deslizamiento *m*; resbalamiento *m* — **i** slittamento *m*; slittare; scivolare

12560 slip — See ream marker.

12561 slip case; slip cover; book case — Box made from paperboard, covered with paper, cloth or leather, to harmonize with book binding, used as a protective cover for de luxe binding. When shelved, the spine of the book is visible.
f gaine *f*; étui *m*; boîte *f* — **d** Schuber *m* (für bessere Bücher); Futteral *n*; Stechfutteral *n*; Schutzhülse *f*; Schutzkarton *m* — **n** doos *fm*; hoes *fm*; huls *fm* — **e** caja-estuche *f*; capa-estuche *f*; estuche *m* — **i** guaina *f* (di cartone); custodia *f*; astuccio *m*

12562 slip compound — In ink manufacture, an ink additive that imparts lubricating qualities to the dried ink film.
f additif *m* facilitant le glissement du film d'encre sèche — **d** Gleitmittel *n* — **n** glijmiddel *n* — **i** additivo *m* scivolante

12563 slip cover — See slip case.

12564 slip, frayed — See frayed slip.

12565 slip galley; column galley — Long narrow galley.

f violon *m* — **d** Spaltenschiff *n*; Kolumnenschiff *n*; Vorteilschiff *n* — **n** kolommengalei *fm* — **e** galera *f* de columna(s) — **i** vantaggio *m* per colonne

12566 slip gauge; slip meter — Instrument to measure the difference in speed between the sheet of paper on the packing of the cylinder and the speed of the printing forme of a flat-bed printing press.

f glissomètre *m* — **d** Schlupfmesser *m* — **n** slipmeter *m* — **e** (aparato) medidor *m* del deslizamiento; dispositivo *m* para la determinación del desarrollo — **i** strumento *m* per misurare lo slittamento

12567 slip meter — See slip gauge.

12568 slip of the pen — See clerical error.

12569 slippage — Slippage can be caused between rollers and plate, plate and blanket, or blanket and paper.

f glissement *m*; dérapage *m* — **d** Schlupf *m* — **n** slip *m* — **e** deslizamiento *m* — **i** scorrimento *m*; slittamento *m*; rotolamento *m* difettoso

12570 slippage flow — A phenomenon which can take place only with suspensions. It occurs at the walls of a capillary-tube viscometer and, to a negligible extent, at the inner surface of the cup and the surface of the bob in rotational viscometers. The particles are held together (by the force of flocculation) as a solid plug which is lubricated by a thin layer of the vehicle. When flow starts, the plug moves as a single piece. Its flow (slippage flow) is not laminar.

f écoulement *m* par glissement — **d** Schlupf *m* — **n** slip *fm* — **e** deslizamiento *m* — **i** flusso *m* difettoso; flusso *m* non laminare

12571 slippery book; loosened binding — Book of which the sections are not firmly bound together.

f livre *m* délié; livre *m* déboîté — **d** zu lose gehefteter Buchblock *m* — **n** slap gebonden boek *n* — **i** libro *m* scartellato; libro *m* sciolto

12572 slipping core — Loose paper near the reel centre which allows the reel to move on the core. Also reel centre rotating on the cones.

f cône *m* glissant — **d** schlupfender Rollenkonus *m* — **n** slippende conus *m* — **e** cono *m* deslizante — **i** anima *f* slittante

12573 slip proof; galley proof; string proof — Rough proof taken of a column of type matter in a galley, printed on a slip about 45 cm (18 in) long, for checking and correcting, and for editing and planning page make-up. In phototypesetting a strip of photographic film or paper of the same length.

f épreuve *f* en placard; épreuve *f* en première — **d** Korrekturfahne *f*; Fahne *f*; Fahnenabzug *m*; Spaltenkorrektur *f*; Fahnenkorrektur *f*; Abzugbogen *m*; Abzugstreifen *m*; Probeabzug *m*; Spaltenabzug *m* — **n** proef *fm* aan stroken; slip *fm* — **e** prueba *f* de galera; prueba *f* de galerada; prueba *f* en galera; galera *f*; galerada *f*; galerada *f* suelta; primera prueba *f* — **i** bozza *f* in colonna; bozzone *m*

12574 slip-ring motor

f moteur *m* à bagues — **d** Schleifringmotor *m*; Schleifringläufermotor *m* — **n** sleepringmotor *m*; motor *m* met sleepringanker — **e** motor *m* de anillos; motor *m* con anillo de frotamiento — **i** motore *m* ad anelli

12575 slip sheeting; interleaving — Slipping of clean unprinted pieces of paper between each of the printed sheets as they are delivered from the press to prevent the freshly printed sheets from setting-off.

f intercalage *m* — **d** Durchschießen *n* (mit Makulatur); Einschießen *n* — **n** tussenschieten *n* — **e** interfoliado *m* (con papel); intercalado *m* — **i** intercalare; interfogliare

12576 slip sheets; interleaves; offsheets; set-off paper — Sheets of coarse paper, kraft or newsprint inserted between freshly printed sheets to prevent offsetting from one sheet to the next one.

f intercalaires *fpl*; macule *f* — **d** Einschießbogen *mpl*; Durchschußbogen *mpl*; Makulaturbogen *mpl*; Abschmutzmakulatur *f*; Abschmutzpapier *n* — **n** smetvellen *npl*; tussenschietvellen *npl* — **e** descargas *fpl*; maculaturas *fpl* — **i** fogli *mpl* antiscartini; fogli *mpl* di separazione; carta *f* per scartino

12577 slips, in — See in slips.

12578 slip stick — See slide rule.

12579 slit diaphragm; slotted diaphragm; slit stop

f diaphragme *m* à fentes — **d** Spaltblende *f*; Schlitzblende *f* — **n** spleetdiafragma *n* — **e** diafragma *m* de ranura — **i** diaframma *m* a fessura; diaframma *m* a feritoia

12580 slit stop — See slit diaphragm.

12581 slitter — On a printing press, a mechanism to automatically cut a sheet into two or more parts after printing and before delivery.

f coupeuse *f* en long; molette *f* coupeuse; refendeuse *f* — **d** Längsschneider *m*; Längsschneidevorrichtung *f* — **n** langssnijmes *n*; langssnij-inrichting *f* — **e** cortadora *f* longitudinal de hojas — **i** taglierina *f* longitudinale

12582 slitter cheeck — Surface against which the slitter operates in a rotary press.

f contre-partie *f* (de molette de refente) — **d** Schneidring *m* — **n** tegenmes *n*; contrames *n* — **e** anillo *m* de corte; contracuchilla *f* circular — **i** controlama *f* di una taglierina

12583 slitter dust — Dust, primarily filler, fibre or coating, thrown off during slitting and remaining in roll.

f poussière *f* de refente; poussière *f* de coupe — **d** Schnittstaub *m*; Schneidstaub *m* — **n** snijstof *n* — **e** polvo *m* de corte — **i** polvere *f* di tagliare

12584 slitter/rewinder — The cutting and rewinding of reels of paper on cones.

f bobineuse-refendeuse *f* — **d** Rollenschneidmaschine *f* — **n** bobineuse *f* — **e** cortadora *f* y rebobinadora — **i** taglierina-ribobinatrice *f*

12585 slitters; cutting disks; circular knives; disk knives — Circular rotating knives to divide the web into narrow ribbons or strips.
f molettes *fpl* de coupe; molettes *fpl* coupeuses; molettes *fpl* de refente; lames *fpl* rotatives; couteaux *mpl* circulaires — **d** Kreismesser *mpl*; Rundmesser *mpl*; Tellermesser *mpl*; Schneidrollen *fpl* — **n** rondmessen *npl* — **e** cuchillos *mpl* circulares; cortadoras *fpl* rotativas — **i** lame *fpl* circolari; coltelli *mpl* circolari

12586 slitter/scorer — Slitter with a set of shafts to carry creasing wheels. This allows scoring and slitting in one operation.
f refouleuse-traceuse *f* — **d** Rill- und Ritzapparat *m* — **n** riller-ritser *m* — **e** cortadora-estriadora *f* — **i** taglierina-cordonatrice *f*; tagliacordona *f*

12587 sliver — Small splinter of wood, longer than a shive and of smaller diameter in proportion to its length. Found more frequently in mechanical than in chemical wood pulp. See also shives.
f bûchette *f* — **d** Ast *m*; Splitter *m* — **n** houtslijpsplinter *m*; houtschilfer *m* — **e** astillita *f* de madera — **i** incotto *m*

12588 slivers — Defects of aluminium foil. Elongated pieces of metal, partly attached to the main body.

12589 sliver screen — Flat type screen or classifier to remove slivers from pulp stock in papermaking.
f épurateur *m* de bûchettes — **d** Sortierer *m* — **n** sorteerzeef *fm* — **e** filtro *m* de astillas — **i** epuratore *m* degli incotti

12590 slop — Composed matter for a newspaper left over when making up and all formes are filled.
f débordé *m*; marbre *m* — **d** Übersatz *m*; Satzrest *m* nach Umbrechen — **n** teveel *n* aan zetsel — **e** material *m* sobrante; composición *f* sobrante — **i** composizione *f* restante; composizione *f* rimanente

12591 slope of a curve
f pente *f* d'une courbe — **d** Neigung *f* einer Kurve — **n** helling *f* van een kromme; helling *f* van een curve — **e** inclinación *f* de una curva — **i** inclinazione *f* di una curva

12592 sloping fraction — See also piece fraction.
f fraction *f* avec barre diagonale — **d** Bruchziffer *f* mit Schrägstrich; Bruchziffer *f* mit schrägem Strich — **n** breukcijfer *n* met schuine deelstreep — **e** fracción *f* con diagonal — **i** frazione *f* con barra diagonale; frazione *f* con barra inclinata

12593 slotted diaphragm — See slit diaphragm.
12594 slow *v* down — To retard the forward speed of the sheet.
f ralentir; diminuer de vitesse — **d** verzögern; verlangsamen — **n** vertragen; afremmen — **e** retardar; retrasar; dilatar; demorar; disminuir la velocidad — **i** rallentare

12595 slowing down device
f dispositif *m* de ralentissement — **d** Ver-

zögerungsvorrichtung *f* — **n** vertragingsmechanisme *n*; vertragingsinrichting *f* — **e** dispositivo *m* retardador — **i** dispositivo *m* ritardatore

12596 sludge — Precipitated solid matter.
f boues *fpl* — **d** Schlamm *m* — **n** slib *n* — **e** lodo *m*; sedimento *m* — **i** morchia *f*; fango *m*; sedimento *m*

12597 sludge build-up; sludging — Defect of blade coated paper. Accumulation of high viscosity and/or paste-like sludge along the inside of the blade at the nip, which occasionally breaks through and is stuck to the surface of the sheet. Found by visual inspection of the blade and/or the sheet.
f encrassement *m* de la lame de couchage — **d** Schlammbildung *f* — **n** slibvorming *f* — **e** formación *f* de lodo — **i** addensamento *m* di patina; accumulo *m* di patina sulla lama

12598 sludging — See sludge build-up.
12599 slug; type slug — Metal bar of the length and width of a line of type, having on its upper edge the type characters to print an entire line. The bar is in fact a solid line of type.
f ligne-bloc *f* — **d** Setzmaschinezeile *f*; gegossene Zeile *f* — **n** zetmachineregel *m* — **e** línea *f* de composición; línea *f* de tipos; lingote *m* de composición — **i** riga *f* di macchina da comporre; riga *f* fusa

12600 slug — Pieces of lead, lower than type high, usually 6 points thick, used for spacing betweeen lines of type.
f lingot *m*; plomb *m* — **d** Reglette *f* — **n** reglet *fmn* — **e** lingote *m*; regleta *f* — **i** lingotto *m*

12601 slug — Hole or scratch in a negative or printing plate.
f égratignure *f* — **d** Schramme *f*; Riß *m*; Kratzer *m*; Kritzer *m* — **n** kras *fm* — **e** arañazo *m*; raspadura *f* — **i** graffiatura *f*; foro *m*

12602 slug — Anchor on a printing plate. See also slugging.
12603 slug caster — See slug-composing machine.
12604 slug-casting machine — See slug-composing machine.
12605 slug clipper — See lead cutter.
12606 slug-composing machine; slug-casting machine; slug caster; line-composing machine; line-casting machine; line caster — Line-casting machine, such as the Intertype, Linotype, and Ludlow.
f machine *f* lignes-blocs; composeuse-fondeuze *f* à lignes-blocs — **d** Zeilensetzmaschine *f*; Zeilengießmaschine *f* — **n** regelzetmachine *f* — **e** compositora *f* para líneas; fundidora *f* de líneas-bloque — **i** (macchina *f*) compositrice *f* monolineare; compositrice *f* di righe; macchina *f* linocompositrice

12607 slugging — Anchoring plate to metal base by studs soldered to the back of the plate to fit holes in the base and soldering the studs into place.
f ancrage *m* à boutons — **d** Verstiften *n*;

Befestigen *n* von Druckplatten auf Metallunterlagen durch Löt- und Stiftverbindung *f* — **n** bevestigen *n* met stiften — **e** soldadura *f* por puntos — **i** ancoraggio *m* mediante spine saldate

12608 slug saw
f scie *f* à lignes-blocs — **d** Zeilensäge *f* — **n** regelzaag *fm* — **e** sierra *f* cortalíneas; sierra *f* cortalingotes; sierra *f* para cortar filetes — **i** sega *f* per linee; sega *f* taglialingotti

12609 slum — See slum *v*.

12610 slum *v* — To hide away for one's own use movable type sorts or other material when these are very short. Place of concealment is a slum.
f amasser; accaparer — **d** hamstern; marderen — **n** hamsteren — **e** acaparar; abastecer — **i** sgraffignare

12611 slur; drag — Blurred or dragged printing detail of an inked image on the press sheet, caused by a movement during impression between paper and type. It may result from excessive underpacking, play between meshing gears, loose blanket.
f papillotage *m*; frisotage *m*; maculage *m*; regiflage *m* — **d** Schmitz *m* — **n** veeg *f*; smet *f* — **e** borrón *m*; ensuciado *m* — **i** doppieggiatura *f*; sbaveggio *m*

12612 slurry — See coating slip.

12613 slushed pulp — See slush pulp.

12614 slush, metallic — See metallic slush.

12615 slush pulp; slushed pulp — Suspension of paper pulp in water containing 1-6% dry stock so that it can be pumped. When a paper mill has its own pulp department, the pulp is often slushed from bleachers direct to beaters.
f pâte *f* liquide — **d** pumpfähiger Papierstoff *m* — **n** waterige stof *fm* — **i** pasta *f* spappolata

12616 small — A size of cut card. See app. no. 3.

12617 small capitals; small caps — Usually about two-thirds size of large caps or equal to the x-height of the fount, indicated in the manuscript by two lines underneath. Abbrev. s.caps or s.c.
f petites capitales *fpl* — **d** Kapitälchen *npl* — **n** klein kapitaal *n* — **e** versalitas *fpl*; versalillas *fpl*; mayúsculas *fpl* pequeñas; letras *fpl* versalitas; letras *fpl* versalillas — **i** maiuscolette *fpl*

12618 small caps — See small capitals.

12619 small caps, accentuated with — See accentuated with small caps.

12620 small double loaf — A size of wrapping paper. See app. no. 3.

12621 small double post — A size of printing paper. See app. no. 3.

12622 small foolscap — A British standard size of writing paper. See app. no. 3.

12623 small half imperial — A size of board. See app. no. 3.

12624 small half royal — A size of board. See app. no. 3.

12625 small hand — A size of wrapping paper. See app. no. 3.

12626 small medium — A size of writing paper. See app. no. 3.

12627 small offset press — Small printing machine using the offset lithographic principle.
f machine *f* offset de bureau — **d** Kleinoffsetmaschine *f* — **n** kleinoffsetpers *f*; kantoor-offsetmachine *f* — **e** máquina *f* pequeña offset; prensa *f* pequeña offset; copiadora *f* offset — **i** macchina *f* offset di piccolo formato

12628 small pica — Name for an old size of type. See app. no. 1.

12629 small post — See pinched post.

12630 small royal — A British standard size of writing paper. See app. no. 3.

12631 small run — See short run.

12632 smalls, accentuated with — See accentuated with small caps.

12633 small single — A size of wrapping paper. See app. no. 3.

12634 small whole royal — A size of board. See app. no. 3.

12635 smash *v*; **nip** *v* — To press the book block after sewing.
f endosser — **d** abpressen — **n** afpersen — **e** apretar — **i** pressare

12636 smearing — Condition in which the impression is slurred and unclear because too much ink was run, or sheets were handled or rubber before it was dry.
f empâtement *m*; maculage *m* — **d** Abschmieren *n*; Schmieren *n*; Verschmieren *n* — **n** overzetten *n*; smetten *n*; smeren *n* — **e** efecto *m* grasiento — **i** controstampa *f*; imbrattamento *m*; insudiciamento *m*; macchiatura *f*

12637 smearing — Spreading ink over areas of the plate where not wanted; wiping the ink off the image areas and onto the non image areas of a plate.
f graissage *m* — **d** Schmieren *n*; Tonen *n* — **n** smeren *n* — **i** velatura *f*; imbrattamento *m*

12638 smoke *v* — To blacken with soot the face of a plate before engraving.
f noircir — **d** schwärzen — **n** zwart maken; inroeten; beroeten; inzwarten — **e** ennegrecer — **i** annerire con fuliggine

12639 smoke proof — Impression from a punch obtained by putting the punch into the flame of a glaring gas burner until its face is covered with soot, and after breathing repeatedly on a piece of paper to moisten it, firmly pressing the punch on the paper.
f fumé *m*; épreuve *f* fumée — **d** Rußabdruck *m*; Rauchabzug *m* — **n** roetafdruk *m* — **e** ahumada *f*; prueba *f* de humo — **i** prova *f* alla fuliggine

12640 smooth breathing — See spiritus lenis.

12641 smooth-file
f lime *f* douce; lime *f* fine — **d** Schlichtfeile *f*; Feinfeile *f*; Glättfeile *f* — **n** zoetvijl *fm* — **e** lima *f* fina — **i** lima *f* per spianare

12642 smoothing press; offset press — Pair of unfelted rolls between the last main press of a paper or board machine and the first drying cylinder, to improve the surface of the paper or board, make it more even-sided and remove the felt mark before drying.

f presse *f* offset — **d** Offsetpresse *f*; Einführpresse *f* — **n** satineerpers *f* — **e** prensa *f* brunidora — **i** pressa *f* ontante

12643 smoothness — Nature of the surface of a sheet of paper. Term preferred to roughness or rugosity.

f lissé *m* (du papier) — **d** Glätte *f* (des Papiers); Papierglätte *f* — **n** gladheid *f* (van papier); papiergladheid *f* — **e** lisura *f* (del papel) — **i** liscio *m* (di carta)

12644 smoothness — The nature of a surface.

f planéité *f* — **d** Ebenheit *f* — **n** vlakheid *f* — **e** llanura *f*; igualdad *f* — **i** levigatezza *f*

12645 smoothness test

f mesure *f* de lissé — **d** Glättemessung *f* — **n** gladheidsmeting *f* — **e** medición *f* de la lisura — **i** misura *f* del liscio; misura *f* di liscio; determinazione *f* del liscio

12646 smoothness tester; roughness tester — Instrument to determine the relative smoothness of a paper surface by measuring the permeability of paper to air sucked through under the influence of a mean vacuum of 0.5 atm. The permeability figure is taken as the time measured in seconds required by 100 cm^3 of air to pass through (Beck, Bendtsen, Gurley, Sheffield).

f essayeur *m* pour le fini; appareil *m* de mesure du lissé; appareil *m* d'essai du lissage — **d** Glätteprüfer *m*; Glättemesser *m* — **n** gladheidsmeter *m* — **e** comprobador *m* del satinado; probador *m* de tersura; satinómetro *m* — **i** apparecchio *m* per prova di liscio

12647 smooth roller

f rouleau *m* lisse — **d** glatte Walze *f* — **n** gladde rol *fm* — **e** rodillo *m* liso — **i** rullo *m* liscio

12648 snakeslip; snake stone; ayr stone; Scotch stone — Very fine-grained stone, first found in Ayrshire (SW Scotland), widely employed in polishing, as a whetstone, and for removing dirt spots or unwanted objects from the lithographic press-plate.

f gomme ponce *f* — **d** Korrekturschiefer *m*; Schleifschiefer *m*; Polierstab *m* — **n** corrigeersteentje *n*; slijpsteentje *n* — **e** lápiz *m* abrasivo; raspadora *f*; piedra *f* de Ayr; piedra *f* de Escocia — **i** pietra *f* pomice; pietra *f* pomice in barrette

12649 snake stone — See snakeslip.

12650 snap — Crispness or spring-back of paper or bristols when bent.

f carteux *m*; sonnant *m* — **d** Klang *m*; Rascheln *n* — **n** klank *m* — **e** carteo *m*; tacto *m* — **i** sonorità *f* (della carta)

12651 snap; snappy — Brilliant, representing a wide range of contrast and middletones in a negative, print or reproduction.

f piquant — **d** brillant; kontrastreich — **n** contrastrijk; briljant; pittig — **i** vivace; ricco di contrasti

12652 snap-out sets — Set of printed forms, bound into a pad with others and perforated near the binding for easy separation, often interleaved with carbon paper.

f formulaires *mpl* snapout; jeux *mpl* de formulaires — **d** Snap-out Durchschreibesätze *mpl*; Schnelltrennsätze *mpl* — **n** scheurstellen *npl* — **e** formularios *mpl* juego múltiple; formularios *mpl* múltiples — **i** moduli *mpl* snap-out

12653 snappy — See snap.

12654 snorkel — Measuring scale on a paper cutter, standing up like a periscope and directly connected to the back gauge. It reflects the actual markings (including a vernier scale) on the back gauge scale.

f périscope *m* — **d** optischer Schnittandeuter *m* — **n** periscoop *m* — **e** periscopio *m* — **i** periscopio *m*; indicatore *m* ottico del taglio

12655 snowflaking — Fine white specks in an area printed by lithography caused by water droplets remaining in the printing nip during impression.

f neigeux *m*; floconneux *m* — **d** Schneeflockeneffekt *m*; schneeflockenartiges Aussehen *n* des Druckes — **i** fioccatura *f*

12656 snowflaking — See skipping.

12657 soap resistance

f résistance *f* aux savons — **d** Seifenechtheit *f* — **n** bestendigheid *f* tegen zeep — **e** resistencia *f* al jabón — **i** resistenza *f* al sapone

12658 soapstone — Impure variety of *steatite.

12659 soap wrappers — Alkali-proof book paper, with high finish, made from soda and sulphite pulps, used for wrapping of soap.

f enveloppes *fpl* à savon — **d** Seifenpackungen *fpl* — **n** zeepwikkels *mpl* — **e** envoltorios *mpl* de jabón — **i** imballagi *mpl* per sapone

12660 socket — Socket to receive a plug that makes an electrical connection with supply wiring.

f prise *f* de courant — **d** Steckdose *f* — **n** stopcontact *n* — **e** enchufe *m*; caja *f* de enchufe — **i** presa *f* di corrente

12661 soda — See sodium carbonate.

12662 soda ash — See sodium carbonate.

12663 soda-chlorine pulp — Unbleached chemical pulp obtained by treating vegetable matter successively with alkali and chlorine.

f pâte *f* au chlore; pulpe *f* au chlore — **d** Natronchlorzellstoff *m* — **n** chloornatroncelstof *fm* — **e** pasta *f* de sodio cloro; pulpa *f* de sodio cloro — **i** pasta *f* al cloro-soda; cellulosa *f* al cloro-soda

12664 soda niter — See sodium nitrate.

12665 soda nitre — See sodium nitrate.

12666 soda process — Process for paper making, in which the cooking liquor is made up of a caustic soda solution.

f procédé *m* à la soude — **d** Natronverfahren *n*; Sodaverfahren *n* — **n** natronprocédé *n* — **e** procedimiento *m* a la sosa — **i** procedimento *m* alla soda

12667 soda pulp — Short-fibred pulp made by cooking wood chips with caustic soda. Used in printing and litho papers principally as a filler to give smoothness, bulk, opacity, and uniform formation.

f pâte *f* à la soude — **d** Natronzellstoff *m* — **n** natroncelstof *fm* — **e** pasta *f* a la soda; pulpa *f* a la sosa — **i** pasta *f* chimica alla soda

12668 sodium; natrium — Light, soft, ductile, malleable, silver white metal, oxidizing rapidly in air. Decomposes in water on contact with vigorous evolution of hydrogen and forms sodium hydroxide. Insoluble in benzene, kerosine and naphtha. Symbol: Na.

f sodium *m*; soude *f* — **d** Natrium *n* — **n** natrium *n* — **e** sodio *m* — **i** sodio *m*

12669 sodium acetate; acetate of soda — Colourless crystals, very soluble in water and ether, slightly soluble in alcohol. Used in photography and manufacture of dry colours. Latin: acetas natricus; natrium aceticum.

Formula: $NaC_2H_3O_2$.

f acétate *m* de sodium; acétate *m* de soude — **d** Natriumazetat *n*; essigsaures Natron *n*; essigsaures Natrium *n* — **n** natriumacetaat *n*; azijnzuur natron *n* — **e** acetato *m* de sodio; acetato *m* sódico; acetato *m* de sosa — **i** acetato *m* di sodio; acetato *m* sodico

12670 sodium acid carbonate — See sodium bicarbonate.

12671 sodium acid sulphate — See sodium bisulphate.

12672 sodium acid sulphite — See sodium bisulphite.

12673 sodium alum; sodium aluminium sulphate; aluminium sodium sulphate — Colourless crystals, soluble in water, not in alcohol, used in papermaking.

Formula: $AlNa(SO_4)_2 \cdot 12H_2O$.

f sulfate *m* (double) de sodium et d'aluminium — **d** Natriumalaun *m*; Aluminium-Natriumsulfat *n* — **n** natriumaluin *m*; natrium-aluminiumsulfaat *n* — **e** sulfato *m* alumínico-sódico — **i** solfato *m* di sodio di alluminio

12674 sodium aluminium sulphate — See sodium alum.

12675 sodium benzoate — White, amorphous, crystalline or granular powder, soluble in water and alcohol, slightly soluble in ethanol.

Formula: $NaC_7H_5O_2$.

f benzoate *m* de sodium; benzoate *m* de soude — **d** Natriumbenzoat *n*; benzoesaures Natrium *n* — **n** natriumbenzoaat *n* — **e** benzoato *m* de sodio; benzoato *m* sódico; benzoato *m* de sosa — **i** benzoato *m* di sodio; benzoato *m* sodico

12676 sodium bicarbonate; sodium acid carbonate; bicarbonate of soda; hydrosodic carbonate; baking soda — White powder or crystalline lumps, soluble in water, insoluble in alcohol. Latin: bicarbonas natricus.

Formula: $NaHCO_3$.

f bicarbonate *m* de sodium; bicarbonate *m* de soude; carbonate *m* acide de sodium; sel *m* de Vichy — **d** Natriumbikarbonat *n*; doppeltkohlensaures Natrium *n*; saures Natriumkarbonat *n*; Natriumhydrogenkarbonat *n* — **n** natriumbi-

carbonaat *n*; dubbelkoolzure soda *fm*; zuiveringszout *n* — **e** bicarbonato *m* de sodio; bicarbonato *m* sódico; bicarbonato *m* de sosa — **i** bicarbonato *m* di sodio; bicarbonato *m* sodico

12677 sodium bichromate — See sodium dichromate.

12678 sodium bisulphate; sodium acid sulphate — Colourless crystals or white, fused lumps. Aqueous solution is strongly acid. Soluble in water.

Formula: $NaHSO_4$.

f bisulfate *m* de sodium; sulfate *m* acide de sodium — **d** Natriumbisulfat *n*; Natriumhydrosulfat *n*; saures Natriumsulfat *n*; Natriumhydrogensulfat *n* — **n** natriumbisulfaat *n*; zuur natriumsulfaat *n* — **e** bisulfato *m* de sodio; bisulfato *m* sódico; bisulfato *m* de sosa — **i** bisolfato *m* di sodio; bisolfato *m* sodico

12679 sodium bisulphite; sodium acid sulphite; sodium hydrogen sulphite; bisulphite of soda — White crystals or crystalline powder, soluble in water, insoluble in alcohol. Used in copper plating and as a photographic reducing agent. Latin: bisulfis natricus; natrium bisulfurosum.

Formula: $NaHSO_3$.

f bisulfite *m* de sodium; bisulfite *m* de soude; sulfite *m* acide de sodium — **d** Natriumbisulfit *n*; saures schwefligsaures Natrium *n*; doppeltschwefligsaures Natron *n*; Natriumhydrogensulfit *n*; Bisulfit *n* — **n** natriumbisulfiet *n* — **e** bisulfito *m* de sosa; bisulfito *m* sódico; bisulfito *m* de sosa — **i** bisolfito *m* di sodio; bisolfito *m* sodico

12680 sodium borate; sodium tetraborate; sodium pyroborate; biborate of soda; borax *trade name* — White crystals or powder, soluble in water and glycerol, insoluble in alcohol. Used in film developing solutions as a restrainer. Latin: biboras natricus.

Formula: $Na_2B_4O_7 \cdot 10H_2O$.

f borate *m* de sodium; borate *m* de soude; tétraborate *m* de sodium; biborate *m* de sodium — **d** Natriumborat *n*; Natriumtetraborat *n*; borsaures Natron *n*; Borax *m* — **n** natriumboraat *n*; natriumtetraboraat *n*; borax *m* — **e** borato *m* de sodio; borato *m* sódico; borato *m* de sosa; atíncar *m*; bórax *m* — **i** borato *m* di sodio; biborato *m* di sodio; biborato *m* sodico; borace *m*

12681 sodium bromide — White crystalline powder or granules, soluble in water, moderately soluble in alcohol. Used in photography. Latin: brometum natricum.

Formula: $NaBr$.

f bromure *m* de sodium; bromure *m* de soude — **d** Natriumbromid *n* — **n** natriumbromide *n* — **e** bromuro *m* de sodio; bromuro *m* sódico; bromuro *m* de sosa — **i** bromuro *m* di sodio; bromuro *m* sodico

12682 sodium carbonate; carbonate of soda; calcined soda; soda ash; soda; anhydrous sodium carbonate — Commercial anhydrous sodium carbonate used in papermaking and as an accelerator for photographic developer, soluble in water, insoluble in alcohol. Latin: carbonas natricus, natrium carbonicum.

Formula: Na_2CO_3.

f carbonate *m* (neutre) de sodium anhydre; carbonate *m* de soude — **d** Natriumkarbonatanhydrid *n*; wasserfreies Natriumkarbonat *n*; kalziniertes Soda *n*; wasserfrei kohlensaures Natrium *n*; kohlensaures Natron *n*; normales Natriumkarbonat *n* — **n** natriumcarbonaat *n*; soda *fm* — **e** carbonato *m* de sodio; carbonato *m* sódico; carbonato *m* de sosa; sosa *f* calcinada; sosa *f* anhidra — **i** carbonato *m* di sodio; carbonato *m* sodico; soda *f*

12683 sodium chloride; table salt; common salt — Colourless, transparent crystals or white, crystalline powder, soluble in water and glycerol, very slightly soluble in alcohol. Latin: chloretum natricum; natrium chloratum.

Formula: NaCl.

f chlorure *m* de sodium; chlorure *m* de soude; chlorure *m* soudique; muriate *m* de sodium; muriate *m* de soude; sel *m* de cuisine; sel *m* blanc; sel *m* marin; sel *m* gemme; sel *m* de gabelle — **d** Natriumchlorid *n*; Chlornatrium *n*; Kochsalz *n* — **n** natriumchloride *n*; chloornatrium *n*; keukenzout *n* — **e** cloruro *m* de sodio; cloruro *m* sódico; cloruro *m* de sosa; sal *f* de cocina — **i** cloruro *m* di sodio; cloruro *m* sodico; sale *m* di cucina

12684 sodium chlorite — White crystals or crystalline powder, soluble in water. Used in papermaking.

Formula: $NaClO_2$.

f chlorite *m* de sodium — **d** Natriumchlorit *n* — **n** natriumchloriet *n* — **e** clorito *m* de sodio; clorito *m* sódico; clorito *m* de sosa — **i** clorito *m* di sodio; clorito *m* sodico

12685 sodium chromate — Yellow crystals, soluble in water, slightly soluble in alcohol. Used for inks, leather tanning, etc.

Formula: $Na_2CrO_4 \cdot 10H_2O$.

f chromate *m* neutre de sodium — **d** Natriumchromat *n*; chromsaures Natrium *n* — **n** natriumchromaat *n* — **e** cromato *m* sódico — **i** cromato *m* di sodio

12686 sodium citrate; trisodium citrate — White crystals or granular powder, soluble in water, insoluble in alcohol. Used in photography.

Formula: $C_6H_5O_7Na_3 \cdot 2H_2O$.

f citrate *m* de sodium; citrate *m* de soude — **d** Natriumzitrat *n*; zitronensaures Natrium *n* — **n** natriumcitraat *n* — **e** citrato *m* de sodio; citrato *m* sódico; citrato *m* de sosa — **i** citrato *m* di sodio; citrato *m* sodico

12687 sodium cyanide; cyanide of sodium — White deliquescent, crystalline powder. Poisonous. Soluble in water, slightly soluble in alcohol. Used in the manufacture of dyes and pigments.

Formula: NaCN.

f cyanure *m* de sodium; cyanure *m* de soude — **d** Natriumcyanid *n*; Cyannatrium *n* — **n** natrium-

cyanide *n* — **e** cianuro *m* de sodio; cianuro *m* sódico; cianuro *m* de sosa — **i** cianuro *m* di sodio; cianuro *m* sodico

12688 sodium dichromate; sodium bichromate; bichromate of soda — Red or orange crystalline solid, soluble in water, insoluble in alcohol. Used as an oxidizing agent in the manufacture of dyes and inks, in the tanning of leather and in electroplating.

Formula: $Na_2Cr_2O_7 \cdot 2H_2O$.

f bichromate *m* de sodium; bichromate *m* de soude — **d** Natriumbichromat *n*; Natriumdichromat *n*; doppeltkohlensaures Natrium *n*; rotes chromsaures Natrium; Natriumpyrochromat *n* — **n** natriumbichromaat *n* — **e** dicromato *m* de soda; bicromato *m* de soda; dicromato *m* sódico — **i** bicromato *m* di sodio; bicromato *m* sodico

12689 sodium dioxide — See sodium peroxide.

12690 sodium formate — White, slightly hygroscopic, crystalline powder, soluble in water, slightly soluble in alcohol, insoluble in ether.

Formula: $NaCHO_2$.

f formiate *m* de sodium — **d** Natriumformiat *n*; ameisensaures Natrium *n* — **n** natriumformiaat *n* — **e** formiato *m* de sodio; formiato *m* sódico; formiato *m* de sosa — **i** formiato *m* di sodio; formiato *m* sodico

12691 sodium hexametaphosphate — Term often improperly used to refer to certain sodium polyphosphates, used in water softening, leather tanning.

Formula: $(NaPO_3)_6$.

f hexamétaphosphate *m* de sodium — **d** Natriumhexametaphosphat *n* — **n** natriumhexametafosfaat *n* — **e** hexametafosfato *m* de sodio; hexametafosfato *m* sódico; hexametafosfato *m* de sosa — **i** esametafosfato *m* di sodio; esametafosfato *m* sodico

12692 sodium hydrate — See sodium hydroxide.

12693 sodium hydrogen sulphite — See sodium bisulphite.

12694 sodium hydrosulphite; hydrosulphite — Light lemon-coloured powder, soluble in water, insoluble in alcohol. Used as a ready compound for bleaching mechanical pulp.

Formula: $Na_2S_2O_4$.

f hydrosulfite *m* de sodium — **d** Natriumhydrosulfit *n* — **n** natriumhydrosulfiet *n* — **e** hidrosulfito *m* de sodio; hidrosulfito *m* sódico; hidrosulfito *m* de sosa — **i** idrosolfito *m* di sodio; idrosolfito *m* sodico

12695 sodium hydroxide; sodium hydrate; caustic soda; white caustic; caustic lye — White, deliquescent pieces, soluble in water, alcohol and glycerol. Used to remove old images from offset pressplates. Latin: natrium hydricus; natrium causticum; hydras natricus. TL-value: 2 mg/m³.

Formula: NaOH.

f hydroxyde *m* de sodium; hydrate *m* de sodium; hydrate *m* de soude; soude *f* caustique; soude *f* à l'alcool; lessive *f* de soude — **d** Natriumhydroxyd *n*; Natriumhydrat *n*; kaustische Soda *f*;

kaustisches Natron *n*; Natronlauge *f*; Ätznatron *n*
— **n** natriumhydroxyde *n*; natronhydraat *n*;
caustische soda *fm*; natronloog *fmn*; etsnatron *n*
— **e** hidróxido *m* de sodio; hidróxido *m* sódico;
hidróxido *m* de sosa; sosa *f* cáustica — **i** idrossido
m di sodio; idrossido *m* sodico; idrato *m* di sodio;
soda *f* caustica

**12696 sodium hypochlorite; Labarraque's
solution; eau de Labarraque** — Salt unstable in
air unless mixed with sodium hydroxide. Soluble
in cold water, decomposed by hot water; used for
bleaching of paper pulp. Latin: hypochloris
natricus.
Formula: NaOCl.
f hypochloride *m* de sodium; chlorure *m* de soude;
eau *f* de Labarraque — **d** Natriumhypochlorit *n*;
unterchlorigsaures Natrium *n*; Labarraquesche
Flüssigkeit *f* — **n** natriumhypochloriet *n*; onder-
chlorigzuur natron *n*; Labarraque-water *n* — **e**
hipoclorito *m* de sodio; hipoclorito *m* sódico; hipoclorito *m* de sosa; agua *f* de Labarraque; solución
f de Labarraque — **i** ipoclorito *m* di sodio; ipoclorito *m* sodico; acqua *f* di Labarraque

12697 sodium hyposulphite — See hypo.

12698 sodium indigo tindisulphonate — The USP
XVII name for *indigo carmine.

12699 sodium iodate
f iodate *m* de sodium — **d** Natriumjodat *n* — **n**
natriumjodaat *n* — **e** yodato *m* de sodio; yodato
m sódico; yodato *m* de sosa — **i** iodato *m* di
sodio; iodato *m* sodico

12700 sodium iodide; anhydrous sodium iodide —
White cubical crystals or powder, soluble in water,
alcohol and glycerol. Used in photography. Latin:
iodetum natricum.
Formula: $NaI \cdot 2H_2O$.
f iodure *m* de sodium — **d** Natriumjodid *n*; Jod-
natron *n* — **n** natriumjodide *n* — **e** yoduro *m* de
sodio; yoduro *m* sódico; yoduro *m* de sosa — **i**
ioduro *m* di sodio; ioduro *m* sodico

12701 sodium kraft paper — See kraft paper.

**12702 sodium metabisulphite; sodium pyro-
sulphite** — Chief constituent of commercial dry
sodium bisulphite, with which most of its properties and uses are practically identical. Latin:
natricum metabisulfurosum.
Formula: $Na_2S_2O_5$.
f métabisulfite *m* de sodium; pyrosulfite *m* de
sodium — **d** Natriummetabisulfit *n*; Natriumpyrosulfit *n* — **n** natriummetabisulfiet *n* — **e** metabisulfito *m* de sodio; metabisulfito *m* de sosa; metabisulfito *m* sódico — **i** metabisolfito *m* di sodio;
metabisolfito *m* sodico

12703 sodium metasilicate — See sodium silicate.

12704 sodium nitrate; soda nitre *GB*; **soda niter**
US — Colourless, transparent crystals, soluble in
water and glycerol, slightly soluble in alcohol.
Latin: nitras natricus.
Formula: $NaNO_3$.
f nitrate *m* de sodium; nitrate *m* de soude — **d**
Natriumnitrat *n*; Natronsalpeter *m* — **n** natrium-
nitraat *n* — **e** nitrato *m* de sodio; nitrato *m*

sódico; nitrato *m* de soda — **i** nitrato *m* di sodio;
nitrato *m* sodico

12705 sodium oxalate — White crystalline
powder, soluble in water, insoluble in alcohol.
Formula: $Na_2C_2O_4$.
f oxalate *m* de sodium — **d** Natriumoxalat *n* — **n**
natriumoxalaat *n* — **e** oxalato *m* de sodio; oxalato
m sódico; oxalato *m* de sosa — **i** ossalato *m* di
sodio; ossalato *m* sodico

12706 sodium paraminophenol — See rodinal.

12707 sodium perborate — White, crystalline
solid, moderately soluble in water (with decomposition) and glycerol. Used as a bleaching agent.
Formula: $NaBO_3 \cdot H_2O_2 \cdot 3H_2O$ or
$NaBO_3 \cdot 4H_2O$.
f perborate *m* de sodium — **d** Natriumperborat *n*
— **n** natriumperboraat *n* — **e** perborato *m* sódico
— **i** perborato *m* di sodio

12708 sodium peroxide; sodium dioxide —
Yellowish white powder, turning yellow when
heated. Soluble in cold water.
Formula: Na_2O_2.
f peroxyde *m* de sodium; bioxyde *m* de sodium —
d Natriumperoxyd *n* — **n** natriumperoxyde *n* — **e**
peróxido *m* de sodio; peróxido *m* sódico; peróxido
m de sosa — **i** perossido *m* di sodio; perossido *m*
sodico

**12709 sodium phosphate; tribasic sodium phos-
phate; trisodium phosphate; trisodium orthophosphate** — Colourless crystals, soluble in water.
Latin: phosphas natricus.
Formula: $Na_3PO_4 \cdot 12H_2O$.
f phosphate *m* trisodique; phosphate *m* tribasique
de sodium — **d** (tertiäres) Natriumphosphat *n*;
normales Natriumphosphat *n*; Trinatrium-
phosphat *n*; dreibasisches Natriumphosphat *n* —
n (driebasisch) natriumfosfaat *n*; trinatrium-
fosfaat *n* — **e** fosfato *m* sódico tribásico; fosfato
m trisódico — **i** fosfato *m* di sodio tribasico

**12710 sodium-potassium tartrate; Rochelle salt;
Seignette salt** — Colourless or white solid, soluble
in water, used in the gravure process.
Formula: $KNaC_4O_6 \cdot 4H_2O$.
f tartrate *m* de sodium et de potassium; sel *m* de
Seignette — **d** Natriumkaliumtartrat *n* — **n**
natrium-kaliumtartraat *n*; seignettezout *n* — **e**
tartrato *m* potásico de sodio — **i** tartrato *m* di
sodio e potassio

12711 sodium pyroborate — See sodium borate.

12712 sodium pyrosulphite — See sodium
metabisulphite.

12713 sodium rhodanate — See sodium
thiocyanate.

12714 sodium rhodanide — See sodium
thiocyanate.

12715 sodium sesquicarbonate; sesqui *sl.* —
Water softening agent and detergent.
f sesquicarbonate *m* de sodium — **d** Natriumsesquikarbonat *n* — **n** natriumsesquicarbonaat *n*
— **e** sesquicarbonato *m* sódico — **i** sesquicarbonato *m* di sodio

12716 sodium silicate; sodium metasilicate;

soluble glass; liquid glass; water glass; silicate of soda — Lumps of greenish glass soluble in steam under pressure, white powders of varying degrees of solubility or liquids cloudy or clear and varying from highly fluid to extreme viscosity. Miscible with some polyhydric alcohols, partially miscible with primary alcohols and ketones. Formulas varying in ratio from $Na_2O \cdot 3.75SiO_2$ to $2Na_2O \cdot SiO_2$ and with various proportions of water.
f silicate *m* de sodium; silicate *m* de soude; verre *m* soluble; verre *m* à eau — **d** Natriumsilikat *n*; Natronsilikat *n*; Wasserglas *n*; Natronwasserglas *n* — **n** natriumsilicaat *n*; waterglas *n*; natronwaterglas *n* — **e** silicato *m* de sodio; silicato *m* sódico; silicato *m* de sosa; vidrio *m* soluble — **i** silicato *m* di sodio; silicato *m* sodico

12717 sodium stearate — White powder, soluble in hot water and hot alcohol, slowly soluble in cold water and cold alcohol, insoluble in many organic solvents.
Formula: $NaC_{18}H_{35}O$.
f stéarate *m* de sodium — **d** Natriumstearat *n* — **n** natriumstearaat *n* — **e** estearato *m* sódico — **i** stearato *m* sodico

12718 sodium sulphantimonate — See sodium-thioantimonate.

12719 sodium sulphate; anhydrous sodium sulphate; exsiccated sodium sulphate; Glauber's salt — White crystals or powder, soluble in water and glycerol, insoluble in alcohol. Used in manufacturing of kraft paper and paperboard. Latin: sulfas natricus; natrium sulfuricum. Glauber's salt: name derived from Johann R. Glauber, German chemist (1604-1668).
Formula: Na_2SO_4.
f sulfate *m* de sodium; sulfate *m* neutre de sodium; sulfate *m* de soude; sel *m* de Glauber — **d** Natriumsulfat *n*; schwefelsaures Natrium *n*; Glaubersalz *n* — **n** natriumsulfaat *n*; Glauberzout *n* — **e** sulfato *m* de sodio; sulfato *m* sódico; sulfato *m* de sosa; sal *f* de Glauber — **i** solfato *m* di sodio; solfato *m* sodico; sale *m* di Glauber

12720 sodium sulphide; sodium sulphuret; sulphide of soda — Yellow or brick-red lumps or flakes, soluble in water, slightly soluble in alcohol, insoluble in ether. Latin: natrium sulfuratum; sulfidum natricum.
Formula: Na_2S.
f sulfure *m* de sodium; monosulfure *m* de sodium; protosulfure *m* de sodium; sulfhydrate *m* de soude — **d** Natriumsulfid *n*; Schwefelnatrium *n*; Schwefelnatron *n* — **n** natriumsulfide *n*; zwavelnatrium *n* — **e** sulfuro *m* de sodio; sulfuro *m* sódico; sulfuro *m* de sosa — **i** solfuro *m* di sodio; solfuro *m* sodico

12721 sodium sulphite; sulphite of soda — White crystals or powder, soluble in water, sparingly soluble in alcohol. Used as a preservative for photographic developers (in anhydrous state). Latin: natrium sulfurosum; sulfis natricus.
Formula: Na_2SO_3 or $Na_2SO_3 \cdot 7H_2O$.

f sulfite *m* de sodium — **d** Natriumsulfit *n*; schwefligsaures Natrium *n*; schwefligsaures Natron *n* — **n** natriumsulfiet *n*; zwaveligzuur natron *n* — **e** sulfito *m* de sodio; sulfito *m* sódico; sulfito *m* de sosa — **i** solfito *m* di sodio; solfito *m* sodico

12722 sodium sulphocyanate — See sodium thiocyanate.

12723 sodium sulphocyanide — See sodium thiocyanate.

12724 sodium sulphuret — See sodium sulphide.

12725 sodium superoxide — Term mistakenly used to refer to sodium peroxide. Formula: NaO_2.

12726 sodium tetraborate — See sodium borate.

12727 sodium-thioantimonate; sodium sulphantimonate; Schlippe's salt; liver of antimony — Double sulphide of antimony.
Formula: $Na_3SbS_4 \cdot 9H_2O$.
f sulfoantimoniate *m* de sodium; sel *m* de Schlippe — **d** Natriumsulfoantimonat *n*; Schlippesches Salz *n* — **n** natriumthioantimoniaat *n* — **e** tioantimoniato *m* de sodio; tioantimoniato *m* sódico; tioantimoniato *m* de sosa — **i** solfoantimoniato *m* di sodio; solfoantimoniato *m* sodico; sale *m* di Schlippe

12728 sodium thiocyanate; sodium sulphocyanate; sodium sulphocyanide; sodium rhodanate; sodium rhodanide — Colourless, deliquescent crystals or white powder, soluble in water and alcohol. Poisonous.
Formula: $NaCNS$.
f sulfocyanure *m* de sodium; thiocyanate *m* de sodium; rhodanate *m* de sodium — **d** Natriumrhodanid *n*; Rhodannatrium *n* — **n** natriumrhodanide *fm* — **e** tiocianato *m* de sodio; tiocianato *m* sódico; tiocianato *m* de sosa — **i** tiocianato *m* di sodio; tiocianato *m* sodico

12729 sodium thiosulphate — See hypo.

12730 soft — Pertaining to a photographic image showing detail and gradation, but lacking proper contrast.
f doux — **d** weich; kontrastlos; grau — **n** zacht; contrastloos; vlak — **e** sin contraste — **i** senza contrasti; morbido

12731 soft copy — Representation of text, or a portion thereof, electronically generated on a screen, such as self-scan display or a TV monitor.

12732 soft dot — Dot where the halation around the edge is excessive and almost equal in area to the dot itself. When the amount of halation is barely noticeable and the dot is very sharp, the dot is called hard. Sometimes improperly used to indicate the case with which the dots yield to chemical reduction. A contact negative has a harder dot than a negative made in the camera.
f point *m* mou — **d** unscharfer Punkt *m*; weicher Punkt *m* — **n** zachte punt *m* — **e** punto *m* débil; punto *m* fofo — **i** punto *m* sfumato; punto *m* morbido

12733 softener — See plasticizer.

12734 softening agent — See plasticizer.

12735 softening point — Temperature at which a

substance becomes plastic.

f point *m* de ramollissement; température *f* de ramollissement — **d** Erweichungspunkt *m*; Erweichungstemperatur *f* — **n** verwekingspunt *n* — **e** punto *m* ablandador; punto *m* de ablandamiento; punto *m* de reblandecimiento; temperatura *f* de reblandecimiento — **i** punto *m* di rammollimento; punto *m* di ammorbidimento; temperatura *f* di ammorbidimento

12736 soften *v* **the ink** — To reduce, to thin, to make the ink more fluid.

f assouplir l'encre — **d** die Farbe geschmeidiger machen — **n** de inkt zacht (smeuig) maken — **e** suavizar la tinta; hacer más fluida la tinta — **i** rendere l'inchiostro scorrevole

12737 soft focus — Effect produced by the use of a special lens to create soft outlines where light areas tend to spread into dark areas.

f flou *m* artistique — **d** Weichzeichnung *f* — **n** soft focus *m* — **e** enfoque *m* suave — **i** fuoco *m* morbido

12738 soft fold — See lapped.

12739 soft-ground etching; vernis mou — Etching method to produce textured and crayon-like effects. The etching ground does not become hard nor does it adhere strongly to the plate. It is sticky and can be easily lifted by drawing on paper put over it, by textured objects, or by the etcher's fingers.

f vernis-mou *m*; gravure *f* au vernis-mou — **d** Weichgrundverfahren *n*; Vernismou-Radierung *f*; Vernismou *f*; Durchdruckverfahren *n* — **n** vernismou *n* — **e** grabado *m* vernis-mou; grabado *m* al barniz blando — **i** incisione *f* alla vernice molle; incisione *f* su vernice molle; vernice *f* molle

12740 soft ink *coll.* — Printing ink exhibiting low tack and a very low resistance to flow (without thixotropy).

f encre *f* douce; encre *f* molle — **d** weiche Druckfarbe *f*; wenig viskose Farbe *f* — **n** zachte inkt *m*; boterachtige inkt *m*; weinig viskeuze inkt *m* — **e** tinta *f* líquida; tinta *f* fluida — **i** inchiostro *m* senza tiro; inchiostro *m* scorrevole

12741 soft lumps — Agglomerates of fibres improperly disintegrated, broken up by treatment in the beater or at the jordan. See also lumps, hard lumps, colour lumps.

f pâtons *mpl* mous — **d** weiche Knoten *mpl* — **n** zachte stofknopen *mpl*

12742 soft packing — Covering of printing cylinder with resilient bulky paper for printing formes containing old, worn types and stereotypes, used for newspaper and similar work.

f habillage *m* mou — **d** weicher Aufzug *m* — **n** zachte legger *m* — **e** cama *f* blanda; padrón *m* blando — **i** rivestimento *m* morbido

12743 soft sized — See slack sized.

12744 software — Manufacturer's programs supplied with the hardware of a system. Subsequently extended by user's own programs. Includes compilers, assemblers, library routines, diagnostic routines and executive programs, and sometimes manuals and systems diagrams. See also composition software.

f programmerie *f*; substance *f*; software *m* — **d** Grundprogramme *npl*; Programmausrüstung *f* — **n** programmatuur *f*; programmapakket *n*; software *m* — **e** programas *mpl* y sistemas de programación; software *m* — **i** programmatura *f*; software *m*

12745 software module — Segment of a program. Can be assembled into different programs to meet a variety of different processing requirements.

12746 soft water

f eau *f* douce — **d** weiches Wasser *n* — **n** zacht water *n* — **e** agua *f* dulce — **i** acqua *f* dolce

12747 softwood; coniferous wood — See also conifer.

f bois *m* résineux; bois *m* conifère — **d** Nadelholz *n* — **n** naaldhout *n*; zacht hout *n* — **e** madera *f* de (las) coníferas; madera *f* de pino — **i** conifere *m*; legno *m* di conifere

12748 softwood pulp; coniferous wood pulp

f pâte *f* de conifères; pâte *f* de bois résineux — **d** Nadelholzzellstoff *m*; Nadelholzhalbstoff *m*; Halbstoff *m* aus Nadelholz — **n** naaldhoutcelstof *fm*; naaldhoutpulp *fm*; halfstof *fm* uit naaldhout — **e** pasta *f* de coníferas; pulpa *f* de coníferas — **i** pasta *f* di conifere; pasta *f* di aghifoglie

12749 sol — Liquid colloidal suspension or solution. Colloidal dispersion that is a liquid.

f sol *m* — **d** Sol *m* — **n** sol *m* — **e** sol *m* — **i** sol *m*

12750 Solander case — Case for fine books, made of two parts, one to slide over the collar of the other. The spine is executed in the same way as that of the book. Named after Daniel Charles Solander (1736-1782), Swedish naturalist. Sometimes called a shoulder case.

f boîte *f* en forme de livre — **d** Schachtel *f* in Buchform — **n** Solander-doos *fm*; doos *fm* met hals — **e** caja *f* Solander; estuche *m* Solander; estuche *m* con cuello — **i** scatola *f* Solander; scatola *f* a forma di libro

12751 solarization — The *reversal of a photographic image by overexposure, fogging of the negative during development, etc.

f solarisation *f* — **d** Solarisation *f*; Bildumkehr *f* — **n** solarisatie *f*; beeldomkering *f* — **e** solarización *f* — **i** solarizzazione *f*

12752 solar spectrum — Band of rainbow colours produced by the ray of sunlight, bent or refracted by passing through a glass prism.

f spectre *m* solaire — **d** Sonnenspektrum *n* — **n** zonnespectrum *n* — **e** espectro *m* solar — **i** spettro *m* solare

12753 solder *v* — To unite two metallic objects or surfaces with molten solder or fusible alloys of lead and tin.

f braser; souder (au cuivre) — **d** löten — **n** solderen — **e** soldar — **i** brasare

12754 solenoid — Electric conductor wound as a helix with small pitch, or as two or more coaxial helices, so that current through the conductor

establishes a magnetic field within the conductor.
f solénoïde *m* — **d** Solenoid *n*; Magnetspule *f* — **n** solenoïde *f* — **e** solenoide *m* — **i** solenoide *m*

12755 solid — Full colour area printed from a smooth block unbroken by dots or other texture patterns.
f aplat *m*; fond *m*; teinte *f* de fond — **d** Vollfläche *f*; Volltonfläche *f* — **n** volvlak *n* — **e** fondo *m*; tinta *f* plana — **i** fondo *m* pieno

12756 solid — See solid matter.

12757 solid — See unleaded.

12758 solid block — See flat-tint plate.

12759 solid fibreboard — Pasted board of minimum substance of 1000 g/m², generally comprising a strong lining (kraft or similar material) suitable for the manufacture of packing cases. Definition according to ISO.
f carton *m* compact — **d** Starkpappe *f* — **n** hardbord *n* — **e** cartón *m* duro; cartón *m* compacto — **i** cartone *m* compatto

12760 solidify *v*
f se solidifier; se prendre en masse — **d** erstarren; fest werden — **n** stollen; vast worden — **e** solidificar — **i** solidificare

12761 solid logic technology — Miniaturized modules used in computers, which result in faster circuiting because of reduced distance for current to travel. Abbrev. SLT.
f technique *f* des circuits intégrés — **d** Technik *f* der integrierten Schaltkreise — **n** techniek *f* der geïntegreerde schakelingen — **e** tecnología *f* de estado sólido — **i** tecnica *f* dei circuiti integrati

12762 solid matter; close matter; solid — Type matter set without leads between the lines, also type matter with few quads in.
f composition *f* pleine; matière *f* pleine; plein *m*; composition *f* compacte — **d** kompresser Satz *m*; undurchschossener Satz *m* — **n** plat zetsel *n*; compact zetsel *n* — **e** composición *f* compacta; composición *f* corriente; composición *f* maciza; composición *f* apretada; composición *f* sin interlíneas; composición *f* corrida; composición *f* seguida; composición *f* llena; composición *f* serrada; composición *f* cargada; composición *f* metida; composición *f* mazorral; composición *f* apiñada; composición *f* amontonada — **i** composizione *f* compatta; composizione *f* non-interlineata; composizione *f* piena

12763 solid plate — Plate with an even printing surface and bearing no etched or engraved design. Used for printing solid tints or uniform depositions of ink in any colour.
f cliché *m* d'aplat — **d** Tonplatte *f* — **n** (ongerasterde) tintplaat *fm* — **e** clisé *m* plano; clisé *m* a masa llena — **i** cliscè *m* di fondo; cliché *m* do fondo

12764 solids content; dry solids content — Percentage of solid material contained in an ink formulation, e.g., pigment, extender, binder, plasticizer, wax, etc.
f teneur *f* en matière sèche; teneur *f* en substance solide — **d** Trockenstoffgehalt *m*; Feststoff-

gehalt *m*; Trockengehalt *m* — **n** gehalte *n* aan droge stof; droge-stofgehalte *n* — **e** contenido *m* de materias sólidas; contenido *m* de sólidos — **i** contenuto *m* in sostanza solida; contenuto *m* in secco; contenuto *m* in solido

12765 solid-state computer — See second-generation computer.

12766 solid state element — Component in a computer that depends on electric or magnetic phenomena in solids, e.g., in semiconductors and ferrites. They perform the same functions as vacuum tubes (valves).
f semiconducteur *m* — **d** Halbleiter *m* — **n** halfgeleider *m* — **e** semiconductor *m* — **i** semiconduttore *m*

12767 solidus; scratch comma; scratch; diagonal; oblique stroke; virgule; shilling mark; shilling stroke; slant; slash — Type sort consisting of a slant line (/), used between the parts of a fraction to separate numerator and denominator, to separate lines of poetry when quoted in run-in fashion and in places as %, c/o, etc.
f barre *f* transversale; barre *f* de fraction oblique; solidus *m* — **d** Schrägstrich *m*; Bruchstrich *m*; Schillingstrich *m* — **n** schuine streep *fm*; schuine breukstreep *fm*; Duitse komma *fmn* — **e** rasgo *m* diagonal; diagonal *m* para quebrados; línea *f* diagonal de quebrados — **i** sbarretta *f* inclinata; barra *f* oblique; barra *f* di frazione; stanghetta *f*

12768 solubility
f solubilité *f* — **d** Löslichkeit *f*; Lösbarkeit *f* — **n** oplosbaarheid *f* — **e** solubilidad *f* — **i** solubilità *f*

12769 soluble — Dissolvable in a liquid solvent.
f soluble — **d** löslich; auflösbar — **n** oplosbaar — **e** soluble; disoluble — **i** solubile

12770 soluble glass — See sodium silicate.

12771 solus — See full position.

12772 solution — Liquid containing dissolved solids.
f solution *f* — **d** Lösung *f* — **n** oplossing *f* — **e** solución *f* — **i** soluzione *f*

12773 solvent — Medium, usually liquid, to dissolve a substance to convert it to a liquid and place it in solution. Frequently an organic solvent. Also a thinner for certain adhesives, inks, paints, etc.
f solvant *m*; dissolvant *m* — **d** Lösungsmittel *n*; Lösemittel *n*; Löser *m* — **n** oplosmiddel *n* — **e** disolvente *m*; solvente *m LA* — **i** solvente *m*

12774 solvent coated paper — Paper coated by a solvent coating process.
f papier *m* couché au solvant — **d** mit Lösungsmittelstrich versehenes Papier *n* — **n** met een oplosmiddel gestreken papier *n* — **e** papel *m* estucado al solvente — **i** carta *f* patinata con miscele a base di solventi

12775 solvent coating — Coating a continuous paper web with resins or plastics dissolved in volatile solvents, and evaporation of the latter. Either the roll or the gravure process may be used.
f couchage *m* par solvant — **d** Lösungsmittel-

streichverfahren *n*; Beschichtung *f* mit in Lösungsmittel gelöstem Stoff — **n** strijkprocédé *n* met oplosmiddel — **e** estucado *m* con mesclas a base de disolventes — **i** patinatura *f* con miscele a base di solventi

12776 solvent naphtha; coal-tar naphtha — Mixture of hydrocarbons derived from coal with a boiling range of approx. 160-195 °C (sometimes called 160 ° benzol; strictly speaking not correct and hence not recommended).
f solvant-naphta *m* — **d** Solventnaphta *fn*; Lösungsbenzol *n*; Schwerbenzol *n* — **n** solvent-nafta *m* — **e** nafta *f* disolvente — **i** benzina *f* solvente

12777 solvent reclamation unit — See solvent recovery plant.

12778 solvent recovery — Recovery of evaporated solvent vapours. They are collected and led over absorptive carbon to which the solvent attaches itself and from which it can be detached in liquid form by the application of steam for collection or distillation for subsequent use.
f recouvrement *m* de solvants; récupération *f* du solvant — **d** Lösemittelwiedergewinnung *f*; Lösemittelgewinnung *f* — **n** terugwinning *f* van oplosmiddelen — **e** recuperación *f* de solventes; recuperación *f* de disolventes — **i** ricupero *m* dei solventi

12779 solvent recovery plant; solvent reclamation unit
f installation *f* de récupération des solvants — **d** Wiedergewinnungsanlage *f*; Rückgewinnungsanlage *f* — **n** terugwinningsinstallatie *f* — **e** sección *f* de recuperación — **i** impianto *m* di recupero

12780 solvent release — Ability of a binder to influence the rate of evaporation of a solvent.
f rétention *f* de solvant — **d** Lösungsmittelrückhaltevermögen *n*; Lösungsmittelfreisetzung *f*; Lösungsmittelabgabe *f* — **n** vasthouden *n* van het oplosmiddel — **i** rilascio *m* di solvente

12781 solvent resistance
f résistance *f* aux solvants — **d** Lösemittelechtheit *f*; Beständigkeit *f* gegen Lösungsmittel — **n** bestendigheid *f* tegen oplosmiddelen — **e** resistencia *f* a los disolventes — **i** resistenza *f* ai solventi

12782 soot — Product of combustion of coal or wood and consisting basically of carbon with a proportion of tars, salts and ash.
f suie *f* — **d** Ruß *m* — **n** roet *n* — **e** hollín *m* — **i** fuliggine *f*

12783 sorbitol — White, odourless, crystalline powder, soluble in water, slightly soluble in methanol, ethanol, acetic acid, phenol and acetamide. Almost insoluble in most other organic solvents.
Formula: $C_6H_8(OH)_6$.
f sorbite *f*; sorbitol *m* — **d** Sorbit *n* — **n** sorbitol *n*; sorbiet *n* — **e** sorbitol *m*; sorbita *f* — **i** sorbitolo *m*

12784 sort *v* — In computer terminology, to arrange data into logical order, e.g. numeric or alphabetic. Most computer manufacturers offer library routines for sorting which are supplied to users.
f trier — **d** sortieren — **n** sorteren — **e** clasificar — **i** selezionare

12785 sorting; assorting — Examination of sheets of paper or board to reject the defective ones. Also the separation into groups according to the quality of rags or waste paper to be used in the manufacture of paper or board.
f triage *m*; revoyage *m* — **d** Sortierung *f*; Sichtung *f* — **n** sorteren *n* — **e** clasificación *f*; escogido *m*; cribado *m* — **i** cernita *f* della carta; assortimento *m* della carta

12786 sorts — Matrices of types of various size or face, as distinct from a complete fount, held in reserve on a typesetting machine. A single character of type is a sort. See also pie characters.
f matrices *fpl* (spéciales) à la main — **d** Handmatrizen *fpl* (an der Setzmaschine); Einhänger *mpl* — **n** handmatrijzen *fpl* — **e** matrices *fpl* de mano; matrices *fpl* auxiliares; auxiliares *fpl*; auxiliares *fpl* de matriz *Pe*; matrices *fpl* sueltas; matrices *fpl* suplementarias; matrices *fpl* de contracaja — **i** matrici *fpl* a mano; matrici *fpl* speciali

12787 sorts case — See fount case.

12788 sorts channel — Channel in the typesetting machine which receives all matrices which are not combined (*combination) to run in the magazines (technically expressed "do not run keyboard"), such as special characters, signs, borders, etc.
f canal *m* de casseau — **d** Handmatrizenkanal *m* — **n** handmatrijzenkanaal *n* — **e** canal *m* de suertes — **i** canale *m* per matrici a mano; canale *m* per matrici speciali

12789 sorts stacker; pie and sorts mechanism; pie stacker — Device on the typesetting machine for assembling matrices in a holder after they have been passed through the sorts channel.
f assembleur *m* de casseau; casseau *m* — **d** Handmatrizenmechanismus *m*; Handmatrizensammler *m*; Einhängersammler *m* — **n** handmatrijzen-verzamelaar *m* — **e** acomodador *m* de suertes; caja *f* de suertes; caja *f* para matrices sueltas; caja *f* para guardar las matrices suplementarias; contracaja *f* — **i** dispositivo *m* di raccolta matrici a mano; dispositivo *m* di raccolta matrici speciali

12790 sos bag — See self-opening satchel bag.

12791 SOS bag — See automatic self-opening bag.

12792 source language — Language in which a program is originally written by the programmer, as opposed to object or machine language. Designed to make the specification of the problem, procedure or program easier.
f langage *m* original — **d** Ursprungssprache *f* — **n** brontaal *fm* — **e** lenguaje *m* fuente — **i** linguaggio *m* originale

12793 source program — Computer program

written in symbolic or high-level language which is automatically translated into machine code by an assembler or compiler before running.

f programme *m* original — **d** Quellenprogramm *n*; Ursprungsprogramm *n* — **n** uitgangsprogramma *n*; oorspronkelijk programma *n* — **e** programa *m* fuente — **i** programma *m* originale

12794 Soxhlet extractor — See extractor.

12795 soya bean oil; soybean oil; Chinese bean oil — Pale, yellow, fixed oil, soluble in alcohol, ether, chloroform and carbon disulphide, semidrying, used in the preparation of printing ink vehicles and in plate making.

f huile *f* de soja; huile *f* de soya — **d** Sojaöl *n*; Sojabohnenöl *n* — **n** soja-olie *fm* — **e** aceite *m* de semilla de soja; aceite *m* (de habas) de soja — **i** olio *m* di soia

12796 soya-bean plate coating; soybean plate coating — Soya bean, introduced to lithographers about 1949, is a substitute for albumen.

f couche *f* au soja — **d** Plattenbeschichtung *f* mit Soja — **n** kopieerlaag *fm* uit soja — **e** capa *f* de emulsión de soja — **i** strato *m* di copia a base di soia

12797 soybean oil — See soya bean oil.

12798 soybean plate coating — See soya-bean plate coating.

12799 space — Distance between one letter and another, between words or between different areas of print.

f espace *f* (masculin comme intervalle de temps) — **d** Zwischenraum *m*; Wortzwischenraum *m*; Raum *m* — **n** ruimte *f*; tussenruimte *f*; wit *n* — **e** espacio *m* — **i** spazio *m*

12800 space — Piece of type metal of varying thickness, to separate one word from another in a line of type. Spaces are of the same size in points as the types used and usually fractions of an em space, thus an em space is one half, a thick space is one third, a middle space is one quarter, and a thin space is one fifth (terms used in hand setting). In Monotype composition and film setting, spaces are usually referred to by their unit width. The thin space used in line-casting machines is one third or one quarter of the em space. Hair spaces vary in width from one eighth to one twelfth of an em and are used between characters.

f espace *f* (masculin comme intervalle de temps) — **d** Spatium *m* (*pl* Spatien) — **n** spatie *f* — **e** espacio *m* — **i** spazio *m*

12801 space — See also magnetic encoded cheques.

f pas *m* de caractère — **d** Zwischenraum *m* — **n** spatie *f* — **e** espacio *m* — **i** spazio *m*

12802 spaceband — Pair of steel wedges on the typesetting machine used between words of a line of matrices to spread and justify the line to the required width.

f espace-bande *f* — **d** Spatienkeil *m*; Ausschließkeil *m* — **n** wigspatie *f*; uitvulspatie *f*; spatie *f*; keilspatie *f depr.* — **e** espaciador *m*; espacio *m* de

cuña; espacio *m* movible — **i** cuneo *m* di giustificazione; spazio *m* giustificatore; spazio *m* mobile

12803 spaceband box — Part of the line-casting machine.

f boîte *f* des espaces-bandes — **d** Spatienkeilkasten *m* — **n** spatiebox *m* — **e** caja *f* de espacios (de cuña); caja *f* de (los) espaciadores; caja *f* del espaciado; espaciador *m Pe* — **i** scatola *f* degli spazi mobili

12804 spaceband key

f touche *f* d'espacement; touche *f* de justification — **d** Ausschlußtaste *f*; Keiltaste *f* — **n** spatietoets *m*; uitvultoets *m* — **e** tecla *f* de los espaciadores; tecla *f* de espaciado; tecla *f* de espaciación; palanca *f* de los espaciadores — **i** tasto *m* degli spazi

12805 space *v* closely

f composer serré — **d** eng halten — **n** nauw zetten — **e** componer cerrado; estrechar — **i** comporre strettamente

12806 space *v* equally — To space evenly.

f espacer régulièrement — **d** (Zeile) ausgleichen — **n** (regels) gelijk uitvullen — **e** igualar el espaciado — **i** spaziare regolarmente

12807 space mark — A proof correction mark. See app. no. 5.

f signe *m* de séparation — **d** Spatiumzeichen *n* — **n** scheidingsteken *n* — **e** signo *m* de separación — **i** segno *m* di separazione

12808 space *v* out — To increase the spacing between words or lines, to make full length, or to cover specified area.

f espacer; chasser — **d** ausschließen; spationieren (1 oder 1 1/2 P.); sperren (ab 2 P.); aussperren; (Zeilen) ausgleichen; (Zeilen) durchschießen — **n** spatiëren; uitdrijven; (regels) interliniëren — **e** entrerrenglonar; espaciar; dar blancos; distanciar — **i** interlineare; distanziare

12809 spacer ring — Disk with a tapering edge carried on a square shaft, used on the Typograph typesetting machine for spacing words.

f anneau-espace *m* — **d** Spatienring *m* — **n** spatiering *m* — **e** anillo *m* de justificación; espacio *m* anillo — **i** spazio *m* ad anello

12810 spacing — See letter spacing.

12811 spacing material

f blancs *mpl* — **d** Ausschluß *m*; Ausschlußmaterial *n* — **n** wit *n*; uitvulwit *n*; uitvulmateriaal *n* — **e** blancos *mpl*; material *m* de blancos; blanco *m* de relleno — **i** bianchi *mpl*

12812 Spanish — The Spanish alphabet consists of 27 letters including ch after c, ll after l, and ñ after n, k and w are not included. Ch, ll and rr must not be separated. The portion carried over to begin with a consonant. Notes of interrogation and exclamation are inverted before, and upright after their phrases. Caps must never be substituted for ñ. In alphabetical order, words beginning with ch form a chapter by itself after c. Double ll, as for example fallada, comes after falucho. Double ll is a chapter by itself after l.

Also ñ (n with tilde) is a chapter by itself after n and follows all other n's.

f espagnol *m*; espagnol *adj.* — **d** Spanische *n*; spanisch *adj.* — **n** Spaans *n*; Spaans *adj.* — **e** español *m*; español *adj.* — **i** spagnuolo *m*; spagnuolo *adj.*

12813 Spanish case — Type case with special arrangement of the boxes.

f casse *f* espagnole — **d** spanischer Schriftkasten *m*; Schriftkasten *m* mit spanischer Belegung — **n** Spaanse letterkast *fm*; Spaanse kast *fm* — **e** caja *f* española — **i** cassa *f* spagnuola

12814 Spanish grass — See esparto.

12815 Spanish leather — See cordovan.

12816 Spanish ñ — See ñ.

12817 Spanish white; bismuth subnitrate; basic bismuth nitrate — White powder used as an ink pigment.
Formula: $Bi(NO_3)OH \cdot 2H_2O$.

f blanc *m* d'Espagne; sous-nitrate *m* de bismuth — **d** Spanischweiß *n*; Wismutsubnitrat *n*; basisches Wismutnitrat *n* — **n** basische bismutnitraat *n* — **e** blanco *m* de España; subnitrato *m* de bismuto — **i** bianco *m* di Spagna; subnitrato *m* di bismuto

12818 spanner — See wrench.

12819 spare parts — Parts of machinery held in reserve for use when needed.

f pièces *fpl* de rechange; pièces *fpl* de réserve — **d** Ersatzteile *mpl* — **n** reserve-onderdelen *npl*; reserve-delen *npl* — **e** repuestos *mpl*; recambios *mpl*; piezas *fpl* de repuesto; piezas *fpl* de recambio — **i** pezzi *mpl* di ricambio

12820 spatula — See ink knife.

12821 special edition (of a book)

f édition *f* spéciale; édition *f* réservée — **d** Sonderausgabe *f* — **n** speciale uitgave *f* — **e** edición *f* especial; edición *f* reservada — **i** edizione *f* speciale

12822 special edition (of a newspaper)

f édition *f* spéciale; numéro *m* spécial — **d** Extra-Ausgabe *f* — **n** extra editie *f* — **e** edición *f* extraordinaria; edición *f* extra; número *m* especial — **i** edizione *f* straordinaria

12823 special purpose computer; specific purpose computer — Computer to cope with a restricted range of problems, e.g., for computer aided typesetting.

f calculateur *m* à usages spéciaux; ordinateur *m* spécialisé — **d** Spezialcomputer *m*; Spezialrechner *m* — **n** speciale computer *m*; computer *m* voor speciale doeleinden — **e** computadora *f* para usos speciales — **i** calcolatore *m* ad usi speciali

12824 specific address — See absolute address.

12825 specific addressing — See absolute coding.

12826 specific coding — See absolute coding.

12827 specific gravity; specific weight — Ratio of the weight of a given volume of a substance or solution compared with the weight of an equal volume of water at some definite temperature. Sometimes called relative density.

f poids *m* spécifique; gravité *f* spécifique — **d** spezifisches Gewicht *n*; Wichte *f* — **n** soortelijk gewicht *n*; specifiek gewicht *n* — **e** peso *m* específico; gravedad *f* específica; densidad *f* específica — **i** peso *m* specifico; gravità *f* specifica

12828 specific heat — Amount of heat required to raise a unit mass of a substance 1 ° of temperature at either constant pressure or constant volume.

f chaleur *f* spécifique — **d** spezifische Wärme *f* — **n** soortelijke warmte *f*; specifieke warmte *f* — **e** calor *m* específico — **i** calore *m* specifico

12829 specific inductive capacity — See dielectric constant.

12830 specific purpose computer — See special purpose computer.

12831 specific volume — Volume of unit weight of a substance, as cubic feet per pound, or gallons per pound, but more frequently millilitres per gram. The reciprocal of density.

f volume *m* spécifique — **d** spezifisches Volumen *n*; Eigenvolumen *n* — **n** soortelijk volume *n*; specifiek volume *n* — **e** volumen *m* específico — **i** volume *m* specifico

12832 specific volume — See bulk.

12833 specific weight — See specific gravity.

12834 specimen; specimen sheet — Representative sheet of paper or other material taken during sampling, printing or manufacture.

f échantillon *m*; feuille-échantillon *f* — **d** Probemuster *n*; Probestück *n*; Probeblatt *n*; Prüfblatt *n*; Muster *n*; Musterbogen *m* — **n** monster *n*; monstervel *n* — **e** muestra *f*; hoja *f* muestra — **i** campione *m*; foglio *m* campione

12835 specimen copy

f numéro *m* spécimen — **d** Probenummer *f*; Probeheft *n* — **n** proefexemplaar *n*; proefnummer *n* — **e** número *m* espécimen — **i** numero *m* di saggio; numero *m* di prova

12836 specimen page — Proof page which shows type character, number of lines, headings, the width to which the matter is set, etc., sent to the author or the publisher.

f page *f* modèle — **d** Probeseite *f* — **n** proefpagina *fm* — **e** página *f* muestra; página *f* de muestra — **i** pagina *f* di prova

12837 specimen sheet — See clean sheet.

12838 specimen sheet — See specimen.

12839 speckle — In rotogravure printing, the result of missing dots or skipping dots, due largely to surface roughness and to the cells in the photogravure plate failing to print. On uncoated paper, skips or partial skips give rise to a general appearance of unevenness. On smooth-coated paper, the skips can stand out, particularly where a dot process is used. Whereas speckle occurs in highlight regions, *mottle is a characteristic failing of solid areas, to which it imparts an unpleasantly uneven appearance.

f impression *f* galeuse; impression *f* grainée — **d** Nichtausdrucken *n*; unruhiger Ausdruck *m*; gekörntes Aussehen *n* — **n** niet uitdrukken *n*;

rauw uitdrukken *n* — **e** impresión *f* sarnosa — **i** stampa *f* granulosa

12840 spectral characteristics — Properties of light or colour due to the relative proportions and distribution of the wavelengths composing them.

f propriétés *fpl* spectrales — **d** Spektraleigenschaften *fpl*; Spektralbereich *m* — **n** spectrale eigenschappen *fpl* — **e** características *fpl* espectrales — **i** caratteristiche *fpl* spettrali

12841 spectral colour — Colour represented by a point on the chromaticity diagram that lies on a straight line between the spectrum locus and the achromatic point.

f couleur *f* spectrale — **d** Spektralfarbe *f* — **n** spectrale kleur *fm* — **e** color *m* espectral — **i** colore *m* spettrale

12842 spectral energy distribution — Energic composition of light source.

f répartition *f* spectrale de l'énergie; répartition *f* spectrale d'énergie — **d** spektrale Energieverteilung *f*; Strahlungsfunktion *f* — **n** spectrale energie-verdeling *f* — **e** distribución *f* de la energía espectral — **i** distribuzione *f* spettrale dell'energia

12843 spectral sensitivity — 1. Sensitivity of a detector measured for narrow spectral bands throughout the spectrum. 2. Sensitivity of a photoelectric device in relation to the wavelength of the incident radiant energy. 3. Emitted radiant-power wavelength distribution of a luminescent screen under a given condition of excitation.

f sensitivité *f* spectrale; sensibilité *f* spectrale — **d** spektrale Empfindlichkeit *f* — **n** spectrale gevoeligheid *f* — **e** sensibilidad *f* espectral — **i** sensibilità *f* spettrale

12844 spectrophotometer — Instrument to measure print colour by comparing it to a standard white reflecting surface. Recording spectrophotometers employ a photoelectric cell to measure colour intensity and print a continuous curve record on charts.

f spectrophotomètre *m* — **d** Spektrophotometer *m* — **n** spectrofotometer *m* — **e** espectrofotómetro *m* — **i** spettrofotometro *m*

12845 spectrophotometry — Photometric comparison between parts of spectra.

f spectrophotométrie *f* — **d** Spektralphotometrie *f* — **n** spectrofotometrie *f* — **e** espectrofotometría *f* — **i** spettrofotometria *f*

12846 spectrum — Series of images formed when a ray of light is subjected to dispersion, more specifically, the rainbow-like band of colours resulting when a ray of white light is separated into its constituent colours by passing through a prism or diffracting grating. The range of visible wavelengths lies between 400 and 700 millimicrons.

f spectre *m* — **d** Spektrum *n* — **n** spectrum *n* — **e** espectro *m* — **i** spettro *m*

12847 spectrum lines — See Fraunhofer's lines.

12848 specular gloss — See gloss.

12849 specular reflection — Reflection which causes a surface to appear somewhat like a mirror. Specularly reflected light is that reflected from a surface at an angle equal to that at which the incident light strikes the surface.

f réflexion *f* spéculaire — **d** Spiegelreflexion *f* — **n** spiegelreflectie *f*; spiegelende reflectie *f* — **e** reflexión *f* especular — **i** riflessione *f* speculare

12850 speed control

f réglage *m* de vitesse — **d** Geschwindigkeitsregelung *f* — **n** snelheidsregeling *f* — **e** regulación *f* de velocidad — **i** regolazione *f* di velocità

12851 speed of an emulsion; film speed — Degree of sensitivity to light. The time factor in sensitivity of light. The amount of moisture picked up by lithographic plate coating solutions affects their speed of action; a 30 ° change in humidity can make the speed of a solution twice as fast. For this reason platemaking departments are frequently air-conditioned.

f rapidité *f* d'une émulsion — **d** Lichtempfindlichkeit *f* einer Emulsion — **n** snelheid *f* van een emulsie; gevoeligheid *f* van een emulsie — **e** sensibilidad *f* a la luz de una emulsión — **i** grado *m* di sensibilità alla luce di un'emulsione

12852 speed of machine — See machine speed.

12853 speed of rotation

f vitesse *f* de rotation; nombre *m* de tours — **d** Drehzahl *f*; Tourenzahl *f* — **n** toerental *n*; aantal *n* toeren; aantal *n* omwentelingen — **e** velocidad *f* de rotación; número *m* de revoluciones — **i** velocità *f* di rotazione; numero *m* di giri

12854 speed-up *sl.* — Increase in output resulting from changes in methods and conditions.

12855 speed *v* **up** — See accelerate *v*.

12856 spelling; orthography — The art of writing words with the proper letters, according to accepted use. Correct spelling. Unless otherwise authorized, the US Government Printing Office follows Webster's spelling. In Germany use is made of the orthographical method of Konrad Duden (1872).

f orthographie *f* — **d** Rechtschreibung *f* — **n** spelling *f*; schrijfwijze *f* — **e** ortografía *f*; deletreo *m* — **i** ortografia *f*

12857 spelling mistake; mistake in spelling; misspelling — See also clerical error, compositor's error.

f faute *f* d'orthographe — **d** orthographischer Fehler *m*; Schreibfehler *m* — **n** spelfout *fm* — **e** falta *f* ortográfica; falta *f* de ortografía — **i** errore *m* ortografico

12858 spelt in full

f écrit en toutes lettres; composé en toutes lettres — **d** vollständig ausgeschrieben; in Worten ausgeschrieben — **n** voluit geschreven; voluit gezet — **e** componer en letras — **i** scritto per esteso; scritto in tutte lettere; tutto testo

12859 spermaceti oil; sperm oil; sperm whale oil — Light yellow liquid wax, soluble in chloroform, ether and benzene.

f huile *f* de spermacéti; huile *f* de baleine — **d** Spermacetöl *n*; Spermwalöl *n*; Walratöl *n* — **n**

spermolie *fm*; spermaceetolie *fm* — **e** aceite *m* de esperma; aceite *m* de cachalote — **i** olio *m* di spermaceti

12860 sperm oil — See spermaceti oil.

12861 sperm whale oil — See spermaceti oil.

12862 spherical aberration — Inability of a photographic lens to convey marginal (not oblique) rays to a point at the same distance as the central rays, manifesting itself by impairment of contrast and definition in the projected image.
f aberration *f* sphérique — **d** sphärische Abweichung *f*; sphärische Aberration *f* — **n** sferische aberratie *f* — **e** aberración *f* esférica; aberración *f* de esferidad — **i** aberrazione *f* sferica

12863 spider — Multiple arm structure supporting reels of paper.
f trèfle *m*; dérouleur *m* en étoile; dérouleur *m* à trois bras; étoile *f* — **d** Rollenstern *m* — **n** rollenster *fm* — **e** trébol *m*; portabobina *m* en estrella — **i** portabobine *m* a stella; portabobine *m* a tre braccia

12864 spike — Slang term for *bodkin, however, its major use is for a pointed instrument on which rejected copy is impaled. See also copy hook.

12865 spike — Transient of short duration, comprising part of a pulse, during which the amplitude considerably exceeds the average amplitude of the pulse.
f pic *m* — **d** Spitze *f* — **n** piek *fm* — **e** pico *m* — **i** guizzo *m*

12866 spike — See bodkin.

12867 spike *v* — To force the point of a bodkin between spaces or quads to make the forme lift. Not a good or economical practice.
f piquer à la pointe — **d** mit der Ahlenspitze anstechen — **n** met de els omhoog steken — **e** agujerear

12868 spike hygrometer — See sword hygroscope.

12869 spindle — Shaft that extends through the core of a reel of paper for running in a reel stand.
f arbre *m* — **d** Spindel *f*; Rollstange *f* — **n** spil *fm*; rolspil *fm*; rolstang *fm* — **e** mandril *m*; alma *f*; núcleo *m*; huso *m*; vástago *m* — **i** albero *m* portabobine

12870 spindleless reel stand — Reel stand supporting the reel on free-running cones at each side of the reel.
f dérouleur *m* sans mandrin; étoile *f* sans arbre — **d** spindellose Rollenlagerung *f*; achslose Rollenlagerung *f*; spindellose Aufhängung *f* der Papierrolle — **n** asloze rollenstandaard *m* — **e** portabobinas *m* sin mandril — **i** aspo *m* di bobinatore senza mandrino

12871 spindle, roller — See roller stock.

12872 spindle *v* **up** — To prepare a reel of paper for running in a reel stand by inserting the spindle or shaft and engaging the cones.
f engager l'arbre; enfiler l'arbre — **d** die Papierrolle aufspindeln; die Papierrolle aufachsen; die Papierrolle fertigmachen — **n** de rol papier in

gereedheid brengen — **e** introducir el mandril — **i** inserire l'albero

12873 spine; back; backbone; shelfback — Back of a book, on which the title is printed. See also back title.
f dos *m* (du livre) — **d** Rücken *m*; Buchrücken *m* — **n** rug *m* (van het boek); boekrug *m* — **e** lomo *m* (del libro) — **i** dorso *m* (del libro)

12874 spine — The main (curved) stroke of the letter S.
f panse *f* — **d** Abstrich *m* — **n** boog *m* — **e** panza *f* — **i** arco *m*

12875 spiral binding; mechanical binding — Method of binding in which the perforated sheets and cover are held by means of wire or plastics ringlets, which may or may not be wound in spirals.
f couture *f* spirale; reliure *f* spirale; reliure *f* mécanique — **d** Spiralheftung *f*; Spiralbindung *f*; Drahtrundbindung *f*; Spezialbindung *f* — **n** spiraalbinding *f* — **e** encuadernación *f* espiral — **i** legatura *f* a spirale

12876 spiral dandy — See dandy.

12877 spiral roller — Ink roller with spiral channels or grooves to distribute and control the ink supply.
f rouleau *m* à spirales; rouleau *m* nervuré; rouleau *m* cannelé en hélice — **d** Spiralfarbübertragungswalze *f* — **n** gespiraliseerde inktrol *fm* — **e** rodillo *m* (entintador) espiroidal — **i** rullo *m* (d'inchiostrazione) a spirale

12878 spiral roller — Idling roller with a spiral groove to smooth out unevenness in the web.
d Glättwalze *f* — **n** gladwals *fm* — **e** rodillo *m* alisador — **i** rullo *m* lisciatore a spirale

12879 spiral-winding machine — Machine to make rube shaped containers.
f machine *f* à fabriquer des tubes spiralés; machine *f* pour l'enroulement en spirale — **d** Hülsenwickelmaschine *f*; Spiralwickelmaschine *f* — **n** spiraal-wikkelmachine *f* — **e** máquina *f* para hacer mandriles; máquina *f* para seccionar los mandriles — **i** macchina *f* per fabbricare tubi avvolti a spirale; macchina *f* per la fabbricazione a spirale di tubi

12880 spirit ink — See flexographic ink.

12881 spirit of turpentine — See turpentine oil.

12882 spirits of turpentine — See turpentine oil.

12883 spiritus asper; rough breathing — See also breathings.
f esprit *m* rude — **d** Spiritus *m* asper — **n** spiritus asper *m* — **e** espíritu *m* áspero; espíritu *m* rudo — **i** espiritu *m* asper; espiritu *m* rude

12884 spiritus lenis; smooth breathing — See also breathings.
f esprit *m* doux — **d** Spiritus *m* lenis — **n** spiritus lenis *m* — **e** espíritu *m* suave — **i** espiritu *m* soave

12885 spirit varnish — Solution of copal or other gums in denatured alcohol applied on a direct roller coating machine to printed forms, such as can labels, for protection and also to increase gloss. Hard-sized, dense paper is required to resist the penetration of the varnish.

f vernis *m* à l'alcool — **d** Spritlack *m*; Spiritus-lack *m* — **n** spiritus-vernis *mn* — **e** barniz *m* al alcohol; laca *f* a base de alcohol; laca *f* en solución alcoholica — **i** vernice *f* ad alcool

12886 spirit wash — See alcohol wash.

12887 spits — Very small ink splashes in gravure printing caused by a burr that is formed by a fine edge given to the blade when working at a steep angle when it wears down.

12888 spits; coating splash — Defect of blade coated paper. Excess of coating in random spots on surface of coating. They may be short and tear drop shaped or long, narrow and acicular in shape. Low angle light will show spits as sitting on top of the sheet by casting shadows. The use of UV light indicates higher coat weight with the spits being more opaque in transmitted light. After supercalendering, it is not possible to see spits using low angle lighting. They are still discernible using UV light transmitted through the sheet.

f éclaboussures *fpl* — **d** Streichfarbespritzer *mpl* — **n** strijklaagspatten *fmpl* — **i** spruzzi *mpl* di patina

12889 spitting — See ink misting.

12890 splash — Mark made from a drop of corrective or dyeing liquid.

f éclaboussure *f* — **d** Spritzer *m* — **n** spat *fm* — **e** salpicadura *f*; salpicón *m*; salpicado *m* — **i** spruzzo *m*; macchia *f*

12891 splice — Joint accomplished by splicing. See also joint.

f raccordement *m*; joint *m* — **d** Spleiß *m*; Bahnverspleißung *f*; Bandverspleißung *f* — **n** las *fm* — **e** empalme *m*; empalmadura *f*; pegadura *f*; junta *f* — **i** giunta *f*; zona *f* d'incollatura

12892 splice *v* — To unite or join the ends of rolled material or of lengths of material, such as foil, film, paper, etc, by mechanical or electrical means or by an adhesive.

f coller; rabouter — **d** ankleben; kleben — **n** aan elkaar plakken; lassen — **e** empalmar — **i** incollare una bobina

12893 splice break — Result of a faulty joint in a paper web.

f rupture *f* de collure — **d** Klebestellenriß *m* — **n** lasbreuk *fm* — **e** rotura *f* del empalme — **i** rottura *f* alla giunzione

12894 splice tag — See flag.

12895 split backs — See poor fold strength.

12896 split-colour printing; split-fountain printing; rainbow printing

f impression *f* irisée — **d** Irisdruck *m*; regenbogenfarbiger Druck *m* — **n** irisdruk *m* — **e** impresión *f* irisada; impresión *f* de iris — **i** stampa *f* iridata; stampa *f* a iride

12897 split delivery — Arrangement on a rotary press folder to transfer alternate copies to separate deliveries.

f sortie *f* à part — **d** getrennte Auslage *f* — **n** ge-

scheiden uitleg *m* — **e** salida *f* selectora; selector *m* de salida — **i** uscita *f* separata

12898 split fountain — Ink fountain divided into sections so that two or more colours may be printed simultaneously.

f encrier *m* partagé; encrier *m* scindé — **d** unterteilter Farbkasten *m*; geteilter Farbkasten *m* — **n** gescheiden inktbak *m*; verdeelde inktbak *m* — **e** tintero *m* dividido (en secciones); tintero *m* seccionado — **i** calamaio *m* suddiviso

12899 split-fountain printing — See split-colour printing.

12900 split fraction — See piece fraction.

12901 split pin *GB*; **cotter pin** *US* — Locking device which consists of a strip of soft or special steel bent in such a way that it forms a loop at one end and two legs at the other. Fitted into slots in the head of slotted or castellated nuts and through a hole in the bolt, the legs being bent in opposite directions, to prevent the split pin from falling out.

f goupille *f* fendue — **d** Splint *m* — **n** splitpen *fm* — **e** pasador *m* hendido; chaveta *f* — **i** copiglia *f*

12902 split skin — In bookbinding, skin of which part of its undersurface is split off.

f peau *f* sciée; cuir *m* scié — **d** Spaltleder *n* — **n** spouwleer *n*; splitleer *n* — **e** cuero *m* chiflado; cuero *m* rebajado — **i** cuoio *m* spaccato; pelle *f* spaccata

12903 splitter — Tool used by newspaper make-up man, consisting of a rectangular flat piece of steel, column width, heightened at the top centre to separate and move masses of type matter. The bottom is bevelled.

f diviseur *m* — **e** divisor *m* — **i** divisorio *m*

12904 splitting of ink film — See ink splitting.

12905 splitting velocity of ink

f vitesse *f* de séparation du film d'encre — **d** Spaltungsgeschwindigkeit *f* der Farbschicht — **n** snelheid *f* van splitsing van de inktlaag — **e** velocidad *f* de disgregación de la capa de tinta — **i** velocità *f* della scissione in strati della pellicola d'inchiostro

12906 spoilage — Sheets spoiled in printing or converting processes.

f gâche *f* (de papier); déchets *mpl* (de tirage); maculatures *fpl* — **d** Makulaturanfall *m*; Makulaturen *fpl* — **n** misdruk *m*; misdrukken *mpl*; afval *mn*; uitschot *n* — **e** maculaturas *fpl*; pérdidas *fpl* — **i** sprechi *mpl*; scarti *mpl*

12907 spoilage — See misprints.

12908 spoils — See misprints.

12909 spoils — See overs.

12910 sponge

f éponge *f* — **d** Schwamm *m* — **n** spons *fm* — **e** esponja *f* — **i** spugna *f*

12911 sponge *v* **over**

f passer l'éponge — **d** mit dem Schwamm abwischen — **n** afsponsen; met de spons afwassen — **e** lavar con (una) esponja — **i** lavare a spugna; dar di spugna a

12912 sponge rubber; foam rubber — Cellular

rubber made by bubbling carbon dioxide through, or whipping air into latex, resembling a natural sponge in structure and used for cushions, vibration dampeners and gaskets.
f caoutchouc *m* éponge — **d** Schaumgummi *m* — **n** schuimrubber *mn* — **e** goma *f* espuma — **i** gomma *f* spugnosa

12913 spontaneous combustion — See spontaneous inflammation.

12914 spontaneous ignition — See spontaneous inflammation.

12915 spontaneous inflammation; spontaneous combustion; spontaneous ignition — Self-ignition of combustible material.
f combustion *f* spontanée; inflammation *f* spontanée; auto-allumage *m*; auto-inflammation *f* — **d** Selbstentzündung *f*; Selbstzündung *f* — **n** zelfontbranding *f* — **e** autoinflamación *f* — **i** combustione *f* spontanea; autocombustione *f*

12916 spool — Metal bobbin holding unprocessed film to be inserted in a camera.
f bobine *f* — **d** Spule *f*; Filmspule *f* — **n** spoel *fm* — **e** carrete *m* — **i** rocchetto *m*

12917 spot *v*; **spot** *v* **out** — To remove pinholes or other small transparent defects in a negative, a photograph or a printing plate.
f reboucher — **d** ausflecken — **n** uitdekken; wegdekken; afdekken; uitpunten — **e** retocar negativos — **i** ritoccare i negativi; smacchiare i negativi *sl.*

12918 spot colour — Small area printed in a second colour.
d Schmuckfarbe *f* — **n** steunkleur *fm* — **e** color *m* solitario — **i** colore *m* supplementare; colore *m* aggiunto

12919 spot of grease — See grease stain.

12920 spot *v* **out** — See spot *v*.

12921 spots — Opaque or transparent (black or white) formations of regular shape and size occurring as blemishes in photographic images.
f poivrage *m* — **d** Flecken *mpl* — **n** vlekjes *npl* — **e** manchuelas *fpl* — **i** macchie *fpl* bianche; macchie *fpl* con centro scuro

12922 spot sheet; spot-up sheet — Marked and patched sheet which carries the makeready and is placed underneath the top sheet on the printing cylinder.
f feuille *f* de mise — **d** Zurichtebogen *m* — **n** toestel *n*; toestelvel *n* — **e** hoja *f* de arreglo; hoja *f* de alzas — **i** foglio *m* d'avviamento

12923 spotter — Worker doing simple opaquing operations to eliminate defects in negatives.
f plumiste *m*; retoucheur *m* — **d** Abdecker *m* — **n** uitdekker *m*; retoucheur *m* — **e** retocador *m* — **i** ritoccatore *m*

12924 spotter — Piece of white or black paper in which a small hole has been cut. Used for comparing areas on the sketch with tints on the chart.
f cache *m* perforé; cache *m* isolateur — **d** Farbenbrille *f*; Abdeckpapier *n* mit Löchern — **n** maskertje *n* — **e** plantilla *f* de comparación — **i** maschera *f* forata

12925 spot-up sheet — See spot sheet.

12926 sprayer — Mechanical device spraying wax or powder which dispenses with the need for slip-sheeting.
f pulvérisateur-antimaculateur *m* — **d** Druckbestäuber *m*; Bestäubungsapparat *m*; Bestäuber *m* — **n** anti-smetapparaat *n* — **e** rociador *m*; pulverizador *m* antirrepinte; antimaculador *m* — **i** spruzzatore *m* antiscartino; dispositivo *m* antiscartino

12927 spray fountain — See automatic fountain.

12928 spraying — See ink misting.

12929 spray nozzle
f gicleur *m* de pulvérisateur; buse *f* de pulvérisation — **d** Zerstäuberdüse *f* — **n** verstuifkop *m*; verstuiverkop *m*; verstuiver *m* — **e** tobera *f* pulverizadora; boquilla *f* pulverizadora; espolvoreador *m* — **i** ugello *m* di polverizzatore; ugello *m* spruzzatore

12930 spread — See double spread.

12931 spread *v* — To thicken or enlarge (of printed areas) by bleeding or lateral penetration of ink.
f étaler — **d** bluten; auslaufen — **n** bloeden; uitlopen; uitvloeien — **e** derramar — **i** spandersi

12932 spread, centre — See centre spread.

12933 spread coated paper — Paper coated by the spread coating process.
f papier *m* couché étalé — **d** im Rakelstreichverfahren gestrichenes Papier *n* — **n** papier *n* met spreidlaag — **e** papel *m* estucado de raspador vertical — **i** carta *f* patinata a lama fissa; carta *f* patinata per spalmatura

12934 spread coating — Coating a continuous paper web with resins or plastics by means of a vertical plate restraining a pond of viscous coating slip which is drawn through the adjustable gap between the plate and the paper by the forward movement of the web over a horizontal support. Separate coating operation.
f couchage *m* par étalement — **d** stehendes Rakelstreichverfahren *n* — **n** spreidlaag-procédé *n* — **e** procedimiento *m* de estucado de raspador vertical — **i** patinatura *f* a lama fissa; patinatura *f* per spalmatura

12935 spreading — Range between the highest and the lowest result in a series of experiments.
f étalement *m* — **d** Spreitung *f* — **n** spreiding *f* — **i** estensione *f*

12936 spreading capacity — See coverage.

12937 spreading coefficient
f coefficient *m* d'étalement — **d** Spreitungskoeffizient *m* — **n** spreidingscoëfficiënt *m* — **i** coefficiente *m* d'estensione

12938 spreading rate — See coverage.

12939 spring — Elastic body or device which yields under pressure and returns to the original state after the pressure is removed. See also tension spring.
f ressort *m* — **d** Feder *m*; Springfeder *m* — **n** veer *f*; springveer *f* — **e** resorte *m*; muelle *m* — **i** molla *f*

12940 spring back — See hollow back.

12941 spring dividers — See dividers.

12942 spring drive device

f dispositif *m* de ressort; dispositif *m* d'accélération — d Federspannungsansatz *m* — n veeraanzetstuk *n* — e dispositivo *m* de aceleración — i dispositivo *m* d'accelerazione

12943 spring forme — Forme in which the type matter does not rest level on the imposing surface or machine bed, due to being locked up badly or too tightly.

f forme *f* bombée — d federnde Druckform *f* — n verende drukvorm *m*; springvorm *m* — e forma *f* elástica — i forma *f* bombata

12944 springiness — See resiliency.

12945 spring loaded idler — See jockey roller.

12946 spring tension

f tension *f* de ressort — d Federspannung *f* — n veerspanning *f* — e tensión *f* de resorte — i tensione *f* della molla

12947 spring, tension — See tension spring.

12948 spring tongue gauge pin — See gauge pin.

12949 sprinkled edges — The cut edges of a book when covered with fine spots of colour(s).

f tranches *fpl* jaspées; tranches *fpl* mouchetées — d Sprengschnitte *fpl*; gesprenkelte Buchschnitte *fpl* — n gespikkelde sneden *fmpl*; gesprenkelde sneden *fmpl* — e cantos *mpl* salpicados; cantos *mpl* jaspeados; cantos *mpl* chispeados — i tagli *mpl* spruzzati

12950 sprocket; sprocket wheel; chain sprocket; chain wheel — Toothed wheel engaging with a conveyor or power chain.

f pignon *m* à chaîne — d Kettenrad *n* — n kettingwiel *n*; kettingrad *n*; tandwiel *n*; tandrad *n* — e rueda *f* de cadena — i ruota *f* dentata per catena

12951 sprocket holes — See feed holes.

12952 sprocket wheel — See sprocket.

12953 spruce — Resinous wood most commenly used in the production of of sulphite and sulphate pulps.

f épinette *f*; épicéa *m* — d Fichte *f*; Fichtenholz *n* — n spar *m*; sparrehout *n* — e abeto *m* (rojo); madera *f* de abeto — i abete *m*; legno *m* d'abete

12954 spur — Slight excrescences at the top of some capital T's (e.g., Granjon Old Face) and serif and terminal of the lower arm of the capital G.

f ergot *m* — d abgewinkelter Serif *m*; Sporn *m* — e terminal *m* — i sperone *m*

12955 squabbled type — Type which is out of line.

f caractère *m* qui chevauche — d tanzende Schrift *f*; tanzender Satz *m* — n dansende letter *fm* — e composición *f* torcida — i composizione *f* accavallata; carattere *m* non allineato

12956 square — That part of the book case which overlaps the edges.

f chasse *f* — d Deckelkante *f*; überstehende Deckelkante *f* — n overstekende rand *m* — i parte *f* della copertina che copre i tagli

12957 square back; flat back — Back or spine of a book formed by the folded sheets being placed together, the back being square with the sides.

f dos *m* carré — d gerader Buchrücken *m* — n rechte rug *m*; Engelse rug *m* — e lomo *m* cuadrado; lomo *m* rectangular; lomo *m* plano; lomo *m* liso — i dorso *m* piatto; dorso *m* quadro

12958 square bag — See bag.

12959 square brackets — See brackets.

12960 square brackets — See parentheses.

12961 square brackets — See punctuation marks.

12962 squared — Halftone plates having for straight edges which can be mechanically cut or bevelled on straight lines.

f au carré — d viereckig; quadratisch — n vierkant — e escuadrado — i squadrato; tagliato in squadra

12963 squared (finish) halftone — Halftone plate presenting an unbroken screen printing surface, which can be finished in square or rectangular shape with or without a border line.

f simili *f* au carré — d viereckige Autotypie *f* — n vierkante autotypie *f* — e medio tono *m* escuadrado — i autotipia *f* quadrata; lastra *f* rettangolare

12964 squared paper; square-ruled paper; quadrillé ruled paper; quadrillé paper; cross section paper — Writing, drawing, tracing or cartridge paper with ruled or printed squares.

f papier *m* quadrillé; papier *m* réglé carré; papier *m* à carreaux — d kariertes Papier *n*; karriertes Papier *n*; rautiertes Papier *n* — n quadrillé gelijnd papier *n*; carré gelijnd papier *n*; geruit papier *n* — e papel *m* cuadriculado — i carta *f* quadrettata

12965 squared paper — Paper trimmed square on four sides or on one side and one end.

f papier *m* équerré — d Papier *n* mit Winkelschnitt auf zwei oder vier Seiten; rechtwinklig geschnittenes Papier *n* — n haaks gesneden papier *n*; papier *n* met haakse hoek — e papel *m* escuadrado — i carta *f* squadrata

12966 square foot — Unit of area. *pl* square feet. Abbrev. ft^2 (or sometimes sq ft). See app. no. 7 and 7a.

f pied *m* carré — d Quadratfuß *m* — n vierkante (Engelse) voet *m* — e pie *m* cuadrado — i piede *m* quadrato; piede *m* quadro

12967 square graver

f burin *m* (à biseau) carré; échoppe *f* (à biseau) carré — d Vierkantstichel *m*; Vierkantstecher *m* — n vierkantsteker *m* — e buril *m* con cuatro cantos — i bulino *m* quadrato

12968 square inch — Unit of area. Abbrev. in^2 (or sometimes sq in). See app. no. 7 and 7a.

f pouce *m* carré — d Quadratzoll *m* — n vierkante inch *mn* — e pulgada *f* cuadrada — i pollice *m* quadrato; pollice *m* quadro

12969 square root — See root.

12970 square root sign — See root sign.

12971 square-ruled paper — See squared paper.

12972 square-ruled screen — See crossline screen.

12973 squares; overlapping edges — Amount of cover board which overhangs the edges of a book;

should be equal on all three sides.

f chasse *f* — **d** überstehende Kanten *fpl*; überstehende Deckelkanten *fpl* — **n** overstekende kanten *mpl*; overhangende kanten *mpl* — **e** cejas *fpl* — **i** unghiature *fpl*; unghie *fpl* dei quadranti delle coperte

12974 square-up — When a carton has been glued and is shipped to the customer for filling, it must fold square when filled and sealed.

12975 square yard — Unit of area. Abbrev. yd² (or sometimes sq yd). See app. no. 7 and 7a.

f yard *m* carré — **d** Quadratyard *m* — **n** vierkante yard *m* — **e** yarda *f* cuadrada — **i** iarda *f* quadrata; iarda *f* quadra

12976 squash — Halo effect around the edge of printed areas in letterpress printing.

f auréole *f*; voile *m* — **d** Quetschrand *m* — **n** kraalrandje *n* — **e** auréola *f* — **i** aureola *f*

12977 squash — See printing pressure.

12978 squeegee — Mounted rubber strip to spread ink through stencil in screen printing.

f raclette *f* (de caoutchouc) — **d** Quetscher *m*; Gummirakel *f* — **n** rakel *m*; trekspaan *fm* — **e** racleta *f* de caucho; rascleta *f* de caucho; escurridor *m*; alisador *m*; alisadora *f* — **i** spatola *f* di gomma; racla *f* di gomma

12979 squeeze — Amount of impression between the plate and the impression cylinder. See also printing pressure.

12980 squeeze *v* out — Squeezing out of the ink under too heavy printing pressure or too much ink results in filling up of halftone screen.

f écraser — **d** quetschen; ausquetschen — **n** wegpersen — **e** aplastar — **i** debordare dell'inchiostro

12981 squeeze rolls — Small size rolls on multicylinder board machine to press the plies of the web together.

f rouleaux *mpl* presseurs; rouleaux *mpl* essoreurs; rouleaux *mpl* foulon — **d** Vorgautschwalzen *fpl*; Vorgautschpreßwalzen *fpl*; Abquetschwalzen *fpl* — **n** koetswalsen *fmpl*; voorkoetswalsen *fmpl* — **e** rodillos *mpl* exprimidores; rodillos *mpl* (de la) anteprensa del manchón — **i** rulli *mpl* spremitore

12982 squirrel-cage induction motor — See squirrel-cage motor.

12983 squirrel-cage motor *GB*; **squirrel-cage induction motor**

f moteur *m* à cage — **d** Käfigläufermotor *m*; Käfigankermotor *m*; Käfigmotor *m* — **n** kooiankermotor *m*; motor *m* met kortsluitanker — **e** motor *m* (con rotor) de jaula; motor *m* (con rotor) de jaula en corto circuito — **i** motore *m* a gabbia

12984 squirt — Flow of molten metal all over the delicate parts of the typesetting machine.

f jet *m* de plomb; jet *m* de métal — **d** Spritzer *m*; Bleispritzer *m*; Überläufer *m* — **n** spuiter *m* — **e** salpicadura *f*; salpicado *m*; rebose *m*; derrame *m* del metal; emplomada *f*; emplomadura *f*; chisquete *m*; chorretada *f* — **i** spruzzo *m*; getto *m* di piombo

12985 stab — See stab hand.

12986 'stab — Applied to establishment printers,

i.e., those paid a set weekly wage as distinct from piece-work. See also stab hand.

f en conscience — **d** auf Wochenlohn — **n** op weekloon; op vast loon — **e** a sueldo fijo — **i** a stipendio fisso

12987 stab, be *v* on

f être en conscience — **d** im gewissen Geld stehen — **n** op vast loon staan — **e** estar a conciencia; estar a sueldo fijo — **i** stare a coscienza; stare a ore

12988 stabbing — To secure a pile of sections or leaves, the required number of staples is first inserted from one side. The wire feed control is set so that the shank of the staple is not long enough to pass through the underside of the pile. The job is then turned over and the same number of stitches inserted from the opposite side. See also side stitching.

f agrafage *m*; piquage *m* à travers; piquage *m* à plat — **d** Nageln *n*; Querheftung *f*; Klammerheftung *f* — **n** krammen *n* — **e** agujerear; cosido *m* a punzón; cosido *m* a diente de perro — **i** cucitura *f* di traverso (a punti metallici)

12989 stab hand; stab — Term derived from *'stab or establishment.

f compositeur *m* en conscience; consciencieux *m* — **d** Gewißgeldsetzer *m* coll.; nicht im Akkord arbeitender Schriftsetzer *m* — **n** op vast loon staande letterzetter *m* — **e** cajista *m* a sueldo fijo; cajista *f* plantilla (a jornal fijo) — **i** manodopera *f* a stipendio fisso

12990 stability — In general, the tendency to remain in a given state or condition, without spontaneous change.

f stabilité *f* — **d** Stabilität *f* — **n** stabiliteit *f* — **e** estabilidad *f* — **i** stabilità *f*

12991 stabilizer; constant-voltage stabilizer *GB*; **voltage stabilizer** *US* — Unit which by electronic means stabilizes the voltage.

f stabilisateur *m* de tension — **d** Stabilisator *m* — **n** stabilisator *m*; spanningsstabilisator *m* — **e** estabilizador *m* de tensión — **i** stabilizzatore *m* di tensione

12992 stable — Capable of lasting without change; permanent; not easily destroyed or decomposed.

f stable; solide; permanent — **d** beständig — **n** bestendig; stabiel — **e** estable — **i** stabile

12993 stab stitching — See side stitching.

12994 stab wages; established wages

f conscience *f* — **d** Gewißgeld *n* — **n** vast loon *n* — **e** conciencia *f*; sueldo *m* fijo — **i** coscienza *f*

12995 stack; calender stack — Calender, generally situated at the end of the paper machine, of which the rolls are of metal only. Used to impart a finish to the paper and to control caliper.

f lisse *f*; calandre *f* finissure — **d** Glättwerk *n*; Glättkalander *m*; Maschinenkalander *m*; Trockenglättwerk *n* — **n** gladwerk *n*; machinekalander *fm*; satineerkalander *fm* — **e** calandra *f* de satinar; mecanismo *m* abrillantador — **i** liscia *f* di macchina

12996 stack — See pile.

12997 stack *v* — To pile up sheets or reels of paper.
f empiler — **d** aufstapeln; stapeln — **n** opstapelen; stapelen — **e** apilar — **i** accatastare
12998 stacker — Device attached to delivery conveyor of a web fed press to collect, compress and bundle signatures.
f paqueteur *m* — **d** Paketausleger *m*; Paketauslage *f* — **n** pakketuitleg *m*; bundelapparaat *n* — **e** sacador *m* de paquetes — **i** impilatore *m*; formapile *m*; accatastatore *m*; stacker *m*
12999 stacking of paper reels — Storing of reels of paper end on end (end stacking) or in pyramid form.
f gerbage *m* de bobines (en canon; en berceau) — **d** Rollenstapelung *f* (Schornsteinstapelung; Sattelstapelung; Sattellagerung) — **n** opstapeling *f* van rollen; stapeling *f* van rollen (eind op eind; rol op rol; in de tier) — **e** apilado *m* de rollos; estibado *m* de bobinas — **i** impilamento *m* delle bobine (estremità ad estremità; a piramide)
13000 staff — The five lines on which notes of music are written or printed (*pl* staves).
f portée *f* (de musique) — **d** Notenlinie *f* — **n** notenbalk *m* — **e** pentagrama *m* — **i** pentagramma *m*
13001 stagger *v* — To display type matter in zigzag fashion.
f composer en zigzag — **d** in Zickzackform setzen; Satz in Zickzackform anordnen — **n** verspringend zetten; om en om zetten — **e** componer en forma de escalón; componer en forma de escalonada; componer en líneas graduales — **i** comporre a zigzag; sfalzata alternativamente
13002 stagger *v* — To arrange piles of paper in zigzag fashion.
f disposer en zigzag; disposer en quinconce; alterner — **d** versetzt abstapeln — **n** in zigzaglijn plaatsen — **e** disponer en zigzag; alternar en zigzag; saltear en zigzag — **i** collocare a zig-zag; disporre a zig-zag; sfalzando alternativamente
13003 stagger bar; head bar — Bar on the plate cylinder in a rotary press against which the plates are placed when they are staggered across the press.
f règle *f* du cylindre cliché; barrette *f* de tête — **d** Plattenanschlag *m* — **n** aanslag *m* voor de drukplaat — **e** barra *f* de entrada — **i** barra *f* appoggio lastra
13004 stagger marks — Darker printing areas parallel to the cylinder axis, occuring at intervals in the rotation of the cylinder on a rotary press.
f stries *fpl* — **d** Rattermarken *fpl* — **n** drukstrepen *fmpl*; strepen *fmpl* — **e** estrías *fpl* — **i** striature *fpl* di stampa
13005 stagger, plate — See plate stagger.
13006 staging — Controlled area-wise etching of positives or negatives. An etching resist is used to paint out sufficiently etched areas in successive stages. Further etching is then confined to unstaged areas. The staging and the etching operations are repeated until the required densities or

dot sizes are obtained.
f morsure *f* et revernissage *m* — **d** Ätzen *n* und Abdecken *n* — **n** etsen *n* en afdekken *n* — **e** proteger ciertas partes del grabado para remordido local — **i** morsura *f* localmente differenziate
13007 staging ink — Ink or solution used in photoengraving and photo-lithography during halftone etching; consists of Syrian asphalt mixed with unbleached beeswax, which mixture is diluted with sufficient benzene to make it flow easily from a brush.
f vernis *m* de bouchage — **d** Abdecklack *m* — **n** afdeklak *m*; afdekverf *fm* — **e** tinta *f* de retocar; tinta *f* para retocar; barniz *m* tapador — **i** inchiostro *m* resistente agli acidi
13008 stain — Local or general discoloration of photographic negatives and prints. The discoloration formed on zinc and copper plates by short immersion in dilute acid or graining baths.
f tache *f* — **d** Fleck *m* — **n** vlek *fm* — **e** mancha *f*; descoloración *f*; mácula *f*; tacha *f*; mancilla *f*; chafarrinada *f*; chafarrinón *m*; lunar *m* — **i** macchia *f*; decolorazione *f*
13009 stained edges — Edges of a book which are decorated with a dye.
f tranches *fpl* coloriées; tranches *fpl* couleur — **d** gefärbter Schnitt *m*; Farbschnitt *m* — **n** geverfde sneden *fmpl* — **e** cantos *mpl* colorados; cantos *mpl* pintados — **i** tagli *mpl* colorati
13010 stained paper — Paper coloured by surface application of a dye solution.
f papier *m* teinté; papier *m* colorié en surface — **d** oberflächengefärbtes Papier *n* — **n** oppervlaktegekleurd papier *n*
13011 staining — Colouring of edges of books for decorative effect. See also gilding.
f coloriage *m* des tranches — **d** Schnittfärben *n* — **n** sneeverven *n* — **e** pintado *m* de los cortes — **i** coloritura *f* dei tagli
13012 stainless steel — Alloy steel containing 4% or more chromium to be resistant to rust and attack from various chemicals.
f acier *m* inoxydable — **d** rostfreier Stahl *m* — **n** roestvrij staal *n* — **e** acero *m* inoxidable — **i** acciaio *m* inossidabile; acciaio *m* resistente alla corrosione
13013 staling — See dark reaction.
13014 stamping — See blocking.
13015 stamping die — Deeply etched or engraved relief plate on brass or zinc for stamping book covers; brass plates are used when the stamping process requires heat.
f matrice *f* de timbrage — **d** Prägestempel *m*; Prägeplatte *f* — **n** bandstempel *n* — **e** plancha *f* de estampar — **i** stampo *m* per impressioni
13016 stamping ink; die stamping ink — Intaglio ink used for die engraving, made from rapid drying, greaseless, tackless gum varnish.
f encre *f* à gravure; encre *f* à impression en relief — **d** Stahlstichdruckfarbe *f*; Reliefdruckfarbe *f* — **n** staalstempeldrukinkt *m*; stempelinkt *m*; reliëf-

drukinkt *m* — **e** tinta *f* para estampar en relieve; tinta *f* para estampación en relieve — **i** inchiostro *m* per stampa a rilievo; inchiostro *m* per impressione a rilievo; inchiostro *m* per rilievografia

13017 stamping machine — See die-stamping press.

13018 stamping press — See blocking press.

13019 stamp paper — See also gummed paper, label paper.

f papier *m* pour timbres — **d** Briefmarkenpapier *n* — **n** papier *n* voor (post)zegels — **e** papel *m* para timbres; papel *m* para sellos (de correo) — **i** carta *f* per (franco)bolli

13020 standard — Unit of references; form, type or set of conditions, accepted as perfect, and used for comparison.

f étalon *m* — **d** Eichmuster *n*; Vergleichsmuster *n* — **n** vergelijkingsmonster *n*; ijkmaat *fm* — **e** testimonio *m*; aforo *m* — **i** standard *m*; norma *f* di riferimento

13021 standard atmosphere — Unit of pressure defined as the pressure exerted by a column of mercury 760 mm high, (29.9213 in), or having a value of 1013.2 millibars.

f atmosphère *f* normale; atmosphère *f* standard — **d** Normalatmosphäre *f* — **n** standaard-atmosfeer *fm* — **e** atmósfera *f* física — **i** atmosfera *f* fisica

13022 Standard Book Number — Number identifying the book to which it is assigned. It consists of nine digits broken into three parts, usually separated by hyphens. The first part identifies the publisher; the second part identifies the title or edition; the third part, always a single digit (check digit), serves as an arithmetic check against errors in transcription of the whole number. The SBN system originated in England. The SBN and the *Library of Congress Catalog Card Number should appear on the copyright page, on successive lines. The SBN is assigned by the publisher, under a system administered by the R.R. Bowker & Co. The LC number is assigned by the Library of Congress upon application of the publisher.

f numéro *m* pour le dépôt legal; référence *f* du dépôt légal — **d** Standardbuchnummer *f* — **n** standaardboeknummer *n* — **e** número *m* de libro del depósito legal — **i** numero *m* del libro per il deposito legale

13023 standard deviation — In statistics, the square root of the mean of the squares of the deviations of all the observations.

f écart-type *m* — **d** mittlere Abweichung *f*; quadratische Abweichung *f*; Standardabweichung *f* — **n** standaard-afwijking *f* — **e** desviación *f* standard — **i** deviazione *f* standard; deviazione *f* quadratica media

13024 standard error — Standard deviation in repeated samples of the mean standard deviation or other statistical measure. It indicates for example the extent to which the sample mean may differ from the batch or consignment mean. Thus it

measures the error involved in estimating the character of the larger whole from apart.

f erreur-type *m* — **d** Standardfehler *m*; mittlerer Fehler *m*; quadratischer Fehler *m* — **n** standaardfout *fm* — **e** error *m* standard — **i** errore *m* standard; errore *m* tipo

13025 standard inks — See process inks.

13026 standardization; normalization — Simplification of sizes, basis weigts, colours, recommended in trade customs. Also simplification in manufacturing processes, test methods, cost accounting, etc.

f normalisation *f*; standardisation *f* — **d** Normung *f*; Standardisierung *f* — **n** normalisatie *f*; standaardisering *f* — **e** normalización *f*; standardización *f* — **i** normalizzazione *f*; standardizzazione *f*

13027 standard line; point common line — Applied to every fount of type with a lower-case alphabet which has a definite number of points to its beard.

f ligne *f* systématique; alignement *m* normal; alignement *m* standard; ligne *f* de parangonnage; chanfrein *m* unique — **d** Normalschriftlinie *f*; Universalschriftlinie *f*; Einheitschriftlinie *f* — **n** universele letterlijn *fm* — **e** línea *f* normal (de los tipos); línea *f* de parangonado — **i** allineamento *m* normale

13028 standard performance — Work study term for a level of performance not yet standardized. See also normal performance.

13029 standard sizes — See international sizes.

13030 standard time — See also allowed time, target time.

f temps *m* de référence; temps *m* normalisé — **d** Vorgabezeit *f* (auf Zeitstudien aufgebaut) — **n** standaard-tijd *m* — **e** tiempo *m* normal — **i** tempo *m* standard

13031 standard work minute; SWM — Time which the worker applies himself to the task with normal effect, and a portion of rest time.

f minute *f* de travail standard; MTS *f* — **d** Standardarbeitsminute *f*; SAM *f* — **n** standaardarbeidsminuut *fm*; SAM *fm* — **e** minuto *m* trabajo standard — **i** minuto *m* di lavoro standard

13032 standing forme

f forme *f* conservée — **d** Stehform *f* — **n** staande vorm *m*; staande drukvorm *m* — **e** molde *m* conservado; molde *m* guardado; forma *f* pendiente (de reimpresión); molde *m* vigente — **i** forma *f* conservata

13033 standing matter; live matter; alive matter — Type forme retained for a possible reprint.

f composition *f* conservée; composition *f* en attente; matière *f* debout — **d** Stehsatz *m*; stehender Satz *m* — **n** staand zetsel *n*; vaststaand zetsel *n* — **e** composición *f* a usar; composición *f* a guardar; composición *f* guardada; composición *f* fija; composición *f* permanente; composición *f* conservada; composición *f* vigente; material *m* fijo; tipo *m* parado; recado *m* — **i** composizione *f* in piedi; composizione *f* da conservare

13034 standing press — Heavy, screw operated press to press and dry books after casing. See also lying press.

f presse *f* à vis — **d** Stockpresse *f*; Spindelpresse *f* — **n** pletpers *f*; kolommenpers *f* — **e** prensadora *f* vertical — **i** pressa *f* verticale a vite

13035 stand oil — Drying oil, partially refined by allowing certain impurities to settle out after heat treatment. Generally used in GB to describe *linseed oil.

f standolie *f* — **d** Standöl *n*; Dicköl *n*; Drucköl *n* — **n** standolie *fm* — **e** aceite *m* contraído — **i** standolio *m*

13036 stand oil — See linseed oil.

13037 Stanhope press — Iron hand press for letterpress printing applying pressure over a series of levers and toggle movements. Named after its inventor Earl Stanhope (1753-1816).

f presse *f* Stanhope; presse *f* modèle Stanhope; presse *f* à bras — **d** Stanhope-Presse *f*; Kniehebelpresse *f* — **n** Stanhope-pers *f*; kniehevelpers *f* — **e** prensa *f* modelo Stanhope — **i** pressa *f* modello Stanhope; torchio *m* con pressione a leva

13038 stannic chloride

f chlorure *m* stannique — **d** Zinnchlorid *n*; Stannichlorid *n* — **n** tinchloride *n*; stannichloride *n* — **e** cloruro *m* estánnico — **i** cloruro *m* stannico

13039 stannous chloride; tin protochloride; tin crystals; tin salt; tin bichloride *depr.***; tin dichloride** *depr.* — White, crystalline mass, which absorbs oxygen from air, being converted into the insoluble oxychloride. Soluble in water, alkalis, tartaric acid and alcohol.

Formula: $SnCl_2$ or $SnCl_2 \cdot 2H_2O$.

f chlorure *m* stanneux; protochlorure *m* d'étain; sel *m* d'étain; bichlorure *m* d'étain — **d** Zinnchlorür *n*; Stannochlorid *n*; reines Zinnsalz *n*; Zinndichlorid *n* — **n** stannochloride *n*; tinchloruur *f* — **e** cloruro *m* de estaño; cloruro *m* estannoso — **i** cloruro *m* di stagno; cloruro *m* stannoso

13040 staple — Loop of metal or piece of wire bent and formed with two points to bind pamphlets. See also wire stitching.

f agrafe *f*; crampon *m* cavalier; cavalier *m* — **d** Heftklammer *f*; Drahtheftklammer *f* — **n** nietje *n*; krammetje *n* — **e** punto *m* metálico; grapa *f* (de alambre); corchete *m* de alambre; broche *m*; grampa *f* *Ar* — **i** punto *m* metallico; grappa *f*; graffa *f* (in box making)

13041 stapler — See wire stitcher.

13042 stapling — See wire stitching.

13043 star — Device to rapidly centre new or worn roller stocks in a roller mould to ensure casting rollers true on their metal bases (or stocks).

f étoile *f* de coulée — **d** Gießstern *m* — **n** gietkruisje *n* — **e** corona *f* dentada — **i** stella *f* di fusione; corona *f* di fusione

13044 star — See asterisk.

13045 starch; amylum — White, amorphous, tasteless powder, irregular lumps or fine powder. Insoluble in cold water, alcohol and ether, forms a jelly with hot water. The starches commonly used in paper are derived from corn, potatoes, tapioca and wheat.

Formula: $(C_6H_{10}O_5)_x$.

f amidon *m* — **d** Stärke *f*; Amylum *n* — **n** zetmeel *n* — **e** almidón *m* — **i** amido *m*

13046 starch glue; starch paste; paste — Mixture of starch and water, to make paper or other material to adhere to something.

f colle *f* de farine; colle *f* d'amidon; empois *m* — **d** Kleister *m*; Stärkekleister *m*; Stärkeklebstoff *m* — **n** stijfsel *mn*; stijfselpasta *m*; plaksel *n* *coll.* — **e** engrudo *m* de almidón — **i** colla *f* d'amido; colla *f* di fecola; salda *f* d'amido

13047 starch gum — See dextrin.

13048 starch paste — See starch glue.

13049 star-delta connection; Y-delta connection

f montage *m* en étoile-triangle; groupement *m* en étoile-triangle; couplage *m* mixte en étoile — **d** Sterndreieckschaltung *f* — **n** ster-driehoekschakeling *f* — **e** conexión *f* en estrella y triángulo — **i** collegamento *m* a stella-triangolo

13050 starred reel; starred roll — Reel of paper with extreme disruption in the pattern of winding as a result of excessive impact.

f bobine *f* écrasée; bobine *f* détériorée aux extrémités — **d** endbeschädigte Rolle *f* — **n** aan het einde beschadigde rol *fm* — **e** bobina *f* estropeada — **i** bobina *f* danneggiata

13051 starred roll — See starred reel.

13052 starred signature — See asterisk signature.

13053 star signature — See asterisk signature.

13054 start *v*; **start** *v* **the run** — To put in motion a press, a folding machine, etc.

f mettre en marche; mettre en route; démarrer; faire rouler — **d** anlassen; abdrucken — **n** draaien; afdraaien *coll.*; "draaien maar"; afdrukken; met afdrukken beginnen — **e** arrancar; ¡adelante! — **i** avviare

13055 star target — Small circular pattern of solid and clear pie wedges. When printed on a press sheet it gives a quick and effective visual indication of ink spread, slur, doubling, etc. Also used for resolution measurements in camera work and plate-making.

f mire-étoile *f*; étoile *f* Siemens — **d** Testfigur *f*; Farbkontrollstreifen *n*; Siemensstern *m* — **n** testplaatje *n*; controleplaatje *n* — **e** cinta *f* de control cromático; estrella *f* de Siemens — **i** croce *f* di Siemens; striscia *f* per il controllo del colore

13056 start button

f bouton *m* démarreur — **d** Startknopf *m*; Anlaßkontakt *m* — **n** startknop *m* — **e** botón *m* arrancador; botón *m* de arrancar — **i** pulsante *m* d'avviamento; pulsante *m* di comando; pulsante *m* di messa in moto

13057 started — Said of the leaves of a book when this shows uneven or stepped edges, due to faulty bookbinding or careless handling.

f ayant un nez; ayant des ondes — **d** verschoben; vorschießend — **n** verschoven; uitstekend — **e**

salida; saliente — **i** saltato

13058 starter — Last page or printing forme to go to press before beginning the run.

f dernière page *f* — **d** letzte Seite *f* — **n** laatste pagina *fm* — **e** página *f* de cierre — **i** ultima pagina *f*

13059 starter

f démarreur *m*; starter *m* — **d** Anlasser *m*; Starter *m* — **n** aanzetter *m*; starter *m* — **e** arrancador *m* — **i** avviatore *m*

13060 starter controller; controller — Electrical regulator to start electric motors.

f combinateur *m* de démarrage — **d** Anlaßfahrschalter *m* — **n** startcontroller *m* — **e** combinador *m* de arranque — **i** combinatore *m* di avviamento

13061 start *v* **the run** — See start *v*.

13062 start-up time — Time at the beginning of a shift. See also shut-down time.

f temps *m* de mise en route (début de poste) — **d** Anlaufzeit *f*; Aufrüstzeit *f* (beim Arbeitsbeginn) — **n** aanlooptijd *m*; inwerktijd *m* — **e** tiempo *m* de arranque — **i** tempo *m* di messa in marcia

13063 star wheel; assembler star — Small wheel about 2.5 cm (1 in) in diameter with three or four prongs, usually made of fibre. The rapid revolving prongs of the star wheel push the matrices and spacebands into the assembler box in sequence as delivered from their chutes.

f molette *f* d'assemblage — **d** Sammelstern *m*; Sammlerstern *m* — **n** sterwieltje *n* — **e** rueda *f* de estrella; estrella *f* — **i** ruota *f* a stella; stella *f* di fibra

13064 stat — See photostat.

13065 static and dynamic tensile-strength tester

f appareil *m* pour l'essai à la rupture statique et dynamique — **d** statischer und dynamischer Zugfestigkeitsprüfer *m* — **n** statische en dynamische trekbank *fm* — **e** ensayador *m* de la resistencia a la tensión estático y dinámico — **i** apparecchio *m* per prova di trazione statica e dinamica

13066 static charge

f charge *m* électrostatique — **d** statische Aufladung *f* — **n** statische lading *f* — **e** carga *f* estática — **i** carica *f* statica

13067 static electricity; statics — Excessive positive or negative electrical charge in a sheet of paper which causes the sheet to be attracted to other sheets and other materials.

f électricité *f* statique; électrisation *f* du papier; aimantation *f* du papier — **d** statische Elektrizität *f* — **n** statische elektriciteit *f* — **e** electricidad *f* estática; electrostática *f* — **i** elettricità *f* statica

13068 static eliminator — See static neutralizer.

13069 static neutralizer; static eliminator; antistatic device; neutralizer — Printing press attachment to remove the static electricity from the paper to avoid ink offsetting and trouble with the paper.

f désélectriseur *m*; appareil *m* pour l'élimination d'électricité statique; éliminateur *m* d'électricité statique — **d** Entelektrisator *m* — **n** des-

elektrisator *m*; apparaat *n* voor het opheffen van statische elektriciteit; ionisatie-apparaat *n* — **e** eliminador *m* de estática; dispositivo *m* antieléctrico estático; equipo *m* antistático; neutralizador *m* de estática — **i** eliminatore *m* di elettricità statica; neutralizzatore *m* di elettricità statica; dispositivo *m* antistatico

13070 statics — Branch of *mechanics; the mathematical and physical study of the behaviour of matter under the action of forces, dealing with cases where no motion is produced.

f statique *f* — **d** Statik *f* — **n** statica *f* — **e** estática *f* — **i** statica *f*

13071 statics — See static electricity.

13072 static tensile-strength tester

f appareil *m* pour l'essai à la rupture statique — **d** statischer Zugfestigkeitsprüfer *m* — **n** statische trekbank *fm* — **e** aparato *m* de ensayo de la resistencia a la tracción estático; dinamómetro *m* para ensayos de tracción — **i** apparecchio *m* per prova di trazione statica

13073 stationary — Fixed in a station, course or mode, unchanging in condition.

f stationnaire — **d** stationär — **n** stationair — **e** estacionario — **i** stazionario

13074 stationer; stationer's shop; stationery store — Place where *stationery is sold.

f papeterie *f* — **d** Papierwarenhandlung *f*; Papierwarengeschäft *n*; Schreibwarengeschäft *n* — **n** winkel *m* in schrijfbehoeften — **i** cartolaio *m*; cartoleria *f*

13075 stationer's shop — See stationer.

13076 stationery — An inclusive term which, as related to paper products, includes papeteries, typewriter paper, packaged papers, billheads and other papers, pens, pencils, ink, notebooks, etc., sold by stationers.

f fournitures *fpl* de bureau; papeteries *fpl* — **d** Schreibwaren *fpl*; Bürobedarf *m* — **n** schrijfbehoeften *fpl*; kantoorbenodigdheden *pl* — **e** papelería *f*; escribanía *f* — **i** articoli *mpl* di cancelleria; articoli *mpl* occorrente per scrivere

13077 stationery store — See stationer.

13078 statistical methods

f méthodes *fpl* statistiques — **d** statistische Methoden *fpl* — **n** statistische methoden *fpl* — **e** métodos *mpl* estadísticos — **i** metodi *mpl* statistici

13079 statistical quality control — System of sampling and testing a manufactured product which assures representative results, to keep variations within established limits.

f contrôle *m* de la qualité statistique — **d** statistische Qualitätskontrolle *f* — **n** statistische kwaliteitszorg *fm*; statistische kwaliteitscontrole *fm* — **e** control *m* estadístico de la calidad — **i** controllo *m* statistico di qualità

13080 statutory copy; deposit copy — In GB the copy of a book that the publisher is required by law to send within twelve months to some prescribed libraries. See also Library of Congress.

f exemplaire *m* du dépôt légal — **d** Pflichtexemplar *n* — **n** wettelijk verplicht exemplaar *n* — **e** ejemplar *m* de depósito legal — **i** esemplare *m* d'obbligo; esemplare *m* del deposito

13081 stave — The five lines in which notes of music are written or printed.
f portée *f* (de musique) — **d** Notenlinie *f* — **n** notenbalk *m* — **e** pentagrama *m* — **i** rigatura *f* del pentagramma

13082 stayer — See corner stayer.

13083 stayflat — Metal or glass plate bearing a tacky or adhesive coating on which films are squeegeed and held in position during the camera exposure.
f plaque *f* adhésive — **d** Klebeplatte *f* — **n** kleefruit *fm*; kleefplaat *fm* — **e** soporte *m* adhesivo — **i** lastra *f* adesiva per supporto

13084 staying machine — See corner stayer.

13085 steam-set ink; vaposet ink — Printing ink which sets nearly instantaneously under the action of water vapour.
f encre *f* steam-set; encre *f* fixée à la vapeur; encre *f* séchant à la vapeur — **d** Steam-set-Farbe *f*; Dampftrockenfarbe *f* — **n** steam-set inkt *m* — **e** tinta *f* steam-set — **i** inchiostro *m* steam-set

13086 steam spraying device
f dispositif *m* souffleur de vapeur — **d** Dampfsprühvorrichtung *f* — **n** stoomnevelinrichting *f*; stoomnevelaar *m* — **e** soplador *m* de vapor — **i** soffiatore *m* di vapore

13087 stearic acid; stearinic acid; stearophanic acid; cetylacetic acid; octadecanoic acid — Colourless, wax-like material, soluble in alcohol, ether, chloroform, carbon disulphide, carbon tetrachloride, sparingly soluble in water.
Formula: $CH_3(CH_2)_{16}CO_2H$.
f acide *m* stéarique — **d** Stearinsäure *f*; Zetylessigsäure *f*; Talgsäure *f*; Bassiasäure *f* — **n** stearinezuur *n* — **e** ácido *m* esteárico — **i** acido *m* stearico

13088 stearin; tristearin; glycerol tristearate — Colourless crystals or powder, insoluble in water, soluble in alcohol, chloroform, carbon disulphide, insoluble in ligroin and ether.
Formula: $(C_{17}H_{35}COO)_3C_3H_5$.
f stéarine *f* — **d** Stearin *n* — **n** stearine *fm*; stearien *fm* — **e** estearina *f* — **i** stearina *f*

13089 stearinic acid — See stearic acid.

13090 stearin pitch — Organic residue to impart flow and good wetting qualities of ink.
f poix *f* stéarique — **d** Stearinpech *n* — **n** stearinepek *mn* — **e** pez *f* de estearina; pez *f* esteárica — **i** pece *f* di stearina

13091 stearophanic acid — See stearic acid.

13092 steatite — Massive form of *talc, a natural acid metasilicate of magnesium. See also soapstone.
f stéatite *f* — **d** Steatit *m*; Speckstein *m* — **n** steatiet *n*; speksteen *mn* — **e** esteatita *f*; jaboncillo *m*; jabón *m* de sastre — **i** steatite *f*

13093 steel engraving — See steel-plate engraving.

13094 steel facing; steeling; acierage — Electrolytical deposition of a thin layer of steel on printing plates, particularly intaglio plates, to lengthen their useful life. Superseded by chromium plating.
f aciérage *m* — **d** Verstählung *f* — **n** verstaling *f* — **e** acerado *m* — **i** acciaiatura *f*

13095 steeling — See steel facing.

13096 steel-plate engraving; steel engraving — Intaglio printing plate engraved by hand by means of gravers, used mostly in printing of bank notes, bonds, certificates, etc. See also siderography.
f gravure *f* sur acier; taille-dure *f* — **d** Stahlstich *m*; Stahlgravur *f* — **n** staalgravure *fm* — **e** grabado *m* en acero — **i** incisione *f* su acciaio

13097 steel-plate engraving ink — Soft intaglio ink, with oil and water resistant pigments, used for steel plate engraving.
f encre *f* pour gravure sur acier; encre *f* tailledure — **d** Stahldruckfarbe *f* — **n** staaldrukinkt *m* — **e** tinta *f* para grabado de acero; tinta *f* para grabado en acero — **i** inchiostro *m* per stampa in acciaio

13098 steel-plate paper — Paper used for steelplate engravings. It may be a ledger, chart, or bristol type, tub-sized to give uniform surface which is the most significant property.
f papier *m* pour gravure sur acier — **d** Stahldruckpapier *n* — **n** staaldrukpapier *n*; papier *n* voor staalgravures; plaatdrukpapier *n* — **e** papel *m* para estampación con grabados de acero — **i** carta *f* per stampa in acciaio

13099 steel rider; steel rider roller; pyramid roller — See also rider roller.
f chargeur *m* d'acier — **d** Stahlreiber *m*; stählerne Reiterwalze *f*; stählerne Deckwalze *f* — **n** stalen ruiterrol *fm* — **e** rodillo *m* cargador de acero; cargador *m* de acero; rodillo *m* de carga de acero — **i** rullo *m* caricatore in acciaio; rullo *m* cavaliere in acciaio; rullo *m* macinatore superiore d'acciaio

13100 steel rider roller — See steel rider.

13101 steel roller
f rouleau *m* d'acier — **d** Stahlwalze *f* — **n** stalen rol *fm* — **e** rodillo *m* de acero — **i** rullo *m* d'acciaio

13102 steel rule
f filet *m* en acier — **d** Stahllinie *f* — **n** stalen lijn *fm* — **e** pleca *f* de acero — **i** filetto *m* d'acciaio

13103 steel rule die — Die constructed by bending sections of steel cutting rule to the desired shape. Rule sections are then located and supported with accurately cut blocks of wood so that the whole assembly can be locked in a chase. It may be called either a block or jig die depending on whether the wood is in individual pieces or whether it is one piece partially cut on a jig saw to accept the formed cutting rule.
d Stahlband-Stanzform *f* — **n** bandstansvorm *m* — **e** troquel *m* de pleca de acero — **i** fustella *f* in nastro d'acciaio; forma *f* in nastro d'acciaio per fustellare

13104 steel strapping — Thin, narrow band of flexible steel to reinforce boxes, crates, etc., extensively used by paper manufacturers.
f feuillard *m* — **d** Bandeisen *n* — **n** bandijzer *n* — **e** hierro *m* llanta; hierro *m* pasamanos; hierro *m* en flejes; fleje *m* de hierro — **i** reggetta *f* metallica
13105 steel to steel — See metal to metal.
13106 steel water roller
f rouleau *m* dégraisseur; dégraisseur *m* — **d** Stahlreiber *m*; Feuchtreiber *m* — **n** stalen vochtrol *fm* — **e** rodillo *m* mojador en acero; rodillo *m* mojador de acero — **i** rullo *m* umettatore d'acciaio
13107 steepening incentive — See accelerating incentive.
13108 steep ratings — Work study term for a set of ratings in which the variations in the rate of working have been overestimated. See also flat ratings.
13109 stem; main stroke; main line — Main straight stroke of a letter, e.g., the uprights in H.
f haste *f*; fût *m*; plein *m*; jambage *m* — **d** Grundstrich *m*; Abstrich *m*; Säule *f* — **n** stok *m*; neerhaal *m* — **e** palo *m* grueso; perfil *m* grueso; grueso *m*; palo *m*; trazo *m* magistral — **i** gambo *m*; asta *f* di lettera
13110 stencil — Piece of paper, board or sheet metal in which letters or designs are cut out for reproduction on a flat surface with a paint brush.
f pochoir *m* — **d** Schablone *f* — **n** schabloon *n* — **e** patrón *m* — **i** maschera *f* di carta
13111 stencil, duplicating — See duplicating stencil.
13112 stencil paper — See duplicating stencil.
13113 stencil, wax — See duplicating stencil.
13114 step — Brief for *step-and-repeat. See also footboard.
13115 step-and-repeat camera — See step-and-repeat machine.
13116 step-and-repeat machine; step-and-repeat camera *GB*; **photocomposing machine** *US*; **photocomposer** *US* — Machine to locate a negative or positive to predetermined positions for exposure onto a large negative or press-plate. The photocomposing layout is called step-and-repeat when a single subject is exposed repeatedly over the sensitized surface, or combination when two or more different subjects are exposed on the same plate. Differentiate from *photo-typesetting machine.
f machine *f* à reports photo(graphiques); machine *f* à copier en répétition(s); machine *f* de copie en retiration; chambre *f* photo(graphique) à répéter; tireuse *f* — **d** Kopiermaschine *f*; Repetier-Kopiermaschine *f* — **n** kopieermachine *f* — **e** máquina *f* repetidora; repetidora *f*; máquina *f* de copia y repetición; cámara *f* de repetición; copiadora *f* de repetición; fotoimpresora *f* de repetición; transportadora *f* fotomecánica; máquina *f* de transporte fotomecánica — **i** macchina *f* ripetitrice; macchina *f* fotoripetitrice
13117 stepless variable speed
f vitesse *f* réglable en continu — **d** stufenlos regel-

bare Geschwindigkeit *f* — **n** traploos regelbare snelheid *f* — **e** velocidad *f* regulable sin escalonamiento — **i** velocità *f* regolabile in continuo
13118 step wedge — See grey scale.
13119 stere — See cubic metre.
13120 stereo — See stereotype plate.
13121 stereo department — See stereotyping section.
13122 stereotype — See stereotype plate.
13123 stereotype *v* — To cast a plate in stereotype metal from a matrix, as of papier-mâché, reproducing the form of the type, plate, etc., from which the matrix was made. Used generally for newspaper and the cheaper forms of printing.
f stéréotypie *f*; clicherie *f* — **d** Stereotypie *f* — **n** stypen; een styp maken — **e** estereotipar; estereotipia *f*; clisar; clisado *m* — **i** eseguire una stereotipia; duplicare in stereotipia
13124 stereotype alloy — See stereotype metal.
13125 stereotype dry mat — See flong.
13126 stereotype metal; stereotype alloy — Alloy somewhat different from a type metal, with a higher lead content, usually 77% lead, 15% antimony and 8% tin.
f matière *f* à stéréo; matière *f* à cliché; métal *m* de stéréotypie — **d** Stereotypiemetall *n* — **n** stereotypiemetaal *n*; stypmetaal *n* — **e** metal *m* para estereotipia; metal *m* para estereotipos; aleación *f* para estereotipia — **i** metallo *m* per stereotipia
13127 stereotype plate; stereotype; stereo — Printing plate cast of a lead-tin-antimony alloy from a paper composition matrix moulded from the original type or illustrations, the printing surface being a replica of the type or original engraving. Also plastic or rubber duplicate plates moulded from original relief printing material.
f stéréotype *m*; stéréo *m*; cliché *m* — **d** Stereoplatte *f*; Stereo *n* — **n** stereotype *fm*; styp *m*; stereo *m* — **e** estereotipo *m*; estéreo *m*; plancha *f* estereotipia; clisé *m* estereotípico; plancha *f* estereotipada; mono *m* Pe — **i** lastra *f* di stereotipia; stereo *m*
13128 stereotyper; stereotypist — In GB sometimes called foundryman.
f stéréotypeur *m* — **d** Stereotypeur *m*; Druckplattengießer *m* — **n** stereotypeur *m*; stypeur *m* — **e** estereotipador *m*; clisador *m* — **i** stereotipista *m*
13129 stereotyping department — See stereotyping section.
13130 stereotyping section; stereotyping department; stereo department
f section *f* de stéréotypie; département *m* de stéréotypie — **d** Stereotypieabteilung *f*; Stereoabteilung — **n** styp-afdeling *f*; styp *m coll.*; stereo *m coll.* — **e** sección *f* estereotípico; taller *m* de estereotipia; departamento *m* de estereotipia — **i** reparto *m* stereotipia
13131 stereotypist — See stereotyper.
13132 sterilization test — Test where a print is cut in half and one portion retained as a standard.

The other portion is put in a pressure cooker under five pounds of pressure for 20 minutes, removed and compared with the piece retained as a standard. Ink should not run or bleed into the fabric.

f essai *m* de stérilisation — **d** Test *m* auf Sterilisierechtheit; Sterilisierungsechtheitsversuch *m* — **n** sterilisatie-proef *fm* — **e** ensayo *m* de esterilización — **i** prova *f* di sterilizzazione

13133 Sterry effect — Photographic effect to decrease the contrast of light-sensitive layers. The exposed layer is bathed for about 1 min in a 1% solution of potassium dichromate and then rinsed. As a result of the reduction of the latent image the contrast is decreased. Instead of potassium dichromate, a weak solution of potassium persulphate or ammonium persulphate can be used. This process is specially applied for making positives from negatives with a too high contrast.

f effet *m* Sterry — **d** Sterry-Effekt *m* — **n** Sterryeffect *n* — **e** efecto *m* Sterry — **i** effetto *m* Sterry

13134 stet — Latin for "let it stand", written in the margin on proofs to cancel an alteration, dots being placed under the word or words that are to remain. See app. no. 5.

f reste; à maintenir — **d** bleibt; vive — **n** blijft zo; laten staan; zo laten — **e** ¡queda!; ¡vale!; deje como está — **i** vive!

13135 stibium — See antimony.

13136 stibnite — See antimony trisulphide.

13137 stick *v* — To stick bills, posters on walls and the like.

f afficher — **d** anschlagen — **n** aanplakken — **e** fijar — **i** affiggere

13138 stick, composing — See composing stick.

13139 stickers, advertising — See poster stamps.

13140 stickiness — See tack.

13141 stickiness — See tack.

13142 sticking tape — See self-adhesive tape.

13143 sticking-up board — See sticking-up sheet.

13144 sticking-up needle — Fixing needle for transfers.

f pointe *f* à piquer (les reports); aiguille *f* à report; pointe *f* à report — **d** Aufstechnadel *f*; Aufstecknadel *f* — **n** opsteeknaald *fm* — **e** alfiler *m* para picar reportes; aguja *f* para picar reportes — **i** spillo *m* da fissare

13145 sticking-up sheet; patching sheet; sticking-up board

f feuille *f* de piquage; carte *f* à piquer — **d** Aufstechbogen *m*; Aufsteckbogen *m* — **n** opsteekvel *n* — **e** hoja *f* de fijación; hoja *f* para picar reportes — **i** foglio *m* per puntature; foglio *m* di montaggio

13146 sticks — Wedge shaped pieces of wood placed at the sides of the pages for locking-up (hence gutter stick, head stick, foot stick, *side stick); now replaced by mechanically operated quoins.

f cales *fpl* — **d** Holzkeile *mpl* — **n** schenen *fmpl* — **e** calzos *mpl* para formas; cuchillos *mpl* — **i** cunei *mpl* (di legno)

13147 stick, setting — See composing stick.

13148 stick *v* **up** — To pin down a transfer.

f piquer (un report) — **d** aufstechen; aufstecken; aufnadeln — **n** opsteken; opprikken — **e** picar (reportes); atacar; apuntalar — **i** puntare

13149 stick-up initial — See cock-up initial.

13150 stiffener; envelope filler — Stout sheet placed in envelopes to protect the enclosure from creasing or crushing. It may be a corrugated board.

f renfort *m* — **d** Verstärkungseinlage *f*; Pappeinlage *f*; dicke Einlage *f* — **n** versterkingsbordje *n*; ruggetje *n* — **e** relleno *m* de sobre — **i** cartone *m* di rinforzo

13151 stiff ink — Insufficiently viscid ink that needs reducing.

f encre *f* ferme; encre *f* dure; encre *f* épaisse — **d** strenge Farbe *f*; zügige Farbe *f* — **n** strenge inkt *m*; stijve inkt *m* — **e** tinta *f* dura; tinta *f* espesa; tinta *f* viscosa — **i** inchiostro *m* duro; inchiostro *m* rigido

13152 stiffness — See rigidity.

13153 stiffness of an ink — Degree to which ink resists stirring.

i rigidità *f* di un inchiostro

13154 stiffness tester — Instrument to measure the stiffness or resistance of paper to bending (Schopper; Gurley).

f appareil *m* à déterminer la raideur — **d** Steifigkeitsprüfgerät *n* — **n** stijfheidsmeter *m* — **e** ensayador *m* de la rigidez (del papel) — **i** apparecchio *m* per misurare la rigidità

13155 stilb — Unit of brightness of a surface equal to 1 candle/cm^3. Abbrev. sb.

f stilb *m* — **d** Stilb *m* — **n** stilb *m* — **e** stilb *m* — **i** stilb *m*

13156 stillage *GB*; **skid** *US* — Load board comprising a single deck supported on bearers or legs, with an interrupted space between the bearers or legs for the entry of a stillage truck. If required, the deck may be fitted with a superstructure. There is a close similarity between stillages and certain types of pallets. Stillages are not normally intended for standing. While a pallet truck might be suitable for moving a stillage, a stillage truck cannot enter a pallet with centre bearers or a bottom deck.

f plateau *m*; plateau *m* sur patins — **d** Ladegestelle *n* — **n** laadbord *n*; verlaadbord *n* — **e** tablado *m*; tarima *f* — **i** piattaforma *f*

13157 still-bath development — See still development.

13158 still development; still-bath development — Development of a photographic emulsion in a tray without agitation or rocking of the tray.

f développement *m* sans agitation; développement *m* statique — **d** Standentwicklung *f* — **n** stilstaand ontwikkelen *n* — **e** revelado *m* en reposo; revelado *m* sin agitar — **i** sviluppo *m* statico; sviluppo *m* senza agitazione

13159 still etching — Etching of halftone images on copper in a tray without agitation of the ferric chloride solution.

f morsure *f* sans agitation; morsure *f* statique — **d** Standätzen *n* — **n** stilstaand etsen *n* — **e** mordido *m* en reposo; mordido *m* sin agitar — **i** mordenzatura *f* statica; mordenzatura *f* senza agitazione

13160 stipple *v* — To engrave by dots instead of lines. In process engraving, used to arrange dots, regular or irregular in formation.

f pointiller — **d** punktieren — **n** punteren; pointilleren; stippelen — **e** puntear — **i** punteggiare

13161 stipple engraving — Method of copperplate engraving in which light and shade are represented by employing dots or flicks instead of lines. The plate is covered with the usual etching ground and the outlines and principal shadows are indicated by dotting this ground with the etching needle or with the *roulette. The plate is then bitten with acid and the ground removed. The composition is finished by dotting or stippling with the print of the *stipple graver.

f gravure *f* au pointillé — **d** Punktiermanier *f* — **n** stippelgravure *fm*; Engelse prent *fm* — **e** grabado *m* punteado; grabado *m* a puntas — **i** incisione *f* a puntini

13162 stipple graver — See stippling tool.

13163 stippling machine — Machine to apply a stipple finish to a printing plate.

f machine *f* à pointiller — **d** Punktiermaschine *f* — **n** stippelmachine *f* — **e** máquina *f* para trazar puntos — **i** macchina *f* per punteggiare

13164 stippling tool; stipple graver — Graver or tool to stipple metal plates and to repair scratches and dots in halftone etchings.

f burin *m* à pointiller — **d** Punktierstichel *m* — **e** buril *m* para puntear; buril *m* para picar; buril *m* de picado; buril *m* de punteado — **i** punta *f* da punteggiare

13165 stir *v*; **agitate** *v* — To move intermittently or constantly in processing to ensure complete and even action of all chemicals.

f agiter; remuer; mélanger; brasser — **d** rühren; aufrühren; mischen; schütteln; bewegen — **n** roeren; omroeren; mengen; schudden; bewegen — **e** remover — **i** agitare; miscelare

13166 stirrer — See agitator.

13167 stirring rod, glass — See glass rod.

13168 stitch driver — Part of the stitch mechanism which forces the staple through the paper.

f pousseur *m* de l'agrafe — **d** Klammertreiber; Heftklammerstößel *m* — **n** drijver *m* — **i** spingitore *m* della grappa

13169 stitched — Bound in a paper cover.

f broché — **d** broschiert — **n** gebrocheerd — **e** encuadernado a la rústica — **i** legato alla rustica

13170 stitcher anvil — Die on which the wire staple is clenched.

d Klammerschließer *m* — **n** ombuiger *m* — **e** yunque *m* de cosedora — **i** appoggio *m* di chiusura

13171 stitcher feed box — Mechanism which feeds the length of wire necessary to form a staple or stitch.

f boîte *f* de fil métallique — **d** Heftdrahttransporteur *m* — **n** hechtdraadvoeder *m* — **e** caja *f* de hilo metálica — **i** alimentatore *m* del filo metallico

13172 stitcher horn — Steel piece around which the wire staple is formed.

f moule *m* d'agrafe — **d** Hefthorn *n* — **n** hechthoorn *m* — **e** molde *m* de grapa — **i** forma *f* d'alimentazione del filo (metallico)

13173 stitcher plunger — Part of the stitching mechanism which under cam operation rises on either side of the stitcher horn to form the staple and then supports the latter while it is being clenched.

d Heftstempel *m*; Klammerstütze *f* — **n** draadsteun *m* — **e** soporte *m* de grapa — **i** appoggio *m* della grappa

13174 stitcher, wire — See wire stitcher.

13175 stitching, all along — See all along stitching.

13176 stitching cylinder — Cylinder which incorporates the mechanism to form and clench the wire staple in the folder of a rotary press.

f cylindre *m* de piquage — **d** Heftzylinder *m* — **n** nietcilinder *m*; hechtcilinder *m* — **e** cilindro *m* cosedor — **i** cilindro *m* di cucitura

13177 stitching head

f tête *f* piqueuse; tête *f* agrafeuse — **d** Heftkopf *m* — **n** hechtkop *m* — **e** cabezal *m* de cosedora — **i** testa *f* cucitrice

13178 stitching needle; bookbinder's needle

f aiguille *f* à relier; aiguille *f* de relieur — **d** Heftnadel *f* — **n** boekbindersnaald *fm* — **e** aguja *f* de encuadernar — **i** ago *m* da legatore; agone *m* da legatore

13179 stitching wire; binding wire — Steel wire of round, square or rectangular shape, used in stitching (round wire; flat wire).

f fil *m* métallique; fil *m* pour brocher — **d** Heftdraht *m*; Draht *m* zum Heften — **n** hechtdraad *n* — **e** alambre *m* para encuadernación; alambre *m* para máquinas de coser; alambre *m* para coser — **i** filo *m* metallico per cucire

13180 stitching, wire — See wire stitching.

13181 stock; stuff — Aqueous suspension of one or more pulps and other material from the stage of disintegration of the pulp to the formation of the sheet of paper or board.

f pâte *f* à papier — **d** Ganzstoff *m*; Papierstoff *m*; Stoff *m*; Zeug *n* — **n** heelstof *fm*; stof *fm* — **e** pasta *f* refinada — **i** impasto *m*; pasta *f* per carta; sospensione *f* fibrosa

13182 stock — Printer's term for paper.

f papier *m* — **d** Papier *n* — **n** papier *n* — **e** papel *m* — **i** carta *f*

13183 stock — Metal part of a printing roller. See also roller stock.

13184 stock *v* — To lay up in store for future use.

f stocker; emmagasiner — **d** lagern; speichern; auf Lager nehmen — **n** opslaan — **e** almacenar — **i** immagazzinare

13185 stock catalogue; publisher's catalogue *GB*;

publisher's list *US*
f catalogue *m* de livres de fonds; catalogue *m* d'éditeur — **d** Verlagskatalog *m*; Lagerkatalog *m* — **n** fondscatalogus *m* — **e** catálogo *m* de libros de fondo; catálogo *m* de editor — **i** catalogo *m* dei libri di fondo; catalogo *m* d'editore

13186 stock chest; stuff chest; machine chest — Large tank with agitator to receive and hold stock from the beaters, or that going to the paper machine.
f cuvier *m* à pâte; réservoir *m* à pâte; cuvier *m* d'alimentation — **d** Stoffbütte *f*; Rührbütte *f*; Vorratsbütte *f*; Maschinenbütte *f* — **n** stofkuip *fm*; roerkuip *fm* — **e** tina *f* de alimentación; tina *f* de la máquina; tina *f* de mezcla; caja *f* de mezcla; caja *f* provista; depósito *m* desgotador — **i** tina *f* di pasta; tino *m* di pasta; tina *f* di macchina; tino *m* di macchina

13187 stock exchange list — See exchange list.

13188 stock item
f papier *m* en stock; article *m* en stock — **d** Lagerbestand *m* — **n** magazijnsoort *n*; magazijnartikel *n* — **e** papel *m* de almacén; artículo *m* de almacén — **i** carta *f* di magazzino; articolo *m* di magazzino

13189 stock order — Order to be filled directly from warehouse inventory of a standard grade, size, weight, and colour, as opposed to a special *making order.
f commande *f* sur stock — **d** Bestellung *f* ab Lager; Lagerauftrag *m* — **n** levering *f* van een magazijnsoort; magazijnorder *fm* — **e** pedido *m* de almacén — **i** ordine *m* di magazzino

13190 stock preparation — The operations which occur between pulping (or bleaching) and formation of the web on a paper machine. It may include for example, repulping, beating, refining, cleaning, etc.
f préparation *f* des pâtes; traitement *m* de la pâte — **d** Stoffansatz *m*; Stoffbereitung *f*; Stoffaufbereitung *f*; Stoffherstellung *f* — **n** stofbereiding *f* — **e** preparación *f* de la pasta; preparación *f* de pastas

13191 stock room — See storehouse.

13192 stock sizes — Standard sizes of papers and boards which are usually stocked by producers, distributors, or consumers and which are reordered from time to time. See app. no. 3.
f formats *mpl* en stock; formats *mpl* sur stock — **d** Lagerformaten *npl* — **n** magazijninformaten *npl* — **e** tamaños *mpl* de almacén; formatos *mpl* de almacén; tamaños *mpl* corrientes; tamaños *mpl* regulares; tamaños *mpl* standard; tamaños *mpl* de corte — **i** formati *mpl* a magazzino

13193 Stoddard solvent — In US a refined low-aromatic petroleum distillate with volatility, flash point and other properties making it suitable for use as a dry-cleaning solvent. Flash point 38 °C, minimum boiling range approx. 150-205 °C. TL-value: 500 ppm. See also SBP spirit.

13194 stokes — Unit of kinematic viscosity. Abbrev. St. In the SI: 1 St = 1 cm²/s = 10^{-4} m²/s.

Name derived from Sir George G. Stokes (1819-1903), English mathematician and physicist. See also centistokes.
f stokes *f* — **d** Stokes *f* — **n** stokes *fm* — **e** stokio *m* — **i** stokes *m*

13195 Stokes' law of fall — The Stokes' law of fall is:
$$\eta = 2(p_b - p)_g R^2/9 \ V,$$
where η is the viscosity of the test material; p is its density; p_b is the density of the ball, and V is its velocity.
f loi *f* de chute de Stokes — **d** Stokesches Fallgesetz *n* — **n** valwet *m* van Stokes — **e** ley *f* de la caída de Stokes — **i** legge *f* della caduta di Stokes

13196 stone — Unit of weight in the avoirdupois system. See app. no. 7 and 7a.

13197 stone — See imposing stone.

13198 stone engraving — Engraving of a lithographic stone, i.e., a very fine-grained limestone. The print from this stone is made on a hand press. See also lithographic stone.
f gravure *f* sur pierre — **d** Steingravur *f* — **n** steengravure *f* — **e** grabado *m* sobre piedra; grabado *m* litográfico — **i** incisione *f* su pietra

13199 stone, grinding — See grinding stone.

13200 stone hand — See lock-up man.

13201 stone, lithographic — See lithographic stone.

13202 stone man — See lock-up man.

13203 stone *v* **out** — To remove an image design from an offset plate with an abrasive stick, sometimes called a *snakeslip.
f poncer — **d** ausschleifen; wegschleifen — **n** uitslijpen; wegslijpen — **i** cancellare con matita abrasiva

13204 stone planing machine — See levigator.

13205 stone printing — Printing from a specially prepared lithographic stone. See also lithography.
f lithographie *f*; impression *f* lithographique — **d** Steindruck *m* — **n** steendruk *m* — **e** litografía *f*; impresión *f* litográfica — **i** litografia *f*; stampa *f* litografica su pietra

13206 stone work — Work in the composing room which is done on an imposing surface, such as imposition, etc.
f travail *m* au marbre — **d** Arbeit *f* an Schließtisch — **n** werk *n* aan de steen; steenwerk *n* — **e** trabajo *m* en mármol; imposición *f* y acuñado *m* — **i** lavoro *m* al piano di imposizione

13207 stop — See diaphragm.

13208 stop *v* — See throw *v* out of gear.

13209 stop bath; short-stop — Dilute acetic acid solution used as an intermediate bath between development and fixing of photographic negatives and prints, to stop further action of the developer.
f bain *m* d'arrêt — **d** Stoppbad *n*; Unterbrechungsbad *n* — **n** stopbad *n* — **e** baño *m* detenedor; baño *m* de detención; baño *m* de contrarrestante; baño *m* de interrupción; baño *m* de paralizador — **i** bagno *m* d'arresto

13210 stop-cylinder machine; stop-cylinder press — Press whose cylinder stops after each impression as opposed to a two-revolution press.

f presse *f* à arrêt de cylindre; presse *f* à temps d'arrêt; presse *f* à déroulement du cylindre; presse *f* en blanc — **d** Stoppzylinderpresse *f*; Schnellpresse *f* — **n** stopcilinderpers *f*; snelpers *f* — **e** máquina *f* de parada de cilindro; prensa *f* de parada de cilindro — **i** macchina *f* ad arresto del cilindro

13211 stop-cylinder press — See stop-cylinder machine.

13212 stop *v* **down** — To reduce the aperture of a lens; to use a smaller aperture.

f diaphragmer — **d** abblenden; Blende verkleinern; Blende schließen — **n** diafragmeren; een kleiner diafragma kiezen — **e** diafragmar — **i** ridurre l'apertura del diaframma

13213 stopgap — See filler.

13214 stop, halftone — See halftone stop.

13215 stop *v* **out; block** *v* **out** — To stage halftone plates during relief etching; to protect certain areas of deep-etch plates by applying gum solution before application of plate bases, so that no ink attracting medium will be deposited on the protected areas.

f reboucher — **d** abdecken — **n** afdekken; indekken — **e** bloquear — **i** ricoprire con vernice resistente

13216 stopping out varnish — Staging solution of fluid acid resists to stage plates.

f vernis *m* de rebouchage — **d** Abdecklack *m* — **n** afdeklak *mn* — **e** barniz *m* para reserva; solución *f* para reserva — **i** vernice *f* per ricopertura

13217 stop-press — Column in a newspaper reserved for late news.

f colonne *f* de dernières nouvelles; blanc *m* pour dernières nouvelles — **d** Depeschen-Spalte *f* — **n** kolom *fm* voor het laatste nieuws — **e** columna *f* de novedades — **i** colonna *f* riservata alle ultime notizie

13218 stop-press unit — Unit for imprinting late news, e.g. *fudge.

f groupe *m* fudge — **d** Depeschen-Eindruckwerk *n* — **n** indrukwerk *n* voor het laatste nieuws — **e** grupo *m* de novedades — **i** dispositivo *m* per la stampa delle ultime notizie

13219 stop-rack locking bar

f fourrier *m* pour crémaillère — **d** Zahnstangesperriegel *m* — **n** tandreepstopper *m* — **e** cerrojo *m* de la cremallera — **i** chiavistello *m* d'arresto della cremagliera

13220 stops, fountain — See fountain stops.

13221 stopwatch

f montre *f* à arrêt; compte-secondes *mpl*; chronomètre *m* — **d** Stoppuhr *f*; Sekundenzähler *m* — **n** stophorloge *n*; stopwatch *mn depr.*; chronometer *m* — **e** reloj *m* de trinquete; cronometrador *m*; cuentasegundos *m*; reloj *m* de segundos muertos; cronógrafo *m* — **i** cronometro *m*; contasecondi *m*

13222 storage — See store.

13223 storage battery — See accumulator.

13224 storage capacity — See store capacity.

13225 storage cell — See accumulator.

13226 storage cell — See cell.

13227 storage life — See shelf life.

13228 storage of paper; paper storage

f magasinage *m* du papier; stockage *m* du papier — **d** Papierlagerung *f* — **n** papieropslag *m* — **e** almacenaje *m* de papel — **i** immagazzinaggio *m* della carta; immagazzinamento *m* della carta

13229 storage room for unsold copies

f cimetière *m* des invendus — **d** Remittendenfriedhof *m* — **n** magazijn *n* voor onverkochte exemplaren

13230 store *GB*; **storage** *US* — Information file accessible to the computer. It holds information such as rate tables, current inventories, balances etc., and programming instructions. An internal store is a part of the computer itself such as drums, cores or thin fim. An external store is separate from the computer such as paper tapes, magnetic tape or punched cards.

f mémoire *f* — **d** Speicher *m* — **n** geheugen *n*; opslagorgaan *n* — **e** memoria *f*; almacenamiento *m* — **i** memoria *f*

13231 store — See storehouse.

13232 store *v* — To transfer information to a device from which it can be obtained when desired.

f mettre en mémoire; mémoriser; emmagasiner — **d** speichern — **n** opslaan; vastleggen — **e** almacenar — **i** memorizzare

13233 store *v* **and forward** — To interrupt data flow from the originating terminal to the receiver by storing the information en route and forwarding it later.

f mémoriser et faire suivre plus tard — **d** speichern und später folgen lassen — **n** opslaan en later laten volgen — **e** almacenar y conmutar hacia adelante — **i** memorizzare e spedire più tardi

13234 store capacity *GB*; **storage capacity** *US* — In computer terminology, amount of data that can be contained in a store device.

f capacité *f* de mémoire — **d** Speicherkapazität *f* — **n** geheugencapaciteit *f*; opslagcapaciteit *f* — **e** capacidad *f* de memoria — **i** capacità *f* di memoria

13235 store cell — See cell.

13236 stored program — Set of instructions in the memory section that can run the computer or be cut in to take over from a normal program when necessary.

f programme *m* mémorisé — **d** gespeichertes Programm *n* — **n** opgeslagen programma *n* — **e** programa *m* almacenado — **i** programma *m* memorizzato

13237 stored-program computer — Computer capable of storing all or some of its instructions and such that the stored instructions may be altered within the computer.

f ordinateur *m* à programme mémorisée — **d** speicherprogrammierter Computer *m*; speicherprogrammierter Rechner *m*; speicherprogrammierte Datenverarbeitungsanlage *f* — **n**

computer *m* met opgeslagen programma — **e** computadora *f* de programa almacenado — **i** calcolatore *m* a programma memorizzato

13238 storehouse; store; stock room — Building or section in which paper or other goods are kept in stock.
f magasin *m*; dépôt *m* — **d** Papierlager *n*; Lagerhaus *n*; Lagerraum *m*; Warenlager *n*; Schriftmagazin *n* — **n** magazijn *n*; papiermagazijn *n* — **e** almacén *m*; bodega *f* para papel *CA* — **i** magazzino *m*

13239 store keeper
f magasinier *m*; garde-magasin *m* — **d** Magazinverwalter *m*; Lagerverwalter *m* — **n** magazijnmeester *m*; magazijnchef *m* — **e** almacenero *m*; almacenista *m*; guardaalmacén *m*; guardalmacén *m*; jefe *m* de almacén — **i** magazziniere *m*

13240 storing; warehousing — Keeping goods in store or warehouse.
f entreposage *m*; stockage *m* — **d** Einlagerung *f*; Auflagerung *f*; Aufbewahrung *f* — **n** opslag *m* — **e** almacenamiento *m* — **i** immagazzinamento *m*

13241 straight accent — See long accent.

13242 straight-edge; ruler — Piece of steel, 90-120 cm (3-4 ft) long, to test the alignment, etc., of pages, particularly in large formes of several pages.
f règle *f* de précision; règle *f* à araser — **d** Lineal *n*; Richtscheit *n*; Richtlineal *n* — **n** liniaal *fmn* — **e** regla *f* de precisión — **i** regolo *m*; riga *f*

13243 straight-line image reverser — Optical device attached to process lenses for lateral reversal of negatives without turning the camera at right angles to the original or copyboard.
f dispositif *m* de retournement de l'image — **d** Umkehrspiegel *m* — **n** omkeerspiegel *m* — **e** invertidor *m* de derecha a izquierda — **i** specchio *m* di rovesciamento; specchio *m* d'inversione; dispositivo *m* per il rovesciamento laterale

13244 straight-line pen — See drawing pen.

13245 straight-line portion — See curve.

13246 straight-line tint — See one-way tint.

13247 straight matter; text matter; text body matter — Type matter that does not contain tables, cuts, formulea, etc.
f composition *f* simple; composition *f* courante — **d** glatter Satz *m* — **n** plat zetsel *n*; plat werk *n* — **e** composición *f* simple; composición *f* corrida; composición *f* de texto; composición *f* ordinaria; composición *f* seguida; composición *f* corriente; materia *f* corrida; materia *f* de texto — **i** composizione *f* corrente

13248 straight run — Running the rotary press without cylinder collection and sometimes without turner bars to give the maximum number of copies to the cylinder revolution.
f sortie *f* directe — **d** doppelte Produktion *f*; ungesammelte Produktion *f*; Nichtsammeln *n* — **n** dubbele produktie *f*; rechte loop *m* — **e** tirada *f* de doble producción; tiro *m* doble; tiro *m* corriente — **i** uscita *f* diretta

13249 strainer — See knot screen.

13250 strain gauge — See resistance strain gauge.

13251 stratch brushed aluminium foil — Foil with a rough surface, produced by mechanical abrasion with a wire brush.

13252 stratified — See layered.

13253 stratified papers
f papiers *mpl* stratifiés — **d** Schichtpapiere *npl* — **n** gelaagde papieren *npl* — **e** papeles *mpl* estratificados — **i** carte *fpl* stratificate

13254 strawboard — Cheap board made from cooked straw used for corrugating or as filler in pasted and lined boards.
f carton *m* de paille; carton *m* paille — **d** Strohpappe *f*; Strohkarton *m* — **n** strobord *n*; strokarton *n* — **e** cartón *m* de paja; cartón *m* paja — **i** cartone *m* paglia

13255 straw paper
f papier *m* paille — **d** Strohpapier *n* — **n** stropapier *n* — **e** papel *m* de paja — **i** carta *f* paglia

13256 strawpulp — Pulp obtained from straw by various processes.
f pâte *f* de paille — **d** Strohzellstoff *m*; Halbstoff *m* aus Getreidestroh; Halbstoff *m* aus Stroh; Strohstoff *m* — **n** strocelstof *fm* — **e** pulpa *f* de paja; pasta *f* de paja — **i** pasta *f* di paglia

13257 stream feeder — Press feeder which keeps several sheets of paper, overlapping each other, moving toward the grippers.
f margeur *m* à nappe — **d** Staffeleinleger *m*; Schuppenanleger *m*; Schuppenanlegeapparat *m* — **n** staffel-inlegapparaat *n* — **e** ponepliegos *m* por escalas — **i** mettifoglio *m* a squame

13258 street — See gutter.

13259 strengthen *v* the ink
f endurcir l'encre — **d** die Farbe strenger machen — **n** de inkt strenger maken — **e** endurecer la tinta — **i** indurire l'inchiostro

13260 stress — Thickening of the stroke of a curve in a letter, which may be vertical, biased, abrupt or gradual.
f graisse *f* — **d** Verdickung *f*; Anschwellen *n* — **n** verdikking *f*

13261 stress — Force acting on a unit area in a solid, as in the theory of elasticity.
f effort *m* — **d** Beanspruchung *f*; Spannung *f* — **n** spanning *f* — **e** esfuerzo *m* — **i** sollecitazione *f*

13262 stretch — Paper stretch is largely due to moisture absorption from the air or during printing operations, greater across the paper grain than along it. When it occurs between printing colours and cannot be compensated for by plant humidity controls, allowances for subject displacements must be provided for in the stripping procedure. Stretch may also be the result of mechanical pressure. It implies an external pulling force.
f allongement *m* — **d** Dehnung *f*; Ausdehnung *f*; Dehnbarkeit *f* — **n** rek *m*; uiteenrekking *f*; uitrekking *f* — **e** estiramiento *m* del papel — **i** allungamento *m* della carta

13263 stretch *v* — The stretching of metal offset printing plates.

f étirer — **d** strecken — **n** strekken — **e** estirar — **i** stirare; mettere in tensione

13264 stretch at break; stretch at breaking point — Elongation of a test piece of paper or board at the moment of rupture under a tensile load, as defined in the standard method of test; usually expressed as a percentage of the initial length.

f allongement m à la (rupture par) traction; allongement m à la tension — **d** Bruchdehnung f; Zerreißdehnung f — **n** rek m bij breuk — **e** alargamiento m de ruptura; alargamiento m de la tensión; alargamiento m a la tracción — **i** allungamento m alla trazione

13265 stretch at breaking point — See stretch at break.

13266 stretchers — Lateral members fixed between the side frames of a printing press.

f entretoises fpl — **d** Traversen fpl; Zwischenbrücken fpl (im Untergestell der Presse); Spannrahmen mpl — **n** dwarssteunen mpl — **e** travesaños mpl — **i** traverse fpl

13267 stretchers — See stretching rolls.

13268 stretching rolls; stretchers — Small rollers which can be moved back and forth or up and down to increase or reduce the tension of the paper web on a paper machine, or on a converting machine.

f rouleaux mpl tendeurs — **d** Spannrollen fpl — **n** spanrollen fmpl — **e** rodillos mpl tensores; rodillos mpl de tensión — **i** rulli mpl da tensione

13269 strike; drive — In type-founding, the process of driving a punch of hardened steel into a bar of copper or brass, so producing a matrix. Also the impression itself.

f poinçonnage m — **d** Abschlag m — **n** afslag m — **e** punzonado m (de matrices) — **i** spinta f del punzone

13270 strike-in — Penetration of the ink vehicle into paper.

f pénétration f — **d** Einschlagen n — **n** wegslaan n — **e** penetración f — **i** penetrazione f

13271 strike'm stiff — See exclamation mark.

13272 strike v out; cross v out; put v a strike through; cancel v

f biffer; barrer; rayer — **d** durchstreichen; ausstreichen — **n** doorhalen; doorschrappen; schrappen; doorstrepen — **e** borrar; rayar; testar; suprimir; cancelar — **i** cancellare

13273 striker — Device, usually attached to a ruling machine and synchronized with the feed release, to determine the length of the ruled lines by causing the ruling pens to drop and lift at given intervals.

f dispositif m d'arrêtage; dispositif m d'arrêt — **d** Abhebevorrichtung f — **n** strijker m — **e** alzaplumas m; muesca f — **i** dispositivo m per levatura intermittente

13274 striker — Worker on strike.

f gréviste m — **d** Streikende m; Streikender m; Ausständige m; Ausständiger m — **n** staker m — **e** huelguista m — **i** scioperante m

13275 strike sheet — See die sheet.

13276 strike-through — Penetration of the vehicle of a printing ink through the sheet so that it is apparent on the opposite side. The result is frequently a stain on the opposite side of the sheet.

f traversement m; transpercement m — **d** Durchschlagen n — **n** doorslaan n — **e** pasar; traspintarse; atravesar — **i** trapasso m; controstampa f per penetrazione

13277 string — Series of two or more characters, usually in machine-readable form, to be analyzed for their composite meaning.

f chaîne f — **d** Kette f — **n** rij fm — **e** secuencia f; serie f — **i** stringa f

13278 string — See page cord.

13279 string and button envelope — Catalogue envelope in which the flap is fastened by wrapping an attached string around a fibre disk on the envelope.

f enveloppe f à bouton — **d** Umschlag m mit Kordelverschluß; Versandtasche f mit Fadenverschluß — **n** envelop(pe) fm met touwtje en knoopje — **e** sobre m de botón; bolsa f de botón — **i** busta f con chiusura a spago

13280 string proof — See slip proof.

13281 string substitution — Use of hardware or software to substitute for one or several characters input by simple keystroking, a more elaborate series of formatted information, such as complex typographical commands. In some cases text material, such as boiler plate, is substituted in this way to simplify the input.

13282 strip v a forme — To unlock the printing forme on a letterpress machine and remove the furniture.

f abattre la garniture; enlever la garniture — **d** den Rahmen abnehmen; das Format abnehmen; das Format abschlagen — **n** een vorm uitslaan — **e** levantar la imposición; desmontar; deshacer; desarmar — **i** disimporre; sguarnire una forma

13283 stripfilm; stripping film — Process film for line and halftone photography, consisting of a gelatino-silver emulsion coated on a temporary support, which permits stripping or removal of the negative image after fixing, washing and transferring, and adhering it in position on a glass flat or other final support. Substitute for wet collodion plates.

f film m pelliculable; stripfilm m — **d** Abziehfilm m; Abzugsfilm m; abziehbarer Film; Kopierfilm m; Stripfilm m — **n** stripfilm m — **e** película f despegable; film m peliculable — **i** pellicola f di emulsione staccabile

13284 stripfilm cement — Liquid adhesive for holding stripfilm negatives on glass supports to prevent lifting and curling.

f adhésif m pour film pelliculable — **d** Filmklebelack m; Stripfilmzement m — **n** stripfilm-cement mn — **e** adhesivo m para montaje de películas despegables — **i** adesivo m per montaggio di pellicole staccabili

13285 stripper — Person in charge of *stripping.

f strippeur m; ≈ copiste m — **d** ≈ Kopierer m;

≈ Kopist *m* — **n** ≈ kopiïst *m* — **e** peliculador *m* — **i** ≈ copista *f*

13286 stripper — See stripping cylinder.

13287 stripper fingers; strippers — Metal fingers to remove the sheet from the printing cylinder of a flat-bed press.

f décolleurs *mpl* — **d** Bogenabstreicher *mpl* — **n** afnemers *mpl* — **e** separadores *mpl*; puentes *mpl* de salida — **i** separatori *mpl*

13288 strippers — See stripper fingers.

13289 stripping — Originally, the removal of the photographic emulsion with its image from individual negatives and combining them in position on a glass plate. Now the use of stripfilm materials, and the cutting, attachment, and other operations for assembling cut film sections to produce a flat.

f pelliculage *m*; montage *m* — **d** Abziehen *n*; Montieren *n*; Strippen *n* — **n** strippen *n*; monteren *n* — **e** peliculado *m*; inversión *f* de la película — **i** distacco *m* della pellicola

13290 stripping — Removing the outer wrappers and damaged outer layers from a reel of paper.

f enlèvement *m* des biftecks — **d** Rollenabriß *m* — **n** verwijdering *f* van de buitenste laag; weghalen *n* van de buitenlaag — **e** desprendimiento *m* del rezago — **i** rimozione *f* dell'involucro esterno

13291 stripping — Trimming off excess material which remains after a processing operation, e.g., removal of excess board after die cutting cartons, or blank of a set-up paper box, removal of excess metal along side seam after flash welding operation.

f décorticage *m*; éjection *f* des déchets — **d** Ausbrechen *n* der Abfälle — **n** uitbreken *n* van het afval — **e** descartonar; expulsión *f* de los recortes — **i** spoglio *m* dei refili

13292 stripping — See roller stripping.

13293 stripping cylinder; stripper — Unit which removes scrap from the cut web.

f cylindre *m* d'éjection; cylindre-éjecteur *m* découpes — **d** Ausbrechzylinder *m*; Ausstoßzylinder *m*; Ausstoßer *m* — **n** uitbreekcilinder *m* — **e** cilindro *m* expulsor — **i** cilindro *m* per lo spoglio dei ritagli

13294 stripping film — See stripfilm.

13295 stroboscope — Optical instrument which enables rapidly moving parts to be viewed as though stationary.

f stroboscope *m*; appareil *m* stroboscopique — **d** Stroboskop *n*; Beobachtungsgerät *n* — **n** stroboscoop *m* — **e** estroboscopio *m* — **i** stroboscopio *m*

13296 stroke — Computer term. To generate a character image(s) by laying down a consecutive series of strokes either vertically or horizontally, as on a cathode ray tube or photographic film or paper, either by the deflection of an electron beam or a coherent light source (laser).

13297 strong acid; active acid — Acid which breaks apart, or ionizes almost completely in a water solution, e.g., hydrochloric acid, nitric acid,

sulphuric acid.

f acide *m* fort — **d** starke Säure *f* — **n** sterk zuur *n* — **e** ácido *m* concentrado — **i** acido *m* forte

13298 strong ink — Long ink with thixotropy.

f encre *f* ferme — **d** strenge Druckfarbe *f*; stramme Druckfarbe *f* — **n** strenge inkt *m* — **e** tinta *f* compacta; tinta *f* consistente — **i** inchiostro *m* duro; inchiostro *m* rigido

13299 strong varnish

f vernis *m* fort — **d** strenger Firnis *m* — **n** strenge vernis *mn* — **e** barniz *m* fuerte — **i** vernice *f* forte; vernice *f* vigorosa

13300 structural change

f changement *m* de structure — **d** Veränderung *f* des Gefüges; Strukturänderung *f* — **n** structurele verandering *f* — **e** variación *f* estructural — **i** cambio *m* strutturale

13301 structural formula — Formula representing the way in which the atoms in a molecule are bonded together.

f formule *f* de constitution; formule *f* rationnelle — **d** Strukturformel *f*; Wertigkeitsformel *f* — **n** structuurformule *fm* — **e** fórmula *f* estructural; fórmula *f* de estructura; fórmula *f* de constitución — **i** formula *f* di struttura

13302 structural viscosity — Viscosity that depends, at least partly, on the structure of the test material. Applied to non-Newtonians.

f viscosité *f* de structure — **d** Strukturviskosität *f* — **n** structurele viscositeit *f*; structuurviscositeit *f* — **e** viscosidad *f* estructural — **i** viscosità *f* strutturale

13303 structure — Arrangement of parts or elements in a substance or object.

f structure *f*; constitution *f* — **d** Struktur *f*; Aufbau *m*; Bau *m*; Beschaffenheit *f*; Konstitution *f*; Gefüge *n* — **n** structuur *f*; bouw *m* — **e** estructura *f*; construcción *f*; formación *f*; constitución *f* — **i** struttura *f*; costituzione *f*

13304 Struwwelpeter — *Picture book for children of 3-6 years, of the physician H. Hoffman, Frankfurt (1845).

13305 stub; counterfoil — Inner end of a duplicate blank in which a memorandum or record is entered corresponding to that of the detached blank.

f souche *f* — **d** Souche *f*; Abschnitt *m* — **n** souche *fm*; strookje *n* — **e** talón *m* (de cheque); talonario *m* — **i** madre *f*; matrice *f* (di un bollettario)

13306 stub — Descriptive column of a *table.

13307 stub — See butt end.

13308 stubs — See compensation guards.

13309 stud horse — Large black type display such as is issued for auction bills, horse sales, etc. A type bolder and bigger than normal in newspapers and other work where small type is necessary. See also poster type, wood letter.

13310 stuff — See stock.

13311 stuff chest — See stock chest.

13312 stuffer; envelope stuffer — Printed piece of

paper or cardboard enclosed with a letter in an envelope.

f encart *m* — **d** Briefbeilage *f*; Briefanlage *f* — **n** bijpakker *m*; bijsluiter *m* — **e** anuncio *m* de incluir (con la correspondencia) — **i** allegato *m* stampato

13313 stump — Roll of paper or leather to tone shades and lights of crayon or charcoal drawing.

f estompe *f*; tortillon *m* — **d** Wischer *m* — **n** doezelaar *m* — **e** esfumino *m*; difumino *m* — **i** sfumino *m*

13314 style of the house — See house style.

13315 stylus printer — See matrix printer.

13316 styrene; phenylethylene — Starting material for polystyrene resins. These have excellent clarity and mechanical properties and good resistance to moisture and chemical attack. Used as a plasticizer for nitrocellulose, and in synthetic rubber.

f styrolène *m* — **d** Styrol *n* — **n** styreen *mn* — **e** estireno *m* — **i** stirolo *m*

13317 styrene resin; styrol; styrolene; vinylbenzene — Refractive, colourless or yellowish oily liquid, soluble in alcohol and ether. Formula: $C_6H_5CHCH_2$.

f résine *f* de styrolène — **d** Styrolharz *n* — **n** styreenhars *mn* — **e** resina *f* de estirol — **i** resina *f* stirolica

13318 styrol — See styrene resin.

13319 styrolene — See styrene resin.

13320 sub-division rule

f ligne *f* de séparation; filet *m* de rubrique — **d** Unterteilungslinie *f* — **n** scheidingslijn *fm* — **e** línea *f* separatoria; línea *f* divisoria — **i** linea *f* di separazione; linea *f* divisoria

13321 sub head — See sub heading.

13322 sub heading; sub head — Head of any description making a subdivision of a chapter.

f sous-titre *m* — **d** Untertitel *m* — **n** tussentitel *m* — **e** subtítulo *m* — **i** sottotitolo *m*

13323 subjective brightness — See brightness.

13324 subject program — Original program, written in a higher level language, before it is compiled into a machine language (or object) program.

13325 sublimate *v* — See sublime *v*.

13326 sublimation — Direct conversion of a substance from the solid state to the gaseous state, without passing through the liquid state.

f sublimation *f* — **d** Sublimation *f* — **n** sublimering *f* — **e** sublimación *f* — **i** sublimazione *f*

13327 sublime *v*; **sublimate** *v* — To volatilize from the solid state to a gas, and then condense again as a solid without passing through the liquid state.

f sublimer — **d** sublimieren — **n** sublimeren — **e** sublimar — **i** sublimare

13328 sub-routine — Routine that is part of another computer routine. Machine coded steps required to instruct the processor to undertake an operation, the parameters being clearly defined. Manufacturers supply a wide range of sub-

routines which may be called during a user's own program (e.g., calculations). A user's own sub-routine may be employed when a sub-routine is called from more than one place in the main program.

f sous-programme *m* — **d** Unterprogramm *n* — **n** onderprogramma *n* — **e** subprograma *m* — **i** sottoprogramma *m*

13329 subscriber — One who pays for a book, magazine, or newspaper for a stated term.

f abonné *m*; abonnée *f* — **d** Abonnent *m* — **n** abonné *m*; intekenaar *m* — **e** abonado *m*; abonada *f*; suscriptor *m*; suscriptora *f*; subscriptor *m*; subscriptora *f* — **i** abbonato *m*; socio *m*

13330 subscribe *v* **to** — To enroll in the list of subscribers.

f s'abonner à — **d** abonnieren; ein Abonnement aufgeben — **n** zich abonneren op; een abonnement nemen op — **e** subscribir(se); abonar(se); inscribir en la lista de subscriptores — **i** abbonarsi

13331 subscription

f abonnement *m*; souscription *f* — **d** Abonnement *n* — **n** abonnement *n*; intekening *f* — **e** abono *m*; subscripción *f*; suscripción *f* — **i** abbonamento *m*

13332 subscription book — Book or set of books intended for marketing by door-to-door agents or by mail.

f livre *m* par souscription; ouvrage *m* par souscription — **d** Bestellbuch *n*; Auftragsbuch *n* — **n** reisexemplaar *n*; reizigersexemplaar *n*

13333 subscription form

f bulletin *m* de souscription; bulletin *m* d'abonnement — **d** Bestellschein *m*; Subskriptionsschein *m* — **n** intekenformulier *n*; bestelformulier *n*; intekenbiljet *n* — **e** boletín *m* de subscripción; boletín *m* de abono — **i** modulo *m* per l'abbonamento

13334 subscripts — See inferior characters.

13335 subset — Contraction of the words "subscriber set", a modulation/demodulation device to make business machines compatible with communicating facilities.

f sous-ensemble *m* — **d** Teilsatz *m*; Untersatz *m*; Teilmenge *f* — **n** deelverzameling *f*; subset *m* — **e** subconjunto *m*; subjuego *m* — **i** sottoinsieme *m*

13336 subsidiaries — See end matter.

13337 substance — See basis weight.

13338 substantive dyes — See direct dyes.

13339 substrate; substratum — Base to which organic colouring agents are fixed to form lakes.

f substrat *m*; substratum *m*; support *m* — **d** Substrat *n*; Trägersubstanz *f*; Trägerstoff *m* — **n** substraat *n*; drager *m* — **e** substrato *m* — **i** supporto *m*

13340 substratum — See substrate.

13341 substructure — Part of the rotary press which supports the main printing units and usually carries the reels.

f sous-structure *f* — **d** Unterbau *m*; Untergestell *n* — **n** onderbouw *m*; onderframe *n* — **e** subestructura *f* − **i** sottostruttura *f*; basamento *m* inferiore

13342 sub-title − Additional title or second title, printed on the title page of a book. See also title, back title, catch title, section title.

f titre *m* de départ − **d** Untertitel *m*; Nebentitel *m* − **n** ondertitel *m* − **e** subtítulo *m*; título *m* de sección; título *m* segundario; título *m* de un apartado − **i** sottotitolo *m*

13343 subtractive colours − See complementary colours.

13344 subtractive colour synthesis − See subtractive synthesis.

13345 subtractive process − See subtractive synthesis.

13346 subtractive synthesis; subtractive colour synthesis; subtractive process − Embracing all colour processes in which three-colour images are superposed to form colour-prints or reproductions, the procedure called subtractive because the colours (images) subtract from white light by colour absorption, the equal absorption by the three primary colours tending to the quenching of light or the formation of grey or black. Subtractive synthesis is the basis of three-colour reproduction as carried out in photoengraving, and also is the principle of images made on multilayer colour films. See also complementary colours.

f synthèse *f* soustractive (des couleurs) − **d** subtraktive Farbsynthese *f*; subtraktive Farbmischung *f*; subtraktives Farbverfahren *n*; subtraktive Dreifarbenmethode *f* − **n** subtractieve kleurmenging *f*; subtractieve synthese *f* − **e** procedimiento *m* substractivo; síntesis *f* substractiva; síntesis *f* por substracción − **i** sintesi *f* sottrattiva; procedimento *m* sottrativo

13347 sub-truck − Movable platform on the press to transport the reel into the loading position.

f chariot *m* (de mise en place des bobines) − **d** Schiebebühne *f*; Fahrbühne *f* − **n** onderschuifwagen *m*; rolwagen *m* − **e** carretilla *f* para bobinas − **i** carrello *m* per trasportare le bobine; piattaforma *f* per trasportare la bobina

13348 suckers; suction cups; suction heads − Rubber suction cups which lift and carry sheets in a printing press.

f suceurs *mpl*; ventouses *fpl* − **d** Saugköpfe *mpl* − **n** zuigkoppen *mpl* − **e** succionadores *mpl*; grifos *mpl* espiradores; aspiradores *mpl*; ventosas *fpl*; toberas *fpl*; chupadores *mpl* Ar − **i** aspiratori *mpl*; ventose *fpl*

13349 suction air

f air *m* d'aspiration − **d** Saugluft *f* − **n** zuiglucht *fm* − **e** aire *m* de aspiración − **i** aria *f* d'aspirazione

13350 suction bar

f tuyau *m* d'aspiration − **d** Saugrohr *n*; Saugröhre *f* − **n** zuigbuis *fm* − **e** barra *f* de aspiración; tubo *m* de aspiración; tubo *m* aspirante; tubo *m* de succión − **i** tubo *m* d'aspirazione

13351 suction box − Device to remove water from the sheet being formed on the Fourdrinier wire of the paper machine or from the wet felt of a cylinder machine before pressing. It is a box with a perforated top over which the wire or felt passes. Water is removed by induced suction within the box.

f caisse *f* aspirante; table *f* aspirante − **d** Saugkasten *m* − **n** zuigkast *fm*; zuigdoos *fm* − **e** caja *f* aspirante; aspirador *m* − **i** cassa *f* aspirante

13352 suction couch roll − Rotary suction box which on modern paper machines has replaced the old jacketed type of *couch roll.

f cylindre *m* coucheur aspirant − **d** Zellensaugwalze *f* − **n** zuigwals *fm* − **e** cilindro *m* aspirante − **i** rullo *m* aspirante

13353 suction cups − See suckers.

13354 suction feeder

f margeur *m* à succion; margeur *m* à pile plane − **d** Sauganleger *m* − **n** zuiginlegapparaat *n* − **e** alimentador *m* de succión; marcador *m* de succión − **i** alimentatore *m* ad aspirazione

13355 suction heads − See suckers.

13356 suction pump − See vacuum pump.

13357 suds − See foam.

13358 sugar aquatint − Variation of the aquatint method to paint positive rather than negative images. As a painting medium on the plate a compound is used that contains sugar. The plate is covered with acid resistant varnish and when dry immersed in water. The varnish film is sufficiently porous to permit passage of the water to the image areas. These lift from the plate and carry the varnished areas above them along in the process. Thereby the plate is laid bare in the image areas. It is then treated like an aquatint plate.

f procédé *m* au sucre − **d** Zuckerradierung *f* − **n** suikerwater-ets *fm* − **e** aguatinta *f* al azúcar − **i** acquatinta *f* allo zucchero

13359 sugar cane bagasse − See bagasse.

13360 sugar cane pulp − See bagasse.

13361 sugar of lead − See lead acetate.

13362 sulfite − See sulphite.

13363 sulfonate *v* − See sulphonate *v*.

13364 sulfur − See sulphur.

13365 sulphate kraft paper − See kraft paper.

13366 sulphate of copper − See copper sulphate.

13367 sulphate of zinc − See zinc sulphate.

13368 sulphate process − Modified soda process in which salt cake or sodium sulphate is used with caustic to prepare the cooking liquor.

f procédé *m* au sulfate − **d** Sulfatverfahren *n* − **n** sulfaat-procédé *n* − **e** procedimiento *m* al sulfato − **i** procedimento *m* al solfato

13369 sulphate pulp − Chemical pulp obtained by digestion of the raw material (wood or other vegetable material) with a liquor consisting essentially of a mixture of caustic soda and sodium sulphide, and possibly other sulphur compounds. Bleached sulphate pulp is used to make printing and litho papers.

f pâte *f* au sulfate — **d** Sulfatzellstof *m* — **n** sulfaatcelstof *fm* — **e** pasta *f* al sulfato; pulpa *f* al sulfato — **i** pasta *f* chimica al sulfato

13370 sulphide — Binary compound of an element or group with sulphur. Salt of hydrogen sulphide, H_2S.
f sulfure *m* — **d** Sulfür *n*; Sulfid *n* — **n** sulfide *n* — **e** sulfuro *m* — **i** solfuro *m*

13371 sulphide of soda — See sodium sulphide.

13372 sulphite *GB*; **sulfite** *US* — Salt of sulphorous acid, H_2SO_3.
f sulfite *m* — **d** Sulfit *n* — **n** sulfiet *n* — **e** sulfito *m* — **i** solfito *m*

13373 sulphite acid liquor; sulphite liquor — Aqueous solution of *calcium bisulphite or calcium and magnesium bisulphites containing a large amount of free sulphur dioxide and limestone or dolomite or lime by passing sulphur dioxide gas up towers packed with limestone, down with water flows. Used in the manufacture of sulphite pulp in the paper industry.
f lessive *f* sulfitée — **d** Sulfitlauge *f* — **n** sulfietloog *fmn* — **e** lejía *f* al sulfito; lejía *f* sulfítica; lejía *f* de bisulfito — **i** liscivio *m* di bisolfito

13374 sulphite bond — Beater-sized, or beater and tub-sized paper, made from bleached sulphite pulp, used for office forms, circular letters, etc.
f bond *m* au bisulfite — **d** Hartpostpapier *n* — **n** hardpostpapier *n*; bondpapier *n* met sulfietcelstof — **e** papel *m* bond al sulfito — **i** carta *f* bond al solfito

13375 sulphite liquor — See sulphite acid liquor.

13376 sulphite of soda — See sodium sulphite.

13377 sulphite paper — Paper made wholly or largely from sulphite pulp.
f papier *m* au bisulfite — **d** Sulfitpapier *n* — **n** sulfietpapier *n* — **e** papel *m* (de) sulfito — **i** carta *f* al solfito

13378 sulphite process — Process in which wood is reduced to paper pulp by treatment with sulphurous acid and its calcium and magnesium salts, under controlled conditions of pressure and temperature. Invented in 1866 by Tilghman.
f procédé *m* au bisulfite — **d** Sulfitverfahren *n* — **n** sulfiet-procédé *n* — **e** procedimiento *m* al bisulfito — **i** procedimento *m* di solfito

13379 sulphite pulp — Chemical pulp derived from wood or other vegetable material by digestion with a bisulphite liquor.
f pâte *f* au bisulfite — **d** Sulfitzellstoff *m* — **n** sulfietcelstof *fm* — **e** pasta *f* al bisulfito; pasta *f* al sulfito; pulpa *f* al bisulfito; pulpa *f* al sulfito — **i** pasta *f* chimica al bisolfito

13380 sulphonate *v GB*; **sulfonate** *v US* — To subject to the treatment of sulphonic acid.
f sulfoner — **d** sulfieren; sulfurieren; sulfonieren — **n** sulfoneren — **e** sulfonar — **i** solfonare

13381 sulphonated castor oil — See turkey red oil.

13382 sulphonated oil; sulphurized oil; sulphur treated oil — Vegetable or animal oil treated with sulphuric acid; the excess of acid is washed out

and the oil neutralized with a small amount of caustic soda or ammonia. Widely used for paper, glue, inks, etc. Chemically this trade term, sulphonated, is incorrect since the oils are sulphated (contain the $-OSO_2OH$ group and not the $-SO_2OH$ group).
f huile *f* sulfonée; huile *f* traitée par le soufre — **d** sulfoniertes Öl *n*; geschwefeltes Öl *n*; mit Schwefel behandeltes Öl *n* — **n** gesulfoneerde olie *fm* — **e** aceite *m* sulfonado — **i** olio *m* solfonato

13383 sulphur *GB*; **sulfur** *US*; **flowers of sulphur** — Non-metallic element. Amorphous, soft or hard yellow, soluble in carbon disulphide, sulphur chloride and sulphur monochloride, insoluble in water. Symbol: S.
f soufre *m* — **d** Schwefel *m* — **n** zwavel *m* — **e** azufre *m* — **i** zolfo *m*

13384 sulphur aquatint — Process where the plate is dusted with sulphur on oil-treated areas, which produces a grainy surface. However, it does not hold up for many prints. Used by Rembrandt van Rhijn and other etchers of the 17th century.
f aquatinte *f* au soufre — **n** zwavel-aquatint *fm* — **e** aguatinta *f* al azufre — **i** acquatinta *f* allo zolfo

13385 sulphurated potash — See potassium trisulphide.

13386 sulphur chloride; sulphur monochloride — Amber to yellowish red, oily, fuming liquid, soluble in alcohol, ether, benzene, carbon disulphide and amyl acetate. Decomposes on contact with water. Must be kept well stoppered. Formula: S_2Cl_2.
f chlorure *m* de soufre — **d** Schwefelchlorür *n*; Chlorschwefel *m* — **n** zwavelchloride *n*; chloorzwavel *m* — **e** sulfocloruro *m*; cloruro *m* de azufre — **i** cloruro *m* di zolfo

13387 sulphur dioxide; sulphurous acid anhydride — Chemical used for the bleaching of textile fibres. TL-value: 5 ppm or 13 mg/m³. Formula: SO_2.
f acide *m* sulfureux; gaz *m* sulfureux; anhydride *m* sulfureux — **d** Schwefeldioxyd *n*; schweflige Säure *f*; Schwefligsäureanhydrid *n* — **n** zwaveldioxyde *n* — **e** dióxido *m* de azufre; anhídrido *m* sulfuroso — **i** biossido *m* di zolfo

13388 sulphuretted hydrogen — See hydrogen sulphide.

13389 sulphuric acid; oil of vitriol — Strongly corrosive, dense, oily liquid, miscible with water. Mixing causes evolution of much heat that can cause explosive splattering. Contact with permanganate is dangerous. TL-value: 1 mg/m³. Latin: acidum sulfuricum. Formula: H_2SO_4.
f acide *m* sulfurique; huile *f* de vitriol — **d** Schwefelsäure *f* — **n** zwavelzuur *n* — **e** ácido *m* sulfúrico; aceite *m* de vitriolo — **i** acido *m* solforico

13390 sulphuric anhydride — See sulphur trioxide.

13391 sulphuric ether — See ether.

13392 sulphurized oil — See sulphonated oil.

13393 sulphur monochloride — See sulphur chloride.

13394 sulphurous acid — Colourless solution of sulphur dioxide in water. The acid is known only through its salts. Latin: acidum sulfurosum.
Formula: H_2SO_3.
f acide *m* sulfureux — **d** schweflige Säure *f* — **n** zwavelig zuur *n* — **e** ácido *m* sulfuroso — **i** acido *m* solforoso

13395 sulphurous acid anhydride — See sulphur dioxide.

13396 sulphur treated oil — See sulphonated oil.

13397 sulphur trioxide; sulphuric anhydride — Chemical used as an oxidizing agent.
Formula: SO_3.
f acide *m* sulfurique; trioxyde *m* sulfurique; acide *m* sulfurique anhydre — **d** Schwefeltrioxyd *n*; Schwefelsäureanhydrid *n* — **n** zwaveltrioxyde *n* — **e** trióxido *m* de azufre; anhídrido *m* sulfúrico — **i** triossido *m* di zolfo; anidride *f* solforica

13398 sumac; sumach; sumag; sumac wax; Japan wax; Japanese tallow; vegetable wax — Dried and powdered leaves of certain species of sumac (any of a genus of woody, erect or root-climbing plants), used for tanning and dying.
f sumac *m* — **d** Schmack *m*; Sumach *m* — **n** sumak *m*; smak *m* — **e** cera *f* zumaque; cera *f* vegetal — **i** sommacco *m*

13399 sumach — See sumac.

13400 sumac tanning
f tannage *m* au sumac — **d** Sumachgerbung *f* — **n** looiing *f* met sumak — **e** curtimiento *m* con zumaque; curtiembre *f* con zumaque; curtido *m* con zumaque — **i** conciatura *f* al sommacco

13401 sumac wax — See sumac.

13402 sumag — See sumac.

13403 summary — See abridgement.

13404 Sunday paper
f journal *m* du dimanche — **d** Sonntagsblatt *n* — **n** zondagsblad *n* — **e** diario *m* dominical; periódico *m* dominical; papel *m* dominical; papel *m* de domingos — **i** giornale *m* domenicale

13405 sundries — Miscellaneous articles, details or items of inconsiderable size or amount individually.
f articles *mpl* divers; fournitures *fpl* — **d** Bedarf *m*; Utensilien *pl* — **n** (diverse) benodigdheden *fpl* — **e** géneros *mpl* diversos — **i** attrezzi *mpl* (diversi)

13406 sunk joint — See French groove.

13407 sun spots; light spots; white spots — Spots caused by dust particles which keep the carbon tissue out of contact with the screen. On etching a light coloured circular area is observed where the screen lines have thickened with a small central hole where the light has been stopped by a dust speck.
f soleils *mpl*; puces *fpl* — **d** Sonnen *fpl*; Sonnenflecken *mpl* — **n** zonnetjes *npl* — **e** solecitos *mpl*; soles *mpl* — **i** soli *mpl*

13408 super — See mull.

13409 supercalender — Type of calender with metal rolls and paper or cotton covered rolls, the diameter of the metal and paper (or cotton) rolls being different to secure an ironing action of the paper and thus obtain a higher finish. Supercalenders are not part of the paper machine but auxiliary to it.
f supercalandre *f*; grande calandre *f* — **d** Superkalander *m*; Satinierkalander *m*; Rollenkalander *m* — **n** superkalander *m*; satineerkalander *m*; rollenkalander *m* — **e** supercalandria *f*; supercalandra *f* — **i** supercalandra *f*

13410 super-calendered paper; s.c. paper — Paper highly calendered in a supercalender to obtain a smoother surface and higher gloss than machine finished paper.
f papier *m* supercalandré; papier *m* surcalandré; papier *m* satiné — **d** hochsatiniertes Papier *n*; scharf satiniertes Papier *n* — **n** hooggesatineerd papier *n*; hooggeglansd papier *n* — **e** papel *m* supercalandrado; papel *m* supersatinado — **i** carta *f* supercalandrata

13411 supercrash — See mull.

13412 superfines — Superior grade of flat writing paper of more or less high finish to meet the requirements.
f papiers *mpl* superfins — **d** Feinstpapier *n* — **n** fijn papier *n* — **e** papeles *mpl* superfinos — **i** carta *f* da scrivere sopraffina

13413 superfine writing paper — Writing paper of superior quality including rag content papers.
f papier *m* écriture surfin — **d** Feinschreibpapier *n* — **n** fijn schrijfpapier *n* — **e** papel *m* superfino de escribir — **i** carta *f* da scrivere di qualità superiore

13414 superimpose *v* — To print over some other printing, such as in process printing.
f superposer (des impressions) — **d** übereinanderdrucken — **n** over elkaar drukken; er overheen drukken — **e** sobreimprimir — **i** sovrastampare

13415 superimpose *v* — In photoengraving, to combine by double exposure one element over another such as type matter over a halftone screen. Also called surprinting.
f faire une surimpression — **d** übereinanderkopieren; übereinanderlegen — **n** over elkaar kopiëren; over elkaar leggen — **e** sobreponer; superponer — **i** sovratrasportare

13416 superior characters; superscripts; superior letters; superior figures — Small letters or figures cast to print above the line, used in mathematical and chemical formulae, or as reference notes.
f exposants *mpl*; lettres *fpl* supérieures; chiffres *mpl* supérieurs; indices *mpl* supérieurs — **d** Exponenten *mpl*; hochstehende Zeichen *npl*; hochstehende Buchstaben *mpl*; hochstehende Ziffern *fpl* — **n** exponenten *mpl*; superieure tekens *npl*; superieuren *mpl*; supers *mpl*; superieure letters *fmpl*; superieure cijfers *npl* — **e** voladitas *fpl*; cerillos *mpl*; aecillas *fpl*; letras *fpl* voladas; números *mpl* volados — **i** esponenti *mpl*; caratteri *mpl* superiori; lettere *fpl* superiori; lettere *fpl* posti in alto; numeri *mpl* posti in alto;

numeri *mpl* superiori

13417 superior figures — See superior characters.

13418 superior letters — See superior characters.

13419 supersaturation — Condition of containing an excess of material over the amount required for saturation.

f sursaturation *f* — **d** Übersättigung *f* — **n** over-verzadiging *f* — **e** supersaturación *f*; hipersaturación *f* — **i** soprasaturazione *f*

13420 superscripts — See superior characters.

13421 superstructure; top rail section — Part of a rotary press which is above the level of the main printing units.

f superstructure *f* — **d** Oberbau *m* — **n** boven-bouw *m*; bovenframe *n* — **e** superestructura *f* — **i** parte *f* superiore; sovrastruttura *f*

13422 supervisor — Office manager in commercial or service establishments.

f chef *m* de bureau; chef *m* de service — **d** Büro-chef *m*; Verwalter *m* — **n** bureauchef *m*; chef *m* de bureau — **e** sobrestante *m*; inspector *m* — **i** capoufficio *m*

13423 supervisor; supervisory programme *GB*; **executive program** *US* — Master program or set of routines (sometimes called *operating system) which supervises the system, controls multi-programming priorities, transfer of data to and from peripheral devices, error detection, suspends and allocates programs, and undertake house-keeping functions; communicates with the operator, usually through the medium of a console typewriter.

f superviseur *m*; programme *m* directeur — **d** Leit-programm *n*; Supervisor *m*; Überwacher *m* — **n** stuurprogramma *n*; supervisor *m*; supervisie-programma *n* — **e** supervisor *m*; programa *m* director — **i** supervisore *m*; programma *m* di controllo

13424 supervisory programme — See supervisor.

13425 suppleness

f souplesse *f* — **d** Geschmeidigkeit *f* (einer Farbe) — **n** zachtheid *f* (van een inkt); soepelheid *f* — **e** flexibilidad *f*; elasticidad *f* — **i** pastosità *f* (di un inchiostro)

13426 support — See easel.

13427 surface active — Able to reduce the surface tension when dissolved in a liquid or applied to its surface. Able to increase the wetting power of a liquid.

f tensio-actif — **d** oberflächenaktiv; kapillaraktiv — **n** oppervlakte-actief — **e** tensioactivo; surfactivo — **i** tensio-attivo

13428 surface active agent; surfactant — Material which when used in small amounts modifies the surface properties of liquids or solids, e.g., detergents, wetting agents, emulsifying agents, dispersing agents, foam inhibitors.

f agent *m* tensio-actif; agent *m* de surface; surfactant *m*; produit *m* de mouillage — **d** Netzmittel *n*; grenzflächenaktiver Stoff *m*; oberflächenaktiver Stoff *m* — **n** bevloeiingsmiddel *n*; grensvlak-actieve stof *fm*; oppervlakactieve stof *fm* — **e**

agente *m* activo de superficie; agente *m* tensio-activo — **i** agente *m* tensio-attivo

13429 surface activity

f activité *f* superficielle; tensio-activité *f* — **d** Oberflächenaktivität *f* — **n** oppervlakte-activiteit *f* — **e** actividad *f* superficial — **i** attività *f* superficiale

13430 surface application; surface treatment — Operation consisting of the deposition of a coat of an appropriate material on the surface of paper or board.

f enduction *f*; traitement *m* de surface — **d** Oberflächenbehandlung *f* — **n** oppervlakte-behandeling *f* — **e** tratamiento *m* superficial; tratamiento *m* de la superficie — **i** trattamento *m* superficiale

13431 surface bonding strength — See bonding strength.

13432 surface colour — Colour perceived by the proportion of the illuminating light that is reflected from a body.

f couleur *f* des objets — **d** Körperfarbe *f* — **n** reflectiekleur *fm* — **e** color *m* objetivo — **i** colore *m* di un oggetto; colore *m* riflesso

13433 surface forces

f forces *fpl* de surface — **d** Oberflächenkräfte *fpl* — **n** oppervlakte-krachten *fmpl* — **e** fuerzas *fpl* superficiales — **i** forze *fpl* superficiali

13434 surface of paper

f surface *f* du papier — **d** Papieroberfläche *f* — **n** papieroppervlak *n* — **e** superficie *f* del papel — **i** superficie *f* della carta

13435 surface plate — One of the two types of lithographic press plates, now getting out of use. A colloid image is formed on the light-sensitized metal plate by the action of actinic light through photographic negatives. The *albumen plate was the most common form of surface plate, but there are others, including some of the bimetal or poly-metal plates on which the image acts as a resist for the removal by acid of the surface metal to get down to the base metal which tends to resist ink.

f plaque *f* (offset) sans creux — **d** Oberflächen-druckplatte *f* — **n** oppervlakte-plaat *f*; negatief-kopie-plaat *f* — **e** plancha *f* con imagen en la superficie — **i** lastra *f* offset secondo il procedimento negativo

13436 surface properties

f propriétés *fpl* de surface — **d** Oberflächeneigen-schaften *fpl* — **n** eigenschappen *fpl* van het oppervlak; oppervlakte-eigenschappen *fpl* — **e** propiedades *fpl* superficiales; calidades *fpl* super-ficiales — **i** proprietà *fpl* superficiali

13437 surface roughness

f rugosité *f* de surface — **d** Oberflächenrauheit *f* — **n** oppervlakte-ruwheid *f*; ruwheid *f* van het oppervlak — **e** rugosidad *f* superficial — **i** rugosità *f* superficiale

13438 surface sized — See animal tub sized.

13439 surface sized paper — See tub sized paper.

13440 surface sizing; top sizing — Surface or tub sizing of paper which has been previously beater sized.

f collage *m* en surface; collage *m* à la cuve; collage *m* en bac; collage *m* superficiel — **d** Oberflächenleimung *f* — **n** oppervlakte-lijming *f*; nalijming *f* — **e** encolado *m* de la superficie — **i** collatura *f* in superficie

13441 surface tension — Stress in a liquid surface due to the lateral attraction of molecules for each other; the force that causes drops of liquid to tend to assume a spherical shape. Measured in dynes/mg.

f tension *f* superficielle — **d** Oberflächenspannung *f* — **n** oppervlakte-spanning *f* — **e** tensión *f* superficial — **i** tensione *f* superficiale

13442 surface tension lowering; reduction in surface tension

f abaissement *m* de la tension superficielle — **d** Oberflächenspannungs-Erniedrigung *f* — **n** oppervlakte-spanningsverlaging *f*; verlaging *f* van de oppervlakte-spanning — **e** baja *f* de la tensión superficial — **i** abbassamento *m* della tensione superficiale

13443 surface treatment — See surface application.

13444 surface-uniformity gloss — See gloss.

13445 surfactant — See surface active agent.

13446 surglacé — Term for *imitation art paper, not used in England.

13447 surimono — Very carefully executed Japanese print (or greeting card), issued at special festivities. See also Japanese colour print.

13448 surplus box

f cassetin *m* à réserve — **d** Ausraffefach *n* — **n** reserve-vak *n* — **e** cajetín *m* de reserva — **i** cassettino *m* di riserva; scomparto *m* di riserva

13449 surprint; overprint — Additional printing over the design areas of previously printed matter. Its equivalent in stripping uses overlay positive films on negatives, or photographic contact procedures to produce overprints as "sale", "void", "discontinued", etc.

f surimpression *f* — **d** Überdruck *m* — **n** opdruk *m* — **e** sobreimpresión *f* — **i** sovrastampa *f*

13450 surprinting — See superimpose *v*.

13451 suspended camera — See overhead camera.

13452 suspension — Two-phase system consisting of very small solid particles distributed in a liquid dispersion medium. If the particles are small enough to pass through ordinary filters and do not settle out on standing, the suspension is called a colloidal suspension or colloid. See also sol.

f suspension *f* — **d** Suspension *f*; Aufschlämmung *f*; Aufhängung *f*; Schwebe *f* — **n** suspensie *f* — **e** suspensión *f* — **i** sospensione *f*

13453 Sütterlinschrift — Handwriting designed by the graphic designer L. Sütterlin (1865-1917), Berlin, introduced in schools (1935-1941) as a basic writing method.

13454 swan socket — See bayonet socket.

13455 swash letters — Ornamental characters in the 17th century style with tails and flourishes, as for instance Caslon Old Face italic. They are a florid version of the standard italic capital letters.

f lettres *fpl* à parafe; lettres *fpl* ornées; capitales *fpl* ornées — **d** Schnörkelbuchstaben *mpl*; Zierbuchstaben *mpl*; geschwungene Versalien *mpl* — **n** sierletters *fmpl*; fantasieletters *fmpl*; sierkapitalen *npl*; versierde kapitalen *npl* — **e** letras *fpl* adornadas; letras *fpl* de fantasía — **i** lettere *fpl* ornate

13456 swatch pad — Assortment of small paper samples fastened together, used for display or advertising. See also sample book.

f carnet *m* d'échantillons — **d** (kleines) Musterbuch *n* — **n** monsterboekje *n* — **e** carnet *m* de muestras; muestrario *m* — **i** campionario *m* di carte

13457 sweat drier; sweat roll; cooling cylinder — Paper machine drying cylinder that operates at a temperature sufficiently low to cause condensation of moisture or sweating on its surface, to cool the sheet and slow the drying process. Usually located at the end of a drier section, and chrome coated or otherwise protected against corrosion.

f cylindre *m* refroidisseur; rouleau *m* refroidisseur — **d** Kühlzylinder *m* — **n** koelcilinder *m* — **e** cilindro *m* refrigerador — **i** rullo *m* raffreddatore; cilindro *m* raffreddatore

13458 sweat *v* on — Operation of firmly attaching zinc and copper etchings, as well as electrotypes, to a solid metal base by a soldering procedure involving the use of a sheet of molten tinfoil as the binding medium between the two surfaces.

f souder le cliché sur bloc métal; monter sur matière — **d** auflöten; aufschweißen; aufschmelzen — **n** solderen — **e** soldar — **i** brasare su zoccolo metallico

13459 sweat roll — See sweat drier.

13460 Swedish olein — See tall oil.

13461 sweeping coil — See deflection yoke.

13462 swell *v* — See also bubbling.

f se gonfler (d'un blanchet); s'enfler (d'un rouleau de pâte) — **d** quellen (des Gummituchs); aufquellen (der Walzenmasse) — **n** zwellen; opzwellen — **e** hincharse; abotagarse; inflarse — **i** gonfiarsi

13463 swell dash — See French rule.

13464 swelled rule — See French rule.

13465 swelling — Increase in volume when solid substances absorb liquids.

f gonflement *m* — **d** Quellung *f*; Quellen *n* (der Fasern) — **n** zwellen *n*; zwelling *f* — **e** hinchamiento *m* (de las fibras) — **i** rigonfiamento *m*

13466 swift — Fast compositor or operator. See also whip hand.

f as *m* — **d** Schnellhase *m* — **n** hardloper *m* *coll.* — **e** cajista *m* hábil — **i** compositore *m* svelto

13467 swimming roller — See expander roll.

13468 swing arm — See swing gripper device.

13469 swinger — The last of a series. Usually the last forme going to press.

f dernier *m* — **d** letzte Form *f* — **n** laatste vorm *m* — **e** última forma *f* — **i** ultima forma *f*

13470 swinger card *US* — Width card of a type face, stored in an octagonal shaped width-card holding device, sometimes called *Lazy Susan.

13471 swing gripper — See swing gripper device.

13472 swing gripper device; swing gripper; swing arm — Transfer mechanism intermediate between the feed board and the cylinder grippers.

f prise *f* à balancier; système *m* de pinces à balancier; pinces *fpl* oscillantes — **d** Schwinggreifer *m*; Vorgreiferschwinganlage *f* — **n** zwaaiende voorgrijper *m* — **e** pinza *f* oscilante — **i** pinze *fpl* oscillanti

13473 swing trick — Job or assignment of filling in for others in the news room during vacations or sickness.

f remplacement *m* — **d** Vertretungsstelle *f* — **n** invaller *m*; plaatsvervanger *m* — **e** suplente *m* — **i** sostituto *m*

13474 switch — Device to make, break, or change the connections in an electrical circuit.

f interrupteur *m*; commutateur *m*; contacteur *m* — **d** Schalter *m*; Unterbrecher *m* — **n** schakelaar *m* — **e** interruptor *m* (on/off switch); conmutador *m* (change-over switch); suiche *m Co* — **i** interruttore *m*

13475 switch — Mechanism to separate copies in a rotary press folder. See also split delivery.

d Exemplarableiter *m* — **n** verdeler *m* — **e** separador *m* — **i** deviatore *m* separatore delle copie

13476 switch board — Apparatus consisting of a panel or a frame on which switching, measuring, controlling and protective devices are mounted with connections so arranged that a number of circuits may be connected, combined, controlled, measured and protected.

f pupitre *m* de commande; coffret *m* de manœuvre — **d** Schaltpult *n*; Schaltkasten *m* — **n** schakelbord *n*; schakelkast *fm*; schakeltafel *fm* — **e** cuadro *m* de distribución; tablero *m* de distribución — **i** quadro *m* dei comandi; pannello *m* dei comandi

13477 switch *v* **off**

f éteindre; débrancher; couper le contact — **d** ausschalten; abschalten — **n** uitschakelen — **e** desconectar; abrir el circuito; apagar; cortar la corriente — **i** disinserire

13478 switch *v* **on**

f mettre le contact; brancher; allumer — **d** einschalten; anschalten; zuschalten — **n** inschakelen — **e** conectar; cerrar el circuito; encender — **i** inserire

13479 swivel joint — See ball-and-socket joint.

13480 SWM — See standard work minute.

13481 sword hygroscope; spike hygrometer; paper hygroscope — Hygrometer to measure the interior moisture content of a pile of paper in comparison to that of the air surrounding, to determine whether or not paper conditioning is necessary.

f hygromètre *m* épée; hygromètre *m* sabre; hygromètre *m* sonde; hygromètre *m* à tige plongeante — **d** Stabhygrometer *m*; Stechhygrometer *m*; Stapelhygrometer *m* — **n** steekhygrometer *m*;

zwaardhygrometer *m*; staafhygrometer *m* — **e** higrómetro *m* de vástago; higroscopio *m* de espada — **i** igrometro *m* a lama; igrometro *m* a sonda

13482 swot — See grind.

13483 swung dash — A *dash like ~, sometimes used to represent the stem of a word.

13484 symbolic addressing — See symbolic coding.

13485 symbolic coding; symbolic addressing — Coding that uses machine instructions with symbolic addresses.

f codage *m* symbolique; adressage *m* symbolique — **d** symbolische Adressierung *f* — **n** symbolische codering *f*; symbolische adressering *f* — **e** codificación *f* simbólica; direccionamiento *m* simbólico — **i** codifica *f* simbolica; indirizzamento *m* simbolico

13486 symbolic language — Discipline that treats formal logic by means of a formalized artificial language or symbolic calculus to avoid the ambiguities and logical inadequacies of natural languages.

f langage *m* symbolique — **d** symbolische Sprache *f*; symbolische Programmiersprache *f* — **n** symbolische taal *fm* — **e** lenguaje *m* simbólico — **i** linguaggio *m* simbolico

13487 sympathetic ink — See invisible ink.

13488 synchronous computer — Computer in which the beginning of any operation is dependent upon timed cycles from an internal clock.

f calculateur *m* synchrone — **d** Synchroncomputer *m*; Synchronrechner *m* — **n** synchrone computer *m* — **e** computadora *f* síncrona — **i** calcolatore *m* sincrono

13489 synchronous mode — Method or working in a continuous time cycle to correspond with the data input or output.

13490 synopsis — Condensed statement or outline of a narrative or treatise.

f synopsis *f* — **d** Synopse *f*; Übersicht *f*; Zusammenfassung *f* — **n** overzicht *n*; synopsis *f*; samenvatting *f* — **e** sinopsis *f* — **i** sinossi *f*

13491 syntax — Structure of expressions in a language. Also the rules governing the structure of a language.

f syntaxe *f* — **d** Syntax *f* — **n** syntaxis *f* — **e** sintaxis *f*; sintasis *f* — **i** sintassi *f*

13492 synthesis — Combining of elements or simple compounds to conform complex compounds of material.

f synthèse *f* — **d** Synthese *f*; Synthesis *f* — **n** synthese *f* — **e** síntesis *f* — **i** sintesi *f*

13493 synthetic — Complex, formed by the chemical union of elements or of simpler compounds. Many dyes, resins, and rubber-like materials are made artificially by synthesis.

f synthétique — **d** synthetisch — **n** synthetisch — **e** sintético — **i** sintetico

13494 synthetic fibre

f fibre *f* synthétique; fibre *f* artificielle — **d** synthetische Faser *f*; Kunstfaser *f* — **n** synthetische vezel *fm*; kunstvezel *fm* — **e** fibra *f* sintética; fibra *f* artificial — **i** fibra *f* sintetica; fibra *f* artificiale

13495 synthetic resin; artificial resin — Complex, amorphous, organic solid or semi-solid material (usually a mixture) built up by the chemical reaction of comparatively simple compounds. Usually soluble in one or more organic solvents.

f résine *f* synthétique; résine *f* artificielle — **d** Kunstharz *n* — **n** synthetische hars *mn*; kunsthars *mn* — **e** resina *f* sintética; resina *f* artificial — **i** resina *f* sintetica; resina *f* artificiale

13496 synthetic rubber; artificial rubber — Material such as neoprene, a trade name for a chlorinated butadiene type, or buna-N, made from butadiene and acrylonitrile, of which offset ink rollers are made to prevent swelling.

f caoutchouc *m* synthétique — **d** synthetischer Gummi — **n** synthetische rubber *mn*; kunstrubber *mn* — **e** caucho *m* sintético; caucho *m* artificial; caucho *m* químico; goma *f* sintética; hule *m* sintético *Me* — **i** caucciù *m* sintetico; gomma *f* sintetica

13497 Syrian asphalt — Asphaltite resembling *gilsonite found in Syria, having a black streak instead of brown. Pure asphalt.

f bitume *m* de Judée — **d** syrischer Asphalt *m* — **n** Syrisch asfalt *n* — **e** betún *m* de Judea — **i** bitume *m* giudaico; bitume *m* di Giudea

13498 syrupy — Having the appearance or quality of syrup; thick or sweet.

f sirupeux — **d** sirupartig; sirupös — **n** stroopachtig; stroperig — **e** parecido a jarabe; almibarado — **i** sciropposo

13499 system — Organization of independent procedures, processes, techniques and methods to form a functional entity. In computer techniques an integrated assembly of hardware and software to implement a given application or set of applications.

f système *m* — **d** System *n* — **n** systeem *n* — **e** sistema *m* — **i** sistema *m*

13500 system analysis — Analysis of an activity to determine what must be accomplished and how to accomplish it (in computer terms of software, hardware, data entry, etc.).

f analyse *f* du système — **d** Systemanalyse *f* — **n** systeemanalyse *f* — **e** análisis *f* del sistema — **i** analisi *f* del sistema

13501 system analyst — See programmer.

13502 systematic random sample

f échantillon *m* systématique aléatoire — **d** systematische Zufallsstichprobe *f* — **n** systematische steekproef *fm* — **e** prueba *f* al azar sistemático; muestreo *m* — **i** prova *f* a caso sistematica

13503 Système International d'Unités — The *International System of Units. A coherent system of scientific units, which is divided in 7 basic quantities: length (l), mass (m), time (t), electric current (I), thermodynamic temperature (T), quality of matter (M), light intensity (I). The system is based on the metre (m), kilogramme (kg), second (s), ampere (A), kelvin (K), mol (mol), and candela (cd). From these all other units can be derived, that is: coulomb, farad, henry, hertz, joule, lumen, lux, newton, ohm, pascal, siemens, tesla, volt, watt, weber. See these terms in the basic table of this dictionary. Not to the SI belong: acre, ångström, are, bushel, drachm, dram, erg, foot, gallon, gill, grain, horsepower, hundredweight, inch, minim, ounce, peck, pint, pound, quart, quarter, scruple, stone, ton, yard. For the conversion of these units, see app. no. 7a. The SI is adopted by the General Conference of Weights and Measures (CGPM) and endorsed to the International Standards Organization (ISO). The system is indicated in all languages with SI. Since January 1978 it is obligatory in the countries of the European Common Market.

f système *m* international d'unités — **d** internationales Einheiten-System *n* — **n** internationaal stelsel *n* van eenheden — **e** sistema *m* internacional de unidades — **i** sistema *m* internazionale di unità

13504 system flowchart — Broad outline of a system, using symbols to indicate the various steps of its implementation.

13505 tab — See index tab.

13506 tab breaker — Roller which separates cartons and breaks the connecting tabs.

f séparateur *m* de poses — **d** Nutzentrenner *m* — **n** uitbreker *m*; uitbreekrol *fm* — **e** separadora *f* de recortes; descartonadora *f* de recortes; dispositivo *m* para descartonar — **i** rullo *m* separatore; separatore *m*

13507 tab card — Die-cut index card with protruding tab for indexing.

f fiche *f* indicatrice; carte *f* à onglet; carte-guide *f* — **d** Leitkarte *f* — **n** tabkaart *fm* — **e** tarjeta *f* de pestaña para ficheros — **i** scheda *f* guida

13508 table — Matter in two or more columns of figures usually set between rules.

f tableau *m*; table *f* — **d** Tabelle *f*; Liste *f* — **n** tabel *fm*; lijst *fm* — **e** tabla *f*; cuadro *m* — **i** tabella *f*

13509 table inking arrangement

f dispositif *m* d'encrage à table plate; encrage *m* à table plate — **d** Tischfarbwerk *n*; Tischfärbung *f* — **n** tafel-inktwerk *n*; tafel-inktgeving *f* — **e** tintaje *m* de mesa — **i** inchiostrazione *f* a tavoletta

13510 table of contents — See contents.

13511 table of signatures — The folded sheets of a book are numbered in the foot margin on the pages 1, 17, 33, 49, etc. This number is called the *signature number. On the third page of each sheet, these numbers are repeated, accompanied by an asterisk. See also asterisk signature.

f table *f* des signatures — **d** Primentafel *f*; Primentabelle *f* — **n** signaturenschema *n* — **e** tabla *f* de (las) signaturas — **i** schema *m* delle segnature

13512 table roll

f pontuseau *m* — **d** Registerwalze *f* — **n** registerwals *fm*; zeeftafelrol *fm* — **i** rullo *m* sgocciolatore

13513 table salt — See sodium chloride.

13514 tablet *US*; **pad** *GB*; **note pad; block** — Pad of fifty or more sheets of writing, book or newspaper, glued together at one edge or end, and backed by a piece of cardboard, with or without cover, used for writing with pencil or pen and ink.

f tablette *f*; bloc *m* (de papier); bloc-notes *m*; bloc *m* de correspondance — **d** Block *m*; Schreibblock *m*; Notizblock *m*; Briefblock *m* — **n** schrijfblok *n*; notitieblok *n* — **e** block *m*; taco *m* de papel blanco; bloc *m* de notas — **i** blocco *m* per note; blocco *m* per appunti

13515 tablet blotting — Light-weight blotting paper used as the first sheet in a tablet or inserted in a box of papeterie. The basis weight ranges from 35 to 80 pounds (19 x 24 - 500).

f buvard *m* pour tablette(s) — **d** Löschpapier *n* für Schreibblöcke — **n** vloeipapier *n* voor notitiebloks — **e** papel *m* secante de blocs; secante *m* de blocs — **i** carta *f* asciugante per blocchi

13516 tablet paper — Hard-sized book paper of chemical wood fibre, with English or high machine finish used in manufacture of writing tablets. Also, improperly, writing paper used in tablets.

f papier *m* pour tablettes — **d** Schreibpapier *n* für Notizblöcke — **n** schrijfpapier *n* voor notitiebloks — **e** papel *m* para blocs — **i** carta *f* per blocchi

13517 tabloid fold — See first parallel fold.

13518 tabloid format — Page size of tabloid newspapers: one-half of a standard size page, 38-43 cm (15-17 in) long and approx. 29.5 cm (11.75 in) wide. The decisive point is not the exact page size, but the fact that the page size remains the same for a long time once it is adopted for a given paper.

f format *m* tabloïd; demi-format *m* — **d** Tabloidformat *n* — **n** tabloidformaat *n* — **e** formato *m* tabloide; tamaño *m* tabloide; formato *m* pequeño; tamaño *m* pequeño — **i** formato *m* tabloide

13519 tabular matter

See tabular work.

13520 tabular work; tabular matter — Figures and other matter ranging vertically in columns, with or without rules.

f composition *f* de tableaux; tableautage *m* — **d** Tabellensatz *m*; tabellarischer Satz *m* — **n** tabelwerk *n*; tabellenzetwerk *n*; staatwerk *n* — **e** composición *f* tabular; composición *f* de tablas; composición *f* de estados; trabajo *m* estadístico; tabla *f*; cuadro *m*; estado *m* *Es* — **i** composizione *f* delle tabelle; composizione *f* tabellare

13521 tabulating attachment

f tabulateur *m* — **d** Tabulator *m* — **n** tabelleerapparaat *n* — **e** máquina *f* tabuladora — **i** tabulatore *m*

13522 tabulating equipment — Equipment used in a system utilizing punched cards for sorting, listing and simple arithmetic processes, producing printed results on continuous stationery. Includes sorters, collators and tabulators.

13523 tachometer — Instrument which shows the speed of the press or the number of copies per hour. See also counter.

f tachymètre *m*; compte-tours *m*; compteur *m* de tours — **d** Tourenzähler *m*; Drehzahlmesser *m*; Umdrehungsmesser *m*; Tachometer *m*; Auflagenzähler *m* — **n** toerenteller *m*; teller *m*; telklok *fm*; klok *fm* — **e** taquímetro *m*; tacómetro *m*; contador *m* de revoluciones; contador *m* de velocidad; indicador *m* de velocidad — **i** tachimetro *m*

13524 tack; tackiness; stickiness — Resistance to splitting of an ink film between two separating surfaces. Property of great importance at the moment of printing since it controls picking of the paper and trapping of colours. Flexographic inks, for example, have very little tack as long as they are fluid, that is, contain most of their original solvent. Tack is measured by inkometers. It is sometimes confused with consistency by thinking

an ink with high tack has high consistency and vice versa. This may be true for some inks, but not for others. Tack and consistency describe separate characteristics.
f poisseux *m* (d'une encre) — **d** Zügigkeit *f*; Zähigkeit *f*; Klebrigkeit *f* — **n** ≈ kleefkracht *fm* (van de inkt); ≈ trekkracht *fm* (van de inkt); splitsingskracht *fm* (van de inkt); tack *m* — **e** pegajosidad *f* (de la tinta) — **i** peciosità *f* (dell'inchiostro); tiro *m* (dell'inchiostro)
13525 tack — Attraction of composition rollers for ink.
f amour *m* (d'un rouleau); tirant *m*; touche *f* — **d** Zügigkeit *f*; Ziehfähigkeit *f*; Klebrigkeit *f* — **n** kleefkracht *fm* — **e** amor *m* de los rodillos; mordiente *m* de los rodillos — **i** tiro *m* del rullo
13526 tack; tackiness; stickiness — Quality of adhesion of an offset blanket. When the rubber of the blanket starts to disintegrate, the blanket is said to be tacky.
f poisseux *m* — **d** Klebrigkeit *f* — **n** kleverigheid *f* — **e** pegajosidad *f*; glutinosidad *f*; untuosidad *f* — **i** attaccaticcio *m* (del blanket)
13527 tack — Small nail.
f petit clou *m* — **d** Nägelchen *n* — **n** spijkertje *n* — **e** tachuela *f* — **i** chiodo *m*
13528 tacker
f agrafeuse *f* — **d** Heftpistole *f*; Heftzange *f* — **n** nietapparaat *n*; nietmachientje *n* — **e** engrapadora *f*; grapadora *f* — **i** pistola *f* a punti metallici
13529 tackiness — See tack.
13530 tackiness — See tack.
13531 tackmeter — Instrument to analyze pull resistance of ink, etc., composed of a flat metallic finger tip, a flat plate on which the material is put, a micrometer screw for changing the initial film thickness, and a lever on which different pulling weights are hung. Each of these factors can be changed independently. The effect of each is thus observed. However, there are instruments of different constructure.
f tackmètre *m* — **d** Tackmeter *n* — **n** tackmeter *m* — tackmetro *m* — **i** tackmeter *m*
13532 tacky ink — Sticky ink made with a stiff varnish which exerts an unusual pull on the surface of the paper.
f encre *f* poisseuse; encre *f* poissante; encre *f* amoureuse — **d** klebrige Farbe *f* — **n** kleverige inkt *m* — **e** tinta *f* pegajosa; tinta *f* glutinosa — **i** inchiostro *m* appiccicoso; inchiostro *m* pecioso
13533 tacky rollers — Rollers which have the correct degree of stickiness to take up and carry the ink.
f rouleaux *mpl* amoureux; rouleaux *mpl* sympathiques — **d** klebrige Walzen *fpl*; zügige Walzen *fpl* — **n** kleefkrachtige rollen *fmpl* — **e** rodillos *mpl* con mordiente — **i** rulli *mpl* attaccaticci
13534 tag — Small stringed label.
f étiquette *f* — **d** Etikett *n*; Anhänger *m*; Anhängeetikett *n* — **n** label *m* — **e** marbete *m*; rótulo *m* — **i** etichetta *f*; cartellino *m*
13535 tag board; manila tag; document manila;

document board — Paperboard used for printed forms, envelopes, shipping tags, file holders, etc., manufactured of rope, jute, sulphite, sulphate or mechanical wood pulp or various combinations of these. Usually of a manila colour, with a smooth finish. The basis weights range from 80 to 300 pounds (22.5 x 28.5 - 500). It has good bending or folding qualities, high bursting and tensile strength, high tearing resistance, and a high water finish.
f carton *m* à étiquettes; carton *m* à bulletins; carte *f* manille à dossiers — **d** Karton *m* für Gepäckanhänger; Anhängekarton *n* — **n** labelkarton *n* — **e** cartoncillo *m* grueso para marbetes; papel *m* grueso para marbetes — **i** cartone *m* per etichette; cartone *m* per cartellini
13536 tag stringing and knotting machine; automatic tag stringer and knotter — Machine to punch holes, to string and knot cords on shipping and merchandise tags and booklets.
f noueuse *f*; machine *f* à nouer les étiquettes — **d** Etikettenknüpfmaschine *f*; Anhängerknüpfmaschine *f* — **n** koordjesknoopmachine *f* — **e** nudadora *f* para etiquetas — **i** macchina *f* per forare e munire di cordino i cartellini
13537 tail — End of a stereo plate formed by excess metal in the casting box.
f jet *m* — **d** Anguß *m* — **n** aangietsel *n* — **e** cabo *m* de fundición; cabo *m* de carga — **i** coda *f* (di una stereo)
13538 tail — For an automatic paster of a rotary press, the amount of web between the splice and the cutting of the expiring web.
f jupe *f* — **d** Papierabschnitt *m* zwischen Klebestelle und Abschlagen der auslaufenden Rolle — **i** coda *f* (di un nastro)
13539 tail — See foot.
13540 tail — See tail margin.
13541 tailband — Silk band at the bottom of the spine of a book. See also headband.
f tranchefile *f* inférieure — **d** unteres Kapitalband *n* — **n** onderste kapitaalbandje *n*; onderste besteekbandje *n* — **e** cabezada *f* del pie — **i** capitello *m* inferiore
13542 tail, blanket — See blanket tail.
13543 tailboard — See apron.
13544 tail bolt — See bolts.
13545 tail cap — See cap.
13546 tail dot — See ordinary terminals.
13547 tail edge — See trailing edge.
13548 tail end hook — See tail hook.
13549 tail hook; tail end hook — Troublesome curl across the back of the sheet, preventing good jogging, caused when paper, particularly coated paper, adheres strongly to the blanket and is pulled off.
f gondolage *m* de la queue de la feuille — **n** krulling *f* aan de afloopkant van het papier — **e** abarquillado *m* de la hoja de papel — **i** imbarcamento *m* di coda
13550 tail margin; tail; foot margin; bottom margin

f marge *f* inférieure; marge *f* de pied; queue *f* — **d** Fußrand *m*; unterer Seitenrand *m* — **n** staartmarge *fm*; ondermarge *fm* — **e** margen *fm* inferior; margen *fm* de pie; pie *m* de página — **i** blanco *m* al piede

13551 tail piece — See apron.

13552 tail-piece; end-piece — Ornamental design at the end of a section, chapter or book.
f cul-de-lampe *m*; cabochon *m* — **d** Schluß-stück *n*; Schlußvignette *f*; Finalstock *m* — **n** sluit-vignet *n*; sluitstuk *n*; sluit-ornament *n* — **e** culo *m* de lámpara; viñeta *f* de cierre — **i** rosone *m* di chiusura; fregio *m* finale; vignetta *f* finale; finalino *m*; finale *m*

13553 take — Copy when more than one compositor is handling a large piece of composition. Also copy in general.
f part *f* de copie — **d** Partie *f* des Manuskripts — **n** deel *n* van de kopij — **e** alcance *m*; agregado *m*; fragmento *m*; pala *f* Ch — **i** parte *f* di manoscritto

13554 take *v* **in a line** — To bring thin spaces in a line of type to get in a word or syllable and preventing hyphenation. See also drive *v* out.
f gagner (une ligne) — **d** eine Zeile einlaufen; eine Zeile einbringen — **n** een regel innemen — **e** ganar línea; comer línea — **i** guadagnare una riga

13555 take *v* **off** — To remove a forme from the press before the work is completed to allow the printing of another job.
f relever — **d** ausheben — **n** opbreken — **e** sacar; levantar — **i** levare

13556 take *v* **out turns** — See insert *v* letter.

13557 take slug — See galley slug.

13558 take *v* **time off** — For the compositor to be idle because of lack of copy or dead matter, for the pressman to wait for paper or a forme.
f être envoyé à la balade — **d** aussetzen; feiern müssen — **n** niets te doen hebben — **e** nada que hacer — **i** aspettare in attesa di materiale

13559 taking-away cylinder — Rotating device to transfer the cut-off signatures or copies to the delivery belts or table. See also paddles, fly.
f moulinet *m*; éventail *m*; araignée *f* — **d** Ausgangszylinder *m*; Ablegezylinder *m*; Schaufelrad *n*; Schaufelstern *m*; Auslegestern *m*; Fächer *m* — **n** waaier *m*; uitlegcilinder *m* — **e** cilindro *m* retirador — **i** cilindro *m* d'uscita

13560 talbot — Unit of luminous energy. A luminous flux of one talbot per second is one lumen. One talbot = 10^7 lumergs.

13561 talbotype; calotype — Old, practically obsolete, process of photography invented by William Henry Fox Talbot (1800-1877).
e talbotipia *f*; calotipia *f*

13562 talc; talcum — Common name for hydrated magnesium silicate.
Formula: $Mg_3(SiO_3)_4$.
f talc *m*; poudre *f* de talc — **d** Talk *m*; Talkum *n* — **n** talk *m*; talkpoeder *n* — **e** talco *m*; polvo *m* de talco — **i** talco *m*; polvere *f* di talco

13563 talcum — See talc.

13564 talking print — See talking shop.

13565 talking shop; talking print
f parler batiau; parler métier; parler affaires — **d** Fachsimpeln *n*; Fachsimpelei *f*; schwätzen — **n** aldoor over het vak praten; zwetsen — **e** parlanchín *m* profesional — **i** chiacchierare sulle faccende del mestiere; colloquio *m* professionale

13566 tall oil; tallöl; pinolin; fluid resin; liquid rosin; Swedish olein — Dark brown, viscous, oily liquid obtained as an alkaline waste liquor in the sulphate process of making wood pulp. It consists of fatty and rosin acids and can be used in the production of printing inks.
f huile *f* de tall — **d** Tallöl *n* — **n** talolie *fm* — **e** aceite *m* de pino — **i** talloil *m*

13567 tallöl — See tall oil.

13568 tampon — See ink ball.

13569 tandem reel stand; tandem roll stand — Dual or single stands one behind the other for feeding multiple webs through a press at the same time.
f double support *m* de bobines — **d** Doppelrollenstern *m* — **n** dubbele rollenster *fm*; tandem-rollenstandaard *m* — **e** soporte *m* doble de bobinas — **i** portabobine *m* doppio

13570 tandem roll stand — See tandem reel stand.

13571 tang — See jet.

13572 tangent — In trigonometry the ratio of the side opposite an acute angle of right triangle and the side adjacent to the same acute angle. Also a line just touching, but not cutting the circumference of a circle or an arc of a circle. See also trigonometric functions.
f tangente *f* — **d** Tangente *f*; Berührungslinie *f* — **n** tangens *f*; raaklijn *fm* — **e** tangente *f* — **i** tangente *f*

13573 tannic acid; gallotannic acid; tannin; digallic acid *misn.* — Lustrous, faintly yellowish, amorphous powder, glistening scales or spongy mass, found in nutgalls, tree barks, such as sumac, oak and hemlock. Soluble in water, alcohol and acetone; almost insoluble in benzene, chloroform, ether and petroleum ether, used to tan skin. Formula: $C_{76}H_{52}O_{46}$. Latin: tanninum; acidum tannicum. Classified into two groups: 1. condensed tannins, which yield *catechol; and 2. hydrolyzable tannins, which yield *pyrogallol. Group 2 is again divided into two groups on the basis of its products of hydrolysis, which are *glucose and 1. ellagic acid ($C_{14}H_6O_8$) or *gallic acid.
f acide *m* tannique; tannin *m*; tanin *m*; acide *m* digallique; acide *m* gallotannique — **d** Gerbsäure *f*; Gallusgerbsäure *f*; Tannin *m* — **n** tannine *fmn*; tannien *fmn*; looizuur *n* — **e** ácido *m* tánico; ácido *m* digálico; ácido *m* galotánico; tanino *m* — **i** acido *m* tannico

13574 tannin — See tannic acid.

13575 tanning — Converting raw hides into leather.
f tannage *m* — **d** Gerbung *f*; Gerben *n* — **n** looiing *f*; looien *n* — **e** curtimiento *m*; curtido *m*

— **i** conciatura *f*

13576 tanning agent
f agent *m* de tannage — **d** Gerbungsmittel *n*; Gerbstoff *m* — **n** looimiddel *n*; looistof *fm* — **e** tanino *m*; curtiente *m* — **i** tannino *m*

13577 tape card — See edge-punched card.

13578 tape clutch — Mechanism to start or stop the captive tapes of a rotary press.
f mécanisme *m* d'engagement de bande; mécanisme *m* d'embrayage des cordons d'alimentation — **d** Kupplung *f* für Einziehbänder — **n** koppeling *f* voor de invoerbanden — **e** dispositivo *m* de arranque de banda — **i** meccanismo *m* d'introduzione del nastro

13579 tape conversion — See conversion.

13580 taped — Packed with tapes instead of with string to avoid string cuts.
f fermé à ruban; attaché avec du papier gommé — **d** mit Klebeband verschlossen — **n** met plakband gesloten — **e** cerrado con cinta engomada — **i** chiuso a nastro adesivo

13581 tape delivery
f sortie *f* à cordons; réception *f* à cordons — **d** Bandausleger *m*; Bänderauslage *f* — **n** banduitleg *m* — **e** receptor *m* a cinta; sacapliegos *m* a cinta(s); sacapliegos *m* de cintas — **i** uscita *f* a nastro; uscita *f* con nastri trasportatori

13582 tape feed hole — See advanced sprocket.

13583 tape merging; computer merging
f mélange *m* de bandes perforées — **d** Lochbandmischung *f*; Lochstreifenkombination *f* — **n** ponsbandsamenvoeging *f* — **e** fusión *m* de cintas perforadas — **i** accoppiatura *f* dei nastri perforati

13584 tape perforating unit — See tape punch.

13585 tape pocket — Pouch on the tape on a rotary press into which the web lead is inserted.
f agrafe *f* de bande — **d** Bändertasche *f*; Büchse *f* — **n** baaninvoerklem *fm* — **e** engrape *m* de la banda — **i** attacco *m* alle cinghie d'introduzione

13586 tape pulleys — Wheels on which tapes run in a rotary press.
f poulies *fpl* des cordons d'engagement — **d** Bandrollen *fpl*; Gurtrollen *fpl* — **n** bandrollen *fmpl* — **e** poleas *fpl* de cinta; poleas *fpl* de guía — **i** pulegge *fpl* delle cinghie d'introduzione

13587 tape punch; tape punching device; tape perforating unit — Machine to punch code holes or feed holes in paper tape.
f perforateur *m* de bande — **d** Lochstreifenstanzer *m* (automatisch); Lochstreifenlocher *m* (manuell); Lochbandperforator *m*; Streifenlocher *m* — **n** bandponsmachine *f*; ponsbandperforator *m* — **e** perforador *m* de cinta; perforador *m* de banda — **i** perforatore *m* di nastro; unità *f* di perforazione del nastro

13588 tape punching device — See tape punch.

13589 taper *v* — To remove by etching the shoulders left by previous etchings.
f dépouiller — **d** rundätzen — **n** rondetsen; wegetsen — **e** limpiar — **i** rimuovere uno strato di ricoprimento

13590 tape reader; paper-tape reader; tape reading device — Device to read the holes and the non-hole combinations in paper tape and convert this information into electronic pulses or charges. The reading may be mechanical, as "fingers" sense the presence or absence of holes, or photo-electric, as light penetrates or fails to penetrate the tape opposite light-serving photodiodes.
f lecteur *m* de bande perforée — **d** Lochbandleser *m*; Streifenleser *m*; Lochstreifenabfühler *m* — **n** ponsbandlezer *m* — **e** lectora *f* de cinta perforada; lectora *f* de cinta de papel — **i** lettore *m* di nastro perforato; lettore *m* di nastro di carta

13591 tape reading device — See tape reader.

13592 tapered dash — See French rule.

13593 tapered pin; taper pin
f goupille *f* conique — **d** Kegelstift *m* — **n** borgpen *fm*; borgstift *fm*; contrapen *fm* — **e** pasador *m* cónico — **i** spina *f* conica

13594 tapered slug — Slug with a slight wedge shape caused by a defect of the trimming knives of the typesetting machine. See also bottle-bottomed type, bottle-necked type.
f ligne *f* conique — **d** konische Zeile *f* — **n** tapse regel *m*; scheve regel *m*; konische regel *m* — **i** riga *f* smussata

13595 taper pin — See tapered pin.

13596 tapes; lead tapes — Strips of leather, fabric, plastics or textile in a rotary press to carry paper from one part of the press to another.
f cordons *mpl* d'engagement; cordons *mpl* de guidage — **d** Transportbänder *npl* — **n** transportbanden *mpl* — **e** cintas *fpl* de transporte; cintas *fpl* transportadoras — **i** nastri *mpl* trasportatori

13597 tapes — Narrow strips of strong material (cloth), about 6.5-12.5 mm (0.25-0.5 in) wide, to which the sections of a book are sewn, extending over the back and onto the cover boards.
f bandes *fpl*; rubans *mpl* — **d** Heftbänder *mpl* — **n** banden *mpl* — **e** cintas *fpl* — **i** nastri *mpl*

13598 tap *v* **out; pat** *v* **out** — To apply a spot of ink to paper with a tapping finger, to distribute the ink to approximately printing film thickness.
f essai *m* d'encre par tache — **d** auftupfen — **n** uittippen — **e** hacer un ensayo de tintas por manchas — **i** spalmare (una macchia d'inchiostro)

13599 Tappi opacity — Contrast ratio measured under conditions specified by the Tappi standard method. The ratio, expressed as a percentage, of the reflectance of a single sheet backed by a black cavity to the reflectance of the same sheet backed by a white body having an absolute reflectance of 0.89. See also opacity.
f opacité *f* d'après Tappi — **d** Opazität *f* nach Tappi — **n** opaciteit *f* volgens Tappi — **e** opacidad *f* según Tappi — **i** opacità *f* secondo Tappi

13600 tap water — Water not specially purified, distilled or otherwise treated, obtained directly from a tap.
f eau *f* de robinet; eau *f* de ville — **d** Leitungswasser *n* — **n** leidingwater *n*; kraanwater *n* — **e** agua *f* de grifo — **i** acqua *f* di rubinetto

13601 target language — See object language.

13602 target program — See object program.

13603 target time — Time less than allowed time in which a task should be completed by an operator working under incentive conditions. See also allowed time, standard time.

n richttijd *m* — **i** tempo *m* minimo

13604 tariff commission

f commission *f* du tarif — **d** Tarifausschuß *m* — **n** tariefscommissie *f* — **e** comisión *f* de la tarifa — **i** comitato *m* della tariffa

13605 tarlatan — Sheer cotton fabric in open plain weave, usually heavy sized, to wipe ink from an hand-engraved intaglio plate.

f tarlatane *f* — **d** Tarlatan *m* — **n** tarlatan *n* — **e** tarlatana *f* — **i** tarlatana *f*

13606 tarnishing — Growing dull or discoloured, or losing luster.

f ternissement *m*; perte *f* de brillant; perte *f* d'éclat — **d** Glanzverlust *m*; Wegnehmen *n* des Glanzes — **n** dof worden *n*; verlies *n* van de glans — **e** deslustrar; empañar; deslucir; quitar el lustre; apagamiento *m* — **i** appannamento *m*; perdita *f* di lucido

13607 tar paper — Paper impregnated or coated with coal tar or asphalt. Cheap brown wrapping paper.

f papier *m* goudron — **d** Goudronnépapier *n*; geringwertiges braunes Packpapier *n* — **n** goudronnépapier *n*; goudronné *n* — **e** papel *m* alquitramado; papel *m* embreado — **i** carta *f* catramata

13608 tarred brown paper — Wrapping paper with some degree of waterproofing, consisting of one sheet of paper coated or impregnated with tar (coal or wood), or bitumen, or several such sheets stuck together.

f papier *m* goudronné; papier *m* bitumé — **d** Teerpapier *n*; Bitumenpapier *n*; bitumenbeschichtetes Papier *n*; bitumenimprägniertes Papier *n* — **n** bitumenpapier *n*; teerpapier *n*; geteerd papier *n* — **e** papel *m* (de estraza) alquimatrado — **i** carta *f* catramata; carta *f* bitumata

13609 tarred thread paper — See reinforced union paper.

13610 tartaric acid; dihydroxysuccinic acid — Colourless, transparent crystals, or white, fine to granular, crystalline powder, soluble in water, alcohol and ether. Used in photography (printing and developing; light-sensitive salts). Latin: acidum tartaricum.

Formula: $C_2H_2(OH)_2(COOH)_2$.

f acide *m* tartarique (ordinaire) — **d** Weinsteinsäure *f*; Weinsäure *f*; Rechtsweinsäure *f*; Dioxybernsteinsäure *f* — **n** wijnsteenzuur *n*; dihydroxy-barnsteenzuur *n* — **e** ácido *m* tartárico; ácido *m* dihidroxisuccinico — **i** acido *m* tartarico

13611 tartrazine yellow — Yellow lake pigment for printing inks, noted for its transparency. Tartrazine is a bright orange-yellow powder, soluble in water.

Formula: $C_{16}H_9N_4O_9S_2$.

f jaune *m* de tartrazine — **d** Tartrazingelb *n* — **n** tartrazine-geel *n* — **e** amarillo *m* de tartracina — **i** giallo *m* di tartrazina

13612 tassel — See book marker.

13613 tearing resistance — See tearing strength.

13614 tearing strength; tearing resistance — Resistance of paper or board to tearing running under the same conditions of the proper tear (Brecht-Imset).

f résistance *f* au déchirement; résistance *f* à la déchirure — **d** Durchreißfestigkeit *f*; Durchreißwiderstand *m* — **n** doorscheurweerstand *m* — **e** resistencia *f* al rasgado — **i** resistenza *f* alla lacerazione

13615 tearing strength; initial tearing strength — Resistance of paper or cardboard to strains causing a tear in the material to start from the edge (Bekk).

f résistance *f* au déchirement au bord du papier — **d** Einreißfestigkeit *f*; Einreißwiderstand *m* — **n** inscheurweerstand *m* — **e** resistencia *f* al rasgado inicial — **i** resistenza *f* alla lacerazione iniziale

13616 tearing strength; edge tearing resistance — Resistance of paper or board to tearing continued for a fixed distance of a tear already started (Elmendorf).

f résistance *f* au déchirement; résistance *f* à la déchirure — **d** Weiterreißfestigkeit *f*; Fortreißwiderstand *m* — **n** verderscheurweerstand *m* — **e** resistencia *f* al rasgado interno — **i** resistenza *f* alla lacerazione

13617 tearing test — The tearing strength as indicated in points per pound on a standardized testing apparatus, such as the Elmendorf test.

f essai *m* à la résistance au déchirement; essai *m* à la résistance à la déchirure — **d** Durchreißprüfung *f* — **n** doorscheurproef *fm* — **e** ensayo *m* de rasgado; ensayo *m* de rasgadura; ensayo *m* de desgarramiento — **i** prova *f* di resistenza alla lacerazione

13618 tearing tester

f appareil *m* pour l'essai à la résistance au déchirement — **d** Durchreißprüfer *m*; Durchreißprüfgerät *n* — **n** doorscheurtoestel *n* — **e** ensayadora *f* de desgarre; ensayadora *f* de rasgado — **i** apparecchio *m* per prova di lacerazione

13619 tear-off calendar — See block calendar.

13620 teaspoon — Deprecated measure of volume. See app. no. 7.

f cuillerée *f* à thé — **d** Teelöffel *m* — **n** theelepel *m* — **e** cucharadita *f* — **i** cucchiaino *m* da tè

13621 technical education

f formation *f* professionnelle; éducation *f* professionnelle — **d** Fachbildung *f* — **n** vakopleiding *f* — **e** educación *f* profesional — **i** istruzione *f* professionale

13622 technical magazine — See trade publication.

13623 Tee-piece — See T-piece.

13624 tee-square — See T-square.

13625 teka oils — See tekaols.

13626 tekaols; teka oils; purified stand oils —

Stand oil extracts from which saturated acids, free acids, etc., have been removed. They dry faster and produce harder films than the untreated stand oils.

f huiles *fpl* de lin fortes; mordants *mpl* — **d** Tekaöle *npl*

13627 telecommunication — Transmission, emission, or reception of signs, signals, writings, images, and sounds of intelligence of any nature by wire, radio, visual, or other electro magnetic means.

f télécommunication *f* — **d** Fernmeldetechnik *f* — **n** telecommunicatie *f* — **e** telecomunicación *f* — **i** telecomunicazione *f*

13628 telephone directory paper — Lightweight catalogue paper similar to newsprint to print telephone directories, etc.

f papier *m* pour annuaires de téléphone — **d** Telephonbücherpapier *n* — **n** papier *n* voor telefoongidsen; telefoongidsenpapier *n* — **e** papel *m* para guías telefónicas — **i** carta *f* per elenchi telefonici

13629 teleprinter — See Teletype.

13630 teleprocessing — Information handling in which a data processing system utilizes communication facilities.

f télétraitement *m* — **d** Datenfernverarbeitung *f* — **n** informatieverwerking *f* op afstand; gegevensverwerking *f* op afstand; afstandgegevensverwerking *f* — **e** teleprocesamiento *m* — **i** teletrattamento *m*

13631 telescoped reel (of paper) — Reel edge alignment runs out, starting at the core as the reel rotates. Reel may also telescope during handling leaving reel ends concave and convexe.

f bobine *f* foirée — **d** in sich verschobene Rolle *f*; seitlich verschobene Papierrolle *f* — **n** uitgelopen rol *fm* — **e** bobina *f* corrida — **i** bobina *f* scampanata

13632 telescopic flow — See laminar flow.

13633 telescoping — Transverse slipping of successive layers of a coil so that the edge of the coil is conical rather than flat.

f télescopage *m* — **n** als een telescoop in elkaar schuiven *n* — **e** enchufado *m* — **i** slittamento *m* trasversale

13634 Teletype — Registered trade-mark of the Teletype Corporation. Not to be confused with *teletypesetter. In common use Teletype has become a synonym for *teletypewriter, teleprinter and printing telegraph. A 5-level system for transmitting messages.

13635 teletypesetter — Registered trade-mark for the line-casting machine, manufactured by the Fairchild Graphic Equipment Co. not to be confused with *Teletype.

f télétypesetter *m*; télécompositeur *m*; composeuse *f* à distance — **d** Teletypesetter *m*; Fernsetzmaschine *f* — **n** teletypesetter *m* — **e** teletipo *m*; máquina *f* telecompositora; máquina *f* teletipográfica — **i** telecompositore *m*

13636 teletypesetting — System, introduced in 1932, whereby all the functions of a hot-metal line-casting machine are controlled by 6-level tape, which is read by an operating unit attached to the keyboard of the line-caster. The primary tape is prepared as a preliminary operation on a perforating keyboard. A reduction of keystrokes and/or operator decisions may be achieved by computer processing. The advantages are faster operation of line-casters and improved performance by the keyboard operator because he is not involved in controlling the functions of the linecaster. The abbreviation TTS for the teletypesetter was used until 1969. Since then it is used by the manufacturers (Fairchild) to designate Total Typesetting System. See also teletypesetter.

f télécomposition *f*; composition *f* à distance — **d** Teletypesatz *m*; Fernsatz *m* — **n** zetten *n* per teletype — **e** teletipocomposición *f* — **i** telecomposizione *f*

13637 teletypewriter — Combination of typewriter keyboard and printer, used to provide sending and receiving facilities for telegraph transmissions. See Teletype.

13638 tell-tale — At the top or bottom of a page (usually as a running head or foot) the name or title of the first entry or topic on a left-hand page and the last entry or topic on a right-hand page, or both the first and last entries on the same page. Commonly used in telephone directories to assist the reader to determine the alphabetic range of entries to be found on the page.

13639 tempelstick — Crayon-like material for marking metal and other surfaces intended to be heated, by melting them at a specific temperature and indicating that the required degree of heat has been obtained.

13640 tempera — Painting technique in which an emulsion consisting of water and pure egg yoke or a mixture of egg and oil is used as a binder or medium, characterized by its lean film-forming properties and drying rate.

13641 temperature control — Maintaining the temperature of photographic solutions at some specified level for highest efficiency and uniformity of negatives and prints. Modern devices include temperature controlled dark room sinks.

f réglage *m* de la température — **d** Temperaturregelung *f* — **n** temperatuurregeling *f*; regeling *f* van de temperatuur — **e** regulación *f* de la temperatura — **i** regolazione *f* di temperatura

13642 temporary storage — See temporary store.

13643 temporary store *GB*; **temporary storage** *US* — Store for registering intermediate results.

f mémoire *f* temporaire — **d** Zwischenspeicher *m* — **n** kladgeheugen *n* — **e** memoria *f* temporal; almacenamiento *m* temporal — **i** area *f* di comodo

13644 tensile strength — Limiting resistance of a test piece of paper or board submitted to a breaking force applied to each of its ends (tensile test) under the conditions defined in the standard method of test; generally expressed as *breaking length.

f résistance *f* à la rupture par traction; résistance *f* à la traction — **d** Bruchwiderstand *m*; Zugfestigkeit *f*; Bruchlast *f* depr. — **n** breeksterkte *f*; treksterkte *f*; trekweerstand *m* — **e** resistencia *f* a la tracción; resistencia *f* a la (rotura por) tensión; resistencia *f* de tensión — **i** resistenza *f* alla trazione

13645 tensile-strength tester — Apparatus to determine the tensile strength and stretch of a paper strip under increasing tension.
f dynamomètre *m*; appareil *m* pour l'essai de traction — **d** Zugfestigkeitsprüfer *m* — **n** treksterktemeter *m* — **e** ensayador *m* de la resistencia a la tensión — **i** apparecchio *m* per la misura della resistenza a trazione

13646 tensile stretch — See stretch at break.

13647 tensiometer — Instrument to measure surface and interfacial tension of printing inks. Surface tension refers to the energy relationship between ink and air surfaces. Interfacial tension refers to energy relationship between pigment and vehicle of inks and is thus an index of the extent of dispersion.
f tensiomètre *m* — **d** Tensiometer *n*; Oberflächenspannungs-Meßgerät *n* — **n** tensiometer *m* — **e** aparato *m* de ensayo de la tensión superficial — **i** tensiometro *m*

13648 tension control — Means for regulating the application of braking effort to the reel of paper or controlling the feed of paper into the press.
f réglage *m* de la tension — **d** Spannungsregelung *f*; Papierspannungsregelung *f* — **n** regeling *f* van de papierspanning; spanningsregeling *f* — **e** regulación *f* de la tensión — **i** regolazione *f* di tensione

13649 tension plate lock-up — Arrangement in a rotary press whereby metal fingers grip pockets on the underside of the plate and lock the plate to the plate cylinder. See also compression plate lock-up.
f fixation *f* du cliché par tension; calage *m* par tension (des stéréos); serrage *m* sous tension — **d** Platteninnenspannung *f*; Zugspannungsverschluß *m* — **n** tension lock-up *m* — **e** dispositivo *m* de fijación por tensión — **i** dispositivo *m* a tensione per il fissaggio delle lastre

13650 tension roller — See jockey roller.

13651 tension spring — See also spring.
f ressort *m* de tension; ressort *m* travaillant à la traction — **d** Zugfeder *f* — **n** trekveer *fm* — **e** resorte *m* de tracción — **i** molla *f* di tensione

13652 terminal — Device related to a system or processor, to provide a method of communication with the system; also a "free-standing" or self-contained unit with or without a video screen for editing.
f terminal *m* — **d** Datenstation *f* — **n** terminal *m* — **e** terminal *m* — **i** terminale *m*

13653 terminal — Mechanical device to establish an electric connection to an apparatus.
f borne *f* — **d** Anschlußklemme *f*; Klemme *f* — **n** aansluitklem *f*; klem *f* — **e** borne *m*; terminal *m* — **i** terminale *m*

13654 terminal — See final letter.

13655 ternion — Term originating from old times, when books could be printed only in formes of two pages and the binding be done by inserting a number of sheets one into the other. See also quaternion, quinternion, sexternion.
f ternion *m* — **d** Ternio *m*; Lage *f* von drei Bogen — **n** katern *fm* van drie — **e** terno *m* — **i** terno *m*

13656 terra silicea — See diatomaceous earth.

13657 tertiary colours — Colours produced by mixture of two secondary colours, or by admixture of the three primary colours.
f couleurs *fpl* tertiaires — **d** Tertiärfarben *fpl* — **n** tertiaire kleuren *fmpl* — **e** colores *mpl* terciarios; colores *mpl* ternarios — **i** colori *mpl* terziari

13658 test — Critical examination or trial of materials under standardized conditions to determine one or more characteristics of the materials.
f essai *m* — **d** Prüfung *f*; Prüfungsversuch *m*; Versuch *m*; Test *m*; Messung *f*; Untersuchung *f* — **n** proef *fm*; onderzoek *n* — **e** ensayo *m*; prueba *f* — **i** prova *f*; saggio *m*

13659 test *v*
f essayer; contrôler — **d** prüfen; versuchen; untersuchen; erproben; probieren — **n** onderzoeken; beproeven — **e** ensayar; probar — **i** provare; esaminare

13660 test board — Container board to comply with test specifications for shipping containers, i.e., Rule 41 of the Consolidated Freight Classification (US).

13661 test conditions
f conditions *fpl* opératoires — **d** Versuchsbedingungen *fpl* — **n** onderzoekingsvoorwaarden *fpl*; omstandigheden *fpl* tijdens het onderzoek — **e** condiciones *fpl* de ensayo — **i** condizioni *fpl* di prova

13662 tester; testing machine
f appareil *m* d'essai — **d** Prüfgerät *n*; Versuchsgerät *n*; Prüfer *m* — **n** proeftoestel *n*; onderzoekingstoestel *n* — **e** ensayadora *f*; probador *m* — **i** apparecchio *m* di prova

13663 testing machine — See tester.

13664 test liner — See test linerboard.

13665 test linerboard; test liner — *Linerboard used as the outside layers in the manufacture of corrugated and solid fibre shipping containers and set forth in Rule 41 of the Consolidated Freight Classification (US). There are two types, jute liner and *kraft liner, usually made from sulphate pulp, waste paper, and clippings; usually hard-sized.
f couverture *f* spéciale — **d** Testliner *m*; Decklage *f* — **n** testliner *m* — **i** testliner *m*

13666 test method
f méthode *f* d'essai — **d** Prüfverfahren *n*; Versuchsmethode *f*; Untersuchungsmethode *f* — **n** onderzoekingsmethode *f*; beproevingsmethode *f* — **e** método *m* de ensayo — **i** metodo *m* di prova

13667 test-paper; indicator paper *depr.* — Paper impregnated with a reagent, as litmus, that changes colour in testing for various substances.
f papier *m* indicateur — **d** Indikatorpapier *n* — **n** indicatorpapier *n* — **e** papel *m* indicador; papel *m* reactivo — **i** carta *f* indicatore

13668 test-piece
f éprouvette *f* — **d** Prüfling *f*; Probe *f*; Versuchsstück *n*; Probestück *n*; Muster *n* — **n** proefstuk *n* — **e** barreta *f* de ensayo; probeta *f* de ensayo — **i** provetta *f*

13669 test specifications
f spécification *f* d'essai — **d** Versuchsvorschrift *f* — **n** onderzoekingsvoorschrift *n* — **e** especificaciones *fpl* de ensayo — **i** specifiche *fpl* di prova

13670 test-strip
f bande-éprouvette *f* — **d** Probestreifen *m*; Prüfstreifen *m*; Teststreifen *m* — **n** proefstrookje *n* — **e** muestra *f* de ensayo; probeta *f* — **i** striscia *f* di prova; provetta *f*

13671 tetrachloromethane — See carbon tetrachloride.

13672 tetravalent; quadrivalent — Having a valence of four, that is capable of combining with four atoms. See also valence.
f tétravalent; quadrivalent — **d** vierwertig — **n** vierwaardig — **e** tetravalente; cuatrivalente — **i** tetravalente

13673 Texoprint — Method to make diapositives from type formes to use these as originals for offset and gravure printing. The type is sprayed with a dull black material, the surface polished with a wash leather and then photographed.
f procédé *m* Texoprint — **d** Texoprintverfahren *n* — **n** Texoprint-procédé *n* — **e** procedimiento *m* Texoprint; método *m* Texoprint — **i** procedimento *m* Texoprint

13674 text — Body matter of a page composed in column or paragraph form and differing from display matter, headings, illustrations, or other printed material. Above 14-point, text is normally identified as display or headline type.
f texte *m* — **d** Text *m* — **n** tekst *m* — **e** texto *m* — **i** testo *m*

13675 text body matter — See straight matter.
13676 text figure — See figure.
13677 text finish — Finish intermediate between antique and machine finish, closely akin to vellum finish.

13678 textile printing
f impression *f* de textile — **d** Textildruck *m*; Stoffdruck *m* — **n** textieldruk *m* — **e** impresión *f* textil — **i** stampa *f* su tessuto

13679 textile sack — See bag.
13680 text matter — See straight matter.
13681 text paper — High-grade, bulky book paper, laid or wove, sometimes deckle-edged, with a medium finish, used in de luxe editions, in some cases, as fancy stationary paper. See also book paper.
f papier *m* pour éditions de luxe — **d** Textpapier *n* — **n** tekstdrukpapier *n*; romandruk-

papier *n*; romandruk *n* — **e** papel *m* para obras de lujo; papel *m* texto — **i** carta *f* per edizioni di lusso

13682 text part — See editorial section.
13683 text stream — Computer term. Flow of text which is not formatted into lines or blocks, although it may contain the information which permits a processor (e.g., computer) to do so.
13684 text string — Computer term. Assortment of consecutive characters which comprise a text stream or some part thereof, as differentiated from a string of formats or command codes.
13685 therblig — Reversed writing of the name Gilbreth, Frank Bunker (1868-1924), American engineer and efficiency expert, who designed symbols for work study elements. The 18 therbligs are: 1. search *v*; 2. find *v*; 3. select *v*; 4. grasp *v*; 5. transport *v*; 6. position *v*; 7. assemble *v*; 8. dissemble *v*; 9. inspect *v*; 10. use *v*; 11. preposition *v*; 12. release *v* load; 13. transport empty; 14. rest for overcoming fatigue; 15. avoidable delay; 16. unavoidable delay; 17. plan *v*; 18. hold *v*.
f therblig *m* (1. chercher; 2. trouver; 3. choisir; 4. saisir; 5. transport en charge; 6. placer; 7. assembler; 8. démonter; 9. contrôler; 10. utiliser; 11. prépositionner; 12. lâcher; 13. transport à vide; 14. se reposer; 15. attente évitable; 16. attente inévitable; 17. réfléchir; 18. tenir) — **d** Therblig *m* (1. suchen; 2. finden; 3. auswählen; 4. ergreifen; 5. beiholen; bewegen mit Last; 6. in Position bringen; 7. zusammenfügen; 8. auseinandernehmen; 9. prüfen; 10. gebrauchen; 11. bereitlegen; ordnen; 12. loslassen; 13. hinlangen; 14. ausruhen; 15. vermeidbare Verzögerung; 16. unvermeidbare Verzögerung; 17. überlegen; 18. festhalten) — **n** therblig *m* (1. zoeken; 2. vinden; 3. kiezen; 4. grijpen; 5. verplaatsen; 6. plaatsen; 7. samenvoegen; monteren; 8. uiteen nemen; demonteren; 9. controle; nagaan; 10. gebruiken; 11. voorrichten; 12. loslaten; 13. reiken; 14. uitrusten; rusten; 15. vermijdbaar oponthoud; 16. onvermijdbaar oponthoud; 17. overleggen; 18. vasthouden) — **e** therblig *m* (1. buscar; 2. encontrar; 3. escoger; 4. coger; 5. transportar; 6. poner; 7. ensamblar; 8. desarmar; 9. controlar; 10. utilizar; 12. desprender; 13. alargar; 14. descansar; 15. retardo evitable; 16. retardo inevitable; 17. estudiar; 18. tener) — **i** therblig *m* (1. cercare; 2. trovare; 3. scegliere; 4. pigliare; 5. trasportare; 6. porre; 7. montare; 8. smontare; 9. collandare; controllare; 10. usare; 11. preposizione; 12. lasciare; 14. riposare; 15. ritardo evitabile; 16. ritardo inevitabile; 17. considerare; 18. tenere)

13686 thermal black — *Carbon black made by the thermatomic process. It consists of relatively coarser particles than channel black; used as a pigment.
f noir *m* thermique — **d** Thermalruß *m* — **n** thermal-zwart *n* — **e** negro *m* térmico — **i** nerofumo *m* termico

13687 thermal conductivity — Quantity of heat passing in unit time through a plate of unit area and unit thickness with a temperature differential of 1° between the faces of the plate.

f conductibilité *f* calorifique; conductivité *f* thermique — **d** Wärmeleitfähigkeit *f* — **n** warmtegeleidings-vermogen *n* — **e** conductibilidad *f* térmica; conductividad *f* térmica — **i** conducibilità *f* termica

13688 thermo-blo — Device to cool the mould of a slug composing machine by means of a large volume of air.

13689 thermographic printing — See thermography.

13690 thermography; thermographic printing; imitation relief printing; imitation relief stamping; imitation embossing; imitation steel-die embossing; plateless engraving; raised printing — Letterpress printing with a special ink. While the ink is still wet on the sheet, it is dusted with a resinous powder which adheres to it. The sheets are then put through a baking process which causes the particles of the powder to fuse with the ink, giving a raised effect to the letters which simulate steel-die or copperplate engraving. Used principally for letterheads, business cards, envelopes, announcement and greeting cards and folders.

f thermogravure *f* — **d** Stahlstichimitation *f*; unechter Stahlstich *m*; Stahlstichnachahmung *f* — **n** thermodruk *m*; imitatie-staalstempeldruk *m*; imitatie-reliëfdruk *m* — **e** termografía *f*; impresión *f* termográfica; termograbado *m*; termoimpresión *f* — **i** termografia *f*; stampa *f* termografica

13691 thermoharding — See thermosetting.

13692 thermoplastic — Property in a plastic material which causes the material to soften each time it is heated and to become rigid when cooled.

f thermoplastique — **d** thermoplastisch — **n** thermoplastisch — **e** termoplástico — **i** termoplastico

13693 thermoplastic binding — See adhesive binding.

13694 thermoplastic-bookbinding machine
f machine *f* pour la reliure thermoplastique — **d** Klebebindemaschine *f* — **n** garenloosbindmachine *f* — **i** macchina *f* per la rilegatura senza cucitura; macchina *f* per la rilegatura in colla plastica

13695 thermo-safety switch
f thermo-interrupteur *m* de sécurité — **d** Thermosicherheitsschalter *m* — **n** thermisch beveiligde schakelaar *m* — **e** interruptor *m* de seguridad termoeléctrico — **i** interruttore *m* termico di sicurezza

13696 thermosetting; thermoharding — Property in a plastic material; when it has been softened by heat and hardened, it cannot be resoftened.

f thermodurcissable; durcissant à chaud — **d** wärmehärtend; wärmehärtbar; hitzehärtend; hitzehärtbar — **n** thermohardend; thermohardbaar; door warmte verhardbaar — **e** termofijo; termoestable; termoendurecible — **i** termoindurente; termostabilizzante

13697 thermosetting ink — Ink from plasticized synthetic resin dissolved in volatile vehicle, set by heating the printed sheet.

f encre *f* fixée à chaud — **d** hitzeabbindende Farbe *f* — **n** warmdrogende inkt *m* — **e** tinta *f* de fraguado térmico; tinta *f* de secado térmico; tinta *f* de secado por calor; tinta *f* termosecante — **i** inchiostro *m* termoindurente; inchiostro *m* termostabilizzante

13698 thermostat — Device, including a relay actuated by thermal conduction or convection, to establish and maintain a desired temperature automatically or signal a change in temperature for manual adjustment.

f thermostat *m*; thermorégulateur *m* — **d** Thermostat *n*; Wärmeregler *m*; Temperaturregler *m* — **n** thermostaat *m*; termostaat *m* — **e** termostato *m* — **i** termostato *m*

13699 thermostatic bath — Vessel of water automatically maintained at a given temperature.

f bain *m* thermostatique — **d** thermostatisch kontrolliertes Bad *n*; thermostatisches Bad *n* — **n** thermostatisch bad *n* — **e** baño *m* termostático — **i** bagno *m* termostatico

13700 thesis; dissertation; academic dissertation — Written or printed report about original research and the critical evaluation of it submitted to a learned body.

f thèse *f*; dissertation *f* académique — **d** Dissertation *f* — **n** proefschrift *n*; dissertatie *f* — **e** tesis *f* doctoral; tesis *f* de doctorado; memoria *f* doctoral; disertación *f* — **i** tesi *m* di laurea; tesi *f*; dissertazione *f* di laurea; dissertazione *f*

13701 thiacetic acid — See thioacetic acid.

13702 thickened oils — See blown oils.

13703 thickener — See bodying agent.

13704 thickening agent — See bodying agent.

13705 thick lead — Lead thicker than two points.
f interligne *f* forte; interligne *f* épaisse — **d** dicke Durchschußlinie *f* — **n** dikke interlinie *f* — **e** interlínea *f* gruesa — **i** interlinea *f* grossa; interlinea *f* di forte spessore

13706 thickness — See calliper.

13707 thickness gauge — See micrometer.

13708 thick space; three-to-em space — Space having a width of one-third of its own body.

f espace *f* forte; espace *f* un tiers; espace *f* de trois points — **d** Drittelgeviert *m*; Drittelspatium *n*; Drittelspatie *n*; starkes Spatie *n*; Dreipunktspatium *n* — **n** dikke spatie *f*; brede spatie *f* — **e** espacio *m* grueso; espacio *m* gordo; espacio *m* de tres puntos; tercio *m* de cuadratín — **i** spazio *m* da tre; spazio *m* terziruolo; terziruolo *m*

13709 thin — Low in photographic density.
f faible — **d** dünn — **n** dun — **e** débil — **i** leggero; a bassa densità

13710 thin *v* — See cut *v*.

13711 thin film — High-speed storage device capable of access times measured in nanoseconds. Consists of a layer of ferromagnetic material, about

four millionth of an inch thick, deposited in minute areas on a thin glass sheet.
f pellicule f mince — d Dünnschicht f — n vlies n — e película f delgada — i pellicola f sottile

13712 thin gauge metal — Zinc and copper sheets for photoengraving rolled to either 18-gauge (1.25 mm or 0.049 in), or 21-gauge (0.8 mm or 0.032 in), instead of the normal 16-gauge (1.65 mm or 0.065 in) thickness. Introduced during World War II as a metal conservation measure. The standard thickness after adoption of the metric system in GB are: for original letterpress photoengravings in copper and zinc 1.626 mm (0.064 in), for flat electrotypes and stereotypes 4.064 mm (0.160 in). There are ISO standards for the thickness of litho plates.
f métal m mince (pour clichés) — d dünnes Klischeemetall n; Dünnmetall n — n dun (cliché-)metaal n — e metai m delgado — i metallo m sottile per fotoincisione

13713 thinkerbelle — Nick name for the general-purpose Honeywell 200 computer, used for typesetting, accounting, payroll handling and analyzing advertising trends.

13714 think piece — See dope story.

13715 thin lead — Lead tinner than two points.
f interligne f fine; interligne f mince — d dünne Durchschußlinie f — n dunne interline f — e interlínea f delgada — i interlinea f fine

13716 thin negative — Underexposed and/or underdeveloped negative which appears less dense than a normal negative.
f négatif m faible; négatif m léger — d dünnes Negativ n — n dun negatief n — e negativo m débil; negativo m claro — i negativo m leggero

13717 thinner — See reducer.

13718 thin space — Space having a width of one-fifth of its own body. Also the matrices from which spaces of the same width are cast in the typesetting machine.
f espace f fine — d dünnes Spatie n; Fünftelgeviert n — n dunspatie f; dunne spatie f — e espacio m delgado; espacio m fino — i spazio m fine

13719 thin varnish
f vernis m faible — d dünner Firnis m — n slappe vernis mn; dunne vernis mn — e barniz m flojo — i vernice f debole

13720 thioacetic acid; thiacetic acid; ethanethiolic acid — Clear, yellow liquid, soluble in water, alcohol and ether.
Formula: CH_3COSH.
f acide m thioacétique — d Thioessigsäure f — n thioazijnzuur n — e ácido m tioacético — i acido m tioacetico

13721 thiocarbamide — See thiourea.

13722 thiosinamine — See allyl thiourea.

13723 thiourea; thiocarbamide — White, lustrous crystals, soluble in cold water, ammonium sulphocyanide solution and alcohol, almost insoluble in ether. Used in photography and photocopying papers.

Formula: CH_4N_2S.
f thiourée f; thiocarbamide f; sulfourée f — d Thiokarbamid m; Thioharnstoff m; Schwefelharnstoff m — n thio-ureum n; thiocarbamide n; zwavel-ureum n — e tiourea f; tiocarbamida f; sulfourea f — i tiourea f; solfourea f; solfocarbamide f; tiocarbamide f

13724 third card — See thirds.

13725 third cover — See inside back cover.

13726 third-generation computer — Computer utilizing *solid logic technology, operating in nanoseconds.
f calculateur m de la troisième génération — d Computer m der dritten Generation; Rechner m der dritten Generation — n computer m van de derde generatie — e computadora f de la tercera generación — i calcolatore m di terza generazione

13727 third-generation typesetter — Photocomposition device which substitutes for photographic master images through which light is flashed to expose the character directly onto photographic film or paper, a character image which is generated by electronic bombardment on the face of a cathode ray tube and photographed directly from the tube. Character masters may be stored as digitized information by the typesetting device, or as photographic masters which are scanned to generate a comparable image on the output tube.
f composeuse f de la troisième génération — d Setzmaschine f der dritten Generation — n zetmachine f van de derde generatie — e componedora f de la tercera generación — i compositrice f di terza generazione

13728 third large — See thirds.

13729 thirds; third card; third large — A standard size of cut card. See app. no. 3.

13730 thirteen as twelve; thirteenth copy — Publisher's term for an additional odd copy.
f treize pour douze; exemplaire m en sus; treizième exemplaire m — d dreizehn für zwölf Stück; Freiexemplar n auf 12 Exemplare; dreizehntes Exemplar n beim Partiebezug — n dertien voor twaalf; dertiende exemplaar n — e trece por doce; décimo tercio — i tredici per dodici; tredicesima

13731 thirteenth copy — See thirteen as twelve.

13732 thirty — American newspaper term (or the figure 30 or XXX or 0), denoting the end of a story or service, by extension the end of a day's work.
f fin f — d Schluß m — n einde n — e fin m; cierre m — i fine f

13733 thirty dash — See thirty rule.

13734 thirty rule; thirty dash — Rule or dash line at the end of an article in a newspaper when all the takes have been set.
f ligne f de la fin — d Schlußlinie f — n sluitlijn fm — e línea f de cierre — i filetto m di chiusura

13735 thirty-sixmo — Book in which each sheet forms 64 leaves or 128 pages. Abbrev. 64mo (no point).

f in-trente-six *m*; in-36 *m* — **d** sechsunddreißiger Form *f* (36-er) — **n** in zesendertigen — **e** en treinta-y-seisavo — **i** in trentaseiesimo

13736 thirty-two-mo — Book in which each sheet forms 32 leaves or 64 pages. Abbrev. 32mo (no point).

f in-trente-deux *m*; in-32 *m* — **d** zweiunddreißiger Form *f* (32-er) — **n** in tweeëndertigen — **e** en treinta-y-dosavo; en treintaidosavo — **i** in trentaduesimo

13737 thixotropy — Property of a liquid or plastic material which involves a reversible decrease of viscosity as the material is agitated or worked. Generally attributed to a loose structure of the disperse solid particles. Also referred to as shear thinning. Term created by Freundlich and Peterfi, 1927.

f thixotropie *f* — **d** Thixotropie *f* — **n** thixotropie *f* — **e** tixotropía *f*; tigmotropismo *m* — **i** tixotropia *f*

13738 thongs — Narrow strips of leather to fasten covers to books.

f lanières *fpl* — **d** Riemen *mpl*; Lederriemen *mpl* — **n** riemen *mpl* — **e** cintas *fpl*; lazos *mpl* — **i** strisce *fpl* (di cuoio)

13739 thousand letter price

f prix *m* aux mil caractères — **d** Tausendbuchstabenpreis *m* — **n** duizendletterprijs *m* — **e** precio *m* por mil letras; precio *m* el millar — **i** costo *m* per mille lettere

13740 thrasher — See duster.

13741 thread *v* — See web *v*.

13742 thread, binding — See binding thread.

13743 thread sealing binding — Binding where thread staples are inserted in the gutters during folding on a special folder. The ends of these staples (two or three to a signature) are sealed down with hot-melt adhesive. When gathered, adhesive plus a gauze or mull is applied without milling the backs.

f piqûre *f* par soudure — **d** Fadensiegeln *n* — **n** warm-hollanderen *n* — **e** cosido *m* por soldadura *f* a puntos — **i** legatura *f* a filo termosaldante; legatura *f* a filo saldato

13744 thread sealing device

f agrégat *m* pour poser et souder des fils de brochage — **d** Fadensiegeleinrichtung *f* — **n** warm-hollanderapparaat *n* — **e** fundente *m* para a cosido por soldadura a puntos — **i** attrezzatura *f* per legatura a filo termosaldante

13745 thread, sewing — See binding thread.

13746 three-colour etcher

f graveur *m* chromiste — **d** Dreifarbenätzer *m* — **n** driekleurenetser *m* — **i** autocromista *m*; incisore *m* a tre colori

13747 three-colour photography — Photography in which coloured originals or subjects are reproduced by making colour separation negatives therefrom through proper colour filters, the negatives dividing the hues of the original into the three primary colours, and from which printing plates are made to be printed in yellow, magenta and cyan inks.

f photographie *f* trichrome — **d** Dreifarbenphotographie *f* — **n** driekleurenfotografie *f* — **e** fotografía *f* tricrómica — **i** fotografia *f* a tre colori; fotografia *f* tricromica

13748 three-colour printing — See three-colour process printing.

13749 three-colour process plates — Plates used for three-colour process printing, specifically referring to tricolour halftone etchings.

f plaques *fpl* trichromes — **d** Dreifarbendruckplatten *fpl*; Dreifarbenklischees *npl* — **n** driekleurenautotypieclichés *npl* — **e** medio tonos *mpl* para tricromía; laminas *fpl* para tricromía; placas *fpl* para la impresión tricrómica — **i** autotipie *fpl* a tre colori; autotipie *fpl* tricromiche; lastre *fpl* di stampa per tricromia; lastre *fpl* per tricromia

13750 three-colour process printing; three-colour printing; trichromatic process printing — Method of printing in which all hues of an original are considered possible of reproduction by use of three separate printing plates, each plate for printing or recording one of the printing colours of the original. The plates usually are made by recourse to three-colour photography and the employment of halftone separation negatives, though they may also include plates made from line drawings and shading sheet originals, as well as those produced by bendaying and manual separation of colours on each of the plates.

f impression *f* en trichrome; procédé *m* trichrome; trichromie *f* — **d** Dreifarbendruck *m* — **n** driekleurendruk *m*; driekleurendruk-procédé *n* — **e** impresión *f* a tres colores; procedimiento *m* de tres colores; impresión *f* en tricromía; tricromía *f* — **i** stampa *f* a tre colori; tricromia *f*

13751 three-column page

f page *f* sur trois colonnes — **d** dreispaltige Seite *f* — **n** pagina *fm* op drie kolommen — **e** página *f* a tres columnas — **i** pagina *f* su tre colonne; pagina *f* a tre colonne

13752 three-dimensional printing — See xography.

13753 three-knife cutting machine — See three-knife trimmer.

13754 three-knife trimmer; three-knife cutting machine; three-sided cutting machine — Automatically operated cutting machine with two parallel and one right-angle knife to trim books and magazines.

f massicot *m* (automatique) trilatéral; massicot *m* à trois lames; massicot *m* tri-lame — **d** Dreimesserautomat *m*; Dreischneidemaschine *f*; Dreischneider *m*; Dreimesserschneidmaschine *f* — **n** driesnijder *m* — **e** cortadora *f* trilateral; guillotina *f* de tres cuchillas; guillotina *f* trilateral — **i** tagliacarte *m* trilaterale

13755 three-layer board — See triplex board.

13756 three-line initial — Initial letter having a depth equal to three lines of text matter.

f initiale *f* sur trois lignes; capitale *f* sur trois lignes; lettrine *f* — **d** dreizeiliger Initial *m*; dreizeiliger Anfangsbuchstabe *m*; Initiale *f* über drei Zeilen — **n** initiaal *n* over drie regels — **e** inicial *f* de tres líneas — **i** iniziale *f* a tre righe

13757 three-line nonpareil — Old name for 18-point type. See app. no. 1.

13758 three-line pica — Old name for 36-point type. See app. no. 1.

13759 three-phase alternating current — See three-phase current.

13760 three-phase circuit — Circuit energized by three electromotive forces which differ in phase by one third of a cycle or 120 °
f circuit *m* triphasé — **d** Dreiphasenkreis *m* — **n** driefasensysteem *n* — **e** circuito *m* trifásico — **i** circuito *m* trifase

13761 three-phase commutator motor
f moteur *m* à collecteur triphasé — **d** Drehstrom-Kommutatormotor *m* — **n** driefasen-collector-motor *m* — **e** motor *m* con colector trifásico — **i** motore *m* collettore trifase

13762 three-phase current; three-phase alternating current
f courant *m* alternatif triphasé — **d** Dreiphasen-strom *m*; Dreiphasenwechselstrom *m*; Drehstrom *m* — **n** driefasen-wisselstroom *m*; draai-stroom *m* — **e** corriente *f* alterna trifásica — **i** corrente *f* alternata trifase

13763 three-phase mains
f réseau *m* triphasé — **d** Dreiphasennetz *n*; Drehstromnetz *n* — **n** driefasen-net *n* — **e** red *f* tri-fásica — **i** rete *f* trifase

13764 three-phase motor
f moteur *m* triphasé — **d** Drehstrommotor *m*; Dreiphasenmotor *m* — **n** draaistroommotor *m* — **e** motor *m* trifásico — **i** motore *m* trifase

13765 three-phase slipring motor
f moteur *m* à bagues triphasé — **d** Drehstrom-Schleifringläufermotor *m*; Drehstrom-Schleifring-motor *m* — **n** driefasen-sleepringmotor *m* — **e** motor *m* de anillos trifásico — **i** motore *m* ad anelli trifase

13766 three-plane watermark — Watermark showing a dark and a light image next to one another, the paper thus having three thicknesses. Combination of a true and shaded watermark.
f filigrane *m* clair-obscur; filigrane *m* à deux tons; filigrane *m* clair et foncé — **d** Hell-Dunkel-Wasser-zeichen *n*; dreistufiges Wasserzeichen *n* — **n** drie-tonig watermerk *n* — **e** filigrana *f* de dos tonos — **i** filigrana *f* chiaro-scuro

13767 three-point bar system; three-point register system — Method and device to register images in colour composing, the device consisting of a right-angled metal bar with three movable points or wedges for accurately positioning any number of photographic images on a single plate or film.
f système *m* de repérage sur trois points — **d** Drei-punktsystem *n* — **n** driepunts-registersysteem *n*; driepunts-aanlegsysteem *n* — **e** sistema *m* de registro de tres guías; dispositivo *m* para poner a registro — **i** sistema *m* di messa a registro con tre punti di riferimento

13768 three-point register system — See three-point bar system.

13769 three-quarter binding — Binding style with the spine and a generous portion of the sides covered in leather or cloth, the rest of the sides being in another coloured cloth or paper.

13770 three-quarter frame — Stand to hold one pair of cases, space at the side for two small trays, and rack-room for five pairs of cases.

13771 three-reel machine
f rotative *f* à trois bobines — **d** Dreirollen-maschine *f* — **n** drierollenpers *f* — **e** máquina *f* de tres bobinas — **i** macchina *f* a tre bobine

13772 three roller mill; triple roller machine
f broyeuse *f* tricylindrique; broyeuse *f* à trois cylindres — **d** Dreiwalzenreibmaschine *f*; Drei-walzwerk *n*; Dreiwalzenstuhl *m* — **n** driewals *fm* — **e** molino *m* (de colores) a tres cilindros — **i** raffinatrice *f* a tre cilindri; mulino *m* a tre cilindri

13773 three-sided cutting machine — See three-knife trimmer.

13774 three-to-em space — See thick space.

13775 three-to-two folder — Folder in which the folding cylinder has a circumference of 3 cut-offs and the cutting cylinder 2 cut-offs, thus giving a ratio of cylinder sizes 3:2.
f plieuse *f* 3/2 — **d** 3/2-Falzapparat *m* — **n** drie-op-twee vouwapparaat *n* — **e** plegadora *f* 3/2 — **i** piegatrice *f* 3/2

13776 three-way process — See indirect process.

13777 threshold limit value; TL-value; maximal allowable concentration *obs.* — Maximum concentration of solvent vapours admissable during an 8-hours day without danger to health. The US Threshold Limits Committee publishes each year a comprehensive list of threshold limit values and their values are widely accepted as authoritative. The German and Scandinavian values agree broadly with those of the ACGIH (American Conference of Governmental Industrial Hygienists). In GB, the values of the Imperial Chemical Industries Ltd. are used. In the USSR and other East-European countries values are used which are generally under those of ACGIH. The threshold value of a chemical mentioned in this dictionary is stated at most of the entries in the basic table.
f concentration *f* maximum dans le lieu de travail; concentration *f* maximum admissible; seuil *m* de concentration — **d** maximale Arbeitsplatz-konzentration *f* — **n** maximale concentratie *f* op de plaats van de arbeid; maximaal toelaatbare concentratie *f* — **e** concentración *f* máxima admisible — **i** concentrazione *f* massima tollerata; concentrazione *f* massima nell'ambiente

13778 threshold performance — See break-even performance.

13779 threshold value
f valeur-seuil *f* — **d** Schwellenwert *m* — **n** drempelwaarde *f* — **e** valor *m* umbral — **i** valore-soglia *m*

13780 throat — Part of the metal pot of a casting

machine through which the alloy flows to the pump. Sometimes called gooseneck.

f gorge *f* de coulée — **d** Gießtopfkehle *f* — **n** keel *fm* van de gietpot — **e** garganta *f* del crisol — **i** canale *m* dal crogiolo

13781 throat element — See throat heater.

13782 throat heater; throat element — Part of the metal pot of a composing machine.

d Gießhalsheizung *f* — **n** keel-element *n* — **e** calentador *m* de garganta

13783 throat of ink rail — Slit or opening in the ink rail through which ink is transferred to the inking drum or roller.

f gorge *f* d'encre — **d** Farbschlitz *m* — **n** inktbakspleet *fm* — **e** grieta *f* del tintero — **i** fessura *f* del calamaio

13784 through drying — See drying through.

13785 throughput — Time required to produce a task through a computer system. Constraints may be imposed by some of the elements of the system. It may be reduced because the program is input bound (e.g., a slow paper tape reader), or output bound (e.g., a slow line printer or typesetter) or computer bound.

f débit *m* — **d** Durchflußleistung *f*; Durchlauf *m*; Datendurchlauf *m* — **n** verwerkingscapaciteit *f*; doorvoercapaciteit *f*; doorvoersnelheid *f* — **e** rendimiento *m* total de procesamiento — **i** capacità *f* di trattamento

13786 throwaways — See handbills.

13787 throwing out — See crazed drying.

13788 throw *v* **into gear**

f faire marcher; faire aller; mettre en marche — **d** ansetzen; anstellen; anlassen; in Gang setzen — **n** aanzetten; op gang brengen — **e** engranar; poner en marcha; poner en juego — **i** avviare; mettere in marcia

13789 throw off — Mechanical device to remove one inking roller from contact with the other inking rollers or forme.

f dispositif *m* de mise hors contact — **d** Walzenabstellvorrichtung *f* — **n** rolafsteller *m* — **e** salvapliegos *m*; desconectador *m* — **i** dispositivo *m* di distacco

13790 throw off — Device to declutch a machine.

f débrayage *m* — **d** Ausrückvorrichtung *f* — **n** afsteller *m*; afstelinrichting *f* — **e** disparador *m* — **i** disimbracamento *m*

13791 throw *v* **off the rollers; lift** *v* **off the rollers** — To remove one inking roller from contact with the other inking rollers or forme.

f enlever les rouleaux — **d** die Walzen abstellen — **n** de rollen terugnemen — **e** parar los rodillos; levantar los rodillos; alzar los rodillos — **i** distaccare i rulli

13792 throw-out; pull-out; fold-out; gatefold — Folded leaf in a book or periodical, which when opened extends beyond the page size.

f planche *f* dépliante — **d** eingehängte Falttafel *f*; Ausschlagseite *f*; Ausschlagtafel *f* — **n** uitslaande plaat *fm* — **e** plancha *f* desplegable — **i** tavola *f* pieghevole

13793 throw *v* **out of gear; stop** *v*

f débrayer; arrêter — **d** abstellen; außer Betrieb setzen; anhalten; ausrücken — **n** afzetten; stoppen — **e** parar; desengranar; desembragar — **i** arrestare; fermare

13794 throw *v* **up** — To render prominent by the use of bold face type.

f faire ressortir — **d** hervorgeben; auszeichnen — **n** laten uitkomen — **e** hacer resaltar; dar énfasis; destacar — **i** mettere in risalto

13795 thumb index; flat cut index; gouge index; side index; banks — Row of thumb or flat side-cut index apertures in the foredge of a book.

f index *m* à encoches; table *f* à encoches; registre *m* à encoches; répertoire *m* encoché — **d** Daumenregister *n*; eingeschnittenes Register *n*; Griffregister *n*; Fingerlochregister *n* — **n** duim-index *m*; duimregister *n* — **e** índice *m* de asas; índice *m* pulgado; registro *m* con digitales; digitales *mpl* — **i** registro *m* con intaccature; registro *m* a camme

13796 thumb nut — See wing nut.

13797 thumb tack — See drawing pin.

13798 thymol — White crystals, soluble in alcohol, carbon disulphide, chloroform, glacial acetic acid, ether and fixed or volatile oils, slightly soluble in water and glycerol.
Formula: $(CH_3)_2CHC_6H_3(CH_3)OH$.

f thymol *m*; acide *m* thymique — **d** Thymol *n* — **n** thymol *n* — **e** timol *m* — **i** timolo *m*

13799 ticket of admission — Ticket which gives the right or permission to enter.

f billet *m* d'entrée; carte *f* d'entrée — **d** Eintrittskarte *f* — **n** toegangsbiljet *n*; toegangskaart *fm* — **e** esquela *f* de entrada — **i** biglietto *m* d'entrada; biglietto *m* d'ingresso

13800 ticket printing machine — Machine to print small tickets from narrow reels or precut blanks.

f machine *f* à imprimer des tickets — **d** Billetdruckmaschine *f*; Fahrkartendruckmaschine *f*; Fahrkartenautomat *m* — **n** kaartjes-automaat *m* — **e** máquina *f* impresora de billetes — **i** macchina *f* da stampa per biglietti

13801 tie *v* **up** — To tie up type matter that has been made up into pages, with page cord to secure it until ready for imposition.

f lier — **d** aufbinden; ausbinden — **n** opbinden — **e** atar; liar; amarrar — **i** legare

13802 tight back; fast back — Cover fastened solidly to the back of the book, so that it does not become hollow when open.

f dos *m* serré — **d** am Buchblock angeklebter Deckenrücken *m*; Buch *n* mit angesetzten Deckeln — **n** boek *n* met aangeplakte band — **e** lomo *m* unido; lomo *m* encolado — **i** dorso *m* attaccato; dorso *m* fisso; dorso *m* unito; dorso *m* alla cappucina

13803 tight edged — Dried out at the edges (of a reel), resulting in a web which is slack in the middle.

f en patate — **d** mit Trockenrändern; mit Randschrumpf; mit Randverkürzung — **n** uitgedroogd; met randkrimp — **e** en cucuruchos — **i** allentato nel mezzo

13804 tighten *v* — See bolt *v*.

13805 tightened inspection
f inspection *f* renforcée — **d** verschärfte Prüfung *f* — **n** verscherpte keuring *f*; verzwaarde keuring *f*; verscherpt onderzoek *n* — **e** inspección *f* reforzada — **i** ispezione *f* rinforzata

13806 tight line — Too closely spaced line of type.
f ligne *f* forte — **d** zu volle Zeile *f* — **n** te volle regel *m*; stijve regel *m* — **e** línea *f* apretada — **i** linea *f* stretta (di spaziatura)

13807 tight negative — Negative so overexposed that the lighter tones are very dense and will come through very light on the positive. Also the shadows will come through noticeable lighter than they should.
f négatif *m* surexposé — **d** dichtes Negativ *n* — **n** dicht negatief *n*; gesloten negatief *n* — **e** negativo *m* sobrexpuesto — **i** negativo *m* sovraesposto

13808 tight reel; dense reel
f bobine *f* serrée — **d** hartgewickelte Rolle *f*; feste Rolle *f* — **n** harde rol *fm*; strakke rol *fm* — **e** bobina *f* dura — **i** bobina *f* compatta

13809 tight register work; close register work — Work requiring precision in the superimposition of the image of the stock.
f travail *m* en repérage précis — **d** Arbeit *f* mit hohen Ansprüchen an Paßgenauigkeit *f*; schwierige Paßform *f* — **n** nauwluisterend paswerk *n*; nauwluisterend sluitwerk *n*; moeilijk sluitwerk *n* — **e** registro *m* exacto — **i** lavoro *m* a stretto registro

13810 tight sheet — Web running with too much tension or brake.
f bande *f* trop tendue; excès *m* de tension — **d** straffe Papierbahn *f* — **n** strakke papierbaan *fm* — **e** banda *f* tirante; banda *f* con exceso de tensión — **i** nastro *m* troppo teso

13811 tight sheet — Bottom make-ready sheet tightened to prevent movement.

13812 tilde — Accent (˜), as used in Spanish and Portuguese ñ, õ.
f tilde *m* — **d** Tilde *f* — **n** tilde *fm* — **e** tilde *fm* — **i** tilde *f*

13813 time, "all in" — See "all in" time.

13814 time card — See clock card.

13815 time/gamma curve
f courbe *f* gamma/temps; courbe *f* de gamme en fonction du temps — **d** Zeitgammakurve *f*; Zeitentwicklungskurve *f*; Gammakurve *f* — **n** tijd/gamma-kromme *fm*; tijd/gamma-curve *fm* — **e** curva *f* tiempo/gama — **i** curva *f* tempo/gamma

13816 time of exposure — See exposure time.

13817 time on daywork — See hours per hour.

13818 time sharing — Technique which utilizes a device for more than one purpose during a given time scale. Time sharing systems are in operation which have many remote terminals to enable several users to be effactually in continuous communication with the central processor. Also

ascribed to simultaneous processing and peripheral activity.
f simultanéité *f* par partage du temps machine; utilisation *f* collective — **d** zeitlich verzahnte Verarbeitung *f* (mehrerer Programme); Zeitmultiplexbetrieb *m*; Zeitwechselbetrieb *m* — **n** tijddeling *f*; tijdscharing *f*; time sharing *m* — **e** simultaneidad *f* del tiempo de máquina; tiempo *m* compartido — **i** distribuzione *f* del tempo; suddivisione *f* del tempo

13819 time standard — Result of an exhaustive work study to arrive at a method of operation which most closely approaches the ideal situation from an ethical viewpoint as well as that of industrial engineering and work technique.
f standard *m* de temps — **d** Standardzeit *f* — **n** tijdnorm *fm* — **e** estandar *m* de tiempo — **i** standard *m* dei tempi

13820 time study — Detailed investigation of an operation in the printing industry, in which the average time taken to do each operation of a complete cycle is recorded.
f chronométrage *m* — **d** Zeitstudie *f* — **n** tijdstudie *f* — **e** cronometraje *m* — **i** studio *m* dei tempi

13821 time ticket — See job ticket.

13822 tin — Silver-white metal, soluble in acids and hot potassium hydroxide solution, insoluble in water. Alloyed with lead and antimony in the manufacture of type and stereotype metal. Symbol: Sn. Latin: stannum.
f étain *m* — **d** Zinn *n* — **n** tin *n* — **e** estaño *m* — **i** stagno *m*

13823 tin — Container made from sheet metal.
f boîte *f* en fer blanc — **d** Büchse *f*; Blechbüchse *f*; Blechdose *f* — **n** bus *fm*; blik *n* — **e** lata *f*; bote *m* de lata — **i** latta *f*; lattina *f*

13824 tin bichloride — See stannous chloride.

13825 tin crystals — See stannous chloride.

13826 tinctorial power — See colour strength.

13827 tinctorial strength — See colour strength.

13828 tin dichloride — See stannous chloride.

13829 tin-foil paper — Paper coated with powdered tin foil, suspendend in a casein solution or other adhesive used for decorative wrapping and preservative purposes.
f papier *m* d'étain — **d** Stanniolpapier *n* — **n** stannioolpapier *n* — **e** papel *m* (de) estaño; papel *m* plateado — **i** carta *f* stagnola

13830 tinny paper — Paper with a clear ring or crackle caused by overbeating of the pulp.
f papier *m* carteux; papier *m* sonnant; papier *m* pétillant — **d** Papier *n* mit gutem Klang; raschelndes Papier *n* — **n** papier *n* met harde klank — **e** papel *m* con sonido duro — **i** carta *f* sonora; carta *f* con buona sonorità

13831 tin-plate — Thin sheet iron or steel coated with tin.
f fer *m* blanc — **d** Blech *n*; Weißblech *n* — **n** blik *n* — **e** hoja *f* de lata; hojalata *f*; palastro *m* — **i** lamiera *f* stagnata; latta *f* stagnata

13832 tin-plate printing machine — See metal

decorating machine.

13833 tin printing — See metal decorating.

13834 tin printing ink — See metal decorating ink.

13835 tin protochloride — See stannous chloride.

13836 tin salt — See stannous chloride.

13837 tint — Unwanted background discoloration printed on the press sheet, usually due to ink emulsification on the press.
f voile *m* — **d** Ton *m* — **n** toon *m* — **e** velo *m* de tinta — **i** velo *m*

13838 tint — Mixture of a primary ink with a complementary ink.
f teinte *f* — **d** Farbton *m*; Mischton *m*; Ton *m* — **e** matice *m*; matiz *m*

13839 tint — See value.

13840 tint base — White ink used in the production of tints.
f blanc *m* de mélange — **d** Mischweiß *n* — **n** mengwit *n*; transparantwit *n* — **e** blanco *m* de mezcla — **i** bianco *m* misto

13841 tint block; tint plate; background plate — Printing plate bearing a stippled line or dot surface, to print light colours, usually in conjunction with a plate in bold or heavy key. Sometimes applied to parts of an illustration requiring darkening in colour. See also flat-tint plate.
f cliché *m* de teinte; cliché *m* de fond — **d** Tonplatte *f* — **n** (gerasterde) tintplaat *fm* — **e** bloque *m* para fondo; plancha *f* para fondo; clisé *m* de fondo — **i** cliché *m* per fondo; cliscè *m* per fondo

13842 tint block — See flat-tint plate.

13843 tinting; emulsifying; emulsification — In lithography, the mixing of ink with the damping solution on the plate which causes an all-over tint on the press-sheet. The non-image areas are readily cleaned by wiping with extra damping solution. Not to be confused with scumming which is more local and due to other reasons. Sometimes also called washing.
f voilage *m*; émulsification *f*; graissage *m*; séchage *m* de la plaque — **d** Emulgieren *n*; sich verfärben; ins Wasser gehen; Verschlierung *f* der Farben; Ausspülen *n* der Farben — **n** emulgeren *n*; tonen *n* door emulgeren; in het water gaan *n* — **e** emulsionado *m* (la tinta se corre por emulsionación) — **i** emulsionare; ingrassare

13844 tinting strength — See colour strength.

13845 tint lay *v* — To transfer benday tints to plates, drawings or other mediums.
f emplacer un grisé au benday — **d** Bendayrastern auflegen — **n** een filmraster aanbrengen — **e** sombrear; grisar; aplicar grisados; aplicar sombreados — **i** applicare fondi; applicare tinte benday

13846 tint plate — See tint block.

13847 tint print solution — Resinous mixture employed in double printing and work involving more than one exposure on the sensitized metal.
f couche *f* à griser — **d** Einkopierlösung *f* — **e** solución *f* resinosa — **i** soluzione *f* resinosa

13848 tint sheet — See shading medium.

13849 tint test — See bleach test.

13850 tintype — See ferrotype.

13851 tip-cat folder — Rotary press folder in which the folding blade shaft is actuated by a cam of special shape and forms the transverse fold by pushing the copy between the folding rollers.
d Tuckerfalzapparat *m* — **n** mes- en rollenvouwapparaat *n* — **e** plegadora *f* tucker — **i** dispositivo *m* piegatore sistema tucker

13852 tip, gilder's — See gilder's tip.

13853 tip *v* **on** — To paste an insert, a leaf or a section on another.
f rapporter — **d** einkleben — **n** inplakken; aanplakken — **e** encolar — **i** incollare

13854 tipped in — Illustration printed separately and gummed or pasted in position by one edge. Frequently used when illustrations on coated (art) paper are required in a book printed on a rough-surfaced paper.
f illustration *f* rapportée — **d** eingeklebte Illustration *f*; eingeklebtes Bild *n* — **n** ingeplakte illustratie *f*; ingeplakte afbeelding *f*; ingeplakte plaat *fm* — **e** encarte *m* pegado; pegado *m* — **i** illustrazione *f* incollata; illustrazione *f* riportata

13855 tissue paper — Thin, soft, resistant paper generally used for packaging delicate articles. Its substance is between 12 and 25 g/m^2.

13856 titanic acid anhydride — See titanium dioxide.

13857 titanic anhydride — See titanium dioxide.

13858 titanic oxide — See titanium dioxide.

13859 titanium — Silvery grey, light, strong metallic element used in alloys (as steel). Symbol: Ti.
f titane *m* — **d** Titan *n* — **n** titaan *n*; titanium *n* — **e** titanio *m* — **i** titanio *m*

13860 titanium dioxide; titanic acid anhydride; titanic anhydride; titanic oxide; titanium white — Very opaque, white, inorganic ink pigment, also used as a filler in papermaking, possessing the greatest hiding powder of all white pigments. Soluble in hot concentrated sulphuric acid and alkalis, insoluble in water and cold dilute acids. TL-value: 15 mg/m^3.
Formula: TiO_2.
f bioxyde *m* de titane; bioxyde *m* de titanium; anhydride *m* titanique; blanc *m* de titane; oxyde *m* de titane — **d** Titandioxyd *n*; Titansäureanhydrid *n*; Titanweiß *n* — **n** titaandioxyde *n*; titaanwit *n* — **e** dióxido *m* de titanio — **i** biossido *m* di titanio; bianco *m* di titanio

13861 titanium white — See titanium dioxide.

13862 title — Name of a book on cover or title page. See also back title, catch title, section title, sub-title.
f titre *m* — **d** Titel *m* — **n** titel *m* — **e** título *m* — **i** titolo *m*

13863 title along the back
f titre *m* de dos en long — **d** Längsrückentitel *m* — **n** rugtitel *m* in langsrichting — **e** título *m* del lomo longitudinal — **i** titolo *m* longitudinale del dorso

13864 title compositor — Compositor of the make-up section engaged in inserting headlines.

f titrier *m* — **d** Überschriftensetzer *m* — **n** koppenzetter *m* — **i** compositore *m* di titoli

13865 title face — See titling.

13866 title letters — See titling.

13867 title-page — Page at beginning of a book, which contains the title or subject of the book, sub-title (if any), name of the author, translator, editor, illustrator, publisher, place and date of publication, etc.

f page *f* de titre; titre *m*; grand-titre *m* — **d** Titel *m*; Haupttitel *m*; Titelseite *f*; Titelei *f* — **n** titelpagina *fm*; titel *m* — **e** portada *f*; carátula *f*; título *m* — **i** pagina *f* di titolo; titolo *m* principale

13868 title piece — See panel.

13869 titling; title face; title letters; full-face type — Fount of capitals for headlines or titles; generally occupying the whole of the body.

f caractères *mpl* de titre; caractères *mpl* pour titres; caractères *mpl* plein œil — **d** Titelschriften *fpl*; Auszeichnungsschriften *fpl* — **n** titelkapitalen *npl* — **e** tipos *mpl* para títulos; titulares *mpl* — **i** caratteri *mpl* da titoli

13870 titration — In analytical chemistry, a process to determine volumetrically the concentration of a desired substance in solution by adding a standard solution of known volume and strength until the reaction is completed, usually as indicated by a change in colour due to an indicator or by electrical measurement.

f titrage *m* — **d** Titration *f*; Titrierung *f* — **n** titratie *f* — **e** titraje *m*; titración *f*; titulación *f*; análisis *f* volumétrico — **i** titolazione *f*

13871 TL-value — See threshold limit value.

13872 to be continued

f à suivre — **d** wird fortgesetzt; Fortsetzung folgt — **n** wordt vervolgd — **e** continúa; continuará — **i** continua

13873 toe of the characteristic curve — In photographic sensitometry, part which extends from the threshold to the start of the "straight line portion" in the case of photographic materials which possess a characteristic curve having a straight line portion. Otherwise, the low density end of a characteristic curve.

f talon *m* de la courbe — **d** Fuß *m* der Schwärzungskurve; unterer Durchgang *m* der Schwärzungskurve — **n** voet *m* van de curve; voet *m* van de kromme — **i** punta *f* della curva caratteristica

13874 toggle joint — Joint having a central hinge like an elbow, and operable by applying the power at the junction, thus changing the direction of motion and giving indefinite mechanical pressure.

f joint *m* à genouillère — **d** Kniehebel *m*; Kniegelenk *n* — **n** kniegewricht *n* — **e** junta *f* de codillo — **i** giunto *m* a ginocchiera

13875 toggle joint press; toggle press

f presse *f* à genouillère — **d** Kniehebelpresse *f* — **n** kniehevelpers *f* — **e** prensa *f* con articulación de la rodilla — **i** pressa *f* a ginocchiera

13876 toggle lever; hinged lever

f levier *m* articulé — **d** Gelenkhebel *m*; Kniehebel *m* — **n** kniehevelarm *m* — **e** palenca *f* articulada — **i** leva *f* articolata

13877 toggle press — See toggle joint press.

13878 toggle switch; tumbler switch — Switch in which a projecting knob or arm, moving through a small arc, causes the contacts to open or close an electric circuit suddenly.

f commutateur *m* à basculeur — **d** Kippschalter *m* — **n** tuimelschakelaar *m* — **e** interruptor *m* de palanca — **i** commutatore *m* a levetta; interruttore *m* a levetta

13879 token — Term signifying 250 impressions, which is the rate per hour used as a basis in changing for presswork. Derived from the old practice of giving a pressman a brass counter, or token, for each 250 impressions.

f demi-rame *f* — **d** halbes Ries *n* — **n** halve riem *m* — **e** media resma *m* — **i** mezza risma *f*; tiratura *f* di 250 copie

13880 tolerance — Usually expressed by the symbol .-, signifies that a standard specified measurement having been established no variation beyond the plus or minus allowance will be acceptable. For example, if a metal plate is specified as 3.82 mm (0.152 in) thick, plus or minus 0.025 (0.001 in), the variation above or below 3.82 mm (0.152 in) cannot exceed 0.025 mm (0.001 in).

f tolérance *f* — **d** Toleranz *f*; zulässige Abweichung *f* — **n** tolerantie *f*; toegestane afwijking *f* — **e** tolerancia *f* — **i** tolleranza *f*

13881 toluene; methyl benzene; methyl benzol; phenylmethane — Colourless, refractive, flammable liquid, soluble in alcohol, benzene and ether, insoluble in water. Used in rotogravure ink. Commercially called toluol, not in modern nomenclature. TL-value: 5 ppm or 375 mg/m^3. Formula: $C_6H_5CH_3$.

f toluène *m*; toluol *m*; méthylbenzène *m* — **d** Toluol *n*; Methylbenzol *n* — **n** tolueen *n*; methylbenzeen *n*; toluol *mn depr.* — **e** tolueno *m*; toluol *m* — **i** toluene *m*; toluolo *m*

13882 toluidine yellow

f jaune *m* de toluidine; jaune *m* permanent — **d** Toluidingelb *n* — **n** toluïdine-geel *n* — **e** amarillo *m* de toluidina; amarillo *m* permanente — **i** giallo *m* di toluidina

13883 toluol — See toluene.

13884 tome — *Volume forming a part of a larger work. Also any book especially a very heavy, large or learned one.

13885 tonal rendition — See tone rendering.

13886 tonal value — See tone value.

13887 tone — In originals and illustrations, the effect due to harmonious relation of light and shade. See also tone value.

f ton *m* — **d** Ton *m* — **n** toon *m* — **e** tono *m* — **i** tono *m*

13888 tone *v* — To change the colour of an original black photographic image to other colours by means of metallic salts or toning dyes.

f virer — **d** tonen — **e** virar — **i** virare

13889 tone-correcting mask — Mask to correct tone, not colour.

f masque *m* de correction de contraste — d Tonwertkorrekturmaske *f* — n tooncorrectiemasker *n* — e máscara *f* de corrección de contraste — i maschera *f* di correzione del tono; maschera *f* di correzione del contrasto

13890 tone correction — Process which changes the tone values of the individual images, usually referring to the grey scale.

f correction *f* de contraste — d Tonwertkorrektur *f* — n tooncorrectie *f* — e corrección *f* de contraste — i correzione *f* del tono; correzione *f* del contrasto

13891 tone-fixing salt

f sel *m* de virage-fixage — d Tonfixiersalz *n* — e sal *f* de viraje — i sale *m* virofissatore

13892 toner — Highly concentrated pigment and/or dye to modify the hue or colour strength of an ink.

f toner *m* — d Schönungsmittel *n*; Schöner *m* — n toner *m* — e color *m* modificador — i toner *m*

13893 tone rendering; tonal rendition; tone rendition

f reproduction *f* des valeurs de tons — d Tonwiedergabe *f*; Tonwertwiedergabe *f* — n toonweergave *fm* — e reproducción *f* de tonos — i riproduzione *f* dei valori tonali

13894 tone rendition — See tone rendering.

13895 toners — See organic pigments.

13896 toner test — Test to estimate the presence of toner by diluting 1 g ink in an alcohol-benzene mixture (acetone in the Anpa method).

13897 tone test — See scumming test.

13898 tone value; tonal value — Density of the negative or positive at any specified point. A photograph giving approximately true monochrome rendition of the various colours and atmospheric effects of nature has correct tone values.

f tonalité *f* — d Tonwert *m*; Tönungswert *m* — n toonwaarde *f* — e valor *m* de tono; valor *m* de tonalidad; valor *m* tonal; tonalidad *f*; gradación *f* de tonos — i valore *m* tonale

13899 toning bath

f bain *m* de virage — d Tonbad *n* — n toonbad *n* — e baño *m* virador; baño *m* de viraje; baño *m* de virado — i bagno *m* di viraggio

13900 tooling — Handwork on an engraving or plate to improve its printing qualities, done by means of gravers or burins.

f burinage *m*; traçage *m* à l'outil; travail *m* à l'outil — d Gravieren *n*; Stechen *n*; Sticheln *n*; Wegsticheln *n*; Durchreißen *n*; Nachschneiden *n* — n graveren *n*; steken *n*; bijsteken *n*; wegsteken *n*; nasteken *n* — e burilado *m*; retoque *m* a buril — i ritoccare col bulino

13901 tooling — Making gold-leaf decorations on book covers.

f décoration *f* de la couverture — d Buchdeckenverzierung *f* — n bandversieren *n* — e estampado *m* de las tapas del libro — i decorazione *f* della coperta

13902 tool rack; rack — Board for hanging tools (hammer, pincers, screw driver, etc.).

f porte-outils *m* — d Werkzeuggestell *n*; Werkzeughalter *m*; Werkzeugregal *n* — n gereedschapsrek *n* — e portaútil *m* — i rastrelliera *f* portautensili

13903 tooth — Rough finish, suitable for pencil or crayon drawing.

f grain *m* — d Korn *n*; Körnung *f* — n korn *m* — e grano *m* — i grana *f* (superficiale)

13904 tooth combination — See combination.

13905 tooth gear — See pinion.

13906 tooth wheel — See gear.

13907 tooth wheel — See pinion.

13908 top — Acid resisting image on metal plates intended for relief etching. The ink image produced by the albumen process, especially after reinforcement with topping powder.

f réserve *f* anti-acide — d Ätzwiderstand *m*; Säurewiderstand *m*; säurefestes Bild *n* als Ergebnis der Kopie — n zuurbestendig beeld *n* — e capa *f* resistente al ácido — i immagine *f* acido-resistente

13909 top blanket; top sheet — In rotary printing, the outside dressing for an impression cylinder.

f blanchet *m* de dessus; feuille *f* de dessus — d Obertuch *n*; Schutztuch *n* — n bovendoek *n* — e paño *m* superior; mantilla *f* — i panno *m* di sopra

13910 top draw sheet — See draw sheet.

13911 top edge gilt — Said of a book when only the top edge shows gilding whilst the other three edges are left plain. Abbrev. t.e.g.

f doré sur tranche supérieure — d mit Kopfgoldschnitt -- n aan de kop verguld — e con cabeza dorada — i dorato al taglio superiore; dorato sul taglio di testa; con taglio superiore dorato

13912 topping — Application of an acid resist to a plate, or strengthening the resist with ink or etching powder.

f coulage *m*; couche *f* — d Einwalzen *n*; Einstäuben *n* — n inrollen *n*; oprollen *n* — e aplicado *m* del esmalte; esmaltado *m* — i applicazione *f* di uno strato protettivo

13913 topping powder — See etching powder.

13914 top rail section — See superstructure.

13915 tops and tails — *Prelims and indexes of a book.

13916 top sheet — See draw sheet.

13917 top sheet — See top blanket.

13918 top side; felt side — Face of a sheet of paper or board opposite to the wire side. In some papermaking processes (hand, mould, multi-vat) the face opposite to that in contact with the wire is not necessarily the right side of the finished paper.

f côté *m* supérieur; côté *m* feutre; face *f* supérieure; face *f* feutre; face *f* fleur — d Oberseite *f*; Filzseite *f* — n bovenzijde *fm*; viltzijde *fm* — e lado *m* de fieltro; cara *f* del papel — i lato *m* fieltro; lato *m* ballerino

13919 top sized paper — See tub sized paper.

13920 top sizing — See surface sizing.

13921 top tone — Colour of a thin film of ink when viewed by reflected light.

f couleur *f* superficielle — **d** Vollton *m* — **e** color *m* superficial — **i** colore *m* superficiale; colore *m* riflesso

13922 torn sheets — In a ream or skid of paper, sheets which are not full size, due to tear-outs or broken ends, not removed in sorting operations.

f feuilles *fpl* déchirées — **d** zerrissene Bogen *mpl*; beschädigte Bogen *mpl* — **n** gescheurde vellen *npl*; beschadigde vellen *npl* — **e** hojas *fpl* desgarradas — **i** fogli *mpl* danneggiati

13923 torntape system — System where the output tape is torn off at the computer and placed manually onto the tape reader of a typesetting machine. With newspapers, *allotters are used to eliminate the necessity of torn tape.

f système *m* à bandes coupées — **d** geschnittener Lochstreifen-System *n* — **n** torntape-systeem *n*; scheurband-systeem *n*; afscheursysteem *n* — **e** sistema *m* de cintas cortadas — **i** sistema *m* a nastri tagliati

13924 torque — The product of the force and the perpendicular distance from the line of action of the force to the axis of rotation. Broadly: a turning or twisting force. See also couple.

f moment *m* de torsion; moment *m* de rotation — **d** Torsionsmoment *n*; Drehmoment *n* — **n** torsiemoment *n* — **e** momento *m* de torsión — **i** momento *m* torcente; momento *m* di rotazione

13925 torsion — Twisting about an axis, produced by the action of two opposing couples acting in parallel planes.

f torsion *f* — **d** Torsion *f*; Drehung *f*; Verdrehung *f*; Zerdrehung *f* — **n** torsie *f*; tordering *f*; torderen *n*; verdraaiing *f* — **e** torsión *f*; torcedura *f* — **i** torsione *f*

13926 torsion balance — If a wire is acted upon by a couple the axis of which coincides with the wire, the wire twists through an angle, determined by the applied couple and the rigidity modulus of the wire. The amount of twist produced can be measured.

f balance *f* de torsion — **d** Torsionswaage *f*; Drehwaage *f* — **n** torsiebalans *fm* — **e** balanza *f* de torsión — **i** bilancia *f* di torsione

13927 total white — See newsprint waste.

13928 touching up — See retouching.

13929 toughness of a film

f ténacité *f* d'un film — **d** Zähigkeit *f* eines Films — **n** taaiheid *f* van een film — **e** tenacidad *f* de una película — **i** solidità *f* di una pellicola

13930 tourist printer; tramp printer — In the guild time a wandering craftsman, going from town to town to get work.

f trimardeur *m* — **d** Reisender *m*; Handwerksbursche *m*; Walzbruder *m* — **n** reizend vakman *m* — **e** tipógrafo *m* errante; tipógrafo *m* ambulante; impresor *m* errante; impresor *m* ambulante — **i** stampatore *m* ambulante; stampatore *m* itinerante; itinerante *m*

13931 town — A size of card. See app. no. 3.

13932 town mains; mains

f réseau *m* du courant; réseau *m* de distribution de ville — **d** Stromnetz *n* — **n** lichtnet *n*; stadselektriciteitsnet *n* — **e** red *f* eléctrica — **i** rete *f* elettrica

13933 toxicity — Quantity or degree of being toxic or poisonous.

f toxicité *f* — **d** Giftigkeit *f* — **n** vergiftigheid *f*; vergiftige werking *f* — **e** toxicidad *f*; venenosidad *f* — **i** tossicità *f*

13934 T-piece; Tee-piece

f branchement *m* en T; raccord *m* en T — **d** T-Stück *n* — **n** T-stuk *n*; driewegkoppeling *f* — **e** pieza *f* en T; unión *f* en T — **i** giunzione *f* in T

13935 trace *v*

f calquer; prendre un calque — **d** pausen; durchpausen; kalkieren; durchzeichnen — **n** calqueren; een calque maken; doortrekken; overtrekken — **e** calcar; hacer un calco — **i** decalcare; fare un calco

13936 tracing — Copy made by tracing on transparent paper.

f calque *m* — **d** Pause *f*; Durchzeichnung *f*; durchgepauste Zeichnung *f* — **n** calque *fm* — **e** calco *m*

13937 tracing linen

f toile *f* à calquer — **d** Pausleinwand *f* — **n** calqueerlinnen *n*; kalkeerlinnen *n* — **e** tela *f* para calcar — **i** tela *f* da lucidi

13938 tracing paper; tracing tissue — Paper made from cotton fibre and/or chemical wood pulps with a coating of Canada balsam in turpentine, or a solution of castor oil or linseed oil in alcohol, to make it translucent. Used for tracing and sketching, and for certain types of friskets.

f papier-calque *m*; papier *m* à décalquer — **d** Pauspapier *n*; Abpauspapier *n*; Durchpauspapier *n*; Transparentzeichenpapier *n*; Detailzeichenpapier *n*; Lichtpauspapier *n* — **n** calqueerpapier *n*; kalkeerpapier *n*; transparant tekenpapier *n* — **e** papel *m* de calco; papel *m* caldo; papel *m* de calcar; papel *m* para calcar — **i** carta *f* per lucidi

13939 tracing tissue — See tracing paper.

13940 track — See channel.

13941 track — See magnetic track.

13942 track parity check — See longitudinal parity check.

13943 trade binding — See edition binding.

13944 trade-mark

f marque *f* de fabrique; marque *f* de commerce — **d** Handelsmarke *f*; Fabrikzeichen *n* — **n** handelsmerk *n*; fabrieksmerk *n* — **e** marca *f* de fábrica; marca *f* comercial; signo *m* de fábrica; sello *m* de fábrica; pie *m* de fábrica — **i** marchio *m* di fabbrica; marca *f* di fabbrica

13945 trade-mark paper — See watermark paper.

13946 trade paper — See trade publication.

13947 trade press

f presse *f* professionnelle; presse *f* technique — **d** Fachpresse *f* — **n** vakpers *fm* — **e** prensa *f* profesional — **i** stampa *f* professionale

13948 trade publication; trade paper; technical

magazine — Magazine devoted to trades, industries or business.

f revue *f* technique; revue *f* commerciale; revue *f* professionnelle; revue *f* de métier — **d** Fachzeitschrift *f*; Fachblatt *n*; Gewerbefachzeitschrift *f* — **n** vaktijdschrift *n*; vakblad *n* — **e** revista *f* técnica; revista *f* comercial; revista *f* profesional; revista *f* industrial; órgano *m* técnico; órgano *m* comercial — **i** periodico *m* professionale

13949 trade setting; setting to the trade
f composition *f* à façon — **d** Lohnsatz *m* — **n** zetten *n* in loondienst; loonzetten *n* — **e** composición *f* asalariada — **i** composizione *f* a economia

13950 traffic — Message information in transit.
f trafic *m* — **d** Nachrichtenverkehr *m*; Verkehr *m* — **n** verkeer *n* — **e** tráfico *m* — **i** traffico *m*

13951 tragacanth gum; gum tragacanth — Dull white, translucent plates or spirally twisted, yellowish powder, soluble in alkaline solutions, aqueous hydrogen peroxide solutions, insoluble in alcohol. Used for sizing by gilders and marblers.
f gomme *f* adragante — **d** Tragant-Gummi *mn* — **n** tragacant *fm*; tragant *fm*; tragacantgom *mn* — **e** goma *f* tragacanta; goma *f* adragante; tragacanto *m*; adraganto *m*; goma *f* alquitira — **i** gomma *f* adragante

13952 trailer label — Computer term. Machine readable record at the end of a file containing data identifying the file and control information.
f label-fin *m* — **d** Nachsatz *m* — **n** staartlabel *m* — **e** rótulo *m* final — **i** dicitura *f* finale

13953 trailer record — Record that follows another record or group of records and contains data related to that record or group of records.
f enregistrement *m* complémentaire — **d** Beisatz *m* — **n** aanvullingsrecord *n* — **e** registro *m* secundario; registro *m* final — **i** registrazione *f* dettaglio

13954 trailing blade coating — See blade coating.

13955 trailing edge; tail edge; back edge — Edge of a plate or forme opposite to the leading edge. See also lead and leave edges.
f bord *m* arrière — **d** Druckende *n* — **n** afloopkant *m* — **e** fin *m* de la plancha; fin *m* de la forma — **i** bianco *m* di coda; bianco *m* di uscita

13956 trainee; learner — Apprentice brought up in the works.
f débutant *m* — **d** Anlernling *m* — **n** leerling *m*; halfwas *m* — **e** aprendiz *m* — **i** apprendista *m*

13957 tramp printer — See tourist printer.

13958 tranchefile chapiteau — Double *headband originating from France.

13959 transcoder — See code converter.

13960 transducer — In electronics, a component which converts a physical change into an electrical signal or vice versa.
f transducteur *m* de mesure — **d** Umformer *m* — **n** omzetter *m* — **e** transductor *m* de medida — **i** trasduttore *m* di misura

13961 transfer — Lithographic image transferred from one surface to another, e.g., from a designer's stone to the printing press-plate by means of a proof on transfer paper. A transfer plate has the images transferred to it.
f report *m*; décalque *m* — **d** Umdruck *m* — **n** overdruk *m* — **e** transporte *m*; reporte *m* — **i** trasporto *m*; riporto *m*

13962 transfer *v* — To pull transfers from original designs and transfer them to the surface of a litho plate or stone.
f faire un report; décalquer — **d** umdrucken — **n** overdrukken maken; overdrukken — **e** reportar — **i** trasportare

13963 transfer channel; intermediate channel — Part of the hot-metal composing machine in which the matrices are transferred from the first to the second elevator.
f canal *m* intermédiaire — **d** Zwischenkanal *m* — **e** canal *m* de traslación; canal *m* de transferencia; canal *m* intermediario; canal *m* intermedio — **i** canale *m* intermedio

13964 transfer ink — Ink used in lithographic work to transfer an image from one printing plate to another.
f encre *f* à report — **d** Umdruckfarbe *f* — **n** overdrukinkt *m*; overdrukverf *m* *obs.* — **e** tinta *f* de reporte; tinta *f* para transporte — **i** inchiostro *m* da trasporto; inchiostro *m* per trasporto

13965 transfer of control — See jump.

13966 transfer paper — Paper used in offset and lithography for transfer work on the plate, e.g., everdamp, rolled Scotch, and French transfer papers.
f papier *m* à reports; papier *m* pour reports; papier *m* autographique; pelure *f* pour reports; Chine *m* à report; hydrochine *m*; hydrochine *m* transparent; pelure *f* à report; couché *m* à report — **d** Umdruckpapier *n*; immerfeuchtes Umdruckpapier *n* — **n** overdrukpapier *n*; natgrauw papier *n*; natgrauw *n* — **e** papel *m* para reporte; papel *m* de transporte; papel *m* para transporte litográfico; papel *m* autográfico — **i** carta *f* da trasporto; carta *f* da trasporto litografico; carte *f* di riporto

13967 transfer paper — See decalcomania paper.

13968 transfer press — Hand-operated press to transfer designs from the transfer paper to the printing surface.
f presse *f* à report — **d** Umdruckpresse *f*; Umdruckmaschine *f* — **n** overdrukpers *f* — **e** prensa *f* para transportar; prensa *f* para reportar — **i** pressa *f* per trasporti litografici; torchio *m* per trasporto litografico

13969 transfer printing; decalcomania printing
f impression *f* des décalcomanies; procédé *m* de décalcomanie; décalcomanie *f* — **d** Abziehbilderverfahren *n*; Abziehbilderdruck *m* — **n** drukken *n* van decalcomanieën; drukken *n* van transfers — **e** impresión *f* de calcomanías; procedimiento *m* para hacer calcomanías; calcomanía *f* — **i** stampa *f* di decalcomanie; decalcomania *f*

13970 transfer rate — Relative speed at which data are transferred between devices. Applied to magnetic tape decks it is the speed of the tape in inches per second multiplied by the number of characters recorded to the inch of magnetic tape.

13971 transferred ink
f encre f transférée — d übertragene Farbe f — n overgedragen inkt m — e tinta f transportada; tinta f transmitida — i inchiostro m trasportato

13972 transfer roller
f rouleau m de transfer; rouleau m distributeur — d Farbübertragungszylinder m; Reibzylinder m — n distributierol m — e rodillo m distribuidor — i rullo m distributore; rullo m macinatore

13973 transfer roller — See ductor.

13974 transfers — See decalcomania.

13975 transfer wedge — Part of the Monotype caster.
f coin m report — d Füllkeil m — n platte wig fm — e cuña f de transferencia — i cuneo m di riempimento

13976 transfer wedge cam — Part of the Monotype caster.
f excentrique m pour coin report — d Füllkeil-Exzenter m — n plattewig-excentriek n — e excéntrica f de la cuña de transferencia — i camma f del cuneo di riempimento

13977 transformer — Electric device consisting essentially of two or more windings wound on the same core, which by electromagnetic induction transforms electric energy from one set of one or more circuits to another set such that the frequency of the energy remains unchanged while the voltage and current usually change.
f transformateur m — d Transformator m — n transformator m — e transformador m — i trasformatore m

13978 transformer board — See presspahn.

13979 transient time — See response time.

13980 transistor — Combination of two types of germanium or two types of silicon fused together to provide two junction layers. Has amplification properties, like a triode valve.
f transistron m; triode f à cristal; transistor m obs. — d Transistor m — n transistor m — e transistor m — i transistore m; triodo m a cristallo

13981 Transitionals — Group of modern *type design classification. This is the evolution of monarchical centuries to an architectonic type with systematically distinguished stem. The axis of the curves is vertical or inclined slightly to the left. The serifs are bracketed and those of the ascenders in the lower-case are oblique. To this group belong Còlumbia, Grandjean, Baskerville, Bell, Caledonia, Fournier, Perpetua (of Eric Gill), Times (of Stanley Morison), Van Dijck, etc.
f Réales fpl — d Barock Antiqua f; vorklassizistische Antiqua f — n Realen fmpl — e Reales mpl — i Transizionali mpl

13982 translation — Rendering from one language into another. See also machine translation.
f traduction f — d Übersetzung f — n vertaling f — e traducción f — i traduzione f

13983 translucency — Property of a material which permits it to transmit light with strong scattering of the light. To be distinguished from transparency: a sheet of bond paper is translucent, whereas a sheet of high-grade glassine paper is fairly transparent.
f translucidité f — d Durchscheinen n — n doorschijnendheid f — e translucidez f — i translucidità f

13984 translucent — Said of materials which permit the passage of light, but which scatter the rays so that objects behind them are not distinctly visible.
f translucide — d durchscheinend; lichtdurchlässig — n doorschijnend — e translúcido — i translucido; traslucido

13985 translucent — See ivory board.

13986 translucent paper — Paper which permits the partial transmission of light but through which objects can be distinguished visually only when the paper is in direct contact with them, for example tracing paper.
f papier m translucide — d durchscheinendes Papier n — n doorschijnend papier n — e papel m translúcido — i carta f translucida

13987 transmission — Percentage of light permitted to pass through the negative.
f transmission f — d Durchlaß m — n doorlating f; transmissie f — e transmisión f — i trasmissione f

13988 transmission — Electrical transfer of a signal, message or other form of intelligence from one location to another.
f transmission f — d Transmission f; Übertragung f — n transmissie f — e transmisión f — i trasmissione f

13989 transmission copy — See transparency.

13990 transmission densitometer — See also densitometer.
f densitomètre m par transmission — d Durchsichtsdensitometer n; Durchsichtsschwärzungsmesser m — n transmissie-densitometer m — e densitómetro m de transmisión — i densitometro m per trasmissione

13991 transmission density
f densité f par transmission — d Durchsichtsschwärze f — n transmissie-zwarting f; zwarting f bij doorzicht — e densidad f de transmisión — i densità f in trasparenza

13992 transmission factor
f facteur m de transmission — d Transmissionsgrad m — n doorlatingsfactor m — e factor m de transmisión — i fattore m di trasmissione

13993 transmission speed; line speed — Maximum rate at which signals may be transmitted over a given channel, usually measured in bits or bauds per second.
f vitesse f de transmission — d Übertragungsgeschwindigkeit f — n overdrachtssnelheid f — e velocidad f de transmisión — i velocità f di trasmissione

13994 transmittance — Ratio of the intensity of transmitted light to the intensity of incident light.

f transmittance *f*; coefficient *m* de transmission — **d** Durchlässigkeit *f*; Transparenz *f* — **n** transmissie *f* — **e** transmisión *f* — **i** trasmittanza *f*; fattore *m* di trasmissione

13995 transmitter — Teletype term for *tape reader.

13996 transmitting pellicle mirror; semi-transparent pellicle mirror
f glace *f* semi-réfléchissante — **d** teildurchlässiger Spiegel *m*; Halbspiegel *m* — **n** halfdoorlatende spiegel *m* — **e** placa *f* semitransparente — **i** specchio *m* (a membrana) semitrasparente

13997 transparency — Property of an ink which permits light to pass through. Lack of hiding power.
f transparence *f* — **d** Transparenz *f*; Lasur *f* — **n** transparantie *f* — **e** transparencia *f* — **i** trasparenza *f*

13998 transparency; transmission copy — Monochrome photographic positive or picture on a transparent support, the image intended for viewing and reproduction by transmitted light. See also colour transparency.
f diapositif *m* — **d** Diapositiv *n*; Dia *m* — **n** diapositief *n*; dia *m* — **e** diapositiva *f*; transparencia *f* — **i** diapositiva *f*; diapositivo *m*; trasparente *m*

13999 transparency — In photography, the light-transmitting power of the silver deposit in a negative. The inverse of opacity.
f transparence *f* — **d** Lichtdurchlässigkeit *f*; Transparenz *f* — **n** lichtdoorlatendheid *f*; lichtdoorlating *f*; transparantie *f* — **e** transparencia *f* — **i** trasparenza *f*

14000 transparency attachment — See transparency holder.

14001 transparency holder; transparency attachment — Carrier or arrangement to hold transparencies (and negatives) in position on process cameras so that the images may be illuminated with transmitted light during photographic reproduction.
f châssis *m* porte-diapositif; porte-modèle *m* en transparence — **d** Diahalter *m* — **n** diahouder *m* — **e** portadiapositivo *f* — **i** portadiapositiva *f*

14002 transparent — Permitting the passing of light without scattering it.
f transparent — **d** durchsichtig; transparent; lasierend — **n** doorzichtig; transparant — **e** transparente — **i** trasparente

14003 transparent inks — Inks which lack hiding power and permit light to pass through. They permit previous printing to show through, the two colours blending to produce a third, e.g., a transparent yellow over a blue to produce a green where the two colours are superimposed.
f encres *fpl* transparentes — **d** lasierende Farben *fpl*; Lasurfarben *fpl*; transparente Farben *fpl* — **n** transparante inkten *mpl* — **e** tintas *fpl* transparentes; colores *mpl* transparentes — **i** inchiostri *mpl* trasparenti

14004 transparentize *v* — To render the paper

support of photographic images translucent by treatment with greasy or oily substances.
f rendre transparent — **d** durchsichtig machen; transparent machen — **n** doorzichtig maken — **e** volverse transparente — **i** rendere trasparente

14005 transparent paper
f papier *m* transparent — **d** durchsichtiges Papier *n* — **n** doorzichtig papier *n*; transparant papier *n* — **e** papel *m* transparente — **i** carta *f* trasparente

14006 transpose *v* — To move letters, words, lines, etc., from one place to another.
f transposer — **d** umheben; umstellen; verheben — **n** verwisselen; omwisselen; verplaatsen — **e** trastocar — **i** trasporre

14007 transpose *v* — To change the image area to a non-image area and vice versa, i.e., black to white or white to black. Not to be confused with lateral reversal.
f inverser noir-au-blanc — **d** Tonwert umkehren — **n** zwart/wit omkeren; zwart/wit maken

14008 transposition mark — Correction mark to indicate that one or more letters or words, or sentences must be transposed. See app. no. 5.
f signe *m* de transposition — **d** Umstellungszeichen *n* — **n** omstellingsteken *n* — **e** signo *m* de transposición — **i** segno *m* di trasposizione

14009 transverse camber — Defect of aluminium foil. Departure of the surface from a straight line joining the two edges and perpendicular to them. See also lateral camber.

14010 transverse collection — See cross association.

14011 trapping — Ability of an already printed ink film to accept a succeeding or overprinted ink film.
f prise *f* (de l'encre) — **d** Farbannahmefähigkeit *f*; Farbaufnahmefähigkeit *f* — **n** hechting *f* van de inkt; inktaanname *f* — **e** toma *f* (de la tinta) — **i** sovrastampabilità *f* di un inchiostro

14012 travellers guide
f itinéraire *m* — **d** Reiseführer *m* — **n** reisgids *m* — **e** itinerario *m* — **i** itinerario *m*; guida *f*

14013 tray rocker — Mechanical arrangement to agitate a tray and the solution in it to obtain uniformity of treatment for photographic materials being processed.
f balanceur *m* de cuvette — **d** Einrichtung *f* zum (mechanischen) Bewegen der Entwicklungsschale — **n** badschommelaar *m* — **i** agitatore *m* per vasche di trattamento

14014 treadle-press — Small (platen) machine worked by the foot. See also cropper.
f pédale *f*; presse *f* à pédale; minerve *f* — **d** Tretpresse *f*; Tretmaschine *f*; Drucktiegel *m* mit Fußbetrieb; Maschine *f* mit Fußantrieb; Tiegelmühle *f* coll. — **n** trappers *f* — **e** prensa *f* de pedal; máquina *f* de pedal; minerva *f* a pedal — **i** pedalina *f*; macchina *f* a pedale

14015 treasury — A size of blotting paper. See app. no. 3.

14016 treatise — Systematic exposition or argument in writing or printing, including a methodical discussion of the facts and principles involved and conclusions reached.

f traité *m* — **d** Abhandlung *f* — **n** verhandeling *f* — **e** tratado *m* — **i** trattato *m*

14017 treble case — Type case which holds three founts of capitals.

f casse *f* pour trois caractères (capitales); casse *f* triple — **d** Schriftkasten *m* für drei Versalienschriften — **n** letterkast *fm* voor drie soorten kapitalen; driedelige kast *fm* — **e** caja *f* por tres cuerpos — **i** cassa *f* tripla

14018 treble rule

f filet *m* double gouttière — **d** feinfettfeine Linie *f* — **n** randlijn *fm* — **e** filete *m* de tres líneas — **i** filetto *m* triplo

14019 tri — See trichloro-ethylene.

14020 triacetate film — See acetate film.

14021 trial order

f commande *f* d'essai — **d** Probelieferung *f*; Probeauftrag *m* — **n** proeforder *fm* — **e** pedido *m* de ensayo — **i** commissione *f* di saggio

14022 trial run — In paper mills, a small making order of a new grade of paper to submit samples to the customer for approval, or an experimental run with modifications in the standard furnish formula or method of processing.

f essai *m* de fabrication; fabrication *f* expérimentale — **d** Probelauf *m*; Probeproduktion *f*; Versuchspartie *f* — **n** proefpartij *f* — **e** partida *f* de prueba — **i** fabbricazione *f* di prova

14023 triangle — Closed plane figure having three sides and three angles. Geometric sign: △.

f triangle *m* — **d** Dreieck *n* — **n** driehoek *m* — **e** triángulo *m* — **i** triangolo *m*

14024 tribasic calcium phosphate — See calcium phosphate.

14025 tribasic sodium phosphate — See sodium phosphate.

14026 tricalcium phosphate — See calcium phosphate.

14027 trichloride of gold — See gold trichloride.

14028 trichloro-ethylene; tri *coll.* — Stable, low boiling, colourless, heavy toxic liquid. Non flammable, non explosive, and non combustible. Will not attack the common metals. Miscible with all common organic solvents, practically insoluble in water. Latin: aethylenum trichloratum. TL-value: 35 ppm or 190 mg/m^3.
Formula: C_2HCl_3.

f trichloréthylène *m* — **d** Trichloräthylen *n* — **n** trichloorethyleen *m*; tri *m* — **e** tricloroetileno *m* — **i** tricloroetilene *m*

14029 trichloromethane — See chloroform.

14030 trichromatic coefficients — See tristimulus values.

14031 trichromatic process printing — See three-colour process printing.

14032 trichromatic units — Relative units of stimulus quantity, applicable to stimuli of any colour and such that the quantity of any stimulus, when expressed in these units is equal to the sum

of the *tristimulus values.

f unités *fpl* trichromatiques — **d** Farbvalenzeinheiten *fpl* — **n** trichromatische eenheden *fpl* — **e** unidades *fpl* tricromáticas — **i** unità *fpl* tricromatiche

14033 trick — See shift.

14034 tricolour filters — Filters employed in three-colour photography for exposure of separation negatives.

f filtres *mpl* trichromes — **d** Dreifarbenfilter *mpl* — **n** driekleurenfilters *npl* — **e** filtros *mpl* tricromáticos; filtros *mpl* para tricromía — **i** filtri *mpl* per tricromia

14035 tricolour inks — See process inks.

14036 triglyceride — Chief constituent of fats and oils, naturally occurring ester of normal acids (fatty acids) and glycerol.
Formula: $CH_2(OOCR_1)CH(OOCR_2)CH(OOCR_3)$, where R_1, R_2 and R_3 are usually of different chain length. Refining processes will often yield a commercial product in which R_1, R_2 and R_3 are the same chain length.

f triglicéride *m* — **d** Triglyzerid *n* — **n** triglyceride *n* — **e** triglicerido *m* — **i** tricliceride *m*

14037 trigonometrical ratios — See trigonometric functions.

14038 trigonometric functions; trigonometrical ratios — In a right-angled triangle ABC, in which ACB = 90 °, *sine BAC = BC/AB, *cosine BAC = AC/AB, *tangent BAC = BC/AC, *cotangent BAC = AC/BC, *secant BAC = AB/AC, *cosecant BAC = AB/BC.

f fonctions *fpl* trigonométrique — **d** trigonometrische Funktionen *fpl* — **n** goniometrische verhoudingen *fpl*; trigonometrische verhoudingen *fpl* — **e** funciones *fpl* trigonométricas — **i** funzioni *fpl* trigonometriche

14039 trihydroxybenzene — See pyrogallol.

14040 trimetallic plate; trimetal plate — Lithographic press-plate consisting of a base metal of stainless steel, aluminium, or other metal, on which other metals are plated, usually copper and then chromium. Lithographic platemaking techniques are used to etch the surface metal of image to reach the copper for image base; surface metal remains for the non printing areas. The nature of the surface metal with the damping solution helps in repelling the ink from non printing areas. Such plates have a long life and no mechanically grained surface.

f plaque *f* trimétallique — **d** Trimetallplatte *f* — **n** trimetaalplaat *fm* — **e** plancha *f* trimetálica — **i** lastra *f* trimetallica

14041 trimetal plate — See trimetallic plate.

14042 trim inclusions — Defect of aluminium foil. Accidental winding in of the edge trim with finished coil.

14043 trim marks — Small tick marks photographed on to the negative from the copy or added afterwards in stripping up the flat. Located so as to print along the margins of the press sheet to guide the cutter in trimming each form to size after printing.

f repères *mpl* de rogne — **d** Schnittmarken *fpl* — **n** snijtekens *npl* — **e** marcas *fpl* de corte — **i** marche *fpl* per la rifilatura

14044 trimmed size — Size of a sheet of paper ready for use.

f format *m* fini — **d** Endformat *n*; beschnittenes Format *n* — **n** schoongesneden formaat *n*; afgesneden formaat *n* — **e** tamaño *m* recortado; formato *m* recortado — **i** formato *m* refilato

14045 trimmer — See guillotine.

14046 trimmer, saw — See saw trimmer.

14047 trimming knife — Pair of knives in the slug composing machine to trim the slug on both sides to proper body thickness as it is ejected from the mould. The bottom of the slug is trimmed by the back knife.

f couteau *m* calibreur — **d** Messerblock *m* — **n** mesblok *n* — **e** cuchillas *fpl* laterales — **i** blocco *m* dei coltelli

14048 trimmings — See shavings.

14049 trim saw

f scie *f* rectifieuse de précision — **d** Präzisionssäge *f* — **n** precisie-zaag *fm* — **e** sierra *f* de precisión — **i** sega *f* rettificatrice

14050 trinitroglycerine — See nitroglycerine.

14051 triple reel stand — See roller star.

14052 triple roller machine — See three roller mill.

14053 triple roll stand — See roller star.

14054 triplex board; three-layer board — Board consisting of three furnish layers felted together during manufacture by pressure while still moist without the use of adhesive.

f carton *m* triplex; carton *m* trois couches — **d** Triplex-Karton *m*; Triplex-Pappe *f*; dreischichtige Pappe *f* — **n** triplexkarton *n* — **e** cartulina *f* triple(x); cartón *m* triplex — **i** cartone *m* triplex

14055 trip mechanism — Mechanism of levers, cams, etc., of a printing press permitting throw off.

f mise *f* hors pression — **d** Druckabstellung *f*; Ausrückvorrichtung *f* — **n** drukafstelling *f* — **i** meccanismo *m* di disinnesto

14056 tripolite — See diatomaceous earth.

14057 trip *v* **the press**

f déclencher — **d** auslösen; stillsetzen — **n** de druk afzetten — **i** fermare la macchina; disinnestare la macchina

14058 trisodium citrate — See sodium citrate.

14059 trisodium orthophosphate — See sodium phosphate.

14060 trisodium phosphate — See sodium phosphate.

14061 tristearin — See stearin.

14062 tristimulus values; trichromatic coefficients *depr.* — Amounts of the three reference or matching stimuli required to give a match with the light considered, in a given trichromatic system (CIE). The symbols recommended (1948) for the tristimulus values are: X, Y, Z, in the CIE standard colorimetric system of 1931 and X_{10}, Y_{10},

Z_{10}, in the CIE supplementary colorimetric system of 1964.

f valeurs *fpl* tristimulus; composantes *fpl* trichromatiques — **d** Tristimulus-Werte *mpl*; Normfarbwerte *mpl* — **n** tristimulus-waarden *fpl*; trichromatische componenten *mpl* — **e** valores *mpl* triestímulos; componentes *mpl* tricromáticos — **i** valori *mpl* tristimulus; coefficienti *mpl* tristimulus; componenti *fpl* tricromatiche

14063 trivalent — Having a *valence of three. Capable of combining with three atoms of hydrogen or the equivalent.

f trivalent — **d** dreiwertig — **n** driewaardig — **e** trivalente — **i** trivalente

14064 trouble shooting

f service *m* de conseils techniques — **d** technischer Beratungsdienst *m* — **n** technische adviesdienst *m* — **e** servicio *m* de asesoramiento técnico — **i** servizio *m* di consiglio tecnico

14065 troy ounce — Unit of weight. Abbrev. tr oz (or sometimes oz t). See app. no. 7 and 7a.

f once *f* troy — **d** Troy-Unze *f* — **n** ons *n* troy — **e** onza *f* troy — **i** oncia *f* troy

14066 troy pound — Unit of weight. Abbrev. tr lb (or sometimes lb t). See app. no. 7 and 7a.

f livre *f* troy — **d** Pfund *n* troy; Troy-Pfund *n* — **n** pond *n* troy — **e** libra *f* troy — **i** libbra *f* troy

14067 truck — See double truck advertisement.

14068 truck carriage — See lift truck.

14069 true resin — See resin.

14070 trumpery paper — See rag.

14071 T-square; tee-square — Straight-edge rule with a projecting helve perpendicular to the rule, used by draughtsmen and photoengravers to trace parallel lines or to ascertain parallelism of edges of plate.

f équerre *f* à T; équerre *f* en té; équerre *f* à chapeau; té *m* (à dessin) — **d** Reißschiene *f* — **n** tekenhaak *m* — **e** escuadra *f* en T; doble escuadra *f* — **i** riga *f* a T

14072 TTS equipment — See advanced sprocket.

14073 tub sized paper; top sized paper; vat sized paper; surface sized paper — Paper immersed in a bath of gelatine or starch to increase its water or ink resistance, strength, and erasing qualities.

f papier *m* collé à la cuve; papier *m* collé en bac; papier *m* collé en superficie; papier *m* collé superficiel — **d** oberflächengeleimtes Papier *n* — **n** nagelijmd papier *n*; oppervlakte-gelijmd papier *n*; papier *n* met oppervlaktelijming — **e** papel *m* encolado en tina; papel *m* encolado superficial; papel *m* aderezado en tina — **i** carta *f* collata in superficie

14074 tuck — Folds which comprise the side walls of square and automatic-type bags, which are folded (tucked) in to permit the bag to be packed flat.

f soufflet *m* — **d** Frosch *m*; seitlicher Einschlag *m* — **n** soufflet *n* — **e** fuelle *m* — **i** soffietto *m*

14075 tuck — End portion of the top or bottom flaps of a folding paper box (carton) inserted inside the container to hold the end (top or

bottom) flaps in place. Various types of cuts and shapes of tuck ends have been developed to hold the flaps, the commonest being a pair of notches at the fold which engage the side flaps and hold the end flaps in place.
f extrémité *f* — **d** Einstecklasche *f* — **i** linguetta *f* a incastro; aletta *f*
14076 tuck and grip folder — See jaw folder.
14077 tucker — See folding blade.
14078 tucker blade — See folding blade.
14079 tucker fold — See jaw fold.
14080 tucking blade — See folding blade.
14081 tumble *v* — To turn crosswise the sheet of paper for perfecting. See also work and tumble.
14082 tumble forme — Imposition style in which the same forme is used for printing both sides of a sheet, the sheets being turned over in their short direction, instead of end over end and as in the work-and-tumble forme. It results in two copies of the same forme back to back, with head appearing at top on both sides.
f forme *f* à l'italienne; forme *f* tête à queue — **d** Umstülp-Druckform *f* — **n** stolpvorm *m* — **e** voltereta *f* — **i** forma *f* da voltare
14083 tumbler switch — See toggle switch.
14084 tung oil; Chinese wood oil; China wood oil; wood oil — Yellow drying oil for varnishes and lacquers. Soluble in chloroform, ether, carbon disulphide and oils.
f huile *f* de bois (de Chine); huile *f* de tung; huile *f* de Canton; huile *f* d'abrasin; huile *f* d'aloecocca — **d** chinesisches Holzöl *n*; Tungöl *n*; Ölfirnisbaumöl *n*; Eläokokkaöl *n* — **n** Chinese houtolie *fm* — **e** aceite *m* de madera de China; aceite *m* de palo; aceite *m* de tung — **i** olio *m* di legno della Cina
14085 tungsten — See wolfram.
14086 tungsten filament lamp — Lamp in which light is produced by means of a tungsten filament heated to incandescence by the passage of an electric current.
f lampe *f* à filament de tungstène — **d** Wolframlampe *f*; Wolframdrahtlampe *f* — **n** wolframlamp *f*; wolfraamlamp *f* — **e** lámpara *f* de filamento de tungsten — **i** lampada *f* a filamento di tungsteno
14087 tunnel drier — See drying tunnel.
14088 turbulence; turbulency — In hydraulics, the haphazard secondary motion caused by eddies within a moving fluid.
f turbulence *f* — **d** Wirbelung *f*; Turbulenz *f* — **n** turbulentie *f*; werveling *f* — **e** turbulencia *f*; remolino *m* — **i** turbolenza *f*
14089 turbulency — See turbulence.
14090 turbulent flow — In rheology, disorganized, non laminar flow, sometimes referred to as hydraulic flow. Opposite of streamline flow.
f écoulement *m* turbulent — **d** turbulente Strömung *f*; Turbulenzstrom *m* — **n** turbulente stroming *f*; turbulentie *f* — **e** flujo *m* turbulento — **i** corrente *f* turbolenta
14091 turkey red oil; sulphonated castor oil —

Reaction product of sulphuric acid and castor oil, soluble in water, also known as alizarin assistant and *alizarin oil because of its use in dyeing with alizarin. Used in paper coatings.
f huile *f* rouge de Turquie; huile *f* de ricin sulfonée — **d** Türkischrotöl *n*; geschwefeltes Rizinusöl *n* — **n** Turks rood-olie *fm*; gesulfoneerde castorolie *fm*; gesulfoneerde wonderolie *fm* — **e** aceite *m* rojo de Turquía; aceite *m* de ricino sulfonado; aceite *m* de castor sulfonado — **i** olio *m* di ricino solfonato
14092 turmeric paper — Paper treated with extract of turmeric root, to detect alkalinity of solutions.
f papier *m* curcuma — **d** Kurkumapapier *n* — **n** kurkumapapier *n* — **e** papel *m* de cúrcuma — **i** carta *f* curcuma
14093 turn — See revolution.
14094 turned commas — See quotation mark.
14095 turned letter; inverted letter; upside down letter; black — Type character deliberately placed face downwards so that the foot prints, to indicate where characters (for the time being run short) will be needed or the manuscript was not clear.
f caractère *m* bloqué; lettre *f* bloquée — **d** auf dem Kopf stehende Drucktype *f*; blockierte Schrift *f*; Fliegenkopf *m coll.* — **n** geblokkeerde letter *fm* — **e** letra *f* vuelta; letra *f* invertida; tipo *m* vuelto; tipo *m* invertido; tipo *m* puesto boca abajo; cabeza *f* de muerto; tipo *m* volcado *Ch*; letra *f* volcada — **i** carattere *m* bloccato
14096 turned line — Line of type turned upside down to indicate that a table or illustration is to be inserted; shows as black bar on proof.
f ligne *f* renversée — **d** blockierte Zeile *f* — **n** omgekeerde regel *m*; op zijn kop gezette regel *m*; geblokkeerde regel *m* — **e** línea *f* vuelta; línea *f* invertida; lingote *m* invertido; línea *f* volcada *Ch*; lingote *m* volcado — **i** riga *f* capovolta
14097 turner bar — See angle bar.
14098 turn *v* **for sorts** — To turn a type face down to take the place of a letter which is exhausted. Shows as black dot on proof.
f bloquer — **d** blockieren — **n** blokkeren; omkeren — **e** bloquear; volver; poner una letra al revés — **i** bloccare; far un blocco; capovolgere un carattere
14099 turn in; turn over — Part of the cover leather or cloth or paper that is turned over from the outside to the inside of a book cover.
f rempli *m* — **d** Einschlag *m* — **n** inslag *m* — **e** alforza *f* — **i** ripiegatura *f*; rimboccatura *f*
14100 turning bar — See angle bar.
14101 turnkey system — System consisting of both hardware and software purchased from a supplier to handle a specified application. No in house technical capability is required.
14102 turn over — Type matter which follows on from the bottom of one column or page at the top of the next.
f texte *m* qui continue — **d** auf die nächste Seite umlaufender Text *m* — **n** overlopende tekst *m*; overlopende pagina *fm*; overlopende regel *m* — **e**

texto *m* que continua — **i** testo *m* che continua

14103 turnover — Apprentice who is transferred or "turned over" to another firm to complete his apprenticeship. See also improver.
f apprenti *m* en cours de perfectionnement — **d** ausgetauschter Lehrling *m* — **n** uitgeleende leerling *m* — **e** aprendiz *m* traslado — **i** apprendista *m* traslocado

14104 turn over — See turn in.

14105 turntable — Arrangement on the truck or stand of process cameras whereby the camera can be turned sideways or at right angle to the copyboard when using a mirror for lateral reversal of negatives.
f plaque *f* tournante; table *f* tournante — **d** Drehscheibe *f*; Drehtisch *m* — **n** draaischijf *fm*; draaitafel *fm* — **e** torreta *f* giratoria; mesa *f* giratoria — **i** piattaforma *f* girevole; tavolo *m* rotante

14106 turpentine; gum turpentine; turps — Crude product obtained from coniferous trees. Soluble in alcohol, ether, chloroform and glacial acetic acid. Often erroneously refers to the distillate, whose proper name is *turpentine oil. TL-value: 100 ppm or 560 mg/m³.
f térébenthine *f* — **d** Terpentin *n* — **n** terpentijn *m* — **e** trementina *f*; terebentina *f* — **i** trementina *f*

14107 turpentine oil; oil of turpentine; spirits of turpentine; spirit of turpentine — Volatile oil obtained by distilling crude turpentine. Latin: oleum terebinthinae. TL-value: 100 ppm. Formula: $C_{10}H_{16}$.
f essence *f* de térébenthine — **d** Terpentinöl *n* — **n** terpentijnolie *fm* — **e** aguarrás *m*; aceite *m* de trementina; esencia *f* de trementina — **i** olio *m* di trementina; essenza *f* di trementina

14108 turpentine substitute; turps subs — Cheaper substitute for turpentine, usually white spirit, a petroleum fraction. The term is misleading and should not be used.
f essence *m* de térébenthine artificielle — **d** Terpentinölersatz *m* — **n** vervangingsmiddel *n* voor terpentijn; terpentijn-substituut *n* — **e** substituto *m* de la trementina — **i** succedaneo *m* della trementina

14109 turps — See turpentine.

14110 turps subs — Contraction of turpentine substitute, a misleading term used in relief stamping. It is a substitute for pure turpentine, generally *white spirit.

14111 turps subs — See turpentine substitute.

14112 turret, lens — See lens turret.

14113 turtle plate — In early times of curved printing formes, plate on which newspaper pages were locked up (in fact sections of the cylinder surface). Use was made of bottle-necked type. Nowadays a dummy plate on a rotary letterpress machine.

14114 tusche — Ink emulsion used in lithography to add printing areas to the plate, usually by hand or by brush to blueline outlines that are printed down on a set of press-plates from a key flat.

f encre *f* grasse; encre *f* litho — **d** Tusche *f*; Fettusche *f* — **n** tusche *fm* — **e** tinta *f* grasa; tinta *f* litográfica — **i** inchiostro *m* grasso litografico

14115 tusching — Adding work to the image on a lithographic press plate, correcting lines and lettering, and adding solids by means of a liquid greasy substance, known as *tusche.

14116 tweezers — Small spring nippers to pick up type, etc.
f pinces *fpl*; pincette *f* à corriger; brucelles *fpl obs.* — **d** Korrigierzange *f*; Pinzette *f* — **n** corrigeertang *fm* — **e** pinzas *fpl* — **i** pinzette *fpl*

14117 twelvemo; duodecimo — Book each sheet of which forms 12 leaves or 24 pages. Abbrev. 12mo (no point).
f in-douze *m*; in-12 *m* — **d** Zwölferformat *n*; Duodezformat *n* — **n** in twaalven — **e** en dozavo; duodécimo — **i** in dodicesimo

14118 twelve-to-pica — Old name for 1-point type. See app. no. 1.

14119 "twelve ways" back-up — See work and tumble.

14120 twenty-fourmo; vigesimoquarto — Book each sheet of which forms 24 leaves or 48 pages. Abbrev. 24mo (no point).
f in-vingt-quatre *m*; in-24 *m* — **d** Vierundzwanzigerformat *n* — **n** in vierentwintigen — **e** en veinti-y-cuatroavo; en veinticuatro; en venticuatroavo — **i** in ventiquattresimo

14121 twice a month — See bi-monthly.

14122 twice a month — See semi-monthly.

14123 twice a week — See bi-weekly.

14124 twice a week — See semi-weekly.

14125 twicer — Employee who works both as a compositor and as a pressman.
f compositeur-pressier *m*; typo-minerviste *m*; amphibie *m sl.* — **d** Schweizerdegen *m*; Setzer-Drucker *m* — **n** zetter-drukker *m*; duizendpoot *m sl.* — **e** cajista-prensista *m*; cajista-minervista *m* — **i** compositore-stampatore *m*; compositore-impressore *m*

14126 twin chase — See folding chase.

14127 twin-flute corrugated fibreboard — See double-double face corrugated fibreboard.

14128 twin-wire paper — Duplex paper formed over two separate wires and united with the undersides together. Two layers of different stuff or two layers of different colour.
f papier *m* deux jets — **d** Duplexpapier *n* — **n** duplex-papier *n* — **e** papel *m* doble — **i** carta *f* duplex; carta *f* con due lati feltro

14129 twist gear
f engrenage *m* (à denture) hélicoïdale — **d** Schraubenradgetriebe *n* — **n** aandrijving *f* met schuine vertanding — **e** engranaje *m* helicoidal — **i** ingranaggio *m* a denti elicoidali

14130 two-colour halftone — See two-colour process.

14131 two-colour machine — See two-colour press.

14132 two-colour offset machine; two-colour offset press

f machine *f* offset deux couleurs; presse *f* offset deux couleurs — **d** Zweifarbenoffsetmaschine *f* — **n** tweekleuren-offsetpers *f* — **e** máquina *f* offset de dos colores; prensa *f* offset de dos colores — **i** macchina *f* offset a due colori; macchina *f* offset bicolore

14133 two-colour offset press — See two-colour offset machine.

14134 two-colour press; two-colour machine — Two-cylinder letterpress machine with two beds which prints two colours on one side at the same feeding.
f presse *f* à deux couleurs; machine *f* à deux couleurs — **d** Zweifarbenmaschine *f* — **n** tweekleurenpers *f* — **e** máquina *f* de dos colores; prensa *f* de dos colores; prensa *f* bicolor — **i** macchina *f* a due colori

14135 two-colour printing — See two-colour process.

14136 two-colour process; two-colour printing — Method of reproduction in which two plates, line or halftone, are printed in two practically complementary colours to give a full-colour effect. A two-colour halftone consists of two plates made from a separate negative from the original, the latter either monochrome or coloured, and the plates etched to produce the desired colour effect when printed in two contrasting colours.
f impression *f* en deux couleurs — **d** Zweifarbenverfahren *n*; Zweifarbendruck *m* — **n** tweekleurenprocédé *n*; tweekleurendruk *m* — **e** impresión *f* en dos colores; impresión *f* bicolor; bicromía *f* — **i** impressione *f* di due colori; stampa *f* di due colori; stampa *f* in bicromia

14137 two-cylinder press — See perfecting press.

14138 two-em quad; 2-em quad
f double cadratin *m* — **d** Doppelgeviertstück *n*; Doppelgeviert *n* — **n** kwadraat *n* op twee vierkanten; stuk *n* van twee — **e** doble cuadrado *m* — **i** doppio quadratone; quadrato *m* a due

14139 two-layer board — See duplex cardboard.

14140 two layer paper — See duplex paper.

14141 two-letter matrix — See double-letter matrix.

14142 two-line brevier — Old name for 16-point type. See app. no. 1.

14143 two-line English — Old name for 38-point type. See app. no. 1.

14144 two-line great primer — Old name for 32-point type. See app. no. 1.

14145 two-line initial — Initial letter having a depth of body (or height of letter on printed page) equal to double that of the size specified, as two-line pica, etc.
f initiale *f* sur deux lignes; lettrine *f*; initiale *f* binaire; capitale *f* en deux lignes — **d** zweizeiliger Anfangsbuchstabe *m*; zweizeiliger Initial *m*; Initiale *f* über zwei Zeilen — **n** initiaal *n* over twee regels — **e** inicial *f* de dos líneas; inicial *f* doble; inicial *f* de doble renglón — **i** iniziale *f* a due righe; iniziale *f* binaria

14146 two-line long primer — Old name for 20-point type. See app. no. 1.

14147 two-line pica — Old name for 24-point type. See app. no. 1.

14148 two-months publication — As the term "*bi-monthly" is confusing, it is better to use the phrase "an every two months appearing publication".
f publication *f* bimestrielle; feuille *f* bimestrielle; revue *f* bimestrielle — **d** Zweimonatschrift *f*; alle zwei Monate erscheinende Publikation *f* — **n** tweemaandelijks (verschijnend) blad *n*; om de twee maanden verschijnend tijdschrift *n* — **e** publicación *f* bimestre; publicación *f* bimestral; revista *f* bimestral — **i** pubblicazione *f* bimestrale; rivista *f* bimestrale; periodico *m* bimestrale; rassegna *f* bimestrale

14149 two-on — See two-up.

14150 two-ply board — Multi-ply board consisting of two furnish layers.
f carton *m* biblex — **d** Duplexpappe *f*; zweischichtige Pappe *f* — **n** karton *n* uit twee lagen; tweelagig karton *n* — **e** cartón *m* de dos capas — **i** cartone *m* a due strati; cartone *m* rifinito sui due lati; cartone *m* rifinito sulle due facce

14151 two-points leads; six-to-pica leads — Leads two typographic points thick. See also lead.
f interlignes *fpl* de deux points — **d** Viertelpetit-Durchschuß *m* — **n** twee-punts interlinies *fpl* — **e** interlíneas *fpl* de dos puntos; lujos *mpl* de dos puntos *Pe* — **i** interlinee *fpl* di due punti

14152 two-revolution machine — See two-revolution press.

14153 two-revolution press; two-revolution machine — Press where the cylinder constantly rotates, printing a sheet on its first revolution and delivering the sheet on its second with an even and steady movement. Distinct from a stop-cylinder machine that rotates once only and comes to a stop for each impression.
f presse *f* à deux tours; presse *f* deux tours; presse *f* à double tour; presse *f* à soulèvement — **d** Zweitourenmaschine *f* — **n** tweetoerenpers *f* — **e** máquina *f* de dos vueltas; máquina *f* de doble revolución; máquina *f* de doble juego; prensa *f* de dos vueltas; prensa *f* de doble revolución; prensa *f* de dos revoluciones — **i** macchina *f* a due giri; macchina *f* a doppiogiro

14154 two set; double production; double plating — Plating the cylinder of a rotary press with duplicate sets of plates round the cylinder to produce two copies for each revolution of the cylinder. See also double plate.
f double production *f*; cylindre *m* double plaque — **d** doppelte Produktion *f*; ungesammelte Produktion *f*; doppelte Plattenbestückung *f* — **n** inrichten *n* van de pers voor dubbele produktie — **e** doble producción *f* — **i** produzione *f* doppia

14155 two-sheet detector; two-sheet trip — Device to trip the press when more than one sheet attempts to enter the press at one time.
f déclanchement *m* en cas de deux feuilles; dispositif *m* évitant le passage de deux feuilles;

contrôle *m* de double feuille; sécurité *f* en cas de deux feuilles — **d** Doppelbogenabführung *f*; Doppelbogenkontrolle *f*; Doppelbogenausrüstung *f*; Gerät *n* zur Verhinderung der Anlage zweier Bogen — **n** inrichting *f* om de inleg van twee vellen te verhinderen — **e** detector *m* de dos hojas — **i** dispositivo *m* di controllo del doppio foglio; dispositivo *m* d'immissione contemporanea di due fogli

14156 two-sheet trip — See two-sheet detector.

14157 two-sided coated paper — Paper coated on both sides.

f papier *m* couché deux faces; papier *m* couché de deux côtés; papier *m* couché double face — **d** zweiseitig gestrichenes Papier *n*; beiderseitig gestrichenes Papier *n*; beidseitig gestrichenes Papier *n*; doppelseitig gestrichenes Papier *n* — **n** tweezijdig gestreken papier *n* — **e** papel *m* estucado dos caras; papel *m* esmaltado por las dos caras; papel *m* couché en ambos lados; papel *m* cuché en ambos lados; papel *m* estucado (por) ambas caras — **i** carta *f* patinata d'ambo i lati; carta *f* patinata sulle due facce

14158 two-sidedness — Unintended difference of varying degree in surface texture or shade existing between the two faces of a sheet of paper or board which is inherent in the method of manufacture.

f envers *m* (d'un papier); inégalité *f* des deux côtés — **d** Zweiseitigkeit *f* — **n** tweezijdigheid *f*; ongelijkzijdigheid *f* — **e** bicarado *m*; diferencia *f* entre las dos caras — **i** doppio viso *m*

14159 two-stage mask; double overlay mask *depr.* — See also contact mask.

f masque *m* compensateur; masque *m* à double recouvrement *obs.* — **d** zweistufige Maske *f* — **n** compensatief-masker *n* — **e** máscara *f* en dos etapas; máscara *f* de dos etapas — **i** maschera *f* a due stadi

14160 two-thirder — Apprentice who has finished two-thirds of his apprenticeship.

f apprenti *m* aux deux tiers de son apprentissage — **d** Lehrling *m* nach zwei Drittens einer Ausbildungszeit — **n** ≈ halfwas *m* — **i** apprendista *m* che ha terminato due terzi del periodo di apprendistato

14161 two-to-one folder — Rotary press folder in which the folding cylinder has a circumference of two cut-offs and the cutting cylinder one cut-off, thus giving a ratio of cylinder sizes 2:1.

f plieuse *f* 2/1 — **d** 2/1-Falzapparat *m* — **n** twee-op-één vouwapparaat *n* — **e** (aparato *m*) plegador 2/1 — **i** piegatrice *f* 2/1

14162 two-up; two-on — Two identical printing subjects on a press plate. Usually made by preparing the flat so that it can be exposed successively in the two required locations.

f par deux; en deux compositions — **d** zweimal; zweimal Satz — **n** twee op één vel; twee tegelijk; tweemaal gezet — **e** postura doble; dos a la vez — **i** composto in doppio; composizione in doppio

14163 two-wire system — See half-duplex

working.

14164 tying machine — See bundle press.

14165 tympan — On a hand press, the frame, usually covered with parchment, on which the sheet is placed when printing.

f tympan *m* — **d** Deckel *m*; Preßdeckel *m*; Tympan *m* — **n** timpaan *n*; tympaan *n* — **e** tímpano *m* — **i** timpano *m*

14166 tympan bails — See bails.

14167 tympan clamps — See bails.

14168 tympan paper — Strong durable paper, plain or oiled, used as a packing on cylinder and platen presses; on rotary web perfecting presses kraft paper, plain or oiled, to prevent offsetting.

f papier *m* d'habillage; papier *m* de garnissage; papier *m* tympan; papier *m* à décharge — **d** Aufzugpapier *n*; Preßdeckelpapier *n*; Antimakulpapier *n*; Straffenpapier *n* — **n** leggerpapier *n*; spanvelpapier *n* — **e** papel *m* para tímpanos; papel *m* de cubrir — **i** carta *f* per abbigliamento; carta *f* per rivestimento

14169 tympan sheet — In boxmaking, usually a hard board or fibre material used for the female counter sheet, e.g., spaulding fibre, red fibre pressboard, tag board or hard manilla. Each plant, however, has its own preference. Cut by pressmen for score width, etc.

14170 tympan sheet — See draw sheet.

14171 type; letter — Piece of metal having for its face a letter or character, usually in high relief, adapted for use in letterpress printing. See also character.

f caractère *m* (d'imprimeur); type *m*; lettre *f* — **d** Buchstabe *m*; Type *f*; Drucktype *f*; Letter *f*; Schrift *f*; Schriftzeichen *n* — **n** letter *fm*; drukletter *fm*; schrift *mn* — **e** tipo *m*; letra *f* de imprenta; carácter *m* (de imprenta); carácter *m* tipográfico; carácter *m* (de impresión) — **i** lettera *f*; tipo *m*; carattere *m* (da stampa)

14172 type alignment slip

d Schriftlinienmeßklötchen *n* — **n** letterlijningsplaatje *n* — **e** listón *m* de medida del alineamiento; chapa *f* de medida del alineamiento — **i** listello *m* per l'allineamento dei caratteri

14173 type alloy — See type metal.

14174 type area — Space occupied by type matter.

f surface *f* de la page — **d** Satzspiegel *m* — **n** zetspiegel *m*; paginaspiegel *m*; bladspiegel *m* — **e** página-caja *f* — **i** luce *f* di composizione; pagina *f* base

14175 type bed — See bed.

14176 type brush; cleaning brush; pick brush — Brush to clean intensively type or plates.

f brosse *f* à nettoyer — **d** Wachsbürste *f*; Satzreinigungsbürste *f*; Putzbürste *f* — **n** vormborstel *m*; wasborstel *m*; plaatborstel *m* — **e** bruza *f* para limpiar; cepillo *m* para limpiar — **i** spazzola *f* per pulire

14177 type cabinet — See cabinet.

14178 type carrier cam — Part of the Monotype machine.

f excentrique *m* pour transporter des caractères

— **d** Schriftholer-Exzenter *m* — **n** letterdrager-excentriek *n* — **i** eccentrico *m* del prendicarattere
14179 type case — See case.
14180 type casting — Casting of type in a type foundry in a machine in which the metal is forced against a matrix with a pump.
f fonte *f* de caractères; fonte *f* de lettres — **d** Schriftguß *m* — **n** lettergieten *n* — **e** fundición *f* de tipos; fundición *f* de caracteres; fundición *f* tipográfica — **i** fusione *f* dei caratteri; fonditura *f* dei caratteri
14181 type-casting machine — Machine used in a type foundry to cast movable types.
f fondeuse *f* de caractères — **d** Schriftgieß-maschine *f*; Gießmaschine *f* — **n** lettergiet-machine *f*; gietmachine *f* — **e** máquina *f* de fundir tipos; fundidora *f* de tipos — **i** macchina *f* da fondere caratteri; fonditrice *f* per caratteri
14182 type correcting — See correction on lead.
14183 type cutter — See punch cutter.
14184 type designer
f lettreux *m*; lettriste *m*; dessinateur *m* de lettres — **d** Schriftzeichner *m* — **n** letterontwerper *m* — **e** diseñador *m* de tipos; letrista *m* — **i** disegnatore *m* di caratteri
14185 type dressing — Kinds of type faces used in a book or newspaper setting.
f style *m* typographique — **d** Schriftmischung *f* — **n** typografie *f*; stijl *m* van de typografie *f* — **e** estilo *m* tipográfico — **i** stile *m* tipografico
14186 type dressing — Process of finishing the incomplete work of the type caster.
14187 type face — See face.
14188 type family — Series of type faces which have common characteristics, and differ only by the increase or decrease of set or thickening and thinning of lines. Also a broad group of type classification, such as Transitionals. See also classification of type design.
f famille *f* de caractères — **d** Schriftfamilie *f*; Schriftgattung *f* — **n** letterfamilie *f*; lettersoort *fm* — **e** familia *f* de tipos; familia *f* tipográfica; familia-tipo *f* — **i** famiglia *f* di caratteri
14189 type font — See fount.
14190 type form — See forme.
14191 type forme — See forme.
14192 type founder — One engaged in the design and production of metal printing type for hand composition.
f fondeur *m* de caractères — **d** Schriftgießer *m* — **n** lettergieter *m* — **e** fundidor *m* de tipos — **i** fonditore *m* di caratteri
14193 type foundry; foundry — Foundry in which types or printing characters are cast.
f fonderie *f* de caractères; fonderie *f* typographique — **d** Schriftgießerei *f*; Gießerei *f* — **n** lettergieterij *f*; gieterij *f* — **e** fundición *f* de tipos; fundición *f* tipográfica — **i** fonderia *f* di caratteri
14194 type fount — See fount.
14195 type gage — See type gauge.
14196 type gage — See type gauge.
14197 type gauge; line gauge; pica rule; typom-

eter *US*; **type gage; line gage** — Rule marked off in picas and nonpareils, also centimetres or inches, to measure width of type, page depths, etc.
f typomètre *m*; lignomètre *m* — **d** Ciceromaß *n*; Zeilenmaß *n*; Zeilenmesser *m* — **n** cicero-maatje *n*; augustijn-maatje *n*; meetlat *fm* — **e** tipómetro *m*; escala *f* tipográfica; lineómetro *m*; medida *f* tipográfica; regla *f* tipométrica; medida *f* de líneas — **i** tipometro *m*
14198 type gauge; type-high gauge; type-height gauge; type gage *US* — Device to measure the printing height of blocks and other printing plates.
f calibre *m* de justification — **d** Schrifthöhe-messer *m*; Schrifthöheprüfer *m*; Klischeeprüfer *m* — **n** letterhoogtemeter *m*; clichéhoogtemeter *m* — **e** calibrador *m* de tipos; aparato *m* para medir la altura de tipos; puente *m* nivelador — **i** calibro *m* per il controllo dell'altezza tipografica
14199 type height — See height to paper.
14200 type-height gauge — See type gauge.
14201 type high — Mounted to the proper height to be used on a printing press. See also height to paper.
f à la hauteur typographique — **d** schrifthoch — **n** op letterhoogte — **e** de la altura del tipo; de la altura tipográfica — **i** di altezza tipografica
14202 type-high gauge — See type gauge.
14203 type-high planer — See block leveller.
14204 type-high stereo
f stéréo *m* à hauteur d'œil — **d** schrifthohes Stereo *n* — **n** styp *m* op letterhoogte — **e** estereo-tipia *f* de la altura tipográfica; estéreo *m* de la altura del tipo; estéreo *m* de la altura tipográfica; plancha *f* de altura tipográfica — **i** stereotipia *f* all'altezza di carattere; stereotipia *f* all'altezza tipografica
14205 type holder; type pallet — Implement to hold type used for decoration of book covers by hand.
f composteur *m* — **d** Schriftkasten *m* — **n** letter-haak *m* — **e** componedor *m*; cajetín *m* — **i** cassettina *f* per caratteri
14206 type line; body line
f ligne *f* de lettre — **d** Schriftlinie *f* — **n** letterlijn *fm* — **e** alineamiento *m*; línea *f* del tipo — **i** allineamento *m* (dei caratteri)
14207 type matter — See composition.
14208 type metal; printer's metal; type alloy — Alloy of lead, tin, antimony, and sometimes copper.
f matière *f* à caractères; matière *f* de caractères; métal *m* typographique; métal *m* à caractères; métal *m* de caractères — **d** Schriftmetall *n*; Letternmetall *n* — **n** lettermetaal *n*; letterspijs *fm* — **e** metal *m* tipográfico; metal *m* de imprenta; aleación *f* tipográfica — **i** lega *f* tipografica; lega *f* per caratteri; metallo *m* per caratteri
14209 type pallet — See type holder.
14210 type planer — See planer.
14211 type plate — See bed.
14212 type pusher

f pousseur *m* du caractère — **d** Buchstaben-ausstoßer *m* — **n** letter-uitstoter *m* — **e** expulsor *m* de letras — **i** espulsore *m* dei caratteri

14213 typescript; typewriter face; typewriter type; ribbon face — Type resembling characters obtained with a typewriter.

f caractère *m* machine à écrire — **d** Schreibmaschinenschrift *f* — **n** schrijfmachineletter *fm*; schrijfmachineschrift *n* — **e** tipo *m* de máquina de escribir — **i** carattere *m* per dattilografia

14214 typescript — See type-written copy.

14215 typescript-proof — See hard copy.

14216 type-setter — See compositor.

14217 typesetting — See compose v.

14218 typesetting machine — See composing machine.

14219 type size — See body size.

14220 type slug — See slug.

14221 type specimen book — Book of the type faces in a printing shop.

f épreuve *f* de caractères (petit format); catalogue *m* de caractères (grand format) — **d** Schriftmusterbuch *n*; Schriftprobe *f*; Schriftprobebuch *n* — **n** letterproef *fm* — **e** muestrario *m* de los tipos; catálogo *m* de los tipos — **i** campione *m* dei caratteri (forma piccola); campionario *m* dei caratteri (forma grossa); catalogo *m* dei caratteri

14222 typewriter double cap — A size of writing paper. See app. no. 3.

14223 typewriter face — See typescript.

14224 typewriter paper; typewriting paper — Bond paper cut to typewriter size. It requires regular bond sizing with no special surface other than the usual bond paper structure.

f papier *m* pour machine à écrire; papier *m* dactylotype — **d** Schreibmaschinenpapier *n*; Hartpostpapier *n* — **n** schrijfmachinepapier *n*; bankpostpapier *n*; bankpost *n*; bondpapier *n*; bond *n* — **e** papel *m* para máquina de escribir; papel *m* coquilla — **i** carta *f* per dattilografia

14225 typewriter size — A size of writing paper. See app. no. 3.

14226 typewriter type — See typescript.

14227 typewriting paper — See typewriter paper.

14228 type-written copy; typescript — Copy produced with a typewriter.

f copie *f* à la machine; écriture *f* à la machine — **d** mit der Schreibmaschine abgeschriebenes Manuskript *n* — **n** getypte kopij *f*; getypt manuscript *n*; met de machine geschreven kopij *f* — **e** copia *f* a máquina de escribir; copia *f* mecanografiada; texto *m* mecanografiado; original *m* mecanografiado; original *m* a máquina de escribir — **i** copia *f* dattilografata

14229 typing error — See typing mistake.

14230 typing mistake; typing error

f erreur *f* de dactylo; erreur *f* dactylographique; faute *f* de frappe — **d** Tippfehler *m*; Schreibfehler *m* — **n** tikfout *f*; typefout *f* — **e** error *m* de mecanografía; falta *f* de mecanografía; falta *f* de tecleo — **i** errore *m* di dattilografia

14231 typist — One who typewrites.

f dactylo *fm*; dactylographe *fm* — **d** Typist *m*; Typistin *f*; Maschinenschreiber *m*; Maschinenschreiberin *f*; Schreibkraft *f*; Daktylo *fm* — **n** typist *m*; typiste *f* — **e** mecanógrafo *m*; mecanógrafa *f*; dactilógrafo *m*; dactilógrafa *f* — **i** dactilografo *m*; dactilografa *f*

14232 typo — Unintensional error introduced by the keyboarder of the initial input, or by the typesetting machine or system. Computer term.

14233 typo — See compositor.

14234 Typograph — Slug-casting typesetting machine essentially in two parts, the base containing the spacing and casting mechanism and the movable upper part comprising the keyboard, keyboard wires, matrices and matrix guide rails. For sizes of 6 to 14-point with measures up to 32 ems 12-point. Invented (1885) by two Americans.

f composeuse *f* Typograph — **d** Typograph-Setzmaschine *f* — **n** Typograph-zetmachine *f* — **e** compositora *f* Typograph — **i** macchina *f* da comporre Typograph

14235 typographical error

f erreur *f* typographique — **d** typographischer Fehler *m* — **n** typografische fout *fm* — **e** error *m* tipográfico; yerro *m* tipográfico — **i** errore *m* tipografico

14236 typographic ink — See ink.

14237 typographic ink — See letterpress ink.

14238 typographic point — See point.

14239 typographic quality — See graphic arts quality.

14240 typographic unit — See also point system.

f unité *f* typographique; point *m* typographique — **d** typographischer Punkt *m* — **n** typografische eenheid *f*; typografische punt *m* — **e** unidad *f* tipográfica; punto *m* tipográfico — **i** unità *f* tipografica; punto *m* tipografico

14241 typography — Character and appearance of printed matter.

f typographie *f* — **d** Typographie *f* — **n** typografie *f* — **e** tipografía *f* — **i** tipografia *f*

14242 typometer — See type gauge.

14243 typon — Common French name for a screen positive ready for printing down on a press-plate. Name derived from the Swiss manufacturer. By extension a negative film, all that must be printed down on metal.

f typon *m*; cello-texte *m*; positif *m* offset — **d** Vorlage *f* auf Film — **n** direkt kopieerbare kopij *f* op film — **e** texto *m* en película; original *m* en película — **i** tipone *m*; velina *f*

14244 Typon process — Reflex printing process in which a special paper or film is used, with a yellow filter inserted between the light and the printing material.

f procédé *m* Typon — **d** Typon-Verfahren *n* — **n** Typon-procédé *n* — **e** procedimiento *m* Typon — **i** procedimento *m* Typon

14245 typote — Small printing press.

14246 typotect — See layout man.

14247 ultra-marine blue — Natural or synthetic inorganic blue pigment occasionally used for printing inks and in paper manufacture (neutralizing yellow colour).
f bleu *m* d'outre-mer — **d** Ultramarinblau *n* — **n** ultramarijn *n* — **e** azul *m* (de) ultramar; azul *m* ultramarino — **i** blu *m* oltremare

14248 ultra-marine yellow — See barium chromate.

14249 ultra-violet — Section beyond the violet end of the visible spectrum, the rays of which extend a high degree of photomechanical action. Its wavelengths range from 100 to 3900 Å.
f ultra-violet *n* — **d** Ultraviolett *n* — **n** ultraviolet *n* — **e** ultraviolado *m*; ultravioleta *f* — **i** ultravioletto *m*

14250 ultra-violet filter — Filter which cuts down ultra-violet rays but allows all visible light to pass.
f filtre *m* ultra-violet — **d** Ultraviolettfilter *m* — **n** ultravioletfilter *n* — **e** filtro *m* ultravioleta; filtro *m* ultra violeta — **i** filtro *m* ultravioletto

14251 ultra-violet lamp; UV lamp
f lampe *f* à rayons ultra-violets — **d** Ultraviolettlampe *f*; UV-Lampe *f* — **n** ultraviolet-lamp *fm*; UV-lamp *fm* — **e** lámpara *f* ultravioleta; lámpara *f* UV — **i** lampada *f* ultravioletta; lampada *f* a raggi ultravioletti; lampada *f* UV

14252 ultra-violet light; black light — Light very rich in the ultra-violet region, used in some gravure processes to expose continuous-tone positives on carbon tissue, being especially effective in reproducing the shadow detail.
f lumière *f* ultra-violette; lumière *f* noire — **d** ultraviolettes Licht *n*; Schwarzlicht *n* — **n** ultraviolet licht *n* — **e** luz *f* ultravioleta; luz *f* negra — **i** luce *f* ultravioletta

14253 ultra-violet range — Region of the spectrum which extends from 300 to 400 nm.
f région *f* ultra-violette — **d** Ultraviolettbereich *mn* — **n** ultraviolet gebied *n* — **i** regione *f* dell'ultravioletto; campo *m* dell'ultravioletto

14254 ultra-violet rays — Light-rays shorter than violet rays in the spectrum, but invisible to the eye, used to detect falsification of documents and presence of certain ingredients or foreign matter in paper. Also used to test fastness to light of inks and papers. Wavelength between 10 and 400 mμ.
f rayons *mpl* ultra-violets — **d** ultraviolette Strahlen *mpl* — **n** ultraviolette stralen *mpl* — **e** rayos *mpl* ultraviolados — **i** raggi *mpl* ultravioletti

14255 umber — Brownish pigment, consisting of the hydrated oxides or iron, used in the paper trade and occasionally in printing inks.
f ombre *f*; terre *f* d'ombre — **d** Umbra *f* — **n** omber *fm* — **e** tierra *f* de Umbría; tierra *f* sombra — **i** terra *f* di ombra

14256 Umlaut — Originally German for a sign placed over the letters as in ä, ö, or ü, changing

the sound. Also used in US for a diacritical mark.
f tréma *m*; accent *m* allemand — **d** Umlaut *n* — **n** trema *n*; deelteken *n*; umlaut *m* — **e** Umlaut *m*; metafonía *f* vocálica; acento *m* alemán — **i** metafonia *f*

14257 unaccounted time — Difference between elapsed time and the sum of the separate times recorded during a time study.
f erreur *f* globale de lecture — **n** verloren opnametijd *m* — **e** error *m* de lectura

14258 unadulterated
f non adultéré — **d** unverschnitten — **n** onversneden — **e** no diluído; sin diluir — **i** non adulterato; non sofisticato

14259 unbleached paper — Paper not treated with bleaching agents.
f papier *m* écru — **d** ungebleichtes Papier *n* — **n** ongebleekt papier *n* — **e** papel *m* no blanqueado; papel *m* crudo — **i** carta *f* non bianchita; carta *f* naturale

14260 unbleached pulp — Chemical wood pulp, before bleaching; light brown in colour. Used only in cheaper grades of paper, such as newsprint, kraft and wrapping papers.
f pâte *f* écrue — **d** ungebleichter Halbstoff *m*; ungebleichter Zellstoff *m* — **n** ongebleekte celstof *fm*; ongebleekte pulp *fm* — **e** pulpa *f* cruda; pulpa *f* no blanqueada; pulpa *f* sin blanquear; pasta *f* sin blanquear; pasta *f* no blanqueada — **i** pasta *f* greggia; pasta *f* non bianchita; pasta *f* naturale

14261 uncalendered paper — See unfinished paper.

14262 uncial; uncial letter — Type family of the 3rd to 5th century, superseded by the half uncial, a form of uncial and cursive letters.
f onciale *f*; lettre *f* onciale — **d** Unziale *f*; Uncial *f*; Unzialschrift *f* — **n** unciaal *fm* — **e** uncial *f*; letra *f* uncial — **i** onciale *f*; lettera *f* onciale

14263 uncial letter
See uncial.

14264 uncut edges; untrimmed edges — Edges of a book as the pages are left by the folding of the sheet.
f bords *mpl* non rognés; tranches *fpl* non rognées — **d** unbeschnittene Kanten *fpl*; unaufgeschnittene Kanten *fpl* — **n** niet-afgesneden kanten *mpl*; onafgesneden kanten *mpl*; niet-opengesneden kanten *mpl* — **e** cantos *mpl* intonsos; barbas *fpl* — **i** bordi *mpl* non refilati

14265 undecipherable — See also illegible.
f indéchiffrable — **d** unentzifferbar — **n** niet te ontcijferen; onontcijferbaar — **e** indescifrable — **i** indecifrabile

14266 under asphaltum — Making a deep-etch offset plate by removing the developing ink from the image areas and then coating these areas with asphaltum to make them permanently ink receptive. With a surface plate the ink of the image is removed and the image put under asphaltum if a plate is to stand more than 24 hours, or if it is to be stored for a rerun.

f sous bitume — **d** eingelackt — **n** in de asfalt gezet — **i** sotto asfalto

14267 under-blanket — Dressing next to the impression cylinder of a rotary press.

f blanchet *m* de dessous — **d** Untertuch *n* — **n** onderdoek *n* — **e** paño *m* inferior — **i** rivestimento *m* inferiore

14268 under-colour — See backing-up colour.

14269 under-colour removal — In four-colour printing, the subtraction of colour density from neutral areas leaving the black printer to replace equivalent density of the colours removed.

f correction *f* des sous-couleurs; élimination *f* des sous-couleurs; affaiblissement *m* de couleurs dans les ombres; réduction *f* des couleurs dans les ombres — **d** Tieftonrücknahme *f*; Farbrücknahme *f*; Unterfarbenkorrektur *f*; Unterfarbenentfernung *f*; Unterfarbenaufhellung *f*; Schattenfarbenreduktion *f* — **n** onderkleur-retouche *f*; onderkleur-correctie *f*; verwijderen *n* van de onderkleur; menggrijsverwijdering *f* — **e** atenuación *f* de color en las sombras — **i** rimozione *f* del sottocolore

14270 under-colours — In colour photography and platemaking, those primary colours and black not sufficiently eradicated in separation negatives and prints to produce an accurate multicolour reproduction without extensive correction and re-etching of halftone printing plates.

f sous-couleurs *fpl* — **d** Unterfarben *fpl* — **n** onderkleuren *fmpl* — **e** colores *mpl* primarios no atenuados — **i** sottocolori *mpl*

14271 undercut; under-etching; side etching — Condition in relief etching in which the acid or mordant has penetrated beneath the printing surface, causing thinning or weakening of lines and dots, and interfering with stereotyping and electrotyping.

f minage *m* (on dit que la planche est minée ou loupée) — **d** Unterätzung *f* — **n** onderetsing *f*; ondermijning *f* — **e** socavado *m* (de los puntos); sobremordido *m* — **i** sottoincisione *f*; morsura *f* sotto squadra laterale

14272 undercut — See cylinder bearer height.

14273 undercut *v* — To spread light beyond the transparent design areas of a negative or positive during exposure. Frequently due to local out-of-contact conditions between the film and the plate.

d unterkopieren — **n** onderkopiëren — **e** socavar — **i** sottocopiare

14274 undercut dot; mushroom dot — Result of careless etching or burnishing of halftone dots.

f point *m* miné; point *m* en champignon — **d** unterätzter Punkt *m*; Rasterpunkt *m* mit überhängender Schulter — **n** ondergeëtste punt *f*; punt *f* met onderetsing — **e** punto *m* en forma de seta; punto *m* en forma de sombrilla; punto *m* en forma de hongo — **i** punto *m* sottoinciso

14275 underdevelopment — Insufficient development, due to developing for too short time, use of a weakened developer or, occasionally, too low a temperature.

f sous-développement *m* — **d** Unterentwicklung *f* — **n** onderontwikkeling *f* — **e** subrevelado *m*; subrevelación *f* — **i** sottosviluppo *m*

14276 under-etching — See undercut.

14277 underexposed

f sous-exposé — **d** unterbelichtet — **n** onderbelicht — **e** sub expuesto; poco expuesto; a falto de exposición; no suficiente exposición — **i** sotto-esposto

14278 underexposure — Insufficient action of light on a sensitized photographic surface, resulting in thin or weak images and loss of detail.

f sous-exposition *f* — **d** unterbelichtung *f* — **n** onderbelichting *f* — **e** subexposición *f*; falta *f* de exposición; falta *f* de poca exposición; poca *f* exposición — **i** sottoesposizione *f*

14279 underfold — Reverse of *overfold.

14280 underhue — See undertone.

14281 underlay; backing up *depr.* — Piece of paper or built-up sheets placed or pasted underneath a printing plate to increase or decrease local pressure when printing, thereby improving the quality of the impression.

f mise *f* de dessous; mise *f* de hauteur sous cliché; mise *f* de puissance — **d** Kraftzurichtung *f*; Plattenzurichtung *f*; Unterlegung *f* — **n** pikeersel *n* onder de voet; krachtpikeersel *n*; onderlegsel *n* — **e** calzo *m*; alza *f*; calza *f*; recorte *m* por debajo; arreglo *m* de plancha — **i** taccheggio *m*; avviamento *m*

14282 underlay — Piece of paper or board placed under a printing plate or a forme to bring it up to proper height when printing.

f hausse *f*; tacon *m*; taquon *m* — **d** Unterlage *f*; Papierunterlage *f*; Klischeeunterlage *f* — **n** ophoogpapier *n* — **e** alza *f* debajo el cliché — **i** rialzo *m* per cliscè; rialzo *m* per cliché

14283 underlay paper — Paper to cut underlays for press makeready, most frequently hard book paper, or thin cardboard.

f papier *m* à hausses — **d** Papier *n* für Kraftzurichtung — **n** papier *n* voor krachtpikeersels — **e** papel *m* para alzas — **i** carta *f* da taccheggio

14284 underline — See legend.

14285 underline *v* — To use a line or lines underneath a word or sentence to stress its importance. For italic use a single line; for small caps a double line, for capitals a treble and for special type, such as bold face, a wavy line.

f souligner — **d** unterstreichen — **n** onderstrepen — **e** subrayar — **i** sottolineare

14286 under-packing — See packing.

14287 underrun; short make — Incompletely made or delivered order. Trade customs allow a certain tolerance for under- or overruns.

f déficit *m* de fabrication; poids *m* en dessous — **d** Untergewicht *n*; Mindergewicht *n*; Minderanfertigung *f* — **n** onderwicht *n*; ondergewicht *n*; tekort *n* — **e** falta *f* de peso; fabricación *f* corta — **i** quantitativo *m* prodotto in meno

14288 underscore *v* — To make a line under a character in written copy; u may be underscored in handwritten copy to distinguish it from an n which may be overscored.
f souligner — **d** unterstreichen — **n** onderstrepen — **e** subrayar — **i** sottolineare

14289 undershot duct; undershot fountain *depr.* — Ink duct utilizing a flexible steel blade mounted below the ductor roller. Weight of ink in duct tends to press on doctor blade and may influence adjustment.
d Heberfarbwerk *n* — **n** likrol-inktwerk *n* — **i** calamaio *m* con prenditore; calamaio *m* con rullo prenditore; calamaio *m* con lama flessibile

14290 undershot fountain — See undershot duct.

14291 underside — See wire-side.

14292 undertone; underhue; printone — Colour of a thin film of an ink as seen on a white background or greatly extended with white. The appearance of an ink when viewed by light transmitted through the film. The undertone of a transparent ink approximates its printing tone on white paper.
f ton *m* nuancé — **d** Nuance *f* in der Aufhellung — **n** gedrukte inktkleur *fm* — **e** tono *m* apagado; tono *m* matizado — **i** tono *m* sfumato; tono *m* attenuato

14293 underweight — See short weight.

14294 uneven pages — See odd pages.

14295 unfair competition
f concurrence *f* déloyale — **d** unlauterer Wettbewerb *m* — **n** deloyale concurrentie *f*; oneerlijke concurrentie *f* — **e** competencia *f* desleal; competencia *f* ilegal — **i** concorrenza *f* sleale

14296 unfinished paper; uncalendered paper — Paper with a rough appearance on both sides after manufacture, which has not been submitted to any finishing treatment.
f papier *m* brut; papier *m* sans apprêt — **d** ungeglättetes Papier *n* — **n** ongesatineerd papier *n* — **e** papel *m* sin acabado — **i** carta *f* ruvida di macchina; carta *f* non calandrata

14297 unfit for work — See incapable of work.

14298 ungathered — Said of printed sheets not collated in book order.
f en feuilles — **d** in Bogen — **n** aan vellen — **e** en hojas; sin alzar — **i** in fogli

14299 unglazed greaseproof paper
f papier *m* simili-sulfurisé non glacé — **d** unsatiniertes fettdichtes Papier *n* — **n** ongesatineerd vetdicht papier *n* — **e** papel *m* semisulfurizado no satinado — **i** carta *f* impermeabile ai grassi non satinata

14300 unglazed paper — Term used to distinguish between glazed and unglazed finishes of certain grades of paper which are made with both finishes, such as onion skin paper.
f papier *m* non glacé; papier *m* mat — **d** unsatiniertes Papier *n* — **n** ongesatineerd papier *n* — **e** papel *m* no satinado; papel *m* sin satinar — **i** carta *f* non satinata

14301 unified European colour scale — Standard of primary inks for three-colour or four-colour letterpress printing in defined order of printed colour, and describing the control method, to enable photoengravers and printers to prepare blocks based on the same colours, suitable for application on the same printing form irrespective of origin, and to obtain printing inks which when applied to a reference paper and at standard thickness, give results complying with the specifications of this international standard (not applicable to fluorescent inks; pigments are not included).
f gamme *f* européenne unifiée — **d** europäische Farbskala *f*; Europaskala *f* — **n** uniforme Europese kleurschaal *fm* — **e** gama *f* europea uniforme — **i** norma *f* unificata europea; scala *f* dei colori europea

14302 unified French colour scale
f gamme *f* française unifiée — **d** französische Farbskala *f* — **n** uniforme Franse kleurschaal *fm* — **e** gamma *f* francesa uniforme — **i** norma *f* unificata francesa; scala *f* dei colori francese

14303 uniform chromaticness scale — Variant of the CIE system. In the UCS the three axes are transformed in such a way that equally measured differences are in better accordance with equal visual colour differences.

14304 uniform formation — Uniform distribution of the fibres in a sheet which gives the paper an even translucency.
f épair *m* régulier — **d** gleichmäßige Durchsicht *f*; ruhige Durchsicht *f* — **n** gelijkmatig doorzicht *n* — **e** transparente *m* uniforme; trasluz *m* uniforme — **i** trasparenza *f* uniforme

14305 uniformly accelerated motion; uniformly increased motion
f mouvement *m* uniformément accéléré — **d** gleichförmig beschleunigte Bewegung *f* — **n** eenparig versnelde beweging *f* — **e** movimiento *m* uniformemente acelerado — **i** movimento *m* uniformemente accelerato

14306 uniformly increased motion — See uniformly accelerated motion.

14307 union-made — Paper made in a mill which is a "closed shop", i.e., one with full recognition of a labour union.

14308 union paper — Two sheets of wrapping paper stuck together with bitumen, tar or similar materials.
f papier *m* doublé-bitumé — **d** Doppelpechpapier *n*; kaschiertes Bitumenpapier *n*; bitumenkaschiertes Papier *n* — **n** bitumenpapier *n* — **e** papel *m* dobleasfalto — **i** carta *f* doppia bitumata

14309 unit — 1/8 part of a set em.
f unité *f* — **d** Einheit *f* — **n** eenheid *f* — **e** unidad *f* — **i** unità *f*

14310 unit arch — Perfecting unit on a rotary press arranged to form an arch, or inverted U design.

f groupement *m* en pont; élément *m* en arche; élément *m* en U renversé — **d** U-werk *n*; U-förmige Anordnung *f* — **n** U-vormige drukeenheid *f* — **e** unidad *f* impresora en forma de U; grupo *m* impresor en forma de U — **i** gruppo *m* stampante ad U

14311 unit count founts — Type founts in which the design of individual characters produces width values which are the same for a given "sign" regardless of the appearance of the typeface. All nine-point lower case "a"s will have the same set width, whether italic, roman, or bold, Univers or Bodoni. However, the width of the "a" will not be the same as the width of the "i". Unit count fount values tend to be universal and correspond to the values assumed by justified wire service output.

14312 unit drive — See individual drive.

14313 unit indicator — Part of the Monotype keyboard.

f indicateur *m* d'unités — **d** Einheitenskala *f* — **n** eenhedenschaal *fm* — **e** indicador *m* de unidades — **i** scala *f* delle unità

14314 unitized founts — Founts that permit variation of width values for any character from face to face and from fount to fount. A Times Roman "a" will not necessarily have the same set width as a Baskerville "a". This allows more scope for the esthetic design of type. Computer term.

14315 unit perfecting press — See blanket-to-blanket press.

14316 unit press — See unit type press.

14317 unit rack stop

f arrêt *m* de calibrage de la roue d'unités — **d** Einheitenwiderstände *mpl* — **n** eenhedenstoppers *mpl* — **i** arresto *m* della cremagliera della unità

14318 unit record equipment — Equipment which performs data processing activities without the use of a computer, by mechanical or electro-mechanical means, usually relying on punched cards for input and punched cards or a printout for output. Programs are designed by wiring plug boards or panels which are then inserted into the appropriate devices.

f matériel *m* classique à carte perforée — **d** Lochkartenmaschinen *fmpl*; elektromechanisch gesteuerte Informations-Bearbeitungsmaschine *f* — **n** ponskaartenapparatuur *f* — **e** equipo *m* de máquinas perforadoras — **i** impianto *m* di macchine perforatrici

14319 unit set; point set — Width of type, when each type is cast to a definite number of points. This is a bad practice, as the design of the type is apt to be distorted when squeezed or expanded to fit the exact number of units.

f force *f* systématique; épaisseur *f* systématique — **d** systematische Dickte *f* — **n** systematische letterdikte *f* — **e** grosor *m* sistemático; espesor *m* sistemático — **i** forza *f* del corpo sistematica; spessore *m* sistematico

14320 unit shift — Monotype system to allocate to a character the value in units of the matrix row

above. Thus a matrix in the 12 unit row in the matrix case may be given a value of 12 units.

14321 unit type press; unit press — Press with one or more printing units in line on the bed plate.

f machine *f* à groupes en ligne; presse *f* à groupes alignés — **d** Parterremaschine *f*; Presse *f* in Reihenbauart *f* — **n** parterre-machine *f* — **e** rotativa *f* de cuerpos; rotativa *f* de grupos; rotativa *f* de unidades de impresión — **i** macchina *f* da stampa a gruppi multipli in linea

14322 unit wheel — Part of the Monotype keyboard.

f roue *f* d'unités — **d** Einheitenrad *n* — **n** eenhedenwiel *n* — **e** rueda *f* de unidades — **i** rotella *f* della unità; ruota *f* delle unità

14323 unit wheel driving cylinder

f cylindre *m* moteur; tube *m* du piston moteur — **d** Luftzylinder *m*; Triebzylinder *m* — **n** luchtcilinder *m*; aandrijfcilinder *m* voor het eenhedenwiel — **i** cilindro *m* motore della ruota delle unità

14324 unit width — See proportionally spaced characters.

14325 universal coupling *GB*; **Hooke's joint** *US*; **cardan joint** — Coupling of two shafts enabling the transmitting of rotation from one shaft to another not lying on or passing through the same straight line.

f joint *m* universal; joint *m* de cardan — **d** Gelenkkupplung *f*; Kardangelenk *n*; Hooke'sches Gelenk *n*; Kreuzgelenk *n* — **n** cardankoppeling *fm*; cardanoverbrenging *f*; kruiskoppeling *f* — **e** junta *f* cardan; junta *f* universal — **i** giunto *m* cardanico; cardano *m*

14326 university library

f bibliothèque *f* universitaire — **d** Universitätsbibliothek *f*; Hochschulbücherei *f* — **n** universiteitsbibliotheek *f* — **e** biblioteca *f* universitaria — **i** biblioteca *f* universitaria

14327 unjustified (paper) tape; idiot tape *sl.*; **minimum coded tape; raw tape** — Paper band encoded on a perforating machine, lacking any typographic or line ending instruction codes, for which the operator needs not to have knowledge of typesetting.

f bande *f* (de texte) brute; bande *f* perforée brute; bande *f* non-justifiée — **d** unjustiertes Endlosband *n*; unjustierter Streifen *m* — **n** vuile band *m*; niet-uitgevulde band *m*; kilometerband *m* *sl.* — **e** cinta *f* non justificada; cinta *f* sin ajustar; cinta *f* boba — **i** nastro *m* non giustificato

14328 unjustified setting; ragged-right setting; rag-right setting

f composition *f* en drapeau; composition *f* non justifiée; lignes *fpl* brisées — **d** Flattersatz *m* (linksbündiger Flattersatz *m*; rechtsbündiger Flattersatz *m*; Satz *m* auf Mittelachse) — **n** niet-uitgevuld zetsel *n*; niet-uitgevulde regels *mpl* — **e** composición *f* a pedazos — **i** composizione *f* non giustificata

14329 unleaded; solid — Set without spacing between lines. See also leaded matter.

f non interligné; compact — **d** nicht durch-schossen; kompreß; ohne Durchschuß — **n** niet ge-ïnterlinieerd; zonder interlinies; plat; compres — **e** sin interlíneas; sin interlineado; sin interlinear — **i** non interlineato

14330 unleaded gasoline — Motor gasoline not containing lead compounds. During printing lead may set down on the offset plate and cause scumming.

f essence *f* non-plombée — **d** bleifreies Benzin *n* — **n** loodvrije benzine *fm*; ongelode benzine *fm* — **e** gasolina *f* sin plomo — **i** benzina *f* senza piombo

14331 unlock *v* — To loosen a forme by un-fastening the quoins.

f desserrer; dégager; désosser — **d** aufschließen — **n** loskooien; losmaken; ontsluiten — **e** desacuñar; abrir; aflojar — **i** aprire; spaginare

14332 unmounted plate — Plate not mounted on a base or block of wood or metal.

f cliché *m* non monté; cliché *m* sans support — **d** unmontierte Platte *f* — **n** ongemonteerd cliché *n*; niet-gemonteerd cliché *n* — **e** plancha *f* sin montar; clisé *m* sin zócalo — **i** cliscè *m* non montato; cliché *m* non montato

14333 unoccupied (cycle) time — See balance time.

14334 unopened edges — Edges of a book which have not been cut open with a knife.

f tranches *fpl* non coupées — **d** unaufgeschnittene Kanten *fpl* — **n** niet-opengesneden randen *mpl* — **e** libro *m* sin cortar; libro *m* intonso — **i** pagine *fpl* non tagliate

14335 unprocessed paper — Paper in sheets or reels before being converted by, e.g., the printer or stationer.

f papier *m* en l'état — **d** unverarbeitetes Papier *n* — **n** onbewerkt papier *n* — **e** papel *m* sin labrar; papel *m* nativo — **i** carta *f* non lavorata

14336 unreliable worker

f homme *m* sur lequel on ne peut pas compter — **d** unzuverlässiger Arbeiter *m*; Schuster *m*; Schusterjunge *m* — **n** onbetrouwbare werker *m*; onbetrouwbaar werkman *m* — **e** trabajador *m* de poco fiar — **i** operaio *m* non fidato; lavorante *m* non fidato

14337 unrestricted cycle — See effort-controlled cycle.

14338 unsaleable book; dud; dead stuck — Book that has proved to be a failure.

f garde-boutique *m*; rossignol *m* — **d** Laden-hüter *m* — **n** onverkoopbaar boek *n*; winkel-dochter *f* — **e** libro *m* invendible; libro *m* muerto — **i** libro *m* non vendibile; libro *m* non esitabile

14339 unsaponifiable — Not capable of being saponified. See also saponification.

f insaponifiable — **d** unverseifbar — **n** onverzeep-baar; niet verzeepbaar — **e** insaponificable — **i** in-saponificabile

14340 unsaturated colours

f couleurs *fpl* non saturées — **d** bezogene Farben

fpl — **n** onverzadigde kleuren *fmpl* — **e** colores *mpl* insaciados — **i** colori *mpl* insaturi

14341 unsaturated fatty acid

f acide *m* gras non saturé — **d** ungesättigte Fett-säure *f* — **n** onverzadigd vetzuur *n* — **e** ácido *m* graso insaciado; ácido *m* graso no saturado — **i** acido *m* grasso insaturo

14342 unseriffed — See sans serif.

14343 unsewn binding — See adhesive binding.

14344 unsharp mask; diffused mask — Mask sometimes made in a printing frame by placing a thin sheet of acetate between the separation nega-tive and the material for the positive, with the illumination placed off-centre and the frame whirled during exposure.

f masque *m* flou — **d** unscharfe Maske *f* — **n** on-scherp masker *n* — **e** máscara *f* borrosa — **i** maschera *f* non nitida

14345 unshift code — See shift code.

14346 unsized paper — Paper to which no sizing has been added during or after its manufacture. Not to be confused with waterleaf paper, which is a papermaker's term for unsized paper, that is permeable to moisture and absorbent.

f papier *m* non collé; papier *m* sans colle — **d** un-geleimtes Papier *n* — **n** ongelijmd papier *n* — **e** papel *m* sin cola; papel *m* sin encolar; papel *m* sin aderezo; papel *m* no encolado — **i** carta *f* non collata

14347 unskilled labourer

f manœuvre *m*; ouvrier *m* non qualifié — **d** un-gelernter Arbeiter *m*; Hilfsarbeiter *m* — **n** on-geschoold arbeider *m*; hulparbeider *m* — **e** obrero *m* no calificado — **i** manovale *m*; lavoratore *m* non qualificato

14348 unslaked lime — See calcium oxide.

14349 unsorted rags — See mixed rags.

14350 unstable

f instable — **d** unstabil; unbeständig — **n** in-stabiel; onstabiel; niet stabiel; onbestendig — **e** in-stable; inestable — **i** instabile

14351 untrimmed edges — See uncut edges.

14352 untrimmed paper — Paper not cut to size and squared with a guillotine or other trimmer.

f papier *m* non rogné — **d** unbeschnittenes Papier *n* — **n** niet-nagesneden papier *n*; niet-schoongesneden papier *n* — **e** papel *m* no re-cortado — **i** carta *f* non refilata

14353 untrimmed size — Size of a sheet of paper, untrimmed and not specially squared, sufficiently large to obtain a trimmed size from it, as required.

f format *m* brut — **d** Rohformat *n*; unbe-schnittenes Format *n* — **n** onafgesneden formaat *n* — **e** tamaño *m* no recortado; formato *m* no recortado — **i** formato *m* non refilato

14354 update *v* — In computer controlled type-setting, to revise a master file by accumulating or inserting up-to-date data and deleting obsolete data in accordance with a predetermined pro-cedure.

f mettre à jour — **d** auf den neuesten Stand

bringen; fortschreiben; aktualisieren — **n** bij-werken — **e** actualizar — **i** aggiornare

14355 upper case; cap case — Part of the case in which are kept the capitals. Formerly that part of the case which was the farthest from the compositor, and inclined at a steeper angle.
f haut de casse *m* — **d** Oberkasten *m* — **n** bovenkast *fm* — **e** caja *f* alta — **i** cassa *f* alta; alta cassa *f*

14356 upper deck — See rostrum.

14357 upper magazine — An upper magazine on a linecasting machine. Some models of paper tape perforator keyboards permit tape control of two magazines when used with suitable linecasting equipment. Depression of the upper magazine key on the keyboard inserts a unique code in the paper tape for subsequent control of the line caster. See also lower magazine, magazine, side magazine.
f magasin *m* supérieur — **d** Obermagazin *n* — **n** bovenmagazijn *n*; bovenste magazijn *n* — **e** magacén *m* alto — **i** magazzino *m* alto

14358 upper rail — See rail.

14359 upright size — See portrait size.

14360 upshift — On a typewriter with striking bars, the key depressed to cause the alternate character (a capital letter or a symbol above a number) to print. On a Selectric typewriter, a phototypesetter, or TTS input from a line caster, the key, when depressed, performs a comparable function.

14361 upside down letter — See turned letter.

14362 up-time; serviceable time
f temps *m* utile — **d** Nutzzeit *f* — **n** bruikbare tijd *m* — **e** tiempo *m* útil — **i** tempo *m* utile

14363 uranium acetate — See uranyl acetate.

14364 uranium ferrocyanide
f ferrocyanure *m* d'uranium — **d** Uraneisenzyanid *n*; Uranferrozyanid *n* — **n** ferrocyaanuraan *n*; uraan-ferrocyanide *n* — **e** ferrocianuro *m* de uranio — **i** ferrocianuro *m* d'uranio

14365 uranium nitrate — See uranyl nitrate.

14366 uranium oxychloride — See uranyl chloride.

14367 uranyl acetate; uranium acetate — Small, yellow crystals, soluble in cold water and alcohol. Decomposed by light or in hot water.
Formula: $UO_2(C_2H_3O_2)_2 \cdot 2H_2O$.
f acétate *m* d'urane — **d** Uranazetat *n*; Uranylazetat *n* — **n** uraniumacetaat *n*; uraanacetaat *n*; uranylacetaat *n* — **e** acetato *m* de uranilo; acetato *m* de uranio — **i** acetato *m* d'uranio; acetato *m* d'uranile

14368 uranyl chloride; uranium oxychloride — Yellow crystals, soluble in water, alcohol and ether.
Formula: $UO_2Cl_2 \cdot H_2O$.
f chlorure *m* d'urane; chlorure *m* d'uranium — **d** Uranylchlorid *n*; Uranchlorid *n* — **n** uranylchloride *n*; uraniumchloride *n*; uraanchloride *n* — **e** cloruro *m* de uranilo; cloruro *m* de uranio

14369 uranyl nitrate; uranium nitrate — Yellow, rhombic crystals. Soluble in water, alcohol and ether. Used in photography. Latin: uranium nitricum.
Formula: $UO_2(NO_3)_2 \cdot 6H_2O$.
f nitrate *m* d'uranyle; nitrate *m* d'urane; nitrate *m* d'uranium; azotate *m* d'urane — **d** Uranylnitrat *n*; Urannitrat *n*; salpetersaures Uran *n* — **n** uranylnitraat *n*; uraniumnitraat *n*; uraannitraat *n* — **e** nitrato *m* de uranilo; nitrato *m* de uranio — **i** nitrato *m* d'uranile; nitrato *m* d'uranio

14370 urea-formaldehyde resin — Thermosetting condensation product of urea, NH_2CONH_2, and formaldehyde, $HCHO$, under pressure.
f résine *f* d'urée-formaldéhyde; résine *f* formolurée; résine *f* urée-formol — **d** Harnstoff-Formaldehydharz *n*; Harnstoff-Aldehydharz *n* — **n** ureum-formaldehydehars *mn* — **e** resina *f* de urea-formaldehído — **i** resina *f* di ureaformaldeide

14371 UV lamp — See ultra-violet lamp.

14372 vacuum back — Plate or board to hold light-sensitive film or paper in a camera by means of suction.
f plaque *f* pneumatique — **d** Vakuumplatte *f*; Ansaugplatte *f* — **n** zuigplaat *fm*; vacuümplaat *fm* — **e** placa *f* neumática; lamina *f* neumática — **i** piastra *f* pneumatica

14373 vacuum copyboard — Board or arrangement on process cameras to hold originals in place during exposure by means of atmospheric pressure. See also camera copyboard.
f porte-modèle *m* à vide — **d** Originalhalter *m* mit Ansaugrahmen; Originalhalter *m* mit Vakuumansaugung — **n** modelbord *n* met zuigwand — **e** portaoriginal(es) *m* al vacío; tablero *m* portaoriginales al vacío; portamodelo *m* al vacío — **i** portaoriginale *m* pneumatico

14374 vacuum holder — Perforated or channeled metal plate on which films and negative papers are held in position in the focal plane of darkroom cameras by withdrawing the air between the film and support, the material then held by the pressure of the atmosphere.
f porte-film *m* pneumatique — **d** Ansaugbrett *n* — **n** zuigcassette *fm* — **e** portapelícula *m* al vacío; respaldo *m* neumático — **i** portapellicola *m* pneumatico

14375 vacuum printing frame — Frame in which contact between a photographic image and sensitized surface is maintained by atmospheric pressure, the air being withdrawn from the frame by an electrically driven pump.
f châssis *m* (à copier) pneumatique — **d** pneumatischer Kopierrahmen *m*; pneumatischer Kontaktrahmen *m*; Vakuumkopierrahmen *m* — **n** pneumatisch kopieerraam *n* — **e** chasis *m* neumático; prensa *f* neumática (de copiar); marco *m* neumático (de copiar); neumático *m* de copiar — **i** telaio *m* pneumatico

14376 vacuum pump; suction pump — Mechanism to maintain very low air pressure.
f pompe *f* à vide; pompe *f* d'aspiration — **d** Vakuumpumpe *f*; Absaugpumpe *f* — **n** vacuümpomp *fm*; zuigpomp *fm*; afzuigpomp *fm* — **e** bomba *f* de vacío; bomba *f* de émbolo; bomba *f* de aspiración — **i** pompa *f* a vuota

14377 vade-mecum — Guide or instruction book.
f vade-mecum *m* — **d** Vademekum *n* (Vademekums *npl*); Handbuch *n*; Leitfaden *m* — **n** vademecum *n* (vademecums *npl*); handboek *n* — **e** vademécum *m* — **i** vade mecum *m*

14378 valence; valency — Combining capacity of an atom based on 1 for hydrogen.
f valence *f*; atomicité *f* — **d** Wertigkeit *f*; Valenz *f* — **n** valentie *f*; waardigheid *f* — **e** valencia *f* — **i** valenza *f*

14379 valency — See valence.

14380 value — Short for Munsell value, a term not admitted by CIE, used to distinguish light pale colours from dark ones, e.g., navy blue is a dark value of blue, whereas sky blue is a light value. Light values are known as tints, while dark values are called shades. Value corresponds in other colour systems to *lightness.
f luminosité *f* — **d** Helligkeit *f* — **n** helderheid *f* — **e** luminosidad *f* — **i** chiarezza *f*

14381 valve bank — Part of the Monotype keyboard.
f boîte *f* de valves — **d** Luftventilkasten *m* — **n** luchtkamer *fm* — **e** caja *f* de válvulas — **i** cassa *f* delle valvole d'aria; cassa *f* degli emboli

14382 Van der Waals equation — Modified equation of state for gases to compensate for actual volume of molecules and for attractive forces existing between the molecules. After Johannes Diderik van der Waals, Dutch physicist, 1837-1923.
f équation *f* de Van der Waals — **d** Van der Waalssche Zustandsgleichung *f* — **n** Van der Waalse vergelijking *f* — **e** ecuación *f* de Van der Waals — **i** equazione *f* di Van der Waals

14383 Van der Waals forces — Forces of cohesion, arising from secondary valences. They do not produce true chemical bonds. They are presumably responsible for pigment flocculation and also for adhesion.
f forces *fpl* Van der Waals — **d** Van der Waalssche Kräfte *fpl* — **n** Van der Waalse krachten *fmpl* — **e** fuerzas *fpl* de Van der Waals — **i** forze *fpl* di Van der Waals

14384 vandyke brown; Kassel brown; Kassel earth; Cassel brown *depr.* — Natural brown-black organic pigment, used for artists' colours and stains.
f brun *m* Van Dyck; terre *f* de Cassel; brun *m* foncé — **d** Vandyckbraun *n*; Kasselerbraun *n* — **n** Van Dijck's bruin *n* — **e** pardo *m* de Van Dijck; pardo *m* Cassel — **i** bruno *m* Van Dijck; marrone *m* di Van Dijck; bruno *m* di Cassel

14385 vandyke print — See brownprint.

14386 Van Gelder — Hand-made paper produced by Van Gelder Zonen, Holland.

14387 vaposet ink
See steam-set ink.

14388 vapour permeability — Property of paper or paperboard which allows the passage of vapour; measured under very carefully specified conditions of total pressure, partial pressures of the vapour on the two sides of the sheet, temperature, and relative humidity. Because of the fact that paper has specific affinity for water vapour, vapour permeability should not be confused with air permeability or porosity.
f perméabilité *f* à la vapeur — **d** Wasserdampfdurchlässigkeit *f* — **n** waterdampdoorlatendheid *f* — **e** permeabilidad *f* al vapor de agua — **i** permeabilità *f* al vapore di acqua; permeabilità *f* al vapore acqueo

14389 vapour-phase inhibitor paper; vpi paper — Fairly light weight wrapping paper, usually made from unbleached kraft pulp and treated with vapour-phase inhibitors to provide anti-rust, anti-tarnish, and anti-corrosion qualities. Used for wrapping metal parts, machinery, cutlery etc. to protect these from corrosion during shipment, storage, etc.
f papier *m* inhibiteur volatile de corrosion — **d** Dampfphasenschutzpapier *n*; Dampfphasenpapier *n*; selbsttätiges Korrosionsschutzpapier *n* — **n** vpi-papier *n* — **e** papel *m* protector a la fase de vapor — **i** carta *f* inibitore volatile di corrosione

14390 vapour-proof paper — Paper treated to resist penetration by gases or vapours.
f papier *m* étanche à la vapeur — **d** dampfdichtes Papier *n* — **n** dampdicht papier *n*; waterdampdicht papier *n* — **e** papel *m* estanco al vapor; papel *m* antivapor — **i** carta *f* a prova di vapore

14391 variable dot leaders — Economical way of producing dotted lines with or without reading matter. See also leaders.

14392 variable-length record system — Method of storing data without regard to the number of digits in the record. Data of varying lengths may be stored in certain media in the minimum space.
f enregistrement *m* à longueur variable — **d** Datei *f* mit variabler Satzlänge — **n** stelsel *m* met variabele recordlengte — **e** registro *m* de longitud variable — **i** registrazione *f* a lunghezza variabile

14393 variable opacity screen — See contact screen.

14394 variable-size folder
f plieuse *f* à format variable — **d** formatvariabler Falzapparat *m*; variabelformatiger Falzapparat *m*; variabeler Falzapparat *m* — **n** variabel vouwapparaat *n* — **i** gruppo *m* piegatore a formato variabile

14395 variable space — Word space which is expanded to make all the lines on a page the same length. In the Monotype system the width of the variable space is calculated to fill the line exactly, whereas in line casting operation the *spaceband is expanded mechanically.
f espace *f* justifiante — **d** variabeler Ausschluß *m* — **n** variabele spatie *f* — **e** espacio *m* variable — **i** spazio *m* variabile

14396 variable word length — Number of characters addressed in a computer which is not a fixed number but is varied by the data or instruction.
f longueur *f* de mot variable — **d** variabele Wortlänge *f* — **n** variabele woordlengte *f* — **e** longitud *f* de palabra variable — **i** lunghezza *f* di parola variabile

14397 variance — In statistics, the squared standard deviation.
f variance *f* — **d** Varianz *f*; Streuung *f* — **n** variantie *f* — **e** variante *f* — **i** varianza *f*

14398 variomat — Appliance on a reproduction camera consisting of an oscillating lens to convert screened negatives or positives into continuous-tone images.

14399 varishade ink — See double-tone ink.

14400 Varityper — Typewriter with interchangeable type faces for reproduction work.
f varityper *f* — **d** Vari-Typer *m* — **n** varityper *m* — **e** varityper *m* — **i** varityper *f*

14401 varnish — Oil which has been bodied by heat or by the addition of gums, resin or other materials, with or without driers, often with plasticizers and waxes. The vehicle of an ink. Gravure varnishes are usually composed of resins and/or film formers dissolved in a solvent or a blend of solvents.
f vernis *m* — **d** Firnis *m*; Lack *m* — **n** vernis *mn* — **e** barniz *m*; laca *f* — **i** vernice *f*; lacca *f*

14402 varnishability — Measure or ability of a sheet of paper to accept varnish. A sheet with a high degree of varnishability will generally be smooth, of low absorbency, and have a minimum amount of colour change on varnishing.
f vernissabilité *f*; aptitude *f* au vernissage — **d** Lackierbarkeit *f*; Lackierfähigkeit *f* — **n** vernisbaarheid *f* — **e** barnizabilidad *f* — **i** verniciabilità *f*

14403 varnished paper
f papier *m* verni — **d** lackiertes Papier *n* — **n** gevernist papier *n* — **e** papel *m* barnizado; papel *m* lacado — **i** carta *f* laccata

14404 varnisher — See varnishing machine.

14405 varnishing — See lacquering.

14406 varnishing machine; varnisher; lacquering machine — Machine to varnish printed paper to brighten and protect the colours.
f vernisseuse *f*; machine *f* à enduire; machine *f* à lacquer — **d** Lackiermaschine *f* — **n** vernismachine *f* — **e** barnizadora *f*; máquina *f* a barnizar — **i** macchina *f* per verniciare; macchina *f* laccatrice

14407 varnish-label paper — Paper for labels which is subsequently to be varnished. It may be uncoated or coated on one side and may have a supercalendered finish. Made of bleached chemical wood pulp and when uncoated resists varnish penetration and discoloration largely by virtue of its high density. See also label paper.
f papier *m* à étiquettes vernies — **d** Papier *n* für lackierte Etiketten — **n** vernisbaar papier *n* voor etiketten en wikkels — **e** papel *m* para etiquetas barnizadas — **i** carta *f* per etichette laccate

14408 varying tension — Pull in the direction of web travel in reel-fed equipment; varies in kilograms of pull per centimetre (pounds per inch) of web width.
f variation *f* de tension — **d** veränderliche Spannung *f* — **n** veranderlijke spanning *f* — **e** tensión *f* variable — **i** tensione *f* variabile

14409 vaseline — See petrolatum.

14410 vat — Tank containing paper making material, used for making paper by hand. Also the tank in which the paper is made upon a semi-submerged cylinder mould.

f cuve *f* — **d** Stoffbütte *f*; Schöpfbütte *f*; Bütte *f* — **n** stofkuip *fm* — **e** tina *f*; cuba *f*; pila *f* — **i** tino *m*

14411 vat-lined board — Board lined on the multi-cylinder machine.

f carton *m* doublé sur machine à formes rondes — **d** gegautschter Karton *m* — **n** gekoetst karton *n* — **e** papel *m* doblado de máquina redonda — **i** cartone *m* accoppiato in macchina a più tamburi

14412 vat machine — See cylinder machine.

14413 vatman; dipper — Craftsman who dips the mould into the pulp and moulds the sheet of paper.

f plongeur *m*; puiseur *m*; formeur *m Be* — **d** Schöpfer *m*; Büttenschöpfer *m*; Büttgeselle *m*; Büttknecht *m*; Taucher *m*; Eintaucher *m* — **n** schepper *m*; formeur *m Be* — **e** sacador *m*; oficial *m* de tina; formador *m* de hoja — **i** immergitore *m*; operaio *m* prenditore; prenditore *m*

14414 vat paper — See hand-made paper.

14415 vat sized paper — See tub sized paper.

14416 V-belt — Driving belt with a section in the form of a trapeze running over suitable shaped pulleys for the transmission of power.

f courroie *f* trapézoïdale — **d** Keilriemen *m* — **n** V-snaar *fm* — **e** correa *f* trapezoidal — **i** cinghia *f* trapezoidale

14417 vegetable glue

f colle *f* végétable — **d** vegetabilischer Leim *m*; Pflanzenleim *m* — **n** plantaardige lijm *m* — **e** cola *f* vegetal — **i** colla *f* vegetale

14418 vegetable glue — See agar.

14419 vegetable gum — See dextrin.

14420 vegetable parchment — See greaseproof paper.

14421 vegetable parchment paper; parchment paper — Paper that has acquired, by the action of sulphuric acid, a continuous texture, which gives it a high degree of resistance to penetration by grease and renders it resistant to disintegration by water even at boiling point.

f papier *m* sulfurisé; parchemin *m* végétal — **d** echtes Pergamentpapier *n*; vegetabilisches Pergament *n* — **n** echt perkamentpapier *n* — **e** papel *m* vegetal; papel *m* pergamino; papel *m* apergaminado; papel *m* sulfurizado; pergamino *m* vegetal — **i** carta *f* pergamena vegetale

14422 vegetable wax — See sumac.

14423 vehicle; binding agent; binding varnish; binding medium; grinding medium; carrying medium — Liquid medium in which pigments and particulate substances are ground and suspended. In ink the vehicle is often linseed oil. See also adhesive.

f véhicule *m*; liant *m*; vernis *m* — **d** Bindemittel *n* — **n** bindmiddel *n* — **e** vehículo *m*; mordiente *m*; pegante *m*; ligante *m*; aglutinante *m* — **i** veicolo *m*; legante *m*

14424 veil — See fog.

14425 vellum; calf vellum — Fine, smooth kind of parchment made from calfskin, prepared with lime and not tanned. See also parchment, forrel.

f parchemin *m* (en cosse); parchemin *m* de veau — **d** Pergament *n* (aus Kalbsfell) — **n** perkament *n* (uit kalfsvel) — **e** pergamino *m*; piel *f* pergamino — **i** pergamena *f* da pelle di vitello; velino *m*

14426 vellum binding — See parchment binding.

14427 vellum paper — Rag content paper, similar to high-grade ledger, usually of a creamy-white colour, with smooth surface, imitating vellum, to print diplomas and certificates.

14428 velocity gradient

f gradient *m* de vitesse — **d** Geschwindigkeitsgefälle *n*; Geschwindigkeitsgradient *m* — **n** snelheidsgradiënt *m*; snelheidsverval *n*; snelheidsdaling *f* — **e** gradiente *m* de velocidad; tasa *f* de cortadura — **i** gradiente *m* di velocità

14429 velocity of light

f vitesse *f* de la lumière — **d** Lichtgeschwindigkeit *f* — **n** lichtsnelheid *f*; snelheid *f* van het licht — **e** velocidad *f* de luz — **i** velocità *f* della luce

14430 velours paper

f papier *m* velours — **d** Velourspapier *n*; Samtpapier *n* — **n** velourspapier *n* — **e** papel *m* terciopelo; papel *m* aterciopelado; papel *m* de paño — **i** carta *f* vellutata

14431 velox — Commercial photopaper, but the name is used for a photoengraving process on any chloride paper. A screened photoprint is made from the original, so that line and drop-out effects are produced, which can be retouched and assembled with line work and prints of type.

f copie *f* par contact sur papier photographique — **d** Velox-Druck *m* — **n** contactafdruk *m* op fotopapier — **e** vélox *m*; copia *f* vélox — **i** copia *f* per contatto su carta Velox

14432 velvet finish — A finish suggesting the feel of velvet. It has a dull surface. Same as English finish.

14433 velvet finish — See English finish.

14434 velvet finished paper — See English finish paper.

14435 velvet paper — See English finish paper.

14436 Venetian — See Humanistics.

14437 Venetian red; English red; colcothar — Ferric oxide pigment obtained by calcining *copperas in the presence of lime. See also Indian red.

f rouge *m* vénitien; rouge *m* de Venise; rouge *m* anglais; rouge *m* d'Angleterre; colcothar *m* — **d** venezianisch Rot *n*; Englischrot *n*; Colcothar *m* — **n** Venetiaans rood *n* — **e** rojo *m* veneciano; colcótar *m* — **i** rosso *m* di Venezia; rosso *m* d'Inghilterra

14438 Venetian turpentine; Venice turpentine; larch turpentine; larch gum — Yellowish to green oleoresin, obtained from Pinus larix, the Tyrolean larch. Becomes hard and brittle on exposure to air; soluble in most organic solvents. Used for varnishes and in lithography. Latin: terebinthina veneta.

f térébenthine *f* de Venise; térébenthine *f* du mélèze — **d** venezianischer Terpentin *m*; Lärchenharzöl *n*; Lärchenterpentin *n*; Venezianeröl *n* — **n** Venetiaanse terpentijn *m*; natuurlijke terpentijn *m* — **e** trementina *f* de Venecia; terebentina *f* veneciana — **i** trementina *f* di Venezia

14439 Venice turpentine — See Venetian turpentine.

14440 ventilation
f aération *f* — **d** Belüftung *f*; Ventilation *f* — **n** ventilatie *f* — **e** ventilación *f* — **i** ventilazione *f*

14441 verification — Assuring the accuracy of input by automatic or semi-automatic checking. Double keyboarding is sometimes used: if the second keyboarder strikes different keys than the first, this will be immediately detected (such as by "locking" the keyboard) so that the operator may determine where the mistake lies. Another method is the inspection of two streams of input (from two separately keyboarded sources), usually with software or hardware which arrests the flow of data.
f vérification *f* — **d** Überprüfung *f*; Kontrolle *f*; Verifikation *f* — **n** controle *fm*; verificatie *f* — **e** verificación *f* — **i** verificazione *f*

14442 verifier — Checking device to verify the accuracy of coded representations in punched cards or paper tape; data are keyboarded a second time with a verifier and each paper tape row or card column is automatically compared with the first punching. The equipment locks if there is a difference between the two transcriptions.
f vérificatrice *f* — **d** Prüfgerät *n*; Lochprüfer *m* — **n** controleponsmachine *f*; controlemachine *f* — **e** verificadora *f* — **i** verificatrice *f*

14443 vermilion — Red, water-insoluble mineral pigment of sulphide of mercury, formerly obtained from *cinnabar, now usually produced by the reaction of mercury and sulphur.
f vermillon *m* — **n** vermiljoen *n* — **e** bermellón *m* — **i** vermiglione *m*

14444 vernier — Device to measure subdivisions of a scale. For a scale graduated in centimetres and tenths, it consists of a scale which slides alongside of the main scale, and on which a length of nine-tenths of a centimetre is subdivided into ten equal parts. Each vernier division is thus 0.09 of a centimetre.
f vernier *m*; nonius *m* — **d** Nonius *m* — **n** nonius *m*; vernier *m* — **e** nonio *m*; vernier *m* — **i** nonio *m*; verniero *m*

14445 vernis mou — See soft-ground etching.

14446 versicle — Typographic sign, ℣, used in mass books.
f verset *m* — **d** Versikelzeichen *n*; Verset *n* — **n** verset *n* — **e** versículo *m* — **i** versetto *m*

14447 verso — Reverse side of a sheet from the recto.
f verso *m*; revers *m* — **d** Rückseite *f* — **n** achterkant *m*; achterzijde *fm* — **e** reverso *m*; dorso *m* — **i** volta *f*; lato *m* in volta

14448 verso pages — See even pages.

14449 vertex — See node.

14450 vertical camera; vertical process camera — Camera in which the main movements and structure are in a vertical plane.
f appareil *m* de reproduction vertical; caméra *f* de reproduction verticale — **d** Vertikalreproduktionskamera *f* — **n** verticale camera *f*; verticale reproduktie-camera *f* — **e** cámara *f* (fotomecánica) vertical; cámara *f* de reproducción vertical — **i** macchina *f* fotografica verticale; apparecchio *m* verticale

14451 vertical cylinder press — See vertical press.

14452 vertical jobbing press — See vertical press.

14453 vertical job press — See vertical press.

14454 vertical press; vertical jobbing press; vertical job press; vertical cylinder press — Letterpress cylinder press in which the forme is rising and lowering in a vertical way.
f presse *f* verticale — **d** Vertikaldruckmaschine *f*; Vertikalmaschine *f*; Vertikalpresse *f* — **n** verticale pers *f* — **e** impresora *f* vertical; minerva *f* vertical — **i** pressa *f* verticale; macchina *f* verticale

14455 vertical process camera — See vertical camera.

14456 viaticum — In ancient Rome, the provision of necessaries for an official journey of a magistrate, in later times provisions for any journey.
f viatique *m*; viaticum *m* — **d** Viatikum *n*; Reiseunterstützung *f*; Zehrpfennig *m* — **n** viaticum *n*; reis- of teerpenning — **e** viático *m*; socorro *m* de viaje — **i** viatico *m*; sovvenzione *f* ai viaggiatori

14457 vibrating ink drum — See reciprocating ink drum.

14458 vibrating ink roller — See reciprocating ink drum.

14459 vibrating roller — See ductor.

14460 vibrator — See ductor.

14461 vice *GB*; **vise** *US* — Part of the slug composing machine in which the matrix line is justified laterally.
f étau *m* — **d** Schraubstockmechanismus *m*; Schraubstock *m* — **n** voorstuk *n* — **e** delantera *f*; marco *m* de la delantera; marco *m* de las quijadas; frente *f* SA; acial *m* Ar — **i** morsa *f*

14462 vice *GB*; **vise** *US* — Tool fixed to a work bench for securely gripping material to be worked on.
f étau *m* — **d** Bankschraube *f*; Bankschraubstock *m*; Schraubstock *m* — **n** bankschroef *fm* — **e** tornillo *m* de banco — **i** morsa *f*

14463 vice closing attachment — Part of the slug composing machine.
f dispositif *m* de fermeture des mâchoires — **d** Schraubstock-Schließschraubenvorrichtung *f*; Schließschraube *f* — **n** ontspanningsmechanisme *n* van het voorstuk — **e** dispositivo *m* de cierre de las quijadas — **i** dispositivo *m* di chiusura delle ganasce

14464 vice jaws — Part of the slug composing machine.
f mâchoires *fpl* (de l'étau) — **d** Schraubstock-

backen *fpl*; Formatbacken *fpl* — **n** formaat-bekken *mpl* — **e** quijadas *fpl* de la delantera — **i** ganasce *fpl* della morsa

14465 Victoria green — See malachite green.

14466 video display terminal; visual display terminal — Editing and also input terminal which presents a series of character images on a TV-like monitor. The operator can modify the displayed text by overstriking, deleting or inserting characters, words, phrases or sentences, or moving words, sentences, paragraphs or other blocks of copy from one position in the text stream to another.
n videoscherm *n*; correctiescherm *n* — **e** terminal *m* de pantalla video

14467 viewing cabinet — Illuminated desk or table in a smaal cabinet whereon colour film transparencies and photographic images can be viewed and examined by transmitted light, the illuminant usually having a colour temperature slightly above 3200 °K to show the best images.
f négatoscope *m*; cabine *f* d'examen; cabinet *m* à juger les valeurs de couleurs — **d** Farbwert-Beurteilungskabine *f*; Farbabmusterungskabine *f* — **n** kleurbeoordelingskast *fm* — **e** mesa *f* de análisis de color; pupitre *m* de tapa iluminada — **i** piano *m* illuminato per l'esame di pellicole fotografiche colorate

14468 vigesimoquarto — See twenty-fourmo.

14469 vignette — In an illustration, the gradual fading of the background away from the principal subject so that it finally blends into the unprinted paper. Usually prepared by air-brushing applied to the artwork, but may be added by handwork on continuous-tone negatives or positives, or by dot-etching halftone positives.
f dégradé *m* — **d** Verlauf *m* — **n** verloop *n* — **e** difuminado *m*; desvanecido *m*; borde *m* esfumado — **i** sfumatura *f* del fondo

14470 vignette — Relatively small decorative design or illustration, usually representing a vine or tendril, put on or just before the title page, or at the beginning or end of a chapter of a manuscript or book.
f vignette *f*; cabochon *m*; fleuron *m*; attribut *m* — **d** Vignette *f*; Zierstück *n* — **n** vignet *n*; sluitornament *n*; sluitstuk *n* — **e** viñeta *f*; piececita *f* de adorno; ornamento *m* de fantasía; adorno *m* de fantasía; marmosete *m*; monigote *m*; rombo *m* — **i** vignetta *f*

14471 vignette *v* — To etch or produce a vignetted halftone.
f dégrader — **d** abtönen; abschattieren — **n** verlopend maken; verloop aanbrengen — **e** degradar; desvanecer; matizar; esfumar; difuminar — **i** sfumare

14472 vignetted contact screen — See contact screen.

14473 vignetted halftone — Halftone plate on which the tones at one or more edges of the subject gradually fade away to pure white.
f simili *f* dégradée; dégradé *m* — **d** Autotypie *f*

mit Verlauf; Rasterklischee *n* mit Verlauf; verlaufende Autotypie *f* — **n** autotypie *f* met verloop — **e** medio tono *m* esfumado; medio tono *m* difumado; medio tono *m* difuminado; medio tono *m* desvanecido; medio tono *m* degradado; medio tono *m* viñeteado; grabado *m* de bordes esfumados — **i** illustrazione *f* con fondo sfumato

14474 Villard effect — Effect where a photographic latent image produced by high energy radiation, e.g., x-rays or corpuscular particles, or by high intensity light, can be destroyed by a second, longer exposure to moderate-intensity light. Named after its discoverer, Paul Villard (1907).
f effet *m* Villard — **d** Villard-Effekt *m* — **n** Villard-effect *n* — **e** efecto *m* de Villard — **i** effetto *m* Villard

14475 vinculum; bar — In mathematics, a horizontal line (a piece of metal rule) used with the radical to indicate a root, placed above two or more characters to unite them. Compositors dislike the use of the vinculum. It is therefore to be avoided in the notation of surds, its place being taken where necessary by brackets.
f trait *m* horizontal — **d** waagerechter Strich *m*; Linie *f* für imaginäre Einheiten — **n** horizontaal lijntje *n* — **e** vínculo *m* — **i** tratto *m* orizzontale; asta *f* orizzontale

14476 vinculum — See braces.

14477 vinegar acid — See acetic acid.

14478 vinylbenzene — See styrene resin.

14479 vinyl resins — Series of resins obtained from the polymerization of various vinyl compounds made from acetylene gas.
f résines *fpl* vinyliques — **d** Vinylharze *npl* — **n** vinylharsen *mnpl* — **e** resinas *fpl* vinílicas; resinas *fpl* de vinyl — **i** resine *fpl* viniliche

14480 virgule — See solidus.

14481 virtual image — Image seen at a point from which the rays of light appear to come to the observer, but do not actually do so, e.g., the image seen in a plane mirror. Such an image cannot be obtained on a screen placed at its apparent position, since the rays of light do not pass through that point.
f image *f* virtuelle — **d** virtuelles Bild *n*; scheinbares Bild *n* — **n** virtueel beeld *n* — **e** imagen *f* virtual — **i** immagine *f* virtuale

14482 viscoelastic — Pertaining to a substance having both viscous and elastic properties.
f visco-élastique — **d** visko-elastisch — **n** visco-elastisch — **e** viscoelástico — **i** viscoelastico

14483 viscometer — See viscosimeter.

14484 viscose sponge; cellulose sponge
f éponge *f* de viscose — **d** Viscoseschwamm *m* — **n** viscose-spons *fm* — **e** esponja *f* de viscosa; esponja *f* de celulosa; esponja *f* celulósica — **i** spugna *f* di viscosa

14485 viscosimeter; viscometer — Apparatus to measure shearing stress and the ensuing rate of shear from which viscosity, plastic viscosity, and yield value can be determined; used for materials that can be made to flow.

f viscosimètre *m* — **d** Viskosimeter *mn*; Zähigkeitsmesser *m* — **n** viscosimeter *m*; viscositeitsmeter *m* — **e** viscosímetro *m* — **i** viscosimetro *m*

14486 viscosity — Resistance of a fluid to flow; the opposite of fluidity. Exact measurements give an evaluation of this property in terms of a coefficient of viscosity which is defined as the shearing force divided by the rate of shear in simple fluids. In thixotropic materials such as printing inks the plastic viscosity is more properly used.

f viscosité *f* — **d** Viskosität *f*; Zähflüssigkeit *f*; Zähigkeit *f*; Dickflüssigkeit *f* — **n** viscositeit *f* — **e** viscosidad *f* — **i** viscosità *f*

14487 viscosity-velocity product — Product of multiplication of viscosity and printing velocity, at which picking occurs, in which the viscosity is expressed in kilo-poises and the velocity in cm/sec.

f formule *f* du VVP; produit *m* de viscosité-vitesse — **d** VVP-Formel *f*; Viskosität-Geschwindigkeitsprodukt *n* — **n** VVP-formule *f*; viscositeit/snelheidsprodukt *n* — **e** fórmula *f* VVP; producto *m* viscosidad/velocidad — **i** formula *f* VVP; prodotto *m* viscosità/velocità

14488 viscount — A size of note paper. See app. no. 3.

14489 vise — See vice.

14490 vise — See vice.

14491 visibility factor

f facteur *m* de visibilité d'une radiation — **d** spektrale Empfindlichkeit *f* (des Auges) — **n** helderheidsfactor *m*; spectrale gevoeligheid *f* (van het oog); ooggevoeligheid *f* — **e** factor *m* de visibilidad — **i** fattore *m* di visibilità (d'una radiazione); sensibilità *f* spettrale (dell'occhio)

14492 visible spectrum — Portion of the electromagnetic spectrum to which the retina is sensitive and by which we see; wavelength from about 400 to 750 mμ.

f spectre *m* visible — **d** sichtbares Spektrum *n* — **n** zichtbaar spectrum *n* — **e** espectro *m* visible — **i** spettro *m* visibile

14493 visiting card — See also business card.

f carte *f* de visite — **d** Visitenkarte *f*; Besuchskarte *f* — **n** visitekaartje *n* — **e** tarjeta *f* de visita — **i** biglietto *m* da visita

14494 visual display terminal — See video display terminal.

14495 visual display unit — See cathode ray tube display.

14496 visual examination

f jugement *m* visuel — **d** visuelle Beurteilung *f* — **n** visuele beoordeling *f* — **e** examen *m* visual — **i** esame *m* visivo

14497 visual scanner — See optical scanner.

14498 vitrauphanie paper; diaphanic paper — Thin transparent paper printed with transparent inks, which can be stuck on window panes. It produces the effect of stained glass.

f papier *m* pour diaphanies; papier *m* pour vitrauphanies — **d** Fensterpapier *n*; Diaphaniepapier *n* — **n** vitrauphanie-papier *n*; vensterpapier *n*; glasvensterpapier *n* — **e** papel *m* diáfano — **i** carta *f* per vetrofania; carta *f* da vetrofanie

14499 vivacity factor — Factor to indicate the ratio between the luminance and the degree of saturation of a colour.

f facteur *m* de vivacité — **d** Vivazitätsfaktor *m* — **n** levendigheidsfactor *m* — **e** factor *m* de vivacidad — **i** fattore *m* di vivacità

14500 viz — Abbrev. for Latin "videlicet", i.e., namely or that is to say. In proofreading it is substituted in reading aloud "namely". (Not italic; comma before.)

f c'est-à-dire; c.à.d. — **d** nämlich — **n** namelijk; n.l.; dat is te zeggen — **e** es decir; (es) a saber; s.e.a. — **i** cioè; vale a dire

14501 VM and P naphtha — Abbrev. for "varnish makers' and painters' naphtha", an American name for a petroleum fraction used mainly as a thinner in varnish and paint. More volatile than *white spirit, its distillation range usually between 90 and 165 °C.

14502 VMCH — Combination of polyvinyl chloride, polyvinyl acetate and a small percentage of maleic acid, used in lacquers to be applied to the inside of metal cans such as beer cans, for tight adhering.

14503 void — In magnetic ink printing, the absence of ink within the specified outline of the printed character.

f manque *m* (d'encre) — **d** Leere *f*; Fehlstelle *f* — **n** kale plek *fm*; weglating *f* — **e** defecto *m* de entintado; calva *f* — **i** vuoto *m*; mancanza *f* d'inchiostro

14504 volatile — Easily passing from a liquid into a gaseous state. Subject to rapid evaporation. Having a high vapour-pressure at room temperature.

f volatile — **d** flüchtig — **n** vluchtig; vervliegend — **e** volátil; volatilizable — **i** volatile

14505 volatile constituents — See volatile matters.

14506 volatile matters; volatile constituents — Matters able to evaporate rapidly.

f matières *fpl* volatiles; substances *fpl* volatiles — **d** flüchtige Stoffe *mpl*; flüchtige Bestandteile *mpl* — **n** vluchtige stoffen *mpl*; vluchtige bestanddelen *npl* — **e** constituyentes *mpl* volátiles; materias *fpl* volatilizables — **i** sostanze *fpl* volatili

14507 volatile store — Store in which data cannot be retained without a continuous power source, e.g., delay line memories.

f mémoire *f* non-permanente — **d** Energiespeicher *m*; energieabhängiger Speicher *m* — **n** vluchtig geheugen *n* — **e** memoria *f* no permanente — **i** memoria *f* non permanente

14508 volt — SI unit of electric potential difference. The absolute volt is the steady potential difference which must exist across a conductor which carries a steady current of one absolute ampere and which dissipates thermal energy at the rate of one watt. Symbol: V. In the SI: 1 V = W/A. One international volt = 1 $V_{int} \approx$ 1.0034 V.

f volt *m* — **d** Volt *m* — **n** volt *m* — **e** voltio *m* — **i**

volt *m*
14509 voltage — Electrical pressure which provides the motivation for current flow.
f voltage *m*; tension *f* — **d** Spannung *f*; elektrische Spannung *f* — **n** (elektrische) spanning *f*; voltage *m* — **e** voltaje *m*; tensión *f* (eléctrica) — **i** voltaggio *m*; tensione *f* elettrica
14510 voltage drop; drop in voltage; loss in voltage — Decrease in voltage due to resistance.
f chute *f* de tension; chute *f* de potentiel — **d** Spannungsabfall *m*; Potentialabfall *m* — **n** spanningsval *m* — **e** caída *f* de tensión; caída *f* de potencial — **i** caduta *f* di tensione; caduta *f* di potenziale
14511 voltage of mains — Voltage of the network.
f tension *f* du réseau — **d** Netzspannung *f* — **n** netspanning *f* — **e** tensión *f* de la red — **i** tensione *f* della rete
14512 voltage selector
f dispositif *m* pour le changement du voltage — **d** Spannungswähler *m* — **n** spanningskiezer *m*; netspanningskiesschakelaar *m* — **e** selector *m* de tensiones — **i** cambia *m* tensione
14513 voltage stabilizer — See stabilizer.
14514 voltampere — In an alternating current system, unit to express the product of the rms amperes and the rms volts.
f volt-ampère *m* — **d** Voltampere *n* — **n** volt-ampère *f* — **e** voltamperio *m*; voltiamperio *m* — **i** voltampere *m*
14515 voltmeter — Instrument to measure the electric tension in volts.
f voltmètre *m* — **d** Voltmeter *n*; Spannungsmesser *m* — **n** voltmeter *m* — **e** voltímetro *m* — **i** voltmetro *m*
14516 volume — Bound book, especially one of a series forming a single work. Abbrev. vol.
f volume *m*; livre *m* — **d** Band *m* — **n** band *m*; deel *n* — **e** volumen *m* — **i** volume *m*
14517 voucher copy; checking copy
f exemplaire *m* justificatif; justificatif *m* — **d**

Belegexemplar *n*; Belegnummer *f* — **n** bewijsexemplaar *n*; bewijsnummer *n* — **e** ejemplar *m* comprobante; ejemplar *m* justifiante; justificante *m* del anuncio; comprobante *m* del anuncio; comprobante *m* de inserción; comprobante *m* del aviso — **i** giustificativo *m*
14518 vpi paper — See vapour-phase inhibitor paper.
14519 vulcanite — See hard rubber.
14520 vulcanite fibre — See vulcanized fibres.
14521 vulcanization — Chemical treatment of crude rubber, by which its elasticity, hardness and durability are increased to meet the requirements needed for commercial use. Applied to rubber platemaking materials.
f vulcanisation *f* — **d** Vulkanisierung *f*; Vulkanisation *f* — **n** vulcanisering *f* — **e** vulcanización *f* — **i** vulcanizzazione *f*
14522 vulcanized fibres; vulcanite fibre — Leatherlike substance made by compression of layers of paper or cloth which have been treated with acids or zinc chloride, used chiefly for electric insulation.
f fibres *fpl* vulcanisées — **d** Vulkanfasern *fpl*; Vulkanfibern *fpl* — **n** fiber *mn* — **e** fibras *fpl* vulcanizadas — **i** fibre *fpl* vulcanizzate
14523 vulcanized oil — Product of the chemical reaction between a vegetable oil (linseed oil, rapeseed oil, cottonseed oil, corn oil) and sulphur chloride; used for ink rollers.
f huile *f* vulcanisée — **d** vulkanisiertes Öl *n* — **n** gevulcaniseerde olie *fm* — **e** aceite *m* vulcanizado — **i** olio *m* vulcanizzato
14524 vulcanized rubber — See hard rubber.
14525 vulcanizing press; rubber-stamp moulding press — Hydraulic press, the platen of which can be heated to form rubber plates from matrices.
f presse *f* à vulcaniser — **d** Vulkanisierpresse *f* — **n** vulcaniseerpers *f* — **e** prensa *f* moldeadora — **i** pressa *f* per vulcanizzare; pressa *f* vulcanizzatrice

14526 wafer — Adhesive disk to fasten envelopes.
f disque *m* de papier — **d** Papieroblate *f* — **n** papieren ouwel *m* — **e** oblea *f* de papel — **i** ostia *f* di carta

14527 waiting time *GB*; **latency** *US* — Time required to locate the first character of a location in store, calculated as access time minus work time.
f temps *m* d'attente — **d** Wartezeit *f* — **n** wachttijd *m* — **e** tiempo *m* de espera — **i** latenza *f*

14528 walking; walking off; walking away — Erosion and gradual disappearance of the printing image on the lithographic press-plate, caused by fogged negatives or insufficient density in positives, particularly in the highlights or in fine printing detail. Cannot be corrected.
f griller — **d** Schwinden *n* der Zeichnung; Nichthaften *n*; Abgehen *n* — **n** kaallopen *n* van de plaat — **e** debilitación *f* de la imagen en la plancha (litoffset) — **i** scomparsa *f* graduale dell'immagine

14529 walking away — See walking.

14530 walking off — See walking.

14531 wallet fold — Parallel fold (6 pages) of which the edges of the outer flaps meet each other at the front side.
f pli *m* en portefeuille — **d** Fensterfalz *m*; Altarfalz *m* — **n** venstervouw *fm* — **e** doblado *m* en paralelo; plegado *m* a corbata — **i** piega *f* a portafoglio

14532 wall map
f carte *f* murale — **d** Wandkarte *f* — **n** wandkaart *fm* — **e** mapa *m* mural — **i** carta *f* murale

14533 wall-paper — Paper, usually with printed decorative patterns in colour, to paste on and cover the walls or ceilings of rooms, hallways, etc.
f papier *m* peint; papier *m* tenture — **d** Tapete *f*; Tapetenpapier *n* — **n** behangselpapier *n* — **e** papel *m* pintado; papel *m* para empapelar — **i** carta *f* da parati

14534 wall-paper colour — Web, usually of advertisement matter, pre-printed in more than one colour, on one or both sides in a repeating design, enabling it to be run through presses of different cut-offs.
f bande *f* de papier à couleur de papier de tenture — **d** Tapetenmuster *n* — **n** baan *fm* papier gelijkend op behangselpapier — **e** color *m* de papel de empapelar — **i** nastro *m* di carta somigliante a carta da parati

14535 wall-paper printing — Printing on a web for subsequent insertion in another press, where the design is such that press to press register is unimportant.
f préimpression *f* pour encartage non repéré — **d** Hi-fi-Druck *m*; Druck *m* von Repetiermustern — **i** prestampa *f* su nastro da inserire; stampa *f* Hi-fi

14536 wall-paper printing — Printing process for the production of wall-papers capable of using pasteous inks drying with a distinct relief. Large drying rooms are required in which the freshly printed webs are festooned for drying and embossing or other finishing machines before re-reeling and cutting.
f impression *f* de papiers teints; impression *f* de papiers tenture — **d** Tapetendruck *m* — **n** behangselpapierdruk *m* — **e** impresión *f* de papel pintado; impresión *f* de empapelado — **i** stampa *f* di carta da parati

14537 wall-paper printing machine
f machine *f* pour l'impression papier peint; machine *f* pour l'impression papier tenture — **d** Tapetendruckmaschine *f* — **n** behangseldrukmachine *f* — **e** máquina *f* para imprimir papeles pintados — **i** macchina *f* per stampare carta da parati

14538 wall publicity
f publicité *f* murale — **d** Giebelreklame *f*; Wandreklame *f* — **n** muurreclame *fm* — **e** reclamo *m* mural; publicidad *f* mural — **i** reclame *f* murale; pubblicità *f* murale

14539 wall socket — See socket.

14540 warehousing — See storing.

14541 warm colour — Colour that has a warm psychological effect on the eye (red, orange and yellow), as opposed to blue, green and violet, which are known as cold colours.
f couleur *f* chaude — **d** Wärmfarbe *f*; warme Farbe *f*; wärmeanzeigende Farbe *f* — **n** warme kleur *fm* — **e** color *m* cálido — **i** colore *m* caldo

14542 warming furnace
f four *m* à réchauffer — **d** Anwärmeherd *m* — **n** stoof *fm* — **e** estufa *f* para calentar — **i** forno *m*

14543 warning light — Light to warn the operator of a machine regarding hazardous conditions in the engine.
f lampe *f* avertisseuse; lampe-témoin *f* — **d** Warnlicht *n* — **n** controlelampje *n*; waarschuwingslampje *n*; alarmlichtje *n* — **e** luz *f* indicadora de atención — **i** luce *f* d'allarme; luce *f* di sicurezza

14544 warp — Threads of a fabric running lengthwise or parallel to the selvedge.
f chaîne *f* — **d** Kette *f* — **n** ketting *f*; schering *f* — **e** urdimbre *f*; mallas *f* — **i** catena *f*

14545 warp *v* — To turn into a dome-shaped or disk-shaped form (of book covers).
f gondoler — **d** sich werfen; sich ziehen; sich krümmen; verziehen — **n** kromtrekken; trekken — **e** torcer; encovarse; alabearse — **i** imbarcarsi

14546 wash *v*
See wash *v* up.

14547 wash-drawing — Sketch or representation of an object in which the colour has been applied in flat washes with a single water-soluble pigment, this diluted to produce varying shades or effects.
f gouache *f*; dessin *m* au lavis; épure *f* au lavis; lavis *m* — **d** Tuschzeichnung *f* — **n** gewassen

tekening *f*

14548 washer — Beater in which cooked rags are defibred, bleached and washed. The obtained half-stuff is dropped into drainers for future use or pumped directly to beaters for further refinement.
f pile *f* laveuse — **d** Waschholländer *m* — **n** washollander *m* — **e** pila *f* lavadora — **i** pila *f* lavatrice

14549 washer — Thin, flat metal ring or a perforated plate used in joints or assemblies to ensure tightness, prevent leakage, or relieve friction (as under the head of a nut or bolt).
f rondelle *f* (de blocage) — **d** Scheibe *f*; Unterlegscheibe *f* — **n** sluitring *m*; onderlegring *m* — **e** arandela *f*; roldana *f* *Me*; golilla *f* — **i** rosetta *f*; rondella *f*

14550 washing — See tinting.

14551 washing fluid
f détergent *m* — **d** Waschmittel *n* — **n** wasmiddel *n* — **e** detergente *m* — **i** detergente *m*; detersivo *m*

14552 washing machine — See also roller washing machine.
f machine *f* à laver — **d** Waschmaschine *f* — **n** wasmachine *f* — **e** lavadora *f*; máquina *f* lavadora; máquina *f* de lavar — **i** lavatrice *f*; macchina *f* lavatrice

14553 washing out — Removal of an image coating which has been applied to the plate, such as developing ink or asphaltum, by use of turpentine or other solvents.
f enlevage *m* à blanc — **d** Auswaschen *n* — **n** uitwassen *n* — **e** lavado *m* — **i** lavaggio *m*

14554 wash-out developer
f révélateur *m* de dépouillement — **d** Auswaschentwickler *m*; Abätzentwickler *m* — **n** uitwasontwikkelaar *m* — **e** revelador *m* de mordido — **i** rivelatore *m* di spoglio

14555 washout solution — See also asphaltum (washout) solution.
f solution *f* d'enlevage — **d** Auswaschmittel *m*; Auswaschlösung *f* — **n** uitwasmiddel *n* — **e** líquido *m* de lavar — **i** soluzione *f* di lavaggio

14556 wash *v* **up; wash** *v* — To clean inking rollers, forme, stone, plate, etc., on a printing press.
f laver — **d** waschen — **n** wassen — **e** lavar; limpiar — **i** lavare

14557 waste *v* — To lose paper by carelessness, improper planning or poor machinery.
f gaspiller du papier — **d** verschwenden; verschmieren — **n** verspillen; bederven; verknoeien — **e** desperdiciar; consumir inútilmente; despilfarrar; quemar *Es* — **i** sprecare; produrre scarti

14558 waste paper — Paper discarded as used, not fit for use, or superfluous.
f bardot *m*; déchets *mpl* de papier — **d** Papierabfall *m*; Abfallpapier *n*; Druckausfall *m*; Makulatur *f* — **n** papierafval *n*; uitschot *n* — **e** papel *m* de recortes; papel *m* de desecho; recortaduras *fpl*; recortes *mpl* — **i** cartaccia *f*; carta *f* straccia; carta *f* di recupero; carta *f* da macero

14559 waste-paper dealer
f négociant *m* en vieux papiers — **d** Altpapierhändler *m* — **n** oudpapierhandelaar *m*; oudpapierhandel *m* — **e** comercio *m* de papel viejo — **i** commerciante *m* in carta da macero

14560 wasters — See run-in sheets.

14561 waste sheets — Sheets of paper spoiled during printing and binding.
f déchets *mpl* — **d** Makulaturbogen *mpl*; Schmutzbogen *mpl* — **n** misdruk *m*; bedorven vellen *npl* — **e** desperdicios *mpl*; retal *m* — **i** scarti *mpl*

14562 waste sheets — See overs.

14563 water — See damping solution.

14564 water absorption
f absorption *f* d'eau — **d** Wasseraufnahme *f* — **n** waterabsorptie *f*; wateropname *fm* — **e** absorción *f* de agua — **i** assorbimento *m* d'acqua

14565 water-absorption capacity
f capacité *f* d'absorption d'eau — **d** Wasseraufnahmefähigkeit *f* — **n** wateropnemend vermogen *n* — **e** capacidad *f* de la absorción de agua — **i** capacità *f* di assorbimento d'acqua

14566 water-based gravure ink
f encre *f* hélio à l'eau — **d** Wassertiefdruckfarbe *f*; Tiefdruckfarbe *f* auf Wasserbasis — **n** water-diepdrukinkt *m*; diepdrukinkt *m* op waterbasis — **e** tinta *f* al agua para huecograbado — **i** inchiostro *m* all'acqua per rotocalco

14567 water based ink — Ink containing a vehicle whose binder is water soluble or water dispersible.
f encre *f* à l'eau — **d** Druckfarbe *f* auf Wasserbasis; Wasserfarbe *f* — **n** inkt *m* op waterbasis — **e** tinta *f* al agua — **i** inchiostro *m* a base d'acqua

14568 water bath; bain marie — Device for heating chemical vessels without contact with flame, the heat being transmitted through water.
f bain-marie *m* — **d** Wasserbad *n* — **n** bain-marie *n*; waterbad *n* — **e** bañomaría *m*; baño *m* de María — **i** bagnomaria *m*

14569 water colour — Paint in which water instead of oil is used as a vehicle.
f couleur *f* à l'eau — **d** Wasserfarbe *f* — **n** waterverf *fm* — **e** aguada *f* — **i** acquerello *m*

14570 water-colour drawing; water-colour painting — Painting executed with water colours and differing from a *wash-drawing in that a full colour instead of a monochromatic effect is obtained by using several colours.
f peinture *f* à l'eau — **d** Tuschzeichnung *f* — **n** waterverftekening *f* — **e** dibujo *m* a la guada — **i** disegno *m* acquerellato

14571 water-colour drawing paper — See water-colour paper.

14572 water-colour painting — See water-colour drawing.

14573 water-colour paper; water-colour drawing paper — Hard-sized paper with coarse grain surface, used for painting with water colours.
f papier *m* pour aquarelle — **d** Aquarellpapier *n* — **n** aquarelpapier *n* — **e** papel *m* acuarela — **i** carta *f* per acquerelli

14574 water cooling
f refroidissement *m* par eau — **d** Wasserkühlung *f* — **n** waterkoeling *f* — **e** enfriamiento *m* de(l) agua — **i** raffreddamento *m* ad acqua

14575 water-damaged reel of paper
f bobine *f* détériorée par l'eau — **d** Rolle *f* mit Wasserschaden — **n** rol *fm* met waterschade — **e** bobina *f* con daño causado por el agua — **i** rotolo *m* di carta danneggiato dall acqua

14576 water-finished paper — Paper with a high finish obtained by damping one or both sides of the web of paper with a film of water, usually applied by water doctors, during its passage through the stack.
f papier *m* apprêté à l'eau; papier *m* calandré humide — **d** Papier *n* mit Feuchtglätte; feucht-geglättetes Papier *n* — **n** water-finished papier *n* — **e** papel *m* satinado al agua — **i** carta *f* lisciata ad umido

14577 water fountain — Metal trough on a lithographic press to hold the damping solution.
f bassine *f* de mouillage; bac *m* de mouillage; cuvette *f* de mouillage — **d** Wischwasserkasten *m*; Feuchtkasten *m*; Wischwassertrog *m* — **n** vocht-bak *m* — **e** fuente *f* del agua; depósito *m* del agua; artesa *f* de mojado — **i** vasca *f* per l'acqua di bagnatura

14578 water-fountain roller — In the water motion of a lithographic press, a non-ferrous roller which revolves in the fountain, to meter in conjunction with the doctor roller the water or fountain solution to the press-plate.
f rouleau *m* de bassine (de mouillage) — **d** Wasser-kastenwalze *f* — **n** vochtbakrol *fm*; bakrol *fm* — **e** rodillo *m* de mojado — **i** rullo *m* della vasca di bagnatura

14579 water free — See anhydrous.

14580 water glass — See sodium silicate.

14581 water-glass — See potassium silicate.

14582 Waterhouse stop — Diaphragm for photographic lenses devised by John Waterhouse in 1856, the aperture taking the form of a thin sheet of metal or paper with circular or irregular openings of a definite size, intended for insertion in a slot cut into the barrel of a lens.
f diaphragme *m* à vannes — **d** Waterhouse-Blende *f*; Waterhouse-Diaphragma *n* — **n** Water-house-diafragma *n*; Waterhouse-stop *m* — **e** dia-fragma *m* de láminas — **i** diaframma *m* di Water-house

14583 water-in-ink emulsion — Emulsion of ink and water in which the water is broken up into fine droplets surrounded by ink.
f émulsion *f* eau dans encre — **d** Wasser/Farbe-Emulsion *f* — **n** water-in-inkt emulsie *f* — **e** emulsión *f* agua y tinta — **i** emulsione *f* acqua-in-chiostro

14584 waterleaf paper — See unsized paper.

14585 water lines — The widely spaced parallel lines appearing in a laid paper.
d Wasserlinien *fpl* — **n** waterlijnen *fmpl* — **e** líneas *fpl* de agua — **i** righe *fpl* d'acqua

14586 waterlogged ink — Ink that contains enough emulsified water to reduce its flow and workability.

14587 watermark — Design in paper which can be seen by holding the paper against the light; formed during the formation of the moist sheet on the wire by means of a raised or recessed pattern. See also imitation watermark, rubber-stamp watermark, shaded watermark.
f filigrane *m*; marque *f* d'eau; marque *f* filigranée (trade-mark) — **d** (echtes) Wasserzeichen *n* — **n** (echt) watermerk *n*; lijnwatermerk *n*; egoutteur-watermerk *n* — **e** filigrana *f*; marca *f* de agua — **i** filigrana *f*

14588 watermark dandy — See dandy.

14589 watermark paper; trade-mark paper — Paper with a distinctive mark or design, to identify the source of supply and quality.
f papier *m* filigrané — **d** Wasserzeichenpapier *n*; Papier *n* mit Wasserzeichen — **n** papier *n* met watermerk — **e** papel *m* con marca de agua; papel *m* con filigrana — **i** carta *f* filigranata

14590 watermark roll — See dandy.

14591 water of crystallization — Water chemically combined in many crystallized substances which may be removed at about 100 °C, usually with loss of crystalline properties. Chemicals in crystal form usually contain a definite amount of water. For example, each molecule of crystallized copper sulphate, $CuSO_4$, contains $5H_2O$. When these five molecules of water are driven off by heat, the $CuSO_4$ becomes a white powder (dehydrated).
f eau *f* de cristallisation — **d** Kristallwasser *n* — **n** kristalwater *n* — **e** agua *f* de cristalización; agua *f* cristalina — **i** acqua *f* di cristallizzazione

14592 water pollution
f pollution *f* de l'eau — **d** Wasserverunreinigung *f*; Wasserverschmutzung *f* — **n** waterverontreini-ging *f* — **e** polución *f* del agua; ensuciamiento *m* del agua — **i** inquinamento *m* dell'acqua

14593 waterproof; water resisting
f résistant à l'eau; hydrofuge — **d** Feuchtigkeit ab-haltend; wasserecht — **n** waterbestendig; water-proef — **e** resistente a la humedad — **i** resistente all'acqua

14594 waterproof kraft — Kraft wrapping paper treated with paraffin, asphalt, or other materials to render it highly resistant to penetration by moisture.
f papier *m* kraft imperméable (à l'eau); papier *m* kraft hydrofuge — **d** wasserdichtes Kraftpapier *n* — **n** waterdicht kraft *n*; waterdicht kraftpapier *n* — **e** papel *m* kraft impermeable (al agua) — **i** carta *f* kraft impermeabile (al acqua)

14595 waterproof paper — Water-repellent paper prepared by combining two sheets of paper by means of asphalt, or by impregnating or coating it with micro-crystalline wax or other suitable water-proofing material.
f papier *m* hydrofuge; papier *m* imperméable — **d** wasserdichtes Papier *n* — **n** waterdicht papier *n* — **e** papel *m* impermeable (al agua) — **i** carta *f*

impermeabile (al acqua)

14596 water pump
f pompe *f* de l'eau — **d** Wasserpumpe *f* — **n** water-pomp *fm* — **e** bomba *f* de agua — **i** pompa *f* d'acqua

14597 water receptive — See hydrophilic.

14598 water repellent — See hydrophobic.

14599 water resistance — Papers for lithography require a fairly high degree of water resistance. Writing papers usually are heavily sized to permit the use of aqueous writing inks and to facilitate erasing. For packaging and for maps or posters subject to outdoor use, papers have been developed that do not disintegrate even when exposed to sea water.
f refus *m* de l'eau; résistance *f* à l'eau — **d** Wasserbeständigkeit *f*; Wasserfestigkeit *f*; Wasserechtheit *f*; Wasserwiderstandsfähigkeit *f* — **n** waterbestendigheid *f* — **e** resistencia *f* del agua — **i** resistenza *f* all'acqua; idroresistenza *f*

14600 water resisting — See waterproof.

14601 water resisting area
f région *f* refusant l'eau; région *f* imprimante — **d** wasserabstoßende Fläche *f*; Druckfläche *f* — **n** waterafstotend deel *n*; waterafstotend gedeelte *n*; drukkende partij *f* — **e** zonas *fpl* de rechazo del agua — **i** zone *fpl* idrorepellenti; zone *fpl* stampanti

14602 water softener
f adoucisseur *m* d'eau — **d** Wasserenthärter *m* — **n** waterontharder *m* — **e** suavizador *m* de agua — **i** addolcitore *m* d'acqua

14603 water soluble
f soluble dans l'eau — **d** wasserlöslich — **n** (in) water oplosbaar — **e** soluble al agua — **i** solubile in acqua

14604 water stains
f taches *fpl* d'eau; taches *fpl* d'humidité — **d** Wasserflecken *mpl*; Feuchtflecken *mpl* — **n** watervlekken *fmpl*; vochtvlekken *fmpl* — **e** manchas *fpl* de agua; manchas *fpl* de humedad — **i** macchie *fpl* d'acqua; macchie *fpl* d'umidità

14605 water supply
f débit *m* d'eau — **d** Wasserführung *f* — **n** water-toevoer *m*; watergeving *f*; waterregeling *f* — **e** abastecimiento *m* de agua — **i** portata *f* d'acqua

14606 water tight
f étanche à l'eau — **d** wasserdicht — **n** waterdicht — **e** estanco; a prueba de agua — **i** a prova d'acqua

14607 water tightness
f étanchéité *f* à l'eau — **d** Wasserdichtheit *f* — **n** waterdichtheid *f* — **e** impermeabilidad *f* — **i** tenta *f* all'acqua

14608 water vapour; aqueous vapour — Water particles evaporated into gaseous form.
f vapeur *f* d'eau — **d** Wasserdampf *m* — **n** waterdamp *m* — **e** vapor *m* de agua — **i** vapore *m* d'acqua; vapore *m* acqueo

14609 water-vapour permeability; aqueous vapour permeability — Rate at which moisture or water vapour passes through one square metre of paper in 24 hours at a definite temperature and humidity.
f imperméabilité *f* à la vapeur d'eau — **d** Wasserdampfdurchlässigkeit *f*; Durchlässigkeit *f* für Wasserdampf — **n** waterdampdoorlatendheid *f*; doorlatendheid *f* voor waterdamp — **e** permeabilidad *f* al vapor de agua — **i** permeabilità *f* dal vapore di acqua; permeabilità *f* al vapore acqueo

14610 water-vapour permeability test
f mesure *f* de la perméabilité à la vapeur d'eau — **d** Wasserdampfdurchlässigkeitsprüfung *f* — **n** waterdampdoorlatendheidsmeting *f*; meting *f* van de waterdampdoorlatendheid — **e** ensayo *m* de permeabilidad al vapor de agua — **i** prova *f* di permeabilità al vapore d'acqua

14611 water-vapour resistance — Property of paper or board to resist the passage of water vapour. Waterproof is not necessarily vapour-resistant. To be made vapour resistant, paper or board needs to be coated with a water resistant, film-forming material, such as asphalt, wax, synthetic resins, etc.
f étanchéité *f* à la vapeur d'eau — **d** Wasserdampdichtigkeit *f* — **n** waterdampdichtheid *f* — **e** impermeabilidad *f* al vapor de agua — **i** impermeabilità *f* al vapore d'acqua; impermeabilità *f* al vapore acqueo

14612 water-vapour transmission rate; moisture-vapour transmission rate *obs.* — Actual rate of water-vapour transmission used to compare water-vapour barrier, wrapping, or container materials.
f étanchéité *f* à la vapeur d'eau — **d** Wasserdampfdurchlässigkeit *f* — **n** waterdampdoorlatendheid *f* — **e** permeabilidad *f* al vapor de agua — **i** coefficiente *m* di trasmissione del vapore d'acqua

14613 Watt — Power required to do work at the rate of 1 joule per second. Unit of power in the MKSA system. Symbol: W.
f watt *m* — **d** Watt *n* — **n** watt *m* — **e** watio *m*; vatio *m*; watt *m* — **i** watt *m*

14614 watthour — Product of power in watts and time in hours. 1 Wh = 3600 joules. Abbrev. Wh. See also kilowatt-hour.
f watt-heure *f* — **d** Wattstunde *f* — **n** watt-uur *n* — **e** vatio-hora *m* — **i** wattora *f*

14615 wavelength; wave range — Distance between two successive crests in the undulation or movement of a ray of light, measured and expressed either in ångström units or millimicrons. Symbol: λ (lambda). It determines its colour, the sensitivity range of normal human vision for spectral colours ranging from 4000 å for extreme violet, to 7600 å for the extreme visible red in the spectrum. (Violet 4100, blue 4700, green 5200, yellow 6000, red 6500 å). There is a tendency to replace the ångström by the *nanometre.
f longueur *f* d'onde — **d** Wellenlänge *f*; Wellenbereich *m* — **n** golflengte *f* — **e** longitud *f* de onda — **i** lunghezza *f* d'onda

14616 waver; waver roller — Ink roller to distribute ink on a table or on other rollers by moving back and forth endways in addition to its rotary movement. On some cylinder presses it is placed diagonally on the ink table, the action of the table giving the vibratory motion.

f rouleau *m* bal(l)adeur; ballade *m* d'encrage — **d** Verreibwalze *f* — **n** inktverdeelrol *fm*; verdeelrol *fm*; heen- en weergaande inktrol *fm* — **e** rodillo *m* batidor; rodillo *m* distribuidor — **i** rullo *m* distributore

14617 wave range — See wavelength.

14618 waver levers — Pivoted armes or levers to effect the reciprocating movement of rollers or wavers.

f bras *mpl* des ballades d'encrage; leviers *mpl* de commande de ballade — **d** Steuerungsantrieb *m*; Steuereinrichtung *f*; Mechanismus *m* für die Erzielung der Axialbewegung; Exzenterhebel *m* — **e** palancas *fpl* de excéntrica — **i** leve *fpl* di comando

14619 waver roller — See waver.

14620 wave rule — See wavy rule.

14621 waviness — See cockling.

14622 wavy line — Line under a word to indicate that it should be in bold type.

f trait *m* ondulé — **d** Wellenlinie *f* — **n** gegolfde lijn *fm*; golflijn *fm* — **e** línea *f* ondulada; raya *f* ondulada — **i** linea *f* ondulata

14623 wavy-line rule — See wavy rule.

14624 wavy paper

f papier *m* gondolé — **d** welliges Papier *n* — **n** golvend papier *n* — **e** papel *m* ondulante; papel *m* ondulado — **i** carta *f* ondulata

14625 wavy rule; wavy-line rule; wave rule — Brass rule with a wavy face.

f filet *m* tremblé; filet *m* ondulé — **d** Wellenlinie *f* — **n** tremblélijn *fm*; golflijn *fm*; gegolfde lijn *fm* — **e** raya *f* ondulada; raya *f* serpentina; raya *f* tremble; raya *f* tremente; raya *f* viborita *Ar*; raya *f* trembleque *Cu*; raya *f* serpentinada *Ch* — **i** filetto *m* ondulato

14626 waxed paper — See paraffin paper.

14627 wax engraving — See cerotype.

14628 waxing paper — Base paper for the manufacture of waxed paper. Weight may vary from light-weight tissue to heavy-weight wrapping paper made from bleached or unbleached sulphite or kraft pulps.

f papier *m* à paraffiner — **d** Wachsrohpapier *n* — **n** grondpapier *n* voor geparaffineerd papier — **e** papel *m* soporte para encerar; papel *m* en bruto para encerar — **i** carta *f* supporto per carta cerata

14629 wax pick test — See wax test.

14630 wax seal

f cachet *m* de cire — **d** Lacksiegel *n* — **n** lakzegel *n* — **e** sello *m* de lacre — **i** sigillo *m* di ceralacca

14631 wax stencil — See duplicating stencil.

14632 wax test; wax pick test — Test of surface strength of paper by heating several grades of wax in stick form and pressing them against paper surface. After cooling the wax sticks are pulled away from paper vertically; the first of the grades of adhesiveness to cause paper surface to adhere to it determines the picking (by number) quality of the paper. See also Dennison wax test.

f essai *m* aux cires; essai *m* de résistance à l'arrachage avec cires — **d** Wachstest *m*; Wachsrupffestigkeitsprüfung *f* — **n** wasproef *fm* — **e** ensayo *m* con cera — **i** prova *f* di resistenza al distacco con cere

14633 wayzgoose; goose — Printer's holiday or celebration, the name sometimes given to printer's annual outing.

f fête *f* annuelle d'une maison d'imprimerie — **d** jährlicher Ausflug *m*; jährliche Landpartie *f* — **n** jaarlijks uitstapje *n* — **e** día *m* de fiesta anual — **i** giorno *m* di festa annuale; giorno *m* festivo annuale

14634 weak acid; inactive acid — Acid not completely ionized when dissolved in water, e.g., acetic acid, often used in lithography as a counter-etch.

f acide *m* faible — **d** schwache Säure *f* — **n** zwak zuur *n* — **e** ácido *m* flojo — **i** acido *m* debole

14635 wearing out — Breakdown of type that sinks under pressure.

f usure *f* — **d** Abnutzung *f* (der Schrift); Abquetschung *f* — **n** slijtage *f* — **e** desgaste *m*; aplastamiento *m* — **i** logorio *m*

14636 weather inks — Inks which turn red or blue, depending on the atmospheric humidity.

f encres *fpl* baromètres — **d** feuchtigkeitsanzeigende Farben *fpl* — **n** weer-inkten *mpl* — **e** tintas *fpl* barómetros — **i** inchiostri *mpl* indicatori dell'umidità atmosferica

14637 weather-o-meter — Instrument to pass light from an electric arc through a glass globe to accelerate the fading of colouring matters, which are periodically wet with water sprays to imitate rain.

14638 weave — Side movement of the web while passing through a press.

f flottement *m* latéral de la bande; déplacement *m* latéral de la bande — **d** Seitenbewegung *f* der Papierbahn — **n** zijdelingse beweging *f* van de papierbaan — **e** movimiento *m* de lado de la banda — **i** movimento *m* laterale del nastro di carta

14639 web; paper web; sheet — In reel-fed printing the material to be printed, from the point where it leaves the reel unit reaching the press delivery mechanism.

f bande *f* de papier; feuille *f* continue; nappe *f* — **d** Papierbahn *f*; Bahn *f*; Endlosbahn *f*; Papierstrang *m* — **n** papierbaan *fm*; baan *fm* — **e** banda *f* de papel; tira *f* de papel (continua); tira *f* continua de la bobina; cinta *f* de papel; papel *m* continuo; banda *f* continua; hoja *f* continua — **i** nastro *m* di carta; nastro *m*

14640 web *v*; **thread** *v* — To insert the paper web into the press.

f passer la bande dans la presse; introduire la bande dans la machine; enfiler le papier — **d** einziehen; einführen — **n** invoeren; doorvoeren — **e** pasar el papel; colocar el papel — **i** introdurre il

nastro
14641 web break — Parting of the paper on a rotary press while the press is running.
f casse *f* de la bande — **d** Papierreißer *m*; Reißer *m* — **n** baanbreuk *m* — **e** rotura *f* de papel; rotura *f* de la banda; ruptura *f* de papel; rompimiento *m* del papel — **i** rottura *f* del nastro
14642 web-break detector — Device to stop automatically the press when the web breaks.
f détecteur *m* de casse (de la feuille continue) — **d** Papierreißschalter *m*; Bahnrißschalter *m*; Stopperschalter *m* — **n** papierbreukschakelaar *m* — **e** detector *m* de rotura de la banda de papel — **i** dispositivo *m* d'arresto automatico in caso di rottura (del nastro)
14643 web cleaner — Rotating brush or suction device located ahead of the first printing unit in a rotary press which, when placed against the moving web, removes paper dust or lint from the web surface.
f nettoyeur *m* de bande; aspirateur *m* de poussière — **d** Entstaubungsanlage *f*; Staubabsaugvorrichtung *f*; Bandreiniger *m* — **n** stof-afzuiginrichting *f* — **e** aspirador *m* de polvo para bobinas — **i** aspirapolvere *m* dal nastro; spazzola *f* spolvatrice del nastro
14644 weber — Unit of magnetic flux. The magnetic flux which when linked with a single turn of wire, generates an electromotive force of one volt in the turn as it decreases uniformly to zero in one second. Abbrev. Wb. In the SI: 10^{-8} Wb = 1 Mx.
f wéber *m* — **d** Weber *m* — **n** weber *m* — **e** weber *m* — **i** weber *m*
14645 web fan out — Expansion of paper in the cross-machine direction rather than in the direction of web travel when it picks up moisture, occuring when the web becomes wider as it progresses through the press.
f extension *f* de la bande (de papier) — **d** Ausdehnung *f* der Papierbahn; Ausweitung *f* der Papierbahn — **n** uitzetten *n* van de papierbaan; breder worden *n* van de papierbaan — **e** ensanchamiento *m* de la banda de papel — **i** aumento *m* della larghezza del nastro
14646 web-fed; reel fed — Said of presses which receive the paper from a reel as distinct from separate sheets.
f alimenté par bobine; à bobine — **d** mit Zuführung des Verarbeitungsgutes in Rollenform — **n** van de rol gedrukt — **e** alimentado por rollos; de alimentación continua — **i** alimentato da bobina
14647 web-fed gravure press — See web-fed gravure rotary press.
14648 web-fed gravure rotary press; web-fed gravure press; web-fed rotogravure press
f rotative *f* en creux à bobines — **d** Tiefdruckrollenrotationsmaschine *f*; Tiefdruckrotationsmaschine *f* — **n** diepdruk-rotatiepers *f* — **e** máquina *f* rotativa de huecograbado; rotativa *f* de huecograbado — **i** rotativa *f* rotocalco da bobina

14649 web-fed machine — See web machine.
14650 web-fed offset rotary (press) — See rotary offset press.
14651 web-fed press — See web machine.
14652 web-fed rotary letterpress machine
f rotative *f* typo à bobines — **d** Buchdruck-Rollenrotationsmaschine *f*; Buchdruck-Rotationsmaschine *f* — **n** boekdrukrotatiepers *f* — **e** máquina *f* rotativa tipográfica; prensa *f* rotativa tipográfica; rotativa *f* tipográfica — **i** rotativa *f* tipografica da bobina
14653 web-fed rotogravure press — See web-fed gravure rotary press.
14654 web lead — Amount of paper in the press when threaded.
f longueur *f* de papier en machine — **d** Papierlänge *f* in der Maschine — **n** lengte *f* van de papierbaan in de machine — **e** longitud *f* de papel en máquina — **i** lunghezza *f* del nastro introdotto nella macchina
14655 web-lead rollers — Pair of grated idler rollers located between printing units on blanket-to-blanket presses in line with lower blanket cylinder to support the web between units, preventing wrinkling and controlling web wrap. Individual rollers are used to guide web when bypassing individual printing units.
f rouleaux *mpl* conducteurs — **d** Leitwalzen *fpl* — **n** geleiderollen *fmpl* — **e** rodillos *mpl* de conducción; rodillos *mpl* de guía — **i** rulli *mpl* di conduzione; rulli *mpl* di guida
14656 web machine; web-fed machine; web press; web-fed press; reel-fed press — Cylinder printing machine in which the paper is fed from a continuous reel.
f rotative *f* à bobine; presse *f* à bobine; presse *f* rotative; presse *f* à papier continu; machine *f* à marge en bobine; machine *f* à papier en bobines — **d** Rollenrotationsmaschine *f*; Rollenpresse *f* — **n** (rollen-)rotatiemachine *f*; rotatiepers *f* — **e** máquina *f* alimentada por bobinas; máquina *f* para imprimir papel continuo; máquina *f* de alimentación continua — **i** macchina *f* alimentata da bobina
14657 web offset press — See rotary offset press.
14658 web paper — Printing paper in a continuous web or reel.
f papier *m* continu; papier *m* en bobines; papier *m* en rouleau; papier *m* sans fin — **d** Rollenpapier *n* — **n** papier *n* aan rollen — **e** papel *m* en bobina; papel *m* en rollo; bobina *f* de papel — **i** carta *f* in bobina; carta *f* continua
14659 web paster — See flying paster.
14660 web press — See web machine.
14661 web tension — Amount of pull or tension applied in the direction of travel of a web by the mechanical components of a web-fed printing or fabricating press.
f tension *f* de la bande — **d** Bahnspannung *f*; Zugspannung *f* — **n** baanspanning *f*; spanning *f* van de papierbaan — **e** tensión *f* de la banda; tensión *f* del papel — **i** tensione *f* del nastro

14662 web tension control
f régulation f de la tension du papier — d Bahn-spannungsregelung f; Papierspannungsregelung f — n regeling f van de papierspanning; regeling f van de baanspanning — e arreglo m de la tensión de la banda — i regolazione f della tensione del nastro

14663 wedding bristol — Superfine, hard sized, pasted bristol, plate, kid, or linen-finished, used for steel-engraved wedding announcements and correspondence.
f bristol m pour billets de mariage — d Bristol-karton m für Vermählungsanzeigen — n karton n voor huwelijksaankondigingen — e cartulina f para participaciones; cartulina f brístol superfina — i cartoncino m per partecipazione di nozze

14664 weddings — Superfine, hard sized paper, plate or kid-finished, with good folding qualities, used for engraved wedding and other social announcements.
f faire-parts mpl; papiers mpl pour faire-part; billets mpl de mariage — d Anzeigenpapiere npl; Heiratsanzeigen fpl — n papier n voor familie-drukwerk; huwelijkscirculaires fmpl — e papeles mpl superfinos para invitaciones; cartulinas fpl superfinas para invitaciones — i carta f per partecipazione di nozze

14665 wedge — Part of the Monotype caster.
f coin m — d Setkeil m — n letterwig fm — e cuña f — i cuneo m

14666 wedge serif — Unbracketed, wedge shaped serif as in some old style and display letters.
f empattement m cunéiforme — d keilförmiger Serif m — n wigvormige schreef fm — e remate m cuneiforme — i terminale m cuneiforme; grazia f cuneiforme

14667 weekly — Publication issued once a week.
f revue f hebdomadaire; hebdomadaire m — d Wochenblatt n; Wochenschrift f — n weekblad n — e semanario m; hebdomadario m; revista f hebdomadaria; periódico m semanal; publicación f semanal; papel m semanal — i settimanale; pubblicazione f settimanale

14668 weft; woof; filler; filling — Cross-wise threads of a fabric, i.e., the threads that run from selvedge to selvedge.
f trame f — d Schuß m; Einschlag m — n in-slag m — e trama f — i trama f

14669 Weigert effect — Photographic effect causing fine grain silver halide emulsions to be double refracting when they first get a pre-exposure with normal light and than with linear polarized light, i.e. in this way can only have an after-exposure with in one direction oscillating light. By this treatment the emulsions are enabled to polarize the light by themselves.
f effet m Weigert — d Weigert-Effekt m — n Weigert-effect n — e efecto m Weigert — i effetto m Weigert

14670 weight fount — See minimum weight fount.

14671 weight of face — Comparative colour value of type faces when printed. The relative weight are extra light, light, semi-light, medium, semi-bold, extra bold, ultra bold.
f couleur f du caractère — d Bild n der Schrift — n vetheid f van de letter; kracht fm van de letter; "kleur" fm van de letter — e grado m de negrura de la letra; negrura f; negror m; grosor m; espesor m; color m; grueso m; mancha f — i tono m dell'occhio dei caratteri

14672 weight of type — Four square inches of type (about 6.45 cm²), composed solid weight approximately one pound (about 0.45 kg).
f poids m du caractère — d Schriftgewicht n — n gewicht n van de letter — e peso m del tipo — i peso m del tipo; peso m del carattere

14673 weights and measures — See app. no. 7.
f poids mpl et mesures fpl — d Gewichte npl und Maße npl — n maten fpl en gewichten npl — e pesas y medidas fpl — i pesi mpl e misure fpl

14674 Weinland effect — Photographic effect by which the sensitivity of a layer for exposures of less luminance can be increased by a short pre-exposure with diffused light of a higher intensity.
f effet m Weinland — d Weinland-Effekt m — n Weinland-effect n — e efecto m Weinland — i effetto m Weinland

14675 well US — Space under the type case in a composing frame to put away copy.
n kopijvak n — i spazio m sotto il banco da composizione

14676 well-closed formation; close formation — Even, regular formation of the fibres, producing a uniform appearance in a sheet, as opposed to a *wild formation which gives a mottled, cloudy appearance.
f épair m fondu — d gut geschlossene Durch-sicht f — n goed gesloten doorzicht n — i formazione f compatta

14677 wet beating — Treatment of paper pulp in which the fibres are strongly fibrillated. Opposite of free beating.
f raffinage m gras — d schmierige Mahlung f — n vette maling f — e refino m graso — i raffina-zione f ad umido

14678 wet board machine — See intermittent board machine.

14679 wet broke — Broke accumulated on the wet end of the paper or board machine.
f cassés mpl de fabrication humides — d Naß-ausschuß m; nasse Fabrikationsabfälle mpl — n natte afval n; afval n bij de natpartij — e des-perdicios mpl de la sección húmeda — i fogliacci mpl umidi

14680 wet-collodion plate
f plaque f au collodion humide; cliché m au collodion humide — d nasse Kollodiumplatte f — n natte collodiumplaat fm — e placa f al colodión húmedo; placa f húmeda — i lastra f al collodio umido

14681 wet collodion process — See collodion process.

14682 wet colour printing; wet-in-wet printing; wet-on-wet printing; wet printing — Procedure of colour printing in which colours are successively and rapidly applied before the previous impression has dried.

f impression *f* à l'humide; impression *f* de fraîche à fraîche; impression *f* simultanée — **d** Naß-in-Naßdruck *m*; Naß-auf-Naßdruck *m* — **n** nat-in-natdruk *m* — **e** impresión *f* húmedo sobre húmedo; impresión *f* en húmedo; impresión *f* a color húmedo — **i** stampa *f* umido-su-umido

14683 wet end — Part of the paper machine between the head box and the drier section. See also Fourdrinier machine.

f section *f* humide — **d** Naßpresse *f*; Naßpartie *f*; Naßpressenpartie *f*; Naßteil *m* — **n** natpartij *f* — **e** lado *m* húmedo; parte *f* húmeda; sección *f* húmeda; amplias zonas *fpl* húmedas; extremo *m* — **i** parte *f* umida

14684 wet felt

f feutre *m* coucheur — **d** Naßfilz *m* — **n** natvilt *n* — **e** fieltro *m* húmedo; fieltro *m* de prensa húmeda — **i** feltro *m* umido

14685 wet-in-wet printing — See wet colour printing.

14686 wet mounting — Placing exposed pigment paper to the copper cylinder in gravure printing by soaking the sheet for a few seconds in water and squeezing it into position by hand with a squeegee, as distinguished from *dry mounting where the water is applied as the paper approaches the cylinder.

f application *f* humide manuelle — **d** nasse Übertragung *f*; Handübertragung *f* — **n** natte overdraging *f* — **i** montaggio *m* ad umido

14687 wet-on-wet printing — See wet colour printing.

14688 wet pick — Occurrence of picking in the presence of water introduced by the printing process. It results from the loss of surface strength of the paper due to damping of the surface before impression.

f arrachage *m* dû à l'eau — **d** Feuchtrupfen *n*; Naßrupfen *n* — **n** vochtpluk *m* — **e** repelado *m* húmedo; picado *m* húmido — **i** strappo *m* (superficiale) a umido

14689 wet plate — Photographic plate produced by the wet collodion process. See also dry plate.

f plaque *f* (au collodion) humide — **d** Naßplatte *f*; (nasse) Kollodiumplatte *f* — **n** natte plaat *fm*; (natte) collodiumplaat *fm* — **e** placa *f* húmeda; placa *f* de colodión (húmedo) — **i** lastra *f* umida; lastra *f* al collodio (umido)

14690 wet-plate process — See collodion process.

14691 wet printing — See wet colour printing.

14692 wet streaks — Dark streaks running parallel to grain direction of sheet, caused by uneven drying across the paper web.

f plages *fpl* humides — **d** Feuchtstreifen *mpl* — **n** vochtstrepen *fmpl* — **e** rayas *fpl* de humedad — **i** fasce *fpl* umide

14693 wet strength paper — Paper treated to increase its wet strength retention.

f papier *m* résistant à l'état humide; papier *m* wet strength — **d** naßfestes Papier *n* — **n** wet-strength papier *n*; natsterk papier *n* — **e** papel *m* resistente a la humedad — **i** carta *f* resistente a umido

14694 wet-strength retention — Ratio between a given strength property of paper or board in the wet state and that of the same paper in the dry state measured under standard conditions.

f indice *m* de résistance à l'état humide — **d** Naßfestigkeitsprozentsatz *m*; Naßfestigkeitszahl *f*; spezifische Naßfestigkeit *f* — **n** natsterktepercentage *n*; natsterkte *f* in procenten — **e** índice *m* de la resistencia en húmedo — **i** resistenza *f* relativa allo stato umido

14695 wet stuff — Papermaker's term for gelatinized or well-hydrated paper stuff.

f pâte *f* grasse — **d** schmierig gemahlener Stoff *m* — **n** vet gemalen stof *fm* — **e** pasta *f* refinada grasa — **i** pasta *f* grassa

14696 wettability — Relative affinity of a liquid for a solid surface. It is measured in terms of the contact angle formed between the liquid and the solid. If the contact angle is zero, complete wettability is said to occur. If the contact angle is greater than 90 °, non-wettability exists.

f mouillabilité *f* — **d** Benetzbarkeit *f* — **n** bevloeiing *f*; bevloeibaarheid *f* — **e** humectabilidad *f*; mojabilidad *f* — **i** umettabilità *f*; bagnabilità *f*

14697 wet-thumb proof — Old-fashioned test consisting in pressing the slightly moistened thumb firmly against the paper surface for a few seconds, then removing it with a jerk. If two papers are compared in the test, the one transferring more dust to the thumb will usually have greater tendency to deposit on the blanket and to cause picking.

f essai *m* au pouce — **d** Daumenprobe *f* — **n** duimproef *fm*; proef *fm* met de natte duim — **i** prova *f* del pollice umido

14698 wetting — Surrounding the minute particles of pigment (or resin) with the "wet" solvent during the ink making procedure. Pigments will wet out easily well, in general, grind easier, form better ink bodies and have a finer dispersion. Also the further thinning of an ink or varnish by the addition of more solvent and/or diluents.

f mouillage *m* — **d** Benetzung *f* — **n** bevloeiing *f* — **e** mojado *m*; mojadura *f* — **i** umettamento *m*; bagnatura *f*

14699 wetting agent — Chemical compound capable of lowering the surface tension of water and water solutions and increasing their wetting power. Often *dispersing agents.

f agent *m* mouillant; mouillant *m*; agent *m* tensioactif — **d** Netzmittel *n*; Befeuchtigungsmittel *n* — **n** bevloeiingsmiddel *n*; bevloeier *m* — **e** agente *m* humector; agente *m* humectante; humectativo *m*; producto *m* humedecedor — **i** agente *m* umettante; agente *m* umidificante; agente *m* bagnante; agente *m* di bagnatura

14700 Whatman('s) paper — English hand-made drawing paper of first quality.

14701 wheat straw
f paille *f* de blé — **d** Weizenstroh *n* — **n** tarwestro *n* — **e** paja *f* de trigo; paja *f* trigaza — **i** paglia *f* di frumento

14702 wheel and pinion folder — See flying tuck.

14703 wheel printer — Line printer which prints its characters from the rim of a wheel around which are disposed the types for the alphabet available. Computer term.
f imprimante *f* à roues — **d** Typenraddrucker *m*; Drucker *m* mit Typenrad — **n** schrijfwieldrukker *m*; wieldrukker *m*; drukker *m* met schrijfwiel — **e** impresora *f* de ruedas — **i** stampatrice *f* a ruote

14704 when room — See punk.

14705 whip — See whip hand.

14706 whip hand; whip; fire eater *obs.* — Hard and accurately working compositor. See also swift.
f piocheur *m* — **d** schnell, gewandt, fehlerfrei arbeitender Setzer *m* — **n** harde werker *m*; werkezel *m* — **e** persona *f* muy trabajadora; buscavida(s) *m*; burro *m* de carga — **i** compositore *m* accurato en veloce

14707 whipping — See whip stitching.

14708 whip stitching; whipping; overcasting; overcast binding; oversewing — Oversewing single or damaged leaves of a book, that have no fold at the back. The leaves are stitched closely together in sections; these sections can then be sewed like folded sheets in the usual way. When bound the leaves of such a book will not open perfectly flat.
f reliure *f* en surjet — **d** Überheften *n*; seitliche Fadenheftung *f* — **n** driegen *n* — **e** sobrehilado *m*; sobrecostura *f*; cosido *m* a diente de perro — **i** cucitura *f* a punto saltato; cucitura *f* a con file refe in piano

14709 whirler; plate whirler; plate coating machine — Apparatus to coat metal and glass plates with light-sensitive solutions by revolving the plate horizontally under the influence of heat.
f tournette *f* — **d** Schleuder *f*; Plattenschleuder *f*; Schleuderapparat *m* — **n** slingerapparaat *n* — **e** aparato *m* giraplanchas; torniquete *m* (de planchas); turneta *f*; centrífuga *f*; giraplanchas *m*; máquina *f* para sensibilizar planchas — **i** tournette *f*; centrifuga *f*

14710 whiskers — See lint.

14711 white; white space — Unprinted part of paper.
f blanc *m* — **d** weiße Stelle *f*; blankgeschlagene Stelle *f* — **n** wit *n*; witte plek *f*; onbedrukte plek *f* — **e** blanco *m*; espacio *m* (en) blanco; gran blanco *m* — **i** bianco *m*; spazio *m* in bianco

14712 white caustic — See sodium hydroxide.

14713 white lead — Name primarily applied to basic lead carbonate, but also used for basic lead sulphate (white lead sulphate) and basic lead silicate (white lead silicate). See also flake white.

14714 white lead — See flake white.

14715 white light — Light approaching daylight in spectral quality, containing all the constituent rays of solar light.

f lumière *f* blanche — **d** weißes Licht *n* — **n** wit licht *n* — **e** luz *f* blanca — **i** luce *f* bianca

14716 white line; blank line — Line of quads, or a line consisting mostly of quads.
f ligne *f* de cadrats; ligne *f* de blanc — **d** Blindzeile *f*; blinde Zeile *f*; freie Zeile *f*; weiße Zeile *f*; Quadratenzeile *f*; Unterschlag *m* (blank line at the end of a page) — **n** blanco regel *m*; regel *m* wit — **e** línea *f* de blancos; línea *f* en blanco — **i** linea *f* bianca; riga *f* bianca; riga *f* morte

14717 white liquor
f lessive *f* neuve — **d** Weißlauge *f*; Kochlauge *f* — **n** witloog *fm* — **e** lejía *f* blanca; lejía *f* nueva — **i** liscivio *m* bianco

14718 white metal — See babbitt metal.

14719 whiteness
f blancheur *f* — **d** Weiße *f*; Weißgehalt *m*; Weißgrad *m* — **n** witheid *f* — **e** blancura *f* — **i** bianchezza *f*

14720 white *v* **out** — To space out composed matter to fill the allotted space, or to improve the typographic effect.
f blanchir; jeter du blanc — **d** austreiben — **n** uitdrijven — **e** distanciar — **i** allargare; allontanare

14721 white-out lettering — See reverse lettering.

14722 white page — See blank page.

14723 white press waste — See newsprint waste.

14724 white print — See diazoprint.

14725 white size — See acid size.

14726 white space — See white.

14727 white spirit — *SBP spirit boiling approx. 150-200 °C, sometimes called mineral turpentine.
f white spirit *m* — **d** White Spirit *m*; Terpentinölersatz *m* — **n** terpentine *f*; kunstterpentijn *m*; lakbenzine *fm*; peut *m* *fam.* — **e** aguarrás *m*; espíritu *m* blanco; white spirit *m* — **i** acqua *f* ragia minerale

14728 white spots — See sun spots.

14729 white sulphite paper — Paper with good strength and appearance, made from bleached sulphite pulp.
f papier *m* au bisulfite blanchi — **d** gebleichtes Sulfitpapier *n* — **n** gebleekt sulfietpapier *n* — **e** papel *m* de bisulfito blanqueado — **i** carta *f* bianchita al solfito

14730 white water; save-all water; pulp water; back water; size water; loose water; free water — Water of a paper mill separated from the stock or pulp suspension, either on the paper machine or accessory equipment, such as thickeners, washers, and save-alls, and also from pulp grinders. It carries an amount of fibre and varying amounts of fillers, dyestuffs, etc.
f eau *f* blanche; eau *f* de retour; eau *f* collée — **d** Weißwasser *n*; Siebwasser *n*; Kreidewasser *n*; Rückwasser *n*; Seitenwasser *n*; Abwasser *n* — **n** witwater *n* — **e** agua *f* blanca; agua *f* de retorno; agua *f* colada — **i** acqua *f* bianca; acqua *f* di sottotela

14731 whiting; Paris white — Finely ground, naturally occurring *calcium carbonate, used with water to clean plates. Not to be confused with precipitated chalk.

f blanc *m* de Meudon — **d** Schlämmkreide *f* — **n** krijt *n* — **e** blanco *m* de Meudón — **i** gesso *m* in polvere

14732 whole-bound — See full-bound.

14733 whole frame — Stand to hold two pairs of type cases side by side, with space between for trays, rack for five pairs, and a small cupboard at side.

f rang *m* carcasse — **e** comodín *m* — **i** banco *m* da composizione con due casse

14734 whole imperial — A size of board. See app. no. 3.

14735 whole stuff — Pulp ready for use in paper-making.

f pâte *f* raffinée — **d** Ganzstoff *m*; Ganzzeug *n* — **n** heelstof *fm* — **e** pasta *f* refinada — **i** pasta *f* raffinata

14736 widow — Left-over line at the bottom of a column, where a paragraph is started with an indent. Sometimes a *wrong overturn or reference to a brief including line or paragraph.

f ligne *f* à voleur — **d** Hängezeile *f*; Waisenkind *n* — **n** weeskind *n*; weesjongen *m* — **e** viúda *f*; línea *f* trunca; línea *f* corta; línea *f* quebrada; fin *m* de párrafo — **i** capoverso *m* al piede di una colonna

14737 width; reel width — Dimension of a web of paper or board measured in the direction across the machine.

f largeur *f*; laize *f* des bobines (de grand format) en centimètres — **d** Rollenbreite *f* — **n** rolbreedte *f*; baanbreedte *f*; breedte *f* van de rol papier — **e** ancho *m* de la banda de papel; anchura *f* de bobina; ancho *m* de bobina — **i** larghezza *f* (della bobina); larghezza *f* del nastro

14738 width; set width; setwise; set size — Horizontal dimension of type. The relative widths of type character are: ultra-condensed, extra-condensed, condensed, semi-condensed, medium, semi-expanded, expanded, extra-expanded, ultra-expanded.

f chasse *f*; largeur *f*; épaisseur *f*; valeur *f* de set — **d** Kegelweite *f*; Dickte *f*; Schriftbreite *f* — **n** dikte *f*; breedte *f*; wijdte *f*; letterbreedte *f* — **e** ancho *m* del tipo; anchura *f* del tipo; espesor *m* del tipo; grosor *m* del tipo — **i** larghezza *f* del fusto del carattere; spessore *m* del carattere

14739 width card — See width plug.

14740 width of column — See column width.

14741 width of paper

f largeur *f* de papier — **d** Papierbreite *f* — **n** papierbreedte *f* — **e** anchura *f* del papel — **i** larghezza *f* della carta

14742 width plug; width card — Card, panel or box to be inserted into a keyboard or typesetter, to ascertain the width (in units) for each particular character.

14743 wild formation — See cloudy formation.

14744 willow; willowing machine; rag willow; rag thrasher; devil — Machine, usually combined with the duster, consisting of two rotating drums with fixed spikes to tease out the rags before they pass to the duster.

f batteuse *f* à chiffons; loup *m*; diable *m* — **d** Haderndrescher *m*; Hadernwolf *m*; Lumpendrescher *m*; Zerreißwolf *m* — **n** lompenwolf *m* — **e** batidor *m* de trapos; batán *m* de trapos; lobo *m*; diablo *m*; rasgador *m*; cernedor *m*; cedazo *m* para trapos

14745 willowing machine — See willow.

14746 wind *v* — To force air into a lift of paper by shaking.

f aérer — **d** aussetzen — **n** aan de lucht blootstellen — **e** exponer al aire; airear — **i** dar aria

14747 winder — Machine to wind paperboard on a mandrel to the desired number of plies for a drum side-wall. Generally, any device to wind material into coils or reels.

f bobineuse *f*; enrouleuse *f* — **d** Rollapparat *m*; Umroller *m*; Aufwickler *m* — **n** bobineuse *f*; opwikkelaar *m*; opwikkelmachine *f* — **e** bobinador *m*; bobinadora *f* — **i** arrotolatore *m*

14748 winder — See wrap round.

14749 window envelope — Envelope with a part made transparent by oil treatment, or with a portion cut out over which a piece of transparent material, like glassine, is pasted, so that an address written on letterhead may be seen, to avoid typing of address on envelope.

f enveloppe *f* à fenêtre (vernissée); enveloppe *f* vitrifiée; enveloppe *f* à panneau transparent; enveloppe *f* à fenêtre (rapportée) — **d** Fensterumschlag *m*; Fensterkuvert *n*; Fensterbriefumschlag *m*; Fensterbriefhülle *f* — **n** vensterenvelop(pe) *fm*; venstercouvert *n*; envelop(pe) *fm* met ingeplakt venster — **e** sobre *m* con ventana; sobre *m* de ventana; sobre *m* con ventanilla; sobre *m* transparente — **i** busta *f* con finestra

14750 wind up — See wrap round.

14751 wing dividers — See dividers.

14752 wing nut; butterfly nut; fly nut; thumb nut

f écrou *m* à oreilles; écrou *m* ailé; écrou *m* papillon — **d** Flügelmutter *f* — **n** vleugelmoer *fm* — **e** tuerca *f* de aletas; tuerca *f* de mariposa — **i** dado *m* ad alette

14753 wink — Work study term for 0.0005 minute, the smallest time unit.

f wink *m* — **d** Wink *m* — **n** wink *m* — **e** wink *m* — **i** wink *m*

14754 winkle bag printer — Under-equipped and inefficient printing house.

f boîte *f* — **d** Quetsche *f*; unvollkommen eingerichtete Druckerei *f* — **n** slecht ingerichte drukkerij *f* — **e** pequeña imprenta *f*; casa *f* de segunda fila — **i** stamparia *f* male attrezzata; stamparia *f* insufficientemente attrezzata

14755 wipe *v* — In rotogravure printing, to clean the engraved cylinder of surplus ink.

f racler — **d** rakeln — **n** rakelen; afrakelen — **e** quitar con racleta — **i** asportare con la racla

14756 wipe *v*; **wipe** *v* **off**

f essuyer; enlever — **d** abwischen — **n** afwissen; wissen; afvegen — **e** limpiar; enjugar — **i** asciugare

14757 wipe *v* off — See wipe *v*.

14758 wipe-off paper — See plate-wiping paper.

14759 wipe-on offset plate — See wipe-on plate.

14760 wipe-on plate; wipe-on offset plate — Aluminium offset plate with an unsensibilized coating which is sensibilized by hand with a sponge immediately before exposure of the plate.
f plaque *f* offset wipe-on; plaque *f* wipe-on — **d** Wipe-on-Platte *f*; handsensibilisierte Platte *f* — **n** wipe-on-plaat *fm*; met de hand gesensibiliseerde plaat *fm*; met de hand lichtgevoelig gemaakte plaat *fm* — **e** plancha *f* sensibilizada a mano — **i** lastra *f* offset wipe-on

14761 wiper, knife — See knife wiper.

14762 wiping blade — See doctor blade.

14763 wiping-off paper — See plate-wiping paper.

14764 wiping paper — See plate-wiping paper.

14765 wire, binding — See stitching wire.

14766 wire bobbin; bobbin — Core to hold stitching wire (thread, etc.).
f bobine *f* (à fil de fer) — **d** Drahtspule *f* (für Heftmaschinen); Spule *f* — **n** spoel *fm* (voor hechtdraad) — **e** carrete *m* (para alambre) — **i** rocchetto *m* (di filo metallico); bobina *f* (di filo metallico)

14767 wire brush — Brush with stiff metal bristles for cleaning.
f brosse *f* métallique — **d** Drahtbürste *f* — **n** staalborstel *m* — **e** cepillo *m* de alambre; broza *f* de alambre; escobilla *f* de alambre; carda *f* metálica — **i** spazzola *f* metallica

14768 wired program computer — Certain in special purpose computers which are hardware programmed by the manufacturers for the range of typographic format and justification operations required. Changes of measure and type size are under operator control and the set width of characters is altered by a replaceable width plug.

14769 wire gauze — See machine wire.

14770 wire lines — See chain lines.

14771 wire lines — See laid lines.

14772 wire lines — See water lines.

14773 wire, machine — See machine wire.

14774 wire mark — Impression left in paper or board by the mesh of the wire on which the sheet was formed.
f marque *f* de la toile; marque *f* de toile — **d** Siebmarkierung *f*; Siebmarke *f* — **n** zeefmarkering *f*; zeefmerk *n* — **e** marca *f* de la tela (metálica) — **i** marcatura *f* della tela; segno *m* della tela

14775 wire printer — See matrix printer.

14776 wire screen — See marbling grill.

14777 wire service output — Information (news stories) received from the press wire services in machine-readable form, as via a paper tape punch. The signals are usually transmitted over a telephone or teletype line as bit serial data.

14778 wire-side; underside; wrong side — Face of a sheet of paper or board which was in contact with the wire during manufacture. See also top side.

f côté *m* toile; côté *m* de la toile; face *m* toile; ì *m* tamis — **d** Siebseite *f* — **n** zeefzijde *fm*; ond\ zijde *fm*; doekzijde *fm* — **e** cara *f* de la tela; ca\ *f* de la malla; cara *f* inferior; lado *m* de la tela lado *m* metálico; lado *m* inferior; lado *m* de abajo; revés *m* del papel — **i** lato *m* tela (della carta)

14779 wire stitcher; wire stitching machine; stapler — Machine to fasten sheets, sections or signatures by means of wire staples.
f brocheuse *f* au fil métallique; piqueuse *f* au fil de fer; piqueuse *f* au fil métallique; agrafeuse *f*; machine *f* à piquer — **d** Drahtheftmaschine *f*; Heftmaschine *f* — **n** nietmachine *f*; draadhechtmachine *f*; hechtmachine *f* — **e** engrapadora *f*; cosedora *f* de alambre; máquina *f* de coser con grapas; máquina *f* de coser con alambre; máquina *f* de coser con hilo metálico; máquina *f* a brochadora; encuadernadora *f* con alambre — **i** cucitrice *f* a filo metallico; cucitrice *f* a punto metallico; aggraffatrice *f* (in box making)

14780 wire stitching; stapling — In bookbinding, to fasten together sheets, signatures or sections with wire staples. There are three methods, *saddle stitching, *side stitching, and *stabbing.
f piquage *m* (au fil) métallique; agrafage *m* — **d** Drahtheftung *f*; Heften *n* mit Draht; Heftung *f* — **n** nieten *n*; hechten *n* — **e** coser con alambre; engrapado *m* con alambre; coser con cochetes; coser con grapas; coser con hilo metálico; abrochar con alambre — **i** cucitura *f* a punto metallico; cucitura *f* a filo metallico; aggraffatura *f* (in converting)

14781 wire, stitching — See stitching wire.

14782 wire stitching machine — See wire stitcher.

14783 wiring diagram — See diagram of connection.

14784 with cross rules
f avec réglure intérieure — **d** mit eingesetzten Querlinien — **n** met (ingebouwde) contralijnen — **e** con rayado interior; pauta interior — **i** con rigatura interna

14785 with separate cross rules
f avec réglure séparée — **d** mit Quersatz für sich — **n** met contralijnen in aparte vorm; met contralijnen apart — **e** con rayado separado; pauta aparte — **i** con rigatura separata

14786 with the letter, impression — See proof with the letter.

14787 with the letter, proof — See proof with the letter.

14788 wolfram; tungsten — Very heavy, hard, brittle, grey metal. Not found native. Soluble in a mixture of nitric acid and hydrofluoric acid. An international official ruling recognizes the use of tungsten for English-speaking countries, but wolfram was approved as official by the International Union of Chemistry in September 1949. Symbol: W.
f wolfram *m*; tungstène *m* — **d** Wolfram *n* — **n** wolfram *n*; wolfraam *n* — **e** wolframio *m*; volframio *m*; tungsteno *m*; tunsteno *m* — **i** wolframio *m*; tungsteno *m*

14789 wood alcohol — See methanol.

14790 wood base; wooden base; wood block — Block used as a mount to make printing plates type high.

f bloc *m* de bois; patin *m* de bois; pied *m* de bois — **d** Holzklotz *m*; Holzfuß *m* — **n** houten voet *m* — **e** base *f* de madera; piso *m* de madera; zócalo *m* de madera — **i** zoccolo *m* di legno; supporto *m* di legno; blocco *m* di legno

14791 wood block — See wood base.

14792 woodblock printings — Books of which the sheets were printed from wooden blocks in which characters or figures were cut, and not from loose types.

f incunables *mpl* tabellaires; incunables *mpl* xylographiques — **d** Holztafeldrucke *mpl* — **n** blokboeken *npl* — **e** impresos *mpl* tabularios; libros *mpl* xilográficos — **i** stampe *fpl* tabellari

14793 Woodbury type; photoglyphy — Print taken from an intaglio plate made from a film of gelatine in relief from a negative. Named after the inventor Walter Bentley Woodbury (1872).

f photoglyptie *f*; photoplastographie *f*; héliogravure *f* au grain de résine — **d** Woodbury-Druck *m* — **n** Woodbury-druk *m* — **e** impresión *f* Woodbury — **i** fotoglittia *f*; stampa *f* Woodbury

14794 woodcut — Printing plate of wood, carrying an illustration in relief, the background carved out with the grain. Also a print obtained from such a wood block. See also wood engraving.

f gravure *f* en taille d'épargne — **d** Holzschnitt *m* — **n** houtsnede *fm* — **e** entalladura *f*; entallamiento *m* — **i** incisione *f* in legno

14795 woodcutter — See xylographer.

14796 wooden base — See wood base.

14797 wood engraver

f graveur *m* sur bois; dominotier *m* *obs.* — **d** Holzschneider *m* — **n** houtgraveur *m* — **e** grabador *m* en madera; xilógrafo *m* — **i** incisore *m* in legno; intagliatore *m* in legno; silografo *m*

14798 wood engraving; xylography — Art of cutting printing formes on wooden blocks.

f xylographie *f* — **d** Holzschneidekunst *f*; Xylographie *f* — **n** houtsnijkunst *f*; xylografie *f* — **e** arte *m* de grabar en madera; xilografía *f* — **i** silografia *f*; xilografia *f*

14799 wood engraving; xylotype — Design cut in the end grain of a type-high block of wood (usually boxwood), or a print from it. See also woodcut.

f gravure *f* sur bois; gravure *f* sur buis; xylogravure *f* — **d** Holzstich *m* — **n** houtgravure *fm* — **e** grabado *m* al boj; grabado *m* en boj; grabado *m* en madera; grabado *m* xilográfico; xilograbado *m*; xilotipo *m* — **i** incisione *f* su legno; incisione *f* xilografica

14800 wooden head *US* — Meaningless headline in a daily newspaper.

14801 wood fibers — See wood fibres.

14802 wood fibres *GB*; **wood fibers** *US*

f fibres *fpl* de bois — **d** Holzfasern *fpl* — **n** houtvezels *fmpl* — **e** fibras *fpl* de madera — **i** fibre *fpl* di legno

14803 woodfree paper — See groundwood-free paper.

14804 wood letter; wood types — Large types cut on wood, used in poster work.

f caractères *mpl* en bois; caractères *mpl* sur bois — **d** Holzschrift *f*; Holzbuchstaben *mpl* — **n** houten letter *fm*; biljetletter *fm* — **e** tipos *mpl* de madera; letras *fpl* de madera; caracteres *mpl* tipográficos en madera — **i** caratteri *mpl* di legno

14805 wood oil — See tung oil.

14806 wood paper — See mechanical paper.

14807 woodpulp; paper pulp — Wood is composed principally of 40-60% cellulose and 20-40% lignin, together with gums, resins, water, and inorganic matter left as ash when the wood is burned. Woodpulp is produced for its cellulose content and used to make various kinds of paper, paperboard, rayon and nitrocelluloses.

f pâte *f* de bois — **d** Holzstoff *m*; Holzzellstoff *m*; Halbstoff *m* aus Holz — **n** houtcelstof *fm*; houtpulp *fm* — **e** pasta *f* de madera; pulpa *f* de madera; pasta *f* mecánica; pulpa *f* leñosa — **i** pasta *f* da legno; pastalegno *m*

14808 woodpulp board — Paperboard made of woodpulp or a combination of woodpulp and waste papers.

f carton *m* de pâte mécanique — **d** Holzpappe *f* — **n** houtbord *n* — **e** cartón *m* de pasta mecánica — **i** cartone *m* da pastalegno

14809 wood rosin — Rosin obtained from pine stumps or other resinous woods by solvent extraction.

f résine *f* naturelle — **d** Wurzelharz *n* — **n** houthars *mn* — **e** resina *f* de madera — **i** resina *f* di legno

14810 woods — See knife cheeks.

14811 wood types — See wood letter.

14812 woof — See weft.

14813 wool fat; wool grease — Fat occurring in wool up to 2% used in the preparation of *lanolin.

f suint *m* de laine; graisse *f* de laine — **d** Wollfett *n* — **n** wolvet *n* — **e** grasa *f* de lana — **i** grasso *m* di lana

14814 wool grease — See wool fat.

14815 wool scale

f échelle *f* de laine (anglais) — **d** Wollskala *f* — **n** (Engelse) wolschaal *fm* — **e** escala *f* de lana — **i** scala *f* di lana

14816 word — A computer word consists of a string of bits (one byte or two or more) which can be manipulated by the computer's instruction set at one time. A word may, for example, consist of 16 bits (two bytes), or of 32 bits (four bytes). There are some 12-bit-word computers.

f mot *m* — **d** Wort *n* — **n** woord *n* — **e** palabra *f* — **i** parola *f*

14817 word division — See division of words.

14818 word division — See hyphenation.

14819 word length; machine word length — Number of bits or characters in a word.

f longueur *f* de mot machine — **d** Wortlänge *f*;

Maschinenwortlänge *f* — **n** woordlengte *f*; machinewoordlengte *f* — **e** longitud *f* de palabra máquina — **i** lunghezza *f* di parola di macchina

14820 word processing — Use of automated typewriters or similar devices for initial input, for subsequent revision, and for final output.

14821 word wrap — Feature, implemented by hardware or software, whereby an entire word must fit on a line (as on a VDT) rather than to be broken arbitrarily between two lines. When additional characters are inserted into a line, causing text to the right of the insertion to move over, word wrap will push any word off one line and onto the next when one or more characters of that word will not fit. A domino effect will thus be observed with respect to succeeding lines. Word wrap may be implemented on a character-by-character basis at the time of the initial input or when insertions are made or, in some cases, only when the material on the screen is "refitted" by specific command. A similar feature is offered by the software driving some line printers, such as a Versatec, so that the hard copy or printout so provided will correspond line-for-line with the otherwise unformatted text displayed on the VDT screen.

14822 work and back; sheetwise forme; sheetwork — Method of working where a number of pages are imposed in two formes, one for printing on one side of the paper and the other on the reverse of the paper, the result being one perfect copy.
f imposer en feuille; imposition *f* en feuille — **d** mit zwei Formen für einen Bogen drucken — **n** schoondruk- en weerdrukvorm apart; binnen- en buitenvorm apart — **e** imposición *f* a blanco y retiración; impresión *f* a blanco y retiración — **i** imposizione *f* con bianca separata dalla volta (in due forme)

14823 work and roll — See work and tumble.

14824 work and tumble; work and roll; "twelve ways" back-up — Printing where the trailing edge of the sheet during first side printing becomes the gripper edge during backing up.
f retourner; impression *f* tête à queue — **d** umstülpen; stülpen (Seitenmarke bleibt, Vordermarke wechselt) — **n** stolpen — **e** retiro *m* de pie con cabeza; impresión *f* a blanco y vuelta; impresión *f* a blanco y voltereta; retirar a la voltereta — **i** capovolgere; imposizione *f* bianca e volta unite

14825 work and turn; half sheetwork — Work printed on both sides of the paper from the same forme, whether a 2, 4, 8 or 16-page, etc. To illustrate, a half sheet of 16's is printed on one side of the paper from a forme of 16 pages, the paper is then turned over, fed to the same gripper lay edge, and printed on reverse side of sheet from the same forme. When cut in half two copies will result.
f imposition *f* en demi-feuille; imposition *f* identique — **d** umschlagen (Vordermarke bleibt,

Seitenmarke wechselt) — **n** keren *n*; keervorm *m*; binnen- en buitenvorm in één vorm; schoon- en weerdruk in één vorm — **e** retiración *f* normal — **i** voltare; imposizione *f* bianca e volta insieme

14826 work and twist; work and whirl — When a sheet is run through the machine once, then turned round, keeping the same side up, but bringing a different edge of the sheet to the grippers. It is then run through again on the same forme.
f basculer en ailes de moulin; imposition *f* en ailes de moulin; impression *f* en ailes de moulin — **d** umdrehen (Vordermarke und Seitenmarke wechseln) — **n** draaien — **e** retiración *f* en molinete — **i** rovesciare

14827 work and whirl — See work and twist.

14828 work bench — See bench.

14829 work content; work standard; work value — Amount of work in a task expressed in work units.

14830 working aperture — Diameter of the actually used part of the lens.
f ouverture *f* utile; ouverture *f* relative; diamètre *m* utile — **d** relative Blende-Öffnung *f* — **n** werkzame opening *f* — **e** abertura *f* práctica — **i** apertura *f* relativa dell'obiettivo

14831 working clothes — See also Saint-Jean.
f vêtements *mpl* de travail; blouse *f* — **d** Arbeitskleidung *f*; Overall *m* — **n** werkkleding *f*; overall *m* — **e** vestidos *mpl* de trabajo; traje *m* de trabajo — **i** abiti *mpl* da lavoro; vesti *mpl* da lavoro; vestiario *m* da lavoro

14832 working conditions
f conditions *fpl* de travail — **d** Arbeitsbedingungen *fpl* — **n** arbeidsvoorwaarden *fpl* — **e** condiciones *fpl* de trabajo — **i** condizioni *fpl* di lavoro

14833 working day
f jour *m* ouvrable — **d** Arbeitstag *m* — **n** werkdag *m*; arbeidsdag *m* — **e** día *m* de trabajo; día *m* de labor; día *m* laboral — **i** giornata *f* di lavoro; giorno *m* di lavoro

14834 working director
f prote *m* à tablier — **d** Faktor *m* "vom Leder" *fam.* (der mitarbeitet) — **n** medewerkend chef *m* — **e** regente *m*

14835 working drawing; block drawing — Perfected drawing of a crudely executed design or preliminary sketch made for reproduction from originals unsuitable in their actual condition.
f dessin *m* d'exécution; maquette *f* d'exécution — **d** Klischeevorlage *f*; Original *n* — **n** werktekening *f*; clichétekening *f* — **e** dibujo *m* de ejecución — **i** disegno *m* definitivo

14836 working scores — Scores that require severe (usually 180 °) bends in either gluing or in box make-up. More cracking is encountered because all the board fibres have to be broken and the score has to be cut proper width to get a good bead.

14837 working up — See rising.

14838 work in shifts — See shift.

14839 workman; labourer
f ouvrier *m* — **d** Arbeiter *m* — **n** arbeider *m*; werkman *m* (*pl* werklieden) — **e** obrero *m*; trabajador *m* — **i** operaio *m*; lavoratore *m*

14840 work v overtime
f faire des heures supplémentaires; veiller — **d** Überstunden machen; Übertich machen; Überstunden arbeiten — **n** overwerken; overuren maken — **e** trabajar fuera de las horas reglementarias — **i** fare lo straordinario

14841 work shop — Building or room where work is carried on.
f atelier *m* — **d** Werkstatt *f* — **n** werkplaats *f* — **e** taller *m*; obrador *m*; aparador *m* — **i** bottega *f*

14842 work simplification
f simplification *f* du travail — **d** Arbeitsvereinfachung *f* — **n** werkmethode-verbetering *f*; methode-verbetering *f* — **e** simplificación *f* del método de trabajo; mejora *f* del método de trabajo; mejoría *f* del método de trabajo; mejoramiento *m* del método de trabajo — **i** semplificazione *f* del lavoro

14843 works manager; factory manager; plant manager
f directeur *m* d'usine — **d** Betriebsführer *m*; Werksleiter *m* — **n** bedrijfsleider *m*; bedrijfschef *m* — **e** director *m* de fábrica; director *m* de usina *SA* — **i** direttore *m* tecnico

14844 work standard — See work content.

14845 work study — Technique and analyzing methods used in performing an operation and of measuring the work involved, to ensure better use of materials, plant and man power, and therefore higher productivity.
f étude *f* du travail — **d** Arbeitsstudie *f*; Arbeitsstudium *n* — **n** werkstudie *f*; arbeidsanalyse *f* — **e** estudio *m* del trabajo — **i** studio *m* del lavoro

14846 work table — See bench.
14847 work ticket — See jacket.
14848 work v up — To form the bands on the backs of books.
f pincer les nerfs — **d** Bünde herausarbeiten — **n** ribben opwerken — **i** sollevare i nervi

14849 work-up — Making of gold leaf decorations on hand-made book covers, without having applied a drawing on it.
f dorure *f* sans tracé — **d** freie Verzierung *f* von Handeinbänden ohne Vorzeichnung

14850 work value — See work content.

14851 world of the press
f monde *m* de la presse; monde *m* du papier — **d** Zeitungswelt *f* — **n** krantenwereld *fm* — **e** mundo *m* de periódicos — **i** mondo *m* giornalaio

14852 worm; endless screw — Gear having at its teeth a specially cut thread in the shape of the threads on a bolt. The worm meshes with a worm wheel.
f vis *f* sans fin — **d** Gewindestift *m*; Schnecke *f*; Madenschraube *f* — **n** worm *m* — **e** tornillo *m* sinfín — **i** vite *f* senza fine

14853 worm gear — Gear of a worm and a worm wheel working together.

f transmission *f* à vis sans fin — **d** Schneckegetriebe *n* — **n** wormwieloverbrenging *f* — **e** engranaje *m* de tornillo sinfín — **i** ruotismo *m* a vite senza fine

14854 worm wheel — Gear or wheel having teeth on a special shape cut around it. It meshes with a worm.
f roue *f* à vis sans fin; roue *f* hélicoïdale — **d** Schneckenrad *n* — **n** wormwiel *n* — **e** rueda *f* serpentina; rueda *f* helicoidal; rueda *f* de tornillo sinfín — **i** ruota *f* elicoidale

14855 worn-out type — Flattened out type. See also batter.
f caractère *m* usagé; caractère *m* écrasé; têtes *fpl* de clous *fam.* — **d** abgequetschte Schrift *f*; abgenutzte Schrift *f*; Nägelköpfe *mpl fam.* — **n** afgereden letter *fm*; plat gereden letter *fm*; afgesleten letter *fm*; versleten letter *fm*; kopspijkers *mpl fam.* — **e** letra *f* gastada; tipo *m* gastado; tipos *mpl* machacados; barba *f* — **i** caratteri *mpl* schiacciati; lettere *fpl* ammaccate; carattere *m* mobile schiacciato; carattere *m* mobile logoro

14856 wove dandy — See dandy.

14857 wove mould — Mould with a woven wire cover. The first wove hand mould was invented by John Baskerville (about 1750) to eliminate laid marks and produce a smoother surface.
f moule *m* vélin — **d** Schöpfbütte *f* mit gewebtem Metallsieb; gewebtes Drahtsieb *n* — **n** schepvorm *m* voor velijn papier — **e** molde *m* para papel avitelado; forma *f* para papel avitelado — **i** forma *f* per velino

14858 wove paper — Paper which is not laid. "Wove" refers to the imprint left on the surface of the paper by the woven wire on which the paper is made.
f papier *m* vélin — **d** Velinpapier *n* — **n** velijn papier *n* — **e** papel *m* avitelado; papel *m* vitela; vitela *f*; papel *m* sin verjuras — **i** carta *f* velina; velino *m*

14859 wrap — Length of contact of paper round a cylinder.
f enveloppement *m* — **d** Umschlingung *f* — **n** contactlengte *f* — **e** abrazo *m* — **i** lunghezza *f* di sviluppo; lunghezza *f* d'avvolgimento

14860 wraparound — See outsert.
14861 wrap-around plate — See wrap round plate.
14862 wrapper — See envelope.
14863 wrapper, dust — See jacket.
14864 wrappers — See newsprint waste.
14865 wrapping — Protective covering put on paper reel at mill. See also mill wrapping.
f emballage *m* de la bobine — **d** Verpackung *f* der Papierrolle — **n** verpakking *f* van de papierrol; rolverpakking *f* — **e** embalaje *m* de la bobina (de papel) — **i** imballo *m* della bobina (de carta)

14866 wrapping paper — Paper made of a large variety of furnishes on a Fourdrinier, cylinder, or Yankee machine, used for wrapping. Strength and toughness are predominant qualities.
f papier *m* d'emballage — **d** Packpapier *n*;

Verpackungspapier *n*; Einschlagpapier *n*; Einwickelpapier *n* — **n** pakpapier *n*; verpakkingspapier *n* — **e** añafea *f*; papel *m* de añafea; papel *m* de envolver; papel *m* para envolver; papel *m* de embalar; papel *m* de embalaje; papel *m* para embalaje — **i** carta *f* da imballaggio; carta *f* da imballo; carta *f* per avvolgere; carta *f* da impacco

14867 wrapping tissue — Thin, soft, relatively tough paper for packaging delicate articles. Its substance is between 12 and 25 g/m^2.

f papier *m* mousseline — **d** Einwickelseidenpapier *n*; Seidenpapier *n* — **n** zijdevloei *n* (een speciaal zijdevloei is juwelierszijde, juweliersvloei dat o.m. chloor- en zuurvrij is) — **e** papel *m* seda (para envolver) — **i** carta *f* velina (da imballo)

14868 wrap round; winder; wind up — Paper accidentally wrapped round the impression cylinder, plate cylinder or inking rollers, usually occurring from a web break.

f enroulement *m* — **d** Papierwickler *m* — **n** wikkelaar *m* — **e** enrollamiento *m* — **i** arrotolamento *m* del nastro

14869 wrap round — See outsert.

14870 wrap round plate *GB*; **wrap-around plate** *US* — Flexible relief printing plate, e.g., of rubber or plastic to replace metal, which is clamped round the cylinder.

f cliché *m* wrap-around; plaque *f* typo wraparound; cliché *m* cintré en relief; typoplaque *f*; plaque *f* enveloppante — **d** Wickelplatte *f* — **n** wikkelplaat *fm*; wikkelbare plaat *fm* — **e** plancha *f* arrollable; plancha *f* adosable; plancha *f* (flexible) rototipográfica; plancha *f* flexible adosable al cilindro portaplanchas — **i** lastra *f* avvolgibile; lastra *f* wrap-around

14871 Wratten filter — Trade name for the dyed gelatine filter of George Eastman, American inventor in the field of photography (1854-1932).

f filtre *m* Wratten; écran *m* Wratten — **d** Wratten-Filter *m* — **n** Wratten-filter *n* — **e** filtro *m* Wratten — **i** filtro *m* Wratten

14872 wrench; spanner — Tool to grip and turn or twist the head of a bolt or a nut. See also Allen wrench, monkey wrench, open-end wrench.

f clef *f* (à écrous); clé *f* — **d** Schraubenschlüssel *m*; Schlüssel *m* — **n** moersleutel *m*; sleutel *m*; schroefsleutel *m* — **e** llave *f* de tuerca — **i** chiave *f* per viti; chiave *f* di serraggio

14873 wrinkles — Creases in paper produced during manufacture. A wet wrinkle occurs at the presses; a dry wrinkle is produced on the calender.

f rides *fpl*; plis *mpl*; faux plis *mpl*; froncis *mpl*; becs *mpl*; bées *fpl* — **d** Falten *fpl*; Quetschfalten *fpl*; Runzeln *fpl* — **n** plooien *fmpl* — **e** arrugas *fpl*; agujetas *fpl* — **i** grinze *fpl*; false pieghe *fpl*

14874 wrinkling

f plissage *m*; formation *f* de faux plis — **d** Faltenschlagen *n*; Faltenbildung *f* — **n** plooien *n* — **e** hacer arrugas; formarse pliegues; formarse agutejas; arrugar — **i** raggrinzamento *m*; formazione *f* di pieghe

14875 wrinkling — See crazed drying.

14876 write *v* — To record data on a magnetic card, type or disk for storage or future use.

f écrire — **d** schreiben — **n** schrijven; opnemen in — **e** escribir — **i** scrivere

14877 writing demy — A size of board. See app. no. 3.

14878 writing ink — See ink.

14879 writing paper — Better, harder, and more highly finished paper than that for printing.

f papier *m* écriture; papier *m* à écrire; papier *m* d'écriture — **d** Schreibpapier *n* — **n** schrijfpapier *n* — **e** papel *m* de escribir — **i** carta *f* da scrivere; carta *f* per scrivere

14880 writing royal — A size of board. See app. no. 3.

14881 wrong fount; wrong letter; distribution fault — Character not in the same type face as the rest of the fount, mostly occurring in hot-metal typesetting.

f œil *m* étranger; coquille *f*; erreur *f* de sorte — **d** Fisch *m*; Zwiebelfisch *m* (Fische durch Unachtsamkeit werden aus einer anderen Schriftsorte zu Zwiebelfischen); falscher Buchstabe *m*; falsche Schrift *f*; Ablegefehler *m* — **n** kastfout *fm*; verkeerde letter *fm*; valfout *fm* (zetmach.) — **e** letra *f* confundida; letra *f* de otro tipo; falta *f* de distribución; falta *f* en la distribución; errata *f* de distribución; pastel *m*; letra *f* falsa; letra *f* equivocada — **i** errore *m* di scomposizione; errore *m* di distribuzione; lettera *f* errata; refuso *m*

14882 wrong lead — See lead of the web.

14883 wrong letter — See wrong fount.

14884 wrong overturn; bad break — Half line at the head of a page. See also widow.

f ligne *f* boiteuse; ligne *f* creuse; fausse ligne *f* — **d** Hurenkind *n* — **n** valse regel *m*; onechte regel *m*; hoerenjong *n depr.* — **e** gazapo *m* garrafal — **i** righino *m*

14885 wrong side — See wire-side.

14886 x-axis — See axis of the abscisse.

14887 xenon — Gaseous element, occurring in extremely small quantities in the atmosphere. Solidifies at a very low temperature; very slightly soluble in water. Symbol: Xe.

f xénon *m* — **d** Xenon *n* — **n** xenon *n* — **e** xenón *m* — **i** xeno *m*

14888 xenon arc lamp — See xenon lamp.

14889 xenon flash — Light source in discharge lamps to give an intense light approaching that of a carbon arc lamp and very short exposure (one microsecond) for high speed photographic applications.

f flash *m* au xénon — **d** Xenonimpulslicht *n* — **n** xenon flash *m*

14890 xenon lamp; xenon arc lamp — Discharge lamp in which the discharge takes place in xenon gas, its light having a colour temperature of 5500-6000 °K.

f lampe *f* au xénon (haute pression); lampe *f* au xénon pulsé; lampe *f* à gaz xénon — **d** Xenonlampe *f* — **n** (hogedruk-)xenonlamp *fm* — **e** lámpara *f* xenón; lámpara *f* xenón de arco; lámpara *f* de arco xenón — **i** lampada *f* allo xeno; lampada *f* intermittente allo xeno

14891 xerography — Electro-physical reproduction process in which electrically charged pigments are transferred from a selen-coated printing forme onto paper or an other substratum, where they are melted.

f xérographie *f* — **d** Xerographie *f* — **n** xerografie *f* — **e** xerografía *f*; jerografía *f* — **i** xerografía *f*

14892 xeronic printer — High-speed printer, almost entirely electronic. Does not require preprinted stationery as it projects and prints the background form at the same time as the information. An image from cathode ray tubes is projected on to a rotating charged selenium drum, which is cascaded with a thermoplastic printing powder electrostatically attracted to the image area. The powder is then transferred and fused to a web of paper.

14893 xerox — Trade-mark for a dry-copying process to reproduce printed, written, or pictorial matter by xerography.

14894 x-height; z-height — Relative height allowed for that part of the lower case fount from the *base line to the top of the lower case x. This affects the relative legibility of the fount size. Recommended in preference to "appearing size"; "small x" and "medium x" are acceptable for comparison.

f œil *m* de la lettre — **d** Mittellänge *f* — **n** x-hoogte *f* — **e** altura *f* media; altura *f* de letra equis — **i** altezza *f* dell'occhio mediano; altezza *f* x dell'occhio

14895 xography; three-dimensional printing — Three-dimensional printing process developed in 1966. Essentially a printed image broken up into stereoscopic elements over which a plastic screen is superimposed to divide the picture into hundreds of tiny vertical strips. The screen focuses on the tiny strips in the picture and gives the viewer the illusion of depth.

f impression *f* tridimensionnelle — **d** dreidimensioneller Druck *m* — **n** xografie *f*; drie-dimensionele druk *m*; drie-dimensionale druk *m* — **e** impresión *f* tridimensional — **i** stampa *f* tridimensionale

14896 xylene; dimethylbenzene; xylol — Clear liquid, toxic and flammable, soluble in alcohol and ether, insoluble in water. Used in intaglio inks. Xylol is used commercially, not in modern nomenclature. TL-value: 100 ppm or 435 mg/m^3. Formula: $C_6H_4(CH_3)_2$.

f xylène *m*; xylol *m* — **d** Xylol *n* — **n** xyleen *n*; dimethylbenzeen *n*; xylol *n depr.* — **e** xileno *m*; xilol *m* — **i** xilene *m*; xilolo *m*

14897 xylographer; woodcutter

f xylographe *m*; graveur *m* sur bois — **d** Holzschneider *m*; Xylograph *m* — **n** houtsnijder *m*; xylograaf *m* — **e** xilógrafo *m*; grabador *m* en madera — **i** xilografo *m*; incisore *m* in legno

14898 xylographic book
See block book.

14899 xylography — See wood engraving.

14900 xylol — See xylene.

14901 xylotype — See wood engraving.

14902 xyz system — See colour triangle.

14903 Yankee cylinder *GB*; **Yankee dryer** *US*
f sécheur gros-frictionneur *m*; frictionneur *m* — **d** Glättzylinder *m* — **n** Yankee-cilinder *m* — **e** secador *m* yanqui; cilindro *m* satinador — **i** cilindro *m* yankee

14904 Yankee dryer — See Yankee cylinder.

14905 Yankee machine; single-cylinder machine — Fourdrinier machine with a single drying cylinder of large circumference with highly polished surface, to make machine-glazed papers.
f machine *f* à cylindre frictionneur; machine *f* frictionneuse; machine *f* à prise automatique — **d** Selbstabnahmemaschine *f* — **n** Yankee-machine *f*; eenzijdig-glad machine *f*; zelfafname-machine *f* — **e** máquina *f* Yanqui; máquina *f* monocilindro — **i** macchina *f* yankee

14906 yapp binding; divinity binding — Binding with edges overlapping the edges of the book block. Bibles are often bound in this way. Named after William Yapp (1854-1875), bookseller, especially of bibles, London.
f yapp *m* — **d** Einband *m* mit übergreifenden Kanten; Kasteneinband *m* — **n** yapp-band *m*; slappe boekband *m* met overhangende randen — **e** encuadernación *f* con bordes cerrados — **i** rilegatura *f* a tagli protetti

14907 yard — Unit of length. Abbrev. yd (no point). See app. no. 7 and 7a.
f yard *m* — **d** Yard *m* — **n** yard *m* — **e** yarda *f* — **i** iarda *f*

14908 yawning boards — Cover boards of a book that open wide away from the book block.
f plats *mpl* qui bâillent — **d** sich werfende Buchdeckel *mpl* — **n** kromtrekkende platten *npl* — **e** tapas *fpl* abarquilladas hacia fuera — **i** quadranti *mpl* che s'arricciano

14909 y-axis
See axis of ordination.

14910 Y-connection
f montage *m* en étoile; groupement *m* en étoile; couplage *m* en étoile — **d** Sternschaltung *f* — **n** sterschakeling *f* — **e** acoplamiento *m* en estrella; conexión *f* en estrella — **i** collegamento *m* a stella

14911 Y-delta connection — See star-delta connection.

14912 year-book; annual — Book or pamphlet published once a year, giving a record of events, statistics, and general information on some subject for reference.
f annuaire *m* — **d** Jahrbuch *n* — **n** jaarboek *n* — **e** anuario *m* — **i** annuario *m*

14913 year of publication
f année *f* de publication — **d** Erscheinungsjahr *n*; Jahrgang *m* — **n** jaar *n* van verschijnen; verschijningsjaar *n*; jaargang *m* — **e** año *m* de publicación — **i** anno *m* di pubblicazione; annata *f*

14914 yellow arsenic sulphide — See orpiment.

14915 yellowed copy

f modèle *m* jauni — **d** vergilbte Vorlage *f* — **n** vergeeld model *n* — **e** modelo *m* amarillento; original *m* amarillento — **i** copia *f* ingiallita

14916 yellow filter
f filtre *m* jaune; écran *m* jaune — **d** Gelbfilter *n* — **n** geelfilter *n* — **e** filtro *m* amarillo — **i** filtro *m* giallo; schermo *m* giallo

14917 yellowing — Gradual change from the original colour of paper as a result of environment or ageing; especially in mechanical paper but in varying degrees in all types of all vegetable fibre. Sometimes called colour reversion.
f jaunissement *m* — **d** Vergilbung *f*; Gelbfärbung *f*; Vergilben *n* — **n** vergeling *f*; vergelen *n*; geel worden *n* — **e** amarilleo *m* — **i** ingiallimento *m*

14918 yellow journal — See yellow paper.

14919 yellow lead oxide — See litharge.

14920 yellow paper; yellow journal; scandal sheet; gutter paper — Newspaper or periodical dealing to a large extent in scandal and gossip. Sensational newspaper. The term is said to originate from 1898 in which year some newspapers tried to incite people for war against Spain.
f journal *m* explosif; journal *m* de chantage; journal *m* à sensation; journal *m* à scandales; feuille *f* à scandales; feuille *f* à ragots; brûlot *m* — **d** Skandalblatt *n*; Hetzblatt *n*; Revolverblatt *n* — **n** sensatieblad *n*; schimpblad *n* — **e** diario *m* amarillisto; diario *m* amarillo; gaceta *f* sensacionalista; órgano *m* sensacionalisto; periódico *m* amarillisto; periódico *m* sensacionalisto; periódico *m* amarillo; periódico *m* del arroyo — **i** giornale *m* scandalistico; foglio *m* scandalistico

14921 yellow plate — See yellow printer.

14922 yellow potassium chromate — See potassium chromate.

14923 yellow potassium prussiate — See potassium ferrocyanide.

14924 yellow printer; yellow plate — Printing plate for which copy is photographed through a blue (or violet) filter.
f cliché *m* du jaune — **d** Gelbdruckplatte *f*; Gelbplatte *f* — **n** gele drukplaat *fm*; geelplaat *fm* — **e** plancha *f* del amarillo — **i** lastra *f* del giallo

14925 yellow prussiate of potash — See potassium ferrocyanide.

14926 yellow strawboard — See yellow strawpaper.

14927 yellow strawpaper; yellow strawboard — Paper and board made solely from unbleached strawpulp, generally yellow.
f papier *m* pure paille; carton *m* pure paille — **d** (reines) Strohpapier *n*; (reine) Strohpappe *f* — **n** stropapier *n*; strobord *n*; strokarton *n* — **e** papel *m* de paja; cartón *m* de paja; cartón *m* amarillo — **i** carta *f* gialla di paglia; cartone *m* giallo di paglia

14928 yellow strawpulp — Pulp obtained by

cooking straw with a chemical agent, generally milk of lime or other weak alkaline solution.

f pâte *f* de paille lessivée — **d** Gelbstrohzellstoff *m*; Gelbstrohstoff *m* — **n** strostof *fm* — **e** pasta *f* amarilla de paja — **i** pasta *f* gialla di paglia; pasta-paglia *f* gialla

14929 yield — Amount or quantity resulting often expressed as the percentage of what is theoretically possible. In papermaking for instance, the percentage which denotes the number of pounds of pulp which are obtained from a cord or a cunit of pulpwood, the number of pounds of paper resulting from a ton of pulp, etc.

f rendement *m* — **d** Ausbeute *f*; Ertrag *m* — **n** rendement *n* — **e** rendimiento *m* — **i** rendimento *m*; resa *f*

14930 yield value — Minimum shearing stress necessary to produce flow in fluid systems such as printing inks. The minimum force necessary to produce movement of a pasty material.

f rigidité *f*; force *f* d'écoulement minimum — **d** Fließgrenze *f*; Streckgrenze *f* — **n** vloeigrens *fm* — **e** límite *m* de fluencia; límite *m* de stiraje; límite *m* de fluidez — **i** rigidità *f* reologica

14931 Zahn cup — Device to measure viscosity of flexographic or gravure inks.
f coupe f de Zahn; coupe f consistométrique de Zahn — d Zahn-Auslaufbecher m — n Zahn-uitloopbeker m — e copa f (de efusión) de Zahn; probeta f de Zahn — i coppetta f di Zahn; bicchierino m di Zahn

14932 zapon lacquer; zapon varnish — Uncoloured solution of cellulose nitrate and amyl acetate with acetone.
f laque f de zapon; laque f zapon; vernis m zapon — d Zaponlack m — n zaponlak mn — e laca f de zapón; laca f zaponica; barniz m zapón — i lacca f zapon; vernice f zapon; vernice f giapponese

14933 zapon varnish — See zapon lacquer.

14934 zero elimination — See zero suppression.

14935 zero v out — To hold to close tolerances, generally within thousandths of a centimetre or inch.
d feinstellen — n fijn stellen — e mantener en pequeña tolerancia — i mantenere tolleranza molto strette

14936 zero suppress — See zero suppression.

14937 zero suppression; zero suppress; zero elimination — Computer term. Deletion of all zeros to the left of the significant part of a number before the print-out operation.
f suppression f des zéros — d Nullenunterdrückung f — n nullenonderdrukking f — e supresión f de (los) ceros — i soppressione f degli zeri

14938 z-height — See x-height.

14939 zigzag folded form
f formule f pliée (en) zig-zag — d Zickzackformular n — n zigzag gevouwen formulier n — e formulario m en zigzag; formulario m en ziszas; formulario m en zigsás; formulario m zigzaguerado; formulario m en línea quebrada — i formulario m a zigzag; modulo m a zigzag

14940 zinc — Shining, white metal with bluish-grey lustre, soluble in acids and alkalis, insoluble in water. Symbol: Zn. Latin: zincum. TL-value of zinc fumes: 15 mg/m³.
f zinc m — d Zink mn — n zink n — e cinc m; zinc m — i zinco m

14941 zinc baryta white
See lithopone.

14942 zinc chloride; butter of zinc — White, granular, deliquescent crystals, soluble in water, alcohol, glycerol and ether. Used as a constituent of galvanizing and electroplating baths and as a tinning solution. Latin: chloretum zincicum.
Formula: $ZnCl_2$.
f chlorure m de zinc; muriate m de zinc; beurre m de zinc — d Zinkchlorid n; Chlorzink n — n zinkchloride n — e cloruro m de cinc — i cloruro m di zinco

14943 zinc chrome — See zinc yellow.

14944 zinc etching; zincograph; zinco — Relief line or halftone etching on zinc, more commonly line etchings on zinc. It is inferior to copper as a material for making halftone blocks.
f cliché m sur zinc; cliché m zinc; zincogravure f — d Zinkätzung f; Zinkklischee n; Zinkdruckplatte f — n zinkcliché n; zinken cliché n; cliché n in zink; zinco fm — e fotograbado m en cinc; grabado m en cinc; grabado m de cinc; cincograbado m; cincografía f; cincotipia f — i zincotipia f; fotoincisione f su zinco; cliscè m di zinco; zinco m

14945 zinc etching — Etching on zinc with nitric acid.
f morsure f du zinc — d Zink ätzen — n etsen n van zink — e grabar en cinc — i morsura f dello zinco

14946 zinc finish — Finish obtained by plating paper when sandwiched between sheets of zinc.
f gaufrure f avec planches de zinc — d mit Metallplattenstruktur geprägtes Papier n — n plaatpersing f — e gofrado m con planchas de cinc — i goffrata tra due fogli di zinco

14947 zinc halftone
f simili f sur zinc — d Zinkätzung f — n autotypie f in zink — e medio tono m de cinc; grabado m directo en cinc; autotipia f de cinc — i fotoincisione f retinata su lastra di zinco; cliscè m di zinco

14948 zinc hydroxide — Colourless crystals, slightly soluble in water. Forms both zinc salts and zincates.
Formula: $Zn(OH)_2$.
f hydroxyde m de zinc — d Zinkhydroxyd n — n zinkhydroxyde n — e hidróxido m de cinc — i idrossido m di zinco

14949 zinco — See zinc etching.

14950 zincograph — See zinc etching.

14951 zinc oxide; Chinese white — Inorganic white pigment used in inks, particularly in semiopaque tint bases and mixing whites. Also used in colour photography, as a photoconductor in office copying machines, and as a vulcanization accelerator for rubber. Soluble in acids and alkalis, insoluble in water and alcohol. Latin: oxydum zincicum.
Formula: ZnO. See also zinc white.
f oxyde m de zinc; blanc m de zinc; blanc m de Chine; blanc m permanent; blanc m couvrant; blanc m à couvrir — d Zinkoxyd n; Zinkweiß n; Deckweiß n — n zinkoxyde n; zinkwit n; dekwit n — e óxido m de cinc; blanco m de cinc; blanco m de China; blanco m opaco; blanco m de cubrir — i ossido m di zinco; bianco m cinese; bianco m opaco; biacca f all'ossido di zinco

14952 zinc plate
f plaque f de zinc — d Zinkplatte f; Zinkblech n — n zinkplaat fm — e lámina f de cinc; plancha f de cinc — i lastra f di zinco

14953 zinc potassium chromate — See zinc yellow.

14954 zinc print — Image in the nature of a photoresist made on zinc with bichromated albumen or cold enamel.
f copie *f* sur zinc — **d** Zinkkopie *f* — **n** zinkkopie *f* — **e** copia *f* sobre cinc; pasado *m* sobre cinc — **i** copia *f* su zinco

14955 zinc sulphate; sulphate of zinc — Colourless crystals, small needles or granular crystalline powder, soluble in water and glycerine, insoluble in alcohol. Used as a bleaching agent for paper. Latin: sulfas zincicus.
Formula: $ZnSO_4 \cdot 7H_2O$.
f sulfate *m* de zinc — **d** Zinksulfat *n* — **n** zinksulfaat *n* — **e** sulfato *m* de cinc — **i** solfato *m* di zinco

14956 zinc sulphide — Opaque yellowish-white powder used both alone and as a component of lithopone in printing inks. Stable if kept dry. Soluble in acids, insoluble in water.
Formula: ZnS.
f sulfure *m* de zinc — **d** Zinksulfid *n* — **n** zinksulfide *n* — **e** sulfuro *m* de cinc — **i** solfuro *m* di zinco

14957 zinc sulphide white — See lithopone.

14958 zinc white — Zinc oxide pastes used by commercial artists. Alternative name for *zinc oxide.

14959 zinc yellow; zinc potassium chromate; zinc chrome; citron yellow — Yellow pigment of comparatively low tinting strength. Partially water-soluble. Consist principally of zinc potassium chromate, about
$4ZnO \cdot K_2O \cdot 4Cr_2O_3 \cdot 3H_2O$.
f jaune *m* de zinc — **d** Zinkgelb *n* — **n** zinkgeel *n* — **e** amarillo *m* de cinc — **i** giallo *m* di zinco

14960 Zip-a-tone — See benday process.

14961 zip code *US* — System to facilitate the delivery of mail, consisting of a five-digit code written directly after the address the first three digits indicating the State and place of delivery, the last two digits the post office or postal zone.

14962 zip fastener; zipper; slide fastener — Fastener consisting of two rows of metal or plastic teeth on strips of tape and a sliding piece which closes an opening by drawing the teeth together. Invented by the American W. Judson (1891).
f fermeture *f* éclair; fermeture *f* glissière; fermeture *f* à curseur — **d** Reißverschluß *m* — **n** treksluiting *f*; ritssluiting *f* — **e** cierre *m* de cremallera; cremallera *f*; cierre *m* relámpago — **i** chiusura *f* lampo

14963 zipper — See zip fastener.

14964 zirconium — Tetravalent metallic element, chemically resembling titanium, prepared as a black amorphous powder, as steel-gray shining scales, or as crystalline laminae, used in special alloys, as an opacifier of lacquers, as an abrasive, in the production of catalysts, and in textile finishing compounds. Symbol: Zr.
f zirconium *m* — **d** Zirkonium *n*; Zirkon *n* — **n** zirkonium *n* — **e** zirconio *m*; circonio *m* — **i** zirconio *m*

14965 zodiacal signs; signs of the zodiac — See app. no. 10.
f signes *mpl* du zodiaque; zodiaques *mpl* — **d** Tierkreiszeichen *npl* — **n** tekens *npl* van de dierenriem; dierenriemtekens *npl* — **e** signos *mpl* zodiacales; signos *mpl* de zodíaco — **i** segni *mpl* dello zodiaco

14966 zonal error — Defect in photographic lenses which causes the size of the image to change with use of apertures or stops at different size.
f aberration *f* locale — **n** plaatselijke aberratie *f* — **e** aberración *f* local — **i** aberrazione *f* locale

14967 zone test — Test for the absorbent property of blotting paper, performed by a drop of ink which spreads itself by capillary action, showing zones of fainter and fainter coloration.
f mesure *f* du pouvoir absorbant — **d** Randzone-Prüfung *f*; Prüfung *f* auf Löschfähigkeit; Prüfung *f* auf Saugfähigkeit — **n** vloeiproef *fm*; onderzoek *n* naar het opzuigend vermogen — **e** ensayo *m* del poder de absorción — **i** prova *f* di assorbilità

14968 zoom lens — Lens system with variable focal distance. Used in some phototypesetters instead of a series of discrete lenses in a turret.

FRANÇAIS

abaissement de la tension superficielle 13442
abaque (de conversion) 9391
abat-feuille 5915
abat-jour 7896
abattre la garniture 13282
abécédaire avec images et figures 10313
abécédaire illustré 10313
aberration 5
aberration astigmatique 852
aberration chromatique 2761
aberration locale 14966
aberration optique 9643
aberration sphérique 12862
abîmer 4178
abonné 13329
abonnée 13329
abonnement 13331
abonner à, s'~ 13330
abrasif 14
abrasimètre 11766
abrasiomètre 11766
abrasion 10
abrégé 20, 3260
abréger 1
abréviation 2
abréviations médicales 8747
abscisse 22
absolument sec 9744
absorbant l'eau 7122
absorber 31
absorption 34, 35
absorption d'eau 14564
absorption de la lumière 8107
absorption d'huile 9550
absorption d'humidité 9037
absorption (du papier) 7327
absorptivité 34
accaparer 12610
accélérateur 53
accélération 52
accélération angulaire 603
accélérer 48
accent 54
accent aigu 175
accent allemand 14256
accent circonflexe 2860
accent grave 1798
accent long 8326
accent séparé 5838
accent superposé 5838
accent tonique 10781
accentuation 62
accentuer 57, 4459
accès 67
accès multiple 9179
accès sélectif 11231
accessibilité 68
accessible 193
accessoires 69
accolade combinée 3300
accolades 1840
accolures 1113

accouplement 3670
accouplement parallèle 9978
accourci 20
accrochage des encres 583
accumulateur 83, 84
accumulation 3078
acétaldéhyde 86
acétate 88
acétate d'alumine 472
acétate d'aluminium 472
acétate d'ammoniaque 513
acétate d'ammonium 513
acétate d'amyle 563
acétate de butyle 2127
acétate de cellulose 2509
acétate de cuivre 3498
acétate de plomb cristallisé 7972
acétate de polyvinyle 10566
acétate de potasse 10623
acétate de potassium 10623
acétate de sodium 12669
acétate de soude 12669
acétate d'éthyle 5300
acétate d'isoamyle 563
acétate d'urane 14367
acétate (neutre) de plomb 7972
acétate polyvinylique 10566
acéto-acétanilide 101
acéto-cellulose 2509
acétone 102
acétyl-acétanilide 101
acétyl-cellulose 2509
achat au comptant 2423
acheminement 11738
achevé d'impression 3102
achevé d'imprimer 3102
achromatique 109
achromatopsie 3123
aciculaire 112
acide 113, 114
acide abiétique 6
acide acétique 95
acide acétique concentré 6418
acide acétique cristallisable 6418
acide acétique glacial 6418
acide acrylique 157
acide arabique 745
acide azotique 9366
acide borique 1798
acide carbolique 10189
acide carbonique 2315
acide chlorhydrique 7102
acide chlorhydronitrique 734
acide chloroazotique 734
acide chromique 2785
acide citrique 2864
acide concentré 3355
acide cyanhydrique 7105
acide cyanique 3909
acide (de morsure) 5286
acide dichloracétique 4292
acide digallique 13573
acide faible 14634

acide fluorhydrique 7106
acide formique 6078
acide fort 13297
acide gallique 6289
acide gallotannique 13573
acide glycérique 6478
acide glycolique 6488
acide gras 5493
acide gras non saturé 14341
acide gras saturé 11880
acide hydroacétique 6488
acide hydrochlorhydrique 7102
acide hydrofluorique 7106
acide hypochloreux 7148
acide iodique 7549
acide lactique 7869
acide libre 6154
acide linoléique 8217
acide linolénique 8218
acide monobasique 9074
acide muriatique 7102
acide nitreux 9380
acide nitrique 9366
acide nitrique fumant 6247
acide nitromuriatique 734
acide oléique 9579
acide oxalique 9797
acide palmitique 9880
acide phénique 10189
acide phosphorique 10205
acide phtalique 10284
acide prussique 7105
acide pyrogallique 11085
acide pyrogallolcarbonique 6289
acide pyroligneux 95
acide salicylique 11851
acide silicique 12414
acide stéarique 13087
acide sulfhydrique 7117
acide sulfureux 13387, 13394
acide sulfurique 13389, 13397
acide sulfurique anhydre 13397
acide tannique 13573
acide tartarique (ordinaire) 13610
acide thioacétique 13720
acide thymique 13798
acidification 130
acidifier 131
acidimétrie 132
acidité 133
acidulation profonde 4081
aciduler 4204
aciérage 13094
acier inoxydable 13012
aciérographie 12385
à-côté 9503
à-coups (de tension) 7656
actinique 159
actinomètre 164
action 12232
action capillaire 2279
action continuatrice 3425

activation 167
activité interfaciale 7493
activité superficielle 13429
acutance 174
adaptateur 180, 7810
adaptation 179
adaptation à l'obscurité 3987
adaptation de l'homme au travail 7067
adapter 178
addende 183
addendum 184
additif 188
additif facilitant le glissement du film d'encre sèche 12562
addition (marginale) 8629
adhérence 204, 3046
adhérence de contact entre feuilles 1598
adhérence mutuelle 1598
adhérer 203
adhésif 213
adhésif à chaud 7048, 7054
adhésif double face pour montage 9146
adhésif hot-melt 7048
adhésif pour film pelliculable 13284
adhésion 205
adjuvant 188, 3317, 4559
adoucisseur d'eau 14602
adressage en absolu 28
adressage symbolique 13485
adresse 192
adresse absolue 25
adresse indirecte 7291
adsorbant 236
adsorption 237
aération 14440
aérer 9100, 14746
aérographe 314
aérosol 284
affaiblir 11340
affaiblissement 11352
affaiblissement de couleurs dans les ombres 14269
affaiblisseur 11344
affaiblisseur cyanure 3916
affaiblisseur de Farmer 5471
affaiblisseur surproportionnel 6991
affection saturnine 8000
affichage 4458
affiche 10609
afficher 13137
affinité 287
affirmation de connexion 6847
affleurage 2005
affleureuse 868
affûteuse 4529, 6643, 7826
agalithe 293
agar-agar 294
agate 296
agate à polir 296

agave 300

agence de coupures de presse 2917

agence de publicité 263

agence d'information(s) 9299

agenda 4279, 9437

agent additif 188

agent adsorbant 236

agent anti-bloquage 657

agent anti-collant 657

agent anti-écume 660

agent antimousse 660

agent antipeaux 689

agent d'azurage optique 9646

agent de blanchiment 1544

agent de blanchiment optique 9646

agent de dispersion 4452

agent de lubrification 8394

agent d'enlevage de voile 12023

agent d'épaississement 1681

agent de surface 13428

agent de tannage 13576

agent dispersant 4452

agent durcissant 6862

agent émulsif 5128

agent émulsifiant 5128

agent fongicide 6251

agent mouillant 14699

agent optique de blanchiment 9646

agent protecteur 689

agents antistatiques 691

agents de surface anioniques 628

agents de surface cationiques 2484

agents de surface non-ioniques 9403

agent tensio-actif 13428, 14699

agglomérat 303

agglomérat de molécules 2967

agglomérats de fibres 5595

agitateur 309, 4241

agitateur en verre 6437

agitation 308

agitation par bullage d'azote 9375

agiter 13165

agrafage 12988, 14780

agrafage à cheval 11831

agrafage à plat 12389

agrafage des coins 3587

agrafe 2877, 9906, 13040

agrafe de bande 13585

agrafe de courroie 1303

agrafeuse 13528, 14779

agrafeuse pour boîtes 1838

agrandir 5198

agrandissement 5201, 8532

agrandisseur 5202

agrégation 304

agrégat pour poser et souder des fils de brochage 13744

aide-conducteur 8456

aide-conducteur de machine à papier 1049

aide-conductrice 8456

aide-mémoire 6816, 8771

aigle 2148

aigreurs 6884

aiguillé 112

aiguille à relier 13178

aiguille à report 13144

aiguille (de graveur) 5281

aiguille de relieur 13178

aiguille (indicatrice) 10524

aiguilles, en forme d'~ 112

aiguiser 7031

aimantation du papier 13067

air comprimé 3322

air d'aspiration 13349

aire de la bouche d'hystérésis 759

aire d'exploration 2906

ais 10740

ais à distribuer 1658

aisément soluble 8425

ais pour forme 8055

ajour 7453, 9107

ajouter 182

ajouter d'adjuvants 4560

ajustage 226

ajustement 226

ajuster 218

ajuster le composteur 12157

ajusteur-monteur (de machines) 5237

album 369

album à colorier 3149

album d'échantillons 11855

album de coupures 11944

albumine bichromatée 4299

albumine des œufs 4988

albumine d'œuf 4988

albumine du blanc d'œuf 371

albumine en paillettes 5748

alcali 414

alcalin 417

alcalinité 418

alcali volatil 533

alcanes 424

alcool 380

alcool allylique 443

alcool amylique 565

alcool anhydre 608

alcool butylique 2128

alcool de bois 8855

alcool dénaturé 4162

alcool dénaturé industriel 7297

alcool de vinaigre 95

alcool d'isopropyle 7619

alcool éthylique 5301

alcool isopropylique 7619

alcool méthylique 8855

alcool ordinaire 5301

alcool polyatomique 10557

alcool polyvinylique 10567

alcool propylique 10973

alcool pur 608

alcools 382

Alde 388

aldéhyde 385

aldéhyde acétique 86

aldéhyde éthylique 86

aldéhyde formique 6048

aléatoire 11230

alène 1672

alfa 5252

alginate 397

algol 398

algorithme 399

algraphie 490

alignement 405

alignement moléculaire 9052

alignement normal 13027

alignement standard 13027

aligner 404, 7739, 8156

aligner, s'~ 7739

alimentateur de métal 8825

alimenté à feuilles 12267

alimenté à la main 6825

alimenté par bobine 14646

alimenté par feuilles 12267

alimenter 8849

alimenteur automatique du creuset 917

alinéa 1875, 9972

alinéa à fleur de marge 5908

alinéa à supprimer 11811

alinéa en sommaire 6854

alkyde 425

alléguer 2861

aller à Saint-Jacques 11493

aller chou pour chou 6000, 8176

aller en Germanie 9783

aller ligne pour ligne 8176

alliage 437

alliage de plomb 7973

alliage de remise au titre 11577

alliage de retirage 11577

alliage d'imprimerie 437

alliage léger 8122

allocation pour besoins personnels 3263

allongement 5361, 11345, 13262

allongement à la rupture 5095

allongement à la (rupture par) traction 13264

allongement à la tension 13264

allongement à l'humidité 7135

allongement de rupture 5095

allongement du blanchet 1521

allongement du point (en offset) 1649

allongement du point (vers l'arrière) 4685

allongement en éventail 5467

allonger 3856, 4372

allonger l'encre 11343

allumer 13478

allure normal 9427

almanach 451

aloès 452

alpha- 453

alphabet 454

alphabet à huit unités 4994

alphabet allemand 6384

alphabet à six unités 12477

alphabet français 6167

alphabet grec 6618

alphabet latin 7940

alphanumérique 461

alsing 1100

altération 1619

alterner 13002

alternomoteur 466

altimètre d'aspiration 42

alude 3138

alumine hydratée 477

aluminium 471

aluminographie 490

aluminotypie 491

alun 468

alun ammoniacal 473

alun d'ammonium 473

alun de chrome 2808

alun de fer ammoniacal 5544

alun de fer potassique 5554

alun de potassium 483

alun ferrique ammoniacal 5544

alun ordinaire 483

alute 3138

alvéole (en héliogravure) 7331

amalgame 494, 6299

amasser 12610

ambiance 864

améliorant 4559

améliorer 4560

amide 500

amidol 4262

amidon 13045

amidon grillé 4243

amine 503

ammoniac 509

ammoniaque 533

ammonio-citrate de fer 5542

ammonium bichromate 526

amorce 1788

amorphe 552

amortisseur à air 318

amortisseur à piston 2032

amortisseur de chocs 2032

amour (d'un rouleau) 13525

amoureux de l'encre 7384

amovible 4214

ampérage 553

ampère 554

ampère-heure 555

ampèremètre 556
amphibie 14125
amplificateur 560
amplitude 561
ampoule électrique 5012
anachromat 568
anaglyphe 569
anagramme 570
analyse 575
analyse de papier 9952
analyse dispersoïdale 10015
analyse du système 13500
analyse granulométrique
    10015
analyseur de couleurs 3115
analyste 576
anastigmat 581
ancrage à boutons 12607
andouilles 8417
anépigraphe 586
angle de contact 3405
angle de coupe 3890
angle de la trame 11973
angle de phase 10185
angle de quatre-vingt-dix
    degrés 11605
angle de réflexion 597
angle de réfraction 598
angle d'incidence 595
angle (d'ornement) 3581
angle droit 11605
angle droit, à ~ 11315
angle du triangle 6068
angle optique 599
anglet 8995
ångström 601
anhydre 607
anhydride 605
anhydride acétique 98
anhydride basique 1199
anhydride carbonique 2315
anhydride chromique 2785
anhydride d'acide 115
anhydride d'acide acétique
    98
anhydride sulfocarbonique
    2316
anhydride sulfureux 13387
anhydride titanique 13860
anilide de l'acide acétyl-
    acétique 101
aniline 613
anion 627
anisotropie 629
annaline 630
anneau de benzène 1329
anneau-espace 12809
anneau-joint 6318
anneaux de Newton 9329
année de publication 14913
annexe 716
annonce 246
annonce avantageuse d'un
    livre sur le point de
    paraître 1647
annonce en plaine page
    6237
annonce matrimoniale 8643
annonces anglaises 2881

annonces groupées 2881
annonce sur le couvre-livre
    1647
annonceur 261
annoncier 250
annotation 635
annoter 633
annuaire 451, 14912
annuaire (des téléphones)
    4408
anode 641
anonyme 652
anopistographe 653
anthologie 654
anthracène 655
anthracine 655
antichlore 658
antihalo 666
antilambda 4393
antimoine 669
antimousse 660, 5940
antioxydant 674
antipeaux 689
antipelliculeur 689
anti-pourriture 6251
antiquaire 675
anti-siccative 659
antivoile 662
aperçu 20
aphorisme 701
apices 12144
aplanat 703
aplanir au rabot 10392
aplat 12755
apochromat 705
apostilb 707
apostille 706
apostrophe 708
appareil 4236
appareil à centrer et à
    cadratiner 11093
appareil à déterminer la
    raideur 13154
appareil à déterminer le
    pouvoir abrasif 11960
appareil à froisser 3780
appareil à humecter 7072
appareil à poudre 4823
appareil à sécher les flans
    8690
appareil automatique
    d'insolation 908
appareil compteur 3653
appareil d'agrandissement
    5202
appareil de chauffage 6920
appareil de climatisation
    321
appareil de collage à main
    10030
appareil de condition-
    nement de papier 9907
appareil de laboratoire
    7856
appareil de lecture 11265
appareil de marge 5508
appareil de mesure 8847
appareil de mesure du
    lissé 12646

appareil de mesure optique
    de profondeur 9654
appareil de reproduction
    10882
appareil (de reproduction)
    horizontal 7040
appareil de reproduction
    vertical 14450
appareil de sélection élec-
    tronique 11910
appareil d'essai 13662
appareil d'essai
    d'arrachage 10302
appareil d'essai
    d'imprimabilité 10797
appareil d'essai du lissage
    12646
appareil d'impression
    supplémentaire 7233
appareil Fenchel 5538
appareil humidificateur
    7072
appareil Mullen 9167
appareil pare-main 6843
appareil pour benday 12215
appareil pour déterminer
    la flexibilité des cartons
    1324
appareil pour déterminer
    la résistance au déchire-
    ment 7324
appareil pour déterminer
    la résistance au rainage
    3719
appareil pour échanger des
    ions 7567
appareil pour examiner la
    solidité des couleurs à la
    lumière 3143
appareil pour fondre les
    rouleaux 6729
appareil pour gros corps
    4460
appareil pour la mesure
    des creux 5277
appareil pour la mesure du
    temps de séchage 11762
appareil pour l'élimination
    d'électricité statique
    13069
appareil pour l'essai à
    l'aptitude à absorber 42
appareil pour l'essai à la
    résistance à la
    perforation 11051
appareil pour l'essai à la
    résistance au déchire-
    ment 13618
appareil pour l'essai à la
    rupture statique 13072
appareil pour l'essai à la
    rupture statique et
    dynamique 13065
appareil pour l'essai de
    froissement 3780
appareil pour l'essai de
    résistance à la perfora-
    tion selon Mullen 9167

appareil pour l'essai de
    résistance au pliage 5980
appareil pour l'essai de
    traction 13645
appareil pour mesurer la
    finesse de broyage 5677
appareil pour mesurer la
    liaison 1721
appareil pour mesurer
    l'étalement 5874
appareil protège-main 6843
appareil soumis à un
    travail très dur 6936
appareil stroboscopique
    13295
appareil suspendu 9759
appareil vérificateur 6334
appels de note 11381
appel sélectif 10548
appendice 716
applicateur d'encre 719
application à sec 4808
application de la feuille de
    papier charbon insolée
    sur la forme de cuivre
    2332
application humide
    manuelle 14686
apprenti 721, 3796
apprenti aux deux tiers de
    son apprentissage 14160
apprenti en cours de
    perfectionnement 7234,
    14103
apprenti (imprimeur) 10809
apprentissage 724
apprenti typographe 722
apprêt 5686
apprêt à la plaque 10437
apprêt coquille 4990
apprêté au feutre 5533
apprêté sur machine 8446
apprêt fantaisie 5458
apprêt grosse toile 3702
apprêt mat 4853
apprêt repsé 11468
apprêt satiné 5179
apprêt similitoile 8184
apprêt toile 8184
approbation 7228
approche 5719, 12364
approches rectifiées 9658
appui-main 6841
apte au travail 8
aptitude à la flexion 5819,
    10482
aptitude à la floculation
    5843
aptitude au pliage 5819,
    10482
aptitude au roulage 11790
aptitude au vernissage
    14402
aptitude du papier à
    l'impression 10796
aquafortiste 5269
aquatinte au soufre 13384
aquatone 737
arabesques 744

araignée 5916, 13559
arbre 968, 12225, 12869
arbre à cames 2254
arbre à papier 9163
arbre de commande 4748
arbre de commande
    principale 8542
arbre de couche 4748
arbre (de) fudge 6227
arbre de serrage de
    blanchet 1514
arbre de transmission 4748
arbre de transmission
    unique 3000
arbre du volant 5936
arbre feuillu 1930
archives pour photos et
    matériel de rédaction
    9101
arc voltaïque 756
ardillons 9269, 10371
aréomètre 761, 7120
aréomètre Baumé 1233
aréopagite 762
arête du couteau 7823
argent 12424
argent halogénique 6806
argent libre 6163
argentomètre 763
argile 2711
argument 764
armite 8017
armoire à sécher 4786
armoire de climatisation
    7074
armoire d'humidification
    7078
armoire tropicale 7074
armoiries 3001
armorial 774
arpète 10809
arquer 1820
arrachage 10295
arrachage à sec 4813
arrachage dû à l'eau 14688
arrêt automatique du
    programme 919
arrêt de calibrage de la
    roue d'unités 14317
arrêt de l'encrage d'une
    page 9856
arrêt de pompe 12330
arrêt de prise d'encre 3892
arrêt de sécurité 8303
arrêter 12359, 13793
arrêter la publication 4424
arrière-plan 1002
arrondi 4550
arrondisseuse 1047
arrondissure des coins
    11723
arrondissure du dos 1046
arrondissure et endorsure
    11827
arsenic jaune 9691
art de graver à l'eau-forte
    802
art de la reliure 801
art de la xylographie 805

art de reliure 801
art du livre 804
article 787
article camouflé 5441
article de fond 7989
article de remplissage 5622
article de rigueur 9217
article de tête 7989
article en stock 13188
article explétif 5622
article faux 5441
article illustré 7163
article maquillé 5441
article principal 7989
articles divers 13405
articles publicitaires 273
article truqué 5441
articulation à genouillère
    1094
artiste graphique 6567
artiste lithographe 8256
art nouveau 800
art qui préserve tous les
    autres, l'~ 781
art typographique 803
as 13466
ascendante 813
ascenseur 8093
ascension capillaire 2283
asphalte 821
asphalte en poudre 823
aspirail 357
aspirateur 335, 6092
aspirateur de poussière
    14643
assemblage de cahiers 8038
assemblage (photo-
    graphique) 10261
assemblage sur chaîne 844
assemblage transversal des
    bandes 3748
assembler 830, 846, 3070,
    6325
assembleur de casseau
    12789
assembleuse 6330
assombrissement 3989
assortiment complet (d'un
    certain œil) 6103
assouplir l'encre 12736
astérisque 850
astigmatisme 853
astralon 855
atelier 4185, 10403, 14841
atelier de coloration 3175
atelier de composition
    (manuelle) 3292
atelier de machinerie 8465
atelier de photographie
    10242
atelier de photogravure
    10889
atelier de reliure 1412, 1421
atelier de reliure
    industrielle 4971
atelier des annonces 248
atelier des machines à
    composer 3311

atelier des travaux de ville
    7667
atelier d'imprimerie 10829
atlas 861
atlas de poche 10507
atmosphère 863
atmosphère normale 13021
atmosphère standard 13021
atomicité 14378
attaché avec du papier
    gommé 13580
attache de courroie 1303,
    1306
attache-lettres 2877
attache-papiers 9906
attaque 3615
attaquer (le métal) 3613
atténuation 867
attrape-science 722
attribut 14470
augmentation de la
    consistance 4368
aulne 387
aune 387
auramine 874
auréole 4569, 12976
aurichlorure 6517
aurines 877
auteur 881
auteur, à compte d'~ 11004
auteur, chez l'~ 11004
auteur des articles de fond
    7985
auteur-éditeur 885
auto-allumage 12915
autobiographie 891
autocode 895
autocolleur 5927
auto-inflammation 12915
automation 929
automatisation 929
autonome 9513
autoplate 932
autoplatine 916
autopolymérisation 934
autoshaver 937
autotypie 6774
autoxydation 940
avant-propos 6043
avant-titre 1217
avertissement au lecteur
    9444
avertissement (préface de
    peu d'étendue) 6043
avis 247, 636
avis au lecteur 9444
avis de copyright 3555
avoir de la main 2046
avoir son bœuf 10086
axe 963
axe des abscisses 22, 967
axe des ordonnées 966
axe des x 967
axe des y 966
axe optique 9644
ayant de consistance 1669
ayant des ondes 13057
ayant un nez 13057
azotate 9361

azotate d'argent 12432
azotate de plomb 7996
azotate de potassium 10660
azotate d'urane 14369
azote 9374
azurant optique 9646
azureur 9646

bac à colle 10029, 12501
bac à encre 7343
bac de morsure 5290
bac de mouillage 14577
bac de tête 6898
bac-évier à aspiration par
    en bas 4674
bacholle 11891
bagasse 1067
bague de bronze 1961
bagues du cylindre 1252
baguettage 8840
baguette 2183
baguette en verre 6437
baguetteuse 8824
bain 1223
bain acidifère 116
bain bichromaté 2774
bain chromaté 2774
bain d'argent 12425
bain d'arrêt 13209
bain de blanchiment 1545
bain de chromage 2773
bain de chrome 2773
bain de cuivrage 3500
bain de décapage 6549
bain de développement
    4224
bain de durcissement 6864
bain de fixage 5736
bain de fixage acide 124
bain de fixage tannant
    acide 128
bain de fixation 5736
bain de morsure
    biodégradable 1436
bain de saturation 11885
bain de teinture 3121
bain de virage 13899
bain électrolytique 5034
bain galvanoplastique 5064
bain-marie 14568
bain pour galvanoplastie
    5064
bain révélateur 4224
bain thermostatique 13699
baissage 11352
baisser 12234
baisser une espace 11063
bakélite 1082
baladeur 4503
balai de charbon 2313
balai de moteur 2313
balance 11898
balance à cendres 818
balance analytique à pesée
    rapide 578
balance à papier 11107
balance à secteur 11107
balance de laboratoire 7855
balance de torsion 13926

balanceur de cuvette 14013
balancier 5933
balayer 11907
ballade d'encrage 14616
bal(l)adeur 9707, 11302
balle 1090
balle (d'encrage) 7329
balle (d'imprimeur) 7329
ballot 1090
ballottement 11255
bamboché 9726
bambou 1109
banc 1311
banc de reproduction 10882
bande 1110
bande à galets 1825
bandeau 6912, 9689, 11002
bandeau général 1404
bande code finale 2897
bande corrigée 2897
bande d'absorption 36
bande de collage 10040
bande de contact 5770
bande de correction 3606
bande de lancement 11002
bande de nouveauté 11002
bande de papier 9949,
  14639
bande de papier à couleur
  de papier de tenture
  14534
bande de publicité 11002
bande dessinée 3233
bande d'essuyage (de
  bronzeuse) 4902
bande (de texte) brute
  14327
bande-éprouvette 13670
bande gommée 12026
bande magnétique 8524
bande mal tendue 12537
bande non-justifiée 14327
bande perforée 11046
bande perforée brute 14327
bande préimprimée 10731
bande publicitaire 11002
bandes 13597
bandes à glissières 1825
bandes du cylindre 1252
bandes du marbre 1275
bande transporteuse 3482
bande trop tendue 13810
banjo 1119
banque de données 4006
banque, (jour de) ~ 10068
baptême 6341
baquet 8626
barbe 2090, 4065
barboter 10355
barboteur 4396, 4847, 10357
bardeau 6117
bardot 14558
barème 11287
barème d'exposition 5375
barreau 1136
barre de calage 2872
barre de distribution 4499
barre de fraction oblique
  12767

barre (de lettre) 1135
barre de pinces 6652
barre de renversement 588
barre de retournement 588
barre de retournement
  pony 10572
barrer 13272
barres conductrices du
  courant 2101
barres de châssis 2370
barres de traverse 3750
barre transversale 3749,
  12767
barrette d'appui 8630
barrette de tête 8630, 13003
baryte anhydre 1152
baryte caustique 1146, 1152
baryte (hydratée) 1146
barytine 1182
baryum 1142
basane 1183
bascule 5836
basculer en ailes de
  moulin 14826
bas de casse 8379, 11065
bas de page 6022
base 1184
base du bloc de pistons
  10377
base en bois contreplaqué
  7890
basicité 1204
basin de morsure 5274
basse tension 8389
bassine de mouillage 14577
bâti 6144
bâti de côté 12368
bâti de fond 1278
bâti d'encrier 7382
bâti du triangle 6069
batiste (de lin) 2232
batteuse à chiffons 14744
battre le briquet 5452
baudruche 6495
baume du Canada 2255
bavure 1248, 6747
bavures 6743
b.d.c. 11065
bec de cornet 6072
bec d'entonnoir 6072
becs 5970, 14873
bées 14873
belle page 11319
belles pages 9495
benday 12216
benzène 1326
benzine cristallisable 1326
benzine de houille 1326
benzoate de sodium 12675
benzoate de soude 12675
benzol 1326
benzotriazole 1336
béquet 6821, 9768, 10045,
  11597
berçage 6397
berceau 3695, 11626
bestseller 1341
béton armé 11425
beurre 5492

beurre de zinc 14942
bibasique 4288
bibelotier 7673, 7679
bibelots 2460
bibliocaste 1749
bibliographe 1360
bibliographie 1361
bibliomane 1362
bibliophile 1364
bibliothèque 8083
bibliothèque de location de
  livres 8036
bibliothèque de prêt 8036
bibliothèque de
  programmes 10917
bibliothèque privée 10872
bibliothèque royale 11745
bibliothèque universitaire
  14326
biborate de sodium 12680
bicarbonate d'ammonium
  517
bicarbonate de sodium
  12676
bicarbonate de soude 12676
bichlorure de mercure 8792
bichlorure d'étain 13039
bichromate 4298
bichromate d'ammoniaque
  526
bichromate d'ammonium
  526
bichromate de potasse
  10648
bichromate de potassium
  10648
bichromate de sodium
  12688
bichromate de soude 12688
bidon d'huile 9552
biffer 13272
bifluorure d'ammonium 519
bifolioter 1381
bilboquets 2460, 7680
bilingue 1383
billes (de grainoir) 3037
billes de verre pour
  grainage 6430
billes pour le grainage 3037
billet de banque 1126
billet d'entrée 13799
billets de mariage 14664
bimanuel 1391
binoculaire 1435
biographie 1437
bioxalate de potasse 10631
bioxalate de potassium
  10631
bioxyde de baryum 1153
bioxyde de carbone 2315
bioxyde de cuivre 3518
bioxyde de sodium 12708
bioxyde de titane 13860
bioxyde de titanium 13860
biseau 1344, 1346
biseautage 1354
biseauteuse 1355
bismuth 1442
bisonne 6627

bisulfate de potasse 10632
bisulfate de potassium
  10632
bisulfate de sodium 12678
bisulfite de calcium 2162
bisulfite de magnésium
  8497
bisulfite de sodium 12679
bisulfite de soude 12679
bisulfure de carbone 2316
bit 1448
bitartrate de potasse 10633
bitartrate de potassium
  10633
bit de contrôle 2664
bit de parité 10008
bitume 1458
bitume à couvrir 828
bitume de Judée 13497
bitume, sous ~ 14266
bivalent 1460
blaireau à épousseter 4898
blaireau à morsure 5275
blanc 1247, 1500, 3651,
  14711
blanc à atténuer 5384
blanc à couvrir 14951
blanc anglais 8276
blanc-bec 3796
blanc brillant 6456
blanc couvrant 10905, 14951
blanc d'alumine 477
blanc d'antimoine 671
blanc d'argent 5749
blanc de baryte précipité
  sec 1156
blanc de blanc 9880
blanc de Chine 14951
blanc de Clichy 5749
blanc de couture 985
blanc de Crems 5749
blanc de dos 985
blanc de fond 6735
blanc de Gênes 5749
blanc de Hambourg 5749
blanc de Hollande 5749
blanc de lithopone 8276
blanc de marge 12387
blanc de mélange 9014,
  13840
blanc de Meudon 14731
blanc de perle 10074
blanc de petit fond 985
blanc de pied 6012, 6024
blanc de pinces 6656
blanc de plomb 5749
blanc de prise 6656
blanc d'Espagne 12817
blanc de tête 6889
blanc de tête et de queue
  d'impression 7974
blanc de titane 13860
blanc de zinc 14951
blanc d'œuf 4988, 6402
blanc d'œufs 371
blanc du doreur 6402
blanc fixe 1156
blanchet 1508

blanchet de caoutchouc 1509
blanchet de dessous 14267
blanchet de dessus 13909
blanchet de la régleuse 1510
blanchet de liège 3576
blanchet de prise d'empreinte 3724
blanchet (en caoutchouc) 1509
blanchet gravé 5183
blanchets automatiques 898
blancheur 14719
blanchiment 1542, 1543
blanchiment optique 9645
blanchir 1538, 14720
blanchir à l'excés 7741
blanchissant optique 9646
blanchissement des tables 11651
blanc opaque 10905
blancophore 9646
blanc permanent 1156, 14951
blanc pour dernières nou- velles 13217
blancs 1531, 9401, 12811
blanc satin 11878
blancs levés 1493
blancs qui lèvent 4716
blanc transparent 7881
bleu 1642, 3917
bleu acier 10992
bleu alcalin 415
bleu alcalin reflexe 415
bleu de Berlin 10992
bleu de bromophénol 1955
bleu de Chine 10992
bleu de cobalt 3006
bleu de Paris 10992
bleu de phtalocyanure 9065
bleu de Prusse 1337, 10992
bleu d'outre-mer 14247
bleu milori 8945
bleu monastral 9065
bleu paon 10069
bleu-vert 3908
bloc 1583
bloc de bois 14790
bloc de correspondance 13514
bloc de montage 9136
bloc de montage en contre- plaqué 7887
bloc (de papier) 13514
bloc de pistons 348, 10376
bloc de remplissage 5627
blocking 1598
bloc matière 8817
bloc nid d'abeilles 7033
bloc-notes 8771, 13514
bloc-notes de caisse 2421
bloc-notes éphéméride 4210
bloc pour calendrier 4018
bloc pour filets-glissières 8683
blocs d'anglets 590

bloc systématique 10056
blond 1561
bloqué 1593
bloquer 14098
blouse 14831
blutoir (à chiffons) 4891
bobine 11354, 12916
bobine, à ~ 14646
bobine (à fil de fer) 14766
bobineau 11354
bobine brute 8942
bobine déformée 1062
bobine desserrée 12535
bobine détériorée aux extrémités 13050
bobine détériorée par l'eau 14575
bobine écrasée 13050
bobine étroite 4385
bobine excentrée 9727
bobine foirée 13631
bobine mère 8942
bobine molle 1070, 8367
bobine petit format 4385
bobine peu serrée 8367, 12535
bobine préimprimée 10730
bobine serrée 13808
bobinette 11354
bobineuse 14747
bobineuse-refendeuse 12584
bobinier 1853, 11369
bois 1582
bois à pâte 11034
bois conifère 12747
bois contreplaqué 10505
bois de bout 3755
bois de fil 8336
bois de montage (de clichés) 9144
bois de pulpe 11034
bois de sapin 5693
bois de scie 7820
bois feuillu 6881
bois non-résineux 6881
bois résineux 11505, 12747
boîte 1826, 1827, 2378, 2396, 3028, 12561, 14754
boîte à calotte 2290
boîte à chocolat 2749
boîte à cigarettes 2830
boîte à défets 6958
boîte à encre 7330
boîte à fromage 2677
boîte à gateaux 2149
boîte à glissière 12552
boîte à monter 12182
boîte à poudrer 4896
boîte à rabattues 2290
boîte à résine 4896
boîte carton 2344
boîte coulissante 12552
boîte de boutons-poussoirs 3451
boîte de distribution 4500
boîte de fil métallique 13171
boîte des éjecteurs 5003

boîte des espaces-bandes 12803
boîte de valves 14381
boîte en carton 2378
boîte en carton ondulé 3620
boîte en fer blanc 13823
boîte en forme de livre 12750
boîte montée 12182
boîte pâtissière 2149
boîte pliante 5975
boîte pliante à ceinture rapportée 2384
boîte pliante à coulisse et tiroir 12291
boîte pliante à fond automatique 2385
boîte pliante à pattes droites 2386
boîte pliante à pattes droites opposées 2882
boîte pliante à pattes rentrantes alternés 2389
boîte pliante à pattes rentrantes opposées à fenêtre pelliculée 2390
boîte pliante à verrouillage sur parois latérales 2381
boîte pliante mixte 2391
boîte pliante presentoir 4461
boîte pliante type coffret en seul pièce 2387
boîte rigide 12182
boîtier 3451
bol d'Arménie 772
bombé 4550
bon à tirer 11286
bond au bisulfite 13374
bon état, en ~ 9259
bonne copie 5440
bonne épreuve 2895
bonne feuille 2896
borate de sodium 12680
borate de soude 12680
bord abîmé 11365
bord arrière 13055
bord avant du cliché 7990
bord de collage 6472
bord de deuil 9149
bord guide 6695
bord pince 6911
bord pinces 6654
bord rogné à vif 5902
bords détendus 12533
bords non rognés 14264
bords ondulés 12534
bordure 1792, 9689
bordure du blanchet 1710
borne 13653
bornes à vis 1426
bossage 9256
bosse 1297
bosselage des clichés 2061
bottin 4408
bouchage 4790, 5633
bouche (de coulée) 10684
bouche du creuset 10684

boucher 4850, 9613
bouche-trou 5621
bouchon 5621
bouchons 12310
boucle 8354, 8355
boucle de régulation 2931
boucle d'hystérésis 7151
boucle fermée 2931
boucler 5687
boudineuse 5411
boues 12596
bouffant 2045
bougie 2266
bougie-pied 6016
bougran 2030
bouillon 11546
bouillons 9322
bouleau 1438
boulon 1714
boulonner 1715
bouquin 11203
bouquineur 1785
bouquiniste 675, 12045
bourdon 9584
bourrage 5621
bourrelet d'encre 4569
bourrelet de refoulage 1243
bourreur de lignes 3313
bourriquet 8081
boursoufflure 1577
boustrophédon 1818
bout de ficelle effiloché 6152
bout de ligne 1889
boutique 3028
boutique de librairie 1778
bouton d'arrêt (en cas d'urgence) 11323
bouton de contact 11060
bouton démarreur 13056
bouton poussoir 11060
boutons 8417
brai 10381
braie 6204
branchement 1854, 9384
branchement en série 12143
branchement en T 13934
brancher 13478
branches du moulinet 9848
branlement 12227
bras (de porte-bobine) 11355
bras des ballades d'encrage 14618
braser 12753
bras preneur 12040
brasser 13165
brayon 1871
brèche 9433
bridage 10774
bride 5752
bride de galée 6281
bride de suppression de jeu 1030
bride réglable 220
brillance 1909
brillancemètre 6454
brillant 6452, 6457

bristol 1921
bristol fabriqué sur forme
    ronde 8933
bristol Fourdrinier 6126
bristol pour billets de
    mariage 14663
brocard 7895
broché 9900, 13169
brocher 1401
brocheuse au fil métallique
    14779
brocheuse automatique
    sans couture 10119
brochure 1933
brochure à points 6183
brome 1949
bromhydrate
    d'ammoniaque 520
bromocollographie 1953
bromoil 1953
bromoléotypie 1953
bromure 1941, 1947
bromure cuivrique 3501
bromure d'ammonium 520
bromure d'argent 12426
bromure de cadmium 2140
bromure de mercure 8790,
    8791
bromure de méthyle 8862
bromure de potassium
    10634
bromure de sodium 12681
bromure de soude 12681
bromure d'éthyle 5303
bromure d'or 6498
bromure d'or et de
    potassium 10651
bromure ferreux 5574
bromure ferrique 5545
bromure glacé 6462
bromure mercureux 8791
bronzage 1968
bronze 1956, 1957
bronze d'aluminium 474
bronze d'or 6499
bronze en poudre 1965
bronzeur 1967
bronzeuse 1969
bronzeuse-essuyeuse 1969
brosse 1981
brosse à bobines 11359
brosse à colle 10023
brosse à développer 4225
brosse à épreuves 1267
brosse à graphiter 1483
brosse à jasper 8622
brosse à lessive 8422
brosse à mouler 1268
brosse à nettoyer 14176
brosse à poudrer 4897
brosse d'artiste 9870
brosse du cylindre 3935
brosse métallique 14767
brosse pour faire des
    épreuves 1267
brosser 1986
brouillard d'encre 7373
broussonétie à papier 9163
broyage à l'huile 5904

broyer 6637
broyeur 4503, 11029
broyeur à copeaux 2727
broyeur muller 9168
broyeuse à billes 1100
broyeuse à rouleaux 11642
broyeuse à trois cylindres
    13772
broyeuse colloïdale 3092
broyeuse d'encre 7371
broyeuse tricylindrique
    13772
brucelles 14116
brûleur à gaz 6312
brûlot 14920
brun foncé 14384
brunissage 2081
brunissoir 2078
brun Van Dyck 14384
bûchette 12587
bûchettes 12310
budget 725
buis 1839
bullage d'azote 9375
bulles d'air 311
bulletin d'abonnement
    13333
bulletin de cours 5343
bulletin de souscription
    13333
bulletin financier 5343
buna 2063
bureau de coupures de
    journeaux 2917
bureau de publicité 263
bureau du prote 11692
burette 9552
burin 2072, 6578
burin (à biseau) carré
    12967
burinage 6848, 13900
burin à pointiller 13164
buse de pulvérisation 12929
buse (de soufflage) 9445
butadiène 2108
butanol 2128
butée mobile 12555
butées réglables de mise
    en registre 11410
buvard 1621
buvard chromo 5143
buvard émaillé 5143
buvard pour tablette(s)
    13515
buvards de publicité 266

CA 465
cabine d'examen 14467
cabine sécheuse 4786
cabinet à juger les valeurs
    de couleurs 14467
cablage fixe, à ~ 6879
câble bowden 1821
câble coaxial 3003
cabochon 13552, 14470
cabochons 1800
cache 6205, 8646, 11834
cache en papier 6504
cache isolateur 12924

cache perforé 12924
cachet de cire 14630
cachet en caoutchouc 11757
cachou 2474
c.à.d. 14500
cadmium 2139
cadrage par filet décollé
    1464
cadrage par filet détaché
    1464
cadran 4259
cadrat 11108
cadratin 5120
cadratinage 11094
cadratinage automatique
    12085
cadrats creux 11175
cadrats évidés 11175
cadre 1792, 1793, 5690, 6145,
    6146
cadre à claire-voie 3703
cadre auquel paraît le
    dernier mot 4470
cadrer en ovale 9742
cadre support des leviers
    de pistons 10378
cadre volant 4063
cage 2398
cahier 12399
cahier encarté 7443
cahier (grand format) 9437
cahiers à plis non coupés
    2933
cahiers, en ~ 7448
caisse 1827
caisse à claire-voie 3703
caisse aspirante 13351
caisse carton 2344
caisse d'arrivée 6898
caisse d'arrivée (de pâte) à
    gravité 6580
caisse d'égouttage 4695
caisse de mélange 9011
caisse en carton ondulé
    3620
caisse pour la refonte 6958
caisse régulatrice 11422
calage instantané (de
    stéréos) 7460
calage par compression des
    stéréos 3325
calage par tension (des
    stéréos) 13649
calage rapide (des plaques)
    7460
calandrage 2191
calandrage à chaud 7053
calandre 2184
calandre à feuilles 12261
calandre à friction 6193
calandré en feuilles 12262
calandre finisseuse 12995
calandre frictionneuse 6193
calandre pour feuilles
    12261
calcination 2153
calcium 2160
calculateur 5265
calculateur analogique 572

calculateur asynchrone 859
calculateur à usages
    spéciaux 12823
calculateur de composition
    3283
calculateur de la deuxième
    génération 12041
calculateur de la troisième
    génération 13726
calculateur de première
    génération 5708
calculateur digital 4359
calculateur électronique
    3330
calculateur numérique 4359
calculateur synchrone
    13488
calculateur universel 6368
calculatrice 2177
calcul du prix de revient
    3629
calculer 2457
cale 11169
cale d'ajustage 225
cale de justification (et
    d'épaisseur) 8193
cale de réglage de pression
    7225
calendrage à chaud 7046
calendrier 2178
calendrier à effeuiller 1590
calendrier bloc 1590
calendrier en effeuilles
    1590
calendrier en feuilles 1590
calendrier éphéméride
    1590, 4210
calepin 9437
caler 8309
caler (la forme) sur le
    marbre 11067
caler les plaques 10464
cales 13146
calibrage de la copie 2629
calibrage (d'un tube) 2204
calibre 2221, 6334
calibre à patins 11599
calibre de justification
    14198
calicot 2206
calligraphe 2217
calligraphie 2218
calomel 8793
calorie 2225
calorifère 6920
calque 13936
calquer 13935
camaïeu 4869
came 2227
came de clavier 7766
came de commande de
    lame plieuse 5926
came de commande du
    cylindre assembleur 3073
came de coupe du fil
    métallique 3882
came d'espacement 7789
came des picots 9268

caméra (de reproduction) horizontale 7040
caméra de reproduction verticale 14450
campagne de publicité 267
campagne publicitaire 267
camphre 2251
canal 2620, 2621, 2825
canal d'assemblage 835
canal de casseau 12788
canal d'information 7308
canal duplex à sens unique 6759
canal intermédiaire 13963
canal simple 12071
canard 7013, 11199
canardier 9303
candela 2265
canevas 2269
canif 11950
cannelé 12146
cannelé (supérieur) de triangle 11652
cannelure 5911
caoutchouc chloré 2741
caoutchouc durci 6874
caoutchouc éponge 12912
caoutchouc (naturel de plantation) 11747
caoutchouc synthétique 13496
caoutchoutage 8284
cap 11068
capable de travailler 8
capacité calorifique 6918
capacité d'absorption d'eau 14565
capacité de mémoire 13234
capacité de transfert 8100
capacité thermique 6918
capillaire 2278
capillarité 2277
capitale 8557
capitale en deux lignes 14145
capitales 2287
capitales bâtons 1609
capitales, en ~ 11068
capitales et bas de casse 2293
capitales et petites capitales 2295
capitales ornées 13455
capitale sur trois lignes 13756
capteur de nœuds 7830
capucin 6335, 6338
caput mortuum 2300
caractère 2628, 8052
caractère à entaille 9109
caractère allongé 5094
caractère anglais 12012
caractère aplati 1809
caractère à rentrure 9109
caractère à vedette 4471
caractère bloqué 14095
caractère cerné 11615
caractère cogné 1228
caractère d'annonce 201

caractère de commande 3452
caractère de composition à la main 6102
caractère d'écriture 12012
caractère demi-gras 12095
caractère (d'imprimeur) 14171
caractère dit de Schwabach 11924
caractère écrasé 1809, 14855
caractère effilé 5094
caractère empaté 2753
caractère encadré 11615
caractère en cuivre 1868
caractère endommagé 1228
caractère en laiton 1868
caractère étroit 3364
caractère évasé 1808
caractère fonctionnel 3452
caractère gothique 6528
caractère gras 1707
caractère large 5358
caractère machine à écrire 14213
caractère maigre 8111
caractère mal croché 5448
caractère mi-gras 12095
caractère qui chevauche 12955
caractère romain 11670
caractères adhésifs 12074
caractères à fractures 1939
caractères à jour 9722
caractères arabes 749
caractères bâtons 1609
caractères brisés 1939
caractères collés 2151
caractères couchés 9511
caractères cunéiformes 3813
caractères d'édition 1751
caractères de labeur 1688
caractères de titre 13869
caractères d'imprimerie 10851
caractères en bois 1608, 14804
caractère serré 3364
caractères étrangers 6038
caractères évidés 3569
caractères fantaisie 5461
caractères grecques 6619
caractères grotesques 6675
caractères inférieurs 7301
caractères italiques 7625
caractères mobiles 9154
caractères noir-au-blanc 11556
caractères par heure 2642
caractères par pouce 2643
caractères plein œil 13869
caractères pour affiches 10612
caractères pour journaux 9316
caractères pour titres 13869
caractères spéciaux 10080

caractères sur bois 14804
caractères taqués 9511
caractères travaux de ville 4471
caractère usagé 14855
caractéristique 5502
caractéristiques 2638
caractéristiques mécaniques 8730
caractéristiques physiques 10286
carbinol 8855
carbonate 2305
carbonate acide de sodium 12676
carbonate cuivrique 3503
carbonate d'ammoniaque 521
carbonate d'ammonium (neutre) 521
carbonate de calcium 2163
carbonate de chaux 2163
carbonate de cuivre 3503
carbonate de fer 5575
carbonate de magnésie 8499
carbonate de magnésium 8499
carbonate de plomb 7976
carbonate de plomb (basique) 5749
carbonate de soude 12682
carbonate (neutre) de potassium 10636
carbonate (neutre) de sodium anhydre 12682
carbone 2304
carborundum 2334
carboxyméthylcellulose 2336
carbure de silicium 12416
carbures d'hydrogène 7100
caricature 2392
carmin d'indigo 7290
carnet d'échantillons 13456
carnet de chèques 2696
carnet (de poche) 9437
carragheen 7574
carré, au ~ 12962
carte à fenêtre 698
carte à gratter 11952
carte à menus 7628
carte à onglet 13507
carte à perforation marginale 4959
carte à piquer 13145
carte astronomique 2497
carte binaire par colonne 3202
carte binaire par ligne 11741
carte bristol 1921
carte bristol collée 10025
carte céleste 2497
carte collection de teintes 3129
carte d'adresse 2105
carte d'affaires 2105
carte d'artiste 798

carte de commerce 2105
carte de correspondance 3610
carte de couleurs 3129
carte de Lyon 10756
carte d'entrée 13799
carte de pointage 2921
carte de recommandation 2105
carte de réponse 11469
carte de visite 14493
carte du ciel 2497
carte (du jour) du Noël 2759
carte en relief 11444
carte géographique 8612
carte-guide 13507
carte hollerith 7021
carte ivoirine 7627
carte magnétique 8503
carte manille à dossiers 13535
carte marine 2648
carte murale 14532
carte perforée 11045
carte porcelaine 7627
carte postale 10605
carte postale de Noël 2759
carte postale illustrée 10312
carte pour albums 378
carte pour jeux de cartes 10478
carte publicitaire 8539
carter 2398
carte routière 11622
cartes 2105
cartes à jouer 10477
cartes découpées à la forme 4315
cartes de vœux 6626
carteux 11254, 12650
cartographe 2377, 8613
carton 1826, 2342, 2393, 7443
carton à bulletins 13535
carton à chocolat 2749
carton à coupes 3816
carton à dessin 4705
carton à étiquettes 13535
carton à l'enrouleuse 8932
carton à matrice 5848
carton amiante 810
carton baryté 1178
carton biblex 14150
carton bois 11027
carton bois blanc 8738
carton bois brun 1975
carton bristol à la machine Fourdrinier 6126
carton buvard chromo 2814
carton collé 10022
carton compact 12759
carton couché 2981
carton couchémachine 8434
carton couché sur machine 8434
carton cuir 8013
carton d'apprêt 10756

carton de paille 13254
carton de pâte brune mixte 1977
carton de pâte chimique 2687
carton de pâte de cuir 8018
carton de pâte mécanique 8738, 14808
carton de pâte mécanique brun 1975
carton de pointage 2921
carton diélectrique 5009
carton doublé 8170
carton doublé sur machine à formes rondes 14411
carton duplex 4875
carton entoilé 2956
carton feutre pour soucoupes pour verres 2979
carton frictionné 8449
carton frictionné d'une côté 8449
carton gris 2726, 9301
carton gris de vieux papiers recouvert brun 1974
carton huilé 9553
carton isolant 10756
carton ivoire 7627
carton mince 9899
carton multijet 9174
carton multijets 9191
carton multiplex 9191
cartonnage à la Bradel 1845
cartonnage anglais 3685
cartonnage moulé 11030
cartonné 7240
cartonnier 5619, 9902
carton ondulé 3619
carton ondulé double cannelure 4598
carton ondulé double-double (face) 4598
carton ondulé double face 4605
carton ondulé simple cannelure 4605
carton ondulé simple face 12455
carton paille 13254
carton paille mixte 9006
carton pâte 9957
carton pierre 9957
carton pliant 5975
carton pour boîte pliante 5974
carton pour boîtes (pliantes) 1831
carton pour carte postale 10608
carton pour cartonnages 1831
carton pour cartothèques 7270
carton pour emboutissage 1659
carton pour fiches 7270

carton pour flans (de clicherie) 5848
carton pour jeux de cartes 10478
carton pour matrices 5848
carton pour montage 9143
carton pour registres 1408
carton (pour) reliure 1408
carton pure paille 14927
carton revêtu de papier 8170
cartons de couvertures 3676
carton silicaté 12411
carton simili-vulcanisé 10756
cartons-pancartes 1534
cartons pour affiches 1534
carton support pour la protection des aliments congelés et surgelés 1189
carton toilé 2956
carton triplex 14054
carton trois couches 14054
cartothèque 2352
cartouche 2394
caséine 2403
cassant 1925
casse 2396
casse à filets 11774
casse à fractions 6142
casse à nombres rompus 6142
casseau 12789
casseau à barrettes mobiles 2419
casseau à chiffres 5608
casseau à espaces 1139
casseau de caractères accentuées 56
casseau pour espaces 1139
casse californienne 2209
casse d'assortiments 6117
casse de caractères pour travaux de ville 7670
casse de la bande 14641
casse (de la bande) 4675
casse de réserve(s) 6117
casse d'espaces 1139
casse double 4585
casse épuisée 435
cassées 2431
casse espagnole 12813
casse mélangée 6098
casse pour trois caractères (capitales) 14017
casser 1891
cassés de fabrication 1934
cassés de fabrication humides 14679
cassés (de fabrication) secs 4769
cassetin 1829
cassetin à réserve 13448
cassetin au diable 6958
casse triple 14017
casse trop pleine 1140
cassette 2429
cassure 1874

catalogue 2462
catalogue de caractères 14221
catalogue d'éditeur 13185
catalogue de livres de fonds 13185
catalogue de vente aux enchères 11847
catalyseur 2463
cataphorèse 5061
catégorie un 9457
cathode 2480
cathode photoélectrique 10214
cation 2483
cavalier 13040
cavitation 2494
cédille 2495
cédule de marche 11806
cello 94
cellophane 2501
cello-texte 14243
cellule 7331
cellule au sélénium 12073
cellule de mémoire 2498
cellule photoélectrique 10215
cellule photoélectrique à couche d'arrêt 1173
cellule photorésistante 10215
cellule photovoltaïque 1173
celluloïd(e) 2507
cellulose 2508
cellulose nitrique 2517
censure de la presse 2526
centigrades 2530
centimètre 2532
centimètre cube 3799
centipoise 2533
centrage automatique 12079
centre 3565
centre éployé 2549
centre optique 9648
centrer la ligne 2551
centrifugal 2552
cent-vingt huit, en ~ 7079
cercle 2835
cercle chromatique 2762
cercle d'accrochage 2546
cercle de confusion 2836
cercle primitif 10383
cérésine 9818
cérotype 2567
certificat de baptême d'apprenti allemand 6340
céruléine 7290
céruse en lamelles 5749
césium 2572
cessé de paraître 4423
cesser de paraître 4424
cesser le travail 3885
c'est-à-dire 14500
c'est connu 6389
cétone 7753
cétopropane 102
chagrin 6179, 12226
chaîne 13277, 14544

chaîne automatique 913
chaîne d'assemblage 843
chaîne de données fermée 2091
chaîne de fabrication 10910
chaîne de réception 4138
chaîne porteuse 3482
chaînette 2581, 7756
chaînettes 2589
chalcographie 2594
chaleur de fusion 6926
chaleur de vaporisation 6928
chaleur latente 7930
chaleur spécifique 12828
chalumeau 1630
chambre claire 2246
chambre de blanchiment 10679
chambre laboratoire 3993, 6277
chambre noire 3992
chambre obscure 3992
chambre photographique 10882
chambre photographique à distorsion 11491
chambre photo(graphique) à répéter 13116
chambre photographique tout métal 433
chambre pour film en rouleaux 11657
chambre pour pellicule en bobines 11657
chambre séchoir 4791
chambre suspendue 9759
chamois 2612
champ d'intensité 5604
champ électrique 5017
champ électrostatique 5066
champlever 3857
champ magnétique 8517
chanfrein d'accrochage de cliché 1345
chanfreiner 2609
chanfrein unique 13027
changement 463, 12295
changement de bobine 11361
changement de couleur 4420
changement de forme 2617
changement de la justification 2616
changement de structure 13300
changement du corps 2615
changement du format 2616
changements dans la forme 2619
changements sous presse 8441
changer 462, 2614
changeur de bobine en marche 5927
chanvre 6963
chanvre de sisal 12476

chapeau 6903
chapeau de palier 1257
chapelle 2625
chaperon 9784
chaperons 9784
chapitre 2626
charbon actif 169
charbon activé 169
charbon à polir 6642
charbon (de bois) 2644
charbon pour lampe à arc 755
charbons 2328
charbons à âme métallique 3568
charbons métallisés 3568
charbons minéralisés 3568
charge 5383, 5623
charge de rupture 8291
charge électrique 5013
charge électrostatique 13066
charger 8289
charger et exécuter 8290
chargeur 8292, 11600
chargeur bal(l)adeur 11305
chargeur commandé 6348
chargeur d'acier 13099
chargeur de caoutchouc 11754
chargeur de disques 4444
chargeur en caoutchouc 11754
chargeurs en pâte 3308
chariot 2364
chariot à bobine 11356
chariot à lingots 10328
chariot (de mise en place des bobines) 13347
chariot d'encrage mobile 9153
chariot d'envoi des matrices 4148
chariot de sortie 4137
chariot de transport de formes 6075
chariot élévateur 8104
chariot élévateur (à fourche) 6045
chariot porte-formes 6075
charnière 6667
charte de couleurs 3157
chasse 12956, 12973, 14738
chasser 12808
chasser un doublure 4741
chasser une ligne 4740
châssis 2650
châssis à bilboquets 7671
châssis à copie 10828
châssis à copier 10828
châssis (à copier) pneumatique 14375
châssis à feuillure 5977
châssis à négatif 2651
châssis à volet pivotant 7009
châssis de copie 10828
châssis déformé 1357

châssis de la distribution 4502
châssis de magasin 8482
châssis de montage 2651
châssis des magasins 8482
châssis numérateur 9451
châssis oblong 6904
châssis porte-diapositif 14001
châssis porte-film 10442
châssis porte-matrices 8685
châssis porte-plaque 10442
châssis porte-trame 11979
châssis pour pellicule en bobine 11658
châtaignier 2701
chaudière 1697, 7755
chaudière à fusion 8770
chauffage au calorifère 2536
chauffage au gaz 6317
chauffage central 2536
chauffage de la machine à composer 3288
chauffage électrique 5020
chaufferette 2075
chauffer les tampons (d'encrage) 5759
chaux 8145
chaux anhydre 2171
chaux caustique 2171
chaux chlorée 2168
chaux délitée 2167
chaux éteinte 2167
chaux hydratée 2167
chaux vive 2171
chef de bureau 13422
chef de chapelle 5490
chef de conscience 7672
chef de service 13422
chef des presses 10762
chef-typo 2914, 3294, 7672
cheminée 6734
chemins de marbre 2367
chemise 7629, 9918
chèque 2695
chèques à marquage magnétique 8516
chercher 12029
cheval-vapeur 7044
chevreau 6493
chèvre du Cap 9466
chicane d'accès 8121
chicane d'entrée 8121
chicanes 1064
chien d'arrêt 10067
chien écrasé 5635
chiffon de nettoyage 2891
chiffons 11210
chiffraison 9449
chiffre 4356, 5607, 9458
chiffre barré 2263
chiffre binaire 1397, 1448
chiffre d'affaires annuel 640
chiffre de contrôle 2665
chiffre de la signature 12401
chiffre Didot 8206

chiffre en marge 11800
chiffres alignants 8206
chiffres arabes 747
chiffres de bulletin des changes 5345
chiffres de bulletin des cours 5345
chiffres Didot 8206
chiffres elzéviriens 6853
chiffres inférieurs 7301
chiffres romains 11671
chiffres supérieurs 13416
chiffreur 5265
chimigraphie 2694
chimiquement pur 2684
Chine à report 13966
chlorate de potasse 10637
chlorate de potassium 10637
chlore 2742
chlore actif 949
chloréthane 5306
chlorhydrate 7104
chlorhydrate d'ammoniaque 522
chlorhydrate de diamino-phénol 4262
chlorhydrate de paramino-phénol 9962
chlorhydroquinone 2747
chlorite de sodium 12684
chloroforme 2746
chloroplatinite de potassium 10640
chlorure 2737
chlorure allylique 444
chlorure chromique 2787
chlorure cuivreux 3832
chlorure cuivrique 3504
chlorure d'allyle 444
chlorure d'aluminium 475
chlorure d'ammonium 522
chlorure d'argent 12427
chlorure de cadmium 2141
chlorure de calcium 2164
chlorure de chaux 2168
chlorure de cuivre 3504
chlorure de phosphoryle 10208
chlorure de platine 10473
chlorure de polyvinyle 10568
chlorure de polyvinylidène 11875
chlorure de potassium 10638
chlorure de sodium 12683
chlorure de soude 12683, 12696
chlorure de soufre 13386
chlorure d'éthyle 5306
chlorure de zinc 14942
chlorure de zinc iodé 6971
chlorure d'or 6517
chlorure d'or et de potassium 10652
chlorure double de platine et de potassium 10640
chlorure d'urane 14368

chlorure d'uranium 14368
chlorure ferreux 5576
chlorure ferrique 5546
chlorure mercureux 8793
chlorure mercurique 8792
chlorure polyvinylique 10568
chlorure soudique 12683
chlorure stanneux 13039
chlorure stannique 13038
choc 7186
choix de la mise en pages 2751
choix des (styles de) caractères 2750
choix du caractère 2750
chromage 2798
chromate d'ammonium 523
chromate de baryum 1143
chromate (neutre) de plomb 7977
chromate neutre de potassium 10641
chromate neutre de sodium 12685
chromate rouge de potassium 10648
chromaticité 2763
chromatographie sur papier 9905
chrome 2796
chromie 2766
chromiste 3164
chromiste benday 1315
chromolithographe 2817
chromolithographie 2818
chromotypographie 10904
chromoxylographie 2821
chronométrage 13820
chronomètre 13221
chute de potentiel 14510
chute de tension 14510
chutes 9504, 12238
cicéro 2827
ciment armé 11425
cimenter 2525
cimetière des invendus 13229
cinabre 2834
cintrage du cliché 10431
cintré 3847
cintrer 1313, 3846
cintreuse 1587
circonférence 2854
circonférence du cylindre d'impression 2855
circonférence minimum et maximum du cylindre gravé 3949
circuit 2837
circuit imprimé 10800
circuit intégré 7477
circuit microminiaturisé 8904
circuit multipoint 9193
circuit porte 6322
circuit triphasé 13760
circulaire 2844
circulaires 6815

cire à cacheter 12027
cire cérésine 9818
cire d'abeilles 1288
cire de carnauba 2360
cire de lignite 9095
cire de montan 9095
cire de paraffine 9969
cire d'Espagne 12027
cire micro-cristalline 8894
cire minérale 9818
cisaille 12242
cisaille à carton 2345
cisaille (à main) 1666
cisaille (à métal) 6705
cisaille circulaire 2848
ciseaux 11929
ciseaux à papier 9940
ciselet 6578
ciseleur 2656
citation 11172
citer 2861
citrate d'ammonium 524
citrate de fer 5548
citrate de fer ammoniacal 5542
citrate de potassium 10644
citrate de sesquioxyde de fer 5548
citrate de sodium 12686
citrate de soude 12686
citrate ferrique 5548
claie 4797
clair-obscur 2707
claquement 2659
claquement de la planche 10461
clarté 8124
classement alphabétique 456
classement chronologique 2822
classer 5613
classer par ordre alphabétique 458
classeur 2352, 5610, 7830
classeur à copeaux 2730
classeur à galées 6284
classeur de lettres 8057
classeur pour feuillets mobiles 1403
classification décimale 4054
classification des caractères 2879
clavette 7760, 12555
clavier 7765
clavier à perforateur 10134
clavier auxiliaire 187
clavier Monotype 9091
clavier non renseigné 1569
clavier perforateur 10134
claviste 7771, 8439, 9639
clé 7759, 7762, 14872
clé Allen 428
clé à molette 9072
clef 7759, 7762
clef à béquille 11170
clef (à écrous) 14872
clef de renversement 11545
clef de serrage 10444

clef de serrage instantané 2056
clef de serrage rapide 2056
clef du calage 10444
clé plate 9620
clé réglable 9072
clichage 10448
cliché 10418, 10907, 13127
cliché à fenêtre 10324
cliché à relief de gélatine 6363
cliché au collodion humide 14680
cliché au trait 8159
cliché chromé 2800
cliché cintré 3848
cliché cintré en relief 14870
cliché-clé 7757
cliché composite 3299
cliché couleur au trait 8165
cliché d'aplat 5809, 12763
cliché de base 7757
cliché de caoutchouc 11751
cliché de fond 13841
cliché (de photogravure) 1582, 5192
cliché de presse zur zinc 9320
cliché de similigravure 6774, 10233
cliché de teinte 5809, 13841
cliché de trait 8159
cliché du bleu 3919
cliché du cyan 3919
cliché du jaune 14924
cliché du magenta 8491
cliché du noir 1491
cliché du rouge 8491
cliché en caoutchouc 11751
cliché en (matière) plastique 10411
cliché en relief 11445
cliché fractionné 12058
cliché intercalé dans le texte 11793
cliché inversé 11557
cliché magnésium 8498
cliché matrice 8658
cliché multiplié 6301
cliché noir au blanc 11557
cliché non monté 14332
cliché nylon 9465
cliché panné 2706
cliché pas assez creux 12229
cliché plat sans contraste 5799
cliché pour fudge 6226
cliché pré-cintré 10718
cliché qui a glissé 9545
clicherie 10889, 13123
clichés à fond perdu 1552
cliché sans support 14332
clichés couleurs pour journaux 9302
clichés en couleurs 10903
clichés exécutés par benday 1316
cliché simili 6774, 10233

cliché simili et trait 8179
clichés simili en couleurs 3165
cliché sur cuivre 3512
cliché sur zinc 14944
cliché tramé 6774, 10233
cliché tramé en lignes 9596
clicheur 10441, 10888
clicheuse journal 932
cliché wrap-around 14870
cliché zinc 14944
client 74, 2915
climatisation 322
cliquet d'arrêt 10067
cloison (de trame) 2523
cloisonné 2924
cloisons de trame 4524
cloquage 1579
cloque 1577
cloqué 4550
clou 1341, 9225
clouage 9226
coagulation 2971
coaguler, se ~ 6353
coaguler, (se) ~ 2970
coauteur 3002
cobalt 3005
cobol 3019
coccine 3021
cochenille 3022
codage 3036
codage en absolu 28
codage symbolique 13485
code 3030
code binaire 1395
code binaire-cyclique 11388
code converteur d'erreurs 5245
code de Baudot 1232
code de Gray 11388
code détecteur d'erreurs 5246
code fonctionnel 6249
code Hollerith 7022
code machine 3331
coder 5154
codeur 3033, 5155
codeur digital 4366
codeur numérique 4366
codifier 5154
coefficient d'absorption 37
coefficient de Callier 2216
coefficient de dilatation 3038
coefficient de filtre 5658
coefficient de friction 3039
coefficient de frottement 3039
coefficient d'étalement 12937
coefficient de transfer d'encre 7406
coefficient de transmission 13994
coefficient de variation 3040
coefficient de vitesse d'écoulement 3041

coefficients trichromatiques 2764
cœur 12196
coffret de manœuvre 13476
cogneur 8311
cohérence 3046
cohésion 3046
cohésion des fibres 1720
coiffe 2272
coin 3578, 14665
coin de cadre 3581
coin de cuivre jaune 1862
coin de justification 7712
coin de laiton 1862
coin de réglage de pression 7225
coin de serrage 11169
coin report 13975
coins arrondis, à ~ 11725
colcothar 14437
collaborateur anonyme 6395
collage 7689, 12503
collage acide 147
collage à la cuve 13440
collage à la main 6838
collage à la résine 11502
collage de fabrication 8941
collage en bac 13440
collage en pile 1265
collage en surface 13440
collage en surface, avec ~ 626
collage superficiel 13440
collation 3068
collationner 3063
colle 6465, 12486
collé 1080, 10024
colle acide 147
collé à la cuve à la gélatine 626
colle animale 623
colle auto-adhesive 10768
colle bichromatée 4303
colle caoutchouc à papier 9904
collecte de données 4008
collecteur 3075
collection d'échantillons 11855
collectionneur de livres 1747
colle d'amidon 13046
colle de copie 10892
colle de farine 13046
colle (de pâte) 213
colle de poisson 5141, 5716, 10892
colle de relieur 1736
colle de résine 11689
colle d'or 6402
colle d'os 1728
colle-émail 5141, 5716
colle forte 623, 6465
colle froide 3056
colle hot-melt 7048
coller 203, 12892
coller des béquets au blanchet 10053

coller sur 10034
coller sur onglets 6691
collé sur toile 9139
colleuse 10035
colle végétale 14417
collier de déviation 4101
collimater 3079
collimateur 3081
collision 7186
collodion 3082
collodion au bromure d'argent 3083
collodion au chlorure d'argent 3084
collodion iodé 7560
colloïdal 3090
colloïde 3089
colloïde protecteur 10979
colloïdes bichromatés 4301
colloïdes hydrophobes 7125
collotypie 3099
colombelle 3209
colombier (entre les mots) 6734
colonne 3199
colonne de dernières nouvelles 13217
colonne de journal 9313
colonnes dans le sens de la rotation 3211
colonnes en sens traverse 3210
colophane 11684
colophon 3102
colorant 4922
colorant d'alizarine 412
colorant d'indanthrène 7258
colorant organique 9677
colorants 4928
colorants acides 122
colorants azoïques 973
colorants basiques 1202
colorants d'aniline 2973
colorants dérivés du goudron de houille 2973
colorants directs 4402
coloration au verso 3105
coloriage des tranches 13011
colorié à la main 6817
colorimètre 3107
colorimètre à cuivre 3505
colorimétrie 3111
colorimétrique 3108
coma 3214
combinaison 3219
combinaison aléatoire de sujets 12137
combinaison chimique 3316
combinateur de démarrage 13060
combustion spontanée 12915
comètes 3231
commande 4736, 9672
commande à bouton-poussoir 11061

commande à compte-ferme 5699
commande à distance 11458
commande à main 6820
commande automatique 902
commande automatique par bouton 11061
commande coaxiale 3004
commande de diaphragme 4274
commande de disque 4439
commande de fabrication 8579
commande de lames engageantes 5929
commande de l'encrage 7348
commande d'entrée/sortie 7427
commande d'essai 14021
commande de vis micrométriques 1119
commande du marbre 1276
commande électrique 5016
commande ferme 5699
commande individuelle par groupe 7295
commande par chaînes 2583
commande par groupe 6685
commander 3449, 9673
commande sur stock 13189
commencer à l'alignement 1289
commencer sans enfoncement 1289
commentaires 635
commenter 632
commettant 2915
commission 9672
commission du tarif 13604
commodité d'accès 68
commutateur 13474
commutateur à basculeur 13878
commutation de circuit 2840
compact 14329
compagnon de la même rangée 1041, 12380
comparateur de couleurs 3129
comparateur (en pression) 1595
comparer 3256
compas à pointes sèches 4511
compas de mesure 4511
compas d'épaisseur 2222
compas droit (à pointes) 4511
compatibilité 3258
compatible 3259
compendium 3260
compensation 5222
compilateur 3269, 3270
compilation 3267

compiler 3268
composant 3281
composantes trichromatiques 14062
composé 3316
composé à la machine 8734
composé à la main 6845
composé complex 3275
composé diazoïque 4283
composé en toutes lettres 12858
composé glycérophtalique 425
composé organique 9679
composer 3282
composer à l'aveugle 12173
composer à nouveau 11493
composer en chiffres 11066
composer en zigzag 13001
composé résineux 11503
composer plein 11812
composer serré 12805
composeuse 3286
composeuse à distance 13635
composeuse de la deuxième génération 12042
composeuse de la troisième génération 13727
composeuse de première génération 5709
composeuse de quatrième génération 6135
composeuse de titres 6906
composeuse-fondeuse à lignes-blocs 12606
composeuse photographique 10279
composeuse Typograph 14234
compositeur 3313
compositeur à la main 6819
compositeur de labeurs 3314
compositeur de titres 6907
compositeur de travaux de ville 7673
compositeur en conscience 12989
compositeur en paquets 3314
compositeur linotypiste 8223
compositeur-pressier 14125
composition 3301, 8702
composition à distance 13636
composition à distribuer 4035
composition à façon 13949
composition à froid 3061
composition à la main 6818
composition alignée 12159
composition automatique (des textes) 3334
composition aux pièces 10322

composition avec beaucoup d'abréviations 3305
composition compacte 12762
composition complétée 2934
composition compliquée 3276, 8009
composition conservée 13033
composition courante 13247
composition de fabrication 5594
composition de mathématiques 8673
composition de musique 9207, 9215
composition des abréviations 12174
composition des annonces 249
composition des formules scientifiques 3307
composition de(s) tableaux 3208
composition de tableaux 13520
composition d'intercalations 4469
composition du grand titre 3306
composition en attente 13033
composition en cul-de-lampe 12334
composition en drapeau 14328
composition en langues étrangères 6037
composition en paquets 10322
composition en sommaire 6854
composition en vedette 4473
composition fibreuse 5592, 5594
composition gros corps 4469
composition interlignée 7980
composition lardée 9003
composition linotypique 8222
composition manuelle 6818
composition mécanique 8438, 8732
composition Monotype 9090
composition non justifiée 14328
composition par ordinateur 3334
composition photo (du texte) 10278
composition photo-mécanique 3061
composition pleine 12762

écumer 12511
écumoire 12514
éditer 4966, 11003
éditeur 11005
éditeur de musique 9214
édition 4970
édition abrégée 19
édition améliorée 3593
édition amplifiée et
    corrigée 558
édition annotée 634
édition à tirage limité 8150
édition augmentée 5200
édition corrigée 3593
édition corrigée et
    amplifiée 11572
édition corrigée et
    augmentée 11572
édition de luxe 4973, 7919
édition de poche 10510
édition du matin 2052
édition en cahiers 4974
édition en fascicules 4974
édition étrangère 8537
édition expurgée 5378
édition grand-format 7919
édition hors commerce
    10871
édition illustrée 7164
édition originale 5702
édition particulière 10871
édition populaire 2660
édition postale 8537
édition princeps 5702
édition privée 10871
édition réduite 19
édition réservée 12821
édition revisée 11573
édition revue et augmentée
    11572
édition spéciale 12821,
    12822
édition sur grands papiers
    7919
éditorial 4977
éditorialiste 7985
éducation professionnelle
    13621
effacer 2899, 5232, 10547
effect Becquerel 1270
effectif moyen de
    l'échantillon 960
effet abrasif 10
effet actinique 160
effet Albert 362
effet Cabannes-Hofmann
    2136
effet Callier 2215
effet Clayden 2885
effet (dans les blancs) 7758
effet (dans les ombres)
    7758
effet de bordure 1794
effet Debot 4039
effet de filtre 5656
effet de halo 6802
effet de proximité 216
effet d'intermittente 7513
effet Doppler 4562

effet Eberhard 4946
effet Herschel 6969
effet Kostinsky 7836
effet Lainer 7879
effet Nyblin 9463
effet photo-électrique 10228
effet Purkinje 11057
effet Ross 11691
effet Russell 11820
effet Sabattier 11825
effet Schwarzschild 11925
effet Sterry 13133
effet Villard 14474
effet Weigert 14669
effet Weinland 14674
effilocher 6151
effort 13261
effort de cisaillement 12241
égalisateur de feuilles 7686
égalisation 5222
égaliser 7685
égouttoir 4694
égratignure 12601
éjecteur 5001
éjection des déchets 13291
élaguer 3741
élargissement de point
    4566
élasticité 5005, 11500
électricité statique 13067
électrisation du papier
    13067
électro 5071
électro-aimant 5037
électrode 5027
électrode à la quinhydrone
    11159
électrode au calomel 2224
électrode de verre 6428
électrode négative 2480
électrodéposition 5063
électrode positive 641
électrolyse 5032
électrolyte 5033
électro-moteur 5042
électron 5043
électronique 5044, 5052
électrophorèse 5061
électrophotographie 5062
électrostatique 5069
élément 5074
élément acyclique 7247
élémentaire 5075
élément chauffant 6923
élément chauffant à
    immersion 7184
élément constitutif 3281
élément cyclique 3924
élément de chauffage 6923
élément de repiquage
    12028
élément de scie 7827
élément d'impression 4060,
    10852
élément d'impression à
    l'aniline 615
élément d'impression flexo-
    graphique 615
élément en arche 14310

élément en U renversé
    14310
élément intermittent 7247
élément occasionnel 7247
élévateur 8093
élévateur de pile 10342
élider un mot 8021
éliminateur d'électricité
    statique 13069
élimination des sous-
    couleurs 14269
élision 5083
ellipse 5085, 5086
ellipsographe 5087
em 5098
émail à froid 3054
émail (à froid) 5138
émail à sec 4773
émail champlevé 2613
emballage 9827
emballage à alvéoles 3257
emballage de la bobine
    8944, 14865
emballage de produits
    alimentaires 6003
emballage pliant 5992
emballer dans une caisse à
    claire-voie 3704
emboîter 2404
emboîteur 2415
embouchure 9151, 9684
embouti du papier 9915
embrayage 3670
embrayage à friction 6194
embrayage
    électromagnétique 5038
emmagasiner 13184, 13232
émousser 1645
empagement 8569, 9868
empâtement 5633, 12636
empattement cunéiforme
    14666
empattement en forme de
    plaque 12531
empattements 12144
empattement, sans ~
    11868
empattements obliques
    9472
empattement (triangulaire)
    1841
empiler 12997
empirique 5118
emplacement 8302
emplacement de l'analyse-
    mémoire 4862
emplacement de l'image-
    mémoire 4862
emplacer un grisé au
    benday 13845
emploi du calculateur
    électronique pour l'auto-
    matisation de la
    composition 3334
emploi simultane de deux
    ouvriers 4671
empois 13046
empoisonnement par le
    chrome 2780

empoisonnement par le
    plomb 8000
empreinte à chaud 1563
empreinte à froid 3051
empreinte à la plaque 1596
empreinte au plomb 365
empreinte dorée 6496
empreinte sur plomb 7993
empreints digitales 5685
émulsier 5128
émulsif 5128
émulsification 5126, 13843
émulsion 5132
émulsion à grain fin 5673
émulsion au collodion 3086
émulsion eau dans encre
    14583
émulsion encre dans eau
    7364
émulsionnage 5126
émulsionner 5129
émulsion pour plaques
    photographiques 10243
encadrement 1792
encadrer 9718
encart 7433, 7442, 7443,
    13312
encartage à l'extérieur 9735
encarté 9735
encarter 7436
encarteuse 7438
encarteuse-piqueuse 6304,
    11829
encastré 2043
enchaînement 3349
enclume d'agrafeuse 696
encoche 9107, 9431, 9432,
    9736
encoche de réglage 227
encocher 9108
encocheuse 11048
encolleuse 6475
encombrement 5859
encrage 7356
encrage à table plate 13509
encrage excessif 3771
encrage pelliculaire
    supérieur 9790
encrage trop chargé 3771
encrassage du blanchet
    offset 3418
encrassement de la lame
    de couchage 12597
encrassement du blanchet
    par le papier 1519
encre 7325, 7326
encre absorbée 10097
encre à dessin 7280
encre à développer 4226
encre à écrire 7326
encre à écriture 7326
encre à gravure 13016
encre à impression en
    relief 13016
encre à impression sur fer-
    blanc 8820
encre à impression sur
    métal 8820

entrée 2470, 7422, 7423
entrée de magasin 2623
entrée en chicanes 8121
entre-enregistrement 7526
entrée/sortie 7425
entrefeuilleter 7498
entrenerfs 9889
entreposage 13240
entretien 8553
entretoises 13266
enveloppe 5209
enveloppe à agrafe 2878
enveloppe à bouton 13279
enveloppe à bouton à
    pression 10744
enveloppe à bouton
    fermoir 10744
enveloppe à fenêtre
    (rapportée) 14749
enveloppe à fenêtre
    (vernissée) 14749
enveloppe américaine 1125
enveloppe à panneau trans-
    parent 14749
enveloppe à soufflet 6732
enveloppe gommée 210
enveloppement 14859
enveloppe postale 3611
enveloppe poste aérienne
    340
enveloppes à savon 12659
enveloppes de rames 11295
enveloppe vitrifiée 14749
envers (d'un papier) 14158
envoi à condition 4146
envoi à l'examen 4146
envoi d'auteur 10734
envoi en communication
    4146
envoyer un livre à une
    adresse et la facture à
    une autre 4762
éosine 5215
épair 8353
épair fondu 14676
épair irrégulier 2962
épair nuageux 2962
épair régulier 14304
épaisseur 9768, 14738
épaisseur de la couche
    3000
épaisseur de la pellicule
    d'encre 7351
épaisseur d'habillage 9834
épaisseur du cliché 10463
épaisseur du papier 2220
épaisseur systématique
    14319
épaississage 8284
épaississant 1681
épaississement 8284
épargner les majuscules
    7737
épaulement 4737, 12343
épicéa 12953
épigramme 5217
épilogue 5219
épinette 12953
épitre dédicatoire 4074

épitre liminaire 4074
éponge 12910
éponge de viscose 14484
époussiéreuse 4891
épreuve 10936
épreuve à la brosse 2012
épreuve à la main 6830
épreuve à report 8266
épreuve avant la lettre
    10940
épreuve avec la lettre grise
    10958
épreuve beurrée 9161
épreuve bromure 1948
épreuve brute 5792
épreuve chargée 4415
épreuve d'auteur 889, 2895
épreuve de caractères
    14221
épreuve de photogravure
    5190
épreuve d'état 5801
épreuve empâtée 9161
épreuve en couleurs 3168
épreuve en creux 7472
épreuve en noir 1492
épreuve en placard 12573
épreuve en première 5801,
    11270, 12573
épreuve en seconde 6062
épreuve en tierce 10759
épreuve faible 8127
épreuve floue 8127
épreuve foncé 3990
épreuve fumée 12639
épreuve galeuse 9161
épreuve nature 5801
épreuve par contact 3414
épreuve peu chargé 2895
épreuve premier état 10940
épreuves en page 9863
épreuves en première 9863
épreuves gammes 10928
épreuves photographiques
    10250
épreuve sur couleurs
    sèches 4819
épreuve sur machine 10759
épreuve sur papier couché
    10948
éprouvette 13668
épuisé 9725
épurateur de bûchettes
    12589
épuration 2889
épure au lavis 14547
équation 5226
équation de Van der Waals
    14382
équerre 1000
équerre à chapeau 14071
équerre à fondre 2453
équerre à T 14071
équerre de coulée 2446
équerre de montage 11416
équerre du moule 2453
équerre en té 14071
équerre en triangle 4715
équerre réglable 219

équilibré 1085
équilibre 5227
équilibre chromatique 3119
équilibre des couleurs 3119
équilibre des mouvements
    1086
équilibre encre-eau 7408
équipe 3255, 12296
équipe de nuit 9346
équipe du jour 4025
équipement de sortie 9733
équipement de traitement
    10895
équipement périphérique
    10143
éraflé 6191
éraillé 6191
éreintement 3967
erg 5238
ergologie 5239
ergot 12954
errata 5241
erratum 5242
erreur 2040
erreur accidentelle 71
erreur dactylographique
    14230
erreur d'auteur 888
erreur de collimation 3080
erreur de composition 3315
erreur de dactylo 14230
erreur de plume 2911
erreur de repérage 11407
erreur de sorte 14881
erreur d'étalonnage 2205
erreur globale de lecture
    14257
erreur probable 10876
erreur-type 13024
erreur typographique 8979,
    14235
érythrosine 5248
esculine 5251
espace 2904, 12799, 12800
espace-bande 12802
espace chromatique 3188
espace de cuivre 3532
espace de trois points
    13708
espace de un point 6752
espace en laiton 1867
espace fine 6752, 13718
espace forte 13708
espace justifiante 14395
espace levée 11226
espacement 8070, 11167
espace mince 6752
espace moyenne 8919
espacer 4740, 12808
espacer régulièrement
    12806
espaces hautes 6997
espace un tiers 13708
espagnol 12812, 12812
esperluète 557
esprit de bois 8855
esprit de vinaigre 95
esprit doux 12884
esprit rude 12883

esprits 1900
esquisse 9717
essai 13658
essai accéléré 49
essai à la plume 10091
essai à la résistance à la
    déchirure 13617
essai à la résistance au
    déchirement 13617
essai au pouce 14697
essai aux cires 14632
essai aux cires Dennison
    4165
essai Cobb 3018
essai d'absorption de
    l'huile 9551
essai de cloquage 1581
essai de corrosion 3616
essai de décoloration 5434
essai de délaminage 10081
essai de fabrication 14022
essai de flexion 1325
essai de flottage 5837
essai de froissement 3782
essai de gondolage 3836
essai de graissage 12022
essai de laboratoire 7859
essai de l'angle de contact
    594
essai de la résistance à
    l'adhérence en surface
    1606
essai d'encre à la spatule
    4703
essai d'encre par tache
    10057, 13598
essai de porosité 10582
essai de porosité à l'encre
    38
essai de pouvoir couvrant
    6978
essai de rayage 11961
essai de rayure à l'ongle
    5684
essai de rayures 11961
essai de résistance à la
    colle 9349
essai de résistance à la
    crevaison 10578
essai de résistance à la
    perforation 10578
essai de résistance à l'ar-
    rachage avec cires 14632
essai de résistance à
    l'éclatement 10578
essai de résistance au
    frottement 13
essai de résistance aux
    acides 140
essai de rigidité H & D
    6824
essai de séchage 4802
essai de stérilisation 13132
essai de vieillissement 302
essai du papier 9952
essai K & N 7727
essai pliagraphe MIT 9001
essai porométrique 10582
essayer 13659

filet (de vis) orthogonal 11316
filet (de vis) rectangle 11316
filet (de vis) rectangulaire 11316
filet double 9987
filet double-fin 4607
filet double gouttière 14018
filet double-maigre 4607
filet double-quart-gras 4632
filet en acier 13102
filet en cuivre 1866
filet fin 6750
filet gouttière 9987
filet gouttière quart-gras 4632
filet gras 6232
filet gras-maigre 9801
filet maigre 6750
filet métrique 8876
filet mi-gras 8757
filet œil de côté 1352
filet ondulé 14625
filet orné 9690
filet perforateur 10133
filet plein 6232
filet pointillé 4577
filet raineur 3720
filet rond 11731
filets assemblés sans biseautage 45
filets systématiques 7861
filet tremblé 14625
filigrane 14587
filigrane à deux tons 13766
filigrane à la molette 7207
filigrane (à) sec 7180
filigrane clair et foncé 13766
filigrane clair-obscur 13766
filigrane enfoncé 12214
filigrane localisé 8301
filigrane ombré 12214
filigrane par pression 7207
film 5636, 10891
film acétate 90, 2510
film à émulsions super-posés 9175
film à modelé continu 6782
filmatrice 6096
film d'encre 7350
film de sécurité 90
film dit à effet lith 8249
film en bobine 11656
fil métallique 13179
film lith 8249
film multi-couches 9175
filmogène 5642, 5646
film orthochromatique 9697
film pelliculable 13283
film photographique 5636
film prétramé 936
film sur la photo en couleurs 3144
film triacétate 90
fil pour brocher 13179
filtrat 5662
filtre 5654

filtre à bande (de trans-mission) étroite 9242
filtre bleu 1636
filtre coloré 3145
filtre coulé sur glace 9657
filtre de gélatine 6360
filtre de sélection 12129
filtre jaune 14916
filtre magenta 8488
filtre orange 9668
filtre polarisant 10542
filtrer 5663
filtre rouge 8488
filtres trichromes 14034
filtre ultra-violet 14250
filtre vert 6623
filtre Wratten 14871
fin 13732
fin d'alinéa 1875
fin de la bobine 2116
finesse de broyage 5676
finesse de grain 5676
finesse de la trame 11992
fini 5686
fini à l'anglaise 5179
fini à la plaque 10437
fini coquille 4990
fini fantaisie 5458
fini grosse toile 3702
fini mat 4853
finir 3885
fini repsé 11468
fini satiné 5179
finissage 5689
finissage à la brosse 2002
fini sur machine 8446
fini toile 8184
fini toile gros grain 3702
fixage 5724
fixateur tannant 6865
fixatif 5725
fixation des clichés par collage 10422
fixation du cliché 3325
fixation du cliché par tension 13649
fixer 5722, 5723
fixer le blanchet 5738
fixer les clichés 10464
flammage 5913
flammèche 4690
flan 5846
flanelle 5754
flan sans garnissage 9835
flan sec 4805
flans fournis par des agences d'information 1699
flash au xénon 14889
fléchissement 4100
fleurdelisé 4073
fleuron 5816, 14470
fleurons 5870
flexe 2860
flexible 5820
flexion 4100
flexographie 5828
flint 5833
flockage 5844

floconneux 12655
floculation 1886, 5842
florilège 654
flôtre 3643
flottage 5852
flottement (latéral) de la bande 12282
flottement latéral de la bande 14638
flou artistique 12737
fluidité 5885
fluidmètre 5874
fluidogène 11345
fluorescence 5888
fluorescent 5890
fluorure chromique 2789
fluorure d'ammonium 530
fluorure de chrome 2789
flushing 5904
flux 3840, 5912
flux de chaleur 6921
flux de rayonnement 11192
flux énergétique 11192
flux lumineux 8414
flux newtonien 9327
flying paster 5927
focale 5945
foehn 5456
foie de soufre 10677
foire de librairie 1752
foire du commerce du livre 1752
foire du livre 1752
folio 5995
folio français 6172
folios de la préface 10724
foliotage 5993
folioter 5997
fonçage 2080
foncer 4077
fonceuse 2995
fonction cumulative 3810
fonctionnaire de bois 8571
fonctionnement 9633
fonctions trigonométrique 14038
fond 1001, 1002, 12755
fond détouré 3075
fonderie de caractères 14193
fonderie typographique 14193
fondeur de caractères 14192
fondeur Monotype 2441, 9075
fondeuse 2440, 11456
fondeuse de caractères 14181
fondeuse d'interlignes 8161
fondeuse display 4467
fondeuse gros corps 4467
fondeuse Monotype 9088
fond perdu 1553
fondre 2435
fonds perdu 5380
fongicide 6251
fonte 6103
fonte bâtarde 1216

fonte de caractères 14180
fonte de lettres 14180
fonte des rouleaux 2452
fonte poreuse 10584
forage 4732
forcé 9217
force adhésive 214
force capillaire 2281
force centrifuge 2553
force coercitive 3043
force de cisaillement 12240
force de corps 1686
force d'écoulement minimum 14930
force du papier 1209
force électromotrice 5041
force en chevaux 7044
forces de surface 13433
forces Van der Waals 14383
force systématique 14319
Ford coupe consisto-métrique 6030
foreuse 4733
formaldéhyde 6048
formaline 6050
format 4375, 6052, 8720
format à l'italienne 7898
format allongé 5093
format atlas 7898
format brut 14353
format de base 1206
format de journaux 9319
format de la presse 11239
format de la rotative 11239
format de livre 1779
format de marbre 1281
format de poche 10516
format d'une feuille 12487
format en hauteur 10590
format en largeur 7898
format en travers 7898
format fini 14044
format intérieur du châssis de serrage 7416
formation de bourrelets 11601
formation de bulles 1579
formation de faux plis 14874
formation de la feuille 6055
formation de micelles 8888
formation de petits cratères (circulaires) 10366
formation de piqûres d'aiguilles 10366
formation de points 4568
formation de soufflures 1579
formation d'une peau 12519
formation professionnelle 13621
format maximum du papier 8710
format minimum 8961
format oblong 7898
formats de cartes 12497
formats de carton 12495
formats de livres 12496

jaune permanent 13882
jaunissement 14917
jet 7657, 10504, 13537
jet d'acide 117
jet de métal 12984
jet de plomb 12984
jeter 2434
jeter du blanc 1856, 7971, 14720
jeton 4724
jeu 1028, 2904, 5361, 9453, 9837
jeu de caractères 2640
jeu de clichés 12166
jeu de données 4015
jeu de rouleaux 12167
jeune rédacteur 3796
jeux de formulaires 12652
joint 7691, 9828, 12891
joint à brides 5753
joint à genouillère 13874
joint à rotule 1094
joint bout à bout 2115
joint de cardan 14325
joint de fabrication 8941
joint en bout 2115
joint presse-étoupe 6318
joint universel 14325
jonction 7492
Jordan 7693
jouer aux (six) cadratins 7653
joule 7696
jour de paie 10068
jour de paiement 10068
journal 7698, 9309
journal anglais grand format 1520
journal à scandales 14920
journal à sensation 14920
journal de chantage 14920
journal de midi 9420
journal de modes 5478
journal d'entreprise 7060
journal du bord 12309
journal du dimanche 13404
journal du matin 9104
journal du soir 5323
journal explosif 14920
journalier 6574
journaliste 7700
journal officiel 6343
journal quotidien 3958
journaux 10738
jour ouvrable 14833
jugement visuel 14496
jupe 13538
jus 1647
justificatif 2569, 14517
justificatif de tirage 871
justification 7708, 8720
justification courte 9243
justification de colonne 3213
justification et coupure des mots 7144
justifier une ligne 7714
jute 7720
juxtaposé 7723

juxtaposition, en ~ 7723

kakémono 7724
kaolin 2711
kauri 7732
kérosène 7751
kérosine 7751
kieselguhr 4280
kilogramme 7797
kilogramme-force 7798
kilohertz 7795
kilovolt-ampère 7799
kilowatt 7799
kilowatt-heure 7800
kiosque à journaux 9317
kraftliner 7839

label de tête 6901
label-fin 13952
labeur 1683, 1783
labeurier 3314
labeur monotone et continu 6635
laboratoire noir 3992
laboratoire obscur 3992
lactate d'ammonium 537
lactate de potassium 10657
laisser en blanc 8020
lait de couchage 2998
laiton 1859
laize des bobines (de grand format) en centimètres 14737
lambert 7882
lame de couteau 7817
lame d'éjecteur 5002
lame de l'encrier 4839, 7366
lame de pli 2754
lame de scie 11894
lame éjectrice 5002
lamelle de guidage 2826
lamelle guidante 2826
lamelles de conduite des matrices 837
lame plieuse 5973
lames-ressorts 7965
lames rotatives 12585
laminé 7886
laminé à la plaque 10437
laminer 7885
lampe à arc 754
lampe à arc à copier 10818
lampe à arc (électrique) fermé 5149
lampe à arc en vase clos 5149
lampe à arc libre 9617
lampe à chambre obscur 3995
lampe à décharge gazeuse 4419
lampe à filament de tungstène 14086
lampe à gaz xénon 14890
lampe à incandescence 7241
lampe à rayons ultra-violets 14251

lampe (à tube) de quartz 11144
lampe au xénon (haute pression) 14890
lampe au xénon pulsé 14890
lampe à vapeur de mercure 8807
lampe à vapeur de mercure à haute pression 6995
lampe avertisseuse 14543
lampe de copie 10834
lampe de néon 9281
lampe éclair électronique 5049
lampe flash 5764
lampe fluorescente 5892
lampe halogène 6805
lampe inactinique 3995
lampe infra-rouge 7313
lampes témoins 11838
lampe-témoin 10347, 14543
lampomètre 6454
langage 7899
langage à instructions 3238
langage algorithmique 400
langage artificiel 790
langage commun machine 3250
langage d'assemblage 842
langage de procédure 10880
langage de programmation 10924
langage d'exécution 9469
langage indépendant de la machine 8451
langage machine 8453
langage naturel 9246
langage orienté vers le problème 10878
langage original 12792
langage symbolique 13486
lanières 13738
lanoléine 7901
lanoline 7901
lanterne de laboratoire 3995
lapis-lazuli 7906
laquage 7866
laque 7863, 7868, 7880
laque à l'asphalte 1459
laque brillante 1914
laque cellulosique 2515
laque de nitrocellulose 9372
laque de zapon 14932
laque écarlate 11914
laque nitrocellulosique 9372
laque offset en creux 4084
laque pour plaque offset 4084
laque zapon 14932
largeur 14737, 14738
largeur de bande 1116
largeur de contact 5770
largeur de machine 8468
largeur de matrice 1869
largeur de papier 14741

largeur moyenne des caractères 5137
larron 1452, 4543
larrons 6976
laser 7928
latitude de développement 4232
latitude de pose 5372
latitude d'exposition 5372
laver 14556
laver le blanchet 2887
laver, se ~ 1556
laveur (de gaz) 12013
lavis 14547
layette 2414
leatheroïd 8017
lecteur 3546, 11265
lecteur (dans une maison d'édition) 11010
lecteur de bande magnétique 8528
lecteur de bande perforée 13590
lecteur de caractères en clair 9651
lecteur de cartes (perforées) 2355
lecteur optique de caractères 9649
lecteur ultrarapide 7002
lecture avant 6093
lecture de marques sensibles 8640
lecture destructive 4212
lecture en tierce 11272
lecture non destructive 9398
lecture optique de caractères 9650
légende 8031
légèrement collé 12538
lentille 8041
lentille biconcave 3350
lentille biconvexe 3478
lentille composée 3319
lentille concave 3350
lentille condensatrice 3367
lentille convergente 3464
lentille convexe 3478
lentille divergente 4507
lentille negative 4507
lentille plan-concave 10399
lentille plan-convexe 10400
lentille positive 3464
lésion du bord de la bobine 11365
lessivage 3486
lessive 8421
lessive de bisulfite 1446
lessive de cuisson 1701
lessive de potasse 10655
lessive de soude 12695
lessive neuve 14717
lessive noire 1487
lessive sulfitée 13373
lessiveur 4352
lettrage avec lettres à dé calquer par pression 7461

longueur focale 5945
longueur focale postérieure 998
losangé 8391
losange, en ~ 8391
lot 1221
lot désassorti 7678
louche à fondre 2447
louche à main 2447, 7871
loup 14744
loupe 8534
loupe de mise au point 5953, 11419
loupe sur pied 2059
loupier 8356
low key 8385
lubrifiant 8394
lubrification 9557
lubrifier 8395, 9548
Ludlow 8401
luisance 6452
lumen 8407
lumen-erg 8408
lumière 8106
lumière actinique 161
lumière arasante 6600
lumière artificielle 793
lumière blanche 14715
lumière diffuse 4346, 4347, 11915
lumière diffusée 4346
lumière du jour 4023
lumière inactinique 9394
lumière incidente 7248
lumière monochromatique 9077
lumière noire 14252
lumière parasite 4346, 11915
lumière polarisée 10541
lumière ponctuelle 10528
lumière punctiforme 10528
lumière rasante 6600
lumière ultra-violette 14252
luminance 8409
lumination 7160
luminescence 8411
luminosité 8413, 14380
lundi perdu 8377
lunette de visée 11419
lustre 6452, 8419
lux 8420
lyophile 8425
lyophobe 8427

macérer 8429
machinal 8725
machine à adresser 196
machine à affranchir 6149
machine à affûter les couteaux 7826
machine à affûter les racles 4529
machine à aiguiser 7826
machine à appliquer le papier charbon 2333
machine à arrondir le dos 1047

machine à arrondir les coins 3583
machine à assembler 6330
machine à baguetter 8824
machine à bronzer 1969
machine à caractères séparés 12447
machine à cintrer 1319
machine à cintrer des clichés 1587
machine à cintrer le carton 2343
machine à clouer (les clichés) 9228
machine à coins ronds 3583
machine à coller des boîtes pliantes 5976
machine à coller les gardes 5167
machine a composer 3286
machine à copier en répé tition(s) 13116
machine à coucher 2995
machine à coudre 12197
machine à couper les anglets 8998
machine à courber du carton 1320
machine à cylindre frictionneur 14905
machine à découper 4318
machine à détruire 12352
machine à deux couleurs 14134
machine à écraser les plis 5972
machine à écrire à (avec) tête imprimante 6519
machine à écrire électrique 5025
machine à égaliser 7687
machine à emballer 9832
machine à emboîter 2425
machine à empaqueter 9832
machine à encarter 7438
machine à encoller 6475
machine à endosser 1015
machine à enduire 14406
machine à enveloppes 5211
machine à épousseter le bronze 4899
machine à équerrer 11896
machine à étiqueter 7850
machine à fabriquer des sacs 1073
machine à fabriquer des tubes de carton 3485
machine à fabriquer des tubes en carton rondes 3485
machine à fabriquer des tubes spiralés 12879
machine à fabriquer les couvertures 2412
machine à faire les couver tures 2412

machine à faire les emboîtages 2412
machine à fermer les sacs 1077
machine à feuilles 12269
machine à forer 4733
machine à forme ronde 3944
machine à formes rondes 9173
machine à fort débit 6936
machine à gommer 6721
machine à grainer des plaques offset 6550
machine à grand rendement 6936
machine à grand vitesse 7000
machine à graver 5195
machine à graver (à l'eau forte) 5280
machine à graver électronique 10229
machine à graver par bullage 2020
machine à groupe satellite 3249
machine à groupes en ligne 14321
machine à guillocher 6703
machine à huit bobines 4997
machine à impression sur fer-blanc 8821
machine à impression sur métal 8821
machine à impression sur tôle 8821
machine à imprimer à l'aniline 5829
machine à imprimer des tickets 13800
machine à imprimer et découper les étiquettes 7852
machine à imprimer le fer blanc 8821
machine à imprimer les blocs-caisses 2422
machine à imprimer sur tissus 2208
machine, à la ~ 8725
machine à lacquer 14406
machine à laminer 7891
machine à laver 14552
machine à l'impression à l'écran de soie 11991
machine à l'impression à trame de soie 11991
machine à l'impression au pochoir sur soie 11991
machine à marge en bobine 14656
machine à marge en feuilles 12269
machine à mettre sous enveloppe 5214
machine à meuler 6643

machine à monter les clichés 9145
machine à mordancer 5280
machine à morsure 5280
machine à morsure continue 10704
machine à morsure rapide 10704
machine à morsure sans poudrage 10704
machine à multiplier sur film 6686
machine à nettoyer les matrices 8687
machine à nouer les éti quettes 13536
machine à numéroter 9452
machine à paginer 9869
machine à papier 9923
machine à papier à table plate 6127
machine à papier en bobines 14656
machine à papier Fourdrinier 6127
machine à perforer 10132
machine à piquer 14779
machine à plier 5966
machine à pointiller 13163
machine à poser des œillets 5417
machine à poudrer 10701
machine à prise auto matique 14905
machine à quatre couleurs 6123
machine à rainer 3718
machine à recouvrir les boîtes 1836
machine à régler 11784
machine à régler de table 1312
machine à relier les coins 3586
machine à report du papier charbon 2333
machine à reports photo (graphiques) 13116
machine à sac à marge en bobine 11367
machine à tambour 3249
machine à titre 4467
machine à tracer 12006
machine automatique à cylindre 904
machine (automatique) à développer 10900
machine automatique de découpe et de rainage 903
machine auxiliaire 942
machine de copie en retira tion 13116
machine découpeuse refouleuse 3889
machine de réglure à disques 4445
machine de remplissage dosage 5631

machine d'impression flexographique 5829
machine Duplex 4879
machine frictionneuse 14905
machine hélio à plaque 10440
machine lignes-blocs 12606
machine lignes-blocs à double distribution 9008
machine octuple 4997
machine offset 9532
machine offset à retiration 9530
machine offset de bureau 12627
machine offset deux couleurs 14132
machine offset une couleur 12450
machine planétaire 3249
machine pour gros corps 4467
machine pour impression sérigraphique 11991
machine pour la reliure thermoplastique 13694
machine pour l'enroulement en spirale 12879
machine pour l'impression en taille-douce 3525
machine pour l'impression papier peint 14537
machine pour l'impression papier tenture 14537
machine rotative 11699
machine rotocalcographique 9532
machines de reliure 1424
machines pour la reliure 1739
mâchoire à découper 8180
mâchoire de la tête d'élévateur 7650
mâchoire de serrage d'habillage 4723
mâchoires 5981
mâchoires (de l'étau) 14464
machurat 1806
macro-instruction 8477
maculage 9540, 12611, 12636
maculage au recto 5712
maculage des marges 1713
maculage en retiration 12047
maculatures 8981, 12906
macule 12576
macule pour estampage 10467
maculer 12162
madrure 9114
magasin 8479, 13238
magasinage du papier 13228
magasin à papiers 9946
magasin auxiliaire 943, 12375
magasinier 13239
magasin inférieur 8382

magasin latéral 12375
magasin supérieur 14357
magenta 8486
magnésie 8500
magnésie calcinée 8500
magnésium 8496
maigre 6153, 8111
maillet de serrage 8586
main 5718
main de papier 11163
maintenance préventive 10777
maintenir, à ~ 13134
maison de l'impression 10829
maison éditrice 11005
maître de danse 2222
maître imprimeur 8659
maître machiniste (de la rotative) 11700
majoration d'opportunité 10546
majuscule 8557
majuscules 2287
majuscules et médiuscules 2295
malachite artificielle 3503
maladie de peau 4200
malaxeur (d'encre) 9007
malléable 8585
manche 1673
manche de la pointe à correction 6831
manchette 8629, 12348
manchon en feutre 3644
manchon mouilleur 3979
mandrin 8587, 11363
mandrin déformé 3783
mandrin de rouleau 11649
mandrin écrasé 3783
manganèse 8589
mangeurs d'eau 6115
manière anglaise 8883
manière criblée 8597
manière de crayon 3708
manière de lavis 735
manière noire 8883
manifold 6172
manipulations du papier 6832
manivelle 11719
manœuvre 8456, 14347
manque de feuille 8983
manque d'encrage 6192
manque d'encre 1876
manque (d'encre) 14503
manque de poids 12341
manteau 1719
mantisse 8607
Manuaires 8609
manuel 6816
manuscrit 3534
manuscrit imprimée 11478
maquette 808, 4208, 7961
maquette collée de mise en pages 10041
maquette collée en pages 10041
maquette complète 3321

maquette de composition des annonces 253
maquette de mise en pages 7197
maquette de pliage 7963
maquette d'exécution 14835
maquette élaborée 3321
maquette papier 4857
maquette truffée 4859
maquettiste 7962
marbrage 8621
marbre 1271, 12590
marbre à refroidir 10429
marbre de mise en pages 8581
marbre de montage 1605
marbre de serrage 7198
marbre d'imposition 7198
marbres 3037
marbré sur tranche 8615
marbreur 8618
marbrure 8621
marche lente 9602
marche-pied 5514
marcher au ralenti 3705
marcher coup par coup 7245
marcher par à-coups 7245
marge 8297, 8627
marge de fond 1035
marge de petit fond 1035
marge de pied 13550
marge de pinces 6656
marge de tête 6213, 6909
marge extérieure 6033
marge inférieure 13550
marge intérieure 1035
marge latérale 12376
marger 5503
marge supérieure 6909
margeur 4940, 5507
margeur à friction 6195
margeur à nappe 13257
margeur à pile plane 13354
margeur à succion 13354
margeur automatique 909
margeur (automatique) 5508
margeuse 5507
marginale 8629
mariage 6299
maroquin 9106
maroquin à grains écrasés 3784
maroquin chagrin 6179
maroquin de Nigeria 9344
maroquin écrasé 3784
maroquin grainé artificiellement 10155
marque (à la) molette 7207
marque d'eau 14587
marque de commerce 13944
marque d'éditeur 7230
marque de fabrique 13944
marque de feutre 5534
marque de la toile 14774
marque de longueur (d'une voyelle) 8326

marque de multiplication 8639
marque déposée 11406
marque de rouleau de soutien 1018
marque de toile 14774
marque d'imprimerie 7229
marque d'imprimeur 7229
marque du fondeur 10369
marque filigranée 14587
marquer les fautes 8633
marques de blanc 1493
marques de calandrage 2192
marques de collage 10031
marques de dents 6351
marques de gomme 6726
marques de rouleaux 11641
marqueur 5741, 11293
marqueur lumineux 8125
marteau de relieur 1014
marteau d'imposer 12248
masquage 8647
masquage magenta 8490
masque 8645
masque à double recouvrement 14159
masque à l'appareil 2241
masque à simple recouvrement 10595
masque compensateur 14159
masque dans la chambre noir 2241
masque de contact 3410
masque de correction de contraste 13889
masque de correction de couleur 3132
masque de détourage 4540, 10861
masque de réduction de contraste 3445
masque des lumières 6989
masque des ombres 12222
masque flou 14344
masque intégral 7475
masque net 12236
masque par contact 3410
masque par projection 10933
masque réserve 11515
masse d'équilibrage 1088
masse molaire 9054
masse volumique apparente 712
massicot 6704, 8652, 12263
massicot à (coupe) programme 920
massicot à levier 8080
massicot à trois lames 13754
massicot (automatique) tri-latéral 13754
massicot à volant 6850
massicoter 3855
massicot rapide 6998
massicot tri-lame 13754
massif 1272

massiquot 6704
mat 8663
matelas d'air 325
matelas de feutres 10339
matériaux bruts 11258
matériaux polaires 10543
matériel auxiliaire 584
matériel classique à carte
  perforée 14318
matériel de traitement
  6877
matériel photosensible à
  trame incorporée 936
matériel publicitaire 273
matière 8702
matière à caractères 14208
matière à cliché 13126
matière à distribuer 4035
matière à galvano 5072
matière à stéréo 13126
matière blanchie 9625
matière bloquée 1593
matière colorante 4922
matière debout 13033
matière de caractères
  14208
matière de charge 5383,
  5623
matière en attente 1593
matière en excès 5338
matière espacée 9625
matière filmogène 5642
matière gluante 213
matière levée 8095
matière morte 4035
matière pleine 12762
matières colorantes 4928
matières colorantes
  basiques 1202
matières plastiques 10412
matières premières 11258
matières volatiles 14506
mâton 2994
mâtons 3153, 5595, 8417
matrice 5846, 8677, 8678,
  8679
matrice à deux œils 4626
matrice de timbrage 4312,
  13015
matrice de verre 6432
matrice duplex 4626
matrice en laiton 1863
matrice fotografique 6096
matrice grésillée 1578
matrice pour gros corps
  4468
matrice sans revêtement
  9835
matrices (spéciales) à la
  main 12786
mattolin 8707
maturation 8705
mauvais calage des
  cylindres 10462
mauvaise aptitude au
  pliage 10573
mauvaise presse 1063
maxwell 8711
Mécanes 8745

mécanique 8744
mécanisme de changement
  de magasin 8484
mécanisme de distribution
  4491
mécanisme d'embrayage
  des cordons d'alimen-
  tation 13578
mécanisme d'engagement
  de bande 13578
mèche 11733
médiane 8746
még(a)- 8763
megahertz 8764
mélange 1560, 9015
mélange de bandes
  perforées 13583
mélange des caractères
  9012
mélange (des couleurs)
  9010
mélanger 13165
mélangeur 9011
mélangeur (d'encre) 9007
mêler les lettres dans une
  casse 10317
mémoire 13230
mémoire à accès direct
  4399
mémoire à bande
  magnétique 8530
mémoire à disque
  magnétique 8510
mémoire à entrée par
  coïncidence 3048
mémoire à ferrites 3574
mémoire à libre accès
  11233
mémoire à noyaux
  magnétiques 3574
mémoire à tambour
  magnétique 8514
mémoire à tores
  magnétiques 3574
mémoire auxiliaire 947
mémoire externe 5389
mémoire intermédiaire
  2033
mémoire interne 7519
mémoire magnétique 8523
mémoire morte 5732
mémoire non-permanente
  14507
mémoire permanente 9418
mémoire principale 8551
mémoire tampon 2033
mémoire temporaire 13643
mémoriser 13232
mémoriser et faire suivre
  plus tard 13233
ménisque 8773
mention de copyright 3555
mention de réserve 3555
mention d'interdiction 3555
merceriser 8775
mercure 8788
mercure doux 8793
mercureux 8784
mercurique 8777

mesure 8722
mesure de la finesse de
  broyage 5678
mesure de la perméabilité
  à la vapeur d'eau 14610
mesure de lissé 12645
mesure d'un pH 10178
mesure du pouvoir
  absorbant 14967
mesureur 8847
mesureur de brillant 6454
mesureur de l'adhésion des
  feuilles de reliures
  collées 6893
mesureur de la
  perméabilité à l'air 345
mesureur de la planéité
  (du marbre) 5797
mesureur de l'aptitude au
  rainage 3719
mesureur de la résistance
  à l'abrasion 11766
mesureur de temps de
  frottage d'une encre
  11762
mesureur de temps de
  séchage 11762
métabisulfite de potassium
  10658
métabisulfite de sodium
  12702
métadioxybenzène 11519
métal à caractères 14208
métal antifriction 982
métal blanc 982
métal de caractères 14208
métal de doublage 1016
métal d'endossage 1016
métal de stéréotypie 13126
métal épais (de
  photogravure de hauteur
  de 11 points) 6940
métal épais pour clichés
  (onze points) 5082
métal mince (pour clichés)
  13712
métal pour machines à
  composer 3209
métal typographique 14208
métamère 8842
métamérisme 8844
métaoxyphénol 11519
méthacrylate de méthyle
  8866
méthanal 6048
méthanol 8855
méthode d'essai 13666
méthode d'impression
  électrostatique 5068
méthodes statistiques
  13078
méthylacétyle 102
méthylbenzène 13881
méthylcellulose 8863
méthylglycol 5310
méthylviolet 8869
métol 8867
mètre 8871
mètre cube 3804

mètre-ruban 6416
metteur des annonces 254
metteur en pages 8573
mettre à jour 14354
mettre à neuf 6255
mettre à zero 5232
mettre de hauteur 1920
mettre du zèle 12170
mettre en caisse 3704
mettre en casse 4484, 7964
mettre en éventail 5466
mettre en forme 4966
mettre en italiques 7623
mettre en macules 7499
mettre en marche 12160,
  13054, 13788
mettre en mémoire 13232
mettre en pages 8570
mettre en pâte 10317
mettre en route 13054
mettre en séquence 12134
mettre en tablette 9843
mettre entre crochets 1843
mettre entre guillemets
  11177
mettre entre parenthèses
  1843
mettre en vedette 4459
mettre la garniture 8561
mettre le blanchet en place
  5738
mettre le contact 13478
mettre les blancs 6387
mettre sous presse 6529
meuble à casses 2414
meuble à galées 6284
meuble-classeur 5619
meuble pour formes
  serrées 6067
meule 6647
meule courante 11801
meule de dessous 1282
meule de fond 1282
meuler 6636
meuleton 1282
meuleton broyeur 7835
mezzo-tinto 8883
micelle 0007
mi-collé 6768
micro-arrachage 2997
micro-contour test 8893
microfiche 8895
microfilm 8896
microfilm en feuille 8895
micromètre 2221, 6334, 8898
micromètre pour mesure
  des épaisseurs de
  blanchet 1523
micron 8905
micro-ondes 8912
microphotographie 8906
microprogrammation 8908
microscope 8910
microscope de poche 10514
microscope électronique
  5058
microscope pour mesure
  des creux 4193
microseconde 8911

objectif anachromatique 568
objectif anastigmatique 581
objectif aplanétique 703
objectif apochromatique 705
objectif composite 3319
objectif de reproduction 10898
objectif interchangeable 7491
objectif Petzval 10175
objectif traité 2982
obligation 1718
obturateur 12361
obturateur à rideau 5947
occasion, d'~ 12043
ocre 9481
octal 9484
octoate 9486
oculaire 5418, 8724
odeur 9498
œil de la lettre 14894
œil du caractère 5423
œil de la lettre 5423
œil électrique 10215
œil étranger 14881
œilleter 5416
œilleteuse 5417
œilleton 5418
œillets 5415
oersted 9502
oeser 3146
office de contrôle des tirages 870
office de justification des tirages 870
office de publicité 274
officine 9505
offset 9525
offset à sec 4812
offset creux 4085
offset en creux 4085, 9521
offset sec 4812
offsettiste 10754
ohm 9547
oléobromie 1953
oléographie 9581
oléophile 9582
oléophobe 9583
ombre 14255
ombres 12223
omettre 9585
omettre un mot 8021
once anglaise 9711
once avoirdupois 9711
once du commerce 9711
once fluide 5886
once troy 14065
onciale 14262
ondulation 3026
onglet (débordant) 7277
onglet en toile 2954
onglets 6692
onglets (de remplissage) 3265
onglette 6578
opacifiant 9612
opacimètre 9603

opacité 9604
opacité d'après Tappi 13599
opacité d'impression 10836
opacité sur fond papier 9606
opacité sur un fond blanc 9608
opalescence 1650
opaline 9610
opaque 9611
opérande 9628
opérateur 9638, 9639, 9640
opérateur de la Linotype 8223
opérateur Ludlow 8402
opération 9634
opération en série 12141
opération parallèle 9985
optique 9665
opuscule 1933
orange de molybdène 9061
or coquille 12292
ordinateur 3330
ordinateur analogique 572
ordinateur à programme mémorisée 13237
ordinateur de la deuxième génération 12041
ordinateur de réserve 1054
ordinateur de secours 1054
ordinateur spécialisé 12823
ordinateur universel 6368
ordinogramme 5867
ordonnée 9676
ordre 9672
ordre alphabétique 456
ordre chronologique 2822
ordre d'appel 8559
ordre de dates 2822
ordre de passage des couleurs 3185
ordre des plaques 10470
oreille 4933, 9331
oreille de la matrice 4932
or en coquille 12292
organe OU 9666
organigramme 5867
organosol 9682
orientation de trame 11973
orifice 9684
orifice de coulée 9684
orifice de graissage 8398
original 3536, 9685
original au trait 8168
original de trait 8168
original prêt à la reproduction 2248
orné de fleurs de Lys 4073
ornement 9688
orpiment 9691
orpin doré 9691
orpin jaune 9691
orseille 9671
ortho- 9694
orthochromatique 9696
ortho-dioxybenzène 11082
orthographie 12856
ortho-phénylène-diamine 9702

oscillation 9706, 9708
oscillation de la racle 4532
oscilloscope 9709
ouate de cellulose 2521
ours 10753
outil 4236
outil à ton vélo 6578
outil de diamant 4269
ouverture de la maille 8813
ouverture de l'objectif 8043
ouverture de piquage 12224
ouverture de trame 11974
ouverture effective 4981
ouverture relative 11430, 14830
ouverture utile 14830
ouvrage à la pièce 10321
ouvrage à la tâche 10321
ouvrage par souscription 13332
ouvrage pressant 11818
ouvrages de ville 7680
ouvrier 9638, 10753, 14839
ouvrier non qualifié 14347
ouvrier qualifié 7701
ouvrir les ficelles 6151
ovale 9741
oxalate chromique 2805
oxalate d'ammonium 541
oxalate de chrome 2805
oxalate de fer 5551
oxalate de fer ammoniacal 5543
oxalate de potassium 10661
oxalate de sodium 12705
oxalate ferreux 5577
oxalate ferrique 5551
oxalate ferrique d'ammonium 5543
oxychlorure de bismuth 10074
oxychlorure de phosphore 10208
oxydant 9809
oxydation 9803
oxydation anodique 649
oxydation par points 7340
oxyde 9805
oxyde cuivrique 3518
oxyde d'aluminium 480
oxyde d'antimoine 671
oxyde de baryum 1152
oxyde de calcium 2171
oxyde de chrome hydraté 2791
oxyde de cuivre 3518
oxyde de magnésium 8500
oxyde de titane 13860
oxyde de zinc 14951
oxyde di-éthylique 5297
oxyde d'or 6511
oxyde ferrique 5552
oxyde magnétique artificiel 1480
oxyde métallique 8829
oxyder 9807
oxyde rouge de fer 5552
oxyde (salin) de plomb 11330

oxydes de fer 7596
oxydes de fer jaunes 7597
oxydes de fer rouges 7595
oxygène 9810
oxylithe 1153
ozachrome 9814
ozalide 1642
ozobrome 2339
ozocérite 9818
ozokérite 9818
ozone 9819

page 9850
page à la française 4092
page à l'italienne 1932, 9476
page blanche 1532
page d'annonces 256
page de départ 4761
page de fin 12334
page de titre 13867
page d'illustrations 10311
page du grand-titre 6244
page en hauteur 4092
page en largeur 9476
page en regard 9641
page en travers 1932
page impaire 11319
page longue 8343
page modèle 12836
page oblongue 9476
page sans faute 10126
pages de fin 5165
pages de gauche 5327
pages impaires 9495
pages interfoliées 7500
pages liminaires 10724
pages paires 5327
page sur deux colonnes 4588
page sur trois colonnes 13751
pageux 8573
page verso 1040
pagination 5993, 9866
pagination non indiquée 1565
paginer 5997, 9865
pagineuse 9869
paille de blé 14701
paille de riz 11595
paillettes 12304
palan à bobines 11370
paléographie 9874
palette 5628, 9878
pâli 4421
palier 1254
palier à billes 1096
palier d'axe 969
palier de cylindre 3931
palier lisse 10390
palimpseste 9876
palladium 9877
palmer 2221, 8898
pamphlet 1933
pancarte 12350
panchromatique 9887
panne 1880
panneau de commande 3459

papier de Hollande 4917, 6834
papier d'emballage 2408, 14866
papier d'emballage séché à l'air 328
papier d'émeri 5111, 11861
papier de musique 9211
papier de pâte chimique 2692
papier d'épicerie 6663
papier de pulp chimique 2692
papier de riz 11594
papier de sécurité 11840
papier de soie du Japon 7638
papier de soie pour la mise en train 8567
papier de sparte 5254
papier de sûreté 2698
papier d'étain 13829
papier d'étiquettes 7851
papier de tirage par contact 3412
papier de tournesol 8280
papier deuil 9150
papier deux jets 14128
papier de verre 6434, 11861
papier d'habillage 9826, 14168
papier diazo 4284
papier diélectrique 4323, 5009
papier d'impression 10838
papier d'impression couleur 6442
papier doré 6512
papier doublé 7888
papier doublé-bitumé 14308
papier double émail 4586
papier doublure 3017, 8207
papier d'Oxford 1359
papier duplex 4881
papier duveteux 10283
papier écriture 14879
papier écriture fin 3963
papier écriture surfin 13413
papier écru 14259
papier écu 10163
papier émaillé 5145
papier émeri 17, 5111
papier émerisé 5111
papier en blanco 1533
papier en bobines 11373, 14658
papier en feuilles 9922
papier en l'état 14335
papier en rouleau 14658
papier en stock 13188
papier entoilé 9958, 11426
papier entoilé (une face) 2957
papier équerré 12965
papier essuie-planche 10467
papier étanche à la vapeur 14390

papier exempt d'acide 127
papier extérieur pour boîtes 1832
papier façonné 11165
papier fantaisie 5460
papier ferrogallique 1644
papier fiduciaire 3839
papier filigrané 14589
papier filtre 5660
papier fin 5679
papier fini toile 8185
papier floqué 5845
papier fluorescent 5893
papier fortement collé 6875
papier frais 6625
papier frictionné d'un côté 8449
papier frictionné (sur machine) 8449
papier garniture 1034, 7026
papier gaufré 5101
papier genre cuve 9131
papier glacé 6445
papier glacé à la calandre frictionneuse 6197
papier glacé albumineux 374
papier glacé par friction 6197
papier glacé sur calandre à friction 6197
papier gommé 6715
papier gommé inerte 9396
papier gondolé 14624
papier goudron 13607
papier goudronné 13608
papier grainé 6543
papier grainé toile 2233
papier granulé 6543
papier gris pour emballages 11982
papier hélio 11707
papier héliographique 1644
papier huilé 9554
papier hydrofuge 14595
papier hydrophile 33
papier ignifuge 5697
papier imitation cuir 792
papier imperméable 14595
papier imperméable à la graisse 6605
papier imprégné 7206
papier incombustible 5697
papier indicateur 13667
papier indien 1359, 2713
papier inerte 5798
papier ingraissable 6605
papier inhibiteur volatile de corrosion 14389
papier ininflammable 5697
papier irisé 9112
papier Japon 7645
papier jeune 6625
papier Joseph 5660
papier journal 9321
papier journal pour vignettes 6787
papier jute 7722
papier kraft 7840

papier kraft blanchi 1539
papier kraft hydrofuge 14594
papier kraft imperméable (à l'eau) 14594
papier labeur 8728
papier laminé 7888
papier lissé une face 5835
papier mâché 9957
papier machine à écrire 1129
papier marbré 8617, 9115
papier mat 8676, 14300
papier maturé 8706
papier métallisé 8830
papier mi-chiffons 11206
papier millimétré 6573
papier millimétrique 6573
papier miméographe 8949
papier mince 5832
papier mince d'emballage 2291
papier ministre 10163
papier monnaie 3839
papier mousseline 14867
papier mousseline pour serviettes 9239
papier multigraphe 8949
papier nacré 9112
papier noir 1488
papier non apprêté 682
papier non carboné pour copie 2325
papier non collé 14346
papier non conforme aux spécifications 7678
papier non glacé 14300
papier non mûri 6625
papier non pelucheux 5879
papier non rogné 14352
papier offset 9528
papier offset couché (classique) 2984
papier opaline 6444
papier opaque 9615
papier ordinaire 8728
papier ozalid 9815
papier paille 13255
papier paille mixte 9006
papier paraffiné 9967
papier parcheminé 7175
papier parchemin imité 6444
papier parfait 10118
papier parfumé 11917
papier peint 14533
papier pelucheux 5882, 6269
papier pelure 6172, 8599, 9597
papier Perse 2725
papier pétillant 13830
papier photographique 10266
papier photographique au trait bleu 1644
papier plano 5798
papier plié par mains 11165

papier plume 5501
papier porcelaine 2603
papier poste 1129
papier pour affiche(s) 10610
papier pour albums 378
papier pour annuaires de téléphone 13628
papier pour aquarelle 14573
papier pour argenterie 12436
papier pour bagues de cigare 2828
papier pour bibles 1359
papier pour billets de banque 3839
papier pour boulangerie 1873
papier pour cartes géographiques 8614
papier pour cartes maritimes 2649
papier pour cartes routières 11623
papier pour cartouches 2395
papier pour chèques 2698
papier pour clavier 9092
papier pour collage 10043
papier pour collotypie 3096
papier pour copie par contact (au gélatino-chlorure d'argent) 6319
papier pour copie sans développement 10837
papier pour copies multiples 8599
papier pour cornets 1075
papier pour coutellerie 694, 3867
papier pour couvertures 3681
papier pour couvertures collé 10026
papier pour décalcomanie 4042
papier pour décalcomanies céramiques 2557
papier pour découpage à la craie 2603
papier pour diaphanies 14498
papier pour documents 8024
papier pour documents de longue conservation 8347
papier pour doublage de saisses 2408
papier (pour) duplicateur-cyclostyle 8949
papier pour éditions de luxe 13681
papier pour enveloppes (postales) 5212
papier pour épreuves 10946
papier pour étagères 12287
papier pour étiquettes 7851
papier pour gardes 1750

papier pour graphiques 6573

papier pour gravure 10457

papier pour gravure sur acier 13098

papier pour héliogravure 11707

papier pour illustrations 6789

papier pour impression de livres 1770

papier pour impression (en) offset 9528

papier pour impression genre cotonnades 2207

papier pour impressions 10838

papier pour livres 1770

papier pour livres comptables 8024

papier pour machine à écrire 14224

papier pour Monotype 9092

papier pour offset 9528

papier pour panneaux-réclames 10610

papier pour patrons (découpe) 10063

papier pour périodiques 8485

papier pour phototypie 3096

papier pour programmes 9386

papier pour registre(s) 8024

papier pour reports 13966

papier pour revues 8485

papier pour rotogravure 11707

papier pour sachets 1075

papier pour similigravure 6789

papier pour tablettes 13516

papier pour taille-douce 10457

papier pour tickets de caisses enregistreuses 11041

papier pour timbres 13019

papier pour titres 8296

papier (pour) valeurs 8296

papier pour vignettes 6789

papier pour vitrauphanies 14498

papier procédé 11952

papier pur chiffon 11205

papier pure paille 14927

papier pur fil 11205

papier quadrillé 12964

papier réactif 8280

papier recoquillé 4432

papier réfractaire 5697

papier réglé 11775

papier réglé carré 12964

papier renforcé de tissus 11426

papier renforcé (de toile) 2957, 11426

papier repsé 11473

papier résistant à l'acide 139

papier résistant à l'état humide 14693

papier résistant aux acides 139

papier résistant aux alcalis 420

papier sablé 11861

papier saisonné 3368

papier sans acide 127

papier sans acidité 127

papier sans apprêt 14296

papier sans bois 2685, 6681

papier sans carbone 9382

papier sans cendre 817

papier sans chlore 2743

papier sans colle 14346

papier sans fin 14658

papier sans pâte mécanique 6681

papiers asphaltés 822

papier satiné 2190, 6445, 13410

papier satiné semi-mat 5180

papiers bitumés 822

papier sec à l'air 332

papier séché à l'air 329, 8312

papier séché à la machine 8444

papier séché en feuilles 8312

papier sensibilisé 12117

papier sensible 12117

papier sens machine 8335

papier sens travers 12326

papier silicaté 12418

papier silicatisé 12418

papier similicouché 7170

papier similicuir 792

papier similiforme 4066

papier simili Japon 7173

papier simili-sulfurisé 6603

papier simili-sulfurisé non glacé 14299

papiers mélangés 9004

papiers minces 8140

papier soie d'emballage 2291

papier soie pour fleurs artificielles 5872

papier sonnant 13830

papier spongieux 10283

papiers pour faire-part 638, 14664

papiers stratifiés 13253

papiers superfins 13412

papier suédé 5845

papier sulfurisé 14421

papier supercalandré 6445, 13410

papier support 1685

papier support pour couchage 1685

papier surcalandré 13410

papier surfacé à la calandre 2194

papier surfacé sur calandre 2194

papier surglacé 7170

papier teinté 13010

papier tenture 14533

papier toilé 8185

papier tournesol 8280

papier transfert 2331

papier translucide 13986

papier transparent 14005

papier très collé 6875

papier tympan 14168

papier vélin 14858

papier vélin blanc 3713

papier vélin Japon 7644

papier velours 5845, 14430

papier vergé 7876

papier vergé crème 3711

papier verni 14403

papier verré 6434

papier volumineux 2051

papier wet strength 14693

papillon 7442

papillons 1019, 6815

papillotage 12611

papillotage typo 1649

papyrolin 9958

papyrus 9959

paquet 9837, 9851

paqueteur 12998

paquetier 3314

para-diaminobenzène 9992

para-dichlorobenzène 9964

para-dioxybenzène 7127

parafes 5863

paraffine 9969

paraffines 424

paraformaldéhyde 9970

parage 9999

paragramme 2911

paragraphe 1875, 9972

paragraphe carré 5908

paraître 714

paraître prochainement, pour ~ 715

parallaxe 9977

parallélisme 9982

paramètre 9988

paraminophénol 9961

paraminophénol-chlor-hydrate 9962

parangonner 8200, 11237

para-oxyphénylglycine 6485

para-phénylène-diamine 9992

parchemin 9994

parchemin d'agneau 7883

parchemin de peau de mouton 6039

parchemin de veau 14425

parchemin (en cosse) 14425

parchemin végétal 14421

parenthèses 9998

parfait 10118

parité 10007

parité horizontale 8339

parler affaires 13565

parler batiau 13565

parler métier 13565

paroir 10000

paroissien 10708

part de copie 13553

partie annonces 257

partie imprimante 7167, 10819

partie publicitaire 257

partie rédactionnelle 4978

parties non imprimées 1710

pas de caractère 12801

pas en magasin 9728

pasquinade 7895

passage, (bon) ~ 11790

passage d'air 343

passage de papier 7997

passage (du courant) 3840

passe 10019

passe (du papier) 9784

passe-partout 7453

passer 9585

passer à la pierre ponce 11036

passer au blanc de magnésie 2597

passerelle 4061

passer en blanc 11816

passer la bande dans la presse 14640

passer l'éponge 12911

passer les feuilles dans la machine 11816

passer les nerfs en carton 7862

passer une annonce 245

passer une commande 9673

passes 9784

passoire longue 6127

passoire ronde 3944

passure en colle 6477

pastel 10033

patate, en ~ 13803

pâté 1620

pâté 10315

pâté 10315, 11026

pâte à haut rendement 7006

pâte à la soude 12667

pâte à nettoyer l'alliage 5912

pâte à papier 13181

pâte à papier blanchie 1540

pâte à papier semi-blanchie 10017

pâte à polir 11069

pâte à rouleaux 11638

pâte au bisulfite 13379

pâte au chlore 12663

pâte au sulfate 13369

pâte au sulfite neutre 9293

pâte blanchie 1540

pâte chimique 2686

pâte chimique mi-blanchie 10017

pâte d'alfa 5255

pâte de bois 14807

pâte de bois brune mécanique 1976

pâte de bois chimique 2686
pâte de bois feuillus 6882
pâte de bois résineux 12748
pâte de chiffons 11208
pâte de conifères 12748
pâte de feuillus 6882
pâte de fibres 5597
pâte de monocotylédones 6575
pâte de paille 13256
pâte de paille lessivée 14928
pâte de paille macérée 8430
pâte de tremble 820
pâte de vieux papiers désencrés 4122
patée 10061
pâte écrue 14260
pâte épurée 65
pâte grasse 14695
pâte kraft 7841
pâte légèrement blanchie 4943
pâte liquide 2998, 12615
pâte maigre 6164
pâte mécanique 8743
pâte mécanique brune 1976
pâte mécanique, sans ~ 6680
pâte mi-chimique 12097
pâte Mitscherlich 8999
pâte nitrosodique 9367
pâte raffinée 14735
pâte satin 11878
pâte sèche à l'air 333
pâte semi-blanchie 12094
pâte semi-chimique 12097
pâte siccative 4784
patin de bois 14790
patine 6441
pâton 2994
pâtons 8417
pâtons durs 6869
pâtons mous 12741
patron 6065, 11967
patte 9212
patté 10061
patte d'araignée 9555
patte de développement 4227
patte d'enveloppe 5755
patte d'index 7277
pattes de mouche 11964
pavillon 5741, 11293
payer sa bienvenue 6614
peau 5638, 8011
peau chamoisée 2612
peau de chagrin 12226
peau de chamois 2612
peau de chèvre 6493
peau de daim 2031, 2612
peau de mouton teintée 3138
peau de phoque 12025
peau de truie 10337
peau d'orange 9670
peau sciée 12902
peau sciée de vache 2038

pectine 10079
pédale 3743, 14014
pédalier 10456
peigne 11637
peigne à marbrer 8624
peigne de perforation 10131
peigne (pour recueillir des matrices) 8695
peille 11210
peinture à la gouache 6530
peinture à l'eau 14570
peinture au pinceau 2017
pelliculage 13289
pellicule 2987, 5637, 5638, 7950, 12517
pellicule cellulosique 2501
pellicule d'acétate 91
pellicule de gomme 6725
pellicule d'encre 7350
pellicule d'oxyde 9806
pellicule mince 13711
pellicule panchromatique 9888
pellicule photographique 5636
peluchage 5880
peluche 5878, 8226
peluche à bronzer 1971
pelure à report 13966
pelure d'ognon 9597
pelure d'oignon 9597
pelure japonaise 7638
pelure pour reports 13966
pendrier à papier 1105
pendule de laboratoire 3994
pénétration 10098, 13270
pénétration de l'encre dans le papier 7327
pénombre 4569, 10107
pensum 10088
pentachlorophénol 10105
pentane 10106
pente 6535
pente d'une courbe 12591
peptisation 10109
perborate de sodium 12707
perçage 4732
percale (de coton) 2232
percarbonate de potassium 10662
perception au verso 10865
perceuse 4733, 11048
perchlorate de potassium 10663
perchlorure de chrome 2787
perchlorure de fer 5546
perçoir 1672
père virgule 3235
perforage 10130
perforage à trous ronds 11727
perforateur 7784
perforateur au kilomètre 9395
perforateur de bande 13587

perforateur de bande de papier 9950
perforateur en bloc 6302
perforateur non justifiant 9395
perforation de zone 9781
perforation transversale 3763
perforeuse 10132
perhydrol 7115
périmé 9724
période de prise 7296
périodique 7514, 7698
périodique de propagande 11000
périscope 12654
perlage 9114
perluète 557
permanence 3397
permanent 12992
permanganate 10148
permanganate de potassium 10664
perméabilité 10151
perméabilité à l'air 344
perméabilité à la vapeur 14388
perméabilité à l'huile 9564
peroxyde de baryum 1153
peroxyde de sodium 12708
peroxyde d'hydrogène 7115
perpendiculaire 10153
persiennes 1064
persulfate d'ammoniaque 543
persulfate d'ammonium 543
persulfate de fer 5556
persulfate de potassium 10666
perte 12357
perte de brillant 13606
perte d'éclat 13606
perte de densité 8373
perte de papier 8375
pèse-argent 763
pèse-bain 761, 763
pèse-lettre(s) 8054
pèse-liqueur 761
petit châssis 7671
petit clou 13527
petite cale en bois à l'intérieur d'une forme de mobile 4915
petites capitales 5329, 12617
petites capitales, avec des ~ 60
petit fond 985
petit sac 10689
petits clous 1688
petits fers 10164
petits fers, aux ~ 948
petit tirage 12336
pétrolatum 10167
pétrole 10168
pétrole (lampant) 7751
peu collé 11139, 12538
peuplier 10577
pH 10177
phase 10184

phase de dispersion 4451
phase d'induction 7296
phase liquide 8238
phénol 10189
phénol-phtaléine 10192
phénomène 10193
phénomène de catoptrie 2486
phénomène d'interférence 7495
phénylamine 613
phloroglucine 10200
phloxine 10201
pH-mètre 10179
phoque 12025
phosphate tribasique de sodium 12709
phosphate tricalcique 2173
phosphate trisodique 12709
phosphore 10202
phosphorescence 10204
phot 10211
photo 10240
photo aérienne 281
photocathode 10214
photochimie 10216
photocollographie 3099
photocomposeuse 10279
photocomposition manuelle 10256
photocompositrice 10279
photocopie 10273
photodiode 10224
photo-électricité 10230
photoélectrique 10225
photo en couleurs 3159
photoglyptie 14793
photographe 10241
photographie 10240, 10252
photographie de trait 8189
photographie directe 4409
photographie en couleurs 3159, 3160
photographie tramée 6790
photographie trichrome 13747
photograveur 10231, 10888
photogravure 10253, 10906
photogravure en couleurs 3167
photogravure (typo-graphique) 10232, 10890
photolettrage 10256
photolitho 10257
photolithographie 10257
photomécanique 10258
photomètre 5373
photo-montage 10261
photo par contact 3414
photoplastographie 14793
photopolymère 10268
photostat 10273
photo-titreuse 6906
phototypie 3099
phototypie en couleurs 3127
phototypographie 10281
phtalate de butyle 4290
phtaléine du phénol 10192

pic 12865
pica 10288
picots 9269, 10371
picots de blanchet 1518
pie (anglais) 6014
pièce de bœuf 6635
pièce défectueuse 4094
pièce de titre 9889
pièce jointe 5153
pièce moulée 2442
pièces de rechange 12819
pièces de réserve 12819
pièces liminaires 10724
pied 6013
pied carré 12966
pied cube 3801
pied-de-biche 1623, 6578
pied de bois 14790
pied de chambre
  photographique 2249
pied de mouche 9973, 11552
pied de plomb 7975
pieds de mouche 8641,
  11381
pied souffleur 1623
pierre à aiguiser Arkansas
  769
pierre à encre 7393
pierre à huile 9568
pierre à parer 10001
pierre gisante 1282
pierre lithographique 8265
pierre ponce 11037
piger 2457
pigment 10330
pigment-coloré 4926
pigment de molybdène
  9060
pigment de phtalocyanine
  10285
pigment minéral 8954
pigment minéral naturel
  4939
pigment naturel 9248
pigments colorés 3161
pigments d'oxyde de fer
  7594
pigments flushés 5901
pigments inorganiques
  7420
pigments nacrés 9224
pigments organiques 9680
pigment terreux 4939
pignon 6346, 10368
pignon à chaîne 12950
pignon à taille hélicoïdale
  6954
pignon conique 1349
pignon du tambour de
  justification 7716
pile à mêler 9011
pile blanchisseuse 10679
pile de disques 4444
pile défileuse 1881
pile de marge 5505
pile (de papier) 10340
pile de recette 4147
pile désagrégante 1881
pile de sortie 4147

pile hollandaise 7020
pile laveuse 10679, 14548
pile raffineuse 11385
pin 10362
pinacryptol jaune 10353
pinacryptol vert 10352
pinacyanol 10350
pinaflavol 10351
pinatypie 10354
pince 6650
pince à chauffer 2077
pince à film 5641
pince à nerfs 1112
pinceau 9870
pinceau à air 314
pinceau à colle 10023
pinceau à épousseter 4898
pinceau à glairer 6424
pinceau à retouche 11535
pinceau chinois 2718
pinceau pneumatique 314
pinceau pour retouche
  11535
pinceau-vaporisateur 314
pince de relieur 1112
pince de sortie 4143
pince-nerfs 1112
pincer les nerfs 14848
pinces 14116
pince selon Finch 5667
pinces oscillantes 13472
pincette à corriger 14116
pinte anglaise 10373
pinte liquide 8239
pinte sèche 4814
piocheur 14706
pipette 10375
pipette à encre 7377
piquage à plat 12988
piquage à travers 12988
piquage (au fil) métallique
  14780
piquage des coins 3587
piquage de tablettes 9845
piquage en coin 3587
piquant 12651
piquer à la pointe 12867
piquer à plate (métallique)
  12389
piquer à travers 12389
piquer (un report) 13148
piquetage 10295
piqueuse à cheval 11830
piqueuse au fil de fer 14779
piqueuse au fil métallique
  14779
piqûre à cheval 11831
piqûre à plat 1617
piqûre par soudure 13743
piqûres 6139, 8928
piqûres d'épingle 10364
pirlouète 557
piste magnétique 8531
pistolet 314, 866
pistolet (de dessinateur)
  6169
pistolet sableur 334
piston à air 347
piston du creuset 10687

placage 10469
placard (d'affiche) 10609
placards, en ~ 7455
placement de tranchefile
  12172
placement en quinconces
  9285
placer des onglets 6690
placer gratuitement 4029
placer interlignes de
  Limoges 2340
placer interlignes en
  carton 2340
plage d'absorption 36
plage d'identification de
  couleur 3190
plages de calandrage 2197
plages de calandre 2192
plages humides 14692
plagiat 10389
planche 1092, 10419
planche à cadrer 12314
planche à dessin 4706
planche à dessiner 4706
planche composite 3299
planche d'aluminium 482
planche (de métal) copiée
  5768
planche dépliante 13792
planche en couleurs 3162
planche en paravent 73
planche maîtresse 8658
planche photopolymère
  10269
planche simili 6774
planchette de galée à
  coulisse 6286
plan de câblage 4258
plan du clavier 7770
plan du diaphragme 4276
planéité 12644
planer 10392
planeuse 1048, 10395
planeuse-rectifieuse 1611
plan focal 5946
planning 10908
planning de marche 11806
plaquage 10345, 10469
plaque 10420
plaqué 10433
plaque adhésive 13083
plaque à dorer 1584
plaque à embosser 5104
plaque à empreindre de
  laiton 1860
plaque à gaufrer 5104
plaque albumine 375
plaque antihalo 993
plaque au collodion
  humide 14680
plaque (au collodion)
  humide 14689
plaque autochrome 892
plaque bimétallique 1393
plaque complétant (un jeu)
  12168
plaque d'aluminium 482
plaque d'aluminium
  anodisée 650

plaque de cuivre 3521
plaque de magnésium 8501
plaque d'essorage 5570
plaque d'étain 10176
plaque de verre 6435, 10420
plaque de zinc 14952
plaque de zinc offset 9541
plaque d'offset en creux
  4080
plaque en laiton 1864
plaque enveloppante 14870
plaque machine 10758
plaque offset 9531
plaque offset en papier
  9932
plaque (offset) sans creux
  13435
plaque offset wipe-on
  14760
plaque perforée 10128
plaque photographique
  10248
plaque pneumatique 14372
plaque précouchée 10732
plaque précoulée 10732
plaque présensibilisée
  10732
plaque process 10891
plaque sèche 4816
plaque sensibilisée à
  l'avance 10732
plaques trichromes 13749
plaque tournante 14105
plaque trimétallique 14040
plaque typo wrap-around
  14870
plaque wipe-on 14760
plastifiant 10410
plastiline 752
plastique 10407
plastiques 10412
plastisol 10415
plastomètre 10416
plat 5769, 5796
plat arrière 991
plat avant 6207
plat de derrière 991
plat de devant 6207
plat d'encrage 7337
plateau 9878, 13156
plateau à claire-voie 3703
plateau de balle 1092
plateau (de galée) à
  coulisse 6286
plateau d'emballage (pour
  papier) 1092
plateau de sortie 4152
plateau sur patins 13156
plate obtenue par nitration
  9364
platine 10451, 10452, 10472
platine automatique 916
platine de pile 1279
plat inférieur 991
plâtre 3774, 6738
plat recto 6207
plats capitonnés 9844
plats de couverture 3676
plats qui bâillent 14908

positif infiniment variable 10593
positif multiplié 9194
positif offset 14243
positif par contact 3413
positif pour projection 12550
positif sur verre 6436
positif tramé 6792
position 7946
postcalculation 3629
poste par tubes pneumatiques 10506
postface 5219
pot 3777
pot à colle (forte) 6469
pot à fusion 8770
potasse 10618, 10622
potasse caustique 10655
potassium 10622
poteaux 1426
potentiel de grille 6634
potentiel d'oxydoréduction 11334
potentiel d'oxyréduction 11334
potentiel rédox 11334
potentiomètre 10681
potin 10176
pouce 7244
pouce carré 12968
pouce cube 3802
poudrage 2599, 10698
poudrage exécuté sur les quatre côtés 6138
poudre à blanchir 1548, 2168
poudre à bronzer 1957
poudre à flan 5634
poudre à garnir 5634
poudre anti-maculage 687, 4824, 9539
poudre d'aluminium 484, 12433
poudre d'amiante 293
poudre de bronze 1957, 1965, 6515
poudre d'émeri 5113
poudre de talc 13562
poudre insecticide 7432
poudrer 4892
poulie 1308
poulies des cordons d'engagement 13586
poupée 10085
pourcentage de défectueux 10111
pourcentage de point 10112
pourcentage de résine 11685
pourpre de bromocrésol 1951
poussette 11060
pousseur de l'agrafe 13168
pousseur du caractère 14212
poussiérage 4894, 6268
poussière de coupe 3896, 12583

poussière (de papier) 5878, 6267, 8226, 9914
poussière de refente 12583
poussière métallique 8822
poussoir 11060
poussoir des matrices 5081
pouvoir absorbant 32
pouvoir adhésif 214
pouvoir colorant 3189, 6977
pouvoir couvrant 6977
pouvoir couvrant en surface 3675
pouvoir inducteur spécifique 4322
pouvoir masquant 6977
pouvoir opacifiant 6977
pouvoir résolvant 11517
pouvoir séparateur 11516
préchargeur pour margeur 10727
précipitation 10717
précipité 10710
précipiter 10711
précis 20, 3260
précision 85
précontrainte 10774
préface 6043
préfixes pour unités 10721
préimpression pour encartage non repéré 14535
prélèvement 11854
prélèvement d'échantillons 11856
préliminaires 10724
première correction 7057
première couleur 5701
première d'auteur 889
première édition 5702
première épreuve 5792
première forme 5707
première impression 5710
premier élévateur 5703
première ligne 7322
première morsure 5787
première page 6215
premier plan 6035
premier pli parallèle 5713
premier tirage 5710
pré-mise des clichés 2062
pré-mise en train 10728
prendre 12154
prendre d'épreuves 10937
prendre empreinte 9121
prendre empreinte à la brosse 1260
prendre en gelée, se ~ 6353
prendre en masse, se ~ 12760
prendre son bœuf 10086
prendre un calque 13935
prendre une analyse-mémoire 4863
preneur 4841
preneur de pétouilles 3686
préparation à la gomme arabique 6711

préparation de la copie 3551
préparation des originaux 3552
préparation des pâtes 13190
préparation du manuscrit 3551
préparation du papier charbon 10333
préparation du travail 10397
préparation scientifique des décisions 9636
préparer 4204
préparer l'original 8642
préposé aux devis 5265
prescription concernant l'échantillonnage 11857
presse 10738
presse à arrêt de cylindre 13210
presse à balancier 5933
presse à ballots 1093
presse à billot 8423
presse à bobine 14656
presse à bras 6839, 13037
presse à bras lithographique 8260
pressé à chaud 7052
presse à contre-épreuves 10951, 10990
presse à copier 3549
presse à cylindre 3948
presse à déroulement du cylindre 13210
presse à deux couleurs 14134
presse à deux tours 14153
presse à dorer 1604, 6407
presse à double tour 14153
presse à empreindre 8699, 9129
presse à empreinte 9129
presse à encoller 12492
presse à endosser 1015, 1745, 6813
presse à épreuves 6282, 10950
presse à estamper 4337
presse à galée 6282
presse à gaufrer 5105
presse à genouillère 368, 13875
presse à groupes alignés 14321
presse (à imprimer) 10736
presse allemande 8423
presse à main 6839
presse à papier continu 14656
presse à paqueter 2064
presse à pédale 14014
presse à percussion 10117
presse à plate 5775
presse à platine 10452
presse à platine à découper 3888
presse à platine Albion 368

presse à platine automatique 916
presse à quatre couleurs 6123
presse à relier 1427
presse à report 13968
presse à retiration 10122
presse à rogner 10484
presse à soulèvement 14153
presse à temps d'arrêt 13210
presse à timbrer 4336
presse à trois cylindres 9626
presse automatique à cylindre 904
presse à vis 1245, 13034
presse à vulcaniser 14525
presse blanchet sur blanchet 1524
presse de découpe 4318
presse (de) Gally 6292
presse de relieur 1427
presse de relieur et pointe à rogner 10484
presse deux tours 14153
presse en blanc 3948, 13210
presse en blanc un tour 12466
presse encolleuse 12492
presse en creux 6587
presse en taille-douce 3525
presse-étoupe 9828
presse Gordon 6526
presse hélio à plaque 10440
presse hydraulique 7097
presse libre 6162
presse litho(graphique) 8263
presse lithographique à bras 8260
presse modèle Stanhope 13037
presse mono-tour 12466
presse offset 9532, 12642
presse offset à bobines 11698
presse offset à plat 9522
presse offset deux couleurs 14132
presse offset une couleur 12450
presse plate 5775
presse pour taille-douce 3525
presse privée 10874
presse professionnelle 13947
presse quotidienne 3959
presse recto-verso 10122
presse recto-verso à quatre couleurs 6125
presse rotative 11699, 14656
presse rotative à plat 5776
presse rotocalco 9532
presses 10737
presse, sous ~ 6521, 7537
presse Stanhope 13037
presse taille-douce 3525

presse technique 13947
presse typo 8063
presse typographique 8063
presseur 1055, 7213
presse verticale 14454
pressier 10753, 10945
pression d'impression 1044, 7219, 10843
pression du cylindre presseur 7218
pression hydraulique 7098
pression minimale 7805
pression osmotique 9710
pression préalable 10774
pressoir 2873
presspahn 10756
prêt pour l'impression 11286
pré-traitement 10776
prime 10781
prime de production très stimulante 50
primeur d'une grosse nouvelle 11932
primitif 10384
prise 8097
prise à balancier 13472
prise de courant 10488, 12660
prise de la plaque en pinces 12158
prise de l'encre 12176
prise (de l'encre) 14011
prise d'encre 4920, 7401
prise de pince 6656
prisme 10869
prisme de retournement 11563
prix à la feuille 10780
prix aux mil caractères 13739
prix-courant 10779
prix coûtant 3631
prix de la composition 3303
prix de revient 3631
prix de vente 12089
prix faible 9286
prix net 9286
probabilité de rejet 10875
procédé à émail à sec 4773
procédé à la colle 6467
procédé à la colle synthétique 10570
procédé à l'albumine 376
procédé à l'albumine bichromatée 2770
procédé à la soude 12666
procédé Alco 379, 384
procédé à l'écran de soie 11989
procédé à l'émail à froid 3055
procédé Alltone 441
procédé analytique 577
procédé aquatone 737
procédé à racle-traineuse 1496
procédé au benday 1318
procédé au bisulfite 13378

procédé au collodion (humide) 3087
procédé au sang (de) dragon 4689
procédé au sucre 13358
procédé au sulfate 13368
procédé autochrome 894
procédé aux trois diaphragmes 10306
procédé Ballard 1095
procédé Bassani 1210
procédé bois en couleurs 2821
procédé brunak 1980
procédé chalcographique 7471
procédé Claybourn 2883
procédé cronak 3737
procédé de blanchiment 1550
procédé de blanchissage 1550
procédé de conversion 3471
procédé de copie 10860
procédé de copie au bitume 824
procédé de couchage 2988
procédé de cuivrage brillant 1095
procédé de décalcomanie 13969
procédé de dessin de maquette aux couleurs fluorescentes 5889
procédé de gillotage 11667
procédé de reproduction 11482
procédé de retournement 11550
procédé de sélection indirecte 7294
procédé de transformation 3477
procédé d'impression 10844
procédé d'interprétation de couleurs par bromure 5442
procédé d'inversion 11550
procédé Dultgen 4855
procédé Henderson 6965
procédé Kemart 7745
procédé lithographique 8270
procédé Manul 8610
procédé mercurographique 11460
procédé offset à sec 4812
procédé offset creux 4088, 9521
procédé ozalid 9816
procédé Pantone 9893
procédé Perry-Higgins 10154
procédé polyvinilique 10570
procédé Renck 11460
procédé semi-chimique 12096
procédé Texoprint 13673
procédé trichrome 13750

procédé Typon 14244
procédure 10879
processus de séchage 4796
prodiguer les majuscules 7743
production de masse 8653
production en ligne 5873
production horaire 9729
produit d'addition 4942
produit d'adsorption 235
produit de blanchiment 1544
produit de mouillage 13428
produit de viscosité-vitesse 14487
produits chimiques 2690
profondeur de foyer 4195
profondeur de gravure 5276
profondeur du logement de l'habillage 3929
programmateur 10922
programme 10914, 11734
programme d'application 718
programme d'assemblage 833
programme de bibliothèque 8088
programme de coupure logique 8319
programme diagnostic 4252
programme directeur 13423
programme du calculateur 3343
programme fixe, à ~ 6879
programme mémorisé 13236
programme original 12793
programmer 10915
programme résultant 9470
programmerie 12744
programmeur 920
projection 10934
projection de sable 11860
projection du caractère 2632
propane-triol 6482
propanol 10973
propanone 102
proportion 10967
proportion de mélange 9013
proportion de pièces défectueuses tolérées dans le lot 8378
proportion du mélange 9013
propriété 5502
propriété photogénique 10237
propriétés de surface 13436
propriétés diélectriques 4324
propriétés rhéologiques 11584
propriétés spectrales 12840
propylène-glycol 10974
prorata, au ~ 10975
prospectus 6815, 10976

prospectus plié 5967
prote 2914, 6041, 7672
prote à tablier 14834
prote aux gosses 723
prote aux machines 10762
protecteur 6689
prote (de la salle des compositeurs) 3294
prote de la salle des presses 10762
protège-courroie 1305
protéine 10981
protochlorure de cuivre 3832
protochlorure de fer 5576
protochlorure de mercure 8793
protochlorure d'étain 13039
protométrie 10983
proton 10984
protosulfate de fer 5580
protosulfure de sodium 12720
protoxyde de plomb 8248
pruche 6961
prussiate de potasse 10647
prussiate jaune (de potassium) 10650
prussiate rouge (de potassium) 10649
pseudonyme 10101
psychromètre 10996
psychromètre tourbillon 12558
psychrométrie 10997
publication bihebdo-madaire 12104
publication bimensuelle 12101
publication bimestrielle 14148
publication mensuelle 9096
publication périodique 7698
publication semi-hebdomadaire 12104
publication semi-mensuelle 12101
publicité murale 14538
publicité par la presse 9311
publier 11003
puces 6976, 13407
puff 11013
puiseur 14413
puissance colorante 3189
puissance couvrante 6977
puissance électrique 5024
puissance en chevaux 7044
pulpe 11026
pulpe à haut rendement 7006
pulpe au chlore 12663
pulpe chimique 2686
pulpe légèrement blanchie 4943
pulpe mécanique 8743
pulp-engine 11385
pulpe nitrosodique 9367
pulp obtenue par nitration 9364

pulvérisateur 866
pulvérisateur-
  antimaculateur 12926
pulvérisateur à poudre
  4823
punaise 4712
pupitre à retouches 11536
pupitre de commande 3395,
  3459, 13476
pupitre de retouche 11536
pureté colorimétrique 3110
pureté (colorimétrique)
  11056
purificateur 12013
pycnomètre 11075
pyrite 11078
pyrocatéchine 1-2 11082
pyrogallol 11085
pyromètre 11086
pyrosulfite de sodium
  12702
pyroxyline 11087

quadrichromie 6124
quadrichromie, en ~ 11109
quadrivalent 13672
quai d'embarquement 8294
qualité de conservation
  7740
qualité de pliure 10482
qualité moyenne 959
qualité moyenne après
  inspection 955
quantifier 11125
quantité de chaleur 11123
quantité de lumière 11124
quantité d'encre 7381
quart de gallon 8240
quarte 8240
quaternion 9485, 11145
quatre-douzes 6130
quatre fois 11088
quatre-vingt-seize, in-~
  9351
questionnaire 11148
queue 13550
queue de morue 5275, 5778
queue de page 12334
queue de rat 6335
quinone 11161
quinternion 11162
quotidien 3958

rabais 4425
rabaisien 3028
rabattre une encre 1879
raboter 10392
rabotoir 10395
rabouter 12892
raccordement 12891
raccord en T 13934
raccorder 10047
racine 11677, 11679
racine carrée 11679
racle (d'essuyage) 4525
racler 14755
raclette 4525
raclette (de caoutchouc)
  12978

radial 11190
radiation 11194
radical 11195, 11679
radioactif 11197
raffinage (en pile) 1266
raffinage (en raffineur)
  11386
raffinage gras 14677
raffinage maigre 6155
raffineur 11385
raffineur conique 7693
raffineuse 11385
raffineuse à meuletons
  7835
raffineuse Jordan 7693
raies de Fraunhofer 6150
raies d'impression 11218
raies noires du spectre
  6150
rainer (peu profond) 3716
rainure 3714
rainure de clavetage 7786
rainure de clavette 7786
rainures de logement des
  griffes 4546
raison inverse de, en ~
  7540
ralenti, au ~ 9602
ralentir 12594
ramasse-pâte 11891
rame 11291
rame de 480 feuilles 12335
ramette 7671
ramie 11228
rampe 2227
rang 2137, 3284, 11182
rang carcasse 14733
rangée de la densité 4175
ranger par ordre alpha-
  bétique 458
rang plat 11229
râpage 6641
râperie 6645
rapidité d'une émulsion
  12851
rapport 11252
rapport de résistance à
  l'éclatement 2100
rapporter 13853
rapporteur 10986
rapport revêtu de
  signatures en rond, ou
  en cercle pour ne pas
  révéler le chef de bande
  11729
rapprocher 2943
raquette (de sortie) 5915
raquette porte-rouleau
  11644
raquettes 5915
raseur 1942
rassemblement de données
  4008
rat de bibliothèque 1785
râteau à matrices 8695
râteau de pression 11951
râteau (d'une presse lithographique à bras) 11951
râtelier à châssis 6067

râtelier à formes 6067
râtelier de séchage 4797
rayer 13272
rayeur 11777
rayon 1664, 11198
rayon-layette 1664, 2414,
  4797
rayon lumineux 8128
rayonnement 11194
rayons actiniques 162
rayons bêta 1343
rayons chimiques 162
rayons ultra-violets 14254
réacidulation 5669
réactif 11288
réactif de Herzberg 6971
réaction 11261
réaction à l'obscurité 3991
réaction au noir 3991
Réales 13981
rebobineuse 11580
rebobineuse à tension
  constante 3400
reboucher 9871, 12917,
  13215
rebuts 8981
réception 4134
réception à chaîne 2580
réception à cordons 13581
réception à l'avant 6208
réception automatique 906
réception frontale 6208
réceptivité à l'encre 7385
recevant les corps gras
  9582
receveur 5924
receveur automatique 906
réchauffeur 6920, 10722
recherche binaire 1399
recherche opérationnelle
  9636
recherche scientifique
  11928
recherche scientifique
  appliquée 720
réciproque 11301
réclame 2471
réclame tapageuse 11013
recomposé 11814
recomposer 11493
reconnaissance
  automatique 925
reconnaissance de
  caractères 2639
reconnaissance de
  configuration 10065
reconnaissance magnétique
  de caractères 8520
reconnaissance optique du
  caractère 9650
recoquillé 4431
recoquillement 3026, 3833
recoupe de hachures 3756
recouvrement 9766, 9770,
  11542
recouvrement de solvants
  12778
rectangle 11315
rectangulaire 11315

rectificateur 12371
rectificateur (de courant)
  11317
rectifier 6636
rectifieuse 1048
recto 11318, 11319
recueil 7698
récupération 11314, 11496
récupération du solvant
  12778
rédacteur 4976
rédacteur au marbre 8572
rédacteur chargé de la
  mise en pages 8572
rédacteur de la rubrique
  financière 2866
rédacteur en chef 4979
rédacteur gérant 11526
rédacteur qui fait la
  marbre 8572
rédacteur responsable
  11526
rédactrice 4976
rédiger 4965
redresseur à cristal 3788
redresseur (de courant)
  11317
réducteur 11344
réducteur de Farmer 5471
réducteur proportionnel
  10969
réduction 11349, 11350
réduction à Fougeadoire
  11753
réduction de point 4574
réduction des couleurs
  dans les ombres 14269
réduction des données 4014
réduction du bruit 9385
réduire 3740, 11341
réduire la consommation
  d'encre 3892
réduire le débit d'encre
  3892
réédition 11429
réfection 8562
refendeuse 12581
référence 11380
référence de source 3721
référence du dépôt légal
  13022
références 1361
réflectomètre 11393
reflet 11387
réflexion 11387, 11389
réflexion spéculaire 12849
refondre 11454
refoulage 3714
refouler (profond) 3716
refouleuse-traceuse 12586
réfractaire aux acides 145
réfraction 11395
réfractomètre 11397
refroidissement 3487
refroidissement à l'air 324
refroidissement par eau
  14574
refroidisseur de nappe
  3490

service des annonces 252
service des commandes
  9674
service des impressions
  10761
service des ventes 11848
serviette en papier 9929
servomoteur 12150
servomoteur de correction
  11412
se solidifier 12760
sesquicarbonate de sodium
  12715
sesquichloride de fer 5546
sesquichlorure de chrome
  2787
sesquioxalate de fer 5551
sesquioxyde de chrome
  2792
sesquisulfate de fer 5556
set 12153
seuil d'allure 1884
seuil de concentration
  13777
seuil de prime 1884
sexternion 12204
siccatif 4784
siccatif au cobalt 3007
siccatif au manganèse 8590
siccatif au naphténate 9237
siccatif au plomb 7979
siccatif cobaltique 3007
siccatif en pâte 10028
siccatif Japon 7636
siccatif liquide 8235
siccité 4772
siccité de l'encre 7342
sidérographie 12385
Sienne 12394
signal analogique 573
signal de commande 3236
signal d'entrée 7428
signal digital 4361
signal d'interruption 7530
signal numérique 4361
signature 12402
signature encartée 7443
signature seconde 851
signe ¢ 2535
signe brève 12323
signe d'alinéa 9973
signe de barres parallèles
  9983
signe de degré 4108
signe de fraction 6143
signe d'égalité 5224
signe de la livre sterling
  10691
signe de la prononciation
  brève 12323
signe de multiplication
  8533, 8639
signe de paragraphe 9973,
  12059
signe de pourcentage 10113
signe de répons 11300
signe de séparation 12807
signe de suppression 4131

signe de transposition
  14008
signe diacritique 4250
signe d'infini 12403
signe d'intégration 7476
signe d'intercalage 2357
signe d'intercalation 7440
signe d'omission 2357
signe graphique 6564
signe minute 8972
signe naturel 750
signe radical 11679
signes algébriques 393
signes arithmétiques 766
signes astronomiques 857
signes botaniques 1802
signes conducteurs 7984
signes d'aspiration 1900
signes de chimie 2691
signes de correction 3600
signes de géométrie 6383
signes de mathématiques
  8674
signes de météorologie
  8846
signes de pharmacie 8747
signes de ponctuation
  11050
signes de référence 11381
signes de répétition 4505
signes du zodiaque 14965
signes hiéroglyphiques
  6979
signes médicaux 8747
signes pour jeu d'échecs
  2699
signes pour jeu de dames
  4699
signet 1762
silicate d'aluminium 485
silicate de potasse 10670
silicate de potassium 10670
silicate de sodium 12716
silicate de soude 12716
silice 12409
silicium 12415
siliciure de carbone 12416
simili 6774, 10233
simili à blancs purs 1628
simili à claire-voie 6788
simili agrandie 1627
simili (à) grosse trame 2978
simili à grosse trame
  imprimé comme cliché
  au trait 11140
simili à hautes lumières
  6987
simili au carré 12963
simili-bristol 1692
similicuir 791, 8014·
simili découpée 9721
simili dégradée 14473
simili de presse à grosse
  trame 9324
simili détourée 9721
simili détourée et dégradée
  9720
simili deux tons 4869
simili-duplex 4877

simili grand creux 4091
similigravure 6774, 6793
similigravure à blancs purs
  6987
similigravure à grain 6545
similigravure en creux
  6783
simili-kraft 1693
similiste 6781
simili-sulfurisé 6603
simili sur zinc 14947
simili veau-marin 359
simogramme 12437
simplification du travail
  14842
simuler 12440
simultanéité par partage
  du temps machine 13818
sinus 12443
sinusoïdale 12444
sirupeux 13498
size presse 12492
s long 8346
slotteur-imprimeur 10812
slotteur-imprimeuse 10812
socle 1272, 1278
sodium 12668
software 12744
soif de lecture 7083
sol 12749
solarisation 12751
solde (d'édition) 11450
soleil 10315
soleils 13407
solénoïde 12754
solide 12992
solide des couleurs 3187
solidifier, se ~ 12760
solidité à la lumière 5484
solidité de la couleur 3142
solidus 12767
solubilité 12768
soluble 12769
soluble dans l'eau 14603
solution 12772
solution alcoolique de
  lavage 383
solution aqueuse 739
solution d'albumine 377
solution d'albumine
  bichromatée 2771
solution d'asphalte 827
solution de blanchiment
  1547
solution de décapage 10298
solution de gomme
  adragante 1273
solution de hypochlorite de
  potasse 7649
solution de lavage anhydre
  609
solution de lavage des
  clichés 10425
solution de lavage pour
  blanchet 1526
solution délayée 739
solution de morsure 5286
solution de morsure en
  creux 4089

solution de mouillage 3980
solution d'enlevage 827,
  14555
solution de protection des
  talus 12382
solution d'iode dans
  l'iodure de potassium
  7564
solution moléculaire 9053
solution saturée 11881
solution tampon 2034
solvant 12773
solvant (de reliure) 209
solvant-naphta 12776
solvant organique 9681
solvant rapide 5487
solvants aliphatiques 409
solvants aromatiques 777
solvants polaires 10544
sommaire 20
somme algébrique 396
sommet 8773
sonnant 11254, 12650
sonnette 5448
sorbite 12783
sorbitol 12783
sortante 9723
sortie 4145, 9730, 9731
sortie à chaîne 2580
sortie à cordons 13581
sortie à la main 9622
sortie à moulinet 5462
sortie à part 12897
sortie à plat 5786
sortie automatique 906
sortie (de feuilles) 4134
sortie directe 13248
sortie en paquets 7788
sortie frontale 6208
souche 13305
soucoupe de verres 1286
soudage par haute
  fréquence 6983
soude 12668
soude à l'alcool 12695
soude caustique 12695
souder (au cuivre) 12753
souder le cliché sur bloc
  métal 13458
souffler une casse 1629
soufflet 1292, 1293, 6730,
  14074
soufflure 336
soufflure de fonte 5810
soufflures 12475
soufre 13383
souiller 3417
souillure du blanchet par
  le papier 1519
souligner 14285, 14288
souple 5820
souplesse 13425
source de lumière 8133
source lumière 8133
sous-bock 1286
souscopie 8970
sous-copie 8970
sous-couche anti-halo 667
sous-couleurs 14270

teneur en eau d'équilibre 5228
teneur en eau (sur brut) 9033
teneur en eau (sur sec) 9035
teneur en fibres de coton 3637
teneur en humidité 9033
teneur en matière sèche 4772, 12764
teneur en substance solide 12764
ténor 7985
tensio-actif 13427
tensio-activité 13429
tensiomètre 13647
tension 4701, 14509
tension anodique 648
tension constante de la bande 3399
tension de grille 6634
tension de la bande 14661
tension d'électrode 5030
tension de ressort 12946
tension du réseau 14511
tension interfaciale 7494
tension préalable 10774
tension superficielle 13441
tenue de fichier 5614
térébenthine 14106
térébenthine de Venise 14438
térébenthine du mélèze 14438
terminaison 6746
terminal 13652
terminer 5687
ternion 13655
ternissement 13606
terre à infusoires 4280
terre à porcelaine 2711
terre de Cassel 14384
terre de diatomées 4280
terre de Sienne 12394
terre de Sienne brûlée 2088
terre d'infusoires 4280
terre d'ombre 14255
terre d'ombre brûlée 2089
terre pourrie (d'Angleterre) 4280
terre siliceuse 12409
test au Scotch 11942
test du micro-contour 8893
tête agrafeuse 13177
tête à graver 5193
tête chercheuse 11912
tête d'analyse 11912
tête de chapitre 2627, 6912
tête (de la page) 6890
tête de lecture 11912
tête de lecture/écriture 11284
tête de l'élévateur 5704
tête de lettre 8058
tête de machine 6898
tête de page 6912
tête en balle de golf 6520

tête encolleuse 10036
tête en sphère 6520
tête piqueuse 13177
têtes de clous 14855
têtière 6889
tétraborate de sodium 12680
tétrachlorométhane 2330
tétrachlorure de carbone 2330
tétravalent 13672
texte 13674
texte en trop 3772
texte préliminaire 10724
texte qui continue 14102
théorie de la diffraction 4345
théorie des ouvertures de la trame 10365
therblig 13685
thermodurcissable 13696
thermogravure 13690
thermo-interrupteur de sécurité 13695
thermomètre à point d'ébullition 1703
thermoplastique 13692
thermoplongeur 7184
thermorégulateur 13698
thermoscellage 6931
thermosoudage 6931
thermostat 13698
thèse 13700
thiocarbamide 13723
thiocyanate d'ammonium 550
thiocyanate de mercure 8806
thiocyanate de potassium 10676
thiocyanate de sodium 12728
thiosulfate d'ammonium 551
thiosulfate de baryum 1162
thiosulfate de sodium 7146
thiourée 13723
thiourée d'allyle 450
thixotropie 13737
thymol 13798
tierce 6062, 10759
tierce des blancs 11022
tige 12231
tiges 1426
tilde 13812
tilleul 1211
timbrage à sec 4825
timbre à date 4021
timbre en caoutchouc 11757
timbres de publicité 10611
timbre sec 4333
timbres-réclames 10611
timbres-vignettes 10611
tirage 2852, 7208, 10817, 11787
tirage à froid 1576
tirage à l'heure 9729
tirage à part 9516

tirage de la chambre (photographique) 2243
tirage du ... 10803
tirage du soufflet 1294
tirage limité 8150
tirage préliminaire 10729
tirant 13525
tiré à part 9516
tire-ligne 4711, 11786
tire l'œil 5414
tirer 10794, 10795
tirer d'épreuves 10937
tirer les épreuves d'essai 11018
tirer une épreuve 11018
tirer une impression 11018
tiret 4001
tiret sur demi-cadratin 4002
tireur d'épreuves 10945
tireuse 3406, 10828, 13116
tiroir de caisse aspirante 4067
tissu imprimé 10801
titane 13551
titrage 13870
titre 6902, 12232, 13862, 13867
titre à gros caractères 1131
titre courant 11805
titre de chapitre 2298
titre de colonne 2298
titre de dédicace 4076
titre de départ 13342
titre de dos 1050
titre de dos en long 13863
titre de la signature 12402
titre de message 6900
titre de page 2298
titre divisionnaire 12060
titre d'obligation 4038
titre encadré 1835
titre en caractères d'affiches 1131
titre général 1404
titre rentrant 3863
titreuse 6906
titrier 13064
titulaire de la rubrique locale 2867
toile 1735
toilé 2955
toile à calquer 13937
toile à reliure 1735
toile artificielle 785
toile-ballon 1102
toile cirée 2559
toile de la machine 8469
toile métallique 8469
tolérance 13880
tolérance de dilution 4374
toluène 13881
toluol 13881
tomber en ligne 5160
tome 10011
ton 13887
tonalité 13898
tonalité chromatique 7066
ton chair 5815

ton continu, à ~ 3433
toner 13892
toner rouge feu 5698
ton moyen 8920
tonne forte 8350
ton nuancé 14292
ton opposé 3442
ton rabattu 12209
ton rompu 12209
tons neutres 9295
tore en ferrite 5561
tore magnétique 8504
torsion 13925
tortillon 13313
touche 7356, 7761, 13525
touche à spatule 4703
touche d'abréviation 7535
touche d'annulation 4130
touche de couleur 3190
touche d'effaçage 4130
touche de justification 7709, 12804
touche d'espacement 12804
touche étalée 4703
toucher 5524
toucheurs 6073
toupillage 8939, 11737
toupilleuse 11732
tour 7938, 11578
tour à guillocher 6703
tour à papier 9953
tour du cylindre 3950
tourelle porte-lentille 8048
tourie 2337
tourillon 7697
tourillons des rouleaux 11640
tourner 5860
tournette 14709
tournevis 12001
tournez s'il vous plaît 10479
tours à la minute 11579
tour sans feuille 8983
tous les droits de traduction (de reproduction, de représentation, d'exécution) réservés 439
tout composé 442
tout droit de reproduction réservé 439
toxicité 13933
traçage à l'outil 13900
tracé 1562, 6698, 7961
tracé (coupant) 11935
tracé (de contours) 7779
tracé de contours 10090
tracer (au couteau) 11936
tracé refoulé 3714
tracé sur papier inactinique 6503
tracette 11937
traduction 13982
traduction algorithmique 401
traduction à machine 8467
traduction autorisée 883
traduction littérale 8247
traduction mécanique 8467
trafic 13950

traîneau du composteur 839
trainées de couchage 11220
trainées de développement 4223
trait 8155
trait de l'épaisseur d'un cheveu 6748
trait d'union 4003, 7143
traité 14016
traitement 10893, 10894
traitement anodique 649
traitement automatique des données 905
traitement d'après 10616
traitement de la pâte 13190
traitement de l'information 7309
traitement des données 4013
traitement de surface 13430
traitement de surface de l'objectif 8045
traitement du fichier 5615
traitement électronique des données 5046
traitement par lot 1222
traitement préparatoire 10776
traitement reversible 11550
traitement séquentiel 12136
traitement ultérieur 10616
trait fin 5674, 6748
trait horizontal 14475
trait latéral 1135
trait ondulé 14622
trait ponctué 4575
traître 1484
traits de plume 5863
traits de repère 11411
trait simili combiné 8179
trame 6145, 14668
trame à grain 6552
trame à points elliptiques 2582
trame à points en chaîne 2582
trame apparente 11981
trame brique 1903
trame circulaire 2847
trame cristal 3759, 6794
trame de contact 3416
trame de similigravure 6794
trame duplex 4882
trame en damier 2667
tramé en lignes 12459
trame en nid d'abeilles 7034
trame en verre 6438
trame fine 5680
trame grossière 2976
trame hélio 6591
trame lignée 6763, 8195
trame magenta 8487
trame metzographe 6552
trame mezzographe 6552
trame mur de briques 1903

trame par contact 3416
trame pour similigravure 6794
trame quadrillée 3759
tramer 11972
trame réticulée 11531
trame rhomboédrique 11923
trame ronde 2847
trame Schulze 11923
trame serrée 5680
trame simili 6794
trame (simili) Pola 10545
tranchante de la lame 7823
tranche dorée 6412
tranche extérieure 6031
tranchefile inférieure 13541
tranchefile (supérieure) 6894
tranches 4961
tranches ciselées 6494
tranches coloriées 3135, 13009
tranches couleur 13009
tranches coupées 3860
tranches ébarbées 3860
tranches jaspées 8615, 12949
tranches marbrées 8615
tranches mouchetées 12949
tranches non coupées 14334
tranches non rognées 14264
tranches orientales 6494
tranches peignées 3217, 8615
tranches rognées 3860
tranches rouges 11324
tranche supérieure dorée 6413
transcodeur 3032
transducteur de mesure 13960
transférable 11448
transférer 11449
transfert d'encre 7405
transfert d'information 7310
transformateur 13977
transformation 3476
transformation automatique des données 905
transformation électronique des données 5046
transistor 13980
transistor à effet photo-électrique 10276
transistron 13980
translucide 13984
translucidité 13983
transmission 13987, 13988
transmission à vis sans fin 14853
transmission de données 4016

transmission entre deux terminaux 10535
transmission par courroie 1302
transmission par fac-similé 5427
transmittance 13994
transparence 12351, 13997, 13999
transparent 14002
transpercement 13276
transporteur 3481
transporteur aérien 9760
transporteur à papier 8104
transporteur de bobines 9936
transporteur de sortie à cordons 3726
transporteur de sortie à courroie 3726
transposer 14006
transvaser 4047
transvision 10865
travail à l'aérographe 316
travail à la main 6851
travail à l'outil 13900
travail au marbre 13206
travail au trait 8205
travail bousillé 1804
travail d'artiste 808
travail de bureau 2912
travail de trait 8205
travail en couleurs 3197
travail en (grande) série 8653
travail en repérage précis 13809
travailler à la roulette 11716
travail mal fait 1804
travail par équipes 12297
travail payé d'avance 7043
travail supplémentaire 5410
travail sur chaîne d'assemblage 844
travail sur presse 10773
travail urgent 11818
travaux d'édition 1783
travaux de ville 7680
travers de la page, en ~ 155
travers de la page, tout en ~ 440
traversement 13276
traverses 3750
trèfle 11648, 12863
treillis 4731
treize pour douze 13730
treizième exemplaire 13730
tréma 4251, 14256
trempage 4395
tremper 631
triage 12785
triage des chiffons 11212
triangle 6063, 14023
triangle chromatique 3193
triangle des couleurs 3193
triangle rectangle 11606
triangle supérieur 1101

trichloréthylène 14028
trichlorométhane 2746
trichlorure de phosphore 10209
trichlorure d'iode 7557
trichromie 13750
trier 12784
trier les pâtes 10303
trieur à copeaux 2730
trieur de bûchettes 2058
trieur de nœuds 7830
triglicéride 14036
trimardeur 13930
trimestriel 11137
tringle à blanchet 1527, 2871
tringle d'échappement 7769
tringle de pinces 6652
tringles de tension de l'habillage 1078
trinitrine 9376
trinitroglycérine 9376
triode à cristal 13980
trioxybenzène 10200, 11085
trioxyde chromique 2785
trioxyde d'antimoine 671
trioxyde sulfurique 13397
tripoli (silicieux) 4280
trisulfure d'antimoine 672
trisulfure d'arsenic 9691
triturateur 11029
trivalent 14063
troisième couverture 7449
troisième page de couverture 7449
trois points 6137
trolley 11654
tronc 12231
tronçonneur 12544
trop grand débit de papier 5337
trop mordu 9751
trou 7017
trou central d'entraînement 7414
trou de graissage 8398
trou graisseur 8398
trous d'aiguille 10364
trous d'entraînement 5517
trous d'épingle 10364
trous et goujons de registre 11409
truelle 3297
truffer 6560
truie 10337
t.s.v.p. 10479
tube à rayons cathodiques 2481
tube capillaire 2284
tube de carton 2348
tube de Nessler 9283
tube de quartz 11144
tube de tension (du papier) non commandé 7155
tube du piston moteur 14323
tube électronique 5056
tube en carton enroulé 3483

voile marginal 4955
voiler 5958
voir ce mot 11168
volant 5935
volatile 14504
volet (de châssis porte-
  plaque) 3998
volt 14508
voltage 14509
voltamètre 3648
volt-ampère 14514
voltige 7373
voltmètre 14515
volume 14516
volume avec dos arrondi

  3684
volume avec dos plat 3685
volume dépareillé 9496
volume spécifique 12831
volumètre de pénétration
  10100
voy cop. 12066
voyelle brève 12340
voyelle longue 8352
voyez copie 12066
vulcanisation 14521
vulcanite 6874

watt 14613
watt-heure 14614

wéber 14644
white spirit 14727
wink 14753
wolfram 14788

xénon 14887
xérographie 14891
xylène 14896
xylographe 14897
xylographie 805, 14798
xylogravure 14799
xylol 14896

yapp 14906
yard 14907

yard carré 12975
yard cube 3805
yeux 5715
youfte 11821

zéro supérieur 4108
zinc 14940
zinc de machine offset
  9541
zincogravure 14944
zirconium 14964
zodiaques 14965
zone 1110, 5602
zone du bord d'impression
  10827

DEUTSCH

absolute Adresse 25
absolute Adressierung 28
absolute Feuchtigkeit 29
absolute Luftfeuchtigkeit 29
absolute Temperatur 30
absolut trocken 9744
absorbieren 31
Absorption 34, 35
Absorptionsbande 36
Absorptionsbereich 36
Absorptionsfähigkeit 32
Absorptionskoeffizient 37
Absorptionsspektrum 41
Absorptionstest 38
Absprung 8101
abspülen 11619
Abstaubband 4902
Abstäuben 4893
Abstaubmaschine (an der Bronziermaschine) 4899
Abstaubpinsel 4898
abstecken 4484
abstellen 13793
abstoßen 11464
Abstoßen der Druckfarbe auf Metallwalzen 11651
Abstoßen (nachfolgender Farben) 3792
Abstreichmesser 7366
Abstrich 4703, 12874, 13109
Abstriche 3742
Abstufung 6534
abstumpfen 1645
Abszisse 22
Abszissenachse 22, 967
abtasten 11907
Abtasten 11911
Abtaster 11910, 11912
Abtaster mit bewegendem Lichtfleck 5928
Abtastkopf 11912
Abtastung 11911
Abtastvorrichtung 11910
Abteilung 4185
Abteilung Photographie 10242
Abteilungsleiter 6041
Abteilungslinie 4514
Abteilungstitel 12060
Abteilungszeichen 4003, 7143
Abteilvorrichtung 7788
abtönen 14471
Abtropfständer 4694
Abwärtsbewegung 4680
Abwasser 14730
abwedeln 4537
Abweichung 5, 4234
Abwickelung 11659
Abwicklung 11659
Abwicklungsfehler 11660
Abwicklungsgeschwindigkeit 2859
abwischen 14756
abwischen, mit dem Schwamm ~ 12911
Abwischpapier 10467
Abzählen 3666

Abzähler 7788
Abzählexemplar 7787
Abzählung 3666
abziehbarer Film 13283
Abziehbilder 4041
Abziehbilderdruck 13969
Abziehbilderpapier 4042
Abziehbilderverfahren 13969
Abziehbürste 1267
abziehen 7031, 11018, 12162
Abziehen 13289
Abziehen der Schöndruckseite 5712
Abzieher 6279, 10945
Abziehfilm 13283
Abziehfolie 1601
Abziehpapier 10946, 12163
Abziehpresse 6282, 10950
Abzug 4425, 10936
Abzugbogen 12573
Abzug machen, einen ~ 11018
Abzug ohne Zurichtung 5792
Abzugpapier 8599, 10946
Abzugpresse 10950
Abzugsfilm 13283
Abzugstreifen 12573
Abzug vor der Schrift 10940
Accent aigu 175
Accent grave 6577
Achatglättrolle 296
Achatstift 296
Achatzeile 298
Achromat 111
achromatisch 109
achromatische Linse 111
Achse 963, 968, 12225
Achsel 4737, 12343
Achselfläche 12343
Achselflächenbreite 12364
Achselhöhe 6949
achselhoher Ausschluß 6997
Achsellager 969
Achselschrägung 1346
Achsenlager 969
Achslager 969
Achslagerung 969
achslose Rollenlagerung 12870
Achtelcicero 4999
Achtelcicero-Durchschuß 5000
Achtelcicero-Regletten 5000
Achtelgeviert 6752
Achtelgeviertspatium 6752
Achtelpetit 10520
Achtelpetitspatium 6752
Achtelalphabet 4994
acht Punkte 4995
Achtrollenmaschine 4997
Achtundvierzigerformat 6090
Achtung, der Fax 2493
Achtzehnerformat 4993

Adapter 180
adaptieren 178
Addiermaschine 6686
Addiermaschinenrollen 186
additive Dreifarbenmethode 191
additive Farbmischung 191
additive Farbsynthese 191
additives Farbverfahren 191
Adhäsion 205
Adreßbuch 4408
Adresse 192
Adressenregister 199
Adressenverzeichnis 4408
adressierbar 193
Adressiermaschine 196
Adreßkarte 2105
Adreßmodifizierung 197
Adsorbat 235
Adsorbens 236
adsorbierte Substanz 235
Adsorption 237
Adsorptionsmittel 236
Adsorptionswert 238
Adsorptiv 235
Aerograph 314
Aerosol 284
Affinität 287
Agalit 293
Agar-Agar 294
Agave 300
Agglomerat 303
Aggregation 304
Ahle 1671
Ahlengriff 1673
Ahlenheft 1673, 6831
Ahlenspitze 1674
Ahlheft 6831
Akkoladen 1840
Akkordarbeit 10321
Akkordzeit 7243
Akkumulator 83, 84
Akontozeichen 81
Akrylatharz 158
Akrylharz 158
Akrylsäure 157
Akrylsäureharz 158
Aktendeckel 1403
Aktenvernichter 12352
Aktenvernichtungsmaschine 12352
Aktenzerreißwolf 12352
Aktie 12232
Aktienpapier 8296
aktinisch 159
aktinisches Licht 161
aktinische Strahlen 162
Aktinometer 164
aktives Chlor 949
Aktivieren 167
Aktivierung 167
Aktivkohle 169
aktualisieren 14354
Akut 175
Akzent 54
Akzentbuchstaben 55
Akzentbuchstabenkasten 56

Akzentkasten 56
akzentuieren 57
Akzentuieren 62
Akzent zum Ansetzen 5838
Akzent zum Übersetzen 5838
Akzidenzabteilung 7667
Akzidenzarbeiten 7680
Akzidenzdruck 7680
Akzidenzdrucker 7679
Akzidenzdruckfarbe 7677
Akzidenzen 7680
Akzidenzfarbe 7677
Akzidenzkasten 7670
Akzidenzlinie 1796
Akzidenzregal 2137
Akzidenzsatz 4469
Akzidenzschließrahmen 7671
Akzidenzschriften 4471
Akzidenzsetzer 7673
Akzidenzsetzerei 7667
Akzidenzzimmer 7667
Alaun 468
Albardinzellstoff 5255
Albert-Effekt 362
Albert-Fischer-Galvano 365
Albertochromie 3099
Albertypie 3099
Albionpresse 368
Album 369
Albumin 371
Albuminkopie 372
Albuminpapier 374
Albuminverfahren 376
Albumkarton 378
Albumpapier 378
Alco-Verfahren 379, 384
Aldehyd 385
Aldehydharz 386
Aldine 388
Aldine-Ausgabe 388
Aldusblätter 5870
Alfapapier 5254
algebraische Summe 396
algebraische Zeichen 393
Alginat 397
Algol 398
algorithmische Sprache 400
algorithmische Übersetzung 401
Algorithmus 399
Algraphie 490
Alinea 1875, 9972
Alinea ohne Einzug 5908
Alineazeichen 9973, 11552
alineieren 404
alinieren 404
aliphatische Lösungsmittel 409
Alizarinfarbstoff 412
Alizarinöl 413
Alkali 414
alkalibeständig 416
alkalibeständiges Papier 420
alkalibeständigkeit 421
Alkaliblau 415
alkaliecht 416

Alkaliechtheit 421
alkalifest 416
alkalifestes Papier 420
Alkalinität 418
alkalisch 417
Alkalität 418
Alkaliviolet 423
Alkane 424
Alkohol 380
Alkohole 382
Alkydharz 425
alleinstehend 6239
alle Rechte vorbehalten 439
Alltone-Verfahren 441
Allylalkohol 443
Allylchlorid 444
Allylharz 447
Allylisothiozyanat 446
Allylthioharnstoff 450
Allzweckrechner 6368
Almanach 451
Aloe 452
Alpha- 453
Alphabet 454
alphabetische Ordnung 456
alphabetischer Ordnung einreihen, in ~ 458
alphabetisches Verzeichnis 455
alphabetisieren 458
Alphabetlänge 459
Alphagras 5252
alphanumerisch 461
Altarfalz 14531
Alte 1799
alte Ausgabe 1038
alte Kamellen 6389
alte Nummer 1038
alter Einband 9571
ältere römische Kursiv 4937
Altern 301
Alterung 301
Alterungsprüfung 302
Alterungsversuch 302
altes Buch 11203
ältester Setzerlehrling 723
altmodisch 9724
Altpapier 9573
Altpapierhändler 14559
Altschriftkunde 9874
Alterwerden 301
Aluminium 471
Aluminium-Ammonium- sulfat 473
Aluminiumätzung 491
Aluminiumazetat 472
Aluminiumbronze 474, 484, 12433
Aluminiumchlorid 475
Aluminiumdruck 490
Aluminiumdruckfarbe 479
Aluminium-Duplikatdruck- form von Ätzungen oder Satzformen 491
Aluminiumfolie 476
Aluminiumhydroxyd 477

Aluminium-Natriumsulfat 12673
Aluminiumoxyd 480
Aluminiumpapier 481
Aluminiumplatte 482
Aluminiumpulver 484
Aluminiumsilikat 485
Aluminiumsulfat 468, 487
Amalgam 494
Amalgam-Druckverfahren 11460
Amboß 696
Ameisensäure 6078
Ameisensäurealdehyd 6048
ameisensaures Natrium 12690
amerikanische Höhe 498
amerikanische Schrifthöhe 498
Amid 500
Amide 500
Amidol 4262
Amin 503
Amine 503
Aminobenzol 613
Aminoharz 507
Aminoplastharz 507
Ammeter 556
Ammonalaun 473
Ammoniak 509
Ammoniakalaun 473
Ammoniaklösung 533
Ammoniakwasser 533
Ammoniumalaun 473
Ammoniumazetat 513
Ammoniumbichromat 526
Ammoniumbifluorid 519
Ammoniumbikarbonat 517, 521
Ammoniumbromid 520
Ammoniumchlorid 522
Ammoniumchromat 523
Ammoniumdichromat 526
Ammonium-Eisenalaun 5544
Ammoniumferrizitrat 5542
Ammoniumfluorid 530
Ammoniumhydrogen- karbonat 526
Ammoniumhydroxyd 533
Ammoniumhyposulfit 551
Ammoniumjodid 535
Ammoniumkarbonat 521
Ammoniumlaktat 537
Ammonium-Nickelsulfat 9336
Ammoniumnitrat 540
Ammoniumoxalat 541
Ammoniumpersulfat 543
Ammoniumrhodanid 550
Ammoniumsulfat 544
Ammoniumsulfid 546
Ammoniumsulfit 547
Ammoniumsulfozyanat 550
Ammoniumsulfozyanid 550
Ammoniumsulfür 546
Ammoniumthiosulfat 551
Ammoniumthiozyanat 550
Ammoniumzitrat 524

Ammonsalpeter 540
amorph(isch) 552
Ampere 554
Amperemesser 556
Amperemeter 556
Amperestunde 555
Amplitude 561
Amplituden-Modulation 562
Amtsblatt 6343
Amylalkohol 565
Amylazetat 563
Amyloxyhydrat 565
Amylum 13045
Anachromat 568
anachromatisches Objektiv 568
Anaglyph 569
Anagramm 570
Analisierung 575
Analogcomputer 572
Analog-Digital-Umsetzer 4366
analoges Signal 573
Analogkanal 571
Analogrechner 572
Analogübertragungskanal 571
Analyse 575
Analysenschnellwaage 578
Analyse-Schnellwaage 578
Analytiker 576
analytisches Verfahren 577
anastatischer Druck 579
anastatisches Druck- verfahren 579
Anastigmat 581
anastigmatisches Objektiv 581
Anätzen 5787
Anätzung 1450, 5787
ändern 462, 2614
Änderung 463, 12295
Änderungen des Verfassers in der Korrektur 887
Änderungen in der Druck- maschine 8441
Änderungen in der Form 2619
Andruck 10936
Andruck auf Azetatfolie 94
Andruck der Klischee- anstalt 5190
andrucken 10937, 11018
Andruckgeschwindigkeit 9602
Andruckmakulatur 11797
Andruck mit der Schrift 10958
Andruckpapier 10946
Andruckpresse 10950
Andruck vor der Schrift 10940
Anepigraphon 586
Anfangslinie 10385
Anfangszeile 7322
anfärben 11666
Anfärben 4924
Anfertigung 8579, 11788

Anfertigungsmenge 8579
anfressen 3613
Anfressen 3615
anführen 11177
Anführung 3239, 11172
Anführungszeichen 3239, 11174
Anführungszeichen bei Bibelzitaten 4393
Anführungszeichen ver- sehen, mit ~ 11177
angebaut 2043
angeschlossen an das übrige System 9599
angetriebene Walze 4739
angewandte naturwissen- schaftliche Forschung 720
Ångström 601
Ångström-Einheit 601
Anguß 7657, 13537
anhaften 203
Anhaften 204
Anhaftung 204
anhalten 12359, 13793
Anhang 184, 716, 5165
Anhängeetikett 1069, 13534
Anhängekarton 13535
anhängen 11811
Anhänger 13534
Anhängerknüpfmaschine 13536
Anhängezettel 1069
Anhäufen 10345
Anhydrid 605
Anilin 613
Anilindruck 5828
Anilindruckfarbe 5825
Anilindruckmaschine 5829
Anilineindruckwerk 615
Anilinfarbe 5825
Anilinfarbstoffe 2973
Anilingummidruck 5828
Anilinkopie 619
Anilinpunkt 617
Anilinschwarztest 9349
Anilinsulfat 622
Anilinsulfatlösung 622
animalischer Leim 623
Anion 627
anion-aktive Verbindungen 628
Anisotropie 629
Anker 771
ankleben 12892
Ankleben des Kapital- bands 12172
Anklebevorrichtung für Stumpfläche 2126
Anklebezettel 7846
Ankörnbad 6549
Ankündigung 247
Anlage 2909, 5153, 10403
Anlagekante (des Bogens) 7949
Anlagesteg 6656
anlassen 631, 13054, 13788
Anlasser 13059
Anlaßfahrschalter 13060

ausbleichen 1538, 5433
Ausbleichen 1542
Ausbleichverfahren 1550
ausblenden 5232
Ausbluten 1557
ausblutende Farbe 1559
Ausbohren 4732
Ausbrechen der Abfälle 13291
Ausbrechzylinder 13293
ausbringen, (eine Zeile) ~ 4740
ausdecken 9613, 12407
Ausdecken kleiner Löcher 10367
Ausdeckmaske 10861
Ausdehnung 4370, 5361, 5362, 13262
Ausdehnung bei Feuchtigkeitseinwirkung 7135
Ausdehnung der Papierbahn 14645
Ausdehnung des Gummituchs 1521
Ausdehnungskoeffizient 3038
Ausdehnungszahl 3038
ausfächern 5466
Ausfall einer Silbe 5083
ausfällen 10711
Ausfallmuster 9740
Ausfallprozentsatz 10111
Ausfallzeit 4679
ausflecken 12917
Ausflecken 10367
Ausflockbarkeit 5843
Ausflocken 5842
Ausflockung 5842
Ausflußviskosimeter 4985
Ausfräsen 11737
Ausführapparat 4145
Ausführbänder 4135, 4150
ausführen 5351
Ausführpunkte 7984
Ausführung 4134
Ausführungspunkte 7984
Ausführungszeichen 2940
ausfüllen 3229
Ausgabe 4970, 9731
Ausgabegerät 9733
Ausgabeliste 10859
Ausgabe mit Anmerkungen 634
Ausgang 1875, 1889, 4134, 9730
Ausgangskolumne 12334
Ausgangsseite 12334
Ausgangszeile 1875, 1889
Ausgangszylinder 13559
ausgeklinkter Buchstabe 9109
ausgeklinktes Klischee 10324
ausgelassene Stellen, (beim Streichen) ~ 12524
ausgesetzt 442, 2934
ausgestanzte Karten 4315
ausgetauschter Lehrling 14103

ausgewölbt 4550
ausgewuchtet 1085
Ausgiebigkeit 3675
ausgleichen 8200
ausgleichen, (Zeile) ~ 12806
ausgleichen, (Zeilen) ~ 12808
Ausgleichsgewicht 1088
Ausgleichswalze 7683
Ausgleichung 2005, 5222
Ausgleichungsstreifen 3265
Ausguß 9684
Aushang 10609
Aushängebogen 244, 2896
Aushänger 244
ausheben 13555
aushöhlen 6687
Auskopierpapier 10837
Auskopierprozeß 10860
Auslagebänder 4135
Auslagekasten 4152
Auslagetrog 4152
auslassen 9585
Auslassung 5083, 9584
Auslassungspunkte 10532
Auslassungszeichen 708
Auslaufbecher 4983
auslaufen 3723, 12931
Auslaufen 1557
auslaufende Farbe 1559
auslauffähiges System 9028
Auslauffarbe 4666
Auslaufviskosimeter 4985
Auslaufzeit 4984
Auslegebänder 4135
Auslegefilz 1013
auslegen 1053
Auslegepapier 12287
Auslegepappe 1013
Auslegerkette 4138
Auslege(r)rechen 5915
Auslegerstäbe 5923
Auslegestern 13559
Auslegetisch 4149
Auslegewagen 4137
Ausleihbibliothek 8036
Ausleiter 4143
auslösen 14057
Ausmalebuch 3149
Ausmaleheft 3149
Ausmalung 7161
Ausmaß 4375
Ausnahmenverzeichnis 5336
ausquetschen 12980
Ausquetschen 3707
Ausraffefach 13448
Ausraffekasten 6117
Ausraffkasten 6117
ausreißen 11017
Ausrichtemarke 6696
ausrichten 218, 404
Ausrichtung 405
Ausrichtung einhalten 7739
ausrücken 13793
Ausrückvorrichtung 13790, 14055
Ausruf(e)zeichen 5348

Ausrufungszeichen 5348
Ausrüstung zur Magazinrahmenbewegung 8484
ausschalten 13477
Ausschärfen 9999
Ausschießbrett 7198
ausschießen 7196
Ausschießen 7202
Ausschießen in Lagen 7201
Ausschießplatte 7198
Ausschießschema 7197
ausschlachten 1894
Ausschlagseite 13792
Ausschlagtafel 13792
ausschleifen 13203
ausschließen 12808
Ausschließen 7708
Ausschließen und Abkürzen von Wörtern 7144
Ausschließhebel 7710
Ausschließkeil 12802
Ausschließtaste 7709
Ausschluß 12811
Ausschlußkasten 1139
Ausschlußkeil 7712
Ausschlußmaterial 12811
Ausschlußstück 5627
Ausschlußtaste 7709, 12804
Ausschlußtrommel 7715
Ausschluß- und Zentriervorrichtung 11093
Ausschlußzeiger 7717
ausschneiden nach der Gehrung 8996
Ausschnitt 7453, 9107, 9304, 9431, 9736
Ausschnittbüro 2917
Ausschuß 1934, 2431, 4418
Ausschwimmen 5852
Außenschattierung bei Ornamenttypen 1249
Außenstelle 11459
äußere Druckform 9714
äußere Form 9714
äußerer Papierrand 6033
äußerer Seitenrand 6033
aussetzen 5687, 13558, 14746
aussetzend 7514
aussparen 9108
Aussparung 9107
aussperren 2340, 7971, 12808
Aussprache-Zeichen 4250
Ausspritzöffnung 9445
Ausspülen der Farben 13843
Ausständige 13274
Ausständiger 13274
ausstanzen 4317
Ausstattungspapier 5460
ausstechen 3857
Ausstechen 4355, 11736
Aussteller 12350
ausstopfen 9842
Ausstoßer 5001, 7790, 13293
Ausstoßmechanismus 5001
Ausstoßplatte 5002

Ausstoßplatten-Kassette 5003
Ausstoßplatten-Magazin 5003
Ausstoßwalze 7790
Ausstoßzylinder 13293
Ausstrahlung 11194
ausstreichen 4128, 13272
Ausströmungsgeschwindigkeits-Koeffizient 3041
Austauschbarkeit 3258
Austauschexemplar 5342
Austiefung 11931
austreiben 2340, 4740, 14720
austrocknen 4206
Austüpfel 10057
auswalzen 11281
auswalzen der Druckfarbe 1870
auswaschen 11619
Auswaschen 14553
Auswaschentwickler 14554
Auswaschlösung 14555
Auswaschmittel 14555
Auswaschreliefdruckstock 9465
Auswaschtinktur 827
auswechselbare Farbwanne 10585
auswechselbares Farbwerk 10587
auswechselbares Objektiv 7491
Auswechselbarkeit 7490
Auswechseln der Papierrolle 11361
Ausweitung der Papierbahn 14645
Auswerfer 5001
Auswerfgreifer 4143
auszeichnen 4459, 13794
Auszeichnungssatz 4473
Auszeichnungsschrift 4471
Auszeichnungsschriften 13869
Auszeichnungszeile 4466
Ausziehen 5397
Ausziehtusche 7280
Ausziehung 5397
Auszug 20
Auszugsnegativ 12130
Autobiographie 891
Autochromplatte 892
Autochromverfahren 894
Autocode 895
Automation 929
automatische Abtastung 925
automatische Anklebevorrichtung 5927
automatische Auslage 906
automatische Bildeinstellung 911
automatische Bremsvorrichtung 899
automatische Datenverarbeitung 905
automatische Drucktücher 898

Bitdichte 1449
Bit/S 1456
Bittersalz 8502
bitte, wenden 10479
Bitumen 1458
bitumenbeschichtetes
   Papier 13608
bitumenimprägniertes
   Papier 13608
bitumenkaschiertes Papier
   14308
Bitumenpapier 13608
Bitzahlprüfung 10009
Blanc fixe 1156
blankgeschlagene Stelle
   14711
Blanklaufen der Walzen
   11651
blanko Papier 1533
blankschlagen 8020
Blankseite 1532
Blankwerden der Walzen
   11651
Blasdüse 9445
Blase 1577
Blasebalg 1293
Blasenätzmaschine 2020
Blasenbilden 1579
Blasenbildung 1579
Blasenprobe 1581
Blasenversuch 1581
blasige Oberfläche 3049
Blasöle 1626
Blasrohr 1630
Blatt 8007, 9309
Blattaluminium 476
Blattbildner 12272
Blattbildung 6055
Blättern 8008
Blätterwald 10738
Blattfedern 7965
Blattgold 6509
Blattschreiber 9854
Blattzeichen 1762
Blaudruckplatte 3919
Blauer Montag 8377
Blaufilter 1636
blaugrün 3908
Blauholz 8324
Blaulack 3054
Blaulackkopierverfahren
   3055
Blaupause 1642, 3917
Blaupauspapier 1644
Blauplatte 3919
Blausäure 7105
blausaures Kalium 10647
Blauschlüssel 1638
Blauschwarz 1635
Blech 13831
Blechbeleistung 8840
Blechbüchse 13823
Blechdose 13823
Blechdruck 8819
Blechdruckfarbe 8820
Blechdruckmaschine 8821
Blechleisten anbringen
   8840
Blei 7968

Bleiasche 4763
Bleiazetat 7972
Bleibarren 10327
bleibende Wärme 7930
bleibt 13134
Bleichbad 1545
bleichen 1538
Bleichen 1542, 1543
bleichendes Mittel 1544
Bleicher 10679
Bleichflüssigkeit 1547
Bleichholländer 10679
Bleichkalk 2168
Bleichlauge 1547
Bleichlösung 1547
Bleichmittel 1544
Bleichpulver 1548, 2168
Bleichromat, (neutrales) ~
   7977
Bleichung 1543
Bleichwasser 1547
Bleidämpfe 7986
bleifreies Benzin 14330
Bleifuß 7975
Bleifuß montiert, auf ~
   9141
Bleiglanz 6272
Bleiglätte 8248
Bleikarbonat 7976
Bleikorrektur 3603
Blei korrigieren, auf
   dem ~ 3591
Bleilegierung 7973
Bleilinie 8003
bleiloser Satz 3061
Bleimater 7993
Bleimatrize 7993
Bleimennige 11330
Bleimonoxyd 8248
Blei montiert, auf ~ 9141
Bleinitrat 7996
Bleioxyd 8248, 11330
Bleiprägung 7994
Bleisalpeter 7996
Bleisatzkorrektur 3603
Bleisatzschrift bei der
   Schriftgrad und Kegel-
   größe nicht überein-
   stimmen 1216
Bleischnitt 7981
Bleisikkativ 7979
Bleispritzer 12984
Bleistange 10327
Bleistege 6262
Bleistiftretusche 3709
Bleistiftzeichnung 10092
Bleisulfat 8005
Bleisulfid 6272
Bleitopfpumpe 10686
Bleitrockner 7979
Bleiunterlage 7975
Bleivergiftung 8000
Bleiverstärker 7991
Bleiweiß 5749
Bleizucker 7972
Blende 4273, 6796
Blendeeinrichtung 4274
Blendenebene 4276

Blendeneinstellscheibe
   4275
Blendenkontrollsystem
   4274
Blendenreihe 8047
Blendenskala 8047
Blende schließen 13212
Blende verkleinern 13212
Blendezahlen 5937
Blendlinie 7691
Blickfang 5414
blind 1561
Blindband 4857
Blindband mit
   eingeklebten Korrektur-
   abzügen 4859
Blindbild 1566
Blinddruck, mit ~ 1562
Blindenschrift 1846
Blindenschriftpapier 1847
blinder Bogen 6192
blindes Material 1531
blinde Zeile 14716
Blindlaufen 1567
Blindmaterial 1531, 6262
Blindmuster 4857
Blindpaginierung 1565
Blindplatten 4860
Blindprägung 1563, 1574,
   1576
Blindprägung, mit ~ 1562
blindschlagen 8020
Blindsteg 8630
Blindwerden 1567
Blindzeile 14716
Blitzröhre 5049
Blochbeutel 1589
Blochbuch 1588
Block 1583, 13514
Blockbodenbeutel 1589
Blockdiagramm 1591
Blocken 1598
Blockfestigkeitsprobe 1606
Blockheftung 1617, 9845
blockieren 14098
blockierte Schrift 1593,
   14095
blockierte Zeile 14096
Blocklänge 1607
Blockpunktprobe 1606
Blockschaltbild 1591
Blockschaltplan 1591
Blockschrift 1609
Blocks leimen 9843
Blocktest 1606
Blockungsfaktor 1599
Blockzwischenraum 7489
Blumenlese 654
Blumenseidenpapier 5872
bluten 12931
Bluten 1557
Bock 1842
Bodenbeutel 11876
Bodenfläche 5859
Bodensatz 4186
Bodenstein des Koller-
   gangs 1282
Bodley-Einband 1675
Bogen 12250

Bogenabstreicher 13287
Bogenanlage, mit ~ 12267
Bogenanlegeapparat 5508
Bogenanleger 909, 5508
Bogenausgang 4134
Bogenauslage 4134
Bogenauswerfer 5915
Bogendruckmaschine 12269
Bogenfänger 5915
Bogenformat 12487
Bogengeradleger 7686
Bogengeradstoßmaschine
   7687
Bogenglattstoßmaschine
   7687
Bogenglattstoßvorrichtung
   7686
Bogen im Blinddruck
   durch die Maschine
   laufen lassen 11816
Bogen, in ~ 14298
Bogen je Stunde 12280
Bogenkalander 12261
Bogenkalandriert 12262
Bogenlampe 754
Bogenlampe mit ein-
   geschlossenem Licht-
   bogen 5149
Bogenlampe mit offenem
   Lichtbogen 9617
Bogenlampenkohle 755
Bogenlampenkohlen 2328
Bogenlicht 756
Bogenlichtaufnahme 2073
Bogenpapier 9922
Bogenpreis 10780
Bogen pro Stunde 7223
Bogensatzstempel 6532
Bogenschneidemaschine
   12263
Bogenschneider 12263
Bogenschneidvorrichtung
   12266
Bogenschüttelmaschine
   7687
Bogensignature 12402
Bogenstempel 6532
Bogentrockner 12265
Bogenzahl 12401
Bogenzähler 3653
Bogenziffer 12401
Bohren 4732
Bohrmaschine 4733
Bollstichel 12017
Bologneser Kreide 2163
Boltstichel 12017
Bolzen 1714
Bolzenschraube 1714
Bolzer (für Kupferstecher)
   5281
Bondpapier 1723
Borax 12680
Bordzeitung 12309
Borsäure 1798
borsaures Natron 12680
Börsenbericht 5343
böse sein 10086
botanische Zeichen 1802
Bowdenzug 1821

Bowdenzugkabel 1821
Bradel-Einband 1845
Braillepapier 1847
braune Holzpappe 1975
braune Mischpappe 1977
brauner Holzschliff 1976
braungedeckte Graupappe 1974
Braunpause 1978
Braunschliff 1976
Braunschliffpappe 1975
brechen 3691
Brechen 1886
Brechungsindex 11396
Brechungsquotient 11396
Brechungswinkel 598
Brechungszahl 11396
Brechzahl 11396
Breitbahnpapier 12326
breite Buchstaben 5358
Breite des Matrizenkörpers 1869
breite Schrift 5358
breites Format 7898
breitlaufende Schrift 5358
Breitstreckwalze 5359
Bremse 1848
Bremstrommel 1852
brennbar 7305
Brennebene 5946
Brennen 2153
Brennerrampe 6312
Brennpunkt 5950
Brennweite 5945
Brenzgallussäure 11085
Brenzkatechin 11082
Brett für Ablegesatz 1658
Brettregal 1664
Briefanlage 13312
Briefbeilage 13312
Briefblock 13514
Briefhülle 3611, 5209
Briefhüllenformaten 12498
Briefhüllenpapier 5212
Briefkopf 8058
Briefmarkendruck 10601
Briefmarkenpapier 13019
Briefordner 8057
Briefpapier 8060, 9442
Brieftasche 5209
Briefumschlag 3611, 5209
Briefumschlagmaschine 5211
Briefumschlagpapier 5212
Briefwaage 8054
brillant 12651
Brillantfarblack 1914
Brillantgrün 1913
Brillantrot 5698
Brinell-Härte 1918
bringen, auf den neuesten Stand ~ 14354
Bristolersatzkarton 1692
Bristolkarton 1921
Bristolkarton, auf der Rundsiebmaschine hergestellter ~ 8933

Bristolkarton für Vermählungsanzeigen 14663
Britische Hitzeeinheit 1924
Brom 1949
Bromammonium 520
Bromäthan 5303
Bromäthyl 5303
Bromethan 8862
Bromide 1941
Bromkalium 10634
Bromkresolgrün 1950
Bromkresolpurpur 1951
Bromkupfer 3501
Bromkupferverstärker 3502
Brommethyl 8862
Bromölverfahren 1953
Bromphenolblau 1955
Bromquecksilber 8790, 8791
Bromsilber 12426
Bromsilberdruck 1947
Bromsilberkollodium 3083
Bromsilberkopie 1947
Bromsilberpapier 1946
Bromsilberpapierprobe 1948
Bronze 1956
bronzeblaue Farbe 1960
Bronzebuchse 1961
Bronzedruck 1966
Bronzedruckfarbe 1964
Bronzefarbe 1964
Bronzeflecke 4164
Bronzelüster 1958
Bronzepigment 1965
Bronzepulver 1957, 1965
Bronzestaub 1957, 1965
Bronzestich 1958
Bronzeunterdruckfarbe 1970
Bronzierapparat 1969
Bronzierdruck 1966
Bronzieren 1968
Bronzierer 1967
Bronziermaschine 1969
Bronzierplüsch 1971
Bronzieruntergrundfarbe 12485
broschiert 9900, 13169
Broschüre 1933
Broschüre mit Blockheftung 5806
Broschürenschnitt 3861
Broteinschlagpapier 1873
Broteinwickelpapier 1873
Broteinwickler 1873
Brotschrift 1688
Bruch 1874, 3692, 5962, 6141
Bruchdehnung 5095, 13264
brüchig 1925
Bruchlast 8291, 13644
Bruchstrich 12767
Bruchwiderstand 13644
Bruchzeichen 6143
Bruchziffer 6141
Bruchzifferkasten 6142
Bruchziffer mit Horizontalstrich 10320

Bruchziffer mit schrägem Strich 12592
Bruchziffer mit Schrägstrich 12592
Bruchziffer mit waagerechtem Strich 10320
Bruchziffern ohne Querstrich 9306
Bruchziffern ohne Schrägstrich 9306
Brücke 1904
Brückenkamera 9759
Brunak-Verfahren 1980
Brustwalze 1899
Bruttogewicht 6674
Buch 1729
Buch aus Besitz einer berühmten Persönlichkeit 848
Buch aus persönlichem Besitz des Autors 848
Buchausstattung 1768
Buch aus zweiter Hand 12044
Buchbeschläge 1800
Buch besprechen, ein ~ 11567
Buchbesprecher 1774
Buchbesprechung 1773
Buch beurteilen, ein ~ 11567
Buchbinde 11002
Buchbindehobel 10484
Buchbindekunst 801
Buchbindeleinen 1735
Buchbinden 1414
Buchbinder 1732
Buchbinderei 1412, 1421
Buchbindereiabteilung 1421
Buchbindereiarbeiter der Bücher einhängt 2415
Buchbindereimaschinen 1424, 1739
Buchbinderhammer 1014
buchbinderische Weiterverarbeitung 6091
Buchbinderleim 1736
Buchbinderpappe 1408
Buchbinderprägepresse 1604
Buchbinderpresse 1427
Buchbinderstempel 1411
Buchblock 1741
Buchbolzen 1426
Buchdecke 2397
Buchdeckel 2397, 3673
Buchdeckel, sich werfende ~ 14908
Buchdeckenherstellmaschine 2412
Buchdeckenmaschine 2412
Buchdeckenverzierung 13901
Buchdruck 8061, 8065, 11446
Buchdrucker 8064
Buchdruckereibesitzer 8659
Buchdruckerkunst 803

Buchdruckerlehrling 10809
Buchdruckerprinzipal 8659
Buchdruckfarbe 8062
Buchdruckfarben 1756
Buchdruckmaschine 8063
Buchdruckpresse 8063
Buchdruck-Rollenrotationsmaschine 14652
Buchdruck-Rotationsmaschine 14652
Bucheignerzeichen 1771
Bucheinbandbeschläge 1800
Bucheinband mit Kameeprägung 2236
Bucheinhängemaschine 2425
Buchenholz 1284
Bücherbörse 1752
Bücherei 8083
Bucherfolg 1341
Bücherfreund 1364
Bücherliebhaber 1364
Büchernarr 1362
Bücherpapier 1770
Bücherrevisor 75
Bücherrücken-Rundmaschine 1047
Büchersammler 1747
Büchersammlung 8083
Büchertrödler 12045
Bücherwurm 1784, 1785
Buchfalz 11591
Buchformat 1779, 6052
Buchformaten 12496
Buchgestalter 1748
Buchgestell 1754
Buchgewerbe 1755
Buchhändler 1775
Buchhändlerbörse 1752
Buchhandlung 1778
Buchhandlungsreisender 1742
buchherstellende Industrie 1755
Buchhobel 10484
Buchhülle 6040
Buch in loser Blattform 8364
Buchkritiker 1774
Buchkunst 804
Buchkünstler 1748
Buchleinen 1735
Buchmalerei 7161
Buchmesse 1752
Buch mit angesetzten Deckeln 13802
Buch mit Autor-Annotationen 848
Buch mit Eselsohren 4544
Buchrevisor 75
Buchrezensent 1774
Buch rezensieren, ein ~ 11567
Buchrezension 1773
Buchrücken 983, 12873
Buchrücken machen, den ~ 988
Buchsbaumholz 1839

Dimensionsstabilität 4376
Dimethylketon 102
DIN-Formate 7521
Dioptrie 4389
Dioxybernsteinsäure 13610
Diphthong 4392
Direktaufnahme 4409
Direktauszug 4412
direkte Beleuchtung 4406
direkter Rasterauszug-
    Prozess 4404
direkter Zugriff 11231
direktes Rasternegativ
    4403
Direktfarbstoffe 4402
dispergieren 4450
Dispergiermittel 4452
Dispergierungsmittel 4452
Dispersion 4454
Dispersionsmittel 4452
Dispersionsphase 4451
Dispersoidanalyse 10015
Dissertation 13700
Distorsion 4481
Distorsionskamera 11491
divergierende Linse 4507
Divis 7143
Divisorium 4520
DK-Meter 4046
D-log-E-Krumme 2637
D-log-E-Kurve 2637
Dochtkohlen 3568
Dochtkohlestifte 3568
Dokumentenpapier 8296,
    8347
Dollarzeichen 4547
Dolomit 4549
Donat 4553
Doppelantrieb 4883
Doppelauslage 4595
Doppelausschluß 4619
Doppelbelichtung 4603
Doppelbesetzung 4671
Doppelbindung 4627
Doppelbogenabführung
    14155
Doppelbogenausrüstung
    14155
Doppelbogenkontrolle
    14155
doppelbreite Maschine
    4670
doppelbreite 3-Einheiten-
    maschine 12207
doppelbreite 6-Einheiten-
    maschine 4654
doppelbreite 8-Einheiten-
    maschine 4635
Doppelbuchstabe 8105
doppelchromsaures
    Ammonium 526
Doppel-Doppel-Wellpappe
    4598
Doppeldruckwerke 4070
Doppelfalzer 5980
Doppelfalzung 5979
Doppelfalzzahl 5979
doppelfarbiges Papier 4881

doppelfeine Linie 4607,
    9987
doppelgestrichenes Papier
    4586
Doppelgeviert 14138
Doppelgeviertstück 14138
Doppelklebeband 4606
Doppelkreuz 4594
Doppellaut 4392
Doppelpechpapier 14308
Doppelpunkt 3101
Doppelrechenantrieb 8596
Doppelregal 4612
Doppelrollenstern 13569
Doppelsalz 4653
Doppelsatz 4579
Doppelschließrahmen 5977
doppelseitiges Klebeband
    4606
doppelseitige Wellpappe
    4605
doppelseitig gestrichenes
    Papier 14157
Doppelsetzkasten 4585
doppelstumpffeine Linie
    4632
doppeltchromsaures
    Kali(um) 10648
doppelte Platten-
    bestückung 14154
doppelte Präzision 4645
doppelte Produktion 13248,
    14154
doppelter Durchschuß 4624
doppelte Rollenlagerung
    4830
doppelter Plattensatz 4641
doppelter Rollenständer
    4830
doppeltkohlensaures
    Ammonium 517
doppeltkohlensaures
    Natrium 12676, 12688
Doppeltondruckfarbe 4666
Doppeltonfarbe 4666
Doppeltrichter 1101
doppeltschwefligsaures
    Kalzium 2162
doppeltschwefligsaures
    Natron 12679
Doppelvokal 4392
doppelwandiger Papiersack
    4873
Doppelzeile 4579
Doppler-Effekt 4562
dos-à-dos-Einband 4563
Doublieren 4646
Drachenblut 4688
Drachenblutverfahren 4689
Drachme 4697
Drahtbürste 14767
Drahtheftklammer 13040
Drahtheftmaschine 14779
Drahtheftung 14780
Drahtrundbindung 12875
Drahtspule (für Heft-
    maschinen) 14766
Draht zum Heften 13179
Dram 4697

draufstechen 12170
Drehbank 7938
Drehdiamant 4269
Drehmoment 13924
Drehraster 2847
Drehscheibe 14105
Drehstift 8587
Drehstrom 13762
Drehstrom-Kommutator-
    motor 13761
Drehstrommotor 13764
Drehstromnetz 13763
Drehstrom-Schleifring-
    läufermotor 13765
Drehstrom-Schleifring-
    motor 13765
Drehtisch 14105
Drehung 13925
Drehwaage 13926
Drehzahl 9455, 12853
Drehzahlmesser 13523
Dreibahnen-Wellpappe
    4605
dreibasisches Natrium-
    phosphat 12709
Dreiblenden-Exposition
    10306
dreidimensioneller Druck
    14895
Dreieck 4715, 14023
Dreifarbenätzer 13746
Dreifarbendruck 13750
Dreifarbendruckfarben
    10897
Dreifarbendruckplatten
    13749
Dreifarbenfilter 14034
Dreifarbenklischees 10903,
    13749
Dreifarbenphotographie
    13747
Dreifarbenplatten 10903
Dreimesserautomat 13754
Dreimesserschneid-
    maschine 13754
Dreiphasenkreis 13760
Dreiphasenmotor 13764
Dreiphasennetz 13763
Dreiphasenstrom 13762
Dreiphasenwechselstrom
    13762
drei Punkte 6137
Dreipunktspatium 13708
Dreipunktsystem 13767
Dreirollenmaschine 13771
dreischichtige Pappe 14054
Dreischneidemaschine
    13754
Dreischneider 13754
dreispaltige Seite 13751
dreistufiges Wasserzeichen
    13766
Dreiwalzenreibmaschine
    13772
Dreiwalzenstuhl 13772
Dreiwalzwerk 13772
dreiwertig 14063
dreizehn für zwölf Stück
    13730

dreizehntes Exemplar
    beim Partiebezug 13730
dreizeiliger Anfangsbuch-
    stabe 13756
dreizeiliger Initial 13756
Dreizylindermaschine 9626
Drell 4731
Drilch 4731
Drill 4731
Drillich 4731
Driographie 4734
Drittelgeviert 13708
Drittelspatie 13708
Drittelspatium 13708
Druckabsteller 7224
Druckabstellung 14055
Druckanfang 10385
Druckanfang der Platte
    7990
Druckanfang und Druck-
    ende 7974
Druckanzeiger 7215
Druckarbeit 10773
Druckarbeit aus dem Haus
    geben, eine ~ 5472
Druckaufbereitung 4967
Druckausfall 14558
Druckautomat 916
Druckbestäuber 688, 4823,
    12926
Druckbestäubungsflüssig-
    keit 686, 9539
Druckbestäubungsmittel
    4824, 9539
Druckbestäubungspuder
    687, 9539
Druckbestäubungspulver
    4824
Druck des Presseurs 7218
Druckeignung 10796
Druckeinheit 4060, 10852
Druckeinstellung 7222
drucken 10794
Drucken 10817
Druckende 13955
drucken mit hoher Druck-
    spannung 8595
Drucken mit magnetische
    Druckfarbe 8521
Drucken mit zu starkem
    Anpreßdruck 6868
drucken, mit zwei Formen
    für einen Bogen ~
    14822
Drucker 10753, 10805
Druckerballen 7329
Druckerei 10829
Druckereibetrieb 10829
Druckereifaktor 10762
Druckereileiter 10762
Druckereiversehen 3315
Druckerfaktor 10762
Drucker für fliegenden
    Abdruck 7012
Drückerfuß 1623
Druckergebnis 10845
Druckerlaubnis 7228
Druckermarke 5816, 7229

Farbplatte, (ungerasterte) ~ 8165
Farbpumpe 7379, 7380
Farbpumpenaggregat 7379
Farbradierung 3140
Farbrasterplatte 892
Farbreagenz 7286
Farbrechen 5850
Farbregister 3169
Farbregulierschrauben 4838
Farbregulierungspult 1119
Farbreiber 9168
Farbreihenfolge 3185
Farbreproduktion 3167, 3171
Farbreproduktionsqualität 3172
Farbretusche 3173, 4927
Farbrücknahme 14269
Farbsättigung 2760, 11882
Farbschicht 7350
Farbschlitz 13783
Farbschnitt 13009
Farbsensibilisierung 3182
Farbskala 3157
Farbskalen 10928
Farbspachtel 7367
Farbspritzer auf der Hinterseite 5405
Farbspritzer auf der Vorderseite 5406
Farbspritzgerät 314
Farbstärke 3189
Farbstein 7393
Farbstellpult 1119
Farbstoff 4922
Farbstoffe 3176, 4928
Farbstofflack 7880
Farbtank 7402
Farbteller 7337
Farbtemperatur 3191
Farb-, Tinten- oder Bleistift-Annahmefähigkeit 1451
Farbtisch 7393, 7400
Farbton 7066, 12209, 13838
Farbtonabschwächer 7881
Farbtonabweichung 9542
Farbton der Farbe im Farbehälter 8654
farbtongleiche Wellenlänge 4552
Farbtönung 7066
Farbtonwiedergabe 3170
Farbtransparenz 3192
Farbtransport 7405
Farbtrennungsnegativ 3184
Farbtrommel 7341
Farbüberstand 1554, 7903
Farbübertragung 7405, 8100
Farbübertragungsmechanismus 7359
Farbübertragungszahl 7406
Farbübertragungszylinder 13972
Farbumlaufpumpe 7380
Farbumwälzsystem 2849
Farbvalenzeinheiten 14032

Farbveränderung beim Trocknen 4768
Farbverbrauch 7334
Farbverteiler 719
Farbverteilerleiste 7383
Farbverteilung 7339
Farbverteilungsbalken 7382
Farbvertiefung 1475
Farbwalzen 7360
Farbwalzen abziehen, die ~ 12281
Farb/Wasser-Abgleich 7408
Farb/Wasser-Balance 7408
Farbwerk 7359
Farbwerk mit Duktorlineal 4534
Farbwerk mit Farbpumpen 912
Farbwert-Beurteilungskabine 14467
Farbwiedergabe 3170, 3172
Farbwiedergabequalität 3172
Farbwischer 4690
Farbzufuhr 7348
Farbzufuhr-Einstellung 3114
Farbzuführung 7348
Farbzusatz 3317
Farbzylinder 4846, 7335, 7341
Farmerscher Abschwächer 5471
Faser 5588
Faserbindung 5590
Fasergußbehälter 11030
Faserlauf 6537
Faserlaufrichtung 8443
Faserrichtung 6537
Faserriß 5593
Faserstaub 8226
Faserstoff 5597
Faserstoffaufschwemmung 5592
Faserstofflage 10504
Faserstoff-Suspension 5592
Faserstoffzusammensetzung 5594
Faserzusammensetzung 5592
Faßikelausgabe 4974
fassoniertes Papier 11165
Fax 3294, 10762
Fazette 5425
Feder 12939
Federleichtpapier 5501
Federliniermaschine 10103
Federliniiermaschine 10103
federnde Druckform 12943
Federspannung 12946
Federspannungsansatz 12942
Federzeichnung 10089
Federzüge 5863
Fehlbogen 8983
Fehlbogenmelder 9430
Fehlbogenschalter 9430
Fehldruck 8981

fehlende Rasterpunkte 8986
Fehler anzeichnen 8633
Fehlererkennungskode 5246
fehlerfreier Abzug 2895
fehlerhaft 4093
fehlerhafte Blätter auswechseln 2260
fehlerhafter Buchstabe 8245
fehlerhaftes Exemplar 4094
Fehlerhäufigkeit 5247
Fehlerkorrekturkode 5245
Fehlerpaket 5243
Fehlstelle 14503
fehlt an Manuskript, es ~ 435
feiern müssen 13558
Feile 5612
Feinätzung 5669
feine Linie 6750
Feinentwurf 3321
feiner Passer 6749
feines Kalbleder 7944
Feinfeile 12641
feinfettfeine Linie 14018
Feingraupappe 9301
Feinheit 5676
Feinheitsanalyse 10015
Feinkorn 5671
Feinkornemulsion 5673
Feinkornentwickler 5672
Feinlayout 3321
Feinmesser 8898
Feinpapier 5679
Feinpostpapier 1723
Feinraster 5680
Feinschreibpapier 13413
feinstellen 14935
Feinstpapier 13412
Feld 5602
Feldstärke 5605
Felleisen 11842
Fenchale-Apparat 5538
Fensterbriefhülle 14749
Fensterbriefumschlag 14749
Fensterfaltschachtel mit gegenüberliegenden Klappen 2390
Fensterfalz 14531
Fensterkuvert 14749
Fensterpapier 14498
Fensterumschlag 14749
Fernbedienung 11458
Fernbetätigung 11458
Fernmeldetechnik 13627
Fernsatz 13636
Fernschaltung 11458
Fernschreibkode 1232
Fernsetzmaschine 13635
Fernsteuerung 11458
Ferri- 5540
Ferri-Ammoniumoxalat 5543
Ferri-Ammoniumsulfat 5544

Ferri-Ammoniumzitrat 5542
Ferribromid 5545
Ferrichlorid 5546
Ferrihydroxyd 5550
Ferrioxalat 5551
Ferrioxyd 5552
Ferrisalz 5555
Ferrisulfat 5556
Ferritkern 5561
Ferrizitrat 5548
Ferrizyankalium 10649
Ferro- 5572
Ferroammonsulfat 5573
Ferrochlorid 5576
Ferroferrioxyd 1480
Ferromagnetismus 5564
Ferrooxalat 5577
Ferrosalz 5579
Ferrosulfat 3499, 5580
Ferrotypie 5568
Ferrozyanid 5562
Ferrozyankalium 10650
Fertigmacherei 6091
Fertigungsablaufstudie 2823
Fertigungslinie 10910
Fertigungsplanung 10908
Fertigungsstraße 10910
feste Bestellung 5699
feste Rolle 13808
feste Wortlänge 5733
festformatiger Falzapparat 5726
Festkomma 5730
festprogrammiert 6879
festschrauben 1715
Festspeicher 5732
Feststoffgehalt 12764
festverdrahtet 6879
fest werden 12760
Festwertspeicher 5732
festziehen 1715
fett 1705, 5394
fettabstoßend 6604, 9583
fettabstoßende Fläche 6607
fettbeständige Druckfarbe 6602
fettbeständiges Papier 6605
fettdichtes Papier 6603
fettechte Druckfarbe 6602
fettechtes Papier 6605
fette Farbe 6612
fette Linie 6232
fettempfänglich 9582
fetter Buchstabe 1707
fetter mittestehender Punkt 2541
fette Schrift 1707
fettfeine Linie 9801
Fettfleck 6608
Fettkreide 8258
Fettsäure 5493
Fettusche 14114
Feuchtaufnahme 9037
Feuchtauftragwalzen 3976
Feuchtdehnung 5361, 7135
Feuchtegradeerhöhung 7071

feuchten 3968
Feuchtfilz 9057
Feuchtflecken 14604
feuchtgeglättetes Papier 14576
Feuchtgehalt 9033
Feuchtigkeit 7073, 9033
Feuchtigkeit abhaltend 14593
feuchtigkeitsabbindende Farbe 9039
feuchtigkeitsanzeigende Farben 14636
Feuchtigkeitsanzeiger 7138
Feuchtigkeitseffekt 7075
feuchtigkeitsempfindlich 7139
Feuchtigkeitsgehalt 5228, 9033
Feuchtigkeitsgehalt berechnet auf absolutem Trockengewicht 9035
Feuchtigkeitsgrad 4113, 9033
Feuchtigkeitsmesser 7137
Feuchtkasten 7078, 14577
Feuchtreiber 13106
Feuchtrupfen 14688
Feuchtschwielen 9041
Feuchtstabilität 7141
Feuchtstreifen 14692
Feuchtung 3977
feuchtungsloser Flachdruck 4734
Feuchtwalzen 3976
Feuchtwalzen reinigen, die ~ 2886
Feuchtwalzenschlauch 3979
Feuchtwalzenstoff 9057
Feuchtwasser 3980
Feuchtwerk 3981
Feuer (der Farbe) 6452
feuerfestes Papier 5697
Feuerzeug 3028
Feuilleton 5585
Feuilleton, im ~ 5586
Fiberleisten 7820
Fibrillen 5598
fibrillieren 5599
Fibrillierung 5600
Fichte 12953
Fichtenholz 12953
figürlichen Schmuck belebt, durch ~ 7011
Filete 5628
Film 2987, 5636, 5637, 7950
filmbildend 5646
Filmbildner 5642
Filmfarbwerk 9790
Filmklammer 5641
Filmklebelack 13284
Filmkorrektur 3602
Filmlochkarte 698
Filmsatzkorrektur 3602
Filmsetzmaschine 10279
Filmspule 12916
Filmträger 5644
Filter 5654
Filterfaktor 5658

Filterfaktorbestimmungsgerät 5661
Filterhalter 5659
filtern 5663
Filterverhältnisskala 5661
Filterwirkung 5656
Filtrat 5662
filtrieren 5663
Filtrierpapier 5660
Filzer 3642
Filzmanchon 3644
Filzmarke 5534
Filzmarkierung 5534
Filzprägung, mit ~ 5533
Filzschlauch 3644
Filzseite 13918
Filztrockner 5532
Finalbuchstabe 5665
Finalstock 13552
Finanzredakteur 2866
Finanzschriftleiter 2866
Finch-Gerät 5667
Fingerabdrücke 5685
Fingerlochregister 13795
Fingernagelprobe 5684
Firnis 14401
Fisch 10315, 14881
Fischhaufen 10315
Fischleim 5141, 5716
Fischöl 5717
Fitzbund 7756
Fixativ 5725
Fixierbad 5736
fixieren 5722, 5723
Fixieren 5724
Fixiernatron 7146
Fixiersalz 5737
Fixierung 5724
Flachbeutel 5773
Flachbürste 5778
Flachdruck 9525, 10401
flache Beleuchtung 5794
Fläche der Hysteresis-Schleife 759
Flächendeckungsgrad 10112
Flächengewicht 1209
flächenvariabler Tiefdruck 760
flache Platte 5799
flaches Bild 5791
flaches Negativ 5795
Flachformdruckpresse 5775
Flachformmaschine 5775
Flachformrotationsmaschine 4879, 5776
flachgefalteter Karton 2380
Flachliegen 5796
flachliegendes Papier 5798
Flachoffsetpresse 9522
Flachs 5811
Flachstereo 5805
Flachstichel 5790
Flammpunkt 5765
flammsicheres Papier 5697
Flanell 5754
Flank 12390
Flankenschutzmittel 12382

Flansch 5752
Flanschverbindung 5753
Flash-Belichtung 5762
Flatteränder 12533
Flattermarken 3066
Flattersatz 14328
Fleck 13008
Flecken 12921
Fleisch 1247
Fleischeinwickelpapier 4781
Fleischerpapier 4781
Fleischfarbe 5815
Fleischton 5815
flexibler Buchumschlag 8153
flexibler Einband 8153
Flexodruck 5828
Flexodruckmaschine 5829
Flexoeindruckwerk 615
Flexofarbe 5825
Flexographie 5828
Flexographiefarbe 5825
flexographische Farbe 5825
Fliegen 7373
fliegendes Blatt 5930
fliegendes Vorsatzblatt 5930
Fliegenkopf 14095
Fliegenköpfe berichtigen 7441
Fliegenköpfe herausnehmen 7441
Fliegenköpfe richtigstellen 7441
Fließarbeit 844
Fließband 843
Fließbandfertigung 844
Fließeigenschaft 5864
Fließen 5864
Fließfertigung 5873
Fließgrenze 14930
Fließgrenze eines Bingham'schen Körpers 1434
Fließkunde 11585
Fließkurve 5868
Fließlehre 11585
Fließpapier 1622
Fließproduktion 5873
Fließprüfer 5874
Flintglas 5833
Flintpapier 5835
Flipflop 5836
Flockenbildung 5842
Flockeneiweiß 5748
Flockung 5842
Florilegium 654
Florpost 6444, 9597
Florpostpapier 6444
Flosse 5718
flüchtig 6229, 14504
flüchtige Bestandteile 14506
flüchtige Stoffe 14506
Flugblätter 6815
Flügelmutter 14752
Flugpostpapier 341
Flugzettel 6815

Fluidität 5885
Fluorammonium 530
Fluoreszenz 5888
Fluoreszenzfarbe 5891
Fluoreszenzfarbstoff 9646
Fluoreszenzlampe 5892
Fluoreszenzverfahren 5889
Fluoreszieren 5888
fluoreszierend 5890
fluoreszierende Farbe 5891
Fluorwasserstoffsäure 7106
Fluschen 5904
Flushing 5904
Flushing-Verfahren 5904
Flush-Montage 5902
Flußbild 5867
Flußdiagramm 5867
flüssige Phase 8238
flüssiges Sikkativ 8235
Flüssigkeit 5885
Flüssigkeitszahl 5885
Flußmittel 5912
Flußsäure 7106
Flute 5911
Föhn 5456
Fokus 5950
folgestanzen 6303
Folgestanzer 6302
Folgezeiger 2471
Folie 5961
Foliendruckmaschine 5647
Folienkaschiermaschine 7891
foliieren 1381
Foliieren 5993
Foliierung 5993
Folioformat 5994
Fond 1001
Footcandle 6016
Ford-Auslaufbecher 6030
Fordbecher 6030
Förderbahn 3481
Förderband 3482
Fördergurt 3482
Förder- und Verteilmechanismus für die Ablegung 4491
Förderung 5397
Fordtopf 6030
Form 6053, 6056
Formaldehyd 6048
Formaldehydharz 6049
Formalin 6050
Formänderung 4105
Format 6052, 6053
Format abnehmen, das ~ 13282
Format abschlagen, das ~ 13282
Formatangabe 3557
Formatausrichtung 7203
Formatbacken 14464
Formatbreite 8720
Formaten van Registerpapier 12499
Format machen 8561
Formatpapier 9922
Formatskala 5124
Formatstege 6262

gebrannte Umbra 2089
Gebrauch eines Elektronenrechners für das Schriftsetzen 3334
Gebrauchsgraphiker 3240
gebraucht 12043
gebrochene Farbe 4414
gebrochene Schriften 1939
gebrochene Ziffer 10320
gebundenes Buch 1815,
 2401
gebundene Wärme 7930
gebürstetes Kunstdruck-
 papier 2001
Geburten, Hochzeiten und
 Todesfälle 1440
gedruckte Kopie 6859
gedruckte Leiterplatte
 10800
gedruckte Schaltung 10800
gedrucktes Manuskript
 11478
Gefach 1829
Gefälle 6535
gefältet, in der Längs-
 richtung ~ 8334
gefärbte Buchschnitte 3135
gefärbte Fasern 3136
gefärbter Schnitt 13009
gefärbtes Schafleder 3138
gefeuert werden 2054
geflushte Farbstoffe 5901
geflushte Pigmente 5901
gefrieren 6166
Gefrierrohkarton 1189
Gefüge 13303
gegautschter Karton 14411
Gegendruck 1044, 7219
Gegendruckzylinder 7212
Gegengewicht 1088
Gegenlaufwalzen-Streich-
 verfahren 11559
Gegenlicht 1031
Gegenmaske 3660
Gegenmuster 3662
Gegenmutter 2672
Gegenseite 9641
gegenseitige Verriegelung
 7506
gegenüberliegende Seite
 9641
gegenüberstehend 7723
geglättetes Papier 2190
geglättetes Saffianleder
 3784
gegossene Zeile 12599
gehämmertes Papier 6809
Gehäuse 2398
geheftet, auf aufgedrehte
 Schnüre ~ 12200
geheftet, auf Bänder ~
 12201
geheftet, (mit Faden) auf
 Gaze ~ 12202
geholländert (mit Buch-
 fadenheftmaschine) 6184
Gehrung 1346, 8995
Gehrung schneiden 8996

Gehrungschneidmaschine
 8998
Gehrungslinie 8997
Gehrungsstück 3581
Geistereffekt 6391
Geistererscheinung 6391,
 6392
Geistersalz 521
gekapselter Motor 5151
geklebt 10024
geklebte Montage 10041
geklebter Bristolkarton
 10025
geklebter Karton 10022
geklebter Umbruch 10041
geklebtes Umschlagpapier
 10026
gekochtes Leinöl 1695
gekörntes Aussehen 12839
gekörntes Papier 6543
gekürzte Ausgabe 19
Gel 6352
Gelatine 6356
gelatineartig 6366
Gelatinefilter 6360
Gelatineleim 623
Gelatine, mit Bichromat
 sensibilisierte ~ 4302
Gelatinereliefplatte 6363
Gelatineschicht 6357
gelatinös 6366
gelbblausaures Kali 10650
Gelbdruckplatte 14924
gelbes Arsensulfid 9691
gelbes Blutlaugensalz
 10650
gelbes chromsaures
 Kalium 10641
Gelbfärbung 14917
Gelbfilter 14916
gelblichweißes geripptes
 Papier 3711
Gelbplatte 14924
Gelbstrohstoff 14928
Gelbstrohzellstoff 14928
Gelegenheitsaufträge 2460
geleimtes Papier 12490
geleimtes Papier, in der
 Masse ~ 1264
Geleitwort 6043
Gelenkhebel 13876
Gelenkkupplung 14325
gelieren 6353
gelochte Karte 11045
gelochter Streifen 11046
gelöschter Kalk 2167
Gemeine 8380, 8967
Gemeinen setzen, in ~
 7737, 11065
Gemeinkosten 9758.
gemeinsamer Antrieb 3004
gemessene Schärfe 174
gemischter Satz 9003
genarbtes Papier 6543
Genauigkeit 85, 4098
Generierer 6377
Generierprogramm 6377
Genosse 12380
geometrische Zeichen 6383

geordnete Datenmenge
 4015
Gepäckanhänger 1069
geperlter Druck 9116
gepolsterter Beutel 7659
geprägtes Papier 5101
geprägtes Pergamin 5100
geprägtes Pergaminpapier
 5100
gequirlter Satz 10315
gerader Buchrücken 12957
geraderichten 404
gerade Seiten 5327
geradestehende Schrift
 11670
geradestoßen 7685
gerasterter Photo-Abzug
 11978
Gerät 4236
Geräteausstattung 6877
Gerät zur Bestimmung der
 Adhäsion 1721
Gerät zur Bestimmung der
 Saughöhe 42
Gerät zur Verhinderung
 der Anlage zweier
 Bogen 14155
gerauhte Druckwalze 9240
gerauhte Lederwalze 9240
geräuschloses Papier 9386
gerben 6861
Gerben 13575
Gerbsäure 13573
Gerbstoff 13576
Gerbung 13575
Gerbungsmittel 13576
gereppt 11468
geringe Druckspannung
 7805
geringwertiges braunes
 Packpapier 13607
gerinnen 2970
Gerinnen 2971
Gerinnung 2971
geripptes Papier 7876
geripptes Schreibpapier
 7878
Geruch 9498
geruchfrei 9499
geruchfreie Farben 9501
geruchlos 9499
geruchlose Farben 9501
Geruchlosmachen 4182
Geruchlosmachung 4182
gerundet 3847
gesammelte Produktion
 3078
gesättigte Farbe 11879
gesättigte Fettsäure 11880
gesättigte Lösung 11881
gesäuberte Ausgabe 5378
gesäubertes Wollfett 7901
Geschäftsbuch 1506
Geschäftsbücherfabrik 77

Geschäftsbücherpapier
 8024
Geschäftsbuchpapier 8024
Geschäftsdrucksache 2107
Geschäftsformulare 2106
Geschäftskarte 2105
geschältes Rundholz 8323
geschleiert 8136
geschleiertes Negativ 5959
geschlossene Datenkette
 2091
geschlossener Kreis 2931
geschlossener Motor 5151
geschmacklose Verzierung
 6414
geschmeidig 5820
Geschmeidigkeit (einer
 Farbe) 13425
Geschmier 1059
geschnitten, auf halbe
 Bahnbreite ~ 12038
geschnittener Loch-
 streifen-System 13923
geschwänzter Buchstabe
 4201
geschwefeltes Öl 13382
geschwefeltes Rizinusöl
 14091
geschweifte Klammern
 1840
Geschwindigkeitsgefälle
 14428
Geschwindigkeitsgradient
 14428
Geschwindigkeitsregelung
 12850
geschwungene Versalien
 13455
Geselle 7701
gesetzliche Pflichtlieferung
 3554
gesetzt aus fetter oder Auszeichnungsschrift 4464
gesetzter Satz, im
 Akkord ~ 10322
gesetzt mit der Maschine
 8734
gespaltenes Schafleder
 12527
Gespan 12380
gespeichertes Programm
 13236
gesprenkelte Buchschnitte
 12949
gesprühte Öle 1626
gesprühtes Leinöl 1625
gestaltlos 552
Gestaltung 7961
Gestaltung (einer Druck-
 sache) 4208
gestanzter Ausschnitt 3872
Gestehungspreis 3631
Gestell 6144
gesteuertes Rundschneid-
 messer 12526
gestiegenes Ausschluß-
 stück 11226

gestrichenes Glacé-Papier, satiniertes gebürstet 1998
gestrichenes Lithopapier 2983
gestrichenes Offsetpapier 2984
gestrichenes Papier 2986
gestürzte Seite 1932
geteilter Farbkasten 12898
getränktes Papier 7206
getrennte Auslage 12897
getrennt (vom übrigen System) 9513
Getriebe 6345
Getriebe laufende Reiterwalze, im ~ 6348
getrübte Farbe 12209
Geviert 5120
Geviertführungspunkte 9623
Geviertskala 5124
Geviertstrich 4001
Gewebedecke 2949
Gewebeeinband 2950, 2951
Gewebefalz 2953
Gewebenummer 8812
Gewebepack 11428
Gewebepackpapier 11428
Gewebepapier 2957, 11426
Gewebezahl 8812
gewebtes Drahtsieb 14857
Gewerbefachzeitschrift 13948
Gewichte und Maße 14673
gewickelte Tube 3483
Gewindestift 14852
Gewinnung 5397
gewissen Geld stehen, im ~ 12987
Gewißgeld 12994
Gewißgeldsetzer 12989
gewöhnlich 11670
gewöhnlicher Äther 5297
gewöhnlicher Kalk 2163
gewölbt 4550
gezählt, jedoch nicht paginierte Seite 1565
Giebelreklame 14538
Gießapparat 2440, 2443
Gießbach 6734
Gießbacke 2453
gießen 2435, 10694
Gießer 2441
Gießerei 14193
Gießform 9118
Gießformeinsatzstück 8193
Gießformkühlung 9124
Gießformwischer 9135
Gießgerät für Flachstereotypie 2450
Gießhalsheizung 13782
Gießhülse 11643
Gießkelle 2447
Gießloch 9684
Gießlöffel 2447
Gießmarke 10369
Gießmaschine 2440, 14181
Gießmund 10684

Gießrad 9125
Gießradschlitten 9126
Gießrahmen 2446
Gießschale 3695
Gießstern 13043
Gießtemperatur 2454
Gießtopf 8836
Gießtopfkehle 13780
Gießwinkel 2453
Gießzettel 6120
Giftigkeit 13933
Gilbert 6398
Gilsonit 6411
Gilsonitasphalt 6411
Gips 3774, 6738
Gitterschachtel 3703
Gitterspannung 6634
Gitterstrom 6632
Glacépapier 5835
Glanz 6452, 8419
Glanzdecke 10756
Glanzdruckfarbe 6453
glänzend 6457
glänzende Fehlerstellen 12304
Glanzfarbe 6453
Glanzfirnis 6455
Glanzkalander 6193
Glanzkupferverfahren 1095
Glanzlack 6455
Glanzlicht 5758, 6425
Glanzmesser 6426, 6454
Glanzmeßgerät 6454
Glanzpapier 5835, 6445
Glanzpappe 10756
Glanzplatte 5570
Glanzprüfer 6426
Glanzschwarz 1915
Glanzverlust 13606
Glasballon 2337
Glaselektrode 6428
Glasfaser 6429
Glashalbzelle 6428
glasige Flecke 5715
glasige Oberfläche 6441
Glaskugeln 3037, 6430
Glaslinse 8041
Glasmärbel 3037, 6430
Glasmater 6432
Glasnegativ 6433
Glaspapier 6434, 11861
Glasplatte 6435, 10420, 12528
Glasplatte (für Sammelmontage) 5766
Glaspositiv 6436
Glasraster 6438
Glasscheibe 5956
Glasskugel 2060
Glasstab (zum Rühren) 6437
Glätte 5686, 8248
Glätte (des Papiers) 12643
Glättemesser 12646
Glättemessung 12645
Glätten 2081
Glätteprüfer 12646
glatter Farbschnitt 3135
glatter Satz 1683, 13247

glatte Walze 6447, 12647
Glättfeile 12641
glatt gewordene Walze 6446
Glättkalander 12995
Glättpresswalze 6448
Glättschaberstreichverfahren 1496
Glättstahl 2078
glattstoßen 7685
Glattstoßmaschine 7687
Glättwalze 6448, 12878
Glattwerden des Drucktuches 6449
Glättwerk 2184, 12995
Glättzahn 2079
Glättzylinder 6448, 14903
Glaubersalz 12719
gleichförmig beschleunigte Bewegung 14305
Gleichgewichtausgleich 5227
Gleichgewichtslage 5227
Gleichheitsstrich 5224
Gleichheitszeichen 5224
gleichlaufende Bewegung 9984
gleichmäßige Durchsicht 14304
Gleichmäßigkeit der Schwärzung über die Fläche eines Films 757
Gleichrichter 11317
Gleichstrom 4400
Gleichstrommotor 4401
Gleichung (im Formelsatz) 5226
gleichwertige Belichtung 5229
Gleitkomma 5839
Gleitlager 10390
Gleitmittel 12562
Glimmerblättchen 12304
Glimmerverunreinigungen 12304
Glöckchen 5448
Glühen 2153
Glühlampe 5012, 7241
Glührückstand 816
Glühung 2153
Glukose 4246
Glykol 5309
Glykolen 6489
Glykolmonomethyläther 5310
Glykolsäure 6488
Glykose 4246
Glyzerid 6479
Glyzerin 6482
Glyzerinsäure 6478
Glyzin 6485
Goldauftrager 9839
Goldbromid 6498
Goldbronze 6499
Goldbronzepuder 6515
Goldchlorid 6517
Golddruck 6496
Golddruckfarbe 6507
goldener Schnitt 6505

Goldfolie 6509
Goldkäferglanz 1958
Goldkaliumbromid 10651
Goldkaliumchlorid 10652
Goldkissen 6501
Goldmesser 6508
Goldoxyd 6511
Goldpapier 6512
Goldprägepresse 1604
Goldprägung 6496
Goldscheidewasser 734
Goldschlägerhaut 6495
Goldschnitt 6412
Goldschnitt, mit ~ 6412
Goldunterdruckfarbe 12485
Goldwarenseidenpapier 12436
Gordon-Maschine 6526
Gordon-Presse 6526
Gotisch 6528
gotische Schrift 6528
Gouache 6530
Goudronnépapier 13607
Gradation 6534
Gradationskurve 2637
Gradbogen 10986
Grade Celsius 2530
Grade Fahrenheit 5435
Gradeinteilung 4518
Gradient 6535
Gradierwaage 761
Grad Kelvin 7744
Gradzeichen 4108
Gramm 6556
Grammgewicht 1209
Grammol 9042
Grammolekül 9042
Graniierstahl 11626
Graphiker 6567
graphische Anstalt 10829
graphische Darstellung 4257
graphischer Betrieb 10829
graphisches Gewerbe 6565, 10831
Graphit 6571
Graphitierbürste 1483
Graphitieren 1482
Graphitschwärze 6571
grappeln 2732
Grat 2090
Gratisexemplar 10733
Gratulationskarten 6626
grau 12730
graue Leinwand 6627
graues Buchbinderleinen 6627
Graukeil 6630
Grauleiter 6629
Graupappe 2726, 9301
Grauschleier 5957
Grauskala 6629
Graveur 2656, 5186
gravieren 5182
Gravieren 13900
Gravierkissen 5189
Gravierkopf 5193
Graviermaschine 5195
Gravis 6577

Kaliumdichromat 10648
Kalium-Eisenalaun 5554
Kalium-Eisensulfat 5554
Kaliumeisenzyanid 10649
Kaliumeisenzyanür 10650
Kaliumferrizyanid 10649
Kaliumferrozyanid 10650
Kaliumhydrat 10655
Kaliumhydrogensulfat 10632
Kaliumhydrosulfat 10632
Kaliumhydroxyd 10655
Kaliumhypermanganat 10664
Kaliumhypochloritlauge 7649
Kaliumhypochloritlösung 7649
Kaliumjodid 10656
Kaliumkarbonat 10636
Kaliumlaktat 10657
Kaliummetabisulfit 10658
Kaliumnitrat 10660
Kaliumoxalat 10661
Kaliumperchlorat 10663
Kaliumperkarbonat 10662
Kaliumpermanganat 10664
Kaliumpersulfat 10666
Kaliumplatinchlorür 10640
Kaliumpolysulfid 10677
Kaliumpyrochromat 10648
Kaliumrhodanid 10676
Kaliumsilikat 10670
Kaliumsulfat 10671
Kaliumsulfid 10672, 10677
Kaliumsulfit 10673
Kaliumthiozyanat 10676
Kaliumthiozyanid 10676
Kaliumzitrat 10644
Kaliumzyanoferrat-II 10650
Kaliumzyanoferrat-III 10649
Kaliwasserglas 10670
Kalk 8145
Kalkhydrat 2167
kalkieren 13935
Kalksalpeter 2169
Kalkstein 2163
Kalkulation 5264
Kalkulationssystem 3630
Kalkulator 5265
Kalligraph 2217
Kalligraphie 2218
Kalomel 8793
Kalomel-Elektrode 2224
Kalorie 2225
kalte Farbe 3052
Kaltemail 3054, 5138
Kaltemailverfahren 3055
kaltgepreßtes Papier 3057
Kaltleim 3056
Kaltnadelradierung 4817
Kaltprägung 3051
Kaltsatz 3061, 10278
kalttrockenfarben 3058
kalttrocknende Farben 3058
Kalzination 2153
Kalzinieren 2153

kalziniertes Soda 12682
Kalzinierung 2153
Kalzium 2160
Kalziumbisulfit 2162
Kalziumchlorid 2164
Kalziumhydroxyd 2167
Kalziumhypochlorit 2168
Kalziumkarbonat 2163
Kalziumnitrat 2169
Kalziumoxyd 2171
Kalziumphosphat 2173
Kalziumsulfat 2174, 3774
Kalziumsulfit 2175
Kambrikleinen 2232
Kambrikpapier 2233
Kamee-Einband 2236
Kamera 10882
Kameraabteilung 6276
Kameraauszug 1294, 2243
Kamerabalgen 1292
Kamerablende 4273
kamerafertige Vorlage 2248
Kamerahinterkasten 2240
Kameralinse 8042
Kamerastativ 2249
Kamerastütze 2249
Kammarmorschnitt 3217
Kampfer 2251
Kampferöl 2252
Kanadabalsam 2255
kanadischer Balsam 2255
Kanal 2620
Kanalruß 2622
kann laufen 11810
Kanten 4961
Kantenabschrägmaschine 1355
Kanteneffekt 1794
Kantensteuerung 12373
Kanvas 2269
Kanzleipapier 8024, 10163
Kaolin 2711
kapillaraktiv 13427
Kapillare 2278
Kapillarität 2277
Kapillarkraft 2281
Kapillarrohr 2284
Kapillarsteigung 2283
Kapillarströmung 2280
Kapillarviskosimeter 2285
Kapillarwirkung 2279
Kapitalband, (oberes) ~ 6894
Kapitälchen 12617
Kapitälchen ausgezeichnet, mit ~ 60
Kapitälchen mit grossen Anfangsbuchstaben 2295
Kapitälchen ohne Initiale 5329
Kapital mit Unterkasten 2293
Kapitalsteg 6656
Kapitel 2626
Kapitelüberschrift 2298, 2627
Kappazahl 7729
Kappenschachtel 2290

Kappvorrichtung 12266, 12277
Kapuziner 6335, 6338
Karagheenmoos 7574
Karbinol 8855
Karbolsäure 10189
Karbonat 2305
Karbondruckfarbe 2312
Karbonfarbe 2312
karbonfreier Formularsatz 2324
karbonfreies Durch-schreibepapier 2325
Karbonisierfarbe 2312
Karbonpapier 2326
Karboxylgruppe 2335
Karboxymethylzellulose 2336
Kardangelenk 14325
Kardinalpunkte 2351
kariertes Papier 12964
Karikatur 2392
Karnaubawachs 2360
Karolingische Minuskeln 2362
Karrageenmoos 7574
Karren 2364
Karrenantrieb 1276
Karrenbahn 1283, 1825
Karrenbewegung 2366
Karrenleisten 2367
Karrenrückgang 2369
karriertes Papier 12964
Karte 8612
Karte binär pro Spalte 3202
Karte binär pro Zeile 11741
Kartei 2352, 5619
Karteikarte 7271
Karteikarton 7270
Karteikasten 5619
Karten 2356
Kartenabfühleinheit 2355
Kartenabfühler 2355
Kartenbild 2350
Kartenformaten 12497
Kartenkarton 2356
Kartenleser 2355
Kartenmischer 3069
Kartensatz 9837
Kartenstapel 9837
Kartentasche 1772
Kartenzeichner 2377
Kartograph 2377, 8613
Karton 2342, 2344, 2393, 5975, 9899
Kartonagenfabrik 9902
Kartonagenglacé 5145
Kartonagenkarton 1831
Kartonagenpappe 1831
Kartonfabrik 1662
Kartonformaten 12495
Karton für Gepäck-anhänger 13535
Karton für Kartonagen 1831
kartoniert 7240

Karton, mit Wasserglas gestrichener ~ 12411
Kartonrolle 11372
Kartonschere 1666, 2345
Kartonschneider 1666, 2345
Kartothekkarte 7271
Kartothekkarton 7270
Kartusche 2394
Kartuschenpapier 2395
kaschieren 7885
Kaschiermaschine 7891
Kaschierpapier 8207, 10043, 12273
kaschiert 7886
kaschierte Pappe 8170
kaschiertes Bitumenpapier 14308
kaschiertes Papier 7888
Kaschierungspapier 8207
Käseblatt 11199
Käseblättchen 11199
Käseechtheit 2679
Kasein 2403
Käseschachtel 2677
Kasselerbraun 14384
Kassenblock 2421
Kassenblockdruck-maschine 2422
Kassenblockmaschine 2422
Kassette 2429, 10442
Kassetteninitial 5429
Kastanienholz 2701
Kasten 2396
Kasteneinband 14906
Kasteneinteilung 7960
Kastenfach 1829
Kastenregal (ohne Setz-pult) 2414
Kastenschema 7960
Kasten verfischen, den ~ 10317
Katalog 2462
Katalysator 2463
Kataphorese 5061
Katechu 2474
Kathedral-Einband 2478
Kathode 2480
Kathodenstrahlröhre 2481
Kation 2483
kationische Reagentien 2484
Katode 2480
Kattun 3636
Kattundruckpapier 2207
Kaudersprache 7648
Kauderwelsch 7648
Kauri 7732
Kauri-Butanolwert 7733
Kauriharz 7732
kaustisches Kali 10655
kaustisches Natron 12695
kaustische Soda 12695
Kautschuk 11747
Kautschukstempel 11757
Kavitation 2494
KB-Wert 7733
Kegel 1686
kegelförmig 3381
Kegelgetriebe 1348

Kegelmühle 7693
Kegelrad 1349
Kegelradantrieb 1348
Kegelstärke 1686
Kegelstift 13593
Kegelstoffmühle 7693
Kegelwechsel 2615
Kegelweite 14738
Kehrbild 7542
Kehrwendestange 10572
Kehrwert 11301
Keil 7760
Keilexzenter 7776
keilförmiger Serif 14666
Keilitzfarbe 11537
Keilnute 7786
Keilriemen 14416
Keilschrift 3813
Keiltaste 12804
Keiltreiber 12318
Kemart-Verfahren 7745
Kenntnisse, Regeln und
    Meßverfahren die dem
    Faktor zur Verfügung
    stehen 10983
Kennzahl 7762
Kennzeichen 2638, 5742
keramische Farbe 2556
keramisches Abziehbild
    2558
Kern 3565, 8504
Kernexzenter 9122
Kernspeicher 3574
Kerosin 7751
Kerze 2266
Kessel 7755
Keton 7753
Ketopropan 102
Kette 13277, 14544
Kettenantrieb 2583
kettenartig verschlungene
    Verzierung 2581
Kettenauslage 2580
Kettenausleger 2580
Kettenbücher 2579
Kettendrucker 2590
Ketteneinband 2584
Kettengreifer 2587
Kettengreiferauslage 2580,
    2587
Kettengreifersystem 2587
Kettenpunktraster 2582
Kettenrad 12950
Kettentrieb 2583
Kettpunktraster 2582
Kiefer 10362
Kiefernholz 10362
Kienruß 7894
Kiesel 12409
Kieselerde 4280, 12409
Kieselglas 5833
Kieselguhr 4280
Kieselsäure 12414
Kilogramm 7797
Kilohertz 7795
Kilopond 7798
Kilowatt 7799
Kilowattstunde 7800
Kinderbuch 2709

kinematische Viskosität
    7801
kinetische Energie 7802
Kiosk 9317
Kippschalter 13878
Kirchenbibel 10708
Kissen 9838
kissenförmige Verzerrung
    10361
Kiste 1827
Kistenauskleidepapier 2408
Kistenauslegepackpapier
    2408
Kistenausschlagpapier 2408
Kistenfutterpapier 2408
Kladdepapier 12005
Klammer 2877, 9906
Klammerheftung 12389,
    12988
Klammern 2920
Klammern, (runde) ~ 9998
Klammerschließer 2910,
    13170
Klammerstütze 13173
Klammertreiber 13168
Klang 11254, 12650
Klappdeckel 5757
Klappe 5755
Klappenfalz 7651
Klappenfalzapparat 7652
Klappentext 1647
klappern, beim Satzen ~
    5452
klären 2901
Klarschriftleser 9651
Klassification der Druck-
    schriften 2879
klassische Druckfarbe 9424
klassizistische Antiqua
    4310
Klatsch 1640, 7757
Klatschdruck 1640
Klatschdrucke 2601
Klauenhammer 12248
Klauenöl 9260
Klaviatur 7765
Klaviaturauslösestab 7769
Klaviaturexzenter 7766
Klebeapparat 10035
Klebeband 215, 6718
Klebeband für Druckstock-
    montage 9146
Klebebindemaschine 10119,
    13694
klebebinden 8405
Klebebinder 10119
Klebebindung 208
Klebefolie 211
Klebekarton 10022
Klebekasten 10029
Klebekopf 10036
Kleb(e)kraft 214
Klebelasche 6472, 7689
Klebemarke 10040
Klebemaschine 6475
Klebemittel 213
Klebemontage 10041
kleben 12892
Klebenaht 6472

Klebepapier 6715, 8207,
    12273
Klebeplatte 13083
Kleberand 6472
Klebescheibe 10036
Klebestärke 4243
Klebestelle 7689
Klebestellenriß 12893
Klebestelle von der Papier-
    fabrik 8941
Kleb(e)stoff 213
Klebestreifen 6718, 10040
Klebeumbruch 10041
Klebevorrichtung 10035
klebrige Farbe 13532
klebrige Walzen 13533
Klebrigkeit 13524, 13525,
    13526
Klebstoffkasten 10029
Klebstoffpumpe 10039
Kleesalz 10631
Kleesäure 9797
Kleinanzeigen 2881
Kleinauflage 12336
Kleinbuchstabe 8967
Kleinbuchstaben 8380
Kleinbuchstaben setzen,
    in ~ 7737
klein drucken 7737
kleine Anzeigen 2881
kleine Ausgabe 19
kleine Buchbinderstempel
    10164
kleine Buchstaben 8380
kleine Löcher in Papier-
    oberfläche 10364
kleiner ätzen 12234
kleines Ries 12335
Kleinoffsetmaschine 12627
kleinste Blende 12224
kleinste Papiermenge für
    Lastwagentransport 2358
kleinste Wagenladung 2358
Kleister 13046
Kleistergraf 1732
Kleisterpinsel 10023
Kleisterstreifen 10031
Kleistertopf 6469
Klemmbacke 1745
Klemme 13653
Klemmklinke 10067
Klemmschraube 2874
Klemmstange 2872
Klimaanlage 321, 9907
Klimakammer 7074
Klimaraum 7074
Klimaschrank 7074
Klimatisieren des Papiers
    3370
klimatisiertes Papier 3368
Klimatisierung des Papiers
    3370
Klimatisierungsanlage 9907
Klischee 1582, 5192, 10418,
    10907, 11445
Klischeeandruck 5190
Klischeeanstalt 10889
Klischeeätzmaschine 5280
Klischeebiegeapparat 1587

Klischeefuß 9136
Klischeehersteller 10888
Klischeeherstellung 10232,
    10890
Klischeehöhenmesser 1595
Klischeeholz 7887, 9144
Klischeemacher 10231
Klischee mit Linienraster
    9596
Klischeemontierholz 9144
Klischeemontiermaschine
    9145
Klischeemontiernägel 1603
Klischeenägel 1603
Klischeeprüfer 14198
Klischeeprüfgerät 1595
Klischeereinigungsmittel
    10425
Klischees mit Tangier-
    raster 1316
Klischeesperrholz 10505
Klischeeunterlage 9136,
    14282
Klischeevorlage 14835
Klischeezurichtung 2061
Klischeur 10888
klopfen, mit der Bürste ~
    1260
Klopfholz 10394
Kloster-Einband 9064
Klotz 9136
Klotzbeutel 1589
Klotzbodenbeutel 1589
Klötze 8323
Klotzpresse 1745, 8423
Knattern der Platte 10461
Knebelgesetz 6271
Kniegelenk 13874
Kniehebel 13874, 13876
Kniehebelpresse 368, 13037,
    13875
Knippvorrichtung 9356
Knitterprobe 3782
Knitterprüfgerät 3780
Knochenleim 1728
Knochenschwarz 1725
Knopfsteuerung 11061
Knoten 5595, 8417
Knotenfänger 7830
Knotenpunkt 9383
K & N-Versuch 7727
koachsialer Antrieb 3004
Koagulation 2971
koagulieren 2970
Koagulieren 2971
Koagulierung 2971
Koautor 3002
koaxiales Kabel 3003
Kobalt 3005
Kobaltblau 3006
Kobaltgrün 3008
Kobaltlinoleat 3009
Kobaltnaphtenat 3010
Kobaltresinat 3015
Kobaltsikkativ 3007
Kobalttrockner 3007
Kochen 3486
Kocher 4352
Kochgebiet 1704

logische Silbentrennung 8319
Logotype 8322
Lohnsatz 13949
Lohntag 10068
Löhnungstag 10068
Lokalredakteur 2867
Lokalredaktion 2868
Lösbarkeit 12768
Löschblatt 1621
löschen 2899, 4129, 5232, 7792
löschendes Lesen 4212
Löschpapier 1622
Löschpapier für Schreibblöcke 13515
Löschtaste 4130
lose aufgewickelte Rolle 8367
Lose-Blätter-Buch 8364
Lose-Blätter-Geschäftsbuch 8365
lose gewickelte Rolle 8367, 12535
Lösemittel 12773
Lösemittelechtheit 12781
Lösemittelgewinnung 12778
Lösemittelwiedergewinnung 12778
lösen 4478
Löser 12773
loser Buchstabe 5448
loses Blatt 8362
löslich 12769
Löslichkeit 12768
loslösen, sich ~ 1580
Lösung 12772
Lösungsbenzin 11897
Lösungsbenzol 12776
Lösungsmittel 12773
Lösungsmittelabgabe 12780
Lösungsmittelfreisetzung 12780
Lösungsmittelrückhaltevermögen 12780
Lösungsmittelstreichverfahren 12775
Lösungsstärke 3356
löten 12753
Lowkey 8385
Lückenbüßer 5621
Ludlow-Setzer 8402
Ludlow-Setzmaschine 8401
Luftabsaugung 335
Luftabzugrohr 5355
Luftbefeuchter 7072
Luftbild 281
Luftbildaufnahme 281
Luftbläschenstreichverfahren 2019
Luftbläschenstrich 2019
Luftblase 336
Luftblasen 311
Luftblasenviskosimeter 2022
Luftbürstenstreichverfahren 338
luftdicht 356
Luftdurchlaß 343

Luftdurchlässigkeit 344
Luftdurchlässigkeitsprüfer 345
Luftdurchlässigkeitsprüfgerät 4176
Luftdüse 9445
Luftdüsenstreichverfahren 338
Luftentfeuchtung 4119
Luftfeuchtigkeit 7077
Luftfeuchtigkeitsmessung 10997
luftförmig 6316
luftgetrocknetes Papier 329, 8312
luftgetrocknetes Verpackungspapier 328
Luftkanal 343
Luftkissen 325
Luftkissentisch 326
Luftkolben 347
Luftkolbenblock 10376
Luftkolbenhebelblock 10378
Luftkompressor 3327
Luftkonditionierung 322
Luftkühlung 324
Luftleitungen 349
Luftmesserstreichverfahren 338
Luftpinsel 314
Luftpinselstreichverfahren 338
Luftpolster 325
Luftpolstertisch 326
Luftpostpapier 341
Luftpostumschlag 340
Luftpresser 354, 3327
Luftpuffer 318
Luftpumpe 354
Luftstift 347
Luftstiftblock 348
Lufttisch 326
lufttrockener Halbstoff 333
lufttrockener Zellstoff 333
lufttrockenes Papier 332
Lufttrockner 330
Lufttrocknung 331
Luftventil 357
Luftventilblock 10377
Luftventilkasten 14381
Luftverunreinigung 351
Luftzylinder 14323
lumbecken 8405
Lumen 8407
Lumen-erg 8408
Lumineszenz 8411
Lumineszenzdruckfarbe 8412
lumineszierende Farbe 8412
Lumpen 11210
Lumpendrescher 14744
Lumpenpapier 11205, 11206
Lumpenschüttelmaschine 4891
Lumpenstoff 11208
Lunker 5810
Lupe 8187, 8534

Lux 8420
Luxusausgabe 4973, 7919
Luxusbuch 4973
Luxuseinband 7919
Lyoneser Einband 8424
lyophil 8425
lyophob 8427

Mäander-Einband 4179
machen, einen Ausschnitt ~ 9108
Madenschraube 14852
Magazin 7698, 8479
Magazineintritt 2623
Magazinrahmen 8482
Magazinrahmen-Umschaltmechanismus 8484
Magazinverwalter 13239
Magenta 8486
Magentadruckfarbe 8489
Magentadruckplatte 8491
Magentafarbe 8489
Magentafilter 8488
Magentakontaktraster 8487
Magentamaskierverfahren 8490
Magentaplatte 8491
mager drucken 10798
magere Schrift 8111
magische Druckfarbe 8492
Magnesia 8500
Magnesium 8496
Magnesiumbisulfit 8497
Magnesium einreiben 2597
Magnesiumkarbonat 8499
Magnesiumklischee 8498
Magnesiumoxyd 8500
Magnesiumplatte 8501
Magnesiumsulfat 8502
Magnetband 8524
Magnetbandlesegerät 8528
Magnetbandspeicher 8530
magnetische Feldstärke 8518
magnetischer Speicher 8523
magnetisches Feld 8517
magnetische Spur 8531
magnetische Zeichenerkennung 8520
magnetisch kodierte Schecks 8516
magnetisierbare Druckfarbe 8519
Magnetkarte 8503
Magnetkern 5561, 8504
Magnetkernspeicher 3574
Magnetplatte 8508
Magnetplattenspeicher 8510
Magnetscheibe 8508
Magnetscheibenspeicher 8510
Magnetspeicher 8523
Magnetspule 12754
Magnettrommel 8512
Magnettrommelspeicher 8514
Mahlarbeit 1266

mahlen 6637
Mahlen 1266
Mahlfeinheit 5676
Mahlfeinheitsmeßgerät 5677
Mahlfeinheitsmessung 5678
Mahlgerät 11385
Mahlgrad 4109, 6161
Mahlgradprüfer 6160
Mahlholländer 7020
Mahlung 1266
Mahlungsgrad 6161
Mahlungsgradprüfgerät 6160
Mahlvorgang 11386
Maioli-Einband 8555
Majuskeln 2287
Makrobefehl 8477
Makroinstruktion 8477
Makulatur 4418, 14558
Makulaturanfall 12906
Makulaturbogen 12576, 14561
Makulaturbogen einschießen 7499
Makulaturen 8981, 12906
Malachitgrün 8582
Maleinharz 8584
Maleinsäureharz 8584
Malerpinsel 9870
Malkarton 798
Malpappe 798
Malzeichen 8639
Manchon 3644
Mangan 8589
Manganosulfat 8591
Mangansikkativ 8590
Mangansulfat 8591
Mangantrockner 8590
Manila-farbiges Schreibpapier 11219
Manila-Imitation 1693
Männchen auf Männchen (setzen) 6000
Männchen auf Männchen setzen 8176
Mann-Maschinenkarte 8606
Mantel 1719
Mantisse 8607
manuelle Steuerung 6820
Manuldruck 8610
Manulverfahren 8610
Manuskript 3534
Manuskript abschätzen 2457
Manuskriptbearbeitung 3551
Manuskript berechnen 2457
Manuskriptberechnung 2629
Manuskriptberechnungstabelle 3542
Manuskriptentschädigung 5395
Manuskripthaken 3547
Manuskripthalter 3544, 4520

Manuskript, mit der Schreibmaschine abgeschriebenes ~ 14228
Manuskriptverbesserung 3551
Manuskriptvorbereitung 3551
Märbel 3037
Marder 10357
marderen 12610
mardern 10355
Marginale 8629
Marginalnote 8629
Marginalziffer 11800
Marke 2909
Markierfilz 8636
Markierung der Stützwalze 1018
Markierungslesung 8640
Markierungslinie 4034
Marmorbecken 8626
Marmorierer 8618
Marmorierfarbe 8623
Marmoriergrund 1273
marmorierte Buchschnitte 8615
marmoriertem Schnitt, mit ~ 8615
marmoriertes Papier 8617
Marmorierung 8621
Marmorkamm 8624
Marmorpapier 8617
Marmorschnitt 8615
Marokkoleder 9106
Marokkoleder aus Nigeria 9344
Maroquin 9106
Maroquin écrasé 3784
Martinusfest 11845
Martinustag 11845
Maschenweite 8813
Maschine (gegangen), in die ~ 6521
Maschine in Trommelbauweise 3249
maschinell 8725
maschinelles Lesen 2639
maschinelle Übersetzung 8467
maschinell gesetzt 8734
maschinell hergestellt 8454
Maschine mit Bogenanlage 12269
Maschine mit einem gemeinsamen Gegendruckzylinder 3249
Maschine mit Fußantrieb 14014
Maschinenätzung 8445
Maschinenaufsteller 5237
Maschinenausrüstung 6877
Maschinenausschuß 1934
Maschinenbefehl 3337
Maschinenbefehlskode 3331
Maschinenbreite 8468
Maschinenbütte 6898, 8432, 13186

Maschinenbüttenpapier 9131
Maschinendefekt 1880
Maschinendruckplatte 10758
Maschinenfett 3394
Maschinenformat 11239
maschinengemacht 8454
Maschinengeschwindigkeit 8466
Maschinengestell 6144
maschinengestrichener Karton 8434
maschinengestrichenes Papier 8435
maschinengetrocknetes Papier 8444
maschinenglatt 8446
maschinenglattes Papier 8447
Maschinenglättung 2191
Maschinenheftung 8464
Maschinenkalander 2184, 12995
maschinenlesbare Information 8459
Maschinenmeister 3456, 9638, 10753
Maschinenmonteur 5237
Maschinenöl 8397
Maschinenpappe 8932
Maschinenplatte 10758
Maschinenprogramm 9470
Maschinenrevision 8441
Maschinenrichtung 8443
Maschinenrolle 8942
Maschinensaal 10761
Maschinensatz 8438, 8732
Maschinenschiff 4152
Maschinenschreiber 14231
Maschinenschreiberin 14231
Maschinensetzer 8439, 9639
Maschinensetzerei 3311
Maschinensieb 8469
Maschinensprache 8453
Maschinenstillstandszeit 4678
Maschinenstreichverfahren 8436
Maschinenstrich 8436
Maschinenübertragung 4808
maschinenunabhängige Programmiersprache 8451
Maschinenwort 3348
Maschinenwortlänge 14819
Maserpapier 6543
Maske 8645
Maskenverfahren 8647
Maskerverfahren 8647
Maskieren 8647
Maskierung 8647
Maskierverfahren 8647
Maß 6334, 8847
Maßbeständigkeit 4376
Massenfabrikation 8653
Massenfertigung 8653

Massenherstellung 8653
Massenleimung 1265
Masse-Reiterwalzen 3308
Massewalze 3309
Massey-Streichverfahren 8651
Maßhaltigkeit 4376
Massicot 8652
Materialdieb 10357
Materialhamsterer 10357
Materialkasten 1139
Materialprüfung 11122
Materialwarenpack 6663
Matern geliefert von Korrespondenten 1699
Maternkarton 5848
Maternpappe 5848
Maternprägepresse 9129
Materntrockenapparat 8690
Materntrockenkasten 8690
Materntrockner 8690
Mater ohne Zurichtung 9835
Mater (vor dem Prägen) 5846
mathematische Formel 8672
mathematischer Satz 8673
mathematische Zeichen 8674
Matrix 8679
Matrixdrucker 8694
Matrize 5846, 8677, 8678
Matrizenbreite 1869
Matrizenführung 2826, 8688
Matrizen in Setzmaschinenmagazin einlaufen lassen 11798
Matrizenkorb 2148
Matrizenohr 4932
Matrizenpappe 5848
Matrizenprägepresse 8699, 9129
Matrizenpulver 5634
Matrizenrahmen 2634, 8685
Matrizenrechen 8695
Matrizenreinigungsgerät 8687
Matrizensatz 12165
Matrizenträger 2634
matt 8663
matte Oberfläche 4853, 8700
mattes Papier 8676
Mattfarbe 4854
Mattfarben 8675
Mattfirnis 8707
mattgeglättet 4853
mattgestrichenes Papier 4852, 8665
Mattglanz 4853
Mattglas 6217
Mattierung 1650
Mattkunstdruckpapier 8665
Mattlack 8707
Mattoir 11626
mattsatiniert 4853, 5179
mattsatiniertes Papier 5180
Mattscheibe 5956

Mattschicht 8700
Maulbeerbaum 9163
Maulbeere 9163
Maulschlüssel 9620
Mauresken 744
maximale Arbeitsplatzkonzentration 13777
Maximalformat des Papiers 8710
Maxwell 8711
Mechanik 8744
mechanisch 8725
mechanische Eigenschaften 8730
mechanische Einstellung 8733
mechanische Farbverstellung 5158
mechanische Kreidereliefzurichtung 8729
mechanische Kreidezurichtung 2602
mechanischer Widerstand 8739
mechanische Stabilität 8740
Mechanismus für die Erzielung der Axialbewegung 14618
Medianwert 8746
Mediävalziffern 6853
Mega- 8763
Megahertz 8764
Mehlen 2599
Mehrarbeit 5410
Mehrbadätzung 9181
Mehrbuchstabenletter 8322
Mehrfachschalter 8189
Mehrfachzugriff 9179
Mehrfarbendruck 3163, 3197, 10904
Mehrfarbendruckfarben 10897
Mehrfarbendruck mit Farbauszügen 3167
mehrfarbig 9169, 10553
Mehrförderung 5337
Mehrkosten 5407
Mehrlagensack 9198
mehrlagiger Karton 9174
mehrlagiger Papiersack 9198
mehrlinsiges Objektiv 3319
mehr Raum 9100
Mehrrechneranlage 9195
Mehrrundsiebmaschine 9173
Mehrschichtenfilm 9175
mehrschichtige Pappe 9191
Mehrstationenschaltung 9193
mehrwandiger Papiersack 9198
mehrwertiger Alkohol 10557
mehrzeilige Überschrift 11815
mehr Zwischenraum 9100
Meinungsfreiheit 6157

Melamin-Formaldehydharz 8766
Melaminharz 8767
Melierfasern 3136
Melierpapier 9115
meliertes Papier 9115
Meltauflecken 6139, 8928
Meniskus 8773
Mennige 11330
menschliche Beziehungen 7070
mercerisieren 8775
Merkantillithograph 3243
Merkmal 5502
Merkmale 2638
Merkuri- 8777
Merkurichlorid 8792
Merkurirhodanid 8806
Merkuro- 8784
Merkurobromid 8790, 8791
Merkurochlorid 8793
merzerisieren 8775
Meßband 6416
Meßbügel 8898
Messen 8722
Messer 7817
Messerbalken 7819
Messerblock 7818, 14047
Messerexzenter 3882
Messerfalzmaschine 7824
Messerleisten 7820
Messerniedergang 4677
Messerputzer 7828
Messerputzfahne 7828
Messerschleifmaschine 7826
Messersektion 7827
Messerstreichverfahren 7821
Messerzylinder 3893
Meßgerät 6334, 8847
Meßgerät zur Bestimmung der Abreibungstrockenzeit 11762
Messing 1859
Messingausschuß 1867
Messingdurchschuß 1867
Messingecke 1862
Messingeinfassung 1861
Messinglinie 1866
Messingmatrize 1863
Messingplatte 1864
Messingprägeplatte 1860
Messingschrift 1868
Messingspatie 1867
Messingwalze 1865
Meßlatte 11899
Meßokular 8724
Meßringe 1252
Meßschleife 8355
Messung 8722, 13658
Meßvorrichtung 8847
Metallasche 4763
Metalldruckfarbe 8828
Metallegierung 437
Metalleiste für Kalender 2183
Metallfuß 8817
Metallkantmaschine 8824

Metallklotz 8817
Metallkopie 8838
Metallkrätze 8832
Metall montiert, auf ~ 9141
Metalloxyd 8829
Metallpapier 8830
Metallreaktivierungsmittel 11576
Metallsalz 8831
Metallschere 6705
Metallschienenandrückmaschine 8824
Metallschmelzofen 8826
Metallschnitt 8597
Metallseife 8833
Metallzuführer 8825
Metallzufuhrstange 10327
Metallzuführung 8825
Metall zum Hintergießen von Galvanos 1016
metamer 8842
metamere Farben 8843
Metamerie 8844
metaschwefligsaures Kali 10658
meteorologische Zeichen 8846
Meter 8871
Methanol 8855
Methansäure 6078
Methylaldehyd 6048
Methylalkohol 8855
Methylaminophenolsulfat 8867
Methylbenzol 13881
Methylbromid 8862
Methylglykol 5310
Methylmethakrylat 8866
Methylviolett 8869
Methylzellulose 8863
Metol 8867
metrisches Gewinde 8876
Metteur 2914, 8573
Metteur-Regal 11229
Mezzotinto 8883
Mezzotintodruck 8883
Mezzotintostich 8883
Mikrobild 8906
Mikrofilm 8896
Mikrokonturtest 8893
mikrokristallinisches Wachs 8894
Mikrometer 2221, 8898
Mikrometerlehre 8898
Mikrometerschraube 8902
Mikrometerskala 8901
mikrominiaturisierte Schaltung 8904
Mikron 8905
Mikroplanfilm 8895
Mikroprogrammierung 8908
Mikrorupfen 2997
Mikrosekunde 8911
Mikroskop 8910
Mikrowellen 8912
Milchglas 9609
Milchsäure 7869

milchsaures Ammonium 537
Milliliter 8937
Millimeterpapier 6573
Millimikron 8938
Millisekunde 8940
Miloriblau 8945
Milorigrün 8946
Minderanfertigung 14287
Minderförderung 7465
Mindergewicht 14287
Mindestfrachtmenge an Papier 2358
Mineralfarbstoff 8954
Mineralöl 8953
Mineralpigment 8954
Mineralwachs 9818
minimaler Druck 7805
Minimalformat 8961
Minimum 2349, 8961
Minuskel 8967
Minuszeichen 8969
Minute 8971
Minutenzeichen 8972
mischbar 8977
Mischbarkeit 8976
Mischbütte 9011
mischen 3064, 8808, 13165
Mischen 9010
Mischer 3069, 9007
Mischfarben 12033
Mischkasten 9011
Mischmaschine 9007
Mischpolymerisat 3495
Mischsatz 9003
Mischton 13838
Mischung 1560, 9010, 9015
Mischungsverhältnis 9013
Mischweiß 6456, 9014, 13840
Mitautor 3002
MIT-Falz 9001
Mitläuferwalze 7155
Mitscherlich-Zellstoff 8999
Mittagsausgabe 9420
Mittagsblatt 9420
Mittel 951
mittelalterlicher Einband 8749
Mittelfacette 2546
Mittelfarben 7510
Mittellänge 14894
Mittellinie 8717
Mittelschicht 5624
Mittelton 8920
Mitteltonblende 8922
Mittelwert 951, 8719
mittlere Abweichung 8715, 13023
mittlere Blendenöffnung 4218
mittlere Druckform 8923
mittlere Geschwindigkeit 961
mittlere Güte 959
mittlerer Fehler 13024
mittlerer Stichprobenumfang 960
mittlere Teilchengröße 957
Mitverfasser 3002

Mixermodell 9008
Mizellbildung 8888
Mizelle 8887
mnemonisch 9016
Mobilometer 9018
Modeblatt 5478
Modell 6065, 6698
Modellpapier 10063
Modezeitschrift 5478
Modul 9031
Modularsystem 9028
Modulation 9029
Modulator 9030
Modus 9021
Moiré 9032
Mol 9042
Molekül 9055
molekular-disperse Lösung 9053
molekulare Ausrichtung 9052
molekulare Lösung 9053
Molekulargewicht 9054
Molekularlösung 9053
Molekülkomplex 2967
Moleskin 9056
Molette-Wasserzeichen 7207
Molton 9057
Molybdatorange 9061
Moment 3668
Monastralblau 9065
Monastralgrün 9066
monatliche Zeitschrift 9096
Monatsheft 9096
Monatsschrift 9096
Mönch 9192
Mönchsbogen 6192
Mönchschlag 6192
Monobrommethan 8862
monochromatisch 9076
monochromatisches Licht 9077
Monogießer 2441, 9075
monoklin 9080
monoklinisch 9080
Monomer 9083
monomolekulare Schicht 9081
Monotype 9087
Monotypegießer 2441
Monotypegießmaschine 9088
Monotypepapier 9092
Monotypesatz 9090
Monotypetastenbrett 9091
Monotypetaster 9091
Montage 1597, 5767
Montage auf nichtaktinischem Papier 6503
Montagelinie 843
Montagelupe 11419
Montageplatte 5766
Montagerahmen 2651
Montagescheibe 2134
Montagestrecke 843
Montagetisch 11417
Montageunterlagepapier 6504

Montanwachs 9095
Montierblock 1605
montieren 9137
Montieren 1597, 13289
Montieren ohne Facette
  5899
Montiermaschine 9145
montierte Vorlageform
  5767
Morgenausgabe 2052
Morgenblatt 9104
Morgenzeitung 9104
Mosaik-Einband 9110
Mount-Hope-Walze 5359
Mudéjar-Einband 9162
Mull 9216
Mullenprüfer 9167
Mullens Berstdruckprüfer
  9167
Müllern 8822
Multiplexer 9189
Multiplex-Karton 9174
Multiplex-Pappe 9174, 9191
Multiplikationszeichen
  8533, 8639
Multiprogramming 9196
Mündung 9684
Munsell-System 9202
Musche 10045
Muschelgold 12292
musierte Schrift 11615
Musikdruck 9213
Musiknotensatz 9215
Musikverlag 9214
Musikverleger 9214
Musselin 9216
Muster 9740, 12834, 13668
Musterband 4857
Musterbeutel 6732
Musterbogen 4331, 12834
Musterbuch 11855
Musterbuch, (kleines) ~
  13456
Musterentnahme 11856
Mustersammlung 11855
Musterzeichnung 4208
Mutter 9459
Mutterbild 8656
Muttergewinde 7520

Nabe 9256
nachahmen 3657, 12440
Nachahmung 3658
Nachätzung 5669
Nachbar-Effekt 216
Nachbehandlung 10616
Nachbelichtung 5762
Nachdruck 11477
Nachdruck verboten 439
Nachdunkeln 3425, 3989
Nachdunkeln der Pigmente
  im Licht 10237
Nacheilung 7465
Nachkalkulation 3629
Nachkleben 289
Nachkopiereffekt 3425
nachlassen 631
Nachleuchten 10204
Nachrichtenagentur 9299

Nachrichtenbüro 9299
Nachrichtenverkehr 13950
Nachruf 5219
Nachsatz 13952
Nachschneiden 5689, 6848,
  11808, 13900
Nachschneider 5688
nachsehen 2663
Nachstechen 6848
Nachstellmutter 222
Nachstellschraube 223
Nachstellung von Farben
  3155
Nachtrag 184
Nachtschicht 9346
Nachwort 5219
Nacktzylinder 7341
nadelförmig 112
Nadelholz 12747
Nadelholzbaum 3382
Nadelholzhalbstoff 12748
Nadelholzzellstoff 12748
Nadellager 9267
Nadeln 9270
Nadelstiche 10363, 10366
Nagel 9225
Nagelapparat 9228
Nägelchen 13527
nagelfest 9414
nagelfeste Farbe 9415
Nägelköpfe 14855
Nagelmaschine 9228
Nageln 12988
Nähen 12192
Nahrungsmittel-
  (ver)packung 6003
Namenregister 9229
nämlich 14500
Nano- 9231
Nanometer 9232
Nanosekunde 9233
Näpfchen 3819
Naphtha 9235
Naphthenat 9236
Naphthenattrockner 9237
Narbung 9234
Nasenklammer aus Mittel-
  und Endstücken 3300
Nasenklammern 1840
Naß-auf-Naßdruck 14682
Naßausschuß 14679
nasse Fabrikationsabfälle
  14679
nasse Kollodiumplatte
  14680
nasses Kollodiumverfahren
  3087
nasse Übertragung 14686
naßfestes Papier 14693
Naßfestigkeitsprozentsatz
  14694
Naßfestigkeitszahl 14694
Naßfilz 14684
Naß-in-Naßdruck 14682
Naßpartie 14683
Naßplatte 14689
Naßpresse 14683
Naßpressenpartie 14683
Naßrupfen 14688

Naßteil 14683
Naßverfahren 3087
Natrium 12668
Natriumalaun 12673
Natriumazetat 12669
Natriumbenzoat 12675
Natriumbichromat 12688
Natriumbikarbonat 12676
Natriumbisulfat 12678
Natriumbisulfit 12679
Natriumborat 12680
Natriumbromid 12681
Natriumchlorid 12683
Natriumchlorit 12684
Natriumchromat 12685
Natriumcyanid 12687
Natriumdichromat 12688
Natriumformiat 12690
Natriumhexametaphosphat
  12691
Natriumhydrat 12695
Natriumhydrogenkarbonat
  12676
Natriumhydrogensulfat
  12678
Natriumhydrogensulfit
  12679
Natriumhydrosulfat 12678
Natriumhydrosulfit 12694
Natriumhydroxyd 12695
Natriumhypochlorit 12696
Natriumjodat 12699
Natriumjodid 12700
Natriumkaliumtartrat
  12710
Natriumkarbonatanhydrid
  12682
Natriummetabisulfit 12702
Natriumnitrat 12704
Natriumoxalat 12705
Natriumperborat 12707
Natriumperoxyd 12708
Natriumphosphat,
  (tertiäres) ~ 12709
Natriumpyrochromat 12688
Natriumpyrosulfit 12702
Natriumrhodanid 12728
Natriumsesquikarbonat
  12715
Natriumsilikat 12716
Natriumstearat 12717
Natriumsulfat 12719
Natriumsulfid 12720
Natriumsulfit 12721
Natriumsulfoantimonat
  12727
Natriumtetraborat 12680
Natriumthiosulfat 7146
Natriumzitrat 12715
Natronchlorzellstoff 12663
Natronlauge 12695
Natronpapier 7840
Natronsalpeter 12704
natronsalpetersaurer Zell-
  stoff 9367
Natronsilikat 12716
Natronverfahren 12666
Natronwasserglas 12716
Natronzellstoff 7841, 12667

Naturgummi 11747
Naturharze 9250
Naturkarton 6126
Naturkautschuk 11747
Naturkunstdruckpapier
  7170
natürliche Größe 9253
natürliche Mineralfarbe
  4939
natürliche Reihenfolge
  bringen, in ~ 12134
natürlicher Farbstoff 4926
natürliches Albumin 4988
natürliches Pigment 9248
natürliche Sprache 9246
Nebenanlage 12545
Nebeneinanderschaltung
  9978
nebeneinanderstehend
  7723
Nebenlicht 5758
Nebenprodukt 2132
Nebentitel 2468, 13342
Negativ 9274
negative Elektrode 2480
negative Linse 4507
negative Schrift 11556
negatives Glasbild 6433
Negativfarbauszug 12130
Negativhalterrahmen 2651
Negativklischee 11557
Negativ mit mehreren
  gleichen Bildern
  (Nutzen) 6300
Negativnutzen 3414
Negativnutzenfilm 3414
Negativpapier 9277
Negativschutzschicht 9279
Negativträger 9275
Neigung 6535
Neigung einer Kurve 12591
Neonlampe 9281
Nessler-Rohr 9283
Nettopreis 9286
Netz 9288
Netzmittel 13428, 14699
Netzspannung 14511
Netzstecker 10488
Netz versehen, mit
  einem ~ 6576
Netzwerk 9288
Neuausgabe 9297, 11429
Neudruck 9297, 11477
neuer Absatz 11552
neu herausgekommen 7719
Neuries 11291
Neusatz 2655, 11814
neu setzen 11493
neutrale Farbe 110
neutrale Töne 9295
neutralisieren 9289
Neutralsulfitzellstoff 9293
Neutraltinten 9295
Neutron 9296
neu umbrechen 9783
Newton 9325
Newtonsche Flüssigkeit
  9328
Newtonsche Ringe 9329

Newtonsches Fließen 9327
Newtonsches Gesetz 9326
Newtonsches Prinzip 9326
Newtonsche Strömung
　9327
nicht aktinisch 9393
nicht-aktinisches Licht
　9394
nicht auf Lager 9728
nicht-ausdrucken 12839
nicht-ausdruckende
　Punkte 8986
nicht-ausgereiftes Papier
　6625
nicht-ausschließender
　Perforator 9395
nicht-beschriftete Fläche
　2906
nicht-beschriftete Ober-
　fläche 2906
nicht beständig 6229
nicht da 4557
nicht-druckende Fläche
　6607
nicht-druckende Flächen
　9401
nicht durchschossen 14329
nicht durch und durch
　trocken 12154
nicht entzündbar 9400
nicht fortgesetzt 4423
Nichthaften 14528
nicht im Handel 10873
nicht-ionogene
　Verbindungen 9403
nicht-kräuselndes
　gummiertes Papier 9396
nicht kursiv 11670
nicht Linie haltend 9723
nicht-metallischer Satz
　3061
nicht mischbar 7185
Nichtmitgehen der Farbe
　im Farbkasten 1011
nicht-Newtonsche Flüssig-
　keit 9407
nicht registerhaltig 9726
nicht-rollendes gummiertes
　Papier 9396
Nichtsammeln 13248
nicht-scharfer Falz 7907
nicht-scharf gebrochener
　Falz 7907
nicht spürbar 7188
nicht-staubendes Papier
　5879
nicht-trocknendes Öl 9399
nicht vorrätig 9728
nicht-zählender Perforator
　9395
nicht-zerstörendes Lesen
　9398
nicht-zügige Farbe 12328
nicht zum Verkauf 10873
nicht-zünftiger Arbeiter
　1484
Nickel 9335
Nickel-Ammoniumsulfat
　9336

Nickelgalvano 9342
Nickelüberzug 9340
Niedergang 4680
niederländisch 4908
Niederländische 4908
Niederschlag 4186, 10710,
　10717
Niederschlag bilden 10711
niederschlagen 10711
Niederspannung 8389
niedrige Auflage 12336
Nigrosin 9348
Nigrosintest 9349
Nit 9359
Nitrat 9361
Nitratzellulose 2517
Nitrierzellstoff 9364
Nitrobenzol 9368
Nitroglyzerin 9376
Nitrolack 9372
Nitronaphthalin 9379
Nitrozellulose 2517, 9371
Nitrozelluloselack 9372
ñ mit Tilde 9223
Nocken 2227
Nockenwelle 2254
Nomogramm 9391
Nonius 14444
Nopak-Mater 9835
Norm 12402
normal- 9423
Normalatmosphäre 13021
Normalblende 9428
Normaldruckfarben 10897
normale Leistung 9427
normales Natriumkarbonat
　12682
normales Natriumphosphat
　12709
Normalfarben 10897
Normalformate 7521
normalisierte Formate 7521
Normal-Kuvertformate
　7522
Normalschriftlinie 1196,
　13027
Normalziffern 6853
Normdruckfarben 10897
Normfarbwertanteile 2764
Normfarbwerte 14062
Normung 13026
Notendruck 9213
Notendruckpapier 9211
Notenlinie 13000, 13081
Notenpapier 9211
Notensatz 9207, 9215
Notenstecher 9209
Notensternchen 850
Notenstich 9210
Notizblock 8771, 13514
Notizbuch 9437
Notizkalender 4279
Not-Rechner 1054
Notverbindung 10046
Nuance 12209
Nuance in der Aufhellung
　14292
Nullenunterdrückung 14937
Numerieren 9449

Numeriermaschine 9452
Numerierrahmen 9451
numeriertes Exemplar 9448
Numerierung 9449
Numerierwerk 9450
Numeroteur 9450
Nummer 3535, 5995
Nummerndruckwerk 9452
Nüsse 11175
nuten 6668
Nutenzahl 9456
Nutenzylinder 6669
Nutzen 3872, 9453
Nutzenfilm 6684
Nutzenkopiermaschine
　6686
Nutzentrenner 13506
Nutzzeit 14362
Nyblin-Effekt 9463
Nylon 9464
Nylonklischee 9465

Oasenziegenleder 9466
Oberbau 13421
obere Linie 2288
oberer Seitenrand 6909
oberes Tuch einer Liniier-
　maschine 1510
oberes Zurichten 9771
Oberfläche 5686
oberflächenaktiv 13427
oberflächenaktiver Stoff
　13428
Oberflächenaktivität 13429
Oberflächenbehandlung
　13430
Oberflächenbeschaffenheit
　1451
Oberflächendruckplatte
　13435
Oberflächeneigenschaften
　13436
oberflächengefärbtes
　Papier 13010
oberflächengeleimtes
　Papier 14073
Oberflächenkräfte 13433
Oberflächenleimung 13440
Oberflächenleimung,
　mit ~ 626
Oberflächenrauheit 13437
Oberflächenspannung
　13441
Oberflächenspannungs-
　Erniedrigung 13442
Oberflächenspannungs-
　Meßgerät 13647
Oberkasten 14355
Oberlänge des Buch-
　stabens 813
Oberlinie 2288
Obermagazin 14357
Obermaschinenmeister
　10762
oberschlächtiger Farb-
　kasten 9790
Oberseite 13918
Obersetzer 2914
Obertrichter 1101

Obertuch 13909
Objektiv 8042
Objektivbeschichtung 8045
Objektivdeckel 8044
Objektivfassung 1170
Objektivöffnung 8043
Objektivskala 8047
Objektivvergütung 8045
Objektivverschluß 12361
Objektprogramm 9470
Obligation 1718, 4038
Ochsengalle 9802
Ochsenklaueöl 9260
Ocker 9481
Oder-Glied 9666
Oersted 9502
Oeser-Folie 3146
Ofenruß 6260
offene Maschenfläche 8813
offener Exemplarausgang
　9622
offene Siebfläche 8813
offizielle Aufforderung
　eine Meldung nicht zu
　bringen 4522
Offizin 9505
off-line 9513
Öffner 11029
Öffnung 9684
Öffnung des Objektivs 8043
Offsetandruckpresse 10951
Offsetblech 9531
Offsetbogenrotations-
　maschine 12270
Offsetdruck 9525, 9536
Offsetdrucker 10754
Offsetdruckfarbe 9523
Offsetdruckmaschine 9532
Offsetdruckplatte 9531
Offsetdruckverfahren 9536
Offsetfarbe 8271, 9523
Offsetmaschine 9532
Offsetmaschinenmeister
　10754
Offsetpapier 9528
Offsetplatte 9531
Offsetpresse 9532, 12642
Offsetrollenrotations-
　maschine 11698
Offsettiefdruck 4085, 9521
Offsettiefverfahren 4088,
　9521
Offsettuchverschmutzung
　1519, 3418
Offsettuchwaschmittel 1526
Offsetzinkplatte 9541
OH-Gruppe 7133
Ohm 9547
ohne Durchschuß 14329
ohne Einzug 12159
ohne Kohle-Papier 9382
ohne Ort und Jahr 9422
Ohr 9331
Oitikaöl 9569
Oktadezensäure 9579
Oktal 9484
Oktav 9485
Oktavformat 9485
Oktoat 9486

Oktodez 4993
Oktodezformat 4993
Okular 5418
Ölabsorption 9550
Ölabsorptionsprobe 9551
ölabstoßend 9566
ölabweisend 9566
Ölaufnahme 9550
Ölbeständigkeit 9567
ölbildendes Gas 5307
Öldurchlässigkeit 9564
Ölechtheit 9567
Oleinsäure 9579
ölen 8395, 9548
Ölen 9557
Oleographie 9581
oleophob 9583
Öler 9552
Ölfirnisbaumöl 14084
Olivenöl 8399
Ölkännchen 9552
Ölkanne 9552
Ölkarton 9553
Ölloch 8398
Ölnut 9555
Ölpapier 9554
Ölsäure 9579
Ölsikkativ 8215
Ölstein 9568
Ölsüß 6482
on-line 9599
opak 9611
opakes Papier 9615
Opalglas 9609
Opalinekarton 9610
Opazität 9604
Opazität gegen Papier-
    unterlage 9606
Opazität gegen weiße
    Unterlage 9608
Opazität nach Tappi 13599
Opazitätsmeßgerät 9603
Operand 9628
Operation 9634
Operations-Charakteristik
    9629
Operationssymbol 9640
Operment 9691
Oppositionsblatt 9642
Optik 9665
Optimalfarbe 11879
optisch ausgeglichen 9658
optische Aberration 9643
optische Abweichung 9643
optische Achse 9644
optische Achse bringen, in
    die ~ 3079
optische Anzeige 4458
optische Aufhellung 9645
optische Brechung 11395
optische Dichte 9653
optische Dichte des Druck-
    bildes 10799
optische Mitte 9648
optisch empfindlich
    machen 9662
optischer Abbildungsfehler
    9643
optischer Abtaster 9660

optischer Aufheller 9646
optischer Kontakt 9652
optischer Mittelpunkt 9648
optischer Schnittandeuter
    12654
optisches Bleichmittel 9646
optisches Lesegerät 9649
optisches Lesen 9650
optisches Schriftlesegerät
    9649
optisches System 8048
optisches Tiefenmeßgerät
    9654
optische Umkehrung 9659
optische Verzeichnung
    9655
optische Verzerrung 9643,
    9655
optische Zeichen-
    erkennung 9650
optisch flacher Farbfilter
    9657
Orangefilter 9668
Orangenschaleneffekt 9670
Ordinate 9676
Ordinatenachse 966
Ordner 1403, 5610, 8057
organische Pigmente 9680
organischer Farbstoff 9677
organisches Lösungsmittel
    9681
organische Substanz 9679
organische Verbindung
    9679
Organosol 9682
Original 808, 3536, 9685,
    14835
Originalausgabe 5702
Originalbild 8656
Originalentwurf 808
Originalgalvano 8658
Originalhalter 2242
Originalhalter mit Ansaug-
    rahmen 14373
Originalhalter mit Vakuum-
    ansaugung 14373
Originalplatte 8658
Originalvorlage 808
Ornament 9688
Orseillenfarbe 9671
Ortho- 9694
orthochromatisch 9696
orthochromatischer Film
    9697
ortho-Dioxybenzol 11082
Orthofilm 9697
orthographischer Fehler
    12857
Ortho-Phenylendiamin
    9702
örtliche Korrekturen 8300
ösen 5416
Ösen 5415
Ösenapparat 5417
Ösen einsetzen 5416
Öseneinsetzmaschine 5417
Ösenhefter 5417
Ösenmaschine 5417
Ösmaschine 5417

osmotischer Druck 9710
Oszillation 9706, 9708
Oszilloskop 9709
Oval 1823, 9741
oval anreißen 9742
oval fräsen 9742
Overall 14831
Oxalsäure 9797
oxalsaures Eisenoxyd 5551
oxalsaures Kali 10631
Oxyd 9805
Oxydation 9803
Oxydationsmittel 9809
Oxydationsverzögerer 674
Oxydhäutchen 9806
oxydieren 9807
Oxydierung 9803
Oxytricarballylsäure 2864
Ozachrom 9814
Ozalidpapier 9815
Ozalidverfahren 9816
Ozobrom-Druck 2339
Ozokerit 9818
Ozon 9819

paarige Matrize 4626
packen 9820
Packmaterial 9827
Packmenge an Bogen-
    papier die eine Kiste
    füllt 2409
Packpapier 14866
Packpresse 2064
Packseidenpapier, (holz-
    haltiges) ~ 2291
Packung 9828
Packungsdichte 9829
Pagina 9850
Paginakopf 6890
paginieren 5997, 9865
Paginiermaschine 9869
Paginierung 4751, 5993,
    9866
Paketauslage 12998
Paketausleger 7788, 12998
Paketsatz 1683, 10322
paketsetzen 11812
Paketsetzer 3314
Paläographie 9874
Palette 9878
Palimpsest 9876
Palladium 9877
Palmholz 1839
Palmitinsäure 9880
Palmkernöl 9882
panchromatisch 9887
panchromatischer Film
    9888
Panfilm 9888
Panne 1880, 2040
Panoramadruck 9891
Panoramazeichnung 9890
Pantograph 9892
Pantonedruck 9893
Pantoneverfahren 9893
Papier 9894, 13182
Papierabfall 14558
Papierabschnitt 12238
Papierabschnitte 9504

Papierabschnitt zwischen
    Klebestelle und Ab-
    schlagen der aus-
    laufenden Rolle 13538
Papier, an der Luft
    getrocknetes ~ 8312
Papier, auf der Leimpresse
    gestrichenes ~ 12493
Papieraufhängeklammern
    2920
Papieraushängevorrichtung
    mit Bogenhalterung
    durch Glaskugeldruck
    1105
Papier aus Leinenlumpen
    8186
Papierbahn 14639
Papierband 9949
Papierbeutel 1066, 9898
Papierbogen 12250
Papierbohrmaschine 4733
Papierbreite 14741
Papierbruch 4675
Papierchromatographie
    9905
Papierdicke 2220
Papierer 9924
Papier erster Wahl 10118
Papierfabrik 9927
Papierfabrikation 9926
Papierfasern 9917
Papierfaserstoff 11026
Papierform 9119
Papierformaten 12500
Papierführung 7997
Papier für keramische
    Abziehbilder 2557
Papier für Kraftzurichtung
    14283
Papier für lackierte
    Etiketten 14407
Papier für langdauernde
    Belege 8347
Papier für Stahlwaren 3867
Papier für Zigarren-
    etiketten 2828
Papiergelddruck 1128
Papierglätte 12643
Papierhalbstoff 6770
Papierherstellung 9926
Papierholz 11034
Papier, im Rakelstreich-
    verfahren
    gestrichenes ~ 12933
Papier in Breitbahn 12326
Papier, in der Streich-
    maschine
    gestrichenes ~ 12127
Papier, in Lagen
    gefalztes ~ 11165
Papier in Schmalbahn 8335
Papierklammer 9906
Papierkleber aus Gummi-
    lösung 9904
Papierklimatisierung 3370
Papierlack 7863
Papierlager 9946, 13238
Papierlagerung 13228

Persönlichkeitsbewertung 8810
Petroläther 10169
Petrolatum 10167
Petroleum 7751
Petroleumäther 10169
Petzval-Objektiv 10175
Pfaublau 10069
Pfeilradgetriebe 6967
Pferdestärke 7044
Pflanzenleim 14417
Pflege (von Maschinen) 8553
Pflichtexemplar 3554, 13080
Pfund 10690
Pfund Sterlingzeichen 10691
Pfund troy 14066
Pfundzeichen 10691
Pfusch 1804
Pfuscher 1806
pH 10177
Phänomen 10193
Phantasiemuster 5458
Phantasiepapier 5460
Phantasieschriften 5461
Phase 10184
Phasemodulation 10186
Phasenwinkel 10185
Phasenzahl 9454
Phenol 10189
Phenol-Formaldehydharz 10190
Phenolharz 10191
Phenolphtalein 10192
Phenylamin 613
Phloroglucin 10200
Phloroglucinlösung 10200
Phloxine 10201
pH-Meßgerät 10179
pH-Messung 10178
pH-Meter 10179
Phosphor 10203, 10207
Phosphoreszenz 10204
Phosphoroxychlorid 10208
Phosphorsäure 10205
Phosphortrichlorid 10209
Phosphorylchlorid 10208
Phot 10211
Photo 10240
Photoabteilung 10242
Photo attraktiver Frauenkörper oder Frauenkörperteile 2678
Photochemie 10216
Photodiode 10224
Photodrucke 10250
photoelektrisch 10225
photoelektrischer Effekt 10228
photoelektrische Registersteuerung 5051
photoelektrischer Registerregler 5051
photoelektrische Zelle 10215
Photoelektrizität 10230
Photograph 10241
Photographie 10240, 10252

photographische Abteilung 10242
photographische Emulsion 10243
photographische Platte 10248
photographische Retusche 10251
photographischer Film 5636
photographischer Film (Platte) mit Lichthofschutzschicht 993
photographisches Mutterbild 10246
photographisches Originalbild 10246
photographisches Papier 10266
Photokathode 10214
Photokopie 10270, 10273
Photolithographie 10257
Photomatrize 6096
photomechanisch 10258
photomechanischer Satz 10278
Photomontage 10261
Photopapier 10266
Photoplatte 10248
Photopolymer 10268
Photopolymerplatte 10269
Photosatz 10278
Photosetzmaschine 10279
Phototypie 3099
Photowiderstandszelle 1173
Photozelle 10215
pH-Papierstreifen 10180
Phthalozyaninpigment 10285
Phthalsäure 10284
pH-Wert 10181
physikalische Eigenschaften 10286
Pica 10288
Pickelbad 10298
Pigment 10330
Pigmentdruck 2327
Pigmentfarbstoffe 9680
pigmentierte Flexodruckfarbe 10331
Pigmentkopie 2327
Pigmentpapier 2331
Pigment, mit Molybdänsäure gefälltes ~ 9060
Pigmentpapier-Übertragungsmaschine 2333
Pilotanlage 10348
Pilotbetrieb 10348
Pilzbefall 6252
pilztötender Stoff 6251
Pilzflecke 12556
Pilzvertilgungsmittel 6251
Pinaflavol 10351
Pinakryptolgelb 10353
Pinakryptolgrün 10352
Pinatypie 10354
Pinazyanol 10350
pinnen 12173
Pinsel 6424, 9870

Pinselätzung 1999
Pinselschrift 2018
Pinseltechnik 2017
Pinte 10373
Pinzette 14116
Pipette 10375
PIV-Getriebe 10387
placiertes Wasserzeichen 8301
Plagiat 10389
Plakat 10609
Plakatbuchstaben 10612
Plakatpapier 10610
Plakatschrift 10612
Plaketteneinband 2236
Plan 4208, 4257
Planetenbewegung 10396
Planfräser 1611
Planfräsmaschine 1611
Planheitsmesser 5797
plankonkave Linse 10399
plankonvexe Linse 10400
Planoauslage 5786
Planobogen 1931
Planoformat 6241
Planopapier 5771
Planschaben und Geraderichten 11710
planschleifen 10392
Planschleifen der Rückseite 1043
Plansumme 725
Planung 10397
Plaste 10412
Plastikheftung 10408
plastisch 10407
plastische Viskosität 10414
Plastisol 10415
Plastometer 10416
Plaststereo 10411
Platin 10472
Platinchlorid 10473
Platinchlorürkalium 10640
Platte 4434, 9531, 10418, 10420
Platte in Umfang, eine ~ 12464
Platte mit zu seichten Punzen 12229
Plattenanschlag 13003
Plattenantrieb 4439
Plattenaufbiegung 1297
Plattenaufwölbung 2029
Plattenbeschichtung 10427
Plattenbeschichtung mit Soja 12796
plattenbreite Farbabstellung 9856
Plattenfolge 10470
plattenförmiger Serif 12531
Plattenfuß 9136
Plattengreifhaken 4542
Plattenhaken 4542
Plattenhalter 2135, 10442
Plattenhalterleisten 2370
Plattenhersteller 10447
Plattenherstellung 10448
Plattenherstellungsabteilung 10449

Platteninnenspannung 13649
Plattenkassette 10442
Plattenklemme 2869
Plattenkornmaschine 6550
Plattenkühler 10429
Plattenkühlvorrichtung 10429
Plattenreinigungsmittel 10425
Plattenrippen 10460
Plattensatiniertes Papier 10437
Plattensatz 12166
Plattenschabemaschine 1611
Plattenschere 12242
Plattenschlag 10461
Plattenschleuder 14709
Plattenschlüssel 10444
Plattenschnellverschluß 7460
Plattenschuh 9136
Plattensitz 10453
Plattenspannschlüssel 2056
Plattenspannvorrichtung mit Ausgleich 3262
Plattenspeicher 8508
Plattenstapel 4444
Plattenstärke 10463
Plattentiefdruckmaschine 10440
Plattenton 12018
Plattenträger 2135
Platten umlegen 10464
Plattenverschiebung 9545
Plattenverschluß 8307
Plattenversetzung 10462
Plattenverspannung 2029
Plattenwandern 9545
Plattenwölbung 1297
Plattenzurichtung 1023, 7497, 14281
Plattenzylinder 10432
Plattenzylinder mit eingefrästen Nuten 6669
Platterton 7340
Plattform 4061
Plattführungsmarken 3943
Platzstelle 2092
Plüschpapier 5845
Plüschtampon 4086
Pluspol 641
Pluszeichen 10499
Plyadhäsivpapier 6932
pneumatische Bremse 313
pneumatischer Kontaktrahmen 14375
pneumatischer Kopierrahmen 14375
Poise 10537
Poiseuillesches Fließen 10538
Poiseuillesche Strömung 10538
Pola-Autotypieraster 10545
Pola-Raster 10545
polare Lösungsmittel 10544

pyroschwefligsaures Kali 10658
Pyroxylin 11087

Quadrantenwaage 11107
Quadrant-Papierwaage 11107
Quadrat 11108
Quadrat auf vier Gevierte 6130
Quadrätchen 5120
Quadrate einteilen, in ~ 6576
quadräteln 7653
Quadratenzeile 14716
Quadratfuß 12966
quadratisch 12962
quadratische Abweichung 13023
quadratischer Fehler 13024
Quadratmetergewicht 1209
Quadratwurzelzeichen 11679
Quadratyard 12975
Quadratzoll 12968
quälende Arbeit 6635
Qualitätskontrolle 11121
quantisieren 11125
Quart 11143
Quartalschrift 11136
Quartformat 11143
Quarzlampe 11144
Quaternio 11145
Quecksilber 8788
Quecksilberbromid 8790, 8791
Quecksilberchlorid 8792, 8793
Quecksilberchlorür 8793
Quecksilberdampflampe 8807
Quecksilberhochdruck-lampe 6995
Quecksilberjodid 8796
Quecksilberrhodanid 8806
Quecksilbersublimat 8792
Quecksilbersulfozyanat 8806
Quecksilberverstärker 8795
quellen 2024, 13462
Quellen 13465
Quellenangabe 1361, 3721
Quellen (des Gummituchs) 5103
Quellenprogramm 12793
Quellung 13465
Querbalken 1135
Querfalz 2756
Querfolio 1932
Querformat 7898
Querformatseite 9476
Querheftung 12389, 12988
Querlaufrichtung 3752
Querlinien 3765, 7041
Querperforation 3763
Querrichtung 3752
Quersatz für sich, mit ~ 12125, 14785
Querstrich 1135

Quetsche 3028, 14754
quetschen 12980
Quetscher 12978
Quetschfalten 2186, 14873
Quetschmarkierung 1979
Quetschrand 4569, 12976
Quinternio 11162
quirlen 10317
qwert-Einteilung 11180

Rabatt 4425
Räderfalzmesser 5929
Räderstempel 1176
radial 11190
Radialbeschleunigung 603
Radierbarkeit 5231
radieren 5182
Radierer 5269
Radierfestigkeit 5231
Radiergummi 5233
Radiermesser 5234, 11950
Radiernadel 5281
Radierung 5270
Radikal 11195
radioaktiv 11197
Radius 11198
Radschneider 6850
Rähmchen (an der Hand-presse) 6204
Rahmen 2650, 6145
Rahmen abnehmen, den ~ 13282
Rakel 4525
Rakelanstellwinkel 4526
Rakelbalken 4530
Rakelbewegung 4532
Rakelblech 4525
Rakeleinschnitt 1497
rakelführendes Rastersteg-netz 4524
Rakelhalter 4530
Rakelhub 4532
Rakelmesser-Schleif-maschine 4529
rakeln 14755
Rakelschleifmaschine 4529
Rakelstreichverfahren 1496, 7821
Rakelstreifen 4531, 11218
Rakeltiefdruck 11705
Rakelträger 4530
Rakelwinkel 600
Ramie 11228
Ramiefaser 11228
Ramsch 11450
Rand 1792, 8627
Randbemerkung 706, 8629
Rande abfallende Bilder, am ~ 1552
Rande angeschnittene Bilder, am ~ 1552
rändeln 9718
Rändelschraube 7831
Ränder 4961
rändern 9718
Randglosse 706, 8629
Randleiste (am Schiff) 6281
Randleisten anbringen 8840

Randleistenandrück-maschine 8824
Randleistenmaschine 8824
Randlinie 1793, 1796, 5690
Randlochkarte 4959
randlose Montage 5902
Randomspeicher 11233
Randschleier 4955
Randschrumpf, mit ~ 13803
Randsteg 12387
Randverkürzung, mit ~ 13803
Rand versehen, mit ~ 9718
Randwinkel 3405
Randwinkelmessung 594
Randziffer 11800
Randzone-Prüfung 14967
Rapsöl 11243
Rascheln 11254, 12650
raschelndes Papier 13830
Raspel 5612
Raste 227
Raster 6794, 11245
Rasterabstand 11977
Rasterätzung 6774, 10233
Rasterätzung mit Verlauf 9720
Rasterauszug 4412
Rasterbelichtung vor-nehmen 11972
Rasterblende 6796
Rasterdiapositiv 6778
Rasterdicken-Ausgleichs-scheibe 11976
Rasterdruck mit zu-gesetzten Rastertönen 9161
Rasterfeinheit 11992
Rasterfenster 11974
Rasterfilm 936
Rasterhalter 11979
Rasterkassette 11979
Rasterklischee 6774, 10233
Rasterklischee mit Verlauf 14473
Rasterkopie 6777
Rasterkopie vornehmen 11972
Rasterlinienzähler 11980
rastern 11972
Rasternegativ 6786
Rasternetz 11245
Rasteröffnung 11974
Rasteröffnung-Theorie 10365
Rasterphotographie 6790
Rasterplatte 6774
Rasterpositiv 6792
Rasterprojektion 1627
Rasterpunkt 6779
Rasterpunkt mit über-hängender Schulter 14274
Rasterstufenkeil 6629
Rastertiefenmesser 5277
Rastertonwert 10112
Rasterträger 11979
Rasterverfahren 6793

Rastervorlage 6777
Rasterwalze 5185
Rasterweite 11992
Rasterwinkel 11973
Rasterwink(e)lung 11973
Rastral 9212
Ratsche 11249
Rattermarken 13004
Rattern 11255
Rauchabzug 12639
rauchende Salpetersäure 6247
Rauheit 11711
rauhe Kante 2090
Rauhigkeit 11711
Raum 12799
Raumbedarf 5859
Raum freilassen 8020
Raumgewicht 712
Raumtemperatur 11676
Rauswerfer 12161
Rauten 8939
rautenförmig 8391
Rautenleder 4291
rautiertes Papier 12964
Reagenz 11288
Reagenzpapier 8280
Reaktion 11261
Rechen 5915, 8695
Rechengerät 2177
Rechenmaschine 2177
Rechenmaschinenpapier 186
Rechenschaltzeit 12186
Rechenscheibe 10970
Rechenschieber 2176, 12554
Rechentabelle 11287
Rechentafel 11287
Rechenwerk 768
Rechner 3330
Rechner der dritten Generation 13726
Rechner der ersten Generation 5708
Rechner der zweiten Generation 12041
Rechnerprogramm 3343
Rechnungskopf 1385
Rechnungsordner 5610
Rechnungsrevisor 75
rechteckiges Dreieck 11606
rechter Winkel 11605
rechte Seite 11319
rechte Seiten 9495
Rechtschreibung 12856
rechtsgängig 2922
rechtsgängiges Gewinde 11609
Rechtsgewinde 11609
Rechtsweinsäure 13610
rechtwinkelig 11315
rechtwinklig 11315
rechtwinkliges Gewinde 11316
rechtwinklig geschnittenes Papier 12965
Redakteurin 4976
redaktioneller Teil 4978
Redaktionsschluß 4032

Rohgummi 11747
Rohleinöl 11257
Rohöl 10168
Rohpapier 1685
Rohrpost 10506
Rohstoffe 11258
Rohwichte 712
Rollapparat 14747
Rolle 1176, 11354
Rolleisen 1176
Rolle Karton 11372
Rolle mit Wassersäcken
  1070
Rolle mit Wasserschaden
  14575
Rollenabriß 13290
Rollenanhebevorrichtung
  11370
Rollenaufzug 11370
Rollenbahn 1825
Rollenbreite 14737
Rollenbürste 11359
Rollenhülse 11363
Rollenhunt 11356
Rollenkalander 13409
Rollenkonus 11362
Rollenlager 11634
Rollenlagerung 11375
Rollenlin(i)iermaschine
  4445
Rollenmixer 11369
Rollenoffsetmaschine 11698
Rollenpaar 12167
Rollenpapier 11373, 14658
Rollenpresse 14656
Rollenrest 2116
Rollenrotationsmaschine
  14656
Rollenschneidmaschine
  12584
rollen, sich ~ 3834
Rollenständer 11375
Rollenstapelung 12999
Rollenstern 11648, 12863
Rollensternarme 11355
Rollensternraum 11374
Rollenstern, seitlich
  angestellter ~ 12383
Rollenstuhl 11375
Rollenwagen 11356
Rollenwechsel 11361
Rolle Papier 11372
Rolle Pappe 11372
Roller-black 11635
Rollfilm 11656
Rollfilmkassette 11658
Rollfilmreproduktions-
  kamera 11657
Rollkassette 11658
Rollkone 11362
Rollstange 12869
Romantiker-Einband 11673
römische Zahlen 11671
römische Zahlenzeichen
  11671
römische Ziffern 11671
Rop-Farbdruck 11680
rösch 6153
rösche Mahlung 6155

röscher Stoff 6164
rösch gemahlener Stoff
  6164
Röslein 5870
Ross-Effekt 11691
Rostflecken 11823
rostfreier Stahl 13012
rostschützendes Papier 694
Rostschutzpapier 694
Rotationer 8950, 11696
Rotationsbuchdruck 11695
Rotationsdruck 11701
Rotationsdrucker 11696
Rotationsmaschine 11699
Rotationsmaschinenführer
  11700
Rotationsmaschinen-
  meister 8950, 11700
Rotationsmaschine, 24-
  seitige ~ 12207
Rotationsoffsetmaschine
  11698
Rotationsstanzformen
  11693
Rotationstiefdruck 11705
Rotationsviskosimeter
  11702
rotativer Buchdruck 11695
Rotdruckplatte 8491
Rötel 11867
Röteldruck 11867
Rötelzeichnung 11867
roter Druckknopf (bei Not-
  fällen) 11323
roter Klebstreifen 8650
roter Kreidestift 11867
roter Titelkopf 11768
rotes Bleioxyd 11330
rotes Blutlaugensalz 10649
rotes chromsaures Natrium
  12688
rotes Klebeband 8650
Rotfilter 8488
rot Gedruckte 11768
Rotplatte 8491
Rotschnitt 11324
Rotwelsch 7648
Roulett 1176
Roulette 11717
roulettieren 11716
Routine 11734
Rüböl 11243
Rubrik 3200, 11768
rubrizierte Anzeigen 2881
Rucke 7656
Rücken 983, 984, 1422,
  12873
Rückenbeklebepapier 1034,
  7026
Rückenfeld 9889
Rückenfräser 1048
Rücken in Feldern 1026
Rücken mit (erhabenen)
  Bünden 11589
Rückenrundemaschine
  1047
Rückenrunden 1046
Rückenrunden und Falzein-
  pressen 11827

Rückenrundklopfen 1046
Rückenschild 9889
Rückenschutzschicht 667
Rückenstütze 4941
Rückentitel 1050
Rückgewinnung 11314
Rückgewinnungsanlage
  11321, 12779
Rückkehrinformation 8211
Rückseite 984, 1040, 14447
Rückseitenbearbeitung
  1043
rücksetzen 11494
Rückstand 11495, 11498
Rückstandgewinnung
  11496
Rückstelltaste 11545
Rückstichheftmaschine
  11830
Rückstichheftung 11831
Rückstichheftung mit drei
  Stichen 2550
Rückübertragung von
  Druckfarbe 1051
Rückwand für Kalender
  2179
Rückwasser 14730
Rückweisegrenze 8378
rückweiser Lauf 7684
Rückweisewahrscheinlich-
  keit 10875
Rüffel 11476
Rüge 11476
Ruhen 4920
ruhige Durchsicht 14304
Rührapparat 309
Rührbütte 9011, 13186
rühren 13165
Rühren 308
Rührholz 4241
Rührstock 4241
Rührwerk 309
rundätzen 13589
runden 1313
runden Ecken, mit ~
  11725
runder Einbandrücken
  11720
Rundgang 4061
Rundgewinde 11731
Rundgießwerk 2449
rundgotische Schrift 8074
Rundlochperforation 11727
Rundlochung 11727
Rundmesser 12585
Rundplatte 3848
Rundraster 2847
Rundschreiben 2844
Rundsiebmaschine 3944
Rundstereo 3849
Rundstereo-Bearbeitungs-
  maschine 937
Rundstereo-Gießwerk 2449
Rundstichel 6531
Rundung 1823
Runzelbildung 6199
Runzelkorn 3094
Runzelkornbildung 11532
Runzelkornraster 11531

Runzeln 3710, 14873
Runzelung 6199
Rupfen 10295
Rupfen ohne Einwirkung
  von Wasser 4813
Rupffestigkeit 10300
Rupffestigkeitsmeßgerät
  10302
Rupffestigkeitsprüfer 10302
Rupffestigkeitsprüfgerät
  10302
Ruß 2311, 12782
Rußabdruck 12639
Russell-Effekt 11820
Rüstzeit 2618
Rutsche 2825
Rüttelgewicht 711
Rütteltisch 7687
Rüttler 7687

Saalmeister 10762
Sabattier-Effekt 11825
Sack 11826
Sack bekommen, den ~
  5694
Sackbildung 12537
Sackpapier 1075
Safawid-Einband 11832
Saffian 9106
Saffianleder 9106
Sägeaggregat 11896
Sägeblatt 11894
Salband 12090
Salizylsäure 11851
Salkante 12090
Salleiste 12090
Salmiakgeist 533
Salpetersalzsäure 734
Salpetersäure 9366
salpetersaurer Natronzell-
  stoff 9367
salpetersaures Blei 7996
salpetersaures Kali(um)
  10660
salpetersaures Kalzium
  2169
salpetersaures Silber 12432
salpetersaures Uran 14369
salpetrige Säure 9380
Salpetrigsäure 9380
Salz 11852
Salzsäure 7102
salzsaures Ammoniak 522
salzsaures Diaminophenol
  4262
salzsaures Eisen 5546
salzsaures Paraminophenol
  9962
SAM 13031
Sämischleder 2612
Sammelbuch 369
Sammeldrahtheftmaschine
  6304, 11829
Sammelelevator 832
Sammelform 6299
Sammelhefter 6304
Sammelheftmaschine 6304
Sammellinse 3367, 3464
sammeln 6325

schwarze Farbe 1479
schwärzen 12638
schwarzer Körper 1467
schwarzer Strahler 1467
Schwarzkunst 8883
Schwarzlauge 1487
Schwarzlicht 14252
Schwarzmischung 12209
Schwarzpapier 1488
Schwarzplatte 1490, 1491
Schwarzschild-Effekt 11925
Schwärzung 1475, 7483,
  9653
Schwärzungsbereich 4175
Schwärzungskurve 2637
Schwärzungsmesser 4167
Schwärzungsumfang 3443,
  4175
Schwärzungsverlust 8373
Schwarz verlaufende
  Rasterätzung, in ~ 1495
Schwarzvignette 1495
schwarz-weiß 1463
Schwarz-Weiß-Abdruck
  1465
schwarz-weiße Randlinie
  1464
Schwarz-Weiß-Original
  8168
schwätzen 13565
Schwebe 13452
Schwebeförderer 9760
Schwefel 13383
Schwefelammonium 546
Schwefelantimon 672
Schwefeläther 5297
Schwefel behandeltes Öl,
  mit ~ 13382
Schwefelblei 6272
Schwefelchlorür 13386
Schwefeldioxyd 13387
Schwefelharnstoff 13723
Schwefelkadmium 2145
Schwefelkalium 10672,
  10677
Schwefelkies 11078
Schwefelkohlenstoff 2316
Schwefelleber 10677
Schwefelnatrium 12720
Schwefelnatron 12720
Schwefelsäure 13389
Schwefelsäureanhydrid
  13397
schwefelsaure Mangan
  8591
schwefelsaurer Kalk 2175
schwefelsaures Aluminium
  487
schwefelsaures Ammonium
  544
schwefelsaures Eisenoxyd
  5556
schwefelsaures Eisen-
  oxydul 5580
schwefelsaures Eisen-
  oxydulammoniak 5573
schwefelsaures Kalzium
  2174, 3774

schwefelsaures Kupfer
  3533
schwefelsaures Magnesium
  8502
schwefelsaures Mono-
  methyl-Paraminophenol
  8867
schwefelsaures Natrium
  12719
schwefelsaure Tonerde 487
Schwefeltrioxyd 13397
Schwefelwasserstoff 7117
Schwefelzinkweiß 8276
Schwefelzyanammonium
  550
Schwefelzyankalium 10676
schweflige Säure 13387,
  13394
Schwefligsäureanhydrid
  13387
schwefligsaures Natrium
  12721
schwefligsaures Natron
  12721
Schweinfürter Grün 5580
Schweinsleder 10337
Schweizerdegen 14125
Schwellenwert 8152, 13779
Schwenkkassette 7009
schwer abzusetzendes
  Manuskript 10088
Schwerbenzin 6941
Schwerbenzol 12776
schwere Druckform 6943
schwere Farbe 6939
schwerer Satz 8009
schweres (Klischee-)Metall
  6940
Schwerestoffauflauf-
  (kasten) 6580
Schwerspat 1182
Schwertfalzmaschine 7824
schwierig abzusetzender
  Satz 8009
schwierige Paßform 13809
schwieriger Satz 3276
schwimmende Walze 5359
Schwimmen von Metall-
  pigmenten auf dem
  Bindemittel 8008
Schwimmprobe 5837
Schwinden der Zeichnung
  14528
Schwinden der Zeichnung
  auf der Druckplatte 1567
Schwinden (des Metalls
  beim Erkalten) 12356
Schwinggreifer 13472
Schwingpsychrometer
  12558
Schwingung 9708
Schwingwalze 7683
Schwund 12356
Schwungkraft 2553
Schwungrad 5935
Schwungradwelle 5936
Schwungscheibe 5935
Scotch-tape-Versuch 11942
Sechsenalphabet 12477

Sechserformat 12205
Sechsformengießrad 12480
Sechs-Gießformen-Gießrad-
  mechanismus 12480
sechsunddreißiger Form
  13735
Sechsundneunzigerformat
  9351
Sedezformat 12481
Sediment 4186
Sedimentation 12065
Seehundleder 12025
segmentieren 12067
Seidenpapier 6172, 14867
Seidenrasterdruck-
  verfahren 11989
Seifenechtheit 12657
Seifenpackungen 12659
Seigerung 12069
seihen 5663
Seite 9851
Seite auf Handtuchformat
  8343
Seite (eines Blattes) 9850
Seite im Hochformat 4092
Seite im Querformat 9476
Seite mit Vorschlag
  beginnen 4761
Seitenabzüge 9863
Seitenanlegemarke 12371
Seitenanschlag 12371
Seitenbewegung der
  Papierbahn 14638
Seitenbreite, über die ~
  155
Seitendrucker 9854
Seitenfalte 6730
Seitenfaltenbeutel 6731
Seitengestell 12368
Seitenhöhe 4196
Seitenkopf 6890
Seitenmagazin 943, 12375
Seitenmarke 12371
Seitennummer 5995
Seitenöffnungen 1238
Seitenpaar in der Mitte
  eines Bogens, aus einem
  Blatt bestehend 2549
Seitenrand 12376
Seitenregister 12372, 12384
Seitenregister-Steuerung
  12373
Seitenregister-Verstellung
  12373
seitenrichtiges Bild 11282
seitenverkehrtes Bild 11285
seitenverkehrtes Negativ
  11551
Seitenwasser 14730
Seitenzahl 5995
Seitenzahl am Fuß (der
  Seite) 4751
48-seitige Rotations-
  maschine 4654
64-seitige Rotations-
  maschine 4635

seitlich bewegbare Reiterwalze 11305
seitliche Begrenzungs-
  lineale 4064
seitliche Fadenheftung
  14708
seitlicher Einschlag 14074
seitlicher Passer 12384
seitliches Verlaufen 7085
seitliches Verlaufen der
  Papierbahn 12282
seitliche Verschiebung
  7933
seitliche Verstellung 12373
seitlich verschobene Papierrolle 13631
Seitwärtsbewegung 7935
Sekante 12032
Sekundärfarben 12033
Sekunde 851
Sekundenzähler 13221
Selbstabnahmemaschine
  14905
Selbstankleber für
  fliegenden Rollen-
  wechsel 5927
Selbstanleger 909
Selbstbiographie 891
Selbstdurchschreibe-
  formularsatz 2324
Selbstdurchschreibepapier
  2325, 9382
Selbstentzündung 12915
Selbstklebeband 12077
selbstklebende Lettern
  12074
selbstklebender Klebstoff
  10768
selbstklebendes Papier
  12076
Selbstklebepapier 12076
Selbstkopierpapier 12081
Selbstkostenpreis 3631
Selbstöffnerbeutel 12083
Selbstpolymerisierung 934
Selbstschalter 926
Selbststart 1788
selbsttätige Auslage 906
selbsttätiger Bogenausleger
  906
selbsttätiger Einleger 909
selbsttätiger Rollenwechsel
  5927
selbsttätiger Wieder-
  holungsruf 923
selbsttätiges Korrosions-
  schutzpapier 14389
Selbstverlag, im ~ 11004
Selbstverleger 885
Selbstzündung 12915
Selektions- und Maskier-
  maschine 11910
Selektorkanal 12071
Selen 12072
Selenzelle 12073
Semikolon 12098
Sender-Signalumsetzer
  9030
Senkkopfschraube 3663

Staubabsaugvorrichtung 14643
staubdicht 4906
Stauben 4894, 5880, 6268
staubendes Papier 5882, 6269
staubgeschützt 4906
Staubkasten 4896
Staubkorn 6553
Staubrette 4064
Staubungsneigung 5881
Stauchfalzmaschine 2027
Stauplatte 4064
Steam-set-Farbe 13085
Stearin 13088
Stearinpech 13090
Stearinsäure 13087
Steatit 13092
Stechen 13900
stechen (in Kupfer usw.) 5182
Stechfutteral 12561
Stechhygrometer 13481
Stechkarte 2921
Stechkissen 9838
Stechschlüssel 9620
Stechzirkel 4511
Steckdose 12660
Stecker 10488
Steckfeld 10490
Steckschriftkasten 2419
Stecktafel 10490
Steg 2523, 2964, 6735
Stege anlegen, die ~ 4722
Stegregal 6263
Stehenbleiben der Farbe im Farbkasten 1011
stehender Satz 13033
stehendes Rakelstreich-verfahren 12934
stehen lassen 6522, 7742
Stehform 13032
Stehsatz 13033
Stehsatzbrett 8055
Stehsatz wiederverwenden 10305
Steife 11613
steife Druckfarbe 6935
steife Schachtel 12182
Steifigkeit 11613
Steifigkeitsprüfgerät 13154
Steifleinen 2030, 2206
Steilheit 6534
Stein 7393, 8265
Steindruck 8270, 13205
Steindruckhandpresse 8260
Steindruckmaschine 8263
Steindruckpresse 8263
Steindruckverfahren 8270
Steindruckwalze 8264
Steinglättung 5834
Steingravur 13198
Steinkohlenteer 2972
Steinschleifer 8081
Steinzeichner 8256
Stellkeil 225
Stellmotor 11412
Stellmutter 222
Stellplakat 12350

Stellschraube 223
Stellschrauben 4838
Stempel 1584, 11040
Stempelbild in punktierten Linien, mit ~ 5582
Stempelsatz, mit ~ 948
Stempelschneiden 11044
Stempelschneider 11043
Stencilpapier 8949
Ster 3804
Stereo 13127
Stereoabteilung 13130
Stereomater 5846
Stereoplatte 13127
Stereotypeur 13128
Stereotypie 13123
Stereotypieabteilung 13130
Stereotypiemetall 13126
Stereotypiepappe 5848
Sterilisierungsechtheits-versuch 13132
Stern 850
Sternchen 850
Sterndreieckschaltung 13049
Sternkarte 2497
Sternschaltung 14910
Sternzeichen 850
Sterry-Effekt 13133
Stetigholländer 1881
Stetigschleifer 3430
Steuereinheit 3461
Steuereinheit für Daten-vermittler 9189
Steuereinrichtung 14618
Steuerkasten 3451
Steuerkurve 2227
Steuerkurve für Abzähler 7789
Steuermotor 3458, 11412
steuern 3449
Steuernocken 2227
Steuernocken für Abzähler 7789
Steuerpult 3459
Steuerstrom 3453
Steuerungsantrieb 14618
Steuerzeichen 3452
Stibnit 672
Stich 8172
Stichel 2072, 2731, 6578
Sticheln 13900
Stichprobe 11854
Stichproben-Entnahme-vorschrift 11857
Stichwort 2469, 2470
Stichwörterverzeichnis 2472
Stichzeile 2466
Stickstoff 9374
Stickstoffstoß 9375
Stift 7234
Stiftdrucker 8694
Stiftschlüssel 428
Stil 8052
Stilb 13155
stillsetzen 14057
Stillwand 4067
Stimmungsbild 9097

stocken 2970
Stockflecken 6139, 8928
Stockpresse 13034
Stoff 13181
Stoffänger 11891
Stoffansatz 13190
Stoffaufbereitung 13190
Stoffauflaufkasten 6898
Stofflöser 11029
Stoffbatzen 2994
Stoffbereitung 13190
Stoffbütte 13186, 14410
Stoffdichteregler 11422
Stoffdruck 13678
Stoffdruckfarbe 2958
Stoffdruckmaschine 2208
Stoffegalisator 868
Stoffeinlauf 6898
Stoffende am Drucktuch, angenähtes ~ 1522
stoffgeleimtes Papier 1264
Stoffherstellung 13190
Stoffleimung 1265
Stoffmühle 7693
Stoffregulierkasten 11422
Stoffzusammenstellung 5592
Stokes 13194
Stokesches Fallgesetz 13195
Stopfer 1880
Stoppbad 13209
Stopper 1880
Stopperschalter 14642
Stoppuhr 13221
Stoppzylinderpresse 13210
Storchschnabel 9892
Störung 1880, 2040, 5494
störungsfreies Laufen 11790
Störungszeit 4679
Stoß 7186, 8091
Stoßdämpfer 2032
Stöße 7656
Stoßlinie 7691
Straffen 4719
Straffenpapier 14168
straffe Papierbahn 13810
Straffer 4719
Strahlenwirkung 160
Strahlteiler 1246
Strahlung 11194
Strahlungsenergie 11191
Strahlungsfluß 11192
Strahlungsfunktion 12842
stramme Druckfarbe 13298
Strammheit 11613
Strangpresse 5411
Strasse 6734
strecken 4372, 13263
Strecken eines Offset-bogens beim Durchgang durch die Presse 5467
Streckenstrich 4002
Streckgang 11816
Streckgrenze 14930
streichen 4128, 4129, 7792
Streichen 2988

Streichen außerhalb der Papiermaschine 3469
Streichen im Tiefdruck-verfahren 6584
Streichen in der Leim-presse 12494
Streichen in der Maschine 9600
Streichfarbespritzer 12888
Streichmaschine 2995
Streichmasse 2998
Streichpapier 1685
Streichräder 3218
Streichrakel 5223
Streichrohpapier 1685
Streichschicht 2989
Streichverfahren 2988
Streichverfahren mit der Luftrakel 338
Streifband mit Werbetext 11002
Streifen 5741
Streifenbildung durch unregelmäßiges Gummieren 6726
Streifenleser 13590
Streifenlocher 9950, 10134, 13587
Streifenspeicher 2498
Streikbrecher 1484
Streikende 13274
Streikender 13274
strenge Druckfarbe 13298
strenge Farbe 12328, 13151
strenger Firnis 13299
Streulicht 4346, 11915
Streumuster-Einband 12092
Streuung 4455, 14397
Strich 2989, 8155, 9234
Strichätzung 8159, 8174
Strichaufnahme 8189
Strich-Auto-Kombination 8179
Strichelung 6740
Strichfarbenreproduktion 8166
Strichklischee 8159
Strichklischee mit ein-kopiertem Raster 8179
Strichkrater 1019
Strichnegativ 8183
Strichoriginal 8205
Strichplatte 8195
Strichplatte, (un-gerasterte) ~ 8165
Strichpositiv 8191
Strichpunkt 12098
Strichraster 8195
Strich-Rasterätzung, (kombinierte) ~ 8179
Strichreproduktion 8205
Strichrupfen 2997
Strichstärkenwandler 8204
Strichstärkeumwandler 8204
Strich, unter dem ~ 5586
Strich, unterm ~ 5586
Strichvorlage 8168

Strichzeichnung 8168, 8171
Stripfilm 13283
Stripfilmzement 13284
Strippen 13289
Stroboskop 13295
Strohkarton 13254
Strohmischpapier 9006
Strohmischpappe 9006
Strohpapier 13255
Strohpapier, (reines) ~
  14927
Strohpappe 13254
Strohpappe, (reine) ~
  14927
Strohstoff 13256
Strohzellstoff 13256
Strom 3840
Stromabnehmer 2313
Stromart 3840
Stromdichte 3842
Stromkreis 2837
stromlos 4028
Stromnetz 13932
Stromstärke 553, 3843
Stromstärkenmesser 556
Strömungsmesser 5874
Stromverbrauch 3841
Stromzuführungsstangen
  2101
Struktur 13303
Strukturänderung 13300
Strukturerkennung 10065
Strukturformel 13301
Struktur in den Tiefen des
  Drucks 11981
Strukturspeicher 9418
Strukturviskosität 13302
Struwwelpeter 10313
Stücklinien 7861
Stücksetzer 3314
Stufengraukeil 6629
stufenloses Regelgetriebe
  7303
stufenlos regelbare
  Geschwindigkeit 13117
stufenlos regelbares
  Getriebe 7303
stülpen 14824
stumpf anfangen 1289
stumpf anfangender
  Absatz 5908
stumpf ausgehen lassen
  5160
stumpfer Stoß 2115
stumpfer Zeilenanfang
  5903, 6235
stumpffeine Linie 8757
stumpf halten 5160
Stumpflasche 2115
Stumpfstoß 2115
Stundenbuch 1767
Stundenleistung 9729
Stundenzettel 7681
stürzender Satz 9511
Stützpresseur 1055
Stützrakel 4533
Stützwalze 1055
Styrol 13316
Styrolharz 13317

Sublimation 13326
sublimieren 13327
Submission begeben, in ~
  5472
Subskriptionsschein 13333
Substrat 13339
Substratfarbe 7880
Subtraktionszeichen 8969
subtraktive Dreifarben-
  methode 13346
subtraktive Farbmischung
  13346
subtraktive Farbsynthese
  13346
subtraktives Farbverfahren
  13346
suchen 12029
Suchen von Fehlern 4040
Suchwort 2469
Sulfatverfahren 13368
Sulfatzellstoff 13369
Sulfid 13370
sulfieren 13380
Sulfit 13372
Sulfitlauge 13373
Sulfitpapier 13377
Sulfitverfahren 13378
Sulfitzellstoff 13379
sulfonieren 13380
sulfoniertes Öl 13382
Sulfür 13370
sulfurieren 13380
Sumach 13398
Sumachgerbung 13400
Summand 183
Superkalander 13409
Supervisor 13423
Suspension 13452
Süß 5492
Swanfassung 1237
symbolische Adressierung
  13485
symbolische Programmier-
  sprache 13486
symbolische Sprache 13486
sympathetische Druckfarbe
  7546
Synchroncomputer 13488
Synchronrechner 13488
Synopse 13490
Syntax 13491
Synthese 13492
Synthesis 13492
synthetisch 13493
synthetische Faser 13494
synthetischer Gummi
  13496
syrischer Asphalt 13497
System 13499
Systemanalyse 13500
systematische Dicke 14319
systematische Zufallsstichprobe 13502
systematisch geschnittene
  Linien 7861
Systemübersicht 3379

tabellarischer Satz 13520
Tabelle 13508

Tabellensatz 13520
Tableau 5768
Tabloidformat 13518
Tabulator 13521
Tachometer 13523
Tackmeter 13531
Tadel 11476
tadelloses Exemplar 3550
Tageblatt 3958
Tagesleuchtfarbe 5891
Tageslicht 4023
Tagespresse 3959
Tagessehen 10267
Tageszeitung 3958
Tagschicht 4025
Talgsäure 13087
Talk 13562
Talkum 13562
talkumieren 4892
Tallöl 13566
Tampon 3954, 7329
Tampon (für Kupferdruck),
  (kleiner) ~ 10085
Tangente 13572
Tangierfell 12216
Tangierfilm 12216
Tangiermaschine 12215
Tangierverfahren 1318
Tanne 5693
Tannenbaumkristalle 4163
Tannin 13573
tanzender Satz 12955
tanzende Schrift 12955
Tänzerwalze 7683
Tapete 14533
Tapetendruck 14536
Tapetendruckmaschine
  14537
Tapetenmuster 14534
Tapetenpapier 14533
Tarif 11900
Tarifausschuß 13604
Tarlatan 13605
Tartrazingelb 13611
Tasche 1772, 1828
Taschenatlas 10507
Taschenausgabe 10510
Taschenbuch 9437, 9895,
  10508
Taschenfalzmaschine 2027
Taschenformat 10516
Taschenmikroskop 10514
Tastapparat 7765
Tastapparat ohne Klartext-
  wiedergabe 1569
Tastatur 7765
Tastaturlocher 7784
Tastaturplan 7770
Taste 7761
Tastenanschlag 7785
Tastenbläser 1623
Tastenbrett 7778
Tastenknopf 7761
Taster 7765, 7771
Tasterin 7771
Taster ohne Klarschrift
  1569
Tasterzirkel 2222
Tastkopf 11912

Tauchbad gestrichenes
  Papier, im ~ 4390
Tauchbeschichtung 4391
Taucher 14413
Tauchsieder 7184
Tauchstreichverfahren
  4391
Tauchstreichverfahren
  gestrichenes Papier,
  im ~ 4390
Tauchwalze 4396, 4847
Tauenpackpapier,
  (braun) ~ 8605
Tauenpapier 8605
Taupunkt 4242
Tauschexemplar 5342
Tauschliste 5344
Tausendbogengewicht 9390
Tausendbuchstabenpreis
  13739
Tausendstel 8926
Technik der integrierten
  Schaltkreise 12761
technischer Beratungs-
  dienst 14064
Teelöffel 13620
Teer 10380
Teerfarbe 5825
Teerfarben 2973
Teerfarbstoffe 2973
Teerpapier 13608
Teerpapier mit Gewebe
  11428
Teil 10011
Teilauszugnegativ 12130
Teilauszugsfilter 12129
teilbreite Papierbahn 4385
Teilchenform 10013
Teilchengröße 10014
teildurchlässiger Spiegel
  13996
Teilfarbenandrücke 10928
Teilkreis 10383
Teilkreisdurchmesser
  10384
Teilmenge 13335
Teilsatz 13335
Tekäöle 13626
Telephonadreßbuch 4408
Telephonbücherpapier
  13628
Telephonverzeichnis 4408
Teletypesatz 13636
Teletypesetter 13635
Teller 4433
Tellerfarbwerk 4442
tellerförmig 4431
tellerförmiges Papier 4432
Tellermesser 12585
Tellerreiber 9168
Temperaturregelung 13641
Temperaturregler 13698
tempern 631
Tenakel 3544
Tensiometer 13647
Terminjäger 10926
Terminkontrolle 10927
Ternio 13655
Terpentin 14106

unterer Seitenrand 13550
untere Schriftlinie 1196
unteres Druckwerk 8381
unteres Kapitalband 13541
unteres Magazin 8382
Unterfarben 14270
Unterfarbenaufhellung 14269
Unterfarbenentfernung 14269
Unterfarbenkorrektur 14269
unterführen 4364
Unterführungszeichen 4505
untergedruckte Farbe 1025
Untergestell 13341
Untergewicht 12341, 14287
Untergrund 1001
Untergrundfarbe 1025
Untergrundplatte 5809
Unterguß 1022
Unterkasten 8379
unterkopieren 14273
Unterlagbogen 1009
Unterlage 9136, 14282
Unterlagen 4005
Unterlänge des Buch-stabens 4201
unterlegen 1920, 11237
Unterlegen 2061
Unterlegscheibe 14549
Unterlegung 14281
Unterprogramm 13328
Unterrubrik 2468
Untersatz 9136, 13335
Unterschlag 14716
unterschließen 7038
unterschnittener Buch-stabe 7748
Unterschrift 8031
unterschwefligsaures Natrium 7146
Unterstrahlung 12374
unterstreichen 14285, 14288
untersuchen 13659
Untersuchung 13658
Untersuchungsmethode 13666
unterteilter Farbkasten 12898
Unterteilungslinie 13320
Untertitel 2468, 13322, 13342
Untertuch 14267
Unterweisungskarte bei Maschinenarbeit 8606
unverarbeitetes Papier 14335
unverbrennbares Papier 5697
unverschnitten 14258
unverseifbar 14339
unvollkommen ein-gerichtete Druckerei 14754
unvollständig 4093
Unze 5886
unzerstörbare Farbe 7259
Unziale 14262

Unzialschrift 14262
unzuverlässiger Arbeiter 14336
Uranazetat 14367
Uranchlorid 14368
Uraneisenzyanid 14364
Uranferrozyanid 14364
Urannitrat 14369
Uranylazetat 14367
Uranylchlorid 14368
Uranylnitrat 14369
Urheberrecht 3553
Urhebervermerk 3721
Ursprungsdaten 11256
Ursprungsprogramm 12793
Ursprungssprache 12792
Utensilien 13405
UV-Lampe 14251
U-werk 14310

Vademekum 14377
Vakantseite 1532
Vakuumkopierrahmen 14375
Vakuumplatte 14372
Vakuumpumpe 14376
Valenz 14378
Van der Waalssche Kräfte 14383
Van der Waalssche Zustandsgleichung 14382
Vandyckbraun 14384
variabele Geschwindigkeit 221
variabeler Ausschluß 14395
variabeler Falzapparat 14394
variabele Wortlänge 14396
variabelformatiger Falz-apparat 14394
Varianz 14397
Variationskoeffizient 3040
Variationsspreite 11236
Variationsweite 11236
Vari-Typer 14400
Vaselin 10167
Vaseline 10167
Vaselinöl 9966
vegetabilischer Leim 14417
vegetabilisches Pergament 14421
Velinpapier 14858
Velourspapier 5845, 14430
Velox-Druck 14431
Venezianeröl 14438
venezianische Renaissance-Antiqua 7068
venezianischer Terpentin 14438
venezianisch Rot 14437
Ventilation 14440
Ventilator 335
veraltet 9724
veränderliche Spannung 14408
Veränderung 463
Veränderung des Gefüges 13300

Verantwortlicher für Papierrollenbremse 1853
verantwortlicher Redakteur 11526
Verarbeitung 3476, 10894
Verarbeitungsteil 10895
Verascher 7249
verätzt 9751
verätztes Klischee 2706
verballhornen 1822
verballhornisieren 1822
verbesserte Auflage 11573
verbesserte Ausgabe 3593, 11573
verbesserte und erweiterte Auflage 11572
verbesserte und vermehrte Auflage 11572
verbiegen 1820
Verbiegung 4481
Verbindung 3316, 3669, 7492, 8212
Verbindungssteg 1905
verblassen 3141, 5433
verblassene Farbe 5431
verblaßt 4421
Verblocken 1598
verblockungshemmendes Mittel 657
verbolzen 1715
Verbreitung 4350
Verbreitung von Informationen 4009
Verbundlinse 3319
Verchromen 2798
Verchromung 2798
Verchromungsanlage 2779, 2795
Verchromungsbad 2773
Verdampfen 5320
Verdampfer 5321
Verdampfung 5320
Verdampfungswärme 6928
Verdichtbarkeit 3324
Verdichtung 3361
verdicken 3353
Verdickung 13260
Verdickungsmittel 1681
Verdrehung 13925
Verdruckbarkeit 11790
Verdrücken 3785
verdrückte Rollenhülse 3783
Verdunklung 3989
verdünnen (Flüssigkeit) 4372
Verdünner 11345
Verdünnungsmittel 11345
Verdünnungsverhältnis 4374
Verdünstung 5320
verestern 5262
Veresterung 5261
Veresterungsgrad 4110
verfahrensorientierte Programmiersprache 10880
verfärben, sich ~ 3141, 5433, 13843

verfärbt 4421
Verfärbung 2083, 4420
Verfärbungserscheinungen auf der Rückseite 3105
Verfasser 881
Verfasserhonorar 890
Verfasserin 881
Verfasserkorrektur 889
Verfasserkorrekturen 887
Verfasser von Leitartikeln 7985
Verfestigung 4368
verfischter Setzkasten 6098
Verflüssigung 8232
Verformung 4105, 4481
verfügbare Zeit 950
vergällen 4161
vergällter Alkohol 4162
vergällter Spiritus 4162
Vergasung 5320
Vergilben 14917
vergilbte Vorlage 14915
Vergilbung 14917
vergleichen 3063, 3256
Vergleichen 3068
Vergleichsmuster 11384, 13020
Vergleichsprobe 11384
Vergoldeeiweiß 6402
Vergoldemesser 6508
Vergoldepolster 6501
Vergoldepresse 1604, 6407
Vergolder 6399
Vergoldrolle 1176
Vergoldung 6496
Vergrauung 1475
vergriffen 9725
vergrößern 5198
Vergrößerung 5201, 8532
Vergrößerung der Um-schlingung 9099
Vergrößerungsapparat 5202
Vergrößerungsgerät 5202
Vergrößerungsglas 8187, 8534
Vergrößerungsglas mit Ständer 2059
Vergrößerungskamera 5202
Vergrößerungskassette 4940
Vergrößerungspapier 5204
Vergrößerungsrahmen 4940
vergütetes Objektiv 2982
Verhältnis 10967, 11252, 11905
Verhältnis, im ~ 10975
Verharzung 11504
verheben 14006
verhobene Zeile 8978
Verifikation 14441
Verkalkung 2153
Verkaufsabteilung 11848
Verkaufsbedingungen 3372
Verkaufspreis 12089
Verkaufsschlager 1341
Verkehr 13950
Verkettung 3349
verkitten 2525

Winkelhacken ausheben,
den ~ 5119
Winkelhaken 3297
Winkelhaken stellen,
den ~ 12157
Winkelrad 1349
Winkelstücke 590
Winkel von neunzig Grad
11605
Winkel von 30 ° (dreißig
Grad), unter einem ~
7251
Wipe-on-Platte 14760
Wirbelstrom 4952
Wirbelung 14088
Wirkleitwert 3374
wirkliches Gewicht 173
wirksame Öffnung 4981
Wirkungskreis 2837
Wirtschaftsredakteur 2866
Wirtschaftsschriftleiter
2866
Wischer 13313
Wischpapier 10467
Wischwalzen 3976
Wischwalzenschlauch 3979
Wischwalzenstoff 9057
Wischwasser 3980
Wischwasserkasten 14577
Wischwassertrog 14577
Wismut 1442
Wismutoxychlorid 10074
Wismutsubnitrat 12817
wissenschaftliche
Forschung 11928
wissenschaftlicher
Formel(n)satz 3307
Witterungszeichen 8846
Wochenblatt 14667
Wochenlohn, auf ~ 12986
Wochenschrift 14667
Wölbung 2028
Wolfram 14788
Wolframdrahtlampe 14086
Wolframlampe 7241, 14086
wolkige Durchsicht 2962
Wollfett 14813
Wollskala 14815
Woodbury-Druck 14793
Wort 14816
Wort am Ende der Zeile
brechen, das ~ 4509
Wort auslassen, ein ~ 8021
Worten ausgeschrieben,
in ~ 12858
Wörterbuch 4306
Wörterbuch-Trenn-
programm 4308
wortgetreue Übersetzung
8247
Wortlänge 14819
Wortteilung 4519
Wort trennen, ein ~ 4509
Worttrennung 4519, 7144
Wort weglassen, ein ~
8021
Wortzwischenraum 12799
Wratten-Filter 14871

wulstartiges Aufbauen von
Druckfarbe auf den Farb-
walzen 11601
Wunderbaumöl 2458
Wurstblatt 11199
Wurstblättchen 11199
Wurzel 11677
Wurzelharz 14809
Wurzelzeichen 11679

x-Achse 967
Xenon 14887
Xenonimpulslicht 14889
Xenonlampe 14890
Xerographie 14891
Xylograph 14897
Xylographie 14798
Xylol 14896

y-Achse 966
Yard 14907

Zähflüssigkeit 14486
Zähigkeit 13524, 14486
Zähigkeit eines Films
13929
Zähigkeitsmesser 14485
Zahl 5607
Zählabstand 11167
Zählapparat 3653
Zähler 3652, 3653
Zählerin 3652
Zählgerät 3653
Zahl in Ziffern ausdrücken
11066
Zahltag 10068
Zählung 3666
Zahlungstag 10068
Zählvorrichtung 3653
Zahn-Auslaufbecher 14931
Zahnrad 6346, 10368
Zahnspiel 1028
Zahnstange 11183
Zahnstangerahmen 11189
Zahnstangesperriegel 13219
Zahnstange und Ritzel
11186
Zahnstreifen 6351
Zahnung 3219
Zapfen 7697
Zapfenlager 2103
Zaponlack 14932
Zäsium 2572
Zauberfarbe 8492
Zehrpfennig 14456
Zeichen 2628
Zeichenabfühlung 8640
Zeichenabmessung 2630
Zeichenband 1762
Zeichenbrett 4706
Zeichendreieck 4715
Zeichenerkennung 2639
Zeichenfeder 4710
Zeichen für Hauchlaute
1900
Zeichen/h 2642
Zeichenkarton 4705
Zeichenkohle 2645
Zeichenmenge 2640

Zeichenpapier 4709
Zeichenrepertoire 11466
Zeichenschräge 2641
Zeichensetzung 11049
Zeichentusche 7280
Zeichenvorrat 2640, 11466
Zeichner 4698
Zeichnikus 4698
Zeichnung 4099
Zeiger 10524
Zeile 8154
Zeile auf die Mitte ein-
stellen, die ~ 2551
Zeile-auf-Zeile gehen 8176
Zeile auf Zeile (setzen)
6000
Zeile ausschließen, eine ~
7714
Zeile brechen 1893
Zeile einbringen, eine ~
13554
Zeile einlaufen, eine ~
13554
Zeile für sich 8160
Zeile halten 7739
Zeile mit der Autoren-
angabe 2131
Zeilen Austreiben 7505
Zeilenbreite 3213
Zeilendrucker 8192
Zeilendurchschuß 7969
Zeilenformat 8720
Zeilenfüllen 11094
Zeilenfüll- und Zentrier-
vorrichtung 11093
Zeilengießmaschine 12606
Zeilenlänge 8720
Zeilenmaß 14197
Zeilenmesser 14197
Zeilensäge 12608
Zeilenschiff 6278
Zeilenschneideapparat 7978
Zeilenschwinder 8193
Zeilensetzmaschine 12606
Zeilenzähler 11800
Zeile trennen 1893
Zeile zentrieren, die ~
2551
Zeile zur Mitte stellen,
die ~ 2551
Zeit der Außerbetrieb-
nahme 4678
Zeitentwicklungskurve
13815
Zeitgammakurve 13815
zeitliche Reihenfolge 2822
zeitlich verzahnte Ver-
arbeitung 13818
Zeitlohnstundenanteil 7056
Zeitmultiplexbetrieb 13818
Zeitschrift 7698, 9309
Zeitschriftenpapier 8485
Zeitstudie 13820
Zeitung 3958, 6344, 9309
Zeitungsausschnitt 9304
Zeitungsausschnittbüro
2917
Zeitungsdruckfarbe 9308

Zeitungsdruckmaschine
9318
Zeitungsdruckpapier 9321
Zeitungsente 7013
Zeitungsfalz 12038
Zeitungsfarbe 9308
Zeitungsformat 9319
Zeitungsgrossist 9300
Zeitungshändler 9300
Zeitungsimpressum 8661
Zeitungskiosk 9317
Zeitungsklischee 9320, 9324
Zeitungskolumne 9313
Zeitungspapier 9321
Zeitungspapiermakulatur
9322
Zeitungspresse 9318
Zeitungsraster 2976
Zeitungsrotationsmaschine
9318
Zeitungsschreiber 7700
Zeitungsschrift 9316
Zeitungssetzer 9303
Zeitungsspalte 9313
Zeitungsüberschriftzeile
1121
Zeitungsvolontär 3796
Zeitungswald 10738
Zeitungswelt 14851
Zeitungswerbung 9311
Zeitungswesen 10738
Zeitungswissenschaft 11927
Zeitung überdurchschnitt-
lichen Formats 1520
Zeitwechselbetrieb 13818
zeitweise 7514
Zelle 2498
Zellensaugwalze 13352
Zellenschmelz 2924
Zellglas 2501
Zellglasbeutel 2502
Zellglasfenster 2505
Zellglasfolie 91, 2503
Zellhorn 2507
Zellophan 2501
Zellstoff 2508, 2686
Zellstoffharz 10381
Zellstoffkarton 2687, 11027
Zellstoffkocher 4352
Zellstoffpapier 2692
Zellstoffpappe 11027
Zellstoffwatte 2521
Zelluloid 2507
Zellulose 2508
Zelluloseazetat 2509
Zelluloseester 2512
Zellulosefasern 2513
Zellulosegummi 2514
Zelluloselack 2515
Zellulosenitrat 2517, 9371
Zellwatte 2521
Zensur der Presse 2526
Zenterlinie 8717
Zentimeter 2532
zentimeterweise vorrücken
7245
Zentipoise 2533
Zentraleinheit 2539

zentrales Transportloch 7414
Zentralheizung 2536
Zentralschmierung 2537
Zentralspeicher 7519
Zentralwert 8746
Zentrierstift 2555
zentriertes Wasserzeichen 8301
zentrifugal 2552
Zentrifugalkraft 2553
Zerdrehung 13925
Zeresin 9818
Zeresinwachs 9818
Zerfaserer 11029
Zerfasern 1266
Zerfaserung 1266
Zerfressen 3615
zerfressene Schicht 3614
Zerlegung von Endlos-formularen 2095
zerquetschte Rollenhülse 3783
Zerrbild 2392
Zerreißdehnung 13264
Zerreißwolf 14744
zerrissene Bogen 13922
zerschnitten 6191
Zerstäuberdüse 866, 12929
zerstörendes Lesen 4212
zerstreutes Licht 4346
Zerstreuung 4350
Zerstreuungskreis 2836
Zerstreuungslinse 4507
Zettel 7846
Zettelpapier 10610
Zetylessigsäure 13087
Zetylsäure 9880
Zeug 13181
Zeugdruckmaschine 2208
Zeugkiste 6958
Zickzackfalz 73
Zickzackform setzen, in ~ 13001
Zickzackformular 14939
Ziegenleder 6493
Ziehdreieck für Farb-auftrag 5223
ziehen, sich ~ 14545
Ziehfähigkeit 13525
Ziehfeder 4711, 11786
Ziehpappe 1659
Zielprogramm 9470
Zielsprache 9469
Zierbuchstaben 13455
Zierleiste 6912
Zierlinie 6182, 9690
Zierrand 9689
Zierschriften 5461
Zierstück 9688, 14470
Ziffer 4356, 5607, 9458
Zifferndarsteller 4366
Ziffernkasten 5608
ziffernmäßig 4358
Ziffernrechner 4359
Ziffernwerk 9450, 9452
Zigarettenkästchen 2830
Zigarettenpapier 2831
Zigarettenschachtel 2830

Zigarettenumhüllung 2830
Zink 14940
Zink ätzen 14945
Zinkätzung 14944, 14947
Zinkblech 14952
Zinkchlorid 14942
Zinkdruckplatte 14944
Zinkgelb 14959
Zinkhydroxyd 14948
Zinkklischee 14944
Zinkkopie 14954
Zinkoxyd 14951
Zinkplatte 14952
Zinksulfat 14955
Zinksulfid 14956
Zinksulfidweiß 8276
Zinkweiß 14951
Zinn 13822
Zinnchlorid 13038
Zinnchlorür 13039
Zinndichlorid 13039
Zinnober 2834
Zinnplatte für Notenstich 10176
Zinsschein 3671
Zirkon 14964
Zirkonium 14964
Zirkular 2844
Zirkumflex 2860
Zirkumflexakzent 2860
Ziseleur 2656
ziselierter Buchschnitt 6494
ziselierter Schnitt 6494
Zitat 11172
zitieren 2861
Zitronensäure 2864
zitronensaures Eisenoxyd-ammonium 5542
zitronensaures Kali(um) 10644
zitronensaures Natrium 12686
Zitskattunpapier 2725
Zoll 7244
Zollangabe 3852
Zolldeklaration 3852
Zonenfarbeinstellung 5158
Zonenschraube 224, 3212
Zonenschrauben 4838, 5158
Zubehör 69
Zubehörteile 69
Zuckerradierung 13358
Zueignung 4074
Zueignungsschrift 4074
zuerst gedruckte Farbe 5701
zufällig 11230
zufälliger Fehler 71
Zufallstichprobe 11235
Zuführbänder 2299
Zuführung des Verarbeitungsgutes in Rollenform, mit ~ 14646
Zuführwalzen 5521
Zug 4701
Zugang 67
Zugänglichkeit 68
zugeklappt 7907

zugeschlagen 7907
Zugfeder 13651
Zugfestigkeit 13644
Zugfestigkeitsprüfer 13645
zügige Farbe 13151
zügige Walzen 13533
Zügigkeit 13524, 13525
Zugriff 67
Zugriffszeit 70
Zugringe 10963
Zugrollen 10963
Zugspannung 14661
Zugspannungsverschluß 13649
Zugwalzen 4713, 4714
zulässige Abweichung 13880
zulässiges Qualitätsniveau 63
zu lose gehefteter Buch-block 12571
zum Geleit 6043
zunehmende Geschwindig-keit 51
Zungenhobel 10484
Zungenschiff 12549
zuordnen 3064
Zurechtweisung 11476
Zurichtebogen 8566, 9773, 12922
Zurichtemesser 3902
zurichten 8564
Zurichtepapier 9768, 9826
Zurichteseidenpapier 8567
Zurichtungsausschnitt 9768
Zurückätzen 996
zurück halten 31
Zurückhalten 8970
Zurückhaltung 11530
Zurücktreiben 11471
Zusammenbacken 1598
zusammenbackende Typen 2151
Zusammenballung 304
Zusammendrückbarkeit 3324
Zusammenfassung 20, 3349, 13490
zusammenführen 846
zusammengebacken 1080
zusammengefallener Satz 10315
zusammengefügte Zeilen 2114
zusammengesetzte Linse 3319
zusammengesetztes Klischee 3299
zusammengesetzte Spalten 3208
zusammengestoßene Linien 45
Zusammenhang 3046
zusammenhängende Rasterpunkte 3388
zusammenhängende Struktur 3389

Zusammenkleben bedruckter Bogen im Stapel 1598
Zusammenlauf der Papier-stränge 3748
zusammenrücken 2943
Zusammenschluß 304
Zusammensetzung 3316
Zusammenstoß 7186
Zusammentraganlage 6330
zusammentragen 3070, 6325
Zusammentragmaschine 6330
Zusammentragtisch 3067
Zusammentrag- und Heft-maschine 6304
Zusatz 716
Zusatzbestellung 1389
Zusätze machen 4560
zusätzliche Bilder er-weitern, durch ~ 6560
zusätzlicher Rollenständer 944
zusätzliche Zeichen 3787
Zusatzmagazin 943
Zusatzmetall 11577
Zusatzmittel 188, 4559
Zusatzspeicher 2033
Zusatztastatur 187
zuschalten 13478
Zuschlag 5395
Zuschnitt 3872
Zuschuß 9784
Zuschußbogen 9784
zu schwache Matrizenzeile 8366
zu schwache Zeile 8366
Zusetzen 5633
zu stark beschneiden 3740
zu starke Quer-schrumpfung 5340
Zuverlässigkeit 11440
zu volle Zeile 13806
Zwangsbewegung 6028
zwangsläufige Bewegung 6028
zwangsweise Bewegung 6028
zweiaugiges Betrachtungs-gerät 1435
Zweibahnen-Wellpappe 12455
zweibasisch 4288
Zweibuchstabenmatrize 4626
Zweifarbendruck 14136
Zweifarbenmaschine 14134
Zweifarbenoffsetmaschine 14132
Zweifarbenverfahren 14136
zweifarbiger Schleier 4295
zweifarbiges Papier 4881
Zweikörpermatrize 4626
Zweilagenbeutel 4873
zweilagiger Sack 4873
Zweilaut 4392
zweimal 14162
zweimal Parallelfalz 12038
zweimal Satz 14162

NEDERLANDS

Aldine 388
aldoor over het vak praten
   13565
alfa- 453
alfabet 454
alfabetische inhoudsopgave
   455
alfabetische volgorde 456
alfabetische volgorde
   plaatsen, in ~ 458
alfabetisch register 455
alfabetiseren 458
alfabetizeren 458
alfabetlengte 459
alfagras 5252
alfanumeriek 461
algebraïsche som 396
algebraïsche tekens 393
algebratekens 393
algemene kosten 9758
alginaat 397
algol 398
algorithmische taal 400
algorithmische vertaling
   401
algoritme 399
algrafie 490
alifatische oplosmiddelen
   409
alinea 1875, 9972
alineateken 9973, 11552
alizarinekleurstof 412
alizarineolie 413
alkali 414
alkalibestendig 416
alkalibestendigheid 421
alkalibestendig papier 420
alkaliblauw 415
alkali-echt 416
alkaliniteit 418
alkalisch 417
alkalivast 416
alkalivast papier 420
alkaliviolet 423
alkanen 424
alkydhars 425
alle rechten voorbehouden
   439
alliage 437
Alltone-procédé 441
allylalcohol 443
allylchloride 444
allylhars 447
allylisothiocyanaat 446
allylthioureum 450
almanak 451
aloë 452
altijddurende kalender
   2178
aluin 468, 483
aluinaardehydraat 477
aluminium 471
aluminiumacetaat 472
aluminium-ammonium-
   sulfaat 473
aluminiumbrons 474
aluminiumchloride 475
aluminiumdruk 490
aluminiumfo(e)lie 476

aluminiumhydroxyde 477
aluminiuminkt 479
aluminiumoxyde 480
aluminiumpapier 481
aluminiumplaat 482
aluminiumpoeder 484
aluminiumsilicaat 485
aluminiumsulfaat 468, 487
aluminiumtypie 491
amalgaam 494
amalgaam-drukprocédé
   11460
amarilpapier 5111
Amerikaanse hoogte 498
Amerikaanse letterhoogte
   498
Amerikaanse verzendbrief
   6171
Amerikaanse vouwbrief
   6171
amide 500
amidol 4262
amine 503
aminobenzeen 613
aminobenzol 613
aminohars 507
ammonia 533
ammoniak 509
ammoniakaluin 473
ammoniumacetaat 513
ammonium-aluminium-
   sulfaat 473
ammoniumbicarbonaat 517
ammoniumbichromaat 526
ammoniumbifluoride 519
ammoniumbromide 520
ammoniumcarbonaat 521
ammoniumchloride 522
ammoniumchromaat 523
ammoniumcitraat 524
ammoniumdichromaat 526
ammoniumfluoride 530
ammoniumjodide 535
ammoniumlactaat 537
ammonium-nikkelsulfaat
   9336
ammoniumnitraat 540
ammoniumoxalaat 541
ammoniumpersulfaat 543
ammoniumrhodanide 550
ammoniumsulfaat 544
ammoniumsulfide 546
ammoniumsulfiet 547
ammoniumsulfocyanide
   550
ammoniumthiocyanaat 550
ammoniumthiosulfaat 551
amorf 552
ampas 1067
amperage 553
ampère 554
ampèremeter 556
ampère-uur 555
amplitude 561
amplitudemodulatie 562
amplitudo 561
amylacetaat 563
amylalcohol 565
anachromaat 568

anachromatische lens 568
anaglyf 569
anagram 570
analist 576
analiste 576
analoge computer 572
analoog-digitaal omzetter
   4366
analoog signaal 573
analoog transmissiekanaal
   571
analyse 575
analytische methode 577
analytische balans 578
anastatisch drukprocédé
   579
anastigmaat 581
anastigmatische lens 581
anderhalf punt 4999
anderhalfpunts interlinies
   5000
anderhalve punt 4999
anepigraaf 586
ångström 601
ångström-eenheid 601
anhydride 605
anhydrisch 607
aniline 613
anilinedruk 5828
aniline-indrukwerk 615
aniline-inkt 5825
anilinekopie 619
anilinepers 5829
anilinepunt 617
anilinesulfaat 622
anion 627
anion-actieve stoffen 628
anisotropie 629
anker 771
annalinwit 630
annotatie 635
annoteren 633
anode 641
anodespanning 648
anodestroom 643
anodestroomrendement 646
anodezak 642
anodische oxydatie 649
anodiseren 649
anodisering 649
anoniem 652
anonieme test 11729
anoniem medewerker 6395
anopistografie 653
anopistografische druk 653
anorganische pigmenten
   7420
anorganisch pigment 8954
anthologie 654
anthraceen 655
anthraceenolie 656
anti-blokmiddel 657
antichloor 658
anti-droger 659
anti-droogstof 659
antieke band 680, 9571
antifrictiemetaal 982
antihalo 666
antihalo-laag 667

antihalo-plaat 993
antimonium 669
antimoniumoxyde 671
antimoniumtrioxyde 671
antimoniumtrisulfide 672
antimoon 669
antimoontrisulfide 672
antimoonwit 671
antioxydant 674
antiquaar 675
antiquariaat 677
antiquariaatsboekhandel
   677
antireflex-laag 667
anti-schuimmiddel 660,
   5940
anti-sluiermiddel 662, 11528
anti-smetapparaat 688,
   12926
anti-smetmiddel 9539
anti-smetpoeder 687, 4824,
   9539
anti-smetvloeistof 9539
anti-static middelen 691
anti-velmiddel 689
anti-velvormer 689
antwoordkaart 11469
antwoordtijd 11523
aparte regel 8160
aplanaat 703
aplanatische lens 703
aplanatisch objectief 703
apochromaat 705
apochromatische lens 705
apostil 706
apostilb 707
apostille 706
apostrof 708
apparaat 4236
apparaat voor het opheffen
   van statische elektri-
   citeit 13069
apparatuur 6877
appendix 716
aquarelpapier 14573
aquatint 735
aquatone-procédé 737
arabesken 744
arabinezuur 745
Arabische cijfers 747
Arabische gom 6707
Arabische letters 749
Arabisch schrift 749
arbeider 14839
arbeidsanalyse 14845
arbeidsdag 14833
arbeidsgeschikt 8
arbeidsongeschikt 7242
arbeidsvoorwaarden 3373,
   14832
arceren 6885
arcering 6740
arcering langs de buiten-
   zijde 1249
areometer 761, 7120
areometer van Baumé 1233
argentometer 763
argument 764
Arkansas-slijpsteen 769

danserrol 7683
data 4005
dataset 4015
datatransmissie 4016
dat is te zeggen 14500
datumregel 4019
datumstempel 4021
dauwpunt 4242
deblokkeren 7441
Debot-effect 4039
decalcomaniepapier 4042
decameter 4046
decanteren 4047, 4048
decibel 4050
decimaal stelsel 8875
decimaalteken 4055
decimale classificatie 4054
decimale komma 4055
decisie 4059
decoderen 4072
decoratieve ligatuur 11120
decoupeerzaag 7663
dedicatie 4074
deel 5474, 10011, 14516
deeldrukken 10928
deelfilter 12129
deelnegatief 3184, 12130
deelteken 4251, 14256
deeltitel 12060
deeltjesgrootte 10014
deeltjesgrootte-bepaling
  10015
deeltjesvorm 10013
deel van de kopij 13553
deelverzameling 13335
defect 4093
defecte exemplaren 11451
defecte letter 1228
defecten 11451
defectenpolis 1389
deflectiespoel 4101
deformatie 4105
degel 10451, 10452
degelautomaat 916
degeldrukker 10456
degelpers 10452
degelpersdrukker 10456
dehydratatie 4121
dehydratie 4121
dekinkt 3680, 9614
dekkelriemen 4068
dekkende inkt 9614
dekkend vermogen 6977
dekkracht 6977
dekkrachtonderzoek 6978
dekplank 6841
deksel 8090
dekvellen 9737
dekverf 9612, 10905
dekvermogen 6977
dekvilt 10339
dekwit 10905, 14951
delaminatie 4124
deloyale concurrent 10778
deloyale concurrentie
  14295
demijohn 2337
demodulatie 4158
demonteerbaar 4214

demonteren 4449
demping 867
den 10362
denatureren 4161
dendrieten 4163
dennehars 6274
dennehout 10362
densiteit 9653
densitometer 4167
densitometrie 4168
densometer 4176
dentelle à l'oiseau-band
  4180
dentelle-band 4181
depot 4186
derivaat 4199
dertiende exemplaar 13730
dertien voor twaalf 13730
deselektrisator 13069
desensibilisatie 4203
desensibilisator 4205
desorptie 4211
desoxydatiemiddel 4184
desoxyderen 4183
detailtekenpapier 12506
detailweergave 4215, 4217
devies 4237
dextrine 4243
dextrose 4246
dia 4278, 12550, 13998
diacritisch teken 4250
diaeresis 4251
diafragma 4273, 6796
diafragmagetallen 5937
diafragmainstelring 4275
diafragmaregeling 4274
diafragmaschaal 8047
diafragmavlak 4276
diafragma voor de midden-
  tonen 4218, 8922
diafragma voor de sluit-
  belichting 6993
diafragmeren 13212
diagnostisch programma
  4252
diagonaal gesneden papier
  591
diagonaalsnijmachine 592
diagram 4257, 5606
diahouder 14001
diahouder met vaste maten
  7810
diamant 4269
diameter van de cilinder
  3938
diameter van de druk-
  cilinder 4260
diameter van de steek-
  cirkel 10384
diaminofenol-hydro-
  chloride 4262
diapositief 4278, 13998
diapositief op glas 6436
diatomeeënaarde 4280
diatomiet 4280
diazokopie 4285
diazopapier 4284
diazotypie 4286
diazoverbinding 4283

dibutylftalaat 4290
dichlorazijnzuur 4292
dichroïsme 4296
dichroïtische sluier 4295
dichtgeslagen, (eenmaal) ~
  7907
dichtheid 4169
dicht negatief 13807
Didonen 4310
Didot-systeem 4311
dieengetal 4328
diëlektricum 4326
diëlektrische constante
  4322
diëlektrische eigen-
  schappen 4324
diëlektrisch papier 4323
diepdruk 7471, 11705
diepdrukcilinder 5184, 6585
diepdruketsing 7467
diepdrukinkt 11706
diepdrukinkt op waterbasis
  14566
diepdrukmachine 6587
diepdrukpapier 11707
diepdrukpers 6587
diepdrukraster 6591
diepdruk-rotatiepers 14648
diepdruk-vellenrotatiepers
  12268
diepetsen 4078
diepetsing 7467
diepetsinkt 4083
diepetslaag 4090
diepetslak 4084
diepetsoplossing 4089
diepetsprocédé 4088
diepgeëtst autotypiecliché
  4081, 4091
diepgelegde offsetplaat
  4080
diepleggen 4082
dieplegprocédé 4088
diepte 5276, 11931
dieptemeetklok 4190
dieptemeter 4190
dieptescherpte 4195
diepvrieskarton 1189
diep zwart 7658
dierenriemtekens 14965
dierlijke lijm 623
diethylether 5297
diffractie 4344
diffractietheorie 4345
diffuseur 4349
diffuse verlichting 4347
diffusiecirkel 2836
diffuus licht 4346
diftong 4392
digitaal 4358
digitaal-analoog 4363
digitaal signaal 4361
digitale computer 4359
digitale opteller 183
digitale rekenautomaat
  4359
dihydroxybarnsteenzuur
  13610
dik 1669

dik clichémetaal 6940
dikke interlinie 13705
dikke lijnen 6884
dikke spatie 13708
dikte 14738
diktemeter 2221, 8898
dikte van de strijklaag
  3000
dikte van het papier 2220
dilatant 4369
dilatatie 4370, 5362
dimensionele stabiliteit
  4376
dimethylbenzeen 14896
dimethylketon 102
DIN-formaten 7521
dinol 11082
dioptrie 4389
direct deelnegatief 4412
directe kleurstoffen 4402
directe rasterfotografie
  4409
directe rastermethode 4404
directe rasteropname 4409
directe verfstoffen 4402
directe verlichting 4406
direct positief 4410
direct rasternegatief 4403,
  4412
direct toegankelijk
  geheugen 4399
direkte toegang 11231
direkt kopieerbare kopij op
  film 14243
dispergeermiddel 4452
dispergeren 4450
disperse fase 4451
dispersie 4454, 4455
dispersie-analyse 10015
dissertatie 13700
distorsie 4481
distorsie-camera 11491
distribueerder 4496
distribueren 2900, 4484,
  7793
distributie 4492
distributiebox 4500
distributiemechanisme
  4491, 4497
distributierol 4503, 13972
distributierol, (heen- en
  weergaande) ~ 9707
distributietafel 4036
distributievorm 4495
distributiezetsel 4035
distributorbar 4499
distributorframe 4502
divergerende lens 4507
dividendbewijs 3671
divisie 7143
D log E-curve 2637
D log E-kromme 2637
dobbelen met kwadraten
  7653
documentpapier 8347
dodekop 2300
doekklemlijst 1513
doek-op-doek-pers 1524
doekpennen 1518

drukken met magnetische inkt 8521

drukken met te grote drukkracht 6868, 8595

drukken met te zware drukspanning 8595

drukken op de achterzijde van een transparante film 1045

drukken, over elkaar ~ 13414

drukken van boekwerken 1783

drukken van decalcomanieën 13969

drukken van postzegels 10601

drukken van transfers 13969

drukken van waardepapieren 12063

drukker 8659, 10753, 10805

drukkerij 9505, 10761, 10829

drukkerijbedrijf 10829

drukker met schrijfwiel 14703

drukkersleerling 10809

drukkersmerk 5816, 7229

drukkerspatroon 8659

drukkerswerkplaats 9505

drukklaar 11286

drukknop 11060

drukknopbesturing 11061

drukknopschakelaar 11062

drukknopstation 3451

drukkosten 10822

drukkracht 7219, 10843

drukkracht-aangever 7215

drukkunst 803

drukletter 14171

drukletters 10851

druklijn 7227, 9352

drukopaciteit 10836

druk over twee pagina's 9891

drukpapier 10838

drukpasta 3317

drukpers 10736

drukpers voor vlakke vormen 5775

drukplaat 5192, 9531, 10418, 12528

drukprijs 10822

drukprocédé 10844

drukproef 2012, 10936

drukproefpapier 10946

drukresultaat 10845

druksmering 6027

druksnelheid 10853

drukspanning 7219, 10843

druks per uur 7223

drukstrepen 13004

druktechniek 10844

druk van ... 10803

druk van de presseur 7218

drukveer 3326, 10770

drukviscositeit 10772

drukvorm 6056

drukvorm voor kleurwerk splitsen, de ~ 4474

drukwerk 10802

dry pint 4814

dubbel-bolle lens 3478

dubbelchroomzure ammoniak 526

dubbelchroomzure kali 10648

dubbel-concave lens 3350

dubbel-convexe lens 3478

dubbel-dubbelzijdig beplakt golfkarton 4598

dubbel-dubbelzijdig beplakt welbord 4598

dubbele belichting 4603

dubbele binding 4627

dubbele drukeenheden 4070

dubbele kast 4585

dubbele nauwkeurigheid 4645

dubbele pers 4670

dubbele plaat 4641

dubbele produktie 13248

dubbele punt 3101

dubbele regel 4579

dubbele rollenster 13569

dubbele stompfijne lijn 4632

dubbele uitleg 4595

dubbele uitvulling 4619

dubbele zak 4873

dubbelfijne lijn 4607, 9987

dubbelgeïnterlinieerd 4624

dubbelgestreken papier 4586

dubbelgezet woord 4579

dubbelgezet woord weghalen, een ~ 4741

dubbel-holle lens 3350

dubbel insluitraam 5977

dubbelkoolzure soda 12676

dubbelkruisje 4594

dubbel raam 5977

dubbel regaal 4612

dubbeltooninkt 4666

dubbel vormraam 5977

dubbelvouwgetal 5979

dubbelvouwproef volgens MIT 2741

dubbelvouwtoestel 5980

dubbelvouwtoestel volgens Schopper 11921

dubbelwandige zak 4873

dubbelzijdig beplakt golfkarton 4605

dubbelzijdig beplakt welbord 4605

dubbelzout 4653

Ducali-band 4833

duim 7244

duimindex 13795

duimproef 14697

duimregister 13795

Duitse alfabet 6384

Duitse hoogte 6385

Duitse komma 12767

duizendletterprijs 13739

duizendpoot 14125

duizendste 8926

Dultgen-procédé 4855

dummy 4857

dun 13709

dun (cliché-)metaal 13712

dundrukpapier 1359, 8140

dunmes 10000

dun negatief 13716

dunne interlinie 13715

dunnen 9999

dunne plek in een offsetrubberdoek 4177

dunne spatie 13718

dunne vernis 13719

dun papier 5832, 8140

dunspatie 13718

dunsteen 10001

duplex-aandrijving 4883

duplex-autotypie 4877

duplex-cliché 4877

duplex-karton 4875

Duplex-machine 4879

duplex-matrijs 4626

duplex-papier 14128

duplex-raster 4882

duplex-reproductie 4869

duplicaat-cliché 4884

duurzaamheid 4886

dwarsdoorsnede van de cilinder 3751

dwarse pagina 1932

dwarskolommen 3210

dwarslijnen 7041

dwarspagina 1932

dwarsperforatie 3763

dwarsregister 12384

dwarsrichting 3752

dwarssteunen 13266

dwarsstreepje 1135

dwarsvouw 2756

dynamische breukenergie 7187

dynamische gladheid van papier 4929

dynamometer 4930

dyne 4931

Eberhard-effect 4946

eboniet 6874

echte oplossing 9053

écrasé 4951

écrasé leer 4951

écrasé maroquin 3784

eczeem 4200

Edison-sluiting 4964

editie 4970

editie-ster 12024

eenbad-etsing 12445

eenbad-etsmethode 12445

eenbasisch zuur 9074

eenfase-etsmachine 10704

eenfase-etsprocédé 10703

eenhedennaald 10523

eenhedenschaal 14313

eenhedenstoppers 14317

eenhedenwiel 14322

eenheid 14309

eenheidsformaten 7521

eenheidsformaten voor enveloppen 7522

eenkleur-drukeenheid 12452

eenkleurig 9078

één kleur, in ~ 9078

eenkleur-offsetpers 12450

één kolom, over ~ 12453

eenmaal gestreken papier 12448

eenmalig carbonpapier 9594

eenparig versnelde beweging 14305

eenpunts interlinie 6744

eenrichtingsduplexkanaal 6759

eens per drie maanden verschijnend blad 11136

eentoerpers 12466

eentonig en langdurig Werk 6635

eenwaardig 9093

eenzijdig bedrukt vel 1573

eenzijdig beplakt golfkarton 12455

eenzijdig beplakt welbord 12455

eenzijdig gekleurd papier 9591

eenzijdig gestreken papier 9590

eenzijdigglad-cilinder 6448

eenzijdig glad karton 8449

eenzijdig-glad machine 14905

eenzijdig glad papier 8449

eenzijdig kleefband 12456

eenzijdig zelfklevend band 12456

eerste druk 5710

eerste elevator 5703

eerste elevatorhaak 7650

eerste keus 10118

eerste kleur 5701

eerste kwaliteit 9457

eerste kwaliteit papier 10118

eerste pagina van het omslag 6207

eerste parallelvouw 5713

eerste proef 5801, 11270

eerste regel 7322

eerste regel over de volle breedte gezet, de volgende regels ingesprongen 11813

eerste regel voluit, de volgende regels ingesprongen 6854

eerste uitgave 5702

effectieve berstdruk 2096

effectieve berststerkte 2096

effectieve berstweerstand 2096

effectieve (lens)opening 4981

effectieve waarde 4982

effening 5222

egoutteur 3985
egoutteur-watermerk 14587
Egyptische Maandag 8377
ei-albumine 371
eierschaal-oppervlak,
  met ~ 4990
eigendomsrecht 3553
eigen drukkerij 10874
eigenhandig geschreven opdracht van de schrijver
  10734
eigen pers 10874
eigenschap 5502
eigenschappen 2638
eigenschappen van het
  oppervlak 13436
eigen uitgave 10871
einde 13732
eindigen 3885
eindigen met zetten 5687
eindschakelaar 8151
eiwit 4988, 6402
eiwit aanbrengen 6420
eiwitbichromaat 4299
eiwitkopie 372
eiwitoplossing 377
eiwitplaat 375
eiwitprocédé 376
eiwitvlokken 5748
elasticiteit 5005, 11500
elastisch 9619
elastische deformatie 5004
elastische vervorming 5004
elektrisch dompelelement
  7184
elektrische aandrijving
  5016
elektrische dompelaar 7184
elektrische energie 5024
elektrische hygrometer
  5021
elektrische kracht 5024
elektrische lading 5013
elektrische schrijfmachine
  5025
elektrische stroom 3840,
  5014
elektrisch etsen 5007
elektrische veldsterkte
  5605
elektrische verwarming
  5020
elektrisch veld 5017
elektrisch vermogen 5024
elektrode 5027
elektrodespanning 5030
elektroforese 5061
elektrofotografie 5062
elektrolyse 5032
elektrolyt 5033
elektrolytisch bad 5034
elektrolytisch etsen 5035
elektrolytisch grein 5036
elektromagneet 5037
elektromagnetische
  eenheid 5040
elektromagnetische
  koppeling 5038

elektromagnetische (laag)-
  diktemeter 5039
elektromotor 5042
elektromotorische kracht
  5041
elektron 5043
elektronenbuis 5056
elektronenflitslamp 5049
elektronenmicroscoop 5058
elektronenstraalbuis 2481
elektronica 5052
elektronisch 5044
elektronische clicheer-
  machine 10229
elektronische data-
  verwerking 5046
elektronische gegevens-
  verwerking 5046
elektronische graveer-
  machine 10229
elektronische regeling van
  het dwarsregister 5054
elektronische regeling van
  het register 5051
elektronische register-
  regeling 5051
elektronische reken-
  machine 3330
elektronische verwerking
  van gegevens 5046
elektronisch graveren 907,
  5048
elektrostatica 5069
elektrostatisch druk-
  procédé 5068
elektrostatische eenheid
  5070
elektrostatische kopieer-
  machine 5067
elektrostatisch veld 5066
element 5074
elementair 5075
elevatorkop 5704
elisie 5083
elkaar verdragend 3259
ellips 5085, 5086
ellipsograaf 5087
ellipspasser 5087
els 1671
elsklos 1673, 6831
elspunt 1674
elzenhout 387
em 5098
email 5138
emailkopie 5146
emaillaag 5147
emaillelaag 5147
emballage 9827
emissiespectrum 5115
emk 5041
em-liniaal 5124
empirisch 5118
em-rek 5122
emulgator 5128
emulgeermiddel 5128
emulgeren 1556, 5126, 5129,
  13843
emulsie 5132
emulsieinkt 5135

emulsielaag gestreken
  papier, met een ~ 5133
emulsiestrijkprocédé 5134
emulsiezijde 5136, 5422
energieverbruik 10705
Engelse band 3685
(Engelse) grein 6539
Engelse lijn 6182
Engelse pint 8239, 10373
Engelse pond 10690
Engelse prent 13161
Engelse rug 12957
Engelse sleutel 9072
(Engelse) voet 6014
Engelse warmte-eenheid
  1924
Engels ons 9711
Engels schrift 12012
Engels zout 8502
enkelbrede pers 12471
enkel deel 9496
enkele pers 12471
enkellijns raster 6763
enkellijnsraster-effect
  12459
en-teken 557
entree betalen 6614
envelop 5209
envelopinsteekmachine
  5214
enveloppe 5209
envelop(pe) 3611
envelop(pe) met drukknoopsluiting 10744
envelop(pe) met gegomde
  klep 210
envelop(pe) met ingeplakt
  venster 14749
envelop(pe) met pen-
  sluiting 2878
envelop(pe) met soufflet
  6732
envelop(pe) met touwtje en
  knoopje 13279
enveloppenformaten 12498
enveloppenmachine 5211
enveloppenpapier 5212
eosine 5215
epigram 5217
epikotehars 5220
epiloog 5219
epoxyhars 5220
erboven zetten en achterin
  houden 7038
erg 5238
ergonomie 5239
ernstige fout 8556
eronder zetten en achter
  inhouden 7038
errata 5241
erratum 5242
erytrosine 5248
esculine 5251
esparto 5252
espartocelstof 5255
espartogras 5252
espartopapier 5254
espehoutstof 820
essence 5257

ester 5259
estergetal 5263
esterhars 5260
esthetisch 286
etalagekaart 12350
ethanal 86
ethanol 5301
etheen 5307
etheenglycol 5309
ether 5297
etherische olie 5257
ethoxylinehars 5220
ethylacetaat 5300
ethylalcohol 5301
ethylbromide 5303
ethylcellulose 5305
ethylchloride 5306
ethyleen 5307
ethyleenglycol 5309
ethyleenglycol-monomethyl-
  ether 5310
ethyljodide 5314
ethylrood 5316
etiket 7846
etiketteermachine 7850
etiketten-automaat 7852
etikettendruk- en stans-
  machine 7852
etikettenlak 7853
etiketten opplakken 7848
etikettenpapier 7851
etikettenvernis 7853
etiketteren 7848
Etruskische band 5317
ets 5270, 5272
etsbad 5274
etsbak 5273, 5290
etsdiepte 5276
etsdieptemeter 5277
etsduur 5288
etsen 1450, 2898, 3900, 4204,
  5271
etsen en afdekken 13006
etsen in koper 3513
etsen van zink 14945
etser 5269, 6781
etsfactor 5285
etsgrond 5278
etsing 1450
etsing met verschillende
  baden 9181
etskali 10655
etskunst 802
etskwast 5275
etsmachine 5280
etsnaald 5281
etsnatron 12695
etspers 3525
etsplaat 5272
etstijd 5288
etsvloeistof 5266, 5286, 9366
et-teken 557
eutecticum 5318
evaporator 5321
Eve-band 5322
even/oneven bit 10008
even/oneven controle 10009
even/oneven toets 10009
even pagina's 5327

éventail-band 5468
evenwicht 5227
evenwichtstoestand 5227
evenwicht tussen inkt en
  water 7408
evenwijdige beweging 9984
evenwijdigheid 9982
excenter 2227
excenter voor de telvinger
  7789
excentriek 2227, 4948
excentriekrol 2253
excentriekschijf 2227, 4949
exemplaar 3535
exemplaar ter bespreking
  240, 11568
ex-libris 1771
expeditie-afdeling 8541
exponenten 13416
exponeren 5367, 5368
exsiccator 4207
extender 5383
extern geheugen 5389
extra cliché 4884
extract 5396
extractie 5397
extractie-apparaat 5398
extra editie 12822
extra kosten 5395, 5407
extra rollenhouder 944
extra rollenstandaard 944
extruder 5411
extrusiecoating 5413
ezel 4941
ezelsoor 4543

fabelboek 1765
fabricagefout 3403
fabricagepartij 1221
fabricageprogramma 11806
fabriek 10403
fabrieksgebouw 10403
fabriekslas 8941
fabrieksmerk 13944
fabrieksmonteur 5237
fabrieksrol 8942
fabrikatieprogram 11806
facet 1344, 1345, 1346
facet-afdruk 5425
facetschaaf 1355, 12314
facetteermachine 1355
facettenbevestiging 3325
facetteren 1354
facsimile 5426
facsimile-overbrenging
  5427
fadensteker 8208
fake-colour-procédé 5442
familieberichten 1440
familiedrukwerk 2460
fanaalviolet 10145
fanfare-band 5464
fantasieletter 5461
fantasieletters 13455
fantasieoppervlak 5458
fantasiepapier 5460
fantasiepreging 5458
farad 5469
faraday 5470

farmaceutische tekens 8747
Farmerse afzwakker 5471
fase 10184
fasehoek 10185
fasemodulatie 10186
faze 10184
fazehoek 10185
feestdag van Sint Jan
  11843
feestdag van Sint Maarten
  11845
felicitatiekaarten 6626
Fenchel-vochtrektoestel
  5538
fenol 10189
fenolformaldehyde hars
  10190
fenolftaleïne 10192
fenolhars 10191
fenylamine 613
ferri- 5540
ferri-ammoniumcitraat
  5542
ferri-ammoniumoxalaat
  5543
ferri-ammoniumsulfaat
  5544
ferribromide 5545
ferrichloride 5546
ferricitraat 5548
ferricyaankali 10649
ferrietkern 5561
ferrihydroxyde 5550
ferri-oxalaat 5551
ferri-oxyde 5552
ferrisulfaat 5556
ferrizout 5555
ferro 5572
ferro-ammonium-sulfaat
  5573
ferrobromide 5574
ferrocarbonaat 5575
ferrochloride 5576
ferrocyaan 5562
ferrocyaankali 10650
ferrocyaankoper 3516
ferrocyaan-uraan 14364
ferromagnetisme 5564
ferro-oxalaat 5577
ferrosulfaat 3499, 5580
ferrotypie 5568
ferrozout 5579
fers pointillés, à ~ 5582
festoendroger 5584
feuilleton 5585
feuilleton-gedeelte, in
  het ~ 5586
fiat geven 10020
fiatteren 10020
fiat voor afdrukken 7228
fiber 14522
fibrillen 5598
fibrilleren 5599
fibrillering 5600
figuren versierd, met ~
  7011
fijne korrel 5671
fijne lijn 6750
fijnetsing 5669

fijnheid 5676
fijnheidsanalyse 10015
fijnheidsmeter 5677
fijnheidsmeting 5678
fijn kalfsleer 7944
fijnkorrel-emulsie 5673
fijnkorrelige emulsie 5673
fijnkorrel-ontwikkelaar
  5672
fijn lijntje 5674, 6748
fijn papier 5679, 13412
fijn raster 5680
fijn schrijfpapier 3963,
  13413
fijn stellen 14935
fileet 6182
filet 5628
film 2987, 5636, 5637, 7950
filmapparaat 12215
filmclip 5641
filmcorrectie 3602
filmdrager 5644
filmdrukmachine 5647
filminktwerk 9790
filmklem 5641
filmlegger 1315
filmpje 5638, 12517
filmplaat 5644
filmraster 8741, 12216
filmraster aanbrengen,
  een ~ 13845
filmraster-procédé 1318
filmvormend 5646
filmvormer 5642
filter 5654
filterfactor 5658
filterfactormeter 5661
filterhouder 5659
filterwerking 5656
filtraat 5662
filtreerpapier 5660
filtreren 5663
financieel redacteur 2866
Finch-klem 5667
fixatief 5725
fixeerbad 5736
fixeermiddel 5725
fixeerzout 5737
fixeren 5722, 5723, 5724
flan 5846
flanel 5754
flaptekst 1647
flauwlijnen 5526
flens 5752
flensverbinding 5753
fletse afdruk 5436
flexibel 5820
flexibele band 8153
flexodruk 5828
flexografie 5828
flexografie-inkt 5825
flexografische inkt 5825
flexo-indrukwerk 615
flexo-inkt 5825
flexopers 5829
flintglas 5833
flip-flop 5836
floorpost 6444, 9597
floorpostpapier 6444

floroglucine 10200
floroglucine-oplossing
  10200
floroglucinol 10200
floxine 10201
fluorescentie 5888
fluorescentiebuis 5892
fluorescentielamp 5892
fluorescentieprocédé 5889
fluorescerend 5890
fluorescerende inkt 5891
fluorescerend papier 5893
fluorwaterstofzuur 7106
flushing-procédé 5904
flying paster 5927
focus 5950
foedraal, (perkamenten) ~
  6040
fo(e)lie 5961
foezelolie 565
föhn 5456
folder 5967
foliëren 1381
folioformaat 5994
fond 1001
fondscatalogus 13185
font 6103
foot 10211
footcandle 6016
Ford-beker 6030
Ford-cup 6030
formaat 4375, 6052
formaatbanden 4064
formaatbekken 14464
formaatbepaling 11906
formaatberekening 11906
formaat maken 8561
formaatstrippen 4064
formaat van een vel
  (papier) 12487
formaatverandering 2616
formaatwit 6262
formaldehyd(e) 6048
formaldehydhars 6049
formaline 6050
formaten van enveloppen
  12498
formaten van kaarten
  12497
formaten van karton 12495
formaten van register-
  papier 12499
formeur 14413
formol 6050
formule 6083
formuleren 6084
formulier 6046
formulieren 2106
formulieren zonder carbon
  2324
fortran 6088
fosfor 10203, 10207
fosforescentie 10204
fosforoxychloride 10208
fosforoxylchloride 10208
fosfortrichloride 10209
fosforzuur 10205
foto 10240
fotoafdeling 6276, 10242

fotoafdruk 10270
fotocel 10215
fotochemie 10216
fotodiode 10224
foto-elektriciteit 10230
foto-elektrisch 10225
foto-elektrische cel 10215
foto-elektrische diode
   10224
foto-elektrisch effect 10228
foto-elektrische kathode
   10214
fotograaf 10241
fotografie 10240, 10252
fotografie met vol-
   uitdrukkend materiaal
   10860
fotografische emulsie 10243
fotografische film 5636
fotografische koppenzetter
   6906
fotografische plaat 10248
fotografische proeven
   10250
fotografische retouche
   10251
fotografische zetmachine
   10279
fotografische zwartplaat
   1490
fotografisch moederbeeld
   10246
fotografisch papier 10266
fotografisch smoutzetsel
   10256
fotografisch zetsel 3061
fotografisch zetten 10278
fotogravure 10253
fotokathode 10214
fotokatode 10214
fotokopie 10270, 10273
fotolithografie 10257
fotomateriaal voor de
   reproduktietechniek
   10891
fotomatrijs 6096
fotomechanisch 10258
fotomontage 10261
fotopagina 10311
fotopapier 10266
fotopisch zien 10267
fotopolymeer 10268
fotopolymeerplaat 10269
fotoproeven 10250
fototypografie 10281
fotozetmachine 10279
fotozetsel 3061
fotozetten 10278
Fourdrinier-machine 6127
fouten-aangevende code
   5246
fouten aanstrepen 8633
fouten-corrigerende code
   5245
foutenfrequentie 5247
fouten-herstellende code
   5245
foutenpakket 5243

fouten-signalerende code
   5246
fouten-verbeterende code
   5245
foutief exemplaar 4094
foutieve pagina's ver-
   wijderen en vervangen
   2260
foutloos zetwerk 10126
foutloze pagina 10126
fout van de auteur 888
fout van de schrijver 888
frame 6144
frankeermachine 6149
Frankforter hoogte 6948
Franse alfabet 6167
Franse lelies versierd,
   met ~ 4073
Franse titel 1217
Frans marokijn 6179
Fraunhoferse lijnen 6150
frees 11733
freesmachine 1048, 11732
frequentie 6186
frequentiemodulatie 6188
frequentieverdeling 6187
frezen 8939, 11737
frictie 6194
frictiekalander 6193
frictiekoppeling 6194
frisket 6204
frontispice 6211
frontplaat 5424
front-uitleg 6208
ftaalzuur 10284
ftalocyanine-pigment 10285
fumaarhars 6246
functiecode 6249
fundament 1272
fungicide 6251
furfural 6257
furfurol 6257
furol 6257
fysische eigenschappen
   10286

gaas 6342, 9164
gaffel voor de cilinder 3936
galei 6278
galeien-regaal 6284
galeirand 6281
galeiregel 6287
galeischuif 6286
gallon 6290
galluszuur 6289
Gally-pers 6292
galvanisatie 10469
galvanisch bad 5064
galvanische neerslag 5028
galvaniseren 5063
galvanisering 10469
galvaniseur 5073
galvano 5071
galvanometer 6293
gamma 6296
gang 10019
gang brengen, op ~ 12160,
   13788
gangetje 429

gappen 10355
Garalden 6308
garenloosbindmachine
   13694
garenloze binding 208
garenloze bindmachine
   10119
garennaaimachine 12197
gas 6310
gasachtig 6316
gasbrander 6312, 11703
gaslichtpapier 6319
gasontladingslamp 4419,
   5892
gasroet 2622, 6311
gasverwarming 6317
gasvormig 6316
gat 7017
gatenspoor 2621
gauffreerpers 5105
gauss 6339
gazet 6344
geabsorbeerde inkt 10097
geadsorbeerde stof 235
geannoteerd exemplaar 848
geanodiseerde aluminium-
   plaat 650
gearceerd stempel 978
geautoriseerde versie 884
geautoriseerde vertaling
   883
gebaksdoos 2149
gebedenboek 10708
gebichromateerde colloïden
   4301
gebichromateerde gelatine
   4302
gebichromateerde gom
   4304
gebichromateerde lijm 4303
gebitumeerde papieren 822
geblazen lijnolie 1625
geblazen oliën 1626
gebleekte celstof 1540
gebleekt kraft(papier) 1539
gebleekt sulfietpapier
   14729
geblokkeerd 11833
geblokkeerde letter 14095
geblokkeerde regel 14096
geblokkeerd zetsel 1593
gebluste kalk 2167
gebogen 3847
gebogen plaat 3848
gebonden boek 1815, 2401
geboorten, huwelijken,
   overlijden 1440
gebrande omber 2089
gebrande siënna 2088
gebrocheerd 9900, 13169
gebroken kleur 4414, 12209
gebroken letters 1939
gebruik maken van staand
   zetsel 10305
gebruiksaanwijzing 7463
gebruikt 12043
gecalculeerde tijd 432
gecapitonneerde platten
   9844

gecarboniseerde
   formulieren 2321
geciseleerd 11472
geciseleerde sne(d)e 6494
gecoate lens 2982
gecompliceerd zetwerk
   3276
gecomprimeerde lucht 3322
geconcentreerd azijnzuur
   6418
geconcentreerd zuur 3355
geconditioneerd papier
   3368
geconjugeerde brand-
   punten 3386
gecorrigeerde band 2897
gecorrigeerde regel 3594,
   3599
gecorrodeerde laag 3614
gecoucheerd offsetpapier
   2984
gecoucheerd papier 2986
gecrêpt papier 3729
gedeeltelijk afgeschermd
   licht 4382
gedeeltelijk ontsloten celstof 12097
gedeeltelijk uitgedekte
   autotypie met verloop
   9720
gedekt grijsbord 2726
gedenatureerde alcohol
   4162
gedeponeerd handelsmerk
   11406
gedifferentieerd stukloon
   4343
gedistilleerd water 4479
gedreven 11472
gedrukt bij ... 10803
gedrukte bedrading 10800
gedrukte inktkleur 14292
gedrukte kopij 6859, 11478
gedrukte schakeling 10800
gedrukt, het worde ~ 7228
gedrukt op de persen van
   ... 10803
gedwongen beweging 6028
geel bloedloogzout 10650
geelfilter 14916
geel koper 1859
geelplaat 14924
geel worden 14917
geëmailleerde band 5139
géén nieuwe alinea 11811
geflushte pigmenten 5901
geforceerde droging 6026
gefrictioneerd papier 6197
gegalvaniseerd 10433
gegevens 4005
gegevens aanbrengen 8642
gegevensbank 4006
gegevenstransmissie 4016
gegevensverbindingen 4011
gegevensverwerking 4013
gegevensverwerking op af-
   stand 13630
gegevensverzameling 4008,
   4015

gegolfde lijn 14622, 14625
gegolfde randen 12534
gegolfd karton 3619
gegomd band 6718
gegomde envelop(pe) 210
gegomde klep, met ~ 6713
gegomd etiket 6714
gegomd papier 6715
gehalte aan droge stof 12764
gehamerd papier 6809
geheel automatisch 6245
geheel doornaaien 426
geheel gesloten inktbak 5150
geheel leer gebonden 6231
geheel leren band 8012
geheel linnen gebonden 6231
geheel metalen camera 433
geheimtekens 3787
gehele breedte van de pagina, over de ~ 440
gehele manuscript in de zetterij 430
geheugen 13230
geheugencapaciteit 13234
geheugenelement 2498
geheugenplaats 8302
gehollanderd 6184
geïllustreerd abc-boek 10313
geïllustreerd artikel 7163
geïllustreerde uitgave 7164
geïmpregneerd papier 7206
geïndexeerd sequentieel 7275
geïntegreerd circuit 7477
geïntegreerde densiteit 7478
geïntegreerde dichtheid 7478
geïntegreerde schakeling 7477
geïnterlinieerd zetsel 7980
geiteleer 6493
gejodeerd collodium 7560
gekartelde schroef 7831
gekartonneerd 7240
gekleurde pigmenten 3161
gekleurde rotatieboekdruk-inkt 3232
gekleurde vezels 3136
gekleurd papier 3137
gekleurd pigment 4926
gekleurd schapeleer 3138
gekneusde plek in het papier 1979
geknoei 1059, 11964
gekoekte letters, aan elkaar ~ 2151
gekoetst karton 14411
gekookte lijnolie 1695
gekoppeld 9599
gekornd papier 6543
gekorreld papier 6543
gekrabbel 11964
gekristalliseerd 3793
gel 6352

gelaagd 7952
gelaagde film 9175
gelaagde papieren 13253
gelamineerd 7886
gelamineerd papier 7888
gelatine 6356
gelatinefilter 6360
gelatinelaag 6357
gelatinelijm 623
gelatinereliëfplaat 6363
gelatineus 6366
geldkolom 3246
gele drukplaat 14924
gelegenheidsdrukwerk 2460
gelei 7654
gelei-achtig 6366
geleidbaarheid 3374
geleidebanden 11807
geleidegaten 5517
geleiderol 12546
geleiderollen 1883, 3375, 14655
geleiders 6700
geleider van de matrijzen-schuif van de distributor 4504
geleidesignaal 3236
geleidestangen 6700
geleiding 3374
geleren 6353
"gelijk-af" monteren 1600
gelijk-afsnijden 3861
gelijke en gelijktijdige beweging 1086
gelijkmaken 8200
gelijkmaking 5222
gelijkmatig doorzicht 14304
gelijkmatigheid van het oppervlak 757
gelijkrichter 11317
gelijkstootmachine 7687
gelijkstoten 7685
gelijkstoter 7686
gelijkstroom 4400
gelijkstroommotor 4401
gelijkteken 5224
gelijktijdige verwerking 3360
gelijk uitvullen, (regels) ~ 12806
gelijkwaardige belichting 5229
gelijmd papier 12490
gelijmd papier, in de stof ~ 1264
gelijnd papier, (batonné) ~ 11775
gelinieerd papier 11775
gelukwensen 6626
gemacereerde strocelstof 8430
gemakkelijk zetsel 9625
gemakkelijk zetwerk 5492
gemarbreerd papier 9115
gemarmerde sneden 8615
gemarmerd papier 8617
gemeenschappelijke aan-drijving 3004

gemeenschappelijke machinetaal 3250
gemengd zetsel 9003
gemeten scherpte 174
gemiddeld doorgelaten uitvalpercentage 955
gemiddelde 951
gemiddelde afwijking 8715
gemiddelde deeltjesgrootte 957
gemiddelde letterbreedte 5137
gemiddelde produktie 956
gemiddelde snelheid 961
gemiddelde steekproef-grootte 960
gemiddelde waarde 8719
gemiddeld kwaliteitsniveau 959
gemonteerd op lood 9141
gemonteerd op metaal 9141
gemonteerd zonder facet 9138
genaaid op band 12201
genaaid op gaas 12202
genaaid op touw 12200
generator 6377
generator van programma's voor het lijsten van gegevens 11470
genereerprogramma 6377
genol 8867
genormaliseerde formaten 7521
genummerd exemplaar 9448
geolied karton 9553
geolied papier 9554
geparaffineerd papier 9967
geparelde afdruk 9116
geparfumeerde drukinkten 10138
geparfumeerd papier 11917
gepenetreerde inkt 10097
geperforeerde plaat 10128
geperst papier 5101
geperst pergamijn 5100
geperst watermerk 7207
gepigmenteerde aniline-inkt 10331
gepigmenteerde flexo-inkt 10331
geplaatst, ten opzichte van elkaar ~ 7723
geplakt 10024
geplakt bristolkarton 10025
geplakt karton 1921, 10022
geplakt omslagpapier 10026
gepolariseerd licht 10541
gepreegd papier 5101
gepresensibiliseerde druk-plaat 10732
geprogrammeerd snijden 10920
geraadpleegde literatuur 1361
gerasterde foto-afdruk 11978
gereedschapsrek 13902

gereinigd wolvet 7901
geretoucheerde foto 5443
gerezen spatie 11226
gerezen wit 1493, 4716
geribbelde rol 12146
geribbelde trekrol 12146
geribbeld papier 11468
gerief 6262
geriefkast 6263, 8002
geruisloos papier 9386
geruisloos programma-papier 9386
geruit leer 4291
geruit papier 12964
gesatineerd, aan vellen ~ 12262
gesatineerd courantdruk 6787
gesatineerd papier 2190, 6445
gescheiden 9513
gescheiden inktbak 12898
gescheiden uitleg 12897
gescheurde vellen 13922
geschiktheid tot vouwen 5819, 10482
geschiktheid van het papier tot bedrukken 10796
geschikt om werk te doen 8
geschikt tot werken 8
geschrift zonder titel 586
gesigneerd exemplaar 896
gesiliconiseerd karton 12411
gesiliconiseerd papier 12418
gesloten booglamp 5149
gesloten gegevensketen 2091
gesloten kring 2931
gesloten lus 2931
gesloten, met plakband ~ 13580
gesloten negatief 13807
gesluierd 8136
gesluierd negatief 5959
gesneden, op halve papier-baan ~ 12038
gespiegelde binaire code 11388
gespikkelde sneden 12949
gespiraliseerde inktrol 12877
gespiraliseerde koker 3483
gespouwen rundleer 2038
gespouwen schapeleer 12527
gesprenkelde sneden 12949
gesteldheid van het papier-oppervlak 1451
gestookte lijnolie 1695
gestreken biljet-litho(-papier) 2983
gestreken offsetpapier 2984
gestreken papier 2986
gesulfoneerde castorolie 14091

gesulfoneerde olie 13382
gesulfoneerde wonderolie 14091
getah pertsja 6733
getal 5607
getarifieerde arbeid 8721
geteerd papier 13608
getekend exemplaar 896
getijdenboek 1767
getypte kopij 14228
getypt manuscript 14228
geveegd beeld 1648
geverfde sneden 3135, 13009
gevergeerd papier 7876
gevergeerd schrijfpapier 7878
gevernist papier 14403
gevoeligheid 12112
gevoeligheidskromme 2637
gevoeligheid van een emulsie 12851
gevoelig papier 12117
gevouwen, in katerns ~ 5964
gevouwen, in langs-richting ~ 8334
gevulcaniseerde olie 14523
gevuld papier 5620
gewapend beton 11425
gewassen tekening 14547
gewatteerde envelop(pe) 7659
gewatteerde platten 9844
gewatteerde zak 7659
gewicht per duizend vel 9390
gewicht per riem 1209, 9390, 11294
gewicht van de letter 14672
gewone drukinkt 9424
(gewone) gloeilamp 7241
gewone uiteinden 9675
gewone zwarte inkten 1756
gezangenboek 2752
gezegeld papier 10163
gezel 7701
gezet op de machine 8734
gezuiverde aardwas 9818
gezuiverde stof 65
gezuiverde uitgave 5378
gids 6698
gietapparaat 2443
gieten 2435, 10694
gieten van rollen 2452
gieter 2441
gieterij 14193
gietgal 336, 5810
gietgallen 12475
gietijzer 2455
gietkoker voor rollen 11643
gietkruisje 13043
gietlaag 2437
gietlepel 2447
gietmachine 2440, 14181
gietmerk 10369
gietmond 10684
gietpolis 6120
gietpot 8836

gietprop 7657
gietprop afbreken, de ~ 1892
gietraam 2446, 2453
gietrad 9125
gietschaal 3695
gietsel 2442
giettemperatuur 2454
gietverpakking 11030
gietvorm 9118
gietvormblad-excentriek 9122
gietvormkoeling 9124
gietvorm voor loodstaven 10334
gietvormwisser 9135
gietwiel 9125
gietwiel met zes giet-monden 12480
gietwielslede 9126
gilbert 6398
gilsoniet 6411
gips 3774, 6738
glacé 5835
glacépapier 5835
gladde rol 6447, 12647
glad geworden rol 6446
gladheid 5686
gladheidsmeter 12646
gladheidsmeting 12645
gladheid (van papier) 12643
glad maken 2037, 2081
gladmaking met de (agaat)-steen 5834
glad oppervlak 6441
gladtand 2079
gladwals 12878
gladwerk 12995
gladworden van het rubber-doek 6449
glans 6452, 8419
glansafdruk 6462
glansbord 10756
glansinkt 6453
glanskalander 6193
glans-koperprocédé 1095
glanslak 6455
glansmeter 6426, 6454
glansplaat 5570
glansvernis 6455
glanszwart 1915
glanzen 5569
glanzend 6457
glasbol 2060
glaselektrode 6428
glaskogels 6430
glasmatrijs 6432
glasnegatief 6433
glasplaat 6435, 10420, 12528
glaspositief 6436
glasraster 6438
glasstaaf 6437
glasvensterpapier 14498
glasvezels 6429
Glauberzout 12719
glazen bol 2060
glazen knikkers 6430
glazen raster 6438
(glazen) roerstaaf 6437

glazige plekken 5715
glazig oppervlak 6441
glijbaan 2825
glijlager 10390
glijmiddel 12562
glimmerdeeltjes 12304
gloeilamp 5012
glucose 4246
glyceride 6479
glycerine 6482
glycerinezuur 6478
glycerol 6482
glyceroltrinitraat 9376
glycine 6485
glycol 5309
glycolen 6489
glycolzuur 6488
glykose 4246
goede kopij 5440
goederenlift 8093
goede staat, in ~ 9259
goed gesloten doorzicht 14676
goedkeur-criterium 64
goedkope uitgave 2660
golf 5911
golfkarton 3619
golflengte 14615
golflijn 14622, 14625
golven 2028, 3026
golvende randen 12534
golvend papier 14624
gom 6706
gom-ets 6711
gomlaag 6712, 6725
gommeermachine 6721
gommen 6719, 6720
gomrand 6472
gomstrepen 6726
gom zetten, in de ~ 6720
goniometrische verhoudingen 14038
goot 2825
gootje 6734
gootsteen 4228
gordijnsluiter 5947
Gordon-pers 6526
Gothische letter 6528
gouache 6530
goud-afval 12508
goudbromide 6498
goudbrons 6499
goudbronspoeder 6515
goudchloride 6517
gouden opdruk 6496
goudfo(e)lie 6509
goudinkt 6507
goudkussen 6501
goudmes 6508
goud-opdraagkwastje 6404
goudoplegger 9839
goud op snee 6412, 6413
goudoxyde 6511
goudpapier 6512
goudronné 13607
goudronnépapier 13607
goudslagershuid 6495
goudslagersvlies 6495
goudstempel 6496

goudvoordrukinkt 12485
graadboog 10986
graad Engler 5177
graad Kelvin 7744
graadteken 4108
graadverdeling 4518
graaf 6577
gradatie 6534
gradatiecurve 2637
gradatiekromme 2637
gradenboog 10986
graden Celsius 2530
graden Fahrenheit 5435
gradient 6535
grafiek 4257
grafiekpapier 6573
grafiet 6571
grafische industrie 6565, 10831
grafische voorstelling 4257
grafisch ontwerper 3240, 6567
grafisch tekenaar 3240
grafiteerborstel 1483
grafiteren 1482
gram 6556
gramgewicht 1209
grammolecule 9042
gratis exemplaar 6156
gratis plaatsen 4029
graveerkop 5193
graveerkussen 5189, 9838
graveermachine 5195
graveernaald 5281
graveren 5182, 13900
graveren in een laag 2992
graveur 5186, 5688
gravis 6577
gravure 5191, 8172
Gray-code 11388
greep 4704, 6826, 6829
greep van de letterkast 4704
grein 6538
greinbad 6549
greinen 6546
greinen met een zandstraal 11860
greinknikkers 3037
greinkogels 3037
greinmachine 6550
grendel 8305
grenskwaliteit voor de af-nemer 8378
grenslaagfotocel 1173
grensvlakactieve stof 13428
grensvlakactiviteit 7493
grensvlakspanning 7494
grenswaarde 8152
Griekse alfabet 6618
Griekse letters 6619
Griekse lettertekens 6619
Griekse rand 6620
grijper 6650
grijperkant 6654, 6911
grijperstang 6652
grijperwit 6656
grijs boekbinderslinnen 6627

grijsbord 1408, 2726
grijs bord 9301
grijs linnen 6627
grijstrap 6629
grijswig 6630
grijze afdruk 5436
groef 6666
groefcilinder 6669
groef maken, een ~ 6668
groenfilter 6623
groepopstelling 3379
groepsaandrijving 6685
groepsgewijze verwerking 1222
groeven 6668
grof grein 2975
grof raster 2976
grofrasterautotypie 2978
grofrasterautotypie als lijn-cliché gedrukt 11140
grofrastercliché 2978, 9324
Grolier-band 6664
grondeerkwastje 6424
grondeermiddel 209
gronderen met eiwit 6420
grondformaat 1206
grondinkt 1185
grondkleur 1025, 1190
grondkoper 1191
grondlijn 1196
grondpapier 1685
grondpapier voor ge-paraffineerd papier 14628
grondplaat 1278
grondsteen van de koller-gang 1282
grondstoffen 11258
grondwerk 1279
grootkorps-apparaat 4460
grootkorps-gietmachine 4467
grootkorpsletter 4471
grootkorpsmatrijs 4468
grootste papierformaat 8710
grootte van het druk-fundament 1281
grootte van het vormraam binnenwerks 7416
grote kop 1131
grote kopregel 1131
grote papierzak 6613
grotere omvatting 9099
Grotesk 6675
grove den 10362
grove lijnen 6884
grove linnenpersing 3702
GSA-systeem 6688
guilloche 6701
guillocheermachine 6703
guillocheren 6702
gulden snede 6505
guttapercha 6733
guttegom 6294

haak 3297
haakjes, (ronde) ~ 9998

haak leeg maken, de ~ 5119
haaknaald 8356
haaks 10153
haaks gesneden papier 12965
haarhygrometer 6742
haarklover 3235
haarlijn 6750
haarlijntje 5674, 6746, 6748
haarspatie 6752
haartje 1248, 6747
haartjes 6267
haastkarweitje 11818
hakselmachine 2578, 2728
halfband 6755
halfchemische stof 12097
halfchemisch procédé 12096
halfcontinu inktwerk 3431
halfdoorlatende spiegel 13996
halfdrogende olie 12100
halfduplexkanaal 6759
halffranse band 6755
halfgebleekte celstof 10017, 12094
halfgebonden 6756
halfgeleider 12099, 12766
halfgelijmd 6768
halfjaarlijks 12093
half kastlijntje 4002
halfleer 11134
halfleer gebonden 11129
halfleren band 6755, 11135
halfleren band, in ~ 11129
halflinnen band 6758, 11130
halfmaandelijks (verschijnend) blad 12101
halfmat papier 4989, 5180
halfperkamenten band 11141
halfschaduw 10107
halfstof 6770, 11026
halfstof uit loofhout 6882
halfstof uit naaldhout 12748
halftint 6773
halftoon 6773
halftoonbeeld 6784
halftoon-model 6777
halftoon-origineel 6777
halfunciaal 6798
halfvet, met ~ 58
halfvette letter 12095
halfvette lijn 8757
halfwas 13956.
14160
halfwekelijks blad 12104
halo 6754, 6801
halo-effect 6802
halogeen 6804
halogeenlamp 6805
halogeenzilver 6806
halovorming 6754
halve kast 6757
halve regel 1889
halve riem 13879

halveringsmethode 1399
halveringsselectie 1399
hamsteraar 10357
hamsteren 10355, 12610
hand 5718
handaandrijfinrichting 1174
handaandrijving 4746
handaanlijmapparaat 10030
handafdruk 6830
handarbeid 6851
handbediening 6820
handbeschermer 6689, 6843
handbesturing 6820
handboek 3260, 6816, 14377
handbord 8932
handbordmachine 7515
handdroog worden 12154, 12176
handel 8078
handelaar in oude boeken 12045
handelaar in tweedehands boeken 12045
handelsdrukkerij 7679
handelsdrukwerk 2107, 7680
handelsmerk 13944
handfotozetsel 10256
handgekleurd 6817
hand gekleurd, met de ~ 6817
handgeschept papier 6834
hand gezet, met de ~ 6845
handgietvorm 6836
handgreep 6829
hand ingelegd, met de ~ 6825
handinleg, met ~ 6825
handje 5718
handletter 6102
handlijmapparaat 10030
handmatrijzen 12786
handmatrijzenkanaal 12788
handmatrijzen-verzamelaar 12789
handpers 6839
handpikeersel 6821, 10045
handrol 1871, 8264
handslijper 8081
handstempelen 6849
handvergulden 6405
hand vervaardigde afdruk, met de ~ 6830
handvol 6826
handwerk 6851
handwiel-snijmachine 6850
handzetsel 6818
handzetter 6819
handzetterij 3292
hangende camera 9759
hangende transportbaan 9760
hangende transportband 9760
Hansageel 6856
harceren 6885
harcering 6740
hard 3447

hard bankpost 1723
hard bankpostpapier 1723
hardbord 12759
harde legger 6873
harden 6861
hardend fixeermiddel 6865
hardend zuurfixeerbad 128
harde punt 6860
harder 6862
harde rol 13808
harde stofknopen 6869
harde werker 14706
hard gelijmd papier 6875
hardheid 4112, 6871
hardheid in Shore 12321
hardheidsgraad 4112
hardheidsmeter 6872, 10417
hardheid volgens Brinell 1918
hardheid volgens Shore 12321
hardhout 6881
hardingsbad 6864
hardingsmiddel 6862
hardloper 13466
hard negatief 6870
hardpostinkt 1722
hardpostpapier 13374
hard rubber 6874
hardware 6877
hardworden van het rubberdoek 6866
harend papier 6269
haren van matrijzen 6743
harig papier 5882
Harley-band 6883
harmonika-vouw 73
hars 10381, 11501
harsachtig hout 11505
harsester 5260
harsgehalte 11685
harsgelijmd papier 11690
harshoudend hout 11505
harslijm 11689
harslijming 11502
harsolie 11687
harspoeder 5283
harsverbinding 11503
hartje 12196
H & C-stijfheidsmeting 6824
H & D-curve 6822
H & D-kromme 6822
H & D-snelheid 6823
H & D-stijfheidsproef 6824
heat-seal-papier 6932
heat-set inkten 6934
hebben we niet 4557
Hebreeuws 6944
hechtcilinder 13176
hechtdraad 13179
hechtdraadgeleiders 7412
hechtdraadvoeder 13171
hechten 8097, 14780
hechten van bloks 9845
hechten van de inkt 583
hechtgaas 9164
hechthoorn 13172
hechting 204

kaliumbromide 10634
kaliumcarbonaat 10636
kaliumchloraat 10637
kaliumchloride 10638
kaliumchromaat 10641
kalium-chroomaluin 2808
kalium-chroomsulfaat 2808
kaliumcitraat 10644
kaliumcyanide 10647
kaliumferricyanide 10649
kaliumferrocyanide 10650
kaliumhydraat 10655
kaliumhydroxyde 10655
kalium-hypochloriet-
  oplossing 7649
kalium-ijzeraluin 5554
kalium-ijzersulfaat 5554
kaliumjodide 10656
kaliumlactaat 10657
kalium-metabisulfiet 10658
kaliumnitraat 10660
kaliumoxalaat 10661
kaliumpercarbonaat 10662
kaliumperchloraat 10663
kaliumpermanganaat 10664
kaliumpersulfaat 10666
kalium-platinachloride
  10640
kaliumpolysulfide 10677
kaliumrhodanide 10676
kaliumsilicaat 10670
kaliumsulfaat 10671
kaliumsulfide 10672
kaliumsulfiet 10673
kaliumsulfocyanide 10676
kaliumthiocyanaat 10676
kaliwaterglas 10670
kalk 8145
kalkeerlinnen 13937
kalkeerpapier 13938
kalkhydraat 2167
kalkspaat 2163
kalligraaf 2217
kalligrafie 2218
kalomel 8793
kam 2227
kamertemperatuur 11676
kamfer 2251
kamferolie 2252
kanaal 2620
kanaalingang 2623
kan afgedrukt worden
  11810
kanten 4961
kantnoot 8629
kantoorbenodigdheden
  13076
kantoorboek 76, 1506
kantoorboekenfabriek 77
kantoor-offsetmachine
  12627
kantoortje van de chef
  11692
kantoorwerk 2912
kanttekening 635, 8629
kant van de gevoelige laag
  5136
kaolien 2711
kap 2860

kapel 3819
kapinrichting 12277
kapitaal 8557
kapitaalbandje,
  (bovenste) ~ 6894
kapitaalhoogte 6947
kapitaal met klein kapitaal
  2295
kapitaal met onderkast
  2293
kapitaal zetten 7743, 11068
kapitalen 2287
kapje 2272, 2860
kapmes 7814
kapot buigen 1820
kapot geëtst 9751
kapot geëtst cliché 2706
kapotte letter 1228
kappagetal 7729
kappen 3885, 11295
kar 2364
karaandrijving 1276
karakter 8052
karbeweging 2366
kardinale punten 2351
kardoes 2395
kardoespapier 2395
karikatuur 2392
karmijnrood 7290
Karolingische minuskels
  2362
karrebaan 1283
karrebanen 2367
karton 2342, 9899
kartonbuigmachine 1320,
  2343
kartonfabriek 1662
kartonformaten 12495
karton met linnen achter-
  zijde 2956
karton, met linnen
  beplakt ~ 2956
kartonnagebord 1831
kartonnagefabriek 9902
kartonnagekarton 1831
kartonnen doos 2344, 2378
kartonnen koker 2348
karton uit twee lagen
  14150
karton voor bierviltjes 2979
karton voor huwelijks-
  aankondigingen 14663
karwei 6635
kas 1934
kassablok 2421
kassablokmachine 2422
kasstuk 1341
kast 2396, 2398
kastanjehout 2701
kastfout 14881
kastgreep 4704
kastindeling 2405, 7960
kast inleggen, een ~ 7964
kastletter 6102
kastlijntje 4001
kastpapier 12287
kastvakje 1829
kast voor een kaartsysteem
  5619

katalysator 2463
katern 12399
katerns, in ~ 7448
katern van drie 13655
katern van vier 11145
katern van vijf 11162
katern van zes 12204
kathedraalband 2478
kathode 2480
kathodestraalbuis 2481
kation 2483
kationactieve stoffen 2484
katoen 3636
katoenhaartjes 3638
katoenlinters 3638
katoenstofgehalte 3637
kauri 7732
kauri-butanolwaarde 7733
kaurigom 7732
kb-getal 7733
kb-waarde 7733
keel-element 13782
keel van de gietpot 13780
keep 9432
keerlaagfotocel 1173
keerstang 588
keervorm 14825
kegelmolen 7693
kegelstofmolen 7693
kegelvormig 3381
keilspatie 12802
Kemart-procédé 7745
kenmerk 2635, 5502
kenmerkende eigenschap
  5502
kennis en middelen ten
  dienste van de chef voor
  een efficiente bedrijfsvoering 10983
kennisgeving 247, 636
keramische inkt 2556
keramisch(e) transfer 2558
keren 14825
keren, over de kop ~ 5860
kerf 9333, 9433
kerf aan de zijde van de
  accenten 9334
kerfzijde 1296
kerkboek 10708
kern 3565, 8504
kernengeheugen 3574
kerosine 7751
kerstkaart 2759
ketel 1697, 7755
keten 2586
keton 7753
ketsen 3792
ketting 14544
kettingaandrijving 2583
kettingband 2584
kettingboeken 2579
kettingdrukker 2590
kettingformulieren 3428,
  5163
kettingformulieren,
  scheiding van ~ 2095
kettinggrijper 2587
kettinggrijperuitleg 2580
kettinglijnen 2589

kettingpuntraster 2582
kettingrad 12950
kettingsteek 7756
kettinguitleg 2580
ketting van de uitleg 4138
kettingvormige versiering
  2581
kettingwiel 12950
keukenzout 12683
keuring 7458
keuringskarakteristiek
  9629
keuze van de bladspiegel
  2751
kieskanaal 12071
kiezelgoer 4280
kiezelzuur 12414
kijk uit, de chef 2493
kilo 7797
kilogram 7797
kilogramkracht 7798
kilohertz 7795
kilometerband 14327
kilowatt 7799
kilowattuur 7800
kinderboek 2709
kinematische viscositeit
  7801
kinetische energie 7802
kiosk 9317
kist 1827
kitten 2525
klaar met zetten 442, 2934
kladgeheugen 13643
kladpapier 12005
klank 11254, 12650
klant 74, 2915
klapdeksel 5757
klassieke strijkmethode
  3469
klats 1640, 7757
klatsen 2601
klauwhamer 12248
klavier 12555
kleefband 215
kleefkracht 214, 13525
kleefkrachtige rollen 13533
kleefkracht (van de inkt)
  13524
kleefmiddel 213
kleefplaat 13083
kleefrand 6472
kleefruit 13083
kleefstof 213
kleefstofbak 10029
kleefstofpomp 10039
kleefstofschijf 10036
kleine advertenties 2881
kleine boekbinderstempels
  10164
kleine letter 8967
kleine letters 8380, 12329
kleine oplaag 12336
kleine oplage 12336
kleiner diafragma kiezen,
  een ~ 13212
kleiner etsen van de rasterpunt 3886

koperdrukpapier 10457
koperen hoek 1862
koperen hoekstuk 1862
koperen kaderlijn 1861
koperen letter 1868
koperen lijn 1866, 11772
koperen matrijs 1863
koperen plaat 1864
koperen preegstempel 1860
koperen spatie 1867, 3532
koperen vochtrol 1865
kopergehalte 3506
kopergraveerkunst 2594
kopergraveur 3522, 10435
kopergravure 3523
koperhuidje 12288
koperneerslag 3508
koperoxyde 3518
koperplaat 3521
koperspatie 3532
kopersulfaat 3533
kopie 2314, 10793
kopieerafdeling 8839, 10449
kopieerinkt 3548
kopieerlaag 6725, 11507
kopieerlaag aanbrengen,
    een ~ 2980
kopieerlaag uit soja 12796
kopieerlaag verwijderen
    2902
kopieerlamp 10818, 10834
kopieerlijm 10892
kopieermachine 6686, 13116
kopieerpers 3549
kopieerraam 10828
kopie maken, een ~ 3538
kopiëren, over elkaar ~
    13415
kopiïst 10447,
    13285
kopij 3534
kopijhaak 3547
kopijhouder 3544
kopij, met de machine
    geschreven ~ 14228
kopijvak 14675
kopijvoorbereiding 3551
kopje, vóórin gehouden ~
    12348
koplabel 6901
koplijn 6913
kopmarge 6909
kop over twee regels 11815
koppel 3668
koppeling 3669, 3670, 8212
koppeling voor de invoer-
    banden 13578
koppelteken 4003, 7143
koppelwoord 3320
koppenzetter 6906, 6907,
    13864
Koppermaandag 8377
kopregel 1121, 6905, 7322
kopschroef 2296
kopshout 3755
kopspijkers 14855
koptitel 6900
kop (van de pagina) 6890

kop van het aanlijm-
    apparaat 10036
kop volgt 6917
kopwit 6889
korn 6538, 13903
kornen 6546
kornraster 6552, 6553, 11531
korrel 6536
korrelraster 6552, 11531
kort accent 12323
kort begrip 3260
korte inkt 12328
korte inleiding onder de
    kop 6903
korte klinker 12340
korte letters 12329
korte regel uitvullen met
    vierkanten, een ~ 5157
kortheid 8037
korting 4425
kortsluiting 12325
korund 3624
kosten van het drukken
    10822
kosten van vervaardiging
    3631
Kostinsky-effect 7836
kostprijs 3631
kostprijssysteem 3630
koude kleur 3052
koudemail 3054, 5138
koudemailprocédé 3055
koudgeperst papier 3057
koudlakprocédé 3055
koudlijm 3056
koud stempelen 3051
koud zetsel 3061
kozo 9163
kraal 1243
kraalrandje 4569, 12976
kraanwater 13600
krabbelaar 11963
krabnaald 9266
krachteloos beeld 5791
krachteloze afdruk 5436
krachtpikeersel 14281
krachtplaat 1491
krachtstroom 5024
kracht van de letter 14671
krachtveld 5604
krachtverbruik 10705
kraft 7840
kraftcelstof 7841
kraftliner 7839
kraftpapier 7840
krammen 12988
krammetje 13040
krans 6801
krant 3958, 9309
kranteknipsel 9304
krantencliché 9320, 9324
krantenformaat 9319
kranteninkt 9308
krantenkiosk 9317
krantenkolom 9313
krantenletter 9316
krantenpapier 9321
krantenpapier voor
    illustratiewerk 6787

krantenpers 9318
krantenraster 2976
krantenreclame 9311
krantenrotatiepers 9318
krantenvouw 12038
krantenwereld 14851
krantenzetter 9303
krantformaat 9319
krantinkt 9308
krantletter 9316
krant van buitengewoon
    formaat 1520
kras 12601
krasnaald 4720, 12007
krasproef 11961
krasproeftoestel 11960
krat 3703, 6146
kreeften 11546
Kremserwit 5749
kreukelproef 3782
kreukeltoestel 3780
krijt 14731
krijtachtig 2606, 3732
krijtachtige inkt 2607
krijtlitho 2598
krijtpikeersel 2602, 8729
krijttekening 2598
krijttekening, op de wijze
    van een ~ 3708
krimp 12355
krimpen 12354
krimpen (van het metaal)
    12356
kring 6801
kringkeuze 2840
kringloop 3921
kristaldetector 3788
kristaldiode 3788
kristalgelijkrichter 3788
kristallijn 3789
kristallijne structuur 3790
kristallisatie 3791
kristalwater 14591
kroes 3777
kromme 3845
kromme regels 3739
krompasser 2222
krom staande regels 3739
kromtrekken 14545
kromtrekkende platten
    14908
kroonmoer 2456
kruipbanduitleg 3726
kruipen 3722
kruipen van het rubber-
    doek 1515
kruis 3749
kruisarcering 3756
kruisbodemzak 11682
kruisdiafragma 3769
kruisharcering 3756
kruisje 3746, 3955
kruiskop 3757
kruiskoppeling 14325
kruislijnraster 3759
kruisstop 3769
kruisverwijzing 3764
kruisvouw 2756, 11607
kruk 11719

krukas 3700
krul 3833
krullen 3834
krulling 3833
krulling aan de afloopkant
    van het papier 13549
krulproef 3836
kubieke centimeter 3799
kubieke decimeter 3800
kubieke (Engelse) voet
    3801
kubieke inch 3802
kubieke meter 3804
kubieke yard 3805
kuil 7265, 8387, 10379, 12472
kuil in een offset-rubber-
    doek 4177
kunst der kopergravure
    2594
kunst die alle andere
    kunsten bewaart 781
kunstdrukkarton 2981
kunstdrukpapier 2986
kunsthars 13495
kunstleer 791, 8014
kunstleerpapier 792
kunstlicht 793
kunstlinnen 785
kunstrubber 2063, 13496
kunststof-cliché 10411
kunststoffen 10412
kunsttaal 790
kunstterpentijn 14727
kunst van het graveren in
    koper 2594
kunstvezel 13494
kurkdoek 3576
kurkumapapier 14092
kursief 7621
kussen 6501, 9838
kussenvormige vertekening
    10361
kuut 175
kwaad zijn 10086
kwadraat 11108
kwadraat op twee vier-
    kanten 14138
kwadraat op vier vier-
    kanten 6130
kwaliteitszorg 11121
kwartaalblad 11136
kwartaal, per ~ 11137
kwartaaluitgave 11136
kwart gallon 8240
kwarto 11143
kwarto-formaat 11143
kwartslamp 11144
kwast 9870
kwastenvanger 7830
kwastetsing 1999
kwastje 6424
kwik 8788
kwikbromide 8790, 8791
kwikchloride 8792
kwikchloruur 8793
kwikdamplamp 8807
kwikjodide 8796
kwikversterker 8795
kwikzilver 8788

luchtbel-viscosimeter 2022
luchtblaasjes 311
luchtbuffer 318
luchtbuis 343
luchtcilinder 14323
luchtcirculatiesysteem 319
luchtdicht 356
luchtdoorlatendheid 344
luchtdoorlatendheidsmeter
  345, 4176
luchtdroging 331
luchtdroog papier 332
luchtdrukrem 313
luchtfoto 281
luchtgedroogd bruinpak
  328
luchtgedroogd bruin pak-
  papier 328
luchtgedroogde celstof 333
luchtgedroogd papier 329
luchtkamer 14381
luchtkanaal 343
luchtklep 347, 357
luchtkoeling 324
luchtkussen 325
luchtleiding(en) 349
luchtpen 347
luchtperspomp 354
luchtpomp 354
luchtpostenvelop(pe) 340
luchtpostpapier 341
luchttafel 326
luchtverontreiniging 351
luchtvochtigheid 7077
luchtvochtigheidsregeling
  4119
Ludlow 8401
Ludlow-zetter 8402
lukrake koppeling van
  onderwerpen 12137
lumbecken 8405
lumen 8407
lumen-erg 8408
luminantie 8409
luminantie-factor 8410
luminescentie 8411
luminescerende inkt 8412
lus 8354, 8355
lux 8420
luxe band 4155
luxe uitgave 4973, 7919
lyofiel 8425
lyofoob 8427

maalbak 7020
maalfijnheid 5676
maalgraad 4109, 6161
maalgraad-bepalingstoestel
  6160
maalsteen 6647
maalteken 8639
maalwerk 11385
maandblad 9096
maandelijks verschijnend
  blad 9096
maar raak zetten 12173
maaswijdte 8813
maat 4375
maatlat 11899

maatvastheid 4376
machinaal 8725
machinaal etsen 8445
machinaal geschept papier
  9131
machinaal gezet 8734
machinaal opzetbare vouw-
  doos 2381
machinaal vervaardigd
  8454
machinale vertaling 8467
machinebreedte 8468
machinecode 3331
machinecorrectie 8441
machine-etsing 8445
machineformaat 11239
machinefundering 1278
machinegalei 4152
machinegeschept papier
  9131
machinegestreken karton
  8434
machinegestreken papier
  8435
machineglad 8446
machineglad papier 8447
machinekalander 2184,
  12995
machinemonteur 5237
machine-olie 8397
machine-onafhankelijke
  taal 8451
machineopdracht 3337
machineopdrachtcode 3331
machineplaat 10758
machinerichting 8443
machinerol 8942
machinesnelheid 8466
machinestilstandtijd 4678
machinestrijkprocédé 8436
machinetaal 8453
machine voor het maken
  van kassabloks 2422
machine voor het
  vernietigen van docu-
  menten 12352
machine voor zwaar werk
  6936
machinewerkplaats 8465
machinewoord 3348
machinewoordlengte 14819
machinezeef 8469
machinezetsel 8438, 8732
machinezetter 8439, 9639
machinezetterij 3311
macro-instructie 8477
macro-opdracht 8477
magazijn 8479, 13238
magazijnartikel 13188
magazijnchef 13239
magazijnformaten 13192
magazijnframe 8482
magazijnframe-verstelling
  8484
magazijnfreem 8482
magazijnmeester 13239
magazijnorder 13189
magazijnsoort 13188
magazijnuitlaat 5249

magazijn voor onverkochte
  exemplaren 13229
magenta 8486
magenta-contactraster 8487
magenta-filter 8488
magenta-inkt 8489
magenta-maskermethode
  8490
magenta-maskerprocédé
  8490
magenta-plaat 8491
magenta-raster 8487
magere letter 8111
magneetband 8524
magneetbandgeheugen
  8530
magneetbandlezer 8528
magneetkaart 8503
magneetkern 8504
magneetkerngeheugen
  3574
magneetschijf 8508
magneettrommel 8512
magnesia 8500
magnesium 8496
magnesiumbisulfiet 8497
magnesiumcarbonaat 8499
magnesiumcliché 8498
magnesiumoxyde 8500
magnesiumplaat 8501
magnesiumsulfaat 8502
magnetische band 8524
magnetische inkt 8519
magnetische kaart 8503
magnetische plaat 8508
magnetische veldsterkte
  8518
magnetisch gecodeerde
  cheques 8516
magnetisch geheugen 8523
magnetisch kerngeheugen
  3574
magnetisch schijfgeheugen
  8510
magnetisch schriftlezen
  8520
magnetisch spoor 8531
magnetisch trommel-
  geheugen 8514
magnetisch veld 8517
magnetiseerbare inkt 8519
Maioli-band 8555
maken 11462
maken van bloks 9845
mal 6169
malachietgroen 8582
maleïnehars 8584
malen 6637
maling 1266, 11386
malingsgraad 4109, 6161
man aan de pers 10753
manchet 11002
manchon 3644
mandfles 2337
mangaan 8589
mangaandroger 8590
mangaansulfaat 8591
mangel 8699
mangelen 8595

manière criblée 8597
manière noire 8883
manilla-gekleurd schrijf-
  papier 11219
manilla-papier 8605
man/machine kaart 8606
mantel 1719
mantisse 8607
Manuaren 8609
manul-procédé 8610
manuscript 3534
manuscripthouder 4520
man van de planning 10926
map 9918
marge 8627
marge aan de kop 6909
marge voor de paskruizen
  8297
marginale 8629
markeervilt 4729, 8636
marmeraar 8618
marmerbad 1273
marmerbak 8626
marmeren 8621
marmerkam 8624
marmerpapier 8617
marmersnee 8615
marmerverf 8623
marokaan 9106
marokijn 9106
marokijnleer 9106
maroquin 9106
masker 8645, 8646
maskeren 8647
masker-methode 8647
masker-procédé 8647
maskertje 12924
massa-produktie 8653
Massey-strijkprocédé 8651
mat 8663
maten en gewichten 14673
materiaaldief 10357
materiaal uit een andere
  vorm 8095
mate van verestering 4110
mate van vochtigheid 4113
mat gecoucheerd papier
  4852
mat gesatineerd 4853, 5179
mat gesatineerd papier
  5180
mat gestreken papier 4852,
  8665
matglas 5956, 6217
mat kunstdrukpapier 8665
matlak 8707
mat oppervlak 8700
mat papier 8676
matrijs 5846, 8677, 8678
matrijzenbreedte 1869
matrijzencontrolestift 4501
matrijzendroger 8690
matrijzendrooginrichting
  8690
matrijzendroogkast 8690
matrijzengeleiders 837,
  8688
matrijzen geleverd door
  derden 1699

moduul 9031
moederbeeld 8656
moederplaat 8658
moeilijk sluitwerk 13809
moeilijk te zetten kopij 10088
moeilijk zetsel 8009
moeilijk zetwerk 3276
moer 9459
moerbeiboom 9163
moerbesboom 9163
moerbout 1714
moersleutel 14872
moet 5106, 6868, 7266
moet in het papier 9915
moet van het facet 5425
moet worden opgenomen 9217
moiré 9032
moiré-effect 9032
moisture-set inkt 9039
mol 9042
molair gewicht 9054
moleculair-disperse op-lossing 9053
moleculaire oriëntering 9052
moleculair gewicht 9054
molecule 9055
moleculenopeenhoping 2967
molekuul 9055
moleskin 9056
molette-watermerk 7207
molton 9057
molybdaat-oranje 9061
molybdeen-oranje 9061
monastraalblauw 9065
monastraalgroen 9066
mondstuk 9151, 9684
monochloorethaan 5306
monochromatisch 9076
monochromatisch licht 9077
monochroom 9078
monoclien 9080
monojoodethaan 5314
monomeer 9083
monomethyl-paramino-fenolsulfaat 8867
monomoleculaire laag 9081
Monotype 9087
Monotype-gieter 2441, 9075
Monotype-gietmachine 9088
Monotype-papier 9092
Monotype-toetsenbord 9091
Monotype-zetsel 9090
monster 11854, 12834
monsterboek 11855
monsterboekje 13456
monstercollectie 11855
monstername 11856
monsterneming 11856
monster uit de partij 9740
monstervel 12834
montaanwas 9095
montage 1597, 5767, 8736
montageband 843

montage op niet-actinisch papier 6503
montagepapier 6504
montageplaat 5766
montageraam 2651
montageruit 5766
montagetafel 11417
monteermachine 9145
monteren 1597, 9137, 13289
monteren met schroef-bouten 582
monteren zonder facet 1600, 5899
montuur van de lens 1170
moordenaar 6993
mordantvernis 1680
motor met kortsluitanker 12983
motor met sleepringanker 12574
mousseline 9216
mozaïek 7411
mozaïekband 9110
mozaïekdrukker 8694
mudéjar-band 9162
muller 9168
multiplex 10505
multiplexer 9189
multiplex karton 9174
multiplex voet 7890
multiplex voor clichévoeten 7887
multi-programmering 9196
Munsell-systeem 9202
muurreclame 14538
muziekdruk 9213
muziekgraveur 9209
muziekgravure 9210
muziekpapier 9211
muziekuitgever 9214
muziekzetsel 9207, 9215

naadje 6734
naaf 9256
naaibank 12195
naaien 12192
naaien met de hand 6846
naaien op de machine 8464
naaien over de gehele lengte 426
naaimachine 12197
naald 10524
naaldboom 3382
naalden 9269, 9270
naalden-excentriek 9268
naaldhout 12747
naaldhoutcelstof 12748
naaldhoutpulp 12748
naaldlager 9267
naaldvormig 112
naamlijst 4408
naamloos 652
naamregister 9229
naam van de drukker 7229
naam van de uitgever(ij) 7230
nabehandeling 10616
nabelichting 5762
nabelichtingslamp 5764

naberekening 3629
nabootsen 12440
nabuureffect 216
nacalculatie 3629
nachtploeg 9346
nadonkeren 3425, 3989
nadruk verboden 439
nafta 9235
naftenaat 9236
naftenaat-droger 9237
nagelijmd met dierlijke lijm 626
nagelijmd papier 14073
nagelvast 9414
nagelvaste inkt 9415, 11765
nagelvastheidsproef 5684, 11961
nakijken 3063, 11571
nakleven 289
nakopieereffect 3425
nalijming 13440
namaak 1691, 3658
namaak-watermerk 7180
namaken 3657
namelijk 14500
nano- 9231
nanometer 9232
nanoseconde 9233
narede 5219
naschrift 5219
nasteken 6848, 11808, 13900
natgrauw 13966
natgrauw papier 13966
nat-in-natdruk 14682
natpartij 14683
natrium 12668
natriumacetaat 12669
natriumaluin 12673
natrium-aluminiumsulfaat 12673
natriumbenzoaat 12675
natriumbicarbonaat 12676
natriumbichromaat 12688
natriumbisulfaat 12678
natriumbisulfiet 12679
natriumboraat 12680
natriumbromide 12681
natriumcarbonaat 12682
natriumchloride 12683
natriumchloriet 12684
natriumchromaat 12685
natriumcitraat 12686
natriumcyanide 12687
natriumformiaat 12690
natriumfosfaat, (drie-basisch) ~ 12709
natriumhexametafosfaat 12691
natriumhydrosulfiet 12694
natriumhydroxyde 12695
natriumhypochloriet 12696
natriumjodaat 12699
natriumjodide 12700
natrium-kaliumtartraat 12710
natriummetabisulfiet 12702
natriumnitraat 12704
natriumoxalaat 12705
natriumperboraat 12707

natriumperoxyde 12708
natriumrhodanide 12728
natriumsesquicarbonaat 12715
natriumsilicaat 12716
natriumstearaat 12717
natriumsulfaat 12719
natriumsulfide 12720
natriumsulfiet 12721
natriumtetraboraat 12680
natriumthioantimoniaat 12727
natriumthiosulfaat 7146
natroncelstof 7841, 12667
natronhydraat 12695
natronkraft 7840
natronkraftpapier 7840
natronloog 12695
natronprocédé 12666
natronwaterglas 12716
natsterk papier 14693
natsterkte in procenten 14694
natsterkte-percentage 14694
natte afval 14679
natte collodiumplaat 14680
natte-collodiumprocédé 3087
natte overdraging 14686
natte plaat 14689
natte-plaatprocédé 3087
natuurkarton 6126
natuurkunstdruk 7170
natuurkunstdrukpapier 7170
natuurlijke grootte 9253
natuurlijke harsen 9250
natuurlijke taal 9246
natuurlijke terpentijn 14438
natuurlijk gedroogd papier 329
natuurlijk pigment 9248
natuurlijk teken 750
natuurrubber 11747
natvilt 14684
nauwer zetten 6386
nauw gespatieerde regel 2938
nauw gespatieerd zetsel 9244
nauwkeurigheid 85
nauwkeurig register 6749
nauwluisterend paswerk 13809
nauwluisterend sluitwerk 13809
nauw spatiëren 7738
nauw zetsel 9244
nauw zetten 7738, 12805
nawoord 5219
nazien 2663
Nederlands 4908, 4908
neergaande beweging 4680
neergang van het mes 4677
neerhaal 13109
neerslaan 10711, 10717, 12065

neerslag 4186, 10710
neerslag vormen 10711
neerwaartse beweging 4680
negatief 9274
negatief beeld 7542
negatief cliché 11557
negatiefhouder 2651, 9275
negatiefkopie-plaat 13435
negatieflak 9279
negatief op glas 6433
negatiefpapier 9278
negatief schrift 11556
negatieve lens 4507
neiging tot stuiven 5881
neiging tot uitvlokken 5843
neonlamp 9281
Nessler-buis 9283
net 9288
netspanning 14511
netspanningskies-
  schakelaar 14512
nettogewicht 9287
nettoprijs 9286
netwerk 3389, 9288
neutrale kleur 110
neutrale sulfietcelstof 9293
neutrale tinten 9295
neutrale tonen 9295
neutraliseren 9289
neutron 9296
neven-apparatuur 12545
neven-element 7247
nevenprodukt 2132
newton 9325
newtonse ringen 9329
newtonse stroming 9327
newtonse vloeistof 9328
niet actinisch 9393
niet-actinisch filter 7236
niet-actinisch licht 9394
niet-afgesneden kanten
  14264
nietapparaat 13528
niet-bedrukt vel 1535
niet-bestorven papier 6625
niet brandbaar 9400
nietcilinder 13176
niet-cyclische gebeurtenis
  7247
niet-drogende olie 9399
niet-drukkend deel van het
  rubberdoek 1710
niet-drukkende partij 6607
niet-drukkende partijen
  9401
nieten 14780
nieten door de rug 11831
nieten door het plat 1617,
  12389
nieten van bloks 9845
niet geïnterlinieerd 14329
niet gekoppeld 9513
niet-gelezen kopij door-
  geven 11217
niet-gemonteerd cliché
  14332
niet-genummerde pagina
  1565
niet haaks gesneden 2477

niet houdbaar 6229
niet in de handel 10873
niet in de lijn 9723
niet ingesprongen 6235,
  12159
niet-ingesprongen alinea
  5908
niet-ingesprongen regel
  5903
niet in register 9726
niet inspringen 1289
niet-inspringende alinea
  5908
niet in voorraad 9728
niet-ionogene stoffen 9403
nietje 13040
niet kristallijn 552
niet-krullend gegomd
  papier 9396
nietmachientje 13528
nietmachine 14779
niet mengbaar 7185
niet-nagesneden papier
  14352
niet-newtonse vloeistof
  9407
niet-opengesneden kanten
  1716, 14264
niet-opengesneden katerns
  2933
niet-opengesneden randen
  14334
niet-opgezette doos 2380
niet oplosbaar 7457
niet op magazijn 9728
niet passend 9726
niet-ronde rol 9727
niet-rond zijnde rol 1062
niet-schoongesneden
  papier 14352
niet smeltbaar 7314
niet stabiel 7459, 14350
niets te doen hebben 13558
niet sterk 6229
niet-strak gewikkelde rol
  12535
niet-stuivend papier 5879
niet-tellende ponsband-
  perforator 9395
niet te mengen 7185
niet te ontcijferen 14265
niet-toegestelde proef 5792
niet uitdrukken 12839
niet-uitdrukkende punten
  8986
niet-uitgevulde band 14327
niet-uitgevulde regels
  14328
niet-uitgevuld zetsel 14328
niet-uitwissend lezen 9398
niet-variabel vouwapparaat
  5726
niet verzeepbaar 14339
niet-vluchtig geheugen
  9418
niet voelbaar 7188
nieuwe alinea maken,
  een ~ 1893
nieuwe druk 11429, 11477

nieuwe uitgave 9297
nieuws-agentschap 9299
nieuwsblad 9309
Nigeriaans marokijn 9344
nigrosine 9348
nijdig zijn 10086
nikkel 9335
nikkel-ammoniumsulfaat
  9336
nikkelgalvano 9342
nikkel-styp 9339
nit 9359
nitraat 9361
nitrobenzeen 9368
nitrocellulose 2517, 9371
nitrocelluloselak 9372
nitroglycerine 9376
nitroglycerol 9376
nitronaftaleen 9379
nitronaftaline 9379
n.l. 14500
ñ met tilde 9223
ñ met tildeteken 9223
nog bruikbaar afsnijsel
  9503
nog leverbaar 7421
nog verkrijgbaar 7421
nok 2227
nokkenas 2254
nomogram 9391
non-ionogene stoffen 9403
nonius 14444
noodverbinding 10046
nootteken 11552
noottekens 8641, 11381
no-pack matrijs 9835
normaal 9423
normaal diafragma 9428
normaalformaten 7521
normaalinkten 10897
normaal insteekdiafragma
  9428
normale prestatie 9427
normalisatie 13026
noten 11381
noten aan de voet van de
  pagina 6021
notenbalk 13000, 13081
notenbalkentrekker 9212
noten onder aan de pagina
  6021
notenpapier 9211
noten tussen de kolommen
  2547
notering 11173
notitieblok 8771, 13514
notitieboekje 9437
nullenonderdrukking 14937
numerieke volgorde
  brengen, in ~ 12134
numeroteur 9450, 9452
nummer 3535
nummeraar 4366
nummeren 5997
nummering 9449
nummerklokje 9450
nummermachine 9452
nummerraam 9451
Nyblin-effect 9463

nylon 9464
nylon cliché 9465

objectief 8042
obligatie 1718, 4038
oblong formaat 7898
ochtendblad 9104
ochtend-editie 2052
octal 9484
octavo 9485
octavo-formaat 9485
octoaat 9486
oculair 5418
oersted 9502
offsetdiepdruk 9521
offsetdiepprocédé 4085,
  4088, 9521
offsetdoek 1509
offsetdoekvervuiling 1519,
  3418
offsetdoekwasmiddel 1526
offsetdruk 9525, 9536
offsetdrukker 10754
offsetdrukplaat 9531
offsetdruktechniek 9525
offsetinkt 8271, 9523
offsetmachine 9532
offsetpapier 9528
offsetpers 9532
offsetplaat 9531
offsetprocédé 9536
offsetproefpers 10951, 10990
offset-rotatiepers 11698
offset-vellenrotatiepers
  12270
offsetzinkplaat 9541
OH-groep 7133
ohm 9547
oiticica-olie 9569
oker 9481
oleïen 9579
oleïne 9579
oleofiel 9582
oleofoob 9583
oleografie 9581
olieaannemend 9582
olieabsorptie 9550
olieabsorptiebepaling 9551
olieabsorptieproef 9551
olieafstotend 9566, 9583
oliebestendigheid 9567
oliedoorlatendheid 9564
oliekan 9552
oliekannetje 9552
oliën 8395, 9548, 9557
oliepapier 9554
oliesiccatief 8215
oliespuit 9552
oliesteen 9568
oliezuur 9579
olijfolie uit lucca 8399
omber 14255
ombuigen 1820
ombuiger 2910, 13170
omgekeerde regel 14096
omgekeerd evenredig 7540
omgekeerd-leesbaar beeld
  11285
omgekeerd negatief 11551

papiermagazijn 9946, 13238

papiermaker 9924

papier met dierlijke lijming 625

papier, met een oplosmiddel gestreken ~ 12774

papier met een "zak" 4432

papier met eierschaaloppervlak 4989

papier met extrusiecoating 5412

papier met extrusie-couche 5412

papier met extrusie-laag 5412

papier met gegoten glanslaag 2436

papier met grote opdikking 5501

papier met haakse hoek 12965

papier met harde klank 13830

papier met harslijming 11690

papier met inwendige lijming 1264

papier met kalanderlijming 2194

papier met linnenpersing 2233, 8185

papier met oppervlaktelijming 14073

papier met parelmoerglans 9112

papier met persing 11473

papier met plaatpersing 10437

papier met rolcoating 11630

papier met rolcouche 11630

papier met rolstrijklaag 11630

papier met scheprand 4066

papier met smeltlaag 7049

papier met spreidlaag 12933

papier met stoflijming 1264

papier, met textielgaas versterkt ~ 11426

papier, met textielweefsel beplakt ~ 2957

papier met vulstof 5620

papier met watermerk 14589

papier, met weefsel versterkt, geteerd ~ 11428

papiermolen 9927

papier, niet op de papiermachine gestreken ~ 12127

papieronderzoek 9952

papier, op de lijmpers gestreken ~ 12493

papier, op de machine gedroogd ~ 8444

papier, op de strijkmachine gestreken ~ 12127

papieroppervlak 9948, 13434

papieropslag 13228

papierporiën 9933

papierrol 11354, 11372

papierrollenkelder 11374

papierrollen-transportinrichting 9936

papierschaar 9940

papiersnijder 3884

papiersnippers 11889

papierstapel 10340

papierstof 5878, 6267, 8226, 9914

papierstrookje 5741

papiertafel 1120

papier, tegen zuur bestand ~ 139

papiertoren 9953

papiertoren-excentriek 9954

papier-uithanginrichting 1105

papier uit houtcelstof 2692

papier uit linnen lompen 8186

papier uit zuiver lompen 11205

papierverbruik 9910

papierverlies 8375

papierverwerking 3477

papiervezels 9917

papier voor aandelen 8296

papier voor albums 378

papier voor bekleding van de binnenzijde van kisten 2408

papier voor blindenschrift 1847

papier voor boterwikkels 2120

papier voor broodwikkels 1873

papier voor buswikkels 7851

papier voor drukproeven 10946

papier voor familiedrukwerk 14664

papier voor gelegenheidsdrukwerk 638

papier voor het beplakken 10043

papier voor keramische transfers 2557

papier voor knippatronen 10063

papier voor krachtpikeersels 14283

papier voor krijtpikeersels 2603

papier voor (post)zegels 13019

papier voor rouwdrukwerk 9150

papier voor servetten 9239

papier voor sigarebandjes 2828

papier voor staalgravures 13098

papier voor telefoongidsen 13628

papier voor tijdschriften 8485

papier voor wegenkaarten 11623

papier voor zakkenvoering 1071

papier voor zeekaarten 2649

papierwals 1824

papierweegschaal 11107

papierweger 11107

papier zonder carbon 2325, 9382

papyrolin 9958

papyrus 9959

para-aminofenol 9961

para-dichloorbenzeen 9964

para-dioxybenzeen 7127

para-fenyleendiamine 9992

paraffine 9969

paraffine-koolwaterstoffen 424

paraffinen 424

paraffine-olie 9966

paraformaldehyde 9970

paragraafteken 12059

parallax 9977

parallelle beweging 9984

parallelle operatie 9985

parallellisme 9982

parallelschakeling 9978

parallelteken 9983

parallelverbinding 9978

parallelvouw 9979

parallelvouwen 9981

parameter 9988

paraminofenol-hydrochloride 9962

paraminofenolontwikkelaar 11628

para-oxyfenyl glycine 6485

pararood 9993

parelen 3706

parelen (van de inkt) 9114

parelmoerpapier 9112

parenthesen 9998

parentheses 9998

pariteit 10007

pariteitsbit 10008

pariteitscontrole 10009

pariteitstoets 10009

parterre-machine 14321

particuliere bibliotheek 10872

particuliere drukkerij/uitgeverij 10874

particuliere uitgave 10871

partij 1221

pasfout 11407

pasgaten en paspennen 11409

pasje 5205

paskijker 11419

paskruizen 11411

pasloep 11419

pasloupe 11419

pas maken 1919

pasmerken 11411

paspen 4673

passen 11402, 11420

passen, (precies) ~ 3229

passen van de plaat 10453

pastei 10315

pastei maken 10317

pastei uitzoeken 10303

pastelinkten 10032

pastelkleuren 10033

pasteltinten 10033

pasverschil 11407

pasverschillen 11405

patentvoet 10056

patronenpapier 10063

patroonpapier 10063

patsen 5943

patzen 2994

pauwblauw 10069

peauter 10176

pectine 10079

pek 10380

pelproef 10081

pelure 6172, 9597

penblok 348

penetratie 10098

penetratie van de inkt in het papier 7327

penetratievolumeter 10100

pennekrullen 5863

pennemesje 11950

pennenlinieermachine 10103

penseel 9870

penseeletsing 1999

penseel getekend lijkende letter, op met het ~ 2018

penseelwerk 2017

pentaan 10106

pentachlorofenol 10105

pentekening 10089

peptisatie 4102, 10109

percentage rasteroppervlak 10112

percentteken 10113

perchloorzure kali 10663

perehout 10075

perforatie 10130

perforatie met ronde gaatjes 11727

perforator 10132

perforeerkam 10131

perforeerlijn 10133

perforeermachine 10132

perforeren 10130

pergamijn 6431

periferie-apparatuur 10143

perilla-olie 10140

periode per seconde 3922

periscoop 12654

perkament 9994

perkamenten band 9995

perkament-ersatz 7175

perkament-ersatzpapier 7175

perkamentpapier, echt ~ 14421

perkament sigaretpapier 8450

perkament (uit kalfsvel) 14425

perkament uit schapevel 6039

perkament van lamshuid 7883

permanent-wit 1156

permanganaat 10148

permeabiliteit 10151

Perry-Higgins-procédé 10154

pers 10736, 10738

persbalk 2873

persbreidel 6271

persbureau 9299

perscensuur 2526

perscorrectie 8441

persdelict 10752

perse nemen, ter ~ 6529, 11067

perse, ter ~ 6521, 7537

perslucht 3322

persmerk 7207

pers nemen, op de ~ 11067

personeelsraad 2625

personenlift 8093

perspartij 10737

perspectieftekening 9890

persplaat 9531, 10758

persproef 6062, 10759

pers, reeds op de ~ 6521

persslijper 10513

persvel 244, 10764

persvergulden 6406, 10749

persvergulder 10748

pers volgens etagebouw 1089

persvrijheid 6158

perswatermerk 7207

perszwengel 1136

pers, 24-zijdige ~ 12207

petits fers, aux ~ 948

petrolatum 10167

petroleum 7751

petroleumether 10169

petroleumfractie 10170

Petzval-objectief 10175

peut 14727

pH 10177

pH-bepaling 10178

pH-meter 10179

pH-meting 10178

pH-onderzoek 10178

pH-strookje 10180

pH-waarde 10181

piauter 10176

pica 10288

piek 12865

pigment 10330

pigmentkleur 1679

pigmentkopie 2327

pigment, met molybdeen- zuur neergeslagen ~ 9060

pigmentpapier 2331

pijnhars 6274

pikeermesje 3902

pikeersel 9768, 10045

pikeersel in het cliché 2062

pikeersel onder de voet 14281

pikeersel tussen cliché en voet 7497

pikeren 9771

pinacyanol 10350

pinaflavol 10351

pinakryptolgeel 10353

pinakryptolgroen 10352

pinatypie 10354

pint 10373

pipet 10375

pittig 12651

PIV-aandrijving 10387

plaat 10419, 10420

plaatborstel 4225, 14176

plaatbuigmachine 1319

plaatcilinder 10432

plaatdiepdrukmachine 10440

plaatdikte 10463

plaatdruk 3527, 10458

plaatdrukinkt 3524, 10254

plaatdrukker 3526

plaatdrukkunst 2594

plaatdrukpapier 10457, 13098

plaatdrukpers 3525

plaatfreesmachine 937

plaathaakgroeven 4546

plaathaken 4542

plaathouder 2135, 10442

plaathouderrails 2370

plaat in één kleur 9079

plaat in kleuren 3162

plaatklem 2869

plaatklemlijst 2872

plaatkoeler 10429

plaatmerken 3943

plaat, met de hand ge- sensibiliseerde ~ 14760

plaat, met de hand licht- gevoelig gemaakte ~ 14760

plaat met het laatste nieuws 6226

plaatopsluitlijst 2872

plaatopsluitmechanisme 8307

plaatpersing 14946

plaatproef 7472

plaatribben 10460

plaatschaar 6705, 12242

plaatselijk blaadje 11199

plaatselijke aberratie 14966

plaatselijke correcties 8300

plaatsen, tussen aanhalings- tekens ~ 4364

plaatsen, tussen haakjes ~ 1843

plaatsen van leestekens 11049

plaatsleutel 2056, 10444

plaatsruimte 5859

plaatsvervanger 13473

plaattoon 7340, 12018

plaatverschuiving 9545

plaatvervaardiging 10448

plaatvolgorde 10470

plagiaat 10389

plakband 211, 215, 6718, 10040, 12026

plakboek 11944

plakkaat 10609

plakkaatpapier 10610

plakken 203

plakken, aan elkaar ~ 12892

plakken van clichés op voeten 10422

plakletters 12074

plakmiddel 213

plakrand 6472

plaksel 13046

plakselpot 6469

plakstrepen 10031

planbord 10398

planconcave lens 10399

planconvexe lens 10400

planeetwiel-beweging 10396

plank 10740

plankstansmes 7660

planner 10926

planning 10397

planningbord 10398

plano 6241

plano formaat 1931

plano papier 5771, 5798

plano-uitleg 5786

plano vellen 5771

plano vellen, aan ~ 7448

plantaardige lijm 14417

plastic band 10408

plastic cliché 10411

plastics 10412

plastic styp 10411

plastisch 10407

plastische viscositeit 10414

plastisol 10415

plastometer 10416

plat 14329

platenman 10441

platen opleggen 10464

plat gedeelte 1684

platgedrukte rol 9727

plat geniet boekje, door het ~ 5806

plat geniete brochure, door het ~ 5806

plat gereden letter 14855

platina 10472

platinachloride 10473

platinochloruur 10473

platte kwast 5778

platten 3676

platte tekst 1683, 1684

platte wig 13975

plattewig-excentriek 13976

platte zak 5773

platvorm 4061

plat werk 13247

plat zetsel 1683, 12762, 13247

plat zetten 11812

platzetter 3314

pletbord 10756

pletpers 13034

ploeg 3255, 10484, 12296

ploegendienst 12297

plooi 3715

plooien 3717, 5970, 14873, 14874

pluche borstel 4086

pluche tampon 4086

plukbeginbeoordelingskast 10301

plukken 10295

plukweerstand 10300

plukweerstandsmeter 10302

plusteken 10499

pneumatisch kopieerraam 14375

pocketboek 9895, 10508

poederen 2599, 4892, 10698

poederen met talkpoeder 10699

poedering 10698

poederkwast 4897

poedermachine 10701

poelie 1308

poetsdoek 2891

poetslap 2891

poetspommade 11069

pointillé-band 10526

pointilleren 13160

poise 10537

Pola-autotypieraster 10545

polaire oplosmiddelen 10544

polaire stoffen 10543

Pola-raster 10545

polarisatie 10539, 10540

polarisatiefilter 10542

polijsten 2037, 2080

polijstmachine 2039

polijstpapier 17

polijststaal 2078

polyacrylaat 10549

polyamide 10550

polyamide-hars 10551

polychroom 9169, 10553

polyester 10555

polyester-hars 10555

polyethyleen 10556

polymeer 10558

polymerisatie 10559

polystyreen 10561

polystyreenhars 10561

polytheen 10556

polyurethaan 10564

polyurethaanhars 10564

polyvinylacetaat 10566

polyvinylalcohol 10567

polyvinylchloride 10568

polyvinylideenchloride 11875

polyvinyl-procédé 10570

rugoverlijmpapier 1034, 7026
rugoverplakpapier 7026
rugschildje 9889
rugtitel 1050
rugtitel in langsrichting 13863
rug (van het boek) 12873
rugwit 985
ruilexemplaar 5342
ruillijst 5344
ruimte 2904, 12364, 12799
ruiterrol 11600
ruiterrollen van specie 3308
ruitvormig 8391
ruitvorm, in ~ 8391
rukken 7656
rundleer 3687
Russell-effect 11820
ruwe leerrol 9240
ruwe leren rol 9240
ruwe olie 10168
ruwe proef 5792
ruwheid 11711
ruwheid van het oppervlak 13437

Sabattier-effect 11825
Safawid-band 11832
saffiaan 9106
saffiaanleer 9106
salicylzuur 11851
salmiak 522
salpeterig zuur 9380
salpeterzure potas 10660
salpeterzuur 9366
salpeterzuur lood 7996
salpeterzuur zilver 12432
SAM 13031
samendrukbaarheid 3324
samengeperste lucht 3322
samengesteld breukcijfer 10320
samengesteld cliché 3299
samengestelde accolade 3300
samengestelde lens 3319
samengesteld objectief 3319
samenhangende structuur 3389
samenklonteren van de rasterpunten 10494
samenlopen der papier-banen 3748
samenpakken 9820
samentrekken 3706, 12354
samenvatting 20, 13490
samenvoegen 830, 846, 8808
samenvouwbare doos 5975
sandarak 6724
sandrak 6724
saraceense band 11874
saran 11875
satellietmachine 3249
satijnwit 11878
satinage 5686
satinage met borstels 2002

satinageplooien 2186
satineercilinder 6448
satineerkalander 12995, 13409
satineerpers 12642
satire 7895
scala-drukken 10928
scala's 10928
scannen 11911
scanner 11910
scanner met bewegende lichtvlek 5928
schaafkarton 11952
schaakbordraster 2667
schaakspeltekens 2699
schaal 4259, 11905
schaaldrukken 10928
schaapsleer 11624
schaar 9433, 11929
schaarde 9433
schabloon 6205, 8646, 11967, 13110
schacht 12231
schaduwbelichting 5762
schaduwmasker 12222
schaduwmerk 12220
schaduwpartijen 12223
schaduwpunten 12219
schaduwwatermerk 12214
schakel 7492
schakelaar 13474
schakelbord 10490, 13476
schakelkast 3395, 13476
schakelpaneel 10490
schakelschema 4258
schakeltafel 3395, 13476
schapeleer 11624, 12249
schapevet 9222
scharlakrood 11914
scharnierende cassette 7009
schavotje 11692
scheef gesneden 2477
scheef sluitraam 1357
scheepskrant 12309
scheidend vermogen 4098, 11516, 11517
scheidingslaag 4513, 11438
scheidingslijn 3871, 4514, 13320
scheidingsteken 12807
scheidingstoets 7535
scheiding van ketting-formulieren 2095
schellak 12289
schellakvernis 12290
schelpgoud 12292
schenen 13146
schepper 14413
schepraam 4063
scheprand 4065
schepvorm 9119
schepvorm voor velijn papier 14857
schering 14544
scherm 11968
scherp 7823
scherp beeld 12235
scherp masker 12236

scherpstelling met behulp van een schaal 11902
scherpte 4099
scherptediepte 4195
scherpte, (visuele) ~ 12237
schets 4208, 9717, 11715
schetspapier 12506
scheur 1874
scheurband-systeem 13923
scheurkalender 1590, 2178
scheurkalenderblok 4018
scheurstellen 12652
scheve belichting 9471
scheven (bij vlas) 12310
scheve regel 13594
schietkatoen 9371
schijf 4433, 4434
schijfaandrijving 4439
schijf-inktwerk 4442
schijnbaar soortelijk gewicht 712
schijnbare dichtheid 711
schijnbare viscositeit 713
schijvenlinieermachine 4445
schijvenpakket 4444
schillen 1165
schiller 1164
schilmachine 1164
schiltrommel 1166
schimmel 9120
schimmeldodende stof 6251
schimmelen 6252
schimmelvlekken 8928, 12556
schimmelvorming 6252
schimpblad 14920
schittering 6425
schoepen 9848
schoepenrad-uitleg 5462
schokbreker 2032
schone band 2897
schone ponsband 2897
schone proef 2895
schoolboek 11920
schoondruk 5711
schoondruk aanbrengen, de ~ 10864
schoondruk- en weerdruk-vorm apart 14822
schoondrukken 10864
schoondrukvorm 5707, 9714
schoon- en weerdruk in één vorm 14825
schoon- en weerdruk-offset-pers 9530
schoon- en weerdrukpers 10122
schoongesneden formaat 14044
schoonmaken 2899
schort 727, 728
schot 3703
schotelvormig 4431
schotelvormig papier 4432
schot (van een papierbaal) 1092
schouder 4737, 12343
schouderhoogte 6949

schraal drukken 10798
schraapstaal 11949
schrapen 3901
schraper 11950
schrappen 4129, 13272
schreefloos 11868
schrens 11982
schreven 12144
schrift 14171
schriftlezen 2639
schriftteken 6564
schrifttekenlezen 2639
schrijfbehoeften 13076
schrijfblok 8771, 13514
schrijfdichtheid 11312
schrijffout 2911
schrijfinkt 7326
schrijfkogel 6520
schrijfletter 12012
schrijfmachineletter 14213
schrijfmachinepapier 14224
schrijfmachineschrift 14213
schrijfpapier 14879
schrijfpapier voor notitie-bloks 13516
schrijfster 881
schrijfwieldrukker 14703
schrijfwijze 12856
schrijven 14876
schrijver 881
schrijver, bij de ~ 11004
schrijver-uitgever 885
schrijver van het hoofd-artikel 7985
schroef 11998
schroefbout 1714
schroefdop 11999
schroeffitting 4964
schroef met verzonken kop 3663
schroefpers 1245
schroefsleutel 14872
schroefsluiting 11999
schroevedraaier 12001
schrootblad 8597
schrootmanier 8597
schrupel 12014
schubornament 7169
schudden 13165
schudwerk 12227
schuifdoosje 12291
schuifdoos(je) 12552
schuifgalei 12549
schuifspanning 12241
schuiftransfers 4041
schuif van de galei 6286
schuilnaam 10101
schuim 5938
schuimblaasjes 5943
schuimen 5939
schuimlepel 12514
schuimrubber 12912
schuimspaan 12514
schuimstrijklaag 2019
schuimwerend middel 660
schuine breukstreep 12767
schuine kant 1345, 1346
schuine kant maken, een ~ 2609

schuine schreven 9472
schuine streep 12767
schuin gesneden papier 591
schuin invallend licht 6600
schuiven 12298
Schulze-raster 11923
schutbladen 5162
schutbladenpapier 1750, 5161
schutbladenplakmachine 5167
schuurmiddel 14
schuurpapier 17, 5111, 6434, 11861
schuurpoeder 5113
schuurweerstand 11
Schwabacher 11924
Schwarzschild-effect 11925
Scripten 12011
scrupel 12014
sealen 6931
sealkrachtmeter 1721
secans 12032
secondaire kleuren 12033
sediment 4186
sedimentatie 12065
segmenteren 12067
segregatie 12069
segrijn 12226
segrijnleer 12226
seignettezout 12710
sein 5742
selectieve roep 10548
seleen 12072
seleen-cel 12073
selenium 12072
selenium-cel 12073
semée-band 12092
sensatieblad 14920
sensibilisator 12118
sensibiliseren 9662, 12114
sensitometrie 12121
sequentiële verwerking 12136
serie- 12138
serie-afdrukmachine 12142
serienummer 12140
serie-operatie 12141
serieschakeling 12143
serigrafie 11989
servettenpapier 9239
servomotor 11412, 12150
set 12153
sextern 12204
sferische aberratie 12862
Shore-hardheid 12321
Shore-hardheidsmeter 12320
showcard 12350
siderografie 12385
siënna 12394
sierkapitalen 13455
sierletters 13455
sierlijn 9690
sierrand 9689
sigarettendoos 2830
sigarettendoosje 2830
sigarettenpapier 2831

signaalomzetter 9022, 12397
signaalserie 5243
signaturenschema 13511
signatuur 12402
silica 12409
silicagel 12410
silicium 12415
siliciumcarbide 12416
simokaart 12437
simuleren 12440
sinus 12443
sinusoïdaal 12444
sinusvormig 12444
sisalhennep 12476
skiver 12527
slaan 2659, 11255
slag 7186
slagregel 2466
slak 12539
slaolie-stijl 800
slap 5739
slap gebonden boek 12571
slap overhangende kanten 4517
slap overhangend omslag 4517
slap papier 10283
slappe afdruk 5436
slappe band 5822, 8153
slappe boekband met over-
hangende randen 14906
slappe kanten 12533
slappe papierbaan 12537
slappe regel 8366
slappe rol 8367, 12535
slappe vernis 8351, 12339, 13719
slechte afdruk 9161
slechte beurt 11476
slechte kopij 1059
slechte pers 1063
slechte rol 1062
slecht gewikkelde rol 8367
slecht ingerichte drukkerij 14754
slecht leesbaar manuscript 1061
slecht leesbare kopij 1059, 1061
slecht verloop 6876
slecht werk 1804
sleepringmotor 12574
slepen 4687
sleutel 7759, 7762, 9072, 9620, 10444, 14872
slib 12596
slibvorming 12597
slijm 9160
slijpen 6636, 6641
slijpen met houtskool 2647
slijper 6638
slijperij 6645
slijpkogels 3037
slijpkool 6642
slijpmachine 6643
slijpmiddel 14
slijppapier 17
slijpsteentje 12648
slijpzand 11858

slijtage 10, 12015, 14635
slijtage van de drukvorm 6077
slijttoestel 11766
slijtvast 12
slijtvaste inkt 11765
slijtvastheid 11, 11767
slijtvastheidsmeter 11766
slijtvastheidsproef 13
slijtweerstandsmeter 11766
slinger 11719
slingerapparaat 14709
slingeren van de papier-
baan 12282
slingering 9708
slinger-psychrometer 12558
slingerwijdte 561
slip 12559, 12569, 12570, 12573
slipmeter 12566
slippen 3722
slippende conus 12572
slips, in ~ 7455
slordige afdruk 9161
slot 2876
slotwoord 5219
sluier 5957
sluierbelichting 5762
sluieren 5958
sluitband 12026
sluiter 12361
sluitingstijd voor nieuws 4032
sluitletter 5665
sluitlijn 13734
sluit-ornament 13552, 14470
sluitring 14549
sluit-s 5666
sluitstuk 11169, 13552, 14470
sluitteken 2940
sluittijd van de redactie 4032
sluittijd voor de zetterij 4033
sluitvignet 13552
sluitwit 6262
sluitwit aanbrengen 4722
sluitzegels 10611
smak 13398
smakeloze versiering 6414
smalband-filter 9242
smalle breedte 9243
smalle letter 3364
smalle papierbaan 4385
smal lopende letter 3364
smaragdgroen 5110
smeedbaar 8585
smeergat 8398
smeergroef 9555
smeermiddel 8394
smeerolie 8397
smeervet 3394, 6601
smeltketel 7755, 8770
smeltkroes 3777
smeltlaag-procédé 7050
smeltlijm 7048
smeltoven 8826
smeltpot 7755, 8770

smeltpunt 8769
smeltwarmte 6926
smeren 1649, 8395, 9557, 12636, 12637
smeren met olie 9548
smering 6610, 9557
smering onder druk 6027
smet 12611
smetringen 1252
smetringhoogte 3929
smetringhoogtemeter 3928
smetten 1478, 12636
smetten op niet te
bedrukken plaatsen 1713
smetten van het drukdoek 1713
smetvellen 12576
smoutinkt 7677
smoutkast 7670
smoutletter 4471
smoutlettergietapparaat 4460
smoutlettermachine 4467
smoutwerk 4473
smoutzetsel 4469
smoutzetter 7673
smoutzetterij 7667
snede 4961
snede met kammarmer 3217
sneden 3860
snee gemarmerd, op ~ 8615
sneeuw (in krantenpapier) 1876
sneevergulder 4956
sneeverven 13011
sneewit 12387
sneldrogend 5482
sneldrogende inkt 11152
sneldrogende lak 11153
snel-etsen 10703
snel-etsmachine 10704
snel-etsprocédé 10703
snelhechter 1403
snelheidsdaling 14428
snelheidsgradiënt 14428
snelheidsregeling 12850
snelheidsverval 14428
snelheid van de machine 8466
snelheid van een emulsie 12851
snelheid van het licht 14429
snelheid van splitsing van
de inktlaag 12905
snelle regeldrukker 7001
snellezer 7002
snellopende machine 7000
snelloper 7000
snelpers 3948, 13210
snelsluitgalei 2654
snelsluitinrichting 7460
snelsnijder 6998
snelsnijmachine 6998
snel verdampend oplos-
middel 5487
snel zetten 12170

terminal, op afstand op-
gestelde ~ 11459
termostaat 13698
terpentijn 14106
terpentijnolie 14107
terpentijn-substituut 14108
terpentine 14727
terra di Siena 12394
terrasiënna 12394
tertiaire kleuren 13657
terugdrijven 11471
terugetsen 996
terughouden 8970
terugkaatsing 11389
terugkeeradres 8211
terugkeermodificator 8211
terugkeeropdracht 8211
terugkloppen 11471
terugloop van de kar 2369
terugstellen 11494
terugsteltoets 11545
terugverlopen 11794
terugvinden 11542
terugwinning 11314, 11496
terugwinningsinstallatie
11321, 12779
terugwinning van oplos-
middelen 12778
terzijde geschoven
exemplaar 7787
testliner 13665
testplaatje 13055
tetra 2330
tetrachloorkoolstof 2330
tetrachloormethaan 2330
teveel aan zetsel 3772,
12590
teveel afsnijden 3740
teveel gedrukte
exemplaren 9782
teveel inkt geven 3771, 5851
te volle kast 1140
te volle regel 13806
Texoprint-procédé 13673
textieldruk 13678
textieldrukinkt 2958
textieldrukmachine 2208
theelepel 13620
theorie van de raster-
openingen 10365
therblig 13685
thermal-zwart 13686
thermisch beveiligde
schakelaar 13695
thermodruk 13690
thermohardbaar 13696
thermohardend 13696
thermoplastisch 13692
thermostaat 13698
thermostatisch bad 13699
thioazijnzuur 13720
thiocarbamide 13723
thio-ureum 13723
thixotropie 13737
thymol 13798
tiendelig stelsel 8875
tientallig stelsel 8875
tijddeling 13818
tijdelijke verbinding 10046

tijdelijk zetter 6574
tijd/gamma-curve 13815
tijd/gamma-kromme 13815
tijdkaart 2921
tijdnorm 13819
tijd om handdroog te
worden 12179
tijd om te zetten 12179
tijdscharing 13818
tijdschrift 7698, 9309
tijdschriften-groothandel
9300
tijdschriftenpapier 8485
tijdschrift, om de twee
maanden
verschijnend ~ 14148
tijdstudie 13820
tijd, uit de ~ 9724
tijd van afsluiting 4032
tikfout 14230
tikken, de letters telkens
onnodig tegen de zet-
haak ~ 5452
tikker 7771
tikster 7771
tilde 13812
time sharing 13818
timpaan 14165
tin 13822
tinchloride 13038
tinchloruur 13039
tinplaat 10176
tint 12209
tinten 1556
tintlegger 1315
tintplaat, (gerasterde) ~
13841
tintplaat, (ongerasterde) ~
12763
titaan 13859
titaandioxyde 13860
titaanwit 13860
titanium 13859
titel 6902, 13862, 13867
titelkapitalen 13869
titelpagina 6244, 13867
titelplaat 6211
titel van een hoofdstuk
2298
titratie 13870
tjap van de drukker 5816,
7229
TL-buis 5892
toebehoren 69
toegang 67
toegangsbiljet 13799
toegangskaart 13799
toegangstijd 70
toegankelijk 193
toegankelijkheid 68
toegepast (natuur)weten-
schappelijk onderzoek
720
toegeslagen, (eenmaal) ~
7907
toegestaan bedrag 725
toegestane afwijking 13880
toegestane tijd 7243
toenemende snelheid 51

toepassingsprogramma 718
toer 11578
toeren per minuut 11579
toerental 9455, 12853
toerenteller 13523
toeslag 5395
toeslag voor persoonlijke
verzorging 3263
toestel 4236, 8566, 9773,
12922
toestellen 8564
toestellen met krijt-
pikeersels 2604
toestelmesje 3902
toestelvel 12922
toestelvloei 8567
toestemming tot afdrukken
geven 10020
toets 7761
toetsaanslag 7785
toetsenbank 7778
toetsenbord 7765
toetsenbordbewerker 7771
toetsenbordbewerkster
7771
toetsenbord-excentriek
7766
toetsenbord-indeling 7770
toetsstangraam 7764
toevallig 11230
toevallige fout 71
toevallig ontdekken 12137
toeval ontdekken, bij ~
12137
toevoegen 182
toevoeging 4559
toevoegingen aanbrengen
4560
toevoegingsmetaal 11577
toevoegingsmiddel 188,
3317, 4942
tol betalen 6614
tolerantie 13880
tolueen 13881
toluïdine-geel 13882
toluol 13881
tonen 2464, 6609, 12020,
12021
tonen door emulgeren
13843
toner 13892
tonvormige vertekening
1171
toon 12018, 13837, 13887
toonaardehydraat 477
toonbad 13899
tooncorrectie 13890
tooncorrectiemasker 13889
toonplaat 5809
toonproef 12022
toonwaarde 13898
toonweergave 13893
topwaarde 10070
torderen 13925
tordering 13925
tornen 7245, 7684
tornmotor 7246
torntape-systeem 13923
torsie 13925

torsiebalans 13926
torsiemoment 13924
touwen 1113
touwen doorhalen 7862
toverinkt 8492
traagheidsmoment 9063
tragacant 13951
tragacantgom 13951
tragant 13951
transferpapier 4042
transfers 4041
transformator 13977
transistor 13980
transmissie 13987, 13988,
13994
transmissie-densitometer
13990
transmissie tussen twee
stations 10535
transmissie-zwarting 13991
transparant 14002
transparante inkten 14003
transparantie 13997, 13999
transparant papier 14005
transparant tekenpapier
13938
transparantwit 477, 7881,
13840
transportbaan 3481
transportband 3482
transportbanden 13596
transporteur 4148
transporteurarm 4144
transportgaten 5517
transportriempje 8689
transportrollen 2373
transportslede 5081
transport van het papier
6832
traploos regelbare
aandrijving 7303
traploos regelbare snelheid
13117
trappers 3743, 14014
trappersdrukker 10456
trapsgewijze inspringing
4256
traversen 3750
trechter 6063, 6254
trechterhoek 6068
trechterinvoerrol 11652
trechterinvoerrollen 1322
trechterkar 6069
trechterneus 6072
trechterrol 11652
trechtervouw 6070
trechtervouwapparaat 6071
trechtervouwrollen 1322
trechterwagen 6069
trefwoord 2469, 2470
trefwoordenregister 2472
trekken 14545
trekkracht (van de inkt)
13524
trekpen 4711, 11786
trekrollen 4713, 4714, 10963
treksluiting 14962
trekspaan 12978
treksterkte 13644

vergaren 3070, 6325
vergeeld model 14915
vergeerrol 7874
vergelen 14917
vergelijken 3068, 3256
vergelijking 5226
vergelijkingsmonster 13020
vergeling 14917
vergiftige werking 13933
vergiftigheid 13933
vergrootglas 8187, 8534
vergrootglas met voet 2059
vergroten 5198
vergroting 5201, 8532
vergrotingsapparaat 5202
vergrotingspapier 5204
vergrotingsschaal 11905
vergroting van de punt
   4566
verguld, aan de kop ~
   13911
vergulde kopsnee 6413
vergulder 6399
verguldkwast 6404
verguld op snee 6412
verguldpers 1604, 6407
vergure 7875
verhandeling 14016
verhardbaar, door
   warmte ~ 13696
verharding door licht 8116
verharsing 11504
verhoging van de vochtig-
   heidsgraad 7071
verhouding 10967, 11252,
   11905
verhoudingsgewicht 5230
verhouding tot, in ~ 10975
verhouding van, naar ~
   10975
verificatie 14441
verkeer 13950
verkeerd aangebrachte
   regel 8978
verkleinen 11341
verkleinen met de rubber-
   pantograaf 11753
verkleinglas 11348
verkleining 11349
verkleiningsschaal 11905
verkleurd 4421
verkleuren 3141, 5433
verkleuring 2083, 4420
verkleuring aan de achterzijde 3105
verknoeien 14557
verkoopafdeling 11848
verkoopprijs 12089
verkoopsprijs 12089
verkoopsvoorwaarden 3372
verkoper 11474
verkoperde styp 3514
verkoperen 3509
verkopering 3509
verkoperingsinrichting
   12518

verkoperingsinstallatie
   3510
verkorte uitgave 19
verlaadbord 13156
verlaging van de opper-
   vlakte-spanning 13442
verlengde letter 5094
verlichting 7160
verlies 12357
verlies van de glans 13606
verloop 6535, 14469
verloop aanbrengen 14471
verlooppraster 3416
verlopen 9783, 11795
verlopende tinten 12212
verlopend maken 14471
Verloren Maandag 8377
verloren opnametijd 14257
verluchting 7161
vermeerderde druk 5200
vermeerderde en
   verbeterde druk 558
vermeerderde uitgave 5200
vermengen 9010
vermenging 9010
vermenigvuldigingsteken
   8533, 8639
vermiljoen 14443
vernier 14444
vernietigende kritiek 3967
vernikkeld 9338
vernikkeling 9340
vernis 7863, 14401
vernisbaarheid 14402
vernisbaar papier voor
   etiketten en wikkels
   14407
vernismachine 14406
vernis-mou 12739
vernissen 7866
verontreinigen 3417
verontreinigingen 7235
verouderd 9724
verouderen 301
veroudering 301
verouderingsproef 302
verpakken in kratten (of
   kisten) 3704
verpakking 9827
verpakkingsindustrie 9823
verpakkingsmachine 9832
verpakkingsmateriaal 9827
verpakkingspapier 14866
verpakkingstechniek 9821
verpakking van de papier-
   rol 8944, 14865
verpakking voor levens-
   middelen 6003
verplaatsbaar 11448
verplaatsen 11449, 14006
verscherpte keuring 13805
verscherpt onderzoek 13805
verschet 6204
verschieten 3141, 5433
verschijnen 714
verschijningsdatum 10999
verschijningsjaar 14913
verschijning staken, de ~
   4424

verschijnsel 10193
verschijnt binnenkort 715
verschijnt niet meer 4423
verschoten 4421
verschoten kleur 5431
verschoven 13057
verschroeide matrijs 1578
verschuiven 12298
verschuiving van de plaat
   9545
verscijfer 11800
verset 14446
vers gedrukt 6189
versierde initiaal 5429
versierde kapitalen 13455
versierde rand 9689
versiering 9688
versiering met assuré-
   lijnen 980
versiersel 9688
versleten letter 14855
versnelde droging 6026
versnelde proef 49
versnellen 48
versnelling 52
versnellingsmiddel 53
versnijden 3856, 4372
versnijmiddel 5384
versnijpigment 5383
versnit 5384
versnummer 11800
verspillen 14557
verspringen 12521
verspringend zetten 13001
verspringen van de platen
   10462
verstaling 13094
verstek 8995
verstekken snijden 8996
verstekmachine 8998
verstelbaar zadel 219
verstelbare flens 220
versterken 7483, 7485
versterker 560, 7484
versterking 7483
versterkingsbordje 13150
versterkte band 11424
versterkte boekband 11424
verstrooiingscirkel 2836
verstrooiingslens 4507
verstuifkop 12929
verstuiver 866, 12929
verstuiverkop 12929
vertaalrecht 11610
vertakking 1854
vertakkingspunt 9384
vertaling 13982
vertalingsrecht 11610
vertegenwoordiger 11474
vertekening 4481, 4482
verticale camera 14450
verticale pers 14454
verticale reproduktie-
   camera 14450
vertragen 12594
vertragingsinrichting 12595
vertragingslijn 4125
vertragingsmechanisme
   12595

vertrouwensman 5490
vervaardiging van de drukplaat 10448
vervaardiging van de stansvorm 4327
vervangen 2614
vervangende vellen 2264
vervangingsmiddel voor
   terpentijn 14108
vervelende man 1942
vervilting 6055
vervliegend 14504
vervolgbladen 3424
vervolgd, wordt ~ 13872
vervolgvellen 3424
vervormde koker 3783
vervormde rol 1062
vervorming 4105, 4481, 4482
vervorming van de raster-
   punt 4565
vervuilde letter 2753
vervuiling van het rubberdoek 1519
verwarmingselement 6923
verwarmingsinstallatie
   6920
verwarmingsrollen 6924
verwarmingstoestel 6920
verwekingspunt 12735
verwerkbaarheid op de
   pers, (gemakkelijke) ~
   11790
verwerking 3476, 10894
verwerkingscapaciteit
   13785
verwerkingseenheid 10895
verwijderen 4129
verwijderen van de onder-
   kleur 14269
verwijdering van de
   buitenste laag 13290
verwijdering van het koperhuidje 12293
verwijzing 3764, 11380
verwijzingsmonster 11384
verwijzingsregel 7706
verwijzingsteken 2357,
   11552
verwijzingstekens 8641,
   11381
verwisselbaarheid 7490
verwisselbaar inktwerk
   9153, 10587
verwisselbaar objectief
   7491
verwisselbare inktbak
   10585
verwisselen 14006
verwrijfrol 4503
verzadigde kleur 11879
verzadigde oplossing 11881
verzadigd vetzuur 11880
verzadiging 11883
verzadigingsbad 11885
verzadigingsgraad 5347,
   11886
verzadigingspunt 11887
verzadiging van de kleur
   2760, 11882

ESPAÑOL

abanico 5915
abarquillado 3026, 3833
abarquillado de la hoja de papel 13549
abarquillarse 2028, 3834
abastecer 12610
abastecido 7421
abastecimiento de agua 14605
abastonado 11868
abecedario con figuras 10313
abecedario con imágenes y grabados 10313
abecedario ilustrado 10313
abecedario pictórico 10313
abedul 1438
aberración 5
aberración astigmática 852
aberración cromática 2761
aberración de esferidad 12862
aberración esférica 12862
aberración local 14966
aberración óptica 9643
abertura de la malla 8813
abertura del objetivo 8043
abertura de (los cuadritos de) la retícula 11974
abertura de media de diafragma 4218
abertura eficaz 4981
abertura práctica 14830
aberturas laterales 1238
abertura útil 4981
abertura útil relativa 11430
abeto 5693
abeto americano 6961
abeto (rojo) 12953
abiseladora 1355
abiselar 1354
abocetador 7962
abollado de plancha 1297
abolladura 4177, 12472
abolladura de plancha 1297
abollar 4178
abombado del papel 1298
abonada 13329
abonado 13329
abonar(se) 13330
abono 13331
abotagarse 13462
abotargarse 2024
abrasímetro 11766
abrasión 10
abrasivo 14
abrazadera 7650
abrazaderas 1840
abrazaderas del tímpano 1078
abrazar con corchetes 1843
abrazo 14859
abreviación 20
abreviar 1
abreviatura 2
abridor de láminas 5688
abrigo del polvo, al ~ 4906
abrillantado a cepillo 2002
abrillantadora 5570

abrillantar 5569
abrir 5466, 14331
abrir el circuito 13477
abrir la forma 4750
abrochar con alambre 14780
abscisa 22
absorber 31
absorción 34
absorción de agua 14564
absorción de humedad 9037
absorción del aceite 9550
absorción del papel 7327
absorción de luz 8107
absorción de tinta 7327
absorción (óptica) 35
acaba de aparecer 7719
acabado 5686
acabado acordonado 11468
acabado aterciopelado 5179
acabado de fieltro, con ~ 5533
acabado de máquina 8446
acabado imitación de tela 8184
acabado imitación lona 3702
acabado inglés 5179
acabado novela 4990
acabado pana 11468
acabado semiáspero 4990
acabado semimate 4990, 5179
acabar el trabajo 3885
acabo 5689
acabo de publicarse 7719
acanaladora 3718
acanaladura 3714
acanalar 3716, 6668
acaparador 10357
acaparar 10355, 12610
acápite 7264, 9972
accesibilidad 68
accesible 193
acceso 67
acceso al azar 11231
acceso directo 11231
acceso múltiple 9179
accesorios 69
acción 12232
accionamiento 4736
accionamiento a mano 1174
accionamiento coaxial 3004
accionamiento combinado 4883
accionamiento del carro 1276
accionamiento eléctrico 5016
accionamiento en paralelo 3004
accionamiento manual 1174
acción capilar 2279
aceitar 8395, 9548
aceite antracénico 656
aceite bruto 10168
aceite contraído 13035
aceite de alcanfor 2252

aceite de alizarina 413
aceite de antraceno 656
aceite de cachalote 12859
aceite de castor 2458
aceite de castor sulfonado 14091
aceite de colza 11243
aceite de engrase 8397
aceite de esperma 12859
aceite de espliego 7943
aceite de esquistosa 12228
aceite de fusel 565
aceite (de habas) de soja 12795
aceite de lavanda 7943
aceite de linaza 8224
aceite de linaza cocido 1695
aceite de linaza crudo 11257
aceite de linaza soplado 1625
aceite de lino 8224
aceite de lino crudo 11257
aceite de madera de China 14084
aceite de manitas 9260
aceite de máquina 8397
aceite de nuez de palma 9882
aceite de oiticica 9569
aceite de oliva 8399
aceite de palmacristi 2458
aceite de palo 14084
aceite de parafina 9966
aceite de perilla 10140
aceite de pescado 5717
aceite de pezuña 9260
aceite de pino 13566
aceite de resina 11687
aceite de ricino 2458
aceite de ricino sulfonado 14091
aceite de semilla de soja 12795
aceite de trementina 14107
aceite de tung 14084
aceite de vitriolo 13389
aceite esencial 5257
aceite lubricante 8397
aceite mineral 8953
aceite no secante 9399
aceite para maquinaria 8397
aceitera 9552
aceite rojo de Turquía 14091
aceite secante 4793, 8215
aceite semisecante 12100
aceites inyectados 1626
aceite sulfonado 13382
aceite volátil 5257
aceite vulcanizado 14523
aceleración 52
aceleración angular 603
acelerador 53
acelerar 48
acento 54
acento agudo 175

acento alemán 14256
acento circunflejo 2860
acento circunflexo 2860
acento diacrítico 4250
acento diéresis 4251
acento grave 6577
acento largo 8326
acento postizo 5838
acento separado 5838
acentuación 62
(acentuado) en cursiva 59
(acentuado) en negrilla 58
(acentuado) en seminegra 58
(acentuado) en versalitas 60
acentuar 57
acepilladora 1611, 10395
acepillar 10392
acerado 13094
acero inoxidable 13012
acetaldehído 86
acetato 88
acetato alumínico 472
acetato amónico 513
acetato celulósico 2509
acetato de aluminio 472
acetato de amilo 563
acetato de amonio 513
acetato de butilo 2127
acetato de celulosa 2509
acetato de cobre 3498
acetato de etilo 5300
acetato de isoamilo 563
acetato de plomo 7972
acetato de polivinilo 10566
acetato de potasa 10623
acetato de potasio 10623
acetato de sodio 12669
acetato de sosa 12669
acetato de uranilo 14367
acetato de uranio 14367
acetato etílico 5300
acetato potásico 10623
acetato sódico 12669
acetil acetanilido 101
acetil celulosa 2509
acetoacetanilida 101
acetona 102
acial 14461
acicalar 6255
aciculado 112
acicular 112
acidez 133
ácido 113, 114
ácido abiético 6
ácido abietínico 6
ácido acético 95
ácido acético glacial 6418
ácido acrílico 157
ácido agálico 6289
ácido arábigo 745
ácido azótico 9366
ácido bicloroacético 4292
ácido bórico 1798
ácido carbólico 10189
ácido carbónico 2315
ácido cianhídrico 7105
ácido ciánico 3909

ajustar 218, 8570
ajustar la forma 8308
ajustar la última línea 5157
ajuste 226, 8307
ajuste de exposición 5374
ajuste de la presión 7222
ajuste de las punturas 9271
ajuste del registro lateral
   (de la bobina) 12373
ajuste del tintero 3114
ajuste tipográfico 8569
alabearse 14545
alambre para coser 13179
alambre para encuader-
   nación 13179
alambre para máquinas de
   coser 13179
álamo 10577
alargamiento 5361
alargamiento a la tracción
   13264
alargamiento de la tensión
   13264
alargamiento de rotura
   5095
alargamiento de ruptura
   13264
alargamiento por humedad
   7135
albayalde 5749
albertipia 3099
álbum 369
álbum de recortes 11944
albúmina bicromatada 4299
albúmina de huevo 371,
   4988
albúmina en escamas 5748
álcali 414
alcalinidad 418
alcalino 417
álcalirresistente 416
alcance 3229, 6223, 11236,
   13553
alcances de reflectividad
   1910
alcanfor 2251
alcanos 424
alcanzar con línea llena
   5160
alcohol 380
alcohol alílico 443
alcohol amílico 565
alcohol anhidro 608
alcohol butílico 2128
alcohol de madera 8855
alcohol de polivinilo 10567
alcohol de quemar 4162
alcohol desnaturalizado
   4162
alcoholes 382
alcohol etílico 5301
alcohol isopropílico 7619
alcohol metilado industrial
   7297
alcohol metílico 8855
alcohol polivalente 10557
alcohol polivinílico 10567
alcohol propílico 10973
alcohol puro 608

alcoholresistente 381
aldehído 385
aldehído acético 86
aldehído fórmico 6048
Aldino 388
aleación (de metal de
   imprenta) 437
aleación de plomo 7973
aleación ligera 8122
aleación para estereotipia
   13126
aleación para máquinas de
   componer 3289
aleación tipográfica 14208
aleatorio 11230
aleta 5755
alfa 5252, 453
alfabetizar 458
alfabeto 454
alfabeto alemán 6384
alfabeto de ocho unidades
   4994
alfabeto de seis unidades
   12477
alfabeto francés 6167
alfabeto griego 6618
alfabeto latino 7940
alfanumérico 461
alfiler para picar reportes
   13144
alforza 14099
algarabía 7648
alginato 397
algodón 3636
algodón pólvora 9371
algol 398
algoritmo 399
algrafía 490
alicates de boca chata
   (para ceñir los nervios)
   1112
alilsulfocarbamida 450
aliltiourea 450
alimentación continua,
   de ~ 14646
alimentación de tinta a la
   página 9856
alimentado a mano 6825
alimentado con hojas 12267
alimentado por rollos 14646
alimentador 5507
alimentadora 5507, 5508
alimentador automático
   909
alimentador automático del
   metal 917
alimentador de metal 8825
alimentador de succión
   13354
alimentar 5503, 8849
alineación 405, 11420
alineación inferior 1196
alineación molecular 9052
alineación paralela 1227
alineador 8193
alineamiento 405, 14206
alinear 404, 7739, 8156
alineo 405
alisado de máquina 8446

alisador 12978
alisadora 12978
alisar 10547
aliso 387
alma 3565, 12869
almacén 8479, 13238
almacenaje de papel 13228
almacenamiento 13230,
   13240
almacenamiento magnético
   8523
almacenamiento temporal
   13643
almacenar 13184, 13232
almacenar y conmutar
   hacia adelante 13233
almacén auxiliar 943
almacén bajo 8382
almacén de papel 9946
almacenero 13239
almacenista 13239
alma deformada 3783
alma (de la bobina) 11363
alma del rodillo 11649
almagre 11867
almanaque 451
almanaque de pared 1590
almanaque en hojas 1590
alma para rodillo 11649
almártaga 8248, 11330
almártega 8248
almazarrón 11867
almibarado 13498
almidón 13045
almohada de grabado 5189
almohadilla de algodón
   10480
almohadilla de dorar 6501
aloe 452
áloe 452
alquifol 6272
alquitrán 10380
alquitrán de hulla 2972
altas luces 6992
altas y bajas 2293
alta tensión 7004
alta y baja 2293
alteración 463
alteraciones del autor 887
alterar 462
alterar con algún
   modificante la tinta 4560
alternar en zigzag 13002
alternomotor 466
alto 813
alto de la página 4196
alto de las mayúsculas
   6947
altura alemana 6385
altura americana 498
altura de carga 10341
altura de Francfort 6948
altura de la corona del
   cilindro 3929
altura de la página 4196
altura del árbol 6949
altura de Leipzig 8034
altura de letra equis 14894

altura del hombro (de tipo)
   6949
altura del ojo 6336
altura de los caminos 1274
altura de los tipos 6953
altura del tipo 6953
altura del tipo, de la ~
   14201
altura de ojo 6952
altura media 14894
altura tipográfica 6953
altura tipográfica, de la ~
   14201
alumbrado 7160
alumbrado concentrado
   10528
alumbrado directo 4406
alumbrado indirecto 7293
alumbrado reflejado 7293
alumbre 468
alumbre de amonio 473
alumbre de cromo 2808
aluminio 471
aluminio en polvo 484
aluminografía 490
aluminotipia 491
alumno 721
alveolado 2924
alveolo 7331
alvéolo 7331
alza 6821, 7497, 9768, 9771,
   10045, 14281
alza debajo el cliché 14282
alzadora con alambre 6304
alzadora cosedora 6304,
   11829
alzadora (de hojas) 6330
alzadora de pliegos 6330
alza mecánica de greda
   2602
alzamiento de blancos
   11620
alzaplumas 13273
alzar 6325
alzar los rodillos 13791
alzar, sin ~ 14298
alzar y acoplar 6325
amador de libros 1364
amalgama 494
amarilleado del reverso
   3105
amarilleo 14917
amarillo bárico 1143
amarillo bencénico 1330
amarillo de bario 1143
amarillo de bencidina 1330
amarillo de cadmio 2146
amarillo de cinc 14959
amarillo de cromo 2783,
   7977
amarillo de tartracina
   13611
amarillo de toluidina 13882
amarillo Hansa 6856
amarillo permanente 13882
amarillos de óxido de
   hierro 7597
amarradora automática
   2064

aplicación (de la emulsión acuosa) de gutagamba 6295
aplicación del papel pigmento 2332
aplicación de quitavelo 12019
aplicación en la máquina 8436
aplicación en seco 4808
aplicado del esmalte 13912
aplicador de fondos 1315
aplicar cartivanas 6690
aplicar el barniz 9871
aplicar el mordiente (para dorar) 6420
aplicar escartivanas 6690
aplicar grisados 13845
aplicar sombreados 13845
aplicar tinta de reserva 9871
apomazar 11036
apostilb 707
apostilla 706, 8629
apóstrofo 708
apoyamanos 6841
apoyo 1842
apoyo del fuelle 1295
aprendiz 721, 3796, 7234, 13956
aprendiza 3796
aprendizaje 724
aprendiz de impresor 10809
aprendiz de tipógrafo 722, 10809
aprendiz gráfico 10809
aprendiz mayor de tipógrafo 723
aprendiz traslado 14103
aprestado 5686
aprestado mate 4853
aprestado por inmersión 4391
apretar 8308, 12635
apretar con tornillos 1715
apretar (la composición) 7738
aprobar 10020
apropiar 178
aproximar 2943
aptitud a la floculación 5843
aptitud para tomar tinta 7385
apuntalar 13148
apuntar 11262
apunte 11715
arabescos 744
arabina 745
arandela 14549
arañamiento 12015
arañazo 10, 1228, 12601
árbol 963, 968, 12225
árbol conductor para las lengüetas 6652
árbol de fronda 1930
árbol del cilindro de cambio 6227
árbol de levas 2254

árbol del volante 5936
árbol de transmisión 4748
árbol frondoso 1930
arcilla 2711
arcilla de aluminio 2711
arco voltaico 754
arco voltaico para copiar 10818
archicarse los puntos 12234
archivador 5619
archivador de cartas 8057
archivar 5613
archivo de la redacción 9101
área de exploración 2906
área de impresión 10819
área de la curva histerética 759
arenero 11604
areómetro 761, 7120
areómetro Baumé 1233
areopagita 762
argentómetro 763
argumento 764
armada 8569
armado 8569
armador 8572, 8573
armadura 6144, 8569
armar 8570
armario de secado 4786
armario para imposiciones 6263, 8002
armario para ramas 6067
armario para secar 4786
armario portagaleras 6284
armario portarramas 6067
armario secador 4786
armas 3001
armazón 6144, 8569
armazón del rodillo 11649
armorial 774
arquitecto de libros 1748
arracada 8629, 11793
arrancado de fibras en seco 4813
arrancador 13059
arrancamiento 3691
arrancamiento (de tinta) 10295
arrancar 11810, 13054
arrancar poco a poco 7245
arreglar 218, 2900, 8564
arreglo 8566
arreglo de la tensión de la banda 14662
arreglo de plancha 1023, 14281
arreglo de tintaje 3114
arreglo excesivo 9775
arreglo mecánico con greda 2604, 8729
arreglo previo 2062
arribo insuficiente de tinta 7464
arrollador de matrices 8699
arromar 1645
arrugado 6191
arrugar 14874
arrugas 5970, 14873

arrugas de calandria 2186
arrugas de expansión 5364
arte conservatriz de todas las artes 781
arte de grabar al aguafuerte 802
arte de grabar en madera 14798
arte de la encuadernación 801
arte de la imprenta 803
arte del libro 804
arte de xilografía 805
arte gráfico 803
arte (gráfico) modernista 800
artesa de encolado 12501
artesa de la cola 10029
artesa de mojado 14577
artesa de revelar con aspiración 4674
artesa para grabar 5290
artes gráficas 6565, 10831
arte tipográfico 803
articulación esférica 1094
artículo 787
artículo de almacén 13188
artículo de fondo 7989
artículo editorial 4977
artículo ilustrado 7163
artículo para llenar 5622
artista dibujante 3240
artista grabador en cobre 3522
artista litógrafo 8256
artista publigráfico 271
artotipia 3099
ascensión capilar 2283
ascensor 8093
ascensor de pila 10342
asegurar con tornillos 582
asentado de las planchas 10453
asentador 10394
asentamiento de la tinta 12176
asentar 10393
aserrar 11893
asfalto 821
asfalto en polvo 823
asiento de la plancha 10453
aspirador 13351
aspirador adelantador 6092
aspirador de polvo para bobinas 14643
aspirador de tinta 7377
aspiradores 13348
aspirante (a reportero) 3796
asterisco 850
astigmatismo 853
astilladora 2728
astillas 12310
astillita de madera 12587
astralón 855
atacador 12318
atacar 13148
atadora 2064
atar 13801

atendedor 3546
atenuación 867
atenuación de color en las sombras 14269
aterciopelado 5844
atíncar 12680
atlas 861
atlas de bolsillo 10507
atlas portátil 10507
atmósfera 863
atmósfera física 13021
atomizador 866
atornillar 582, 1715
atravesar 13276
atril 3544, 4520, 4941
augaripola 2206
aumento 5395, 9784
aumento de consistencia 4368
aumento de velocidad 51
auramina 874
aureola 4569, 6754, 6801
auréola 12976
aurinas 877
autobiografía 891
autocarga 1788
autocentrado 12079
autocentrador y cuadrador 12085
autocódigo 895
autocromía 3162
autocubierta 7474
autógrafo 10734
autoinflamación 12915
automación 929
autómata de exposición 908
automatismo 929
automatización 929
autooxidación 940
autopolimerización 934
autopositiva 4410
autopositivo directo 4410
autor 881
autora 881
autor-editor 885
autor, en casa del ~ 11004
autotipia 6774
autotipia de blancos puros 6987
autotipia de cinc 14947
autotipia de grano 6545
autotipia de mordido extra-hondo 4091
autotipia de periódicos de trama gruesa 11140
autotipia de trama gruesa 2978
autotipia duplex 4877
autotipia silueteada 9721
auxiliar del prensista 8456
auxiliar de redacción 3796
auxiliares 12786
auxiliares de matriz 12786
avalorar un manuscrito 2457
avalorar un original 2457
avance 5337
avance (del sueldo) 7043
avisador 261

avisar 245
avisero 250
avisero-cajista 250
aviso 246
aviso a toda página 6237
ayudante de conductor de la máquina de papel 1049
ayudante del compaginador 8571
ayudante del corrector 3546
ayudante mecánico 8456
aziminobenceno 1336
ázote 9374
azúcar de uva 4246
azufre 13383
azul acero 10992
azul Berlín 1337
azul cobalto 3006
azul chino 10992
azul de álcali 415
azul de añil 7288
azul de bromofenol 1955
azul de cobalto 3006
azul de Paris 10992
azul de Prusia 10992
azul (de) ultramar 14247
azul ftalociánico 9065
azul milori 8945
azul pavo real 10069
azul ultramarino 14247
azul verdoso 3908
azurada 6740
azurado 2697, 6740

badana 1183
badana (blanca) 12249
badana colorada 3138
bagazo 1067
bailarín 3985
bajada de las matrices 8689
baja de la tensión superficial 13442
bajar el componedor 5119
bajar un blanco subido 11063
baja tensión 8389
bala 1090, 7329
bala de entintar 3954
balance del color 3119
balancín 2227
balancín de la cuchilla 3882
balanza 11898
balanza de ceniza 818
balanza de cuadrante para papel 11107
balanza de laboratorio 7855
balanza de torsión 13926
balanza para cenizas 818
balanza rápida de análisis 578
baldés 3138
bálsamo del Canadá 2255
bambú 1109
bambuc 1109
bancada 6144
banco de datos 4006

banco de pruebas 4861
banco de taller 1311
banco de trabajo 1311
banda 1110
banda adhesiva 215
banda autoadhesiva 12077
banda con exceso de tensión 13810
banda continua 14639
banda corregida 2897
banda de absorción 36
banda de contacto 5770
banda de corrección 3606
banda de papel 9949, 14639
banda (de papel) con comba 12537
banda de papel para pH 10180
banda (de transmisión) 1301
banda extensimétrica 11509
banda magnética 8524
banda perforada 11046
bandas laterales 1275
banda tirante 13810
bandeja estibadora 9878
bandera 5742
banderilla 11597
banderita 5741, 7828
baño 1223
baño ácido 116, 5274
baño al bicromato 2774
baño cromatado 2774
baño curtiente 6864
baño de ácido 116
baño de blanqueo 1545
baño de cobrizar 3500
baño de color 3121
baño de contrarrestante 13209
baño de cromatado 2773
baño de detención 13209
baño de endurecimiento 6864
baño de fijado 5736
baño de goma tragacanto 1273
baño de graneador 6549
baño de granear 6549
baño de interrupción 13209
baño de María 14568
baño de mordido 5274
baño de paralizador 13209
baño de plata 12425
baño de revelado 4224
baño de saturación 11885
baño detenedor 13209
baño de virado 13899
baño de viraje 13899
baño electrolítico 5034
baño endurecedor 6864
baño fijador 5736
baño fijador ácido 124
baño fijador-endurecedor ácido 128
baño galvanoplástico 5064
bañomaría 14568
baño mordiente 5274
baño para grabar 5274

baño revelador 4224
baño termostático 13699
baño virador 13899
baquelita 1082
barajas 10477
barba 14855
barba (del papel) 4065
barbas 14264
barbujitas 5943
baremo 11287
barillas sacapliegos 5923
bario 1142
barita 1182
baritina 1182
barniz 7863, 14401
barnizabilidad 14402
barnizado 7866
barnizado a calandra 2198
barnizadora 14406
barniz al alcohol 12885
barniz blando dúctil 8351
barniz brillante 6455
barniz celulósico 2511
barniz de aceite de linaza 8225
barniz de asfalto 828, 1459
barniz de brillo 6455
barniz de copal 3494
barniz de goma laca 12290
barniz de secado rápido 11153
barniz flojo 12339, 13719
barniz fuerte 13299
barniz lito(gráfico) 8279
barniz lustroso 6455
barniz mate 6455
barniz para cubrir 8648
barniz para etiquetar 7853
barniz para etiquetas 7853
barniz para negativos 9279
barniz para rebajar 5384
barniz para reserva 13216
barniz para sobreimpresión 9780
barniz para tapar 8648
barniz secante 4804
barniz tapador 13007
barniz zapón 14932
barquín 1293
barquinera 1293
barra de aspiración 13350
barra de cabeza 8630
barra de distribución 4499
barra de distribución de tinta 7382
barra de entrada 3757, 13003
barra de grapa 2872
barra de inflexión 588
barra de inversión 588, 10572
barra de la mantilla 1513
barra de las pinzas 6652
barra de la tela 2871
barra de la trama 2523
barra (del husillo de la prensa) 1136
barra de metal 10327
barra de uña 2872

barra de vidrio 6437
barra distribuidora 4499
barra portacuchilla 7819
barra rascadora 5223
barras de guía 6700
barras de rodillos 11641
barras ómnibus 2101
barra tensora de la mantilla 1527
barrenado 4732
barreta de ensayo 13668
barrilete del objetivo 1170
barrote de la galera 6281
base 1184, 6013, 9136
base de acetato 89
base de celdillas 10056
base del bloque de pistón 10377
base de madera 14790
base de metal 8817
base de plomo 7975
base filiforme 1841
base metálica 8817
bases oblicuas 9472
base triangular 1841
basicidad 1204
bastardilla, de ~ 7621
bastardilla, en ~ 7621
bastardillas 7625
bastidor 3998, 4063, 6204, 10442
bastidor de la cremallera 11189
bastidor del triángulo 6069
bastidor para coser libros 12195
bastidor para cosido 12195
batán de trapos 4891, 14744
batería de rodillos 12167
batería de secadores 4774
batería de tintaje semicontinua 3431
batidor 309
batidora 11385
batidora Jordán 7693
batidor de trapos 4891, 14744
batir a mano 1870
batir la tinta 11666
batir la tinta en el tintero 11817
bautizar 6341
bayeta 1508, 5754
becerillo 2199
becerro 2199
benceno 1326
benzoato de sodio 12675
benzoato de sosa 12675
benzoato sódico 12675
benzol 1326
benzotriazol 1336
bermellón 14443
betún 1458
betún de Judea 1458, 13497
bibásico 4288
bibliocasto 1749
bibliófilo 1364
bibliografía 1361
bibliógrafo 1360

bibliómano 1362
bibliópegia 801
bibliópola 1775
bibliorrato 8057
biblioteca 8083
biblioteca circulante 8036
biblioteca del rey 11745
biblioteca de programas 10917
biblioteca particular 10872
biblioteca real 11745
biblioteca universitaria 14326
bicarado 14158
bicarbonato amónico 517
bicarbonato de amonio 517
bicarbonato de sodio 12676
bicarbonato de sosa 12676
bicarbonato sódico 12676
bicloruro de mercurio 8792
bicromato 4298
bicromato amónico 526
bicromato de amonio 526
bicromato de potasa 10648
bicromato de potasio 10648
bicromato de soda 12688
bicromato potásico 10648
bicromía 14136
bidón de aceite 9552
biela conductora para las lengüetas 6652
bifluoruro amónico 519
bifoliar 1381
bifurcación 1854, 7705, 9384
bigote 4001, 6182
bigotillos 12144
bilingüe 1383
billete bancario 1126
billete de banco 1126
bimano 1391
bímano 1391
binocular 1435
biografia 1437
bioxalato de potasa 10631
bioxalato de potasio 10631
bioxalato potásico 10631
birlí 12334
bisagra de tela 2954
bisel 1344, 8995
biseladora 1355, 12314
biselar 1354, 2609
bisemanario 12104
bismuto 1442
bisulfato de potasa 10632
bisulfato de potasio 10632
bisulfato de sodio 12678
bisulfato de sosa 12678
bisulfato potásico 10632
bisulfato sódico 12678
bisulfito cálcico 2162
bisulfito de calcio 2162
bisulfito de magnesio 8497
bisulfito de sosa 12679, 12679
bisulfito sódico 12679
bisulfuro de carbono 2316
bit 1448
bitartrato de potasa 10633
bitartrato de potasio 10633

bitartrato potásico 10633
bit de comprobación 2664
bit de paridad 10008
bitono 4869
bitumen 1458
bivalente 1460
blanco 1500, 14711
blanco de antimonio 671
blanco de cabecera 6889
blanco de cabeza 6889
blanco de cinc 14951
blanco de cubrir 14951
blanco de China 14951
blanco de entrada 6656
blanco de España 12817
blanco de Krems 5749
blanco de la letra 1247
blanco del tipo 1247
blanco de Meudón 14731
blanco de mezcla 13840
blanco de pie 6012
blanco de pinzas 6654
blanco de plomo 5749
blanco de relleno 12811
blanco fijo 1156
blanco inglés 8276
blanco interior 3651
blanco interno 3651
blanco levantado 1493
blanco lustroso 6456
blanco opaco 14951
blanco para mezclar 9014
blanco para retoque 10905
blanco permanente 1156
blancos 1531, 6992, 9401, 12811
blanco satín 11878
blanco transparente 9014
blanco y negro 1463
blancura 14719
blando 5739, 5740
blandura 5740
blanqueado 1542, 1543
blanqueador 10679
blanqueador óptico 9646
blanqueante óptico 9646
blanquear 1538
blanqueo 1542
blanqueo óptico 9645
blanquillo 12249
bloc de calendario 4018
bloc de notas 13514
block 13514
blocking 1598
bloque 1583
bloquear 8309, 9137, 13215, 14098
bloque de cuchillas 7818
bloque de madera contra-chapeada 7890
bloque de madera laminada 7890
bloque de madera terciada 7890
bloque de metal 10327
bloque de pistón 10376
bloque de pistones de aire 348
bloque de relleno 5627

bloque para fondo 13841
bloque para fundir orlas y filetes 8683
bobina aplastada 1062
bobina con daño causado por el agua 14575
bobina corrida 13631
bobina de ancho reducido 4385
bobina decentrada 9727
bobina de deflexión 4101
bobina de desviación 4101
bobina deformada 1062
bobina de papel 11354, 14658
bobina de papel continuo 11354
bobina de papel de periódico 11354
bobina de papel de una (sola) página 4385
bobinador 14747
bobinadora 14747
bobina dura 13808
bobina estropeada 13050
bobina floja 8367, 12535
bobina no redonda 9727
bobina preimprimida 10730
bobinero 11369
bocacé 2030
boca de fundición 10684
boca de fundir 10684
boca del crisol 10684
boca de molde 10684
bocetista 3240, 7962
boceto 7961
bodega para papel 9946, 13238
boj 1839
bol arménico 772
bolas de vidrio para amolar 6430
bolas graneadoras 3037
bolas para amolar 3037
bolas para granear 3037
bolas para granear planchas litográficas 6430
bol de Armenia 772
boleta de trabajo 7630
boleta-horario 7681
boletín de abono 13333
boletín de la bolsa 5343
boletín de letras incompletas 1389
boletín de subscripción 13333
boletín oficial 6343
boletín para el personal 7060
bolígrafo 1104
bolsa 1066, 1828, 11826
bolsa con fondo cruzado 11682
bolsa con fondo plano 11876

bolsa con fondo rect-angular y fuelles laterales 12083
bolsa con foro múltiple 9198
bolsa con grapa 2878
bolsa con pliegue lateral 6731
bolsa de ánodo 642
bolsa de bloque 1589
bolsa de botón 13279
bolsa de celofán 2502
bolsa de envío 10511
bolsa de expedición 10511
bolsa de fondo cuadrangular 1589
bolsa de papel 1298, 9898
bolsa de papel de doble hoja 4873
bolsa de ruta 7630
bolsa de varias capas 9198
bolsa grande 6613
bolsa para bizocho 1441
bolsa para compras 2372
bolsa (para) libros 1772
bolsa plana 5773
bolsa portátil 2372
bolsilla 1828
bolsillo 10689
bomba alimentadora 5512
bomba de ácido 141
bomba de agua 14596
bomba de aire 354
bomba de aspiración 14376
bomba de circulación 2851
bomba de circulación de la tinta 7380
bomba de cola 10039
bomba de émbolo 14376
bomba del crisol 10686
bomba de tinta 7380
bomba de vacío 14376
bomba para subir la tinta 7379
bombeado 4550
bombilla 10375
bombilla eléctrica 5012
bombilla para cámara oscura 3995
bombo 1647, 11013
bombo de impresión 7211
bombo de presión 7211
bombo descortezador 1166
bombona 2337
boquilla 9151, 9445, 9684
boquilla del crisol 10684
boquilla del soplador 1623
boquilla encoladora 10036
boquilla pulverizadora 12929
boquilla sopladora 1630
borato de sodio 12680
borato de sosa 12680
borato sódico 12680
bórax 12680
borde 1793, 8627
borde antique 4065
borde a sangre 1553
borde de encolado 6472

borde de guía 6695
borde de impresión 10827
borde de la galera 6281
borde de luto 9149
borde esfumado 14469
borde greco 4065
borde guillotinado a sangre 1553
borde perdido 1553
bordes cerrados 4517
bordes flojos 12533
bordón 9584
borla para polvo 4897
borne 13653
borrador 12005
borrador de tinta 7345
borrar 2899, 4128, 5232, 11494, 13272
borrarse 2464, 5633
borras de algodón 3638
borratintas 7345
borrón 1620, 12611
borrón (de tinta) 9069
bosquejo 7961, 9717
bosquejo anuncio 253
botador 5002, 5003, 12318
bote de lata 13823
bote de tinta 7330
botón arrancador 13056
botón de arrancar 13056
botón de contacto 11060
botón (de la tecla) 7761
botón de presión 11323
botonera 3459, 11060
botón pulsador 11060
botón rojo (en caso de emergencia) 11323
bramante 1430, 9855
brazo elevador 12040
brazos (del estante porta-bobinas) 11355
brea 10380, 10381
breve 12323
brida 5752
brida de supresión de juego 1030
brida eliminadora de holguras 1030
brillantadora 5570
brillante 6457
brillo 6452
broca (para taladradora) 11733
brocha 9870
brocha china 2718
brocha de aire 314
brocha de la cola 10023
brocha para agua fuerte 5275
brocha para desempolvar 4898
brocha para despolvorear 4898
brocha plana 5778
broche 2876, 13040
bromo 1949
bromóleo 1953
bromoleotipia 1953
bromuro 1941

bromuro amónico 520
bromuro cúprico 3501
bromuro de amonio 520
bromuro de cadmio 2140
bromuro de cobre 3501
bromuro de etilo 5303
bromuro de hierro 5545
bromuro de mercúrico 8790
bromuro de mercurio 8790, 8791
bromuro de metilo 8862
bromuro de oro 6498
bromuro de oro y potasio 10651
bromuro de plata 12426
bromuro de potasa 10634
bromuro de potasio 10634
bromuro de sodio 12681
bromuro de sosa 12681
bromuro férrico 5545
bromuro ferroso 5574
bromuro mercurioso 8791
bromuro potásico 10634
bromuro sódico 12681
bronce 1859, 1956, 1957
bronceado 1968
bronceador 1967
bronceadora 1969
bronce al aluminio 474
bronce al oro 6499
bronce de estampación 1584
broza de alambre 14767
broza para (limpiar con) lejía 8422
brozar 1986
bruñidora mecánica 2039
bruñidor de ágata 296
bruñidor (de diente de lobo) 2079
bruñir 2081
bruza 1981
bruza de batir 1268
bruza de bobinas 11359
bruza de revelado 4225
bruza para amoldar 1268
bruza para limpiar 14176
bruza para pruebas 1267
bruza para sacar pruebas 1267
bruza para tirar pruebas 1267
bruzar 1986
bucarán 2030
bucle 8355
budinadora 5411
buen original 5440
buen uso, en ~ 9259
bujía 2266
bujía-metro 6016
bujía-pie 6016
bulón 1714
bullones 1800
buna 2063
burbuja 1577, 5810
burbuja de aire 336
burbujas 12475
burbujas de aire 311
burbujas de espuma 311

burbujoso 10583
buril 2072
burilado 13900
burilado a mano 6848
buril agrisado 8208
burilar 5182
buril con cuatro cantos 12967
buril (de grabar) 6578
buril de picado 13164
buril de punteado 13164
buril para picar 13164
buril para puntear 13164
buril plano 5790
buril rascador 9266
buril raspador 11949
buril rayador 8208
buril redondo 12017
burro 1745, 2414, 3284
burro de carga 14706
buscar 12029
buscar el registro 1919
buscavida(s) 14706
busqueda binaria 1399
bustrófedon 1818
butadieno 2108
butanol 2128
byte 2133

caballete 4941
caballete para secar negativos 4694
caballo de energía 7044
caballo de fuerza 7044
caballo de potencia 7044
caballo de vapor 7044
cabecera 6903, 6905, 7322
cabecera a todo plana 1131
cabecera de carta 8058
cabecera de factura 1385
cabecera (del periódico) 8661
cabecera-marca 8661
cabecero 2627, 6907
cabeza 813, 2297, 6905
cabezada 2272, 6894
cabezada del pie 13541
cabeza de capítulo 2627
cabeza de grabar 5193
cabeza (de la página) 6890
cabeza de lectura/grabación 11284
cabeza del primer elevador 5704
cabeza de muerto 14095
cabeza de página 11805
cabeza dorada, con ~ 13911
cabeza en bandera 1131
cabeza encoladora 10036
cabeza escudriñadora 11912
cabeza giratoria 1094
cabeza grabadora 5193
cabeza impresora 6520
cabezal de cosedora 13177
cabeza lectora 11912
cabeza lectora/grabadora 11284

cable bowden 1821
cable coaxial 3003
cabo de carga 13537
cabo de fundición 7657, 13537
cabo del embudo 6072
cabria de bobinas 11370
cabujones 1800
cabuya 300
cachou 2474
cachú 2474
cadena automatizada de producción continua 913
cadena de datos cerrada 2091
cadena del receptor 4138
cadeneta 7756
cadenetas 2589
cadmio 2139
caída 9511
caída de potencial 14510
caída de tensión 14510
caja 1827, 2378, 2398
caja alta 14355
caja aspirante 13351
caja baja 8379
caja blancos 1139
caja californiana 2209
caja con listones móviles 2419
caja corredera 12552
caja corrediza (de cartón) 12291
caja de alimentación 6898
caja de alimentación por gravedad 6580
caja (de cartón) 1826
caja de cartón 2344, 2378
caja de cartón ondulado 3620
caja de cartón semirrigida 5975
caja de cifras quebradas 6142
caja de distribución 4500
caja de enchufe 12660
caja de entrada 6898
caja de entrada por gravedad 6580
caja de espacios (de cuña) 12803
caja de espolvorear 4896
caja de fundir 2443, 2450, 9118
caja de hilo metálica 13171
caja de imprenta 2396
caja de ingletes 8998
caja del distribuidor 4500
caja del espaciado 12803
caja de (los) espaciadores 12803
caja demasiado llena 1140
caja de mezcla 9011, 13186
caja de pastel 2149
caja de resina 4896
caja de suertes 12789
caja de suertes sobrantes 6117
caja de tapa 2290

caja de tipo(s) 2396
caja de tipos para re-
    miendos 7670
caja de válvulas 14381
caja distribuidora 4500
caja empastelada 6098
caja española 12813
caja espolveadora 4896
caja-estuche 12561
caja exposición 4461
caja media 6757
caja-molde 2443
caja montada 12182
caja numeradora 9450
caja para blancos 1139
caja para chocolate 2749
caja para filetes 11774
caja para guardar las
    matrices suplementarias
    12789
caja para letras acentuadas
    56
caja para matrices sueltas
    12789
caja para números 5608
caja para plecas 11774
caja para queso 2677
caja para rayas 11774
caja plegable 5975
caja plegable con
    enganches en los
    costados 2381
caja plegable con fondo
    automático 2385
caja plegable con solapas
    alternas 2389
caja plegable con solapas
    opuestas 2882
caja plegable de una sola
    pieza 2387
caja plegable mixta 2391
caja plegadiza 5975
caja por tres cuerpos 14017
caja posterior de la cámara
    2240
caja provista 13186
caja reguladora 11422
caja rigida 12182
caja Solander 12750
caja solapas abrocha-
    miento con ventana de
    celofán 2390
caja solapas abrocha-
    miento dos piezas con
    pestana inferior a toda
    altura y reborde 2384
caja solapas ranuradas
    2386
caja (tipográfica) 2396
cajera 7786
cajetilla para cigarillos
    2830
cajetín 1829, 14205
cajetín de (las) ánimas
    6958
cajetín del diablo 6958
cajetín del pastel 6958
cajetín de reserva 13448
cajista a sueldo fijo 12989

cajista de anuncios 250
cajista de líneas 3313, 6819
cajista de remiendos 7673
cajista distribuidor 4496
cajista especializado en la
    composición de páginas
    (de libros) 3314
cajista hábil 13466
cajista liniero 3313, 6819
cajista-minervista 14125
cajista montador 8573
cajista plantilla (a jornal
    fijo) 12989
cajista-prensista 14125
cajista remendista 7673
cajista (tipógrafo) 3313
cajita para chocolate 2749
cajo 7690
cajón (de metal) 6958
cajón de polvos 10701
cajón humectante 7072,
    7078
cajón humectativo 7078
cajón para botar el plomo
    6958
cajón para echar el plomo
    6958
cajuela blancos 1139
cajuela para blancos 1139
cal 8145
cala 8193
calado defectuoso 10462
caladora 7663
cala móvil 8193
calandra 2184, 8699, 9129
calandra de satinar 12995
calandrado con calor 7046
calandrado en caliente
    7046, 7053
calandrado en hojas 12262
calandra para hojas 12261
calandrar 2191
calandria 2184, 8699, 9129
calandria a fricción 6193
calandria para hojas 12261
calandria para satinar el
    papel en hojas 12261
calcar 13935
calcinación 2153
calcio 2160
cal clorada 2168
calco 13936
calcografía 2594, 7471
calcomanía 13969
calcomanía cerámica 2558
calcomanías 4041
calculador 5265
calculadora 2177
calculador analógico 572
calculador (de datos) 3330
calculador de exposiciones
    5369
calculador numérico 4359
calcular el espacio de origi-
    nal(es) 2457
calcular el material 2457
calcular el precio de la
    composición 2459

calcular un manuscrito
    2457
calcular un original 2457
cálculo de espacio 3541
cálculo de la composición
    3541
cálculo de las dimensiones
    11906
cálculo de los gastos 3629
cálculo de material 3541
cálculo de original(es) 3541
cálculo (de precios) 5264
cálculo previo 3541
cálculos previos 3541
cálculo tipográfico 3541
caldera 7755
caldera de fundición 8836
caldera (de refundición)
    8770
caldera de refundir 8770,
    8826, 11456
caldera de vapor 1697
caldera para metal 8826
calderico 7755
calderilla 7755
calderón 7755, 9973, 12059
calefacción a gas 6317
calefacción central 2536
calefacción eléctrica 5020
calendario 451, 2178
calendario de bloque de
    notas 4210
calendario de notas 4210
calendario de taco 1590
calendario en bloc 1590
calentado 2074
calentador 6920
calentador de garganta
    13782
calentador de la
    componedora 3288
calentamiento excesivo
    9761
calentar las balas (de entin-
    tado) 5759
calentar los tampones 5759
calibrado 2204
calibrador 8898
calibradora 1048, 1611
calibrador de atrás 1000
calibrador del cilindro 3942
calibrador de tipos 14198
calibrador para la mantilla
    1523
calibre 6334
calibre del cuerpo 1686
calibre del espesor de la
    cama 9830
calibre del espesor de
    revestimiento 9830
calibre de puente 11599
calibre medidor de espesor
    electromagnético 5039
calibre para cilindro 11599
calicó 2206
calidad aceptable 63
calidad de la superficie
    1451

calidades superficiales
    13436
calidad intermedia 959
calidad media después de
    inspección 955
caligrafía 2218
calígrafo 2217
caliza lenta 4549
calomel 8793
calomelanos 8793
calor de evaporación 6928
calor de fusión 6926
calor desarrollado 7930
calor de vaporización 6928
calor específico 12828
caloría 2225
calor latente 7930
calotipia 13561
calva 14503
cal viva 2171
calza 7497, 14281
calzado 9771
calzar 1920, 8564, 9771
calzar la matriz 1053
calzo 7497, 9768, 14281
calzos para formas 13146
calle 6734
callejón 6734
cama 9825
cama blanda 12742
cama de corcho 3576
cama de goma tragacanto
    1273
cama de la cuchilla 3882
cama de las agujas 9268
cama del teclado 7766
cama dura 6873
cama fija 9825
cámara ampliadora 5202
cámara clara 2246
cámara colgante 9759
cámara de acondiciona-
    miento de aire 7074
cámara de cuarto o(b)scuro
    3993
cámara de fotomecánica
    10882
cámara de galería 6277
cámara de repetición 13116
cámara de reproducción
    10882
cámara de reproducción
    horizontal 7040
cámara de reproducción
    vertical 14450
cámara de retratos 6277
cámara de suspensión 9759
cámara fotográfica de
    distorsión 11491
cámara (fotográfica) de
    película en rollo 11657
cámara fotográfica de re-
    producción 10882
cámara (fotomecánica) hori-
    zontal 7040
cámara (fotomecánica)
    vertical 14450
cámara lúcida 2246
cámara obscura 3992

carpeta 1403
carpeta de anillas 11618
carpeta de argollas 11618
carrageen 7574
carrete 12916
carrete (para alambre) 14766
carretilla de elevación 8104
carretilla elevadora 8104
carretilla montacargas 8104
carretilla para bobinas 13347
carretilla para formas 6075
carretilla para transportar bobinas (de papel) 11356
carretilla portaformas 6075
carril 1283
carro 2364
carro conductor 4148
carro de entrega 4148
carro de los rodillos 11654
carro de traslación del elevador 5081
carro para transportar barras de metal 10328
carro para transportar lingotes 10328
carro portaformas 6075
carro portalingotes 10328
carro transportador 4148
cartabón 4715
carta circular 2844
carta de correspondencia 3610
carta de exposición 5369
carta de marear 2648
carta de navegar 2648
carta de pose 5369
carta marina 2648
carta marítima 2648
carta muestra del color 3129
cartas 10477
cartel 10609
cartel de escaparate 12350
carteles 6815
carteo 12650
carteo del papel 11254
cartera (de sobre) 5755
cartivanas 6692
cartivanas de relleno 3265
cartógrafo 2377, 8613
cartón 2393
cartón acanalado 3619
cartón acanalado doble 4605
cartón acanalado doble-doble 4598
cartón acanalado por una cara 12455
cartón ahuesado 7627
cartón aislante eléctrico 5009
cartonaje 9902
cartón amarillo 9006, 14927
cartón brillo 10756
cartoncillo 2342
cartoncillo gris 9301

cartoncillo grueso para marbetes 13535
cartoncillo para cajas plegables 5974
cartoncillo para tarjetas 7270
cartoncillos 1534
cartoncino para reproducciones por grabado 11952
cartón compacto 12759
cartón compacto de pasta mecánica parda 1975
cartón con burbujas 1578
cartón condensador 5009
cartón corrugado 3619
cartón corrugado doble 4605
cartón corrugado doble-doble 4598
cartón corrugado por una cara 12455
cartón cuero 8013
cartón de amianto 810
cartón de asbesto 810
cartón de celulosa 11027
cartón de dibujo 4705
cartón de dos capas 14150
cartón de estereotipia 5848
cartón de fibras de cuero 8018
cartón de madera blanca 8738
cartón de madera parda 1975
cartón (de) matriz 5848
cartón de paja 13254, 14927
cartón de pasta de cuero 8018
cartón de pasta mecánica 8738, 14808
cartón de pasta química 2687
cartón de pulpa química 2687
cartón de reforzar blancos 1013
cartón de reforzar matrices 1013
cartón de varias capas 9191, 10022
cartón duro 12759
cartonear 1053, 9121
cartoné, en ~ 7240
cartón engomado 1013
cartonera 8699
cartonería 1662, 9902
cartón estirado 1659
cartón estucado 2981
cartón estucado a máquina 8434
cartón fieltro 1013
cartón forrado 8170
cartón forrado de tela 2956
cartón glaseado a máquina 8449
cartón gris 1013, 1974, 2726, 9301
cartón gris mixto 1977

cartón grueso 8932
cartón hidráulico 10756
cartón marfil 7627, 7628
cartón moldeado 11030
cartón monolúcido 8449
cartón muelle 1013
cartón multicapa 9174
cartón múltiple 9191
cartón ondulado 3619
cartón ondulado doble cara 4605
cartón ondulado forrado a dos caras 4605
cartón ondulado por una cara 12455
cartón ordinario 2726
cartón paja 9006, 13254
cartón para cajas 1831
cartón para cartonaje 1831
cartón para dibujo 4705
cartón para embutición 1659
cartón para encuadernaciones 1408
cartón para estuches 5974
cartón para matrizar 5848
cartón para monturas 9143
cartón para pinturas 798
cartón para redondelitos de cerveza 2979
cartón para tapas 1408
cartón para tarjetas perforadas 11041
cartón pegado 10022
cartón piedra 9957
cartón prensado 10756
cartón satinado por una cara 8449
cartón siliconizado 12411
cartón triplex 14054
cartucho 1066
cartulina 2342, 9899
cartulina aceitada 9553
cartulina baritada 1178
cartulina bristol 1921
cartulina bristol fabricada en la máquina a forma redonda 8933
cartulina bristol fabricada en la máquina Fourdrinier 6126
cartulina bristol imitación 1692
cartulina bristol pegada 10025
cartulina brístol superfina 14663
cartulina de celulosa 11027
cartulina duplex 4875
cartulina estucada 2981
cartulina marfil 7627, 7628
cartulina multicapa 9174
cartulina opalina 9610
cartulina para álbumes 378
cartulina para cálices 3816
cartulina para ficheros 7270
cartulina para naipes 10478

cartulina para participaciones 14663
cartulina para postales 10608
cartulina para tarjetas 2356
cartulina para tarjetas postales 10608
cartulina pegada 10022
cartulina secante al cromo 2814
cartulinas gruesas 1534
cartulina soporte 1685
cartulinas superfinas para invitaciones 14664
cartulina triple(x) 14054
casa de imprenta 10829
casa de segunda fila 3028, 14754
casado 7202, 7203
casado de colores 3155
casa editora 11005
casa editorial 11005
casa impresora 10829
casar 7196
cascabel 4716, 5448
cascabelas 1493
cascarilla de cobre 3532
cascarilla electrotipada 5071
cascarilla (galvánica) 12288
caseína 2403
caseta 2429
casquillo de bronce 1961, 2103
casquillo de metal antifricción 2103
castaña 2337
castaño 2701
cata 7458
catacú 2474
catalizador 2463
catálogo 2462
catálogo de editor 13185
catálogo de libros de fondo 13185
catálogo de los tipos 14221
catálogo de subasta 11847
catecol 11082
catechu 2474
catión 2483
cátodo 2480
cátodo fotoelectrónico 10214
caucho artificial 2063, 13496
caucho buna 2063
caucho clorado 2741
caucho endurecido 6874
caucho (natural) 11747
caucho químico 13496
caucho sintético 2063, 13496
caucho vulcanizado 6874
cavitación 2494
cazo de cola 6469
cazo de colada 2447
cazo para cola 6469
cazuela 3297

cazuelita (de la retícula) 7331
cedazo para trapos 4891, 14744
cedilla 2495
cegado 1561, 1567
cegarse 2464
ceguedad de los colores 3123
cejas 12973
cejas del cilindro 1252
celofán 2501
celofana 2501
célula de memoria 2498
célula de selenio 12073
célula fotoeléctrica 10215
célula fotoeléctrica de bloqueo 1173
célula fotoeléctrica de detención 1173
celuloide 2507
celulosa 2508
celulosa de álamo (temblón) 820
cementar 2525
cemento armado 11425
censura de (la) prensa 2526
centelleo 5765
centígrados 2530
centímetro 2532
centímetro cúbico 3799
centipoise 2533
centrado automático 12079
centrador 11093
centrar la línea 2551
centrífuga 14709
centrífugo 2552
centro óptico 9648
ceñir 9358
cepilladora 1611, 10395
cepillar 10392
cepillo 1981
cepillo biselador 1355
cepillo de alambre 14767
cepillo de grafitar 1483
cepillo del cilindro 3935
cepillo de revelado 4225
cepillo para batir matrices 1268
cepillo (para biselar) 12314
cepillo para grafitar 1483
cepillo para jaspear 8622
cepillo para lejía 8422
cepillo para limpiar 14176
cepillo para revelar 4225
cera carnauba 2360
cera de abejas 1288
cera de carandai 2360
cera de carnauba 2360
cera de ceresina 9818
cera de parafina 9969
cera microcristalina 8894
cera mineral 9818
cera montana 9095
cera vegetal 13398
cera zumaque 13398
cerco de filetes 1792
cerco de rayas 1792
cerdo 10337

cerillos 13416
cerito 4108
cernedor 4891, 14744
cerotipia 2567
cerrado con cinta engomada 13580
cerrar 8308, 11171
cerrar a su hora 5687
cerrar el circuito 13478
cerrojo de la cremallera 13219
certificación de circulación 871
certificado 2569
certificado de aprendizaje 6340
certificado de circulación 871
cerusa 5749
cesio 2572
cesta de (las) matrices 2148
cetona 7753
cian 3908
cianea 3908
cianógeno 3908
cianotipia 1642
cianuro de cobre 3507
cianuro de plata 12428
cianuro de potasa 10647
cianuro de potasio 10647
cianuro de potasio cúprico 3530
cianuro de sodio 12687
cianuro de sosa 12687
cianuro potásico 10647
cianuro sódico 12687
cibernética 3920
cícero 2827
ciclo 3921
ciclo en trabajo libre 4986
ciclohexanona 3925
ciclo por segundo 3922
ciencia diaria 11927
cientoveinte-y-ochoavo, en ~ 7079
cierre 2876, 4032, 8307, 13732
cierre de cremallera 14962
cierre de filetes 1792
cierre de los rodillos 11637
cierre de planchas compensador 3262
cierre de planchas por compresión 3325
cierre de rayas 1792
cierre lateral móvil de caja aspirante 4067
cierre relámpago 14962
cifra 4356, 5607, 9458
cifra binaria 1397, 1448
cifra de tirada 2852
cifra marginal 11800
cifras alineadas 8206
cifras árabes 747
cifras arábigas 747
cifras bajas 6853
cifras para la nota del cambio 5345
cifras romanas 11671

cigüeñal 3700
cilindrar 2191
cilindro aspirante 13352
cilindro batidor de la tinta 7341
cilindro colector 3072
cilindro con revestimiento 9824
cilindro cosedor 13176
cilindro de acero por cobrear 1192
cilindro de cambio 6225
cilindro de contrapresión 1055
cilindro de corte 3893
cilindro de huecograbado 5184, 6585
cilindro de impresión 7211, 7212
cilindro de inmersión 4847
cilindro de la mantilla 1516
cilindro de la pila holandesa 7822
cilindro de las planchas 10432
cilindro del tintero 4847
cilindro de plancha 10432
cilindro de presión 11750
cilindro de registro 11415
cilindro de retiración 12046
cilindro de una sola plancha 12464
cilindro distribuidor 7341
cilindro doblador 5978
cilindro empresor 7211
cilindro esqueleto 12504
cilindro expulsor 13293
cilindro guía 12546
cilindro impresor 7211
cilindro mandil 1899
cilindro oscilante 9707
cilindro plegador 5978
cilindro portaforma 10432
cilindro portamantilla 1516
cilindro portaplancha(s) 10432
cilindro ranurado 6669
cilindro refrigerador 13457
cilindro retirador 13559
cilindro satinador 6448, 14903
cilindro secador 4788
cilindros secadores 4730
cilindro vestido 9824
cinabrio 2834
cinc 14940
cincel 2731
cincelador 2656
cincograbado 14944
cincografía 14944
cincotipia 14944
cincho 2546
cinchos de freno 1307
cinta adhesiva 211, 215, 12026
cinta autoadhesiva de doble cara 4606
cinta autoadhesiva por una cara 12456

cinta autocolante por una cara 12456
cinta boba 14327
cinta con doble cara adhesiva 4606
cinta continua 843, 3482
cinta correcta 2897
cinta de control cromático 13055
cinta de corrección 3606
cinta de enmascarar 8650
cinta de papel 9949, 14639
cinta (de papel) con comba 12537
cinta engomada 6718, 10040, 12026
cinta indicadora 1762
cinta magnética 8524
cinta non justificada 14327
cinta para despolvorear 4902
cinta para montar grabados 9146
cinta perforada 11046
cintas 1113, 13597, 13738
cintas conductoras 4150, 11807
cintas de arrastre 4150
cintas de conducción 11807
cintas de entrada 2299
cintas de transporte 13596
cinta sin ajustar 14327
cinta sin fin 843
cintas sacapliegos 4150
cintas transportadoras 13596
circonio 14964
circuito 2837
circuito cerrado 2931
circuito impreso 10800
circuito integrado 7477
circuito microminiaturizado 8904
circuito multipunto 9193
circuito puerta 6322
circuito trifásico 13760
circulación 2852
circular 2844
círculo 2835
círculo cromático 2762
círculo de difusión 2836
circunferencia 2854
circunferencia del cilindro impresor 2855
circunferencia máxima y mínima del cilindro 3949
circunferencia primitiva 10383
circunflejo 2860
cirílico 3952
cita 11172
citar 2861
citas 11381
citocromía 6124, 10904
citrato amónico 524
citrato de amonio 524
citrato de hierro 5548
citrato de hierro amoniacal 5542

citrato de potasa 10644
citrato de potasio 10644
citrato de sodio 12686
citrato de sosa 12686
citrato férrico 5548
citrato férrico-amónico 5542
citrato potásico 10644
citrato sódico 12686
cizalla 1666, 6705
cizalla biseladora 8998
cizalla circular 2848
cizalla para cortar cartón 2345
clara de huevo 371, 4988, 6402
claridad 1909, 8124
claros 6992
claroscuro 2707
claroscuros 12223
clasificación 12785
clasificación decimal 4054
clasificación de tipos 2879
clasificación de trapos 11212
clasificador 2352, 5610, 5619, 8057
clasificar 12784
claudátur 1844
clavadora 9228
clavar 9226
clave 7762
clavija de ajuste 4673
clavo 6868, 7266, 9225
clavos 1800
clavos para montar clichés 1603
cliché 1582, 10418, 10907
cliente 74, 2915
climatización 322
climatización del papel 3370
clisado 13123
clisador 13128
clisar 13123
clisé 1582, 10418, 10907
clisé al fotorrelieve 6363
clisé a masa llena 12763
clisé autotípico 10233
clisé calado 10324
clisé combinado 8179
clisé compuesto 8179
clisé con relieve de gelatina 6363
clisé con ventana 10324
clisé de caucho 11751
clisé de fondo 13841
clisé (de fotograbado) 5192
clisé de línea y trama 8179
clisé del negro 1491
clisé de magnesio 8498
clisé de medias tintas 6774
clisé de medios tonos 6774
clisé de medio tono 6774
clisé de nilón 9465
clisé de plástico 10411
clisé de prensa 9324
clisé desplazado 9545
clisé de trama 6774

clisé de trazo 8159
clisé directo 6774
clisé doble 4641
clisé duplicado 4884
clisé elástico 11751
clisé electrotípico 5071
clisé en plástico 10411
clisé estereotípico 2442, 13127
clisé fotograbado 10233
clisé fraccionado 12058
clisé machacado 2706
clisé magullado 2706
clisé mixto 8179
clisé negativo 11557
clisé pisado 2706
clisé plano 12763
clisé pluma 8159
clisé poco profundo 12229
clisés a sangre 1552
clisés con grisado bendéi 1316
clisés de punto para simili-grabados 3165
clisés en serie 6301
clisé sin zócalo 14332
clisés para impresión multi-color 10903
clisés sin margen 1552
clisés (tramado) para tri-cromía 10903
clisé tipográfico en relieve 11445
clisé tramado 6774
cloisonné 2924
clorato de potasa 10637
clorato de potasio 10637
clorato potásico 10637
clorhidrato 7104
clorhidrato de diamino-fenol 4262
clorhidrato de paramino-fenol 9962
clorito de sodio 12684
clorito de sosa 12684
clorito sódico 12684
cloro 2742
cloro activo 949
cloroformo 2746
clorohidroquinona 2747
cloroplatinito de potasa 10640
cloroplatinito de potasio 10640
cloroplatinito potásico 10640
cloruro 2737
cloruro alumínico 475
cloruro amónico 522
cloruro áurico 6517
cloruro cálcico 2164
cloruro crómico 2787
cloruro cúprico 3504
cloruro cuproso 3832
cloruro de alilo 444
cloruro de aluminio 475
cloruro de amonio 522
cloruro de azufre 13386
cloruro de cadmio 2141

cloruro de cal 2168
cloruro de calcio 2164, 2168
cloruro de cinc 14942
cloruro de cobre 3504, 3832
cloruro de estaño 13039
cloruro de etilo 5306
cloruro de hidrógeno 7102
cloruro de hierro 5546, 5576
cloruro de oro 6517
cloruro de oro y potasio 10652
cloruro de plata 12427
cloruro de platino 10473
cloruro de polivinilo 10568
cloruro de potasa 10638
cloruro de potasio 10638
cloruro de sodio 12683
cloruro de sosa 12683
cloruro de uranilo 14368
cloruro de uranio 14368
cloruro de vinilideno 11875
cloruro estánnico 13038
cloruro estannoso 13039
cloruro etílico 5306
cloruro férrico 5546
cloruro ferroso 5576
cloruro mercúrico 8792
cloruro mercurioso 8793
cloruro polivinílico 10568
cloruro potásico 10638
cloruro sódico 12683
coagulación 2971
coagular 2970, 6353
coágulo 2971
coautor 3002
cobalto 3005
cobertura 7356
cobertura de la plancha 10427
cobol 3019
cobre 3497
cobreado 3509
cobreamiento 3509
cobre de base 1191
coccín 3021
cocción 3486
cocido 2074
cocina 3175
cochinilla 3022
codificación 3036
codificación absoluta 28
codificación real 28
codificación simbólica 13485
codificador 3033
codificadora 5155
codificador digital 4366
codificador numérico 4366
codificar 5154
código 3030
código binario 1395
código cíclico binario 11388
código corrector de errores 5245
código de Baudot 1232
código de detección de errores 5246
código de función 6249
código de Gray 11388

código de máquina 3331
código de operación 6249
código de ordenador 3331
código hollerith 7022
coeficiente de absorción 37
coeficiente de Callier 2216
coeficiente de dureza Brinell 1918
coeficiente de expansión 3038
coeficiente de fricción 3039
coeficiente de frotamiento 3039
coeficiente del filtro 5658
coeficiente de rozamiento 3039
coeficiente de traspaso de tinta 7406
coeficiente de variación 3040
coeficiente de velocidad de gasto 3041
coeficiente kappa 7729
coeficientes del diafragma 5937
cogebocados 4143
cohesión 3046
cohesión de fibras 1720
coincidencia 11420
cojín de grabador 9838
cojinete 1254
cojinete a bobillas 1096
cojinete de agujas 9267
cojinete de bolas 1096
cojinete del árbol principal 8549
cojinete del cilindro 3931
cojinete del eje 969
cojinete de rodillos 11634
cojinete de rodillos de alineación automática 12078
cojinete liso 10390
cola 4201, 6465, 12486
cola ácida 147
cola animal 623
cola autoadhesiva 10768
cola bicromatada 4303, 10892
colaborador anónimo 6395
colaborador oculto 6395
colaborador permanente 3207
colacionar 3063
cola con resina libre 147
cola de Bengala 294
cola de caucho 9904
cola de dorar 6402
cola de gelatina 623
cola de huesos 1728
cola de la mantilla 1522
cola de pescado 5141, 5716, 10892
cola de pez 5716
cola de resina 11689
colador 5654
cola en frío 3056
cola esmalte 1084, 5141, 6467

cola fuerte 623
colaje a la resina 11502
cola para fotograbado 5716
cola para libros 1736
colapez 5716
colar 2435, 5663
cola resinosa 11689
cola vegetal 14417
colcótar 2300, 14437
colchón 9825
colchón de aire 325
coleccionador de libros 1747
coleccionar 6325
colección de muestras 11855
colector 3075
colega de hierro 3286
colero 6469
colgadores 2920
colgar (el papel) 6852
cólicos saturninos 8000
colilla 8995
colilladora 8998
colillar 8996
colimador 3081
colimar 3079
colocación 7946
colocación aislada 7439
colocación de alzas 9771
colocación de letras adhesivas 7461
colocación de letras transferibles 7461
colocación de los rodillos 11646
colocación qwert 11180
colocar 1885
colocar el papel 14640
colocar interlíneas 7971
colocar las planchas 10464
colocar los pliegos cosidos en la cubierta 2404
colocar ojetes 5416
colocar por orden alfabético 458
colocar un rodillo 12156
colodio-bromuro de plata 3083
colodio-cloruro de plata 3084
colodión 3082
colodión yodado 7560
colofón 3102
colofonia 11684
coloidal 3090
coloide 3089
coloide de protección 10979
coloide protector 10979
coloides bicromatados 4301
coloides hidrófobos 7125
colón 3101
colón perfecto 3101
color 3113, 14671
color acromático 110
colorante 4922
colorante de alizarina 412
colorante indantrénico 7258

colorante indantreno 7258
colorante lacado 7880
colorante orgánico 9677
colorantes 3176
colorantes ácidos 122
colorantes azoicos 973
colorantes básicos 1202
colorantes de alquitrán de hulla 2973
colorantes de anilina 2973
colorantes directos 4402
color básico 1190
color cálido 14541
color carne 5815
color contrastante 3442
color de fondo 1190
color de la luz 3156
color de papel de empapelar 14534
coloreado a mano 6817
color en la masa 8654
color en polvo 4771
colores binarios 12033
colores complementarios 3272
colores del espectro 3186
colores delicados 10033
colores difumados 12212
colores esfumados 12212
colores fundamentales 10782
colores insaciados 14340
colores intermedios 7510
colores metaméricos 8843
colores normales 10782
colores pastel 10033
color espectral 12841
colores primarios 10782
colores primarios no atenuados 14270
colores primitivos 10782
colores secundarios 12033
colores simples 10782
colores suaves 10033
color estable 5481
colores terciarios 13657
colores ternarios 13657
colores tiernos 10033
colores transparentes 14003
color firme 5481
color frío 3052
colorimetría 3111
colorimétrico 3108
colorímetro 3107
colorímetro de cobre 3505
color indeleble 5481
color inestable 6230
color liso 5808
color modificador 13892
color muerto 5431
color no resistente 6230
color objetivo 13432
color para jaspear 8623
color pasado 5431
color quebrado 5431
color saciado 11879
color saturado 11879
color seco 1679, 4771
color sin trama 5808

color solitario 12918
color superficial 13921
color verde esmeraldo 5110
colotipia 3099
coloxilina 11087
columna 3199, 3200
columna (compuesta) sin errores 10126
columna de novedades 13217
columna periodística 9313
columnaria 3209, 3871
columnas, en ~ 7455
columna sencilla 12453
columnas en el sentido de rotación 3211
columnas transversales 3210
columnista titular de sección 3207
coma 3214, 3234
coma de decimales 4055
coma fija 5730
coma flotante 5839
comas invertidas 11174
comas vueltas 11174
combado de plancha 1297
combadores 3218
combinación al azar de asuntos 12137
combinación de los dientes 3219
combinador de arranque 13060
combinados 8736
combustible 7305
comentar 632
comer 6386
comercio de libros 1778
comercio de papel viejo 14559
comer línea 6386, 13554
cométas 3231
comiéncese sin sangría 1289
comilla de abertura 3239
comilla de escomienzo 3239
comilla final 2940
comillas 4505
comillas de estilo francés 11174
comillas de trinchante 11174
comillas francesas 11174
comillas invertidas 11174
comisión de la tarifa 13604
comodín 2137, 14733
comodín-chibalete 2414, 3284
comodín para composición en galeras 6284
compaginación 8569, 9868
compaginador 2914, 8573
compaginador de anuncios 254
compaginador de platina 8572
compaginar 830, 8570

compañero de la misma calle 1041
compañero de sección 12380
comparación 2204
comparación con, en ~ 10975
comparación de, en ~ 10975
comparador Cleveland del punto de inflamación 2913
comparar 3256
compartimiento 1829
compartir 5472
compás de calibre 2222
compás de división 4511
compás de espesor 2221
compás de puntas curvas 2222
compatibilidad 3258
compatible 3259
compendio 20, 3260
compensación 5222
competencia desleal 14295
competencia ilegal 14295
compilación 3267
compilador 3269, 3270
compilar 3268
completamente automático 6245
componedor 14205
componedora 3286
componedora de la segunda generación 12042
componedora de la tercera generación 13727
componedora de primera generación 5709
componedor de cazuela 3297
componedor para titulares 4460
componente 3281
componentes físicos 6877
componentes tricromáticos 14062
componer 3282
componer a ciegas 12173
componer a plana (y) renglón 8176
componer cerrado 12805
componer con espacios gruesos 7741
componer con espacios medianos 7738
componer con minúsculas 7737
componer de nuevo 11493
componer en forma de escalón 13001
componer en forma de escalonada 13001
componer en letras 12858
componer en líneas graduales 13001
componer progresivo 6000

componer rápidamente 12170

componer sin sangrar 12159

componer sin sangría 12159

composición 3301, 8702

composición abierta 9625

composición a distribuir 4035

composición a guardar 13033

composición a mano 6818

composición a máquina 8732

composición amontonada 9244

composición a pedazos 14328

composición apiñada 9244

composición apretada 12762

composición asalariada 13949

composición a usar 13033

composición cargada 12762

composición compacta 12762

composición completa 2934

composición complicada 3276, 8009

composición con computadora 3334

composición con muchas abreviaturas 3305

composición conservada 13033

composición con titulares 4469

composición corrida 1683, 12762, 13247

composición corriente 1683, 12762, 13247

composición de abreviaturas 12174

composición de ancho de columna 3208

composición de ancho interlineado 9625

composición de anuncios 249

composición de avisos 249

composición de carátula 3306

composición de cola 11638

composición de disposición variada 4469, 4473

composición de epigrafía 4473

composición de epigráfica 4469

composición de estados 11776, 13520

composición de fibras 5592

composición de fórmulas científicas 3307

composición de la pasta 5592

composición de la pasta de fibras 5594

composición de la pulpa 5592

composición de lenguas extranjeras 6037

composición de máquina 8438, 8732

composición de matemáticas 8673

composición de música 9207

composición de pasta 11638

composición de portada 3306

composición de tablas 13520

composición de texto 1683, 13247

composición de tipos grandes 4469, 4473

composición de tipos para música 9215

composición echada 9511

composición en frío 3061

composición en paquete 10322

composición enrevesada 3276

composición epigráfica 4473

composición espaciado ancho 9625

composición estrecha 9244

composición fibrosa 5592, 5594

composición fija 13033

composición fotográfica 10278

composición fotomecánica 10278

composición fototipográfica 10278

composición fría 3061

composición guardada 13033

composición inclinada 9511

composición interlineada 7980

composición intricada 3276

composición ladeada 9511

composición linotípica 8222

composición llena 12762

composición maciza 12762

composición manual 6818

composición mecánica 8438, 8732

composición mecanotípica 8732

composición mezclada 9003

composición mixta 9003

composición monotípica 9090

composición ordinaria 13247

composición permanente 13033

composición programada 3334

composición publicitaria 249

composición quebrada 9625

composición regleteada 7980

composición seguida 1683, 11812, 12762, 13247

composición serrada 12762

composición simple 13247

composición sin interlíneas 12762

composición sin plomo 3061

composición sobrante 12590

composición sobre película 10278

composición sólida 11812

composición tabular 13520

composición titular 4473

composición torcida 12955

composición tumbada 9511

composición versiforme 4473

composición vigente 13033

compositora 5536

compositora fotográfica 10279

compositor a mano 6819

compositora para líneas 12606

compositora Typograph 14234

compositor de anuncios 250

compositor de periódicos 9303

compositor linotipista 8223

compositor pasajero 6574

compositor (tipográfico) 3313

comprador de la permeabilidad al aire 345

compresibilidad 3324

compresor 3327

compresora 2064

compresor de aire 354

comprobación 2661

comprobación automática 901

comprobación de color 3128

comprobación de materiales 11122

comprobación de paridad horizontal 8340

comprobación por redundancia 11353

comprobador de dureza 10417

comprobador del satinado 12646

comprobante 4724

comprobante de inserción 14517

comprobante del anuncio 14517

comprobante del aviso 14517

compuerta 6322

compuerta de luz 8121

compuesto 3316

compuesto a mano 6845

compuesto a máquina 8734

compuesto complejo 3275

compuesto de caja 6845

compuesto de diazo 4283

compuesto diazoico 4283

compuesto resinífero 11503

computadora 3330

computadora analógica 572

computadora asíncrona 859

computadora de emergencia 1054

computadora de la segunda generación 12041

computadora de la tercera generación 13726

computadora de primera generación 5708

computadora de programa almacenado 13237

computadora de reserva 1054

computadora de uso general 6368

computadora digital 4359

computadora para usos especiales 12823

computadora síncrona 13488

computador (electrónico) 3330

computador para composición tipográfica 3283

computador universal 6368

computista 5265

comunicador 4748

concentración 3356

concentración ácida 119

concentración de los iones de hidrógeno 7113

concentración hidrogeniónica 7113

concentración iónica de hidrógeno 7113

concentración máxima admisible 13777

concentrado 3352

concentrar 3353

concéntrico 3357

conciencia 12994

conciencia, estar a ~ 12987

condensación 3361

condensador 2275, 3367

condensadora 3367

condensar 1, 9820

condiciones atmosféricas 864

condiciones de ensayo 13661

condiciones de entrega 4151

condiciones del ambiente 864

condiciones de trabajo 3373, 14832

condiciones de venta 3372
condiciones para la inserción de anuncios 268
conducción de la hoja de papel 7997
conducta de las matrices 8688
conductancia 3374
conductibilidad térmica 13687
conductividad de calor 6919
conductividad térmica 13687
conducto de aire 349
conductor 10753
conductor rotativista 11700
conectar 13478
conexión 8211
conexión abordonado 5753
conexión de bridas 5753
conexión en estrella 14910
conexión en estrella y triángulo 13049
conexión en paralelo 9978
conexión en serie 12143
conexiones de los datos 4011
conexión provisional 10046
confección 7961
confeccionador 7962, 8572
confeccionadora de tapas 2412
confeccionar 8570
confección de planchas 10448
confección de suajes 4327
confección de tapas 2411
confección de troqueles 4327
configuración 3379
congelar 6166
conglutinante 213
cónico 3381
conífera 3382
coniforme 3381
conjunción 3320
conjunto de caracteres 2640
conjunto de datos 4015
conmutación de circuito 2840
conmutador 13474
cono de bobina 11362
cono deslizante 12572
cono portabobinas 11362
cono triangular 6063
consejo obrero 2625
conservabilidad 7740, 12286
conservada 7742
conservado 7742
conservar 7742
¡conserve! 7742
consistencia 1677, 3393, 12286
consistente 1669
consola 3395, 3396
constancia 3397
constante 3398

constante dieléctrica 4322
constitución 13303
constituyentes volátiles 14506
construcción 13303
consumir inútilmente 14557
consumo de corriente 3841
consumo de energía 10705
consumo de energía eléctrica 3841
consumo de fuerza 10705
consumo de papel 9910
consumo de tinta 7334
contacto 3414
contacto absoluto 9652
contacto hermético 9652
contactor de fin de carrera 8151
contado 3666
contado de caracteres 2629
contador 3652
contadora 3652, 3653
contador automático 3653
contador de revoluciones 13523
contador de velocidad 13523
contador (público) 75
contagiar 3417
contaminar 3417
contenido 3421
contenido de algodón 3637
contenido de cenizas 816
contenido de cobre 3506
contenido de humedad 9033
contenido de humedad a peso seco absoluto 9035
contenido de materias sólidas 12764
contenido de resina 11685
contenido de sólidos 12764
contenido en resina 11685
continúa 13872
continuará 13872
contorno 1792, 3437, 9716
contornos para separación de colores 7779
contracaja 12789
contracción (del papel) 12355
contracubierta 991
contracuchilla circular 12582
contraerse 12354
contrafibra, a ~ 292
contrahacer 3657
contrahacer el color 3155
contrahilo, a ~ 292
contramaestre 3456
contramáscara 3660
contramuestra 3662
contrapegar 7885
contrapeso 1088
contraportada 6211
contrapresión 1044
contraprueba 3662, 11570
contrapunzón 3661

contrario al de las agújas del reloj 3655
contrastado 3447
contraste 3439
contraste alto 6981
contraste, sin ~ 12730
contratapa 991
contratipo 3662
contrato de propaganda 269
contrato de publicidad 269
contrato de publicidad de un año 251
contratuerca 2672
contribución 787
control a distancia 11458
controlador de la producción 10926
controlador de plazos 10926
control de calidad 11121
control de la producción 10927
control de paridad 10009
control de plazos 10927
control de recepción de los materiales 11122
control estadístico de la calidad 13079
control local 4539
control manual 6820
control remoto 11458
conversión 3467
conversión de código 3031
convertidor 3475, 7544
convertidor de código 3032
convertidor de señal 12397
convertir 3474
conveyor 3481
coordenadas 3491
coordenadas de cromaticidad 2764
coordenadas tricromáticas 2764
coordinador publicitario 80
copa Cleveland 2913
copa de efusión 4983
copa (de efusión) de Zahn 14931
copa de Ford 6030
copela 3819
copia 3534
copia a diazo 4285
copia a la albúmina 372
copia a la anilina 619
copia a la cola 5146
copia al carbón 2314
copia al esmalte 5146
copia al ferroprusiato 1642
copia a máquina de escribir 14228
copia anticipada 241
copia brillante 6462
copia cianográfica 1642
copiado de planchas 10448
copiadora 10828
copiadora de repetición 6686, 13116
copiadora offset 12627

copia en papel bromuro 1947
copia en papel carbónico 2314
copia en papel pigmento 2327
copia fotográfica 10270
copia (fotográfica) 10793
copia fotográfica tramada 11978
copia fotostática 10273
copia impresa 11478
copia legible 6859
copia lista para cámara 2248
copia mecanografiada 14228
copia por contacto 3414
copiar 10795
copiar por contacto 3415
copiar (por contacto) 3538
copia sobre cinc 14954
copia sobre metal 8838
copia vélox 14431
copista 10447
copolímero 3495
copulación 3669
copyright 3553
corazón para cosido 12196
corchete 2876
corchete de alambre 13040
corchete de tres piezas 3300
corchetes 1840
cordel 9855
cordobán 3561
cordón de la correa 1306
cordones 1113
corindón 3624
corona 3589
corona dentada 13043
coronas del cilindro 1252
corondel 3209
corondeles 2589
corquis 9717
corral 6734
correa del colector 834
correa del reunidor 834
correa (de transmisión) 1301
correa de transmisión 4745
correa portadora 3482
correas de entrega 4135
correas de salida 4135
correa sin fin 843
correa transportadora 3482
correa transportadora de matrices 8689
correa trapezoidal 14416
corrección 463, 3597
corrección con mascarillas 8647
corrección de contraste 13890
corrección de las pruebas finales 11272
corrección del color 3131
corrección del original 3551

debilitamiento de la
densidad 11352
debilitar 3900, 11340
debilitarse 12234
decalco 1640
decalcomanías 4041
decámetro 4046
decantación 4048
decantar 4047
decantos cortados 3860
decapar 2902, 3656
decibel 4050
decibelio 4050
decímetro cúbico 3800
décimoctavo, en ~ 4993
décimosexto 12481
décimo tercio 13730
decisión 4059
decisión lógica 8318
declaración de aduana 3852
declive 6535
decoloración 5433
decoloración al secarse
4768
decoloramiento 4420
decolorar gradualmente
3141
decomponer un molde para
impresión a colores 4474
decoración con líneas
azuradas 980
decorado con flores de lis
4073
dedicación 4074
dedicatoria 4074, 4076
dedicatoria autógrafa 10734
dedo guiador 2826
defecto 2040
defecto de construcción
3403
defecto de entintado 14503
defecto de imagen 4096
defectos 11451
defectos del desarrollo
11660
defectuoso 4093
definición 4098
definido 3447
deflectores 1064
deformación 4105
deformación elástica 5004
deformación por cizalla-
miento 12243
degradar 14471
dejar (en) blanco 8020
dejar espacio 8020
deje como está 13134
delaminación 4124
delantal (de trabajo) 727
delante 6207
delantera 6031, 14461
dele 4128
delear 4129
deleátur 4128
deletreo 12856
delicuescencia 4133
delimitador 4132
delito de prensa 10752
demodulación 4158

demorar 12594
dendritas 4163
densidad 4169
densidad aparente 711
densidad de bits 1449
densidad de corriente 3842
densidad del ácido 119
densidad de registro 9829,
11312
densidad de transmisión
13991
densidad específica 12827
densidad integrada 7478
densidad óptica 9653
densímetro 7120
densitometría 4168
densitómetro 4167
densitómetro de reflexión
11393
densitómetro de trans-
misión 13990
dentillado granoso 6537
departamento 4185
departamento de anuncios
252
departamento de avisos
252
departamento de
composición 3292
departamento de
embarques 8541
departamento de empaque
8541
departamento de estereo-
tipia 13130
departamento de
expedición 8541
departamento de
formación 8576
departamento de fotografía
10242
departamento de máquinas
10761
departamento de obras
7667
departamento de obra
suelta 7667
departamento de pedidos
9674
departamento de prensas
10761
departamento de
preparación de la copia
10449
departamento de publi-
cidad 252
departamento de remen-
dería 7667
departamento de ventas
11848
depósito 4186, 8479
depósito auxiliar 943, 12375
depósito bajo 8382
depósito de cobre 3508
depósito del agua 14577
depósito desgotador 13186
depósito de tinta 7403
depósito electrolítico 5028
depósito legal 3554.

8085
depósito para cuadratines
5122
depósito para material
usado 4036
depresión 8387
depuración 2889
depuración de un
programa 4040
depurador de metal 11576
derecha, hacia la ~ 2922
derecho de traducción
11610
derechos de autor 3553
derechos reservados por ...
439
derivado 4199
dermatitis 4200
dermatosis del cromo 2799
dermitis 4200
derramamiento 5864
derramar 12931
derramarse 3723
derrame del metal 12984
desabonarse a una revista
2262
desacomodar 5694
desacuñar 14331
desacuñar la composición
4750
desarmable 4214
desarmador 12001
desarmar 1894, 4449, 13282
desarrollo 11659
desbarbador 5234, 11949
desbarbar 3858
desbloquear 7441
descarga corona 3590
descarga de los rodillos
metálicos 11651
descargas 12576
descarte 3183
descartonadora de recortes
13506
descartonar 13291
descenso de la cuchilla
4677
descodificar 4072
descoloración 13008
descoloración parcial 3105
descolorado 1543, 4421
descoloramiento 4420
descolorarse 3141
descolorimiento 1557
descomposición en fibrilas
5600
desconchados 4769
desconectador 13789
desconectar 1878, 13477
descortezado 1165
descortezadora 1164
descuento 4425
descuento para colegas
4426
descuñador 12318
desecador 4207
desecante de cobalto 3007
desecar 4206
desembragar 13793

desempastelar 2900, 10303
desempolvar una caja 1629
desencuadernar 11017
desengranar 13793
desengrasar 2898, 4106
desengrasar los mojadores
2886
desensibilización 4203
desensibilizador 4205
desensibilizar 4204
desfibrado 6641
desfibrador 6638
desfibrador continuo 3430
desfibrador de prensas
10513
desfibraduría 6645
desfloculación 4102
desgarro de la banda 4675
desgaste 10, 14635
desgaste de la forma 6077
desgaste por roce 11
desgaste por rozamiento 11
desglosar 4474
deshacer 4484, 7793, 13282
deshacer la forma 1894
deshidratación 4121
deshilachar los clavos 6151
deshilachar los cordeles
6151
deshumectación 4119
deslizadora del colector 839
deslizadora del disco 9126
deslizadora del reunidor
839
deslizador del disco 9126
deslizamiento 12559, 12569,
12570
deslizamiento de la
mantilla 1515
deslucir 13606
deslumbramiento 6425
deslumbre 6425
deslustrar 13606
desmochar 3740
desmoldeo de la cascarilla
(galvánica) 12293
desmontable 4214
desmontar 1894, 4449, 13282
desnaturalizar 4161
desorción 4211
desoxidante 4184
desoxidar 4183
desoxigenar 4183
despachar 5694
despachar un libro a una
dirección y la factura al
otro 4762
despacho 9505
despedida de inmediato
2054
despedida en el acto 2054
despedida incontinenti
2054
despedido en jergo 2054
despedir a una persona
5694
despegamiento 6199
despegar escalonando 5466
despegar esparciendo 5466

desperdiciar 14557
desperdicio 8981
desperdicios 14561
desperdicios de la sección húmeda 14679
despilfarrar 14557
desplazamiento del clisé 9545
desplazamiento lateral 7085, 7933
desplazar 12298
desplazarse 3722
desplegar 4459
despolvado 4893
despolvoreado 4893
desprender el papel del manchón 3641
desprenderse 1580
desprendimiento de letra(s) 4716
desprendimiento del rezago 13290
desprovisto de agua 607
despumar 12511
destacado 4464
destacar 4459, 13794
desteñirse 5433
destornillador 12001
destructor de libros 1749
desuso, en ~ 9724
desvanecer 14471
desvanecer del color 3141
desvanecido 14469
desvelar 2901
desviación 4234
desviación de la banda de papel 12282
desviación de rayos de luz 4235
desviación media 8715
desviación standard 13023
detalle 4215
detector de dos hojas 14155
detector de rotura de la banda de papel 14642
detergente 4219, 14551
deterioración del canto del bobina 11365
determinación 4220
determinación de pH 10178
detersivo 4219
detersorio 4219
detintado 4123
devanecimiento 5433
devastar 9999
devoluciones 11546
dextrina 4243
dextrosa 4246
diablerías de las erratas 8980
diablo 4891, 14744
día de fiesta anual 14633
día de labor 14833
día de paga 10068
dia de trabajo 14833
diafragma 4273
diafragma de láminas 14582

diafragma de los detalles 4218
diafragma de medio tono 6796
diafragma de ranura 12579
diafragma en cruz 3769
diafragma iris 7572
diafragma para acentuar los blancos 6993
diafragma para detalles 9428
diafragma para los medios tonos 8922
diafragma para reforzar los negros 12224
diafragmar 13212
diagonal para quebrados 12767
diagrama 4257, 5606
diagrama de cromaticidad 3193
diagrama de la subdivisión de las cajas 7960
diagramador 7962, 8572
diagrama fotográfico 2134
día laboral 14833
diamante de corto 4269
diámetro del cilindro 3938
diámetro del cilindro impresor 4260
diámetro del círculo primitvo 10384
diapositiva 4278, 13998
diapositiva al contacto 3409
diapositiva cromática 3192
diapositiva de color 3192
diapositiva en color(es) 3192
diapositiva para proyección 12550
diapositiva por contacto 3409
diapositivo 4278
diapositivo reticulado 6778
diario 3958, 4279
diario amarillisto 14920
diario amarillo 14920
diario de anuncios 272
diario de la mañana 9104
diario de la noche 5323
diario de la tarde 5323
diario de tamaño extraordinario 1520
diario dominical 13404
diario matinal 9104
diario matutino 9104
diario oficial 6343
diarios viejos 9322
diario vespertino 5323
diarista 7700
diarucho 11199
diatomita 4280
diazotipia 4286
diazotipo 4286
dibásico 4288
dibujante 4698
dibujante comercial 271, 3240

dibujante de trabajos de publicidad 3240
dibujante gráfico 271, 3240
dibujante maquetista 271, 3240
dibujante publicitario 271, 3240
dibujo 4208, 7165
dibujo acabado 7961
dibujo a la guada 14570
dibujo a lápiz 10092
dibujo a la pluma 8171
dibujo al carbón 2646
dibujo a pincel seco 4770
dibujo a pluma 10089
dibujo de conexiones 4258
dibujo de ejecución 14835
dibujo definitivo 7961
dibujo de lápiz 10092
dibujo de línea 8171
dibujo de líneas 8171
dibujo de pluma 8171
dibujo de pluma (y tinta) 10089
dibujo de trazo 8171
dibujo hecho al carbon 3708
dibujo litográfico a lápiz 2598
dibujo litográfico de artista 8255
dibujo panorámico 9890
diccionario 4306, 4307
diccionario de excepciones 5336
dicroísmo 4296
dicromato 4298
dicromato amónico 526
dicromato de amonio 526
dicromato de potasa 10648
dicromato de potasio 10648
dicromato de soda 12688
dicromato potásico 10648
dicromato sódico 12688
Didodianos 4310
diecciochavo, en ~ 4993
dieciseisavo 12481
dieléctrico 4326
diente 5106, 6868, 7266
diente en el papel 9915
diéresis 4251
diesis 4594
dieziochavo, en ~ 4993
diez-y-seisavo, en ~ 12481
diferencia de matiz 9542
diferencia de tinte 9542
diferencia entre las dos caras 14158
diferencias del registro 11405
diferencias de puntura 11405
difracción 4344
difracción de la luz 1321
difuminado 14469
difuminar 14471
difumino 13313
difusión de luz 4350
difusor 4349

digital 4358
digital-analógico 4363
digitales 13795
dígito 4356
dígito binario 1448
dígito de verificación 2665
dilatación 4370, 5362
dilatador 4369
dilatancia 4368
dilatante 4369
dilatar 12594
diluir 3856
diluir (en agua) 4372
diluir la tinta 11343
diluyente 11345
dimensión 4375
dimetilcetona 102
diminución del papel 12355
diminuir el consumo de tinta 3892
dina 4931
dinamómetro 4930
dinamómetro para ensayos de tracción 13072
diodo de cristal 3788
diodo fotoeléctrico 10224
dioptría 4389
dióxido carbónico 2315
dióxido de azufre 13387
dióxido de carbono 2315
dióxido de titanio 13860
diptongo 4392
dirección 192
dirección absoluta 25
direccionamiento absoluto 28
direccionamiento simbólico 13485
dirección de fibra 8443
dirección de la máquina 8443
dirección de las fibras 6537
dirección indirecta 7291
dirección real 25
dirección transversal 3752
directo 9599
director de fábrica 14843
director de usina 14843
disco 4259, 4433, 4434
disco de moldes 9125
disco de seis moldes 12480
disco excéntrico 4949
disco magnético 8508
discontinuado 4423
discontinuar la publicación 4424
discromatopsia 3123
diseminación de informaciones 4009
diseñador 3240
diseñador de tipos 14184
diseñador publicitario 271
diseño 4208, 7961
diseño anuncio 253
diseño de contorno 10090
disertación 13700
disminuir el entintado 3892
disminuir la velocidad 12594

disoluble 12769
disolvente 12773
disolventes alifáticos 409
disolventes aromáticos 777
disolver 4478
disparador 13790
disparar 12160
dispersar 4450
dispersión 4454
dispersión de la luz 4455,
    8129
dispersivo 4452
disponer en zigzag 13002
disposición de la dis-
    tribución de las cajas
    7960
disposición de los cajetines
    2405
dispositivo 4236
dispositivo alimentador del
    crisol 917
dispositivo antieléctrico
    estático 13069
dispositivo conductor 2370
dispositivo de aceleración
    12942
dispositivo de arranque de
    banda 13578
dispositivo de bloqueo 8305
dispositivo de cierre de las
    quijadas 14463
dispositivo de des-
    embrague 4071
dispositivo de fijación por
    tensión 13649
dispositivo de impresión de
    últimas noticias 6224
dispositivo de impresión
    suplementaria 7233
dispositivo de laboratorio
    7856
dispositivo de refrigeración
    3489
dispositivo de salida 4134,
    5462, 9733
dispositivo de secado 4775
dispositivo impresor
    electrostático 5067
dispositivo lavador 11655
dispositivo mojador 3981
dispositivo para des-
    cartonar 13506
dispositivo para la
    determinación del
    desarrollo 12566
dispositivo para lavar
    rodillos 11655
dispositivo para poner a
    registro 13767
dispositivo portador 2370
dispositivo registrador
    11310
dispositivo retardador
    12595
dispositivo salvapliegos
    7224
dispositivos de pinza 9356
dispuesto para la
    impresión 11286

distancia de (la) trama
    11977
distancia focal 5945
distancia focal imagen 998
distanciar 12808, 14720
distorsión 4481, 4482
distorsión de barrel 1171
distorsión de concavidad
    10361
distorsión de convexidad
    1171
distorsión del punto
    reticular 4565
distorsión esférica 1171
distorsión fortuita 7665
distorsión óptica 9655
distribución 4492
distribución de frecuencia
    6187
distribución de
    informaciones 4009
distribución de la energía
    espectral 12842
distribución de los
    cajetines 2405
distribución de tinta 7339
distribuidor 4496, 4497
distribuidor de la tinta 719
distribuidor manual de
    cola 10030
distribuir 2900, 7793
distribuir (la composición)
    4484
distribuir la tinta 4487
distribuir los tipos 4484
distribuir tipo nuevo en las
    cajas 7964
disulfuro de carbono 2316
ditografía 4579
dividir 1
dividir una palabra 4509
divina proporción 6505
divisa 4237
división 4003, 7143
división de las palabras
    4519
división del tintero 6114
división de palabras 7144
división de sílabas 7144
división en grados 4518
división incorrecta de
    palabra 1057
división silábica en el
    diccionario 4308
divisor 12903
divisor del tintero 6108
divisor de luz 1246
divisorio 3544, 4520
doblado a lo largo 8334
doblado en acordeón 73
doblado en cuadernos 5964
doblado en fuelle 73
doblado en paralelo 9979,
    14531
doblado en tres 7496
doblado en zigzag 7496
doblador 5965
dobladora 5965, 5966
dobladora triangular 6071

doblamiento 11463
doblar 5963
doblar en paralelo 9981
doble caja 4585
doble cruz 4594
doble cuadrado 14138
doble escuadra 14071
doble exposición 4603
doble grupos de impresión
    4070
doble impresión 4646
doble interlínea, con ~
    4624
doble justificación 4619
doble letra 4579
doble página 2549, 4663
doble precisión 4645
doble producción 14154
doble renvío 3764
doble sencillo 9987
doble tono 4869
doble unidades impresoras
    4070
doblez 5962
doblez de acordeón 73
doblez en zigzag 73
doce puntos inglés-
    americanos 10288
dolomía 4549
dolomita 4549
donato 4553
dorado a mano 6405
dorado a prensa 6406
dorado a volante 10749
dorado de prensa 10749
dorador 6399
dorador a prensa 10748
dorador de cantos 4956
dorador de cortes 4956
dorso 984, 1021, 14447
dorso antihalo 667
dorso del calendario 2179
dorso de página 1040
dos a la vez 14162
dosificador de luz 8118
dos interlíneas (entre cada
    renglón), con ~ 4624
dos puntos 3101, 12482
dozavo, en ~ 14117
dracma 4697
drago 4688
dril 4731
driografía 4734
dúctil 8585
duodécimo 14117
duplicado 4579
durabilidad 4886
dureza 4112, 6871
dureza Brinell 1918
dureza según Shore 12321
duro 3447
durómetro 6872, 10417
durómetro de Shore 12320

ebonita 6874
ebuliómetro 1703
ebulioscopio 1703
ecuación 5226

ecuación de Van der Waals
    14382
ecuadernación jansenista
    7635
eczema de cromo 2799
echado 9511
echar forma 6529
echar fundición 7964
edición 4970
edición abreviada 19
edición aldina 388
edición ampliada y
    aumentada 558
edición anotada 634
edición aumentada 5200
edición corregida 3593,
    11573
edición corregida y
    aumentada 558, 11572
edición de bolsillo 10510
edición de la mañana 2052
edición de la tarde 9420
edición de lujo 4973, 7919
edición de mula 2052
edición de pocos
    ejemplares 8150
edición económica 2660
edición emendada y
    ampliada 11572
edición en cuadernos 4974
edición en fascículos 4974
edición enmendada 3593
edición especial 12821
edición espurgada 5378
edición extra 12822
edición extraordinaria
    12822
edición fina 4973
edición fuera de comercio
    10871
edición ilustrada 7164
edición limitada 8150
edición matutina 2052
edición no oficial 10871
edición no venal 10871
edición nueva 11429
edición original 5702
edición para el correo 8537
edición popular 2660
edición príncipe 5702
edición reservada 12821
edición revisada y
    aumentada 11572
editar 4966, 11003
editor 11005
editor de música 9214
editorial 4977
editorialista 7985
educación profesional
    13621
efectivo medio de la
    muestra 960
efecto actínico 160
efecto Becquerel 1270
efecto borde 1794
efecto Cabannes-Hofmann
    2136
efecto continuo 3425
efecto de Albert 362

efecto Debot 4039
efecto de Callier 2215
efecto de Clayden 2885
efecto de Doppler 4562
efecto de Eberhard 4946
efecto de halo 6802
efecto de Herschel 6969
efecto de intermitencia 7513
efecto de Kostinsky 7836
efecto de la aureola 6802
efecto de la humedad 7075
efecto del filtro 5656
efecto de proximidad 216
efecto de tramado lineal 12459
efecto de Villard 14474
efecto en los blancos y en las sombras 7758
efecto fotoeléctrico 10228
efecto grasiento 12636
efecto Lainer 7879
efecto Nyblin 9463
efecto Purkinje 11057
efecto Ross 11691
efecto Russell 11820
efecto Sabattier 11825
efecto Schwarzschild 11925
efecto Sterry 13133
efecto Weigert 14669
efecto Weinland 14674
egoutteur 3985
eje 963, 968, 12225
ejecución simultánea de instrucciones 3360
ejecutar 5351
ejecutivo de cuenta 80
ejecutivo de órdenes 80
eje de abscisas 967
eje de la barra tensora de la mantilla 1514
eje de las ordenadas 966
eje de levas 2254
eje dell'y 966
eje del rodillo 11649
ejemplar 3535
ejemplar adolecido 4094
ejemplar anticipado 240
ejemplar anticipado a la puesta en venta 240
ejemplar atrasado 1038
ejemplar autografiado 896
ejemplar autógrafo 896
ejemplar comprobante 14517
ejemplar de cambio 5342
ejemplar de canje 5342
ejemplar de depósito legal 13080
ejemplar dedicado por el autor 896
ejemplar defectuoso 4094
ejemplar de obsequio 3278
ejemplar de prueba 240
ejemplar de publicidad 11568
ejemplar de regalo 6156, 10733
ejemplar echado 7787

ejemplar enviado a críticos antes de la fecha de publicación 240
ejemplares por hora 7223
ejemplares sobrantes 11546
ejemplar excelente 3550
ejemplar fuera de venta 6156
ejemplar gratuito 6156, 10733
ejemplar impecable 3550
ejemplar justifiante 14517
ejemplar no reciente 1038
ejemplar no venal 6156
ejemplar numerado 9448
ejemplar para la crítica de la prensa 11568
ejemplar preliminar de muestra 240
ejemplar suministrado gratuitamente 6156
eje óptico 9644
elasticidad 5005, 11500, 13425
electricidad estática 13067
electro 5071
electrochapeado 5063
electrodo 5027
electrodo de calomelanos 2224
electrodo de carbón para lámpara de arco 755
electrodo de quinhidrona 11159
electrodo de vidrio 6428
electrodos de carbón para lámpara de arco 2328
electroforesis 5061
electrofotografía 5062
electrogalvanización 5063
electroimán 5037
electrólisis 5032
electrólito 5033
electromotor 5042
electrón 5043
electrónica 5052
electrónico 5044
electrostática 5069, 13067
electrotipador 5073
electrotipia 5071
electrotipia extradura moldeada en plomo 365
electrotipista 5073
electrotipo 5071
elemental 5075
elementar 5075
elemento 5074
elemento acíclico 7247
elemento cíclico 3924
elemento de caldeo 6923
elemento de calefacción 6923
elemento de impresión a la anilina 615
elemento impresor 7167, 10852
elemento impresor reversible 11560
elemento intermitente 7247

elementos superficiales no iónicos 9403
elemento sustentador de la hoja cortante 7827
elevador del aparato colector 832
elevador de pila 10342
elevador-reunidor 832
eliminación de fallas 4040
eliminador de espuma 660
eliminador de estática 13069
eliminar 4129
elipse 5085
elipsis 5086
elipsógrafo 5087
elisión 5083
elogiar 10489
elogio 1647
emanación de tinta 7373
embaladora 1093
embalaje 9827
embalaje de la bobina (de papel) 8944, 14865
embocadura 9151, 9684
émbolo (de la bomba) del crisol 10687
émbolo del aire 347
emborronador de cuartilla 3235
embotar 1645
embrague 3670
embrague de fricción 6194
embrague electromagnético 5038
embuchar 7436
embudo 6063, 6254
embudo plegador 6063
embuste 7013
embutidora 4337
embutir los grabados en el texto 1885
eme 5098
empalmador 10035
empalmadora de unión a tope 2126
empalmador de la banda (de papel) a toda velocidad 5927
empalmadura 12891
empalmar 12892
empalmar a mano 6838
empalme 2115, 12891
empañamiento 1619
empañar 13606
empaque 9825
empaquetadura 9828
empaquetar 9820
emparejador 7687, 10394
emparejadora 7686, 7687
emparejar 7685
empastador 1732, 10030
empastadura 1414
empastar 1414
empastarse 2464, 5633
empastelado 10315
empastelamiento 10315
empastelar 10317
empernar 1715

empírico 5118
emplanado 8569
emplanador 8572, 8573
emplanaje 8569
emplane 8569
emplazamiento 8302
emplazamiento aislado (en el texto) 6239
emplazamiento aislado (entre texto) 6239
emplear abundancia de mayúsculas 7743
empleo simultáneo de dos operarios 4671
emplomada 12984
emplomadura 12984
empolvado 6268, 10698
empolvamiento de la mantilla 1519
empolvar 4892
empolvoramiento 10698
empolvorar 4892
empolvoreado con talco 10699
empolvorizar 4892
emporrado 2043
empuñadura 6829
emulgente 5128
emulsificación 5126
emulsificar 5129
emulsión 5132
emulsionado 13843
emulsión agua y tinta 14583
emulsión al colodión 3086
emulsionante 5128
emulsionar 1556, 5129
emulsión de colodión 3086
emulsión de grano fino 5673
emulsión fotográfica 10243
emulsión tinta en agua 7364
encaballado 10315
encaballado de pliegos 8038
encaballamiento 10315
encabezado 2297, 6905
encabezamiento 2297, 6905
encabezamiento encasillado 1835
encadenamiento 3349
encajado de cuadernos 8038
encajar 7436
encajarse 3748
encaje 7442, 7443
encajonar 3704
encaminado 11738
encargado de armadura 8572
encargado de sección de cajas 3294
encargar 9673
encargo 9672
encarrujarse 3834
encartado 7433
encartaje 7433
encarte 7442, 7443
encarte pegado 13854

encartonado en tapas 7240
encender 13478
encerado 2559
encerrada 5153
enclavamiento mutuo 7506
encobreamiento 3509
encogimiento del papel
  12355
encolado a la resina 11502
encolado de la superficie
  13440
encolado (del lomo) 6477
encolado (del papel) 12503
encolado en la pila 1265
encolado en la superficie
  (con gelatina) 626
encolado en tina 626
encolado ligero 11139
encoladora 6475
encolado sobre cartivanas
  7084
encolado sobre lienzo 9139
encolado sobre tela de lino
  9139
encoladura 12503
encolamiento 12503
encolar 10034, 13853
encolar a mano 6838
encolar sobre cartivanas
  6691
encorvado 3847
encorvadora de planchas
  1319
encorvar 1313
encostradura de la tinta
  12519
encorvarse 14545
encrespadura 3710
encuadernación 1414
encuadernación a la
  americana 10408
encuadernación a la
  antigua 680
encuadernación à la Bradel
  1845
encuadernación a la
  fanfarria 5464
encuadernación a la
  francesa 6755
encuadernación a la
  holandesa 6755
encuadernación a la
  Lionesa 8424
encuadernación al estilo
  romántico 11673
encuadernación a mosaico
  9110
encuadernación a mosaico
  embutido 9110
encuadernación antigua
  9571
encuadernación a toda piel
  6231
encuadernación a toda tela
  6231
encuadernación Bradel
  1845
encuadernación camafeo
  2236

encuadernación Canevari
  2267
encuadernación catedral
  2478
encuadernación cincelada
  3808
encuadernación con bordes
  cerrados 14906
encuadernación con el
  nombre del propietario
  10159
encuadernación con lomo
  de tela 11130
encuadernación con lomo
  plástico 10408
encuadernación con
  mosaico 9110
encuadernación cuero 8012
encuadernación de abanico
  5468
encuadernación de
  aficionado 495
encuadernación de alero
  de cabaña 3634
encuadernación de
  Cambridge 2234
encuadernación de edición
  4972
encuadernación de
  editorial 4972
encuadernación de Eve
  5322
encuadernación de hojas
  movibles 8364
encuadernación de hojas
  sueltas 8364
encuadernación de media
  pasta 11135
encuadernación de medio
  pergamino 11141
encuadernación dentelle
  4181
encuadernación dentelle à
  la grecque 4179
encuadernación dentelle à
  l'oiseau 4180
encuadernación de
  pergamino 9995
encuadernación dos-à-dos
  4563
encuadernación Ducali
  4833
encuadernación en becerro
  2200
encuadernación
  encadenada 2584
encuadernación en
  camafeo 2236
encuadernación encolada
  208
encuadernación en cuir
  bouilli 3807
encuadernación en cuir
  ciselé 3808
encuadernación en esmalte
  5139
encuadernación en
  material plástico 5822

encuadernación en media
  tela 6758
encuadernación en pasta
  francesa 2200
encuadernación en piel
  8015
encuadernación en serie
  4972
encuadernación en tela
  2948
encuadernación espiral
  12875
encuadernación espléndida
  4155
encuadernación estilo
  Bodley 1675
encuadernación etrusco
  5317
encuadernación flexible
  5822, 8153
encuadernación Grolier
  6664
encuadernación Harley
  6883
encuadernación heráldica
  775
encuadernación holandesa
  11129
encuadernación Maioli
  8555
encuadernación manual y
  de esmeralda calidad
  5390
encuadernación media piel
  6755
encuadernación medioeval
  8749
encuadernación monástica
  9064
encuadernación mosaico
  9110
encuadernación mudéjar
  9162
encuadernación pegada 208
encuadernación persa
  11832
encuadernación pointillé
  10526
encuadernación reforzada
  11424
encuadernación romántica
  11673
encuadernación sarracena
  11874
encuadernación sembrado
  12092
encuadernación semitela
  11130
encuadernación sin coser
  208
encuadernación sin cosido
  208
encuadernación sin costura
  208
encuadernación veneciana
  4833
encuadernación vistosa
  5464

encuadernado a la rústica
  13169
encuadernado a media piel
  11129
encuadernado a medio
  6756
encuadernado en rústica
  9900
encuadernado en tela 2950
encuadernador 1732
encuadernadora con
  alambre 14779
encuadernar 1401
encuadernar de nuevo
  11296
encuadernar sin costuras
  Lumbeck 8405
encuadrado 1792
encuadrar 9718
encumbramiento 5859
enchufado 13633
enchufe 12660
enderazamiento del
  gálvano 12529
enderezador de pliegos
  7686
enderezar (el papel) 7685
endurecedor 6862
endurecer la tinta 13259
endurecimiento de la
  mantilla 6866
endurecimiento lustroso
  6441
endurecimiento por la luz
  8116
enegricimiento 1475
energía cinética 7802
energía de radiación 11191
energía eléctrica 5024
energía radiante 11191
enfajilladora 2064
enfardadora 1093
enfocador 5953
enfocar 5952
enfoque 5950
enfoque automático 911
enfoque mecánico 8733
enfoque según escala 11902
enfoque suave 12737
enfrenamiento 1851
enfrenamiento de la
  bobina (de papel) 11358
enfriador 3489
enfriamiento de aire 324
enfriamiento de(l) agua
  14574
engomado 6719
engomadora 6721
engomados (los sobres) a
  toda la solapa 6713
engomadura 6719
engranaje 6345
engranaje a velocidad
  variable 7303
engranaje cónico 1348
engranaje de contragolpe
  1029
engranaje de tornillo sinfín
  14853

engranaje doble angular 6967

engranaje doble helicoidal 6967

engranaje helicoidal 6954, 14129

engranaje rectificador (de holguras) 1029

engranaje sin holguras 1029

engranar 8811, 13788

engrandecimiento del punto 4566

engrapado a la esquina 3587

engrapado con alambre 14780

engrapado de cuaderno 11831

engrapadora 13528, 14779

engrapado talonario 12389

engrape de la banda 13585

engrasado 6610, 12020

engrasado (en offset) 6609

engrasar 8395

engrase 6610

engrase central 2537

engrase forzado 6027

engrase líquido 8397

engrudo 9160

engrudo de almidón 13046

enjuagar 11619

enjugar 14756

enjugar los rodillos 12281

enlace doble 4627

enlomar 988

enmascarado 8647

enmascarado magenta 8490

enmascarar (una parte de un negativo) 12407

enmendar 11262

enmienda 3597

enmiendas 3605

enmiendas en el molde 2619

enmiendas en la forma 2619

ennegrecer 12638

ennegrecimiento 3989

enrolladora de bobinas 10576

enrolladora pope 10576

enrollamiento 14868

ensambladora 6330

ensamblar 830

ensanchamiento de la banda de papel 14645

ensanchamiento de la hoja en abanico 5467

ensayadora 13662

ensayadora a la flexión según Schopper 11921

ensayadora de desgarre 13618

ensayadora (del esfuerzo) de flexión 1324

ensayadora de rasgado 13618

ensayadora plegadora según Schopper 11921

ensayador de doblez 5980

ensayador de la absorción 42

ensayador de la resistencia a la abrasión 11766

ensayador de la resistencia al arrancamiento 10302

ensayador de la resistencia a la tensión 13645

ensayador de la resistencia a la tension estático y dinámico 13065

ensayador de la resistencia al repelado 10302

ensayador de la rigidez (del papel) 13154

ensayador del grado de refino 6160

ensayador de plegado 5980

ensayador de resistencia al doblez 5980

ensayador de resistencia al estallido 2097

ensayador de resistencia al reventamiento 2097

ensayador Mullen 9167

ensayar 13659

ensayo 7458, 13658

ensayo acelerado 49

ensayo al frotamiento 3782

ensayo al reventamiento 10578

ensayo al rozamiento 13

ensayo anónimo 11729

ensayo Cobb 3018

ensayo con cera 14632

ensayo con la uña de dedo 5684

ensayo de abarquillado 3836

ensayo de absorción de aceite 9551

ensayo de arañazo 11961

ensayo de blanqueo 1551

ensayo de bloqueo 1606

ensayo de burbujas 1581

ensayo de ceras Dennison 4165

ensayo de cinta adhesiva al tacto de marca Scotch 11942

ensayo de cinta adhesiva de celulosa 11942

ensayo de corrosión 3616

ensayo de desgarramiento 13617

ensayo de engrasado 12022

ensayo de envejecimiento 302

ensayo de esterilización 13132

ensayo de flexión 1325

ensayo de la absorción de la tinta 38

ensayo de laboratorio 7859

ensayo del ángulo de contacto 594

ensayo del poder cubriente 6978

ensayo del poder de absorción 14967

ensayo de materiales 11122

ensayo de pérdida del color 5434

ensayo de permeabilidad al vapor de agua 14610

ensayo de porosidad 10582

ensayo de rasgado 13617

ensayo de rasgadura 13617

ensayo de resistencia al ácido 140

ensayo de resistencia al estallido 10578

ensayo de resistencia al frote 3782

ensayo de resistencia (del papel) bajo presión 10578

ensayo de rigidez H & D 6824

ensayo de rotura 10578

ensayo de tinta a la pluma 10091

ensayo de tinta por manchas 10057

ensayo en laboratorio 7859

ensayo micro-contor 8893

ensayo MIT 9001

ensayo Mullen 10578

ensayo por flotación 5837

ensayo sobre colores secos 4819

ensombrecimiento 3989

ensuciado 12611

ensuciamiento del agua 14592

ensuciamiento del aire 351

ensuciamiento de la mantilla 1519

ensuciamiento de la mantilla offset 3418

entalladura 7265, 7453, 9107, 14794

entallamiento 14794

entintado 7339, 7356

entintado a la página 9856

entintado del fondo de la plancha 1478

entintaje 3114

entintar 11666

entintar con rodillo de mano 1870

entrada 3229, 6614, 7264, 7422, 7423

entrada de la plancha 7990

entrada del colector 837

entrada del depósito 2623

entrada del reunidor 837

entrada/salida 7425

entradilla 2470

entrar 7262

entrar en prensa 6529, 11067

entrar en tapas 7862

entrar tapas 2404

entrecomillar 11177

entrecomillas 11174

entrega 3391, 4134, 5474

entrega automática 906

entrega de cintas 3726

entrega por cadena 2580

entregar 5687

entremeter 7971

entrenervios 9889

entrerrenglón 7505

entrerrenglonadura 7505

entrerrenglonar 7971, 12808

entretemiento preventivo 10777

entrevista 7534

envarilladora 8824

envase de alimentos 6003

envejecimiento 301

envenenamiento de dicromato 2780

envío a condición 4146

envío a examen 4146

envío condicional 4146

envío para inspección 4146

envoltorios de jabón 12659

envoltorios de las resmas 11295

envoltura plegadizada 5992

eñe con tilde 9223

eosina 5215

epígrafe 4466, 6905, 8031

epigrama 5217

epílogo 5219

epsomita 8502

equilibrado 1085

equilibrio 5227

equilibrio de la humedad relativa 5228

equilibrio tinta/agua 7408

equipo 3255, 12296

equipo antimaculador 688

equipo antistático 13069

equipo auxiliar 10143

equipo central 2539

equipo de máquinas perforadoras 14318

equipo de periférico 10143

equipo de tratamiento 10895

erg 5238

ergología 5239

eritrosina 5248

errata 5242, 8979

errata de copia 2911

errata de distribución 14881

errata de redacción 2911

erratas 5241

error accidental 71

error de aforo 2205

error de imprenta 3315, 8979

error del autor 888

error del cajista 3315, 8245

error del compositor 3315

error de lectura 14257

error de mecanografía 14230

error de pluma 2911

error de probabilidades 10876
error standard 13024
error tipográfico 3315, 14235
(es) a saber 14500
esbozo 7961
esbozo anuncio 253
escala de colores 3157
escala de contraste 3443
escala de distancia 5955
escala de enfoque 5955
escala de filtros de color 5661
escala de grises 6629
escala de lana 14815
escala del objetivo 8047
escala de reproducción 11484, 11905
escala de valuación 11904
escala enfocadora 5955
escala micrométrica 8901
escala para enfocar 5955
escala tipográfica 14197
escamas de pescado 7169
escape 5249
escapes de aire 5355
escarabajo 336, 5810
escarabajos 12475
escartivanas 6692
escartivanas de relleno 3265
escayola 6738
escisión de la tinta 7397
esclavo de galera 3313
escobilla de alambre 14767
escobilla de carbón 2313
escobilla del cilindro 3935
escobilla para golpear 1268
escogido 12785
escogido de trapos 11212
escopladura 9107, 9432
escopladura al borde 9736
escopladura al centro 7453
escopladura exterior 9736
escopleadura 9107, 9432
escopleadura adentro 7453
escopleadura al borde 9736
escopleadura al centro 7453
escopleadura exterior 9736
escopleadura interior 7453
escoplear 2732, 3857, 9108
escoplo redondo 6531
escoria 4763
escorilla 12539
escribanía 13076
escribir 14876
escritor 881
escritora 881
escritor del artículo de fondo 7985
escritura Braille 1846
escritura cuneiforme 3813
escritura de espejo 8975
escritura de pincel 2018
escritura en relieve 1846
escritura inglesa 12012
escritura para ciegos 1846
Escrituras 12011

escrúpulo 12014
escuadra 1000, 7686
escuadra de corredera 219
escuadra de delineante 4715
escuadra de dibujo 4715
escuadrado 12962
escuadra en T 14071
escuadra graduable 219
escuadra lateral 12371
escuadra para fundir 2446, 2453
escuadra regulable 219
escudo de armas 3001
escudriñador electrónico 11910
escudriñamiento electrónico 11911
esculina 5251
escurridor 4694, 12978
es decir 14500
esencia 5257
esencia de lavanda 7943
esencia de petróleo 6320
esencia de trementina 14107
esfera 4259
esfuerzo 13261
esfumado áspero 6876
esfumado en negro 1495
esfumar 14471
esfumino 13313
esmaltado 2074, 13912
esmalte a la cola 5141, 6467
esmalte de pasado 1084
esmalte en frío 3054
esmalte (frío) 5138
esmalte ordinario 5141
esmaragdino 5110
esmeraldino 5110
esmerilar 7031
espaciado de (las) letras 8070
espaciado entre letras 8070
espaciado óptico 9658
espaciador 12802, 12803
espaciar 1856, 7971, 12808
espaciatura 8070
espacio 12364, 12799, 12800, 12801
espacio anillo 12809
espacio cromático 3188
espacio de bronce 1867
espacio de cobre 3532
espacio de cuña 12802
espacio de eme 5120
espacio de latón 1867
espacio delgado 13718
espacio de pelo 6752
espacio de tres puntos 13708
espacio de un punto 6752
espacio en blanco 1500
espacio (en) blanco 14711
espacio entrefino 6752
espacio fino 13718
espacio gordo 13708
espacio grueso 13708
espacio levantado 11226

espacio mediano 8919
espacio medio 8919
espacio movible 12802
espacio muerto 2904
espacios altos 6997
espacios de altura del hombro 6997
espacio variable 14395
español 12812, 12812
esparto 5252
espato pesado 1182
espátula 7367
espátula de dorar 6508
espátula de tinta 7367
espátula para tinta 7367
especificaciones de ensayo 13669
espectro 12846
espectro continuo 3432
espectro de absorción 41
espectro de emisión 5115
espectro de líneas 8196
espectrofotometría 12845
espectrofotómetro 12844
espectro solar 12752
espectro visible 14492
espejo 8974
espejo inversor 11561
espesador 1681
espesamiento 2971, 8284
espeso 1669
espesor 14671
espesor de la cama 9834
espesor de la capa 3000
espesor de la capa de tinta 7351
espesor de la plancha 10463
espesor del papel 2220
espesor del tipo 14738
espesor sistemático 14319
espigas del rodillo 11640
espíritu áspero 12883
espíritu blanco 14727
espíritu de madera 8855
espíritu rudo 12883
espíritus 1900
espíritu suave 12884
espolvoreado 10698
espolvoreado en cuatro direcciones 6138
espolvoreador 12929
espolvorear con blanco de magnesio 2597
esponja 12910
esponja celulósica 14484
esponja de celulosa 14484
esponja de viscosa 14484
es propiedad de ... 439
espuma 5938
espumadera 12514
espumadura 5939
espumar 12511
espúreo 1691
esquadra comprobante 4724
esquela de entrada 13799
esquelas de defunción 9150
esquelas de muerte 9150

esquelas funerarias 9150
esquelas mortuorias 9150
esqueleto 3703, 6046
esqueletos comerciales 2106
esquema 9717
esquema de bloques 1591
esquema de cableado 4258
esquema de conexiones 4258
esquema de la póliza de tipos 6120
esquema del lanzado 7197
esquema de los cajetines 2405
esquema funcional 1591
esquina 3581, 8997
esquina doblada 4543
esquinadores 3218
esquinar 8996
esquinazo 3581
esquinazo de latón 1862
esquinazos 590
esquinero 8998
esquirol 1484
estabilidad 3397, 7740, 12286, 12990
estabilidad a los ácidos 143
estabilidad a los álcalis 421
estabilidad a prueba de luz 5484
estabilidad dimensional 4376
estabilidad frente a los productos químicos 2688
estabilidad mecánica 8740
estabilizador de tensión 12991
estable 12992
establecimiento 10403
establecimiento gráfico 10829
establecimiento tipográfico 10829
estacionario 13073
estación de lectura 11276
estación de mando 3451
estación palpadora 11276
estado 13520
estampación en acero 4334
estampación en bronce 1966
estampación en oro 6496
estampación en plomo 7994
estampación en relieve 4334
estampación en talla dulce 3527
estampación fotocalcográfica 7471
estampado a seco 1563
estampado con bronce 1596
estampado con molde 1596
estampado de las tapas del libro 13901
estampado en frío 3051
estampado en relieve 4334
estampado en seco 1562, 1574, 1576

fotografía tricrómica 13747
fotógrafo 10241
fotohuecograbado 10253
fotoimpresora de
    repetición 13116
fotolitografía 10257
fotomecánica 10889
fotomecánico 10231, 10258
fotómetro 5373
fotomontaje 10261
fotopolímero 10268
fotoquímica 10216
fotostato 10273
fototipia de Albert 3099
fototipia en colores 3127
fototipocromía 3162
fototipografía 10281, 10890
fototransistor 10276
fotozincotipia 8159
fracción 6141
fraccionar 1893, 4509
fracción compuesta 10320
fracción con diagonal 12592
frágil 1925
fragilidad (de una película)
    1926
fragmento 13553
fraguar 12154
fraile 1452, 6192
franela 5754
franela de algodón 9057
franela de caucho 1509
franjas (en la impresión)
    por franquicia de
    engranajes 6351
franqueadora 6149
frasqueta 6204
frecuencia 6186
frecuencia de errores 5247
frecuencia relativa 11432
frenado 1851
freno 1848
freno automático 899
freno de aire comprimido
    313
freno del cilindro 3933
freno neumático 313
frente 1296, 14461
fresa 11733
fresado 8939, 11737
fresadora 937, 11732, 11733
friso 9688
frontis 6211
frontispicio 6211
ftalato de butilo 4290
fuelle 1292, 1293, 6730,
    14074
fuente del agua 14577
fuente de luz 8133
fuente de matrices 12165
fuente (de tipos) 6103
fuera de comercio 10873
fuera de puntura 9726,
    11407
fuera de registro 9726
fuero de línea 9513
fuerte 5394
fuerza 3397
fuerza adhesiva 214

fuerza capilar 2281
fuerza centrífuga 2553
fuerza coercitiva 3043
fuerza colorante 3189
fuerza cortante 12240
fuerza de cizallamiento
    12240
fuerza de empuje 12241
fuerza de impacto 7187
fuerza de impresión 7219
fuerza del cuerpo 1686
fuerza de presión 10843
fuerza de saturación 11886
fuerza eléctrica 5024
fuerza electromotriz 5041
fuerza licuante 8233
fuerza mecánica tensora
    10774
fuerzas de Van der Waals
    14383
fuerzas superficiales 13433
fugitiva 6229
función acumulativa 3810
funcionamiento 9633
funciones trigonométricas
    14038
funda protectora 10980
fundente para a cosido por
    soldadura a puntos 13744
fundición 2442, 8836
fundición bastarda 1216
fundición de caracteres
    14180
fundición de rodillos 2452
fundición (de tipos) 6103
fundición de tipos 14180,
    14193
fundición tipográfica 14180,
    14193
fundido 2074
fundidor 9118
fundidora 2443
fundidora automática para
    planchas curvas 932
fundidora curva 2449
fundidora de líneas-bloque
    12606
fundidora de tejas 2449
fundidora de tejas de
    impresión 932
fundidora de tipos 14181
fundidora (de tipos) 2440
fundidora de titulares 4467
fundidora Monotype 9088
fundidora para estereotipia
    plana 2450
fundidora para estereo-
    tipias semicilíndricas
    2449
fundidor de tipos 14192
fundidor monotipista 2441,
    9075
fundir 2435
fundir rodillos 2960
fundir tejas 2434
fungicida 6251
furfurol 6257
fusionar 8808

fusión de cintas perforadas
    13583

gaceta 6344
gaceta oficial 6343
gaceta sensacionalista
    14920
gacetero 7700
gacetín 1829
gafa (de uña) 2869
galena 6272
galera 6278, 12573
galera con lengüeta 12549
galera con pala 12549
galerada 6278, 12573
galerada suelta 12573
galera de columna(s) 12565
galera (de) volandera 12549
galera para anuncios 202
galera rama 2654
galera receptora 4152
galería fotográfica 6276
galerín 6278
galerón (de pruebas) 4861
galipodio 6274
galón 6290
galvanización 10469
galvanizado 10433, 10469
galvano 5071
galvano Albert-Fischer 365
galvanómetro 6293
galvano niquelado 9342
galvanoplastia 5063
gama 6296
gama de colores 3157
gama de grises 6629
gama europea uniforme
    14301
gamma francesa uniforme
    14302
gamuza 2612
ganar 6386
ganar línea 13554
gancho portaoriginales
    3547
garabateador 11963
garabatos 11964
Garaldos 6308
garganta del crisol 13780
garrafón 2337
garrapatos 11964
gas 6310
gasa 6342
gasa para encuadernar
    9164
gaseoso 6316
gasolina 6320
gasolina a punto de
    ebullición especial 11897
gasolina sin plomo 14330
gastos adicionales 5407
gastos de composición 3303
gastos de impresión 10822
gastos generales 9758
gatillo de trinquete 10067
gato de tornillo 7632
gauss 6339
gazapo garrafal 14884
gel 6352

gelatina 623, 6356
gelatina bicromatada 4302
gelatina cromatada 4302
gelatinocloruro (de plata)
    6365
gelatinografía 6363
gelatinoso 6366
gel de sílice 12410
gelosa 294
generación de caracteres
    2632
generador 6377
generador de caracteres
    2633
generador de programas
    para listar informes
    11470
generador de vapor 1697
géneros diversos 13405
génuli 9691
gilbert 6398
gilsonita 6411
giraplanchas 14709
girar 5860
girar lentamente 3705
giro de ventas anual 640
girones, en ~ 5739
glaseado de la mantilla
    6449
glaseado en hojas 12262
glaseadura 6441
glasear 2191
glasilla 9164
glasímetro 6426
glicérido 6479
glicerina 6482
glicerol 6482
glicina 6485
glicol 5309
glicolas 6489
glucosa 4246
glutinosidad 13526
gobernar 3449
gobierno electrónico 5054
gofrado 5102
gofrado a mano 6849
gofrado con planchas de
    cinc 14946
gofrar 5099
golilla 14549
golpear 10393
golpear innecesariamente
    el componedor con el
    tipo 5452
golpe (periodístico) 11932
golpes 7656
golpeteo 11255
golpeteo de la plancha
    10461
goma 6706
goma adragante 13951
goma alquitira 13951
goma arábiga 6707
goma bicromatada 4304
goma contracorte 3870
goma copal 3493
goma de borrar 752, 5233,
    7345
goma de celulosa 2514

goma de India 4688
goma de kauri 7732
goma desensibilizadora 6711
goma desoxidante 6711
goma elástica 5233, 11747
goma espuma 12912
goma guta 6294
gomaguta 6294
goma laca 12289
goma mordiente 6711
goma sandáraca 6724
goma sintética 13496
goma tragacanta 13951
gordura 9222
gorrón 7697
gorrones del rodillo 11640
gotero 10375
gouache 6530
grabado 1582, 5191, 5192, 5606, 7165, 10419
grabado a gubia 11736
grabado al agua fuerte 5270, 5272
grabado al aguafuerte en cobre 3513
grabado al aguatinta 735
grabado a la maneranegra 8883
grabado al barniz blando 12739
grabado al boj 14799
grabado al buril 8172
grabado al claroscuro 2707
grabado a máquina 8445
grabado a pincel 1999
grabado a puntas 13161
grabado a puntaseca 4817
grabado a un solo baño 12445
grabado blando 5799
grabado calcográfico 7467
grabado colorido con muñeca 3140
grabado con baños graduales 9181
grabado con blancos 6987
grabado con exceso 9751
grabado con poco contraste 5799
grabado continuo 10703
grabado cromado 2800
grabado de base 7757
grabado de bordes esfumados 14473
grabado de cinc 14944
grabado de combinación 3299, 8179
grabado de línea 8174
grabado de línea para colores 8165
grabado de música 9210
grabado de pluma 8174
grabado de punzones 11044
grabado de registro 7757
grabado de trama gruesa 2978
grabado de trazo 8159, 8174

grabado directo degradado 9720
grabado directo de trama gruesa 11140
grabado directo en cinc 14947
grabado electrolítico 5007, 5035
grabado electrónico 907
grabado en acero 13096
grabado en boj 14799
grabado en cinc 14944
grabado en cobre 3523
grabado en el strato 2992
grabado en hueco 7467
grabado en linóleo 8214
grabado en madera 14799
grabado en negro 8883
grabado en plomo 7981
grabado en relieve 11442
grabado en talla dulce 3523
grabado fototipográfico 10890
grabado litográfico 13198
grabado pluma 8159
grabado punteado 13161
grabador 5186, 5688, 10888
grabadora 5195, 5280
grabador a buril 10435
grabadora continua 10704
grabadora de mordido continuo 10704
grabadora electrónica 10229
grabador al agua fuerte 5269
grabado rápido (sin escalar) 10703
grabado rápido sin escalas 10703
grabado rápido (sin espolvorear) 10703
grabadora rápida 10704
grabador de medio tonos 6781
grabador de música 9209
grabador de punzones 11043
grabado recortado 9721
grabador en madera 14797, 14897
grabado reticulado 6774
grabados a sangre 1552
grabados bendéi 1316
grabado silueteado 9721
grabado sobre piedra 13198
grabados para impresión 1699
grabados para policromía 3165
grabados sin margen 1552
grabado vernis-mou 12739
grabado xilográfico 14799
grabar 5182, 5268
grabar (al agua fuerte) 5271
grabar a la ruleta 11716
grabar en cinc 14945
grabar en relieve 5291

gradación 6534
gradación de tonos 13898
gradiente de velocidad 14428
grado de acidez 133
grado de blancor 4115
grado de blancura 4115
grado de claridad 8124
grado de dilución 4374
grado de dureza 4112
grado de dureza según Brinell 1918
grado de engrasamiento 4109
grado de ennegrecimiento 4175
grado de humedad 4113, 9033
grado del albura 4115
grado del cuerpo 1686
grado del esterificación 4110
grado de negrura de la letra 14671
grado de refino 4109, 6161
grado Engler 5177
grado H & D 6823
grado higrométrico 9033
grado Kelvin 7744
grados Celsius 2530
grados Fahrenheit 5435
graduación 226
graduar un rodillo 12156
gráfica hombre/máquina 8606
gráfico 4257
gráfico simultaneado 12437
grafista 6567
grafitar 1482
grafito 6571
gramaje del papel 1209
gramo 6556
grampa 13040
gran blanco 14711
grandes luces 6992
graneado 6546
graneado por chorro de arena 11860
graneador 11626
graneadora (de planchas) 6550
granear químicamente 2683
granillo 6536
granillo del papel 8226
grano 6536, 6538, 13903
grano atravesado 3752
grano de fototipia 3094
grano de heliotipia 3094
grano (de la piel) 9234
grano del papel 6537
grano de resinada 6553
grano electrolítico 5036
grano fino 5671
grano grueso 2975
grano (inglés) 6539
grano químico 2682
granos de arena 6658
granulado 6536
grapa 2869

grapa (de alambre) 13040
grapa de la tela 2871
grapa de unión de correas 1303
grapadora 13528
grapas intermediarias 7511
grasa 6601
grasa consistente 3394
grasa de carnero 9222
grasa de lana 14813
grasa de lana purificada 7901
gravedad específica 12827
greca 6620
gredoso 2606, 3732
grieta del tintero 13783
grifos espiradores 13348
grisado 8741
grisado bendéi 12216
grisado de resina 6552
grisador 1315
grisado rayado 9596
grisar 13845
grosor 14671
grosor del tipo 14738
grosor sistemático 14319
Grotesca 6675
grueso 13109, 14671
grueso pesado 5394
grumo (de estucado) 2994
grumo (de pasta) 2994
grumos 8417
grumos de fibras 12310
grupo 6299
grupo carboxílico 2335
grupo convertidor 7544
grupo de cambio para las últimas noticias 6228
grupo de entintado 7359
grupo de impresión 10852
grupo de impresión inferior 8381
grupo de impresión reversible 11560
grupo de impresión suplementaria 7233
grupo de novedades 13218
grupo hidróxilo 7133
grupo impresor 10852
grupo impresor en forma de U 14310
grupo impresor reversible 11560
grupo OH 7133
grupo oxhidrilo 7133
grupo oxidrilo 7133
guacal 3703
guache 6530
gualdrapeado 7496, 7907
guardaalmacén 13239
guarda de correa 1305
guarda del alma 11362
guardafilo 3904
guardafrenos 1853
guardalmacén 13239
guardamanos 6843
guardapapeles 9918
guardapolvo 7629
guardar, a ~ 6522

guardar la composición 7742

guardas 5162

guarismo 5607, 9458

guarismos 747

guarnecer 4721

guarnecer la forma 4722, 8561

guarnecido de los rodillos de la plegadora 3786

guarnición 6262, 9828

guarniciones 6262

guarniciones de hierro 7590

guata celulósica 2521

guata de celulosa 2521

gubia 6532

guía 2909, 6335, 6697

guía capuchina 5516

guía (de ajuste) 6287

guía de densidades 6629

guía de direcciones 4408

guía de entrada 2623

guía (de entrada) 5516

guía de frente 6212

guía de lado 12371

guía delantera 6212

guía de lengüeta 6338, 7955

guía de resorte 6338, 7955

guía de sensibilidad 6629

guía de tope 12371

guiado de la hoja de papel 7997

guía lateral 12371

guíalateral electrónico 5054

guía para minervas 6338

guía pecho 12371

guía (posterior) 1000

guías 3375, 6700

guías del hilo metálico 7412

guía telefónica 4408

guía trazado para comprobación de registro 6698

guillemets 11174

guilloque 6701

guilloquear 6702

guillotina 6704

guillotina con programa de corte 920

guillotina de palanca 8080

guillotina de tres cuchillas 13754

guillotina de volante 6850

guillotinar 3855

guillotina rápida 6998

guillotina trilateral 13754

guillotinista 3884

guión 4003, 7143

guión de ene 4002

guiones conductores 7984, 9623

guiones de conducción 7984, 9623

guiones suspensivos 9623

gutagamba 6294

gutapercha 6733

gutiambar 6294

hacer arrugas 14874

hacer el arreglo 8564

hacer el cálculo de espacio 2457

hacer el cálculo de la composición 2457

hacer el cálculo de material 2457

hacer el cálculo tipográfico 2457

hacer el registro 1919

hacer inodoro 4182

hacer más fluida la tinta 12736

hacer ranuras 6668

hacer resaltar 13794

hacer tapas 2411

hacer una corrección mal hecha 1822

hacer una impresión secundaria 7231

hacer una nueva composición 11493

hacer un calco 13935

hacer un ensayo de tintas por manchas 13598

halo 4569, 6754, 6801

halógeno 6804

hallar el foco 5952

hardware 6877

haya 1284

haz de luz 1244

haz luminoso 1244

hebdomadario 14667

hebreo 6944

heliograbado 10253

heliotipia 3099

hemicelulosa 6960

hender 3716

hendido 3714

hendido para el doblado 1243

hendidura 6666

henequen 300

henry 6966

hermético 356

hertz 6970

hexametafosfato de sodio 12691

hexametafosfato de sosa 12691

hexametafosfato sódico 12691

hidratación 7095

hidrato 7089

hidrato alumínico 477

hidrato de alúmina 477

hidrato de cal 2167

hidrato de calcio 2167

hidratos carbónicos 2301

hidratos de carbono 2301

hidrocarburos 7100

hidrocarburos aromáticos 776

hidrocarburos parafínicos 424

hidroclorato 7104

hidrocloruro de diaminofenol 4262

hidrófilo 7122

hidrófobo 7124

hidrógeno 7107

hidrógeno nítrico 9366

hidrógeno sulfurado 7117

hidrógrafo 7118

hidrólisis 7119

hidrómetro 7120

hidroquinona 7127

hidrosulfito de sodio 12694

hidrosulfito de sosa 12694

hidrosulfito sódico 12694

hidróxido bárico 1146

hidróxido cálcico 2167

hidróxido crómico 2791

hidróxido cuproamoniacal 3821

hidróxido de aluminio 477

hidróxido de amonio 533

hidróxido de bario 1146

hidróxido de cinc 14948

hidróxido de cromo 2791

hidróxido de litio 8251

hidróxido de potasa 10655

hidróxido de potasio 10655

hidróxido de sodio 12695

hidróxido de sosa 12695

hidróxido férrico 5550

hidróxido potásico 10655

hidróxido sódico 12695

hiel de buey 9802

hiel de toro 9802

hieroglíficos 6979

hierro 7575

hierro colado 2455

hierro de dorar 1411

hierro de fundición 2455

hierro en flejes 13104

hierro fundido 2455

hierro llanta 13104

hierro pasamanos 13104

hierros puntillados, con ~ 5582

higroestabilidad 7141

higrómetro 7137

higrómetro de caballo 6742

higrómetro de vástago 13481

higrómetro eléctrico 5021

higroscopicidad 7140

higroscópico 7139

higroscopio 7138

higroscopio de espada 13481

higrotermógrafo 7142

hilo 1430

hilo bramante 1430

hilo (bramante) 9855

hilo corto 3752

hilo cruzado 3752

hilo de encuadernación 1430

hinchamiento (de las fibras) 13465

hincharse 2024, 13462

hinchazón (de la mantilla) 5103

hipersaturación 13419

hipoclorito cálcico 2168

hipoclorito de calcio 2168

hipoclorito de sodio 12696

hipoclorito de sosa 12696

hipoclorito sódico 12696

hiposulfito amónico 551

hiposulfito de amonio 551

hirma 12090

histéresis 7150

histograma 7010

historiado 7011

historieta cómica 3233

historieta gráfica 3233

hoja 8007, 12250

hoja adhesiva 211

hoja continua 14639

hoja de acetato 91

hoja de aluminio 476

hoja de alzas 12922

hoja de arreglo 9773, 12922

hoja debajo el papel de impresión 1009

hoja de fijación 13145

hoja de guarda exterior 5930

hoja de la mañana 9104

hoja de la noche 5323

hoja de las cotizaciones 5343

hoja de lata 13831

hoja de(l) cuchillo 7817

hoja de máquina 10764

hoja de muestra 4331

hoja de papel preimpresa 10731

hoja de registro 11022

hoja de sierra 11894

hoja de trabajo 7681

hoja en blanco 1535

hoja en color 3146

hoja exterior de la cama 4719

hoja in-plano 1931

hojalata 13831

hoja metálica 5961

hoja modelo 4331, 8202

hoja muestra 12834

hoja no impresa 1535

hoja para picar reportes 13145

hoja periódica 9309

hoja plana 1931

hoja raspadora 4525

hojas adicionales 3424

hojas con manchas de humedad 6139

hojas costeras 9737

hojas de falta 9784

hojas de más 9784

hojas de papel defectuosas 11541

hojas desgarradas 13922

hojas, en ~ 7448, 14298

hojas interiores 7454

hojas intonsas 1716

hojas para interfoliar 7500

hojas por hora 12280

hojas rosáceas 6139

hojas sueltas 6815

hoja suelta 8362

holandés 4908, 4908
holandesa 7020
hollín 12782
hollín de acetileno 106
hollín de lámpara 7894
hombre de confianza 5490
hombre trivial 1942
hombro 4737, 12231, 12342, 12343
hombro saliente 7748
homocéntrico 3357
hondura de la presión 5106
honorarios del autor 890
hora de cerrar 4032
hora de clausura 4032
hora de(l) cierre 4032
horas especiales 9793
horas extraordinarias 9793
horma superior 1101
horma triangular 6063
hormigón armado 11425
hornillo para esmaltar planchas 2075
horno de refundición 8770, 8826, 11456
horno de secado 4794
horno para calentar 2075
horno para metal 8826
horno para refundir 8770
horqueta de rodillo 11639
horquilla de la correa 1304
horquilla del cilindro 3936
horquilla de rodillo 11639
hueco 11931
huecograbado 7471, 11705
huecograbado autotípico 760, 6783
huecograbado convencional 3462
hueco-offset 4085
huelga 1028
huelgo 1028
huelguista 13274
huella 6868, 7266
huella en el papel 9915
hule 1508, 2559, 5233, 11747
hule sintético 13496
Humanísticos 7068
humectabilidad 14696
humectación 7071
humectador 7072
humectadora de chorritos 7072
humectativo 14699
humedad 7073
humedad absoluta 29
humedad del aire 7077
humedad relativa (del aire) 11433
humedecedor 7072
humedecer 3968
humidificación 7071
husillo del rodillo 11649
huso 12869

igualación 5222
igualadora 7687
igualador de pliegos 7686
igualar 7685, 10393

igualar el espaciado 12806
igualdad 12644
ilegible 7157
iluminación 7160, 7161
iluminación a contraluz 1031
iluminación difusa 4347
iluminación lateral 12374
iluminación oblicua 9471
iluminación uniforme 5794
ilustración 7165, 10419
ilustración a doble plana 4663
ilustraciones a sangre 1552
ilustraciones decayentes 1552
ilustraciones voladas 1552
ilustración señuelo 5414
ilustrar 7162
imagen 5606, 7166, 10419
imagen al directo 11282
imagen al revés 11285
imagen borrosa 1648
imagen cegada 1648
imagen con falta de contraste 5791
imagen con poco contraste 5791
imagen de medio tono 6784
imagen de plata 12429
imagen de revés 11285
imagen de tarjeta 2350
imagen empastelada 1648
imagen falta de afinidad para la tinta 1566
imagen fantasma 6391, 10182
imagen invertida 7542
imagen latente 7931
imagen litográfica 8261
imagen maculada 1648
imagen nítida 12235
imagen original 8656
imagen original fotográfica 10246
imagen oscura 8385
imagen pálida 6985
imagen real 11289
imagen remosqueada 1648
imagen sombreada 1648
imagen virtual 14481
imbricación 7169
imitación cuero 791
imitación, de ~ 1691
imitación de cuero de Rusia 499
imitación de papel a mano 9131
imitación de papel estucado 7170
imitación de piel de foca 359
imitación de piel de Rusia 499
impalpable 7188
impedancia 7189
impeledora 4504
impermeabilidad 7194, 14607

impermeabilidad al vapor de agua 14611
imponedor 8311
imponer 7196
imposición 7202
imposición a blanco y retiración 14822
imposición de altura 1251
imposición de pie 6024
imposición en cuadernos 7201
imposiciones de hierro 7590
imposiciones (de metal) 6262
imposición plomada 2964
imposición y acuñado 13206
impositor 8311
impregnación 11884
imprenta 10829
imprenta comercial 7679
imprenta particular 10874
imprenta privada 10874
imprentar 10794
imprentario 8659
imprentero 8659
impresión 7209, 10792, 10817
impresión a blanco y retiración 14822
impresión a blanco y voltereta 14824
impresión a blanco y vuelta 14824
impresión a colores 3163
impresión a color húmedo 14682
impresión a dos o tres colores en la prensa diaria 11680
impresión a la anilina 5828
impresión algráfica 490
impresión a mano 6830
impresión anastática 579
impresión anopistográfica 653
impresión anticipada 10729
impresión a tres colores 13750
impresión a un solo color 9079
impresión autocrómica 894
impresión bicolor 14136
impresión clandestina 12053
impresión con clisés flexibles 5828
impresión con clisés plásticos 5828
impresión con grueso mínimo 7805
impresión con malla de seda 11989
impresión con muñequilla al aguafuerte 3140
impresión con pantalla de seda 11989

impresión con plancha de acero 10458
impresión con plancha de cobre 10458
impresión con planchas de aluminio 490
impresión con tinta magnética 8521
impresión con tintas planas 8166
impresión con trama de seda 11989
impresión débil 5436
impresión de billetes de banco 1128
impresión de blanco 1573, 5711
impresión de calcomanías 13969
impresión de empapelado 14536
impresión de iris 12896
impresión de la faceta 5425
impresión de libros 1783
impresión de música 9213
impresión de papeles de valor 12063
impresión de papel moneda 1128
impresión de papel pintado 14536
impresión de relieve 11446
impresión de sellos de correo 10601
impresión electrostática 5068
impresión empastada 9161
impresión, en ~ 6521
impresión en cobre 3527
impresión en colores 3162, 3163, 3197
impresión en dos colores 14136
impresión en hojalata 8819
impresión en húmedo 14682
impresión en negro 1465
impresión en talladulce 10458
impresión en tricromía 13750
impresiones digitales 5685
impresiones por hora 7223
impresiones reportadas 2601
impresión excesiva 6868
impresión fantasma 6391, 10182
impresión granulosa 6554
impresión grumosa 9116
impresión hectográfica 6945
impresión hueco-offset 9521
impresión húmedo sobre húmedo 14682
impresión irisada 1958, 12896

impresión litográfica 9536, 13205
impresión metalográfica 8819
impresión monocromo 9079
impresión multicolor 10904
impresión offset 9536
impresión plana 10401
impresión planográfica 10401
impresión policroma 3167, 10904
impresión por estarcido 11989
impresión rotativa 11701
impresión sarnosa 12839
impresión serigráfica 11989
impresión sobre hojalata 8819
impresión sobre metal 8819
impresión termográfica 13690
impresión textil 13678
impresión tipográfica 8065
impresión tipográfica indirecta 4812
impresión tridimensional 14895
impresión Woodbury 14793
impreso 10792
impreso comercial 2107
impreso en el taller de ... 10803
impreso monocromo 9079
impresora 10805
impresora al vuelo 7012
impresora de alta velocidad 7001
impresora de barras 1169
impresora de cadena 2590
impresora de ruedas 14703
impresor ambulante 13930
impresora por contacto 3406
impresora por páginas 9854
impresora por puntos 8694
impresora serie 12142
impresora-troqueladora 10812
impresora vertical 14454
impresor errante 13930
impresor por líneas 8192
impresor remendista 7679
impresor tipógrafo 8064
impresos 10802
impresos publicitarios 11001
impresos tabulares 14792
imprimabilidad del papel 10796
imprímase 7228
imprimátur 7228
imprimir 10794
imprimir al revés 1045
imprimir el blanco 10864
imprimir el retiro 1052
imprimir en exceso de huella 8595

imprimir sensibilización 2980
impulsión 4736, 11035
impulsión de cadena 2583
impulsión eléctrica 5016
impulsión separada 7295
impulsión unitaria 7295
impulso 4736, 11035
impulso a mano 4746
impulso del carro 1276
impulso mayor 8542
impulso por correa 1302
impulsora 4504
impurezas 7235
inactínico 9393
inalterable a los ácidos 145
inatacable por el ácido 145
inatacable por los ácidos 145
incapaz de trabajar 7242
incidencia 2040
incinerador 7249
inciso 9972
Incisos 7250
inclinación de carácter 2641
inclinación de la cuchilla 600
inclinación de la racleta 600
inclinación de una curva 12591
inclinado en ángulo de 30 ° (treinta grados) 7251
incluída 5153
incoloro 3151
incompleto 4093
incorporado 2043
incrustación 5156
incunable 7256
indentar 7262
indescifrable 14265
indeseados 3438
indicación 11380
indicación de medidas de reproducción 3557
indicaciones para componer 3285
indicación tipológica de los textos 11180
indicador 7286, 10524
indicador de atrás 1000
indicador de enrutado 11739
indicador de la abertura del diafragma 4275
indicador de (la lineatura de) la retícula 11980
indicador de las unidades 10523
indicador de presión 7215
indicador de prioridad 10867
indicador de profundidad 4190
indicador de unidades 14313
indicador de velocidad 13523

indicar erratas 8633
indicar los colores (con letras) 7763
índice 7269, 7277
índice alfabético 455
índice de acidez 134
índice de asas 13795
índice de contraste 3441
índice de estallido 2100
índice de éster 5263
índice de ilustraciones 8243
índice de kauri butílico 7733
índice de kb 7733
índice de la resistencia en húmedo 14694
índice de nombres propios 9229
índice de refracción 11396
índice de reventamiento 2100
índice de saponificación 11872
índice de yodo 7555
índice kappa 7729
índice pulgado 13795
índigo 7288
indirecto 9513
indisoluble 7457
inducido 771
industria del libro 1755
industria de materiales de embalaje 9823
industria gráfica 10831
inerte 7299
inestable 7459, 14350
inferiores 7301
inficionar 3417
inflamable 7305
inflamación cutánea 4200
inflamación de la piel 4200
inflarse 13462
información 7307
información especulativa 4561
información legible por la máquina 8459
informe sobre circulación 871
infrarrojo 7311
infundible 7314
infundio 5441, 7013
ingenio (para encuadernadores) 10484
inglesa 12012
inglete 8995
ingleteadora 8998
ingletear 8996
inhibición del secante 4728
iniciación 7539
inicial 7321
inicial de doble renglón 14145
inicial de dos líneas 14145
inicial de tres líneas 13756
inicial doble 14145
inicial embutida 4752
inicial encastrada 5429
inicial encuadrada 5429

inicial encuadrada en ornamento 5429
inicial en orla 5429
inicial en racada 3865
inicializar 7320
inicial mayúscula 7321
inicial orlada 5429
inicial saliente 3029
inicial sobre una línea 3029
ininflamable 9400
inmersión 4395
inmiscible 7185
inodoro 9499
insaponificable 14339
inscribir en la lista de subscriptores 13330
insecticida 7432
inserción 7434
inserción publicitaria 246
insertar 7436
insertar gratis 4029
insetter 7446
insolación 5368
insolación doble 4603
insolación principal 8543
insolar 5367
insoluble 7457
insonorización 9385
inspección 7458
inspección, para su ~ 726
inspección reforzada 13805
inspector 13422
instable 14350
instalación de acondicionamiento de aire 321
instalación de acondicionamiento para papel 9907
instalación de aire acondicionado 321
instalación de cromado 2779, 2795
instalación de desagüe 4695
instalación de deshidratación 4695
instalación de pasta desfibrada 6645
instalación de pasta mecánica 6645
instalación para cobrear 3510
instalación para cromado 2795
instituto de investigaciones 11492
instrucción 7462
instrucción de máquina 3337
instrucción de ordenador 3337
instrucciones de manejo 7463
instrucciones para el compositor 3285
instrucciones para mostreo 11857
instrucciones para muestreo 11857

letras voladas 13416
letra uncial 14262
letra volcada 14095
letra vuelta 14095
letrero 12350
letrista 14184
leva 2227
leva de (la) bomba 11038
leva del separador 4144
levantamiento de blancos 11620
levantamiento de espacios 11620
levantamiento de letras 4716
levantamiento de tipo 4716
levantar 13555
levantar (la forma) 8094
levantar la forma de la máquina 8094
levantar la imposición 13282
levantar letra 3282
levantar los rodillos 13791
léxico 4306
ley de Hooke 7037
ley de imprenta 6271
ley de la caída de Stokes 13195
leyenda 8031
lezna 1671
lezna de encuadernación 1672
liar 13801
libelo 7895
libertad de comentario 6157
libertad de expresión 6157
libertad de (la) imprenta 6158
libertad de (la) prensa 6158
libertad de opinar 6157
libertad de opinión 6157
libraco 11203
libracho 11203
libra inglés 10690
libra troy 14066
libre de ácidos 126
librería 1778, 8083
librería anticuaria 677
librería circulante 8036
librería de lance 677
librería de viejo 677
librero 1775
librero de viejo 12045
libreta de apuntes 4279
libreta de cheques 2696
libreta de notas 9437
libretilla para anotaciones 8771
librillo 1758
librito 1758
librito de notas 9437
libro 1729
libro antifonal 2752
libro con anotaciones (autográficos) 848
libro con estampas 10310
libro con orejas 4544

libro con puntas dobladas 4544
libro (con tapa confeccionada) empastado 2401
libro cosido con alambre 5806
libro de apuntes 4279, 9437
libro de bolsillo 9895, 10508
libro de cánticos 2752
libro de comercio 76
libro de contabilidad 1506
libro de coro 2752
libro de cuentas 76
libro de cheques 2696
libro de devociones 10708
libro de extractos 11944
libro de fábulas 1765
libro de grabados 10310
libro de (gran) éxito 1341
libro de hojas cambiables 8365
libro de hojas movibles 8365
libro de hojas sueltas 8365
libro de horas 1767
libro de imágenes 10310
libro de indicaciones 7463
libro de lance 12044
libro de mayor venta 1341
libro de memoria 4279
libro de mucha venta 1341
libro de ocasión 12044
libro de oraciones 10708
libro de recortes 11944
libro de rezo 10708
libro de segunda mano 12044
libro de tapa montada 2401
libro de texto 11920
libro encuadernado 1815
libro escolar 11920
libro ilustrado 10310
libro impreso con grabados xilográficos 1588
libro incunable 7256
libro infantil 2709
libro intonso 14334
libro invendible 14338
libro muerto 14338
libro para niños 2709
libro para pintar 3149
libro rayado de contabilidad 1506
libros encadenados 2579
libro sin cortar 14334
libros xilográficos 14792
licuefacción 8232
ligado 8105
ligado de fibras 5590
ligadura 8105
ligadura doble 4627
ligamiento de las tintas en superposición 583
ligante 14423
lignina 8141
ligroína 8144
lijadora 8081
lijante 14

lima 5612
lima fina 12641
limar 5618
límite de estiraje según Bingham 1434
límite de fluencia 14930
límite de fluencia según Bingham 1434
límite de fluidez 14930
límite de imposición 4034
límite de stiraje 14930
limpiacuchillas 7828
limpiador de (las) cuchillas 7828
limpiador del molde 9135
limpiamantillas 1526
limpiar 13589, 14556, 14756
limpiar la mantilla 2887
limpiar los rodillos 12281
limpiarrodillos 11655
línea 8154, 8155
línea ágata 298
línea apretada 13806
línea básica 1196
línea central 8717
línea con enmienda 3599
línea corregida 3594
línea corta 1889, 14736
línea cumbrera 1404
línea de blancos 14716
línea de cabeza 7322
línea de cierre 13734
línea de composición 12599
línea de contacto 7691, 9352
línea de continuación 7706
línea de división 4514
línea de entrada de presión 4034
línea de fabricación 5873
línea de fecha 4019
línea de fin de párrafo 1889
línea de guía 6287
línea de identificación 6287
línea de imposición 4034
línea de impresión 7227
línea del pie 6018
línea del principio de impresión 10385
línea del tipo 14206
línea de parangonado 13027
línea de presión 7227
línea de procedencia 4019
línea de puntos 4575
línea de reclamo 2466
línea de retardo 4125
línea de sección 4514
línea de tipos 12599
línea de titulares 4466
línea de traslado 7706
línea diagonal de quebrados 12767
línea divisoria 13320
línea en blanco 14716
línea enmendada 3599
línea epigráfica 4466
línea fina 6748
línea floja 8366

línea inicial 7322
línea invertida 14096
Lineales 8157
línea normal (de los tipos) 13027
línea ondulada 14622
línea perdida 2465
línea poco espaciada 2938
línea quebrada 1889, 14736
líneas ásperas 6884
líneas de agua 2589, 14585
línea separatoria 13320
líneas negras del espectro 6150
líneas torcidas 3739
línea superior 2288
línea transversal (de la letra) 1135
línea trunca 14736
lineatura de la trama 11992
lineatura fina 5680
línea volcada 14096
línea vuelta 14096
lineómetro 14197
Lineotipia 8221
lingote 2964, 10327, 12600
lingote de composición 12599
lingote de madera de 48 puntos 1927
lingote indicador 6287
lingote invertido 14096
lingote poroso 10584
lingotero 6263, 6279, 8002
lingotes de plomo 6262
lingotes empalmados 2114
lingotes para unir 2114
lingotes yuxtapuestos 2114
lingote volcado 14096
linguete 10067
lino 5811
linógrafo 8223
linoleato 8215
linoleato cobaltoso 3009
linoleato de cobalto 3009
linóleo 8219
linóleum 8219
Linotipia 8221
linotipista 8223, 9639
Linotipo 8221
línteres 3638
linterna para cuarto oscuro 3995
línters 3638
liofílico 8425
liófobo 8427
líquido anhidro para limpieza de planchas 609
líquido antimaculador 686
líquido antimancha 686
líquido colado 5662
líquido de lavar 14555
líquido filtrado 5662
líquido Newtoniano 9328
líquido non Newtoniano 9407
líquido seudoplástico 10995
lista código 3034
lista de abreviaturas 8242

lista (de duplicados) para
   canje 5344
lista de precios 10779
lista de salida 10859
lista de tipos 6120
lista encadenada 2586
listín 5343
listón de la guillotina 3904
listón de medida del alineamiento 14172
listones de cuchilla 7820
lisura 5686
lisura (del papel) 12643
lisura dinámica del papel
   4929
litargirio 8248
litoffset 9525
litografía 8254, 8255, 8261,
   8270, 13205
litografía sin agua 4734
litógrafo 8256
litógrafo comercial 3243
lito-offset 9525
litopón 8276
litro 8281
lobo 14744
localización 11542
lodo 12596
logaritmo 8315
lógica 8316
logotipo 8322
lomera 1034
lomera despegada 7023
lomera destacada 7023
lomera libre 7023
lomeras 7026
lomo 983, 1422
lomo aserrado 11895
lomo con nervios 11589
lomo cuadrado 12957
lomo (del libro) 12873
lomo despegado 7023
lomo destacado 7023
lomo encolado 13802
lomo en compartimientos
   1026
lomo en tela 2947
lomo libre 7023
lomo liso 12957
lomo plano 12957
lomo rectangular 12957
lomo redondeado 11720
lomo unido 13802
lona 2269, 4731
longitud de bloque 1607
longitud de corte 3906
longitud de foco 5945
longitud de la banda de
   papel entre dos grupos
   impresores 8287
longitud del alfabeto 459
longitud de onda 14615
longitud de onda
   complementaria 3273
longitud de onda
   dominante 4552
longitud de palabra
   máquina 14819

longitud de palabra
   variable 14396
longitud de papel en
   máquina 14654
longitud fija de palabra
   5733
longitud focal 5945
lote 1221
lubricación 9557
lubricación a presión 6027
lubricación central 2537
lubricante 8394
lubricar 8395, 9548
lubrificación 9557
lubrificante 8397
lubrificar 9548
luces 6992
lucha contra el ruido 9385
ludlista 8402
Ludlow 8401
lugar solar 5859
lujo 7969
lujos de dos puntos 14151
lumen 8407
lumen-erg 8408
luminancia 8409
luminescencia 8411
luminiscencia 8411
luminosidad 8413, 14380
luna 6435
luna de montaje 5766
lunar 13008
lunares 6976
lunares transparentes
   12304
lungitud 8037
lupa 8187, 8534
lupa de enfoque 5953
lupa de pie 2059
lupa montada 2059
lustre 6452, 8419
lux 8420
luz 8106
luz actínica 161
luz artificial 793
luz atenuada 4346, 4382
luz blanca 14715
luz de arco voltaico 756
luz de corte 3906
luz de incidencia 7248
luz de incidencia rasante
   6600
luz del día 4023
luz difusa 4346
luz dispersa 11915
luz diurna 4023
luz inactínica 9394
luz incidente 7248
luz indicadora de atención
   14543
luz interior de la rama
   7416
luz monocromática 9077
luz negra 14252
luz parasitaria 11915
luz polarizada 10541
luz puntiforme 10528
luz ultravioleta 14252

llamada 5718, 11552
llamadas 11381
llamadas de corrección
   3600
llamada selectiva 10548
llamadas (indicativas) 8641
llamadas marginales 8641
llanura 12644
llave 7759
llave a botón 11062
llave Allen 428
llave de acuñar 11170
llave de plancha 10444
llave de tuerca 11170, 14872
llave fija 9620
llave inglesa 9072
llaves 1840
llenadora-dosificadora 5631
llenar la línea con cuadratines 5157
llenar la última línea 5160
llenarse 2464, 5633

macerar 8429
maceta 8586
macroinstrucción 8477
mácula 13008
mácula de retiración 12047
maculado 8473, 9540
maculado de los márgenes
   1713
maculatura de descarga
   12163
maculaturas 12576, 12906
machacadora de pulpa
   11029
machacadora (de pulpa)
   11385
machón 8661
machote 6046, 8661
madera a contrafibra 3755
madera a contrahilo 3755
madera compensada 10505
madera contrachapeada
   7887, 10505
madera de abedul 1438
madera de abeto 5693,
   12953
madera de álamo 10577
madera de aliso 387
madera de árboles foli-
   culares 6881
madera de boj 1839
madera de campeche 8324
madera de chopo 387
madera de haya 1284
madera de hilo 8336
madera de (las) coníferas
   12747
madera de peral 10075
madera de pino 10362,
   12747
madera de pino melis 1107
madera de tilo americano
   1211
madera laminada 7887
madera (para la
   fabricación) de papel
   11034

madera para montaje (de
   clisés) 9144
madera para zócalos 9144
madera resinosa 11505
madera, sin ~ 6680
madera terciada 10505
maduración 8705
maestro 1799
maestro de escritura 2217
maestro impresor 8659
magacén alto 14357
magacén auxiliar 12375
magenta 8486
magnesia 8500
magnesio 8496
magnetismo remanente
   11452
magnetismo residual 11452
magro 6153
maguey 300
mala prensa 1063
malbaratador 10778
maleable 8585
maleta con traje de trabajo
   y trebejos 11842
mal obreo 11963
malla de seda 11967
mallas 14544
mallete 8586
manantial de luz 8133
mancilla 13008
manco de peso 12341
mancha 1620, 9069, 13008,
   14671
mancha de grasa 6608
mancha de luz 5758
mancha de pringue 6608
mancha de tinta en el re-
   verso 5405
mancha de tinta en la cara
   5406
manchar 3417
manchas de agua 5715,
   14604
manchas de bronze 4164
manchas de cobre 4164
manchas de herrumbre
   11823
manchas de humedad
   14604
manchas del rodillo de
   sostén 1019
manchas de moho 8928,
   12556
manchas de secado 4792
manchego 9584
manchón 3645, 3979
manchón de fieltro 3644
manchuelas 12921
mandadero de imprenta
   10809
mandar 3449
mandato 3236
mandil 727
mando 4736
mando a botón (pulsador)
   11061
mando a distancia 11458
mando a mano 6820

mando automático 902
mando de entrada/salida 7427
mando del carro 1276
mando del diafragma 4274
mando electrónico 5054
mando individual 7295
mando por botón (pulsador) 11061
mandril 8587, 11363, 12869
mandril deformado 3783
mandril del rodillo 11649
manecilla 2876, 5718
manejo 6820, 6832
manera punteada 8597
manga 9057
manganeso 8589
mango de la punta 6831
mango del punzón 1673
manipulado del papel 6832
manipulador 9638
manipular localmente 4537
manivela 11719
mano 5718
mano de papel 11163
manómetro tensor 10775
mano perdida 9784
mantener en pequeña tolerancia 14935
mantenimiento 8553
mantilla 13909
mantilla de calzar 3724
mantilla de caucho 1509, 3724
mantilla de corcho 3576
mantilla de la prensa 1508
mantilla del cilindro 1508, 9825
mantilla de respaldar 3724
mantilla hundida 5183
mantilla litográfica 1509
mantillas automáticas 898
mantisa 8607
manual 6816
Manuales 8609
manuscrito 3534
manuscrito de difícil lectura 1059, 1061
manuscrito impreso 11478
mapa astronómico 2497
mapa caminero 11622
mapa celeste 2497
mapa en relieve 11444
mapa geográfico 8612
mapa mural 14532
maqueta 808, 4857, 7961
maqueta acabada 4859
maqueta anuncio 253
maqueta de plegada 7963
maqueta detallada 3321
maquetista 7962
máquina, a ~ 8725
máquina a barnizar 14406
máquina a brochadora 14779
máquina afiladora de cuchillas 7826
máquina (a forma) redonda 3944

máquina alimentada por bobinas 14656
máquina astilladora 2728
máquina automática de platina 916
máquina auxiliar 942
máquina calibradora 1048
máquina caucho contra caucho 1524
máquina cepillo 12314
máquina cilíndrica 3948, 5775
máquina cilíndrica automática 904
máquina compositora 3286
máquina curvadora de clichés 1587
máquina de alimentación continua 14656
máquina de alimentación en hojas 12269
máquina de alta velocidad 7000
máquina de alto rendimiento 6936
máquina de amarrar 2064
máquina de aplicar el papel pigmento 2333
máquina de atar 2064
máquina de bolsas en bobina 11367
máquina de broncear 1969
máquina de cajas plegadizas 5976
máquina de calandria 2184
máquina de cama plana 5775
máquina de cartón a la enrolladura 7515
máquina de cerrar bolsas 1077
máquina de cilindro 3948
máquina de cilindro y platina 3948
máquina de clavar 9228
máquina de componer 3286
máquina de componer para titulares (y anuncios) 4467
máquina de composición 3286
máquina de copia y repetición 13116
máquina de cortar paja 2578
máquina de coser 12197
máquina de coser con alambre 14779
máquina de coser con grapas 14779
máquina de coser con hilo metálico 14779
máquina de cuatro colores 6123
máquina de cubrir 7891
máquina de curvar 1319
máquina de descortezar 1164

máquina de doble impresión offset 9530
máquina de doble juego 14153
máquina de doble revolución 14153
máquina de dos colores 14134
máquina de dos vueltas 14153
máquina de encorvar 1319
máquina de encuadernación rústica 10119
máquina de encuadernación sin costura 10119
máquina de encuardernación a la americana 10119
máquina de enrollar y hacer tubos en espiral 3485
máquina de escribir con cabeza impresora 6519
máquina de escribir eléctrica 5025
máquina de estucar 2995
máquina de etiquetar 7850
máquina de forrar 7891
máquina de fotocomposición 10279
máquina de fototipocomposición 10279
máquina de franquear 6149
máquina de fresar 937
máquina de fundir 2443
máquina de fundir tipos 14181
máquina de grabar 5195
máquina de grabar al ácido 5280
máquina de grabar al agua fuerte 5280
máquina de granear 6550
máquina de granular 6550
máquina de gran velocidad 7000
máquina de guilloches 6703
máquina de hacer tapas 2412
máquina de huecograbado 6587
máquina de huecograbado a planchas 10440
máquina de (im)presión planocilíndrica 5775
máquina de imprimir talonarios de caja 2422
máquina de imprimir tejidos 2208
máquina de imprimir y troquelar etiquetas 7852
máquina de lavar 14552
máquina de matrizar 8699
máquina de meter (libros) en tapas 2425
máquina de montaje 9145
máquina de mordido continuo 10704

máquina de mordido rápido 10704
máquina de mordido sin polvo 10704
máquina denatrizar 9129
máquina de numerar 9452
máquina de ocho bobinas 4997
máquina de papel 9923
máquina de parada de cilindro 13210
máquina de pedal 14014
máquina de pegar guardas 5167
máquina de película enrollada 11657
máquina de percusión 10117
máquina de platina 10452
máquina de plegar 5966
máquina de poner ojillos 5417
máquina de poner tapas 2425
máquina de prensar 2064, 8699
máquina de presión plana 10452
máquina de rayar 11784
máquina de reacción offset 9530
máquina de redondear 1047
máquina de redondear esquinas 3583
máquina de retiración 10122
máquina de retiración offset 9530
máquina de taladrar 4733
máquina de tamiz largo 6127
máquina de tamiz redondo 3944
máquina de tapas sueltas 2412
máquina de tiro y retiro de cuatro cilindros 6125
máquina de titulares 4467
máquina de transporte fotomecánica 13116
máquina de tres bobinas 13771
máquina de tres cilindros 9626
máquina dobladora con cuchillo 7824
máquina Duplex 4879
máquina electrónica de grabar 10229
máquina embaladora 1093
máquina empaquetadora 9832
máquina encartadora 7446
máquina encoladora 6475
máquina etiquetadora 7850
máquina fototituladora 6906
máquina Fourdrinier 6127

máquina fresadora 11732
máquina fundidora 2443
máquina fundidora de la
  Monotipia 9088
máquina fundidora del
  Monotipo 9088
máquina Gally 6292
máquina gofradora 5105
máquina Gordon 6526
máquina graneadora 6550
máquina holandesa 7020
máquina impresora de
  billetes 13800
máquina insertadora 7438
máquina lavadora 14552
máquina Linotipia 8221
máquina mezcladora 9008
máquina monocilindro
  14905
máquina numeradora 9452
máquina offset 9532
máquina offset de dos
  colores 14132
máquina offset plana 9522
máquina paginatrice 9869
máquina para despolvorear
  4899
máquina para doblar
  cartón 2343
máquina para embalar
  1093, 9832
máquina para encolar
  12492
máquina para encolar
  cajas plegables 5976
máquina para
  encuardernar sin cosido
  10119
máquina para enderezar
  pliegos 7687
máquina para escribir
  direcciones 196
máquina para fabricar
  bolsas 1073
máquina para foto-
  composición 10279
máquina para fundir por
  líneas 8161
máquina para grabado electrónico 10229
máquina para hacer
  contra-pruebas 10990
máquina para hacer
  ingletes 8998
máquina para hacer
  mandriles 12879
máquina para hacer tubos
  espirales para envases
  3485
máquina para hueco-
  grabado 6587
máquina para impresión
  de películas 5647
máquina para imprimir
  direcciones 196
máquina para imprimir
  hojalata 8821
máquina para imprimir
  papel continuo 14656

máquina para imprimir
  papeles pintados 14537
máquina para juntar
  esquinas de cajas 3586
máquina para la
  confección de sobres
  5211
máquina para la impresión
  en hojalata 8821
máquina para la impresión
  tipográfica 8063
máquina para la re-
  producción 10882
máquina para limpiar
  matrices 8687
máquina para meter en
  sobres 5214
máquina para morder 5280
máquina para offset
  rotativo a bobinas 11698
máquina para ojetear 5417
máquina para paginar 9869
máquina para plegar cajas
  plegables 5976
máquina para rajar 11937
máquina para ranurar 3718
máquina para rayar 11784
máquina para sacar cajos
  1015
máquina para sacar cajos
  de mano 6813
máquina para seccionar los
  mandriles 12879
máquina para sensibilizar
  planchas 14709
máquina para trazar
  puntos 13163
máquina para troquelar
  4318
máquina pequeña offset
  12627
máquina perforadora 10132
máquina plana 5775, 6127
máquina plana de cartón
  7515
máquina planocilíndrica
  3948, 5775
máquina plegadora 5966
máquina prensadora de
  pliegos 5972
máquina pulidora 2039
máquina rápida 7000
máquina rayadora de
  discos 4445
máquina rayadora de mesa
  1312
máquina rebobinadora
  11580
máquina redonda 9173
máquina repetidora 13116
máquina reveladora de
  películas 10900
maquinaria para encuader-
  nación 1424
máquina rotaplana 5776
máquina rotativa 11699
máquina rotativa de huecograbado 14648

máquina rotativa Duplex
  4879
máquina rotativa offset (de
  bobina) 11698
máquina rotativa tipo-
  gráfica 14652
máquina roto offset 11698
máquina sacapruebas
  10950
máquina satélite 3249
máquinas de encuader-
  nación 1739
máquina semirrotativa
  5776
máquinas y equipo 6877
máquina tabuladora 13521
máquina taladradora 4733
máquina telecompositora
  13635
máquina teletipográfica
  13635
máquina tipográfica 8063
máquina tituladora 4467
máquina Yanqui 14905
maquinista de la rotativa
  8950, 11700
maquinista (impresor)
  10753
maquinista offset 10754
maquinista tipógrafo 8064
marbete 7846, 13534
marca comercial 13944
marca comercial registrada
  11406
marca de agua 14587
marca de agua imitada
  7180
marca de agua molette
  7207
marca de bayeta 3785
marca de cabecera 8661
marca de fábrica 13944
marca de imprenta 7229
marca de impresor 5816
marca de la tela (metálica)
  14774
marca del editor 7230
marca del fieltro 5534
marca del fundador 10369
marca del impresor 7229
marca del manchón 5534
marca del rodillo de sostén
  1018
marca de propiedad 1771
marca de separación de
  resma 11293
marca diacrítica 4250
marcado manual, a ~ 6825
marcador 5507, 5508
marcadora 5507
marcador automático 909
marcador de succión 13354
marcador luminoso 8125
marcar 8633
marcar (a mano) 5503
marca registrada 11406
marcas de corrección 3600
marcas de corte 3899, 14043
marcas de dimensión 4377

marcas del pliego 3066
marcas de registro 11411
marcas de registro de la
  plancha 3943
marcas de registro del
  cilindro 3947
marcas de tamaño 4377
marcas guías 3066
marcas por la calandra
  2192
marcas recorte 3742
marca tipográfica 7229
marco 1792, 2650
marco de bronce 1861
marco de copia 4940
marco de imprimir 10828
marco de la delantera
  14461
marco de las quijadas
  14461
marco-frasqueta 11834
marco intermediario 180
marco lateral 12368, 12371
marco neumático (de
  copiar) 14375
marcha lenta 9602
marcha normal 9427
marcha sin dificultades
  11790
margen 8627
margen de cabeza 6909
margen de cosido 1035
margen de entrada de
  pinzas 6654
margen de exposición 5372
margen de fondo 1035
margen de la guía de
  pecho 7949
margen delantero 6213
margen del crucero 1035
margen del lomo 1035
margen de pie 13550
margen de pose 5372
margen de uña(s) 6656
margen exterior 6033, 6213
margen externo 6033
margen inferior 13550
margen interior 1035
margen lateral (exterior)
  6033, 12376
margen lateral (externo)
  12376
margen lateral interior
  1035
margen medianil 1035
margen para las uñas 6656
margen para marcas de
  registro 8297
margen sujetaplanchas
  12158
marginador 4940
marginadora 5507
marginar 5503
marginera 2909
marión azul 3917
marmita 8770
mármol 7198, 8581
marmoración 8621
marmosete 14470

marquesina de la lámina expulsora 5003
marroqué 9106
marroquín 9106
marroquín de cabra 6493
marroquín frances 6179
marroquín nigeriano 9344
marroquín persiano 10155
martillo de encuadernador 1014
martillo para redondear lomos 1014
masa para estucar 2998
máscara 6205, 8645, 8646
máscara acladora 6989
máscara aclaradora 6989
máscara borrosa 14344
máscara correctiva del color 3132
máscara de claros 6989
máscara de contacto 3410
máscara de corrección de contraste 13889
máscara de corrección del color 3132
máscara de dos etapas 14159
máscara de rebaje 4540
máscara de reducción de contraste 3445
máscara de reserva 11515
máscara de sombra 12222
máscara en dos etapas 14159
máscara en la cámara 2241
máscara integral 7475
máscara nítida 12236
máscara para respaldo de cámara 2241
máscara para sombras 12222
máscara por contacto 3410
máscara por proyección 10933
máscara positiva 10595
máscara proyectada 10933
máscara reductora de contraste 3445
mascarilla 6205, 8646
mascarilla por contacto 3410
masicote 8652
mat 5846
mate 8663
materia colorante 4922
materia corrida 13247
materia de texto 13247
materia fibrosa 5597
material a deshacer 4035
material anticipado 241
material auxiliar 584
material clisado 1699
material compuesto sin intención de publicarlo 1694
material de blancos 1531, 12811
material de carga 5383, 5623

material de propaganda 273
material de publicidad 273
material de relleno 5623
materiales polares 10543
material estereotipado 1699
material fijo 13033
material levantado 8095
material muerto 4035
material para futura información 4561
material sobrante 3772, 12590
material suplementario 5165
material usado 4035
materia orgánica 9679
materia prima 11258
materias colorantes 3176, 4928
materias volatilizables 14506
matice 13838
matiz 7066, 12209, 13838
matizar 3155, 14471
matraquar 2659
matricera 8699, 9129
matrices auxiliares 12786
matrices de contracaja 12786
matrices de mano 12786
matrices sueltas 12786
matrices suplementarias 12786
matriz 8677, 8678, 8679
matrizador 8699, 9129
matrizadora 8699
matriz corrediza 8696
matriz de bronce 1863
matriz de cartón 4805
matriz (de cartón) 5846
matriz de dos letras 4626
matriz de estereotipia 5846
matriz de fotocomposición 6096
matriz de latón 1863
matriz de papel 5846
matriz de plomo 7993
matriz de titulares 4468
matriz de vidrio 6432
matriz doblada 4626
matriz estereotípica 5846
matriz fotográfica 6096
matriz que no necesita relleno 9835
matriz seca 4805
matriz sin respaldo 9835
matriz titular 4468
maxwell 8711
mayúscula 8557
mayúsculas 2287, 5329
mayúsculas pequeñas 12617
mayúsculas y mayúsculas pequeñas 2295
mayúsculas y versalillas 2295
mayúsculas y versalitas 2295

mazo 8586
mecánica 8744
mecánica, sin ~ 6680
mecánico 8725
mecanismo 4236
mecanismo abrillantador 12995
mecanismo alimentador 5508
mecanismo cuadrador y centrador 11093
mecanismo de cambio de almacén 8484
mecanismo de conducción 3481
mecanismo de distribución 4491
mecanismo de entintado 7359
mecanismo de entrega 4145
mecanismo de impresión 10852
mecanismo de inmovilización 8305
mecanismo de mojado 3981
mecanismo de salida 4134, 4145
mecanismo de seis moldes (de rueda) 12480
mecanismo sacudidor 12227
mecanismo salvapliegos 7224
mecanógrafa 14231
mecanógrafo 14231
Mecanos 8745
mecanotipista 7771, 8439
mecha deshilada 6152
mechero a gas 6312
media 951
media caja 6757
mediacaña 9801
mediana 8746
medianil 985, 3749, 6735
media pasta 6755
media pasta, de ~ 11134
media piel 11134
media portada 1217
media progresiva 10929
media resma 13879
media tela 6758
media tinta 6773, 8920
medición 8722
medición de la fineza 5678
medición de la lisura 12645
medición de la rigidez H & D 6824
medición de pH 10178
medida 4375, 8720, 8722
medida arrollable 6416
medida corta 9243
medida de la altura del cliché 1595
medida de la columna 3213
medida de la velocidad de secado 4802
medida del cuerpo 1686
medida de líneas 14197
medida de los colores 3111

medida tipográfica 14197
medidora de molturación 5677
medidor a mostrador de profundidad 4190
medidor de brillo 6454
medidor de espesor electro-magnético 5039
medidor de gruesos 8898
medidor de la altura de coronas 3928
medidor de la dureza 6872
medidor del caudal 5874
medidor del deslizamiento, (aparato) ~ 12566
medidor del plano (del mármol) 5797
medidor del secado 11762
medidor del tiempo de frotamiento de la tinta 11762
medidor de luz 5373
medidor de opacidad 9603
medidor de pH 10179
medidor de profundidad de clisés 5277
medidor óptico de hondura 9654
medio acelerador 53
medio aritmético 765
medio cuadratín 5205
medio de adsorción 236
medio de oxidación 9809
medio encolado 6768
medios tonos para poli-cromía 3165
medio tono 6773
medio tono amplificado 1627
medio tono blando 5799
medio tono con partes en blanco 1628
medio tono con poco contraste 5799
medio tono con zonas al trazo 8179
medio tono de cinc 14947
medio tono degradado 14473
medio tono de mordido extrahondo 4091
medio tono desvanecido 14473
medio tono de trama ampliada 1627
medio tono de trama ancha 2978
medio tono de trama gruesa 2978
medio tono difumado 14473
medio tono difuminado 14473
medio tono directo siluetado 1628
medio tono escuadrado 12963
medio tono esfumado 14473
medio tono recortado 1628
medio tono silueteado 9721

molde conservado 13032
molde de estado 11776
molde de filetes 11776
molde de fundición curva 2449
molde de fundición manual 6836
molde de fundir 9118
molde de fusión 10334
molde de grapa 13172
molde de imprenta 6056
molde de impresión 6056
molde de plástico 10411
molde de rueda 9125
molde electrotipado 5071
molde esqueleto 11299
molde guardado 13032
molde hueco 11299
molde para estampar las tapas 4312
molde para estampar tapas 1584
molde para estereotipia curva 2449
molde para estereotipia plana 2450
molde para fundir orlas y filetes 8696
molde para fundir rodillos 6729, 11643
molde para panes de metal 10334
molde para papel avitelado 14857
molde (para papel hecho a mano) 9119
molde tipográfico 6056
molde vigente 13032
moldista de remiendos 7673
molécula 9055
molécula-gramo 9042
moledora de tintas 7371
moler (la tinta) 6637
molesquina 9056
moleta 9168
moletón 9057
molino a rodillos 11642
molino coloidal 3092
molino cónico 7693
molino de afloras 868
molino de bolas 1100
molino de cilindros 11642
molino (de colores) a tres cilindros 13772
molino de muelas 7835
molino de papel 9927
molino de ruedas verticales 7835
molino de rulos 7835
molino de tintas 7371
molturación 1266, 11386
momento de inercia 9063
momento del cierre 4032
momento de torsión 13924
momio 5492
monigote 14470
mono 13127
monoclínico 9080

monocromático 9076
monocromía 9079
monocromo 9078
monómero 9083
monometiléter de etilen-glicol 5310
Monotipia 9087
monotipista 7771, 9639
Monotipo 9087
monovalente 9093
monóxido de plomo 8248
montado adhesivo 10422
montado a flor 9138
montado a ras 9138
montado en metal 9141
montador 8573
montador de máquinas 5237
montaje 9136
montaje al ras de la letra 1600
montaje a ras 5899
montaje de grabados 1597
montaje fotográfico 10261
montaje sin bisel 5899
montaje sin chaflán 5899
montaje sobre papel in-actínico 6503
montar 5236, 9137
montar varillas 8840
montear la plancha 8309
montón 10340
montura 9136
montura a bayoneta 1237
montura de un objetivo 1170
moral 9163
mordante 3544
mordaza 7650, 8180
mordaza de cierre 2056
mordaza para encuadernar 1745
mordaza rápida 2056
mordedura 1450, 5787
mordedura segunda 12037
morder 5271
mordido 1450
mordido a máquina 8445
mordido a un solo baño 12445
mordido con baños graduales 9181
mordido continuo 10703
mordido electrolítico 5007
mordido en hueco 4082
mordido en profundo 4082
mordido en reposo 13159
mordido extrahondo 4081
mordido preliminar 5787
mordido profundo 4081
mordido sin agitar 13159
mordiente 6402, 9366, 14423
mordiente de los rodillos 13525
mordiente para broncear 1970
mordiente para encuader-nación (en falsa piel) 209

mordiente para grabar 5274
mordiente (para grabar) 5286
morera 9163
mortaja 9107, 9432
mortaja al borde 9736
mortaja interior 7453
mortajar 9108
mosaico 7411
motas 5715
mote 4237
motor acorazado 5151
motor blindado 5151
motor con anillo de frota-miento 12574
motor con colector trifásico 13761
motor (con rotor) de jaula 12983
motor (con rotor) de jaula en corto circuito 12983
motor de ajuste del registro 11412
motor de anillos 12574
motor de anillos trifásico 13765
motor de arranque 7246
motor de corriente alterna 466
motor de corriente continua 4401
motor de mando 3458, 4747
motor eléctrico 5042
motor mayor 8546
motor principal 8546
motor trifásico 13764
mover (intermitentemente) 7684
movilidad 9017
moviliómetro 9018
movimiento alternativo de los rodillos 11645
movimiento circular de las lámparas de arco 2845
movimiento de atras para adelante 11304
movimiento de descenso 4680
movimiento de descenso de la cuchilla 4677
movimiento de lado 7935
movimiento de lado de la banda 14638
movimiento del carro 2366
movimiento de planeta 10396
movimiento de retorno del carro 2369
movimiento de vaivén 6397, 11304
movimiento de vaivén de los rodillos 11645
movimiento forzado 6028
movimiento giratorio de las lámparas de arco 2845
movimiento lateral 7935
movimiento oscilante 9706

movimiento paralelo 9984
movimiento recíproco 11304
movimientos equilibrados 1086
movimientos falsos 5453
movimientos innecesarios 5453
movimiento uniforme-mente acelerado 14305
muaré 9032
mucílago 9160
muchacho 5240
mucho contraste, con ~ 3447
muela 6647
muela de fondo 1282
muela yacente 1282
muelle 12939
muelle de compresión 3326
muelle de presión 10770
muesca 6666, 7265, 9107, 9432, 13273
muesca de chaveta 7786
muesca de graduación 227
muesca exterior 9736
muesca interior 7453
muestra 11854, 12834
muestra de ensayo 13670
muestra (de ensayo) de laboratorio 7858
muestra de escogido al alzar 11235
muestra de fabricación 9740
muestra de referencia 9740, 11384
muestra representativa 11475
muestrario 11855, 11856, 13456
muestrario de los tipos 14221
muestreo 13502
muletón 1508, 9057
multicolor 9169, 10553
multicomputadora 9195
multiperforador desde el emisor 6302
multiplexor 9189
multiprogramación 9196
mullido 5740
mundo de periódicos 14851
munición 1096
muñequilla de algodón 10480
muñequilla de eje 7697
muñequilla de revelado 4227
muñón 7697
muñones del rodillo 11640
muselina 9216
musgo de Irlanda 7574
mutilar 1822
muy negro 5394

nacimientos, matrimonios, fallecimientos 1440
nada que hacer 13558

páginas verso 5327
página vertical 4092
paja 5810
paja de arroz 11595
paja de trigo 14701
pajas 12475
paja trigaza 14701
pala 13553
palabra 14816
palabra clave 2469
palabra compuesta 3320
palabra de máquina 3348
palabra de ordenador 3348
palabra repetida 4579
pala (de la galera) 6286
paladio 9877
palanca 8078, 12555
palanca de justificación 7710
palanca de los espacia-dores 12804
palancas de excéntrica 14618
palas 9848
palastro 13831
palatina 5778
palenca articulada 13876
paleografía 9874
paleta 5628
paleta de tinta 7367
paletas 9848
paletina para mordida 5275
palidecido 4421
palillo de levantar (el pan de oro al dorar) 6404
palimsesto 9876
palmeador 10394
palmear 10393
palmetear 10393
palmotear 10393
palo 11932, 13109
palo alto 813
palo (de bastón), sin ~ 11868
palo de campeche 8324
palo fino 6746
palo grueso 13109
palo guardafilo 3904
pana 1880
pancromático 9887
pan de metal 10327
pan de oro 6509
panel 9889
panel de acoplamiento 10490
panel de control 3459
pan para dorar 6509
pantalla 2482, 5420, 11967, 11968
pantalla cerrada 3388
pantalla de contacto 3416
pantalla de lámpara 7896
pantalla del imagen 4470
pantalla para enfocar 5953
pantógrafo 9892
panza 1298, 2029, 12874
pañete 1508
paño 1508
paño inferior 14267

paño superior 13909
paño superior de fieltro 4729
paño superior de la rayadora 1510
papel 9309, 9894, 13182
papel absorbente 33
papel acanillado 7876
papel aceitado 9554
papel aclimatado 3368
papel acondicionado 3368
papel a contrafibra 3752
papel acordonado 11473
papel acuarela 14573
papel aderezado en tina 14073
papel adiáfeno 9615
papel aéreo 341
papel afelpado 5845
papel ahuesado 3712
papel aislante eléctrico 5009
papel a la sosa 7840
papel al bromuro de plata 1946
papel albuminado 374
papel al chlorobromuro 2744
papel alfa 5254
papel (al) ferroprusiato 1644
papel alisado 8447
papel alquitramado 13607
papel aluminio 481
papel ambientado 3368
papel anticorrosivo 694
papel antivapor 14390
papel apergaminado 7175, 14421
papel apergaminado y gofrado 5100
papel apilado 5798
papel a prueba de álcalis 420
papel a prueba de grasa 6603
papel arenado 11861
papel arte 6789
papel artístico bático 1226
papel áspero 682
papel atercíopelado 5845, 14430
papel autoadherente 12076
papel autoadhesivo 12076, 12086
papel autocierre 12086
papel autocolante 12076
papel autocopiador 9382
papel autográfico 13966
papel autopegante 12076
papel avión 341
papel avitelado 14858
papel bañado en gelatina 625
papel baritado 1179
papel barnizado 14403
papel biblia 1359
papel bicolor 4881
papel bond 1723

papel bond al sulfito 13374
papel Braille 1847
papel bromuro 1946
papel calandrado 2190
papel caldo 13938
papel calicó 2207
papel carbón 2326
papel carbón para (usar) una vez 9594
papel carbón para uso una vez 9594
papel cargado 5620
papel cebolla 341, 9597
papel cebolla para cartas 6444
papel cianográfico 1644
papel cícero 6789
papel cigarillo 2831
papel Cobb 3017
papel colorado en la calandra 2188
papel coloreado 3137, 6442
papel coloreado por una cara 9591
papel con filigrana 14589
papel con marca de agua 14589
papel con sonido duro 13830
papel con tejido 11426
papel continuo 14639
papel coquilla 14224
papel cortado al sesgo 591
papel cortado en diagonal 591
papel costero 9737
papel couché en ambos lados 14157
papel crema 3712
papel cremoso 3712
papel crêpe 3729
papel crêpé 3729
papel crespado 3729
papel cristal 6431
papel cromo 2820
papel crudo 14259
papel cuadriculado 6573, 12964
papel cuché en ambos lados 14157
papel charol 5835, 6197
papel China 2713
papel chupón 1622
papel damasco 3963
papel (de acabado) semi-mate 5180
papel (de) albúmina 374
papel de almacén 13188
papel de alto brillo 6197
papel de amianto 812
papel de ampliación 5204
papel de añafea 14866
papel de arena 11861
papel de arroz 11594
papel de asbesto 812
papel de barba 4066
papel (de) barba 4917
papel de barita 1179
papel débil 5832

papel de bisulfito blanqueado 14729
papel de bromuro 1946
papel de calcar 13938
papel decalco 4042
papel de calco 13938
papel de carpetas 8347
papel de carta 8060
papel de carta azul celeste velin 981
papel de carta azul celeste verjurado 979
papel de carta(s) 9442
papel de carteles 10610
papel de cera 9967
papel de ciegos 1847
papel de cigarillos 2831
papel de cloruro (de plata) 2739
papel de color 3137
papel de colores 6442
papel de compuesto diazoico 4284
papel de condensador 5009
papel de contacto 3412
papel de copia 12051
papel de copias 2326
papel de cubiertas de varias capas 10026
papel de cubrir 14168
papel de cúrcuma 14092
papel de China 1359
papel de desecho 14558
papel de desecho sin surtido 9004
papel de dibujo 4709
papel (de) doble faz 4881
papel de domingos 13404
papel de embalaje 14866
papel de embalaje secado al aire 328
papel de embalar 14866
papel de encolado duro 6875
papel de envolver 14866
papel de escribir 14879
papel de escribir (aperga-minado) 1129
papel de escribir verjurado 7878
papel de esmeril 5111
papel (de) esmeril 17
papel de estado 3839
papel (de) estaño 13829
papel (de estraza) alqui-matrado 13608
papel (de) fantasía 5460
papel defectuoso 11541
papel (de) ferroprusiato para copias 1644
papel de filtrar 5660
papel de filtro(s) 5660
papel de forro 1034, 10043
papel de fumar 2831
papel de gelatina 625
papel de hilo 8186
papel de Holanda 4917
papel de imprenta 10838
papel de impresión 10838

papel de la mañana 9104
papel de la noche 5323
papel de la prensa 10838
papel de larga duración 8347
papel de la tarde 5323
papel delgado 5832, 6172
papel de lija 17, 6434, 11861
papel de línea marca 7876
papel de luto 9150
papel de música 9211
papel de paja 13255, 14927
papel de paño 14430
papel de pasta de madera 8735
papel de pasta mecánica 8735
papel de pasta química 2692
papel de periódico para ilustraciones 6787
papel (de) prensa 9321
papel de primera calidad 10118
papel de primer escogido 10118
papel de raso 5845
papel de recortes 14558
papel de reportar 2331
papel de seda 6172
papel de seda para arreglo 8567
papel de seda para hacer flores 5872
papel de seda para recorte 8567
papel de segunda (clase) 11541
papel de seguridad 11840
papel (de) seguridad 2698
papel de straza 11982
papel (de) sulfito 13377
papel de tina 4066, 6834
papel de tornasol 8280
papel de transporte 13966
papel de trapos 11206
papel (de) vidrio 6434
papel de yute 7722
papel diáfano 14498
papel diario(s) 9321
papel diazo 4284
papel dieléctrico 4323
papel doblado de máquina redonda 14411
papel doble 14128
papel dobleasfalto 14308
papel dominical 13404
papel dorado 6512
papel duplex 4881
papel duplo 4881
papel edición 1770
papel (embalaje) ondulado 3619
papel embetunado entelado 11428
papel embreado 13607
papel empaquetado en postetas dobladas 11165
papel en blanco 1533

papel en bloc 8771
papel en bobina 14658
papel en bobinas 11373
papel en bruto para encerar 14628
papel encerado 9554, 9967
papel encolado 12490
papel encolado a la resina 11690
papel encolado animal 625
papel encolado con cola animal 625
papel encolado en la calandra 2194
papel encolado en la pila 1264
papel encolado en tina 14073
papel encolado superficial 14073
papel en forma de platillo 4432
papel en forma de plato 4432
papel engomado 6715
papel engomado inerte 9396
papel en hojas 5771, 5798, 9922
papel enmascar 6504
papel en rollo 14658
papel entelado 9958, 11426
papelera 9918, 9927
papelería 13076
papelero 9924
papeles alquitranados 822
papeles asfaltados 822
papeles bituminados 822
papel escuadrado 12965
papeles de deshecho 4418
papeles de obra 1770
papeles embreados 822
papeles estratificados 13253
papeles ligeros 8140
papel esmaltado 2986
papel esmaltado a cepillo 1998
papel esmaltado de una cara 9590
papel esmaltado doble 4586
papel esmaltado (para cartonaje) 5145
papel esmaltado por las dos caras 14157
papel esmeril 5111
papeles superfinos 13412
papeles superfinos para invitaciones 14664
papel estanco al vapor 14390
papel estriado 7876
papel estucado 2986
papel estucado abrillantado a cepillo 2001
papel estucado a cepillo 1992
papel estucado al solvente 12774

papel estucado a máquina 8435
papel estucado a rodillo 11630
papel estucado brillante 6459
papel estucado de raspador vertical 12933
papel estucado de una cara 9590
papel estucado doble 4586
papel estucado dos caras 14157
papel estucado dos veces 4586
papel estucado en el size press 12493
papel estucado en la prensa encoladora 12493
papel estucado fuera máquina 12127
papel estucado glaseado 2983
papel estucado mate 4852, 8665
papel estucado moldeado 2436
papel estucado (por) ambas caras 14157
papel estucado por emulsión 5133
papel estucado por extrusión 5412
papel estucado por fusión 7049
papel estucado por inmersión 4390
papel estucado por una cara 9590
papel estucado simple 12448
papel estucado y esmaltado 4586
papel exento de ácido 127
papel filtro(s) 5660
papel fino 5679
papel fluorescente 5893
papel fofo 5501, 10283
papel forrado de tela 2957
papel fotográfico 10266
papel glaseado 6445
papel glaseado a máquina 8449
papel glasín 6431
papel gofrado 5101
papel granulado 6543
papel grueso para marbetes 13535
papel hecho a base de pulpa mecánica 8735
papel hecho a mano 6834
papel holandés 4917
papel ilustración 6789
papel imitación antiguo 4917
papel imitación barba 9131
papel imitación cuero 792
papel imitación estucado 7170

papel imitación manila 1693
papel imitación tela 8185
papel imperfecto 11541
papel impermeable a la grasa 6605
papel impermeable (al agua) 14595
papel impregnado 7206
papel incombustible 5697
papel India 1359
papel indicador 13667
papel inerte al ácido 139
papel infalsificable 2698
papel ininflamable 5697
papel inmaduro 6625
papel iridescente 9112
papel irisado 9112
papelista 9924
papel Japón 7645
papel japonés 7644
papel japonés para copiar 7638
papel jaspeado 8617
papel kraft 7840
papel kraft blanqueado 1539
papel kraft impermeable (al agua) 14594
papel kraft para cartón corrugado 7839
papel lacado 14403
papel laminado 7888
papel leñoso 8735
papel libre de ácido 127
papel ligero 5501
papel liviano 5501
papel maduro 8706
papel majado 9957
papel manila 8605
papel manila para escribir 11219
papel martillado 6809
papel mascado 9957
papel mate 8676
papel matutino 9104
papel mecánico 8735
papel metálico 8830
papel metalizado 8830
papel milimetrado 6573
papel ministro 10163
papel moneda 3839
papel monolúcido 8449
papel nacarado 9112
papel nativo 14335
papel negativo 9278
papel negro 1488
papel no blanqueado 14259
papel no encolado 14346
papel no recortado 14352
papel no satinado 14300
papel no transparente 9615
papel obra de acabado ligero 4989
papel offset 9528
papel offset estucado 2984
papel ondulado 14624
papel ondulante 14624
papel opaco 9615

papel ozalid 9815
papel paja 9006
papel paño 5845
papel para acciones 8296
papel para alzar 9826
papel para alzas 14283
papel para alzas mecánicas 2603
papel para anaqueles 12287
papel para arreglo 9826
papel para arreglo (mecánico) con greda 2603
papel para atlas 8614, 11623
papel para autotipias 6789
papel para blocs 13516
papel para bolsas 1075
papel para borradores 12005
papel para calcar 13938
papel para calcomanías 4042
papel para calcomanías cerámicas 2557
papel para carátulas 3681
papel para carnicerías 4781
papel para cartas 9442
papel para cartas marítimas 2649
papel para carteles 10610
papel para cartuchos 2395
papel para ciegos 1847
papel para contabilidad 8024
papel para contralomo 1034
papel para copias 2326, 8599
papel para copias sin carbón 2325
papel para correo aéreo 341
papel para correspondencia aérea 341
papel para croquis 12506
papel para cubiertas 3681
papel para cubrir 6504
papel para cuchillería 3867
papel para cheques 2698
papel para diario 9321
papel para diazotipia 4284
papel para dibujo 4709
papel para documentos 8296
papel para embalaje 14866
papel para empapelar 14533
papel para envolver 14866
papel para envolver pan 1873
papel para estampación con grabados de acero 13098
papel para etiquetas 7851
papel para etiquetas barnizadas 14407
papel para filtrar 5660
papel parafina 9967
papel parafinado 9967

papel para forrar 8207, 10043
papel para forrar cajas 1832, 2408
papel para forrar (cartón) 12273
papel para forros 8207
papel para forros de sacos 1071
papel para fotograbados 6789
papel para fotografía 10266
papel para fototipia 3096
papel para fraguado térmico 6932
papel para grabados 10457
papel para grasas 6603
papel para guardas 1750, 5161
papel para guías telefónicas 13628
papel para heliografía 3096
papel para huecograbado 11707
papel para joyería 12436
papel para la tirada 12276
papel para libros 1770
papel para limpiar 10467
papel para luz de gas 6319
papel para manteca 2120
papel para mantequilla 2120
papel para mapas 2649
papel para mapas camineros 11623
papel para mapas (geográficos) 8614
papel para mapas geográficos 11623
papel para máquina calculadora 186
papel para máquina de escribir 14224
papel para marbetes 7851
papel para medio tono 6789
papel (para) mimeógrafo 8949
papel para monotipia 9092
papel para monotipo 9092
papel para música 9211
papel para negativos 9278
papel para obras 1770, 8728
papel para obras de lujo 13681
papel para offset 9528
papel para patrones de labores 10063
papel para pegar 10043
papel para periódicos 9321
papel para positivado directo 10837
papel para prensa 9321
papel para programas 9386
papel para pruebas (de imprenta) 10946
papel para reclamos 10610
papel para recubrir cajas 2408
papel para reporte 13966

papel para revistas 8485
papel para rótulos 7851
papel para sacos 1075
papel para sellado térmico 6932
papel para sellos (de correo) 13019
papel para servilletas 9239
papel para sobres 5212
papel para sortijas de puro 2828
papel para talla dulce 10457
papel para tapas 3681
papel para timbres 13019
papel para tímpanos 14168
papel para títulos 8296
papel para transporte litográfico 13966
papel para valores 8296
papel para vitolas 2828
papel perfumado 11917
papel pergamino 14421
papel periódico 9321
papel persa 2725
papel picado 2577
papel pigmentado 2331
papel pigmento 2331
papel pintado 14533
papel plano 5771
papel plateado 481, 13829
papel pluma 2051, 5501
papel polvoriente 5882
papel prensado en frío 3057
papel prensa para ilustraciones 6787
papel protector a la fase de vapor 14389
papel protector contra la oxidación 694
papel puro de trapos 11205
papel quebrado 7678, 9737
papel quincenal 12101
papel raspador 17
papel rayado 11775
papel reactivo 13667
papel reforzado 11426
papel refractario 5697
papel registro 8024
papel resistente al ácido 139
papel resistente a la humedad 14693
papel resistente a los álcalis 420
papel revestido 7888
papel satinado 2190, 6445
papel satinado al agua 14576
papel satinado mate 5180
papel satinado por una cara 8449
papel satinado sin estucar 7170
papel secado al aire 329, 8312
papel secado a máquina 8444

papel secante 33, 1622
papel secante de blocs 13515
papel secante esmaltado 5143
papel seco al aire 332
papel seda Japón 7638
papel seda japonés 7638
papel seda (para envolver) 14867
papel semanal 14667
papel semibrillante 4989
papel semisatinado 8447
papel semisulfurizado no satinado 14299
papel sensibilizado 12117
papel sensible 12117
papel servilleta 9239
papel siliconizado 12418
papel similcuero 792, 8016
papel simil pergamino 7175
papel sin acabado 14296
papel sin ácido 127
papel sin aderezo 14346
papel sin carbón 2325
papel sin ceniza 817
papel sin cloro 2743
papel sin cola 14346
papel sin encolar 14346
papel sin fibra de madera 2685
papel sin (fibra de) madera 6681
papel sin impresión 1533
papel sin labrar 14335
papel sin pasta mecánica 2685
papel sin pasta mecánica de madera 6681
papel sin pelusa 5879
papel sin satinar 14300
papel sin verjuras 14858
papel soporte 1685
papel soporte para encerar 14628
papel soporte para estucar 1685
papel sulfito satinado 8450
papel sulfurizado 14421
papel supercalandrado 8447, 13410
papel superfino de escribir 13413
papel supersatinado 13410
papel sutil 5832
papel talco 12436
papel tela 8185
papel tela para escribir 2233
papel terciopelo 14430
papel texto 13681
papel tornasol 8280
papel translúcido 13986
papel transparente 14005
papel vegetal 14421
papel velin cremoso 3713
papel velloso 6269
papel vergé 7876
papel vergueteado 7876

papel verjurado 7876
papel verjurado cremoso 3711
papel veteado 6543
papel vía aérea 341
papel vitela 14858
papel vitela imitación Japón 7173
papel voluminoso 2051
papel zaraza 2725
papiro 9959
papirolina 9958
paquete 8541, 9837
paquetero 3314
par 3668
para-aminofenol 9961
parachoques 2032
parada automática del programa 919
parada (de venta) 9317
para-diclorobenceno 9964
para-fenilendiamina 9992
parafina 9969
parafina líquida 9966
parafinas 424
paraformaldehído 9970
paragolpes 2032
para-hidroxifenilglicina 6485
paralaje 9977
paralelelismo 9982
paraluz 8121
parámetro 9988
paraminofenol 9961
parangonar 8200, 11237
parar 12359, 13793
parar los rodillos 13791
parar tipo 3282
parasol 5420
parche 10045
pardo Cassel 14384
pardo de Van Dijck 14384
parecido a jarabe 13498
paréntesis cuadrados 1844
paréntesis rectangulares 1844
paréntesis (redondos) 9998
paridad 10007
paridad horizontal 8339
parlanchín profesional 13565
paro de seguridad 8303
párrafo 9972
párrafo aparte 1875, 9973
párrafo francés 6854
párrafo sin entrado 5908
párrafo sin sangría 5908
parte editorial 4978
parte húmeda 14683
parte impresora 7167
parte rechazanda la tinta 6607
parte sobresaliente 7747
partición de las palabras 4519
partícula copulativa 3320
partículas ásperas y duras 6658
partida de fabricación 8579

partida de prueba 14022
pasada 10019
pasado a la albúmina 372
pasado a la sala de composición 430
pasado de plancha 8838
pasado en hojas 12262
pasador cónico 13593
pasador de ajuste (repetidora) 4673
pasadores para hojas sueltas 1426
pasador hendido 12901
pasado sobre cinc 14954
pasar 5367, 13276
pasar el papel 14640
pasar en blanco 11816
pasar por la calandra 2191
pasar saltando 12521
pasillo 429
pasta 11026
pasta adhesiva 6465
pasta a la soda 12667
pasta al bisulfito 13379
pasta al bisulfito de calcio 8999
pasta al sulfato 7841, 13369
pasta al sulfito 13379
pasta amarilla de paja 14928
pasta blanqueada 1540
pasta de álamo (temblón) 820
pasta de alfa 5255
pasta de alta rendición 7006
pasta de árboles foliculares 6882
pasta de coníferas 12748
pasta de fibras 5597
pasta de madera 14807
pasta (de) madera parda 1976
pasta de madera preparada mecánicamente 8743
pasta (de madera), sin ~ 6680
pasta de Mitscherlich 8999
pasta de monocotiledóneos 6575
pasta de paja 13256
pasta depurada 65
pasta de recortes detintados 4122
pasta de rodillos 11638
pasta de sodio cloro 12663
pasta de sulfito neutro 9293
pasta detergente para metal de fundición 5912
pasta de trapos 11208
pasta gelatinosa 11638
pasta kraft 7841
pasta magra 6164
pasta mécanica 14807
pasta mecánica (de madera) 8743
pasta (mecánica), sin ~ 6680

pasta mucilaginosa 9160
pasta nitrosódica 9367
pasta no blanqueada 14260
pasta para libros 1736
pasta para pulir 11069
pasta para rodillos 11638
pasta química 2686
pasta química de paja 8430
pasta refinada 13181, 14735
pasta refinada grasa 14695
pasta secada al aire 333
pasta secante 10028
pasta semiblanqueada 10017, 12094
pasta semiquímica 12097
pasta sin blanquear 14260
pasta (suavizante) de tinta 3317
pastel 10315, 14881
pastera 10029
pastilla 10327
pata 6013
pata de araña 9555
pato 10315
patrón 1799, 6065, 8566, 13110
patrón de color 3190
patrón de corcho 3576
patrón duro 6873
pauta 9212
pauta aparte 14785
pauta interior 14784
pectina 10079
pedacites de papel 2577
pedalinista 10456
pedido 9672
pedido de almacén 13189
pedido de ensayo 14021
pedido de entrega diferida 8559
pedido en firme 5699
pedido escalonado 8559
pedido urgente 11818
pedir 9673
pedir para inspección 11488
pegado 1080, 10024, 12274, 13854
pegadura 12891
pegajosidad 1598, 13526
pegajosidad (de la tinta) 13524
pegajosidad relativa 11435
pegamento 213, 9160
pegante 14423
pegar a mano 6838
pegarse 203
pegar sobre cartivanas 6691
pegar una alza 10053
pegatina 12076
peine 11637
peine de perforar 10131
peine para el jaspeado 8624
peine para las matrices 8695
peine perforador 10131
peiñadores 3218

película 2987, 5636, 5638, 7950, 12517
película con trama incorporada 936
película de acetato 90, 91
película de acetato de celulosa 2510
película de aluminio 476
película de batihoja 6495
película de color 3144
película de extremo contraste 10891
película de fotomecánica 10891
película de goma 6725
película de grisar 12216
película (de grisar) bendéi 12216
película delgada 13711
película de medio tono 6782
película despegable 13283
película de tinta 7350
película de tono continuo 6782
película de triacetato 90
película de varias capas 9175
peliculado 13289
peliculador 13285
película en colores 3144
película en rollo 11656
película fotocroma 3144
película fotográfica 5636
película lith 8249
película ortocromática 9697
película pancromática 9888
película para grisado 12216
película para tramado directo 936
película pretramada 936
pelito 2090, 6747
pelo 2090, 6747
peltre 10176
peluche de broncear 1971
pelusa 6267
pelusa del papel 5878, 8226
pendiente 6535
penetración 10098, 13270
pentaclorofenol 10105
pentagrama 13000, 13081
pentano 10106
penumbra 10107
peptización 4102, 10109
pequeña imprenta 14754
pequeños avisos 2881
pequeños hierros 10164
pequeños hierros, a ~ 948
peral 10075
perborato sódico 12707
percal 2232
percarbonato de potasa 10662
percarbonato de potasio 10662
percarbonato potásico 10662
perclorato de potasa 10663
perclorato de potasio 10663

perclorato potásico 10663
percloruro de hierro 5546
pérdida 12357
pérdida de color 5433
pérdida de densidad 8373
pérdida de papel 8375
pérdidas 12906
perdido 9784
perfecto estado, en ~ 9259
perfil 7779
perfil fino 5674, 6746
perfil grueso 13109
perforación 10130
perforación central de arrastre 7414
perforación de zona 9781
perforaciones de arrastre 5517
perforación redonda 11727
perforación transversal 3763
perforador 7784, 10134
perforadora 10132, 10134
perforador de banda 13587
perforador de cinta 13587
perforador de cinta de papel 9950
perforador sin contador 9395
pergamino 9994, 14425
pergamino de cordero 7883
pergamino de piel de carnero 6039
pergamino vegetal 14421
periferia 2854
periódico 7698
periódico amarillisto 14920
periódico amarillo 14920
periódico competencia 9642
periódico cotidiano 3958
periódico de cambiar 5342
periódico (de información) 9309
periódico del arroyo 14920
periódico de provincia 10988
periódico de tamaño extra-ordinario 1520
periódico diario 3958
periódico dominical 13404
periódico provinciano 10988
periódico rival 9642
periódico semanal 14667
periódico sensacionalisto 14920
periodicucho 11199
periodiquito 11199
periodista 7700
período de inducción 7296
período de paro de la máquina 4678
periscopio 12654
perito de contabilidad 75
permanencia 3397
permanencia del color 3142
permanganato 10148
permanganato de potasa 10664

permanganato de potasio 10664
permanganato potásico 10664
permeabilidad 10151
permeabilidad al aceite 9564
permeabilidad al aire 344
permeabilidad al vapor de agua 14388, 14609, 14612
perno 1714
perno de ajuste del tintero 224
peróxido bárico 1153
peróxido de bario 1153
peróxido de hidrógeno 7115
peróxido de sodio 12708
peróxido de sosa 12708
peróxido hidrogenado 7115
peróxido sódico 12708
perpendicular 10153
perseverencia 3397
persianas (de cierre) 1064
persona muy trabajadora 14706
persulfato amónico 543
persulfato de amonio 543
persulfato de potasa 10666
persulfato de potasio 10666
persulfato potásico 10666
pesa-ácidos 761
pesacartas 8054
pesalicores 761
pesas y medidas 14673
pescante de bobinas 11370
peso atómico 865
peso balancín 1088
peso básico (del papel) 1209
peso bruto 6674
peso cúbico 712
peso de base 1209
peso de la resma 11294
peso del tipo 14672
peso de quilibrado 1088
peso elemental (del papel) 1209
peso en gramos 1209
peso en seco 4828
peso equivalente 5230
peso específico 12827
peso mínimo 8963
peso molecular 9054
peso neto 9287
peso por metro cuadrado 1209
peso por mil hojas 9390
peso por resma 1209, 9390, 11294
peso por unidad cúbica 712
peso real 173
peso verdadero 173
pestaña 1344
pestaña de índice 7277
petición automática de repetición 923
petrolato 10167
petróleo 10168
petróleo de lámpara 7751

pez 10380, 10381
pez de estearina 13090
pez esteárica 13090
pez griega 11684
pH 10177
phot 10211
pica 10288
picado 3615
picado de la tinta 1567
picado (del papel) 10295
picado húmido 14688
picado seco 4813
picar composición 10294
picar (reportes) 13148
picar una forma 10294
picar un molde 10294
picnómetro 11075
pico 12865
pie 6013
piececita de adorno 14470
pie cuadrado 12966
pie cúbico 3801
pie de fábrica 13944
pie de grabado 8031
pie (de grabado) 9136
pie de imprenta 7229
pie de la cámara 2249
pie de página 6022, 13550
pie de plomo 7975
piedra aceitada para afilar 9568
piedra amoladera 6647
piedra amoladora Arkansas 769
piedra de Ayr 12648
piedra de Escocia 12648
piedra de imposición 7198
piedra litográfica 8265
piedra para apomazar 11037
piedra para mesclar la tinta 7393
piedra para rebajar (pieles) 10001
piedra pómez 11037
pie (inglès) 6014
piel 5638, 8011, 12517
piel de ante 2031
piel de becerro 2199
piel de cabra 6493
piel de carnero 1183
piel de carnero chiflado 12527
piel de cerdo 10337
piel de foca 12025
piel de gamuza 2612
piel de Rusia 11821
piel losanjeada 4291
piel pergamino 14425
piel (superior) de becerro 7944
pieza colada 2442
pieza en T 13934
pieza fundida 2442
piezas angulares 590
piezas de esquina 590
piezas de recambio 12819
piezas de repuesto 12819
pigmento 10330

pigmento de tierra 4939
pigmento ftalociánico 10285
pigmento inorgánico 8954
pigmento mineral 8954
pigmento molibdeno 9060
pigmento natural 9248
pigmentos de color 3161
pigmentos de óxido de hierro 7594
pigmentos flush 5901
pigmentos inorgánicos 7420
pigmentos nacarados 9224
pigmentos nacarinos 9224
pigmentos orgánicos 9680
pila 10340, 14410
pila batidora 11385
pila blanqueadora 10679
pila de discos 4444
pila de entrada 5505
pila de entrega 4147
pila (de las hojas impresas) 4147
pila del marcador 5505
pila del ponepliego 5505
pila del receptor 4147
pila de revelado 4228
pila de salida 4147
pila desfibradora 1881
pila holandesa 7020
pila Jordán 7693
pila lavadora 14548
pila refinadora 11385
pinabete 6961
pinacianol 10350
pinacriptol amarillo 10353
pinacriptol verde 10352
pinaflavol 10351
pinatipia 10354
pincel 9870
pincel china 2718
pincel de la cola 10023
pincel de retoque 11535
pincel luminoso 8125
pincel neumático 314
pincel para aplicar el mordiente de dorar 6424
pinchos fijadores de la mantilla 1518
pino 10362
pino de Virginia 6961
pinotipia 10354
pintado a mano 6817
pintado a pincel 2017
pintado de los cortes 13011
pinta inglés 10373
pinta líquida 8239
pinta seca 4814
pinza de cadena 2587
pinza (de enganche) 6650
pinza de entrega 4143
pinza de salida 4143
pinza Finch 5667
pinza oscilante 13472
pinza para película 5641
pinzas 14116
pinzas de cierre 4542
piñón 6346, 10368
piñón cónico 1349

plegar 5963
plegar en paralelo 9981
pliego 12399
pliego a la francesa 6171
pliego cruzado 2756
pliego de arreglo 8566
pliego de cama intermedio 7497
pliego de cubrir 4719
pliego de encaje 7443
pliego de máquina 244, 10759
pliego de prensa 10759
pliego de principios 10724
pliego de prueba 10759
pliego longitudinal 6070
pliegos de reposición 2264
pliegos por hora 7223
pliegue 3715, 5962
pliegue a cuchilla 7651
pliegue en abanico 73
pliegue lateral 6730
plombaginar 1482
plomo 7968
plomo del tintero 6108
plumadas 5863
pluma de dibujo 4710
pluma de rayar 11785
plumilla (inglesa) 12012
poca exposición 14278
poco encolado 12538
poco expuesto 14277
poder absorbente 32, 44
poder cubridor 3675
poder cubriente 3675, 6977
poder de adhesión 214
poder de cubrimiento de la tinta 7334
poder de resolución 11517
poder específico inductor 4322
poder lumínico 7487
poder resolutivo 11517
poder resolvente 11517
poder separador 11516, 11517
poise 10537
polarización 10539
polarización electrolítica 10540
polea (de transmisión) 1308
poleas de cinta 13586
poleas de guía 13586
poliacrilato 10549
poliamida 10550
policromático 10553
policromía 3162, 3163, 3167, 6124, 10904
policromía de línea 8166
policromo 9169, 10553
policromotipografía 3163
poliéster 10555
poliestireno 10561
polietileno 10556
polimerización 10559
polímero 10558
polistireno 10561
póliza de tipos 6103
póliza (de tipos) 6120

polo negativo 2480
polonesa 6404
polo positivo 641
polución del agua 14592
polución del aire 351
polvillo 6267
polvillo de papel 8226
polvo 9914
polvo antimaculador 687, 4824
polvo cortante 3896
polvo de amianto 293
polvo de blanqueo 1548, 2168
polvo de bronce 1957, 1965
polvo de corte 3896, 12583
polvo de esmeril 5113
polvo de papel 8226, 9914
polvo de plata 12433
polvo de resina 5283
polvo de talco 13562
polvo dorado 6515
polvo insecticida 7432
polvo metálico 8822
polvo para grabar 5283
polvo para patrices 5634
polvoreado 4894
polvo (resinoso) para reserva 5283
polvorizado 2599
ponedor 5504, 5507
ponedora 5507
ponedor automático 909
ponepliegos 5504, 5508
ponepliegos automático 909, 5508
ponepliegos por escalas 13257
poner alzas 9771
poner con cifras 11066
poner con números 11066
poner de bastardilla 7623
poner de (letras) mayúsculas 11068
poner de mayúsculas 11068
poner de seguido 11811
poner de versales 11068
poner en abanico 5466
poner en código 5154
poner en cursiva 7623
poner en el tendedor 11185
poner en foco 5952
poner en juego 13788
poner en letras bastardillas 7623
poner en (letras) mayúsculas 11068
poner en línea 404, 3079
poner en marcha 12160, 13788
poner en minúsculas 11065
poner en prensa 11067
poner en resalte 4459
poner en secuencia 12134
poner entre comillado 11177
poner entre comillas 4364, 11177

poner entre paréntesis 1843
poner entre vírgulas 11177
poner entre virgulillas 11177
poner en versales 11068
poner los tipos 4484
poner sin párrafo 11811
poner sombras 6885
poner una letra al revés 14098
poner un anuncio 245
poner varillas 8840
porcentaje de cenizas 816
porcentaje de desechos 10111
porcentaje de desperdicios 10111
porcentaje de humedad 9033
porcentaje de piezas defectuosas 8378
porfolio 5967
poro 5810
poros de papel 9933
porosidad 44, 10581
porosímetro (ad aria) 4176
poroso 10583
portaalmacenes 8482
portabobina doble 4830
portabobina en estrella 12863
portabobinas 11375
portabobinas sin mandril 12870
portabobina superpuesto 944
portacarbón 2317
portada 6215, 6244, 13867
portadiapositivo 14001
portadilla 1217
portador de películas enrolladas 11658
portador de películas en rollo 11658
portafiltro 5659
portagaleras 4861
portalámpara de bayoneta 1237
portalámpara(s) de rosca 4964
portalámpara(s) Edison 4964
portalámpara(s) roscado 4964
portalibros 1754
portamodelo 3545
portamodelo al vacío 14373
portamoldes 4861
portanegativos 9275
portaobjetivos giratorio 8048
portaoriginal(es) 2242, 3545
portaoriginal(es) al vacío 14373
portapáginas 9861
portapantalla 11979
portapelícula al vacío 14374
portapelículas 5644, 11658

portaplaca fotográfico 10442
portaplaca fotográfico con bisagra 7009
portaplaca giratorio 7009
portaplacas 10442
portaplanchas 10442
portarramas 6067
portarraspadora 4530
portarretícula 11979
portarrodillos 11639, 11644
portarrollo doble 4830
portarrollos 11375
portarrollos de estrella 11648
portarrollo-tensor-empalmador automático 11648
portatrama 11979
portauñas 6652
portaútil 13902
posa de altas luces 6988
posar 5367
posar por un filtro 5663
pose 5368
pose doble 4603
pose principal 8543
posición 8302
posición adelantada 6387
positiva brillante 6462
positivadora 10828
positivo 10591
positivo brillante 6462
positivo de imágines combinadas 6684
positivo de inversión 4410
positivo de línea 8191
positivo de medio tono 6792
positivo de trama 6792
positivo de trazo 8191
positivo infinitamente variable 10593
positivo por contacto 3413
positivo punteado 6792
positivo sobre vidrio 6436
positivo tramado 6792
posofotómetro 5373
posómetro 5373
posos de cobre 3508
postdata 5219
posteta 8091
postetas alternadas 9285
postizo 1691
postura doble 14162
potasa 10618
potasa cáustica 10655
potasio 10622
potencial de óxidorreducción 11334
potencia lumínica 7487
potencia luminosa 7487
potenciómetro 10681
potrillo 932
preámbulo 7539, 9444
prearreglo 10728
precalentador 10722
precio con descuento 9286
precio de coste 3631

precio de costo 3631
precio de venta 12089
precio el millar 13739
precio neto 9286
precio por hoja 10780
precio por mil letras 13739
precipitación 10717
precipitado 10710
precipitar 10711
precisión 85
precisión larga 4645
preexposición 5762
prefacio 6043, 9444
prefijos para unidades métricas 10721
prelectura 6093
prensa 10738
prensa a brazo 6839
prensa Albion 368
prensa a mano 6839
prensa automática de platina 916
prensa bicolor 14134
prensa brunidora 12642
prensa caucho contra caucho 1524
prensa cilíndrica 3948, 5775
prensa con articulación de la rodilla 13875
prensa cotidiana 3959
prensa de brazo 6839
prensa de cama plana 5775
prensa de cilindro 3948
prensa de cilindro y platina 3948
prensa de copiar 3549, 10828
prensa de cortar y marcar 3889
prensa de cuatro colores 6123
prensa de doble impresión 10122
prensa de doble revolución 14153
prensa de dorador 1604
prensa de dorar 6407
prensa de dos colores 14134
prensa de dos revoluciones 14153
prensa de dos vueltas 14153
prensa de embalar 1093
prensa de encuadernación 1427
prensa de estereotipar 9129
prensa de Gally 6292
prensa de grabado en cobre 3525
prensa de gran cilindro 12466
prensa de hacer balas 1093
prensa (de imprenta) 10736
prensa de (im)presión planocilíndrica 5775
prensa de la tela 3645
prensa de matrices estereotipias 9129

prensa de ocho bobinas 4997
prensa de parada de cilindro 13210
prensa de pedal 14014
prensa de picado 5972
prensa de platina 10452
prensa de pliegos 5972
prensa de positivar 10828
prensa de pruebas 10950
prensa de pruebas offset 10951
prensa de serigrafía 11991
prensa de tambor 12466
prensa de tiro y retiro 10122
prensa de tiro y retiro de cuatro cilindros 6125
prensa de una revolución 12466
prensa de vacio 10828
prensa diaria 3959, 10738
prensado en caliente 7052
prensador 2873, 3642
prensadora vertical 13034
prensa Duplex 4879
prensa embaladora 1093
prensa, en ~ 6521, 7537
prensa encoladora 12492
prensa flexográfica 5829
prensa fotográfica 10828
prensa hidráulica 7097
prensa hidráulica de matrizar 9129
prensa húmeda 3645
prensa independiente 6162
prensa libre 6162
prensa litográfica 8263
prensa litográfica a brazo 8260
prensa modelo Stanhope 13037
prensa moldeadora 14525
prensa neumática 10828
prensa neumática (de copiar) 14375
prensa offset 9532
prensa offset de dos colores 14132
prensa offset de un color 12450
prensa offset plana 9522
prensa para cortar a troquel 4318
prensa para cortar y hender 3889
prensa para dorar 1604, 6407
prensa para el acabado de libros a mano 8423
prensa para encuadernar 1745
prensa para enfardar 2064
prensa para estampar (con grabados de acero) 4336
prensa para estampar matrices 8699, 9129
prensa para impresión con trama de seda 11991

prensa para impresión de grabados en cobre 3525
prensa para imprimir en hojas 12269
prensa para matrizar 9129
prensa para pruebas de galeradas 6282
prensa para pruebas offset 10951
prensa para reportar 13968
prensa para (sacar) pruebas 10950
prensa para timbrar 4336
prensa para tipografía 8063
prensa para transportar 13968
prensa pequeña offset 12627
prensa plana 5775
prensa plana con papel continua 5776
prensa planocilíndrica 3948
prensa profesional 13947
prensa rotaplana 5776
prensa rotativa tipográfica 14652
prensa roto offset 11698
prensa sacapruebas 10950
prensa sacapruebas offset 10951
prensa semirrotativa 5776
prensa serigráfica 11991
prensista (tipográfico) 10753
preparación científica de las decisiones 9636
preparación de la pasta 13190
preparación de originales 3552
preparación de pastas 13190
preparación de planchas 10448
preparación para la imprenta 4967
preparado del papel pigmento 10333
preparar la cama del cilin- dro 4721
preparar (la tinta) 6637
preparar para la imprenta 4966
presión 10843
presión de contacto 7805
presión de impresión 7219
presión de leve contacto 7805
presión de rodillo de presión 7218
presión excesiva 5339
presión hidráulica 7098
presión mínima 7805
presión osmótica 9710
presión suave 7805
prespan 10756
prestación normal 9427
presupuesto 725, 5264
pretratamiento 10776

prima estimuladora a la producción 50
primera calidad 9457
primera calidad, de ~ 10118
primera edición 5702
primera forma 5707
primera hoja de la cama 9773
primera impresión 5710
primera página 6215
primera página de cubierta 6207
primera prueba 12573
primeras 10118
primera tirada 5710
primer cajista 2914
primer color 5701
primer elevador 5703
primer mordido 5787
primero doblado en paralelo 5713
primer plano 6035
primicia informativa 11932
primicia periodística 11932
principiante 3796
principio de impresión 10385
prisma 10869, 11563
prisma inversor 11563
probabilidad de lanza- miento 10875
probador 13662
probador de tersura 12646
probar 13659
probeta 13670
probeta de color 3190
probeta de ensayo 13668
probeta de Zahn 14931
probeta Ford 6030
procedencia 4019
procedimiento 10879
procedimiento aditivo 191
procedimiento a la albú mina 376
procedimiento a la albúmina (cromatada) 2770
procedimiento a la cola esmalte 6467
procedimiento al asfalto 824
procedimiento a la sosa 12666
procedimiento al betún 824
procedimiento al bisulfito 13378
procedimiento al bromóleo 1953
procedimiento Alco 379, 384
procedimiento al cobre brillante 1095
procedimiento al colodión (húmedo) 3087
procedimiento al esmalte en frío 3055
procedimiento al esmalte (ordinario) 6467

procedimiento al esmalte
seco 4773
procedimiento al hueco-
offset 4088
procedimiento Alltone 441
procedimiento al oleo-
bromo 1953
procedimiento al sulfato
13368
procedimiento analítico 577
procedimiento autocrómica
894
procedimiento Ballard 1095
procedimiento Bassani
1210
procedimiento bendéi 1318
procedimiento brunak 1980
procedimiento Claybourn
2883
procedimiento cronak 3737
procedimiento de
blanquear 1550
procedimiento de
conversión 3471
procedimiento de drago
4689
procedimiento de Dultgen
4855
procedimiento de en-
mascarado 8647
procedimiento de entintaje
11667
procedimiento de estucado
2988
procedimiento de estucado
a cuchilla 7821
procedimiento de estucado
con rodillo invertido
11559
procedimiento de estucado
de raspador vertical
12934
procedimiento de estucado
por cepillos 1993
procedimiento de estucado
por inmersión 4391
procedimiento de estucado
por rodillos 11631
procedimiento de fluo-
rescencia 5889
procedimiento de grisado
bendéi 1318
procedimiento de
impresión 10844
procedimiento de
impresión acuatono 737
procedimiento de inversión
11550
procedimiento de medio
tono 6793
procedimiento de medio
tono directo 4404
procedimiento de mordido
rápido 10703
procedimiento de mordido
sin polvo 10703
procedimiento de placa
húmeda 3087

procedimiento de
reproducción 11482
procedimiento de sangre
de dragón 4689
procedimiento de secado
4796
procedimiento de selección
y tramado simultáneos
4404
procedimiento de tres
colores 13750
procedimiento flushing
5904
procedimiento Henderson
6965
procedimiento indirecto
7294
procedimiento Kemart
7745
procedimiento litográfico
8270
procedimiento Manul 8610
procedimiento offset 9525,
9536
procedimiento ozalid 9816
procedimiento Pantone
9893
procedimiento para hacer
calcomanías 13969
procedimiento Parry-
Higgins 10154
procedimiento per lotes
1222
procedimiento rápido de
mordido 10703
procedimiento Renck 11460
procedimiento semiquímico
12096
procedimiento substractivo
13346
procedimiento Texoprint
13673
procedimiento tricromático
directo 4404
procedimiento tricrómico
191
procedimiento Typon 14244
procesado 10893
procesado automático de
datos 905
procesado electrónico de
los datos 5046
procesadora (automática)
10900
procesamiento secuencial
12136
proceso 10894
proceso polivinílico 10570
proceso reversible 11550
producción en cadena 8653
producción en masa 8653
producción en serie 8653
producción/hora 9729
producción media 956
producción por línea de
fabricación 5873
producto accesorio 2132
producto acelerador 53
producto antisecativo 659

producto auxiliar 4559
producto derivado 4199
producto espesante 1681
producto humedecedor
14699
productos químicos 2690
producto
viscosidad/velocidad
14487
profundidad central 3651
profundidad de foco 4195
profundidad de grabado
5276
profundidad de mordedura
5276
profundidad de mordido
5276
profundímetro 4190
programa 10914, 11734
programa almacenado
13236
programación de la
producción 10908
programación del trabajo
10397
programa de aplicación 718
programa de biblioteca
8088
programa de
compaginación 833
programa de corte lógico
8319
programa de ensamblaje
833
programa de fabricación
11806
programa de la
computadora 3343
programa diagnóstico 4252
programa director 13423
programador 10922
programa fuente 12793
programa objeto 9470
programar 10915
programa standard 8088
programas y sistemas de
programación 12744
prohibida la reproducción
439
prólogo 7539, 9444
promedio 951
propaganda periodística
9311
propanol 10973
propiedad absorbente 32
propiedad artística 3553
propiedad de absorción 32
propiedad de la superficie
1451
propiedades dieléctricas
4324
propiedades reológicas
11584
propiedades superficiales
13436
propiedad fotogénica 10237
propiedad intelectual 3553
propiedad literaria 3553
propiedad superficial 1451

propilenglicol 10974
propio a tomar tinta 7384
proporción 10967
proporción a, en ~ 10975
proporcional a 10975
proporción áurea 6505
proporción de mezcla 9013
propulsión 4736
propulsión de cadena 2583
propulsión de engranajes
6345
propulsión por correa 1302
prosa 12153
prospecto 10976
prospecto plegado 5967
protector de las manos
6689
protector de manos 6843
proteger ciertas partes del
grabado para remordido
local 13006
protegido contra el polvo
4906
proteína 10981
protocloruro de mercurio
8793
protón 10984
próxima publicación, de ~
715
proyección 10934
proyectista de libros 1748
proyectista gráfico 7962
proyecto borrador 11715
prueba 10936, 13658
prueba al azar sistemático
13502
prueba antes de la letra
10940
prueba con la letra 10958
prueba con la uña de dedo
5684
prueba con pocos errores
2895
prueba corregida 2895
prueba de, a ~ 11511
prueba de agua, a ~ 14606
prueba de aire, a ~ 356
prueba débil 8127
prueba de blanqueo 1551
prueba de Cobb 3018
prueba de color(es) 3168
prueba de ensayo 5190
prueba de envejecimiento
302
prueba de galera 12573
prueba de galerada 12573
prueba de grabador 5190
prueba de huecograbado
7472
prueba de humo 12639
prueba de imprenta 11270
prueba de K & N 7727
prueba de la absorción de
la tinta 38
prueba de laboratorio 7859
prueba de lanzado 6062
prueba del autor 889
prueba del poder cubriente
6978

prueba de máquina 5792
prueba de plana 2012
prueba de primera 11270
prueba de primeras 5801
prueba de rascado, a ~ 9414
prueba de resistencia al ácido 140
prueba de segunda capilla 889
prueba en acetato 94
prueba en color(es) 3168
prueba en galera 12573
prueba en limpio 2895
prueba en negro 1492
prueba en papel bromuro 1948
prueba fotográfica tramada 11978
prueba hecha con un cepillo 2012
prueba limpia 2895
prueba llena de erratas 4415
prueba llena de errores 4415
prueba llena de moscas 4415
prueba mentirosa 4415
prueba moteada 4415
prueba nítida 2895
prueba plegada de erratas 4415
prueba positiva sobre papel brillante 6462
prueba posterior 11570
pruebas de estado 10928
pruebas de gama 10928
pruebas de la plana entera 9863
pruebas de página 9863
pruebas de reproducción 11483
pruebas de tricromía 10928
pruebas fotográficas 10250
prueba sin arreglo en la forma 5792
prueba sobre papel estucado 10948
prueba sobrexpuesta 3990
pruebas para reproducción 11483
pruebas progresivas 10928
prueba sucia 4415
prueba tipográfica 11270
pruebero 10945
prusiato amarillo 10650
prusiato de potasa 10650
prusiato rojo 10649
psicrometría 10997
psicrómetro 10996
psicrómetro de torbellino 12558
publicación 4970
publicación bimensual 12101
publicación bimestral 14148
publicación bimestre 14148

publicación bisemanal 12104
publicación mensual 9096
publicación quincenal 12101
publicación semanal 14667
publicación trimestral 11136
publicarse 714
publicidad 247
publicidad de prensa 9311
publicidad mural 14538
publicidad periodística 9311
publicista 7700
publicitario 229
publigráfico 3240
puente de la trama 2523
puente nivelador 14198
puentes de salida 13287
puerta 6322
puesta a punto 226
puesta a punto del registro circunferencial 2857
puesta a punto de un programa 4040
puesto de periódicos 9317
pulgada 7244
pulgada cuadrada 12968
pulgada cúbica 3802
pulido 2080
pulido con carbón de tilo 2647
pulido con carbón vegetal 2647
pulidora 2078
pulimentar 10547
pulir 2037
pulpa 11026
pulpa a la sosa 12667
pulpa al bisulfito 13379
pulpa al sulfato 7841, 13369
pulpa al sulfito 13379
pulpa blanqueada 1540
pulpa cruda 14260
pulpa de alfa 5255
pulpa de alta rendición 7006
pulpa de árboles foliculares 6882
pulpa de coníferas 12748
pulpa de madera 14807
pulpa (de) madera parda 1976
pulpa de monocotiledóneos 6575
pulpa de paja 13256
pulpa de sodio cloro 12663
pulpa de trapos 11208
pulpa kraft 7841
pulpa leñosa 14807
pulpa magra 6164
pulpa no blanqueada 14260
pulpa química 2686
pulpa química de paja 8430
pulpa semiblanqueada 10017, 12094
pulpa sin blanquear 14260
pulsación 7785

pulsador 11060, 11062
pulsador rojo 11323
pulverizador-antimaculador 4823
pulverizador antirrepinte 12926
pulverizador (antirrepinte) 4823
pulverizador de aire comprimido 314
pulverizador neumático 314
punta 3578
punta de correcciones 1671
punta de la lesna 1674
punta del fotograbador 5281
punta doblada 4543
puntas para montar clisés 1603
puntas redondeadas 11725
punteado de trama 4568
punteado reticular 4568
puntear 13160
puntera 3578
puntero de espaciado (de la scala indicadora) 7717
puntero de las unidades 10523
puntillado de trama 4568
puntillado en las altas luces 6986
puntillado reticular 4568
puntito 6779
puntitos 10364
puntizones 7875
punto 10518, 10520
punto ablandador 12735
punto admirativo 5348
punto centelleo 5765
punto de ablandamiento 12735
punto de admiración 5348
punto de agarre 1905
punto de anilina 617
punto débil 12732
punto de cadeneta 7756
punto de comprobación 2673
punto de condensación 4242
punto de ebullición 1702
punto de engrase 8398
punto de fusión 8769
punto de ignición 5765
punto de inflamabilidad 5765
punto de inflamación 5765
punto de interrogación 7527
punto de la trama 6779
punto de reblandecimiento 12735
punto de rocío 4242
punto de saturación 11887
punto de sujeción 1905
punto diacrítico 4250
punto duro 6860
punto en forma de hongo 14274

punto en forma de seta 14274
punto en forma de sombrilla 14274
punto estirado 4685
punto fijo 5730
punto final 6238
punto focal 5950
punto fofo 12732
punto interrogante 7527
punto isoeléctrico 7615
punto metálico 13040
punto muerto 4920
punto nodal 9383
punto recortado 6860
punto redondo 6238
punto reticular 6779
puntos 10363
puntos cardinales 2351
puntos cegados 3388
puntos conductores 2936, 7984, 9623
puntos corridos 7984
puntos de conducción 2936, 7984, 9623
puntos de oscuro 12219
puntos de sombra 12219
puntos fallados 8986
puntos (reticulares) unidos 3388
puntos suspensivos 7984, 9623, 10532
punto tipográfico 10520, 14240
punto y acápite 9973
punto y aparte 9973
punto y coma 12098
punto y medio 4999
punto y seguido 11811
puntuación 11049
punturas 9269, 10371
punturas regulables de registro 11410
punzón 1672, 3651, 11040
punzonado (de matrices) 13269
punzonadora 11048
punzonar 6303
punzón de correcciones 1671
puño 6829
pupitre de mando 3395
pupitre de mando a distancia del entintado 1119
pupitre de retoque 11536
pupitre de tapa iluminada 14467
pupitre para retocar 11536
pureza colorimétrica 3110
pureza (colorimétrica) 11056
pureza de excitación 5347
púrpura de bromocresol 1951
purpurina 1957
purpurina oro 1964
purpurina plata 12433

quebradizo 1925
quebrado 6141
quebrado de segunda clase 7678
quebrado de segunda (clase) 11541
quebrado postizo 10320
quebrado seccional 10320
quebrados sin barra 9306
quebrar el cabo 1892
¡queda! 13134
quedar semiseco 12154
quemado 2074
quemar 14557
queroseno 7751
quijada 4737, 7650, 12343
quijada del primer elevador 5704
quijadas de la delantera 14464
quijadas plegaderas 5981
quilo 7797
quilogramo 7797
químicamento puro 2684
quimicografía 10906
químicos 2690
quimigrafía 2694
quincenario 12101
quinol 7127
quinona 11161
quinterno 11162
quiosco de periódicos 9317
quiselgur 4280
quitapresión 7224
quitar con el cepillo 10392
quitar con (el) escoplo 2732
quitar con la lima 5618
quitar con racleta 14755
quitar el lustre 13606
quitar espacio 2943
quitatinta 7345
quitavelo 12023

rabajarse 6687
rabo de la mantilla 1522
racleta 4525
racleta de caucho 12978
radiación 11194
radiador completo 1467
radial 11190
radical 11195, 11679
radio 11198
radioactivo 11197
raedera 12007
raedor 14
ráfaga 5758
ráfagas en la impresión 11641
raíz 11677
raja cortante 11935
rajar 11936
rama 2650
rama dúplex 5977
rama numeradora 9451
rama para encabezados 6904
rama para remiendos 7671
rama sesgada 1357
ramio 11228

ramo de preparación de la copia 10449
ramo experimental 10348
ranudora-impresora 10812
ranura 3714, 6666, 9432
ranuradas de pinzas 4546
ranura de engrase 9555
ranuradora 3718
ranura para chaveta 7786
ranurar 3716
ranuras de calandra 2187
rascador 5223
rascador para dibujo litográfico 9266
rascleta de caucho 12978
rasgador 4891, 14744
rasgo 5502
rasgo ascendente 813
rasgo descendiente 4201
rasgo diagonal 12767
rasgos 12144
rasgos (de pluma) 5863
raspado del negativo 3901
raspador 4525, 11949, 11951
raspadora 4525, 12648
raspadora de retorno 5850
raspador de retoque 11950
raspadura 10, 12601
raspar 11947
raspar un rodillo en el sentido (de la flor) 11946
rasqueta 4525
rastrillo 11951
rata de biblioteca 1785
ratón de biblioteca 1785
rauteadora 11732
raya 3714, 7143, 8155
raya achaflanada 8997
raya cuadratín 4001
raya de adorno 9690
raya de bronce 1866
raya de caña sencilla 9987
raya de columna 3209
raya de contorno 1796
raya de división 4514
raya de latón 1866
raya delgada 5674
raya (de) luto 6232
raya de media caña 9801
raya de medio cuadratín 4002
raya de paleta 4703
raya de pie 6023
raya de separación 3871
rayado 3209, 6740, 9596
rayado interior, con ~ 14784
rayador 11777
rayadora 11784
rayadora de discos 4445
rayadora de mesa 1312
rayado separado, con ~ 14785
rayadura 11781
raya entre columnas 3209
raya fundida al canto 1352
raya ingleteada 8997
raya ondulada 14622, 14625
raya ornamental 9690

raya para columna 3209
raya puntillada 4577
rayar 3716, 11773, 13272
rayas de Fraunhofer 6150
rayas de goma 6726
rayas de humedad 14692
rayas de raspadora 4531
rayas de secado 4792
raya serpentina 14625
raya serpentinada 14625
rayas tenues 5526
rayas transversales 7041
raya tremble 14625
raya trembleque 14625
raya tremente 14625
raya viborita 14625
rayita 4001, 4003, 7143
rayo de luz 8128
rayos actínicos 162
rayos beta 1343
rayos ultraviolados 14254
razón 11252
razón de, a ~ 10975
razón inversa de, en ~ 7540
reacción 11261
reacción de obscuridad 3991
reacción elástica 11500
reactivo 11288
reactivos catiónicos superficiales 2484
realces 6992
Reales 13981
rebaba 2090, 4737, 6747, 9766, 12343
rebaba (de fundición) 1248
rebabas 6743
rebaja 4425
rebajado 11350
rebajado de la densidad 11352
rebajador 11344, 11345
rebajadora 11732
rebajador de altas luces 6991
rebajador (de) Farmer 5471
rebajamiento 996, 11350
rebajar 10547
rebajar (el punto) 3900
rebajar puntos 3886
rebajar (un color) 4372
rebajo 1345, 9432
rebobinador 11580
rebobinadora 11580
rebobinadora a tensión constante 3400
rebobinar 11581
rebordes laterales del cilindro 1252
rebose 12984
recado 13033
recambiabilidad 7490
recambios 12819
recargo 5395, 5407
receptor 4149
receptor a cinta 13581
receptora doble 4595
receptor automático 906

receptor de cadena 2580
recibo de tinta 7385
recién impreso 6189
recipiente de tinta 7403
recíproco 11301
reclamo 247, 2471
reclamo mural 14538
recodo 6063
recodo desviador 588
recoge-astillas 2058
recoge-nudos 7830
recoge pasta 11891
recogida de datos 4008
recomponer 11493
reconocimiento de caracteres 2639
reconocimiento de caracteres de tinta magnética 8520
reconocimiento de configuración 10065
reconocimiento magnético de caracteres 8520
reconocimiento óptico de caracteres 9650
recopilación de datos 4008
recopilador 3269
recorrer 9783
recorrer (une línea) 4740
recortado al registro 3907
recortaduras 9504, 14558
recortar 1822, 3741
recortar a ras 3861
recortar con exceso 3740
recorte 7497, 9768, 10045, 12238
recorte de periódico 9304
recorte de prensa 9304
recorte por debajo 1023, 14281
recortes 1934, 4418, 9503, 9573, 12238, 14558
rectangular 11315
rectángulo 11315
rectángulo de proporciones áureas 6505
rectificación cromática 3131
rectificado de planitud del dorso 1043
rectificador 11317
rectificadora 1048
rectificadora para raspadores 4529
rectificador de cristal 3788
rectificar 11262
recto 11318
recuadrado con raya negra 1464
recuadrar 9719
recuadro 1792, 5690, 9889
recubridor 5278
recubrimiento 7903, 9770
recubrimiento con goma 6720
recubrimiento del cilindro 9825
recubrir con película plástica 7885

recuperación 11314, 11496, 11542
recuperación de disolventes 12778
recuperación de solventes 12778
rechazando la tinta 7387
rechazando los cuerpos grasos 6604
rechazar 11464
rechazo de las sobreimpresiones 3792
rechinar 2659
red 9288
redactar 4965
redactor 4976
redactora 4976
redactor de anuncios 229
redactor de información financiera 2866
redactor de información local 2867
redactor de platina 8572
redactor de textos (para anuncios) 229
redactor en jefe 4979
redactor gerente 11526
redactor publicitario 229
redactor responsable 11526
red eléctrica 13932
redondeamiento del lomo 1046
redondear esquinas 11723
redondelito para cerveza 1286
redondo 11670
red trifásica 13763
reducción 11349
reducción de los datos 4014
reducción de los puntos reticulares 4574
reducir 11341
reducir (el tamaño de los puntos tramadas) 3886
reducir la densidad 3886
reductor 11344
reductor de altas luces 6991
reductor Farmer 5471
reenrollar 11581
referencia 3764
refilón 8995
refinado 5669, 11386
refinador 11385
refinadora 11385
refinadora Jordán 7693
refinaje 11386
refino 1266
refino graso 14677
refino magro 6155
reflectómetro 11393
reflejo 11387
reflexión 11387, 11389
reflexión especular 12849
reforzado 7483
reforzador 7484
reforzador al yoduro mercúrico 8797

reforzador de bromuro cúprico 3502
reforzador de plomo 7991
reforzador de sales de mercurio 8795
reforzador de yodo 7554
reforzador de yoduro de potasio 7556
reforzar 7485
refracción 11395
refracción de la luz 11395
refractómetro 11397
refrenamiento 1851
refrenamiento de la bobina (de papel) 11358
refrigeración 3487
refrigerador del molde 9124
refuerzo 7483
refundir 8808, 11454
regencia 11692
regente 14834
regente de cajas 3294
regente de imprenta 10762
registrador 11310
registrar 8314
registro 1762, 5741, 11309, 11402, 11403
registro circunferencial 2856
registro con digitales 13795
registro (de) dirección 199
registro de longitud fija 5729
registro de longitud variable 14392
registro de los colores 3169
registro exacto 6749, 13809
registro final 13953
registro indicativo 9027
registro lateral 12372
registro secundario 13953
registro sin retorno al estado de referencia 9413
regla de cálculo 2176, 10970, 12554
regla de precisión 13242
regla de registro 11416
regla deslizante 12554
regla graduada 11899
reglaje separado, con ~ 12125
regla plegadora 5973
reglas laterales del carro 1275
regla tipométrica 14197
regleta 3213, 8193, 11421, 12600
regleta de pie 6024
regleta (gruesa) 2964
regleta marginal 12387
regletas 4064
regleta sacalíneas 3295
regletear 7971
regrabado 5669
regrabar 4078
regulación 226
regulación de entintado 7348

regulación de la cantidad de tinta 7348
regulación de la temperatura 13641
regulación de la tensión 13648
regulación de registro electrónica 5051
regulación de velocidad 12850
regulación local 4539
regulador de la cuchilla plegadora 5926
regulador del diafragma 4274
regular 218
régulo 982
rehacimiento 8562
reimponer 11423
reimpresión 11429, 11477
reimpresión aparte 9516
rejalgar 9691
rejilla de caracteres 2634
rejilla para jaspear 8625
rejilla para secar negativos 4694
relación a, en ~ 10975
relación en, en ~ 10975
relaciones humanas 7070
relais 11437
relé 11437
relevador 11437
reloj de cámara obscura 3994
reloj de cuarto obscuro 3994
reloj de segundos muertos 13221
reloj de trinquete 13221
relumbre 5758
rellenar 9842
relleno 5622, 9846
relleno del dorso (de la cascarilla del galvano) 1022
relleno de sobre 13150
remanencia 11452
remanente de ejemplares 11546
remanso 1011
remate cuneiforme 14666
remates 12144
remates oblicuos 9472
remedio 11453
remender 10047
remendería 2460, 7680
remendista 250, 7679
remolino 14088
remosqueamento 4646
remosqueo 1649, 4646, 8473
remover 13165
rendición de colores 3172
rendición en color 3170
rendimiento 14929
rendimiento anódico 646
rendimiento medio 956
rendimiento total de procesamiento 13785
renglón 8154

renglones chuecos 3739
renglones encaballados 3739
renglones torcidos 3739
renvío recíproco 3764
reología 11585
reparar 11462
repasar 3063
repelado de la tinta 1567
repelado del estucado 2997
repelado (del papel) 10295
repelado húmedo 14688
repelado seco 4813
repelente al aceite 9566
repelente al agua 7124
repelente la tinta 7387
repeler 11464
repelón 5880
reperforador 11465
reperforadora 11465
repetido 4579
repetidora 13116
repetir palabras 4580
repintado 289, 8473, 9540
repintado a la cara de blanco 5712
repintar 12162
repinte 8473, 9540
reportar 13962
reporte 11540, 13961
reporte de la mantilla 1051
reporte del papel pigmento 2332
reporte en seco 4808
reporte litográfico 8266
reporter novel 3796
reportero novicio 3796
reportero nuevo 3796
reposición de faltas 8562
reprensión 11476
representación gráfico 6564
representación visual 4458
reprimenda 11476
reproducción 11480
reproducción de detalles 4217
reproducción de tonos 13893
reproducción de un dibujo a lápiz 10094
reproducción en colores 3162, 3171
reproductibilidad 11479
reprografía 11485
repuestos 12819
repujado 11471, 11472
resaltar 4459
resalto en el corte del papel 4700
resbalamiento 12559
reseña de un libro 1773
reseñante 1774
reseñar un libro 11567
reserva 142, 11507
reserva colloidal 3093
reserva de goma 4090
reservados todos los derechos 439
residuo 11450, 11498, 11546

resiliencia 11500
resina 11501
resina acrílica 158
resina alcídia 425
resina alcohílica 425
resina aldehídica 386
resina alílica 447
resina alquídica 425
resina amínica 507
resina artificial 13495
resina copal 3493
resina cumarona 3650
resina damar 3964
resina de acetaldehído 87
resina de acrilo 158
resina de aldehído 386
resina de cumarón 3650
resina de éster 5260
resina de estireno 10561
resina de estirol 13317
resina de fenol-form-
  aldehído 10190
resina de formaldehído
  6049
resina de madera 14809
resina de melamina 8767
resina de melamínica formaldehída 8766
resina de poliéster 10555
resina de poliuretano 10564
resina de urea-form-
  aldehído 14370
resina epiclorhidrina 5220
resina epoxi 5220
resina epoxídica 5220
resina epóxido 5220
resina esterificada 5260
resina fénica 10191
resina fenólica 10191
resina fumárica 6246
resina isociánica 7614
resina maleica 8584
resina melamina form-
  aldehída 8766
resina melamínica 8767
resina poliamida 10551
resina poliéster 10555
resinas de vinyl 14479
resina sintética 13495
resinas naturales 9250
resinas vinílicas 14479
resinato cobaltoso 3015
resinato de cobalto 3015
resinificación 11504
resistencia a la abrasión 11
resistencia al aceite 9567
resistencia al ácido 143
resistencia a la flexión
  1323
resistencia al álcali 421
resistencia a la luz 5484
resistencia a la (rotura por)
  tensión 13644
resistencia al arranca-
  miento 10300
resistencia a la tinta 7388
resistencia a la tracción
  13644
resistencia al calor 6929

resistencia al doblado 5979
resistencia al frote 11
resistencia al jabón 12657
resistencia a los di-
  solventes 12781
resistencia al plegado 5979
resistencia al queso 2679
resistencia al rasgado
  13614
resistencia al rasgado
  inicial 13615
resistencia al rasgado
  interno 13616
resistencia al repelado
  10300
resistencia al rozamiento
  11, 5231, 11767
resistencia del agua 14599
resistencia de placa 647
resistencia de tensión
  13644
resistencia dieléctrica 4325
resistencia efectiva al
  estallidomiento 2096
resistencia efectiva al
  reventamiento 2096
resistencia (eléctrica) 11508
resistencia frente a los
  productos químicos 2688
resistencia insuficiente al
  plegado 10573
resistencia interna 647
resistencia mecánica 8739
resistencia química 2688
resistente a la abrasión
  9414
resistente a la humedad
  14593
resistente al alcool 381
resistente a la tinta 7389
resistente al calor 6930
resistente al desgaste por
  abrasión 12
resistente al desgaste por
  roce 12
resistente al frotamiento
  11764
resistente a los ácidos 145
resistente a los álcalis 416
resma 11291
resma de 480 hojas 12335
resorcina 11519
resorte 12939
resorte de compresión 3326
resorte de presión 10770
resorte de tracción 13651
resortes de hoja 7965
resortes de lámina 7965
resortes de planos 7965
respaldo antihalo 667
respaldo de la cámara 2240
respaldo neumático 14374
resplandor 1909
responsorio 11300
respuesta hablada 869
respuesta transitoria 11525
resto 11498
restregador 10480

resultado de impresión
  10845
resumen 20
retal 14561
retardador 11528
retardar 12594
retén 10067
retención 11529, 11530
retenedor 10067
retícula de contacto 3416
retícula de medio tono
  3759
retícula (de medio tono)
  6794
retícula de puntos elípticos
  2582
reticulado 6773, 11532
retícula en forma de
  ladrillo 1903
retícula lineal 6763
retícula magenta de
  contacto 8487
retícula para autotipia 6794
retiración 1021
retiración en molinete
  14826
retiración normal 14825
retirar 1052
retirar a la voltereta 14824
retirar le escoria 12511
retiro 1021, 7418
retiro de pie con cabeza
  14824
retocado al aerógrafo 316
retocado fotográfico 10251
retocador 11533, 12923
retocar negativos 12917
retoque 11534
retoque a buril 13900
retoque al aerógrafo 316
retoque con lápiz lito-
  gráfico 3709
retoque con tintes 4927
retoque (de la autotipia) a
  buril 11808
retoque de los colores 3173
retoque de los medios
  tonos 4567
retoque de puntos 4567
retoque fotográfico 10251
retorno del carro 2368
retrasar 12594
retraso de papel 7465
retribución pecuniaria 890
reubicable 11448
reubicar 11449
reunidor 832
reunidora de hojas 6330
reunidora de pliegos 6330
reunir 846, 6325
reunirse 3748
revelado 4231
revelado cromógeno 2815
revelado en reposo 13158
revelador 4222
revelador ácido 121
revelador a la hidro-
  quinona 7128
revelador al hierro 7588

revelador de grano fino
  5672
revelador de mordido 14554
revelador de paramino-
  fenol 11628
revelado sin agitar 13158
revelar 4221
revendedor de periódicos
  9300
reventamiento 2092
reversión 11548
reverso 14447
revés del papel 14778
revestimiento de corcho
  3576
revestimiento del cilindro
  9825
revestimiento electrolítico
  de cobre 3509
revestir 4721
revisar 11262, 11571
revisor de libros 75
revista 7698
revista bimestral 14148
revista comercial 13948
revista de comunicación
  (interna) 7060
revista de modas 5478
revista de publicidad 11000
revista hebdomadaria
  14667
revista industrial 13948
revista mensual 9096
revista órgano 7060
revista para uso del
  personal 7060
revista profesional 13948
revista quincenal 12101
revista técnica 13948
revista trimestral 11136
revolución 11578
revolución del cilindro 3950
revoluciones por minuto
  11579
rezago 2116
rezagos 11450
ribeteadora 8824
ribete de metal 2183
riel 1283
riga 8154
rigidez 11613
rivalidad 3420
rizado del papel 3833
robar 10294
rociada de mordiente 117
rociador 12926
rociador antimaculador
  4823
rocío antimaculador 9539
rodamiento a bolas 1096
rodamiento de bolas 1096
rodamiento de rodillos
  11634
rodamina B 11586
rodanuro amónico 550
rodanuro de amonio 550
rodetes 5521
rodillo 11633
rodillo acanalado 12146

rodillo accionado 4739
rodillo alisador 6448, 12878
rodillo batidor 4503, 9707, 14616
rodillo cargador 11600
rodillo cargador accionado 6348
rodillo cargador de acero 13099
rodillo cargador de caucho 11754
rodillo compensador 7683
rodillo dador 4503
rodillo de acero 13101
rodillo de afiligranar 3985
rodillo de ágata 296
rodillo de cabeza 1899
rodillo de carga 11600
rodillo de carga de acero 13099
rodillo de carga de caucho 11754
rodillo de caucho 11756
rodillo de cola 3309
rodillo de cuero 8019
rodillo de distribución 4503
rodillo de excéntricas 2253
rodillo de expulsión 7790
rodillo de flotación 5359
rodillo de gelatina 3309
rodillo de goma 11756
rodillo de goma del teclado 7767
rodillo del tintero 4846, 4847, 7335
rodillo de mano 1871, 8264
rodillo de mojado 14578
rodillo dentellado 12146
rodillo de pasta 3309
rodillo de presión 7213
rodillo desgotador 3985
rodillo de vaivén 9707, 11302
rodillo de verjuras 7874
rodillo distribuidor 11302, 13972, 14616
rodillo endurecido 6446
rodillo (entintador) espiroidal 12877
rodillo expansivo 5359
rodillo filigranador 3985
rodillo flotador 5359
rodillo glaseado 6446
rodillo grabador 5185
rodillo graneado 9240
rodillo guía 12546
rodillo inmersor 4396
rodillo libre 7155
rodillo liso 6447, 12647
rodillo marcador 3985
rodillo mojador de acero 13106
rodillo mojador de bronce 1865
rodillo mojador de latón 1865
rodillo mojador en acero 13106
rodillo oscilador 9707

rodillo oscilante 9707
rodillo oscilatorio 9707
rodillo para marcas de agua 3985
rodillo refrigerador 3490
rodillo refrigerante 3490
rodillo reticulado 5185
rodillos alimentadores 5521
rodillos caballeros de pasta (gelatinosa) 3308
rodillos cargadores de pasta (gelatinosa) 3308
rodillos conductores 1883, 3375
rodillos conductores (de la hoja) 2373
rodillos con mordiente 13533
rodillos dadores 6073
rodillos de agua 3976
rodillos de alimentación 5521
rodillos de avance 5521
rodillos de batido 7360
rodillos de calefacción 6924
rodillos de conducción 3375, 14655
rodillos de entintación 6073
rodillos de entintar 7360
rodillos de entrada 1322, 8850
rodillos de guía 14655
rodillos de introducción 8850
rodillos (de la) anteprensa del manchón 12981
rodillos de pinza 9355
rodillos de tensión 13268
rodillos de tinta 7360
rodillos de tintaje 7360
rodillos de tiro 4713, 10963
rodillos de tracción 10963
rodillos entintadores 6073, 7360
rodillos exprimidores 12981
rodillos guía 3375
rodillos guía (de la hoja) 2373
rodillos guía (del papel) 2373
rodillos humectadores 3976
rodillos intermediarios 7512
rodillos introductores 4714
rodillos llamadores 4713
rodillos mojadores 3976
rodillos plegadores 5986
rodillos tensores 13268
rodillos tiradores 4713
rodillo superior del.triángulo 11652
rodillo tensor 7683
rodillo tomador 4841
rodillo verjurador 7874
rodinal 11628
rodón 10950
rojo Berlín 1339
rojo de cadmio 2144
rojo de etilo 5316

rojo de la India 7281
rojo litol 8274
rojo para 9993
rojo para laca C 11329
rojos de óxido de hierro 7595
rojo veneciano 14437
roldana 14549
rollo de cartón 11372
rollo de papel 11354, 11372
rollos de papel para máquina de calcular 186
rollo tensor 5360
romano 11670
rombo 14470
romboidal 8391
romper el jito 1892
romperse 1874, 1891
rompimiento del papel 14641
rondana 9828
rondeado y formación del lomo 11827
rop-color 11680
rosca a derechas 11609
rosca a izquierdas 8028
rosca (de tornillo) rectangular 11316
rosca hembra 7520
rosca interior 7520
rosca interna 7520
rosca métrica 8876
rosca redonda 11731
rotación a la izquierda, de ~ 3655
rotación inversa, de ~ 3655
rotaplana 5776
rotativa 11699
rotativa a doble ancho 4670
rotativa a un solo ancho 12471
rotativa con 3 grupos de 8 páginas 12207
rotativa de cuerpos 14321
rotativa de grupos 14321
rotativa de huecograbado 14648
rotativa de huecograbado en hojas 12268
rotativa de periódicos 9318
rotativa de unidades de impresión 14321
rotativa de 6 grupos de 4 páginas 4654
rotativa de 8 grupos de 4 páginas 4635
rotativa offset 11698
rotativa offset en hojas 12270
rotativa tipográfica 14652
rotativista 11696
rotocalco 9525
rótula 1094
rótulo 4466, 6905, 7846, 11768, 13534
rótulo de encabezamiento 6900
rótulo final 13952
rótulo inicial 6901

rotura 1874
rotura de la banda 4675, 14641
rotura del empalme 12893
rotura de papel 14641
roturas de fabricación 4769
rúbrica 3200, 11768
rubro 11768
rueda cónica 1349
rueda de cadena 12950
rueda de dorar 1176
rueda de estrella 13063
rueda de fundición 9125
rueda (de) molde 9125
rueda dentada 6346
rueda de palas 5916
rueda de tornillo sinfín 14854
rueda de trinquete 11249
rueda de unidades 14322
rueda fileteadora 1176
rueda fundidora 9125
rueda helicoidal 6954, 14854
rueda motriz 4749
rueda para puntillado 4578
ruedas alimentadoras 3218
rueda serpentina 14854
rueda volante 5935
rugosidad 11711
rugosidad superficial 13437
ruleta 1176, 11717
rulo de mano 1871
ruptor de fin de carrera 8151
ruptura 1874
ruptura de papel 14641
rutina 11734
rutina de biblioteca 8088
rutina standard 8088

sablón 11858
sacaboquear 4317
sacador 4149, 14413
sacador de cadena 2580
sacador de paquetes 12998
sacador de pruebas 10945
sacador en postetas 7788
sacalíneas 3295
sacapliegos 4134
sacapliegos a cadena 2580
sacapliegos a cinta(s) 13581
sacapliegos de abanico 5915
sacapliegos de cadena 2580
sacapliegos de cintas 13581
sacapruebas 10945
sacar 13555
sacar contramoldes 4474
sacar el papel de la tela 3641
sacar pruebas 10937
sacar una prueba 11018
sacatipia 10950
sacatornillos 12001
saco 1066
saco con foro múltiple 9198
saco con válvula de apertura automática 924
saco de papel 9898

saco (de papel) 11826
saco de papel duplicado 4873
saco de varias capas 9198
saco pequeño 10689
sal 11852
sal ácida 146
sala da cajas 3292
sala de clasificación 5692
sala de componer 3292
sala de composición 3292
sala de composición de los anuncios 248
sala de corrección 11274
sala de escogido 5692
sala de insolación 8839
sala de las componedoras 3311
sala de lectura 11275
sala de máquinas 10761
sala de máquinas de componer 3311
sala de prensas 10761
sala de redacción de información local 2868
sal amoniaca 522
sal amónica 522
sal crómica 2782
sal de cocina 12683
sal de Epsom 8502
sal de Glauber 12719
sal de la Higuera 8502
sal de viraje 13891
sal doble 4653
saldos 11450
sal férrica 5555
sal ferrosa 5579
sal fijadora 5737
salida 4134, 9730, 9731, 13057
salida a cadena 2580
salida alternativa 4595
salida de palas 5462, 5916
salida doble 4595
salida frontal (de las hojas) 6208
salida manual 9622
salida plana 5786
salida por cadenas 2580
salida selectora 12897
saliente 13057
salir a luz 714
salir(se) 4716
sal metálica 8831
salón de lectura 11275
sal para fijar 5737
salpicado 12890, 12984
salpicadura 12890, 12984
salpicón 12890
saltar 8021, 12521
saltear en zigzag 13002
salto 997, 7705, 9584
salvamanos 6689
salvapliegos 7224, 13789
salvapliegos de fallo de hoja 9430
sandáraca 6724
sangrado 7264

sangrado a la francesa 6854, 11813
sangrador (automático) 11093
sangrar 7262
sangre de drago 4688
sangre de dragón 4688
sangre drago 4688
sangría 7264
sangría de escalón 4755
sangría diagonal 4256
sangría en gradas 4755
sangría escalonada 4256, 4755
sanguina 11867
saponificable 11870
saponificación 11871
satinado a la piedra (de ágata) 5834
satinado común 5179
satinado en hojas 12262
satinado entre planchas de cinc 10437
satinadora 2184, 5570, 6193
satinar 2191
satinómetro 12646
sátira 7895
saturación 11883
saturación del color 2760, 11882
saturnismo 8000
Schwabacher 11924
s.e.a. 14500
sebo 9222
secadero 4791, 4797
secadero de suspensión 5584
secado 4783
secado acelerado 6026
secado a fondo 4799
secado al aire 331
secado de la tinta 7342
secado de la tinta en las alvéolos del cilindro 4790
secado demasiado rápido de la tinta 11981
secado forzado 6026
secado infrarrojo 7312
secado por aire 331
secador 5456
secadora 4786
secadora abrillantadora 4777
secadora de matrices estereotípicas 8690
secadora (de matriz) 8690
secadora para estereo-matrices 8690
secado rápido, de ~ 5482
secador de aire 330
secador de fieltro 5532
secador de hojas 12265
secador de matrices (estereotípicas) 8690
secador de paño 5532
secador de tendido 5584
secador líquido 8235
secador yanqui 14903
secante 1621, 4784, 12032

secante de barniz (para tintas) 7636
secante de blocs 13515
secante de cobalto 3007
secante de manganeso 8590
secante de naftenato 9237
secante de pasta 10028
secante de plomo 7979
secante Japón 7636
secante líquido 8235
secante naftalino 9237
secantes publicitarios 266
secar a ventilador 5463
secativo 4784
sección 4185
sección de anuncios 248
sección de cajas 3292
sección de componer 3292
sección de emplane 3292
sección de encargos 9674
sección de encuadernación 1421
sección de prensas 10737
sección de preparación de la copia 10449
sección de recuperación 12779
sección de talleres 8465
sección estereotípico 13130
sección fija 3200
sección húmeda 14683
sección transversal del cilindro 3751
seco absoluto 9744
secuencia 13277
secuencia de indicativo 7275
sedimentación 12065
sedimento 4186, 12596
segmentar 12067
segregación 12069
segunda cubierta 7452
segunda mano, de ~ 12043
segunda prueba 11570
segunda signatura 851
segundo elevador 12040
segundo pliegue paralelo 12038
seguridad 11440
seguro 11833
seguro de la bomba de fundición 12330
seguro de la fundición 12330
seleccionadora para la separación tricroma y confección de los clisés 3177
selección cromática 3183
selección de color(es) 3183
selección de la página caja 2751
selección del tipo 2750
selección de probetas 11856
selección en colores 3183
selecciones de colores a mano 8731
selección fotomecánica de colores 3183

selección tricroma 3183
selector de matrices 4501
selector de salida 12897
selector de tensiones 14512
selenio 12072
sellado al calor 6931
sellado en caliente 6931
sello de caucho 11757
sello de fábrica 13944
sello de goma 11757
sello de hule 11757
sello de lacre 14630
sellos de reclamo 10611
semanario 14667
semestral 12093
semicelda de quinhidrona 11159
semicelda de vidrio 6428
semicelulosa 6960
semicolon 12098
semiconductor 12099, 12766
semiencolado 6768
seminegra, (acentuado) en ~ 58
semipasta 6770
semirrotativa 5776
semiuncial 6798
seno 12443
sensibilidad 12112
sensibilidad a la luz de una emulsión 12111
sensibilidad cromática 3181
sensibilidad espectral 12843
sensibilizado cromático 3182
sensibilizado óptico 9662
sensibilizado para deter-minar colores 9662
sensibilizador 12118
sensibilizador óptico 9661
sensibilizar 2980, 12114
sensible a la luz 12111
sensitometría 12121
sentar la forma 6529
sentido atravesado 3752
sentido contrario 3752
sentido de fabricación 8443
sentido de fibra 8443
sentido transversal 3752
señal 4543, 5741
señal analógica 573
señal de apartado 9973
señal de entrada 7428
señal de integración 7476
señal de intercalación 7440
señal de interrupción 7530
señal de lectura 1762
señal de libro 1762
señal de mando 3236
señal de párrafo 9973
señal de quitar (letra) 4131
señal de suprimir (letra) 4131
señal diacrítica 4250
señal digital 4361
señales de atención 3600
señales de corrección 3600
señales de encuadre 3742

sobre de correspondencia 3611
sobre de ventana 14749
sobre engomado 210
sobreexponer 9755
sobreexponer excesivamente 2084
sobreexposición 9754
sobreexpuesto 9753, 9778, 10502
sobrehilado 14708
sobreimpresión 9779, 13449
sobreimpresión fotomecánica 4647
sobreimprimir 13414
sobremordido 9751, 14271
sobre para vía aérea 340
sobrepeso 9794
sobreponer 13415
sobreposición de colores 7903
sobreprecio 5395
sobrerrevelado 9750
sobrestante 13422
sobresueldo por cuidado personal 3263
sobretapas 7629
sobre transparente 14749
socavado (de los puntos) 14271
socavar 14273
socorro a los enfermos 7260
socorro de viaje 14456
sodio 12668
software 12744
sol 12749
solapa 1647, 5755
solarización 12751
soldadura por alta frecuencia 6983
soldadura por calefacción 6931
soldadura por puntos 12607
soldar 12753, 13458
solecitos 13407
solenoide 12754
soles 13407
solidez 7740, 12286
solidez a la luz 5484
solidez a los ácidos 143
solidez a los álcalis 421
solidez al queso 2679
solidez del color 3142
solidificar 12760
solido cromático 3187
soltura del estuco 10082
solubilidad 12768
soluble 12769
soluble al agua 14603
solución 12772
solución (a base) de asfalto 827
solución acuosa 739
solución amoniacal 533
solución amortiguadora 2034
solución compensadora 2034

solución de albúmina 377
solución de albúmina bicromatada 2771
solución de blanquear 1547
solución de casca para curtir 10298
solución de cloruro de cinc yodado 6971
solución de Labarraque 12696
solución de mojador 3980
solución detergente para limpiar 10425
solución de yodo yoduro de potasio 7564
solución humectadora 3980
solución lavamantilla 1526
solución mojadora 3980
solución molecular 9053
solución mordiente para hueco-offset 4089
solución para limpiar mantillas 1526
solución para quitar la capa de cola residual 12023
solución para reserva 13216
solución reguladora 2034
solución resinosa 13847
solución saturada 11881
solución tampón 2034
solvente 11345, 12773
solvente orgánico 9681
solvente rápido 5487
solventes polares 10544
sombra 6740
sombras 12223
sombreado de líneas cruzadas 3756
sombreado en tipos ornamentales 1249
sombrear 6885, 13845
sombrerete de cabeza 1257
sonido del papel 11254
soplador 9445
soplador de aire 5456
soplador de chorro de arena 334
soplador de vapor 13086
sopladura 336, 1577, 5810
sopladuras 12475
soplete 1630
soporte adhesivo 13083
soporte de acetato 89
soporte de cámara 2249
soporte de grapa 13173
soporte de la raspadora 4533
soporte del árbol principal 8549
soporte del fuelle 1295
soporte doble de bobinas 13569
soporte lateral de bobina 12383
soporte para almacenes 8482
sorbita 12783
sorbitol 12783

sosa anhidra 12682
sosa calcinada 12682
sosa cáustica 12695
sótano de las bobinas 11374
standardización 13026
stilb 13155
stokio 13194
stop 4273
stop en cruz 3769
strato interno 5624
suaje 3903, 4313
suavizador de agua 14602
suavizante 4942, 11345
suavizar la tinta 12736
subconjunto 13335
subcontratar 5472
subdividir 12067
subencargar 5472
subestructura 13341
subexposición 14278
subexpuesto 8970
subjuego 13335
subletras 7301
sublimación 13326
sublimado (corrosivo) 8792
sublimar 13327
subnitrato de bismuto 12817
subnúmeros 7301
subproducto 2132
subprograma 13328
subrayar 14285, 14288
subrevelación 14275
subrevelado 14275
subscribir(se) 13330
subscripción 13331
subscriptor 13329
subscriptora 13329
substancia fibrosa 5597
substancia filmógena 5642
substancia incrustante 5156
substituto de la trementina 14108
substrato 13339
subtítulo 467, 13322, 13342
succibilidad 32
succionadores 13348
suciedades 7235
sueldo a horas 7056
sueldo fijo 12994
sueldo fijo, a ~ 12986
sueldo fijo, estar a ~ 12987
suiche 13474
sujeta ajustable 220
sujetacuartillas 3544
sujetacuchillas 12541
sujetador 9906
sujetapapeles 9906
sujeta regulable 220
sujetar (la plancha) 8309
sulfato alumínico 468, 487
sulfato alumínico-amónico 473
sulfato alumínico-potásico 483
sulfato alumínico-sódico 12673
sulfato amónico 544

sulfato amónico de níquel 9336
sulfato cálcico 3774
sulfato cérico 2562
sulfato ceroso 2568
sulfato crómico-potásico 2808
sulfato de aluminio 487
sulfato de aluminio potásico 483
sulfato de amonio 544
sulfato de anilina 622
sulfato de bario (artificial) 1156
sulfato de cal 2174
sulfato de calcio 2174, 3774
sulfato de cinc 14955
sulfato de cobre 3533
sulfato de hierro 3499, 5580
sulfato de hierro amoniacal 5544
sulfato de magnesia 8502
sulfato de manganeso 8591
sulfato de monometil paraminofenol 8867
sulfato de níquel y amonio 9336
sulfato de plomo 8005
sulfato de potasa 10671
sulfato de potasio 10671
sulfato de sodio 12719
sulfato de sosa 12719
sulfato férrico 5556
sulfato férrico-amónico 5544
sulfato férrico-potásico 5554
sulfato ferroso 3499, 5580
sulfato ferroso-amónico 5573
sulfato magnésico 8502
sulfato manganoso 8591
sulfato potásico 10671
sulfato sódico 12719
sulfito 13372
sulfito amónico 547
sulfito bárico 1158
sulfito cálcico 2175
sulfito de amonio 547
sulfito de bario 1158
sulfito de calcio 2175
sulfito de potasa 10673
sulfito de potasio 10673
sulfito de sodio 12721
sulfito de sosa 12721
sulfito potásico 10673
sulfito sódico 12721
sulfocianato de potasa 10676
sulfocianato de potasio 10676
sulfocianato potásico 10676
sulfocianuro amónico 550
sulfocianuro bárico 1161
sulfocianuro de amonio 550
sulfocianuro de bario 1161
sulfocloruro 13386
sulfonar 13380
sulfourea 13723

tarjeta de ventana 698
tarjeta de visita 14493
tarjeta hollerith 7021
tarjeta horaria 2921
tarjeta magnética 8503
tarjeta para avisos 12350
tarjeta perforada 11045
tarjeta postal 10605
tarjeta postal con vistas
10312
tarjeta postal ilustrada
10312
tarjetas cortadas a troquel
4315
tarjetas de felicitación 6626
tarjetas de participación
638
tarjetas troqueladas 4315
tarjetón de anuncio 12350
tarjetón publicitario 12350
tarlatana 9164, 13605
tartrato potásico de sodio
12710
tarugo 11362
tasa 5319
tasa de cortadura 14428
tecla 7761
tecla de abreviaturas 7535
tecla de anulación 4130
tecla de espaciación 12804
tecla de espaciado 7709,
12804
tecla de los espaciadores
12804
tecla de restitución 11545
teclado 7765
teclado adicional 187
teclado auxiliar 187
teclado ciego 1569
teclado Monotype 9091
teclado perforador 10134
teclista 7771
teclista tipográfico 9639
técnica de embalaje 9821
técnica de imprimir por
amalgamación 11460
técnico publicitario 229
tecnología de estado sólido
12761
teja de estereotipia 3849
tejidos estampados 10801
tejo 9889
tejuelo 1050, 9889
tela artificial 785
tela de balón 1102
tela de caucho 1509
tela de encuadernación
1735
tela de máquina 8469
tela Fourdrinier 8469
tela gris para encuader-
nación 6627
tela para calcar 13937
tela para encuadernación
1735
telar 12195
tela sin fin de la máquina
de papel 8469
telecomunicación 13627

teleprocesamiento 13630
teleta 1621
teleta esmaltada 5143
teleterminal 11459
teletipo 13635
teletipocomposición 13636
temperatura absoluta 30
temperatura ambiente 496,
11676
temperatura de ebullición
1702
temperatura de fundición
2454
temperatura de fundir 2454
temperatura de inflama-
ción 5765
temperatura de(l) color
3191
temperatura de reblandecimiento 12735
templar 631
tenacidad de una película
13929
tenazas de fundido 2077
tendalero 4791
tendedero 4791
tendedores 2920
tendencia a la pelusa 5881
tender en el ambientador
11185
tener buen tacto 8092
tensioactivo 13427
tensión anódica 648
tensión constante de la
bande 3399
tensión de ánodo 648
tensión de cizallamiento
12241
tensión de cortadura 12241
tensión de electrodo 5030
tensión de la banda 14661
tensión de la red 14511
tensión de la superficie de
límite 7494
tensión del papel 14661
tensión de rejilla 6634
tensión de resorte 12946
tensión (eléctrica) 14509
tensión interfacial 7494
tensión superficial 13441
tensión variable 14408
teoría de difracción 4345
teoría de la abertura de la
retícula 10365
teoría de Newton 9326
tercera cubierta 7449
tercera página de cubierta
7449
tercio de cuadratín 13708
terebentina 14106
terebentina veneciana
14438
terminado 442
terminal 12954, 13652, 13653
terminal cuadrado 12531
terminal de pantalla video
14466
terminar a final de párrafo
5160

terminar con línea llena
5160
terminar un alcance con
línea llena 5160
término de entrega 4141
termoendurecible 13696
termoestable 13696
termofijo 13696
termograbado 13690
termografía 13690
termoimpresión 13690
termoinmersor eléctrico
7184
termómetro para punto de
ebullición 1703
termoplástico 13692
termostato 13698
terno 13655
tesis de doctorado 13700
tesis doctoral 13700
testar 13272
testera 6889
testimonio 13020
tetracloruro carbónico 2330
tetracloruro de carbono
2330
tetracromía 6124
tetravalente 13672
textiles estampados 10801
texto 13674
texto de aleta 1647
texto en película 14243
texto mecanografiado
14228
texto que continua 14102
therblig 13685
tiempo compartido 13818
tiempo de acceso 70
tiempo de arranque 13062
tiempo de avería 4679
tiempo de caída 4049
tiempo de cambio 2618
tiempo de ciclo 3923
tiempo de composición
3298
tiempo de efusión 4984
tiempo de entrega 4032
tiempo de espera 14527
tiempo de exposición 5376
tiempo de fraguado 12179
tiempo de funcionamiento
9632
tiempo de grabado 5288
tiempo de inducción 7296
tiempo de insolación 5376
tiempo del cierre 4032
tiempo de mordido 5288
tiempo de ordenación
12360
tiempo de pose 5376
tiempo de preparación
12186
tiempo de respuesta 11523,
11524
tiempo de revelado 4233
tiempo de secado 4800
tiempo de semisecado
12179
tiempo disponible 950

tiempo doble 4603
tiempo inactivo 4678
tiempo incentivo 7243
tiempo muerto 1087
tiempo normal 13030
tiempo presupuestado 432
tiempo real 11290
tiempo standard de re-
cambio 464
tiempo suplementario 9793
tiempo útil 14362
tierra arcillosa blanca 2711
tierra blanca 630
tierra de diatomeas 4280
tierra de infusorios 4280
tierra de Siena 12394
tierra de Siena calcinada
2088
tierra de Umbría 14255
tierra de Umbría quemada
2089
tierra infusoria 4280
tierra sombra 14255
tigmotropismo 13737
tijeras 11929
tijeras para cartón 12242
tijeras para cortar papel
9940
tilde 13812
tilo 1211
timbrado a seco 4825
timbrado en relieve 4334
timbrado en seco 1562
timbre a seco 4333
timol 13798
tímpano 14165
tina 14410
tina de alimentación 8432,
13186
tina de cabeza 8432
tina de encolado 12501
tina de (la) máquina 8432
tina de la máquina 13186
tina de mezcla 9011, 13186
tina mezcladora 9011
tinta 7325, 7326
tinta absorbida 10097
tinta a la anilina 5825
tinta al agua 14567
tinta al agua para hueco-
grabado 14566
tinta al alcohol 5825
tinta amortiguadora 1025
tinta antifricción 11765
tinta apagada 4854
tinta argentina 12430
tinta azul bronce 1960
tinta bebida 10097
tinta blanda dúctil 8338
tinta brillante 6453
tinta cerámica 2556
tinta compacta 13298
tinta comunicativa 3548
tinta consistente 13298
tinta corta 12328
tinta co-solvente 3627
tinta cubriente 3680, 9614
tinta China estilográfica
4708

tinta de alta viscosidad 6935
tinta de alto brillo 6453
tinta de alto lustre 6984
tinta de aluminio 479
tinta de anilina 5825
tinta de anilina pigmentada 10331
tinta de base 1185
tinta de bronce 1964
tinta de carbonizar 2312
tinta de color para periódicos 3232
tinta de copiar 3548
tinta de cubrir 9612
tinta (de) China 7280
tinta de doble tono 4666
tinta de dos tonos 4666
tinta de emulsión 5135
tinta de flexografía 5825
tinta de fraguado rápido 11152
tinta de fraguado térmico 13697
tinta de grabar 5279
tinta de imprenta 7325
tinta (de impresión) normal 9424
tinta de imprimir 7325
tinta de luminescencia 8412
tinta de luminiscencia 8412
tinta de opacar 9612
tinta de penetración 10099
tinta de plata 12430
tinta de reporte 4226, 13964
tinta de retocar 13007
tinta de retoque 10905, 11537
tinta de revelar 4226
tinta de secado por calor 13697
tinta de secado rápido 11152
tinta de secado térmico 13697
tinta de seguridad (para cheques) 11837
tinta de serigrafía 11987
tinta duplicativa 8948
tinta dura 13151
tinta encáustica 5148
tinta espesa 12328, 13151
tinta fija 10144
tinta fijandose por humedad 9039
tinta fijandose por vapor 9039
tinta flexográfica 5825
tinta flexográfica pigmentada 10331
tinta fluida 12740
tinta fluorescente 5891
tinta glutinosa 13532
tinta grasa 6612, 14114
tinta gredosa 2607
tinta inalterable 10144
tinta indeleble 7259
tinta invisible 7546

tintaje 7339, 7356
tintaje cilíndrico 4442
tintaje de disco 4442
tintaje de mesa 13509
tinta líquida 12740
tinta lito(gráfica) 8271
tinta litográfica 14114
tinta lustrosa 6453
tinta magenta 8489
tinta mágica 8492
tinta magnética 8519
tinta mate 4854
tinta mecánica 8741, 12216
tinta metálica 8828
tinta metálica de bronce 1964
tinta metálica de oro 1964
tinta moisture-set 9039
tinta mordiente 12485
tinta negra 1479
tinta (negra) para ilustraciones 6785
tinta offset 8271, 9523
tinta opaca 9612, 9614
tinta para calcar 2312
tinta para diáfanos 4272
tinta para dibujar 7280
tinta para escribir 7326
tinta para estampación en relieve 13016
tinta para estampar en relieve 13016
tinta para fotograbado 5279
tinta para fototipia 3095
tinta para grabado de acero 13097
tinta para grabado en acero 13097
tinta para heliotipia 3095
tinta para huecograbado 11706
tinta para huecograbado (a grano de resina) 10254
tinta para impresión con trama de seda 11987
tinta para impresión sobre hojalata 8820
tinta para impresión sobre metal 8820
tinta para imprimir en oro 6507
tinta para la imprenta de tejidos 2958
tinta para marcar 8637
tinta (para) mimeógrafo 8948
tinta para mordido 5279
tinta para mordido profundo 4083
tinta para papel bond 1722
tinta para papel cartas 1722
tinta para periódico(s) 9308
tinta para rayar 11783
tinta para remiendos 7677
tinta para retocar 13007
tinta para revelar 4226
tinta para talla dulce 3524
tinta para transporte 13964

tinta pegajosa 13532
tinta penetrada 10097
tinta pesada 6939
tinta plana 5808, 12755
tinta plateada 479
tinta poco dúctil 12328
tinta porométrica 10099
tinta prueba de rascado 9415
tinta que se corre 1559
tinta que se decolora 1559
tinta resistente a la luz 10144
tinta resistente al frote 11765
tinta resistente al grasa 6602
tinta reveladora 4226
tinta rezagada en la forma 7386
tintas al pastel 10032
tintas apagadas 8675
tintas barómetros 14636
tintas de fraguado al calor 6934
tintas de fraguado al frío 3058
tintas de secado al frío 3058
tintas de secado en caliente 6934
tintas de tricromía 10897
tinta simpática 7546
tintas inodoras 9501
tintas mates 8675
tintas normalizadas 10897
tintas para obras 1756
tintas para policromía 10897
tintas para tricromía 10897
tintas perfumadas 10138
tintas primarias sustractivas 10897
tintas sin olor 9501
tinta steam-set 13085
tintas transparentes 14003
tintas tricromáticas 10897
tinta termosecante 13697
tinta terrona 4939
tinta tipográfica 8062
tinta transmitida 13971
tinta transportada 13971
tinta viscosa 13151
tinte 4922
tintero a inyección (por bomba) 912
tintero cerrado 5150
tintero con cuchilla 4534
tintero de cuchilla entintado pelicular 9790
tintero (de prensa) 7343
tintero dividido (en secciones) 12898
tintero móvil 9153
tintero para impresión posterior 4735
tintero portátil 10587
tintero portátil para colores 10585

tintero seccionado 12898
tintómetro 7375
tintura 4922
tioantimoniato de sodio 12727
tioantimoniato de sosa 12727
tioantimoniato sódico 12727
tiocarbamida 13723
tiocianato amónico 550
tiocianato de amonio 550
tiocianato de potasa 10676
tiocianato de potasio 10676
tiocianato de sodio 12728
tiocianato de sosa 12728
tiocianato mercúrico 8806
tiocianato potásico 10676
tiocianato sódico 12728
tiosulfato amónico 551
tiosulfato bárico 1162
tiosulfato de amonio 551
tiosulfato de bario 1162
tiosulfato de sodio 7146
tiosulfato sódico 7146
tiourea 13723
tipo 14171
tipo abierto 5358
tipo alargado 3364
tipo a mano 6102
tipo ancho 5358
tipo compacto 3364
tipo común 1688
tipo condensato 3364
tipo corriente 1688
tipocromía 3163
tipo cuyo cabeza es mas grueso que el pie 1809
tipo cuyo pie es mas grueso que la cabeza 1808
tipo chupado 3364
tipo de bronce 1868
tipo de caja 6102
tipo de carácter publicitario 201
tipo de cartel 10612
tipo de dibujo publicitario 201
tipo de fundición 6102
tipo de latón 1868
tipo de letra 5423
tipo de mano 6102
tipo de máquina de escribir 14213
tipo de ojo ancho 5358
tipo de texto 1688
tipo en hueco 9109
tipo escritura 12012
tipo esqueleto 3364
tipo estrecho 3364, 5094
tipo estropeado 1228
tipo extendido 5358
tipo extenso 5358
tipoffset 4812
tipo gastado 14855
tipografía 8061, 8065, 14241
tipografia indirecta 4812
tipografia rotativa 11695

tipógrafo ambulante 13930
tipógrafo cajista 3313
tipógrafo compositor 3313
tipógrafo errante 13930
tipo invertido 14095
tipo levantado 4716
tipo machacado 1228
tipo metido 3364
tipómetro 14197
tipómetro (del monotipo) 5124
tipo negrillo 1707
tipo negrito 1707
tipo negro 1707
tipo-offset 4812
tipo parado 13033
tipo para libros 1751
tipo para publicidad 201
tipo publicitario 201
tipo puesto boca abajo 14095
tipos abastonados 1609
tipos ahuecados 3569
tipo saledizo 7748
tipos cursivos 7625
tipos de fantasía 5461
tipos de imprenta 10851
tipos de madera 1608, 14804
tipos de palo seco 1609
tipo seminegro 12095
tipos extranjeros 6038
tipos grandes 4471
tipos griegos 6619
tipos machacados 14855
tipos mechados 3569
tipo sobresaliente 7748
tipos para diarios 9316
tipos para juego de damas 4699
tipos para juego del ajedrez 2699
tipos para periódicos 9316
tipos para remiendos 4471
tipos (para) titulares 4471
tipos para títulos 4471, 13869
tipos pegados 2151
tipos publicitarios 4471
tipos quebrados 1939
tipos sueltos 9154
tipo sucio 2753
tipo suelto 6102
tipo usual 1688
tipo volcado 14095
tipo vuelto 14095
tira cómica 3233
tira continua de la bobina 14639
tirada 2852, 7208, 11787
tirada aparte 9516
tirada baja 12336
tirada combinada 3078
tirada con los productos coleccionados 3078
tirada de blanco 5711
tirada de doble producción 13248
tirada (de la) primera 5710

tirada de primera cara 5711
tirada efectiva 9729
tirada en blanco 5711
tira de muñecos 3233
tira de muñequitos 3233
tira (de papel) con comba 12537
tira de papel (continua) 14639
tira de papel para pH 10180
tiradera (de la caja) 4704
tirador 10753, 10945, 11786
tiraje 2852, 7208, 11787
tiraje bajo 12336
tiraje primero 5710
tiralíneas 4711, 11786
tirar 10794
tirar el blanco 10864
tirar impresión 11018
tirilla cómica 3233
tirilla de muñecos 3233
tirilla de muñequitos 3233
tiro corriente 13248
tiro de la cámara 2243
tiro del fuelle 1294
tiro doble 13248
tiro en blanco 9714
tiro primero 5710
titanio 13859
titración 13870
titraje 13870
titulación 13870
tituladora 4467
titular 6905
titular a toda página 1131
titular a todo despliegue 1131
titular en bandera 1131
titular encuadrado 1835
titulares 13869
titulera 4467
titulero 8402
título 6902, 6905, 13862, 13867
título alineado 5903
título a ras con el texto 12348
título a toda página 1131
título a toda plana 1131
título a todo despliegue 1131
título a todo lo ancho de la plana 1131
título corrido 11805
título de cabacera 11805
título de corrido 4019
título dedicatorio 4076
título de fecha 4019
título del capítulo 2298
título del lomo 1050
título del lomo longitudinal 13863
título del periódico 8661
título de sección 2468, 13342
título desplegado 4466
título de un apartado 13342

título en bandera 1131
título encasillado 1835
título entrado 3863
título grande 4466
título no centrado 12348
título no sangrado 12348
título parcial 12060
título por venir 6917
título principal 6243
título segundario 13342
título sin sangría 5903
tixotropía 13737
tobera pulverizadora 12929
toberas 13348
todo columna, a ~ 155
todo despliegue, a ~ 155
todo el ancho de la página, de ~ 440
todo extensión, a ~ 155
todo plana, a ~ 155
tolerancia 13880
tolueno 13881
toluol 13881
tolva 6254
tomada 6826
toma (de la tinta) 14011
toma de tinta 7401
tomador 4841, 12040
tomador automático 909
tomador de pliegos 5924
tomo 10011
tonalidad 13898
tonalidad cromática 7066
tonelada larga 8350
tono 13887
tono apagado 14292
tono continuo, de ~ 3433
tono matizado 14292
tono medio 8920
tono no reticulado, de ~ 3433
tonos neutros 9295
tope 2032, 6697
torcedura 13925
torcer 1820, 14545
tórculo 3525
torneadora 7938
tornilladores de ajuste automático 5158
tornillo 1714, 11998
tornillo alimentador 224
tornillo avellanado 3663
tornillo de ajuste 223
tornillo de ajuste del tintero 3212
tornillo de apriete 2874
tornillo de banco 14462
tornillo de bloqueo 2874
tornillo de cabeza avellanada 3663
tornillo de cabeza hexagonal 2296
tornillo de cabeza ovalada 9743
tornillo de movimiento 223
tornillo de regulación 223
tornillo micrométrico 8902
tornillo moleteado 7831

tornillos de libros de hojas cambiables 1429
tornillos del registro circunferencial 2858
tornillos graduadores del tintero 4838
tornillo sinfín 14852
tornillos reguladores del tintero 4838, 5158
torniquete (de planchas) 14709
torno 7938
toroide magnético 8504
toro magnético 8504
torre del papel 9953
torreta giratoria 14105
torsión 13925
tostado 2074
toxicidad 13933
trabajador 14839
trabajador de poco fiar 14336
trabajar fuera de las horas reglamentarias 14840
trabajo aburrido 6635
trabajo adicional 5410
trabajo a la cinta continua 844
trabajo a la cinta sin fin 844
trabajo a la correa sin fin 844
trabajo a mano 6851
trabajo a piezas 10321
trabajo a turnos 12297
trabajo comercial 7680
trabajo de desfibrado 1266
trabajo de emergencia 11818
trabajo de impresión 10773
trabajo de línea 8205
trabajo de obra pacotilla 7680
trabajo de oficina 2912
trabajo de remendería 7680
trabajo de remiendo 7680
trabajo en colores 3197
trabajo en mármol 13206
trabajo especulativo 4561
trabajo estadístico 13520
trabajo inferior 1804
trabajo innecesario 5499
trabajo inservible 1804
trabajo inútil 5499
trabajo manual 6851
trabajo por equipos 12297
trabajo teórico 4561
trabajo urgente 11818
trabazón de las fibras 5590
tracción 4701
traducción 13982
traducción algorítmica 401
traducción autorizada 883
traducción literal 8247
traducción mecánica 8467
tráfico 13950
tragacanto 13951
traje de trabajo 14831
trama 6145, 14668

xilotipo 14799

ya en venta 7421
yarda 14907
yarda cuadrada 12975
yarda cúbica 3805
yerro de bulto 8556
yerro de colimación 3080
yerro de imprenta 3315
yerro del autor 888
yerro del cajista 8245
yerro tipográfico 14235
yeso 3774, 6738
yeso rojo 11867
yodato de sodio 12699
yodato de sosa 12699
yodato sódico 12699

yodo 7552
yodobenceno 7561
yodoetano 5314
yodometría 7563
yodurar 7559
yoduro 7550
yoduro amónico 535
yoduro crómico 2804
yoduro de amonio 535
yoduro de cadmio 2143
yoduro de cromo 2804
yoduro de etilo 5314
yoduro de mercurio 8796
yoduro de plata 12431
yoduro de potasa 10656
yoduro de potasio 10656
yoduro de sodio 12700

yoduro de sosa 12700
yoduro mercurioso 8796
yoduro potásico 10656
yoduro sódico 12700
yunque 696
yunque de cosedora 13170
yute 7720
yuxtapuesto 7723

zángala 2030
zapatilla 12555
zapato 6958
zaraza 2206, 2725
zedilla 2495
zinc 14940
zincografía 8159

zincotipia 8159
zirconio 14964
zócalo 1272, 1278, 9136
zócalo de madera 14790
zona de blancos 6992
zona de mancha 10819
zona de negros 12223
zona de sombras 12223
zona interfacial 7492
zona no impresa 2906
zonas de rechazo del agua
  14601
zonas descargadas de
  presión 1710
zonas no impresas 9401
zopilotes 1493

ITALIANO

abbagliamento 6425

abbassamento della
tensione superficiale
13442

abbassare uno spazio 11063

abbecedario illustrato
10313

abbigliamenti 898

abbigliamento duro 6873

abbigliamento eccessivo
9775

abbonamento 13331

abbonarsi 13330

abbonato 13329

abbozzo 4208

abbreviare 1

abbreviatura 2

abbreviazione 2

abbreviazioni di chimica
2691

aberrazione 5

aberrazione astigmatica
852

aberrazione cromatica 2761

aberrazione locale 14966

aberrazione ottica 9643

aberrazione sferica 12862

abete 5693, 12953

abete del Canada 6961

abile al lavoro 8

abiti da lavoro 14831

abrasimetro 11766

abrasione 10, 12015

abrasivo 14

accaparrare 10355

accaparratore 10357

accapo 7322

accartocciamento 3833

accartocciarsi 3834

accartocciato 4431

accatastare 12997

accatastatore 12998

accecamento 1567

accecato 1561

acceleramento 52

accelerante 53

accelerare 48

acceleratore 53

accelerazione 52

accelerazione angolare 603

accentare 57

accentatura 62

accentazione 62

accento 54

accento acuto 175

accento acuto primario
10781

accento circonflesso 2860

accento grave 6577

accento separato 5838

accento sovrapposto 5838

accessibilità 68

accesso 67

accesso diretto 11231

accesso multiplo 9179

accessori 69

accessorio per grandi
caratteri 4460

acciaiatura 13094

acciaio inossidabile 13012

acciaio resistente alla
corrosione 13012

accoppiamento 3669, 3670

accoppiamento di frizione
6194

accoppiare 7885

accoppiato 7886

accoppiato (con materiale
in fogli) 12274

accoppiatura dei nastri perforati 13583

accordo dei colori 3155

accostamento dei colori
3155

accottimare 5472

accumulatore 83, 84

accumulo di depositi d'inchiostro sui rulli 11601

accumulo d'inchiostro
10345

accumulo di patina sulla
lama 12597

accuratezza 85

acetaldeide 86

acetato 88

acetato d'alluminio 472

acetato d'amile 563

acetato d'ammonio 513

acetato d'etile 5300

acetato di butile 2127

acetato di cellulosa 2509

acetato di isoamile 563

acetato di piombo 7972

acetato di polivinile 10566

acetato di potassio 10623

acetato di rame 3498

acetato di sodio 12669

acetato d'uranile 14367

acetato d'uranio 14367

acetato etilico 5300

acetato polivinilico 10566

acetato potassico 10623

acetato sodico 12669

acetilcellulosa 2509

acetoacetoanilide 101

acetone 102

aciculare 112

acidificare 131

acidificazione 130

acidimetria 132

acidità 133

acido 113, 114

acido abietico 6

acido acetico 95

acido acetico cristal-
lizzabile 6418

acido acetico glaciale 6418

acido acrilico 157

acido arabico 745

acido bicloroacetico 4292

acido borico 1798

acido cianico 3909

acido cianidrico 7105

acido citrico 2864

acido cloridrico 7102

acido concentrato 3355

acido cromico 2785

acido debole 14634

acido fenico 10189

acido fluoridrico 7106

acido formico 6078

acido forte 13297

acido fosforico 10205

acido ftalico 10284

acido gallico 6289

acido glicolico 6488

acido grasso 5493

acido grasso insaturo 14341

acido grasso saturo 11880

acido idrossiacetico 6488

acido iodico 7549

acido ipocloroso 7148

acido lattico 7869

acido libero 6154

acido linoleico 8217

acido linolenico 8218

acido monobasico 9074

acido muriatico 7102

acido nitrico 9366

acido nitrico fumante 6247

acido nitroso 9380

acido oleico 9579

acido ossalico 9797

acido palmitico 9880

acido pirogallico 11085

acido prussico 7105

acidoresistente 145

acido salicilico 11851

acido silicico 12414

acido solfidrico 7117

acido solforico 13389

acido solforoso 13394

acido stearico 13087

acido tannico 13573

acido tartarico 13610

acido tioacetico 13720

acqua ammoniacale 533

acqua bianca 14730

acqua di bagnatura 3980

acqua di cristallizzazione
14591

acqua di Javel 7649

acqua di Labarraque 12696

acqua di rubinetto 13600

acqua di sottotela 14730

acqua distillata 4479

acqua dolce 12746

acquaforte 802, 9366

acquafortista 5269

acqua ossigenata 7115

acqua ragia minerale 14727

acqua regia 734

acquatinta 735

acquatinta allo zolfo 13384

acquatinta allo zucchero
13358

acquerello 14569

acquisto a contanti 2423

acromatico 109

acromatopsia 3123

acutezza 174

adattamento 179, 180

adattamento all'oscurità
3987

adattamento delle
condizioni di lavoro
all'uomo 7067

adattare 178

adattatore 180, 7810

addendo 184, 872

addensamento di patina
12597

addetto ai preventivi 5265

addetto al carico delle
bobine 11369

addetto alla preparazione
delle lastre 10447

addetto alla rotativa 11696

additivo 188, 4942

additivo diluente 11345

additivo per inchiostri 4559

additivo per inchiostro
3317

additivo scivolante 12562

addizionatore 183

addolcitore d'acqua 14602

aderente 1080

aderenza 204

aderenza superficiale 1598

aderire 203

adesione 205

adesivo 213

adesivo a caldo 7054

adesivo hot-melt 7048

adesivo per carta in
soluzione gommosa 9904

adesivo per montaggio di
pellicole staccabili 13284

adesivo pressosensibile
10768

adsorbato 235

adsorbente 236

adsorbimento 237

aerografo 314

aerosol 284

affiggere 13137

affilare 6636

affilatoio 6647

affilatrice per lame 7826

affilatrice per racle 4529

affine al grasso 9582

affinità 287

affioramento 5852

affisso 10609

aforisma 701

aforismo 701

agalite 293

agar-agar 294

agata 296

agave 300

agente 11474

agente (additivo) di cottura
1701

agente antiadesivo 657

agente antipellicolare 689

agente antiscartino 686

agente antischiuma 660

agente bagnante 14699

agente di bagnatura 14699

agente di dispersione 4452

agente diluente 5384

agente di sbianca 1544

agente disperdente 4452

agente d'ispessimento 1681

agente emulsionante 5128

agente tensio-attivo 13428

anello di guarnizione a
    tenuta 6318
anello di regolazione 2931
anepigrafo 586
angolazione del retino
    11973
angoli arrotondati 11725
angolo 3578, 8995
angolo del cono piegatore
    6068
angolo di campo 599
angolo di contatto 3405
angolo di fase 10185
angolo d'incidenza 595
angolo di novanta gradi
    11605
angolo di riflessione 597
angolo di rifrazione 598
angolo di taglio 3890
angolo d'ottone 1862
angolo retto 11605
ångström 601
anidride 605
anidride acetica 98
anidride basica 1199
anidride carbonica 2315
anidride di acido 115
anidride solfocarbonica
    2316
anidride solforica 13397
anidro 607
anilina 613
anima d'acciaio dei cilindri
    rotocalco 1192
anima deformata 3783
anima (della bobina) 11363
anima del rullo 11649
anima ovalizzata 3783
anima slittante 12572
anione 627
anisotropia 629
annalina 630
annata 14913
annerimento da calandra
    1475
annerire con fuliggine
    12638
anno di pubblicazione
    14913
annotare 633
annotazione 635
annotazioni marginali ri-
    entranti nel testo 3866
annuario 14912
annullare 5232
annunci 638
annuncio 636
annunziare 245
annunzio a piena pagina
    6237
annunzio di matrimonio
    8643
annunzio (pubblicitario)
    246
annunzi pubblicitari
    classificati 2881
anodo 641
anonimo 652
anopistografo 653

antialo 666
anticipo 7043
anticloro 658
antiessicante 659
antilogia 654
antimonio 669
antiossidante 674
antipellicola 689
antiquario 675
antiscartinatura 10699
antiscartino secco 4823
antisiccativo 659
antivelo 662
antracene 655
anziano 1799
aperto alle estremità 10061
apertura 9107
apertura del'obiettivo 8043
apertura del retino 11974
apertura di diaframma
    media 4218
apertura di diaframma per
    accentuare le ombre
    12224
apertura di maglia 8813
apertura di sfogo 357
apertura effettiva 4981
apertura relativa 11430
apertura relativa dell'ob-
    iettivo 14830
apostilb 707
apostrofo 708
appannamento 1619, 13606
apparato estrattore 5398
apparecchio 4236
apparecchio antiscartino a
    spruzzo 688
apparecchio a ponte 9759
apparecchio ausiliario 942
apparecchio contatore 3653
apparecchio da laboratorio
    7856
apparecchio di Fenchel
    5538
apparecchio di Finch 5667
apparecchio di misura
    6334, 8847
apparecchio di misura
    della pre-tensione 10775
apparecchio di Mullen 9167
apparecchio di prova 13662
apparecchio di riscalda-
    mento 6920
apparecchio di sabbiatura
    334
apparecchio (fotografico) a
    rullo di pellicole 11657
apparecchio fotografico
    metallico 433
apparecchio ingranditore
    5202
apparecchio lettore 11265
apparecchio orrizontale
    7040
apparecchio per
    determinare la flessi-
    bilità 1324

apparecchio per
    determinare la
    resistenza all'abrasione
    11766
apparecchio per
    determinare la
    resistenza alla
    cordonatura 3719
apparecchio per
    determinare la
    resistenza alla flessione
    1324
apparecchio per dosare la
    luce 8118
apparecchio per inchiodare
    9228
apparecchio per incollatura
    a mano 10030
apparecchio per la deter-
    minazione
    dell'assorbimento 42
apparecchio per la misura
    della resistenza a
    trazione 13645
apparecchio per la misura
    del sfregamento del'in-
    chiostro 11762
apparecchio per la misura
    del tempo d'es-
    siccamento 11762
apparecchio per la misura
    ottica della profondità
    9654
apparecchio per la prova di
    scoppio 2097
apparecchio per la
    selezione elettronica
    11910
apparecchio per la
    selezione elettronica dei
    colori 3177
apparecchio per misurare
    l'abbagliamento 6426
apparecchio per misurare
    l'adesione 1721
apparecchio per misurare
    la durezza 6872
apparecchio per misurare
    la profondità d'incisione
    5277
apparecchio per misurare
    la resistenza alla lacera-
    zione iniziale 7324
apparecchio per misurare
    la rigidità 13154
apparecchio per misurare
    la solidità alla luce 3143
apparecchio per misurare
    la tendenza alla raschia-
    tura 11960
apparecchio per prova di
    brillantezza 6454
apparecchio per prova di
    lacerazione 13618
apparecchio per prova di
    liscio 12646
apparecchio per prova di
    perforazione 11051

apparecchio per prova di
    permeabilità all'aria 345
apparecchio per prova di
    resistenza alle doppie
    pieghe 5980
apparecchio per prova di
    resistenza allo scoppio
    9167
apparecchio per prova di
    trazione statica 13072
apparecchio per prova di
    trazione statica e
    dinamica 13065
apparecchio per prove di
    resistenza allo strappo
    10302
apparecchio per prove di
    sgualcimento 3780
apparecchio per prove di
    stampabilità 10797
apparecchio per ri-
    produzione 10882
apparecchio Schopper per
    prove di piegatura 11921
apparecchio sospeso 9759
apparecchio verticale 14450
apparire 714
appena pubblicato 7719
appena uscito 7719
appendice 716
appianare la composizione
    10393
appianare la forma 10393
appiattimento 5769
appicciosità ritardata 289
applicare brachette 6690
applicare el mordente per
    la doratura 6420
applicare fondi 13845
applicare il strato rico-
    prente 2980
applicare polvere di
    magnesia 2597
applicare tinte benday
    13845
applicazione della colla (al
    dorso) 6477
applicazione di uno strato
    protettivo 13912
appoggiamano 6841
appoggio della grappa
    13173
appoggio di chiusura 13170
apprendista 721, 7234,
    13956
apprendista che ha
    terminato due terzi del
    periodo di apprendistato
    14160
apprendista compositore
    722
apprendista compositore
    anziano 723
apprendista incaricato di
    tirare le prove 6279
apprendista tipografico
    10809
apprendista tipografo 10809
apprendistato 724

apprendista traslocado 14103
aprire 14331
aprire la forma 4750
aquatone 737
arabeschi 744
arancio molibdato 9061
archivio 5611
archivio della redazione 9101
archivio principale 8655
arco 12874
area del ciclo d'isteresi 759
area d'esplorazione 2906
area di comodo 13643
area non stampata 2906
aree d'ombra 12223
areometro 761, 7120
areometro di Baumé 1233
argento 12424
argento in polvere 12433
argento libero 6163
argentometro 763
argilla 2711
argomento 764
aria compressa 3322
aria d'aspirazione 13349
armadio 2137
armadio condizionato per flani 7078
armadio di asciugamento 4794
armadio essiccatore 4786
armadio per marginatura 6263
armerista 774
arrestare 12359, 13793
arresti 6115
arresto automatico della pompa 12330
arresto automatico del programma 919
arresto del flusso d'inchiostro 3892
arresto della cremagliera della unità 14317
arresto posteriore regolabile 219
arricciamento 3833
arricciatura 3026
arrotolamento del nastro 14868
arrotolatore 14747
arrotolatore pope 10576
arrotondamento degli angoli 11723
arrotondamento del dorso 1046
arrotondare il dorso e imprimere la piega 11827
arte ad acqua forte 802
arte conservatrice dell'arte 781
arte della legatoria 801
arte della stampa 803
arte del libro 804
arte d'incisione all'acqua-forte 802
arte impresoria 803

arte tipografica 803
articoli di cancelleria 13076
articoli occorrente per scrivere 13076
articolo 787
articolo di fondo 7989
articolo di magazzino 13188
articolo di riempimento 5622
articolo espletivo 5622
articolo illustrato 7163
arti grafiche 6565
artista litografico 8256
ascendente 813
ascensione capillare 2283
ascensore 8093
ascissa 22
asciugamento 4783
asciugare 14756
asciugare ventilando 5463
asciugatrice-smaltatrice 4777
asciugatura 4783
asfalto 821
asfalto polverizzato 823
asfalto, sotto ~ 14266
aspettare in attesa di materiale 13558
aspirapolvere dal nastro 14643
aspiratore a ventosa 6092
aspiratori 13348
aspo di bobinatore senza mandrino 12870
asportare col bulino 11736
asportare con la racla 14755
asportare dai rulli l'eccesso di inchiostro 12281
asprezze 6884
asquaio di sviluppo 4228
asse 963, 968, 10740, 12225
asse a manovella 3700
asse dell'ascissa 22, 967
asse dell'ordinata 966
asse dell'y 966
assegni a marcazione magnetica 8516
assegni numerati con inchiostri magnetici 8516
assegno bancario 2695
asse ottico 9644
assicurare in macchina 11067
assistente del correttore 3546
assistente di macchina 8456
assorbenza 32
assorbimento 34
assorbimento d'acqua 14564
assorbimento della carta 7327
assorbimento della luce 8107
assorbimento dell'in-chiostro 7327
assorbimento d'olio 9550

assorbimento d'umidità 9037
assorbimento (ottico) 35
assorbire 31
assortimento completo (di caratteri) 6103
assortimento della carta 12785
assortimento di caratteri 2640
assortimento di matrici 12165
assortitore di minuzzoli 2730
assottigliare 12234
asta ascendente 813
asta capillare 6746
asta del regolo del calamaio 7383
asta di lettera 13109
asta discendente 4201
asta orizzontale 14475
asta trasversale 1135
asterisco 850
astigmatismo 853
astralon 855
astuccio 12561
astuccio accassetto e manicotto 12291
astuccio con interno scorre-vole 12291
astuccio pieghevole 2378, 5975
astuccio pieghevole con chiusura a lembi aggraffabili 2882
astuccio pieghevole con chiusura automatica a incastro del fondo 2385
astuccio pieghevole con falde laterali ridotte 2386
astuccio pieghevole con falde rientranti alternate 2389
astuccio pieghevole con falde rientranti opposte e finestra trasparente 2390
astuccio pieghevole con falde superiori ridotte e inferiori complete 2391
astuccio pieghevole con striscia di cartone a strappo 2384
astuccio pieghevole tipo cofanetto in un sol pezzo 2387
atlante 861
atlante tascabile 10507
atmosfera 863
atmosfera fisica 13021
atomizzatore 866
attaccaticcio (del blanket) 13526
attacco alle cinghie d'intro-duzione 13585
attacco di colata 7657
attacco Edison 4964
attendibilità 11440

attenersi esattamente al manoscritto 6000
attenuazione 867
attestato della tiratura 871
attinico 159
attinometro 164
attitudine alla piegatura 10482
attitudine alla stampa della carta 10796
attivazione 167
attività all'interfacie 7493
attività superficiale 13429
attraverso tutta la pagina 440
attrazione 5414
attrezzatura per legatura a filo termosaldante 13744
attrezzi (diversi) 13405
attualità 6223
aumentare il contenuto di un libro inserendo altro materiale illustrato 6560
aumento della larghezza del nastro 14645
aumento dell'avvolgimento 9099
aumento dell'esposizione 10502
aumento di prezzo 5395
aureola 6754, 12976
aurine 877
autobiografia 891
autocarica 1788
autocentratura 12079
autocodice 895
autocombustione 12915
autocopertina 7474
autocromista 13746
autoeditore 885
automaschera 7475
automatizzazione 929
automazione 929
autopaster 5927
autoplate 932
autopolimerizzazione 934
autore 881
autore-editore 885
autossidazione 940
autotipia 6774, 10233
autotipia a retino grosso 2978
autotipia a trama grossa 2978
autotipia con fondo eliminato 9721
autotipia con fondo sfumato 9720
autotipia con retinatura grossolana 2978
autotipia decontornata 9721
autotipia di alte luci 6987
autotipia duplex 4877
autotipia quadrata 12963
autotipia schermata 9721
autotipia schermata con vignette 9720
autotipie a tre colori 13749

autotipie tricromiche 13749
autrice 881
avaria 1880
avere la mano svelta 8092
aviso a piena pagina 6237
avvelenamento da cromo
   2780
avvelenamento da piombo
   8000
avviamento 1023, 9771,
   14281
avviamento chimico 8729
avviamento con rilievo in
   creta 8729
avviamento di rilievo in
   creta 2604
avviare 8564, 13054, 13788
avviare a colpi (una
   macchina) 7684
avviare a piccoli impulsi
   7684
avviatore 13059
avvicinamento 5719
avvicinare 2943
avvicinato otticamente 9658
avvisare 245
avviso 247
avviso al lettore 9444
avviso per il lettore 9444
avvitare 582, 1715
avvolgimento ripiegato
   5992
azione 12232
azione capillare 2279
azocoloranti 973
azotato 9361
azoto 9374
azzerare 11494
azzurro di Berlino 1337
azzurrante ottico 9646
azzurro di bromofenolo
   1955
azzurro di indaco 7288
azzurro milori 8945

bacchetta di sgancio 7769
bachelite 1082
bacinella di sviluppo 4230
bacinella per gli acidi 5273
bagassa 1067
bagnabilità 14696
bagnatura 3977, 14698
bagno 1223
bagno d'acido 116
bagno d'argento 12425
bagno d'arresto 13209
bagno di coloritura 3121
bagno di cromatura 2773,
   2774
bagno di fissaggio 5736
bagno di fissaggio acido
   124
bagno di fissaggio acido
   induritore 128
bagno di marmorizzazione
   8626
bagno d'incisione 5274
bagno d'incisione bio-
   degradabile 1436

bagno d'indurimento 6864
bagno di ramatura 3500
bagno d'irruvidimento 6549
bagno di saturazione 11885
bagno di sviluppo 4224
bagno di viraggio 13899
bagno elettrolitico 5034
bagno fissatore 5736
bagno galvanico 5064
bagnomaria 14568
bagno rivelatore 4224
bagno sbiancante 1545
bagno termostatico 13699
balestra 8055, 12549
balla 1090
ballerino filigranatore 3985
balsamo 1107
balsamo del Canada 2255
bambù 1109
banca di dati 4006
banco 1311
banco da carta 1120
banco da composizione
   3284, 4861, 11182
banco da composizione con
   due casse 14733
banco di comando 3395
banco doppio 4612
bancone 1311
bancone da composizione
   11182, 11229
bancone doppio 4612
bancone per forme 6067
banconota 1126
banda 1110
banda d'assorbimento 36
bandierina 5741
baracca 10315
barattolo d'inchiostro 7330
barba 4065
bardotto 8571
bario 1142
baritina 1182
barra 10327
barra appoggio lastra 13003
barra delle pinze 6652
barra diagonale 588
barra diagonale pony 10572
barra di bloccaggio (del
   mecanismo di fissaggio)
   2872
barra di bloccaggio (del rivestimento) 2871
barra di distribuzione 4499
barra di frazione 12767
barra di guida della
   scomposizione 4499
barra d'inversione 588
barra di rovesciamento 588
barra di tensione del
   blanket 1527
barra oblique 12767
barra per volgere 588
barra portapinze 6652
barra premicarta 2873
barra trasversale 3749
barre collettrici 2101
barre regolabili del portalastra 2370

basamento 1272
basamento inferiore 13341
base 1184
base a nido d'ape 7033
base del blocco degli stantuffi 10377
base del blocco dei pistoni
   10377
base di montaggio 9136
base di piombo 7975
base monopigmentata 1185
base per il montaggio 9136
basicità 1204
bassa cassa 8379
bassa densità, a ~ 13709
bassa resistenza, a ~ 6229
bassa resistenza alla piegatura 10573
bassa tensione 8389
bassa tiratura 12336
bastoncino 11868
bastone 11868
battenti di piega 5981
battesimo 6341
battitoio 10394
battuta 7785
bava 6747
bava di fonderia 1248
bava (metallica) 2090
bave 6743
bazzana 1183
bazzana colorata 3138
becco 4543
becher ad efflusso 4983
benzene 1326
benzina 6320
benzina a punto di
   ebollizione speciale
   11897
benzina per pulizie 2894
benzina pesante 6941
benzina senza piombo
   14330
benzina solvente 12776
benzina speciale 11897
benzoato di sodio 12675
benzoato sodico 12675
benzolo 1326
benzotriazolo 1336
betulla 1438
biacca all'ossido di zinco
   14951
biacca di Krems 5749
bianca 1573
bianchezza 14719
bianchi 12811
bianchi di testa e di coda
   7974
bianchi tipografici 1531,
   6262
bianco 1500, 14711
bianco al piede 6012
bianco brillante 6456
bianco cinese 14951
bianco d'allungamento 9014
bianco d'antimonio 671
bianco del dorso 985
bianco di coda 13955
bianco diluente 9014

bianco di piombo 5749
bianco di riduzione 5383
bianco di Spagna 12817
bianco di spalla 1247
bianco di testa 6889
bianco di titanio 13860
bianco di uscita 13955
bianco e nero 1463
bianco fisso 1156
bianco interno dell'occhio
   3651
bianco misto 13840
bianco opaco 14951
bianco per presa pinze
   6656
bianco per ritocchi 10905
bianco satin 11878
bianco tipografico 6735
bibasico 4288
bibliocasta 1749
bibliofilo 1364
bibliografia 1361
bibliografo 1360
bibliomane 1362
biblioteca 8083
biblioteca circolante 8036
biblioteca di programmi
   10917
biblioteca particolare 10872
biblioteca privata 10872
biblioteca reale 11745
biblioteca universitaria
   14326
biborato di sodio 12680
biborato sodico 12680
bicarbonato d'ammonio 517
bicarbonato di sodio 12676
bicarbonato sodico 12676
bicchierino di Zahn 14931
bicchierino Ford 6030
bicloruro di mercurio 8792
bicromato 8083
bicromato d'ammonio 526
bicromato di potassio 10648
bicromato di sodio 12688
bicromato potassico 10648
bicromato sodico 12688
bifluoruro d'ammonio 519
biglie 3037
biglie di vetro per
   granitura 6430
biglietti 2356
biglietti augurali 6626
biglietti d'augurio 6626
biglietto da visita 14493
biglietto d'entrada 13799
biglietto di banca 1126
biglietto d'ingresso 13799
biglietto di visita
   commerciale 2105
bilancia 11898
bilancia analitica rapida
   578
bilancia a settore per carta
   11107
bilancia da cenere 818
bilancia da laboratorio
   7855
bilancia di torsione 13926

bilanciato 1085
bilanciere a mano 5933
bilingue 1383
bimano 1391
binari del carro portaforme 1825
binda a vite 7632
binoccolo 1435
biografia 1437
biossalato di potassio 10631
biossalato potassico 10631
biossido di carbonio 2315
biossido di titanio 13860
biossido di zolfo 13387
bisellare 2609
bisellatura 1354
bisello 1344
bismuto 1442
bisolfato di potassio 10632
bisolfato di sodio 12678
bisolfato potassico 10632
bisolfato sodico 12678
bisolfito calcico 2162
bisolfito di calcio 2162
bisolfito di magnesio 8497
bisolfito di sodio 12679
bisolfito sodico 12679
bisolfuro di carbonio 2316
bit 1448
bitartrato di potassio 10633
bitartrato potassico 10633
bit di controllo 2664
bit di parità 10008
bitume 1458
bitume di Giudea 1458, 13497
bitume giudaico 13497
bivalente 1460
blanco al piede 13550
blanket 1509
blanket abbassato 5183
blindatrice 8824
bloccaggio rapido 7460
bloccaggio reciproco 7506
bloccare 8309, 14098
blocco 1583, 8303
blocco degli stantuffi 10376
blocco dei coltelli 7818, 14047
blocco dei pistoni 348, 10376
blocco del calamaio 6108
blocco del libro 1741
blocco di legno 14790
blocco di montaggio 9136
blocco di riempire 5627
blocco note di cassa 2421
blocco per appunti 8771, 13514
blocco per calendario 4018
blocco per fondere filetti 8683
blocco per note 13514
blocco sistematico 10056
blocco-spazio 5627
blu 1642
blu alcalino 415
blu cobalto 3006
blu di Cina 10992

blu di cobalto 3006
blu di Prussia 10992
blu milori 8945
blu Monastral 9065
blu oltremare 14247
blu pavone 10069
blurb 1647
blu riflesso 415
bobina 11354
bobina compatta 13808
bobina danneggiata 13050
bobina deflettrice 4101
bobina deformata 1062
bobina di carta 11354
bobina di fabbrica 8942
bobina (di filo metallico) 14766
bobina di piccolo formato 4385
bobina lenta 12535
bobina molle 8367
bobina ovalizzata 9727
bobina prestampata 10730
bobina scampanata 13631
bobina stretta 4385
bocca di soffiaggio 9445
bocchello del crogiolo 10684
bocchetta del crogiolo 10684
bocchetta di fusione 10684
bocchino 9684
boccola 2103
boccola in bronzo 1961
Bodoniani 4310
bolla 1577
bolle d'aria 311
bollettino della bolsa valori 5343
bollitore per fibre 4352
bollitura 3486
bolo armenico 772
bolo d'Armenia 772
bombato 4550
bombatura 2028
bombatura di lastra 1297
borace 12680
borato di sodio 12680
borchie 1800
bordi molli 12533
bordi non refilati 14264
bordi non tagliati 1716
bordi ondulati 12534
bordo 1792
bordo a filo 5902
bordo centrale smussato 2546
bordo collato 6472
bordo del lato pinza 6911
bordo d'incollatura 6472
bordo inferiore di un bulino 1299
bordo lato squadra 7949
borsa con abiti di lavoro e arnesi 11842
borsa con vesti di lavoro e arnesi 11842
bosso 1839
bottega 14841

bozzetto 9717
bozza a colori 3168
bozza alla spazzola 2012
bozza corretta 2895
bozza da forma ad incavo 7472
bozza d'autore 889
bozza definitiva 10759
bozza del cliché 5190
bozza del cliscè 5190
bozza di macchina 10759
bozza di stampa 10936
bozza in colonna 12573
bozza in nero 1492
bozza senza avviamento della forma 5792
bozza senza errori 2895
bozza su acetato 94
bozze fotografiche 10250
bozze impaginate 9863
bozzettista 3240
bozzetto 4208, 7961, 11390
bozzetto annunzio 253
bozzetto inserzione 253
bozzetto pubblicitario 253
bozzone 12573
bracci (del portabobine) 11355
braccio di torchio 1136
brachetta in tela 2954
braghette di compensazione 3265
braghette di riempimento 3265
brasare 12753
brasare su zoccolo metallico 13458
brillante 6457
brillantezza 6452
brillanza 1909
bromo 1949
bromografo 3406
bromuro 1941
bromuro d'ammonio 520
bromuro d'argento 12426
bromuro d'etile 5303
bromuro di cadmio 2140
bromuro di ferro 5545
bromuro di mercurio 8790, 8791
bromuro di metile 8862
bromuro di potassio 10634
bromuro di rame 3501
bromuro di sodio 12681
bromuro d'oro 6498
bromuro d'oro e potassio 10651
bromuro etilico 5303
bromuro ferrico 5545
bromuro ferroso 5574
bromuro mercurico 8790
bromuro mercuroso 8791
bromuro potassico 10634
bromuro rameico 3501
bromuro sodico 12681
bronzatrice 1969
bronzatura 1968
bronzina 2103
bronzista 1967

bronzo 1956
bronzo d'alluminio 474
bronzo d'oro 6499
bronzo pesto 1965
brossura 1933
bruciatore a gas 6312, 11703
brunire 2081
brunitoio 2079
brunitore 2078, 2079
bruno di Cassel 14384
bruno Van Dijck 14384
buccia d'arancia 9670
buco da oliare 8398
budget 725
bulinare 5182
bulino 2072, 5186, 6578
bulino piatto 5790
bulino quadrato 12967
bulino rigatore 8208
bulino tondo 6531, 12017
bullone 1714
buna 2063
buona copia 5440
buono a stampare 7228, 11286
buono a tirare 7228, 11286
buono manoscritto 5440
busta all'americana 1125
busta a sacchetto 10511
busta a soffietto 6732
busta con bottone a pressione 10744
busta con chiusura a spago 13279
busta con finestra 14749
busta con lembo chiudibile con fermaglio 2878
busta (da lettere) 5209
busta (da lettere) gommata 210
busta delle istruzioni 7630
busta di corrispondenza 3611
busta di posta aerea 340
busta imbottita jiffy 7659
bustrofedone 1818
butadiene 2108
butanolo 2128
butil acetato 2127
byte 2133

cabina di asciugamento 4786, 4794
cacciacunei 12318
cacciavite 12001
cadmio 2139
caduta di potenziale 14510
caduta di tensione 14510
calamaio 7343
calamaio a lama 4534
calamaio a sezioni 4735
calamaio chiuso 5150
calamaio con lama flessibile 14289
calamaio con prenditore 14289
calamaio con rullo prenditore 14289

calamaio con trasferimento pellicolare continuo 9790
calamaio intercambiabile 10585
calamaio portatile 10587
calamaio suddiviso 12898
calandra 2184
calandra a frizione 6193
calandra per flani 8699
calandra per fogli 12261
calandrato in fogli 12262
calandratura 2191
calandratura a caldo 7046, 7053
calce 8145
calce clorurata 2168
calce viva 2171
calcinazione 2153
calcio 2160
calcografia 2594, 3527, 10458
calcografo 3522, 3526, 10435
calcolare il prezzo di composizione 2459
calcolatore ad usi speciali 12823
calcolatore analogico 572
calcolatore a programma memorizzato 13237
calcolatore asincrono 859
calcolatore della composizione 3283
calcolatore della seconda generazione 12041
calcolatore dell'esposizione 908
calcolatore digitale 4359
calcolatore di prima generazione 5708
calcolatore di terza generazione 13726
calcolatore elettronico 3330
calcolatore numerico 4359
calcolatore sincrono 13488
calcolatore universale 6368
calcolatrice 2177
calcolo dei costi 3629
calcolo preventivo 5264
calcomanie 4041
caldaia 7755
caldaia (a vapore) 1697
caldaia per fondere 8770
calendario 2178
calendario a blocco 1590
calendario blocco da tavolo 4210
calendario da sfogliare 1590
calibro 6334
calibro a cavaliere 11599
calibro di spessore 2221
calibro di spessore elettromagnetico 5039
calibro per il controllo dell'altezza del cilindro 3942
calibro per il controllo dell'altezza tipografica 14198

calibro per il controllo dello spessore del rivestimento 9830
calibro per misurare l'altezza dei clichè 1595
calico 2206
calligrafia 2218
calligrafo 2217
calomelano 8793
calore di evaporazione 6928
calore di fusione 6926
calore latente 7930
calore specifico 12828
caloria 2225
cambiamenti nella forma 2619
cambiamento di corpo 2615
cambiamento di giustezza 2616
cambiare 462, 2614
cambia tensione 14512
cambiatore di bobine automatico 5927
cambio della forma 2617
cambio di bobina 11361
cambio strutturale 13300
camera di asciugamento 4791
camera di combustione 12013
camera di correttori 11274
camera oscura 3992
camma 2227
camma cesoiatrice 3882
camma degli aghi di trascinamento 9268
camma del cuneo di riempimento 13976
camma dell'espulsore 7789
camma di comando della lama piegatrice 5926
camma di comando del tamburo raccoglitore 3073
camma per comando delle lame piegatrice 12541
campagna pubblicitaria 267
campanella 5448
campionamento 11856
campionario 11855
campionario dei caratteri 14221
campionario di carte 13456
campionatura 11856
campione 11854, 12834
campione da laboratorio 7858
campione dei caratteri 14221
campione di colore 3190
campione di riferimento 11384
campione prelevato a caso 11235
campione rappresentativo 11475
campo 5602
campo del bordo di stampa 10827

campo dell'ultravioletto 14253
campo d'intensità 5604
campo elettrico 5017
campo elettrostatico 5066
campo magnetico 8517
canale 2620, 2621, 6666
canale analogico 571
canale dal crogiolo 13780
canale d'aria 343
canale d'informazione 7308
canale di raccolta 835
canale duplex a senso unico 6759
canale d'uscita 4139
canale intermedio 13963
canale per matrici a mano 12788
canale per matrici speciali 12788
canale selettore 12071
canaletto 429, 6734
canapa 6963
canapa sisal 12476
canapa sisalana 12476
canavaccio 2269
cancellabilità 5231
cancellare 2899, 5232, 13272
cancellare con matita abrasiva 13203
cancellaria per uso commerciale 2107
candela 2265, 2266
canfora 2251
cannello ferruminatorio 1630
cannocchiale per messa a registro 11419
cannoncino 1242
canovaccio 2269
canto di guida 6695
caolino 2711
capacità assorbente 44
capacità coprente 3675
capacità di assorbimento d'acqua 14565
capacità di memoria 13234
capacità di trasporto 8100
capacità di trattamento 13785
capacità termica 6918
capillare 2278
capillarità 2277
capitello di testa 6894
capitello inferiore 13541
capitello superiore 6894
capitolo 2626
capo (della pagina) 6890
capo macchinista 10762
capo reparto 6041
caposquadra alla rotativa 11700
capoufficio 13422
capoverso 1875, 9972
capoverso allineato con le linee di testo 5908
capoverso al piede di una colonna 14736
capoverso a sommario 6854

capoverso non rientrante 5908
capoverso senza a capo 5908
capovolgere 14824
capovolgere un carattere 14098
cappello 3757, 6903
cappello del cuscinetto 1257
caput mortuum 2300
carattere 2628, 8052
carattere all'aria 4716
carattere allungato 5094
carattere basso 8967
carattere bloccato 14095
carattere danneggiato 1228
carattere (da stampa) 14171
carattere deteriorato 1228
carattere detto di Schwabach 11924
carattere di comando 3452
carattere discendente 4201
carattere di scrittura (a mano) 12012
carattere funzionale 3452
carattere gotico 6528
carattere grassetto 1707
carattere incorniciato 11615
carattere intasato 2753
carattere largo 5358
carattere magro 8111
carattere mobile logoro 14855
carattere mobile schiacciato 14855
carattere neretto 1707, 12095
carattere non allineato 12955
carattere per annunci pubblicitari 201
carattere per composizione a mano 6102
carattere per dattilografia 14213
carattere più grosso al piede che in testa 1808
carattere più grosso in testa che al piede 1809
carattere rialzato 4716
carattere rientrante 9109
carattere sporco 2753
carattere stretto 3364
caratteri all'ora 2642
caratteri arabici 749
caratteri bastardi 1216
caratteri bastoncini 1609
caratteri bastoni 1609
caratteri bianchi su fondo stampato 11556
caratteri Braille 1846
caratteri cavi 3569
caratteri comuni 1688
caratteri contornati 9722
caratteri correnti 1688
caratteri corsivi 7625
caratteri da giornale 9316
caratteri da stampa 10851

caratteri da titoli 13869
caratteri dei pezzi degli dama 4699
caratteri dei pezzi degli scacchi 2699
caratteri delle carte di gioco 2354
caratteri del testo 1688
caratteri di grande corpo 4471
caratteri di legno 1608, 14804
caratteri di occhio mediano 12329
caratteri di testo 1751
caratteri fantasia 5461
caratteri fratti 1939
caratteri greco 6619
caratteri incollati 2151
caratteri in fascio 10315
caratteri in negativo 11556
caratteri in rilievo per i ciechi 1846
caratteri maiuscoli 2287
caratteri mobili 9154
caratteri per affissi 10612
caratteri per cartelloni 10612
caratteri per libri 1751
caratteri per manifesti 10612
caratteri per pollice 2643
caratteri per titoli 4471
caratteri pubblicitari 4471
caratteri schiacciati 14855
caratteristica 5502
caratteristica operativa 9629
caratteristica superficiale (della carta) 1451
caratteristiche 2638
caratteristiche dielettriche 4324
caratteristiche fisiche 10286
caratteristiche meccaniche 8730
caratteristiche spettrali 12840
caratteri stranieri 6038
caratteri superiori 13416
caratteri trasferibili a secco 12074
carboidrati 2301
carbonato 2305
carbonato calcico 2163
carbonato d'ammonio 521
carbonato di calcio 2163
carbonato di ferro 5575
carbonato di magnesio 8499
carbonato di piombo 7976
carbonato di piombo basico 5749
carbonato di potassa 10636
carbonato di potassio 10636
carbonato di rame 3503
carbonato di sodio 12682
carbonato ferroso 5575

carbonato sodico 12682
carbone assorbente 169
carbone attivo 169
carbone da disegnare 2645
carbone di legna 2644
carbone (di salice) per levigare 6642
carbone per lampada ad arco 755
carboni a miccia 3568
carbonio 2304
carboni per lampade 2328
carborundum 2334
carbossimetilcellulosa 2336
carbrotipia 2339
carburo di silicio 12416
cardano 14325
carica 5383, 5623
carica elettrica 5013
caricamento automatico del metallo 917
caricare 8289
caricare e va 8290
carica statica 13066
carica termoindurente 5634
caricatore 8292
caricatura 2392
carico di carta di un vagone ferroviario 2358
carico di rottura 8291
carminio d'indaco 7290
carrello accatastore 8104
carrello d'uscita 4137
carrello elevatore 8104
carrello elevatore a forca 6045
carrello elevatore a forcella 6045
carrello per il trasferimento delle matrici 4148, 5081
carrello per il trasporto delle bobine 11356
carrello per il trasporto di barre di metallo 10328
carrello per trasportare le bobine 13347
carrello portaforma 6075
carrello portarulli 11654
carrello trasportatore 6075
carro 2364
carro d'inchiostrazione mobile 9153
carta 9894, 13182
carta a bassa grammatura 8140
carta abrasiva 17, 5111
carta accoppiata 7888
carta acidoresistente 139
carta adesiva 6715
carta a fibre longitudinali 8335
carta a fibre trasversali 12326
carta al bromuro 1946
carta al clorobromuro 2744
carta al cloruro (d'argento) 2739

carta al cloruro d'argento 6319
carta alla barite 1179
carta all'albumina 374
carta al solfito 13377
carta al solfito lisciata a macchina 8450
carta a mano 6834
carta a mano-macchina 9131
carta anticorrosiva 694
carta antiruggine 694
carta a pressione 11473
carta a prova di vapore 14390
carta asciugante per blocchi 13515
carta asciugata all'aria 329
carta asciugata stesa all'aria 8312
carta assorbente 33, 1622
carta autoadesiva 12076, 12086
carta autocopiante 2325
carta autoricalcante 12081
carta barbata 4066
carta baritata 1179
carta batik 1226
carta batista 2233
carta bianca 1533
carta bianchita al solfito 14729
carta bibbia 1359
carta bicolore 4881
carta bitumata 13608
carta bond al solfito 13374
carta Braille 1847
carta burro 2120
carta calandrata 2190
carta calibrata 7893
carta carbone 2326
carta carbone per una copia 9594
carta caricata 5620
carta catramata 13607, 13608
cartaccia 9573, 14558
carta cerata 9967
carta che spolvera 5882, 6269
carta China 2713
carta chintz 2725
carta cianografica 1644
carta Cina 2713
carta collata 12490
carta collata alla colofonia 11690
carta collata in liscia di macchina 2194
carta collata in pressa collante 2194
carta collata in superficie 14073
carta collata nell'olandese 1264
carta colorata 3137
carta colorata in calandra 2188
carta comune 12005

carta con buona sonorità 13830
carta con collatura animale 625
carta condizionata 3368
carta con due lati feltro 14128
carta con legno 8735
carta con pasta legno 8735
carta con patinatura per colata 2436
carta con stracci 11206
carta con taglio diagonale 591
carta continua 14658
carta copiativa senza carbonatura 2325
carta crema 3712
carta crespata 3729
carta cromo 2820
carta curcuma 14092
carta da burro 2120
carta da buste 5212
carta da cancelleria 10163
carta da coltelleria 3867
carta da copertina accoppiata con adesivo 10026
carta da copia Giappone 7638
carta da copia senza carbone 9382
carta da disegno 4709
carta da filtrare 5660
carta da filtro 5660
carta da giornale per illustrazioni 6787
carta da giornali 9321
carta da imballaggio 14866
carta da imballo 14866
carta da imballo essiccata all'aria 328
carta da impacco 14866
carta da involgere sottile 2291
carta da lettere 8060, 9442
carta d'alluminio 481
carta da lutto 9150
carta da macero 14558
carta damascata 3963
carta d'amianto 812
carta da minuta 12005
carta da modelli (per abiti) 10063
carta da musica 9211
carta danneggiata 2431
carta da parati 14533
carta da pasta meccanica 8735
carta da protocollo 10163
carta d'asbesto 812
carta da scrivere 14879
carta da scrivere di buona qualità 1129
carta da scrivere di qualità superiore 13413
carta da scrivere manila 11219

carta da scrivere
   sopraffina 13412
carta da scrivere velina
   azzurrina 981
carta da scrivere vergata
   7878
carta da scrivere vergata
   azzurrina 979
carta da sigarette 2831
carta da stampa 10838
carta da stampa calco-
   grafica 10457
carta da stampa offset 9528
carta da stampa per ciechi
   1847
carta da stampa per
   edizioni 1770
carta da stampa per
   illustrazioni 6789
carta da stampa rotocalco
   11707
carta da stracci di lino
   8186
carta da taccheggio 14283
carta da trasporto 13966
carta da trasporto lito-
   grafico 13966
carta da vetrofanie 14498
carta da visita
   commerciale 2105
carta d'avviamento 9826
carta dei colori 3129, 3157
carta del lomo 7026
carta di alfa 5254
carta diazo 4284
carta di Cobb 3017
carta dielettrica 4323
carta di magazzino 13188
carta di Natale 2759
carta di pasta chimica da
   legno 2692
carta di prima scelta 10118
carta di puri stracci 11205
carta di qualità inferiore
   per edizioni 8728
carta di recupero 9573,
   14558
carta di rinforzo sul dorso
   1034
carta di risposta 11469
carta di rivestimento 10043
carta di scarica 12163
carta di seconda scelta
   7678
carta di sicurezza 2698,
   8296, 11840
carta di sparto 5254
carta di straccio 11206
carta (di) supporto 1685
carta d'Olanda 4917
carta doppia bitumata
   14308
carta doppia bitumata rinforzata 11428
carta dorata 6512
carta duplex 4881, 14128
carta esente da acidi 127
carta esente da ceneri 817
carta esente da cloro 2743

carta esente di polvere
   5879
carta essiccata con aria 329
carta essiccata in
   macchina 8444
carta fantasia 5460
carta filigranata 14589
carta filtro 5660
carta fine 5679
carta fluorescente 5893
carta foderata di tela 9958
carta fortemente collata
   6875
carta fotografica 10266
carta (fotografica) che non
   necessita di sviluppa
   10837
carta fotosensibile 12117
carta fotosensibile per
   stampa a contatto 6319
carta geografica 8612
carta (geografica) in rilievo
   11444
carta gialla di paglia 14927
carta Giappone 7645
carta giapponese 7645
carta Giappone vellutata
   7644
carta giornale 9321
carta goffrata 5101
carta gommata 6715
carta gommata inarriccia-
   bile 9396
carta gommata inerte 9396
carta granulata 6543
carta grossa 2051
carta imbarcata 4432
carta imitazione cuoio 8016
carta imitazione
   pergamena giapponese
   7173
carta impacco risme 11295
carta impermeabile ai
   grassi 6603
carta impermeabile ai
   grassi non satinata 14299
carta impermeabile (al
   acqua) 14595
carta impregnata 7206
carta inattinica per
   montaggi 6504
carta in bobina 14658
carta incombustibile 5697
carta India 1359
carta indicatore 13667
carta in fogli 5771, 9922
carta inibitore volatile di
   corrosione 14389
carta in piano 5798
carta in rotoli 11373
cartaio 9924
carta iridescente 9112
carta isolante 5009
carta kraft 7840
carta kraft bianca 1539
carta kraft impermeabile
   (al acqua) 14594
carta laccata 14403
carta lineata 11775

carta lino 8185
carta liscia di macchina
   8447
carta lisciata ad umido
   14576
carta lisciata alla calandra
   a frizione 6197
carta lisciata in macchina
   8447
carta lucidata a spazzola
   2001
carta madreperla 9112
carta manila 8605
carta marina 2648
carta marmorizzata 8617
carta martellata 6809
carta metallizzata 8830
carta millimetrata 6573
carta monocolorata 9591
carta monolucida 8449
carta monolucida patinata
   a macchina 2983
carta monopatinata 9590,
   12448
carta murale 14532
carta naturale 14259
carta nautica 2648
carta nera 1488
carta non bianchita 14259
carta non calandrata 14296
carta non collata 14346
carta non lavorata 14335
carta non lucida 8676
carta non refilata 14352
carta non satinata 14300
carta non stagionata 6625
carta offset 9528
carta offset patinata 2984
carta oleata 9554
carta ondulata 3619, 14624
carta opaca 9615
carta opalina 9610
carta ordinaria per
   edizioni 8728
carta ozalid 9815
carta paglia 9006, 13255
carta paraffinata 9967
carta patinata 2986
carta patinata a emulsione
   5133
carta patinata a lama fissa
   12933
carta patinata a rullo 11630
carta patinata a spazzola
   1992
carta patinata brillante
   6459
carta patinata cast-coated
   2436
carta patinata con miscele
   a base di solventi 12774
carta patinata d'ambo i lati
   14157
carta patinata da un lato
   9590
carta patinata due volte
   4586
carta patinata fuori
   macchina 12127

carta patinata in macchina
   8435
carta patinata in pressa
   collante 12493
carta patinata in size press
   12493
carta patinata lucidata a
   spazzola 1998
carta patinata mat 4852
carta patinata non lucida
   8665
carta patinata per es-
   trusione 5412
carta patinata per
   immersione 4390
carta patinata per spalmatura 12933
carta patinata sulle due
   facce 14157
carta patinata su una
   faccia 9590
carta patinata su una sola
   faccia 12448
carta patinata su una
   superficie 9590
carta patinata su uno solo
   lato 12448
carta pelle d'aglio 9597
carta pelle d'uovo 4989
carta per abbigliamento
   14168
carta per accoppiamento
   10043
carta per acquerelli 14573
carta per affisi 10610
carta per album 378
carta per appoggio pagine
   9861
carta per argenteria 12436
carta per asportare 10467
carta per assegni 2698
carta per autotipie 6789
carta per avviamento 8567
carta per avviamento in
   creta 2603
carta per avviamento
   meccanico 2603
carta per avvolgere 14866
carta per avvolgere il pane
   1873
carta per banche 1723
carta per biglietti di banca
   3839
carta per bisguardie 5161
carta per blocchi 13516
carta per bozze 10946
carta per buste 5212
carta per cartamoneta 3839
carta per carta valori 8296
carta per carte geografiche
   8614
carta per cartelloni 10610
carta per carte stradali
   11623
carta per cartografia 8614
carta per cartucce 2395
carta per ciclostilare 8949
carta per condensatori
   elettrici 5009

carta per copertine 3681
carta per copia a contatto 3412
carta per copie dattiloscritte 12051
carta per copie multiple 8599
carta per dattilografia 12051, 14224
carta per decalcomanie 4042
carta per decalcomanie su ceramica 2557
carta per documenti di lunga conservazione 8347
carta per edizioni di lusso 13681
carta per elenchi telefonici 13628
carta per eliotipia 3096
carta per etichette 7851
carta per etichette laccate 14407
carta per (foderare) armadi 12287
carta per foderare di sacchetti 1071
carta per foderare le scatole 1832
carta per fototipia 3096
carta per (franco)bolli 13019
carta pergamena 6431
carta pergamenacea 7175
carta pergamena goffrata 5100
carta pergamena imitata 6444, 7175
carta pergamena vegetale 14421
carta per generi alimentari 6663
carta per ingrandimento 5204
carta per isolamento elettrico 5009
carta per la stampa calcografica 10457
carta per lavori editoriali 1770
carta per le fascette dei sigari 2828
carta per lucidi 13938
carta per macelleria 4781
carta per manifesti 10610
carta per mappe marittima 2649
carta per modelli 10063
carta per monotipia 9092
carta per negativi 9278
carta per offset 9528
carta per partecipazione di nozze 14664
carta per posta aerea 341
carta per programmi 9386
carta per prove 10946
carta per quotidiani 9321
carta per registri 8024

carta per risguardi 1750
carta per rivestimento 8207, 14168
carta per rivestimento interno di casse 2408
carta per riviste 8485
carta per rotocalco 11707
carta per sacchetti 1075
carta per sacchi 1075
carta per salviette 9239
carta per scartino 12576
carta per schizzi 12506
carta per scrivere 14879
carta per stampa in acciaio 13098
carta per tovaglioli 9239
carta per vetrofania 14498
cartapesta 9957
carta piegata in mazette 11165
carta piegata in quinterni 11165
carta pigmento 2331
carta pressata a freddo 3057
carta priva di cariche 817
carta profumata 11917
carta quadrettata 12964
carta rasata 5835
carta registro 8024
carta resistente agli acidi 139
carta resistente agli alcali 420
carta resistente ai grassi 6603, 6605
carta resistente a umido 14693
carta rigata 11775
carta rinforzata (con tessuto) 11426
carta riso 11594
carta rivestita 7888
carta ruvida di macchina 14296
carta sabbia 11861
carta satinata 2190, 5180, 6445
carta satinata colorata 6442
carta scartata 7678
carta secca all'aria 332
carta semimat 4989
carta sensibile 12117
carta senza acido 127
carta senza carbonatura 9382
carta senza ceneri 817
carta senza legno 6681
carta senza pastalegno 2685
carta senza polvere 5879
carta seta per fiori 5872
carta silenziosa 9386
carta siliconata 12418
carta similpelle 792
carta smaltata 5145
carta smalto asciugante 5143
carta smalto sugante 5143

carta smerigliata 17, 5111
carta smeriglio 5111
carta sonora 13830
carta sottile 1359, 5832
carta spalmata 2986
carta spalmata con hot-melt 7049
carta spalmata hot-melt 7049
carta spugnosa 10283
carta squadrata 12965
carta stagionata 8706
carta stagnola 13829
carta stesa 5798
carta straccia 14558
carta stradale 11622
carta suga 1622
cartasuga 1622
carta supercalandrata 13410
carta supporto alla patinatura 1685
carta supporto per carta cerata 14628
carta tela 9958
carta telata 2233, 2957, 8185, 11426
carta termosaldante 6932
carta tipo antico 682
carta tipo cotone 2207
carta translucida 13986
carta trasparente 14005
carta trattata per estrusione 5412
carta uomo/macchina 8606
carta uso manilla 1693
carta uso mano 9131
carta uso patinata 7170
carta uso pelle 792
carta uso pergamena 7175
carta vecchia 9573
carta velina 14858
carta velina crema 3713
carta velina (da imballo) 14867
carta velina francese 6172
carta vellutata 5180, 5845, 14430
carta vergata 7876
carta vergata crema 3711
carta vetrata 5111, 6434
carta voluminosa 5501
carte asciuganti per messaggi pubblicitari 266
carte asfaltate 822
carte bitumate 822
carte da gioco 10477
carte di iuta 7722
carte di recupero da carta da giornale 9322
carte di riporto 13966
carte leggere 8140
cartellina 9918
cartellino 13534
cartellino di lavoro 7681
cartellino di presenza 2921
cartellino per bagaglio 1069
cartellone 10609

cartello reclamistico 12350
carte non assortite 9004
carte per annunci 638
carte stratificate 13253
cartiera 9927
cartina di tornasole 8280
cartina di tornasole per il pH 10180
cartina per il pH 10180
cartoccio 2394
cartografo 2377, 8613
cartolaio 13074
cartoleria 13074
cartolina di corrispondenza 3610
cartolina illustrata 10312
cartolina postale 10605
cartolina postale pubblicitaria 8539
cartonato 7240
cartoncini d'augurio 6626
cartoncini fustellati 4315
cartoncino 2342, 9899
cartoncino a più strati 9174
cartoncino avorio 7627
cartoncino bristol 1921
cartoncino bristol fabbricato su macchina a tavola piana 6126
cartoncino bristol fabbricato su macchina in tondo 8933
cartoncino da cartoline postali 10608
cartoncino da disegno 4705
cartoncino duplex 4875
cartoncino opalino 7627
cartoncino patinato 2981
cartoncino per carte da gioco 10478
cartoncino per partecipazione di nozze 14663
cartoncino per schedario 7270
cartoncino per schede perforate 11041
cartoncino uso bristol 1692
cartone 2393
cartone accoppiato in macchina a più tamburi 14411
cartone a due strati 14150
cartone a mano 8932
cartone a più strati 9191
cartone a più strati incollati 10022
cartone avorio 7628
cartone baritato 1178
cartone compatto 12759
cartone cromo assorbente 2814
cartone cuoio 8013
cartone da cascami di cuoio 8018
cartone da disegno 798
cartone d'amianto 810
cartone da montaggio 9143
cartone da pastalegno 14808

cartone da pasta
  meccanica 8738
cartone d'asbesto 810
cartone di cellulosa 11027
cartone di pasta bruna
  mista 1977
cartone di pasta chimica
  2687
cartone di pasta meccanica
  bruna 1975
cartone di polpa chimica
  2687
cartone di rinforzo 13150
cartone giallo di paglia
  14927
cartone grigio 2726, 9301
cartone grigio rivestito in
  bruno 1974
cartone monolucido 8449
cartone multistrato 9174
cartone oleato 9553
cartone ondulato 3619
cartone ondulato a due
  onde 4598
cartone ondulato ad una
  onda 4605
cartone ondulato doppio
  4605
cartone ondulato doppio-
  doppio 4598
cartone ondulato semplice
  12455
cartone ondulato triplo
  4598
cartone paglia 9006, 13254
cartone patinato 2981
cartone patinato in
  macchina 8434
cartone per astucci pieghevoli 5974
cartone per bicchierini
  3816
cartone per cartellini 13535
cartone per cartonagi 1831
cartone per coppette 3816
cartone per etichette 13535
cartone per flani 5848
cartone per formatura 1659
cartone per imbutitura
  1659
cartone per isolamento
  elettrico 5009
cartone per raschiare 11952
cartone per rilegatura 1408
cartone per scatole 1831
cartone per sottobicchieri
  di birra 2979
cartone pressato 10756
cartone presspan 10756
cartone rifinito sui due lati
  14150
cartone rifinito sulle due
  facce 14150
cartone rigido per
  pubblicità 1534
cartone rivestito 8170
cartone siliconato 12411

cartone supporto per la
  protezione di cibi
  congelati 1189
cartone telato 2956
cartone triplex 14054
cartonificio 1662
cartoteca 2352, 5619
casa editrice 11005
cascio 4063
caseina 2403
casella del titolo 9889
cassa 1827
cassa alta 14355
cassa aspirante 13351
cassa a trafori 3703
cassa bassa 8379
cassa californiana 2209
cassa con strice in legno
  2419
cassa d'afflusso 6898
cassa d'afflusso a gravità
  6580
cassa da filetti 11774
cassa da numeri 5608
cassa degli emboli 14381
cassa delle valvole d'aria
  14381
cassa del piombo di ri-
  fondere 6958
cassa di regolazione 11422
cassa di riserva 6117
cassa di sgocciolamento
  4695
cassa doppia 4585
cassa esaurita 435
cassa in cartone ondulato
  3620
cassa per composizione
  2396
cassa per frazioni 6142
cassa per la spaziatura
  1139
cassa per lettere accentate
  56
cassa spagnuola 12813
cassa spolverizzatrice 4896
cassa sporca 6098
cassa tripla 14017
cassa troppo piena 1140
cassetta 2429
cassetta a trafori 3703
cassetta di scomposizione
  8479
cassetta per polvere 4896
cassetta posteriore della
  macchina fotografica
  2240
cassettina per caratteri
  14205
cassettino 1829
cassettino degli accordi
  6958
cassettino dei refusi 6958
cassettino del diavolo 6958
cassettino di riserva 13448
cassettino per spaziatura
  1139
casso 4063
castagno 2701

castello 2414
catalizzatore 2463
catalogo 2462
catalogo d'editore 13185
catalogo dei caratteri 14221
catalogo dei libri di fondo
  13185
catalogo di vendita
  pubblica 11847
catena 14544
catena di dati chiusa 2091
catena d'uscita 4138
catione 2483
catodo 2480
catodo fotoelettrico 10214
catrame di carbone 2972
cattiva critica 1063
cattiva recensione 1063
caucciù 1509, 11747
caucciù clorurato 2741
caucciù sintetico 13496
cavallo di forza 7044
cavallo-vapore 7044
cava per chiavetta 7786
cavarighe 3295
cavitazione 2494
cavo bowden 1821
cavo coassiale 3003
cediglia 2495
cedola 3671
cella 7331
cella al selenio 12073
cella di memoria 2498
cella fotoelettrica 10215
cellofan 2501
cellula fotoelettrica 10215
celluloide 2507
cellulosa 2508
cellulosa al cloro-soda
  12663
cellulosa tecnica 2686
cementare 2525
cemento armato 11425
censura di stampa 2526
centigradi 2530
centiguide 4068
centimetro 2532
centimetro cubo 3799
centipoise 2533
centoventottesimo, in ~
  7079
centrare 2551
centratore e aggiustatore
  automatico 11093
centrifuga 14709
centrifugo 2552
centro focale 5950
centro ottico 9648
cera d'api 1288
cera di carnauba 2360
cera di lignite 9095
cera di paraffina 9969
cera di Spagna 12027
ceralacca 12027
cera microcristallina 8894
cera minerale 9818
cera montana 9095
cerchio 2835
cerchio cromatico 2762

ceresina 9818
cernita degli stracci 11212
cernita della carta 12785
certificato 2569
certificato della tiratura
  871
cesellatore 2656
cesello 2731
cesio 2572
cesoia 12242
cesoia a mano per cartone
  1666
cesoia circolare 2848
cesoia per cartoni 2345
cesoia per interlinee 7978
cessare il lavoro 3885
cesto delle matrici 2148
cesto portamatrici 2148
chassis telaio 10442
cheque 2695
cherosene 7751
chetone 7753
chiacchierare sulle
  faccende del mestiere
  13565
chiarezza 8124, 14380
chiaroscuro 2707
chiave 7759, 7762
chiave a boca 9620
chiave di serraggio 14872
chiave di serraggio rapido
  2056
chiave inglese 9072
chiave per lastra 10444
chiave per serrature di
  chiusura 11170
chiave per viti 14872
chiave per viti Allen 428
chiave regolabile 9072
chiavetta 7760
chiavistello d'arresto della
  cremagliera 13219
chiazza 8473
chiazza d'inchiostro 1620
chiazzatura 1649, 9114
chilo 7797
chilogramma 7797
chilogrammo 7797
chilowatt 7799
chilowattora 7800
chimicamente puro 2684
chimigrafia 2694
chinone 11161
chiodatura 9226
chiodi di montaggio 1603
chiodo 9225, 13527
chips 2729
chiudere 8308
chiudere le righe 7714
chiuso a nastro adesivo
  13580
chiusura a vite 11999
chiusura lampo 14962
ciabattino 1806
ciano 3908
cianografia 3917
cianotipia 1642
cianuro d'argento 12428
cianuro di potassio 10647

cianuro di potassio
   rameico 3530
cianuro di rame 3507
cianuro di sodio 12687
cianuro potassico 10647
cianuro sodico 12687
cibernetica 3920
cicero 2827
ciclo 3921
ciclo d'isteresi 7151
cicloesanone 3925
ciclo in lavoro libero 4986
cifra 4356, 5607
cifra binaria 1448
cifra cancellata 2263
cifra con barra diagonale
   2263
cifra con cancellatura 2263
cifra di controllo 2665
cifra marginale 11800
cifre allineate al piede e in
   testa 8206
cifre arabe 747
cifre arabiche 747
cifre discendenti 6853
cifre romane 11671
cilindri essiccatori 4730
cilindro abbigliato 9824
cilindro a coltelli 7822
cilindro a scanalature 6669
cilindro asciugafeltro 5532
cilindro ballerino 3985
cilindro capotela 1899
cilindro con una lastra per
   giro 12464
cilindro del calamaio 7335
cilindro di accumulo 3072
cilindro di carta 1824
cilindro di contropressione
   1055, 7212
cilindro di cucitura 13176
cilindro d'impressione 7211
cilindro d'inchiostrazione
   7341
cilindro di piega 5978
cilindro di pressione 7211,
   7213
cilindro di pressione (della
   stampa) in volta 12046
cilindro di taglio 3893
cilindro d'uscita 13559
cilindro essiccatore 4788
cilindro impressore 7211
cilindro lisciatore 6448
cilindro macinatore 7341
cilindro manicotto 3645
cilindro motore della ruota
   delle unità 14323
cilindro per le ultime
   notizie 6225
cilindro per lo spoglio dei
   ritagli 13293
cilindro piegatore 5978
cilindro portablanket 1516
cilindro portacaucciù 1516
cilindro portalama 3893
cilindro portalastra 10432
cilindro portatessuto
   gommato 1516

cilindro pressore 7213,
   11750
cilindro raffreddatore
   13457
cilindro rivestito 9824
cilindro rotocalco 5184,
   6585
cilindro uscitafogli 12504
cilindro yankee 14903
cimice 4712
cimosa 12090
cinabro 2834
cinghia del compositoio 834
cinghia del raccoglitore 834
cinghia di trasmissione
   1301, 4745
cinghia trapezoidale 14416
cinghia trasportamatrici
   8689
cioè 14500
circolare 2844
circolo di confusione 2836
circonferenza 2854
circonferenza del cilindro
   di pressione 2855
circonferenza periferica del
   cilindro 3949
circonferenza primitiva
   10383
circonflesso 2860
circuito 2837
circuito flip-flop 5836
circuito integrato 7477
circuito micro-miniatu-
   rizzato 8904
circuito multinodale 9193
circuito porta 6322
circuito stampato 10800
circuito trifase 13760
cirillico 3952
citare 2861
citazione 11172
citrato d'ammonio 524
citrato di ferro 5548
citrato di ferro ammo-
   niacale 5542
citrato di potassio 10644
citrato di sodio 12686
citrato ferrico 5548
citrato ferrico ammonia-
   cale 5542
citrato ferrico d'ammonio
   5542
citrato potassico 10644
citrato sodico 12686
classificatore 5610
classificazione decimale
   4054
classificazione dei caratteri
   2879
cliché 1582, 5192, 10418,
   10907, 11445
cliché al tratto 8159
cliché a mezzatinta 6774
cliché a retino 10233
cliché a retino lineare 9596
cliché autotipico inciso in
   profondità 4091
cliché combinato 3299

cliché con feritoia 10324
cliché con retinatura
   grossolana 2978
cliché del cian 3919
cliché del nero 1491
cliché difettoso 2706
cliché di gomma 11751
cliché di magnesio 8498
cliché di nailon 9465
cliché di plastica 10411
cliché di rame 3512
cliché do fondo 12763
cliché doppiatinta 4877
cliché duplicato 4884
cliché inserito nella
   composizione 11793
cliché negativo 11557
cliché non montato 14332
cliché per fondo 13841
cliché precurvato 10718
cliente 74, 2915
climatizzatore 321
climatizzazione 322
cliscè 1582, 5192, 10418,
   10907, 11445
cliscè a colori per giornali
   9302
cliscè al tratto 8159
cliscè a mezzatinta 6774
cliscè a retino 10233
cliscè a retino lineare 9596
cliscè colore al tratto 8165
cliscè composto 3299
cliscè con feritoia 10324
cliscè con retinatura grossolana 2978
cliscè del blue 3919
cliscè del nero 1491
cliscè di fondo 12763
cliscè di gomma 11751
cliscè di magnesio 8498
cliscè di nailon 9465
cliscè di plastica 10411
cliscè di rame 3512
cliscè di zinco 14944, 14947
cliscè doppiatinta 4877
cliscè duplicato 4884
cliscè inserito nella
   composizione 11793
cliscè misto al tratto e
   retinata 8179
cliscè negativo 11557
cliscè non montato 14332
cliscè per fondo 13841
cliscè precurvato 10718
cliscè retinato grossolanamento 11140
cloisonné 2924
clorato di potassio 10637
clorato potassico 10637
clorito di sodio 12684
clorito sodico 12684
cloro 2742
cloro attivo 949
cloroformio 2746
cloroidrato 7104
cloroidrochinone 2747
clorioioduro di zinco 6971
cloropropilene 444

cloruro 2737
cloruro allilico 444
cloruro aurico 6517
cloruro calcico 2164
cloruro cromico 2787
cloruro d'alluminio 475
cloruro d'ammonio 522
cloruro d'argento 12427
cloruro d'etile 5306
cloruro di cadmio 2141
cloruro di calcio 1548, 2164,
   2168
cloruro di ferro 5546, 5576
cloruro di platino 10473
cloruro di polivinile 10568
cloruro di polivinilidene
   11875
cloruro di potassio 10638
cloruro di rame 3504, 3832
cloruro di sodio 12683
cloruro di stagno 13039
cloruro di zinco 14942
cloruro di zolfo 13386
cloruro doppio di platino e
   potassio 10640
cloruro d'oro 6517
cloruro d'oro e potassio
   10652
cloruro etilico 5306
cloruro ferrico 5546
cloruro ferroso 5576
cloruro mercurico 8792
cloruro mercuroso 8793
cloruro polivinilico 10568
cloruro potassico 10638
cloruro rameico 3504
cloruro rameoso 3832
cloruro sodico 12683
cloruro stannico 13038
cloruro stannoso 13039
coagulare 2970
coagularsi 2970
coagulazione 2971
coautore 3002
cobalto 3005
cobol 3019
coccinella 3022
coda del blanket 1522
coda (di una stereo) 13537
coda (di un nastro) 13538
codice 3030
codice autocorrettivo 5245
codice binario 1395
codice ciclico binario 11388
codice di Baudot 1232
codice di Gray 11388
codice d'istruzioni di
   macchina 3331
codice d'operazione 6249
codice hollerith 7022
codice rivelatore d'errori
   5246
codifica in assoluto 28
codificare 5154
codifica simbolica 13485
codificatore 3033, 5155
codificazione 3036
coefficiente d'assorbimento
   37

consumo d'inchiostro 7334
contafili 8187
contaminazione dell'aria 351
contare 3666
contasecondi 13221
contatore 3652, 3653
contatrice 3652, 3653
contatto ottico 9652
conteggio copie con distanziamento gruppi 11167
conteggio dei caratteri 2629
contenitore dell'inchiostro 7403
contenitore in cartone ondulato 3620
contenitore in cellulosa modellata 11030
contenuto d'umidità 9033
contenuto d'umidità calcolato sul secco 9035
contenuto d'umidità d'equi-librio 5228
contenuto in ceneri 816
contenuto in fibre di cotone 3637
contenuto in resina 11685
contenuto in resina naturale 11685
contenuto in secco 4772, 12764
contenuto in solido 12764
contenuto in sostanza solida 12764
contesa 3420
continua 13872
contornare ovale 9742
contorno 1792, 3437, 9716
contorno a greca 6620
contorno d'ottone 1861
contraffare 3657
contraffazione 3658
contrappeso 1088
contrarsi 12354
contrassegni di raccolta 3066
contrastato 3447
contrasti, senza ~ 12730
contrasto 3439
contrasto alto 6981
contratto d'annunzi 269
contratto pubblicitario annuale 251
contrazione della carta 12355
contributo 787
controdado 2672
controlama di gomma 3870
controlama di una taglie-rina 12582
controllare 2663
controllo 2661
controllo automatico 901
controllo dei materiali in entrata 11122
controllo dei termini (di consegna) 10927
controllo del colore 3128

controllo del registro 11404
controllo del registro laterale 12373
controllo di parità 10009
controllo di parità orizzontale 8340
controllo di produzione 10908
controllo di qualità 11121
controllo di qualità dei materiali in entrata 11122
controllo di ridondanza 11353
controllo elettronico del registro laterale 5054
controllore dei termini 10926
controllore della produzione 10926
controllo statistico di qualità 13079
controluce 1031
contromaschera 3660
contropressione 1044
contropunzone 3661
controracla 4533
controstampa 9540, 12636
controstampa per penetra-zione 10865, 13276
controstampa per riflessione 10865
controstampare 12162
controstampa sul cilindro della stampa in volta 12047
controstampo 3661
controtipo 3662
conversione 3467
conversione di codice 3031
convertire 3474
convertitore 3475
convertitore di codice 3032
convertitore di segnale 12397
convogliamento 11738
coordinate 3491
coordinate tricromatiche 2764
coperchio 8090
coperchio cerniera 5757
coperchio del cuscinetto 1257
coperta 2397, 3673
coperta (di pergamena) 6040
coperta incollata 6466
coperta splendida 4155
coperte imbottite 9844
coperte sporgenti 4517
copertina 2397, 3673
copertina con unghiatura 5381
copertina del dorso interno 7449
copertina di protezione 7629
copertina frontale interna 7452

copertina in carta kraft 7839
copertina integrale 7474
copertina in tela 2951
copertina per cartoni 12273
copertina provvisoria a foglio sciolto 1403
copertina splendida 4155
copertina telata 2951
copertura di protezione 10980
copia 3534
copia a contatto 3414, 10270
copia al bromuro 1947
copia all'albumina 372
copia all'anilina 619
copia allo smalto 5146
copia anticipata 241
copia arretrata 1038
copia autografata 896
copia carbone 2314
copia cianografica 1642
copia combinata 4647
copia completa in sala compositori 430
copia dattilografata 14228
copia estratta 7787
copia fotografica 10270
copia (fotografica) 10793
copia fotografica retinata 11978
copia gratuita 6156
copia ingiallita 14915
copia leggibile 6859
copia malamente leggibile 1061
copia numerata 9448
copia omaggio 3278
copia per contatto su carta Velox 14431
copiare 3538, 10795
copiare per contatto 3415
copia sovraesposta 3990
copia stampata 11478
copia su carta pigmento 2327
copia su lastra metallica 8838
copia su zinco 14954
copie invendute 11546
copie/ora 7223
copie stampate in più 9782
copiglia 12901
copista 10447, 13285
copolimero 3495
coppale 3493
coppella 3819
copperosa verde 3499
coppetta di Zahn 14931
coppetta Ford 6030
coppia 3668
copri-obiettivo 8044
copriposta 10339
coprire 9613, 9719
copyright 3553
corda per la legatura delle pagine 9855
cordonare 3716

cordonatrice 3718
cordonatura 3714
cordonatura segnapiega 1243, 3715
coriandoli 2577
corindone 3624
cornice 1792
cornice (della forma) per carta a mano 4063
cornice grecata 6620
corona 3589
corona dentata registrabile per l'eliminazione del gioco 1030
corona di fusione 13043
corone di controllo 1252
corpo completo 1467
corpo della maiuscola 6947
corpo di Bingham 1433
corpo (di un carattere) 1686
corpo falso 5451
corpo nero 1467
corpo otto 4995
corpo perfecto 1467
corpo tre 6137
corpo 12 10288
correggere (bozze) 11262
correggere esageratamente 1822
correggere in piombo 3591
correggere senza lettore 7042
correggere sul piombo 3591
correggia di trasmissione 1301
correggie di frenatura 1307
correlazione 3608
corrente alternata 465
corrente alternata trifase 13762
corrente anodica 643
corrente continua 4400
corrente di comando 3453
corrente di Foucault 4952
corrente di griglia 6632
corrente elettrica 5014
corrente (elettrica) 3840
corrente laminare 7884
corrente parassita 4952
corrente portatore 2371
corrente, senza ~ 4028
corrente turbolenta 14090
correttivo per inchiostri 4559
correttore (di bozze) 10954
correttore in piombo 3596
correttore pedantesco 3235
correzione 3597
correzione cromatica 3131
correzione del contrasto 13890
correzione del tono 13890
correzione di macchina 8441
correzione in pellicola 3602
correzione in piombo 3603
correzione in prima 7057
correzione preliminare 7057

correzione ultima 11272
correzioni 3605
correzioni d'autore 887
correzioni dell'autore 887
correzioni locali 8300
corrodere 3613
corrosione 3615
corrosivo 5266
corsa del carro 1283
corsa del soffietto 1294
corsa discendente della
  lama 4677
corsivo 7621, 7625
corsivo, in ~ 59, 7624
corsivo, (posto) in evidenza
  in ~ 59
corsivo romano antico 4937
corso di pubblicazione,
  in ~ 715
corso di stampa, in ~ 7537
cortocircuito 12325
coscienza 12994
coscienza, stare a ~ 12987
cosecante 3625
coseno 3626
costante 3398
costante dielettrica 4322
costanza 3397
costi di stampa 10822
costi generali 9758
costituzione 13303
costo per mille lettere
  13739
cotangente 3633
cote 6647
cotone 3636
cottimo 10321
cottura 2074, 3486
coulomb 3646
cremagliera 11183
cremagliera del tipometro
  5122
cremagliera e pignone
  11186
crenatura 7747
crepolatura 3694
cresolo 3731
crespatura 3730
creta di Bologna 2163
cretoso 2606, 3732
cristallino 3789
cristallizzato 3793
cristallizzazione 3791
critica d'un libro 1773
critica fulminante 3967
criticare un libro 11567
critico 1774, 3734
croce 3746, 3955, 12161
croce di Siemens 13055
crocetta 12161
crocette di registro 11411
crogiolo 3777, 8836
crogiuolo 3777, 8836
crogiuolo di fusione 8770
crogiuolo di rifusione 11456
cromaticamente corretto
  109
cromaticità 2763
cromato barico 1143

cromato d'ammonio 523
cromato di bario 1143
cromato di piombo 7977
cromato di potassio 10641
cromato di sodio 12685
cromatografia su carta
  9905
cromato potassico 10641
cromatura 2798
cromia 2766
cromista 1315, 3164
cromo 2796
cromolitografia 2818
cromolitografo 2817
cromosilografia 2821
cromoxilografia 2821
cronista principiante 3796
cronometro 13221
cucchiaino da tè 13620
cucchiaio 7871
cucchiaio (da fonditore)
  2447
cucchiaio per scoriare
  12514
cucchiaio schiumatore
  12514
cucito a punti correnti 6184
cucitoio 12195
cucito su corde 12200
cucito su fettucce 12201
cucito su filo con garza
  12202
cucito su nastri 12201
cucitrice a filo metallico
  14779
cucitrice a filo refe 12197
cucitrice a punti metallici
  per scatole 1838
cucitrice a punto metallico
  14779
cucitrice a sella 11830
cucitrice per scatole 1838
cucitura 12192
cucitura a blocco 1617
cucitura a chiusura lampo
  12198
cucitura a con file refe in
  piano 14708
cucitura a filo metallico
  14780
cucitura a macchina 8464
cucitura a mano 6846
cucitura a punto appicci-
  cato 426
cucitura a punto corrente
  6183
cucitura a punto metallico
  14780
cucitura a punto saltato
  14708
cucitura a sella 11831
cucitura a tre punti 2550
cucitura attraverso la
  copertina 1617
cucitura degli angoli 3587
cucitura dei blocchi 9845
cucitura di piatto a punti
  metallici 12389

cucitura di traverso a punti
  metallici 1617, 12389
cucitura di traverso (a
  punti metallici) 12988
cuffia 2272
culla 3695
cunei (di legno) 13146
cuneo 14665
cuneo a tono continuo 3436
cuneo dei grigi 6630
cuneo di giustificazione
  7712, 12802
cuneo di regolazione 225
cuneo di regolazione della
  pressione 7225
cuneo di riempimento
  13975
cuneo di serraggio 11169
cuneo di serrare 11169
cuneo serraforma 11169
cuoio 8011
cuoio artificiale 791
cuoio cilindrato 4951
cuoio con motivo a losanga
  4291
cuoio di foca 12025
cuoio di pecora 1183, 11624
cuoio di porco 10337
cuoio di Russia 11821
cuoio di Spagna 3561
cuoio di vacca 3687
cuoio di vitello 2199
cuoio imitazione foca 359
cuoio spaccato 2038, 12902
cuore della cucitura 12196
cupone 3671
curva 3845
curva caratteristica 2637
curva della sensibilità 6822
curva di annerimento 6822
curva di distribuzione 4493
curva di fluidità 5868
curva di scorrimento 5868
curva D log E 2637
curva H & D 6822
curvalinee 6169
curvare 1313, 3846
curva tempo/gamma 13815
curvato 3847
curvatura della lastra 10431
cuscinetto 1254
cuscinetto ad aghi 9267
cuscinetto a rulli 11634
cuscinetto a rulli oscillanti
  12078
cuscinetto a sfere 1096
cuscinetto d'appoggio per
  incisori 5189
cuscinetto del cilindro 3931
cuscinetto dell'albero 969
cuscinetto dell'albero
  principale 8549
cuscinetto liscio 10390
cuscinetto per incisori 9838
cuscinetto per la doratura
  6501
cuscini 1710
cuscino d'aria 325
cuscino pneumatico 325

custodia 12561

daccapo 1875
dactilografa 14231
dactilografo 14231
dado 9459
dado a corona 2456
dado ad alette 14752
dado ad intagli 2456
dado cieco 2289
dado di regolazione 222
dagherrotipo 3956
daltonico 3122
daltonismo 3123
damigiana 2337
dammara 3964
danneggiamento del canto
  bobina 11365
dar aria 14746
dar di spugna a 12911
dare alla stampa 10020
dare in subappalto 5472
dare lucido 5569
da sopprimere 4128
data di pubblicazione 10999
dati 4005
dati bruti 11256
dattilografo monotipista
  7771
deassorbimento 4211
debordare dell'inchiostro
  12980
decalcare 13935
decalco di guida 1640
decalcomania 13969
decalcomania su ceramica
  2558
decalcomanie 4041
decametro 4046
decantare 4047
decantazione 4048
decappare 3656
decibel 4050
decimetro cubo 3800
decisione 4059
decisione logica 8318
decodificare 4072
decolorazione 13008
decorazione a linee
  parallele 980
decorazione della coperta
  13901
decorazione delle rile-
  gatura con ferri
  azzurrati 980
decorazione del libro 1768
decorazione di cattivo
  gusto 6414
decorazione esagerata 6414
decorazione per floccula-
  zione 5844
dedica 4074
definizione 4098
deflocculazione 4102
deformato a coppa 4431
deformazione 4100, 4105
deformazione al taglio
  12243

deformazione a ventaglio 5467

deformazione del punto di retino 4565

deformazione elastica 5004

deformazione per cesoiamento 12243

degradazione di colore 12209

delaminazione 4124

deleatur 4128

delimitatore 4132

demodulazione 4158

denaturare 4161

dendrite 4163

densimetro 761

densità 4169

densità apparente 711, 712

densità dell'inchiostro 7336

densità di bit 1449

densità di corrente 3842

densità di registrazione 9829, 11312

densità di stampa 10799

densità integrata di un retino 7478

densità in trasparenza 13991

densità ottica 9653

densitometria 4168

densitometro 4167, 7120

densitometro per riflessione 11393

densitometro per trasmissione 13990

denso 1669

dente d'arresto 1277, 10067

deodorizzazione 4182

deossidare 4183

deposito 4186

deposito di fibre di carta 4894

deposito di rame 3508

deposito legale 3554, 8085

deposizione elettrolitica 5028

depressione 8387, 12472

depressione in un blanket 4177

depressioni 1711

dermatite 4200

dermatite da cromo 2799

desensibilizzatore 4205

desensibilizzazione 4203

destra, a ~ 2922

destrina 4243

destrosio 4246

detergente 4219, 14551

determinazione 4220

determinazione della scala di riproduzione 3557

determinazione del liscio 12645

detersivo 4219, 14551

detersivo per blanket 1526

detersivo per cliscè 10425

dettagliato 3321

dettaglio 4215

deumidificazione 4119

deviatore separatore delle copie 13475

deviazione 4234

deviazione di raggi di luce 4235

deviazione media 8715

deviazione quadratica media 13023

deviazione standard 13023

dextrina 4243

diaframma 4273

diaframma a croce 3769

diaframma a feritoia 12579

diaframma a fessura 12579

diaframma a iride 7572

diaframma da inserire 6796

diaframma di Waterhouse 14582

diaframma normale 9428

diaframma per fotografia con retino 6796

diaframma per le alte luci 6993

diaframma per toni intermedi 8922

diagramma 4257, 5606

diagramma di cromaticità 3193

diagramma di flusso 5867

diagramma simultaneo 12437

diametro del cilindro 3938

diametro del cilindro d'impressione 4260

diametro primitivo 10384

diapositiva 4278, 13998

diapositiva a contatto 3409

diapositivo 4278, 13998

diapositivo a colori 3192

diapositivo per proiezione 12550

diapositivo retinato 6778

diario 4279

diatomite 4280

diazocopia 4285

diazotipia 4286

dibutilftalato 4290

dichiarazione doganale 3852

diciottesimo, in ~ 4993

dicitura 8031

dicitura finale 13952

dicroismo 4296

didascalia 8031

dielettrico 4326

dieresi 4251

difetti 11451

difetti di registro 11405

difetti di rotolamento 11660

difetto 2040

difetto di costruzione 3403

difetto grave 8556

difetto principale 8556

difettoso 4093

differenzas di registro 11405

diffrazione 4344

diffrazione della luce 1321

diffusione 1557

diffusione della luce 4350, 8129

diffusore 4349

digitale 4358

digitale-analogico 4363

digitalizzatore 4366

dilatante 4369

dilatanza 4368

dilatazione 4370, 5362

dilatazione d'umidità 7135

diluente 5384, 11345

diluire 3856, 4372

diluire l'inchiostro 11343

dimensione 4375

dimensione delle particelle 10014

dimensione media delle particelle 957

dimensioni del carattere 2630

dimetilchetone 102

diminuire lo spazio 2943

dina 4931

dinamometro 4930

diodo a cristallo 3788

diodo fotoelettronico 10224

diottria 4389

dipinto a pennello 2017

direttore tecnico 14843

direzione antiorario, in ~ 3655

direzione contro-fibra, in ~ 292

direzione della fibra 6537

direzione di fabbricazione 8443

direzione di fibra 6537

direzione di macchina 8443

direzione oraria, in ~ 2922

direzione trasversale 3752

diritti d'autore 3553

diritto d'entrata 6614

diritto di traduzione 11610

discendente 4201

disco 4433, 4434

disco magnetico 8508

disconnettere 1878

disegnatore 4698

disegnatore di caratteri 14184

disegnatore grafico 1748, 3240, 6567

disegnatore pubblicitario 271

disegnatrice 4698

disegno 4208

disegno a carboncino 2646

disegno acquerellato 14570

disegno a lapis 10092

disegno al carbone 2646

disegno al tratto 8171

disegno a matita 10092

disegno a penna 10089

disegno a pennello secco 4770

disegno con gesso litografico 2598

disegno definitivo 14835

disegno di contorno 10090

disegno eseguito a mano 808

disegno litografico dell'artista 8255

disegno originale 808

disegno panoramico 9890

disidratazione 4121

disimbracamento 13790

disimporre 13282

disinchiostrazione 4123

disinnestare la macchina 14057

disinserire 13477

disinseritore di pressione 7224

disossidante 4184

dispensa 10012

disperdere 4450

dispersione 4454

dispersione della luce 4455, 8129

disporre a squama 5466

disporre a zig-zag 13002

disporre fogli di carta sotto il rivestimento di gomma 10053

dispositivo 4236

dispositivo antiscartino 12926

dispositivo antistatico 13069

dispositivo a pinze per sospendere la carta 1105

dispositivo a tensione per il fissaggio delle lastre 13649

dispositivo automatico per taglio in fogli 12277

dispositivo con ruota a palette 5916

dispositivo d'accelerazione 12942

dispositivo d'arresto 7224

dispositivo d'arresto automatico in caso di rottura (del nastro) 14642

dispositivo d'esecuzione d'operazioni concorrenti 9189

dispositivo di arresto 8305

dispositivo di bloccaggio per compressione 3325

dispositivo di chiusura 8305

dispositivo di chiusura delle ganasce 14463

dispositivo di controllo del doppio foglio 14155

dispositivo di disimbracamento 4071

dispositivo di disinnesto 4071

dispositivo di distacco 7224, 13789

dispositivo d'immissione contemporanea di due fogli 14155

dispositivo di presa (del cilindro) 3936

dispositivo di raccolta matrici a mano 12789

dispositivo di raccolta matrici speciali 12789

dispositivo di rientranza e centratura 12085

dispositivo di sicurezza 8303, 8305

dispositivo di staffatura 3325

dispositivo di tensione con compensazione 3262

dispositivo di terza piega 11138

dispositivo d'uscita 9733

dispositivo essiccatore 4775

dispositivo inchiostratore semicontinuo 3431

dispositivo incollatore 10035

dispositivo ottico d'esplorazione 9660

dispositivo per effettuare una giunzione testa a testa 2126

dispositivo per il fissaggio di lastre 8307

dispositivo per il rovesciamento laterale 13243

dispositivo per il taglio in fogli 12266

dispositivo per la centratura e spaziatura 11093

dispositivo per l'arresto 7224

dispositivo per la stampa delle ultime notizie 6224, 13218

dispositivo per le giunte al volo delle bobine 5927

dispositivo per levatura intermittente 13273

dispositivo per l'introduzione di rulli 11653

dispositivo per rimuovere la polvere 9903

dispositivo piegatore sistema tucker 13851

dispositivo raffreddatore della forma 9124

dispositivo refrigerante 3489

dispositivo ritardatore 12595

dispositivo rivelatore foglio mancante 9430

dispositivo stampatore elettrostatico 5067

dispositivo uscitafoglio con catene portapinze 2587

disposizione 7946

disposizione dei tasti 7770

disposizione quinconce 9285

disseminazione d'informazioni 4009

dissertazione 13700

dissertazione di laurea 13700

distaccare i rulli 13791

distacco (della patina) 3691

distacco della pellicola 13289

distanza del retino 11977

distanza focale 5945

distanza tra i cilindri 8841

distanziare 12808

distanzi focali 3386

distorsione 4481, 4482

distorsione a botte 1171

distorsione a forma di cuscino 10361

distorsione anormale 7665

distorsione con incurvamento convesso 1171

distorsione ottica 9655

distribuire 2900, 4484, 7793

distribuire l'inchiostro con rullo a mano 1870

distributore 4497

distributore d'inchiostro 719

distribuzione 4492

distribuzione cumulativa 3810

distribuzione dell'inchiostro 7339

distribuzione del tempo 13818

distribuzione di frequenza 6187

distribuzione d'informazioni 4009

distribuzione spettrale dell'energia 12842

dito di guida 2826

dittongo 4392

dividere il manoscritto in più parti 4510

divisa 4237

divisione 7143

divisione del calamaio 6114

divisione della cassa 2405

divisione della parola 4519

divisione delle parole in sillabe 7144

divisione in gradi 4518

divisione scorretta di parole 1057

divisione sillabica nel dizionario 4308

divisore del calamaio 1904

divisorio 4520, 12903

dizionario 4306, 4307

dizionario delle eccezioni 5336

dodicesimo, in ~ 14117

dodici punti inglese-americano 10288

dolomia 4549

domanda 11147

dopo trattamento 10616

doppia crocetta 4594

doppia esposizione 4603

doppia giustificazione 4619

doppia impressione 4646

doppia pagina nel centro di un quaderno 2549

doppia precisione 4645

doppiare una parola 4580

doppieggiatura 12611

doppi gruppi stampanti 4070

doppio legame 4627

doppione 4579

doppio quadratone 14138

doppio viso 14158

dorato al taglio superiore 13911

doratore 6399

doratore alla pressa 10748

doratore dei tagli 4956

dorato sul taglio di testa 13911

doratura alla pressa 6406, 10749

doratura a mano 6405

doratura di mano 6405

dorso 983, 984

dorso alla cappucina 13802

dorso a nervi 11589

dorso a scomparti 1026

dorso attaccato 13802

dorso con salti 7023

dorso del libro 1422

dorso (del libro) 12873

dorso di calendario 2179

dorso di tela 2947

dorso fisso 13802

dorso grecato 11895

dorso piatto 12957

dorso quadro 12957

dorso tondo 11720

dorso unito 13802

dosatore di luce 8118

dramma 4697

driografia 4734

due punti 3101, 12482

duplicare in stereotipia 13123

duplicato 4884

durabilità 4886

durata 12286

durata d'incisione 5288

durezza 4112, 6871

durezza Brinell 1918

durezza secondo Brinell 1918

durezza secondo Shore 12321

durometro 6872, 10417

durometro Shore 12320

ebanite 6874

ebraico 6944

eccentrica 4948

eccentrico 2227

eccentrico del cilindro di pressione 7214

eccentrico della lama della forma a fondere 9122

eccentrico della pompa 11038

eccentrico della torre della carta 9954

eccentrico dello spazio mobile 7776

eccentrico del prendi-carattere 14178

eccentrico di tastiera 7766

eccessivamente ornato 7011

eccesso di contrazione 5340

eccesso d'inchiostro 5851

eccesso di restringimento 5340

eccesso di ritiro 5340

ecco, il proto 2493

ecco qua 11168

eczema da cromo 2799

edicola 9317

editore 11005

editore di musica 9214

editoriale 4977

edizione 4970

edizione accresciuta 5200

edizione ampliata 5200

edizione ampliata e migliorata 558

edizione (a tiratura) limitata 8150

edizione commentata 634

edizione corretta 3593

edizione dell'estero 8537

edizione di lusso 4973, 7919

edizione di mezzogiorno 9420

edizione economica 2660

edizione emendata 3593

edizione illustrata 7164

edizione in fascicoli 4974

edizione in quaderni 4974

edizione migliorata e ampliata 11572

edizione nuova 9297, 11429

edizione originale 5702

edizione particolare 10871

edizione popolare 2660

edizione privata 10871

edizione purgata 5378

edizione ridotta 19

edizione riveduta 11573

edizione riveduta e accresciuta 11572

edizione speciale 12821

edizione straordinaria 12822

edizione tascabile 10510

effetto alone 4569, 6802

effetto attinico 160

effetto Becquerel 1270

effetto bordo 1794

effetto Cabannes-Hofmann 2136

effetto Debot 4039

effetto dell'umidità 7075

effetto di adiacenza 216

effetto di Albert 362

effetto di Callier 2215

effetto (di) Clayden 2885

effetto di Doppler 4562

effetto di Eberhard 4946

effetto di Herschel 6969

effetto di Kostinsky 7836

effetto d'intermittenza 7513

effetto di retino lineare 12459
effetto fotoelettrico 10228
effetto Lainer 7879
effetto nelle ombre 7758
effetto Nyblin 9463
effetto Purkinje 11057
effetto Ross 11691
effetto Russell 11820
effetto Sabattier 11825
effetto Schwarzschild 11925
effetto Sterry 13133
effetto Villard 14474
effetto Weigert 14669
effetto Weinland 14674
Egiziani 8745
eiettore 5001
elaboratore d'emergenza 1054
elaboratore di dati 3330
elaboratore di riserva 1054
elaboratore universale 6368
elaborazione a blocchi 1222
elaborazione automatica dei dati 905
elaborazione dell'informazione 7309
elaborazione elettronica dei dati 5046
elaborazione sequenziale 12136
elasticità 5005, 11500
elementare 5075
elemento 5074
elemento aciclico 7247
elemento ciclico 3924
elemento colore 3194
elemento di riscaldamento 6923
elemento di somma logica 9666
elemento di stampa 4060, 10852
elemento intermittente 7247
elemento O 9666
elemento OR 9666
elemento per stampa all'anilina 615
elemento per stampa flessografica 615
elemento riscaldante ad immersione 7184
elemento riscaldatore 6923
elenco dei richiami 2472
elenco delle abbreviazioni 8242
elenco delle illustrazioni 8243
elenco (di duplicati) per cambio 5344
elenco telefonico 4408
elettricità statica 13067
elettrodo 5027
elettrodo al chinidrone 11159
elettrodo a vetro 6428
elettrodo di calomelano 2224

elettroforesi 5061
elettrofotografia 5062
elettroincisione 5007
elettrolisi 5032
elettrolito 5033
elettromagnete 5037
elettromotore 5042
elettrone 5043
elettronica 5052
elettronico 5044
elettrostatica 5069
elettrotipista 5073
elevatore della pila 10342
eliminare 7792
eliminare del fondo attorno 12407
eliminare una pagina incorretta ad una buona pagina 2260
eliminatore di elettricità statica 13069
eliotipia 3099, 10253
elisione 5083
ellisse 5085
ellissi 5086
ellissografo 5087
Elzeviri 6308
em 5098
emettere 4863
emicellulosa 6960
empirico 5118
emulsificante 5128
emulsificarsi 1556
emulsificazione 5126
emulsionare 5129, 13843
emulsione 5132
emulsione acqua-inchiostro 7364, 14583
emulsione a grana fine 5673
emulsione al collodio 3086
emulsione fotografica 10243
energia cinetica 7802
energia di radiazione 11191
energia elettrica 5024
entrata a labirinto 8121
entrata del canale 2623
entrata del compositoio 837
eosina 5215
epigramma 5217
epilogo 5219
epuratore degli incotti 12589
epurazione 2889
equazione 5226
equazione di Van der Waals 14382
equilibrato 1085
equilibrazione 5222
equilibrio 5227
equilibrio cromatico 3119
equilibrio dei movimenti 1086
equilibrio di colori 3119
equilibrio inchiostro/acqua 7408
equipaggiamento ausiliario 10143

erg 5238
ergonomia 5239
eritrosina 5248
ermetico 356
errata 5241
errata corrige 5241
erratum 5242
errore dell'autore 888
errore di collimazione 3080
errore di compositore 8245
errore di dattilografia 14230
errore di distribuzione 14881
errore di registro 11407
errore di scomposizione 14881
errore di scrittura 2911
errore di stampa 3315, 8979
errore di taratura 2205
errore occasionale 71
errore ortografico 12857
errore probabile 10876
errore sfuggito all'autore 888
errore standard 13024
errore tipo 13024
errore tipografico 3315, 8979, 14235
errori di composizione 8980
esame di invecchiamento 302
esame, in ~ 726
esametafosfato di sodio 12691
esametafosfato sodico 12691
esame visivo 14496
esaminare 13659
esattezza 85
esaurito 9725
esclusione di una pagina dall'inchiostrazione 9856
esculina 5251
esecuzione simultanea d'istruzioni 3360
eseguire 5351
eseguire una fusione 2434
eseguire una stereotipia 13123
eseguire un getto 2434
esemplare 3535
esemplare anticipato 240
esemplare autografato 896
esemplare del deposito 13080
esemplare difettoso 4094
esemplare di scambio 5342
esemplare d'obbligo 13080
esemplare gratuito 6156, 10733
esemplare numerato 9448
esemplare omaggio 3278
esemplare perfetto 3550
esemplare per la recensione 240
esemplare per recensione 11568
esemplare preliminare 240

esente da acidi 126
espansibile 9619
espansione 5361, 5362
espiritu asper 12883
espiritu rude 12883
espiritu soave 12884
esplorare 11907
esplorazione elettronica 11911
esponente di testa 2469, 2470
esponenti 13416
esporre 5367
esposimetro 5373
esposimetro computerizzato 908
esposimetro integratore 8118
esposizione 5368
esposizione con lampada ad arco 2073
esposizione equivalente 5229
esposizione flash 5762
esposizione in bianco 5762
esposizione per alte luci 6988
esposizione principale 8543
espulsore degli scarti 11953
espulsore dei caratteri 14212
espurgare 1822
essenza di trementina 14107
essere adiarto 10086
esse rotonda 5666
essiccaflani 8690
essiccamento 4783, 4796
essiccamento ad aria 331
essiccamento all'infrarosso 7312
essiccamento a raggi infrarossi 7312
essiccamento dell'inchiostro negli alveoli dei cilindri 4790
essiccamento forzato 6026
essiccamento troppo rapido dell'inchiostro 11981
essiccante al cobalto 3007
essiccante al manganese 8590
essiccante al naftenato 9237
essiccante al piombo 7979
essiccante liquido 8235
essiccare 4206
essiccatoio 4797
essiccatoio a festoni 5584
essiccatoio per flani 8690
essiccatore 4207
essiccatore ad aria 330
essiccatore per carta in fogli 12265
essiccatori 4730
essiccazione a fondo 4799
estendibile 9619
estensimetro 11509
estensione 12935

estensione delle zone stampanti 1554, 7903
estere 5259
estere di cellulosa 2512
esterificare 5262
esterificazione 5261
estetico 286
estrarre 5119
estrarre il riporto 11281
estratto 5396
estrattore 5398
estrattore d'aria 335
estrattore di Soxhlet 5398
estrazione 5397
estremità sfiaccilata 6152
estrusore 5411
etandiolo 5309
etanolo 5301
etere 5297
etere di petrolio 10169
etere etilico 5297
etere monometilico del glicol etilenico 5310
etere solforico 5297
etichetta 7846, 13534
etichetta gommata 6714
etichetta iniziale 6901
etichetta per bagaglio 1069
etichettare 7848
etichettatrice 7850
etil acetato 5300
etilcellulosa 5305
etilene 5307
etilenglicol 5309
eutettico 5318
evaporazione 5320
ex-libris 1771
extra costi 5407

fabbisogno di carta 9910
fabbrica 10403
fabbrica di carta 9927
fabbrica di cartonaggi 9902
fabbrica di cartone 1662
fabbrica di libri in bianco 77
fabbrica d'inchiostri (da stampa) 7370
fabbrica d'inchiostri da stampa 10833
fabbrica di pastalegno 6645
fabbrica di pasta meccanica 6645
fabbricante lastre 10447
fabbricazione 11788
fabbricazione della carta 9926
fabbricazione di prova 14022
faccetta 1344, 5425
faccia anteriore del fusto del carattere 1296
facilità di manipolazione 8233
facsimile 5426
fadeometro 5432
faggio 1284
falda 5755
falsa immagine 6391

false lastre 4860
false pieghe 14873
falsi decalchi litografici 2601
falso frontespizio 1217
falso giornalistico 5441
fame di lettura 7083
famiglia di caratteri 14188
fango 12596
Fantasie medioevali 8609
farad 5469
faraday 5470
fare le bozze 11018
fare lo straordinario 14840
fare presa 203, 12154
fare un calco 13935
farfalle 1019
farina fossile 4280
far un blocco 14098
fasce 1113
fasce umide 14692
fascia d'assorbimento 36
fascicolo 1933, 5474, 10012
fascicolo arretrato 1038
fascio di luce 1244
fase 10184
fase dispersa 4451
fase liquida 8238
fatto a macchina 8454
fattore d'aggruppamento 1599
fattore di Callier 2216
fattore di luminanza 8410
fattore d'incisione 5285
fattore di scoppio 2100
fattore di trasmissione 13992, 13994
fattore di visibilità (d'una radiazione) 14491
fattore di vivacità 14499
fattore filtro 5658
fattore lavoro 2635
fattore relativo di scoppio 2094
fattorino 5240, 10809
fattura intestata 1385
felpa per bronzatura 1971
feltro da taccheggio (per flani) 1013
feltro essiccatore 4729
feltro manicotto 3644
feltro marcatore 8636
feltro per asciugante 1511
feltro ponitore 3643
feltro umido 14684
f.e.m. 5041
fenilamina 613
fenolftaleina 10192
fenolo 10189
fenomeno 10193
fenomeno di luce catodica riflessa 2486
fenomeno d'interferenza 7495
feritoia 9107
fermaglio 2876, 9906
fermare 13793
fermare la macchina 14057
fermata 1880

fermomacchina 1880
ferricianuro di potassio 10649
ferricianuro potassico 10649
ferrico 5540
ferro 7575
ferro azzurrato 978
ferrocianuro 5562
ferrocianuro di potassio 10650
ferrocianuro di rame 3516
ferrocianuro d'uranio 14364
ferrocianuro potassico 10650
ferromagnetismo 5564
ferro per dorare 1411
ferroso 5572
ferrotipia 5568
ferrotipo 5568
fessura del calamaio 13783
festa di San Giovanni 11843
festa di San Martino 11845
feuilleton 5585
fianco 12390
fianco inclinato del rilievo dell'occhio 1346
fibra 5588
fibra artificiale 13494
fibra sintetica 13494
fibre colorati 3136
fibre d'amianto 811
fibre di carta 9917
fibre di cellulosa 2513
fibre di latifoglia 4052
fibre di legno 14802
fibre di vetro 6429
fibre liberiane 1219
fibre sciolte 8361
fibre vulcanizzate 14522
fibrillare 5599
fibrillazione 5600
fibrille 5598
fiele di bovo 9802
fiele di bue 9802
fiera del libro 1752
filettatura di vite rettangolare 11316
filettatura femmina 7520
filettatura metrica 8876
filettatura tonda 11731
filetti senza smusso 45
filetti sistematici 7861
filetti trasversali separati 12125
filetti uniti di testa 45
filetto 11772
filetto a curva 6532
filetto al piede 6023
filetto a paletta 5628
filetto azzurrato 2697
filetto capillare 6750
filetto chiaroscuro 9801
filetto d'acciaio 13102
filetto di chiusura 13734
filetto di contorno 1796
filetto di inquadratura 1796
filetto di piombo 8003

filetto di separazione 3871, 4514
filetto di taglio 3903
filetto doppio 9987
filetto doppio chiaro 4607
filetto doppio filo 4607
filetto doppio semiscuro 4632
filetto d'ottone 1866
filetto finissimo 6750
filetto fra due colonne 3209
filetto inglese 6182
filetto interno 7520
filetto in testa 6913
filetto medio nero 8757
filetto mezzo scuro 8757
filetto nero 6232
filetto ondulato 14625
filetto ornamentale 9690
filetto Oxford 9801
filetto per cordonatura 3720
filetto perforatore 10133
filetto per fustellare 4319
filetto per perforare 10133
filetto per tracciare 11938
filetto punteggiato 4577
filetto scuretto 8757
filetto scuro 6232
filetto sinistro 8028
filetto sinistrorso 8028
filetto smussato 1352
filetto tagliato ad angolo 8997
filetto triplo 14018
filettura destrorsa 11609
filigrana 14587
filigrana alla moletta 7207
filigrana a secco 7180
filigrana chiaro-scuro 13766
filigrana collocata regolare 8301
filigrana imitata 7180
filigrana in chiaroscuro 12220
filigrana in scuro 12214
filigrana localizzata 8301
film 5636
film a colori 3144
film di acetato di cellulosa 2510
film d'inchiostro 7350
film lith 8249
filmogeno 5642, 5646
filo della lama 7823
filo metallico per cucire 13179
filo morto 2090
filoni 2589
filo refe 1430
filtraggio 5656
filtrare 5663
filtrato 5662
filtri per tricromia 14034
filtro 5654
filtro a banda (spettrale) ristretta 9242
filtro arancio 9668

filtro blu 1636
filtro colorato 3145
filtro colorato ottico 9657
filtro colore 3145
filtro compensatore 3261
filtro di correzione 3261
filtro di gelatina 6360
filtro di selezione 12129
filtro giallo 14916
filtro inattinico 7236
filtro magenta 8488
filtro polarizzatore 10542
filtro rosso 8488
filtro ultravioletto 14250
filtro verde 6623
filtro Wratten 14871
finale 13552
finalino 13552
fine 13732
fine bobina 2116
finestra 1500, 7453, 9107
finestra di cellofan 2505
finezza del retino 11992
finir la copia 5687
finitura 5686, 6091
finitura a spazzole 2002
finitura fantasia 5458
finitura mat 4853
finitura mediante feltri
    marcatori, con ~ 5533
finitura semimatta 4990
finto cuoio 791
fioccatura 12655
fiorone 5816
fissaggio 5724
fissaggio dei cliscè allo
    zoccolo mediante
    adesivo 10422
fissaggio dell'inchiostro
    12176
fissare 5722, 5723
fissare le corde 7862
fissare il blanket 5738
fissare mediante viti 582
fissativo 5725
fissatore-induritore 6865
flanella 5754
flangia 5752
flano 5846
flano bruciacchiato 1578
flano in piombo 7993
flano secco 4805
flano senza taccheggio 9835
flessibile 5820
flessibilità 5819, 10482
flessografia 5828
flocculazione 5842, 5844
florilegio 654
floroglucinolo 10200
floscia 5739
floxina 10201
fluidità 5864, 5885
fluorescente 5890
fluorescenza 5888
fluoruro cromico 2789
fluoruro d'ammonio 530
fluoruro di cromo 2789
flusso capillare 2280
flusso di calore 6921

flusso difettoso 12570
flusso di Poiseuille 10538
flusso di radiazione 11192
flusso luminoso 8414
flusso newtoniano 9327
flusso non laminare 12570
focallizzare 5952
foderatrice per scatole 1836
fogliacci 1934
fogliacci secchi 4769
fogliacci umidi 14679
foglia colorata 3146
foglia d'oro 6509
fogli aggiunti 9784
fogli antiscartini 12576
fogli con macchie di muffa
    6139
fogli con macchie
    d'umidità 6139
fogli danneggiati 6139,
    13922
fogli di continuazione 3424
fogli di protezione 9737
fogli di risguardo 5162
fogli di scarto 9737
fogli di separazione 12576
fogli esterni 9737
fogli, in ~ 7448, 14298
fogli intercalati 7500
fogli macchiati 6139
foglio 5961, 8007, 9309,
    12250
foglio adesivo 211
foglio anticipato 244
foglio campione 12834
foglio campione tipo 9740
foglio d'alluminio 476
foglio d'avviamento 8566,
    12922
foglio della carta di
    tiratura 12276
foglio di acetato 91
foglio di avviamento 9773
foglio di carta assorbente
    1621
foglio di cellofan 2503
foglio di macchina 10764
foglio di maestra 1009, 4719
foglio di montaggio 13145
foglio di norma 8202, 11022
foglio di pressa 10764
foglio di registro 8202,
    11022
foglio di stampa 10764
foglio mancato 8983
foglio mobile 8362
foglio modello 4331
foglio non ripiegato 1931
foglio non stampato 1535
foglio non teso 12537
foglio per puntature 13145
foglio per stampa 1601
foglio pubblicitario 272
foglio scandalistico 14920
fogli per ora 12280
fogli rifatti 2264
folio, in ~ 5994
fondente 5912
fondere 2435

fonderia di caratteri 14193
fonditore di caratteri 14192
fonditore monotipista 2441,
    9075
fonditrice 2440
fonditrice di caratteri
    mobili 12447
fonditrice di linee 8161
fonditrice display 4467
fonditrice Monotype 9088
fonditrice per caratteri
    14181
fonditrice per stereotipia
    curva 2449
fonditrice per stereotipia
    piana 2450
fonditura dei caratteri
    14180
fondo 1001
fondo contornado 3875
fondo perduto 5380
fondo pieno 12755
footcandle 6016
foratrice 4733
foratura 4732
forbici 11929
forbici da carta 9940
forca portarulli 11637
forcella del cilindro 3936
forcella della coreggia 1304
forcella portarulli 11639
forchetta di presa 3936
forellini 10364
fori di trascinamento 5517
fori e chiodini di registro
    11409
forma a fondere 2443, 9118
forma a fondere barre di
    metallo 10334
forma a fondere stretta
    11299
forma a mano 6836, 9119
forma bombata 12943
forma cerografica 2567
forma composta 6299
forma con molti filetti
    11776
forma conservata 13032
forma d'alimentazione del
    filo (metallico) 13172
forma da scomporre 4495
forma da voltare 14082
forma della particella 10013
forma di bianca 5707
forma di fusione 2453, 9118
forma di losanga, a ~ 8391
forma di peso medio 8923
forma di scomposizione
    4495
forma di stampa 6056
forma duplicata di plastica
    10411
forma fustellatrice 3897
forma in bianca 9714
forma incavografica 5104
forma in nastro d'acciaio
    per fustellare 13103
forma interna 7418
formaldeide 6048

forma leggera 8139
formalina 6050
forma multipla 6299
forma per fondere filetti
    8696
forma per la fusione dei
    rulli 6729, 11643
forma per la stampa a
    colori 3147
forma per velino 14857
forma pesante 6943
formapile 12998
formare 9121
formare il dorso (di un
    libro) 988
forma stampante 6056,
    10418, 10907
forma stampante in rilievo
    11445
forma stampante pre-
    curvata 10718
formati a magazzino 13192
formati dei libri 12496
formati del cartone 12495
formati della carta 12500
formati della carta per
    registri 12499
formati delle buste 12498
formati delle cartoline
    12497
formati (di carte)
    normalizzati 7521
formati internazionali 7521
formati internazionali delle
    buste 7522
formati unificati (della
    carta) 7521
formato 4375, 6052, 6053
formato allungato 5093
formato base 1206
formato dei classici, in ~
    9485
formato della macchina
    11239
formato del libro 1779
formato del piano
    portaforma 1281
formato di giornali 9319
formato foglio 12487
formato massimo del foglio
    8710
formato minimo 8961
formato non refilato 14353
formato oblungo 7898
formato pieno 6241
formato refilato 14044
formato tabloide 13518
formato tascabile 10516
formato verticale 10590
formazione compatta 14676
formazione dei punti 4568
formazione del falso 1008
formazione del foglio 6055
formazione della carta 6055
formazione del morso 1008
formazione di bolle 1579
formazione di micelle 8888
formazione di pellicola
    12519

formazione di pieghe 3717, 14874
formazione di schiuma 5939
formiato di sodio 12690
formiato sodico 12690
formol 6050
formola chimica 2681
formola matematica 8672
formula 6083
formula del benzene 1329
formula di struttura 13301
formulare 6084
formulario 6046
formulario a zigzag 14939
formula VVP 14487
fornitura ausiliaria 584
forno 14542
forno essiccatore 4794
forno per la cottura dello smalto 2075
forno per la fusione metallica 8826
foro 7017, 12601
foro centrale di trascinamento 7414
fortran 6088
forza adesiva 214
forza capillare 2281
forza centrifuga 2553
forza coercitiva 3043
forza del corpo sistematica 14319
forza di coesione 1720
forza di colore 3189
forza di pressione 7219
forza di saturazione 11886
forza di taglio 12240
forza elettrica 5024
forza elettromotrice 5041
forze di Van der Waals 14383
forze superficiali 13433
fosfato calcico 2173
fosfato di calcio 2173
fosfato di sodio tribasico 12709
fosforescenza 10204
fosforo 10203, 10207
fossa 10379
fot 10211
foto 10240
foto a colori 3159
foto al tratto 8189
fotocatodo 10214
fotocella 10215
fotocellula 10215
fotocellula a strato di sbarramento 1173
fotochimica 10216
fotocompositrice 10279
fotocomposizione 3061, 10278
fotocomposizione manuale 10256
fotocopia 10270, 10273
fotodiodo 10224
fotoelettricità 10230
fotoelettrico 10225

fotoformatura 10906
fotoformatura incavografica 10253
fotoformatura rilievografica 10233
fotoglittia 14793
fotografia 10240, 10252
fotografia a colori 3159, 3160
fotografia aerea 281
fotografia al tratto 8189
fotografia a mezzatinta 6790
fotografia a stampa diretta 10860
fotografia a tre aperture di diaframma 10306
fotografia a tre colori 13747
fotografia a tre esposizioni 10306
fotografia debole 8127
fotografia diretta 4409
fotografia ritoccata 5443
fotografia tricromica 13747
fotografo 10241
fotoincisione 10232, 10233, 10890
fotoincisione retinata su lastra di zinco 14947
fotoincisione su zinco 14944
fotoincisione (su zinco) con retinatura grossolama 9324
fotoincisore 6781, 10231, 10888, 10889
fotolito 10257
fotolitografia 10257
fotomat 6096
fotomatrice 6096
fotomeccanico 10258
fotometro 5373
fotomontaggio 10261
fotopolimero 10268
fototipia 3099
fototipia a colori 3127
fototipografia 10281
fototitolatrice 6906
fototransistor 10276
fragile 1925
fragilità (di una pellicola) 1926
franatura 3785
francobolli pubblicitari 10611
fraschetta 6204
frastagliati 6191
frate 6192
frazione 6141
frazione composta 10320
frazione con barra diagonale 12592
frazione con barra inclinata 12592
frazione con barra orizzontale 10320
frazione della distillazione del petrolio 10170
frazioni senza barra 9306

frazioni senza linea 9306
fregio di testa 6912
fregio finale 13552
frenatura 1851
frenatura della bobina (di carta) 11358
freno 1848
freno ad aria compressa 313
freno al cilindro 3933
freno automatico 899
freno pneumatico 313
frequenza 6186
frequenza d'errori 5247
frequenza di retino 11992
frequenza relativa 11432
fresa 11733
fresatrice 937, 1048, 11732
fresatura 8939, 11737
friabile 1925
frizione 6194
fronte del carattere 1296
frontespizio 6211
fruscio 11254
frusta spingimatrici alla scatola di distribuzione 4504
fuliggine 12782
fulmicotone 9371
fungicida 6251
funzionamento 9633
funzione cumulativa 3810
funzioni trigonometriche 14038
fuochi coniugati 3386
fuoco 5950
fuoco morbido 12737
fuori commercio 10873
fuori linea 9513
fuori registro 9726
fuoritinta 9542
fuori vendita 10873
furfurolo 6257
fusello 9690
fusetto 9690
fusione 2442, 5913
fusione dei caratteri 14180
fustagno 9056
fustella 4313
fustella in nastro d'acciaio 13103
fustellare 4317
fustellato 3872
fustelle rotative 11693
fusto 12231

gabbia (di legno) 3703
galena 6272
galipot 6274
gallone 6290
galoppino 5240
galvanizzazione 10469
galvano 5071
galvano Albert-Fischer 365
galvano madre 8658
galvanometro 6293
galvano nichelato 9342
galvano originale 8658
galvanoplastica 5063

galvanotipista 5073
gambero 4579
gambo 13109
gamma 6296
gamma di contrasto 3443
ganasce della morsa 14464
ganasce di piega 5981
ganascia 7650
gancio 3547, 8354
garza 6342, 9164
gas 6310
gassoso 6316
gauss 6339
gazzetta 6344
gazzetta poco importante 11199
gazzetta ufficiale 6343
gazzettiere 7700
gel 6352
gelare 6353
gelatina 623, 6356, 7654
gelatina al bicromato 4302
gelatina bicromatata 4302
gelatinatrice 12501
gelatino-cloruro (d'argento) 6365
gelatinoso 6366
gel di silice 12410
gelo di silice 12410
generatore 6377
generatore di caratteri 2633
generatore di programmi per listare informazioni 11470
gergo 7648
geroglifici 6979
gesso 3774, 6738
gesso in polvere 14731
gesso litografico 8258
gesso rosso 11867
gessoso 2606
gettare 2434
getto 2442, 10504
getto d'acido 117
getto di piombo 12984
gherone 3581
ghigliottina per il taglio di materiale metallico 6705
ghisa 2455
giaconetta 7633
gialli all'ossido di ferro 7597
giallo di barite 1143
giallo di benzidina 1330
giallo di cadmio 2146
giallo di cromo 2783, 7977
giallo di tartrazina 13611
giallo di toluidina 13882
giallo di zinco 14959
giallo Hansa 6856
gigliato 4073
gilbert 6398
gilsonite 6411
giocare coi quadrati 7653
giocare con nove quadrati 7653
gioco 1028, 2904

gioco fra in fianchi dei denti 1028
giornale 3958
giornale della moda 5478
giornale della sera 5323
giornale del mattino 9104
giornale di bordo 12309
giornale domenicale 13404
giornale d'opposizione 9642
giornale più largo della norma 1520
giornale provinciale 10988
giornale scandalistico 14920
giornali 10738
giornaliero 6574
giornalismo 11927
giornalista 7700
giornalucolo 11199
giornata di lavoro 14833
giorno di festa annuale 14633
giorno di lavoro 14833
giorno d'inaugurazione 8377
giorno di paga 10068
giorno festivo annuale 14633
girare al minimo 3705
giro 8355, 11578
giro controllato 8355
giro d'affari annuale 640
giro del cilindro 3950
giunta 7689, 12891
giunta di fabbricazione 8941
giunto 3670, 7689
giunto a ginocchiera 13874
giunto cardanico 14325
giunto elettromagnetico 5038
giunto sferico 1094
giuntura 7689
giunzione della cinghia 1306
giunzione in T 13934
giunzione testa a testa 2115
giustapposto 7723
giustezza di colonna 3213
giustezza (di riga) 8720
giustificare le righe 7714
giustificare una linea 7714
giustificativo 14517
giustificazione 7708
glassina 6431
glassina goffrata 5100
gliceride 6479
glicerina 6482
glicerolo 6482
glicina 6485
glicol 5309
glicol etilenico 5309
glicoli 6489
glicolo di propilene 10974
glucosio 4246
goffare 3641
goffrare 5099

goffrata tra due fogli di zinco 14946
goffrato 1562
goffratura 5102
goffratura a mano 6849
gomma 6706
gomma adragante 13951
gomma al bicromato 4304
gomma arabica 6707
gomma bicromatata 4304
gomma clorurata 2741
gomma di cellulosa 2514
gomma dura 6874
gomma esterificata 5260
gommagutta 6294
gomma lacca 12289
gomma naturale 11747
gomma pane 752
gomma per cancellare 5233
gomma (per cancellare) 752
gomma per inchiostro 7345
gomma sintetica 2063, 13496
gomma spugnosa 12912
gommata su tutta la falda 6713
gommatura 6719, 6720
gonfiarsi 2024, 13462
goniometro 10986
gradazione 6534
gradi Celsius 2530
gradiente 6535
gradiente di velocità 14428
gradi Fahrenheit 5435
gradino sul fianco 12342
grado di bianco 4115
grado di durezza 4112
grado di esterificazione 4110
grado di macinazione 5676
grado di raffinazione 4109
grado di sensibilità alla luce di un'emulsione 12851
grado d'umidità 4113, 9033
grado Engler 5177
grado H & D 6823
grado Kelvin 7744
graffa 13040
graffa composta 3300
graffa con becco centrale 3300
graffa con naso centrale 3300
graffa di giunzione della cinghia 1303
graffe 1840
graffiamento 12015
graffiatura 12601
grafico 4257
grafico d'attività multipla 9180
grafitatura 1482
grafite 6571
grammatura 1209
grammo 6556
grammo molecola 9042
grana 6536, 6538, 9234
grana chimica 2682

grana d'eliotipia 3094
grana elettrolitica 5036
grana fine 5671
grana fototipica 3094
grana grossolana 2975
grana (superficiale) 13903
grande quantità di materiale ricomposto 11814
grandezza delle particelle 10014
grandezza naturale 9253
granelli di sabbia 6658
granitoio 6550
granitore 11626
granitura 6546
granitura chimica 2683
grano (inglese) 6539
granulazione con sabbia 11860
grappa 13040
grappa composta 3300
grappe 1840
grassetto 1705
grassetto, in ~ 58
grassetto, (posto) in evidenza in ~ 58
grasso 6601
grasso denso 3394
grasso di lana 14813
grasso di lana purificato 7901
gravità specifica 12827
grazia cuneiforme 14666
grazie 12144
grazie inclinate 9472
grazie, senza ~ 11868
greca 6620
grecare 11893
grembiale 727, 728
grembiule 727, 728
griffa di bloccaggio 8310
griglia del portaracla 4524
griglia di caratteri 2634
griglia per marmorizzare 8625
griglia per spruzzare 8625
grinze 14873
grinze d'espansione 5364
grinze di piegatura 5970
gronda 11891
Grotesca 6675
grumi 8417
gruppo carbossilico 2335
gruppo colore 12452
gruppo d'alimentazione 8850
gruppo delle leve degli stantuffi 10378
gruppo di lavoro 3255
gruppo di piega 5965
gruppo di piste 1110
gruppo di stampa 10852
gruppo di stampa a colori 9172
gruppo di stampa colore 9172
gruppo idrossilico 7133
gruppo inchiostratore 7359

gruppo inferiore 8381
gruppo OH 7133
gruppo per la stampa della copertina 3683
gruppo piegatore a formato variabile 14394
gruppo piegatore a ganasce 7652
gruppo stampante 10852
gruppo stampante ad U 14310
gruppo stampante delle ultime notizie 6228
gruppo stampante reversibile 11560
gruppo stampante supplementare 7233
guadagnare spazio 6386
guadagnare una riga 13554
guaina anodica 642
guaina (di cartone) 12561
guance 7820
guardavista 5420
guardia 5930
guardie 5162
guarnire 8561
guarnire la forma 4722
guarnizione 9828
guastamestieri 1806
guasto 1880, 2040, 5494
guazzo 6530
guida 14012
guida commerciale 4408
guida lavoro 7630
guide 6700
guide del carro 2367
guide del filo metallico 7412
guide di rotolamento del piano portaforma 1275
guillosciare 6702
guizzo 12865
guttaperga 6733

hardware 6877
henry 6966
hertz 6970

iarda 14907
iarda cuba 3805
iarda quadra 12975
iarda quadrata 12975
idratazione 7095
idrato 7089
idrato di cromo 2791
idrato di potassio 10655
idrato di sodio 10695
idrato ferrico 5550
idroapritore 11029
idrocarburi 7100
idrocarburi aromatici 776
idrocarburi di paraffina 424
idrochinone 7127
idrocloruro di diammino-fenolo 4262
idrocloruro di para-amminofenolo 9962
idrofilo 7122

inchiostri stabilizzati a freddo 3058
inchiostri termostabilizzanti 6934
inchiostri trasparenti 14003
inchiostro 7325, 7326
inchiostro a base d'acqua 14567
inchiostro a co-solvente 3627
inchiostro ad acqua 5135
inchiostro ad alta brillantezza 6984
inchiostro ad alta viscosità 6935
inchiostro a doppia tonalità 4666
inchiostro a filo corto 12328
inchiostro all'acqua 5135
inchiostro all'acqua per rotocalco 14566
inchiostro all'anilina 5825
inchiostro appiccicoso 13532
inchiostro a rapida essiccazione 11152
inchiostro a rapida presa 11152
inchiostro argento 12430
inchiostro assorbito 10097
inchiostro base 1185
inchiostro brillante 6453
inchiostro bronzo 1964
inchiostro bronzo azzurro 1960
inchiostro burroso 2121
inchiostro calcografico 10254
inchiostro che sanguina 1559
inchiostro ciclostile 8948
inchiostro classico 9424
inchiostro colorato da giornali 3232
inchiostro coprente 3680, 9612
inchiostro da copia 3548
inchiostro da disegno 7280
inchiostro da giornale 9308
inchiostro da illustrazione 6785
inchiostro d'alluminio 479
inchiostro da scrivere 7326
inchiostro da stampa 7325
inchiostro da stampa offset 9523
inchiostro da trasporto 13964
inchiostro deep-etch 4083
inchiostro d'emulsione 5135
inchiostro di China 7280
inchiostro di penetrazione 10099
inchiostro doppia tinta 4666
inchiostro duro 13151, 13298
inchiostro filante 8338

inchiostro flessografico 5825
inchiostro flessografico pigmentato 10331
inchiostro fluorescente 5891
inchiostro gessoso 2607
inchiostro grasso 6612
inchiostro grasso litografico 14114
inchiostro indelebile 5148, 7259
inchiostro invisibile 7546
inchiostro lacca 7868
inchiostro litografico 8271
inchiostro lucido 6453
inchiostro luminescente 8412
inchiostro magenta 8489
inchiostro magico 8492
inchiostro magnetico 8519
inchiostro mat 4854
inchiostro metallizzato 8828
inchiostro mimeografico 8948
inchiostro moisture-set 9039
inchiostro nero 1479
inchiostro offset 8271, 9523
inchiostro opaco 9614
inchiostro oro 6507
inchiostro pecioso 13532
inchiostro penetrato 10097
inchiostro per calcografia 3524
inchiostro per carbonatura 2312
inchiostro per carta dura per corrispondenza 1722
inchiostro per copiare 3548
inchiostro per decalcomanie su ceramica 2556
inchiostro per eliotipia 3095
inchiostro per fototipia 3095
inchiostro per giornali 9308
inchiostro per impressione a rilievo 13016
inchiostro per incisione 5279
inchiostro per incisione in profondità 4083
inchiostro per la stampa di carte di sicurezza 11837
inchiostro per la stampa di tessuti 2958
inchiostro per la stampa su superfici paraffinate 5485
inchiostro per lavori avventizi 7677
inchiostro per lineare 11783
inchiostro per mordenzatura 5279

inchiostro per morsura 5279
inchiostro per penne stilografiche 4708
inchiostro per rigature 11783
inchiostro per rilievografia 13016
inchiostro per rotocalco 11706
inchiostro per scrivere 7326
inchiostro per serigrafia 11987
inchiostro per stampa a rilievo 13016
inchiostro per stampa in acciaio 13097
inchiostro per stampa oro 6507
inchiostro per stampa su latta 8820
inchiostro per stampigliatura 8637
inchiostro per sviluppo 4226
inchiostro per taglio dolce 3524
inchiostro per trasporto 13964
inchiostro per vetrofanie 4272
inchiostro pesante 6939
inchiostro prottetivo 4226
inchiostro quick-set 11152
inchiostrorepellente 7387
inchiostro resistente agli acidi 13007
inchiostro resistente ai grassi 6602
inchiostro resistente alla luce 10144
inchiostro resistente alla raschiatura 9415
inchiostro resistente allo strofinio 11765
inchiostro rigido 13151, 13298
inchiostro rimasto sulla forma 7386
inchiostro scorrevole 12740
inchiostro senza tiro 12740
inchiostro serigrafico 11987
inchiostro sottostampa per bronzatura 1970
inchiostro stabilizzante ad umido 9039
inchiostro steam-set 13085
inchiostro termoindurente 13697
inchiostro termo-stabilizzante 13697
inchiostro tipografico 8062
inchiostro trasferibile a caldo 5107
inchiostro trasportato 13971
incidente meccanico 1880

incidere 3857, 5182, 5268, 5271
incidere col bulino 4355
incidere in profondità 4078, 4082
incidere in rilievo 5291
inceneratore 7249
incisione 1450, 5191
incisione ad acquaforte 5270
incisione ad una fase 10703
incisione a fondo punteggiato 8597
incisione a fondo sporco 11202
incisione al bulino 8172
incisione all'acqua forte 5270
incisione all'acqua forte a colori 3140
incisione alla vernice molle 12739
incisione all'incavo 4081
incisione al tratto 8174
incisione a più bagni 9181
incisione a puntasecca 4817
incisione a puntini 13161
incisione (chimica) in incavo 7467
incisione con bagno unico 12445
incisione con mordente 1450
incisione d'approfondimento 4081
incisione dei punzoni 11044
incisione dello strato 2992
incisione del punto 4567
incisione del rame 3513
incisione di colori in legno 3195
incisione di musica 9210
incisione di note musicali 9210
incisione elettrolitica 5035
incisione elettronica 907, 5048
incisione frazionata 12058
incisione in forma di disegno a matita 3708
incisione in legno 14794
incisione in piombo 7981
incisione in rame al bulino 3523
incisione in rilievo 11442
incisione in una sola fase 12445
incisione (per acqua forte) 5272
incisione preliminare 5787
incisione profonda 4081
incisione senza coperture 10703
incisione senza polvere 10703
incisione su acciaio 13096
incisione su legno 14799
incisione su linoleum 8214
incisione su pietra 13198

lastra di zinco 14952
lastra di 3.865 mm di spessore 6940
lastra doppia 4641
lastra d'ottone 1864
lastra eliotipica 3097
lastra fotografica 10248
lastra fotopolimerica 10269
lastra fotosensibile 10891
lastra fototipica 3097
lastra incisa 5192
lastra incisa in incavo per imprimere in rilievo 5104
lastra madre 8658
lastra metallica di undici punti 5082
lastra offset 9531
lastra offset deep-etch 4080
lastra offset di carta 9932
lastra offset di zinco 9541
lastra offset secondo il procedimento negativo 13435
lastra offset wipe-on 14760
lastra originale 8658
lastra per fondo a tono pieno 5809
lastra perforata 10128
lastra per la stampa delle ultime notizie 6226
lastra per macchina 10758
lastra per rilievo in ottone 1860
lastra precurvata 10718
lastra presensibilizzata 10732
lastra rettangolare 12963
lastra secca 4816
lastra senza contrasti 5799
lastra trimetallica 14040
lastra troppo poco profunda 12229
lastra umida 14689
lastra wrap-around 14870
lastre dei colori 3165
lastre di stampa per tricromia 13749
lastre in due colori 10903
lastre in più colori 10903
lastre ottenute per fotoformatura 10903
lastre per tricromia 13749
latenza 14527
latitudine di posa 5372
lato anteriore della lastra 7990
lato ballerino 13918
lato comandi 4744
lato comando 4744
lato d'avviamento 9630
lato dello strato sensibile 5422
lato di bianca 11318
lato di comando 4744
lato di emulsione (fotosensibile) 5422
lato di marginazione 7949
lato di servizio 9630

lato emulsionato 5136
lato fieltro 13918
lato in volta 14447
lato operatore 9630
lato pinze 6654
lato tela (della carta) 14778
latta 13823
latta d'inchiostro 7330
latta stagnata 13831
lattato d'ammonio 537
lattato di potassio 10657
lattato potassico 10657
lattina 13823
lattografia 8819
lavaggio 14553
lavare 14556
lavare a spugna 12911
lavare il blanket 2887
lavarulli 11655
lavatrice 14552
lavorante non fidato 14336
lavora rappezzato 1804
lavoratore 14839
lavoratore non qualificato 14347
lavorazione 10894
lavorazione con aerografo 316
lavorazione dell'impasto 11386
lavoro a colori 3197
lavoro a cottimo 10321
lavoro alla linea di montaggio 844
lavoro alla macchina 10773
lavoro al piano di imposizione 13206
lavoro al tratto 8205
lavoro a mano 6851
lavoro amministrativo 2912
lavoro a stretto registro 13809
lavoro avventizi 2460
lavoro d'urgenza 11818
lavoro extra 5410
lavoro giornaliero misurato 8721
lavoro in équipe 12297
lavoro inutile 5499
lavoro manuale 6851
lavoro straordinario 9793
lavoro urgente 11818
layout 7961
legaccio della cinghia 1306
lega (di metalli) 437
lega di piombo 7973
lega leggera 8122
legame fibroso 5590
legante 14423
lega per caratteri 14208
lega per galvanotipia 5072
legare 1401, 1414, 13801
legare con il procedimento di Lumbeck 8405
legare di nuovo 11296
lega tipografica 14208
legato alla rustica 13169
legato in carte 9900
legato in mezza pelle 11129

legatore 1732
legatoria 1412
legatoria industriale 4971
legatrice ad anello 11618
legatrice impaccatrice 2064
legatura 1415
legatura a fanfara 5464
legatura a filo saldato 13743
legatura a filo termosaldante 13743
legatura a fogli mobili 8364
legatura a fogli sciolti 8364
legatura alla Bradel 1845
legatura alla cattedrale 2478
legatura alla lionese 8424
legatura al stile a ventaglio 5468
legatura al stile romantico 11673
legatura a mano 5390
legatura a meandri 4179
legatura a merlotti 4180
legatura antica 680, 9571
legatura architettonica 3634
legatura a spirale 12875
legatura a ventaglio 5468
legatura con cammeo 2236
legatura con dorso di materiale plastico 10408
legatura con dorso mezzo tondo 3684
legatura con dorso piatto 3685
legatura con dorso plastico 10408
legatura con lavoro mosaicato 9110
legatura con medaglione 2236
legatura da amatore 495
legatura da Canevari 2267
legatura da Maioli 8555
legatura dentelle 4181
legatura di Bodley 1675
legatura di Cambridge 2234
legatura di Eve 5322
legatura di nome del proprietario 10159
legatura dos-à-dos 4563
legatura Ducali 4833
legatura etrusco 5317
legatura flessibile 5822
legatura giansenista 7635
legatura Grolier 6664
legatura Harley 6883
legatura incatenata 2584
legatura in colla plastica 208
legatura in cuir bouilli 3807
legatura in cuir ciselé 3808
legatura in cuoio 8012, 8015
legatura in cuoio cesellato 3808
legatura in cuoio sbalzato 3807

legatura in mezza pelle 6755, 11135
legatura in mezza pergamena 11141
legatura in mezza tela 6758, 11130
legatura in pelle 8012, 8015
legatura in pelle di vitello 2200
legatura in pergamena 9995
legatura in smalto 5139
legatura in tela 2948
legatura medioevale 8749
legatura mezzatela 11130
legatura monastica 9064
legatura mudéjar 9162
legatura persiana 11832
legatura rinforzata 11424
legatura romantica 11673
legatura saracena 11874
legatura seminata 12092
legatura senza cucitura 208
legatura sile a mille punti 10526
legatura sile filigrana 10526
legatura splendida 4155
legge della caduta di Stokes 13195
legge di Hooke 7037
leggenda 8031
leggere 11263
leggere le bozze 11262
leggero 13709
legge sulla stampa 6271
leggibilità 8032, 11264
legno a fibre longitudinali 8336
legno compensato 7887, 10505
legno d'abete 12953
legno da zoccolatura 9144
legno di abete 5693
legno di alno 387
legno di balsamo 1107
legno di betulla 1438
legno di bosso 1839
legno di campeggio 8324
legno di conifere 12747
legno di faggio 1284
legno di latifoglie 6881
legno di ontano 387
legno di pero 10075
legno di pino 10362
legno di pioppo 10577
legno di testa 3755
legno di tiglio 1211
legno laminato 7887
legno per carta 11034
legno resinoso 11505
lembo sporgente 7277, 9756
lente 8041, 8042
lente acromatica 111
lente anacromatica 568
lente anastigmatica 581
lente aplanatica 703
lente aplanetica 703
lente apocromatica 705
lente biconcava 3350

liquido non newtoniano 9407

liquido pseudo-plastico 10995

liscia di macchina 12995

lisciare 2081, 10392, 10547

lisciatore alla pietra 296

lisciatura a pietra 5834

lisciatura in macchina 2191

liscio (di carta) 12643

liscio dinamico di carta 4929

liscio-macchina 8446

liscivia di bisolfito 1446

liscivio 8421

liscivio bianco 14717

liscivio di bisolfito 13373

liscivio di sbianca 1547

liscivio nero 1487

lista codice 3034

lista concatenata 2586

lista d'uscita 10859

listella del vantaggio 6281

listello d'appoggio 8630

listello per l'allineamento dei caratteri 14172

listello salvafilo 3904

listino dei corsi 5343

listino dei prezzi 10779

litargirio 8248

litografia 8254, 8255, 8270, 13205

litografia a colori 2818

litografo 8256

litografo per lavoro commerciale 3243

litopone 8276

litro 8281

livellare la forma 10393

livello di qualità accettabile 63

locale di pigmenti e coloranti 3175

locazione 8302

locazione di memoria trasferita 4862

logaritmo 8315

logica 8316

logorio 10, 14635

logotipo 8322

lotta contro il rumore 9385

lotto 1221

lubrificante 8394

lubrificare 8395, 9548

lubrificazione 9557

lubrificazione centralizzata 2537

lubrificazione forzata 6027

luccichini 12304

luce 8106

luce artificiale 793

luce attinica 161

luce bianca 14715

luce d'allarme 14543

luce del giorno 4023

luce di composizione 14174

luce diffusa 4346

luce di lampada ad arco 756

luce di sicurezza 14543

luce diurna 4023

luce entro il telaio 7416

luce inattinica 9394

luce incidente 7248

luce incidente radente 6600

luce interna del telaio 7416

luce monocromatica 9077

luce parassita 4346, 11915

luce polarizzata 10541

luce puntiforme 10528

luce ultravioletta 14252

lucidare 5569, 6255

lucidatura 2037

lucidatura a smalto 10437

lucidatura a spazzole 2002

lucidatura del blanket 6449

lucidezza 6452

luci di sicurezza 11838

lucido 6452, 8419

Ludlow 8401

lumen 8407

lumen-erg 8408

luminanza 8409

luminescenza 8411

luminosità 8413

lunghezza 8037

lunghezza d'alfabeto 459

lunghezza d'avvolgimento 14859

lunghezza della pagina 9858

lunghezza del nastro introdotto nella macchina 14654

lunghezza di blocco 1607

lunghezza di parola di macchina 14819

lunghezza di parola variabile 14396

lunghezza di rottura 1888

lunghezza di sviluppo 14859

lunghezza di taglio 3906

lunghezza d'onda 14615

lunghezza d'onda complementare 3273

lunghezza d'onda dominante 4552

lunghezza fissa di parola 5733

lunghezza focale 5945

lunghezza focale posteriore 998

lunghezza nastro fra gruppi stampanti contigui 8287

lupa 8187

lustro 6452

lux 8420

macchia 12890, 13008

macchia di grasso 6608

macchia di luce 5758

macchia d'inchiostro sul lato in bianca 5406

macchia d'inchiostro sul lato in volta 5405

macchia di patina 2994

macchia di prova (spalmata a mano) 10057

macchia di stampa 8473

macchia d'unto 6608

macchia nera di stampa per eccesso d'inchiostro 9069

macchiare 12162

macchiatura 12636

macchie bianche 12921

macchie bronzo 4164

macchie con centro scuro 12921

macchie d'acqua 14604

macchie del cilindro d'appoggio 1019

macchie di asciugamento 4792

macchie di muffa 8928, 12556

macchie di patina 3153

macchie dovute al legante di patina 1410

macchie d'umidità 14604

macchie nere 1493

macchie ruggine 11823

macchina, a ~ 8725

macchina a bronzare 1969

macchina accoppiatrice 7891

macchina a cilindro 3948

macchina a contatto 3406

macchina ad alta velocità 7000

macchina ad arresto del cilindro 13210

macchina a doppiogiro 14153

macchina ad otto bobine 4997

macchina a due colori 14134

macchina a due giri 14153

macchina affilalama 7826

macchina affilaracle 4529

macchina affilatrice 6643

macchina a foglio 12269

macchina a incidere 5280

macchina a incidere senza coperture 10704

macchina alimentata da bobina 14656

macchina, andato in ~ 6521

macchina a paginare 9869

macchina a pedale 14014

macchina a platina 10452

macchina a quattro colori 6123

macchina arrotondadorsi 1047

macchina a tamburo 3249

macchina a tavola piana 6127

macchina a tre bobine 13771

macchina a tre cilindri 9626

macchina ausiliaria 942

macchina automatica fustellatrice 903

macchinabilità 11790

macchina bisellatrice 1355

macchina calcolatrice 2177

macchina compositrice 3286

macchina compositrice di titoli 4467

macchina compositrice per grandi caratteri 4467

macchina confezionatrice di copertine 2412

macchina continua (per carta) 9923

macchina cucitrice a filo refe 12197

macchina cucitrice a sella 11830

macchina curvatrice 1319, 1320

macchina da comporre Typograph 14234

macchina da foglio 12269

macchina da fondere caratteri 14181

macchina da guillosciare 6703

macchina da incidere 5195, 5280

macchina da stampa 10736

macchina da stampa ad alta velocità 7001

macchina da stampa ad un giro 12466

macchina da stampa a gruppi multipli in linea 14321

macchina da stampa calcografia 3525

macchina da stampa litografica 8263

macchina da stampa offset 9532

macchina da stampa per biglietti 13800

macchina da stampa per forme piane 5775

macchina da stampa pianocilindrica 904

macchina (da stampa) pianocilindrica 5776

macchina da stampa pianocilindrica a un giro 12466

macchina da stampa rotativa per forme piane 5776

macchina discontinua in tondo per cartoni 7515

macchina di struttura robusta 6936

macchina dosatrice e riempitrice 5631

macchina Duplex 4879

macchina etichettatrice 7850
macchina fabbricare tubi 3485
macchina flessografica 5829
macchina fonditrice 2440
macchina foratrice 4733
macchina fotoaddizionatrice 6686
macchina fotocompositrice di titoli 6906
macchina fotografica 10882
macchina fotografica a camera oscura 3993
macchina fotografica a distorsione 11491
macchina fotografica da laboratorio 6277
macchina fotografica orizontale 7040
macchina fotografica per pellicole in bobine 11657
macchina fotografica sospesa 9759
macchina fotografica verticale 14450
macchina fotoripetitrice 6686, 13116
macchina fustellatrice 4318
macchina fustellatrice a platina 3888
macchina gomma (contro) gomma 1524
macchina gommatrice 6721
macchina Gordon 6526
macchina granitrice (a biglie) 6550
macchina impaccatrice 9832
macchina in bianca e volta 10122
macchina incassatrice 2425
macchina incollatrice 6475
macchina incollatrice per astucci pieghevoli 5976
macchina ingommatrice 6721
macchina inseritrice 7438
macchina in tondo 3944
macchina in tondo a più tamburi 9173
macchina laccatrice 14406
macchina lavatrice 14552
macchinale 8725
macchina legatrice in colla (plastica) 10119
macchina legatrice senza cucitura 10119
macchina linocompositrice 8221, 12606
macchina Linotype 8221
macchina numeratrice 9452
macchina occhiellatrice 5417
macchina offset a due colori 14132
macchina offset bicolore 14132

macchina offset con forma in piano 9522
macchina offset da bobina 11698
macchina offset di piccolo formato 12627
macchina offset in bianca e volta 9530
macchina offset monocolore 12450
macchina offset rotativa 11698
macchina paginatrice 9869
macchina patinatrice 2995
macchina per affilare le lame 7826
macchina per affrancare 6149
macchina per applicare fondi benday 12215
macchina per arrotondare angoli 3583
macchina per carta 9923
macchina per confezionare in buste 5214
macchina per cordonare 3718
macchina per curvare il cartone 1320
macchina per distruggere documenti 12352
macchina per etichettare 7850
macchina per fabbricare sacchetti 1073
macchina per fabbricare sacchetti alimentata da bobina 11367
macchina per fabbricare tubi avvolti a spirale 12879
macchina per fermare spigoli 3586
macchina per forare e munire di cordino i cartellini 13536
macchina perforatrice 10132
macchina per formare i dorsi 1015
macchina per formare i dorsi a mano 6813
macchina per forme stereotipiche curve 932
macchina per il lavaggio dei rulli 11655
macchina per il trasporto della carta al pigmento 2333
macchina per incisione 5280
macchina per incollare 6475
macchina per incollare i risguardi 5167
macchina per indirizzi 196
macchina per la chiusura di sacchi 1077

macchina per la fabbricazione a spirale di tubi 12879
macchina per la manifattura di buste 5211
macchina per la rilegatura in colla plastica 13694
macchina per la rilegatura senza cucitura 13694
macchina per la stampa all'anilina 5829
macchina per la stampa dei blocchi note di cassa 2422
macchina per la stampa di pellicole 5647
macchina per la stampa in bianca e volta a quattro colori 6125
macchina per la stampa offset 9532
macchina per la stampa su latta 8821
macchina per lavare i rulli 11655
macchina per lavorazioni pesanti 6936
macchina per l'incisione elettronica 10229
macchina per morsura a bolle 2020
macchina per pulire matrici 8687
macchina per punteggiare 13163
macchina per raccogliere fogli singoli 6330
macchina per scrivere a sfera 6519
macchina per scrivere elettrica 5025
macchina per stampa con incisione in acciaio 4336
macchina per stampare carta da parati 14537
macchina per stampare e fustellare etichette 7852
macchina per stampa rilievografica 4336
macchina per stampa su tessuti 2208
macchina per tagliare ad angolo 8998
macchina per verniciare 14406
macchina piallatrice 10395
macchina piana 5775
macchina piegatrice 5966
macchina piegatrice a coltello 7824
macchina pressatrice delle pieghe 5972
macchina punzonatrice 11048
macchina raccoglitrice 6330
macchina raccoglitrice-cucitrice a filo metallico 6304
macchina rapida 7000

macchina rigatrice a penna 10103
macchina rigatrice a tavola 1312
macchina ripetitrice 13116
macchina rotativa 11699
macchina rotativa Duplex 4879
macchina rotocalco 6587
macchina rotocalco a lastre 10440
macchina rotocalco da foglio 12268
macchina satellite 3249
macchina serigrafica 11991
macchina sminuzzatrice 2727
macchina spianatrice 10395
macchina spolveratrice degli stracci 4891
macchina spolverizzatrice 10701
macchina tipografica 8063
macchina tipo planetario 3249
macchina tiralinee 11784
macchina tiraprove 10950
macchina tiraprove offset 10990
macchina tracciatrice 11937, 12006
macchina trasportatrice della carta pigmento 2333
macchina trituratrice 12352
macchina verticale 14454
macchina yankee 14905
macchine curvatrice (di lastre di stampa) 1587
macchine per legatoria 1424, 1739
macchine per montare i clichè 9145
macchinista 10753
macchinista per offset 10754
macchiolina bianca 1452
macchioline vetrose 5715
macerare 8429
macinare l'inchiostro 6637
macroistruzione 8477
maculatura 8473, 9540
maculatura della bianca 5712
maculatura dovuta a cuscini 1713
maculature 4418
madre 13305
madrevirola a baionetta 1237
madrevite di regolazione 222
maestra 1009
magazziniere 13239
magazzino 8479, 13238
magazzino alto 14357
magazzino ausiliario 12375
magazzino carta 9946

magazzino della lama
d'espulsione 5003
magazzino inferiore 8382
magazzino supplementare
943
magenta 8486
maggiorazione per
opportunità 10546
maggiorazioni di costo 5407
magnesia 8500
magnesio 8496
magnetismo residuo 11452
magro 6153
maiuscola 8557
maiuscole 2287
maiuscole e maiuscolette
2295
maiuscolette 5329, 12617
maiuscoletto, in ~ 60
maiuscoletto, (posto) in
evidenza in ~ 60
malatitta cutanea 4200
malatitta della pelle 4200
malleabile 8585
manata 6826, 8091
manca il carattere, ci ~
435
mancanza d'immagine 4096
mancanza d'inchiostro
14503
mancato trascinamento
dell'inchiostro nel
calamaio 1011
mandare in baracca 10317
mandrino 8587
manganese 8589
manico della lesina 1673
manico della punta 6831
manico (di cassa) 4704
manicotto dei rulli
bagnatori 3979
manicotto dei rulli
umidificatori 3979
maniera criblée 8597
maniera nera 8883
manière noire 8883
manifesto 10609
manina 5718
mano 5718
mano della carta 2046
manodopera a stipendio
fisso 12989
manopola 6829
manoscritto 3534
manoscritto anticipato 241
manoscritto cattivo 1059
manoscritto supplementare
inserito nelle bozze
11597
manovale 14347
manovella 11719
mantenere tolleranza
molto strette 14935
mantissa 8607
manuale 6816
manutenzione 8553
manutenzione preventiva
10777
mappa 8612

mappa astronomica 2497
mappa celeste 2497
marca di fabbrica 13944
marca di fabbrica
registrata 11406
marca di registro 6697
marca di riferimento 4034
marca per interpolazione
7440
marcare 8633
marcatore luminoso 8125
marcatura del feltro 5534
marcatura della tela 14774
marcatura del rullo di
sostegno 1018
marche di guida del
cilindro 3947
marche di guida della
lastra 3943
marche di piegatura 5989
marche di registro 11411
marche di taglio 3899
marche per la rifilatura
14043
marchio della fonderia
10369
marchio depositato 11406
marchio di fabbrica 13944
marcia lenta 9602
marciare ad impulsi 7245
marciume 9120
marezzatura 9670
marginare 8561
marginatore 4940
marginatura 6262
marginatura al piede 6024
margine 2964, 6213, 8627
margine del taglio 12387
margine di cucitura 1035
margine di dorso 1035
margine di piega 1035
margine di pinza 6656
margine di sicurezza 11440
margine di taglio 6033
margine di testa 6909
margine laterale 12376
margine per crocette per il
registro 8297
margine principale 6656
margini rifilati 3860
marmorizzatore 8618
marmorizzazione 8621
marocchino 9106
marocchino di levante a
grani schiacciati 3784
marocchino di levante
cilindrato 3784
marocchino di Nigeria 9344
marocchino francese 6179
marocchino persiano 10155
marrone di Van Dijck
14384
martellatura 11471
martelletto per chiudere le
forme 12248
martello da rigelatore 1014
maschera 8645
maschera a contatto 3410
maschera a due stadi 14159

maschera da ritocco 11515
maschera di carta 6205,
13110
maschera di copertura 8646
maschera di correzione del
contrasto 13889
maschera di correzione del
tono 13889
maschera di regolazione
del contrasto 4540
maschera di riduzione del
contrasto 3445
maschera forata 12924
maschera integrale 7475
maschera nella macchina
fotografica 2241
maschera nitidissima 12236
maschera non nitida 14344
maschera per la correzione
dei colori 3132
maschera per le alte luci
6989
maschera per smorzare il
contrasto 3445
maschera per sovra-
posizione 10861
maschera per zone di
ombra 12222
maschera positiva 10595
maschera proiettata 10933
mascheratura 8647
mascheratura magenta
8490
massicot 8652
mat 8663
materiale in eccesso 5338
materiale in fondo al libro
5165
materiale in scomposizione
4035
materiale pubblicitario 273
materiale sollevato 8095
materiali polari 10543
materia prima 11258
materie coloranti 3176,
4928
materie plastiche 10412
matita Conté 3419
matita rossa 11867
matrice 8677, 8678, 8679
matrice a doppia incisione
4626
matrice da piombo 7993
matrice (di un bollettario)
13305
matrice di vetro 6432
matrice d'ottone 1863
matrice per grande
carattere 4468
matrici a mano 12786
matrici speciali 12786
maturazione 8705
maxwell 8711
mazzetta di fogli 11163
mazzuola 8586
mecanismo d'espulsione a
intermittenza 7788
meccanica 8744
meccanico 8725

meccanismo 4236
meccanismo della ruota di
fusione a sei forme
12480
meccanismo di commuta-
zione del magazzino
8484
meccanismo di disinnesto
14055
meccanismo di
distribuzione 4491
meccanismo d'introduzione
del nastro 13578
meccanismo d'uscita 4145
meccanismo per la scompo-
sizione 4491
media 951
media aritmetica 765
media cassa 6757
mediana 8746
media progressiva 10929
medio de adsorbimento 236
mega- 8763
megahertz 8764
membrana intestinale del
bue 6495
memoria 13230
memoria a disco magnetico
8510
memoria a nastro
magnetico 8530
memoria a nuclei di ferrite
3574
memoria a nuclei
magnetici 3574
memoria a tamburo
magnetico 8514
memoria ausiliaria 947
memoria d'accesso casuale
11233
memoria d'accesso diretto
4399
memoria di transito 2033
memoria esterna 5389
memoria intermedia 2033
memoria interna 7519
memoria magnetica 8523
memoria non permanente
14507
memoria permanente 9418
memoria ponte 2033
memoria principale 8551
memoria protetta 5732
memoria selezione di
corrente di coincidenza
3048
memorizzare 13232
memorizzare e spedire più
tardi 13233
menabò 4857
menisco 8773
meno 8969
mensola 1842
mensola della tastiera 7778
mensola per vantaggi 6284
menzione del diritto
d'autore 3555
menzione della fonte 3721
mercurico 8777

montato su tela 9139
montatura 11013
montatura del calendario 2183
montatura dell'obiettivo 1170
morbido 12730
morchia 12596
mordente per bronzare 12485
mordente per bronzatura 1970
mordenzato eccessivamente 9751
mordenzatora 1450
mordenzatura senza agitazione 13159
mordenzatura statica 13159
morsa 14461, 14462
morsa per la formazione del dorso 1745
morsetti 1426, 2920
morsetto 2869, 2877
morsetto per il fissaggio del blanket 1513
morsetto per il rivestimento 4723
morso 6667, 7690
morso in tela 2953
morsura 1450
morsura al pennello 1999
morsura a macchina 8445
morsura dello zinco 14945
morsura del punto 3886
morsura finale 5669
morsura localmente differenziate 13006
morsura preliminare 5787
morsura sotto squadra laterale 14271
mosaico 7411
motore a corrente alternata 466
motore a corrente continua 4401
motore ad anelli 12574
motore ad anelli trifase 13765
motore ad impulsi 7246
motore a gabbia 12983
motore chiuso 5151
motore collettore trifase 13761
motore di comando 3458, 4747
motore elettrico 5042
motore per la regolazione del registro 11412
motore principale 8546
motore trifase 13764
motto 4237
movimento alternativo 11304
movimento alternato laterale della racla 4532
movimento circolare delle lampade ad arco 2845
movimento coatto 6028
movimento del carro 2366

movimento di andirivieni 6397
movimento di ritorno del carro 2369
movimento discendente 4680
movimento di va e veni a doppia cremagliera 8596
movimento di va e vieni dei rulli 11645
movimento forzato 6028
movimento ipocicloidale 10396
movimento laterale 7935
movimento laterale del nastro di carta 14638
movimento non necessario 5453
movimento oscillante 9706
movimento parallelo 9984
movimento planetario 10396
movimento su e giú 11304
movimento uniformemente accelerato 14305
mozzino 12334
mozzo della ruota 9256
mucillagine 9160
muffa 9120
mulino a palle 1100
mulino a rulli 11642
mulino a tre cilindri 13772
mulino per dispersioni colloidali 3092
mulino per la macinazione degli inchiostri 7371
multicolore 9169, 10553
multiprogrammazione 9196
muschio d'Irlanda 7574
mussola 9216
mussolina 9216
muta dei rulli 12167

nafta 9235
naftenato 9236
naftenato cobaltoso 3010
naftenato di cobalto 3010
nailon 9464
nano 9231
nanometro 9232
nanosecondo 9233
nascite, matrimoni, morti 1440
naso 9331
naso del cono di piega 6072
naso del piegatore 6072
nastri 1113, 13597
nastri conduttori 11807
nastri da introduzione 2299
nastri di guida 2299
nastri d'uscita 4135, 4150
nastri metallici per fissare il timpano 1078
nastri trasportatori 13596
nastro 14639
nastro adesivo 215, 6718, 12026
nastro adesivo a doppia faccia 4606

nastro autoadesivo 12077
nastro autoadesivo a doppia faccia 4606
nastro autoadesivo su una sola faccia 12456
nastro di carta 9949, 14639
nastro di carta somigliante a carta da parati 14534
nastro di gomma 6718
nastro di montaggio 9146
nastro di sigillatura 12026
nastro estensimetro 11509
nastro gommato 6718, 10040
nastro graduato 6416
nastrolibro 1762
nastro magnetico 8524
nastro mal teso 12537
nastro non giustificato 14327
nastro perforato 11046
nastro (perforato) corretto 2897
nastro (perforato) delle correzioni 3606
nastro per mascheratura 8650
nastro prestampato 10731
nastro spolveratore 4902
nastro trasportatore 3482
nastro troppo teso 13810
nebbia d'inchiostro 7373
negativa 9274
negativo 9274
negativo a contatto 3411
negativo al tratto 8183
negativo a tono continuo 3434
negativo con immagine rovesciata 11551
negativo di carta 9930
negativo di montaggio 2134
negativo di selezione 3184, 12130
negativo di selezione diretta 4412
negativo di separazione diretta 4412
negativo duro 6870
negativo fortemente contrastato 6870
negativo in mezzatinta 6786
negativo invertito 11551
negativo leggero 13716
negativo multiplo 6300
negativo piatto 5795
negativo retinato 6786
negativo retinato direttamente 4403
negativo senza punti nelle alte luci 6990
negativo sottoesposto 8386
negativo sovraesposto 13807
negativo su lastra di vetro 6433
negativo velato 5959
neretto 1705, 12095

neretto, in ~ 58
neretto, (posto) in evidenza in ~ 58
neri 1493
nero 5394
nero bluastro 1635
nero brillante 1915
nero d'acetilene 106
nero da gas 6311
nero di carbone 2311
nero di ferro 1480
nero d'ossa 1725
nerofumo di gas 2622
nerofumo di lampada 7894
nerofumo fornace 6260
nerofumo termico 13686
nero giaietto 7658
nero roller 11635
nervare i dorsi 9358
nervature 11223
nervature sul dorso (della lastra) 10460
nervi 11223
nervi falsi 5450
nervi finti 5450
neutralizzare 9289
neutralizzatore di elettricità statica 13069
neutrone 9296
newton 9325
nichel 9335
nichelato 9338
nichelatura 9340
nigrosina 9348
nit 9359
nitidezza 4099, 12237
nitidezza dell'immagine 7168
nitidezza misurata 174
nitrato 9361
nitrato barico 1150
nitrato calcico 2169
nitrato d'ammonio 540
nitrato d'argento 12432
nitrato di bario 1150
nitrato di calcio 2169
nitrato di cellulosa 2517, 9371
nitrato di piombo 7996
nitrato di potassio 10660
nitrato di sodio 12704
nitrato d'uranile 14369
nitrato d'uranio 14369
nitrato potassico 10660
nitrato sodico 12704
nitrobenzene 9368
nitrocellulosa 2517, 9371
nitrocotone 9371
nitroglicerina 9376
nitronaftalene 9379
nitronaftalina 9379
nodi 5595
nodo di catenella 7756
nome dell'editore 7230
nome dello stampatore 7229
nomogramma 9391
non adulterato 14258
non ce n'è 4557

non infiammabile 9400
non interlineato 14329
non in vendita 10873
nonio 14444
non più pubblicato 4423
non sofisticato 14258
non tagliato 1717
non tiene il registro 9726
norma di riferimento 13020
normale 9423
normalizzazione 13026
norma unificata europea 14301
norma unificata francese 14302
nota di margine 8629
nota marginale 8629
notazione decimale codificata in binario 1396
note al centro fra le colonne 2547
note a piede di pagina 6021
note a pie di pagina 6021
note marginali rientranti nel testo 3866
notizia falsa 7013
notizia inventata 7013
notizie superate 6389
nottolino (d'arresto) 10067
novantasei, in ~ 9351
novità editoriale 7719
ñ spagnuola 9223
nucleo 3565
nucleo in ferrite 5561
nucleo magnetico 8504
numerare 1381, 5997
numerare le pagine 9865
numeratore 9450
numeratore (automatico) 9452
numerazione 9449
numerazione delle pagine 5993, 9866
numeri arabi 747
numerico 4358
numerico-analogico 4363
numeri del bollettino di cambio 5345
numeri del diaframma 5937
numeri posti in alto 13416
numeri romani 11671
numeri sotto la linea 6853
numeri superiori 13416
numerizzatore 4366
numero 5607, 9458
numero al piede (di pagina) 4751
numero arretrato 1038
numero binario 1397
numero cancellato 2263
numero d'accettazione 64
numero d'acidità 134
numero della segnatura 12401
numero del libro per il deposito legale 13022
numero di commissione 5617

numero di copie (che si ottengono da un foglio) 9453
numero di fasi 9454
numero di giri 9455, 12853
numero di iodio 7555
numero di linee per pollice 8198
numero di pagina 5995
numero di prova 12835
numero di saggio 12835
numero di saponificazione 11872
numero di scanalature 9456
numero di serie 12140
numero kappa 7729
numero medio di campioni 960
nuova riga per composizione in piedi 2655
nuvolosità 2962
nylon 9464

obbligazione 1718, 4038
obelisco 3955
obiettivo 8042
obiettivo acromatico 111
obiettivo anacromatico 568
obiettivo anastigmatico 581
obiettivo aplanatico 703
obiettivo apocromatico 705
obiettivo compuesto 3319
obiettivo da riproduzione 10898
obiettivo intercambiabile 7491
obiettivo Petzval 10175
obiettivo trattato 2982
occhiellare 5416
occhiellatrice 5417
occhielli 5415
occhiello 4933
occhietto 1217
occhio del carattere 5423
ocra 9481
oculare 5418
oculare per misurare 8724
odore 9498
oersted 9502
officina 9505
officina meccanica 8465
officina tipografica 10829
offset a secco 4812
offset inciso 9521
offuscare 4077
ohm 9547
olandese 4908, 4908, 7020
olandese di sbianca 10679
olandese sfilacciatrice 1881
oleofobo 9583
oleografia 9581
oleorepellente 9566
oliatore 9552
olio da macchina 8397
olio d'antracene 656
olio di alizarina 413
olio di canfora 2252
olio di colza 11243

olio di flemma 565
olio di legno della Cina 14084
olio di lino 8224
olio di lino cotto 1695
olio di lino grezzo 11257
olio di noce di palma 9882
olio di paraffina 9966
olio di perilla 10140
olio di pesce 5717
olio di piede di bue 9260
olio di ravizzone 11243
olio di resina 11687
olio di ricino 2458
olio di ricino solfonato 14091
olio di schisto 12228
olio di semi di lino 8224
olio di semi di lino soffiato 1625
olio di soia 12795
olio di spermaceti 12859
olio di trementina 14107
olio d'oiticica 9569
olio d'oliva de Lucca 8399
olio essenziale 5257
olio essenziale di lavanda 7943
olio essiccante 4793
olio essiccativo 4793
oliofilia 9582
olio lubrificante 8397
olio minerale 8953
olio non siccativo 9399
olio semisiccativo 12100
olio solfonato 13382
olio vulcanizzato 14523
oli soffiati 1626
ombre 12223
ombre di incollatura 10031
ombreggiatura 1249, 7266
omettere 9585
omettere una parola 8021
omissione 9584
omogeneizzatore dell'impasto 868
oncia 5886
oncia avoirdupois 9711
onciale 14262
oncia troy 14065
onda 5911
onorari dell'autore 890
opacimetro 9603
opacità 9604
opacità di stampa 10836
opacità secondo Tappi 13599
opacità su fondo bianco 9608
opacità su fondo carta 9606
opacizzare 4850, 9613
opaco 9611
operaia addetta al mettifoglio 5507
operaio 14839
operaio addetto a levare 5924
operaio addetto al mettifoglio 5507

operaio incaricato del montaggio delle lastre in macchina 10441
operaio non fidato 14336
operaio prenditore 14413
operaio qualificato 7701
operando 9628
opera senza titolo 586
operatore 9638, 9640
operatore alla Ludlow 8402
operatore alla rotativa 11696
operatore alla tastiera 7771
operatore di macchina 8439
operazione 9634
operazione parallela 9985
operazione seriale 12141
opuscolo 1933
opuscolo cucito di piatto (a punti metallici) 5806
ora di andare in macchina 4033
orcina 9671
ordinamento alfabetico 456
ordinare 9673
ordinata 9676
ordinatore 3330
ordinatore digitale 4359
ordinazione 9672
ordinazione fissa 5699
ordinazione su appello 8559
ordine alfabetico 456
ordine cronologico 2822
ordine dei colori successivamente 3120
ordine di magazzino 13189
orecchia 4543
orecchia della matrice 4932
orecchietta della matrice 4932
ore pagate all'ora 7056
ore, stare a ~ 12987
ore straordinarie 9793
organo elaboratore 10895
organosolo 9682
orifizio 9684
originale 808, 3536, 9685
originale al tratto 8168
originale difficile 10088
originale per riflessione 11390
originale pronto per la riproduzione 2248
orlo di lutto 9149
ornamenti a svolazzi e ghirigori 5804
ornamenti decorazioni 5870
ornamento 9688
ornamento a catenelle 2581
ornamento a ripetizione 4270
ornamento a squama di pesce 7169
ornamento del libro 1768
oro in conchiglia 12292
oro in foglie 6509
oro in polvere 6515

pie 6013
piede 6013
piede cubo 3801
piede della pagina 6022
piede (inglese) 6014
piede quadrato 12966
piede quadro 12966
piedino 4941
piedino soffiatore 1623
pie di pagina 6022
piega 3715, 5962
piega a croce 2756
piega al cono 6070
piega alla barra diagonale
  11591
piega a portafoglio 14531
piegacarta 5987
piega fra ganasce 7651
piega incrociata 2756, 11607
piega laterale a soffietto
  6730
piega longitudinale 8334
piega parallela 9979
piegare 5963
piegato 7907
piegatore 5965
piega trasversale 2756
piega traversa 2756
piegatrice 5966
piegatrice a castelli 2027
piegatrice a cono 6071
piegatrice a formato fisso
  5726
piegatrice a ganasce 7652
piegatrice a tasche 2027
piegatrice invariabile 5726
piegatrice per cartoncino
  2343
piegatrice 2/1 14161
piegatrice 3/2 13775
piegatura ad organetto 73
piegatura a fisarmonica 73
piegatura a intercalare
  5964
piegatura a Leporello 73
piegatura a organetto 73
piegatura a quinterni 5964
piegatura a soffieto 73
piegatura non marcata
  7907
piegatura parallela 9981
pieghe dovute alla
  calandra 2186
pieghevole 5820
pieghevole alla francese
  6171
pieghevole pubblicitario
  5967
pietra abrasiva 6647
pietra a olio 9568
pietra da affilare 6647
pietra da affilare Arkansas
  769
pietra di fondo nella
  molazza 1282
pietra litografica 8265
pietra per scarnire 10001
pietra pomice 11037, 12648

pietra pomice in barrette
  12648
pigmenti all'ossido di ferro
  7594
pigmenti colorati 3161
pigmenti inorganici 7420
pigmenti madreperlacei
  9224
pigmenti organici 9680
pigmento 10330
pigmento al molibdeno
  9060
pigmento colorato 4926
pigmento di ftalocianina
  10285
pigmento inorganico 8954
pigmento minerale 8954
pigmento naturale 9248
pigmento rosso fuoco 5698
pigmento secco 4771
pigmento terroso 4939
pignone 10368
pignone conico 1349
pignone del tamburo di
  giustificazione 7716
pila (di carta) 10340
pila di dischi 4444
pila di fogli 10340
pila d'uscita 4147
pila lavatrice 14548
pila mettifoglio 5505
pinatipia 10354
pinholing 10366
pino 10362
pinta inglese 10373
pinta liquida 8239
pinza 6650
pinza del mecanismo
  d'uscita 4143
pinze da rilegatore 1112
pinze oscillanti 13472
pinzetta per pellicola 5641
pinzette 14116
piombo 7968
pioppe 10577
pipetta 10375
pipetta per inchiostro 7377
pirite 11078
pirocatechina 11082
pirocatecolo 11082
pirometro 11086
pirossilina 11087
pista di perforazione 2621
pista magnetica 8531
pistola a punti metallici
  13528
pistola pneumatica 314
pistone ad aria 347
pittura a guazzo 6530
più piccolo assortimento
  2349
più spazio 9100
plagio 10389
planarità 5796
plastico 10407
plastico geografico 11444
plastificante 10410
plastisol 10415
plastometro 10416

platina 10451, 10452
platina automatica 916
platina dell'olandese 1279
platina Gally 6292
platino 10472
poise 10537
polarizzazione 10539
polarizzazione elettrolitica
  10540
poliacrilato 10549
poliamide 10550
policromia 3162, 3163, 10904
policromia al tratto 8166
policromo 9169, 10553
poliestere 10555
polietilene 10556
polimerizzazione 10559
polimero 10558
polire 2080, 11036
polisolfuro di potassio
  10677
polisolfuro potassico 10677
polistirene 10561
polistirolo 10561
politore alla pietra 296
politrice 2039
politura 2037
politura con carbone di
  legna 2647
polivinilacetato 10566
polivinilcloruro 10568
polizza (di caratteri) 6120
pollice 7244
pollice cubo 3802
pollice quadrato 12968
pollice quadro 12968
polo negativo 2480
polo positivo 641
polvere 6267, 9914
polvere antiscartino 687,
  4824, 9539
polvere da bronzatura 1957
polvere d'alluminio 484
polvere d'amianto 293
polvere di bronzo 1965
polvere di carta 5878, 6267,
  8226, 9914
polvere di smeriglio 5113
polvere di tagliare 3896,
  12583
polvere di talco 13562
polvere insetticida 7432
polvere metallica 8822
polvere per granitura 11858
polvere resinosa 5283
polverizzatore 866
pomice 11037
pompa a vuota 14376
pompa d'acqua 14596
pompa d'aria 354
pompa della colla 10039
pompa di alimentazione
  5512
pompa di circolazione 2851
pompa di crogiolo 10686
pompa per inchiostro 7379
pompa per la circolazione
  dell'inchiostro 7380
pompa pneumatica 354

pompa resistente agli acidi
  141
ponitore 3642
ponticelli 2589
pori della carta 9933
porosimetro (ad aria) 4176
porosità 44, 10581
poroso 10583
porpora di bromocresolo
  1951
porporina 1957
porre 3641
porre i caratteri nuovi in
  cassa 7964
porre in evidenza 4459
porta 6322
portabobine 11375
portabobine a stella 11648,
  12863
portabobine a tre braccia
  12863
portabobine doppio 4830,
  13569
portabobine laterale 12383
portabobine supplementare
  944
portabobine triplo 11648
portacarbone 2317
portacasse 2137
portacoltelli 7819
portadiapositiva 14001
portafilm 5644
portafiltri 5659
portalama 7819
portalampada a baionetta
  1237
portalastra 2135
portalibri 1754
porta manoscritto 3544
portanegativo 2135, 9275
portaobiettivo multiplo
  8048
portaoriginale 2242, 3545
portaoriginale pneumatico
  14373
portapagina 9861
portapellicola 5644
portapellicola pneumatico
  14374
portaracla 4530
portare alla riga successiva
  11795
portare a registro 1919
portaretino 11979
portare una riga sul
  vantaggio 5119
portasmusso regolabile 220
portata 1236
portata d'acqua 14605
portata di sviluppo 4232
posa 5368
positiva su lastra di vetro
  6436
positivo 10591
positivo a contatto 3413
positivo al tratto 8191
positivo infinitamente
  variabile 10593
positivo in mezzatinta 6792

processo di patinatura 2988
processo di trasformazione
  3477
processor 10900
processo rolup 11667
prodotti chimici 2690
prodotto antiscartino 9539
prodotto derivato 4199
prodotto per ispessire gli
  inchiostri 1681
prodotto secondario 2132
prodotto viscosità/velocità
  14487
produrre scarti 14557
produzione continua 5873
produzione di massa 8653
produzione doppia 14154
produzione in grande
  quantità 8653
produzione media 956
produzione oraria 9729
profondità dell'impronta
  5106
profondità di fresatura
  11931
profondità di fuoco 4195
profondità d'incisione 5276
profondità di passata della
  fresa 11931
profondità focale 4195
progettista di pubblicità
  271
progettista grafico 3240,
  7962
progetto 4208
progetto di massima 11715
progetto (grafico) 7961
programma 10914, 11734
programma assiematore
  833
programma d'applicazione
  718
programma del calcolatore
  3343
programma diagnostico
  4252
programma di biblioteca
  8088
programma di controllo
  13423
programma di divisione
  delle parole in sillabe
  8319
programma di fabbri-
  cazione 11806
programma di marcia
  11806
programma memorizzato
  13236
programma oggettivo 9470
programma originale 12793
programmare 10915
programmatore 10922
programmatura 12744
proiezione 10934
proiezione del carattere
  2632
proliferazione di funghi
  6252

pronto per la stampa 11286
prontuario 6816
prontuario dei conti fatti
  11287
propagandare insistente-
  mente 10489
propanolo 10973
proporzione 10967
proporzione di, in ~ 10975
proprietà fotogenica 10237
proprietà letteraria
  (riservata) 3553
proprietà reologiche 11584
proprietà superficiali 13436
propulsione a cinghia 1302
propulsione elettrica 5016
prospetto 10976
prossima pubblicazione,
  di ~ 715
proteina 10981
protettore dei fianchi 12382
protezione 6689
protezione di sicurezza
  della cinghia 1305
proto 2914, 3294, 6041
protone 10984
prova 13658
prova a caso 11235
prova a caso sistematica
  13502
prova accelerata 49
prova al bromuro 1948
prova alla cera Dennison
  4165
prova alla fuliggine 12639
prova al nastro adesivo
  11942
prova anonima 11729
prova anti-lettere 10940
prova avanti-lettere 10940
prova d'accartocciamento
  3836
prova d'acido, a ~ 145
prova d'acqua, a ~ 14606
prova d'arricciamento 3836
prova d'autore 889
prova della forza di colore
  1551
prova dell'angolo di
  contatto 594
prova della porosità 10582
prova della resistenza
  all'unghia 5684
prova dell'assorbimento
  d'olio 9551
prova delle bolle 1581
prova dell'inchiostro alla
  spatola 4703
prova del pollice umido
  14697
prova del potere coprente
  6978
prova del tratto di penna
  10091
prova d'essiccamento 4802
prova di abrasione 13
prova di abrasività 13
prova di assorbilità 14967
prova di Cobb 3018

prova di corrosione 3616
prova di doppie pieghe
  MIT 9001
prova di flessione 1325
prova di flottazione 5837
prova di invecchiamento
  302
prova di laboratorio 7859
prova di macchina 6062
prova d'imbarcamento 3836
prova d'ingrassamento
  12022
prova di permeabilità al
  vapore d'acqua 14610
prova di raschiatura 11961
prova di resistenza agli
  acidi 140
prova di resistenza al
  distacco con cere 14632
prova di resistenza
  all'acqua della collatura
  di un'etichetta 9349
prova di resistenza alla
  lacerazione 13617
prova di resistenza all'autoadesione 1606
prova di resistenza allo
  scoppio di Mullen 10578
prova di rigidità H & D
  6824
prova di scumming 12022
prova di sgualcimento 3782
prova di solidità alla luce
  5434
prova di stampa 2896,
  10936
prova di stampa su colori
  secchi 4819
prova di sterilizzazione
  13132
prova dopo lettera 10958
prova in nero 1492
prova K & N 7727
prova micro-contour 8893
prova per correzione
  d'autore 889
prova pulita 2895
provare 13659
provare al tatto 5524
prova sopra carta patinata
  10948
prove progressive 10928
provetta 13668, 13670
prussiato giallo 10650
prussiato rosso 10649
pseudonimo 10101
pseudotricromia 5442
psicrometria 10997
psicrometro 10996
psicrometro ad imbraga-
  mento 12558
pubblicare 11003
pubblicato presso l'autore
  11004
pubblicazione 4970
pubblicazione bimestrale
  14148
pubblicazione mensile 9096

pubblicazione quindicinale
  12101
pubblicazione settimanale
  14667
pubblicista 7700
pubblicità a piena pagina
  6237
pubblicità attorniata da
  testo 6239
pubblicità di giornali 9311
pubblicità murale 14538
pulegge delle cinghie
  d'introduzione 13586
puleggia 1308
pulire 2080
pulisci-lama 7828
pulito 9259
pulsante 11060
pulsante d'avviamento
  13056
pulsante di comando 13056
pulsante di messa in moto
  13056
pulsante rosso (di emer-
  genza) 11323
pulsantiera di comando
  3451
punta da punteggiare 13164
punta della curva carat-
  teristica 13873
punta della lesina 1674
punta di centratura 2555
punta di diamante 4269
punta per correzioni 1671
punta per mordenzatura
  5281
punta per tracciare 12007
puntare 13148
puntasecca 4817
punte di spillo 10363
punteggiare 13160
punteggiatura 11049
punteruolo per legatori
  1672
punti cardinali 2351
punti di conduzione 2936,
  7984, 9623
punti di reticenza 10532
punti di retino chiusi 3388
punti di retino nelle zone
  d'ombra 12219
punti di retino uniti 3388
punti mancanti 8986
puntina (da disegno) 4712
puntine di fissaggio 10371
puntine di registro 9270
punti nelle alte luci 6986
puntine per fissare il
  blanket 1518
puntini di guida 7984
puntini sospensivi 10532
punto 6238, 10518, 10520
punto centrato 2541
punto d'anilina 617
punto decimale 4055
punto del retino 6779
punto di ammorbidimento
  12735
punto di catenella 7756

punto di contatto 7691
punto di controllo 2673
punto di ebollizione 1702
punto di fusione 8769
punto d'impressione 7227
punto d'infiammabilità 5765
punto di rammollimento 12735
punto di rugiada 4242
punto di salto 9384
punto di saturazione 11887
punto duro 6860
punto e mezzo 4999
punto esclamativo 5348
punto e virgola 12098
punto interrogativo 7527
punto isoelettrico 7615
punto metallico 13040
punto morbido 12732
punto nodale 9383
punto sfumato 12732
punto sottoinciso 14274
punto tipografico 10520, 14240
punzonare 6303
punzone 11040
punzonista 11043
purezza colorimetrica 3110
purezza (colorimetrica) 11056
purezza d'eccitazione 5347

quaderni, in ~ 7448
quaderno 12399
quadrante 4259
quadranti che s'arricciano 14908
quadratino 5205
quadratino lineato 4002
quadratino rigato 4002
quadrati vuoti 6173, 11175
quadrato 5120, 11108
quadrato a due 14138
quadrato da una 5120
quadrato lineato 4001
quadratone 5120
quadrato rigato 4001
quadrato tondo 5120
quadrettare 6576
quadricromia 6124
quadro 1792, 6145, 6146
quadro dei comandi 13476
quadro di comando 3459
quadruplo 11088
qualità di conservazione 7740
qualità media 959
qualità media del prodotto finito 955
quantità di calore 11123
quantità di luce 11124
quantità d'inchiostro 7381
quantità minima 8963
quantità per imballaggio in casse 2409
quantità per vagone 2358
quantitativo prodotto in meno 14287

quantizzare 11125
quarta liquida 8240
quartino 11143
quarto colla, un ~ 11139
quarto, in ~ 11143
quastatore dei prezzi 10778
quaternio 11145
quattro colori, in ~ 11109
questionario 11148
quindicina 12101
quinterno 11162, 12399
quinterno di carta 11163
quod vide 11168
quotazione 11173
quotidiano 3958

rabuffo 11476
raccogliere 3070, 6325
raccoglitore 5610
raccoglitore di corrispondenza 8057
raccoglitore di libri 1747
raccoglitore (elevatore) 832
raccoglitrice 6330
raccoglitrice-cucitrice a punti metallici 6304, 11829
raccolta dati 4008
raccolta di quinterni 8038
raccolta di ritagli 11944
raccordare 10047
raccordo fra terminale 1841
racla 4525
racla di gomma 12978
racla di Meyer 5223
racla di ritorno 5850
racletta 4525
raddoppio 11463
raddrizzatore 11317
raddrizzatore a cristallo 3788
raddrizzatura dei galvano 12529
radiale 11190
radiazione 11194
radiazioni attiniche 162
radicale 11195, 11679
radice 11677
radioattivo 11197
raffinatore 11385
raffinatore conico 7693
raffinatore Jordan 7693
raffinatrice 11385
raffinatrice a tre cilindri 13772
raffinazione 1266, 11386
raffinazione ad umido 14677
raffinazione magra 6155
raffinometro 6160
rafforzamento 7483
raffreddamento 3487
raffreddamento ad acqua 14574
raffreddamento ad aria 324
raffreddatore per lastre 10429
raggi attinici 162
raggi beta 1343

raggio 11198
raggio di luce 8128
raggio luminoso 8128
raggi ultravioletti 14254
raggrinzamento 6199, 14874
raggruppamento (di negativi; di positivi) 6684
rallentare 12594
ramanzina 11476
ramatura 3509
ramatura Ballard 1095
rame 3497
rame di base 1191
rameoso 3831
ramico 3822
ramié 11228
rapido essiccamento, a ~ 5482
rapportatore 10986
rapporto 11252
rapporto di miscelazione 9013
rapporto di riproduzione 11484
rappresentante 11474
raschiare 11947
raschiare il cuoio dal rullo in direzione contraria alla fibra 11946
raschietto 5234, 11949
raschietto litografico 9266
raspare 5618
rassegare 6166
rassegarsi 6166
rassegna 9309
rassegna bimestrale 14148
rastello 9212
rastello delle matrici 8695
rastrelliera d'essiccazione 4797
rastrelliera per asciugare 4797
rastrelliera per forme 6067
rastrelliera per seccare 4797
rastrelliera portautensili 13902
rastrello (per torchio litografico) 11951
reagente 11288
reattivo di Herzberg 6971
reazione 11261
recensione d'un libro 1773
recensore 1774
recipiente della colla 6469
reciproco 11301
reclame murale 14538
recto 11319
recupero 11314
redattore 4976
redattore capo 4979
redattore della cronaca locale 2867
redattore di impaginazione 8572
redattore finanziario 2866
redattore responsabile 11526
redattrice 4976

redigere 4965
redopotenziale 11334
refilare 3855
refilare a filo della copertina 3861
refili 12238
refuso 8245, 14881
reggetta metallica 13104
reggi-originale 8180
regione dell'ultravioletto 14253
registrare 8314
registratore 11310
registrazione 226, 11309
registrazione a lunghezza fissa 5729
registrazione a lunghezza variabile 14392
registrazione della lastra con striscioline di carta 12303
registrazione dettaglio 13953
registrazione senza ritorno allo stato di riferimento 9413
registro 11402, 11403
registro a camme 13795
registro a fogli mobili 8365
registro circonferenziale 2856
registro con intaccature 13795
registro dei colori 3169
registro indice 9027
registro indirizzo 199
registro laterale 12372, 12384
registro perfetto 6749
regolare 218, 3449
regolare il contrasto 4537
regolare un rullo 12156
regolatore della densità dell'impasto 11422
regolazione 226
regolazione degli aghi di trascinamento 9271
regolazione dei rulli 11646
regolazione del contrasto 4539
regolazione della pressione di stampa 7222
regolazione della tensione del nastro 14662
regolazione dell'esposizione 5374
regolazione dell'inchiostrazione 3114
regolazione del registro circonferenziale 2857
regolazione del registro laterale 12373
regolazione di temperatura 13641
regolazione di tensione 13648
regolazione di velocità 12850

regolazione elettronica del registro 5051
regolazione PIV 10387
regoletto 11421
regoli laterali 4064
regolo 13242
regolo calcolatore 2176, 12554
regolo del calamaio 7382
regolo della cassa aspirante 4067
regolo di piega 5926
regolo di registro 11416
regolo proporzionale 10970
relazioni umane 7070
relè 11437
rendere l'inchiostro scorrevole 12736
rendere più scuro 4077
rendere trasparente 14004
rendimento 14929
rendimento anodico 646
rendimento dell'inchiostro 7334
rendimento medio 956
reologia 11585
reparto 4185
reparto assortimento 5692
reparto composizione 3292
reparto composizione della pubblicità 248
reparto copia 8839
reparto dell'impaginazione 8576
reparto di composizione per lavori commerciali 7667
reparto di macchine compositrices 3311
reparto fotografico 6276, 10242
reparto lavori avventizi 7667
reparto legatura 1421
reparto ordinazioni 9674
reparto preparazione lastre 10449
reparto pubblicità 252
reparto spedizione 8541
reparto stereotipia 13130
reparto vendite 11848
repellente ai grassi 6604
repellente all'olio 9566
reperimento 11542
repertorio 11466
reprocamera 10882
resa 14929
resa dei colori 3170
resa dei dettagli 4217
resa di copie non vendute 11546
residui della doratura 12508
residuo 11495, 11498
residuo bobina 2116
resina 11501
resina acrilica 158
resina alchilica 425
resina aldeidica 386

resina alla melammina 8767
resina allilica 447
resina amminica 507
resina artificiale 13495
resina coppale 3493
resina cumaron-indenica 3650
resina dammara 3964
resina di acetaldeide 87
resina di formaldeide 6049
resina di isocianato 7614
resina di kauri 7732
resina di legno 14809
resina di poliuretani 10564
resina di ureaformaldeide 14370
resina epossidica 5220
resina esterificata 5260
resina fenolformaldeide 10190
resina fenolica 10191
resina fumarica 6246
resina maleica 8584
resina melammino-form-aldeidica 8766
resina poliammidica 10551
resina poliesterica 10555
resina sintetica 13495
resina stirolica 13317
resinato cobaltoso 3015
resinato di cobalto 3015
resine naturali 9250
resine viniliche 14479
resinificazione 11504
resistente a 11511
resistente agli acidi 145
resistente agli alcali 416
resistente al caldo 6930
resistente al calore 6930
resistente all'abrasione 12
resistente all'acqua 14593
resistente all'alcool 381
resistente alla raschiatura 9414
resistente alla scalfittura 9414
resistente all'inchiostro 7389
resistente allo strofinio 11764
resistenza agli acidi 143
resistenza agli agente chimici 2688
resistenza agli alcali 421
resistenza ai solventi 12781
resistenza al calore 6929
resistenza al formaggio 2679
resistenza all'abrasione 11
resistenza alla cancellatura 5231
resistenza all'acqua 14599
resistenza alla doppie pieghe 5979
resistenza alla flessione 1323
resistenza alla lacerazione 13614, 13616

resistenza alla lacerazione iniziale 13615
resistenza alla luce 5484
resistenza alla piegatura 5979
resistenza alla trazione 13644
resistenza all'impatto 7187
resistenza all'inchiostro 7388
resistenza all'olio 9567
resistenza allo scollamento 1720
resistenza allo sfregamento 11, 5231, 11767
resistenza allo strappo 10300
resistenza allo strofinio 11, 11767
resistenza al sapone 12657
resistenza differenziale 647
resistenza effettiva allo scoppio 2096
resistenza (elettrica) 11508
resistenza interna 647
resistenza meccanica 8739
resistenza relativa allo stato umido 14694
resorcina 11519
responsabile dei rapporti con la clientela 80
responsabile del freno bobina 1853
responsorio 11300
resto 11450, 11498
restaurazione della carta 12355
restringimento (della carta) 12355
rete 9288
rete elettrica 13932
rete trifase 13763
reticulazione 11532
retinare 11972
retino a contatto 3416
retino a contatto magenta 8487
retino a grana 6552
retino a linee 8195
retino a linee incrociate 3759
retino a mattoni 1903
retino a nido d'ape 7034
retino a punto ellittico 2582
retino a scacchiera 2667
retino a trama esagonale 11923
retino a trama grossa 2976
retino autotipico 6794
retino (autotipico) Pola 10545
retino circolare 2847
retino con trama a muratura di mattoni 1903
retino di Schulze 11923
retino di vetro 6438
retino duplex 4882
retino fine 5680

retino granulare 11531
retino granulato 6552
retino lineare 6763, 8195
retino per rotocalco 6591
retino romboedrico 11923
retribuzione dell'autore 890
rettangolare 11315
rettifica della forma elettrotipica 1043
rettificare 6636, 7441
rettificatore 11317
rettificatrice 1048
reversione 11548
rialzi 11223
rialzi finti 5450
rialzo per cliché 14282
rialzo per cliscè 14282
riassunto 20
ribasso 4425
ribobinare 11581
ribobinatrice 11580
ribobinatura a tensione costante 3400
ricalcatura 11471
ricapitolazione 20
ricco di contrasti 12651
ricerca binaria 1399
ricerca operativa 9636
ricercare 12029
ricerca scientifica 11928
ricerca scientifica applicata 720
ricettività all'inchiostro 7385
ricettivo all'inchiostro 7384
richiami 11381
richiamo 2471, 11380
richiedere in esame 11488
richiesta automatica di ripetizione 923
richiesta ufficiale di silenzio stampa 4522
ricognizione ottica dei caratteri 9650
ricomporre 11493
riconoscimento automatico 925
riconoscimento dei caratteri ottici 9650
riconoscimento di caratteri 2639
riconoscimento di configurazione 10065
riconoscimento magnetico di caratteri 8520
ricoperto elettroliticamente 10433
ricoprimento della lente 8045
ricoprire 9871
ricoprire a gommagutta 6295
ricoprire con vernice resistente 13215
ricupero 11314, 11496
ricupero dei solventi 12778
riducente 4184
ridurre 3741, 11341

ridurre l'apertura del diaframma 13212
riduzione 4425, 11349, 11350
riduzione dei dati 4014
riduzione dei punti 4574
riduzione del flusso
d'inchiostro 3892
riduzione dell'esposizione
8970
riduzione del rumore 9385
riduzione di densità 11352
riduzione di opacità 11352
riduzione mediante stampa
su foglio teso di gomma
11753
riempimento 5621, 5633,
9846
riempire 9842
riempitivo 5621, 5635
rientranza 7264
rientranza al disotto della
linea 4755
rientranza diagonale 4256
rientrare 3229, 6386, 7262
rifacimento 8562
riferimenti 11381
riferimento 11380
riferimento alla fonti 1361
rifilare 3855
rifilare a filo della
copertina 3861
rifilare eccessivamente
3740
rifilatura della lastra a filo
5910
rifinitura 5689
rifinitura al bulino 6848
rifiuta l'inchiostro 7387
rifiutare 11464
rifiuto dell'inchiostro 3792
rifiuto dell'inchiostro da
parte dei rulli 11651
rifiuto dell'inchiostro di
sovrastampa 3792
rifiuto di sovrastampa 3792
rifiuto di stampa a punta
di spillo 10366
riflessione 11387, 11389
riflessione speculare 12849
riflesso di luce 6425
riflettometro 11393
rifondere 11454
rifornire di matrici i
magazzini 11798
rifrattometro 11397
rifrazione 11395
riga 8154, 13242
riga a T 14071
riga bianca 14716
riga capovolta 14096
riga corretta 3594
riga di correzione 3599
riga di daccapo 7322
riga di macchina da
comporre 12599
riga di un titolo di giornale
1121
riga fusa 12599
riga morte 14716

riga per la data 4019
riga per se stessa 8160
riga porosa 10584
riga principale di un titulo
6905
rigare 11773
riga smussata 13594
riga sotto il titolo
riportante il nome
dell'autore 2131
riga spostata 8978
rigato con linee parallele
1227
rigatrice 11784
rigatrice a dischi 4445
rigatrice a tavola 1312
rigatura 11781
rigatura del pentagramma
13081
rigatura interna, con ~
14784
rigatura per partite
contabili 82
rigatura separata, con ~
14785
rigature di stampa roto-
calco 11218
rigeneratore del metallo
11576
righe d'acqua 2589, 14585
righe di fondo 5526
righe trasversale 3765
righino 1889, 14884
rigidezza 11613
rigidità 11613
rigidità dielettrica 4325
rigidità di un inchiostro
13153
rigidità reologica 14930
rigidità secondo Bingham
1434
rigonfiamento 13465
rigonfiamento (del blanket)
5103
rigonfiamento sul tessuto
gommato 5103
rilascio di solvente 12780
rilegare di nuovo 11296
rilegato in tela 2950
rilegatore 1732
rilegatura 1415
rilegatura araldica 775
rilegatura a tagli protetti
14906
rilegatura commerciale
4972
rilegatura di lusso 7919
rilegatura editoriale 4972
rilegatura flessibile 8153
rilegatura industriale in
serie 4972
rilegatura intera 6231
rilievo della carta 9915
rilocabile 11448
rilocare 11449
rimando (in un libro) 3764
rimando reciproco 3764
rimaneggiare 9783
rimanenza 11452

rimanenza di bobina 2116
rimanenze 11450, 11546
rimboccatura 14099
rimedio 11453
rimozione della conchiglia
12293
rimozione della velatura
12019
rimozione dell'involucro
esterno 13290
rimozione del sottocolore
14269
rimuovere del strato di
riserva 2902
rimuovere la forma della
macchina 8094
rimuovere uno strato di ricoprimento 13589
rinforzare 7485
rinforzatore 7484
rinforzatore al bromuro
rameico 3502
rinforzatore all'ioduro di
mercurio 8797
rinforzatore all'ioduro di
potassio 7556
rinforzatore allo iodio 7554
rinforzatore al mercurio
8795
rinforzatore al piombo 7991
rinforzo 7483
ringraziamento ai collaboratori 150
riordinare in sequenza
12134
riparare 11462
riperforatore 11465
ripiegatura 14099
riportare alla riga pre-
cedente 11794
riporto 11540, 13961
riposta parlata 869
riprensione 11476
ripristinare 11494
riproducibilità 11479
riproduttrice 6302
riproduzione 11480
riproduzione a colori 3167,
3171
riproduzione ad un colore
9079
riproduzione con retino
granulato 6545
riproduzione dei colori
3172
riproduzione dei valori
tonali 13893
riproduzione di un disegno
a matita 10094
riproduzione duotone 4869
riproduzione monocroma
9079
riproduzione vietata 439
riprografia 11485
riquadro 1793
riquadro ornamentale 9689
riscaldamento a gas 6317
riscaldamento centrale
2536

riscaldamento della
compositrice 3288
riscaldamento elettrico
5020
riscaldatore ad immersione
7184
rischiarire 2901
risciacquare 11619
riscontrare 2663
riserva 11507
riserva colloidale 3093
risguardi 5162
risguardo 5930
risma 11291
risma ridotta 12335
risolvenza 11516
risparmiare le maiuscole
7737
risposta transitoria 11525
ristampa 11429, 11477
ristampa anastatica 579
ristampa corretta ed
arricchita di nuove
aggiunte 558, 11572
ristorare 6255
risultato di stampa 10845
ritagli 9504, 11889, 12238
ritagliare illustrazioni da
un libro per inserirle in
un altro 6561
ritaglio a mano 10045
ritaglio di giornale 9304
ritardatore 11528
ritenuto 11529
ritenzione 11530
ritirarsi 12354
ritiro della carta 12355
ritiro (del metallo) 12356
ritiro del nastro in senso
trasversale 5340
ritoccare col bulino 13900
ritoccare i negativi 12917
ritoccare i negativi 12917
ritoccatore 11533, 12923
ritoccatore intagliatore
5688
ritocco 11534
ritocco al bulino (di lastre
retinate) 11808
ritocco a spruzzo 316
ritocco colore 3173
ritocco con coloranti 4927
ritocco con gesso grasso
litografico 3709
ritocco fotografico 10251
ritorno del carro 2368
riunione trasversale dei
nastri 3748
riunire 846, 6325
rivedere 2663, 11571
rivelatore 4222
rivelatore all'idrochinone
7128
rivelatore cromogeno 4923
rivelatore di para-amminofenolo 11628
rivelatore di spoglio 14554
rivenditore di periodicos
9300

sacchetto di cellofan 2502
sacchetto per biscotti 1441
sacchetto piatto 5773
sacco 11826
sacco a parete multipla 9198
sacco di grande capacità 6613
sacco filtro per anodi 642
saggio 13658
saggio al nastro adesivo 11942
saggio della carta 9952
saggio della forza di colore 1551
saggio della porosità 10582
saggio della resistenza all'unghia 5684
saggio del potere coprente 6978
saggio di abrasione 13
saggio di laboratorio 7859
saggio K & N 7727
saggio micro-contour 8893
saggio MIT 9001
saggio porometrico 10582
sagoma protettiva di carta 6205
sala compositori 3292
sala composizione 3292
sala di composizione di annunzi pubblicitari 248
sala di correttori 11274
sala di lettura 11275
sala di macchine compositrices 3311
sala di redazione 2868
sala macchine 10761
salario a cottimo differenziato 4343
sala stampa 10761
salda d'amido 13046
saldatura ad alta frequenza 6983
sale 11852
sale acido 146
sale di cromo 2782
sale di cucina 12683
sale di fissaggio 5737
sale di Glauber 12719
sale di Schlippe 12727
sale doppio 4653
sale ferrico 5555
sale ferroso 5579
sale metallico 8831
sale virofissatore 13891
saltare 12521
saltato 13057
salti di patinatura 12524
salto 997, 1854, 7705
salvamani 6843
salvamano 6689
salvietta di carta 9929
sandracca 6724
sangue di drago 4688
sanguigna 11867
sapone di resina 11689
sapone metallico 8833
saponificabile 11870

saponificazione 11871
saran 11875
satinatura mat 5179
satira 7895
saturazione 11883
saturazione del colore 11882
saturazione di colore 2760
saturnismo 8000
sbaglio dell'autore 888
sbandamento laterale del nastro 12282
sbarbare (uno stereo) 11947
sbarre a pinze per sospendere la carta 1105
sbarre di guida 6700
sbarretta inclinata 12767
sbassare 10392
sbattere i caratteri di continuo contro il compositoio 5452
sbattimento 11255
sbattimento della lastra 10461
sbavare 11947
sbavatura 4690
sbaveggio 12611
sbiadire 5433
sbiadito 4421
sbianca 1543
sbiancante 1544
sbiancante ottico 9646
sbianca ottica 9645
sbiancare 1538
scabrosita 11711
scaffale 1664, 2414
scaffale per interlinee 8002
scala 4259
scala a fuoco 5955
scala controllo 6629
scala dei colori 3157
scala dei colori europea 14301
scala dei colori francese 14302
scala dei grigi 6629, 6630
scala della riproduzione 11484
scala delle unità 14313
scala dell'obiettivo 8047
scala di lana 14815
scala di rapporto dei fattori filtro 5661
scala di riproduzione 11905
scala di valutazione 11904
scala graduata 11899
scala micrometrica 8901
scala per messa a fuoco 5955
scaldare i tampone (d'inchiostro) 5759
scalpellare 2732
scalpello piatto 5790
scambiatore ionico 7567
scambio d'esperienza 5346
scambio d'ioni 7566
scambio ionico 7566
scanalare 6668, 9108
scanalatura 9431

scanalatura per chiavetta 7786
scanalatura per lubrificazione 9555
scanalature per le staffe intermedie 4546
scandire 11907
scanner 3177, 11910
scappamento 5249
scarabocchiatore 11963
scarabocchio 11964
scarica corona 3590
scarnitura 9999
scarti 4418, 12906, 14561
scarti d'avviamento 11797
scarti di carta giornale 9322
scarti di doratura 12508
scartini 7503
scarto 1934, 8981
scatola 1826, 2378
scatola a cappuccio 2290
scatola a forma di libro 12750
scatola a montare 12182
scatola con interno scorrevole 12552
scatola da esposizione 4461
scatola da montare 5975
scatola degli spazi mobili 12803
scatola di cartone 2344, 2378
scatola di distribuzione 4500
scatola d'imballaggio a compartimenti 3257
scatola di protezione 2398
scatola di sigarette 2830
scatola in cartone ondulato 3620
scatola per cioccolato 2749
scatola per dolci 2149
scatola per formaggio 2677
scatola rigida 12182
scatola Solander 12750
scatoletta 1826
scatoletta per cioccolato 2749
scatolificio 9902
scatolina 1826
scatolina per cioccolato 2749
scatto 10067
scelta del carattere 2750
scelta della impaginazione 2751
scheda a finestra 698
scheda a perforazione marginale 4959
scheda binaria in colonna 3202
scheda binaria per riga 11741
scheda (di schedario) 7271
scheda guida 13507
scheda hollerith 7021
scheda magnetica 8503
scheda perforata 11045

scheda perforata sui margini 4959
schedario 2352, 5619
schegge 12310
scheggiatura 9433
schema a blocchi 1591
schema della cassa 2405
schema della tastiera 7770
schema delle piegature 7963
schema delle segnature 13511
schema di cablaggio 4258
schema di cassa per composizione 7960
schema di connessione 4258
schema di flusso 5867
schema d'impostazione 7197
schema funzionale 1591
schermo 11968
schermo giallo 14916
schermo (per la proiezione) di imaggini 4470
schermo per messa a fuoco 5956
schermo scorrevole 3998
schermo serigrafico 11967
schermo smerigliato 5956
schiacciamento 5769
schiarimento di un negativo 3901
schiarire 2901, 9627
schiocco 2659
schiumare la scoria 12511
schiumini 5943
schizzo 9717, 11715
schizzo a penna 10089
sciacquare 11619
scienza e misura al servizio del proto 10983
sciogliere 4478
scioperante 13274
sciropposo 13498
scissione (in strati) della pellicola d'inchiostro 7397
scivolamento del clichè 9545
scivolare 12559
scivolo 2825
scolapiatti 4694
scolorazione 2083
scolorimento 3141, 4420
scolorimento dell'immagine 5433
scomparsa graduale dell'immagine 14528
scomparto 1829
scomparto di riserva 13448
scomporre 2900, 4484, 7793
scomporre una forma 1894
scompositore 4496
scomposizione 4035, 4492
sconto 4425
sconto per colleghi 4426
scoppio 2092
scoppiometro 2097

sfaldamento del blanket
  1580
sfalzando alternativemente
  13002
sfalzata alternativemente
  13001
sfarinatura 2599
sfera 8353
sfera di battuta 6520
sfere per granitoio 3037
sfibratore 6638
sfibratore a pressione
  10513
sfibratore continuo 3430
sfibratura 6641, 6645
sfilacciare 6151
sfondo 1002
sfrangiatura 4065
sfridi 12238
sfumare 14471
sfumatura del fondo 14469
sfumatura troncata 6876
sfumino 13313
sgarzino 3902
sghembo 1345
sgraffignare 12610
sgrassare 2898, 4106
sgrassare rulli bagnatori
  2886
sguardia 5930
sguardie 5162
sguarnire una forma 13282
siccatività dell'inchiostro
  7342
siccativo giapponese 7636
siccativo in pasta 10028
siccativo liquido 8235
siccità 4772
sicuro 11833
siderografia 12385
Siena 12394
sigillatura a caldo 6931
sigillo 12024
sigillo di ceralacca 14630
sigla dello stampatore 5816
silicato d'alluminio 485
silicato di potassio 10670
silicato di sodio 12716
silicato potassico 10670
silicato sodico 12716
silicio 12415
sillabario illustrato 10313
silografia 805, 14798
silografo 14797
simboli astronomici 857
simboli di correzione 3600
simbolo contabile 81
simbolo convenzionale 750
simbolo del dollaro 4547
simbolo di delimitazione
  4132
similcuoio 8014, 8014
similfoca 359
similpelle 8017
simulare 12440
sinistra, a ~ 3655
sinossi 13490
sintassi 13491
sintesi 13492

sintesi additiva 191
sintesi sottrattiva 13346
sintetico 13493
sinusoidale 12444
sistema 13499
sistema a due fori 1400
sistema a nastri tagliati
  13923
sistema bagnatore 3981
sistema BBR 1241
sistema binario 1400
sistema di bagnatura a
  lama d'aria 327
sistema di circolazione
  dell'inchiostro 2849
sistema Didot 4311
sistema di messa a registro
  con tre punti di riferimento 13767
sistema di messa a registro
  delle lastre 11410
sistema di misura in punti
  10534
sistema di ricircolazione
  d'aria 319
sistema di scuotimento
  della tavola piana
  continua 12227
sistema di stampa 10844
sistema essadecimale 6973
sistema GSA 6688
sistema inchiostratore 7359
sistema internazionale di
  unità 13503
sistema metrico decimale
  8875
sistema metrico (di misura)
  8875
sistema modulare 9028
sistema Munsell 9202
sistema operativo 9631
sistema per la deter-
  minazione dei costi 3630
sistema portante 2374
sistema trasportatore per
  bobine di carta 9936
size press 12492
slitta del compositoio 839
slitta della ruota porta-
  forme 9126
slittamento 12559, 12569
slittamento del cliscè 9545
slittamento trasversale
  13633
slittare 12559
smacchiare i negativi 12917
smallatura ad incavo 2613
smaltare 2074
smalto 5138
smalto a freddo 3054
smalto freddo 5138
smeraldo verde 5110
smerigliare 7031
sminuzzatrice 2728
smontabile 4214
smontare 4449
smorgamento 867
smorto 4421

smussare 1645, 2609
smussatura 1354, 8939
smusso 1344, 5425
smusso sui bordi 1345
socio 13329
soda 12682
soda caustica 12695
sodio 12668
soffiare una cassa 1629
soffiatore d'aria 5456
soffiatore di vapore 13086
soffiatura 5810
soffiatura d'aria 336
soffiature 12475
soffietto 1292, 1293, 6730,
  14074
software 12744
soglia 1884
sol 12749
solarizzazione 12751
solenoide 12754
solfato ammonico d'allu-
  minio 473
solfato barico 1156
solfato calcico 2174, 3774
solfato cerico 2562
solfato ceroso 2568
solfato d'alluminio 468, 487
solfato d'ammonio 544
solfato d'anilina 622
solfato di bario 1156
solfato di calcio 2174, 3774
solfato di cromo e potassio
  2808
solfato di ferro 5580
solfato di ferro ammo-
  niacale 5544
solfato di magnesio 8502
solfato di manganese 8591
solfato di metil-paramminofenolo 8867
solfato di nichel e
  ammonio 9336
solfato di piombo 8005
solfato di potassio 10671
solfato di potassio e
  alluminio 483
solfato di rame 3533
solfato di sodio 12719
solfato di sodio di
  alluminio 12673
solfato di zinco 14955
solfato doppio d'alluminio
  e potassio 483
solfato doppio di cromo e
  potassio 2808
solfato ferrico 5556
solfato ferrico ammonia-
  cale 5544
solfato ferrico d'ammonio
  5544
solfato ferrico di potassio
  5554
solfato ferroso 3499, 5580
solfato ferroso di ammonio
  5573
solfato manganoso 8591
solfato potassico 10671
solfato sodico 12719

solfito 13372
solfito barico 1158
solfito calcico 2175
solfito d'ammonio 547
solfito di bario 1158
solfito di calcio 2175
solfito di potassio 10673
solfito di sodio 12721
solfito potassico 10673
solfito sodico 12721
solfoantimoniato di sodio
  12727
solfoantimoniato sodico
  12727
solfocarbamide 13723
solfocianuro barico 1161
solfocianuro d'ammonio
  550
solfocianuro di bario 1161
solfocianuro di mercurio
  8806
solfocianuro di potassio
  10676
solfocianuro potassico
  10676
solfonare 13380
solfourea 13723
solfuro 13370
solfuro barico 1157
solfuro d'ammonio 546
solfuro d'argento 12435
solfuro di bario 1157
solfuro di cadmio 2145
solfuro di carbonio 2316
solfuro di potassio 10672
solfuro di sodio 12720
solfuro di zinco 14956
solfuro potassico 10672
solfuro sodico 12720
soli 13407
solidificare 12760
solidità alla luce 5484
solidità di colore 3142
solidità di una pellicola
  13929
solido cromatico 3187
soli toni, a ~ 6985
sollecitazione 13261
sollecitazione al taglio
  12241
sollevamento degli spazi
  11620
sollevare i nervi 14848
solubile 12769
solubile in acqua 14603
solubilità 12768
soluzione 12772
soluzione acquosa 739
soluzione acquosa di adragante 1273
soluzione anidra per il
  lavaggio delle lastre 609
soluzione d'albumina 377
soluzione d'asfalto 827
soluzione deep-etch 4089
soluzione di albumina
  bicromatata 2771
soluzione di bagnatura
  3980

spruzzo 12890, 12984
spugna 12910
spugna di viscosa 14484
spulciatura d'un
    programma 4040
spuma 5938
spuntare 1645
squadra 2909, 3255, 12296
squadra d'arresto 5516
squadra del giorno 4025
squadra di fusione 2446
squadra di marginatura
    6338, 7955
squadra di notte 9346
squadra frontale 6212
squadra laterale 7955
squadra per fusione 2453
squadra posteriore 1000
squadra posteriore rego-
    labile 219
squadrato 12962
squadre per unione filetti
    ad angolo 590
squadretta verificatrice
    4724
sta bene 11286
stabile 12992
stabilere i bianchi 6387
stabilimento 10403
stabilità 12990
stabilità ad umido 7141
stabilità dimensionale 4376
stabilità meccanica 8740
stabilizzatore di tensione
    12991
stabilizzazione 12176
staccare l'eccedente del
    colore dai rulli 12281
stacker 12998
staffa di bloccaggio 8310
staffare 8309
staffe intermedie 7511
staffe intermedie di
    bloccaggio 4542
stagno 13822
stampa 10738, 10792, 10802
stampa a colori 3163
stampa a fondo pun-
    teggiato 8597
stampa a iride 12896
stampa all'anilina 5828
stampa a mano 6830
stampa anastatica 579
stampa a parte 9516
stampa a (più) colori 10904
stampa a quattro colori
    6124
stampa a secco 1563, 1574,
    4825
stampa a tre colori 13750
stampabilità 10796
stampa bronzo 1966
stampa calcografica 3527
stampa chiazzata 9116
stampa clandestina 12053
stampa colorata al
    tampone 3140
stampa con accumulo 3078

stampa con inchiostri
    magnetizzabili 8521
stampa con lastre d'allu-
    minio 490
stampa con poca pressione
    7805
stampa con pressione
    minime 7805
stampa da lastre d'acciaio
    incise 4334
stampa dei libri 1783
stampa del verso 1021
stampa di banconote 1128
stampa di biglietti di banca
    1128
stampa di carta da parati
    14536
stampa di carte-valori
    12063
stampa di decalcomanie
    13969
stampa di due colori 14136
stampa di fondi 1478
stampa di francobolli 10601
stampa di lavori avventizi
    7680
stampa di lavori commer-
    ciali 7680
stampa di materiali
    trasparenti sul lato
    rovescio 1045
stampa di musica 9213
stampa di note musicali
    9213
stampa doppia 4646
stampa driografica 4734
stampa elettrostatica 5068
stampa flessografica 5828
stampa fotografica 10270
stampa fresca 6189
stampa giapponese 7637
stampa granulosa 6554,
    12839
stampa Hi-fi 14535
stampa impestata 9161
stampa in bianca 5711
stampa in bianco e nero
    1465
stampa in bicromia 14136
stampa incavorilievografica
    4334
stampa in forma di
    disegno a matita 3708
stampa in piano 10401
stampa in rame 10458
stampa in rilievo 11446
stampa in rotativa 11701
stampa in trasparenza 1045
stampa in volta 1021
stampa iridata 1958, 12896
stampa libera 6162
stampa litografica 8270
stampa litografica su
    pietra 13205
stampa mancata 1876
stampa morbida 5436
stampa occasionale a colori
    sui giornali 11680
stampa offset 9525, 9536

stampa offset deep-etch
    4085
stampa offset incisa 4085
stampa panoramica 9891
stampa particolare 10874
stampa perlate 9116
stampa piana 10401
stampa planografica 10401
stampa preliminare 10729
stampa professionale 13947
stampa quotidiana 3959
stampare 10794
stampare con eccesso di
    pressione 8595
stampare con relativa-
    mente piccole quantità
    d'inchiostro 10798
stampare in bianca 10864
stampare in volta 1052
stampa retinata impestata
    9161
stamparia insufficiente-
    mente attrezzata 14754
stamparia male attrezzata
    14754
stampa rotativa 11701
stampa rotocalco 7471
stampa senza bagnatura
    4734
stampa, sotto ~ 7537
stampa su latta 8819
stampa su tessuto 13678
stampa termografica 13690
stampa tipografica 8065
stampa tipografica
    indiretta 4812
stampati pubblicitari 11001
stampato 10802
stampato a colori 3162
stampato commerciale 2107
stampato fustellato 3872
stampato in rilievo a secco
    1562
stampato per i tipi di ...
    10803
stampatore 10753
stampatore ambulante
    13930
stampatore di lavori avven-
    tizi 7679
stampatore in rame 3526
stampatore itinerante
    13930
stampatore offset 10754
stampatrice 10736, 10805
stampatrice a barra 1169
stampatrice a catena 2590
stampatrice a fili 8694
stampatrice a ruote 14703
stampatrice continua 7012
stampatrice in parallelo
    8192
stampatrice pagina per
    pagina 9854
stampatrice rapida 7001
stampatrice serie 12142
stampa tridimensionale
    14895

stampa umido-su-umido
    14682
stampa Woodbury 14793
stampe per riproduzione
    11483
stamperia 10829
stamperia di lavori avven-
    tizi 7679
stamperia di scarsa qualità
    3028
stamperia di seconda mano
    3028
stampe tabellari 14792
stampigliatura 1596
stampo a goffrare 4312
stampo di chiusura 2910
stampo e controstampo
    8583
stampo i legno compensato
    7660
stampone 244
stampo per impressioni
    1584, 13015
standard 13020
standard dei tempi 13819
standard di ricambio 464
standardizzazione 13026
standolio 13035
standolio di lino 8224
stanghetta 12767
stantuffo del crogiolo 10687
statica 13070
stazionario 13073
stazione di esplorazione
    11276
stazione di lettura 11276
stearato sodico 12717
stearina 13088
steatite 13092
stecca per piegare 5973
stecca (piegacarta) 5987
stecche del levafoglio 5923
stella 850
stella crosciata 12161
stella di fibra 13063
stella di fusione 13043
stelloncino 5621
stemma 3001
stemmario 774
stemperamento 2005
stencil 4885
stendere ad aciugare 11185
stereo 13127
stereo curvo 3849
stereo di gomma 11751
stereo di plastica 10411
stereo piano 5805
stereo ramato 3514
stereotipia all'altezza di
    carattere 14204
stereotipia all'altezza tipo-
    grafica 14204
stereotipia nichelata 9339
stereotipia piana 5805
stereotipista 13128
stilb 13155
stile 8052
stile tipografico 7062, 14185
stilo 12007

stima 5264

stimare un originale per il lavoro di composizione 2457

stimolo luminoso 8135

stipendio fisso, a ~ 12986

stirare 13263

stirolo 13316

stokes 13194

storcere 1820

stracci 9005, 11210

strappare 11017

strappi 7656

strappo 1874

strappo superficiale a secco 4813

strappo (superficiale) a umido 14688

strappo (superficiale) della carta 10295

stratificato 7952

stratificazione della lastra 10427

strato 2987, 5637, 7950

strato acidoresistente 142

strato antiacido 142

strato antialo 667

strato bicromatato 4300

strato corroso 3614

strato deep-etch 4090

strato di colla bicromatada 6473

strato di copia a base di soia 12796

strato di gelatina 6357

strato di gomma 6712

strato di gomma (resistente agli acidi) 6725

strato di separazione 4513

strato intermedio 7509

strato monomolecolare 9081

strato per incisione in profondità 4090

strato resistente agli acidi 10272

strato sensibile alla luce 8131

strato sensibilizzato 12116

strato smalto 5147

strettoio di legatore 1427

striature della carta patinata a lama 11220

striature dello sviluppatore 4223

striature di gomma 6726

striature di racla 4531

striature di stampa 13004

stringa 13277

stringere 7738

strisce di calandra 2197

strisce (di cuoio) 13738

strisce testimoni nella stampa del gioco tra i denti degli ingranaggi 6351

striscia anteriore non stampante delle lastre 12158

striscia di copertura 11834

striscia di prova 13670

striscia di rifilatura 3905

striscia di schiacciamento 3786

striscia divisoria 3905

striscia per il controllo del colore 13055

strisci dei rulli 11641

strisciolina di carta 5741, 11293

strisciolina di conteggio 11293

striscioline 6692

striscione pubblicitario 11002

stroboscopio 13295

strofinaccio 2891

strumento di misura (di resistenza allo scoppio) Mullen 9167

strumento di taratura 8657

strumento per misurare lo slittamento 12566

strumento registratore 11310

struttura 13303

struttura cristallina 3790

struttura reticolata 3389

stuccatura di minutissimi fori 10367

studio dei tempi 13820

studio dei tempi di produzione 2823

studio del lavoro 14845

stufa essiccatore 4794

stufa per essiccare 4794

subappaltare 5472

sublimare 13327

sublimato (corrosivo) 8792

sublimazione 13326

subnitrato di bismuto 12817

succedaneo della trementina 14108

suddividere una parola (in sillabe) 4509

suddivisione del tempo 13818

sudiciume 10292

sunto 20

supercalandra 13409

superficie asferica 829

superficie con soffiature 3049

superficie della carta 9948, 13434

superficie di stampa 10819

superficie indurita lucida 6441

superficie non lucida 8700

superficie percentuale di punto 10112

superficie vaiolata 3049

supervisore 13423

supplemento pubblicitario 278

supporto 13339

supporto dell'albero 969

supporto dell'albero principale 8549

supporto del soffietto 1295

supporto di astralon 855

supporto di legno 14790

supporto in acetato 89

surriscaldamento 9761

sussidio per malattia 7260

sviluppare 4221

sviluppatore 4222

sviluppatore acido 121

sviluppatore al ferro 7588

sviluppatore (automatico) 10900

sviluppatore cromogeno 4923

sviluppatore di para-amminofenolo 11628

sviluppatrice automatica 10900

sviluppo 4222, 4231

sviluppo a grana fine 5672

sviluppo cromogeno 2815

sviluppo eccessivo 9750

sviluppo senza agitazione 13158

sviluppo statico 13158

svincoli 1238

svolazzi 3066

tabella 13508

tabella dei tempi d'esposizione 5375

tabella d'esposizione 5369, 5375

tabella di conversione 3472

tabella per tipoconteggio 3542

tabulatore 13521

tacca 9333, 9432, 12401

tacca di registro 6697

tacca di regolazione 227

tacca dorsale 9334

tacca esterna 9736

tacca francese 9334

taccheggiare un flano 1053

taccheggio 2061, 6335, 9768, 14281

taccheggio a mano 10045

taccheggio con rilievo in gesso 2602

taccheggio laterale 12371

taccheggio mediante morsura di una lastra retinata 2062

taccheggio tra cliché e piede 7497

taccuino 9437

tachimetro 13523

tackmeter 13531

tagli 4961

tagliacarte a ghigliottina 6704

tagliacarte a mano 10484

tagliacarte a programma 920

tagliacarte trilaterale 13754

tagliacartone 2345, 12242

tagliacordona 12586

taglia da fibra 5593

tagliafogli 12263

taglialinee 7978

tagliando 3671

tagli a pettine 3217

tagliare 3855

tagliare angoli 8996

tagliare con tinte scure 1879

tagliare interlinee 3878

tagliare un inchiostro 1879

tagliarina diagonale 592

tagliarina per cartone 1666

tagliata non a squadra 2477

tagliato in squadra 12962

tagliatrice-cordonatrice 3889

tagli cesellati 6494

tagli colorati 3135, 13009

tagli dalla calandra 2187

taglierina a leva 8080

taglierina con volantino a mano 6850

taglierina-cordonatrice 12586

taglierina ghigliottina 6704

taglierina longitudinale 12581

taglierina longitudinale e trasversale 12264

taglierina per fogli 12263

taglierina rapida 6998

taglierina-ribobinatrice 12584

tagli goffrati 6494

tagli marmorizzati 3217, 8615

taglio a registro 3907

taglio da un giornale 9304

taglio dolce 3523

taglio dorato 6412

taglio nella carta 1497

taglio programmato 10920

taglio superiore dorato 6413

taglio superiore dorato, con ~ 13911

tagli rifilati 3860

tagli rossi 11324

tagli spruzzati 12949

talco 13562

talloil 13566

tamburo del freno 1852

tamburo di giustificazione 7715

tamburo essiccatore 4788

tamburo magnetico 8512

tamburo raccoglitore 3072

tamburo scortecciatore 1166

tamburo uscitafogli 12504

tampone 3954, 10085

tampone di cotone 10480

tampone di felpa 4086

tampone d'incisione 4086

tampone di peluche 4086

tampone di sviluppo 4227

tampone (inchiostratore) 7329

triangolo rettangolare 11606
tricloroetilene 14028
triclorometano 2746
tricloruro di fosforo 10209
tricloruro di iodio 7557
tricloruro iodico 7557
tricromia 13750
trigliceride 14036
trimestrale 11137
trinciapaglia 2578
triodo a cristallo 13980
triossido d'antimonio 671
triossido di zolfo 13397
trippoli 4280
trisolfuro d'antimonio 672
trituratrice 12352
trivalente 14063
trucioli 2729
truschino 4720
tubazioni dell'aria 349
tubo attorcigliato 3483
tubo capillare 2284
tubo catodico 2481
tubo da raggi catodici 2481
tubo d'aspirazione 13350
tubo dell'obiettivo 1170
tubo di cartone 2348
tubo di Nessler 9283
tubo elettronico 5056
tubo fluorescente 5892
tungsteno 14788
tunnel di essiccamento 4803
tunnel di essiccazione 4803
turbolenza 14088
turno 12296
tutti i diritti riservati 439
tutto composto 442
tutto testo 12858

ufficio controllo delle tirature 870
ufficio del proto 11692
ufficio di ritagli (di giornali) 2917
ufficio pubblicitario 274
ugello 5456, 9151, 9445
ugello di polverizzatore 12929
ugello spruzzatore 12929
ultima forma 13469
ultima pagina 13058
ultravioletto 14249
umettabilità 14696
umettamento 14698
umettare 3968
umettazione 3977
umidificatore d'aria 7072
umidificazione 3977, 7071
umidità 7073
umidità assoluta 29
umidità dell'aria 7077
umidità relativa (dell'aria) 11433
una colonna, su ~ 12453
uncinetto 8356
unghiature 12973

unghie dei quadranti delle coperte 12973
uniformità di superficie 757
unire 2943
unità 14309
unità ångström 601
unità aritmetica 768
unità di calore inglese 1924
unità di controllo 3461
unità di conversione 3475
unità di perforazione del nastro 13587
unità di stampa 10852
unità di sviluppo (automatico) 10900
unità di visualizzazione 2482
unità elaborativa 2539
unità elettromagnetica 5040
unità elettrostatica 5070
unità stampante (a elementi sovrapposti) 4060
unità supplementare per la stampa del sigillo 12028
unità tipografica 14240
unità tricromatiche 14032
uomo di coscienza (degli operai) 5490
urto 7186
usato 12043
uscire alla luce 714
uscita 4134, 9730, 9731
uscita a mano 9622
uscita a mulinello 5462, 5916
uscita a nastro 13581
uscita automatica 906
uscita con catene 2580
uscita con nastri trasportatori 3726, 13581
uscita diretta 13248
uscita fogli 4134
uscita frontale (dei fogli) 6208
uscita in doppio 4595
uscita in piano 5786
uscita separata 12897
uso di due o più tipi di inchiostro nella stampa in policromia 11866
uso telata, (carta) ~ 11468
usura 10
usura della forma 6077

vade mecum 14377
vaiolature 10363
vale a dire 14500
valenza 14378
valore di adsorbimento 238
valore di kauri-butanolo 7733
valore di non-saturazione 4328
valore di punta 10070
valore effettivo 4982
valore limite 8152

valore medio 8719
valore pH 10181
valore-soglia 13779
valore tonale 13898
valori tristimulus 14062
valutazione 5319
valutazione dei colori 3117
valutazione di merito 8810
valvola d'aria 357
valvola elettronica 5056
vantaggio 6278
vantaggio della macchina compositrice 4152
vantaggio per colonne 12565
vantaggio per composizioni pubblicitarie 202
vantaggio serraforma 2654
vapore acqueo 14608
vapore d'acqua 14608
vapori di piombo 7986
vaporizzatore 5321
varianza 14397
variatore dello spessore dei tratti 8204
varityper 14400
vasca con aspirazione 4674
vasca d'incisione 5290
vasca di sviluppo 4229
vasca per l'acqua di bagnatura 14577
vaschetta dell'adesivo 10029
vaselina 10167
vecchio libro 11203
vedi a tergo 10479
vedi manoscritto 12066
vedi originale 12066
vedi retro 10479
vegetazione micelica 6252
veicolo 14423
velarsi 5958
velato 8136
velatura 2464, 6609, 12637
velatura su lastra per ossidazione 7340
velina 2314, 14243
velino 14425, 14858
velo 5957, 13837
velo chimico 2680
velocità accelerante 51
velocità angolare 604
velocità della luce 14429
velocità della macchina 8466
velocità della scissione in strati della pellicola d'inchiostro 12905
velocità di cesoiamento 11250
velocità di lettura 11273
velocità di rotazione 12853
velocità di stampa 10853
velocità di tiratura 10853
velocità di trasferimento di bit 1456
velocità di trasmissione 13993
velocità media 961
velocità periferica 2859

velocità regolabile 221
velocità regolabile in continuo 13117
velo dicroico 4295
velo d'inchiostro 12018
velo marginale 4955
vendita, in ~ 7421
venditore di libri usati 12045
Veneziani 7068
ventilazione 14440
ventiquattresimo, in ~ 6090, 14120
ventosa aspirante 6092
ventose 13348
verdazzurro 3908
verde brillante 1913
verde cobalto 3008
verde di bromocresolo 1950
verde di cadmio 2142
verde di cobalto 3008
verde di cromo 2777
verde malachite 8582
verde milori 8946
verde Monastral 9066
vergatura 7875
verificare 2663
verificatrice 14442
verificazione 14441
verme dei libri 1784
vermiglione 14443
vernice 7863, 14401
vernice ad alcool 12885
vernice ad olio di seme di lino 8225
vernice a finire 9780
vernice alla cellulosa 2515
vernice all'asfalto 828
vernice antiacido di protezione per morsura 5278
vernice a rapido essiccamento 11153
vernice brillante 6455
vernice d'asfalto 828
vernice debole 12339, 13719
vernice di allungamento 8351
vernice di gomma lacca 12290
vernice essiccante 4804
vernice forte 13299
vernice giapponese 14932
vernice litografica 8279
vernice molle 12739
vernice mordente 1680
vernice non lucida 8707
vernice per etichette 7853
vernice per ricopertura 13216
vernice per soprastampa 9780
vernice protettiva 8648
vernice siccativa 4804
vernice smalto 1084
vernice vigorosa 13299
vernice zapon 14932
verniciabilità 14402
verniciatura 7866

APPENDIXES

Appendix 1 | Old type-body names

Pica system

points	inches*	mm	United Kingdom	USA
1	0.013837	0.351457	twelve-to-pica	one point
1½	0.020755	0.527	eight-to-pica	one and a half point
2	0.0276774	0.703	six-to-pica	two points
3	0.041511	1.054	half nonpareil, four-to-pica	excelsior
3½	0.0484295	1.230	minikin	ruby
4	0.055348	1.406	brilliant	brilliant
4½	0.062267	1.581	diamond	diamond, gem
5	0.069185	1.757	pearl	pearl
5½	0.0761035	1.933	ruby	agate
6	0.083022	2.108	nonpareil	nonpareil
6½	0.0899405	2.284	minionette	emerald
7	0.096859	2.460	minion	minion
8	0.110696	2.811	brevier	brevier
9	0.124533	3.163	bourgeois	bourgeois
10	0.13837	3.514	long primer	long primer
11	0.152207	3.865	small pica	small pica
12	0.166044	4.217	pica	pica
14	0.193718	4.920	english	english
16	0.221392	5.622	great primer, two-line brevier	columbian
18	0.249066	6.325	three-line nonpareil	great primer
20	0.27674	7.028	paragon, two-line long primer	paragon
22	0.304414	7.731		double small pica
24	0.332088	8.434	two-line pica	double pica
28	0.387436	9.839		double english, two-line english
32	0.442784	11.245	two-line great primer	double columbian, four-line-brevier, broken bastard
36	0.498132	12.651	three-line pica	double great primer, two-line great primer
40	0.55348	14.056		double paragon
42	0.581154	14.759		seven-line nonpareil
44	0.608828	15.462		canon
48	0.664176	16.867	four-line pica	four-line pica
60	0.83022	21.084	five-line pica	five-line pica
72	0.996264	25.301	six-line pica	meridian**, six-line pica

* See appendix no. 6.   ** See in the basic table.

Continental or Didot system. (For Fournier system see in basic table)

points	m.m.	mm 1-1-'78	France	Germany
1	0.376065	0.375	un point	Achtelpetit
1$^1/_2$			un point et demi	Achtelcicero
2	0.752	0.75	deux points	Viertelpetit, Non Plus Ultra
2$^1/_2$	0.940	0.94	microscope, myope	Microscopic
3	1.128	1.13	diamant	Viertelcicero, Brilliant, Mailänder
4	1.504	1.50	perle	Halbpetit, Diamant
4$^1/_2$				
5	1.880	1.88	Parisienne, sédanaise	Perl, Pariser
6	2.256	2.25	nonpareille, nompareille	Nonpareille, Halbcicero
6$^1/_2$	2.444	2.44		Insertio
7	2.632	2.63	mignonne	Kolonel, Mignon
7$^1/_2$			petit-texte	Brener
8	3.009	3.00	gaillard(e)	Petit, Jungfer
9	3.385	3.38	petit-romain	Borgis, Bourgeois
10	3.761	3.75	philosophie	Korpus, Garamond
11	4.137	4.13	cicéro	Rheinländer, Manitzer, Kleincicero
12	4.513	4.50	Saint Augustin*	Cicero
13				
14	5.265	5.25	gros-texte	Mittel, Media
16	6.017	6.00	gros-romain*	Tertia, Biblia
18	6.769	6.75	petit-paragon*	Paragon, 1$^1/_2$ Cicero
20	7.521	7.50	gros-paragon*	Text, Textur, Secunda
22			palestine*	grober Text
24	9.026	9.00		Doppelcicero, Zweicicero, $^1/_2$ Konkordanz
28	10.530	10.50	(petit) canon*	Doppelmittel, Roman
32	12.034	12.00		Doppeltertia, Kleine Kanon
36	13.538	13.50	trismégiste	Kanon, Dreicicero, $^3/_4$ Konkordanz
40				Doppeltext
42	15.795	15.75	gros-canon*	grobe Kanon
48	18.051	18.00	Sabon	kleine Missal, Viercicero, Konkordanz
54	20.308	20.25		Missal
56			missel, double canon*	
60	22.564	22.50		grobe Missal, Fünfcicero
66	24.820	24.75		kleine Sabon
72	27.077	27.00	triple canon, double trismégiste	Sabon, Sechscicero
84	31.589	31.50		grobe Sabon, Siebencicero
88				
96	36.102	36.00	grosse-nonpareille	Real, Achtcicero
100			moyenne-de-fonte	
138			grosse-sans pareille	

* according to Marius Audin: Saint-Augustin 12 et 13 points, gros-romain 15 et 16 points, petit-paragon 18 et 20 p. gros-paragon 21 et 22 p., palestine 24 p. petit-canon 28 p. et 32 p., gros-canon 40 et 44 p; double canon 48 et 56 p.

p.t.o.

Old type-body names (continued)

points	The Netherlands	Spain	Italy
1	een punt	un punto	uno punto
1½	anderhalve punt	un punto y medio	
2	twee punten	dos puntos	due punti
2½		dos puntos y medio	
3	microscoop	diamante, ala de mosca	diamante, occhio di mosca
4	diamant	perla, ojo de mosca	milanina, perla
4½	robijn	quatro puntos y medio	
5	parel	diamante y perla, ágata, parisina, parisiano, parisiena	parigina, parmigianina normanno
6	nonparel	nomparela, nomparell, Victoria, jubileo	nompariglia, nonpareille
6½	joli	seis puntos y medio	
7	kolonel	miñona, del siete, glosilla	mignona
7½	brevier	texto pequeño, breviario	
8	galjard	gallarda	testino, gallarda
8½			gagliarda, gagliardao
9	garmond	burguesa, romano pequeña, novela	garamoncino
10	dessendiaan	filosofía, entredós	garmone, garamonda
11	mediaan	lecturilla, lectura (chica)	filosofia
12	cicero, augustijn	cícero, lectura gorda**	lettura
13		San Agustín	
14	groot augustijn	gran texto, texto grande, atanasia	silvio
15			soprasilvio
16	tekst	romana (grande), colombia	testo
18	(kleine) paragon	parangona pequeña, gran novela, texto	piccolo parangone, grosso texto
20	groot paragon	misal, parangona grande, gran parangón	parangone, imperiale, ascendonica parangone, imperiale, ascendonica
22		parangona gruesa, parangón, doble lectura chica	grosso parangone
24	(kleine) kanon, dubbel cicero, dubbel augustijn	palestina, doble lectura	palestina
26		pequeño canon	
28	kanon	peticano, peticanon	
			sopracanoncino
32	groot kanon		cannone
36	Parijse kanon	trimegista, trismegista, canon	trerighe, sopra
40	(kleine) Sabon		corale
42	dubbel groot paragon		
44		gran canon	canone, ducale
48	grote sabon	concordancia (tipográfica)	reale
56	missaal	doble canon	imperiale
60	groot missaal		
72		doble tri(s)megista	papale
88		triple canon	
96		gran nonparela, nonparela gorda	papale cancelleresco
100			grosso di fonderia

** according to the Fournier system: San Agustín.

Remark to appendix 1

Since January 1st, 1978, it is in the Countries of the European Common Market officially no longer permissible to make use of other units of measure besides the traditional measure, the typographic point. According to the 'Système International' *qv*, the Didot point is 1000.333/2660 mm $\approx$ 0.376065 $\times$ 10^{-3} m. See Didot system; point; point system. See also preceding pages. At the time of the delivery of the manuscript for this dictionary to the publisher, the metric system was not general in use. Also due to the new typesetting techniques are now in use Didot, pica, inch and metric units. Conversion of measures of one system into another is a source of confusion.

Appendix 2 | Comparative synopsis of type-design classification and current type names

Modern names

Group	GB*	*France (M. Vox)*	*Germany (DIN 16518)*	*Netherlands (M. Vox)*	*Spain (M. Vox)*	*Italy (Aldo Novarese)*
I	Humanistics	Humanes	Venezianische Renaissance Antiqua	Humanen	Humanísticos	Veneziani
II	Garaldics	Garaldes	Französische Renaissance Antiqua	Garalden	Garaldos	Elzeviri
III	Transitionals	Réales	Barock Antiqua (vorklassizistische Antiqua)	Realen	Reales	Transitionali
IV	Didonics	Didones	klassizistische Antiqua	Didonen	Didonianos	Bodoniani
V	Mechanistics	Mécanes	serifenbetonte Linear-Antiqua	Mechanen	Mecanos	Egiziani
VI	Lineals	Linéales	serifenlose Linear-Antiqua	Linearen	Lineales	Lineari
VII	Inciseds	Incises	(sonstige) Antiqua-Varianten	Inciezen	Incisos	Lapidari
VIII	Scripts	Scriptes	Schreibschriften	Scripten	Escrituras	Scritti
IX	Manuals	Manuaires	handschriftliche Antiqua	Manuaren	Manuales	Fantasie Medioevale
X	broken types	caractères brisés, caractères à fractures	gebrochene Schriften	gebroken letters	tipos quebrados	caratteri fratti
XI	foreign types	caractères étrangers	fremde Schriften	vreemde letters	tipos extranjeros	caratteri stranie

* Names according to BS 2961-1967.

Old, current names

GB/US	France	Germany	Netherlands	Spain	Italy
(Venetian) Old Style	—	Mediäval, Antiqua	Mediaevals-	Romana antigua, Mediaeval	Romano antico
—	Elzévir			—	Elzeviriano
Transitional	—	(Antiqua)	Renaissance-overgangs-	—	—
Modern Face	Didot	klassizistische Antiqua	klassicistische	Redondo	Bodoniano
Egyptian, Clarendon Ionic	Egyptienne, Italienne	Aegyptienne Egyptienne	Egyptiennes	Egipcio Egipciano Clarendón	Egiziano Latino
Sans Serif, Gothic Grotesque	Antique	Grotesk	Schreeflozen, Grotesken, Antieken	Grotesca	Grottesco
—	—	—	—	—	—
Scripts	caractères d'écriture	Schreibschrift	schrijfletters	Escritura	Scritti
—	—	—	—	—	—
Black letter	—	Fraktur, Gotisch (Textur), Schwabacher, Rotunda, Kanzlei	—	Gótico	Gotico
—	—	—	—	—	—

(The rows from "Mediaevals-" through "Renaissance-overgangs-" in the Netherlands column are bracketed together with the label **Romeinen**.)

Paper size names in Britain and America

In preparing this dictionary paper size names, according to the American Paper and Pulp Association appeared to be obsolescent. A considerable number of names, used principally abroad, denote various sizes of cards, boards and ledger, writing, printing and wrapping papers. Various cut sizes of paper and paperboard are not generally designated by specific names. Many of the English names are familiar in the United States. However, though these names are out of use, it may be useful to record them, were it merely in reply to occasional questions.

Albert, Albert note	9.8 × 15.2 cm (3⁷/₈ × 6 in)	commercial note	20.3 × 25.4 cm (8 × 10 in)
antiquarian	78.7 × 134.6 cm (31 × 53 in)	correspondence	8.8 × 11.4 cm (3¹/₂ × 4¹/₂ in)
atlas	66 × 86.8 cm (26 × 34 in)	court	8.8 × 11.4 cm (3¹/₂ × 4¹/₂ in)
bag cap	50.7 × 60.9 cm (20 × 24 in)	cover royal	52 × 64.7 cm (20¹/₂ × 25¹/₂ in)
baronet	8.8 × 14.6 cm (3¹/₂ × 5³/₄ in)	crown	38 × 50.7 cm (15 × 20 in)
bath note	20.3 × 35.5 cm (8 × 14 in) or 20.3 × 17.7 cm (8 × 7 in)	crown octavo	12 × 18.4 cm (4³/₄ × 7¹/₄ in) or 12.7 × 19 cm (5 × 7¹/₂ in)
billet note	15 × 20 cm (6 × 8 in)	czarina	11.4 × 15.2 cm (4¹/₂ × 6 in)
boudoir	10.8 × 13.9 cm (4¹/₄ × 5¹/₂ in)	demy	4.44 × 57.1 cm (17¹/₂ × 22¹/₂ in) for boards
broad double demy	53.3 × 81.2 cm (21 × 32 in)		44.4 × 57.1 cm (17¹/₂ × 22¹/₂ in), 36.8 × 46.9 cm
broad double medium	58.4 × 91.4 cm (23 × 36 in)		(14¹/₂ × 18¹/₂ in) and 36.8 × 45.7 cm
business	5.7 × 9.8 cm (2¹/₄ × 3⁷/₈ in)		(14¹/₂ × 18 in)
cabinet	10.7 × 16.5 cm (4¹/₄ × 6¹/₂ in)	diamond	8.8 × 13.3 cm (3¹/₂ × 5¹/₄ in)
cap	49.5 × 60.9 cm (19¹/₂ × 24 in), for writing paper 35.5 × 43.1 cm (14 × 17 in)	double bag cap	60.9 × 101.5 cm (24 × 40 in)
		double cap	60.9 × 99 cm (24 × 39 in)
carte de visite	6.3 × 10.4 cm (2¹/₂ × 4¹/₈ in)	double copy	50.7 × 81.2 cm (20 × 32 in)
casing	91.4 × 116.8 cm (36 × 46 in)	double crown	50.7 × 76.1 cm (20 × 30 in)
check folio	43.1 × 60.9 cm (17 × 24 in)	double demy	57.1 × 88.8 cm (22¹/₂ × 35 in)
check royal	48.2 × 66 cm (19 × 26 in) or 48.2 × 71 cm (19 × 28 in)	double double cap	71.1 × 86.8 cm (28 × 34 in)
		double double imperial	114.3 × 121.8 cm (45 × 48 in), for boards
club	3.6 × 7.3 cm (1⁷/₁₆ × 2⁷/₈ in)		76.1 × 111.7 cm (30 × 44 in) or
colombier	59.6 × 88 cm (23¹/₂ × 34¹/₂ in)[1]		81.2 × 111.7 cm (32 × 44 in)
commercial letter, commercial post	27.9 × 43.1 cm (11 × 17 in) folded to 12.7 × 20.3 cm (5 × 8 in)	double double small hand	76.1 × 101.5 cm (30 × 40 in)
		double elephant	68.5 × 101.5 cm

	(27 × 40 in)	double small hand	35.5 × 55.8 cm
double flat foolscap	40.6 × 66 cm		(14 × 22 in),
	(16 × 26 in)		38 × 58.4 cm
double folio	55.8 × 86.8 cm		(15 × 23 in),
	(22 × 34 in)		40.6 × 60.9 cm
double foolscap, double cap	43.1 × 68.5 cm		(16 × 24 in)
	(17 × 27 in,		43.1 × 68.5 cm
	for writing paper		(17 × 27 in),
	41.9 × 67.3 cm		50.7 × 76.1 cm
	(16$^1/_2$ × 26$^1/_2$ in)		(20 × 30 in),
double globe	71.1 × 96.5 cm		48.2 × 73.6 cm
	(28 × 38 in)		(19 × 29 in),
double hambro	68.5 × 76.1 cm		53.3 × 78.7 cm
	(27 × 30 in)		(21 × 31 in),
double imperial, double	76.1 × 111.7 cm		50.7 × 73.6 cm
imperial cap	(30 × 44 in),		(20 × 29 in)
	for wrapping paper		43.1 × 58.4 cm
	73.6 × 114.3 cm		(17 × 23 in),
	(29 × 45 in)		45.7 × 71.1 cm
double large	11.4 × 15.2 cm		(18 × 28 in), or
	(4$^1/_2$ × 6 in)		33 × 53.3 cm
double large post	53.5 × 83.8 cm		(13 × 21 in)
	(21 × 33 in)	double small post	45.7 × 73.6 cm
double long demy	40.6 × 106.6 cm		(18 × 29 in) or
	(16 × 42 in)		48.2 × 77.8 cm
double long medium	45.7 × 116.8 cm		(19 × 30$^1/_2$ in)
	(18 × 46 in)	double small royal	60.9 × 96.5 cm
double lump	81.2 × 106.6 cm		(24 × 38 in)
	(32 × 42 in)	double super royal	69.8 × 104.1 cm
double medium	58.4 × 91.4 cm		(27$^1/_2$ × 41 in) or
	(23 × 36 in)		69.8 × 101.5 cm
double music	53.3 × 71.1 cm		(27$^1/_2$ × 40 in)
	(21 × 28 in)	duchess	7.6 × 14.9 cm
double nicanee	114.3 × 142.2 cm		(3 × 5$^7/_8$ in)
	(45 × 56 in)	duchess quarto	15.2 × 22.8 cm
double official	13.9 × 17.7 cm		(6 × 9 in)
	(5$^1/_2$ × 7 in)	duke	13.6 × 17.7 cm
double pinched post	46.9 × 73.6 cm		(5$^3/_8$ × 7 in),
	(18$^1/_2$ × 28 in)		for cards
double post	48.2 × 77.8 cm		8.8 × 13.9 cm
	(19 × 30$^1/_2$ in)		(3$^1/_2$ × 5$^1/_2$ in)
double pott	38 × 63.4 cm	elephant	53.3 × 83.8 cm
	(15 × 25 in),		(21 × 33 in)
	for boards	emperor	121.8 × 182.8 cm
	43.8 × 73.6 cm		(48 × 72 in)
	(17$^1/_4$ × 29 in)	empire	10.7 × 15.8 cm
double quad pott	81.2 × 126.9 cm		(4$^1/_4$ × 6$^1/_4$ in)
	(32 × 50 in)	engineering	55.8 × 76.1 cm
double royal	63.4 × 101.5 cm		(22 × 30 in)
	(25 × 40 in)	executive	6.3 × 10 cm
double small	9.2 × 12 cm		(2$^1/_2$ × 3$^{15}/_{16}$ in)
	(3$^5/_8$ × 4$^3/_4$ in)	extra antiquarian	86.8 × 137.1 cm
	or 8.8 × 12.7 cm		(34 × 54 in)
	(3$^1/_2$ × 5 in)	extra atlas	67.3 × 81.9 cm
double small cap	63.4 × 86.6 cm		(26$^1/_2$ × 32$^1/_4$ in)
	(25 × 34 in)	extra double crown	55.8 × 81.2 cm
double small demy	50.7 × 78.7 cm		(22 × 32 in)
	(20 × 31 in)	extra large	10.4 × 14.9 cm
double small foolscap	41.9 × 67.3 cm		(4$^1/_8$ × 5$^7/_8$ in)
	(16$^1/_2$ × 26$^1/_2$ in)	extra large atlas	68.5 × 88.8 cm

(27 × 35 in)
or 66 × 96.5 cm
(26 × 38 in),
for boards
67.3 × 81.9 cm

extra large casing — (26½ × 32¼ in) 101.5 × 121.8 cm

extra large lump — (40 × 48 in) 60.9 × 91.4 cm

extra large post — (24 × 36 in) 45 × 57.1 cm

extra small — (17¾ × 22½ in) 6.3 × 10.4 cm

extra thirds — (2½ × 4⅛ in) 4.4 × 7.6 cm

flat cap — (1¾ × 3 in) 35.5 × 43.1 cm

flat foolscap — (14 × 17 in) 33 × 40.6 cm

folio, folio post — (13 × 16 in) 43.1 × 55.8 cm

folio note — (17 × 22 in) 13.9 × 21.5 cm

foolscap — (5½ × 8½ in) 34.2 × 43.1 cm

foolscap folio — (13½ × 17 in) 20.3 × 33 cm

foolscap long folio — (8 × 13 in) 16.5 × 40.6 cm

foolscap 1½ sheet, foolscap and half — (6½ × 16 in) 34.2 × 64.7 cm

foolscap 1⅓ sheet, foolscap and third — (13½ × 25½ in) 34.2 × 57.1 cm

foolscap 4 to — (13½ × 22½ in) 16.5 × 20.3 cm

gentlemen's card — (6½ × 8 in) 3.8 × 7.6 cm

grocers 4-lb — (1½ × 3 in) 50.7 × 66 cm (20 × 26 in) or 45.7 × 66 cm (18 × 26 in)

grocers 6-lb — 55.8 × 81.2 cm (22 × 32 in)

half large — 5.7 × 7.6 cm (2¼ × 3 in)

half small — 4.6 × 6 cm (1¹³/₁₆ × 2⅜ in)

half super royal — 36.8 × 50.7 cm (14½ × 20 in), for boards 32.3 × 48.2 cm (12¾ × 19 in)

haven cap — 53.3 × 66 cm (21 × 26 in)

imperial — 55.8 × 76.1 cm (22 × 30 in)

imperial cap — 57.1 × 73.6 cm (22½ × 29 in)

imperial card — 9.5 × 14.6 cm

imperial quarto — (3¾ × 5¾ in) 29.2 × 39.3 cm

index — (11½ × 15½ in) 64.7 × 77.8 cm

index royal — (25½ × 30½ in) 52 × 63.5 cm

intimation — (20½ × 25 in) 9.2 × 15.2 cm

Kent cap — (3⅝ × 6 in) 45.7 × 55.8 cm

ladies' cards — (18 × 22 in) 6.3 × 8.8 cm

large — (2½ × 3½ in) 7.6 × 11.4 cm (3 × 4½ in), for board 48.2 × 60.9 cm (19 × 24 in)

large atlas — 68.5 × 88.8 cm (27 × 35 in) for board 68.5 × 86.3 cm (27 × 34 in)

large casing — 101.5 × 121.8 cm (40 × 48 in) and 96.5 × 121.8 cm (38 × 48 in)

large court 8vo — 10.1 × 12.3 cm (4 × 4⅞ in)

large double loaf — 41.9 × 58.4 cm (16½ × 23 in)

large foolscap — 34.2 × 43.1 cm (13½ × 17 in)

large half royal — 35.5 × 53.3 cm (14 × 21 in)

large imperial — 57.1 × 81.2 cm (22½ × 32 in), for wrapping paper 60.9 × 81.2 cm (24 × 32 in), for boards 55.8 × 81.2 cm (22 × 32 in)

large middle — 46.9 × 60.3 cm (18½ × 23¾ in)

large official postcard — 10.4 × 14.9 cm (4⅛ × 5⅞ in)

large post — 41.9 × 53.3 cm (16½ × 21 in)

large post 8vo — 12.7 × 20.3 cm (5 × 8 in)

large post 4to — 20.3 × 25.4 cm (8 × 10 in)

large royal — 52 × 68.5 cm (20½ × 27 in)

large single — 73.6 × 83.8 cm (29 × 33 in)

large size for driers — 60.9 × 91.4 cm (24 × 36 in)

large whole royal — 52.7 × 67.9 cm

	$(20^3/_4 \times 26^3/_4$ in)
law cap	30.4×38 cm
	$(12 \times 15$ in) or
	31.7×40.6 cm
	$(12^1/_2 \times 16$ in)
ledger royal	48.2×60.9 cm
	$(19 \times 24$ in)
ledger super royal	48.2×68.5 cm
	$(19 \times 27$ in)
long	48.2×81.2 cm
	$(19 \times 32$ in)
long colombier	60.9×124.4 cm
	$(24 \times 49$ in)
long double demy	40.6×106.6 cm
	$(16 \times 42$ in)
long double elephant	69.2×126.9 cm
	$(27^1/_4 \times 50$ in)
long double medium	45.7×116.8 cm
	$(18 \times 46$ in)
long elephant (in reels)	55.8 or 57.1 cm
	$(22$ or $22^1/_2$ in)
long royal	53.3×86.8 cm
	$(21 \times 34$ in)
long thin	53.5×76.1 cm
	$(21 \times 30$ in)
lumber hand	45.7×58.4 cm
	$(18 \times 23$ in),
	44.4×57.1 cm
	$(17^1/_2 \times 22^1/_2$ in),
	45.7×57.1 cm
	$(18 \times 22^1/_2$ in),
	or 49.5×57.1 cm
	$(19^1/_2 \times 22^1/_2$ in)
medium	45.7×58.4 cm
	$(18 \times 23$ in)
medium copying	45.7×57.1 cm
	$(18 \times 22^1/_2$ in)
medium 8vo	13.6×21.2 cm
	$(5^3/_8 \times 8^3/_8$ in)
medium post	45.1×57.1 cm
	$(18 \times 22^1/_2$ in)
medium 4to	21.2×27.3 cm
	$(8^3/_4 \times 10^3/_8$ in),
	drawing
	22.8×29.2 cm
	$(9 \times 11^1/_2$ in)
middle	46.9×55.8 cm
	$(18^1/_2 \times 22$ in)
middle demy	46.9×57.1 cm
	$(18^1/_2 \times 22^1/_2$ in)
middle hand	40.6×55.8 cm
	$(16 \times 22$ in) or
	40.6×53.3 cm
	$(16 \times 21$ in)
middles	55.8×81.2 cm
	$(22 \times 32$ in),
	50.7×76.1 cm
	$(20 \times 30$ in),
	48.2×60.9 cm
	$(19 \times 24$ in) or

	44.4×57.1 cm
	$(17^1/_2 \times 32^1/_2$ in)
midget	3.3×6 cm
	$(1^5/_{16} \times 2^3/_8$ in)
miss	5×7.3 cm
	$(2 \times 2^7/_8$ in)
mr.	3.9×7.9 cm
	$(1^9/_{16} \times 3^1/_8$ in)
mr. and mrs.	5.7×8.5 cm
	$(2^1/_4 \times 3^3/_8$ in)
mrs.	5.7×8 cm
	$(2^1/_4 \times 3^3/_{16}$ in)
music demy	39.3×50 cm
	$(15^1/_2 \times 20$ in)
nicanee	71.1×114.3 cm
	$(28 \times 45$ in)
oblong double foolscap	34.2×86.8 cm
	$(13^1/_2 \times 34$ in)
oblong double small foolscap	33.6×83.8 cm
	$(13^1/_4 \times 33$ in)
official	8.8×13.9 cm
	$(3^1/_2 \times 5^1/_2$ in)
old rag	7.4×12 cm
	$(2^{15}/_{16} \times 4^3/_4$ in)
paste royal	52×63.5 cm
	$(20^1/_2 \times 25$ in)
pasting royal	49.5×60.9 cm
	$(19^1/_2 \times 24$ in)
personal	4.7×8.5 cm
	$(1^7/_8 \times 3^3/_8$ in)
pinched post	36.8×46.9 cm
	$(14^1/_2 \times 18^1/_2$ in)
plutarch	66×91.4 cm
	$(26 \times 36$ in)
pocket note	15.2×24.1 cm
	$(6 \times 9^1/_2$ in)
portfolio	68.5×86.8 cm
	$(27 \times 34$ in)
post	38.7×48.2 cm
	$(15^1/_4 \times 19$ in)
post 8vo	10.7×18 cm
	$(4^1/_4 \times 7^1/_8$ in),
	for cards
	11.4×17.7 cm
	$(4^1/_2 \times 7$ in)
post 4to	18.7×22.8 cm
	$(7^3/_8 \times 9$ in)
postal	57.1×72.3 cm
	$(22^1/_2 \times 28$ in)
postal card	8.8×13.9 cm
	$(3^1/_2 \times 5^1/_2$ in)
postcard	8.8×13.9 cm
	$(3^1/_2 \times 5^1/_2$ in)
pott	31.7×38 cm
	$(12^1/_2 \times 15$ in)
powder leaf	46.9×66 cm
	$(18^1/_2 \times 26$ in) or
	45.7×66 cm
	$(18 \times 26$ in)
princeps	10.7×14.2 cm

	(4$^1/_4$ × 5$^5/_8$ in)		(5 × 6$^1/_2$ in)
princess	10.4 × 14.2 cm	rex	12.7 × 16.5 cm
	(4$_1/_8$ × 5$^5/_8$ in) or		(5 × 6$^1/_2$ in),
	10.7 × 14.2 cm		for cards
	(4$^1/_4$ × 5$^5/_8$ in)		8.2 × 12.7 cm
professional	5.7 × 9.8 cm		(3$^1/_4$ × 5 in)
	(2$^1/_4$ × 3$^7/_8$ in)	royal	50.7 × 63.4 cm
Prussian	81.2 × 106.6 cm		(20 × 25 in),
	(32 × 42 in)		for wrapping paper
pulp royal	52 × 63.4 cm		50.7 × 63.4 cm
	(20$^1/_2$ × 25 in)		(20 × 25 in) or
purple no. 3	43.1 × 66 cm		53.3 × 66 cm
	(17 × 26 in)		(21 × 26 in)
purple no. 4	46.9 × 71.1 cm	royal copying	52.7 × 62.8 cm
	(18$^1/_2$ × 28 in)		(20$^3/_4$ × 24$^3/_4$ in)
quad bag cap	101.5 × 121.8 cm	royal hand	50.7 × 63.4 cm
	(40 × 48 in)		(20 × 25 in)
quad crown	76.2 × 101.6 cm	royal 4to	26 × 31.7 cm
	(30 × 40 in)		(10$^1/_4$ × 12$^1/_4$ in)
quad demy	88.8 × 114.3 cm	saddleback	91.4 × 114.3 cm
	(35 × 45 in)		(36 × 45 in)
quad foolscap	68.5 × 86.8 cm	salesman	5 × 9 cm
	(27 × 34 in)		(2 × 3$^9/_{16}$ in)
quad foolscap 1$^1/_2$ sheet	86.8 × 102.8 cm	sheet and half demy usual	45 × 85.7 cm
	(34 × 40$^1/_2$ in)		(17$^3/_4$ × 33$^3/_4$ in)
quad globe	96.5 × 142.2 cm	sheet and half foolscap	34.2 × 64.7 cm
	(38 × 56 in)		(13$^1/_2$ × 25$^1/_2$ in),
quad imperial	114.3 × 147.3 cm		33.6 × 62.2 cm
	(45 × 58 in)		(13$^1/_4$ × 24$^3/_4$ in) or
quad large	15.2 × 22.8 cm		34.2 × 62.2 cm
	(6 × 9 in)		(13$^1/_2$ × 24$^1/_2$ in)
quad large post	83.8 × 109.2 cm	sheet and half imperial	55.8 × 114.3 cm
	(33 × 43 in)		(22 × 45 in)
quad medium	96.5 × 121.8 cm	sheet and half post	49.5 × 59.6 cm
	(38 × 48 in)		(19$^1/_2$ × 23$^1/_2$ in) or
quad post	81.2 × 101.5 cm		50.1 × 59.6 cm
	(32 × 40 in)		(19$^3/_4$ × 23$^1/_2$ in)
quad pott	63.4 × 81.2 cm	sheet and half quad foolscap	86.8 × 102.8 cm
	(25 × 32 in),		(34 × 40$^1/_2$ in)
	63.4 × 76.2 cm	sheet and half small foolscap	33.6 × 62.8 cm
	(25 × 30 in) or		(13$^1/_4$ × 24$^3/_4$ in)
	66 × 81.2 cm	sheet and third foolscap	34.2 × 57.1 cm
	(26 × 32 in)		(13$^1/_2$ × 22$^1/_2$ in),
quad royal	101.5 × 126.9 cm		33.6 × 55.8 cm
	(40 × 50 in)		(13$^1/_4$ × 22 in) or
quad small	12 × 18.4 cm		34.2 × 55.8 cm
	(4$^3/_4$ × 7$^1/_4$ in),		(13$^1/_2$ × 22 in)
	12.7 × 17.7 cm	sheet and third small foolscap	33.6 × 55.8 cm
	(5 × 7 in) or		(13$^1/_4$ × 22 in)
	13.9 × 17.7 cm	short	43.1 × 53.3 cm
	(5$^1/_2$ × 7 in)		(17 × 21 in)
quad small demy	78.7 × 101.5 cm	single hambro	46.9 × 60.9 cm
	(31 × 40 in)		(18$^1/_2$ × 24 in) or
queen's	8.8 × 13.6 cm		41.9 × 58.4 cm
	(3$^1/_2$ × 5$^3/_8$ in)		(16$^1/_2$ × 23 in)
reduced small	5.3 × 8.8 cm	single loaf	60.9 × 86.8 cm
	(2$^1/_8$ × 3$^1/_2$ in)		(24 × 34 in) or
regina	12 × 16.8 cm		58.4 × 83.8 cm
	(4$^3/_4$ × 6$^5/_8$ in) or		(23 × 33 in)
	12.7 × 16.5 cm	single lump	54.5 × 68.5 cm

	(21½ × 27 in)
single small hand	38 × 50.7 cm (15 × 20 in)
small	6 × 9.2 cm (2³/₈ × 3⁵/₈ in) or 6.3 × 8.8 cm (2½ × 3½ in)
small demy	46.9 × 55.8 cm (18½ × 22 in)
small double loaf	41.9 × 53.3 cm (16½ × 21 in)
small double post	48.2 × 73.6 cm (19 × 29 in)
small foolscap	33.6 × 41.9 cm (13¼ × 16½ in)
small half imperial	38 × 56.5 cm (15 × 22¼ in)
small half royal	33 × 51.4 cm (13 × 20¼ in)
small hand	50.7 × 76.1 cm (20 × 30 in), 48.2 × 73.6 cm (19 × 29 in), 48.2 × 71.1 cm (19 × 28 in), 45.7 × 71.1 cm (18 × 28 in), 45.7 × 66 cm (18 × 26 in), 43.1 × 63.4 cm (17 × 25 in), 40.6 × 60.9 cm (16 × 24 in), 38 × 58.4 cm) (15 × 23 in), 38 × 50.7 cm (15 × 20 in), 35.5 × 55.8 cm (14 × 22 in), 35.5 × 50.7 cm (14 × 20 in), 33 × 48.2 cm (13 × 19 in) or 33 × 45.7 cm (13 × 18 in)
small medium	44.4 × 55.8 cm (17½ × 22 in)
small post	36.8 × 46.9 cm (14½ × 18½ in)
small royal	48.2 × 60.9 cm (19 × 24 in)
small single	54.6 × 68.5 cm (21½ × 27 in)
small whole royal	49.5 × 64.7 cm (19½ × 25½ in)
thirds, third card, third large	3.8 × 7.6 cm (1½ × 3 in)
town	5 × 7.6 cm (2 × 3 in)
treasury	48.2 × 60.9 cm (19 × 24 in)
typewriter double cap	40.6 × 66 cm (16 × 26 in)
typewriter size	21.5 × 27.9 cm (8½ × 11 in)
viscount	12.7 × 16.5 cm (5 × 6½ in)
whole imperial	57.1 × 81.2 cm (22½ × 32 in)
writing demy	45.7 × 55.8 cm (18 × 22 in)
writing royal	48.2 × 60.9 cm (19 × 24 in)

Sheet sizes in the United States

The following are some of the common sheet sizes (in inches) in the United States:

bible paper	25 × 38, 28 × 42, 28 × 44, 32 × 44, 35 × 45, 38 × 50
blanks (plain)	22 × 28, 22½ × 24½, 22½ × 28½, 23 × 43, 28 × 44, 34 × 43
blanks (coated):	22 × 28, 22½ × 42½, 28 × 44
blotting paper	19 × 24, 24 × 38
bogus bristol	22½ × 28½
bond paper (wood pulp and cotton content)	8½ × 11, 8½ × 13, 8½ × 14, 10 × 14, 11 × 17, 17 × 22, 17 × 28, 17½ × 22½, 19 × 24, 20 × 28, 22 × 25½

	22×34	label paper (coated one-side book)	20×26
	$22^1/_2 \times 35$		25×38
	23×29		26×40
	24×38		28×42
	28×34		28×44
	34×44		32×44
	35×45		35×45
book paper (uncoated)	$17^1/_2 \times 22^1/_2$		36×48
	19×25		38×50
	$22^1/_2 \times 35$		41×54
	23×29	ledger paper (wood pulp and cotton	
	23×35	content)	17×22
	25×38		17×28
	28×42		19×24
	28×44		22×34
	32×44		$22^1/_2 \times 22^1/_2$
	35×45		$22^1/_2 \times 34^1/_2$
	38×50		24×38
book paper (coated two sides)	$22^1/_2 \times 35$		$24^1/_2 \times 24^1/_2$
	24×36		28×34
	25×38	lithol label paper (coated one side)	25×38
	26×40		28×42
	28×42		28×44
	28×44		32×44
	32×44		35×45
	35×45		36×48
	36×48		38×50
	38×50		41×54
boxboard	25×40	manifold paper	17×22
butchers paper	12×18		17×26
	18×24		17×28
	20×30		19×24
	24×36		21×32
	30×40		22×34
cover paper (wood pulp and cotton			24×38
content)	20×26		26×34
	23×29		28×34
	23×35	manuscript cover	18×31
	26×40	mill banks	22×28
	35×46		28×44
document manila	$22^1/_2 \times 28^1/_2$	mill bristols	$22^1/_2 \times 28^1/_2$
	24×36		(2, 3 and 4 ply,
glassine	24×36		125, 150 and 175
	25×40		pounds)
	30×40	newsprint	21×32
gummed papers	17×22		22×24
	20×25		24×36
	25×38		25×38
index bristol (wood pulp and cotton			28×34
content)	$20^1/_2 \times 24^3/_4$		28×42
	$22^1/_2 \times 28^1/_2$		34×44
	$22^1/_2 \times 35$		36×48
	$25^1/_2 \times 30^1/_2$		38×50
	35×45	offset book paper (uncoated)	$17^1/_2 \times 22^1/_2$
kraft wrapping	18×24		$22^1/_2 \times 35$
	20×30		25×38
	24×36		28×42
	30×40		32×44
	40×48		35×45
	48×60		36×48

	38 × 50		24 × 38
	38 × 52		28 × 34
	41 × 54		28 × 38
offset book paper (coated)	22 × 29	tag board	$22^1/_2 \times 28^1/_2$
	$22^1/_2 \times 35$		24 × 36
	25 × 38		30 × 40
	28 × 42		or double
	28 × 44	text paper	23 × 29
	32 × 44		23 × 35
	35 × 45		25 × 38
	38 × 50		26 × 40
opaque circular	17 × 22		35 × 45
	$17^1/_2 \times 22^1/_2$		38 × 50
	22 × 34	thick China	22 × 28
	23 × 29		or double
	23 × 35	tough check	22 × 28
	25 × 38		or double
	28 × 34	translucents	$22^1/_2 \times 28^1/_2$
	35 × 45		or double
	38 × 50	waxed paper	9 × 12
photomount board	23 × 29		12 × 12
postcard bristol	$22^1/_2 \times 28^1/_2$		$12 \times 13^1/_2$
postcards (coated)	$22^1/_2 \times 28^1/_2$		12 × 18
	or double		18 × 24
railroad manila	17 × 22		24 × 36
	17 × 28	wedding papers	17 × 22
	19 × 24		22 × 34
	22 × 34		35 × 45
	24 × 38	wrapping tissues	10 × 15
	28 × 34		10 × 30
	34 × 44		12 × 18
rotogravure paper	25 × 38		12 × 24
	28 × 42		15 × 20
	28 × 44		18 × 24
	32 × 44		20 × 30
	35 × 45		24 × 36
	38 × 50	writing paper	17 × 22
safety paper	17 × 22		17 × 28
	17 × 28		19 × 24
	19 × 24		22 × 34
	19 × 26		24 × 38
	19 × 28		28 × 34
	22 × 34		

Paper size names in France (in centimetres)

There are three groups, e.g. the main group carré 45 × 56, the secondary groups couronne 36 × 45, raisin 50 × 64, jésus 56 × 72 and the auxiliary groups pot 32 × 40, écu 40 × 50. The most frequently used sizes are indicated by an asterisk. There is a considerable variation in sizes.

aigle (papier)	50 × 80		48 × 61
atlas ou journal	65 × 94		48 × 62
carré (carton)	47 × 62*	cavalier (impressions)	45 × 60
carré (papier)	45 × 56*	cloche	29 × 39
cavalier	46 × 60		30 × 40
	46 × 62	cloche (de Paris)	29 × 30
	47 × 62	colombier (journal, petit colombier	

ou colombier affiche)	57 × 80	grand double raisin	69 × 102
	60 × 80 *	grand jésus (le plus usité)	56 × 72
	62 × 90		56 × 76*
	63 × 90*	grand médian	46 × 60¹/₂
colombier belge	62 × 85	grand monde	90 × 120
colombier (carton)	71 × 100		90 × 126
colombier chromo	68 × 92		93 × 123
coquille	44 × 55		94 × 111
	44 × 56*	grand monde (cartes, dessins)	87 × 119
	45 × 55	grand monde (carton)	90 × 125
	45 × 56	grand raisin	51 × 69
coquille allemand	48 × 58	grand raisin (carton)	52 × 67
coquille (sans colle pour copies de		grand soleil	57 × 80
lettres)	45 × 57		60 × 80*
couronne (édition ou papeterie ou			69 × 100
registre)	36 × 45	grand univers (carton)	100 × 148
	36 × 46*	jésus (carton)	58 × 78
	37 × 47*	jésus (musique)	54 × 70
	37 × 48		55 × 70
demi carré	33¹/₂ × 42	jésus (ordinaire)	55 × 70
demi raisin	37 × 46		56 × 72
double carré	56 × 88	pelure (papier pour fleurs artificielles)	50 × 70
	56 × 90*	petit à la main	20 × 36
	57 × 89	petit aigle (carton)	75 × 105
double cavalier	58¹/₂ × 92	petit aigle (papier)	66 × 100
double cloche (écrits)	39 × 58		70 × 94
double colombier (affiches)	80 × 129*	petit cloche normande	26 × 36
double colombier	96¹/₂ × 127	petit jésus	50 × 70
double coquille	56 × 88*		56 × 72
	57 × 89	petit médian	40 × 53.3
double couronne	46 × 72	petit monde	85 × 107
	47 × 74*	petit raisin	32 × 43
double couronne (carton)	50 × 78	petit soleil	50 × 68
double grand soleil	80 × 120*	pittoresque	57 × 78
double jésus	76 × 112*		62 × 84
double pot	40 × 62*	pot	31 × 39
double pot (carton)	45 × 67		31 × 40*
double raisin	63¹/₂ × 102		32 × 40
	65 × 100*	quadruple carré	89 × 114
double tellière	43 × 69	quadruple colombier	74 × 94
	44 × 64*	quadruple coquille	88 × 112*
écu (carton)	40 × 52	quadruple couronne	74 × 94*
	42 × 57	quadruple raisin	100 × 130*
écu (papier ou français)	40 × 50		102 × 127
	40 × 52	quadruple tellière	69 × 86
écu belge	40 × 53	raisin	50 × 65
éléphant	62 × 77	royal	48 × 63
grand aigle (carton)	80 × 114	soleil	58 × 80
grand aigle (papier)	74 × 105	soleil (carton)	60 × 82
	75 × 103	super royal	50 × 70
	75 × 106	tellière	32 × 44*
grand colombier (ou colombier			33 × 44
d'éditions)	60 × 90		34 × 44
	63 × 90	tellière (comptes)	35 × 45
	66 × 92	univers	100 × 130

Old size names (in centimetres)

almasso	32¹/₂ × 44	charpentier	format in-18 jésus, emprunté
aux armes	35 × 44¹/₂		d'un éditeur spécial.

espagne	57 × 78	pro patria	34.5 × 43
florette	32 × 40	procureur	26 × 33
florette (écrits)	34 × 44		26 × 35
griffen	35 × 45	rosette	27 × 34.7
lys	31.7 × 39.7	ruche	36 × 46.2

Paper size names in Germany

In 1883 in Germany is tried to order size names with roman numbers. Today size names are completely out of use and sizes are now given in centimetres. However, old names with their approximate dimensions were as follows (in centimetres):

alt gross Median	48 × 58	Jesus	52 × 73
alt klein Regal	40 × 50		55 × 72
alt super Royal	50 × 76	Kanzlei	21 × 33
Atlas	83 × 118	klein Median	40 × 56
Billetpost	18 × 23		41 × 51
Bienenkorb	36 × 45		41 × 53
	37 × 46	klein Oktave	41 × 51
Bischof	38 × 48	klein Quart	42 × 52
breit Kanzlei	38 × 50	klein Regal	47 × 60
Colombier	60 × 90		47 × 62
Deutsch Post	46 × 59	klein Register	40 × 50
Deutsch Quart	22 × 28		41 × 51
Doppeltüten	50 × 76		44 × 50
double Elephant	67 × 103	klein Royal	44 × 50
Einfachtüten	37 × 45		48 × 64
Einhorn	38 × 48	kopier alt Regal	50 × 60
Ein-Pfund Beutel	46 × 75	Krönli	37 × 47
Elephant	67 × 92	Lexikon	49 × 64
	70 × 100		50 × 65
extra Regal	56 × 66	Löwen	38 × 48
Federn	35 × 44		40 × 50
Französisch Post	44 × 56	Median	45 × 58
Französisch Quart	21½ × 27	Notenquart	54 × 68
Glocken	41 × 60	Oktav und klein Duodez	43 × 52
gross Adler	70 × 107	Oktavdruck	42 × 53
gross Duodez	47 × 58	Oktave	14¼ × 22½
gross Elephant	67 × 103	Oktavpost	21 × 27
gross Glocken	44 × 60	Postkarton	49 × 59
gross Imperial		Postquart	46 × 59
gross Lexikon	54 × 70	Pro Patria	34 × 43
gross Median	44 × 58	Prospekt	46 × 59
	45 × 58	Quart	47 × 60
	46 × 59	Raisin	48 × 64
gross Oktav und Sedez	45 × 58	Regal	49 × 64
gross (Pro) Patria	36 × 45	Register	42 × 53
	37 × 45	Reichsformat	33 × 42
gross Raisin	50 × 65	Royal	49 × 61
gross Regal	51 × 68	Schreibmaschinenfolio	33 × 43
gross Register		Schulformat (Schulheft)	33 × 42
gross Royal	50 × 65		34 × 42
Halb-Pfund Beutel	40 × 63	Stab	36 × 45
Hochquart	47 × 65		38 × 48
Imperial	55 × 76	super Royal	50 × 70
	57 × 78		54 × 68
	57 × 80		54 × 70
	60 × 84	Vereinsdruck I	49½ × 76

| Vereinsdruck II | 55 × 84 | Vereinsdruck IV | 70 × 100 |
| Vereinsdruck III | 64 × 94 | Weltformat | 50 × 70.7 |

Paper size names in the Netherlands

Standard sizes in centimetres. Extra sizes are indicated by an asterisk. The legislation of the names was withdrawn on 8 February 1977.

achtdubbel bijkorf	94 × 150	dubbel super royaal	70 × 100
achtdubbel klein mediaan	110 × 160	groot mediaan	47 × 62
achtdubbel schrijf	88 × 138	groot post	46 × 59
adelaar	75 × 100	groot register kwadraat	62$^1/_4$ × 62$^1/_4$*
atlas	64 × 75	imperiaal	56 × 75
briefkaartkarton	43 × 61	journaal	52 × 73*
	47 × 57*	klein mediaan (enkel klein mediaan)	40 × 55
bijkorf (enkel bijkorf)	37$^1/_2$ × 47	klein register kwadraat	57 × 57*
carré	100 × 100	klein royaal	52 × 62
colombier	62 × 85	mediaan	47 × 56*
dubbel adelaar	100 × 150	olifants	62 × 75
dubbel atlas	75 × 128	oriënt	63 × 90*
dubbel briefkaartkarton	57 × 94*	post	44 × 56
	61 × 86	register mediaan	42 × 55
dubbel bijkorf	47 × 75	royaal	50 × 65
dubbel colombier	85 × 124	schrijf (enkel schrijf)	34$^1/_2$ × 44
dubbel groot mediaan	62 × 94	super royaal	50 × 70
dubbel groot post	59 × 92	vierdubbel bijkorf	75 × 94
dubbel imperiaal	75 × 112	vierdubbel groot mediaan	94 × 124
dubbel klein colombier	80 × 120*	vierdubbel klein mediaan	110 × 160
dubbel klein mediaan	55 × 80	vierdubbel groot post	92 × 118
dubbel kroon	51 × 76*	vierdubbel imperiaal	112 × 150
dubbel olifants	75 × 124	vierdubbel kroon	76 × 102*
dubbel oriënt	90 × 126*	vierdubbel post	88 × 112
dubbel post	56 × 88	vierdubbel royaal	100 × 130
dubbel register mediaan	55 × 84	vierdubbel schrijf	69 × 88
dubbel royaal	65 × 100	vierdubbel steendruk mediaan	92 × 120*
dubbel schrijf	44 × 69	vierdubbel super royaal	100 × 140

Paper size names in Belgium

Sizes not legislated in this country. Sizes in centimetres.

French name		Flemish name	French name		Flemish name
carton postale	43 × 61	briefkaartkarton	éléphant	61 × 77	olifants
colombier	62 × 85	colombier	emballage	78 × 90	verpakkingspapier
coquille	44 × 56	coquille	grand aigle	75 × 100	adelaar
double couronne	51 × 76	dubbel kroon	grand médian	46 × 60$^1/_2$	groot mediaan
double petit		dubbel klein	grand raisin	50 × 70	groot raisin
médian	53$^1/_2$ × 80	mediaan	grande coquille	46 × 59	groot coquille
double propatria	43 × 69	dubbel propatria	jésus	55 × 73	Jezus
double ruche	46 × 72	dubbel bijkorf	raisin	50 × 65	raisin

Paper size names in Spain

Paper size names in Spain have always been very unimportant. They were never legislated and are now fully out of use. However, old names were as follows (in centimetres):

España (papel a máquina)

marca folio	32×44
marquilla	$38^1/_2 \times 55$
doble marca o marca mayor	44×64
papel seda, para copias	45×57
indianas, fantasías, jaspeados, charolados, mates y chagrin	50×65
doble marquilla	55×77
cuadruple o doble marca mayor	64×88
cartulinas bristol o mate	77×110

Papel para registros (libros rayados)

folio	32×44
corona	36×47
escudo	40×54
carré	44×60
marca mayor	44×64
doble corona	47×72
raisin	48×66

petit sol	50×68
jesus	54×74
doble escudo	54×80
gran sol	57×80

Para cartas

coquil (ordinario)	43×55
coquil (fino)	44×56

Catalunya (papel hecho a mano)

ofici català	23×34
marca foli regular	32×44
marca foli perllongat	$34 \times 46^1/_2$
marquilla regular	$36^1/_2 \times 53$
marquilla perllongat	$39^1/_2 \times 56$
doble marca o marca major	47×66
doble marca perllongat o marca major perllongat	47×68

Paper size names in Italy

These names are not legislated and now out of use.

Carte per atti ufficiali

processo	27×38
comune	29×40
protocollo	32×44
arispetto	34×46
leona o stato	37×49
doppio processo	38×54
bastarda	44×55
doppio protocollo	44×64
reale	46×62
doppio arispetto	46×68
realone	48×64
doppia leona	49×74
imperialino	53×70
quadruplo processo	54×76
doppia bastarda	55×88
imperiale	58×78
colombier	63×90
quadruplo protocollo	64×69
quadruplo arispetto	68×92

Carte da stampa

realino	40×54
mezzano	44×60

reale	48×66
doppia leona	50×76
sott'imperiale	54×74
imperiale	60×80
colombier	63×90
quadruplo protocollo	63×86
	64×88
doppio reale o elefante	66×96
elefante grande	70×110
ottuplo processo	75×110
quadruplo leona o bislunga	76×100
papele	80×116

Carte da scrivere

olandina	36×56
	18×27
processo	38×56
	27×37
notarile	39×58
	28×38
pellegrina	42×62
	30×41
protocollo	43×63
	31×42
quartina	$43 \times 55^1/_2$

conchiglia	21 × 27 43 × 55½	quadrotta	18 × 22 47 × 58
officio o arispetto	27 × 42 45 × 68	leona	28 × 44½ 50 × 76
sestina	33½ × 44 46 × 56		37 × 49

International paper sizes

The basis of the international sizes, or A sizes, is a rectangle having an area of one square metre, the sides of which are in the ratio $1 : \sqrt{2}$ (or $1:1.414$); one side is consequently 841 mm, the other 1189 mm. In order to meet the smaller sizes after trimming the untrimmed sizes for the A-sizes are: $RA_0 = 860 \times 1220$ mm, $RA_1 = 610 \times 860$ mm and $RA_2 = 430 \times 610$ mm. The C and D sizes are practically not in use. The C_0 size is 917×1297 mm; the D_0 size is 771×1090 mm.

A sizes | | | B sizes | | |

designation	millimetres	inches	designation	millimetres	inches
A_0	841 × 1189	33.11 × 46.81	B_0	1000 × 1414	39.37 × 55.67
A_1	594 × 841	23.39 × 33.11	B_1	707 × 1000	27.83 × 39.37
A_2	420 × 594	16.54 × 23.39	B_2	500 × 707	19.68 × 27.83
A_3	297 × 420	11.69 × 16.54	B_3	353 × 500	13.90 × 19.68
A_4	210 × 297	8.27 × 11.69	B_4	250 × 353	9.84 × 13.90
A_5	148 × 210	5.83 × 8.27	B_5	176 × 350	6.93 × 9.84
A_6	105 × 148	4.13 × 5.83	B_6	125 × 176	4.92 × 6.93
A_7	74 × 105	2.91 × 4.13	B_7	88 × 125	3.46 × 4.92
A_8	52 × 74	2.05 × 2.91	B_8	62 × 88	2.44 × 3.46
A_9	37 × 52	1.46 × 2.05	B_9	44 × 62	1.73 × 2.44
A_{10}	26 × 37	1.02 × 1.46	B_{10}	31 × 44	1.22 × 1.73

International sizes of envelopes

E–A sizes

designation	without window millimetres	designation	without window millimetres	with window millimetres
E–A 3	312 × 441	E–A 5	156 × 220	156 × 220
E–A 3/2	156 × 441	E–A 5/6	110 × 220	110 × 220
E–A 4	220 × 312	E–A 6	110 × 156	110 × 156
E–A 4/2	115 × 312	E–A 7	78 × 110	

E–B sizes | | E–C sizes | |

designation	millimetres	designation	millimetres
E–B 4	262 × 371	E–C 4	240 × 340
E–B 4/2	131 × 371	E–C 4/2	120 × 340
E–B 5	185 × 262	E–C 5	170 × 240
E–B 6	131 × 185	E–C 6	120 × 170
E–B 7	92 × 131	E–C 7	85 × 120
E–B 8	65 × 92	E–C 8	60 × 85

Appendix 4 | French classification of the fibre content of paper (AFNOR Q 00-001)

class	type	fibre composition in per cents (tolerance: five per cent)	
P	1	100 mechanical wood pulp (of which 100 of straw)	
	2	95 idem (of which 80 of straw)	5 unbleached chemical pulp
	3	85 idem (of which 50 of straw)	15 idem
O	1	100 mechanical wood pulp	
	2	95 idem	5 idem
I	1	80 idem	20 idem
	2	65 idem	35 idem
II	1	50 idem	50 idem
	2	55 idem	10 idem and 35 bleached chemical pulp
III	1	35 idem	65 idem
	2	35 idem	25 idem and 40 bleached chemical pulp
IV	0	20 idem	80 idem
	1	25 idem	55 idem and 20 bleached chemical pulp
	2	25 idem	35 idem and 40 bleached chemical pulp
	3	25 idem	10 idem and 65 bleached chemical pulp
V	1	100 unbleached chemical pulp (traces of mechanical wood pulp)	
	2	70 idem	30 bleached chemical pulp
	3	40 idem	60 idem
VI		20 idem	80 idem
VII	1	100 bleached chemical pulp	
	2	100 idem	of which 25 of rags
	3	100 idem	of which 50 of rags
	4	100 idem	of which 75 of rags
	5	100 idem	of which 100 of rags

Extracts from BS 1219 : 1958, reproduced by permission of the British Standards Institution, 2 Park street, London W. 1, from whom copies of the complete standard may be obtained. In most countries proof correction marks are standardized. However, there is no or slight correlation between these standards, the British standard being the most extensive and fulfilling the purposes of this dictionary.

no.	instruction	textual mark	marginal mark
1	Correction is concluded	None	/
2	Insert in text the matter indicated in margin	⋏	New matter followed by /
3	Delete	Strike through characters to be deleted	ȹ
4	Delete and close up	Strike through characters to be deleted and use mark 21	ȹ̃
5	Leave as printed	. . . . . . under characters to remain	stet
6	Change to italic	———— under characters to be altered	ital
7	Change to even small capitals	════ under characters to be altered	s.c.
8	Change to capital letters	════ under characters to be altered	caps
9	Use capital letters for initial letters and small capitals for rest of words	════ under initial letters and ════ under the rest of the words	c. & s.c.
10	Change to bold type	⌇⌇⌇ under characters to be altered	bold
11	Change to lower case	Encircle characters to be altered	l.c.
12	Change to roman type	Encircle characters to be altered	rom
13	Wrong fount. Replace by letter of correct fount	Encircle character to be altered	w.f.
14	Invert type	Encircle character to be altered	℧
15	Change damaged character(s)	Encircle character(s) to be altered	x
16	Substitute or insert character(s) under which this mark is placed, in 'superior' position	/ through character or ⋏ where required	⌐ under character (e.g. ⅹ⌐)

no.	instruction	textual mark	marginal mark
17	Substitute or insert character(s) over which this mark is placed, in 'inferior' position	/ through character or ⋏ where required	∧ over character (e.g. ⬧)
18	Underline word or words	_____ under words affected	*underline*
19	Use ligature (e.g. ffi) or diphthong (e.g. œ)	⌣ enclosing letters to be altered	⌣ enclosing ligature or diphthong required
20	Substitute separate letters for ligature or diphthong	/ through ligature or diphthong to be altered	write out separate letters followed by /
21	Close up—delete space between characters	⊃ linking characters	⊃
22	Insert space	⋏	#
23	Insert space between lines or paragraphs	> between lines to be spaced	#
24	Reduce space between lines	(connecting lines to be closed up	less #
25	Make space appear equal between words	/ between words	eq #
26	Reduce space between words	/ between words	less #
27	Add space between letters	ıııııı between tops of letters requiring space	letter #
28	Transpose	⎍⎎ between characters or words, numbered when necessary	trs
29	Place in centre of line	Indicate position with ⌐ ¬	centre
30	Indent one em	⊏	▢
31	Indent two ems	⊏⊐	▢▢
32	Move matter to right	⊑ at left side of group to be moved	⊑
33	Move matter to left	⊒ at right side of group to be moved	⊒
34	Move matter to position indicated	[ ] at limits of required position	move
35	Take over character(s) or line to next line, column or page	⊏	take over

no.	instruction	textual mark	marginal mark
36	Take back character(s) or line to previous line, column or page	⌐	*take back*
37	Raise lines	↑ over lines to be moved / ‿ under lines to be moved	*raise*
38	Lower lines	⌐ over lines to be moved / ↓ under lines to be moved	*lower*
39	Correct the vertical alignment	‖	‖
40	Straighten lines	═ through lines to be straightened	═
41	Push down space	Encircle space affected	⊥
42	Begin a new paragraph	[before first word of new paragraph	*n.p.*
43	No fresh paragraph here	⌒ between paragraphs	*run on*
44	Spell out the abbreviation or figure in full	Encircle words or figures to be altered	*spell out*
45	Insert omitted portion of copy NOTE. The relevant section of the copy should be returned with the proof, the omitted portion being clearly indicated	λ	*out see copy*
46	Substitute or insert comma	/ through character or λ where required	,/
47	Substitute or insert semi-colon	/ through character or λ where required	;/
48	Substitute or insert full stop	/ through character or λ where required	⊙
49	Substitute or insert colon	/ through character or λ where required	⊙
50	Substitute or insert interrogation mark	/ through character or λ where required	?/
51	Substitute or insert exclamation mark	/ through character or λ where required	!/
52	Insert parentheses	λ or λλ	(/)/

no.	instruction	textual mark	marginal mark
53	Insert (square) brackets	λ or $\lambda\lambda$	[/]/
54	Insert hyphen	λ	|-|
55	Insert en (half-em) rule	λ	_en_
56	Insert one-em rule	λ	_em_
57	Insert two-em rule	λ	_2 em_
58	Insert apostrophe	λ	⁊
59	Insert single quotation marks	λ or $\lambda\lambda$	⁊ ⁊
60	Insert double quotation marks	λ or $\lambda\lambda$	⁊ ⁊
61	Insert ellipsis	λ	··· /
62	Insert leader	λ	⊙
63	Insert shilling stroke	λ	⦸
64	Refer to appropriate authority anything of doubtful accuracy	Encircle words, etc. affected	?

MARKS FOR THE CLARIFICATION OF MATHEMATICAL COPY

no.	in margin	meaning	in copy
65	Gk	Use Greek letter	Encircle letter in lead pencil or underline in green
66	Ger	Use German (Fraktur) letter	Encircle letter in green or lead pencil
67	rom	Use roman	Encircle letter or letters in lead pencil or underline in red
68	scr	Use script	Encircle letter in red or lead pencil
69	fig 1/2/etc.	Use figure	Pencil circle round figure
70	2-line fr.	Use fraction made up two lines deep	Pencil circle round fraction
71	11pt.fr. etc. (according to point size)	Use fraction of text size	Pencil circle round fraction
72	dec	Use decimal point	Pencil $\vee$ under point
73	No mark is required	Space to be a hair space or 2 or 3 units according to Style of House	black $\wedge$ or green /

no.	in margin	meaning	in copy
74	*No mark is required*	Space to be either a thick space or 5 units	*black* ↑ *or red* /
75	*No mark is required*	Space to be an en space or 9 units	*black* ⋀̱
76	*No mark is required*	Space to be an em space or 18 units	*black* ⋀
77	*ital*	Set in italic	*Pencil underline of mathematical values in text*

ADDITIONAL MARKS FOR THE CORRECTION OF MATHEMATICAL PROOFS

no.	in margin	meaning	in text
78	*Letter required followed by* Ⓖⓚ	Use Greek letter	/ *through letter*
79	*Letter required followed by* Ⓖⓔⓡ	Use German (Fraktur) letter	/ *through letter*
80	*Letter required followed by* Ⓡⓞⓜ	Use roman	/ *through letter*
81	*Letter required followed by* Ⓢⓒⓡ	Use script	/ *through letter*
82	*x* (showing letter required)	Use superior to superior (e.g. '2' in y^{a2})	/ *through letter*
83	*x* (showing letter required)	Use inferior to inferior (e.g. '2' in y_{a2})	/ *through letter*
84	*x* (showing letter required)	Use superior to inferior (e.g. '2' in y_{a2})	/ *through letter*
85	*x* (showing letter required)	Use inferior to superior (e.g. '2' in y^{a2})	/ *through letter*
86	Ⓕⓘⓖ 1/2/etc.	Use figure	/ *through letter*
87	*2-line fr.*	Use fraction made up two lines deep	*Circle round fraction*
88	*11pt. fr.* (according to point size)	Use text size fraction	*Circle round fraction*
89	Ⓓⓔⓒ	Use a decimal point	∨
90	*hair thick* # Ⓝ2#Ⓝ Ⓝ5#Ⓝ	Space to be hair space or 2 units or either a thick space or 5 units as indicated	∧ *where required*

Appendix 6 | Some rules for hyphenation

There are no hard-and-fast rules for the application of hyphens. It is more or less governed by the view and styles adopted by English and American printers. In general the following notes abstracted from the GPO Style Manual (Washington, 1963), may be of some assistance.

A hyphen must be used:

1. to avoid doubling a vowel or tripling a consonant, *except* after the short prefixes, *co, de, pre, pro,* and *re,* which are generally printed solid. (anti-inflation; micro-organism; brass-smith; thimble-eye; shell-like; hull-less; cooperation; deemphasis; preexisting).

2. between words, or abbreviations of words, combined to form a unit modifier immediately preceding the word modified, *except* as indicated sub 26. This applies particularly to combinations in which one element is a present or past participle. (collective-bargaining talks; English-speaking nation; high speed line; long-term-payment loan; multiple-purpose uses; part-time personnel; 1-inch diameter; a 4-percent increase (*but*: 4 percent of hydrochloric acid; U.S.-flag ship).

3. do not confuse a modifier with the word it modifies. (American flag-ship; well-trained schoolteacher; wooden-shoe maker; light-blue hat; elementary-school teacher; *but*: common stockholder; small businessman).

4. where two or more hyphened compounds have a common basic element and this element is omitted in all but the last term. (2- or 3-em quads, *not* 2 or 3-em quads; 2- to 3- and 4- to 5-ton trucks; 2-by 4-inch boards; long- and short-term money rates, *not* long and short-term money rates; *but*: twofold or threefold, *not* two or threefold; goat, sheep and calf skins, *not* goat sheep and calfskins).

5. to prevent mispronunciation, to insure a definite accent on each element of the compound, or to avoid ambiguity. (non-civil-service position; non-tumor-bearing tissue; re-cover (cover again); re-treat (treat again); un-ionized, un-uniformity).

6. with the prefixes *ex, self,* and *quasi.* (ex-governor; ex-trader; self-control; quasi-academic; quasi-young; *but*: selfhood, selfsame).

7. to join a prefix or combining form to a capitalized word, *unless* usage demands otherwise. The hyphen is retained in words of this class set in caps. (anti-Arab; pro-British; non-Government; post-World War II or post-Second World War; *but*: nongovernmental; transatlantic).

8. between the elements of compound numbers from twenty-one to ninety-one and in adjective compounds with a numeric first element. (twenty-one; 24-inch ruler; 8-hour day; 10-minute delay; 20th-century progress; 3-to-1 ratio; two-sided question; *but* one hundred and twentyone; $ 20 million airfield).

9. between the elements of a fraction, but *not* between the numerator and the denominator when the hyphen appears in either or in both (one-thousandth; two-thirds; two one-thousandths; twenty-three thirtieths; twenty-one thirty-seconds; three-fourths of an inch).

10. in a unit modifier following and reading back to the word or words modified and is always printed in the singular. (motor, alternating-current, 3-phase, 60-cycle, 115-volt; glass jars: 5-gallon, 2-gallon, 1-quart; belts: 2-inch, $1^1/_4$-inch, $^1/_2$-inch, $^1/_4$-inch).

11. with the adjectives *elect* and *designate*, as the last element of a title. (President-elect; Vice-President-elect; minister-designate).

12. between the elements of technical compound units of measurement. (candle-hour; horse-power-hour; kilowatt-hour; passenger-mile).

13. between the elements of an improvised compound. (blue-pencil, *v.*; know-how, *n.*; make-believe, *n.*; how-to-be-beautiful course).

14. in a prepositional-phrase compound noun consisting of three or more words. (man-of-war; mother-in-law; mother-of-pearl; patent-in-fee; *but*: coat of arms; heir at law; next of kin; officer in charge).

15. in verbs when the corresponding noun form is printed as separate words. (cold-shoulder; blue pencil; cross-brace).

16. in a compound formed of repetitive or conflicting terms and in a compound naming the same thing under two aspects. (comedy-ballet; dead-live; devil-devil; farce-melodrama; walkie-talkie; young-old).

17. in a nonliteral compound expression containing an apostrophe in its first element. (assess'-eyes; ass's-foot; bull's eye; cat's-paw; crow's-nest; *but*: The cat's paw is soft. There is the crow's nest).

18. to join a single capital letter to a noun or a participle. (H-bomb; T-shaped; U-boat; V-necked; X-ray; S-iron; T-square).

No hyphen is used:

19. when words appear in regular order and the omission causes no ambiguity in sense or sound. (banking hours; blood pressure; eye opener; living costs; palm oil; patent right; training ship; violin teacher).

20. in words expressing a literal or nonliteral (figurative) unit idea that would not be as clearly expressed in unconnected succession. (afterglow; cupboard; newsprint; forget-me-not; right-of-way).

21. between two nouns that form a third when the compound has only one primary accent, especially when the prefixed noun consists of only one syllable or when one of the elements loses its original accent. (airship; bathroom; bookseller; footnote; locksmith; workman).

22. in a noun consisting of a short verb and an adverb as its second element, *except* when the use of the solid form would interfere with comprehension. (blowout; break down; flareback; holdup; makeready; pickup; runoff, setup; showdown; throwaway; *but*: cut-in; tie-in).

23. usually in compounds beginning with the following nouns: book; eye; horse; house; mill; play, school; shop; snow; way; wood; work).

24. in compounds ending in the following, especially when the prefixed word consists of one syllable. (boat; book; holder; house; maker; man; shop; store; time (not clock); worker; working; writing; etc.).

25. in compass directions consisting of two points, *but* a hyphen is used after the first point when three points are combined. (northeast; southwest; north-northeast; south-southwest).

26. where meaning is clear and readability is not aided. Restraint should be exercised in forming unnecessary combinations of words used in normal sequence. (durable goods industry; civil rights case; high school student; life insurance company; portland cement plant; *but*: no-hyphen rule (readability aided); *not* no hyphen rule).

27. in a compound predicate adjective or predicate noun the second element of which is a present participle. (The duties were price fixing. The effects were far reaching. The area was used for beet raising).

28. in a two-word modifier the first element of which is a comparative or superlative. (better drained soil; higher level decision; larger sized dress; lower income group; *but*: lowercase type (printing); upperclassman; bestseller (noun); lighter-than-air craft: higher-than-market price).

29. in a two-word unit modifier the first element of which is an adverb ending in *ly*, or in a threeword modifier the first two elements of which are adverbs. (eagerly awaited moment; heavily laden ship; unusually well preserved specimen; not too distant future, *but*: ever-rising flood; still-new car; well-known lawyer).

30. in proper nouns used as unit modifiers, either in their basic or derived form, *except* after combining forms. (Latin American countries; United States laws; Red Cross nurse; Afro-American program; Franco-Russian war; *but*: Indochina border; North American-South American sphere; French-English descent).

31. in a unit modifier consisting of a foreign phrase. (ante bellum days; bona fide transaction; per diem employee).

32. in a unit modifier containing a letter or a numeral as its second element. (article 3 provisions; class II railroad; grade A paper).

33. in a unit modifier enclosed in quotation marks, *unless* it is normally a hyphened term. Quotation marks are not to be used in lieu of a hyphen. ('blue sky' law; 'good neighbour' policy; 'tie-in' sale; but: right-to-work law).

34. in combination colour terms, but *with* a hyphen when such colour terms are unit modifiers. (bluish green; dark green; orange red; bluish-green feathers; iron-gray sink; silver-gray body).

35. in words ending on *like*, *except* when tripling a consonant or when the first element is a proper name. (lifelike; lilylike; girllike; bell-like; Florida-like).

36. in a modifier consisting of a possessive noun preceded by a numeral. (I month's layoff; 1 week's pay; 2 hours' work; 3 weeks' vacation).

37. in civil or military titles denoting a single office, *except* in a double title. (assistant attorney general; commander in chief; notary public; secretary general; under secretary, *but*: undersecretaryship: vice president, *but*: vice-presidency; secretary-treasurer; treasurer-manager).

38. in scientific terms (chemicals, diseases, animals, insects, plants) used as unit modifiers if no hyphen appears in their original form. (carbon monoxide poisoning; methyl bromide solution; whooping cough remedy; *but*: Russian-olive planting; white-pine weevil; Douglas-fir tree).

39. in chemical elements used in combination with figures, even as a unit modifier. (polonium 210; uranium 235; *but*: U^{235}; Sr^{90}; $92U^{234}$; Freon 12).

40. in chemical formulas. (9-nitroanthra (1,9, 4, 10) bis (1) oxathiazone-2, 7-bisdioxide; Cr-Ni-Mo; 2,4-D).

41. in idiomatic phrases. (come by; inasmuch as; insofar as; Monday week).

Appendix 7 | Weights and measures

Current US system and former GB system. See also app. no. 7a. The United Kingdom embarked on a ten-year programme in 1965 to achieve substantial completion for a changeover from the imperial system of units to the metric system (SI). Australia, New Zealand and other Commonwealth countries have since adopted similar programmes. (For monetary units the UK decimalized its currency on 15 February 1971. One pound sterling = 100 pence.) Currently the US continues to use the inch-pound system, but legislation to change to the metric system is under consideration. For the equivalent names and their abbreviations in other languages, see in the basic table.

Lengths (GB and US system)

inch. British and American unit of length.
   12 inches = 1 foot
   1 inch (*GB*) = 2.539998 centimetres
   1 inch (*US*) = 2.540005 centimeters
foot. British and American unit of length.
   3 feet = 1 yard
   1 foot (*GB*) = 30.47997 centimetres
   1 foot (*US*) = 30.48006 centimeters
yard. British and American unit of length.
   1 yard = 3 feet = 36 inches.
   1 yard (*GB*) = 91.43992 centimetres
   1 yard (*US*) = 91.440183 centimeters

Lengths (metric system)

centimetre (*GB*), *centimeter* (*US*). Metric unit of length.
   1 centimetre = 0.39370 inch
decimetre (*GB*), *decimeter* (*US*). Metric unit of length.
   1 *decimetre* = 3.937011 inches (*GB*) = 0.328043 foot (*GB*)
                 = 3.937000 inches (*US*) = 0.3280833 foot (*US*)
metre (*GB*), *meter* (*US*). Metric unit of length.
   1 metre = 39.3701 inches (*GB*) = 3.280843 feet (*GB*) = 1.093614 yards (*GB*)
           = 39.3700 inches (*US*) = 3.280833 feet (*US*) = 1.093611 yards (*US*) = 1.198838 rod (*US*).

Area (GB and US system)

square inch. British and American unit of area.
   144 square inches = 1 square foot.
   1 square inch (*GB*) = 6.4515898 square centimetres
   1 square inch (*US*) = 6.4516258 square centimeters
square foot. British and American unit of area.
   1 square foot = 144 square inches.
   1 square foot (*GB*) = 9.29089 square decimetres
   1 square foot (*US*) = 9.290341 square decimeters
square yard. British and American unit of area.
   1 square yard = 9 square feet.
   1 square yard (*GB*) = 0.836126 square metre (or centare)
   1 square yard (*US*) = 0.83613 square meter (or centare)

Area (metric system)

square centimetre (GB), *square centimeter (US)*. Metric unit of area.
   1 square centimetre = 0.15500 square inch (*GB* and *US*)
square decimetre (GB), *square decimeter (US)*. Metric unit of area.
   1 square decimetre = 15.500 square inches (*GB* and *US*)
square metre (GB), *square meter (US)* centare. Metric unit of area.
   1 square metre = 10.76390 square feet (*GB*) = 1.195992 square yards (*GB*)
   1 square metre = 10.76387 square feet (*US*) = 1.195985 square yards (*US*)

square hectometre (GB), *square hectometer (US)* *hectare.* Metric unit of area.
   1 square hectometre = 2.471058 acres *(GB)* = 2.471044 acres *(US)*

Volume (GB and US system)

cubic inch. British and American unit of volume.
   1 cubic inch = 0.5541 fluid ounce = 4.4329 fluid drams
   1 cubic inch *(GB)* = 16.3870253 cubic centimetres
   1 cubic inch *(US)* = 16.387162 cubic centimeters
cubic foot. British and American unit of volume.
   27 cubic feet = 1 cubic yard.
   1 cubic foot *(GB)* = 6.229 gallons *(GB)* = 28.316677 cubic decimetres
   1 cubic foot *(US)* = 7.481 gallons *(US)* = 1728 cubic inches *(US)* = 59.844 liquid pints *(US)* = 29.922
   liquid quarts *(US)* = 25.714 dry quarts *(US)* = 28.31701 cubic decimeters
cubic yard. British and American unit of volume.
   1 cubic yard = 27 cubic feet.
   1 cubic yard *(GB)* = 0.76455285 cubic metre.
   1 cubic yard *(US)* = 0.76455945 cubic meter.

Volume (metric system)

cubic centimetre (GB), *cubic centimeter (US)*. Metric unit of volume.
   1 cubic centimetre = 0.035195 fluid ounce *(GB)* = 0.281157 fluid drachm *(GB)* = 16.894 minims *(GB)*
   1 cubic centimetre = 0.033814 fluid ounce *(US)* = 0.27051 fluid dram *(US)* = 16.231 minims *(US)*
   1 cubic centimetre = 0.061023 cubic inch *(GB* and *US)*
cubic decimetre (GB), *cubic decimeter (US)*. Metric unit of volume.
   1 cubic decimetre = 61.023 cubic inches = 0.035314 cubic foot
cubic metre (GB), *cubic meter (US)*, *stere.* Metric unit of volume.
   1 cubic metre = 35.31477 cubic feet *(GB)* = 1.307954 cubic yards *(GB)*
   1 cubic meter = 35.314445 cubic feet *(US)* = 1.3079428 cubic yards *(US)* = 264.173 gallons *(US)*

Capacity (liquid measure)

gill. British and American liquid measure.
   1 gill *(GB)* = 0.25 liquid pint *(GB* and *US)* = 5 fluid ounces *(GB)* = 0.03125 gallon *(GB* and *US)* = 0.14206
   litre = 142.07 cubic centimeters
   1 gill *(US)* = 0.25 liquid pint *(US* and *GB)* = 4 fluid ounces *(US)* = 32 fluid drams *(US)* = 1920 minims
   (US) = 0.125 liquid quart *(US)* = 0.03125 gallon *(US* and *GB)* = 7.21875 cubic inches = 0.118292 liter
   = 118.295 cubic centimeters
pint, liquid pint. British and American liquid measure.
   1 pint *(GB)* = 0.125 gallon *(GB* and *US)* = 0.5 quart *(GB* and *US)* = 4 gills *(GB* and *US)* = 20 fluid
   ounces *(GB)* = 1.20094 pints *(US)* = 0.56825 litre = 568.26 cubic centimetres.
   1 pint *(US)* = 0.125 gallon *(US* and *GB)* = 0.5 quart *(US* and *GB)* = 4 gills *(US* and *GB)* = 16 fluid
   ounces *(US)* = 128 fluid drams *(US)* = 7680 minims *(US)* = 28.875 cubic inches *(US)* = 0.83268 pints
   (GB) = 0.473167 liter = 473.179 cubic centimeters
quart, liquid quart. British and American liquid measure.
   1 quart *(GB)* = 0.25 gallon *(GB* and *US)* = 2 liquid pints *(GB* and *US)* = 1.13650 litres = 1136.52 cubic
   centimetres
   1 quart *(US)* = 0.25 gallon *(US* and *GB)* = 2 liquid pints *(US* and *GB)* = 8 gills *(US)* = 32 fluid ounces
   = 256.00 fluid drams = 57.749 cubic inches = 0.946333 liter = 946.358 cubic centimeters
gallon, British imperial gallon.
   1 gallon *(GB)* of water at 15 °C weighs 10 pounds avoirdupois. 1 gallon *(US)* of water at 15 °C weighs
   8.337 pounds avoirdupois
   1 gallon *(GB)* = 0.5 peck *(GB)* = 4 liquid quarts *(GB* and *US)* = 8 liquid pints *(GB* and *US)* = 32 liquid
   gills *(GB* and *US)* = 160 fluid ounces *(GB)* = 277.3 cubic inches = 0.16054 cubic foot = 4.54596 liters
   = 4546.1 cubic centimeters
   1 gallon *(US)* = 4 liquid quarts *(US* and *GB)* = 8 liquid pints *(US* and *GB)* = 32 liquid gills *(US* and *GB)*
   = 128 fluid ounces *(US)* = 231.00 cubic inches *(US)* = 0.13368 cubic foot *(US)* = 0.83268 gallons *(GB)*
   = 3.7853 liters = 3785.4 cubic centimeters

Capacity (metric system)

litre (GB), *liter (US)*. Metric unit of capacity.
   1 Liter is the volume of pure water at 4 °C and 760 mm pressure which weighs 1 kilogram(me)
   1 Liter = 1.000027 cubic decimetre
   1 Liter = 0.21998 gallon *(GB)* = 0.10999 peck *(GB)* = 1.7598 pints *(GB)* = 0.87990 quart *(GB)* = 7.0392
   gills *(GB)* = 35.196 fluid ounces *(GB)* = 61.025 cubic inches
   1 Liter = 0.26417762 gallon *(US)* = 0.11351 peck *(US)* = 0.908102 dry quart *(US)* = 1.8162 dry pints
   (US) = 1.056710 liquid quarts *(US)* = 2.1134 liquid pints *(US)* = 8.4538 gills *(US)* = 33.8147 fluid ounces
   = 270.5179 fluid drams *(US)*
millilitre (GB), *milliliter (US)*. Metric unit of capacity.
   1 millilitre = 0.035196 fluid ounce *(GB)*
   1 milliliter = 0.0338147 fluid ounce *(US)* = 0.2705179 fluid dram *(US)* = 16.2311 minims *(US)*
centilitre (GB), *centiliter (US)*. Metric unit of capacity.
   1 centilitre = 0.33815 fluid ounce *(US)* = 2.705179 fluid drams *(US)* = 0.61025 cubic inch
decilitre (GB), *deciliter (US)*. Metric unit of capacity.
   1 decilitre = 0.176 pint *(GB)* = 3.38147 fluid ounces *(US)*
decalitre (GB), *dekaliter (US)*. Metric unit of capacity.
   1 decalitre = 1.3513 pecks *(US)* = 9.08102 dry quarts *(US)* = 18.1620 dry pints *(US)* = 10.00027 cubic
   decimetres
hectolitre. Metric unit of capacity.
   1 hectolitre = 2.7497 bushels *(GB)* = 2.8378 bushels *(US)* = 11.3513 pecks (US)

Apothecaries' fluid measure

minim. Apothecaries' fluid measure.
   1 minim *(GB)* = 0.059194 cubic centimetre
   1 minim *(US)* or fluid minim = 0.061610 millimeter
drachm, fluid drachm. A British measure.
   1 drachm = 0.125 fluid ounce *(GB and US)* = 60 minims = *(GB and US)* = 3.5515 cubic centimetres
dram, fluid dram.
   1 fluid dram *(US)* = 0.125 fluid ounce *(US and GB)* = 60 minims *(US and GB)* = 0.225586 cubic inch =
   3.6967 cubic centimeters
fluid ounce.
   1 fluid ounce *(GB)* = 8 fluid drachms *(GB)* = 480 minims *(GB and US)* = 28.4130 cubic centimetres
   1 fluid ounce *(US)* = 0.03125 liquid quart *(US)* = 0.0625 liquid pint = 0.25 gill *(US)* = 8 fluid drams =
   480 minims *(US and GB)* = 0.29579 deciliter = 29.5737 cubic centimeters

Dry measure

pint, dry pint.
   1 dry pint *(US)* = 0.5 quart *(US)* = 33.600 cubic inches = 0.550599 liter = 550.61 cubic centimeters
quart, dry quart.
   1 dry quart *(US)* = 0.038889 cubic foot *(US)* = 0.125 peck = 2 dry pints = 67.2006 cubic inches = 1.10120
   liters = 1101.23 cubic centimeters
peck.
   1 peck *(GB)* = 2 gallons *(GB)* = 554.6 cubic inches = 9.0919 litres
   1 peck *(US)* = 8 quarts *(US)* = 16 pints *(US)* = 537.605 cubic inches *(US)* = 8.80958 liters
bushel.
   1 bushel *(GB)* = 0.125 quarter *(GB* capacity) = 1.03205 bushels *(US)* = 1.2843 cubic feet = 8 gallons *(GB)*
   = 2219.3 cubic inches = 36.3677 liters.
   1 bushel *(US)* = 4 pecks = 32 quarts (dry) = 64 pints (dry) = 2150.42 cubic inches = 0.35239 cubic meter
   = 35.238 liters

Mass

In *GB* and *US* three systems are in use, i.e. avoirdupois, troy and apothecaries'. Grain is same in all.

Weight avoirdupois (commercial)

grain (GB and US).
   1 grain = 0.0020833 ounce (apothecaries' or troy) = 0.0022857 ounce avoirdupois = 0.01667 dram apothecaries' or troy = 0.03657143 dram avoirdupois = 0.04166667 pennyweight troy = 0.05 scruples = 64.798918 milligrams

dram.
   1 dram avoirdupois = 0.4557292 dram apothecaries' or troy = 0.0625 ounce avoirdupois = 0.056966146 ounce apothecaries' or troy = 1.139323 pennyweight = 1.3671875 scruples = 27.34375 grains = 1.771845 grams

ounce.
   1 ounce avoirdupois = 16 drams avoirdupois = 18.22917 pennyweights = 21.875 scruples apothecaries' = 437.5 grains = 0.9114583 ounce apothecaries' or troy = 7.29166 drams apothecaries' or troy = 0.062500 pound avoirdupois = 0.075954861 pound apothecaries' or troy = 28.349527 grams

pound.
   1 pound avoirdupois (*GB* and *US*) is the mass of 27.681 cubic inches of water weighed in air at 4 °C and 760 mm pressure.
   1 pound avoirdupois = 16 ounces avoirdupois = 256 drams avoirdupois = 291.6667 pennyweights = 350.01 scruples = 7000 grains = 1.2152778 pounds apothecaries' or troy = 14.5833 ounces apothecaries' or troy = 0.4535924 kilogram(me)

stone (GB).
   1 stone = 14 pounds avoirdupois = 6.350 kilograms

quarter (GB).
   ¹/₄ long hundredweight = 28 pounds = 12.70 kilograms
   (*US*) ¹/₄ short hundredweight = 25 pounds = 11.340 kilograms
   (*GB*) ¹/₄ long ton = 560 pounds = 254.01 kilograms
   (*US*) ¹/₄ short ton = 500 pounds = 226.795 kilograms

hundredweight, long hundredweight.
   1 long hundredweight = 112 pounds = 0.05 long ton = 4 quarters (*GB*) = 50.8023 kilograms
   1 short hundredweight = 100 pounds = 0.05 short ton = 4 quarters (*GB*) = 1600 ounces avoirdupois = 0.044643 long ton = 0.0453592 metric ton = 45.3592 kilograms

ton, long ton, gross ton (GB and US).
   1.1200 short tons = 22.400 short hundredweights = 2240 pounds avoirdupois = 2722.22 pounds apothecaries' or troy = 1.0160470 metric tons = 1016.0470 kilograms

ton, short ton, net ton (US).
   1 short ton = 20 short hundredweights = 0.89286 long ton = 2000 pounds avoirdupois = 2430.56 pounds apothecaries' or troy = 0.907185 metric ton = 907.185 kilograms

Troy weight

Grain is same in avoirdupois, troy and apothecaries'. Ounce and pound are same in troy and apothecaries'.

grain. Same as grain avoirdupois (*qv*).
pennyweight.
   1 pennyweight = 24 grains = 0.05 ounce = 0.0548571 ounce avoirdupois = 0.8777143 dram avoirdupois = 1.55517 grams

dram.
   1 dram = 3 scruples = 60 grains = 2.5 pennyweights = 0.12510 ounce apothecaries' = 0.1371429 ounce avoirdupois = 2.194286 drams avoirdupois = 3.8879351 grams

ounce.
   1 ounce = 8 drams = 20 pennyweights = 24 scruples = 480 grains = 0.08333 pound = 0.06857143 pound avoirdupois = 1.09714 ounces avoirdupois = 31.103481 grams

pound.
   1 pound = 12 ounces = 96 drams = 240 pennyweights = 288 scruples = 5670 grains = 13.165714 ounces avoirdupois = 210.6514 drams avoirdupois = 0.3732418 kilogram

Apothecaries weight

grain. Same as grain avoirdupois (*qv*), troy and apothecaries'.
scruple.
    1 scruple = 0.3333 dram = 0.7314286 dram avoirdupois = 0.833333 pennyweight = 1.2959784 grams
dram. Same as dram troy. Same as British drachm
ounce. Same as ounce troy (*qv*)
pound. Same as pound troy (*qv*)

Mass (metric system)

milligram(me) (GB), milligram (US).
    1 milligram = 0.01543236 grain
centigram(me) (GB), centigram (US).
    1 centigram = 0.1543236 gram
decigram(me) (GB), decigram (US).
    1 decigram = 1.543236 grams
gram(me) (GB), gram (US).
    1 gram = 0.0352739 ounce avoirdupois = 0.0321507 ounce apothecaries' and troy = 0.564383 dram avoirdupois = 0.257206 dram apothecaries' and troy = 0.6430149 pennyweight = 0.771618 scruple = 15.4324 grains
decagram(me) (GB), dekagram (US).
    1 decagram = 0.35273957 ounce avoirdupois = 5.64383 drams avoirdupois
hectogram(me).
    1 hectogram = 3.52739 ounces avoirdupois
kilogram(me).
    1 kilogram = 2.2046223 pounds avoirdupois = 2.679.2285 pounds apothecaries' or troy = 32.150742 ounces apothecaries' or troy = 35.273957 ounces avoirdupois = 257.21 drams apothecaries' or troy = 564.38 drams avoirdupois = 643.01 pennyweights = 771.62 scruples
tonne, metric ton, millier.
    1 tonne = 0.984206 long ton = 1.10231 short ton = 22.046223 short hundredweights = 2204.6223 pounds avoirdupois = 2679.23 pounds apothecaries' or troy = 1000 kilograms

Domestic measures (approximately)

    1 teaspoon = 1 fluid dram = 3.5 cm³
    1 dessert spoon = 2 fluid drams = 6 cm³
    1 table spoon = ¹/₂ fluid ounce = 14.5 cm³
    3 pennies = 1 ounce av. = 28 grammes

Conversion factors of the most important units to the SI-units. Since 1 January 1978, the International System of Units (See especially Système International d'Unités) is obligatory in the countries of the European Common Market. See also prefixes for metric units in the basic table. (Remark: 10^{-3} means 0.001. Thus 10^{-3} m = 0.001 meter or 1 mm.)

Lengths
 1 inch — 25.4×10^{-3} m
 1 foot = 0.3048 m
 1 yard = 0.9144 m
 1 centimetre = 0.01 m = 10^{-2} m
Area
 1 square inch = 645.16×10^{-6} m²

 1 square foot $\approx$ 0.092903 m²
 1 square yard $\approx$ 0.836127 m²
Volume
 1 cubic inch $\approx$ 16.3871×10^{-6} m³
 1 cubic foot $\approx$ 28.3168×10^{-3} m³
 1 cubic yard $\approx$ 0.764555 m³

Capacity (liquid measure)

gill
 1 UKgill = $^1/_4$ UKpt $\approx$ 0.142065×10^{-3} m³
 1 USgill = $^1/_4$ USliqpt $\approx$ 0.118294×10^{-3} m³
pint, liquid pint
 1 UKpt = $^1/_8$ UKgal $\approx$ 0.568261×10^{-3} m³
 1 liquid pint = $^1/_8$ USgal $\approx$ 0.473176×10^{-3} m³
quart, liquid quart
 1 UKqt = $^1/_4$ UKgal $\approx$ 1.13652×10^{-3} m³

 1 liquid quart = $^1/_4$ USgal $\approx$ 0.946353×10^{-3} m³
gallon
 1 UKgal $\approx$ 4.54609×10^{-3} m³
 1 USgal = 231 in³ $\approx$ 3.78541×10^{-3} m³
 1 dry gal US = $^1/_2$ USpk $\approx$ 4.40488×10^{-3} m³
litre
 1 liter = 10^{-3} m³
 1 millilitre = 1 mil = 1 ml = 10^{-6} m³

Apothecaries' fluid measure

minim
 1 UK min = $^1/_{60} \times$ UK fl dr $\approx$ 59.1939×10^{-9} m³
 1 US min = $^1/_{60} \times$ US fl dr $\approx$ 61.6115×10^{-9} m³
drachm, fluid drachm
 1 UK fl dr = 1 fluid drachm = $^1/_8$ UK fl oz $\approx$
  3.55163×10^{-6} m³

 1 US fl dr = 1 fluid dram = $^1/_8$ US fl oz $\approx$ 3.69669
  $\times 10^{-6}$ m³
fluid ounce
 1 UK fl oz = $^1/_{20}$ UKpt $\approx$ 28.4131×10^{-6} m³
 1 US fl oz = $^1/_{16}$ liq pt $\approx$ 29.5735×10^{-6} m³

Dry measure

pint, dry pint
 1 UK pt = $^1/_8$ UKgal $\approx$ 0.568261×10^{-3} m³
 1 dry pt = $^1/_2$ US dry qt $\approx$ 0.550610×10^{-3} m³
quart, dry quart
 1 UK qt = $^1/_4$ UKgal $\approx$ 1.13652×10^{-3} m³
 1 dry qt = $^1/_4$ dry gal $\approx$ 1.10122×10^{-3} m³

peck
 1 UKpk = 2 UKgal $\approx$ 9.09218×10^{-3} m³
 1 USpk = $^1/_4$ USbu $\approx$ 8.80977×10^{-3} m³
bushel
 1 UKbu = 8 UKgal $\approx$ 36.3687×10^{-3} m³
 1 USbu = 2150.42 in³ $\approx$ 35.2391×10^{-3} m³

Weight avoirdupois

grain
 1 gr = $^1/_{7000}$ lb $\approx$ 64.7989×10^{-6} kg
 1 gr/UKgal $\approx$ 14.2538×10^{-3} kg/m³
 1 gr/USgal $\approx$ 17.1181×10^{-3} kg/m³
dram
 1 dram avdp = $^1/_{16}$ oz $\approx$ 1.77185×10^{-3} kg
ounce
 1 oz = $^1/_{16}$ lb $\approx$ 28.3495×10^{-3} kg

 See also troy ounce; fluid ounce; apothecaries'
  ounce
pound
 1 pound = 1 lb $\approx$ 0.453592 kg
 See also troy pound
stone
 1 stone = 14 lb $\approx$ 6.35029 kg

quarter
    1 quarter = 28 lb ≈ 12.7006 kg
hundredweight
    1 UKcwt (long hundredweight or gross hundred-
        weight) = 112 lb ≈ 50.8023 kg
    1 UScwt (short hundredweight or net hundred-
        weight) = 100 lb ≈ 45.3592 kg

ton
    1 UK ton (long or gross ton US) = 2240 lb ≈
        1.01605×10^3 kg
    1 US ton (short ton or net ton) = 2000 lb ≈
        907.185 kg
    See also metric ton

Troy weight

grain
    Same as grain avoirdupois
pennyweight
    1 pennyweight = 1 dwt = 24 gr ≈ $1.55517 \times$
        10^{-3} kg
dram
    1 dram US = 1 drachm UK = 1 apothecaries'
        dram = 60 grains ≈ 3.88793×10^{-3} kg
    See also drachm; fluid drachm

ounce
    20 dwt (pennyweight) ≈ 31.1035×10^{-3} kg
    See also ounce avoirdupois
pound
    1 pound troy = 1 t lb = 12 tr oz ≈ 0.37342 kg
    See also pound avoirdupois

Apothecaries' weight

grain
    Same as grain avoirdupois
scruple
    1 scruple = 20 grains ≈ 1.29598×10^{-3} kg
dram
    Same as dram troy
ounce
    Same as ounce troy

pound
    Same as pound troy
tonne
    1 (metric) tonne = 10^3 kg
    See also long ton; short ton

a	are. See app. no. 6.
a	atto. See prefixes for metric units.
å	Obsolete symbol for ångström. See Å.
a.a.O.	am angegebenen Orte, in Fußnoten der Hinweis dass ein Werk schon früher als Quelle genannt wurde. German for 'loco citato', in the place cited.
A	ampere *qv.*
Å	SI-symbol for angström *qv.*
AA	author's alterations. Changes required by the author or his representative, after type has been correctly set to the original specifications. Sometimes called author's correction (AC).
AB	Aktiebolag(et). Swedish for joint-stock company.
ABC	Audit Bureau of Circulation *qv.*
AC	alternating current *qv.*
AC	author's correction. See also AA.
AD	anno Domini. Latin for 'in the year of our Lord'. Should be placed in small caps before the figures. See also BC.
adj.	adjective.
ADP	automatic data processing. See also EDP.
Adpro	Official American name for the process inks for newspaper colour printing.
ADS	autograph document signed.
A-flute	See flute.
Ag	(no point) argentum. See silver.
AG	Aktiengesellschaft. German for shareholders' corporation.
Ah	ampere-hour *qv.*
Al	(no point) aluminium.
a.m.	ante meridiem. Latin for 'before midday' (lower case, points).
AM	amplitude modulation *qv.*
ANG	American newspaper guide.
anon.	anonymous (author).
AOQ	average outgoing quality *qv.*
ap	(no point) apothecaries'. See app. no. 7 and 7a.
a.p.	author's proof.
APL	A programming language used for problem solving, usually at remote terminals, especially useful for mathematical functions and extensive arrays of data.
AQL	acceptable quality level *qv.*
ARQ	automatic repetition *qv.*
ARRA	Dutch for automatic relais rekenmachine Amsterdam.
ASA	American Standards Association.
asb	apostilb *qv.*
ASCII	American National Standard Code for Information Interchange. It is based upon a seven-level code which provides for 128 discrete characters.
ASN	average sample number *qv.*
ASR	Automatic send-receive set. A combination teletypewriter transmitter and receiver with transmission capability from either keyboard or paper tape.
at	atmosphere. See in the basic table.
AT & T	American Telegraph & Telephone.
atm	atmosphere. See in the basic table.
ATS	administrative terminal system *qv.*
at wt	(no point) atomic weight.
Au	(no point) aurum. See gold.
av	average.
avdp	avoirdupois. See app. no. 7 and 7a.
B	A mark on paper, indicating a secondary quality.
Ba	(no point) barium *qv.*
BAL	British anti-Lewisite. In some cases a treatment for chromic dermatitis *qv.*
BASIC	A programming language used for problem solving and intended for persons who are not professional programmers.
BB	double rough. A mark indicating a degree of roughness in the finish of superfine drawing papers.
BB	British term applied to note papers implying black bordered.

B bristol	See bogus bristol.
BBR	See in the basic table.
BC	before Christ. Should be placed in small caps after the figures of the year. See also AD.
BCD	binary coded decimal *qv*.
b.d.c.	bas de casse. French for 'put down' *qv*.
BDH	British drug houses.
Be	(no point) beryllium.
Bé	Baumé *qv*.
BEP	brevet d'enseignement professionnelle (replaces CAP).
bf	bold-faced type. A proofreader's mark. See app. no. 5.
B-flute	See flute.
Bi	(no point) See bismuth.
BIPC	bulletin of the Institute of Paper Chemistry.
BMT	basic motion time.
bpi	bits per inch.
BP chromo	Trade name for proofing chromo paper.
Br	(no point) bromine *qv*.
BS	British standard.
Btu	British thermal unit *qv*.
BTW	belasting op de toegevoegde waarde. Dutch for value added tax.
bu	(no point) bushel. See app. no. 7 and 7a.
BUGRA	Internationale Ausstellung für Buchgewerbe und Graphik (Leipzig).
BGW	Birmingham wire gauge.
©	To satisfy the requirements of international copyright a book must contain the name of the copyright owner, date and place of initial publication, and the symbol ©.
c	centi. See prefixes for metric units.
C	(no point) See carbon.
C	Celsius. See centigrades.
C	coulomb.
C	100. See Roman numerals.
ca	centiare. See app. no. 7.
Ca	(no point) calcium *qv*.
c.a.	caja alta. Spanish proofreader's mark meaning 'put in uppercase letters'.
CA	Italian for corrente alternata.
cal	(no point) calorie *qv*.
CAO	Dutch for collectieve arbeidsovereenkomst.
cap	See foolscap.
CAP	The former term for Certificat d'Aptitude Professionnelle. See BEP.
c.b.	caja baja. Spanish proofreader's mark meaning 'put in lowercase letters'.
CBS	Columbia Broadcasting System.
cc	cubic centimetre. See app. no. 7.
CC	Italian for 'corrente continua'. See direct current.
ccs	hundred call seconds.
CCT	computer controlled typesetting *qv*.
cd	(no point) See candela.
Cd	(no point) See cadmium.
cf	confer. Latin term meaning 'compare'.
C-flute	See flute.
cg	(no point) centigram(me). See app. no. 7.
CGO	Can go over. Copy that may be used at a later time. See punk.
CGS	centimetre/gramme/second.
CI	colour index.
cic	cicero *qv*.
CIE	Commission Internationale d'Eclairage.
cl	centilitre. See app. no. 7.
Cl	(no point) See chlorine.
cm	(no point) centimetre. See app. no. 7.
cm²	(no point) square centimetre. See app. no. 7.
cm³	(no point) cubic centimetre. See app. no. 7.

CMC	carboxymethylcellulose *qv*.
CMC 7	magnetic character code. See in the basic table.
CMT	concora medium test. A method for testing the rigidity of paperboard flutes.
Co	(no point) cobalt *qv*.
COBOL	See in the basic table.
c/o	care of.
colloq.	colloquially.
COM	computer output microfilm.
c 1 s	coated one side.
context	See in the basic table.
cos	(no point) See cosine.
cosec	(no point) See cosecant.
cot	(no point) See cotangent.
cP	(no point) See centipoise.
CP	chemically pure.
CP	computer keyboard perforator.
CPGM	Conférence Générale des Poids et Mesures.
cph	characters per hour *qv*.
cpi	characters per inch *qv*.
cpl	characters per line.
cpm	characters per minute, more often used for cards (punch) per minute.
cps	(no point) characters per second.
cps	See c/s.
cps	cycles per second *qv*.
CPU	central processing unit *qv*.
CPV	French for 'chlorure de polyvinyle' *qv*.
Cr	(no point) See chromium.
CR	compensating relaxation allowance *qv*.
CRT	cathode ray tube *qv*.
cs	or cps, cycles per second. 1 cs = 1 Hertz *qv*.
Cs	(no point) See cesium.
cSt	centistokes *qv*.
CT	computer typesetting installation.
CTR	computer tape reader. Normally attached to a photographic typesetting device, when needed by the device to read computer tape.
CTS	computer typesetting.
c 2 s	coated two sides.
Cu	(no point) cuprum. See copper.
cu ft	(no point) cubic foot *qv*.
cu in	(no point) cubic inch *qv*.
cu yd	(no point) cubic yard *qv*.
CUIM	continuously undulating inking mechanism *qv*.
cV	cheval-vapeur. French for 'horsepower' *qv*.
CWo	cream wove.
CWP	chemical wood pulp.
cwt	hundredweight. See app. no. 7 and 7a.
CZ 13	impression with magnetic ink designed by SEL *qv*.
d	deci. See prefixes for metric units.
d	dextro.
D	500. See Roman numerals.
da	deca. See prefixes for metric units.
dal	decalitre. See app. no. 7.
dam	decametre. See prefixes for metric units.
DAS	deutsche Auslegeschrift.
dB	decibel *qv*.
DBPC	ditertiary butyl para-cresol, an anti-oxydant for paraffine.
d.c.	double crown *qv*, or double cap *qv*.
DC	direct current *qv*.
DDT	dichloro-diphenyl-trichloroethane. An insecticide harmless to humans when used externally

	in diluted forms.
DEC	digital equipment system.
del	delineavit. Latin for 'drawn by', preceding or following on the name of the designer.
dg	decigram(me). See app. no. 7.
dil	dilute.
DIN	Deutsche Industrie Normen (also: 'das ist Norm').
DI-plate	direct image plate.
DK	Dezimal Klassifikation.
dl	decilitre. See app. no. 7.
dm	decimetre. See app. no. 7.
dm²	square decimetre. See app. no. 7.
dm³	cubic decimetre. See app. no. 7.
DMD	direct mat drop.
DOD-Farben	Druck-ohne-Durchschuß-Farben (Ermöglichen Illustrationsdruck ohne Durchschuß oder Bestäubung).
DOS	disc operating system.
DPLG	diplomé par le gouvernement.
dpt	dioptre.
dr	(no point) dram. See app. no. 7 and 7a.
dr ap	(no point) dram apothecaries. See app. no. 7 and 7a.
dr fl	(no point) fluid drachm. See app. no. 7 and 7a.
DRGM	Deutsches Reichsgebrauchsmuster (German protection of patents).
dry pt	(no point) dry pint. See app. no. 7 and 7a.
dry qt	(no point) dry quart. See app. no. 7 and 7a.
d to a	digital to analog.
DTP	directory tape processor.
dwt	pennyweight. See app. no. 7 and 7a.
dyn	dyne *qv*.
E	exa. See prefixes for metric units.
EAN	European article numbering. A system of symbols for the specification of products sold in food packages.
EBCDIC	extended binary coded decimal interchange code. An 8-bit code which can represent 256 characters.
EDP	electronic data processing *qv*. See also ADP.
EDSAC	electronic delayed storage automatic calculator.
edu	edition.
EDVA	See composing computer.
Ef-paper	English finished paper *qv*.
E-flute	See flute.
e.g.	exempli gratia, that is 'for example'. Put in lower case, not italic, and preceded by comma. Use 'for example', rather than 'e.g.'.
Elhi	elementary and high school (textbook publishing).
emf	electromotive force *qv*.
EM language	Common machine language, used for magnetic ink printing.
emu	electromagnetic unit.
ENIAC	electronic numerical integrator and calculator.
eod	every other day. Blind date abbreviation in newspaper advertisements.
EOM	end of message.
EPC-process	See in the basic table.
EPM	electronic photocomposing machine.
ERA	electronic reading apparatus.
ES-paper	engine-sized paper *qv*.
esu	electrostatic unit.
et seq	et sequens. Latin for 'and the following'.
E13B	See in the basic table.
eV	eingetragener Verein.
exc	excudit. Latin for 'he printed' or 'engraved' (this).
f	femto. See prefixes for metric units.

F	Fahrenheit *qv*.
F	farad *qv*.
F	faraday *qv*.
F	(no point) fluorine.
fac	facsimile *qv*.
fam	familiar.
FCT	flat crush test.
FDC	food, drug and cosmetic.
Fe	(no point) ferrum. See iron.
fec	fecit or fecerunt (plural). Latin for '(so-and-so) made this picture'. Used with artist's signature.
FEM	fuerza electromotriz. Spanish for 'electromotive force' *qv*.
fifo	first in, first out. Term used with the storage of newspaper reels.
fl	floruit date. The approximate date used during which a person was alive or flourished because the exact date is not known. Abbreviation placed before the date used.
fl dr	fluid drachm. See app. no. 7 and 7a.
fl oz	fluid ounce. See app. no. 7 and 7a.
FM	frequency modulation *qv*.
FOC	father of the chapel *qv*.
FOS	Film ohne Schicht. German for 'film without layer'.
FRED	figure reading electronic device.
ft	foot, or feet. See app. no. 7 and 7a.
ft²	square foot, or square feet. See app. no. 7 and 7a.
ft³	cubic foot, or cubic feet. See app. no. 7 and 7a.
ft L	foot-lambert *qv*.
FTP	folded, trimmed and packed (books).
FyI	for your information.
g	gram(me). See app. no. 7.
G	giga. See prefixes for metric units.
G	gauss *qv*.
gal	gallon. See app. no. 7 and 7a.
g.a.o.f.	gummed all over flap *qv*.
Gb	(no point) gilbert *qv*.
GDT	graphic display terminal. A device on which the composed text is made visible.
g.e.	gilt edge.
GERS	General Electric Recording Spectrophotometer.
GEU	gamme Européenne unifiée.
GFU	gamme Française unifiée (d'encres primaires typographiques).
g.h.	For this abbreviation, see in the basic table.
gi	(no point) gill *qv*. See also in app. no. 7 and 7a.
gigo	garbage in, garbage out. Programming slang for bad input produces bad output. See garbage.
GIP	glazed imitation parchment.
GmbH	Gesellschaft mit beschränkter Haftung. German for 'limited (liability) company'.
gr	(no point) grain. See app. no. 7 and 7a.
GRACE	graphic arts composing equipment.
GSA	See GSA system in the basic table.
gt	gilt.
gt.t.	gilt top.
g.t.e.	gilt top edge.
h	hecto. See prefixes for metric units.
h	hour.
H	(no point) See hydrogen.
H	henry *qv*.
ha	hectare. See app. no. 7.
H & D curve	Hunter and Driffield curve *qv*.
hg	hectogram(me). See app. no. 7.
Hg	(no point) hydrargyrum. See mercury.
hl	hectolitre. See app. no. 7.

hm	hectometre. See app. no. 7.
hm²	square hectometre. See app. no. 7.
HMSO	Her (or His) Majesty's Stationery Office.
hp	(no point) horsepower *qv.*
HP	hot pressed.
HR	humidité relative. French for relative humidity *qv.*
HS	hard sized (paper) *qv.*
HT	high-speed typesetter.
HTC, HTK, H 2 K	head to come *qv.*
Hz	(no point) hertz *qv.*

I	See Roman numerals.
I	(no point) iodine *qv.*
IAL	international algebraic language. See algol.
IAS	immediate access storage *qv.*
ib.	ibidem. Latin term meaning 'in the same place'.
IBM	International Business Machines.
IC	integrated circuit *qv.*
id.	idem. Latin term meaning 'the same'.
i.e.	id est. Latin term for 'that is'. Not italic, no capitals, comma before.
IMPACT	inventory management program and control techniques.
IMS	industrial methylated spirit *qv.*
in	(no point) inch. See app. no. 7 and 7a.
in²	(no point) square inch. See app. no. 7 and 7a.
in³	(no point) cubic inch. See app. no. 7 and 7a.
I/O	input/output *qv.*
IOCS	input/output control system *qv.*
IPA	international phonetic alphabet.
iph	impressions per hour *qv.*
ISBN	international standard book number.
ISCC	Inter-Society Colour Council.
ISO	International Standards Organization.
ISO-code	The International Standards Organization's 7-level code for computer output and information interchange.
ital.	italic *qv.*
IUE	Intensitätsumkehreffekt.
IVA	imposta sul valore aggiunto. Italian for VAT *qv.*
IW	interword knob.

J	joule *qv.*
JES	Japanese Engineering Standards.

k	kilo. See prefixes for metric units.
K	(no point) kalium. See potassium.
K	(no point) Kelvin *qv.*
K	A symbol which stands for 1,000, more precisely in describing a computer's memory it is the binary number 1,024. However, in popular parlance, computer 'types' refer to prices as $ 100 K for $ 100,000; or salary of $ 18 K.
kb-value	kauri-butanol value *qv.*
kc	kilocycle *qv.*
kg	(no point) kilogram(me). See app. no. 7.
KG (a.A.)	Kommandit-Gesellschaft (auf Aktien).
kgf	(no point) kilogramme-force.
kHz	(no point) See hertz.
KPR	Kodak photoresist. A kind of light-sensitive coating, in 1953 announced by the Eastman Kodak Company.
KTS	keep type standing.
kV	(no point) kilovolt.
kW	(no point) kilowatt.
kWh	(no point) kilowatt-hour *qv.*

l	(no point) litre. See app. no. 7.
L	Lineaniert. German on Apo-Ronar process lenses focal length 24.36 or 48 cm.
L	50. See Roman numerals.
L	(no point) Lambert *qv*.
Lat	Latin.
lb	(no point) pound. See app. no. 7 and 7a.
lb av	(no point) pound avoirdupois. See app. no. 7 and 7a.
lb/in²	(no point) pounds per square inch.
lb t	(no point) troy pound. See app. no. 7 and 7a.
l.c.	lower case.
l.c.	loco citato, that is 'in the place cited'.
LC number	Library of Congress catalog card number *qv*.
LCD	liquid crystal display. The reading of fluid crystals.
LDR	light dependent resistance. The inner resistance, for example in a cadmium-sulphide cell, which decreases according to as it receives more light.
lf	long fold.
LGCP	lexical graphic composer printer.
LID	low-intensity desensitization *qv*.
Linasec	a line a second.
liq pt	(no point) liquid pint. See app. 7 and 7a.
liq qt	(no point) liquid quart. See app. no. 7 and 7a.
lit.	literally.
lm	(no point) lumen *qv*.
loc. cit.	See l.c.
log	(no point) logarithm *qv*.
logo	See logotype.
lp	long primer.
lp	large post.
lph	lines per hour.
lpi	lines per inch.
lpm	lines per minute.
LS	lectori salutem.
LS	locus sigilli. Latin for 'in the place of the seal' as on a document.
LSI	large-scale integrated (circuitry) *qv*.
LTPD	lot tolerance per cent defective *qv*.
lx	(no point) lux *qv*.
m	(no point) metre. See app. no. 7.
m-	meta- *qv*.
m²	(no point) square metre. See app. no. 7.
m³	(no point) cubic metre. See app. no. 7.
mA	(no point) milliampere.
M	mega. See prefixes for metric units.
M	1,000. See Roman numerals.
MAC	maximum allowable concentration *qv*.
MAK	German abbreviation for maximum allowable concentration *qv*.
mc	megacycle *qv*.
mc paper	machine-coated paper *qv*.
MCA	matrix case arrangement.
MEDLARS	medical literature analysis and retrieval system. See also data bank.
MEMO	maintenance electronic monitor and observation, a term used in computer controlled type-setting.
mf paper	machine-finished paper *qv*.
MF	modern face.
mg	milligram(me). See app. no. 7.
Mg	magnesium *qv*.
mg paper	machine-glazed paper *qv*.
m-Höhe	Höhe der Mittellängen.
MHR	machine-hour rate (system).
MICR	magnetic ink character recognition *qv*.

min	minim. See app. no. 7.
mired	micro reciprocal degree.
MIS	matt one side *qv*.
MIT	Massachussets Institute of Technology, Boston, US.
MKSA	metre/kilogramme/second/ampere. See also SI
MKZ	mechanische Kreiderelief-Zurichtung. German for mechanical chalk overlay.
ml	(no point) millilitre. See app. no. 7.
mμ	millimicron
μA	microampere.
μm	micron *qv*. micrometre.
mm	(no point) millimetre. See app. no. 7.
mm^2	(no point) square millimetre. See app. no. 7.
mm^3	(no point) cubic millimetre. See app. no. 7.
Mn	manganese *qv*.
MNA	master negative arrangement.
MO	making order *qv*.
MODEM	See in the basic table.
MOF	meilleur ouvrier de France.
MΩ	megaohm
MOVaR	moisture vapour resistance (board).
mp	melting point.
M paper	Paper slightly defective yet salable as 'seconds'. The bundles of such paper are stencilled 'Ms'. Perfect sheets of the same grades are sometimes referred to as 'Ps'.
MPY	multiply.
ms	manuscript. See copy.
ms	millisecond.
Ms	See M paper.
msec.	millisecond.
mss	manuscripts.
MTBF	mean time between failures. See down time.
MTM	methods time measurement.
MTS	See 'standard work minute'.
MVTR	moisture vapour transmission rate.
MWP	mechanical wood pulp.
MWS	Mehrwertsteuer. German for VAT.
Mx	(no point) maxwell *qv*.
n	nano. See prefixes for metric units.
n-	normal *qv*.
N	(no point) Newton *qv*.
N	(no point) nitrogen *qv*.
Na	(no point) natrium. See sodium.
n-Berechnung	Satzpreisberechnung nach Anzahl der n in einer Zeile.
NCR	no carbon required *qv*. See also OKP.
NCR	National Cash Register.
NdlR	note de la rédaction.
neg	negative.
NEN	Nederlandse norm (Dutch standard).
NEP	not easy pulpable.
n.f.	A compositor's slang term meaning to ignore completely.
Ni	(no point) nickel *qv*.
nm	(no point) nanometre. See prefixes for metric units.
NOB	not on bonus. See also time on daywork.
not.	not glazed, i.e., semi-rough. A mark indicating a degree of roughness of superfine drawing papers. See also B.
n.p.	new paragraph.
n paper	Paper inferior in quality to M paper. It usually contains specks and wrinkles.
ns	(no point) nanosecond. See prefixes for metric units.
nsec	(no point) See ns.
NSCC	Neutralsulfit-Semichemical Verfahren.

nt	(no point) nit *qv*.
#	number. An arbitrary sign representing a quantity.
o-	ortho *qv*.
O	(no point) oxygen *qv*.
O	printer's term for overseer *qv*.
ob.	obiit. That is 'He (she or it) died'.
obit.	obituary. A notice of a person's death, usually with a short biographical account, in a newspaper.
OC	operating characteristic *qv*.
OCR	optical character recognition *qv*.
o'er	to be closed up.
Oe	(no point) oersted *qv*.
OEM	original equipment manufacturer.
Ω	(no point) ohm *qv*.
OJT	office de justification des tirages des organes quotidiens et périodiques.
o.k.	oll korrect, alteration of 'all correct'. See pass for press.
OKP	ohne Kohle-Paper. See also NCR.
op.	opus, that is 'work'.
op. cit.	opere citato, that is 'in the work quoted'.
O.P.	out of print *qv*.
org.	organic.
o.s.	out of stock *qv*. Sometimes: o/s.
O.S.	Old Style
oz	(no point) ounce. See app. no. 7 and 7a.
oz ap	(no point) apothecaries' ounce. See app. no. 7 and 7a.
oz fl	(no point) fluid ounce.
oz t	(no point) troy ounce. See app. no. 7 and 7a.
p	(no point) pico. See prefixes for metric units.
p	page.
p-	para-
P	perfect.
P	peta. See prefixes for metric units.
P	(no point) See poise.
P	(no point) See phosphorus.
Pa	(no point) Pascal *qv*.
pat.	patented.
PAM	pulse amplitude modulation *qv*.
PAO	Preisanordnung.
par.	paragraph.
PAX	private automatic exchange.
Pb	(no point) plumbum. See lead.
PB report	publication board report.
PCM	pulse code modulation *qv*.
PCMI	photochromic micro image. See microfilm.
Pd	(no point) See palladium.
PERT	project evaluation and review technique.
PFM	pulse frequency modulation *qv*.
pH	See in the basic table.
pinx.	pinxit, or pinxerunt (plural). Latin form used in signing pictures, meaning 'he (they) painted'. Used with author's signature.
PIV	positive infinitely variable *qv*.
pk	paardekracht. Dutch for horsepower *qv*.
pk	(no point) peck. See app. no. 7 and 7a.
PL/l	programming language one; a high-level programming language.
PLV	publicité sur le lieu de vente.
p.m.	post meridiem, i.e. after midday (lower case; points).
PM	phase modulation *qv*.
PMA	Abbreviation used to describe a pigment which has been precipitated with phosphomolybdic

	acid to give it permanence and insolubility.
PMH	production per man hour.
PMS	pantone matching ststem.
PMTA	Colours made of complex compounds of basic dyestuffs with phosphomolybdic acid.
pnxt.	See pinx.
P.O. box	Post Office box.
p.o.p.	printing out paper *qv*.
POP	post office preferred.
pos.	positive.
ppH	prints per hour.
ppm	parts per million.
PPR	printed paper rate.
PPS	page printing system.
PROM	programmable read-only memory. See read-only memory.
ps	(no point) picosecond. See prefixes for metric units.
Ps	See M paper.
psec	picosecond. See ps.
psi	(no point) pounds per square inch, gauge pressure, rather than absolute pressure. Used in stating air pressure, bursting strength of paper, etc.
PS	Pferdestärke. See horsepower.
pt	(no point) pint; liquid pint as well as dry pint. See app. no. 7 and 7a.
pt	point *qv*.
Pt	(no point) platinum *qv*.
PT	plain transparent (Said of regenerated cellulose film).
PT	electronic phototypesetting.
PTA	Abbreviation used to describe a pigment which has been precipitated with phosphotungstic acid to give it permanence and insolubility.
PTM	pulse time modulation *qv*.
p.t.o.	please turn over (Remark at the bottom of a page).
PU	pick up *qv*.
PVaC	polyvinylacetate *qv*.
PVal	polyvinylalcohol *qv*.
PVC	polyvinylchloride. See saran.
PVDC	polyvinylidene chloride
PVOH	polyvinylalcohol
PVP	publicité par voie postale.
PVVC	polivinyl-vinylidene- chloride.
qy	query.
qt	(no point) quart. Liquid quart as well as dry quart. See app. no. 7 and 7a.
qv	quod vide. See in the basic table.
r	(no point) revolution.
®	registered trade-mark
R	retrees *qv*.
RAM	random access store *qv*.
rb	See run-back.
RCA	Radio Corporation of America.
rd	See run-down.
redox	reduction and oxydation *qv*.
r°	recto.
rev	revolution.
rev/min	See rpm.
RF	relative Feuchtigkeit. German for 'relative humidity' *qv*.
RH	relative humidity *qv*.
RHD	regenerated halide desensitization.
rms	root mean square.
ROM	read-only memory *qv*.
ROP	See in the basic table.
RP	initials which stand for reprint.

RPG	report programme generator *qv*.
rpm	revolutions per minute. In scientific work rev/min (no points) is preferred.
rps	revolutions per second. In scientific work rev/s (no points) is preferred.
rsc	regular slotted container. A container of which outer flaps meet. Inner flaps do not meet unless length and width happen to be the same.
RSVP	répondez s'il vous plait. Placed on invitation printing. May be followed by the address of the sender.
RTF	roller top of former *qv*.
RUG	red under gold
RV	relatieve vochtigheid. Dutch for 'relative humidity' *qv*.
rv/s	revolutions per second.
s	(no point) second.
s	(no point) scruple. See app. no. 7 and 7a.
S	(no point) sulphur *qv*.
SAE	Society of Automotive Engineers. The use of these letters with any materials or articles means that they meet the standards of the society.
SAM	See standard work minute.
S & SC	sized and supercalandered *qv*.
S.a.r.l.	société à responsabilité limitée.
sb	(no point) stilb *qv*.
Sb	(no point) stibium. See antimony.
SBN	standard book number *qv*.
SBR	styrene-butadiene rubber.
sc	See sculp.
SC	single column.
SC	small capitals *qv*.
SC paper	supercalandered paper *qv*.
s caps	small capitals *qv*.
SCR	silicon controlled rectifier.
sculp	sculpsit. Latin for 'engraved by', preceding or following on the name of the engraver.
sec	(no point) secant *qv*.
SEL	Standard Elektrik Lorenz.
S/F	store and forward *qv*.
sh cwt	(no point) short hundredweight *qv*. See app. no. 7 and 7a.
sh t	(no point) short ton. See app. no. 7 and 7a.
SI	(no point) See Système International d'Unités.
s.l.e.a.	sine loco et anno. Latin for 'without place and date'
SLT	solid logic technology *qv*.
SM	siehe Manuskript. German for 'see copy' *qv*.
SMIG	salaire minimum interprofessionnel garanti.
Sn	(no point) stannum. See tin.
soat	surface oil absorbency test.
SOS bag	self-opening satchel bag *qv*.
sp. gr.	specific gravity *qv*.
sq ft	square foot (or square feet). See app. no. 7 and 7a.
sq in	square inch. See app. no. 7 and 7a.
sq yd	square yard. See app. no. 7 and 7a.
SR	Schopper-Riegler. See degree of beating.
ss	same size. Instruction for block makers.
St	stokes *qv*.
St	Saint. In alphabetical arrangement always place under Saint, not under St.
stet	See in the basic table.
sub	substract.
SWM	standard work minute *qv*.
syn.	synonymous.
t	tome *qv*.
t	tonne. See app. no. 7 and 7a.
T	tera. See prefixes for metric units.

tan	(no point) See tangent.
TC	tape combiner.
TEAM	top European advertising media.
teg.	top edge gilt *qv*.
TGL	Technische Normen. Gütevorschriften und Lieferbedingungen. (verbindliche Staatliche Standart in Ost Deutschland).
3-D printing	See xograph.
Ti	(no point) titanium *qv*.
TL-lamp	tubular luminescent lamp.
TL-value	threshold limit value *qv*.
TNO	toegepast natuurwetenschappelijk onderzoek. Dutch for applied scientific research.
tn sh	short ton. See app. no. 7 and 7a.
TPR film	thermoplastic recording film.
tr	troy. See app. no. 7 and 7a.
TR	tape reading device.
tr lb	troy pound. See app. no. 7 and 7a.
tr oz	troy ounce. See app. no. 7 and 7a.
trs	proofreader's mark for transpose *qv*.
TS	tub sized.
TSAD	tub sized and air dried.
tsp	teaspoon. See app. no. 7.
t.s.v.p.	tournez s'il vous plait. French for 'please turn over' *qv*.
TTP	tabular tape processor.
TTS	Total Typesetting Systems. Name used since 1969 by the manufacturers (Fairchild) for their original typesetter. See teletypesetting.
TTU	teletypewriter terminal unit *qv*.
TVA	taxe à la valeur ajoutée. French for VAT *qv*.
2 s B	two sides bright. See matt one side.
UAC	Uniforme Artikel Code (Dutch for EAN, European Article Numbering).
u.c.	upper case *qv*.
UCR	under-colour removal *qv*.
UCS	uniform chromaticness scale *qv*.
UDC	universal decimal classification.
UNE	norma española
UPC	universal product code (The term used in US for EAN, European Article Numbering).
USASCCII	See ASCII.
USP	United States Pharmacopoeia.
V	See Roman numerals.
V̄	vertatur. Latin for p.t.o. (Please turn over).
V	(no point) volt *qv*.
VA	(no point) voltampere *qv*.
VAT	value added tax.
VCE	visual colour efficiency.
VDT	video display terminal *qv*.
VDU	visual display unit.
VE	volume equivalent. A draft standard for the volumetric measurement of gravure cells depth
viz.	See in the basic table.
VM & P naphtha	See in the basic table.
VMCH	See in the basic table.
v°	verso.
VO	vedi originale. Italian for 'see copy'.
vol.	volume *qv*.
VPC	vente par correspondance.
VPI-paper	vapour-phase inhibitor paper *qv*.
VU	volume unit *qv*.
VVP	viscosity/velocity product *qv*.
W	(no point) watt *qv*.

W	(no point) wolfram *qv*.
WATS	wide area telephone service *qv*.
Wb	(no point) weber *qv*.
wf	wrong fount *qv*.
wf	water finished paper.
WF	In time studywork, abbreviation for work factor.
Wh	watt hour *qv*.
wpm	words per minute.
WS	work simplification.
w.v.p.	water-vapour permeability.
w.v.t.r.	water-vapour transmission rate.
X	See Roman numerals.
x paper	See pattern paper.
xx paper	Retrees or slightly damaged paper of different reams, marked xx and invoiced at 10 per cent. Less than good. See retree.
xxx paper	Broken or outsides *qv*.
yd	(no point) yard. See app. no. 7 and 7a.
yd^2	(no point) square yard. See app. no. 7 and 7a.
yd^3	(no point) cubic yard. See app. no. 7 and 7a.
Zn	(no point) See zinc.
Zr	(no point) See zirconium.

In the Middle Ages and up to the 17th century, in Latin many abbreviations were in use, as:

aliqñ = aliquando = at some time or other
alr = aliter = by another way
añ = ante = before
aplice = apostolice = apostolico (adverb)
aploy = apostolorum = of the apostles
bñ = bene = well
cãm = causam = cause, reason (accusative)
corpibus = corporibus = to, for the bodies (dative or ablative)
dc̄m = dictum = saying, proverb
dño = domino = Lord (dative or ablative)
dñs = dominus = Lord (nominative)
ds = Deus = God
ẽ = est = is
eēt = esset = let him, her, it, be (subjunctive 3rd person)
em̃ = enim = just then, just so, just as you say, yes indeed
ẽt = etiam = 1) also, 2) yet, still, 3) certainly
etem̃ = etenim = indeed, and in fact for
ffem = fratrem = brother (accusative)
g̓ = ergo = hence
gr̄a = gratia = pleasantness, favour, gratitude
hēat = habeat = may he, she, it have
.i. = id est = that is
igr̄, igr, g = igitur = then, therefore, accordingly, consequently
ip̄ius = ipsius = the same
ip̄oy = ipsorum = ipsorum, of the same (plural)
īr (perhaps for ir̄) = ire = to go
lrās, lrīs = litteras, litteris = letters, despatches epistles
mō = modo = by measure, with a limit

mr̄is = matris = of the mother
nr̄ = noster = our, our own, ours
nr̄is = nostris = of our . . . .
oēs = omnes = all men, all persons, everybody
oĭm = omnium = of all
or̄o = oratio = a speaking, speech, prayer
orōnē = orationem = prayer (accusative)
pñtes = praesentes = those who are present
pp = propter = near, hardby, at hand
pr̄ = pater = a father, sire
pr̄is = patris = of the father
p̄t = potest = can, is able to
qđ = quod = at, as for, in the fact, that is that, the fact that, that because
qm̃ = quoniam = seeing that, whereas, since, because
rō = ratio = a reckoning, account calculation, computation
r̄furr̄ctōem = resurrectionem = resurrection (accusative)
.ſ. = scilicet = one may know
ſc̄us = sanctus = holy
ſñia = sententia = opinion, meaning
ſp = semper = always
ſpm̄ = spiritum = spirit (accusative)
ſp̄us = spiritus = spirit (nominative)
ſc̄e = sancte = holy (vocative)
tm̃ = tantum = so great a . . . . (accusative)
tñ = tamen = however, nevertheless
tpe = tempore = at the right time
tp̄s = tempus = time
uñ = unde = whence
xp̄s = Christus

Abbreviations of names and their addresses of non-profit institutes, organisations, federations, unions, etc., which may be useful for the international exchange of experience and technical information. List checked January 1979. Addresses subject to change.

AAPA	All Asian Printers' Association.
ABEJ	Association Belge des Editeurs de Journaux, 20 rue Belliard, Bruxelles 4e.
ABMPM	Association of British Manufacturers of Printers' Machinery. Changed its name (1966) to British Printing Machinery Association. See BPMA.
ACANOR	Association des Cadres de Normalisation.
ACEPP	Association des Cadres et Employés de Presse et de Publicité.
ACGIH	American Conference of Governmental Industrial Hygienists.
ACIMGA	Associazione Costruttori Italiani Macchine Grafiche e Affini, Lungo Po Antonelli 45, 10153 Torino.
ACP	Association of Correctors of the Press. See NGA.
ACS	American Chemical Society. 1155 Sixteenth Street NW, Washington (DC 20036).
ACTFP	Association des Chefs et Techniciens de Fabrication de la Publicité Maîtres-Artisans du Livre et petites Imprimeries, 30 rue Claude-Decaen, Paris 12e.
ADERP	Association pour le Développement de l'Enseignement et des Recherches de Paris, avenue Pierre Derbie de Serbie, Paris 16e.
ADETEM	Association pour le Développement des Techniques de Marketing.
ADICEP	Amicale des Dirigeants des Centres de Productivité.
AELE	Association Européenne de Libre Echange.
AFAP	Association Française pour l'Accroissement de la Productivité. See CNAP.
AFC	Association Française de Colorimétrie, 57 rue Cuvier, Paris 5e.
AFCIQ	Association Française pour le Contrôle Industriel de la Qualité, 6 rue Royale, Paris 1e.
AFDEL	Association Française pour le Développement de la Lecture, 117 boulevard Saint-Germain, Paris 6e.
AFIRO	Association Française d'Informatique et de Recherche Operationnelle.
AFMAC	Association Française des Constructeurs de Machines Automatiques de Conditionnement.
AFNOR	Association Française de Normalisation, 92 Courbevoie, Paris 1.
AFORP	Association pour la Formation et le Perfectionnement du Personnel des Entreprises Industrielles de la Région Parisienne.
AFPA	Association Nationale pour la Formation Professionnelle des Adults. Ministère du Travail, 13 Place de Villier, 93 Montreuil.
AGI	Alliance Graphique International, Dufourstraße 107, 8008 Zürich.
AGIRC	Association Générale des Institutions de Retraites des Cadres.
AGV	Arbeitsgemeinschaft der Graphischen Verbände des Deutschen Bundesgebiets.
AIERI	Association Internationale des Etudes et Recherches sur l'Information.
AIFEC	Associazione Italiana Fotocompositori, Editoriali e Commerciali.
AIFMP	All India Federation of Master Printers, 133 C. Vakola Pipe Lane, Bombay 55.
AIGA	American Institute of Graphic Arts, 1059 Third Avenue, New York (N.Y. 10016).
ALA	Amalgamated Lithographers of America. See LPIU.

All-Japan-Federation of Printing and Bookbinding Machinery Manufacturers' Associations, 7 Azumabashi 3-chome, Sumida-ku, Tokyo.

All-Union Research Institute for the Polygraphic Industries; and All-Union Research Institute for Printing Machines, Ul. Čechova 18 ᵃ, Moscow K 6.

AMA	American Management Association, 135 West 50th street, New York (N.Y. 10020).
ANC	Australian Newspapers Council, 100 Bathurst street, Sydney, N.S.W.
ANFACI	Association Nationale des Fabricants de Colles Industrielles.
ANFOPPE	Association Nationale pour la Formation et le Perfectionnement du Personnel d'Encadrement.
ANGB	Algemene Nederlandse Grafische Bond, Koninginneweg 20, Amsterdam.
ANIMA	Association Nationale Italienne des Industries Mécaniques et Similaires.
ANPA	American Newspaper Publishers Association
ANPARI	American Newspaper Publishers Association Research Institute.
APA	American Photoplatemakers Association, 166 West Van Buren street, Chicago (Ill. 60604) (The former APEA).
APEC	Association pour l'Emploi des Cadres.
API	Associated Printing Industries, 610, 16th street, Room 206, Oakland (Cal. 94612).
API	American Paper Institute, 260 Madison Avenue, New York.

APIG   Associazione Periti Industriali Grafici.

APP   Association Professionnelle de la Papeterie.

APPA   American Paper & Pulp Association, 122 East 42nd street, New York (N.Y. 10017).

APPITA   Australian and New Zealand Pulp and Paper Industry Technical Association, 476 Collins street, P.O. box 1643, Melbourne C, 1 (Vic.).

APRKT   Archives for Printing, Repro and Kindred Trades, Schillerstraße 18, Postschließfach 121, 322 Alfeld/Leine, West Germany.

APS   Association Patronale des Sérigraphes.

APT   Association of Printing Technologists. See IOP.

ASA   See USASI.

ASCELF   Association Syndicale des Cadres de l'Edition et de la Librairie Française.

ASF   Association Française de la Sérigraphie, 2 rue du Colonel Driant, Paris 1e.

ASLP   Amalgamated Society of Lithographic Printers, Senefelder House, 137 Dickinson Road, Rusholme, Manchester 14.

ASMIC   Association pour l'Organisation des Missions de Coopération Technique.

Asociación de Investigación de la Industria Gráfica, Pedro Martinez Artola 4–10, Bilbao 12.

Asociación de Investigación Técnica de la Industria Papelera Española, Plaza del Marqués de Salamanca 9, 3e derecha, Madrid 6.

Asociación Española para el Progreso de las Artes Gráficas, Madera 10, 2e, Madrid 13.

ASSCO   European Solid Fibreboard Case Manufacturers' Association, Bleicherweg 33, Zürich.

ASSEDIC   Association pour l'Emploi dans l'Industrie et le Commerce.

Associated Third Class Mail Users, 100 Indiana Avenue NW, Washington (DC 20001).

Association des Compagnons de Lurs, Chancellerie, Lure en Provence (Basses Alpes).

Association des Maîtres Imprimeurs du Grand-Duché de Luxembourg, 18 rue de la Poste, Luxembourg.

Association Professionnelle des Représentants et Employés de la Papeterie et Imprimerie, 69 rue de Bretagne, Paris 3e.

Associazione Nazionale Italiana Industrie Grafiche Cartotecniche e Trasformatrici, Piazza della Conciliazione 1, 20123 Milano.

ASTEF   Association pour l'organisation de Stages en France.

ASTEGI   Associazione Tecnici Grafici Giornali Italiani.

ASTM   American Society for Testing Materials, 1916 Race street, Philadelphia (Pa. 19103).

ATA   Advertising Typographers Association of America, Inc., 461 Eight Avenue, New York (N.Y. 10001).

ATF   American Type Founders, Inc., 200 Elmora Avenue, Elisabeth (N.J.).

ATICELCA   Associazione Tecnica Italiana per la Cellulosa e la Carta, Via Sandro Botticelli 19, 20133 Milano.

A. Typ. I.   Association Typographique Internationale (Founded in 1957 to foster international contact and cooperation among persons engaged in or otherwise interested in typographic technology) Kattowitzer strasse 57, D-6230 Frankfurt 80.

BAG   Buchhändler-Abrechnungs-Genossenschaft, Leipzig.

BAIE   British Association of Industrial Editors Ltd, 17 Church street, London SW 13.

BASF   Badische Anilin und Soda Fabrik.

Battelle Memorial Institute, 505 King Avenue, Columbus (Ohio 43201).

BCA   British Carton Association, 35 New Bridge street, London EC 4.

BDBI   Bund Deutscher Buchbinder-Innungen, Postschließfach 194, Hedwigstraße 35, 41 Duisburg 1.

BDG   Bund Deutscher Gebrauchsgraphiker e.V., Ottostraße 9, 8 München 2.

BDZV   Bundesverband Deutscher Zeitungsverleger e.V., Kölner Straße 135, 532 Bad Godesberg.

BELCA   Bureau Européen de Liaison des Colles et Adhésifs.

BEMA   Business Equipment Manufacturers Association of America, 235 East 42nd street, New York (N.Y. 10017).

BESA   British Engineering Standards Association.

BFI   Beutel- und Flexodruck-Industrie, Zürich.

BFMP   British Federation of Master Printers, 11 Bedford Row, London W.C. 1.

BGV   Bundesvereinigung der Deutschen Graphischen Verbände e.V. (Fachverbände; Buchdruck; Zeitungsdruck; Flachdruck; Tiefdruck; Chemigraphie; Galvanoplastik und Stereotypie; Stempel- und Graveurgewerbe; Flexographie; Kleinoffset; Siebdruck), Postfach 503, 6200 Wiesbaden.

BGV   Belgische Vereniging van Dagbladuitgevers, Belliardstraat 20, Brussels IV.

BICO   Bureau International d'Informations et de Cooperation des Editeurs de Musique.

BMI	Book Manufacturers' Institute, Inc., 25 West 43rd street, New York (N.Y. 10036).
BN	Bibliothèque Nationale, 58 rue de Richelieu, Paris 7e.
	Bond van Handelaren in Machines en Materialen voor de Grafische en Aanverwante Industrieën, Zeestraat 78, The Hague.
BPA	Business Publications Audit.
BPBIRA	British Paper and Board Industry Research Association, St. Winifred's Laboratories, Welcomes Road, Kenley (Surrey) (Merged in 1967 with Patra *qv.*).
BPBMA	British Paper and Board Makers' Association, Plough Place, Fetter Lane, London EC 4.
BPMA	British Printing Machinery Association, 12 Cliffords Inn, Fetter Lane, London EC 4. See also ABMPM.
BRAEC	Bureau de Recherche et d'Action Economique, Paris.
BSA	Bank Stationers Association, National Head Quarters, 230 West 41st street, New York (N.Y. 10036).
BSA	Bibliographical Society of America.
BSI	British Standards Institution, British Standards House, 2 Park Street, London W. 1.
BTG	British Typographers Guild.
BUMA	Bureau voor Muziek-Auteursrecht, Marius Bauerstraat 30, Amsterdam-C.
CAAN	Continental Advertising Agency Network.
CAB	College Algrafisch Beraad, Van Eeghenstraat 70, Amsterdam-Z.
Cadippe	Comité d'Action pour Développer l'Intéressement du Personnel à la Productivité des Entreprises.
CAE	Compagnie Européenne d'Automatisme Electronique, 14 rue de la Baume, Paris 10e.
CANCAVA	Caisse Autonome Nationale de Compensation de l'Assurance Vieillesse Artisanale, 28 boulevard de Grenelle, Paris 15e.
	Carnegie Institute of Technology, Schenley Park, Pittsburgh 13 (Pa.) (Merged in 1966 with Melton Institute).
CCFI	Compagnie des Chefs de Fabrication de l'Imprimerie et des Industries annexes, 35 rue des Grands-Champs, Paris 20e.
CCIP	Chambre de Commerce et de l'Industrie de Paris.
CEAFC	Centre d'Etudes et d'Application pour la Formation des Cadres.
CEBUCO	Centraal Bureau voor Couranten-Publiciteit van de Nederlandse Dagbladpers, Weesperstraat 89, Amsterdam.
CECIAG	Congrès Européen de Coordination des Industries et Arts Graphiques. Organisé par le Centre Parisien de Congrès Internationaux (CEPACI), 120 Avenue Emile-Zola, Paris 15e.
CEE	Communauté Economique Européenne (Les Pays du Marché Commun).
CEFE	Comité d'Etudes du Financement des Entreprises.
CEGE	Comité d'Etudes de la Gestion des Entreprises.
CEI	Comité Européen des Associations de Fabricants de Peintures et d'Encres d'Imprimerie, Bruxelles.
CEN	Comité Européen des Normes.
	Centralne Laboratorium Poligraficzne, Ul. Minska 69, Warszawa.
CEOP	Confédération Européenne des Organisations de Publicitaires.
CEPAC	Confédération Européenne de l'Industrie des Pâtes, Papiers et Cartons, Bruxelles.
CEPACI	See CECIAG.
CESP	Centre d'Etudes des Supports de Publicité.
CFDT	Confédérations Française Démocratique du Travail, 26 rue Montholon, Paris 9e.
CFPC	Centre chrétien des Patrons et Dirigeants d'entreprise français.
CFRO	Centre français de Recherche Opérationnelle.
CFTC	Confédération Française des Travailleurs Chrétiens, 26 rue Montholon, Paris 9e.
CGC	Syndicat des Cadres de la Maîtrise du Livre, 30 rue de Grammont, Paris 2e.
CGT	Confédération Générale du Travail, 213 rue Lafayette, Paris, 10e.
CGT-FO	Ingénieurs et Cadres et Ouvriers du Livre, 198 avenue du Maine, Paris 14e.
CIE	Commission Internationale de l'Eclairage, 25 rue Pépinière, Paris 8e. See also ICI.
CIESJ	Centre International d'Enseignement Supérieur du Journalisme.
CIGA	Société des Maîtres-Imprimeurs de l'Argentine.
CIH	Club International d'Héliogravure.
CIS	Composition Information Services, 1605 North Cahuenga Boulevard, Los Angeles (Cal. 90028).
CISCO	Centro Italiano Studi sul Cartone Ondulatto.
CJP	Centre des Jeunes Patrons.

CLA   Canadian Lithographers Association, The Arcade Building, 74 Victoria street, Toronto 1
                      (Ontario).
CNAH   Centre National pour l'Amélioration de l'Habitation, 31 avenue Pierre Derbie de Serbie,
                      Paris 16e.
CNAM   Confédération Nationale de l'Artisanat et des Métiers, 292 rue Saint-Martin, Paris 3e.
CNAP   Centre National pour l'Amélioration de la Productivité, 6 rue Royale, Paris 8e, (the former,
                      AFAP).
CNCE   Centre National du Commerce Extérieur.
CNET   Centre National d'Etude des Télécommunications.
CNIT   Centre National des Industries et des Techniques, 2 Place de la Défense, 92 Puteaux. See
                      also TPG.
CNOF   Comité National de l'Organisation Française, 3 rue Cassette, Paris 6e.
CNP   Comité National Permanent.
CNPF   Conseil National du Patronat Français, 31 avenue Pierre Derbie de Serbie, Paris 8e.
CNRS   Centre National de la Recherche Scientifique.
COBELPA   Association des Fabricants de Pâtes, Papiers et Cartons de Belgique, 14 rue de Crayer,
                      Bruxelles 5e.
COID   Council of Industrial Design, 28 Haymarket, London S.W. 1.
COP   Commissie Opvoering Productiviteit. Dutch for Committee for Increased Productivity.
COPACEL   Confédération Française de l'Industrie des Papiers, Cartons et Celluloses.
CPI   Council of Printing Industries, 181 Bay street, Toronto (Ontario).
CPIMA   Chicago Printing Ink Manufacturers Association, 608 South Dearborn street, Chicago
                      (Ill. 60605).
CPPA   Canadian Pulp and Paper Association, 2280 Sun Life Building, Montreal 2.
CPTA   Canadian Paper Trade Association, 55 York street, 13th floor, Toronto 1 (Ontario).
CQFD   Centre Qualitatif Français de la Diffusion.
CRCDIPE   Caisse de Retraite de non salariés de la Presse.
CRPILIC   Caisse de Retraite et de Prévoyance de l'Imprimerie de Labeur et des Industries Graphiques.
CSAPE   Central States Association of Photo-Engravers, 750 Prospect avenue, Cleveland (Ohio 44115).
CSG   Centro di Sperimentazione Grafica, Rome, See ENCC.
CST   Compagnie Générale du Télégraphie sans Fil, 7 rue de Madrid, Paris 8e.
CTIP   Centre Technique de l'Industrie des Papiers, Cartons et Celluloses.

DAMW   Deutsches Amt für Material- und Warenprüfung.
D&ADA   Designers and Art Directors Association.
Dansk Litograflang. Landemaerket 11, Kobenhavn.
Dansk Papaeskefabrikant Forening. Vesterbrogade 101, Kobenhavn V.
Dansk Provins Bogtrykkerforening, Bogtryggernes Hus, Helgajev 26, Odense.
Danske Presses Faellesindkøbs Forening, Radhuspladsen, 55 Kobenhavn.
DIN   Deutsches Institut für Normung e.V., Burggrafenstraße 4–7, Postfach 1107, 1000 Berlin 30
                      (Namensänderung ab 1-9-'75 für Deutscher Normenausschuß DNA).
DMAA   Direct Mail Advertising Association, Inc., 230 Park avenue, Rm 666, New York (N.Y. 10017).
DNA   See DIN.
DRN   Drukinkt Research Nederland (i.e. Dutch Printing Ink Research Group), Groot Haese-
                      broekseweg 1, Wassenaar (the Netherlands).
DRUPA   Internationale Messe Druck und Papier. Organisation: Nordwestdeutsches Ausstellungs-
                      Gesellschaft m.b.H. (NOWEA), Ehrendorf 4, Düsseldorf.

EAAA   Association Européenne des Agences de Publicité, 27 Arosastraße, ch. 8000, Zürich.
ECGAI   Educational Council of the Graphic Arts Industry, Inc. Has merged into the GATF (1966).
ECMA   European Carton Makers Association, Raamweg 13, The Hague.
ECMA   European Computer Manufacturers Association, 11–1204 rue d'Italie, Geneva.
EFR   Editeurs Français Réunis.
EMA   Envelope Manufacturers Association, 1 Rockefeller Plaza, New York (N.Y. 10020).
EMBALPACK   Groupement Européen des Fabricants de Papiers d'Emballage, Brussels.
EMPA   Eidgenössische Materialprüfungs- und Versuchsanstalt für Industrie, Bauwesen und Gewerbe,
                      Unterstraße 11, Sankt Gallen (Switzerland).
ENCC   Ente Nazionale per la Cellulosa e per la Carta (Centro di Sperimentazione Grafica), Viale
                      Regina Margherita 262, Roma.
Ente Italiano Cellulosa e Carta, Via Assisi 163, Roma.

EO	Euroffset, Via G. V. Zeviani 2, Verona (The offset equivalent of ERA *qv.*).
EOQC	European Organization for Quality Control.
EOST	Ecole d'Organisation Scientifique du Travail.
EPAM	Employing Printers Association of Montreal, Inc., 1509 Sherbrooke street West, Montreal 25, Quebec.
EPCF	European Plastic Coaters Federation.
EPF	European Packaging Federation.
ERA	European Rotogravure Association, Kopernikusstraße 9, 8 München 80.
ERIC	Etudes et Recherches Industrielles et Commerciales, Paris.
ESMA	Engraved Stationary Manufacturers Association, 233 Tower Building, Washington (DC 20005).
EUpagraph	Association Européenne de l'Industrie des Papiers Graphiques, Brussels.
FAC	Franklin Association of Chicago, Inc., 12 East Grand avenue, Chicago (Ill. 60611).
FATIPEC	Fédération d'Associations de Techniciens des Industries de Peintures, Vernis, Emaux et Encres d'Imprimerie de l'Europe Continental, 28 rue Saint-Dominique, Paris 7.
FCMA	Finnish Carton Makers Association, c/o G. A. Serlachius, Oy Tako, Tampere.

Fédération de Journaux Belges, 20 rue Belliard, Bruxelles 4e.
Fédération Nationale des Hebdomadaires d'Informations, 20 rue Belliard, Bruxelles 4e.
Fédération Française des Syndicats Patronaux de l'Imprimerie et des Industries Graphiques, 115 boulevard Saint-Germain, Paris 8e.
Fédération des Chambres Syndicales des Fabricants de Cartonnages de France, 182 rue de Rivoli, Paris 1e.

FEDES	Fédération Européenne des Fabricants de Sacs en Papier et d'Emballages Similaires.
FEFCO	Fédération Européenne des Fabricants de Carton Ondulé, 90 rue d'Amsterdam, Paris 9e.
FEGRAB	Federatie der Grafische Bedrijven in België.
FESPA	Federation of European Screen Printers Associations, Van Eeghenstraat 70, Amsterdam.
FETRA	Fédération des Industries Transformatrices de Papier et Carton, 93 Avenue Louise, Brussels 4e.
FFI	Fachverband Faltschachtel-Industrie e.V., Postfach 220, 605 Offenbach 1.
FFITP	Fédération Française des Industries Transformatrices des Plastiques.
FFTL	Fédération Française des Travailleurs du Livre, 7 rue Jules Breton, Paris 6e.
FGE	Organisatie van Fabrikanten van Grafische Eindprodukten (the former Bond van Boekbinderspatroons), Vondelstraat 172, Amsterdam.
FGI	Fédération Graphique Internationale.
FIB	Fachgemeinschaft Industrielle Buchbinderei e.V., Arndtstraße 47, Frankfurt a/Main 33529.
FID	Fédération Internationale de Documentation, Hofweg 7, The Hague.
FIEJ	Fédération Internationale des Editeurs de Journaux et Publications, 6 bis rue Gabriel Laumain, Paris 10e.
FINAT	Fédération Internationale des Fabricants et Transformateurs d'Adhésifs et Thermocollants sur papier et autres supports.
FIPP	Fédération Internationale de la Presse Périodique.
FMPNZ	Federation of Master Printers of New Zealand, Box 1422, Wellington.
FMPSA	Federation of Master Printers of South Africa, United Buildings, Co. Fox & Eloff street, Johannesburg.
FNMAL	Fédération Nationale des Maîtres Artisans du Livre, 5 rue Crussol, Paris 11e.
FNPF	Fédération Nationale de la Presse Française.
FNSIC	Fédération Nationale des Syndicats d'Ingénieurs et de Cadres Supérieurs, 30 rue de Grammont, Paris 2e.
FO	Fédération Force Ouvrière.
FOGRA	Deutsche Gesellschaft für Forschung im Graphischen Gewerbe, Streitfeldstraße 19, 8 München 80.

Forschungsgesellschaft Druckmaschinen e.V., Lyonerstraße, Postfach 109, 6 Frankfurt a/Main-Niederrad.

FPBAA	Folding Paper Box Association of America, 222 West Adams street, Chicago (Ill. 60606).
FPCMA	Fibreboard Packing Case Manufacturers Association.
FTA	Flexographic Technical Association, 1416 Avenue M, Brooklyn (N.Y. 11230).
FTL	Fédération Française des Travailleurs du Livre. See FFTL.
FUGRA	Association des Jeunes et Futurs Patrons des Entreprises Graphiques, Vleminckveld 18, Antwerp 1.
FULPC	Federazione Unitaria dei Lavoratori Poligrafice e Cartai.
GAA	Graphic Arts Association, Inc., 1003 Omaha National Bank Building, Omaha (Nebraska 68102).

GAABC	Graphic Arts Association of British Columbia, 475 Howe street, Vancouver (British Columbia).
GAAC	Graphic Arts Advertisers Council, 230 West 41st street, New York City.
GAAE	Graphic Arts Association Executives, 321 Tower Building, Washington (DC 20005).
GAESD	Graphic Arts Equipment and Supply Dealers.
GAIA	Graphic Arts Industries Association, 1509 Sherbrooke W., Montreal 25, Quebec (Canada).
GAIU	Graphic Arts International Union. The amalgamation of the Lithographers and Photo-engravers International Union (LPIU), the International Stereotypers and Electrotypers Union (ISEU), the International Printing Pressmen and Assistants Union (IPP & AU), the Amalgamated Lithographers of America (ALA) and the International Brotherhood of Bookbinders (IBB).
GAMIS	Graphic Arts Marketing and Information Service (In 1966 newly organized division of PIA *qv*.).
GARC	Graphic Arts Research Centre, 65 Plymouth avenue South, Rochester (N.Y. 14608) (the former Rochester Institute of Technology RIT).
GASB	Gemeinsamer Ausschuß des Schweizerischen Buchbindergewerbes, 1 rue du Temple, Geneva.
GASI	Group Advertising Services International.
GATF	Graphic Arts Technical Foundation, Inc., 4615 Forbes Avenue, Pittsburgh (Penn. 15213) (The former LTF).
GEA	Gravure Engravers Association, Inc., 166 West Van Buren street, Chicago (Ill. 60604).
GEA	Grafisch Economisch Adviescentrum, Ter Gouwstraat 1, Amsterdam.
GEEC	Groupement Européen d'Etiquetage et de Conditionnement.
GFL	Grafiska Forskning-Laboratoriet, Drottning Kristina Väg 61, Stockholm-O.
GFR	Groupe Fraternel des Représentants et Agents Commerciaux des Industries Graphiques, 30 rue des Boulets, Paris 11e.
GFT	Gesellschaft zur Förderung des Tiefdrucks m.b.H.
GIAG	Groupement des Industries des Arts Graphiques.
GIENNEP	Groupement Interprofessionnel des Entreprises du Nord, Nord-Est Parisien.
GIFCO	Gruppo Italiano Fabbricanti Cartone Ondulato.
GJC	Graphischer Juniorenclub des Hauptverbandes der Graphischen Unternehmungen Österreichs, 4 Grünauergasse, 1010 Vienna.
GKf	Gebonden Kunsten Federatie (Vereniging van Beoefenaars der Gebonden Kunsten. See GVN).
GLV	Graphische Lehr- und Versuchsanstalt.
GPO	Government Printing Office, Washington (DC 20025).
	Graafinen Keskuslito, P. Esplanaadinkatu 25 a, Helsinki.
	Graafinen Tutkimuslaitos (Graphic Arts Institute) (Lönnrotinkatu 37, Helsinki 18.
	Grafiske Höjskole (The Graphic College of Denmark), Julius Thomsensgade 3B, Copenhagen V.
	Graphic Arts Institute, 8955 St. Hubert, Montreal.
GRI	Gravure Research Inc., 22 Manhasset Avenue, Manorhaven, Port Washington, Long Island (N.Y. 11050).
GSPO	Government Security Printing Office, Djalan Trunodjo 4 B, Blok K V, Kebajoran Baru, Djakarta.
GTA	Gravure Technical Association, Lincoln Building, 60 East 42nd street, New York (N.Y. 10017).
GVN	Grafisch Vormgevers Nederland (Beroepsvereniging) (Amalgamation of GKf and VRI), Nieuwe Keizersgracht 58, 1018 DT Amsterdam.
GVR	Genootschap voor Reclame, Koningslaan 2, Amsterdam.
HGUO	Hauptverband der Graphischen Unternehmungen Österreichs, Grünauergasse 4, Vienna I.
HMSO	Her (His) Majesty's Stationary Office, 49 High Holborn, London W.C.1.
	Höhere Graphische Bundeslehr- und Versuchsanstalt, Vienna.
HPV	Hauptverband der Papier- und Pappeverarbeitenden Industrie e.V., Arndtstraße 47, Frankfurt a/Main. 33529.
	Hungarian Technical Association of Paper Industry & Graphic Arts, Fontca 68, Budapest.
IAA	International Advertising Association.
IAEE	Institut d'Administration et d'Economie des Entreprises
IAES	International Association of Electrotypers and Stereotypers, Inc., 758 Leader Building, Cleveland (Ohio 44114).

IAPHC	International Association of Printing House Craftsmen, 7599 Kenwood road, Cincinnati (Ohio 45236).
IARIGAI	International Association of Research Institutes for the Graphic Arts Industry,
IBB	International Brotherhood of Bookbinders, 1612 K Street N.W., Washington, D.C. See GAIU.
IBBI	International Boards on Books for Young People.
IBE	Institut Belge de l'Emballage, 15 rue Picard, Bruxelles 2e.
IBFI	International Business Forms Industries, 1730 N. Lynn street, Arlington. (Virginia 22209).
IBM	International Business Machines Corp.
IBN	Institut Belge de Normalisation, 29 avenue de la Brabançonne, Bruxelles 4.
IBPA	International Business Press Associates.
ICC	International Container Corporation.
ICCA	International Corrugated Cartonmakers Association.
ICCAS	International Center for the Communication Arts and Sciences, New York.
ICI	International Commission on Illumination.
ICOGRADA	International Council of Graphic Design Associations, 12 Blendon Terrace, Plumstead Common, London SE 18 7RS.
ICSID	International Council of Societies of Industrial Design, Avenue Legrand 45, 1050 Brussels.
ICT	International Computers and Tabulators.
ICTA	International Centre for Typographic Arts, Inc., P.O. box 2438, Grand Central Station, New York 17, N.Y.
IDD	Institut für Druckmaschinen und Druckverfahren, Technische Hochschule, Alexanderstraße 22, Darmstadt.
IDET	Institut pour le Développement Economique et Technique.
IEC	International Technical Commission.
IEPLV	Institut Européen de la Publicité sur le Lieu de Vente, Paris.
IESTO	Institut d'Etudes Supérieures et Techniques d'Organisation.
IfD	Institut für Dokumentation der Deutschen Akademie der Wissenschaften, Berlin.
IFEC	Institut Français de l'Emballage et du Conditionnement, 40 rue du Colisée, Paris 8e.
IFIP	International Federation for Information Processing.
IFMP	International Federation of Master Printers. See IMPA.
IFOSA	International Federation of Stationary Associations.
IFRA	Inca-Fiej Research Association, Washingtonplatz 1, D-6100 Darmstadt (the former INCA).
IFRI	Inca-Fiej Research Institute. See IFRA.
IFTIM	Institut de Formation aux Techniques d'Implantation et de Manutention, 46 rue Troyon, 92 Sèvres.
IGF	International Graphical Federation.
IgT	Institut für Grafische Technik, Inselstraße 20, 701 Leipzig
IGT	Instituut voor Grafische Techniek TNO, Ter Gouwstraat 1, PO box 4150 1009 AD Amsterdam-O. Discontinued 1983. See Reproset.
III	Istituto Italiano Imballagio, Padua.
IIT	Illinois Institute of Technology, Chicago.
IMPA	International Master Printers Association, 57 Wharfdale Road, London N1 (the former IFMP).
IN	Imprimerie Nationale, 39 rue de la Convention, Paris 15e.
INCA	International Newspaper and Color Association. See IFRA.
INIAG	Institut National des Industries et Arts Graphiques, 51 Boulevard St. Michel, Paris 5e.
INPA	International Newspaper Promotion Association.
INS	Institut National de Sécurité.
INSEE	Institut de la Statistique et des Etudes Economiques, 29 quai Branly, Paris 7e.

Institut der Papiertechnische Stiftung, Lothstraße 34, 8000 München 2.
Institut für Angewandte Geodäsie, Fortshausstraße 151, Frankfurt a/Main.
Institut für Verpackung, Riesaer Strasze 7, 8023, Dresden.
Institute of Newspapers Controllers & Finance Officers, P.O. box 68, Fairhaven (N.J. 07701).
International Association of Blue print & Allied Industries, 33 East Congress Parkway, Chicago (Ill. 60605).
International Graphic Arts Education Association, 85 Plymouth Avenue South, Rochester (N.Y. 14608).

IOP	Institute of Printing, 8 Lousdale Gardens, Tunbridge Wells, Kent (TNI INU).
IPA	International Publishers Association.
IPC	Institute of Paper Chemistry, P.O. box 1048, Appleton (Wisc. 54922).
IPEAC	Institut pour la Promotion Economique par l'Action Commerciale.

IPEX	International Printing Machine and Allied Trade Exhibition, Organisation: F. W. Bridges & Sons Ltd., Grand Buildings, Trafalgar Square, London W.C. 2.
IPG	Independent Packaging Group, Ltd., 26 rue Lachenal, 1200 Geneva.
IPI	Interchemical Corporation, Printing Ink Division. New York.
IPM	Institut für Polygrafische Maschinen, Schönbachstraße 66, 7027 Leipzig.
IPPAU	International Printing Pressmen and Assistants Union. See GAIU.
IPREIG	Institut Professionnel de Recherches et d'Etudes des Industries Graphiques, 17 rue des Reculettes, Paris 13e.
IPU	International Photoengravers Union. See LPIU.
IRANOR	Instituto Nacional de Racionalización y Normalización, Serrano 150, Madrid 6.
IRCHA	Institut National de Recherche Chimique Appliquée, 12 quai Henri IV, Paris.
IRD	Institut für Rationalisierung in der Druckindustrie e.V., Frankfurt/Main.
IRSEC	Institut de Recherche, Sondage et Etude de Commercialisation, 20 rue de la Michodière, Paris 2e, et 39 rue d'Amsterdam, Paris 8e.
ISA	Instrument Society of America, 530 William Penn Place, Pittsburgh (Penn. 15219).
ISA	International Federation of the National Standardizing Association.
ICC	Inter-Society Color Council, c/o Eastman Kodak Company, Color Technology Division, Building 65, Rochester, N Y. (See also in basic table).
ISEU	International Stereotypers and Electrotypers Union. See GAIU.
ISG	Institut Supérieur de Gestion. Paris.
ISO	International Standards Organization, 32 route de Malagnon, Geneva.
ISPA	International Small Printers' Association.
	Istituto Rizzoli per l'Insegnamento delle Arti Grafiche, Via Botticelli N 19, Milan.
ITCA	International Typographic Composition Association, The Georgetown Building, 2233 Wisconsin avenue NW, Washington (DC 200037).
ITU	International Typographic Union. See A. Typ. I.
	Japanese Government Printing Bureau Research Institute, 1-400 Oji Kita-ku, Tokyo.
JPA	Japan Printers' Association, 23, 2-chome Shintomicho, chuo-ku, Tokyo.
KDT	Kammer der Technik (Deutsche Demokratische Republik).
KIO	Kring Industriële Ontwerpers, Keizersgracht 321, Amsterdam.
KNUB	Koninklijke Nederlandse Uitgevers Bond, Keizersgracht 391, Amsterdam.
	Københavns Bogtrykkerforening, Landemaerket 11, Kobenhavn.
KVGO	Koninklijk Verbond van Grafische Ondernemingen, Van Eeghenstraat 70, Amsterdam.
	Lak- og Farveindustries Forsknings-Laboratorium.
LEPA	Lithographic Engravers and Platemakers Association. See LPGA.
	Library Institute of Paper Chemistry, Appleton (Wisc.).
LMNA	Label Manufacturers National Association, Shoreham Building, 15th and H streets, NW, Washington (DC 20005).
LNA	Lithographers National Association, 420 Lexington avenue, New York 17 (N.Y.).
	London College of Printing, Elephant and Castle, London EC 1.
LPGA	Lithographic Plate Grainers Association, Inc., (LPGA) and Lithographic Engravers and Platemakers Association, Inc., (LEPA), West 42nd street, New York, N.Y.
LPIU	The amalgamation (1964) of Amalgamated Lithographers of America (ALA) and International Photoengravers Union (IUP). See GAIU.
LTF	Lithographic Technical Foundation. The former name of GATF qv.
LTS	London Typographical Society. See NGA.
MASAI	Mail Advertising Service Association International, 815 – 17th street N.W., Washington (DC 20006).
MBA	Master Bookbinders Association (Merged with London Master Printers Association (1967)).
MBAL	Master Bookbinders Alliance of London, 11 Bedford Row, London W.C. 1.
MCVA	Machine Compositors' Vigilant Association (Popularly known as the Vig).
	Member & Association Services Department. Printing Industries of America, Inc., 20 Chevy Chase Circle NW, Washington (DC 20015).
	Metropolitan Lithographers Association, 250 West 57th street, New York (N.Y. 10019).
MIT	Massachussetts Institute of Technology, Cambridge (Mass. 02140).
MLA	Metropolitan Lithographers Association.

MPS	Master Printers Section, Printing Industries of America, Inc., 20 Chevy Chase Circle NW, Washington (DC 20015).
Mugecia	Mutuelle Générale du Commerce de l'Industrie et de l'Artisanat, Paris.
NAAP	National Association of Advertising Publishers.
NAGRA	Normenausschuß für das Graphische Gewerbe. See DNA.
NALC	National Association of Litho Clubs, Inc., 230 West 41st street, New York (N.Y. 10036).
NAPIM	National Association of Printing Ink Manufacturers, 39 West 55th street, Suite 904, Washington (DC 10019).
NAPL	National Association of Photo-Lithographers, 230 West 41st street, New York (N.Y. 10036).
NAPM	National Association of Paper Merchants, 27 Chancery Lane, London W.C. 2.
	National Association of Blueprint & Diazotype Coaters, 1925 K street NW, Suite 402, Washington (DC 20006).
NATSOPA	National Society of Operative Printers and Assistants, Caxton House 13–16 Borough road, St. George's Circus, London, S.E. 1. See SOGAT.
NBC	National Book Committee.
NBCK	Nederlandse Bond van Copieerders en Kleinoffsetdrukkers, Alexander Numankade 14, Utrecht.
NBFA	National Business Forms Association, 1522 K street NW, Washington (DC 20005).
NBL	National Book League, 7 Albemarle street, London W.1.
NBS	National Bureau of Standards, Washington, (DC) 25.
NCGB	Nederlands Christelijke Grafische Bond, Valeriusplein 30, Amsterdam.
NCR	National Cash Register Co., Fundamental Research Department, Main and K streets, Dayton (Ohio 45409).
NDP	Nederlandse Dagbladpers, Johan Vermeerstraat 14, Amsterdam.
NEDC	National Economic Development Council, London.
	Nederlandse Vereniging van Drukinktfabrikanten en Fabrikanten van Rollenspecie, The Hague.
NEPEA	Netherlands Photoengravers Export Association u.a., Churchill-laan 35a, Amsterdam-Z.
	Newspaper Society, Whitefriars House, 6 Carmelite street, London E.C. 4.
NFPA	National Flexible Packaging Association, 1170 Shaker Boulevard, Cleveland (Ohio 44120).
NGA	National Graphical Association (The amalgamation (1964) of the London Typographical Society (LTS), the Typographical Association (TA), the National Union of Press Telegraphists and the Association of Correctors of the Press (ACP). The LTS was formed by the amalgamation of the London Society of Compositors and the Printing Machine Managers Trade Society), Bromham Road, Bedford (GB).
NGDR	Nederlands Genootschap voor Documentreproductie.
NIDER	Nederlands Instituut voor Documentatie en Registratuur. See NOBIN.
NIIPM	Naučno Issledoratel'sky Institut Poligrafičeskogo Masinostroenija (Institute for Scientific Research), Moscow M, Kaluzcky per., 4.
NIVE	See NVM – NIVE.
NKGB	Nederlandse Katholieke Grafische Bond, P.C. Hooftstraat 170, Amsterdam-Z .
NMDA	National Metal Decorators Association, 2300 Sixth street, Rockford, Ill.
NMPP	Nouvelles Messageries de la Presse Parisienne.
NNA	National Newspaper Association, 491 National Press Building, 14th and F streets, Washington (DC 20004).
NNI	Nederlands Normalisatie-Instituut, Polakweg 5 , Rijswijk (Z.H.) (the Netherlands).
NNP	Nederlandse Nieuwsblad Pers, Van Blankenburgstraat 74, The Hague.
NOBIN	Nederlands Orgaan voor de Bevordering van de Informatieverzorging, Octrooi- en Informatiedienst voorheen NIDER, Willem Witsenplein 6, P.O. box 2647, The Hague.
	Norske Bokbindermestres Forbund, H. Heyersdalsgt. 1, Oslo.
	Norske Bogtrykkerförening, St. Olavsgate 28, Oslo.
	Norske Eskefabrikkers Landsforening, P.O. box 1731, Vika, Oslo.
NOTU	Nederlandse Organisatie van Tijdschrift-Uitgevers, Herengracht 257, Amsterdam.
NOWEA	Nordwestdeutsches Ausstellungs-Gesellschaft mbH. See Drupa.
NPBMA	National Paper Box Manufacturers Association, Suite 910, City Centre Building, 121 N. Broad street, Philadelphia (Pa. 19107).
NPEA	National Printing Equipment Association, Inc., room 500, 217 Broadway, New York, (N.Y. 10007).
NPIRI	National Printing Ink Research Institute, Lehigh University, Bethlehem (Pa. 18015).
NPTA	National Paper Trade Association, Inc., 220 – 42nd street, New York (N.Y. 10017).
NSES	National Society of Electrotypers & Stereotypers (Amalgamated with the National Graphical

	Association (NGA).
NUJ	National Union of Journalists.
NUPB & PW	National Union of Printing, Bookbinding and Paper Workers, 74 Nightingale Lane, London S.W. 12.
NVC	Nederlands Verpakkingscentrum, Lange Voorhout 58a, The Hague.
NVM-NIVE	Nederlandse Vereniging voor Management, Van Alkemadelaan 700, The Hague (the former NIVE, i.e. Nederlandse Instituut voor Efficiency, since Sept. '72. To avoid the loss of the initials NIVE after 45 years, the new initials are placed before NIVE).

Nyomdaipari Egyesülés, Budapest V.

OAAA	Outdoor Advertising Association of America, 24 West Erie street, Chicago, Ill.
OCDE	Organisation de Coopération et de Développement Economique, 2 rue André-Pascal, Paris 16e. See OECD.
OECD	Organization for Economic Co-operation and Development. See OCDE.
OECQ	Organisation Européenne pour le Contrôle de la Qualité.
ONSEP	Office National de Statistiques et d'Etudes de Presse.
ORTF	Office de Radiodiffusion Télévision Française, 116 avenue du Président Kennedy, Paris 16e.
OSA	Optical Society of America, Inc., 1155 – 16th street N.W., Washington (D.C. 20036).

Oy Keskuslaboratorio (Centrallaboratorium AB).

PAC	Packaging Association of Canada.

Packaging Institute, Inc., 342 Madison Avenue, New York (N.Y., 10017).
Papierindustriens Forskningsinstitut (the Norwegian Pulp and Paper Research Institute), P.O. Box 250, Vinderen.
Pappersbrukens Centrallaboratoriet, Drottning Kristina Väg 61, Stockholm.

P & KTF; PKTF	Printing and Kindred Trades Federation, 60 Doughty street, London W.C. 1. See SOGAT.
PATRA	Printing, Packaging and Allied Trades Research Association, Patra House, Randalls road, Leatherhead (Surrey). See also PIRA.
PCA	Packaging Corporation of America, Evanston, Ill.
PCMAA	Printers Cost and Management Accounting Association.
PEN	Poets, Playwrights, Editors, Essayists, Novelists (founded 1921).
PERI	Photoengravers Research Institute, Park Forest, Ill.
PIA	Printing Industries of America, 1730 Lynn street, Rosslyn (Virginia).
PIMA	Printing Ink Manufacturers' Association of Australia.
PIMNY	Printing Industries of Metropolitan New York.
PIRA	Printing Industries Research Association, (The Amalgamation of PATRA *qv* and BPBIRA *qv*), Randalls Road, Leatherhead (Surrey).
PMA	Paper Makers Association of Great Britain and Ireland, Melbourne House, Aldwyck, London W.C. 2.
PMAA	Paper Makers Advertising Association.
PM & OA	Printing Managers and Overseers Association, 50–51 Temple Chambers Avenue, London E.C. 4.
PME	Petites et Moyennes Entreprises, 18 rue Forluny, Paris 17e.
PMMI	Packaging Machinery Manufacturers Institute, Inc., 2000 K street N.W., Washington (DC 20006).

Polygrafické Závodyn, p. Vyvojovy Závod, Ul. Ceskoslovenskej armády 18, Bratislava.
Polygraphisches Institut, N. Rakitin 2, Sofia.

PPIC	Pulp and Paper Institute of Canada, 3420 University street, Montreal 2 (Quebec).
PPN	De Periodieke Pers in Nederland, Burgemeester Reigerstraat 89, Utrecht.

Printing Bureau of the Ministry of Finance (serves as the Japanese Government Printing Office), Oji, Kita ward, Tokyo.
Printing Paper Manufacturing Association, 122 East 42nd street, New York (N.Y. 10017).
Printing and Allied Trades Employers' Federation of Australia, 136 Jolimont Road, East Melbourne.

PROGRA	Promotie-Centrum voor Grafische Bedrijven en Aanverwante Bedrijven, Handelsstraat 54, Brussel 4.
PTA	Printing Trades Alliance, Alliance House, 50–51 Fetter Lane, London EC 4.

Pulp & Paper Research Institute of Canada, 3420 University street, Montreal.

R & E	Research and Engineering Council of the Graphic Arts Industry, 1515 Wilson boulevard, Arlington (VA 2209).

RCA	Radio Corporation of America, Graphic Systems Division, Princeton, N.J.
REA	Reunion Européenne d'Automatisme.
REFA	Reichsausschuss für Arbeitsstudien, Havelstrasse 16, Darmstadt 16.
Reproset	Ter Gouwstraat 1, Amsterdam-O. See also IGT.
RIT	Rochester Institute of Technology, Graphic Arts Research Department, P.O. box 3409, Rochester (N.Y. 14614) (Name changed into Graphic Arts Research Centre, See GARC).

Rolled Label Manufacturers Association, 333 Tower Building, Washington (DC 20005).
Rotary Business Forms Section of Printing Industries of America, Inc. See PIA.

RPS	Royal Photographic Society of Great Britain, 16 Princess Gate, London S.W. 7.
SAE	Society of Automotive Engineers.
SBMSV	Schweizerische Buchdruck-Maschinenmeister und Stereotypeure-Verband.
SBPIM	Society of British Printing Ink Manufacturers, 5–11 Theobalds House, London W.C. 1.

Scandinavian Paint and Printing Ink Research Institute, Hörsholm.
Schweizerische Buchdruckerverein, Postfach 121, 8030, Zürich.

SCIPAG	Syndicat de Constructeurs de Machines pour les Industries du Papier, du Carton et des Arts Graphiques, 10 avenue Hoche, Paris 8e.
SEPFA	Société d'Edition et de Publicité Française et Américaine, 11 rue Royale, Paris 8e.
SGAA	Southern Graphic Arts Association, 1000 – 17th Avenue, Nashville (Tenn. 37212).
SIA	Society of Industrial Artists & Designers, 12 Carlton House Terrace, London SW 1.
SIAG	Société Interprofessionnelle Artisanale de Garantie, 13 rue Jacques Bigen, Paris 17e.
SICOB	Salon International de l'Equipement de Bureau, 6 Place de Valois, Paris 1e.

Sindicato Nacional del Papel, Prensa y Artes Gráficas, Plaza Callao 4, Madrid.

SMAG	Spécialistes du Matériel des Arts Graphiques, Paris.
SMIC	(label de chambre) Syndicale des négociants, constructeurs et réparateurs de Machines d'Imprimerie et de Cartonnage.
SNAD	Syndicat National des Artistes Dessinateurs, 9–11 rue Berryer, Paris.
SNAPO	Syndicat National de la Publicité par l'Objet.
SNCML	Syndicat National des Cadres et Maîtrise du Livre, 30 rue Notre-Dame-des Victoires, Paris 2e.
SNCTL	Syndicat National des Cadres Technique du Livre, 7 rue Jules Breton, Paris 13e.
SNEP	Société Nationale des Enterprises de Presse.
SNEPP	Syndicat National des Extrudeurs de Profilés Plastiques.
SNPQR	Syndicat National de la Presse Quotidienne Régionale, 6 bis Gabriel Laumain, Paris 10e.
SNV	Schweizerische Normenvereinigung.
SOFIMA	Société de Financement de Matériel d'Imprimerie.
SOGAT	Society of Graphical and Allied Trades. Amalgamation of Printing and Kindred Trades Association (P & KTF) and National Society of Operative Printers and Assistants (NATSOPA).
SPPA	Screen Process Printing Association, 549 W. Randolph street, Chicago 6 (Illinois).
SPPAI	Screen Process Printing Association International, 708 Associations Building, 1145 Nineteenth street NW, Washington (DC 20006).
SPPP	Société Professionnelle des Papiers de Presse.
SQR	Syndicat des Quotidiens Régionaux, 6 bis Gabriel Laumain, Paris 10e.
SRE	Society of Reproductions Engineers, 18307 Detroit (Mich. 48235).
SSMI	Section Suisse des Maîtres Imprimeurs.
SSPL	Société Suisse des Patrons Lithographes, Effingerstraße 14, Bern. See also VSLB.
STA	Scottish Typographical Association, 136 West Regent street, Glasgow C 2.

Stanford Research Institute, Menlo Park, Calif.

STD	Society of Typographic Designers, Busby House, 33 Aldrington road, London S.W. 16.

Sveriges Grafiska Arbetsgivare och Industriorganisationer, P.O. box 16383, Blasicholmsgate 4a, Stockholm.

SVW	Staatliche Autorisierte Versuchsanstalt der höheren graphischen Bundes-, Lehr- und Versuchanstalt, Vienna 7.
SYFACAR	Chambre Syndicale des Fabricants de Cartonnages, 93 Avenue Louise, Bruxelles 5.
SYNAP	Syndicat National des Attachés de Presse Parisienne.
TA	Typographical Association. See NGA.
TAGA	Technical Association of the Graphic Arts, P.O. box 3064, Federal Station, Rochester (N.Y. 14614).
TALI	Technical Association of the Lithographic Industry (Founded Chicago 1948, since 1950 TAGA *qv*).

TAPPI   Technical Association of the Pulp & Paper Industry, 360 Lexington Avenue, New York (N.Y. 10017).

Technical Association of Graphic Arts, 4 Ginza 5, Chuo-ku, Tokyo.

Tidningspappersbrukens Forskning-Laboratorium, Drottning Kristina Väg 55, Stockholm-O.

TNO   Central Organisation for Applied Scientific Research in the Netherlands. (See list of abbreviations, *app. 8*), Juliana van Stolberglaan 148, P.O. box 297, The Hague. (TNO is divided in 5 main groups; the Central Organisation, the Organisation for Industrial Research, the Organisation for Nutrition and Food Research, the Organisation for Health Research, the National Defense Research Organisation, and comprises together 58 research institutes.)

Tokyo Institute of Technology, Meguro-ku, Tokyo.

TPG   Techniques Papetières et Graphiques. Organisation: Société pour l'Organisation de Salons Industriels et Techniques, 40 rue du Colisée, Paris 8e. Exhibition Hall: Rond-Point de la Défense. See also CNIT.

Type Directors Club, 122 East 42nd street, New York (N.Y. 10017).

Typomundus International Centre for the Typographic Arts, P.O. box 2438, Grand Central Station, New York (N.Y. 17).

UBPP   Unie van de Belgische Periodieke Pers, Belliardstraat 20, Brussel 4.

UFAG   Union Française des Associations Graphiques, 69 rue de Bretagne, Paris 3e.

UGRA   Verein zur Förderung Wissenschaftlicher Untersuchungen im Graphischen Gewerbe, Postfach 121, 8030 Zürich.

UIAA   Union Internationale des Associations d'Annonceurs.

UIP   Union Internationale de la Photogravure, 117 boulevard Saint-Germain, Paris 6e.

UIP   Union Internationale des Publicitaires.

UNEEPF   Union Nationale des Editeurs-Exportateurs des Publications Françaises.

UNIDIG   Union Nationale Intersyndicale des Industries Graphiques.

UNIDO   United Nations Industrial Development Organization.

UNIGRA   Union des Industries Graphiques et du Livre (Unie der Grafische Bedrijven en het Boek), 76 rue Renkin, Bruxelles 1e.

Union de la Presse Périodique Belge, 20 rue Belliard, Bruxelles 4e.

UNIPAC   Union Industrielle des Fabricants de Papiers et Cartons.

UPSPI   Union Parisienne des Syndicats Patronaux de l'Imprimerie, 117 boulevard Saint-Germain, Paris 6e.

USASI   United States of America Standards Institute (the former name for ASA *qv*).

USFO   Union Syndicale Française du Carton Ondulé.

UTA   United Typothetae of America.

VBM   Verband deutschschweizerischer Buchbindermeister, Postfach 8023, Zürich.

VDL   Vereinigung Deutscher Layoutsetzereien e.V.

VDMA   Verein Deutscher Maschinenbau-Anstalten e.V. (Fachgemeinschaft Druck- und Papiermaschinen), Lyonerstrasse, P.O. box 109, 6 Frankfurt a/Main-Niederrad.

VDP   Verband Deutscher Papierfabriken e.V. (Vereinigung der Papier- und Pappen-, Zellstoff- und Holzstofferzeugung), (the former Treuhandstelle der Zellstoff- und Papierindustrie e.V.).

VEA   Nederlandse Vereniging van Erkende Reclame-Adviesbureaux, A. J. Ernststraat 169, 1083 GT Amsterdam.

Verband der Faltschachtelindustrie Österreichs, Hintere Zollamtstraße 1, 1030 Vienna.

Vereeniging ter Bevordering van de Belangen des Boekhandels, Lassusstraat 9, Amsterdam.

Verein Schweizerischen Cartonnage-Fabriken, Gutenbergstraße 21, Postfach 2456, 3001 Bern.

Verein Schweizerischer Lithographiebesitzer, Schlosshalden-straße 20, 3000 Bern 15.

Verein Zellcheming, Verein der Zellstoff- und Papier-Chemiker und Ingenieurs, Rheinstraße 51, 61 Darmstadt.

Vereinigung der Heimat- und Standortpreße e.V., Korbach.

Vereniging van Kartonnagefabrikanten in Nederland, Laan Copes van Cattenburch 79, 2585 EW The Hague.

Vereniging van Verf-en Drukinktfabrikanten, Groot Haesebroekseweg 1, 2240 AB Wassenaar (the Netherlands).

Vereniging van Nederlandse Golfkartonfabrikanten, Julianastraat 30, 2001 DA Haarlem (the Netherlands).

Vereniging van Papiergroothandelaren, Zeestraat 78, 2518 AD The Hague.

VFG   Verein zur Förderung und Forschung für die graphischen Gewerbe und Verwandten Wirtschaftszweige, p/a Hohere Graphische Bundes-Lehr und Versuchsanstalt, Vienna 7.

VGJO   Vereniging van Grafische Jonge Ondernemers, Van Eeghenstraat 70, Amsterdam (the former VJAG, Vereniging van Jonge en Aanstaande Patroons in de Grafische Bedrijven).

VGRO   Vereniging van Grafische Reproductie-Ondernemingen. See VNCI.

VI   Vezelinstituut TNO (Fibre Research Institute TNO), Schoemakerstraat 97, Delft (the Netherlands).

Vig   See MCVA.

VNCI   Vereniging van Nederlandse Chemigrafische Inrichtingen (Changed its name into VGRO).

VINITI   Vscsojuznyj Institut Naučnoj I Tehničeskoj Informacii, Baltijskaja ulica 14, Moscow A 219.

VNIIPP   Vscsojuznyj Naučno Issledovatel'skij Institut Poligraficeskoj Promyšlennosti, Cvetnoj bul'var 30, Moscow I 51.

VSLB   Verein Schweizerischer Lithographiebesitzer. See SSPL.

VSS   See APS

Vyzkumny Ustav Polygraficky (Forschungsinstitut für Polygrafie), Prague.

Wissenschaftliches Forschungsinstitut für den Polygrafischen Maschinenbau, m. Kalusjeskijper 4, Moscow W.71

WONA   Web-Offset Newspaper Association (Division of the Newspaper Society of the UK).

WPUC   Waste Paper Utilization Council.

YMP   Young Master Printers.

YNA   Young Newspapermen's Association, Whitefriars House, 6 Carmelite street, London EC 4.

Yngre Bogtrykkers under Dansk Provins Bogtrykkersforening, Kochsgade 23, Odense, Denmark.

Young Copenhagen Master Printers, Landskrona gade 70, Kobenhavn.

ZFA   Zentral-Fachausschuß für das Graphische Gewerbe des Deutschen Bundesgebietes und Berlin, Herforderstraße 28, Bielefeld.

Astronomical signs and symbols

sun and moon:

⊙ sun   ● new moon   ○ full moon
◑ moon   ☾ first quarter moon   ☽ last quarter moon

major planets:

☿ Mercury   ♂ Mars   ♅ Uranus
♀ Venus   ♃ Jupiter   ♆ Neptune
♁ Earth   ♄ Saturn   ♇ Pluto

asteroids (in order of discovery):

① ⚳ Ceres   ③ Juno
② ⚴ Pallas   ④ Vesta

stars, nubelae, etc.:

✳ star   ○ planetary nubela
≪ comet   ◉ annular nubela
⊕ globular cluster

aspects and nodes:

☌ conjunction   □ quadrature   Ω ascending node
⚹ sextile   △ semi-square   ℧ descending node
⚺ semi-sextile   ⧄ sesqui-quadrate   ♈ vernal equinox
⚻ quincunx   ☍ opposition   ♎ autumnal equinox
   quintile

positions:

⊖ sun's centre   ☉ sun's lower limb   ☾ moon's lower limb
☉ sun's upper limb   ☽ moon's upper limb   ⊠ station mark

Botanical signs and symbols

⊙① annual   ? doubtful   ⊙ monocarpus
♂ antheridia   △ evergreen   ♂-♀ monoecious
① autumn flowering   ♀ female   ∞ number indefinite
② biennial   ☿ hermaphrodite   ♀ oogamia
∧ climbing plant   ✕ hybrid   ○ one or absent
♂♀ dioecious   ♂ male   ♃ perennial

!   personally verified

♂♀☿   polygamous

§   section (of a genus)

☾   spring flowering

☽   summer flowering

♄   tree

☽   winter flowering

Chemical signs

$+$ together with

$=$ are equal to

$\rightarrow$ and $\leftarrow$ denotes a reaction in the direction indicated

$\rightleftarrows$ denotes a reaction which proceeds simultaneously in both directions

$\downarrow$ after a symbol or formula, denotes the precipitation of a specified substance

$\uparrow$ after a symbol or formula, denotes that a specified substance passes off as a gas

$\equiv$ or $\rightleftharpoons$ denoting the quantities of specified substances which will enter into a reaction, without leaving excess material

Geometrical signs and symbols

Most frequently occurring symbols are:

$\angle$ angle

$\cap$ arc of a circle

$\bigcirc$ circle

$\equiv$ or $\simeq$ congruent to

⬡   cube

$\#$ equal and parallel

$\veebar$ equal angles

$\parallel$ parallel to

$\perp$ perpendicular to

▭   rectangle

$\diamondsuit$ regular pentagon

▱   rhomboid

$\llcorner$ right angle

◁   right-angled triangle

$\triangledown$ sector

$\frown$ semi circle

$\sim$ similar

$\because$ since

⊲   spherical angle

$\square$ square

$\therefore$ therefore

⏢   trapezoid

$\triangle$ triangle

$-$ vinculum

Mathematical signs

The most frequently occurring signs are listed below. See also algebraical signs; arithmetical signs

$\approx$ approximately equal to

$\because$ because, or sine

$\varpropto$ between

$\sqrt[3]{}$ cube root

$\bigcirc$ degree

$\div$ divided by

$=$ equal to

$\geqq$ or $\gtreqless$ or $\geqslant$ equal to or

  greater than

$\leqq$ or $\lesseqgtr$ or $<$ equal to or less than

!   factorial of

$>$ greater than

$\equiv$ identical with

∞ infinity

$\int$ integral

$<$ less than

$-$ minus

$\mp$ minus or plus

$'$ minute or prime

$|$ modulus

$\gg$ much greater than

$\ll$ much less than

$\times$ multiplied by

$\sqrt[n]{}$	nth root	$+$	plus	$\sim$	similar to
$\neq$ or $\pm$	not equal to	$\pm$	plus or minus	$/$	solidus
$\ngtr$	not greater than	$\propto$	proportional to	$\sqrt[2]{}$	square root
$\nless$	not less than	$''$	second, or double	Σ	summation
$\parallel$	parallel to		prime	$\therefore$	therefore, hence

Medical and pharmaceutical (apothecaries') signs and symbols

Quantities are written in roman lower-case numerals: the final i becoming j, e.g.: vij ($= 7$).

℈	scruple	ℨ iss	a dram and a half	f ℥	fluid dram
℈ ss	half a scruple	ℨ ij	two drams	f ℥	fluid ounce
℈ i	one scruple	ℨ	ounce	O	pint (octarius)
℈ iss	a scruple and a half	℥ ss	half an ounce	C	gallon (congius)
℈ ij	two scruples	℥ i	one ounce	℞	recipe (take)
ℨ	dram	℥ iss	an ounce and a half	aa	of each
ℨ ss	half a dram	℥ ij	two ounces	fs	semi
ℨ i	one dram	♏	minim		

Meteorological signs and symbols

The most frequently occurring symbols are:

⋎	aurora (borealis)	⟷	ice needles and crystals	⁂	sleet
⌓	dew	$=$	light fog	⁂	sleet shower
⋖	distant lightning (i.e., thunder inaudible)	⌣	lunar corona	△	small hail
		⌣	lunar halo	✳	snow
↓→	drifting snow (near ground)	⤜	mirage	✳	snow shower
			mist, same as light fog	✳ ⤳	snowstorm
,	drizzle	0	pure air (abnormal visibility)	✳	soft hail
S	dust			V	soft rime
⚡	duststorm	o	rain	①	solar corona
$\equiv$	fog	▽	rain showers	⊕	solar halo
⊔	gale	⌒	rainbow	⊙	sunshine
$\sim$	glazed frost		sandstorm, same as duststorm	⚡	thunder and lightning
▲	hail	≡	shallow fog		
∞	haze (dry haze)	⋖	sheet lightning		
△	ice grains				

Zodiacal signs

Signs which indicate the twelve parts of an imaginary belt encircling the heavens and extending about 8° on each side of the ecliptic, within which are the orbits of the moon, sun and larger planets. The signs of the zodiac are :

spring signs:
    ♈ Aries, the Ram
    ♉ Taurus, the Bull
    ♊ Gemini, the Twins

summer signs:
    ♋ Cancer, the Crab
    ♌ Leo, the Lion
    ♍ Virgo, the Virgin

autumn signs:
    ♎ Libra, the Balance
    ♏ Scorpio, the Scorpion
    ♐ Sagittarius, the Archer

winter signs:
    ♑ or ♉ Capricornus, the Goat
    ♒ Aquarius, the
        Water Bearer
    ♓ Pisces, the Fishes.

Indexes to Appendixes 1, 2 and 3

Dutch

Spanish

Italian

Dictionaries, glossaries and vocabularies

American Paper and Pulp Association. The Dictionary of Paper. New York, 1965.

Anpa R.I. bulletin. Communication Terms. New York, 1967.

Archambeaud, P. Dictionnaire Anglais-Français et Français-Anglais des Industries Graphiques. Paris.

Bargilliat, A. Vocabulaire pratique Anglais-Français et Français-Anglais des termes techniques concernant la cartographie. Paris. 1944.

Born, Ernst. Lexikon für das Graphische Gewerbe. Frankfurt a/Main. 1958.

Clason, W. E. Lexicon of international and national units. Amsterdam. 1964.

Collins, F. Howard, Authors and Printers Dictionary. London. 1973.

Commission Internationale d'Eclairage. International lighting vocabulary. Paris. 1957.

Computer Terms for the Typographic Industry. Washington DC. 1973.

Comte, René; et André Pernin. Lexique des industries graphiques. Paris. 1963.

Condensed Chemical Dictionary. New York. 1966.

Dizionario Cartotecnico. Rome. 1969.

Encyclopædia Britannica. Chicago (Ill.). 1962.

European Carton Makers Association. Dictionary. Offenbach-Main. 1963.

European Productivity Agency of the OEEC. Glossary of work study terms. Paris. 1958.

Fouchier, J.; and H. Billet. Chemical Dictionary. Amsterdam. 1961.

Glaister, G. A. Glossary of the book. London. 1960.

Glossaries of the Association Française de Normalisation. Paris.

Glossaries of the British Standards Institution. London.

Glossaries of the Fachnormenausschuß Graphisches Gewerbe. Berlin.

Glossaries of the International Organization for Standardisation. Geneva.

Glossary of Automatic Typesetting & Related Computer Terms. Los Angeles. 1966.

Handbook of Chemistry and Physics. Cleveland (Ohio). 1960.

Harper's Dictionary for the Graphic Arts. New York. 1963.

Harrap's standard French and English dictionary. London, 1960.

Hellwig, Wilhelm. Wörterbuch der Fachausdrücke des Buch- und Papiergewerbes mit besonderer Berücksichtigung der wichtigsten Druckverfahren. Frankfurt a/Main. 1917.

Hofmann, C. Typografen ABC. 's-Gravenhage. 1944.

Hostettler, R.; E. Kopley, H. Strehler. The Printer's Terms. St. Gallen. 1958.

Hoyer, Fritz. Papiersortenlexikon. Stuttgart. 1929.

Institute of Printing Ltd. Computer Typesetting Conference. Report of Proceedings. London. 1965.

Interchemical Ink Corporation. Flexographic Ink Handbook.

Jong, G. J. Verklarend woordenboekje voor de reproductietechniek in de grafische vakken. Amsterdam. 1940.

Kenneison, W. C.; and A. J. B. Spilman. Dictionary of printing, papermaking and bookbinding. London. 1963.

King. M. K. Photographic Dictionary. Düsseldorf.

Klimschs Wörterbuch der Fachausdrücke des Druckgewerbes, der Reproduktionstechnik, der Buchbinderei und Papierverarbeitung. Frankfurt a/Main. 1941.

Krug, Karl. Fachwörter ABC für graphische Berufe. Krefeld.

Labarre, E. J. Dictionary and encyclopædia of paper and papermaking. Amsterdam. 1952.

Lafontaine, Gérard H. Dictionary of terms used in the paper, printing and allied industries. Toronto. 1954.

Lexikon der graphischen Technik. Leipzig. 1962.

Linotype & Machinery Ltd. Dictionary of printing terms. London. 1962.

Ljungberg, Hans. Typografisk Ordlista. Stockholm. 1948.

Martinot. E. Woordenboek Nederlands-Engels en Engels-Nederlands fotografie en acoustiek. Hilversum. 1949.

Materiallexikon für die graphische Industrie. Leipzig. 1955.

Mengel, W. Druckschriften der Gegenwart. Stuttgart. 1966.

Mohrberg. W. Technisches Wörterbuch Zellstoff und Papier. Darmstadt. 1955.

Morin, Edmond. Dictionnaire de l'Imprimerie. Brussels. 1963–1965.

National Printing Ink Research Institute, Printing ink definitions. New York. 1952.

Nederlandsche Bond van Boekbinders-Patroons, Codelist of binding operations. Amsterdam. 1960.

Palmer, H. Kleines ABC der Druckerei. Stuttgart.

Pepper, W. Dictionary of newspapers and printing terms. New York. 1959.

Phillips, A. Computer peripherals and typesetting. London. 1968.

Pocket encyclopedia of papers & graphic arts terms. Kaukauna (Wisc.). 1960.

Polygraph Dictionary. Fachausdrücke der graphischen Industrie. Frankfurt a/Main. 1958.

Proost en Brandt N.V. Papier op papier. Verklaring van gewone en ongewone uitdrukkingen. Amsterdam. 1959.

Rabaté, Henri. Glossaire trilingue, spécial aux industries des cires, huiles, gommes, résines, pigments, vernis, encres, peintures, produits d'entretien et préparations assimilées. Paris. 1948.

Random House Dictionary of the English Language. New York, 1967.

Research and Engineering council of the graphic arts industry, Inc. Rubber and plastics used in the printing industry. Washington. 1953.

Rodríguez, César. Diccionario Bilingüe de las Artes Gráficas, New York, 1966.

Rupp, E. Materiallexikon für die graphische Industrie. Leipzig. 1955.

Schlemmiger, Johann. Fachwörterbuch des Buchwesens. Darmstadt. 1954.

Stoeckhert, K. Kunststoff-Lexikon. München. 1953.

Seybold Report. Glossary of terms pertaining to computerized text processing and photocomposition. Haddon-field (N.J.). 1974.

Uvarov, E. B.; and D. R. Chapman. A dictionary of Science. Harmondsworth (Middlesex). 1951.

Van Dale's nieuw groot Woordenboek der Nederlandse taal. The Hague. 1970.

Wall, E. J. Dictionary of Photography. New York.

Webster's Seventh New Collegiate Dictionary. Springfield (Mass.). 1970.

Wheelwright, William Bond. Paper trade terms. Boston. 1947.

Winkler, T. T. Geïllustreerd handwoordenboek voor de grafische vakken, Baarn, 2e druk.

Text books

American Bankers Association. The common machine language for mechanized check handling. New York. 1959.

American Photoengravers Association. The art of photoengraving. Chicago. 1952.

Apps, E. A. Ink technology for printers and students. London. 1963.

Archambeaud, Pierre. Encres d'imprimerie noir et couleurs. Paris. 1956.

Audin, Marius. Le Livre, son architecture, sa technique. Paris. 1924.

Bargilliat, Alain. Impression offset. Paris. 1960.

Bargilliat, Alain. Offset-litho. Procédés manuels. Paris. 1945.

Bargilliat. Alain. Photo litho. Photo et copie. Paris. 1958.

Bargilliat, Alain. Typographie impression. Paris. 1945.

Bargilliat, Alain; et Jacques Campbell. Eléments de sciences pour les industries graphiques. Paris. 1956.

Baudry, G. Hélio, Gravure et tirage. Paris. 1953.

Baudry, G.; et R. Marange. Comment on imprime. Paris. 1966.

Bekk, J. Het Papier. Amsterdam. 1942.

Blottiau, Félicien. Colorimétrie. Encyclopédie photométrique. Paris. 1951.

Bowles, Reg. F. Printing ink manual. Cambridge. 1961.

Braun, Alexander. Atlas der Zeitungs- und Illustrationsdruck. Frankfurt a/Main. 1960.

Bustanoby, J. A. Principles of color and color mixing. New York. 1947.

Cartwright, H. H.; and Robert Mackay. Rotogravure. A survey of European and American methods. Lyndon (Kentucky). 1956.

Cermak, Werner. Handbuch für den Siebdruck. Leipzig. 1961.

Champlain Company, Inc. Commercial rotogravure-printing with Champlain rotogravure press equipment. Bloomfield (New Jersey). 1957.

Champlain Company, Inc. Web-fed cutting and creasing of folding cartons. Roseland (New Jersey). 1962.

Clerc, L. P. La technique photographique. Paris. 1950.

Cox, Arthur. Optics. The technique of definition. London. 1946.

Day, Fred. T. Paper converting and usage. Mechanical materials handling. London. 1954.

Diehl, Edith. Bookbinding. Its background and technique. New York. 1946.

Ellis, C. Printing inks. Their Chemistry and technology. New York. 1940.

Erbs, H. Der Rotationsdruck. Leipzig. 1952.

Gravure Technical Association, Inc. Technical guide for the gravure industry. New York. 1975.

Green, H. Industrial rheology and rheological structures. New York, 1949.

Halpern, Bernhard R. Colour stripping for offset lithography. New York. 1955.

Halpern, Bernard R. Offset stripping (Black and white). New York. 1958.

Handbuch für Reproduktionstechnik. Frankfurt a/Main. 1954.

Henningsen, Th. Das Handbuch für den Buchbinder. St. Gallen. 1950.

Judd, Deane B. Color in business, science and industry. New York. 1959.

Kowaliski, Paul. Problèmes photographiques de la photogravure. Paris. 1958.

Laborderie, F. de; et J. Boisseau. Toute l'imprimerie. Les techniques et leurs applications. Paris. 1960.

Lange, M. J. de. Eenheid in Eenheden. The Hague. 1957.

Larsen, Louis M. Industrial printing inks. New York. 1962.

Lawrence, John, R. Polyester resins. New York. 1960.

Letouzey, Victor. Pourquoi et comment nous voyons les couleurs. Paris. 1959.

Lithographic Technical Foundation. Shop manuals.

Martin, Gérard. La physico-chimie des encres. Paris. 1961.

Mosher, Robert H. Specialty papers. Their properties and applications. Brooklyn (N.Y.). 1950.

Murail, Maurice. Typographie. Typogravure. Paris. 1954.

Optical Society of America. The Science of color. New York. 1953.

Papermaking. A general account of its history, processes and application. Kenley (Surrey). 1950.

Parson, E. S. The application of statistical methods to industrial standardisation and quality control. London. 1935.

Printing ink handbook. New York. 1958.

Reed, Robert F. What the lithographer should know about ink. New York. 1960.

Reed, Robert F. What the lithographer should know about paper. New York. 1957.

Salomon. Louis. Steréo-galvano. Paris. 1950.

Schröder, Willy. Die stereotypie. Frankfurt a/Main. 1950.

Strauss, Victor. The Printing Industry. New York. 1967.

Valette, Georges. Typographie. Composition. Paris. 1945.

Vareine, Jean. Aspects modernes de la photomécanique. Paris. 1961.

Whetton, Harry. Practical printing and binding. London. 1948.

Wolfe, Herbert J. Pressmen's ink handbook. Caldwell (New Jersey). 1952.

Wolfe, Herbert J. Printing and litho inks. New York. 1957.

COLOPHON

Electronic data processing:
Büro für Satztechnik W. Meyer KG, Weissensberg, W. Germany
Printed by:
N.V. Noord-Nederlandse Drukkerij, Meppel, the Netherlands
Bound by:
Paardekooper-Wöhrmann B.V., Zutphen, the Netherlands